Nanostructures for Oral Medicine

Nanostructures in Therapeutic Medicine Series

Nanostructures for Oral Medicine

Edited by

Ecaterina Andronescu
University Politehnica of Bucharest, Bucharest, Romania

Alexandru Mihai Grumezescu
University Politehnica of Bucharest, Bucharest, Romania

ELSEVIER

Elsevier
Radarweg 29, PO Box 211, 1000 AE Amsterdam, Netherlands
The Boulevard, Langford Lane, Kidlington, Oxford OX5 1GB, United Kingdom
50 Hampshire Street, 5th Floor, Cambridge, MA 02139, United States

Library of Congress Cataloging-in-Publication Data
A catalog record for this book is available from the Library of Congress

British Library Cataloguing-in-Publication Data
A catalogue record for this book is available from the British Library

ISBN: 978-0-323-47720-8

Publisher: Matthew Deans
Acquisition Editor: Simon Holt
Editorial Project Manager: Sabrina Webber
Production Project Manager: Julie-Ann Stansfield
Designer: Greg Harris

Typeset by Thomson Digital

Contents

List of Contributors xxv
Foreword of the Series xxxii
Preface xxxv

CHAPTER 1 Bacterial Polyester Nanoparticles for Drug Delivery 1
Mohsin Shah, Sung C. Yoon
1 Introduction 1
2 Amphiphlic Block Copolymers 2
3 Polymeric Nanoparticles in Drug-Delivery Applications 3
4 Biodegradable Bacterial Polyesters Poly(R)-Hydroxyalkanoic Acids 4
4.1 Properties of PHA 5
4.2 Biodegradation of PHA 6
4.3 General Applications of PHA 6
4.4 PHA as Drug Carrier 7
5 Targeting Drug Delivery using Ligand-Conjugated PHA Nanoparticles (Active Targeting) 9
6 Chemical Modification of PHA Polymers for Drug Delivery (Passive Targeting) 10
7 Synthesis of Surface Modified PHB-mPEG Diblock Copolymer 11
8 Self-Assembly of Diblock Copolymer Into Nanoparticles 12
9 Characterization of PHA–mPEG Nanoparticles 13
9.1 Core–Shell Topology in Aqueous Solution 13
9.2 Morphological Characteristics of PHA–mPEG Nanoparticles 15
9.3 Drug Loading in PHA–mPEG Nanoparticles 16
9.4 Drug Release from PHA–mPEG Nanoparticles 17
9.5 Cellular Uptake of PHA–mPEG Nanoparticles 18
10 Drawbacks and Challenges 19
11 Conclusions and Future Perspectives 20
References 20

CHAPTER 2 A Novel Approach to the Oral Delivery of Bionanostructures for Systemic Disease 27
Rosa Martha Perez Gutierrez, Juan Vicente Mendez Mendez, Israel Arzate Vazquez
1 Introduction 28
2 NanoStructures for Oral Drug Delivery 29

3 The Gastrointestinal Mucus Barriers 30
3.1 Mucus in the GI Tract 30
3.2 Absorption in the Gastrointestinal Tract 32
3.3 Transport Mechanisms in the GI Tract 33
3.4 Mucosal Barrier 34
3.5 Mucus Interactions 34
4 Strategies for Effective Oral Delivery of Drugs 35
5 Challenges in the Oral Delivery Route 36
6 Characteristics of the Transporter Used in Oral Delivery Nanostructures 37
6.1 Oral Delivery Nanostructures 37
6.2 Hydrogels 37
6.3 Dendrimers 38
6.4 Micelles 40
6.5 Liposomes 40
6.6 Nanoparticles 41
6.7 Carbon Structures 42
6.8 Polymers Used in Drug-Delivery Systems 43
7 Therapeutic Applications of Oral Delivery Nanostructures 45
7.1 Cerebrovascular Disease 45
7.2 Hypercholesterolemia 46
7.3 Arthritis 46
7.4 Nanostructures in Cancer 47
7.5 Diabetes 48
7.6 Hepatitis B 49
8 Synthesis of Nanoparticles 49
9 Characterization of Nanoparticles 49
9.1 Light Microscopy 49
9.2 Scanning Electron Microscopy 49
9.3 Transmission Electron Microscopy 50
9.4 Dynamic Light Scattering 50
9.5 Zeta Potential (ζ) Analysis 50
9.6 Raman Spectroscopy 50
9.7 Atomic Force Microscopy 51
9.8 X-Ray Photoelectron Spectroscopy or Electron Spectroscopy for Chemical Analysis 51
9.9 Fourier Transform Infrared Spectrometry 51
10 Conclusions 52
References 52

CHAPTER 3 Biodegradable Polymeric Nanostructures: Design and Advances in Oral Drug Delivery for Neurodegenerative Disorders **61**
Andreza Maria Ribeiro, Francisco Veiga, Ana Figueiras
1 Introduction 61
1.1 Neurodegenerative Disorders: an Overview 63
2 Biodegradable Polymeric Materials 64
2.1 Natural Polymers 64
2.2 Synthetic Polymers 65
3 Nanocarriers Technology and Recent Modifications to Improve Their Specificity 66
3.1 Polymeric Nanoparticles 68
3.2 Polymeric Micelles 69
3.3 Dendrimers 71
3.4 Liposomes 73
3.5 Functional Surface Building 75
4 Advances in Polymeric Nanocarriers as Vehicles for Oral-Delivery Systems 76
5 New Approaches and Clinical Applications in Neurodegenerative Disorders 78
6 Future Perspectives 79
References 80

CHAPTER 4 Nanostructured Systems for Transbuccal Drug Delivery **87**
Michelle Franz-Montan, Daniele Ribeiro de Araújo, Lígia Nunes de Morais Ribeiro, Nathalie Ferreira Silva de Melo, Eneida de Paula
1 Introduction 88
2 Oral Mucosa 89
2.1 Concept and Functions 89
2.2 Organization and Structure 89
2.3 Permeability 93
2.4 Permeation Pathways 94
3 Buccal Mucosa Chemical Permeation Enhancers: Chemical Structure, Mechanisms, and Biopharmaceutical Applications 94
3.1 Surfactants and Bile Salts 95
3.2 Fatty Acids 98
3.3 Cyclodextrins 98
3.4 Chelators 98
3.5 Polymers: The Case of Chitosan and Its Derivatives and Associations 99
4 Nanoparticles and Nanostructures for Transbuccal Drug Delivery 100
4.1 Nanostructures 102
4.2 Nanoparticles 104

5 Limitations....113
6 Conclusions....114
References....114

CHAPTER 5 **Trends in Orally Viral Vector Gene Delivery and Therapy**....123
Jingqi Xie, Ruian Xu
1 Introduction....124
2 Nanoparticles and Microparticles Applications in Oral Delivery of Drug Molecules and Therapeutic Genes....124
3 The Crucial Intake Site of the Absorption of Nanoparticles and Microparticles in the Gastrointestinal Tract....125
4 The Characteristics of Particles and the Effect of Intake In Vivo....126
4.1 The Performance of the Particles....126
4.2 The Surface Modification and Charge of the Particles....127
4.3 The Size of the Particles....127
4.4 The Age of the Animals....128
5 Protein Transduction and the Absorption of Nanoparticles....128
6 Ways to Promote the Absorption of Particles....129
7 The Feasibility of Oral Gene Delivery....130
8 The Oral Gene Delivery Vector System....130
9 The Uptake of Oral Viral Vector....131
10 The Selection of Viral Vectors Suitable for Oral Delivery....132
10.1 Retrovirus....133
10.2 Adenovirus....133
10.3 Adeno-Associated Virus....135
10.4 Other Viral Vector Systems....137
11 The Digestion Circumstance of Oral Gene Delivery....138
12 The Transmission, Distribution, and Transgenic Product Pharmacokinetics of Oral Viral Vectors....138
13 Potential, Problems, and Prospects....139
14 Conclusions....140
References....140

CHAPTER 6 **Nanostructures for Oral Delivery of Therapeutic Nucleic Acids**....147
Virginie Busignies, Christine Charrueau, Pierre Tchoreloff, Virginie Escriou
1 Introduction....147
2 Oral Delivery of Therapeutic Nucleic Acids....149
2.1 Advantages of the Oral Route of Administration....149
2.2 Obstacles to Nucleic Acid Oral Delivery and Potential Targets....150
2.3 Experimental Models for Studying the Stability and Efficiency of Oral Nucleic Acid Delivery Systems....152

3 Oral Delivery Systems for Nucleic Acid Administration....159
3.1 Naked Molecules (Essentially Oligonucleotides) 159
3.2 Cationic Lipid-Based Nanostructures....159
3.3 Condensing Polymer-Based Nanostructures 162
3.4 Other Nanostructures....166
3.5 Micrometric Delivery Systems and Minitablets....167
4 Conclusions 168
References....169

CHAPTER 7 Challenges in Oral Drug Delivery: A Nano-Based Strategy to Overcome 173
Pragasam Viswanathan, Yuvashree Muralidaran, Gokulakannan Ragavan
1 Introduction 174
2 Biopharmaceutical Classification of Drugs 175
3 Measurement of Parameters Determining Bioavailability....176
3.1 Dissolution 176
3.2 Solubility 178
3.3 Permeability 178
4 Transport Mechanism in GIT....179
5 Gastrointestinal Luminal Milieu and Drug Absorption 179
5.1 pH Variation in GIT....181
5.2 Gastrointestinal Microbiota....182
5.3 Gastric Emptying Rate 183
6 Improving Drug Bioavailability....183
7 Pharmaceutics Approach Enabling Physical Modification of Drugs....183
7.1 Particle Size Reduction 183
7.2 Modification of Crystal Habit 184
7.3 Drug Dispersion in Carriers 184
7.4 Complexation 185
7.5 Solubilization by Surfactants 185
7.6 Chemical Modification....185
8 Improving Stability of Oral Drugs....186
9 Nanostructures in Oral Drug Delivery....186
9.1 Polymeric Formulation....186
9.2 Lipid-Based Formulations....194
9.3 Metal and Inorganic-Based Nanocarriers....194
10 Conclusions 196
Conflict of Interest....197
References....197

CHAPTER 8 Oral Pellets Loaded with Nanoemulsions 203
Thi Trinh Lan Nguyen, Nicolas Anton, Thierry F. Vandamme
1 Pellets 203
1.1 Introduction to Pellets 203
1.2 General Description of the Extrusion Spheronization Process (Wet Mass Extrusion) 205
1.3 Process and Equipment 207
1.4 Formulation 211
1.5 Evaluation of Pellets 213
2 Nanoemulsions 213
2.1 Introduction 213
2.2 Method 215
2.3 Materials Used in Nanoemulsions Production 216
3 Biopharmaceutical Classification System (BSC) and Applications of Pellet to Improve Dissolution Rate of Drugs in BCS Class II 217
4 Case Study 221
4.1 Coating of Pellets 221
4.2 Dissolution Tests 222
4.3 Reconstitution Study 222
4.4 Results 222
5 Conclusions 223
References 223

CHAPTER 9 Oral Drug Delivery Potential of Dendrimers 231
Anurag K. Singh, Ashok K. Sharma, Iliyas Khan, Avinash Gothwal, Lokesh Gupta, Umesh Gupta
1 Introduction 232
2 Oral Route and its Implication 232
3 Nanocarriers in Oral Delivery 234
3.1 Poly(Lactic-*co*-Glycolic Acid) Nanoparticles 234
3.2 Solid Lipid Nanoparticles 238
3.3 Hydrogel NPs (Nanogels) 238
3.4 Polymeric Micelles 238
3.5 Liposomes 239
3.6 Nanoemulsion and Self-Nanoemulsifying System 239
3.7 Nanocrystals 241
3.8 Dendrimers 241
3.9 Nanospheres 241
3.10 Carbon Nanotubes 242

4 Dendrimers in Oral Drug Delivery......242
5 Synthetic Approaches of Dendrimer......243
5.1 Divergent Approach......243
5.2 Convergent Approach......243
5.3 Miscellaneous Approaches......244
6 Unique Properties of Dendrimers......247
7 Host Guest Interactions of Dendrimers and Bioactives......248
8 Drugs Delivered Through Oral Route Using Dendrimers as a Nanocarrier......248
8.1 Propranolol......249
8.2 Triclosan......250
8.3 Nitric Oxide......251
8.4 Clotrimazole......251
8.5 Ketoconazole......251
8.6 Sulfamethoxazole......252
8.7 Silicone Urinary Catheters......252
9 Conclusions and Future Prospects......252
References......255

CHAPTER 10 Nanovaccines for Oral Delivery-Formulation Strategies and Challenges......263
Sayali Karandikar, Amit Mirani, Vishal Waybhase, Vandana B. Patravale, Sandeep Patankar
1 Introduction......263
2 Oral Vaccine Delivery......264
3 Intestinal Immune Response......266
3.1 Antigen Uptake......266
3.2 Immune Response......268
4 Challenges in Oral Vaccine Delivery......268
5 Nanocarrier Based Oral Vaccine Delivery......268
5.1 Antigen-Loading Mechanism in Nanocarriers......269
5.2 Nanocarrier Properties Affecting Antigen Uptake......270
6 Nanocarrier Formulation Strategies......271
6.1 Vaccine Composition......271
6.2 Type of Nanocarrier......272
7 Regulatory Aspects of Oral Nanovaccines......286
7.1 Considerations for Regulating Nanocarrier-Based Vaccine Delivery System......286
8 Future Prospects......288
References......289

CHAPTER 11 Nanosystems for Oral Delivery of Immunomodulators **295**
Prachi B. Kharkar, Swapnil S. Talkar, Namrata A. Kadwadkar, Vandana B. Patravale
1 Introduction 295
1.1 Summary of Immune System 296
1.2 Overview of Immunomodulators 299
1.3 Current Routes of Administration 300
2 Oral Route for Immunomodulators 301
2.1 Conventional Immunomodulators 301
2.2 Drawbacks of Conventional Oral Systems 302
3 Need of Nanosystems for Immunomodulators 302
4 Oral Nanosystems for Immunomodulators 303
4.1 Oral Nanosystems for Synthetic Immunomodulators 305
4.2 Oral Nanosystems for Herbal Immunomodulators 306
4.3 Oral Nanosystems for Vaccines and Adjuvant 310
4.4 Oral nanosystems for Toxins and Venoms 313
5 Methods for Characterization of Nanosystems 315
6 Fate of Immunomodulators After Oral Delivery 320
7 Future Prospects 323
7.1 Oral 323
7.2 Other Preferred Routes of Administration 324
8 Conclusions 326
References 328

CHAPTER 12 Tannic Acid Modification of Metal Nanoparticles: Possibility for New Antiviral Applications **335**
Malgorzata Krzyzowska, Emilia Tomaszewska, Katarzyna Ranoszek-Soliwoda, Karolina Bien, Piotr Orlowski, Grzegorz Celichowski, Jaroslaw Grobelny
1 Introduction 335
2 Syntheses and Characterization of Metallic Nanoparticles Colloids (AuNPs, AgNPs, Ag/CuNPs) with Tannic Acid 336
2.1 Tannins 336
2.2 Synthesis of mNPs With Tannins 336
2.3 Detailed Synthesis of Metallic Nanoparticles (mNPs) 342
2.4 Characterization Techniques 344
3 Toxicity and Antiviral Properties of Tannic Acid Modifies Metallic Nanoparticles Colloids (AuNPs, AgNPs, Ag/CuNPs) 345
3.1 Toxicity of Tannic Acid Modified Metallic Nanoparticles Colloids 345
3.2 Herpes Simplex Infection 351

3.3 Metallic Nanoparticles in Treatment of Virus-Mediated STDs 352
3.4 Antiviral Activities of Tannic Acid–Modified Metallic Nanoparticles Colloids 354
4 Conclusions 360
References 360

CHAPTER 13 Polyelectrolyte-Drug Ionic Complexes as Nanostructured Drug Carriers to Design Solid And Liquid Oral Delivery Systems 365
María E. Olivera, Ruben H. Manzo, Fabiana Alovero, Alvaro F. Jimenez-Kairuz, María V. Ramírez-Rigo
1 Introduction 366
2 Pharmaceutical Polyelectrolytes 366
3 Polyelectrolyte-Drug Complexes in Aqueous Dispersion 371
3.1 Polyelectrolyte-Drug Counterionic Condensation 371
3.2 Species Distribution in Polyelectrolyte-Drug Complexes 373
3.3 Affinity Constants for Counterionic Condensation 374
3.4 Electrokinetic Potentials and Hydrodynamic Diameters of Polyelectrolyte-Drug Complexes 375
3.5 Rheological Properties of Polyelectrolyte-Drug Complexes 377
3.6 Pharmaceutical Relevance of the Polyelectrolyte-Drug Interactions 378
3.7 Interaction with Biological Membranes 385
3.8 A Model to Interpret the Properties of PE-D Complexes 385
3.9 Summary of Technological and Biopharmaceutical Properties That Determine the Potential Applications of PE-D Dispersions 387
4 Polyelectrolyte-Drug Complexes in the Solid State 388
4.1 Wetting of Particulate and Compacted PE-D Complexes 389
4.2 Carbomer-D Matrices 389
4.3 Carboxymethylcellulose-D Matrices 393
4.4 Carrageenan-D Matrices 399
4.5 EE-D Matrices 400
5 Conclusions 402
References 405

CHAPTER 14 Polysaccharide Based Nano/Microformulation: An Effective and Versatile Oral Drug Delivery System 409
Md Nurunnabi, Vishnu Revuri, Kang Moo Huh, Yong-Kyu Lee
1 Introduction 409
2 Chitosan 415
3 Hyaluronic Acid 418

4 Chondroitin Sulfate ... 420
5 Heparin ... 420
6 Cyclodextrins ... 423
7 Other Polysaccharides ... 424
8 Polysaccharide Blends (PSB) in Oral Drug Delivery ... 429
9 Conclusion and Future Prospects ... 429
References ... 430

CHAPTER 15 Recent Advancements in Oral Delivery of Insulin: From Challenges to Solutions ... 435
Raje Chouhan, Shilpi Goswami, Anil Kumar Bajpai
1 Introduction ... 436
2 Diabetes Mellitus ... 436
2.1 Type 1 Diabetes ... 437
2.2 Type 2 Diabetes ... 439
3 Diagnosis Test for Diabetes ... 440
4 Insulin Therapy ... 440
5 Plants Used in the Treatment of Diabetes ... 441
6 Diabetes in World Scenario ... 441
7 Insulin ... 442
7.1 Insulin Analogs ... 443
8 Action of Insulin ... 445
9 Oral Delivery of Insulin ... 447
10 Problems in Oral Delivery ... 448
10.1 Chemical Barrier ... 448
10.2 Enzymatic Barrier ... 449
10.3 Absorption Barrier ... 450
10.4 Stability Barrier ... 451
11 Strategies to Improve Oral Delivery of Insulin ... 451
11.1 Absorption Enhancers ... 451
11.2 Enzyme Inhibitors ... 453
11.3 Mucoadhesive Polymers ... 453
11.4 Chemical Modifications of Proteins and Peptides ... 454
11.5 Formulation Vehicles ... 455
12 Recent Breakthroughs in Oral Insulin Delivery ... 456
13 Commercialization ... 457
13.1 Oral-Lyn ... 457
13.2 Eligen ... 457
13.3 CODES (Colon Specific Delivery System) technology ... 459
13.4 Capsulin ... 459
13.5 Octreolin ... 459

13.6 ORMD-0801 459
13.7 AI-401 460
13.8 Oradel 460
13.9 Macrulin 460
14 Recent Challenges and Future Prospects 460
References 461

CHAPTER 16 Nanotechnology for Oral Delivery of Anticancer Drugs: An Insight Potential 467
Madhu Gupta, Vikas Sharma, Nagendra S. Chauhan
1 Introduction 467
2 Remarkable Addressing Issues Concerns for Challenging of Oral Chemotherapy 469
2.1 Patient's Adherence 475
2.2 Economic Factor 475
3 Specific Consideration of Anticancer Drugs Appropriate for Oral Delivery 476
3.1 Poor Drug Solubility 476
3.2 Intestinal Transit Time 477
3.3 Drug Metabolism and Intestinal Efflux Pumps 477
4 Nanocarrier System for Oral Delivery of Cancer Therapeutics 478
4.1 Drug Nanocrystals 480
4.2 Polymeric Nanosystems 481
4.3 Lipidic Nanoplatform Systems 499
4.4 Dendrimers 504
5 Concluding Remarks 505
References 505

CHAPTER 17 Nanomaterials: Promising Structures for the Management of Oral Cancer 511
Görkem Eskiizmir, Aylin T. Ermertcan, Kerim Yapici
1 Oral Cancer 512
1.1 Oral Cancer: Overview 512
1.2 Epidemiology 512
1.3 Etiopathogenesis 512
1.4 Clinical Features and Diagnosis 513
1.5 Treatment 514
1.6 Prognosis 515
2 Nanoparticles in Cancer Treatment and Diagnosis 516
2.1 Magnetic Nanoparticles 516
2.2 Gold Nanoparticles 517

2.3 Graphene Oxide and Graphene NPs 519
2.4 Carbon Nanotubes 520
3 Nanomaterials in Early Diagnosis of Oral Cancer 523
4 Nanotherapeutics in Oral Cancer 524
4.1 Nanomaterials for Drug Delivery Systems 525
4.2 Nanocarriers in Oral Cancer 529
4.3 Nanomaterials for Combination Treatment Modalities in Oral Cancer 529
5 Nanomaterials in Tumor Imaging of Oral Cancer 531
5.1 Optical Imaging 531
5.2 Magnetic Resonance Imaging 531
5.3 Positron Emission Tomography 532
5.4 Nanotheranostics in Oral Cancer 533
6 Conclusions 533
References 533

CHAPTER 18 Therapeutic Potential of Solubilized Nanolignin Against Oral Diseases 545
Hiroshi Sakagami, Hong Sheng, Toshikazu Yasui, Kunihiko Fukuchi, Takaaki Oizumi, Hirokazu Ohno, Masaji Yamamoto, Toshiyuki Fukuda, Kayoko Kotohda, Hidenobu Yoshida, Taisei Kanamoto, Shigemi Terakubo, Hideki Nakashima
1 Studies of Nanostructure of Lignin 546
1.1 Structure of Lignin 546
1.2 Encounter to LCC 547
2 Biological Activity 549
2.1 In Vivo Study 549
2.2 In Vitro Study with Cultured Cells 550
2.3 Antiviral Activity 557
2.4 Synergism with Vitamin C 560
2.5 Anti-UV Activity 561
3 Efficient Utilization of Plant Resources by Alkaline Extraction 562
3.1 Alkaline Extraction of Leaf of *Sasa senanensis* Rehder 562
3.2 Alkaline Extraction of Licorice Root 563
3.3 Alkaline Extraction of Green Tea Leaves 563
3.4 Effect of $FeSO_4$ on Anti-HIV Activity of LCC 564
4 Utilization of Lignin and Its Components 564
4.1 Industrial Use: Dental Application 564
4.2 Utilization of Cellulose 566
4.3 Manufacture of Metal Nanoparticles 566
5 Clinical Application 567
5.1 Effect of lignin-vitamin C tablet on HSV-1-infection 567

5.2 Effect of SE on Oral Lichenoid Dysplasia 567
5.3 Antihalitosis Effect of SE-Containing Toothpaste 567
6 Future Direction 568
References 569

CHAPTER 19 Novel Lipid Nanostructures for Delivery of Natural Agents with Antioxidant, Antiinflammatory and Antistroke Potential: Perspectives and Outcomes 577
Hafsa Ahmad, Abhishek Arya, Satish Agrawal, Anil K. Dwivedi
1 Introduction 578
2 Stroke 579
2.1 Types of Stroke on the Basis of Pathogenesis 580
2.2 Risk Factors 580
2.3 Pathophysiology of Stroke: (Mechanism of Ischemia) 581
2.4 Cell Death in Cerebral Ischemia 581
3 Oxidative Stress Inflammation, Antioxidants, and Their Mechanism of Action 582
3.1 Mechanisms of Antioxidants 584
3.2 Natural Antioxidants 584
4 Bioactives with Potential Antioxidant, Antiinflammatory, and Antistroke Effects 584
5 Limitations of Bioactives to Elicit the Desired Therapeutic Response 589
6 Novel Lipid Nanostructures 589
6.1 Advantages of LNs-Based Delivery 590
6.2 Mechanism of Drug Delivery via LNs 591
6.3 Basic Ingredients of Lipid Nanocarriers 592
6.4 Types of LNs 592
6.5 Insights into SLNs and Liposomes with Their Methods of Preparation 595
6.6 Characterization of LNs 597
6.7 Challenges of LNs-based Delivery Systems 599
6.8 Role of LNs in Delivery of Bioactives for the Brain 599
7 Conclusions: Way Ahead with LNs 600
References 601

CHAPTER 20 Chitosan-Based Nanoparticulate Systems for Oral Drug Delivery 607
Sougata Jana, Sabyasachi Maiti
1 Introduction 607
2 Novel Technologies for Chitosan-Based Nanoparticles 611
2.1 Synthesis of CS-NPs by Ionic Gelation 611
2.2 Solid–Liquid Phase Separation Method 613
2.3 Solvent Injection Method for CS-SLNs 613
2.4 Emulsification and Ultrasonication Method for TMC-*g*-PA SLNs 613

2.5 In Situ Precipitation Method for Magnetic Nanohydrogels 614
2.6 Self-Assembled CS-Tocol Nanoparticles 616
2.7 Spinning Disk Processor for CS–PMA NPs 617
2.8 Dialysis Method for Chitosan-Based Nanohydrogels 618
2.9 Amide Coupling Reaction for CS-Drug Nanoconjugates 618
2.10 Sol–Gel Method for TiO_2-Coated CS-Based Nanohybrid Composites 619
3 Oral Chitosan Nanosystems: Key Findings 619
4 Conclusions 631
References 631

CHAPTER 21 Phytonanoconjugates in Oral Medicine 639
L. Bharathi Priya, R. Baskaran, V. Vijaya Padma
1 Introduction 640
2 Mechanism of Polyphenol Protection 641
3 Nanostructures in Oral Delivery 642
3.1 Liposomes 642
3.2 Micelles/Polymeric Micelles 644
3.3 Dendrimers 644
3.4 Self-Emulsifying Drug-Delivery Systems 645
3.5 Inorganic Nanoparticles 645
3.6 Organic Polymeric Nanoparticles 646
3.7 Solid Lipid Nanoparticles and Nanostructured Lipid Carriers 646
3.8 Nanoemulsions 647
3.9 Emulsomes 647
3.10 Cerasomes 648
3.11 Nanocrystals 648
3.12 Multifunctionalized carrier systems 648
4 Mechanism of Absorption of Phytonanoconjugates in Epithelial Cells Via Endocytotic Cellular Uptake 648
4.1 Paracellular Pathway 649
4.2 Transcellular Pathway 649
5 Biofate and Pharmacokinetics of the Nanocarriers in the GI Tract After Oral Administration 651
6 Phytonanoparticles in Oral Medicine 651
6.1 Epigallocatechin-3-gallate 651
6.2 Resveratrol 653
6.3 Curcumin 654
6.4 Silymarin 656
7 Other Phytonanoconjugates 657
8 Remarks and Future Perceptions 657
References 658

CHAPTER 22 Design and Development of Pharmaceutical Microprocesses in the Production of Nanomedicine 669
Norma A. Noguez Méndez, Carlos T. Quirino Barreda, Abraham F. Vega, Jorge E. Miranda Calderon, César G. Urioste, Xochitl C. Falomec, Alejandro R. Martínez, Máximo P. Díaz
1 Microfluidics: Origins, Present, and Future 670
1.1 Physical Foundations of Microfluidics 672
1.2 Applications of Microfluidics in Pharmaceutical Processes and Nanomedicine 673
2 Lyotropic Liquid Crystals as Drug Delivery Systems 676
2.1 Introduction 676
2.2 Liquid Crystals (LCs) 676
2.3 Thermotropic Liquid Crystals (TLCs) 676
2.4 Lyotropic Liquid Crystals (LLCs) 677
3 Lipid-Carrying Systems 679
3.1 Introduction 679
3.2 Solid Lipid Nanoparticles 680
4 Microspheres 683
4.1 Introduction 683
4.2 Criteria for the Design and Development of Microspheres 683
4.3 Physicochemical Characteristics of Chitosan 684
4.4 Microsphere Preparation Methods 685
4.5 Application of Microspheres in the Transport and Vectorization of Pharmaceutical Compounds 685
5 Polymerosomes 687
5.1 Introduction 687
5.2 Requirements for the Production of Polymerosomes 687
5.3 Synthesis of a Polymerosome from Hydrophilic-Hydrophobic Copolymers 688
5.4 Polymerosome Applications 689
6 Conclusions 690
References 691

CHAPTER 23 Mesoporous Materials and Technologies for Development of Oral Medicine 699
Shou-Cang Shen, Yuan-Cai Dong, Kumaran Letchmanan, Wai Kiong Ng
1 Introduction 700
2 Development of Mesoporous Materials as Potential Drug Carriers 701
2.1 Mesoporous Silica 701
2.2 Mesoporous Carbon 703
2.3 Mesoporous Composites 705
2.4 Other Mesoporous Materials 706

3 Drug-Loading Techniques 708
3.1 Solution Immersion....708
3.2 Impregnation and Solvent Evaporation 709
3.3 Cospray Drying....710
3.4 Other Loading Methods....711
4 Characterization and Stabilities Studies of Poorly Soluble Oral Drug Formulation 712
4.1 Effect of Pore Size of Mesoporous Drug Carriers 713
4.2 Stability of Amorphous Formulation and Supersaturation....716
4.3 Sustained Release of Poorly Soluble Drugs....718
4.4 Enhancement of In Vivo Bioavailability....719
4.5 Cytotoxicity of Mesoporous Drug Delivery System 720
5 Application of Functional Mesoporous Materials in Smart Oral Dosage Form 721
5.1 Slow Sustained Release of APIs....721
5.2 Stimuli Responsive Smart Drug Delivery 723
5.3 Development of Tablet Form 731
6 Summary and Outlook....734
Reference 735

CHAPTER 24 Nanodentistry: Novel Approaches 751
Nehi Sinha, Niha M. Kulshreshtha, Meenal Dixit, Indrani Jadhav, Divya Shrivastava, Prakash S. Bisen
1 Introduction....752
1.1 The Bottom-Up Approach 752
1.2 The Top-Down Approach....755
1.3 The Functional Approach....757
1.4 The Biomimetic Approach....757
2 Types and Properties of Nanostructures Used in Dentistry 757
2.1 Nanocomposites....758
2.2 Nanoclusters....760
2.3 Nanorobots....761
2.4 Nanosolutions 761
2.5 Nanolight Curable Glass Ionomers....762
2.6 Nanoimpression Materials 762
2.7 Nanoencapsulated Materials....762
2.8 Other Nanostructures 763
3 Nanotechnology in Dentistry....763
3.1 Nanodiagnostics....763
3.2 Nanotechnology in Oral and Maxillofacial Surgery 767
3.3 Nanotechnology in Orthodontic Treatment....768
3.4 Nanotechnology in Prosthodontics....768

3.5 Nanotechnology in Periodontics 768
3.6 Nanotechnology in Conservative Dentistry and Endodontics 769
4 Preventive Nanodentistry 770
4.1 Dental Caries 770
4.2 Antibacterial Nanotherapy 770
4.3 Biomimetic Remineralization—Reversing an Incipient Caries 770
4.4 Biomimetic Remineralization—Recurrent Decay 770
4.5 Caries Vaccine 771
4.6 Periodontal Diseases 771
5 Challenges Faced by Nanodentistry 771
6 Conclusions 772
References 772

CHAPTER 25 The Role of Nanomedicine, Nanotechnology, and Nanostructures on Oral Bone Healing, Modeling, and Remodeling 777
Mohamadreza Baghaban-Eslaminejad, Ahmad Oryan, Amir Kamali, Ali Moshiri
1 Introduction 778
2 Nanomedicine 779
3 TERM and Nanomedicine 780
4 Fabrication Methods for Nanomaterials and Nanostructures 781
4.1 Physical and Chemical Synthesis of Nanomaterials 782
4.2 Biological Synthesis of Nanoparticles 782
5 Application of Nanotechnology in Dentistry 788
5.1 Nanomaterials 788
5.2 Nanodiagnostics 805
5.3 Nanorobotics 807
6 Application of Nanomedicine in Oral Bone Regeneration 808
6.1 Nanoceramics 808
6.2 Nanometals 811
6.3 Nanopolymers 811
6.4 Nanocomposites and Nanoscaffolds 811
7 Nanotechnology and Cell Therapy in Oral Bone Regeneration 814
8 Current Challenges and Future Directions 816
References 818

CHAPTER 26 Therapeutic Applications of Nanotechnology in Dentistry 833
Muhammad S. Zafar, Zohaib Khurshid, Shariq Najeeb, Sana Zohaib, Ihtesham Ur Rehman
1 Introduction 833
2 Origin of Nanotechnology Concept and History 835

3 Nature and Timeline of Nanomaterials 835
4 Synthesis, Characteristics, and Classifications of Nanomaterials 836
5 Nanobiomaterials in Dentistry 838
5.1 Nanobiomaterials in Restorative Dentistry 839
5.2 Nanobiomaterials for Preventive Dentistry 847
5.3 Nanobiomaterials for Endodontics 848
5.4 Nanobiomaterials in Oral Implantology 852
6 Conclusive Remarks 855
References 855

CHAPTER 27 Oral Nanomedicine and the Emergent Process of Clinical Translation 863
Gabriel H. Hawthorne, Marcelo P. Bernuci, Mariza Bortolanza, Ana C. Issy, Elaine Del-Bel
1 Introduction 863
2 Understanding Clinical Trials 864
3 Overview of Oral Nanomedicine and Clinical Trials 867
4 Analyzing Selected Trials 868
4.1 Bleaching Efficacy 869
4.2 Desensitizing Efficacy 871
4.3 Dental Implants 873
4.4 Response to Pulpotomy 875
4.5 Alveolar Socket Healing 876
4.6 Gingival Invaginations 876
4.7 Antimicrobial Effect 877
4.8 Ongoing Trials 878
4.9 A Note on Indirect Approaches 879
5 Final Considerations 879
6 Appendix 880
References 880

CHAPTER 28 Carbon-Based Nanostructures for Electrochemical Analysis of Oral Medicines 885
Sevinc Kurbanoglu, Bengi Uslu, Sibel A. Ozkan
1 Introduction 886
2 Oral Drugs 887
2.1 Solids and Semisolid Oral Dosage Forms 887
2.2 Liquid Oral Dosage Forms 891
3 Voltammetric Methods for Electrochemical Analysis of Oral Medicines Using Carbon-Based Nanostructures 892
3.1 Potential Sweep Voltammetry 893

3.2 Alternating Current Voltammetry 895
3.3 Stripping Voltammetry 896
4 Carbon-Based Nanostructures in Electrochemical Analysis of Oral Medicines 898
4.1 The Unique Element Carbon 898
4.2 Graphene and Graphene Oxide 899
4.3 Carbon Nanotubes 901
4.4 Carbon Nanofibers 905
4.5 Carbon Nanohorns 915
4.6 Fullerenes 917
4.7 Carbon Nanodiamond 919
5 Conclusions 923
References 925

CHAPTER 29 Scientometric Overview Regarding Oral Cancer Nanomedicine 939
Ozcan Konur
1 Overview 939
1.1 Issues 939
1.2 Methodology 940
1.3 The Cancer Research in General: Overview 941
1.4 The Oral Cancer Research in General: Overview 943
1.5 The Nanoresearch in General: Overview 944
1.6 The Research on the Oral Cancernanomedicine in General: An Overview 945
2 The Citation Classics in Oral Cancer Nanomedicine 946
2.1 The Citation Classics in Oral Cancer Nanomedicine: Overview 946
2.2 The Citation Classics in Head and Neck Cancer Nanomedicine 951
2.3 The Citation Classics in Oral Cancer Nanomedicine 952
2.4 The Citation Classics in Pharyngeal Cancer Nanomedicine 954
2.5 The Citation Classics in Thyroid and Tracheal Cancer Nanomedicine 955
3 Conclusions 956
References 956

Index 963

List of Contributors

Satish Agrawal
CSIR-Central Drug Research Institute, Lucknow, India

Hafsa Ahmad
CSIR-Central Drug Research Institute, Lucknow, India

Fabiana Alovero
National University of Cordoba (UNC), Córdoba, Argentina

Nicolas Anton
University of Strasbourg, Strasbourg, France

Abhishek Arya
CSIR-Central Drug Research Institute, Lucknow, India

Mohamadreza Baghaban-Eslaminejad
Royan Institute for Stem Cell Biology and Technology, ACECR, Tehran, Iran

Anil Kumar Bajpai
Government Autonomous Science College, Jabalpur, India

R. Baskaran
National Health Research Institutes, Zhunan Taiwan

Marcelo P. Bernuci
University Center of Maringa, Maringá, Brazil

Karolina Bien
Military Institute of Hygiene and Epidemiology, Warsaw, Poland

Prakash S. Bisen
Jaipur National University, Jaipur; Jiwaji University; Tropilite Foods Pvt Ltd., Gwalior, India

Mariza Bortolanza
University of Sao Paulo, São Paulo, Brazil

Virginie Busignies
University of Bordeaux; CNRS, Bordeaux INP, ENSAM, I2M, Talence, France

Grzegorz Celichowski
University of Lodz, Lodz, Poland

Christine Charrueau
UTCBS, CNRS; INSERM; University of Paris Descartes, Sorbonne Paris City;
Chimie ParisTech, PSL Research University, Paris, France

Nagendra S. Chauhan
Government Ayurvedic College Campus, Raipur, India

Raje Chouhan
Government Autonomous Science College, Jabalpur, India

Máximo P. Díaz
Instituto Tecnológico de Querétaro, Santiago de Querétaro, Mexico

Daniele Ribeiro de Araújo
Federal University of ABC, Santo André, Brazil

Nathalie Ferreira Silva de Melo
São Leopoldo Mandic Institute and Research Center, Campinas, Brazil

Lígia Nunes de Morais Ribeiro
University of Campinas—UNICAMP, Campinas, Brazil

Eneida de Paula
University of Campinas—UNICAMP, Campinas, Brazil

Elaine Del-Bel
University of Sao Paulo, São Paulo, Brazil

Meenal Dixit
Jaipur National University, Jaipur, India

Yuan-Cai Dong
Institute of Chemical and Engineering Sciences, A*STAR, Jurong Island, Singapore

Anil K. Dwivedi
CSIR-Central Drug Research Institute, Lucknow, India

Aylin T. Ermertcan
Manisa Celal Bayar University, Manisa, Turkey

Virginie Escriou
UTCBS, CNRS; INSERM, University of Paris Descartes, Sorbonne Paris City; Chimie ParisTech, PSL Research University, Paris, France

Görkem Eskiizmir
Manisa Celal Bayar University, Manisa, Turkey

Ana Figueiras
REQUIMTE/LAQV, Group of Pharmaceutical Technology, Faculty of Pharmacy, University of Coimbra, Coimbra; Health Sciences Centre, Faculty of Health Sciences, University of Beira Interior, Covilhã, Portugal

Michelle Franz-Montan
University of Campinas—UNICAMP, Piracicaba, Brazil

Kunihiko Fukuchi
Showa University, Tokyo, Japan

Toshiyuki Fukuda
Satoen Co., Ltd., Shizuoka, Japan

Shilpi Goswami
Government Autonomous Science College, Jabalpur, India

Avinash Gothwal
Central University of Rajasthan, Ajmer, India

Jaroslaw Grobelny
University of Lodz, Lodz, Poland

Lokesh Gupta
Central University of Rajasthan, Ajmer, India

Madhu Gupta
Shri Rawatpura Sarkar Institute of Pharmacy, Datia, India

Umesh Gupta
Central University of Rajasthan, Ajmer, India

Rosa Martha Perez Gutierrez
National Polytechnic Institute, Mexico City, Mexico

Gabriel H. Hawthorne
University Center of Maringa, Maringá, Brazil

Kang Moo Huh
Chungnam National University, Daejeon, Republic of Korea

Ana C. Issy
University of Sao Paulo, São Paulo, Brazil

Indrani Jadhav
Jaipur National University, Jaipur, India

Sougata Jana
Gupta College of Technological Sciences, Asansol, India

Alvaro F. Jimenez-Kairuz
National University of Cordoba (UNC), Córdoba, Argentina

Namrata A. Kadwadkar
Institute of Chemical Technology, Mumbai, India

Amir Kamali
Shiraz University, Shiraz, Iran

Taisei Kanamoto
St. Marianna University School of Medicine, Kanagawa, Japan

Sayali Karandikar
Bharati Vidyapeeth's College of Pharmacy, Navi-Mumbai, India

Iliyas Khan
Central University of Rajasthan, Ajmer, India

Prachi B. Kharkar
Institute of Chemical Technology, Mumbai India

Zohaib Khurshid
King Faisal University, Al-Hofuf, Saudi Arabia

Ozcan Konur
Yildirim Beyazit University, Ankara, Turkey

Kayoko Kotohda
Cascade Resources Laboratory Co., Ltd., Nagano, Japan

Malgorzata Krzyzowska
Military Institute of Hygiene and Epidemiology, Warsaw, Poland

Niha M. Kulshreshtha
Jaipur National University, Jaipur, India

Sevinc Kurbanoglu
Ankara University, Ankara, Turkey

Yong-Kyu Lee
Korea National University of Transportation, Chungju, Republic of Korea

Kumaran Letchmanan
Institute of Chemical and Engineering Sciences, A*STAR, Jurong Island, Singapore

Sabyasachi Maiti
Gupta College of Technological Sciences, Asansol, India

Ruben H. Manzo
National University of Cordoba (UNC), Córdoba, Argentina

Alejandro R. Martínez
Instituto Tecnológico de Querétaro, Santiago de Querétaro, Mexico

Juan Vicente Mendez Mendez
National Polytechnic Institute, Mexico City, Mexico

Jorge E. Miranda Calderon
Laboratory of Molecular Pharmacy and Controlled Release, México D.F., Mexico

Amit Mirani
Institute of Chemical Technology, Mumbai, India

Ali Moshiri
Iran University of Medical Sciences, Tehran, Iran

Yuvashree Muralidaran
VIT University, Vellore, India

Shariq Najeeb
Al-Farabi Colleges, Riyadh, Saudi Arabia

Hideki Nakashima
St. Marianna University School of Medicine, Kanagawa, Japan

Wai Kiong Ng
Institute of Chemical and Engineering Sciences, A*STAR, Jurong Island, Singapore

Thi Trinh Lan Nguyen
University of Strasbourg, Strasbourg, France

Norma A. Noguez Méndez
Laboratory of Molecular Pharmacy and Controlled Release,
México D.F., Mexico

Md Nurunnabi
Chungnam National University, Daejeon; Korea National University of Transportation, Chungju, Republic of Korea

Hirokazu Ohno
Maruzen Pharmaceuticals Co. Ltd., Tokyo, Japan

Takaaki Oizumi
Daiwa Biological Research Institute Co., Ltd., Kanagawa, Japan

María E. Olivera
National University of Cordoba (UNC), Córdoba, Argentina

Piotr Orlowski
Military Institute of Hygiene and Epidemiology, Warsaw, Poland

Ahmad Oryan
Shiraz University, Shiraz, Iran

Sibel A. Ozkan
Ankara University, Ankara, Turkey

V. Vijaya Padma
Bharathiar University, Coimbatore, India; China Medical University; Asia University, Taichung, Taiwan

Xochitl C. Palomec
Laboratory of Molecular Pharmacy and Controlled Release, México D.F., Mexico

Sandeep Patankar
Bharati Vidyapeeth's College of Pharmacy, Navi-Mumbai, India

Vandana B. Patravale
Institute of Chemical Technology, Mumbai, India

L. Bharathi Priya
Bharathiar University, Coimbatore, India

Carlos T. Quirino Barreda
Laboratory of Molecular Pharmacy and Controlled Release, México D.F., Mexico

Gokulakannan Ragavan
VIT University, Vellore, India

María V. Ramírez-Rigo
National University of the South, Bahía Blanca, Argentina

Katarzyna Ranoszek-Soliwoda
University of Lodz, Lodz, Poland

Ihtesham Ur Rehman
The University of Sheffield, Sheffield, United Kingdom

Vishnu Revuri
Korea National University of Transportation, Chungju, Republic of Korea

Andreza Maria Ribeiro
Faculty of Pharmacy, University of Coimbra, Coimbra, Portugal; Department of Engineering and Material Sciences, University of Federal of Paraná (UFPR), Curitiba, Brazil

Hiroshi Sakagami
Meikai University School of Dentistry, Saitama, Japan

Mohsin Shah
Khyber Medical University, Peshawar, Pakistan

Ashok K. Sharma
Central University of Rajasthan, Ajmer, India

Vikas Sharma
Shri Rawatpura Sarkar Institute of Pharmacy, Datia, India

Shou-Cang Shen
Institute of Chemical and Engineering Sciences, A*STAR, Jurong Island, Singapore

Hong Sheng
Meikai University School of Dentistry, Saitama, Japan

Divya Shrivastava
Jaipur National University, Jaipur, India

Anurag K. Singh
Central University of Rajasthan, Ajmer, India

Nehi Sinha
Jaipur National University, Jaipur, India

Swapnil S. Talkar
Institute of Chemical Technology, Mumbai, India

Pierre Tchoreloff
University of Bordeaux; CNRS, Bordeaux INP, ENSAM, I2M, Talence, France

Shigemi Terakubo
St. Marianna University School of Medicine, Kanagawa, Japan

Emilia Tomaszewska
University of Lodz, Lodz, Poland

César G. Urioste
Laboratory of Molecular Pharmacy and Controlled Release, México D.F., Mexico

Bengi Uslu
Ankara University, Ankara, Turkey

Thierry F. Vandamme
University of Strasbourg, Strasbourg, France

Israel Arzate Vazquez
National Polytechnic Institute, Mexico City, Mexico

Abraham F. Vega
Laboratory of Molecular Pharmacy and Controlled Release, México D.F., Mexico

Francisco Veiga
REQUIMTE/LAQV, Group of Pharmaceutical Technology, Faculty of Pharmacy, University of Coimbra, Coimbra, Portugal

Pragasam Viswanathan
VIT University, Vellore, India

Vishal Waybhase
Bharati Vidyapeeth's College of Pharmacy, Navi-Mumbai, India

Jingqi Xie
Huaqiao University, Quanzhou; Engineering Research Center of Molecular Medicine, Ministry of Education; Key Laboratory of Molecular Medicine in Fujian Province, Xiamen, China

Ruian Xu
Huaqiao University, Quanzhou; Engineering Research Center of Molecular Medicine, Ministry of Education; Key Laboratory of Molecular Medicine in Fujian Province, Xiamen, China

Masaji Yamamoto
Maruzen Pharmaceuticals Co. Ltd., Tokyo, Japan

Kerim Yapici
Cumhuriyet University, Sivas, Turkey

Toshikazu Yasui
Meikai University School of Dentistry, Saitama, Japan

Sung C. Yoon
Gyeongsang National University, Jinju; ReSEAT Program, KISTI, Daejeon, Republic of Korea

Hidenobu Yoshida
Cascade Resources Laboratory Co., Ltd., Nagano, Japan

Muhammad S. Zafar
Taibah University, Al Madinah, Saudi Arabia; Riphah International University, Islamabad, Pakistan

Sana Zohaib
King Faisal University, Al-Hofuf, Saudi Arabia

Foreword of the Series

Material science and engineering at the nanoscale have brought revolutionary advances to the biomedical sciences, overturning many of the known traditional approaches. Nanotechnology has driven many of the most successful innovative technologies, and their impressive record of accomplishment has made nanostructures promising candidates for future therapy. The advantages that nanomaterials have already provided to therapeutics, such as targeted and controlled delivery, wide accessibility, high specificity, low side effects, improved efficiency, and impressive versatility are currently considered key elements in designing personalized medicine approaches for prophylaxis, diagnosis, and therapy.

Therapeutic nanostructures can be highly diverse, and their unique properties have led to the development of highly specialized biosensors, more efficient drug delivery vehicles, and controlled release targeting systems to fight severe or incurable diseases, such as cancer, infections, and cardiovascular disease.

In view of the astounding progress made in the field of therapeutic nanotechnology, and its rapidly progressing expansion, this book aims to collect together in one place all the most recent and innovative aspects regarding the impact of nanomaterials in both current and future therapy. The book is organized in five volumes, covering the main areas that are relevant for the design and implementation of nanostructures in medical therapies.

The first volume, *Nanostructures for Novel Therapy: Synthesis, Characterization, and Applications* describes methods to obtain and characterize nanosystems, emphasizing their biomedical applications. Special attention in this volume was paid to modern synthesis methods to reduce side effects and limit the toxicity of nanomaterials in biomedical applications. Numerous examples of nanostructures designed for therapy, as well as the most efficient synthesis and characterization routes for these materials are clearly described and critically analyzed.

The second volume, entitled *Nanostructures for Drug Delivery* covers one of the most widely utilized and investigated applications of nanomaterials in the biomedical field, namely drug delivery. The design of nanoscale agents in order to specifically and safely carry therapeutic agents to their final destination is an intriguing approach to future targeted therapeutics. This approach could provide a treatment for previously incurable diseases, as well as reducing the side effects of current drugs. Many highly active drugs are severely limited by side effects related to their unspecific sites of action. This book introduces the readers to the amazing field of nanomedicine by discussing the versatility and variety of nanovehicles for drug delivery and targeting. Moreover, readers will find numerous examples, and will learn about the currently used or investigational drug delivery agents for therapy, prophylaxis, and diagnosis.

The third volume, *Nanostructures for Antimicrobial Therapy* highlights the impressive progress made by nanotechnology in the design of novel antimicrobial approaches. Since microbial resistance to antibiotics is a very real and worrying issue growing throughout the world, the development of more efficient antimicrobial agents has a high priority to allow control of infections in the future. Antimicrobial nanosystems have proved to be highly efficient against drug-resistant microorganisms, are able to fight biofilm-associated infections and control the social behavior of microbial communities. Nanostructures can also reduce microbial virulence factors and reduce pathogenesis mechanisms thus offering a promising alternative for future therapy.

The fourth volume, entitled *Nanostructures for Cancer Therapy* covers the applications of nanomedicine in cancer diagnosis and treatment. The use of nanoparticles for cancer therapy is not in itself a new approach, but numerous advances have been recently made in this area, and the aim of this book is to cover the most interesting new approaches in the management of this deadly disease. Nanosized drugs are currently believed to represent the most efficient approach in cancer chemotherapy, and this volume provides coverage of the latest novel findings, while also discussing possible improvements in more established types of nanosystems to increase the efficiency of cancer therapy.

Last but not least, the fifth volume of this series entitled *Nanostructures for Oral Medicine* covers the progress made in applications of nanotechnology in treating various diseases of the oral cavity and in dentistry. Readers will have the chance to learn about the most efficient modern materials used to treat or to prevent widely encountered oral diseases, such as gingivitis, periodontitis, caries, and dental plaque. Moreover restorative dentistry also now makes wide use of nanomaterials.

Overall, this book series will provide a state-of-the-art compendium of knowledge, and a crystal ball to see into the future of biomedical nanotechology and nanomedicine. It will appeal to researchers, clinicians, engineers, pharmacologists, pharmacists, oncologists, infectious disease experts, and dentists. More many interested general readers may discover the impact, current progress and future applications of nanotechnology in therapeutics and diagnosis. Taken together, nanoscale approaches will improve the efficiency of personalized medicine for better management of diseases in the 21st century.

Michael R. Hamblin

Wellman Center for Photomedicine, Massachusetts General Hospital, Boston, MA, United States;
Harvard Medical School, Boston, MA, United States;
Harvard-MIT Division of Health Sciences and Technology, Cambridge, United States

Preface

ABOUT THE SERIES (VOLUMES I–V)

In our permanently changing world, novel therapeutics are constantly required to manage health and well-being of population. Although numerous diseases are currently considered incurable, massive progress made in biomedicine but also associated fields, such as chemistry, physics, engineering, pharmacology, and materials science, offers a new light to the therapeutics domain. In this context, most physicists and researchers believe that a personalized and adequate treatment may significantly improve the outcome of severe diseases and ensure a faster healing. Nanotechnology offers great perspectives for personalized medicine, since nanostructured therapeutics proved their efficiency and amazing impact in improving therapy, prophylaxis, and diagnosis. The emerging field of nanosized materials have numerous applications in the biomedical field, especially in therapy. This series of five volumes came out by the need of learning about recent progress of the science of nanostructured materials in order to improve current therapy and lead to the next level. The books offer an interesting and updated perspective regarding applications of nanomaterials in therapy of most investigated and difficult to treat diseases, such as cancer and severe infections. The presentation approach of each chapter contained in those five volumes is clear and easy to understand by most readers and for biomedical specialists, researchers, and engineers. The series is organized in an attractive manner for students and academics on the field, starting with a volume dealing with synthesis, characterization, and main applications of nanostructures, emphasizing on their impact in therapy. Next volume reveals the most recent progress made on a very investigated field, considered a key element in personalized medicine and future therapy, namely nanostructured drug delivery systems. Their impact in antimicrobial therapy is also widely discussed and suggestive examples are given and explained. Moreover, a whole volume is dedicated to the management of the disease of the century—cancer—revealing the huge value added by the utilization of nanosystems in the therapy of this deadly disease. Important aspects related to improved diagnosis and prophylaxis are highlighted. In the last volume, the progress and novel applications of nanotechnology in oral medicine are dissected. The field of oral diseases represents a wide-interest and priority field since both physicists and researchers believe that they can be prevented and treated more easily with targeting systems and nanofunctional prosthetics. All chapters are clearly illustrated to highlight most important or difficult to understand aspects and suggestive examples are often enumerated in organized tables, which are explained and discussed. Overall, the series contain very recent but accessible information regarding the progress of nanostructures in therapeutics and give a novel perspective about future therapy of severe diseases.

Alexandru Mihai Grumezescu
University Politehnica of Bucharest, Bucharest, Romania

ABOUT VOLUME V

The fifth volume of the series *Therapeutic Nanostructures* is entitled *Nanostructures for Oral Medicine*. This volume reveals numerous applications of nanostructures in oral medicine. Various key approaches, such as drug design and delivery, oral vaccines, and therapeutic systems for oral applications, are discussed and exemplified in this volume. Oral medicine represents a widely investigated field, where nanotechnology has found numerous applications. From improved implantable materials to specialized drug-delivery systems, nanomaterials have offered solutions to serious clinical conditions from diagnosis to therapy. The book describes not only the preparation of oral nanomedicines but also their main properties, applications and principal advantages and drawbacks. Applications and impact on human life quality of recently developed nanomaterials aimed at oral therapy are also presented here. Similar to general medicine, the trend in oral medicine seems to follow technological progress, the specificity and sensibility of nanostructured drugs playing an important role in the design of tailored approaches for oral therapy. Aspects related to oral cancers, oral drug delivery, personalized therapy, regenerative local medicine, and disease-preventing oral nanoagents are widely dissected in this book.

Volume V contains 29 chapters, prepared by outstanding international researchers from Turkey, India, Pakistan, South Korea, Brazil, Portugal, Iran, China, Poland, France, Argentina, Japan, Taiwan, Mexico, Singapore, Iran, Saudi Arabia, and the United Kingdom.

In Chapter 1, entitled *Bacterial Polyester Nanoparticles for Drug Delivery*, Mohsin Shah and Sung Chul Yoon describe the methods of preparation and characterization of drug-encapsulated polymeric nanoparticles formulated with various copolymers, drug release kinetics, and evaluation of their localization in cancer cells in vitro and in vivo.

Rosa Martha Perez Gutierrez et al., in Chapter 2, entitled *A Novel Approach to the Oral Delivery: Nanostructures and Biomaterials for Systemic Disease*, shed light on oral delivery nanostructures researched in medicine, including synthesis techniques and materials that can be effectively used for controlled oral drug delivery application used in disease treatments.

Chapter 3, prepared by Andreza Maria Ribeiro et al., entitled *Biodegradable Polymeric Nanostructures: Design and Advances in Oral Drug Delivery for Neurodegenerative Disorders*, highlights the main advances in polymeric nanocarriers as vehicles for oral delivery systems used in the therapy of neurodegenerative disorders. First, a brief introduction into the field of biodegradable polymeric materials used in nanostructures formulation is presented. Considering their structure and characteristics, the next section offers a detailed description of the technologies and recent modifications to improve their specificity. Finally, new approaches are presented with clinical applications in neurodegenerative disorders.

Chapter 4, entitled *Nanostructured Systems for Transbuccal Drug Delivery*, prepared by Michelle Franz-Montan et al., outlines the use of nanostructures as tools for the development of new formulations for topical use at the oral mucosa in dental clinical practice. Topics covered in this chapter include the structure, organization, and permeability of oral mucosa; permeation pathways at oral mucosa; chemical permeation enhancers for oral mucosa topical formulations; mechanisms of permeation enhancers; and the use of the following nanostructures to improve transbuccal drug delivery for the treatment of oral mucosa infections, inflammation, and pain conditions: liposomes, polymeric nanoparticles/micelles, nanostructured lipid carriers, and cyclodextrins.

Chapter 5, entitled *Trends in Orally Viral Vector Gene Delivery and Therapy*, prepared by Jingqi Xie and Ruian Xu, highlights the merits and disadvantages of novel applications and their relationship between sizes and structures of nanoparticles and the efficacy of oral gene delivery and therapy. In addition, this chapter reveals orally viral vector gene delivery timing, and systematically discusses the effect of age, delivery route, cell preference, organ distribution, and so on on orally viral vector delivery and trends in oral viral vector gene delivery and therapy.

Virginie Busignies et al., in Chapter 6, entitled *Nanostructures for Oral Delivery of Therapeutic Nucleic Acids*, discuss the barriers to oral delivery of nucleic acids and review the various technologies already tested in preclinical study and the experimental models available to assay oral delivery systems.

Pragasam Viswanathan et al., in Chapter 7, entitled *Challenges in Oral Drug Delivery: A Nano-Based Strategy to Overcome*, offer an overview about the strategy to override such drawbacks by ushering in various nanotechnology platforms, such as polymer-based nanocarriers, lipid-based nanocarriers, and metal- and inorganic-based nanoparticles.

Thi Trinh Lan Nguyen et al., in Chapter 8, *Oral Pellets Loaded With Nanoemulsions*, present an up-to-date review about nanoemulsion and its subsequent transformation into pellet systems that offer several advantages for improving solubility and dissolution rates of poorly water-soluble drugs for oral delivery in order to overcome the poor absorption of some active pharmaceutical ingredients.

Anurag Kumar Singh et al., in Chapter 9, entitled *Oral Drug Delivery Potential of Dendrimers*, focus on the oral potential applicability of dendrimers for delivering various categories of drugs, such as antimicrobial agents, antifungal drugs, and certain catheter-related substances.

Chapter 10, prepared by Sayali Karandikar et al., entitled *Nanovaccines for Oral Delivery-Formulation Strategies and Challenges*, summarizes nanoformulation strategies for oral delivery of vaccines, which can potentially help to overcome existing problems. Also, this chapter briefly presents their challenges and regulatory aspects.

Chapter 11, prepared by Prachi B. Kharkar et al., entitled *Nanosystems for Oral Delivery of Immunomodulators*, gives an insight into nanostructurated immunomodulators, their routes of administration, and newer sophisticated methods of their formulation for better therapeutic effect.

Chapter 12, entitled *Tannic Acid Modification of Metal Nanoparticles: Possibility for New Antiviral Applications*, prepared by Malgorzata Krzyzowska et al., presents an up-to-date overview about tannic-acid-modified silver nanoparticles consisting of a novel potential microbicide for herpes virus infection in the mucosal tissues.

Chapter 13, *Polyelectrolyte-Drug Ionic Complexes as Nanostructured Drug Carriers to Design Solid and Liquid Oral Delivery Systems*, prepared by Maria E. Olivera et al., deals with the description of main pharmaceutical applications of drug-delivery systems based on the unique properties of complexes of polyelectrolytes with ionizable drugs.

Chapter 14, prepared by Md Nurunnabi et al., entitled *Polysaccharide Based Nano/Micro Formulation: An Effective and Versatile Oral Drug Delivery System*, focuses on potential applications of polysaccharides for oral drug delivery, their biocompatibility, as well as their advantages and disadvantages over conventional strategies. The authors also highlight recent advances in oral drug delivery based on specific polysaccharides, including chitosan, heparin, hyaluronic acid, beta glucan, and chondroitin sulfate.

Raje Chouhan et al., in Chapter 15, entitled *Recent Advancements in Oral Delivery of Insulin: From Challenges to Solutions*, present an overview of diabetes highlighting various aspects of insulin therapy, including insulin analogs, oral delivery of insulin, mechanistic action of insulin, challenges of oral administration of insulin, and breakthroughs in insulin delivery at the global level. The chapter also discusses economical aspects related to the commercialization level of insulin.

Madhu Gupta et al., in Chapter 16, *Nanotechnology for Oral Delivery of Anticancer Drugs: An Insight Potential*, review the challenges encountered in oral delivery with various nanoconstructs employed for oral anticancer drug delivery comprehensively as well as their applications.

Chapter 17, entitled *Nanomaterials: Promising Structures for the Management of Oral Cancer*, prepared by Görkem Eskiizmir et al., presents the role and potential effectiveness of using nanomaterials in different aspects of oral cancer. Recently, nanotechnology has offered new strategies in cancer management: (1) early detection of cancers via specialized probes and biosensors, (2) better diagnosis and staging via improving the imaging systems, (3) nanoscale drug-delivery systems that may provide a significant decrease in adverse effects and toxicities of chemotherapeutics, and (4) nanotheranostics that integrate both diagnosis and therapy simultaneously.

Hiroshi Sakagami et al., in Chapter 18, entitled *Therapeutic Potential of Solubilized Nanolignin Against Oral Diseases*, offer relevant information about the lignin-carbohydrate complex and its excellent anti-HIV activity and anti-UV activity, in contrast to its weak tumor-selective cytotoxicity.

Chapter 19, prepared by Hafsa Ahmad et al., *Novel Lipid Nanostructures for Delivery of Natural Agents With Antioxidant, Antiinflammatory, and Antistroke Potential: Perspectives and Outcomes*, presents many bioactive nanocombinations with antioxidant, anti-inflammatory, and neuroprotective effects that might promote alleviation of neuronal and tissue damage following stroke and reverse pathophysiological changes in ischemia-reperfusion injury that fail clinically due to degradation, poor aqueous solubility, and subsequent bioavailability.

Sougata Jana and Sabyashachi Maiti, in Chapter 20, entitled *Chitosan-Based Nanoparticulate Systems for Oral Drug Delivery*, discuss various methods of preparation of chitosan-based nanoparticles, structural modifications of chitosan, and the latest developments in the field of oral drug delivery.

Chapter 21, prepared by Lohanathan Bharathi Priya and Viswanadha Vijaya Padma, entitled *Phytonanoconjugates in Oral Medicine*, focuses on the role of nanostructures conjugated to phytochemicals (phytonanoconjugates) in oral medicine by extensively reviewing of the latest research findings.

Chapter 22, entitled *Design and Development of Pharmaceutical Microprocesses in the Production of Nanomedicine*, prepared by Norma Angélica Noguez Méndez et al., presents an up-to-date overview about the microfluidics and examples of nano- and microparticles involved in this field: liquid crystals (LCs) and solid lipid nanoparticles (SLNs). Also, the authors present details about synthetic polymersomes that imitate biological membrane functions.

Chapter 23, prepared by Shou-Cang Shen et al., entitled *Mesoporous Materials and Technologies for Development of Oral Medicine*, presents recent progress in the development of mesoporous materials as drug carriers for applications in oral medicine formulation according to the types of mesoporous excipients, drug loading techniques, characterization and stability studies, and applications in oral medicine.

Nehi Sinha et al., in Chapter 24, entitled *Nanodentistry: Novel Approaches*, focus on the role of nanostructures to keep teeth and oral tissues healthy and functioning. The challenges faced by this technology for the betterment of human health are also discussed.

Chapter 25, entitled *The Role of Nanomedicine, Nanotechnology, and Nanostructures on Oral Bone Healing, Modeling, and Remodeling*, prepared by Mohamadreza Baghaban-Eslaminejad et al., covers the most important information regarding manufacturing methods, available and future options, and the role of nanostructures on bone health, healing, and repair based on their basic to clinical evidence.

Muhammad Sohail Zafar et al., in Chapter 26, entitled *Therapeutic Applications of Nanotechnology in Dentistry*, highlight recent developments regarding therapeutic applications of nanomaterials in dentistry. In addition, chemistry, synthesis, properties, and benefits of therapeutic nanomaterials over conventional materials in relation to dentistry are discussed.

Chapter 27, prepared by Gabriel Henrique Hawthorne et al., *Oral Nanomedicine and the Emergent Process of Clinical Translation*, discusses the translational research from laboratory approaches to clinical applications inside oral nanomedicine. Analyzing the clinical trials conducted in a field is the only way of predicting the potential for clinical impact of the new approaches, and oral nanomedicine stands in a promising position.

Chapter 28, prepared by Sevinc Kurbanoglu et al., entitled *Carbon-Based Nanostructures for Electrochemical Analysis of Oral Medicines*, gives an up-to-date overview about electrochemical assay of oral drugs using carbon-based nanostructures such as fullerene polymers, carbon nanotubes, carbon nanohorns, carbon nanodiamonds, and grapheme. Also, advantages and disadvantages of producing techniques are detailed.

Chapter 29, entitled *Scientometric Overview Regarding Oral Cancer Nanomedicine*, by Ozcan Konur, highlights important papers influencing the development of the research field as well as in determining the key research fronts in the field of cancer nanomedicine. This book chapter presents the first bibliometric study focusing on the 25 citation classics in oral cancer nanomedicine following a brief bibliometric overview of the research in these underlying research areas.

Ecaterina Andronescu
University Politehnica of Bucharest, Bucharest, Romania

Alexandru Mihai Grumezescu
University Politehnica of Bucharest, Bucharest, Romania

CHAPTER

BACTERIAL POLYESTER NANOPARTICLES FOR DRUG DELIVERY

1

Mohsin Shah*, Sung C. Yoon,†**

**Khyber Medical University, Peshawar, Pakistan*

***Gyeongsang National University, Jinju, Republic of Korea*

†ReSEAT Program, KISTI, Daejeon, Republic of Korea

CHAPTER OUTLINE

1 Introduction 1
2 Amphiphlic Block Copolymers 2
3 Polymeric Nanoparticles in Drug-Delivery Applications 3
4 Biodegradable Bacterial Polyesters Poly(R)-Hydroxyalkanoic Acids 4
4.1 Properties of PHA 5
4.2 Biodegradation of PHA 5
4.3 General Applications of PHA 5
4.4 PHA as Drug Carrier 7
5 Targeting Drug Delivery using Ligand-Conjugated PHA Nanoparticles (Active Targeting) 9
6 Chemical Modification of PHA Polymers for Drug Delivery (Passive Targeting) 10
7 Synthesis of Surface Modified PHB-mPEG Diblock Copolymer 11
8 Self-Assembly of Diblock Copolymer Into Nanoparticles 12
9 Characterization of PHA–mPEG Nanoparticles 13
9.1 Core–Shell Topology in Aqueous Solution 13
9.2 Morphological Characteristics of PHA–mPEG Nanoparticles 15
9.3 Drug Loading in PHA–mPEG Nanoparticles 16
9.4 Drug Release from PHA–mPEG Nanoparticles 17
9.5 Cellular Uptake of PHA–mPEG Nanoparticles 18
10 Drawbacks and Challenges 19
11 Conclusions and Future Perspectives 20
References 20

1 INTRODUCTION

Over the past few decades, there has been considerable progress in developing biodegradable nanoparticles as effective drug delivery devices (Jendrossek and Handrick, 2002) Different types of biodegradable polymers both synthetic and natural have been utilized in the preparation of nanoparticles

(Anderson and Dawes, 1990; Doi, 1990). Nanoparticles are submicron-sized polymeric colloidal particles with a size range 1–1000 nm and drugs may be encapsulated, adsorbed, or dispersed in them (Jendrossek et al., 1996; Jendrossek and Handrick, 2002; Verlinden et al., 2007). In the preparation of particles for drug-delivery systems, biodegradable materials are preferably used in order to prevent the side effects associated with nonbiodegradable polymers. The products result after the degradation of the biopolyesters that must be biocompatible and nontoxic. In general, nanoparticles prepared from polymers can be divided into three different categories, depending on their structure. Capsules are the first category. Nanocapsules are submicroscopic colloidal drug carrier systems composed of an oily or an aqueous core surrounded by a thin polymer membrane (Madison and Huisman, 1999). Another example of a vesicular system is the liposomes. In an aqueous environment, phospholipid molecules form a lipid bilayer, resulting in the formation of liposomes. Several types of liposomes can be prepared, including unilaminar, multilaminar, and multivesicular (Sun et al., 2007). The disadvantages of liposomes are the unreliable reproducibility of liposomes and exchange of phospholipids with certain blood components (Sudesh and Doi, 2000). The second category is polymeric micelles. Polymeric nanoparticles are nanosized, supramolecular core/shell structures formed by self-aggregation of individual amphiphilic macromolecules comprised of inner concealed hydrophobic and outer exposed hydrophilic domains (Jendrossek and Handrick, 2002; Jendrossek et al., 1996; Verlinden et al., 2007). This core–shell structure enables the solubilization of hydrophobic drugs in the core hydrophobic domain (Kumagai and Doi, 1993). The micelles are formed spontaneously above a certain copolymer concentration: the critical micelle concentration (CMC) (Kim et al., 1999; Kumagai and Doi, 1993). The third category of nanoparticles is particles that consist of a more or less homogeneous polymeric matrix. Analogous to microspheres, these particles are generally referred to as nanospheres. These particles are larger than micelles and may be more polydispersed in terms of size (Pötter and Steinbüchel, 2005).

2 AMPHIPHLIC BLOCK COPOLYMERS

Amphiphilic copolymers can be synthesized by introducing hydrophilic groups, such as hydroxyl, carboxyl, amine, glycol, and hydrophilic polymers, such as polyethylene glycol (PEG), poly(vinyl alcohol), poly(acryl amide), poly(acrylic acid), poly(hydroxyethyl methacrylate), poly(vinyl pyridine), and poly(vinyl pyrrolidone) to a hydrophobic central moiety (Mergaert et al., 1994). In the amphiphilic block copolymers, the hydrophilic part is covalently linked with strongly hydrophobic polymers. On exposing to the aqueous environments, the hydrophobic core of the polymer is buried inside with exposed hydrophilic domains (Klingbeil et al., 1996). In such amphiphilic structure of polymeric micelles, hydrophobic drug molecules are solubilized within the hydrophobic core, which acts as a cargo for loading high amounts of drugs, whereas the shell maintains a hydration barrier that protects the integrity of each micelle (Kumagai and Doi, 1993). Moreover, the hydrophilic moiety used as surface coating also protect the nanoparticles from the induction of immune response. The use of biodegradable amphiphilic nanoparticles as drug carriers is advantageous because it releases the drug via two steps; first, diffusion release of the mPEG segment, and second, degradation release of the biodegradable polymer segment (Jendrossek et al., 1995). Several studies have been performed over the past few years with micelle-forming block copolymers. Most of these have been conducted on loading of hydrophobic drugs in the amphiphilic copolymers with core–shell morphology (Bhatt et al., 2008; Cui et al., 2002; Kim et al., 1999). Various hydrophobic drugs, including griseofulvin (Cui et al., 2002), doxorubicin (DOX) (Preusting et al., 1993; Wong et al., 2002), and amphotericin B (Martin and Williams 2003) have been successfully loaded into the micelle compartments of these copolymers. Some of these have been used in clinical trials and the in vitro and in vivo efficacies of these micelles have been evaluated.

3 POLYMERIC NANOPARTICLES IN DRUG-DELIVERY APPLICATIONS

With the development of new materials and a combination of nanotechnology and biotechnology it could be possible to make artificial organs and implants through cell growth, which could repair damaged nerve cells, replace damaged skin, tissue, or bone (Williams et al., 1999). Another application of bionanotechnology in medicine is drug delivery where research is especially intensive on the possibility of manipulating nanoparticles to deliver drugs because nanoparticles can have a better solubility and absorption potential than bigger particles (Verlinden et al., 2007). The nanoparticles can carry the drug and perhaps release it in fine-tuned doses over an extended time period to a targeted area, reducing the side effects of the traditional drugs.

Polymeric nanoparticles in pharmaceutical applications have gained plenty of research attention during recent decades (Zhang et al., 1997). Although, the research concerning formulation of polymer based nanoparticles into drug delivery devices has been extensive, only a few polymeric nanoparticulate products have reached the market. One known product, Abraxane, consist of intravenously administered 130-nm nanoparticles prepared from the protein albumin bound with paclitaxel, a drug used in cancer therapy (Zhang et al., 1995). Another cancer drug, DOX transdrug, consisting of DOX-loaded poly(isohexylcyanoacrylate) nanoparticles is currently at the Phase II/III clinical trials (Xiong et al., 2010). In particular among the drugs used in nanoparticle formulations, cancer therapeutics are widely studied because their formulation might reduce toxicity of the drug while improving efficacy of the treatment (Mei et al., 2006). In addition to drug molecules, other candidates to be encapsulated in or coupled with nanoparticles include macromolecules like proteins, peptides, and genes (nucleic acids) (Zhao et al., 2003). These kinds of molecules tend to be inactivated in the body by enzymatic degradation. In terms of controlled release, nanoparticles provide protection against the body conditions resulting in sustained release and maintenance of bioactivity before the drug reaches the target. The most effective strategy adopted by these nanoparticulated drug carriers to selectively target the solid tumors is the exploitation of anatomical and pathophysiological abnormalities of the tumor vasculature, utilizing the principle of enhanced permeability and retention (EPR) effect (Williams et al., 1999). The difference between normal inflammatory tissues and tumor tissues is reflected in their clearance velocities. Usually the macromolecules like nanoparticles, nanocapsules; lipid nanoparticles delivered into the interstitial space of normal inflammatory tissue is cleared more rapidly. On the other hand tumor tissues have leaky vessels and poor aligned lymphatic vessels as results the macromolecules leak out more easily into the tumor tissues and retain longer due to their defective lymphatic system. This phenomenon is called as EPR effect as explained by Maeda et al. (2001). The size and surface of the nanoparticulate carrier plays a very crucial role with respect to uptake by the tumor. Particles <200 nm in size with hydrophilic surfaces tend to exhibit improved EPR effect, which has been attributed to the increased residence time of the carrier in blood (Williams et al., 1999). Moreover, this increased plasma half-life of particulate drug carrier, as well as accumulation in tumor is strictly correlated to the molecular weight of the macromolecule or polymer used for making the nanoparticles long circulating.

The benefits of drugs loaded in the polymeric nanoparticles include protection of the encapsulated pharmaceutical substance, improved efficacy, fewer adverse effects, controlled release and drug targeting to tumor tissues by the EPR effect resulting from defective tumor vascular architecture and impaired lymphatic drainage. Polymeric micelles provide a "shielding effect" in intracellular drug delivery, wherein drugs remain inside micelles upon endocytosis.

4 BIODEGRADABLE BACTERIAL POLYESTERS POLY(R)-HYDROXYALKANOIC ACIDS

Poly(R)-hydroxyalkanoic acids (PHA) are a group of storage compounds of carbon and energy that are accumulated during unbalanced growth by many bacteria, that is, in the presence of an excess of a carbon source and if growth is limited by another nutrient, such as nitrogen (Jendrossek and Handrick, 2002; Jendrossek et al., 1996; Verlinden et al., 2007). When the supply of limiting nutrient is restored, the PHA can be depolymerized and subsequently metabolized as carbon and energy sources (Anderson and Dawes, 1990; Doi, 1990). PHA in bacterial cells is deposited intracellularly in the form of inclusion bodies (granules) and it forms up to 90% of the cellular dry weight (Anderson and Dawes, 1990; Doi, 1990; Jendrossek et al., 1996) as shown in Fig. 1.1A. The many different types of PHA isolated to date are primarily linear; head-to-tail polyesters composed of 3-hydroxy fatty acids monomers (Madison and Huisman, 1999). In such types of

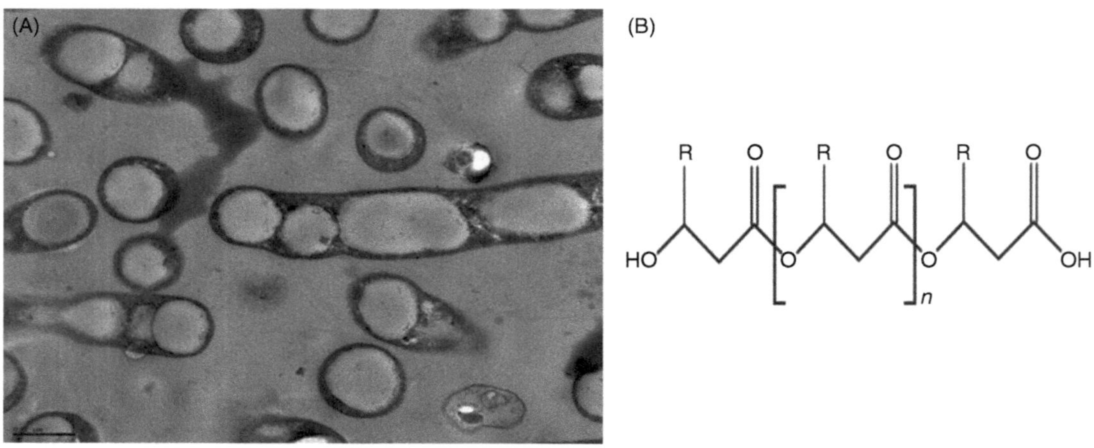

FIGURE 1.1

(A) Transmission electron microscopic image of accumulated of PHA granules in *Hydrogenophaga pseudoflava*. About 70%–90% of bacterial weight consists of accumulated PHA granules in bacterial cytoplasm. Scale bar = 0.5 μm. (B) Chemical structure of polyhydroxyalkanoates (PHAs)

Part B: Reprinted with permission from Sun, Z., Ramsay, J.A., Guay, M., Ramsay, B.A., 2007. Fermentation process development for the production of medium-chain-legth poly-3-hydroxyalkanoates. Appl. Microbiol. Biotechnol. 75, 475–485.

polyesters, the carboxyl group of one monomer forms an ester bond with the hydroxyl group of the neighboring monomer (Fig. 1.1B). In all PHAs that have characterized so far, the hydroxyl-substituted carbon atom is of R configuration, except in some cases where there is no chirality (Madison and Huisman, 1999). Polyhydroxalkanoates (PHAs) isolated from bacterial cells show material properties that are similar to petrochemically produced polymers. Many microorganisms secrete enzymes, which have the ability to degrade these polyesters enzymatically. Apart from their biodegradable nature, PHA-based polyesters are natural, renewable, and biocompatible. The wide range of physical properties of PHA-based materials and the extended performance, obtainable by chemical modification of the surface or blending provide a broad range of potential end-use applications (Verlinden et al., 2007). The key enzymes utilized by microorganisms for the degradation of PHA-based materials are both extracellular and intracellular PHA depolymerases. Another important property of PHAs is that they can be produced on renewable resources (Braunegg et al., 2004; Madison and Huisman, 1999). Fermentative production of PHAs is based on agricultural products, such as sugars and fatty acids as carbon and energy sources. The origin of this agricultural feedback is CO_2 and water, and upon degradation in the environment, it is again converted into CO_2 and water.

4.1 PROPERTIES OF PHA

The physical properties of PHA are different, and two physical states of PHA granules can be distinguished represented by intracellular "native PHA granules" and partially "crystalline PHA granules" (Jendrossek and Handrick, 2002). The native PHA granules are in amorphous state, the surface of which is surrounded with a layer consisting of phospholipids and granule-associated proteins (Hocking and Marchessault, 1994; Pötter and Steinbüchel, 2005). After extraction from the cells using some organic solvents, the phospholipid and protein layer is damaged or lost and the amorphous nature is converted into complete crystalline structure and the PHA after isolation is tough and brittle in nature; such a PHA is called crystalline PHA (Doi, 1995). Many researchers have shown that crystalline PHA is unsuitable for design of different products due to its high crystallinity, thermosensitivity, and brittle in nature (Kim et al., 1999; Kumagai and Doi, 1993). Copolymerization and blending with other degradable polymers were also suggested to enhance the physical properties of the polymer (Doi, 1990; Kim et al., 1999). Copolymers of PHA can be produced by cofeeding of mixed substrates, such as a mixture of glucose and valerate or butyrate, which results in the formation of polymers containing 3-hydroxyvalerate (3HV) or 4-hydroxybutyrate (4HB) monomers (Doi, 1990; Madison and Huisman, 1999). The high number of monomers and the variable monomeric composition of PHA result in an enormous variation of the physical and chemical characteristics of different PHA (Jendrossek and Handrick, 2002). Based on the carbon chain length, PHA is generally divided into two groups, that is, short-chain-length (SCL) and medium-chain-length (MCL) PHA. SCL-PHA consists of (R)-hydroxyalkanoates with carbon number range C3–C5, on the other hand MCL-PHA consist of aliphatic or aromatic (R)-hydroxyalkanoates of C6–C14 (Madison and Huisman, 1999; Sudesh and Doi, 2000). Polyhydorxybutyrate (PHB), poly(3-hydroxyvalerate) (PHV), poly(3-hydroxybutyrate-*co*-3-hydroxyvalerate) (PHBV), and poly(3-hydroxybutyrate-*co*-4-hydroxybutyrate) (PHB4HB) form a class of PHAs polyesters typically referred as SCL-PHA.

4.2 BIODEGRADATION OF PHA

Besides the typical properties described earlier, an important characteristic of PHA-based materials is their biodegradability in natural environment. In nature, a vast consortium of microorganisms is able to degrade PHAs by secreting PHA hydrolases and PHA depolymerases (Jendrossek et al., 1996; Madison and Huisman, 1999; Verlinden et al., 2007). The activities of these enzymes may vary and depend on the composition of the polymer and also on the environmental conditions (Doi, 1990; Verlinden et al., 2007). Although PHA is water-insoluble, hydrophobic, and partially crystalline polymers, they can be degraded by a large variety of microorganisms (Jaeger et al., 1995). PHA can be degraded either intracellularly by intracellular PHA depolymerases of the accumulating strain, or extracellularly by extracellular PHA depolymerases. Intracellular PHA depolymerases can only degrade native amorphous PHA and cannot degrade crystalline denatured PHA. On the other hand, extracellular PHA depolymerases can degrade only crystalline denatured PHA while it cannot degrade native amorphous PHA (Jendrossek et al., 1996; Jendrossek and Handrick, 2002). Enzymes catalyzing the extracellular degradation of denatured PHA are called extracellular PHA depolymerases. It is well known that microorganisms degrade wide varieties of PHA homopolyesters and copolyesters by excreting extracellular PHB depolymerases into the environment and utilizing the decomposed components as a nutrient source (Abe et al., 1994; Doi, 1990). However, the structural composition of monomer unit in the copolyesters is an important parameter that determines the mobilization of copolyesters by PHB depolymerases (Doi, 1990). PHA-degrading microorganisms are present in diverse habitat, such as soil, activated sludge, laboratory atmosphere, and seawater (Kasuya et al., 1994; Mukai et al., 1993). The enzymatic degradation usually proceeds by the surface erosion of the polymer and the resulting products after enzymatic degradation are water-soluble monomers or oligomers or both.

4.3 GENERAL APPLICATIONS OF PHA

The majority of expected applications of PHAs are as replacement for petrochemical plastics. The plastic materials in current use for packaging and coating applications can be replaced partially or entirely by PHA-based biodegradable materials (Verlinden et al., 2007). The wide range of applications of PHA-based materials and the new and improved products obtained from the copolymerization and blending provide a broad range of potential end-use applications (Bhatt et al., 2008; Cui et al., 2002). PHA has emerged as potential material utilized widely in varieties of medical applications (Martin and Williams, 2003; Williams et al., 1999; Xiong et al., 2010). PHA-based materials are nontoxic and well tolerated by the human body (Mei et al., 2006; Williams et al., 1999; Zhao et al., 2003). PHB and its copolymer PHBV have widely been used for medical applications, such as surgical sutures, wound dressings, lubricating powders, orthopedic uses, and pericardial substrate, as well as in tissue engineering materials (Hocking and Marchessault, 1994; Williams et al., 1999). PHA-based materials are also widely used in the preparation of microparticles and nanospheres for the delivery of varieties of drugs to their site of action without inducing any toxic inflammatory responses for the cells (Chen et al., 2006; Lionzo et al., 2007). Microspheres of PHB loaded with rifampicin were investigated for their use as chemoembolizing agent (Zhang et al., 1997). Thus, functionalized PHA granules seem to be excellent candidates for targeted drug delivery as they combine the biocompatible polymer properties with the properties of formation nano- or microbased carriers.

With technology now in place to allow the properties of PHA polymers to be tailored to specialized application, coupled with a significant increase in the need for new absorbable biomaterials, this class of polymers now appears to have a bright future in medicine and pharmacy. Indeed, a wide range of applications for PHA-based polymers have been described, which include their use as sutures, meniscus repair devices, bone plates and bone plate systems, ligament and tendon grafts, surgical mesh, skin substitutes, bulking and filling agents, formation of cardiac stents, and ocular cell implants (Williams, 2000; Williams et al., 2000).

4.4 PHA AS DRUG CARRIER

Biomaterials have been widely used in medical applications, such as drug delivery, tissue engineering, device-based therapies, and medical imaging (Langer and Tirrell, 2004; Mitragotri and Lahann, 2009). Recently, biodegradable polymeric nanoparticles as drug carriers have attracted great research interest because of their biocompatibility, biodegradability, and sustained release of drugs (Panyam and Labhasetwar, 2003). Synthetic and naturally occurring polymers have played important role in the treatment of diseases and the improvement of health care. Among them, PHAs are promising materials for biomedical applications in tissue engineering and drug-delivery systems because they are natural, renewable, biodegradable, and biocompatible thermoplastics (Hazer, 2010). Different types of PHA homopolymers and copolymers produced by microorganisms have been suggested to be utilizable as tissue engineering materials and controlled release drug vectors (Khang et al., 2001; Köse et al., 2003; Timbart et al., 2004). However, only few PHA-based materials to date have been studied and their pharmacological applications have been evaluated. Microparticles prepared from either PHBV or blends of PHBV and polycaprolactone were used for the sustained release of drugs (Khang et al., 2001; Lionzo et al., 2007). Similarly, wafers produced from PHBV were also utilized for the release of antibiotics (Khang et al., 2000). Nanoparticles prepared from poly(3-hydroxybutyrate) (PHB) are getting much attention because it provides a safe, nontoxic, and biodegradable surface for coating with other materials (Pötter and Steinbüchel, 2005). Apart from that, according to ISO 10993, toxicity experiment on PHB indicated that the substance is safe to be used as nanoparticles in animals (Pötter and Steinbüchel, 2005). However, direct administration of unmodified microbial polyesters as therapeutic agents were found to induce inflammatory responses in animal tissues (Qu et al., 2006; Valappil et al., 2006).

Literature review currently shows two main methods of delivering drugs using PHA as a drug carrier system to a specific site particularly to cancer cells. The first is active targeting of the drug molecule, involving specific surface attached ligands to the polymer surface and then recognition by the specific receptors, which are expressed on the surface of cancer cells. In this method the drug molecules are conjugated with the surface attached ligands by making chemical interactions. The second approach is passive targeting involving no surface-attached ligands. The nanoparticles are simply internalized by simple endocytosis without involving any surface attached ligands. In some cases, the surface is modified with nontoxic and inert substances, which prevent nanoparticles upsonization by the phagocytic cells and also gives dispersibility to the hydrophobic polymer core in aqueous solution. The PHA-based nanoparticles used in drug delivery system is summarized in Table 1.1.

Table 1.1 PHA Nanoparticles in Drug Delivery System

Type of Polymer	Encapsulated Substance	Biological Applications	References
Controlled drug delivery (passive targeting)			
P(3HB)	Rhodamine B isothiocyanate as a drug model	Slow and sustained release of anticancer drug intracellularly	Xiong et al. (2010)
P(3HB)	TGX221, a PI3K inhibitor	Sustain release of anticancer drug improving stability of the systemic anticancer drug	Lu et al. (2011)
PHB–PEG–PHB	Pyrene-loaded nanoparticles	Slow and sustained release of pyrene	Chen et al. (2006)
PHV-*b*-mPEG	Thymoquinone as a drug model	Slow and sustained release of anticancer drug intracellularly	Shah et al. (2011)
P(4HB)-mPEG	Cisplatin as drug model	Bioavailability of anticancer drugs	Shah et al. (2014)
P(3HB-*co*-3HHx)	Rhodamine B isothiocyanate as a drug model		Xiong et al. (2010)
P(3HB-*co*-3HHx)-*b*-PEG	Rapamycin as a drug model	Intracellular delivery release of anticancer drug	Lu et al. (2014)
P(3HB-*co*-3HV)	Ellipticine as a drug model	Improvement of bioavailability of anticancer drug	Masood et al. (2013)
P(3HB-*co*-3HV)	Paclitaxel as a drug model	Parenteral delivery of anticancer drug to endometrial and primary ovarian cancer cells	Vilos et al. (2013)
P(3HV-*co*-4HB)-*b*-PEG	Cisplatin as a drug model	Bioavailability of anticancer drug	Shah et al. (2012)
Targeted drug delivery (active targeting)			
P(3HB-*co*-3HV)/FA-PEG–NH_2 conjugate	DOX as a drug model	Folate-mediated intracellular delivery of anticancer drug	Zhang et al. (2010)
P(3HB-*co*-3HV)/magnetite composite	Alexa Fluor 488 dye as imaging agent	Targeted delivery to intestinal tissue	Vilos et al. (2013)
P(3HB-*co*-3HHx)	Phasin-mediated immobilization of costimulator B7	Complementary anticancer therapy	
P(3HB-*co*-3HHx) with immobilized epidermal growth factor receptor fused with phasin	Rhodamine B isothiocyanate as a drug model	Targeted intracellular delivery carcinoma cells	Yao et al. (2008)
P(3HB) with immobilized tumor-specific ligand, RGD4C, fused with PHB synthase	Nile Red as a drug model	Targeted intracellular delivery to breast cancer cells	Lee et al. (2011); Kim et al. (2009)
P(3HB)	*Mycobacterium* spp. antigens; Ag85A and ESAT-6	Tuberculosis protein vaccine	Parlane et al. (2011)
P(3HB-*co*-3HHx) with immobilized human αl-acid glycoprotein fused with phasin	Rhodamine B isothiocyanate as a drug model	Targeted delivery to liver tumor	Yao et al. (2008)
P(3HB-*co*-3HHx)	Etoposide as a drug model	Folate-mediated intracellular delivery of anticancer drug	Kilicay et al. (2011)

5 TARGETING DRUG DELIVERY USING LIGAND-CONJUGATED PHA NANOPARTICLES (ACTIVE TARGETING)

Different types of surface attached ligands have been exploited using PHA nanocarriers. For example, folate-conjugated P(3HB-*co*-3HO) nanoparticles encapsulating DOX (DOX/FA-P(3HB-*co*- 3HO) were synthesized by Zhang et al. (2010) for targeted drug delivery of cancer. The study confirmed that folate conjugated nanoparticles were efficiently internalized by HeLa cells as compared to nonfolate conjugated cells due to competition for the folate receptor, which increased the internalization specificity of folate expressed on cancer cells. Another study done by Kilicay et al. (2011) on delivering etoposide using P(3HB-*co*-3HHx) nanoparticles conjugated to folate (ETO/FA-P(3HB-*co* 3HHx)) in mouse fibroblast L929 cell lines. It was observed that ETO/FA-P(3HB-*co*-3HHx) nanoparticles exhibited much greater cytotoxicity (approximately 58% remaining viable cells) than nonfolate-conjugated particles (approximately 70% remaining viable cells), which were more cytotoxic than free etoposide drug in solution (approximately 88% remaining cell viability).

Apart from folate-based surface-attached ligands, PHA-associated surface granules, such as (PhaP) and PHA synthase (PhaC) have been widely used as surface attachment ligand. Phasin (PhaP) is an amphiphilic protein responsible for the regulation of the size of polyhydroxyalkanoic acid (PHA) granules and their number density in bacterial cells (Pötter and Steinbüchel, 2005; Prieto et al., 1999). It was reported previously that phasins form a layer on the surface of PHA granules, thereby stabilizing them in the aqueous intracellular environment of the bacteria (Nigmatullin et al., 2015). It has been confirmed that PhaP binds with higher affinity to hydrophobic materials (Pötter et al., 2002). Phasin proteins are thermally stable up to 121°C for 20 min (Handrick et al., 2004). They are also stable upon exposure to acid and alkali, as well as organic solvents, such as C1–C8 alkanol, acetone, acetonitrile, and chloroform (Handrick et al., 2004). On the other hand, side PhaC (PHA synthase) required for polymerization is linked to the active site cysteine residue (Pötter et al., 2005; York et al., 2002). It was suggested that the insertion of a desired gene sequence to the PhaP or PhaC genes can produce fusion proteins that could be used for wide range applications (Hooks et al., 2014; Li et al., 2015). Rho et al. (2014) prepared $(His)_6$-tagged or GST-fusion recombinant phasin and utilized for the artificial granule preparation. In addition, they also coloaded P(3HB) depolymerase with the recombinant phasin to prepare self-degradable phasin-coated P(3HB) granules. Yao et al. (2008) fused PhaP with recombinant human epidermal growth factor (rhEGF) and produced rhEGF-Phap fusion proteins. These fusion proteins were then linked to presynthesized P(3HB-*co*-3HHx) nanoparticles. In order to find intracellular accumulation of the receptor conjugated nanoparticles, Rhodamine-B was loaded into the particles and transplantable murine hepatoma 22 (H22) model cells were implanted into BALB/c mice, which were then treated with Rhodamine-B loaded particles. A variety of tissue types were then sampled to assess particle distribution. Fluorescent microscopy studies demonstrated that Rhodamine-B loaded rhEGF-PhaP-PHBHHX nanoparticles were found selectively accumulated within the tumor in large amounts.

Many studies have also described PhaC as a linker region to conjugate it to PHA nanoparticles. Studies conducted by Kim et al. (2009); Lee et al. (2011) described the use of PhaC to links targeting moieties like RGD4C, a ligand-binding $\alpha v\beta 3$ integrin overexpressed in many cancers and preprepared P(3HB) nanoparticles and block copolymers. They found that RGD4C was overexpressed in MDA-MB 231 breast cancer cells showing its higher affinity for targeted particles. Similarly, Kim et al (2009) produced functionalized P(3HB) granules and linked an A33scFv (targeting moiety) and GFP (imaging moiety) to PhaC. Microscopic observations showed that functionalized particles were taken up by A33 expressing cells, and not by A33 nonexpressing cells.

A variety of different particles were prepared by Jahns et al. (2008) for theranaustic purposes while linking PhaC with inorganic molecules, such as silica and gold-bound polyester particles to PHA granules. Vilos et al. (2013) also described polymer-loaded superparamagnetic iron oxide nanoparticles (SPIONs) by encapsulating SPIONs within P(3HB-*co*-3HV) (SPION-P(3HB-*co*-3HV)), forming theranostic nanoparticles for a wide variety of theranostic applications.

6 CHEMICAL MODIFICATION OF PHA POLYMERS FOR DRUG DELIVERY (PASSIVE TARGETING)

The surface modification of a polymer with nontoxic and blood compatible material is essential in order to avoid recognition by macrophages, to prolong blood circulation time and sustained release of the encapsulated drugs (Duan et al., 2006; Mosqueira et al., 2001). Amphiphilic PHAs can be synthesized by introducing hydrophilic groups, such as hydroxyl, carboxyl, amine, sulfonic acid, ethylene glycol, and PEG (Qu et al., 2006). In chemical grafting reactions, some hydrophilic groups have been attached in the PHA chain to obtain amphiphilic polymer. PEG is widely used as hydrophilic nontoxic segment in combination with hydrophobic biodegradable aliphatic polyesters. Incorporation of a hydrophilic mPEG group on the surface of nanoparticles was found to show resistance against opsonization and phagocytosis and showed prolonged residence time in blood compared to the nanoparticles prepared without mPEG (Knop et al., 2010). Microbial PHB was found to be tolerable in animal tissues without causing any inflammatory reactions. However, in some cases, chemically unmodified microbial PHB was observed to aggravate inflammatory responses in animal tissues (Qu et al., 2006). Therefore, finding an additional nontoxic candidate to PHA is warranted in order to deliver drugs without inducing a reaction from the body's immune system.

PEG has already been widely used as a hydrophilic segment in combination with hydrophobic biodegradable aliphatic polyesters, such as PHB, poly(L-lactide), and poly(D,L-lactide-*co*-glycolide) due to its low toxicity and nonimmunogenic nature (Knop et al., 2010). The prolonged circulation time of the mPEG-coated particles was attributed to the increased thickness of high molecular weight mPEG, which protect the nanoparticles from opsonization and phagocytosis (Gref et al., 1994). Surface modification of a polymer with mPEG as nontoxic blood compatible material reduces the risk of side effects associated with unmodified polymers (Kumar and Homme, 2008). Polysorbates and PEG on the particle surface are known to enable the transport across the blood–brain barrier (Garcia-Garcia et al., 2005). It was reported previously that mPEG interacts with phospholipids' head groups in cell surface membranes, and this promotes an enhanced in membrane fluidity, which results in ions and small metabolites permeation intracellularly (Ingram and Buttke, 1984; Yamazaki and Ito, 1990). PEGylation modifies the physiochemical properties including chain conformation, electrostatic binding, and most noticeably, hydrophobicity (Foster, 2010). Moreover, PEG has been approved by the Food and Drug Administration for human use and the mPEG released from the degradation of polyester–mPEG block polymers can be easily removed from the body via renal system (Knop et al., 2010). A wide variety of PEGylated nanoparticles have recently been used for passive targeting of drug molecules especially for cancer therapy (Shah et al., 2010, 2012).

7 SYNTHESIS OF SURFACE MODIFIED PHB-mPEG DIBLOCK COPOLYMER

The synthesis of carboxyl terminal end-caped samples PHB-block-mPEG, P(3HB-*co*-3HV)-block-mPEG and P(3HB-*co*-4HB)-block-mPEG are mostly carried out by transesterification method in the melt (Ravenelle and Marchessault, 2002). In this method a weighted amount of the PHA polymer and mPEG are mixed in the presence of catalyst and heated up at high temperature for a short time, which melts the high molecular weight PHA into low molecular weight PHAs (approximately 2300–7300). After completion of the reaction, the product formed is usually in the form of a waxy product that contains a mixture of end-capped and nonend-capped polymers. A colloidal suspension of the final product is formed by dissolving it in organic solvent like chloroform, which is dropped in to aqueous solution and then dialized against distilled water. The dialysis step removes the unbound free mPEG if any in the mixture. Lyophilization of the product results in the formation of light whitish powder (yield 70%–80%). The final product is then purified by precipitation in methanol followed by filtration (final yield 50%–60%). In the transesterification reaction, the hydroxyl group from mPEG attacks on the terminal carboxyl group of PHA polymers resulting in the formation of PHA-block-mPEG diblock copolymers. The high molecular weight PHA homopolymers or copolymers are depolymerized in the reaction into low molecular weight polymers. The terminal hydroxyl group of PHA substrate(s) is removed due to thermal dehydration leaving behind an alkenic group but having an intact terminal carboxyl group (62). It was suggested that the hydroxyl terminal group of the PHA polyesters are the most sensitive to thermal temperature as result such group disappear in the reaction at high temperature (190°C) and usually is removed in the form of water molecules (Ravenelle and Marchessault, 2002). Moreover, the presence majority of hydroxyl groups were reported previously to induce immune response (Knop et al., 2010). Therefore, complete removal of hydroxyl group makes the nanoparticles prepared from such polymers less immunogenic. Moreover, this transesterification method is much useful as compared to others because in this method the heating step result in the formation of low molecular weight polymers from higher molecular weight PHA polymers. The promising characteristics of the low molecular weight polymers are many, such as they can be easily molded into any type of nanostructures with least aggregation and sedimentation in aqueous solution as opposed to higher molecular weight polymers. Furthermore, the high temperature used during isolation of the polyesters from bacterial sources and during the preparation of diblock, copolymer inactivates almost all the bacterial toxins if present. A chemical transesterification reaction of PHA-block-mPEG diblock copolymers synthesis scheme is shown in Fig. 1.2. The product formed as a result of transesterification reaction contained bulk amount of diblock copolymers, in addition to small amount of unreacted free polymers (Ravenelle and Marchessault 2002). The purified products could be easily characterized by technique, such as GPC and ^{1}H-NMR analysis to confirm the formation of diblock copolymers. Another method called ring-opening polymerization is also used for surface modification of many PHA-based materials (Foster, 2010). Kumagai and Doi (1993); Shuai et al. (2000) synthesized triblock copolymers of PHB-*co*-PEG-*co*-PHB using ring-opening polymerization of [R,S]-β-butyrolactone with PEG based macroinitiators and produced atactic PHA segment with improved solubility in aqueous solution.

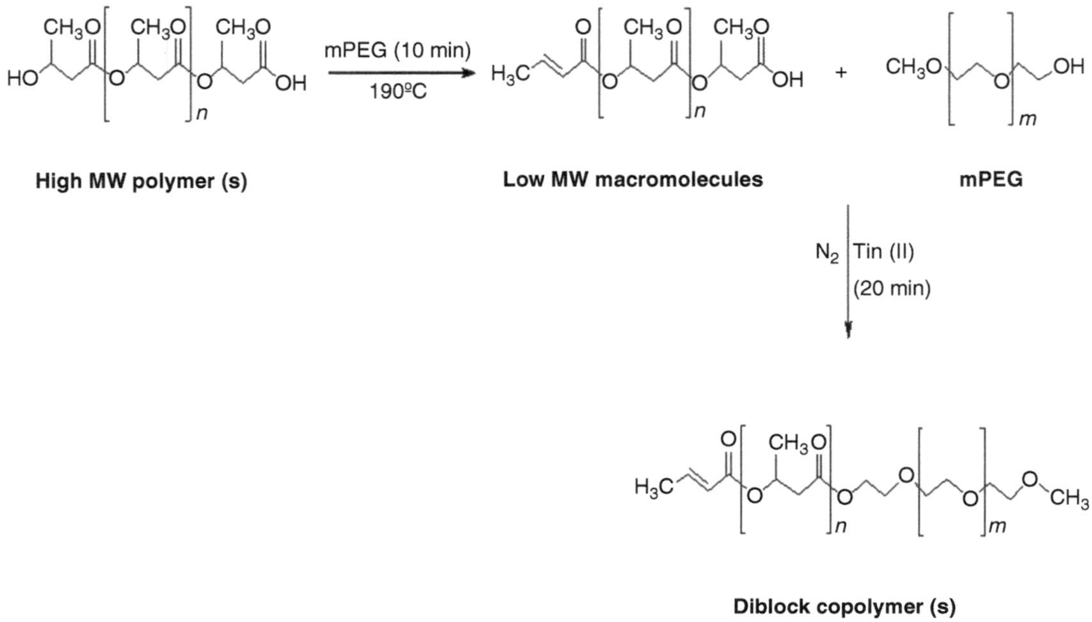

FIGURE 1.2 Chemical Coupling Reaction between PHA Polymers and mPEG, Catalyzed by Bis(2-ethylhexanoate) Tin Catalyst at 190°C

8 SELF-ASSEMBLY OF DIBLOCK COPOLYMER INTO NANOPARTICLES

PHA–mPEG nanoparticles are mostly prepared by emulsification/solvent evaporation method. Different types of PHA, such as PHB-mPEG, PHV-mPEG, P(4HB)-mPEG, P(3HV-*co*-4HB)-*b*-mPEG, P(3HB-*co*-12 mol.% 3HV)–mPEG, P(3HB-*co*-33 mol.% 3HV)–mPEG, P(3HB-*co*-6 mol.% 4HB)–mPEG, P(3HB-*co*-20 mol.% 4HB)–mPEG, PHBHH × PEG hybrid copolymers were already reported using this method (Lu et al., 2014; Shah et al., 2010, 2011, 2012, 2014). In this method, the diblock copolymer is first dissolved in a water-immiscible organic solvent like dichloromethane or chloroform under continuous stirring. The drug particle is also added during this preparation step. This mixture is added to a detergent solution in water, such as sodium deoxycholate, sodium cholate or poly(vinyl alcohol), and is probe sonicated or homogenized. This method is useful in the encapsulation of drug particles in the nanodroplet of the polymer solution and is called oil-in-water emulsion. The emulsion formed is stirred gently in a fume hood at room temperature until complete evaporation of the organic phase leaving behind the drug-loaded nanoparticles in aqueous solution stabilized by the detergents. The nanoparticles are purified by centrifugation and then washed with fresh water to remove the excessive emulsifier and unentrapped surface-attached free drug particles. In a modified emulsion/solvent evaporation method, pouring of the organic phase into water result in the self-assembly of the diblock copolymer into nanosized droplets, probably caused by the decreased interfacial tension between the organic and the aqueous phase and interfacial turbulence induced by the rapid evaporation of organic solvent. This method is useful in the preparation of 100–300 nm drug-loaded PHA–mPEG

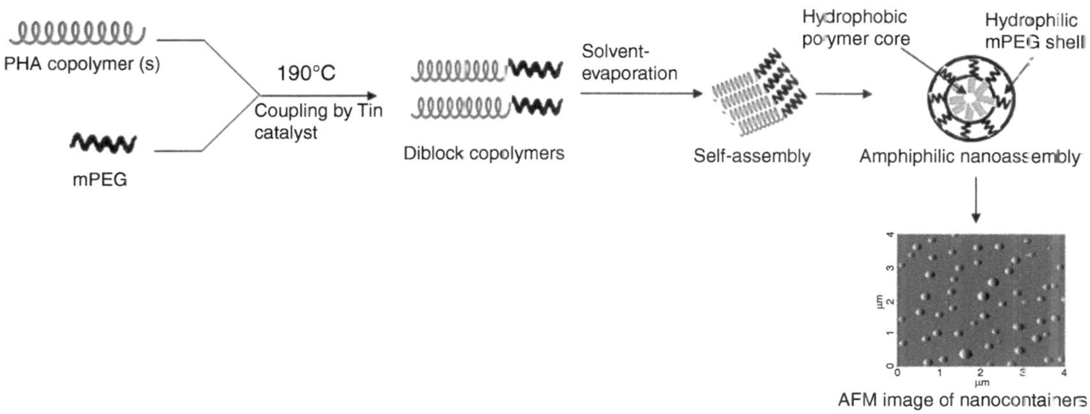

FIGURE 1.3 Synthesis Scheme for the Formation of Amphiphilic Copolymeric Nanocontainers from PHA–mPEGdiblock Copolymers, Catalyzed by Tin Catalyst

nanoparticles. Drug-free nanoparticles can also be prepared by the same method. This method is also used for the preparation of poly(lactic acid) (PLA) and poly(lactic-*co*-glycolic acid) (PLGA) nanoparticles. The formation of diblock copolymer and then its subsequent self-assembly is shown diagrammatically in Fig. 1.3.

Another method widely used for the preparation of PHA-based drug-loaded nanoparticles is solvent-displacement method. In this method, the polymer (diblock copolymer) and drug mixture are dissolved in a water-miscible liquid, such as acetone or ethanol. This mixture is poured into a solution of surfactant, such as sodium cholate or poly(vinyl alcohol) for the stabilization and formation of nanodroplets in aqueous solution. The organic phase is evaporated leaving behind nanoparticles with a size range from 100 to 200 nm. This method is widely used for the preparation of polymer nanoparticles, such as PLA, PLGA, PCL, and so on (Lammers et al., 2008).

9 CHARACTERIZATION OF PHA–mPEG NANOPARTICLES

9.1 CORE–SHELL TOPOLOGY IN AQUEOUS SOLUTION

The most promising characteristics of PHA–mPEG nanoparticles are the biodegradation of the polymer core (PHA moiety) by its enzyme PHA depolymerases. This property could be exploited to check the core–shell morphology of surface modified PHA with any surface-modifying agent in vitro. To understand the core–shell topography of the PHA–mPEG nanoparticles, a comparative enzymatic degradation was carried out against PHA–mPEG and PHA particles in the presence of PHA depolymerase (Shah et al., 2010). The degradation reaction was followed by measuring turbidity of the reaction solution at 660 nm using purified enzyme from *Ralstonia pickettii* T1. Fig. 1.4 shows the degradation profiles of two mPEGylated nanoparticles P(3HB-*co*-12 mol.% 3HV)–mPEGand P(3HB-*co*-20 mol.% 4HB)–mPEG and two non-mPEG counterpart particles P(3HB-*co*-12 mol.% 3HV) and P(3HB-*co*-20 mol.% 4HB). The PEGylated sample was degraded very slowly compared to the counterpart non-mPEG control sample (Fig. 1.4A). A plausible explanation for the slower degradation rate of all

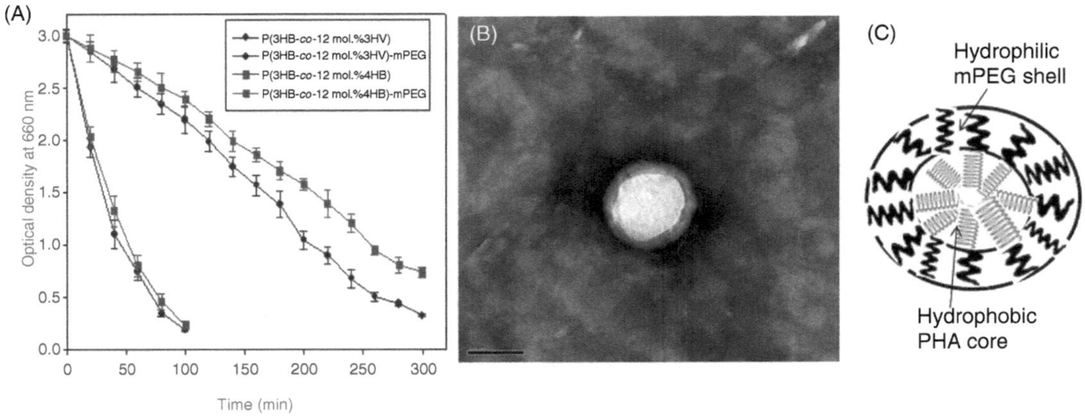

FIGURE 1.4

(A) Relative enzymatic degradation of copolymer nanoparticles to determine the localization of PHA and mPEG segment in nanoparticles: open symbols for controls, (○) P(3HB-*co*-12 mol.% 3HV) and (□) P(3HB-*co*-20 mol.% 4HB) nanoparticles; closed symbols, (●) P(3HB-*co*-12 mol.% 3HV)-mPEG and (■) P(3HB-*co*-20 mol.% 4HB)-mPEG nanoparticles. (B) Enlarged TEM image of PHA–mPEG shell structure morphology of single nanoparticle; scale bar = 100 nm. (C) Diagrammatic representation of nanoparticle with central PHA core and outer mPEG shell.

Modified from Shah, M., Ullah, N., Choi, M.H., Kim, M.O., Yoon, S.C., 2012. Amorphous amphiphilic P(3HV-co-4HB)-b-mPEG block copolymer synthesized from bacterial copolyester via melt transesterification: nanoparticle preparation, cisplatin-loading for cancer therapy and in vitro evaluation. Europ. J. Pharm. Biopharm. 80, 518–527.

mPEGylated systems was due to the coverage of the enzymatically inactive mPEG segment on the surface of enzymatically active PHA copolymer core. The accumulation of a thick surface layer of mPEG on the surface f PHA polymer provides the steric hindrance to the PHA depolymerase to penetrate to the surface and to degrade its respective substrate in to degradable units. This type of study has been performed by Shah et al. using different types of PHA copolymers end-capped with mPEG, which was compared with nonend-capped PHA copolymers as shown in Fig. 1.4A (Shah et al., 2010).

Apart from degradation study, detailed TEM observation could also clearly demonstrate the two separate layers formed by the nanoparticles. TEM observation further confirmed the core–shell morphology of the nanoparticles. A clear boundary between the inner hydrophobic polymer core and outer mPEG shell is indicated in the magnified photograph of the TEM (Fig. 1.4B). A diagrammatic representation of core–shell structure of PHA–mPEG particle is also shown in Fig. 1.4C. Although TEM observations were not found effective for the determination of core–shell topology of the particulate carriers and only dense core of the NPs was visualized without any contribution from the hydrated hydrophilic segment (Shahin and Lavasanifer, 2010). However, in our studies, the clear boundary between the two segments might be due to the transmittance of electron beam from the heavy metal of platinum in cisplatin in the dense polymer core, which reflect effectively the electron beam and the clear boundary between the two phases might be more effective. A similar TEM observation of the core–shell morphology with outermost crystalline shell and inner amorphous polymer core was conducted previously for aliphatic copolyester nanoparticles (Papadimitriou and Bikiaris, 2009).

Furthermore, it can be clearly elucidated from our previous work (Shah et al., 2011) of zeta potential measurements of the two types of particles prepared from simple PHV and PHV-*b*-mPEG that the surface of native PHV was much more highly negatively charged (–25 mV) than that of PHV-mPEG (–15 mV) particles, indicating that mPEG covered most of the surface exposed carboxyl group because of the neutral nature of mPEG segment on the surface of PHA polymer.

9.2 MORPHOLOGICAL CHARACTERISTICS OF PHA–mPEG NANOPARTICLES

Most of the matrix studies to date focused on the utilization of PHA copolymers at the level of micron-sized particles using high molecular weight unmodified PHA copolymers as a component or a blend form mixed with other polyesters (Lionzo et al., 2007; Sendil et al., 1999). However, the microparticles prepared from such copolymers are bigger in size and have difficulty crossing cellular membranes of the body (Panyam and Labhasetwar, 2003). A decreased rate of clearance and extended circulation time has also been observed with smaller diameter nanoparticles as compared to those with a larger diameter (Moghimi et al., 1993). The extremely fine size of nanoparticles facilitates their penetration across various biological barriers, such as blood brain barriers for drug delivery. Moreover, phagocytosis of particles is highest for particles between 1 and 2 μm in size, and gradually drops as particle size decreases (Moghimi et al., 1993). Therefore, nanoparticles preparation with smaller diameter from the low molecular weight PHA–mPEG diblock copolymers may enhance the therapeutic application of these materials for drug delivery.

The PHA–mPEG nanoparticles prepared by oil-in-water solvent-evaporation method are usually spherical in shape and uniformly distributed in size with a size range of 100–300 nm in diameter. The shape of the nanoparticle plays an important role on the drug release mechanism. Nanoparticles prepared using only PHA block without capping with PEG usually result in bigger particles size usually in the micrometer range. It is because of the aggregation of PHA block in aqueous solution owing to strong hydrophobicity. Therefore, surface modification with strongly hydrophilic material is essential to increase the water dispersibility of the nanosized PHA nanoparticles in aqueous milieu. However, the nanoparticles prepared by using different PHA block copolymers with varying the comonomer ratios in the polymer chain show variations in surface morphology of the formed nanoparticles. The nanoparticles prepared with increasing the comonomer ration show some nonuniformed and larger sized nanoparticles. For example, the P(3HB-*co*-33 mol.% 3HV)-mPEG and P(3HB-*co*-20 mol.% 4HB)-mPEG nanoparticles sample exhibited nonuniformly distributed particles in the AFM images (Fig. 1.5). However, when the level of 3HV or 4HB is decreased like P(3HB-*co*-12 mol.% 3HV)-mPEG and P(3HB-*co*-6 mol.% 4HB)-mPEG relatively uniform distributed particles was observed. Such nonuniform distribution of the nanoparticles by increasing the comonomer ratios of 3HV or 4HB might be due to some aggregation of the nanoparticles, especially the nanoparticles in the composition. A similar comonomer-ratio dependent behavior was also reported for PLGA copolymer nanoparticles (Papadimitriou and Bikiaris, 2009).

Another important property of polymeric nanoparticles is their surface charge called zeta potential. Zeta potential represents an index for particle stability in aqueous environment. As the zeta potential of the particles in aqueous solution increases, the repulsive interactions between particles also increases leading to the formation of more stable particles with a more uniform size distribution. More negative zeta potential of the nanoparticles was suggested for more dispersion stability of the nanoparticles in the medium (Jeong et al., 2009). This stability is important in preventing aggregation and subsequent

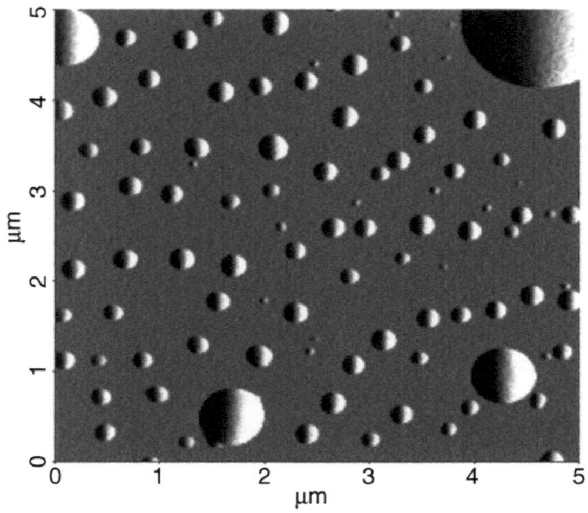

FIGURE 1.5 A 3-D AFM Image of PHA–mPEG Nanoparticles with a Scan Size of 5 μm × 5 μm

precipitation leading a stable colloidal dispersion of the particles. The repulsion among the highly negatively charged NPs was suggested to provide extra stability in aqueous solution (Liu et al., 2000). It is suggested in literature that particles with zeta potentials more positive than +30 mV or more negative than −30 mV are more stable (Müller et al., 2001). Surface modification is the most commonly used approach for controlling the zeta potential of nanoparticles and thereby improving their stability (Vila et al., 2002). Thus, the PHA–mPEG nanoparticles form a stable colloidal suspension in aqueous medium. Moreover, the amphiphilic nature of the nanoparticles causes them to repel them and to stabilize each other and to prevent aggregation.

9.3 DRUG LOADING IN PHA–mPEG NANOPARTICLES

The drugs (especially anticancer drugs, such as DOX, thymoquinon, etc.) isolated from natural resources and marketed are mostly hydrophobic in nature. The main problem associated with such types of drugs is their low bioavailability and precipitation is nonspecific targets parts of the body resulting in severe side effects and patient's compliance. Therefore, loading of these drugs in nanoparticles results in the protection of drugs and also decreases their nonspecific accumulation. PHA used in drug delivery system is mostly crystalline, brittle, and strongly hydrophobic in nature. Therefore, the PHA-based nanoparticles in current research are mostly used for hydrophobic drug delivery system.

Drug particles loaded in to PHA nanoparticles are usually done by two methods. In the first method a concentrated solution of solid drug particles (e.g., thymoquinon, DOX, cisplatin) are dispersed into the liquid medium that entrapped the drug into the polymer matrix of nanoparticles followed by continuous stirring and sonication during preparation steps. Second, the drug particles are adsorbed directly onto the inner cavities of the preformed nanoparticles in aqueous solution. The amount of drug encapsulated is higher when incorporated during the formation of nanoparticles; however, there is a possibility that drug stability gets affected by processing parameters, such as the method of preparation and presence and types of detergents used for the preparation (Yoo et al., 2000). The drug particles are either chemically conjugated to the polymer core matrix or entrapped physically in the polymer matrix. The latter approach involves physical adsorption of the drug on the nanoparticle surface.

The central core domain (PHA) acts as cargo for loading the highest amount of drugs when the drug particles are incubated in aqueous solution prior to nanoparticle formation to allow drug–PHA binding. The main mechanism involved in drug encapsulation in PHA–mPEG nanoparticles are either physical entrapment of the solid drug particles in the cavities of the core PHA domain or attachment of the drugs to the hydrophobic polymer core by hydrophobic interactions with the drug particles, which usually completely solubilize the drug inside their core domains. It is well known that most of the PHA polymers are crystalline in nature, therefore, not suitable for the preparation of amphiphilic diblock copolymers (Chen et al., 2006). Polymer nanoparticles prepared from such copolymers form dense crystalline cores with strong intermolecular forces, which prevent the encapsulation of appreciably high amount of hydrophobic drugs (Karanikolopoulos et al., 2010). The increased drug-loading efficiency with decreasing the crystallinity of the copolymers was also reported previously (Papadimitriou and Bikiaris, 2009). Moreover, rapid release of the drug was reported for polymers having higher crystallinity in the matrix compared to amorphous polymers (Papadimitriou and Bikiaris, 2009), which was suggested to be due to the formation of microchannels in the highly crystalline polymers (Sinha et al., 2004). Therefore, amorphous polymer was suggested to improve the drug loading efficiency and drug release profile (Papadimitriou and Bikiaris, 2009; Shah et al., 2012).

9.4 DRUG RELEASE FROM PHA–mPEG NANOPARTICLES

The mechanism of drug release from most of the polymeric nanoparticles is either by degradation (surface erosion), diffusion, or a combination of both. Both of the mechanisms depend on the characteristics of the polymer used for the preparation of nanoparticles. In PHA-based polymers the drug release is by diffusion because of the longer degradation time for most of the PHA polymers. However, in polymers, such as PLGA the release is due to surface erosion or degradation, because in PLGA polymers degradation is only a few hours. It has also been reported that for particles with a compact structure, drug release is controlled by degradation of the polymer matrix, whereas for particles with pores and channels, release mainly depends on the diffusion of the drug (Perez et al., 2001).

The diffusion release from PHA nanoparticles is biphasic. In this biphasic release pattern the drug firstly is released at burst manner followed by slow and sustained release. P(3HB) nanoparticles containing prednisolone were prepared using high pressure homogenization by Koosha et al. (1988). They observed biphasic release pattern of up to 50% of drug loading, with an initial burst effect followed by slow release of drug, and the completion was achieved in 1–2 days. The initial fast release of the drug from nanoparticles was suggested that some amount of the drug was absorbed on the surface or loosely bound to the inner polymer core and was lost during the initial stage of incubation (Allen et al., 1999), while the strongly encapsulated drug in the core domains followed slow and sustained release kinetics. The slow release of the drug from the core domains of the nanoparticles was attributed to the hydrophobic interactions between the hydrophobic drugs and hydrophobic polymer core (Liu et al., 2000). The release of the encapsulated drug in the matrix of the polymer core was also suggested to be dependent on the comonomer ratio used in the copolymer preparation steps. For example, the relatively rapid release rate of the nanoparticles containing high amount of 3HV or 4HB comonomer units might be due to the higher encapsulation efficiency of the nanoparticles compared to that of low comonomer units in the composition. A similar trend of higher encapsulation efficiency of drug-loaded nanospheres with

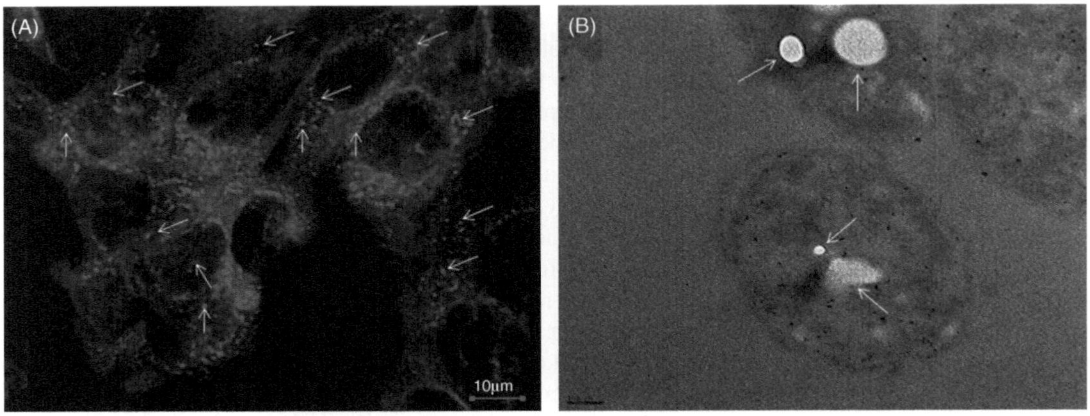

FIGURE 1.6 Microscopic Study Showing the Cellular Internalization of the P(3HV-*co*-4HB)-*b*-mPEG NPs

(A) CLSM images of the DU145 cells treated with rhodamine-loaded NPs for 12 h (scale bar 10 μm); (B) TEM images showing the uptake of NPs in DU145 cells incubated for 6 h (scale bar = 100 nm); *arrows* in the figure represent intracellular localization of NPs.

higher 3HV comonomer unit in the composition was observed by Shah et al. (2010). Moreover, the utilization of amorphous nanoparticles instead of crystalline one was also reported to decrease the burst release pattern (Papadimitriou and Bikiaris, 2009; Shah et al., 2012).

9.5 CELLULAR UPTAKE OF PHA–mPEG NANOPARTICLES

The major dilemma to the clinician is the multidrug resistance of malignant tumors to most chemotherapeutic agents, which remains the major cause of cancer therapy failure. One reason is the increased activity of an efflux pump, such as P-glycoprotein for most of the low molecular weight cytotoxic agents. However, being drug loaded or conjugated inside the polymeric nanocontainers was suggested to bypass this resistance mechanism (Gottesman et al., 2002; Kopeček and Kopečková, 2003; Larsen et al., 2000). The major advantage of polymeric nanoparticles is that they not only protect encapsulated drugs from degradation but also prolong the residence time of drugs in the absorption site. This result increased bioavailabilities of drugs and also bypassed the efflux pump in some of the resistant tumors. Shah et al. (2012) studied the cellular uptake of Rhodamine-loaded PHA–mPEG nanoparticles using confocal laser scanning microscopy and TEM analysis. They observed that amorphous PHA–mPEG nanoparticles loaded with Rhodamine was taken up by sufficient number by the prostate cancer DU145 cells, and the red fluorescence, which is characteristic of Rhodamine, was concentrated clearly in the cytoplasm of the cells (Fig. 1.6A). The time dependent uptake and cytoplasmic localization were further supported by transmission electron microscopic observation (Fig. 1.6B). These results suggest that amorphous nanoparticles were internalized presumably via endocytosis like most of the other PEGylated drug conjugates (Bareford and Swaan, 2007). Most of the PHA–PEG nanoparticles internalization also takes through endocytotic mood of action as shown in Fig. 1.7.

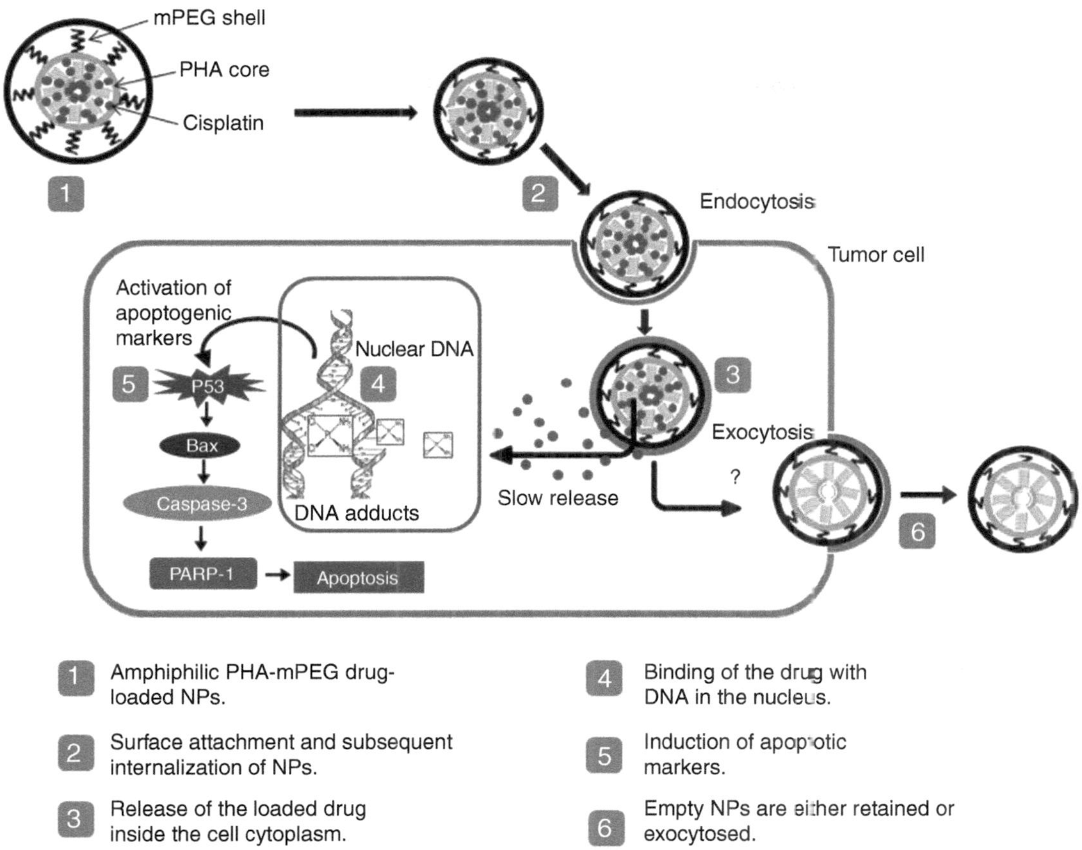

FIGURE 1.7 Schematic Representation of Drug Loaded PHA–mPEG Nanoparticles and Internalization and Activation of Apoptotic Markers

10 DRAWBACKS AND CHALLENGES

There is considerable interest nowadays in PHA-based nanoparticles because of their biodegradable and biocompatible nature. However, several problems are still associated with using PHA-based materials as a drug delivery system. Direct administration of unmodified microbial polyesters as therapeutic agents can induce inflammatory responses in animal tissues (Qu et al., 2006). As a consequence, surface modification was suggested to improve their efficacy. Another disadvantage of PHA polymers is their lengthy degradation time. It is well known that the degradation rate of a piece of PHA is typically in the order of a few months to years (Jendrossek et al., 1996; Mergaert et al., 1994). This longer degradation time renders PHA-based nanoparticles of lesser appeal in the area where drug release is achieved by surface erosion. However, this can be counteracted by using low molecular weight polyesters with a short degradation time which are proving more useful as a drug delivery system. It is well known that most of the PHA polymers are crystalline and brittle in nature, therefore, not suitable for the preparation of amphiphilic diblock copolymers (Chen et al., 2006). Polymer nanoparticles prepared from such copolymers form dense crystalline cores with strong intermolecular forces, which prevent the encapsulation of appreciably high amounts of hydrophobic drugs (Saigusa et al., 2002). Moreover, rapid release of the drug has been reported for polymers having higher crystallinity in the matrix, as compared to amorphous polymers (Shinomiya et al., 1997), which may be due to the formation of microchannels in highly crystalline polymers (Kasuya et al., 2003). These properties can be modified by copolymerization with hydroxyvalerate to yield more amorphous polymers, such as P(HB-*co*-HV). Consequently, using amorphous polymers instead of crystalline ones is highly suggested in order to reduce the burst release pattern associated with most of the polymer-loaded drugs.

11 CONCLUSIONS AND FUTURE PERSPECTIVES

The ability to deliver drugs locally is of great benefit in the treatment of many diseases. Nanoparticles can function as a drug carrier, providing controlled delivery of the drug. For optimum drug delivery, nanoparticles should readily be biocompatible (not harmful for humans) and biodegradable (deteriorate and expulse in the body conditions). Biodegradable polymers containing the entrapped drug can be placed in the body and are used for localized drug delivery and/or the controlled release of a drug over an extended period of time. These properties, as well as targeting and controlled release, can be affected by nanoparticle material selection and by surface modification. Materials, such as synthetic polymers, proteins, and other natural macromolecules are used in the preparation of nanoparticles. The use of biodegradable polyester copolymers offers the possibility of tuning properties, such as the rate of degradation and the mechanical characteristics, by changing the molecular weight and/or the molar composition of these copolymers. For these reasons, PHA–PEG nanoparticles provide the optimum platform for drug delivery of toxic hydrophobic drugs.

REFERENCES

Abe, H., Matsubara, I., Doi, Y., Hori, Y., Yamaguchi, A., 1994. Physical properties and enzymatic degradability of poly(3-hydroxybutyrate) sterioisomers with different sterioregulatories. Macromolecules 27, 6018–6025.

Allen, C., Maysinger, D., Eisenberg, A., 1999. Nano-engineering block copolymer aggregates for drug delivery. Colloids Surf. B 16, 3–27.

Anderson, A.J., Dawes, E.A., 1990. Occurrence, metabolism, metabolic role, and industrial use of bacterial polyhydroxyalkanoates. Microbiol. Rev. 54, 450–472.

Bareford, L.M., Swaan, P.W., 2007. Endocytic mechanisms for targeted drug delivery. Adv. Drug Del. Rev. 59, 748–758.

Bhatt, R., Shah, D., Patel, K.C., Trivedi, U., 2008. PHA–rubber blends: synthesis, characterization and biodegradation. Biores. Technol. 99, 4615–4620.

Braunegg, G., Bona, R., Koller, M., 2004. Sustainable polymer production. Polym. Plast. Technol. 43, 1779–1793.

Chen, C., Yu, C.H., Cheng, Y.C., Yu-Peter, H.F., Cheung, M.K., 2006. Biodegradable nanoparticles of amphiphilic triblock copolymers based on poly(3-hydroxybutyrate) and poly(ethylene glycol) as drug carriers. Biomaterials 27, 4804–4814.

Cui, Y., Wang, X., Hong, L., Tang, X., 2002. Crystallization behavior of PHA/PEDEA oligoesterblends. J. Appl. Polym. Sci. 84, 1363–1368.

Doi, Y., 1990. Microbial Polyesters. VHC, New York.

Doi, Y., 1995. Microbial synthesis, physical properties, and biodegradability of polyhydroxyalkanoates. Macromol. Symp. 98, 585–599.

Duan, Y., Sun, X., Gong, T., Wang, Q., Zhang, Z., 2006. Preparation of DHAQ-loaded mPEG-PLGA-mPEG nanoparticles and evaluation of drug release behaviors in vitro/in vivo. J. Mater. Sci. Mater. Med. 17, 509–516.

Foster, L.J.R., 2010. PEGylation and bioPEGylation of polyhydroxyalkanoates: synthesis, characterisation and applications, biopolymers. In: Elnashar, M. (Ed.), In Tech. Available from: http://www.intechopen.com/books/biopolymers/pegylation-and-biopegylation-of-polyhydroxyalkanoates-synthesis-characterisation-and-applications

Garcia-Garcia, E., Andrieux, K., Gil, S., Couvreur, P., 2005. Colloidal carriers and blood–brain barrier (BBB) translocation: a way to deliver drugs to the brain? Int. J. Pharm. 298, 274–292.

Gottesman, M.M., Fojo, T., Bates, S.E., 2002. Multidrug resistance in cancer: role of ATP dependent transporters. Nat. Rev. Cancer 2, 48–58.

Gref, R., Minamitake, Y., Peracchia, M.T., Trubetskoy, V., Torchilin, V., Langer, R., 1994. Biodegradable long-circulating polymeric nanospheres. Science 263, 1600–1603.

Handrick, R., Technow, U., Reichart, T., Reinhardt, S., Sander, T., Jendrossek, D., 2004. The activator of the *Rhodospirillum rubrum* PHB depolymerase is a polypeptide that is extremely resistant to high temperature (121 degree C) and other physical or chemical stresses. FEMS Microbiol. Lett. 230, 265.

Hazer, B., 2010. Amphiphilic poly(3-hydroxy alkanoate)s: potential candidates for medical applications. Article ID 423460, 8 p.

Hocking, P.J., Marchessault, R.H., 1994. In: Griffin, G.J.L. (Ed.), Chemistry and Technology of Biodegradable Polymers. Chapman and Hall, London, p. 48, (Chapter 4).

Hooks, D.O., Venning-Slater, M., Du, J., Rehm, B.H.A., 2014. Polyhydroyxalkanoate synthase fusions as a strategy for oriented enzyme immobilisation. Molecules 19, 8629–8643.

Ingram, L.O., Buttke, T.M., 1984. Effects of alcohols on microorganisms. Adv. Micro. Phys. 25, 253–300.

Jaeger, K.E., Steinbuchel, A., Jendrossek, D., 1995. Substrate specificities of bacterial polyhydroxyalkanoates depolymerases and lipases: bacterial lipases hydrolyze poly(ω-hydroxyalkanoates). Appl. Environ. Microbiol. 61, 3113–3118.

Jahns, A.C., Haverkamp, R.G., Rehm, B.H.A., 2008. Multifunctional inorganic-binding beads self-assembled inside engineered bacteria. Bioconjugate Chem 19, 2072–2080.

Jendrossek, D., Handrick, R., 2002. Microbial degradation of polyhydroxyalkanoates. Annu. Rev. Microbiol. 56, 403–432.

Jendrossek, D., Schirmr, A., Schlegel, H.D., 1996. Biodegradation of polyhydroxyalkanoic acids. Appl. Microbiol. Biotechnol. 46, 451–463.

Jendrossek, D., Frisse, A., Behrends, A., Andermann, M., Kratzin, M.H.D., Stanislawski, T., Schlegel, H.G., 1995. Biochemical and molecular characterization of the Pseudomonas lemoignei polyhydroxyalkanoate depolymerase system. J. Bacteriol. 177, 596–607.

Jeong, I., Kim, B.S., Lee, H., Lee, K.M., Shim, I., Kang, S.K., Yin, C.S., Hahm, D.H., 2009. Prolonged analgesic effect of PLGA-encapsulated bee venom on formalin induced pain in rats. Int. J. Pharm. 380, 62–66.

Karanikolopoulos, N., Zamurovic, M., Pitsikalis, M., Hadjichristidis, N., 2010. Poly(D,L lactide)-*b*-poly(*N,N*-dimethylamino-2-ethylmethacrylate): synthesis, characterization, micellization behavior in aqueous solutions, and encapsulation of the hydrophobic drug dipyridamole. Biomacromolecules 1 , 430–438.

Kasuya, K., Doi, Y., Yao, T., 1994. Enzymatic degradation of poly[(R)-3-hydroxybutyrate] by *comamonas testosteroni* ATSU of soil bacterium. Polym. Degrad. Stab. 45, 379–386.

Kasuya, K., Takano, T., Tezuka, Y., Hsieh, W.C., Mitomo, H., Doi, Y., 2003. Cloning, expression and characterization of a poly(3-hydroxybutyrate) depolymerase from *Marinobacter* sp. NK-1. Int. J. Biol. Macromol. 33, 221–226.

Khang, G., Choi, H.S., Rhee, J.M., Yoon, S.C., Cho, J.C., Lee, H.B., 2000. Controlled release of gentamicin sulfate from poly(3-hydroxybutyrate-*co*-3-hydroxyvalerate) wafers for the treatment of osteomyelitis. Korean Polym. J. 8, 253–260.

Khang, G., Kim, S.W., Cho, J.C., Rhee, J.M., Yoon, S.C., Lee, H.B., 2001. Preparation and characterization of poly(3-hydroxybutyrate-*co*-3-hydroxyvalerate) microspheres for the sustained release of 5-fluorouracil. Bio-Med. Mater. Eng. 11, 89–103.

Kilicay, E., Demirbilek, M., Turk, M., Guven, E., Hazer, B., Denkbas, E.B., 2011. Preparation and characterization of poly(3-hydroxybutyrate-*co*-3-hydroxyhexanoate) (PHBHHX) based nanoparticles for targeted cancer therapy. Eur. J. Pharm. Sci. 44, 310–320.

Kim, M.N., Lee, A.R., Chin, I.J., Yoon, J.S., 1999. Biodegradability of poly(3-hydroxybutyrate) blended with poly(ethylene-*co*-vinyl acetate) or poly(ethylene oxide). Euro. Polym. J. 35, 1153–1158.

Kim, H., Lee, J., Kim, H., Kim, Y., 2009. Enzymatic synthesis of a drug delivery system based on polyhydroxyalkanoate-protein block copolymers. Chem. Commun. 46, 7104–7106.

Klingbeil, B., Kroppenstedt, R., Jendrossek, D., 1996. Taxonomical identification of *Streptomyces exfoliates* K10 and characterization of its poly(3-hydroxybutyrate) depolymerase genes. FEMS Microbiol. Lett. 142, 215–221.

Knop, K., Hoogenboom, R., Fischer, D., Schubert, U.S., 2010. Poly(ethylene glycol) in drug delivery: pros and cons as well as potential alternatives. Angew. Chem. Int. Ed. 49, 6288–6308.

Koosha, F., Muller, R.H., Davis, S.S., 1988. A continuous flow system for in vitro evaluation of drug-loaded biodegradable colloidal barriers. J. Pharm. Pharmacol. 40, 131.

Kopeček, J., Kopečková, P., 2003. Macromolecular therapeutics: state-of-the-art and future potential. Bull. Tech. Gattefossé 96, 9–21.

Köse, G.T., Ber, S., Korkusuz, F., Hasirci, V., 2003. Poly(3-hydroxybutyric acid-*co*-3-hydroxyvaleric acid) based tissue engineering matrices. J. Mater. Sci. Mater. Med. 14, 121–126.

Kumagai, Y., Doi, Y., 1993. Synthesis of a block copolymer of poly(3-hydroxybutyrate) and poly(ethyelene glycol) and its application to biodegradable polymer blends. J. Polym. Environ. 1, 81–87.

Kumar, V., Homme, R.K.P., 2008. Thermodynamic limits on drug loading in nanoparticle cores. J. Pharm. Sci. 97, 4904–4914.

Lammers, T., Hennink, W., Storm, G., 2008. Tumor-targeted nanomedicines: principles and practice. Br. J. Cancer 99 (3), 392–397.

Langer, R., Tirrell, D.A., 2004. Designing materials for biology and medicine. Nature 428, 487–492.

Larsen, A.K., Escargueil, A.E., Skladanowski, A., 2000. Resistance mechanisms associated with altered intracellular distribution of anticancer agents. Pharmacol. Ther. 85, 217–229.

Lee, J., Jung, S., Park, C., Kim, H., Batt, C.A., Kim, Y., 2011. Tumor-specifichybrid polyhydroxybutyrate nanoparticle: surface modification of nanoparticle by enzymatically synthesized functional block copolymer. Bioorg. Med. Chem. Lett. 21, 2941–2944.

Li, X., Chang, H., Luo, H., Wang, Z., Zheng, G., Lu, X., He, X., Chen, F., Wang, T., Liang, J., Xu, M., 2015. Poly (3-hydroxybutyrate-*co*-3-hydroxyhexanoate) scaffolds coated with PhaP-RGD fusion protein promotes the proliferation and chondrogenic differentiation of human umbilical cord mesenchymal stem cells in vitro. J. Biomed. Mater. Res. A 103, 1169–1175.

Lionzo, M.I.Z., Re, M.I., Guterres, S.S., Pohlmann, A.R., 2007. Microparticles prepared with poly(hydroxybutyrate-*co*-hydroxyvalerate) and poly(ε-caprolactone) blends to control the release of a drug model. J. Microencapsul. 24, 175–186.

Liu, M., Kono, K., Fréchet, J.M.J., 2000. Water-soluble dendritic unimolecular micelle: their potential as drug delivery agents. J. Control. Release 65, 121–131.

Lu, X., Ciraolo, E., Stefenia, R., Chen, G., Zhang, Y., Hirsch, E., 2011. Sustainedrelease of PI3K inhibitor from PHA nanoparticles and in vitro growth inhibition of cancer cell lines. Appl. Microbiol. Biotechnol. 89, 1423–1433.

Lu, X.Y., Li, M.C., Zhu, X.L., Fan, F., Wang, L.L., Ma, J.G., 2014. Microbial synthesized biodegradable PHBHHxPEG hybrid copolymer as an efficient intracellular delivery nanocarrier for kinase inhibitor. BMC Biotechnol. 14 (1), 4.

Madison, L.L., Huisman, G.W., 1999. Metabolic engineering of poly(3-hydroxyalkanoates): from DNA to plastic. Microbiol. Mol. Biol. Rev. 63, 21–53.

Maeda, H., Sawa, T., Konno, T., 2001. Mechanism of tumor-targeted delivery of macromolecular drugs, including the EPR effect in solid tumor and clinical overview of the prototype polymeric drug SMANCS. J. Control. Release 74, 47–61.

Martin, D.P., Williams, S.F., 2003. Medical applications of poly(4-hydroxybutyrate): a strong flexible absorbable biomaterials. Biochem. Eng. J. 16, 97–105.

Masood, F., Chen, P., Yasin, T., Fatima, N., Hasan, F., Hameed, A., 2013. Encapsulation of ellipticine in poly-(3-hydroxybutyrate-*co*-3-hydroxyvalerate) based nanoparticles and its in vitro application. Mater. Sci. Eng. C 33, 1054–1060.

Mei, N., Zhou, P., Pan, L.F., Chen, G., Wu, C.G., Chen, X., Shao, Z.Z., Chen, G.Q., 2006. Biocompatibility of poly (3-hydroxybutyrate-*co*-3-hydroxyhexanoate) modified by silk fibroin. J. Mat. Sci Mat. Med. 17, 749–758.

Mergaert, J., Anderson, C., Wouters, A., Swings, J., 1994. Microbial degradation of poly(3-hydroxybutyrate) and poly(3-hydroxybutyrate-*co*-3-hydroxyvalerate) in compost. J. Environ. Polym. Degrad. 2, 177–183.

Mitragotri, S., Lahann, L., 2009. Physical approaches to biomaterial design. Nat. Mater. 8, 15–23.

Moghimi, S.M., Hedeman, H., Muir, I.S., Illum, L., Davis, S.S., 1993. An investigation of the filtration capacity and the fate of large filtered sterically stabilized microspheres in rat spleen. Biochim. Biophys. Acta 1157, 233–240.

Mosqueira, V.C.F., Legrand, P., Gulik, A., Bourdon, O., Gref, R., Labarre, D., Barratt, G., 2001. Relationship between complement activation, cellular uptake and surface physicochemical aspects of novel PEG-modified nanocapsules. Biomaterials 22, 2967–2979.

Mukai, K., Yamada, K., Doi, Y., 1993. Enzymatic degradation of poly(hydroxyalkanoates) by a marine bacterium. Polym. Degrad. Stab. 41, 85–91.

Müller, R.H., Jacobs, C., Kayser, O., 2001. Nanosuspensions as particulate drug formulations in therapy: rationale for development and what we can expect for the future. Adv. Drug. Deliv. Rev. 47, 3–19.

Nigmatullin, R., Thomas, P., Lukasiewicz, B., Puthussery, H., Roy, I., 2015. Polyhydroxyalkanoates, a family of natural polymers, and their applications in drug delivery. J. Chem. Technol. Biotechnol. 90, 1209–1221.

Panyam, J., Labhasetwar, V., 2003. Biodegradable nanoparticles for drug and gene delivery to cells and tissue. Adv. Drug. Del. Rev. 55, 329–347.

Papadimitriou, S., Bikiaris, D., 2009. Novel self-assembled core–shell nanoparticles based on crystalline amorphous moieties of aliphatic copolyesters for efficient controlled drug release. J. Control. Release 138, 177–184.

Parlane, N.A., Grage, K., Mifune, J., Basaraba, R.J., Wedlock, D.N., Rehm, B.H.A., Buddle, B.M., 2011. Vaccines displaying mycobacterial proteins on biopolyester beads stimulate cellular immunity and induce protection against tuberculosis. Clin. Vaccine Immunol. 19, 37–44.

Perez, C., Sanchez, A., Putnam, D., Ting, D., Langer, R., Alonso, M.J., 2001. Poly(lactic acid)-poly(ethylene glycol) nanoparticles as new carriers for the delivery of plasmid DNA. J. Control. Release 75 (1), 211–224.

Pötter, M., Steinbüchel, A., 2005. Poly(3-hydroxybutyrate) granule-associated proteins: impacts on poly(3-hydroxybutyrate) synthesis and degradation. Biomacromolecules 6, 552–560.

Pötter, M., Madkour, M.H., Mayer, F., Steinbuchel, A., 2002. Regulation of phasin expression and polyhydroxyalkanoate (PHA) granule formation in *Ralstonia eutropha* H16. Microbiology 148, 2413–2426.

Pötter, M., Muller, H., Steinbüchel, A., 2005. Influence of homologous phasins (PhaP) on PHA accumulation and regulation of their expression by the transcriptional repressor PhaR in *Ralstonia eutropha* H16. Microbiology 151, 825–833.

Preusting, H., Kingma, J., Huisman, G., Steinbüchel, A., Witholt, B., 1993. Formation of polyester blends by a recombinant strain of *Pseudomonas oleovorans:* different poly(3-hydroxyalkanoates) are stored in separate granules. J. Environ. Polym. Degrad. 1, 11–21.

Prieto, M., Buhler, B., Jung, K., Witholt, B., Kessler, B., 1999. PhaF a polyhydroxyalkanoate-granule-associated protein of *Pseudomonas oleovorans* GPo1 involved in the regulatory expression system for *pha* genes. J. Bacteriol. 181, 858–868.

Qu, X.H., Wu, Q., Zhang, K.Y., Chen, G.Q., 2006. In vivo studies of poly(3-hydroxybutyrate-*co*-3-hydroxyhexanoate) based polymers: biodegradation and tissue reactions. Biomaterials 27, 3540–3548.

Ravenelle, F., Marchessault, R.H., 2002. One-step synthesis of amphiphilic diblock copolymers from bacterial poly([R]-3hydroxybutyric acid). Biomacromolecules 3, 1057–1064.

Rho, J.K., Choi, M.H., Gutierrez, M., Tian, B., Yoo, T., Baek, J.E., Shah, M., Yoon, S.C., 2014. Amorphous poly(3-hydroxybutyrate) nanoparticles prepared with recombinant phasins and PHB depolymerase. J. Appl. Polym. Sci. 131, Available from: http://onlinelibrary.wiley.com/wol1/doi/10.1002/app.41074/full.

Saigusa, H., Shiraki, M., Saito, T., 2002. Cloning of an intracellular D(−)-3-hydroxybutyrate-oligomer hydrolase gene from *Ralstoniaeutropha* H16 and identification of the active site serine residue by site-directed mutagenesis. J. Biosci. Bioeng. 94, 106–112.

Sendil, D., Gürsel, I., Wise, D.L., Hasirci, V., 1999. Antibiotic release from biodegradable PHBV microparticles. J. Control. Release 59, 207–217.

Shah, M., Naseer, M.I., Choi, M.H., Kim, M.O., Yoon, S.C., 2010. Amphiphilic PHA-mPEGcopolymericnanoconatianers for drug delivery: preparation, characterization and in vitro evaluation. Int. J. Pharm. 400, 165–175.

Shah, M., Choi, M.H., Ullah, N., Kim, M.O., Yoon, S.C., 2011. Synthesis and characterization of PHV-block-mPEGdiblock copolymer and its formation of amphiphilic nanoparticles for drug delivery. J. Nanosci. Nanotech. 11, 5702–5710.

Shah, M., Ullah, N., Choi, M.H., Kim, M.O., Yoon, S.C., 2012. Amorphous amphiphilic P(3HV-*co*-4HB)-*b*-mPEG block copolymer synthesized from bacterialcopolyester via melt transesterification: nanoparticle preparation, cisplatin-loading for cancer therapyand in vitro evaluation. Eur. J. Pharm. Biopharm. 80, 518–527.

Shah, M., Choi, M.H., Ullah, N., Kim, M.O., Yoon, S.C., 2014. Nanoscale poly (4-hydroxybutyrate)-mPEG carriers for anticancer drugs delivery. J. Nanosci. Nanotech. 14, 1–6.

Shahin, M., Lavasanifer, A., 2010. Novel self-associating poly(ethylene oxide)-bpoly(epsilon-caprolactone) based drug conjugates and nano-containers for paclitaxel delivery. Int. J. Pharm. 389, 213–222.

Shinomiya, M., Iwata, T., Kasuya, K., Doi, Y., 1997. Cloning of the gene for poly(3-hydroxybutyric acid) depolymerase of *Comamonas testosterone* and functional analysis of its substrate-binding domain. FEMS Microbiol. Lett. 154, 89–94.

Shuai, X.T., Jedlinski, Z., Luo, Q., Nozirov, F., 2000. Synthesis of novel block copolymers of poly(3-hydroxybutyric acid) with poly(ethyelene glycol) through anionic polymerization. Chin. J. Polym. Sci. 18, 19–23.

Sinha, V.R., Bansal, K., Kaushic, R., Kumria, R., Trehan, A., 2004. Poly-ε-caprolactone microspheres and nanospheres: an overview. Int. J. Pharm. 278, 1–23.

Sudesh, K., Doi, Y., 2000. Molecular design and biosynthesis of biodegradable polyesters. Polym. Adv. Technol. 11, 865–872.

Sun, Z., Ramsay, J.A., Guay, M., Ramsay, B.A., 2007. Fermentation process development for the production of medium-chain-length poly-3-hydroxyalkanoates. Appl. Microbiol. Biotechnol. 75, 475–485.

Timbart, L., Renard, E., Langlois, V., Guerin, P., 2004. Novel biodegradable copolyesters containing blocks of poly(3-hydroxyoctanoate) and poly(ε-caprolactone): synthesis and characterization. Macromol. Biosci. 4, 1014–1020.

Valappil, S.P., Misra, S.K., Boccaccini, A., Roy, I., 2006. Biomedical applications of polyhydroxyalkanoates, an overview of animal testing and in vivo responses. Expert. Rev. of Med. Dev. 3, 853–868.

Verlinden, R.A.J., Hill, D.J., Kenward, M.A., Williams, C.D., Radecak, I., 2007. Bacterial synthesis of biodegradable polyhydroxyalkanoates. J. Appl. Microbiol. 102, 1437–1449.

Vila, A., Sanchez, A., Tobo, M., Calvo, P., Alonso, M.J., 2002. Design of biodegradable particles for protein delivery. J. Control. Release 78, 15–24.

Vilos, C., Gutierrez, M., Escobar, R., Morales, F., Denardin, J., Velasquez, L., Altbir, D., 2013. Superparamagnetic poly (3-hydroxybutyrate-*co*-3 hydroxyvalerate) (PHBV) nanoparticles for biomedical applications. Electr. J. Biotechnol. 16, 8–18.

Williams, S.F., 2000. Bioabsorbable, biocompatible polymers for tissue engineering. PCT Patent Application No. WO 00/51662.

Williams, S.F., Martin, D.P., Horowitz, D.M., Peoples, O.P., 1999. PHA applications: addressing the price performance issue: I. Tissue engineering. Int. J. Biol. Macromol. 25, 111–121.

Williams, S.F., Martin, D.P., Skraley, F., 2000. Medical devices and applications of polyhydroxyalkanoate polymers. PCT Patent Application No. WO 00/56376.

Wong, P.A., Chua, H., Lo, W., Lawford, H.G., Yu, P.H., 2002. Production of specific copolymers of polyhydroxyalkanoates from industrial waste. Appl. Biochem. Biotechnol., 655–662, 98–100.

Xiong, Y.C., Yao, Y.C., Zhan, X.Y., Chen, G.Q., 2010. Application of polyhydroxyalkanoates nanoparticles as intracellular sustained drug-release vectors. J. Biometer. Sci. Polym. Ed. 21, 127–140.

Yamazaki, M., Ito, T., 1990. Deformation and instability in membrane structure of phospholipid vesicles caused by osmophobic association-mechanical stress model for the mechanism of poly(ethylene glycol) induced membrane fusion. Biochemistry 29, 1309–1314.

Yao, Y., Zhan, X., Zhang, J., Zou, X., Wang, Z., Xiong, Y.C., Chen, J., Chen, G.Q., 2008. A specific drug targeting system based on polyhydroxyalkanoate granule binding protein PhaP fused with targeted cell ligands. Biomaterials 29, 4823–4830.

Yoo, H.S., Lee, K.H., Oh, J.E., Park, T.G., 2000. In vitro and in vivo anti-tumor activities of nanoparticles based on doxorubicin_PLGA conjugates. J. Control. Release 68, 419–431.

York, G.M., Stubbe, J., Sinskey, A.J., 2002. The *Ralstonia eutropha* PhaR protein couples synthesis of the PhaPphasin to the presence of polyhydroxybutyrate in cells and promotes polyhydroxybutyrate production. J. Bacteriol. 184, 59–66.

Zhang, L.L., Xiong, C.D., Deng, X.M., 1995. Biodegradable polyester blends for biomedical application. J. Appl. Polym. Sci. 56, 103–112.

Zhang, L.L., Deng, X.M., Zhao, S.J., Huang, Z.T., 1997. Biodegradable polymer blends of poly(3-hydroxybutyrate) and hydroxyethyl cellulose acetate. Polymer 38 (24), 6001–6007.

Zhang, C., Zhao, L., Dong, Y., Zhang, X., Lin, J., Chen, Z., 2010. Folate-mediated poly(3-hydroxybutyrate-*co*-3-hydroxyoctanoate) nanoparticles for targeting drug delivery. Eur. J. Pharm. Biopharm. 76, 10–16.

Zhao, K., Deng, Y., Chen, J.C., Chen, G.Q., 2003. Polyhydroxyalkanoate (PHA) scaffolds with good mechanical properties and biocompatibility. Biomaterials 24, 1041–1045.

CHAPTER

2

A NOVEL APPROACH TO THE ORAL DELIVERY OF BIONANOSTRUCTURES FOR SYSTEMIC DISEASE

Rosa Martha Perez Gutierrez, Juan Vicente Mendez Mendez, Israel Arzate Vazquez

National Polytechnic Institute, Mexico City, Mexico

CHAPTER OUTLINE

1 Introduction 28
2 NanoStructures for Oral Drug Delivery 29
3 The Gastrointestinal Mucus Barriers 30
3.1 Mucus in the GI Tract 30
3.2 Absorption in the Gastrointestinal Tract 32
3.3 Transport Mechanisms in the GI Tract 33
3.4 Mucosal Barrier 34
3.5 Mucus Interactions 34
4 Strategies for Effective Oral Delivery of Drugs 35
5 Challenges in the Oral Delivery Route 35
6 Characteristics of the Transporter Used in Oral Delivery Nanostructures 37
6.1 Oral Delivery Nanostructures 37
6.2 Hydrogels 37
6.3 Dendrimers 38
6.4 Micelles 40
6.5 Liposomes 40
6.6 Nanoparticles 41
6.7 Carbon Structures 42
6.8 Polymers Used in Drug-Delivery Systems 43
7 Therapeutic Applications of Oral Delivery Nanostructures 45
7.1 Cerebrovascular Disease 45
7.2 Hypercholesterolemia 46
7.3 Arthritis 46
7.4 Nanostructures in Cancer 47
7.5 Diabetes 48
7.6 Hepatitis B 49

Nanostructures for Oral Medicine

8 **Synthesis of Nanoparticles** 49
9 **Characterization of Nanoparticles** 49
9.1 Light Microscopy 49
9.2 Scanning Electron Microscopy 49
9.3 Transmission Electron Microscopy 50
9.4 Dynamic Light Scattering 50
9.5 Zeta Potential (ζ)Analysis 50
9.6 Raman Spectroscopy 50
9.7 Atomic Force Microscopy 51
9.8 X-Ray Photoelectron Spectroscopy or Electron Spectroscopy for Chemical Analysis 51
9.9 Fourier Transform Infrared Spectrometry 51
10 **Conclusions** 52
References 52

1 INTRODUCTION

Nanomedicine uses nanotechnology to improve human health at the molecular scale. These combine the effect therapeutic drugs themselves and the abilities to deliver drugs to specific tissues or cells and also creates the release under certain physiological conditions. Nanoparticles are small, with a increase in the surface area with a large proportion of atoms/molecules, so that they have a larger surface area per unit/mass compared with other large particles that may increase biological activity.

Due to the importance of effective use and the safety drugs, the controlled drug delivery system has been developed. During the past decades, important studies had been made in the research of drug-delivery systems (DDS) for their great potential to administer local anesthesia, antioxidants, chemotherapy, antiinflammatories, which are attributed to the developments in nanotechnology involving the preparation of carriers of different structures for protecting or delivering several pharmaceutical substances. The synthesis of nanostructures and novel nanomaterials have received special interest because it makes it possible to control geometry, surface charge, physicochemical properties, and surface tailoring with bioactive compounds.

A variety of nanoplatforms, such as nanoparticles (Hu et al., 2012), superparamagnetic iron oxide nanoparticles (Shi et al., 2012), graphene (Yang et al., 2013), biodegradable polymers (Feng et al., 2009), metal nanostructures (Narayanan et al., 2008), dendrimers (Cloninger, 2002), and liposomes (Cevc, 2003), silicon-based structures (Li et al., 2004) have been developed for drug delivery. Release of the drugs to specific areas within the body can reduce systemic side effects and allow a more efficient use of the drug. Nanostructures are candidates that efficiently allow targeted delivery of drugs. A sustained drug release and the intracellular entry capability are properties of the nanostructures that can minimize side effects and allow the direct treatment of the cause of the disease. Nanocarriers allow the delivery of small-molecule drugs, proteins, and nucleic acids, and also improve the solubility of drugs. This is helpful for drug candidates that did not pass through trial phases and can be reformulated with new carrier systems.

The pathophysiological changes of the diseases offer opportunities for the delivery of the drugs loaded in nanostructures. Bioactive targeting can be achieved by taking advantage of these specific characteristics of abnormal tissues (Vasir et al., 2005). An ideal targeting system should be present at sufficient concentrations at the target site, should not lose its activity, and should have long circulation time.

Oral delivery is an important route of administration due to patient compliance and ease of administration to deliver therapeutics via nanoparticles. Oral formulations are easy to produce because they do not need to be manufactured under sterile conditions.

A good nanoparticle designed to obtain the desired results is in the function of the capability to deliver the drug at the right place and at the right time in an amount adequate enough to yield a therapeutic response (Thanki et al., 2013). The most important characteristics for oral via applications are the biocompatibility and biodegradability of the material, which has to be biocompatible and easy to biodegrade by the organism. For this reason some of the most used materials are biopolymers and lipids; however, other materials with good biocompatibility are used, such as metals and carbon nanotubes. The size of the particles is important because the carrier has to pass gastric and immune system barriers, because the appropriatesize of the nanoparticle helps to improve the efficacy (Wina and Feng, 2005). The loaded drug, the surface properties of the nanoparticules, and the designed nanoparticles can be coated with different substances to improve its efficacy (Wina and Feng, 2005).

Drug release, in the case of toxic drugs, is desired to release the medication slowly in order to reduce the toxicity. For this reason an important characteristic of the nanoparticles is that they can control the drug liberation rate to increase the efficacy and at the same time extend the effects of the drug resulting in patient compliance. This chapter highlights nanostructures in the oral route of drug administration and the interaction mechanisms between nanostructures and drug molecules.

2 NANOSTRUCTURES FOR ORAL DRUG DELIVERY

The oral route for drug delivery is regarded as the optimal route for achieving therapeutic benefits owing to increased patient compliance. Although other rouers of administration as intranasal, transdermal, injectable, rectal have had a great advances the oral drug delivery remains well ahead of the pack as the preferred delivery route (Taratula et al., 2013). The development of new encapsulation methods for drug carriers allows the delivery of not only small-molecule drugs but also the delivery of nucleic acids and proteins to specific areas within the body, and controls drug rate liberation, which will reduce systemic side effects and allow for more efficient use of the applied drug (Gareth and Hughes, 2005).

Oral application has some advantages, like applications safety, convenience, and discomfort elimination compared with other methods. Also, economically, the elimination of specialized applicants is interesting in the development of this method. The treatment cost for the patient can be highly reduced due to avoidance of hospitalization, sterile manufacturing, and trained personnel assistance. Oral chemotherapy reduces the discomfort of injection and can be conducted at home. Also, this eliminates risks of infection compared with intravenous methods (Thanki et al., 2013).

Nanotubes, nanocells, nanoparticles, lipid emulsions, nucleic acid nanostructures (Singh and Lillard 2009), lipid–drug conjugates, nanocrystals (Gulsun et al., 2009), and dendrimers (Menjoge et al., 2010) are structures tested to deliver drugs. However, one of the most investigated are micelles and liposomes, which can be used to entrap and release drug molecules (Gareth and Hughes, 2005). Several structures have been tested in vitro and/or in vivo, in order to determine parameters like efficiency of the encapsulation, correct targeting, toxicity, and nanoestructura shape (size).

In vivo experiments are generally conducted on rodents. This type of study should take into consideration how the results translate to humans. Mice, rats, and rabbits are most commonly used, but these rodents do not produce as much mucin as humans, thus reducing the barrier properties to nanoparticle distribution and drug delivery in the gastrointestinal (GI) tract (Kreuter et al., 1989). These have a high density of Peyer's patches as compared to humans (Gruber et al., 1987). In addition, rodents have less acidic stomachs, which can produce a significant alteration in the stability. Also, they have significant differences in bacterial colonization, stomach emptying times, mucus thicknesses, and intestinal transit times, among others.

3 THE GASTROINTESTINAL MUCUS BARRIERS

3.1 MUCUS IN THE GI TRACT

Mucosae is a protective layer of fluid called mucus, which acts as a physical barrier to biological and chemical attack, as well as a natural lubricant opposing shear damaging, which have homeostatic functions, such as in regulating ion transport and water balance, clearing of cellular debris, mucosal immune-regulation (Fig. 2.1). Mucus is composed of carbohydrates, proteins, salts, cellular debris, lipids, bacteria, and antibodies. The term mucins includes a heterogeneous group of glycoproteins that are coded by the MUC gene family but differ in their glycosylation and polypeptide sequences (Dekker et al., 2002).

Mucins with high molecular weight (106–107 g/mol) are formed by glycoproteins ≈75%, carbohydrate/25% aminoacid residues linked via o-glycosidic bonds between serine or threonine residues and *N*-acetylgalactosamine (Cone and Mucus, 2005). The low isoelectric point of mucin determines its negative charge at most physiological pH values.

Mucus is a complex hydrogel with mucins as the main protein component, which can be either secreted or cell-bound of the which, MUC6, MUC5AC, and MUC2, are the secreted mucin types found throughout the GI tract (Corfield et al., 2001). The 2%–5% of mucus by wet weight, mucin molecules entangle and crosslink adhesively and reversibly to form a dynamic viscoelastic gelwith shear-thinning properties. Large amounts of threonine, proline, and serine are found in the protein backbone linked to oligosaccharides.

These oligosaccharides vary in branching and size depending on the location; when they have a long, branched oligosaccharide side chains on mucins, they protect against proteolytic degradation (Macadam, 1993). The GI tract presents adherent mucus layers that include the collective glycolipids, cell-bound mucins (MUC1, 3, 4, 12, and 13) and glycoproteins that make up the glycocalyx.

To protect the epithelium, there is constant movement of the adherent layer and the glycocalyx in the GI tract that serves to eliminate microorganisms introduced in the diet and damaging compounds. In addition, GI mucus secretes antimicrobial molecules as defensins, immunoglobulins, and lysozymes (McGuckin et al., 2011). Also, it presents a balance between nutrient and absorption rate and protective capability (Corfield et al., 2001). Therefore, a regulated balance occurs among degradation, removal, and mucin synthesis, which affect the thickness of the mucus barrier. The viscoelastic properties of GI mucus are essential for lubrication and protective properties, helping food to be peristaltically transported through the GI tract without damaging the epithelium. Mucus is continuously secreted to remove pathogens and to lubricate the epithelium, assisting material that passes through it, decreasing the permanency time of nanoparticles that penetrate in adherent layer of mucus. Mucins present in mucus are involved in adhesion phenomena.

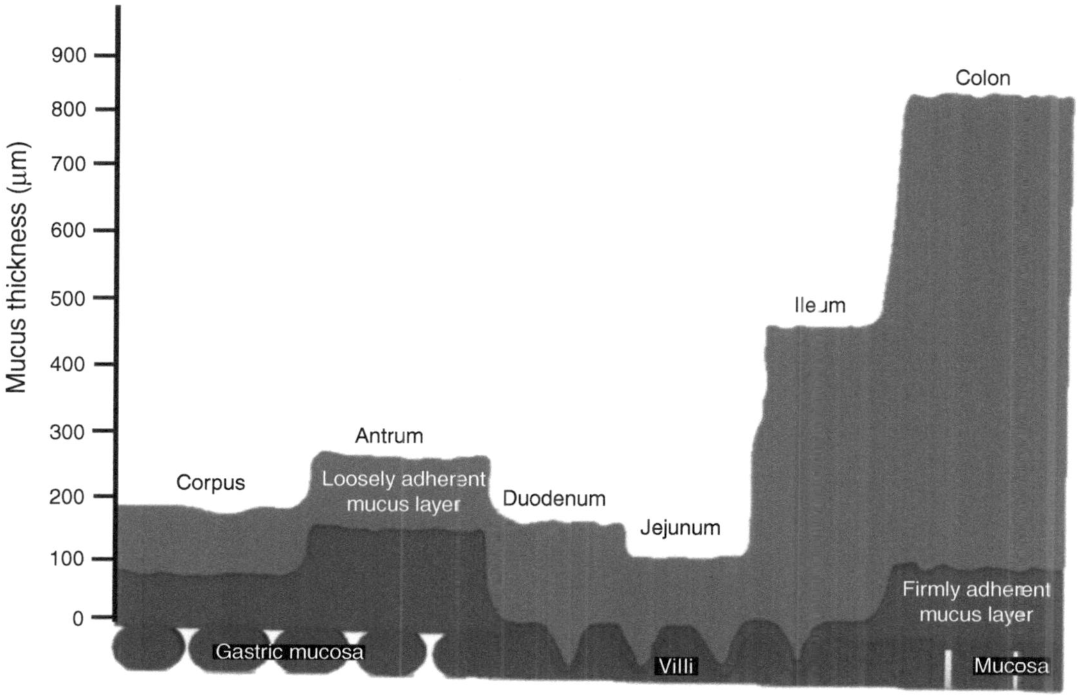

Thickness (μm)	Corpus	Antrum	Duodenum	Jejunum	Ileum	Colon
Total	189±11	274±41	170±38	123±4	480±47	830±110
Firmly adherent	80±5	154±16	16±3	15±2	29±8	116±51
Loosely adherent	109±12	120±38	154±39	108±5	447±47	714±109
n	6	6	7	6	6 (8 Firm)	11

FIGURE 2.1 Thicknesses of the Mucus Gel Layers in the Antrum, Corpus, Distal Ileum Proximal Jejunum, Midduodenum, and Proximal Colon of the Rat GI

In vitro techniques are useful in determining the ability of the nanocarrier system to overcome the barriers in the GI tract in the in vivo environment. Models that simulate membrane analysis and gastric fluids are used to correlate in vivo environment without the use of human cell lines. The GI tract presents a unique microenvironment of ionic strength and enzymes that affect the colloidal stability and chemicals of the nanocarrier.

Another important consideration is that the barrier properties of mucus are altered in disease states (Fig. 2.2). The GI tract is confronted with the important mission of allowing uptake and the transfer of fluids and nutrients, while being confronted with pathogenic bacteria and harmful substances. Even so the GI tract is able to maintain a healthy homeostasis, considering that it is colonized by ∼1014

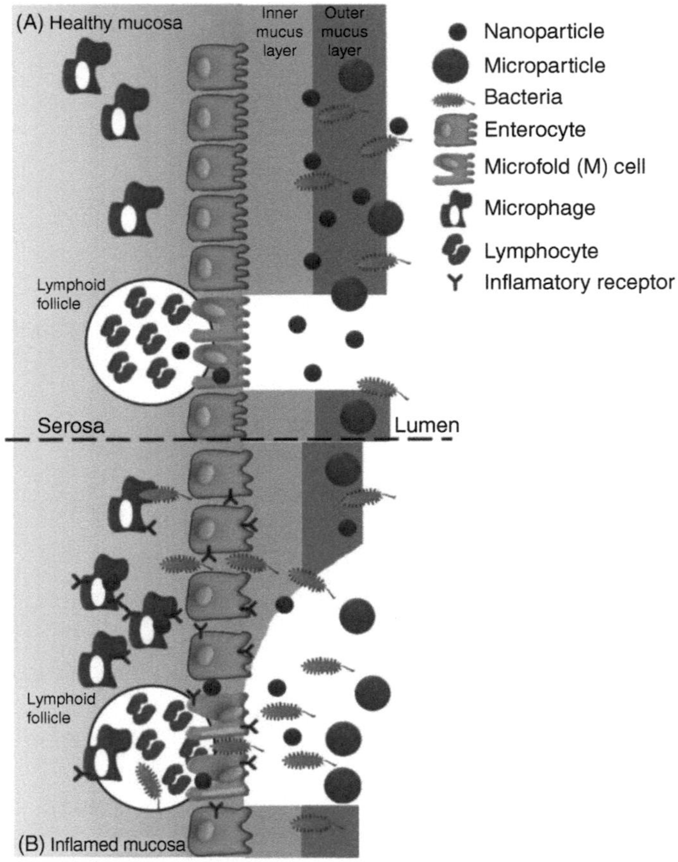

FIGURE 2.2 Changes in Inflammatory Bowel Disease in the Mucosal Barrier

The healthy mucosa (A) is protected by an outside mobile mucus-gel layer and inner adherent both acts as a barrier to hydrophobic compounds and large molecules. Mucosal inflammation (B) is associated with a loss of epithelial barrier integrity, outside mobile mucus-gel layers, and inner adherent. In addition, increased of intestinal bacteria leading to a hitch of immune cells to the mucosal. These changes lead to a uptake of nanoparticles, accumulation by macrophages and enterocytes increasing the exposure of inflammatory receptor targets to nanoparticles.

bacteria of more than 500 different species and the enormous surface area (300–400 m^2 in humans) (Schenk and Mueller, 2008). Although it is unclear if a disease is first established, followed by an alteration in the mucosal barrier, it is known that histochemical alterations in mucins occur in many GI diseases (Owen and Reid, 1995).

3.2 ABSORPTION IN THE GASTROINTESTINAL TRACT

The absorption of a drug depend mainly on the physiology of the GI tract, solubility of the drug, type of dosage formulation, chemical form, pK_a particle, size, hydrophobicity, molecular weight, ionization constants, pH stability of absorbing molecules, and biological barriers that restrict absorption from the GI tract. Some drugs are absorbed by passive diffusion in the GI tract, where the drug concentration gradient across the barrier is proportional to the rate of absorption. Other drugs are absorbed by a combination of passive and active transport that, depending on their location, can increase and/or decrease absorption. Among the most common limitations in the absorption by the GI there are a low bioavailability, a low solubility, and the mucus barrier can prevent drug penetration and a poor stability in the gastric environment. Poor water solubility of a drug is a main challenge in drug delivery research and limits pharmacokinetic parameters and bioavailability. To overcome these problems the nanostructure formulations are being developed to protect drugs and release them in a controlled manner.

3.3 TRANSPORT MECHANISMS IN THE GI TRACT

There are different mechanisms for molecules to cross the cell membrane: via transcellular, paracellular, receptor-mediated transport, and carrier-mediated.

Paracellular transport is the passage of a substance through the intercellular spaces in between epitelial cells, which is passive and results from diffusion and is under the control of tight junctions. These constitute the main rate limiting barriers toward the paracellular transport for permeation of ions and larger substances (Madara, 1998). The tight junctions have biophysical characteristics including size and charge selectivity, conventional ion channels, competition between permeant molecules, dependency of permeability on the ion concentration, and sensitivity to pH and anomalous mole-fraction effects (Tang and Goodenough, 2003). Paracellular transport varies among epithelia in terms of electrical resistance and shows small differences in ionic selectivity, complementing the transcellular mechanism by defining the degree and selectivity of reverse leaks for solutes and ions, making an important tissue-specific contribution to overall transport (Powell, 1981). In addition it is not largely determined by lipophilicity and the hydrogen-bonding capacity.

Transcellular transport is the pathway of substances through the intestinal epithelial cells by transcytosis, a process by which particles are taken up by cells, depending on various physicochemical properties of particles, such as lipophilicity, size, hydrogen bond potential, surface hydrophobicity, charge, the physiology of the GI tract, and the presence of a ligand at the particle surface (Florence, 2004).

M cells and enterocytes are the main intestinal cells for transport. Enterocytes represent the majority of cells lining the GI tract. M cells are located within the epithelium of Peyer's patches and represent a small proportion of the intestinal epithelium (1% of the total intestinal surface, i.e., 5% of the human follicle-associated epithelium [FAE]) (Giannasca et al., 1999). M cells have a high endocytosis ability and a high transcytotic capacity and are able to transport a wide variety of materials, including nanoparticles (Clark et al., 2000). M cells by adsorptive endocytosis viaclathrin-coated pits, phagocytosis, fluid phase endocitosis, and vesicles, take particles, microorganisms, and macromolecules. In addition, they are also used by a series of pathogens as a route for host invasion (Gebert et al., 1996). However, this route is limited to the transport of low molecular-weight lipophilic drugs.

In carrier-mediated transport the drugs are transferred across the cell membrane or entire cell and then released into circulation from the basal surface of the enterocyte (Russell-Jones, 1999). The process is mainly used for small hydrophilic molecules (Barthe et al., 1999). The carriers transport them across the membranes into the GI epithelium, and recognize target molecules through membrane receptors.

In receptor-mediated transport, drugs function as receptors for surface-attached ligands or specific ligand for surface-attached receptors (Russell-Jones, 1996). This transportation entails cell invagination,which leads to formation of a vesicle. This transportation is known as endocytosis, which includes pinocytosis, phagocytosis, potocytosis (nonclathrin-mediated), and receptor-mediated endocytosis (clathrin-mediated) (Swaan, 1998). Initially in this process there is a specific cell-surface receptor, receptor clustering, and internalization through coated vesicles into endosomal acidic compartments binding of the ligand. The follow pathway is dependent on the low endosomal pH and the type of the receptor/ligand pair. After protein drugs are transported to the GI tract, they enter the circulation through two different absorption pathways: the intestinal lymphatics and portal blood. In the intestinal, lymphatics are transported in highly lipophilic ligands that cross the epithelial barrier, which directly delivers them to the vena cava. Portal blood is the main pathway for the orally administered protein drugs. Hydrophilic ligands are carried to the liver through the hepatic portal vein, and then by the hepatic artery to the circulation, for subsequent delivery to their sites of action.

3.4 MUCOSAL BARRIER

Mucus determines the bioavailability and absorption of orally administered drugs. The mucosal barrier is a epithelial cell lining, which is bound by tight junctions that repel harsh fluids that may injure the mucosal lining. The blanket is secreted by surface epithelial cells and mucosal neck cells. This insoluble mucus forms a protective gel-like coating over the entire surface of the gastric mucosa (Meyer et al., 1984). Glycocalyx is the main mucosal barrier against delivery proteins in the oral route. The glycocalyx is a fibrous coat that is weakly acidic and contains sulfated mucopolysaccharides. The glycocalyx and mucus layers form barriers to proteins and peptides. Due to the interactive nature of these layers and viscosity, it offers resistance to the protein diffusion, which must first diffuse through these layers to reach the epithelial surface. The interaction of mucus and particles involves hydrogen bonding, van der Waals interactions, electrostatic/ionic interactions, and hydrophobic forces, which can influence the nanoparticle retention at the mucosal surface (Lai et al., 2009). Electrostatic adhesive interactions with particle aggregation and mucin affect the nanoparticle transport rate. Negatively charged groups, such as sulfate-and carboxylate showed a higher transport rate than neutral or positively groups. The amine nanoparticle transport is severely limited, likely by electrostatic adhesive interaction and particle aggregation with mucin (Crater and Carrier, 2010). In addition, drugs delivered to mucosal surfaces are usually removed by mucus clearance mechanisms. Mucus is one of the main obstacles to orally administrated drugs. Nanoparticles as drug carriers are an alternative to diffuse into the mucus layer and avoids the mucilliary clearance. On the other hand the mesh-pore spacing of the mucus is 50–1800 nm, so that there is a size limit to cross the intestinal mucosal barrier (Lai et al., 2010). Studies have shown that nanoparticles under 200 nm size effectively diffuse through the mucus (Primard et al., 2010). The transport of nanoparticles in different mucosal sites is dependent on its size. The pore network accommodates the movement of a number of particles whenever electrostatic mucoadhesive forces and hydrophobic can be minimized (Fig. 2.3).

3.5 MUCUS INTERACTIONS

Polymers can present different mucoadhesive depending on the bulk. On the other hand size and the available surface area determine the ability of particles to fit in the low viscosity aqueous mesh spaces and channels formed within the mucin matrix that composes mucus. The surface-area-to-volume ratio of nanosystems is very important as the interface available to establish bonding. Mucoadhesive

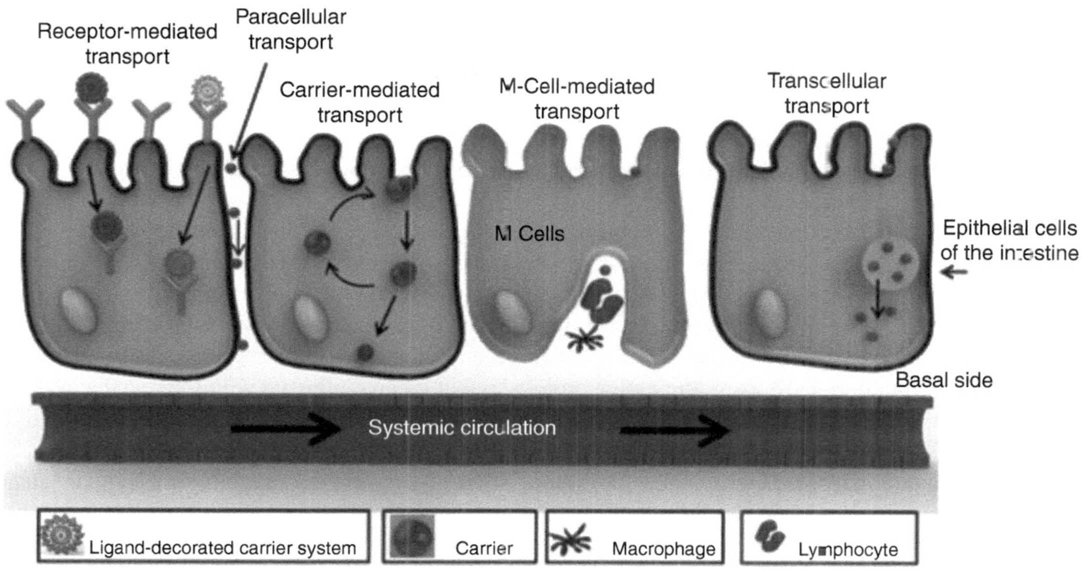

FIGURE 2.3 Drug Uptake Mechanisms in the Intestinal Epithelium

particles in the micrometric range due to the inability to fit in these channels tend to remain at the top layers of the mucus (Ponchel et al., 1997). The correct size adjustment is a fundamental characteristic in the modulation of the mucoadhesive under pathophysiological conditions The enhanced ability to penetrate mucus has been seen that it is not linear with size reduction. The nanosystems size may be a method that can be used to regulate mucoadhesion together. So that size, surface chemistry, and surface charge of nanosystems determine their mucoadhesive behavior.

4 STRATEGIES FOR EFFECTIVE ORAL DELIVERY OF DRUGS

Besides the existing barriers to oral delivery, there are also some advantages in the GI which include a sticky mucus lining, presence of microfold cells in the Peyer's Patch, and a large surface area, which is a highly beneficial property. The large surface area of 250 m^2 of the small intestine (Lozier et al., 1997), increasing the opportunity for transport and uptake that can occur through facilitated mechanisms or passive, active. Typically, nanoparticles tend to be transported by the transcellular route due to their size, functional groups, and surface (Rieux et al., 2005). The mucus lining consists in a viscous hydrated gel that reduces the shear effect from the movement of gastric juices. When a nanoparticle adheres to the mucus layer can increase the opportunity of transcellular migration into the lower layers of the epithelial layer, the lamina propia, and the epidermal wall (Ensign et al., 2012). Finally, it is the immunological organ, the Peyer's Patch, which contains specialized microfold cells (M cells) that can increase the transcytotic ability of the nanoparticles. M cells are highly efficient antigen deliverers that assist the transport of nanoparticles from the small intestine through the Peyer's Patch. Protection of peptide, drugs, or nucleic acid contents from degradative enzymes increased retention in the GI tract and increased mucoadhesion are some advantages of using nanoparticles.

Some factors should be taken into consideration in the development an oral delivery system, as the modification of epithelial permeability by small molecules (absorption enhancers), properties of itself, such as enzyme susceptibility and lipophilicity, the use of particulate delivery carrier system to traverse to the site of absorption as formed micelle or drugs loaded to nanoparticles and the addition of novel functionality capable of receptor recognition.

Absorption enhancers contain a hydrophobic moiety (polar charged) that associate with the drug molecules by charge-charge interaction or create a more lipophilic drug/carrier complex or intermolecular forces, or enabling the transport across the epithelial membrane. Enhancers fall into two categories: those that enable the systemic absorption of the drug via transcellular absorption without compromising the integrity of the intestinal epithelium, and those that induce chemical modification. The addition of functional groups to the native protein so that it can conjugate with the reactive groups without affecting receptor binding and activation might also improve the lipophilicity of the native protein. The resulting groups would readily be cleaved during or after absorption and promote the oral absorption of intact macromolecules. The addition of polar groups and conjugation of polymers, such as poly(ethylene glycol) (PEG) or chitosan (CS) are some common strategies to improve drug solubility (Lee et al., 2008). Polymer drug conjugates are widely used as DDS by improving solubility.

Production of pharmaceutical proteins for oral delivery plant-base is an excellent way to protect the proteins from the hard environment of the GI tract via encapsulation within the plant cell (Limaye et al., 2006). This system offers good manufacturing practice, low cost, no human pathogen risks, free from toxins, and easy scale up. Expression of proteins in chloroplasts has been found at high levels of expression.

5 CHALLENGES IN THE ORAL DELIVERY ROUTE

Generally, the peptides, vaccines, and proteins are administered parenterally because of their poor bioavailability from other different routes of administration, including the oral route. The enzymatic hydrolysis, its susceptibility to acid, and its physicochemical properties, including charge, size, and hydrophilicity, produce a poor intestinal absorption of these drugs. In addition to the hydrolysis in the stomach and GI tract, protein and peptide drugs must also survive degradation via bacterial fermentation in the colon (Chourasia and Jain, 2003). Oral routes of administration for biologic drugs have been studied in recent years, which can be enhanced to targeting diseased tissues, improved compliance, and controlled release rate. Development of an effective oral delivery system for biopharmaceuticals requires understanding the mechanisms involved in their absorption across targeted tissues, as well as the several barriers of the digestive tract. The challenges to oral delivery of biologic drugs are formulations of the components, processing conditions, and optimizing delivery approaches. The actual route for the delivery of pharmaceuticals most used is oral. However, attempts to deliver peptides and large molecular weight proteins orally are limited.

Other challenges of orally delivered drugs are that after drug transportation across the epithelium and subsequent absorption into the blood system and reaching the specific tissue or cell so it must also overcome additional intracellular and extra barriers, such as endocytosis into the target cells and entry into the nucleus (Belting et al., 2005).

6 CHARACTERISTICS OF THE TRANSPORTER USED IN ORAL DELIVERY NANOSTRUCTURES

6.1 ORAL DELIVERY NANOSTRUCTURES

In order to improve the efficiency of DDs, researchers have drawn on the use of nanostructures synthesized through nanotechnology techniques. Fig. 2.4 shows several nanostructures have been used as vehicles for oral drug delivery, such as nanoemulsions, micelles, nanoparticles, dendrimers, liposomes polymersomes, carbon structures, and hydrogels.

Oral delivery of polymer therapeutics is prevented by barriers in the GI tract. These include their hydrophilicity and large size limiting transepithelial transport. However, enzymatic activity and variations in pH in the GIT can affect the stability of polymer therapeutics leading to degradation of the matrix and premature release of the drug (Ponchel and Irache, 1998).

6.2 HYDROGELS

Hydrogels offer excellent potential as oral therapeutic systems due to biocompatibility and diversity of natural and synthetic materials and physiochemical and biological properties provide a variety of potential for pharmaceutical applications. In addition, stimuli-responsive hydrogels take advantage of physiological changes of the intestinal tract to achieve controlled release of protein, peptide, and are site-specific.

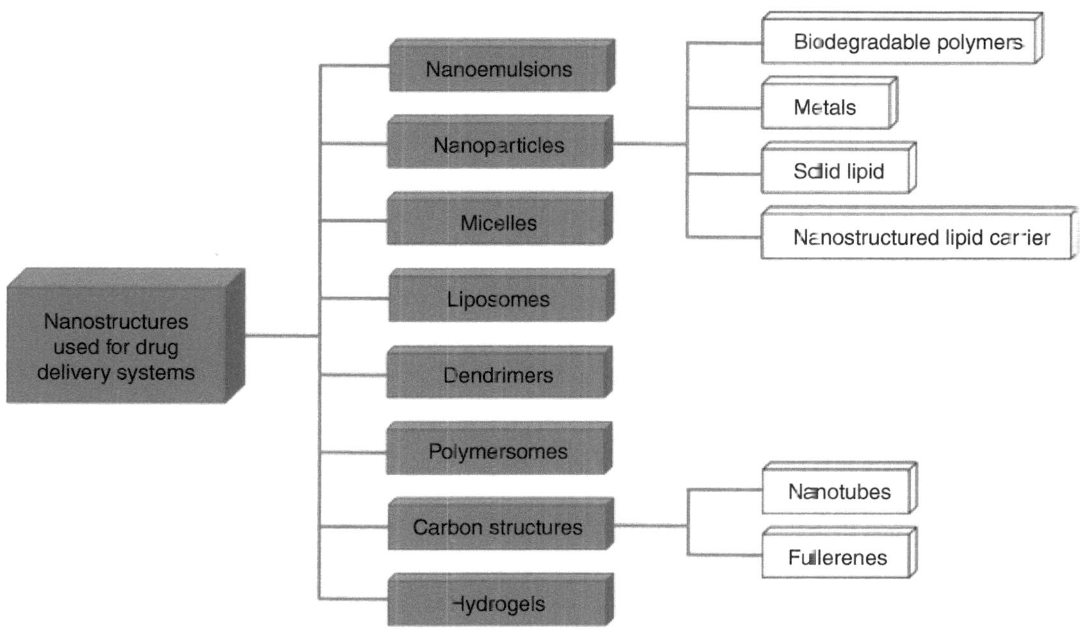

FIGURE 2.4 Diagram of Main Nanostructures Used as Drug Delivery Systems

Hydrogel nanoparticles, also called nanogels, are systems composed of hydrophobic polysaccharides, which can encapsulate different compounds, such as drugs, proteins, and antigens (Singh and Lillard, 2009). The synthesis of nanoparticles is relatively simple and based on the establishments of polymers crosslinked via chemical bonds, ionic interactions, and hydrogen or physical bonds (Gui et al., 2006). Because the nanogels are made mainly of biomolecules, such as polysaccharides and proteins, this feature provides a high degree of biocompatibility. Additionally, another advantage possessed by such nanostructures are that the release and encapsulation properties can be modified depending on the pH.

For instance, nanogels based on pullulan (polysaccharide) and cholesterol (compound encapsulated) have been developed and it has been found that these particles stimulate the immune system and they are quickly consumed by dendritic cells (Singh and Lillard, 2009).

Moreover, nanogels also have been intended for the treatment of chronic diseases, such as diabetes. Hidrogel nanoparticles were used by Kazunari et al. (1998) to encapsulate insulin to prevent the self-aggregation and the enzymatic degradation in oral DDS.

Natural polymers, such as CS and alginate are attractive matrices for oral drug delivery due to their physiochemical properties, biocompatibility, and mild gelation conditions (George and Abraham, 2006). Alginate is extracted from brown seaweed and is a linear polysaccharide composed of alternating blocks of β-D-mannuronic (M-block) acid and 1–4 linked α-L-guluronic (G-block) residues. Their ability to form gels under mild conditions is important for the applications of alginates as drug oral delivery systems. Alginate as coated beads, plain beads, and microcapsules for entrapping various biological molecules, including hemoglobin (Rasmussen et al., 2003), heparin (Edelmana et al., 2000), vaccines (Kim et al., 2002), and melatonin (Lee et al., 2006), have been studied. Also microspheres coated in serum albumin-alginate for the intestinal delivery based on pH variations of probiotic yeast were investigated (Hebrard et al., 2010). Coated beads and microcapsules are more effective as oral delivery systems (George and Abraham, 2006) modifying the alginate matrices allows the control of drug release.

Hyaluronic acid (HA) is another natural polymer used in oral drug delivery, which is an anionic glycosaminoglycan. The presence of repeat units with one carboxylic group produced a pH-responsiveness, and such behavior is enhanced in crosslinked hydrogel networks (Fiorica et al., 2013). pH-responsive HA nanostructures were used for oral insulin delivery systems, showing enhanced delivery via transcellular pathway (Han et al., 2012). Another study developed with increased carboxylic groups with increased carboxylic groups a derivative of HA to optimize the system for delivery to the colon using α-chymotrypsin demonstrated pH-sensitive release (Fiorica et al., 2013).

6.3 DENDRIMERS

Dendrimers are structures that have been proposed as oral DDS. They are called arborols because they have a tree-like architecture. They are hyperbranched macromolecules well defined; this feature is distinguished from hyperbranched polymers that exhibit random branches (Duncan and Izzo, 2005). Due to the manufacturing process it is possible to have a precise control of dendrimers; sizes of 10–20 nm in diameter are the most common. Features, such as well-defined structure, monodispersity of size, high stability, and surface functionalization capability make these structures potential candidates for use in DDS (Hughes, 2005). In recent years, simple dendrimers have been combined with other molecules (polymers and cyclodextrins) in order to improve the drug release process. Fig. 2.5 shows some typical dendritic architectures used as DDS. Also, a variety of polymers have been used for the synthesis of dendrimers, among them: polyamidoamine (PAMAM), PEG, poly(propylene imine), melamine, and

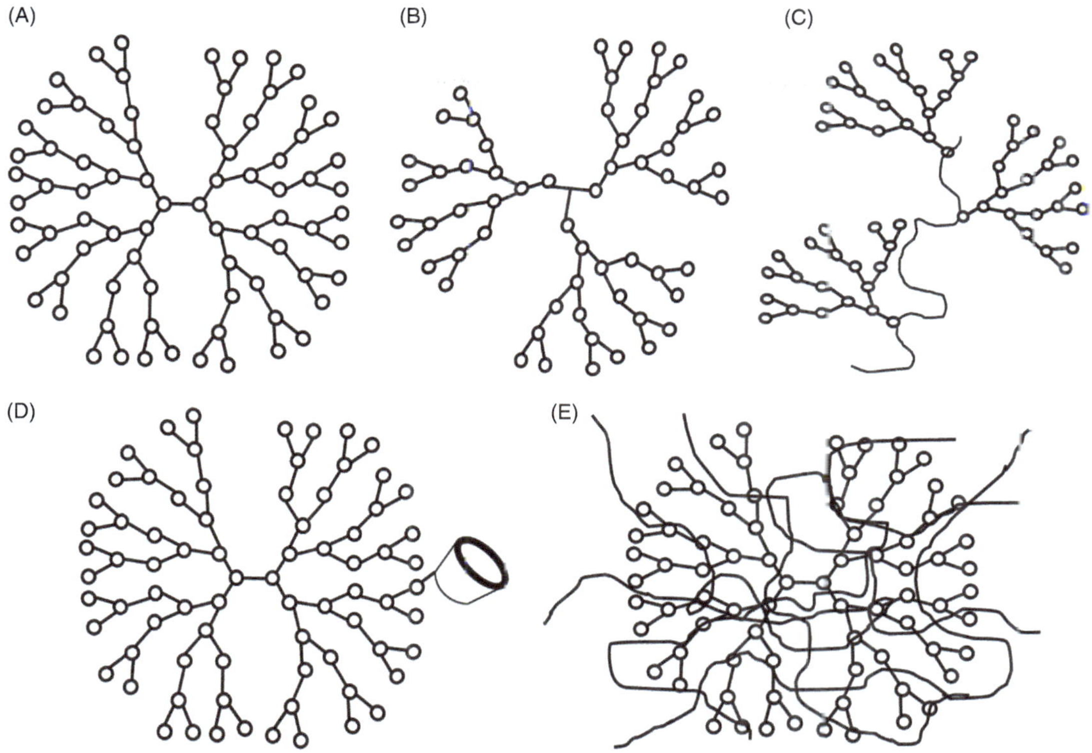

FIGURE 2.5 Dendritic Architectures Used as Drug Delivery Systems

Divergent dendrimer (A), convergent dendrimer (B), dendronized polymers (C), complex hybrids (dendrimer-cyclodextrins) (D), and dendrimer gels polymers (E).

chitin derivates. Basically, drugs are delivered by two mechanisms: encapsulation and establishment of covalent bonds.

PAMAM has been widely used because of its high solubility in water and large number of functional groups available on the surface. The macromolecular nature and hydrophilic tendency of the PAMAM dendrimers cause low oral bioavailability—and the charged PAMAM dendrimers are toxic. However, this can be solved by surface modifications to reduce toxicity, optimize the dosage, and increase permeability. A combination of transepithelial translocation and penetration produces beneficial oral drug delivery using PAMAM dendrimers.

An asymmetric doxorubicin-functionalized dendrimer was produced by PEGylation of one side of a third-generation 2,2-bis(hydroxymethyl)propionic acid dendrimer and the attachment of the acyl hydrazone linkage to the other side (G4) with the drug resulting in 8–10 wt.% of doxorubicin (Lee et al., 2006). No healing was observed with doxorubicin-alone treatments whereas a single injection of the dendrimer conjugate of doxorubicin caused complete tumor recession and 100% survival of mice.

Another study prepared conjugates of PAMAM dendrimers with cisplatin, a potent anticancer drug which is used to treat several types of cancers including lymphomas, carcinomas, sarcomas, and testicular cancer (Malik et al., 1999). This nanostructure has poor aqueous solubility and nonspecific toxicity. However, dendrimer platinum complex shows increased solubility, decreased toxicity, has selective accumulation in tumors, increasing the ability of the cisplatin in the treatment of B16F10 melanoma.

6.4 MICELLES

Polymeric micelles are characterized by core–shell structure, which is formed from self-aggregation of amphiphilic polymers (Kedar et al., 2010). The amphiphilic polymers contain a hydrophobic part that is located on the inside (core), while the hydrophilic part is located on the outside (shell) (Letchford and Burt, 2007). With this architecture, polymeric micelles can be loaded with hydrophobic drugs or with poor aqueous solubility in the core. The hydrophilic shell provides protection to the drug in limiting opsonin adsorption, which contributes to a longer blood circulation time (Lu and Park, 2013). Although typically the micelles are composed of polymers having a hydrophilic (A) and hydrophobic (B) part, it is possible to use copolymers with three parts (A–B–A) (Sosnik and Raskin, 2015). Polymeric micelles are quite stable and have values with lower critical micelle concentration (CMC) compared with surfactant micelles. Some of the molecules constituting the hydrophilic portion of the micelles include PEG, poly(ethylene oxide) (PEO), poly(acryloylmorpholidone), poly(trimethylene carbonate), and poly(vinylpyrrolidone). Moreover, the hydrophobic core is made of molecules, such as poly(L-amino acid), polyesters, and Pluronics (Kedar et al., 2010).

Biodegradable micelles loaded with insulin–phospholipid complex (entrapment efficiency was up to 90%) were obtained by a reverse micelle-solvent evaporation method, in which soybean phosphatidylcholine (SPC) was used to improve the liposolubility of insulin, and to control drug release. Intragastric administration of the 20 IU/kg nanoparticles reduced 57.4% fasting plasma glucose levels within the first 8 h of administration and this continued for 12 h (Cui et al., 2006).

Taxol, or paclitaxel, a classical antitumor drug shows poor oral bioavailability. To improve the oral bioavailability of Taxol, glycyrrhizic acid was used as the carrier (Yang et al., 2015). The Taxol-loaded micelles were prepared with an ultrasonic dispersion method, displayed spherical shapes and small particles with an encapsulation efficiency of 90%. The increase of oral absorption of Taxol from Taxol-loaded glycyrrhizic acid micelles could be due to the increased absorption in colon intestine and jejunum.

6.5 LIPOSOMES

Liposomes are spherical vesicles composed of lipid bilayers, which confine water inside (Fig. 2.5A). Phospholipids are the main molecule forming these bilayers, but also may contain cholesterol (Allen and Cullis, 2013). Cholesterol has the function of increasing the mechanical stiffness and stability (Gharib et al., 2015). The sizes of liposomes depend on the number of bilayers composed; therefore, unilamellar vesicles have sizes from 20 to 100 nm, while multilamellar vesicles show values up to 500 nm. The methods of preparation of liposomes are thin-film hydration, reverse-phase evaporation, freeze and thaw, ethanol injection and freeze-drying (Gharib et al., 2015). The liposomes can encapsulate hydrophilic drugs in the aqueous core and lipophilic drugs in the membrane. Different types of drugs and molecules have been encapsulated efficiently using liposomes: lipids, insulin, vaccines, antibiotics, cyclodextrins, nucleic acids, anticancer, and antifungal drugs (Cui et al., 2015). Liposomes have adequate properties of encapsulation and drug release, due to the chemical and physical characteristics of the lipid membranes. Among the variables affecting the encapsulation efficiency of drugs includes the following: membrane composition, membrane charge, and encapsulation method used (Eloy et al., 2014).

The use of liposomes as oral DDS is problematic due to the poor stability of the vesicles under the physiological conditions found in the GI tract. Nevertheless, it has been reported that phospholipid-based liposomes improved low-bioavailability drugs, including peptides and proteins.

Cyclosporin A (CyA) was formulated in lecithin vesicles with an incorporation of >98% and was compared with CyA-formulation Sandimmun Neoral in rat intestinal tissues using everted gut sacs and in situ circulation methods. Although the mucus is a barrier blocking the diffusion of the liposomal system, Sandimmun-Neoral-andCsA-containing vesicles showed equal absorption levels in the intestine (Guo et al., 2001). In another study, CyA was formulated in proliposomes with egg lecithin CyA, and Cremophor EL in a methanol/chloroform mixture onto lactose and liposomes were formed after hydrate with water. The absorption for this liposomal was nine times higher than for free drugs and four times higher than for a marketed sample of microemulsion (Shah et al., 2006).

A recent study showed that after administration of calcitonin-loaded CS-coated liposomes in rats an increase in the effect of oral calcitonin compared to calcitonin in solution about 11-fold. Coated liposomes could permeate the mucus layer in the small intestine of rats (Werle and Takeuchi, 2009).

Oral delivery of oxymatrine, a natural quinolizidine alkaloid found in sophora roots, used clinically for treating hepatitis B (Cao et al., 2009) shown in a pharmacokinetic study in rats of multivesicular liposomes coated with *N*-trimethyl chitosan (TMC) about 3.26 times that of a simple drug solution, indicating that *N*-trimethyl chitosan-coated liposomes may be interesting as carriers for oral drug administration.

In another study the blood glucose level was significantly decreased by orally administered liposomal insulin. Three types of insulin-loaded liposomes were prepared: phosphatidylcholine/dipalmitoyl phosphatidylethanol (1:1 w/w), dipalmitoyl, dipalmitoyl phosphatidyl etanol/dipalmitoyl phosphatidylcholine/palmitoyl-stearoyl sucrose (1:1:0.2) liposomes and liposomes composed of natural phosphatidylinositol and phosphatidylcholine (1:1). Oral administration of all liposomal species to rats resulted in hyperinsulinemia (Kisel et al., 2001).

Liposomes (HSPC, DPPC, DMPC, and DMPG) can make stable the oral vaccine preserving their stability against decomposition (Ogue et al., 2006). Mice orally vaccinated with ovalbumin (OVA) (model antigen) showed significantly raised immunoglobulin A values (IgA) when liposomes were administered, in comparison with the administration of OVA solution. The formulation with phospholipids resulted in an improvement of the immune response. Another study showed that liposomal OVA-induced suppression of the proliferative responses of popliteal lymph node cells from the treated mice to OVA, suggesting that a developed tolerance occurs (Masuda et al., 2002).

6.6 NANOPARTICLES

Nanoparticles are nanostructures that have been widely used in medical and pharmaceutical applications, specifically in the diagnosis and DDS. The use of nanoparticles as DDS is associated with excellent surface properties. These structures have a remarkable surface-to-volume ratio, which makes good candidates for uses as drug carriers. The nanoparticles can be classified into two types: nanospheres and nanocapsules (Fig. 2.5B–C). Nanospheres are a matrix material, where the drug is dispersed homogeneously over the matrix. On the other hand, the nanocapsules are hollow vesicles, where the drug is located inside and is surrounded by a membrane material (Alai et al., 2015). The nanoparticles offer the possibility to functionalize their surfaces to link various drugs, as well as to be used as targets. Different types of materials have been used for the synthesis of nanoparticles for use as drug carriers. The most common nanoparticles used in DDS include: biodegradable polymer nanoparticles, metal nanoparticles, SLNs, and nanostructured lipid carriers. Each of the mentioned nanoparticles has different chemical and physical properties, which have the capacity of encapsulation of various types of drugs. For instance, SLNs consist of biodegradable and compatible lipids. Considering the composition of these nanostructures are ideal for encapsulating lipophilic or poor aqueous solubility drugs (Neves et al., 2015). An important feature that should have lipids used in the manufacture of these nanoparticles is that they must remain in solid state at room temperature (Makwana et al., 2015). Recently, nanostructured lipid carriers have been synthesized and these offer a better encapsulation of several drugs, as well as drug release (Zhang et al., 2014). Another type of nanoparticles used as drug carriers are those that are made from biodegradable polymers. The advantages of these nanoparticles are biocompatibility and good encapsulation efficiency (Sahoo et al., 2015). Natural (CS, alginate, dextran) and synthetic (polylactic acid [PLA], poly(lactic-*co*-glycolic acid [PLGA], PCL) polymers can be used for their synthesis (Fonte et al., 2015). Finally, metal nanoparticles also have been used as DDS (Pooja et al., 2015). These are made of elements, such as gold, silver, platinum, and palladium. The drugs may be linked or embedded in the metal nanoparticles. Possible interactions between the drug and the metal nanoparticles include: lipophilic interaction, silanization, and electrostatic attraction (Hughes, 2005).

A study encapsulated in poly(D,L-lactide-*co*-glycolide) (PLG) nanoparticles three antitubercular drugs, isoniazid (INH), pyrazinamide (PZA), and rifampicin (RIF). Drug encapsulation efficiencies were 66.3% ± 5.8% for INH, 68% ± 5.6% for PZA and 56.9% ± 2.7% for RIF. After on oral administration of drug-loaded nanoparticles to *Mycobacterium tuberculosis*-infected mice at every 10th day, no tubercle bacilli could be detected in the tissues after five oral doses of treatment. Based on these, datananoparticle was reduce in dosing frequency for better management of tuberculosis (Pandey et al., 2003).

Lectins are highly specific proteins that bind to carbohydrate moieties of the glycocalyx of the mucus layer and intestinal enterocytes. They are involved in adhesion processes and several cell recognition. Their conjugation to polymeric nanoparticles efficiently increased interactions with the mucus, increasing their transport across the intestinal mucosa (Irache et al., 1994).

Oral administration of lectin conjugated nanoparticles loaded with insulin, improved the intestinal absorption of insulin decreasing the glucose level in blood (Zhang et al., 2006). Insulin as a hydrophilic peptide can be incorporated with high efficiency about 98% in nanoparticles, showing sustained drug release behavior and a good stability (Liu et al., 2007).

6.7 CARBON STRUCTURES

Other types of nanomaterials that have been proposed as potential drug carriers are carbon-based structures (Raissi and Mollania, 2014). There are several carbon configurations that can be obtained as nanotubes (CNT), fullerenes, nanohorns, and so on (Fig. 2.5D–E). These structures are made of carbon atoms; in the case of CNT, this can be single-wall (SCNT) or multiwall (MCNT) and the CNT can be found with open or closed ends. The diameters of these structures include values of 1–10 nm, which may readily penetrate the cells. Carbon nanotubes and fullerenes have been made through various methods, such as electric arc discharge, laser ablation, chemical vapor deposition, and combustion process (Hughes, 2005). The main advantage of the carbon-based structures is that their surfaces may be functionalized; this fact allows the links of some drugs via covalent or noncovalent interactions. Besides these structures have the ability to be loaded within the interior with drugs, improving the protection and release process (Wong et al., 2013). Finally, another interactive way between the drugs and CNT is as nanocatheters (Foldvari and Bagonluri, 2008).

Anticancer agents are small molecules and can be loaded into or onto CNTs through p–p stacking interactions between the drug molecules and pseudoaromatic double bonds of the graphene sheet or an covalent immobilization onto the reactive functional groups present on the sidewalls of CNTs.

Cisplatin, a small molecule, can be loaded into SWCNTs with a diameter of 1.3 to 1.6 nm (Tripisciano et al., 2009). Using the dialysis membrane method the cisplatin was continually released for almost a week, with maximum release during 72 h and up to 1 week.

Doxorubicin was loaded on CNT based on p–p stacking interactions by simply mixing the drug with an aqueous dispersion of CNTs stabilized by Pluronic F127 (nonionic surfactant). Doxorubicin-MWCNT complex shows improved cytotoxity in comparison with both doxorubicin-Pluronic complex-es and doxorubicin on MCF-7 human breast cancer cell line. These results indicate that MWCNTs can increase the delivery of doxorubicin and improve the cellular uptake of doxorubicin (Zhang et al., 2009).

Many cancers metastasize through lymphatic canal. Polyacrylic acid (PAA) can be incorporated onto CNTs, making them highly hydrophilic and gemcitabine was loaded into the nanosystem with a loading efficiency of 62%. It was found that CNTs were seen only in the lymphatic nodes and were absent in the major organs, such as kidney, liver, spleen, lungs, and heart, after 3 h of subcutaneous injection. Without the help of this nanostructures, gemcitabine cannot be exclusively distribute in the lymphatic system (Yang et al., 2009).

6.8 POLYMERS USED IN DRUG-DELIVERY SYSTEMS

Polymer materials have several desirable properties for drug carrier use as biodegradability, functionalization (the process of adding new functions, features, capabilities, or properties to a material by changing the surface chemistry of the material) biocompatibility and capability. Through structural manipulation and functionalization of polymer, drugs can be incorporated within the polymer. Entrapping or encapsulating the drug within a polymer allows for greater control of the pharmacokinetic behavior of the active drug molecule. The drug can be released with a more ideal, near zero-order kinetic profile, which establishes a more constant flow of the drug out of the carrier. This pharmacokinetic behavior maintains more appropriate steady levels of the drug at the site of delivery. In contrast, conventional oral drug delivery typically follows first-order release kinetics where the drug release rate is proportional to the amount of drug remaining in the drug carrier.

Peptide transporter PEPT1 (SLC15A1) has been studied in several clinical trials using valacyclovir and beta-lactam antibiotics, which are absorbed by PEPT1 due to their structural similarities with native peptides. The data obtained show that utilization of this transporter as the tool for oral delivery is by mainly prodrug approaches (Kikuchi et al., 2008).

In addition, monocarboxylate transporters (MCTs) (SLC16A family) are a peptide transporter, accept monocarboxylates as nicotinic acid and lactic acid as substrates, which have an activity pH-dependent transport to improve the absorption/uptake of monocarboxylic acid-type drugs (Tsuji et al., 1994).

Organic anion transporting polypeptide (OATP) is the anion transporter in liver and function as drug absorptive transporter in the small intestine. OATPs show a broader substrate selectivity than PEPT1, leading to greater use as transporters for oral delivery (Jacquemin et al., 1994). The utilization of OATPs, PEPT1, transporters, and other uptake/influx transporters may be the next strategy for oral delivery carrying compounds through the intestine.

CS and its derivatives are a linear polysaccharide of random β-(1→4)-linked to deacetylated unit of D-glucosamine and an acetylated unit *N*-acetyl-D-glucosamine (Muzzarelli and Muzzarelli, 2005). This acetylation and deacetylated gives to CS different properties, such as biodegradability and biocompatibility (Grenha et al., 2007). CS is insoluble in water, their solubility is pH-dependent at pH value of approximately 5.8 due to the progressive protonation of pendant amine moieties. Due to the ability of CS to establish hydrogen, hydrophobic, and ionic bonds, with the negatively charged of mucin, CSt is considered an important mucoadhesive agent for a broad variety of administration routes, such as oral (Thanou et al., 2001).

The polyester PLGA is a copolymer of PLA and PGA. It is the best defined biomaterial available for drug delivery with respect to design and performance. PLA contains an asymmetric α-carbon, which is typically described as the D or L form. The enantiomeric forms of the polymer PLA are poly-D-lactic acid (PDLA) and poly-L-lactic acid (PLLA). PLGA is generally an acronym for poly D,L-lactic-coglycolic acid where both enantiomeric forms are in equal ratio. PLGA is biodegradable, biocompatible, and is able to deliver its payload with appropriate duration, concentration, and biodistribution for the intended therapeutic effect.

Poly(methacrylic acid), poly(acrylic acid), and derivatives (PAA), also usually referred to as carbomer, are made of a high molecular–weight polymer of acrylic acid. Due to the good biocompatibility and the presence of salient carboxylic acid units (one per repeating unit), PAA display a high adhesive bond strength in contact with tissues, enhancing the mucosal penetration of drugs; in the strong interaction with mucus the presence of a unionized carboxyl group is very important. These interactions are the result of the hydrogen bonds between PAA and the proton-accepting groups of the mucin (Gu et al., 1988). The presence of numerous carboxyl groups make that PAA adopts a more favorable macromolecular conformation and an increased accessibility of its hydrogen-bonding groups improving the reduction in mucus viscosity, transport rates, and the delayed release of hydrophilic compounds make proteolytic enzyme functionalized NPs improving the targeted drug delivery of drugs at different mucosal surfaces (Müller et al., 2012). PAA NPs are able to cross the mucus layer and remained in the jejunum, and duodenum where the absorption of the drug principally occurs (Müller et al., 2012).

The copolymers PEG/poly(ethylene oxide) (PEO) depending on the synthetic method, is the type of product obtained with different properties. This can specifically interact with components of mucin due to the lack of side functional groups as carboxylic acid and amine (Patel and Patel, 2010), though the mechanism would rely on the fast interpenetration of PEG chains with the lining mucus layer (De Ascentiis et al., 1995). In an attempt to improve the adhesiveness of PEG a modification has been made with PAA or PLGA (Cleary et al., 2004).

Poly(vinyl pyrrolidone) (PVP), also known as povidone, is a nonionic synthetic linear polymercontaininggroups of 1-vinyl-2-pyrrolidinone.These polymer are chemically inert and essentially nontoxic for nonparenteral exposure used for its mucoadhesive properties due to their mild-mucoadhesive properties (Salamat-Miller et al., 2005). The mild-adhesive properties of PVP seem to limit its usefulness for the development of mucoadhesive dosage forms containing mixtures of different polymers (Chun et al., 2013).

Among the polymers used in DDS, polylactides (PLAs) are widely used because they are compostable, biodegradable, and biocompatible with degradation products nontoxic to living organisms (Garlotta, 2001). However, PLAs is limited for biomedical uses, due to presence of hydrophobic side groups with low reactivity and a low rate of degradation, compared to other polyesters. In an attempt to improve the chemical, biological, and physical properties of PLA, is blending with different polymers that may or may not be degradable (Xu et al., 2009). Among the polymers blended with PLA, PVP is one of the most common (Zereshki et al., 2011) because it is a homopolymer of amphiphilic nature arising from the presence of amide groups in its chemical structure that possesses properties, such as strong adsorption capacity, good biocompatibility, high surface activity, and low toxicity (Wang et al., 2012)

The mechanism for intestinal absorption of drugs and transporters has been studied during the past decades. Various studies in vitro and in vivo suggested a contribution of the carrier-mediated absorption of drugs via the nutrient transport systems for oligopeptides, amino acids, water-soluble vitamins, monocarboxylic acids, bile acids, sugars, nucleosides, and amines (Tamai and Tsuji, 1996).

7 THERAPEUTIC APPLICATIONS OF ORAL DELIVERY NANOSTRUCTURES

7.1 CEREBROVASCULAR DISEASE

Vinpocetine (ethyl-apovincamine-22-oate, VP), is a derivative of the alkaloid vincamine (commercial name Cavinton), used in the treatment of ischemic stroke and other cerebrovascular disease (Lin et al., 2014), also was found that when vinopepsin was encapsulated shown positive influence in the aggregation blood and plasms and also changes in the viscosity of the whole blood. However, one of the disadvantages of VIN is its poor solubility in water and in the intestinal tract. Look into different lipid-based synthesis mixtures to improve the viability and obtain promising results. Also see (Zhuang et al., 2010), a study of nanostructured lipid carriers (NLC) formulation for VIN as an oral delivery system. The preparation was made by the high pressure homogenization method; the results were promising for oral delivery of poorly water-soluble drugs.

HIV and other lentiviruses have the ability to infect and replicate in nondividing cells, including monocyte/macrophage lineage. HIV infection produces a reduction in immune defense causing extensive destruction of macrophages, dendritic cells, T-helper cells, and other cellular components associated with cell-mediated immunity (Wyatt and Sodroski, 1998). HIV invade the central nervous system (CNS) early in the course of the infection and primarily target the brain, perivascular macrophages, microglia, and mononuclear macrophages (Martin et al., 2001).

Exist barriers to antiretroviral (ARV) penetration into the brain that treatment hindering HIV-associated CNS complications. This can be solved using drug carriers of submicron scale or nanometer (Ojewole et al., 2008). Depending on the type of nanocarriers is the efficiency of delivery and of load. There is a broad range of nanocarriers available, including polymeric systems, liposomes, micelles, and nanoparticles; unfortunately only a few of them are suitable for brain delivery. Due to complexity, the CNS must choose carefully the biomaterials particularly in safety considering the duration of ARV therapy.

Nanocarriers must be fully biodegradable and nontoxic. Characteristics covered by polymers include the polyesters and acrylic polymers that contain lactide units and pluronic block copolymers/surfactants for preparing brain-targeting micellar systems (Spitzenberger et al., 2007). Nanoparticles of poly(butyl cyanoacryalate) (PBCA), an acrylic polymer, has been the most studied polymer for brain delivery. PBCA have a good accumulation in both cerebrospinal fluid and brain tissues without physical disruption of the blood–brain barrier (BBB) integrity. PBCA is biodegradable and its lipophilicity facilitates encapsulation of diverse types of compounds as loperamide, dalargin, amitriptyline, loperamide, doxorubicin, and methotrexate. However, in vivo biodegradation produced formaldehyde, which is toxic (Alyaudtin et al., 2001).

Asmethylmethacrylate-sulfopropylmethacrylate (MMA-SPM) is an acrylic copolymer in which its negative charges exhibit a higher loading capacity for polar compounds when compared to PBCA. This MMA-SPM nanocarrier system increases the permeability of zidovudine that was the first ARV compound commercially available for the treatment of AIDS (Kuo and Chen, 2006). Also, it improved the permeability of lamivudine in vitro BBB model of bovine brain-microvascular endothelial cells aiding delivery to infected brain mononuclear macrophages (Svenson, 2009).

Carbon dioxide and water are degradation products of polyesters, such as PLA and poly(D,L-lactide-*co*-glycolide) (PLGA). They are relatively nontoxic, conferring properties that make them interesting for brain delivery. Because of their safety profiles, PLA and PLGA are of the few polymers officially approved for clinical use (Dechy-Cabaret et al., 2004).

7.2 HYPERCHOLESTEROLEMIA

Nanostructured lipid carriers (NLCs) from a mixture of squaleno and precirol were chose to load lovastatin, which is a cholesterol-lowering agent for oral administration via. NLCs increased oral absorption and produced a prolonged efficacy (Chen et al., 2010). Different emulsifiers have been used to prepare NLCs, including Pluronic F68 (PF68), Myverol 18–04 K, and soybean phosphatidylcholine, improving the oral bioavailability and the short half-life of lovastatin, which is critical in the efficacy of the oral administration. The findings show that lovastatin loading in NLCs increased oral absorption and sustained drug release. Also, it was shown that it is possible to modify the release profiles as a function of the lipid matrix and emulsifier type, which can be used to increase the efficacy of lovastatin in the control of hypercholesterolemia.

7.3 ARTHRITIS

Triptolide (TP) nanostructures carriers (TP-NLCs) are used for oral administration. TP is a compound that has activity as an antiinflammatory, immunosuppressive, antifertility, anticystogenesis, and anticancer agent. However, this compound is toxic. It is known to cause hepatic, renal, digestive, reproductive, and hematological systems damage. This is due to fluctuation of the TP concentration in the plasm for oral administration. The idea of using nanostructures is to reduce the drug release rate improving its safety uses. The results demonstrate that TP-NLCs increased drug encapsulation efficiency and stable drug incorporation for more than 1 month. TP-NLCs manifested a better in vitro sustained-release pattern compared to TP-SLNs. Furthermore, it was associated with reduced subacute toxicity in male rats (Zhang et al., 2014). TP is a drug with antiinflammatory, autoinmune, antifertility, and antineoplastic activities; however, this is poorly soluble in water and has high toxic limits in its applications to external uses (Mei et al., 2005). Nanoparticules encapsulation for oral application is a good alternative to control release rates and to reduce the toxicity. SLN were synthetized to encapsulated triptolide, tristearin glyceride (TSG), poloxamine 908, and medicinal grade of soybean lecithin. The nanoencapsulation suggest that SLN delivery systems can enhance the antiinflammatory activity and the protective effect against triptolide-induced hepatotoxicity.

Floating microparticles system are made to encapsulated ketoprofen, which is a nonsteroidal antiinflammatory agent used in arthritis disease. The objective is to design a structure to resist in the stomach without contact with the mucosa, two polymers were used Eudragit S100 (ES) and Eudragit RL (ERL), in four relations to form the floating microparticles. The results show good floatability characteristics, which make this approach a possible solution to oral DDS (El-Kamel et al., 2001).

7.4 NANOSTRUCTURES IN CANCER

Oral chemotherapy is ideal in the clinical practice of chemotherapy improving the quality of life of the patients. Actually, the anticancer drugs are administered through i.v. injection causing high peak above the maximum tolerable concentration (MTC) of the drug in the plasma producing serious side effects. Thus, oral administration keeps a sustained concentration of the drug in the circulation to maintain a prolonged exposure to the drug, as well as to avoid high peaks above MTC decreasing the side effects and increasing the therapeutic efficacy. Moreover, also oral chemotherapy can improve their quality of life, reduce their medical expenses, and in case of cancer patients at the latest stage, who are too weak to resist hardness in medical treatment.

Currently known anticancer drugs with excellent anticancer effects, such as Taxanes (docetaxel and aclitaxel) are not absorbable/interactive in the GI tract; therefore, they are not orally bioavailable. Various solutions, including pharmaceutical nanotechnology, have thus developed to cross drug barriers used in the molecular basis of the various physiological drug barriers that have been intensively investigated.

The GI problems are the most frequent side effects of cancer chemotherapy, which lead to diarrhea, ulceration of the GI tract, and mucositis. Because this has increased the nanoparticulate carriers composed of proteins, peptides, polysaccharides, probiotics, and synthetic polymers for oral administration of nutraceuticals and pharmaceuticals (Lopes et al., 2010). Drug delivery in cancer is important for optimizing the effect of drugs and reducing toxic side effects. Several nanotechnologies, mostly based on nanoparticles, can facilitate drug delivery to tumors. Structures, such as micelles, liposomes, hydrogels, dendrimers, nanotubes, nanocells, quantum dots, XPclad have been used to improve the effects of the drugs used to treat this disease.

An NLC with Tamoxifen (TAM) is prescribed as an endocrine therapy to combat breast cancer. The use of TAM has shown significant reduction of cancer recurrence (How et al., 2013). However, TAM has low solubility and low selectivity, resulting in thromboembolic events and uterine malignancies in long-term usage of the drug, a disadvantage that can be overcome by improving specific targeting and solubility using NLC.

It is known that orally administered anticancer treatments have little chance to get into the blood system, and there are two factors that can help to overcome this problem. The nanoparticles size and the appropriate coating can contribute to crossing the intestinal barrier and immune system to reach the tumor site. The nanoparticles with the right size and appropriate surface coating can be used to deliver the drugs to the blood system.

Another alternative to chemotherapies is the use of nanotubes to deliver the drug without lateral side effects and take advantage of the transporting capabilities of carbon nanotubes coated with the appropriate surface modifications in the in vivo experiment multiwall nanotubes that were delivered orally to mice. The results showed that for a short time, fewer toxic effects were found; however, changes due to immune system were present (Golokhvast et al., 2013).

Other structures used in the treatment of cancer are the dendrimers as the PAMAM to transport of propranolol across monolayers of the human colon adenocarcinoma cell line, propranolol was selected model due to its poor water solubility, improving the efficacy of low solubility. (D'Emanuele et al., 2004).

Chemotherapy is associated with unpleasant side effects many chemotherapeutic drugs are, in themselves, carcinogenic; treatment with these drugs carries a risk of secondary cancer. Many cancer researchers today are therefore committed to producing a new generation of more specific, better-targeted cancer therapies, to improve the quality of life of cancer patients, and to minimize the risks of their treatment. There has been considerable interest in drug delivery for enhancing therapeutic efficacy and minimizing systemic side effects (Murphy et al., 2008).

7.5 DIABETES

The delivery of proteins via the oral route is a challenge that emerged from the importance of diabetes as a disease, considering insulin as the most commonly employed model therapeutic protein. After many attempts, oral delivery of insulin remains as a challenge for many researchers. The main obstacle is the GI tract, which has the function of breaking down peptides and proteins into their amino acids. However, to enzymatic barriers, there are physical barriers that make the oral delivery of a peptide or protein difficult (Uchiyama et al., 1999). Also lining of the GI tract is composed of a thick wall of epithelial cells covered by a layer of mucus and polysaccharides, which can inhibit the oral delivery of a peptide as insulin via the hepatic portal.

New techniques have been developed to improve the oral bioavailability of therapeutic proteins. These include the use of permeation enhancers, protease inhibitors, microsphere encapsulation, and enteric coatings (Damge et al., 2007). Among these methods microsphere encapsulation is an oral drug delivery vehicle that has the potential to outstrip physical and enzymatic barriers of the GI tract. Microparticles composed of poly(ethylene glycol) and poly(methacrylic acid) containing insulin administered orally diabetic rats (type 1) produced an inhibition of postprandial blood glucose levels (Morishita et al., 2006).

The development of biodegradable and biocompatible poly(lactic acid)–bpoly(ethylene glycol) (PLA–PEG) polymeric nanoparticles for oral administration of insulin were functionalized to target the neonatal Fc receptor (FcRn), to facilitate the crossing of the intestinal epithelium with IgG, the results found shown that it is possible to transport the drug efficiently across the intestinal epithelium (Alexis et al., 2013). In another study conjugation of the insulin with ethylene glycol tetracetic acid, poly(γ-glutamic acid), and CS-produced nanoparticles with a prolonged decrease of the blood glucose levels in a diabetic rat (Chuang et al., 2012).

The nature of the polymer increases the residence time of a drug delivery system within the GI tract. A drug encapsulated within the hydrophobic bioadhesive polymer p(FASA) prolong the residence time within the GI tract, resulting in an increase intestinal drug absorption (Mathiowitz et al., 1997). To prolong the residence time within the GI tract, it can be used as a bioadhesive polymer that may increase the probability of a particle uptake by the enterocytes of the small intestine for thus microsphere size should not exceed 5 μm in diameter and that particle uptake increases as sphere size decreases. In another study insulin microspheres encapsulated in the bioadhesive polymer p(FASA) administered to type 1 diabetic rats and insulin concentration was determined that at 2 h it was 74.2 μIU/mL and 4 h was 36.6 μIU/mL. The relative bioavailability was determined to be 5.5 ± 1.7% (Furtado et al., 2008).

7.6 HEPATITIS B

The materials used for the encapsulation of hepatitis B vaccine for oral application were the poly(vinyl alcohol)-coated and CS. The results shown similar response compared with intramuscular vaccination (Shrestha and Rath, 2013). In another study that used PLA, PLGA, and PEG to encapsulating hepatitis B surface antigen (HBsAg) to form copolymeric blocks of PLA and PEG. The results demonstrated good stability of nanoparticles in GI environments (Jain et al., 2010).

8 SYNTHESIS OF NANOPARTICLES

Nanoparticle synthesis refers to methods for creating nanoparticles. Nanoparticles can be derived from larger molecules, or synthesized by "bottom–up" methods that, for example, nucleate and grow particles from fine molecular distributions in liquid or vapor phase. Synthesis can also include functionalization by conjugation to bioactive molecules. Synthesis of nanomaterials in high yield and low cost has been a great challenge since the very early development of nanoscience. Application of the nanoparticles in medicine depends on the ability to synthesize particles with different shape, monodispersity, chemical composition, and size.

9 CHARACTERIZATION OF NANOPARTICLES

9.1 LIGHT MICROSCOPY

Light microscopy has been used for a long time, but yet has limited magnification. The long focal length allows the observation of great samples, unlike other methods, where it is only possible to display a small region. Optical microscope involves the passage of visible light transmitted or reflected by the sample through single or multiple lenses, which allow an enlarged samples view and the magnification range is 10×–1000× with a resolving power of about 0.2 μm.

9.2 SCANNING ELECTRON MICROSCOPY

Scanning electron microscopy (SEM) is based on an electron beam on the sample to form a magnified image of a material and increased up to 1000 times the resolving power. When electrons interact with the sample, a series of signals as secondary electrons, backscattered electrons, transmitted electrons, characteristic X-rays, among others that can be captured for generated imaging. Within the scanning electron microscopes we can find different operating modes. One is the environmental mode, which compared to convention, allows observation of samples with high humidity, nonconductive, sensitive, without being conductive or require the gold film.

Determine the imaging of the sample surface by scanning with a beam of electrons in high vacuum in a scan pattern, when the beam of electrons strikes the surface of the sample, it interacts with the atoms to yield signals in the form of secondary electrons, back scattered electrons and generate characteristic X-rays that contain information about the samples as topography, surface, composition.

It is possible to detect particles >10 nm, direct imaging technique, and is visible the morphology of the particle. Their composition can be determined by energy dispersive X-ray and spectroscopy (EDX).

Agglomeration may be caused by drying. The analysis of agglomeration in the dispersion is not possible, the electron beam may influence the sample, limited resolution in size. Only the solid core of a nanoparticle is observed, precipitation of other dissolved compounds from the dispersion is possible and only a small number of particles is examined.

9.3 TRANSMISSION ELECTRON MICROSCOPY

The image of a sample is obtained by the transmission of an electron beam at high vacuum. In this technique the sample is prepared in very thin layers in which the electron beam is focus traveling through the sample, In this microscope it is possible to reach high resolution thousands times better than optical light. The images contrast is generated by the crystallinity of the material, variations in the sample thickness, material density, etc.

TEM is a technique of direct imaging of higher resolution than the SEM. It is possible to detect particles >1 nm. The particle composition can be determined by energy dispersive X-ray spectroscopy (EDX) and the crystallinity and morphology can be observed. Agglomeration may be caused by drying, the analysis of agglomeration in the dispersion is not possible, electron beams may influence the sample, limited resolution in size, and only the solid core of a nanoparticle is observed. Precipitation of other dissolved compounds from the dispersion is possible and only a small number of particles is examined.

9.4 DYNAMIC LIGHT SCATTERING

The intensity of scattered light fluctuates when a laser light beam at the surface to emulsions and molecules in suspension that undergo Brownian motion. The technique is applicable from about 0.001 to several microns. The measurements are fast, from seconds to minutes, and a very small amount of sample is required to be measured. The shape information is not easily obtained. The technique does not produce a high-resolution histogram of the size distribution, dust may affect measures, making interpretation difficult and the multiple scattering affects the data analysis.

9.5 ZETA POTENTIAL (ζ)ANALYSIS

Zeta potential (ζ)analysis is a technique for determining the surface charge of nanoparticles in solutions. Nanoparticles have a surface charge that attracts a thin layer of ions of opposite charge to the nanoparticle surface. Zeta potential is an important tool for understanding the state of the nanoparticle surface and predicting the long-term stability of the nanoparticle.

9.6 RAMAN SPECTROSCOPY

Raman spectroscopy (RS) is nondestructive and rapid technique with molecular specificity that can be implemented for real-time analysis. RS is based on the inelastic scattering of light, in which an incident excitation radiation interacts with most of the molecules inducing a red shift of the energy of the incident photon in an amount equal to the internal vibrational energy. Energy quantity that is lost or increased is observed as a change in the irradiated photon energy; this energy loss is characteristic of each functional group of the molecules. Raman is a technique that produces a single precise spectral fingerprint of a molecule or molecular single structure. RS is involved as a light source, a monochromator, sample holder, and the detector. Factors affecting Raman spectrum analysis could include high-signal noise relationship, the instruments stability, and sufficient resolution. Actual development using Raman spectrometer for NIR (near infrared) laser excitation; resolve the problem to avoid affecting the Raman fluorescence signals. Recently, the Raman scattering technique has been used for food analysis, for archaeology, and nanotechnology.

9.7 ATOMIC FORCE MICROSCOPY

Atomic force microscopy (AFM) is one kind of scanning probe microscopes (SPM). SPMs are designed to measure local properties, such as friction, height, and magnetism to acquire an image with a probe, the SPM raster scans the probe over a small area of the sample, measuring the local property simultaneously. Simultaneous mapping of chemical composition and topography is the main goal of microscopic techniques. The capacity of AFM to imaging with nanometer-scale resolution a variety of surfaces provides information about the chemical composition of the simple. A selective method to study chemical properties is functionalizing the AFM probe to sense of biological interaction and a specific chemical. Imaging material properties by AFM methods is a young topic, so it is likely to find difficulties and challenges in this application. AFM is a powerful method for mapping variations in the viscoelasticity, composition, adhesion, and friction of the sample surface. This method is called simply phase imaging or phase-contrast imaging.

9.8 X-RAY PHOTOELECTRON SPECTROSCOPY OR ELECTRON SPECTROSCOPY FOR CHEMICAL ANALYSIS

This technique is widely used in surface characterization and analysis of biomedical polymers. The main advantages are methods nondestructive, have a sensitivity of parts/1000, analyzes all elements except hidrogen and helium, have a surface sensitive of 10–200 A°, is a method quantitative, provides chemical bonding information, interpretation, and straightforward theory , and have a high information content. Among the disadvantages are to make the analysis that is required as high vacuum (10^{-8}–10^{-10}), slow 0.5–8 h/sample, and is an expensive technique and have low resolution of ~ 0.5–1.0 ev.

9.9 FOURIER TRANSFORM INFRARED SPECTROMETRY

The main disadvantage of the conventional method of IR spectrum is its inherent slowness. In IR each point of the spectrum has to be recorded separately, while the frequency is swept smoothly across the whole span of the spectrum. IR spectrum of a compound, has few peaks, most of the time is spent in recording the background noise. Using the technique FTIR simultaneous and almost instantaneous recording of the whole spectrum in the magnetic resonance, microwave, and infrared region is possible. FTIR was developed to overcome the limitations encountered with dispersive instruments. The main difficulty was the slow scanning process for which was developed a simple optical device called an interferometer. The interferometer produces a unique type of signal, which has all of the infrared frequencies encoded into it. The signal is acquired instantaneously which allows making simultaneous measurements by the FT-IR. Sensitivity is improved with FT-IR resulting in lower noise levels This makes FT-IR an invaluable tool for quality assurance applications. This quantitative method can be easily calibrated and developed, which can be incorporated into simple procedures for routine analysis. Thus, the Fourier transform infrared technique has brought important advantages to infrared spectroscopy. It has made possible the development of many new sampling techniques which were designed to tackle challenging problems that were impossible by older technology.

10 CONCLUSIONS

Oral delivery is the route of administration most used and readily accepted. Many drugs are administered by the oral route because of high absorption in the GI tract. However, many drugs are not suitable for oral administration due to poor stability, bioavailability, and solubility. Consequently, encapulation of these drugs in nanostructures can avoid these limitations. This chapter describes the nanocarriers across GI barriers as absorption in intestinal epithelium, mucus glycan layer, pH, enzymatic degradation, and abundant microbial population. Increasing the relevance of nanocarriers in oral drug delivery system is focused mainly on solving problems related to the bioavailability and solubilization of drugs in the GI system. Polymer materials for oral application is a good alternative to control release rates and targeting to reduce the toxicity to improve the oral biocompatibility, capability, and bioavailability of therapeutic drugs, permitting incorporation of new functions, features, capabilities, or changing the surface chemistry of the material to increase the effectiveness of the active drug. Also, innovative strategies are described for enhancing the effectiveness of nanostructure delivery to the GI system.

REFERENCES

Alai, M.S., Lin, W.J., Pingale, S.S., 2015. Application of polymeric nanoparticles micelles and in insulin oral delivery. J. Food Drug Anal. 23, 351–358.

Alexis, F., Kuo, T.T., Levy-Nissenbaum, E., Karnik, R., Blumberg, R.S., Langer, R., Farokhzad, O.C., 2013. Transepithelial transport of Fc-targeted nanoparticles by the neonatal fc receptor for oral delivery. Sci. Transl. Med. 5doi: 10.1126/scitranslmed.3007049, 213ra167.

Allen, T.M., Cullis, P.R., 2013. Liposomal drug delivery systems: from concept to clinical applications. Adv. Drug Deliv. Rev. 65, 36–48.

Alyaudtin, R.N., Reichel, A., Löbenberg, R., Ramge, P., Kreuter, J., Begley, D.J., 2001. Interaction of poly(butylcyanoacrylate) nanoparticles with the blood–brain barrier in vivo and in vitro. J. Drug Target 9, 209–221.

Barthe, L., Woodley, J., Houin, G., 1999. Gastrointestinal absorption of drugs: methods and studies. Fundam. Clin. Pharmacol. 13, 154–168.

Belting, M., Sandgren, S., Wittrup, A., 2005. Nuclear delivery of macromolecules: barriers and carriers. Adv. Drug Deliv. Rev. 57, 505–527.

Cao, J., Sun, J., Wang, X., Li, X., Deng, Y., 2009. N-trimethyl chitosan-coated multivesicular liposomes for oxymatrine oral delivery. Drug Dev. Ind. Pharm. 35, 1339–1347.

Cevc, G., 2003. Hydration-driven transport of deformable lipid vesicles through fine pores and the skin barrier. Biophys. J. 84, 1010–1024.

Chen, C.C., Tsai, T.H., Huang, Z.R., Fang, J.Y., 2010. Effects of lipophilic emulsifiers on the oral administration of lovastatin from nanostructured lipid carriers: physicochemical characterization and pharmacokinetics. Eur. J. Pharm. Biopharm. 74, 474–482.

Chourasia, M.K., Jain, S.K., 2003. Pharmaceutical approaches to colon targeted drug delivery systems. J. Pharm. Sci. 6, 33–66.

Chuang, E.Y., Lin, K.J., Su, F.Y., Chen, H.L , Maiti, B., Ho, Y.C., Yen, T.C., Panda, N., Sung, H.W., 2012. Calcium depletion-mediated protease inhibition and apical–junctional complex disassembly via an EGTA-conjugated carrier for oral insulin delivery. J. Control. Release 27, 00803–00807.

Chun, M.K., Bhusal, P., Choi, H.K., 2013. Application of carbopol/PVP inter-polymer complex to prepare mucoadhesive floating granule. Arch. Pharm. Res. 36, 745–751.

Clark, M.A., Hirst, B.H., Jepson, M.A., 2000. Lectin-mediated mucosal delivery of drugs and microparticles. Adv. Drug Deliv. Rev. 43, 207–223.

Cleary, J., Bromberg, L., Magner, E., 2004. Adhesion of polyether-modifiedpoly(acrylic acid) to mucin. Langmuir 20, 9755–9762.

Cloninger, M.J., 2002. Biological applications of dendrimers. Curr. Opin. Chem. Biol. 6, 742–746.

Cone, R.A., Mucus, 2005. In: Mestecky, J., Lamm, M.E., Strober, W., Bienenstock, J. McGhee, J.R., Mayer, L. (Eds.), Mucosal Immunology. third ed. Academic Press, San Diego, CA, pp. 49–72.

Corfield, A.P., Carroll, D., Myerscough, N., Probert, C.S.J., 2001. Mucins in the gastrointestinal tract in health and disease. Front. Biosci. 6, D1321–D1357.

Crater, J.S., Carrier, R.L., 2010. Barrier properties of gastrointestinal mucus to nanoparticle transport. Macromol. Biosci. 10, 1473–1483.

Cui, F., Shi, K., Zhang, L., Tao, A., Awashima, Y., 2006. Biodegradable nanoparticles loaded with insulin–phospholipid complex for oral delivery: preparation, in vitro characterization and in vivo evaluation. J. Control. Release 114, 242–250.

Cui, M., Wu, W., Hovgaard, L., Lu, Y., Chen, D., Qi, J., 2015. Liposomes containing chlolesterol analogues of botanical origin as drug delivery systems to enhance the oral absorption of insulin. Inter. J. Pharm. 489, 277–284.

Damge, C., Maincent, P., Ubrich, N., 2007. Oral delivery of insulin associated to polymeric nanoparticles in diabetic rats. J. Control. Release 117, 163–170.

De Ascentiis, A., deGrazia, J.L., Bowman, C.N., Colombo, P., Peppas, N.A., 1995. Mucoadhesion of poly(2-hydroxyethyl methacrylate) is improved when linear poly(ethylene oxide) chains are added to the polymer network. J. Control. Release 33, 197–201.

Dechy-Cabaret, O., Martin-Vaca, B., Bourissou, D., 2004. Controlled ring-opening polymerization of lactide and glycolide. Chem. Rev. 104, 6147–6176.

Dekker, J., Rossen, J.W., Buller, H.A., Einerhand, A.W., 2002. The MUC family: anobituary. Trends Biochem. Sci. 27, 126–131.

D'Emanuele, A., Jevprasesphant, R., Penny, J., Attwood, D., 2004. The use of a dendrimer-propranolol prodrug to bypass efflux transporters and enhance oral bioavailability. J. Control. Release 95, 447–453.

Duncan, R., Izzo, L., 2005. Dendrimer biocompatibility and toxicity. Adv. Drug Deliv. Rev. 57, 2215–2237.

Edelmana, E.R., Nathan, A., Katada, M., 2000. Perivascular graft heparin delivery using biodegradable polymer wraps. Biomaterials 21, 2279–2286.

El-Kamel, A.H., Sokar, M.S., Gamal, A.L., Naggar, V.F., 2001. Preparation and evaluation of ketoprofen floating oral delivery system. Int. J. Pharm. 220, 13–21.

Eloy, J.O., De Souza, M.C., Petrilli, R., Abriata, J.P., Lee, R.J., Marchetti, J.M., 2014. Liposomes as carriers of hydrophilic small molecule drugs: strategies to enhance encapsulation and delivery. Colloids Surf. B 123, 345–363.

Ensign, L.M., Cone, R., Hanes, J., 2012. Oral drug delivery with polymeric nanoparticles: the gastrointestinal mucus barriers. Adv. Drug Deliv. Rev. 19, 165–170.

Feng, J., Su, W., Wang, H., Huang, F., Zhang, X., Zhuo, R., 2009. Facile fabrication of diblock methoxy polyethylene glycol-poly-(trimethylene carbonate) and its self-assembled micelles as drug carriers. ACS Appl. Mater. Interfaces 12, 2729–2737.

Fiorica, C., Pitarresi, G., Palumbo, F.S., Di Stefano, M., Calascibetta, F., 2013. A new hyaluronic acid pH sensitive derivative obtained by ATRP for potential oral administration of proteins. Int. J. Pharm. 457, 150–157.

Florence, A.T., 2004. Issues in oral nanoparticle drug carrier uptake and targeting. J. Drug Target 12, 65–70.

Foldvari, M., Bagonluri, M., 2008. Carbon nanotubes as functional excipients for nanomedicines: II. Drug delivery and biocompatibility issues. Nanomed. Nanotech. Biol. Med. 4, 183–200.

Fonte, P., Araùjo, F., Silva, C., Pereira, C., Reis, S., Santos, H.A., Sarmento, B., 2015. Polymer-based nanoparticles for oral insulin delivery: revisited approaches. Biotech. Adv. 33, 1342–1354.

Furtado, S., Abramson, D., Burrill, R., Olivier, G., Gourd, C., Bubbers, E., Mathiowitz, E., 2008. Oral delivery of insulin loaded poly(fumaric-*co*-sebacic) anhydride microspheres. Int. J. Pharm. 347, 149–155.

Gareth, A., Hughes, P., 2005. Nanostructure-mediated drug delivery. Nano. Nano., Biol. Med. 1, 22–30.

Garlotta, D., 2001. A literature review of poly(lactic acid). J. Polym. Environ. 9, 63–84.

Gebert, A., Rothkötter, H.J., Pabst, R., 1996. M cells in Peyer's patches of the intestine. In: Kwang, W.J. (Ed.), International Review of Cytology. Academic Press, San Diego, California, pp. 91–159.

George, M., Abraham, T.E., 2006. Polyionic hydrocolloids for the intestinal delivery of protein drugs: alginate and chitosan: a review. J. Control. Rel. 114, 1–14.

Gharib, R., Greige-Gerges, H., Fourmentin, S., Charcosset, C., Auezova, L., 2015. Liposomes incorporating cyclodextrin-drug inclusion complexes: current state of knowledge. Carbohydr. Polym. 129, 175–186.

Giannasca, P.J., Giannasca, K.T., Leichtner, A.M., Neutra, M.R., 1999. Human intestinal M cells display the sialyl Lewis A antigen. Infect. Immun. 67, 946–953.

Golokhvast, K.S., Chaika, V.V., Kuznetsov, L.V., Elumeeva, K.V., Kusaikin, M.I., Zakharenko, A.M., Kiselev, N.N., Panichev, A.M., Reva, G.V., Usov, V.V., Reva, I.V., Yamamoto, T., Gulkov, A.N., 2013. Effects of multiwalled carbon nanotubes received orally during 6 days on the gastrointestinal tract. Bull. Exper. Biol. Med. 155, 788–792.

Grenha, A., Grainger, C.I., Dailey, L.A., Seijo, B., Martin, G.P., Remunan-Lopez, C., Forbes, B., 2007. Chitosan nanoparticles are compatible withrespiratory epithelial cells in vitro. Eur. J. Pharm. Sci. 31, 73–84.

Gruber, P., Longer, M.A., Robinson, J.R., 1987. Some biological issues in oral, controlled drug delivery. Adv. Drug Deliv. Rev. 1, 1–18.

Gu, J.M., Robinson, J.R., Leung, S.H., 1988. Binding of acrylic polymers to mucin/epithelial surfaces: structure–property relationships. Crit. Rev. Ther. Drug. Carrier Syst. 5, 21–67.

Gui, G.F., Zhuo, Y., Chai, Y.Q., Xiang, Y., Yuan, R., 2006. The Ru complex and hollow gold nanoparticles branched-hydrogel as signal probe for construction of electrochemiluminescent aptasensor. Bios. Bioelect. 77, 7–12.

Gulsun, T., Gursoy, R., Neslihan, O.L., 2009. Nanocrystal technology for oral delivery of poorly water-soluble drugs. J. Pharm. Sci 34, 11–15.

Guo, J., Ping, Q., Chen, Y., 2001. Pharmacokinetic behavior of cyclosporin A in rabbits by oral administration of lecithin vesicle and Sandimmun Neoral. Int. J. Pharm. 216, 17–21.

Han, L.N., Zhao, Y.F., Yin, L.F., Li, R.M., Liang, Y., Huang, H., Pan, S.R., 2012. Insulin-loaded pH-sensitive hyaluronic acid nanoparticles enhance transcellular delivery. AAPS Pharm. Sci. Tech. 13, 836–845.

Hebrard, G., Hoffart, V., Beyssac, E., Cardot, J.M., Alric, M., Subirade, M., 2010. Coated whey protein/alginate microparticles as oral controlled delivery systems for probiotic yeast. J. Microencapsul. 27, 292–302.

How, C.W., Rasedee, A., Manickam, S., 2013. Tamoxifen-loaded nanostructured lipid carrier as a drug delivery system: characterization, stability assessment, and cytotoxicity. Colloids Sur. B. Biointerfaces 112, 393–399.

Hu, J., Wu, T., Zhang, G., Liu, S., 2012. Efficient synthesis of single gold nanoparticle hybrid amphiphilic triblock copolymers and their controlled self-assembly. J. Am. Chem. Soc. 134, 7624–7627.

Hughes, G.A., 2005. Nanostructure-mediated drug delivery. Nanomed. Nanotech. Biol. Med. 1, 22–30.

Irache, J.M., Durrer, C., Duchene, D., Ponchel, G., 1994. Preparation and characterization of lectin-latex conjugates for specific bioadhesion. Biomaterials 15, 899–904.

Jacquemin, E., Hagenbuch, B., Stieger, B., Wolkoff, A.W., Meier, P.J., 1994. Expression cloning of a rat liver Na(+)-independent organic anion transporter. Proc. Natl. Acad. Sci. USA 91, 133–137.

Jain, A.K., Goyal, A.K., Mishra, N., Vaidya, B., Mangal, S., Vyas, S.P., 2010. PEG-PLA-PEG block copolymeric nanoparticles for oral immunization against hepatitis B. Int. J. Pharm. 387, 253–262.

Kazunari, A., Seiichi, K., Shozo, S., Don, M., Miroslav, B., Sung, W.K., Junzo, S., 1998. Self assembled hydrogel nanoparticle of cholesterol-bearing pullulan as a carrier of protein drugs: complexation and stabilization of insulin. J. Control. Release 54, 313–320.

Kedar, U., Phutane, P., Shidhaye, S., Kadam, V., 2010. Advances in polymeric micelles for drug delivery and tumor targeting. Nanomed: Nanotech. Biol. Med. 6, 714–729.

Kikuchi, A., Tanaka, M., Tomoyasu, T., Maeda, M., Tamai, I., 2008. Improvement of intestinal absorption by peptide-derivation of nonabsorbable drug via peptide transporter PEPT1. J. Pharm. Sci. 98, 1775–1787.

Kim, B., Bowersock, T., Griebel, P., Kidane, A., Babiuk, L.A., Sanchez, M., Attah-Poku, S., Kaushik, R.S., Mutwiri, G.K., 2002. Mucosal immune responses following oral immunization with rotavirus antigens encapsulated in alginate microspheres. J. Control. Release 85, 191–202.

Kisel, M.A., Kulik, L.N., Tsybovsky, I.S., Vlasov, A.P., Vorob'yov, M.S., Kholodova, E.A., 2001. Liposomes with phosphatidylethanol as a carrier for oral delivery of insulin: studies in the rat. Int. J. Pharm. 216, 105–114.

Kreuter, J., Muller, U., Munz, K., 1989. Quantitative and microautoradiographic study on mouse intestinal distribution of polycyanoacrylate nanoparticles. Int. J. Pharm. 55, 39–45.

Kuo, Y.C., Chen, H.H., 2006. Effect of nanoparticulate polybutylcyanoacrylate and methyl methacrylate sulfopropyl methacrylate on the permeability of zidovudine and lamivudine across the in vitro blood–brain barrier. Int. J. Pharm. 327, 160–169.

Lai, S.K., Wang, Y.Y., Hanes, J., 2009. Mucus-penetrating nanoparticles for drug and gene delivery to mucosal tissues. Adv. Drug Deliv. Rev. 61, 158–171.

Lai, S.K., Wang, Y.Y., Hida, K., Cone, R., Hanes, J., 2010. Nanoparticles reveal that humancervicovaginal mucus is riddled with pores larger than viruses. Proc. Natl. Acad. Sci. U S A 7, 598–603.

Lee, C.C., Gillies, E.R., Fox, M.E., Guillaudeu, S.J., Fréchet, J.M., Dy, E.E., Szoka, F.C., 2006. A single dose of doxorubicin-functionalized bow-tie dendrimer cures mice bearing C-26 colon carcinomas. Proc. Natl. Acad. Sci. USA 103, 16649–16654.

Lee, E.J., Lee, I.H., Lee, M., Yu, H., Kim, S.Y., Chae, S.Y., Jon, S., 2008. Conjugated chitosan as a novel platform for oral delivery of paclitaxel. J. Med. Chem. 51, 6442–6449.

Letchford, K., Burt, H., 2007. A review of the formation and classification of amphiphilic block copolymer nanoparticulate structures: micelles, nanospheres, nanocapsules and polymersomes. Eur. J. Pharm. Biopharm. 65, 259–269.

Li, Z., Wen, L., Shao, L., Chen, J., 2004. Fabrication of porous silica nanoparticles and their applications in release control. J. Control. Release 98, 245–254.

Limaye, A., Koya, V., Samsam, M., Daniell, H., 2006. Receptor-mediated oral delivery of a bioencapsulated green fluorescent protein expressed in transgenic chloroplasts into the mouse circulatory system. FASEB J. 20, 959–961.

Lin, C., Chen, F., Ye, T., Zhang, L., Zhang, W., Liu, D., Xiong, W., Yang, X., Pan, W., 2014. A novel oral delivery system consisting in drug-in cyclodextrin-in nanostructured lipid carriers for poorly water-soluble drug: vinpocetine. Int. J. Pharm. 465, 90–96.

Liu, J., Gong, T., Wang, C., Zhong, Z., Zhang, Z., 2007. Solid lipid nanoparticles loaded with insulin by sodium cholate-phosphatidylcholine-based mixed micelles: preparation and characterization. Int. J. Pharm. 340, 153–162.

Lopes, J.C.M., Martins-Lopes, P., Souto, E.B., 2010. Nanoparticulate carriers (NPC) for oral pharmaceutics and nutraceutics. Pharmazie 65, 75–82.

Lozier, J.N., Yankaskas, J.R., Ramsey, W.J., Chen, L., Berschneider, H., Morgan, R.A., 1997. Gut epithelial cells as targets for gene therapy of hemophilia. Hum. Gene Ther. 8, 1481–1490.

Lu, Y., Park, K., 2013. Polymeric micelles and alternative nanonized delivery vehicles for poorly soluble drugs. Inter. J. Pharm. 453, 198–214.

Macadam, A., 1993. The effect of gastrointestinal mucus on drug absorption. Adv. Drug Deliv. Rev. 11, 201–220.

Madara, J.L., 1998. Regulation of the movement of solutes across tight junctions. Annu. Rev. Physiol. 60, 143–159.

Makwana, V., Jain, R., Patel, K., Nivsarkar, M., Joshi, A., 2015. Solid lipid nanoparticles (SLN) of Efavirenz as lymph targeting drug delivery systems: elucidation of mechanism of uptake using chylomicron flow blocking approach. Inter. J. Pharm. 495, 439–446.

Malik, N., Evagorou, E.G., Duncan, R., 1999. Dendrimer–platinate: a novel approach to cancer chemotherapy. Anticancer Drugs 10, 767–776.

Martin, J., LaBranche, C.C., Gonzalez-Scarano, F., 2001. Differential CD4/CCR5 utilization, gp120 conformation, and neutralization sensitivity between envelopes from a microglia-adapted human immunodeficiency virus type 1 and its parental isolate. J. Virol. 75, 3568–3580.

Masuda, K., Horie, K., Suzuki, R., Takayoshi, Y., Hirano, K., 2002. Oral delivery of antigens in liposomes with some lipid compositions modulates oral tolerance to the antigens. Microbiol. Immunol. 46, 55–58.

Mathiowitz, E., Jacob, J.S., Jong, Y.S., Carino, G.P., Chickering, D.E., Chaturvedi, P., Santos, C.A., Vijayaraghavan, K., Montgomery, S., Bassett, M., Morrell, C., 1997. Biologically erodable microspheres as potential oral drug delivery systems. Nature 386, 410–414.

McGuckin, M.A., Linden, S.K., Sutton, P., Florin, T.H., 2011. Mucin dynamics and enteric pathogens. Nat. Rev. Microbiol. 9, 265–278.

Mei, Z., Li, X., Wu, Q., Hu, S., Yang, X., 2005. The research on the anti-inflammatory activity and hepatotoxicity of triptolide-loaded solid lipid nanoparticle. Pharmacol. Res. 51, 345–351.

Menjoge, A.R., Kannan, R.M., Kim, C.J., Romero, R., 2010. Dendrimer-based drug and imaging conjugates: design considerations for nanomedical applications. Drug Disc. Today 15, 171–185.

Meyer, R.A., McGinley, D., Posalaky, Z., 1984. The gastric mucosal barrier: structure of intercellular junctions in the dog. J. Ultrastruct. Res. 86, 192–201.

Morishita, M., Goto, T., Nakamura, K., Lowman, A.M., Takayama, K., Peppas, N.A., 2006. Novel oral insulin delivery systems based on complexation polymer hydrogels: single and multiple administration studies in type 1 and 2 diabetic rats. J. Control. Release 110, 587–594.

Müller, C., Leithner, K., Hauptstein, S., Hintzen, F., Salvenmoser, W., Bernkop-Schnürch, A., 2012. Preparation and characterization of mucus-penetrating papain/poly(acrylic acid) nanoparticles for oral drug delivery applications. J. Nanopart. Res. 15, 1–13.

Murphy, E.A., Majeti, B.K., Barnes, L.A., Makale, M., Weis, S.M., Lutu-Fuga, K., Wrasidlo, D.A., Cheresh, 2008. Nanoparticle-mediated drug delivery to tumor vasculature suppresses metastasis. Proc. Natl. Acad. Sci. U SA 105, 9343–9348.

Muzzarelli, R.A.A., Muzzarelli, C., 2005. Chitosan chemistry: relevance to the biomedical sciences. Adv. Polym. Sci. 186, 151–209.

Narayanan, R., Tabor, C., El-Sayed, M.A., 2008. Can the observed changes in the size or shape of a colloidal nanocatalyst reveal the nanocatalysis mechanism type: homogeneous or heterogeneous? Top. Catal. 48, 60–74.

Neves, A.R., Queiroz, J.F., Costa-Lima, A.A., Figueiredo, F., Fernandes, R., Reis, S., 2015. Cellular uptake and transcytosis of lipid-based nanoparticles across the intestinal barrier: relevance for oral drug delivery. J. Colloid Interf. Sci. 463, 258–265.

Ogue, S., Takahashi, Y., Onishi, H., Machida, Y., 2006. Preparation of double liposomes and their efficiency as an oral vaccine carrier. Biol. Pharm. Bull. 29, 1223–1228.

Ojewole, E., Mackraj, I., Naidoo, P., Govender, T., 2008. Exploring the use of novel drug delivery systems for antiretroviral drugs. Eur. J. Pharm. Biopharm. 70, 697–710.

Owen, D.A., Reid, P.E., 1995. Histochemical alterations of mucin in normal colon, inflammatory bowel-disease and colonic adenocarcinoma. Histochem. J. 27, 882–889.

Pandey, R., Zahoor, A., Sharma, S., Khuller, G.K., 2003. Nanoparticle encapsulated antitubercular drugs as a potential oral drug delivery system against murine tuberculosis. Tuberculosis 83, 373–378.

Patel, D.J., Patel, J.K., 2010. Mucoadhesive effect of polyethyleneoxideon famotidine nanosuspension prepared by solvent evaporation method. Int. J. Pharm. Pharm. Sci. 2, 122–127.

Ponchel, G., Irache, J.M., 1998. Specific and non-specific bioadhesive particulate systems for oral delivery to the gastrointestinal tract. Adv. Drug. Delivery Rev. 34, 191–219..

Ponchel, G., Montisci, M.J., Dembri, A., Durrer, C., Duchêne, D., 1997. Mucoadhe-sion of colloidal particulate systems in the gastro-intestinal tract. Eur. J. Pharm. Biopharm. 44, 25–31.

Pooja, D., Panyaram, S., Kulhari, H., Reddy, B., Rachamalla, S.S., Sistla, R., 2015. Natural polysaccharide functionalized gold nanoparticles as biocompatible drug delivery carrier. Intern. J. Biol. Macromol. 80, 48–56.

Powell, D.W., 1981. Barrier function of epithelia. Am. J. Physiol. 241, G275–G288.

Primard, C., Rochereau, N., Luciani, E., Genin, C., Delair, T., Paul, S., Verrier, B., 2010. Traffic of poly(lactic acid) nanoparticulate vaccine vehicle from intestinal mucus to sub-epithelial immune competent cells. Biomaterials 31, 6060–6068.

Raissi, H., Mollania, R., 2014. Immunosuppressive agent leflunomide: a SWNTs-immobilized dihydroortate dehydrogenase inhibitory effect and computational study of its adsorption properties on zigzag single walled (6,0) carbon and boron nitride nanotubes as controlled drug delivery devices. Eur. J. Pharm. Scie. 56, 37–54.

Rasmussen, M.R., Snabe, T., Pedersen, L.H., 2003. Numerical modelling of insulin and amyloglucosidase release from swelling Ca–alginate beads. J. Control. Release 91, 395–405.

Rieux, A.D., Ragnarsson, E.G.E., Gullberg, E., Préat, V., Schneider, Y.J., Artursson, P., 2005. Transport of nanoparticles across an in vitromodel of the human intestinal follicle associated epithelium. Europ. J. Pharm. Sci., 455–465.

Russell-Jones, G.J., 1996. The potential use of receptor-mediated endocytosis for oral drug delivery. Adv. Drug Deliv. Rev. 20, 83–97.

Russell-Jones, G.J., 1999. Carrier-mediated transport, oral drug delivery. In: Mathiowitz, E. (Ed.), Encyclopedia of Controlled Drug Delivery. John Wiley & Sons, NewYork, pp. 173–184.

Sahoo, N., Sahoo, R.K., Biswas, N., Guha, A., Kuotsu, K., 2015. Recent advancement of gelatin nanoparticles in drug and vaccine delivery. Inter. J. Biol. Macromol. 81, 317–331.

Salamat-Miller, N., Chittchang, M., Johnston, T.P., 2005. The use of mucoadhesive polymers in buccal drug delivery. Adv. Drug. Deliv. Rev. 57, 1666–1691.

Schenk, M., Mueller, C., 2008. The mucosal immune system at the gastrointestinal barrier. Best Pract. Res. Cl Ga. 22, 391–409.

Shah, N.M., Parikh, J., Namdeo, A., Subramanian, N., Bhowmick, S., 2006. Preparation, characterization and in vivo studies of proliposomes containing Cyclosporine A. J Nanosci. Nanotechnol. 6, 2967–2973.

Shi, P., Qu, K., Wang, J., Li, M., Ren, J., Qu, X., 2012. pH-responsive NIR enhanced drug release from gold nanocages possesses high potency against cancer cells. Chem. Commun. 48, 7640–7642.

Shrestha, B., Rath, J.P., 2013. Poly(vinyl alcohol)-coated chitosan microparticles act as an effective oral vaccine delivery system for hepatitis B vaccine in rat model. IET Nanobiotechnol. 4, 1–7.

Singh, R., Lillard, Jr., J.W., 2009. Nanoparticle-based targeted drug delivery. Exper. Mol. Pathol. 86, 215–223.

Sosnik, A., Raskin, M.M., 2015. Polymeric micelles in mucosal drug delivery: challenges towards clinical translation. Biotech. Adv. 33, 1380–1392.

Spitzenberger, T.J., Heilman, D., Diekmann, C., Batrakova, E.V., Kabanov, A.V., Gendelman, H.E., Elmquist, W.F., Persidsky, Y., 2007. Novel delivery system enhances efficacy of antiretroviral therapy in animal model for HIV-1 encephalitis. J. Cereb. Blood Flow Metab. 27, 1033–1042.

Svenson, S., 2009. Dendrimers as versatile platform in drug delivery applications. Eur. J. Pharm. Biopharm. 71, 445–462.
Swaan, P.W., 1998. Recent advances in intestinal macromolecular drug delivery via receptor-mediated transport pathways. Pharm. Res. 15, 826–834.
Tamai, I., Tsuji, A., 1996. Carrier-mediated approaches for oral drug delivery. Adv. Drug Deliv. Rev. 20, 5–32.
Tang, V.W., Goodenough, D.A., 2003. Paracellular ion channel at the tight junction. Biophys. J. 84, 1660–1673.
Taratula, O., Kuzmov, A., Shah, M., Garbuzenko, O.B., Minko, T., 2013. Nanostructured lipid carriers as multifunctional nanomedicine platform for pulmonary co-delivery of anticancer drugs and siRNA. Tenth Internat. Nano. Drug Deliv. Symp. 171, 349–357.
Thanki, K., Gangwal, R.P., Sangamwar, A.T., Jain, S., 2013. Oral delivery of anticancer drugs: challenges and opportunities. J. Control. Release 170, 15–40.
Thanou, M., Verhoef, J.C., Junginger, H.E., 2001. Oral drug absorption enhance-ment by chitosan and its derivatives. Adv. Drug. Deliv. Rev. 52, 117–126.
Tripisciano, C., Kraemer, K., Taylor, A., Borowiak-Palen, E., 2009. Single-wall carbon nanotubes based anticancer drug delivery system. Chem. Phys. Lett. 478, 200–205.
Tsuji, A., Takanaga, H., Tamai, I., Terasaki, T., 1994. Transcellular transport of benzoic acidacross Caco-2 cells by a pH-dependent and carrier-mediated transport mechanism. Pharm. Res. 11, 30–37.
Uchiyama, T., Sugiyama, T., Quan, Y.S., Kotani, A., Okada, N., Fujita, T., Muranishi, S., Yamamoto, A., 1999. Enhanced permeability of insulin across the rat intestinal membrane by various absorption enhancers: their intestinal mucosal toxicity and absorption enhancing mechanism of n-lauryl-beta-D-maltopyranoside. J. Pharm. Pharmacol. 51, 1241–1250.
Vasir, J.K., Reddy, M.K., Labhasetwar, V., 2005. Nanosystems in drug targeting: opportunities and challenges. Curr. Nanosci. 1, 47–64.
Wang, B.L., Liu, X.S., Ji, Y., Ren, K.F., Ji, J., 2012. Fast and long-acting antibacterial properties of chitosan–Ag/polyvinylpyrrolidone nanocomposite films. Carbohydr. Polym. 90, 8–15.
Werle, M., Takeuchi, H., 2009. Chitosan-aprotinin coated liposomes for oral peptide delivery: development, characterisation and in vivo evaluation. Int. J. Pharm. 370, 26–32.
Wina, K.Y., Feng, S.S., 2005. Effects of particle size and surface coating on cellular uptake of polymeric nanoparticles for oral delivery of anticancer drugs. Biomaterials 26, 2713–2722.
Wong, B.S., Yoong, S.L., Jagusiak, A., Panczyk, T., Ho, H.K., Ang, W.H., Pastorin, G., 2013. Carbon nanotubes for delivery of small molecule drugs. Adv. Drug Deliv. Rev. 65, 1964–2015.
Wyatt, R., Sodroski, J., 1998. The HIV-1 envelope glycoproteins: fusogens, antigens, and immunogens. Science 280, 1884–1888.
Xu, F., Cui, F.Z., Jiao, Y.P., Meng, Q.Y., Wang, X.P., Cui, X.Y., 2009. Improvement of cytocompatibility of electrospinning PLLA microfibers by blending PVP. J. Mater. Sci. Mater. Med. 20, 1331–1338.
Yang, D., Yang, F., Hu, J., Long, J., Wang, C., Fu, D., Ni, Q., 2009. Hydrophilic multi-walled carbon nanotubes decorated with magnetite nanoparticles as lymphatic targeted drug delivery vehicles. Chem Commun. 29, 4447–4449.
Yang, K., Feng, L., Shi, X., Liu, Z., 2013. Nano-graphene in biomedicine: theranostic applications. Chem. Soc. Rev. 42, 530–547.
Yang, F., Zhang, Q., Liang, Q., Wang, S., Zhao, B., Wang, Ya., Cai, Y., Li, G., 2015. Bioavailability enhancement of paclitaxel via a novel oral drug delivery system: paclitaxel-loaded glycyrrhizic acid micelles. Molecules 20, 4337–4356.
Zereshki, S., Figoli, A., Madaeni, S.S., Galiano, F., Drioli, E., 2011. Pervaporation separation of ethanol/ETBE mixture using poly(lactic acid)/poly(vinyl pyrrolidone) blend membranes. J. Membr. Sci. 373, 29–35.
Zhang, N., Ping, Q., Huang, G., Xu, W., Cheng, Y., Han, X., 2006. Lectin-modified solid lipid nanoparticles as carriers for oral administration of insulin. Int. J. Pharm. 327, 153–159.

Zhang, X., Meng, L., Lu, Q., Fei, Z., Dyson, P.J., 2009. Targeted delivery and controlled release of doxorubicin to cancer cells using modified single wall carbon nanotubes. Biomaterials 30, 6041–6047.

Zhang, C., Peng, F., Liu, W., Wan, J., Wan, C., Xu, H., Lam, C.W., Yang, X., 2014. Nanostructured lipid carriers as a novel oral delivery system for triptolide: induced changes in pharmacokinetics profile associated with reduced toxicity in male rats. Int. J. Nanomed. 9, 1049–1063.

Zhuang, C.Y., Li, N., Wang, M., Zhang, X.N., Pan, W.S., Peng, J.J., Pan, Y.S., Tang, X., 2010. Preparation and characterization of vinpocetine loaded nanostructured lipid carriers (NLC) for improved oral bioavailability. Int. J. Pharm. 394, 179–185.

CHAPTER

3 BIODEGRADABLE POLYMERIC NANOSTRUCTURES: DESIGN AND ADVANCES IN ORAL DRUG DELIVERY FOR NEURODEGENERATIVE DISORDERS

Andreza Maria Ribeiro*,, Francisco Veiga*,†, Ana Figueiras*,†,‡**

**Faculty of Pharmacy, University of Coimbra, Coimbra, Portugal*

***Department of Engineering and Material Sciences, University of Federal of Paraná (UFPR), Curitiba, Brazil*

†REQUIMTE/LAQV, Group of Pharmaceutical Technology, Faculty of Pharmacy, University of Coimbra, Coimbra, Portugal

‡Health Sciences Centre, Faculty of Health Sciences, University of Beira Interior, Covilhã, Portugal

CHAPTER OUTLINE

1. Introduction 61
 1.1 Neurodegenerative Disorders: an Overview 63
2 Biodegradable Polymeric Materials 64
 2.1 Natural Polymers 64
 2.2 Synthetic Polymers 65
3 Nanocarriers Technology and Recent Modifications to Improve Their Specificity 66
 3.1 Polymeric Nanoparticles 68
 3.2 Polymeric Micelles 69
 3.3 Dendrimers 71
 3.4 Liposomes 73
 3.5 Functional Surface Building 75
4 Advances in Polymeric Nanocarriers as Vehicles for Oral-Delivery Systems 76
5. New Approaches and Clinical Applications in Neurodegenerative Disorders 78
6. Future Perspectives 79
References 80

1 INTRODUCTION

Nanotechnology is a science that has gained attention in recent years, particularly in the medical, pharmaceutical, and chemical fields, although this concept has already been introduced more than 40 years ago (Nath and Banerjee, 2013). This science allows the manipulation of matter in atomic and molecular scale, creating novel materials and developing new products in nanoscale range (1–100 nm).

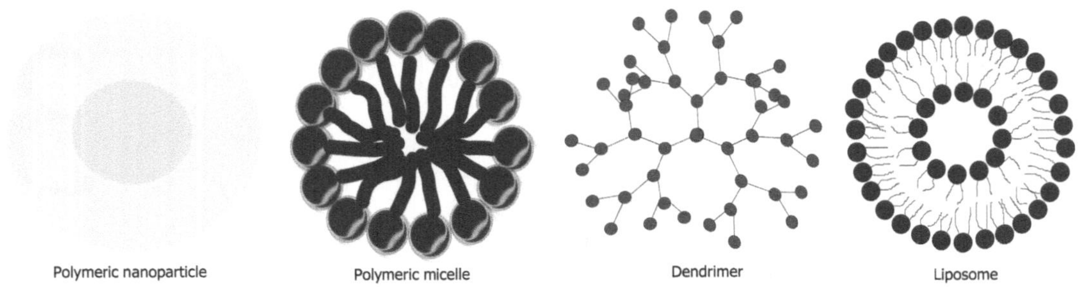

FIGURE 3.1 Different Types of Nanoparticles for Active Molecules Transport

In the medical-pharmaceutical field, nanotechnology mainly deals with the characterization, manipulation, preparation, and application of biological structures, in particular nanoparticles (NPs) (Fig. 3.1) (Ochekpe et al., 2009).

Polymers are the most prominent among the materials used for controlled release formulations and drug-targeting systems due their diversity in structure and chemistry. Biodegradable polymers have the potential to exceed the drawbacks associated with conventional, nondegradable polymers. The primary concern is the problem associated with the cell toxicity. Other issues include the benefits of biocompatibility and reproducibility when biodegradable polymers are used. With the development of several nanosystems, the importance of polymer structure and property relationships has emphasized their use in pharmaceutical sciences (Smith, 2005; Vroman and Tighzert, 2009).

Nanotechnology has been widely applied, especially in the area of drug-delivery systems (DDS), because the NPs are able to interact with biological systems on a molecular level and at the same time stimulate, respond, and interact with target cells and tissues, inducing desired physiological responses (Modi et al., 2010). Therefore, nanotechnology has been used therapeutically for several levels of disorders, and also for diagnosis and imaging. Currently, the most promising application is the use of nanosystems to transport drugs across the blood–brain barrier (BBB) (Kreuter, 2007).

The incidence of neurological diseases, such as Alzheimer's disease, Parkinson's disease, glioblastoma, and other dementia disorders, has rapidly increased with the progressive growth in average life expectancy. However, formulations targeting the central nervous system (CNS) remain a challenge due to the difficulty that the drugs have for reaching the brain (Ramos-Cabrer and Campos, 2013). The CNS is a complex and vulnerable system protected by a biological barrier, the BBB, which is a good defense mechanism of the brain against external molecules that might cause any kind of brain damage and is, therefore, quite selective on the passage of molecules between the blood and the brain (Ramos-Cabrer and Campos, 2013). Yet this defensive mechanism complicates attempts to carry out therapeutic strategies in order to reach the brain and exert their pharmacological actions in neurological targets.

The BBB is formed by a monolayer of endothelial cells connected to each other through the tight junctions, having direct or indirect contact with the terminals of astrocytes, neurons perivascular, and pericytes (Ramos-Cabrer and Campos, 2013). Endothelial brain cells are highly polarized and they have low pinocytic activity, but they contain different mechanisms of active transport in order to ensure homeostasis of the brain. Moreover, they also have high concentrations of P-glycoprotein, which is an

ATP-dependent protein, responsible for the active transport involved in the multiple drug efflux, including chemotherapeutic agents (Neha et al., 2013). NPs, due to their small size, have been considered possible therapeutic strategies to pass through the BBB and reach the brain. However, the modification of their structures has become essential to increasing specificity and selectivity, thereby promoting a more effective and targeted release of drugs. One of the most common modifications is the addition of binders to the surface of NPs, so they can directly interact with endothelial cells, such as transferrin, insulin, and thiamine, which have specific receptors on the BBB (Tosi et al., 2012). Researchers have found that during the interaction between the NPs with the corresponding receptors and ligands, endocytic vesicles are created, which can promote the NPs transcytosis through the BBB cells and eventually release the NPs into the brain by exocytosis (Tosi et al., 2012).

This chapter aims to make an approach to the principal nanoparticle systems that are considered promising in oral transport of drugs to the brain. The goal is to focus on the main characteristics of each one, and the relevance that nanotechnology can provide in order to improve the treatment and prevention of different neurological pathologies. For this purpose the following subsections are explored.

1.1 NEURODEGENERATIVE DISORDERS: AN OVERVIEW

"Neurodegenerative diseases are defined as hereditary and sporadic conditions which are characterized by progressive nervous system dysfunction." Alzheimer's disease (AD) is the most common neurodegenerative disorder (ND). Currently, there are 36 million of people stricken with NDs in the world. It is projected that there will be 65.7 million people with dementia worldwide in the year 2030 and the prevalence of affected will live in developing countries (Duthey, 2013). AD is characterized by progressive deposition of β-amyloid protein in the space between neurons, forming senile plaques. On the other hand, deposition of Tau protein within neurons, forming neurofibrillary tangles, has also been observed. This pathology is responsible for dementia and loss of cognitive functions (Akula, 2013). Other kinds of NDs include brain cancer, degenerative nerve diseases, encephalitis, epilepsy, genetic brain disorders, head and brain malformations, hydrocephalus, stroke, multiple sclerosis, amyotrophic lateral sclerosis (ALS or Lou Gehrig's disease), Huntington's disease, and prion disease. (Korsak and Kozyreva, 2015).

Gliomas are primary brain tumors originating from glial cells. They can be astrocytomas, glioblastomas, oligodendrogliomas, and ependymomas. Glioblastoma multiforme (GBM) is an astrocytoma of grade IV and represents 70% of all malignant gliomas (Chung et al., 2014). Physiological characteristics of a solid tumor, such as glioblastoma, can be used for the benefit of therapy when using NPs. In fact, around the tumor it's possible to observe fibrin deposits and the formation of new blood vessels that allow the penetration of the NPs in the tumor neovascularization by the so-called enhanced and retention effect (Oliveira et al., 2012).

Cerebral paralysis, which refers to a group of disorders, targets the motor control and posture as a result of a nonprogressive injury upon the CNS development. So it is a disease that mostly affects babies and may appear in the pre- or postnatal stages (Balakrishnan et al., 2013). Worldwide, there is an incidence of 0.2% in newborns, increasing to 4%–10% in premature newborns and those of low birth weight. This disease is a permanently disabling condition and it makes for a major consumer of health care resources in developed countries (Sewell et al., 2014).

Parkinson's disease, another degenerative disorder of the CNS, is chronic and progressive. It is caused by a strong decrease in the production of dopamine, particularly in a small brain region called the substantia nigra. The patient's motor control is lost, causing characteristic signs and symptoms that allow clinical diagnosis of Parkinsonism. The clinical presentation basically consists of four main signs: tremors, slowness and poverty of voluntary movement (akinesia or bradykinesia), stiffness (of muscles, particularly at level of the joints), and postural instability (difficulties related to balance, with frequent falls). Presently, the therapy for Parkinson's disease is the oral administration of dopamine agonists, such as levodopa (Modi et al., 2010). Another drug used is amantadine, which reduces rigidity, tremor, and akinesia. Additionally, a variety of compounds are currently studied, such as cell lines derived of neurotrophic factors, rasagiline, creatine, coenzyme Q-10, and riluzole (Leyva-Gómez et al., 2015). Therapeutic treatment has been focused on restoring dopaminergic neurotransmission (Meissner et al., 2011). Deep brain stimulation and transplantation of fetal dopamine neurons have been explored to complement the drug treatment (Freed et al., 2001).

2 BIODEGRADABLE POLYMERIC MATERIALS

The term biodegradable defines all the materials capable of undergoing decomposition into carbon dioxide, water, methane, inorganic compounds, or biomass, being the enzymatic action of microorganisms the predominant mechanism of decomposition (Smith, 2005). Biodegradable polymeric materials were first introduced in the 1980s. The worldwide interest in these polymers has grown substantially in recent times. Nevertheless, the high cost of their production is still a great disadvantage compared to conventional polymers. These materials are currently used in a number of industrial applications and have attracted an increasing amount of attention over the last decades, predominantly due to environmental concerns. In general, the biodegradable polymers used in medicine must have a high degree of biocompatibility, that is, the material can interact with the body without inducing toxic response, immunogenicity, thrombogenic, or carcinogen (Lam et al., 1992). Generally, the degradation of polymers for biomedical applications is often a desirable goal after their introduction in the body and the performance of their function. However, biodegradable polymeric materials become very attractive in nanotechnology processes for drug release control and therapeutic improvement. The biodegradability of polymers takes place through the action of enzymes and chemical deterioration associated with corporeal organism fluids (Vroman and Tighzert, 2009). Biodegradable polymers undergo more easily the action of microorganisms whose organic functions contain aliphatic chains: carbonyl, hydroxyl, ester, and hydroxy acids, which are more susceptible to enzymatic action (Franchetti and Marconato, 2006). Several sources of biodegradable polymers were observed, made from natural and synthetic materials (Vroman and Tighzert, 2009).

2.1 NATURAL POLYMERS

Natural polymers are formed during the growth cycle of living organisms. Their synthesis generally involve enzyme-catalyzed reactions and chain growth reactions of activated monomers, forming within the cells by complex metabolic processes (Franchetti and Marconato, 2006). They can be derived from natural sources, such as corn, cellulose, potato, sugarcane, or they can be synthesized by bacteria from small molecules, such as butyric acid or valeric acid yielding polyhydroxybutyrate and polyhydroxybutyrate-*co*-valerate, respectively, or they can be derivatives of animal sources, such as chitin, chitosan, or proteins (Nair and Laurencin, 2007).

One useful strategy to improve the characteristics of natural polymers is their modification, using proper reagents, to obtain semisynthetic polymers (Zhao et al., 2012) Modified dextran, cyclodextrins (CDs), starch, cellulose, alginate, and chitosan derivatives are the largest semisynthetic polymers used in biomedical devices, pharmaceutical technology, and drug delivery (Mogoşanu and Grumezescu, 2015). A common example is the modification of different chemical derivatives of natural CDs (α, β, and γ glucopyranose molecules), by the substitution of some hydrogen bonds that form chemical groups, improving some properties of natural CDs like solubility, stability, bioavailability, and decreasing toxicity and hemolytic activity (Veiga et al., 2006). Thus, a variety of nanocarriers can be created by exploiting and modifying natural polymers in order to treat dementia.

2.2 SYNTHETIC POLYMERS

Synthetic biodegradable polymers can be obtained from fossil sources, oil, or the mixture of biomass and petroleum. The most popular biodegradable polymers derived from petroleum are polycaprolactones, polyesteramides, and aliphatic or aromatic copolyesters (Ray and Bousmina, 2005). The polymers (Table 3.1) most widely used, especially in the clinical field, are poly(lactic acid) (PLA), poly(glycolic acid) (PGA), poly(lactic glycolic acid) (PLGA), and poly(ε-caprolactone) (PCA). Degradable polyesters have achieved success in drug delivery. The biodegradable synthetic polymers have advantages over natural origin ones because this type of material is easily manipulated by reducing variations between produced batches (Middleton and Tipton, 2000). Synthetic biodegradable polymers can be modified,

Table 3.1 The Most Commonly Used Synthetic Polymers in the Preparation of Polymeric Nanocarriers

Synthetic Polymer	Active Molecule	Nanocarrier System	References
Polylactides (PLA)	Rhodamine	Nanoparticles	Crawford (2015)
Polyglycolides (PGA)	Lidocaine	Nanoparticles	Wang (2012)
Poly(lactide-*co*-glycolides) (PLGA)	Celecoxib	Nanoparticles	Kim et al. (2011)
Polyanhydrides	Curcumin	Micelles	Wang et al. (2014)
Polyorthoesters	Celocoxib	Nanoparticles	Palamoor and Jablonski (2013)
Polycyanoacrylates	Horseradish peroxidase	Nanoparticles	Lin et al. (2012)
Polycaprolactone	Cyclosporine	Micelles	Wilson et al. (2014)
Polyglutamic acid	Antigens	Nanoparticles	Wang et al. (2008)
Polymalic acid	Doxorubicin	Nanoparticles	Huang et al. (2012)
Poly(*N*-vinyl pyrrolidone)	Paclitaxel	Nanoparticles	Zheng et al. (2015)
Poly(vinyl alcohol)	Amphotericin B	Nanoparticles	Yamamoto et al. (2013)
Poly(acrylic acid)	Papain	Nanoparticles	Müller et al. (2013)
Poly(ethylene glycol)	Quercetin	Nanoparticles	El-Gogary et al. (2014)
Poly acrylamide	Hyaluronic acid	Nanoparticles	Sun et al. (2012)
Poly(methyl methacrylate)	Ibuprofen	Nanoparticles	Saade et al. (2013)

functionalized easily, and they allow production of tailor-made nanocarriers (Dubruel, 2014), plus their process and time of degradation can be changed. Biodegradable polymers are used extensively in the pharmaceutical field because of their safety and biocompatibility. The versatility provided by synthetic polymers allows obtaining microparticles with specific release profiles that can be administered through different routes, such as parenteral, oral, or ophthalmic (Brannon-Peppas, 1995).

3 NANOCARRIERS TECHNOLOGY AND RECENT MODIFICATIONS TO IMPROVE THEIR SPECIFICITY

Recently, several types of nanosystems with the potential for drug release in the brain have emerged (Modi et al., 2010). These systems are developed at the nanoscale (less than 1 mm), and they can be constituted of different biodegradable materials, such as natural (alginate) or synthetic polymers (PLGA), lipids (phospholipids), and organometallic compounds (Rawat et al., 2006).

In general, NPs are considered excellent vehicles for the transport and release of drugs due to the many advantages offered (Rawat et al., 2006). Among these are:

1. High chemical and biological stability and the absence of toxicity and immunogenicity;
2. Greater therapeutic efficacy, since they promote an increase in drug solubility, protection to the enzymatic degradation and alteration of its pharmacokinetic and administration profile: favoring a major distribution, specificity, constant and controlled drug release, thereby decreasing the dose administered;
3. The ability to incorporate (higher concentration) hydrophilic and hydrophobic substances, such as drugs, siRNA, DNA, enzymes, and proteins, minimizing the side effects associated with free drug, and therefore their cytotoxicity;
4. Administration capacity by different routes, either oral, inhaled, or parenterally, as well as increasing specificity to achieve certain biological targets; they can be used in tissue engineering;
5. Reduced size, which allows them to reach deep areas of the organism and crossing different anatomical structures, such as the BBB;
6. Ease of manipulation (Masserini, 2013; Ochekpe et al., 2009; Rawat et al., 2006).

All these advantages are significant; however, there are also some limitations that must be taken into consideration in the development of these systems. The possible toxicity of products resulting from their degradation, biological incompatibility of the used materials, and higher costs (depending on the material and the production process) compared to conventional pharmaceutical formulations and low circulation time in the bloodstream due to the rapid uptake by the reticuloendothelial system (RES), providing constant challenges to achieve strategies that can overcome the evident limitations and improve their characteristics (Cho et al., 2008; Prausnitz et al., 2004).

To control the properties of NPs in vivo, some modifications in their structure have been carried out (Torchilin, 2009) with the goal of adjusting their characteristics according to the pathological site where they will act and, according to their purpose, that is, diagnostic or therapeutic, with these objectives:

1. Increase the stability and circulation time of the NPs in the blood flow (Torchilin, 2009). PEGylation is one of the strategies most used in this sense. For this purpose the structure of the NPs is modified by the addition of the synthetic polymer polyethylene glycol (PEG), resulting in the formation of a polymeric layer on the surface of the NPs, which are impervious to other

solutes, even at relatively low polymer concentrations (Torchilin, 2009). On the other hand, PEG prevents the interaction and binding of blood components with the surface of the NPs through steric repulsion, preventing opsonization of the NPs and their capture by the RES, which is not more than one mononuclear phagocyte system (MPS) comprised of a diverse group of cells located in different tissues. This strategy allows the NP circulation in the bloodstream without being recognized, thereby preventing their rapid clearance from the organism (Husseini and Pitt, 2008; Morshed et al., 2013).

2. Direct NPs to the target pathological tissue by incorporating ligands to the surface thereof, which exhibits selectivity for certain receptor or target cell existing in the surface of the molecules (Torchilin, 2009). These ligands allow controlling the NPs biodistribution and their specific catchment by cells in the active site. Some ligands normally used are sugars, peptides, proteins, aptamers, monoclonal antibodies (Mourya et al., 2011). However, some problems can be considered when these systems are designed: the ligand to the NP surface can increase the uptake by the RES, despite being sterically protected, promoting the development of an unwanted immune response, depending on the ligand type and the NP composition; the amount of ligand can become a critical factor for the success of binding to the target, and for maintaining an NP prolonged circulation in the blood stream (Torchilin, 2009).
3. Give the NPs sensitivity to certain intrinsic stimuli characteristics of the pathological area, such as temperature and pH, or external stimuli applied on the same area, such as ultrasound (Mourya et al., 2011). Depending on the stimulus, different responses can be observed, such as breakage of the structure of the NP and changes in shape, volume, permeation rate, hydration status, and changes in conformational states (Mourya et al., 2011). This strategy allows that the drug release from the NP becomes dependent on the physiological characteristics of the pathologic tissue, which are not observed in normal tissues (Mourya et al., 2011). Accordingly, the so-called intelligent polymers, or stimuli-responsive polymers have been widely used in the preparation of NPs, once they have an active response to small signals and changes in the surrounding environment, suffering drastic changes in microstructure and leading to desirable modification in their physicochemical properties, which are reversible transitions (Zhang et al., 2013).

This mechanism has been presenting particular relevance in the antineoplastic therapy. In this situation, the NPs have to pass from the bloodstream into the tumor interstitium, and in many cases, they have to enter the endosomes. However, each compartment has its own pH: the pH of the endosomal system (between 5.0 and 6.0) is substantially lower than the physiological pH (pH 7.4) and the extracellular pH of solid tumors (pH 6.8) (Provenzale and Silva, 2009). Apparently, these pH changes cause instability in NPs and even affect their transport through the blood stream until they reach the therapeutic target. However, previous studies have demonstrated that a coating of NPs with PEG is reversibly removed at pH 6.8 and reconstituted at pH 7.4. It is noteworthy that a NP of this form can be modified to provide information about the pH of a particular biological compartment. This fact can be important in several aspects, such as understanding drug resistance to chemotherapeutic agents, which are sensitive to pH (Provenzale and Silva, 2009). Another example is the occurrence of cerebral ischemia after lactic acidosis, wherein the pH decreases dramatically, this type of NP is a good strategy to reach the affected area and is able to act (Ramos-Cabrer and Campos, 2013).

4. Add a contrast agent to the NPs, providing visual information in real time on their distribution and accumulation in the target tissue (Torchilin, 2009). This strategy, known as theranostic, is promising because it makes use of NPs to build a processing platform that combines a diagnosis in situ, by imaging with a specific therapy by drug encapsulation and monitoring the response to this therapy (Torchilin, 2009). The most commonly contrasted agents used are particles of iron oxide and gadolinium chelates, while nanosystems with greater success in theranostic are micelles, liposomes, and more recently dendrimers (Ramos-Cabrer and Campos, 2013; Torchilin, 2009).

Most neurological disorders are specific and therefore only affect a particular organ, tissue, or even a single specific cell line, such as the dopaminergic neurons of the substantia nigra in Parkinson's disease (Ramos-Cabrer and Campos, 2013). Thus, the theranostic is based on the following: systemic administration of NPs that are distributed in the body; their ability to reach the site of pathology, through strategic modifications in their structure and ease to concentrate the drug at the desired site; use of imaging technologies, such as magnetic resonance imaging or tomography to view the route of NPs in the body, as well as to locate, delineate, and quantify the injured area and to monitor the progression of the disease in response to the treatment (Ramos-Cabrer and Campos, 2013).

3.1 POLYMERIC NANOPARTICLES

Polymeric nanoparticles (PNs) are defined as particulate dispersions or solid particles with a size between 10 and 1000 nm (typically 200 nm) (Nagavarma et al., 2012). Developed in the 1990s, PNs have been important not only for the ability to transport drugs in general, but because they have been considered as promising carriers for the CNS (Neha et al., 2013). PNs are colloidal systems composed of biocompatible and biodegradable natural, synthetic, or semisynthetic polymers (Rawat et al., 2006; Torchilin, 2009). Nowadays, the use of polymers in NPs preparation is outstanding with well-established methods and extensive information about the toxicity of these materials (Leyva-Gómez et al., 2015).

PNs is a term that encompasses two types of different structures, nanocapsules and nanospheres, which differ in their composition and structural organization (Schaffazick et al., 2003). Nanospheres (Fig. 3.2) are systems formed from a dense polymer matrix on which the drug is dispersed evenly and after released by

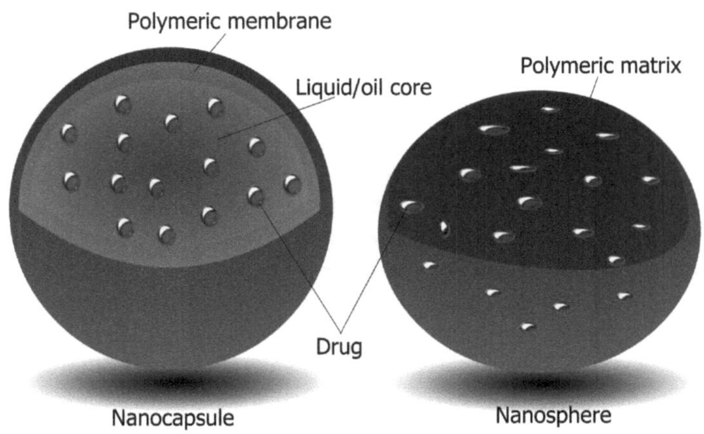

FIGURE 3.2 Illustration of Two Types of Polymeric NPs

diffusion. Nanocapsules are thus systems with a reservoir liquid core, generally oily, where the drug is dissolved and this core is surrounded by a thin polymeric membrane (Modi et al., 2010; Steichen et al., 2013). The advantages of nanocapsules relative to nanospheres are due to the fact that the nanocapsules have the ability to incorporate drugs in solid or liquid forms as a molecular dispersion; the low content of polymers, reducing potential toxicities; and the increased ability to incorporate hydrophobic drugs, due to their lipophilic core (Blouza et al., 2006). Some drawbacks associated with PNs include the presence of organic solvents residues after their preparation, onset of polymerization, and production of toxic monomers.

The release of the drug from the PNs is influenced by the properties of the polymer used, like the biodegradation profile, the distribution of molecular mass, and the hydrophobicity, as well as by the properties of the drug (Des Rieux et al., 2006). Thus, in order to achieve a controlled drug release, the use of natural polymers, such as chitosan, cellulose, sodium alginate, and collagen has been substituted for biodegradable synthetic polymers, such as PLA, PGA, poly(caprolactone) (PCL), and PLGA (Neha et al., 2013). In this point of view polymer selection becomes a decisive factor because natural polymers provide a relatively rapid drug release from the PNs, and because they decompose in a few hours, while synthetic polymers favor a prolonged drug release because they can resist degradation in the body for days or even weeks (Des Rieux et al., 2006; Schaffazick et al., 2003). Thus, it is possible to control the properties of the PNs through their composition, such as the type of polymer used, the respective molecular weight and the drug-polymer ratio, as well as the terminal functional groups added to the structure to confer particular specificity (Dikpati et al., 2012). Moreover, various factors can influence the stability of NPs, like the chemical stability of the polymers under different storage conditions. Furthermore, also the zeta potential, drug content, particle size, distribution of molecular weight and pH are physicochemical parameters that must be taken into account during stability monitoring (Schaffazick et al., 2003).

The preparation of PNs can be performed by different methods. In general, it can be classified into two different categories. They can be directly prepared by in situ polymerization reactions of monomers dispersed or by precipitation of preformed polymers, synthetic or natural (Quintanar-Guerrero et al., 1998). The preparation always requires the use of a polymer, one or more solvents, a nonsolvent, one or more surfactants and oil in the case of nanocapsules (Quintanar-Guerrero et al., 1998). In addition to the general advantages of PNs, they can stand out for more ease in modification of their structure and the fact that PNs have higher stability in gastrointestinal fluids when orally administered, unlike micelles and liposomes. This fact is due to the establishment of strong molecular interactions, such as covalent bonds, which make their structure more rigid and stable (Ramos-Cabrer and Campos, 2013). Furthermore, they have the ability to incorporate hydrophilic or hydrophobic drugs, instead of polymeric micelles (PMs). Moreover, s have a large capacity for loading drugs; they can be prepared in large quantities and by a variety of methods (Neha et al., 2013). The major disadvantage of this kind of systems is their uptake by the RES, leading to a therapeutic failure due to insufficient concentration of the drug present in the blood that passes through the BBB. However, the use of PEG, hydrophilic surfactants, such as polaxamers, polysorbates, and ligands on their surface can help to overcome this limitation (Neha et al., 2013).

3.2 POLYMERIC MICELLES

In recent years, PMs (Fig. 3.3) have emerged as promising DDS (Alvarez-Lorenzo et al., 2010; Ribeiro et al., 2012; Sosnik and Menaker Raskin, 2015), with main emphasis on oncology because of their peculiar structure, with important applications in the transport of proteins, genetic material and hydrophobic drugs (Alvarez-Lorenzo et al., 2011; Francis et al., 2004).

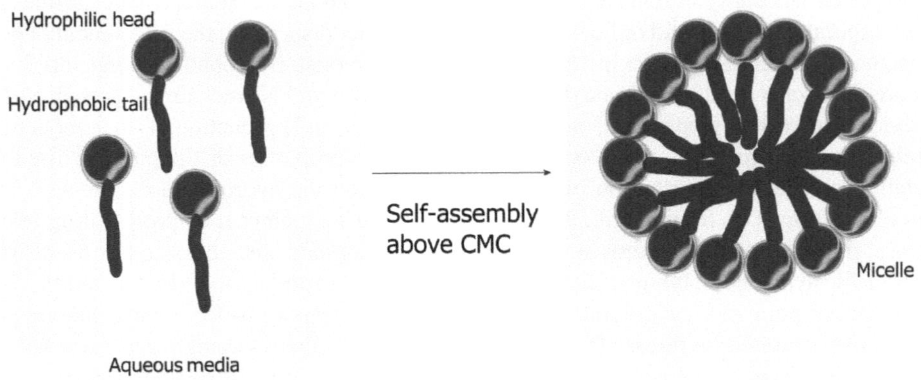

FIGURE 3.3 Micelles Formation Process

These systems are obtained from amphiphilic block copolymers, which can aggregate spontaneously in aqueous environment above a certain concentration, called the critical micellar concentration (CMC), thereby forming a spherical structure with a hydrophobic inner core surrounded by a hydrophilic outer layer, and with a diameter 10–100 nm (Husseini and Pitt, 2008; Mourya et al., 2011). The micellization process is influenced by several factors, namely, the size of the hydrophobic portion, the concentration of amphiphilic compounds, the temperature and the surrounding solution. Moreover, the formed micelle size depends on the molecular weight of the amphiphilic copolymers, the number of aggregates formed between them, the ratio between the number of hydrophilic and hydrophobic segments, as well as the preparation process (Mourya et al., 2011). The hydrophobic interactions are the main driving forces behind the formation of micelles. However, other intermolecular forces, such as hydrogen bonds and electrostatic interactions are also present, to promote greater stability of the structure (Morshed et al., 2013).

The hydrophobic core of PMs has a high ability to encapsulate hydrophobic drugs (Ribeiro et al., 2012) via covalent bonds to copolymers or physical incorporation within the core, increasing the aqueous solubility thereof (Husseini and Pitt, 2008; Singh et al., 2012). On the other hand, the hydrophilic outer stabilizer serves as the interface between the hydrophobic core and the external aqueous medium due to repulsive forces (Francis et al., 2004; Mourya et al., 2011). Accordingly, the PMs are systems with a high potential to carry drugs with poor aqueous solubility, improving their oral bioavailability and pharmacokinetics in the organism (Mourya et al., 2011). Unlike the conventional surfactant micelles, the PMs have low CMC values and, also, the establishment of covalent bonds between the monomers that constitute the hydrophobic core, avoid dynamic exchange of monomers between the free and micellar pseudosolution-phase, conferring rigidity and stability to the micelle in aqueous media (Morshed et al., 2013; Mourya et al., 2011). Moreover, PMs do not exhibit cytotoxicity (Ribeiro et al., 2012) and allow conjugation with a multiplicity of functional groups providing therapeutic advantages (Francis et al., 2004).

Various copolymers have been used in the formation of micelles (Alvarez-Lorenzo et al., 2010; Gaucher et al., 2010). However, the use of biodegradable and biocompatible polymers limits their application (Sutton et al., 2007). For the hydrophobic segment, the most common is the use of derivatives of polyesters, polyethers, and polypeptides. The polyesters, such as PLA, PCL, PLGA, and PGA have the advantage of being approved by the Food and Drug Administration (FDA) for biomedical applications (Morshed et al., 2013). Moreover, PEG is used as the hydrophilic component in most of produced PMs, because it does not show immunogenicity or toxicity. In addition to the advantage already mentioned earlier, PEG increases the time circulation in the bloodstream (Husseini and Pitt, 2008; Morshed et al., 2013).

The aforementioned polymers can be in the form of diblock, triblock, or more complex structures. Among the amphiphilic materials most widely used are the hydrophilic copolymers derivatives from the polyethylene oxide (PEO) and hydrophobic derivatives of polypropylene oxide (PPO), with particular focus on poloxamers, which are characterized by a linear and bifunctional trilayer PEO–PPO–PEO structure. Previous studies have demonstrated that the use of poloxamers can be advantageous in oncological situations, because at low concentrations they have the ability to counteract the multiple drug resistance developed by cancer cells that were repeatedly exposed to chemotherapeutic agents (Husseini and Pitt, 2008).

In addition to the several advantages already mentioned in respect of NPs in general, and thus also suited to the PMs, the latter also have some peculiarities. The preparation of PMs is readily accomplished in large quantities and reproducibly, in addition to the flexibility in the modification of morphology, allowing easy adjustment of structure of the amphiphilic copolymers when necessary (Francis et al., 2004; Mourya et al., 2011).

The main PMs limitation is the instability associated to their reversible structure composed by noncovalent bonds, apart from the fact that this kind of system only demonstrates advantage in the transport of hydrophobic drugs. On the other hand, the PMs as a large part of the nanoparticulate systems are subject to rapid clearance (Husseini and Pitt, 2008). However, their low toxicity and rapid rate of clearance make them suitable for intravenous administration (Mourya et al., 2011).

3.3 DENDRIMERS

Although dendrimers were synthesized for the first time in 1978, and termed as "cascade molecules," the development of this class of systems only occurred in 1985 with the synthesis of dendrimers of poly(amidoamine) (PAMAM)—having today numerous applications in the field of diagnosis and therapy (Boas et al., 2006).

Dendrimers (Fig. 3.4) are new molecular architectures with 3–10 nm highly branched formed from a central atom, such as nitrogen (Balakrishnan et al., 2013; Puvvada, 2011). After, a series of chemical reactions occur with the addition of carbon and other elements, resulting in repetition and branching chains that give rise to radial concentric layers, each branch has a binary disjunction (Boas et al., 2006; Masserini, 2013; Puvvada, 2011).

Active molecules encapsulation in dendrimers requires that both the dendrimer structure and the hydrophobic compound be soluble in a volatile organic solvent, which is miscible with water.

The surface of these three-dimensional macromolecules can be modified by the introduction of functional groups at their terminal monomers, changing the physicochemical properties, such as stiffness, chemical reactivity, and solubility (Newkome et al., 2006). Therefore, the diversity of dendritic

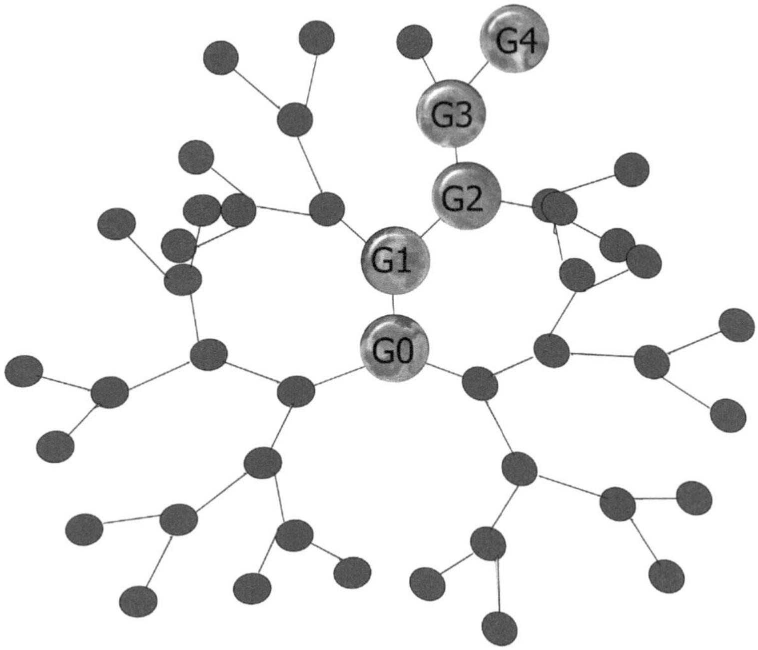

FIGURE 3.4 Generation 4 Dendrimer Structure

structures depends on the choice of the surface group and the type of ligands that are used. For each layer of branching units, the number of modifiable groups on the surface doubles or even triples while the dendrimer diameter only increases 1 or 2 nm (Sarin et al., 2008).

The synthesis of dendrimers is performed step-by-step, through branched polymer units and can be performed by two different methods: a divergent or a convergent method. In the first case, the dendrimer is constructed from a center by sequential addition of monomers in the core. In the second case, first the branches are synthesized, and only subsequently they are all linked to a central core. However, the divergent synthesis is preferred, since it generates pure products and it presents larger yield (Esfand and Tomalia, 2001; Langereis et al., 2007).

Therefore, the most common dendrimers are composed of PAMAM, polyesters, and PEGs, taking into account that the main cleavable groups are amides, esters, and carbamates (Gingras et al., 2007). Choosing different polymers with different terminal functional groups, it is possible to accurately control the size of the dendrimer, the shape, polarity, solubility, and flexibility (Langereis et al., 2007).

Dendrimers, due to the many advantages offered, are excellent candidates for the transport of drugs (Kannan et al., 2008; Xu et al., 2013), and the drug can be incorporated by different ways: it can be incorporated into the dendritic structure through electrostatic interactions, hydrophobic and hydrogen bonds; the monomers can contain the drug, providing an exponential increase in drug content with the addition of each subsequent generation; interactions with branched structures by means of spacers; and the drug attached directly to the surface of the dendrimer through covalent and electrostatic interactions (D'Emanuele and Attwood, 2005; Liu and Fréchet, 1999; Masserini, 2013). Often, the dendrimers can be conjugated with PEG in the surface in order to create a hydrophilic layer around the hydrophobic dendrimer, forming the so called "micelle unimolecular" capable of carrying hydrophobic drugs inside and hydrophilic drugs in the surface layer of the PEG (Crampton and Simanek, 2007).

The disintegration process of a cleavable dendrimer or depolymerization can occur in three forms: drug or smaller chemical species not covalently bound to the dendrimer can be released; consecutive cleavage of covalent bonds with initial removal of functional groups on the periphery, then breaking the branches (or part thereof) to the core; and a combined mode in which the hydrolysis of covalent bonds of the periphery is critical to the release of noncovalently bounded molecules (D'Emanuele and Attwood, 2005; Gingras et al., 2007).

Dendrimers are definitely systems with great potential to carry active molecules and contrast agents, the numerous advantages offered are: highly branched and regular architecture; low polydispersity, which facilitate their passage through biological barriers, such as the BBB; large number of peripheral functions; ability to make multiple copies of surface groups to a biological recognition process; ease of preparation; small size when compared with micelles, liposomes, and PNs; and ability to incorporate hydrophilic and hydrophobic drugs (Akula, 2013; Singh, 2013). However, they have the same limitation as other polymeric particles, namely the possibility of some toxicity associated to the used polymer (Boas et al., 2006) and their positively charged surface (Ziemba et al., 2012). In this way, for biomedical functions, the modification of the surface of dendrimers should be performed in order to improve their pharmacokinetic and safety properties (low immunogenicity and hemotoxicity) (Onoue et al., 2014).

3.4 LIPOSOMES

Liposomes were developed in the 1960s and so they are the first generation of nanosystems in DDS (Masserini, 2013). However, their structure has not stagnated, and various studies have been responsible for the developments of this type of system, giving rise to the most recent cationic liposomes, immunoliposomes, virosomes, among others (Spuch and Navarro, 2011).

Liposomes are spherical vesicles comprised by one or more lipid bilayers that enclose an internal aqueous compartment, varying in size from nanometers to micrometers (diameter) (Masserini, 2013). These nanosystems can be classified into three types: small unilamellar vesicles (SUVs) (Fig. 3.5), when they have a single lipid bilayer and a size of 25–100 nm; large unilamellar vesicles (LUVs), when they have a single lipid bilayer and a size of 200–800 nm; and multilamellar vesicles (MLVs), when they have multiple concentric lipid bilayers and a maximum of 14 bilayers with a size 500–5000 nm (Spuch and Navarro, 2011).

Due to their amphiphilic composition, liposomes can encapsulate hydrophilic or hydrophobic drugs: the first are incorporated into the internal aqueous core while the second are inserted into or adsorbed on lipid bilayers (Sharma and Sharma, 1997). However, the size and number of lipid bilayers and the proportion of lipid-drug concentrations are important factors influencing the efficiency of encapsulation of the drugs into the liposomes (Sharma and Sharma, 1997; Spuch and Navarro, 2011). In the case of the hydrophilic drugs, their percentage of encapsulation depends on the composition of the lipid bilayer and the liposome preparation method (Spuch and Navarro, 2011). Moreover, the vesicle size is a critical factor in determining their half-life and transport of molecules, whereas MLVs are the first to be eliminated from the blood circulation because of the phagocytosis (Chapman and Pascu, 2012; Spuch and Navarro, 2011).

Liposomes can be made up of different types of lipids. However, phospholipids are the most commonly used, such as phosphatidylcholine, extracted from egg yolk or soybeans, phosphatidylserine,

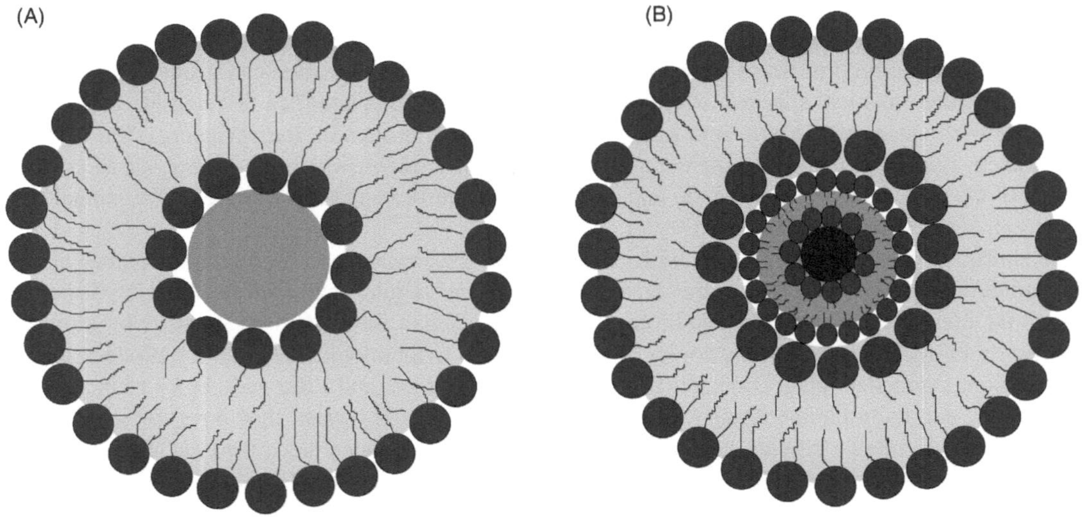

FIGURE 3.5

Illustration of (A) unilamellar liposome and (B) multilamellar liposome

phosphatidylglycerol, and sphingomyelin (Batista et al., 2007; Masserini, 2013; Sharma and Sharma, 1997).

The formation of liposomes occurs spontaneously when phospholipids are in contact with an aqueous medium, due to the hydrophilic interactions of the polar head of the phospholipid with water, leading to formation of unilamellar and multilamellar structures (Sharma and Sharma, 1997; Vemuri and Rhodes, 1995).

The cholesterol, an important constituent of cell membrane, is often included in the liposome formulations in order to increase their stability (Masserini, 2013). Thus, this sterol favors a better fluidity of the lipid bilayer, reducing its permeability and also stabilizes this bilayer in the presence of biological fluids (Vemuri and Rhodes, 1995).

Featuring the general advantages of NPs, liposomes can be distinguished by their structural similarity to the cell membrane (lipid bilayer with cholesterol), allowing it to establish a deep interaction with the cells, or even with individual cellular compartments. On the other hand, their formulation requires a small number of excipients. The preparation is easy and, in addition, the liposomes present the ability to incorporate hydrophobic and hydrophilic drugs (Neha et al., 2013; Vemuri and Rhodes, 1995).

Although liposomes are nanosystems with beneficial characteristics for the transport of drugs, they also have some disadvantages related to the rapid uptake by the RES. These systems show a low encapsulation efficiency of drugs and a high instability in the body, because they are very sensitive to the pH and temperature. These limitations become particularly significant in the oral administration, taking into account the low stomach pH and the presence of bile salts that can destabilize the liposomal structure (Chapman and Pascu, 2012; Spuch and Navarro, 2011).

As mentioned earlier, liposomes have undergone structural changes in recent decades in order to respond to therapeutic needs. First, liposomes were developed with PEG. However, several studies revealed that these structures were captured in an undifferentiated form or by pathological cells, as well as by healthy cells, emerging the need to develop liposomes with specificity for certain therapeutic targets (Vemuri and Rhodes, 1995). Currently, some types of liposomes are considered promising strategies in the transport of drugs: cationic liposomes in which a strong affinity between the positive charge of the lipid nanosystems is created with the negative charge of the genetic material to be transported; immunoliposomes, which are based on a combination of monoclonal antibodies deposited on the surface of the system; virosomes, in which a linkage is created between the viral cover and the liposome (Masserini, 2013; Spuch and Navarro, 2011).

3.5 FUNCTIONAL SURFACE BUILDING

NPs have an advantage over microparticles due to their nanosizes. NPs, however, exhibit a very short circulating life duration within the body after intravenous administration, because of their rapid clearance from the body by phagocytic cells. In recent times, this problem has been solved by the functionalization of NPs surface (Mahapatro and Singh, 2011). The NPs can be modified on their surface by adding functional groups. These modifications on their surfaces can make them biosensors, fluorescent markers, and drug carriers with defined targets for several medical applications (Rotomskis et al., 2006). Even having the propensity of being collected by macrophages and considering that they can accumulate in the spleen and liver, strategies, such as coating PEG can increase blood circulation time and to increase their accumulation in tumors (Chatterjee et al., 2008). PEG offers advantages, such as high solubility in water and organic solvents, biocompatibility, lack of toxicity and immunogenicity (Wattendorf and Merkle, 2008). Depending on the system, the material and the solvent in which the particles are dispersed, the choice of the linker can produce stable suspensions. First, the molecules must be bound to the particle surface by any attractive interaction, chemisorption, electrostatic attraction or hydrophobic interactions, most commonly provided by a functional group of the ligand molecule. Various chemical functional groups have an affinity with inorganic surfaces; for example, the thiol group has an affinity for gold. Due to interactions between the ligand molecule and the solvent, polar linker, or ionic provides solubility in polar solvents. NPs with apolar ligands, such as hydrocarbon chains are stable only in nonpolar organic solvents (Sperling and Parak, 2010). Particularly marked polymeric particles are capable of being recognized by receptors on target cells, enhancing the efficiency of DDS. Therefore, to increase the efficiency of delivery systems, target ligands can be linked to the nanoparticle drug loaders (Cho et al., 2008). Such ligands include antibodies, carbohydrates (galactose, and mannose), aptamers (DNA or RNA oligonucleotides, or peptide molecules that can bind to a specific target antigen) and other small molecules (Fig. 3.6). Monoclonal antibodies have been used as binders in the DDS with enough frequency (Kim et al., 2009), as well as the vectors based on folate (Moghimi et al., 2001). The use of aptamers offers many advantages, such as nonimmunogenicity, stability over wide ranges of pH and temperature, and tumor specificity (Kim et al., 2009).

The polymers, besides conferring mechanical and chemical stability for nanomaterials, can be used for surface modification of NPs during synthesis or subsequent steps (Vladkova, 2010). The silicon oxide film with organic functions (amine, carboxyl, and hydroxyl) helps bind molecules with biological activity on the surface of the NPs (Lydon et al., 1985). Thiolate NPs have been widely used in biomedical applications due to the presence of the thiol (SH) groups in their free surfaces, which allow the combination of molecules with biological interest and promote their differential migration for specific targets (Valois et al., 2010).

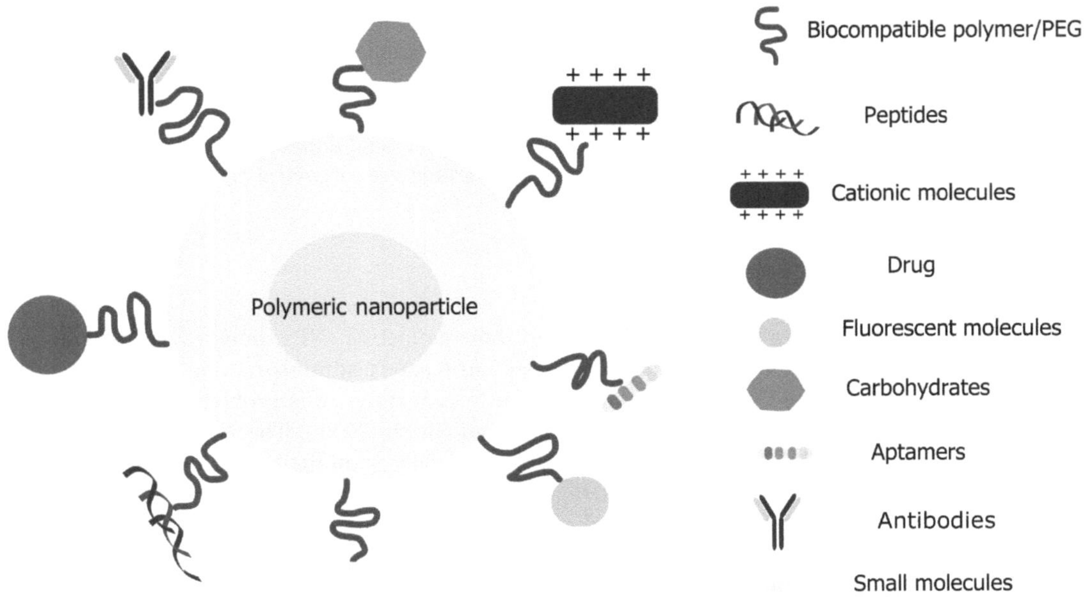

FIGURE 3.6 Illustration of Surface-Modified NPs With Different Targeting Molecules

The binding of NPs with biomolecules can facilitate transduction processes (electronic or optical systems) of biological events in the development of biosensors (Penn et al., 2003). The NPs functionalized with enzymes allow the biomolecules to maintain similar activity of free enzymes in solution, also improves the interaction with the substrate, and avoids aggregation of free proteins (Jia et al., 2003).

The NPs can be comprised of various metals (Au, Ag, Pt, and Cu) or semiconductors (PbS, Ag_2S, CdS, CdSe, and TiO_2) (Fedlheim and Foss, 2001) and they can be stabilized by anionic ligands, such as derivatives of carboxylic acids (citrate and lipoic acid) and by the adsorption of positively charged proteins from electrostatic interactions (Meziani et al., 2002). Thus, the particle surface properties are key factors that determine their permanence in the systemic circulation and increase the bioavailability of the active ingredient in order to obtain a specific pharmacologic answer with an adequate quantity in the target site.

4 ADVANCES IN POLYMERIC NANOCARRIERS AS VEHICLES FOR ORAL-DELIVERY SYSTEMS

The oral route for drugs administration is the preferred by the patients due its versatility and increased patient compliance. Nanocarriers often have been explored for the delivery of a wide number of active biomolecules, drugs, proteins, or enzymes by this route (Galindo-Rodriguez et al., 2005; Woitiski et al., 2009, 2010). The nanocarriers have the ability to protect drugs from the degradation in the gastrointestinal tract (GIT), and this technology can allow target delivery of drugs to various parts of the organism. Nanosystems enable the delivery of hydrophobic drugs and can provide means of bypassing the liver, thereby preventing the first pass metabolism (Simões et al., 2015). This technology increases oral bioavailability of drugs due to their specialized uptake mechanisms, such as absorptive endocytosis and their permanence in the blood circulation for a long time, releasing the drug in a controlled way, taking to less plasma fluctuations and minimizing side effects (Muhamad12 et al., 2014). The inactivation and digestion by proteolytic enzymes in the luminal cavity and poor permeability across intestinal epithelium of molecules with high molecular weight are other problems that have overtaken in the oral route (Carino and Mathiowitz, 1999). Due to the increased drug loading capacity of nanostructured lipid carriers (NLCs), most of these studies have been focused on the ability of these systems to improve the oral bioavailability of poorly water-soluble drugs (classified according Biopharmaceutics Classification System [BSC] as class II or IV) (Beloqui et al., 2015).

Tsai et al. (2011) used solid lipid NPs of glyceryl monostearate in order to incorporate apomorphine, a dopamine receptor agonist. The aim of the study was to increase oral drug bioavailability and take data about brain regional distribution in rats with 6-hydroxydopamine-induced lesions. The NPs were able to increase (12-fold) the drug bioavailability when compared with the drug in solution, and the total rotation number increased from 20 to 94. In another study, Ngwuluka et al. (2011) used methacrylate copolymer–lipid NPs for oral drug delivery of L-DOPA. The formulation was made using multicrosslinking technology and the properties of the system surface were characterized. The study shown sustained release of L-DOPA when incorporated in the NPs and concluded that their systems could catch applications for oral drug delivery.

Huanbutta et al. (2013) demonstrated that the spinning disk processing for preparation of chitosan NPs loaded with diclofenac sodium was able to obtain 88% of drug entrapment efficiency and high systemic outreach, giving proof of concept in noninvasive, oral delivery for brain targeting. Dai et al. (2004) prepared pH-sensitive NPs with two polymers, methylacrylic acid, and methacrylate and they showed that these NPs increased the oral bioavailability of drugs, such as cyclosporine-A by releasing their load at a specific pH within the GIT.

Chitosan, a cationic polysaccharide, has been widely employed with biopharmaceutical and biomedical applications. Lalatsa et al. (2012) prepared quaternary ammonium palmitoyl glycol chitosan NPs for the delivery of a therapeutic peptide (leucine5-enkephalin) to the brain. They identified that 3% of the drug was able to reach the systemic circulation and brain 30 min after oral administration. Also, the new NPs were able to protect the peptide from enzymatic degradation in the GIT. When they compared with other studies using chitosan with different molecular weights (Chae et al., 2005) they concluded that it is critical to select chitosan with suitable physical properties in order to prepare NPs containing anti-Alzheimer drugs to be delivered by the oral route (Sarvaiya and Agrawal, 2015b).

Debajan and coworkers (Das and Lin, 2005) evaluated the oral administration of poly(butylcyanoacrylate) nanoparticulate delivery systems (PBCA-NDSs), double-coated with Tween 80 and polyPEG 20000 for brain delivery of hexapeptide dalargin, an antinociceptive peptide that does not cross BBB by itself. Dalargin-loaded PBCA-NDS were successively double-coated with Tween 80 and PEG 20000 in varied concentrations of up to 2% each. The measurement of in vivo central antinociceptive effect of dalargin along with a dose response curve was obtained by the tail flick test following the oral administration of PBCA-NDSs to mice. Results from the tail flick test indicated that significant dalargin-induced analgesia was observed when administered orally. In other studies researchers investigated the potential of solid lipid NPs in improving the oral bioavailability of paclitaxel. Paclitaxel-loaded SLNs (PTX-SLNs) were prepared by modified solvent injection method using stearylamine as lipid, soya lecithin, and poloxamer 188 as emulsifiers. The studies of pharmacokinetic and bioavailability were conducted in male Swiss albino mice after oral administration of PTX-SLNs. Paclitaxel showed a slow and sustained in vitro release profile. After oral administration of PTX-SLNs, drug exposure in plasma and tissues was 10- and twofold higher, respectively, when compared with free paclitaxel solution. PTX-SLNs produced a high mean concentration (C_{max}) (10.274 ng/mL) compared with the free paclitaxel solution (3.087 ng/mL). The absorbed drug was found to be distributed in liver, lungs, kidneys, spleen, and brain. The results suggested that PTX-SLNs dispersed in an aqueous environment are promising novel formulations that can increase the oral bioavailability of hydrophobic drugs, like paclitaxel and they were quite safe for oral delivery of this drug as observed by in vivo toxicity studies (Pandita et al., 2011). Also, Yang et al. (1999) investigated the specific changes in the body distribution of camptothecin through incorporation into solid lipid NPs by peroral route with promising results in sustained release and targeting system for camptothecin delivery in the brain.

5 NEW APPROACHES AND CLINICAL APPLICATIONS IN NEURODEGENERATIVE DISORDERS

Several strategies have been used for drug administration in order to treat dementia of the brain. In this way, different approaches in nanotechnology, inspired by recent advances in biomaterials, have been applied to neurodegenerative diseases treatment. Consequently, advances in new nanocarriers for neurologic drugs have focused on the development of biodegradable NPs that can cross the BBB, being noninvasive treatments with prolonged release, sufficient loading capacity, and low toxicity (Leyva-Gómez et al., 2015). Given the complexity of the brain, depth, and comprehensive toxicological studies of brain targeting nanomedicines are essential (Wong et al., 2012). Several studies have shown that nanocarriers, such as polymeric or lipid nanoparticles, liposomes, dendrimers, and micelles can increase the local drug concentration gradients, facilitating drug transport into the brain via endocytotic pathways and inhibiting the ATP-binding cassette (ABC) transporters expressed at the barrier sites (Wong et al., 2010), constituting an attractive strategy.

Several studies show that chitosan and its derivatives as nanocarrier materials have shown good end-result expectations in drug delivery to the brain with clinical application in Alzheimer's disease (Sarvaiya and Agrawal, 2015a).

In another interesting study, Mittal et al. (2011) developed Tween 80 (T-80) coated PLGA NPs to deliver estradiol to the brain upon oral administration. The NPs were evaluated in an ovariectomized rat model of Alzheimer's disease that mimics the postmenopausal conditions. After the oral administration of T-80-coated NPs significantly higher brain estradiol levels after 24 h were observed when compared with uncoated NPs. The authors showed that the brain estradiol levels were almost similar to those obtained after administration of the same dose of drug suspension via intramuscular route, indicating an increased fraction of bioavailable drug reaching the brain when administered orally. Also, the group treated with NPs was successful in preventing the expression of amyloid beta-42 (Aβ42) immunoreactivity in the hippocampus region of brain.

Researchers have developed DDS using nanostructured materials, which cross the BBB and show low toxicity. In one study, authors synthesized a nanostructured conjugated chelator prototype nano-N2PY, which demonstrated ability to protect human cortical neurons from oxidative toxicity induced by amyloid-β peptide and, additionally, inhibited the formation of aggregates of this peptide. Overall, the results indicated that this nanostructured chelating transport system could significantly improve the therapeutic efficacy and decreased the toxicity constituting an alternative therapy for AD and other neurodegenerative diseases associated with an excess of transition metals (Liu et al., 2009).

Dendrimers have the surface modified in order to inhibit the efflux effect of the P-glycoprotein, which is one of the major drug efflux pumps in the GIT (Liu et al., 2015). Oral absorption enhancement of probucol (an antioxidant drug) was achieved by PEGylated G5 PAMAM dendrimer modified nano-liposomes allowing a significant improvement in drug pharmacodynamic effects. The uptake mechanism was explored in vitro in Caco-2 cells with the results suggesting that the absorption improvement of G5-PEG modified PB-liposome was related with the P-glycoprotein efflux pump but was independent of caveolae endocytosis pathways (Ma et al., 2015). In other study, Patel et al. (2006) showed a reduction in the deposition of A peptides when sialic acid functionalized dendritic polymers were used. These vectors were found to mimic the cell surface sialic acid clusters and minimized amyloid protein related neurotoxicity. Formulations based on NLCs (nanostructured lipid carriers), are highlighted for oral drug delivery due to their physicochemical properties per improving their bioavailability and their ability to circumvent P-gp drug efflux. Beloqui et al. (2015) studied the potential of NLCs to enhance the oral bioavailability of saquinavir, a BCS 456 class IV drug and P-gp substrate, and they analyzed NLC transport mechanisms across the intestinal barrier. The modification of some critical physicochemical parameters of the NLC formulation, allowed to overcome the P-gp drug efflux and altered the transcytosis mechanism of the NPs (Beloqui et al., 2013).

6 FUTURE PERSPECTIVES

An increase in the incidence of neurological disorders has been observed in the last decades. On the other hand the oral route for drug delivery is regarded as a good route for therapeutic benefits owing to major patient compliance. Despite big advances in nasal, injectable, transdermal, and other routes of administration, the oral drug delivery remains well ahead as the preferred delivery route. The nanotechnology can overcome the challenges associated with this route of administration as stability, poor solubility, and biocompatibility of drugs. Still, the BBB impermeability remains a limitation on the development of diagnostic methods and therapeutic targets for neurological therapies, because this barrier exerts a protective action against external molecules in the brain. The suggestion for future research in NDs is to head therapeutics in order to use noninvasive treatments, systems with the capacity to cross the BBB, prolonged drug release, satisfactory drug loading, and the use of biodegradable materials with low toxicity (Leyva-Gómez et al., 2015).

The need to develop new strategies to improve the transport of drugs to the brain has been a constant challenge and nanotechnology is contributing significantly to the success of these investigations in neuroscience. However, efficient therapeutic options to treatment are limited for many patients with CNS disorders. Micelles, polymeric NPs, liposomes, and dendrimers are the most studied nanosystems in this area and begin to demonstrate in clinical and nonclinical trials that they can be used as DDS in the CNS. In general, these nanovectors are biodegradable, without immunogenicity; they increase the drugs solubility and in addition they show small sizes, which allows them to interact on a molecular level with biological systems.

Despite these, nanosystems present stability in the bloodstream protecting the drug from enzymatic degradation and promoting prolonged drug release, which are fundamental prerequisites in nanomedicine, these are no longer considered sufficient requirements for the development of effective new therapies.

Thus, assigning specificity to the nanovectors and directing them toward certain pathological cells constitutes the latest challenge, given the difficulties in understanding some of the mechanisms of nanochemistry and above all difficulties in the technological and physiological characterization of these systems. Surface modifications of NPs can prolong the residence time of nanovectors in blood circulation.

The number of clinical and nonclinical assays with nanosystems for neurological targets grew in the last decade and some of the results are quite promising, verifying that a significant proportion of the system reaches the brain when compared to the administered dose, showing that these systems are potential carriers for drugs and other molecules to the brain (Tosi et al., 2012).

There are already nanosystems patented as DDS to the brain. These data are positive indicators that reveal the progress of studies in the field of neuroscience and nanotechnology, and show that there are good prospects in the near future with regard to diagnosis and therapy in neurological diseases. Since PD therapy is chronic, drug oral administration is extremely needed. Consequently, different nanocarrier systems should be able to have appropriate oral drug absorption and then overcome the BBB, producing a drug-prolonged release after oral administration. However, it becomes necessary to conduct further studies in this area, namely, the molecular and cellular mechanisms involved in the BBB and the interaction of nanosystems with neurological target tissues. On the other hand, the design of controlled preclinical studies, becomes a challenge and can contribute for the development and application of nanotechnology in clinical neuroscience.

REFERENCES

Akula, K.K., 2013. Revisiting Neuroscience with Nanomedicine, a Renaissance in Remedy. Innovations in Pharmaceutical and Pharmacotherapy (IPP). Innovations in Pharmaceuticals and Pharmacotherapy 1 (4), 1–60.

Alvarez-Lorenzo, C., Rey-Rico, A., Sosnik, A., Taboada, P., Concheiro, A., 2010. Poloxamine-based nanomaterials for drug delivery. Front. Biosci. 2, 424–440.

Alvarez-Lorenzo, C., Sosnik, A., Concheiro, A., 2011. PEO-PPO block copolymers for passive micellar targeting and overcoming multidrug resistance in cancer therapy. Current Drug Targets 12, 1112–1130.

Balakrishnan, B., Nance, E., Johnston, M.V., Kannan, R., Kannan, S., 2013. Nanomedicine in cerebral palsy. Int. J. Nanomed. 8, 4183.

Batista, C.M., Carvalho, C.M.D., Magalhaes, N.S., 2007. Lipossomas e suas aplicações terapêuticas: estado da arte. Braz. J. Pharm. Sci. 43, 167–179.

Beloqui, A., Solinís, M.Á., Gascón, A.R., Del Pozo-Rodríguez, A., Des Rieux, A., Préat, V., 2013. Mechanism of transport of saquinavir-loaded nanostructured lipid carriers across the intestinal barrier. J. Control. Release 166, 115–123.

Beloqui, A., Solinís, M.Á., Rodríguez-Gascón, A., Almeida, A.J., Préat, V., 2015. Nanostructured lipid carriers: promising drug delivery systems for future clinics. Nano.: Nano., Biol. Med. 12 (1), 143–161.

Blouza, I.L., Charcosset, C., Sfar, S., Fessi, H., 2006. Preparation and characterization of spironolactone-loaded nanocapsules for paediatric use. Int. J. Pharm. 325, 124–131.

Boas, U., Christensen, J.B., Heegaard, P.M., 2006. Dendrimers in Medicine and Biotechnology: New Molecular Tools. Royal Society of Chemistry, London.

Brannon-Peppas, L., 1995. Recent advances on the use of biodegradable microparticles and nanoparticles in controlled drug delivery. Int. J. Pharm. 116, 1–9.

Carino, G.P., Mathiowitz, E., 1999. Oral insulin delivery1. Adv. Drug Deliv. Rev. 35, 249–257.

Chae, S.Y., Jang, M.-K., Nah, J.-W., 2005. Influence of molecular weight on oral absorption of water soluble chitosans. J. Control. Release 102, 383–394.

Chapman, M., Pascu, S.I., 2012. Nanomedicines design: approaches towards the imaging and therapy of brain tumours. J. Nano. Nanotechnol. S4:006, 4.
Chatterjee, D.K., Fong, L.S., Zhang, Y., 2008. Nanoparticles in photodynamic therapy: an emerging paradigm. Adv. Drug Deliv. Rev. 60, 1627–1637.
Cho, K., Wang, X., Nie, S., Shin, D.M., 2008. Therapeutic nanoparticles for drug delivery in cancer. Clin. Cancer Res. 14, 1310–1316.
Chung, E.J., Cheng, Y., Morshed, R., Nord, K., Han, Y., Wegscheid, M.L., Auffinger, B., Wainwright, D.A., Lesniak, M.S., Tirrell, M.V., 2014. Fibrin-binding, peptide amphiphile micelles for targeting glioblastoma. Biomaterials 35, 1249–1256.
Crampton, H.L., Simanek, E.E., 2007. Dendrimers as drug delivery vehicles: non-covalent interactions of bioactive compounds with dendrimers. Polym. Int. 56, 489–496.
Crawford, L., 2015. Exploitation of P-glycoprotein at the blood brain barrier for targeted drug delivery. 3730419, 155.
Dai, J., Nagai, T., Wang, X., Zhang, T., Meng, M., Zhang, Q., 2004. pH-sensitive nanoparticles for improving the oral bioavailability of cyclosporine A. Int. J. Pharm. 280, 229–240.
Das, D., Lin, S., 2005. Double-coated poly (butylcynanoacrylate) nanoparticulate delivery systems for brain targeting of dalargin via oral administration. J. Pharm. Sci. 94, 1343–1353.
D'Emanuele, A., Attwood, D., 2005. Dendrimer–drug interactions. Adv. Drug Del. Rev. 57, 2147–2162.
Des Rieux, A., Fievez, V., Garinot, M., Schneider, Y.-J., Préat, V., 2006. Nanoparticles as potential oral delivery systems of proteins and vaccines: a mechanistic approach. J. Control. Release 116, 1–27.
Dikpati, A., Madgulkar, A., Kshirsagar, S.J., Bhalekar, M., Chahal, A.S., 2012. Targeted drug delivery to CNS using nanoparticles. JAPS J. 2, 179–191.
Dubruel, P., 2014. Cationic Polymers in Regenerative Medicine. Royal Society of Chemistry, London.
Duthey, B., 2013. Background paper 6.11: Alzheimer disease and other dementias. 1–74.
El-Gogary, R.I., Rubio, N., Wang, J.T.-W., Al-Jamal, W.T., Bourgognon, M., Kafa, H., Naeem, M., Klippstein, R., Abbate, V., Leroux, F.D.R., 2014. Polyethylene glycol conjugated polymeric nanocapsules for targeted delivery of quercetin to folate-expressing cancer cells in vitro and in vivo. ACS Nano 8, 1384–1401.
Esfand, R., Tomalia, D.A., 2001. Poly (amidoamine)(PAMAM) dendrimers: from biomimicry to drug delivery and biomedical applications. Drug Discov. Today 6, 427–436.
Fedlheim, D.L., Foss, C.A., 2001. Metal Nanoparticles: Synthesis, Characterization, and Applications. CRC Press, Boca Raton, FL.
Franchetti, S.M.M., Marconato, J.C., 2006. Biodegradable polymers-a partial way for decreasing the amount of plastic waste. Química Nova 29, 811–816.
Francis, M.F., Cristea, M., Winnik, F.M., 2004. Polymeric micelles for oral drug delivery: why and how. Pure Appl. Chem. 76, 1321–1335.
Freed, C.R., Greene, P.E., Breeze, R.E., Tsai, W.-Y., Dumouchel, W., Kao, R., Dillon, S., Winfield, H., Culver, S., Trojanowski, J.Q., 2001. Transplantation of embryonic dopamine neurons for severe Parkinson's disease. N. Engl. J. Med. 344, 710–719.
Galindo-Rodriguez, S.A., Allemann, E., Fessi, H., Doelker, E., 2005. Polymeric nanoparticles for oral delivery of drugs and vaccines: a critical evaluation of in vivo studies. Crit. Rev. in Therap. Drug Carr. Sys. 22 (5), 419–464.
Gaucher, G., Satturwar, P., Jones, M.-C., Furtos, A., Leroux, J.-C., 2010. Polymeric micelles for oral drug delivery. Eur. J. Pharm. Biopharm. 76, 147–158.
Gingras, M., Raimundo, J.M., Chabre, Y.M., 2007. Cleavable dendrimers. Angew. Chem. Int. Ed. 46, 1010–1017.
Huanbutta, K., Sriamornsak, P., Luangtana-Anan, M., Limmatvapirat, S., Puttipipatkhachorn, S., Lim, L.-Y., Terada, K., Nunthanid, J., 2013. Application of multiple stepwise spinning disk processing for the synthesis of poly(methyl acrylates) coated chitosan–diclofenac sodium nanoparticles for colonic drug delivery. Eur. J. Pharm. Sci. 50, 303–311.

Huang, Z.W., Laurent, V., Chetouani, G., Ljubimova, J.Y., Holler, E., Benvegnu, T., Loyer, P., Cammas-Marion, S., 2012. New functional degradable and bio-compatible nanoparticles based on poly (malic acid) derivatives for site-specific anti-cancer drug delivery. Int. J. Pharm. 423, 84–92.

Husseini, G.A., Pitt, W.G., 2008. Micelles and nanoparticles for ultrasonic drug and gene delivery. Adv. Drug Deliv. Rev. 60, 1137–1152.

Jia, H., Zhu, G., Wang, P., 2003. Catalytic behaviors of enzymes attached to nanoparticles: the effect of particle mobility. Biotechnol. Bioeng. 84, 406–414.

Kannan, R., Iezzi, R., Rajaguru, B., Kannan, S., 2008. Dendrimers for sustained release of compounds. Google Patents.

Kim, S., Kim, J.-H., Jeon, O., Kwon, I.C., Park, K., 2009. Engineered polymers for advanced drug delivery. Eur. J. Pharm. Biopharm. 71, 420–430.

Kim, T.-H., Jeong, Y.-I., Jin, S.-G., Pei, J., Jung, T.-Y., Moon, K.-S., Kim, I.-Y., Kang, S.-S., Jung, S., 2011. Preparation of polylactide-*co*-glycolide nanoparticles incorporating celecoxib and their antitumor activity against brain tumor cells. Int. J. Nanomed. 6, 2621.

Korsak, M., Kozyreva, T., 2015. Beta amyloid hallmarks: from intrinsically disordered proteins to Alzheimer's disease. Springer, New York.

Kreuter, J., 2007. Nanoparticles: a historical perspective. Int. J. Pharm. 331, 1–10.

Lalatsa, A., Garrett, N., Ferrarelli, T., Moger, J., Schatzlein, A., Uchegbu, I., 2012. Delivery of peptides to the blood and brain after oral uptake of quaternary ammonium palmitoyl glycol chitosan nanoparticles. Mol. Pharm. 9, 1764–1774.

Lam, K.H., Schakenraad, J., Nieuwenhuis, P., 1992. Biocompatibility of degradable biomaterials. Thesis, State University of Groningen, Groningen, The Netherlands.

Langereis, S., Dirksen, A., Hackeng, T.M., Van Genderen, M.H., Meijer, E., 2007. Dendrimers and magnetic resonance imaging. New J. Chem. 31, 1152–1160.

Leyva-Gómez, G., Cortés, H., Magaña, J.J., Leyva-García, N., Quintanar-Guerrero, D., Florán, B., 2015. Nanoparticle technology for treatment of Parkinson's disease: the role of surface phenomena in reaching the brain. Drug Discov. Today 20, 824–837.

Lin, Y., Pan, Y., Shi, Y., Huang, X., Jia, N., Jiang, J.-Y., 2012. Delivery of large molecules via poly (butyl cyanoacrylate) nanoparticles into the injured rat brain. Nanotechnology 23, 165101.

Liu, M., Fréchet, J.M., 1999. Designing dendrimers for drug delivery. Pharm. Sci. Technol. Today 2, 393–401.

Liu, G., Men, P., Kudo, W., Perry, G., Smith, M.A., 2009. Nanoparticle–chelator conjugates as inhibitors of amyloid-β aggregation and neurotoxicity: a novel therapeutic approach for Alzheimer disease. Neuro. Lett. 455, 187–190.

Liu, Y., K Tee, J., NC Chiu, G., 2015. Dendrimers in oral drug delivery application: current explorations, toxicity issues and strategies for improvement. Current Pharm. Design 21, 2629–2642.

Lydon, M., Minett, T., Tighe, B., 1985. Cellular interactions with synthetic polymer surfaces in culture. Biomaterials 6, 396–402.

Ma, Q., Han, Y., Chen, C., Cao, Y., Wang, S., Shen, W., Zhang, H., Li, Y., Van Dongen, M.A., He, B., 2015. Oral absorption enhancement of Probucol by PEGylated G5 PAMAM dendrimer modified nanoliposomes. Mol. Pharm. 12, 665–674.

Mahapatro, A., Singh, D.K., 2011. Biodegradable nanoparticles are excellent vehicle for site directed in-vivo delivery of drugs and vaccines. J. Nanobiotechnol. 9, 1–11.

Masserini, M., 2013. Nanoparticles for brain drug delivery. ISRN Biochem. 2013.

Meissner, W.G., Frasier, M., Gasser, T., Goetz, C.G., Lozano, A., Piccini, P., Obeso, J.A., Rascol, O., Schapira, A., Voon, V., Weiner, D.M., Tison, F., Bezard, E., 2011. Priorities in Parkinson's disease research. Nat. Rev. Drug Discov. 10, 377–393.

Meziani, M.J., Rollins, H.W., Allard, L.F., Sun, Y.-P., 2002. Protein-protected nanoparticles from rapid expansion of supercritical solution into aqueous solution. J. Phys. Chem. B 106, 11178–11182.

Middleton, J.C., Tipton, A.J., 2000. Synthetic biodegradable polymers as orthopedic devices. Biomaterials 21, 2335–2346.
Mittal, G., Carswell, H., Brett, R., Currie, S., Kumar, M.R., 2011. Development and evaluation of polymer nanoparticles for oral delivery of estradiol to rat brain in a model of Alzheimer's pathology. J. Control. Release 150, 220–228.
Modi, G., Pillay, V., Choonara, Y.E., 2010. Advances in the treatment of neurodegenerative disorders employing nanotechnology. Annal. N Y Acad. Sci. 1184, 154–172.
Moghimi, S.M., Hunter, A.C., Murray, J.C., 2001. Long-circulating and target-specific nanoparticles: theory to practice. Pharmacol. Rev. 53, 283–318.
Mogoşanu, G.D., Grumezescu, A.M., 2015. Pharmaceutical natural polymers: structure and chemistry. Handbook of Polymers for Pharmaceutical Technologies: Structure and Chemistry Scrivener Publishing–Wileypp. 477–519.
Morshed, R.A., Cheng, Y., Auffinger, B., Wegscheid, M.L., Lesniak, M.S., 2013. The potential of polymeric micelles in the context of glioblastoma therapy. Front. Pharmacol. 4 (157), 1–15.
Mourya, V., Inamdar, N., Nawale, R., Kulthe, S., 2011. Polymeric micelles: general considerations and their applications. Ind. J. Pharm. Ed. Res. 45, 128–138.
Muhamad12, I.I., Selvakumaran, S., Lazim, N.A.M., 2014. Designing polymeric nanoparticles for targeted drug delivery system. Nanomedicine, 287–313.
Müller, C., Leithner, K., Hauptstein, S., Hintzen, F., Salvenmoser, W., Bernkop-Schnürch, A., 2013. Preparation and characterization of mucus-penetrating papain/poly (acrylic acid) nanoparticles for oral drug delivery applications. J. Nano. Res. 15, 1–13.
Nagavarma, B., Yadav, H.K., Ayaz, A., Vasudha, L., Shivakumar, H., 2012. Different techniques for preparation of polymeric nanoparticles—a review. Asian J. Pharm. Clin. Res. 5, 16–23.
Nair, L.S., Laurencin, C.T., 2007. Biodegradable polymers as biomaterials. Progress Polym. Sci. 32, 762–798.
Nath, D., Banerjee, P., 2013. Green nanotechnology: a new hope for medical biology. Environ. Toxicol. Pharmacol. 36, 997–1014.
Neha, B., Ganesh, B., Preeti, K., 2013. Drug delivery to the brain using polymeric nanoparticles: a review. Int. J. Pharm. Life Sci. 2, 107–132.
Newkome, G.R., Kotta, K.K., Moorefield, C.N., 2006. Design, synthesis, and characterization of conifer-shaped dendritic architectures. Chem. Eur. J. 12, 3726–3734.
Ngwuluka, N.C., Pillay, V., Choonara, Y.E., Modi, G., Naidoo, D., Toit, L.C.D., Kumar, P., Ndesendo, V.M.K., Khan, R.A., 2011. Fabrication, modeling and characterization of multi-crosslinked methacrylate copolymeric nanoparticles for oral drug delivery. Int. J. Mol. Sci. 12, 6194.
Ochekpe, N.A., Olorunfemi, P.O., Ngwuluka, N.C., 2009. Nanotechnology and drug delivery part 1: background and applications. Trop. J. Pharm. Res. 8 (3), 265–274.
Oliveira, L., Taveira, E., Souza, L., Marreto, R., Lima, E., Taveira, S., 2012. Aplicações das nanopartículas lipídicas no tratamento de tumores sólidos: revisão de literatura. Revista Brasileira de Cancerologia 58, 695–701.
Onoue, S., Yamada, S., Chan, H.-K., 2014. Nanodrugs: pharmacokinetics and safety. Int. J. Nano. 9, 1025.
Palamoor, M., Jablonski, M.M., 2013. Synthesis, characterization and in vitro studies of celecoxib-loaded poly (ortho ester) nanoparticles targeted for intraocular drug delivery. Colloids Surf. B 112, 474–482.
Pandita, D., Ahuja, A., Lather, V., Benjamin, B., Dutta, T., Velpandian, T., Khar, R., 2011. Development of lipid-based nanoparticles for enhancing the oral bioavailability of paclitaxel. AAPS PharmSciTech 12, 712–722.
Patel, D., Henry, J., Good, T., 2006. Attenuation of β-amyloid induced toxicity by sialic acid-conjugated dendrimeric polymers. Biochim. Biophys. Acta 1760, 1802–1809.
Penn, S.G., He, L., Natan, M.J., 2003. Nanoparticles for bioanalysis. Curr. Opin. Chem. Biol. 7, 609–615.
Prausnitz, M.R., Mitragotri, S., Langer, R., 2004. Current status and future potential of transdermal drug delivery. Nat. Rev. Drug Discov. 3, 115–124.

Provenzale, J., Silva, G., 2009. Uses of nanoparticles for central nervous system imaging and therapy. Am. J. Neuroradiol. 30, 1293–1301.

Puvvada, N.A., 2011. Nanomedical platform for drug delivery. J. Nano. Nanotechnol 2, 122.

Quintanar-Guerrero, D., Allémann, E., Doelker, E., Fessi, H., 1998. Preparation and characterization of nanocapsules from preformed polymers by a new process based on emulsification-diffusion technique. Pharm. Res. 15, 1056–1062.

Ramos-Cabrer, P., Campos, F., 2013. Liposomes and nanotechnology in drug development: focus on neurological targets. Int. J. Nanomed. 8, 951.

Rawat, M., Singh, D., Saraf, S., Saraf, S., 2006. Nanocarriers: promising vehicle for bioactive drugs. Biol. Pharm. Bull. 29, 1790–1798.

Ray, S.S., Bousmina, M., 2005. Biodegradable polymers and their layered silicate nanocomposites: in greening the 21st century materials world. Prog. Mater. Sci. 50, 962–1079.

Ribeiro, A., Sosnik, A., Chiappetta, D.A., Veiga, F., Concheiro, A., Alvarez-Lorenzo, C., 2012. Single and mixed poloxamine micelles as nanocarriers for solubilization and sustained release of ethoxzolamide for topical glaucoma therapy. J.Royal Soc. Interface 9, 2059–2069.

Rotomskis, R., Streckytė, G., Karabanovas, V., 2006. Nanodariniai diagnostikai ir gydymui: nanomedicinos link. Medicina (Kaunas) 42 (7), 542–558.

Saade, H., Barrera, C., Espinoza, A., López-Quintanilla, M.L., Fernández, S., López, R.G., 2013. Ultrafine nanoparticles of ibuprofen-poly (methyl methacrylate) by a polymerization-loading method. Drug Deliv. Lett. 3, 54–60.

Sarin, H., Kanevsky, A.S., Wu, H., Brimacombe, K.R., Fung, S.H., Sousa, A.A., Auh, S., Wilson, C.M., Sharma, K., Aronova, M.A., 2008. Effective transvascular delivery of nanoparticles across the blood-brain tumor barrier into malignant glioma cells. J. Trans. Med. 6, 80.

Sarvaiya, J., Agrawal, Y., 2015a. Chitosan as a suitable nanocarrier material for anti-Alzheimer drug delivery. Int. J. Biol. Macromol. 72, 454–465.

Sarvaiya, J., Agrawal, Y.K., 2015b. Chitosan as a suitable nanocarrier material for anti-Alzheimer drug delivery. Int. J. Biol. Macromol. 72, 454–465.

Schaffazick, S.R., Guterres, S.S., Freitas, L.D.L., Pohlmann, A.R., 2003. Physicochemical characterization and stability of the polymeric nanoparticle systems for drug administration. Química Nova 26, 726–737.

Sewell, M.D., Eastwood, D.M., Wimalasundera, N., 2014. Managing common symptoms of cerebral palsy in children. BMJ 349, g5474.

Sharma, A., Sharma, U.S., 1997. Liposomes in drug delivery: progress and limitations. Int. J. Pharm. 154, 123–140.

Simões, S.M., Figueiras, A.R., Veiga, F., Concheiro, A., Alvarez-Lorenzo, C., 2015. Polymeric micelles for oral drug administration enabling locoregional and systemic treatments. Expert Opin. Drug Deliv. 12, 297–318.

Singh, S.B., 2013. Novel approaches for brain drug delivery system: review. Int. J. Pharm. Res. Rev. 2, 8.

Singh, A.V., Khare, M., Gade, W., Zamboni, P., 2012. Theranostic implications of nanotechnology in multiple sclerosis: a future perspective. Autoimmune Dis. 2012, 160830.

Smith, R., 2005. Biodegradable Polymers for Industrial Applications. CRC Press, Boca Raton, FL.

Sosnik, A., Menaker Raskin, M., 2015. Polymeric micelles in mucosal drug delivery: challenges towards clinical translation. Biotechnol. Adv. 33, 1380–1392.

Sperling, R.-A., Parak, W., 2010. Surface modification, functionalization and bioconjugation of colloidal inorganic nanoparticles. Philos. Trans. R Soc. 368, 1333–1383.

Spuch, C., Navarro, C., 2011. Liposomes for targeted delivery of active agents against neurodegenerative diseases (Alzheimer's disease and Parkinson's disease). J. Drug Deliv. 2011, 469679.

Steichen, S.D., Caldorera-Moore, M., Peppas, N.A., 2013. A review of current nanoparticle and targeting moieties for the delivery of cancer therapeutics. Eur. J. Pharm. Sci. 48, 416–427.

Sun, H., Benjaminsen, R.V., Almdal, K., Andresen, T.L., 2012. Hyaluronic acid immobilized polyacrylamide nanoparticle sensors for CD44 receptor targeting and pH measurement in cells. Biocon. Chem. 23, 2247–2255.

Sutton, D., Nasongkla, N., Blanco, E., Gao, J., 2007. Functionalized micellar systems for cancer targeted drug delivery. Pharm. Res. 24, 1029–1046.

Torchilin, V., 2009. Multifunctional and stimuli-sensitive pharmaceutical nanocarriers. Eur. J. Pharm. Biopharm. 71, 431–444.

Tosi, G., Ruozi, B., Belletti, D., 2012. Nanomedicine: the future for advancing medicine and neuroscience. Nanomedicine 7, 1113.

Tsai, M.J., Huang, Y.B., Wu, P.C., Fu, Y.S., Kao, Y.R., Fang, J.Y., Tsai, Y.H., 2011. Oral apomorphine delivery from solid lipid nanoparticleswith different monostearate emulsifiers: pharmacokinetic and behavioral evaluations. J. Pharm. Sci. 100, 547–557.

Valois, C.R., Braz, J.M., Nunes, E.S., Vinolo, M.A., Lima, E.C., Curi, R., Kuebler, W.M., Azevedo, R.B., 2010. The effect of DMSA-functionalized magnetic nanoparticles on transendothelial migration of monocytes in the murine lung via a β 2 integrin-dependent pathway. Biomaterials 31, 366–374.

Veiga, F., Pecorelli, C., Ribeiro, L., 2006. As Ciclodextrinas em Tecnologia Farmacêutica. Minerva Coimbra, Coimbra.

Vemuri, S., Rhodes, C., 1995. Preparation and characterization of liposomes as therapeutic delivery systems: a review. Pharm. Acta Helvetiae 70, 95–111.

Vladkova, T.G., 2010. Surface engineered polymeric biomaterials with improved biocontact properties. Int. J. Poly. Sci. 2010, 22pp..

Vroman, I., Tighzert, L., 2009. Biodegradable polymers. Materials 2, 307–344.

Wang, M., 2012. Application of layer-by-layer assembly for controlled drug release from polyglycolide nanoparticles. Oklahoma State University. Dissertation or Thesis.

Wang, X., Uto, T., Akagi, T., Akashi, M., Baba, M., 2008. Poly(γ-glutamic acid) nanoparticles as an efficient antigen delivery and adjuvant system: potential for an AIDS vaccine. J. Med. Virol. 80, 11–19.

Wang, J., Yang, G., Guo, X., Tang, Z., Zhong, Z., Zhou, S., 2014. Redox-responsive polyanhydride micelles for cancer therapy. Biomaterials 35, 3080–3090.

Wattendorf, U., Merkle, H.P., 2008. PEGylation as a tool for the biomedical engineering of surface modified microparticles. J. Pharm. Sci. 97, 4655–4669.

Wilson, D.R., Zhang, N., Silvers, A.L., Forstner, M.B., Bader, R.A., 2014. Synthesis and evaluation of cyclosporine A-loaded polysialic acid–polycaprolactone micelles for rheumatoid arthritis. Eur. J. Pharm. Sci. 51, 146–156.

Woitiski, C.B., Veiga, F., Ribeiro, A., Neufeld, R., 2009. Design for optimization of nanoparticles integrating biomaterials for orally dosed insulin. Eur. J. Pharm. Biopharm. 73, 25–33.

Woitiski, C.B., Neufeld, R.J., Veiga, F., Carvalho, R.A., Figueiredo, I.V., 2010. Pharmacological effect of orally delivered insulin facilitated by multilayered stable nanoparticles. Eur. J. Pharm. Sci. 41, 556–563.

Wong, H.L., Chattopadhyay, N., Wu, X.Y., Bendayan, R., 2010. Nanotechnology applications for improved delivery of antiretroviral drugs to the brain. Adv. Drug Deliv. Rev. 62, 503–517.

Wong, H.L., Wu, X.Y., Bendayan, R., 2012. Nanotechnological advances for the delivery of CNS therapeutics. Adv. Drug Deliv. Rev. 64, 686–700.

Xu, L., Zhang, H., Wu, Y., 2013. Dendrimer advances for the central nervous system delivery of therapeutics. ACS Chem. Neurosci. 5, 2–13.

Yamamoto, S., Kaneo, Y., Maitani, Y., 2013. Hydrophobized poly(vinyl alcohol) for encapsulation of amphotericin B in nanoparticles. J. Drug Deliv. Sci. Technol. 23, 129–135.

Yang, S., Zhu, J., Lu, Y., Liang, B., Yang, C., 1999. Body distribution of camptothecin solid lipid nanoparticles after oral administration. Pharm. Res. 16, 751–757.

Zhang, N., Wardwell, P.R., Bader, R.A., 2013. Polysaccharide-based micelles for drug delivery. Pharmaceutics 5, 329–352.

Zhao, D., Zhuo, R.-X., Cheng, S.-X., 2012. Alginate modified nanostructured calcium carbonate with enhanced delivery efficiency for gene and drug delivery. Mol. BioSys. 8, 753–759.

Zheng, D., Ye, H., Luo, C., Xu, H., Meng, L., 2015. The efficient apoptotic induction of paclitaxel-loaded poly (*N*-vinylpyrrolidone)-block-poly (-caprolactone) nanoparticles in the in vitro study of lung cancer cell lines. J. Nanomater. 501, 791063.

Ziemba, B., Matuszko, G., Bryszewska, M., Klajnert, B., 2012. Influence of dendrimers on red blood cells. Cell. Mol. Biol. Lett. 17, 21–35.

CHAPTER

NANOSTRUCTURED SYSTEMS FOR TRANSBUCCAL DRUG DELIVERY

4

Michelle Franz-Montan*, Daniele Ribeiro de Araújo, Lígia Nunes de Morais Ribeiro†, Nathalie Ferreira Silva de Melo‡, Eneida de Paula†**

**University of Campinas—UNICAMP, Piracicaba, Brazil*

***Federal University of ABC, Santo André, Brazil*

†University of Campinas—UNICAMP, Campinas, Brazil

‡São Leopoldo Mandic Institute and Research Center, Campinas, Brazil

CHAPTER OUTLINE

1 Introduction 88
2 Oral Mucosa 89
- 2.1 Concept and Functions 89
- 2.2 Organization and Structure 89
- 2.3 Permeability 93
- 2.4 Permeation Pathways 94

3 Buccal Mucosa Chemical Permeation Enhancers: Chemical Structure, Mechanisms, and Biopharmaceutical Applications 94
- 3.1 Surfactants and Bile Salts 95
- 3.2 Fatty Acids 98
- 3.3 Cyclodextrins 98
- 3.4 Chelators 98
- 3.5 Polymers: The Case of Chitosan and Its Derivatives and Associations 99

4 Nanoparticles and Nanostructures for Transbuccal Drug Delivery 100
- 4.1 Nanostructures 102
- 4.2 Nanoparticles 104

5 Limitations 113
6 Conclusions 114
Acknowledgments 114
References 114

1 INTRODUCTION

Several diseases affect the oral cavity, including mucositis, periodontitis, recurrent aphthous stomatitis, ulcers, lichen planus, herpes simplex, cancer, vesiculobullous diseases, salivary dysfunctions, and candidiasis. Although topical treatment is an attractive therapeutic approach, few formulations have so far been developed for this purpose (Paderni et al., 2012; Sankar et al., 2011).

The oral mucosa lines the palate (hard and soft), tongue, buccal, gingival, and floor regions of the mouth. There are three different types of mucosa, according to their function: masticatory, lining, and specialized (Hassan et al., 2010; Squier and Brogden, 2011). There are differences in mucosal structure among these types, which can influence drug delivery due to their varying degrees of permeability (Morales and McConville, 2014).

Compelling benefits of the buccal route include safety, increased drug bioavailability at the affected site, fast onset of action, avoidance of hepatic first-pass effects, and destruction in the gastrointestinal tract, less invasive procedures, reduced side effects, easy access, and better patient compliance (Chinna Reddy et al., 2011; Sanz et al., 2015; Shakya et al., 2011). These unique features make it worthwhile to try to achieve sustained drug release at the mouth for the treatment of oral diseases. In addition, due to the rich blood supply, transbuccal delivery is an attractive route of administration that can provide systemic effects.

Similar to the skin, the oral mucosa acts as a protective barrier for deeper layers, limiting the entry of substances from the exterior (Lesch et al., 1989; Squier and Hopps, 1976). Nevertheless, its permeability is known to be 4–4000 times greater than that of skin (Galey et al., 1976; Squier and Hall, 1985b), which often leads to the erroneous notion that the oral mucosa is a permeable tissue. On the contrary, the oral mucosa presents an effective barrier to penetration, and successful transbuccal drug delivery remains a great challenge. Other limitations on transmucosal drug delivery that have been identified are the presence of saliva, with continuous washing of the mucosa sites, accompanied by the constantly applied mechanical forces induced by swallowing, speaking, and chewing (Chinna Reddy et al., 2011; Paderni et al., 2012). In addition, the mucus can disrupt the penetration of drug-delivery systems, and the use of mucoadhesive structures may be necessary to overcome this problem (Holpuch et al., 2010; Madhav et al., 2009).

For these reasons, the low therapeutic efficacy of formulations applied topically at the oral mucosa is of great concern, and new strategies are required to increase drug retention and penetration using this route (Patel et al., 2011; Şenel and Hincal, 2001).

Several approaches have been proposed with the aim of improving transmucosal drug delivery. New technologies have already been used in novel commercially available tablets, wafer/film, spray, mouthwash, gel, and paste formulations, which have been reviewed elsewhere (Hearnden et al., 2012). Mucoadhesive dosage forms, including tablets, patches, films, and semisolid and liquid systems, have also been suggested as promising tools for the treatment of oral diseases (Paderni et al., 2012). Similarly, the development of drug-delivery systems, such as polymeric nanoparticles (PN), liposomes, nanostructured lipid carriers (NLCs), cyclodextrins (CDs), and buccal delivery films are techniques that can be applied to improve transbuccal delivery.

The aim of this work is to review nanostructured systems for transbuccal drug delivery in order to achieve local effects in the treatment of diseases that commonly affect the oral mucosa. Although several other formulations for topical application at the oral mucosa have been developed, only nanostructured DDS are considered in this review.

The following topics are covered: (1) organization, structure, and permeability of the oral mucosa; (2) permeation pathways at the oral mucosa; (3) buccal mucosa chemical permeation enhancers: chemical structures, mechanisms, and biopharmaceutical applications; and (4) the use of the following nanostructured systems to improve transbuccal delivery for the treatment of oral mucosa infections, inflammation, and pain conditions: CDs; liposomes; solid lipid nanoparticles (SLNs); NLCs; PN and micelles.

2 ORAL MUCOSA

2.1 CONCEPT AND FUNCTIONS

Considering its anatomical position, the oral cavity connects the exterior and the gastrointestinal environment. It is lined by a moist mucous membrane, also called mucosa (Squier and Brogden, 2011). The moist condition is ensured by the constant production by the salivary glands of saliva, whose composition is very similar to that of an ultrafiltrate of plasma, with 95%–99% of water, proteins, enzymes, electrolytes, and nucleic acids (Dawes et al., 2015; Shakya et al., 2011). Key functions of the salivary film include reduction of the susceptibility of oral mucosa to abrasion, as well as cleaning effects. It is essential for the sensation of taste, because it dissolves and distributes substances near the taste buds; it acts to protect the teeth due to its remineralization and buffering abilities; and it promotes the initial digestion of starches (by α-amylases), among other effects. Detailed information about the composition and functions of saliva is provided in a recent review by Dawes et al. (2015). On the other hand, the presence of saliva can negatively influence drug delivery, because it constantly washes the oral mucosa, dilutes the drug, and can reduce the contact and bioavailability of topically applied formulations, in a phenomenon known as "saliva wash out" (Chinna Reddy et al., 2011; Paderni et al., 2012; Patel et al., 2011).

During daily activities, such as swallowing, speaking, and chewing, the oral cavity is exposed to many chemical compounds and microorganisms, and is submitted to mechanical forces with different intensities. For this reason, one of the most important functions of the oral mucosa is protection (Squier and Brogden, 2011), which is conferred by its action as a permeability barrier, due to the structure and organization of its outermost layer, the epithelium (discussed in next section). In addition to its role as a permeability barrier, some specific lipids present have important antimicrobial properties and contribute to its immunoprotective function (Dawson et al., 2013).

Furthermore, the oral mucosa has sensory functions involving taste, tactile, thermal, proprioception, and pain sensations that are essential in life. All these sensory functions are necessary to start the reflexes of chewing, masticatory strength, swallowing, gagging, retching, and salivating, among others (Squier and Brogden, 2011).

2.2 ORGANIZATION AND STRUCTURE

The oral mucosa covers the inner lips, gingiva, palate, floor of the mouth, cheeks, and dorsum of the tongue. These regions present different mucosal structures. Essentially, three types of mucosa are found in the oral environment, based on their function: masticatory, lining, and specialized. The distribution of different oral mucosa types is illustrated in Fig. 4.1.

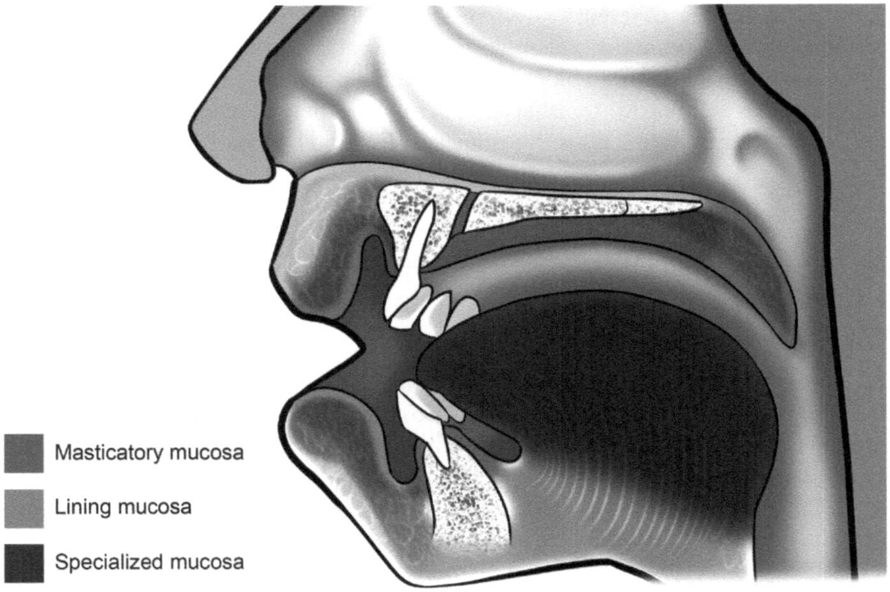

FIGURE 4.1 Schematic Illustration of the Anatomical Locations of the Three Different Oral Mucosa Types

Considering a total surface area inside the mouth of about 200 cm^2, the relative occupation of each type of mucosa has been calculated to be approximately 25% for the masticatory, 60% for the lining, and 15% for the specialized mucosa (Collins and Dawes, 1987).

Structurally, the oral mucosa is divided into two distinct layers: the epithelium, which is a stratified squamous layer, and the underlying connective tissue, called the lamina propria. The submucosa region, located under the lamina propria of the nonkeratinized tissue, comprises a loose, fatty, glandular tissue containing the major blood vessels and nerves that supply the mucosa (Squier and Brogden, 2011). This layer is absent in the masticatory and specialized mucosa, which are directly attached to the periosteum of the underling bone (maxilla or alveolar bone), and the tongue muscle, respectively. The structural components of the oral mucosa are illustrated in Figs. 4.2 and 4.3.

The masticatory mucosa comprises the hard palate (the upper surface of the mouth) and the gingiva (gums), which are covered by a keratinized stratified squamous epithelium, as detailed in Fig. 4.2A. The lining mucosa is found at the buccal mucosa (inner cheeks), sublingual region (floor of the mouth), underside of tongue, soft palate, and upper and lower lips (inner region), and is covered by a nonkeratinized stratified squamous epithelium, as detailed in Fig. 4.3A.

Keratinocytes constitute the main cell type of the oral epithelium. One of the factors that contributes to the important barrier property of the epithelium is that these cells are physicochemically adhered to each other due to the presence of desmosomes (Squier and Brogden, 2011).

As a result of its lining function, the epithelium in all the oral mucosa regions constantly proliferates in order to maintain its integrity. This is achieved by cells from the basal layer (the *stratum basale*) undergoing mitotic division, with the cells then migrating to the surface and replacing those that desquamate. The process occurs according to functional demands related to the location of the cells.

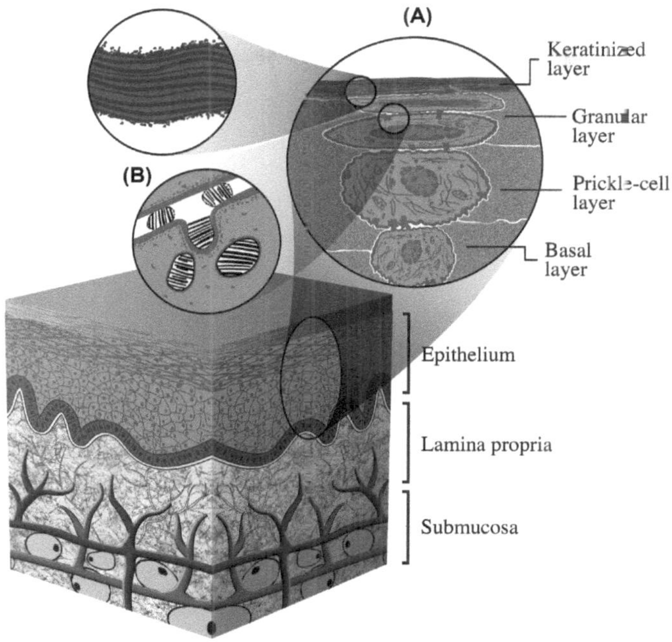

FIGURE 4.2 Illustration of Anatomical Structures That Comprise the Masticatory Oral Mucosa

Details are shown of the keratinized epithelium layers during differentiation (A) and the membrane-coating granules during extrusion of the lamellar content to the intercellular space (B).

As the cells move toward the superficial layers, they undergo several differentiation processes, which together are known as epithelium proliferation and turnover (Katchburian and Arana, 2012; Squier and Brogden, 2011).

Despite the differences between the layers of the keratinized and nonkeratinized epithelia during differentiation, there are changes that are common to both, involving cell size and shape, with the production of structural proteins (tonofilaments and tonofibrils), new organelles, and intercellular material. The basal (*stratum basale*) and prickle-cell (*stratum spinosum*) layers are common to both epithelia.

At the prickle-cell layer, small organelles called membrane-coating granules start to appear. In the keratinized epithelium, these are elongated and filled with a lamellae-like material (Fig. 4.2B), while in the nonkeratinized epithelium they are rounded and filled with an amorphous material (Fig. 4.3B) (Squier and Brogden, 2011). As the cells move toward the next cell layer (the granular and intermediate layers for the keratinized and nonkeratinized epithelia, respectively), they start to become more flattened and the membrane-coating granules fuse with the cell membrane, discharging their contents to the intercellular space, which forms a barrier to many compounds (Hayward, 1979; Squier, 1991; Squier and Brogden, 2011).

Another common change observed during the differentiation process, most clearly observed in the keratinized epithelium, is the development of a cornified envelope, with thickening on the inner part of the cell membrane attributed to the presence of the involucrin protein, which confers additional barrier resistance to both epithelia (Squier and Brogden, 2011).

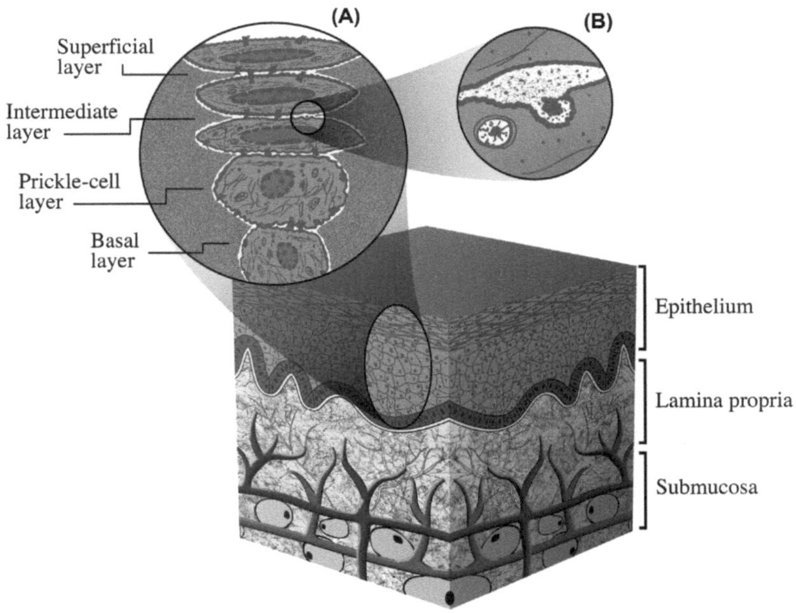

FIGURE 4.3 Illustration of the Anatomical Structures That Comprise the Lining Oral Mucosa

Details are shown of the nonkeratinized epithelium layers during differentiation (A) and the membrane-coating granules during extrusion of the amorphous content to the intercellular space (B).

The main differences between the two epithelia are in the superficial layers, as can be seen in Fig. 4.2A and Fig. 4.3A. Following the prickle-cell layer, the next superficial layers at the keratinized epithelium are known as the granular (*stratum granulosum*) and keratinized (*stratum corneum*) layers, while at the nonkeratinized epithelium they are known as the intermediate (*stratum intermedium*) and superficial (*stratum superficiale*) layers (Squier and Brogden, 2011; Katchburian and Arana, 2012).

The granular layer is characterized by the presence of intracellular keratohyalin granules and, as they reach the keratinized layer, the cells lose all the organelles and nuclei and remain filled with keratin alone, becoming flattened and dehydrated, and starting to desquamate. In contrast, few changes occur at the nonkeratinized epithelium, where the cells of the intermediate layer contain significantly fewer granules, compared to the granular layer. The cells become more flattened and the organelle contents decrease, and these features are intensified as the cells reach the superficial layer, although the nuclei are retained (Katchburian and Arana, 2012; Squier and Brogden, 2011).

The specialized mucosa is found at the dorsum of the tongue. It is composed of a mixture of masticatory and lining mucosa, with regions of keratinized and nonkeratinized epithelium (Squier, 1991; Squier and Brogden, 2011; Squier and Rooney, 1976). In addition, in the region of the taste buds, this mucosa also presents fungiform, foliate, and circumvallate papillae, responsible for the sensation of taste (Squier and Brogden, 2011).

2.3 PERMEABILITY

Since the beginning of the 1970s, the permeability of oral mucosa has been the focus of work conducted by Squier and colleagues (Lesch et al., 1989; Squier, 1973, 1984, 1991; Squier and Hall, 1984, 1985a,b; Squier and Hopps, 1976; Squier and Rooney, 1976; Wertz and Squier, 1991). The fact that the oral mucosa is a highly permeable tissue was acknowledged several years ago, and there has been increased interest in oral mucosa permeability and ways to get through this barrier. Transbuccal delivery has become a useful administration route, due to the various advantages mentioned earlier.

The main factors affecting the permeability of oral mucosa are the characteristics of the epithelium, such as its structure (thickness and keratinization) and the lipid composition of intercellular material derived from the membrane-coating granules described earlier (Squier, 1984 Squier and Hall, 1984; Squier and Hopps, 1976). Considering the variety of oral mucosa structures, differences in permeability are expected (Lesch et al., 1989; Squier, 1991; Squier and Hall, 1985b). It is worth noting that it is not only the superficial layers of the epithelium that play an important role in reducing permeability, but also the basal membrane (Alfano et al., 1977). Furthermore, mucus and saliva also contribute to reducing drug penetration (Sanz et al., 2015).

Among all the different factors, the intercellular material derived from the membrane-coating granules is considered the most important (Squier, 1984; Squier and Hall, 1984; Squier and Hopps, 1976). Besides the differences in organization described above and detailed in Fig. 4.2B and Fig. 4.3B, the components of the intercellular material are also different. The lipid components of the keratinized epithelium comprise sphingomyelin, glucosylceramides, ceramides, and other nonpolar lipids organized in a lamellar phase. However, for the nonkeratinized epithelium, the lipid components are cholesterol esters, cholesterol, and glycosphingolipids, organized in a nonlamellar lipid phase (Sanz et al., 2015; Wertz and Squier, 1991). In addition, the presence of ceramides increases lipid organization, with keratinized epithelium presenting about 75% of ceramides and nonkeratinized epithelium about 5% (Squier and Brogden, 2011). The characteristics of such lipids and their organization result in the observed differences in permeability (Squier and Hall, 1984, 1985b; Squier and Hopps, 1976).

Studies of the differences in permeability among the different regions of the oral mucosa have shown that the permeability of the mucosal epithelium increases in the order: palate < buccal < floor of the mouth (Harris and Robinson, 1992; Lesch et al., 1989; Shojaei et al., 1998). The lateral border of the tongue has been found to present permeability intermediate between the buccal and floor of the mouth regions (Lesch et al., 1989).

Besides the characteristics of the barrier membrane, account should also be taken of the physicochemical features of the studied drug, which also play an important role in permeability (Paderni et al., 2012) and is discussed further. It was recently demonstrated that the dorsum of the tongue is significantly more permeable toward lidocaine hydrochloride, compared to the buccal and palatal epithelia, and that there is no difference between these two barriers that could explain this observation (Franz-Montan et al., 2016).

It is important to mention that other parameters can also affect the permeability of oral mucosa, such as aging (with minor thinning and flattening of the epithelium cells as people become older), inflammation, and the presence of extrinsic factors (e.g., tobacco, smoke, alcohol, toothpastes, and mouth rinses). These situations can either increase thickness, while stimulating cell proliferation, or decrease thickness, while suppressing proliferation (Squier and Brogden, 2011).

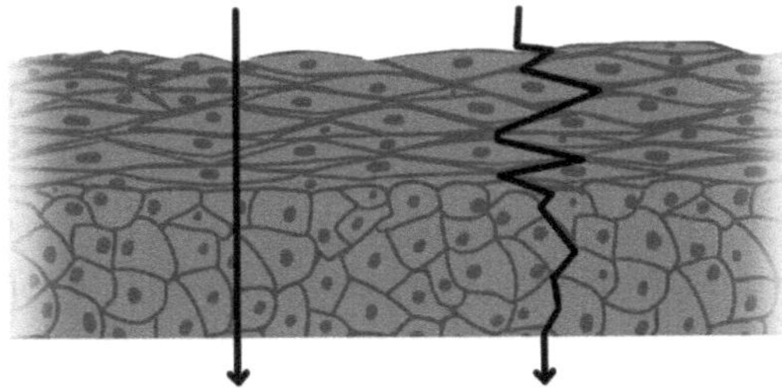

FIGURE 4.4 Permeation Pathways of Drugs Across Oral Epithelium

Transcellular (intracellular) (left) and paracellular (intercellular) (right).

2.4 PERMEATION PATHWAYS

The literature reports several parameters that can affect transbuccal permeation, including biological factors (age, oral mucosa area, blood flow, and tissue condition), physicochemical properties of the drug (size, partitioning, diffusion coefficient, and solubility), and aspects related to the final formulation and application method (occlusion ability, viscosity, drug concentration, frequency of application, and mucoadhesive property) (Paderni et al., 2012; Sanz et al., 2015).

Transbuccal permeation depends on release of the drug from the final formulation to the mucosa surface, followed by partitioning of the drug into the epithelium layers, and finally reaching its site of action, whether local or systemic (Sanz et al., 2015; Şenel and Hincal, 2001).

Although transbuccal permeation can occur by means of active transport by a carrier, or due to other specialized mechanisms (e.g., endocytosis/exocytosis), it generally occurs by passive diffusion in two possible routes: paracellular (intercellular) and/or transcellular (intracellular) (Hearnden et al., 2012; Mashru et al., 2005; Patel et al., 2011; Sanz et al., 2015; Şenel and Hincal, 2001), as illustrated in Fig. 4.4. It has been found that for the majority of drugs, the paracellular route is most important for permeation (Şenel and Hincal, 2001).

Despite all the possible drug permeation pathways described earlier, active substances can often still fail to permeate through oral mucosa, and transbuccal drug delivery remains a great challenge. Several approaches have been investigated in order to achieve an effective therapeutic effect, such as the use of mucoadhesive properties to increase drug retention, and the use of permeation enhancers and drug-delivery systems to increase drug penetration, as discussed later in this chapter.

3 BUCCAL MUCOSA CHEMICAL PERMEATION ENHANCERS: CHEMICAL STRUCTURE, MECHANISMS, AND BIOPHARMACEUTICAL APPLICATIONS

The search for an ideal pharmaceutical formulation for buccal drug delivery should consider several factors: physicochemical properties of the formulation components (solubility, partitioning coefficient, compatibility with the other constituents of the formulation, concentration, molecular weight, and charge); mechanism of action; mucoadhesive capacity and the ability to interact with the mucus layer (interfacial phenomena involving the mucus surface and its glycoproteins); formulation-related factors (flexibility and swelling of films, cross-linking density, pH, rheological profile, drug release rate, organoleptic properties, and biocompatibility); physiological factors (buccal epithelium morphology and its variations, composition and flow of the saliva, and area of application), and patient compliance (Chinna Reddy et al., 2011; Salamat-Miller et al., 2005; Sanz et al., 2015; Sudhakar et al., 2006).

Several studies have described the development and mechanisms of action of chemical substances able to alter the penetration of different drugs and/or bioactive substances across the buccal mucosa. In this section, strategies are described that are based on traditional and new chemical permeation/penetration enhancers proposed as formulation additives for buccal drug delivery.

The main mechanisms responsible for the penetration of molecules across the buccal mucosa include simple diffusion, active transport, and pinocytosis or endocytosis, with passive diffusion being the primary mechanism, as already discussed. Considering these mechanisms, different chemical substances can be incorporated into formulations in order to increase the penetration flux of drugs across the buccal mucosa.

By definition, penetration enhancers, permeation promoters, and absorption enhancers are chemical substances able to promote drug permeation through the buccal epithelium using different mechanisms of action, depending on their chemical structures and physicochemical properties (Chinna Reddy et al., 2011; Nicolazzo et al., 2004; Şenel and Hincal, 2001).

The biopharmaceutical properties of an ideal penetration enhancer should include low toxicity (systemic and local); it needs to be nonallergenic, nonirritating, and pharmacologically inert so that the tissue can recover its normal morphological/functional characteristics after removal of the enhancer (Chinna Reddy et al., 2011). Several compounds have been investigated as additives with penetration enhancing properties in order to obtain more effective formulations for buccal delivery. Some of these compounds and their mechanisms of action are presented in Table 4.1 (Sudhakar et al., 2006).

3.1 SURFACTANTS AND BILE SALTS

Among the penetration enhancers that have been used, surfactants (sodium dodecyl sulfate, cetyltrimethyl ammonium bromide, poloxamers, etc.) and bile salts (sodium glycocholate, sodium fusidate, etc.) have been shown to be effective as buccal penetration enhancers, especially for lipophilic drugs. The main mechanism of action attributed to surfactants and bile salts is related to their ability to extract lipids from the membrane, and/or to disturb the epithelial membrane packing (Attia et al., 2004; Chinna Reddy et al., 2011; Nicolazzo et al., 2005).

Many studies have described the effectiveness of these chemicals for the enhancement of drug permeation through buccal mucosa. In the case of triamcinolone acetonide, for example, sodium taurocholate, sodium cholate, and sodium deoxycholate, at 5%, showed the highest enhancement factor values for permeation through porcine buccal mucosa (Shin and Kim, 2000). In another study, Hu et al. (2011) showed that dodecyl 2-(*N,N*-dimethyl amino) propionate hydrochloride was able to significantly enhance the permeation of ondansetron hydrochloride across porcine buccal tissue. Similar results were obtained by Wei et al. (2012) for lidocaine and nicotine buccal permeation, and use of 5% (w/v) dodecyl 2-(*N,N*-dimethyl amino) propionate hydrochloride did not induce severe local toxic effects in porcine buccal tissue. On the other hand, sodium dehydrocholate was not able to enhance the permeation rate of naltrexone hydrochloride through buccal epithelium, at three different concentrations tested (0.1, 0.5, and 1%) (Giannola et al., 2007).

While the above studies employed the addition of surfactants or bile salts to water, other strategies have used their incorporation in hydrogels and films. Shidhaye et al. (2010) developed a mucoadhesive

Table 4.1 Chemical Structures and Mechanisms of Action of Penetration Enhancers Discussed in This Chapter

Penetration Enhancers	Example and Chemical Structure	Mechanism of Action
Surfactants	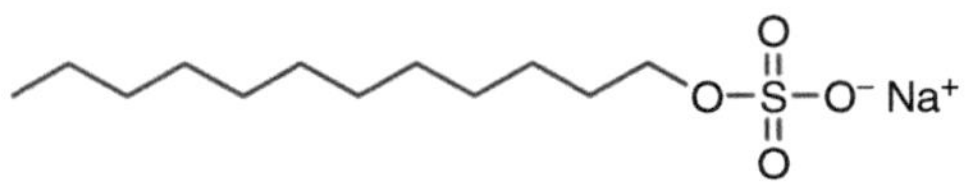 Sodium dodecyl sulfate	Extraction of lipids and disturbance of the membrane lipid bilayer
	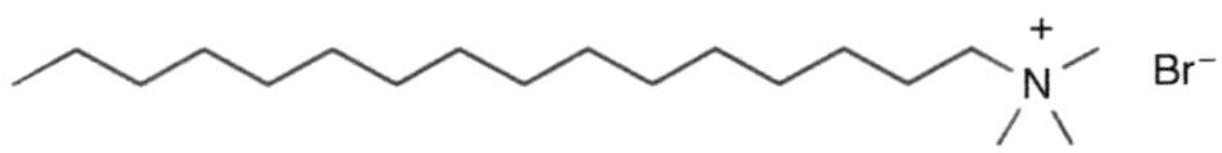 Cetyltrimethyl ammonium bromide	
Bile salts	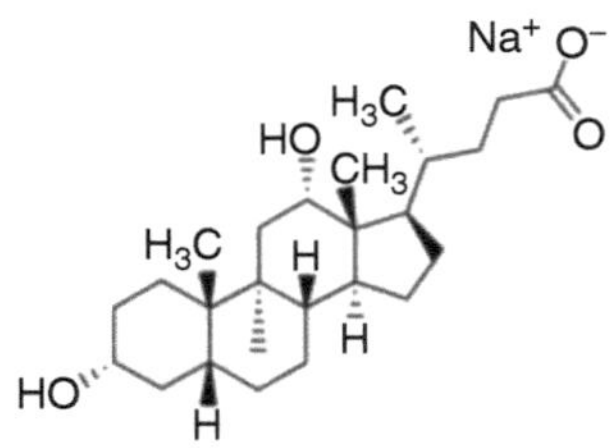 Sodium deoxycholate	Disturbance of intercellular lipids and protein domain integrity
	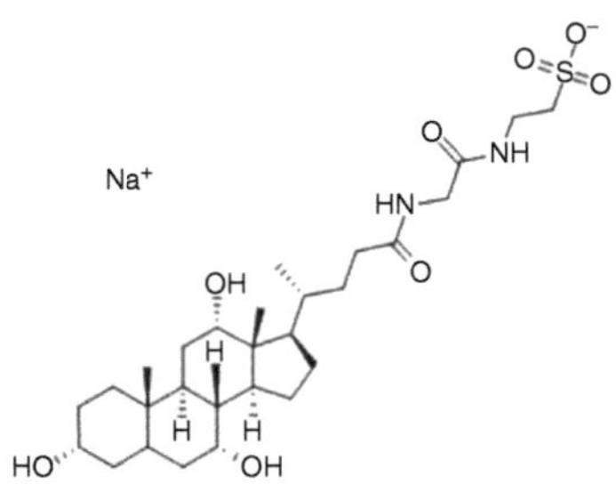 Sodium tauroglycocholate	
Fatty acids	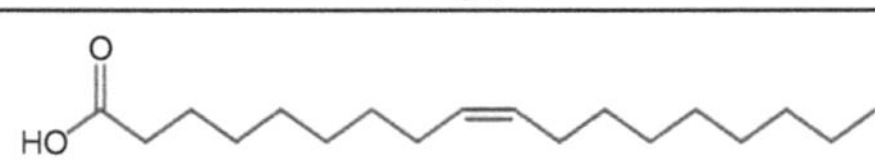 Oleic acid	Increased fluidity of phospholipid domains
Chelators	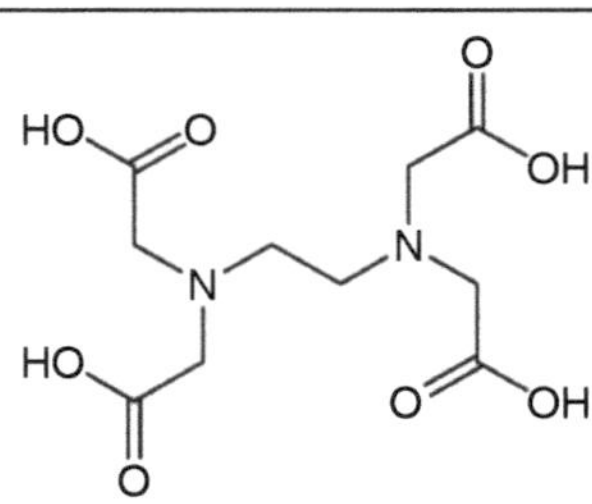 EDTA	Interaction with divalent cations and enzyme inhibition

Table 4.1 Chemical Structures and Mechanisms of Action of Penetration Enhancers Discussed in This Chapter (*cont.*)

Penetration Enhancers	Example and Chemical Structure	Mechanism of Action
Polymers	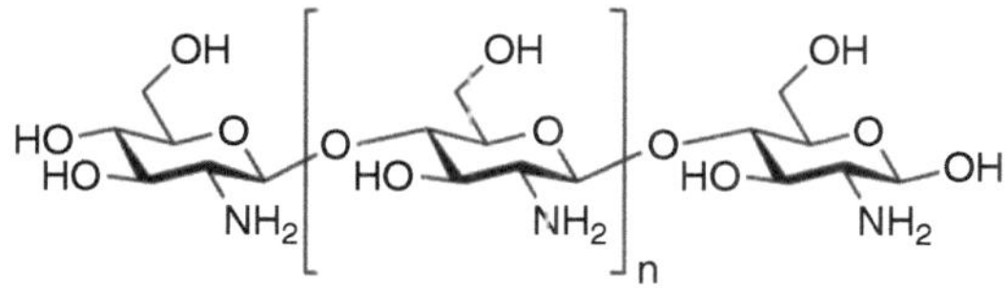 Chitosan	Enzyme inhibition and ionic interaction with negative charges on the mucosal surface
	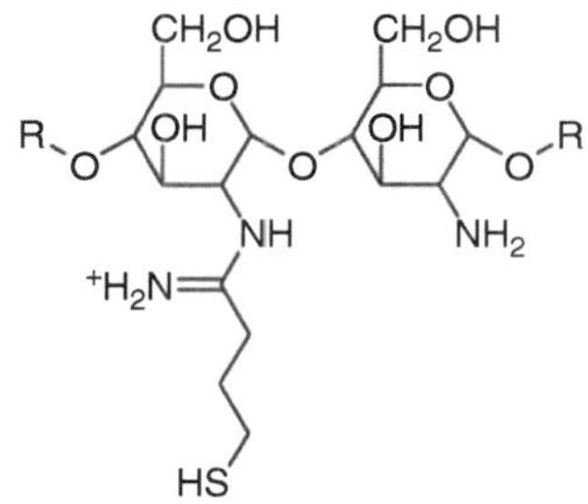 Thiolated chitosan	

tablet composed of polyvinyl pyrrolidone K 30, poloxamer 407 (Pluronic F-127), and a magnesium oxide matrix containing 1% sodium lauryl sulfate, which presented satisfactory mucoadhesive strength and enhanced permeation of the drug through buccal mucosa, without causing local toxicity. Other work has reported the development of piroxicam hydrogels using different matrices (hydroxypropylmethylcellulose, hydroxypropylcellulose, sodium alginate, methylcellulose hydroxyethylcellulose, Carbopol 934, sodium carboxymethylcellulose, Pluronic F-127, or polyvinyl alcohol) containing 1% sodium lauryl sulfate and 3% sodium deoxycholate or 3% sodium tauroglycocholate. The highest fluxes across the mucosa were observed for the formulations containing 1% sodium lauryl sulfate, highlighting their potential therapeutic use in both animal and human inflammation models (Attia et al., 2004).

Despite the substantial effects of surfactants and bile salts as penetration enhancers, these compounds can potentially cause concentration-dependent side effects, such as swelling, irritation, and tissue ulceration, associated with histological changes (formation of vacuoles, reduced desmosome numbers, and loss of the upper cell layers) after application of bile salts, although the effects have been found to be reversible within 24 h (Ikinci et al., 2000). The use of these compounds also resulted in a bitter taste, which reduced patient compliance. For this reason, new penetration enhancers have been proposed, such as lysalbinic acid, a nonionic surfactant obtained from the alkaline hydrolysis of albumin or casein. The work of Starokadomskyy and Dubey (2006) described the synthesis, physicochemical characterization, and performance of this compound as a penetration enhancer, showing that the administration of 1% lysalbinic acid led to a fivefold increase in α-interferon permeation across the buccal mucosa.

3.2 FATTY ACIDS

One of the first studies using fatty acids investigated the permeation profiles of insulin from Pluronic F-127 hydrogel containing unsaturated fatty acids, such as oleic (18:1), eicosapentaenoic (20:5), or docosahexaenoic acid (22:6). After buccal administration in rats, the oleic acid formulation showed the most pronounced hypoglycemic effects (Morishita et al., 2001). In another study, Sakata et al. (2011) reported the development of a semisolid formulation for clonazepam, with a matrix containing polyethylene glycol (1500) and oleic acid. This association resulted in enhanced absorption by the buccal mucosa, suggesting it as an alternative to intravenous injection of clonazepam. A similar study also investigated glycol-fatty acid association, showing that the combination of oleic acid (1%) and polyethylene glycol 400 (5%) was able to control the buccal absorption of donepezil hydrochloride (Caon et al., 2014).

These promising results encouraged the development of new pharmaceutical formulations containing synthetic oleic acid derivatives, such as dicarboxylic ester or its acid derivatives, and its bicephalous dianionic surfactant. The optimal penetration-enhancing effect was observed for oleic acid bicephalous dianionic surfactant, but was associated with more severe tissue morphological changes, such as vacuole formation and increased intercellular spaces (Ojewole et al., 2014).

Although fatty acids have been used as penetration enhancers, physicochemical studies of their mechanisms of action only exist for skin delivery. For mucosa models, some studies have reported the enhancement effects of these compounds, but it was assumed that increased fluidity of phospholipid domains was the mechanism responsible for the effectiveness of fatty acids. Fluorescence anisotropy studies with the 1,6-diphenyl-1,3,5-hexatriene probe demonstrated that oleic acid disrupted the organization of membrane lipids in the polar head region, suggesting interaction with the membrane lipid domains (Turunen et al., 1994). The literature is lacking in specific physicochemical studies designed to elucidate the mechanisms of action and acute/chronic effects of fatty acids in buccal tissue.

3.3 CYCLODEXTRINS

Some penetration enhancers act to alter the aqueous solubility of drugs. This category includes the natural CD (β-cyclodextrin) and its hydroxypropyl-β-cyclodextrin derivative. The low solubility of a drug in the saliva can limit its buccal absorption and bioavailability. The ability of CDs to form inclusion complexes with drugs alters both their physicochemical and biopharmaceutical properties. The benefits of drug-CD inclusion complexes are considered later in this chapter (Section 4.1), in the discussion of nanostructures used for buccal delivery systems.

3.4 CHELATORS

Chelation agents (EDTA, citric acid, sodium salicylate, methoxy salicylates, etc.) have been described in the literature as buccal penetration enhancers. The initial strategy was to evaluate their performance as single additives. One of the first studies showed that Na_2EDTA was ineffective in enhancing insulin penetration across the buccal mucosa (Aungst and Rogers, 1988). In contrast, it was found elsewhere that EDTA was effective in reducing the degradation of Leu-enkephalin by peptidases in buccal tissue. Evidence was provided of a mechanism of action based on the chelation activity of calcium and the possible inhibition of peptidases and proteases present within buccal mucosa (Walker et al., 2002). More recent applications of chelation agents have been based on their combination with polymers such as chitosan, as described in the next section.

3.5 POLYMERS: THE CASE OF CHITOSAN AND ITS DERIVATIVES AND ASSOCIATIONS

In addition to the numerous studies reporting the penetration enhancing activities of low molecular weight agents (surfactants, fatty acids, chelators, etc.), some works have reported formulation designs based on polymeric absorption enhancers, together with their main properties such as mechanism of action, structure–activity relationships, and biocompatibility (Di Colo et al., 2008).

Different types of polymers, such as polyacrylic acid (Carbopol) (Shin and Kim, 2000), hydroxypropyl cellulose (Abdel-Hamid et al., 2006), carboxymethyl cellulose (Bahri-Najafi et al., 2013), and pectins (Mura et al., 2015) have been used as matrices for transbuccal drug delivery. However, chitosan has been shown to be one of the most effective penetration enhancers, due to its mucoadhesive properties (Dhawan et al., 2014; Lv et al., 2015; Sandri et al., 2004; Şenel and Hincal, 2001). The use of high molecular weight enhancers can have advantages, since some are nonabsorbable and present mucoadhesive properties that act to reduce systemic toxicity, while at the same time prolonging contact of the formulation with the buccal mucosa (Di Colo et al., 2008).

The permeation mechanism of action of polymers is, in part, a result of their enzymatic inhibition properties, which results in reduced enzymatic degradation of protein and peptide drugs. For example, polyacrylic acid polymers (Carbopol and other carbomers) are able to interact with enzyme cofactors such as divalent cations (calcium and zinc), leading to loss of the functional activity of the enzyme (Bernkop-Schnürch et al., 1997). In the case of chitosan, the combination with chelators (e.g., EDTA) causes the inhibition of metalloproteinase activity, and the insertion of thiol groups on the carbomer chemical structure (thiomers) enhances the enzyme inhibition capacity (Bernkop-Schnürch et al., 2004).

In relation to the mechanisms of mucoadhesion, it is necessary to consider interfacial phenomena and forces such as van der Waals interactions, hydrogen bonding, disulfide bridging, hydration forces, electrostatic double-layer forces, hydrophobic interactions, covalent bonds, and steric forces (Sudhakar et al., 2006). In general terms, theories concerning the possible mechanisms of mucoadhesion have been developed considering: (1) adsorption, where primary and secondary covalent and noncovalent chemical bonds are formed when contact between the mucus and the polymer occurs; (2) diffusion of the polymer chain into the mucosal environment; (3) electronic, related to interfacial phenomena between the polymer and the mucus surface; (4) fracture, describing the force applied to evaluate the strength of the adhesive bonds; and (5) wetting, describing the ability of polymers to adhere and spread on the buccal tissue (Salamat-Miller et al., 2005). In the case of chitosan, for example, its mucoadhesive property is due to the formation of hydrogen bonds or ionic interactions between the positively charged amino groups and the negatively charged sialic acid residues from mucus glycoproteins or mucins (Meng-Lund et al., 2014).

Chitosan is a positively charged high molecular weight polysaccharide that has been widely used in the development of new pharmaceutical formulations for transbuccal drug delivery, due to its mucoadhesive capacity, biocompatibility, and biodegradability (Sandri et al., 2004). Although these properties are well described in the literature, several works have described a new generation of chitosan-based formulations produced using changes in its chemical structure, conjugation with other enhancers, and association with other enhancers in the same formulation. Some of these important results are described in this section.

The development of bioadhesive formulations has been the objective of several studies. Sander et al. (2013) reported the development of chitosan discs and the permeation enhancing effects of aqueous chitosan solution in TR146 cells (a human buccal epithelium model), using metformin as a drug model. The results demonstrated that the permeation of metformin was increased by the presence of chitosan, due to its bioadhesive and permeation enhancement effects. Another study proposed a chitosan-based mucoadhesive film for increasing the retention of 5-aminolevulinic acid in the buccal epithelium and its application in the treatment of tumors of the oral cavity (Costa et al., 2014). In similar work, Abruzzo et al. (2012) described the development and characterization of films composed of chitosan and gelatin for buccal delivery of propranolol hydrochloride, showing a relationship between the chitosan concentration and a high in vivo residence time in the buccal cavity.

Chitosan-based hydrogels have also been used for buccal drug delivery. Studies have evaluated the influence of the chitosan concentration on physicochemical properties and biopharmaceutical performance. In a recent work, Graciano et al. (2015) incorporated the photosensitizer toluidine blue O into chitosan hydrogels intended for use in buccal cancer photodynamic therapy. Promising results were obtained in terms of physicochemical properties (pseudoplastic behavior, mucoadhesion, pH, etc.), extended drug release, and the ability to induce cell apoptosis.

Other studies have used the simple association of chitosan with different polymers or low molecular weight enhancers, in attempts to modify rheological properties or to combine the mechanisms of action of different penetration enhancers. Nerkar and Gattani (2013) reported the association of linseed mucilage gel with chitosan, Carbopol 934P, and carboxymethyl cellulose, showing that the hydrogel containing mucilage and chitosan presented high mucoadhesion, satisfactory gel strength, sustained venlafaxine diffusion, and increased drug bioavailability. Cid et al. (2012) described another delivery strategy involving the combination of chitosan with laurocapram (a lauric acid-derivative low molecular weight enhancer) in hydrogels for the local buccal delivery of celecoxib, and investigated their mucoadhesion properties and ability to retain the drug in the buccal epithelium layers. Pignatello et al. (2009) described the development of a chitosan-glutamate hydrogel containing lidocaine hydrochloride, for use in painful oral cavity conditions.

Promising results have also been obtained in studies of chitosan structure–activity relationships. Notably, a buccal delivery system composed of catechol-functionalized chitosan was able to control lidocaine release for 3 h (Xu et al., 2015). Elsewhere, thiolated chitosan-based hydrogels were found to present enhanced mucoadhesive characteristics and a satisfactory protein delivery profile (Ayensu et al., 2012). In a similar application, the delivery of insulin across excised rabbit buccal mucosa was enhanced by thiolated dimethyl ethyl chitosan films, which were suggested as potential new formulations for buccal delivery of high molecular weight compounds (Mortazavian et al., 2014).

A different chitosan derivative, carboxymethyl chitosan, was found to be effective as a penetration enhancer in combination with phospholipid-bile salt mixed micelles for delivery of cucurbitacin B, a potent anticancer agent (Lv et al., 2015). Also, for cancer therapy, another study showed the greater antimicrobial effects of an oral gel composed of *N*-succinyl chitosan, eugenol, honey, and sodium hyaluronate, compared to a commercially available formulation. The oral gel induced tissue regeneration for 15 days, in assessment using an oral mucositis animal model (Dhawan et al., 2014).

4 NANOPARTICLES AND NANOSTRUCTURES FOR TRANSBUCCAL DRUG DELIVERY

Traditional dosage forms such as creams, gels, and ointments have several limitations in terms of efficacy and safety, especially concerning topical application at skin or mucosa. The main limitations have already been discussed. Besides the permeation enhancers, nanotechnology can be used in the pharmaceutical field to develop products that are more effective and less toxic to patients and the environment, and that can be economically viable in scaled-up production. These objectives can be achieved by manufacturing products using biocompatible and biodegradable materials.

Due to the lack of effective commercially available products for topical treatment of oral mucosa disease, inflammation, and pain conditions, nanostructured DDS have been studied since the last quarter of the 20th century. These DDS, usually composed of polymeric and lipid-based materials, provide smart systems with multifaceted properties.

In 1960, the first nanotechnological carriers for use in DDS were described. These systems, called liposomes, have become one of the most widely used lipid-based DDS. Thirty years later, SLN (1991) and NLC (2000) systems were created, with higher physicochemical stability and the ability to encapsulate highly hydrophobic drugs (Attama et al., 2012; Ribeiro et al., 2016).

Among the PN, CD has the important property of forming inclusion complexes with different molecules. Almost 100 years after its isolation, CD was first used as a carrier for pharmaceuticals at the end of the 1970s (Loftsson and Duchene, 2007). Limitations of natural CDs, such as limited aqueous solubility, have led researchers to focus on the development of other PN with different morphologies and properties. These include nanocapsules (NC), nanospheres (NS), and polymeric micelles (PM) (Hoffman, 2008).

Although both lipid and PN are still under development, aiming to overcome specific limitations in their use, several important advances have already been achieved. Fig. 4.5 illustrates the evolution of the development of the lipid and PN that are discussed in detail in this chapter, which offer promising DDS for local treatment using transbuccal delivery.

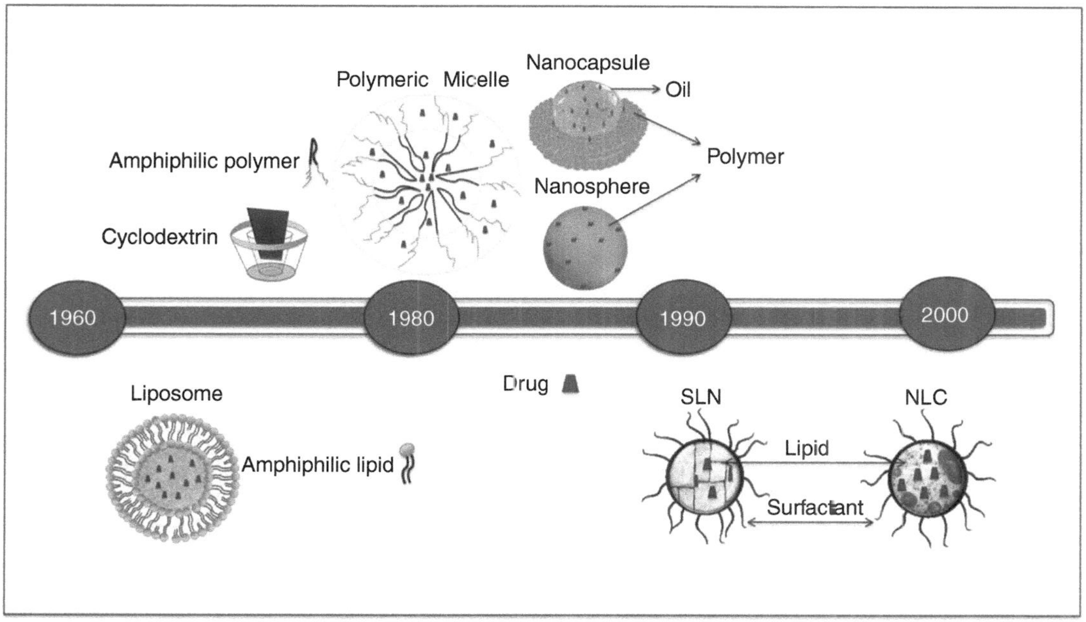

FIGURE 4.5 Schematic Representation of the Timeline of Development of Nanostructured DDS Used in Transbuccal Drug Delivery

4.1 NANOSTRUCTURES

4.1.1 Cyclodextrins

CDs are cyclic oligosaccharides that are able to form inclusion complexes in their macrocyclic ring cavities, which have become very important carriers in drug-delivery systems. They are able to improve the water solubility of hydrophobic drugs (Loftsson and Brewster, 2011; Loftsson and Duchene, 2007), including those used for buccal topical delivery (Hassan et al., 2010; Sudhakar et al., 2006).

Although the natural CDs were discovered in 1890, it was almost a hundred years later (Fig. 4.5) that their application as a DDS was first reported for the sublingual delivery of prostaglandin E2 (Loftsson and Duchene, 2007). In addition, CDs have also been studied as permeation enhancers, as discussed previously.

Many works in the literature have described the advantages of using CDs as drug-delivery systems for local anesthetics (for a review, see de Paula et al., 2012). However, only a few studies have described their use in transbuccal delivery. Recent work has investigated the complexation of tetracaine (TTC) with βCD and hydroxypropyl-βCD (HP-βCD) (de Lima et al., 2012), and bupivacaine (BVC) with HP-βCD (Serpe et al., 2014), for infiltrative anesthesia in rats. In the first case (de Lima et al., 2012), rats were each given an injection close to the infraorbital foramen, and blockade of the infraorbital nerve was assessed by the aversive response of the rat to upper lip pinching. Complexation with TTC:β-CD and TTC:HP-β-CD resulted in a 36% improvement in the total analgesic effect elicited by a 0.5% TTC solution. In the second case (Serpe et al., 2014), the onset, success, and duration of pulpal anesthesia were assessed after blockade of the inferior alveolar nerve by electrical stimulation. A bupivacaine-HP-βCD complex induced better anesthesia than BVC solution, and the anesthetic properties were similar to those of bupivacaine with epinephrine.

For topical delivery to the mouth, Arakawa et al. (2005) prepared a βCD polymer with low molecular weight (MW 5000, equivalent to 4–5 CDs, crosslinked by the use of epichlorohydrin), which formed 1:1 molar ratio (drug:cyclodextrin) inclusion complexes with lidocaine and ketoprofen, in a mucoadhesive buccal film of hydroxypropyl cellulose or polyvinyl alcohol. The release of lidocaine from the system into artificial saliva (pH 5.7) was not retarded by more than 40%, which was attributed to the low association constant (ca. 10 M^{-1}) determined for lidocaine:βCD, in agreement with a previous study ($K_a = 35\ M^{-1}$, lidocaine:HP-βCD; Moraes et al., 2007). Although complexation did not to improve the solubility of lidocaine (as observed for lipophilic ketoprofen), the complexation of the drug with the polymeric CD acted to prolong the release of lidocaine to the mouth.

The same strategy was adopted by Mura and colleagues (Jug et al., 2010). For the delivery of bupivacaine, 1:1 inclusion complexes were prepared either with βCD or with crosslinked βCD polymers (MW = 4500) to form mucoadhesive buccal films. The in vitro rate of dissolution of bupivacaine in artificial saliva decreased when the drug was complexed with βCD, while complexation of bupivacaine with the βCD polymer increased the dissolution rate. The findings suggested that the tailoring of drug release rates could provide ways of providing a rapid onset of anesthetic action, as in the case of simple dental procedures, or prolonged effects, as required in postsurgical procedures or in the treatment of oral mucositis. Considering that these soluble polymeric CDs are not adsorbed, due to their high molecular mass and high hydrophilicity, they merely serve as temporary carriers that are highly suitable for delivery to the buccal mucosa (Jug et al., 2010).

A sustained delivery, CD-based, local buccal delivery system for the nonsteroidal antiinflammatory drug (NSAID) flufenamic acid was proposed for the treatment of oral mucosa inflammations

(Mura et al., 2010). The antiinflammatory drug was complexed with HP-βCD (1:1 mol.%) in order to improve its dissolution and release rates, and buccal films were prepared with the mucoadhesive polymer chitosan, a copolymer of polyvinyl alcohol and polyethylene, plus plasticizers. The formulation was characterized and presented satisfactory swelling behavior and good mucoadhesive and mechanical properties for the topical administration of flufenamic acid in the oral cavity. It presented a neutral pH and its soft, elastic, and flexible properties enabled it to protect the inflamed tissue, with prolonged in situ permanence.

Miro et al. (2009) proposed the use of CDs in hydrophilic matrices for the oral delivery of diclofenac, due to their modulatory effects not only on drug dissolution rates but also on the aqueous medium-membrane interface environment. NSAID formulations were prepared in polyoxyethylene (PEO) tablets, with and without HP-βCD, and it was shown that the CD increased the release rate of the neutral form of diclofenac, as well as its cumulative flux through artificial and porcine buccal mucosa. It was proposed that CDs in PEO polymeric systems could be used for the delivery of triamcinolone acetonide to the oral cavity. An increase was observed in the release rate of the corticosteroid present in simulated buccal fluids, and it was concluded that the strategy of combining CDs with PEO films provided effective buccal drug delivery (Miro et al., 2013).

In work by Choi et al. (2014), the anticancer drug Paclitaxel was complexed with methyl-βCD and prepared in a hydrogel form for the treatment of oral cancers. Thermoreversible (Pluronic F127) and mucoadhesive (PEG) polymers were used to prepare hydrogels. The topical formulation prolonged the release of the anticancer drug and reduced the intrinsic toxicity of Paclitaxel in cultured KB cells (derived from an epidermal carcinoma of the mouth). In this combined approach, methyl-βCD acted as a solubility enhancer for the relatively insoluble Paclitaxel in the development of a topical oral delivery system.

The transmucosal buccal delivery of the antiviral acyclovir as a 1:1 inclusion complex in HP-βCD, using buccal patches, was proposed for the treatment of herpes simplex infections. Different polymers were tested, and higher cumulative release of acyclovir was observed from patches containing the inclusion complex, which increased the bioavailability of acyclovir (Saxena et al., 2011).

Antimycotics intended to act against fungal infections in the oral cavity were complexed with HP-βCD and prepared as a chewing gum (Jacobsen et al., 1999). Release studies conducted using a mastication device showed that complexation with the CD favored the release of both miconazole and econazole from chewing gum. The miconazole:HP-βCD formulation presented higher antimycotic activity against *Candida albicans*, compared to the chewing gum containing miconazole alone.

Two other studies have described CD-based antimycotic systems with great potential for buccal delivery. dos Santos et al. (2010) tested the association of CDs with hydrogels for the delivery of miconazole. Polymers of 2-hydroxyethyl methacrylate were functionalized with α-, β-, and γ-cyclodextrin. The CDs were then able to form inclusion complexes with miconazole. The γ-CD:miconazole-loaded hydrogels were able to completely prevent *C. albicans* biofilm formation, as shown in in vitro microbiological tests, with inhibition following the order: γ-CD > α-CD > β-CD. Also, for the treatment of oropharyngeal candidiasis, Udgirkar Dattatreya et al. (2009) proposed the complexation of ketoconazole with β-CD (1:1 mol.%), for administration as a buccoadhesive tablet of hydroxypropyl methylcellulose and Carbopol. Permeation experiments revealed that the strategy was suitable for optimizing the release of the antimycotic to the excised oral mucosa of sheep, probably due to the solubilizing effect of complexing ketoconazole with β-CD.

4.2 NANOPARTICLES

4.2.1 Liposomes

Liposomes are lipid-based vesicles that are able to encapsulate hydrophobic drugs (between the lipids that form the lamella) and hydrophilic drugs (in the internal aqueous phase) (Fig. 4.5). They are probably the most well-known and widespread drug delivery carriers (Allen and Cullis, 2013), including for oral applications (Erjavec et al., 2006; Hearnden et al., 2012; Smistad et al., 2007).

The encapsulation of local anesthetics in extruded (400 nm) liposomes composed of egg phosphatidylcholine (PC), cholesterol (Chol), and α-tocopherol (α-Toc), at a ratio of 4:3:0.07 mol.%, was investigated by Franz-Montan et al. (2007). In a double-blinded and randomized crossover study, liposomal ropivacaine (1%) in a gel was topically applied to the maxillary-buccal fold of the right canine, reducing the pain caused by needling (as evaluated using the visual analog scale, VAS), as well as increasing the duration of soft tissue anesthesia in the oral mucosa of 30 healthy human volunteers, compared to an equivalent ropivacaine solution, EMLA cream (a mixture of 2.5% lidocaine and 2.5% prilocaine), or 20% benzocaine gel.

Similar results were observed for a 10% benzocaine liposomal formulation, which elicited the same response (using the VAS scale), when applied in a Carbopol gel, as a commercial 20% benzocaine gel (Franz-Montan et al., 2010a). In another randomized, single-blinded, placebo-controlled (2.5% lignocaine/2.5% prilocaine) crossover study, liposome-encapsulated 2% ropivacaine gel was applied (for 30 min) in the buccal fold of the maxillary lateral incisors region of 40 volunteers, followed by insertion of a dental needle until periosteum contact (at the same topical application site), and the pain was rated using the VAS. Use of the liposomal ropivacaine resulted in lower pain response to needle insertion, and longer soft tissue anesthesia, compared to ropivacaine gel, with performance similar to that of the commercial 20% benzocaine formulation (Franz-Montan et al., 2010b). Nevertheless, neither the liposomal anesthetic formulations nor any of the other tested formulations were able to reduce the pain caused by the insertion of a needle in the palatal mucosa (Franz-Montan et al., 2012).

Zed et al. (1996) found that the use of liposome-encapsulated tetracaine (5%), topically applied in the oral mucosa of volunteers prior to injection of a local anesthetic, provided more pain relief (reducing the discomfort of needle penetration and injection of 4% prilocaine), compared to the use of 20% benzocaine gel.

In a double-blinded, crossover, placebo-controlled study, Franz-Montan et al. (2015) evaluated liposomal lidocaine (2.5 and 5%) gel formulations in terms of their anesthetic efficacy in the palatal mucosa of 40 healthy volunteers, as well as their in vitro permeation across pig palatal mucosa. Compared to 5% lidocaine ointment and EMLA, the liposomal lidocaine gel presented better in vivo topical anesthetic efficacy, and the permeation parameters (flux and permeability coefficient) showed that it was suitable for anesthesia of the oral mucosa.

Paphangkorakit et al. (2012) described buccal anesthesia using a formulation containing 2% lidocaine plus epinephrine in liposomes (egg PC:Chol, 1:1 mol.%) prepared using 1 min sonication with an ultrasonic dental scaler. The gel was applied prior to palatal injection in 22 volunteers, and the pain-relieving effect was compared to that of a control (18% benzocaine/2% tetracaine gel), using the VAS method. A split-mouth design was adopted, so that the degree of pain could be compared for the same subject, but the protocol required repeated pinpricks, and the tests could not be blinded. The success of the liposomal formulation was reflected in the pain score (on a scale of 0–10), which decreased from 4.8 with the lidocaine control gel to 4.1 with the liposomal 2% lidocaine + epinephrine formulation.

Niosomes are a special type of liposome prepared with nonionic surfactants. The only recent report describing the use of niosomes for oral transmucosal drug delivery concerned the analgesic and antiinflammatory treatment of pain related to oro-dental diseases (Abdelbary and Aburahma, 2014). A mucoadhesive proniosomal gel containing the NSAID lornoxican was prepared by dissolving phosphatidyl choline, cholesterol, and surfactant (using a PC:Chol:surfactant ratio of 1:0.1:1 mol.%) in a minimum amount of water plus ethanol. The provesicles were spontaneously converted to niosomes following hydration with saliva. Nonionic detergents from the Span, Tween, Brij, Cremophor, and Pluronic series were tested. Span 60–based niosomes (~250 nm) presented higher encapsulation efficiency (>99%), satisfactory rheological and mucoadhesive behavior, and better performance in the delivery of lornoxicam through bovine buccal mucosal membrane, compared to Carbopol gel with lornoxicam at the same dosage (8 mg).

Corticosteroids are used to treat the discomfort of mouth sores. Harsanyi et al. (1986) tested the potential of liposomes as carriers of drugs to ulcerated oral mucosa, with encapsulation of radioactive triamcinolone acetonide palmitate in large multilamellar lipid vesicles (dipalmitoyl PC:Chol, 2:1 mol.%). Hamsters with multiple confluent ulcers in both cheek pouches were treated by topical application or intramucosal injection of the liposome formulation. Overall, use of the liposomes increased the local drug concentration and decreased the systemic drug concentration, but the mean drug concentrations in the ulcerated mucosa were higher in animals that received the intramucosal injection, compared to topical application.

Triamcinolone has also been encapsulated in hydrogenated soybean multilamellar liposomes (PC:Chol = 1:0.5, 450 nm) in an acrylic acid copolymer mucoadhesive ointment for the treatment of aphthous ulcers (canker sores) in hamster cheek pouches (Sveinsson and Holbrook, 1993; Sveinsson and Mezei, 1992). The local drug concentration was significantly higher than that achieved for the control (free corticoid applied in ointment form). Dexamethasone was encapsulated in such liposomes and then applied in three groups of rats: topical application, intramucosal injection, and control groups. Buccal mucosa ulcers were induced by silver nitrate, and the rats were killed at 3 and 24 h after application. Ulcerated buccal mucosa, intact adjacent mucosa, and distant mucosa were excised and analyzed, and liposomes were found to increase the local drug concentration and decrease the systemic concentration. Furthermore, the liposomes acted to maintain the drug in the ulcerated area, demonstrating their utility in the treatment of oral ulcers (Farshi et al., 1996).

Antibiotics such as doxycycline and minocycline have been encapsulated in liposomal formulations designed for topical treatment of periodontitis in rats (Jin et al., 2010; Liu et al., 2013). Gingival index, probing depth, and histopathological analysis of mononuclear and broken bone cells revealed that liposome-encapsulated antibiotics were more effective, compared to control groups, in the control of periodontitis. Di Turi et al. (2012) described the development of extruded (200–300 nm) stealth liposomes, empty or containing $FeCl_2$ and $FeCl_3$ (magnetic liposomes), designed for the delivery of antibiotics into dentinal tubules. It was reported that these liposomes were able to penetrate deeply into the tubules, sometimes reaching their terminal ends.

Another treatment proposed for periodontitis involved the use of liposome-encapsulated sensitizers for photodynamic therapy (Ko et al., 2013). Positively charged extruded liposomes (dioleoyl trimethylammonium propane:palmitoyloleoyl PC, 1:1 mol.%) containing zinc phthalocyanine were designed to act against bacteria. Since bacterial membranes are highly anionic, cationic liposomes are expected to attach to the bacterial biofilm, increasing the photosensitizer concentration at the site of action over a prolonged period. Following illumination, the colony-forming units of *Porphyromonas gingivalis* decreased, showing that localization of the photosensitizer at the bacteria surface (by means of liposome binding) enhanced the photodynamic cytotoxicity of the phthalocyanine. This approach could be exploited for localized delivery to the periodontal pocket and to plaque bacteria.

Lactoferrin and lactoperoxidase have been liposome-encapsulated for the control of development of oral microbiota and caries in rats (Martínez-Gomis et al., 1999). While lactoperoxidase is known to reduce the adherence of bacteria, lactoferrin has effects that are both bacteriostatic (attributed to iron sequestration, reducing its availability for microorganisms) and bactericidal (by direct interaction between bacterial walls and apolactoferrin). The drugs encapsulated in small unilamellar vesicles of soybean phosphatidylcholine were topically applied in each side of the mouth of rats. The incidence of caries in the animals treated with liposome-encapsulated lactoferrin and lactoperoxidase was significantly lower than in those treated with lactoferrin and lactoperoxidase in solution, or with NaF.

4.2.2 Lipid carriers

Thanks to a memorable meeting between Professor Rainer H. Müller and Rosa M. Gasco, these lipid nanosystems subsequently became widely studied. Important work on this theme was also conducted by the research group led by Professor J. S. Lucks (Müller et al., 2011). Based on these efforts, at the end of the 1990s SLNs and NLCs were invented in order to address limitations in DDS. The new materials included polymeric micro- and nanoparticles, fat emulsions, and liposomes (Muller et al., 2000), whose applications as nanocarriers in DDS have grown exponentially, including in approaches using topical release.

SLN, considered the first generation of lipid nanoparticles, consist of a solid lipid phase dispersed in an aqueous phase containing one or more surfactants (charged or anionic) (Bunjes, 2011). This ordered internal arrangement hinders efficient encapsulation and causes drug expulsion during storage, due to the reduced mobility of molecules. The second generation of lipid carriers, NLC, has a different lipid phase, which is composed of a blend of solid and liquid lipids at room temperature, providing a less-ordered internal organization that provides higher encapsulation capacity and long-term stability (Morales et al., 2015).

There are several methods that have been described for the production of lipid nanoparticle systems, such as solvent injection, evaporation, and diffusion, microemulsion dilution and cooling, coacervation, phase inversion temperature techniques, gas-assisted melted atomization, supercritical fluid extraction of emulsions, membrane contactors, electrospraying, cryogenic micronization, spray-drying, spray-congealing, ultrasonication, and high pressure homogenization (HPH) (Battaglia and Gallarate, 2012). The last technique has been more widely employed, due to its many desirable features, such as simplicity and short preparation time, easy scaling-up, and no use of organic solvents (Pardeike et al., 2009). Nonetheless, the HPH method has certain drawbacks, such as the use of high temperatures and cavitation forces, which can cause mechanical and thermodynamic stress in the resulting colloidal system.

The physicochemical characterization of SLN/NLC is mainly based on the analysis of structure-dependent properties. These include particle size, polydispersity index (PDI), and particle surface charge expressed as zeta potential (ZP) values, which can be determined using photon correlation spectroscopy or nanoparticle tracking analysis. The morphology of such systems can be characterized using transmission electronic microscopy (TEM). Particle sizes in these systems are generally in the range between 50 and 1000 nm. A monodisperse distribution (PDI < 0.2) and ZP values of around ± 15–50 mV (depending on the stabilizer used) are desirable, because they indicate good to excellent

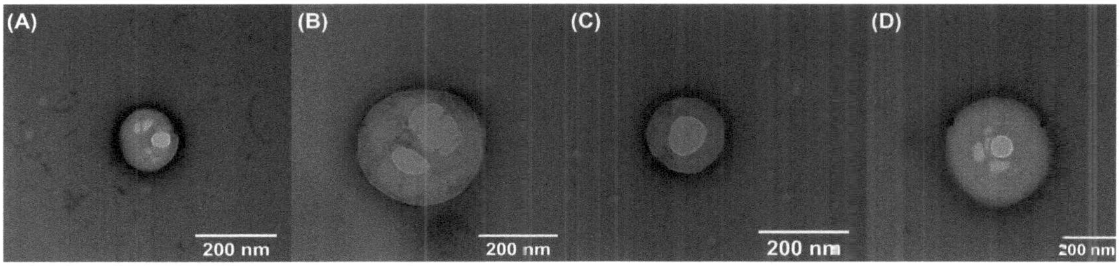

FIGURE 4.6

(A) TEM images of different formulations composed of cetyl palmitate-SLN, (B) myristyl myristate-SLN, (C) cetyl palmitate and caprylic triglycerides-NLC, and (D) myristyl myristate and caprylic triglycerides-NLC as the lipid phase, and poloxamer 188 as the aqueous phase in all the systems (magnification of 100,000×; scale bar: 200 nm).

physical stability (Attama et al., 2012). The morphology of SLN/NLC is generally spherical, with a clearly delimited surface (Barbosa et al., 2013; Ribeiro et al., 2016), as can be seen in the TEM images (Fig. 4.6) of four different formulations of SLN (A and B) and NLC (C and D).

These lipid nanoparticles are potentially useful as nanocarriers for topical applications at the oral mucosa. They can improve permeation due to their nanometric sizes and consequently high-contact surfaces, ability to encapsulate hydrophobic drugs, and good physicochemical stability. They also offer biocompatibility and biodegradability, enabling sustained therapeutic action. However, colloidal systems are very fluid, which can be a disadvantage when it is desired to maintain contact at a specific site (such as the oral mucosa). The incorporation of appropriate matrices can be used in order to circumvent this potential limitation and increase viscosity, firmness, and stability (Souto et al., 2004). Special properties have been obtained by the incorporation of these colloidal dispersions into different (bio) polymeric matrices, with the preparation of pharmaceutical forms including films, sponges, patches, and hydrogels (Wong et al., 2004), which can act as smart drug-delivery systems.

To date, few works have described the use of SLN/NLC systems for topical application at the oral mucosa. Lipid nanoparticles loaded with the antifungals miconazole and clotrimazole have been successfully used as topical delivery systems. Mendes et al. (2013) proposed a colloidal system based on miconazole-loaded NLC for the treatment of oral candidiasis. In terms of the capacity to solubilize miconazole, the most suitable excipients were found to be Gelucire 43/01 as a solid lipid and Capryol PGMC and Mygliol 812 as liquid lipids, plus Tween 80 as nonionic surfactant. Particle size, PDI, and encapsulation efficiency (%EE) were around 200 nm, 0.230, and >87%, respectively, indicative of good physicochemical properties. The in vitro antifungal activity was evaluated using the agar well-diffusion method, with comparison of the minimum inhibitory concentrations (MIC) of the pure drug, a commercial miconazole gel, and miconazole-loaded NLC dispersions (incorporation in hydrogels), which confirmed the antifungal activity of the latter. The dialysis bag diffusion technique was used to determine in vitro drug release, with sustained diffusion observed for miconazole-loaded NLC prepared as a hydrogel, indicating that the formulation was a potential candidate for the treatment of oral fungal disorders.

Esposito et al. (2013) described the in vitro analysis of clotrimazole-loaded NLC and other formulations for the treatment of fungal infections. Dispersions of NLC were prepared by the ultrasonication method, with tristearin and miglyol employed as structural lipids and poloxamer 188 as surfactant. The colloidal system presented suitable particle size (~198 nm), PDI (~0.200), and drug recovery (~97%). Viscosization of the system with poloxamer 407 resulted in satisfactory characteristics. In vitro evaluation of the diffusion of clotrimazole, using the Franz-cell diffusion method, revealed a suitable sustained release profile, with the NLC and polymeric 3D network matrices acting as barriers to prevent fast release of the drug. Determination of in vitro anticandidal activity, employing *C. albicans* as a test strain, revealed effective activity against the microorganism, with a substantial microbial decrease of around 65% for clotrimazole-NLC dispersions. The formulation also presented good physicochemical stability.

Recurrent aphthous stomatitis is a disease characterized by one or more ulcers, commonly present in nonkeratinized oral mucosa, which cause considerable pain and can be related to autoimmune conditions. Karavana et al. (2012) developed a potential candidate formulation for the treatment of recurrent aphthous stomatitis at the oral mucosa, based on cyclosporine A-loaded SLN incorporated into different polymeric matrices to form biocompatible bioadhesive gels. The optimal formulation found was composed of Compritol 888 ATO blended with cyclosporine A as the lipid phase, and poloxamer 188 and Tween 80 as the aqueous phase. The size, PDI, ZP, and EE% values were 203 nm, 0.374, −26.5 mV, and >94%, respectively. Investigation of the in vitro drug release from the gels was carried out using Franz diffusion cells with cow buccal tissue, and the results revealed that 72% of the initial drug dose was present in the extracted tissues, indicating the suitability of this DDS. Evaluation of in vivo efficacy using an oral ulcer model revealed the healing of buccal ulcers in rabbits, and histological analysis showed that there was a significant decrease in ulcer extent, with the rate of tissue repair increasing following treatment with the gel containing cyclosporine-SLN. The formulation remained on the rabbit buccal mucosa for up to 6 h after administration.

Polymyxins are a class of gold standard antibiotics against resistant Gram-negative infections, and are commonly used in the treatment of oral mucositis. Given the increase in bacterial resistance against antibiotics, the development of DDS is essential in trying to solve this global public health problem. A versatile strategy to overcome the major limitation of lipid nanoparticles, namely the difficulty in achieving efficient encapsulation of hydrophilic molecules, has been proposed by Severino et al. (2015). Using ionotropic gelation between polymyxin B sulfate and sodium alginate loaded in SLN (composed of cetyl alcohol as the lipid phase and polysorbate 20 and trioleate sorbitan as the aqueous phase), it was possible to prepare an optimized system with high %EE (~83%). Elevated antimicrobial activity was observed in in vitro susceptibility tests, and cytotoxicity was lower than for the pure drug. The optimal SLN formulation presented size and PDI of around 439 nm and 0.241, respectively, with a slight decrease of ZP from −69 to −58 mV in the presence of polymyxin B sulfate-loaded sodium alginate.

In other work with modified SLN/NLC systems, Rao (2007) patented a SLN system that included therapeutic ligands or peptides at the surfaces of lipid nanoparticles, which in turn were covered with carbohydrates or lectins. The authors claimed that these conjugated systems optimized uptake by the mucosal route, and that they were suitable for the delivery of drugs and vaccines at oral, sublingual, and buccal sites. The resulting polymerized SLNs showed the ability to treat or prevent diseases, with reproducible composition and methods.

SLNs have shown potential for the treatment of neoplasms. In pioneering work, Holpuch et al. (2010) proved the feasibility of SLN internalization by monolayer-cultured human oral squamous cell carcinoma, and demonstrated in vitro penetration of the SLN in the basal and connective tissue layers of healthy mucosa. For these purposes, four human cell lines obtained from squamous cell carcinomas (which retained features of the oral mucosa) were employed for the evaluation of internalization using luminescent microscopy. The penetration of healthy mucosa was analyzed using human oral tissues collected from 25 normal individuals that were submitted to elective oral surgical procedures. Ten samples were used for histological analysis of SLN penetration, and the remaining samples were used for Western blot and immunohistochemistry analyses (phase III efflux enzyme expression). The results revealed that the SLN systems presented excellent hemocompability, the ability to be internalized and penetrate mucosal tissues, and suitability for use as carriers to deliver chemopreventive agents at the oral mucosa. The SLN were based on emulsifying wax and Brij 78 as structural lipid and stabilizer, respectively, and incorporated idarubicin or BODIPY FL C_{12} as model fluorescent probes. The average size, PDI and ZP values were 90 nm, 0.100 and −14 mV, respectively.

An emerging approach in DDS is the incorporation of phytochemicals in nanostructured systems. Curcumin (Cur), 1,7-*bis*(4-hydroxy-3-methoxyphenyl)-1,6-heptadiene-3,5-dione, an active compound obtained from the rhizomes of *Curcuma longa*, exhibits many useful pharmacological properties, including antimicrobial, antitumor, antioxidant, and antiinflammatory activities. However, its clinical success has been hindered by its hydrophobicity, high sensitivity to neutral and basic pH, and photochemical instability, as well as by its rapid metabolization and systemic elimination. The encapsulation of curcumin in SLN/NLC systems, therefore, offers a way of improving its bioavailability and physicochemical stability.

Hazzah et al. (2015a) reported the development of curcumin-loaded SLN, targeting the oral mucosa. Three recent papers described the stages from preformulation up to short-term clinical evaluation of these colloidal systems. The first work concerned the development and characterization of optimized Cur-SLN systems and their antimicrobial properties. The preformulation studies analyzed the effects of differences in lipid concentration and type, as well as the use of two different surfactants, in lipid nanoparticles prepared by a hot homogenization method. The optimum formulation for loading with curcumin employed Gelucire 50/13 as structural lipid and poloxamer 407 as stabilizer. The mean particle diameter, PDI, ZP, and %EE were 290 nm, 0.470, −15 mV, and 89%, respectively. Evaluation of mucoadhesion, using the modified mucin particle method, showed that the best adhesion was achieved with Cur-SLN, corroborating the in vitro mucoadhesion, permeation, and retention tests (carried out using chicken buccal mucosa). These findings highlighted the role of SLN as a protective host that stabilized and improved the bioavailability of curcumin, enhancing its resistance in the oral cavity medium. In vitro dialysis experiments confirmed the sustained release profile of curcumin, and the in vitro antimicrobial activity was higher for Cur-SLN than for the controls (Hazzah et al., 2015a). The optimized Cur-SLN formulation possessed suitable physicochemical properties (such as bactericidal effects) for preparation in the forms of sponges and gels.

Hazzah et al. (2015b) also described the preparation of Cur-SLN loaded in a mucoadhesive gel for the in vivo treatment of local precancerous lesions at the oral mucosa. Use of the mucoadhesive gel enabled a significant amount of curcumin to be retained in chicken mucosa, as confirmed by histological analysis. Good mucoadhesion properties were observed, but the residence time was fairly short (25 min). In addition, short-term clinical effects were evaluated in 10 patients presenting oral erythroplakia. The results showed a substantial reduction of pain, followed by complete healing after 6 weeks, even though the curcumin concentrations employed were lower than the usual oral dose.

Finally, Hazzah et al. (2015c) demonstrated the successful incorporation of the Cur-SLN formulation described earlier in several polymeric matrices designed as sponges, in order to improve the in vivo residence time. The best results were found for Cur-SLN-loaded sponges composed of hydroxypropyl methylcellulose (HPMC 4000) or polycarbophil, with glycerol as a plasticizer and mannitol as a cryoprotectant, which resulted in an effective sponge-like structure. The mucoadhesion residence time was assessed in vivo in six healthy human volunteers, with placement of the selected sponge on the gingiva (upper canine region) until absorption was complete. The residence time was almost four times longer for Cur-SLN-loaded sponges produced with a polycarbophil matrix, compared to those produced with HPMC 4000. This was probably due to differences in the swelling properties of the materials, with greater hydration increasing the interaction with mucin. Curcumin release was evaluated in vivo with five healthy human volunteers. At predetermined intervals, aliquots of saliva were collected for spectrophotometric quantification of curcumin. The most mucoadhesive sponge showed the best sustained release, and elevated levels of curcumin were detected in the saliva after 14–15 h, even after destruction of the sponge. It was concluded that use of this optimized Cur-SLN-loaded formulation resulted in excellent short-term clinical effects, contributing to scientific advancement in the treatment of microbial infections at the oral mucosa.

4.2.3 Polymeric nanoparticles

Various types of nanocarriers have been employed for drug delivery, but micelles and PN, especially, have potential for use in mucosa (Holpuch et al., 2010; Liu et al., 2015; Sosnik and Menaker Raskin, 2015).

PN are one of the types of nanocarriers that can be produced using biodegradable and biocompatible materials, with sizes between 10 and 1000 nm (Samiei et al., 2016). They can divided into NC and NS (Fig. 4.5). NC are vesicular systems comprising an oily core surrounded by a polymeric wall, while NS consist of a polymer matrix structure. The materials used for NP production, which may be of natural or synthetic origin, include chitosan, alginate, gelatin, polycyanoacrylate, PEO/polyethylene glycol (POE/PEG), poly-ε-caprolactone (PCL), poly-D,L-lactide (PLA), and poly-D,L-lactide-*co*-glycolide (PLGA), among others (Calixto et al., 2014; Estanqueiro et al., 2015).

It is difficult to characterize these systems, due to the colloidal nature of the particles. Characterizations generally consider morphology (analyzed using atomic force microscopy or TEM), suspension pH, ZP, hydrodynamic diameter, and size distribution (using photon correlation spectroscopy), nanoparticle concentration (using nanoparticle tracking analysis), encapsulation efficiency and drug loading, release kinetics, and physicochemical stability. Many of the techniques used enable assessment of the interactions between drugs and nanoparticles, providing valuable information during the preformulation stage (Malloy, 2011; Mora-Huertas et al., 2010; Pathak and Preuss, 2013).

Recent studies have aimed at understanding the dynamics of interactions between the buccal mucosa and PN (Mazzarino et al., 2014; Roblegg et al., 2012), providing information important in the development of PN for drug delivery in buccal mucosa. There is widespread use of PN in the buccal cavity for treatment of periodontal disease, prevention of biofilm formation, reducing hypersensitivity of the teeth, local anesthesia, and cancer treatment. The use of these nanocarriers for drug release in the buccal cavity has gained attention due to the advantages they can provide, such as modulation of the release profile, protection of labile drugs, improved bioavailability, and mucoadhesion, among others (Mazzarino et al., 2014).

In an investigation of the dynamics of PN transport through the buccal mucosa, Roblegg et al. (2012) conducted a study employing nanoparticles of carboxyl polystyrene (20 and 200 nm, negative surface charge) and amino polystyrene (200 nm, positive surface charge). Permeation through excised porcine buccal mucosa was measured using a vertical Franz diffusion apparatus. The cellular route was evaluated in the H376 cell line. The results indicated that the biological medium influenced the dispersion of PN, with all the particles becoming aggregated in the presence of simulated saliva. The amino polystyrene nanoparticles were able to permeate the mucosal layer and reach the deepest regions, while the 200 nm carboxyl polystyrene nanoparticles were aggregated and retained in the mucosal lamina propria. On the other hand, the 20 nm nanoparticles were able to permeate mucus, but were unable to traverse the superficial layers of the epithelium. In vitro experiments performed using cell cultures and excised buccal epithelium showed that the PN permeation route followed a transcellular pathway, and that both surface charge and size influenced the permeation. Hence, particles with positive charge and diameters in the region of 200 nm presented improved permeation ability in the buccal mucosa.

Piñón-Segundo et al. (2005) developed and characterized NS produced with PLA, PLGA, and cellulose acetate phthalate, containing triclosan as an antimicrobial agent at different concentrations. Physicochemical characterization showed that NS with sizes between 176 and 458 nm were obtained, with %EE of triclosan between 63% and 89%. In vitro tests revealed that drug release was significantly faster for all the NS tested, with diffusion being the main release mechanism. In vivo studies in beagle dogs showed that PLGA nanoparticles loaded with triclosan were able to reduce the gingival inflammation induced by periodontitis.

Sandri et al. (2006) studied the penetration ability of chitosan in excised porcine buccal mucosa, using three different forms: chitosan solution, chitosan nanoparticles (CN), and trimethyl chitosan. The systems were loaded with fluorescein isothiocyanate dextran, as a model molecule. Evaluation was made of the morphology and histology of tissues after contact with the formulations, using optical and TEM techniques. The location of fluorescein in the tissues was determined by confocal laser scanning microscopy. The permeation pathway of the drug was assessed in Franz diffusion cells, using isolated buccal epithelial cells. The mean diameter of the CN was around 233 nm. The penetration pathway study suggested that permeation of all three systems tested followed the paracellular route. It was observed that chitosan could destabilize the lipid organization in the epithelium, facilitating permeation, without causing any cell damage, as already discussed. It was also evident that CN and trimethyl chitosan increased fluorescein permeation through buccal epithelium, compared to the plain chitosan solution. Hence, the combination of increased permeation and the mucoadhesive properties of chitosan makes these nanoparticles promising carriers for the development of formulations for buccal use (Sandri et al., 2006).

Other studies have described the use of PN in buccal mucosa, but most have focused on the effects of surfactants present in the PN on the mechanisms of interaction among the PN, mucus, and the epithelial layer. McCarron et al. (2004, 2007) carried out two in vitro studies of the effects of PN in modulating the adherence of *C. albicans* blastospores to human buccal epithelium. In the first study, PNs composed of poly(propylcyanoacrylate) were produced and incubated with the stationary phase of blastospores, which were then incubated with mucosal epithelial cells. The particle size range was between 80 and 7000 nm, depending on the surfactant used. The results showed that adsorption of PN by the blastospores increased their hydrophobicity, reducing their adhesion to epithelial cells by 73% in the case of nanoparticles stabilized with sodium docusate (~183 nm), compared to untreated blastospores. The antiadherent effect could also be explained by steric hindrance of the adsorbed nanoparticles (McCarron et al., 2004).

In a second study, the antiadherent and antifungal activities of PN produced with poly(ethylcyanoacrylate) were evaluated. Human buccal epithelium was collected from 10 healthy volunteers and the cells were pretreated with PN, with the particles being incubated in cell suspensions.

The adhesion of *C. albicans* was assessed after the incubation process. Subsequently, evaluation was made of the antifungal activity of the PN due to the peripheral release of surfactants present on the nanoparticle surface. The PN were produced with cationic, anionic, and nonionic surfactants. The antifungal effect was evaluated employing two methods (zone of inhibition and viable cell counts). The cationic surfactant cetrimide was most effective in preventing adherence (~52%) and presented the greatest antifungal activity, with larger zones of inhibition. This DDS could be promising for mouthwash use, in view of the antiadhesion and antifungal effects of the nanoparticles (McCarron et al., 2007).

Mazzarino et al. (2014) evaluated mucoadhesive films containing PN coated with chitosan for the buccal release of curcumin, as described earlier. PCL nanoparticles coated with chitosan containing curcumin were prepared and impregnated in films of chitosan with different molecular weights. The PN and films were both characterized, and in vitro curcumin release tests were performed using simulated saliva solution. The mean diameter of the nanoparticles ranged between 218 and 256 nm, with %EE of 98%. Homogeneous films with uniform dispersion of PN and excellent physicochemical properties were obtained. The in vitro release of curcumin was greater for the films containing PN, compared to the films containing the free drug, probably due to the low solubility of the drug in the film. The nanoencapsulation was advantageous in terms of the increased amount of drug available for permeation through the film, and was suitable for use in the case of periodontal diseases requiring prolonged drug release (Mazzarino et al., 2014).

Wang et al. (2014) studied nanoparticles composed of monomethoxy poly(ethylene glycol)-poly(ε-caprolactone) containing naringenin, embedded in mucoadhesive tablets manufactured using hydroxypropylmethylcellulose and a milk protein concentrate. The aim of the encapsulation was to increase drug solubility for the treatment of inflammatory and ulcerative diseases of the oral cavity, such as periodontitis. The nanoparticles were prepared and characterized, and incorporated into the tablets by freeze-drying. In vitro release was evaluated using a USP I apparatus to simulate the in vivo buccal cavity conditions. The mean diameters of the NP before and after lyophilization were 53.9 and 54.6 nm, respectively, and the %EE was up to 95%, indicating good interaction between the nanoparticles and naringenin. The release of naringenin from the buccal nanoparticle tablets was influenced by pH (faster at pH 7.0 than at pH 5.5), and was faster and more complete, compared to control tablets containing free naringenin. The findings showed that encapsulation was able to increase the solubility and diffusion of naringenin, enhancing its bioavailability (Wang et al., 2014).

Yao et al. (2015) studied the antimicrobial agent minocycline in normal PEG-PLA and in a surface modified with RGD peptide nanoparticles, with the aim of treating periodontitis in dogs. The cellular uptake and internalization of nanoparticles were evaluated in Calu-3 cells. The nanoparticles were applied directly in the bottom of the periodontal pocket, and the efficacy was evaluated using clinical periodontal parameters such as plaque index, gingival index, clinical attachment level, and periodontal pocket depth. The evaluations were conducted on the first day, and then 7 and 14 days after treatment. The animals were divided into five groups (saline, free minocycline, nanoparticles loaded with the drug with and without peptide, and the commercial product Periocline). The NP with and without peptide showed mean diameters of 98.3 and 106 nm, respectively, with satisfactory colloidal parameters and monocycline %EE of ~45%. In vitro release was slower and more sustained for the RGD-modified NP (14 days), compared to the free drug (2 days). Inclusion of the peptide increased the uptake of the nanoparticles by epithelial cells, suggesting intracellular transport by lysosomes and the Golgi apparatus. The RGD-modified NP were able to maintain an effective drug concentration in the gingival crevicular fluid for more than 2 weeks, which was longer than achieved with the controls. The dogs showed reduced symptoms of periodontitis, compared to the control group (Yao et al., 2015).

4.2.4 Polymeric micelles

PM (Fig. 4.5) are spherical colloidal particles with diameters between 20 and 100 nm (Cabral and Kataoka, 2014; Kwon, 2003), formed by a structure of amphiphilic block copolymers, above their critical micelle concentration, surrounding a hydrophobic core that carries the active substance (Thipparaboina et al., 2015). They are typically composed of copolymers of polyethylene glycol (PEG, PEO), polyoxypropylene, poly(isopropyl acrylamide), and PCL, among others (Sosnik and Menaker Raskin, 2015). In recent years, PM have emerged as carriers for drugs that are weakly soluble in water molecules, because their hydrophobic cores present high loading capacities (Lu and Park, 2013; Thipparaboina et al., 2015).

Due to their unique characteristics, PM have been identified as candidates for drug delivery to mucous membranes, but their low residence time imposes limits on their use, making it necessary to change their structure and method of administration (Liu et al., 2015; Sosnik and Menaker Raskin, 2015). The interaction of PM with the mucus layer is one way to prolong their presence at the application site, and mucoadhesive PM have shown promise for use in buccal mucosa (Sosnik and Menaker Raskin, 2015).

Finally, Wayakanon et al. (2013) reported the preparation of another type of polymer-based nanostructured carrier, namely polymersomes, which are polymeric vesicles that contain an internal aqueous core different to that of the traditional PM. Polymersomes were prepared with the poly(2-methacryloyloxyethyl phosphorylcholine-poly(2-diisopropylaminoethyl methacrylate) (PMPC25-PDPA70) copolymer and were loaded with the antibiotics metronidazole and doxycycline. The system was proposed for eradication of the intracellular pathogen *P. gingivalis* in oral mucosa models (monolayer and organotypic), using normal oral keratinocyte (H357) and fibroblast (TR146) cell lines. The cells were previously infected with *P. gingivalis*. The polymersomes showed an average diameter of 130 nm, and the concentrations of the antibiotics used were 0.102 μg/mL (metronidazole) and 0.19 μg/mL (doxycycline). This study showed that the incorporation of polymersomes into endosomes was pH-dependent, with release of the active drugs at low pH. The use of polymersomes resulted in a significantly greater reduction in levels of the intracellular pathogen, compared to the free antibiotics, opening new perspectives in the treatment of periodontitis (Wayakanon et al., 2013).

5 LIMITATIONS

Although DDS show considerable promise for improving therapeutic effects, when applied topically at the oral mucosa in the treatment of oral conditions, there are a number of limitations that remain to be overcome in order for the methodology to achieve its full potential. Patient acceptance and adherence to the treatment depend on the ease of application of the DDS. Furthermore, the formulations should have pleasing taste and odor, and should not disturb daily activities such as speaking, chewing, and swallowing. They must not be irritating to the oral mucosa, or be dangerous if unintentionally swallowed. They should also be easily removed when undesired side effects occur (Hearnden et al., 2012; Shakya et al., 2011). Finally, stability issues and costs are always a concern in the development of ideal formulations.

6 CONCLUSIONS

The studies described in this review reflect the importance of pharmaceutical nanotechnology as an emerging field in DDS. Manipulation of materials in the nanometer size range provides these systems with innovative properties that cannot be obtained at larger dimensions. Particular attention should be given to mucoadhesive nanostructured pharmaceutical formulations, composed of drug-loaded nanocarriers associated with (bio)polymeric matrices, specifically designed for oral mucosa applications. The submolecular design of materials can provide solutions to difficulties, including the transfer of drugs through relatively impermeable barriers (such as the stratified squamous oral epithelium), the low availability of hydrophobic drugs, their rapid degradation, and retention of active agents at the desired sites of action, protecting them from the aqueous environment. Despite the considerable promise of these systems, the literature remains sparse concerning the development of CDs, liposomes, SLNs, NLCs, PN, and micellar formulations for use in transbuccal applications. We believe that the findings of the research described in the present review should stimulate the development of new DDS formulations and lead to the provision of even more effective topical delivery systems for use in dentistry.

ACKNOWLEDGMENTS

We are grateful to Andrew George Allen for English revision and to Felipe Blanco for figure preparation. Financial support was provided by the São Paulo Research Foundation (FAPESP, grants 2012/06974-4 and 2014/14457-5).

REFERENCES

Abdelbary, G.A., Aburahma, M.H., 2014. Oro-dental mucoadhesive proniosomal gel formulation loaded with lornoxicam for management of dental pain. J. Liposome Res. 2104, 1–15, Available from: http://informahealthcare.com/doi/abs/10.3109/08982104.2014.941861.

Abdel-Hamid, S.M., et al., 2006. Formulation of an antispasmodic drug as a topical local anesthetic. Int. J. Pharm. 326, 107–118.

Abruzzo, A., et al., 2012. Mucoadhesive chitosan/gelatin films for buccal delivery of propranolol hydrochloride. Carbohydrate Polym. 87 (1), 581–588, Available from: http://dx.doi.org/10.1016/j.carbpol.2011.08.024.

Alfano, M.C., Chasens, A.I., Masi, C.W., 1977. Autoradiographic study of the penetration of radiolabelled dextrans and inulin through non-keratinized oral mucosa in vitro. J. Periodon. Res. 12 (5), 368–377, Available from: http://www.ncbi.nlm.nih.gov/pubmed/143519.

Allen, T.M., Cullis, P.R., 2013. Liposomal drug delivery systems: from concept to clinical applications. Adv. Drug Deliv. Rev. 65 (1), 36–48, Available from: http://dx.doi.org/10.1016/j.addr.2012.09.037.

Arakawa, et al., 2005. Oromucosal delivery of venlafaxine by linseed mucilage based gel: in vitro and in vivo evaluation in rabbits. Arch. Pharm. Res. 36, 846–853.

Attama, A.A., Momoh, M.A., Builders, P.F., 2012. Lipid nanoparticulate drug delivery systems: a revolution in dosage form design and development. In: Sezer, Ali Demir (Ed.), Recent Advances in Novel Drug Carrier Systems. InTech, pp. 107–140, Available from: http://www.intechopen.com/books/recent-advances-in-novel-drug-carrier-systems/lipid-nanoparticulate-drug-delivery-systems-a-revolution-in-dosage-form-design-and-development.

Attia, M., et al., 2004. Transbuccal permeation, anti-inflammatory activity and clinical efficacy of piroxicam formulated in different gels. Int. J. Pharm. 276 (1–2), 11–28, Available from: http://linkinghub.elsevier.com/retrieve/pii/S0378517304000663.

Aungst, B.J., Rogers, N.J., 1988. Site dependence of absorption-promoting actions of laureth-9, Na salicylate, Na2EDTA and aprotinin on rectal, nasal and buccal insulin delivery. Pharm. Res. 5 (5), 305–308.

Ayensu, I., Mitchell, J.C., Boateng, J.S., 2012. Development and physico-mechanical characterisation of lyophilised chitosan wafers as potential protein drug delivery systems via the buccal mucosa. Colloids Surf. B 91 (1), 258–265.

Bahri-Najafi, R., et al., 2013. Formulation of nicotine mucoadhesive tablet for smoking cessation and evaluation of its pharmaceuticals properties. Adv. Biomed. Res. 2, 88.

Barbosa, R.M., et al., 2013. Cytotoxicity of solid lipid nanoparticles and nanostructured lipid carriers containing the local anesthetic dibucaine designed for topical application. J. Phys.Confer. Ser. 429 (1), 012035, Available from: http://www.scopus.com/inward/record.url?eid=2-s2.0-84876853791&partnerID=tZOtx3y1.

Battaglia, L., Gallarate, M., 2012. Lipid nanoparticles: state of the art, new preparation methods and challenges in drug delivery. Exp. Opin. Drug Deliv. 9 (5), 497–508.

Bernkop-Schnürch, A., Paikl, C., Valenta, C., 1997. Novel bioadhesive chitosan-EDTA conjugate protects leucine enkephalin from degradation by aminopeptidase N. Pharm. Res. 14 (7), 1997.

Bernkop-Schnürch, A., Guggi, D., Pinter, Y., 2004. Thiolated chitosans: development and in vitro evaluation of a mucoadhesive, permeation enhancing oral drug delivery system. J. Control. Release 94 (1), 177–186.

Bunjes, H., 2011. Structural properties of solid lipid based colloidal drug delivery systems. Curr. Opin. Colloid Interface Sci. 16 (5), 405–411, Available from: http://linkinghub.elsevier.com/retrieve/pii/S1359029411000835.

Cabral, H., Kataoka, K., 2014. Progress of drug-loaded polymeric micelles into clinical studies. J. Control. Release 190, 465–476.

Calixto, G., et al., 2014. Nanotechnology-based drug delivery systems for treatment of oral cancer: a review. Int. J. Nanomed. 9, 3719–3735.

Caon, T., et al., 2014. Exploiting the buccal mucosa as an alternative route for the delivery of donepezil hydrochloride. J. Pharm. Sci. 103 (6), 1643–1651, Available from: http://www.ncbi.nlm.nih.gov/pubmed/24687969.

Chinna Reddy, P., Chaitanya, K.S.C., Madhusudan Rao, Y., 2011. A review on bioadhesive buccal drug delivery systems: current status of formulation and evaluation methods. Daru 19 (6), 385–403, Available from: http://www.pubmedcentral.nih.gov/articlerender.fcgi?artid=3436075&tool=pmcentrez&rendertype=abstract.

Choi, S.G., et al., 2014. Thermosensitive and mucoadhesive sol-gel composites of paclitaxel/dimethyl-β-cyclodextrin for buccal delivery. PloS one 9 (9), e109090, Available from: http://www.pubmedcentral.nih.gov/articlerender.fcgi?artid=4183572&tool=pmcentrez&rendertype=abstract.

Cid, Y.P., et al., 2012. In vitro characterization of chitosan gels for buccal delivery of celecoxib: influence of a penetration enhancer. AAPS PharmSciTech 13 (1), 101–111, Available from: http://link.springer.com/10.1208/s12249-011-9725-8.

Collins, L.M., Dawes, C., 1987. The surface area of the adult human mouth and thickness of the salivary film covering the teeth and oral mucosa. J. Dent. Res. 66 (8), 1300–1302.

Costa, I.D.S.M., et al., 2014. Chitosan-based mucoadhesive films containing 5-aminolevulinic acid for buccal cancer's treatment. J. Photochem. Photobiol. B 140, 266–275, Available from: http://www.ncbi.nlm.nih.gov/pubmed/25190225.

Dawes, C., et al., 2015. The functions of human saliva: a review sponsored by the world workshop on oral medicine VI. Arch. Oral Biol. 60 (6), 863–874, Available from: http://linkinghub.elsevier.com/retrieve/pii/S0003996915000692.

Dawson, D.V., et al., 2013. Organization, barrier function and antimicrobial lipids of the oral mucosa. Int. J. Cosmet. Sci. 35 (3), 220–223.

de Paula, E., et al., 2012. Micro and nanosystems for delivering local anesthetics. Exp. Opin. Drug Deliv. 9 (12), 1505–1524, Available from: http://www.ncbi.nlm.nih.gov/pubmed/23140102.

Dhawan, N., et al., 2014. *N*-succinyl chitosan as buccal penetration enhancer for delivery of herbal agents in treatment of oral mucositis. Curr. Drug Deliv. 11, 415–425.

Di Colo, G., Zambito, Y., Zaino, C., 2008. Polymeric enhacers of mucosal epithelia permeability: synthesis, transepithelial penetration-enhancing properties, mechanism of action, safety issues. J. Pharm. Sci. 97 (5), 1652–1680.

Di Turi, G., et al., 2012. Sub-micrometric liposomes as drug delivery systems in the treatment of periodontitis. Int. J. Immunopathol. Pharmacol. 25 (3), 657–670, Available from: http://www.ncbi.nlm.nih.gov/pubmed/23058016.

dos Santos, J.-F.R., et al., 2010. Functionalization of acrylic hydrogels with alpha-, beta- or gamma-cyclodextrin modulates protein adsorption and antifungal delivery. Acta Biomater. 6 (10), 3919–3926, Available from: http://www.ncbi.nlm.nih.gov/pubmed/20417319.

Erjavec, V., et al., 2006. In vivo study of liposomes as drug carriers to oral mucosa using EPR oximetry. Int. J. Pharm. 307 (1), 1–8, Available from: http://www.ncbi.nlm.nih.gov/pubmed/16257157.

Esposito, E., et al., 2013. Clotrimazole nanoparticle gel for mucosal administration. Mater. Sci. Eng. C 33 (1), 411–418, Available from: http://linkinghub.elsevier.com/retrieve/pii/S0928493112004390.

Estanqueiro, M., et al., 2015. Nanotechnological carriers for cancer chemotherapy: the state of the art. Colloids Surf. B 126, 631–648.

Farshi, F.S., et al., 1996. In-vivo studies in the treatment of oral ulcers with liposomal dexamethasone sodium phosphate. J. Microencapsul. 13 (5), 537–544.

Franz-Montan, M., et al., 2007. Liposome-encapsulated ropivacaine for topical anesthesia of human oral mucosa. Anesth. analges. 104 (6), 1528–1531, Available from: http://www.ncbi.nlm.nih.gov/pubmed/17513653.

Franz-Montan, M., Silva, A.L.R., et al., 2010a. Liposomal encapsulation improves the duration of soft tissue anesthesia but does not induce pulpal anesthesia. J. Clinic. Anesth. 22 (5), 313–317, Available from: http://www.ncbi.nlm.nih.gov/pubmed/20542419.

Franz-Montan, M., De Paula, E., et al., 2010b. Liposome-encapsulated ropivacaine for intraoral topical anesthesia. Oral Surg., Oral Med., Oral Pathol., Oral Radiol. Endodontol. 110 (6), 800–804, Available from: http://dx.doi.org/10.1016/j.tripleo.2010.07.005.

Franz-Montan, M., et al., 2012. Liposomal delivery system for topical anaesthesia of the palatal mucosa. Br. J. Oral Maxil. Surg. 50 (1), 60–64.

Franz-Montan, M., et al., 2015. Liposomal lidocaine gel for topical use at the oral mucosa: characterization, in vitro assays and in vivo anesthetic efficacy in humans. J. Liposome Res. 2104, 1–9, Available from: http://www.ncbi.nlm.nih.gov/pubmed/24807821.

Franz-Montan, M., et al., 2016. Evaluation of different pig oral mucosa sites as permeability barrier models for drug permeation studies. Eur. J. Pharm. Sci. 81, 52–59, Available from: http://linkinghub.elsevier.com/retrieve/pii/S0928098715300294.

Galey, W.R., Lonsdale, H.K., Nacht, S., 1976. The in vitro permeability of skin and buccal mucosa to selected drugs and tritiated water. J. Invest. Dermatol. 67 (6), 713–717, Available from: http://www.ncbi.nlm.nih.gov/pubmed/1033956.

Giannola, L.I., et al., 2007. Diffusion of naltrexone across reconstituted human oral epithelium and histomorphological features. Eur. J. Pharm. Biopharm. 65 (2), 238–246, Available from: http://linkinghub.elsevier.com/retrieve/pii/S0939641106001640.

Graciano, T.B., et al., 2015. Using chitosan gels as a toluidine blue O delivery system for photodynamic therapy of buccal cancer: in vitro and in vivo studies. Photodiag. Photodyn. Ther. 12 (1), 98–107, Available from: http://linkinghub.elsevier.com/retrieve/pii/S1572100014001392.

Harris, D., Robinson, J.R., 1992. Drug delivery via the mucous membranes of the oral cavity. J. Pharm. Sci. 81 (1), 1–10.

Harsanyi, B.B., Hilchie, J.C., Mezei, M., 1986. Liposomes as drug carriers for oral ulcers. J. Dent. Res. 65 (9), 1133–1141.

Hassan, N., et al., 2010. Chemical permeation enhancers for transbuccal drug delivery. Exp. Opin. Drug Deliv. 7 (1), 97–112, Available from: http://www.ncbi.nlm.nih.gov/pubmed/20017661.

Hayward, A.F., 1979. Membrane coating granules. Int. Rev. Cytol. 59, 97–124.

Hazzah, H.A., et al., 2015a. Gelucire-based nanoparticles for curcumin targeting to oral mucosa: preparation, characterization, and antimicrobial activity assessment. J. Pharm. Sci. 104 (11), 3913–3924, Available from: http://doi.wiley.com/10.1002/jps.24590.

Hazzah, H.A., et al., 2015b. A new approach for treatment of precancerous lesions with curcumin solid-lipid nanoparticle-loaded gels: in vitro and clinical evaluation. Drug Deliv. 00 (00), 1–11, Available from: http:// www.tandfonline.com/doi/full/10.3109/10717544.2015.1065524.

Hazzah, H.A., et al., 2015c. Lyophilized sponges loaded with curcumin solid lipid nanoparticles for buccal delivery: development and characterization. Int. J. Pharm. 492 (1), 248–257, Available from: http://www. sciencedirect.com/science/article/pii/S0378517315005426.

Hearnden, V., et al., 2012. New developments and opportunities in oral mucosal drug delivery for local and systemic disease. Adv. Drug Deliv. Rev. 64 (1), 16–28, Available from: http://linkinghub.elsevier.com/ retrieve/pii/S0169409X11000433.

Hoffman, A.S., 2008. The origins and evolution of "controlled" drug delivery systems. J. Control. Release 132 (3), 153–163, Available from: http://dx.doi.org/10.1016/j.jconrel.2008.08.012.

Holpuch, A.S., et al., 2010. Nanoparticles for local drug delivery to the oral mucosa: proof of principle studies. Pharm. Res. 27, 1224–1236.

Hu, L., et al., 2011. Enhanced in vitro transbuccal drug delivery of ondansetron HCl. Int. J. Pharm. 404 (1–2), 66–74, Available from: http://www.ncbi.nlm.nih.gov/pubmed/21056647.

Ikinci, G., et al., 2000. In vitro/in vivo studies on a buccal bioadhesive tablet formulation of carbamazepine. Pharmazie 55, 415–425.

Jacobsen, J., Bjerregaard, S., Pedersen, M., 1999. Cyclodextrin inclusion complexes of antimycotics intended to act in the oral cavity: drug supersaturation, toxicity on TR146 cells and release from a delivery system. Eur. J. Pharm. Biopharm. 48 (3), 217–224.

Jin, H., et al., 2010. Doxycycline nano-liposome slow-release gel improves rat periodonttis. Shangai J. Stomatol. 19, 508–511.

Jug, M., et al., 2010. Preparation and solid-state characterization of bupivacaine hydrochloride cyclodextrin complexes aimed for buccal delivery. J. Pharm. Biomed. Anal. 52 (1), 9–18.

Karavana, S.Y., et al., 2012. A new approach to the treatment of recurrent aphthous stomatitis with bioadhesive gels containing cyclosporine A solid lipid nanoparticles: in vivo/in vitro examinations. Int. J. Nanomedicine 7, 5693–5704, Available from: http://www.pubmedcentral.nih.gov/articlerender.fcgi?artid=3497893&tool=p mcentrez&rendertype=abstract.

Katchburian, E., Arana, V., 2012. Histologia e Embriologia Oral, third ed. Guanabara Koogan, Rio de Janeiro.

Ko, A., et al., 2013. Photodynamic therapy of *Porphyromonas gingivalis* via liposome-encapsulated sensitizers. J. Calif. Dent. Assoc. 41 (11), 827–830, Available from: http://www.ncbi.nlm.nih.gov/pubmed/24341134.

Kwon, G.S., 2003. Polymeric micelles for delivery of poorly water-soluble compounds. Crit. Rev. Ther. Drug Carrier Syst. 20 (5), 357–403.

Lesch, C.A., et al., 1989. The permeability of human oral mucosa and skin to water. J. Dent. Res. 68 (9), 1345–1349.

de Lima, R.A.F. et al., 2012. Improvement of tetracaine antinociceptive effect by inclusion in cyclodextrins. J. Drug Target., 20(1), pp. 85–96. Available from: <http://www.scopus.com/inward/record.url?eid=2-s2.0-83155183254&partnerID=40&md5=c1dbb5e1fffeb3052090589232eec35f>

Liu, D., et al., 2013. Minocycline hydrochloride liposome controlled-release gel improves rat experimental periodontitis. West China J. Stomatol. 31, 592–596.

Liu, M., et al., 2015. Developments of mucus penetrating nanoparticles. Asian J. Pharm. Sci. 10 (4), 275–282.

Loftsson, T., Brewster, M.E., 2011. Pharmaceutical applications of cyclodextrins: effects on drug permeation through biological membranes. J. Pharm. Pharmacol. 63 (9), 1119–1135.

Loftsson, T., Duchene, D., 2007. Cyclodextrins and their pharmaceutical applications. Int. J. Pharm. 329 (1–2), 1–11, Available from: http://linkinghub.elsevier.com/retrieve/pii/S0378517306009665.

Lu, Y., Park, K., 2013. Polymeric micelles and alternative nanonized delivery vehicles for poorly soluble drugs. Int. J. Pharm. 453 (1), 198–214.

Lv, Q., et al., 2015. Mucoadhesive buccal films containing phopholipid-bile salts-mixed micelles as an effective carrier for curcurbitacin B delivery. Drug Deliv. 22, 351–358.

Madhav, N.V.S., et al., 2009. Orotransmucosal drug delivery systems: a review. J. Control. Release 140 (1), 2–11.

Malloy, A., 2011. Count, size and visualize nanoparticles. Mater. Today 14 (4), 170–173.

Martínez-Gomis, J., et al., 1999. Effects of topical application of free and liposome-encapsulated lactoferrin and lactoperoxidase on oral microbiota and dental caries in rats. Arch. Oral Biol. 44 (11), 901–906.

Mashru, R., et al., 2005. Transbuccal delivery of lamotrigine across porcine buccal mucosa: in vitro determination of routes of buccal transport. J. Pharm. Pharm. Sci. 8 (1), 54–62, Available from: http://www.ncbi.nlm.nih.gov/pubmed/15946598.

Mazzarino, L., Borsali, R., Lemos-Senna, E., 2014. Mucoadhesive films containing chitosan-coated nanoparticles: a new strategy for buccal curcumin release. J. Pharm. Sci. 103 (11), 3764–3771.

McCarron, P.A., et al., 2004. Bioadhesive, non-drug-loaded nanoparticles as modulators of candidal adherence to buccal epithelial cells: a potentially novel prophylaxis for candidosis. Biomaterials 25 (12), 2399–2407.

McCarron, P.A., et al., 2007. Anti-adherent and antifungal activities of surfactant-coated poly(ethylcyanoacrylate) nanoparticles. Int. J. Pharm. 340 (1–2), 182–190.

Mendes, a.I., et al., 2013. Miconazole-loaded nanostructured lipid carriers (NLC) for local delivery to the oral mucosa: improving antifungal activity. Colloids Surf. B 111, 755–763, Available from: http://dx.doi.org/10.1016/j.colsurfb.2013.05.041.

Meng-Lund, E., et al., 2014. A mechanistic based approach for enhancing buccal mucoadhesion of chitosan. Int. J. Pharm. 461 (1–2), 280–285, Available from: http://www.sciencedirect.com/science/article/pii/S0378517313009617.

Miro, A., et al., 2009. Modulation of release rate and barrier transport of diclofenac incorporated in hydrophilic matrices: role of cyclodextrins and implications in oral drug delivery. Eur. J. Pharm. Biopharm. 72 (1), 76–82, Available from: http://dx.doi.org/10.1016/j.ejpb.2008.12.006.

Miro, A., et al., 2013. Engineering poly(ethylene oxide) buccal films with cyclodextrin: a novel role for an old excipient? Int. J. Pharm. 452 (1–2), 283–291, Available from: http://dx.doi.org/10.1016/j.ijpharm.2013.05.030.

Moraes, C.M., et al., 2007. Characterization of lidocaine:hydroxypropyl-β-cyclodextrin inclusion complex. J. Incl. Phenom. Macrocycl. Chem. 57 (1–4), 313–316.

Mora-Huertas, C.E., Fessi, H., Elaissari, A., 2010. Polymer-based nanocapsules for drug delivery. Int. J. Pharm. 385 (1–2), 113–142, Available from: http://www.ncbi.nlm.nih.gov/pubmed/19825408.

Morales, J.O., McConville, J.T., 2014. Novel strategies for the buccal delivery of macromolecules. Drug Develop. Ind. Pharm. 40 (5), 579–590, Available from: http://www.ncbi.nlm.nih.gov/pubmed/24611816.

Morales, J.O.B., et al., 2015. Lipid nanoparticles for the topical delivery of retinoids and derivatives. Nanomedicine 10 (2), 253–269, Available from: <http://www.scopus.com/inward/record.url?eid=2-s2.0-84921528965&partnerID=40&md5=0ad64c93a3b2e6a29741a032a9465ed6>.

Morishita, M., et al., 2001. Pluronic F-127 gels incorporating highly purified unsaturated fatty acids for buccal delivery of insulin. Int. J. Pharm. 212 (2), 289–293, Available from: http://www.ncbi.nlm.nih.gov/pubmed/11165086.

Mortazavian, E., et al., 2014. Preparation, design for optimization and in vitro evaluation of insulin nanoparticles integrating thiolated chitosan derivatives. J. Drug Deliv. Sci. Technol. 24 (1), 40–49, Available from: http://www.sciencedirect.com/science/article/pii/S1773224714500068.

Muller, R.H., Mader, K., Gohla, S., 2000. Solid lipid nanoparticles (SLN) for controlled drug delivery ± a review of the state of the art. Eur. J. Pharm. Biopharm. 50, 161–177.

Müller, R.H., Shegokar, R., Keck, C.M., 2011. 20 years of lipid nanoparticles (SLN and NLC): present state of development and industrial applications. Curr. Drug. Discov. Technol. 8 (3), 207–227, Available from: http://www.ncbi.nlm.nih.gov/pubmed/21291409.

Mura, P., et al., 2010. Development of mucoadhesive films for buccal administration of flufenamic acid: effect of cyclodextrin complexation. J. Pharm. Sci. 99 (10), 4215–4227.

Mura, P., et al., 2015. Amidated pectin-based wafers for econazole buccal delivery: formulation optimization and antimicrobial efficacy estimation. Carbohyd. Polym. 121, 231–240, Available from: http://dx.doi.org/10.1016/j.carbpol.2014.11.065.

Nerkar, P.P., Gattani, S.G., 2013. Oromucosal delivery of venlafaxine by linseed mucilage based gel: in vitro and in vivo evaluation in rabbits. Arch. Pharm. Res. 36 (7), 846–853, Available from: http://www.ncbi.nlm.nih.gov/pubmed/23588682.

Nicolazzo, J.A., Reed, B.L., Finnin, B.C., 2004. Assessment of the effects of sodium dodecyl sulfate on the buccal permeability of caffeine and estradiol. J. Pharm. Sci. 93 (2), 431–440, Available from: http://dx.doi.org/10.1002/jps.10559\nhttp://onlinelibrary.wiley.com/store/10.1002/jps.10559/asset/10559_ftp.pdf?v=1&t=hkn0nuaz&s=8fcd9b620a09f02bf4f7a988cd65d304c9d384de.

Nicolazzo, J.A., Reed, B.L., Finnin, B.C., 2005. Buccal penetration enhancers: How do they really work? J. Control. Release 105 (1–2), 1–15, Available from: http://linkinghub.elsevier.com/retrieve/pii/S0168365905001409.

Ojewole, E., et al., 2014. Novel oleic acid derivatives enhance buccal permeation of didanosine. Drug Dev. Ind. Pharm. 40 (5), 657–668, Available from: http://www.ncbi.nlm.nih.gov/pubmed/24592892.

Paderni, C., et al., 2012. Oral local drug delivery and new perspectives in oral drug formulation. Oral Surg.. Oral Med., Oral Pathol. Oral Radiol. 114 (3), e25–e34, Available from: http://linkinghub.elsevier.com/retrieve/pii/S2212440312001927.

Paphangkorakit, J., Sangsirinakagul, C., Priprem, A., 2012. Relief of palatal injection pain by liposome-encapsulated 2% lignocaine prepared by ultrasonic dental scaler. Br. J. Oral Maxil. Surg. 50, 784–787, Available from: http://www.ncbi.nlm.nih.gov/pubmed/16141539.

Pardeike, J., Hommoss, A., Müller, R.H., 2009. Lipid nanoparticles (SLN, NLC) in cosmetic and pharmaceutical dermal products. Int. J. Pharm. 366 (1–2), 170–184.

Patel, V.F., Liu, F., Brown, M.B., 2011. Advances in oral transmucosal drug delivery. J. Control. Release 153 (2), 106–116, Available from: http://dx.doi.org/10.1016/j.jconrel.2011.01.027.

Pathak, Y., Preuss, C., 2013. Orally delivered nanoparticle drug delivery systems for dental applications and their toxicity on systemic organs. In: Hartsfield, K.S.A.K. (Ed.), Nanobiomaterials in Clinical Dentistry. Elsevier Inc, Amsterdam, The Netherlands, pp. 497–508.

Pignatello, R., Basile, L., Puglisi, G., 2009. Chitosan glutamate hydrogels with local anesthetic activity for buccal application. Drug Deliv. 16 (3), 176–181.

Piñón-Segundo, E., et al., 2005. Preparation and characterization of triclosan nanoparticles for periodontal treatment. Int. J. Pharm. 294 (1–2), 217–232.

Rao, K.K., 2007. Polymerized solid lipid nanoparticles for oralor mucosal delivery of therapeutic proteins and peptides, US 2007/0,237,826 A1, 1(19).

Ribeiro, L.N.M., et al., 2016. Nanostructured lipid carriers as robust systems for lidocaine-prilocaine transbuccal release. Eur. J. Pharm. Sci. 93, 192–202.

Roblegg, E., et al., 2012. Evaluation of a physiological in vitro system to study the transport of nanoparticles through the buccal mucosa. Nanotoxicology 6 (4), 399–413.

Sakata, O., Machida, Y., Onishi, H., 2011. Semi-solid dosage form of clonazepam for rapid oral mucosal absorption. Drug Dev. Ind. Pharm. 37, 809–814.

Salamat-Miller, N., Chittchang, M., Johnston, T., 2005. The use of mucoadhesive polymers in buccal drug delivery. Adv. Drug Deliv. Rev. 57 (11), 1666–1691, Available from: http://linkinghub.elsevier.com/retrieve/pii/S0169409X05001444.

Samiei, M., et al., 2016. Nanoparticles for antimicrobial purposes in endodontics: a systematic review of in vitro studies. Mater. Sci. Eng. C 58, 1269–1278.

Sander, C., Nielsen, H.M., Jacobsen, J., 2013. Buccal delivery of metformin: TR146 cell culture model evaluating the use of bioadhesive chitosan discs for drug permeability enhancement. Int. J. Pharm. 458 (2), 254–261, Available from: http://dx.doi.org/10.1016/j.ijpharm.2013.10.026.

Sandri, G., et al., 2004. Assessment of chitosan derivatives as buccal and vaginal penetration enhancers. Eur. J. Pharm. Sci. 21 (2–3), 351–359.

Sandri, G., et al., 2006. Histological evaluation of buccal penetration enhancement properties of chitosan and trimethyl chitosan. J. Pharm. Pharmacol. 58 (10), 1327–1336.

Sankar, V., et al., 2011. Local drug delivery for oral mucosal diseases: challenges and opportunities. Oral Dis. 17, 73–84.

Sanz, R., et al., 2015. Enhancing topical analgesic administration: review and prospect for transdermal and transbuccal drug delivery systems. Curr. Pharm. Des. 21, 2867–2882.

Saxena, A., Tewari, G., Saraf, S.A., 2011. Formulation and evaluation of mucoadhesive buccal patch of acyclovir utilizing inclusions phenomenon. Int. J. Pharm.Tech. Res. 47 (4), 887–897.

Şenel, S., Hincal, A.A., 2001. Drug permeation enhancement via buccal route: possibilities and limitations. J. Control. Release 72 (1–3), 133–144.

Serpe, L., et al., 2014. Anaesthetic efficacy of bupivacaine 2-hydroxypropyl-β-cyclodextrin for dental anaesthesia after inferior alveolar nerve block in rats. Br. J. Oral Maxil. Surg. 52 (5), 452–457, Available from: http://www.ncbi.nlm.nih.gov/pubmed/24673836.

Severino, P., et al., 2015. Sodium alginate-cross-linked polymyxin B sulphate-loaded solid lipid nanoparticles: antibiotic resistance tests and HaCat and NIH/3T3 cell viability studies. Colloids Surf. B 129, 191–197, Available from: http://linkinghub.elsevier.com/retrieve/pii/S0927776515001927.

Shakya, P., et al., 2011. Sublingual mucosa as a route for systemic drug delivery. J. Control. Release 151, 2–9, Available from: http://dx.doi.org/10.1016/j.jconrel.2010.11.003.

Shidhaye, S.S., et al., 2010. Buccal drug delivery of pravastatin sodium. AAPS PharmSciTech 11 (1), 416–424.

Shin, S.C., Kim, J.Y., 2000. Enhanced permeation of triamcinolone acetonide through the buccal mucosa. Eur. J. Pharm. Biopharm. 50 (2), 217–220.

Shojaei, A.H., Zhuo, S.L., Li, X., 1998. Transbuccal delivery of acyclovir (II): feasibility, system design, and in vitro permeation studies. J. Pharm. Pharm. Sci. 1 (2), 66–73.

Smistad, G., Jacobsen, J., Sande, S., 2007. Multivariate toxicity screening of liposomal formulations on a human buccal cell line. Int. J. Pharm. 330 (1–2), 14–22, Available from: http://www.ncbi.nlm.nih.gov/pubmed/16997516.

Sosnik, A., Menaker Raskin, M., 2015. Polymeric micelles in mucosal drug delivery: challenges towards clinical translation. Biotechnol. Adv. 33 (6 Part 3), 1380–1392.

Souto, E.B., et al., 2004. Evaluation of the physical stability of SLN and NLC before and after incorporation into hydrogel formulations. Eur. J. Pharm. Biopharm. 58 (1), 83–90, Available from: http://www.ncbi.nlm.nih.gov/pubmed/15207541.

Squier, C.A., 1973. The permeability of keratinized and nonkeratinized oral epithelium to horseradish peroxidase. J. Ultrastruct. Res. 43 (1), 160–177.

Squier, C.A., 1984. Effect of enzyme digestion on the permeability barrier in keratinizing and non-keratinizing epithelia. Br. J. Dermatol. III, 253–264.

Squier, C.A., 1991. The permeability of oral mucosa. Crit. Rev. Oral Biol. Med. 2 (1), 13–32.

Squier, C.A., Brogden, K.A., 2011. Barrier functions of oral mucosa. Human Oral Mucosa: Development, Structure, and FunctionWiley-Blackwell, New Jersey, pp. 113-144.

Squier, C.A., Hall, B.K., 1984. The permeability of mammalian nonkeratinized 0ra. l epithelia to horseradish peroxidase applied in vivo and in vitro. Arch. Oral Biol. 29 (I), 45–50.

Squier, C.A., Hall, B.K., 1985a. In-vitro permeability of porcine oral mucosa after epithelial separation, stripping and hydration. Arch. Oral Biol. 30 (6), 485–491.

Squier, C.A., Hall, B.K., 1985b. The permeability of skin and oral mucosa to water and horseradishperoxidase as related to the thickness of the permeability barrier. J. Invest. Dermatol. 84 (3), 176–179.

Squier, C.A., Hopps, R.M., 1976. A study of the permeability barrier in epidermis and oral epithelium using horseradish peroxidase as a tracer in vitro. Br. J. Dermatol. 95, 123–129, 0007-0963.

Squier, C.A., Rooney, L., 1976. The permeability of keratinized and nonkeratinized oral epithelium to lanthanum in vivo. J. Ultrastruct. Res. 54 (2), 286–295.

Starokadomskyy, P.L., Dubey, I.Y., 2006. New absorption promoter for the buccal delivery: preparation and characterization of lysalbinic acid. Int. J. Pharm. 308 (1–2), 149–154, Available from: http://www.sciencedirect.com/science/article/pii/S0378517305007490.

Sudhakar, Y., Kuotsu, K., Bandyopadhyay, A.K., 2006. Buccal bioadhesive drug delivery: a promising option for orally less efficient drugs. J. Control. Release 114 (1), 15–40, Available from: http://linkinghub.elsevier.com/retrieve/pii/S0168365906002021.

Sveinsson, S.J., Holbrook, W.P., 1993. Oral mucosal adhesive ointment containing liposomal corticosteroid. Int. J. Pharm. 95 (1–3), 105–109.

Sveinsson, S.J., Mezei, M., 1992. In vitro oral mucosal absorption of liposomal triamcinolone acetonide. Pharm. Res. 9 (10), 1359–1361.

Thipparaboina, R., et al., 2015. Micellar carriers for the delivery of multiple therapeutic agents. Colloids Surf. B 135, 291–308.

Turunen, T.M., et al., 1994. Effect of some penetration enhancerrs on epithelial membrane lipid domains: evidence from fluorescence spectroscopy studies. Pharm. Res. 11 (2), 288–294.

Udgirkar Dattatreya, B., Hiremath, S.N., Rao, K., Sreenivasa, P.D., 2009. Formulation and in-vitro evaluation of buccoadhesive tablets containing ketoconazole inclusion complex with β-cyclodextrin. Res. J. Pharm. Tech. 2 (2), 396–404.

Walker, G.F., Langoth, N., Bernkop-Schnürch, A., 2002. Peptidase activity on the surface of the porcine buccal mucosa. Int. J. Pharm. 233 (1–2), 141–147, Available from: http://www.ncbi.nlm.nih.gov/pubmed/11897418.

Wang, K., et al., 2014. Preparation and in vitro release of buccal tablets of naringenin-loaded MPEG-PCL nanoparticles. RSC Adv. 4 (64), 33672–33679.

Wayakanon, K., et al., 2013. Polymersome-mediated intracellular delivery of antibiotics to treat *Porphyromonas gingivalis*-infected oral epithelial cells. FASEB J. 27 (11), 4455–4465.

Wei, R., et al., 2012. Effects of iontophoresis and chemical enhancers on the transport of lidocaine and nicotine across the oral mucosa. Pharm. Res. 29 (4), 961–971.

Wertz, P.W., Squier, C.A., 1991. Cellular and molecular basis of barrier function in oral epithelium. Crit. Rev. Ther. Drug Carrier Syst. 8, 237–269.

Wong, H.L., et al., 2004. Development of solid lipid nanoparticles containing ionically complexed chemotherapeutic drugs and chemosensitizers. J. Pharm. Sci. 93 (8), 1993–2008.

Xu, J., et al., 2015. Biomaterials Genipin-crosslinked catechol-chitosan mucoadhesive hydrogels for buccal drug delivery. Biomaterials 37, 395–404, Available from: http://dx.doi.org/10.1016/j.biomaterials.2014.10.024.

Yao, W., et al., 2015. RGD functionalized polymeric nanoparticles targeting periodontitis epithelial cells for the enhanced treatment of periodontitis in dogs. J. Colloid Interface Sci. 458, 14–21.

Zed, C.M., Epstein, J., Donaldson, D., et al., 1996. Topical liposome encapsulated tetracaine versus benzocaine: a clinical investigation. J. Dent. Res. 75, 247.

CHAPTER

TRENDS IN ORALLY VIRAL VECTOR GENE DELIVERY AND THERAPY

5

Jingqi Xie*,,†, Ruian Xu*,**,†**

*Huaqiao University, Quanzhou, China

**Engineering Research Center of Molecular Medicine, Ministry of Education, Xiamen, China

†Key Laboratory of Molecular Medicine in Fujian Province, Xiamen, China

CHAPTER OUTLINE

1 Introduction 124
2 Nanoparticles and Microparticles Applications in Oral Delivery of Drug Molecules and Therapeutic Genes 124
3 The Crucial Intake Site of the Absorption of Nanoparticles and Microparticles in the Gastrointestinal Tract 125
4 The Characteristics of Particles and the Effect of Intake In Vivo 126
4.1 The Performance of the Particles 126
4.2 The Surface Modification and Charge of the Particles 127
4.3 The Size of the Particles 127
4.4 The Age of the Animals 128
5 Protein Transduction and the Absorption of Nanoparticles 128
6 Ways to Promote the Absorption of Particles 129
7 The Feasibility of Oral Gene Delivery 130
8 The Oral Gene Delivery Vector System 130
9 The Uptake of Oral Viral Vector 131
10 The Selection of Viral Vectors Suitable for Oral Delivery 132
10.1 Retrovirus 133
10.2 Adenovirus 133
10.3 Adeno-Associated Virus 135
10.4 Other Viral Vector Systems 137
11 The Digestion Circumstance of Oral Gene Delivery 138
12 The Transmission, Distribution, and Transgenic Product Pharmacokinetics of Oral Viral Vectors 138
13 Potential, Problems, and Prospects 139
14 Conclusions 140
References 140

1 INTRODUCTION

By the statistics, the account of oral medicine makes up about half of the total each year. Oral medicine is one of the main methods for disease prevention and treatment because of its noninvasive nature. It is popular in health care due to its convenience, efficiency, comfort, and safety. Moreover, oral formulations are also less expensive to produce for the reason that they do not need to be manufactured under sterile conditions (Yun et al., 2013). When compared with injections, oral routes avoid pain and discomfort, as well as eliminate contaminations (des Rieux et al., 2006). The potential benefits of the oral administration of nucleic acids for the treatment of a wide variety of diseases that involve the gastrointestinal (GI) tract (Kriegel et al., 2013) and viruses represent highly evolved natural vectors for the transfer of foreign genetic information into cells and therefore show high transfection efficiencies (Hearnden et al., 2012). This attribute has led to extensive attempts to engineer recombinant viral vectors for the delivery of therapeutic genes into diseased tissues (Kay et al., 2001; Lim et al., 2010; Mingozzi and High, 2011; Nathwani et al., 2011). Nanotechnology has solved many drug delivery problems (Cooper et al., 2014; Sun et al., 2014). Nanoparticles have lots of advantages, such as small size, high surface area, and modification using functional groups for high capacity or selectivity (Yun et al., 2013). However, there is still a lack of qualitative and quantitative information about how macromolecules and nanoparticles in particular, were taken up and absorbed in the GI tract (Xu et al., 2006a). Oral gene therapy provided a convenient way for genetic maladies and those diseases difficult to be treated using conventional methods (During et al., 2000; Xu, 2014).

2 NANOPARTICLES AND MICROPARTICLES APPLICATIONS IN ORAL DELIVERY OF DRUG MOLECULES AND THERAPEUTIC GENES

For the past few decades, there has been plenty of research in nanopolymers and micron particles used as physical approaches to alter and improve the pharmacokinetic and pharmacodynamic properties of various types of drug molecules (Mohanraj and Chen, 2007). Polylactic-*co*-glycolic acid (PLGA) and polylactic acid (PLA), granted by the Food and Drug Administration (FDA), can be used for oral delivery, which is used widely for biodegradable of biocompatibility nano- and microparticles (Garinot et al., 2007). Most studies used polystyrene rubber (PS) on the bowel absorption of particulate matters. PS was a useful polymer for particle uptake research (Dorney, 2013) and can carry charges on its surface. Therefore, it can be covered by proteins and adhesion factors easily. In addition, the covalent binding to the fluorescence probe made PS an effective tool for the study of oral particles. Nevertheless, the biodegradation limited its clinical application. Recent studies demonstrated that PEG–PLA–PEG nanoparticles depict enhanced mucosal uptake leading to effective immune response as compared to other copolymeric nanoparticles (Jain et al., 2010). PLA has been copolymerized with PEG to enhance stability and bioavailability of these nanoparticles.

The particle-coating technology had been used in the wrap of hydrophily and hydrophobicity fluorescein, charged compounds, and radioactive molecular. Probe inclusion needed to achieve minimum disclosure for the reason why the unwittingly releasing of probe can cause an exorbitant quantitative evaluation and deep microscopic analysis background. Polymer can be radiolabeled (Arshady, 2001) or integrated to fluorescein directly (Ahmed et al., 2001). The latter has not yet been applied to the research of oral absorption. Fluorescence can make the particles more clearly visible in tissues and improve the sensibility of quantities analysis. Thus, the epipolic nano- and macroparticles are the most popular histology qualitative and quantitative tool for intestinal uptake research. By using different fluorescent substances to mark particles, researchers can make a difference between tracking the way of uptake and comparing the absorption of particles with different coatings (Morgan et al., 2008). Data from experiments showed that after taking orally PLGA-coating DNA, Balb/c mice produced mucosal and systemic immune responses. With the development of PLA-coating technology, 60% of packing efficiency is realized, the release efficiency of DNA varies from prescriptions. In addition, it is found that the DNA that was released from PLGA compounds showed better stability by the determination of the DNA superhelix degree. However, in the PLGA autohydrolysis environments, DNA occurs with some degree of hydrolysis (Page and Cudmore, 2001).

3 THE CRUCIAL INTAKE SITE OF THE ABSORPTION OF NANOPARTICLES AND MICROPARTICLES IN THE GASTROINTESTINAL TRACT

In the absorptive and nonabsorptive epithelia of the small intestine, it was hypothesized that the trafficking of particles occurred primarily by three routes: (1) through the densely clustered M cells of the Peyer's patches (PPs); (2) between the microvilli (brush border) located at the apical surfaces of the enterocytes; (3) between two adjacent enterocytes (Bakhru et al., 2013). Particles enter the guts through three routes, including receptor-mediated, clathrin-mediated, or caveolae-mediated endocytosis. Furthermore, particle entry by these routes was generally referred to as phagocytosis, transcellular uptake, and paracellular transport. Nevertheless, the majority of the literature surveyed suggests that particle uptake predominantly occurs in the M-cell bearing PPs of the small intestine (Bakhru et al., 2013; Xu et al., 2008).

Nano- and microparticles offer considerable potential for drug and vaccine delivery via the GI tract (Shakweh et al., 2004). Intestinal epithelial cells were combined with different types of cells, including absorbability intestinal cells, goblet cells, and M cells of PPs. M cells were distinctive epithelial cells that occurred only in the follicle-associated epithelia that overlie organized mucosa-associated lymphoid tissues (Kraehenbuhl and Neutra, 2000). A large body of evidences showed that M cell was the main access; it could not only intake the particles with a diameter of about 500 nm, but also allowed a number of pathogens, including bacteria, viruses and prions present in the gut-associated lymphoid tissue. Efficient transport of antigens and pathogens requires adherence to M-cell apical surfaces (Lelouard et al., 2012). This was the direct indicator that lactide copolymer particles by transcellular transport via M cells to cross the epithelial cell for the first time. These observations assisted the potential utility of such vehicles for oral vaccine delivery. Utilizing scanning electron microscope (SEM), natural ester copolymer microspheres could not target M cells on the surface; it can also hold a great promise to get an improvement for those particles intakes by modifying their surface (Jepson et al., 1996). Gullberg et al. (2000) established a simplified phenotypic model of the human M cell and showed that the induction of an M-cell-like morphology, particle transport, and expression of M-cell-specific adhesion molecules in the model can occur independently of migration of the B-lymphocytes into the epithelial cell monolayer.

Recent studies suggest that M cells and PPs may not be the only sites for intake. The intestinal mucosa epithelial cells have played an important role in the inversion process between particles and intestinal epithelial cells. By using colloidal probes with easily traceable markers, researchers discovered that polyethylene particles can transfer in transcelluar and paracellular pathways through mucosal epithelial cells and follicle-associated epithelium cells, which are covered with the PPs after oral administration in one hour. Microsome has also been observed in the spleen and liver. The degree of intestinal epithelial cells absorbing peptide parcel particles remains controversial. Because the number of M cells is far less than the intestinal epithelial cell, the relationship between particles absorbing and immunohistochemical result continued to be almost impossible to define (Artursson et al., 2012). Previous studies implied that microvillus get very few PLG microsphere and lipidosome intake (Brayden and Baird, 2004). However, there are experiments that showed that the normal intestinal mucosa epithelial cells in rats had a significant absorption with cyano-acrylate nanoparticle with a diameter of about 300 nm (Pinto-Alphandary et al., 2003). There were gangliosides glycolipid receptors existing on all nucleated cell surfaces, and cholera toxin B subunit (CTB) had a strong affinity for gangliosides glycolipid receptors. The combination of cholera toxin and antigens-coated lipidosome surface and researchers observed the level of mucus membrane IgA and serum IgM in mice increased after oral administration. Although gangliosides glycolipid receptors were less specific to M cells, with thinner extracellular polyose it can be more easily accessible than adjacent intestinal mucosa cells. At last, pathogens can also use enterocoel non-PPs lymph node methods, which were discovered earlier, like dendritic cells (Rescigno et al., 2001). Those methods were also the potential targets of pathogens and vaccine modification.

4 THE CHARACTERISTICS OF PARTICLES AND THE EFFECT OF INTAKE IN VIVO

It has been considered that the main obstacle for particle uptake might be the GI barriers and mucosal barriers (Yun et al., 2013). Currently, the key factor for control particles absorbed in the gut has been widely studied, including solubility in the gut lumen, surface charge and hydrophobicity, the presence of specific ligands, such as lectins and invasions, the properties and size of particles (During et al., 1998; Powell et al., 2010; Xu et al., 2008), absorb channel, zeta potential, medium, surface modification, such as covered by hemagglutinin or other adherence factor, stomach content, and the age of the animal are also considered to be one of the important factors influencing the absorbtion of particles (Xu et al., 2006b).

4.1 THE PERFORMANCE OF THE PARTICLES

Encapsulating or incorporating peptides in particles should at least protect these substances against degradation and, in some cases, also enhance their absorption (Li et al., 2012). Combining viral gene delivery with engineered biomaterials could overcome some limitations related to nonlocalized gene delivery, insufficient transduction, and long-term transgene expression (Jang et al., 2011; Xu et al., 2008). The encapsulation of insulin into mucoadhesive nanoparticles was a key factor in the improvement of its oral absorption and oral bioactivity (Brayden and Baird, 2004). Chitosan is a biocompatible and slowly biodegradable polymer that has been widely used as a delivery vehicle for gene therapy and vaccine design owing to its inherent muco-adhesive feature and the ability to modulate the integrity of epithelial tight junctions reversibly (Chen et al., 2013). It could also increase the transcellular and paracellular transport across mucosal epithelium (Pan et al., 2002), a further indication of its potential in oral gene delivery and in generating protective mucosal immune responses (Van der Lubben et al., 2001). Oral chitsotan has been widely applied to reduce the fat and cholesterol uptake from food, chitsotan has been used in oral gene delivery system of intestinal epithelial cells, PPs, and the lymphonodus of mesentery (Kadiyala et al., 2010). Chitosan–DNA microparticles were prepared by using a complex coacervation process and stability of plasmid DNA was investigated in this complex. The chitosan–DNA microparticles could also protect the encapsulated plasmid DNA from nuclease degradation (Guliyeva et al., 2006). Ataman-Onal et al. (2015) use a colloidal synthetic bioresorbable vector comprising protein substances and a viral vector comprising the nucleotide sequences, which code for the corresponding protein substances, and makes it possible to increase the effectiveness of the vaccination associated with the use of a protein substance, while decreasing the antivector immunity. Other biodegradable polymerization anhydride, such as fumaric acid and sebacic acid, has also been used in oral DNA delivery.

4.2 THE SURFACE MODIFICATION AND CHARGE OF THE PARTICLES

Some studies have tried to confirm under the physiological in vivo conditions the change of the particle surface to influence the capacity absorbed by M cells. Smith et al. (1987) found the particle surface materials combined with M cells in that order of preference: bGH < bSA < hIgG (in mice) and bGH < bGH–Ab (in mice and rats). This performance can also be observed in the intake of microspheres in M cells and the presence of particles that coated with protein in rat mesenteric lymph. The transmission selectivity of the M-cell transcytosis emphasized that the M cell surface was the primary site to identify what kind of antigen can be presented at the intestinal immune system (Page and Cudmore, 2001). Kulkarni and Feng (2013) showed that nanoparticles after TPGS modification got enhanced cellular uptake efficiency. On the other hand, the development of mucus-penetrating particles, by rendering the surfaces of particles nonmucoadhesive, provided a powerful strategy for overcoming the mucus barrier (Lai et al., 2009). Positive charges that appeared to improve the efficacy of gene transfer and drug delivery have been reported (Fröhlich, 2012). Previous studies revealed that in all cell lines the surface charge influenced the cellular uptake rate and showed positive correlations. Yue et al. (2011) suggested that nanoparticles positively charged could escape from lysosome after being internalized and exhibit perinuclear localization while the negatively and neutrally charged particles prefer to colocalize with lysosome.

4.3 THE SIZE OF THE PARTICLES

Studies have shown that the size of the particles was the major parameter for absorption (Hussain et al., 2001). Translocation was largely dependent on particle size: smaller particles are more readily absorbed (Hillyer and Albrecht, 2001). Although, 10 μm diameter particles could be swallowed by the M cell, the smaller, more efficiently absorption. Lamprecht et al. (2001) demonstrated that the 100 nmsize particles have a significantly higher efficiency of uptake by the intestinal tissue in terms of their total mass and the number compared to larger size particles. He et al. (2012) have been using rhodamine B-labeled carboxylated chitosan grafted nanoparticles with different particle sizes (300, 600, and 1000 nm) and similar zeta potentials to form drug-loaded polymeric nanoparticles; as a result, the smallest one (300 nm) got the highest absorptivity. Awaad et al. (2012) investigated specific sizes of different groups (namely, 95, 130, 200, 340, 695, and 1050 nm) and had a similar trend. Kulkarni and Feng (2013) pointed out that particles smaller than 200 nm might be more similar to low-density lipoprotein than the other sizes and more possible to cross GI barriers to have a longer half-life in the blood system.

4.4 THE AGE OF THE ANIMALS

Some investigation showed that the age of the animals may affect the uptake rate. After the inject with rhodamine B-labeled synthetic latex particles, 1.8 μm in diameter, aged mice exhibited significantly more fluorescent particle accumulation in PPs. Seifert et al. (1996) confirmed the uptake dynamics in rat thoracic lymph with diverse ages. Doyle-McCullough et al. (2007) showed that particle uptake being significantly greater in young adult males (7 weeks) than in younger (3 weeks) and older (17 and 52 weeks) age groups, which is consistent with previous works (Mandir et al., 2005). The uptake occurred in PPs and guts-associated lymphoid tissue (GALT). The absorption states with the greatest number of particles were rats, hamsters, and mice, in that order, implying that this was caused by more than one factor. PPs were considered the principal sites for particle absorption, it therefore could be assumed that those animals that had more PPs, such as rabbits, may have higher uptake rates than other animals. The high rate of uptake in aged mice may stand for the function of PPs that get better with age. For aged animals, the slower particles transport rates and the longer duration of exposure with epithelial cells made the difference. The animal immune and pathological state may also be having an effect, because when the mucosal system was triggered, the amount of PP will increase (Jepson et al., 1996).

5 PROTEIN TRANSDUCTION AND THE ABSORPTION OF NANOPARTICLES

Previous studies found the Tat protein separated from HIV-1could enter the cell-by-cell membrane and reported the protein transduction phenomenon respectively (Frankel and Pabo, 1988; Green and Loewenstein, 1988). It was so-called protein transduction that protein transduction domains—composed by some short small alkaline peptide—mediated could improve proteins to cross over the cell membrane to achieve protein delivery between cells. Studies found that this structural domain transduction did not depend on energy, endocytosis, receptor, and ion channel but through lipid bilayer directly to achieve transmembrane transport and it could also carry several compounds, such as other DNA, polypeptide, and heterologous protein via paracellular transport and did not make damaged cells. Protein transduction without obvious separation cells selectivity could be used in different types of cultured cells (Xiong et al., 2006). Up to now, there were many native proteins with transduction potential found, mainly including the polypeptides with a structurally related to HIV-1 transactivator Tat (48–60), the polypeptides derived from drosophila homeobox transcription factors ANTP(43–58), the membrane translocation hydrophobic sequence peptide came from Karpori fibroblast growth factors and the chimera combined with neuropeptide, glycerin peptide, wasps toxic peptides, and MCD peptides, the optimal efficiency of transduction come from HSV-VP22, ANTP, and HIV-1 Tat (Xiong et al., 2006).

Different experiments supported different results so more and deeper studies were needed in order to understand the mechanism of how Tat proteins enter a cell. Some investigations suggested that the Tat protein transduces into cells via residues from 49 to 57 sites combined in low affinity with cell surface 90kU protein, but more studies confirmed that the transduction of Tat protein may not need cell receptors. Eguchi et al. (2001) made a nystatin test in exogenous gene delivery mediated by Tat protein and found that nystatin could inhibit the process of exogenous substances through the fossa approach to cross the cells. Similar results were obtained to demonstrate that Tat protein may use this pathway through the plasma membrane by using fillipin for inhibition study (Efrat et al., 1991). Moore and Payne (2004) found that the transduction of Tat protein caused the rearrangement of membrane phospholipids through the lack of cell surface phosphatidylserine and positively polylysine could prevent the process. Thus, the transduction of Tat protein caused phosphatidylserine transferred from plasma membrane inside to outside not fully by electrostatic interactions.

Previous investigators used fluorescent protein and PTD to transfect cells but did not observe any existence of PTD in the cytoplasm or nucleus. Only a strong adsorption on the surface of living cells could be observed. When samples were subjected to the fixative, like formaldehyde, then the PTD could be observed. Therefore, the phenomenon of PTD could be considered using traditional endocytosis to form an inclusion body after sticking to the cytomembrane surface to get into the cytoplasm and endonuclear as other positive macromoleculars (Xiong et al., 2006). Whether the pathogen using the PTD mechanism in enteric non-PPs lymphatic channels to cross the cell or not remains to be solved by further exploration to make a comprehensive explanation.

6 WAYS TO PROMOTE THE ABSORPTION OF PARTICLES

For many researchers, the failure of the study proceeds in biodegradable lactide and PLG particles on orally polypeptide to promote the research of receptor targeted PLGA (Eguchi et al., 2001). Additional studies chose polymer oral preparations (Lambkin and Pinilla, 2002) or nasal delivery (Thanos et al., 2003).

Normal M cells could pass cargo to the immune response organization under the epithelial cell directly. The direct transmission path was of obvious advantage for vaccine delivery, especially in the pathogen antigen expressed in natural epithelial cell and M cell. PP lymph node follicle-linked epithelial cell was the site for particles delivery to lymphoid tissue directly. Eldridge et al. (1991) studied the PLG microparticles with entrapped protein or adsorbed DNA as a vaccine delivery system caused mucosal immunity. Apical membrane targeting, which was determined by orally peptide (those have been experiment on human) and human M-cell model, mainly based on a combination of peptide and carrier surfactant or the genes, which was certified a high expression in follicle-associated epithelium by the experiment of peptide coupling with sorbefacient, including protocadherin, laminin, matrix metalloproteinase 15, and perfringocin enterotoxin receptor (Giannasca et al., 1999). The intestine epithelia were composed of a mass of constant cycle cells, which are derived from small intestinal gland (with large areas of therapeutic gene transduction of the surface). In order to improve the absorption of M cells, researchers used a new Ulex europaeus agglutinin (UEA) analogue, which was derived from combinatorial libraries and picked out by binding affinities screening, but the limitations of the composite particles delivery system itself restricted the pervasive application in clinical. First, (UEA) was not the pervasive receptor of human M cell. Second, the targeting of high-affinity aglucon caused a low level for particles adhered to M cells. Third, in the selection of antigen processing, relationships of dose with reactions are still unclear (Brayden and Baird, 2004). Last, although the transduction efficiency of those systems kept promoting, it still could not compare with viral systems. Thus, it is worth further exploration.

A set of experiments and analysis demonstrated that the drug with particle size 20–50 nm could be most likely absorbed. Thus, the orally nanocrystallization for medicine became a hot spot in the research of nanophase carrier materials. The oral gene medicine should meet the requirements of DNA, RNA, or protein carries with a size at the optimal ranges (20–50 nm). After oral administration, DNA genes or protein medicine could be delivered to the accurate target cells in order to achieve the final goal–disease prevention.

7 THE FEASIBILITY OF ORAL GENE DELIVERY

Intestines can provide nearly 300 m^2 mucosal surfaces, which contain rich immune-mediated tissue like intraepithelial lymphocyte, lamina propria, and PPs (Page and Cudmore, 2001). Plenty of blood capillaries near the intestinal epithelial cell layer make the protein encoded by foreign genes in epithelial cell and can not only enter the systemic circulation but also make exogenous DNA or vector carries exogenous DNA via transcellular transport by endocytosis or paracellular route by rupture lacuna at the tight junction moment between cells enter many organs via the systemic circulation after expression. After oral delivery, gene expression protein products persist on human intestinal cells (Palfi et al., 2014; Xu et al., 2008). Therapeutic protein products can enter the blood of patients without injection and it is a more convenient and effective way in protein delivery system (Xu et al., 2008). Oral gene therapy provided genetic diseases and diseases that cannot be treated in conventional methods in a convenient way with low risk in dosing and safety.

In addition to the digestive gene therapy, through the digestive tract gene vaccine immunization was also possible. Except for feasibilities and low cost, another advantage of nucleic acid vaccine was that it was an immune process that mimics the natural infection. The form of the encoded antigen protein presented by major histocompatibility complex I and II, it could not only get general immunization by PP but also build mucosal immunity. This is the first line of defense for the pathogen infected by intestinal, urinary tract, and respiratory mucosa. The main advantage was the fact that it could provide a GI and systemic treatment. For DNA vaccine, the digestive tract pathway could also stimulate a mucosal immune response. Nowadays, digestive gene therapy and immunity is one of the hottest subjects in the field of oral gene therapy.

8 THE ORAL GENE DELIVERY VECTOR SYSTEM

Up to now, oral gene delivery system has developed physical method chemical synthesis vector (including polymer and lipidosome carrier), viral vector, and bacteria delivery systems (Xu et al., 2006b). Takamura et al. (2004) found that plasmid DNA could be packaged in vitro into a virus-like particle (VLP) composed of open reading frame 2 of hepatitis E virus, which is an orally transmissible virus. These VLPs can deliver this foreign DNA to the intestinal mucosa in vivo. These findings suggest that VLPs derived from orally transmissible viruses can serve as vectors for delivery of genes to mucosal tissue by oral administration for the purpose of DNA vaccination and gene therapy.

Recombinant bacteria are efficient vehicles for the delivery and display of heterologous antigens for oral vaccination. Recently, recombinant strains of *Salmonella* and *Shigella* have been used to deliver DNA vaccines to the GALT (Bai et al., 2011; Shata et al., 2000). After infection, *Shigella* vectors underwent programmed autolysis within the host cell, delivering the DNA vaccine cargo into the cytoplasm (Shata and Hone, 2001). *Salmonella* vectors were understood to act by inducing apoptosis of the infected host cell, resulting in delivery of the DNA vaccine to the nearby dendritic cells (Page and Cudmore, 2001; Shata et al., 2000).

Years of observation and experiments have indicated that the viral vector, the most efficient tool for gene transmission that has its own special features and characters, such as easy for regulation and tissue specificity. Some viral systems could prompt the transfer genes integrated into the chromosome during the transmission. Thus, it has the potential for long-term gene expression. Viral vectors utilize the intrinsic function of wild-type viral, which were susceptible to infection. However, after the processing by genetic engineering, the function of replication and cytolysis could be minimized or totally disappear. Various viral vectors have been applied to transfer DNA to intestinal structure and be used to gene correction and induce mucosal immunity (Xu, 2014).

9 THE UPTAKE OF ORAL VIRAL VECTOR

Although the uptake site of the most common recombinant viral vector has not been clearly identified, previous studies on pathogens and prions infected in the intestinal tract provide proof that virus combined with the M-cell receptors in intestinal PPs and then turned into epithelial cells. Reovirus is a kind of virus that enters the host via M cells combined with protein σ-1. This ligand, reovirus protein σ-1, when conjugated to polylysine, can bind the apical surface of M cells from nasal-associated lymphoid tissues (Wu et al., 2001). Reovirus can combine with normal intestinal epithelial cell receptors like the adhesion molecule of epithelial cell tight junction as other pathogens with a fondness for M cells (Tyler et al., 2001).

The intestinal mucosa is a major barrier to overcome oral drug delivery (Deli, 2009). Once the barrier of mucus is crossed, drug-loaded nanoparticles have to be transported across the intestinal epithelium via the paracellular pathway, transcytosis, and receptor-mediated transcytosis by enterocytes or M cells. Whether drugs are absorbed and delivered in the systemic circulation free or encapsulated would strongly influence their pharmacokinetics and their biodistribution (Plapied et al., 2011).

The cocultured model proved that PrPsc got from infected sheep brain homogenate across M cells get a better result than Caco-2 cells. When the PrPsc injected into mouse brains, the transport particles still keeped infectivity (Heppner et al., 2001). While scrapie spread by genetically sensitive sheep's mouth, the PP absorption happened to prion (Heggeb et al., 2003). There was evidence that the M cell of normal PP played a crucial role in the prion oral delivery, which is related to bovine spongiform encephalopathy (BSE) (Terry et al., 2003). Some researchers considered that the M cell was the important incursion approach in PrPsc spreading. The histology of sheep PPs showed that the intestine nervous system presented a high innervation level in the circular region of PP. Nerve transport and lymphatic drainage affected together, making the PrPsc being transported rapidly from intestine lymph nodes. PrPsc agents usually were replicated in lymphoid tissues, before they spread into the central nervous system. Huang et al. (2002) extended this theory and suggested that dendritic cells were a cellular bridge between the gut lumen and lymphoid transmissible spongiform encephalopathies replicative machinery. PrPsc shifted from guts to the follicular dendritic cell from the B cell follicles in PP circular region through this mechanism. Normal dendritic cells can uptake prion from guts directly, just as what M cell does. Ghosh (2002) thought that prion may cross the brush border of the epithelial cell in the same way as they cross the PPs and respectively represent the enteric nervous system and lymphatic system.

Experimental data revealed that laminin receptor precursor (LRP) could be the prion's receptor in the brush border. If it can be identified that whether the LRP in human PPs and recently developed human M-cell tissue culture models increased or not, that would be very significant. But how prion and M cell and all the dendritic cell subsets interact with each other still needs more investigations in order to define why they have no effect in prion interspecific orally infection. Species specificity degradation pathway and the genetic diversity of PrP may affect the susceptibility. Recently, yeast two-hybrid screening identified the 37-kD LRP as a putative receptor for prion proteins (PrP) (Shmakov et al., 2000). The major implication of intestinal epithelial brush border expression of the LR may be an increased susceptibility to oral infection with prion proteins. The research on the gene expression and hybridization in situ in mouse follicle-associated epithelium confirmed the laminin β3 was an inducible FAE-specific gene and it was easy to imagine that this acrosomal membrane protein may interact with prion (El Bahi et al., 2002). Maybe the infection in the mammalian host by prion could occur in several ways.

There was evidence that M cells had evolved from mucosal epithelial cells of PPs. As its temporary phenotype, the ability to indentify the human M-cell receptor had been fully developed. When the enterocyte cell line Caco-2 clone was cultured with freshly isolated murine PP lymphocytes, the B cell/Caco-2 cell mixture showed the consistent feature with M-cell phenotype, inducing the transposition of particles that have been enhanced and the anchoring strength of *Vibrio cholerae* was reinforced (Kernéis et al., 1997). When it was applied to the detection of pathogens, including HIV (Fotopoulos et al., 2002) and poliovirus (Ouzilou et al., 2002), primary coculture models to be accepted generally, it became a method for human model M-cell biological observation. The data obtained from the coculture model agreed with the conclusion from recent research, which showed that the CD155 receptor of PP may play an important role in the absorption of poliovirus (Iwasaki et al., 2002).

So far, the cell surface structures that could divide the human M cell and other epithelial cells, such as intestinal mucosa epithelial cells and goblet cells, which were so little described. Lelouard et al. (2001) identified as the rabbit homologue of human Muc2 (a dome enterocyte membrane-associated mucin, which is always absent from M-cell glycocalyx) using monoclonal antibodies to recognize intestinal mucin-related epitopes. It is proposed that the presence, on the surface of M cells, of carbohydrates also expressed on Muc2, together with the absence of an enterocyte-associated mucin, could favor pathogen attachment and accessibility to the M-cell luminal membrane.

10 THE SELECTION OF VIRAL VECTORS SUITABLE FOR ORAL DELIVERY

The selection of suitable viral vectors depends on treatment purposes and the vectors' own characteristics. The transient expression of adenovirus could be used as vaccine, antiallergic agents, and the vector to kill cancer cells. But the treatment for diabetes and antiangiogenesis needs vectors with a long-term stable expression like retrovirus (Rv) and adeno-associated virus (AAV). The specific targeting for cells and transduction efficiency is also important (Giacca and Zacchigna, 2012). As of July 2015, there are 2210 gene therapy clinical trials worldwide and among them adenovirus vector is the most often used at a rate of 21.7%, higher than Rv, 18.9% (Anon, 2015). The natural tropism of the vector must often be ablated or diminished to avoid toxic side effects and might also need to be engineered to infect target cells that it does not infect naturally (Waehler et al., 2007). Viral vectors fall into one of two main categories: integrating vectors, which insert themselves into the recipient's genome, such as gamma-retroviral vectors and lentiviral vectors, are generally used to transfect actively dividing cells, as they are stably inherited; nonintegrating vectors, which usually (although not always) form an extrachromosomal genetic element, such as adenoviral vectors and AAV vectors, can be used to transfect quiescent or slowly dividing cells, but they are quickly lost from cells that divide rapidly (Sheridan, 2011; Xu, 2014). Later mentioned are the most common viral vector types that have been widely used in orally experimental studies and potential candidates for oral therapy.

10.1 RETROVIRUS

Rv is the earliest viral vector proved through animal model experiments that can transmit the DNA to GI epithelial cells. Rv was chosen because it penetrates the nucleus only at mitosis (Wang and Zheng, 2002), thus transfection is restricted to proliferating cells and GI cells have continuously proliferating characteristics.

The recombinant retroviral vector, which was based on murine leukemia virus, is a commonly used vector in gene therapy. Its biology characteristics have been well studied and the transgene can be integrated into host chromosome genomics effectively. Furthermore, its virus promoter in the long terminal repeat (LTR) sequence existing in many culture cellular types. Poor expression in vivo has often been attributed to a specific decrease in transcription of the transferred gene once the transduced cells are grafted to the animal and generally below the detection level. Loss of gene expression from the transduced gene has been observed in several systems, including hematopoietic cells, hepatocytes, fibroblasts, and keratinocytes (Ghazizadeh et al., 1997; Strauss, 1994). The reason why low expression in vivo traditionally can be construed as once the transducer was transferred to animals, the transcription of transgene would reduce specifically. Selective methylation of proviral sequence was considered a plausible approach for inactivation. Vector design, characteristics of transduction, and host immune response can relate to the transgene expression (Riviere et al., 1995). Many studies have been reported about cleaner productive process and more efficient Rv producing. And the vector system derived from lentivirus has been proven to transduct dividing and nondividing cells successfully. It was still unknown whether the recent vector technology development could overcome the limitation of intestinal gene transfer by Rv and made it back on a potentially candidate of intestinal gene therapy (Page and Cudmore, 2001). The poor transfection efficiency in vivo still limits the pervasive application of Rv in oral gene therapy so far.

10.2 ADENOVIRUS

Adenoviruses (Ads) are a family of more than 50 viral mammalian pathogens, whose nonenveloped protein capsids embody a single copy of a double-stranded DNA genome. Based on their ability to agglutinate blood erythrocytes and the homology of their genomes, Ads have been classified into species A through F (Korokhov et al., 2003). Adenovirus has obvious epithelial cells targeting, high-transfection efficiency, persist long for free and without the risk of insertional mutagenesis. It is clear that for a wide variety of cell types, Ad gave more-efficient gene transfer compared with other systems, especially in vivo. Ad vectors could transfer genes to both proliferating and quiescent cells (Mountain, 2000). The well-understood life cycle of these viruses, combined with relatively simple methods for the generation, propagation, and purification of recombinants derived from Ad2 and Ad5, made them attractive candidates as gene delivery vectors for human gene therapy (Korokhov et al., 2003). A number of systems have been developed for producing adenoviral vectors devoid of all viral coding sequences (Sakhuja et al., 2003).

Ads are the most commonly used vector. In 2003, China became the first country to approve a gene therapy–based product for clinical use (Wirth et al., 2013). Gendicine, developed by SiBiono Gene Tech Co., is an adenoviral vector, wherein the E1 gene is replaced with a human p53 cDNA. Gendicine is a nonreplicative virus and received approval for the treatment of head- and neck squamous cell carcinoma (Peng, 2005). The Ads has received general attention and is attributes to the following points: (1) as a eukaryotic gene regulation model, its genomic structure and function have been studied widely and deeply and provided a solid foundation for develop gene vectors; (2) large packing capacity, a wide range of host, high infection efficiency, and the infectious to nonproliferative splinter cells; (3) stable with physical and chemical properties, easy to prepare, purification and enrichment; (4) lower genetic toxicity. The infection and reproduction process of Ads with no need for integrate to host genome and the foreign gene transducted by Ads express as an attachment; (5) Ads could be used as live vaccine and no discernible ill effects have shown up to inoculators, Ads were generally relatively safe when compared to other vectors; (6) Ad 2 and Ad 5 have nononcogenic function to both human and animals.

When Ads infect targeting cells, the knob domain in virus capsid fibrous protein recognize and combine with the coxsackie and adenovirus receptor (CAR) on the surface of cells at first, then through the interactions of the coat neighboring protein and the integrin family molecules, which is on the surface of cells to internalize into the cell. Ads are integrated vectors and in proliferative cell there will be a loss of genome. The globular knob domains at the tips of the fibers directly contact the CAR can determine the cell tropism of Ads (Douglas, 2007). CAR was a component of the tight junction and of the functional barrier to paracellular solute movement (Cohen et al., 2001). The elimination of CAR combines was not enough to change the tropism of Ads normally and another kind of serotype fiber camouflage Ad 5 could be used (Einfeld and Roelvink, 2002). Ogawara et al. (2004) conjugated bifunctional polyethylene glycol (PEG) onto the adenoviral capsid to inhibit the interaction between viral knob and coxsackie-adenovirus receptor (CAR). Subsequently, they introduced an integrin-specific RGD peptide or E-selectin–specific antibody to the other functional group of the PEG molecule for the retargeting of the adenovirus to activate endothelial cells. In vitro studies showed that this approach resulted in the elimination of transgene transfer into CAR-positive cells, while at the same time specific transgene transfer to activate endothelial cells was achieved. The approach described here could form the basis for further development of adenoviral gene therapy vectors with improved pharmacokinetics and increased efficiency and specificity of therapeutic gene transfer into endothelial cells in disease.

As the dispersing agent of virus, cyclodextrin can enhance the absorption and endocytosis of Ad in differentiated Caco-2 monolayer and mice jejunum. Cyclodextrin could prevent the aggregation of the virus and increase the amount of virus particles that have contact with enterocyte. Croyle et al. (1998) identified a small intestine serotype Ad 41. Compared with Ad 5, Ad 41 has a stronger tendency to differentiate intestinal epithelial cell and has more potential to be a highly efficient vector for delivery of transgenic proteins to differentiate enterocytes—probable targets for oral gene delivery (Ko et al., 2009).

In studies of using the Ads transduct enterocyte in vivo, Foreman et al. (1998) have demonstrated that secreted alkaline phosphatase (SEAP) can be detected in the circulation following intraluminal administration of the vector. Recent research revealed that oral Ad4 vector priming might enhance the efficacy of poorly immunogenic vaccines, such as H5N1 (Gurwith et al., 2013). It should determine that the GI tract cells served as the sites of synthetic genes via the mediated by Ads to achieve the gene therapy. The dilution effect of the GI tract asks for high titer viruses and the production of Ads make it a promising vector. It has been proved that after Ads transduct to intestinal epithelial cells, it can secrete transgenosis-coded protein (alanine phosphorylase) to blood in vivo and secrete interleukin-1 receptor antagonist through lateral basilar membrane in vitro. Those phenomena determined the potential for intestinal epithelial cells to be a synthesis factory of exogenous gene expression product. Wirtz et al. (2002) experimented with colonic epithelial cells transduction mediated by Ads in vivo and showed that the effects on colitis mice were much better than untreated mice. Inflammation and immunogenicity side effects, which often stem from Ads, were reduced greatly. Those discoveries account for the use of packaging Ads in a colon and have the value of controlled release.

However, two decades of the extensive use of Ad-based vectors as prototypes of future gene therapeutics have revealed a number of limitations of this vector system, which have hampered its rapid transition into the clinic. One of these drawbacks is the relative inefficiency of gene delivery by Ad vectors to certain types of diseased human tissues. On the other hand, the susceptibility of various normal tissues to Ad infection makes them random targets for Ad vectors and results in the suboptimal distribution of the viruses upon administration to patients (Korokhov et al., 2003). In contrast with the Rv vector, Ads vector has higher transfection efficiency to intestinal epithelial cells. But because its proper genome cannot integrate into the chromosome of the host cell, the expression time of foreign gene, which is mediated by Ads, will be shorter and need repeated dosing. It would influence the lead to immunogenicity side effects.

The development of gutless Ads offers hope for involving this question. Except two reversed LTR and packaging signal sequence, gutless Ads lack all viral coding regions and consequently exhibit low toxicity and immunogenicity, thus permitting prolonged expression of the transgene. In addition, the elimination of all the adenoviral genes leaves room to allocate large expression cassettes (Wang et al., 2004).

10.3 ADENO-ASSOCIATED VIRUS

AAV vectors are currently among the most frequently used viral vectors for gene therapy. At recent meetings of the American Society for Gene Therapy, nearly half of the presentations involved the use of AAV (Daya and Berns, 2008). AAV is a kind of single-strand DNA virus, which has not yet caused human disease. AAV is defective and relies on the helper virus. Wild-type AAV is a nonpathogenic parvovirus, which is incapable of autonomous replication and spread (Maheshri et al., 2006). Because AAV vectors are administered directly to the patient, the likelihood of a host immune response is high, as shown by human studies (Mingozzi and High, 2013). Twelve types of AAVs have been separated and studies on antibody cross-reactivity determine there are seven serotypes. Among them, AAV6 is the recombinant of AAV1 and 2. The seroepidemiology research point out that AAV2, 3, and 5 are present in the human body, AAV4 infect nonhuman primates chiefly. The host of AAV1 and 6 is still not clear because they had not been separated, but their antibody had been found in human and nonhuman primates. Recent studies using vectors derived from alternative AAV serotypes, such as AAV1, 4, 5, and 6, have shown improved potency and broadened tropism of the AAV vector by packaging the same vector genome with different AAV capsids (Gao et al., 2005). The targeting of every type made them suitable for application in vivo (Gao et al., 2002). AAV1 have been demonstrated that they could be used in advanced heart failure (Jessup et al., 2011), AAV6 and AAV9 performed better for motor neuron diseases treatment than the other serotypes (Snyder et al., 2011). Furthermore, some serotypes may generally be similar in their ability to target and transduce rhesus hepatocytes, such as AAV2, AAV5, and AAV8 (Davidoff et al., 2005; Xu, 2014); they showed different prevalence of immunity (Allay et al., 2011; Xu, 2014). Using a recombinant adeno-associated virus (rAAV) carrying the RPE65 gene (AAV2-hRPE65v2) for the treatment with congenital blindness have been proved to be a safe and stable procedure (Bennett et al., 2012). AAV5 vector is for high efficacy and mutually exclusive expression in neurons or astrocytes (Drinkut et al., 2012). Existing AAV vectors cannot be duplicated but keep the capability of infecting dividing and nondividing cells. Compared with adenovirus and Rv, AAV vectors have lower immunogenicity and can be produced without helper virus. In the rAAV vector system, almost all of the parental genome has been deleted such that only the terminal repeats remain, containing recognition signals for DNA replication and packaging (Maheshri et al., 2006). Because this kind of vector did not contain all the virus genes makes the possibility of recombinant and virus gene expression reduced to minimum. Although AAV may cause an immune response, when compared with early inflammation that was caused by Ads the reaction seems very little. A significant advantage of AAV is its long-term gene expression. The reason may be that the AAV vector, including foreign genes, could integrate to the host chromosome.

Another characteristic of AAV that made it particularly suitable for acting as orally gene transmission vector is its tolerance. AAV can resist temperature, is extremely pH and solvent. After infection by AAV in the human body, the wild-type can be found in respiratory organs and GI tract secretions, so the intestinal tract is the common host tissue (Xu et al., 2006b). When delivered in vivo, AAV vectors can have a long-term expression in terminal differentiated cells (Xu, 2014). Early studies have determined that the expression of transgenosis in primordial stem cell, intestinal cell, nerve cell, and lamina propia cell can last for months. Because of the limited capacity, AAV is not being used in the treatment of disease that needs transferring large DNA fragments. However, recent studies show that this limitation can be solved. Sun et al. (2000) used the hybridizing dimerization of DNA to overcome the vector packaging limitation. This can resolve the problem for packaging large genes into AAV vectors, and can also provide a new paradigm of including gene regulatory elements that are now too large to be accommodated. More importantly, oral gene therapy can transfer various vectors; multiple genes can be transferred at the same time without the limit of size and length of insert DNA. This is a significant advantage when long fragment gene for treatment is needed (Xu, 2004). Lopes et al. (2013) make the MYO7A cDNA packaged in single AAV2 and AAV5 vectors and as two overlapping halves in dual AAV2 vectors. Those single vector preparations successfully transduced photoreceptor and retinal pigment epithelium cells, resulting in functional, full-length MYO7A protein and correction of mutant phenotypes, suggesting successful homologous recombination of gene fragments. Flotte et al. (2011) determined the feasibility and safety of AAV gene therapy for $\alpha 1$-antitrypsin deficiency, and indicating that serum levels of vector-derived normal human AAT > 20 lg/ml would be achieved. Bowles et al. (2012) performed the first clinical trial using a chimeric AAV capsid variant for successful treatment of Duchenne muscular dystrophy. There was no cellular immune response to this rationally designed vector and the safe and well-tolerated lays the foundation of customizing AAV vectors that best suit the clinical objective. Nathwani et al. (2011) demonstrated successful conversion of severe hemophilia B to mild or moderate disease in six adult males who underwent intravenous infusion of an AAV vector expressing factor IX. AAV used in different gene encoding got excellent effects (Hao et al., 2012; Maclaren et al., 2014). In a word, AAV vector, as an efficient orally transfer carrier, has great potential for genetic modification treatment and gene vaccine. Up to now there are no direct comparison research of AAV and Ads transfer DNA through intestinal structure, but correlational studies showed that AAV vector with lower immune response and longer transgenosis expression should be a more effective DNA transmission tool.

Most experts consider that the AAV is a more secure vector, proved by many studies and a commonly used medium in gene therapy and genetic engineering. This virus was used for carrying genes to the specific organ and tissue of human and organism in many countries. In the autumn of 2005, the FDA approved the multidose gene therapy for rheumatic arthritis in the United States more than 20 places and recruited two groups of 201 patients in total (32 of them using placebo), no one at a serious side effect. However, in 2007, an American woman Jolee Mohr, at age 36, who appeared with high fever, vomiting, general extensive bleeding, and hepatic failure, died three weeks after secondary administration. This accident sparked widespread controversy for the security of AAV. Terry Flotte, the president of the University of Massachusetts medical college, considered that because TNF profilin can restrain the immune system, it may be related to the death of Jolee, such as lead to pyemia and bacterial infection. On the other hand, Carte suggested that TNF profilin may not be pathogenic because it had not been detected in primates and human serum and only observed in the joint. And Harvard University molecular plastic surgeon Christopher Evans believed the crux may consist in the number of gene therapy trials because Jolee accepted more than one injection than previous patients. It therefore may allow patients to be more sensitive to AAV and cause adverse reactions. It had happened before, in the state of Pennsylvania (United States), two patients using the AAV as the transporter received liver administration and produced moderate immune response.

10.4 OTHER VIRAL VECTOR SYSTEMS

Although many viral vectors could generate mucosal immune response, only one part of them can be used in the foreign gene transmission through the digestive tract. Lentiviral vectors have been repeatedly shown to have low immunogenicity and an ability to transduce neuronal cells with high efficiency and carry a larger therapeutic cargo than the popular AAV vectors. Palfi et al. (2014) have generated a tricistronic lentiviral vector named ProSavin, which was safe and well tolerated in patients and can be used as a long-term treatment for Parkinson's disease, while Gherardi and Esteban (1999) demonstrated that vaccinia virus (VV) recombinants administered by the oral route generate mucosal and systemic immune responses against antigens of the virus vector and the recombinant products. In this study they found that after oral administration of a VV recombinant expressing the luciferase reporter gene, VV gene expression took place for several days in gut-associated lymphoid (GALT) tissues, as well as in the spleen of mice. Hoyng et al. (2014) have developed a fully functional lentivirus vector system capable of regulating transgene expression in vivo via a tetracycline inducible transactivator that possesses enhanced immune-evasive properties. In addition, plant-specific virus can also express exogenous antigen products to cause an immune response in the way of gene be transferred by the digestive tract. Generally, this kind of virus enters the digestive tract for gene delivery in the way of wrapped by plant cells. Therefore, it can defend the interference effect from intestine microenvironment and produce a better immune effect. What's more, baculovirus- mediated gene delivery have the same potential being used in viral vector production (Airenne et al., 2013). Other viral vectors that can be used in digestive tract gene delivery include poliomyelitis cells (have natural tropism of enterocytes), Sindbis virus, and Venezuelan equine encephalomyelitis virus and so on (Xu et al., 2006a).

Over the past years, improved manufacturing processes have been developed with reduced replication-competent virus and increased yield for Ad, Rv, and AAV. This is an important development because it means the feasibility of manufacturing these vectors in adequate quality and quantity for clinical studies need no longer be a major factor in vector selection (Mountain, 2000; Xu, 2014).

11 THE DIGESTION CIRCUMSTANCE OF ORAL GENE DELIVERY

Information of orally viral transmission development was limited. However, the TNO Stomach and Small Intestine Model (TIM-1) was a suitable model to evaluate the viability of recombination microorganisms and the activity of exogenous gene (Xu et al., 2008). Compared with animal models, this model has advantages of high accuracy, good reproducibility, easy to operate and gather samples from the alimentary tract in any time and sites without the ethic limits. The computer program has been designed to accept parameters and data obtained from in vivo studies in animals or human volunteers, such as the quantity and duration of a meal, the pH curves for the stomach and duodenum, secretion rates into the different compartments, water absorption from the small intestine and gastric and ileal delivery into the duodenum and colon (Minekus et al., 1995). A variety of studies have shown that this system is available, including digestive and absorptive of nutrition, the survival rate of microorganisms, and effectiveness of absorbing the mutagen of mineral substance, vitamins, and food (Drouault et al., 1999).

Taïeb et al. (2004) reported that the oligosaccharide and glycosphingolipid of bacteria vector surface can provide a special ground substance for the pathogen targeting the brush border. It equally shows that the lipid vector can be used as the port for drug stress endocytosis (Taïeb et al., 2004). However, the conditions necessary for viral vectors was still not clear. Secretion efficiency of recombination microorganisms had not been quantified directly. We proved for the first time that rAAV vector can cause synthesis and secretion of different sizes of proteins in animal digestive tract environment. This research revealed that after oral administration of rAAV, there was a long-term stable expression of β-galactosidase enzyme in intestinal epithelial cells and lamina propria cells (During et al., 1998, 2000). This work proved that oral gene delivery can secrete compounds in the digestive tract directly and open the door for developing innovation drug delivery vectors.

Recent studies showed that baculovirus displaying hemagglutinin vaccine enhanced the mucosal IgA level significantly. Furthermore, immunohistochemistry results demonstrated that the live vaccine was able to transduce and express hemagglutinin in the intestinal epithelial cells both in vitro and in vivo (Prabakaran et al., 2010). Gref et al. (2012) demonstrated that PEG-coated nanospheres exhibited increased blood circulation times and reduced liver accumulation. Despite the recent attention to intranasal administration to enhance mucosal immune responses, oral vaccination is still considered as the best approach to increase patient compliance.

12 THE TRANSMISSION, DISTRIBUTION, AND TRANSGENIC PRODUCT PHARMACOKINETICS OF ORAL VIRAL VECTORS

The research on the distribution and elimination in different tissues of viral vectors will provide insight into the pharmacokinetic parameters of pesticide effect and gene expression levels. Evidences collected from the Wistar rat after orally treated with rAAV for 6 h and used PCR to measure the distribution of AAV revealed that AAV only existed in the guts and liver and not found in the spleen, testis, brain, and lung (Xu et al., 2008).

The research on viral vector delivery pathway, distribution, and gene expression after oral administration of rAAV can provide strong proof for oral gene delivery and gene therapy. Using PCR to determine undigested AAV genomic DNA in the blood sample of portal and caudal vein at different time courses, we found that AAV vectors can be passed to the liver by the portal vein directly (Xu et al., 2008). The positive PCR results were detected in all portal vein samples but little or nearly not in caudal vein samples. The trace amounts of positive fragments can be detected after 1 h. The intestinal epithelial was a highly vascularized tissue with capillary very close to the epithelial layer. The AAV peak appeared at the 3 h after oral and then decreased gradually. PCR data also showed that AAV vector still existed in the liver after 8 h. Real-time PCR can further expand the detection range and determined the amount and specific location of viral vectors in different cells. All the results indicated clearly that viral vectors, such as rAAV could cross the GI epithelial cells by way of transcellular and intercellular then enter the liver by the portal vein. According to the experiments of the time needs of transgenosis expressed in GI tract, we discovered that the gene product began to appear at 6 h after oral admission in guts. Drawing blood from portal and caudal vein at the same time, we used the IRMA, which was high specificity for human insulin, less cross-reactive than 0.005% in proinsulin, and no cross reactions with C-peptide, glucagon, or rat insulin to measure the insulin concentration. The insulin concentration of the blood sample in portal and caudal vein were (6.2 ± 2.8) μU mL ($n = 6$) and (1.8 ± 0.9) μU mL respectively. However, there was no insulin detected in the control group. This implied that the protein encoded by intestinal epithelial tissue may secrete into portal circulation and make the transgenosis insulin level in portal vein reach the peak. There was also the possibility that the open-ended lymph pathway in intestinal villus could let the insulin enter into the lymph and then transfer the insulin into blood by pleural canals, which are connected by venous circulation. After 6 h of medication, it could be observed that there was a high level of insulin in the blood taken from experimental animals. Excitingly, the mature insulin generated can cross the membrane of the epithelial cell and capillary membrane then enter into the circulatory system. It can be the principal channel of mature insulin transmission. However, the levels of mature transgenosis insulin drop below 0.7 μU mL on day 3 and below the detection bottom on day 7. The reduced level of blood transgenosis insulin agreed with the defluvium of those transducted intestinal epithelial cells.

Oral delivery of insulin and other bioactive proteins and peptides continue to receive much attention, despite the significant obstacles presented by the harsh environmental conditions within the GI tract. The next steps forward would require engineering of the delivery vehicle to interact with or respond to this harsh, variable environment. The vehicle's material constitution must ensure safe passage of the protein cargo to the blood or other destinations within the body, protecting it throughout its transit through the digestive circuitry intent on breaking it down (Bakhru et al., 2013).

13 POTENTIAL, PROBLEMS, AND PROSPECTS

Compared with traditional ways, the oral viral vector transmission system has the following advantages: (1) it is allowed to use the receptor known hypersensitivity to peptic secretion. Because its gene expression pathway can be adjusted, therefore, it can target the specific loci in the digestive tract and the release would be managed. It can achieve the same effect at a lower dose and avoid the active substance degraded by acid and protease in absorption or reaction sites; (2) the biotransformation of AAV generated compounds that were directly acting on the digestive tract. For example, the AAV can modify the metabolic disorder, which is caused by the lack of enzymes in the stomach and intestine or organ failure (Alexander et al., 2008); (3) the active product can be released by cell lysis or expressed in GI condition (Steidler et al., 2000); (4) we can use a variety of oral viral vectors at the same time; (5) we can transport a variety of genes at the same time without the capacity limit (Xu, 2004); (6) reduce the immunity side effects caused by repeated intake of medicines.

Oral gene vectors also have potential in the development of oral DNA vaccine. This kind of vaccine can present protein antigen in GALT (Rosenkranz et al., 2003), which can limit the distribution of transferred DNA in guts and retain the advantages of nonviral gene transmission, including repeated intake of medicines.

The main limiting factor for oral gene vector is the poor gene expression efficiency. How to express the right amount of gene medicine in specific targeting tissue at the right time is the most important key for future research.

To explore fully the potential of oral transmission, it is necessary to determine the way of transport from guts to liver. Whether the transgenic product in liver was released to the circulatory system directly or transmitted by biliary drainage still remain unclear (Xu et al., 2006b). There are questions that still need systematic investigation, such as the huge individual difference of transduction efficiency, potential cytotoxicity, and the GI circumstance effect on viral vector absorption kinetics.

14 CONCLUSIONS

It is extremely important to select a suitable gene medicine for treatment. Despite the fact that there was great progress in orally viral vector delivery, it is far from optimized. Compared with the nonviral vectors, such as lipid, the viral vector has more complicated biological characteristics and higher difficulty in studying. Applying the oral vector to gene therapy still faces challenges and obstacles in the technology of transmission, gene, and cell engineering, including the degradation of DNA enzymes in guts, the body fluid barrier, the stable synthesis and release for protein products, the lack of specific targeting, the organic response generated by organism physiological needs, and the safety of the medication.

The perfection of knowledge will certainly enable us to establish ideal treatments of transmission technology. With the increasing development of this technology, those obstacles would be solved gradually. Oral nanoparticle gene therapy can become a clinical practicable solution. The application of nanomaterials in oral medicine research can accelerate such a process.

REFERENCES

Ahmed, F., Alexandridis, P., Neelamegham, S., 2001. Synthesis and application of fluorescein-labeled Pluronic block copolymers to the study of polymer-surface interactions. Langmuir 17, 537–546.

Airenne, K.J., Hu, Y.-C., Kost, T.A., Smith, R.H., Kotin, R.M., Ono, C., Matsuura, Y., Wang, S., Yl-Herttuala, S., 2013. Baculovirus: an insect-derived vector for diverse gene transfer applications. Mol. Ther. 21, 739–749.

Alexander, I., Cunningham, S., Logan, G., Christodoulou, J., 2008. Potential of AAV vectors in the treatment of metabolic disease. Gene Ther. 15, 831–839.

Allay, J.A., Sleep, S., Long, S., Tillman, D.M., Clark, R., Carney, G., Fagone, P., Mcintosh, J.H., Nienhuis, A.W., Davidoff, A.M., 2011. Good manufacturing practice production of self-complementary serotype 8 adeno-associated viral vector for a hemophilia B clinical trial. Human Gene Ther. 22, 595–604.

Anon, 2015. Gene Therapy Clinical Trials Worldwide. Available from: http://www.wiley.com/legacy/wileychi/genmed/clinical/

Arshady, R., 2001. Radiolabeled and Magnetic Particulates in Medicine & Biology. Citus Books, London.

Artursson, P., Palm, K., Luthman, K., 2012. Caco-2 monolayers in experimental and theoretical predictions of drug transport. Adv. Drug Deliv. Rev. 64, 280–289.

Ataman-Onal, Y., Delair, T., Verrier, B., 2015. Composition comprising a colloidal synthetic bioresorbable vector and a viral vector. Google Patents.

Awaad, A., Nakamura, M., Ishimura, K., 2012. Imaging of size-dependent uptake and identification of novel pathways in mouse Peyer's patches using fluorescent organosilica particles. Nanomedicine 8, 627–636.

Bai, Y., Gong, H., Li, H., Vu, G.-P., Lu, S., Liu, F., 2011. Oral delivery of RNase P ribozymes by *Salmonella* inhibits viral infection in mice. Proc. Nat. Acad. Sci. 108, 3222–3227.

Bakhru, S.H., Furtado, S., Morello, A.P., Mathiowitz, E., 2013. Oral delivery of proteins by biodegradable nanoparticles. Adv. Drug Deliv. Rev. 65, 811–821.

Bennett, J., Ashtari, M., Wellman, J., Marshall, K.A., Cyckowski, L.L., Chung, D.C., Mccague, S., Pierce, E.A., Chen, Y., Bennicelli, J.L., 2012. AAV2 gene therapy readministration in three adults with congenital blindness. Sci. Trans. Med. 4, 120ra15–1120ra.

Bowles, D.E., Mcphee, S.W., Li, C., Gray, S.J., Samulski, J.J., Camp, A.S., Li, J., Wang, B., Monahan, P.E., Rabinowitz, J.E., 2012. Phase 1 gene therapy for Duchenne muscular dystrophy using a translational optimized AAV vector. Mol. Ther. 20, 443–455.

Brayden, D.J., Baird, A.W., 2004. Apical membrane receptors on intestinal M cells: potential targets for vaccine delivery. Adv. Drug Deliv. Rev. 56, 721–726.

Chen, M.-C., Mi, F.-L., Liao, Z.-X., Hsiao, C.-W., Sonaje, K., Chung, M.-F., Hsu, L.-W., Sung, H.-W., 2013. Recent advances in chitosan-based nanoparticles for oral delivery of macromolecules. Adv. Drug Deliv. Rev. 65, 865–879.

Cohen, C.J., Shieh, J.T., Pickles, R.J., Okegawa, T., Hsieh, J.-T., Bergelson, J.M., 2001. The coxsackievirus and adenovirus receptor is a transmembrane component of the tight junction. Proc. Nat. Acad. Sci. 98, 15191–15196.

Cooper, D.L., Conder, C.M., Harirforoosh, S., 2014. Nanoparticles in drug delivery: mechanism of action, formulation and clinical application towards reduction in drug-associated nephrotoxicity. Expert Opin. Drug Deliv. 11, 1661–1680.

Croyle, M.A., Stone, M., Amidon, G.L., Roessler, B.J., 1998. In vitro and in vivo asessment of adenovirus 41 as a vector for gene delivery to the intestine. Gene. Ther. 5, 645–655.

Davidoff, A.M., Gray, J.T., Ng, C.Y., Zhang, Y., Zhou, J., Spence, Y., Bakar, Y., Nathwani, A.C., 2005. Comparison of the ability of adeno-associated viral vectors pseudotyped with serotype 2, 5, and 8 capsid proteins to mediate efficient transduction of the liver in murine and nonhuman primate models. Mol. Ther. 11, 875–888.

Daya, S., Berns, K.I., 2008. Gene therapy using adeno-associated virus vectors. Clin. Microbiol. Rev. 21, 583–593.

Deli, M.A., 2009. Potential use of tight junction modulators to reversibly open membranous barriers and improve drug delivery. Biochim. Biophys. Acta 1788, 892–910.

des Rieux, A., Fievez, V., Garinot, M., Schneider, Y.-J., Pr At, V., 2006. Nanoparticles as potential oral delivery systems of proteins and vaccines: a mechanistic approach. J. Control. Release 116, 1–27.

Dorney, J., 2013. Polystyrene: A Potential Standard For Developing In Vitro Cellular Tracking Methods For Nanotoxicology. Ph.D. Thesis. Dublin Institute of Technology, Dublin.

Douglas, J.T., 2007. Adenoviral vectors for gene therapy. Mol. Biotechnol. 36, 71–80.

Doyle-McCullough, M., Smyth, S., Moyes, S., Carr, K., 2007. Factors influencing intestinal microparticle uptake in vivo. Int. J. Pharm. 335, 79–89.

Drinkut, A., Tereshchenko, Y., Schulz, J.B., B Hr, M., K Gler, S., 2012. Efficient gene therapy for Parkinson's disease using astrocytes as hosts for localized neurotrophic factor delivery. Mol. Ther. 20, 534–543.

Drouault, S., Corthier, G., Ehrlich, S.D., Renault, P., 1999. Survival, Physiology, and Lysis of Lactococcus lactis in the Digestive Tract. Appl. Environ. Microbiol. 65, 4881–4886.

During, M.J., Xu, R., Young, D., Kaplitt, M.G., Sherwin, R.S., Leone, P., 1998. Peroral gene therapy of lactose intolerance using an adeno-associated virus vector. Nat. Med. 4, 1131–1135.

During, M.J., Symes, C.W., Lawlor, P.A., Lin, J., 2000. An oral vaccine against NMDAR1 with efficacy in experimental stroke and epilepsy. Science 287, 1453.

Efrat, S., Surana, M., Fleischer, N., 1991. Glucose induces insulin gene transcription in a murine pancreatic beta-cell line. J. Biol. Chem. 266, 11141–11143.

Eguchi, A., Akuta, T., Okuyama, H., Senda, T., Yokoi, H., Inokuchi, H., Fujita, S., Hayakawa, T., Takeda, K., Hasegawa, M., 2001. Protein transduction domain of HIV-1 Tat protein promotes efficient delivery of DNA into mammalian cells. J. Biol. Chem. 276, 26204–26210.

Einfeld, D., Roelvink, P., 2002. Advances towards targetable adenovirus vectors for gene therapy. Curr. Opin. Mol. Thera. 4, 444–451.

El Bahi, S., Caliot, E., Bens, M., Bogdanova, A., Kernéis, S., Kahn, A., Vandewalle, A., Pringault, E., 2002. Lymphoepithelial interactions trigger specific regulation of gene expression in the M cell-containing follicle-associated epithelium of Peyer's patches. J. Immunol. 168, 3713–3720.

Eldridge, J.H., Staas, J.K., Meulbroek, J.A., McGhee, J.R., Tice, T.R., Gilley, R.M., 1991. Biodegradable microspheres as a vaccine delivery system. Mol. Immunol. 28, 287–294.

Flotte, T.R., Trapnell, B.C., Humphries, M., Carey, B., Calcedo, R., Rouhani, F., Campbell-Thompson, M., Yachnis, A.T., Sandhaus, R.A., Mcelvaney, N.G., 2011. Phase 2 clinical trial of a recombinant adeno-associated viral vector expressing α1-antitrypsin: interim results. Hum. Gene Ther. 22, 1239–1247.

Foreman, P.K., Wainwright, M.J., Alicke, B., Kovesdi, I., Wickham, T.J., Smith, J.G., Meier-Davis, S., Fix, J.A., Daddona, P., Gardner, P., 1998. Adenovirus-mediated transduction of intestinal cells in vivo. Hum. Gene Ther. 9, 1313–1321.

Fotopoulos, G., Harari, A., Michetti, P., Trono, D., Pantaleo, G., Kraehenbuhl, J.-P., 2002. Transepithelial transport of HIV-1 by M cells is receptor-mediated. Proc. Nat. Acad. Sci. 99, 9410–9414.

Frankel, A.D., Pabo, C.O., 1988. Cellular uptake of the tat protein from human immunodeficiency virus. Cell 55, 1189–1193.

Fröhlich, E., 2012. The role of surface charge in cellular uptake and cytotoxicity of medical nanoparticles. Int. J. Nanomedicine 7, 5577.

Gao, G.-P., Alvira, M.R., Wang, L., Calcedo, R., Johnston, J., Wilson, J.M., 2002. Novel adeno-associated viruses from rhesus monkeys as vectors for human gene therapy. Proc. Nat. Acad. Sci. 99, 11854–11859.

Gao, G., Vandenberghe, L.H., Wilson, J.M., 2005. New recombinant serotypes of AAV vectors. Curr. Gene Ther. 5, 285–297.

Garinot, M., Fi Vez, V., Pourcelle, V., Stoffelbach, F., Des Rieux, A., Plapied, L., Theate, I., Freichels, H., J R Me, C., Marchand-Brynaert, J., 2007. PEGylated PLGA-based nanoparticles targeting M cells for oral vaccination. J. Control. Release 120, 195–204.

Ghazizadeh, S., Carroll, J.M., Taichman, L.B., 1997. Repression of retrovirus-mediated transgene expression by interferons: implications for gene therapy. J. Virol. 71, 9163–9169.

Gherardi, M.M., Esteban, M., 1999. Mucosal and systemic immune responses induced after oral delivery of vaccinia virus recombinants. Vaccine 17, 1074–1083.

Ghosh, S., 2002. Intestinal entry of prions. Z. Gastroenterol. 40, 37–39.

Giacca, M., Zacchigna, S., 2012. Virus-mediated gene delivery for human gene therapy. J. Control. Release 161, 377–388.

Giannasca, P.J., Giannasca, K.T., Leichtner, A.M., Neutra, M.R., 1999. Human intestinal M cells display the sialyl Lewis A antigen. Infect. Immun. 67, 946–953.

Green, M., Loewenstein, P.M., 1988. Autonomous functional domains of chemically synthesized human immunodeficiency virus tat trans-activator protein. Cell 55, 1179–1188.

Gref, R., Domb, A., Quellec, P., Blunk, T., Müller, R., Verbavatz, J., Langer, R., 2012. The controlled intravenous delivery of drugs using PEG-coated sterically stabilized nanospheres. Adv. Drug. Deliv. Rev. 64, 316–326.

Guliyeva, Ü., Öner, F., Özsoy, Ş., Haziroğlu, R., 2006. Chitosan microparticles containing plasmid DNA as potential oral gene delivery system. Eur. J. Pharm. Biopharm. 62, 17–25.

Gullberg, E., Leonard, M., Karlsson, J., Hopkins, A.M., Brayden, D., Baird, A.W., Artursson, P., 2000. Expression of specific markers and particle transport in a new human intestinal M-cell model. Biochem. Biophys. Res. Comm. 279, 808–813.

Gurwith, M., Lock, M., Taylor, E.M., Ishioka, G., Alexander, J., Mayall, T., Ervin, J.E., Greenberg, R.N., Strout, C., Treanor, J.J., 2013. Safety and immunogenicity of an oral, replicating adenovirus serotype 4 vector vaccine for H5N1 influenza: a randomised, double-blind, placebo-controlled, phase 1 study. Lancet Infect. Dis. 13, 238–250

Hao, Z.M., Cai, M., Lv, Y.F., Huang, Y.H., Li, H.H., 2012. Oral administration of recombinant adeno-associated virus-mediated bone morphogenetic protein-7 suppresses CCl4-induced hepatic fibrosis in mice. Mol. Ther. 20, 2043–2051.

He, C., Yin, L., Tang, C., Yin, C., 2012. Size-dependent absorption mechanism of polymeric nanoparticles for oral delivery of protein drugs. Biomaterials 33, 8569–8578.

Hearnden, V., Sankar, V., Hull, K., Juras, D.V., Greenberg, M., Kerr, A.R., Lockhart, P.B., Patton, L.L., Porter, S., Thornhill, M.H., 2012. New developments and opportunities in oral mucosal drug delivery for local and systemic disease. Adv. Drug Deliv. Rev. 64, 16–28.

Heggeb, R., Gonz Lez, L., Press, C.M., Gunnes, G., Espenes, A., Jeffrey, M., 2003. Disease-associated PrP in the enteric nervous system of scrapie-affected Suffolk sheep. J. Gen. Virol. 84, 1327–1338.

Heppner, F.L., Christ, A.D., Klein, M.A., Prinz, M., Fried, M., Kraehenbuhl, J.-P., Aguzzi, A., 2001. Transepithelial prion transport by M cells. Nat. Med. 7, 976–977.

Hillyer, J.F., Albrecht, R.M., 2001. Gastrointestinal persorption and tissue distribution of differently sized colloidal gold nanoparticles. J. Pharm. Sci. 90, 1927–1936.

Hoyng, S., Gnavi, S., De Winter, F., Eggers, R., Ozawa, T., Zaldumbide, A., Hoeben, R., Malessy, M., Verhaagen, J., 2014. Developing a potentially immunologically inert tetracycline-regulatable viral vector for gene therapy in the peripheral nerve. Gene Ther. 21, 549–557.

Huang, F.-P., Farquhar, C.F., Mabbott, N.A., Bruce, M.E., Macpherson, G.G., 2002. Migrating intestinal dendritic cells transport PrPSc from the gut. J. Gen. Virol. 83, 267–271.

Hussain, N., Jaitley, V., Florence, A.T., 2001. Recent advances in the understanding of uptake of microparticulates across the gastrointestinal lymphatics. Adv. Drug Deliv. Rev. 50, 107–142.

Iwasaki, A., Welker, R., Mueller, S., Linehan, M., Nomoto, A., Wimmer, E., 2002. Immunofluorescence analysis of poliovirus receptor expression in Peyer's patches of humans, primates, and CD155 transgenic mice: implications for poliovirus infection. J. Infect. Dis. 186, 585–592.

Jain, A.K., Goyal, A.K., Mishra, N., Vaidya, B., Mangal, S., Vyas, S.P., 2010. PEG–PLA–PEG block copolymeric nanoparticles for oral immunization against hepatitis B. Int. J. Pharm. 387, 253–262.

Jang, J.-H., Schaffer, D.V., Shea, L.D., 2011. Engineering biomaterial systems to enhance viral vector gene delivery. Moll Ther. 19, 1407–1415.

Jepson, M.A., Clark, M., Foster, N., mason, C.M., Bennett, M., Simmons, N., Hirst, B., 1996. Targeting to intestinal M cells. J. Anat. 189, 507.

Jessup, M., Greenberg, B., Mancini, D., Cappola, T., Pauly, D.F., Jaski, B., Yaroshinsky A., Zsebo, K.M., Dittrich, H., Hajjar, R.J., 2011. Calcium upregulation by percutaneous administration of gene therapy in cardiac disease (CUPID) a phase 2 trial of intracoronary gene therapy of sarcoplasmic reticulum Ca2+-ATPase in patients with advanced heart failure. Circulation 124, 304–313.

Kadiyala, I., Loo, Y., Roy, K., Rice, J., Leong, K.W., 2010. Transport of chitosan—DNA nanoparticles in human intestinal M-cell model versus normal intestinal enterocytes. Eur. J. Pharm. Sci. 39, 103–109.

Kay, M.A., Glorioso, J.C., Naldini, L., 2001. Viral vectors for gene therapy: the art of turning infectious agents into vehicles of therapeutics. Nat. Med. 7, 33–40.

Kernéis, S., Bogdanova, A., Kraehenbuhl, J.-P., Pringault, E., 1997. Conversion by Peyer's patch lymphocytes of human enterocytes into M cells that transport bacteria. Science 277, 949–952.

Ko, S.Y., Cheng, C., Kong, W.P., Wang, L., Kanekiyo, M., Einfeld, D., King, C.R., Gall, J.G., Nabel, G.J., 2009. Enhanced induction of intestinal cellular immunity by oral priming with enteric adenovirus 41 vectors. J. Virol. 83, 748–756.

Korokhov, N., Mikheeva, G., Krendelshchikov, A., Belousova, N., Simonenko, V., Krendelshchikova, V., Pereboev, A., Kotov, A., Kotova, O., Triozzi, P.L., 2003. Targeting of adenovirus via genetic modification of the viral capsid combined with a protein bridge. J. Virol. 77, 12931–12940.

Kraehenbuhl, J.-P., Neutra, M.R., 2000. Epithelial M cells: differentiation and function. Ann. Rev. Cell Dev. Biol. 16, 301–332.
Kriegel, C., Attarwala, H., Amiji, M., 2013. Multi-compartmental oral delivery systems for nucleic acid therapy in the gastrointestinal tract. Adv. Drug Deliv. Rev. 65, 891–901.
Kulkarni, S.A., Feng, S.S., 2013. Effects of particle size and surface modification on cellular uptake and biodistribution of polymeric nanoparticles for drug delivery. Pharm. Res. 30, 2512–2522.
Lai, S.K., Wang, Y.-Y., Hanes, J., 2009. Mucus-penetrating nanoparticles for drug and gene delivery to mucosal tissues. Adv. Drug Deliv. Rev. 61, 158–171.
Lambkin, I., Pinilla, C., 2002. Targeting approaches to oral drug delivery. Exp. Opin. Biol. Ther. 2, 67–73.
Lamprecht, A., Sch Fer, U., Lehr, C.-M., 2001. Size-dependent bioadhesion of micro-and nanoparticulate carriers to the inflamed colonic mucosa. Pharm. Res. 18, 788–793.
Lelouard, H., Reggio, H., Roy, C., Sahuquet, A., Mangeat, P., Montcourrier, P., 2001. Glycocalyx on rabbit intestinal M cells displays carbohydrate epitopes from Muc2. Infect. Immun. 69, 1061–1071.
Lelouard, H., Fallet, M., De Bovis, B., M Resse, S., Gorvel, J.P., 2012. Peyer's patch dendritic cells sample antigens by extending dendrites through M cell-specific transcellular pores. Gastroenterology 142, 592–601, e3.
Li, P., Nielsen, H.M., Müllertz, A., 2012. Oral delivery of peptides and proteins using lipid-based drug delivery systems. Exp. Opin. Drug Deliv. 9, 1289–1304.
Lim, S.T., Airavaara, M., Harvey, B.K., 2010. Viral vectors for neurotrophic factor delivery: a gene therapy approach for neurodegenerative diseases of the CNS. Pharm. Res. 61, 14–26.
Lopes, V.S., Boye, S.E., Louie, C.M., Boye, S., Dyka, F., Chiodo, V., Fofo, H., Hauswirth, W.W., Williams, D.S., 2013. Retinal gene therapy with a large MYO7A cDNA using adeno-associated virus. Gene Ther. 20, 824–833.
Maclaren, R.E., Groppe, M., Barnard, A.R., Cottriall, C.L., Tolmachova, T., Seymour, L., Clark, K.R., During, M.J., Cremers, F.P., Black, G.C., 2014. Retinal gene therapy in patients with choroideremia: initial findings from a phase 1/2 clinical trial. Lancet 383, 1129–1137.
Maheshri, N., Koerber, J.T., Kaspar, B.K., Schaffer, D.V., 2006. Directed evolution of adeno-associated virus yields enhanced gene delivery vectors. Nat. Biotechnol. 24, 198–204.
Mandir, N., Fitzgerald, A.J., Goodlad, R.A., 2005. Differences in the effects of age on intestinal proliferation, crypt fission and apoptosis on the small intestine and the colon of the rat. Int. J. Exp. Pathol. 86, 125–130.
Minekus, M., Marteau, P., Havenaar, R., 1995. A multi-compartmental dynamic computer-controlled model simulating the stomach and small intestine. Alt. Lab. Anim. 23, 197–209.
Mingozzi, F., High, K.A., 2011. Therapeutic in vivo gene transfer for genetic disease using AAV: progress and challenges. Nat. Rev. Gen. 12, 341–355.
Mingozzi, F., High, K.A., 2013. Immune responses to AAV vectors: overcoming barriers to successful gene therapy. Blood 122, 23–36.
Mohanraj, V., Chen, Y., 2007. Nanoparticles-a review. Trop. J. Pharm. Res. 5, 561–573.
Moore, V.D.G., Payne, R.M., 2004. Transactivator of transcription fusion protein transduction causes membrane inversion. J. Biol. Chem. 279, 32541–32544.
Morgan, T.T., Muddana, H.S., Altinoglu, E.I., Rouse, S.M., Tabakovic, A., Tabouillot, T., Russin, T.J., Shanmugavelandy, S.S., Butler, P.J., Eklund, P.C., 2008. Encapsulation of organic molecules in calcium phosphate nanocomposite particles for intracellular imaging and drug delivery. Nano Lett. 8, 4108–4115.
Mountain, A., 2000. Gene therapy: the first decade. Trends Biotechnol. 18, 119–128.
Nathwani, A.C., Tuddenham, E.G., Rangarajan, S., Rosales, C., Mcintosh, J., Linch, D.C., Chowdary, P., Riddell, A., Pie, A.J., Harrington, C., 2011. Adenovirus-associated virus vector–mediated gene transfer in hemophilia B. N. Engl. J. Med. 365, 2357–2365.
Ogawara, K.-I., Rots, M.G., Kok, R.J., Moorlag, H.E., Van Loenen, A.-M., Meijer, D.K., Haisma, H.J., Molema, G., 2004. A novel strategy to modify adenovirus tropism and enhance transgene delivery to activated vascular endothelial cells in vitro and in vivo. Hum. Gene Ther. 15, 433–443.

Ouzilou, L., Caliot, E., Pelletier, I., Pr Vost, M.-C., Pringault, E., Colb Re-Garapin, F., 2002. Poliovirus transcytosis through M-like cells. J. Gen. Virol. 83, 2177–2182.

Page, D.T., Cudmore, S., 2001. Innovations in oral gene delivery: challenges and potentials. Drug Disc. Today 6, 92–101.

Palfi, S., Gurruchaga, J.M., Ralph, G.S., Lepetit, H., Lavisse, S., Buttery, P.C., Watts, C., Miskin, J., Kelleher, M., Deeley, S., 2014. Long-term safety and tolerability of ProSavin, a lentiviral vector-based gene therapy for Parkinson's disease: a dose escalation, open-label, phase 1/2 trial. Lancet 383, 1138–1146.

Pan, Y., Li, Y.-J., Zhao, H.-Y., Zheng, J.-M., Xu, H., Wei, G., Hao, J.-S., 2002. Bioadhesive polysaccharide in protein delivery system: chitosan nanoparticles improve the intestinal absorption of insulin in vivo. Int. J. Pharm. 249, 139–147.

Peng, Z., 2005. Current status of gendicine in China recombinant human Ad-p53 agent for treatment of cancers. Hum. Gene Ther. 16, 1016–1027.

Pinto-Alphandary, H., Aboubakar, M., Jaillard, D., Couvreur, P., Vauthier, C., 2003. Visualization of insulin-loaded nanocapsules: in vitro and in vivo studies after oral administration to rats. Pharm. Res. 20, 1071–1084

Plapied, L., Duhem, N., Des Rieux, A., Pr At, V., 2011. Fate of polymeric nanocarriers for oral drug delivery. Curr Opin. Colloid Interf. Sci. 16, 228–237.

Powell, J.J., Faria, N., Thomas-Mckay, E., Pele, L.C., 2010. Origin and fate of dietary nanoparticles and microparticles in the gastrointestinal tract. J. Autoimm. 34, J226–J233.

Prabakaran, M., Madhan, S., Prabhu, N., Qiang, J., Kwang, J., 2010. Gastrointestinal delivery of baculovirus displaying influenza virus hemagglutinin protects mice against heterologous H5N1 infection. J. Virol. 84, 3201–3209.

Rescigno, M., Urbano, M., Valzasina, B., Francolini, M., Rotta, G., Bonasio, R., Granucci, F., Kraehenbuhl, J.-P., Ricciardi-Castagnoli, P., 2001. Dendritic cells express tight junction proteins and penetrate gut epithelial monolayers to sample bacteria. Nat. Immun. 2, 361–367.

Riviere, I., Brose, K., Mulligan, R.C., 1995. Effects of retroviral vector design on expression of human adenosine deaminase in murine bone marrow transplant recipients engrafted with genetically modified cells. Proc. Nat. Acad. Sci. 92, 6733–6737.

Rosenkranz, C.D., Chiara, D., Agorio, C., Baz, A., Pasetti, M.F., Schreiber, F., Dematteis, S., Martinez, M., Sztein, M.B., Chabalgoity, J.A., 2003. Towards new immunotherapies: targeting recombinant cytokines to the immune system using live attenuated *Salmonella*. Vaccine 21, 798–801.

Sakhuja, K., Reddy, P.S., Ganesh, S., Cantaniag, F., Pattison, S., Limbach, P., Kayda, D.B., Kadan, M.J., Kaleko, M., Connelly, S., 2003. Optimization of the generation and propagation of gutless adenoviral vectors. Human Gene Ther. 14, 243–254.

Seifert, J., Haraszti, B., Sass, W., 1996. The influence of age and particle number on absorption of polystyrene particles from the rat gut. J. Anat. 189, 483.

Shakweh, M., Ponchel, G., Fattal, E., 2004. Particle uptake by Peyer's patches: a pathway for drug and vaccine delivery. Exp. Opin. Drug Deliv. 1, 141–163.

Shata, M.T., Hone, D.M., 2001. Vaccination with a Shigella DNA vaccine vector induces antigen-specific CD8+ T cells and antiviral protective immunity. J. Virol. 75, 9665–9670.

Shata, M.T., Stevceva, L., Agwale, S., Lewis, G.K., Hone, D.M., 2000. Recent advances with recombinant bacterial vaccine vectors. Mol. Med. Today 6, 66–71.

Sheridan, C., 2011. Gene therapy finds its niche. Nat. Biotechnol. 29, 121–128.

Shmakov, A.N., Bode, J., Kilshaw, P.J., Ghosh, S., 2000. Diverse patterns of expression of the 67-kD laminin receptor in human small intestinal mucosa: potential binding sites for prion proteins? J. Pathol. 191, 318–322.

Smith, M., James, P., Tivey, D., 1987. M cell numbers increase after transfer of SPF mice to a normal animal house environment. Am. J. Pathol. 128, 385.

Snyder, B.R., Gray, S.J., Quach, E.T., Huang, J.W., Leung, C.H., Samulski, R.J., Boulis, N.M., Federici, T., 2011. Comparison of adeno-associated viral vector serotypes for spinal cord and motor neuron gene delivery. Hum. Gene Ther. 22, 1129–1135.

Steidler, L., Hans, W., Schotte, L., Neirynck, S., Obermeier, F., Falk, W., Fiers, W., Remaut, E., 2000. Treatment of murine colitis by *Lactococcus lactis* secreting interleukin-10. Science 289, 1352–1355.

Strauss, M., 1994. Liver-directed gene therapy: prospects and problems. Gene Ther. 1, 156–164.

Sun, L., Li, J., Xiao, X., 2000. Overcoming adeno-associated virus vector size limitation through viral DNA heterodimerization. Nature Med. 6, 599–602.

Sun, T., Zhang, Y.S., Pang, B., Hyun, D.C., Yang, M., Xia, Y., 2014. Engineered nanoparticles for drug delivery in cancer therapy. Angew. Chem. Int. Ed. 53, 12320–12364.

Taïeb, N., Yahi, N., Fantini, J., 2004. Rafts and related glycosphingolipid-enriched microdomains in the intestinal epithelium: bacterial targets linked to nutrient absorption. Adv. Drug Deliv. Rev. 56, 779–794.

Takamura, S., Niikura, M., Li, T., Takeda, N., Kusagawa, S., Takebe, Y., Miyamura, T., Yasutomi, Y., 2004. DNA vaccine-encapsulated virus-like particles derived from an orally transmissible virus stimulate mucosal and systemic immune responses by oral administration. Gene Ther. 11, 628–635.

Terry, L., Marsh, S., Ryder, S., Hawkins, S., Wells, G., Spencer, Y., 2003. Detection of disease-specific PrP in the distal ileum of cattle exposed orally to the agent of bovine spongiform encephalopathy. Vet. Rec. 152, 387–392.

Thanos, C., Liu, Z., Reineke, J., Edwards, E., Mathiowitz, E., 2003. Improving relative bioavailability of dicumarol by reducing particle size and adding the adhesive poly (fumaric-*co*-sebacic) anhydride. Pharm. Res. 20, 1093–1100.

Tyler, K.L., Clarke, P., Debiasi, R.L., Kominsky, D., Poggioli, G.J., 2001. Reoviruses and the host cell. Trends Microbiol. 9, 560–564.

Van der Lubben, I., Verhoef, J., Borchard, G., Junginger, H., 2001. Chitosan for mucosal vaccination. Adv. Drug Deliv. Rev. 52, 139–144.

Waehler, R., Russell, S.J., Curiel, D.T., 2007. Engineering targeted viral vectors for gene therapy. Nat. Rev. Genet. 8, 573–587.

Wang, H., Zheng, W.L., 2002. The current research status of the digestive tract pathway gene therapy. Chem. Life 22, 353–356.

Wang, L., Hernández-Alcoceba, R., Shankar, V., Zabala, M., Kochanek, S., Sangro, B., Kramer, M.G., Prieto, J., Qian, C., 2004. Prolonged and inducible transgene expression in the liver using gutless adenovirus: a potential therapy for liver cancer. Gastroenterology 126, 278–289.

Wirth, T., Parker, N., Yl -Herttuala, S., 2013. History of gene therapy. Gene 525, 162–169.

Wirtz, S., Becker, C., Blumberg, R., Galle, P.R., Neurath, M.F., 2002. Treatment of T cell-dependent experimental colitis in SCID mice by local administration of an adenovirus expressing IL-18 antisense mRNA. J. Immunol. 168, 411–420.

Wu, Y., Wang, X., Csencsits, K.L., Haddad, A., Walters, N., Pascual, D.W., 2001. M cell-targeted DNA vaccination. Proc. Nat. Acad. Sci. 98, 9318–9323.

Xiong, F., Zhang, C., Yu, M.J., 2006. The application of protein transduction in gene therapy. J. Biomed. Eng. 29, 250–254.

Xu, R.A., 2004. Oral gene delivery for gene therapy. Cancer Gene Ther. 11, 843.

Xu, R.A., 2014. Adeno-Associated Virus: From Virus to Vector. Science Press, Peking.

Xu, R.A., Diao, Y., Xie, H.T., 2006a. Latest research and development of gene drug in clinical application. Chin. J. Clin. Pharmacol. Ther. 10, 1092–1097.

Xu, R.A., Ma, H., Li, H., 2006b. Trends in Gene Therapy Research. NOVA Science Publisher, New York.

Xu, R.A., Chen, L., Xiao, W.D., 2008. Molecular Gene Medicine. Peking University Medical Press, Beijing.

Yue, Z.G., Wei, W., Lv, P.P., Yue, H., Wang, L.Y., Su, Z.G., Ma, G.H., 2011. Surface charge affects cellular uptake and intracellular trafficking of chitosan-based nanoparticles. Biomacromolecules 12, 2440–2446.

Yun, Y., Cho, Y.W., Park, K., 2013. Nanoparticles for oral delivery: targeted nanoparticles with peptidic ligands for oral protein delivery. Adv. Drug Deliv. Rev. 65, 822–832.

CHAPTER 6

NANOSTRUCTURES FOR ORAL DELIVERY OF THERAPEUTIC NUCLEIC ACIDS

Virginie Busignies*, Christine Charrueau, Pierre Tchoreloff*, Virginie Escriou****

*University of Bordeaux; CNRS, Bordeaux INP, ENSAM, I2M, Talence, France

**UTCBS, CNRS; INSERM; University of Paris Descartes, Sorbonne Paris City; Chimie ParisTech, PSL Research University, Paris, France

CHAPTER OUTLINE

1 **Introduction** 147
2 **Oral Delivery of Therapeutic Nucleic Acids** 149
 2.1 Advantages of the Oral Route of Administration 149
 2.2 Obstacles to Nucleic Acid Oral Delivery and Potential Targets 150
 2.3 Experimental Models for Studying the Stability and Efficiency of Oral Nucleic Acid Delivery Systems 152
3 **Oral Delivery Systems for Nucleic Acid Administration** 159
 3.1 Naked Molecules (Essentially Oligonucleotides) 159
 3.2 Cationic Lipid-Based Nanostructures 159
 3.3 Condensing Polymer-Based Nanostructures 162
 3.4 Other Nanostructures 166
 3.5 Micrometric Delivery Systems and Minitablets 167
4 **Conclusions** 168
References 169

1 INTRODUCTION

Since the demonstration that gene therapy can actually treat patients (Wang and Gao, 2014) and the multiplication of clinical trials based on RNA interference (RNAi) (Ozcan et al., 2015), the concept of therapeutic nucleic acid, that is, the possibility to use a nucleic acid as a drug, has become a reality. Gene therapy involves local or systemic administration of a nucleic acid construct that will result in expression of the gene-encoded protein. The expected effect can be prevention, treatment, and even curing of diseases. RNAi-based drugs harness a small RNAs-based endogenous process of gene silencing to treat disease. Both strategies rely on the efficient administration of nucleic acid molecules. Thus, therapeutic nucleic acid molecules can be distinguished (Fig. 6.1) in terms of function on the one hand, and in terms of structure on the other hand. In terms of function, nucleic acid molecules can either switch on expression of a protein (gene therapy), or switch off expression of a protein (siRNA, shRNA, or DNA-directed RNAi

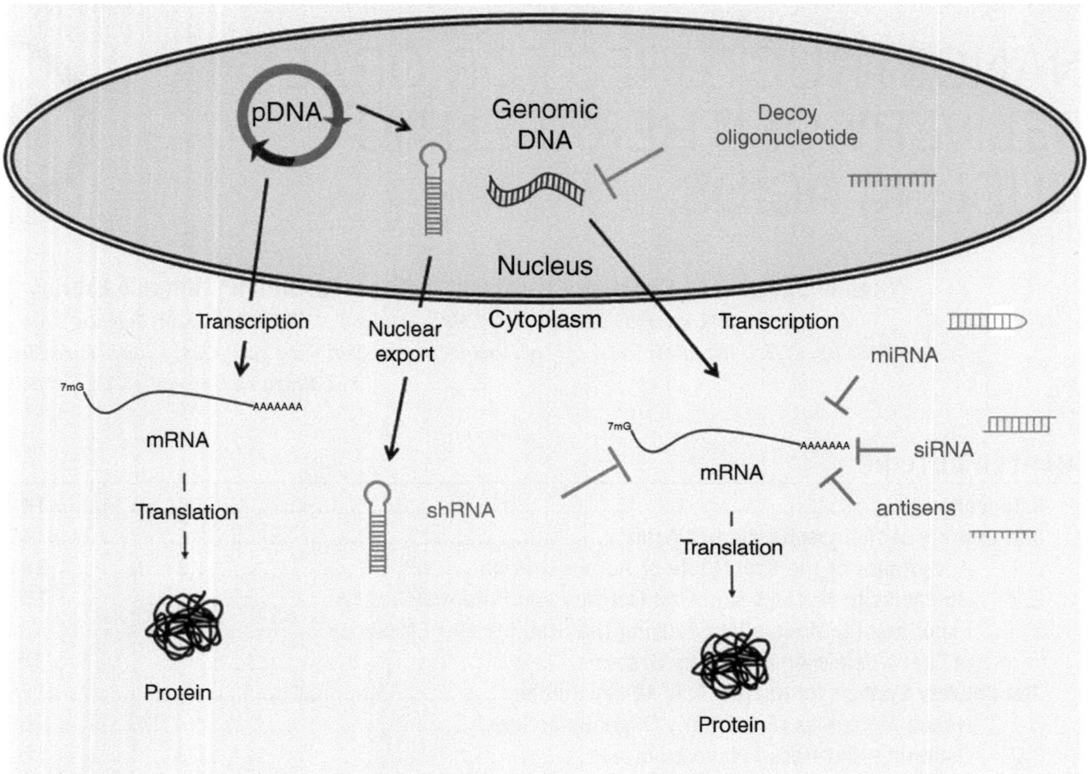

FIGURE 6.1 Overview of Therapeutic Nucleic Acids Pathway and Sites of Action

therapeutics, antisense oligonucleotide, decoy oligonucleotide). In terms of structure, gene and shRNA are based on DNA coding sequences that will trigger either the production of a protein via transcription of cDNA then translation of mRNA in the case of gene, or the production of interfering duplex shRNA. Antisense or decoy oligonucleotides and siRNA are small synthetic DNA- or RNA-based oligonucleotides. Nucleic acids are susceptible to degradation, particularly by nucleases, so efficacy may be enhanced by delivery system strategies that protect the drug. Moreover, in order to fulfill their function, the nucleic acids must penetrate the cells where the machineries required for the implementation of their function are located. Cellular delivery of all types of nucleic acid molecules is challenging because these negatively charged, water soluble molecules are high molecular weight compounds with very low membrane permeability. Because they are unable to cross cell membranes on their own, they have to be vectorized. Viral and nonviral vectorization strategies exist. We focus here on nonviral methods (Yin et al., 2014). To date, most nonviral nucleic acids delivery strategies have concentrated on the parenteral route of delivery and oral administration has been less investigated. Yet, oral drug products account for half of the annual drug delivery market, so oral route for therapeutic nucleic acid has to be considered.

The two main obstacles for oral delivery strategies to achieve this goal are the harsh environment of the digestive tract, in particular the acidic milieu of the stomach, and subsequently the crossing of the intestinal epithelia to reach the circulation (O'Neill et al., 2011). Nucleic acid modification and formulation into pH-responsive slow-release capsules can be used for stability purposes, while tight junction modulators, such as caprate can be used to penetrate the epithelium. Several methods and delivery systems have been developed for parenteral delivery of nucleic acids (Yin et al., 2014). Numerous in vivo studies have used cationic lipid-based vectors but they are not very popular for oral gene delivery. This is mainly due to their instability in the gut environment and could explain why polymer-based approaches have been used more commonly. Both condensing and noncondensing polymers have been utilized (Bhavsar and Amiji, 2007a). The cationic polysaccharide, chitosan, is among the most widely used condensing polymers for intestinal gene therapy research. It is a natural biodegradable mucoadhesive polysaccharide derived from crustacean shells, already available in pill form as an alternative therapy to reduce dietary fat and cholesterol absorption. This slowly degradable polymer has been shown to increase transcellular and paracellular transport of macromolecules across intestinal epithelial monolayers. It forms nanoparticulate complexes with nucleic acids that are biodegradable, biocompatible, and relatively nontoxic in nature. Chitosan also has mucoadhesive and tight junction destabilizing effects. Poly(lactide-*co*-glycolide) (PLGA) has been one of the most commonly used noncondensing polymers in intestinal gene delivery. It has generally been formulated as microparticles to deliver DNA vaccines.

This chapter reviews the various delivery systems that have been developed for oral administration of therapeutic nucleic acids, with a focus on nanostructures. It also discusses the physiological and cellular barriers to this delivery, as well as the experimental models available to evaluate the stability of the delivery systems and the efficiency of the delivery.

2 ORAL DELIVERY OF THERAPEUTIC NUCLEIC ACIDS

2.1 ADVANTAGES OF THE ORAL ROUTE OF ADMINISTRATION

Drugs can be administered by many routes but the most convenient route and then the most frequently used is the oral route. Several reasons may explain it. The first one is the patient's comfort. Compared to parenteral routes which consist in injecting the drug via a needle, this is a noninvasive and painless route for patients. This ease of administration results in a high patient compliance even for repeatable administration associated with a reduced health-care cost (Bhavsar and Amiji, 2007a; Raoof et al., 2004). Several therapeutic strategies are possible. It is first possible to cure local diseases, such as, for example, food allergies, inflammatory bowel diseases (IBD), familial adenomatous polyposis (FAP), intestinal cancers, or the intestinal symptoms of cystic fibrosis. Localized and targeted therapies make it possible to reduce side effects. Second, considering that the gut epithelium is highly vascularized, the nucleic acid can access the systemic circulation and can be delivered to a distant site of action after transport across the intestinal epithelium. This can be promising in the treatment of disease like hemophilia (Bhavsar and Amiji, 2008b; Kriegel et al., 2013; O'Neill et al., 2011).

The other advantages of the oral route are linked to the physiology of the gastrointestinal tract. The gut is characterized by a large surface of absorption composed of a high number of cells available for nucleic acid uptake. The crypts of Lieberkuhn are in stem cells. The transfection of these cells could make possible the long-term expression of therapeutic genes. The gut is also an entry portal for pathogens. GALT, antigen-sampling M-cells in the follicle-associated epithelium (FAE) of lymphoid follicles and Peyer's patches can be targeted in the perspective of orally delivered DNA vaccines (Bhavsar and Amiji, 2007a; O'Neill et al., 2011). The last point to consider is the high turnover of intestinal epithelial cells (Kriegel et al., 2013) because the enterocyte lifetime is estimated to be 5–7 days (O'Neill et al., 2011). Considering this fast proliferation, repeated dosing can be considered to avoid an unreasonable cytotoxicity.

Nevertheless, over all the described advantages, the efficiency of gene delivery is compromised by the different barriers exhibited by the gastrointestinal tract. To better understand the obstacles encountered by nucleic acids after oral administration, the following section describes extracellular, intercellular, and cellular barriers present in the gastrointestinal tract and the potential of targeting in therapeutic applications.

2.2 OBSTACLES TO NUCLEIC ACID ORAL DELIVERY AND POTENTIAL TARGETS

The nucleic acids are efficient only inside the cytosol of the cell. Nevertheless, it could be difficult to reach due to the several barriers that exist in the gastrointestinal tract. Fig. 6.2 shows these barriers to oral gene delivery. As a consequence, nucleic acids must be protected during transit to enable intestinal cellular uptake and then increase their transfection efficiency (Forbes and Peppas, 2012).

2.2.1 Physiological barriers

The first barrier is the pH in the gastrointestinal tract. The nucleic acids first reach the stomach known as a harsh acidic environment (pH 1.5–1.9). This low pH causes denaturation and depurination of nucleic acids leading to a decrease of their efficiency (Kriegel et al., 2013). Moreover, stability over a

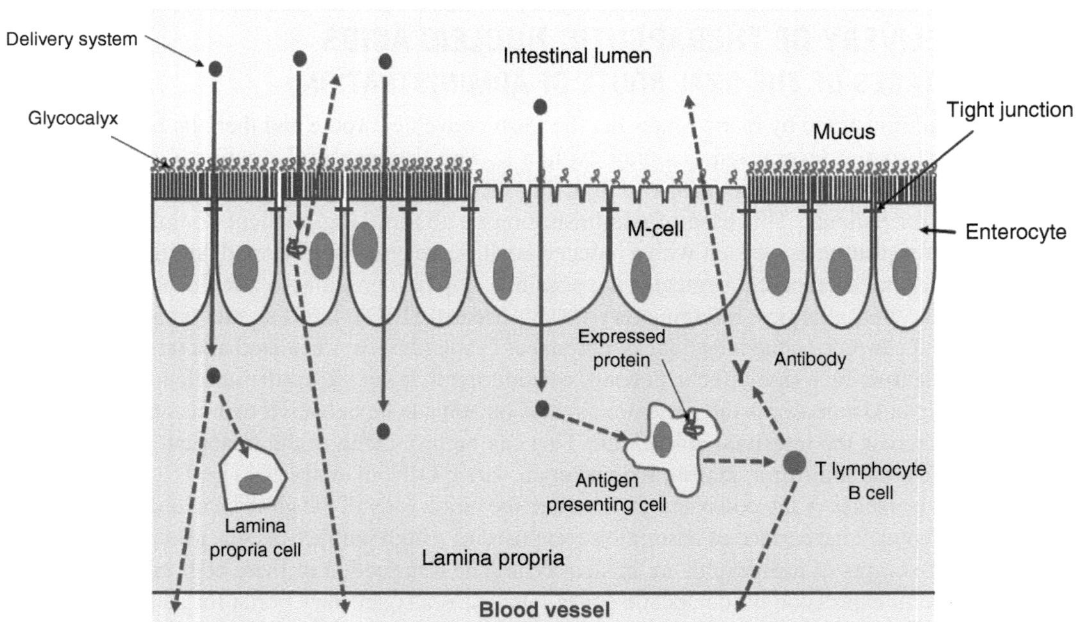

FIGURE 6.2 Fate of Nucleic Acids Delivery System in the Gut Following an Oral Administration With the Description of the Different Routes, Barriers, and Targets

wide range of pH is essential because the pH in the intestine can be variable and different from values observed in the stomach (pH 5–9 in the lower intestine).

Degradative enzymes are also found in gastrointestinal tracts in order to degrade nucleic acid in food. We can cite pepsin in the stomach and amylase, trypsin, or lipase in the gut lumen and on the brush border membrane. Obviously, the presence of these enzymes can lead to the rapid degradation of nucleic acid (Akhtar, 2009; Bhavsar and Amiji, 2007a; O'Neill et al., 2011).

Furthermore, the uptake of vectors containing nucleic acids can be reduced due to the fluid flow and peristaltic activity of the gastrointestinal tract because the contact time with the epithelial layer is reduced.

Physical barriers also exist (Kriegel et al., 2013; O'Neill et al., 2011). The gastrointestinal tract is covered by a sticky layer of mucus under the epithelium. The role of the mucus is to catch foreign agents. Its thickness changes along the gastrointestinal tract, being thicker in the colon than in the ileum. Its turnover is quite short (50–270 min) and then can cause the excretion of the nucleic acids before reaching the epithelium. Another barrier is the glycocalyx. This is a glycoprotein and polysaccharide layer with a thickness of 400–500 nm present over the apical membrane of enterocytes. This layer acts as a size-selective diffusional barrier for viruses, bacteria, and also vectors that contain nucleic acids. Finally, the negatively charged cell surfaces can also be a barrier as nucleic acids are negatively charged and repelled by the cell surfaces.

2.2.2 Cellular barriers and potential cellular targets

Three layers compose the intestinal mucosa. The first one is a single layer of epithelial cells. Under this first layer, the lamina propria is vascularized and composed of numerous immune cells. The last layer is called the muscularis mucosa. At the surface of intestinal mucosa is also found immune inductive tissue (GALT), including lymphoid follicles and Peyer's patches. These cells are important for mucosal and systemic immunity and are below the epithelial layer between villi.

When arriving in the gut after administration, nucleic acids first encounter an epithelial layer, which is composed of enterocytes, goblet cells, M cells of the Peyer's patches, Paneth cells, intraepithelial lymphocytes, and hormone-secreting endocrine cells (O'Neill et al., 2011). The epithelial layer is mostly composed of enterocytes. Their apical membrane is covered by the glycocalyx. Beyond enterocytes, goblet cells are the next most common cells. These are the cells that produce mucus. M cells are the specialized cells in the small intestine. These cells take up bacterial and viral antigens to make possible phagocytose by macrophages and finally initiate immune response in lamina propria. However, M cells cover a small surface of the gut and when they are targeted, this can lead to patient-to-patient variability in bioavailability (Akhtar, 2009).

Depending on the therapeutic strategy, the cells that compose the epithelial layer can be targeted or they can just represent a barrier to cross in order to access the vascularized lamina propria. As shown on Fig. 6.2, the crossing of epithelial layer can occur between the cells (it corresponds to paracellular route) or through the cells (it is called transcellular route).

The single way in the paracellular route is through the tight path junctions between cells. Because it represents less than 1% of surface area in intestines, it limits the possibility of crossing. In addition, in the human intestine the paracellular pores are characterized by a range diameter of 0.5–3 nm. This can restrain the paracellular way because the vectors used for the delivery of nucleic acids can be characterized by a higher size (O'Neill et al., 2011).

The transcellular route is more interesting than the previous one because the surface area for crossing is important. The first step is in the endocytosis of the delivery system of nucleic acids at the apical pole of the epithelial cell. In the cytoplasm, the degradation of nucleic acids can occur due to the presence of nucleases and lysosomes (Kriegel et al., 2013). The nucleic acids may also be captured by endosome and ejected to the lumen. Nuclear uptake is also possible. It is interesting if the goal is the transfection in the epithelial cell. Nevertheless, the lifetime of epithelial cells is short, about 5–7 days. As a consequence, repeated administration is essential for efficacy. Moreover, even if enterocytes are the most common cells of the epithelial layer, the access to enterocytes can be limited by the glycocalyx layer and by a lower endocytic activity of these cells. The glycocalyx layer on goblet cells is reduced compared to epithelial cells, but for the moment these cells are not identified as a target in gene therapies. Because of their role in immune response, the endocytic activity of M cells is higher than the previous ones. Contrary to epithelial cells, they are covered with reduced glycocalyx and a mucus layer. Furthermore, M cells present other advantages. Lysosomes are lesser than in other cells leading to a reduced degradation of nucleic acids. Their basolateral membrane is like a pocket that limits the travel way for nucleic acids. They are also closed to antigen-presenting cells that justify their potential in DNA vaccine delivery (Roy et al., 1999). Nevertheless, as previously mentioned, the main restraint is their poor proportion in the gastrointestinal tract.

Lamina propria is richly vascularized and composed of immune cells with T and B cells, dendritic cells, and macrophages. These cells can be targeted in therapy strategies. In inflammatory diseases, such as inflammatory bowel diseases (IBD), macrophages produce proinflammatory cytokines that influence the disease state. It explains why these cells have been chosen as target in several studies (Bhavsar and Amiji, 2008a; Kriegel and Amiji, 2011a,b; Wilson et al., 2010,). In vaccine delivery and food allergy treatment (Jones et al., 1997; Roy et al., 1999), it can be interesting to target dendritic cells. In fact, these cells are the starting point of the immune response. They also can participate to the uptake of nucleic acids or their vectors across the epithelial layer thanks to their projection into the intestinal lumen. Because the endocytic activity of dendritic cells is equal to that of M cells, they can contribute to the uptake of nucleic acid vectors (O'Neill et al., 2011).

To conclude, Fig. 6.2 summarizes the delivery of nucleic acids and/or their vectors in the gut after oral administration with the description of uptake and transport across the epithelium. The different barriers and targets can also be visualized. The delivery system of nucleic acids can make possible the transport from the intestinal lumen to lamina propria and bloodstream by paracellular route through tight junctions or by transcellular routes through enterocytes or M cells. The nucleic acids can transfect three kinds of cells: the epithelial cells, the lamina propria cells, or antigen-presenting cells. In the case of the transfection of epithelial cells, the expressed proteins can enter the lumen or the bloodstream. The transfection of antigen-presenting cells is desired in the case of vaccination. The transfection results in stimulation of B cells and T lymphocytes. The produced antigen can be secreted in the intestinal lumen or in the bloodstream leading to systemic immunity.

2.3 EXPERIMENTAL MODELS FOR STUDYING THE STABILITY AND EFFICIENCY OF ORAL NUCLEIC ACID DELIVERY SYSTEMS

The evaluation of the stability and efficiency of oral formulations of nucleic acids calls to several methods, including in vitro, in cellulo, ex vivo, in situ, and in vivo methods whose advantages and limitations are presented in Table 6.1. The first three types of experimental methods preceding the in situ and in vivo assays are implemented with the aim of predicting the efficiency of nucleic acid delivery systems in a whole living organism. Considering the successive obstacles that these systems have to

Table 6.1 Advantages and Limitations of Experimental Methods for Evaluating Stability and Efficiency of Oral Formulations of Nucleic Acids

Experimental Methods	Advantages	Limitations
In vitro methods		
• Stability assessment • Release of nucleic acids	Easy implementation; low cost	Results highly dependent on the medium composition; distant from in vivo conditions
In cellulo methods		
• Cellular uptake • Intracellular trafficking • Permeation across cells • Transfection efficiency • Biological effect detection	Implementation relatively easy; closer to biological conditions than in vitro methods; methods of choice for testing nucleic acid biological effects; moderate cost	Results highly dependent on the cell line and on the testing conditions; necessity to extract the nucleic acids from cells in uptake and intracellular distribution studies
Ex vivo methods		
• Methods of transport across gastrointestinal membranes	Closer to in vivo conditions than in vitro and in cellulo methods	Difficulty of implementation; limited survival of the isolated intestine; higher cost than above-cited methods
In situ methods		
• Intestinal transfection efficiency in anesthetized animals	Intermediate between ex vivo and in vivo methods; possibility to test conventional oral forms like tablets that can neither be crushed before administration nor be swallowed by rodents	Difficulty of implementation; high cost
In vivo methods		
• Biodistribution in healthy or diseased animals • Translocation in target tissues • Therapeutic efficacy in diseased animals	Closest conditions to future utilization in medicine; determination of the nucleic acid delivery system fate in a whole organism	Difficulty of implementation; high cost; anatomical and physiological differences between animal and human gastrointestinal tracts

overcome before delivering their load to its intracellular site of action, several complementary models are available to mimic each challenging step, from the in vitro ones to methods using biological supports like cells in culture or intestinal mucosa. In situ and in vivo models are the most elaborated ones with endoscopic/surgical intestinal administration and oral administration, respectively.

2.3.1 *In vitro models*

These are the simplest models employed. They allow assessing (1) the stability of the formulations and their ability to protect the nucleic acids from degradation either by the gastrointestinal fluids (Ballarin-Gonzales et al., 2013; O'Neill et al., 2011) or by nucleases (Bhavsar and Amiji, 2007b; Han et al., 2014); and (2) the release of the nucleic acids from their delivery systems (Bhavsar and Amiji, 2007b, 2008b).

2.3.1.1 Stability

The stability of the formulations in extracellular conditions of the gastrointestinal tract is essential to their further activity. In vitro methods studying stability aim at mimicking the gastrointestinal environment in terms of pH, temperature, agitation, length of exposure, and biological composition. The integrity of the formulations and nucleic acids as well is determined after their exposure to the challenging medium. Structural stability of nanoparticulate formulations can be monitored in terms of particle size and zeta potential (Han et al., 2014; He et al., 2015). Polyacrylamide gel electrophoresis is also used to evaluate the biological stability (Ballarin-Gonzales et al., 2013). The pH value may vary from 1.2 to 1.6 for gastric fluids, to 6.0–6.8 for intestinal fluids (O'Neill et al., 2011). The temperature of 37°C is usually used as body temperature (Han et al., 2014; He et al., 2015) but room temperature is also employed (Ballarin-Gonzales et al., 2013). Mechanical or magnetic agitation mimics peristaltism. The length of exposure reflects the transit time in each segment of the gastrointestinal tract. In practice, it varies from 1 h (Ballarin-Gonzales et al., 2013) to 2 h (Han et al., 2014; He et al., 2015). As for composition, several media can be used including artificial ones like simulated gastric fluid and intestinal fluids (O'Neill et al., 2011) prepared according to US Pharmacopoeia formulas, and natural ones, such as intraluminal fluids collected in animals (He et al., 2015). The composition of the medium is a critical parameter that may greatly influence the results of in vitro stability assays, especially depending on the presence or absence of bile salts, digestive enzymes (lipase, amylase, and trypsin), or mucus. In artificial media, the composition is well controlled since it reproduces formulas from the US Pharmacopoeia; while in natural intraluminal fluids, the composition may vary from one species to another, and within an animal species from one individual to another.

The ability of the formulations to protect their load from endogenous nucleases can also be determined through incubation with DNases or RNases. The principle is close to that of above-mentioned experiments in gastrointestinal conditions. The formulations are exposed to DNase-I (Bhavsar and Amiji, 2007b; Han et al., 2014) or to RNase-A (Han et al., 2014) at 37°C, for 20 min (Bhavsar and Amiji, 2007b) or 2 h (Han et al., 2014), and the biological stability of nucleic acids is determined by gel electrophoresis (Bhavsar and Amiji, 2007b; Han et al., 2014).

2.3.1.2 Release

Besides stability studies, in vitro methods also comprise release studies. From a general point of view, the release of the therapeutic load from its pharmaceutical form is a necessary step for conventional drugs that activity depend on the absorption of their free soluble form. In such cases, drug release is classically evaluated in vitro by dissolution and dialysis assays. However, nucleic acid-based therapies are very different from conventional drugs because they cannot cross the cellular membrane unless carried by an adapted nanoscale delivery system. For that reason, nucleic acid release from their nanoparticulate carrier is not a key point in the development of such formulations. Nonetheless, some authors have implemented release studies to characterize oral formulations of nucleic acids because of the complexity of the oral system they have designed as a vehicle: gelatin nanoparticles, encapsulating the plasmid DNA, have been next included into poly(ε-caprolactone) microspheres (Bhavsar and Amiji, 2007b, 2008b). It was therefore necessary to determine whether these microspheres could release the nanoparticles carrying the nucleic acid. For that purpose, PBS containing protease was used as a release medium at 37°C, for 2 h exposure, followed by 5 h exposure in PBS with protease and lipase, the latter degrading the poly(ε-caprolactone) matrix and the former degrading the gelatin nanoparticles. The release was evaluated by fluorescence detection of the plasmid DNA whose stability was further studied by agarose gel electrophoresis. The polymer degradation was also visualized by scanning electron microscopy (Bhavsar and Amiji, 2007b).

In summary, the results of in vitro stability testing often allow to screen several formulations and to choose the most protective one. Advantages of in vitro methods are their easy implementation and low cost. Nonetheless, they present limitations because their results are highly dependent on the experimental conditions, in particular the composition of the tested media. In addition, they are relatively distant from in vivo conditions.

2.3.2 In cellulo models

In cellulo methods for evaluating oral formulations mainly use intestinal cultured cells as a biological support. These techniques either study the uptake of nucleic acid into cultured cells by the direct mean of intracellular measurement, or explore the mechanisms involved in the uptake process by using endocytic inhibitors, or determine intracellular trafficking through examination with transmission electron microscopy, or assess permeation across cells, or evaluate transfection efficiency by detecting a biological effect within the cells.

Besides rapidly growing immortalized cells that are poor models of the intestinal epithelium, a variety of more physiologically relevant differentiated cell culture models are available, including Caco-2 cells, Ht29 cells, cocultures of Caco-2 with Ht29 cells, or M cells, or Raji-B cells, multilayer cell models of intestinal epithelial cells (Caco-2 or T84 cells) plus basolateral macrophages (O'Neill et al., 2011), and other cell culture models.

- Caco-2 cells established from a human colon adenocarcinoma represent a well-characterized model for assessing oral delivery of drugs. Caco-2 cell monolayers consist in highly polarized cells possessing the characteristics of mature enterocytes, that is, tight junctions, a glycocalyx, and apical microvilli. They also express several membrane and cytosolic enzymes common to enterocytes in vivo. The Caco-2 Transwel system that separates apical and basolateral compartments allows measuring the transcellular transport of nucleic acid delivery systems. However, Caco-2 cells present an increasing resistance to transfection in relation to their differentiation; this phenomenon may be partially due to decrease in particle uptake, as well as poor nuclear accessibility resulting from the non-dividing nature of these differentiated cells (O'Neill et al., 2011).

  It is also possible to quantify the level of expressed proteins in the basolateral compartment (Martien et al., 2008); and to evaluate potential tight junction-destabilization or monolayer integrity disruption through the measurement of the transepithelial electrical resistance of Caco-2 monolayers (O'Neill et al., 2011). However, Caco-2 cells are devoid of mucus layers and represent only enterocytes. For those reasons, other cell lines can be preferred.
- Among the clones of Ht29 cells coming from a human colon carcinoma, the Ht29GlucH clone forms monolayers including mucin-secreting goblet cells able to produce a mucus layer (Wikman et al., 1993). These cells as well as the coculture of Caco-2 and Ht29GlucH cells present the advantages to allow assessing the effect of mucus on nucleic acid transport (Cryan and O'Driscoll, 2003). However, Ht29GlucH present certain limitations, notably the multilayer growth of cells (Schipper et al., 1999). This phenomenon is overcome by the subclone Ht29MTXE12, which can provide a more reliable model for gene delivery assessment (O'Neill et al., 2011).

- Follicle-associated epithelium (FAE) models associate Caco-2 monolayers and M cells obtained by murine lymphocyte-stimulated conversion of enterocytes, or human Raji-B cells (O'Neill et al., 2011). These models possess some of the characteristics of the FAE of Peyer's patches, that is, fewer or no microvilli and reduced alkaline phosphatase activity. They permit to assess the contribution of M cells to transfection and/or transepithelial transport, and may be of particular interest in evaluating M-cell-targeted oral formulations (O'Neill et al., 2011).
- Multilayer cell culture systems associating an epithelial monolayer of Caco-2 or T84 cells to a basolateral cell type, are of special interest when the target cells are located beneath the intestinal epithelium. This is the case of IEC (intestinal epithelial cells)-macrophage cocultures. The stimulation of macrophages by LPS (lipopolysaccharide) can induce a barrier dysfunction characteristic of inflammatory bowel diseases, through the production of TNFα by macrophages and IL-8 by IECs (Clayburgh et al., 2004).
- Other cell culture models.

Besides intestinal cultured cells, other models are used to evaluate oral formulations of nucleic acids. For example, mouse peritoneal exsudate cell macrophages (PECs) were employed to test β1,3-D-glucan-encapsulated siRNA (Aoudi et al., 2009); and rat PECs were used to evaluate chitosan-siRNA nanoparticles (He et al., 2015). Human hepatoma BEL-7402 cells allowed assessing uptake, intracellular distribution, and gene silencing by shRNA and siRNA encapsulated in multifunctional polymeric nanoparticles (Han et al., 2014). Mouse monocyte macrophages Raw 264.7 cells were used to study the cellular uptake and intracellular trafficking of polymeric siRNA nanoparticles (He et al., 2013a).

The critical parameters of in cellulo methods include the cell line employed, the conditions of their culture, and the composition of the culture medium in which the formulation are tested, the concentrations at which the cells are exposed and the duration of exposure. The extraction method of the nucleic acids from cells, and the detection limit of the measurement method are critical as well in uptake studies. Finally, the sensitivity and specificity of detection of a biological effect are also of major importance.

In summary, in cellulo methods are very useful to test nucleic acid formulations in terms of uptake and biological efficacy. Their advantages are a relatively easy implementation and a moderate cost. They are also closer to biological conditions that in vitro methods. However, their results are highly dependent on the cell line and the testing conditions employed.

2.3.3 Ex vivo models

Ex vivo methods consist in studying the permeation of nucleic acid delivery systems across the intestine. A part of the intestine is collected in euthanized animals, rinsed, tied at both ends to form a bag, and maintained at 37°C in an oxygenated buffer under smooth shaking. The formulation to be tested is placed either inside the bag in the noneverted intestine method or in the external buffer bath in the everted intestine method. Measurement of the transport across the gut membrane allows calculating the apparent permeability coefficient. A few studies have been implemented on either everted rat jejunum (Hughes et al., 1995) or on noneverted rat ileum (He et al., 2015; Zhang et al., 2013b). In the first case, the study suggested that antisense oligonucleotide transport across the jejunum occurred through slow passive diffusion (Hughes et al., 1995). In the second case, ternary polymeric nanoparticles enhanced siRNA transport across the ileum as compared to naked siRNA (Zhang et al., 2013b). In the third case, the ileal segments were selected to contain either no Peyer's patch or 5–6 Peyer's patches in order to represent nonfollicle-associated epithelia and follicle associated epithelia, respectively. The apparent permeability of mannose-modified chitosan-siRNA nanoparticles was enhanced by targeting normal enterocytes and M cells that express mannose receptors (He et al., 2015).

Although more physiological than in vitro and in cellulo methods, ex vivo methods present limitations linked to the difficulties of implementation and to limited survival of the isolated intestine. They are also costly.

2.3.4 In situ models

In situ methods take place at the limit between ex vivo and in vivo methods. They consist in endoscopically or surgically delivering the oral formulation directly into an isolated intestinal segment of anesthetized animals, and in further studying the nucleic acid effect. In situ methods have been rarely employed in the literature to test oral nucleic acid delivery systems. The colon epithelium could be transiently transfected by cationic liposomes of recombinant DNA (Liptay et al., 1998; Schmid et al., 1997). The model of ligated rat intestinal loop allowed assessing efficient intestinal transfection by various synthetic vectors of plasmid DNA (Cryan and O'Driscoll, 2003).

Such methods may be of interest when the pharmaceutical form cannot be administered to the animal by the oral route. This is the case for solid dosage forms like tablets that could neither be swallowed by small animals like rodents, nor be crushed and administered by gavage as a suspension without losing a matrix effect or a coating protection. Nonetheless, these methods are difficult to implement and invasive since they necessitate an endoscopic or a surgical delivery in anesthetized animals; they are also costly.

2.3.5 In vivo models

Above-cited models are very useful but none of them fully reproduces the complex conditions of in vivo environment. In vivo models represent the most physiologically relevant approaches to study nucleic acid delivery via the oral route of administration. Most of these models are implemented in rodents (Ballarin-Gonzales et al., 2013; Bhavsar and Amiji, 2007b, 2008a; Han et al., 2014; He et al., 2015; Roy et al., 1999; Zhang et al., 2013a) but dogs are also used (Raoof et al., 2004). Oral gavage represents the unique way of administration of liquid suspensions in rodents while tablets could be administered orally to dogs (Raoof et al., 2004). Experiments are carried out either in healthy animals or in diseased animals. Healthy animals are used for biodistribution studies (He et al., 2013a; Raoof et al., 2004; Zhang et al., 2013b) or translocation studies (Ballarin-Gonzales et al., 2013), or in vivo protein expression (Bhavsar and Amiji, 2008b, or in vivo transgene expression (Bhavsar and Amiji, 2007b) after euthanasia.

Animal disease models are necessary to assess the ability of nucleic acid delivery systems to induce a biological response, either protein expression, transgene expression, gene silencing, or therapeutic effect. Local as well as distant biological responses can be detected depending on the disease to be treated. Therapies to the intestine are intended for the treatment of local intestinal diseases. These pathologies include inflammatory bowel diseases, that is, Crohn's disease and ulcerative colitis, colorectal cancer, and duodenal ulcers, which are good candidates for gene therapy intervention (O'Neill et al., 2011). Local therapies also cover oral DNA administration for vaccination purposes because the intestinal mucosal surface offers an opportunity for generation of both mucosal and systemic immunity (Bhavsar and Amiji, 2007a). Therapies via the intestine are intended for the systemic treatment of distant pathologies like cancers, cystic fibrosis, hemophilia, diabetes (O'Neill et al., 2011), hepatic tumor (Han et al., 2014), and hepatic injury (Aoudi et al., 2009; He et al., 2013b, 2015; Yin et al., 2013).

- Models of local intestinal diseases

  Crohn's disease is one of the two major forms of inflammatory bowel diseases. It mainly occurs in the distal ileum or the proximal colon while the second major form of inflammatory bowel diseases, ulcerative colitis, affects only the large intestine. More than 40 models of intestinal inflammation are described in rodents. They can be chemically induced by oral administration of dextran sodium sulfate (Kriegel and Amiji, 2011a,b; Wilson et al., 2010; Zhang et al., 2013a), or by rectal administration of trinitrobenzene sulfonic acid (Bhavsar and Amiji, 2008a), or by intraperitoneal injection of lipopolysaccharide (Laroui et al., 2011). Intestinal inflammation can also result from genetic manipulation like IL-10 knockout (Kühn et al., 1993), or depend on transfer of cells to immunodeficient animals (Wirtz and Neurath, 2007). These models are characterized by alterations close to human ones, including epithelium disruption, immune cell infiltration, and luminal bacterial translocation (O'Neill et al., 2011). The severity indicators are body weight loss, stool consistency, and rectal bleeding as clinical signs; colon macroscopic score, colon length and weight as macroscopic markers; proinflammatory cytokines, colon histology and myeloperoxidase activity as inflammation markers.

  Colorectal cancer is one of the most common forms of intestinal cancer. It can be chemically induced in rodents by the carcinogen azoxymethane, or its precursor 1,2-dimethylhydrazine, reproducing histological and morphological features of human colon cancer (O'Neill et al., 2011). Both earlier cited chemicals used in combination can induce the development of colitis-associated colon tumor (Tanaka et al., 2003). In addition, genetic mouse models spontaneously developing intestinal cancer can be obtained by mutations in the Apc gene, like the ApcMin mouse with multiple intestinal neoplasia, for example (Moser et al., 1990).

  Finally, experimental chronic duodenal ulceration can be chemically induced by cysteamine in rats (Szabo et al., 2001).
- Models for vaccination study
  These models study the induction of immunity in healthy mice against pathogen agents like rotavirus (Chen et al., 1998; Herrmann et al., 1999), bovine adenovirus type-3 (Mittal et al., 2001), or pathologies like HIV (Kaneko et al., 2000), and hepatitis B (He et al., 2005).
- Models of pathologies distant from the digestive tract
  Several models of nonintestinal diseases have been used to evaluate oral nucleic acid formulations intended for systemic effect. They include cystic fibrosis model (Alton et al., 1993), hemophilia model (Bowman et al., 2008), diabetes model (Yoon and Jun, 2002), hepatic tumor model (Han et al., 2014), and lipopolysaccharide/D-galactosamine-induced hepatic injury (Aoudi et al., 2009; He et al., 2013b, 2015; Yin et al., 2013).
  The critical parameters of in vivo models are the modalities of administration, mostly by gavage in rodents, the relevance of the experimental model compared to human diseases, and the choice of indicators of disease severity. In vivo methods offer the closest conditions to future utilization in medicine and allow the determination of the nucleic acid delivery system fate in a whole organism. Their limitations include difficulties of implementation and high cost. Of

note, differences in anatomy and physiology between animal models and humans may hamper prediction of oral nucleic acid delivery performance in future clinical application. Hence, several parameters like, for example, gastrointestinal transit time, intraluminal pH under fasted or fed conditions, bile concentration, metabolic activity of phase I and phase II enzymes, membrane transporters, may notably vary in animals compared to human gastrointestinal characteristics (Sjögren et al., 2014).

3 ORAL DELIVERY SYSTEMS FOR NUCLEIC ACID ADMINISTRATION (TABLE 6.2)

3.1 NAKED MOLECULES (ESSENTIALLY OLIGONUCLEOTIDES)

The base sequence of antisense oligonucleotide is complementary to its targeted gene's mRNA. Since unmodified DNA oligonucleotide structure is based on a phosphodiester linkage sensitive to naturally occurring nucleases, antisense drug candidates are generally modified to decrease their sensitivity to nuclease. They exhibit modified backbones, where the classical phosphodiester backbone is replaced by phosphorothioate backbone, or else sugar moieties of the nucleotides are altered. Modified oligodeoxynucleotides of a novel class, defined as MBOs (novel hybrid DNA/RNA mixed-backbone oligonucleotides) have shown a significant improvement of pharmacokinetic properties in vivo compared with phosphorothioate-oligodeoxynucleotides. MBOs administered p.o. are absorbed in the gastrointestinal tract and distributed to major organs of mice and rats. GEM 231, an MBO with a hybrid DNA/RNA structure containing 2'-O-methyl-ribonucleosides at the 5' and 3' ends targets the protein kinase A type I (PKAI) subunit RIα. This protein transduces mitogenic signals from different growth factors and oncogenes and is overexpressed in the majority of human cancers. When administered p.o. to nude mice bearing GEO human colon carcinoma xenografts, GEM 231 has antitumor activity (Tortora et al., 2000). Similar results were obtained in SCID or nude mice with xenografts of human cancers of the colon (LS174T), breast (MDA-MD-468), and lung (A549) (Wang et al., 1999). The mechanism by which modified oligonucleotide reaches its targeted cells is still unclear.

3.2 CATIONIC LIPID-BASED NANOSTRUCTURES

Cationic lipids are amphiphilic molecules able to interact with nucleic acid molecules. They form complex upon mixing via electrostatic interactions between the positive charges of the cationic lipid and the negative charges of the nucleic acid. Once complexed with cationic lipids, nucleic acid molecules are protected against nuclease activity and are readily taken up by cells by endocytosis or other mechanisms. Since the complexes formation lies on electrostatic interactions, cationic lipids can interact with both DNA and RNA and this irrespective of the size of the molecule, even if a collapse event is not expected when smaller molecules like siRNA or oligonucleotides are used.

3.2.1 *Per se*

Only two publications show oral administration of particles formed with plasmid and cationic lipids. Liu et al. (2014) used commercial cationic liposome Lipofectamine to prepare self-assembled complexes with plasmid DNA containing influenza M1 gene and administered intragastrically by

Table 6.2 Delivery Systems for Oral Delivery of Nucleic Acids

Molecule	Vector	Particle Size	Gene	Specie	Dose	Model	References
siRNA							
	GeRP	2–4 μm	Map4k4	Mouse	20 μg/μg	LPS/DGalN-ALI	Aoudi et al. (2009)
	TKN	600 nm	TNFα	Mouse	0.23 and 2.3 mg/kg	DSS-UC	Wilson et al. (2010)
	NiMOS	2–3 μm	Cyclin D1/ TNFα	Mouse	1.2 mg/kg	DSS-UC	Kriegel and Amiji (2011a,b)
	(PEI/PLA) Cs modified - alginate	380 nm	TNFα	Mouse	Not given	LPS/DGalN-ALI	Laroui et al. (2011)
	Cs NP	150 nm	no target	Mouse	28 and 78 μg/mouse	Biodistribution	Ballarin-Gonzales et al. (2013)
	T-Cs NP	150 nm	TNFα	Mouse		LPS/DGalN-ALI	He et al. (2013a,b)
	HTP-Cs NP	130 nm	TERT	Mouse	0.2 and 6.5 μg/kg	tumor	Wei et al. (2013)
	SSNP	130 nm	TNFα	Mouse	200 μg/kg	LPS/DGalN-ALI	Yin et al. (2013)
	TTG- Cs NP	150 nm	Map4k4	Mouse	250 μg/kg	DSS-UC	Zhang et al. (2013a,b)
	TTG- Cs NP	130 nm	VEGF (+shRNA Survivin)	Mouse	200 μg/kg (1 mg pDNA/kg)	Hepatoma	Han et al. (2014)
	TTM-Cs NP	150 nm	TNFα	Mouse	30 μg/kg	LPS/DGalN-ALI and DSS-UC	Chu et al. (2015)
	TTM-Cs NP	150 nm	TNFα	Rat	20 and 50 μg/kg	LPS/DGalN-ALI	He et al. (2015)
Oligonucleotide							
	Naked		PKA I	Mouse	0.5–10 mg/kg	Tumor	Wang et al. (1999)
	Tablet		TNFα	Dog	80 mg/dog	Biodistribution	Raoof et al. (2004)
	Tablet/Capsules		TNFα	Man	100 and 140 mg	Toxicity/pharmacokinetics	Tillman et al. (2007)
	(PLA/PLGA) Chitosan	370 nm	NF κB	Rat	2 mg/rat	DSS-UC	Tahara et al. (2011)
	Naked		PKA I	Mouse	1–10 mg/kg	Tumor	Tortora et al. (2000)
Plasmid DNA							
	PLGA	2 μm	Luciferase	Mouse	50 μg/mouse	Vaccination	Jones et al. (1997)
	Cs NP	150–300 nm	Peanut arah2 allergen	Mouse	50 μg/mouse	Allergy	Roy et al. (1999)
	Cs NP	70–150 nm	LacZ/mEPO	Mouse	50 μg/mouse	Hemophilia	Chen et al. (2004)

	NiMOS	5 μm	LacZ/GFP	Rat	100 μg/rat	Gene expression	Bhavsar and Amiji (2007b)
	NiMOS	2–5 μm	IL-10	Mouse	100 μg/mouse	TNBS-induced colitis	Bhavsar and Amiji (2008a)
	Cs NP	200–400 nm	FVIII	Mouse	50 μg/mouse	Hemophilia	Dhadwar et al. (2010)
	PLGA	0.5 μm	rotavirus VP6	Mouse	50 μg/mouse	Vaccination	Chen et al. (1998)
	N-acetylated Cs NP	not given	LacZ/IL-10	Mouse	6 mg/mouse	Gene expression	Kai and Ochiya (2004)
	Alginate	5–10 μm	LacZ	Mouse	not given	Vaccination	Mittal et al. (2001)
	PLGA	2–9 μm	env HBV	Mouse	20–200 μg/mouse	Vaccination	He et al. (2005)
	PLGA	1–10 μm	env HIV	Mouse	10 μg/mouse	Vaccination	Kaneko et al. (2000)
	Alginate	45 μm	CtxB	Mouse	50–150 μg/mouse	Vaccination	Rosli et al. (2013)
	Lipofecamine	not given	Influenza M1 gene	Mouse	100 μg/mouse	Vaccination	Liu et al. (2014)
	Cs NP	100–400 nm	Dust mite allergen Der p2	Mouse	100 μg/mouse	Vaccination	Li et al. (2009)
	PLA-PEG-PLA/ PLGA– PEG–PLGA micelles	Not given	HCV MA gene	Mouse	100 μg/mouse	Vaccination	Yang et al. (2011)
	Alginate	50 μm	GFP	Mouse	50–150 μg/mouse	Gene expression	Nograles et al. (2012)
	Silver nanoparticle	25 nm	Avian virus H5 gene	Chick-en	not given	Vaccination	Jazayeri et al. (2012)
	PEO-PPO-PEO micelles	170 nm	Parathyroid hormone	Rat	720 μg/rat	Parathyroidect-omised rat	Chou et al. (2009)
	Cs NP in Jello	300 nm	FVIII	Mouse	50–600 μg/mouse	Hemophilia	Bowman et al. (2008)
	Lipofectin	Not given	APC	Mouse	20 μg/mouse	Tumor	Lew et al. (2002)

Cs NP, Chitosan nanoparticle; DGalN-ALI, D-galactosamine-induced acute liver injury; DSS-UC, dextran sulfate sodium-induced ulcerative colitis; GeRP, β1,3-D-glucan-encapsulated siRNA particles; HTP, N-*((2-hydroxy-3-trimethylammonium) propyl); LPS, lipopolysaccharide; Map4k4, mitogen-activated protein kinase kinase kinase kinase 4; NF κB, nuclear factor κB; NiMOS, nanoparticles in microspheres oral system; pDNA, plasmid DNA; PEI, polyethylenimine; PKAI, protein kinase A type I; PLA, poly(D,L-lactic acid); shRNA, small hairpin RNA; siRNA, small interfering RNA; SSNP, supramolecular self-assembled nanoparticles; T-Cs NP, trimethyl chitosan nanoparticle; TERT, telomerase reverse transcriptase; TKN, thioketal nanoparticles; TNFα, tumor necrosis factor α; TTG, thiolated trimethyl galactose; TTM, thiolated trimethyl mannose; VEGF, vascular endothelial growth factor; APC, adenomatous polyposis coli; Cs NP, chitosan nanoparticle; CtxB, cholera enterotoxin B subunit; FVIII, coagulation factor VIII; GFP, green fluorescent protein; HBV, hepatitis B virus; HCV, hepatitis C virus; HIV, human immunodeficiency virus; IL-10, interleukin-10; MA, multiple-epitope antigen; NiMOS, nanoparticles in microspheres oral system; PEG, polyethyleneglycol; PEO-PPO-PEO, poly(ethylene oxide)-poly(propylene oxide)-poly(ethylene oxide); PLA, poly(D,L-lactic acid); PLGA, poly(D,L-lactic-*co-*glycolic acid).*

gavage 100 μg of plasmid to mice, three times at weekly intervals. One week after the final booster vaccination, mice were challenged with influenza virus to examine the efficiency of oral vaccination. Similarly, Lew et al. (2002) associated plasmid DNA expressing human APC (adenomatous polyposis coli) gene with cationic liposome Lipofectin and administered transorally by gavage 20 μg of plasmid to Min (Multiple Intestinal Neoplasia) mouse, that exhibit a mutation of the APC gene and consequently develop neoplastic intestinal polyps with elevated COX-2 levels similar to patients with familial adenomatous polyposis and sporadic colorectal cancer. The effects of APC gene replacement on intestinal neoplasia were observed 2 months after the oral administration. Although these publications show some efficacy following oral administration of these relatively simple formulations, most teams prefer to include the complex in a polymer matrix to provide better protection against the harsh conditions in the stomach.

3.2.2 *Inside a polymeric matrix*

Wilson et al. (2010) prepared formulations of siRNA termed TKN (thioketal nanoparticles) generated by first preparing siRNA-DOTAP complexes then using a single-emulsion protocol to encapsulate these complexes into nanoparticles using a polymer, poly-(1,4-phenyleneacetone dimethylene thioketal) or PPADT, that degrades selectively in response to reactive oxygen species (ROS). DOTAP, or *N*-[1-(2,3-dioleoyloxy)propyl]-*N,N,N*-trimethylammonium methyl, is a quaternary ammonium salt lipid capable of binding nucleic acid via electrostatic interactions and forming complexes or particles in which the nucleic acid is protected from nucleases. Once siRNA/DOTAP complexes are formed, they are added to an organic solution containing PPADT to obtain nanoparticles. The authors showed that PPADT is composed of thioketal linkages that are stable to acid-, base-, and protease-catalyzed degradation and ROS-sensitive. Therefore, upon oral administration, the siRNA-particles remain stable in the gastrointestinal tract, thereby protecting siRNA and preventing its release and, at sites of intestinal inflammation, where high levels of ROS are produced, ROS-induced degradation of PPADT triggers the release of siRNA to inflamed intestinal tissues. Tahara et al. (2011) described the preparation of nanospheres of chitosan-modified PLGA encapsulated complexes of oligonucleotides and DOTAP. Chitosan modified and unmodified PLGA provided 4 h of protection for the DNA cargo from DNase I degradation and the chitosan modified provided protection from acidic pH for 2 h. Once formulated, the efficiency of this decoy oligonucleotide targeting the nuclear factor κB was tested on oral administration in DSS-induced acute colitis in rats. Daily oral administration of high doses (2 mg oligonucleotide/rat) of nanospheres significantly improved DSS-induced diarrhea, bloody feces, shortening of colon length, and myeloperoxidase activity. Nanospheres exhibited more adhesion and penetration in inflamed tissues than in normal tissues.

3.3 CONDENSING POLYMER-BASED NANOSTRUCTURES

Besides cationic lipids, other cationic molecules of polymeric structure are frequently used to vectorize therapeutic nucleic acids. For oral administration the most popular is chitosan, a biodegradable polysaccharide modified from chitin, abundant in crustacean shells. Chitin is converted to chitosan by alkaline deacetylation. The major method for the formulation of chitosan-based nucleic acid delivery systems is complex coacervation in which the anionic nucleic acid complexes with cationic chitosan with the aid of either stirring or vortex force. Chitosan behaves as a linear polyelectrolyte in solution with the *N*-deacetylated subunits conferring a highly positive charge density.

3.3.1 *Per se*

3.3.1.1 Chitosan

In 1999, Roy et al. (1999) published a proof-of-concept study using chitosan-plasmid DNA nanoparticles containing a dominant peanut allergen gene to demonstrate transduced gene expression in the epithelium of the gut upon oral administration of the DNA vaccine. Nanoparticles, fairly spherical and 150–300 nm in size were prepared by complex coacervation of high-molecular weight (about 390,000 Da) chitosan and plasmid, by simply mixing preheated (55°C) solutions of plasmid in sodium sulfate and chitosan in sodium acetate-acetic acid buffer, pH 5.7. Mice were immunized orally with a dose of 50 µg plasmid by oral gavage. Mice immunized with nanoparticles showed a substantial reduction in allergen-induced anaphylaxis, demonstrating that mucosal as well as systemic immunity was successfully achieved upon DNA vaccine delivery via oral mucosal surfaces. The same preparation protocol was used by Chen et al. (2004) to produce chitosan-based nanoparticles with plasmids containing either LacZ gene or mEPO under inducible promoter (tetOn system). They obtained nanoparticles fairly spherical with a size of 70–150 nm. They observed expression of β-galactosidase in stomach and small intestine 3 days after oral gavage of mice with 50 µg LacZ gene-containing plasmid. They also administered once a week nanoparticles with plasmid expressing mEpo and observed an inducible modification of hematocrit during the first four days of treatment. A second administration, 6 days after the first one, was still able to promote a second hematocrit increase.

Following the same preparation protocol, Li et al. (2009) were able to produce chitosan-based nanoparticles with plasmid expressing house dust mite allergen Der p2. The obtained nanoparticles were found fairly spherical with a diameter size of 100–400 nm. Mice were immunized with nanoparticles administered by oral gavage (100 µg plasmid). It was shown that chitosan-DNA complexes can generate high expression of allergen in stomach and small intestine, 3 days postadministration, and prevent subsequent sensitization of Th2 cell-regulated specific IgE responses in a complete immunization study. Dhadwar et al. (2010) examined the influence of chitosan-DNA plasmid ratio on the size, morphology, and transfection abilities of nanoparticles. They identified the best ratio yielding the smallest particle sizes (200–400 nm) without producing large aggregates and exhibiting transfection efficiency on cells. Nanoparticles were prepared with plasmid containing canine FVIII cDNA. Hemophilia A mice (C57Bl/6$^{FVIII-/-}$) were administered by oral gavage with chitosan nanoparticles (150 µg DNA plasmid expressing FVIII) and demonstrated transient canine FVIII expression 1 day after treatment together with partial phenotypic correction. The delivered plasmid was detected in the intestine and, to a lesser extent, in the liver. Repeated weekly administrations restored FVIII activity. In order to enhance resistance toward acids of chitosan-based plasmid nanoparticles, Kai and Ochiya (2004) proposed a preparation method in which nanoparticles were *N*-acetylated once formed and combined with mouse food. They showed that the *N*-acetylated particles were more efficient to deliver gene to the intestines than particles prepared without *N*-acetylation; this result was most significant in the case of the duodenum. Chitosan was also used to prepare nanoparticles with siRNA (Ballarin-Gonzales et al., 2013) at various chitosan to siRNA ratios. After oral gavage in mice, northern and quantitative PCR detection were used 1 and 5 h postadministration to study structural integrity and localization of siRNA. High levels of intact siRNA were detected in the stomach, small intestine and colon 1 and 5 h after oral gavage of nanoparticles. Intact siRNA was also detected in the liver, spleen, and kidney 1 h after administration, with an organ distribution pattern influenced by chitosan to plasmid ratio used to prepared nanoparticles. These observations suggest a systemic translocation to peripheral organs after oral gavage in mice.

The earlier publications used unmodified chitosan, that produces, upon gentle mixture with nucleic acid solutions, chitosan/nucleic acid particulate complexes with broad size distributions (mean sizes from 100 to 600 nm), depending on the molecular weight of chitosan and chitosan to nucleic acid ratio. Because chitosan is soluble in an acidic pH range, the standard method of nanoparticle preparation comprises dissolving chitosan in acetic and/or hydrochloric acid.

3.3.1.2 Modified chitosan

Chitosan as a carrier for oral delivery of nucleic acids has many advantages, including its strong nucleic acid affinity, bioadhesion, and biodegradability. However, its insolubility in neutral and alkaline environments limits its application. To improve the solubility of chitosan over a wide pH range, trimethylation modification of chitosan has been proposed. Zhang et al. (2013b) have synthesized TMC (*N*-trimethyl chitosan) with 34% trimethylation degree and used it to prepare fluorescent nanoparticles encapsulating TAMRA-labeled siRNA to perform a biodistribution study upon oral administration to mice (4 μg siRNA/mouse). Nanoparticles were shown to be able to promote the intestinal absorption of siRNA and subsequent entry into the systemic circulation and other reticuloendothelial tissues (liver and spleen). To achieve nanoparticles with compact and stable structures, a new preparation method has been proposed, based on ionic gelation in the presence of an anionic component, tripolyphosphate (TPP). This method has been proposed because siRNA does not carry sufficient negative surface charge to form stable nanoparticles with cationic modified chitosan via simple complexation initiated by electrostatic interaction. TPP is a known food additive approved by the FDA. Ionic gelation occurs spontaneously in aqueous solution through electrostatic interaction between cationic polymers and anionic TPP without sonication or heating, which can preserve biological activity of siRNA (Zhang et al., 2013b). Furthermore, more modifications of methylated chitosan have been proposed to enhance its bioadhesion capacity through covalent binding with mucin glycoproteins and improve its permeation-enhancing potency. First, the synthesis of a partially quaternized derivative of chitosan, namely *N*-((2-hydroxy-3-trimethylammonium) propyl) chitosan chloride (HTCC) has been described. Wei et al. (2013) have prepared HTCC-based nanoparticles of fluorescent siRNA and followed the biodistribution of siRNA thanks to in vivo fluorescence imaging after oral administration to mice bearing LLC (Lewis Lung cells) tumors. The fluorescence intensity in gastrointestinal tract remained visible even 12 h after administration, suggesting an efficient retention of nanoparticles in the intestinal tract. Meanwhile, the fluorescence signal was visualized and gradually increased in the LLC tumor after 12 h administration, suggesting entry into the circulation and probably tumor localization through the enhanced permeability and retention (EPR) effect. Thiolated modification of chitosan with thiol-bearing compounds can also improve bioadhesion properties of nanoparticles. Thiolated trimethyl chitosan (TTMC) has been synthetized and used to prepare nanoparticles of siRNA through ionic gelation with sodium TPP (He et al., 2013b; Zhang et al., 2013b). Compared to unmodified chitosan-based nanoparticles, TAMRA-siRNA/TTMC particles showed the highest distribution percentages in most organs and plasma several hours after oral gavage administration in mice (Zhang et al., 2013b). In addition, orally delivered TTMC particles with siRNA targeting TNFα at a siRNA dose of 200 μg kg^{-1} significantly blocked LPS/D-GalN-triggered serum TNFα production in an acute hepatic injury model (He et al., 2013b). The last modification of chitosan that has been suggested is the addition of targeting groups able to bind specific extracellular receptors in order to target nanoparticles to specific cellular types. Mannose- (Chu et al., 2015; He et al., 2013a, 2015) or galactose- (Han et al., 2014; Zhang et al., 2013a) modified trimethyl chitosan-cystein conjugates have been synthesized and used to prepare nanoparticles with siRNA via ionic gelation with addition of TPP. Compared with nanoparticles without galactose, galactosylated nanoparticles were shown to more efficiently promote the distribution of fluorescent siRNA in ulcerative colon (in dextran-sulfate sodium (DSS)-induced ulcerative colitis mouse model) following oral administration. In addition, daily oral administration of galactosylated nanoparticles containing siRNA targeting Map4ka (mitogen-activated protein kinase kinase kinase kinase 4) significantly improved DSS-induced body weight loss, colon length shortening and increase of myeloperoxidase activity (Zhang et al., 2013a). Nanoparticles of galactosylated chitosan with various degrees of modification have been used to encapsulate siRNA and shRNA within the same nanoparticles (Han et al., 2014). Nanoparticles modified with a moderate degree (17%) outperformed the other particles, with no galactosylation or with higher (27%) or lesser (7%) degree of modification, in terms of siRNA accumulation in organs including tumors after oral administration to mice bearing hepatoma BEL-7402 tumors. Likewise, chitosan has also been modified with mannose and mannosylated nanoparticles of siRNA have been administered orally to mice to evaluate the biodistribution of fluorescent siRNA (He et al., 2013a). Mannose is expected to be bound and taken up by macrophages. Mannosylated nanoparticles promote siRNA distribution in most organs (liver, spleen, lung, and intestine), suggesting epithelial absorption of siRNA and subsequent trafficking to systemic circulation and other reticuloendothelial tissues. A single gavage of nanoparticles of siRNA targeting TNFα blocked LPS/DGalN-triggered serum TNFα production in both time- and dose-dependent manners in mice (He et al., 2013a) and in rats (He et al., 2015). As for galactosylated nanoparticles, the degree of modification has to be controlled because the ligand grafting density affects the efficiency of the mannosylated nanoparticles with TNFα siRNA orally administered to mice bearing DSS-induced colitis (Chu et al., 2015). Finally, modified chitosan has also been incorporated into supramolecular self-assembled nanoparticles (SSNP) designed to deliver siRNA and able to overcome the absorption and transfection barriers posed by intestine. These nanoparticles were composed of (1) oleyl conjugated trimethyl chitosan (OTMC) expected to enhance permeation and transfection, (2) poly(γ)-(4((2-(piperidin-1-yl)ethyl)-amino)methyl)benzyl-L-glutamate (PVBLG-8), a cationic α-helical polypeptide designed to promote cellular internalization and endosomal escape, (3) oleyl-PEG-mannose (OPM) designed to target the nanoparticles to enterocytes and macrophages that express mannose receptors, thus improving the intestinal absorption and macrophage uptake, and (4) oleyl-PEG-cysteamine expected to improve the mucoadhesion of the nanoparticles by forming disulfide bonds with mucin glycoproteins enriched in the intestinal mucosa. In contrast with previous nanoparticles (see earlier), in this strategy, each group (ligand, SH-, …) is provided by separate components instead of being grafted on the same chitosan molecule. SSNP were used to encapsulate siRNA targeting TNFα and were given by gavage to mice suffering from LPS/D-GalN-induced acute hepatic injury. SSNP orally administered 24 h before LPS/D-GalN treatment significantly improved the survival rate and inhibited the elevation of transaminase serum levels (Yin et al., 2013).

3.3.2 Inside a micrometric matrix

Nanoparticles formed by condensing polymers, like chitosan or PEI, have also been incorporated inside a micrometric matrix before oral administration. Bowman et al. (2008) mixed unmodified chitosan-based nanoparticles of plasmid DNA expressing FVIII with strawberry Jell-O brand gelatin to avoid gavage and get mice to eat nanoparticles contained inside gelatin. Functional FVIII protein was detected in plasma following oral administration of the chitosan nanoparticles to hemophilia A mice. More sophisticated multicomponent systems consisting of precondensing of nucleic acid with cationic polymer and subsequent incorporation of the obtained nanoparticles into micrometric matrix have been described for oral delivery of plasmid DNA or siRNA. Laroui et al. (2011) prepared complexes of PEI and siRNA, subsequently loaded into polylactide (PLA) in a single emulsion, followed by a double emulsion covered with polyvinylalcohol (PVA). The final particles of 300–500 nm in diameter were further encapsulated into a biomaterial composed of alginate and chitosan and given by gavage to mice before induction of acute inflammation by LPS injection. Nanoparticles loaded with siRNA targeting TNFα were shown to specifically reduce the TNFα expression in colonic tissue in LPS-treated mice. Aoudi et al. (2009) also reported oral delivery of siRNA with the so-called GeRP (β1,3-D-glucan-encapsulated siRNA particles). These particles are prepared following a layer by layer approach where 2–3 μm large porous hollow outer β1,3-D-glucan shells from baker's yeast were first filled with tRNA and PEI to build the cationic core that will encapsulate the siRNA, associated with a cationic peptide, endoporter, and followed by another PEI layer (Aoudi et al., 2009). Mice were given by oral gavage 10–20 μg/kg of siRNA targeting Map4k4 and were injected with LPS to induce acute inflammation. Oral delivery of GeRP containing siRNA anti-Map4k4 blocked the increase of serum TNFα and lethality. GeRPs were suggested to target the M cells of the Peyer's patches in the intestine in order to reach the GALT located below. Over time, the ingested GeRPs may migrate throughout the lymphatic tissue away from the gut. The last system called NiMOS (for nanoparticles in microspheres oral system) consists of pDNA (Bhavsar and Amiji, 2007a,b, 2008a,b) or siRNA (Kriegel and Amiji, 2011a,b) physically encapsulated by type B gelatin nanoparticles, which are further encapsulated within poly(ε-caprolactone) (PCL) microparticles. This system has been designed to facilitate protection of the nanoparticles from the harsh environment of the stomach, and then upon reaching the intestine, PCL is expected to be degraded by intestinal lipases, releasing the nanoparticles, which are then available for uptake by cells of the small and large intestine, where gelatin can be degraded by proteases. NiMOS were used to orally deliver plasmid DNA expressing reporter genes (Bhavsar and Amiji, 2007b) or IL-10 (Bhavsar and Amiji, 2008a) to mice or rats. A 100 μg dose given by oral gavage lead to transgene expression 5 days postadministration in intestine and even therapeutic efficiency in TNBS- (2,4,6-trinitrobenzenesulfonic acid) induced acute colitis model in mice. Similarly, NiMOS encapsulated siRNA targeting TNFα (Kriegel and Amiji, 2011a) or Cyclin D1 (Kriegel and Amiji, 2011b) were orally administered to mice suffering from DSS-induced acute colitis and successful gene silencing led to decreased expression of proinflammatory cytokines, increase of body weight and reduced tissue myeloperoxidase activity.

3.4 OTHER NANOSTRUCTURES

Jazayeri et al. (2012) have reported silver-based nanoparticles (AgNP) for oral DNA vaccine delivery. They used a green method to synthesize AgNP from $AgNO_3$, β-D-glucose and PEG as reduction agent and stabilizer, respectively. Although some concerns about their toxicity have been raised, in this study AgNP were conjugated to plasmid DNA expressing the avian influenza virus H5 gene and orally administered to 1-day old chicks. No toxicity induced by AgNP delivery was observed and rapidly increasing antibodies against H5 were detected in serum samples collected on days 14 and 35 postimmunization. Polymeric micelles have also been reported to orally deliver plasmid DNA. Nonionic PEO-PPO-PEO (poly(ethylene oxide)-poly(propylene oxide)-poly(ethylene oxide) triblock copolymer has been used to condense pDNA to form micelles with a diameter of 170 nm and nearly neutral (ζ potential −4 mV). Six doses of micelles containing LacZ encoding plasmid were orally administered to mice (Chang et al., 2004) and led to expression of the encoded protein in the villi, crypts, and goblets cells of the duodenum and in the crypt cells of the stomach. LacZ gene activity was also seen in the liver. The same polymeric micelles were also administered to parathyroidectomized rats to deliver plasmid DNA expressing parathyroid hormone (PTH) (Chou et al., 2009) and led to expression of PTH in gastrointestinal tissues. Triblock copolymer micelles were prepared with copolymers poly(D,L-lactic acid)-*co*-poly(ethylene glycol)-*co*-poly(D,L-lactic acid) (PLA-PEG-PLA) and poly(D,L-lactic-*co*-glycolic acid)-*b*-poly(ethylene glycol)-*b*-poly(D,L-lactic-*co*-glycolic acid) (PLGA-PEG-PLGA) and used to entrap plasmid DNA. The hydrophilic PEG segment has been incorporated into polylactide and glycolic acid to improve the hydrophilicity of the polymers. Micelles of PLA-PEG-PLA/PLGA-PEG-PLGA with encapsulated plasmid expressing multiple-epitope antigen gene of hepatitis C virus were orally administered to mice with 100 μg pDNA per vaccination. The results indicated that oral immunization achieved satisfying immune responses (Yang et al., 2011) with scarcely any side effects.

3.5 MICROMETRIC DELIVERY SYSTEMS AND MINITABLETS

Besides nanostructures, larger particles of rather micrometric size and minitablets have also been reported for oral delivery of nucleic acids. Although they do not fall within the nanostructure category in the strict sense, we chose to present them as they are much used in particular for genetic vaccination.

3.5.1 PLGA micrometric formulations

PLGA, which has FDA approval for drug implants, is widely used for the production of biodegradable and biocompatible micro- and nanoparticles. PLGA has been one of the most commonly used non-condensing polymers in intestinal gene delivery. It has generally been formulated as microparticles to deliver DNA vaccines. Although PLGA-based microparticles cannot themselves transduce cells, they are phagocytosed by M cells in the Peyer's patches, making them a useful polymer system for oral vaccination.

The first report of systemic and mucosal antibody responses to encoded protein upon oral administration of the plasmid expressing the antigen has been published in 1997 (Jones et al., 1997). PLGA-based microparticles of plasmid expressing photinus luciferase (50 μg pDNA) were orally administered to mice by gavage. Luciferase specific IgG and IgM antibodies were detected in serum samples 3 and 6 weeks postadministration. In addition, significant levels of specific mucosal IgA antibodies were also detectable, suggesting that PLGA encapsulation has the ability to protect pDNA against degradation after administration, and to facilitate its uptake into appropriate cells for the subsequent expression and presentation of antigen, in such a way as to elicit both systemic and mucosal antibody responses. Afterward, similar studies were conducted to induce immune responses toward various antigens, rotavirus VP6 protein (Chen et al., 1998), HIV envelope glycoprotein (Kaneko et al., 2000), hepatitis B virus small envelope protein (He et al., 2005), by oral immunization with PLGA-based microparticles containing antigen-coding DNA plasmid. In these three reports, one or several doses of vaccine orally given to mice elicited both antigen-specific serum antibodies and induced resistance to mucosal viral transmission in one model (Kaneko et al., 2000).

3.5.2 Alginate micrometric formulations

Sodium alginate is a naturally occurring polysaccharide that can be easily crosslinked into a solid matrix with the addition of divalent cations, crosslinking in a water-in-oil emulsion resulting in the formation of microspheres. Alginate microparticles are safe to use in animals and have been used to encapsulate purified proteins, bacterial outer membrane proteins, and virus. Mittal et al. (2001) encapsulated plasmid encoding the bacterial β-galactosidase gene inside alginate microspheres and compared immune responses following various routes of administration, oral, intranasal, intramuscular, subcutaneous, or intraperitoneal. They reported that oral immunization of mice induced slightly higher titers of LacZ specific IgA antibodies at mucosal sites compared to those obtained by systemic routes, namely intramuscular, subcutaneous and intraperitoneal. The ability of orally administered alginate microspheres to deliver plasmid DNA into intestinal cells of mice and induce the expression of the encoded protein was reported by Nograles et al. (2012) and Rosli et al. (2013). Using expression plasmid carrying gene-coding sequences for the cholera enterotoxin B subunit (ctxB) immunostimulatory antigen, the latter have observed IgA antibody responses at local mucosal sites following oral administration of alginate microspheres containing plasmid DNA (Rosli et al., 2013).

3.5.3 Tablets

ISIS 104838, a 2'-O-methoxy ethyl partially modified phosphorothioate antisense oligonucleotide targeting human TNFα mRNA was incorporated into minitablets containing sodium caprate permeation enhancer (Raoof et al., 2004). Sodium caprate enhances drug absorption, causes dilatation of tight junctions, enhances paracellular transport and is FDA approved as an emulsifier or binder food additive. An Opadry subcoat was applied to the minitablets that were then further coated with hydroxy-propyl methylcellulose (HPMC-50) and orally administered to beagle dogs to investigate tolerability and tissue distribution of ISIS 104838. The tissue distribution profile showed that the highest concentrations of ISIS 104838 were observed in the kidney, followed by the liver, lymph nodes, and spleen. In addition, the study showed that multidose administration was well tolerated during the in-life phase of the study as well as through clinical pathology (standard clinical chemistry and haematology tests) and postmortem examination at the end of the study. Oral absorption of ISIS 104838 was also investigated in human healthy volunteers (Tillman et al., 2007). Four different capsule formulations were prepared, one containing a single population of minitablets designed to perform an immediate releasing of oligonucleotide, the others containing two populations of minitablets, including the one permitting immediate release of ISIS 104838 and another minitablet population. The other minitablet populations were either a short- or long-delayed release of sodium caprate and were coated with different levels of methacrylate polymer Eudragit RS 30D (poly(ethyl acrylate-*co*-methyl methacrylate-*co*-trimethylammonioethyl methacrylate chloride). Minitablets were then encapsulated in gelatin capsules, and then an Eudragit L30D55-based (poly(methacrylic acid-*co*-ethyl acrylate)) enteric outer coating of the capsules was added. These various formulations were well tolerated in both fasted and fed volunteers. Oral absorption of ISIS 104838 was demonstrated with an average 9.5% plasma bioavailability.

4 CONCLUSIONS

Increasing nucleic acid delivery systems are developed for the oral route. Although therapeutic nucleic acids molecules exhibit unique properties in terms of stability and characteristics like global charge and size, an effective delivery through oral administration has been demonstrated in various preclinical models. However, at present, no system has reached clinical trials in humans, apart from the delivery of naked antisense oligonucleotides. It is therefore still necessary to improve the efficiency of these vectors as well as to characterize their toxicity, in particular for the most recent discovered molecules, such as RNAi-trigger molecules.

REFERENCES

Akhtar, S., 2009. Oral delivery of siRNA and antisense oligonucleotides. J. Drug Target. 17 (7), 491–495.

Alton, E.W., Middleton, P.G., Caplen, N.J., Smith, S.N., Steel, D.M., Munkonge, F.M., Jeffery, P.K., Geddes, D.M., Hart, S.L., Williamson, R., Fasold, K.I., Miller, A.D., Dickinson, P., Stevenson, B.J., McLachlan, G., Dorin, J.R., Porteous, D.J., 1993. Non-invasive liposome-mediated gene delivery can correct the ion transport defect in cystic fibrosis mutant mice. Nat. Genet. 5, 135–142.

Aoudi, M., Tesz, G.J., Nicoloro, S.M., Wang, M., Chouinard, M., Soto, E., Ostroff, G.R., Czech, M.P., 2009. Orally delivered siRNA targeting macrophage Map4k4 suppresses systemic inflammation. Nature 458, 1180–1184.

Ballarin-Gonzales, B., Dagnaes-Hansen, F., Fenton, R.A., Gao, S., Hein, S., Dong, M., Kjems, J., Howard, K.A., 2013. Protection and systemic translocation of siRNA following oral administration of chitosan/siRNA nanoparticles. Mol. Ther. Nucleic Acids 2, e76.

Bhavsar, M.D., Amiji, M.M., 2007a. Polymeric nano- and microparticle technologies for oral gene delivery. Exp. Opin. Drug Deliv. 4 (3), 197–213.

Bhavsar, M.D., Amiji, M.M., 2007b. Gastrointestinal distribution and in vivo gene transfection studies with nanoparticles-in-microsphere oral system (NiMOS). J. Control. Release 119, 339–348.

Bhavsar, M.D., Amiji, M.M., 2008a. Oral IL-10 gene delivery in a microsphere-based formulation for local transfection and therapeutic efficacy in inflammatory bowel disease. Gene Ther. 15 (17), 1200–1209.

Bhavsar, M.D., Amiji, M.M., 2008b. Development of novel biodegradable polymeric nanoparticles-in-microsphere formulation for local plasmid DNA delivery in the gastrointestinal tract. AAPS PharmSciTech 9, 288–294.

Bowman, K., Sarkar, R., Raut, S., Leong, K.W., 2008. Gene transfer to hemophilia A mice via oral delivery of FVIII-chitosan nanoparticles. J. Control. Release 132, 252–259.

Chang, S.-F., Chang, H.-Y., Tong, Y.-C., Chen, S.-H., Hsaio, F.-C., Lu, S.-C., Liaw, J., 2004. Nonionic polymeric micelles for oral gene delivery in vivo. Hum. Gene Ther. 15, 481–493.

Chen, S.C., Jones, D.H., Fynan, E.F., Farrar, G.H., Clegg, J.C.S., Greenberg, H.B., Herrmann, J.E., 1998. Protective immunity induced by oral immunization with a rotavirus DNA vaccine encapsulated in microparticles. J. Virol. 72 (7), 5757–5761.

Chen, J., Yang, W.-L., Li, G., Qian, J., Xue, J.-L., Fu, S.-K., Lu, D.-R., 2004. Transfection of mEpo gene to intestinal epithelium in vivo mediated by oral delivery of chitosan-DNA nanoparticles. World J. Gastroenterol. 10 (1), 112–116.

Chou, F.-F., Huang, S.-C., Chang, S.-F., Liaw, J., Hung, P.-H., 2009. Oral gene therapy for hypoparathyroidism: a rat model. Hum. Gene Ther. 20, 1344–1350.

Chu, S., Tang, C., Yin, C., 2015. Effects of mannose density on in vitro and in vivo cellular uptake and RNAi efficiency of polymeric nanoparticles. Biomaterials 52, 229–239.

Clayburgh, D.R., Shen, L., Turner, J.R., 2004. A porous defense: the leaky epithelial barrier in intestinal disease. Lab. Invest. 84, 282–291.

Cryan, S.A., O'Driscoll, C.M., 2003. Mechanistic studies on nonviral gene delivery to the intestine using in vitro differentiated cell culture models and an in vivo rat intestinal loop. Pharm. Res. 20, 569–575.

Dhadwar, S.S., Kiernan, J., Wen, J., Hortelano, G., 2010. Repeated oral administration of chitosan/DNA nanoparticles delivers functional FVIII with the absence of antibodies in hemophilia A mice. J. Thromb. Haemost. 8, 2743–2750.

Forbes, D., Peppas, N., 2012. Oral delivery of small RNA and DNA. J. Control. Release 162 (2), 438–445.

Han, L., Tang, C., Yin, C., 2014. Oral delivery of shRNA and siRNA via multifunctional polymeric nanoparticles for synergistic cancer therapy. Biomaterials 35, 4589–4600.

He, X.-W., Wang, F., Jiang, L., Li, J., Liu, S.-K., Xiao, Z.-Y., Jin, X.-Q., Zhang, Y.-N., He, Y., Li, K., Guo, Y.-J., Sun, S.-H., 2005. Induction of mucosal and systemic immune response by single-dose oral immunization with biodegradable microparticles containing DNA encoding HBsAg. J. Gene Virol. 86, 601–610.

He, C., Yin, L., Tang, C., Yin, C., 2013a. Multifunctional polymeric nanoparticles for oral delivery of TNF-α siRNA to macrophages. Biomaterials 34, 2843–2854.

He, C., Yin, L., Tang, C., Yin, C., 2013b. Trimethyl chitosan-cysteine nanoparticles for systemic delivery of TNF-α siRNA via oral and intraperitoneal routes. Pharm. Res. 30, 2596–2606.

He, C., Yin, L., Song, Y., Tang, C., Yin, C., 2015. Optimization of multifunctional chitosan-siRNA nanoparticles for oral delivery applications, targeting TNF-α silencing in rats. Acta Biomater. 17, 98–106.

Herrmann, J.E., Chen, S.C., Jones, D.H., Tinsley-Bown, A., Fynan, E.F., Greenberg, H.B., Farrar, G.H., 1999. Immune responses and protection obtained by oral immunization with rotavirus VP4 and VP7 DNA vaccines encapsulated in microparticles. Virology 259, 148–153.

Hughes, J.A., Avrutskaya, A.V., Brouwer, K.L., Wickstrom, E., Juliano, R.L., 1995. Radiolabeling of methylphosphonate and phosphorothioate oligonucleotides and evaluation of their transport in everted rat jejunum sacs. Pharm. Res. 12, 817–824.

Jazayeri, S.D., Ideris, A., Zakaria, Z., Shameli, K., Moeini, H., Omar, A.R., 2012. Cytotoxicity and immunological responses following oral vaccination of nanoencapsulated avian influenza virus H5 DNA vaccine with green synthesis silver nanoparticles. J. Control. Release 161, 116–123.

Jones, D.H., Corris, S., McDonald, S., Clegg, J.C., Farrar, G.H., 1997. Poly(DL-lactide-*co*-glycolide)-encapsulated plasmid DNA elicits systemic and mucosal antibody responses to encoded protein after oral administration. Vaccine 15 (8), 814–817.

Kai, E., Ochiya, T., 2004. A method for oral DNA delivery with N-acetylated chitosan. Pharm. Res. 21 (5), 838–843.

Kaneko, H., Bednarek, I., Wierzbicki, A., Kiszka, I., Dmochowski, M., Wasik, T.J., Kaneko, Y., Kozbor, D., 2000. Oral DNA vaccination promotes mucosal and systemic immune responses to HIV envelope glycoprotein. Virology 267, 8–16.

Kriegel, C., Amiji, M., 2011a. Oral TNF-α gene silencing using a polymeric microsphere-based delivery system for the treatment of inflammatory bowel disease. J. Control. Release 150 (1), 77–86.

Kriegel, C., Amiji, M., 2011b. Dual TNF-α/cyclin D1 gene silencing with an oral polymeric microparticle system as a novel strategy for the treatment of inflammatory bowel disease. Clin. Transl. Gastroenterol. 2 (3), 1–12.

Kriegel, C., Attarwala, H., Amiji, M., 2013. Multi-compartemental oral delivery systems for nucleic acid therapy in the gastrointestinal tract. Adv. Drug Deliv. Rev. 65, 891–901.

Kühn, R., Löhler, J., Rennick, D., Rajewsky, K., Müller, W., 1993. Interleukin-10-deficient mice develop chronic enterocolitis. Cell 75, 263–274.

Laroui, H., Theiss, A.L., Yan, Y., Dalmasso, G., Nguyen, H.T.T., Sitaraman, S.V., Merlin, D., 2011. Functional TNFα gene silencing mediated by polyethyleneimine/TNFα siRNA nanocomplexes in flamed colon. Biomaterials 32, 1218–1228.

Lew, J.I., Guo, Y., Kim, R.K., Vargish, L., Michelassi, F., Arenas, R.B., 2002. Reduction of intestinal neoplasia with adenomatous polyposis coli gene replacement and COX-2 inhibition is additive. J. Gastrointest. Surg. 6, 563–568.

Li, G., Liu, Z., Liao, B., Zhong, N., 2009. Induction of TH1-type immune response by chitosan nanoparticles containing plasmid DNA encoding house dust mite allergen Der p2 for oral vaccination in mice. Cell. Mol. Immunol. 6 (1), 45–50.

Liptay, S., Weidenbach, H., Adler, G., Schmid, R.M., 1998. Colon epithelium can be transiently transfected with liposomes, calcium phosphate precipitation and DEAE dextran in vivo. Digestion 59, 142–147.

Liu, J., Wu, J., Wang, B., Zeng, S., Qi, F., Lu, C., Kimura, Y., Liu, B., 2014. Oral vaccination with a liposome-encapsulated influenza DNA vaccine protects mice against respiratory challenge infection. J. Med. Virol. 56, 886–894.

Martien, R., Loretz, B., Sandbichler, A.M., Schnürch, A.B., 2008. Thiolated chitosan nanoparticles: transfection study in the Caco-2 differentiated cell culture. Nanotechnology 19, 045101.

Mittal, S.K., Aggarwal, N., Sailaja, G., van Olphen, A., HogenEsch, H., North, A., Hays, J., Moffatt, S., 2001. Immunization with DNA, adenovirus or both in biodegradable alginate microspheres: effect of route of inoculation on immune response. Vaccine 19, 253–263.

Moser, A.R., Pitot, H.C., Dove, W.F., 1990. A dominant mutation that predisposes to multiple intestinal neoplasia in the mouse. Science 247, 322–324.

Nograles, N., Abdullah, S., Shamsudin, M.N., Billa, N., Rosli, R., 2012. Formation and characterization of pDNA-loaded alginate microspheres for oral administration in mice. J. Biosci. Bioeng. 113 (2), 133–140.

O'Neill, M., Bourre, L., Melgar, S., O'Driscoll, C., 2011. Intestinal delivery of non-viral gene therapeutics: physiological barriers and preclinical models. Drug Discov. Today 16 (5–6), 203–218.

Ozcan, G., Ozpolat, B., Coleman, R.L., Sood, A.K., Lopez-Berestein, G., 2015. Preclinical and clinical development of siRNA-based therapeutics. Adv. Drug Deliv. Rev. 87, 108–119.

Raoof, A.A., Chiu, P., Ramtoola, Z., Cumming, I.K., Teng, C., Weinbach, S.P., Hardee, G.E., Levin, A.A., Geary, R.S., 2004. Oral bioavailability and multiple dose tolerability of an antisense oligonucleotide tablet formulated with sodium caprate. J. Pharm. Sci. 93 (6), 1431–1439.

Rosli, R., Nograles, N., Hanafi, A., Shamsudin, M.N., Abdullah, S., 2013. Mucosal genetic immunization through microsphere-based oral carriers. Hum. Vaccin. Immunother. 9 (10), 2222–2227.

Roy, K., Mao, H.Q., Huang, S.K., Leong, K.W., 1999. Oral gene delivery with chitosan-DNA nanoparticles generates immunologic protection in a murine model of peanut allergy. Nat. Med. 5 (4), 387–391.

Schipper, N.G., Vårum, K.M., Stenberg, P., Ocklind, G., Lennernäs, H., Artursson, P., 1999. Chitosans as absorption enhancers of poorly absorbable drugs. 3: influence of mucus on absorption enhancement. Eur. J. Pharm. Sci. 8, 335–343.

Schmid, R.M., Weidenbach, H., Liptay, S., Adler, G., 1997. Direct gene transfer into the colon using a double-balloon catheter. Endoscopy 29, 39–43.

Sjögren, E., Abrahamsson, B., Augustijns, P., Becker, D., Bolger, M.B., Brewster, M., Brouwers, J., Flanagan, T., Harwood, M., Heinen, C., Holm, R., Juretschke, H.P., Kubbinga, M., Lindahl, A., Lukacova, V., Münster, U., Neuhoff, S., Nguyen, M.A., Peer, A.V., Reppas, C., Hodjegan, A.R., Tannergren, C., Weitschies, W., Wilson, C., Zane, P., Lennernäs, H., Langguth, P., 2014. In vivo methods for drug absorption - comparative physiologies, model selection, correlations with in vitro methods (IVIVC), and applications for formulation/API/excipient characterization including food effects. Eur. J. Pharm. Sci. 57, 99–151.

Szabo, S., Deng, X., Khomenko, T., Yoshida, M., Jadus, M.R., Sandor, Z., Gombos, Z., Matsumoto, H., 2001. Gene expression and gene therapy in experimental duodenal ulceration. J. Physiol. Paris 95, 325–335.

Tahara, K., Samura, S., Tsuji, K., Yamamoto, H., Tsukada, Y., Bando, Y., Tsujimoto, H., Morishita, R., Kawashima, Y., 2011. Oral nuclear factor-κB decoy oligonucleotides delivery system with chitosan modified poly(D,L-lactide-*co*-glycolide) nanospheres for inflammatory bowel disease. Biomaterials 32, 870–878.

Tanaka, T., Kohno, H., Suzuki, R., Yamada, Y., Sugie, S., Mori, H., 2003. A novel inflammation-related mouse colon carcinogenesis model induced by azoxymethane and dextran sodium sulfate. Cancer Sci. 94, 965–973.

Tillman, L.G., Geary, R.S., Hardee, G.E., 2007. Oral delivery of antisense oligonucleotides in man. J. Pharm. Sci. 97 (1), 225–236.

Tortora, G., Bianco, R., Damiano, V., Fontanini, G., De Placido, S., Bianco, A.R., Ciardello, F., 2000. Oral antisense that targets protein kinase A cooperates with Taxol and inhibits tumor growth, angiogenesis, and growth factor production. Clin. Cancer Res. 6, 2506–2512.

Wang, D., Gao, G., 2014. State-of-the-art human gene therapy: Part I. Gene delivery technologies. Discov. Med. 18 (97), 67–77.

Wang, H., Cai, Q., Zeng, X., Yu, D., Agrawal, S., Zhang, R., 1999. Antitumor activity and pharmacokinetics of a mixed-backbone antisense oligonucleotide targeted to the RIα subunit of protein kinase A after oral administration. Proc. Natl. Acad. Sci. USA 96 (24), 13989–13994.

Wei, W., Lv, P.-P., Chen, X.-M., Yue, Z.-G., Fu, Q., Liu, S.-Y., Yue, H., Ma, G.-H., 2013. Codelivery of mTERT siRNA and paclitaxel by chitosan-based nanoparticles promoted synergistic tumor suppression. Biomaterials 34, 3912–3923.

Wikman, A., Karlsson, J., Carlstedt, I., Artursson, P., 1993. A drug absorption model based on the mucus layer producing human intestinal goblet cell line HT29-H. Pharm. Res. 10, 843–852.

Wilson, D.S., Dalmasso, G., Wang, L., Sitaraman, S., Merlin, D., Murthy, N., 2010. Orally delivered thioketal-nanoparticles loaded with TNFα-siRNA target inflammation and inhibit gene expression in the intestines. Nat. Mater. 9 (11), 923–928.

Wirtz, S., Neurath, M.F., 2007. Mouse models of inflammatory bowel disease. Adv. Drug Deliv. Rev. 59, 1073–1083.

Yang, Y., Kuang, Y., Liu, Y., Li, W., Jiang, Z., Xiao, L., Li, M., 2011. Immunogenicity of multiple-epitope antigen gene of HCV carried by novel biodegradable polymers. Comp. Immunol. Microbiol. Infect. Dis. 34, 65–72.

Yin, L., Song, Z., Qu, Q., Kim, K.H., Zheng, N., Yao, C., Chaudhury, I., Tang, H., Gabrielson, N.P., Uckun, F.M., Cheng, J., 2013. Supramolecular self-assembled nanoparticles mediate oral delivery of therapeutic TNF-α siRNA against systemic inflammation. Angew. Chem. 52, 5757–5761.

Yin, H., Kanasty, R.L., Eltoukhy, A.A., Vegas, A.J., Dorkin, J.R., Anderson, D.G., 2014. Non-viral vectors for gene-based therapy. Nat. Rev. Genet. 15, 541–555.

Yoon, J.W., Jun, H.S., 2002. Recent advances in insulin gene therapy for type 1 diabetes. Trends Mol. Med. 8, 62–68.

Zhang, J., Tang, C., Yin, C., 2013a. Galactosylated trimethyl chitosan-cysteine nanoparticles loaded with Map4ka siRNA for targeting activated macrophages. Biomaterials 34, 3667–3677.

Zhang, J., He, C., Tang, C., Yin, C., 2013b. Ternary polymeric nanoparticles for oral siRNA delivery. Pharm. Res. 30, 1228–1239.

CHAPTER

7

CHALLENGES IN ORAL DRUG DELIVERY: A NANO-BASED STRATEGY TO OVERCOME

Pragasam Viswanathan, Yuvashree Muralidaran, Gokulakannan Ragavan
VIT University, Vellore, India

CHAPTER OUTLINE

1 Introduction 174
2 Biopharmaceutical Classification of Drugs 175
3 Measurement of Parameters Determining Bioavailability 176
3.1 Dissolution 176
3.2 Solubility 178
3.3 Permeability 178
4 Transport Mechanism in GIT 179
5 Gastrointestinal Luminal Milieu and Drug Absorption 179
5.1 pH Variation in GIT 181
5.2 Gastrointestinal Microbiota 182
5.3 Gastric Emptying Rate 183
6 Improving Drug Bioavailability 183
7 Pharmaceutics Approach Enabling Physical Modification of Drugs 183
7.1 Particle Size Reduction 183
7.2 Modification of Crystal Habit 184
7.3 Drug Dispersion in Carriers 184
7.4 Complexation 185
7.5 Solubilization by Surfactants 185
7.6 Chemical Modification 185
8 Improving Stability of Oral Drugs 186
9 Nanostructures in Oral Drug Delivery 186
9.1 Polymeric Formulation 186
9.2 Lipid-Based Formulations 194
9.3 Metal and Inorganic-Based Nanocarriers 194
10 Conclusions 196
Acknowledgment 196
References 197

Nanostructures for Oral Medicine

1 INTRODUCTION

A drug-delivery system should show rapid absorption with minimal adverse effects, which are determined by the route of administration. In clinical arenas, noninvasive rotes of administration like oral, intranasal, and topical are chosen over various other parenteral routes to avoid toxicity and to promote better quality of disease management. Among the noninvasive routes, oral specific drug delivery is highly preferable due to its reliability, ease of use, and therapeutic efficiency.

However, to achieve and maintain the therapeutics range, targeted drug delivery for an extended period of time along with a better control of systemic drug delivery is needed. Drug delivery systems are framed depending on the drug release rate. Sustained release systems facilitate the slow release of drug over an extended period of time, and if it maintains constant drug levels in blood or the target tissue, it is considered as controlled-release system. Further, a prolonged release system is attributed to drugs that extend the duration of action to achieve constant drug levels in blood (Neetika et al., 2012; Pundir et al., 2013).

Efficacy of any drug delivery system is determined by the extent of bioavailability of the drug. Good oral bioavailability takes place when the drug has maximal permeability and solubility at the site of absorption. The fundamental relationship between the rate of drug absorption measured as a permeability coefficient and extent of absorption, is the pioneer for experimental and mathematical models, which predicts the absorption of oral drug products (Volpe, 2010). The mathematical theories applied for quantifying drug dissolution are the Noyes–Whitney equation, Nernst–Brunner equation and Hixson–Crowell equation, for drug solubility, solubility constant, and other empirical methods are used (Dash et al., 2010).

Mostly, oral drugs are administered in the form of tablets, wherein the drug disintegration and dissolution relies on mass transport phenomena that includes (1) water uptake, (2) swelling, (3) disintegration, and (4) dissolution. The hydration rate and wettability of tablets largely depends on the particle size distribution of disintegrant that influences the water transfer rate, and its shape that determines the rate of swelling (Zhao and Augsburger, 2005). Further swelling of the tablet depends on its porosity; tablets with high porosity do not provide enough swelling force for the tablet to disintegrate. However, tablets with low porosity provide sufficient swelling force. The deformation of disintegrants on compression is ascertained to elevate its swelling capacity in aqueous environment. In addition, hydrophilicity of the tablet also promotes its disintegration by capillary action termed wicking, which facilitates the formation of hydrophilic network between drug particles by replacing air by water molecules. In case of nonswellable disintegrants, electric repulsive force among particles, mediated by water molecules has shown to facilitate disintegration (Bala et al., 2012).

However, in conventional drug delivery systems, the major limitation is nonspecificity of drug action due to inefficient drug formulation. Thus, designing drugs with greater target specificity is very important, which can be achieved by altering the pharmacokinetics and pharmacodynamics of drugs employed (Surendiran et al., 2009). The controlled targeted release (CTR) is made possible with size reduction to nanometer range that helps in improving solubility and bioavailability by increasing surface area, rapid onset of therapeutic action, decreased fed/fasted variability and patient-to-patient variability (Duggal, 2011).

The nanotechnology-based systems are classified into nanomaterials and nanodevices, where nanomaterials further segregated as nanocrytalline and nanostructured materials. The nanostructured materials are used in oral drug delivery, which are polymer-based like dendrimers, nanoparticles, and micelles, and nonpolymers based like carbon nanotubes (CNTs), metallic nanoparticles, and quantum dots. This chapter discusses the challenges faced in oral drug delivery that hampers the bioavailabilty of drugs, and the possible measures to overcome this difficulty by acquainting nanotechnology in oral drug delivery.

2 BIOPHARMACEUTICAL CLASSIFICATION OF DRUGS

Biopharmaceutical classification system (BCS) is intended for oral immediate release products that are absorbed in the intestine, and aims at replacing certain studies by in vitro dissolution tests that could minimize the drug exposure to healthy human subjects, which reduce the cost and time in drug development process. Besides, it facilitates in identifying the rate limiting step, the critical determinant in drug bioavailability (Lennernas and Abrahamsson, 2005). Accordingly, the drug substances have been classified into four classes based on its dissolution, solubility, and permeability in the intestine. Class II and Class IV are further divided into subclasses based on acidic (1), basic (2), or neutral (3) characteristics of the drug in the physiological pH range (~pH < 7.5) (Fig. 7.1).

Dissolution: A drug product is considered rapidly dissolving when 85% or more of the labeled amount of drug substance dissolves within 30 min using USP apparatus 1 or 2 in a volume of 900 mL or less of buffer solutions.

Solubility: A drug substance is considered highly soluble when the highest dose strength is soluble in 250 mL or less of water over a pH range of 1–7.5 at 37°C.

Permeability: A drug substance is considered highly permeable when the extent of absorption in humans is greater than 90% of an administered dose, based on mass-balance or compared with an intravenous reference dose (Reddy and Karunakar, 2011; Tsume et al., 2014).

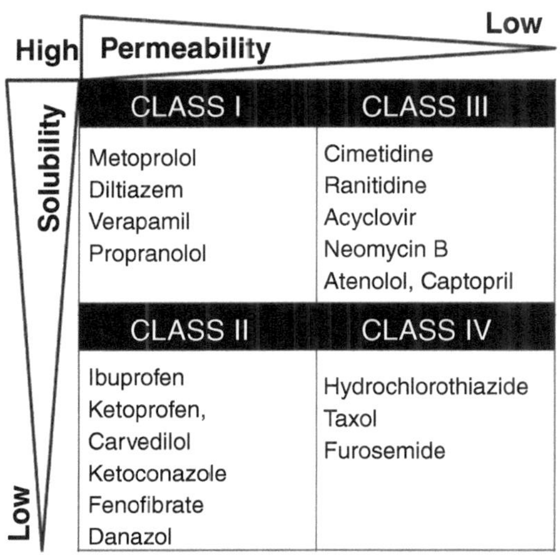

FIGURE 7.1 Biopharmaceutical Classification System (BCS) of Drugs: Classification of Drugs into Four Classes According to its Permeability and solubility

Class I–High Permeability, High Solubility: These compounds are well absorbed and their absorption rate is usually higher than the excretion rate. These drugs are formulated as immediate release products. If dissolution rate is very rapid, then gastric emptying becomes the rate-determining step. The examples for this Class I category include metoprolol, diltiazem, verapamil, and propranolol (Siya et al., 2015).

Class II–High Permeability, Low Solubility: This class of drugs possesses divergent biopharmaceutical properties that have an enormous impact on their formulation, gastrointestinal solubility, dissolution rate, and, hence, in vivo oral absorption and bioavailability profiles. Class IIa are weak acids that exhibit low solubility at gastric pH and dissolve quickly at intestinal pH and behave as a Class I drug. Alternatively, Class IIb are weak bases that exhibit high solubility at gastric pH and precipitate at intestinal pH. However solubility of Class IIc neutral drugs does not depend on in vivo pH change, whereas in vivo ambience, such as surfactant and lipids, played a vital role. The examples for this category include Class IIa ibuprofen and ketoprofen (weak acids); for IIb, carvedilol and ketoconazole (weak bases); and for IIc, fenofibrate and danazol (neutral) (Tsume et al., 2014).

Class III–Low Permeability, High Solubility: Drug permeability is the rate-limiting step for drug absorption, but the drug is solvated very quickly. These drugs exhibit a high variation in the rate and extent of drug absorption. Because the dissolution is rapid, the variation is attributable to alteration of physiology and membrane permeability rather than the dosage form factors. Class I criteria can be applied if the permeability or gastrointestinal duration time remains unchanged.

The examples for this category include cimetidine, ranitidine, acyclovir, neomycin B, atenolol, and captopril.

Class IV: The drugs of this class are problematic for effective oral administration. These compounds have poor bioavailability. They are usually not well absorbed through the intestinal mucosa, and a high variability is expected. Fortunately, extreme examples of Class IV compounds are the exception rather than the rule, and these are rarely developed and marketed. Nevertheless, several Class IV drugs do exist. The examples for this category include hydrochlorothiazide, taxol, and furosemide (Siya et al., 2015).

3 MEASUREMENT OF PARAMETERS DETERMINING BIOAVAILABILITY

3.1 DISSOLUTION

The bioavailability of drug is determined by its degree of solubility that directly influences the dissolution rate. Dissolution is the process by which a solid substance enters the solvent phase to yield a solution, that is, mass transfer from solid surface to liquid phase. As given by IUPAC, the term "dissolution" is defined as "The mixing of two phases with the formation of one new homogeneous phase (i.e., the solution)." Dissolution rate is the amount of drug substance that goes in solution (dc) in the time interval (dt) under standardized conditions of temperature and solvent composition (Siepmann and Siepmann, 2013).

$$\text{Dissolution rate} = \mathrm{d}c/\mathrm{d}t$$

Further, this dissolution rate also depends on the nature of drug, its polymorphic form, crystallinity, particle size, solubility, and amount in the pharmaceutical dosage form (Costa and Lobo, 2001).

Drug dissolution occurs by five major physical phenomena:

1. Wetting of drug.
2. Breaking of attractive electrostatic force between cations and anions in the drug.
3. Solvation of drug molecules by water molecules.
4. Diffusion of drug particles through the liquid, unstirred boundary layer surrounding the system into the well-stirred bulk fluid.
5. In well-stirred bulk fluid, the drug molecules/ions/atoms are transported by convection in the liquid.

Process of dissolution is of two consecutive stages. The first stage is instantaneous, which constitutes the interfacial reaction between solute and solvent, in which the solute molecules are displaced by solvent molecules. The second stage is relatively the slowest process, which involves diffusion of solute molecules through the boundary layer. This boundary layer is formed between the two phases, which disappears when homogenous solution is obtained. Because the overall rate of dissolution depends on the slowest step, the latter is considered to be the rate limiting step (Mohanachandran et al., 2011).

Mathematical expression of dissolution has been given by the Noyes–Whitney equation, Nernst–Brunner and Hixson–Crowell equation.

Fick's law of diffusion states that change in concentration is directly proportional to the concentration difference over the diffusion layer.

$$J = D.\mathrm{d}c / \mathrm{d}x$$

According to the Noyes–Whitney equation, the dissolution of the drug at constant agitation and temperature is proportional to the difference between the instantaneous concentration, C at time t, and the saturation solubility C_s.

$$\mathrm{d}C / \mathrm{d}t = D(C_s - C)$$

where D is the diffusion constant, and the mechanism of dissolution was related to the diffusion of solute through the thin diffusion layer.

Later, Nernst and Brunner in 1904 developed the Nernst–Brunner equation based on Fick's law of diffusion and the Noyes–Whintey equation.

$$\mathrm{d}C / \mathrm{d}t = DS(C_s - C) / Vh$$

where D is the diffusion coefficient, h is the thickness of the diffusion layer, S is the surface area and V is the volume of the dissolution medium. The rate of dissolution is considered to be dependent on the exposed surface, the rate of stirring, temperature, structure of the surface, and the arrangement of the apparatus.

In the aforementioned studies, time-dependent changes on the surface of a dissolving substance were not considered. However, decades previously, Hixson and Crowellin designed a mathematical equation addressing the fact that the surface of a dissolving substance often changes with time in practice, for example, if spherical particles dissolve in well-agitated bulk fluids the radius of the spheres continuously decreases. The following equation

$$\mathrm{d}C / \mathrm{d}t = -K \cdot St\,(C_s - Ct)$$

denotes that the rate of drug dissolution is proportional to the available surface area (which is a function of time), *St*, and the difference in concentration. Where, the negative sign the mass change of nondissolved substance.

Cubic root law is designed by letting *S* be proportional to $w^{2/3}$, relating time to the cubic-root of weight and in the special case of sink conditions, where small concentrations are considered and the difference ($C_s - C$) can be considered as constant.

$$w_0^{1/3} - w^{1/3} = k_2 t$$

where M_0 and M_t denote the mass of the particle at time $t = 0$ and at time *t*, respectively (Dokoumetzidis and Macheras, 2006).

3.2 SOLUBILITY

Solubility occurs under dynamic equilibrium, which means that solubility results from the simultaneous and opposing processes of dissolution and phase joining (e.g., precipitation of solids). Solubility equilibrium occurs when the two processes proceed at a constant rate. Under certain conditions equilibrium solubility may be exceeded to give a so-called supersaturated solution, which is metastable (Ojha and Prabhakar, 2013).

IUPAC defines solubility as the analytical composition of a saturated solution expressed as a proportion of a designated solute in a designated solvent. Solubility may be stated in units of concentration, molality, mole fraction, mole ratio, and other units.

Solubility is the important parameter to achieve desired concentration of drugs in the systemic circulation that could exert desired physiological response. Any drug to be absorbed must be present in the form of an aqueous solution at the site of absorption. Since, water remains the solvent of choice for liquid pharmaceutical formulations, poor water soluble drugs have slow absorption leading to inadequate bioavailability. Most of the drugs are either weakly acidic or weakly basic, having poor aqueous solubility, especially in BCS Class II drugs, the rate-limiting step is drug release from dosage form and its solubility in gastric pH than absorption (Savjani et al., 2012).

However, the disadvantage is that solubility strongly depends on the presence of other species in the solvent (e.g., the common ion effect). Saturated solution of ionic compounds with low solubility is expressed in terms of solubility constants, which describe the balance between undissolved and dissolved salts. Other than solubility constant, which is temperature-dependent, and independent of the presence of other species in the solvent, other physical constants, such as enthalpy of fusion and empirical methods, namely, Hansen solubility parameters and Hildebrand solubility parameters are also employed in predicting solubility. In addition, solubility of polymers is described by Flory–Huggins solution theory, and the differential solubility of a compound in a hydrophobic and hydrophilic solvent is measured by partition coefficient (log P) (Savjani et al., 2012).

Various solubilization techniques, including cosolvency, addition of surfactants, salt formation, and solid dispersion are employed into solubilized poorly soluble drugs (Jouyban et al., 2014; Seo et al., 2015).

3.3 PERMEABILITY

With the advent of drug discovery, evaluation of permeability characteristics of drug molecules has become vital in drug development. Currently, there exists several permeability assays including in situ perfusion through isolated intestinal segments, ex vivo diffusion across tissues, and in vitro permeation through cell monolayers or artificial membranes (Motz et al., 2007a,b).

The cell culture–based models make the study of permeability mechanisms feasible because it displays structural and biological properties characteristic of differentiated intestinal epithelia (e.g. microvilli, tight junctions) that express different drug transporter systems and metabolic enzymes However, cellular models have several disadvantages: cellular monolayers comprise just one type of cell that fails to mimic the exact in vivo cellular ambience, and they are associated with an inherent degree of variability; with culture time, passage number and composition of culture media capable of affecting cellular properties.

Several other cellular models designed include $CaCO_2$ model, modified HT29–MTX and Madin–Darby canine kidney (MDCK) cells, an alternative to $CaCO_2$ model. In an attempt to minimize heterogeneity, a clone from a late passage of the parental $CaCO_2$ cell line has been produced, termed TC7. The rat-derived 2/4/A1 cell lines are reported to better reflect the epithelial barrier in vivo, with regard to passive drug transport (Awortwe et al., 2014).

Alternatively, considering the cost and time consumed in cellular models, noncellular models based on artificial membranes were established that offers higher throughout capacity compared to their cellular-based counterparts. These models include parallel artificial membrane permeability assay (PAMPA), phospholipid vesicle-based permeation assay (PVPA), which are most effective for modeling passive, transcellular permeability of poorly soluble drug compounds transported via this route (Buckley et al., 2012).

4 TRANSPORT MECHANISM IN GIT

Transportation of drugs across the biological membrane in the gut is the crucial step in determining the bioavailability, where permeability of the membrane plays the critical role. This membranous barrier consists of lipid bilayer with hydrophilic head facing outward and inward lipophilic tails facing each other. The physicochemical property of the drug influences its entry into the cell, in specific, hydrophilicity and lipophilicity of the drug ascertains the transport pathway across the membrane. Such transport pathways include paracellular transport, transcellular transport, endocytosis, transcytosis, and membrane proteins mediated influx and efflux of drug molecules (Buckley et al., 2012). Generally, the low molecular weight hydrophobic molecules, which can diffuse through the membrane are transported transcellularily, while on the contrary, hydrophilic molecules are transported paracellularily across the membrane. In some instances drugs are transported inside a membrane vesicle by fluid-phase endocytosis (pinocytosis), an energy-dependent saturable process (Xu et al., 2013) (Fig. 7.2).

5 GASTROINTESTINAL LUMINAL MILIEU AND DRUG ABSORPTION

The absorption of drugs is dependent on the intrinsic factors like drug stability at gastric pH, mucosal permeability, and solubility in mucosal fluid, and the physiological factors, such as gastrointestinal transit time, pH, surface area, enzymatic activity, and colonic microflora (Xu et al., 2013).

The primary port for oral drug delivery is the buccal cavity. The mucosa in the mouth forms the tough layer and protective layer that guard the gut by moistening the food into a soft bolus. The composition

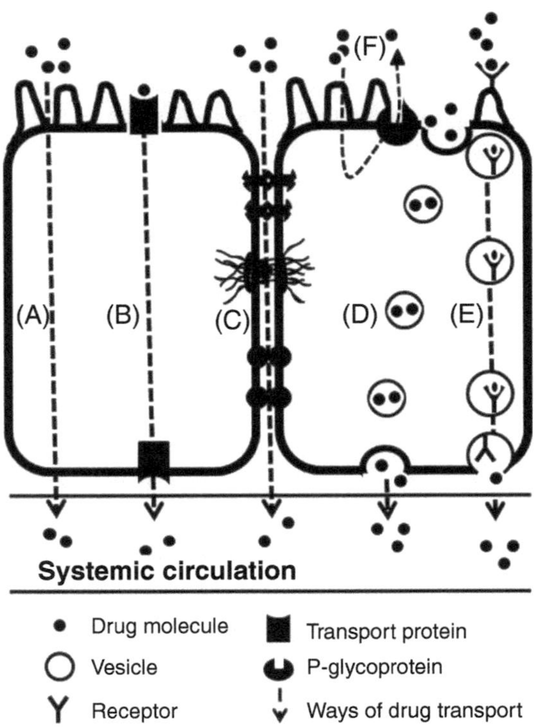

FIGURE 7.2 Schematic Representation of Preferred Ways of Drug Transport to Systemic Circulation, and Drug Efflux via P-glycoprotein (P-gp)

(A) Transcellular absorption of lipid soluble drugs, (B) transport protein mediated absorption of drugs, (C) paracellular transport of water soluble drugs, (D) pinocytosis, (E) receptor mediated endocytosis of drug molecule.

and pH of saliva varies with the rate of secretion, the pH range varies from 7.4 to 6.2. After the drug leaves the buccal cavity, transit through the esophagus is complete within 10 s. In the stomach 3 L of fluid is secreted every 24 h, the fluid consists of mucus secreted by goblet cells, enzymes pepsinogen and lipase, hydrochloric acid and intrinsic factor secreted by parietal cells that facilitates absorption of vitamin B12 in distal portion of small intestine (ileum). The small surface area, gastric mucus, and gastric pH serve as a barrier for effective drug absorption (Schubert, 2014). In addition, pepsin and pancreatic proteases are the major threat for the stability of proteins and peptides, and lipases hinders with the release of drug from fat/oil containing dosage forms.

Villi in the intestinal epithelium play a pivotal role in drug absorption that increases the total surface area in the gastrointestinal tract (GIT) to 300–400 m^2. Mucus secreting goblet cells and absorptive enterocytes interspersed with follicle associated epithelium (FAE) embraces villi. The mucosal surface provides an effective barrier to macromolecules and microorganisms present in the intestinal lumen. In favor of drug absorption, intestines possess specialized epithelial cells termed microflod cells (M cells) overlying organized lymphoid follicle, which naturally exists to transport wide range of microorganisms and macromolecules to the intestinal immune system (Ensign et al., 2012; Nicoletti, 2000).

A crucial factor that acts as a barrier to drug absorption is the intestinal mucus secreted by goblet cells. Mucin is the major protein component that is either secreted or cell-bound. Secreted mucin monomers linked by di-sulfide bonds forms a large molecule of 0.5–40 MDa size. These mucin molecules entangle and crosslink adhesively and reversibly to form a dynamic viscoelastic gel with shear-thinning properties. Further, mucin consists of protein backbone, O-linked oligosaccharides (2%–3%) and *N*-linked oligosaccharides (40%–80%). The structure and function of mucus vary depending on its location in the body. Long-branched oligosaccharide side chains on mucins protects against proteolytic degradation by digestive proteases. The viscoelastic property of mucus is contributed by hydrophobic, lipid-coated domains in mucins that govern adhesive trapping interactions with mucus and mucin–mucin interactions (Lai et al., 2009).

The thickness of the mucus varies depending on the nutrient absorption rate and protective capability against varying pH along the GIT. It further acts as a lubricant and facilitates colonization of commensal bacteria aiding protection from pathogens. Lipids in mucus contribute to its viscoelastic property and hydrophobic nature that governs mucoadhesion of drug molecules. However, nonspecific mucoadhesion in loosely adherent mucus layers is very common in that it minimizes the drug absorption (Nicoletti, 2000). In addition, pancreatic lipase and proteases are secreted into the lumen with pancreatic fluid that influences drug absorption. The digestion process of fatty acids by lipase-mediated hydrolysis is completed by bile salts secreted in duodenum. This further facilitates micellization of lipophilic compounds that governs drug absorption.

In the large intestine, mucosal thickness and epithelial interaction with the drug compounds varies markedly from the small intestine. In addition, the distal part of the colon contains highest number of bacterial population 10^{11}–10^{12} bacteria/gram of colon content compared to the upper parts of gut with 10^{3}–10^{4} bacterial/gram of jejunum content. Though, the extensive luminal metabolism by microflora minimizes the drug absorption rate and the prolonged transit time of 17.2 h in colon facilities significant drug absorption (Druart et al., 2014; Gopalsamy et al., 2015).

5.1 pH VARIATION IN GIT

An important limitation to drug absorption in the GIT is the variation in pH that influences the dissolution of dosage forms. Further, dissolution profile of enteric-coated polymers is also highly dependent on pH change. An increase in pH by 1 or 2 units over a critical pH range results in rapid drug dissolution and release. In the case of lipophilic drugs pH of half maximal absorption rather than pKa should be considered, wherein, extraction occurs even when the concentration of nonionized form is less (Dressman, 1986).

5.1.1 Gastric pH

The stomach is the port of entry into the GIT, where the immediate-release (IR) dosage forms disintegrate, facilitating the dissolution of active pharmaceutical ingredients (API) in a timely manner that gets absorbed in small intestine. However, the proximal and distal region of stomach functions differentially in response to food intake. The pH response to meals in proximal region was of greater magnitude to that in the distal region, and the peak pH varied depending on the meal constituents solid, semisolid, or liquid. The fasted gastric pH is highly acidic that lies between 0.3 and 1.9 and in response to buffering effects of meals pH rises to a plateau pH of 5.5, the plateau period is very short followed by most variable period of decline in pH toward baseline pH over a period of 60–90 min (Clark et al., 1993; Jantratid and Dressman, 2009). Thus, poorly soluble weak bases are preferred mostly under fasted conditions and the Median gastric residence time is reported to be 1.1 h.

5.1.2 Intestinal pH

The small intestine is the region where the drug is best absorbed and the transit time is 7.5 h. The dissolution of poorly soluble, weakly acidic compounds and lipophilic compounds is highly favored in this segment. Challenges related to delivery targeting the small intestine include the development of formulations of poorly soluble and high molecular weight drugs, and delivering of protein peptides due to their instability in the GIT environment (Gupta et al., 2009).

The luminal pH gradient was steepest in the proximal 10 cm of the duodenum, where acidity was reduced from pH 2 to pH 6.4 in the fasting state and from pH 1.7 to pH 4.3 in the second and third postprandial hour. The intestinal post prandial pH can reach 7.6. The initial drop in pH during postprandial state is due to the influx of acidic chyme. The chyme is buffered by bicarbonate secretion as it moves distally. The mean pH in proximal small intestines for the first hour of drug transit is 6.6. Acidity was further reduced in the distal duodenum to a pH between 5 and 6 at the duodenojejunal junction (Ovesen et al., 1986).

5.1.3 Colonic pH

In colon-specific drug delivery, the drug has to traverse the whole alimentary canal to reach the target site. The colonic pH is mainly influenced by short chain fatty acids (SCFA); these are weak acids of pKa 4.8, present as organic anions in the normal colonic lumen. A fall in intraluminal SCFAs along the distal colon contributes to colonic mucosal bicarbonate secretion, leading to a rise in pH near neutral. The solubilization of drug is the rate-limiting factor in colonic drug absorption due to low colonic luminal fluid volume, higher viscosity, and neutral pH. Further, nonspecific interactions of the drug with the colonic content, for example, dietary residues, intestinal secretions, mucus, or fecal matter can have a negative influence on the stability of the drug (Amidon et al., 2015; Nugent et al., 2001).

5.2 GASTROINTESTINAL MICROBIOTA

The human GIT harbors a diverse range of bacterial community with increasing gradient from stomach to jejunum, ileum and colon, where the concentration peaks to 10^{11}–10^{12} bacterial cells per gram of stool (Biagi et al., 2012). Stomach consists of 10^2–10^3, duodenum $<10^{4-5}$, ileum 10^3–10^7, and colon 10^9–10^{12} of bacterial population (Lin et al., 2014).

Orally administered drugs before being absorbed in the GIT is metabolized by gut microbial population that hinders with drug bioavailability. Once these compounds are orally administered, they are transformed to bioactive, bioinactive, or toxic metabolites by intestinal microbiota before their absorption into the blood. However, modulation of microbial enzyme activity in gut also leads to another major problem termed drug–drug interaction (DDI). Further, DDI occurs due to inhibition or induction of drug metabolizing enzymes, such as cytochrome P450 enzymes or transporters involved in absorption or excretion processes that alter the drug pharmacokinetics (Yoo et al., 2014). Location and diet highly influence microbial growth and its metabolism. Short-chain fatty acids, such as acetic, propionic, and butyric acids produced by microbial fermentation help in reducing the colonic pH and provide energy sources for colonic epithelium (Walker et al., 2005).

Indeed, intestinal microflora have also been associated with the production of toxic or genotoxic metabolites from drugs, food components, and detoxification of other chemicals, such as chromium complexes.

The gut microbes act in many ways that could impact on drug toxicity and carcinogenicity. They possess the capability to metabolize numerous drug-related compounds and increase the half-life of xenobiotics by hydrolysis and consequent reabsorption. Further, microbes constantly dose the GIT with drug-like substances that are cometabolised, and lead to biotransformation of the drug into a toxic metabolite (Holmes and Nicholson, 2005; Jin et al., 2014). Contrarily, gut microflora also has the beneficiary role of enhancing drug bioavailability by removing polar moiety from derived conjugates, and by promoting biliary recycling of compounds, such as ouabain, digoxin, and steroid hormones (Martinez and Amidon, 2002).

5.3 GASTRIC EMPTYING RATE

Gastric transit time also strongly influences drug bioavailability. Animal species with rapid GI transit (e.g., dogs) may be more sensitive to changes in the in vivo dissolution rate as compared to species with a longer transit time, such as swine. In Class II drugs of low solubility and high permeability, rate limiting step is determined by dissolution rate, which is directly related to the factors entailed in gastric transit time (Martinez and Amidon, 2002). A controlled gastric retention of solid dosage forms can be achieved by mucoadhesion, floatation, and sedimentation (Shashank et al., 2013).

6 IMPROVING DRUG BIOAVAILABILITY

As per the definition of bioavailability, a drug with poor bioavailability is one with poor aqueous solubility and/or slow dissolution rates in the biologic fluids, poor stability of the dissolved drug at the physiologic pH, inadequate partition coefficients, and, thus, poor permeation through the biomembrane and extensive presystemic metabolism. Thus, to improve the bioavailability of a drug, multiple phenomena have been designed. Such approaches include:

1. Pharmaceutics approach: modification of formulation, manufacturing processes or physiochemical properties of the drug without changing the chemical structure.
2. Pharmacokinetic approach: modification of chemical structure, but it has numerous hitches, such as being expensive, time-consuming, and requires repetition of chemical studies, risk of precipitation, and adverse effects. In addition, it cannot be assured that the new entity will not lead to another pharmacokinetic problem.
3. Biological approach: in this, route of drug administration may be changed, such as parenteral form instead of oral form (Bajaj et al., 2011).

7 PHARMACEUTICS APPROACH ENABLING PHYSICAL MODIFICATION OF DRUGS

7.1 PARTICLE SIZE REDUCTION

Since dissolution is directly proportional to the surface area of the drug, to increase the dissolution rate, the surface area has to be increased by reducing the particle size. This can be achieved by the following techniques:

1. Micronization: increases the surface area by reducing the particle size that overcomes the mechanical stress employed in conventional methods, such as communition and spray drying.

2. Nanosuspension: the submicron colloidal dispersion of pure particles of drug, which are stabilized by surfactants. Techniques for production of nanosuspension include homogenization and wet milling.

In homogenization the suspension is forced under pressure through a valve that has nanoaperture. This causes bubbles of water to form, which collapse as they come out of valves. This mechanism cracks the particles. Three types of homogenizers are commonly used, such as conventional homogenizers, sonicators, and high shear fluid processors.

In wet milling process the active drug in the presence of surfactant is defragmented by milling. Drying of nanosuspensions can be done by lyophilisation or spray drying. However, the major concern in particle size reduction is that the high-energy molecule is converted to low-energy crystalline form, which may not be therapeutically active. Other techniques include sonocrystallization, which utilizes ultrasound power characterized by a frequency range of 20–100 kHz that promotes nucleation rate for crystal formation, size reduction, and controlling size distribution of API and supercritical fluid (SCF) process in which the drugs are solubilized within SCF that recrystallizes at a reduced particle size (Geetha et al., 2014).

However, particle size reduction has its own drawbacks; this includes adsorption of air on the surface of hydrophobic drugs, which inhibits the wetting and agglomeration fine particles to minimize the surface energy, which all leads to a decrease in the effective surface area for dissolution (Nitin et al., 2014; Patel et al., 2012).

7.2 MODIFICATION OF CRYSTAL HABIT

Crystal habit can be altered by the use of different solvents and processing conditions concomitant with altering the polymeric state. Polymorphism is the ability of an element or compound to crystallize in more than one crystalline form. Though the polymorphs are chemically identical, they exhibits different physicochemical properties, including solubility, melting point, density, texture, and stability (Tiwary, 2001).

7.3 DRUG DISPERSION IN CARRIERS

Drug dispersion employs dispersion of one or more active ingredients in amorphous particles or crystalline particles by melting, solvent, or fusion-solvent methods. Further, the dissolution characteristics of the dispersed drug depend on the selection of the carrier. A water soluble carrier promotes faster release of drug from matrix, and a poorly soluble or insoluble carrier causes slower release of drug from matrix. Examples of hydrophilic carriers include polyvinylpyrrolidone, polyethylene, glycols, and plasdone-S630, surfactants like Tween 80, docusate sodium, Myrj-52, pluronic-F68, and sodium lauryl sulfate are also used (De et al., 2011).

Solid dispersions enhance the dissolution rate by reducing particle size, improving wettability, and increasing particle porosity. However, these characteristics are achieved when the parameters, such as carrier molecular weight and composition, drug crystallinity, and particle porosity and wettability are successfully controlled. Besides, solid dispersions also possess certain disadvantages, including conversion of amorphous state to crystalline state, phase separation due to moisture absorption, crystal growth, or conversion from a metastable crystalline form to a more stable structure during storage. Some of the solid dispersion types include eutectics, amorphous precipitation in crystalline matrix, solid solutions, glass suspensions, and glass solution (Sridhar et al., 2013).

7.4 COMPLEXATION

Complexation is the association between two or more molecules to form a nonbonded entity with a well-defined stiochiometry. Complexation relies on relatively weak forces, such as london forces, hydrogen bonding, and hydrophobic interactions. Types of complexing agents include inorganic, coordination, chelates, metal-olefin like ferrocene, inclusion and molecular complexes (Jatwani et al., 2014) Formation of soluble complexes improves drug dissolution, for example, efficiency of ergotamine tartarate is improved by simultaneous intake of caffeine that forms soluble and absorbable complexes with ergotamine. Caffeine improves dissolution of ergotamine by a factor of 3 in gastric juice, and at intestinal pH it increases its partitioning in lipid phase (Paradkar, 2008)

7.5 SOLUBILIZATION BY SURFACTANTS

Surfactants are amphipathic compounds that contain both hydrophobic and hydrophilic groups, thus soluble in both organic solvents and water. When small nonpolar molecules are added they can accumulate in the hydrophobic core of the micelles. When water, oil, and a surfactant are mixed, the surfactant rests at the water–oil interface. These systems, depending on their stability, are called emulsions or microemulsions (thermodynamically stable).

There are four kinds of surfactants: anionic, cationic, zwitterionic, and nonionic, wherein anionic surfactants impart a negative charge on the surface, cationic surfactants impart a positive charge. Since the likely charges repel, it promotes electrostatic stabilization by reducing the rate of droplet coalescence. Zwitterionic surfactants can possess both positive and negative charges, depending on the environment present. In addition, nonionic surfactants promotes steric stabilization, the hydrophobic heads are polymeric chains dissolved in water that prevents other droplet coalescence (Mishra et al., 2009; Sekhon, 2013).

A microemulsion is a four-component system composed of external phase, internal phase, surfactant, and cosurfactant that forms internal or dispersed phase of $<0.1\ \mu$ droplet diameter. The addition of surfactant, which is predominately soluble in the internal phase unlike the cosurfactant, results in the formation of an optically transparent, low viscosity, isotropic, thermodynamically and kinetically stable emulsion that improves absorption of poorly water soluble drugs. Microemulsions are highly preferred for oral drug delivery systems as it governs a better drug solubilization, protection against enzymatic hydrolysis and augments absorption rate due to a surfactant-induced improvement in the permeability (Gundogdu, 2013). Physiological surfactant bile acids and lyso-lecithin also promote drug absorption.

7.6 CHEMICAL MODIFICATION

Solubility of ionizable organic solutes can be increased by adjusting the pH of the solution. A drug that can be efficiently solubilized by pH control should be either weak acid with a low pKa or a weak base with a high pKa. On the other hand, solubility of nonionizable, hydrophobic substances can have improved solubility by changing the dielectric constant (a ratio of the capacitance of one material to a reference standard) of the solvent by the use of cosolvents rather than the pH of the solvent.

Other techniques that improve drug solubility include use of cocrystallizing agents saccharin, nicotinamide, and acetic acid, solubilization in cosolvents, hydrotrophy, which utilizes large amounts of additives that act by complex formation and use of solubilizing agents (Patel et al., 2012).

8 IMPROVING STABILITY OF ORAL DRUGS

When the stability of the oral drug is concerned, its intactness in GIT until it reaches the target area of action is pivotal. This is facilitated by polymeric coating of active compounds that could disintegrate at the target region, such phenomena involve enteric coating of active ingredients, which facilitates protection from gastric acidic environment and promotes its disintegration in the neutral part of the small intestine. The solubility profile highly depends on the choice of polymer and the thickness of the coated layer. A thick coat causes delays in drug release, which is both pH and time-dependent. Most of the enteric-coated drug delivery systems targeting colon delivery utilizes Eudragit (L-100/S-100) or cellulose acetatephthalate (CAP), shellac (ES), and ethyl cellulose (EC). The reason for enteric coating includes:

1. Protection of acid-labile drugs from gastric fluid.
2. Prevention of gastric distress or nausea due to irritation from a drug.
3. To deliver drugs intended for local action in the intestines.
4. To deliver drugs that are optimally absorbed in the small intestine to their primary absorption site in their most concentrated form.
5. To provide a delayed-release component for repeat action tablets.

In addition, complexation also improves drug stability. In drug-cyclodextrin complex synthesis, no covalent bonds are formed or broken down; however, dissolution occurs in aqueous ambience. Further, to overcome the enzymatic barriers of certain drugs, metabolism inhibitors are administered along with the drug (Bozdag et al., 1999; Sachdeva and Kataria, 2013).

9 NANOSTRUCTURES IN ORAL DRUG DELIVERY

As the GIT forms a formidable barrier for passage of larger drug molecules like proteins and peptides, this necessitates the breakdown of these molecules into smaller molecules. Further, as discussed earlier, gastric and intestinal enzymes also regulate the drug absorption. Considering these facts, wide variety of delivery strategies like enteric-coated capsules, tablets, liposomes, use of bio-adhesive agents and permeability enhancers are being employed along with particle size reduction, but these approaches failed to achieve CTR, enabling multiple drug administration. To facilitate CTR, polymers were employed as drug-carrying matrices that governed the sustained release of drugs within the therapeutic window (Chirra and Desai, 2012). In addition, the versatility of using polymers in combination with nanotechnology has led to several innovative ideas of miniaturization of oral drug delivery carriers for controlled release and targeting. Though there are various methodology, namely, micronization to improve drug dissolution, such techniques fail for very low solubility drugs. Several particulate carriers like liposomes, microspheres, micelles, and nanoparticles, solid lipid nanoparticles have been designed to overcome such problems (Table 7.1) (Sagadevan and Periasamy, 2014).

9.1 POLYMERIC FORMULATION

9.1.1 Polymeric Nanoparticle

Polymeric nanoparticles are prepared from biocompatible and biodegradable polymers with size between 10 and 1000 nm, where the drug is dissolved, entrapped, encapsulated, or attached to a polymer

Table 7.1 Summarizes the Properties, Advantage, Disadvantage and Application of Various Forms of Nanostructures in Oral Drug Delivery

No	Nanostruc-tures	Properties	Advantage	Disadvantage	Application in Oral Drug Delivery	Structure
1. Polymeric formulations						
1.1 Nanoparticles						
a.	Nanospheres	10–200 nm in diameter; can be biodegradable, nonbiodegradable, immune or magnetic nanospheres	Desired release profile; organ-targeted release; reduced toxicity	Hydrophobic surfaces are highly susceptible to opsonization and clearance by reticulo-endothelial system; limited drug loading; particle aggregation	Tumor-targeted delivery; oral delivery of bioactive molecules (proteins and peptides)	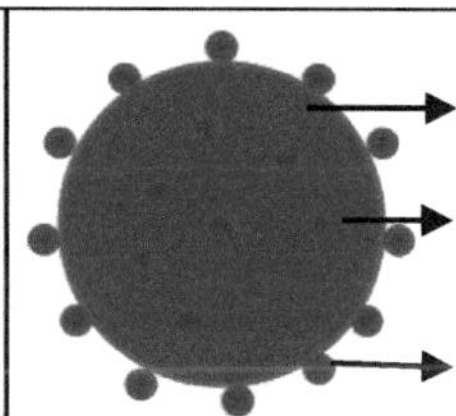
b.	Nanocapsules	10–1000 nm in size; made of active material (core) and protective shell	Sustained release; incremental drug selectivity and effectiveness; improve drug bioavailability; reduced drug toxicity	Low drug-loading capacity of carrier material	Tumor-targeted delivery; oral delivery of bioactive molecules (proteins and peptides especially insulin); hormone-targeted delivery of SiRNA; liver cell type delivery of plasmids	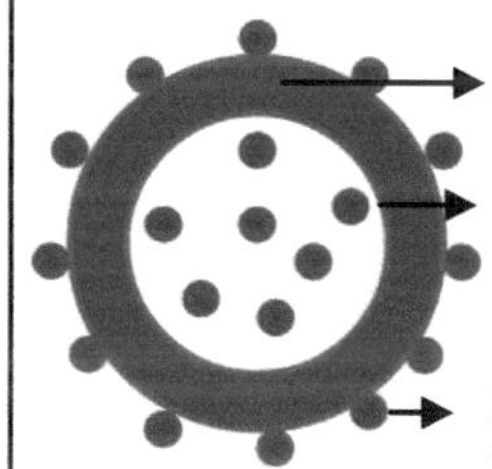
c.	Polymer-drug conjugate	6–15 nm in size; consists of soluble polymer backbone, a biodegradable linker, and a covalently linked deactivated drug	Prevents premature degradation in biological environment; targeted delivery; mucoadhesive property	Some carriers reduces or inactivates the protein activity and lack biodegradability, for example, PEG	Conjugates accumulated in solid tumors, reducing systemic toxicity	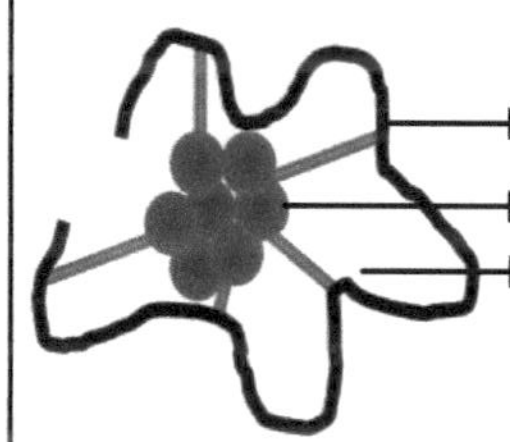

(Continued)

Table 7.1 Summarizes the Properties, Advantage, Disadvantage and Application of Various Forms of Nanostructures in Oral Drug Delivery (*cont.*)

No	Nanostructures	Properties	Advantage	Disadvantage	Application in Oral Drug Delivery	Structure
1.2	Polymeric micelles	10–100 nm in size; consists of hydrophilic shell and hydrophobic core encapsulating drug molecules	Increase drug stability; increase rate of absorption of hydrophobic drugs; low toxicity	Difficult polymer synthesis; immature drug-incorporation technology; slow extravazation; possible chronic liver toxicity due to slow metabolic process	Receptor mediated targeting for cancer therapy by passive targeting, active targeting, or by ligand coupling	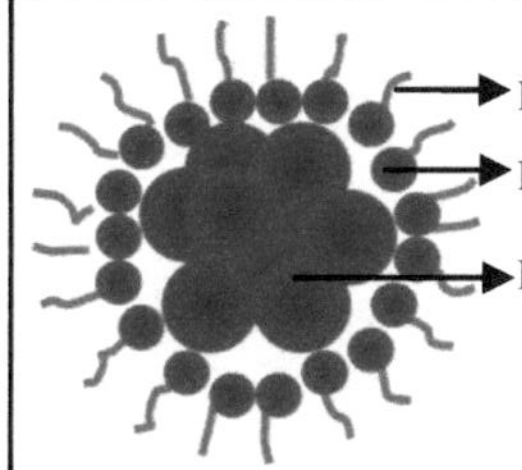
1.3	Nanocrystals	>100 nm in size; stabilized by surface active agent; crystalline or amorphous structure	100% drugs without carriers;increase saturation solubility, dissolution velocity; increased mucoadhesiveness; limited toxicity	Stability depends on molecular structure of drug; only certain categories of drugs are suitable; need of expensive equipments	Increased bioavailability in cryptosporidiosis patients by administering bupravaquone in nanosuspension form to cure diarrhea; naproxen nanosuspension showed faster onset of action and reduced gastric irritancy	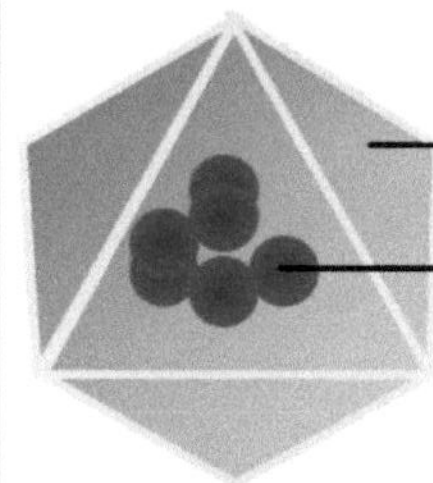
1.4	Nanoemulsion	Mean droplet diameter 50–1000 nm; exists as oil-in-water or water-in-oil form	Increased rate of absorption, and bioavailability; eliminates variability in absorption; helps solubilize lipophilic drug and masks unpleasant taste of some drugs; enhance stability of drug	Short period of stability; expensive process; need of large concentration of surfactants and cosurfactants	Increased systemic circulation of hydrophobic drugs like paclitaxel in oral drug delivery	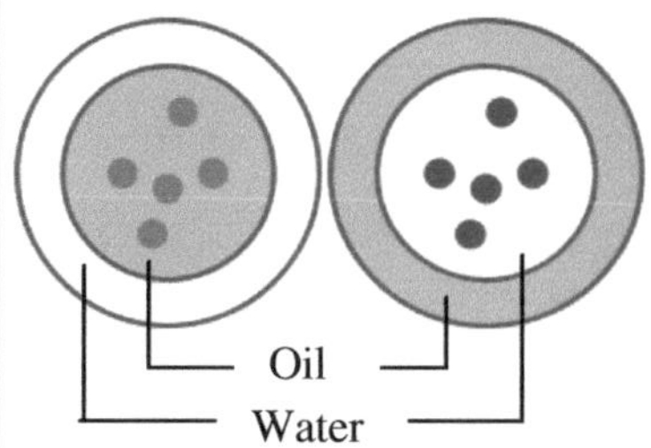

1.5	Dendrimers	Stable, well-defined structure monodispersity of size, surface functionalization capability	Modulate tight junctions and the integrity of cellular membranes; reduce toxicity and increased stability	Need of adjuvants to induce immunogenic response of coated antigens	Surface modified to inhibit the efflux effect of P-glycoprotein; enhanced the drug absorption of 5-FU anticancer drug; increased drug solubility of nonsteroidal antiinflammatory drugs	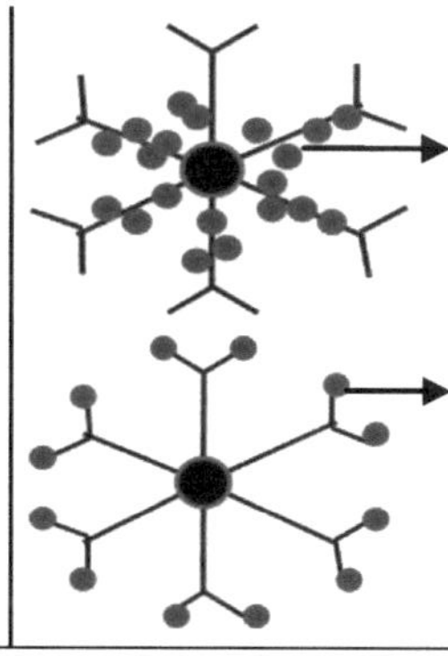
2. Lipid based formulations						
2.1	Liposomes	90–150 nm in size	Increased pharmacokinetics of drug; target specificity; decreased toxicity	Poor mucus penetrating properties of liposomes; easily trapped or wiped by the mucus blanket; cationic liposomes are cytotoxic; cost of production is high	Not usually used in oral drug delivery	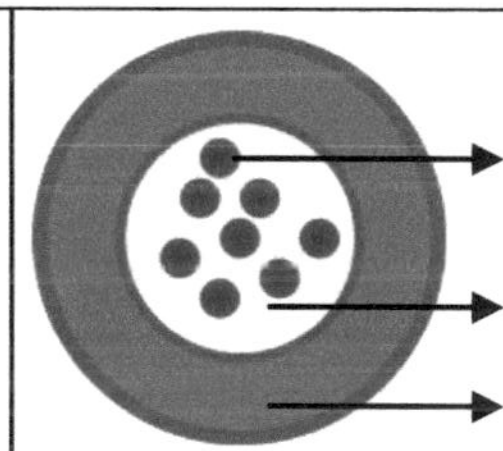
2.2	Solid lipid nanoparticles (SLNs)	50–100 nm in size; physiochemical diversity; biocompatibility and proven ability to enhance; oral bioavailability of poorly water soluble lipophilic drug uptake through selective lymphatic uptake	Increased stability; biocompatible and biodegradable; controlled drug delivery and specific targeting	Poor drug loading capacity; drug expulsion after polymeric transition during storage; relatively high water content of the dispersions (70–99.9%). The low capacity to load water soluble drugs due to partitioning effects during the Production process.	Used in treatment of neoplasms and breast cancer; antitubercular chemotherapy; sleath nanoparticles for increased drug absorption	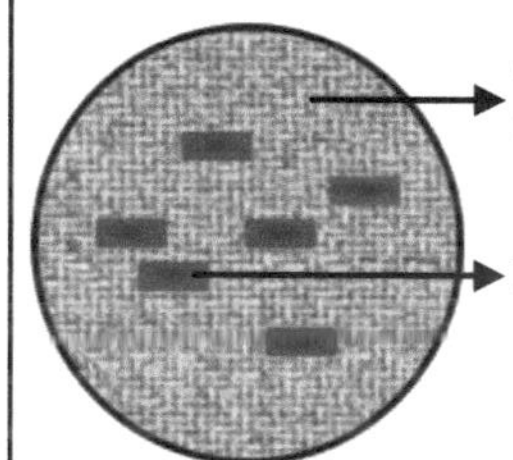

(Continued)

Table 7.1 Summarizes the Properties, Advantage, Disadvantage and Application of Various Forms of Nanostructures in Oral Drug Delivery (*cont.*)

No	Nanostructures	Properties	Advantage	Disadvantage	Application in Oral Drug Delivery	Structure
2.3	Nanostructured lipid carriers (NLCs)	10–500 nm in diameter; influenced by the type of surfactant used; smaller size and superior particle surface to volume ratio	Increased solubility and storage stability; improved permeability and bioavailability; reduced adverse effect; prolonged half-life, and tissue-targeted delivery	—	Improved bioavailability of lovastatin, cholesterol lowering drug, repaglinide, an antidiabetic agent with poor water solubility and a short half-life and chemotherapeutic agent etoposide	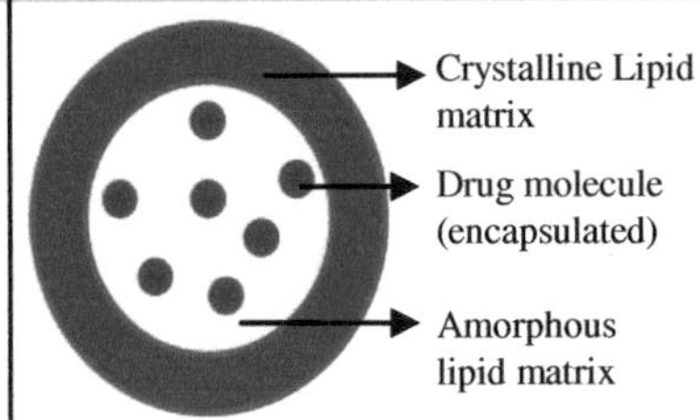
3. Metal and inorganic based nanocarriers						
3.1	Carbon nanotubes	Mean diameter <100 nm; large surface-to-volume ratio	Controlled lipophilicity enhanced permeability, distribution and retention in targeted organ; enhanced intracellular penetration	Lack of solubility; difficulty in maintaining high quality with identical characteristics and minimal impurities	Targeting Amphotericin B to cells; slow release of cisplatin, effective in terminating growth of human lung cancer cells	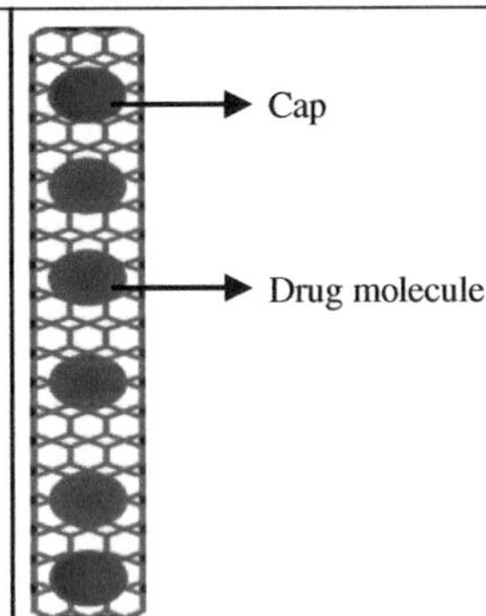

3.2	Nanoshells	1–20 nm nanometers thick; properties can be modified by changing either the constituting materials or core-to-shell ratio	Drug can be encapsulated or adsorbed onto the shell surface; possess highly favorable optical and chemical properties for biomedical imaging and therapeutic applications	Synthesis is complex and time consuming	Nanoshells can be tagged with specific antibodies for diseased tissues or tumors Gold nanoshell in detection of colloids	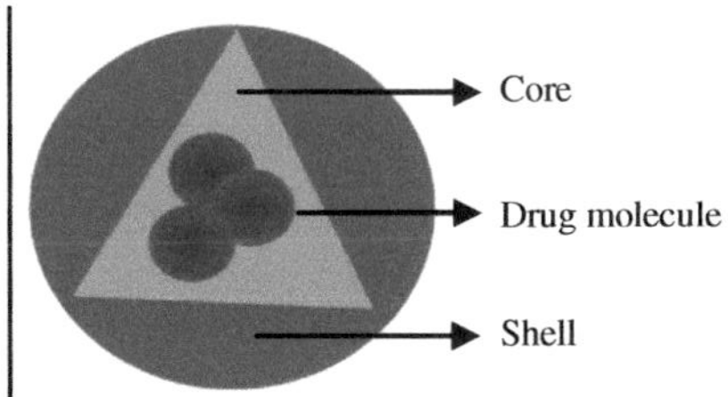
3.3	Quantum dots	2–10 nm, which, after polymer encapsulation, generally increases to 5–20 nm in diameter	Targeted delivery of small molecule hydrophobic drugs and hydrophilic therapeutic agents including small interfering RNA (siRNA) and antisense oligodeoxynucleotide (ODN) and biomolecules	Some quantum dots are toxic to human and environment, for example, proton-resistant quantum dots; they are not uniform: each design is a unique combination of physicochemical properties that influence biological activity and toxicity	Targeted drug delivery to tumor cells; medical imaging and disease detection	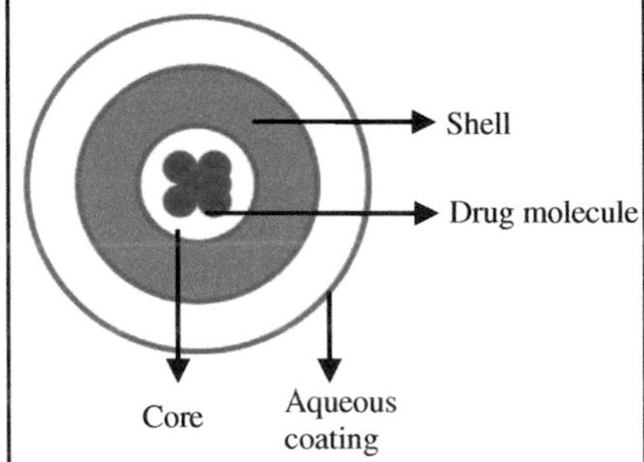

nanoparticle matrix. Various natural polymers like gelatin, albumin, and alginate are used to prepare the nanoparticles, however, they have some inherent disadvantages like poor batch-to-batch reproducibility, prone to degradation and potential antigenicity. Synthetic polymers used for nanoparticles preparation may be in the form of preformed polymer, for example, polyesters like polycaprolactone (PCL), polylactic acid (PLA), or monomers that can be polymerized in situ, for example, poly(alkylcyanoacrylate) (Kumari et al., 2010; Jayasree et al., 2014).

Polymeric nanoparticles can be divided into two main families: nanospheres, which have a homogeneous structure in the whole particle, and nanocapsules, which exhibit a typical core-shell structure.

1. Nanospheres are the spherical particle, which have the size between 10 and 200 nm in diameter and exhibits some new enhanced size dependent properties in comparison of larger spheres of the same material. Nanospheres can be amorphous or crystalline in nature and also they have the ability to protect the drug from enzymatic and chemical degradation (Singh et al., 2012; Yun et al., 2013).
2. Nanocapsules, due to their micronizing size, can be used as carriers for oral administration of peptides and proteins, enzymes, organic or inorganic catalysts, oils, adhesives, surface polymers, inorganic microparticles and nanoparticles, latex particles, or even biological cells particularly biodegradable poly(isobutylcyanoacrylate) nanocapsules. However, they are restricted due to the GI barriers of the epithelium and by the degradation of digestive enzymes (Kothamasu et al., 2012).
3. Polymer drug conjugate is a conjugation of low molecular weight drug with a polymer that could improve the pharmacokinetics and pharmacodynamics of therapeutics. The major advantage is that it facilitates retention of drugs in cancer cells through enhanced permeation and retention effect using passive delivery approach. Poly(ethylene glycol) (PEG) is the commonly used carrier because of a lack of immunogenicity, antigenicity, and toxicity. However, PEGylation can lead to a decrease in protein activity, which is compensated by increased circulation time. To overcome this, a strategy called polymer-masking-unmasking protein therapy (PUMPT) is established, which involves conjugating drugs to a biodegradable polymer, dextrin (Liechty et al., 2010).

9.1.2 Polymeric Micelles

Among the different polymer-based drug delivery systems, PMs represent a promising delivery vehicle especially intended for poorly water-soluble pharmaceutical active ingredients in order to improve their oral bioavailability. PMs are formed from amphiphilic copolymers that have a hydrophilic shell and a hydrophobic core. Micellar morphology is determined by choice of amphiphilic copolymer architecture, namely, diblock copolymers, triblock copolymers, multiblock copolymers, star-shaped, graft, and brush copolymer (Mori and Muller, 2003; Yokoyama, 2011). The PMs, such as crosslinked polyacrylic acids (PAA), carboxypolymethylene, carboxymethyl cellulose, alginate, chitosan (CS), and their derivatives are employed in enhancing mucoadhesive property. In addition, PM due to its stability in acidic milieu, facilitates safe transport of API and P-glycoprotein inhibitor leading to drug accumulation, thus increasing bioavailability (Xu et al., 2013).

The advantages of this system include their small size, lower toxicity, and advent of adverse reactions, and potentially long blood circulation times. One of the key issues of PMs is their stability in the physiological environment that may lead to side effects and adverse reactions (Lua and Park, 2013).

9.1.3 Nanocrystals

Drug nanocrystals are essentially nanoscopic crystals of the parent compound. By definition, the dimensions of nanocrystals are less than 1 μm, but for practical purposes they are often less than 500 nm Preparation of nanocrystals takes place either via top–down or bottom–up techniques. After nanonization, drug nanocrystals are often formulated into conventional dosage forms, such as tablets, capsules, pellets, and suspensions for IV administration. Advantages of nanocyrstals include excellent reproducibility and applicability to a wide range of drugs with various solubility profiles. However, the nanocrystal approach sometimes requires high-energy input, which drives up the cost of production (Junghanns and Muller, 2009; Katteboinaa et al., 2009).

9.1.4 Nanoemulsion

Nanoemulsions consist of two immiscible liquids (usually an oil phase and an aqueous phase) where one liquid acts as a continuous medium and the other as a dispersion medium of droplets in the other liquid. Nanoemulsions for solubilization of poorly water-soluble drugs usually consist of oil phase dispersed in aqueous medium, where the lipophilic drug is in oil phase. The most widely used components of the oil phase are saturated and unsaturated fatty acids, fatty acid esters, and soybean oils. In order to stabilize these drug-loaded oil droplets, nonionic or amphoteric surfactants, such as poloxamers, lecithin, and Tween 80 are commonly used.

Advantages of nanoemulsion include high drug-loading capacity and the potential to increase permeability in GI tract, thus increasing bioavailability. However, lack of stability, with flocculation and coalescence often take place during storage, and the lack of controlled release mechanism is also a limitation if delivering cytotoxic compounds, such as anticancer drugs (Lua and Park, 2013).

9.1.5 Dendrimers

Dendrimers consist of hydrophilic surface and hydrophobic core produced by convergent or divergent polymerization of branching units. These nanocarriers are formed by a central core with branching polymers and peripheral functional groups. It facilitates modulation of tight junctions and the integrity of cellular membranes, a major benefit for oral drug delivery. Further, to govern more desirable application in oral delivery, dendrimers are surface-modified. Cationic dendrimers are commonly conjugated with neutral or negatively charged ligands, such as PEG, to reduce potential toxicity in the GIT. The muco-adhesive property of dendrimers enhanced the residence time of drug in intestinal epithelium, in addition, dendrimers themselves are capable of penetrating intestinal membrane preferable through lymphoid tissues, thus enhancing drug absorption (Cheng et al., 2008).

In order to prevent the removal of drug from GIT and to improve its retention time, the efflux effect of P-glycoprotein is inhibited by surface modified dendrimers. Another approach to enhance drug bioavailability is to directly conjugate dendrimer with drugs either to form a dendrimer–drug conjugation or complex (Liu et al., 2015).

9.1.6 Self-assembled chitosan (CS) nanoparticles

Self-assembly is the most cost-effective and efficient method for CS nanocarriers that are employed in delivery of proteins and peptides, genes, small molecules, and combinational drugs. The mucoadhesive and absorption-enhancing property of CS governs opening of cellular tight junctions making it efficient for oral administration (Mukhopadhyay et al., 2013).

9.2 LIPID-BASED FORMULATIONS

9.2.1 Liposomes

Liposomes consist of self-assembled lipid bilayer membranes with size ranging from 90 to 150 nm, which are formed by phospholipids and cholesterol that enclose an aqueous core. Phospholipids consist hydrophilic heads and hydrophobic long tails that facilitate incorporation of hydrophilic molecules within the inner compartments and hydrophobic compounds within the hydrophobic bilayer. However, liposomes are not preferred for oral drug delivery due to its susceptibility to digestion in the GIT. This exhorts the need for coating liposomes with biopolymer, such as Eudragit S100 coat that allowed an appropriate pH-dependent drug release, but it could not prevent the ingress of bile salts that lead to premature release of drugs. Hence liposome-in-microsphere (LIM) comprising CS-coated liposome encapsulated by Eudragit S100 was employed, governing both pH and enzyme mediated release of drug (Barea et al., 2012). Pectin coated liposomes enhanced the mucoadhesive property and the residence time in GIT mucosa (Hua, 2014; Patel and Panda, 2012).

9.2.2 Solid Lipid Nanoparticles

SLNs are an efficient and nontoxic lipophilic colloidal drug carrier prepared either with physiological lipids or lipid molecules and surfactants. The advantages of SLNs include avoidance of chemical crosslinking and toxic solvents. Further, SLNs can incorporate hydrophilic or lipophilic drug with high payload than biopolymers-based nanoparticles and are employed as proteins and peptide delivery system (Almeida and Souto, 2007).

In order to increase the absorption rate and improve the delivery efficacy of SLN, surface modification and biopolymer coating with enhanced mucoadhesive properties and sustained release profile have been developed. This includes covalent conjugation of PEG onto stearic acid to fabricate SLN with better mucoadhesive property. In addition, biopolymers like CS and alginate coating also improves the mucoadhesive property with avoidance of potential toxicity of chemical residues during conjugation. Due to enhanced mucoadhesive property of CS-coated SLN, prolonged retention time promoting enhanced absorption of encapsulated insulin was observed. However, long-term storage may lead to particle aggregation and case formation, due to collision of dispersed lone particles (Luo et al., 2015).

9.2.3 Nanostructured Lipid Carriers

NLCs are a second generation of lipid nanoparticles that provide better water solubility and prevent GI irritation and toxicity on normal tissue. Thus, this class of drug carriers is preferable for anticancer therapy. NLC is a blend of solid lipid with fluid lipid and surfactant that facilitates higher loading capability, chemical stability, and controlled release of oral drug at the targeted site. However, selection of optimal emulsifiers is vital for governing these advantages. NLCs amend the disadvantage of SLNs, wherein particles disperse in a "pearl-like network" pattern that prevents collision and flocculation facilitating long-term storage. NLCs are more advantageous than other lipid carriers, and are involved in several applications as discussed in Table 7.1 (Fang et al., 2012; How et al., 2011; Liu and Wu, 2010; Selvamuthukumar and Velmurugan, 2012; Zhou et al., 2015).

9.3 METAL AND INORGANIC-BASED NANOCARRIERS

Nanoparticles are considered to be a multifunctional platform because they can be used for many imaging and therapeutic functions. These are synthesized by different organic, inorganic, or hybrids of organic and inorganic materials. However, inorganic platforms are very important for diagnosis and simultaneous therapy due to their easy modification, stability, and high drug-loading capacity. However, the extrinsic property of a nanoparticle (small size) rather than its intrinsic property (nanoscale property) features its increase in residence time facilitating absorption (Auan et al., 2009).

9.3.1 Layer-by- Layer Structured Nanocarriers

Layered coating of drugs imparts protection and sustained release, but it is not suited for intestinal drug delivery as it is micro-sized and is unstable in the stomach. Thus, layer-by-layer (LBL) drug carriers are employed for targeted delivery to the intestine that possess good stability at low stomach fluid and enhanced mucosal/membrane penetration. Recent studies have reported that LBL comprising casein-coated iron oxide nanoparticles core loaded with hydrophobic drug doxorubicin (DOX)–enhanced cellular uptake, mucus penetration and could overcome enzymatic and pH mediated degradation in acidic environment (Huang et al., 2015).

On the other hand, beta-caesin (β-CN) nanoparticles facilitate delivery of drugs in the stomach as the proline residues lead to an open tertiary structure, which are easily accessible to gastric proteases. Another important feature of β-CN making it highly suitable for gastric delivery is its pronounced amphiphilic structure, which enables it to self-assemble in aqueous solution, there by forming stable micellar structures (Shapira et al., 2012).

9.3.2 Silica-Based Nanocarriers

Silica nanotubes derived from $Gd(OH)_3$ nanorods are highly resistant to acidic pH, and faster dissolution rate is achieved at pH 8. This degradation pattern indicates that the silica nanotubes are highly preferable for oral drug delivery that could overcome the harsh acidic environment. In addition, the nanotubes dissolution can be altered based on thickness of the silica nanotubes (Hu et al., 2010).

Mesoporous silica nanoparticles (MSNs), an effective oral drug delivery system, especially for water insoluble drugs govern improved oral bioavailability. The fabrication of MSNs is simple, cost-effective, and controllable. In addition, the high surface area and large pore volume of this material allow it to be used for encapsulating various therapeutic agents with high loading capacity and delivering them to desired locations (Li et al., 2015).

9.3.3 Gold Nanoparticles

Gold NPs are widely used for oral drug delivery because it exhibits low cytotoxicity. However, toxicity depends on the nature of gold NPs, very small Au-cluster fits into the grooves of DNA-molecules, induce oxidative stress, and thereby cause cytotoxic effects. Gold NPs form several shapes like nanospheres, nanorods, branched, nanorod, nanostar, nanocube, and nanoshell. Due to the small size and large surface, shape, and crystallinity, gold NPs can easily travel into the target cells and can bear high drug load (Khan et al., 2014).

9.3.4 Carbon Nanotubes

CNTs are allotropes of carbon with a cylindrical nanostructure of elongated fullerenes, with small mean diameters of <100 nm. The large surface-to-volume ratio provides a good platform for efficient transportation of active drug components. Absorption is the primary measure after oral administration of CNTs, absorbed CNTs are transported from the administration sites to the effect-relevant sites by blood or lymphatic circulation. It has also been established that physically shortened CNTs that are orally administered can be absorbed through the columnar cells of intestinal mucous membrane (Kumar et al., 2012).

9.3.5 Nanoshells

Nanoshell comprises a spherical core of a compound surrounded by a shell or outer coating of thin layer of another material of 1–20 nm thick. Nanoshell materials can be synthesized from semiconductors, metals, or insulators. The property of nanoshells is determined by the materials used and core-to-shell ratio. Metal nanoshell comprise dielectric core enclosed by metallic shell, especially gold (AuNSs). In these cases, drug is encapsulated or adsorbed onto the shell surface via specific functional groups or by electrostatic stabilization. AuNS are employed to deliver antitumor drugs (e.g., doxorubicin, paclitaxel, small interfering RNA, and single-stranded DNA) into cancer cells, which enhance the efficacy of treatment. AuNSs can also be functionalized with active targeting ligands, such as antibodies, aptamers, and peptides to increase the particles' specific binding to the desired targets (Mudshinge et al., 2011).

9.3.6 Quantum Dots

QDs consist of an inorganic core, an inorganic shell and aqueous organic coating. Commonly used coatings include amphiphilic polymers, amphiphilic lipids, hydrophilic thiols, and multidentate coordinating ligands that are stable only in neutral and basic pH. To meet the acidity of stomach in oral drug delivery acid stable coatings termed proton sponge polymers are being used (Mohs et al., 2009). The size of QDs can be continuously altered between 2 and 10 nm. Various studies on efficacy of QD-mediated oral drug delivery concluded that absorption is size-dependent, QDs enhanced the absorption rate by 250-fold compared to microparticles. Regarding surface modification, positively charged particles show better absorption than neutral or negatively charged particles. Contrarily, a recent study using gold nanoparticles has reported that negatively charged particles show higher absorption rates than positively charged particles. Thus surface modification is independent of absorption rate (Rodriguez-Fragoso et al., 2014; Tsoi et al., 2013).

10 CONCLUSIONS

Nanobased strategy has been rising at a high rate in orally controlled drug delivery, which improves the drug retention time in GIT, eventually ameliorating its bioavailability. As the physiological barriers in GIT hinders with the pharmacokinetics and pharmacodynamics of drug, strategies like structural modification of drugs are being enforced to achieve maximal drug absorption. Various other efficient approaches for formulation development of oral delivery of therapeutics have been developed and are implemented in large-scale production. However, studies focusing on pharmacokinetics and pharmacodynamics of nanostructured drugs have to be widely exercised to obviate its toxic effects on normal functioning of human organs, and to ensure its safety effects in oral drug delivery.

ACKNOWLEDGMENT

The authors are also thankful to VIT University for providing the resources necessary in preparing this chapter.

CONFLICT OF INTEREST

The authors declare that there was no conflict of interest in writing this chapter.

REFERENCES

Almeida, A.J., Souto, E., 2007. Solid lipid nanoparticles as a drug delivery system for peptides and proteins. Adv. Drug Deliv. Rev. 59, 478–490.

Amidon, S., Brown, J.E., Dave, V.S., 2015. Colon-targeted oral drug delivery systems: design trends and approaches. AAPS PharmSciTech 16, 731–741.

Auan, M., Rose, J., Bottero, J.-Y., Lowry, G.V., Jolivet, J.-P., Wiesner, A.M.R., 2009. Towards a definition of inorganic nanoparticles from an environmental, health and safety perspective. Nat. Nanotechnol. 4, 1–8.

Awortwe, C., Fasinu, P.S., Rosenkranz, B., 2014. Application of Caco-2 cell line in herb-drug interaction studies: current approaches and challenges. J. Pharm. Pharm. Sci. 17, 1–19.

Bajaj, H., Bisht, S., Yadav, M., Singh, V., 2011. Bioavailability enhancement: a review Int. J. Pharm. Bio. Sci. 2, 202–216.

Bala, R., Kanna, S., Pawar, P., 2012. Polymers in fast disintegrating polymers: a review. Asian J. Pharm. Clin Res. 5, 8–14.

Barea, M.J., Jenkins, M.J., Lee, Y.S., Johnson, P., Bridson, R.H., 2012. Encapsulation of liposomes within pH responsive microspheres for oral colonic drug delivery. Int. J. Biomater. 2012, 1–9.

Biagi, E., Candela, M., Fairweather-Tait, S., Franceschi, C., Brigidi, P., 2012. Aging of the human metaorganism: the microbial counterpart. Age (Dordr) 34, 247–267.

Bozdag, S., Çalis, S., Sumnu, M., 1999. Formulation and stability evaluation of enteric-coated omeprazole formulations. STP Pharm. Sci. 9, 321–327.

Buckley, S.T., Fischer, S.M., Fricker, G., Brandl, M., 2012. In vitro models to evaluate the permeability of poorly soluble drug entities: challenges and perspectives. Eur. J. Pharm. Sci. 45, 235–250.

Cheng, Y., Xu, Z., Ma, M., Xu, T., 2008. Dendrimers as drug carriers: applications in different routes of drug administration. J. Pharm. Sci. 97, 123–143.

Chirra, H.D., Desai, T.A., 2012. Emerging microtechnologies for the development of oral drug delivery devices. Adv. Drug Deliv. Rev. 64, 1569–1578.

Clark, G.W.B., Jamieson, J.R., Hinder, R.A., Polishuk, P.V., Demeester, T.R., Gupta, N., Cheng, S.C., 1993. The relation between gastric pH and the emptying of solid,semisolid and liquid meals. J. Gastrointest. Mot. 5, 273–279.

Costa, P., Lobo, J.M.S., 2001. Modeling and comparison of dissolution profiles. Eur. J. Pharm. Sci. 13, 123–133.

Dash, S., Murthy, P.N., Nath, L., Chowdhury, P., 2010. Kinetic modeling on drug release from controlled drug delivery systems. Acta Pol. Pharm. 67, 217–223.

De, P.K., Sahana, B., Rakshit2, S., 2011. Enhancement of dissolution rate and stability study of ofloxacin solid dispersion. Der Pharm. Sinica 2, 169–181.

Dokoumetzidis, A., Macheras, P., 2006. A century of dissolution research: from Noyes and Whitney to the biopharmaceutics classification system. Int. J. Pharm. 321, 1–11.

Dressman, J.B., 1986. Comparison of canin and human gastrointestinal physiology. Pharm. Res. 3, 124–131.

Druart, C.L., Neyrinck, A.M., Vlaeminck, B., Fievez, V., Cani, P.D., Delzenne, N.M., 2014. Role of the lower and upper intestine in the production and absorption of gut microbiota-derived PUFA metabolites. Plos one 9 1–9.

Duggal, D., 2011. Role of nanotechnology in new drug delivery systems. Int. J. Drug Dev. Res. 3, 4–8.

Ensign, L.M., Cone, R., Hanes, J., 2012. Oral drug delivery with polymeric nanoparticles: the gastrointestinal mucus barriers. Adv. Drug Deliv. Rev. 64, 557–570.

Fang, C.-L., A. Al-Suwayeh, S., Fang, J.-Y., 2012. Nanostructured lipid carriers (NLCs) for drug delivery and targeting. Re. Pat. Nanotechnol. 7, 41–55.

Geetha, G., Poojitha, U., Khan, K.A.A., 2014. Various techniques for preparation of nanosuspension: a review. Int. J. Pharm. Res. Rev. 3, 30–37.

Gopalsamy, G.L., Alpers, D.H., Binder, H.J., Tran, C.D., Ramakrishna, B.S., Brown, I., Manary, M., Mortimer, E., Young, G.P., 2015. The relevance of the colon to zinc nutrition. Nutrients 7, 572–583.

Gundogdu, E., 2013. A microemulsion for the oral drug delivery of pitavastatin. Pharma. Anal. Acta 04, 3–5.

Gupta, H., Bhandari, D., Sharma, A., 2009. Recent trends in oral drug delivery: a review. Rec. Pat. Drug Deliv. Formul. 3, 162–173.

Holmes, E., Nicholson, J., 2005. Variation in gut microbiota strongly influences individual rodent phenotypes. Toxicol. Sci. 87, 1–2.

How, C.W., Abdullah, R., Abbasalipourkabir, R., 2011. Physicochemical properties of nanostructured lipid carriers as colloidal carrier system stabilized with polysorbate 20 and polysorbate 80. Afr. J. Biotechnol. 10, 1684–1689.

Hu, K.W., Hsu, K.C., Yeh, C.S., 2010. pH-Dependent biodegradable silica nanotubes derived from Gd(OH)3 nanorods and their potential for oral drug delivery and MR imaging. Biomaterials 31, 6843–6848.

Hua, S., 2014. Orally administered liposomal formulations for colon targeted drug delivery. Front. Pharmacol. 5, 138.

Huang, J., Shu, Q., Wang, L., Wu, H., Wang, A.Y., Mao, H., 2015. Layer-by-layer assembled milk protein coated magnetic nanoparticle enabled oral drug delivery with high stability in stomach and enzyme-responsive release in small intestine. Biomaterials 39, 105–113.

Jantratid, E., Dressman, J., 2009. Biorelevant dissolution media simulating the proximal human gastrointestinal tract: an update. Dissol. Tech. 16, 21–25.

Jatwani, S., Rana, A.C., Singh, G., Aggarwal, G., 2014. An overview on solubility enhancement techniques for poorly soluble drugs and solid dispersion as an eminent strategic approach. Int. J. Pharm. Sci. Res. 3, 942–956.

Jayasree, J., Sivaneswari, S., Hemalatha, G., Preethi, N., Mounika, B., Murthy, S.V., 2014. Role of various natural, synthetic and semi-synthetic polymers on drug release kinetics of losartan potassium oral controlled release tablets. Int. J. Pharm. Invest. 4, 183–188.

Jin, M.J., Kim, I.S., Kim, D.H., Yoo, H.H., 2014. Effects of intestinal microbiota on the bioavailability of geniposide in rats. J. Agric. Food Chem. 62, 9632–9636.

Jouyban, A., Panahi-Azar, V., Fakhree, M.A.A., Ahmadian, S., 2014. Solubility of phenytoin in aqueous mixtures of ethanol and sodium dodecyl sulfate at 298 K. Rev. Colomb. Cienc. Quím. Farm. 43, 153–161.

Junghanns, J.-U.A.H., Muller, R.H., 2009. Nanotechnology, drug delivery and clinical applications. Int. J. Nanomed. 2, 295–309.

Katteboinaa, S., P. Chandrasekar, V.S.R., Balaji, S., 2009. Drug nanocrystals: a novel formulation approach for poorly soluble drugs. Int. J. PharmTech Res. 1, 682–694.

Khan, A., Rashid, R., Murtaza, G., Zahra1, A., 2014. Gold nanoparticles: synthesis and applications in drug delivery. Trop. J. Pharm. Res. 13, 1169–1177.

Kothamasu, P., Kanumur, H., Ravur, N., Maddu, C., Parasuramrajam, R., Thangavel, S., 2012. Nanocapsules: the weapons for novel drug delivery systems. Bioimpacts 2, 71–81.

Kumar, S.P., Prathibha, D., Shankar, N.L.G., Parthibarajan, R., Mastyagiri, L., Shankar, M., 2012. Pharmaceutical applications of carbon nanotube-mediated drug delivery systems. Int. J. Pharm. Sci. Nanotech. 5, 1685–1696.

Kumari, A., Yadav, S.K., Yadav, S.C., 2010. Biodegradable polymeric nanoparticles based drug delivery systems. Colloids Surf. B 75, 1–18.

Lai, S.K., Wang, Y.Y., Hanes, J., 2009. Mucus-penetrating nanoparticles for drug and gene delivery to mucosal tissues. Adv. Drug Deliv. Rev. 61, 158–171.

Lennernas, H., Abrahamsson, B., 2005. The use of biopharmaceutic classification of drugs in drug discovery and development: current status and future extension. J. Pharm. Pharmacol. 57, 273–285.

Li, L., Liu, T., Fu, C., Tan, L., Meng, X., Liu, H., 2015. Biodistribution, excretion, and toxicity of mesoporous silica nanoparticles after oral administration depend on their shape. Nanomedicine 11, 1915–1924.

Liechty, W.B., Kryscio, D.R., Slaughter, B.V., Peppas, N.A., 2010. Polymers for drug delivery systems. Annu. Rev Chem. Biomol. Eng. 1, 149–173.

Lin, C.S., Chang, C.J., Lu, C.C., Martel, J., Ojcius, D.M., Ko, Y.F., Young, J.D., Lai, H.C., 2014. Impact of the gut microbiota, prebiotics, and probiotics on human health and disease. Biomed. J. 37, 259–268.

Liu, C.-H., Wu, C.-T., 2010. Optimization of nanostructured lipid carriers for lutein delivery. Colloids Surf. A 353, 149–156.

Liu, Y., Tee, J.K., Chiu, G.N.C., 2015. Dendrimers in oral drug delivery application: current explorations, toxicity issues and strategies for improvement. Curr. Pharm. Des. 21, 2629–2642.

Lua, Y., Park, K., 2013. Polymeric micelles and alternative nanonized delivery vehicles for poorly soluble drugs. Pure Appl. Chem. 453, 198–214.

Luo, Y., Teng, Z., Li, Y., Wang, Q., 2015. Solid lipid nanoparticles for oral drug delivery: chitosan coating improves stability, controlled delivery, mucoadhesion and cellular uptake. Carbohydr. Polym. 122, 221–229.

Martinez, M.N., Amidon, G.L., 2002. A Mechanistic approach to understanding the factors affecting drug absorption: a review of fundamentals. J. Clin. Pharmacol. 42, 620–643.

Mishra, M., Muthuprasanna, P., Prabha, K.S., Rani, P.S., Babu, I.A.S., Chandiran, I.S., Arunachalam, G., Shalini, S., 2009. Basics and potential applications of surfactants: a review. Int. J. Pharm.Tech. Res. 1, 1354–1365.

Mohanachandran, P.S., Sindhumol, P.G., Kiran, T.S., 2011. Superdisintegrants: an overview. Int. J. Pharm. Sci. Rev. Res. 6, 105–109.

Mohs, A.M., Duan, H., Kairdolf, B.A., Smith, A.M., Nie, S., 2009. Proton-resistant quantum dots: stability in gastrointestinal fluids and implications for oral delivery of nanoparticle agents. Nano Res. 2, 500–508.

Mori, H., Muller, A.H.E., 2003. New polymeric architectures with (meth)acrylic acid segments. Prog. Polym. Sci. 28, 1403–1439.

Motz, S.A., Klimundova, J., Schaefer, U.F., Balbach, S., Eichinger, T., Solich, P., Lehr, C.M., 2007a. Automated measurement of permeation and dissolution of propranolol HCl tablets using sequential injection analysis. Anal. Chim. Acta 581, 174–180.

Motz, S.A., Schaefer, U.F., Balbach, S., Eichinger, T., Lehr, C.M., 2007b. Permeability assessment for solid oral drug formulations based on Caco-2 monolayer in combination with a flow through dissolution cell. Eur. J. Pharm. Biopharm. 66, 286–295.

Mudshinge, S.R., Deore, A.B., Patil, S., Bhalgat, C.M., 2011. Nanoparticles: emerging carriers for drug delivery. Saudi Pharm. J. 19, 129–141.

Mukhopadhyay, P., Sarkar, K., Chakraborty, M., Bhattacharya, S., Mishra, R., Kundu, P.P., 2013. Oral insulin delivery by self-assembled chitosan nanoparticles: in vitro and in vivo studies in diabetic animal model. Mater. Sci. Eng. C 33, 376–382.

Neetika, B., Arsh, D., Manish, G., 2012. An overview on variousapproaches to oral controlled drug delivery system via gastroretentive drug delivery system. Int. Res. J. Pharm. 3, 128–133.

Nicoletti, C., 2000. Unsolved mysteries of intestinal M cells. Gut 47, 735–739.

Nitin, B.B., Adhikrao, V.Y., Sachin, S.M., Rohan, A.K., Ashok, A.H., Sachin, S.S., Sameer, J.N., 2014. A review on development of biorelevant dissolution medium. J. Drug Deliv. Ther. 4, 140–148.

Nugent, Rampton, Evans, 2001. Intestinal luminal pH in inflammatory bowel disease: possible determinants and implications for therapy with aminosalicylates and other drugs. Gut 48, 571–577.

Ojha, N., Prabhakar, B., 2013. Advances in solubility enhancement techniques. Int. J. Pharm. Sci. Rev. Res. 21 (2), 351–358.

Ovesen, L., Bendtsen, F., Tage-Jensen, U., Pedersen, N.T., Gram, B.R., Rune, S.J., 1986. Intraluminal pH in the stomach, duodenum, and proximal jejunum in normal subjects and patients with exocrine pancreatic insufficiency. Gastroenterology 90, 958–962.

Paradkar, A.R., 2008. Biopharmaceutics & Pharmacokinetics. Pragathi Books Pvt. Ltd.

Patel, N., Panda, S., 2012. Liposome drug delivery system: a critic review. J Pharm. Sci. Bio-sci. Res. 2, 169–175.
Patel, J.N., Rathod, D.M., Patel, N.A., Modasiya, M.K., 2012. Techniques to improve the solubility of poorly soluble drugs. Int. J. Pharm. Life Sci. 3, 1459–1469.
Pundir, S., Badola, A., Sharma, D., 2013. Sustained release matrix technology and recent advance in matrix drug delivery system: a review. Int. J. Drug Res. Tech. 3, 12–20.
Reddy, B.B.K., Karunakar, A., 2011. Biopharmaceutics classification system: a regulatory approach. Dissol. Tech. 18, 31–37.
Rodriguez-Fragoso, L., Gutiérrez-Sancha, I., Rodríguez-Fragoso, P., Rodríguez-López, A., Reyes-Esparza, J., 2014. Pharmacokinetic properties and safety of cadmium-containing quantum dots as drug delivery systems. In: PhD. Demir Sezer, A. (Ed.), Application of Nanotechnology in Drug Delivery. InTech, Available from: http://www.intechopen.com/books/application-of-nanotechnology-in-drug-delivery/pharmacokinetic-properties-and-safety-of-cadmium-containing-quantum-dots-as-drug-delivery-systems.
Sachdeva, V., Kataria, M.K., 2013. Enhancement of bioavailability through increase in drug permeation, stability and retention time. Int. Res. J. Pharm. 4, 59–66.
Sagadevan, S., Periasamy, M., 2014. A review on role of nanostructures in drug delivery system. Rev. Adv. Mater. Sci. 36, 112–117.
Savjani, K.T., Gajjar, A.K., Savjani, J.K., 2012. Drug solubility: importance and enhancement techniques. ISRN Pharm. 2012, 1–10.
Schubert, M.L., 2014. Gastric secretion. Curr. Opin. Gastroenterol. 30, 578–582.
Sekhon, B.S., 2013. Surfactants: pharmaceutical and medicinal aspects. J. Pharm. Technol. Res. Man. 1, 11–36.
Selvamuthukumar, S., Velmurugan, R., 2012. Nanostructured lipid carriers: a potential drug carrier for cancer chemotherapy. Lipids Health Dis. 11, 2–8.
Seo, J.H., Park, J.B., Choi, W.K., Park, S., Sung, Y.J., Oh, E., Bae, S.K., 2015. Improved oral absorption of cilostazol via sulfonate salt formation with mesylate and besylate. Drug Des. Devel. Ther. 9, 3961–3968.
Shapira, A., Davidson, I., Avni, N., Assaraf, Y.G., Livney, Y.D., 2012. β-Casein nanoparticle-based oral drug delivery system for potential treatment of gastric carcinoma: stability, target-activated release and cytotoxicity. Eur. J. Pharm. Biopharm. 80, 298–305.
Shashank, C., Prabha, K., Sunil, S., Kumar, A.V., 2013. Approaches to increase the gastric residence time: floating drug delivery systems: a review. Asian J. Pharm. Clin. Res. 6, 1–9.
Siepmann, J., Siepmann, F., 2013. Mathematical modeling of drug dissolution. Int. J. Pharm. 453, 12–24.
Singh, A., Garg, G., Sharma, P., 2012. Nanospheres: a novel approach for targeted drug delivery system. Int. J. Pharm. Sci. Rev. Res. 5, 84–88.
Siya, D., Kunde, S., Bhilegaonkar, S., Godbole, A.M., Gajre, P., 2015. Biopharmaceutical classification system: a brief account. IJRM 1, 20–46.
Sridhar, I., Doshi, A., Joshi, B., Wankhede, V., Doshi, J., 2013. Solid dispersions: an approach to enhance solubility of poorly water soluble drug. J. Sci. Innov. Res. 2, 6685–6694.
Surendiran, A., Sandhiya, S., Pradhan, S.C., Adithan, C., 2009. Novel applications of nanotechnology in medicine. Indian J. Med. Res. 130, 689–701.
Tiwary, A.K., 2001. Modification of crystal habit and its role in dosage form performance. Drug Dev. Ind. Pharm. 27, 699–709.
Tsoi, K.M., Dai, Q., Alman, B.A., Chan, W.C.W., 2013. Are quantum dots toxic? Exploring the discrepancy between cell culture and animal studies. Acc. Chem. Res. 46, 662–671.
Tsume, Y., Mudie, D.M., Langguth, P., Amidon, G.E., Amidon, G.L., 2014. The biopharmaceutics classification system: subclasses for in vivo predictive dissolution (IPD) methodology and IVIVC. Eur. J. Pharm. Sci. 57, 152–163.
Volpe, D.A., 2010. Application of method suitability for drug permeability classification. AAPS J. 12, 670–678.
Walker, A.W., Duncan, S.H., McWilliam Leitch, E.C., Child, M.W., Flint, H.J., 2005. pH and peptide supply can radically alter bacterial populations and short-chain fatty acid ratios within microbial communities from the human colon. Appl. Environ. Microbiol. 71, 3692–3700.

Xu, W., Ling, P., Zhang, T., 2013. Polymeric micelles, a promising drug delivery system to enhance bioavailability of poorly water-soluble drugs. J. Drug Deliv. 2013, 1–15.

Yokoyama, M., 2011. Clinical applications of polymeric micelle carrier systems in chemotherapy and image diagnosis of solid tumors. J. Exp. Clin. Med. 3, 151–158.

Yoo, D.-H., Kim, I.S., Le, T.K.V., Jung, I.-H., Yoo, H.H., Kim, D.-H., 2014. Gut microbiota-mediated drug interactions between lovastatin and antibiotics. Drug Metab. Dispos. 42, 1508–1513.

Yun, Y., Cho, Y.W., Park, K., 2013. Nanoparticles for oral delivery: targeted nanoparticles with peptidic ligands for oral protein delivery. Adv. Drug Deliv. Rev. 65, 822–832.

Zhao, N., Augsburger, L.L., 2005. The influence of swelling capacity of superdisintegrants in different Ph media on the dissolution of hydrochlorothiazide from directly compressed tablets. AAPS PharmSciTech 6, e120–e126.

Zhou, X., Zhang, X., Ye, Y., Zhang, T., Wang, H., Ma, Z., Wu, B., 2015. Nanostructured lipid carriers used for oral delivery of oridonin: an effect of ligand modification on absorption. Int. J. Pharm. 479, 391–398.

CHAPTER

ORAL PELLETS LOADED WITH NANOEMULSIONS

8

Thi Trinh Lan Nguyen, Nicolas Anton, Thierry F. Vandamme
University of Strasbourg, Strasbourg, France

CHAPTER OUTLINE

1 Pellets 203
1.1 Introduction to Pellets 203
1.2 General Description of the Extrusion Spheronization Process (Wet Mass Extrusion) 205
1.3 Process and Equipment 207
1.4 Formulation 211
1.5 Evaluation of Pellets 213
2 Nanoemulsions 213
2.1 Introduction 213
2.2 Method 215
2.3 Materials Used in Nanoemulsions Production 216
3 Biopharmaceutical Classification System (BSC) and Applications of Pellet to Improve Dissolution Rate of Drugs in BCS Class II 217
4 Case Study 221
4.1 Coating of Pellets 221
4.2 Dissolution Tests 222
4.3 Reconstitution Study 222
4.4 Results 222
5 Conclusions 223
References 223

1 PELLETS

1.1 INTRODUCTION TO PELLETS

Historically, the term *pellet* has been used by a number of industries to describe a variety of agglomerates produced from diverse raw materials, using different pieces of manufacturing equipment. These agglomerates include fertilizers, animal feeds, iron ores, and pharmaceutical dosage units and thus do not only differ in composition but also encompass different sizes and shapes. In the pharmaceutical industry, pellets can be defined as spherical or nearly spherical, free-flowing particles with a narrow size distribution, varying between 500 and 1500 μm manufactured by the agglomeration of fine powders or granules of drug substances and excipients using appropriate processing equipment (Ghebre-Sellassie, 1989; Ghebre-Sellassie and Knoch, 2007). Pellets are commonly placed into hard gelatine capsules but can also be compressed to tablets; furthermore, modified release dosage forms can be created (Conine and Hadley, 1970; Csobán et al., 2015; Ghanam and Kleinebudde, 2011; Hosseini et al., 2013; Malinowski and Smith, 1975; Mehta et al., 2012; Millili and Schwartz, 1990; Xu et al., 2015).

Pellets can be produced through different procedures. A variety of techniques is available for pellet manufacturing. Layering processes have been used over the years. These processes have some limitations, such as nonuniformity in the size of the pellets and less drug loading. In recent years extrusion spheronization, cryopelletization, freeze pelletization, and hot melt extrusion have been used to produce spherical pellets:

1. *Solution and suspension layering*: Coating processes are performed in either a rotated drum pan coater or fluidized bed with conventional top spray, bottom-spray (Wurster apparatus) (Priese and Wolf, 2013; Sinchaipanid et al., 2004) and side (tangential) spray to apply drug and/or binder solution or suspension to solid cores that can be inert materials (Lust et al., 2013; Priese and Wolf, 2013; Suhrenbrock et al., 2011) or granules of the same drug.
2. *Dry powder layering*: This involves the use of modified rotating pans, rotor-granulators/rotor tangential spray fluid-bed to deposit successive layers of dry powder (drug and excipients) on inert materials with the help of an adhesive solution/binding liquid (Nastruzzi et al., 2000; Park et al., 2016).
3. *Direct powder pelletization*: The technique uses high shear mixers (Hamdani et al., 2002; Kristensen et al., 2000a,b; Schæfer, 1996) and centrifugal fluid-bed or rotary fluid-bed granulators (Pašić et al., 2010; Vilhelmsen et al., 2004; Zema et al., 2008) to apply agglomeration liquid direct to a powder mixture of a drug and expients followed by pelletization by means of a rotating disc. A binder can be added as a liquid (wet pelletization) or added as a molten binder before or during the process (melt pelletization).
4. *Hot-melt extrusion*: The drug's substance and excipients (as polymers and waxes) are blended and extruded at a predetermined temperature. The extrusion temperature must be high enough to melt at least one or more of the formulation components. The molten extrudate is cut at the die-face pelletization into uniform cylindrical segments and/or not are spheronized in a jacketed spheronizer to generate uniformly size pellets (Bialleck and Rein, 2011; Follonier et al., 1994, 1995; Radl et al., 2010; Yeung and Rein, 2015; Young et al., 2002).
5. *Cryopelletization*: Pellets can be produced by allowing droplets of liquid formulation, such as solution, suspension, or emulsion to come in contact with liquid nitrogen at −160°C in which liquid nitrogen is used as solidifying medium. The procedure permits instantaneous and even freezing of the material being processed due to the rapid heat transfer that occurs between the droplets and the liquid nitrogen. The required amount of liquid nitrogen for manufacturing quantity depends on the solids content and temperature of the solution or suspension being processed. The pellets are dried in conventional freeze dryers to remove water or organic solvents (Erber and Lee, 2015a,b).
6. *Freeze pelletization*: A simple technique for producing spherical pellets for pharmaceutical use. In this technique, a molten-solid carrier/matrix is introduced as droplets into an inert column of liquid in which the molten solid is immiscible. The molten solid moves in the liquid column as droplets and solidifies into spherical pellets (Cheboyina et al., 2004; Cheboyina and Wyandt, 2008a,b).

7. *Extrusion-spheronization*: Can be defined as the process of forcing a material through an orifice or die under controlled conditions thus forming cylinders or strands called extrudates. During spheronization, these extrudates are broken into small cylinders and consequently rounded into spheres (pellets). Hence, extrusion/spheronization is a multiple-step process capable of making uniformly sized spherical particles referred to as pellets and involving the following sequential steps: (1) dry blending, (2) wet granulation, (3) extrusion, (4) spheronization, (5) drying, and (6) optional screening (Erkoboni, 2003).

Compared with other drug-delivery systems, such as a single unit dosage form, pellets have several advantages:

- Pellets can improve the bioavailability, easier to administer, and provide for dose flexibility.
- The gastrointestinal transit time, especially the residence time in the stomach, is more uniform and less sensitive to food ingestion compared to single unit dosage forms because the small pellets can pass the pylorus even in the closed state. This leads to smaller intra- and interindividual variations in the pharmacokinetic parameters.
- The local concentration of a drug is relatively low because the particles are well dispersed after swallowing; irritation of the gastrointestinal mucosa is therefore minimized.
- Pellets with a coating for modified release have a lower risk of dose dumping than coated tablets.
- Pellets with different coatings can be mixed and filled into capsules to enable the desired dissolution profile to be achieved.
- Pellets containing different drugs can be filled into capsules without the risk of interaction of the substances during preparation and storage (Knop and Kleinebudde, 2005; Krämer and Blume, 1994).

Disadvantages of pellets:

- The volume per dose is usually higher than for tablets because of the lower bulk densities of pellets compared to compressed tablets.
- Compared to larger single unit dosage forms, the specific surface area per dose is higher and more coating material is necessary to obtain coatings of the same thickness and the same functionality.
- The preparation of pellets and the subsequent filling into capsules and the compaction of pellet containing tablets are more complicated and time-consuming than the production of compressed tablets made from granules or powder mixtures (Knop and Kleinebudde, 2005).

1.2 GENERAL DESCRIPTION OF THE EXTRUSION SPHERONIZATION PROCESS (WET MASS EXTRUSION)

In 1970, Conine and Hadley were the first to describe the production of pellets via extrusion spheronization in the pharmaceutical industry. Reynolds (1970) elaborated on this work through the further description of the equipment and process mechanisms. Extrusion spheronization is a multistep process consisting of consecutive stages (Fig 8.1).

The extrusion-spheronization process typically involves a specific sequence of steps (Fig. 8.2). The characteristics and quality of the final pellets produced by the extrusion spheronization process are dependent on the physicochemical properties of the active pharmaceutical ingredient (API) and excipients, and process variables. Specifically, the physicochemical properties of the API and excipients that affect the types of pellets produced by the process include particle size and surface area, moisture

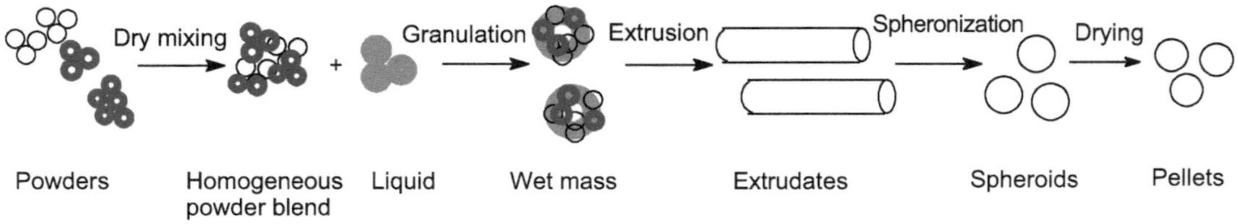

FIGURE 8.1 Schematic Overview of the Steps Involved in an Extrusion-Spheronization Process

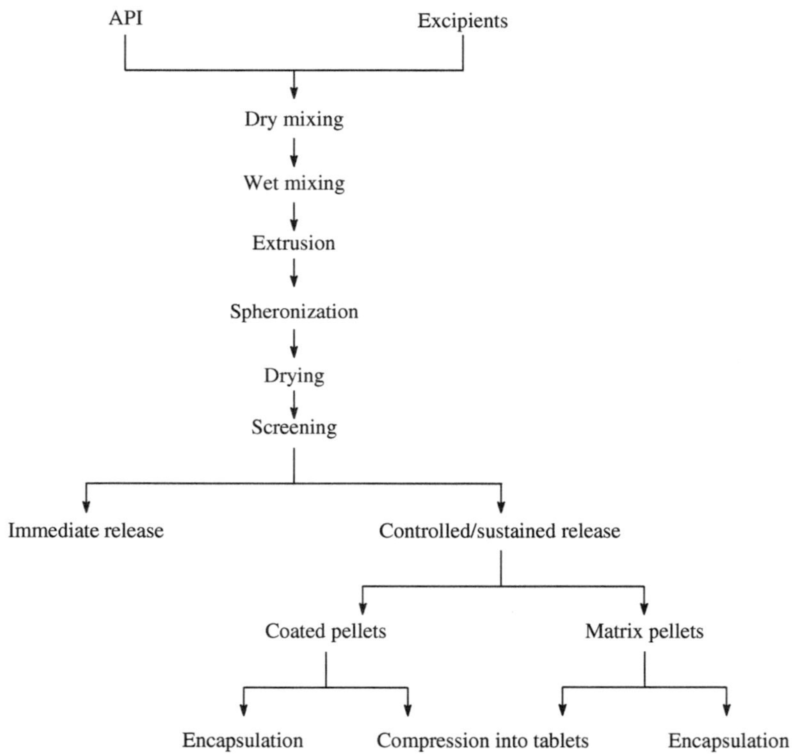

FIGURE 8.2 The Extrusion Spheronization Process (Wet Mass Extrusion)

sorption capacity, polymorphism of the API, aqueous solubility, and density. The various process variables that affect the quality of the pellets include mixing time and speed, type and amount of binder/solvent added to wet the powder mixture, type of extruder, diameter of the orifice, and thickness of the die or screen, rate of extrusion, spheronization time, speed, load, and plate design, and drying time and rate (Hellén et al., 1993a,b).

The main steps of the process are:

1. *Dry mixing of ingredients* to achieve homogeneous powder dispersion.
2. *Wet massing* to produce a sufficiently plastic wet mass.

3. *Extrusion* to form rod-shaped particles of uniform diameter.
4. *Spheronization* to round off these rods into spherical particles.
5. *Drying* to achieve the desired final moisture content.
6. *Screening* (optional) to achieve the desired narrow size distribution.

The disadvantages of the extrusion spheronization process are that it is a multistep, labor-intensive, and time-consuming batch process. Each step in the process has its own variables that have major effects on the quality of the final product. Consequently, it is necessary to optimize the processing parameters for each formulation. Furthermore, the process requires the use of a relatively large amount of water compared to the traditional granulation process, uniform distribution of the water in the wetted mass, and appropriate drying conditions, which may not be suitable for moisture and heat sensitive drugs (Trivedi et al., 2007).

1.3 PROCESS AND EQUIPMENT

- *Dry mixing and wet granulation*: A uniform blend has to be obtained before wet granulation can take place, as it affects the pellets' content uniformity. An uneven distribution may also result in local over wetting during granulation, for materials with large differences in size and solubility properties (Parikh, 2009). Dry mixing and wet granulation are usually carried out in the same equipment (e.g., planetary mixer, high shear mixer, sigma blade mixer).
- *Extrusion*: During extrusion, the plastic wet mass is forced through a die and shaped into cylindrical particles with an uniform diameter. The resultant extrudates diameter is determined by the diameter of the die, and its length depends on the properties of the wet mass and the extruder type. A variety of extruder types has been developed, and can be divided into three classes based on their feed mechanism (i.e., the way by which the wet mass is transported through the extrusion zone): screw feed extruders, gravity feed extruders, and piston feed extruders, shown in Fig. 8.3.

  According to Reynolds (1970) and Rowe (1985) an axial screw extruder produces a denser material than a radial screw extruder. The latter has a higher output but also produces but shows greater heat production during the processing. Pellet quality is dependent on the thickness of the screen and the diameter of the perforations (Hellén et al., 1992). A thinner screen produced a rough and loosely bound extrudate, whereas a thicker screen forms smooth and well-bound extrudate because of the higher densification of the wet mass. Similarly, the diameter of the perforations determines the size of the pellets: a larger diameter in the perforations will produce pellets with a larger diameter under similar processing conditions (Hellén et al., 1993b).

  A number of studies for comparison of the extrusion and spheronization behavior of wet powder masses processed by a ram extruder and a cylinder extruder (Fielden et al., 1992b), comparison between the extrusion forces and sphere quality of a gravity feed extruder and a ram extruder (Baert et al., 1992), comparison between a gravity feed extruder and a twin screw extruder and a ram extruder (Baert et al., 1993a,b), comparison between a twin-screw extruder and a rotary ring die press (Schmidt et al.,1997; Schmidt and Kleinebudde, 1998). Vervaet et al. (1994) reported on the monitoring of the extrusion process using a basket extruder and on the influence of the extrusion screen specifications on the pellet quality.
- *Spheronization*: A spheronizer consists of a bowl with fixed sidewalls and a bottom plate or disk that rotates at a high speed (Fig. 8.4). The spheronizer is filled with extrudates, and due to

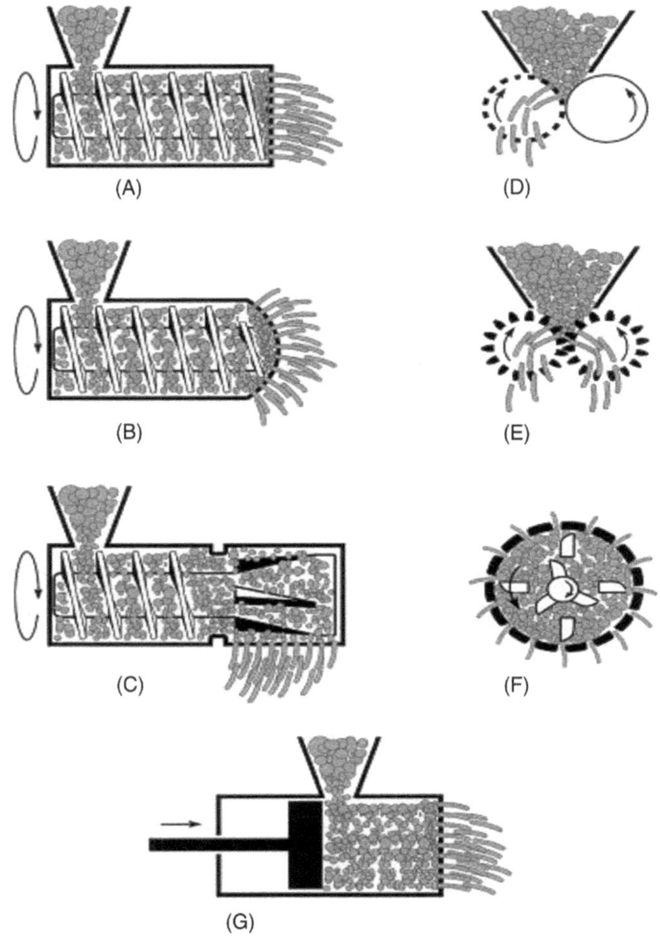

FIGURE 8.3 Schematic Overview of the Extruder Types Applied in Extrusion-Spheronization

Screw feed extruders with axial (A), dome (B), and radial (C) type, gravity feed extruders with cylinder (D), gear (E), and radial type (F) and piston feed ram extruders (G).

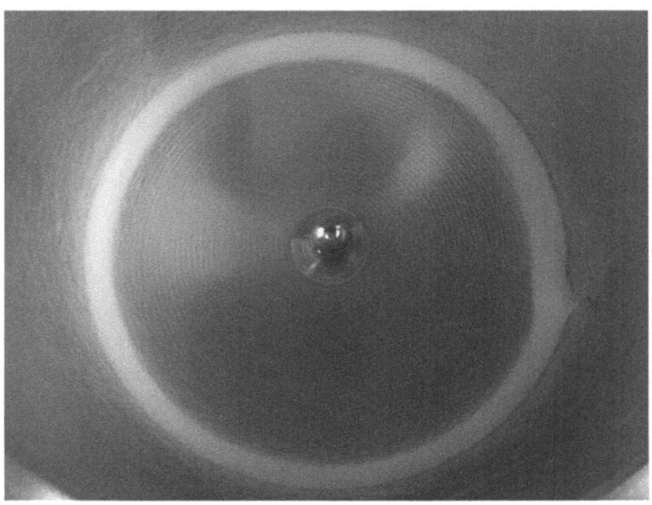

FIGURE 8.4 Photograph (Plan View) of Spheroniser Plate in Motion

frictional forces generated by particle–particle and particle–equipment interaction, the extrudates are initially broken into smaller cylinders and then rounded into spheres. To enhance the forces generated as particles move across its surface, the bottom plate generally exhibits a grooved surface (cross-hatched pattern or radial pattern).

Different manufacturers produce different types of plates. Most spheronizer plates have grooved surfaces, which are designed to increase the frictional forces (Vervaet et al., 1995).

Pellet quality is also dependent on spheronizer load, which affects the particle size distribution, bulk, and tap density of the final pellets. The increase in the spheronizer speed and a low spheronizer load will result in wider particle size distribution with a smaller yield of pellets, whereas it increases with extended spheronization time at a higher spheronizer load. Hellén et al. (1992) reported that the bulk and tap density increased and the size of the pellets decreased with an increasing spheronizer load.

Four different spheronizer friction plate patterns (i.e., cross-hatch, radial, striated edge pattern) have been used (Fig. 8.5A–D) (Michie et al., 2012; Podczeck and Newton 2014; Caleva).

The transformation from a cylindrical extrudate in a spherical particle occurs in various stages, and two models have been proposed to describe this process (Fig. 8.5A–B). According to Rowe (Fig. 8.6A), the cylinders are rounded off at the edges, followed by the formation of dumbbell-shaped particles, ellipsoids, and finally spheres. The model proposed by Baert et al. (Fig. 8.6B) suggests that the cylinders are rounded off at the edges, but are additionally bent, resulting in the formation of a rope-shaped particle. Next, a dumbbell with a twisted middle is formed, initiating particle breakage into two spherical particles with a cavity on their flat side. Further rounding in the spheronizer creates completely spherical articles. Recently, Koester and Thommes (2010) reported that these generally adopted pelletization mechanisms need to be extended, to account for the material transfer between pellet particles (Fig. 8.6C). In addition to plastic deformation, the authors also observed a material transfer between pellet particles for different formulations. Herewith, regional distinctions in the amount of mass transfer and an influence of the water content were observed.

The duration of spheronization is usually 2–10 min (Gamlen, 1985) and a rotational speed between 200 and 400 rpm of the friction plate is optimum to obtain a highly spherical pellet (West and Rowe, 1988).

- *Drying*: This is the final step of extrusion/spheronization process. The spherical particles are then dried at room temperature or at an elevated temperature (in an oven, a fluidized bed or microwave oven) until the desired residual moisture level is achieved. As the different process steps are influenced by a number of interrelated process and formulation variables, the extrusion-spheronization process requires the necessary control to obtain the required pellet quality.

  Wet pellets are mostly dried in an oven or fluid-bed, although microwave and freeze drying have been also used to study the influence of drying method on pellet properties. The main differences between oven and fluid-bed drying are the rate of granulation liquid evaporation and the way how the material is handled during drying: during oven drying in a static bed liquid evaporates from the material over a longer period of time, while during fluid-bed drying the turbulent motion of dried material in a heated air stream promotes significantly faster drying (Lieberman and Rankell, 1970). Several authors studied the influence of different drying techniques on pellet characteristics (Airaksinen et al., 2004; Bashaiwoldu et al., 2004; Bataille et al., 1993 Kleinebudde, 1994; Römer et al., 2007; Scusa et al., 1992, 2002; Wlosnewski et al., 2010).

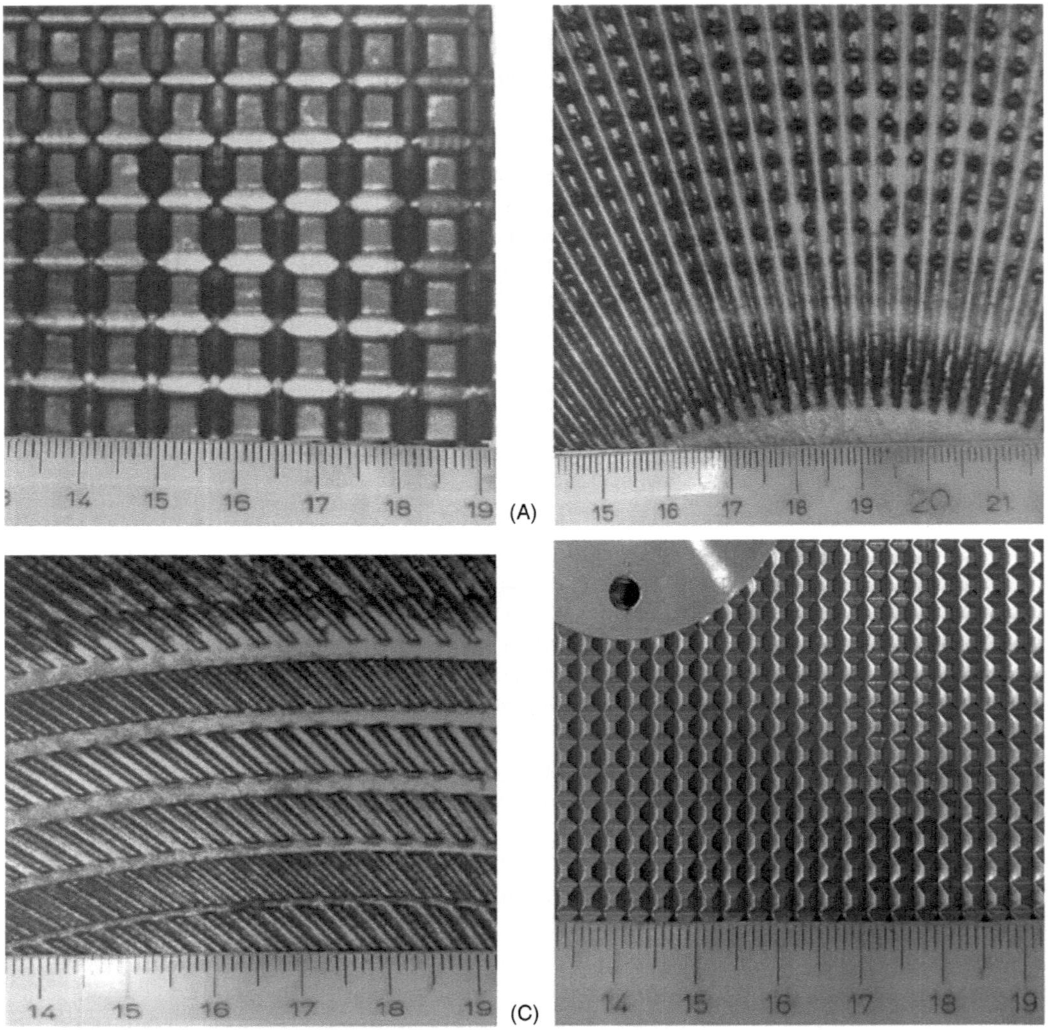

(A) (B) (C) (D)

FIGURE 8.5 Friction Plate Design

(A) Cross-hatch pattern; (B) radial pattern; (C) striated edge pattern; (D) cross-hatched friction plate (pyramidal elements on a square pattern, spacing 1.40 mm, height 0.86 mm and width at top 0.50 mm).

It was shown that the drying conditions have an impact of the physico-mechanical properties of pellets, but they did not lead to disintegration of the pellets.

- *Screening*: This may be necessary to achieve the desired size distribution, and for this purpose sieves are used. In case of pellets prepared by extrusion-spheronization, screening is essentially required after manufacture, in order to avoid pellets having high size distribution (Koo and Heng, 2001).

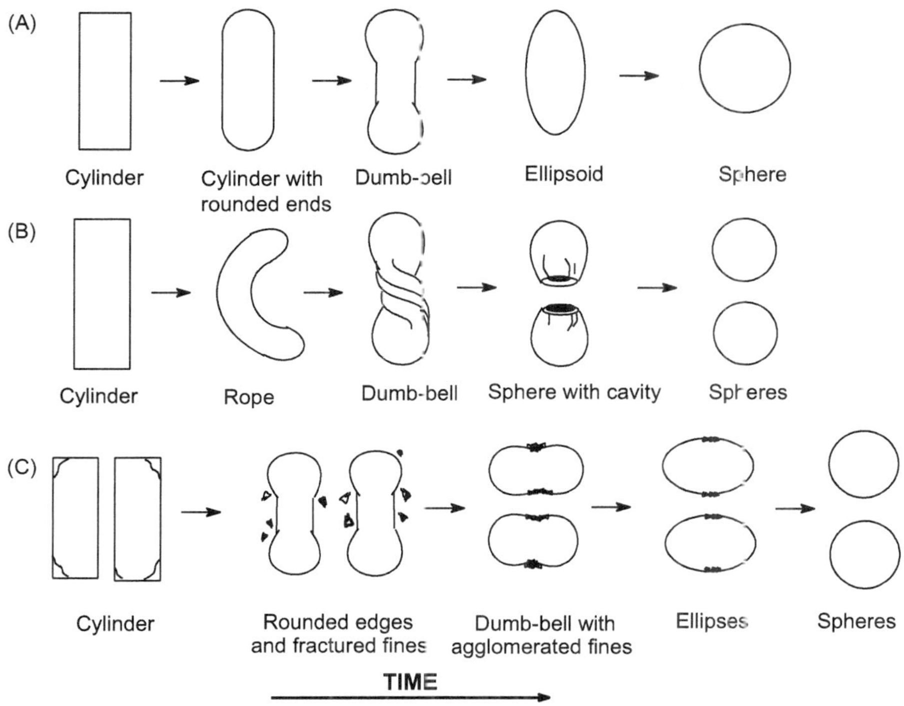

FIGURE 8.6

Schematic presentation of the different pellet formation stages during spheronization according to Rowe (1985) (A), Baert and Remon (1993) (B), and Koester and Thommes (C).

1.4 FORMULATION

Selection of suitable excipients for any formulation is one of the major tasks during formulation development. In a multistep process, the quality of the final product depends on the quality of intermediate products obtained at the end of each step, and that depends significantly on the type of the excipients used. The formulation must have certain qualities in order to produce extrudates that can be spheronized into appropriate particle-size pellets with desired surface characteristics. The following section describes various excipients that have been used to prepare pellets using the extrusion-spheronization process (Trivedi et al., 2007).

1.4.1 Microcrystalline cellulose as spheronization aid

Microcrystalline cellulose (MCC) is the most commonly used spheronizing aid in a formulation undergoing extrusion spheronization. It is available in different grades and particle sizes. Of all the different brands and grades of MCC, Avicel PH 101 or Emcocel 50 has been the most widely used. MCC helps in the formation of spheres because of its unique properties. Like other cellulosic materials, MCC is a filamentous material with a large surface area, high internal porosity, and high moisture retaining property (Shah et al., 1995).

MCC is the golden standard as an extrusion spheronization aid based on its good binding properties that provide cohesiveness to a wetted mass containing MCC. Furthermore, it is able to absorb and retain a large quantity of water due to its large surface area and high internal porosity (Sonaglio et al., 1995), thus facilitating extrusion, improving wetted mass plasticity and enhancing spheronization. Moreover, by controlling the movement of water through the plastic mass, it prevents phase separation during extrusion or spheronization. Due to these properties MCC-based pellets produced via extrusion spheronization have a good sphericity, low friability, high density, and smooth surface properties.

Its capacity to retain very large quantities of water internally means that wet masses made with MCCs have rheological properties that are very suitable for extrusion spheronization (Fielden et al., 1992a, 1995).

1.4.2 Alternative excipients for microcrystalline cellulose

To date, MCC was considered a golden standard for extrusion spheronization. However, the use of MCC has some disadvantages. Because MCC is water insoluble, pellets made with MCC take a long time to disintegrate and, therefore, release of drug from the pellets is not immediate.

However, several investigators have attempted to replace MCC as a spheronizing agent, but with limited success. For example, Agrawal et al. (2004) studied the effect of formulation and process variables on the physical properties of the pellets prepared with varying concentrations of chitosan, ethyl cellulose, and hydroxypropyl methylcellulose (HPMC), without using MCC. Results of the study indicated that process variables, such as spheronizer and extruder speed, and formulation variables, such as chitosan, HPMC, and water content significantly affected the physical properties of the pellets. Pellet size decreased with an increase in chitosan content. Glyceryl monostearate has also been used as an appropriate replacement for MCC for producing spherical pellets, using the extrusion spheronization process in two studies (Basit et al., 1999; Newton et al., 2001). Other excipients that have been used to prepare spherical pellets include: hydroxypropyl cellulose (Kleinebudde, 1994), HPMC, and hydroxyl ethyl cellulose (HEC) (with isopropanol as a solvent) (Chatlapalli and Rohera, 1998), pectinic acid (Tho et al., 2002), powdered cellulose (Alvarez et al., 2003), starch-dextrin (Prieto et al., 2005), crosslinked polyvinylpyrrolidone (Liew et al., 2005), K-carrageenan (Ghanam and Kleinebudde, 2011; Koester and Thommes, 2010; Thommes and Kleinebudde, 2006), modified MCC (Podczeck, 2008; Podczeck et al., 2008), and isomalt (Antal et al., 2013).

1.4.3 Use of other excipients in the extrusion-spheronization process

Other excipients, such as water soluble/swellable polymers (e.g., polyvinylpyrrolidone, hydroxypropyl methylcellulose, polyethylene oxide), ionic polymers (e.g., carboxypolymethylene, chitosan, Eudragit), hydrophobic/water insoluble polymer (e.g., ethylcellulose), waxes (e.g., glyceryl monostearate, carnuba wax), disintegrants (e.g., croscarmellose sodium, sodium starch glycolate, crospovidone), and polysaccharides (e.g., starch, dextrin), are also added to the formulations containing MCC in order to attain the desired drug release characteristics from the pellets (Trivedi et al., 2007).

Some of the excipients used in combination with MCC to improve pellet disintegration and/or drug release from MCC-based pellets include *surface active agents*. Surfactants are added to the liquid to improve wettability by lowering the interfacial tension between the liquid and drug particles. Examples are sodium lauryl sulphate, polyethylene glycol, polysorbate 80, glyceryl and sorbitan mono-oleate, sorbitan mono-palmitate, glycerol monostearate, and self-emulsifying drug delivery systems (SEDDS) (Dukić et al., 2009).

Table 8.1 Examples of Commonly Used Excipients	
Filler	MCC, starch and derivatives, sucrose, lactose, mannitol, dicalcium diphosphate, β-cyclodextrine, glucose
Binder	Gelatin, HPC, HPMC, MC, PVP, sucrose, starch, water
Lubricant	Calcium stearate, glycerin, PEG, magnesium stearate
Separating agent	Kaolin, talc, silicon dioxide
Disintegrant	Alginates, croscarmellose sodium, sodium starch glycolate
pH adjuster	Citrate, phosphate
Surfactant	Polysorbate 80, Solidum lauryl sulphate, polyethylene glycol, glyceryl and sorbitan monooleate, sorbitan monopalmitate, glycerol monostearate, self-emulsifying systems
Spheronization enhancer	MCC, sodium carboxymethyl cellulose
Glidant	Talc, starch, magnesium stearate
Release modifier	Ethyl cellulose, carnauba wax, shellac

The types and concentrations of specific excipients incorporated in the formulations depend on the specific need and the desired drug-release rates from the pellets. For example, in spite of the large surface area of pellets that provides faster absorption from this dosage form, poorly soluble drugs might require super disintegrants to compensate for the delay in drug dissolution due to the high density of pellets (Souto et al., 2005) (Table 8.1).

1.5 EVALUATION OF PELLETS

A number of techniques are used to evaluate the physical properties of pellets prepared by the extrusion-spheronization process (Table 8.2) (Aulton and Taylor, 2013; Eriksson et al., 1997; Knop and Kleinebudde, 2005; Trivedi et al., 2007).

2 NANOEMULSIONS

2.1 INTRODUCTION

Nanoemulsions are defined as the structures having typically 20 to 200 nm droplet diameters (Solans et al., 2005). Nanoemulsion term has been first used by Nakajima et al. (1993) or "ultrafine emulsion" (Nakajima, 1997), "submicron emulsion" (Amselem and Friedman, 1998), and called by different terms like "miniemulsion" (El-Aasser and Sudol, 2004). Until today, it is recommended by Solans et al. (2005) to use the term "nanoemulsion" among the other terms, because it clearly describes that droplets are at nanolevel and it differs completely from the term of microemulsion or conventional emulsions (of micrometric size). Nanoemulsions are particular types of emulsions because they can have a transparent and semitranslucent appearance, they have very small droplet size, they are stable against sedimentation creaming and they tend to increase bioavailability (Shakeel et al., 2008; Solans

Table 8.2 List of Different Evaluation Methods for Determination of Pellet Properties

Properties	Requirement	Evaluation Methods
Particle shape, sphericity or roundness	Uniform spherical shape	Two-dimensional methods: Image analysis (aspect ratio) OPCS (one-plane critical stability) Shape factor (e_R) Three-dimensional methods Heywood's shape coefficient Permeametry shape factor
Particle-size distribution	Narrow, uniform size	Image analysis
Mean particle size	Between 0.5 and 1.5 mm	Sieve or image analysis
Density	High	Bulk density; tapped density
Flow property	Very good	angle of repose; dynamic flow; Carr's index; Hausner's ratio
Porosity	Low	Surface-area analysis; porosimetry; true and bulk densities
Surface roughness	Smooth	Microscopic method (SEM-scanning electron microscopy)
Strength of pellets	High	Friability Crushing strength
Drug content	From very low to more than 95%	Assay
Drug release	Have desired drug release characteristics	Dissolution

and Solé, 2012; Wang et al., 2008). In addition to their high colloidal stability, nanoemulsions need less than 10% surface-active agent in preparation stages (while this rate is 20% or higher in microemulsions). Furthermore, since nanoemulsions provide a wide surface area, they lead active components to be penetrated quickly (Laouini et al., 2012). Due to their long-term stability over storage and an improved high bioavailability, they have attained particular interest as delivery systems for bioactive substances, such as carotenoids, phytostetol, polyunsaturated fatty acids, goryzanol, lipophilic vitamins, and numerous other compounds. Contrary to thermodynamically stable microemulsions, nanoemulsions are unstable systems and can be exposed to environmental degradation (Solans and Solé, 2012). The formation and manufacturing control of nanoemulsions require high shear force applications in order to cope with the surface tension of droplets (Mason et al., 2006). Since nanoemulsions are non-equilibrium systems, they cannot be formed spontaneously and typically energy input of a mechanical device is required. Thus, in general, high-energy methods are used for preparation of nanoemulsions [high pressure homogenizers (Sanguansri and Augustin, 2006; Solans et al., 2005)], microfluidizers (Dalgleish et al., 1997; Guraya and James, 2002; Jafari et al. 2006, 2007; Kwon et al., 2002; Swientek, 1990; Thompson and Singh, 2006), and ultrasonic generators (Chandrapala et al., 2012; Ghosh et al., 2013; Li and Chiang, 2012; Sanguansri and Augustin, 2006; Xia et al., 2001). There are also a number of studies applying low-energy methods for production of nanoemulsions, by using the chemical potential of the components under special conditions (Acosta, 2009; Calderó et al., 2011; Rao and McClements, 2010; Solans and Solé, 2012).

In general oil-in-water (O/W) nanoemulsions have started to be investigated for a while and their role in such polymerization reactions as nanoreactors has been researched (Solans et al., 2005). In previous studies, O/W nanoemulsion systems have been used as a carrier and combiner in food in order to encapsulate w-3 fatty acids in yogurt (Chee et al., 2005) and ice cream (Chee et al., 2007). Therefore, it could be concluded that O/W nanoemulsions offer a good potential in drug applications especially for encapsulation of lipophilic compounds.

2.2 METHOD

2.2.1 High-energy emulsification methods

The formation of nanoemulsions by high-energy methods is governed by the selected composition (i.e., surfactants and functional compounds) and by the quantity of energy applied. Therefore, nanoemulsions produced through high energy methods present a natural predisposition to preserve the nanoemulsions formation against formulation modification, for example, addition of monomer, surfactant and cosurfactant (Anton et al., 2008).

The mechanical processes generating nanoemulsions can be divided into three major groups based on the used devices (Anton et al., 2008; Sanguansri and Augustin 2006):

1. High-pressure homogenization in high-pressure homogenization, the mixture is subject to very high pressures and is pumped through a restrictive valve. The very high shear stress causes the formation of very fine emulsion droplets (Quintanilla-Carvajal et al., 2010; Sanguansri and Augustin, 2006).
2. Ultrasound is when two immiscible liquids are submitted to high-frequency sound waves in the presence of a surfactant, emulsion droplets are formed by cavitation. This causes intense shock waves in the surrounding liquid and the formation of liquid jets at high speed is responsible for the formation of emulsion droplets. However, this technology has not yet been proved as efficient for industrial-scale applications (Quintanilla-Carvajal et al., 2010; Sanguansri and Augustin, 2006).
3. High-speed devices rotor/stator devices (such as Ultra-Turrax) when compared with the other high energy approaches do not provide a good dispersion in terms of droplet sizes. With the energy provided mostly being dissipated, generating heat (Anton et al., 2008; Walstra, 1993).

Regarding the formation of nanoemulsions by high-energy approaches, most of the energy provided is dissipated in the form of heat. Though this idea is generalized, only very few authors report energy calculations in their publications.

2.2.2 Low-energy emulsification methods

In low-energy approaches, nanoemulsions are obtained as a result of phase transitions produced during the emulsification process, which is carried out, generally, at constant composition and changing the temperature (Izquierdo et al. 2001; Morales et al. 2003) or at constant temperature and changing the composition (Usón et al. 2004).

The methods used more often are:

- Membrane emulsification is a low-energy process that requires less surfactant (when compared with high energy methods) and produces emulsions with a narrow size distribution range. This method involves formation of a dispersed phase (droplets) through a membrane into a continuous phase. Nevertheless, this method has as limitation the low flux of the dispersed phase through the membrane, this being an issue during scale-up (Sanguansri and Augustin, 2006).

- Spontaneous emulsification is a mechanism that occurs when an organic phase and an aqueous phase are mixed, with the organic phase being a homogeneous solution of oil, lipophilic surfactant and water-miscible solvent, and the aqueous phase consisting of water and hydrophilic surfactant (Bouchemal et al., 2004). The spontaneous features of this method result of the initial nonequilibrium states of two bulks liquids when they are brought into contact without stirring. It is only under specific conditions that spontaneous emulsification occurs. Spontaneous emulsification is produced by different mechanisms (e.g., diffusion of solutes between two phases, interfacial turbulence, surface tension gradient, dispersion mechanism, condensation mechanism), which seem to be affected by the systems' compositions and their physicochemical characteristics like the physical properties of the oily phase and nature of the surfactants (Bouchemal et al., 2004). This process itself increases entropy and thus decreases the Gibbs free energy of the system (Anton et al., 2008).
- Solvent displacement is a method that consists of mixing a water-miscible organic solvent containing lipophilic functional compounds in an aqueous phase containing an emulsifier. The rapid diffusion of the organic solvent in the aqueous phase promotes the formation of nanoemulsions enabling their preparation in one step at low-energy input with high yield of encapsulation. Finally, the organic solvent is removed from the nanodispersion under reduced pressure. Nevertheless, the use of this technique is limited to water-miscible solvents (Chu et al., 2007; Yin et al., 2009).
- Emulsion inversion point varies the composition of the system at a constant temperature. The structures are formed through a progressive dilution with water or oil in order to create kinetically stable nanoemulsions (Anton et al., 2008; Sadtler et al., 2010).
- Phase inversion point uses the specific ability of surfactants (nonionic) to alter their affinities to water and oil in function of temperature at a fixed composition (Shinoda and Saito, 1968, 1969). It consists in suddenly breaking up the microemulsions maintained at the phase inversion point by a rapid cooling (Izquierdo et al., 2001, 2004; Sadurní et al., 2005) or by a dilution in water or oil (Anton et al., 2007, 2008). The nanoemulsions immediately formed are kinetically stable and can be considered as irreversible.

This process is relatively simple, prevents the encapsulated drug being degraded during processing, consumes low amounts of energy and allows an easy industrial scale-up (Anton et al., 2008; Anton and Vandamme, 2009).

2.3 MATERIALS USED IN NANOEMULSIONS PRODUCTION

In general, the O/W nanosized emulsion should be formulated with compatible vehicles and additives. The components of internal and external phases of nanosized emulsion should be chosen to confer enhanced solubility and stability to the incorporated lipophilic drug. In addition, the selected excipients should also preferably be chosen to favorably influence the biofate or therapeutic index of the incorporated drug following administration via parenteral, ocular, percutaneous, and nasal routes. This section is a comprehensive presentation of the general considerations concerning excipient selection and their optimum concentrations. Common excipients used for formulation of O/W nanosized emulsions are summarized in Table 8.3 (Tamilvanan et al., 2010).

Table 8.3 Excipients Used for Formulation of O/W Nanosized Emulsions

Oils	Sesame oil, castor oil, soyabean oil, paraffin oil, paraffin light, lanolin, vaseline, corn oil, glycerin monostearate, medium-chain monoglycerides, medium-chain triglycerides, squalen
Emulsifiers	Cholesterol, phospholipids (lipoid), polysorbate 80 and 20 (Tween 80 and 20), Transcutol P, Cremophor RH, Poloxamer 407, Poloxamer 188, Miranol C_2 M and MHT, Tyloxapol (α-tocopheryl polyethylene glycol succinate)
Cationic lipids and polysaccharide	Stearylamine, oleylamine, chitosan
Miscellaneous	α-tocopherol, glycerin, xylitol, sorbitol, thiomersal, ethylenediamine tetraacetic acid, methyl paraben, propyl paraben

3 BIOPHARMACEUTICAL CLASSIFICATION SYSTEM (BSC) AND APPLICATIONS OF PELLET TO IMPROVE DISSOLUTION RATE OF DRUGS IN BCS CLASS II

Biopharmaceutical classification system (BCS) divides all active pharmaceutical ingredients (API) into four classes based on drug dissolution rate and gastrointestinal permeability. BCS Class II is defined by drugs of high permeability and low solubility (Amidon et al., 1995).

The poor solubility of a drug is a major problem that limits the development of highly potent pharmaceutics. The drugs with low solubility lead to low oral bioavailability and erratic absorption, which is particularly pertinent to drugs within Class II of the BCS.

Examples of such Class II drugs include glibenclamide, phenytoin, danazol, ketoconazole, mefenamic acid, nifedipine, rifampicin, ethambutol, pyrazinamide, isoniazid, quinidine, chloroquine, mebendazole, niclosamide, prasiquantel, atenolol, piroxicam, and amitriptyline (Amidon et al., 1995; Dahan et al., 2009; Takagi et al., 2006).

Various strategies have been widely investigated to enhance the bioavailability of poorly absorbed drugs in order to increase their clinical efficacy when administered orally. It is estimated that between 40 and 70% of all new chemical entities identified in drug discovery programs are insufficiently soluble in aqueous media (Lindenberg et al., 2004; Lipinski, 2002; Lipinski et al., 2001).

Generally, the rate-limiting step for absorption of the drugs in this class is the dissolution velocity arising from low solubility. Although the drugs have a high permeability, the poor solubility results in a low concentration gradient between gut and blood vessel consequent to a limitation of drug transport and oral absorption.

IUPAC has defined solubilization of a drug molecule as a process by which an agent increases the solubility or the rate of dissolution of a solid or liquid solute (Gamsjäger et al., 2008). The solubility of drug molecules is an important factor to take into account during the development and design of new drug products.

Nowadays, 38% of drug compounds in drug development have a poor aqueous solubility. Therefore, one of the most challenging tasks in drug development is to improve the drug solubility in order to enhance the bioavailability of these drugs. Several strategies have been employed to overcome these limitations. The approaches to increase the solubility and the available surface area for dissolution are classified as physical and chemical modifications. For the physical modification, the techniques include decreasing particle size (micronization, nanonization), formation of polymorphs/pseudopolymorphs (including solvates), complexation/solubilization (by means of using surfactants or cyclodextrins, conjugation to dendrimers, and an addition of cosolvents), and preparation of drug dispersions in carriers (eutectic mixtures, nonmolecular solid dispersions, solid solutions). Pouton (2006) provides a brief indication of the main formulation options and advantages and disadvantages of some approaches (Table 8.4).

Particle size reduction has been a much smarter approach that can be applied to nonspecific formulation for many years. The micronization of drugs leads to an increase in their surface area, which proportionally increases in rate of dissolution and rate of diffusion (absorption). However, for very low solubility compounds, the micronization fails to improve the saturation solubility and increase the bioavailability of the drug. Therefore, the further step to reduce the particle dimension to nanometer size range has been invented. Recently, particle diminution to the submicron range has emerged to be

Table 8.4 Options For Formulation of Poorly Water-Soluble Drugs

Technology	Potential Advantage	Potential Disadvantage
Conventional micronization	Known technology, freedom to operate, solid dosage form	Insufficient improvement in dissolution rate
Nanocrystals obtained by ball-milling	Established products on the market, experienced technology provider (Elan), solid dosage form possible	Available only under license, secondary process required to avoid aggregation of nanocrystals
Nanocrystals obtained by dense gas technology	Alternative nanocrystal processing method, still room to develop new IP	Unproven technology, secondary process required to avoid aggregation of nanocrystals
"Solid solutions"—drug immobilized in polymer	Freedom to operate, new extrusion technology offers solvent-free continuous process	Physical stability of product questionable—drug or polymer may crystallize
Self-dispersing "solid solutions" with surfactants	Steric hindrance to aggregation built into product, amenable to extrusion	Physical stability of product questionable—drug or polymer may crystallize
Lipid solutions (LFCS Type I lipid systems)	Freedom to operate, safe and effective for lipophilic actives, drug is presented in solution avoiding the dissolution step	Limited to highly lipophilic or very potent drugs, requires encapsulation
Self-emulsifying drug delivery systems (SEDDS) and SMEDDS (LCFS Type II or Type III lipid systems)	Prior art available, dispersion leads to rapid absorption and reduced variability, absorption not dependent on digestion	Surfactant may be poorly tolerated in chronic use, soft gel or hard gel capsule can be used in principle but seal must be effective
Solid or semisolid SEDDS	Could be prepared as a free flowing powder or compressed into tablet form	Surfactant may be poorly tolerated in chronic use, reduced problem of capsule leakage, physical stability of product questionable—drug or polymer may crystallize
Surfactant-cosolvent systems (LFCS Type IV "lipid" systems)	Relatively high solvent capacity for typical APIs	Surfactant may be poorly tolerated in chronic use, significant threat of drug precipitation on dilution

a powerful formulation approach that can increase the dissolution rate and the saturation solubility, subsequently improve the bioavailability of poorly water-soluble drugs and may also decrease systemic side effects. For the chemical modification, the used technique is the synthesis of soluble prodrugs and salts (Chen et al., 2011; Gao et al., 2008; Junghanns and Müller, 2008; Magdalene, 2001; Möschwitzer and Müller, 2006).

Over the last decade, drug nanocrystals are considered as a novel approach to improve the solubility of hydrophobic drugs because the technique is simple and effective, which can quickly launch product to the market. The nanocrystals were invented at the beginning of the 1990s and the first products appeared very fast on the market from the year 2000 onward. Additionally, drug nanocrystals are a universal approach generally applied to all poorly soluble drugs for the reason that all drugs can be disintegrated into nanometer-sized particles (Müller et al., 2011).

To date, only few data are available on the incorporation of drug nanocrystals into solid dosage forms, although one multiparticulate dosage form containing drug nanocrystals has already entered the market (Emend by Merck, pellets containing aprepitant approved by the FDA and introduced to the market in 2003). Emend capsules contain spray-coated pellets with the nanocrystalline aprepitant, sucrose, MCC, hyprolose, and sodium dodecylsulfate (Merck, Drug Information Emend, 2004).

The formulation of ternary solid dispersions of ketoprofen with macrogol and kollagen hydrolizate derivative (KLHT) as carriers was elaborated on the basis of the results of the experiments in which different methods of solid dispersion preparation (melting, solvent method, different coolings, different concentrations of drug/carriers and molecular weight of macrogol were tested. The best solid dispersion consisted of: ketoprofen–macrogol 6000–KLHT: 1–8.9–0.1. The increase in the amount of released ketoprofen from solid dispersion pellets was 3.8-times greater than from the pellets containing the drug alone (Jachowicz et al., 2000).

The self-emulsifying (SE) pellets combine both advantages of self-emulsifying drug delivery systems (SEDDS) and pellets, and the extrusion/spheronization technique has been firstly introduced to prepare the SE pellets by Newton et al. (2001, 2007). In addition, Tuleu et al. (2004) has made a comparative bioavailability study of progesterone self-emulsifying formulations presented in a pellet and liquid form in dogs and the bioavailability of the two dosage forms was equivalent.

Dukić-Ott et al. (2007) evaluated modified starch (high-amylose, crystalline, and resistant starch) as the main excipient for immediate-release pellets containing poorly soluble drugs (hydrochlorothiazide and piroxicam) and prepared via extrusion spheronization.

Möschwitzer and Müller (2006) used a mucoadhesive nanosuspension of poorly soluble hydrocortisone acetate produced by high-pressure homogenization as layering dispersion in a fluidized bed process, followed by the application of an enteric coating to achieve a controlled drug release.

Abdalla et al. (2008) investigated the influence of physiological dilution media and enzymatic digestion on the solubilization capacity of the formulation for the model drug progesterone from self-emulsifying pellet and used Solutol HS 15 and medium chain glycerides as lipid mixtures.

Iosio et al. (2008) were to produce by coextrusion-spheronization pellets with two cohesive layers, one of them containing a self-emulsifying system for vinpocetine, a poorly water soluble model drug.

Wang et al. (2010) demonstrated that the self-emulsifying nitrendipine pellets with 30% of the liquid SEDDS were successfully developed by the use of extrusion-spheronization technique and the SE pellets could improve oral absorption of poorly soluble drug, such as nitrendipine.

Lei et al. (2011) studied solid self-nanoemulsifying cyclosporin A pellets prepared by fluid-bed coating.

Franceschinis et al. (2011) prepared a solid self-emulsifying drug delivery systems (solid-SEDDS) by means of a wet granulation process was optimized in a lab-scale high shear mixer in order to improve the dissolution rate of piroxicam, a poorly water-soluble model drug. With this aim, the classic liquid granulation binder was replaced with an O/W microemulsion, loaded with the drug.

Wang et al. (2012) were increasing the oral bioavailability of poorly water-soluble carbamazepine using immediate-release pellets supported on SBA-15 mesoporous silica. The aim of this study was to develop an immediate-release pellet formulation with improved drug dissolution and adsorption. Carbamazepine (CBZ), a poorly water-soluble drug, was adsorbed into mesoporous silica (SBA-15) via a wetness impregnation method and then processed by extrusion spheronization into pellets. Flow ability and wettability of the drug-loaded silica powder were evaluated by bulk and tapped density and by the angle of repose and contact angle, respectively. The drug-loaded silica powder was formulated into pellets to improve flowability. With maximum drug loading in SBA-15 matrices determined to be 20 wt.%., in vitro release studies demonstrated that the carbamazepine dissolution rate was notably improved from both the SBA-15 powder and the corresponding pellets as compared with the bulk drug. Correspondingly, the oral bioavailability of SBA-15-CBZ pellets was increased considerably by 1.57-fold in dogs ($P < 0.05$) compared with fast-release commercial carbamazepine tablets.

Vaassen et al. (2012) studies lipid-based pellets of the poorly soluble drug NXP 1210 were successfully produced by the solid lipid extrusion spheronization at room temperature. The dissolution profiles depicted a short lag time (about 2 min) and then fast and complete dissolution of NXP 1210 by increasing the amount of crospovidone. The initial release was more delayed with an increased amount of PVA–PEG-graft copolymer. Dissolution rate could also be influenced by changing the lipid binder from pure hard fat into a mixture of hard fat, glycerol distearate, and glycerol trimyristate. The formulations are feasible for taste-masked granules or pellets containing poorly soluble drugs.

Cao et al. (2012) developed novel mucoadhesive pellets containing valsartan (VAL) with enhanced oral bioavailability. Two types of VAL-loaded core pellets were prepared by an extrusion-spheronization method, and further dry-coated with a mixture of hydroxypropyl methylcellulose (HPMC) and carbomer (CB) at different ratios. The core pellets containing poloxamer 188 as a solubilizer and NaOH as a pH modulator showed a significantly higher drug release rate than common pellets or drug powder in vitro.

Tian et al. (2013) studied nanostructured lipid carriers (NLCs) to improve the oral bioavailability of fenofibrate from pellet.

Lust et al. (2013) studied drug-layer coated pellets of piroxicam. These exhibited improved dissolution rate compared to the powder forms of piroxicam. The presence of hydrophilic HPMC plays a major role in improving the dissolution characteristics of piroxicam solid-state forms in drug-layer coated pellets.

Recently, much attention has been paid to lipid-based formulations, especially self-emulsifying and self-nanoemulsifying drug delivery systems (SNEDDSs). In general, SNEDDS is an isotropic mixture of drug, oil, surfactants, and cosurfactants, which forms an O/W nanoemulsion upon contact with aqueous medium under gentle agitation (Pouton and Porter, 2008; Zvonar et al., 2010). SNEDDSs are commonly formulated in a liquid state and further encapsulated into soft gelatine capsules before use. However, the soft capsule formulations have some disadvantages, including low drug compatibility, poor stability, drug leakage and precipitation, capsule aging, and high production costs. Therefore, it is desirable to formulate SNEDDSs into solid dosage forms.

Fluid-bed coating technique is a process to prepare pellet-based solid SNEDDS formulations (Tian et al., 2013), such as solid dispersion and cyclodextrin inclusion complexes, from liquids or solidied nanostructured lipid carriers onto pellets (Lei et al., 2011). This technique is based on removal of the solvent from the bulk liquid, while the solid precipitates and deposits on the surface of nonpareil cores simultaneously.

Up until now, melt extrusion spheronization has been employed as the only technique to prepare SNEDDS pellets. The formulation process depends highly on liquid SNEDDSs as moistening agents or adhesives to prepare paste for extrusion, which makes formulation development less flexible.

4 CASE STUDY

The case study involves the use of Cremophor ELP (polyoxiethylated-35 castor oil, hydrophilic–lipophilic balance approximately 12–14) and Labrafil M 1944 CS (Oleoyl macrogol-6 glycerides) to improved dissolution and enhanced oral bioavailability of a typical BCS II drug.

Drug A is a crystalline compound characterized by a low water solubility so that it can cause problems in formulating and limiting the bioavailability.

Powders were blended in different proportions and wetted by mixing of Cremophor ELP and Labrafil M 1944 CS in mixers UMC5 Stephan (Germany), which was fitted with knife blade impellers of 0.14 m in diameter. Mixing rate was 3000 rpm. Mixing time was set to 10 min. Further addition of water was added until a mass suitable for extrusion was obtained and then the plastic mass was being mixed for a further 3 min. During granulation, the material was repeatedly scraped from the mixing bowl walls, to ensure uniform water distribution

The wet mass was extruded at an extrusion speed of 20 rpm using a single screw extruder (Variable Density Extrusion Caleva 120, UK, for the laboratory) equipped with an axial extrusion in a single bench-top unit (extrusion screen: thickness: 1.2 mm, perforation diameter: 1 mm). The extrudates were spheronized for 6 min at 1200 rpm in a 250 mm radial plate spheroniser using a "cross-hatch" frictional plate of 3x3 mm pitch and 1.2 mm depth (Caleva MBS 250, Caleva, Sturminster Newton, Dorset, UK). The resulting pellets were dried in the ventilated oven Memmert (Germany) at 45°C until a constant weight had been reached. The pellets were stored in sealed bags.

4.1 COATING OF PELLETS

Pellet coating was performed with a lab-scale air flow technology coater (Innojet VENTILUS V-2.5-Herbert Hüttlin) with the assembled product bowl IPC 1 and the nozzle IRN 2 (Innojet Rotojet spraying nozzle).

As the goal was to prevent the release of drug in the stomach, the pellets were enteric-coated using Acryl–EZE. The coating suspension was prepared by dispersed Acryl–EZE (10% w/v) in water for 1 h.

The suspension was then filtered through 200 μm sieve and agitated continuously throughout the coating process. The weighed quantity (100 g) of pellets was charged into the glass coating chamber. The pan kept rotating, and coating dispersion was sprayed on the surface of pellet under the processing conditions listed here:

- Inlet air temperature = 60C.
- Air flow rate = 60 m^3/h.
- Spray rate of coating solution = 2.6 g/min.
- Atomization pressure = 5 psi.

4.2 DISSOLUTION TESTS

The dissolution studies were carried out according to the USP (XXXVI) using reciprocating-cylinder method (USP apparatus 3) at 15 dpm in 230 mL of simulated gastric fluid (pH 1.2) and intestinal fluid (pH 6.8) without enzyme (dissolution media at 37 ± 0.5°C).

Uncoated pellets equivalent to 25 mg drug A were subjected to the dissolution studies in 230 mL 0.1 N HCl or pH 6.8 phosphate buffer, respectively.

For the drug release studies of the enteric-coated pellets, 0.1 N HCl (230 mL) was used as a dissolution medium for the first 2 h and then pH 6.8 phosphate buffer (230 mL).

Samples (5 mL) were withdrawn at regular time intervals and filtered using a Whatman filter paper (Cat. No. 1001 090). An equal volume of the respective dissolution medium was added to maintain the volume constant. Drug A content of the samples was analyzed by UV spectrophotometer (UV2401PC, SHIMADZU, Kyoto, Japan) at suitable nanometer. All measurements were performed in triplicate from three independent samples.

4.3 RECONSTITUTION STUDY

One gram of uncoated pellets or enteric-coated pellets were gently agitated with a magnetic stirrer in 30 mL of deionized water or pH 6.8, respectively. Samples were taken after 30 min and filtered through 0.45 m micropore filters. The droplet size of resultant emulsions was determined by Zetasizer Nano ZS (Malvern Instruments, UK) at a wavelength of 635 nm with the scatter angle fixed at 173°C, and the temperature was maintained at 25°C.

4.4 RESULTS

In vitro release profiles of drug A from coated pellets 17.5% w/w with Acryl–EZE revealed that no drug was released for 2 h in acidic dissolution media (Fig. 8.7). The dissolution rate of the drug A from

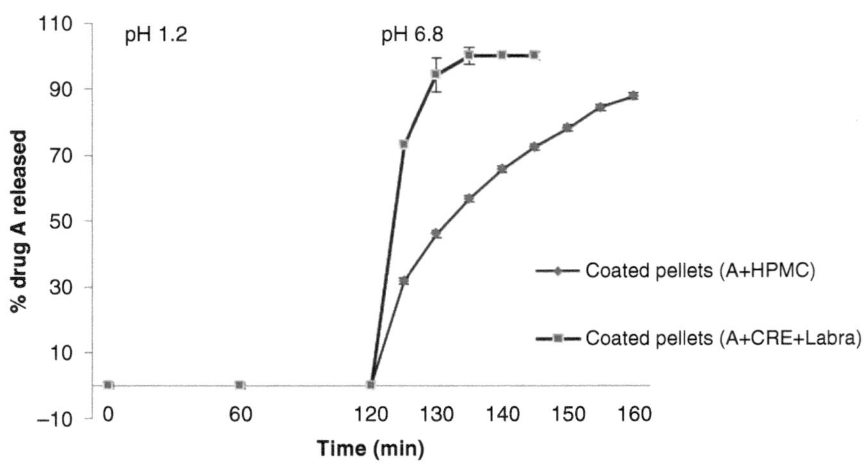

FIGURE 8.7

Comparison of in vitro release profiles of drug A from pellets coated (with Acryl–EZE ratio at coating level of 15% w/w) containing HPMC and pellets coated containing Cremophor ELP and Labrafil M 1944 CS.

solid dispersion pellet is greater than from pellets containing physical mixture (hydroxypropyl methylcellulose E6 (HPMC) as a pellet binder). Addition of Cremophor ELP and Labrafil M 1944 CS in the formulation indeed was found to improve the dissolution rates substantially in pH 6.8. In the pH 6.8, after 10 min, the percent dissolution was improved from 32% to 73%. The release of drug A increased further up to 90% within 15 min due to the addition of Cremophor ELP and Labrafil M 1944 CS.

The average droplet size of the nanoemulsion formed from uncoated pellet and enteric-coated pellets were 153.2 ± 13.06 and 121.0 ± 12.98 nm, respectively, with narrow distributions.

On the basis of these results, it was suggested that formulation of pellet containing Cremophor ELP and Labrafil M 1944 CS can be promising as an enteric-coated pellet of drug A with practical expectation for rapid pharmacological actions.

5 CONCLUSIONS

Nanoemulsion and its subsequent transformation into pellet systems offer several advantages for improving solubility and dissolution rates of poorly water-soluble drugs for oral delivery are developed. This technology could be developed to overcome the poor absorption of some active pharmaceutical ingredients.

REFERENCES

Abdalla, A., Klein, S., Mäder, K., 2008. A new self-emulsifying drug delivery system (SEDDS) for poorly soluble drugs: characterization, dissolution, in vitro digestion and incorporation into solid pellets. Eur. J. Pharm. Sci. 35 (5), 457–464.

Acosta, E., 2009. Bioavailability of nanoparticles in nutrient and nutraceutical delivery. Curr. Opin. Colloid Interface Sci. 14, 3–15.

Agrawal, A.M., Howard, M.A., Neau, S.H., 2004. Extruded and spheronized beads containing no microcrystalline cellulose: influence of formulation and process variables. Pharm. Dev. Tech. 9, 197–217.

Airaksinen, S., Karjalainen, M., Räsänen, E., Rantanen, J., Yliruusi, J., 2004. Comparison of the effects of two drying methods on polymorphism of theophylline. Int. J. Pharm. 276, 129–141.

Alvarez, L., Concheiro, A., Gomez-Amoza, J.L., Souto, C., Martinez-Pacheco, R., 2003. Powdered cellulose as excipient for extrusion-spheronization pellets of a cohesive hydrophobic drug. Eur J. Pharm. Biopharm. 55, 291–295.

Amidon, G.L., Lennernas, H., Shah, V.P., Crison, J.R., 1995. A theoretical basis for a biopharmaceutic drug classification: the correlation of in vitro drug product dissolution and in vivo bioavailability. Pharm. Res. 12, 413–420.

Amselem, S., Friedman, D., 1998. Submicron emulsions as drug carriers for topical administration. In: Benita, S. (Ed.), Submicron Emulsions in Drug Targeting and Delivery. ninth ed. Harwood Academic Publishers, London, pp. 153–173.

Antal, I., Kállai, N., Luhn, O., Bernard, J., Nagy, Z.K., Szabó, B., Klebovich, I., Zelkó, R., 2013. Supramolecular elucidation of the quality attributes of microcrystalline cellulose and isomalt composite pellet cores. J. Pharm. Biomed. Anal. 84, 124–128.

Anton, N., Vandamme, T.F., 2009. The universality of low-energy nano-emulsification. Int. J. Pharm. 377 (1–2), 142–147.

Anton, N., Gayet, P., Benoit, J.-P., Saulnier, P., 2007. Nanoemulsions and nanocapsules by the PIT method: an investigation on the role of the temperature cycling on the emulsion phase inversion. Int. J. Pharm. 344 (1–2), 44–52.

Anton, N., Benoit, J.P., Saulnier, P., 2008. Design and production of nanoparticles formulated from nano-emulsion templates—a review. J. Control. Release 128 (3), 185–199.

Aulton, M.E., Taylor, K.M.G., 2013. Aulton's pharmaceutics: the design and manufacture of medicines, fourth ed., Churchill Livingstone, London. (Chapter 28), p. 481.

Baert, L., Remon, J.P., 1993. Influence of amount of granulation liquid on the drug release rate from pellets made by extrusion spheronisation. Int. J. Pharm. 95, 135–141.

Baert, L., Remon, J.P., Knight, P., Newton, J.M., 1992. A comparison between the extrusion forces and sphere quality of a gravity feed extruder and a ram extruder. Int. J. Pharm. 86, 187–192.

Baert, L., Remon, J.P., Elbers, J.A.C., Van Bommel, E.M.G., 1993a. Comparison between a gravity feed extruder and a twin screw extruder. Int. J. Pharm. 99, 7–12.

Baert, L., Vermeersch, H., Remon, J.P., Smeyers- Verbeke, J., Massart, D.L., 1993b. Study of parameters important in the spheronization process. Int. J. Pharm. 96, 225–229.

Bashaiwoldu, A.B.B., Podczeck, F., Newton, M., 2004. A study on the effect of drying techniques on the mechanical properties of pellets and compacted pellets. Eur. J. Pharm. Sci. 21, 119–129.

Basit, A.W., Newton, J.M., Lacey, L.F., 1999. Formulation of ranitidine pellets by extrusion spheronization with little or no microcrystalline cellulose. Pharm. Dev. Tech. 4, 499–505.

Bataille, B., Ligarski, K., Jacob, M., Thomas, C., Duru, C., 1993. Study of the influence of spheronization and drying conditions on the physico-mechanical properties of neutral spheroids containing Avicel PH 101 and lactose. Drug Det. Ind. Pharm. 19, 653–671.

Bialleck, S., Rein, H., 2011. Preparation of starch-based pellets by hot-melt extrusion. Eur. J. Pharm. Biopharm. 79 (2), 440–448.

Bouchemal, K., Briançon, S., Perrier, E., Fessi, H., 2004. Nanoemulsion formulation using spontaneous emulsification: solvent, oil and surfactant optimisation. Int. J. Pharm. 280 (1–2), 241–251.

Calderó, G., García-Celma, M.J., Solans, C., 2011. Formation of polymeric nano-emulsions by a low-energymethod and their use for nanoparticle preparation. J. Colloid Interf. Sci. 353, 406–411.

Cao, Q.R., Liu, Y., Xu, W.J., Lee, B.J., Yang, M., Cui, J.H., 2012. Enhanced oral bioavailability of novel mucoadhesive pellets containing valsartan prepared by a dry powder-coating technique. Int. J. Pharm. 434 (1–2), 325–333.

Chandrapala, J., Oliver, C., Kentish, S., Ashokkumar, M., 2012. Ultrasonics in food processing. Ultrason. Sonochem. 19, 975–983.

Chatlapalli, R., Rohera, B.D., 1998. Physical characterization of HPMC and HEC and investigation of their use as pelletization aids. Int. J. Pharm. 161, 179–193.

Cheboyina, S., Wyandt, C.M., 2008a. Wax-based sustained release matrix pellets prepared by a novel freeze pelletization technique I: formulation and process variables affecting pellet characteristics. Int. J. Pharm. 359 (1–2), 158–166.

Cheboyina, S., Wyandt, C.M., 2008b. Wax-based sustained release matrix pellets prepared by a novel freeze pelletization technique II. In vitro drug release studies and release mechanisms. Int. J. Pharm. 359 (1–2), 167–173.

Cheboyina, S., Chambliss, W.G., Wyandt, C.M., 2004. A novel freeze pelletization technique for preparing matrix pellets. Pharm. Tech. 28, 98–108.

Chee, C.P., Djordjevic, D., Faraji, H., Decker, E.A., Hollender, R., McClements, D.J., Peterson, D.G., 2007. Sensory properties of vanilla and strawberry flavored ice cream supplemented with omega-3 fatty acids. Milchwissenschaft. 62, 66–69.

Chee, C.P., Gallaher, J.J., Djordjevic, D., Faraji, H., McClements, D.J., Decker, E.A., Hollender, R., Peterson, D.G., Roberts, R.F., Coupland, J.N., 2005. Chemical and sensory analysis of strawberry flavored yogurt supplemented with an algae oil emulsion. J. Dairy. Res. 72, 311–316.

Chen, H., Khemtong, C., Yang, X., et al., 2011. Nanonization strategies for poorly water-soluble drugs. Drug Discov. Today 16 (7/8), 354–360.

Chu, B.-S., Ichikawa, S., Kanafusa, S., Nakajima, M., 2007. Preparation of protein-stabilized β-carotene nanodispersions by emulsification–evaporation method. J. Amer. Oil Chem. Soc. 84 (11), 1053–1062.
Conine, J.W., Hadley, H.R., 1970. Preparation of small solid pharmaceutical spheres. Drug Cosmet. Ind. 106, 38–41.
Csobán, Z., Kállai-Szabó, B., Kállai-Szabó, N., Sebe, I., Gordon, P., Antal, I., 2015. Improvement of mechanical properties of pellet containing tablets by thermal treatment. Int. J. Pharm. 496 (2), 489–496.
Dahan, A., Miller, J.M., Amidon, G.L., 2009. Prediction of solubility and permeability class membership: provisional BCS classification of the world's top oral drugs. AAPS J. 11 (4), 740–746.
Dalgleish, D.G., West, S.J., Hallett, F.R., 1997. The characterization of small emulsion droplets made from milk proteins and triglyceride oil. Colloids Surf. A 123, 145–153.
Dukić, R., Mens, A., Adriaensens, P., Foreman, P., Gelan, J.P., Remon, J., Vervaet, C., 2009. Production of pellets via extrusion–spheronization without the incorporation of microcrystalline cellulose: a critical review. Eur. J. Pharm. Biopharm. (71), 38–46.
Dukić-Ott, A., Remon, J.P., Foreman, P., Vervaet, C., 2007. Immediate release of poorly soluble drugs from starch-based pellets prepared via extrusion/ spheronization. Eur. J. Pharm. Biopharm. 67 (3), 715–724.
El-Aasser, M.S., Sudol, E.D., 2004. Miniemulsions: overview of research and applications. J. Coat. Technol. Res. 1, 21–31.
Erber, M., Lee, G., 2015a. Cryopellets based on amorphous organic calcium salts: production, characterization and their usage in coagulation diagnostics. Powd. Technol. 280, 10–17.
Erber, M., Lee, G., 2015b. Development of cryopelletization and formulation measures to improve stability of *Echis carinatus* venum protein for use in diagnostic rotational thromboelastometry. Int. J. Pharm. 495 (2), 692–700.
Eriksson, M., Alderborn, G., Nyström, C., Podczeck, F., Newton, J.M., 1997. Comparison between and evaluation of some methods for the assessment of the sphericity of pellets. Int. J. Pharm. 148 (2), 149–154.
Erkoboni, D.F., 2003. Extrusion/spheronization. Ghebre-Sellassie, I., Martin, C. (Eds.), Pharmaceutical Extrusion Technology, Vol. 133, Marcel Dekker, New York, pp. 277–322.
Fielden, K.E., Newton, J.M., Rowe, R.C., 1992a. The influence of lactose particle size on spheronization of extrudate processed by a ram extruder. Int. J. Pharm. 81, 205–212.
Fielden, K.E., Newton, J.M., Rowe, R.C., 1992b. A comparison of the extrusion and spheronization behaviour of wet powder masses processed by a ram extruder and a cylinder extruder. Int. J. Pharm. 81, 225–233.
Fielden, K.E., Newton, J.M., Rowe, R.C., 1995. The influence of lactose particle size on spheronization of extrudate processed by a ram extruder. Int. J. Pharm. 81, 205–224, 1992.
Follonier, N., E. Doelker, E., Cole, E.T., 1994. Evaluation of hot-melt extrusion as a new technique for the production of polymer-based pellets for sustained-release capsules containing high loadings of freely soluble drugs. Drug Dev. Ind. Pharm. 20, 1323–1339.
Follonier, N., E. Doelker, E., Cole, E.T., 1995. Various ways of modulating the release of diltiazem hydrochloride from hot-melt extruded sustained-release pellets prepared using polymeric materials. J. Control. Release 36, 243–250.
Franceschinis, E., Bortoletto, C., Perissutti, B., Dal Zotto, M., Voinovich, D., Realdon, N., 2011. Self-emulsifying pellets in a lab-scale high shear mixer: formulation and production design. Powd. Technol. 207 (1–3), 113–118.
Gamlen, M.J., 1985. Pellet manufacture for controlled release. Manuf. Chem. 56, 55–59.
Gamsjäger, H., Lorimer, J.W., Scharlin, P., Shaw, D.G., 2008. Glossary of terms related to solubility. IUPAC-Pure Appl. Chem. 80, 233–276.
Gao, L., Zhang, D., Chen, M., 2008. Drug nanocrystals for the formulation of poorly soluble drugs and its application as a potential drug delivery system. J. Nano. Res. 10, 845–862.
Ghanam, D., Kleinebudde, P., 2011. Suitability of κ-carrageenan pellets for the formulation of multiparticulate tablets with modified release. Int. J. Pharm. 409 (1–2), 9–18.
Ghebre-Sellassie, I., 1989. Pellets: a general overview. In: Pharmaceutical Pelletization Technology, Marcel Dekker, Inc., New York, Basel, pp. 1–13.

Ghebre-Sellassie, I., Knoch, A., 2007. Pelletization Techniques. In: James, Swarbrick (Ed.), Encyclopedia of Pharmaceutical Technology. third ed. Informa Healthcare, New York, pp. 2651–2663.
Ghosh, V., Mukherjee, A., Chandrasekaran, N., 2013. Ultrasonic emulsification of food-grade nanoemulsion formulation and evaluation of its bactericidal activity. Ultrason. Sonochem. 20, 338–344.
Guraya, H.S., James, C., 2002. Deagglomeration of rice starch-protein aggregates by high pressure homogenization. Starch 54 (3–4), 108–116.
Hamdani, J., Moës, A.J., Amighi, K., 2002. Development and evaluation of prolonged release pellets obtained by the melt pelletization process. Int. J. Pharm. 245 (1–2), 167–177.
Hellén, L., Ritala, M., Yliruusi, J., Palmroos, P., Kristoffersson, E., 1992. Process variables of the radial screen extruder. Part II. Size and distribution of pellets. Pharm. Tech. Int. Biophys. 4, 50–57.
Hellén, L., Yliruusi, J., Merkku, P., Kristoffersson, E., 1993a. Process variables of instant granulator and spheroniser: I. Physical properties of granules, extrudate and pellets. Int. J. Pharm. 96, 197–204.
Hellén, L., Yliruusi, J., Kristoffersson, E., 1993b. Process variables of instant granulator and spheroniser: II. Size and size distributions of pellets. Int. J. Pharm. 96, 205–216.
Hosseini, A., Körber, M., Bodmeier, R., 2013. Direct compression of cushion-layered ethyl cellulose-coated extended release pellets into rapidly disintegrating tablets without changes in the release profile. Int. J. Pharm. 457, 503–509.
Iosio, T., Voinovich, D., Grassi, M., Pinto, J.F., Perissutti, B., Zacchigna, M., Quintavalle, U., Serdoz, F., 2008. Bi-layered self-emulsifying pellets prepared by co-extrusion and spheronization: influence of formulation variables and preliminary study on the in vivo absorption. Eur. J. Pharm. Biopharm. 69 (2), 686–697.
Izquierdo, P., Esquena, J., Tadros, T.F., Dederen, C., Garcia, M.J., Azemar, N., 2001. Formation and stability of nanoemulsions prepared using the phase inversion temperature method. Langmuir 18 (1), 26–30.
Izquierdo, P., Esquena, J., Tadros, T.F., Dederen, J.C., Feng, J., Garcia-Celma, M.J., 2004. Phase behavior and nanoemulsion formation by the phase inversion temperature method. Langmuir 20 (16), 6594–6598.
Jachowicz, R., Nürnberg, E., Pieszczek, B., Kluczykowska, B., Maciejewska, A., 2000. Solid dispersion of ketoprofen in pellets. Int. J. Pharm. 206 (1–2), 13–21.
Jafari, S.M., He, Y., Bhandari, B., 2006. Nano-emulsion production by sonication and microfluidization: a comparison. Int. J. Food Prop. 9, 475–485.
Jafari, S.M., He, Y., Bhandari, B., 2007. Optimization of nano- emulsions production by microfluidization. Eur. Food Res. Technol. 225, 733–741.
Junghanns, J.U.A.H., Müller, R.H., 2008. Nanocrystal technology, drug delivery and clinical appications. Int. J. Nanomed. 3 (3), 295–309.
Kleinebudde, P., 1994. Shrinking and swelling properties of pellets containing microcrystalline cellulose and low substituted hydroxypropyl cellulose: I. Shrinking properties. Int. J. Pharm. 109 (3), 209–219.
Knop, K., Kleinebudde, P., 2005. Pharmaceutical pellets, definition, properties, production. ExAct Exp. Act. Pharm. 15, 2–5.
Koester, M., Thommes, M., 2010. New insights into the pelletization mechanism by extrusion/spheronization. AAPS Pharm. Sci. Technol. 11, 1549–1551.
Koo, O.M.Y., Heng, P.W.S., 2001. The influence of microcrystalline cellulose grade on shape and shape distributions of pellets produced by extrusion-spheronization. Chem. Pharm. Bull. 49, 1383–1387.
Krämer, J., Blume, H., 1994. Biopharmaceutical aspects of multiparticulates. In: Swarbrick, J. (Ed.), Multiparticulate Oral Drug Delivery. first ed. Marcel Dekker, New York, pp. 307–332, 1994.
Kristensen, J., Schæfer, T., Kleinebudde, P., 2000a. Direct pelletization in a rotary processor controlled by torque measurements. I. Influence of process variables. Pharm. Dev. Technol. 5, 247–256.
Kristensen, J., Schæfer, T., Kleinebudde, P., 2000b. Direct pelletization in a rotary processor controlled by torque measurements. II: Effects of changes in the content of microcrystalline cellulose. AAPS PharmSci. 2, Article 24.
Kwon, S.S., Nam, Y.S., Lee, J.S., Ku, B.S., Han, S.H., Lee, J.Y., Chang, I.S., 2002. Preparation and characterization of coenzyme Q10-loaded PMMA nanoparticles by a new emulsification process based on microfluidization. Colloids Surf. A 210, 95–104.

Laouini, A., Fessi, H., Charcosset, C., 2012. Membrane emulsification: a promising alternative for vitamin E encapsulation within nano-emulsion. J. Membr. Sci. 423–424, 85–96.
Lei, Y., Lu, Y., Qi, J., Nie, S., Hu, F., Pan, W., Wu, W., 2011. Solid self-nanoemulsifying cyclosporin A pellets prepared by fluid-bed coating: preparation, characterization and in vitro redispersibility. Int. J. Nanomed. 6, 795–805.
Li, P., Chiang, B., 2012. Process optimization and stability of D-limonene-in-water nanoemulsions prepared by ultrasonic emulsification using response surface methodology. Ultrason. Sonochem. 19, 192–197.
Lieberman, Herbert A., Rankell, Albert, 1970. Drying. In: Leon, Lachman, Lieberman Herbert A., Kanig, Joshep L. (Eds.), The Theory and Practice of Industrial Pharmacy. Lea & Febiger, Philadelphia, pp. 22–48.
Liew, C.V., Gu, L., Soh, J.L., Heng, P.W., 2005. Functionality of cross-linked polyvinylpyrrolidone as a spheronization aid: a promising alternative to microcrystalline cellulose. Pharm. Res. 22 (8), 1387–1398.
Lindenberg, M., Kopp, S., Dressman, J.B., 2004. Classification of orally administered drugs on the World Health Organization model list of essential medicines according to the biopharmaceutics classification system. Eur. J. Pharm. Biopharm. 58 (2), 265–278.
Lipinski, C.A., 2002. Poor aqueous solubility: industry wide problem in drug discovery. Am. Pharm. Rev. 5, 82–85.
Lipinski, C.A., Lombardo, F., Dominy, B.W., Feeney, P.J., 2001. Experimental and computational approaches to estimate solubility and permeability in drug discovery and development settings. Adv. Drug Deliv. Rev. 46 (1–3), 3–26.
Lust, A., Lakio, S., Vintsevits, J., Kozlova, J., Veski, P., Heinämäki, J., Kogermann, K., 2013. Water-mediated solid-state transformation of a polymorphic drug during aqueous-based drug-layer coating of pellets. Int. J. Pharm. 456 (1), 41–48.
Magdalene, R., 2001. Pure drug nanoparticles for the formulation of poorly soluble drugs. New Drugs 3, 62–68.
Malinowski, H.J., Smith, W.E., 1975. Use of factorial design to evaluate granulations prepared by spheronization. J. Pharm. Sci 64, 1688–1692.
Mason, T., Wilking, J., Meleson, K., Chang, C., Graves, S., 2006. Nanoemulsions: formation, structure, and physical properties. J. Phys. Cond. Matt. 18, 635–666.
Mehta, S., De Beer, T., Remona, J.P., Vervaet, C., 2012. Effect of disintegrants on the properties of multiparticulate tablets comprising starch pellets and excipient granules. Int. J. Pharm. 422, 310–317.
Michie, H., Podczeck, F., Newton, J.M., 2012. The influence of plate design on the properties of pellets produced by extrusion and spheronization. Int. J. Pharm. 434 (1–2), 175–182.
Millili, G.P., Schwartz, B., 1990. The strength of microcrystalline cellulose pellets: the effect of granulating with water-ethanol mixtures. Drug Dev. Ind. Pharm. 16, 1411–1426.
Morales, D., Gutiérrez, J.M., García-Celma, M.J., Solans, Y.C., 2003. A study of the relation between bicontinuous micro- emulsions and oil/water nano-emulsion formation. Langmuir 19 (18), 7196–7200.
Möschwitzer, J., Müller, R.H., 2006. Spray coated pellets as carrier system for mucoadhesive drug nanocrystals. Eur. J. Pharm. Biopharm. 62 (3), 282–287.
Müller, R.H., Gohla, S., Keck, C.M., 2011. State of the art of nanocrystals e special features, production, nanotoxicology aspects and intracellular delivery. Eur. J. Pharm. Biopharm. 78, 1–9.
Nakajima, H., 1997. Microemulsions in cosmetics. In: Solans, C., Kunieda, H. (Eds.), Industrial Applications of Microemulsions. Marcel Dekker, New York, pp. 175–197.
Nakajima, H., Tomomasa, S., Okabe, M., 1993. Preparation of nanoemulsions: proceedings first world congress on emulsions, Paris, France, pp. 19–22.
Nastruzzi, C., Cortesi, R., Esposito, E., Genovesi, A., Spadoni, A., Vecchio, C., Menegatti, E., 2000. Influence of formulation and process parameters on pellet production by powder layering technique. AAPS PharmSciTech 1 (2), E9.
Newton, J.M., Petersson, J., Podczeck, F., Clarke, A., Booth, S., 2001. The influence of formulation variables on the properties of pellets containing a self-emulsifying mixture. J. Pharm. Sci. 90, 987–995.
Newton, J.M., Pinto, M.R., Podczeck, F., 2007. The preparation of pellets containing a surfactant or a mixture of mono- and di-gylcerides by extrusion/spheronization. Eur. J. Pharm. Sci. 30, 333–342.

Park, H.J., Lee, G.H., Jun, J.H., Son, M., Choi, Y.S., Choi, M.K., Kang, M.J., 2016. Formulation and in vivo evaluation of probiotics-encapsulated pellets with hydroxypropyl methylcellulose acetate succinate (HPMCAS). Carbohydr. Polym. 136, 692–699.

Parikh, Dilip M., 2009. Handbook of pharmaceutical granulation technology, drugs and the pharmaceutical sciences. Informa Health.

Pašić, M., G. Betz, G., Hadžidedić, Š., Kocova El-Arini, S., Leuenberger, H., 2010. Investigation and development of robust process for direct pelletization of lansoprazole in fluidized bed rotary processor using experimental design. J. Drug Deliv. Sci. Technol. 20 (5), 367–376.

Podczeck, F., 2008. A novel aid for the preparation of pellets by extrusion/spheronization. Pharm. Technol. Eur. 20, 26–31.

Podczeck, F., Newton, J.M., 2014. Influence of the standing time of the extrudate and speed of rotation of the spheroniser plate on the properties of pellets produced by extrusion and spheronization. Adv. Powd. Technol. 25, 659–665.

Podczeck, F., Knight, P.E., Newton, J.M., 2008. The evaluation of modified microcrystalline cellulose for the preparation of pellets with high drug loading by extrusion/spheronization. Int. J. Pharm. 350 (1–2), 145–154.

Pouton, C.W., 2006. Formulation of poorly water-soluble drugs for oral administration: physicochemical and physiological issues and the lipid formulation classification system. Eur. J. Pharm. Sci. 29 (3–4), 278–287.

Pouton, C.W., Porter, C.J.H., 2008. Formulation of lipid-based delivery systems for oral administration: materials, methods and strategies. Adv. Drug Deliv. Rev. 60, 625–637.

Priese, F., Wolf, B., 2013. Development of high drug loaded pellets by design of experiment and population balance model calculation. Powd. Technol. 241, 149–157.

Prieto, S.M., Mendez, J.B., Oreto Espinar, F.J., 2005. Starch-dextrin mixtures as base excipients for extrusion spheronization pellets. Eur. J. Pharm. Biopharm. 59, 511–521.

Quintanilla-Carvajal, M., Camacho-Díaz, B., Meraz-Torres, L., Chanona-Pérez, J., Alamilla-Beltrán, L., Jimenéz-Aparicio, A., 2010. Nanoencapsulation: a new trend in food engineering processing. Food Eng. Rev. 2 (1), 39–50.

Radl, S., Tritthart, T., Johannes, G., Khinast, J.G., 2010. A novel design for hot-melt extrusion pelletizers. Chem. Eng. Sci. 65 (6), 1976–1988.

Rao, J.J., McClements, D.J., 2010. Stabilization of phase inversion temperature nanoemulsions by surfactant displacement. J. Agric. Food Chem. 58, 7059–7066.

Reynolds, A.D., 1970. A new technique for the production of spherical particles. Manuf. Chem. Aer. News 44, 40–44.

Römer, M., Heinämäki, J., Miroshnyk, I., Sandler, N., Rantanen, J., Yliruusi, J., 2007. Phase transformations of erythromycin A dihydrate during pelletisation and drying. Eur. J. Pharm. Biopharm. 67, 246–252.

Rowe, R.C., 1985. Spheronization: a novel pill-making process? Pharm. Int. 6, 119–123.

Sadtler, V., Rondon-Gonzalez, M., Acrement, A., Choplin, L., Marie, E., 2010. PEO-covered nanoparticles by emulsion inversion point (eip) method. Macromol. Rapid Comm. 31 (11), 998–1002.

Sadurní, N., Solans, C., Azemar, N., García-Celma, M.J., 2005. Studies on the formation of O/W nano-emulsions, by low-energy emulsification methods, suitable for pharmaceutical applications. Eur. J. Pharm. Sci. 26 (5), 438–445.

Sanguansri, P., Augustin, M.A., 2006. Nanoscale materials development: a food industry perspective. Trends Food Sci. Technol. 17, 547–556.

Schæfer, T., 1996. Melt pelletization in a high shear mixer. X. Agglomeration of binary mixtures. Int. J. Pharm. 139, 149–159.

Schmidt, C., Kleinebudde, P., 1998. Comparison between a twin-screw extruder and a rotary ring die press. Part II: influence of process variables. Eur. J. Pharm. Biopharm. 45 (2), 173–179.

Schmidt, C., Lindner, H., Kleinebudde, P., 1997. Comparison between a twin-screw extruder and a rotary ring die press. Part I. Influence of formulation variables. Eur. J. Pharm. Biopharm. 44, 169–176.

Shah, R.D., Kabadi, M., Pope, D.G., Augsburger, L.L., 1995. Physico-mechanical characterization of the extrusion-spheronization process. Part II: rheological determinants for successful extrusion and spheronization. Pharm. Res. 12 (4), 496–507.

Shakeel, F., Baboota, S., Ahuja, A., Ali, J., Shafiq, S., 2008. Accelerated stability testing of celecoxib nanoemulsion containing cremophor-EL. Afr. J. Pharm. Pharmacol. 2, 179–183.

Shinoda, K., Saito, H., 1968. The effect of temperature on the phase equilibria and the types of dispersions of the ternary system composed of water, cyclohexane, and nonionic surfactant. J. Colloid Interf. Sci. 26 (1), 70–74.

Shinoda, K., Saito, H., 1969. The stability of o/w type emulsions as a function of temperature and the hlb of emulsifiers: the emulsification by pit-method. J. Colloid Interf. Sci. 30 (1), 258–263.

Sinchaipanid, N., Chitropas, P., Mitrevej, A., 2004. Influences of layering on theophylline pellet characteristics. Pharm. Dev. Technol. 9, 163–170.

Solans, C., Solé, I., 2012. Nano-emulsions: formation by low-energy methods. Curr. Opin. Colloid Interf. Sci. 17, 246–254.

Solans, C., Izquierdo, P., Nolla, J., Azemar, N., Garcia-Celma, M.J., 2005. Nanoemulsions. Curr. Opin. Colloid Interf. Sci. 10, 102–110.

Sonaglio, D., Bataille, B., Ortigosa, C., Jacob, M., 1995. Factorial design in the feasibility of producing Microcel MC 101 pellets by extrusion/spheronisation. Int. J. Pharm. 115, 53–60.

Sousa, J.J., Sousa, A., Podczeck, F., Newton, J.M., 1992. Influence of process conditions on drug release from pellets. Int. J. Pharm. 144, 159–169.

Sousa, J.J., Sousa, A., Podczeck, F., Newton, J.M., 2002. Factors influencing the physical characteristics of pellets obtained by extrusion-spheronization. Int. J. Pharm. 232, 91–106.

Souto, C., Rodriguez, A., Parajes, S., Martinez-Pacheco, R., 2005. A comparative study of the utility of two superdisintegrants in microcrystalline cellulose pellets prepared by extrusion-spheronization. Eur. J. Pharm. Biopharm. 61 (1–2), 94–99.

Suhrenbrock, L., Radtke, G., Knop, K., Kleinebudde, P., 2011. Suspension pellet layering using PVA-PEG graft copolymer as a new binder. Int. J. Pharm. 412 (1–2), 28–36.

Swientek, R.J., 1990. Microfluidizing technology enhances emulsion stability. Food Proc., 152–153.

Takagi, T., Ramachandran, C., Bermejo, M., Yamashita, S., Yu, L.X., Amidon, G.L., 2006. A provisional biopharmaceutical classification of the top 200 oral drug products in the United States, Great Britain, Spain, and Japan. Mol. Pharm. 3, 631–643.

Tamilvanan, S., Senthilkumar, R.S., Baskar, R., Sekharan, T.R., 2010. Manufacturing techniques and excipients used during the formulation of oil-in-water type nanosized emulsions for medical applications. J. Excip. Food Chem. (1), 11–29.

Tho, I., Sande, S.A., Kleinebudde, P., 2002. Pectinic acid, a novel excipient for production of pellets by extrusion/spheronization: preliminary studies. Eur. J. Pharm. Biopharm. 54, 95–99.

Thommes, M., Kleinebudde, P., 2006. Use of κ-carrageenan as alternative pelletization aid to microcrystalline cellulose in extrusion/spheronization. I. Influence of type and fraction of filler. Eur. J. Pharm. Biopharm. 63, 59–67.

Thompson, A.K., Singh, H., 2006. Preparation of liposomes from milk fat globule membrane phospholipids using a microfluidizer. J. Dairy Sci. 89, 410–419.

Tian, Z., Yi, Y., Yuan, H., Han, Y., Zhang, X., Xie, Y., Lu, Y., Qi, J., Wu, W., 2013. Solidification of nanostructured lipid carriers (NLCs) onto pellets by fluid-bed coating: preparation, in vitro characterization and bioavailability in dogs. Powd. Technol. 247, 120–127, 2013.

Trivedi, N.R., Rajan, M.G., Johnson, J.R., Shukla, A.J., 2007. Pharmaceutical approaches to preparing pelletized dosage forms using the extrusion–spheronisation process. Crit. Rev. Ther. Drug Carrier Syst. 24, 1–40.

Tuleu, C., Newton, M., Rose, J., Euler, D., Saklatvala, R., Clarke, A., Booth, S., 2004. Comparative bioavailability study in dogs of a self-emulsifying formulation of progesterone presented in a pellet and liquid form compared with an aqueous suspension of progesterone. J. Pharm. Sci. 93, 1495–1502.

Usón, N., Garcia, M.J., Solans, C., 2004. Formation of water-in- oil (W/O) nano-emulsions in a water/mixed non-ionic surfactant/oil systems prepared by a low-energy emulsification method. Colloids Surf. A 250 (1–3), 415–421.

Vaassen, J., Bartscher, K., Breitkreutz, J., 2012. Taste masked lipid pellets with enhanced release of hydrophobic active ingredient. Int. J. Pharm. (429), 99–103.

Vervaet, C., Baert, L., Risha, P.A., Remon, J.P., 1994. The influence of the extrusion screen on pellet quality using an instrumented basket extruder. Int. J. Pharm. 107, 29–39.

Vervaet, C., Baert, L., Remon, J.P., 1995. Extrusion-spheronisation: a literature review. Int. J. Pharm. 116, 131–146.

Vilhelmsen, T., Kristensen, J., Schaefer, T., 2004. Melt pelletization with polyethylene glycol in a rotary processor. Int. J. Pharm. 275 (1–2), 141–153.

Walstra, P., 1993. Principles of emulsion formation. Chem. Eng. Sci. 48 (2), 333–349.

Wang, X., Jiang, Y., Wang, Y., Huang, M., Ho, C.T., Huang, Q., 2008. Enhancing anti-inflammation activity of curcumin through O/W nanoemulsions. Food Chem. 108, 419–424.

Wang, Z., Sun, J., Wang, Y., Liu, X., Liu, Y., Fu, Q., Meng, P., He, Z., 2010. Solid self-emulsifying nitrendipine pellets: preparation and in vitro/in vivo evaluation. Int. J. Pharm. 383 (1–2), 1–6.

Wang, Z., Chen, B., Quan, G., Li, F., Wu, Q., Dian, L., Dong, Y., Li, G., Wu, C., 2012. Increasing the oral bioavailability of poorly water-soluble carbamazepine using immediate-release pellets supported on SBA-15 mesoporous silica. Int. J. Nanomed. 7, 5807–5818.

West, A.J., Rowe, R.C., 1988. Production spheronisers ten years on. Manuf. Chem. 59, 79–80.

Wlosnewski, J.C., Kumpugdee-Vollrath, M., Sriamornsak, P., 2010. Effect of drying technique and disintegrant on physical properties and drug release behaviour of microcrystalline cellulose-based pellets prepared by extrusion/spheronization. Chem. Eng. Res. Des. 88, 100–108.

Xia, H., Zhang, C., Wang, Q., 2001. Study on ultrasonic induced encapsulating emulsion polymerization in the presence of nanoparticles. J. Appl. Polym. Sci. 80, 1130–1139.

Xu, M., Liew, C.V., Heng, P.W.S., 2015. Evaluation of the coat quality of sustained release pellets by individual pellet dissolution methodology. Int. J. Pharm. 478, 318–327.

Yeung, C.W., Rein, H., 2015. Hot-melt extrusion of sugar-starch-pellets. Int. J. Pharm. 493 (1–2), 390–403.

Yin, L.-J., Chu, B.-S., Kobayashi, I., Nakajima, M., 2009. Performance of selected emulsifiers and their combinations in the preparation of β-carotene nanodispersions. Food Hydrocolloids. 23 (6), 1617–1622.

Young, C.R., John, J., Koleng, J.J., McGinity, J.W., 2002. Production of spherical pellets by a hot-melt extrusion and spheronization process. Int. J. Pharm. 242 (1–2), 87–92.

Zema, L., Palugan, L., Maroni, A., Foppoli, A., Sangalli, M.E., Gazzaniga, A., 2008. The use of ß-cyclodextrin in the manufacturing of disintegrating pellets with improved dissolution performances. AAPS PharmSciTech 9 (2), 708–717.

Zvonar, A., Berginc, K., Kristl, A., et al., 2010. Microencapsulation of self-microemulsifying system: Improving solubility and permeability of furosemide. Int. J. Pharm. 388, 151–158.

FURTHER READINGS

Möschwitzer, J., Müller, R.H., 2007. Drug nanocrystals: the universal formulation approach for poorly soluble drugs. In: Thassu, D., Deleers, M., Pathak, Y. (Eds.), Nanoparticulate Drug Delivery Systems. Informa Healthcare, New York, pp. 71–88.

CHAPTER

9 ORAL DRUG DELIVERY POTENTIAL OF DENDRIMERS

Anurag K. Singh, Ashok K. Sharma, Iliyas Khan, Avinash Gothwal, Lokesh Gupta, Umesh Gupta

Central University of Rajasthan, Ajmer, India

CHAPTER OUTLINE

1 **Introduction** 232
2 **Oral Route and its Implication** 232
3 **Nanocarriers in Oral Delivery** 234
- 3.1 Poly(Lactic-*co*-Glycolic Acid) Nanoparticles 234
- 3.2 Solid Lipid Nanoparticles 238
- 3.3 Hydrogel NPs (Nanogels) 238
- 3.4 Polymeric Micelles 238
- 3.5 Liposomes 239
- 3.6 Nanoemulsion and Self-Nanoemulsifying System 239
- 3.7 Nanocrystals 241
- 3.8 Dendrimers 241
- 3.9 Nanospheres 241
- 3.10 Carbon Nanotubes 242

4 **Dendrimers in Oral Drug Delivery** 242
5 **Synthetic Approaches of Dendrimer** 243
- 5.1 Divergent Approach 243
- 5.2 Convergent Approach 243
- 5.3 Miscellaneous Approaches 244

6 **Unique Properties of Dendrimers** 247
7 **Host Guest Interactions of Dendrimers and Bioactives** 248
8 **Drugs Delivered Through Oral Route Using Dendrimers as a Nanocarrier** 248
- 8.1 Propranolol 249
- 8.2 Triclosan 250
- 8.3 Nitric Oxide 251
- 8.4 Clotrimazole 251
- 8.5 Ketoconazole 251
- 8.6 Sulfamethoxazole 252
- 8.7 Silicone Urinary Catheters 252

9 **Conclusions and Future Prospects** 252
10 **Abbreviations** 255
References 255

1 INTRODUCTION

Among other routes, oral administration is the preferred route for delivery of bioactives. The benefits point of view is that it is easy for self-medication, painless, and patient-friendly. Generally, this route suppresses risk of disease transmission, increases patient compliance, and reduces cost that usually occurs with parenteral route. Oral routes of drug administration provide stretchy and controlled dosing regimen. It is mainly suitable for long-lasting therapy (Bromberg, 2008; Gaucher et al., 2010; Ruenraroengsak et al., 2010; Yamanaka and Leong, 2008). The human intestinal epithelium that is absorptive and is made up of villi that surge the total absorptive surface area (300–400 m^2) in the gastrointestinal tract (GIT) (Schenk and Mueller, 2008). Apart from these potential advantages, an oral formulation faces numerous common drawbacks, particularly in the case of peptides and proteins delivery: like low solubility and/or bioavailability, poor stability in the gastric environment, and the mucus barrier that can preclude drug penetration and successive absorption. Bioavailability of orally administered drugs is mainly influenced by their properties. Permeability and solubility are two important parameters for their absorption via passive diffusion. The biopharmaceutic classification system (BCS) describes four categories of drugs that basically founded on their solubility and their permeability (Shugarts and Benet, 2009). Orally administered drugs suffer through the chemical and enzymatic gastrointestinal (GI) liquids, cross the mucus layer before being absorbed. If utmost insignificant or small molecules are resistant to the harsh environment of the GITand can be absorbed, the intestinal barrier limits the oral absorption of macromolecules, such as nucleic acids, proteins, and vaccines. Hence, protective vehicles to circumvent destruction in the GIT and possibly enhance oral absorption are needed. To solve the aforementioned drawbacks that are associated with conventional delivery vehicles, nanoparticulate formulations are being established that may encapsulate and protect drugs and release them in a temporally or spatially controlled manner. It is easy to increase the surface area of nanoparticles (NPs) to enhance or reduce bioadhesion to target specific cells. The mucus layers that guard the epithelial surfaces have been underlined as major barriers to NP penetration (Cone, 2009; Lai et al., 2009; Suk et al., 2009, 2011; Wang et al., 2008a; Yang et al., 2011b). Nanotechnology holds some merits to the drug delivery area in general and oral drug delivery in particular. It allows (1) the targeting of drugs to specific part of the GIT, (2) the delivery of poorly water-soluble drugs, (3) the transcytosis of bioactives transversely to the tight intestinal barrier, and (4) the intracellular and transcellular delivery of large macromolecules (Farokhzad and Langer, 2009). A nanomedicine eventually increases the specificity, efficacy, tolerability, and therapeutic index of drugs (Hall et al., 2007). Ideally, nanocarriers must be stable, less toxic or nontoxic, nonthrombogenic, nonimmunogenic, noninflammatory, biodegradable, avoid uptake by reticulo-endothelial system, and should have applicability in the delivery of various molecules, such as small drugs, proteins, vaccines, or nucleic acids (des Rieux et al., 2006; Kumari et al., 2010). Among the overall nanocarriers dendrimers have provided an auspicious approach to obtain the required pharmacokinetic and biopharmaceutical properties for medicines.

2 ORAL ROUTE AND ITS IMPLICATION

Oral route is one of the most conventional/traditional method because of its suitability and patient acceptance especially for pediatriatics and geriatric patients (Date and Nagarsenker, 2007; Kim et al., 2013; Plapied et al., 2011; Shinkar et al., 2012). However, many biological therapeutics that have poor solubility, poor permeability, and/or poor stability in the GI environment have poor oral bioavailability due to formidable barriers posed by the GIT and are therefore rarely used for oral drug delivery (Ensign et al., 2012). Thus, understanding the characteristics and functions of these barriers is essential in designing an effective carrier for oral administration of drugs. The pH in the GIT varies greatly according to location, with a pH of 1.2 in the empty stomach, 5–7 in the small intestine and 6–7.5 in the colon (Gaucher et al., 2010). Intestinal epithelium is an efficient physical barrier that covers the surfaces of the GIT. This epithelial cell layer comprises absorptive cells (enterocytes), secretory cells (goblet cells and Paneth cells), and M cells, which are specialized epithelial cells that reside predominantly in the Peyer's patches (Fig. 9.1). Physical integrity of the intestinal epithelium can be attributed to the presence of tight

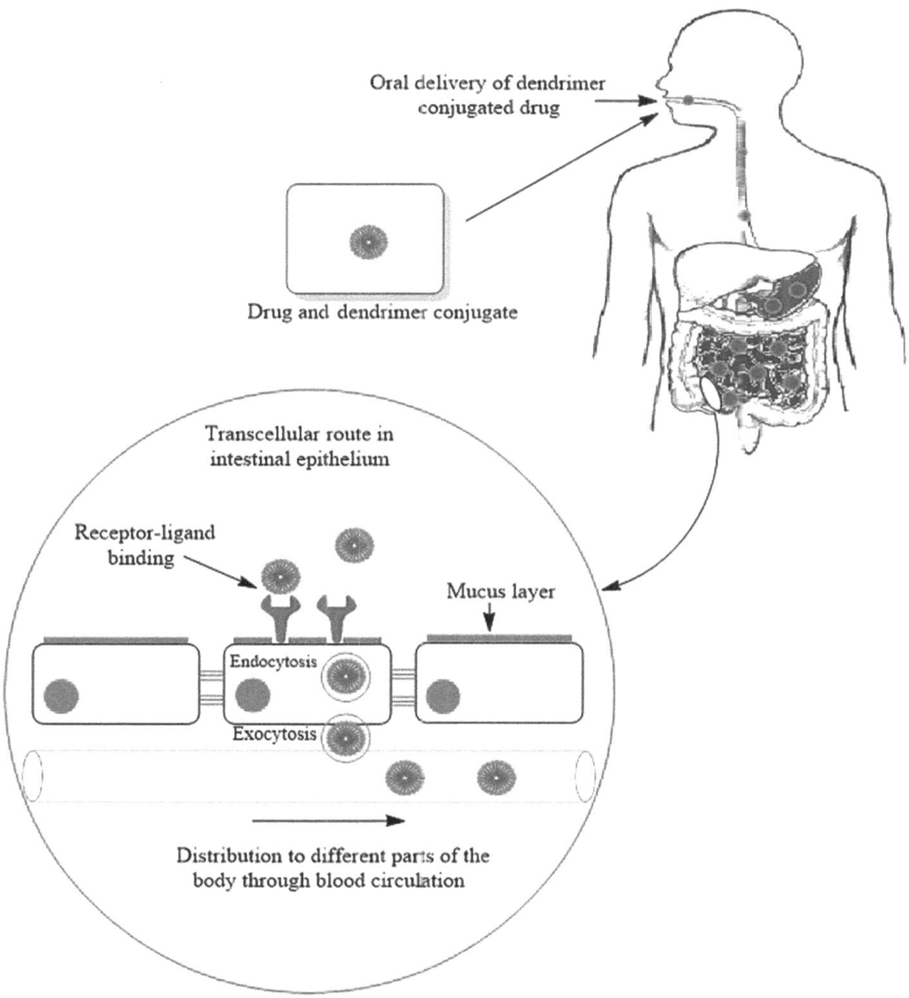

FIGURE 9.1 Schematics Diagrammatic Uptake Mechanism of Various Drug through Oral Route

Table 9.1 BCS Drug Classification

Class	Permeability	Solubility
Class I	High permeability	High solubility
Class II	High permeability	Low solubility
Class III	Low permeability	High solubility
Class IV	Low permeability	Low solubility

junctions between contiguous epithelial cells (Chen et al., 2011). As shown in Table 9.1, therapeutic agents may be classified in four classes BCS, according to their permeability and solubility. These drugs that are administered orally must survive transit through the chemical and enzymatic GI liquids, across the epithelium and mucus layer before being absorbed. These drugs can be taken up by interests or M cells of Peyer's patches via the transcellular pathway or between adjacent cells (paracellular pathway; Fig. 9.2) (Desai et al., 2012). Various factors that limit drug absorption through the GI mucosa contribute to restricting the bioavailability of the drug (Mrsny, 2012). Among these factors, poor drug permeability through the GI mucosa and/or low aqueous solubility is/are of central importance. For drugs belonging to biopharmaceutical classification system II or IV, poor aqueous solubility and intrinsic dissolution rate are the major factors leading to lower bioavailability. Poor GI permeability is another factor that significantly affects oral bioavailability of many drugs, such as drugs belong to BCS Class III.

3 NANOCARRIERS IN ORAL DELIVERY

NPs are colloidal particles, in which a drug is dissolved, entrapped, encapsulated, or attached to the NPs matrix. There are many methods to prepare NPs and the choice of methods depends on the nature of the polymers as well as on the drug to be encapsulated. NPs have been extensively applied to oral delivery of drugs. Because they can protect the drug against enzymatic and hydrolytic degradation in the GIT, they prolong the residence time in the gut by mucoadhesion and notably improve the absorption, bioavailability of drugs (Chen et al., 2011; Khalil et al., 2013; Sarparanta et al., 2012). Increased bioavailability has been achieved following oral administration of these drug-entrapped NPs. For example, in the case of the anticancer drugs that are delivered through nanocarriers, the in vivo bioavailability of the notoriously insoluble drug paclitaxel was reported to be increased by 10 times in comparison to orally administered Taxol (Zhao and Feng, 2010). Polymer materials have the best combination of characteristics: they are stable and allow high loading of many agents, they provide control over drug release kinetics, they can be readily modified to display a variety of surface-attached ligands, and many polymers have a long history of safe use in humans (Patel et al., 2012). A variety of biodegradable and nonbiodegradable polymeric nanocarriers have been employed for oral drug delivery (Table 9.2).

3.1 POLY(LACTIC-*CO*-GLYCOLIC ACID) NANOPARTICLES

Particles consisting of poly(lactic-*co*-glycolic acid) (PLGA) have been extensively studied as therapeutic delivery vehicles for many bioactives owing to their biodegradability and biocompatibility. PLGA microparticles have received approval from the Food and Drug Administration (FDA)

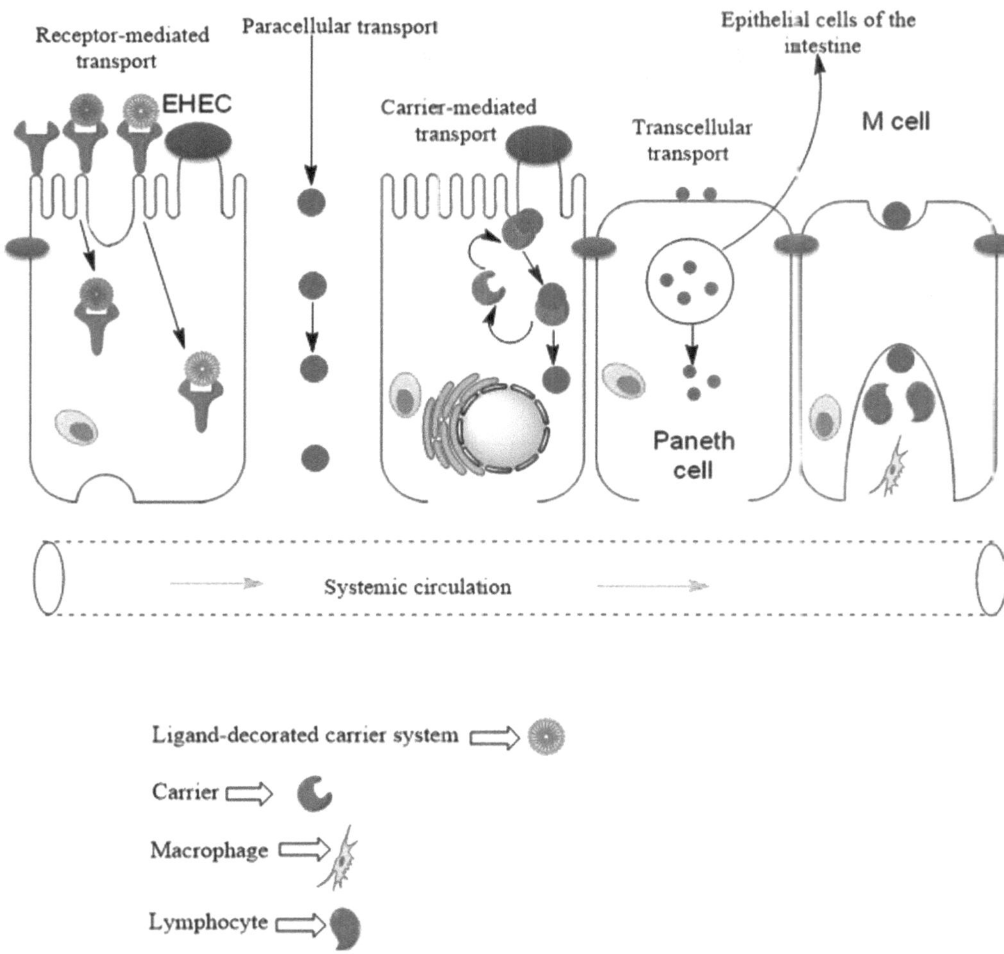

FIGURE 9.2 Drug Uptake Mechanisms in the Intestinal Epithelium Via Paracellular Pathway

for establishing sustained release of leuprolide (Lupron Depot) and triptorelin (Trelstar) (Xie and Smith, 2010). PLGA NPs have shown the ability to improve oral bioavailability of drugs with low solubility such as paclitaxel and curcumin (Cur) (Khalil et al., 2013; Roger et al., 2012). Xie et al. (2011) prepared Cur-loaded PLGA NPs (Cur-PLGA NPs) by a solid-in-oil-in-water (S/O/W) solvent evaporation method. The entrapment efficiency and loading rate of Cur were 91.96% and 5.75%, respectively. Cur-PLGA NPs showed ~640-fold in water solubility relative to that of native Cur. After oral administration of Cur-PLGA NPs, the relative bioavailability was 5.6-fold and had a longer half-life compared with that of native Cur. These results showed that the effect in improving oral bioavailability of Cur may be associated with improved water solubility. PLGA NPs can also offer sustained in vivo release of the drug after oral administration, for example, cyclosporine-loaded PLGA NPs showed sustained release of the drug in the plasma with concomitant decrease in

Table 9.2 Types of Nanocarriers Conjugated With Drug and Their Effect on Pharmacological Property

Nanocarriers	Drug Used	Effect on Pharmacological Property	Size	Composition	References
Polymeric nanoparticles	Curcumin	Curcumin-PLGA NPs showed ~640-fold in water solubility relative to that of native Cur. After oral administration of Cur-PLGA NPs, the relative bioavailability was 5.6-fold and had a longer half-life compared with that of native Cur.	—	PLGA	Roger et al. (2012)
SLN	Puerarin	The relative bioavailability of puerarin improved (43-fold) when incorporated into the SLNs. These data collectively support that SLNs are a promising delivery system for the enhancement of oral absorption of puerarin, a poorly water-soluble drug.	—	—	Luo et al. (2011)
Hydrogel NPs (nanogels) for oral drug delivery	Silymarin	The resulting alginate-PLGA nanogel were found to be new biodegradable carriers that could enhance the dissolution and oral bioavailability of silymarin and at the same time conferred sustained oral release of the drug.	10–100 nm	PLGA	El-Sherbiny et al. (2011)
Polymeric micelles	Paclitaxel	The formulation was administered intravenously or orally to rats. The oral bioavailability was estimated at 12.4%.	85–90 nm	PEG-*b*-P(VBODENA-*co*-AA)y	Kim et al. (2008); Lee et al. (2007)

Liposomes in oral drug delivery	Somatostatin analog octreotide	According to their study, oral administration of these liposomes elicited mild and prolonged hypoglycemic effect in parallel with increase in blood insulin levels for a duration of ~20 h, peaking at ~8–12 h.	—	Tetraether lipids (TELs) derived from archaea bacteria	Parmentier et al. (2011)
Nanoemulsion and self-nanoemulsifying drug delivery system for oral drug delivery	Curcumin	The preparation increased the solubility of Cur in water for 46,000,000-fold.	<100 nm	A folate-modified SNEDDS (FSNEDDS)	Kohli et al. (2010)
Nanocrystals	Coenzyme Q10(CoQ10)	CoQ10 dissolution was enhanced by the preparation of nanocrystals and the in vivo study in rats demonstrated that the oral bioavailability of nanocrystals was ~2.5-fold higher than that of coarse suspensions.	<500 nm	—	Piao et al. (2011)
Dendrimers	Short hairpin RNA	These studies suggest that RNAi-mediated hTERT gene silencing, coupled with dendrimer delivery, could provide a promising approach for the treatment of oral cancer in which hTERT is abundantly expressed.	107–315 nm	PAMAM	Liu et al. (2011)
Nanospheres	Cyclosporine-A	It was found that the formulation of Cyclosporine-A into Eudragit–S100 nanospheres improved its bioavailability compared to the Neoral–microemulsion system.	100–200 nm	Eudragit-based nanospheres	Wang et al. (2008b)
CNTs	Cyclo-(D-tryptophan-tyrosine) peptide	Which can deliver gene to duodenum, stomach, liver and kidney through oral administration.	—	Carbon cylinders composed of benzene rings	Hsieh et al. (2012)

the cyclosporine-associated nephrotoxicity, whereas commercial formulation Sandimmune Neoral failed to achieve both (Italia et al., 2007).

3.2 SOLID LIPID NANOPARTICLES

As they are derived from physiologically compatible lipids, solid lipid nanoparticles (SLNs) represent a safe and effective alternative, which include additional advantages and are devoid of the potential toxicities of conventional polymeric NPs (Chakraborty et al., 2010). Stability of SLNs is increased compared to other lipid nanocarriers and they can withstand the conditions in the upper GIT. Therefore, SLNs have been extensively studied as delivery systems to improve oral absorption and bioavailability of poorly soluble drugs (bufalin, risperidone, and puerarin) and hydrophilic drug such as insulin (Liu et al., 2011; Luo et al., 2011; Sarmento et al., 2007; Silva et al., 2012). Research data indicates that the oral bioavailability of poorly hydrophilic drugs can be enhanced when they are incorporated into SLNs. Recently, puerarin-loaded SLNs (Pue-SLNs) were prepared by Luo et al. (2011). Orally administered Pue-SLNs were rapidly absorbed, as evidenced with a shorter T_{max} for Pue-SLNs than for Pue suspension. The relative bioavailability of Pue improved (>threefold) when incorporated into the SLNs. These data collectively support that SLNs are a promising delivery system for the enhancement of oral absorption of puerarin, a poorly water-soluble drug.

3.3 HYDROGEL NPS (NANOGELS)

As a family of nanoscale particulate materials, hydrogel NPs (nanogels) have been showing its benefits in oral drug delivery. Interestingly, hydrogel nanoparticulate materials demonstrate the simultaneous characteristics of hydrogels and NPs, such as hydrophilicity, flexibility, versatility, and high water absorptivity (Bardajee et al., 2011; Kim et al., 2011). Nanogels with a three-dimensional network structure and a diameter of 10–100 nm have some smart functional properties that are pH-sensitive and/or temperature-sensitive, and so on (Kurnia et al., 2012; Murdan 2003; Wenceslau et al., 2012; Zhang et al., 2009). As a representative for oral drug delivery of nanogels, El-Sherbiny et al. (2011) prepared a biodegradable pH-responsive alginate-PLGA nanogel (silymarin-loaded PLGA nanogels) for oral delivery of silymarin, a hydrophobic natural therapeutic drug. Silymarin-loaded PLGA NPs were prepared using the "single-emulsion solvent evaporation" method through a procedure similar to that described by Cartiera et al. (2009). The suspensions of these silymarin-loaded PLGA NPs in aqueous sodium alginate solutions were then employed to produce pH-sensitive hydrogel microparticles via the ionotropic crosslinking in mild aqueous medium. The resulting alginate-PLGA nanogel were found to be a new biodegradable carriers that could enhance the dissolution and oral bioavailability of silymarin and at the same time conferred sustained oral release of the drug.

3.4 POLYMERIC MICELLES

Surfactants and amphiphilic polymers can self-assemble above the critical micellar concentration (CMC) in colloidal dispersions of molecular aggregates of ~20–100 nm known as micelles. Polymeric micelles present an abundant potential as drug delivery systems for compounds that are hydrophobic and exhibit poor bioavailability. The unique core–shell structure of micelles enables incorporation of poorly soluble drugs into the inner hydrophobic core, thus enhancing their stability and bioavailability (Coimbra et al., 2012). Micelles have been extensively investigated as a promising drug carrier by intravenous administration, primarily because they can reach the lesion sites by avoiding biological barriers in the human body such as limited GI absorption and high hepatic first-pass effect (Huang et al., 2012). In recent years, polymeric micelles have also been studied as an oral drug delivery system as a means to improve the oral delivery of hydrophobic compounds (Kim et al., 2008). However, most micelles were either studied on cell models or not tested thoroughly enough to prove their efficacy. In vivo data pertaining to orally administered micelles are scarce. The most interesting results were obtained with the polymeric micelles as carriers for paclitaxel. The formulation was administered intravenously or orally to rats. The oral bioavailability was estimated at 12.4% (Lee et al., 2007). Another study has reported an oral bioavailability of 6.5% for Taxol (Cremophor EL micelles) (Peltier et al., 2006). The use of polymeric micelles for oral drug delivery is still fairly new, and few systems have been extensively investigated.

3.5 LIPOSOMES

Liposomes are microscopic vesicles composed of membrane-like lipid bilayer vesicles that have been studied extensively as potential drug carriers (Sudimack et al., 2002). They can be made of natural phospholipids and cholesterol in the form of one or multiple concentric bilayers capable of encapsulating hydrophilic and hydrophobic drugs. Liposomal delivery systems have been adopted as a promising approach to improve the unfavorable physicochemical properties of drugs. The system can significantly improve the membrane permeability and cellular absorption of highly hydrophilic drugs, such as acyclovir as well as peptide and protein drugs, which were degraded in the GIT and were poorly absorbed due to the high molecular weight and hydrophilicity (Law et al., 2000; Li et al., 2012). Liposomes have many advantages, including the similarity with biological membrane, entrapping ability of structurally diverse drugs, and low toxicity. However, liposomal drug-delivery system suffers from the instability of lipid vesicles in intestines, including chemical and enzymatic destabilization, which limits their application and development as an oral drug delivery carrier. Polymer-coated and modified liposomes have been actively adopted to overcome the instability of liposomal carriers and to improve the stability and efficiency of liposomal delivery systems, as well as to extend the residence time in intestines (Takeuchi et al., 2005).

3.6 NANOEMULSION AND SELF-NANOEMULSIFYING SYSTEM

Recently, much attention has been focused on nanoemulsions and self-nanoemulsifying drug delivery system (SNEDDS) to improve the oral bioavailability of poorly water-soluble drugs. These systems improve the oral bioavailability of poorly water-soluble drugs by different mechanisms, including improved drug solubilization and protection against physicochemical and enzymatic degradation. In addition, the small size of these droplets increases the interphase between the lipophilic droplet and the aqueous medium of the gut facilitating a homogeneous and wide distribution of the drug along the GIT. However, these systems usually require a relatively large amount of surfactants, which can be of interest to modify the permeability of the membrane and they may induce GI side effects at the same time.

3.6.1 Nanoemulsion for oral drug delivery

Nanoemulsions (also called as microemulsions in many publications) are thermodynamically stable systems with the droplet size usually <100 nm. Nanoemulsions have been widely studied to increase the bioavailability of water-insoluble drugs (Rao and Shao, 2008). Silymarin was solubilized into oil-in-water (O/W)-based nanocarrier to increase its oral bioavailability (Parveen et al., 2011). These systems result in improved solubility, dissolution, and the area under the curve, *C*max of nanoemulsion after oral administration were fourfold and sixfold higher than those of drug suspension of silymarin. This study showed that the nanoemulsion being a versatile technology has the potential to improve the biopharmaceutical properties of silymarin. Surfactants and cosurfactants were grouped in two different combinations to construct the pseudoternary phase diagrams by Bali et al. (2010). A nanoemulsion formulation of ezetimibe containing Capryol 90 (10%, v/v), Tween-20 (33.33%, v/v), PEG400 (16.67%, v/v) and double-distilled water (40%, v/v) were successfully optimized. The in vivo studies revealed a significantly greater extent of absorption than the conventional tablet (Ezedoc10, ezetimibe 10 mg) formulation. Cremophor-free oral microemulsion of paclitaxel was developed to enhance its permeability and oral absorption. The developed microemulsion systems increased both the permeability and area under the curve of paclitaxel as compared to cremophor EL (Nornoo et al., 2009).

3.6.2 Self-nanoemulsifying drug-delivery system

Emulsion systems are linked with their own complications, including stability and manufacturing associated with their commercial production. Self-emulsification drug delivery system (SEDDS) is one formulation technique that can solve such problems. "SEDDS" is a wide term, archetypally producing emulsions with a droplet size ranging from a few nanometers to several microns. "Self-microemulsifying drug delivery systems" (SMEDDS) specifies the formulations forming transparent microemulsions with oil droplets ranging between 100 and 250 nm. "Self-nanoemulsifying drug delivery systems" (SNEDDS) is a recent term construing the globule size range <100 nm (Kohli et al., 2010). At present, there are four drug products, Sandimmune and Sandimmun Neoral (cyclosporine A), Norvir (ritonavir), and Fortovase (saquinavir) on the pharmaceutical market, the active compounds of which have been formulated into specific SEDDS (Gursoy and Benita, 2004). Although there are several SEDDS formulations approved by the FDA, predicting in vivo performances of SNEDDS formulations from in vitro data are still in its infancy. The performances of SNEDDS formulations are different in different types of in vitro studies. Because of SNEDDS's multiple absorption mechanisms, such as passive diffusion, active transport, paracellular, endocytosis, and P-glycoprotein activity, the single-mechanism in vitro studies (such as dissolution or permeability) might not be appropriate for evaluating the in vivo–in vitro correlation of SNEDDS formulations. The present study demonstrates that care should be taken when prioritizing which SNEDDS to use in vivo. Recently, much attention is given for SNEDDS to improve the solubility, dissolution, and oral absorption for poorly water-soluble drugs, such as apigenin, gemfibrozil, and coenzyme Q10 (CoQ10) and zedoary essential oil (Nepal et al., 2010; Villar et al., 2012; Zhao et al., 2010, 2013). SNEDDS is thermodynamically and isotropic stable solution consisting of oil, surfactant, cosurfactant, and drug, which can spontaneously produce O/W nanoemulsion when mixed with water under mild stirring (Hong et al., 2006; Shen et al., 2010). The nanoemulsion represents the drug in a dissolved form, and the small droplet size (5100 nm) provides a large interfacial surface area favoring drug absorption (Cui et al., 2009; Liu et al., 2009). Tamoxifen citrate is a highly lipophilic drug, which is an antiestrogen for peroral breast cancer treatment. Tamoxifen citrate-loaded SNEDDS comprised of tamoxifen citrate (1.6%), maisine 35–1 (16.4%), caproyl 90 (32.8%), cremophor RH40 (32.8%), and propylene glycol (16.4%) was prepared to improve oral bioavailability (Elnaggar et al., 2009). In vitro release in phosphate buffer (pH 7.4) revealed a gradual release pattern with 100% release at 6 h. SNEDDS of tamoxifen citrate exhibited a significant increase in release rate compared to the drug suspension under the identical conditions. These results proposed that the prepared system could be promising to improve oral efficacy of the tamoxifen citrate.

3.7 NANOCRYSTALS

Drug nanocrystals are essentially nanoscopic crystals of the parent compound. By definition, the dimensions of nanocrystals are <1 μm but for practical purposes they are often ~500 nm in dimensions. Preparation of nanocrystals takes place either via top-down or bottom-up techniques. In contrast to the previously discussed carriers, nanocrystals could attain complete drug loading (Mauludin et al., 2009). There is no carrier material and it may be formulated as suspensions or more frequently as dry dosage forms (tablets and capsules) (Kesisoglou et al., 2007). A range of oral formulations are available for use in the clinic and include Rapamune(rapamycin), Emend (aprepitant), TriCor (fenofibrate), Megace ES (Megestrol), and Triglide (fenofibrate). Panzem nanocrystal colloidal dispersion in combination with fixed-dose temozolomide to patients with recurrent glioblastoma multiforme is investigated in Phase II. A stated advantage of nanocrystal systems is that they can be formulated without the need of problem surfactants such as cremophor EL. Nanocrystals achieve therapeutically relevant bioavailability by increased rates of dissolution and saturation velocity as well as improving adhesiveness to cell membranes. Piao et al. (2011) prepared CoQ10-loaded nanocrystals. The dissolution profiles of CoQ10 in nanocrystals and its oral absorption in rats compared with a coarse suspension of CoQ10 were investigated. CoQ10 dissolution was enhanced by the preparation of nanocrystals and the in vivo study in rats demonstrated that the oral bioavailability of nanocrystals was ~2.5-fold higher than that of coarse suspensions. Nowadays, it is clear that drug nanocrystals used for oral administration can only lead to better bioavailability when the compounds show dissolution rate limited bioavailability. However, this approach can add tremendous dosing flexibility for other administration routes especially when highly concentrated formulations are needed.

3.8 DENDRIMERS

Dendrimers are highly complex molecules with a core, branches, and end groups. Dendrimers become progressively larger with the addition of each generation (shell) of branches and can be built from the molecular level to nanoscale size (1–10 nm) (D'Emanuele et al., 2004; Kateb et al. 2011). Drug molecules can be attached to functional groups on the dendrimer surface or shielded in the dendritic channels of the interior environment of the sphere. Poly(amidoamine) (PAMAM) dendrimers and its potential use in oral drug delivery had been evaluated by Sadekar and Ghandehari (2012). It has been proved that PAMAM dendrimers were transported by a combination transcellular and paracellular routes. Depending on surface chemistry, PAMAM dendrimers can open the tight junctions of epithelial barriers. Ke et al. (2008) developed a complex of drug and PAMAM for oral administration and demonstrated that a significant increase in oral bioavailability (200-fold) over that free drug. However, studies have followed on the extent of drug (and not dendrimer) absorption, and most of them have only been conducted in vitro. No direct evidence of dendrimer permeability after oral administration was reported (Kaminskas et al., 2011).

3.9 NANOSPHERES

These NPs are spherical structures composed of a matrix system in which drug is distributed by entrapment, attachment, or encapsulation. Nanospheres are typically larger than micelles having diameters between 100 and 200 nm, which can be used for oral applications (Zhang et al., 2011). Recently, the oral delivery of cyclosporine A by using Eudragit based nanospheres was examined. It was found that the formulation of cyclosporine A into Eudragit S100 nanospheres improved its bioavailability compared to the neoral microemulsion system (Wang et al., 2008a). The increase in absorption was attributed to a shorter transit time in the stomach, greater bioadhesiveness and better protection against degradation.

3.10 CARBON NANOTUBES

Carbon nanotubes (CNTs) are carbon cylinders composed of benzene rings and they have attracted a tremendous amount of attention regarding their use in biomedical applications. Application of CNTs for the oral delivery of drugs has attracted great interest. Hsieh et al. (2012) prepared cyclo-(D-tryptophan-tyrosine) peptide nanotubes, which can deliver gene to duodenum, stomach, liver, and kidney through oral administration.

4 DENDRIMERS IN ORAL DRUG DELIVERY

Oral administration of chemotherapeutics has gained the advantages in terms of patient preference, cost-effectiveness, ease of administration, and improving quality of life in palliative care (Aisner, 1992; DeMario and Ratain, 1998). Conclusively, in terms of treatment and its outcome, oral chemotherapeutics are advantageous for protracted dosage regimens as is the case for schedule-dependent cytotoxic drugs (DeMario and Ratain, 1998; Terwogt et al., 1999).

Oral drug delivery is the preferred route for polymeric drug carriers and has been used for the administration of many therapeutically active agents. However, the oral bioavailability of polymeric drug carriers is frequently limited by their large size range and molecular weight. Therefore, the difference in the mechanism and net permeability of polymeric drug carriers has been ascribed to the difference in their structural features such as molecular weight, molecular geometry, hydrodynamic volume, and charge (Tajarobi et al., 2001). Dendrimers as drug delivery carriers are of pronounced interest due to their highly controllable structure as well size and the terminal functional groups of dendrimers show higher chemical reactivity compared with other polymers (Frechet, 1994). The multifuctionality property of dendrimers leads to conjugation to various biologically active molecules. In this affection, PAMAM dendrimers have been broadly explored for oral drug delivery (El-Sayed et al., 2002; Kitchens et al., 2006; Kolhatkar et al., 2008; Najlah et al., 2007; Vijayalakshmi et al., 2010), because they belong to water-soluble polymers that are characterized by unique tree-like branching architecture and a compact spherical shape in solution (Tomalia, 1994). Its potential use as a drug carrier arises from the large number of arms and surface amine groups that can be utilized to immobilize enzymes, drugs, antibodies, or other bioactive agents (Peppas, 1995; Wilbur et al., 1998). PAMAM dendrimers have also been widely investigated in vitro for potential oral drug delivery applications as intestinal penetration enhancers as well as vehicle for transepithelial transport of small molecules (Sadekar and Ghandehari, 2012). Cationic, amine terminated as well as anionic, carboxylic acid-terminated PAMAM dendrimers are known to modulate tight junctions and increase paracellular transport of small molecules (El-Sayed et al., 2002, 2003; Kitchens et al., 2006). These are also known to be translocated transversely the intestinal epithelial barrier by endocytic mechanisms (Goldberg et al., 2010, 2011; Kitchens et al., 2007, 2008). Along with the cationic charge of PAMAM dendrimers are liable to interact with lipid bilayers that are present on cell membranes then increase their cellular uptake (Lin et al., 2010). Solubilizing potential of dendrimer pooled with their transepithelial transport make PAMAM dendrimers attractive for oral delivery of BCS Class IV drugs, such as camptothecin and its derivatives: 7-ethyl-10-hydroxy-camptothecin (SN38). PAMAM-SN38 complexes and conjugates have previously been evaluated for their transepithelial transport in vitro across Caco-2 monolayers (Goldberg et al., 2011; Thiagarajan et al., 2010; Vijayalakshmi et al., 2010). The permeability of SN38 was seen to increase when associated with PAMAM, which was attributed to increased solubility, decreased toxicity, saturable transcellular process, such as endocytic uptake and concentration-driven paracellular diffusion, such as tight junction modulation. These studies did not evaluate the in vivo oral absorption of SN38 when complexed or conjugated to the PAMAM dendrimer. There have been very limited studies in vivo to evaluate PAMAM dendrimers for oral drug delivery (Huang et al., 2011; Ke et al., 2008). In this study, camptothecin was formulated and codelivered with cationic, amine-terminated PAMAM dendrimer generation 4 (G4.0) and anionic, carboxylateterminated generation 3.5 (G3.5) in CD-1 mice. Wiwattanapatapee et al. (2000) evaluated the effect of size, concentration, and charge of PAMAM dendrimer on uptake and transport across the adult rat intestine in vitro using the everted rat intestinal sac system. Researchers studied the cationic PAMAM dendrimers (generations 3 and 4) and anionic PAMAM dendrimers (generations 2.5, 3.5, and 5.5). Mullin (2003) showed that I-labeled dendrimers proposed that transport across the intestinal membrane was charge-dependent, and they demonstrated that cationic dendrimers tissue uptake was higher than serosal transport. Finally, 2.5G and 3.5G dendrimers were assumed to be transported through third-phase endocytosis. In contrast, 5.5G and cationic dendrimers were observed to be taken up by specific or nonspecific adsorptive endocytosis.

5 SYNTHETIC APPROACHES OF DENDRIMER

Basically, dendrimers are synthesized using central polyfunctional cores through repetitive addition of branching units or monomers to every functional group, which results in the next generation of dendrimer and increases in end groups for further reaction. Tomalia and coworkers, developed two main synthetic approaches for the dendrimers, namely divergent and convergent approaches.

5.1 DIVERGENT APPROACH

In divergent approach there is two steps; first is the activation of functional surface groups, and second is the addition of branching units (Fig. 9.3). It includes development of dendrimers begins from a core, which reacts with multiple moles of reagent having minimum two protecting branching positions, abide by elimination of protecting groups, results in first-generation dendrimers. The procedure is repetitive until the outcome of the required size/generation of dendrimer. Starbust or polyamidoamines (PAMAMs) dendrimers were synthesized the first time by using this technique (Kandpal and Saxena, 2015; Patri et al., 2002; Svenson and Tomalia, 2005).

5.2 CONVERGENT APPROACH

An alternative approach for dendrimer synthesis proposed by Hawker and Frechet (1990). Only one type of functional group of outermost generations is the main limitation of the divergent approach. The convergent method is also included in two steps, one is repetitive coupling of protected or deprotected branch to create a central point functioned dendron; and next is divergent interior fixing steps, which

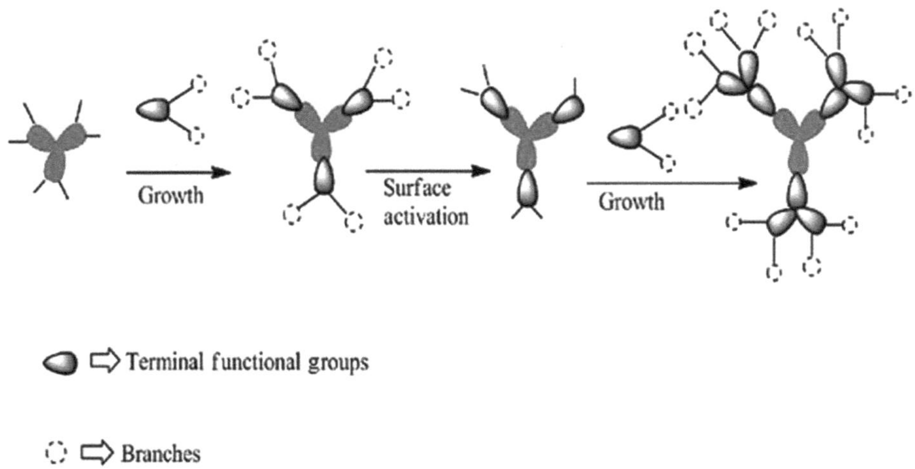

FIGURE 9.3 Divergent Synthetic Approach for Dendrimer

results several multidendron dendrimers (Fig. 9.4). This synthetic approach provides precise molecular weight and production of dendrimers with exact functionalities (Hawker and Frechet, 1990). This approach holds the unique advantages like a somewhat simple purification of needed product, incidence of shortcoming in concluding structure is reduced (Khoee and Hemati, 2013).

5.3 MISCELLANEOUS APPROACHES

Three new approaches used for the synthesis of dendrimer, which are described in the following sections.

5.3.1 Lego chemistry approach

"Lego" chemistry uses highly functionalized cores and branched monomers to generate phosphorus-based dendrimers (Svenson and Tomalia, 2005). A number of variants made afterward allows multiplying various end-facade groups from 48 to 250 in a single step. They require only one step in each generation, use a small amount of solvent, and permit simplistic purification (Maraval et al., 2003a,b). (Fig. 9.5)

5.3.2 Click chemistry approach

The "click" chemistry technique by which several authors reported the use of Cu (I)-catalyzed reaction of 1,2,3-triazoles from azides and alkynes to make dendrimers with a variety of facade groups in high yield and purity (Lee et al., 2006a,b; Wu et al., 2004) (Fig. 9.6).

5.3.3 Double exponential approach

The "double exponential" approach permits the formation of monomers via divergent and convergent growth using only one substrate, then the resultant two products react to produce an orthogonally protected trimer that can be employed in a further step. The main advantage is rapid synthesis and applied to either divergent or convergent methods (Svenson and Tomalia, 2005) (Fig. 9.7).

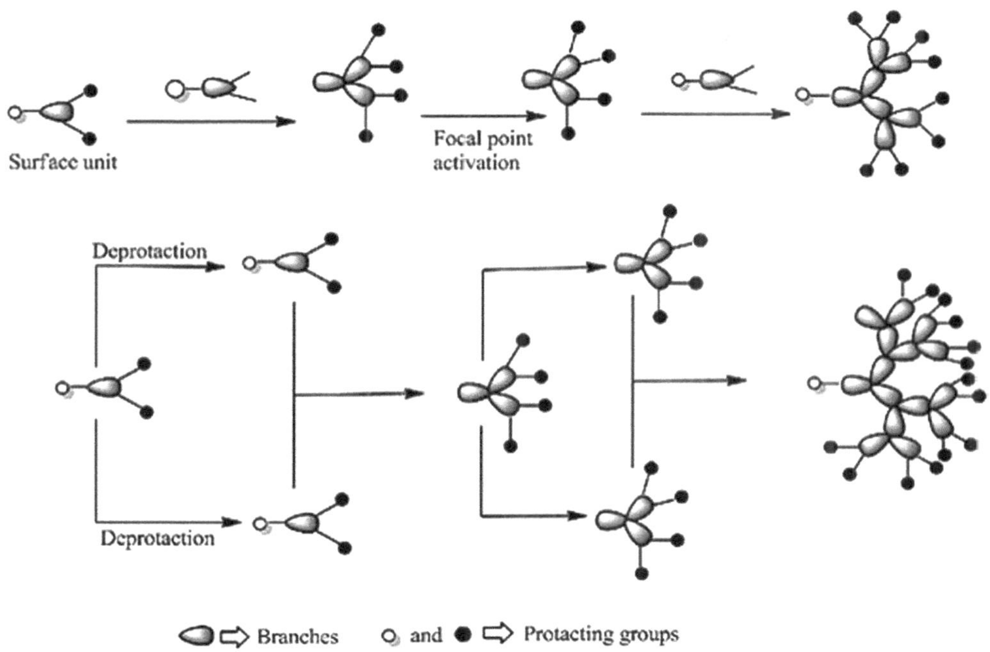

FIGURE 9.4 Convergent Synthetic Approach for Dendrimer

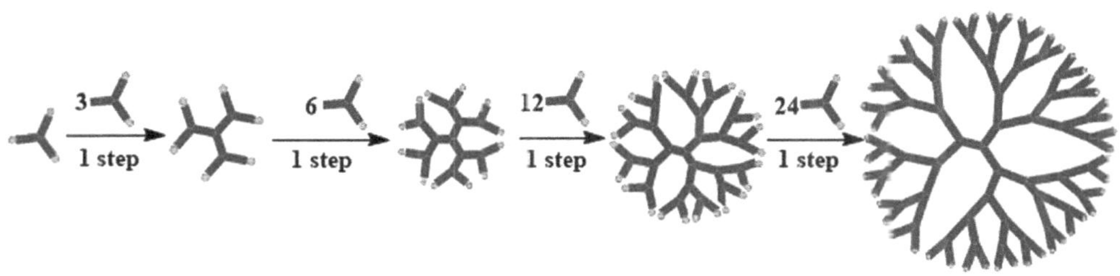

FIGURE 9.5 Lego Chemistry Synthetic Approach for Dendrimer

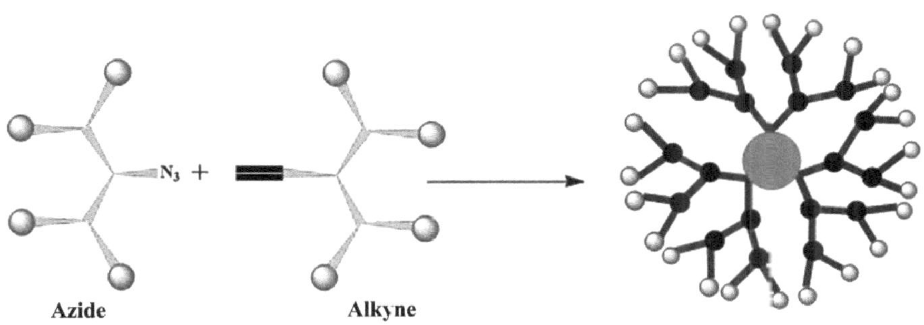

FIGURE 9.6 Click Chemistry Synthetic Approach for Dendrimer

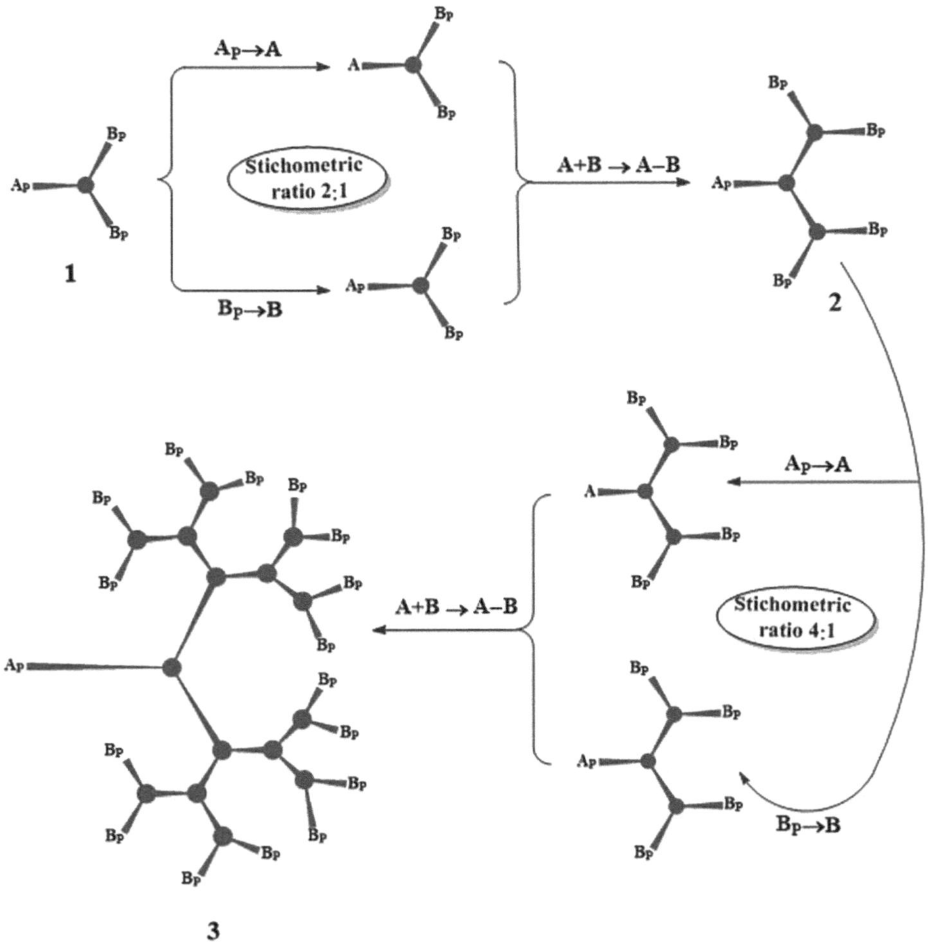

Repeat unit ⟹ •

Oligomeric sequence of monomers ⟹ **A and B**

Diprotected monomer ⟹ **Ap Bp**

Selective deprotection ⟹ **Ap → A**

Selective deprotection ⟹ **Bp → B**

Coupling ⟹ **A + B → A–B**

FIGURE 9.7 First Two Generations of Double Exponential Dendrimer Growth

6 UNIQUE PROPERTIES OF DENDRIMERS

The unique structure and properties of dendrimers are important in making it suitable for biomedical applications. Physico-chemical properties of the dendrimers are based on its generation, surface functionalities, and different types of core structure (Table 9.3A). Dendrimeric solutions exhibit minor viscosity in comparison to linear polymers, which do not enhance linearly with molecular weight, but exhibit highest at the fourth generation and turn down thereafter (Mourey et al., 1992). It happens because of the shape changes according to the generations developed. Lower generation dendrimers are spherical and have huge surface areas in relation to volume while the higher generation dendrimers have more compact spherical shapes and are tightly packed. Due to the presence of hydrophobic core and hydrophilic surface they have the ability to solubilize hydrophobic molecules in aqueous media. Dendrimers offer diverse benefits as compared to other polymeric nanocarriers, which makes them not as much prone to reticulum endothelium uptake. Dendrimers have a lower polydispersity index because of rigorous managed synthesis. With increasing density of branches the exterior branches organize themselves around a lesser density core in spheres shape of consequent more periphery density and the majority of the place remains hollow in the direction of the core. The void space may be used for drug loading. Multiple functional groups (Table 9.3B) are present on the outer surface of dendrimers, which can be used to attach vectors, devices, or ligands for targeting to a particular site in the body.

Table 9.3A Influence of Different Cores of Generation 4 Dendrimers With Primary Amines As Terminal Functional Groups

Core	Log K_{ow}
EDA	−2.38 ± 0.02
DAB	−2.35
DAH	−2.25
DAD	−2.33
Cyst	−.62

EDA, Ethylene diamine; DAB, diaminobutane; DAH, diaminohexane; DAD, diaminododecane; Cyst; cystamine.
Barcelo, D., Farre, M., 2012. Analysis and Risk of Nanomaterials in Environmental and Food Samples. Elsevier Publication, Newnes.

Table 9.3B Influence of Different Terminal Functionality of EDA Core PAMAM Dendrimers

Generation	Terminal Group	Log K_{ow}
4	Amine	−2.38 ± 0.02
4	Amidoethanol	−2.47
4	Succinamic acid	−2.53 ± 0.06
3.5	Sodium carboxylate	−2.33
4	Tris(hydroxymethyl amino methane)	−1.39

Barcelo, D., Farre, M., 2012. Analysis and Risk of Nanomaterials in Environmental and Food Samples. Elsevier Publication, Newnes.

Dendrimers can be modified as stimuli responsive to releasing the drug. Dendrimers might show enhanced permeability retention (EPR) effect, which allows them to target tumor cells more effectively than small molecules. These can be synthesized and designed for specific applications. Due to their feasible topology, functionality, and dimensions, their size is very close to various important biological polymers and assemblies, such as DNA and proteins which are physiologically ideal. This property, along with the cationic surface charge, makes these carriers as ideal transfecting agents in gene delivery (Morgenroth et al., 1997; Nanjwade et al., 2009; Patri et al., 2002).

7 HOST GUEST INTERACTIONS OF DENDRIMERS AND BIOACTIVES

In this section we discuss supramolecular chemistry, which deals with the molecular interaction except for covalent, and provides information about specific interaction and the relationship between geometrical structure and binding sites. Host guest interaction involves the binding of a substrate within a host molecule. As we have discussed earlier, dendrimer has a good capability of holding the hydrophobic as well as hydrophilic drugs/bioactives inside it due to its structure. There are two requirements for the host guest interaction: first is the precise control over geometry; another is interactional complementary structure; and both can be achieved in dendrimer due to its tailor made functionality. There can be several types of interactions, that is, ionic, electrostatic, hydrogen-bonding, hydrophobic. Host guest chemistry of dendritic molecules starts attracting the researchers in the last decade of the 20th century. Buhleier et al. (1978) and Vogtle coworkers reported the first document on host guest interaction. Recently Yang et al. (2011a) reported characterization of host guest chemistry of fully acetylated PAMAM and dexamethasone 21-phosphate (Ac-G5/Dp21) and cationic dendrimer-dexamethasone 21-phosphate (G5/Dp21) complex using H1 NMR and 2D NOE NMR at different pH. At neutral pH no interaction was observed between Ac-G5 and Dp21 because Dp21 was unable to locate the cavities of dendrimer, while at acidic pH interaction was observed because at lower pH enhances polarity of cavities. This leads to ionic interaction between dendrimer-drugs and cationic dendrimers that can encapsulate Dp21 at neutral pH. Upon these results the authors concluded that surface functionality of the dendrimer influences the host–guest interaction. At the same time period similarly Shao et al. (2011) investigated the host–guest interaction of G3 PAMAM and G4 PPI with phenylbutazone using 1H NMR and 2D NOESY. Spatial arrangement of phenylbutazone in pocket of PAMAM dendrimers observed less interaction as compared to PPi dendrimer and this led to higher loading capacity of drug in G3 PAMAM dendrimer than PPI dendrimer at same feed ratio. Host–guest chemistry is still in infancy even after three decades after the first dendrimer synthesis. This has to be explored further.

8 DRUGS DELIVERED THROUGH ORAL ROUTE USING DENDRIMERS AS A NANOCARRIER

As per well-known reported literature, many therapeutic active agents, like anticancer drugs, antihypertensive agents, and antimicrobial agents, have been delivered through dendrimer as nanocarriers. Here we discuss in detail the oral delivery of antimicrobial, antifungal, and dentifrice agents, as well catheters by this nanocarriers (Fig. 9.8).

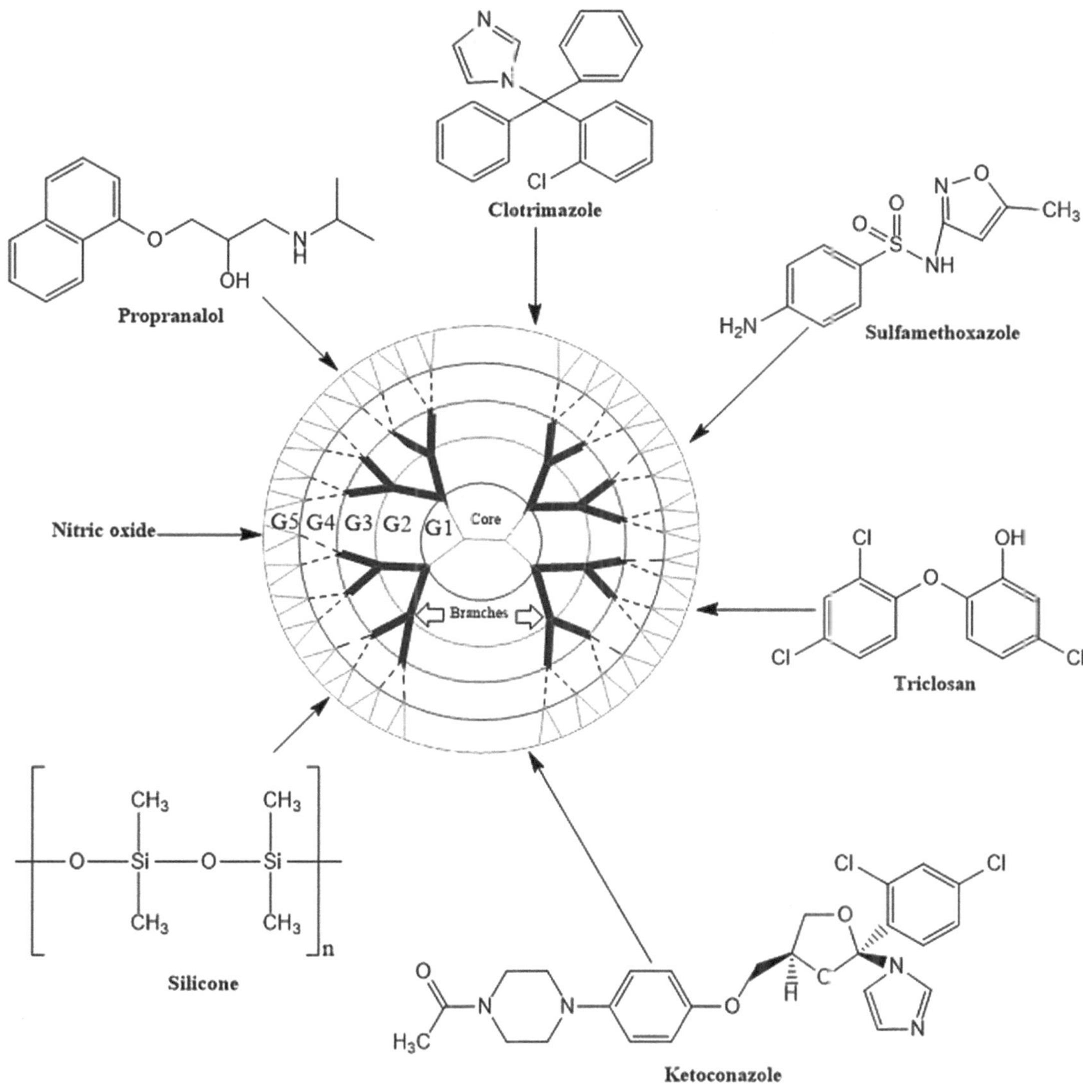

FIGURE 9.8 Drug and Dendrimer Conjugation

8.1 PROPRANOLOL

Propranolol is a sympatholytic nonselective beta blocker. It is used to treat high blood pressure, a number of heart dysrhythmias, thyrotoxicosis, and essential tremors. D'Emanuele et al. (2004) reported dendrimer–propranolol conjugates to bypassing the efflux transporters. They used PAMAM G3 dendrimers for conjugation with propranolol. They analyzed propranolol, propranolol-G3 dendrimers, and propranolol-lauroyl-G3 dendrimer conjugates across Caco-2 cell monolayers. They analyzed basolateral to apical direction; propranolol was significantly greater than apical to basolateral. In the presence of Cyclosporin A, propranolol was increased from apical to basolateral (A→B) because Cyclosporine A is a P-gp inhibitor. PAMAM G3 dendrimer does not significant change in Apical to basolateral direction because PAMAM G3 does not have P-gp inhibitor property, but it produces significant changes in basolateral to apical (B→A) direction. They also discovered that the propranolol, when conjugated with PAMAM G3 dendrimer then significant increments, produces in terms of apical to basolateral (A→B) transport direction and decreases in basolateral to apical (B→A) transport direction. Hence, they observed that when propranolol was covalently attached with PAMAM G3 dendrimers it was able to cross the P-gp efflux transporter. They also conjugated propranolol with lauroyl chains and PAMAM G3 dendrimer. In this study they observed much more increase in apical to basolateral (A→B) transporter than simple propranolol-PAMAM G3 dendrimer conjugates. The authors also analyzed the effects of cyclosporin A, colchicine, and temperature on the G3-P6 conjugates (propranolol-PAMAM G3 dendrimers) and they found that the colchicine caused a decrease in apical to basolateral (A→B) and basolateral to apical transporter (B→A) from 11% to 20% of true data value and temperature also caused significant decrement (A→B), which was 2% of true data value at 3 h and no significant difference in the permeability of propranolol conjugates in the presence and absence of cyclosporin A. They observed that the permeation of dendrimer and drug conjugates was temperature-sensitive and reduced by the presence of colchicine, which is an endocytosis inhibitor suggesting that the transport through Caco-2 cell monolayers involves endocytosis mediated transepithelial transporter through a transcellular route. In this study the authors concluded that propranolol, when covalently conjugated with PAMAM G3 dendrimer, it will significantly increase apical to basolateral (A→B) transporter and when conjugated with lauoryl chains, propranolol also significantly increased in apical to basolateral (A→B) transport. It is all due to covalently linked PAMAM G3 dendrimer cases, not a simple mixture of propranolol and PAMAM G3 dendrimer.

8.2 TRICLOSAN

Triclosan is most commonly used as an antiplaque agent. It is nonionic in nature and lipid soluble bisphenol antibacterial. It has broad spectrum antimicrobial activity against both Gram-positive and Gram-negative bacteria, yeasts, and some fungi. Gardiner et al. (2008) reported PAMAM dendrimers used as a delivery carrier for triclosan. They used HPLC method for analysis of triclosan solubility using naphthalene as an internal standard. The authors prepared different concentration ranges (1–5 mM) of unsubstituted PAMAM G3 dendrimer and they observed that triclosan solubility was significantly increased with increasing PAMAM G3 dendrimer concentration in a linear manner. They also prepared different pH range buffer solutions containing PAMAM G3 dendrimer from pH ranges 5 to 12 and observed that there was no significant improvement in solubility. Other than this study they also prepared PAMAM G3 dendrimer conjugated with phenylalanine to investigate the solubility of triclosan. They concluded that there was no increasing in solubility of triclosan due to dendrimers conjugated with phenylalanine. From this study the authors concluded that triclosan solubility increased with using PAMAM G3 dendrimer concentrations due to ionization effects, not due to encapsulation.

8.3 NITRIC OXIDE

Nitric oxide (NO) is a reactive, gaseous radical, generally produced endogenously during the immune response to occupying organisms. Oral mucosal epithelial cells produce NO in response to microbes and proinflammatory incitements initiated upon the deposition of dental plaque. Backlund et al. (2014) prepared PAMAM G1 generation dendrimer and silica particle mediated exogenous NO formulation and evaluated its antibacterial efficacy on periodontal pathogens. The effectiveness of macromolecular NO release was proven by a 3-log reduction at 4 mg/mL in periodontopathogenic *Aggregatibacter actinomycetemcomitans* and *Porphyromonas gingivalis* viability. Both NO-releasing dendrimers and silica showed reductions in HGF-1 viability (75 ± 3 and 61 ± 12% compared with untreated cells) at the maximum concentrations required to kill periodontopathogens (2 and 4 mg/mL, respectively), but only minimally when compared with chlorhexidine. In contrast, PROLI/NO was more toxic to HGF-1 at the maximum concentration required to kill periodontopathogens (18 ± 2% compared with untreated cells for 4 mg/mL). Likewise, 0.12 and 0.20% (w/w) chlorhexidine (CHX) negatively affected HGF-1 viability (17 ± 2 and 22 ± 3% compared with untreated cells, respectively). All these results propose the potential usefulness of macromolecular NO-release scaffolds as a novel platform for the effective treatment of periodontal disease.

8.4 CLOTRIMAZOLE

Clotrimazole is an azole type antifungal agent. It is generally used in the treatment of vaginal and skin infections. Clotrimazole has poor solubility, which is a disadvantage of its clinical use. Winnicka et al. (2011) reported dendrimers has the ability to solubilize the water-insoluble drug and passage of drug through cell membranes. The authors used PAMAM dendrimers to enhance the solubility and antifungal activity of clotrimazole. They used PAMAM G2 and G3 generation dendrimers with amine and hydroxy end groups. They analyzed solubility of clotriamazole using the shake flask method. They reported that the solubility of clotriamazole using PAMAM-NH_2 G3 dendrimers exhibit highest solubility among PAMAM-NH_2 G2, PAMAM-OH G3, and PAMAM-OH G2 generation. PAMAM-NH_2 G3 dendrimers enhanced solubility from 0.38 to 2.5 μg/mL, which is almost 6.7-fold increased from pure clotriamazole. Antifungal activity was analyzed against various species of *Candida* cultures. The authors reported that the PAMAM-NH_2 G2 dendrimers was more potent antifungal than PAMAM-NH_2 G3 dendrimers. Clotrimazole-PAMAM-NH_2 G2 dendrimers showed enhanced antifungal activity approximately 4-32-fold more potent than pure clotriamazole.

8.5 KETOCONAZOLE

Ketoconazole is an imidazole-type antifungal agent, which is used in both systemic and topical fungal infections. Ketoconazole inhibits the ergosterol synthesis, which is a constituent of cell membrane of fungi. It is a lipophilic drug and it has poor aqueous solubility. So, solubility of ketoconazole was attempted by using PAMAM dendrimer. Winnicka et al. (2012) prepared hydrogels of ketoconazole using PAMAM dendrimer of NH_2 and OH terminated of G2 and G3 generation. They performed solubility study using the shake flask method. They analyzed that PAMAM-NH_2 dendrimer of G2 generation showed increasing solubility from 0.98 μg/mL to 11.12 μg/mL and PAMAM-NH_2 G3 dendrimer showed solubility from 0.98 μg/mL to approximately 7.00 μg/mL, which is interestingly increment on solubility. On the other side, PAMAM-OH G2 dendrimer showed solubility increment from 0.98 μg/mL to 5.84 μg/mL and PAMAM-OH G3 dendrimer showed solubility increased from 0.98 μg/mL to approximately 5.70 μg/mL. They also performed antifungal activity against various strains of *Candida* strains. They reported that PAMAM-NH_2 dendrimer effectively increased the antifungal action against all *Candida* strains compared to pure ketoconazole. The authors conclude that PAMAM-NH_2 G2 dendrimer showed 2-16-fold more potent against *Candida albicans* than pure ketoconazole, which is 0.008 μg/mL and 0.064 μg/mL, which is MIC (Minimum Inhibitory Concentration). PAMAM-NH_2 G2 also showed MFC (Minimum Fungicidal Concentration) for *Candida albicans* strains, which is 2 μg/mL and 32 μg/mL for pure ketoconazole.

8.6 SULFAMETHOXAZOLE

Sulfamethoxazole is a sulfonamide derivatives. It is low soluble in water, rapid elimination through blood, and it produces several side effects, including skin rashes, fever, hepatotoxicity, haematological disorder, and lymphadenopathy. So poor solubility restricts their use in topical and parenteral applications. In order to improve solubility of sulfamethoxazole, PAMAM dendrimers was used to attempt increasing solubility. Dendrimers are hyperbranched, monodisperse, and three-dimensional macromolecules. Polyamidoamine (PAMAM) dendrimers with an ellipsoidal or spheroidal shape is one of the most important studied starburst macromolecules. Ma et al. (2007) reported evaluation of PAMAM dendrimers as drug carriers of antibacterial drugs, using sulfamethoxazole as a model drug. They investigate the PAMAM dendrimers (G2–G4) as a solubility enhancer; effects of concentration, generation, ion concentration, and temperature on the solubility and in vitro release behavior and antibacterial effects in the presence of PAMAM. They reported that the solubility of sulfamethoxazole was successfully improved by using PAMAM dendrimers of G3 generation approximate 40-fold increment in solubility in 10 mg/mL G3 PAMAM dendrimers compared to distilled water. The solubility of sulfamethoxazole was enhanced by increasing dendrimer generation from lower to higher. The authors also reported antibacterial activity against *Escherichia coli.* They found that the sulfamethoxazole-PAMAM (G3 generation) dendrimers showed a 4- or 8-fold increase in antibacterial activity than pure sulfamethoxazole.

8.7 SILICONE URINARY CATHETERS

Lopez et al. (2011) reported a new approach to prevent the growth of bacterial flora in the medical device–related infections. The authors claimed that they conjugated biofunctinalization of silicone (PDMS) with PAMAM G5 generation dendrimer that is a generated amino terminated surface and further covalently attached a carboxy-terminated mannose derivatives to promote the adherence of benign *E. coli* 83972. The authors investigated the stability of the device and effectively observed that the biofilm was able to reduce the uropathogenic *Enterococcus faecalis* by 104-fold and the benign *E. coli* 83972 was reduced by only 5.5-fold. (Table 9.4).

9 CONCLUSIONS AND FUTURE PROSPECTS

The selection of the route of administration along with its delivery device technologies remains a significant challenge, concerning the frequency of dosing, dosing, dose volume, and number of treatments. Ideally, a single pill administration in one time for chronic diseases is currently a dream.

Table 9.4 List of Drugs and Their Effect on Solubility and Stability With Dendrimer

Drug	Basic Moiety	Effect on Solubility/ Stability	Effect on Pharmacological Activity	Target Organism/Cell Line	Dendrimer	References
Sulfamethoxazole	Sulfonamide derivatives	The solubility of sulfamethoxazole was successfully approx. 40-fold increment by using PAMAM dendrimers.	Anti-bacterial activity increase 4–8-fold.	*E. coli*	PAMAM (G3)	Ma et al. (2007)
Clotrimazole	Azole derivatives	PAMAM-NH2 G3 dendrimers enhanced solubility 0.38–2.5 µg/ml, which is almost 6.7-fold increased from pure clotriamazole.	Antifungal activity increase approximately 4.32-fold.	Various species of *Candida*	PAMAM-NH2 (G3) and (G4)	Winnicka et al. (2011)
Propranolol	—	PAMAM G3 doesn't have P-gp inhibitor property but it produces significant changes in basolateral to apical (B A) direction.	Propranolol when covalently conjugated with PAMAM G3 dendrimer then it will significantly increase apical to basolateral (A B) transporter.	Caco-2 cell monolayers	PAMAM (G3)	D'Emanuele et al. (2004)
Triclosan	Bisphenol	Triclosan solubility increases with using PAMAM G3 dendrimer concentrations due to ionization effects.	—	—	PAMAM (G3)	Gardiner et al. (2008)
Ketoconazole	Imidazole derivative	PAMAM-NH2 dendrimer of G2 generation showed increasing solubility from 0.98 µg/mL to 11.12 µg/mL.	PAMAM-NH2 G2 dendrimer showed 2–16-fold more potent against *C. albicans* than pure ketoconazole, which is 0.008 µg/mL and 0.064 µg/mL, which is MIC (minimum inhibitory concentration).	*C. albicans*	PAMAM-NH2 (G2) and (G3)	Winnicka et al. (2012)

(*Continued*)

Table 9.4 List of Drugs and Their Effect on Solubility and Stability With Dendrimer (*cont.*)

Drug	Basic Moiety	Effect on Solubility/ Stability	Effect on Pharmacological Activity	Target Organism/Cell Line	Dendrimer	References
Nitrous oxide	—	NO-releasing dendrimers and silica showed reductions in (human gingival fibroblasts) HGF-1 viability (75 ± 3 and 61 ± 12% compared with untreated cells).	Both NO-releasing dendrimers and silica showed reductions in (human gingival fibroblasts) HGF-1 viability (75 ± 3 and 61 ± 12%).	*Aggregatibacter actinomycetemcomitans* and *Porphyromonas gingivalis*	PAMAM (G1)	Backlund et al. (2014)
Silicone urinary catheters	—	Stability of the device against uropathogenic *Enterococcus faecalis* by 104-fold and the benign *E. coli* 83972 one was reduced by only 5.5-fold.	—	*E. coli* 83972	PAMAM (G5)	Lopez et al. (2011)

Pharmacokinetics (PK), exposure–response relationship, and the PK/pharmacodynamic (PD) index are important tools that probably suggest the best route of administration in order to maximize the therapeutic effect and minimize the toxic effects. It is well known that polymeric nanocarrier-based drug delivery systems can greatly enhance bioavailability as well therapeutic efficacies relative to plain drug and decrease the outcome side effects of drugs. In this track, the dendrimers emerge as a unique nanocarrier comparison with other classical nanocarriers due to their biocompatibilities, tuneable physicochemical properties, and tailor-made surface functionalities. All these properties of dendrimers make them ideal nanocarriers and choice of nanocarriers for the huge varieties of drugs for their oral route of administration. To date, the foremost way of administration of dendrimers comprise intravenous and, to a lesser extent, intraperitoneal routes. Recently, other routes have been described, such as transdermal, ocular, oral, pulmonary, intranasal, and intravaginal for the treatment of life-threatening diseases requiring parenteral and nonparenteral routes of administration. Both encapsulation (noncovalent) and conjugation (covalent attachment) properties of dendrimer have been intensively pointed out for drug delivery concerns. Here in this chapter the focus area is oral potential applicability of dendrimer for delivering various categories of drugs like antimicrobial agents, antifungal drugs, and certain catheter-related substances.

10 ABBREVIATIONS

BCS Biopharmaceutics classification system
CNTs Carbon nanotubes
Cur Curcumin
FDA Food and Drug Administration
GIT Gastrointestinal tract
NO Nitric oxide
NPs Nanoparticles
PAMAM Poly(amidoamine)
PLGA Poly(lactic-*co*-glycolic acid)
SEDDS Self-emulsification drug delivery system
SLNs Solid lipid nanoparticles
SNEDDS Self-nanoemulsifying drug delivery system

REFERENCES

Aisner, J., 1992. Current approaches to small cell lung cancer. Hematol. Oncol. 10, 7–20.
Backlund, C., Sergesketter, A., Offenbacher, S., Schoenfisch, M., 2014. Antibacterial efficacy of exogenous nitric oxide on periodontal pathogens. J. Dent. Res. 93 (11), 1089–1094.
Bali, V., Ali, M., Ali, J., 2010. Study of surfactant combinations and development of a novel nanoemulsion for minimising variations in bioavailability of ezetimibe. Colloids Surf. B 76, 410–420
Bardajee, G.R., Pourjavadi, A., Soleyman, R., 2011. Novel nano-porous hydrogel as a carrier matrix for oral delivery of tetracycline hydrochloride. Colloids Surf. A 392, 16–24.
Bromberg, L., 2008. Polymeric micelles in oral chemotherapy. J. Control. Release 128, 99–112.
Buhleier, E., Wehner, W., Vögtle, F., 1978. Cascade-and Nonskid-Chain-like Syntheses of Molecular Cavity Topologies. Synthesis 1978, 155–158.

Cartiera, M.S., Ferreira, E.C., Caputo, C., Egan, M.E., Caplan, M.J., Saltzman, W.M., 2009. Partial correction of cystic fibrosis defects with PLGA nanoparticles encapsulating curcumin. Mol. Pharm. 7, 86–93.

Chakraborty, S., Shukla, D., Vuddanda, P.R., Mishra, B., Singh, S., 2010. Utilization of adsorption technique in the development of oral delivery system of lipid based nanoparticles. Colloids Surf. B 81, 563–569.

Chen, M.-C., Sonaje, K., Chen, K.-J., Sung, H.-W., 2011. A review of the prospects for polymeric nanoparticle platforms in oral insulin delivery. Biomaterials 32, 9826–9838.

Coimbra, M., Rijcken, C.J., Stigter, M., Hennink, W.E., Storm, G., Schiffelers, R.M., 2012. Antitumor efficacy of dexamethasone-loaded core-crosslinked polymeric micelles. J. Control. Release 163, 361–367.

Cone, R.A., 2009. Barrier properties of mucus. Adv. Drug Deliv. Rev. 61, 75–85.

Cui, J., Yu, B., Zhao, Y., Zhu, W., Li, H., Lou, H., Zhai, G., 2009. Enhancement of oral absorption of curcumin by self-microemulsifying drug delivery systems. Int. J. Pharm. 371, 148–155.

Date, A.A., Nagarsenker, M., 2007. Design and evaluation of self-nanoemulsifying drug delivery systems (SNEDDS) for cefpodoxime proxetil. Int. J. Pharm. 329, 166–172.

D'Emanuele, A., Jevprasesphant, R., Penny, J., Attwood, D., 2004. The use of a dendrimer-propranolol prodrug to bypass efflux transporters and enhance oral bioavailability. J. Control. Release 95, 447–453.

DeMario, M.D., Ratain, M.J., 1998. Oral chemotherapy: rationale and future directions. J. Clin. Oncol. 16, 2557–2567.

des Rieux, A., Fievez, V., Garinot, M., Schneider, Y.-J., Préat, V., 2006. Nanoparticles as potential oral delivery systems of proteins and vaccines: a mechanistic approach. J. Control Release 116, 1–27.

Desai, P.P., Date, A.A., Patravale, V.B., 2012. Overcoming poor oral bioavailability using nanoparticle formulations—opportunities and limitations. Drug Disc. Today 9, e87–e95.

Elnaggar, Y.S., El-Massik, M.A., Abdallah, O.Y., 2009. Self-nanoemulsifying drug delivery systems of tamoxifen citrate: design and optimization. Int. J. Pharm. 380, 133–141.

El-Sayed, M., Ginski, M., Rhodes, C., Ghandehari, H., 2002. Transepithelial transport of poly (amidoamine) dendrimers across Caco-2 cell monolayers. J. Control. Release 81, 355–365.

El-Sayed, M., Rhodes, C., Ginski, M., Ghandehari, H., 2003. Transport mechanism(s) of poly(amidoamine) dendrimers across Caco-2 cell monolayers. Int. J. Pharm. 265, 151–157.

El-Sherbiny, I.M., Abdel-Mogib, M., Dawidar, A.-A.M., Elsayed, A., Smyth, H.D., 2011. Biodegradable pH-responsive alginate-poly(lactic-*co*-glycolic acid) nano/microhydrogel matrices for oral delivery of silymarin. Carbohy. Polym. 83, 1345–1354.

Ensign, L.M., Cone, R., Hanes, J., 2012. Oral drug delivery with polymeric nanoparticles: the gastrointestinal mucus barriers. Adv. Drug Deliv. Rev. 64, 557–570.

Farokhzad, O.C., Langer, R., 2009. Impact of nanotechnology on drug delivery. ACS Nano. 3, 16–20.

Frechet, J.M., 1994. Functional polymers and dendrimers: reactivity, molecular architecture, and interfacial energy. Sci. AAAS 263, 1710–1714.

Gardiner, J., Freeman, S., Leach, M., Green, A., Alcock, J., D'Emanuele, A., 2008. PAMAM dendrimers for the delivery of the antibacterial triclosan. J. Enz. Inhib. Med. Chem. 23, 623–628.

Gaucher, G., Satturwar, P., Jones, M.-C., Furtos, A., Leroux, J.-C., 2010. Polymeric micelles for oral drug delivery. Eur. J. Pharm. Biopharm. 76, 147–158.

Goldberg, D.S., Ghandehari, H., Swaan, P.W., 2010. Cellular entry of G3. 5 poly (amido amine) dendrimers by clathrin-and dynamin-dependent endocytosis promotes tight junctional opening in intestinal epithelia. Pharm. Res. 27, 1547–1557.

Goldberg, D.S., Vijayalakshmi, N., Swaan, P.W., Ghandehari, H., 2011. G3. 5 PAMAM dendrimers enhance transepithelial transport of SN38 while minimizing gastrointestinal toxicity. J. Control. Release 150, 318–325.

Gursoy, R.N., Benita, S., 2004. Self-emulsifying drug delivery systems (SEDDS) for improved oral delivery of lipophilic drugs. Biomed. Pharmacother. 58, 173–182.

Hall, J.B., Dobrovolskaia, M.A., Patri, A.K., Mcneil, S.E., 2007. Characterization of nanoparticles for therapeutics. Nanomedicine (London) 2(6), 789–803.

Hawker, C.J., Frechet, J.M., 1990. Preparation of polymers with controlled molecular architecture. A new convergent approach to dendritic macromolecules. J. Am. Chem. Soc. 112, 7638–7647.

Hong, J.-Y., Kim, J.-K., Song, Y.-K., Park, J.-S., Kim, C.-K., 2006. A new self-emulsifying formulation of itraconazole with improved dissolution and oral absorption. J. Control. Release 110, 332–338.

Hsieh, W.-H., Chang, S.-F., Chen, H.-M., Chen, J.-H., Liaw, J., 2012. Oral gene delivery with cyclo-(D-Trp-Tyr) peptide nanotubes. Mol. Pharm. 9, 1231–1249.

Huang, X., Wu, Z., Gao, W., Chen, Q., Yu, B., 2011. Polyamidoamine dendrimers as potential drug carriers for enhanced aqueous solubility and oral bioavailability of silybin. Drug Dev. Ind. Pharm. 37, 419–427.

Huang, X., Xiao, Y., Lang, M., 2012. Micelles/sodium-alginate composite gel beads: a new matrix for oral drug delivery of indomethacin. Carbohy. Polym. 87, 790–798.

Italia, J., Bhatt, D., Bhardwaj, V., Tikoo, K., Kumar, M.R., 2007. PLGA nanoparticles for oral delivery of cyclosporine: nephrotoxicity and pharmacokinetic studies in comparison to Sandimmune Neoral. J. Control. Release 119, 197–206.

Kaminskas, L.M., Boyd, B.J., Porter, C.J., 2011. Dendrimer pharmacokinetics: the effect of size, structure and surface characteristics on ADME properties. Nanomedicine 6, 1063–1084.

Kandpal, S., Saxena, A., 2015. Synthesis & characterization of novel silicone dendrimers as base stock for high performance lubricants. J. Org. Chem. 791, 232–237.

Kateb, B., Chiu, K., Black, K.L., Yamamoto, V., Khalsa, B., Ljubimova, J.Y., Ding, H., Patil, R., Portilla-Arias, J.A., Modo, M., 2011. Nanoplatforms for constructing new approaches to cancer treatment, imaging, and drug delivery: what should be the policy? Neuroimage 54, S106–S124.

Ke, W., Zhao, Y., Huang, R., Jiang, C., Pei, Y., 2008. Enhanced oral bioavailability of doxorubicin in a dendrimer drug delivery system. J. Pharm. Sci. 97, 2208–2216.

Kesisoglou, F., Panmai, S., Wu, Y., 2007. Nanosizing—oral formulation development and biopharmaceutical evaluation. Adv. Drug Deliv. Rev. 59, 631–644.

Khalil, N.M., Do Nascimento, T.C.F., Casa, D.M., Dalmolin, L.F., De Mattos, A.C., Hoss, I., Romano, M.A., Mainardes, R.M., 2013. Pharmacokinetics of curcumin-loaded PLGA and PLGA–PEG blend nanoparticles after oral administration in rats. Colloids Surf. B 101, 353–360.

Khoee, S., Hemati, K., 2013. Synthesis of magnetite/polyamino-ester dendrimer based on PCL/PEG amphiphilic copolymers via convergent approach for targeted diagnosis and therapy. Polymer 54, 5574–5585.

Kim, S., Kim, J.Y., Huh, K.M., Acharya, G., Park, K., 2008. Hydrotropic polymer micelles containing acrylic acid moieties for oral delivery of paclitaxel. J. Control. Release 132, 222–229.

Kim, Y., Thapa, M., Hua, D.H., Chang, K.-O., 2011. Biodegradable nanogels for oral delivery of interferon for norovirus infection. Antiviral Res. 89, 165–173.

Kim, G.G., Poudel, B.K., Marasini, N., Lee, D.W., Hiep, T.T., Yang, K.Y., Kim, J.C., Yong, C.S., Choi, H.-G. 2013. Enhancement of oral bioavailability of fenofibrate by solid self-microemulsifying drug delivery systems Drug Dev. Ind. Pharm. 39, 1431–1438.

Kitchens, K.M., Kolhatkar, R.B., Swaan, P.W., Eddington, N.D., Ghandehari, H., 2006. Transport of poly (amidoamine) dendrimers across Caco-2 cell monolayers: influence of size, charge and fluorescent labeling Pharm. Res. 23, 2818–2826.

Kitchens, K.M., Foraker, A.B., Kolhatkar, R.B., Swaan, P.W., Ghandehari, H., 2007. Endocytosis and interaction of poly (amidoamine) dendrimers with Caco-2 cells. Pharm. Res. 24, 2138–2145.

Kitchens, K.M., Kolhatkar, R.B., Swaan, P.W., Ghandehari, H., 2008. Endocytosis inhibitors prevent poly (amidoamine) dendrimer internalization and permeability across Caco-2 cells. Mol. Pharm. 5, 364–369.

Kohli, K., Chopra, S., Dhar, D., Arora, S., Khar, R.K., 2010. Self-emulsifying drug delivery systems: an approach to enhance oral bioavailability. Drug Discov. Today 15, 958–965.

Kolhatkar, R.B., Swaan, P., Ghandehari, H., 2008. Potential oral delivery of 7-ethyl-10-hydroxy-camptothecin (SN-38) using poly (amidoamine) dendrimers. Pharm. Res. 25, 1723–1729.

Kumari, A., Yadav, S.K., Yadav, S.C., 2010. Biodegradable polymeric nanoparticles based drug delivery systems. Colloids Surf. B 75, 1–18.

Kurnia, J.C., Birgersson, E., Mujumdar, A.S., 2012. Analysis of a model for pH-sensitive hydrogels. Polymer 53, 613–622.

Lai, S.K., Wang, Y.-Y., Hanes, J., 2009. Mucus-penetrating nanoparticles for drug and gene delivery to mucosal tissues. Adv. Drug Deliv. Rev. 61, 158–171.

Law, S., Huang, K., Chiang, C., 2000. Acyclovir-containing liposomes for potential ocular delivery: corneal penetration and absorption. J. Control. Release 63, 135–140.

Lee, J.W., Kim, B.-K., Kim, H.J., Han, S.C., Shin, W.S., Jin, S.-H., 2006a. Convergent synthesis of symmetrical and unsymmetrical PAMAM dendrimers. Macromolecules 39, 2418–2422.

Lee, S., Han, S.Y., Lee, T.G., Chung, G., Lee, D., Oh, H.B., 2006b. Observation of pronounced b•, y cleavages in the electron capture dissociation mass spectrometry of polyamidoamine (PAMAM) dendrimer ions with amide functionalities. J. Am. Soc. Mass Spectrom. 17, 536–543.

Lee, S.C., Huh, K.M., Lee, J., Cho, Y.W., Galinsky, R.E., Park, K., 2007. Hydrotropic polymeric micelles for enhanced paclitaxel solubility: in vitro and in vivo characterization. Biomacromolecules 8, 202–208.

Li, P., Nielsen, H.M., Müllertz, A., 2012. Oral delivery of peptides and proteins using lipid-based drug delivery systems. Exp. Opin. Drug Deliv. 9, 1289–1304.

Lin, Y.-L., Khanafer, K., El-Sayed, M.E., 2010. Quantitative evaluation of the effect of poly (amidoamine) dendrimers on the porosity of epithelial monolayers. Nanoscale 2, 755–762.

Liu, Y., Zhang, P., Feng, N., Zhang, X., Wu, S., Zhao, J., 2009. Optimization and in situ intestinal absorption of self-microemulsifying drug delivery system of oridonin. Int. J. Pharm. 365, 136–142.

Liu, Y., Wang, P., Sun, C., Zhao, J., Du, Y., Shi, F., Feng, N., 2011. Bioadhesion and enhanced bioavailability by wheat germ agglutinin-grafted lipid nanoparticles for oral delivery of poorly water-soluble drug bufalin. Int. J. Pharm. 419, 260–265.

Lopez, A.I., Kumar, A., Planas, M.R., Li, Y., Nguyen, T.V., Cai, C., 2011. Biofunctionalization of silicone polymers using poly (amidoamine) dendrimers and a mannose derivative for prolonged interference against pathogen colonization. Biomaterials 32, 4336–4346.

Luo, C.-F., Yuan, M., Chen, M.-S., Liu, S.-M., Zhu, L., Huang, B.-Y., Liu, X.-W., Xiong, W., 2011. Pharmacokinetics, tissue distribution and relative bioavailability of puerarin solid lipid nanoparticles following oral administration. Int. J. Pharm. 410, 138–144.

Ma, M., Cheng, Y., Xu, Z., Xu, P., Qu, H., Fang, Y., Xu, T., Wen, L., 2007. Evaluation of polyamidoamine (PAMAM) dendrimers as drug carriers of anti-bacterial drugs using sulfamethoxazole (SMZ) as a model drug. Eur. J. Med. Chem. 42, 93–98.

Maraval, V., Caminade, A.M., Majoral, J.P., Blais, J.C., 2003a. Dendrimer design: how to circumvent the dilemma of a reduction of steps or an increase of function multiplicity? Angew. Chem. Int. Ed. 42, 1822–1826.

Maraval, V., Pyzowski, J., Caminade, A.-M., Majoral, J.-P., 2003b. Lego chemistry for the straightforward synthesis of dendrimers. J. Org. Chem. 68, 6043–6046.

Mauludin, R., Müller, R.H., Keck, C.M., 2009. Development of an oral rutin nanocrystal formulation. Int. J. Pharm. 370, 202–209.

Morgenroth, F., Reuther, E., Müllen, K., 1997. Polyphenylene dendrimers: from three-dimensional to two-dimensional structures. Angew. Chem. Int. Ed. 36, 631–634.

Mourey, T.H., Turner, S., Rubinstein, M., Fréchet, J., Hawker, C., Wooley, K., 1992. Unique behavior of dendritic macromolecules: intrinsic viscosity of polyether dendrimers. Macromolecules 25, 2401–2406.

Mrsny, R.J., 2012. Oral drug delivery research in Europe. J. Control. Release 161, 247–253.

Mullin, D., 2003. Prometheus in Gloucestershire: Edward Jenner, 1749-1823. J. Allergy Clin. Immune. 112, 810–814.

Murdan, S., 2003. Electro-responsive drug delivery from hydrogels. J. Control. Release 92, 1–17.

Najlah, M., Freeman, S., Attwood, D., D'Emanuele, A., 2007. In vitro evaluation of dendrimer prodrugs for oral drug delivery. Int. J. Pharm. 336, 183–190.

Nanjwade, B.K., Bechra, H.M., Derkar, G.K., Manvi, F., Nanjwade, V.K., 2009. Dendrimers: emerging polymers for drug-delivery systems. Eur. J. Pharm. Sci. 38, 185–196.

Nepal, P.R., Han, H.-K., Choi, H.-K., 2010. Preparation and in vitro–in vivo evaluation of Witepsol® H35 based self-nanoemulsifying drug delivery systems (SNEDDS) of coenzyme Q 10. Eur. J. Pharm. Sci. 39, 224–232.

Nornoo, A.O., Zheng, H., Lopes, L.B., Johnson-Restrepo, B., Kannan, K., Reed, R., 2009. Oral microemulsions of paclitaxel: in situ and pharmacokinetic studies. Eur. J. Pharm. Biopharm. 71, 310–317.

Parmentier, J., Thewes, B., Gropp, F., Fricker, G., 2011. Oral peptide delivery by tetraether lipid liposomes. Int. J. Pharm. 30, 150–157.

Parveen, R., Baboota, S., Ali, J., Ahuja, A., Vasudev, S.S., Ahmad, S., 2011. Oil based nanocarrier for improved oral delivery of silymarin: in vitro and in vivo studies. Int. J. Pharm. 413, 245–253.

Patel, K., Padhye, S., Nagarsenker, M., 2012. Duloxetine HCl lipid nanoparticles: preparation, characterization, and dosage form design. AAPS PharmSciTech 13, 125–133.

Patri, A.K., Majoros, I.J., Baker, J.R., 2002. Dendritic polymer macromolecular carriers for drug delivery. Curr. Opin. Chem. Biol. 6, 466–471.

Peltier, S., Oger, J.-M., Lagarce, F., Couet, W., Benoît, J.-P., 2006. Enhanced oral paclitaxel bioavailability after administration of paclitaxel-loaded lipid nanocapsules. Pharm. Res. 23, 1243–1250.

Peppas, N., 1995. Star polymers and dendrimers: prospects of their use in drug delivery and pharmaceutical applications. Control. Release Soc. 12, 12–13.

Piao, H., Ouyang, M., Xia, D., Quan, P., Xiao, W., Song, Y., Cui, F., 2011. In vitro–in vivo study of CoQ10-loaded lipid nanoparticles in comparison with nanocrystals. Int. J. Pharm. 419, 255–259.

Plapied, L., Duhem, N., Des Rieux, A., Préat, V., 2011. Fate of polymeric nanocarriers for oral drug delivery. Curr. Opin. Colloid Interf. Sci. 16, 228–237.

Rao, S.V.R., Shao, J., 2008. Self-nanoemulsifying drug delivery systems (SNEDDS) for oral delivery of protein drugs: I. Formulation development. Int. J. Pharm. 362, 2–9.

Roger, E., Kalscheuer, S., Kirtane, A., Guru, B.R., Grill, A.E., Whittum-Hudson, J., Panyam, J., 2012. Folic acid functionalized nanoparticles for enhanced oral drug delivery. Mol. Pharm. 9, 2103–2110.

Ruenraroengsak, P., Cook, J.M., Florence, A.T., 2010. Nanosystem drug targeting: facing up to complex realities. J. Control. Release 141, 265–276.

Sadekar, S., Ghandehari, H., 2012. Transepithelial transport and toxicity of PAMAM dendrimers: implications for oral drug delivery. Adv. Drug Deliv. Rev. 64, 571–588.

Sarmento, B., Ribeiro, A., Veiga, F., Sampaio, P., Neufeld, R., Ferreira, D., 2007. Alginate/chitosan nanoparticles are effective for oral insulin delivery. Pharm. Res. 24, 2198–2206.

Sarparanta, M.P., Bimbo, L.M., Mäkilä, E.M., Salonen, J.J., Laaksonen, P.H., Helariutta, A.K., Linder, M.B., Hirvonen, J.T., Laaksonen, T.J., Santos, H.A., 2012. The mucoadhesive and gastroretentive properties of hydrophobin-coated porous silicon nanoparticle oral drug delivery systems. Biomaterials 33, 3353–3362.

Schenk, M., Mueller, C., 2008. The mucosal immune system at the gastrointestinal barrier. Best Pract. Res Clin. Gastroenterol. 22, 391–409.

Shao, N., Su, Y., Hu, J., Zhang, J., Zhang, H., Cheng, Y., 2011. Comparison of generation 3 polyamidoamine dendrimer and generation 4 polypropylenimine dendrimer on drug loading, complex structure, release behavior, and cytotoxicity. Int. J. Nano. 6, 3361.

Shen, Q., Li, X., Yuan, D., Jia, W., 2010. Enhanced oral bioavailability of daidzein by self-microemulsifying drug delivery system. Chem. Pharm. Bull. 58, 639–643.

Shinkar, D.M., Dhake, A.S., Setty, C.M., 2012. Drug delivery from the oral cavity: a focus on mucoadhesive. PDA J. Pharm. Sci. Technol. 66(5), 466–500.

Shugarts, S., Benet, L.Z., 2009. The role of transporters in the pharmacokinetics of orally administered drugs. Pharm. Res. 26, 2039–2054.

Silva, A., Kumar, A., Wild, W., Ferreira, D., Santos, D., Forbes, B., 2012. Long-term stability, biocompatibility and oral delivery potential of risperidone-loaded solid lipid nanoparticles. Int. J. Pharm. 436, 798–805.

Sudimack, J.J., Guo, W., Tjarks, W., Lee, R.J., 2002. A novel pH-sensitive liposome formulation containing oleyl alcohol. Biochim. Biophys. Acta 1564, 31–37.

Suk, J.S., Lai, S.K., Wang, Y.-Y., Ensign, L.M., Zeitlin, P.L., Boyle, M.P., Hanes, J., 2009. The penetration of fresh undiluted sputum expectorated by cystic fibrosis patients by non-adhesive polymer nanoparticles. Biomaterials 30, 2591–2597.

Suk, J.S., Lai, S.K., Boylan, N.J., Dawson, M.R., Boyle, M.P., Hanes, J., 2011. Rapid transport of muco-inert nanoparticles in cystic fibrosis sputum treated with *N*-acetyl cysteine. Nanomedicine 6, 365–375.

Svenson, S., Tomalia, D.A., 2005. Dendrimers in biomedical applications—reflections on the field. Adv. Drug Deliv. Rev. 57, 2106–2129.

Tajarobi, F., El-Sayed, M., Rege, B., Polli, J., Ghandehari, H., 2001. Transport of poly amidoamine dendrimers across Madin–Darby canine kidney cells. Int. J. Pharm. 215, 263–267.

Takeuchi, H., Matsui, Y., Sugihara, H., Yamamoto, H., Kawashima, Y., 2005. Effectiveness of submicron-sized, chitosan-coated liposomes in oral administration of peptide drugs. Int. J. Pharm. 303, 160–170.

Terwogt, J.M.M., Schellens, J.H., Wim, W., Beijnen, J.H., 1999. Clinical pharmacology of anticancer agents in relation to formulations and administration routes. Cancer Treat. Rev. 25, 83–101.

Thiagarajan, G., Ray, A., Malugin, A., Ghandehari, H., 2010. PAMAM-camptothecin conjugate inhibits proliferation and induces nuclear fragmentation in colorectal carcinoma cells. Pharm. Res. 27, 2307–2316.

Tomalia, D.A., 1994. Starburst/cascade dendrimers: fundamental building blocks for a new nanoscopic chemistry set. Adv. Mater. 6, 529–539.

Vijayalakshmi, N., Ray, A., Malugin, A., Ghandehari, H., 2010. Carboxyl-terminated PAMAM-SN38 conjugates: synthesis, characterization, and in vitro evaluation. Bioconj. Chem. 21, 1804–1810.

Villar, A.M.S., Naveros, B.C., Campmany, A.C.C., Trenchs, M.A., Rocabert, C.B., Bellowa, L.H., 2012. Design and optimization of self-nanoemulsifying drug delivery systems (SNEDDS) for enhanced dissolution of gemfibrozil. Int. J. Pharm. 431, 161–175.

Wang, X.-Q., Dai, J.-D., Zhang, H., Zhang, X., Wang, J.-C., Zhang, Q., 2008a. Absorption mechanism of cyclosporine A loaded pH-sensitive nanoparticles in rats. J. Nanosci. Nanotechnol. 8, 2422–2431.

Wang, Y.Y., Lai, S.K., Suk, J.S., Pace, A., Cone, R., Hanes, J., 2008b. Addressing the PEG mucoadhesivity paradox to engineer nanoparticles that "slip" through the human mucus barrier. Angew. Chem Int. Ed. 47, 9726–9729.

Wenceslau, A.C., Dos Santos, F.G., Ramos, É.R., Nakamura, C.V., Rubira, A.F., Muniz, E.C., 2012. Thermo-and pH-sensitive IPN hydrogels based on PNIPAAm and PVA-Ma networks with LCST tailored close to human body temperature. Mater. Sci. Eng. C 32, 1259–1265.

Wilbur, D.S., Pathare, P.M., Hamlin, D.K., Buhler, K.R., Vessella, R.L., 1998. Biotin reagents for antibody pretargeting. 3. Synthesis, radioiodination, and evaluation of biotinylated starburst dendrimers. Bioconj. Chem. 9, 813–825.

Winnicka, K., Sosnowska, K., Wieczorek, P., Sacha, P.T., Tryniszewska, E., 2011. Poly (amidoamine) dendrimers increase antifungal activity of clotrimazole. Biol. Pharm. Bull. 34, 1129–1133.

Winnicka, K., Wroblewska, M., Wieczorek, P., Sacha, P.T., Tryniszewska, E., 2012. Hydrogel of ketoconazole and PAMAM dendrimers: formulation and antifungal activity. Molecules 17, 4612–4624.

Wiwattanapatapee, R., Carreño-Gómez, B., Malik, N., Duncan, R., 2000. Anionic PAMAM dendrimers rapidly cross adult rat intestine in vitro: a potential oral delivery system? Pharm. Res. 17, 991–998.

Wu, P., Feldman, A.K., Nugent, A.K., Hawker, C.J., Scheel, A., Voit, B., Pyun, J., Frechet, J.M., Sharpless, K.B., Fokin, V.V., 2004. Efficiency and fidelity in a click-chemistry route to triazole dendrimers by the copper (I)-catalyzed ligation of azides and alkynes. Angew. Chem. Int. Ed. 43, 3928–3932.

Xie, H., Smith, J.W., 2010. Fabrication of PLGA nanoparticles with a fluidic nanoprecipitation system. J. Nanobiotechnol. 8, 18.

Xie, X., Tao, Q., Zou, Y., Zhang, F., Guo, M., Wang, Y., Wang, H., Zhou, Q., Yu, S., 2011. PLGA nanoparticles improve the oral bioavailability of curcumin in rats: characterizations and mechanisms. J. Agri. Food Chem. 59, 9280–9289.

Yamanaka, Y.J., Leong, K.W., 2008. Engineering strategies to enhance nanoparticle-mediated oral delivery. J. Biomater. Sci. 19, 1549–1570.

Yang, K., Weng, L., Cheng, Y., Zhang, H., Zhang, J., Wu, Q., Xu, T., 2011a. Host—guest chemistry of dendrimer—drug complexes. 6. Fully acetylated dendrimers as biocompatible drug vehicles using dexamethasone 21-phosphate as a model drug. J. Phys. Chem. B 115, 2185–2195.

Yang, M., Lai, S.K., Wang, Y.Y., Zhong, W., Happe, C., Zhang, M., Fu, J., Hanes, J., 2011b. Biodegradable nanoparticles composed entirely of safe materials that rapidly penetrate human mucus. Angew. Chem. Int. Ed. 50, 2597–2600.

Zhang, J., Xie, R., Zhang, S.-B., Cheng, C.-J., Ju, X.-J., Chu, L.-Y., 2009. Rapid pH/temperature-responsive cationic hydrogels with dual stimuli-sensitive grafted side chains. Polymer 50, 2516–2525.

Zhang, Y., Wei, W., Lv, P., Wang, L., Ma, G., 2011. Preparation and evaluation of alginate–chitosan microspheres for oral delivery of insulin. Eur. J. Pharm. Biopharm. 77, 11–19.

Zhao, L., Feng, S.S., 2010. Enhanced oral bioavailability of paclitaxel formulated in vitamin E-TPGS emulsified nanoparticles of biodegradable polymers: in vitro and in vivo studies. J. Pharm. Sci. 99, 3552–3560.

Zhao, Y., Wang, C., Chow, A.H., Ren, K., Gong, T., Zhang, Z., Zheng, Y., 2010. Self-nanoemulsifying drug delivery system (SNEDDS) for oral delivery of Zedoary essential oil: formulation and bioavailability studies. Int. J. Pharm. 383, 170–177.

Zhao, L., Zhang, L., Meng, L., Wang, J., Zhai, G., 2013. Design and evaluation of a self-microemulsifying drug delivery system for apigenin. Drug Devel. Ind. Pharm. 39, 662–669.

CHAPTER

NANOVACCINES FOR ORAL DELIVERY-FORMULATION STRATEGIES AND CHALLENGES

10

Sayali Karandikar*, Amit Mirani, Vishal Waybhase*, Vandana B. Patravale**, Sandeep Patankar***

**Bharati Vidyapeeth's College of Pharmacy, Navi-Mumbai, India*

***Institute of Chemical Technology, Mumbai, India*

CHAPTER OUTLINE

1 **Introduction** 263
2 **Oral Vaccine Delivery** 264
3 **Intestinal Immune Response** 266
3.1 Antigen Uptake 266
3.2 Immune Response 268
4 **Challenges in Oral Vaccine Delivery** 268
5 **Nanocarrier Based Oral Vaccine Delivery** 268
5.1 Antigen-Loading Mechanism in Nanocarriers 269
5.2 Nanocarrier Properties Affecting Antigen Uptake 270
6 **Nanocarrier Formulation Strategies** 271
6.1 Vaccine Composition 271
6.2 Type of Nanocarrier 272
7 **Regulatory Aspects of Oral Nanovaccines** 286
7.1 Considerations for Regulating Nanocarrier-Based Vaccine Delivery System 286
8 **Future Prospects** 288
References 289

1 INTRODUCTION

Infectious diseases continue to take a heavy toll on human life and are the predominant cause of mortality all over the globe. The first scientific accomplishment to save the lives of millions from infectious disease was witnessed in 1798 due to the arduous efforts of Edward Jenner, who used cowpox virus vaccine for prevention of smallpox infection (Marasini et al., 2014; Riedel, 2005; Tennant et al., 2014). In the past few decades, research focus has been radically oriented toward the development of vaccines that will comply with the needs of patients/consumers in terms of safety, efficacy, and, more

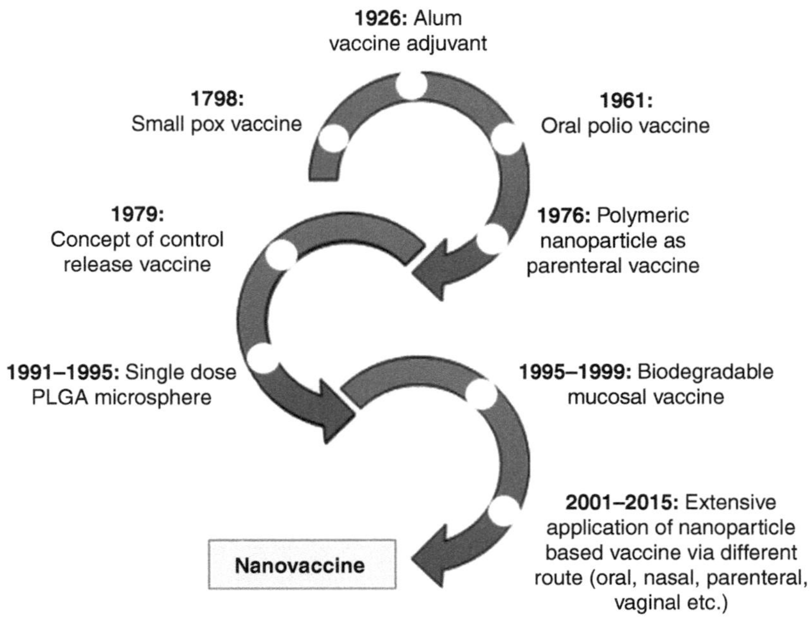

FIGURE 10.1 Milestones Achieved in Vaccine Development

judiciously, the economy. The milestones achieved in vaccine development are depicted in Fig. 10.1 (Gonzalez-Aramundiz et al., 2012).

Vaccines primarily comprise an antigen (live attenuated, killed, inactivated organism), which triggers the immune system of body so as to guard it from the hostile attack of similar antigens in the future. A vaccine containing a live attenuated organism as an antigen induces strong innate and long-lasting immune responses without the need of additional adjuvants. However, the use of live attenuated vaccines is limited due to its instability, safety issues regarding patient health (allergic reaction), and difficulties in pathogen culturing. To circumvent this problem, a vaccine containing inactivated whole organism came into existence in the nineteenth century, which possessed a weaker immunity level and required additional immune modulators (i.e., adjuvants) for effectiveness. In the past few decades, vaccine research has been steered toward the development of novel and safe adjuvant-based vaccines, such as subunit vaccines, DNA vaccines, and conjugate vaccines, which exhibit stronger immunogenicity (Gilligan et al., 1991; Marasini et al., 2014; Tennant et al., 2014).

2 ORAL VACCINE DELIVERY

To date, the most commonly employed route for the delivery of vaccines is the parenteral route (subcutaneous or intramuscular), which imparts effective cellular and systemic immune responses. However, the parenteral route has some inherent limitations, for example, low patient compliance due to its pain during administration, inability for self-medication, and lack of mucosal immune responses. Thus, there is a compelling need to develop a noninvasive route for vaccine administration. Several noninvasive routes,

Table 10.1 Commercially Available Oral Vaccines

Brand	Disease	Available Formulations	Composition	Mechanism of Action
RotaTeq/Rotarix	Gastroenteritis	Solution/suspension	Monovalent live-attenuated human rotavirus and multivalent animal-human reassortant rotavirus	Mucosal IgA and systemic neutralizing IgG antibodies specific for homotypic or heterotypic VP4 and VP7 antigens
Dukoral/Shanchol	Cholera	Suspension	Whole killed *Vibrio cholerae* O1 strains (Classical and El Tor biotypes) with CTB (Dukoral) or without CTB (Shanchol)	Gut antitoxin- and CTB-specific IgA and antibacterial and LPS-specific antibodies
Orochol	Cholera	Lyophilized product	Live recombinant vaccine lacking CTA (CVD 103-HgR)	Vibriocidal antibodies (possibly not the main effector mechanism but correlates well with protection)
Vario	Polio	Liquid	Trivalent, bivalent, and monovalent vaccines	Mucosal IgA and systemic IgG neutralizing antibodies

such as oral, nasal, vaginal, and rectal routes have been exploited in the past. Among all, the oral route is regarded as an ideal means of delivering vaccines, offering significant advantages over parenteral, nasal, rectal, and vaginal routes like noninvasive, ease of administration, capacity for mass immunization, broad safety profile, ease of production, and storage. Complexity in the manufacture and scale-up of parenteral vaccines due to their requirement of stringent environmental control and purity comparatively renders the preference to oral vaccines. The broad immune response (i.e , mucosal and systemic) associated with the oral route is propelling researchers to explore vaccine delivery for therapeutic as well as prophylactic purposes (D'Souza et al., 2014; Gamazo and Irache, 2014; Rieux et al., 2006; Wang and Coppel, 2008). To date, several studies have been reported for oral vaccines, for example, oral delivery of recombinant or attenuated strains of bacteria (such as *Salmonella*, *Escherichia coli*, *Listeria*, *Shigella*, and *Lactobacilli*) and live attenuated recombinant strain of virus (such as adenoviruses, poxviruses, influenza) as a vector antigen showing promising immune responses. The commercially available oral vaccines and those in clinical trials are listed in Tables 10.1 and 10.2 (Marasini et al., 2014).

Table 10.2 Oral Vaccines Under Clinical Trials

Brand	Disease	Formulation	Status
PXVX0103	Bird flu (avian influenza)	Capsules	Phase I clinical trial completed
PXVX0200	Cholera	Not disclosed	Phase II clinical trial completed

3 INTESTINAL IMMUNE RESPONSE

The intestine provides major immunological barriers for access of pathogenic antigens. For efficient designing of an oral vaccine, a thorough mechanism of immune response associated with orally administered antigens needs to be understood.

The gut associated lymphoid tissue (GALT) has a role to play in regulation of the intestinal immune system. It consists of two sites—inductive sites and effector sites. The antigens are first encountered at the inductive sites to initiate an immune response that is subsequently expressed at effector sites. The effector sites are loosely organized and divided into lamina propria lymphocytes and intraepithelial lymphocytes. The lamina propria lymphocytes (IgA secreting B-cells and memory T effector cells) are scattered in lamina propria, while the intraepithelial lymphocytes (cytotoxic T cells) are positioned in basolateral spaces between luminal epithelial cells. The intraepithelial lymphocytes are mainly of two types (i.e., Type A and Type B) based on their mode of antigen recognition. Type A lymphocytes are major histocompatibility complexes (MHCs) directed against the peptides from invading pathogens. Type B lymphocytes show limited T-cell receptor diversity. The inductive sites are highly organized mucosal tissue. They comprise of Peyer's patches (PPs), antigen presenting cells (APCs), and lymphoid follicles (Gilligan et al., 1991; Marasini et al., 2014; Shakweh et al., 2004; Tennant et al., 2014; Wang and Coppel, 2008). PPs are collections of lymphoid follicles and distributed throughout the small intestine. PPs mainly comprise four regions, namely, germinal center (large lymphocytes), follicular area surrounding the germinal center (small lymphocytes), parafollicular area surrounding follicular area ($CD8^+$ suppressor cells and $CD4^+$ helper T-cells), and last, the dome region just above the follicular area projecting into the intestinal lumen. The dome region comprises lymphocytes, macrophages, and dendritic cells (DCs). This area is mainly used for antigen processing and presentation. PPs are separated from lumen by follicle-associated epithelium (FAE) (Gilligan et al., 1991; Shakweh et al., 2004; Wang and Coppel, 2008).

The intestinal antigen uptake and associated immune responses are summarized in Fig. 10.2 and are briefly discussed in the next sections:

3.1 ANTIGEN UPTAKE

The tightly adhered epithelial cells of intestines facilitate the antigen uptake preferably by two mechanisms, namely, microfold cell (M cell)-mediated antigen uptake and enterocytes-mediated antigen uptake. M cells and enterocytes predominantly reside in the FAE. FAE are single layered–specialized epithelial cells overlying the PPs. The mechanisms involved in antigen uptake and their transport across intestinal epithelial cells are summarized in the next sections (Gilligan et al., 1991; Marasini et al., 2014; Shakweh et al., 2004; Tennant et al., 2014; Wang and Coppel, 2008).

3.1.1 M cell mediated antigen uptake

M cells are characterized by short irregular microvilli, with flattened apical surfaces and a low number of cytoplasmic lysosomes. M cells express surface receptors (ganglioside [GM1], b1 integrins, cellular prion protein, and C5a receptors) in luminal regions, which facilitate antigen uptake by absorptive or nonabsorptive endocytosis. M cells enable the uptake of antigens from apical surfaces of the intestinal lumen to the underlying immune system of the mucosae (macrophages/GALT). The M cell–mediated antigen uptake involves three different mechanisms (actin dependent phagocytosis by macrophages, DCs)–mediated luminal sampling, and transepithelial DC expanding dendrites to the lumen).

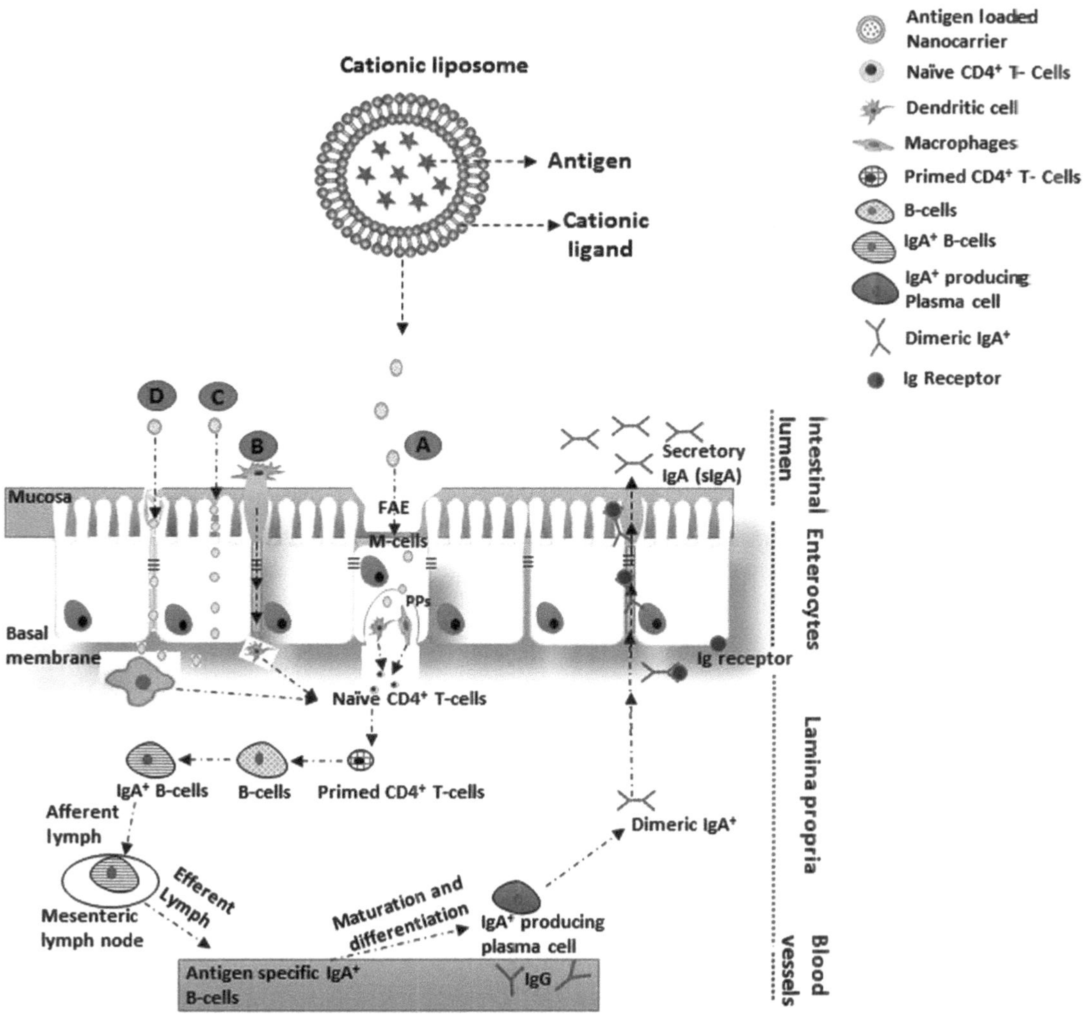

FIGURE 10.2 Nanocarrier Uptake Mechanism and Intestinal Immune Response

Nanocarrier uptake mechanism: (A) Macrophages and DCs mediated uptake via M-cell in enterocytes; (B) DCs -mediated sampling into the luminal region of the intestine; (C) transcellular uptake of macromolecules across the enterocytes; (D) paracellular uptake of large particles from the passage between the enterocytes. Intestinal immune response: (1) The antigen from intestinal lumen are taken up by local lymphocytes into inductive sites of gut associated mucosal tissue (PPs, lymphoid follicles, and APCs); (2) the antigen at inductive sites are presented to DCs. DCs process the antigen into small fragments. These antigen-loaded DCs migrate to the subepithelial dome in PPs or to the mesenteric lymph nodes and provide simulatory signals to activate naive $CD4^+$ T-cells. The activated $CD4^+$ T-cells further trigger B cells and undergo isotype switching to generate antigen-specific IgA^+ B-cells. This IgA^+ B cells leave the PPs through afferent lymphatics to the regional mesenteric lymph node and finally reach the systemic circulation. (3) Similarly, antigen presenting DCs actively migrate to mesenteric lymph nodes for further $CD4^+$ T-cell activation and subsequent IgA^+ B cell production. The antigen IgA^+ specific B cells move to effector sites in the GALT (lamina propria), where they undergo differentiation and maturation to generate high-affinity IgA^+ producing plasma cells to produce a polymeric form of IgA^+. This polymeric form of IgA binds to Ig receptors expressed on the basolateral surface of epithelial cells to form secretory IgA (sIgA). sIgA in the intestinal lumen is responsible for protection against invading organisms.

3.1.2 Enterocytes-mediated antigen uptake

Enterocytes-mediated antigen uptake involves two mechanisms, that is, paracellular (between adjacent cells) and transcellular (through the cell) mechanism. The paracellular mechanism involves uptake of large particles (>1 μ) from the passage between the enterocytes. The transcellular mechanism involves uptake of macromolecules at enterocytes lining the intestinal epithelia. This is facilitated by endocytosis where the molecule is endocytosed at apical surface and exocytosed at basolateral membrane followed by their entry into macrophages or GALT.

3.2 IMMUNE RESPONSE

The antigen from intestinal lumen are taken up by local lymphocytes into inductive sites of gut associated mucosal tissue (PPs, lymphoid follicles, and APCs. The antigen at inductive sites are presented to DCs. DCs process the antigen into small fragments. These antigen-loaded DCs migrate to the subepithelial dome in PPs or to the mesenteric lymph nodes and provide simulatory signals to activate naive $CD4^+$ T-cells. The activated $CD4^+$ T-cells further trigger B cells and undergo isotype switching to generate antigen specific IgA^+ B cells. These IgA^+ B cells leave the PPs through afferent lymphatics to the regional mesenteric lymph node and finally reach the systemic circulation. Similarly, antigen-presenting DCs actively migrate to mesenteric lymph nodes for further $CD4^+$ T-cell activation and subsequent IgA^+ B cell production. The antigen IgA^+ specific B cells move to effector sites in the GALT (lamina propria), where they undergo differentiation and maturation to generate high affinity IgA^+ producing plasma cells to produce a polymeric form of IgA^+. This polymeric form of IgA binds to Ig receptors expressed on the basolateral surface of epithelial cells to form secretory IgA (sIgA). sIgA in the intestinal lumen is responsible for protection against invading organisms (Marasini et al., 2014).

4 CHALLENGES IN ORAL VACCINE DELIVERY

The possibility of achieving mucosal and systemic immunization, in case of oral route delivery, renders it an obvious preference for vaccine administration. However, it exhibits several formulation hurdles, namely, (1) the antigen is prone to degradation in the gastrointestinal tract (GIT), due to susceptibility of the former to the harsh gastrointestinal mileu; (2) the short exposure time of antigen in mucosal tissue is responsible for its lower intestinal uptake. Owing to this, the high doses of antigen are required to elicit immune response. However, the high dose does not always ensure enhanced immune response. (3) Dilution of antigen due to the large surface area of the GIT creates a need for higher quantities of antigen/adjuvant formulation and consequently, higher cost; (4) conventional subunit oral vaccine (Gamazo and Irache, 2014; Marasini et al., 2014; Rieux et al., 2006; Wang and Coppel, 2008; Wilkhu et al., 2011).

Thus, the aforementioned drawbacks suggest the need for the development of a novel oral delivery system, which not only improves immunogenicity and efficacy but also reduces toxicity.

5 NANOCARRIER BASED ORAL VACCINE DELIVERY

To combat the aforementioned problems associated with oral vaccines, nanocarrier-based oral vaccine delivery systems were introduced. Nanocarriers refer to particulate systems with sizes within the range of 1–1000 nm. Nanocarrier-based systems offer several advantages, such as improving oral bioavailability at GALT (by improving stability of antigen in the gastric environment), site-specific delivery and reduced side effects and enhancement of immunogenicity due to their self-adjuvant properties. Nanocarrier-based systems induce both cellular and humoral immune response, which may be attributed to rapid uptake of nanoparticle systems across intestinal lumen by M cells leading to efficient antigen recognition and presentation to APCs (DCs and macrophages). The APCs further activate T-cells and B cells to activate systemic and mucosal immune response (Gonzalez-Aramundiz et al., 2012; Rieux et al., 2006; Zaman et al., 2013; Zhao et al., 2014). To achieve prompt immune response, it is necessary to understand factors, such as antigen-loading mechanisms in nanocarrier and nanocarrier properties affecting their intestinal uptake.

5.1 ANTIGEN-LOADING MECHANISM IN NANOCARRIERS

The antigen can be loaded in nanocarrier by several mechanisms, namely, physical adsorption, encapsulation, encapsulation with coating, encapsulation with targeting, chemical conjugation, and conjugation with targeting mechanism (Fig. 10.3). The physical adsorption of antigen onto a nanocarrier can be achieved either by charge or hydrophobic interaction. Therefore, the interaction between nanocarrier and antigen is relatively weak, which may lead to rapid dissociation of antigen and nanocarrier in vivo Similarly, in encapsulation, antigens are mixed with nanocarrier precursors during synthesis, resulting in the encapsulation of antigen into a nanocarrier. Antigen is released only when nanoparticles are dissembled in vivo or inside the cell. The nanocarrier encapsulated with antigen can be targeted to a specific site by surface coating of targeting ligand. On the other hand, for chemical conjugation, antigen is chemically crosslinked to the surface of a nanoparticle. Antigen is taken up by the cell together with the nanoparticle and is then released inside the cell. Similarly, the nanocarrier conjugated with antigen

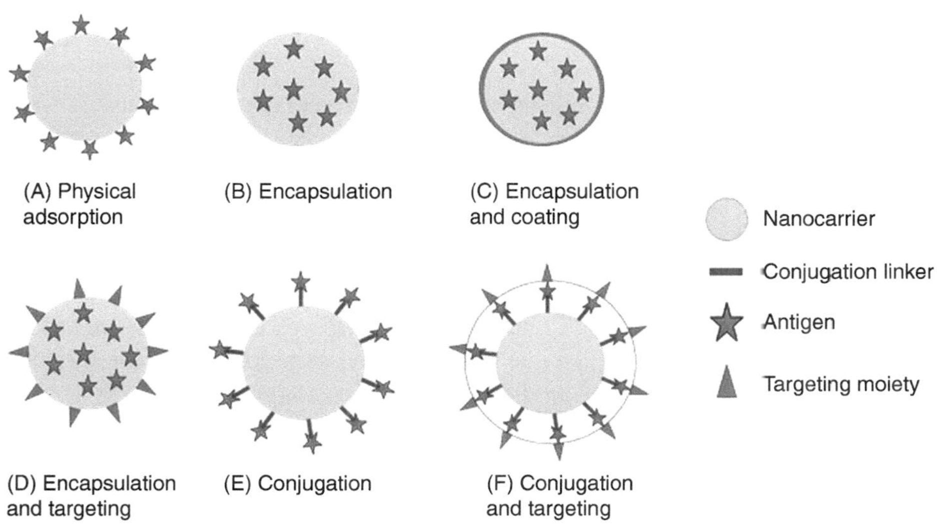

FIGURE 10.3 Antigen-Loading Mechanism in Nanocarriers

(A) Physical adsorption; (B) encapsulation; (C) encapsulation and coating; (D) encapsulation and targeting; (E) conjugation; (F) conjugation and targeting.

can be targeted to the specific site by surface coating with targeting ligand (Oyewumi et al., 2010; Wendorf et al., 2006; Zhao et al., 2014).

5.2 NANOCARRIER PROPERTIES AFFECTING ANTIGEN UPTAKE

Nanocarrier uptake by intestinal M cells and enterocytes involves binding to the surface receptor in the intestine followed by their uptake by specific endocytosis pathway as explained in Section 3.1. The antigen uptake is influenced by several physicochemical parameters of nanocarriers, such as size, charge, shape, hydrophilicity, and encapsulation efficiency. The nanocarrier uptake in turn is significantly influenced by the immune response (Marasini et al., 2014; Pattani et al., 2014; Shakweh et al., 2004; Wilkhu et al., 2011). These parameters are briefly discussed in the next section.

5.2.1 Size

The particle size of nanovaccine plays a pivotal role in its uptake of antigen by M cells and enterocytes. A delivery system with particle size greater than 5 μm can be readily taken up by M cells but retained for longer periods in PPs (34 h). Particles with size range within 5–0.5 μm can be readily taken up by the M cells and phagocytosed resulting in a humoral immune response. The systems with medium particle size 150–500 nm are mostly carried into the lymphatic system by DCs. The systems with particle sizes in the range of 20–50 nm particles are directly delivered to peripheral lymph node via enterocytes. However, study results comparing the effect of particle size on immune responses are inconsistent. Gutierro et al. (2002) reported a higher IgG titer obtained with 1 μm particles compared with 200–500 nm particles. Awaad et al. (2012) studied the uptake of particles with different size ranges (95, 130, 200, 340, 695, and 1050 nm) and concluded that the particles with a size range of 95–200 nm represent optimum M cells uptake compared to antigens with particle sizes >300 nm.

5.2.2 Particle charge

Nanocarrier surface charge plays a significant role in particle uptake by PPs because M cells and mucosal lining have a negative charge, which permits electrostatic interaction with nanocarriers carrying a positive charge, hence favoring ready uptake of nanocarriers with a cationic charge. Foged et al. (2005) concluded that the cationic surface could significantly enhance the uptake of particles by macrophages and DCs. The surface charge of particles also influences intracellular localization, for example, negatively and neutrally charged nanoparticles colocalize with lysosomes as against positively charged nanoparticles, which could manage to escape from lysosomes after being internalized (Foged et al., 2005). Contrary to this, Shakweh et al. (2005) studied the nano- and microparticles of poly(lactic-*co*-glycolic acid) (PLGA) exhibiting different surface charges and prepared them by using different stabilizing agents. The stabilizers used were polyvinyl alcohol (negatively charged), hydrophobically modified hydroxyethyl cellulose (neutral), polyethyeneimine (positively charged). The results showed that the positively charged nano/microparticles exhibit less uptake compared to negative and neutral charged particles. This may be attributed to the strong electrostatic interaction between the positively charged nanocarriers and negatively charged membrane, leading to the reduction in their uptake. The increase in uptake of negatively charged molecules could be due to their interaction with the hydrophilic mucus layer, thus favoring internalization(Shakweh et al., 2005).

5.2.3 Particle shape

Particle shapes of nanocarriers may have an influence on the immune response as specific shapes of particles facilitates uptake and phagocytosis by M cells. The particles, such as spherical, cylindrical, or rod shaped are endocytosed more efficiently compared to disc-shaped particles. For example, Nikura et al. (2013) revealed that the spherical gold nanoparticles were more efficiently taken up by APC uptake compared to cube- and rod-shape nanoparticles. Champion and Mitragotri (2009) revealed that worm-like particles exhibit negligible phagocytosis compared to spherical particles. Yoo et al. (2010) studied the effect of particle shape (spherical and elliptical disk) on rate of endocytosis by endothelial cells. The study showed elliptical particles were slowly endocytosed compared to spherical particles (Yoo et al., 2010).

5.2.4 Hydrophobicity

Nanocarriers with high-surface hydrophobicity, result in higher PPs uptake, hence, they show better immune response than those containing hydrophilic ones. The nanocarriers with hydrophobic polymers such as PLA, PLGA showed higher PPs uptake compared to those with hydrophilic polymers, such as hydroxyl propyl methyl cellulose and so on. Eldridge et al. (1990) studied the effect of different PLGA polymers for their uptake by PPs. The data revealed that the microparticles with hydrophobic PLGA (60 and 100 kDa) exhibit more uptake compared to particles with less hydrophobic PLGA (20 kDa) (Eldridge et al., 1990).

6 NANOCARRIER FORMULATION STRATEGIES

6.1 VACCINE COMPOSITION

The vaccine composition plays a crucial role in eliciting a substantial immune response. The vaccine primarily comprises an antigen (to provoke an immune response), and may have in conjunction an adjuvant (to induce or enhance immune response) and excipients. The excipients must be necessarily nontoxic and nonreactive with the antigen.

6.1.1 Antigen

As discussed in the previous sections, different types of antigen (live attenuated, killed organism, subunit, etc.) have been harnessed to develop vaccines. To develop an efficient vaccine, antigen must possess high purity, inherent immunogenicity, nontoxicity, and be proven to be safe for human use.

6.1.2 Adjuvant

Adjuvants are the chemical substances that induce or enhance the antigen specific immune response without precipitating any toxicity. The rationale for selection of adjuvant depends on factors, such as the type of antigen, nature of immune response, and route of vaccination. Adjuvanticity can be achieved by using chemical entity, which activates the immune system (immunomodulator) through its interaction with specific receptors (Toll-like receptors, TLR, agonists, NOD-like receptors, agonist, etc.) of the innate immune system. Adjuvant must be nontoxic, non antigenic, biocompatible, biodegradable, stable, and cost effective, and should be able to induce both humoral and cell-mediated immunity.

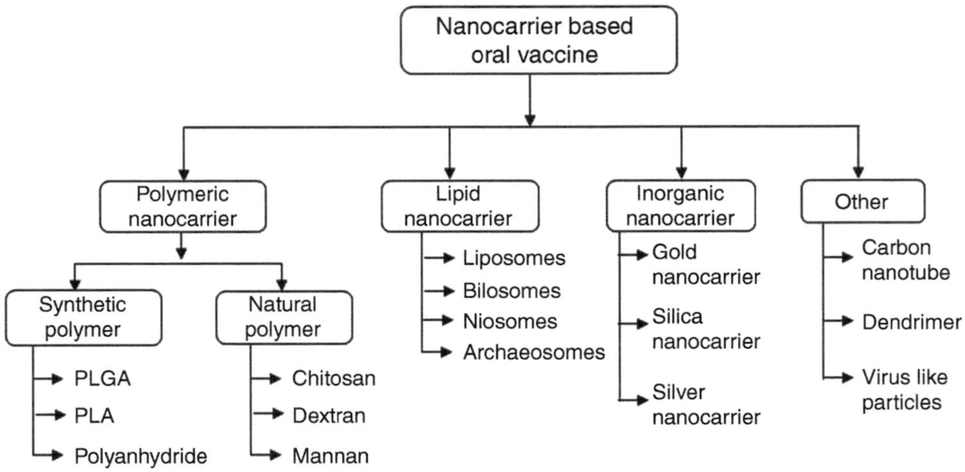

FIGURE 10.4 Classification of Nanocarrier-Based Oral Vaccine

6.2 TYPE OF NANOCARRIER

Nanocarriers for oral vaccine delivery have a noteworthy influence on immune response. Nanocarriers of each class exhibit varying stability profiles and cellular uptake mechanism, which correspondingly affects the immune response. Strikingly, the polymers (PLGA, PLA, Dextran, etc.) employed for fabrication of nanocarriers also act as self-adjuvants, thus exhibiting additive effect when loaded with antigen. This emphasizes the need to critically consider the type of nanocarrier and its fabrication while devising strategies for effective oral vaccine delivery. The key parameters to be considered for the former are antigen properties, such as antigen solubility, antigen dose, pH sensitivity of antigen, and formulation properties like ease in fabrication, self-adjuvant properties, no requirement for special storage conditions. Different types of nanocarriers, namely, polymeric nanocarriers (synthetic and natural source), lipidic nanocarriers, and inorganic nanocarriers have been exploited for antigen delivery (Kim et al., 2014). These are depicted in Fig. 10.4 and are briefly discussed in Table 10.3

6.2.1 Polymeric nanocarrier

Polymeric nanocarriers are colloidal systems exhibiting a size range 10–1000 nm (Fig. 10.5). Over the past few decades, polymeric nanoparticles have been explicitly tapped for their potential in antigen delivery basically owing to their high immunogenicity and stability after oral administration. The immunogenicity potential of polymeric nanocarriers is mainly attributed to their self-adjuvant properties and inherently smaller particle size together contributing favorably in facilitating increased APCs uptake. The enhanced stability features are imparted due to the feature of antigen encapsulation. Numerous polymers from both natural (alginate, starch, dextran, chitosan) and synthetic (PLGA, polyanhydride) origin have been utilized in the past for antigen delivery. Among these, the natural source polymers seem to be beneficial due to their low cost, biocompatibility, and aqueous solubility. However, they do come in conjunction with certain disadvantages like the presence of extraneous matter, an aspect unavoidable due to their natural origin source with consequential batch-to-batch variability. As against this, the synthetic polymers can be manufactured with more reproducibility and can be fine-tuned so as

Table 10.3 Nanocarrier-Based System for Oral Vaccine Delivery

Nanosystem	Polymer	Composition	Antigen Loading	Size	Observations	References
Polymeric Nanocarrier advantages: 1. Readily surface modified for enhanced immunogenicity. 2. Effectively used for targeting. Disadvantages: 1. Limited aqueous solubility. 2. Premature release. 3. Low loading. 4. Require organic solvent.	PLGA	a. Ovalbumin antigen and MPLA as adjuvant	9.6%	320.0 nm	Increased uptake by M-cells of Peyer's patches. Showed both systemic and mucosal immune response.	Sarti et al. (2011)
		b. Ulex europaeus lectin anchored nanoparticle loaded with HBsAg antigen, trehalose and magnesium hydroxide	45.35%	380.3 nm	Trehalose and magnesium hydroxide prevented degradation is gastric environment. Facilitated M cell mediated uptake and induced systemic and mucosal immunity.	Gupta et al. (2007)
	Dextran	a. Imiquimod loaded in acetylated dextran nanoparticles	100%	233 nm	Imiquimod loaded nanoparticles increased Toll like receptor (TLR-7) and thus induced more immune response compared to conventional Imiquimod formulation. Increased MHC-I and MHC-II presentation.	Bachelder et al. (2010)
Lipid Nanocarrier Advantages: 1. Self-adjuvants. 2. Encapsulate hydrophilic and lipophilic drug. 3. Increase permeation across intestine. Disadvantages: 1. Poor GI stability. 2. Premature antigen release.	Liposomes	a. Lectin-anchored liposomes	—	100 nm	Enhanced lectin mediated uptake across M-cell.	Chen et al. (1996a)
	Bilosomes	a. HBsAg loaded in bile salts vesicles	20%–24%	220.0 nm	Increased stability due to presence of bile salts. Rapid uptake across intestinal lumen. Induced both systemic and mucosal immunity.	Shukla et al. (2010)
		b. Glucomannosylated bilosomes loaded with BSA antigen	7.13%	154.0 nm	Increased GI stability due to presence of bile salts. Induced systemic and mucosal immune response.	Indulkar et al. (2014)
	Niosomes	a. Ovalbumin loaded in non-ionic surfactant, that is, Wasag 7 (stearic sucrose ester 70% and palmitate sucrose ester 30%)	Not available	Not available	Better stability compared to conventional liposomes. Presence of surfactant facilitated enhanced uptake across intestine. Induced both systemic and mucosal immune response.	Rentel et al. (1999)
	Archaeosomes	a. Bovine serum albumin antigen loaded in polar lipid extracted from *Sulfobus acidocaldarius*	Not available	500 nm	Increased stability in GI tract and showed enhanced immune response.	Li et al. (2011)

(*Continued*)

Table 10.3 Nanocarrier-Based System for Oral Vaccine Delivery (*cont.*)

Nanosystem	Polymer	Composition	Antigen Loading	Size	Observations	References
Inorganic Nanocarrier Advantages: 1. Better protection of adsorbed antigen. 2. No premature release. 3. Easy surface modification. Disadvantages: 1. Low biodegradability. 2. No role as self-adjuvant. 3. Poor aqueous solubility.	Gold nanoparticle	a. *Tetanus toxoid* and quillaja saponin loaded gold nanoparticle coated with chitosan	60%	40 nm	Stable in GI system and showed higher loading. Induced both systemic and mucosal immune response. Quillaja saponin served as immunomodulator.	Barhate et al. (2013)
	Silica nanoparticle	a. Bovine serum albumin antigen loaded silica nanoparticles	Not available	Not available	Provided slow release. Non-toxic and showed higher stability in GI tract. Induced both systemic and mucosal immune response.	Wang et al. (2011)
	Silver nanoparticle	a. Plasmid encoding hemagglutinin gene of avian influenza virus loaded in silver nanoparticle coated with polyethylene glycol	Not available	4–18 nm	Showed higher stability in GI tract. Showed enhanced uptake across intestine.	Jazayeri et al. (2012)
ISCOMs Advantages: 1. Self-adjuvants. 2. High loading of lipophilic antigens. Disadvantages: 1. No depot release. 2. Not suitable for hydrophilic antigens.	Quillaja saponin extract	a. Influenza virus loaded in ISCOMs	Not available	Not available	High immune response due to presence of saponin; induced both systemic and mucosal immune response; better stability	Ghazi et al. (1995)
Virus Like Particles (VLPs) Advantages: 1. Mimics original virus. 2. Self-adjuvants. 3. Higher GI stability. Disadvantages: 1. Polydispersed particle size. 2. Batch to batch variations.	Virus like particle	a. HBsAg loaded in VLPs	Not available	20–100 nm	Mimic virus like entry; more stable in GI tract; exhibited better uptake by M cells, enterocytes, dendritic cells across the intestine and showed stronger immune response	Shah et al. (2015)

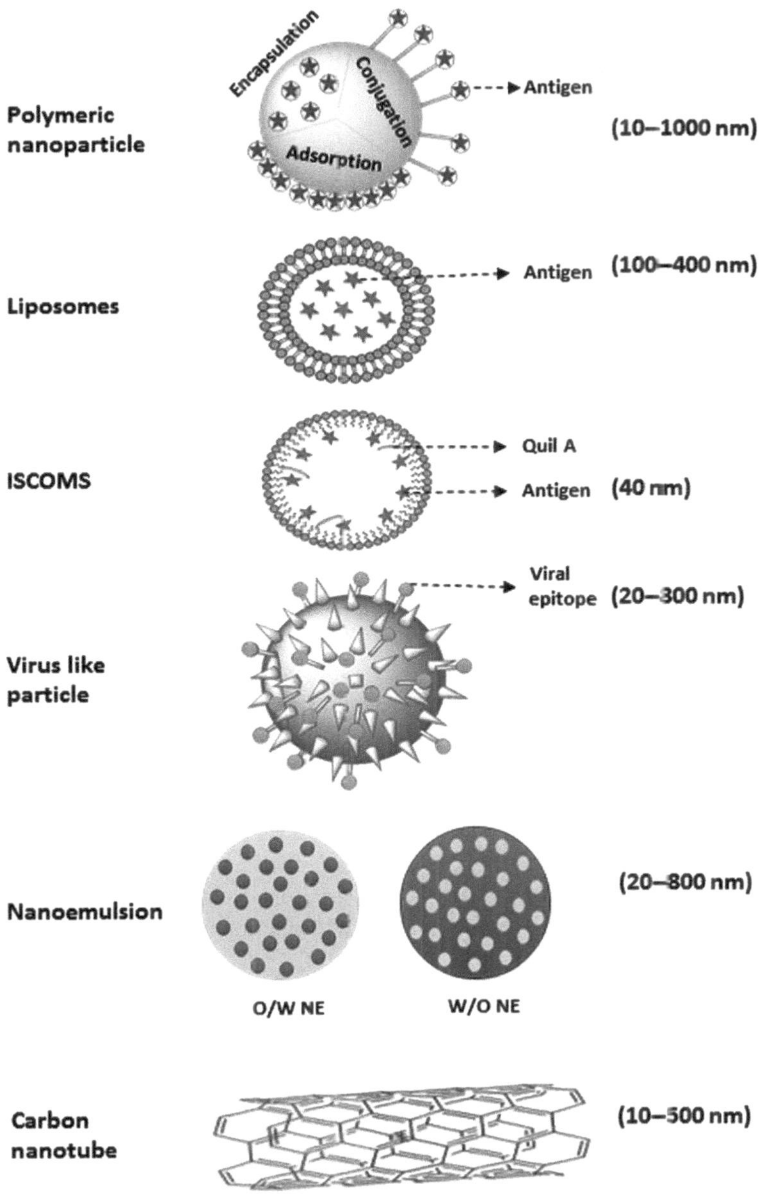

FIGURE 10.5 Illustration of Nanocarrier-Based Oral Vaccine

to achieve controlled molecular weight compositions of polymer and copolymer-desired degradation rates. Different synthetic and natural source-based polymeric nanocarriers are briefly discussed in the following section.

6.2.1.1 Poly(lactic-*co*-glycolic acid)

PLGA is a biodegradable copolymer ester of two α-hydroxyacids (lactic and glycolic acids) and has been approved by the FDA for antigen delivery. PLGA nanocarriers undergo degradation by bulk erosion mechanism, thereby eliciting sustained release of antigen from the matrix. PLGA nanoparticles are efficiently taken up by M cells across the intestinal epithelial cell, thus promoting a strong immune response. It does not interfere in the normal functioning of DC like maturation, migration, and cytokine secretion. The PLGA polymer is reported to possess self-adjuvant properties, consequently their nanocarrier with a specific antigen may display an additive immune response (Allahyari and Mohit, 2016). Several parameters may affect the PLGA nanocarrier mediated immune response, namely, polymer properties (hydrophilicity, crystallinity, molecular weight), nanocarrier size, charge, encapsulation efficiency, and antigen release. The PLGA nanocarriers with significantly smaller particle sizes (~200nm) effectively cross the intestinal epithelial barrier and are carried toward the B cell by DC cells of PPs. PLGA nanoparticles effectively encapsulate antigens and counter their degradation by proteolytic enzymes. However, the polymer is susceptible to degradation at gastric pH. To overcome this hurdle, the nanoparticles need to be surface modified using polyethylene glycol (PEG), thus preventing their degradation and facilitating their uptake across the intestinal epithelium (Garinot et al., 2007). Recently, several studies have been conducted using PLGA-based oral nanovaccines and are discussed in the following section.

Sarti et al. (2011) developed PLGA nanoparticles loaded with ovalbumin antigen (OVA) and monophosphoryl lipid A (MPLA) as an adjuvant for oral immunization in mice. The PLGA nanoparticles were spherical in shape, 320 nm in size, and negatively charged (−20 mV) with a payload of around 8.6% of ovalbumin and 0.96% of MPLA. The developed nanocarriers were slowly hydrolyzed and effectively taken by APCs across the intestinal epithelium cells, thus inducing time-dependent systemic and mucosal immune response. The ovalbumin/MPLA loaded PLGA nanoparticles showed an 8.5 time increase in serum IgG levels compared to soluble antigen. The increased immune response is attributed to encapsulation of antigen in the nanosystem. The developed nanoparticulate system also exhibits an increase in IgA titer, which could be possibly due to induction of T helper lymphocytes (Th1) and cytolytic T lymphocytes (CTL T) responses.

Gupta et al. (2007) developed hepatitis B antigen (HBsAg) encapsulated PLGA nanoparticles for oral delivery. PLGA nanoparticles are, however, prone to degradation in acidic environment and loose antigenicity in the presence of organic solvent. To circumvent this limitation, trehalose and magnesium hydroxide were used in combination with antigen. Trehalose is a protein stabilizer so it afforded protection from degradation of antigen in organic solvents whereas magnesium hydroxide complemented by creating basic microenvironment and thereby maintaining stability of antigen. The said HBsAg-loaded PLGA nanoparticles further anchored with lectin ulex europaeus so as to target M cells of PPs and showed enhanced immunogenicity (systemic and mucosal immunity) as compared to unmodified PLGA nanoparticles.

Ma et al. (2014) developed ulex europaeus agglutinin-1 (UEA-1) anchored PLGA nanoparticles loaded with OVA and MPLA as an adjuvant for oral immunization in mice. The PLGA nanoparticles were spherical in shape, 220 nm in size, and showed loading capacity of 15.9% and encapsulation efficiency of 95%. The developed nanocarriers slowly hydrolyzed in gastric environment and were readily uptaken by APCs via M cells in intestinal lumen and were carried toward the DCs, which subsequently activated the T-cell and B cell–mediated immune response. This resulted in induced enhanced systemic and mucosal immune response by the developed nanocarriers as compared to unmodified PLGA nanocarriers.

6.2.1.2 Poly(D,L-lactic acid)

PLA is a biodegradable and biocompatible polymer, which undergoes erosion to form monomeric units of lactic acid. PLA nanoparticles are considered to be safe and devoid of any severe toxicity. PLA has been widely explored for antigen delivery in the past due to its self-adjuvant properties. PLA nanoparticles loaded with oral antigen are readily taken up by M cells across the intestinal epithelium and induce both mucosal and systemic immune response. Recently, several studies have been performed using PLA-based nanocarriers for oral vaccine delivery and are discussed in following section.

Makadia and Siegel (2011) developed polyethylene glycol conjugated PLA nanoparticles loaded with hepatitis B antigen for oral vaccine delivery. The data revealed that the PEG-PLA nanoparticles showed enhanced intestinal uptake and induced antigen specific humoral and cellular immune responses without requiring any booster dose. The conjugation of PEG to PLA resulted in lowering of encapsulation efficiency of hepatitis B surface antigen within the nanoparticles.

Primard et al. (2013) studied the uptake capacity of PLA-based nanocarrier for oral antigen delivery. The PLA nanoparticles, being 200 nm in size, were effectively taken up from intestinal mucosa to PPs compared to soluble antigen and thus induced enhanced mucosal and systemic immune response.

Jain et al. (2014) developed hepatitis B surface antigen-loaded nanoparticles prepared by using tricoblock polymer of PLA and polyethylene for oral vaccine delivery. The results showed better stability of nanoparticles in gastric environment and hence maintained antigenicity of the encapsulated hepatitis B antigen. PEG conjugation at both ends of PLA provided efficient cellular uptake and induced enhanced cellular, humoral, and mucosal immunity compared to unmodified nanoparticles.

6.2.1.3 Polyanhydride

Polyanhydride is a copolymer of methyl vinyl and maleic anhydride, which provides hydrophobic backbone with hydrolytically labile anhydride linkages. Polyanhydride is noncytotoxic, nonmutagenic, biocompatible, and has excellent sustained release properties (Gómez et al., 2007). Recently, polyanhydride-based nanocarriers has been approved by the FDA for its application on cancer chemotherapy. Polyanhydride nanoparticles have been widely explored for antigen delivery due to their self-adjuvant properties. The physicochemical properties of polyanhydride, such as hydrophilicity, crystallinity affect the immune responses. The surface modification of polyanhydride with PEG resulted in increased hydrophilicity and hence showed an increase in release of encapsulated antigen from nanoparticles. The polyanhydride-based nanoparticles fabricated by solvent displacement method required lesser amounts of organic solvents, hence minimal effect of organic solvent on antigen degradation was observed. Polyanhydride possessing mucoadhesive property was rapidly taken up by intestinal lumen and facilitated mucosal immune response. Recently, several studies have been performed using polyanhydride-based oral nanovaccine and are discussed in following section.

Salman et al. (2007) developed ovalbumin-loaded polyanhydride nanoparticles (300–400 nm) coated with mannose or *Salmonella enteritidis* derived flagellin for oral delivery. Mannose interacted with mannose receptors and facilitated enhanced intestinal lumen uptake compared to uncoated antigen loaded nanoparticles. The developed nanoparticles also revealed increased systemic immune response

(Th1 and Th2 response) as compared to uncoated nanocarriers. In addition to this, oral immunization using the surface coated or uncoated nanoparticles elicited higher mucosal immunity compared to their administration by subcutaneous route.

Ojer (2013) studied the cytotoxicity and cell interaction pattern of antigen-loaded polyanhydride conventional nanoparticles, nanoparticles containing 2-hydroxypropyl-β-cyclodextrin (NP-HPCD) and nanoparticles coated with PEG 6000. The nanoparticles displayed a size of about 175 nm with negative surface charge. The results showed that all three nanoparticles showed cytoadhesion on the cell surface and also showed no cytotoxicity.

Gómez et al. (2007) developed polyanhydride based nanoparticles containing OVA and lipopolysaccharide (LPS) as an adjuvant for oral immunization in mice. The result revealed that the OVA-LPS-NPs exhibited higher immune response compared to OVA-NPs. OVA-LPS-NPs showed higher affinity for proximal parts of the intestine and associated with Th-2-like responses.

6.2.1.4 Chitosan

Chitosan is a linear polysaccharide composed of randomly arranged β-linked-D-glucosamine and *N*-acetyl-D-glucosamine. Chitosan has been widely explored for antigen delivery due to its biocompatibility, biodegradability, and self-adjuvant properties. Additionally, it has bioadhesive properties that not only enhance its intestinal uptake but also provide antigen protection from the gastric environment (Sayın et al., 2008). Furthermore, chitosan can be customized depending on solubility profile over a wide pH range, providing more flexibility for antigen encapsulation and release (Jabbal-Gill et al., 2012). Recently, several studies have been performed using chitosan nanocarrier-based oral vaccine delivery and these are briefly described as follows.

Li et al. (2009) performed a study to investigate the effect of chitosan-DNA nanoparticles on immune response in mice after oral administration. The nanoparticles were prepared by complexing chitosan with plasmid DNA. DNA was completely encapsulated into chitosan-DNA nanoparticles. Encapsulation in nanoparticles offered protection to DNA from degradation in the GIT at different concentrations and pH. Cytotoxicity of chitosan was also evaluated and nontoxicity was confirmed. Gene expression studies were performed and the results showed that oral gene delivery with chitosan-DNA nanoparticles can generate higher-level expression of gene in vivo.

Oliveira et al. (2012) developed a novel approach of oral vaccination for helminth infection schistosomiasis with DNA-chitosan nanoparticles. Schistosomiasis is a severe helminth infection with high levels of morbidity and mortality. Chitosan nanoparticles were loaded with plasmid DNA encoding Rho1-GTPase protein of *Schistosoma mansoni.* The prepared nanoparticles were found to be stable in simulated intestinal and gastric fluids. The results of in vivo studies performed in mice suggested that oral immunization was not able to induce high levels of specific antibodies but rather induced high levels of the modulatory cytokine IL-10. Mice immunized with plain chitosan nanoparticles presented only 47% protection against parasite infection compared to chitosan-encapsulated nanoparticles. Thus, the study concluded that chitosan plays an important role in inducing a protective immune response against schistosomiasis.

Like encapsulation, antigens can also be adsorbed onto the surface of chitosan microparticles via electrostatic interaction. These microparticles can be further coated with alginate. The alginate coating over chitosan microparticles possess two advantages—they provide stability to the antigen adsorbed on chitosan microparticle surface and prevents immediate desorption of antigen from the particle surface in the GIT.

Li et al. (2008) prepared alginate-coated chitosan microparticles using bovine serum albumin (BSA) as a model antigen. A layer-by-layer technology was employed for coating the BSA-loaded chitosan microparticles with sodium alginate. As demonstrated by the results, the release of BSA from the nanoparticles was modulated by the alginate coating and the latter also provided protection to the antigen from degradation in acidic medium, as proved by the in vitro studies.

6.2.1.5 Dextran

Dextrans are polymers mainly consisting of α-1,6-glycosides with small α (1,3) branches and obtained from bacteria, particularly from *Lactobacillus*, *Leuconostoc*, and *Streptococcus* species. Dextran is high molecular weight, water soluble, and hydrolyzed only under highly acidic or alkaline conditions. Dextran has been widely explored for antigen delivery due to their self-adjuvant, biocompatibility, biodegradability, ease in availability, and low cost (Cordeiro et al., 2015). Dextran can be further customized to make it more stable and possess better adjuvanticity. For example, high molecular weight dextran derivative, that is,. diethylaminoethyl (DEAE) dextran, a polycationic polymer, showed enhanced adjuvant properties against foot-and-mouth disease and Venezuelan equine encephalomyelitis in mammals (Adams and Mallapragada, 2014). Bachelder et al. (2010) explored acetylated dextran based microparticles as delivery platforms for vaccine adjuvants. Acetylated dextran microparticles were loaded with imiquimod so as to increase Toll-like receptor (TLR-7), a potent inducer of immune response for oral delivery. Acetylated dextran is water soluble and biocompatible. Acetylated dextran showed rapid release of antigen in acidic environments as it shows pH-dependent degradation by acetyl hydrolysis. This pH-sensitivity facilitates a significant increase in both MHC I and MHC II presentation relative to other carrier materials. The degradation of acetylated dextran microparticles does not lead to acidic by-products and thus it is suitable to be used for pH sensitive antigens. The results showed that the imiquimod-induced higher immune response compared to conventional delivery of imiquimod. The encapsulation of the imiquimod in acid-degradable particles can significantly enhance the in vitro activation of macrophages and DCs.

6.2.1.6 Mannan

Chemically, mannans are polymers consisting of D-mannose monomers connected by β-(1,4) bonds. These are storage polysaccharides that have the ability to interact with C-type lectins, which are a group of receptors expressed in APCs, that is, DCs and macrophages. These receptors, which include the mannose receptor and the mannose-binding lectin, are responsible for the recognition of different carbohydrates present in the cell wall of certain pathogens. Harde et al. (2014) developed glucomannosylated (GM) chitosan nanoparticles for oral immunization using BSA as a model antigen. Glucomannan was used as a site-directed ligand to target macrophages. The developed nanoformulation was found to have improved stability in gastric environments and provide specificity toward the M cell. This facilitated rapid uptake across the intestinal lumen to APCs, which then activate immune response. The GM chitosan nanoparticle induced stronger immune response compared to non-GM ones.

6.2.2 Lipid nanocarriers

Lipid nanocarriers are colloidal system with a size range 10–1000 nm. Over the past few decades, polymeric nanoparticles have been widely explored for antigen delivery owing to their ability to incorporate antigen and protect it from harsh gastric environment. Several lipid-based systems that have been explored in past for antigen delivery are liposomes, bilosomes, niosomes, archaeosomes, nanoemulsion, and so on, which are briefly discussed in the following section.

6.2.2.1 Liposomes

Liposomes are vesicular structures composed of one or more phospholipid bilayers surrounding an aqueous core (Fig. 10.5). Liposome represents an ideal nanocarrier for both hydrophilic and lipophilic entities (Torchilin, 2005). Liposomes exhibit enhanced intestinal uptake, better encapsulation efficiency of both antigen and adjuvants, and are available in varied size ranges. These characteristics make them ideal for antigen delivery (Henriksen-Lacey et al., 2011). Antigens are loaded in liposomes via covalent lipid conjugation (either pre- or postvesicle formation), noncovalent surface attachment (biotin/antibody-epitope interactions), encapsulation, and surface adsorption. Liposomes are readily taken up across intestinal lumen via M cells and enterocytes to the PPs and there they activate APC, DCs, macrophages to induce humoral and cell-mediated immune responses (Fujii et al., 1993). Several parameters of liposomes, such as vesicle size, bilayer structure, lamellarity, charge, and lipid transition temperature may affect the immune response and hence need to be considered while designing oral liposomal based vaccine. Large vesicles (250–750 nm) showed better Th1 response. Further, unilamellar vesicles induce more immune response compared to multilamellar vesicles. In terms of the charge of liposomes, cationic liposomes promote stronger immune response compared to neutral or anionic one (Kraajeeld et al., 1984). The liposomes with lipid transition temperature slightly higher than the body temperature elicit higher immune response, which may be due to formation of less rigid lipid and thereby enhance mucosal adherence. Recently, several studies have been performed using liposome based oral nanovaccine and are discussed in the following section.

Chen et al. (1996b) developed antigen-loaded polymeric liposomes for oral vaccine delivery. The results showed that the polymeric liposomes are more stable in gastric environment compared to conventional liposomes. The polymeric liposomes are readily taken up across the intestinal epithelium to macrophages, DCs owing to their size and induce stronger immune responses.

Chen et al. (1996a) developed lectin-modified liposomes loaded with antigen for oral vaccine delivery. The lectins used were ulex europaeus agglutinin I and wheat germ agglutinin modified with polymer *N*-glutaryl-phosphatidylethanolamine. The results revealed that the lectin coated liposomes are readily taken up across the intestinal lumen to PPs. This may be attributed to the affinity of lectin-modified liposomes toward the intestinal lectin receptors. Higher affinity to PPs led to a higher uptake of these liposomes. The modified liposomes showed enhanced immune responses compared to conventional liposomes.

Watarai and Sasaki (2014) studied the effect of cationic liposomes for oral vaccine delivery. Stearylamine was used to make cationic liposomes. The modified liposomes are readily taken up across the intestinal lumen due to presence of positive charge on surface and presented to macrophages and DC cells, which further induced antigen specific Th1 and Th2 immune response in mice.

Minato et al. (2003) developed polyethylene glycol coated liposomes loaded with ovalbumin as an antigen for oral delivery. The PEG-modified liposomes (1.7–3.3 μm multilamellar vesicles) induce a higher immune response compared to low dose to unmodified liposomes when orally administered in mice. The enhanced immune response is attributed to their intestinal uptake by M cells in PPs. After the uptake, the antigen is then carried forward to DC and induces mucosal immune response. The PEG-coated formulation, being hydrophilic, suppressed their uptake from macrophages and thus showed lower systemic immune responses.

6.2.2.2 Bilosomes

Bilosomes are liposomes containing biodegradable and biocompatible bile salts (e.g., sodium deoxycholate). Incorporation of bile salts in liposome not only enhanced their stability in the gastric environment but also facilitate increased permeability across the intestinal lumen. Unlike liposomes, bilosomes do not require special storage conditions. Bilosomes have been widely explored for antigen delivery because of their self-adjuvant properties, high antigen encapsulation, different size vesicle formation, and, importantly, their rapid and efficient uptake via M cells and enterocytes of intestinal lumen (Conacher et al., 2001; Patravale and Prabhu, 2014). Bilosomes are reported to induce mucosal immunity not only at the site of induction but also at the other distant mucosal sites. Recently, several studies have been performed using bilosomes-based oral nanovaccine and are discussed in following section.

Indulkar et al. (2014) developed GM bilosomes loaded with bovine serum albumin antigen for oral delivery. The GM bilosomes have vesicle sizes of 157 nm and surface charges of −21.8 mV with entrapment efficiency of 71.3 ± 4.3%. The results revealed that the GM bilosomes are stable in a gastric environment and provide sustained release of antigen. The GM bilosomes are readily taken up across the intestinal lumen and carried to APCs, that is, macrophages that induce both systemic and mucosal immune response. The GM bilosomes induce higher immune responses compared to that obtained with ovalbumin-alum adsorbed after oral immunization. In addition, the GM bilosomes-based oral immunization showed a similar immune response to that of ovalbumin-alum by intramuscular route.

Premanand et al. (2013) developed bilosomes loaded with recombinant baculovirus antigen for oral delivery in mice. The developed bilosomes formulation showed enhanced mucosal and humoral immune response, which may be attributed to their ability to remain stable in a gastric environment, more permeability across intestine, and self-adjuvant properties.

Wilkhu et al. (2013) studied the effect of bilosomes formulation parameters as vaccine nanocarrier The study results showed that the large vesicles (~6 μm) exhibit rapid uptake by PPs resulting in higher immune response compared to lower size vesicles (~2 μm). The size of vesicles and their stability was influenced by the concentration of bile salts, high concentration of bile salts showed better results in smaller sized vesicles with more gastric stability.

Singh et al. (2004) developed bilosomes loaded with cholera toxin B subunit vaccine for oral delivery. The developed formulation showed stability at gastric environment and was rapidly taken up across intestinal lumen via M cells, which later elicit both systemic and mucosal immune responses. The oral immunization of cholera toxin B conjugated bilosomes exhibit equivalent immune response compared to parenteral administration of antigen with Freund's complete adjuvant.

Shukla et al. (2010) developed bilosomes loaded with hepatitis B antigen for oral delivery. The developed vaccine (vesicle size 220 nm and entrapment efficiency 20%–24%) was found to be stable in a gastric environment and exhibited rapid uptake by M cells across the intestinal lumen in PPs. The results showed that the antigen loaded bilosomes induce both systemic and mucosal immune response at very low dose compared to conventional liposomes formulation.

Mann et al. (2006) studied the effect of vesicle size of antigen loaded bilosomes on oral immune response. The two different-sized vesicles (lower 60–100 nm and higher 400–2500 nm) were prepared and their results showed that the vesicles with size ranges 400–2500 nm induced higher systemic and mucosal immune response. This higher immune response with large vesicles might be attributed to their high antigen loading capacity ,which enables their release for prolonged periods of time.

6.2.2.3 Niosomes

Niosomes are vesicular systems comprising of bilayer made up of nonionic surfactants. Niosomes act as an excellent carrier for loading both hydrophilic and lipophilic drugs. Niosomes have been widely explored for antigen delivery because of their biocompatibility, nontoxic, high antigen encapsulation efficiency, high stability in the gastric environment and high permeability across the intestine

(Shilpa et al., 2011). Recently, several studies have been performed using polyanhydride-based oral nanovaccine and are discussed in following section.

Rentel et al. (1999) developed ovalbumin-loaded niosomes for oral delivery and studied the effect of nonionic surfactant on immune response. The niosomes were prepared using two different nonionic surfactants, that is, Wasag7(stearate sucrose ester 70% and palmitate sucrose ester 30%) and Wasag15 (stearate sucrose ester 30% and palmitate sucrose ester 70%). Wasag7 being more hydrophobic showed ready uptake across intestinal lumen to PPs, which further induced enhanced immune response. However, Wasag15 being hydrophilic was unable to cross the lumen and hence showed no enhancement of immune response compared to soluble antigen.

Jain et al. (2005) developed hepatitis B antigen-loaded mannosylated niosomes for oral delivery to study the effect of humoral, cellular, and mucosal immune response. The niosomes were mannosylated with *o*-palmitoylmannan, which protected the niosome from degradation in gastric environment. The modified niosomes also exhibited enhanced uptake across intestinal lumen to PPs and induced higher humoral, cellular, and mucosal immune responses compared to unmodified niosomes and soluble antigen alone.

6.2.2.4 Archaeosomes

Archaeosomes are vesicular systems enclosing specialized polar lipid extracted from microbes of the class *Archaea*. The archaeosomal lipids have membrane-spanning properties, which provide rigidity to the vesicle membrane. These lipids are less sensitive to oxidation and leakage problem compared to liposomes. Archaeosomes have been widely explored for antigen delivery, due to their self-adjuvant properties. Archaeosomes exhibit enhanced uptake by macrophages and thereby activate APCs, which then carry the antigen to T-cells, resulting in induction of potent humoral and cellular immune responses. Archaeosomes exhibit 3–53 times greater uptake by phagocytes than conventional liposomes. Li et al. (2011) developed ovalbumin-loaded archaeosomes for oral vaccine delivery. The polar lipid employed for preparation of archaeosomes composed of lipid fraction E of *Sulfobus acidocaldarius*. The results showed that the antigen-loaded archaeosomes exhibit enhanced stability in gastric pH. It also exhibited better systemic IgG and mucosal IgA response compared to liposomes.

6.2.2.5 Nanoemulsion

Nanoemulsion is an isotropic system of two immiscible liquids stabilized by surfactant. Nanoemulsion can be either in the form of water-in-oil or oil-in-water–based system with droplet sizeS in the range of 100–1000 nm (Fig. 10.5). Nanoemulsion has been widely explored for antigen delivery mainly due to the enhanced permeability across intestinal lumen, higher antigen loading, stability in gastric environment, ease of preparation, and low cost. Nanoemulsion is readily taken up across the intestinal lumen due to the presence of surfactants in system. From lumen antigens are transferred to APCs in PPs and then activate T-cells and B cells to induce immune response. Recently, several studies have been performed using polyanhydride-based oral nanovaccine and are discussed in the following section.

Ge et al. (2009) developed antigen-loaded nanoemulsion-based oral vaccine for tumor therapy. The incorporated antigen is a complex mixture of melanoma antigen, heat shock protein, and *Staphylococcal* toxin A. The developed nanoemulsion showed high antigen encapsulation (87%) and particle size in range of 15–25 nm. The antitumor activity of orally administered nanoemulsion showed an immune response similar to that produced by subcutaneous route. In addition, the nanoemulsion-loaded antigen showed a higher cellular immune response compared with conventional formulations.

Shahiwala and Amiji (2008) developed squalene oil containing water in oil in water multiple emulsion loaded with OVA for oral delivery in mice. Squalene was used as an adjuvant. The results showed that the developed formulation induced both systemic and mucosal immune response when compared to nasal immunization. The increased systemic response (increased IgG levels) may be attributed to the increased uptake of negatively charged emulsion droplets by the lymphatic system. OVA in emulsion formulations was protected from degradation and endocytosed intact by M cells. This leads to efficient antigen presentation and thus increase in IgA levels to provide mucosal immunity.

6.2.3 Inorganic nanocarriers

Inorganic nanocarriers are the potential antigen delivery systems due to their unique physical and chemical properties (high surface area, high ligand binding affinity due to functional group at surface), better stability at gastric pH, high antigen loading and more importantly due to enhanced immune response (Fujita and Taguchi, 2011). Several inorganic materials, such as gold, silver, and silica have been widely explored for antigen delivery and are discussed in following section.

6.2.3.1 Gold nanocarrier

Gold nanocarrier is the colloidal carrier generated from gold salt [H(AuCl4)] in water with a particle size in the range of 1–100 nm. The size-dependent toxicity of gold nanocarriers can be reduced by modifying their surfaces. Gold nanocarriers have been widely explored for antigen delivery due to their biocompatibility, lack of immunogenicity, and high antigen loading. In addition to this, gold nanocarriers provide protection of entrapped antigens from harsh gastric environment and facilitate their intracellular uptake by endocytosis. Barhate et al. (2013) developed chitosan coated gold nanocarrier loaded with tetanus toxoid antigen and quillaja saponaria (QS) extract for oral delivery. QS extract was used as an adjuvant, which is known to enhance Th1 immune responses. The synthesized gold nanocarrier were spherical in shape, around 40 nm in size, positively charged (around +35 mV) with 60% antigen encapsulation. The results revealed that the modified gold nanocarrier remained stable in gastric environment compared to conventional/soluble antigen The systemic and mucosal immune responses produced by adjuvant containing chitosan-coated gold nanocarriers were found to be higher compared to plain chitosan-coated gold nanocarrier. This proved that incorporation of QS exhibits immunomodulation by triggering the receptors present on DCs and thus improved antigen recognition and processing. Thus, the chitosan-coated gold nanocarrier comprising QS as a coadjuvant prove to be an efficient for oral immunization of TT.

6.2.3.2 Silver

Silver (Ag) has been used mostly for the synthesis of stable dispersions of nanoparticles. Recently, Ag nanoparticle-based oral antigen delivery was studied by Jazayeri et al. The authors developed polyethylene glycol-coated Ag nanocarriers loaded with plasmid encoding hemagglutinin gene of avian influenza virus for oral vaccine delivery. The Agnanocarriers were synthesized by green technology using Ag nitrate, D-glucose (reducing agent) and PEG (stabilizer). The developed nanoparticles showed particle sizes in range of 4–18 nm with positive charges facilitating higher uptake across the intestine. The results revealed that the PEG-conjugated nanocarriers induce higher antibodies and cell-mediated responses compared to unconjugated nanocarrier.

6.2.3.3 Silica nanocarrier

Silica-based nanocarriers have been widely used for antigen delivery due to some properties, such as biocompatibility, high surface area, high antigen loading, nontoxic, and possess high stability at gastric pH. Silica being insoluble in nature provides the slow release of antigens creating a depot phenomenon. Silica-based nanocarriers for oral antigen delivery provide better uptake across the intestinal lumen and carried APCs, which further induces both systemic and mucosal immune response. Wang et al. (2011) developed bovine serum albumin-loaded silica nanoparticles for oral vaccine delivery. The developed nanocarriers were studied to identify the effect of pore shape and size on immune response. The results revealed that the nanocarrier with honeycomb shaped pores and size of 430 nm induces higher systemic and mucosal immune response compared to nanoparticles of smaller size. Thus, the results conclude the dependence of immune response on the size and architecture of silica nanoparticles.

6.2.3.4 Superparamagnetic iron oxide nanocarrier

Superparamagnetic iron oxide nanocarriers (SPIONs) is a paramagnetic nanocarrier mainly utilized for targeted delivery of drug/antigen to the required site. SPIONs have been explored for antigen delivery due to their high stability, efficient targeting, and so on. Al-Deen et al. (2011) developed polyethyleneimine coated SPIONs subunit oral malarial vaccine loaded with DNA. Nanocarrier-provided active targeting to vector and thus increased efficiency of vaccine. SPIONs/PEI complexes produced at acidic conditions showed the best DNA binding and gene transfection efficiency compared to those generated under neutral pH, possibly due to the protonated structure of the branched polymer that entrapped and protected the DNA.

6.2.4 Other

6.2.4.1 Carbon nanotube

Carbon nanotubes belong to the fullerenes family and are defined as a scaffold (sphere, ellipsoid, or a tube) made entirely of carbon atoms. Carbon nanotubes are mainly categorized as single-walled carbon nanotube and multiwall carbon nanotube-based on their geometry and size (Fig. 10.5). Carbon nanotubes are relatively inert, nonimmunogenic, and nontoxic (oral route). They are highly stable particulate systems, which provide depot release being insoluble in nature. Carbon nanotubes make an ideal carrier for antigen delivery by oral route. The antigen entrapped within the scaffold are taken up across the lumen by M cells-mediated phagocytosis by macrophages. The phagocytosed antigens are transferred to DCs, which further stimulate T-cell and B cells–mediated immune response. Carbon nanotubes activate both MHC Class I and MHC Class II, and suggests that they induce preferably a humoral immune response. Carbon nanotube-mediated immune response was mainly affected by several parameters, such as size. For example, single-walled carbon nanotubes, being small in size, rapidly internalize by DCs or macrophages via micropinocytosis and hence induce stronger immune response. Wang et al. (2011) studied carbon nanotubes as an adjuvant for oral vaccine delivery. Hydrophobic carbon nanotubes with the size of 470 nm were synthesized using sucrose as carbon source. The oral immunization data of developed carbon nanotubes suggest an effective stimulation of mucosal immune response (increase in intestinal IgA). It also induces Th1 and Th2–mediated immune response.

6.2.4.2 Virus-like particles

Virus-like particles (VLPs) are self-assembled viral protein complexes with sizes ranging 20–800 nm (Fig. 10.5). VLPs mimic the native virus genetic material, making them noninfectious and safe for use. VLPs is an ideal carrier for antigen delivery, as it mimics the native virus, hence, can be readily recognized by the host immune system. Several reports suggest that VLPs exhibit both humoral and cellular immune response even in the absence of an adjuvant (Crisci et al., 2012). VLPs-based systems are also reported to have better stability, fewer side effects, and, most importantly, they form polydispersed systems. The polydispersed system exhibits better uptake by M cells, enterocytes, and DCs across the intestine and induce stronger immune response. VLPs-based vaccine currently available in the regulatory market are Recombivax (hepatitis virus), Gardasil, and Cervarix, (human papillomavirus infection) vaccine. Recently, several studies have been performed using VLPs-based oral nanovaccine and are discussed in the following section.

Shah et al. (2015) developed a VLPs-based oral subunit vaccine using maize-derived hepatitis B antigen. The data showed that the antigen-loaded VLPs-based nanocarrier with sizes ranging 20–100 nm induce both systemic and mucosal immune response when orally administered to mice.

Takamura et al. (2004) developed VLPs-based oral subunit vaccines loaded with HIV envelope cDNA. The developed VLPs showed enhanced stability of encapsulated cDNA in gastric environment and facilitated higher concentration of antigen across intestinal lumen.

6.2.4.3 Dendrimers

Dendrimers are the three-dimensional, monodisperse, globular macromolecules with a perfectly branched tree-like structure. Dendrimers comprise a well-defined inner core surrounded by many branches attached to it. The branching in dendrimer signifies the "dendrimer generation," where the "first-generation" dendrimer is comprised of only one layer of branching points while that of fifth-generation dendrimer comprises five branching units. Higher generation dendrimers possess multiple functional groups on their surface, which provide coupling with biologically active moiety. Dendrimers have been widely explored nanocarriers for antigen delivery due to their biocompatibility, biodegradability, and self-adjuvant properties. Dendrimers facilitate rapid uptake of antigen across the intestinal epithelium and express strong immunogenicity. The immune response induced by dendrimers is mainly affected by size, charge, and antigen loading. Dendrimers with particle sizes of <200 nm, spherical shape, and high antigen-loading facilitate stronger immune responses compared to conventional oral vaccine. The commonly employed polymer for fabrication of dendrimers are polyethyleneimine, polyamidoamine (PAMAM), poly-(*N*-isopropylacrylamide), and polypropyleneimine. Recently, several studies have been performed using dendrimer-based oral nanovaccines and are discussed in the following section.

García-Vallejo et al. (2013) developed antigen-encapsulated PAMAM dendrimers for targeting DCs. The data showed better internalization and enhanced immune response, which may be attributed to their multivalency and high generation characteristic of dendrimers. Lower generation dendrimers led to linear increases in binding efficacy and internalization, to a lesser extent, as it results in cluster formations of DC signaling tetramers. In contrast, higher generation dendrimers possess higher binding capabilities, antigen processing, and presenting abilities. These dendrimers, though bigger in size, have spherical shapes that enable them to prevent cluster formation and undergo conformational changes required to induce cell signaling and immune responses.

Madhusudana et al. (2014) developed a fourth-generation plasmid loaded poly(ether imine) (PETIM) dendrimer-based rabies vaccine. The data revealed that the dendrimer-based vaccine showed enhanced immunogenicity and efficacy compared to conventional systems. The dendrimer with 500 nm facilitates increased uptake by APCs, leading to efficient antigen release and generations of higher levels of antibody responses.

6.2.4.4 ISCOMs

ISCOMs (immunostimulatory complexes) are spherical, hollow, rigid, cage-like structures of 40 nm size and exhibit a strong negative charge (Fig. 10.5) (Rimmelzwaan and Osterhaus, 1995). It comprises of cholesterol, phospholipid, saponin, and immunogen (antigen), usually a protein. ISCOMs have been widely explored for antigen delivery as it mimics a virus particle in terms of size and shape (Barr, 1998). ISCOMs high immune response is mainly associated to the presence of QS extract (Quil A), which exhibits strong immunostimulatory activity (Boyaka et al., 2001). ISCOMs have been reported to have both humoral and cellular immune responses (Sun et al., 2009). ISCOMs without antigen is available in the market under trademark ISCOMATRIX and is reported to have potential immunomodulatory properties, being a mucosal adjuvant. Similarly, ISCOMs derivative with cationic charge (PLUSCOMs) provide high-antigen loading and therefore enhanced T-cell response (Nielsen et al., 2015). Ghazi et al. (1995) studied the ISCOMs-based oral immunization against the influenza virus. The results revealed that oral immunization with ISCOMs showed increased systemic and humoral immune response compared to the same by intramuscular route. In addition, ISCOMs showed higher immune response compared to soluble antigen irrespective of the route of immunization.

7 REGULATORY ASPECTS OF ORAL NANOVACCINES

Vaccine has proven to be the most effective prophylactic aid for infectious disease for many years. In order to get regulatory approval for commercialization of any vaccine, it must comply with the regulatory requirements and risk management of scale-up and manufacturing process so as to assure quality, safety, and efficacy to the patients. With the advent of nanocarrier-based vaccine systems, meticulous understanding of the nanocarrier manufacturing process and the components used for vaccine fabrication, such as antigens and adjuvants, have become the need of the hour. To date, no specific guidelines for nanocarrier-based oral vaccine delivery are available. Hence, this section attempts to summarize various issues and regulatory challenges associated with both the nanocarrier-based systems and vaccine manufacturing process at their individual front and also presents the respective regulatory approaches and various methods to overcome related shortcomings for better regulatory practices.

7.1 CONSIDERATIONS FOR REGULATING NANOCARRIER-BASED VACCINE DELIVERY SYSTEM

7.1.1 Environmental, health, and safety risk

Until now, evaluation and estimation of environmental, health, and safety risks associated with nanocarrier-based systems have been overlooked. Thus, to address this issue the nanocarrier-based vaccine delivery must be investigated for various aspects, such as their fate in air, water, and soil, biodegradation, bioaccumulation, their probable conversion to toxic metabolites, and their interaction with environmental contaminants. In addition to this, the probable human exposure anticipated during manufacturing of nanovaccine must also be considered.

7.1.2 Ethical, legal, and social issues

There have been numerous ethical, legal, and social issues associated with nanocarrier-based systems, such as disparity in control measures on these systems in developed and underdeveloped countries. Also, there is overall lack of harmonized regulations, clarity in intellectual property rights for nanosystems, public awareness, and so on. All these issues must be dealt with while regulating nanocarrier-based vaccine delivery systems.

7.1.3 Regulatory challenges associated with nanoparticulate systems

There have been several reports briefing regulatory hurdles associated with nanocarrier-based systems, namely:

- Lack of harmonized regulatory and policy courses, disparity in basic scientific knowledge, and societal risk based on assessment of exposure, hazard detection, and their characterization.
- Lack of specific tools to estimate human and environmental nanocarrier-based system exposure.
- Lack of harmonized regulation to ensure safe and efficacious nanosystem development.

7.1.4 Regulatory challenges associated with vaccine systems

Regulatory issues must be emphasized while developing vaccine delivery systems. As the vaccine delivery system is evolving and advancing, the manufacturer has to overcome many challenges to address regulatory compliance during development, preclinical, and clinical studies. Currently, the topic of major concern is the use of adjuvants and their characterization. In this regard, several regulatory agencies have provided generalized guidelines for vaccine adjuvant and adjuvanted vaccine for their nonclinical and clinical evaluation. Vaccine adjuvants exist as components in licensed vaccines with specific antigen in combinations. Adjuvanted vaccine is considered a new entity that will require independent physicochemical characterization and nonclinical and clinical evaluation. Hence, it is mandatory for the manufacturers of adjuvanted vaccines to provide scientific rationale supporting the benefit of adding the adjuvant and the choice of specific adjuvant(s). Before evaluating a particular adjuvant in combination with an antigen in a clinical trial, it is recommended that data from in vitro and/or in vivo studies be generated to support corresponding rationale for including the specific adjuvant in the vaccine formulation and also for selecting the dose range of adjuvant to be tested. It is recommended that proof-of-concept studies to support the use of an adjuvant be carried out to evaluate vaccine formulations with and without the adjuvant. The preclinical toxicity studies of the final adjuvanted vaccine formulation should be adequate to identify and characterize potential adverse effects of the vaccine in order to conclude that it is reasonably safe to proceed further for clinical investigations. The aforementioned issues are individually considered in several regulatory polices framed by countries around the globe, namely the United States (USFDA), Europe (EMEA), and Harmonized Agency (WHO) and these provide guidelines for manufacturing and clinical trial pertaining to nanocarrier delivery systems and vaccine development at their individual front (Tables 10.4 and 10.5).

Thus, considering all this regulatory aspects, there is compelling need for harmonized regulations for nanocarrier-based vaccine delivery, which will not only ensure risk mitigation during manufacturing but also provide safety, efficacy, and availability of licensed vaccines for betterment of the masses.

Table 10.4 Agencies Under Initiatives for Nanoparticulate Drug Delivery Systems

Country	Agencies Under Initiatives for Regulating Nanoparticulate Drug Delivery
USA	Government Initiatives • Nanotechnology Initiatives • The Director of the Office of Science and Technology Policy • US Environmental Protection Agency • The USFDA • The Consumer Product Safety Commission • The Occupational Safety and Health Administration • US Department of Labor Nongovernment Initiatives • The Project on Emerging Nanotechnologies • The Environmental Defense Fund • The Natural Resources Defense Council • NANO Risk Framework
EU	Government Initiatives • The Communication from the Commission on the Precautionary Principle Towards a European Strategy for Nanotechnology • Nanosciences and Nanotechnologies: An action plan for Europe 2005–2009 • The European Group on Ethics in Science and New Technologies • Code of Conduct for Responsible Nanosciences and Nanotechnologies Research • Motion for a European Parliament Resolution on Regulatory Aspects of Nanomaterial • The Scientific Committee on Emerging and Newly Identified Health Risks Nongovernment Initiatives • Nanotechnology Engagement Group

Table 10.5 Regulatory Bodies for Vaccine Manufacturing

Country	Regulatory Body
United States	The Center for Biologics Evaluation and Research (CBER) of the US Food and Drug Administration (FDA) is the federal agency responsible for regulating vaccines and other biological products
Europe	The European pharmaceutical legislation
WHO	WHO Collaborating Centers and WHO Expert Committee on Biological Standardization (ECBS) in close collaboration with the regional and national regulatory authorities (NRAs) control manufacturing of biological product, such as vaccines

8 FUTURE PROSPECTS

Though oral nanovaccine delivery appears to be a promising approach due to its inherent advantages, such as patient compliance, invoking desired mucosal immune responses, as well as the convenience of mass vaccination, certain key aspects need to be taken into consideration. The comprehensive review of literature available in this regard reflects contradictory research findings in many cases. These are in the context of the effect observed due to varying size ranges of the nanocarrier-based systems, sometimes advocating particle sizes beyond nanometric size for specific and better immune responses. We feel the possible reasons for this disparity may be due to the observed assessment of these responses in the animal models selected. Suitability of the animal models and the responses obtained due to inherent variations in them, need to be rationally evaluated and duly validated. Also, relevant animal models simulating response in clinical conditions need to be identified and employed for testing, before drawing inferences on the efficacy of the formulations.

With respect to the approach of potential use of adjuvants in enhancing immune responses in the vaccine delivery systems, certain aspects should be given careful consideration. Since adjuvants have to be considered in combination with specific antigen and are regarded as a new chemical entity, their use has to be with support of extensive safety data and caution from regulatory perspective. The toxicity of adjuvants and their characterization issues in some cases will be a major hurdle. Hence, one has to assess risk-benefit ratios for the same. In this context nanocarrier-based vaccines with some of the carrier systems acting as self-adjuvants will be beneficial to avoid delays due to regulatory considerations. Also, one cannot overlook the fact that nanocarrier-based systems and their issues are yet to be thoroughly investigated for their complexity, especially pertaining to their toxicity, stability, and scalability aspects. Since research on this front is still in the developmental stage, batch-to-batch variation in manufacturing and scale-up issues have to be assessed critically. Thorough validation measures for raw materials (especially of natural origin), for bulk drug synthesis and process validation during manufacturing should be supplemented with comprehensive statistical analysis. Nevertheless, with the advances in technology and emerging sophisticated instrumentation for critical issues of characterization and analysis of these materials, their prevalent issues will shortly be resolved. With focused efforts in this direction, soon the hurdles being faced currently by nanotechnology-based delivery systems will be positively circumvented, both from manufacturing and regulatory point of view.

The purpose of this chapter was to provide comprehensive data of nanovaccines for oral delivery by giving ready reference for their formulation strategies and challenges. This serves to encourage and propel research efforts in this direction and helps this delivery system "see the light of the day" in the near future, for the betterment of the masses.

REFERENCES

Adams, J., Mallapragada, S., 2014. Enhancing the immune response through next generation polymeric vaccine adjuvants. Technology 2 (1), 1–12.

Al-Deen, F., Ho, J., Selomulya, C., Ma, C., Coppel, R., 2011. Superparamagnetic nanoparticles for effective delivery of malaria DNA vaccine. Langmuir 27 (7), 3703–3712.

Allahyari, M. and Mohit, E., 2016. Peptide/protein vaccine delivery system based on PLGA particles. Hum Vaccin. Immunother. 12(3), 806–823.

Awaad, A., Nakamura, M., Ishimura, K., 2012. Imaging of size-dependent uptake and identification of novel pathways in mouse Peyer's patches using fluorescent organosilica particles. Nanomedicine 8 (5), 627–636.

Bachelder, E., Beaudette, T., Broaders, K., Fréchet, J., Albrecht, M., Mateczun, A., Ainslie, K., Pesce, J., Keane-Myers, A., 2010. Vitro analysis of acetylated dextran microparticles as a potent delivery platform for vaccine adjuvants. Mol. Pharm. 7 (3), 826–835.

Barhate, G., Gautam, M., Gairola, S., Jadhav, S., Pokharkar, V., 2013. Quillajasaponaria extract as mucosal adjuvant with chitosan functionalized gold nanoparticles for mucosal vaccine delivery: stability and immunoefficiency studies. Int. J. Pharm. 441 (1–2), 636–642.

Barr, I., 1998. ISCOMs and other saponin based adjuvants. Adv. Drug Deliv. Rev. 32 (3), 247–271.

Boyaka, P.N., Marinaro, M., Jackson, J.J., et al., 2001. Oral QS-21 requires early IL-4 help for induction of mucosal and systemic immunity. J. Immunol. 166 (4), 2283–2290.

Champion, J.A., Mitragotri, S., 2009. Shape induced inhibition of phagocytosis of polymer particles. Pharm. Res. 26, 244–249.

Chen, H., Torchilin, V., Langer, R., 1996a. Lectin bearing polymerized liposomes as potential oral vaccine carriers. Pharm. Res. 13 (9), 1378–1383.

Chen, H., Torchilin, V., Langer, R., 1996b. Polymerized liposomes as potential oral vaccine carriers: stability and bioavailability. J. Control. Release 42 (3), 263–272.

Conacher, M., Alexander, J., Brewer, J., 2001. Oral immunization with peptide and protein antigens by formulation in lipid vesicles incorporating bile salts (bilosomes). Vaccine 19 (20–22), 2965–2974.

Cordeiro, A., Alonso, M., de la Fuente, M., 2015. Nanoengineering of vaccines using natural polysaccharides. Biotechnol. Adv. 33 (6), 1279–1293.

Crisci, E., Bárcena, J., Montoya, M., 2012. Virus-like particles: the new frontier of vaccines for animal viral infections. Vet. Immunol. Immunopathol. 148 (3–4), 211–225.

D'Souza, M.J., Tawde, S.A., Akalkotkar, A., Chablani, L., D'Souza, M., Chiriva-Internati, M., 2014. Nanotechnology in vaccine delivery. Giese, M. (Ed.), Molecular Vaccines: From Prophylaxis to Therapy, Vol. 2, Springer International Publishing, Switzerland, p. 727.

Eldridge, J.H., Hammond, C.J., Meulbroek, J.A., et al., 1990. Controlled vaccine release in the gut associated lymphoid tissue. I. Orally administered biodegradable microsphere target the Peyer's patches. J. Control. Release 11, 205–214.

Foged, C., Brodin, B., Frokjaer, S., et al., 2005. Particle size and surface charge affect particle uptake by human dendritic cells in an in vitro model. Int. J. Pharm. 298, 315–322.

Fujii, Y., Aramaki, Y., Hara, T., Yachi, K., Kikuchi, H., Tsuchiya, S., 1993. Enhancement of systemic and mucosal immune responses following oral administration of liposomes. Immunol. Lett. 36 (1), 65–69.

Fujita, Y., Taguchi, H., 2011. Current status of multiple antigen-presenting peptide vaccine systems: application of organic and inorganic nanoparticles. Chem. Central J. 5 (1), 48.

Gamazo, C., Irache, J.M., 2014. Antigen delivery systems as oral adjuvants. Giese, M. (Ed.), Molecular Vaccines: From Prophylaxis to Therapy, 2, Springer International Publishing, Switzerland, p. 603.

García-Vallejo, J., Ambrosini, M., Overbeek, A., van Riel, W., Bloem, K., Unger, W., Chiodo, F., Bolscher, J., Nazmi, K., Kalay, H., van Kooyk, Y., 2013. Multivalent glycopeptide dendrimers for the targeted delivery of antigens to dendritic cells. Mol. Immunol. 53 (4), 387–397.

Garinot, M., Fiévez, V., Pourcelle, V., Stoffelbach, F., des Rieux, A., Plapied, L., Theate, I., Freichels, H., Jérôme, C., Marchand-Brynaert, J., Schneider, Y., Préat, V., 2007. PEGylated PLGA-based nanoparticles targeting M cells for oral vaccination. J. Control. Release 120 (3), 195–204.

Ge, W., Li, Y., Li, Z., Zhang, S., Sun, Y., et al., 2009. The antitumor immune responses induced by nanoemulsion-encapsulated MAGE1-HSP70/SEA complex protein vaccine following peroral administration route. Cancer Immunol. Immunother. 58, 201–208.

Ghazi, H., Potter, C., Smith, T., Jennings, R., 1995. Comparative antibody responses and protection in mice immunised by oral or parenteral routes with influenza virus subunit antigens in aqueous form or incorporated into ISCOMs. J. Med. Microbiol. 42 (1), 53–61.

Gilligan, C.A., Wan, P.O., Li, A., 1991. Oral vaccines: design and delivery. Int. J. Pharm. 75, 1–24.

Gómez, S., Gamazo, C., Roman, B., Ferrer, M., Sanz, M., Irache, J., 2007. Gantrez AN nanoparticles as an adjuvant for oral immunotherapy with allergens. Vaccine 25 (29), 5263–5271.

Gonzalez-Aramundiz, J., Cordeiro, A., Csaba, N., et al., 2012. Nanovaccines: nanocarriers for antigen delivery. Biol. Aujourd'hui 206 (4), 249–261.

Gupta, P., Khatri, K., Goyal, A., Mishra, N., Vyas, S., 2007. M-cell targeted biodegradable PLGA nanoparticles for oral immunization against hepatitis B. J. Drug Target. 15 (10), 701–713.

Gutierro, I., Hernandez, R.M., Igartua, M., et al., 2002. Size dependent immune response after subcutaneous, oral and intranasal administration of BSA loaded nanospheres. Vaccine 21, 67–77.

Harde, H., Agrawal, A., Jain, S., 2014. Development of stabilized glucomannosylated chitosan nanoparticles using tandem crosslinking method for oral vaccine delivery. Nanomedicine 9 (16), 2511–2529.

Henriksen-Lacey, M., Korsholm, K., Andersen, P., Perrie, Y., Christensen, D., 2011. Liposomal vaccine delivery systems. Exp. Opin. Drug Deliv. 8 (4), 505–519.

Indulkar, A., Jain, S., Harde, H., Agrawal, A., 2014. Oral mucosal immunization using glucomannosylated bilosomes. J. Biomed. Nanotechnol. 10 (6), 932–947.

Jabbal-Gill, I., Watts, P., Smith, A., 2012. Chitosan-based delivery systems for mucosal vaccines. Exp. Opin. Drug Deliv. 9 (9), 1051–1067.

Jain, S., Singh, P., Mishra, V., Vyas, S., 2005. Mannosylated niosomes as adjuvant–carrier system for oral genetic immunization against hepatitis B. Immunol. Lett. 101 (1), 41–49.

Jain, A., Reddy, A.V., Muntimadugu, E., Khan, W., 2014. Nanotechnology in vaccine delivery. Curr. Trends Pharm. Sci. 1, 17–27.

Jazayeri, S., Ideris, A., Zakaria, Z., Shameli, K., Moeini, H., Omar, A., 2012. Cytotoxicity and immunological responses following oral vaccination of nanoencapsulated avian influenza virus H5 DNA vaccine with green synthesis silver nanoparticles. J. Control. Release 161 (1), 116–123.

Kim, M., Park, J., Shon, Y., Kim, G., Shim, G., Oh, Y., 2014. Nanotechnology and vaccine development. Asian J Pharm. Sci. 9, 227–235.

Kraajeeld, C., Schilham, M., Jansen, J., Benaissa-Trouw, B., Harmsen, M., van Houte, A., Snippe, H., 1984. The effect of liposomal charge on the neutralizing antibody response against inactivated encephalomyocarditis and Semliki forest viruses. Clin. Exp. Immunol. 56, 509–514.

Li, X., Kong, X., Shi, S., Zheng, X., Guo, G., Wei, Y., Qian, Z., 2008. Preparation of alginate coated chitosan microparticles for vaccine delivery. BMC Biotechnol. 8 (1), 89.

Li, G., Liu, Z., Liao, B., Zhong, N., 2009. Induction of Th1-type immune response by chitosan nanoparticles containing plasmid DNA encoding house dust mite allergen Der 2 for oral vaccination in mice. Cell Mol. Immunol. 6 (1), 45–50.

Li, Z., Zhang, L., Sun, W., Ding, Q., Hou, Y., Xu, Y., 2011. Archaeosomes with encapsulated antigens for oral vaccine delivery. Vaccine 29 (32), 5260–5266.

Ma, T., Wang, L., Yang, T., Ma, G., Wang, S., 2014. M-cell targeted polymeric lipid nanoparticles containing a toll-like receptor agonist to boost oral immunity. Int. J. Pharm. 473, 296–303.

Madhusudana, S., Padinjaremattathil Thankappan, U., Desai, A., Sagar, B., Govindasamy, J., Yamajala, B., Narayanaswamy, J., 2014. Enhancement of immunogenicity and efficacy of a plasmid DNA rabies vaccine by nanoformulation with a fourth-generation amine-terminated poly(ether imine) dendrimer. Int. J. Nanomed. 9, 627–634.

Makadia, H., Siegel, S., 2011. Poly lactic-*co*-glycolic acid (PLGA) as biodegradable controlled drug delivery carrier. Polymers 3 (4), 1377–1397.

Mann, J., Scales, H., Shakir, E., Alexander, J., Carter, K., Mullen, A., Ferro, V., 2006. Oral delivery of tetanus toxoid using vesicles containing bile salts (bilosomes) induces significant systemic and mucosal immunity. Methods 38 (2), 90–95.

Marasini, N., Skwarczynski, M., Toth, I., 2014. Oral delivery of nanoparticle based vaccines. Exp. Rev. Vaccines 13 (11), 1361–1376.

Minato, S., Iwanaga, K., Kakemi, M., Yamashita, S., Oku, N., 2003. Application of polyethyleneglycol (PEG)-modified liposomes for oral vaccine: effect of lipid dose on systemic and mucosal immunity. J. Control. Release 89 (2), 189–197.

Nielsen, H., Hubschmann, H., Rades, T., 2015. ISCOMS as a vaccine delivery system. Subunit Vaccine Delivery vol. 8Springer, NY, pp. 141-158.

Nikura, K., Matsunaga, T., Suzuki, T., et al., 2013. Gold nanoparticles as a vaccine platform: influence of size and shape on immunological responses in vitro and in vivo. ACS Nano 7, 3926–3938.

Ojer, P., 2013. Cytotoxicity and cell interaction studies of bioadhesive poly(anhydride) nanoparticles for oral antigen/drug delivery. J. Biomed. Nanotechnol. 9 (11), 1891–1903.

Oliveira, C., Rezende, C., Silva, M., Borges, O., Pêgo, A., Goes, A., 2012. Oral vaccination based on DNA-chitosan nanoparticles against *Schistosoma mansoni* infection. Sci. World J. 2012, 1–11.

Oyewumi, M.O., Kumar, A., Cui, Z., 2010. Nano-microparticles as immune adjuvants: correlating particle sizes and the resultant immune responses. Exp. Rev. Vacc. 9, 1095–1107.

Patravale, V., Prabhu, P., 2014. Potential of nanocarriers in antigen delivery: the path to successful vaccine delivery. Nanocarriers 1 (1), 10–45.

Pattani, A., Gupta, P., Curran, R., Malcolm, R., 2014. Vaccine delivery systems: roles, challenges and recent advances. Giese, M. (Ed.), Molecular Vaccines: From Prophylaxis to Therapy, 2, Springer International Publishing, Switzerland, p. 743.

Premanand, B., Prabakaran, M., Kiener, T., Kwang, J., 2013. Recombinant baculovirus associated with bilosomes as an oral vaccine candidate against HEV71 infection in mice. PLoS One 8 (2), e55536.

Primard, C., Poecheim, J., Heuking, S., Sublet, E., Esmaeili, F., Borchard, G., 2013. Multifunctional PLGA-based nanoparticles encapsulating simultaneously hydrophilic antigen and hydrophobic immunomodulator for mucosal immunization. Mol Pharm. 10 (8), 2996–3004.

Rentel, C., Bouwstra, J., Naisbett, B., Junginger, H., 1999. Niosomes as a novel peroral vaccine delivery system. Int. J. Pharm. 186 (2), 161–167.

Riedel, S., 2005. Edward Jenner and the history of smallpox and vaccination. Baylor Univ. Med. Center Proc. 18, 21–25.

Rieux, A., Virginie, V., Garinot, M.a., Schneider, Y., Préat, V., 2006. Nanoparticles as potential oral delivery systems of proteins and vaccines: a mechanistic approach. J. Control. Release 116, 1–27.

Rimmelzwaan, G., Osterhaus, A., 1995. A novel generation of viral vaccine based on the ISCOM matrix. In: Powell, M. (Ed.), Vaccine Design: The Subunit and Adjuvant Approach. Plenum Press, New York, p. 543.

Salman, H., Gamazo, C., Agüeros, M., Irache, J., 2007. Bioadhesive capacity and immunoadjuvant properties of thiamine-coated nanoparticles. Vaccine 25 (48), 8123–8132.

Sarti, F., Perera, G., Hintzen, F., Kotti, K., Karageorgiou, V., 2011. In vivo evidence of oral vaccination with PLGA nanoparticles containing the immunostimulant monophosphoryl lipid A. Biomaterials 32, 4052–4057.

Sayın, B., Somavarapu, S., Li, X., Thanou, M., Sesardic, D., Alpar, H., enel, S., 2008. Mono-*N*-carboxymethyl chitosan (MCC) and *N*-trimethyl chitosan (TMC) nanoparticles for non-invasive vaccine delivery. Int. J. Pharm. 363 (1–2), 139–148.

Shah, S., Hayden, C., Fischer, M., Howard, J., Rao, A., 2015. Biochemical and Biophysical Characterization of Maize-derived HBsAg for the development of an Oral Vaccine. Arch. Biochem. Biophys. 588, 41–49.

Shahiwala, A., Amiji, M., 2008. Enhanced mucosal and systemic immune response with squalane oil-containing multiple emulsions upon intranasal and oral administration in mice. J. Drug Target. 16 (4), 302–310.

Shakweh, M., Ponchel, G., Fattal, 2004. Particle uptake by Peyer's patches: a pathway for drug and vaccine delivery. Exp. Opin. Drug Deliv. 1 (1), 141–163.

Shakweh, M., Besnard, M., Nicolas, V., et al., 2005. Poly (lactide-*co*-glycolide) particles of different physicochemical properties and their uptake by peyer's patches in mice. Eur. J. Pharm. Biopharm. 61, 1–13.

Shilpa, S., Srinivasan, B., Chauhan, M., 2011. Niosomes as vesicular carriers for delivery of proteins and biologicals. Int. J. Drug Deliv. 3 (1), 14–24.

Shukla, A., Katare, O., Singh, B., Vyas, S., 2010. M-cell targeted delivery of recombinant hepatitis B surface antigen using cholera toxin B subunit conjugated bilosomes. Int. J. Pharm. 385 (1–2), 47–52.

Singh, P., Prabakaran, D., Jain, S., Mishra, V., Jaganathan, K., Vyas, S., 2004. Cholera toxin B subunit conjugated bile salt stabilized vesicles (bilosomes) for oral immunization. Int. J. Pharm. 278 (2), 379–390.

Sun, H., Xie, Y., Ye, Y., 2009. Advances in saponin-based adjuvants. Vaccine 27 (12), 1787–1796.

Takamura, S., Niikura, M., Li, T., Takeda, N., Kusagawa, S., 2004. DNA vaccine-encapsulated virus-like particles derived from an orally transmissible virus stimulate mucosal and systemic immune responses by oral administration. Gene Ther. 11, 628–635.

Tennant, S.M., Muhsen, K., Pasetti, M.F., 2014. Gut immunology and oral vaccination. Giese, M. (Ed.), Molecular Vaccines: From Prophylaxis to Therapy, 1, Springer, Verlag-Wien, pp. 59–84, 2013.

Torchilin, V., 2005. Recent advances with liposomes as pharmaceutical carriers. Nat. Rev. Drug Disc. 4 (2), 145–160.

Wang, L., Coppel, R., 2008. Oral vaccine delivery: can it protect against the non-mucosal pathogen? Exp. Rev. Vacc. 7 (6), 729–738.

Wang, T., Zou, M., Jiang, H., Ji, Z., Gao, P., Cheng, G., 2011. Synthesis of a novel kind of carbon nanoparticle with large mesopores and macropores and its application as an oral vaccine adjuvant. Eur. J. Pharm. Sci. 44 (5), 653–659.

Watarai, S., Sasaki, Y., 2014. Evaluation of stearylamine-,modified liposomes for the oral vaccine adjuvant. J. Infect. Dis. Ther. 2 (3), 1–6.

Wendorf, J., Singh, M., Chesko, J., et al., 2006. A practical approach to the use of nanoparticles for vaccine delivery. J. Pharm. Sci. 95, 2738–2750.

Wilkhu, J., McNeil, S., Kirby, D., Perrie, Y., 2011. Formulation design considerations for oral vaccines Ther. Deliv. 2 (9), 1141–1164.

Wilkhu, J., McNeil, S., Anderson, D., Perrie, Y., 2013. Characterization and optimization of bilosomes for oral vaccine delivery. J. Drug Target. 21 (3), 291–299.

Yoo, J., Doshi, N., Mitragotri, S., 2010. Endocytosis and intracellular distribution of PLGA particles in endothelial cells: effect of particle geometry. Macromol. Rapid Comm. 31, 142–148.

Zaman, M., Good, M., Toth, I., 2013. Nanovaccines and their mode of action. Methods 60 (3), 226–231.

Zhao, L., Seth, A., Wibowo, N., Zhao, C., Mitter, N., Yu, C., Middelberg, A., 2014. Nanoparticle vaccines Vaccine 32 (3), 327–337.

FURTHER READING

Chen, Y., Mohanraj, V., Parkin, J., 2003. Chitosan-dextran sulfate nanoparticles for delivery of an anti-angiogenesis peptide. Lett. PeptSci. 10 (5–6), 621–629.

CHAPTER

NANOSYSTEMS FOR ORAL DELIVERY OF IMMUNOMODULATORS

11

Prachi B. Kharkar, Swapnil S. Talkar, Namrata A. Kadwadkar, Vandana B. Patravale
Institute of Chemical Technology, Mumbai, India

CHAPTER OUTLINE

1 Introduction 295
1.1 Summary of Immune System 296
1.2 Overview of Immunomodulators 299
1.3 Current Routes of Administration 300
2 Oral Route for Immunomodulators 301
2.1 Conventional Immunomodulators 301
2.2 Drawbacks of Conventional Oral Systems 302
3 Need of Nanosystems for Immunomodulators 302
4 Oral Nanosystems for Immunomodulators 303
4.1 Oral Nanosystems for Synthetic Immunomodulators 305
4.2 Oral Nanosystems for Herbal Immunomodulators 306
4.3 Oral Nanosystems for Vaccines and Adjuvant 310
4.4 Oral nanosystems for Toxins and Venoms 313
5 Methods for Characterization of Nanosystems 315
6 Fate of Immunomodulators After Oral Delivery 320
7 Future Prospects 323
7.1 Oral 323
7.2 Other Preferred Routes of Administration 324
8 Conclusions 326
References 328

1 INTRODUCTION

Immunomodulators (IMs) are one of the most vital therapeutic agents that not only help to modulate the immune response to specific antigen, but also cause least toxicity to the host system. This chapter gives an insight into IMs, their routes of administration, and newer sophisticated methods of their formulation for better therapeutic effect. Prior to the detailed discussion on IMs, the first section gives an overview of the immune system for valuable perception and understanding of the same and, further, the section covers types of IMs and their current routes of administration.

1.1 SUMMARY OF IMMUNE SYSTEM

Immunology is one of the most rapidly developing thrust areas of medical biotechnology research. It not only deals with physiological functions of immune system in healthy individuals but also with malfunctions of immune responses associated with progression of neoplastic diseases, organ transplantation, and several autoimmune diseases. In addition to this, it has a profound influence on prevention and treatment of inflammatory diseases of skin, gut, respiratory tract, joints, and central organs (Patil et al., 2012). Hence, it is crucial to understand the immune system and its responses in depth.

The immune system is the most complex biological host defense system that has evolved to protect the organism from the foreign pathogens, microorganisms, viruses, as well as the diseases like cancer and AIDS. It is capable of producing a variety of cells and molecules capable of precisely recognizing and excluding the foreign invaders, all of which act together in a dynamic cascade network. Basically, the immune system discriminates between nonself and self-antigens. Nonself- antigens could be derived from infectious organisms, transplanted organs, and skin or endogenous cells whereas self-antigens are derived from the body's own tissue. It was assumed earlier that the immune system does not react to self-antigens. However, Paul Ehrlich realized that the immune system could go askew and it not only reacts against foreign antigens but also attacks self-antigens. This condition, which he termed "horror autotoxicus," can result in a number of chronic and acute diseases, including rheumatoid arthritis, multiple sclerosis, Grave's disease, and certain types of diabetes (Kindt et al., 2007). The immune system has two mechanisms that help to eliminate the foreign organism or altered self-cell:

- Cell-mediated immunity.
- Humoral immunity.

Cell mediated immunity: This immunity is generally activated by effector T cells in response to the foreign antigen, as well as altered self-cell. Both helper T cells (T_H) and cytotoxic T cells (T_C) function as effector cells in cell-mediated immune reactions. After activation, T_H cells secrete cytokines which are responsible for the stimulation of various phagocytic cells, enabling them to phagocytose and kill the foreign bodies in an effective manner. T_C cells are involved in the killing of altered self-cells; they help in the killing of tumor or virally infected cells.

Humoral immunity: This immunity is activated by interaction of B cells with the antigen, subsequently leading to their proliferation and differentiation into antibody secreting plasma cells. Antibody functions by binding to antigen and neutralizing or eliminating it. Antigens from the body can be eliminated by several ways. For example, antibody can be crosslinked with many antigens and form clusters that are easily engulfed by the phagocytic cells, antigen–antibody complexes can activate the complement system for elimination of organisms by lysis, or antibody can also neutralize toxin or viruses by coating them.

Defense by immune system can be divided into two parts—recognition and response. Recognition is extremely important for its ability to discriminate between self- and nonself-components. The immune system recognizes specific molecular patterns that characterize species of common pathogens, it can also detect the slight chemical differences that differentiate one foreign organism from another, for example, Toll like receptors (TLR), 11 TLRs have been found in human beings to date. Each TLR represents an individual repertoire of highly conserved pathogen molecules. They detect a broad variety of bacteria, viruses, fungi, and even some protozoa (Table 11.1).

Table 11.1 Examples of Toll Like Receptors in Human

TLR	Ligand	Target Microbe
TLR1	Triacyl lipopeptides	Mycobacterium
TLR2	Peptidoglycan Zymosan GPI-linked protein	Gram-positive bacteria yeast and other fungi trypansomes
TLR3	Double-stranded DNA (dsDNA)	Viruses
TLR4	LPS	Gram-negative bacteria
TLR5	Flagellin	Bacteria
TLR6	Lipoproteins	Bacteria
TLR7, TLR8	Single-stranded RNA (ssRNA)	Viruses
TLR9	CpG DNA	Bacteria and viruses
TLR10	Unknown	Unknown
TLR11	Profilin	Bacteria

CpG DNA, cytosine and guanine rich DNA; DNA, deoxy ribonucleic acid; GPI, glucose phosphate isomerase; LPS, lipopolysaccharide; TLR, Toll like receptor; RNA; ribonucleic acid.

Recognition of foreign substances by multiple components of the immune system triggers a variety of effect or responses in order to eliminate or neutralize a particular type of invader. Certain responses lead to the memory response, characterized by a more rapid and intensified immune response, which precludes us from contacting a particular disease for the second time. The immunological memory is the foundation for vaccination, which is a means of "educating" the immune system to prepare it for later attacks.

There are two systems of immunity: innate immunity and adaptive immunity, which work together to protect the body. Both these responses have two components each, that is, cellular and humoral as mentioned earlier in this section. Innate immunity includes molecular and cellular mechanisms reassigned before an infection and composed/designed to prevent or eliminate it. This highly effective first line of defense prevents most infections at the onset or eliminates them within hours of encounter. The recognition system of innate immunity discriminates between self- and nonself ones, but they are not efficient enough to distinguish small differences in foreign molecule. Innate immunity lacks the specificity to produce the memory cells. On the other hand the second form of immunity, known as adaptive immunity, develops in response to infection and is able to recognize, eliminate, and then remember the invading pathogen (by producing memory cells). Adaptive immunity depends on the innate immunity and begins a few days after the infection. It requires time to marshal and provides the comprehensive second line defense that eliminates pathogens that evade the innate response or persist in spite of them. An important consequence of adaptive immunity is the memory (Kindt et al., 2007). Fig. 11.1 elaborates different components of immune system.

Both immune systems function in a highly cooperative and interactive manner by producing a combined response, which is more effective than the individual response. An important innate defense

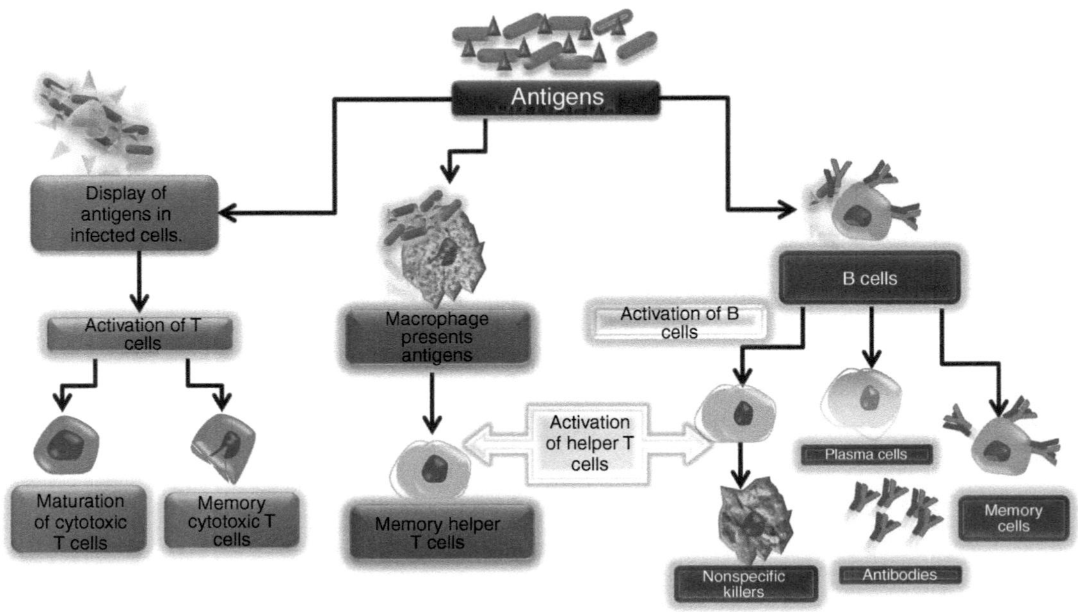

FIGURE 11.1 Components of Immune System

mechanism is the engulfment of extracellular particulate material by phagocytosis. Most phagocytosis is conducted by the specialized cells, such as blood monocytes, neutrophils, and tissue macrophages. Similarly, a variety of soluble factors, such as the lysosome, interferon, and the components of the complement system contribute to innate immunity.

An adaptive immune response involves lymphocytes and antigen presenting cells (APCs). First, lymphocytes are the type of white blood cells formed in the bone marrow by the process of hematopoiesis. Lymphocytes leave the bone marrow, circulate in the blood and lymphatic system, and reside in the various lymphoid organs. There are two major populations of lymphocytes: B lymphocytes (B cells), and T lymphocytes (T cells). B lymphocytes mature in bone marrow and each expresses a unique antigen-binding receptor on the membrane. The binding of antigen to the antigen-binding receptor of B cell causes the cell to differentiate rapidly into memory B cells and effector B cells (plasma cells). Memory B cells have a longer life span. Plasma cells have a short life span and they secrete generous amounts of antibody. Single plasma cells can generate hundreds to thousands of antibody per second. T lymphocytes also arise in the bone marrow. Unlike B cells, which mature in bone marrow, T cells migrate to the thymus gland for maturation. Mature T cells express a unique antigen binding molecule, the T cell receptor (TCR) on their membrane. T cells displaying CD4 (membrane glycoprotein) function as T_H cells, whereas those displaying CD8 function as T_C cells. T cell receptors recognize antigen that is bound to cell membrane proteins called major histocompatibility complex (MHC) molecules. There are two major types of MHC molecules: MHC class I, which are expressed on all the nucleated cells, and MHC class II, which are expressed only by the APCs. After T_H cells recognize the antigen presented on MHC class II molecule complex, the cell activates and secretes cytokines, which eventually activates the B cells, T_C cells, macrophages, and various other cells that participate in the immune response. APCs include B lymphocytes, macrophages, and dendritic cells. APCs internalize the antigen, process it, and present it on the membrane with help of MHC class II molecules. Then this antigen is recognized by the T_H cells for the activation of the immune response.

Further, T_H cells evolve in two subpopulations, which are distinguished by the different set of cytokines they secrete. One population, called T_H1 subset, secretes IL-2, IFN-γ, TNF-β. This subset helps in cytotoxic T cell activation, production of the opsonization-promoting IgG antibodies, and a delayed type of hypersensitivity, which is responsible for cell-mediated immunity. This subset is also associated with excessive inflammation and tissue injury. Another subset, called T_H2 subset, secretes IL-4, IL-5, IL-6, and IL-10. This subset acts as a helper for B cell activation and promotes the production of relatively large amount of IgM, IgE, and some IgG isotypes, which are responsible for humoral immunity. Apart from these two subsets, there is another type T_H3 subset, which is responsible for production of antiinflammatory antibodies (IgA) that are important in secretory immunity (Patil et al., 2012).

1.2 OVERVIEW OF IMMUNOMODULATORS

IMs are prophylactic or therapeutic agents, which are of biological or synthetic origin, can modulate, suppress, or stimulate any of the components of the immune system including both innate and adaptive immunity (Agarwal and Singh, 1999). Immunomodulation means all the therapeutic interventions used for modification of immune system. Augmentation of immune response is necessary for the prevention of infections in various conditions likes immunodeficiency, transplants, autoimmunity, allergy, and cancer (Gea-Banacloche, 2006). Cytokines play an important role in immunomodulation. The availability of purified cytokines or soluble cytokine receptors offer great therapeutic activity in immunotherapy by the means of immunomodulation. Research indicates inclusion of both IFN-γ and IFN-β into clinical phase. Cytokines have been used as immunoadjuvants along with vaccines but are at an experimental stage. Most remarkable progress in this field is exemplified by the discovery of cyclosporine, a potent immunosuppressant that has been widely used in graft rejection. This drug is also explored in cases of several autoimmune diseases where prednisone, dexamethasone, and methotrexate have shown failure.

1.2.1 Classification of immunomodulators

According to clinical perspective, IMs can be classified into three categories:

1. Immunoadjuvants

   These agents are generally used in the vaccine development for the generation of greater immune response (Agarwal and Singh, 1999). They act as an immune enhancer of antigenic component of vaccine. Adjuvants augment, stimulate, potentiate, enhance, activate, or modulate the immune response at either cellular or humoral level. Best known examples include Freund's adjuvant, Bacillus Calmette–Guérin (BCG), *Corynebacterium parvum*; all of which contain bacterial antigen. Endogenous adjuvants include histamine, IL-1, tuftsin, transfer factor, and interferon. Although, their mode of action may be nonspecific but it results in increased immune responsiveness to a variety of antigens or group of antigens.
2. Immunostimulants

   These agents also stimulate or boost either cellular or humoral immune responses. Some of them are specific immunostimulants, such as vaccines, which stimulate an immune response against specific antigens contrary to nonspecific stimulants. These types of stimulants are being used widely in the cases of autoimmunity, allergy, immunodeficiency, and cancer. In healthy individuals, stimulants act as a prophylactic agent, that is, they potentiate the basal levels of immune system. When an individual is exposed to pathogen, it elicits the heightened immune response that allows the rapid clearance of pathogen and its products that eventually prevent disease. In individuals with impaired immune response, stimulants act as immunotherapeutic agents (Agarwal and Singh, 1999). Immunocompromised condition includes patients with primary and secondary immunodeficiency. A condition resulting from a genetic or developmental defect in immune system is called as primary immunodeficiency. For example, severe combined immunodeficiency (SCID), Wiskott–Aldrich syndrome (WAS), and X-linked agammaglobulinemia. Secondary immunodeficiency is the loss of immune function, which results from exposure to various agents, for example, AIDS, malignancy, different types of immunostimulants like cytokines (IL-2, interferons, G-CSF), microbial toxins/fragments, herb, venom and so on.

3. Immunosuppressants

   These agents weaken or suppress the activation of the immune system. In general, immunosupression is of two types, deliberate induced and nondeliberate immune suppression. In deliberately induced immunosuppression, there is a need for the same because it is performed mainly to prevent body from rejecting an organ transplant, treating graft-versus-host disease after a bone marrow transplant, or the treatment of autoimmune diseases like rheumatoid arthritis, Grave's disease, and myasthenia gravis. Cortisone was the first immunosuppressant identified but its use is limited because of side effects, but discovery of cyclosporine led to the remarkable choice and is being used in various transplantations (including kidney, liver, and heart). Tacrolimus and sirolimus antibiotics are also used in transplantation and graft rejection. Nondeliberate immunosuppression can occur in aging, malnutrition, cancer, and chronic infections like HIV. In this case, unwanted immunosuppression leads to the increase in susceptibility to various pathogens, such as bacteria, fungi, viruses, or protozoans (Abbas et al., 2012).

Thus, IMs function by stimulating or suppressing natural and adaptive immune mechanism, which helps the body to tackle various medical conditions. IMs are derived from various sources like natural (include plant and animal origin, such as venom or toxin) or synthetic. These IMs are administered orally, parenterally, or topically as per the requirement and need of the disease in question. The next subsection gives a brief overview of the current routes of administration of IMs (Fig. 11.2).

1.3 CURRENT ROUTES OF ADMINISTRATION

Though IMs are a well-explored class of therapeutics, the need for absolute administration of IMs remains an issue to be addressed. IMs are available from different sources like synthetic, herbal, bacterial, and animal; in the current scenario, they are administered by all possible routes. IMs are majorly dependent on the desired biological activity whether they are used for a particular type of disease condition like rheumatoid arthritis, Grave's disease, or in a situation like graft rejection. In cases of vaccines, mostly parenteral and oral route is preferred whereas IMs from synthetic source are administered via topical, oral, and parenteral routes. Similarly, herbal IMs, venoms, and toxins are administered via oral

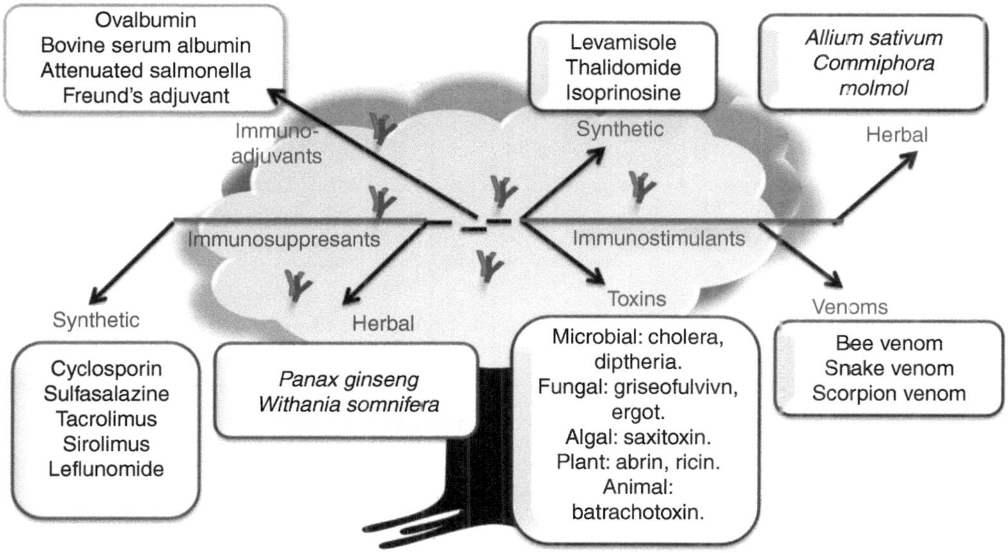

FIGURE 11.2 Classification and Examples of Immunomodulators

and parenteral routes. Among all the earlier listed routes for IMs delivery, the oral route is the most preferred because it offers several advantages:

- Most convenient, as it can be self administered, pain-free, easy to take and hence, offers patient compliance.
- The absorption takes place along the whole length of the GIT (Gastrointestinal tract).
- It is economical as compared to parenteral route.
- It does not require any sterilization.
- It is very convenient to sustain, extend or control the release of active from several dosage forms, such as tablet, capsule through oral delivery (Verma et al., 2010).

In autoimmune diseases like inflammatory bowel diseases (IBD) where inflammation of GIT is the main cause, the oral delivery of immunosuppressant is the choice over others as it shows a targeted action.

The next section focuses on the oral delivery of the IMs and the marketed preparations available to date. Further, it also cites the drawbacks of the current oral delivery systems and hence the need to explore the newer novel methods of efficacious IM delivery via the oral route.

2 ORAL ROUTE FOR IMMUNOMODULATORS

2.1 CONVENTIONAL IMMUNOMODULATORS

IMs are not only used in autoimmune diseases, graft rejections, but they are also well explored for therapy of life-threatening diseases like cancer to improve immune response. Taking into account this

Table 11.2 Conventional Marketed Products for Immunomodulators

Brand name	Company	IM	Dosage Form
Neoral	Novartis	Cyclosporin	Microemulsion filled capsule (25, and 100 mg)
Asacol	Zeria Pharmaceutical	Mesalazine	Tablet (400 mg)
Imuran	Aspen	Azathioprine	Tablet (25, 50, and 75 mg)
Prograf	Astellas Pharma	Tacrolimus	Capsule (0.5, 1, and 5 mg)
Rapamune	Pfizer	Sirolimus	Tablet and solution (2 and 5 mg)
Rotarix	GlaxoSmithKline	Human rotavirus	Oral suspension 1.5 mL
Vivotip	Berna Biotech	Typhoid vaccine	Capsules (2–10 × 10^9 CFU)

CFU, Colony forming units.

fact, the current market for IMs is huge. However, oral delivery of IMs is challenging due to several hurdles related to GIT, like first pass metabolism, variance of pH in GIT environment, instability of IMs and so on.

For oral delivery of any active, tablet and capsules are the choice of dosage forms as they offer patient compliance. Similarly, in the case of IM oral delivery these dosage forms are already available in market. Some examples of the marketed preparations for oral delivery of IMs are listed in Table 11.2.

As depicted in Table 11.2, enteric-coated tablets, soft gelatin capsules, suspensions, and microemulsions are the currently available oral dosage forms of IMs in market. However, these dosage forms are not able to address hurdles related to the oral delivery of IMs, like targeting to particular disease and variance in GIT environment.

2.2 DRAWBACKS OF CONVENTIONAL ORAL SYSTEMS

The incompetency of the conventional oral systems is due to indiscriminate reasons. These include instability of the conventional oral system resulting in degradation of the therapeutic drug on or before administration, rapid metabolism after administration, poor distribution to the targeted sites, binding to plasma proteins thus restricting the amount of free drug available for absorption. Furthermore, other reasons include degradation by intestinal enzymes, high adherence to the mucus but insufficient penetration through the intestinal mucosa, higher incidence of untoward effects of the therapeutic drug as compared to the pharmacological effects (Gaspar et al., 2008), higher dosing and dose frequency (Singh and Chandra, 2012). A few examples of toxicity include renal toxicity caused by oral delivery of cyclosporine (Singh et al., 2009), cardiac damage, or hypertension by tacrolimus tablets (Mihatsch et al., 1998).

3 NEED OF NANOSYSTEMS FOR IMMUNOMODULATORS

Therapeutic agents currently on the market can be positioned into either one of the four categories: chemically synthesized compounds, which are small molecules; botanically available molecules that are isolated from plants, fungi, and molds; biotherapeutic macromolecules, which could be either naturally occurring or can be engineered from a known biological template; for example, monoclonal antibodies designed to treat cancer; and nucleic acid–based therapeutics that are designed to interfere with natural translation process by mRNA.

Isolation of naturally occurring small molecules is a herculean task, which leads to a call on developing commercially feasible large-scale chemical synthesis of complex natural products.

The therapeutic agents available show shortcomings in their properties as an ideal agent, moreover, the current conventional dosage forms lack in oral bioavailability, once-a-day treatment, minimal or no side effects, and effectiveness in all the patients. Not only the improvement of existing drug delivery systems is important but also it is a valuable method to generate new therapeutics.

Nanosystems will provide target-based delivery. Nanosystems will not only enhance efficacy and potency but will also improve specificity without any untoward effects of the drug delivery system. It would also help to modulate the release profile of drugs either by sustained or controlled effect. The fate of the proteins in the gut is well established and known; the presence of cellular proteases and peptidases together can degrade both the proteins and peptides (Woodley, 1993). Hence, the therapeutic agents, such as proteins or peptides, can barely reach the site of absorption thereby lowering the bioavailability. Nanosystems have a capability to be incorporated into variety of dosage forms like tablets, capsules, liquids, suspensions, gels, pellets, and so on.

Modifications of conventional dosage forms by introducing the nanosystems, such as vesicular systems, colloidal systems, polymer-based systems, nanoparticles (NPs), lipid-based systems and so on, will help enhance the potential of the currently available therapeutic agents.

4 ORAL NANOSYSTEMS FOR IMMUNOMODULATORS

It is crucial to understand the interactions of the nanosystems with the IMs in terms of its effect as whole drug conjugated nanosystems on the targeting cell surface. The drug release pattern of the nanosystems, stability of the therapeutic IMs, and molecular mechanism of the cell signaling involved in the pathophysiology of disease in consideration are the key points to be considered while selecting an appropriate nanosytem.

Oral nanosystems that are established include:

- Vesicular systems: liposomes, proliposomes, phytosomes
- Biphasic systems: micro/nanoemulsions
- Particulate systems: NPs, microspheres, micropellets

This section highlights the techniques involved in the preparation of the nanosystems in brief. Furthermore, the section also reviews individual classes of the IMs with the nanosystems that have been tried and worked out to date.

Liposomes can help in passive targeting to the cells of the immune system, especially the cells of the mononuclear phagocytic system. Liposomes can potentially preserve the entrapped substance and thus increase the bioavailability (Akbarzadeh et al., 2013; Maherani et al., 2011) (Table 11.3).

Proliposomes are a dry form of the liposomes, which show improved stability as compared to the liposomes. Proliposomes form liposomal suspension when in contact with water (Thapa et al., 2013).

The key difference between liposomes and phytosomes is that in case of liposomes the drug is incorporated in the cavity formed by the phospholipids; whereas, in case of phytosomes, the drug becomes the integral part in forming the lamelle complex structure.

Table 11.3 Methods for Liposome Preparation

Mechanical dispersion method (for hydrophilic drugs)	Solvent dispersion method (for hydrophobic drugs)
1. Lipid hydration method	1. Ethanol injection method
2. Sonication	2. Ether injection method
3. Microfluidization/microemulsion method	3. Double emulsion method
4. Membrane extraction method	4. Reversed phase evaporation of vesicles
5. Dried reconstituted vesicles	5. Stable plurilamellar vesicles
6. Ethanol injection method	6. Detergent removal method

Preparation of phytosomes is simple, which involves a film hydration method. Aprotic solvent is added to a phospholipid followed by a phytoconstitutent. Next, they are subjected to drying on a rotary evaporator and finally hydrated using water or buffer.

Microemulsion/Nanoemulsion is an optically isotropic and thermodynamically stable system containing oil, water, and an amphiphile. These can be in varied size-ranges ranging from few nanometers to microns (Zolnik et al., 2010). Self-emulsifying techniques give better oral delivery as the active moiety is not subjected to the biochemical degradation owing to GIT (bile salts and lipases) (Li et al., 2015). There are two methods of preparation: phase titration and phase inversion.

Solid lipid nanoparticles (SLNs) are NPs ranging from 50 to 1000 nm, made from the lipids that remain solid at room temperature and are stabilized using surfactants. These show promising drug targeting, protection of active moiety from chemical degradation, and reduction of drug toxicity. Currently, these nanosystems are explored for transdermal application. However, with appropriate selection of lipid and surfactant they can be made acceptable for oral delivery, which can prevent GI enzymatic degradation of drug.

Methods of SLN preparation include (Almeida and Souto, 2007; Ekambaram et al., 2012; Thapa et al., 2013).

1. Microemulsion-based SLN
2. High-pressure homogenisation SLN
3. Solvent emulsification evaporation
4. Solvent emulsification diffusion
5. Lipid particles from super critical fluid technology (RESS)
6. Hot melt fusion method
7. High-speed homogenization
8. Precipitation technique

Nanostructured lipid carriers (NLCs) have been developed to overcome the drawbacks associated with SLN. NLCs in contrast to SLN possess a higher drug-loading capacity as they are prepared by blending a liquid lipid with the solid one. The incorporation of liquid lipid helps to increase the loading of active into the core structure and hence the probability of drug expulsion during storage is less compared to SLN. NLCs have benefits like controlled release and an increased chemical stability of the incorporated actives. Furthermore, they can be produced easily on a large scale (Mehnert and Mäder, 2001; Selvamuthukumar and Velmurugan, 2012).

Polymeric nanoparticles (NPs) are colloidal systems with particle size ranging from 10 to 1000 nm. These have benefits of enhanced solubility and reduced toxicity. Biocompatibility and biodegradability of the polymers used are of paramount importance. Poly (lactide-*co*-glycolide) (PLGA) or polylactide (PLA) polymeric particles are used for oral immunotherapy not only for their special carrier properties but also because of their adjuvant effects in case of vaccine formulations. Also, these polymers are eliminated through the body by Krebs cycle making its application in humans admissible with ease (Thapa et al., 2013).

Inorganic nanoparticles reveal unique physical and chemical properties as their size approaches in nanometer range. The surface morphology and particle size of the inorganic NPs provides exceptionally good colloidal stability and high antigen encapsulation efficiency, which makes them good carriers for drug/vaccine deliveries. Due to these properties, they are explored as potential oral vaccine carriers. The high surface area of NPs provides good rigidity and stability to the adsorbed antigens against harsh GI environments and prevents premature leaking of the antigen. These NPs are biocompatible and inert, that is, immune response is not induced against them. For example, gold NP, iron NP, and silica NP (Marasini et al., 2014).

Microspheres are spherical particles of 1–1000 μm size. A uniform dispersion of drug in the polymer matrix is noted with release following first-order kinetics. Various natural and synthetic polymers have been explored for preparation of microspheres and include gelatin, modified starches, dextrans, and polypropylene. The microspheres have large surface-to-volume ratio and hence the interfacial characteristics of microspheres often dictate their activity. These can be administered by either oral route or by injection. Emulsion–solvent evaporation, spray drying, and chemical cross-linking are the major techniques used for preparation of microspheres.

Micropellets are solid particles with a size range of 1–1000 μm. Spray drying is the technique used for preparation of micropellets. Here the drug can be dispersed or dissolved in the polymeric solution, which would be subjected to spray drying. Spray drying process can be used for thermolabile drugs and also can be scaled up with ease. Micropellets prevent dose dumping, which is a common problem with the conventional dosage forms (Kulkarni, 2011).

The next subsection gives a detailed look at different classes of IMs and the nanosystems in research and development.

4.1 ORAL NANOSYSTEMS FOR SYNTHETIC IMMUNOMODULATORS

Actives derived synthetically have always proven their potency in the management of different kinds of diseases. On the same virtue, IMs derived synthetically are well explored as stimulants, suppressants, and adjuvants. Currently, IMs from synthetic sources have established a market in treatment of various autoimmune diseases, for example, rheumatoid arthritis and IBD. Numerous studies were carried out by the researchers on different types of nanosystems like polymeric, liposomes, and other lipidic systems. Immunosuppressant drugs like cyclosporine, tacrolimus, sulfasalazine, and dexamethasone are well explored using nanoformulations.

Shin et al. (2010) have evaluated tacrolimus-loaded NPs using PLGA and PEG-PLGA polymers for the treatment of autoimmune diseases. The particle size and encapsulation efficiency of PLGA and PEG–PLGA NPs were found to be almost similar, that is, around 220 nm and 60%, respectively. The in vitro release studies were carried out at various pH of GIT (1.2, 4.0, 6.8, and 7.4). The release rate of tacrolimus was found to be higher at pH 1.2 as compared to others; this suggests that the release of tacrolimus from NPs is a pH-dependent phenomenon. The lymphatic targeting efficiency of NPs was compared with marketed Prograf injection by intra-venous (IV) administration and it was found to be higher. Rather, PEGylated NPs showed better targeting than non-PEGylated one. The area under the curve (AUC) for PEGylated NPs was found to be 39,526.18 ± 3,411.35 ng min/mL whereas that of Prograf injection was 28,058.66 ± 4,531.07 ng min/mL. Concentration of NPs in mesenteric and axillary lymph nodes were found to be higher than Prograf injection; this indicated that the formulated NPs have better lymphatic targeting ability than the marketed formulation (Shin et al., 2010).

Wang et al. (2014) have studied three different types of nanosystems (PLGA NPs, NLCs, and SMEDDs) loaded with cyclosporine. The particle size and encapsulation efficiency of PLGA NPs was 182.2 ± 12.8 nm and 87.6% ± 1.6%, of NLCs 89.7 ± 9.0 nm and 80.3 ± 0.6%, respectively. The particle size for SMEDDS was found to be 26.9 ± 1.9 nm. All NPs were negatively charged. In vitro release studies were carried out and cumulative release of drug in phosphate buffer saline (PBS) from all carriers was extremely low within 12 h. Oral bioavailability of drug was evaluated in Beagle dogs and it was found that for C_{max}, NLCs showed the maximum value among all vehicles (2.32 ± 1.05 μg/mL) but was comparable with Sandimmune Neoral (1.76 ± 0.77 μg/mL). This suggests that lipid-based drug carriers have promising results due to their preferential uptake by the lymphatic system. But, the C_{max} of PLGA NPs was found to be lower than that of others (Wang et al., 2014). Table 11.4 depicts some more examples.

4.2 ORAL NANOSYSTEMS FOR HERBAL IMMUNOMODULATORS

Since the development of human civilization, mineral, plant, and animal products have been the main sources of drugs. These sources of drugs have shown promising curative and remedial properties. But over the time, the therapeutic effect of the natural IMs has been compromised owing to their inefficient bioavailability and also their untoward actions.

The nanosystems for drug delivery are well established and have shown promising results in clinical trials to name a few, Marqibo [vincristine sulphate] liposomal formulation, LEP-ETU [Paclitaxel] liposomal formulation (Fan and Zhang, 2013). This subsection enables us to understand the potential outcome of nanosystem for herbal IMs.

Sou et al. (2008) developed an intravenous injection that could be delivered to the tissue macrophages. The intravenous injection was prepared from a modified nanolipid vehicle loaded with curcumin. Curcumin was encapsulated into a phospholipid vehicle comprising 1,2-dimyristoylsn-glycero-3-phosphocholine; 1,5-dihexadecyl ester; 1,2-distearoylsn-glycero-3-phosphoethanolamine-*N*-[monomethoxypoly(ethylene glycol) (5000)] in a molar ratio of 10:1:0.06 (Thapa et al., 2013). Liposomes tend to show instability, which could be overcome by proliposomes without changing intrinsic characteristics. Silymarin proliposomes were prepared and evaluated for their physicochemical properties, pharmacokinetic characteristics, and bioavailability of silymarin proliposomes in rats after oral administration by Xiao et al. (2005). They showed encapsulation efficiency of 90% and an average particle size of 238.8 nm with good stability. Also, they reported an improved bioavailability of proliposomes following oral administration (Xiao et al., 2005). Herbal drug delivery has a promising value and is reflected by the development in the area of phytosomes. Echinacea phytosome containing echinacosides from *Echinacea angustifolia*, *Echinacea purpurea* roots stimulate cellular and humoral

Table 11.4 Nanosystems for Synthetic Immunomodulators

Immunomodulator Agent	Pharmacological Activity	Formulation Type	Particle Size (nm)	EE (%)	Effect of Formulation	References
Tacrolimus	IBD (colitis)	PLGA nanoparticles Eudragit nanoparticles	455 469	52 64	Compared to conventional oral tacrolimus solution, both the nanoformulations were found to reduce the adverse effects of tacrolimus, (inflammatory activity and marked reduction in nephrotoxicity)	Meissner et al. (2006)
5ASA	IBD	PCL	200–350	—	5ASA covalently bound to PCL polymer slowed down the drug release, this allowed an efficient colonic targeting	Pertuit et al. (2007)
Sulfasalazine	Inflammatory paw edema	Inorganic nanoparticles (MgAl–NO_3)	100–150	—	Intercalation of sulfasalazine into the interlayer of a layered double hydroxide and further coating with a pH-sensitive eudragit copolymer lead to a controlled release system, reduced dosing frequency, protective effect for scavenging of superoxide free radicals as well as DEM induced lipid peroxidation	Kankala et al. (2015)
Sulfasalazine	IBD	Mesoporous silica nanoparticles	100	—	pH-responsive controllable drug release system allowed high loading of anionic drugs	Lee et al. (2008)
Tacrolimus	Graft rejection	SMEDDs	<25	—	Therapeutic efficacy of drug was improved and toxicity was reduced than marketed formulation	Borhade et al. (2008)
Tacrolimus	Autoimmune diseases	PLGA nanoparticles PEG-PLGA nanoparticles	218 220	60 60.3	PLGA NPs showed significantly higher lymphatic targeting efficiencies than Prograf injection, Higher distribution of PEG-PLGA NPs.	El-Shabouri (2002)

(Continued)

Table 11.4 Nanosystems for Synthetic Immunomodulators (*cont.*)

Immunomodulator Agent	Pharmacological Activity	Formulation Type	Particle Size (nm)	EE (%)	Effect of Formulation	References
Cyclosporin-A	Graft rejection, autoimmune diseases	Polymeric [chitosan-HCl, gelatin A and sodium glycocholate (SGC)] nanoparticles	104–148	21–24	Positively charged Chitosan NPs showed improved absorption rate and overall bioavailability than marketed formulation Neoral (microemulsion), reduced dosing frequency and side effects	Italia et al. (2007)
Cyclosporin-A	Autoimmune diseases	PLGA nanoparticles	143	71.9	Didodecylmethylammonium bromide (DMAB) as stabilizer increased EE of drug. Significantly improved intestinal uptake and oral bioavailability as well as lower nephrotoxicity as compared to Neoralwas seen	Dai et al. (2004)
Cyclosporin-A	Auto immune diseases	PLGA NPs NLCs SMEDDs	182 89 26	87.6 80 –	Lipid-based drug delivery systems proved to be advantageous than others in improving oral absorption, their oral bioavailability were comparable with that of the marketed formulation	Jaiswal et al. (2004)
Cyclosporin-A	Autoimmune diseases	Poly(methacrylic acid and methacrylate) copolymer NPs	37.4–106.7	90.9–99.9	pH-sensitive NPs exhibitedpromising results compared to marketed formulations in terms of oral bioavailability was observed.	Müller et al. (2006)
Cyclosporin-A	Autoimmune diseases	PLGA NPs	< 250	88–91	Monodisperse NPs with high EE and reproducibility were produced by high-pressure emulsification-solvent evaporation technique. The technique can be automated and scaled up.	Wang et al. (2004)

Cyclosporin-A	Autoimmune diseases	SLN nanocrystals	157 962	96.1 100	Nanocrystals did not show promising results whereas plasma concentrations of drug loaded SLNs were similar to that of marketed formulation; this might be because of lymphatic uptake of lipidic carriers.	Zhang et al. (2000)
Cyclosporin-A	Autoimmune diseases	HPMCP NPs 1. HP50 2. HP55	50–60	>95	pH sensitive polymer HPMCP NPs showed oral bioavailability comparable tothe marketed formulation.	Ubrich et al. (2005)
Cyclosporin-A	Autoimmune diseases	Stearic acid NPs	316	88	Oral bioavailability of NPs was significantly lower themarketed formulation.	Gulbake and Jain (2013)
Mesalazine	Ulcerative colitis	Chitosan NPs	157	—	Eudragit-coated NPs showed protection against SGF and targeted delivery to colon	del Giudice et al. (2014)

5ASA, 5-amino salicylic acid; DEM, dimethyl maleate; HPMCP, hydroxypropyl methylcellulose phthalate; IBD, inflammatory bowel disease; MgAl-NO_3, magnesium aluminium nitrate; NLC, nanostructured lipid carrier; PCL, polycaprolactone; PEG, polyethylene glycol; PLGA, poly (lactide-co-glycolide); SLN, solid lipid nanoparticles; SMEDDS, self-microemulsifying drug delivery systems.

immune defense, activates B and T lymphocytes, and stimulates tissue necrosis (Karimi et al., 2015). Ginseng phytosome containing 37.5% ginsenosides from *Panax ginseng* increases catalase, superoxide dismutase, glutathione peroxidase, and gluthathione reductase activities and prevents depletion of these antioxidant enzymes (Jain et al., 2010).

In case of a study done in rat intestines for silymarin microemulsion, Yuan et al. (2004) observed absorption of the silymarin microemulsion in the middle and lower segments of the intestine. The absorption was observed to be a first-order process with a passive diffusion mechanism (Yuan et al., 2004).

Triptolide obtained from *Tripterygium wilfondil* used in the treatment of autoimmune disease was formulated into SLN by Mei et al. (2003). They showed decreased toxicity of triptolide SLN as compared to free triptolide (Thapa et al., 2013). Jin et al. (2006) prepared the SLN of crytoptanshinone, which inhibits the activity of COX-2 but not its expression. Also, it shows cytotoxic activity. Incorporating crytoptanshinone in SLN showed a marked increase in the bioavailability when compared against crytoptanshinone suspension. A significant enhancement in bioavailability of crytoptanshinone suggests SLNs being a powerful tool for improving oral absorption (Jin et al., 2006; Zhou et al., 2013).

Liu et al. developed triptolide-loaded poly (DL-lactic acid) NPs. Triptolide illustrated the problem of poor solubility and toxicity. This was overcome by loading it in NPs using a biocompatible and biodegradable polymer which were uniform in size, spherical in shape with smooth surface, with improved solubility and reduced toxicity (Thapa et al., 2013). Bao et al. (2002) developed NPs of cuscutachinensis by nanosuspension method to improve absorption of the poorly water soluble constituent. These NPs were used to prevent liver and kidney senescence and regulate the immune system. Some studies showed anticancer, antiaging, and immune-stimulatory effects. Cuscutachinesis NPs showed an entrapment efficiency of 90% with improved water solubility and improved hepatoprotective and antioxidant effects (Bao et al., 2002). Bhatia et al. (2011) have developed chitosan NPs for the extract of *Ziziphus mauritiana* and checked their immunomodulatory activity. The in vivo study indicated elevated and maximum delayed type hypersensitivity response in case of extract loaded NPs as compared to plant extract. The GI instability of plant extract has been overcome due to extract-loaded NPs, which are stable in elevated pH conditions and do not degrade by the digestive enzymes (Bhatia et al., 2011).

Silymarin solution versus silymarin microspheres was studied by Abrol et al. They showed an improved synergistic hepatoprotective effect of silymarin microspheres as compared to silymarin solution (Dixit et al., 2007). Tong et al. (2003) prepared camptothecin (CPT)-loaded PLGA microspheres and studied them for antitumor activity. The CPT-PLGA microspheres created an acidic microenvironment, which provides the pH for the active lactone form CPT for its activity. Cytotoxicity assay showed CPT-PLGA microspheres retained antitumor activity with quick uptake by B16 cells. The microsphere approach stabilized and showed effective delivery of CPT for the treatment of cancer(Tong et al., 2003).

Andrographolide-loaded sodium alginate micropellets were prepared and studied for its entrapment efficiency, which was 94.89% with drug release of 86% in 8 h (Shariff et al., 2007).

A few more reported nanosystems for herbal IMs are given in Table 11.5.

4.3 ORAL NANOSYSTEMS FOR VACCINES AND ADJUVANT

Vaccines are for prophylactic use. They lead to the activation of both innate and acquired immune systems to induce a specific immune response against a particular disease. There are three types of vaccines; killed, live, and subunit vaccines. They are able to recognize and activate PRRs (pattern recognition receptors), including TLRs, on the surface of APCs. This activation induces the development

Table 11.5 Nanosystems for Herbal Immunomodulators

Herbal Medicine	Source and Chemical Constituents	Pharmacological Activity	Formulation Type	Particle Size (nm)	EE (%)	Benefit of Formulation	References
Triptolide	*T. wilfondil* Hook F; diterpenoid triepoxide	Autoimmune diseases, such as rheumatoid arthritis, psoriasis, antineoplastic activity leukemia,	SLN	<200	–	Enhanced anti-inflammatory activity	Mei et al. (2003)
Curcuminoids	*Curcuma longa*; curcuminoids	Antitumor	SLN	447	–	Enhanced stability of curcuminoids	Tiyaboonchai et al. (2007)
Cryptotanshinone	*Salvia miltiorrhiza Bunge*	Immunomodulator COX-2 inhibitor	SLN	127.3	93.7	Enhanced bioavailability of cryptotanshinone	Hu et al. (2010)
Silymarin	*Silybum mari anum*	Hepatoprotective Im munomodulator	Proliposomes	238.8	90	Improved hepatoprotective activity	Xiao et al. (2005)
Triptolide	*T. wilfondil* Hook F; Diterpenoid triepoxide	Autoimmune diseases, such as rheumatoid arthritis, psoriasis, antineoplastic activity leukemia,	Polymeric nanoparticles {Poly (DL-lactic acid) nanoparticles}	149.7	85.7	Enhanced solubility and reduced toxicity	Liu et al. (2005)
Ziziphusmauritiana	*Ziziphus mauritiana whole plant extract*	Immunomodulatory activity	Chitosan based nanoparticles	–	–	Enhanced activity of the extract	Bhatia et al. (2011)
Cuscutachinensis	*Cuscutachinensis Lam*; (active constituents–flavonoids and lignans, such as quercetin, kaempferol	Used as tonic for the liver and kidney, prevent senescence and regulate the immune system. Some studies showed anticancer, antiageing and immune-stimulatory effects.	Nanoparticles nanosuspension	267	90	Enhanced solubility	Bao et al. (2002)

(Continued)

Table 11.5 Nanosystems for Herbal Immunomodulators (*cont.*)

Herbal Medicine	Source and Chemical Constituents	Pharmacological Activity	Formulation Type	Particle Size (nm)	EE (%)	Benefit of Formulation	References
Paclitaxel	*Taxus brevifolia*	Anticancer	Nanoparticles	—	—	Reduction of side effects	Trickler et al. (2008)
Ginseng root	*Panax quinque-folium*	Immunostimulant activity	Nanoparticles	20		Enhance immuno-stimulant activity	Akhter et al. (2016)
Berberine	*Coptischinensis franch*	Anticancer	Nanoemulsion	—	—	Improved resi-dence time hence absorption and anticancer activity	Tan et al. (2011)
Camptothecin	*Camptotheca acuminata*	Cytotoxic, anticancer	Microspheres	–	–	Dose reduction, enhancement of cytotoxicity	Chao et al. (2010)
Silymarin	*Silybum mari-anum*	Hepatoprotective Im-munomodulator	Microspheres	1.3 μm	94.1	Enhanced hepato-protective activity	Dixit et al. (2007)
Andrographolide	*Andrographis-paniculata*	Hepatoprotectant	Micropellets	—	91.83	Improved hepato-protectant activity	Arshia et al. (2007)
Curcumin	*Curcuma longa;* natural poly-phenol	Antiinflammatory	Microemulsion	61.8–79.35	—	Enhanced antiin-flammatory effects	Wang et al. (2008)
Curcumin	*Curcuma longa;* polyphenols	Antitumor	Liposomes	—	—	Improved drug de-livery of curcumin	Sou et al. (2008)
Zerumbone	isolated from rhizomes of *Zingiber zerum-bet*	Antitumor	Nanostructure lipid carriers	–	–	Increased levels of IL-2, IFN-γ (antitumor related cytokine) and IL-1β, IL-6 (inflam-matory related cytokines).	Kong et al. (2013)

COX, Cyclooxygenase; IFN, interferon; IL, interleukin; SLN, solid lipid nanoparticles.

of a T cell specific response and also B cell antibody response (del Giudice et al., 2014). Oral administration of vaccines induces an immune response through gut-associated lymphoid tissue (GALT) in the small intestine, ascending colon, and distant tissues, such as mammary glands and salivary glands (Patravale and Prabhu, 2014). The effective delivery of vaccines using nanosystems has been meticulously studied by researchers.

Garinot et al. (2007) have carried out studies on radio-labeled vaccines where ovalbumin was labeled with ^{3}H by reductive alkylation method using sodium boro [^{3}H] hydride (100 mCi). Encapsulation of vaccine was carried out using polymers like PLGA, PLGA-PEG, and PCL-PEG. Three different nanosystems were formulated: A(PLGA:PCL-PEG, 5:5); B (PLGA:PLGA-PEG:PCL-PEG, 4.5:4.5:1); and, C (PLGA:PLGA-PEG:PCL-PEG, 7:1.5:1.5). Targeting of NPs was achieved by surface labeling with RGD peptide to target human M cells where formulation C was selected for targeting. Particle size of all three formulations was found to be around 200 nm and encapsulation efficiency about 30%–50%. When formulations were tested for stability in biological media, after 2 h of incubation with gastric fluid (0.1 N HCl), 5% of ^{3}H-labeled ovalbumin was observed in supernatant after centrifugation of NP suspension. During the stability experiment not less than 10% release of [^{3}H]-labeled ovalbumin was observed. In vitro transport of NPs by M cells was evaluated and transport of formulation A was not significantly high but on the other hand transport for formulation C was found to be 4-fold higher and for B was 10-fold higher. RGD-labeled formulation containing PCL-PEG-GRGDS showed 3.5-fold higher transport compared to nonlabeled one. Immunization was studied for oral administration of vaccine NPs in NMRI mice. Though the mean production of IgG was not increased with targeting, the number of mice producing IgG was slightly higher with targeted formulations compared to nontargeted ones. Uptake of targeted fluorescent NPs given by oral gavage in mouse Peyer's patches (PPs) was studied using confocal microscopy and it was found that NPs were localized mainly in the PPs and also they were often colocalized with M cells (Garinot et al., 2007).

Guittre et al. have formulated (bovine serum albumin) BSA-loaded PLGA NPs of three different particle sizes (200, 500, and 1000 nm). These nanoformulations were evaluated by administering them via different routes like intranasal, oral, and subcutaneous to Balb/c mice. The in vivo antibody response was studied and it was found that the subcutaneous injection of NPs with particle sizes of 1000 nm showed higher IgG levels compared to other two NPs with lower sizes. These differences in the immune response could be due to the reason that distribution in the lymphoid tissue is size-dependent and particles having size 1–2 μm undergo preferential phagocytosis. Nanoparticles with 1000 nm size have shown similar results even in case of oral administration. This might be because particles in lower size ranges could bring a possible reduction on the immunogenicity of the administered NPs. Similar observations were seen in case of intranasal administration. The highest capability of 1000 nm size particles to stimulate IgG responses in mice by all three routes, compared to smaller sized particles, might be an indication that the stimulation mechanisms of the immune response are depending on the size of administered NPs (Gutierro et al., 2002). Table 11.6 depicts various vaccines formulated as nanosystems.

4.4 ORAL NANOSYSTEMS FOR TOXINS AND VENOMS

Toxins are synthesized by various organisms, invertebrates, even higher plants and animals. It is the metabolic product that evolved as a defense mechanism for the purpose of killing or repelling from pathogens or predators. Toxins may be small molecules, peptides, glycoproteins, or proteins that interact

Table 11.6 Nanosystems for Vaccines and Adjuvants

Vaccine	Activity	Formulation Type	Particle Size (nm)	EE(%)	Effect of Formulation	References
Der p 2 encoding DNA	Allergen-induced airway eosinophil dominant inflammation	Chitosan NPs	100–400	100	Elevated IgG2 level indicating improved T_H1 response and declined IgE level indicating prevention of T_H2 response	Chew et al. (2003)
Rho1-GTPase Protein encoding DNA	Protection against *S. mansoni* infection	Chitosan NPs	150–500	—	Immunization against schistosomiasis	Oliveira et al. (2012)
Attenuated *Samonella*	Cancer immunotherpay	β-cyclodextrin-PEI600 NPs	—	—	Remarkable T cell activation and cytokine production, inhibition of tumor growth	Hu et al. (2015)
OVA	Autoimmune disease and allograft rejection	Chitosan NPs	—	—	Reduction of antigen-specific delay type hypersensitivity response and antibody formation	Goldmann et al. (2012)
OVA and adjuvant MPLA	Autoimmune disease	PLGA NPs	296–375	66–86	OVA/MPLA loaded NPs stimulated higher antigen specific IgG and IgA antibodies, induction of systemic and mucosal immune responses	Sarti et al. (2011)
HbsAg	Immunoadjuvant	PLGA NPs	380		M-cell targeted vaccine with plant lectin UEA1-loaded NPs elicited significantly higher immune response as compared to nontargeted NPs.	Gupta et al. (2007)
OVA	Autoimmune disease	PLGA NPs Lipidic NPs	558 215		Lipidic NPs were superior as compared to PLGA NPs, even they had higher affinity for M cells. OVA-lipid NPs effectively induced high level of IgG and IgA antibody responses.	Ma et al. (2014)
BSA antigen	Immunoadjuvant	Silica NPs	430	—	Stimulated humoral and secretory IgA response, induction of both T_H1 and T_H2 response	Wang et al. (2014)

DNA, Deoxyribose nucleic acid; Ig, immunoglobulins; MPLA, monophosphoryl lipid A; NPs, nanoparticles; OVA, ovalbumin; PEI, polyethylenimine; T_H, T helper; UEA, Ulex europaeus.

with biomolecules like receptors, channels, or enzymes for their activity (Hodgson, 2012). Over the period, attempts have been made to understand the mechanism of action of toxins. For example, a variety of research is going on for discovery of the important pathways in which these toxins interact at various levels. These studies revealed how toxins interact with their targets at cellular or molecular level and have helped rapid expansion of the use of toxin in various medical conditions

Venoms are secretory products of animal species, which are being synthesized and stored in specific area in the body so whenever there is a need it is delivered to the target. Species have evolved with different mechanisms for delivering venom. For example, bee and wasp venom is delivered by sting, centipedes that use forcipules (modified legs), and the venom of snake delivered by fangs (Hodgson, 2012). Generally, venom is composed of several bioactive molecules, such as enzymes, proteins, toxin, peptides, or polypeptides, which are highly specific in their action and have great affinity for different functional organization of cells and tissues. Today, several components of venom have been isolated with a known mode of action; these components are widely used in several applications like pharmaceutical agents, diagnostic reagents, or preparative tools in neurobiology, cancer biology, and complement research. Table 11.7 cites a few examples of toxins and venoms.

Although various kinds of toxins have been identified from different sources, only some of them are explored for therapeutic activity. Most of the venoms are fatally harmful for human beings but some of them have proven to be beneficial for mankind. Venom and its components provide great opportunity in drug designing, because their three-dimensional structures allow the interaction of small molecules, having therapeutic and pharmacological properties (Pal et al., 2002). Table 11.8 cites a few nanosystems for toxins and venoms.

Nanoformulations of venom and toxin would not only offer insights to newer drugs, but also good drug delivery systems, which would have better bioavailability and greater therapeutic efficacy. Research on nanoconjugated venom–toxins shows good promise in the field of drug discovery and drug delivery, but thorough research is required before it can be taken to clinical trials.

5 METHODS FOR CHARACTERIZATION OF NANOSYSTEMS

This section addresses the characterization of nanosystems and its impact on its delivery thus helps us understand the fate of the nanosystem after oral delivery.

In general the nanosystems are characterized for their physical attributes, that is, size, shape, percentage drug entrapped, percentage drug release, and chemical composition (Saraf, 2010). The response of the nanosystems is a major attribute of membrane permeability, chemical constituents, quantity, and purity of materials and physical size.

Characterization techniques:

1. *Imaging*: this can be done by transmission electron microscopy (TEM), scanning electron microscopy(SEM), cryo transmission electron microscopy(CryoTEM), Cryo Field Emission Scanning Electron microscopy(cryoFESEM). These techniques help understand changes in structures with time and also help understand stability changes (Almgren et al., 2000; Dayan and Touitou, 2000; Dubes et al., 2003; Rizwan et al., 2007).
2. *Particle size and zeta potential*: nanosizing is an important parameter to understand whether the therapeutic agent is in the desired window of uptake. An example in the case of bypassing the GI

Table 11.7 Examples of Toxins and Venoms

Origin	Toxin/Venom	Source	Action
Microbial	Cynobacterial (anatoxin, microcystin, and nodularin)	*Anabaena, Aphanizomenon, Nodularia, Oscillatoria,* and *Microcystis*	—
	Botulinum	*Clostridium botulinum*	Interfere with cholinergic neuro-transmission
	Cholera	*Vibrio cholera*	Cause massive diarrhea
	Diphtheria	*Corynebacterium diphtheriae*	RNA translation inhibitor
	Exotoxin A	*Pseudomonas aeruginosae*	Inhibits elongation factor 2
	Tetanus	*Clostridium tetani*	Neurotoxic
Mycotoxin (Fungal toxin)	Ergot alkaloids	*Claviceps purpurea*	Affect nervous system and vasocon-strictor
	Aflatoxins	*Aspergillus flavus*	Affects cytochrome P450
	Amantin	*Amanita phalloides*	Affects RNA polymerase II
	Griseofulvin	*Penicillium griseofulvum*	Affects microtubules
Algal toxin	Saxitoxin	*Gonyaulax catenalla*	Block sodium channels of nerve membrane
Plant toxin	Abrin	Castor bean (*Ricinus communis*)	Affect gastrointestinal tract
	Ricin	Castor bean (*R. communis*)	Antitumor activity
	Buckthorn toxins	*Karwinskia humboldtiana* fruit	demyelination of peripheral nerves
	Urushiols	*Toxicodendron* species oil	Dermatitis and hypersenstivity
	Picrotoxin	*Anamirta cocculus*	Block GABA receptor
	Strychnine	*Strychnos nux vomica*	Hypersensitivity to sensory stimuli and convulsion
	Grayanotoxins	Leaves and flowers of plants of the *Ericaceae* family	Affects both cardiovascular and respiratory system
	Belladonna	*Atropa belladonna*	Affects nervous system
Animal toxin	Tetrodotoxin	Ovary and liver of puffer fish, in the eggs of the California newt	Affect nerve and muscle by block-ing sodium channel
	Batrachotoxin	Columbian arrow poison frog *Phyllo-bates aurotaenia*	Blocks neuromuscular transmission and evokes muscular contracture

Coelenterates	Palytoxin	*Palythoa toxica*	Intense vasoconstrictor
Arachnids	Scorpion toxin	Scorpionidae and buthidae family *Centruroides* *Sculpturatus*	Neurotoxic
	Widow spider venom	*Latrodectus mactans*	Neurotoxic
	Recluse spider venom	*Loxosceles recluse*	Cytotoxic and responsible for skin lesion
Insects	Bee venom	*Apis mellifera mellifera* and *A.m. ligustica*	Inhibits protein kinase C, myosin light chain kinase
Fish	Stonefish venom	*Synanceia horrida*, *Synanceia verrucosa*	Affects respiratory and cardiac system
Snake	Cobra venom	Cobra	Neurotoxic
	α-Bungarotoxin	*Bungarus multicinctus* and *B. caeruleus*	Block the acetylcholine receptor
	β Bungarotoxin	*B. multicinctus* and *B. caeruleus*	Block neuromuscular transmission
	Mamba venom	*Dendroaspis angusticeps*	Acetylcholinesterase an voltage gated K^+ channels blocker
	Rattlesnake venom	*Crotalidae* family	Cause pulmonary congestion and hemorrhage, hemolysis

GABA, γ-Amino butyric acid; RNA, ribo nucleic acid.

Table 11.8 Nanosystems for Toxins and Venoms

Toxin	Pharmacological Activity	Formulation Type	Particle Size (nm)	Benefits of Formulation	References
Tetanus toxoid	Prevention against tetanus	PVAL-PLGA Nanoparticles	100 or 500	Induction of IgG and IgA antibody responses strongly influenced by the NP size distribution. Highest antibody titers found in case of 100 nm NP.	Jung et al. (2001)
Cholera toxin B subunit (CTB)	Prevention against cholera	PLGA nanoparticles	420	Microencapsulated CTB induced Serum antibodies against intact CTB.	O'Hagan et al. (1995)
Protective antigen from *B. pertussis*	Prevention against pertussis	PLGA nanoparticles	200–600	Elicited IgG response, Induction of spleen cells (secretion of IL-5) and of antigen-specific IgA in the lungs.	Conway et al. (2001)
Toxoplasma gondii GRA1 protein	Prevention against toxoplasmosis	Chitosan nanoparticles	400	Elicited mixed IgG2a/IgG1 response.	Bivas-Benita et al. (2003)
Cholera toxin (CT) or its B subunit (CTB)	Prevention against cholera	Liposomes dipalmitoylphosphatidyl-ethanolamine (DPPE), DPPE-maleimidomethylcyclohexane-1-carboxylate (DPPE-MCC); Distearoylphosphatidylcholine (DSPC)	100	Induction of serum IgG anti-CT salivary IgA anti-CT response.	Harokopakis et al. (1995)
Tetanus toxoid	Prevention against tetanus	Stabilized chitosan nanoparticles (sChNPs)	116	Elicited serum anti-TT IgG response and increase levels of IL-2 and IFN-γ.	Harde et al. (2015)
		Mannosylated chitosan nanoparticles (Ch-GM-NPs)	123	Elicited serum anti-TT IgG response as well as secretary IgA response and increase levels of IL-2 and IFN-γ.	
Diphtheria toxin	Prevention against diphtheria	Chitosan microparticles	4700	Higher levels of specific anti-DT IgA and IgG in the gastrointestinal tract.	van der Lubben et al. (2003)

Bufalin (toad venom)	Antitumor activity	Wheat germ agglutinin (WGA)-grafted lipid nanoparticles	170	Activity against hepatocellular and lung carcinoma.	Liu et al. (2010)
Naja naja oxiana venom	Immunoadjuvant	Chitosan nanoparticles	120	Acts as a carrier to increase immune response against specific antigen.	Mohammadpour-dounighi et al. (2010)
Mesobuthus eupeus scorpion venom	Immunoadjuvant	Chitosan nanoparticles	300–400	Acts as a carrier to increase immune response against specific antigen.	Mohammadpour Dounighi et al. (2012)
Monophosphoryl lipid A (endotoxin from Salmonella)	Immunoadjuvant	PLGA nanoparticles	325	Promotes both systemic and mucosal immune responses.	Sarti et al. (2011)
Tetnus toxoid	Prevention against tetanus	Gold nanoparticles	40	Significant TT-specific IgG and IgA immune responses.	Barhate et al. (2013)

CTB, Cholera toxin B subunit; DT, diphtheria toxin; IFN, interferons; Ig, immunoglobulins; IL, interleukins; NPs, nanoparticles; PLGA, poly (lactide-co-glycolide); PVAL; polyvinyl alcohol.

barrier, a size of range of 15–30 nm is desired. In the case of nanosystems, in order to promote the humoral response the particle size needs to be in the range of 1–10 μm, and for induction of cellular response size below 1 μm is required. Mastersizer for nanosystems above 10 μm and Zetasizer for nanosystems 10 nm–6 μm can be used for size determination. Zetasizer can also be used for determining the zeta potential of the nanosystem. Zeta potential helps understand the stability of the nanosystems. Typically, the zeta potential should be between +30mV and −30 mV (Akhter et al., 2016; Dayan and Touitou, 2000).

3. *Transition temperature*: differential scanning calorimetry (DSC) is a thermoanalytical technique utilized to determine drug–lipid compatibility and their interactions, if any. The temperature is kept in a definite range with predetermined heating rate (Pandita and Sharma, 2013).
4. *Surface tension measurement*: in case of vesicular systems (liposomes, phytosomes, virosomes) the measurement of the surface tension of the drug in the aqueous solution can be measured by Du Nouy ring tensiometer (Dayan and Touitou, 2000).
5. *Drug content stability*: quantification of drug amount can be done by appropriate spectroscopic or HPLC methods.
6. *Spectroscopic evaluations*: to compare the formation of complex and to study the interaction between the immunomodulator and platform nanosystems (lipids, phospholipids, etc.).
   6.1 *Hydrogen-nuclear magnetic resonance (H^1 NMR)*: when the nanosystem is subject to H^1NMR marked changes in signal are observed due to complex formed between the immunomodulator and nanosystems. Also broadening of signal can be observed in case of phospholipid based system.
   6.2 *Carbon-nuclear magnetic resonance (C^{13} NMR)*: with this method the conformation of the carbon skeleton can be understood, which is depicted by the signal changes.
   6.3 *Fourier transform infrared spectroscopy (FTIR)*: spectrum of complex versus spectrum of individual components and its physical mixtures confirms complex formation by infrared.
7. *In vitro and in vivo evaluation*: selection of appropriate models needs to be done as per the therapeutic activity of the constituents. For example, Cefodizime can be evaluated for models of infection (Labro, 1990). *Astragalus membranaceus* extract can be evaluated for models of cytotoxicity (Cho and Leung, 2007). Dissolution study is a quality control parameter employed to estimate drug release studies. For example, study of immunostimulant monophosphoryl lipid A (MPLA), in vitro study of GantrezAN NPs as adjuvants for oral immunotherapy (Gómez et al., 2007; Sarti et al., 2011).

6 FATE OF IMMUNOMODULATORS AFTER ORAL DELIVERY

Oral administration is the preferred route for drug delivery. It is patient-friendly, painless, and easy for self-medication. Compared to parenteral delivery, it suppresses risk of disease transmission and increases patient compliance. It allows flexible and controlled dosing schedule. It is particularly convenient for chronic therapy (Ruenraroengsak et al., 2010; Yamanaka and Leong, 2008).

The log P values of IMs lies in the range of 0–3, hence, oral route of delivery is facile to formulate. Oral, intestinal, colonic absorption is observed to be best with logP values in the range of 0–3, for example, Oxamisole log P 1.74, allicin log P 2.02, silymarin log P 2.63. Therapeutic agents in this range are hydrophilic and would not be assumed to follow transcellular route of absorption through passive absorption. The paracellular space lies between 10 and 30–50 Å. The paracellular route is restricted to relatively small hydrophilic molecules that can fit in these dimensions; absorption of macromolecules is not easy (Morishita and Peppas, 2006).

Oral bioavailability of drug is decisively inveigled by its properties and properties of the carrier system. A drug that is administered orally must thrive through the transit of the GIT. Small molecules can withstand the harsh environment presented by the GIT, and hence can be absorbed. In case of large molecules, such as proteins, vaccines, peptides, and nucleic acid, the intestinal barrier limits their absorption and hence a suitable protective carrier is required to avoid the destruction of the large molecules and enhance the oral absorption (Plapied et al., 2011). This section gives an understanding of immunomodulator loaded nanosystems. Fig. 11.3 is the representation of uptake of IM-loaded nanosystems (Lamprecht, 2015; Moghimipour et al., 2015; Yun et al., 2013).

Liposomes are lipidic vesicles formed by one or several phospholipid bilayers encapsulating an aqueous core. These are rapidly cleared through the bloodstream by macrophages of the reticuloendothelial system (RES). Liposomes with neutral vesicles show a lower binding affinity and hence phagocytosis as opposed the negatively charged surface, generally containing the negatively charged phospholipids phosphatidylserine (PS) and phosphatidylglycerol (PG), which tend to show higher binding. The same is true for positively charged liposomes or polyplexes. Liposomes have also been developed along this line. For example, doxorubicin-loaded liposomes constituting of phosphatidylglycerol, phosphatidylcholine, and cholesterol, in a range of 300–500 nm showed an increased liver uptake, and reduced cardiac toxicities in a phase I clinical study. Here it was also suggested that after capture by Kupffer cells, the liposome matrix became leaky, and the drug was released in free form to the tumor

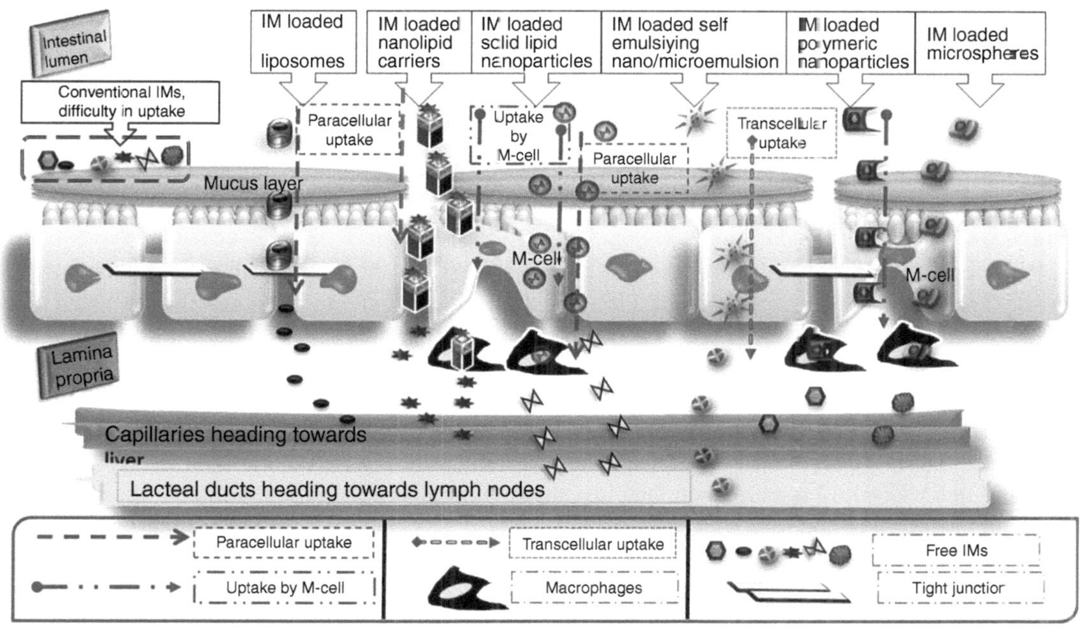

FIGURE 11.3 Uptake of Immunomodulators Nanosystem in Intestinal Lumen

(Gabizon et al., 1989). AmBisome showed a 50- to 70-fold decrease in toxicity and 5-fold increase in dose loading owing to 60–80 nm SUV liposomes formulated. The small size of these liposomes helps them escape immediate clearance by the macrophages of the liver and spleen and thus reach the other infected tissue. pH sensitive liposomes undergo a transition from lamellar to inverted micelle structure at low pH resulting in fusion between the liposomal and the endosomal membranes, and the destabilization of endosomes and also destabilization of the liposomes making them leaky and hence the release in the liposomal contents (Adler-Moore and Proffitt, 2002; Drummond et al., 2000). Glycosylphosphatidylinositol-linked folate receptor (FR), is generally overexpressed on the surface of cancer cells, but highly restricted in normal tissue. FR can conveniently help the folic acid–linked nanosystem escape into the cytosol through the receptor mediated endocytosis (RME). Liposomes linked with the phospholipid-anchored folic acid or PEG-folic acid phospholipid conjugate showed preferential uptake by FR-expressing cells (Gabizon et al., 2004; Hilgenbrink and Low, 2005).

Polymeric micro/nanoparticles: in case of polymeric micelles, the polymer gets partly absorbed as unimers and micelles. In case of PNPs and targeted polymeric nanoparticles (TPNPs), the stability of the PNPs/TPNPs affects the absorption, transcytosis of the PNPs/TPNPs or drug may occur (Plapied et al., 2011). To quote an example, in case of vaccine delivery via oral route, an increase in uptake in rats by NPs-loaded vaccine through the intestine due to the presence of villus enterocytes was observed. Microencapsulated vaccines are formulated using aliphatic polyesters the poly (lactide-*co*-glycolides) [PLGs]. 4 μm particle size was effective for induction of serum antibody response and 7 μm particle size was found to show strong mucosal IgA response. Data in rodents suggest that targeting ligands may be effective in case of microparticles, but the low levels of particle uptake from the gut is a major drawback. Whether the low uptake of the microparticles in humans can cause effective oral vaccine is still in question. A continued work can help benefit the development of polymeric nanosystem system (Lavelle and O'Hagan, 2006). Noscapine, an opium alkaloid, is in clinical trials for the treatment of low-grade Hodgkin's lymphoma or chronic lymphocytic leukemia refractory to chemotherapy and hematological malignancies. Noscapine has been studied for its bioavailability, metabolic fate, and pharmacokinetics because it has been used as an antitussive for past 40 years. Singh and Chandra (2012) synthesized lung-targeting gelatin microparticles of noscapine by improving its plasma concentration. The gelatin is a natural polymer with low immunogenicity and biodegradability. The cytotoxic drug noscapine reaches the target with enhanced bioavailability and the gelatin carrier get degraded naturally.

Polymeric NPs coated with poly ethylene glycol (PEG) have shown promising results compared to non-PEGylated NPs. Tacrolimus-loaded PEGylated NPs were found to be more efficient when compared to marketed formulation. PEG is a flexible, hydrophilic, nonionic polymer; it not only makes NPs a long circulating one (stealth effect) but also prolongs their plasma half-life. This helps NPs in effective biodistribution and improves oral bioavailability (Shin et al., 2010).

Curcumin SMEDDS prepared by Cui et al. (2009) showed passive transfer of curcumin by diffusion across the lipid membranes in rat intestines, thus contributing to the intestinal absorption of the curcumin in SMEDDS. Further, when they investigated curcumin SMEDDS in mice it showed 93.8% absorption, which was 3.86 times more than the curcumin suspension. The reason for marked improved absorption being the perfunctory formation of microemulsion in the intestine presents curcumin in dissolved form thus aiding absorption. Also, surfactants used in preparation of the SMEDDS aid and abide by reducing the interfacial tension and enhance penetration of curcumin into the epithelial cells (Cui et al., 2009).

Mucus adhesion and mucus penetration are the two key factors for the efficient oral delivery of IMs. Nanosystems should be small to avoid significant stearic inhibition offered by the fiber mesh of mucus and should also dodge adhesion to mucin fibers. Concurrently, the nanosystems should be mucoadhesive such that the retention time and thus the contact time with the intestinal mucosa is higher. Since nanosystems are cleared from the mucosal tissue, the elaboration of mucus penetrating systems should be a priority. NPs offer better protection to the encapsulated therapeutic agent and have better stability as opposed to the liposomal vesicular structures (Lai et al., 2009).

7 FUTURE PROSPECTS

Nanotechnology platforms have been researched as adjuvants, carriers, and drug delivery systems to target various disorders, such as immunodeficiencies, cancer, and inflammatory diseases. Though some formulations are in clinical trials and many of them are in various phases of preclinical trials, our understanding of nanosystem interaction with components of immune system has improved a lot. Although it is explored well, in the future many questions need to be answered for a deeper understanding of the mechanism of immunomodulation by these nanosystems.

7.1 ORAL

The virtues of oral route have made it a route of choice for the administration of different conventional systems. IMs administered via oral route need to be exposed to the gut, which is already full of commensal bacteria, enzymes, nutrients, and lumens. However, the oral route, being the most preferred one, comes with several hurdles for effective delivery to the immune cells. First, oral IM nanosystems have to be exposed to an extremely acidic pH, proteolytic enzymes, and bile salts leading to their degradation in the GIT. Second, they need to cross various biological barriers (tight epithelial cellular junctions and a thick mucus layer) in the intestinal lumen to reach to the systemic circulation (Marasini et al., 2014). In addition to this, lower uptake of the IMs is also related to their short exposure time to mucosal membranes. Consequently, for a potent and sustained immunogenic effect, IMs given by oral means might require larger and frequent dosing compared with their systemic administration (Bergqvist et al., 2013). However, repetitive oral administration of larger oral doses of IMs might stimulate systemic nonresponsiveness, that is, oral tolerance (Pabst and Mowat, 2012). Further, lack of availability of potent IMs with low toxicity is another major obstruction in the development of oral IMs (Premanand et al., 2013).

To date, oral routes for delivery of various nanosystems have proved its efficacy to a good extent in delivery of IMs as well as other actives. However, nanosystems are not 100% effective to improve oral bioavalability of IMs. Hence, more focused research is needed in the field of nanotechnology research, which would make it a targeted, effective, and more bioavailable delivery system. Table 11.9 depicts some of the advantages and disadvantages associated with the use of various nanosystems in the area of oral IM delivery.

It can be implicated with reasonable certainty that in future NP-based IMs will overcome the barriers of GIT and will selectively target intestinal immune cells without toxic effects. The only difficulty remaining is if they can be produced by economically feasible processes to allow low-cost oral immunization in the treatment of various diseases.

Table 11.9 Advantages and Disadvantages of Nanosystems Used for Immunomodulator Oral Delivery

Nanosystems	Advantages	Disadvantages
• Liposomes	• Both hydrophilic and lipophilic IMs can be loaded • Modified liposomes, such as bilosomes or arachesomes are relatively stable in GI fluids	• Hydrophillic IMs have limited loading • Poor GI stability of naked liposomes • Lower mucus penetrating property
• Nanoemulsion	• Both hydrophilic and lipophilic IMs can be loaded • Addition of surfactant provides better permeation across intestine	• Chances of premature active leakage • Poor GI stability
• Polymeric NPs	• Surface properties can be easily modified for better activity • Release of IMs at targeted site	• Limited aqueous solubility • Burst release • Low IM loading • Formulation might involve use of organic solvents
• Lipidic nanocarriers	• Sufficient entrapment of hydrophobic IMs • Preffered uptake by lymphatic system improves bioavailability	• Very low entrapment of hydrophillic IMs • Poor aqueous solubility
• Inorganic NPs	• Better protection of adsorbed IMs • Less chances of burst release • Easy surface alteration	• Low biodegradability • Limited aqueous solubility

GI, Gastro intestinal; IMs, immunomodulators; NPs, nanoparticles.

7.2 OTHER PREFERRED ROUTES OF ADMINISTRATION

Other preferred routes for delivery of nanoformulated IMs include parenteral, topical, buccal, sublingual, pulmonary, nasal, vaginal, and rectal. These routes provide rapid onset of action of drugs and better bioavailability. Parenteral route consists of delivery through intraperitoneal, intramuscular, intravenous, and subcutaneous. Although oral route is found to be most convenient by all age groups, parenteral route has been preferred in various clinical conditions, such as toxic shocks and venomous sting. Other routes are preferred for those IMs that are not stable orally, such as venom, synthetic drugs, and antibiotics. Nasal, pulmonary, vaginal, and rectal route are generally opted for induction of site-specific mucosal immunity. One advantage of these routes is that they avoid first pass metabolism. Previously, a number of various IMs has been explored via above routes; a few examples are quoted in Table 11.10.

Parenteral route provides immediate onset of action, that is, intravenous in 15–30 s, intramuscular in 10–20 min and subcutaneous in 15–30 min. These routes are promising for poorly bioavailable drugs. Although, results have shown promising activity on administration of IMs via parenteral routes but this route has some disadvantages like pain, irritation, or discomfort to the patient, requirement of trained personnel and IV injection may lead to the risk of overdose.

Table 11.10 Other Routes for Immunomodulators Delivery

Immunomodulators	Formulation Type	Particle Size (nm)	Route of Administration	Benefits of Formulation	References
Bee venom phospholipase A2 (PLA2)	PLGA microspheres	5000	Intraperitoneal	Elicited IgG response	Guerin et al. (2002)
Echis carinatlrs (snake venom)	Liposomes	100–1000	Intravenous and subcutaneous	Higher humoral and oral immune response	Laing et al. (1987)
Echis carinatlrs (snake venom)			Intravenous	Higher immune response	Theakston (1989)
Crotalus dunssus (Rattle snake venom)			Intravenous	Protective and persistent immune response elicited	Freitas et al. (1989)
Crotoxm from *C. dunssus ternficus* (snake venom)			Intravenous and subcutaneous	Protective IgG immune response increased	Freitas and Frézard (1997)
Tryus senulatlrs (scorpion venom)			Subcutaneous	Higher immune response	Chavez-Olortegui et al. (1991)
Hepatitis B	Liposomes	710	Intramuscular	Protective IgG immune response increased	Jain et al. (2009)
Matrix2 protein of Influenza A virus	Gold nanoparticles	12	Intranasal	Higher level of IgG (anti-M2e)	Tao et al. (2014)
Pneumococcal surface protein A (PspA) of *Streptococcus pneumoniae*	Cationic cholesteryl group-bearing pullulan nanogel	40	Intranasal	Induced protective immunity against lethal *S. pneumoniae* and induced systemic and nasal T_H17 PspA-specific IgG, and IgA antibody responses.	Kong et al. (2013)
Clostridium botulinum type A neurotoxin BoHc/A	Cationic cholesteryl group-bearing pullulan nanogel	10	Intranasal	Induced serum IgG and secretary IgA response	Nochi et al. (2010)
Triptolide (isolated from a shrub *T. wilfondil*)	Solid lipid nanoparticles	147	Topical	Inhibit the expression of IL-8, TNF-α and various inflammatory cytokines	Mei et al. (2003)
Tetanus toxoid	PVAL-PLGA nanoparticle	100 or 500	Intranasal	Induction of IgG and IgA antibody responses	Jung et al. (2001)
Bee venom	PVAL–chitosan hydrogel		Topical	Reduction in IL-6 level (responsible for wound healing)	Amin and Abdel-Raheem (2014)

Ig, Immunoglobulins; IL, Interleukins; PLGA, poly (lactide-co-glycolide); PVAL, polyvinyl alcohol; TNF, tumor necrosis factor; T_H, T helper.

8 CONCLUSIONS

In the past few years, the oral delivery of IMs has been studied at the preclinical level to build a strong proof of concept for oral administration of the IMs. Oral delivery of the therapeutic agents (herbals, proteins, monoclonal antibodies, vaccines, synthetics) by nanosystems should offer a felicitous opportunity for patient friendly, needle free delivery.

These nanosystems involve numerous hurdles, such as scale up, synthesis of polymers, and so on even though these nanosystems improve the bioavailability whether it reaches the therapeutic levels, the patient-to-patient variation in pharmacokinetics can be high for therapeutic delivery of the IMs. The recent more sophisticated methods for developing nanosystems should help go past the hurdles and yield products for the market and finally the patients.

The better understanding of the cellular uptake mechanisms of the nanosystems, modulation of release according to chrono-pharmacokinetics of nanosystems, study of the fate of nanocarriers in the body, targeted delivery through tailoring the properties of nanosystems is required. A thorough, in-depth understanding and investigation is required of the pathophysiological changes in the GIT in each disease condition for better absorption of the therapeutic drugs. The use of immunomodulatory lipids would make nanosystems more effective as they have preferential uptake in addition to their immunomodulation effect. PEGylation of NPs also helps in targeting, as well as increasing their half-life. These novel nanosystems might be useful to address specific needs. Nanosystems could be targeted to M cells or immunocompetent cells in the GIT for oral immunisation. If preclinical data reports a low but significantly enhanced bioavailability of protein, the feasibility of oral nanosystem delivery of proteins remains open and should be investigated with optimized systems. Enhanced permeability and retention effect can be passively targeted in case of anticancer drugs; oral polymeric micellar system can come to an aid for efficacious delivery. A number of current and new IMs are investigated for their conventional oral delivery; they should be further prompted and supported to study them for nanosystems for obtaining better bioavailability. An approach toward determined efforts for large-scale manufacturing and production of the novel systems is yet another aspect that needs to be addressed. Large-scale parameters and designing of equipments for the continuous production of nanosytems opens an arena of research. Future research should consider cost and benefits in routine patient care, such that current and future patients receive the full benefits of their treatment.

ABBREVIATIONS

5-ASA	5-Amino salicylic acid
AIDS	Acquired immunodeficiency syndrome
APCs	Antigen presenting cells
AUC	Area under curve
BCG	Bacillus calmette–guérin
BSA	Bovine serum albumin
CD	Cluster of differentiation
CFU	Colony forming unit
Ch-GM-NPs	Mannosylated chitosan nanoparticles
COX	Cyclooxygenase
CpG DNA	Cytosine and guanine rich DNA

Cryo FESEM	Cryo field emission scanning electron microscopy
Cryo TEM	Cryo transmission electron microscopy
CT	Cholera toxin
CTB	Cholera toxin B subunit
DEM	Diethylmaleate
DMAB	Didodecylmethylammonium bromide
DNA	Deoxy ribonucleic acid
DPPE	Dipalmitoylphosphatidyl-ethanolamine
DPPE-MCC	DPPE-maleimidomethylcyclohexane-1-carboxylate
DSC	Differential scanning calorimetry
DsDNA	Double stranded deoxy ribonucleic acid
DSPC	Distearoylphosphatidylcholine
DT	Diphtheria toxin
EE	Encapsulation efficiency
FR	Folate receptor
FTIR	Fourier transform infrared spectroscopy
GABA	γ-aminobutyric acid
GALT	Gut-associated lymphoid tissue
G-CSF	Granulocyte-colony stimulating factor
GI	Gastrointestinal
GIT	Gastrointestinal tract
GPI	Glucose phosphate isomerase
HCL	Hydrochloric acid
HIV	Human immunodeficiency virus
HPLC	High pressure liquid chromatography
HPMCP	Hydroxypropyl methylcellulose pthalate
IBD	Inflammatory bowel disease
IFN	Interferons
Ig	Immunoglobulins
IL	Interleukins
IMs	Immunomodulators
IV	Intravenous
LPS	Lipopolysaccharide
M2e	Matrix2 protein of influenza A virus
MHC	Major histocompatibility complex
MPLA	Mono phosphoryl lipid A
Mrna	Messenger single stranded ribonucleic acid
μ	Micron
μg/mL	Microgram per millilitre
MgAl-NO_3	Magnesium aluminium nitrate
mg/mL	Milligram per millilitre
ng min/mL	Nanogram minute per millilitre
NLC	Nanostructured lipid carrier
Nm	Nanometer
NMR	Nuclear magnetic resonance
NPs	Nanoparticles
OVA	Ovalbumin
PBS	Phosphate buffer saline

PCL	Poly capro lactone
PEG	Poly ethylene glycol
PEI	Polyethyleneimine
PG	Phosphatidylglycerol
PLA	Polylactide
PLA2	Phospholipase A2
PLGA	Poly (lactide-*co*-glycolide)
PNPs	Polymeric nanoparticles
PPs	Peyer's patches
PRRs	Pattern recognition receptors
PS	Phosphatidylserine
Psp A	Pneumococcal surface protein A
PVLA	Poly vinyl alcohol
RNA	Ribonucleic acid
RES	Reticulo endothelial system
RESS	Rapid expansion of super critical solvent
RGD	Arginylglycylaspartic acid
sChNPs	Stabilized chitosan nanoparticles
SCID	Severe combined immunodeficiency
SEM	Scanning electron microscopy
SGC	Sodium glycocholate
SLN	Solid lipid nanoparticle
SMEDDs	Self-microemulsifying drug delivery systems
ssRNA	Single stranded ribonucleic acid
SUV	Small unilamellar vesicle
T_C cell	T cytotoxic cell
TCR	T-cell receptor
TEM	Transmission electron microscopy
T_H cell	T helper cell
TLR	Toll like receptor
TNF	Tumor necrosis factor
TPNPs	Targeted polymeric nanoparticles
TT	Tetanus toxoid
UEA	*Ulex europaeus* 1
WAS	Wiskott–Aldrich syndrome
WGA	Wheat germ agglutinin

REFERENCES

Abbas, A.K., Lichtman, A.H., Pillai, S., 2012. Basic Immunology: Functions and Disorders of the Immune System. Elsevier Health Sciences, New York.

Adler-Moore, J., Proffitt, R.T., 2002. AmBisome: liposomal formulation, structure, mechanism of action and pre-clinical experience. J. Antimicrob. Chemo. 49 (Suppl 1), 21–30.

Agarwal, S.S., Singh, V.K., 1999. Immunomodulators: a review of studies on Indian medicinal plants and synthetic peptides. Part-I: medicinal plants. Proc. Indian Nat. Sci. Acad. B 65 (3–4), 179–204.

Akbarzadeh, A., Rezaei-Sadabady, R., Davaran, S., Joo, S.W., Zarghami, N., Hanifehpour, Y., Samiei, M., Kouhi, M., Nejati-Koshki, K., 2013. Liposome: classification, preparation, and applications. Nanoscale Res. Lett. 8 (1), 102.

Akhter, K.F., Mumin, M.A., Lui, E., Charpentier, P.A., 2016. Microfluidic synthesis of ginseng polysaccharide nanoparticles for immunostimulating action on macrophage cell lines. ACS Biomater. Sci. Eng. 2 (1), 95–103.

Almeida, A.J., Souto, E., 2007. Solid lipid nanoparticles as a drug delivery system for peptides and proteins. Adv. Drug Deliv. Rev. 59 (6), 478–490.

Almgren, M., Edwards, K., Karlsson, G., 2000. Cryo transmission electron microscopy of liposomes and related structures. Colloids Surf. A 174 (1), 3–21.

Amin, M.A., Abdel-Raheem, I.T., 2014. Accelerated wound healing and anti-inflammatory effects of physically cross linked polyvinyl alcohol–chitosan hydrogel containing honey bee venom in diabetic rats. Arch. Pharm. Res. 37 (8), 1016–1031.

Arshia, S., Manna, P.K., Paranjothy, K.L.K., Manjula, M., 2007. Entrapment of andrographolide in cross-linked alginate pellets: II.physicochemical characterization to study the pelletization of andrographolide. Pak J. Pharm. Sci. 20 (1), 1–9.

Bao, X., Wang, Z., Fang, J., Li, X., 2002. Structural features of an immunostimulating and antioxidant acidic polysaccharide from the seeds of *Cuscuta chinensis*. Planta. Medica. 68 (3), 237–243.

Barhate, G., Gautam, M., Gairola, S., Jadhav, S., Pokharkar, V., 2013. Quillaja saponaria extract as mucosal adjuvant with chitosan functionalized gold nanoparticles for mucosal vaccine delivery: stability and immunoefficiency studies. Int. J. Pharm. 441 (1), 636–642.

Bergqvist, P., Stensson, A., Hazanov, L., Holmberg, A., Mattsson, J., Mehr, R., Bemark, M., Lycke, N.Y., 2013. Re-utilization of germinal centers in multiple Peyer's patches results in highly synchronized, oligoclonal, and affinity-matured gut IgA responses. Mucosal. Immunol. 6 (1), 122–135.

Bhatia, A., Shard, P., Chopra, D., Mishra, T., 2011. Chitosan nanoparticles as carrier of immunorestoratory plant extract: synthesis, characterization and immunorestoratory efficacy. Int. J. Drug Deliv. 3 (2), 381.

Bivas-Benita, M., Laloup, M., Versteyhe, S., Dewit, J., De Braekeleer, J., Jongert, E., Borchard, G., 2003. Generation of *Toxoplasma gondii* GRA1 protein and DNA vaccine loaded chitosan particles: preparation, characterization, and preliminary in vivo studies. Int. J. Pharm. 266 (1), 17–27.

Borhade, V., Nair, H., Hegde, D., 2008. Design and evaluation of self-microemulsifying drug delivery system (SMEDDs) of tacrolimus. AAPS PharmSciTech 9 (1), 13–21.

Chao, P., Deshmukh, M., Kutscher, H.L., Gao, D., Rajan, S.S., Hu, P., Laskin, D.L., Stein, S., Sinko, P.J., 2010. Pulmonary targeting microparticulate camptothecin delivery system: anti-cancer evaluation in a rat orthotopic lung cancer model. Anticancer Drugs 21 (1).

Chavez-Olortegui, C., Amara, D.A., Rochat, H., Diniz, C., Granier, C., 1991. In vivo protection against scorpion toxins by liposomal immunization. Vaccine 9 (12), 907–910.

Chew, J.L., Wolfowicz, C.B., Mao, H.Q., Leong, K.W., Chua, K.Y., 2003. Chitosan nanoparticles containing plasmid DNA encoding house dust mite allergen, Der 1 for oral vaccination in mice. Vaccine 21 (21), 2720–2729.

Cho, W.C.S., Leung, K.N., 2007. In vitro and in vivo immunomodulating and immunorestorative effects of *Astragalus membranaceus*. J. Ethnopharmacol. 113 (1), 132–141.

Conway, M.A., Madrigal-Estebas, L., McClean, S., Brayden, D.J., Mills, K.H., 2001. Protection against *Bordetella pertussis* infection following parenteral or oral immunization with antigens entrapped in biodegradable particles: effect of formulation and route of immunization on induction of Th1 and Th2 cells. Vaccine 19 (15), 1940–1950.

Cui, J., Yu, B., Zhao, Y., Zhu, W., Li, H., Lou, H., Zhai, G., 2009. Enhancement of oral absorption of curcumin by self-microemulsifying drug delivery systems. Int. J. Pharm. 371 (1), 148–155.

Dai, J., Nagai, T., Wang, X., Zhang, T., Meng, M., Zhang, Q., 2004. pH-sensitive nanoparticles for improving the oral bioavailability of cyclosporine A. Int. J. Pharm. 280 (1), 229–240.

Dayan, N., Touitou, E., 2000. Carriers for skin delivery of trihexyphenidyl HCl: ethosomes vs. liposomes. Biomaterials 21 (18), 1879–1885.

del Giudice, M.M., Leonardi, S., Galdo, F., Allegorico, A., Filippelli, M., 2014. Probiotics and vaccination in children. J. Vaccines Vaccin. 5 (226), 2.

Dixit, N., Baboota, S., Kohli, K., Ahmad, S., Ali, J., 2007. Silymarin: a review of pharmacological aspects and bioavailability enhancement approaches. Indian J. Pharmacol. 39 (4), 172.

Drummond, D.C., Zignani, M., Leroux, J.C., 2000. Current status of pH-sensitive liposomes in drug delivery. Prog. Lipid Res. 39 (5), 409–460.

Dubes, A., Parrot-Lopez, H., Abdelwahed, W., Degobert, G., Fessi, H., Shahgaldian, P., Coleman, A.W., 2003. Scanning electron microscopy and atomic force microscopy imaging of solid lipid nanoparticles derived from amphiphilic cyclodextrins. Eur. J. Pharm. Biopharm 55 (3), 279–282.

Ekambaram, P., Sathali, A.A.H., Priyanka, K., 2012. Solid lipid nanoparticles: a review. Sci. Rev. Chem. Commun. 2 (1), 80–102.

El-Shabouri, M.H., 2002. Positively charged nanoparticles for improving the oral bioavailability of cyclosporin-A. Int. J. Pharm. 249 (1), 101–108.

Fan, Y., Zhang, Q., 2013. Development of liposomal formulations: from concept to clinical investigations. Asian J. Pharm. Sci. 8 (2), 81–87.

Freitas, T.V., Frézard, F., 1997. Encapsulation of native crotoxin in liposomes: a safe approach for the production of antivenom and vaccination against Crotalus durissus terrificus venom. Toxicon 35 (1), 91–100.

Freitas, T.V., Tavares, A.P., Theakston, R.D.G., Laing, G., New, R.R.C., 1989. Use of liposomes for protective immunisation against *Crotalus durissus* (tropical rattlesnake) venom. Toxicon 27 (3), 341–347.

Gabizon, A., Peretz, T., Sulkes, A., Amselem, S., Ben-Yosef, R., Ben-Baruch, N., Catane, R., Biran, S., Barenholz, Y., 1989. Systemic administration of doxorubicin-containing liposomes in cancer patients: a phase I study. Eur. J. Cancer Clin. Oncol. 25 (12), 1795–1803.

Gabizon, A., Shmeeda, H., Horowitz, A.T., Zalipsky, S., 2004. Tumor cell targeting of liposome-entrapped drugs with phospholipid-anchored folic acid–PEG conjugates. Adv. Drug. Deliv. Rev. 56 (8), 1177–1192.

Garinot, M., Fiévez, V., Pourcelle, V., Stoffelbach, F., des Rieux, A., Plapied, L., Theate, I., Freichels, H., Jérôme, C., Marchand-Brynaert, J., Schneider, Y.J., 2007. PEGylated PLGA-based nanoparticles targeting M cells for oral vaccination. J. Control. Release 120 (3), 195–204.

Gaspar, M.M., Cruz, A., Fraga, A.G., Castro, A.G., Cruz, M.E.M., Pedrosa, J., 2008. Developments on drug delivery systems for the treatment of mycobacterial infections. Curr. Topics Med. Chem. 8 (7), 579–591.

Gea-Banacloche, J., 2006. Immunomodulation. In: Runge, M., Patterson, C. (Eds.), Principles of Molecular Medicine. Humana Press, Inc., Totowa, NJ.

Goldmann, K., Ensminger, S.M., Spriewald, B.M., 2012. Oral gene application using chitosan-DNA nanoparticles induces transferable tolerance. Clin. Vacc. Immunol. 19 (11), 1758–1764.

Gómez, S., Gamazo, C., San Roman, B., Ferrer, M., Sanz, M.L., Irache, J.M., 2007. Gantrez AN nanoparticles as an adjuvant for oral immunotherapy with allergens. Vaccine 25 (29), 5263–5271.

Guerin, V., Dubarryr, M., Robic, D., Brachet, F., Rautureau, M., Andre, C., Bourbouze, R., Tome, D., 2002. Microsphere entrapped bee-venom phospholipase A2 retains specific IgE binding capacity: a possible use for oral specific immunotherapy. J. Microencaps. 19 (6), 761–765.

Gulbake, A., Jain, S.K., 2013. colon specific delivery of mesalazine using biocompatible polymeric nanoparticles. Int. J. Pharmacol. Pharm. Technol. 2 (1), 17–24.

Gupta, P.N., Khatri, K., Goyal, A.K., Mishra, N., Vyas, S.P., 2007. M-cell targeted biodegradable PLGA nanoparticles for oral immunization against hepatitis B. J. Drug Target. 15 (10), 701–713.

Gutierro, I., Hernandez, R.M., Igartua, M., Gascon, A.R., Pedraz, J.L., 2002. Size dependent immune response after subcutaneous, oral and intranasal administration of BSA loaded nanospheres. Vaccine 21 (1), 67–77.

Harde, H., Agrawal, A.K., Jain, S., 2015. Tetanus toxoids loaded glucomannosylated chitosan based nanohoming vaccine adjuvant with improved oral stability and immunostimulatory response. Pharm. Res. 32 (1), 122–134.

Harokopakis, E., Childers, N.K., Michalek, S.M., Zhang, S.S., Tomasi, M., 1995. Conjugation of cholera toxin or its B subunit to liposomes for targeted delivery of antigens. J. Immunol. Methods 185 (1), 31–42.

Hilgenbrink, A.R., Low, P.S., 2005. Folate receptor-mediated drug targeting: from therapeutics to diagnostics. J. Pharm. Sci. 94 (10), 2135–2146.

Hodgson, E., 2012. Toxins and venoms. Prog. Mol. Biol. Transl. Sci. 112, 373–415.
Hu, L., Xing, Q., Meng, J., Shang, C., 2010. Preparation and enhanced oral bioavailability of cryptotanshinone-loaded solid lipid nanoparticles. AAPS PharmSciTech 11 (2), 582–587.
Hu, Q., Wu, M., Fang, C., Cheng, C., Zhao, M., Fang, W., Chu, P.K., Ping, Y., Tang, G., 2015. Engineering nanoparticle-coated bacteria as oral DNA vaccines for cancer immunotherapy. Nano Lett. 15 (4), 2732–2739.
Italia, J.L., Bhatt, D.K., Bhardwaj, V., Tikoo, K., Kumar, M.R., 2007. PLGA nanoparticles for oral delivery of cyclosporine: nephrotoxicity and pharmacokinetic studies in comparison to Sandimmune Neoral. J. Control. Release 119 (2), 197–206.
Jain, N., Gupta, B.P., Thakur, N., Jain, R., Banweer, J., Jain, D.K., Jain, S., 2010. Phytosome: a novel drug delivery system for herbal medicine. Int. J. Pharm. Sci. Drug Res. 2 (4), 224–228.
Jain, V., Vyas, S.P., Kohli, D.V., 2009. Well-defined and potent liposomal hepatitis B vaccines adjuvanted with lipophilic MDP derivatives. Biol. Med. 5 (3), 334–344.
Jaiswal, J., Gupta, S.K., Kreuter, J., 2004. Preparation of biodegradable cyclosporine nanoparticles by high-pressure emulsification-solvent evaporation process. J. Control. Release 96 (1), 169–178.
Jin, D.Z., Yin, L.L., Ji, X.Q., Zhu, X.Z., 2006. Cryptotanshinone inhibits cyclooxygenase-2 enzyme activity but not its expression. Eur. J. Pharmacol. 549 (1), 166–172.
Jung, T., Kamm, W., Breitenbach, A., Hungerer, K.D., Hundt, E., Kissel, T., 2001. Tetanus toxoid loaded nanoparticles from sulfobutylated poly (vinyl alcohol)-graft-poly (lactide-*co*-glycolide): evaluation of antibody response after oral and nasal application in mice. Pharm. Res. 18 (3), 352–360.
Kankala, R.K., Kuthati, Y., Sie, H.W., Shih, H.Y., Lue, S.I., Kankala, S., Jeng, C.C., Deng, J.P., Weng, C.F., Liu, C.L., Lee, C.H., 2015. Multi-laminated metal hydroxide nanocontainers for oral-specific delivery for bioavailability improvement and treatment of inflammatory paw edema in mice. J. Colloid Interf. Sci. 458, 217–228.
Karimi, N., Ghanbarzadeh, B., Hamishehkar, H., Keivani, F., Pezeshki, A., Gholian, M.M., 2015. Phytosome and liposome: the beneficial encapsulation systems in drug delivery and food application. Appl. Food Biotechnol. 2 (3), 17–27.
Kindt, T.J., Goldsby, R.A., Osborne, B.A., Kuby, J., 2007. Kuby Immunology. Macmillan, New York.
Kong, I.G., Sato, A., Yuki, Y., Nochi, T., Takahashi, H., Sawada, S., Mejima, M., Kurokawa, S., Okada, K., Sato, S., Briles, D.E., 2013. Nanogel-based PspA intranasal vaccine prevents invasive disease and nasal colonization by *Streptococcus pneumoniae*. Infect. Immun. 81 (5), 1625–1634.
Kulkarni, G.T., 2011. Herbal drug delivery systems: an emerging area in herbal drug research. J. Chronot. Drug Deliv. 2 (3), 113–119.
Labro, M.T., 1990. Cefodizime as a biological response modifier: a review of its in-vivo, ex-vivo and in-vitro immunomodulatory properties. J. Antimicrob. Chemother. 26 (Suppl C), 37–47.
Lai, S.K., Wang, Y.Y., Hanes, J., 2009. Mucus-penetrating nanoparticles for drug and gene delivery to mucosal tissues. Adv. Drug Deliv. Rev. 61 (2), 158–171.
Laing, G., Theakston, R.D.G., New, R.R.C., 1987. Use of Liposomes Incorporating Immunostimulant for Parenteral and Oral Immunization Against Snake Venom. In: Progress in Venom and Toxin Research. Faculty of Medicine, University of Singapore, Singapore, p. 283.
Lamprecht, A., 2015. Nanomedicines in gastroenterology and hepatology. Nat. Rev. Gastroenterol. Hepatol. 12 (4), 195–204.
Lavelle, E.C., O'Hagan, D.T., 2006. Delivery systems and adjuvants for oral vaccines. Expert Opin. Drug Deliv. 3 (6), 747–762.
Lee, C.H., Lo, L.W., Mou, C.Y., Yang, C.S., 2008. Synthesis and characterization of positive-charge functionalized mesoporous silica nanoparticles for oral drug delivery of an anti-inflammatory drug. Adv. Funct. Mater. 18 (20), 3283–3292.
Li, Q., Zhai, W., Jiang, Q., Huang, R., Liu, L., Dai, J., Gong, W., Du, S., Wu, Q., 2015. Curcumin–piperine mixtures in self-microemulsifying drug delivery system for ulcerative colitis therapy. Int. J. Pharm. 490 (1), 22–31.

Liu, M., Dong, J., Yang, Y., Yang, X., Xu, H., 2005. Anti-inflammatory effects of triptolide loaded poly (D, L-lactic acid) nanoparticles on adjuvant-induced arthritis in rats. J. Ethnopharmacol. 97 (2), 219–225.

Liu, Y., Wang, P., Sun, C., Feng, N., Zhou, W., Yang, Y., Tan, R., Chen, Z., Wu, S., Zhao, J., 2010. Wheat germ agglutinin-grafted lipid nanoparticles: preparation and in vitro evaluation of the association with Caco-2 monolayers. Int. J. Pharm. 397 (1), 155–163.

Ma, T., Wang, L., Yang, T., Ma, G., Wang, S., 2014. Homogeneous PLGA-lipid nanoparticle as a promising oral vaccine delivery system for ovalbumin. Asian J. Pharm. Sci. 9 (3), 129–136.

Maherani, B., Arab-Tehrany, E., Mozafari, M.R., Gaiani, C., Linder, M., 2011. Liposomes: a review of manufacturing techniques and targeting strategies. Curr. Nanosci. 7 (3), 436–452.

Marasini, N., Skwarczynski, M., Toth, I., 2014. Oral delivery of nanoparticle-based vaccines. Exp. Rev. Vaccines 13 (11), 1361–1376.

Mehnert, W., Mäder, K., 2001. Solid lipid nanoparticles: production, characterization and applications. Adv. Drug Deliv. Rev. 47 (2), 165–196.

Mei, Z., Chen, H., Weng, T., Yang, Y., Yang, X., 2003. Solid lipid nanoparticle and microemulsion for topical delivery of triptolide. Eur. J. Pharm. Biopharm. 56 (2), 189–196.

Meissner, Y., Pellequer, Y., Lamprecht, A., 2006. Nanoparticles in inflammatory bowel disease: particle targeting versus pH-sensitive delivery. Int. J. Pharm. 316 (1), 138–143.

Mihatsch, M.J., Kyo, M., Morozumi, K., Yamaguchi, Y., Nickeleit, V., Ryffel, B., 1998. The side-effects of ciclosporine-A and tacrolimus. Clin. Nephrol. 49 (6), 356–363.

Moghimipour, E., Ameri, A., Handali, S., 2015. Absorption-enhancing effects of bile salts. Molecules 20 (8), 14451–14473.

Mohammadpour Dounighi, N., Eskandari, R., Avadi, M.R., Zolfagharian, H., Mir Mohammad Sadeghi, A., Rezayat, M., 2012. Preparation and in vitro characterization of chitosan nanoparticles containing *Mesobuthus eupeus* scorpion venom as an antigen delivery system. J. Venom. Anim.Toxins Incl. Trop. Dis. 18 (1), 44–52.

Mohammadpourdounighi, N., Behfar, A., Ezabadi, A., Zolfagharian, H., Heydari, M., 2010. Preparation of chitosan nanoparticles containing *Naja naja oxiana* snake venom. Nanomedicine 6 (1), 137–143.

Morishita, M., Peppas, N.A., 2006. Is the oral route possible for peptide and protein drug delivery? Drug Disc. Today 11 (19), 905–910.

Müller, R.H., Runge, S., Ravelli, V., Mehnert, W., Thünemann, A.F., Souto, E.B., 2006. Oral bioavailability of cyclosporine: solid lipid nanoparticles (SLN®) versus drug nanocrystals. Int. J. Pharm. 317 (1), 82–89.

Nochi, T., Yuki, Y., Takahashi, H., Sawada, S.I., Mejima, M., Kohda, T., Harada, N., Kong, I.G., Sato, A., Kataoka, N., Tokuhara, D., 2010. Nanogel antigenic protein-delivery system for adjuvant-free intranasal vaccines. Nat. Mater. 9 (7), 572–578.

O'Hagan, D.T., McGee, J.P., Lindblad, M., Holmgren, J., 1995. Cholera toxin B subunit (CTB) entrapped in microparticles shows comparable immunogenicity to CTB mixed with whole cholera toxin following oral immunization. Int. J. Pharm. 119 (2), 251–255.

Oliveira, C.R., Rezende, C.M., Silva, M.R., Borges, O.M., Pêgo, A.P., Goes, A.M., 2012. Oral vaccination based on DNA-chitosan nanoparticles against *Schistosoma mansoni* infection. Sci. World J. 2012, 11.

Pabst, O., Mowat, A.M., 2012. Oral tolerance to food protein. Mucosal Immunol. 5 (3), 232–239.

Pal, S.K., Gomes, A., Dasgupta, S.C., Gomes, A., 2002. Snake venom as therapeutic agents: from toxin to drug development. Indian J. Experm. Biol. 40 (12), 1353–1358.

Pandita, A., Sharma, P., 2013. Pharmacosomes: an emerging novel vesicular drug delivery system for poorly soluble synthetic and herbal drugs. ISRN Pharm. 2013, 10.

Patil, U.S., Jaydeokar, A.V., Bandawane, D.D., 2012. Immunomodulators: a pharmacological review. Int. J. Pharm. Pharm. Sci. 4 (Suppl 1), 30–36.

Patravale, V., Prabhu, P., 2014. Potential of nanocarriers in antigen delivery: the path to successful vaccine delivery. Nanocarriers 1 (1), 2299–3991.

Pertuit, D., Moulari, B., Betz, T., Nadaradjane, A., Neumann, D., Ismaïli, L., Refouvelet, B., Pellequer, Y., Lamprecht, A., 2007. 5-amino salicylic acid bound nanoparticles for the therapy of inflammatory bowel disease. J. Control. Release 123 (3), 211–218.

Plapied, L., Duhem, N., des Rieux, A., Préat, V., 2011. Fate of polymeric nanocarriers for oral drug delivery. Curr. Opin. Colloid Interf. Sci. 16 (3), 228–237.

Premanand, B., Prabakaran, M., Kiener, T.K., Kwang, J., 2013. Recombinant baculovirus associated with bilosomes as an oral vaccine candidate against HEV71 infection in mice. PloS One 8 (2), e55536.

Rizwan, S.B., Dong, Y.D., Boyd, B.J., Rades, T., Hook, S., 2007. Characterisation of bicontinuous cubic liquid crystalline systems of phytantriol and water using cryo field emission scanning electron microscopy (cryo FESEM). Micron 38 (5), 478–485.

Ruenraroengsak, P., Cook, J.M., Florence, A.T., 2010. Nanosystem drug targeting: facing up to complex realities. J. Control. Release 141 (3), 265–276.

Saraf, S., 2010. Applications of novel drug delivery system for herbal formulations. Fitoterapia 81 (7), 680–689.

Sarti, F., Perera, G., Hintzen, F., Kotti, K., Karageorgiou, V., Kammona, O., Kiparissides, C., Bernkop-Schnürch, A., 2011. In vivo evidence of oral vaccination with PLGA nanoparticles containing the immunostimulant monophosphoryl lipid A. Biomaterials 32 (16), 4052–4057.

Selvamuthukumar, S., Velmurugan, R., 2012. Nanostructured lipid carriers: a potential drug carrier for cancer chemotherapy. Lipids Health Dis. 11, 159.

Shariff, A., Manna, P.K., Paranjothy, K.L.K., Manjula, M., 2007. Entrapment of andrographolide in cross-linked alginate pellets: I. Formulation and evaluation of associated release kinetics. Pakistan J. Pharm. Sci. 20 (1), 1–9.

Shin, S.B., Cho, H.Y., Kim, D.D., Choi, H.G., Lee, Y.B., 2010. Preparation and evaluation of tacrolimus-loaded nanoparticles for lymphatic delivery. Eur. J. Pharm. Biopharm. 74 (2), 164–171.

Singh, P., Chandra, R., 2012. Noscapine-encapsulated nanopolymers: a novel drug-carrier system to target the tumor cells. Int. J. Green Nanotechnol. 4 (2), 80–92.

Singh, R., Srivastava, A., Kapoor, R., Sharma, R.K., Mittal, R.D., 2009. Impact of *CYP3A5* and *CYP3A4* gene polymorphisms on dose requirement of calcineurin inhibitors, cyclosporine and tacrolimus, in renal allograft recipients of North India. Naunyn Schmiedebergs Arch. Pharmacol. 380 (2), 169–177.

Sou, K., Inenaga, S., Takeoka, S., Tsuchida, E., 2008. Loading of curcumin into macrophages using lipid-based nanoparticles. Int. J. Pharm. 352 (1), 287–293.

Tan, W., Li, Y., Chen, M., Wang, Y., 2011. Berberine hydrochloride: anticancer activity and nanoparticulate delivery system. Int. J. Nanomed. 6, 1773.

Tao, W., Ziemer, K.S., Gill, H.S., 2014. Gold nanoparticle-M2e conjugate coformulated with CpG induces protective immunity against influenza A virus. Nanomedicine 9 (2), 237–251.

Thapa, R.K., Khan, G.M., Parajuli-Baral, K., Thapa, P., 2013. Herbal medicine incorporated nanoparticles: advancements in herbal treatment. Asian J. Biomed. Pharm. Sci. 3 (24), 7–14.

Theakston, R.D.G., 1989. Snake venoms in science and clinical medicine 2. Applied immunology in snake venom research. Trans. Royal Soc. Trop. Med. Hyg. 83 (6), 741–744.

Tiyaboonchai, W., Tungpradit, W., Plianbangchang, P., 2007. Formulation and characterization of curcuminoids loaded solid lipid nanoparticles. Int. J. Pharm. 337 (1), 299–306.

Tong, W., Wang, L., D'Souza, M.J., 2003. Evaluation of PLGA microspheres as delivery system for antitumor agent-camptothecin. Drug Dev. Ind. Pharm. 29 (7), 745–756.

Trickler, W.J., Nagvekar, A.A., Dash, A.K., 2008. A novel nanoparticle formulation for sustained paclitaxel delivery. AAPS PharmSciTech 9 (2), 486–493.

Ubrich, N., Schmidt, C., Bodmeier, R., Hoffman, M., Maincent, P., 2005. Oral evaluation in rabbits of cyclosporin-loaded Eudragit RS or RL nanoparticles. Int. J. Pharm. 288 (1), 169–175.

van der Lubben, I.M., Kersten, G., Fretz, M.M., Beuvery, C., Verhoef, J.C., Junginger, H.E., 2003. Chitosan microparticles for mucosal vaccination against diphtheria: oral and nasal efficacy studies in mice. Vaccine 21 (13), 1400–1408.

Verma, P., Thakur, A.S., Deshmukh, K., Jha, A.K., Verma, S., 2010. Routes of drug administration. Int. J. Pharm. Studies Res. 1 (1), 54–59.

Wang, K., Qi, J., Weng, T., Tian, Z., Lu, Y., Hu, K., Yin, Z., Wu, W., 2014. Enhancement of oral bioavailability of cyclosporine A: comparison of various nanoscale drug-delivery systems. Int. J. Nanomed. 9, 4991.

Wang, X., Jiang, Y., Wang, Y.W., Huang, M.T., Ho, C.T., Huang, Q., 2008. Enhancing anti-inflammation activity of curcumin through o/w nanoemulsions. Food Chem. 108 (2), 419–424.

Wang, X.Q., Dai, J.D., Chen, Z., Zhang, T., Xia, G.M., Nagai, T., Zhang, Q., 2004. Bioavailability and pharmacokinetics of cyclosporine A-loaded pH-sensitive nanoparticles for oral administration. J. Control. Release 97 (3), 421–429.

Woodley, J.F., 1993. Enzymatic barriers for GI peptide and protein delivery. Crit. Rev. Thera. Drug Carrier Sys. 11 (2–3), 61–95.

Xiao, Y.Y., Song, Y.M., Chen, Z.P., Ping, Q.N., 2005. Preparation of silymarin proliposomes and its pharmacokinetics in rats. Acta Pharm. Sinica 40 (8), 758–763.

Yamanaka, Y.J., Leong, K.W., 2008. Engineering strategies to enhance nanoparticle-mediated oral delivery. J. Biomater. Sci. 19 (12), 1549–1570.

Yuan, Q., Li, X.R., Wang, H.J., Li, X.Y., Liu, Y., 2004. The absorption kinetics of silymarin microemulsion in rat intestine. Acta Pharm. Sinica 39 (8), 631–634.

Yun, Y., Cho, Y.W., Park, K., 2013. Nanoparticles for oral delivery: targeted nanoparticles with peptidic ligands for oral protein delivery. Adv. Drug Deliv. Rev. 65 (6), 822–832.

Zhang, Q., Yie, G., Li, Y., Yang, Q., Nagai, T., 2000. Studies on the cyclosporin A loaded stearic acid nanoparticles. Int. J. Pharm. 200 (2), 153–159.

Zhou, J., Xu, X.Z., Hu, Y.R., Hu, A.R., Zhu, C.L., Gao, G.S., 2013. Cryptotanshinone induces inhibition of breast tumor growth by cytotoxic CD4+ T cells through the JAK2/STAT4/perforin pathway. Asian Pacific J. Canc. Prev. 15 (6), 2439–2445.

Zolnik, B.S., Gonzalez-Fernandez, A., Sadrieh, N., Dobrovolskaia, M.A., 2010. Minireview: nanoparticles and the immune system. Endocrinology 151 (2), 458–465.

FURTHER READING

Mohamad, N.E., Abu, N., Rahman, H.S., Ky, H., Ho, W.Y., Lim, K.L., How, C.W., Rasedee, A., Alitheen, N.B., Yeap, S.K., 2015. Nanostructured lipid carrier improved in vivo anti-tumor and immunomodulatory effect of Zerumbone in 4T1 challenged mice. RSC Adv. 5 (28), 22066–22074.

CHAPTER

TANNIC ACID MODIFICATION OF METAL NANOPARTICLES: POSSIBILITY FOR NEW ANTIVIRAL APPLICATIONS

12

Malgorzata Krzyzowska*, Emilia Tomaszewska**, Katarzyna Ranoszek-Soliwoda**, Karolina Bien*, Piotr Orlowski*, Grzegorz Celichowski**, Jaroslaw Grobelny**

*Military Institute of Hygiene and Epidemiology, Warsaw, Poland

**University of Lodz, Lodz, Poland

CHAPTER OUTLINE

1 Introduction 335
2 Syntheses and Characterization of Metallic Nanoparticles Colloids (AuNPs, AgNPs, Ag/CuNPs) with Tannic Acid 336
2.1 Tannins 336
2.2 Synthesis of mNPs With Tannins 336
2.3 Detailed Synthesis of Metallic Nanoparticles (mNPs) 342
2.4 Characterization Techniques 344
3 Toxicity and Antiviral Properties of Tannic Acid Modifies Metallic Nanoparticles Colloids (AuNPs, AgNPs, Ag/CuNPs) 345
3.1 Toxicity of Tannic Acid Modified Metallic Nanoparticles Colloids 345
3.2 Herpes Simplex Infection 351
3.3 Metallic Nanoparticles in Treatment of Virus-Mediated STDs 352
3.4 Antiviral Activities of Tannic Acid–Modified Metallic Nanoparticles Colloids 354
4 Conclusions 360
Acknowledgments 360
References 360

1 INTRODUCTION

It has been shown that silver nanoparticles (NPs) can interact with several viruses to inhibit their ability to bind host cells in vitro. However, little is known about possible use of AgNPs as efficient microbicides in vivo and their side effects. Herpes simplex virus type 1 (HSV-1) and type 2 (HSV-2) infections are the most common cause of oral herpes and genital herpes, respectively, affecting approximately 60% to 95% of adults worldwide. HSV-2–caused infections heightens the risk of sexually transmitted diseases (STDs), such as human immunodeficiency virus (HIV). Therefore, there is a need to define an effective antiherpes microbicide, which should work by binding a cell-free virus, blocking its further spread and inactivate free virions, being at the same time safe and nontoxic when applied locally on the skin and mucosal tissue.

Hydrolysable tannins, such as tannic acid, have been shown to exert antiviral effects against HSV-1/2 and several other viruses as well as antiinflammatory properties. Tannins have also been used in the synthesis of noble metal NPs allowing for synthesis of NPs of precise size and shape. The combination of tannic acid with metallic NPs may allow preparation of a new type of microbicide with improved antimicrobial activity and the antiinflammatory properties.

2 SYNTHESES AND CHARACTERIZATION OF METALLIC NANOPARTICLES COLLOIDS (AuNPs, AgNPs, Ag/CuNPs) WITH TANNIC ACID

2.1 TANNINS

Tannins are water-soluble phenol derivatives naturally synthesized and accumulated by higher plants as secondary metabolic products. From a chemical point of view, tannins are polyphenols with molecular weights between 500 and 3000 Da. In complexes with saccharides, alkaloids, and proteins its molecular weight can increase even up to 20,000 Da (Haslam Edwin, 1998) that exhibit characteristic reactions for phenols. The chemical structure of tannic acid depends on the plant species producing the compound. Currently, more than 8000 different tannins have been isolated and chemically characterized. However, there are certainly many more tannins with chemical structures that have not been precisely determined yet.

All tannins have some common features, which enable classification of these types of compounds in two main groups, three types of hydrolysable tannins: gallotannines, ellagitannines, and complex tannins (sugars derivatives—mainly glucose, gallic acid, and ellagic derivatives) and condensed tannins (nonhydrolysable) called procyanidins containing condensed carbon chain typical for flavonoids (Khanbabaee and Van Ree, 2001). Condensed tannins are much more resistant to microbial degradation than hydrolysable tannins and exhibit stronger antibacterial, antiviral, and antifungal activity.

The chemical structure of tannins determines its biological activity. Moreover, tannins can act as potential: metal ion chelating agent, biological antioxidant, or, depending on its concentration, as a complexing or precipitating agent (in low concentrations as a complexing, and in high as a precipitating, agent). Because of the variety of possible tannin structures and diverse biological effects, it is difficult to predict the effects of tannins on any living organism. The combination of tannins and metal NPs is also very weakly presented in literature. However, the combination of tannins with metallic NPs allows it to prepare new types of nanomaterials with improved antimicrobial activity of NPs and the antiinflammatory properties of tannic acid.

2.2 SYNTHESIS OF mNPs WITH TANNINS

The procedures of synthesis metallic NPs, including gold (AuNPs) and silver (AgNPs), are already well known and precisely described in the literature (Shenashen et al., 2014; Tan and Cheong, 2013; Zhao et al., 2013). The methods of AuNPs and AgNPs synthesis can be based on both physical and chemical processes (Njoki et al., 2010; Wuithschick et al., 2013). Those synthesis procedures can also be adopted to obtain bimetallic NPs systems (e.g., AgNPs doped with Cu–Ag/CuNPs) with enhanced bioactivity.

The aim of this study was to investigate the influence of the metallic NPs modified with tannins on the biological processes of viruses and cell lines. Before all biological tests that the reagents could affect, the biological response of viruses or cell lines were exposed to NPs colloids, but the usage of which was crucial at the synthesis stage to obtain NPs with specific size and size distribution has been removed by colloids purification.

Another important point affecting the biological activity of mNPs is its size and size distribution. It is known that small NPs with the size below 10 nm are more toxic than the bigger ones (Gliga et al., 2014; Ivask et al., 2014). Moreover, only with the use of monodisperse NPs with defined sizes is it possible to find the correlation between the size of nanoobjects and its biological activity. Hence, the use of NPs with specific and precisely controlled size and narrow size distribution seems to be fully justified.

The preparation of NPs colloid with specific and strictly defined size and narrow size distribution is possible only with the precise control over the synthesis process. In the synthesis of mNPs two reagents already known for its reducing and stabilizing properties can be used: sodium citrate and tannic acid. The change of the amount and molar ratio of these reagents allows to obtain NPs in a wide range of sizes (from 10 to 60 nm) without the necessity of usage any additional reagents. The only exception is silver NPs colloid with the size below 10 nm, the synthesis of which requires the usage of a stronger reducing agent (sodium borohydride).

It is already known that both tannic acid and citric ions are involved in the reduction of metal ions to metallic NPs as well as in the stabilization process of resulted colloidal systems (Ahmad, 2014; Bastus et al., 2011; Dadosh, 2009). Those compounds, as well as its reaction products, can interact not only between each other but also with metal ions and form complexes with different stability constants (ionic interactions and/or hydrogen bonds formation) (Kumar et al., 2007). Therefore, the NPs preparation conditions have to be optimized to obtain stable mNPs colloids. The structure of citrate ion and tannic acid as well as the scheme of its oxidation process, which occurs during the NPs synthesis processes, is presented in Fig. 12.1 (Ojea-Jimenez et al., 2010).

The overall redox reaction occurring during the synthesis of AuNPs with gold (III) chloride and citrate ions is presented in Fig. 12.2.

Tannic acid is a natural compound that contains a glucose part that binds to five branches of condensed gallic acids (each branch can contain from one to three gallic acid molecules) schematically illustrated in Fig. 12.3B. Statistically commercial tannic acid contains 10 gallic acid molecules (Fig. 12.3A).

Hence, in all of mNPs synthesis procedures described herein, the molar ratios of reagents refer to one molecule of gallic acid. The schemes of the reactions occurring during the synthesis of mNPs with the use of tannic acid are shown in Fig. 12.4.

The optimum reaction conditions to obtain mNPs with specific and defined size and narrow size distribution were found by determination of the influence of tannic acid and citrate ions on the resulted mNPs. NPs syntheses were carried out with different molar ratios of reagents in order to find the correlation between its amount and the size of obtained NPs.

The changes of the molar ratio of tannic acid and citrate ions with the constant ratio of gold (III) chloride result in only small changes of the size of synthesized AuNPs (Fig. 12.5). This effect may be caused by the high oxidizing potential of $AuCl_4^{3-}$ ion. With the increase of the tannic acid concentration the reducing potential and reduction rate, compared to the rate of stable nuclei formation, also increases.

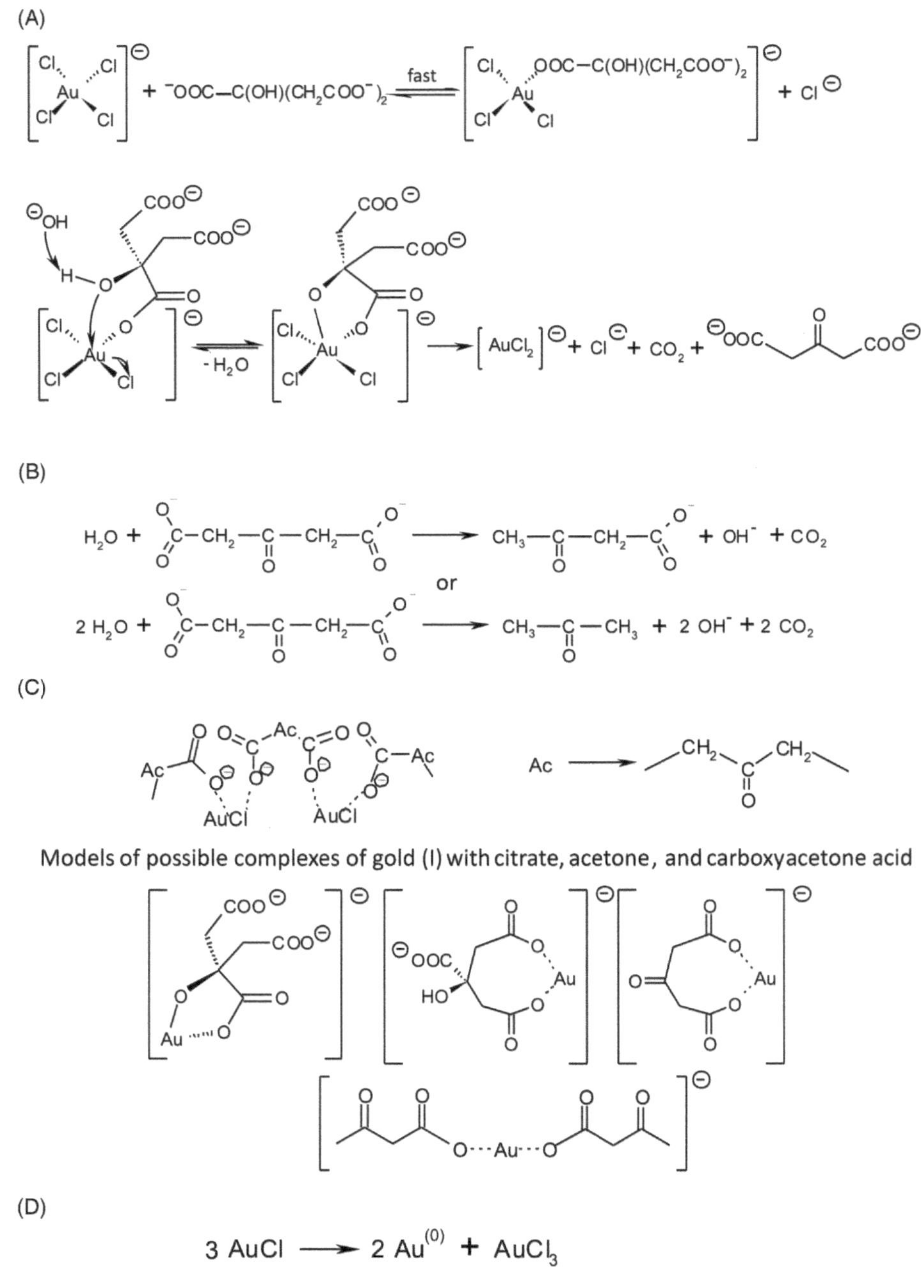

FIGURE 12.1 The Structure of Citrate Ion and Its Oxidation Processes That Occurs During the NPs Synthesis

(A) Exchange of the chloride ligand to citrate, which allows the reduction of gold (III) to gold (I). Oxidation of sodium citrate to the acetonedicarboxylic acid. (B) At higher temperature acetonedicarboxylic acid decomposes to carboxyacetone acid or acetone. (C) Complexation of acetonedicarboxylic acid and AuCl. (D) Disproportination of AuCl.

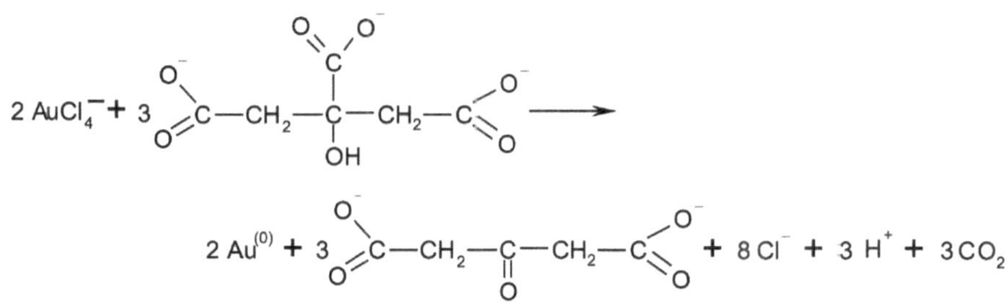

FIGURE 12.2 The Overall Redox Reaction in the Synthesis of Gold Nanoparticles With Citrate Ions

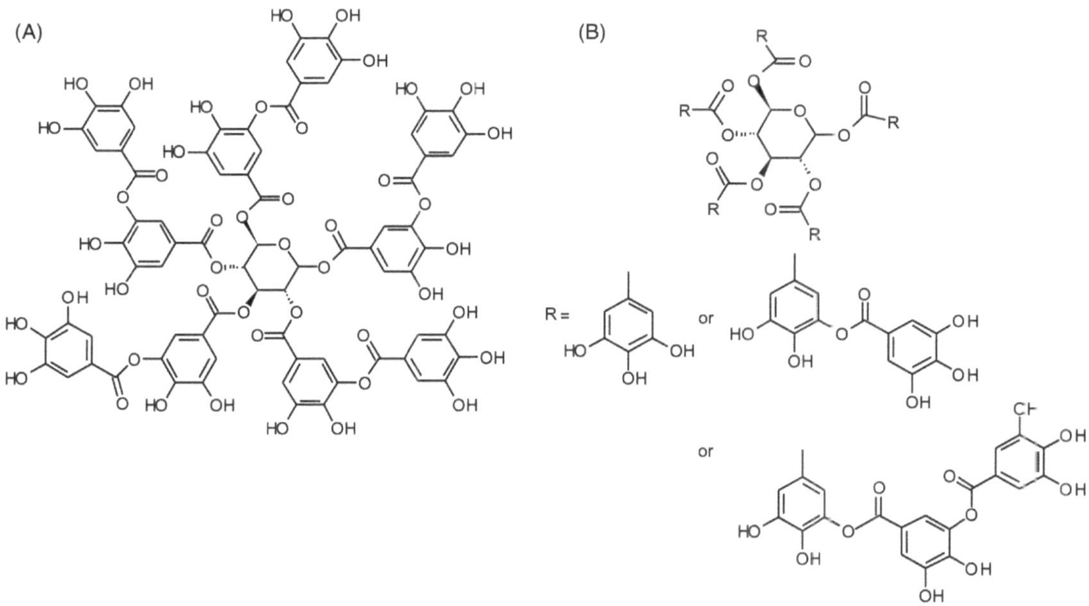

FIGURE 12.3

(A) The conventional formula of tannic acid; (B) condensed gallic acid structural fragments presenting the structure of tannic acid.

On the contrary, the increase of the citrate concentration with strong stabilizing properties results in formation of smaller AuNPs (amount of reducing agent constant). Moreover, the increase of the sodium citrate concentration (which is a salt of a weak acid and a strong base) causes the increase of the solution pH, and hence increases the reducing potential of tannic acid. Therefore, at the first step of the synthesis process, an increased number of nuclei, which may be stabilized by sodium citrate, are being formed. The rate of the new nuclei formation exceeds the rate of the diffusion of metal atoms formed in the reduction process, hence AuNPs with the smaller size are obtained.

The results clearly indicate that the optimization of the synthesis conditions and the reagents ratio (tannic acid and sodium citrate) allow to obtain AuNPs with specific size (from 9 to 18) and narrow size

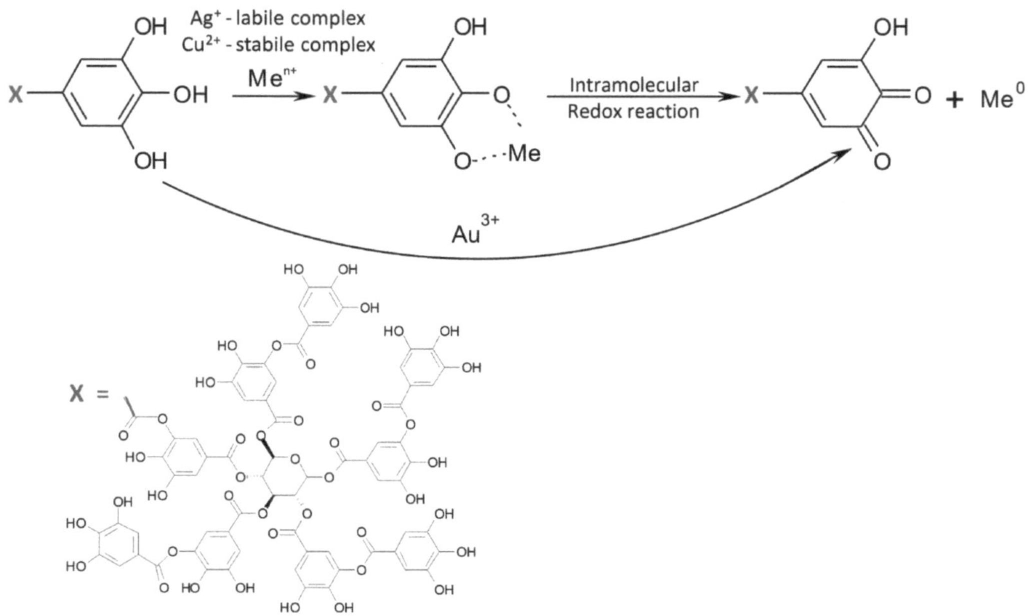

FIGURE 12.4 The Scheme of the Reactions Occurring During the Synthesis of mNPs With the Use of Tannic Acid

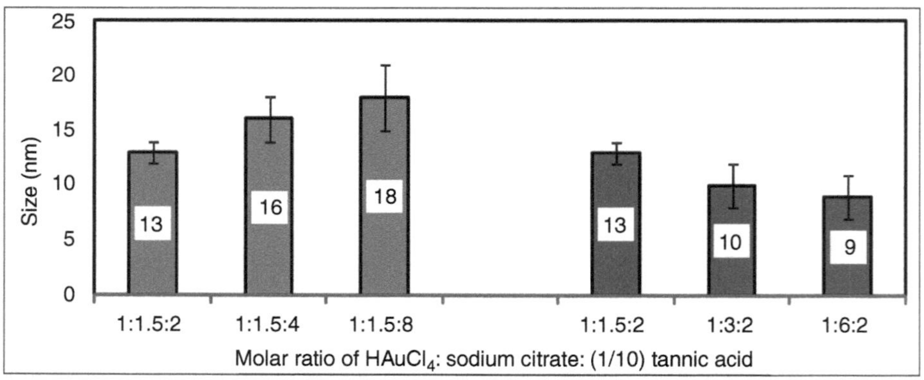

FIGURE 12.5 The Mean Size of AuNPs (DLS Measurements) for Different Molar Ratios of Sodium Citrate and Tannic Acid With a Constant Ratio of Gold (III) Chloride

distribution. Larger AuNPs with the size of about 30 and 60 nm were prepared with the use of only one reagent: tannic acid or sodium citrate, respectively. After the NPs formation the concentrations of all reagents in colloid were adjusted to those in synthesis of AuNPs with the size of about 10 nm.

The use of a mixture of reducing agents (sodium citrate and tannic acid) in the synthesis of AgNPs results in formation of monodisperse NPs with a larger size range compared to AuNPs. AgNPs syntheses are reproducible for all investigated reagents ratios and particles are homogenous in size and shape. Moreover, dynamic light scattering (DLS) and zeta potential measurements confirmed colloids stability in time.

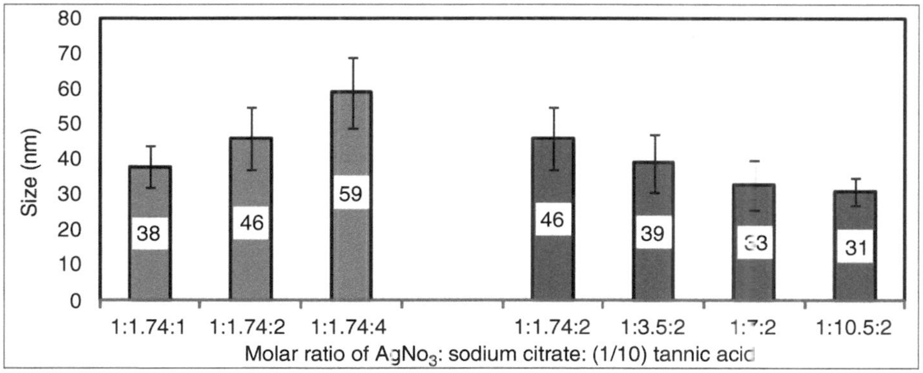

FIGURE 12.6 The Mean Size of AgNPs (DLS measurements) for Different Molar Ratios of Sodium Citrate and Tannic Acid With a Constant Ratio of Silver Nitrite

The molar ratio of reagents used in the synthesis procedures of AgNPs significantly influences the size of formed particles. The mean size of AgNPs (DLS measurements) obtained for different molar ratio of citrate ions and tannic acid with a constant ratio of silver nitrite is shown in Fig. 12.6.

The increase of the citrate ions concentration in case of AgNPs syntheses results in formation of smaller particles compared to results obtained for AuNPs. However, this phenomenon for AgNPs is observed only for specific contents of sodium citrate (up to 3.5 molar excess of sodium citrate to tannic acid). When the molar ratio exceeds the value of 3.5 no further increase of NPs size is observed. The size of synthesized AgNPs remains stable and is equal to about 30 nm. The reducing potential of tannic acid increases with the pH value, which is influenced by the citrate ion concentration. The citrate ions increase the pH value only up to 6.5 until the citrate buffer is formed. Further addition of citrate ions does not influence the pH value significantly, thereby the reduction of potential of tannic acid remains stable and the rate of the reduction reaction also remains constant. As a consequence, the number of nuclei formed at the initial stage of the reduction is constant, and the size of the NPs remains unchanged.

With the increase of the tannic acid concentration its reducing potential also increases and thereby the rate of the reduction reaction also increases. This, in turn, causes the diffusion rate of generated metal atoms to exceed the rate of stable nuclei formation (the stabilization of more nuclei is not possible because of the constant citrate ions concentration). The initial number of nuclei centers determines the final number of formed NPs (new nuclei centers are not formed and reduced metal atoms adsorbed only on already existing nuclei centers and results in formation of larger NPs). On the other hand, for lower concentration of tannic acid the diffusion rate of metal atoms and the rate of nuclei formation remain on the same level. Hence, the nucleation process lasts longer, more nuclei centers are being formed and consequently, more NPs with the smaller size can be obtained.

Among all mNPs syntheses carried out with citric ions and biologically active tannic acid a series of colloids with increasing size of metallic NPs were selected: AgNPs with the size of about 10, 33, 46, and (10–65 nm); AuNPs with the size of about 10, 29, 62 nm; and bimetallicAg/CuNPs with the size of about 29 and 63 nm as the aim of this research was to investigate the influence of the size of mNPs on its biological activity (toxicity and antiviral activity). Hence, for the research, the concentration of biologically active tannic acid to metals was maintained at constant level. Moreover, the concentrations of other compounds present in the colloids were also standardized (the ratio between components was constant). The descriptions of the mNPs synthesis procedures of all colloids used in biological tests are presented in the next section.

2.3 DETAILED SYNTHESIS OF METALLIC NANOPARTICLES (mNPs)

Gold (AuNPs), silver (AgNPs), and bimetallic (Ag/CuNPs) NPs were synthesized according to chemical reduction method, which is the most common approach for the synthesis of mNPs. The method is based on the use of organic and/or inorganic agents for the reduction of metallic ions (e.g., Ag^+) in aqueous or nonaqueous solutions. The agents reduce metallic ions and lead to the formation of metallic silver (Ag^0), followed by formation of oligomeric clusters that lead to the metallic colloidal particles. The protective agents stabilize dispersive NPs during the preparation procedure, and protect NPs against agglomeration, sedimentation, or losing surface properties. Small modifications in synthesis conditions can lead to dramatic changes in final product structure, average size, size distribution, or stability.

2.3.1 Synthesis of gold nanoparticles (AuNPs)

AuNPs with the size of about 10, 29, and 62 nm were synthesized by gold (III) chloride hydrate ($HAuCl_4{\cdot}H_2O$, Sigma-Aldrich, ≥49% Au basis) reduction with sodium citrate ($C_6H_5Na_3O_7{\cdot}2H_2O$, Sigma-Aldrich, ≥99%) or ammonium citrate tribasic ($C_6H_{17}N_3O_7$, Sigma, ≥97%) and tannic acid ($C_{76}H_{52}O_{46}$, Fluka, pure). Before biological tests all AuNPs colloids were filtered through a 0.1 μm polyvinylidene fluoride (PVDF) membrane.

2.3.1.1 10 nm AuNPs

To the aqueous solution of $HAuCl_4$ (94.4 g, 0.018%) heated to the boiling point under reflux, a warm mixture of tannic acid (1.7 g, 1%) and sodium citrate (3.9 g, 1%) was added. The mixture was continuously stirred during the reagents addition and immediately after the solution colored to ruby-red resulting from synthesized gold NPs. The solution was stirred for additional 15 min and next cooled down to room temperature. The molar ratio of gold chloride, sodium citrate, and tannic acid was 1:3:2, respectively. The concentration of gold in AuNPs colloid was 100 ppm.

2.3.1.2 29 nm AuNPs

AuNPs with the size of about 29 nm and concentration of gold in NPs of about 100 ppm, were synthesized with $HAuCl_4$ and tannic acid with the molar ratio 1:22, respectively. Aqueous solution of $HAuCl_4$ (96.2 g, 0.018%) was stirred and heated to the boiling point under reflux and next a warm aqueous solution of tannic acid (3.8 g, 5%) was added. After a few minutes the solution colored to ruby-red indicating the formation of AuNPs. The solution was stirred for next 15 min under reflux and after that cooled down to room temperature. Before the biological tests, the colloid was purified with the excess amount of tannins with centrifugation (RCF = 7900 g, 20 min). Each time the removed supernatant was replaced with the aqueous solution of sodium citrate (0.037%). The final concentration of tannic acid was 200 ppm.

2.3.1.3 62 nm AuNPs

To the 95.7 g of gold (III) chloride (0.018%), stirred and heated to the boiling point under reflux a 4.3 g of aqueous solution of tribase ammonium citrate (1%) was added. Next, a solution was heated for additional 15 min and afterward cooled down to room temperature. The concentration of tannic acid in the final product was changed and was equal to 200 ppm. The molar ratio of $HAuCl_4$ to ammonium citrate was 1:3.5.

2.3.2 Synthesis of silver nanoparticles (AgNPs)

AgNPs with the size of about 10, 33, 46, and 10–69 nm were synthesized by silver nitrite ($AgNO_3$, Aldrich, 99.999% metals basis) reduction with sodium citrate ($C_6H_5Na_3O_7{\cdot}2H_2O$, Sigma-Aldrich, ≥99%) and tannic acid ($C_{76}H_{52}O_{46}$, Fluka, pure) or sodium borohydride ($NaBH_4$, Fluka, ≥99%). All AgNPs colloids were stored in darkness and before the biological tests filtered through a 0.1 μm polyvinylidene fluoride (PVDF) membrane.

2.3.2.1 10 nm AgNPs

To the aqueous solution of $AgNO_3$ (94.5 g, 0.0166%) a mixture of aqueous solutions of tannic acid (0.63 g, 5%) and sodium citrate (4.2 g, 4%) was added. After about 10 s an aqueous solution of $NaBH_4$ (0.7 g 2%) was added to the reaction mixture and the solution was stirred for additional 15 min. The molar ratio of silver nitrite to sodium citrate, tannic acid and sodium borohydride was 1:7:2:4. The concentration of silver in AgNPs was 100 ppm.

2.3.2.2 33 nm AgNPs

A warm mixture of aqueous solution of tannic acid (0.63 g, 5%) and sodium citrate (4.2 g, 4%) was added to an aqueous solution of silver nitrite (95.2 g, 0.0165%) stirred under reflux to the boiling point. Immediately after the reducing agents addition, the solution colored to brown, which indicated the formation of AgNPs. The solution was stirred under reflux for an additional 15 min and then cooled down to room temperature. The molar ratio of silver nitrite to sodium citrate, tannic acid and was 1:7:2.

2.3.2.3 46 nm AgNPs

To the aqueous solution of silver nitrite (98.3 g, 0.0160%) stirred and warmed under reflux a mixture of tannic acid (0.63 g, 5%) and sodium citrate (1.04 g, 4%) was added. After the reducing mixture addition the solution colored to brown indicating the formation of AgNPs. The solution was stirred for additional 15 min and cooled down to room temperature. The molar ratio of silver nitrite to sodium citrate and tannic acid was 1:1.74:2.

2.3.2.4 10–69 nm AgNPs

An aqueous solution of silver nitrite (95.8 g, 0.0154%) was stirred and heated to the boiling point under reflux. Next, warmed aqueous solution of sodium citrate (4.2 g, 2%) was added to the reaction mixture. After a few minutes the solution colored to yellow indicating the formation of AgNPs. The solution was stirred for next 15 min and cooled down to room temperature.

2.3.3 Synthesis of bimetallic nanoparticles (Ag/CuNPs)

Ag/CuNPs with the size of about 29 nm and 63 nm were synthesized by reduction of silver ($AgNO_3$, Aldrich, 99.999% metals basis) and copper (II) nitrite [$Cu(NO_3)_2{\cdot}2.5H_2O$, Aldrich, 99.99+% metals basis] with a mixture of sodium citrate ($C_6H_5Na_3O_7{\cdot}2H_2O$, Sigma-Aldrich, ≥99%) and tannic acid ($C_{76}H_{52}O_{46}$, Fluka, pure). The weight concentration of silver and copper in Ag/CuNPs was 100 and 15 ppm, respectively.

2.3.3.1 29 nm Ag/CuNPs

94.406 g of a mixture of aqueous solutions of $AgNO_3$ (0.017%) and $Cu(NO_3)_2$ (0.005%) was stirred and heated to the boiling point under reflux. Next, a warm mixture of sodium citrate (5.247 g, 4%) and tannic acid (0.790 g, 5%) was added and the mixture was continuously stirred. Immediately after the reagents addition, the solution colored first to yellow and after a few seconds to bright brown, which indicated the formation of NPs. The mixture was stirred for an additional 15 min and next cooled down to room temperature. The molar ratio of $AgNO_3$ to $Cu(NO_3)_2$ and reductive mixture was 1:7:2, respectively. The molar ratio of sodium citrate to tannic acid in reductive mixture was 1:0.255, respectively.

2.3.3.2 63 nm Ag/CuNPs

Bimetallic Ag/CuNPs with the size of about 63 nm were synthesized according to procedure described earlier but with different reagents amount. Briefly, 97.558 g of a mixture of aqueous solutions of $AgNO_3$ (0.017%) and $Cu(NO_3)_2$ (0.005%) was stirred and heated to the boiling point under reflux. Next, a warm mixture of sodium citrate (1.304 g, 4%) and tannic acid (1.581 g, 5%) was added and the solution was continuously stirred. The molar ratio of sodium citrate to tannic acid in reductive mixture was 1:0.255, respectively. After the reductive mixture addition, the solution colored first to yellow and after a few seconds to bright brown, which indicated the formation of NPs. The mixture was stirred for additional 15 min and next cooled down to room temperature. The molar ratio of $AgNO_3$ to $Cu(NO_3)_2$ and reductive mixture was 1:1.74:4, respectively. The weight concentration of silver and copper in Ag/CuNPs was 100and 15 ppm, respectively.

2.4 CHARACTERIZATION TECHNIQUES

There are number of methods for NPs size characterization; for example, scanning transmission electron microscopy (STEM), atomic force microscopy (AFM), analytical ultracentrifugation (AUC), DLS, flow field fractionation (FFF) and so on. The precise characterization of particles requires the usage of few of them. However, although there is a range of methods available for the particle size determination, users should remember about the strengths and weaknesses of applied methods to recognize its limitations.

Aqueous colloids of AuNPs, AgNPs, Ag/CuNPs were examined using a DLS and STEM techniques. Results acquired by DLS and STEM techniques give complementary information about the size, size distribution, shape, and colloidal stability of NPs.

2.4.1 Dynamic light scattering

DLS is one of the most popular, fast, and easy-to-operate sizing techniques, especially for colloidal suspensions. There are several advantages of DLS technique: automated operation, short time of measurement, sensitivity to aggregates, noninvasive for colloids, no calibration and no specific sample preparation requirements. Therefore, this technique is increasingly used for NPs characterization in many fields of science and industry. In spite of many advantages, the precision of the determined NP size is not entirely understood due to the number of unevaluated effects, such as concentration of NP suspension, scattering angle, shape of NPs, and polydispersity of colloids (Tomaszewska et al., 2013).

All DLS measurements were performed using a nano ZS zetasizer system (Malvern Instruments, Worcestershire, United Kingdom). The measured parameters were: a laser wavelength of 633 nm (He-Ne), a scattering angle of 173 degree, a measurement temperature of 25°C, a medium viscosity of 0.8872 mPa*s, and a medium refractive index of 1.330. The samples were loaded into quartz cuvette and five measurements were performed for which the mean result was recorded.

Table 12.1 The Characterization Results Obtained for Gold (AuNPs), Silver (AgNPs), and Bimetallic (Ag/CuNPs) Nanoparticles

	DLS		STEM
NPs Type	**Hydrodynamic Size (nm)**	**Zeta (mV)**	**Size (nm)**
AuNPs	10 ± 2	−33 ± 12	5 ± 1
	29 ± 6	−32 ± 1	20 ± 3
	62 ± 12	−49 ± 3	59 ± 7
AgNPs	10 ± 3	−31 ± 7	4.5 ± 1.2
	33 ± 7	−58 ± 2	28 ± 7
	46 ± 9	−55 ± 2	33 ± 8
	10 ± 1; 69 ± 13	−61 ± 1	10-65
Ag/CuNPs	27 ± 9	−49 ± 6	18 ± 4
	63 ± 10	−36 ± 9	49 ± 8

2.4.2 Electron microscopy

Electron microscopy is an excellent technique available, which can give the direct structural and size information about NPs. Modern, high-resolution microscopes are capable of investigating the structure of NPs down to nanometer level. Quantitative information analyses based on micrograph acquired and size histogram created, give the size distribution result of investigated NPs. An obvious disadvantage of microscopy analysis is the limited number of examined objects (usually not more than several hundred). Therefore, the sample preparation procedures should quarantine a random population of selected NPs.

The shape, size, and size distribution of NPs were investigated using scanning electron microscopy (SEM, Nova NanoSEM 450, FEI, accelerating voltage of 30 kV) equipped with a STEM II detector located under the sample for detection of electrons passing through the sample as in the case of transmission electron microscopy (TEM). The samples for STEM measurements were prepared as: a drop of vigorously shacked colloid was deposited onto carbon-coated copper grids (300 mesh) and the suspension was left for 1 h for solvent evaporation before measurements.

2.4.3 mNPs characterization results

The mNPs characterization results are mentioned in (Table 12.1) (Figs. 12.7–12.9)

3 TOXICITY AND ANTIVIRAL PROPERTIES OF TANNIC ACID MODIFIES METALLIC NANOPARTICLES COLLOIDS (AuNPs, AgNPs, Ag/CuNPs)

3.1 TOXICITY OF TANNIC ACID MODIFIED METALLIC NANOPARTICLES COLLOIDS

We previously tested toxic and antiinflammatory properties of tannic acid modified 13, 33, and 46 nm AgNPs in murine 291.03C keratinocyte and RAW 264.7 monocyte cell lines (Orlowski et al., 2013). Both cell lines exposed to 1–10 μg/mL of 13, 33, and 46 nm AgNPs exhibited a dose-dependent

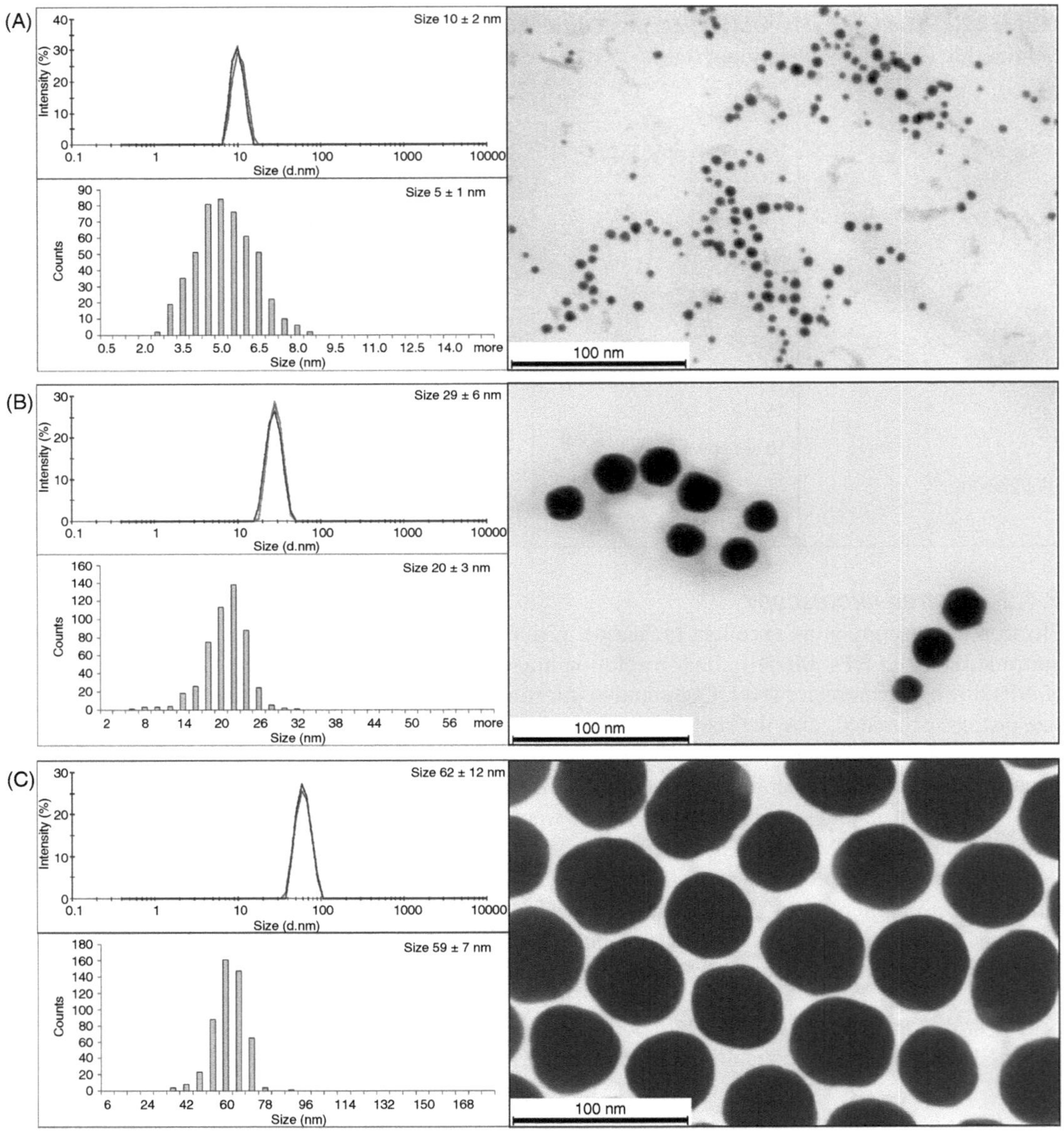

FIGURE 12.7

DLS size distribution graph, size distribution histogram, and STEM image of AuNPs with the following approximate size: A: (A) 10 nm, (B) 29 nm, (C) 62 nm.

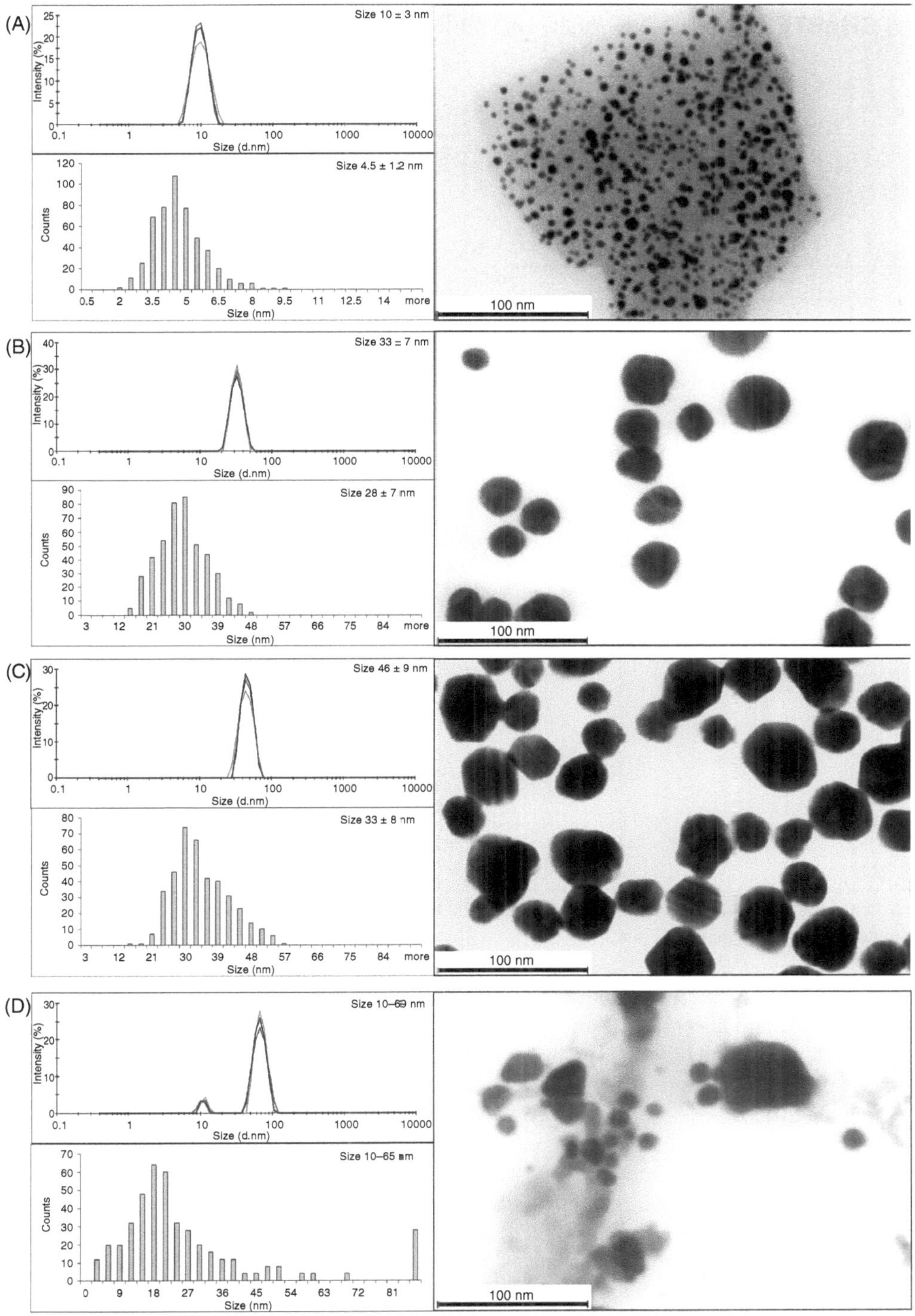

FIGURE 12.8

DLS size distribution graph, size distribution histogram and STEM image of AgNPs with the following approximate size: (A) 13 nm, (B) 33 nm, (C) 46 nm, (D) 10–69 nm.

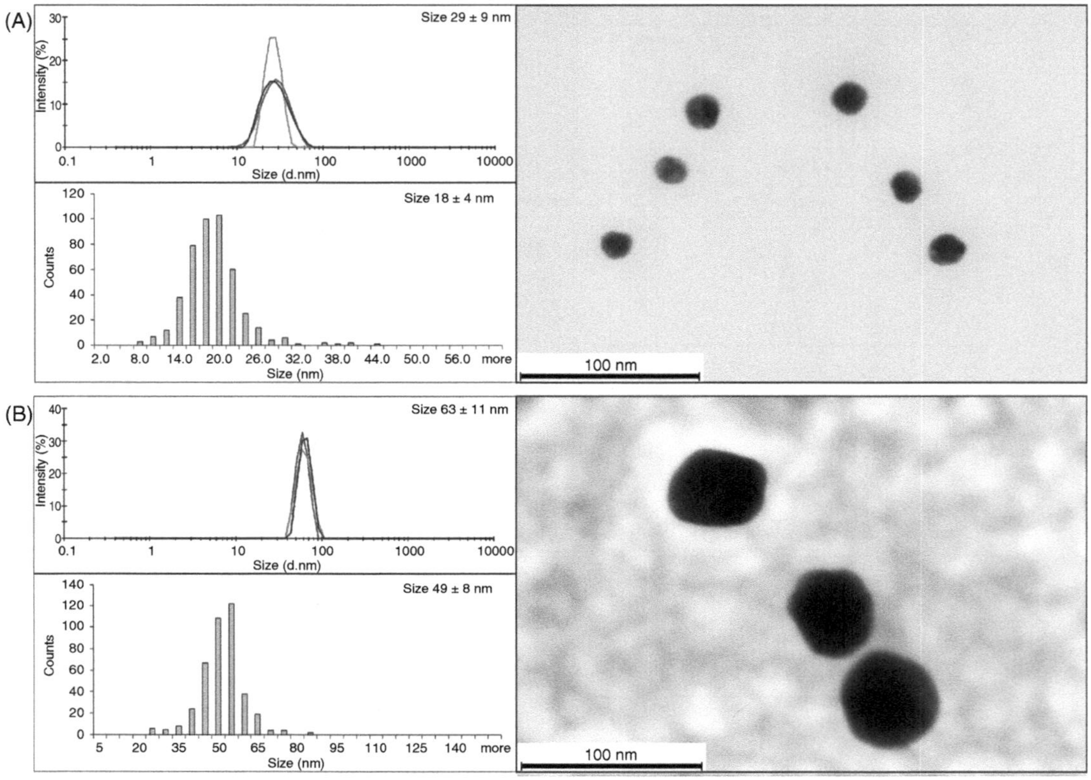

FIGURE 12.9

DLS size distribution graph, size distribution histogram and STEM image of Ag/CuNPs with the following approximate size: (A) 29 nm, (B) 63 nm.

toxicity expressed as the decrease of mitochondrial potential, neutral red uptake, induction of caspase-9 activity, apoptosis, and necrosis; 13 nm tannic-acid modified AgNPs but also 10–65 nm unmodified AgNPs induced production of ROS only by monocytes but not by keratinocytes (Orlowski et al., 2013). Higher sizes of AgNPs were relatively nontoxic and did not produced reactive oxygen species (ROS) (Orlowski et al., 2013).

Here, we tested toxicity of tannic acid modified and unmodified NPs of different sizes (Fig. 12.10) to compare the effect of size, material, and tannic acid modification upon the toxicity of silver and silver/copper bimetallic NPs using staining with a cationic dye, 5,5′,6,6′-tetrachloro1,1′,3,3′-tetraethyl-benzimidazolylcarbocyanine iodide (JC-1). This staining allows it to identify cells with decreased mitochondrial potential resulting from disturbances in the mitochondrial functioning (Salvioli et al., 1997). The following sizes of tannic acid modified and unmodified AgNPs were used: modified 10, 33, and 46 nm, unmodified 10–69 nm, as well as Ag/CuNPs: 29 and 63 nm. To assess the toxicity of Ag and Ag/CuNPs we used the mouse monocyte RAW 265.7 cell line as the in vitro model because we previously showed that this cell line is prone to fast and abundant internalization of NPs, reflected by toxicity (Orlowski et al., 2013). Using concentration of 5 μg/mL, we found that only tannic acid modified silver

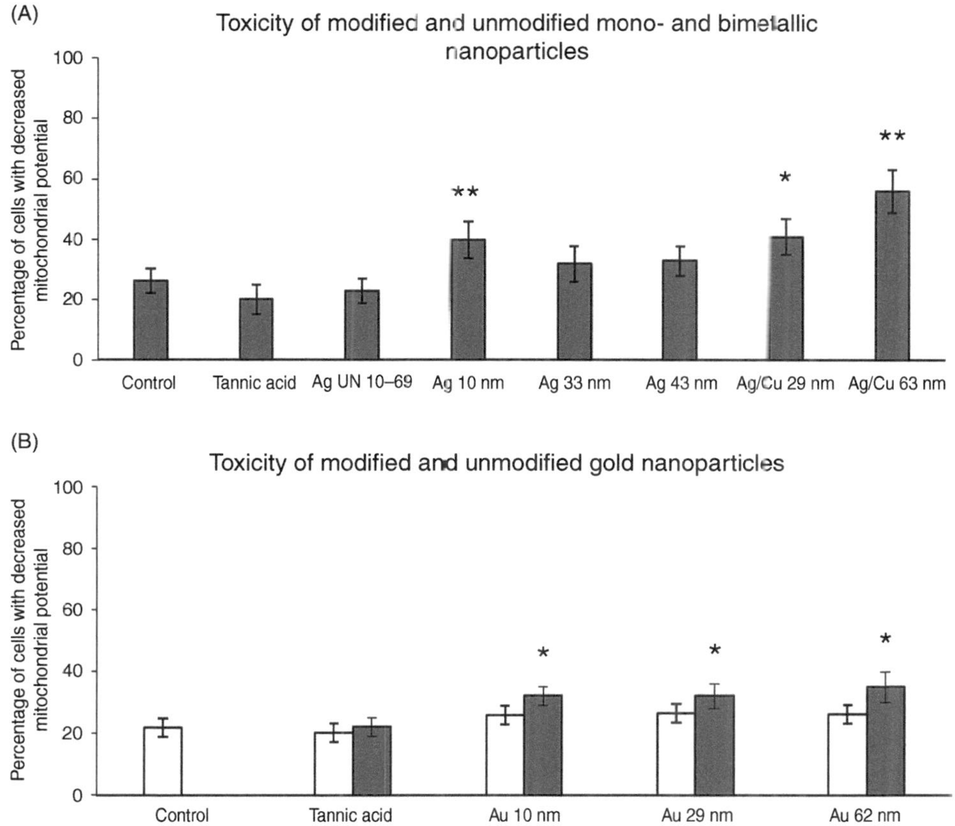

FIGURE 12.10

Cytotoxicity assays in RAW 264.7 cell line at 24 h of exposition to: (A) tannic acid modified 10, 33, and 46 nm AgNPs, unmodified 10–69 nm AgNPs, 29 and 63 nm Ag/CuNPs at 5 μg/mL, (B) 10, 29, and 62 nm AuNPs at 5 and 10 μg/mL. Mitochondrial potential expressed as the percentage of cells with green JC-1 fluorescence. Each bar represents the mean from three experiments ±SEM, * represents significant differences with $P \leq 0.05$, while ** means $P \leq 0.001$.

NPs sized 10 nm showed significant decrease of mitochondrial potential in contrast to unmodified AgNPs, showing no toxicity ($P \leq 0.01$) (Fig. 12.10A). Tannic acid modified silver/copper bimetallic NPs sized 29 and 63 nm showed significant toxicity ($P \leq 0.05$) (Fig. 12.10A). On the other hand, using the same concentration of tannic acid modified AuNPs of similar sizes: 10, 29, and 62 nm we found no significant induction of toxicity (Fig. 12.10B).

Carlson et al. (2008) determined the ability of silver NPs (15, 30, 55 nm) to induce oxidative stress within NR8383 rat alveolar macrophages, at concentrations up to 100 μg/mL and for a duration of up to 24 h. Particles were internalized by cells and cell morphology was detrimentally affected by NP exposure. All particles elicited an increase in ROS production, and depletion of cellular GSH, thereby indicating that AgNPs influence an oxidative response (Carlson et al., 2008).

Studies have shown AgNP toxicity to be both size- and shape-dependent; for example, the study by Carlson et al. (2008) with alveolar macrophages indicated that AgNPs with a mean size of 15 nm induced the greatest loss in mitochondrial activity. Liu et al. (2010) showed that toxicity of silver NPs in four human cell lines derived from different tissues (A549, HepG2, SGC7901, and MCF-7) depended on their size, with 5 nm particles being more toxic than 20 and 50 nm. On the other hand, human epidermal keratinocytes (HEKs) did not show different cytotoxicity when exposed to 20, 50, and 80 nm AgNPs (Samberg et al., 2010).

The smaller the NPs, the more easily they become internalized by the cells, especially by phagocyting cells, such as macrophages. A number of studies suggested that cytotoxic effects of silver NPs were mediated by the generation of oxidative stress (Park et al., 2010). Oxidative stress is induced when the generation of ROS exceeds the cell's antioxidant capacity. Apart from the damaging effects to cellular proteins, lipids, and DNA, an increasing level of ROS triggers the cell to respond by activating pro-inflammatory signaling cascades, and ultimately induces cell death (Park et al., 2010; Ryter et al., 2007). Mitochondria also appear to be sensitive targets for AgNP toxicity. Asharani et al. (2009) suggested that the disruption of the mitochondrial respiratory chain by AgNP increased ROS production and interruption of ATP synthesis, thus leading to DNA damage.

Recent studies have demonstrated the antimicrobial activity of CuNPs against methicillin-resistant *Stapholococcus aureus* and *Escherichia coli* (Kruk et al., 2015; Singh et al., 2015). However, copper NPs are limited by their widely demonstrated toxic properties. A study by Yang et al. (2010) has shown that CuNPs with greater surface oxidation have higher ROS-generating capacity. Cuprous NPs can selectively induce apoptosis and suppress the proliferation of tumor cells, showing great potential as a clinical cancer therapy. In a study by Wang et al. (2012) treatment with CuNPs caused a G1/G0 cell cycle arrest in tumor cells. Furthermore, CuNPs enclosed in vesicles entered or were taken up by mitochondria, which damaged their membranes, thereby inducing apoptosis (Wang et al., 2012).

Since gold has traditionally been considered inert and biocompatible, and due to its physicochemical properties and high surface area, AuNPs are more and more used in biomedical research such as site-specific drug delivery (Arvizo et al., 2010). Gold NPs have been found to be "nontoxic" according to many reports. Fraga et al. (2013) investigated the cytotoxic effects of AuNPs in human liver HepG2 cells, utilizing ~20 nm spherical AuNPs (0–200 μM Au) with two surface coatings [citrate (Cit) compared with 11-mercaptoundecanoic acid (11-MUA)]. TEM revealed that both AuNPs were internalized in a concentration-dependent manner and did not induce significant cytotoxicity. Also, Brown et al. (2014) investigated how AuNPs of different sizes (20 and 100 nm) influenced macrophage cytokine responses when dispersed in lung or blood relevant fluids and found that AuNPs had no significant impact on macrophage viability. However, other groups have found that gold NPs are "toxic". For example, Pan et al. (2009) found that 1.4 nm, but not 15 nm gold NPs triggered necrosis, mitochondrial damage, and induced an oxidative stress. This result highlights possible size-dependent toxicity of gold NPs (Pan et al., 2009). In our study, we noticed toxicity of AuNPs only at higher concentrations, irrespectively of the size (Fig. 12.10).

Because the mechanism underlying induction of apoptosis via mitochondrial pathway by AgNPs and Ag/CuNPs may be through the generation of ROS, we measured ROS generation using 2′,7′-dichlorofluorescin diacetate (DCFH-DA) fluorescence in monocytes (Fig. 12.11). The results showed significant production of ROS by RAW 264.7 monocytes exposed to 10 nm AgNPs, 29, and 63 nm Ag/CuNPs ($P \leq 0.05$). Altogether, these findings indicate that small-sized tannic acid–modified AgNPs are potent inducers of ROS production in monocytes, while the presence of Cu in bimetallic

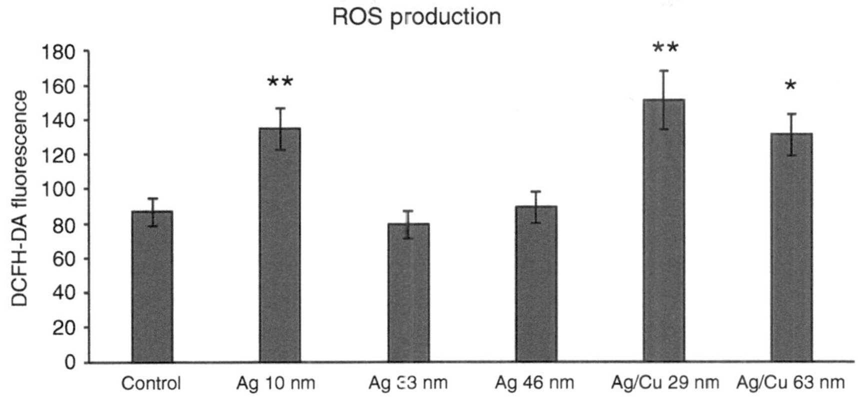

FIGURE 12.11 The Relative Levels of Intracellular ROS in RAW 264.7 Monocyte Culture After 24 h of Incubation With AgNPs or Ag/CuNPs Expressed as the Mean DCFH-DA Fluorescence

Cells were treated with 5 μg/mL of 10, 33, and 46 nm AgNPs and 29 or 63 nm Ag/CuNPs. Each bar represents the mean from three experiments ± SEM, * represents significant differences with $P \leq 0.05$, while ** means $P \leq 0.001$.

NPs increases their toxicity, irrespective of their size due to ROS production (Fig. 12.11). We did not find any ROS production by AuNPs (data not shown).

The increased toxicity of small-sized tannic acid–modified NPs can be also explained by the insitric properties of tannic acid—its ability to form insoluble complexes with proteins, nucleic acids, and carbohydrates. Therefore, tannic acid modification may help to bind NPs to the cell surface and external receptors, enabling more efficient uptake of NPs. Adding the known toxicity of small-sized NPs we can therefore expect that tannic acid modification does not help to reduce toxicity of small NPs, while for bigger sizes it may help to internalize NPs by the cells. In a previous paper we showed that tannic acid–modified NPs accumulate inside intracellular vacuoles with the characteristic "corona effect," further indicating binding to extracellular receptors and receptor-driven internalization (Orlowski et al., 2013).

3.2 HERPES SIMPLEX INFECTION

HSV causes a contagious infection that affects approximately 60% to 95% of adults worldwide. HSV-1 is associated mainly with infections of the mouth, pharynx, face, eye, and central nervous system (CNS), while HSV-2 is associated with infections of the anogenital region, although both serotypes may infect both areas. HSV-1 and -2 persist in the body by becoming latent in the cell bodies of nerves and after the initial or primary infection. People infected with HSV-1 or HSV-2 can expect to have several (typically four or five) outbreaks (symptomatic recurrences) within a year (Gupta et al., 2007; Tobian and Quinn, 2009). The World Health Organization (WHO) estimates that every year 20 million people become newly infected and 400 million people are already carrying HSV-2 virus, which makes the HSV-2 infection a worldwide problem. Genital herpes is the most characteristic sign of HSV-2 infection, sometimes occurring with fever, body aches, and swollen lymph nodes, but most infections give no symptoms (World Health Organization, 2015). Genital HSV-2 infection is more common in women (approximately one out of five women 14–49 years of age) than in men (about one out of nine men 14–49 years of age). Therefore, we may assume that ulcerations resulting from recurrent HSV-2 infection consists of one of the most common factors influencing integrity and functioning of vaginal mucosa (Gupta et al., 2007; Halioua and Malkin, 1999). There is a strong association between bacterial and viral STDs and both the acquisition and transmission of HIV infection. Most of the evidence about the role of STDs in the transmission and control of HIV is focused on HSV-2 infection, causing genital ulcers (Fox and Fidler, 2010). Dissemination of HIV is higher for people carrying HSV-2, and they are 3–4 times more likely to acquire HIV. Similarly, HIV-positive patients more frequently acquire HSV-2. For this reason, successfully developed material, protecting from HSV-2, would also decrease susceptibility to HIV infection.

Recurrent herpes infections are also a problem in persons with immunodeficiencies, such as the patients with malignancies. Frequent and extensive local inflammatory changes the influence arrangement of medical procedures in patients with malignancies, causing a change of schedules for necessary medical procedures until the infection heals (Gupta et al., 2007). This may influence the ultimate results of anticancer treatments. Any type of antiherpes treatment that can shorten the period of acute infection will add to the total outcome of treatment (Halioua and Malkin, 1999). The vaccines against HSV-1 or HSV-2 infection are still not ready, although several experimental models have been proposed.

Therefore, there is an urgent need to develop an effective antiherpesviral microbicide, which can be applied locally by persons suspecting recurrent HSV-1 or HSV-2 infection to block development of lesions but also for the use by persons suspecting a new infection. The only way of herpes treatment is the local or systemic use of antiviral drugs, blocking viral replication. In the case of recurring herpes infection, the use of acyclovir and its analogues, blocking viral replication is limited by the necessity of inhibitors to penetrate infected cells, and by development of resistance to inhibitors (Gupta et al., 2007). Inhibitors of viral replication do not work upon the cell-free virus released from the infected cells and cannot block further spread of the cell-free virus. Ideally, an antiviral microbicide should work by binding the cell-free virus and thus blocking its further spread to neighboring uninfected cells, but it should also "disinfect" the body fluids containing the infectious virions. At the same time, it should possess a potential ability to be able to remain in the tissue long enough to provide protection from infection by the residual active virus.

Treatment of initial genital herpes takes about 10 days but chronic suppressive therapy for recurrent disease may last up to 1 year. The global cost of such treatment is difficult to estimate because many countries do not collect statistical data about HSV-2 cases. It should not be forgotten that present forms of treatment means only suppressing symptoms. Regardless financial aspect, avoiding HSV-2 infection could diminish physiological problems associated with STDs, such as depression, diminution of self-esteem and feeling of negative stigma (Gupta et al., 2007).

3.3 METALLIC NANOPARTICLES IN TREATMENT OF VIRUS-MEDIATED STDs

The use of NPs-based antiviral agents has several advantages. NPs have the characteristics of high surface-to-volume ratios, enabling the packaging of multiple antiviral agents onto the surface of NPs. This type of approach provides the possibility of targeting specific biological sites actively or passively by means of controlling hydrophobicity/lipophilicity, or modification of nanocarrier surfaces with desired compounds.

Historically, soluble silver compounds, such as silver salts, have been used for treating STDs, including syphilis and gonorrhoea (Frost, 1924). Credés prophylaxis of gonococcal ophthalmia neonatorum was the standard care in many countries until the end of the 20th century (Hoyme, 1993).

Currently, silver sulfadiazine is listed by WHO as an essential antiinfective topical medicine (Rashaan et al., 2014). Silver particles (AgNPs), with a long history of general use as an antiseptic and disinfectant, are able to interact with disulfide bonds of the glycoprotein/protein contents of microorganisms such as viruses, bacteria, and fungi (Gajbhiye et al., 2009; Lara et al., 2011a).

HIV is an enveloped virus. Derived from the membrane of the host cell, it is equipped with virus encoded proteins: gp 120 and gp 41. Gp120 is needed to attach to the host cell, and gp 41 is critical for the cell fusion process. The primary targets of HIV-1 are T-lymphocytes and macrophages, but the virus can infect other cells, such as monocytes and dendritic cells (Miller et al., 1991; Schuitemaker et al., 1992). Elechiguerra et al. (2005) were the first to demonstrate that polyvinylpyrrolidone-coated silver NPs (PVP-coated AgNPs) undergo a size-dependent interaction with HIV-1 and only the particles within the range of 1–10 nm were able to bind to the virus The precise mechanism of action is not completely understood, but PVP-coated AgNPs inhibited HIV-1 infection (regardless of viral tropism or resistance profile) by binding to gp 120 in a manner that prevents CD4-dependent virion binding, fusion, and infection. Therefore, PVP-coated AgNPs blocked HIV-1 cell-free and cell-associated infection and acted as a virucidal agent, also in the form of a gel (Lara et al., 2010a,b). Furthermore, addition of AgNPs to HIV-1 neutralizing antibodies significantly increased the neutralizing potency of antibodies in prevention of the cell-associated HIV-1 transmission and infection (Lara et al., 2011b).

Gold NPs (Au NPs) have also been shown as promising effective inhibitors of HIV fusion (Bowman et al., 2008). These NPs have the advantage of allowing the formation of multivalent conjugates that effectively inhibit HIV-1 fusion to human cells. Bowman et al. (2008) were the first to demonstrate the inhibition of HIV fusion with multivalent Au NPs. The 2.0 nm diameter mercaptobenzoic acid coated gold NPs were conjugated to SDC-1721, a derivative of TAK-779, a known CCR5 antagonist. The SDC-1721–gold NP conjugates displayed anti-HIV-1 activity comparable to that of TAK-779 (Bowman et al., 2008). Martinez-Avila et al. (2009) proposed another approach of multivalent AuNPs, in which water-soluble gold glyconanoparticles (manno-GNPs) presenting truncated (oligo)mannosides of the high-mannose undecasaccharide Man(9)GlcNAc(2) completely inhibited the binding of dendritic cell-specific intracellular adhesion molecule-3-grabbing nonintegrin (DC-SIGN) binding to gp 120 of HIV-1. Similarly, modification of AuNPs with polyethylene glycol showed to be efficient in inhibition of M-tropic, T-tropic, dual tropic, and resistant HIV-1 isolates (Vijayakumar and Ganesan, 2012). The proposed mechanism was blocking viral entry by binding with gp 120 (Vijayakumar and Ganesan, 2012).

The HSV virion consists of double-stranded DNA enclosed in an icosadeltahedral capsid composed of four structural proteins. The outermost layer of the virion is an envelope constructed from the lipid bilayer, derived from the infected host cell. The surface of the envelope is covered with 10–12 glycoproteins (Wald, 1998). The penetration of both HSV-1 and HSV-2 into the host cell involves at least five viral envelope glycoproteins (gC, gB, gD, gH, gL), together with three classes of host cell membrane receptors (herpes virus entry mediator, nectin-1, nectin-2, and cell surface glycosaminoglycans, preferentially heparan sulfate) (Akhtar and Shukla, 2009; Wald, 1998). Heparan sulfate thus becomes a very attractive target to generate NPs able to block HSV interaction with the host cell surface. Baram-Pinto et al. (2009) constructed silver NPs that were capped with mercaptoethane sulfonate (Ag-MES). The mercaptoethanesulfonate (MES)-decorated AgNPs of 4 nm in size mimicked heparan sulfate on the host cell and blocked the virus attachment to the cells (Baram-Pinto et al., 2009). In the same manner, sulfonate-decorated Au NPs inhibited the entrance of the virus into the cells and prevented the cell-to-cell spread of HSV-1 (Baram-Pinto et al., 2010). The proposed mode of antiviral action of Ag-MES and Au-MES NPs involved effective engagement of the viral envelope glycoproteins by mimicking the cell surface HS by MES on the surface of NPs (Baram-Pinto et al., 2010). Mishra et al. (2011) proposed a similar mode of targeting the initial steps of HSV-1 infection with zinc oxide (ZnO) micro–nanostructures (MNSs) that were capped with multiple nanoscopic spikes mimicking cell-induced filopodia. The MNSs were predicted to target the virus to compete for its binding to cellular heparin sulfate through their partially negatively charged oxygen vacancies on their nanoscopic spikes (Mishra et al., 2011). Antoine et al. (2012) also showed that ZnO tetrapod structures significantly block HSV-2 entry into target cells.

Condoms are only 85% effective in preventing HIV and other STDs. There is a strong need to develop new nanotechnology-based condoms that are more potent than the first-generation products. Mohammed Fayaz et al. (2012) developed an Ag-NPs-coated polyurethane condom (PUC) and investigated its antimicrobial activity toward HIV-1 and HSV. Their results showed that Ag-NPs-coated PUC exhibited strong inactivating activity against both M-tropic and T-tropic HIV-1 strains but also against HSV-1 and HSV-2. In addition, Ag-NPs-coated PUC was able to inhibit the growth of bacteria and fungi (Mohammed Fayaz et al., 2012).

Copper has long been used as an antibacterial material, and several copper compounds have been reported to exhibit viral inactivation (Borkow and Gabbay, 2005). Recently, the inactivation of avian influenza virus by copper metal (Noyce et al., 2007) and the inactivation of HIV by copper ions (Sagripanti and Lightfoote, 1996) and copper oxide (Borkow and Gabbay, 2005) have been reported.

The studies using copper alloys as antiviral surfaces against the 2009 pandemic H1N1 influenza virus (CuI NPs) (Fujimori et al., 2012), GI.1 Norwalk virus–like particles (Au/CuS core/shell NPs) (Broglie et al., 2015) and nonenveloped feline calicivirus (CuI NPs) (Shionoiri et al., 2012) suggested that the Cu component may be a key factor in controlling the antiviral activity leading to capsid protein degradation and the loss of viral RNA. So far, no antiviral activity of Cu NP alloys against DNA viruses has been shown.

3.4 ANTIVIRAL ACTIVITIES OF TANNIC ACID–MODIFIED METALLIC NANOPARTICLES COLLOIDS

Tannins are polyphenolic derivatives naturally synthesized by plants as a metabolic product. For centuries tannins have been used in medicine as agents supporting wound healing, due to its antipathogen properties resulting from the precipitation of surface proteins. In plants, tannins act as protecting agents against pathogens, such as bacteria, viruses, and fungi, as well as against other organisms, particularly insects.

Tannin toxicity for fungi, bacteria, and yeasts has been known for many years (Buzzini et al., 2008). Different mechanisms of tannin antimicrobial activity have been proposed: inhibition of extracellular microbial enzymes, deprivation of the substrates required for microbial growth, or direct action on microbial metabolism through inhibition of oxidative phosphorylation. The antimicrobial toxicity of tannins have been shown against, among others, *S. aureus*, *Clostridium perfringens*, *Salmonella typhimurium*, *Bacillus subtilis*, *E. coli* (Akiyama et al., 2001; Buzzini et al., 2008; Chung et al., 1998; Payne et al., 2013).

Lin et al. (2011) showed that hydrolysable tannins isolated from fruits of *Terminalia chebula* Retz. interact with HSV-1 glycoproteins to prevent virus attachment, entry, and cell-to-cell spread. The authors identified chebulagic acid (CHLA) and punicalagin (PUG), two hydrolysable tannins, which mediated their antiviral activities by targeting HSV-1 viral glycoproteins that interact with cell surface heparin sulphate (Lin et al., 2011). Later, Lin et al. (2013) explored the antiviral potential of these two tannins against a number of viruses known to interact with glycosaminoglycans, such as heparan sulfate. Viral models included human cytomegalovirus (HCMV), hepatitis C virus (HCV), dengue virus (DENV), measles virus (MV), and respiratory syncytial virus (RSV). CHLA and PUG were effective in abrogating infection in dose-dependent manners without significant cytotoxicity (Lin et al., 2013). Several tannins of various sources were also described to inhibit HIV-1 infection (Bedoya et al., 2010; Martino et al., 2004; Notka et al., 2004).

Identification of multiple drugs that can act on different phases of the viral life cycle can be particularly useful in managing HSV-1 and -2 infection or reactivation in either immunocompromised individuals or cases of acyclovir resistance. So far, there have not been any studies showing the use of silver NPs functionalized with tannic acid.

In our previous work (Orlowski et al., 2014), we demonstrated that tannic acid modified silver NPs sized 13, 33, and 46 nm are capable of reducing HSV-2 infectivity both in vitro and in vivo. The antiviral properties of AgNPs were tested in mouse keratinocyte 291.03C cell culture by standard plaque forming assay (PFU/mL). Incubation of HSV-2 with different sizes and concentrations of AgNPs for 60 min prior to infection showed a dose-dependent virus inactivation by tannic acid–modified AgNPs sized 13, 33, and 46 nm and much less efficient virus inhibition by unmodified AgNPs sized 10–65 nm (Orlowski et al., 2014). Of all tested sizes, 33 nm showed the highest antiviral activity already at 0.5 and 1 μg/mL (Orlowski et al., 2014).

The antiviral activity of tannic acid–modified silver NPs also required direct interaction and blocked virus attachment, penetration, and further spread, albeit inhibition of attachment was clearly size-dependent with larger sizes being more effective in blocking the HSV-2 attachment to the host cells (Orlowski et al., 2014). Although we tested tannic acid-modified AuNPs of similar sizes (13, 20, 30, and 40 nm) for possible antiviral effect against HSV-2, we found no significant antiviral activity (data not published).

Therefore, we can assume that the antiviral activities of tannic acid–modified NPs are the effect of the combined activities of silver and tannic acid. The surface of HSV-2 is composed of glycoproteins with N- or O-linked sugars (Wald, 1998), which can readily interact with tannic acid moieties on the surface of NPs and help the nanosilver to interact with the virus envelope and destroy it (Fig. 12.12). By binding to virions, NPs also play a role of a physical obstacle, making virion-receptor interaction difficult (Fig. 12.12).

Taking into account promising antiviral properties of copper alloys we tested antiviral activities of tannic acid modified silver–copper alloys in the form of 29 and 63 nm NPs and compared with the tannic acid–modified AgNPs of similar sizes in 291.03C mouse keratinocyte cell cultures. We found that tannic acid–modified 29 nm Ag/CuNPs at 5 μg/mL inhibited HSV-2 333 strain adhesion to the host cells much better than tannic acid–modified 33 nm AgNPs ($P \leq 0.001$) (Fig. 12.13A). Similarly, Ag/CuNPs of both tested sizes inhibited cell penetration significantly better that AgNPs of the similar sizes ($P \leq 0.001$) (Fig. 12.13B).

To confirm our in vitro findings, we used the well-established murine model of HSV-2 genital infection (Orlowski et al., 2014). For in vivo experiments, we incubated HSV-2 for 1 h prior to infection with 5 μg/mL of tannic acid–modified 13, 33, and 46 nm AgNPs, then the mixtures were used for intravaginal infection of C57BL6 mice. At the peak of local vaginal infection (48 h p.i.), we observed that incubation of HSV-2 with all tested AgNPs significantly reduced viral titers measured by real time PCR in comparison to HSV-2 infected control (Orlowski et al., 2014). We also assessed the potential use of AgNPs as an antiviral agent in animals with ongoing intravaginal

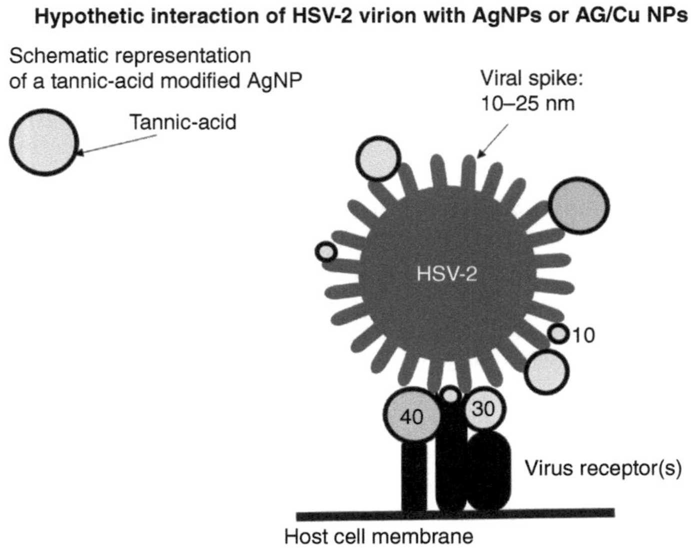

FIGURE 12.12 Schematic Representation of Interaction Between Tannic Acid–Modified AgNPs or Ag/CuNPs and HSV-2 Virion

HSV-2 infection. The animals were infected with untreated HSV-2 and then the vaginal tissues were irrigated 3 times with 33 nm AgNPs at 3 or 18 h p.i. (Orlowski et al., 2014). Postinfection treatment with 33 nm AgNPs resulted in a significant decrease of the viral titers for both time points (Orlowski et al., 2014). Following results obtained for Ag-Cu alloys in vitro, we repeated in vivo experiments, using tannic acid modified 29 and 63 nm Ag/CuNPs for preincubation with HSV-2, as described previously, and compared with tannic acid–modified AgNPs of similar sizes (Fig. 12.14). Silver–copper NPs showed the same antiviral activity in vivo as tannic acid–modified AgNPs of the similar sizes (Fig. 12.14).

To test inflammatory properties of tannic acid–modified AgNPs or Ag/CuNPs with or without infection, we performed staining with the antigranulocyte receptor-1 (Gr-1) mAb, RB6-8C5 clone in the vaginal tissues treated once a day for 2 days with 100 μL of tannic acid–modified 10, 33, and 46 nm AgNPs or 30 nm Ag/CuNPs or unmodified 10–69 nm AgNPs at 48 h after start of treatment but also in mice infected with HSV-2 treated with modified and unmodified AgNPs or Ag/CuNPs, as in the infectious experiments described earlier. Gr-1 antibody binds to Ly6G, which is present on neutrophils, and to Ly6C, which is expressed on neutrophils, and subpopulations of monocytes (Ly6G+ and Ly6C+). Both cells populations actively participate in early inflammation during HSV-2 infection (Krzyzowska et al., 2014). We observed that all tested AgNPs and Ag/CuNPs did not induce significant inflammatory reaction when applied to uninfected mice at 100 μg/mL (Fig. 12.15A). Interestingly, all tested AgNPs and Ag/CuNPs significantly decreased the numbers of Gr-1-positive cells (neutrophils and inflammatory monocytes) within the vaginal tissue during HSV-2 infection at 48 h (Fig. 12.15B).

We have previously described that monocytes exposed in vitro to tannic acid–modified AgNPs strongly down-regulated production of TNF-α, while keratinocytes showed an opposite effect of

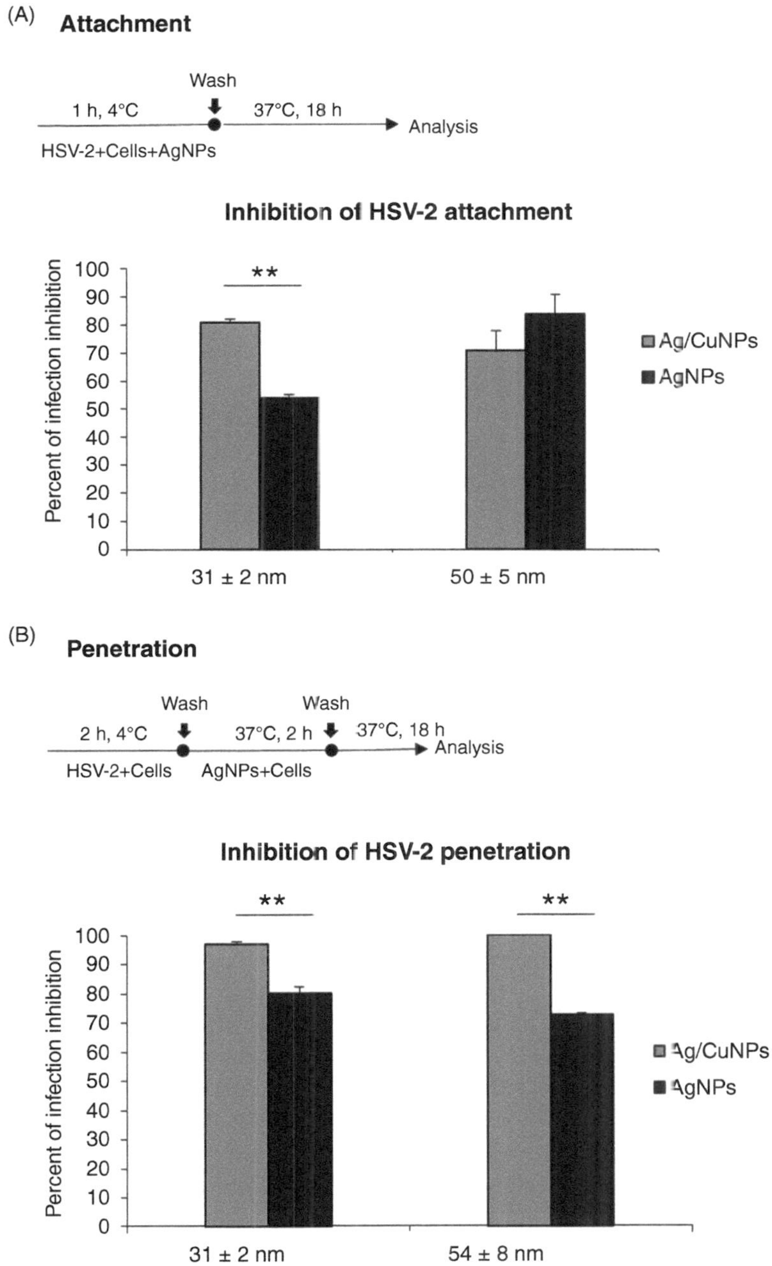

FIGURE 12.13

Tannic Acid–Modified AgNPs and Ag/CuNPs Block HSV-2 Attachment (A) and Penetration (B). Viral inhibition (%) for virus attachment and penetration experiments in 291.03C cell cultures with the use of 31 ± 2 nm and 54 ± 8 nm AgNPs or Ag/CuNPs (5 μg/mL). At 24 h p.i. cells and supernatants were collected and titrated to determine PFU/mL in comparison to HSV-2 infected cultures with standard virological method. The data are shown as means from three independent experiments ± SEM. *Asterisk* represents significant differences with $P \leq 0.001$.

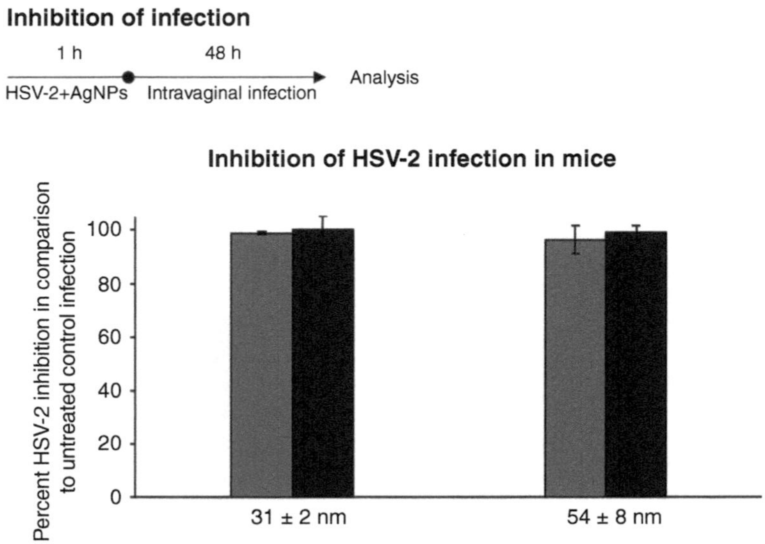

FIGURE 12.14 Tannic-Acid Modified AgNPs and Ag/CuNPs Reduce HSV-2 Infection In Vivo

C57BL/6 mice were infected with HSV-2 preincubated or not with 33 and 46 nm AgNPs or 29 and 63 Ag/CuNPs (5 μg/mL). Viral inhibition (%) for virus infection in comparison to HSV-2 infected control measured as HSV-2 DNA titers (copies/μg DNA) in the whole vaginal tissues determined by real-time PCR at 48 h p.i.

increased TNF-α production (Orlowski et al., 2013). Therefore, we must take into account the fact that depending on size, toxic effects, and cell type, silver NPs influence inflammatory response in a different manner. In our previous paper (Orlowski et al., 2014), cytokine and chemokine production during HSV-2 infection showed size-related differences for each NP-type presence of 13 nm AgNPs at the time of infection led to production of CCL2, IFN-γ, and IL-10, while 33 nm AgNPs induced CCL2, TNF-α, and IL-10. Larger-sized AgNPs, 46 nm, influenced production of only IL-6. The time of postinfection treatment also influenced the profile of produced cytokines.

The early anti-HSV-2 response in vivo involves a plethora of cells, such as neutrophils, macrophages, monocytes, natural killer cells, and dendritic cells, which upon infection produce different cytokines and chemokines. The balance of the cytokine network can regulate the enhancement or inhibition of the inflammatory and adaptive immune responses. As mentioned earlier, the influence of tannic acid–modified AgNPs or Ag/CuNPs may be critical but also a complex issue, therefore, it needs further studies performed both in different cell types in vitro and in vivo, both in control conditions and during viral infection. However, taking into account our in vivo results, we can presume that tannic acid–modified silver and silver–copper alloy NPs are good candidates for NPs combining antiviral activity with a lack of pro-inflammatory potential.

Since both HSV-1 and HSV-2 possess similar structures of surface proteins, tannic acid–modified AgNPs or Ag/CuNPs can be used not only as a microbicide to treat HSV-2 infections of the anogenital area, but also to treat oral herpes infections in the form of a protective gel or cream to be applied topically.

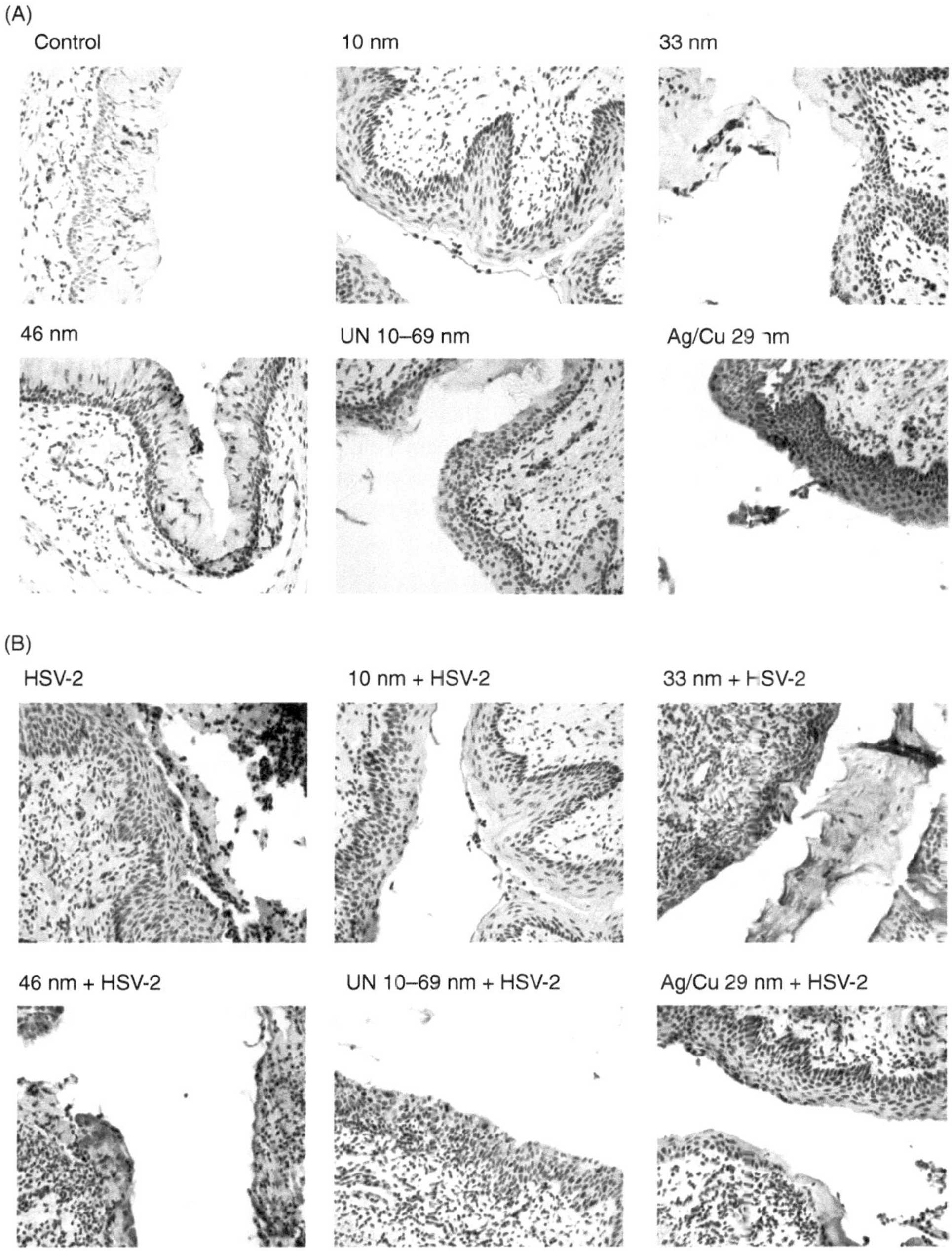

FIGURE 12.15 Tannic Acid–Modified AgNPs Reduce Inflammation During HSV-2 Infection of C57BL/6 Mice

Immunohistochemical staining for Gr-1+ cells in the vaginal tissue of: (A) control mice 48 h after treatment with tannic acid modified 10, 33 and 46 nm AgNPs or 30 nm Ag/CuNPs or unmodified 10–69 nm AgNPs (100 μg/mice); (B) mice at 48 h p.i. with HSV-2 preincubated with tannic acid modified 10, 33, and 46 nm AgNPs or 30 nm Ag/CuNPs or unmodified 10–69 nm AgNPs (5 μg/mice).

4 CONCLUSIONS

Here, we described methods of synthesis, characterization, and toxicity of the metallic NPs modified with tannic acid. Furthermore, we also showed the antiviral effects of tannic acid–modified AgNPs or Ag/CuNPs against HSV-2 both in vitro and in vivo. The preparation of silver, gold, and bimetallic (silver–copper) NPs with specific and strictly defined size and narrow size distribution was possible with the use of tannic acid. The latter was involved in the reduction of silver, gold, and copper ions to metallic NPs, as well as in the stabilization process of the resulting colloidal systems. Toxicity of the obtained NPs depended on the metal and size of the obtained NP. Tannic acid modification did not help to reduce toxicity of small NPs. However, it could help to bind NPs to the cell surface and external receptors, enabling more efficient uptake of NPs. Taking into account the fact that the surface of HSV-2 is built from the eukaryotic cell membrane incrusted with viral glycoproteins, tannic acid modification also helps to bind NPs to the viral surface and blocks further binding of the virion with its membrane receptors. This study indicates that by direct blocking of viral attachment and penetration, tannic acid–modified AgNPs or Ag/CuNPs show good antiviral properties both in vitro and in vivo. At the same time, lack of significant inflammatory reaction in vivo suggests that AgNPs and Ag/CuNPs are safe when applied onto mucosal tissue. Altogether, we show that tannic acid–modified silver NPs are a novel potential microbicide for herpes virus infection in the mucosal tissues. Further studies are required to test the use of AgNPs or Ag/CuNPs as microbicides in the pharmaceutically accepted forms.

ACKNOWLEDGMENTS

This work was supported by National Science Centre grant no. 2014/13/B/NZ5/01356.

REFERENCES

Ahmad, T., 2014. Reviewing the tannic acid mediated synthesis of metal nanoparticles. J. Nanotechnol. 2014, 1–14.

Akhtar, J., Shukla, D., 2009. Viral entry mechanisms: cellular and viral mediators of herpes simplex virus entry. FEBS J. 276, 7228–7836.

Akiyama, H., Fujii, K., Yamasaki, O., Oono, T., Iwatsuki, K., 2001. Antibacterial action of several tannins against *Staphylococcus aureus*. J. Antimicrob. Chemother. 48, 487–491.

Antoine, T.E., Mishra, Y.K., Trigilio, J., Tiwari, V., Adelung, R., Shukla, D., 2012. Prophylactic, therapeutic and neutralizing effects of zinc oxide tetrapod structures against herpes simplex virus type-2 infection. Antiviral Res. 96, 363–375.

Arvizo, R., Bhattacharya, R., Mukherjee, P., 2010. Gold nanoparticles: opportunities and challenges in nanomedicine. Exp. Opin. Drug Deliv. 7, 753–763.

Asharani, P.V., Low Kah Mun, G., Hande, M.P., Valiyaveettil, S., 2009. Cytotoxicity and genotoxicity of silver nanoparticles in human cells. ACS Nano 3, 279–290.

Baram-Pinto, D., Shukla, S., Gedanken, A., Sarid, R., 2010. Inhibition of HSV-1 attachment, entry, and cell-to-cell spread by functionalized multivalent gold nanoparticles. Small 6, 1044–1050.

Baram-Pinto, D., Shukla, S., Perkas, N., Gedanken, A., Sarid, R., 2009. Inhibition of herpes simplex virus type 1 infection by silver nanoparticles capped with mercaptoethane sulfonate. Bioconjug. Chem. 20, 1497–1502.

Bastus, N.G., Comenge, J., Puntes, V., 2011. Kinetically controlled seeded growth synthesis of citrate-stabilized gold nanoparticles of up to 200 nm: size focusing versus Ostwald ripening. Langmuir 27 (17), 11098–11105.

Bedoya, L.M., Abad, M.J., Sanchez-Palomino, S., Alcami, J., Bermejo, P., 2010. Ellagitannins from *Tuberaria lignosa* as entry inhibitors of HIV. Phytomedicine 17, 69–74.

Borkow, G., Gabbay, J., 2005. Copper as a biocidal tool. Curr. Med. Chem. 12, 2163–2175.

Bowman, M.C., Ballard, T.E., Ackerson, C.J., Feldheim, D.L., Margolis, D.M., Melander, C., 2008. Inhibition of HIV fusion with multivalent gold nanoparticles. J. Am. Chem. Soc. 130, 6896–6897.

Broglie, J.J., Alston, B., Yang, C., Ma, L., Adcock, A.F., Chen, W., Yang, L., 2015. Antiviral activity of gold/copper sulfide core/shell nanoparticles against human norovirus virus-like particles. PLoS One 10, e0141050.

Brown, D.M., Johnston, H., Gubbins, E., Stone, V., 2014. Cytotoxicity and cytokine release in rat hepatocytes, C3A cells and macrophages exposed to gold nanoparticles—effect of biological dispersion media or corona. J. Biomed. Nanotechnol. 10, 3416–3429.

Buzzini, P., Arapitsas, P., Goretti, M., Branda, E., Turchetti, B., Pinelli, P., Ieri, F., Romani, A., 2008. Antimicrobial and antiviral activity of hydrolysable tannins. Mini. Rev. Med. Chem. 8, 1179–1187.

Carlson, C., Hussain, S.M., Schrand, A.M., Braydich-Stolle, L.K., Hess, K.L., Jones, R.L., Schlager, J.J., 2008. Unique cellular interaction of silver nanoparticles: size-dependent generation of reactive oxygen species. J. Phys. Chem. B 112, 13608–13619.

Chung, K.T., Lu, Z., Chou, M.W., 1998. Mechanism of inhibition of tannic acid and related compounds on the growth of intestinal bacteria. Food Chem. Toxicol. 36, 1053–1060.

Dadosh, T., 2009. Synthesis of uniform silver nanoparticles with a controllable size. Mater. Lett. 63, 2236–2238.

Elechiguerra, J.L., Burt, J.L., Morones, J.R., Camacho-Bragado, A., Gao, X., Lara, H.H., Yacaman, M.J., 2005. Interaction of silver nanoparticles with HIV-1. J. Nanobiotechnol. 3, 6.

Fox, J., Fidler, S., 2010. Sexual transmission of HIV-1. Antiviral Res. 85, 276–285.

Fraga, S., Faria, H., Soares, M.E., Duarte, J.A., Soares, L., Pereira, E., Costa-Pereira, C., Teixeira, J.P., De Lourdes Bastos, M., Carmo, H., 2013. Influence of the surface coating on the cytotoxicity, genotoxicity and uptake of gold nanoparticles in human HepG2 cells. J. Appl. Toxicol. 33, 1111–1119.

Frost, A.T., 1924. The present-day trend of treatment of gonorrhoea, with special reference to cataphoretic application of colloid silver. Proc. Royal Soc. Med. 17, 1–6.

Fujimori, Y., Sato, T., Hayata, T., Nagao, T., Nakayama, M., Nakayama, T., Sugamata, R., Suzuki, K., 2012. Novel antiviral characteristics of nanosized copper(I) iodide particles showing inactivation activity against 2009 pandemic H1N1 influenza virus. Appl. Environ. Microbiol. 78, 951–955.

Gajbhiye, M., Kesharwani, J., Ingle, A., Gade, A., Rai, M., 2009. Fungus-mediated synthesis of silver nanoparticles and their activity against pathogenic fungi in combination with fluconazole. Nanomedicine 5, 382–386.

Gliga, A.R., Skoglund, S., Wallinder, I.O., Fadeel, B., Karlsson, H.L., 2014. Size-dependent cytotoxicity of silver nanoparticles in human lung cells: the role of cellular uptake, agglomeration and Ag release. Part. Fibre Toxicol. 11, 11.

Gupta, R., Warren, T., Wald, A., 2007. Genital herpes. Lancet 370, 2127–2137.

Halioua, B., Malkin, J.E., 1999. Epidemiology of genital herpes: recent advances. Eur J. Dermatol. 9, 177–184.

Haslam Edwin, 1998. Plant Polyphenols—Vegetable Tannins Revisited. Cambridge University Press, Cambridge

Hoyme, U.B., 1993. Clinical significance of Crede's prophylaxis in Germany at present. Infect. Dis. Obstet Gynecol. 1, 32–36.

Ivask, A., Kurvet, I., Kasemets, K., Blinova, I., Aruoja, V., Suppi, S., Vija, H., Kakinen, A., Titma, T., Heinlaan, M., Visnapuu, M., Koller, D., Kisand, V., Kahru, A., 2014. Size-dependent toxicity of silver nanoparticles to bacteria, yeast, algae, crustaceans and mammalian cells in vitro. PLoS One 9, e102108.

Khanbabaee, K., Van Ree, T., 2001. Tannins: classification and definition. Nat. Prod. Rep. 18, 641–649.

Kruk, T., Szczepanowicz, K., Stefanska, J., Socha, R.P., Warszynski, P., 2015. Synthesis and antimicrobial activity of monodisperse copper nanoparticles. Colloids Surf. B 128, 17–22.

Krzyzowska, M., Baska, P., Grochowska, A., Orlowski, P., Nowak, Z., Winnicka, A., 2014. Fas/FasL pathway participates in resolution of mucosal inflammatory response early during HSV-2 infection. Immunobiology 219, 64–77.

Kumar, S., Gandhi, K.S., Kumar, 2007. Modeling of formation of gold nanoparticles by citrate method. Ind. Eng. Chem. Res. 46, 3128–3136.

Lara, H.H., Ayala-Nunez, N.V., Ixtepan-Turrent, L., Rodriguez-Padilla, C., 2010a. Mode of antiviral action of silver nanoparticles against HIV-1. J. Nanobiotechnol. 8, 1.

Lara, H.H., Garza-Trevino, E.N., Ixtepan-Turrent, L., Singh, D.K., 2011a. Silver nanoparticles are broad-spectrum bactericidal and virucidal compounds. J. Nanobiotechnol. 9, 30.

Lara, H.H., Ixtepan-Turrent, L., Garza-Trevino, E.N., Rodriguez-Padilla, C., 2010b. PVP-coated silver nanoparticles block the transmission of cell-free and cell-associated HIV-1 in human cervical culture. J. Nanobiotechnol. 8, 15.

Lara, H.H., Ixtepan-Turrent, L., Garza Trevino, E.N., Singh, D.K., 2011b. Use of silver nanoparticles increased inhibition of cell-associated HIV-1 infection by neutralizing antibodies developed against HIV-1 envelope proteins. J. Nanobiotechnol. 9, 38.

Lin, L.T., Chen, T.Y., Chung, C.Y., Noyce, R.S., Grindley, T.B., Mccormick, C., Lin, T.C., Wang, G.H., Lin, C.C., Richardson, C.D., 2011. Hydrolyzable tannins (chebulagic acid and punicalagin) target viral glycoprotein-glycosaminoglycan interactions to inhibit herpes simplex virus 1 entry and cell-to-cell spread. J. Virol. 85, 4386–4398.

Lin, L.T., Chen, T.Y., Lin, S.C., Chung, C.Y., Lin, T.C., Wang, G.H., Anderson, R., Lin, C.C., Richardson, C.D., 2013. Broad-spectrum antiviral activity of chebulagic acid and punicalagin against viruses that use glycosaminoglycans for entry. BMC Microbiol. 13, 187.

Liu, W., Wu, Y., Wang, C., Li, H.C., Wang, T., Liao, C.Y., Cui, L., Zhou, Q.F., Yan, B., Jiang, G.B., 2010. Impact of silver nanoparticles on human cells: effect of particle size. Nanotoxicology 4, 319–330.

Martinez-Avila, O., Bedoya, L.M., Marradi, M., Clavel, C., Alcami, J., Penades, S., 2009. Multivalent manno-glyconanoparticles inhibit DC-SIGN-mediated HIV-1 trans-infection of human T cells. ChemBioChem 10, 1806–1809.

Martino, V., Morales, J., Martinez-Irujo, J.J., Font, M., Monge, A., Coussio, J., 2004. Two ellagitannins from the leaves of *Terminalia triflora* with inhibitory activity on HIV-1 reverse transcriptase. Phytother. Res. 18, 667–669.

Miller, A.D., Garcia, J.V., Von Suhr, N., Lynch, C.M., Wilson, C., Eiden, M.V., 1991. Construction and properties of retrovirus packaging cells based on gibbon ape leukemia virus. J. Virol. 65, 2220–2224.

Mishra, Y.K., Adelung, R., Rohl, C., Shukla, D., Spors, F., Tiwari, V., 2011. Virostatic potential of micro-nano filopodia-like ZnO structures against herpes simplex virus-1. Antiviral Res. 92, 305–312.

Mohammed Fayaz, A., Ao, Z., Girilal, M., Chen, L., Xiao, X., Kalaichelvan, P., Yao, X., 2012. Inactivation of microbial infectiousness by silver nanoparticles-coated condom: a new approach to inhibit HIV- and HSV-transmitted infection. Int. J. Nanomed. 7, 5007–5018.

Njoki, P.N., Luo, J., Kamundi, M.M., Lim, S., Zhong, C.-J., 2010. Aggregative growth in the size-controlled growth of monodispersed gold nanoparticles. Langmuir 26 (16), 13622–13629.

Notka, F., Meier, G., Wagner, R., 2004. Concerted inhibitory activities of *Phyllanthus amarus* on HIV replication in vitro and ex vivo. Antiviral Res. 64, 93–102.

Noyce, J.O., Michels, H., Keevil, C.W., 2007. Inactivation of influenza A virus on copper versus stainless steel surfaces. Appl. Environ. Microbiol. 73, 2748–2750.

Ojea-Jimenez, I., Romero, F.M., Bastus, N.G., Puntes, V., 2010. Small gold nanoparticles synthesized with sodium citrate and heavy water: insights into the reaction mechanism. J. Phys. Chem. C 114, 1800–1804.

Orlowski, P., Krzyzowska, M., Zdanowski, R., Winnicka, A., Nowakowska, J., Stankiewicz, W., Tomaszewska, E., Celichowski, G., Grobelny, J., 2013. Assessment of in vitro cellular responses of monocytes and keratinocytes to tannic acid modified silver nanoparticles. Toxicol. In Vitro 27, 1798–1808.

Orlowski, P., Tomaszewska, E., Gniadek, M., Baska, P., Nowakowska, J., Sokolowska, J., Nowak, Z., Donten, M., Celichowski, G., Grobelny, J., Krzyzowska, M., 2014. Tannic acid modified silver nanoparticles show antiviral activity in herpes simplex virus type 2 infection. PLoS One 9, e104113.

Pan, Y., Leifert, A., Ruau, D., Neuss, S., Bornemann, J., Schmid, G., Brandau, W., Simon, U., Jahnen-Dechent, W., 2009. Gold nanoparticles of diameter 1.4 nm trigger necrosis by oxidative stress and mitochondrial damage. Small 5, 2067–2076.

Park, E.J., Yi, J., Kim, Y., Choi, K., Park, K., 2010. Silver nanoparticles induce cytotoxicity by a Trojan-horse type mechanism. Toxicol. In Vitro 24, 872–878.

Payne, D.E., Martin, N.R., Parzych, K.R., Rickard, A.H., Underwood, A., Boles, B.R., 2013. Tannic acid inhibits *Staphylococcus aureus* surface colonization in an IsaA-dependent manner. Infect. Immun. 81, 496–504.

Rashaan, Z.M., Krijnen, P., Klamer, R.R., Schipper, I.B., Dekkers, O.M., Breederveld, R.S., 2014. Nonsilver treatment vs. silver sulfadiazine in treatment of partial-thickness burn wounds in children: a systematic review and meta-analysis. Wound Repair Regen. 22, 473–482.

Ryter, S.W., Kim, H.P., Hoetzel, A., Park, J.W., Nakahira, K., Wang, X., Choi, A.M., 2007. Mechanisms of cell death in oxidative stress. Antioxid. Redox. Signal. 9, 49–89.

Sagripanti, J.L., Lightfoote, M.M., 1996. Cupric and ferric ions inactivate HIV. AIDS Res. Hum. Retroviruses 12, 333–337.

Salvioli, S., Ardizzoni, A., Franceschi, C., Cossarizza, A., 1997. JC-1, but not DiOC6(3) or rhodamine 123 is a reliable fluorescent probe to assess delta psi changes in intact cells: implications for studies on mitochondrial functionality during apoptosis. FEBS Lett. 411, 77–82.

Samberg, M.E., Oldenburg, S.J., Monteiro-Riviere, N A., 2010. Evaluation of silver nanoparticle toxicity in skin in vivo and keratinocytes in vitro. Environ. Health Perspect. 118, 407–413.

Schuitemaker, H., Kootstra, N.A., Groenink, M., De Goede, R.E., Miedema, F., Tersmette, M., 1992. Differential tropism of clinical HIV-1 isolates for primary monocytes and promonocytic cell lines. AIDS Res. Hum. Retroviruses 8, 1679–1682.

Shenashen, M.A., El-Safty, S.A., Elshehy, E.A., 2014 Synthesis, morphological control, and properties of silver nanoparticles in potential applications. Part. Part. Syst. Char. 31, 293–316.

Shionoiri, N., Sato, T., Fujimori, Y., Nakayama, T., Nemoto, M., Matsunaga, T., Tanaka, T., 2012. Investigation of the antiviral properties of copper iodide nanoparticles against feline calicivirus. J. Biosci. Bioeng. 113, 580–586.

Singh, A., Ahmed, A., Prasad, K.N., Khanduja, S., Singh, S.K., Srivastava, J.K., Gajbhiye, N.S., 2015. Antibiofilm and membrane-damaging potential of cuprous oxide nanoparticles against *Staphylococcus aureus* with reduced susceptibility to vancomycin. Antimicrob. Agents Chemother. 59, 6882–6890.

Tan, K.S., Cheong, K.Y., 2013. Advances of Ag, Cu, and Ag–Cu alloy nanoparticles synthesized via chemical reduction route. J. Nanopart Res. 15, 1537.

Tobian, A.A., Quinn, T.C., 2009. Herpes simplex virus type 2 and syphilis infections with HIV: an evolving synergy in transmission and prevention. Curr. Opin. HIV AIDS 4, 294–299.

Tomaszewska, E., Soliwoda, K., Kadziola, K., Tkacz-Szczesna, B., Celichowski, G., Cichomski, M., Szmaja, W., Grobelny, J., 2013. Detection limits of DLS and UV-Vis spectroscopy in characterization of polydisperse nanoparticles colloids. J. Nanomaterials 2013, 1–10.

Vijayakumar, S., Ganesan, S., 2012. Gold nanoparticles as an HIV entry inhibitor. Curr. HIV Res. 10, 643–646.

Wald, A., 1998. Herpes. Transmission and viral shedding. Dermatol. Clin. 16, 795–797, xiv.

Wang, Y., Zi, X.Y., Su, J., Zhang, H.X., Zhang, X.R., Zhu, H.Y., Li, J.X., Yin, M., Yang, F., Hu, Y.P., 2012. Cuprous oxide nanoparticles selectively induce apoptosis of tumor cells. Int. J. Nanomed. 7, 2641–2652.

World Health Organization, 2015. WHO Model of Essential Medicines. Available from: http://www.who.int/medicines/publications/essentialmedicines/en/index.html

Wuithschick, M., Paul, B., Bienert, R., Sarfraz, A., Vainio, U., Sztucki, M., Kraehnert, R., Strasser, P., Rademann, K., Emmerling, F., Polte, J., 2013. Size-controlled synthesis of colloidal silver nanoparticles based on mechanistic understanding. Chem. Mater. 25, 4679–4689.

Yang, Z., Liu, Z.W., Allaker, R.P., Reip, P., Oxford, J., Ahmad, Z., Ren, G., 2010. A review of nanoparticle functionality and toxicity on the central nervous system. J. Royal Soc. Interf. 7 (Suppl 4), S411–S422.

Zhao, P., Li, N., Astruc, D., 2013. State of the art in gold nanoparticle synthesis. Coord. Chem. Rev. 257, 638–665.

CHAPTER

13

POLYELECTROLYTE-DRUG IONIC COMPLEXES AS NANOSTRUCTURED DRUG CARRIERS TO DESIGN SOLID AND LIQUID ORAL DELIVERY SYSTEMS

María E. Olivera*, Ruben H. Manzo*, Fabiana Alovero*, Alvaro F. Jimenez-Kairuz*, María V. Ramírez-Rigo**

**National University of Cordoba (UNC), Córdoba, Argentina*

***National University of the South, Bahía Blanca, Argentina*

CHAPTER OUTLINE

1 **Introduction** 366
2 **Pharmaceutical Polyelectrolytes** 366
3 **Polyelectrolyte-Drug Complexes in Aqueous Dispersion** 371
3.1 Polyelectrolyte-Drug Counterionic Condensation 371
3.2 Species Distribution in Polyelectrolyte-Drug Complexes 373
3.3 Affinity Constants for Counterionic Condensation 374
3.4 Electrokinetic Potentials and Hydrodynamic Diameters of Polyelectrolyte-Drug Complexes 375
3.5 Rheological Properties of Polyelectrolyte-Drug Complexes 377
3.6 Pharmaceutical Relevance of the Polyelectrolyte-Drug Interactions 378
3.7 Interaction with Biological Membranes 385
3.8 A Model to Interpret the Properties of PE-D Complexes 385
3.9 Summary of Technological and Biopharmaceutical Properties That Determine the Potential Applications of PE-D Dispersions 387
4 **Polyelectrolyte-Drug Complexes in the Solid State** 388
4.1 Wetting of Particulate and Compacted PE-D Complexes 389
4.2 Carbomer-D Matrices 389
4.3 Carboxymethylcellulose-D Matrices 398
4.4 Carrageenan-D Matrices 399
4.5 EE-D Matrices 400
5 **Conclusions** 402
References 405

1 INTRODUCTION

This chapter deals with the description of main pharmaceutical applications of drug delivery systems (DDS) based on the unique properties of complexes of polyelectrolytes (PE) with ionizable drugs (D). It contains the contributions of recent years in the area of PE-D and covers molecular aspects relevant to know the factors that govern their behavior, as well as the potentialities in oral pharmacotherapy of the systems obtained both, in dispersion and solid forms. Interpolyelectrolyte complexes are not covered by this chapter.

PE are defined as hydrophilic polymers containing ionizable groups, which can dissociate generating charged polymer chains (macroions) and small counterions when dispersed in a polar solvent (Barrat and Meyer, 1996; Dobrynin and Rubinstein, 2005).

In the field of pharmacy and pharmaceutical technology, some PE have been used empirically, based on properties that are common to other macromolecules. However, in the past decades, several applications emerged related to their specific properties (Guo et al., 1998; Hugerth, 2001; Radeva, 2001; Thünemann et al., 2004). In this way, it is possible to find early applications of the PE-D systems (Bonferoni et al., 1998; Holgado et al., 1995; Lu et al., 1991). However, a systematic study of their properties was not available, and many aspects of their behavior were recently addressed.

In recent years interesting revisions about some properties of PE and PE-D complexes have been published (De Robertis et al., 2015; Lankalapalli and Kolapalli, 2009; Manzo et al., 2014; Müller, 2014).

The interaction of water-soluble PE with inorganic or organic counterions, such as ionizable D usually generates stable colloidal dispersions in an aqueous medium. In such dispersions, a high proportion of D is electrostatically attached to the PE, forming PE-D complexes that behave as drug carrier systems. In some cases, the PE-D interaction can also originate insoluble products. Besides, solid-state PE-D complexes are prepared by several methods and exhibit interesting properties for the development of oral DDS.

Ideally, an oral DDS should protect D from degradation along the gastrointestinal tract and allow its release in a specific area of the gut where it is best absorbed or where it has to exert its activity. In this context, PE-D systems can help meet the above requirements and can also interact favorably with gastrointestinal tract membranes, promoting the absorption of D or protecting the gut from damage action of D.

2 PHARMACEUTICAL POLYELECTROLYTES

There is an interesting number of PE approved for use as excipients in pharmaceutical formulations (Table 13.1).

The ionizable moiety of most of anionic PE is a carboxylic group. Such is the case of carbomers (C), also known as carbopol or carboxyvinyl polymers, alginic acid (AA), carboxymethylcellulose (CMC), and hyaluronic acid (HA), also known as hyaluronan. Other anionic PE, such as dextran sulfate, cellulose sulfate, carrageenans (CG), and nucleic acids are characterized by the presence of sulfate of phosphate ionizable groups.

Among cationic PE, the nitrogen atom is the protonable center. Eudragit E100 (EE), chitosan and its derivative trimethyl chitosan (a PE having a quaternary ammonium as a permanent cationic group) are representative members of this group.

Table 13.1 Some Polyelectrolytes Currently Used as Pharmaceutical Excipients

Substance	Structure	Functional Category
(a) Anionic PE		
Acrylic acid derivatives		
Carbomer	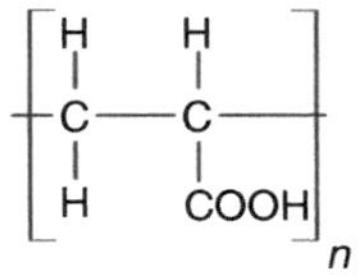	Bioadhesive substance; release-modifying agent; emulsion stabilizer; suspending agent; tablet binder
Polycarbophil	Acrylic acid polymer crosslinked with divinyl glycol	Bioadhesive substance; release-modifying agent; tablet binder; suspending agent
Anionic polymethacrylates		
Eudragit (-S, -L)	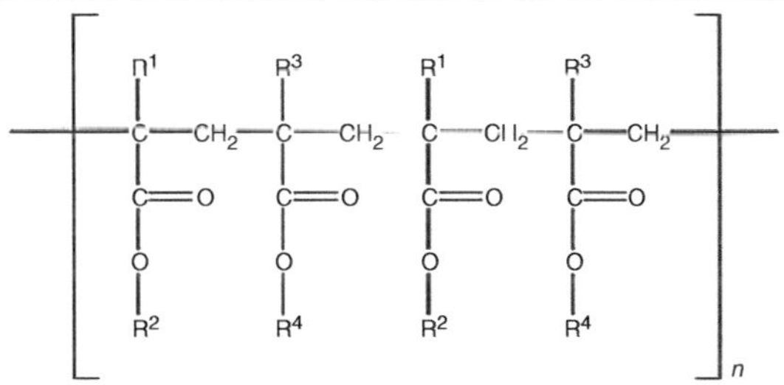 For Eudragit L and S: R^1, $R^3 = CH_3$ $R^2 = H$ $R^4 = CH_3$	Coating agent; tablet binder or diluents
Anionic polysaccharides		
Alginic Acid	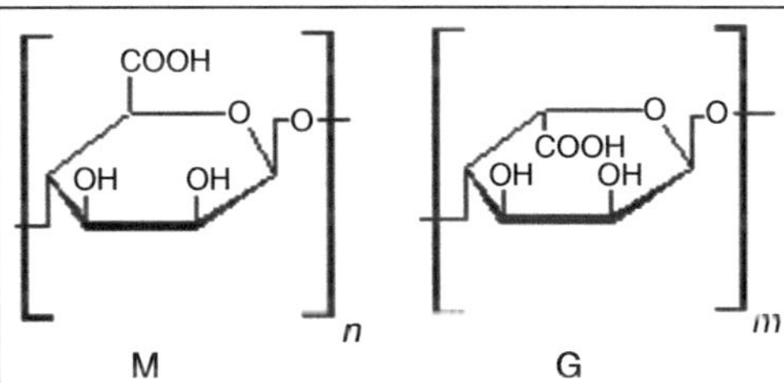 Mannuronic (M) and guluronic (G) units	Release-modifying agent; tablet binder; tablet disintegrant; viscosity-increasing agent

(*Continued*)

Table 13.1 Some Polyelectrolytes Currently Used as Pharmaceutical Excipients (*cont.*)

Substance	Structure	Functional Category
Carboxymethyl-cellulose	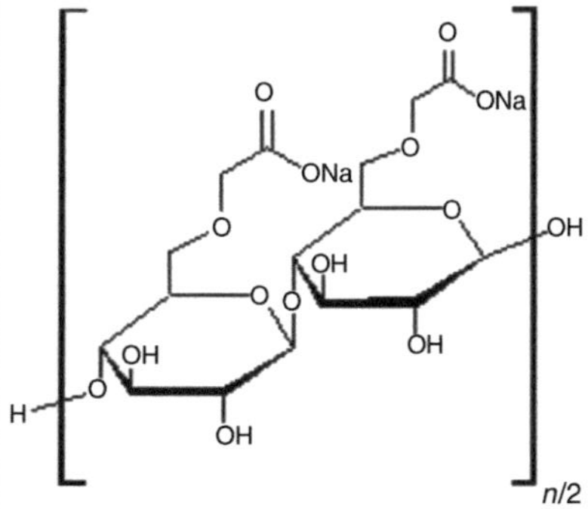 For example, Sodium carboxymethylcellulose with a substitution degree of 1.0	Stabilizing agent; tablet and capsule disintegrant; viscosity-increasing agent; release-modifying agent
Carboxymethyl-starch	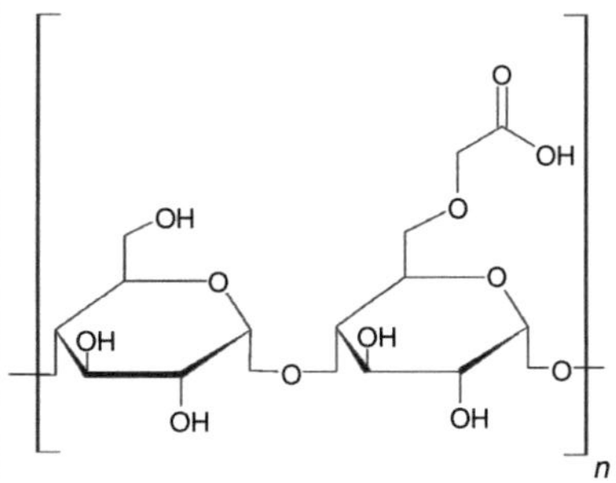	Tablet and capsule disintegrant
Hyaluronic acid	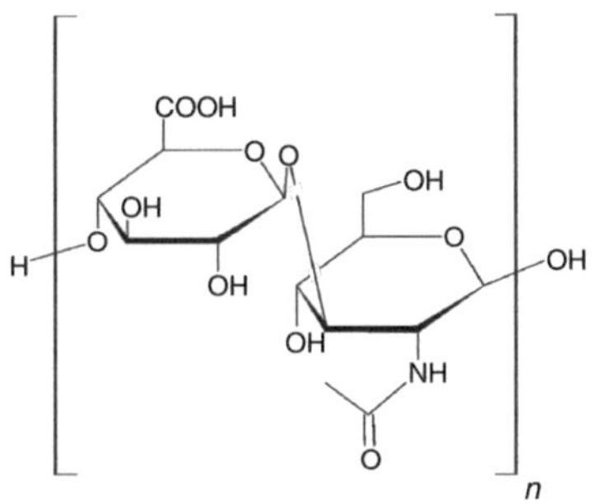	Humectant; lubricant; release-modifying agent; viscosity-increasing agent

Carrageenan (λ^-, κ^-, ι^-)	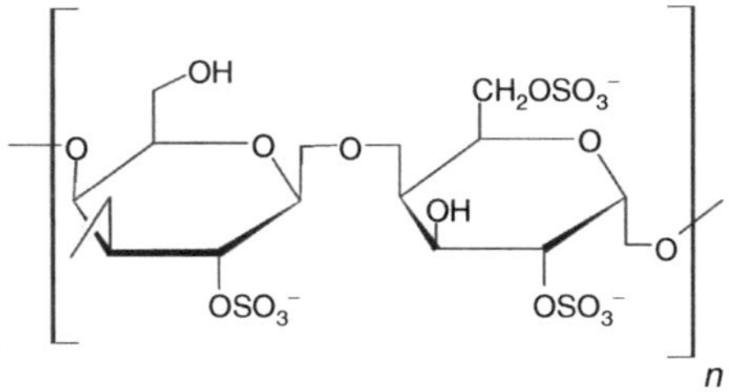 For example, Lambda carrageenan	Emulsifying agent; gel base; release-modifying agent; viscosity-increasing agent; suspending agent
Dextran sulfate	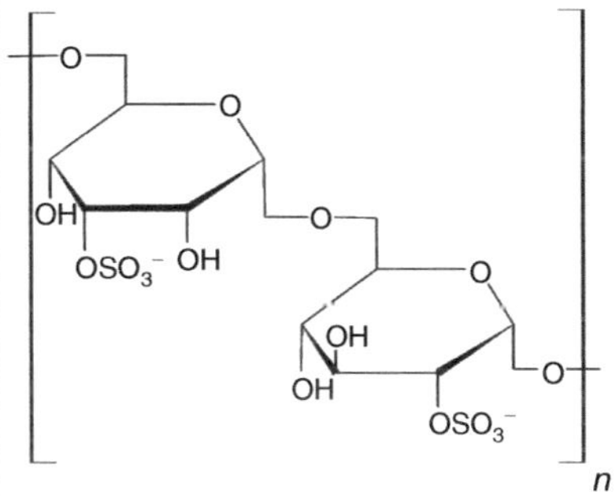	Encapsulation agent; drug delivery agent
Cellulose sulfate	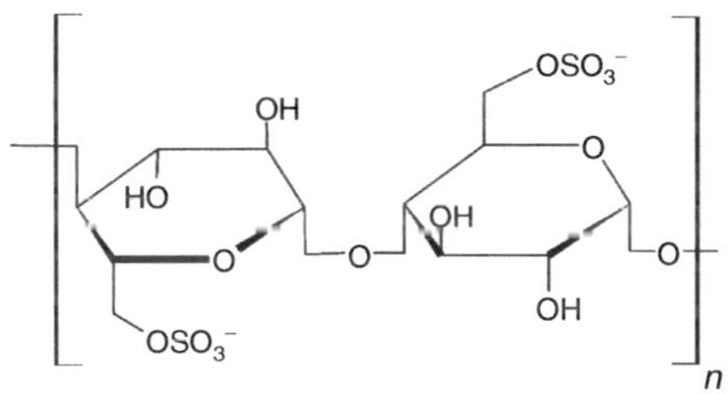	Encapsulation agent; drug delivery agent

(*Continued*)

Table 13.1 Some Polyelectrolytes Currently Used as Pharmaceutical Excipients (*cont.*)

Substance	Structure	Functional Category
(b) Cationic PE		
Cationic polymethacrylates		
Eudragit (-E, -RL, -RS)	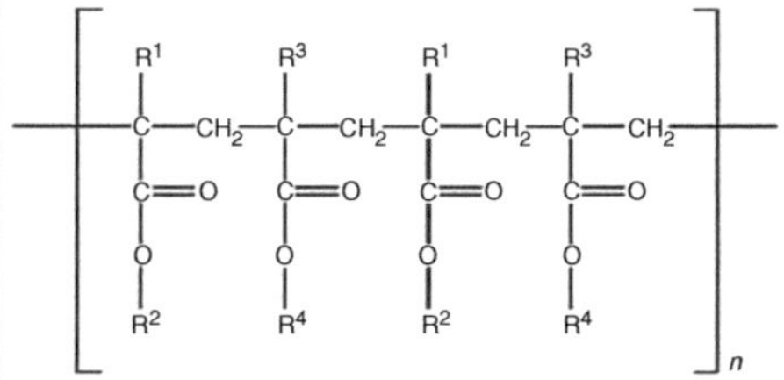 For Eudragit E: R^1, $R^3 = CH_3$ $R^2 = CH_2CH_2N(CH_3)_2$ $R^4 = CH_3$, C_4H_9 For Eudragit RL and RS: $R^1 = H$, CH_3 $R^2 = CH_3$, C_2H_5 $R^3 = CH_3$ $R^4 = CH_2CH_2N(CH_3)_3^+Cl^-$	Coating agent; tablet binder or diluents
Polysaccharides		
Chitosan	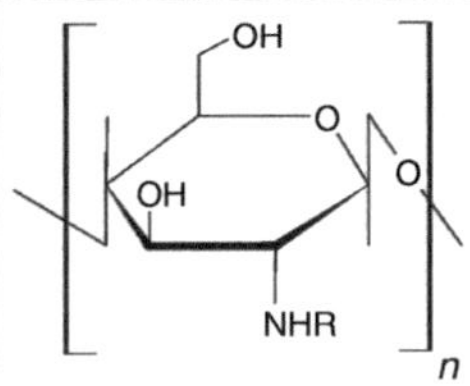 $R = H$ or $COCH_3$	Coating agent; disintegrant; mucoadhesive; permeability enhancer; viscosity-increasing agent
Trimethyl chitosan	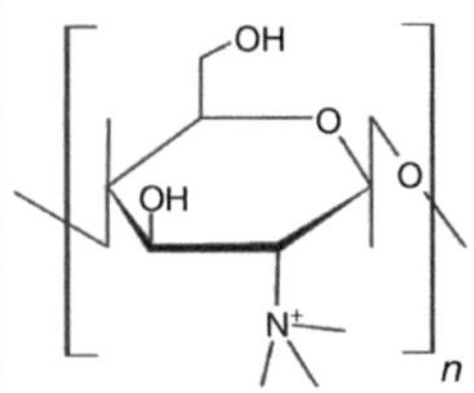	Permeability enhancer; drug delivery agent

From: Mourya, V.K., Inamdar, N.N., 2009. Trimethyl chitosan and its applications in drug delivery. J. Mater. Sci. Mater. Med. 20(5), 1057–1079; Rowe, R.C., Sheskey, P.J., Quinn, M.E., 2009. Handbook of Pharmaceutical Excipients. 6th ed, Pharmaceutical Press and American Pharmacists Association, USA.

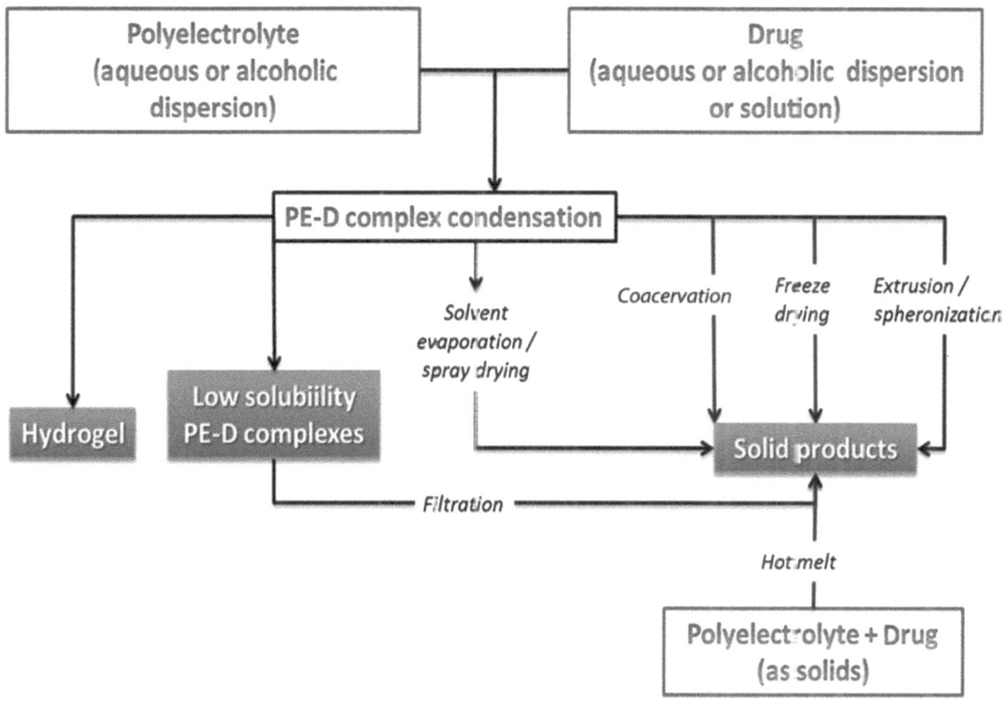

FIGURE 13.1 Procedures Used to Obtain Polyelectrolyte-Drug Complexes

Fig. 13.1 shows the procedures for obtaining PE-D complexes by acid-base reaction. PE-D complexes can be obtained both, as aqueous dispersions (hydrogels) or in solid state (García et al., 2015; Guzmán et al., 2012; Jimenez-Kairuz et al., 2003; Kindermann et al., 2011; Quinteros, 2008). Solid state complexes are characterized by an ionic interaction between the PE and the D. In many cases, they spontaneously revert to the original nanosized colloidal dispersion upon contact with an aqueous medium.

3 POLYELECTROLYTE-DRUG COMPLEXES IN AQUEOUS DISPERSION

The study of physical and chemical properties of PE-D complexes in aqueous dispersions and their behavior in fluids and biological systems is of great importance not only to rationalize the development of liquid DDS but also to provide the basis for the development of oral solid DDS.

3.1 POLYELECTROLYTE-DRUG COUNTERIONIC CONDENSATION

The interaction of anionic or cationic PE with model D having basic or acid groups, respectively (Fig. 13.2), yields a high proportion of counterionic condensation and yields stable colloidal dispersions. However, in some cases, the complementary addition of an inorganic counterion contributes to the required degree

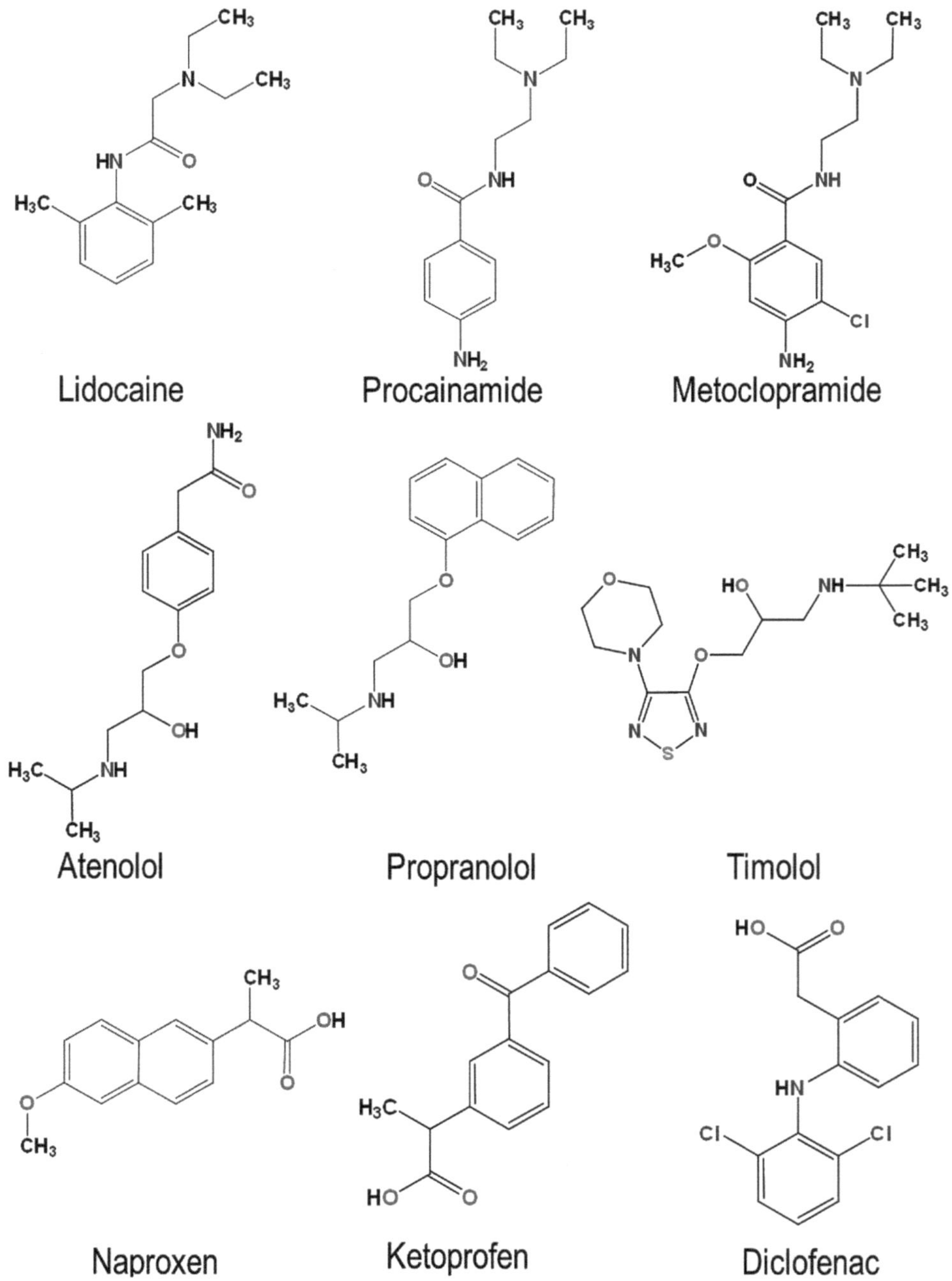

FIGURE 13.2 Molecular Structure of Some Model Drugs Used in Several Studies

of dispersibility. For instance, the interaction of the linear polymethacrylate EE with organic acids yields stable aqueous dispersions when an inorganic anion (e.g., Cl^-) is incorporated in the system.

Eq. (13.1) depicts the reaction between acid groups of a PE (*RAH*) with a basic D (*B*).

$$RAH + B \rightleftarrows RA^- + BH^+ \rightleftarrows [RA^-BH^+] \tag{13.1}$$

Where B and BH^+ are the neutral and protonated species, *RAH* and RA^- represent the undissociated and dissociated fractions of pending groups of the PE and $[RA^- BH^+]$ represents the counterionic condensed fraction.

In the same way, PE having protonable amino groups (RNR_1R_2) can react with an acid D (AH) generating an analog process of counterion condensation (Radeva, 2001), in which *AH* and A^- are the neutral and anionic species, and RNR_1R_2, $RNR_1R_2H^+$, and $[RNR_1R_2H^+A^-]$ have the same meaning of Eq. (13.1).

$$RNR_1R_2 + AH \rightleftarrows RNR_1R_2H^+ + A^- \rightleftarrows RNR_1R_2H^+A^- \tag{13.2}$$

Eqs. (13.1) and (13.2) describe the main interaction between the ionized groups of the macromolecule and its opposite charged partners. Both equations can also take place from salified PE with salts of D. In such cases, the corresponding salts will be formed as the byproduct of the reaction.

The knowledge about the factors that determine the interaction between ionic or ionizable D and PE is relevant in several fields. At present, a detailed description about the factors governing such interaction is not fully available.

The classical description of ion–ion interaction recognizes two relatively stable regions: one referred to as a solvent-separated or loose ion pair and the other referred to as a contact or tight ion pair (Grant and Higuchi, 1990). Similarly, the theory of counterionic condensation with PE recognizes two extreme modes referred to as loose and covalent-like association. The former is the delocalized confinement of the counterions within a condensation volume in the immediate vicinity of the PE, due only to long-range interactions, while the latter is a short-range, site-specific interaction (Benegas et al., 1999; Porasso and Benegas, 2001; Radeva, 2001).

Theoretical treatments mainly address the interaction of acid linear PE with inorganic cations. With organic counterions, nonelectrostatic contributions, such as hydrogen bonding and hydrophobic interactions, among others, would also play a role in the association process (Guzmán et al., 2012). However, the main contribution to the overall interaction arises from the electrostatic attraction.

3.2 SPECIES DISTRIBUTION IN POLYELECTROLYTE-DRUG COMPLEXES

Since D speciation produces the free forms *B* and BH^+ or *AH* and A^- and also ion pairs with the ionizable groups of the PE, $[RA^-BH^+]$ or $[RNR_1R_2H^+A^-]$, the stoichiometric concentration of D, B_{st}, or AH_{st} is distributed as Eqs. (13.3) and (13.4):

$$B_{st} = (B) + (BH^+) + ([RA^-BH^+]) \tag{13.3}$$

$$AH_{st} = (AH) + (A^-) + ([RNR_1R_2H^+A^-]) \tag{13.4}$$

Species distribution has been determined through selective extraction of the neutral species with organic solvent, dialysis, ultrafiltration, or NMR spectroscopy (Ardusso, 2012; Guzmán, 2012; Jimenez-Kairuz et al., 2002, 2003; Jimenez-Kairuz, 2004; Ramírez Rigo, 2006). Typical results of species distribution of model D are shown in Fig. 13.3.

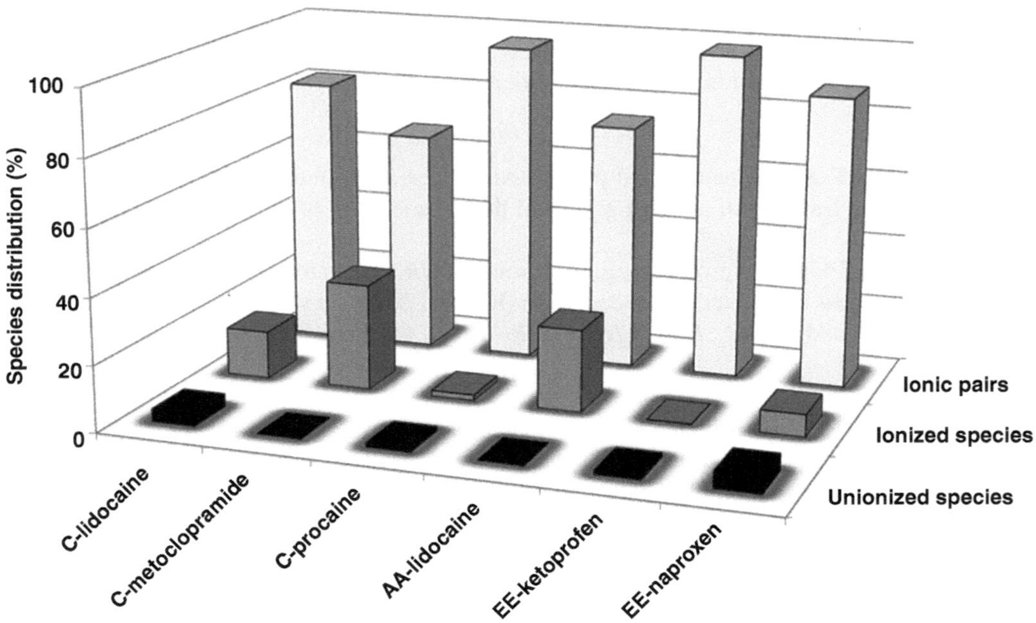

FIGURE 13.3 Distribution of Species from Various PE-D Hydrogels

Adapted from Jimenez Kairuz, A.F., Allemandi, D.A., Manzo, R.H., 2004. The improvement of aqueous chemical stability of a model basic drug by ion pairing with acid groups of polyelectrolytes. Int. J. Pharm. 269(1), 149–156; Ramírez Rigo, M.V., 2006. Preparación y estudio de sistemas portadores de fármacos. Ph D Thesis. Facultad de Ciencias Químicas, Universidad Nacional de Córdoba, Córdoba, Argentina; Quinteros, D.A., 2010. Desarrollo de nuevas estrategias de formulación de fármacos mediante el acomplejamiento con polielectrolitos. Ph D Thesis. Facultad de Ciencias Químicas, Universidad Nacional de Córdoba, Córdoba, Argentina.

An early attempt to get PE-D affinity parameters was based on plotting mgPE/µmol of D against 1/(D) based on data obtained from dialysis equilibrium (Bonferoni et al., 2000). More recently, starting from the data of species distribution at equilibrium, two approaches were reported to render affinity PE-D parameters. One approach based on the correlation of $([RA^{-}BH^{+}])$ against (B) was developed with a set of complexes of C loaded with lidocaine giving a linear isotherm expressed as:

$$([RA^{-}BH^{+}]) = 13.5(B) + 1.9.10^{-4}$$

The second approach is based on the mass law as described by Eqs. (13.1) and (13.2).

3.3 AFFINITY CONSTANTS FOR COUNTERIONIC CONDENSATION

According with Eq. (13.1), the affinity constant for the counterionic condensation K_{cc} for an anionic PE and a basic drug B is expressed as:

$$K_{cc} = ([RA^{-}BH^{+}])/(RAH).(B) = ([RA^{-}BH^{+}])/(RA^{-}).(BH^{+}) \quad (13.5)$$

The following approach was used to solve Eq. (13.5) in an aqueous dispersion:

$$(RA^-)+(HO^-)=(BH^+)+(H^+) \tag{13.6}$$

Under conditions in which $(RA^-) >> (HO^-)$ and $(BH^+) >> (H^+)$, Eq. (13.6) can be reduced to $(RA^-) = (BH^+)$.

The same treatment can be applied to a cationic PE and an acid drug AH through Eqs. (13.7) and (13.8):

$$K_{cc} = ([RNR_1R_2H^+A^-])/(RNR_1R_2)(AH) = ([RNR_1R_2H^+A^-])/(RNR_1R_2H^+)(A^-) \tag{13.7}$$

To solve Eq. (13.7) the approach that equilibrium 8 was essentially shifted to the right was considered.

$$RNR_1R_2 + Cl^- + H^+ \rightleftarrows RNR_1R_2 + RNR_1R_2H^+ + Cl^- \tag{13.8}$$

Therefore $HCl_{st} = (Cl^-)$ and the charge balance is

$$(RNR_1R_2H^+)+(H^+)=(HO^-)+(A^-)+(Cl^-) \tag{13.9}$$

Under conditions in which $(Cl^-) >> (A^-)$, Eq. (13.9) is reduced to $(RNR_1R_2H^+) = (A^-) + (Cl^-)$ or $(RNR_1R_2H^+) = (Cl^-)$ that led to solve Eq. (13.7).

Table 13.2 shows the K_{cc} obtained for a set of PE-D complexes.

3.4 ELECTROKINETIC POTENTIALS AND HYDRODYNAMIC DIAMETERS OF POLYELECTROLYTE-DRUG COMPLEXES

The acid-base interaction described by Eqs. (13.1) and (13.2) originates dispersions with high electrokinetic potentials (ζ) that contribute to the physical stability of the systems. As expected, the dispersions obtained with anionic PE yield a negative ζ while those obtained with cationic PE display a positive ζ, as it is reported in Table 13.3.

Table 13.2 Affinity Constants (K_{cc}) of Several Polyelectrolyte-Drug Complexes

PE-D Complex[a]	PE Concentration (%)	Log K_{cc}	References
EL-lidocaine$_{100}$	1	3.17	Ardusso (2012)
ES-lidocaine$_{100}$	1	3.70	Ardusso (2012)
C-lidocaine$_{100}$	0.1	3.56	Jimenez-Kairuz (2004)
EL-metoclopramide$_{50}$	1	4.55	Ardusso (2012)
ES-metoclopramide$_{50}$	1	4.59	Ardusso (2012)
C-metoclopramide$_{50}$	0.1	4.89	Jimenez-Kairuz (2004)
HANa-lidocaine$_{100}$	0.15	4.36	Battistini (2015)
HANa-procainamide$_{100}$	0.15	6.14	Battistini (2015)
HANa-atenolol$_{100}$	0.15	4.94	Battistini (2015)

[a]The subscripts indicate the percentage of functional groups of the polyelectrolytes neutralized with drug.

Table 13.3 Electrokinetic Potential of Aqueous Dispersions of Complexes at Concentrations of 1.0% of PE

PE-D complex[a]	ζ (mV)	References
HA-atenolol$_{30}$-Na$_{50}$	−22	Battistini (2015)
EL-metoclopramide$_{30}$-Na$_{50}$	−31	Ardusso (2012)
EL-propranolol$_{30}$-Na$_{50}$	−40	Ardusso (2012)
EL-atenolol$_{30}$-Na$_{50}$	−35	Ardusso (2012)
C-ciprofloxacin$_{30}$-Na$_{50}$	−25	Vilches et al. (2002)
C-erythromicin$_{30}$-Na$_{50}$	−67	Jimenez-Kairuz (2004)
C-metoclopramide$_{30}$-Na$_{50}$	−69	Jimenez-Kairuz (2004)
C-ofloxacin$_{30}$-Na$_{50}$	−72	Jimenez-Kairuz (2004)
C-erythromicin$_{50}$	−65	Jimenez-Kairuz (2004)
C-azithromycin$_{50}$	−74	Jimenez-Kairuz (2004)
C-metoclopramide$_{50}$	−67	Jimenez-Kairuz (2004)
EE-ofloxacin$_{50}$-Cl$_{50}$	+23	Quinteros (2010)
EE-indometacin$_{50}$-Cl$_{50}$	+51	Quinteros (2010)
EE-diclofenac$_{50}$-Cl$_{50}$	+38	Quinteros (2010)
EE-flurbiprofen$_{50}$-Cl$_{50}$	+43	Quinteros (2010)
EE-naproxen$_{50}$-Cl$_{50}$	+37	Quinteros (2010)
EE-ketoprofen$_{50}$-Cl$_{50}$	+38	Quinteros (2010)
EE-salicylic acid$_{50}$-Cl$_{50}$	+35	Quinteros (2010)
EE-benzoic acid$_{50}$-Cl$_{50}$	+43	Quinteros (2010)
EE-enalapril$_{30}$-maleate$_{60}$	+28	Ramírez-Rigo et al. (2014)

[a]*The subscripts indicate the percentage of functional groups of the polyelectrolytes neutralized with drug and other counterion.*

The characterization of the surface charge of the PE-D systems, may also be useful to explain other interesting phenomena as their interaction with biological membranes, such as the gastrointestinal mucosa or the outer membrane of bacteria. This information is an interesting tool for the rational design of bioadhesive systems to increase intrabacterial D uptake or drive them to a specific absorption site or cellular target.

On the other hand, there is an important amount of reports dealing with dynamic light scattering (DLS) of flexible linear PE with acid pending groups neutralized with monovalent inorganic cations (mainly Na^+) that show a multimodal spectrum of relaxation times (Oosawa, 1997; Radeva, 2001). In particular, two clearly differentiated diffusion modes were identified, one with diffusion coefficient ranging from 10^{-7} to 10^{-5} cm^2/s, and the other ranging from 10^{-9} to 10^{-8} cm^2/s, which were regarded as "fast" and "slow" modes, respectively.

DLS data from aqueous dispersions of two linear flexible polymethacrylic acid derivatives, Eudragit L100 and S100 of average molecular weight approx. 135,000, loaded with lidocaine and atenolol

Table 13.4 pH and Dynamic Light Scattering Data of Eudragit L and Eudragit S Complexes With Model Drugs in Aqueous Dispersions at 1.0% of Polyelectrolyte

Eudragit-Drug$_x$	pH	Diffusion Coefficient (10^9 cm^2/s)	HD (nm)[a]
EL-lidocaine$_{100}$	7.31	3.7 ± 1.1	1334 ± 673
ES-lidocaine$_{100}$	7.50	3.9 ± 0.7	1313 ± 257
EL-atenolol$_{100}$	7.32	8.8 ± 2.4	597 ± 174
ES-atenolol$_{100}$	7.75	5.3 ± 0.1	927 ± 8

[a]*HD; hydrodynamic diameter.*
Adapted from Ardusso, M., 2012. Utilización de materiales portadores polielectrolito-fármaco en el desarrollo de sistemas de liberación de fármacos. Ph D Thesis. Facultad de Ciencias Químicas, Universidad Nacional de Córdoba, Córdoba, Argentina.

as model D were reported (Ardusso, 2012). Under the experimental conditions used there, a unimodal distribution was only observed, rendering diffusion coefficients in the interval of 4 to 9 × 10^{-9} cm^2/s and virtual hydrodynamic diameters of 600–1300 nm (Table 13.4).

Such diffusion coefficients clearly lie within the range of the slow mode early defined. This slow mode of diffusion has been associated to the presence of multichain domains (clusters) with dimensions appreciably exceeding the size of single chains. The origins of these domains as well as the mechanism by which macromolecules of the same charge interact between them are not satisfactorily understood

3.5 RHEOLOGICAL PROPERTIES OF POLYELECTROLYTE-DRUG COMPLEXES

Anionic PE, such as the branched C, CMC, and AA, are currently used on salification with inorganic counterions, for their ability of increasing the viscosity of aqueous dispersion media, acting as viscosifying agents and emulsifiers, among other applications.

In the same way, their complexes with D partially maintain this property exhibiting a rheological behavior typical of this group of anionic macromolecules.

Some of the properties of PE-D dispersions are unique to this group of macromolecules (Manzo et al., 2014):

1. They display non-Newtonian flux (plastic or pseudoplastic).
2. The viscosity of the dispersions increases exponentially with concentration.
3. The larger the molecular weight of a linear PE, the higher the viscosity generated.
4. Branched PE generates higher viscosity than their linear counterparts.
5. PE containing monomeric units of carbohydrates are less sensitive to salt effect than those with hydrocarbon backbone, for which the viscosity of the dispersions is drastically reduced.
6. The dispersions of cationic PE exhibit lower viscosity than anionic PE counterparts.
7. In general, the dispersions of PE-D complexes display lower viscosities than those obtained with inorganic counterions.
8. Complexes of the same PE with different D exhibit a considerable variation in their viscosities. Table 13.5 shows the viscosity of some PE-D aqueous dispersions.

Table 13.5 Apparent Viscosity (25°C) of PE-D Aqueous Dispersions

PE-D Complex	PE Concentration (%)	Shear Rate (s^{-1})	Viscosity (cP)
C-atenolol$_{50}$[a]	0.25	100	666.9
C-procainamide$_{50}$[a]	0.25	100	435.0
C-pilocarpine$_{50}$[a]	0.25	100	438.7
C-lidocaine$_{50}$[a]	0.20	100	199.6
C-metoclopramide$_{50}$[a]	0.25	100	31.58
C-nafazolin$_{50}$[a]	0.25	100	17.03
C-erythromycin$_{50}$[a]	0.25	100	14.29
C-azithromycin$_{50}$[a]	0.25	100	11.99
AA-lidocaine$_{50}$[b]	5.0	50	229.0
AA-atenolol$_{50}$[b]	5.0	50	80.0

[a]*Adapted from Jimenez Kairuz, A.F., Allemandi, D.A., Manzo, R.H., 2004. The improvement of aqueous chemical stability of a model basic drug by ion pairing with acid groups of polyelectrolytes. Int. J. Pharm. 269(1), 149–156.*
[b]*Adapted from Ramírez Rigo, M.V., 2006. Preparación y estudio de sistemas portadores de fármacos. Ph D Thesis. Facultad de Ciencias Químicas, Universidad Nacional de Córdoba, Córdoba, Argentina.*

3.6 PHARMACEUTICAL RELEVANCE OF THE POLYELECTROLYTE-DRUG INTERACTIONS

The PE-D dispersions exhibit a set of properties that could be favorably exploited to increase the compatibility and stability of D, as well as to modulate their release.

In fact, as previously mentioned, the phenomenon of counterionic condensation imparts particular properties to PE-D dispersions, which are relevant for their applications in DDS. Among them, the increase in apparent solubility of poorly soluble D (Baena et al., 2011; Quinteros, 2008, 2010; Vilches et al., 2002; Vilches, 2003) the improvement in chemical stability of D (Esteban et al., 2009; Jimenez-Kairuz, 2004; Zoppi et al., 2012) and the release properties of the D from the complexes are worth to be described.

3.6.1 Increase in apparent solubility of drugs by complexation with polyelectrolytes

Table 13.6 shows the apparent solubility increase of poorly soluble D in aqueous PE-D dispersions, in the range of pH where solubility is determined by the neutral species of the drugs (*B* or *AH*).

The additional introduction of an inorganic ion with a charge of the same sign as the D, for example, Na^+ in anionic PE-D complexes or Cl^- in cationic PE-D ones, generally increases the quality of the dispersions. Thus, the transparency of the dispersions can be increased when an inorganic ion is incorporated (Guzmán, 2012; Quinteros, 2008).

3.6.2 Increase in chemical stability of drugs by complexation with polyelectrolytes

The negative ζ of anionic PE-D complexes provides a microenvironment whose pH is lower than that of the bulk. Moreover, molecules involved in the interaction with the ionized groups of the PE have

Table 13.6 Increased Aqueous Compatibility (25°C) for Acid, Basic, and Zwitterionic Drugs of Low Solubility by Complexation With Polyelectrolytes

PE-D Complex	pH	$[D]_{hydrogel}/[D]_{water}$ [a,b]
C-metoclopramide$_{50}$ [c]	7.6	60
C-erythromycin$_{50}$ [c]	7.6	10
C-norfloxacin$_{30}$-Na_{45} [d]	7.5	7.2
C-ciprofloxacin$_{30}$-Na_{45} [d]	7.0	34
EE-indometacin$_{50}$-Cl_{50} [e]	3.8	318
EE-diclofenac$_{50}$-Cl_{50} [e]	3.7	200
EE-furosemide$_{50}$-Cl_{50} [e]	3.9	121
EE-dexamethasone phosphate$_{25}$-Cl_{75} [f]	2.0	53

[a]*Ratio between D concentration in PE dispersions and D solubility at the same pH.*
[b]*The compatibility limits of drug in PE aqueous dispersions were not reached because the complexes generated a viscous environment that prevents the addition of the solid in order to saturate the system.*
Data were adapted from [c]*Jimenez Kairuz, A.F., Allemandi, D.A., Manzo, R.H., 2004. The improvement of aqueous chemical stability of a model basic drug by ion pairing with acid groups of polyelectrolytes. Int. J. Pharm. 269(1), 149–156;* [d]*Vilches, A.F., Jimenez-Kairuz, A.F., Alovero, F., Olivera, M.E., Allemandi, D.A., Manzo, R.H., 2002. Release kinetics and up-take studies of model fluoroquinolones from carbomer hydrogels. Int. J. Pharm. 246, 17–24;* [e]*Quinteros, D.A., 2010. Desarrollo de nuevas estrategias de formulación de fármacos mediante el acomplejamiento con polielectrolitos. Ph D Thesis. Facultad de Ciencias Químicas, Universidad Nacional de Córdoba, Córdoba, Argentina;* [f]*Guzmán, M.L., 2012. Desarrollo de nuevos sistemas farmacoterapéuticos y evaluación de sus propiedades biofarmacéuticas. PhD Thesis. Facultad de Ciencias Químicas, Universidad Nacional de Córdoba, Córdoba, Argentina.*

less kinetic energy than those that are free in solution. On this basis, it was proposed that the aforementioned factors may increase the chemical stability of D susceptible to specific base catalysis, a degradation mechanism that affects the stability of many D in solution.

Experiments with procaine as a model D showed an increase in stability compared with free procaine, when it was complexed as C-procaine, at a similar pH (Jimenez-Kairuz et al., 2004). Similar results were obtained in C-pilocarpine (Zoppi et al., 2012), C-azithromycin (Esteban et al., 2009), and polyvinylpyrrolidone-poly(acrylic acid)-pilocarpine systems (Abd El-Rehim et al., 2013). Such results are resumed in Table 13.7.

Table 13.7 Stability Increase of Drugs by Complexation With Carbomer

PE-D Complex	pH	Temperature (°C)	$k_{water}/k_{hydrogel}$ [a]
C-procaine$_{100}$	8.5	20	6.1
C-procaine$_{100}$	8.5	40	7.2
C-procaine$_{50}$	7.5	40	4.2
C-azithromycin$_{50}$	8.0	25	28.0
C-pilocarpine$_{100}$	7.0	25	5.4

[a]*Ratio between degradation constants of D in an aqueous dispersion of PE and in water at the same pH.*

3.6.3 Release of drugs from polyelectrolyte-drug complexes

3.6.3.1 The reversibility of the ionic interaction in PE-D complexes

The study of the behavior of PE-D dispersions shows that the equilibrium of counterionic condensation is not affected by the presence of nonelectrolytes. However, as expected, it is progressively displaced by the addition of increasing concentrations of neutral salts (Guzmán et al., 2012; Jimenez-Kairuz et al., 2002, 2003; Quinteros, 2008; Ramírez Rigo, 2006; Vilches et al., 2002).

As can be observed in Fig. 13.4, the addition of increasing percentages of NaCl to an anionic PE-D complex produces a protogenic effect generated by the shift to the right of the following equilibria Eqs. (13.10) and (13.11):

$$[RA^{-}BH^{+}] + \mathrm{NaCl} \rightleftarrows [RA^{-}Na^{+}] + BH^{+} + Cl^{-} \tag{13.10}$$

$$BH^{+} \rightleftarrows B + H^{+} \tag{13.11}$$

In contrast, cationic PE-D complexes generate a proton uptake effect (Fig. 13.5), which can be explained by Eqs. (13.12) and (13.13).

$$[\mathrm{RNR_1R_2H^{+}A^{-}}] + \mathrm{NaCl} \rightleftarrows [\mathrm{RNR_1R_2H^{+}Cl^{-}}] + \mathrm{A^{-}} + \mathrm{Na^{+}} \tag{13.12}$$

$$\mathrm{A^{-}} + \mathrm{H^{+}} \rightleftarrows \mathrm{AH} \tag{13.13}$$

3.6.3.2 Kinetic release of drugs from polyelectrolye–drug complexes

3.6.3.2.1 Polyelectrolye–drug complexes with acid or basic model drugs

The reversibility of the PE–D interactions described in the previous section determines the rate and extent of the release of the loaded D from the complexes in aqueous dispersions.

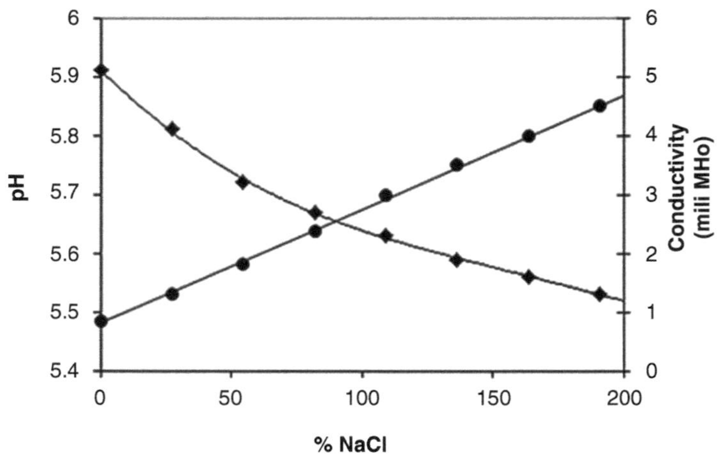

FIGURE 13.4

Effect of addition of increasing amounts of NaCl to aqueous dispersions of AA-lidocaine$_{100}$ on (◆) pH and (●) aonductivity. The content of NaCl is expressed as percentage of the carboxylic groups of AA.

Adapted from Ramírez Rigo, M.V., 2006. Preparación y estudio de sistemas portadores de fármacos. Ph D Thesis. Facultad de Ciencias Químicas, Universidad Nacional de Córdoba, Córdoba, Argentina.

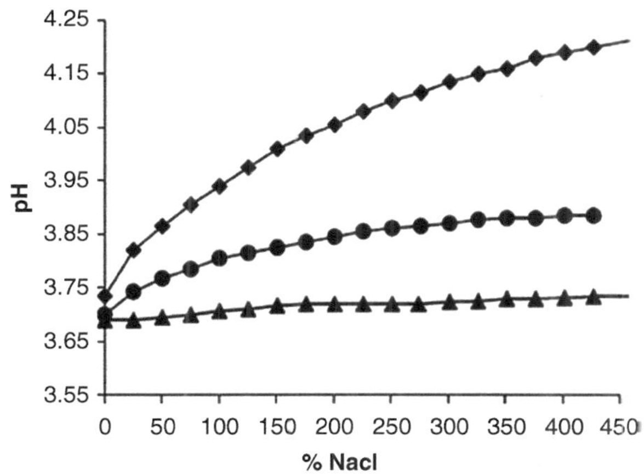

FIGURE 13.5 Effect of the Addition of NaCl to Aqueous Dispersions of Eudragit E100 50% Neutralized with Different Drugs

(◆) EE-diclofenac$_{50}$-Cl$_{50}$, (●) (EE-salicylic acid$_{50}$-Cl$_{50}$, (▲) (EE-Cl$_{50}$) as control. The content of NaCl is expressed as percentage of cationic groups of EE.

Adapted from Quinteros, D.A., 2010. Desarrollo de nuevas estrategias de formulación de fármacos mediante el acomplejamiento con polielectrolitos. Ph D Thesis. Facultad de Ciencias Químicas, Universidad Nacional de Córdoba, Córdoba, Argentina.

In fact, experiments performed in bicompartimental cells limited by a semipermeable membrane that prevents the diffusion of the complex demonstrated that the release of D toward water as receptor medium is mainly produced by the diffusion of the neutral species *B* or *AH* able to freely diffuse since the fraction of free charged ones (A^- or BH^+) are compromised with the electrical gradient of the PE. In such cases Fickian diffusion and a slow release rate are observed, either for anionic or cationic PE (Fig. 13.6).

The diffusion of the neutral species promotes the dissociation of ionic pairs according with the equilibria described in Eqs. (13.1) and (13.2) producing a feedback mechanism. In other words, the fraction of ionic pairs acts as a reservoir of D. It has been shown that the release rate is proportional to such fraction (Fig. 13.7).

As shown in Fig. 13.6 the addition of neutral salts (e.g., NaCl) in the receiving compartment increases the release rate of the D since the diffusion of ions from the receptor compartment toward the dispersion promotes ion exchange with PE-D complex, as was described in Eqs. (13.5)–(13.8). It was observed in EE-benzoic acid complexes that the release rate increases with increasing NaCl concentration (Fig. 13.8).

3.6.3.2.2 Polyelectrolyte–drug complexes with other drugs

Dexamethasone phosphate was used as a model D containing phosphate groups able to interact with EE (Guzmán et al., 2012). The complexes obtained present low solubility, which is increased by the addition of Cl^- as counterion. EE-dexamethasone phosphate-Cl complexes showed an affinity remarkably higher than that observed with drugs containing carboxylic acid groups. Drug delivery in bicompartimental cells was very slow in water and increased as it was replaced by NaCl solution. A sustained release of the D with zero order kinetics was ascribed to the ionic attraction between EE and the phosphate groups of D, which involves the two acid groups of the molecule.

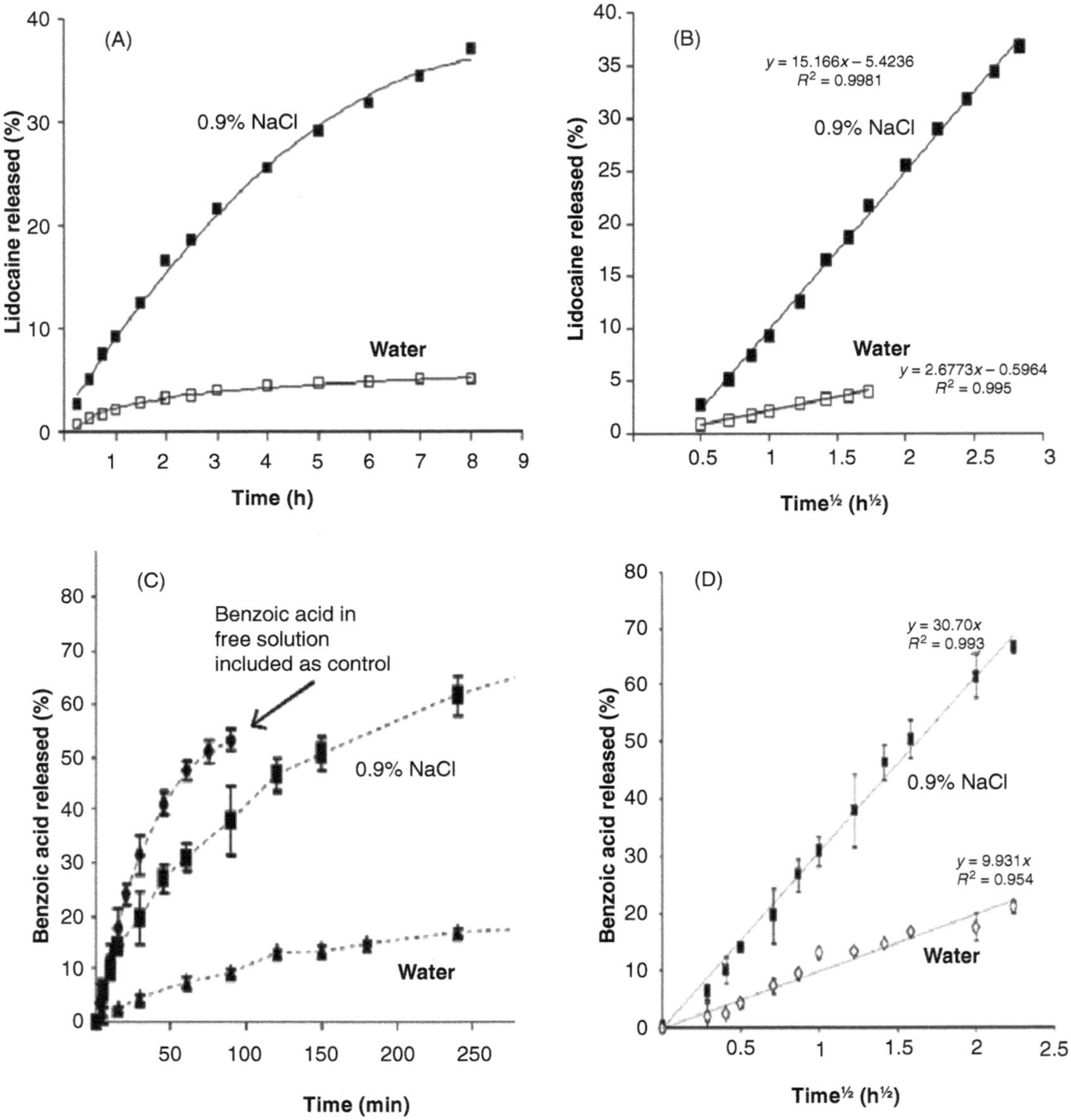

FIGURE 13.6 Release of Model D from Anionic and Cationic Polyelectrolyte-D Complexes Using Water or 0.9% NaCl Solution as Receptor Media

Lidocaine release profile from AA-lidocaine$_{75}$ dispersion (A) and their fitting according to Higuchi's equation (B); Benzoic acid release profiles from EE–benzoic acid$_{50}$ dispersion (C); and their fitting according to Higuchi's equation (D).

Adapted from Ramírez Rigo, M.V., 2006. Preparación y estudio de sistemas portadores de fármacos. Ph D Thesis. Facultad de Ciencias Químicas, Universidad Nacional de Córdoba, Córdoba, Argentina; Quinteros, D.A., 2010. Desarrollo de nuevas estrategias de formulación de fármacos mediante el acomplejamiento con polielectrolitos. Ph D Thesis. Facultad de Ciencias Químicas, Universidad Nacional de Córdoba, Córdoba, Argentina.

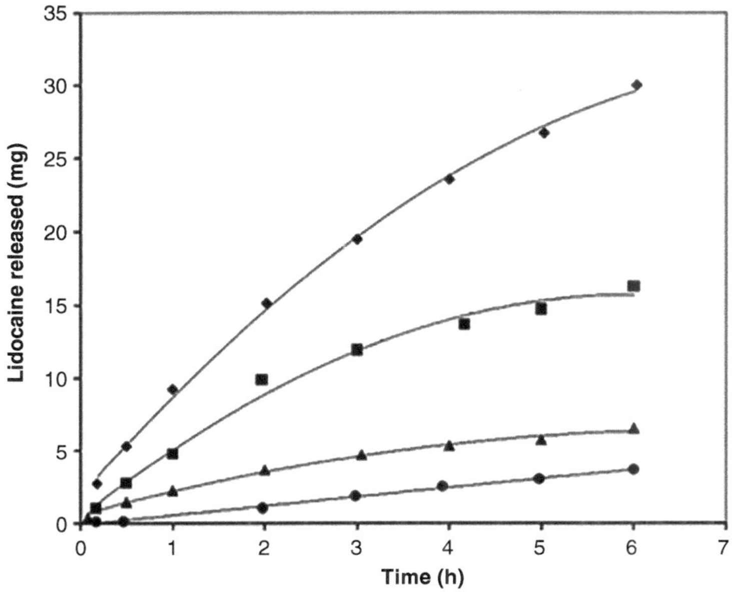

FIGURE 13.7 Lidocaine Release Profiles From Hydrogels of Carbomer in Water

(◆) C-lidocaine$_{100}$, (■) C-lidocaine$_{75}$, (▲) C-lidocaine$_{50}$, (●) C-lidocaine$_{25}$.

Adapted from Jimenez Kairuz, A.F., Allemandi, D.A., Manzo, R H., 2004. The improvement of aqueous chemical stability of a model basic drug by ion pairing with acid groups of polyelectrolytes. Int. J. Pharm. 269(1), 149–156.

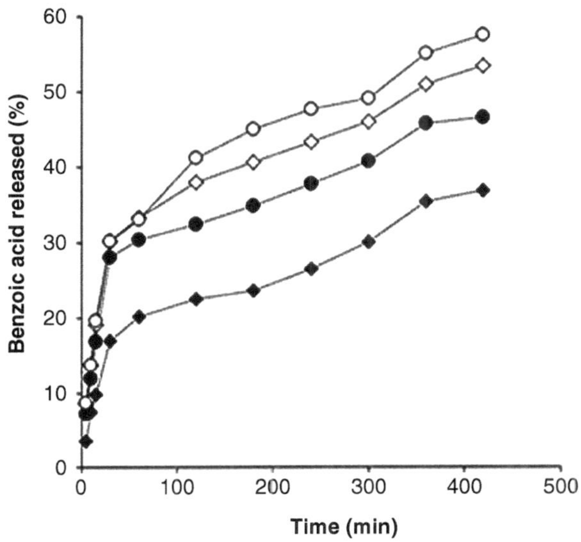

FIGURE 13.8

Release of benzoic acid from ee-benzoic acid$_{50}$-Cl$_{25}$ dispersions added with (◆) 50%, (●) 100%, (◇) 200%, and (○) 300% of NaCl in the donor compartment. NaCl is referred to as the molar percentage of in the complexes.

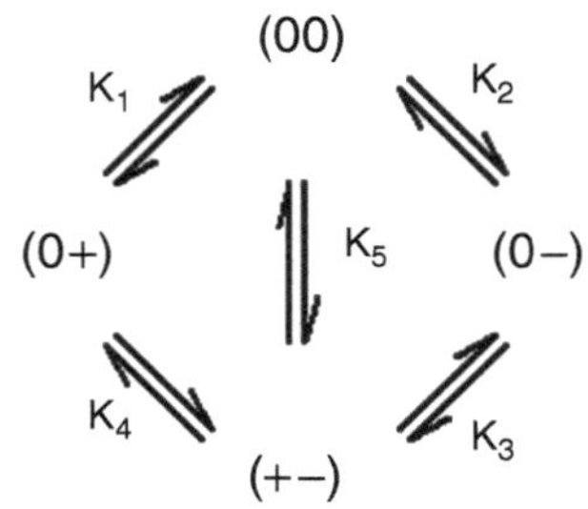

FIGURE 13.9 Acid-Base Equilibria Speciation of Zwitterionic Compounds

Zwitterions are molecules containing ionizable acid and basic groups capable of ionizing in aqueous solution, according to Fig. 13.9.

These molecules can interact with both anionic and cationic PE, forming ion pairs with the counter-ionic species. However, the main difference with monofunctional molecules is that, under conditions where D is released from the PE, the species that can freely diffuse are both (00) and (+−). This fact produces a different kinetic behavior to that previously described.

Indeed, it has been observed that the release kinetic of D from these complex systems approaches to zero order when water or saline solution are used as receiving medium (Figs. 13.10 and 13.11).

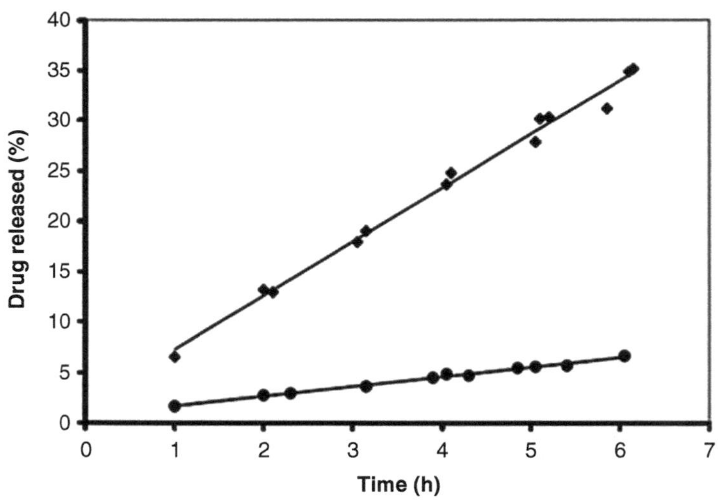

FIGURE 13.10

Release profile of ciprofloxacin from C–ciprofloxacin$_{50}$-Na$_{50}$, aqueous dispersion toward water (●) or 0.9% NaCl solution (■) used as receptor media.

Adapted from Vilches, A.P., 2003. Polielectrolitos solubles como portadores de fármacos ionizables. Preparación y estudio de sus propiedades farmacotécnicas. Ph D thesis. Facultad de Ciencias Químicas, Universidad Nacional de Córdoba, Córdoba, Argentina.

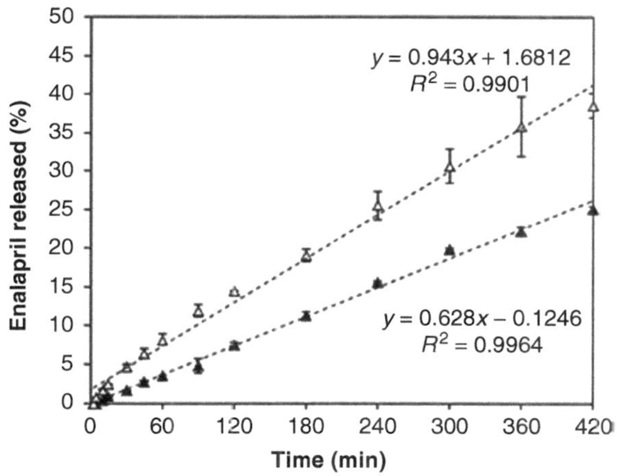

FIGURE 13.11

Enalapril release from EE-enalapril$_{30}$-maleate$_{60}$ aqueous dispersion toward water (▲) and 0.9% NaCl solution (Δ).

Adapted from Ramírez-Rigo, M.V., Olivera, M.E., Rubio, M., Manzo, R.H., 2014. Enhanced intestinal permeability and oral bioavailability of enalapril maleate upon complexation with the cationic polymethacrylate Eudragit E100. Eur. J. Pharm. Sci. 55, 1–11.

3.7 INTERACTION WITH BIOLOGICAL MEMBRANES

As informed previously, the characterization of the surface charge of the PE-D systems may be useful to explain their interaction with biological membranes, such as the gastrointestinal mucosa or the outer membrane of bacteria. In fact, the ex vivo and in vivo evaluation of the complex between Eudragit E100 and enalapril maleate led to interesting results. Indeed, oral administration of the complex resulted in higher oral bioavailability in rats, compared to the free D, due to an increased intestinal permeability of enalapril when it is complexed with the cationic polymethacrylate (Fig. 13.12). Such improvement was associated with the interaction between the positively charged systems and the intercellular junctions in biological membranes (Ramírez-Rigo et al., 2014).

Besides, the Gram-positive and Gram-negative intrabacterial uptake of fluoroquinolone antimicrobials is greater when they are complexed with C than when they are in solution (Fig. 13.13) (Romero, 2012). This behavior can be associated with the carrier system's capacity to interact with the bacterial walls.

3.8 A MODEL TO INTERPRET THE PROPERTIES OF PE-D COMPLEXES

The models that interpret the electrostatic interactions in colloidal systems provide the basis for analyzing the properties of the PE-D dispersions in equilibrium (Sinko, 2005).

On this basis, a model that considers the effect of high negative ζ values observed in the C–D hydrogels was proposed.

In the C–D dispersions, under equilibrium conditions, the concentration of the cationic species BH^+ and H^+ in the vicinity of the PE should be greater than in the fluid monomeric phase due to the electrostatic attraction of the negative charges on the macromolecular phase. Similarly, OH^- concentration in the fluid phase should be greater than in the vicinity of the PE.

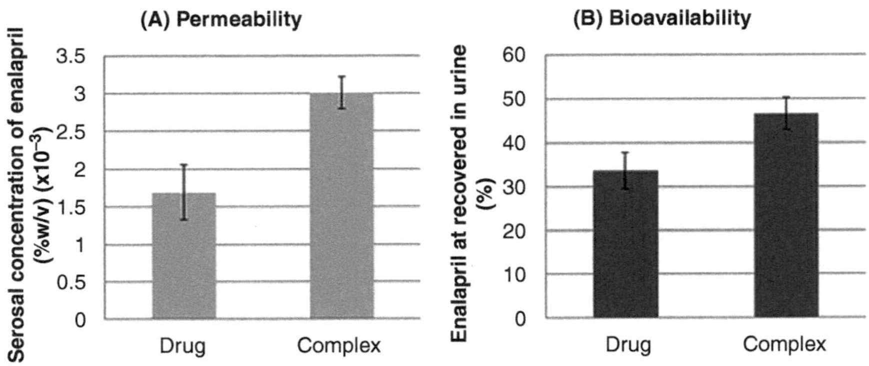

FIGURE 13.12 Intestinal Permeability and Oral Bioavailability of Enalapril in Rats Using Enalapril Maleate Alone or Complexed With Eudragit E100 (EE-enalapril$_{30}$-maleate$_{60}$ Complex Was Used)

(A) Cumulative serosal concentration of the prodrug determined at $t = 85$ min; (B) cumulative enalaprilat percentage recovered in urine at $t = 36$ h.

Adapted from Ramírez-Rigo, M.V., Olivera, M.E., Rubio, M., Manzo, R.H., 2014. Enhanced intestinal permeability and oral bioavailability of enalapril maleate upon complexation with the cationic polymethacrylate Eudragit E100. Eur. J. Pharm. Sci. 55, 1–11.

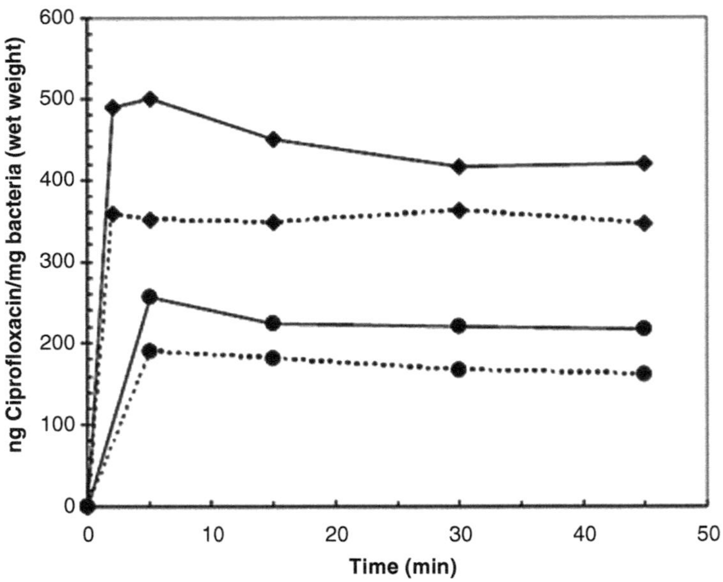

FIGURE 13.13

Ciprofloxacin intrabacterial uptake from C-ciprofloxacin$_{30}$-Na$_{45}$, against *escherichia coli* (♦) and *staphylococcus aureus* (●). The dotted line shows data obtained from a solution of free ciprofloxacin.

Adapted from Romero, V.L., 2012. Evaluación de los efectos de polímeros aniónicos y catiónicos en el desempeño y la eficacia de agentes antimicrobianos Fluoroquinolónicos. Ph D Thesis. Facultad de Ciencias Químicas, Universidad Nacional de Córdoba, Córdoba, Argentina.

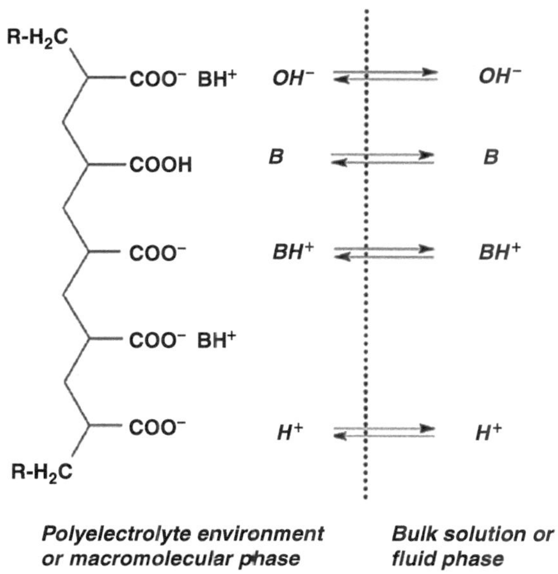

FIGURE 13.14 Model Representing the Distribution of All Species in Equilibrium in an Aqueous PE-D Dispersion

In the monomeric phase, electroneutrality requirements in absence of other species are as shown in Eq. (13.14):

$$[BH^+] + [H^+] = [OH^-] \tag{13.14}$$

In this context, Fig. 13.14 was proposed as a model that describes the anionic PE-D system, considering the relevant interactions that occur both in the vicinity of the PE and the fluid monomeric phase (Jimenez-Kairuz et al., 2003).

According to this model, the vicinity of the PE is characterized by a lower pH than the bulk solution and a high proportion of BH^+ present in the form of ion pairs. In addition, the model assumes that the concentration of neutral species [B] is equal in both environments.

This model allowed to explain the results observed in the release kinetics of D at different pH conditions.

3.9 SUMMARY OF TECHNOLOGICAL AND BIOPHARMACEUTICAL PROPERTIES THAT DETERMINE THE POTENTIAL APPLICATIONS OF PE-D DISPERSIONS

The results presented in the preceding paragraphs allow the following aspects:

1. The PE-D systems form colloidal dispersions of nanometric dimensions and behave as a reservoir of D, in which a high proportion of D is reversibly attached to the PE, as ionic pairs (RA^-BH^+) or ($RR_1R_2NH^+A^-$).
2. The ion-pair equilibrium is not modified by the addition of nonelectrolytes and is partially displaced by the addition of salts.

3. The release of D in water responds to a kinetic model proportional to the square root of time and is characterized by the following aspects:
   a. The release rate of D in water is slow due to the high degree of PE–D ionic condensation.
   b. The release rate of D can be increased by the diffusion of neutral salts (such as NaCl) toward the dispersion, due to ion exchange.
4. PE-zwitterionic D dispersions that have been reported, exhibited release kinetics that approximates zero order in both water and in salt solutions.
5. The chemical stability of D, which hydrolyze in aqueous solution, is significantly increased upon PE complexation due to counterionic condensation.

In summary, the PE-D systems in dispersion allow:

- Improved critical properties of D, such as low water compatibility or, in some cases, chemical instability, if they contain an appropriate ionizable group.
- Sustained release of D in the presence of the salinity of biological fluids.
- Regulated and adapted their properties through (1) the proportion of D incorporated into the PE, (2) the addition of inorganic counterions, (3) the salinity and pH of the dispersion.

4 POLYELECTROLYTE-DRUG COMPLEXES IN THE SOLID STATE

Solid dosage forms for oral administration are the most prescribed and dispensed medicines, not only as immediate release, but also as controlled release formulations.

In the field of oral DDS, hydrophilic polymeric matrices are among the most represented and studied because of their versatility, simplicity, and cost-effective methods for productions (Gallo et al., 2013; Nokhodchi and Asare-Addo, 2014). In this way, the introduction of swellable PE-D matrices (SPDM) represents an interesting contribution toward the development of novel and more effective DDS. The characterization of their relevant properties is the main point addressed in the following sections.

As reported in Fig. 13.1, the PE-D complexes can be obtained in solid state by several methods. In addition to those in which the reaction between PE and D occurs in dispersion, it has been reported the obtaining of PE-D complexes in solid state; for example, by hot-melt extrusion (Kindermann et al., 2011).

The PE-D complexes in solid state are easily obtained from either anionic or cationic PE loaded with D in different percentages. They can also be added with inorganic counterions, allowing them to modify their properties. In general, amorphous and stable solids are obtained, whose properties are different from that of their precursors (Jimenez-Kairuz et al., 2005; Ramírez Rigo et al., 2006).

The characterization of the solid state using FTIR spectroscopy, X-powder ray diffraction and thermal analysis (DSC-TG) indicates that the PE-D bond is ionic and the systems are constituted by a single phase (Bermúdez et al., 2008; Guzmán et al., 2012; Quinteros, 2008).

The literature reports a variety of solid PE-D complexes using anionic PE (C, AA, CMC, and λ-CG) or cationic PE (EE, chitosan, and cationic derivatives of cellulose) with D containing basic or acid groups, respectively (Bettini et al., 1994; Bonferoni et al., 2000, 2008; Kim and Park, 2010; Kindermann et al., 2011; Ramírez Rigo et al., 2004).

In general, these complexes exhibit vapor sorption-desorption isotherms, which are typical of amorphous materials (Fig. 13.15). However, this property is not critical to the formulation of solid dosage

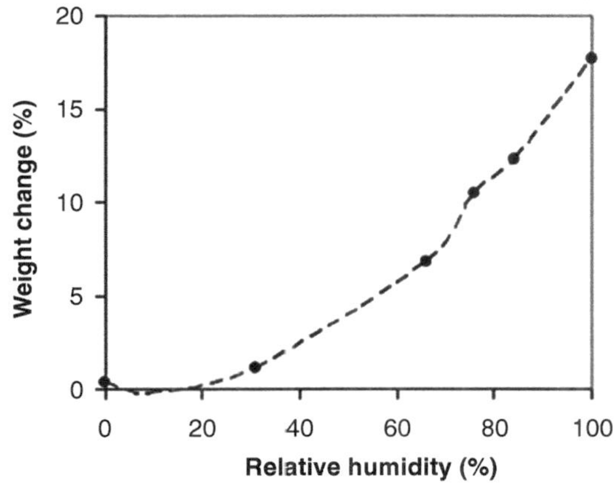

FIGURE 13.15 Water Vapor Sorption-Desorption Isotherm of CMC-atenolol$_{50}$

Adapted from Ramírez Rigo, M.V., 2006. Preparación y estudio de sistemas portadores de fármacos. Ph D Thesis. Facultad de Ciencias Químicas, Universidad Nacional de Córdoba, Córdoba, Argentina.

forms. In fact, it was described that the water present in the complexes confers a plasticizing effect, which favorably affected the modulation of D release from the matrix (Bonferoni et al., 2000).

Moreover, available data on compressibility, such as Hausner indexes and angles of repose show flow properties, which are acceptable and comparatively better than the respective physical mixtures of polymer and free D.

4.1 WETTING OF PARTICULATE AND COMPACTED PE-D COMPLEXES

One property that characterizes the PE-D solid materials is their behavior in an aqueous medium. For instance, complexes of C, EL, and AA with model D are easily wetted and have an important water uptake capacity (Fig. 13.16). When put in contact with aqueous media they increase in volume and can also revert to the hydrogel form. In contrast, CMC-D and EE-D complexes capture a limited proportion of water and retain their morphology.

The compaction of C-D and AA-D complexes results in SDPM. They are a particular type of hydrophilic matrices characterized by uniform distribution of D in the mass matrix (Jimenez-Kairuz et al., 2005; Ramírez Rigo et al., 2006). These matrices may also contain a high proportion of D. Table 13.8 reports the proportion of D contained in matrices prepared with PE-D complexes of different compositions.

4.2 CARBOMER-D MATRICES

4.2.1 Water uptake

As shown in Fig. 13.17, C-D SDPM can absorb water in a quick and sustained way (Jimenez-Kairuz et al., 2005). The water uptake coefficient is inversely proportional to the lipophilicity of the D loaded, expressed as its octanol-water partition. When saline is used instead of water, the uptake is slower but still sustained over time.

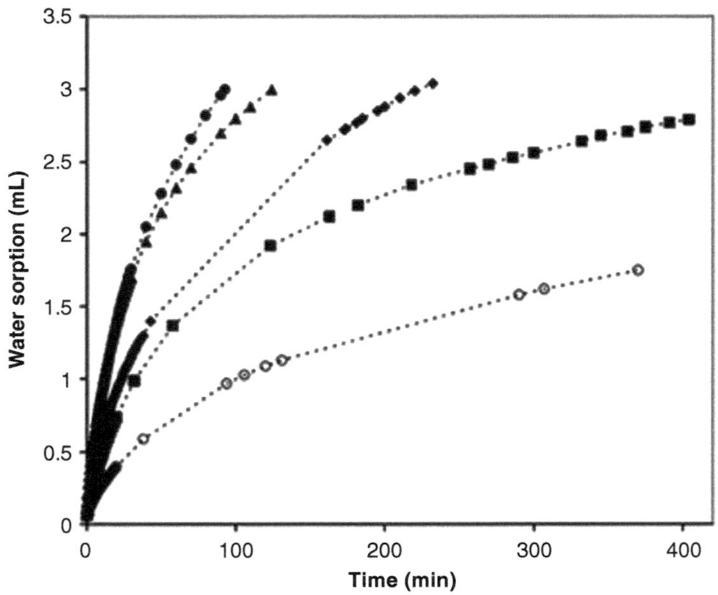

FIGURE 13.16 Water Sorption Isotherms of Different Particulate Solid Complexes (50 mg)

(○) C, (■) C-lidocaine$_{100}$, (◆) C-lidocaine$_{75}$, (▲) C-lidocaine$_{50}$, (●) C-lidocaine$_{25}$.

Adapted from Jimenez Kairuz, A.F., Allemandi, D.A., Manzo, R.H., 2004. The improvement of aqueous chemical stability of a model basic drug by ion pairing with acid groups of polyelectrolytes. Int. J. Pharm. 269(1), 149–156.

Table 13.8 Amount of D Contained in 200 mg-SPDM of PE-D Complexes Prepared at Different Percentages of Neutralization of the PE

	mg (% w/w)			
PE-D Complex	***x**** **= 100%**	***x**** **= 75%**	***x**** **= 50%**	***x**** **= 25%**
C-atenolol$_x$	137.0 (68.5%)	120.0 (60.0%)	109.0 (54.5%)	87.0 (43.5%)
C- metoclopramide$_x$	156.5 (78.3%)	145.9 (73.3%)	128.5 (64.3%)	94.7 (47.4%)
C-lidocaine$_x$	147.5 (73.8%)	136.0 (68.0%)	117.0 (58.5%)	83.0 (41.5%)
C-erythromycin$_x$	179.6 (89.8%)	173.7 (86.9%)	163.0 (81.5%)	137.5 (68.8%)
C-azithromycin$_x$	180.0 (90.0%)	174.2 (87.1%)	163.6 (81.8%)	138.4 (69.2%)
AA-atenolol$_x$	—	115.5 (57.8%)	77.0 (38.5%)	—
AA-metoclopramide$_x$	—	—	82.8 (41.4%)	—

**Percentage of ionizable groups of the polyelectrolyte neutralized by the drug.*

Adapted from Jimenez Kairuz, A.F., Allemandi, D.A., Manzo, R.H., 2004. The improvement of aqueous chemical stability of a model basic drug by ion pairing with acid groups of polyelectrolytes. Int. J. Pharm. 269(1), 149–156; Ramírez Rigo, M.V., 2006. Preparación y estudio de sistemas portadores de fármacos. Ph D Thesis. Facultad de Ciencias Químicas, Universidad Nacional de Córdoba, Córdoba, Argentina.

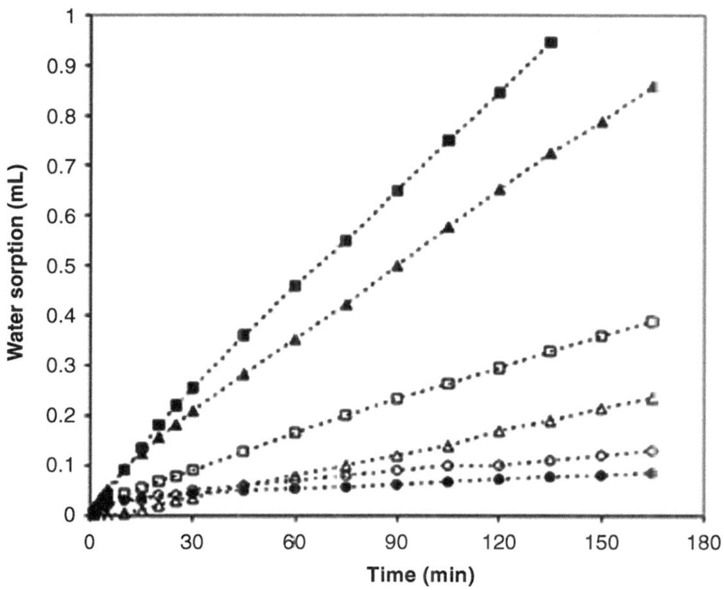

FIGURE 13.17 Water Sorption of Some Swellable Drug-Polyelectrolyte Matrices (SDPM)

C-atenolol$_{50}$ (■); C-lidocaine$_{50}$ (▲) and C-metoclopramide$_{50}$ (●). Filled symbols: water; empty symbols: 0.9% NaCl.

Adapted from Jimenez Kairuz, A.F., Allemandi, D.A., Manzo, R.H., 2004. The improvement of aqueous chemical stability of a model basic drug by ion pairing with acid groups of polyelectrolytes. Int. J. Pharm. 269(1), 149–156.

The study of the dynamics of aqueous medium penetration into the SDPM provides an interesting insight into the mechanisms governing the release of the D or PE-D from the compacted material

4.2.2 Drug diffusion front movement in SDPM

The gradual penetration of the aqueous medium in the matrix generates an outer layer of hydrogel that expands. Its outer and inner edges are designated as the erosion and swelling fronts, respectively (Fig. 13.18) (Bettini et al., 1994; Colombo et al., 1995, 2000).

The evolution of these fronts was studied using the methodology proposed by Bettini et al. (1994) for the study of conventional hydrophilic matrices, which allows visualizing the interfaces between dissolution medium and gel layer (erosion front) and between gel layer and not yet hydrated core of the matrix (swelling front). A cylindrical matrix is clamped between two transparent disks that are immersed in water or an aqueous solution at 37°C. The introduction of an acid-base indicator to the dissolution media allows to identify a third front in the hydrogel layer called diffusion front, which is associated to the free D released from the PE. Fig. 13.19 is representative of the dynamic fronts of a SDPM in contact with water and saline (Jimenez-Kairuz et al., 2005).

4.2.3 Kinetics and release mechanism

4.2.3.1 Carbomer-D SDPM

Studies about D release from SDPM of C-D were performed in the dissolution equipment described in pharmacopoeias. Fig. 13.20 shows the release rate of some model D (atenolol, lidocaine, and

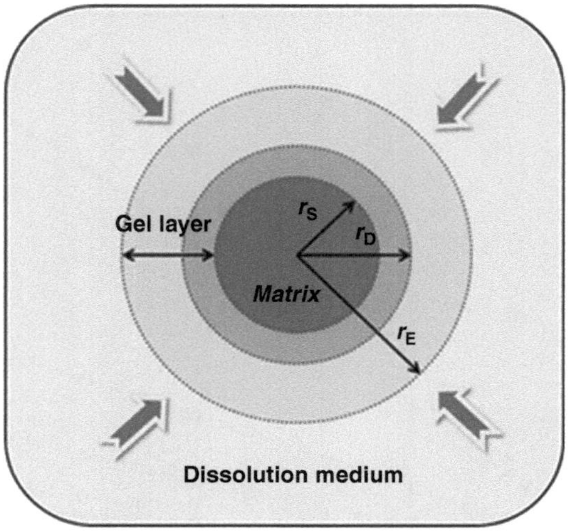

FIGURE 13.18 Schematic Representations of the Swelling Fronts for a SDPM Observed Upon Contact With the Dissolution Medium

r_S, Swelling radius; r_D, diffusion radius; r_E, erosion radius.

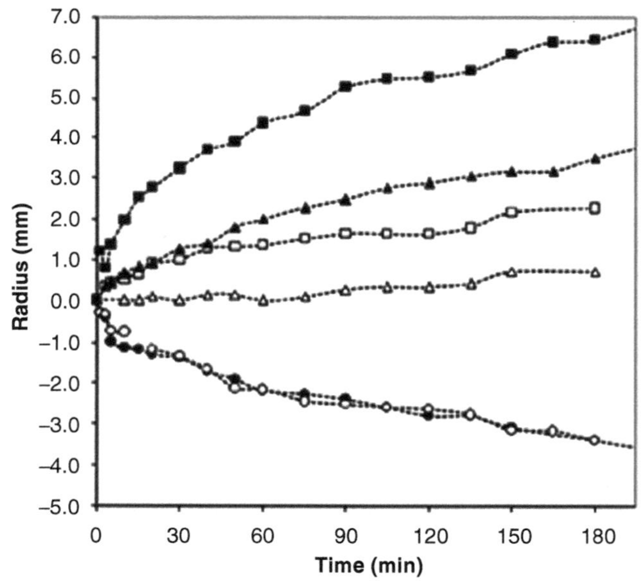

FIGURE 13.19 Front Movements of a SDPM Based on C-atenolol$_{75}$ in Contact With Water (Filled Symbols) and 0.9% NaCl Solution (Open Symbols)

(■) Erosion, (▲) diffusion, (●) swelling.

Adapted from Jimenez Kairuz, A.F., Allemandi, D.A., Manzo, R.H., 2004. The improvement of aqueous chemical stability of a model basic drug by ion pairing with acid groups of polyelectrolytes. Int. J. Pharm. 269(1), 149–156.

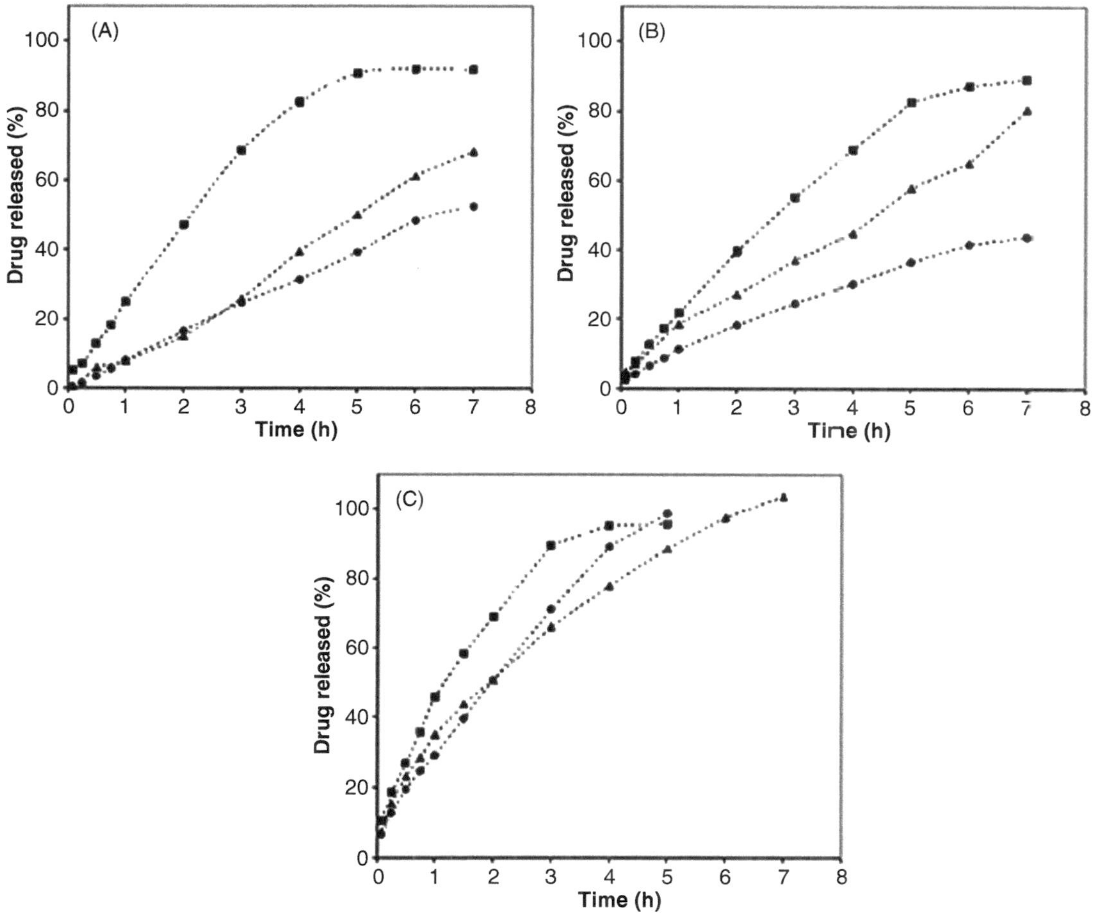

FIGURE 13.20

Release profile of atenolol (■), lidocaine (▲) and metoclopramide (●). From C–D_{50} SDPM in (A) Water, (B) NaCl 0.9%, (C) HCl 0.1M.

Adapted from Jimenez Kairuz, A.F., Allemandi, D.A., Manzo, R.H., 2004. The improvement of aqueous chemical stability of a model basic drug by ion pairing with acid groups of polyelectrolytes. Int. J. Pharm. 269(1), 149–156.

metoclopramide) in water, saline, and HCl solution. Table 13.9 reports the release kinetics adjusted according to Korsmeyer's model.

It can be seen in Fig. 13.20 that the release of D in water and saline occurs with similar rates, showing that the release behavior of the matrices is different from that seen with the C-D aqueous dispersions. Indeed, while in aqueous dispersion the presence of salts considerably increases the rate of release promoting ion exchange between inorganic cation/organic cation, the kinetic control in the SDPM would be associated with the availability of water in the system (Jimenez-Kairuz et al., 2005)

However, the release rate of D is higher in HCl solution, probably associated with the regeneration of the undissociated form of the carboxyl groups of C.

Table 13.9 Kinetic Data of Various Materials in Different Media Calculated by Korsmeyer's Model

	Drug Release								
	Water			NaCl 0.9%			HCl 0.1M		
SDPM	***n***	r^2 **(N)**	k **(h^{-n})**	***n***	r^2 **(N)**	k **(h^{-n})**	***n***	r^2 **(N)**	k **(h^{-n})**
C-atenolol$_{100}$	0.97	0.994 (7)	0.16	0.61	0.997 (7)	0.23	0.59	0.999 (6)	0.49
C-atenolol$_{75}$	1.14	0.991 (7)	0.23	0.70	0.996 (7)	0.24	—	—	—
C-atenolol$_{50}$	0.92	0.999 (6)	0.25	0.76	0.996 (7)	0.23	0.66	0.994 (5)	0.50
C-atenolol$_{25}$	0.89	0.999 (11)	0.15	0.68	0.992 (7)	0.21	—	—	—
C-lidocaine$_{75}$	1.17	0.993 (6)	0.08	0.66	0.992 (7)	0.18	—	—	—
C-lidocaine$_{50}$	1.24	0.995 (6)	0.07	0.78	0.993 (5)	0.17	0.58	0.999 (6)	0.35
C-lidocaine$_{25}$	0.89	0.984 (5)	0.18	0.71	0.991 (4)	0.24	—	—	—
C-metoclopramide$_{100}$	1.09	0.997 (11)	0.04	0.65	0.999 (11)	0.13	—	—	—
C-metoclopramide$_{75}$	1.03	0.998 (10)	0.11	0.66	0.998 (10)	0.14	0.83	0.999 (4)	0.52
C-metoclopramide$_{50}$	1.08	0.998 (11)	0.07	0.70	0.997 (11)	0.12	0.79	0.998 (6)	0.25
n average	1,04 (±0,12)			0,69 (±0,05)			0,69 (±0,11)		

(N), Number of experimental data used for regression analysis.
Adapted from Jimenez Kairuz, A.F., Allemandi, D.A., Manzo, R.H., 2004. The improvement of aqueous chemical stability of a model basic drug by ion pairing with acid groups of polyelectrolytes. Int. J. Pharm. 269(1), 149–156.

The data set provided by sorption studies, dynamic fronts, and release kinetics of this new type of hydrophilic matrices, are interpreted in Fig. 13.21, which contains the sequence of events that occur when the matrix comes into contact with the release media.

As noted earlier, the results suggest that, unlike C-D dispersions the kinetic control in the matrices is related to the availability of water in the system (Jimenez-Kairuz et al., 2005), and thus, the concentration of cationic species DH^+ released by ion exchange in NaCl solution would not be the main controlling factor. Moreover, in these matrices the release occurs essentially by diffusion of D and DH^+ free forms due to the viscous gel layer formed (Fig. 13.21, sequence A).

4.2.3.2 Alginic acid-D matrices

AA SDPM prepared using model D are characterized by a lower water uptake compared to C-D ones. As shown in Fig. 13.22, water uptake can be increased by the addition of NaAA to the matrices (Ramírez Rigo et al., 2006).

As observed C-D SDPM, those with AA-D also develop swelling and erosion fronts. However, the hydrogel layer generated in the last ones is less extensive and significantly less viscous than that of C-D systems (Jimenez-Kairuz et al., 2005; Ramírez Rigo et al., 2006) (Fig. 13.23).

The release kinetics of the AA-D SDPM was determined in water and both simulated intestinal and gastric fluids (Fig. 13.24). The release occurs with a remarkable zero order in all cases, with similar rates in water and simulated intestinal fluid while in simulated gastric fluid the release rate is slower.

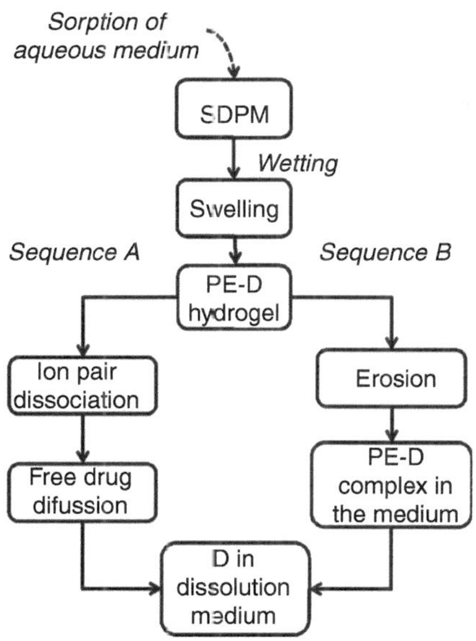

FIGURE 13.21 Drug Release Events of Swellable Drug-Polyelectrolyte Matrices (SDPM)

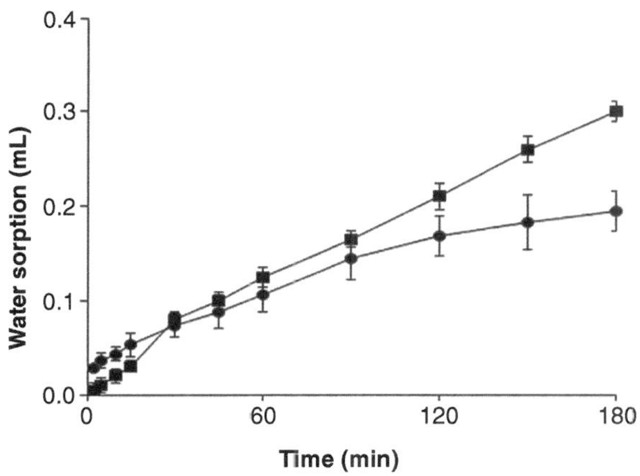

FIGURE 13.22

Water sorption of (●) AA-atenolol$_{50}$, (■) 1:2 AA-atenolol$_{50}$: NaAA SDPM.

Adapted from Ramírez Rigo, M.V., 2006. Preparación y estudio de sistemas portadores de fármacos. Ph D Thesis. Facultad de Ciencias Químicas, Universidad Nacional de Córdoba, Córdoba, Argentina.

In mixed AA: NaAA SDPM the release rate decreases as the proportion of NaAA increases, which allows to modulate it. Interestingly, in experiments simulating the gastrointestinal transit of such matrices (2 h in acid solution and then 3 h at pH 6.8), the release rate remained constant over the entire range, as illustrated in Fig. 13.25.

In AA-D SDPM the release occurs mainly by erosion of the hydrogel layer (Fig. 13.26) (Benegas et al., 1999; Radeva, 2001; Ramírez Rigo et al., 2006).

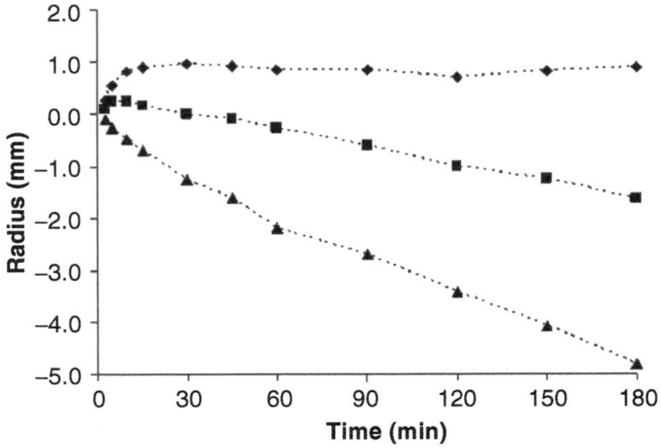

FIGURE 13.23 Front Dynamics in AA-atenolol$_{50}$ SDPM in Water

(♦) Erosion front, (■) diffusion front, and (▲) swelling front.

Adapted from Ramírez Rigo, M.V., 2006. Preparación y estudio de sistemas portadores de fármacos. Ph D Thesis. Facultad de Ciencias Químicas, Universidad Nacional de Córdoba, Córdoba, Argentina.

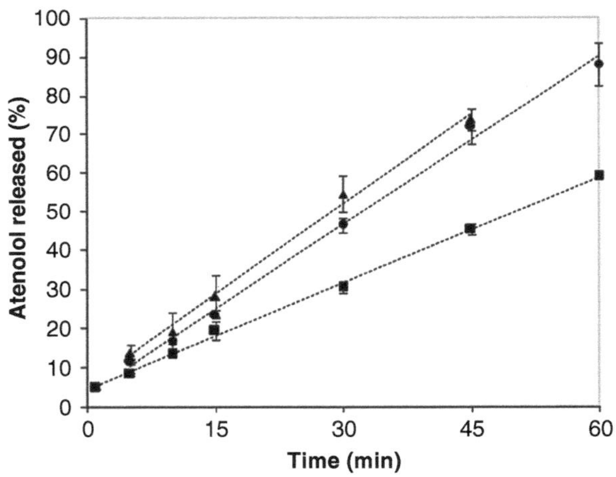

FIGURE 13.24

Atenolol release from AA-atenolol$_{50}$ matrix in: (●), water; (■) HCl 0.01N; (▲), simulated intestinal fluid without enzyme.

Adapted from Ramírez Rigo, M.V., 2006. Preparación y estudio de sistemas portadores de fármacos. Ph D Thesis. Facultad de Ciencias Químicas, Universidad Nacional de Córdoba, Córdoba, Argentina.

In the same way, the erosion of the hydrogel layer is the main route of release in AA-D:NaAA matrices, that is, the sequence B described in Fig. 13.21 (Ramírez Rigo et al., 2006). This mechanism was established by the observed reduction in the mass of the matrix during the dissolution process and the variation in the release rate observed with changes in the stirring speed of the dissolution equipment (Kavanagh and Corrigan, 2004), which showed a sensitivity typical of erodible matrices

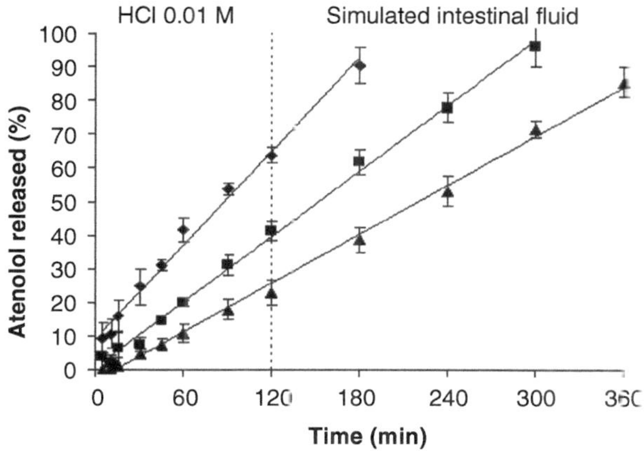

FIGURE 13.25 Atenolol Release Profiles From AA-atenolol$_{50}$

NaAA mixed matrices at different proportions: (◆), 1:1; (■) 1:2; (▲), 1:3 in HCl 0.01N and simulated intestinal fluid without enzyme.

Adapted from Ramírez Rigo, M.V., 2006. Preparación y estudio de sistemas portadores de fármacos. Ph D Thesis. Facultad de Ciencias Químicas, Universidad Nacional de Córdoba, Córdoba, Argentina

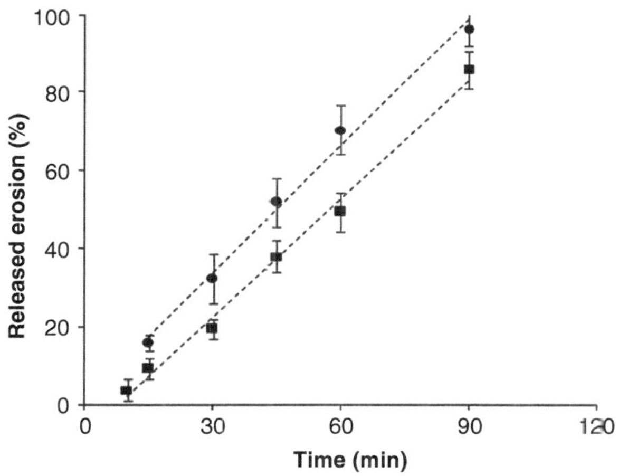

FIGURE 13.26

Atenolol release (●) from AA-atenolol$_{50}$. NaAA mixed matrix (1:1) and erosion (■) in simulated intestinal medium without enzyme.

Adapted from Ramírez Rigo, M.V., 2006. Preparación y estudio de sistemas portadores de fármacos. Ph D Thesis. Facultad de Ciencias Químicas, Universidad Nacional de Córdoba, Córdoba, Argentina.

(Fig. 13.27). Such behavior is different from C-D SDPM and may be associated with the characteristics of the carrier polymers because C yields high viscosity hydrogels with significant entanglement between chains (Jimenez-Kairuz, 2004) whereas AA produces a lower viscosity and therefore more easily eroded hydrogel (Ramírez Rigo, 2006)

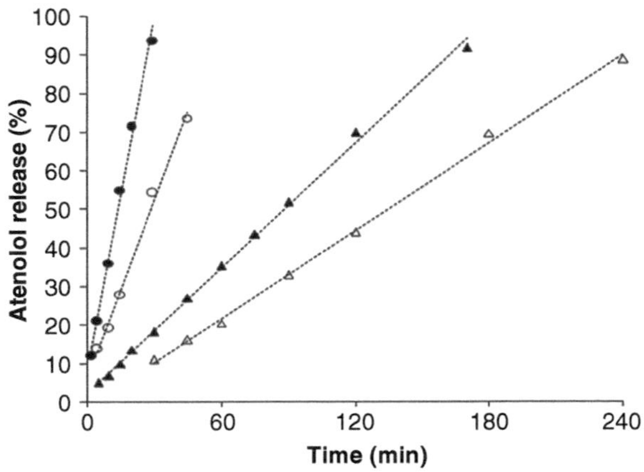

FIGURE 13.27 Atenolol Release Profiles from AA-atenolol$_{50}$

NaAA mixed matrix: (●) 1:0, (▲) 1:1 at different rotation speeds. 50 rpm (*open symbols*) and 100 rpm (*filled symbols*).

Adapted from Ramírez Rigo, M.V., 2006. Preparación y estudio de sistemas portadores de fármacos. Ph D Thesis. Facultad de Ciencias Químicas, Universidad Nacional de Córdoba, Córdoba, Argentina.

4.3 CARBOXYMETHYLCELLULOSE-D MATRICES

By acid-base reaction of the acid form of CMC, as a particulate material, with a solution containing a basic D (or a soluble salt $BH^{+}A^{-}$), insoluble solid CMC-D complexes with different degrees of substitution are obtained (Ramírez Rigo et al., 2004). Figs. 13.28 and 13.29 show the water sorption behavior and the dissolution profiles of some CMC-D matrices.

As can be seen there, CMC-D compacted materials rapidly sorb a limited amount of water maintaining its morphology. In these systems the formation of a hydrogel layer on the surface of the matrix

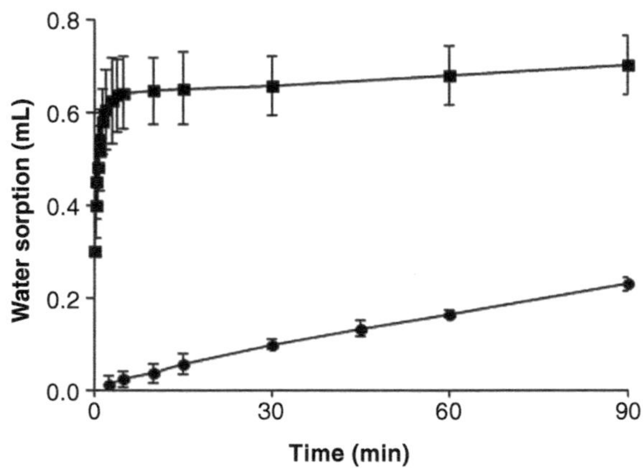

FIGURE 13.28

Water sorption profiles of matrices (200 mg) (■) CMC-atenolol$_{50}$, (●) CMC-atenolol$_{50}$:NaCMC (1:2).

Adapted from Ramírez Rigo, M.V., 2006. Preparación y estudio de sistemas portadores de fármacos. Ph D Thesis. Facultad de Ciencias Químicas, Universidad Nacional de Córdoba, Córdoba, Argentina.

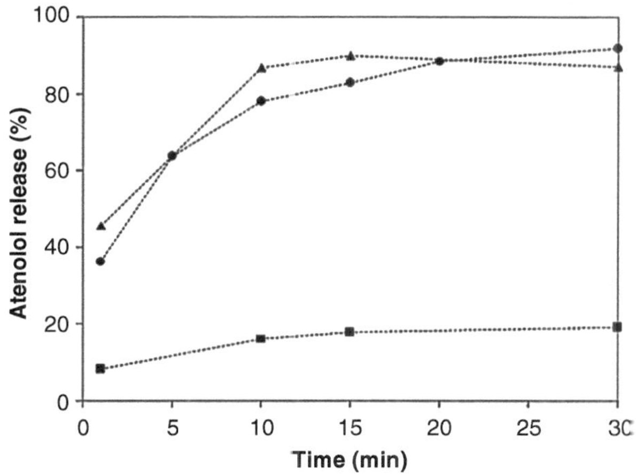

FIGURE 13.29

Release profile of atenolol from CMC-atenolol$_{50}$ matrix in (■) Water (▲) NaCl 0.9% and (●) HCl 0.1M.

Adapted from Ramírez Rigo, M.V., 2006. Preparación y estudio de sistemas portadores de fármacos. Ph D Thesis. Facultad de Ciencias Químicas, Universidad Nacional de Córdoba, Córdoba, Argentina.

is not observed (Ramírez Rigo et al., 2004, 2009). When water is used as a dissolution medium only a limited amount of D is released, quickly reaching the equilibrium depending on the affinity constant. However, in saline or acid media a rapid release is observed. This behavior is a consequence of the porosity of the matrix, which allows the rapid penetration of the dissolution medium and the subsequent release of D since a gel layer is not formed (Ramírez Rigo, 2006; Ramírez Rigo et al., 2009).

Incorporation of NaCMC into the matrix originates SDPM that develops a hydrogel layer similar to those of C and AA, which retards D release (Fig. 13.30) and also the water sorption (Fig. 13.28) (Ramírez Rigo et al., 2009). It can be seen there that the decreased rate of release correlates with the proportion of NaCMC incorporated. This allows the modulation of the release of D depending on the design requirements of the formulations. The CMC-based system behaves similarly to those described previously (Ramírez Rigo, 2006).

4.4 CARRAGEENAN-D MATRICES

Due to its half-ester sulfate moieties, carrageens (CG) exhibit a high capacity to react with positively charged substances. In particular λ-CG having the highest proportion of sulfate pending groups have attracted the attention to be used as carrier for basic D.

Because in commercial CG the sulfate moieties are salified with alkaline cations, they react with D salts (i.e., hydrochlorides, tartrates) according with Eq. (13.1) to give CG-D complexes.

An early study showed that CG could control the initial release of salbutamol sulfate, regardless of the dissolution media, which was attributed to the ionic interaction between the D and the polymer (Bonferoni et al., 1994).

Later, the ionic interaction between λ-CG and basic D in oral controlled release matrix tablets was characterized (Bonferoni et al., 2008). It was observed that different D give complexes with quite different characteristics of solubility and D release kinetics (Aguzzi et al., 2002). In fact, as it was

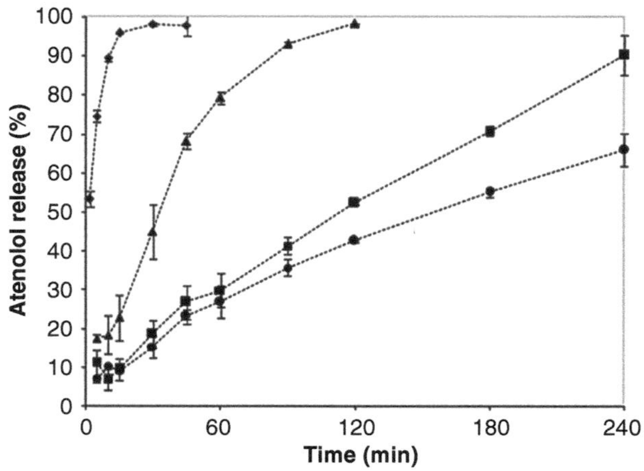

FIGURE 13.30 Atenolol Release Profiles From Different CMC-atenolol$_{50}$

NaCMC mixed matrices: (◆) 1:0, (▲) 1:1, (■) 1:2, (●) 1:3 in simulated intestinal fluid without enzyme.

Adapted from Ramírez Rigo, M.V., 2006. Preparación y estudio de sistemas portadores de fármacos. Ph D Thesis. Facultad de Ciencias Químicas, Universidad Nacional de Córdoba, Córdoba, Argentina.

observed with CMC-D complexes, matrices of the low soluble complex of λ-CG-diltiazem was characterized by a fast water uptake due only to capillarity, but no further gelation was observed.

On the other hand, matrices of the more soluble complex λ-CG-metoprolol developed a thick gel layer around the matrix tablets that modulate the diffusive D release profiles.

Therefore, the different solubility of the complexes result in differences in water uptake and gelation properties.

4.5 EE-D MATRICES

Several EE-D complexes were prepared with a set of nonsteroidal anti-inflammatory drugs. As mentioned earlier, such complexes are insoluble and difficult to disperse in water and consequently exhibited an almost negligible release rate in water when assayed as particulate powders or as matrices. Although the release rate is increased when the medium is pH 7.4 buffer or saline, it is still very limited. Moreover, no erosion was observed in the matrices, which remained intact during the assays.

In contrast, EE-D complexes prepared with mesalazine exhibit good water solubility and their matrices dissolve rapidly without the generation of a gel layer (Fig. 13.31). The same behavior was observed with EE-benzoic acid complexes.

On the other hand, as previously reported, the addition of Cl^- as a second counterion in the complexes of EE-nonsteroidal anti-inflammatory drugs increases their aqueous compatibility. As expected, their matrices exhibit a progressive increase in the rate of dissolution in water and buffer solutions as a function of the proportion of Cl^- incorporated (Fig. 13.32). However, a gel layer was not produced and D release occurs by erosion of the complex from the matrix (Quinteros, 2010).

The strategy of introducing a gelling agent in the matrix was used successfully in the CMC-D and AA-D systems, where the incorporation of NaCMC or NaAA generated a hydrogel layer that modulates the release (Ramírez Rigo et al., 2006, 2009). In a similar way, mixed matrices of EE-mesalazine

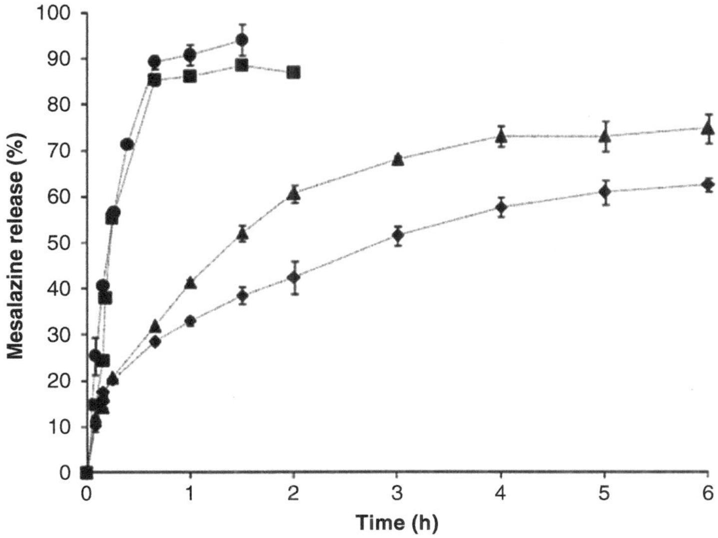

FIGURE 13.31 Mesalazine Dissolution Profiles From EE-mesalazine$_x$ Matrices Using Buffer pH 7.4 as Dissolution Medium

(♦) EE-mesalazine$_{25}$, (▲) EE-mesalazine$_{50}$, (■) EE-mesalazine$_{75}$, (●) EE-mesalazine$_{100}$.

Adapted from Quinteros, D.A., 2010. Desarrollo de nuevas estrategias de formulación de fármacos mediante el acomplejamiento con polielectrolitos. Ph D Thesis. Facultad de Ciencias Químicas, Universidad Nacional de Córdoba, Córdoba, Argentina.

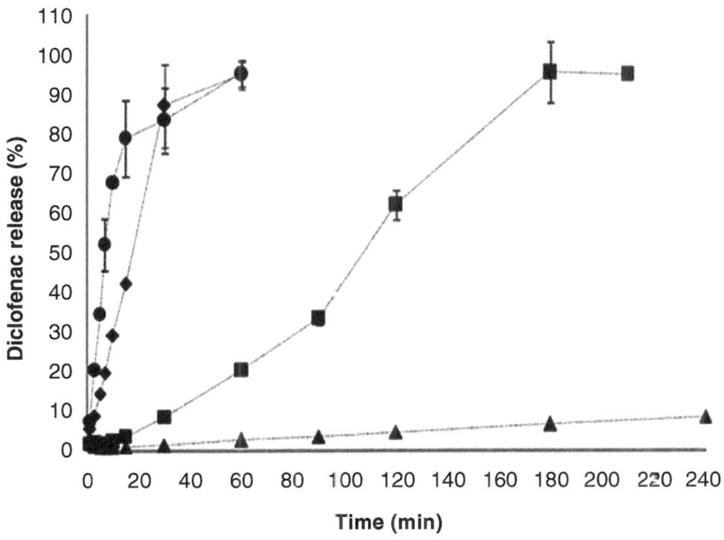

FIGURE 13.32 Diclofenac Dissolution Profiles From Eudragit E100 Matrices EE-diclofenac$_{50}$-Cl$_x$ using NaCl 0.9% as Dissolution Medium

(▲), EE-diclofenac$_{50}$-Cl$_{10}$, (■) EE-diclofenac$_{50}$-Cl$_{25}$, (♦) EE-diclofenac$_{50}$-Cl$_{35}$, (●) EE-diclofenac$_{50}$-Cl$_{50}$.

Adapted from Quinteros, D.A., 2010. Desarrollo de nuevas estrategias de formulación de fármacos mediante el acomplejamiento con polielectrolitos. Ph D Thesis. Facultad de Ciencias Químicas, Universidad Nacional de Córdoba, Córdoba, Argentina.

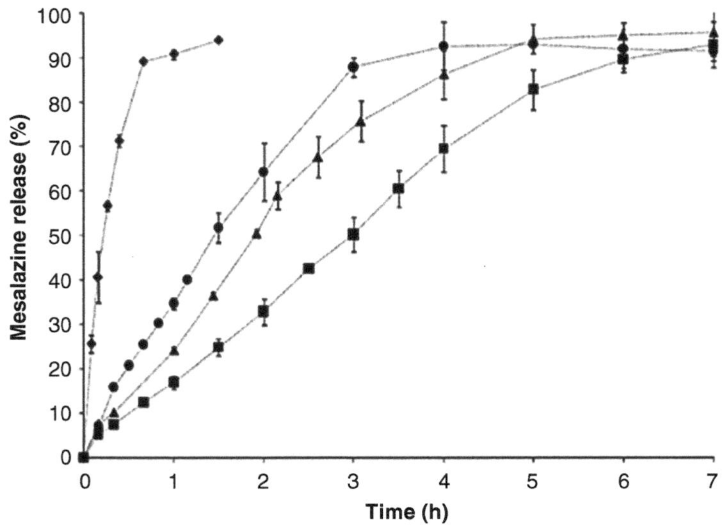

FIGURE 13.33 Mesalazine Dissolution Profiles From EE-mesalazine$_{100}$

C_x Matrices in buffer pH 7.4. (■) EE-mesalazine$_{100}$:C_{50}, (▲) EE-mesalazine$_{100}$: C_{30}, (●) EE-mesalazine$_{100}$:C_{10} (♦), EE-mesalazine$_{100}$.

complexes with increasing amounts of C as gelling agent resulted in the formation of a gel layer that modulates the D release (Fig. 13.33). This system was used to develop a DDS for the colonic release of mesalazine (Quinteros et al., 2010).

5 CONCLUSIONS

Table 13.10 contains a summary of the release properties of the matrices obtained with C, AA, CMC, λ-CG, and EE.

The findings on the behavior of these matrices, in addition to the information provided by the studies of these systems in aqueous dispersion allow establishing some general conclusions:

- The SDPM act as versatile release systems wherein the release rate and kinetics of D delivery can vary according to the design objectives of each system, releasing the molecules or the PE-D complexes as nanocarriers able to interact with biological systems.
- Changes in the rate of release of D can be generated by considering the phenomena involved at molecular scale.
- Unlike conventional extended release systems, consisting of the physical mixture of D and polymeric carriers, in PE-D complexes D molecules are homogeneously distributed in the mass of the PE allowing multiparticulate systems being formulated readily.
- The PE-D complexes may contain a high proportion of D so that the loading capacity of the PE used is considerable.

Table 13.10 Comparative Drug Release Rate and Mechanisms of Matrices Based on PE-D Complexes

PE-D Complex	Hydrogel Layer Extensión	Drug Release Rate	Release Mechanism
C-D	Large	Slow[a]	Ionic exchange/diffusion
AA-D	Narrow	Moderate[a]	Erosion
AA-D: NaAA	Intermediate	Slow[a]	Erosion
CMC-D	Absent	Fast	Ionic exchange
CMC-D: NaCMC	Intermediate	Slow[a]	Ionic exchange/diffusion/erosion
EE-$D_{50}Cl_{50}$	Absent	Fast[a]	Erosion/Ionic exchange
EE-D:C	Intermediate	Slow[a]	Ionic exchange/diffusion
CG-D	Intermediate	Slow[a]	Ionic exchange/diffusion

[a]*Able to be modulated.*

Moreover, it should be mentioned that SDPM have been prepared from PE-D complexes, which were processed using operations and equipments commonly used in the pharmaceutical industry. For instance, Bermúdez et al. (2008) studied the release of a series of complexes C-ciprofloxacin-Na. Extended release SDPM tablets of such complexes can be satisfactorily prepared through the current processes of tableting technology. Ciprofloxacin release in simulated gastric fluid from such matrices can be widely modulated by changes in the complex composition (Fig. 13.34). On these bases

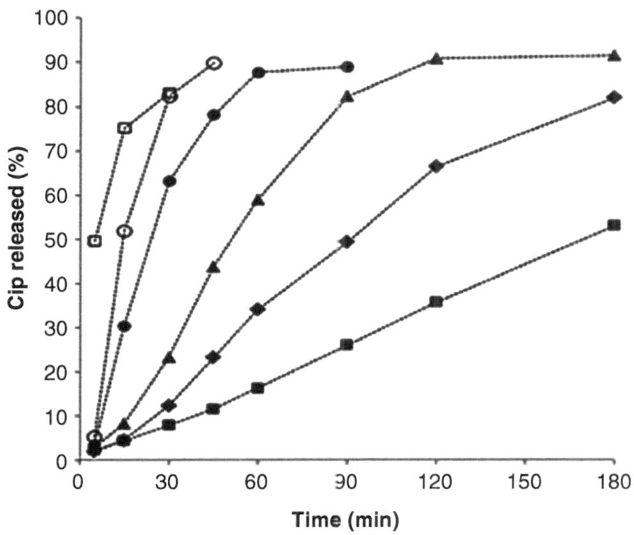

FIGURE 13.34 In Vitro Release Profiles of Ciprofloxacin in Simulated Gastric Fluid From C-ciprofloxacin$_{50}$Na$_x$ SDPM Tablets, Containing the Equivalent at 500 mg of Ciprofloxacin

(□) Na = 14%, (○) Na = 13%, (●) Na = 12.5%, (▲) Na = 12.25%, (◆) Na = 12%, (■) Na = 10%.

an extended-release SDPM tablet that releases 90% of the dose in simulated gastric fluid in about 120 min was developed.

Pellets and spheronized beads were also obtained from C, AA, and carrageen complexes (Bommareddy et al., 2006; Paker-Leggs and Neau, 2009; Rajesh et al., 2014). Accordingly, matrix systems described are presented as interesting formulation options that can be exploited favorably in the area of DDS.

ABBREVIATIONS

AA	Alginic acid
AH	Acid drug
$\mathbf{AH_{st}}$	Stoichiometric concentration of an acid drug
A^-	Dissociated species of an acid drug (*)
AH	Neutral species of an acid drug (*)
$\mathbf{B_{st}}$	Stoichiometric concentration of a basic drug
B	Neutral species of a basic drug (*)
BH^+	Protonated species of a basic drug (*)
C	Carbomer 934
CG	Carrageens
CMC	Carboxymethylcellulose
D	Drug
DDS	Drug delivery systems
EE	Eudragit E 100
EL	Eudragit L 100
ES	Eudragit S 100
HA	Hyaluronic acid/hyaluronan
K_{cc}	Affinity constant for counterionic condensation
NaAA	Sodium alginate
NaCMC	Sodium carboxymethylcellulose
PE	Polyelectrolyte
RAH	Acid pending groups of a polyelectrolyte
RAH	Undissociated fraction of the acid pending groups of a polyelectrolyte (*)
RA^-	Dissociated fraction of the acid pending groups of a polyelectrolyte (*)
$[RA^-BH^+]$	Counterionic condensed fraction between the dissociated acid pending groups of a polyelectrolyte and a protonated basic drug (*)
$\mathbf{RNR_1R_2}$	Basic pending groups of a polyelectrolyte
RNR_1R_2	Nonprotonated fraction of the amino pending groups of a basic polyelectrolyte (*)
$RNR_1R_2H^+$	Protonated fraction of the amino pending groups of a basic polyelectrolyte (*)
$[RNR_1R_2H^+A^-]$	Counterionic condensed fraction between the protonated basic pending groups of a polyelectrolyte and a dissociated acid drug (*)
SDPM	Swellable Drug-Polyelectrolyte Matrix

(*) Italics indicate the dissociated or protonated species of a polyelectrolyte or a drug.

REFERENCES

Abd El-Rehim, H.A., Swilem, A.E., Klingner, A., Hegazy, E.-S.A., Hamed, A.A., 2013. Developing the potential ophthalmic applications of pilocarpine entrapped into polyvinylpyrrolidone-poly(acrylic acid) nanogel dispersions prepared by γ radiation. Biomacromolecules 14 (3), 688–698.

Aguzzi, C., Bonferoni, M.C., Fortich, M.R., Rossi, S., Ferrari, F., Caramella, C., 2002. Influence of complex solubility on formulations based on lambda carrageenan and basic drugs. AAPS Pharm. Sci. Tech. 3 (3), E27.

Ardusso, M., 2012. Utilización de materiales portadores polielectrolito-fármaco en el desarrollo de sistemas de liberación de fármacos. Ph D Thesis. Facultad de Ciencias Químicas, Universidad Nacional de Córdoba, Córdoba, Argentina.

Baena, Y.A., Manzo, R.H., Ponce D'Leon, L.F.Q., 2011. Preparation and physicochemical characterization of some polyelectrolyte-diclofenac complexes. Vitae 18 (3), 305–311.

Barrat, J.-L., Meyer, M.C., 1996. Theory of polyelectrolyte solution. Prigonine, I (Ed.), Polymeric Systems: Advances in Chemical Physics, 94, John Wiley & Sons. Inc., New York, pp. 1–65.

Battistini FD. 2015. Desarrollo de complejos polielectrolito-fármaco y su utilización en el diseño de sistemas de liberación modificada. Ph D Thesis. Facultad de Ciencias Químicas, Universidad Nacional de Córdoba, Córdoba, Argentina.

Benegas, J.C., Paoletti, S., Van Den Hoop, M.A.G.T., 1999. Affinity interactions in counterion-polyelectrolyte systems: competition between different counterions. Macromol. Theor. Simul. 8 (1), 61–64.

Bermúdez, J.M., Jimenez-Kairuz, A.F., Olivera, M.E., Allemandi, D.A., Manzo, R.H., 2008. A ciprofloxacin extended release tablet based on swellable drug polyelectrolyte matrices. AAPS Pharm. Sci. Tech. 9 (3), 924–930.

Bettini, R., Colombo, P., Massimo, G., Catellani, P.L., Vitali, T., 1994. Swelling and drug release in hydrogel matrices: polymer viscosity and matrix porosity effects. Eur. J. Pharm. Sci. 2, 213–219.

Bommareddy, G.S., Paker-Leggs, S., Saripella, K.K.. Neau, S.H., 2006. Extruded and beads containing Carbopol 974P to deliver nonelectrolytes and salts of weakly basic drugs. Int. J. Pharm. 321 (1–2), 62–71.

Bonferoni, M.C., Rossi, S., Tamayo, M., Pedraz, J.L., Dominguez-Gil, A., Caramella, C., 1994. On the employment of lambda carrageenan in a matrix system. II. λ-carrageenan and hydroxypropylmethylcellulose mixtures. J. Control. Release. 30, 175–182.

Bonferoni, M.C., Rossi, S., Ferrari, F., Bertoni, M., Bolhuis, G.K., Caramella, C., 1998. On the employment of lambda carrageenan in a matrix system. III Optimization of a lambda carrageenan-HPMC hydrophilic matrix. J. Control. Release 51, 231–239.

Bonferoni, M.C., Rossi, S., Ferrari, F., Bettinetti, G.P., Caramella, C., 2000. Characterization of a diltiazem-lambda carrageenan complex. Int. J. Pharm. 200, 207–216.

Bonferoni, M.C., Sandri, G., Rossi, S., Ferrari, F., Caramella, C., 2008. Polyelectrolyte–Drug complexes of Lambda carrageenan and basic drugs: relevance of particle size and moisture content on compaction and drug release behavior. Drug Devel. Ind. Pharm. 34, 1188–1195.

Colombo, P., Bettini, R., Massimo, G., Catellani, P.L., Santi, P., Peppas, N.A., 1995. Drug diffusion front movement is important in drug release control from swellable matrix tablets. J. Pharm. Sci. 84 (8), 991–997.

Colombo, P., Bettini, R., Santi, P., Peppas, N.A., 2000. Swellable matrices for controlled drug delivery: gel-layer behaviour, mechanisms and optimal performance. PSTT 3 (6), 198–204.

De Robertis, S., Bonferoni, M.C., Elviri, L., Sandri, G., Caramella, C., Bettini, R., 2015. Advances in oral controlled drug delivery: the role of drug-polymer and interpolymer non-covalent interactions. Exp. Opin. Drug Deliv. 12 (3), 441–453.

Dobrynin, A.V., Rubinstein, M., 2005. Theory of polyelectrolytes in solutions and at surfaces. Prog. Polym. Sci. 30, 1049–1118.

Esteban, S.L., Manzo, R.H., Alovero, F.L., 2009. Azithromycin loaded on hydrogels of carbomer: chemical stability and delivery properties. Int. J. Pharm. 366, 53–57.

Gallo, L., Piña, J., Bucalá, V., Allemandi, D., Ramírez-Rigo, M.V., 2013. Development of a modified-release hydrophilic matrix system of a plant extract based on co-spray-dried powders. Powder Tech. 241, 252–262.

García, A., Leonardi, D., Piccirilli, G.N., Mamprin, M.E., Olivieri, A.C., Lamas, M.C., 2015. Spray drying formulation of albendazole microspheres by experimental design. In vitro-in vivo studies. Drug Dev. Ind. Pharm. 41 (2), 244–252.

Grant, D.W, Higuchi T., 1990. Ion pairs and solubility behavior. Solubility behavior of organic compounds, vol. XXI. In: Techniques of chemistry. Wiley Interscience, NewYork, USA.

Guo, J.-H., Skinner, G.W., Harcum, W.W., Barnum, P.E., 1998. Pharmaceutical applications of naturally occurring water-soluble polymers. Pharm. Sci. Technol. Today 1, 254–260.

Guzmán, M.L., 2012. Desarrollo de nuevos sistemas farmacoterapéuticos y evaluación de sus propiedades biofarmacéuticas. PhD Thesis. Facultad de Ciencias Químicas, Universidad Nacional de Córdoba, Córdoba, Argentina.

Guzmán, M.L., Manzo, R.H., Olivera, M.E., 2012. Eudragit E100 as a drug carrier: the remarkable affinity of phosphate ester for dimethylamine. Mol. Pharm. 9 (9), 2424–2433.

Holgado, M.A., Fernandez-Arevalo, M., Alvarez-Fuentes, J., Caraballo, I., Llera, J.M., Rabasco, A.M., 1995. Physical characterization of carteolol: Eudragit® L binding interaction. Int. J. Pharm. 114, 13–21.

Hugerth, A.M., 2001. Micropolarity and microviscosity of amitriptyline and dextran sulfate/carrageenan-amitriptyline system: the nature of polyelectrolyte-drug complexes. J. Pharm. Sci. 90 (10), 1665–1677.

Jimenez-Kairuz, A.F., 2004. Investigación y desarrollo de nuevos materiales con utilidad en tecnología farmacéutica para el diseño de sistemas terapéuticos. Ph D thesis, Facultad de Ciencias Químicas, Universidad Nacional de Córdoba, Córdoba, Argentina.

Jimenez-Kairuz, A.F., Allemandi, D.A., Manzo, R.H., 2002. Mechanism of lidocaine release from carbomer-lidocaine hydrogels. J. Pharm. Sci. 91, 267–272.

Jimenez-Kairuz, A.F., Allemandi, D.A., Manzo, R.H., 2003. Equilibrium properties and mechanism of kinetics release of metoclopramide from carbomer hydrogels. Int. J. Pharm. 250, 129–136.

Jimenez-Kairuz, A.F., Allemandi, D.A., Manzo, R.H., 2004. The improvement of aqueous chemical stability of a model basic drug by ion pairing with acid groups of polyelectrolytes. Int. J. Pharm. 269 (1), 149–156.

Jimenez-Kairuz, A.F., Allemandi, D.A., Manzo, R.H., 2005. Swellable drug-polyelectrolyte matrices (SDPM): characterization and delivery properties. Int. J. Pharm. 288, 87–99.

Kavanagh, N., Corrigan, O.I., 2004. Swelling and erosion properties of hydroxypropyl methylcellulose (Hypromellose) matrices—influence of agitation rate and dissolution medium composition. Int. J. Pharm. 279, 141–152.

Kim, K.-S., Park, S.-J., 2010. Characterization and release behaviors of porous PCL/Eudragit RS microcapsules containing tulobuterol. Colloids Surf. B 76 (2), 404–409.

Kindermann, C., Matthée, K., Strohmeyer, J., Sievert, F., Breitkreutz, J., 2011. Tailor-made release triggering from hot-melt extruded complexes of basic polyelectrolyte and poorly water-soluble drugs. Eur. J. Pharm. Biopharm. 79 (2), 372–381.

Lankalapalli, S., Kolapalli, V.R.M., 2009. Polyelectrolyte complexes: a review of their applicability in drug delivery technology. Indian J. Pharm. Sci. 71, 481–487.

Lu, M.Y., Borodkin, S., Woodward, L., Li, P., Diesner, C., Hernandez, L., Vadnere, M., 1991. A polymer carrier system for taste masking of macrolide antibiotics. Pharm. Res. 8, 706–712.

Manzo, R.H., Jimenez-Kairuz, A.F., Olivera, M.E., Alovero, F., Ramirez Rigo, M.V., 2014. Thermodynamic and Rheological Properties of Polyelectrolyte Systems, Polyelectrolytes, Thermodynamics and Rheology. Springer International Publishing, Switzerland, 215–244.

Müller, M., 2014. Sizing, shaping and pharmaceutical applications of polyelectrolyte complex nanoparticles. Adv. Polym. Sci. 256, 197–260.

Nokhodchi, A., Asare-Addo, K., 2014. Drug release from matrix tablets: physiological parameters and the effect of food. Exp. Opin. Drug Deliv. 11 (9), 1401–1418.

Oosawa, F., 1997. Polyelectrolytes. Marcel Dekker Inc., New York.

Paker-Leggs, S., Neau, S.H., 2009. Pellet characteristics and drug release when the form of propranolol is fixed as moles or mass in formulations for extruded and spheronized Carbopol-containing pellets. Int. J. Pharm. 369 (1–2), 96–104.

Porasso, R.D., Benegas, J.C., Van den Hoop, M.A.G.T., Paoletti, S., 2001. Chemical bonding of divalent counterions to linear polyelectrolytes: theoretical treatment within the counterion condensation theory. Phys. Chem. Chem. Phys. 3 (6), 1057–1062.

Quinteros, D.A., Ramírez Rigo, M.V., Jimenez Kairuz, A.F., Olivera, M.E., Manzo, R.H., Allemandi, D.A., 2008. Interaction between a cationic polymethacrylate (Eudragit E100) and anionic drugs. Eur. J. Pharm. Sci. 33 (1), 72–79.

Quinteros, D.A., 2010. Desarrollo de nuevas estrategias de formulación de fármacos mediante el acomplejamiento con polielectrolitos. Ph D Thesis. Facultad de Ciencias Químicas, Universidad Nacional de Córdoba, Córdoba, Argentina.

Quinteros, D.A., Manzo, R.H., Allemandi, D.A., 2010. Design of a colonic delivery system based on cationic polymethacrylate (Eudragit E100)-mesalamine complexes. Drug Deliv. 17 (4), 208–213.

Radeva, T., 2001. Physical chemistry of polyelectrolytes. Ed. Marcel Dekker Inc, New York.

Rajesh, A.M., Bhatt, S.A., Venkata Ramana, G., Brahmbhatt, H., Gajjar, K., Tripathi, C.B., Anand, P.S., Popat, K.M., 2014. Taste masking by functional cross-linked copolymers and sustain release of drug through interpenetrating polymer network with sodium alginate and κ-carreganeen biopolymers. Int. J. Pharm. Pharm. Sci. 6 (11), 352–359.

Ramírez Rigo, M.V., 2006. Preparación y estudio de sistemas portadores de fármacos. Ph D Thesis. Facultad de Ciencias Químicas, Universidad Nacional de Córdoba, Córdoba, Argentina

Ramírez Rigo, M.V., Allemandi, D.A., Manzo, R.H., 2004. A linear free energy relationship treatment of the affinity between carboxymethylcellulose and basic drugs. Mol. Pharm. 1 (5), 383–386.

Ramírez Rigo, M.V., Allemandi, D.A., Manzo, R.H., 2006. Swellable drug–polyelectrolyte matrices (SDPM) of alginic acid: characterization and delivery properties. Int. J. Pharm. 322, 36–43.

Ramírez Rigo, M.V., Allemandi, D.A., Manzo, R.H., 2009. Swellable drug-polyelectrolyte matrices of drug-carboxymethylcellulose complexes. Characterization and delivery properties. Drug Deliv. 16, 108–115.

Ramírez-Rigo, M.V., Olivera, M.E., Rubio, M., Manzo, R.H., 2014. Enhanced intestinal permeability and oral bioavailability of enalapril maleate upon complexation with the cationic polymethacrylate Eudragit E100. Eur. J. Pharm. Sci. 55, 1–11.

Romero, V.L., 2012. Evaluación de los efectos de polímeros aniónicos y catiónicos en el desempeño y la eficacia de agentes antimicrobianos Fluoroquinolónicos. Ph D Thesis. Facultad de Ciencias Químicas, Universidad Nacional de Córdoba, Córdoba, Argentina.

Sinko, P.J., 2005. Colloids. Martin's Physical Pharmacy. Fifth ed. Lippincott Williams & Wilkins Ed, New York, pp. 469–498.

Thünemann, A.F., Müller, M., Dautzenberg, H., Joanny, J., Löwen, H., 2004. Polyelectrolyte Complexes. Advances in Polymer Sciences. vol. 166, Springer-Verlag Berlin, Heidelberg, pp. 113–171.

Vilches, A.P., 2003. Polielectrolitos solubles como portadores de fármacos ionizables. Preparación y estudio de sus propiedades farmacotécnicas. Ph D thesis. Facultad de Ciencias Químicas, Universidad Nacional de Córdoba, Córdoba, Argentina.

Vilches, A.P., Jimenez-Kairuz, A.F., Alovero, F., Olivera, M.E., Allemandi, D.A., Manzo, R.H., 2002. Release kinetics and up-take studies of model fluoroquinolones from carbomer hydrogels. Int. J. Pharm. 246, 17–24.

Zoppi, A., Linck, Y.G., Monti, G.A., Genovese, D.B., Jimenez Kairuz, A.F., Manzo, R.H., Longhi, M.R., 2012 Studies of pilocarpine: carbomer intermolecular interactions. Int. J. Pharm 427 (2), 252–259.

FURTHER READING

Higashi, K., Yamamoto, K., Pandey, M.K., Mroue, K.H., Moribe, K., Yamamoto, K., Ramamoorthy, A., 2014. Insights into atomic-level interaction between mefenamic acid and eudragit epo in a supersaturated solution by high-resolution magic-angle spinning NMR spectroscopy. Mol. Pharm. 11 (1), 351–357.

Sedlák, M., 2002. Mechanical properties and stability of multimacroion domains in polyelectrolyte solutions. J. Chem. Phys. 116 (12), 5236–5245.

Sinha Roy, D., Rohera, B.D., 2002. Comparative evaluation of rate of hydration and matrix erosion of HEC and HPC and study of drug release from their matrices. Eur. J. Pharm. Sci. 16, 193–199.

CHAPTER

14 POLYSACCHARIDE BASED NANO/MICROFORMULATION: AN EFFECTIVE AND VERSATILE ORAL DRUG DELIVERY SYSTEM

Md Nurunnabi*,, Vishnu Revuri**, Kang Moo Huh*, Yong-kyu Lee****

*Chungnam National University, Daejeon, Republic of Korea

**Korea National University of Transportation, Chungju, Republic of Korea

CHAPTER OUTLINE

1 Introduction 409
2 Chitosan 415
3 Hyaluronic Acid 418
4 Chondroitin Sulfate 420
5 Heparin 420
6 Cyclodextrins 423
7 Other Polysaccharides 424
8 Polysaccharide Blends (PSB) in Oral Drug Delivery 429
9 Conclusion and Future Prospects 429
References 430

1 INTRODUCTION

The oral route has been the much-preferred route of drug administration among all of the routes due to the patient convenience and, most importantly, requires no qualified personalities for the administration (Khatun et al., 2013; Patel et al., 2012; Sastry et al., 2000). However, the delivery of biomolecules like peptides, proteins, siRNA, or DNA has been critically impeded due the stumbling barriers imposed by the gastrointestinal (GI) system (Guliyeva et al., 2006; He et al., 2008; Morishita and Peppas, 2006). The poor stability of the biomolecules in the acidic gastric environment, proteolytic enzyme degradations, the mucosal barriers, variable pH along the different intestinal regions, and the lymphatic clearance impedes their transcendence that results in the poor drug/biomolecules bioavailability. In order to develop robust drug carrier systems that can endure these impediments and enhance the bioavailability of the drugs, it is very important to learn about the GI pathophysiology and the impediments in detail. The digestive organ, the stomach—with small intestine, aids in the digestion and absorption of nutrients from the ingested food. However, stomach modulates the pH blow 3.0 in order to facilitate active proteolytic enzyme activity (Nurunnabi et al., 2016). This might limit the stability and solubility of the biomolecules that can end up toward degeneration and degradation. Moreover, the stomach secretes diverse proteolytic enzymes like pepsin, chymotrypsin and so on, which can specifically degrade these peptides and have a critical impact over their bioavailability (Guliyeva et al., 2006; Garcia-Fuentes et al., 2005). In order to prevent the orally administered biomolecules from the acid degradations, thick enteric coatings have been one of the plausible solutions where these coatings prevent the inner cores from stomach acids and proteolytic enzymes (Carino and Mathiowitz, 1999). The intestinal mucosal secretion from goblet cells lining the small intestines is the next barrier that prevents the drug transport. Mucus is a viscoelastic hydrogel that grants defensive and lubricating properties in the GI tract. Moreover, the thickness of this mucus varies throughout the GI tract (Vasir et al., 2003; Xu et al., 2004). The mucus layers materialize into a protective sieve that restricts the transport of larger particles and foreign pathogens through hydrophobic or nonspecific interactions. This limits the residence time of the orally targeted nanomaterials in the GI tract. Recent advances include the use of mucoadhesive polymers that can improve the residence time as well as provide transport of these materials into the blood circulations (Hearnden et al., 2012; Paliwal et al., 2012). The third barrier is encountered from the epithelial tight junctions that are impervious and inflexible, which makes it hard to channel the drug delivery systems into the blood circulations. However, transient transcellular and paracellular transport mechanisms facilitate the nonspecific passive diffusion of the molecules (Khatun et al., 2014). The uptake and transport through these passive drug transport mechanisms are very meager, which results in insignificant therapeutic efficacies of the drug delivery systems. Moreover, the variance in the GI pH ranging from 1 to 8 can result in oxidation, hydrolysis, or deamidation of the molecules that can result in the loss of the therapeutic activity of the targeted molecules (Allard et al., 2009). Even if the drug delivery systems traverse these barriers, they encounter the biological filtering portal circulatory system, the liver that screens, metabolizes these molecules and makes them hard to attain the desired therapeutic levels (Fig. 14.1).

Diverse strategies have been employed to expand the significance of the oral delivery. Different synthetic and natural polymers based nanocarrier systems have been developed as drug delivery depots to confront the impediments and improve the therapeutic efficacy (Rastogi et al., 2007). For example, enteric coatings over the conventional drugs have been one of the oldest strategies where these drugs are coated with the enteric polymers in order to protect the drug/biomolecules from the GI acidic degradations and facilitate pH dependent release of the cargo at different fragments of the small intestines (Sarmento et al., 2007). Numerous enteric coating materials have been listed in Table 14.1. However, these enteric coatings could not address the issues associated with the mucosal barriers and some of the enteric coatings can induce the toxicity. Moreover some of these enteric-coated drug delivery systems fail to correlate between in vitro and in vivo studies.

Synthetic polymers have been one of the promising strategies in the development of oral drug delivery systems (Hejazi and Amiji, 2003; Semete et al., 2010). Using the synthetic polymers helps to control the size, charge, and other physiochemical properties of the nanocarriers, which can regulate the levels of drug release and transport into the enterohepatic circulation. Moreover, it is easy to synthesize and conjugate the biological molecules to achieve the active targeting of these polymeric nanomaterials. In general, the conjugation of the polymers with the nanoparticles or drugs would be via self-assembly where the cores of the polymeric materials are loaded with the drug molecules. Moreover, by optimizing the self-assembly parameters and selection of appropriate polymers, one could modulate the activity and tune the

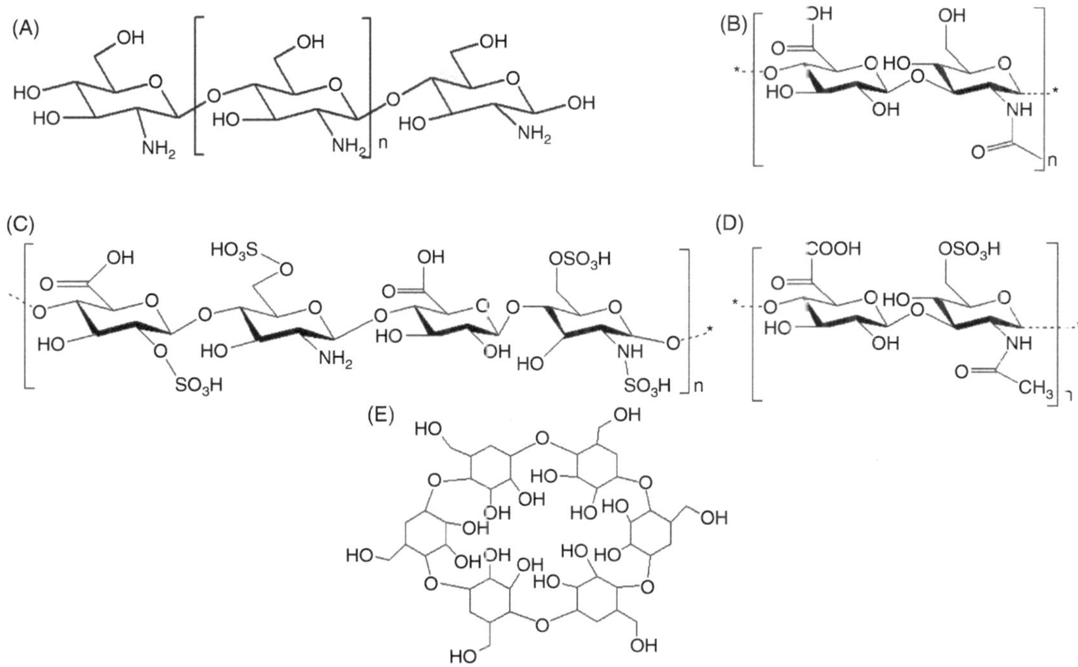

FIGURE 14.1 Typical Chemical Structure of Commonly Used Polysaccharides for Oral Drug Delivery

(A) Chitosan, (B) hyaluronic acid, (C) heparin, (D) chondroitin sulfate, and (E) cyclodextrin.

Table 14.1 List of Different Enteric Coating Polymers and Their Intestinal Dissolution pH

Enteric Polymers	Intestinal dissolution pH
Cellulose acetate phthalate	6.2
Cellulose acetate trimellitate	5
Hydroxypropyl methylcellulose phthalate	4.5–5.5
Poly(methacrylic acid-co-methyl methacrylate)	5.5–7.0
Poly(vinyl acetate phthalate)	5
Shellac (esters of aleurtic acid)	7

From Hussan et al. (2012).

drug release from these polymeric nanosystems. The mucoadhesion improves the bioavailability of the drugs and facilitates improved therapeutic efficacy with minimal doses. These polymers form a weak intermolecular bonding (hydrogen bond, hydrophobic interactions, physisorption, or in a combination) with the mucus layers and aid in the transport of the nanosystems. For example, conjugating the surface of the PLGA nanoparticles with poly(ethylene glycol) (PEG) could favor them to traverse the mucosal barriers. Other polymers like polyacrylic acid, poly (fumeric-*co*-sebacic) anhydride, poly lactic acid (PLA), poly lactic-*co*-glycolic acid (PLGA) are some of the mucoadhesive polymers. However, the quantitative levels of polymeric nanomaterials that can cross the epithelial tight junctions are very low (Gao et al., 2013). Though some of these polymeric nanomaterials acts as paracellular transport permeation enhancers, they lack the selectivity, that is, they can channel the transport of foreign pathogens through the tight junction openings leading to unnecessary complications (Rastogi et al., 2007; Venishetty et al., 2012). In order to cover these barriers, ligands specific to the receptors expressed on the intestinal walls have been designed. Some of these examples include Fc peptides, vitamin B12 analogs, lectins, invasins, or bile acid derivatives (Huang, 2009; Kochut and Dersch, 2013; Petrus et al., 2009; Radaev and Sun, 2001; Smart, 2004). They could improve the bioavailability (Gao et al., 2013) of these drug targets. Different strategies have been implemented to orally deliver the drug/biological molecules and improve their in vivo stability, absorption and therapeutic efficacy. For example, recently a pH responsive PLGA nanoparticles loaded with insulin were formulated. The use of sodium oleate to induce the hydrophobic ion pairing technology for improving the loading of hydrophilic insulin into the hydrophobic cores of PLGA nanoparticles and further enteric coating by Eudragit aided in sustained released of insulin in the GI tract.

Thioketal nanoparticles are a class of nanoparticle system where they release the cargo in response to the reactive oxygen species (ROS) present in the cells (Kim et al., 2015). For example, Wilson et al. (2010) synthesized poly-(1,4-phenyleneacetone dimethylene thioketal) polymers-based nonviral vectors for the oral delivery of TNF-α siRNA to treat colon cancer. These nanoparticles could endure the acidic, basic, or enzymatic degradations in the GI tract. Since the thiketal linkages are very specific to the ROS that are specifically present in the inflamed tissues, they could induce the localized release of siRNA at the targeted site.

Dendrimers-based technologies are a class of polymeric nanomaterials that have been investigated for oral drug delivery (Duncan and Izzo, 2005; Najlah et al., 2007). The precision in regulating the size and physiochemical properties, their surface hydrophilicity, their hallow interiors incorporated in the drug molecules, their ability to endure the GI degradations, their ability to act as a permeation enhancers, and aid in paracellular transport made dendrimers one of the plausible solutions in improving the oral availability of the drug molecules. Different cationic, anionic, and nonionic dendrimers have been fabricated to study the GI stability, for comparing their uptake and transport of the drug molecules. However, the toxicity issues based on the generations of the dendrimers and poor efficacy compared to other carriers limited their therapeutic applications.

Biological polymers have gained significant attention in recent times to fabricate oral drug delivery systems. This significance is due to their biocompatibility, biodegradability, nontoxicity with negative immunogenicity. Proteins and polysaccharides are widely used natural polymers for the drug delivery applications. One of the classical examples of protein-based biopolymers is gelatin. Gelatin is a functional protein that is water soluble and forms a gel-type structure under specific processing parameters. Gelatin has played a predominant role in pharmaceutical and food industries due to their abundance and relatively easy extraction (Kuijpers et al., 2000). However, gelatin could easily disintegrate in the acidic gastric pH (Kumari et al., 2010). Several polymers-coated gelatin capsules have been fabricated in order to overcome the above-mentioned impediments and achieve the oral delivery of biological molecules. For example, a pH sensitive hydrogel composed of gelatin and poly acrylic acid was fabricated to study the pH dependent release of ketoprofen from the hydrogel matrices. Very recently the mucoadhesive property of the gelatin was explored to fabricate semiinterpenetrating scaffolds for the oral mucosal delivery. The controlled release of the nystatin from these hydrogels indicates their importance in oral drug delivery. Other proteins like immunoglobulin G (IgG) targeting neonatal Fc receptors in the GI tract, bovine serum albumin have also been explored for oral drug delivery. For example, Pridgen et al. (2013) fabricated PLA-PEG nanoparticles coated with IgG Fc peptides for the oral delivery of insulin (Pridgen et al., 2013). The Fc receptors resulted in enhanced uptake of the IgG Fc–coated nanoparticles and had a significant effect over the insulin release and regulated blood glucose levels compared to the pristine nanoparticles group.

Polysaccharides (PS) are a class of natural polymers that plays a predominant role in the biological system. They are composed of monosaccharide monomers that are conjugated by the glycosidic linkages. They are the basic components of the extra cellular matrix (ECM) that can regulate the integrity of the ECM. They possess multiple functional groups to favor the conjugation and transport of drug/biological molecules. They can be easily degraded to the monomers of monosaccharides without initiating any systemic toxicity and immunogenicity. They can easily be recognized by the receptors expressed on the surface of specific cell type and aid in active targeting of biological molecules. The amine, carboxyl, or hydroxyl groups of the polysaccharides can favor the noncovalent interactions with the biological tissues and improve the bioadhesion of the nanomaterials. Moreover the sources for the extraction are relatively abundant and cost of extraction, purifications are economical. The molecular weight, charge, and monomer composition of the polysaccharides govern their physiochemical properties like degrees of hydrophilicity, water sorption, degradation, solubility, and have a marked effects over their biocompatibility and biological activity. Polysaccharides can be divided in homopolysacharides and heteropolysaccharides based on the monomer composition. Some of the typical examples of homopolysaccharides include cellulose, glycogen, xylan, and chitin and so on, where the polysaccharides are polymers of single monomer units. In case of heteropolysaccharides, the repeating units are composed of different monomer subunits. For example, cellulose a homopolysaccharide is a polymer of D-glucose subunits with $\beta(1\rightarrow4)$ glycosidic linkages, whereas chondroitin sulfate is a sulfated heteropolysaccharide that is composed of composed of *N*-acetylgalactosamine and glucuronic acid. In a polyelectrolyte perspective, the PS can be divided in polyelectrolyte PS and nonpolyelectrolyte polysaccharides. The polyelectrolyte polysaccharides can further be divided into cationic polysaccharides (chitosan) and anionic polysaccharides (hyaluronic acid, heparin chondroitin sulfate, etc.) based on the presence of positive and negative charges. The conformation of the polysaccharides principally depends on the ionic strength, pH of the biological media, temperatures, and the local receptors/proteins that initiate the interactions with these polysaccharides. These chemical/physical stimuli can be exploited to trigger the phase transition of these polysaccharides. For example, by applying external stimuli like heat to alter the local temperatures or electric field, the charge distributions within the polysaccharides can be easily altered. Moreover, conjugating metallic nanoparticles like gold or iron oxide, a synergistic effect over the behavior of the polysaccharides can be seen. The ability of the polysaccharides to respond the local stimuli can assist in the development of stimuli responsive drug delivery systems and improve the targeted delivery of the biomolecules.

Several strategies have been employed to fabricate polysaccharide-based drug delivery systems. Covalently crosslinked nanoparticles have been principally used for the conjugation of the drug or biomolecules. For example, initially glutaraldehyde crosslinkers were used to covalently conjugate and stabilize the chitosan nanoparticles. However, the toxicity profiles of glutaraldehyde over the cell viability restrained their use for crosslinking. Recently, the biologically compatible amide-based crosslinking have been widely used to conjugate the biological molecules. The carboxyl/amine groups of the polysaccharides were principally employed in the covalent conjugation (EDC-NHS chemistry) of the polysaccharides. The covalent conjugation was stable under low, neutral, and mild basic pH conditions. However, the size of the particles may vary depending on the pH of the biological systems. Ionic crosslinks-stabilized polysaccharides nanoparticles have gained significant attention in recent years due to the ease of conjugation and preparation procedures. Lower molecular weight cationic or anionic polymers could engage in the conjugation and synthesis of ionically crosslinked polysaccharides nanoparticle systems. Tripolyphoshate (TPP) is one of the most widely used anionic crosslinkers that can induce ionotropic gelation with the positively charged chitosan and aid in the formation of stabilized nanoparticles. The size of the nanoparticles can be varied by altering the pH, chitosan to TPP ratio, and the molecular weight of the chitosan. This method has been widely used for the delivery of hydrophobic nanomaterials, peptides, and biomolecules. For example, Amidi et al. (2006) have fabricated trimethyl chitosan-TPP nanoparticles for the nasal delivery of ovalbumin. These particles had 95% of protein-loading efficiency without losing the structural integrity of ovalbumin. Calcium-assisted ionic crosslinked nanoparticles have also been fabricated as drug delivery systems. The negatively charged PS with carboxyl groups can interact with the bivalent calcium ions to form calcium crosslinked nanoparticles. Nesamony et al. (2012) used ultrasound-assisted microemulsion technique for the synthesis of bovine serum albumin (BSA)-loaded calcium alginate nanoparticles. The particles had a mean particle size of 350 nm with 40% loading efficacy of BSA and took 16 h to completely release the loaded biomolecules. Several researchers have explored the use of chitosan as a potential carrier for oral gene therapy. The electrostatic forces-driven conjugation of negatively charged siRNA/DNA/biomolecules to the positively charged chitosan PS and subsequent synthesis of nanoparticle is one of the promising and efficient modes of delivering the biomolecules. For example, Choi et al. (2015) fabricated siRNA loaded-calcium phosphate nanoparticles and further decorated to glutamine conjugated oligochitosan to study the osteogenic differentiation of mesenchymal stem cells. The results showed that the synthesized siRNA loaded nanoparticles had a mean diameter of 172 nm with a mean loading efficacy of about 82%. Further, coating of glutamine conjugated oligochitosan over the calcium phosphate nanoparticles resulted in improving the stability of these nanoparticles. The hydrophilic polysaccharide can result in amphiphilic polysaccharide upon conjugation with the hydrophobic molecules. The functional groups over the surface of the polysaccharide can favor covalent conjugation of hydrophobic molecules. Upon contact with water, these drug-conjugated PS would self-assemble to form micelles in order to decrease the interfacial free energy. Zhang et al. (2013) briefed the importance of polysaccharides in designing micellar-based drug delivery systems. For example, Kesharwani et al. (2015) synthesized 3,4-difluorobenzylidene curcumin loaded hyaluronic acid nanomicelles to target cancer stem cells. Styrene-*co*-maleic anhydride (SMA) was conjugated to hyaluronic acid in order to facilitate the formation of micellar structures. The ability of hyaluronic acid to target the CD44 receptors was explored to target the pancreatic cancer stem cells. The drug-loaded nanomicelles had a mean particle size of 114 nm. The release of drugs from the nanomicelles was restrained in acidic pH compared to the burst release in basic pH. This is because at the acidic pH the polymers tend be protonated and attain hydrophobic (HPB) configurations. The HPB configurations of hyaluronic acid-SMA polymers could result in strong interactions with the drugs and aid in regulating the drug release. Different core-sell-based polysaccharide-based nanoformulations have also been formulated for the drug delivery of desired molecules. For example, Bagre et al. (2013)

synthesized alginate coated chitosan-based core-shell nanoparticles for the oral delivery of enoxaparin. Initially, enoxaparin loaded chitosan nanoparticles were fabricated by ionic gelation method using sodium tripolyphosphate, which was followed by dropwise addition of sodium alginate under mild stirring. The chitosan-alginate core-shell nanoparticles had a mean diameter of 335 nm with enoxaparin loading efficiency of 71%. Recently, hydrogel-based drug delivery systems have gained significant attention in the oral delivery of biological molecules. Hydrogels can effectively sense the local pH and subsequently swell/deswell in order to protect or release the drug from the hydrogel matrices. Recently, Mukhopadhyay et al. (2014) synthesized succinyl chitosan-grafted polyacrylamide (FAA) porous hydrogels for the controlled release of insulin. The pores within the hydrogels were dependent on the concentrations of the crosslinker methylenebisacrylamide used. The hydrogels had a loading efficiency of 38% and resulted in regulating the blood glucose levels by delivering insulin around 4.8%. Other crosslinked PS hydrogel networks, injectable hydrogel networks, semiinterpenetrating polysaccharide networks have been explored as novel polysaccharide based drug delivery systems.

Here we discuss the biological importance and advantages of some of the polysaccharides used in oral drug delivery applications. We focus on the mechanisms by which the PS could endure and assist in oral drug delivery.

2 CHITOSAN

Chitosan is a positively charged linear PS with a random arrangement of β-(1–4)-linked D-glucosamine and *N*-acetyl-D-glucosamine. Chitosan is a derivative of chitin that results from the deamination of chitin. The presence of ample number of amino groups on the surface of chitosan aids in generating positive zeta potentials. As the pK_a of amino groups on the chitosan is around 6.5, they tend to remain protonated at acidic and neutral pH and result in stabilizing the nanoparticles. However, the physiochemical properties of the chitosan can be varied based on the pH, molecular weights and the degree of deacetylation. Chitosan is one of commonly used polysaccharides for diverse applications such as food processing, cosmetics, fabrics, and water treatment. The presence of amino functional groups on the surface of the chitosan helps them chemically or physically interact with the nanoparticles, drugs, polymers, and cells. Chitosan has been one of the most widely explored polysaccharide in the pharmaceutical and biomedical sectors. As these materials are biocompatible, biodegradable, nontoxic, and possess wound healing and antimicrobial properties they are widely investigated for myriad of biological applications. Orally administered chitosan can reduce the absorption of dietary fat and cholesterol. Their ability to prevent the loaded therapeutic molecules from GI and enzymatic degradations and facilitate mucoadhesion of the loaded therapeutic molecules to the secreted mucus aids in prolonging their residual time in the small intestines. Moreover, their propensity to open the tight junctions and aid in the paracellular transport of the drug-loaded chitosan molecules made them one of the promising materials for oral drug delivery. Several derivatives of chitosan have been synthesized in order to derive/modulate the physiochemical properties of chitosan polysaccharide. For example, the amino groups in the chitosan can be quaternized in order to regulate the positive charge and improve the aqueous solubility of chitosan. Trimethyl chitosan (TMC), triethyl chitosan (TEC), diethylmethyl chitosan (DEMC), and dimethylethyl chitosan are some of the typical quaternized derivatives of chitosan. The degree of quaternization determines the extent of mucoadhesive property and the improved tight junction opening. Increased quaternization and reduced length of alkyl groups substituted can enhance the tight junction opening and the permeability of the nanoparticles. Hence, the efficacy in opening the tight junctions and mediating the transport is in the following order.

TMC > DMEC > DEMC > TEC > chitosan

This is because the driving force shielding the positive charge of nitrogen substituted ethyl groups is steric hindrance. This subsequently lowers the efficacy of ethyl quaternized chitosan in the transport enhancement. Thiolated chitosan derivatives possess enhanced mucoadhesive property compared to the pristine chitosan. This is because the thiolated chitosan molecules can form a disulfide bonds with the glycoproteins in the mucus. As the disulfide bond strengths are stronger than the ionic interactions that are seen by the chitosan molecules, they can improve the retention times compared to the pristine chitosan. TMC-cysteine derivatives have a synergistic effect compared to the TMC alone. This is because the cysteine domains can enhance the mucoadhesion and TMC can assist in a tight junction opening and subsequent permeation compared to TMC alone. Amphiphilic chitosan derivatives by grafting the hydrophobic moieties like bile acids; fatty acids have been explored to enhance the mucoadhesion of the chitosan nanoparticles. As the sulfonated residues, charged acidic groups on sialic acid and hydrophobic methyl groups on the fucose residues of the mucus are the preliminary modes to enhance the mucoadhesion, incorporating a hydrophobic moiety can significantly improve the mucoadhesion and prolong the residence time of the chitosan drug conjugates. Other chitosan derivatives include carboxyl chitosan, chelating agents bearing chitosan derivatives are designed to prevent the enzyme-assisted degradations and PEGylated chitosan derivatives to improve the stability of the nanoparticles with minimal toxicity. The mechanism by which the chitosan nanoparticles could be reversible opens the tight junctions and facilitates the paracellular transport of the chitosan nanoparticles has been critically reviewed. The expression of claudin-4, a transmembrane protein regulating the tight junction integrity, can be modulated by the presence of chitosan at the tight junctions. The presence of chitosan on the surface of the epithelial cells reduced the expression of claudin-4 at the membrane and the cytosol. This resulted in impaired membrane integrity and aided in paracellular transport of the chitosan nanomaterials. However, removing the chitosan from the extracellular fluids resulted in improving the expression of claudin-4 and helped in maintaining the integrity of intestinal tight junctions. Chitosan nanoparticles can also reach the enterohepatic circulations via transcellular pathways. M cells present in the Peyer's patches and the enterocytes can assist in uptake and transport of chitosan molecules similar to the transport of pathogens. As the chitosan polysaccharide has a mucoadhesive property they can retain in the mucus layers and preferentially aid in improved transport compared to the other materials. The stability of the chitosan nanoparticles primarily depends on pH of the biological media. Transmission electron microscopy helped to study the stability and integrity of the chitosan nanoparticles at different biological pH. At pH <7, the particles remain intact and stable. However, at pH >7.4, deprotonation, subsequent aggregation and disintegration of the particles were observed, which resulted in the release therapeutic cargo materials from the chitosan cores. Several different strategies based on chitosan and chitosan derivatives have been formulated for the delivery of biomolecules and drugs. Recently, a mannose modified TMC nanoparticle loaded with siRNA was used to knock down the TNF-α expressions in the gut associated macrophages. The use of mannose resulted in improving the internalization of the nanoparticles by clatherin-independent mannose targeting and uptake in the macrophages. Moreover, the thiol and trimethyl groups in the chitosan resulted in prolonging the intestinal retention time and improved the bioavailability of siRNA. Single oral gavage of the siRNA nanocomplex with a concentration of 50 μg/kg in Sprague Dawley rats resulted in a 65% reduction in the serum TNF-α expression. The final nanocomplex was able to reduce the expressions of TNF-α in liver, spleen, and lung tissues with enriched macrophage concentrations and prevents the incidence of acute hepatic injuries. In another study, Huang et al. (2014)

formulated polycaprolactone (PCL) conjugated hyaluronic acid (HA) nanoparticles and further wrapped with chitosan to improve the oral delivery of anticancer drugs. Paclitaxel (PTX) and doxorubicin (DOX) were the model drugs that were loaded into the cores of the nanoparticles (Huang et al., 2014). Further coating with chitosan resulted in shielding and stabilizing the particles in acidic pH. However, at pH 7.4, the enfolding was released that assisted in release of the drug loaded hyaluronic acid-*g*-PCL into the systemic circulation. The orally delivered PTX-loaded nanoparticles resulted in 100% survival of the tumor-bearing mice. Moreover, the relative tumor volumes and the tumor burden were regulated in the chitosan coated hyaluronic acid-*g*-PCL nanoparticles compared to the pristine IV administered PTX and hyaluronic acid-*g*-PCL nanoparticles. These studies indicate the importance of chitosan in improving the oral delivery of anticancer drugs. Iqbal et al. (2012) fabricated chitosan-thioglycolic acid oral gel for the delivery of synthetic nanopeptide, Leuprolide acetate. The fabricated oral gels contained glutaraldehyde (GSH), which helped in improving the permeation and transport of the Leuprolide. The addition of 0.5% (w/v) GSH resulted in improving the Leuprolide transport 3.7 times compared to the pristine Leuprolide. Loading GSH into the gels resulted in inhibiting the activity of protein tyrosine phosphatase (PTP), an enzyme that is abundant at mucosal membranes. The presence of GSH results in forming a disulfide bond with the cysteine residues of PTP. This results in improving the number of phosphorylated tyrosine moieties on the extracellular loops of occludin and subsequently opened the tight junctions. The presence of thiol groups in the oral gel could assist in reducing the disulfide GSH (GSSG) to GSH and aid in improving the interactions with PTP. A 10-fold improvement in the bioavailability of oral gel–loaded Leuprolide was observed compared to the control group in vivo. However, these formulations were partially effective resulting in only 1% of bioavailability of orally administered Leuprolide, which is suboptimal compared to the other peptide-based formulations (Fig. 14.2).

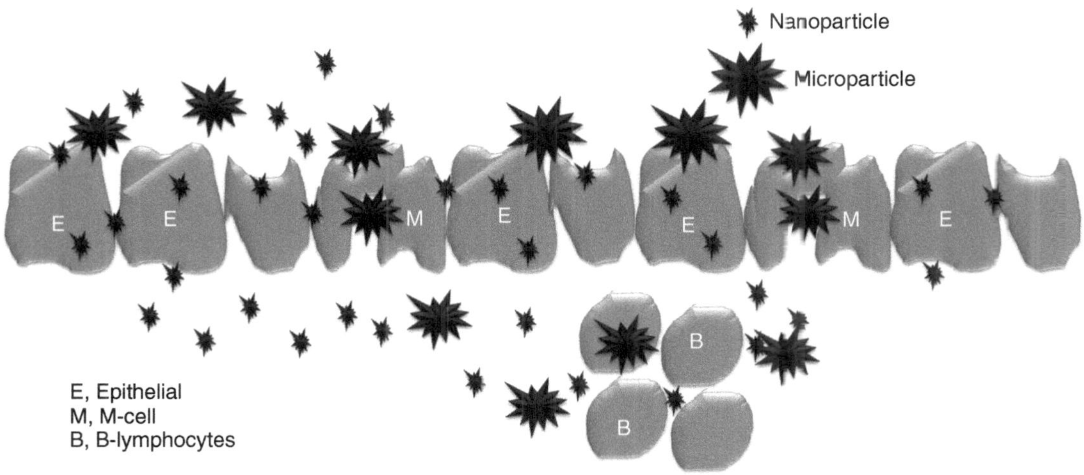

FIGURE 14.2 Scheme of Intestinal Epithelium. E, M, and B Represents M Cell, Epithelial Cell, and B Lymphocytes, Respectively

M cells located between the epithelial cells in the lymphoid. Microparticles usually taken up by M cells, whereas nanoparticles taken up by epithelial cells. For vaccine delivery microparticulate formulation supposed to be more effective since they should transport into the Peyer's patches. However, most often microparticles take up by the phagocytotic cells.

3 HYALURONIC ACID

Hyaluronic acid is a nonsulfated, nonimmunogenic, and nonadhesive glycosaminoglycan that is highly abundant in the biological systems. They belong to the class of hetero-PS with repeating units of D-glucuronic acid and *N*-acetylglucosamine and a molecular weight distribution ranging from 10^3 to 10^7 daltons. The predominant biological functions like wound healing, initiating the signal transduction to improve the cell proliferations, and regulating the physiochemical properties of pericellular matrices made them one of the widely investigated biomaterial for drug delivery applications. Their viscoelastic, electrostatic, and networking properties play a vital role in many biological activities. Having such abundance and biological relevance, hyaluronic acid has been one of the highly investigated materials for biomedical applications. Ease of surface modifications on hyaluronic acid favored conjugation and crosslinking with other functional moieties that could be employed for drug delivery, wound healing, and other applications. The ability of hyaluronic acid to preferentially target the CD44 receptors that are abundant on the cancer cells made them a promising material in cancer research. Moreover, the selective degradations by hyaluronidase enzymes made them an interesting material in drug delivery perspective. As hyaluronic acid has very short plasma half-life of 2.5–5 min, different chemical derivatives have been formulated. For example, hyaluronic acid has been crosslinked with divinyl sulfone, dialdehyde, or glycidyl ether to improve the bioavailability of hydrophobic drugs like PTX, DOX, and biomolecules. However, most of these formulations are widely investigated for intravenous delivery. A very few formulations have been explored for oral delivery. For example, Piao et al. (2007) fabricated hyaluronic acid–PEG microspheres by a spray-drying method in order to improve the bioavailability of Piroxicam. A 3-fold increases in the dissolution rates resulted in improved plasma concentrations of Piroxicam. Moreover, a 2-fold increase in the area under the curve was observed in Piroxicam-loaded microspheres compared to pristine Piroxicam. The spray-drying technique was also used to fabricate the hyaluronic acid microspheres to transport cyclosporine A. However, these formulations suffer from poor liposolubility and high molecular weight. Hsieh et al. (2014) fabricated adipic acid dihydrazide (ADH) modified hyaluronic acid and studied the effect of molecular weights and degrees of HA modifications. Their results claim that the low molecular weight hyaluronic acid (50 kD) with reduced degrees of ADH substitution can easily be metabolized and results in rapid clearance from the biological system. However, excessive chemical modifications could result in reduced uptake due to steric hindrance and lack of recognition by the specific homing receptors like CD44 or hyaluronan receptor for endocytosis (HARE). Han et al. (2012) used ADH-modified HA for designing the pH responsive insulin delivery system. The low ionization of HA at acidic pH resulted in preventing the release of insulin in acidic pH. However, the degree of ionization changes at pH 6.8 and can assist in the release of insulin from the hyaluronic acid matrices. The insulin-loaded hyaluronic acid nanoparticles were able to induce the glycemic effects that resulted in 24% reduction in blood (Fig. 14.3) glucose levels within 2 h and about 35% reduction in blood glucose levels in during the next 6 h. Fiorica et al. (2013) fabricated a pH-sensitive hyaluronic acid derivative using ATRP polymerization in order to conjugate the pH-responsive sodium methacrylate to the hyaluronic acid. These hyaluronic acid derivatives were insoluble at gastric pH and assisted in protecting the peptide α-chymotrypsin from degradation. Though hyaluronic acid represents one of the promising candidates for oral drug delivery, more extensive research in terms of understanding the mechanism of drug transport through the epithelial tight junctions and comparative efficacy over other PS-based drug delivery systems is required.

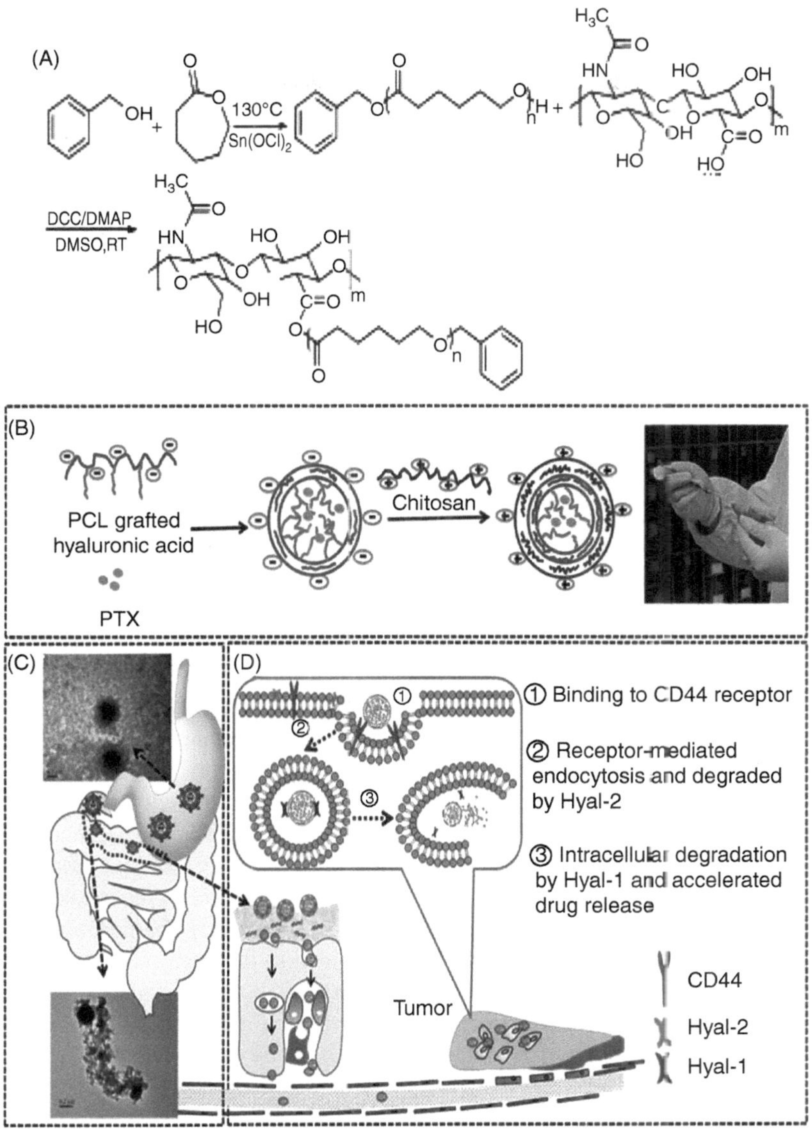

FIGURE 14.3 (A) Synthesis process of hyaluronic acid-*g*-PCL copolymer. (B) Formation of paclitaxel loaded nanoparticle into PCL grafted hyaluronic acid. (C) Mechanism of transportation of the nanoparticle in GI tract and (D) schematic presentation of cellular uptake pathways and subcellular release behavior.

From Huang, P., et al., 2014. Improving the oral delivery efficiency of anticancer drugs by chitosan coated polycaprolactone-grafted hyaluronic acid nanoparticles. J. Mater. Chem B, 2, 4021–4033.

4 CHONDROITIN SULFATE

Chondroitin sulfate is one of the sulfated GAGs with repeating disaccharide units of β-1,4-linked D-glucuronic acid (GlcA) and β-1,3-linked *N*-acetyl galactosamine (GalNAc). Based on the position of sulfur groups substituted on the chondroitin sulfate backbone different chondroitin sulfate has been designated. Chondroitin sulfate is one of the primary components of ECM and abundant in various biological tissues like skin, bone, cartilage, blood vessels, nerve tissues. The antioxidant, antiartherosclerosic, anticoagulant, antithrombosis, and negative immunogenic properties of chondroitin sulfate made one of the promising materials in the biomedical research. They have been one of the most widely used materials in the treatment of osteoarthritis. Moreover, they play a predominant role in neural signal transduction and improve the activity of central nervous system. Furthermore, tight packing of chondroitin sulfate with highly charged sulfated groups could restrict the compression of cartilage and aid in cartilage regeneration. Moreover, the ability of chondroitin sulfate to target the CD44 receptors on the cancer cells made them to be explored in active targeting for cancer therapy. The presence of carboxyl and hydroxyl functional groups allows chondroitin sulfate to involve in deluge of covalent/ionic interactions with other molecules. For example Xiao et al. (2014) fabricated α-linolenic acid–conjugated chondroitin sulfate micellar oral drug delivery systems in order to improve the intestinal absorptions and oral bioavailability of low molecular weight chondroitin sulfate. The extent of hydrophobicity and the molecular weights of the polysaccharide were the deciding factors in improving the oral availability of chondroitin sulfate. The results showed that the micellar structures of the chondroitin sulfate-α-linolenic acid micelles could induce an effect over the F-actin cytoskeleton and subsequently improve the paracellular uptake and transport of the chondroitin sulfate derivatives into the enterohepatic circulation. Cho et al. (2014) synthesized chondroitin sulfate-wrapped gold nanoparticles for improving the oral delivery of insulin. Chondroitin sulfate was used as a reducing and stabilizing agent for the synthesis of gold nanoparticles. The mean concentrations of insulin in the blood plasma were 6-fold higher than the pristine insulin administered orally and resulted in 32% reduction in the blood glucose levels. Hu et al. (2012) fabricated FITC-BSA-loaded chondroitin sulfate-chitosan nanoparticles by ion-gelation method and investigated the effect of the charge over the drug administration. By modulating the surface charge of the nanoparticles they were able to see the variance in the uptake and bioavailability of the nanoparticles in the human Caco-2 monolayers. The positively charged nanoparticles showed enhanced uptake with a 42% increase in the uptake compared to the negatively charged nanoparticles. Though studies confirmed the effect of charge on the transport, a conclusive study on the mechanism of chondroitin sulfate-mediated transport (paracellular or transcellular or CD44 mediated endocytosis) and comparison in between different transport studies needs to be addressed.

5 HEPARIN

Heparin is a long-chain polysaccharide best known for its anticoagulant effect. There is both low and high molecular weight heparin used for different purposes. Heparin usually is a highly water soluble and negatively charged surface that cannot be absorbed through the GI tract when administered orally. However, due to its stability in acidic conditions, such as stomach and most importantly biocompatibility properties, heparin has been widely considered for oral drug delivery formulation and carrier (Fig. 14.4).

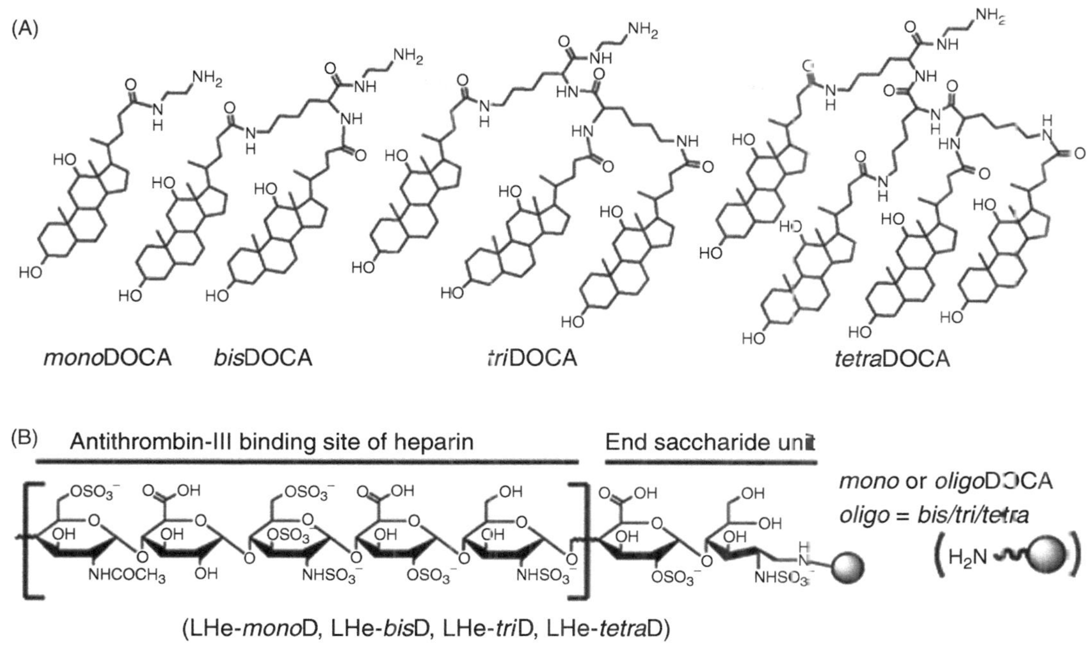

FIGURE 14.4 (A) Chemical structure of deoxycholic acid derivatives are showing monoDOCA, bisDOCA, triDOCA and tetraDOCA. (B) Chemical conjugation between low molecular weight heparin and oligoDOCA.

From Al-Hilal, T.A., et al., 2014. Oligomeric bile acid-mediated oral delivery of low molecular weight heparin. J. Control. Release 175 (1), 17–24.

Heparin is one of the highly sulfated GAGs that has played a prominent role in various biological activities. They are composed of pyranosyluronic acid and glucosamine. Heparin has the highest anionic charge density compared to any other biological molecules. The affinity of heparin to bind to antithrombin III enzymes results in the inactivation of various blood-clotting proteins like thrombin or Factor Xa and aids in the anticoagulation of blood. Moreover, heparin can interact and inhibit the activity of growth factors like VEGF and bFGF and help in inhibiting the angiogenesis of the cancer cells. Recent studies proved the ability of heparin to induce caspase-dependent apoptosis and suppress the cancer growth and metastasis. As heparin bears high negative charges, they tend to provoke electrostatic repulsion with the mucus layers and assist in poor oral absorption.

Since heparin itself does not absorb when orally administered, they have been chemically modified by conjugation oral absorption enhancers like protamine or bile acids to their backbone. Byun and coworkers have accomplished a series of works regarding oral delivery of heparin and its derivatives (Lee et al., 2003; Nurunnabi et al., 2012). The conjugation of bile acids to the heparin backbone aided in the bile acid transporter proteins–assisted active transport of the heparin-bile acid derivatives across the epithelial tight junctions. Initially, they conjugated a couple of deoxycholic acid (DOCA), a secondary bile acid on the back-bone of low molecular weight heparin. DOCA is a hydrophobic bile acid that enhances oral absorption of the carrier heparin as the heparin-DOCA conjugates become lipophilic and form self-assembled micelles in an aqueous solution. However, in the case of heparin-DOCA conjugation systems, hydrophobic DOCA are located in the core while it forms micelles and therefore the expected absorption enhancement does not meet the demand. This is because DOCA, located in the cores, can practically not come in contact with the bile acid transporter though more than 50% of heparin was absorbed as quantified. The reason of absorption of heparin was assumed due to increasing their lipophilic properties. Since hydrophobic bile acid are rarely or hardly exposed outside the micelles, we assumed that conjugating hydrophilic bile acid, such as taurocholic acid might overcome the limitation and enhance oral absorption.

The taurocholic acid (TCA)-conjugated heparin derivatives are expected to expose the TCA that are conjugated on the surface of heparin to the bile acid transporter of the small intestinal membrane and aid in improved oral absorption. Our group has designed LMWH-TCA–based nanocarriers for the oral delivery of doxorubicin. Each LMWH molecules were conjugated with four TCA molecules and were further loaded with the anticancer drug doxorubicin in order to develop oral drug delivery platforms of heparin nanocarriers. The studies and results have demonstrated that heparin can protect internal drugs from the harsh GI environment and the conjugated taurocholic acid greatly enhances absorption and bioavailability. The orally administered formulation significantly inhibited growth of tumors compared to the saline-treated control model. Taslim et al. has latter on introduced another model of oral heparin delivery which shows a new mechanism of oral absorption through apical sodium dependent bile acid transporter (ASBT) (Al-Hilal et al., 2014). They have evaluated that self-conjugation of four DOCA (tetraDOCA) has greater oral absorption potentiality than single DOCA conjugation. This tetraDOCA-conjugated heparin formulation absorbs through vesicle formation in the intestinal membrane where a singular DOCA may absorb through bile acid transporter (Fig. 14.5).

Zehedina and coworkers has also considered heparin as a carrier for oral anticancer drug delivery as particulate formulation (Khatun et al., 2013). Being a stable polysaccharide, heparin can shield the loaded drug from the harsh acidic environment of the GI tract. For this oral delivery system, taurocholic acid was considered as a potent bile acid, which is hydrophilic in nature and known as a potent absorption enhancer that interacts with bile acid transporters of the small intestinal membrane in the GI tract. In an earlier project our group has also developed a platform formulation for oral delivery of optical contrast agent such as semiconductor quantum dots. The purpose of developing this hybrid nanoparticle was for the diagnosis of the GI tract through noninvasive imaging. The study has revealed that highly hydrophobic quantum dots loaded into the core of lipophilic micelles (heparin-DOCA), which was highly stable in acidic stomach and became able to reach to ileum without any damage (Khatun et al., 2012). This result supports the hypothesis that heparin shielding can protect the nanoparticle from the harsh conditions in the stomach. The bile acid-linked nanoparticle was absorbed through bile acid transporter of ileum; as a result we were able to image the entire small intestine.

In this strategy, both taurocholic acid and docetaxel was conjugated on the backbone of heparin through chemical modification. The heparin-docetaxel-taurocholic acid conjugates formed self-assemble micelles that are in nanoparticle range (less than 200 nm in diameter). During self-assemble hydrophilic taurocholic acid locates on the outer surface the nanoparticle while hydrophobic docetaxel locates inside the core. In vitro studies demonstrated that the nanoparticle is highly stable in biological media, such as fasted state simulated intestinal fluid (FaSSIF) and fed state simulated intestinal fluid (FeSSIF) because the size of the particle was almost constant. In vivo studies demonstrated that the nanoparticles absorption occurred through the bile acid transporter in ileum part of small intestine and significant amount of docetaxel was measured from biodistribution observation. Most importantly, the orally administered formulation was potent for inhibiting tumor growth with a single oral dosage within 3 weeks of administration.

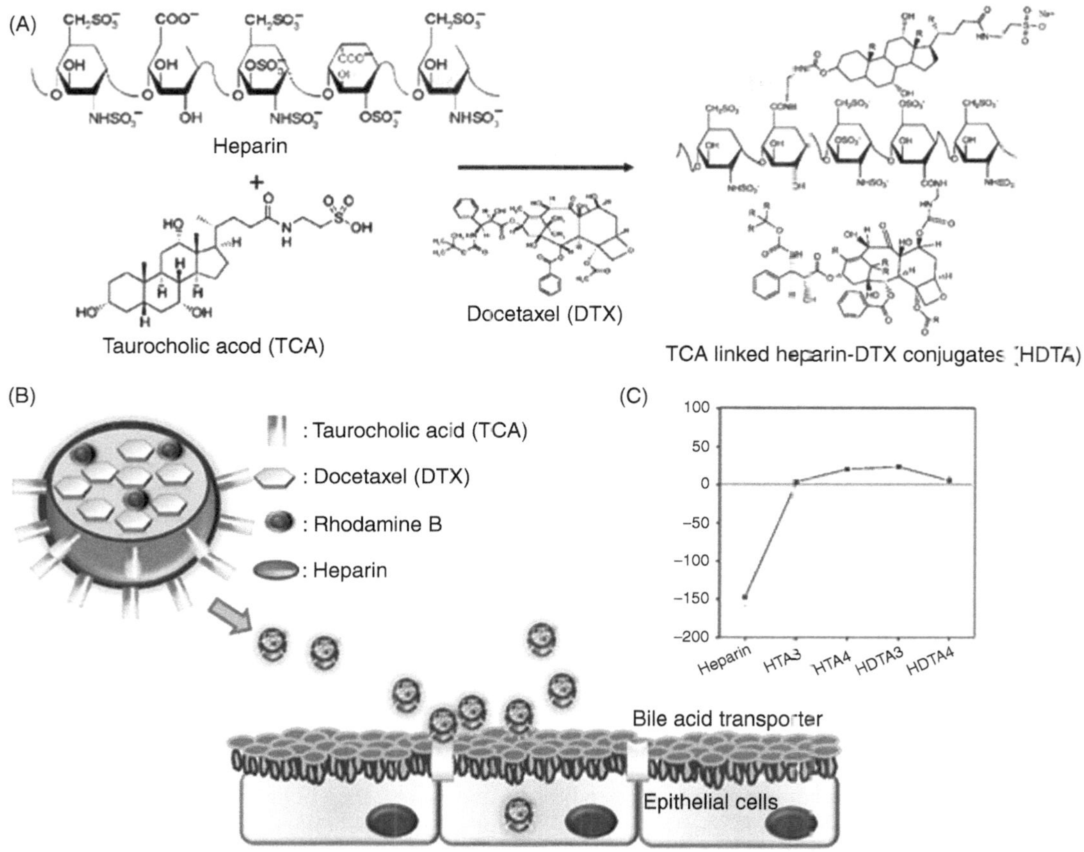

FIGURE 14.5 Synthesis of Taurocholic Acid–Linked Heparin–Docetaxel Conjugates (HDTX) and the Oral Absorption Mechanism of HDTX through Epithelial Cells of Small Intestine

(A) Chemical structure of heparin, docetaxel and TCA and chemical illustration of HDTX conjugates.
(B) Schematic illustration of HDTA nanoparticles for oral delivery of anticancer drug using bile acid transporter.
(C) Zeta potential of HDTA after conjugation with docetaxel and TCA (data represent mean ± SEM, $n = 3$).

From Khatun, Z., et al., 2013. Oral delivery of taurocholic acid linked heparin-docetaxel conjugates for cancer therapy. J. Control Release 170(1), 74–82.

6 CYCLODEXTRINS

Cyclodextrin is a group of oligosaccharides that are formed by cyclic arrangement of glucopyranose units conjugated by β 1,4 glycosidic linkages. In nature, cyclodextrins are available in three different forms α, β, and γ based on the number of number of glucose monomers. The number of glucose monomers in α, β, and γ cyclodextrin is 6, 7, and 8, respectively. Cyclodextrins have a lipophilic central core with hydrophilic outer surfaces. These lipophilic cores helped to improve the loading of hydrophobic drugs and led them to act as a potential biological character. β form of cyclodextrins are widely used in the pharmaceutical industry. Different derivatives of β-cyclodextrins are fabricated in order to improve the oral bioavailability and solubility of the cyclodextrins. For example, by modifying the hydroxyl groups of cyclodextrins with OH groups drastically improved the solubility of cyclodextrins. Some of the potential derivatives include randomly methylated β-cyclodextrin, branched β-cyclodextrin, and so on. Several factors determine the drug/cyclodextrin interactions. For example, ionic interactions between the drug molecules and cyclodextrin can improve the loading efficacy and stability of the complexes compared to the negatively charged or uncharged drug conjugates. Second, increased temperatures can assist in destabilization of the drug/cyclodextrin conjugation. This is because the hydrophobic and/or vanderwall forces, which stabilize the drug interactions with cyclodextrin and would be debilitated at increased temperatures. The degree of substitution also plays a crucial role in solubility, complexation, and release of the cargo from these molecular cages. Different strategies and formulations have been designed in order to deliver therapeutic molecules with high-precision orally. Recently Zhang and coworkers fabricated oridonin loaded hydroxypropyl-β-cyclodextrin surface stabilized with polaxamer to improve the oral availability of oridonin (Yan et al., 2008). The resultant studies showed improvement in the oral availability of oridonin by 214% compared to control. They believed that cyclodextrins-dependent intestinal absorption and the antidrug efflux property of cyclodextrin molecules resulted in such improvement. Ijaz et al. (2015) fabricated thiolated β-cyclodextrins in order to improve the mucoadhesive properties of miconazole nitrate-loaded β-cyclodextrin. The degree of thiolation had marked effects over improving the mucoadhesive property of β-cyclodextrins. The number of thiol groups increased from 561 to 1783 and increased the mucoadhesion from 2.86% to 49% compared to pristine polysaccharide. Moreover, a 150-fold increase in the solubility of miconazole nitrate was observed. Trapani et al. (2010) compared the therapeutic efficacy of chitosan and chitosan-cyclodextrin in small peptides. Efficacy of different cyclodextrin derivatives conjugated with chitosan were compared with chitosan in the transport of glutathione peptide was studied. The selection of suitable cyclodextrin (sulfobutyl ether-β-cyclodextrin) played an important role in improving the glutathione uptake and peptide release from the nanoparticles. Though an in vivo study needs to be conducted, they believe chitosan-cyclodextrin derivatives could be a promising candidate for oral peptide delivery. Other studies based on the development sulfobutyl ether-β-cyclodextrin–loaded nanolipid carriers showed a 592% increase in the bioavailability of vinpocetine over cyclodextrin-vinpocetine-loaded nanolipid carriers. Though the overall studies showed the feasibility of using the cyclodextrin as a potential candidates for oral drug delivery, several key factors relating the toxicity, hemophilic profiles, and concrete evidence regarding the efficacy of the carriers over other conventional PS based nanocarriers is required to accentuate their applications in biomedical and pharmaceutical industries.

7 OTHER POLYSACCHARIDES

Besides chitosan, hyaluronic acid, heparin, and chondroitin sulfate A, as discussed earlier, there are many other polysaccharides also considered for oral drug delivery systems. Konjac glucomannan composed of β-1,4 linked with D-mannose and D-glucose has a quite high molecular weight ranges from 500,000 to 2,000,000, such as $9 \pm 1 \times 10^5$ g/mol^{-1}. This polysaccharide contains a huge number of hydrogen bonds and are responsible for water absorption and the swelling ability of Konjac glucomannan. But, because the colon contains a very low amount of water, taking an excess amount of konjac glucomannan through orally may lead to diarrhea. A lack of adequate water in the colon also inhibits

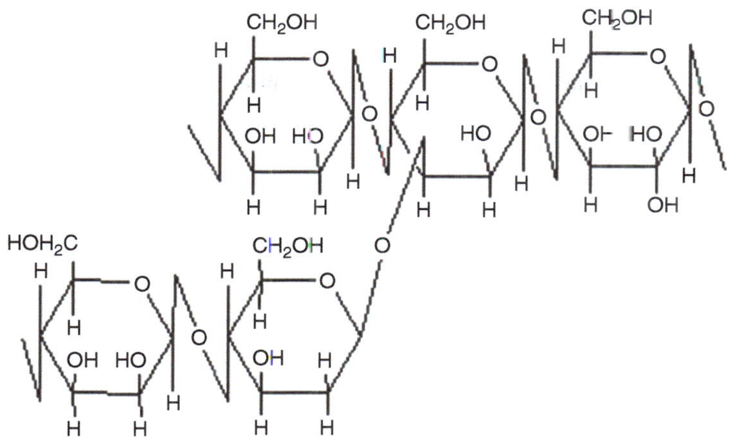

FIGURE 14.6 Typical Chemical Structure of Konjac Glucomannan, Which Contains β-1,3 and β-1,4 Repeating Units

drug release, therefore chemical and physical modifications were required to improve water-retaining ability (Fig. 14.6).

Psyllium polysaccharides obtained from husks of *Plantago ovata* seeds are considered to be one of the effective plant-based polysaccharides for oral delivery. They are glucoarabinoxylane compounds composed of glucuronic acid, arabinoxylan, glucuronic acid, xylose, and galactose sidechains attached to D-xylp backbone. The significant features, like the instant ability to form a viscous gel in the presence of water and ability to restrict the hydrolysis and prevent bacterial degradations, made them one of the promising candidates for oral therapy. Moreover, their ability to promote healthy bowel activity, regulate the blood glucose levels in type II diabetes mellitus patients, and reduce the blood cholesterol and low-density lipoprotein levels made them a highly investigated material in oral drug delivery (Singh et al., 2006a). As the colon is enriched with polysaccharides, derivatives of psyllium polysaccharides with their ability to endure the bacterial degradations have been critically investigated as a carrier for colon-specific drug delivery. They could easily be grafted or crosslinked with other synthetic or natural polymers in order to improve their therapeutic efficacy. For example, Singh and Chauhan (2009) employed free-radical polymerization techniques to crosslink methylacrylamide with psyllium husk and form a three-dimensional hydrogel network. They used a swelling equilibrium technique to load insulin into the hydrogel networks and studied the pH-dependent drug release profiles of the hydrogels. Their results conclude the efficient release of insulin at pH 7.4 compared to pH 2.2 and followed a non-Fickian diffusion pattern. They also studied the release of tetracyclin hydrochloride from these hydrogels (Singh et al., 2006b). Similar results were observed where the drug release patterns were strictly governed by the pH of the media with which they were incubated. The polymer swellings were higher in basic mediums compared to acidic mediums due the lack of ionization of $CONH_2$ groups, which resulted in the collapse of the polymer chains and the prevention of water diffusion and the drug release from the polymer matrix. However, at higher pH, the carboxyl groups of the PS gets ionized followed by repulsion and subsequent swelling of the polymers that favor the drugs to be dissolved and diffused out from the polymer matrices. They also observed that the rate of drug release was faster in the earlier stages compared to the later stages. In another study, Kharia et al. (2010) has treated psyllium husk to fabricate floating drug delivery systems to improve the oral bioavailability of Acyclovir to treat viral infections. The tablets were prepared by a wet granulation method using sodium bicarbonate as a gas-generating agent. The increase of the polysaccharide concentrations reduced the drug release rates from the swollen polymer matrices. The release profiles followed Higuchi kinetics followed by diffusion of the drug from the swollen matrices. Their observations proved that formulations delivered with a time delay of 24 h postadministrations would be efficient to treat viral diseases. Moreover, Kaialy et al. (2013) fabricated hydroxypropyl methyl cellulose (HPMC)-loaded psyllium tablets to study the sustained release of theophylline from the tablet matrices. As HPMC reduced the release of the drug, hydrophilic filler, lactose was further loaded to achieve zero order kinetics release profiles of theophylline. The overall results conclude the efficient use of psyllium polysaccharides as prominent materials to control the release of orally delivered drugs. As psyllium polysaccharides are nontoxic, cost effective, and FDA-approved for drug delivery applications, diverse strategies can be designed considering psyllium for oral therapeutic delivery.

Alginate has been one of the most explored polysaccharides for oral delivery. Diverse strategies, including microspheres, micellar structures, core–shell nanostructures, hydrogels, are sensitive to biological stimuli like pH, temperature has been developed to show their feasibility in oral drug delivery. Alginate polysaccharides are composed of glucoronate and mannuronate subunits arranged in either blocks or consecutive positions. The glucoronate blocks are expected to interact with the calcium ions to form interconnected physically crosslinked hydrogels. The rich abundance of carboxyl and hydroxyl groups on the PS backbone aids in the synthesizing different alginate derivatives with diverse properties for miscellaneous applications (Lee and Mooney, 2012). For example, different hydrophobic alkyl chains like dodecyl or octadecyl groups conjugated via ester/amide bonds can result in the formation of micellar structures that can assist in the delivery of hydrophobic drugs. Other polymers, like poly(έ-caprolactone) or poly butyl methacrylate, were also conjugated to the alginate backbones to regulate the release of drugs from the alginate micelles. Moreover, the rate of release strictly depended on the chain length of the hydrophobic polymers. Peptide-conjugated alginate derivatives have been fabricated to improve the interactions with the extracellular matrices and enhance the drug transport. For example, RGD peptides that are specific to the integrin receptors are abundant on the cell surface. The abundance of carboxyl groups on the alginate backbones aids in the synthesis of pH responsive drug delivery systems. The carboxyl groups protonate in the acidic pH that assists in the formation of hydrogen bonding with the neighboring molecules that result in shrinking and preventing the core materials from gastric degradations. However, one of the major disadvantages of the alginate nanoparticles in the drug delivery applications is their lower encapsulation and swift release of drugs from the PS matrices (Sosnik, 2014). This is due to the higher porosity of the alginate nanoparticles that hastens the drug leakage from the alginate-based nanocarriers. Moreover, presence of $CaCl_2$ solutions accelerates the release of drugs from the alginate cores networks. In order to prevent the rapid drug dissolution, different polymers like chitosan or dextran sulfate polymers have been conjugated to alter the drug release patterns. Chitosan with a positive charge, mucoadhesive and gastric stability properties have been most explored in the alginate-based micro/nanosystems. Different chitosan and chitosan derivatives have been physically, covalently crosslinked with alginate PS to obtain stable drug delivery systems. For example, physical crosslinker Ca^{2+} ions crosslinked carboxymethyl chitosan, and alginate microsphere hydrogels that are pH responsive and impervious to acid degradations have been designed for the oral delivery of bovine serum albumin (BSA). As the microsphere formulations are fabricated in aqueous solutions at neutral pH, they aid in improving the stability of biological peptides. Only 15% of BSA was released at pH 1.2 while 85% was released at pH 7.4, which confirms the tissue (small intestines) specific release of orally delivered biological peptides (Lin et al., 2005). Later, chitosan-coated alginate beads–based floating drug delivery system has been fabricated for the sustained release of orally administered metronidazole to treat Helicobacter pylori (Ishak et al., 2007). Coating the surface of alginate beads with chitosan improved the drug conjugation to 89% from 77%. Moreover, 4 h were required for the complete dissolution of the drugs from the polymer matrices with a 100% clearance of *Helicobacter pylori* bacteria from the stomach. In another study, chitosan and dextran coating over insulin-loaded alginate spheres were compared with the terms of peptide stability and gastric protection properties. Dextran-coated alginate beads showed enhanced insulin stability over chitosan coating (Martins et al., 2007). Coating the alginate beads with dextran improved the insulin encapsulation efficiencies and reduced the burst release of insulin from the alginate beads. The enhancement in insulin loading was due to the interactions between the negatively charged dextrin moieties with the amino acid residues of insulin. Drying techniques like freeze-drying is much preferable over oven-based drying of the alginate spheres. This is because the heat could aid in collapsing of alginate structures and aid in altering the morphology of the microspheres. Later, a dual crosslinked (Ca^{2+} and SO^{42-}) *N*-α-glutaric acid chitosan/alginate hydrogels have been fabricated for the oral delivery of peptide drugs (Gong et al., 2011). Only 18% of BSA was released in acidic pH while almost 100% drug release was observed at intestinal pH (pH 6.8–7.2). Synthetic polymers have also been conjugated for developing oral drug delivery systems. For example, PLGA nanospheres loaded into alginate microspheres have been fabricated for the oral delivery of antioxidant, silymarin (El-Sherbiny et al., 2011). Initially silymarin was loaded into the PLGA nanospheres via single emulsion-solvent evaporation technique followed by loading these nanoparticles into calcium crosslinked alginate matrices. This hybrid system was pH sensitive and improved the bioavailabilty, stability of silymarin. Furthermore, a new micellar system coated with alginate was fabricated for the oral delivery of indomethacin (Huang et al., 2012). These systems could result in uniform coating of alginate over the micellar structures, aid in preventing the six-arm block copolymer, poly (ε-caprolactone)-block-(dimethylamino) ethyl methacrylate from acidic degradations and control the indomethacin release from the micellar structures. This fabrication technique helps in fabricating materials with uniform thickness and effectively regulates the drug release. Mukhopadhyay et al. (2015) fabricated chitosan core alginate shell-based nanoparticles for the oral delivery of insulin. The fabricated nanoparticles had a mean diameter of 150 nm with 85% encapsulation of insulin in the chitosan cores (Mukhopadhyay et al., 2015). As less as 8% bioavailability of orally delivered insulin resulted in regulating the blood glycemic levels to a minimum of 85 mg/dL at 2 h postadministration. Moreover, in vivo toxicity studies validated the safety of chitosan alginate nanoparticles for oral delivery of insulin. Aikawa et al. (2015) fabricated liposome-loaded manganese-porphyrin nanoparticles that were further coated with sodium alginate to inhibit the superoxide activity in cancer cells. Alginate helped in improving the stability of the particles and also retained the structural integrity in acidic pH. However, lower therapeutic efficacy and lower encapsulation efficiencies were observed in the cancer cells. To establish the mechanisms involved in controlling the cancer further results are needed. As alginate is less toxic, biocompatible, biodegradable, and FDA-approved as food additive, they could be seen as a promising material for oral therapeutics delivery research. Moreover, their pH sensitive, mucoadhesive, and gastric protection properties could assist them in developing ideal oral drug delivery systems.

Dextran is one of the widely used polysaccharides in oral delivery systems. These biological polymers are branched glucans where the main chain glucose molecules are connected with (1–6) glycosidic links and branched at (1–3) position. Their ability to induce antithrombotic effects in order to decrease the blood viscosity made them an interesting material in medical fields. Bearing a negative charge they can easily conjugate with the positively charged materials and aid in improving their therapeutic properties (Varshosaz, 2012). For example, conjugating polyethyleneimine with dextran improved their serum stability and also reduced the toxicity induced by polyethyleneimine (Tiyaboonchai et al., 2015). Advancements in the science and engineering helped to synthesize various derivatives of dextrans. The hydroxyl groups on the glucose backbone principally aids in developing various dextran derivatives. For example, many hydrophobic moieties like polyalkyl chains, bile acids, and other hydrophobic materials were conjugated to dextran backbone to form micellar structures. For example, dextran was grafted with polyethylene glycolalkylether chains to form micelles of 30 nm to deliver cyclosporine A (CsA) (Francis et al., 2003). The size and the properties of the micelles were dependent on the molecular weight of dextran, number of hydrophobic chains substituted, and the length of the hydrophobic chains. Moreover, lower critical micellar concentrations were observed when the hydrophobic chain lengths were higher. However, the drug loading efficiencies were poor in this system and had no active targeting of the orally delivered molecules. Later, vitamin B12 was conjugated over the surface of dextran-polyethyleneoxide cetyl ether to deliver CsA (Francis et al., 2005). An increase in the drug loading to 8.3% was observed with this system. As the GI epithelia have vitamin B12 receptors, they could actively be transported and cross the ilial monolayer tight junctions. Moreover, the ability of dextran to endure gastric degradations aided in improving the stability of these nanoparticles. These systems had negligible effects over membrane integrity and took a minimum of 4 h to attain enhanced drug transport. This delay of 4 h was explained due to the time taken by the intrinsic factor proteins to form complex and aid in the active transport of the drug. However, no drug was retained within the cells after 24 h. Probing through the current literature it is evident that the dextran has been used as an efficient carrier in oral delivery. The selective degradations by the enzymes like dextranase that are vividly expressed in colon can aid in colon specific release of dextran. For example, Rai et al. (2015) fabricated epichlorohydrin crosslinked dextran microspheres that were further coated with enteric polymers for the colon specific oral delivery of 5-fluorouracil (5-FU). The drug release was enhanced at pH 7.4 and in the presence of dextranase. The mean residence time of erudite-dextran loaded 5-FU was 14-fold higher than 5-FU with a peak recovered concentrations of around 66% at the colon. In another study, the solid lipid nanoparticles containing ibuprofen were loaded into dextran hydrogels in order to formulate the oral drug delivery systems (Casadei et al., 2006). The solid lipid nanoparticles obtained by O/W emulsion technique was further loaded into the methacrylated dextran via photochemical method. The coating of methacrylated dextran hydrogels resulted in 60% retainment of drug molecules inside the hydrogels in acidic pH. Saquinavir (Saq)-loaded nanostructured lipid structures (NLC) coated with dextran-protamine structures have been designed in order to improve the mucoadhesive and release of Saq-NLC (Beloqui et al., 2014). By conjugating protamine and dextran over the NLC improvement in the cell penetration and gastric protection of the NLC was observed. Moreover, the release of Saq was observed only in intestinal fluids rather than in stomach. The stability of the particles was observed only with the coating of dextran-protamine over NLC. A 2-fold increase in the NLC was observed in mucus secreted Caco-2 monolayers. However, a 9-fold increase was observed in Caco-2 monolayers. In another study, HT29 cancer cells were treated with cysteine trimethyl chitosan-carboxymethyl dextran nanoparticles that contained histone methyl transferase enzyme, hSET1 (Dinarvand et al., 2015). They have also included an anticancer drug SN38 in order to improve the therapeutic efficiencies of these nanoparticles. The cysteine-assisted disulfide bondings resulted in shielding the drug release. SN38 release was enhanced in the presence of 10 mM glutathione. This synergistic nanosystem resulted in a 78% reduction in the cancer cell death. These studies conclude the efficiency of Dextran coatings in oral drug delivery systems.

8 POLYSACCHARIDE BLENDS (PSB) IN ORAL DRUG DELIVERY

Polymer or polysaccharide blending system has an advantage over plain microsphere system in terms of controlled release. Usually drugs from a plain microsphere system show faster release profile as the microspheres easily get in contact with the aqueous medium in vitro and in vivo condition and start swelling due to absorption of solvent thus the microspheres start to disintegrate and diffuse. For example, plain alginate microsphere showed the fast release of the drug due to swelling and disintegration in biomimetic GI tract medium, such as a fasted state simulated intestinal fluid (FaSSIF) and fed state simulated intestinal fluid (FeSSIF) media. But the release was controlled by blending the alginate microspheres with polymers that prevent swelling and disintegration as well. Lee et al. (2003) has demonstrated that when alginate was coated with chitosan or blended with HPMC they provided an extended release profile of drug molecule. Besides sustained release of drugs, the blended system also proved as a feasible method for encapsulating protein, peptide, microorganisms, and plasmid DNA.

Due to the feasibility of the blending method compared to other chemical modification methods, it has attracted great attention for industrial translation through reproducibility and scale-up production. Konjac glucomannan is one of the promising polysaccharides, which was considered for oral delivery of drug for colon targeting through absorbing in colon that to treat colonic diseases specifically.

9 CONCLUSION AND FUTURE PROSPECTS

Harnessing the biological significance of polysaccharide to develop efficient nanocarriers to orally deliver the therapeutic molecules has gained substantial attention in the pharmaceutical and biomedical industries. Ease of extraction and purification steps made them more accessible for biomedical applications. Chitosan derivatives have extreme significance in oral delivery due to their ability to overcome the GI barriers, generate mucoadhesion, and aid in transient openings of tight junctions. They have been a promising material in siRNA, DNA, and peptide delivery. Though chitosan derivatives are well explored, issues related to chitosan nanoparticle instability during the extreme acidic conditions (like gastric emptying) and achieve active targeting of these molecules to a specific cell types still needs to be achieved. Heparin-bile acid conjugate nanocarriers have recently gained significant attention by the research community due to their ability to generate site specific ASBT-mediated active transport of the nanocarriers. This could significantly reduce IC50 of the orally delivered drugs and aid in improving the therapeutic efficacy of the nanocarrier systems. PSB have gained a significant attention due their ability to tune the drug release at determined pH levels. However, a concrete study is required to establish a stable PS-based nanocarrier systems to orally transport the biological molecules. Further studies are required to compare the therapeutic efficacies of different polysaccharide derivative-based nanocarrier systems, study the molecular mechanisms related to the role of polysaccharide in improving the transport of the nanocarriers and a thorough pharmacokinetic and pharmacodynamics profiles of the drug-loaded carriers. Moreover, the limitations associated to the impurities, batch reproducibility, and toxicity profiles need to be surpassed in order to transcend their biomedical applications to clinical levels. Rather confining to the current polysaccharide nanosystems, research in terms of developing polysaccharide derivatives is required to achieve the barriers imposed by current systems and accomplish the impossible oral delivery of therapeutic molecules possible.

REFERENCES

Aikawa, T., et al., 2015. A drug formulation using an alginate hydrogel matrix for efficient oral delivery of the manganese porphyrin-based superoxide dismutase mimic. Biomater. Sci. 3 (6), 861–869.

Al-Hilal, T.A, et al., 2014. Functional transformations of bile acid transporters induced by high-affinity macromolecules. Sci. Rep. 4, 4163.

Allard, S., et al., 2009. Oxidation of iodide and iodine on birnessite (δ-MnO_2) in the pH range 4–8. Water Res. 43 (14), 3417–3426.

Amidi, M., et al., 2006. Preparation and characterization of protein-loaded *N*-trimethyl chitosan nanoparticles as nasal delivery system. J. Control. Release 111 (1–2), 107–116.

Bagre, A.P., et al., 2013. Alginate coated chitosan core shell nanoparticles for oral delivery of enoxaparin: in vitro and in vivo assessment. Int. J. Pharm. 456 (1), 31–40.

Beloqui, A., et al., 2014. Dextran–protamine coated nanostructured lipid carriers as mucus-penetrating nanoparticles for lipophilic drugs. Int. J. Pharm. 468 (1–2), 105–111.

Carino, G., Mathiowitz, E., 1999. Oral insulin delivery. Adv. Drug Deliv. Rev. 35 (2–3), 249–257.

Casadei, M.A., et al., 2006. Solid lipid nanoparticles incorporated in dextran hydrogels: a new drug delivery system for oral formulations. Int. J. Pharm. 325 (1–2), 140–146.

Cho, H.J., et al., 2014. Chondroitin sulfate-capped gold nanoparticles for the oral delivery of insulin. Drug Des. Devel. Ther. 9, 3109–3118.

Choi, B., et al., 2015. Glutamine-chitosan modified calcium phosphate nanoparticles for efficient siRNA delivery and osteogenic differentiation. J. Mater. Chem. B 3, 6448–6455.

Dinarvand, M., et al., 2015. Oral delivery of nanoparticles containing anticancer SN38 and hSET1 antisense for dual therapy of colon cancer. Int. J. Biol. Macromol. 78, 112–121.

Duncan, R., Izzo, L., 2005. Dendrimer biocompatibility and toxicity. Adv. Drug Deliv. Rev. 57 (15), 2215–2237.

El-Sherbiny, I.M., et al., 2011. Biodegradable pH-responsive alginate-poly (lactic-*co*-glycolic acid) nano/micro hydrogel matrices for oral delivery of silymarin. Carbohydr. Polym. 83 (3), 1345–1354.

Fiorica, C., et al., 2013. A new hyaluronic acid pH sensitive derivative obtained by ATRP for potential oral administration of proteins. Int. J. Pharm. 457 (1), 150–157.

Francis, M.F., et al., 2003. Solubilization of cyclosporin A in dextran-*g*-polyethyleneglycolalkyl ether polymeric micelles. Eur. J. Pharm. Biopharm. 56 (3), 337–346.

Francis, M.F., Cristea, M., Winnik, F.M., 2005. Exploiting the vitamin B12 pathway to enhance oral drug delivery via polymeric micelles. Biomacromolecules 6 (5), 2462–2467.

Gao, X., et al., 2013. Biodegradable pH-responsive polyacrylic acid derivative hydrogels with tunable swelling behavior for oral delivery of insulin. Polymer 54 (7), 1786–1793.

Garcia-Fuentes, M., et al., 2005. A comparative study of the potential of solid triglyceride nanostructures coated with chitosan or poly(ethylene glycol) as carriers for oral calcitonin delivery. Eur. J. Pharm. Biopharm. 25 (1), 133–143.

Gong, R., et al., 2011. A novel pH-sensitive hydrogel based on dual crosslinked alginate/*N*-α-glutaric acid chitosan for oral delivery of protein. Carbohydr. Polym. 85 (4), 869–874.

Guliyeva, Ü., et al., 2006. Chitosan microparticles containing plasmid DNA as potential oral gene delivery system. Eur. J. Pharm. Biopharm. 62 (1), 17–25.

Han, L., et al., 2012. Insulin-loaded pH-sensitive hyaluronic acid nanoparticles enhance transcellular delivery. AAPS PharmSciTech 13 (3), 836–845.
He, C., Kim, S.W., Lee, D.S., 2008. In situ gelling stimuli-sensitive block copolymer hydrogels for drug delivery. J. Control. Release 127 (3), 189–207.
Hearnden, V., et al., 2012. New developments and opportunities in oral mucosal drug delivery for local and systemic disease. Adv. Drug Deliv. Rev. 64 (1), 16–28.
Hejazi, R., Amiji, M., 2003. Chitosan-based gastrointestinal delivery systems. J. Control. Release 89 (2), 151–165.
Hsieh, C.M., et al., 2014. Biodistribution profiling of the chemical modified hyaluronic acid derivatives used for oral delivery system. Int. J. Biol. Macromol. 64, 45–52.
Hu, C., et al., 2012. Influence of charge on FITC-BSA-loaded chondroitin sulfate-chitosan nanoparticles upon cell uptake in human Caco-2 cell monolayers. Int. J. Nanomed. 7, 4861–4872.
Huang, C., 2009. Receptor-Fc fusion therapeutics, traps, and MIMETIBODY technology. Curr. Opin. Biotechnol. 20 (6), 692–699.
Huang, X., Xiao, Y., Lang, M., 2012. Micelles/sodium-alginate composite gel beads: a new matrix for oral drug delivery of indomethacin. Carbohydr. Polym. 87 (1), 790–798.
Huang, P., et al., 2014. Improving the oral delivery efficiency of anticancer drugs by chitosan coated polycaprolactone-grafted hyaluronic acid nanoparticles. J. Mater. Chem. B 2, 4021–4033.
Hussan, D.S., et al., 2012. A review on recent advances of enteric coating. IOSR J. Pharm. 2 (6), 5–11.
Ijaz, M., et al., 2015. Synthesis and characterization of thiolated β-cyclodextrin as a novel mucoadhesiveexcipient for intra-oral drug delivery. Carbohydr. Polym. 132, 187–195.
Iqbal, J., et al., 2012. Thiolated chitosan: development and in vivo evaluation of an oral delivery system for leuprolide. Eur. J. Pharm. Biopharm. 80 (1), 95–102.
Ishak, R.A.H., et al., 2007. Preparation, in vitro and in vivo evaluation of stomach-specific metronidazole-loaded alginate beads as local anti-Helicobacter pylori therapy. J. Control. Release 119 (2), 207–214.
Kaialy, W., et al., 2013. Psyllium: a promising polymer for sustained release formulations in combination with HPMC polymers. Pharm. Dev. Technol. 19 (3), 269–277.
Kesharwani, P., et al., 2015. Hyaluronic acid engineered nanomicelles loaded with 3,4-difluorobenzylidene curcumin for targeted killing of CD44+ stem-like pancreatic cancer cells. Biomacromolecules 16 (9), 3042–3053.
Kharia, A., et al., 2010. Design and optimization of floating drug delivery system of acyclovir. Indian J. Pharm. Sci. 72 (5), 599–606.
Khatun, Z., et al., 2012. Oral delivery of near-infrared quantum dot loaded micelles for noninvasive biomedical imaging. ACS Appl. Mater. Interf. 4 (8), 3880–3887.
Khatun, Z., et al., 2013. Oral delivery of taurocholic acid linked heparin-docetaxel conjugates for cancer therapy. J. Control. Release 170 (1), 74–82.
Khatun, Z., et al., 2014. Oral absorption mechanism and anti-angiogenesis effect of taurocholic acid-linked heparin-docetaxel conjugates. J. Control. Release 177 (1), 64–73.
Kim, J.S., et al., 2015. ROS-induced biodegradable polythioketal nanoparticles for intracellular delivery of anti-cancer therapeutics. J. Ind. Eng. Chem. 21, 1137–1142.
Kochut, A., Dersch, P., 2013. Bacterial invasion factors: tools for crossing biological barriers and drug delivery? Eur. J. Pharm. Biopharm. 84 (2), 242–250.
Kuijpers, A.J., et al., 2000. In vitro and in vivo evaluation of gelatin-chondroitin sulphate hydrogels for controlled release of antibacterial proteins. Biomaterials 21 (17), 1763–1772.
Kumari, A., Yadav, S.K., Yadav, S.C., 2010. Biodegradable polymeric nanoparticles based drug delivery systems. Colloids Surf. B 75 (1), 1–18.
Lee, K.Y., Mooney, D.J., 2012. Alginate: properties and biomedical applications. Prog. Polym. Sci. 37 (1), 106–126.

Lee, D.W., et al., 2003. Preparation and release characteristics of polymer-coated and blended alginate microspheres. J. Microencapsul. 20 (2), 179–192.
Lin, Y.-H., et al., 2005. Physically crosslinked alginate/N,O-carboxymethyl chitosan hydrogels with calcium for oral delivery of protein drugs. Biomaterials 26 (14), 2105–2113.
Martins, S., et al., 2007. Insulin-loaded alginate microspheres for oral delivery: effect of polysaccharide reinforcement on physicochemical properties and release profile. Carbohydr. Polym. 69 (4), 725–731.
Morishita, M., Peppas, N.A., 2006. Is the oral route possible for peptide and protein drug delivery? Drug Discov. Today 11 (19–20), 905–910.
Mukhopadhyay, P., et al., 2014. pH sensitive *N*-succinyl chitosan grafted polyacrylamide hydrogel for oral insulin delivery. Carbohydr. Polym. 112, 627–637.
Mukhopadhyay, P., et al., 2015. pH-sensitive chitosan/alginate core-shell nanoparticles for efficient and safe oral insulin delivery. Int. J. Biol. Macromol. 72, 640–648.
Najlah, M., et al., 2007. In vitro evaluation of dendrimer prodrugs for oral drug delivery. Int. J. Pharm. 336 (1), 183–190.
Nesamony, J., et al., 2012. Calcium alginate nanoparticles synthesized through a novel interfacial cross-linking method as a potential protein drug delivery system. J. Pharm. Sci. 101 (6), 2177–2184.
Nurunnabi, M., et al., 2016. Design and strategies for bile acid mediated therapy and imaging. RSC Adv. 6 (78), 73986–74002.
Nurunnabi, M., et al., 2012. Heparin based nanoparticles for cancer targeting and imaging. Quant, Imaging Med. Surg. 2 (3), 219–226.
Paliwal, R., et al., 2012. Chitosan nanoconstructs for improved oral delivery of low molecular weight heparin: in vitro and in vivo evaluation. Int. J. Pharm. 422 (1–2), 179–184.
Patel, D.A., et al., 2012. Buccal mucosa as a route for systemic drug delivery: a review. Int. J. Drug Dev. Res. 4 (2), 99–116.
Petrus, A.K., Fairchild, T.J., Doyle, R.P., 2009. Traveling the vitamin B12 pathway: oral delivery of protein and peptide drugs. Angew. Chem. Int. Ed. Engl. 48 (6), 1022–1028.
Piao, M.G., et al., 2007. Enhanced oral bioavailability of piroxicam in rats by hyaluronate microspheres. Drug Dev. Ind. Pharm. 33 (4), 485–491.
Pridgen, E.M., et al., 2013. Transepithelial transport of Fc-targeted nanoparticles by the neonatal Fc receptor for oral delivery. Sci. Transl. Med. 5 (213), 213ra167.
Radaev, S., Sun, P.D., 2001. Recognition of IgG by Fcgamma receptor. The role of Fc glycosylation and the binding of peptide inhibitors. J. Biol. Chem. 276 (19), 16478–16483.
Rai, G., et al., 2015. Enteric-coated epichlorohydrin crosslinked dextran microspheres for site-specific delivery to colon. Drug Dev. Ind. Pharm. 41 (12), 2018–2028.
Rastogi, R., et al., 2007. Alginate microspheres of isoniazid for oral sustained drug delivery. Int. J. Pharm. 334 (1–2), 71–77.
Sarmento, B., et al., 2007. Alginate/chitosan nanoparticles are effective for oral insulin delivery. Pharm. Res. 24 (12), 2198–2206.
Sastry, S., Nyshadham, J., Fix, J., 2000. Recent technological advances in oral drug delivery: a review. Pharm. Sci. Tech. Today 3 (4), 138–145.
Semete, B., et al., 2010. In vivo uptake and acute immune response to orally administered chitosan and PEG coated PLGA nanoparticles. Toxicol. Appl. Pharmacol. 249 (2), 158–165.
Singh, B., Chauhan, N., 2009. Modification of psyllium polysaccharides for use in oral insulin delivery. Food Hydrocolloids 23 (3), 928–935.
Singh, B., Chauhan, G.S., Bhatt, S.S., et al., 2006a. Metal ion sorption and swelling studies of psyllium and acrylic acid based hydrogels. Carbohydr. Polym. 64 (1), 50–56.
Singh, B., Chauhan, G.S., Sharma, D.K., et al., 2006b. The release dynamics of model drugs from the psyllium and *N*-hydroxymethylacrylamide based hydrogels. Int. J. Pharm. 325 (1–2), 15–25.

Smart, J.D., 2004. Lectin-mediated drug delivery in the oral cavity. Adv. Drug Deliv. Rev. 56 (4), 481–489.
Sosnik, A., 2014. Alginate particles as platform for drug delivery by the oral route: state-of-the-art. ISRN Pharm. 2014, 1–17.
Tiyaboonchai, W., Rodleang, I., Ounaroon, A., 2015. Mucoadhesive polyethylenimine–dextran sulfate nanoparticles containing *Punica granatum* peel extract as a novel sustained-release antimicrobial. Pharm. Dev. Technol. 20 (4), 426–432.
Trapani, A., et al., 2010. A comparative study of chitosan and chitosan/cyclodextrin nanoparticles as potential carriers for the oral delivery of small peptides. Eur. J. Pharm. Biopharm. 75 (1), 26–32.
Varshosaz, J., 2012. Dextran conjugates in drug delivery. Exp. Opin. Drug Deliv. 9 (5), 509–523.
Vasir, J.K., Tambwekar, K., Garg, S., 2003. Bioadhesive microspheres as a controlled drug delivery system. Int. J. Pharm. 255 (1–2), 13–32.
Venishetty, V.K., et al., 2012. Design and evaluation of polymer coated carvedilol loaded solid lipid nanoparticles to improve the oral bioavailability: A novel strategy to avoid intraduodenal administration. Colloids Surf. B 95, 1–9.
Wilson, D.S., et al., 2010. Orally delivered thioketal nanoparticles loaded with TNF-α-siRNA target inflammation and inhibit gene expression in the intestines. Nat. Mater. 9 (11), 923–928.
Xiao, Y., et al., 2014. Enhancing the intestinal absorption of low molecular weight chondroitin sulfate by conjugation with α-linolenic acid and the transport mechanism of the conjugates. Int. J. Pharm. 465 (1–2), 143–158.
Xu, W., et al., 2004. Intranasal delivery of chitosan-DNA vaccine generates mucosal SIgA and anti-CVB3 protection. Vaccine 22 (27–28), 3603–3612.
Yan, Z., et al., 2008. Characterization and in vivo evaluation of an inclusion complex of oridonin and 2-hydroxypropyl-β-cyclodextrin. Drug Dev. Ind. Pharm. 34 (6), 632–641.
Zhang, N., et al., 2013. Polysaccharide-based micelles for drug delivery. Pharmaceutics 5 (2), 329–352.

FURTHER READINGS

Chalasani, K.B., et al., 2007. Effective oral delivery of insulin in animal models using vitamin B12-coated dextran nanoparticles. J. Control. Release 122 (2), 141–150.
García-González, C.A., Alnaief, M., Smirnova, I., 2011. Polysaccharide-based aerogels—Promising biodegradable carriers for drug delivery systems. Carbohydr. Polym. 86 (4), 1425–1438.

CHAPTER

15

RECENT ADVANCEMENTS IN ORAL DELIVERY OF INSULIN: FROM CHALLENGES TO SOLUTIONS

Raje Chouhan, Shilpi Goswami, Anil Kumar Bajpai

Government Autonomous Science College, Jabalpur, India

CHAPTER OUTLINE

1 **Introduction** ... 436
2 **Diabetes Mellitus** ... 436
2.1 Type 1 Diabetes ... 437
2.2 Type 2 Diabetes ... 439
3 **Diagnosis Test for Diabetes** ... 440
4 **Insulin Therapy** ... 440
5 **Plants Used in the Treatment of Diabetes** ... 441
6 **Diabetes in World Scenario** ... 441
7 **Insulin** ... 442
7.1 Insulin Analogs ... 443
8 **Action of Insulin** ... 445
9 **Oral Delivery of Insulin** ... 447
10 **Problems in Oral Delivery** ... 448
10.1 Chemical Barrier ... 448
10.2 Enzymatic Barrier ... 449
10.3 Absorption Barrier ... 450
10.4 Stability Barrier ... 451
11 **Strategies to Improve Oral Delivery of Insulin** ... 451
11.1 Absorption Enhancers ... 451
11.2 Enzyme Inhibitors ... 453
11.3 Mucoadhesive Polymers ... 453
11.4 Chemical Modifications of Proteins and Peptides ... 454
11.5 Formulation Vehicles ... 455
12 **Recent Breakthroughs in Oral Insulin Delivery** ... 456
13 **Commercialization** ... 457
13.1 Oral-Lyn ... 457
13.2 Eligen ... 457

13.3 CODES (Colon Specific Delivery System) technology 459
13.4 Capsulin 459
13.5 Octreolin 459
13.6 ORMD-0801 459
13.7 AI-401 460
13.8 Oradel 460
13.9 Macrulin 460
14 Recent Challenges and Future Prospects 460
References 461

1 INTRODUCTION

Diabetes is one of the major afflictions of modern Western society. To date, blood-sugar levels are controlled by diabetic patients via insulin. It is introduced directly into the bloodstream using injections. Subcutaneous insulin injections cause itching, pain, allergic reaction, hyperinsuliemia, and lipodistrophy around the injection area (Krishnan and Sudhakar, 2015). This unpleasant method is undesirably required because stomach acid destroys protein-based substances, such as insulin, making oral insulin consumption useless. There are different new systems that are based on inhaling the insulin and on a controlled release of insulin into the bloodstream. The proper delivery of insulin in the blood stream for the treatment of diabetic patients can be achieved by nanotechnology. Production of pharmaceutically active proteins, such as insulin, in large quantities has become feasible. Pulmonary delivery (Shaikh et al., 2010), buccal delivery (George and Roy, 2013), transdermal delivery (Pandey, 2010), nasal delivery (Rajeswari, 2011), and oral delivery (Olya et al., 2014) are some different delivery routes considered for delivery of insulin. The oral route is considered to be the most convenient and comfortable means for administration of insulin for less invasive and painless diabetes management, leading to a higher patient compliance. Nevertheless, the intestinal epithelium is a major barrier to the absorption of hydrophilic drugs, as they cannot diffuse across epithelial cells through lipid-bilayer cell membranes to the bloodstream, because of the large size, hydrophilicity, and instability of these macromolecules. There are different approaches, such as microsphere, liposomes, permeation enhancer, enzyme inhibitors, and nanoparticles (Pandit and Joshi, 2015) have been used for protecting proteins. These approaches are attractive and able to protect active degradation, improving drug transmucosal transport and provide controlled release properties for encapsulated drugs.

2 DIABETES MELLITUS

In diabetes mellitus the body does not produce or properly use insulin. Insulin is essential for the cellular uptake of glucose into most tissues in the body. Insulin opens the doors of the body's cells to glucose circulating in the blood. The glucose enters the cells and is used as the cells' fuel for energy. Insulin binds to a spot on the cell surface called a receptor. Likened to a lock and key, insulin is the key that opens up the lock (receptor) so that glucose can pass through the door into the cell. Using this analogy in type 1 diabetes, the keys have been stolen (no insulin is made by the pancreas). In type 2, the door won't open fully even with the right key. The lock and key model for diabetes is shown in

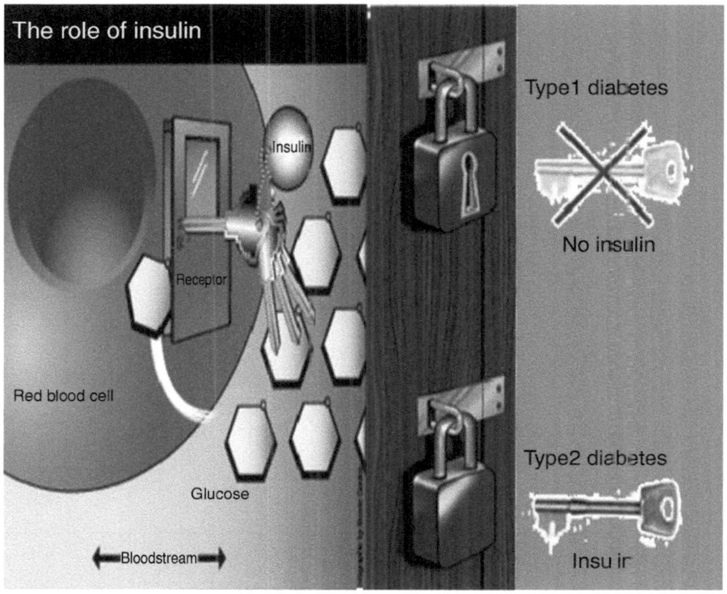

FIGURE 15.1 Lock and Key Model for Diabetes

Fig. 15.1. The central identifying feature of diabetes mellitus is a chronic and substantial elevation of the circulating blood glucose concentration. In addition to blood glucose concentration, glycosylated hemoglobin is used to measure blood glucose levels over a long period of time. The chronic effects of diabetes mellitus contribute to different types of medical problems, such as retinopathy, neuropathy, coronary heart disease, gastrointestinal disorders, kidney disease, impaired sexual function, bone and rheumatic disorders, endocrine diseases, central nervous system problems, and psychological and psychiatric conditions.

1. *Neuropathy*: loss of peripheral nerve function, tingling sensations, numbness, loss of function, pain, and muscle weakness
2. *Retinopathy*: a serious eye disease that can result in blindness
3. *Nephropathy*: a common complication and a leading cause of death in diabetes
4. *Foot ulcers*: lack of blood supply and peripheral neuropathy are the key factors in the development of diabetic foot ulcers

Diabetes is generally divided into two major categories: type 1 diabetes, or insulin dependent diabetes mellitus (IDDM), type 2 diabetes or noninsulin dependent diabetes (NIDDM) (Ozougwu et al., 2013).

2.1 TYPE 1 DIABETES

Type 1 diabetes is known as dependent, juvenile, or childhood onset. In this type, the diabetes patient daily requires administration of insulin. In this disease the body attacks the pancreatic β-cells

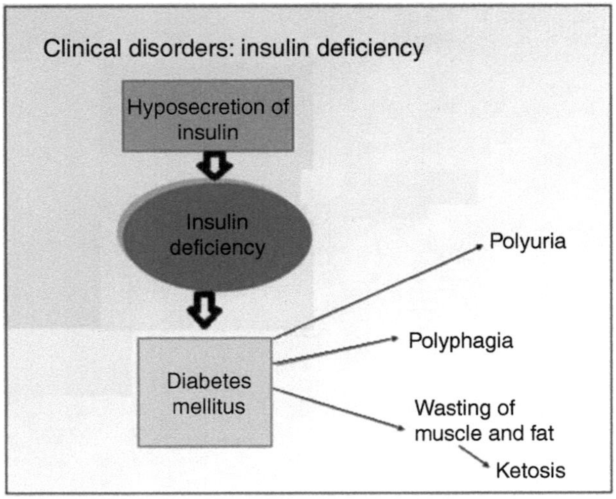

FIGURE 15.2 Symptoms of Diabetes

that produce insulin. About 12% of people with type 1 diabetes have clinical depression (Roy and Lloyd, 2012). Type 1 diabetes is usually developed due to an autoimmune disorder. This happens when the body's immune system behaves inappropriately and starts seeing one of its own tissues as foreign. In this case, the islet cells of the pancreas, which produce insulin, are seen as the "enemy" by mistake. The body then creates antibodies to fight the "foreign" tissue and destroys the islet cells' ability to produce insulin. The lack of sufficient insulin thereby results in diabetes. It is, however, still unknown why this autoimmune diabetes develops.

Most often it is a genetic tendency. Sometimes it follows a viral infection such as mumps, rubella, cytomegalovirus, measles, influenza, encephalitis, polio, or Epstein-Barr virus. Without the treatment of insulin, type 1 diabetics experience acute effects of the disease, such as ketoacidosis, a condition where the body begins to burn fat and muscle tissues for energy, causing the release of ketone bodies into the bloodstream. Ketoacidosis is extremely taxing on the kidneys and endothelial cells in the capillaries and eyes, causing extreme damage to those tissues and leading to kidney failure, blindness, and amputation of tissues of the extremities.

The classical symptoms of type 1 diabetes include: polyuria (excessive urination), polydipsia (increased thirst), xerostomia (dry mouth), polyphagia (increased hunger), fatigue, and weight loss (Chen et al., 2012). All symptoms for diabetes are shown in Fig. 15.2. Almost exclusively, the treatment for type 1 diabetes is insulin therapy. Patients with type 1 diabetes should have a diet that has approximately 35 calories per kg of body weight per day (or 16 calories per pound of body weight per day). Type 1 diabetes tends to have fewer tendencies to have other family members affected with diabetes than type 2. In the first large family study of diabetes, less than 4% of parents and 6% of siblings of a person with diabetes also had diabetes. In studies with identical twins less than 50% of the siblings of a person with diabetes also had diabetes versus almost 100% of siblings of people with type 2 diabetes. Children of type 1 diabetic fathers are more likely to develop type 1 autoimmune diabetes than children of type 1 diabetic mothers.

2.2 TYPE 2 DIABETES

Type 2 diabetes is known as adult onset diabetes. Type 2 diabetes mellitus (T2DM) and prediabetes are increasingly observed among children, adolescents, and younger adults. It is generally caused by a combination of low production of insulin from the pancreas and insulin resistance in the tissues that require insulin. It is the most common type of diabetes affecting approximately 90% of diabetic patients (Ding et al., 2014). This type of diabetes produces some insulin, which significantly reduces the possibility of ketoacidosis, although the risk is still much higher than a nondiabetic person. A type 2 diabetes patient is overweight although there are some exceptions. Type 2 diabetes is associated with insulin resistance. This type of diabetes is obtained as a hereditary tendency from one's parents Insulin levels in these patients are usually normal or higher than average but the body's cells are rather sluggish to respond to it. Type 1 and type 2 diabetes are shown in Fig. 15.3.

There are different factors that are responsible for development of type 2 diabetes. The strongest factor for its development seems to be genetic predisposition and other factors are obesity and high caloric intake. Twenty percent of people with type 2 diabetes have antibodies to their islet cells, which are detectable in their blood resulting in the expected low levels of insulin. Patients with type 2 diabetes generally are put on a 1500–1800 calorie diet per day to promote weight loss and then the maintenance of ideal body weight. However, this may vary depending on the person's age, sex, activity level, current

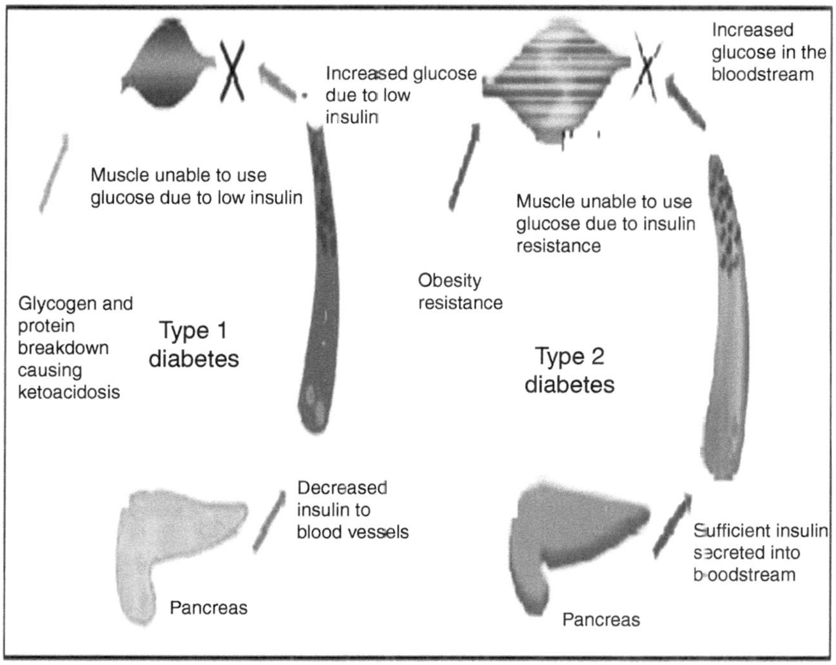

FIGURE 15.3 Type 1 and Type 2 Diabetes Mellitus

weight, and body style. More obese individuals may need more calories initially until their weight is less. Type 2 diabetes tends to be fairly hereditary in contrast to type 1 diabetes.

Approximately 38% of siblings and one third of children of people with type 2 diabetes will develop diabetes or abnormal glucose metabolism at some point. The degree of obesity also seems to be a factor with a larger percentage of diabetes developing in those who are more obese. Studies with identical twins showed that 90%–100% of the time when diabetes developed in one it would also develop in the other compared with 50% in type 1 diabetes.

3 DIAGNOSIS TEST FOR DIABETES

In the normal cases, that is, a healthy person, the urine will not contain sugar but within diabetes patients glucose will overflow through the kidneys and as a result of that urine will include glucose and as a result the urine test will show a positive result. This test (GTT) and oral glucose tolerance test (OGTT), used to detect whether the body is having problems processing glucose, will indicate whether the patient is suffering from impaired glucose tolerance or diabetes (Bassam and Rasool, 2013). There are several tests used to detect this disease, including:

- *Fasting plasma glucose (FPG)*: Fasting means after not having anything to eat or drink (except water) for at least 8 h before the test.
- *Random (also called casual) plasma glucose test*: Checks blood glucose levels at any time of the day.
- *A1C*: The A1C test measures average blood glucose for the past 2 to 3 months.
- *Oral glucose tolerance test (OGTT)*: Checks blood glucose levels before and 2 h after patient drink a special sweet drink.

4 INSULIN THERAPY

Type 1 diabetes must be treated with insulin shots. This involves injecting insulin under the skin in the fat for it to get absorbed into the blood stream where it can then access all the cells of the body that require it. Insulin cannot be taken as a pill because the juices in the stomach would destroy the insulin. Insulin therapy by multiple daily injections attempts to mimic the natural production of insulin in response to blood glucose levels. Prior to insulin analogs, normal insulin or insulin formulations were used, but they poorly matched normal glucose control. Insulin therapy remains as the most rigorous and physiological treatment for diabetes and is the only option for patients suffering from type 1 diabetes. Insulin administration is generally performed by subcutaneous injection, while insulin production and insulin analog technologies have made parenteral delivery much more effective. When injected, insulin enters the circulatory system from the periphery and has the largest effect on muscle and adipose tissues. These tissues have a delayed onset to insulin and a sustained duration of action. In healthy individuals, insulin produced by the β-cells in the pancreas is released into the portal circulation in response to the presence of glucose and its effect on peripheral tissues occurs after first pass hepatic clearance. Nonetheless, insulin therapy based on subcutaneous insulin delivery remains the primary means of treatment for diabetes.

5 PLANTS USED IN THE TREATMENT OF DIABETES

Plants play a major role in controlling diabetes from early times. Compared to allopathic medications, it does not show any side effects, so recently patients are returning to natural medications. Some of the plants used in the treatment are: *Aegle marmelos* Corr (baelfruit), *Azardirachta indica* Ajuss (neem tree), *Emblica officinalis* S Gaertn (gooseberry), *Mimosa pudica* Linn (touch me not plant). *Momordica charanti* Linn (bitter guard), *Jatropha curcas* Linn (biodiesel plant), *Acacia catechu* W&A (Leguminosae), *Aerva lanata* Juss (Amarantaceae), *Alpinia calcarata* Rosc (zingiberaceae), *Benincasa hispida* Thunb(Cucurbitaceae), *Capsicum annum* Linn (Solanaceae), *Cedrus deodara* Roxb(Coniferae), *Coccinia indica* W&A (Cucurbitaceae), *Eugenia jambolana* Lam (Myrtaceae), *Ficus gibosa* BI (Moraceae), *Ficus glomerata* Roxb (Moraceae), *Helicteres isora* Linn (Sterculiaceae), *Holostemma annulare* K.Schum (Asclepiadaceae), *Plumbago rosea* Linn (Plumbaginaceae), *Rubia cordifolia* Linn (Rubiaceae), *Salacia oblonga* Wall (Hippocrateaceae), *Saraca indica* Linn(Leguminosae), *Stroblanthus hyneanus* Nees (Acanthaceae), *Tinospora cordifolia* Miers (Menispermaceae), *Tragia involucrate* Linn (Euphorbiaceae), *Tribulus terrestris* Linn (Zygophyllaceae), and so on. These are the plants that have the activity of controlling diabetes (Shivashankar and Mani, 2011).

6 DIABETES IN WORLD SCENARIO

Diabetes mellitus, or simply diabetes, is a very common chronic disease, whose prevalence continues to increase, especially in the Western world. However, diabetes and its complications are often diagnosed late due to the long subclinical evolution, unclear pathogenesis, and poor medical facilities in some regions over the world. According to recent estimates, approximately 285 million people worldwide (6.6%) in the 20–79 year age group will have diabetes in 2010 and by 2030, 438 million people (7.8%) of the adult population, is expected to have diabetes. The largest increases will take place in the regions dominated by developing economies (Ramachandran et al., 2010). Today, approximately 113 million Asians are thought to be diabetic and more than 95% of these people have type 2 diabetes. The World Health Organization has predicted that the major burden will occur in the developing countries. There will be a 42% increase from 51 to 72 million in the developed countries and 170% increase from 84 to 228 million, in the developing countries. India has a population of 1.1 billion and has an estimated 40 million people with diabetes, the largest in the world. This is followed by China with 39 million with diabetes, Japan 6.9 million, Thailand 3.1 million, Philippines and Korea both with 3.0 million, and Indonesia with 2.8 million. The disorder is spreading more rapidly in the Asia Pacific region than anywhere else. By 2025 the number of diabetics in Asia is expected to hit 170 million, with India and China together accounting for more than 100 million sufferers. The islands in the Pacific that have high rates of type 2 diabetes include Tonga, Fiji, and French Polynesia, but it is the small island of Nauru with a population of 14,019 that has the highest prevalence of type 2 diabetes per population in the world. Approximately one-third of the adult population of Nauru, of which the average age is 21 years old, has type 2 diabetes. In Australia over the past 2 decades the prevalence rates of type 2 diabetes has more than doubled. More than 6% of the 22 million people living in Australia have type 2 diabetes, and it is estimated by 2030 that 3.3 million Australians will have developed the disorder (Piero et al., 2014).

7 INSULIN

The discovery of insulin is one of the greatest discoveries of modern medical sciences. It is the most studied protein and it was the first ever naturally produced protein to have its amino acid sequence determined, a feat for which Frederick Sanger won the Nobel Prize in Chemistry in 1958. Insulin is composed of two polypeptide chains and has a molecular weight of 5808 Da (Saima et al., 2012). The two chains are connected to one another by two disulfide bonds; as seen in Fig. 15.4. It is produced and secreted from the β-cells in the islets of Langerhans in the pancreas.

The primary role of insulin is to facilitate the transport of glucose from the bloodstream to cells in the body. The secretion of insulin is a carefully regulated process that is meant to maintain healthy levels of blood glucose in the bloodstream during periods of both feeding and fasting. Insulin has polypeptide structure, which is synthesized in pancreatic B cells, have many important function in the mammalian body. It can bind to receptors in surface of cells. The receptors located in cell membranes bind with each other with a disulfide bond and consist of four subunits. Two of these subunits, located in the cell surface membrane, named beta, the other two of these subunits located out of the cell surface named alpha. Insulin binds the first alpha subunits and leads to autophosphorilation of beta subunits, which extend cytoplasm by having bonds of two subunits, and induces converting active protein kinase. Therefore, insulin triggers many enzymes phosphorylation and shows its intracellular effects. In brief, insulin is necessary for glucose carrying in tissues, so the tissues are more permeable to glucose. Except for endocytosis of insulin, the cells are more permeable for K^+ and PO-4. Intracellular enzymes activation level with phosphorylation change in approximately in 10–15 s (Ünal et al., 2012).

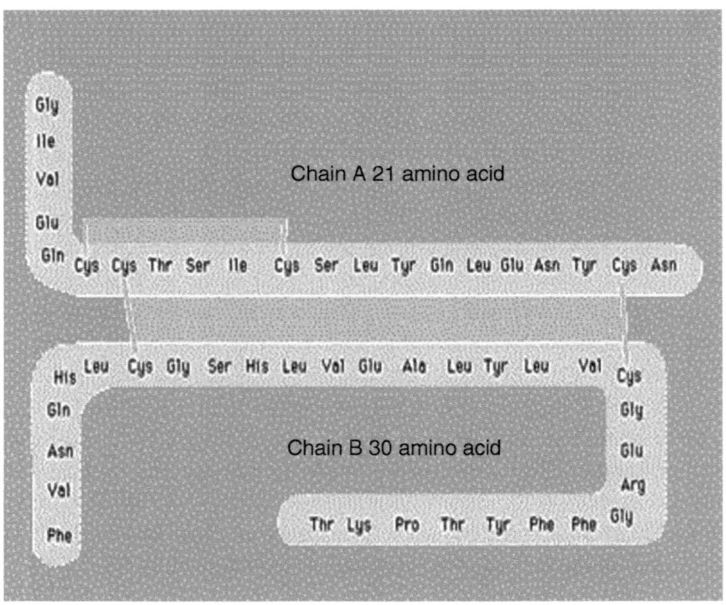

FIGURE 15.4 Structure of Insulin

7.1 INSULIN ANALOGS

Insulin analogs are synthetically produced variations of insulin that are different in amino acid sequence than that of native human insulin. A diagram of the amino acid sequence of native human insulin is given in all cases, insulin analogs vary from this structure. So the pharmacokinetics of insulin formulation can be altered by bioengineered modifications of the natural amino acid sequence to produce "insulin analogs." Similar to formulator insulin variations, insulin analogs change the timing and duration for the effect of insulin and they are separated accordingly. Insulin analogs figures are shown in Fig. 15.5.

- *Fast acting analogs*: Aspart, Lispro, Glulisine
- *Long acting analogs*: Glargine and Detemir

7.1.1 Fast-acting insulin analogs

Fast-acting insulin analogs were designed to give more freedom to the diabetic patient so that they could administer insulin near mealtime.

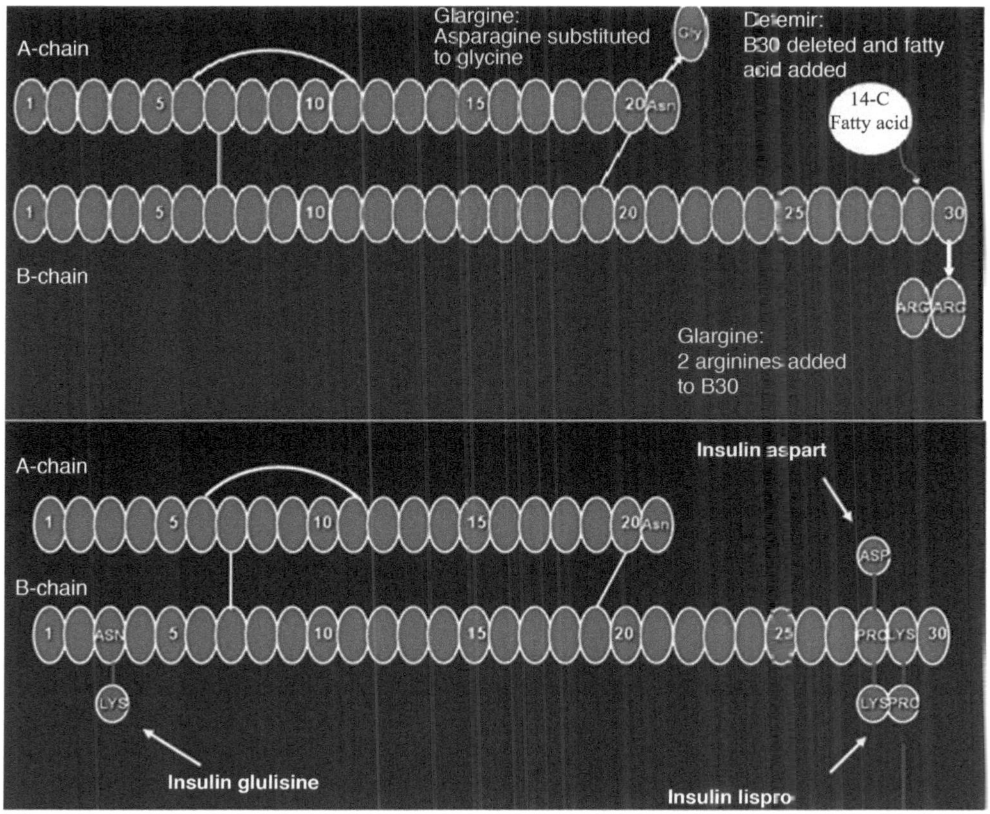

FIGURE 15.5 Analogs of Insulin

- *Lispro*: The first available insulin was insulin lispro. In insulin lispro the two residues in the B-chain of insulin are substituted for one another, the lysine at position 29 and the proline at position 28 (Varewij and Janssen, 2012). The change causes a reduced tendency of the analog to self-associate into dimmers by 200–300 times. Because insulin is only biologically active as a monomer, the hypoglycemic effect of insulin Lispro is more rapid than that of native insulin, which exhibits some self-associative behavior.
- *Aspart*: It is another fast-acting insulin analog, in which the B-chain of insulin is again modified. In insulin as part of the proline at position 28 is replaced with an aspartic acid residue (Negrato et al., 2012). The effect is similar to that of insulin Lispro in terms of reducing self-association of the protein.
- *Glulisine*: It is the third fast-acting insulin analog. Insulin glulisine has substitutions in the B-chain where glutamate replaces lysine at the 29th position and lysine replaces asparagine at the 3rd position. This analog exhibits improved monomer stability, reduced self-association, and an increased rate of absorption while maintaining an unaltered affinity for the insulin receptor. These types of insulin are designed to take care of the sugar a person eats. They include:
    - *Aspart*: Starts to act in 5 to 15 min, peaks in 90 min, lasts 3 h
    - *Lispro*: Starts to act in 5 to 15 min, peaks in 90 min, lasts 3 h
    - *Regular*: Starts to act in 30 to 60 min, peaks in 2 h, lasts 4 to 6 h

7.1.2 Long-acting insulin analogs

Long-acting insulin analogs were originally designed for the need of diabetics to maintain a healthy blood glucose levels throughout the night.

- *Glargine*: Insulin glargin is also known as Lantus. It is a long-acting basal insulin analog. It is given once daily to control the blood sugar level. It consists of microcrystals that slowly release insulin, giving a long duration of action of 18 to 26 h, with a "peakless" profile (according to the insulin glargine package insert) (Hilgenfeld et al., 2014). Pharmacokinetically, it resembles basal insulin secretion of nondiabetic pancreatic beta cells. Sometimes, in type 2 diabetes and in combination with a short-acting sulfonylurea (drugs that stimulate the pancreas to make more insulin), it can offer moderate control of serum glucose levels. In the absence of endogenous insulin—type 1 diabetes, depleted type 2 (in some cases) or latent autoimmune diabetes of adults in late stage—insulin glargine needs the support of fast-acting insulin taken with food to reduce the effect of prandially derived glucose.
- *Detemir*: It is a long-acting human insulin analog for maintaining the basal level of insulin. The trade name of detemir is Levemir. It is an insulin analogue in which a fatty acid (myristic acid) is bound to the lysine amino acid at position B29. It is quickly absorbed after which it binds to albumin in the blood through its fatty acid at position B29. It then slowly dissociates from this complex
- *Glargine*: Starts to act about 2 h after taking, and can last 20 to 24 h
- *Detemir*: Starts to act about 2 h after taking, and lasts between 14 and 24 h
  The peaks of insulin analogs are shown in Fig. 15.6.

7.1.3 Premixed analog

Insulin glargine cannot be mixed with any other insulin by virtue of its acidic pH and high Zn content, that is, 30 μg/mL. Premixed preparations are available with protamine insulin. In these preparations the

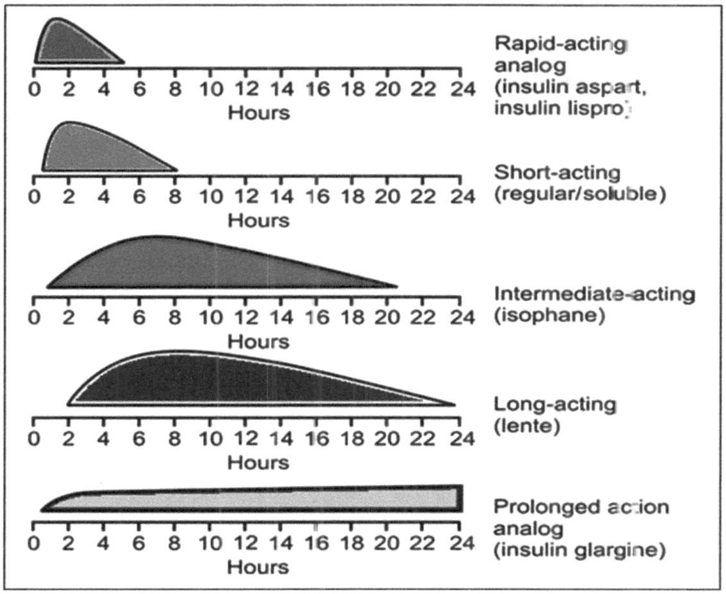

FIGURE 15.6 Peak Time for Insulin Analogs

protamine-insulin part has to be formulated with the same insulin analog like lispro or aspart. These preparations are available in 25:75, 30:70, 50:50 proportions (Seshiah, 2012). The premixed insulin analog provides better postprandial coverage and less hypoglycemic tendencies between meals.

8 ACTION OF INSULIN

Insulin acts as specific receptors. It is located on the cell membrane. The insulin receptor is a heterotetrameric glycoprotein, which consists of 2 extracellular α and 2 transmembrane β subunits linked together by disulfide bonds. Binding of insulin to α subunits, induce aggregation and internalization of the receptor along with the bound insulin molecules. This activates tyrosine kinase activity of the β subunits. In turn, a cascade of phosphorylation and dephosphorylation reactions is set into motion resulting in stimulation or inhibition of enzymes involved in the rapid metabolic actions of insulin. Certain second messengers like phosphatidyl inositol triphosphate (PIP3), which are generated through activation of a specific PI3-kinase, also mediate the action of insulin on metabolic enzymes. Insulin stimulates the uptake of amino acids and synthesis of proteins in muscle tissues, regulates gene transcription, increases glycogen and fatty acid synthesis, and inhibits the breakdown of glycogen, fat, and protein for energy. Insulin is the central actor in homeostasis. The actions of insulin are shown in Fig. 15.7 and the pathway of insulin is shown in Fig. 15.8.

The actions of insulin on cells include:

1. *Increased glycogen synthesis*: insulin forces storage of glucose in liver (and muscle) cells in the form of glycogen; lowered levels of insulin cause liver cells to convert glycogen to glucose and

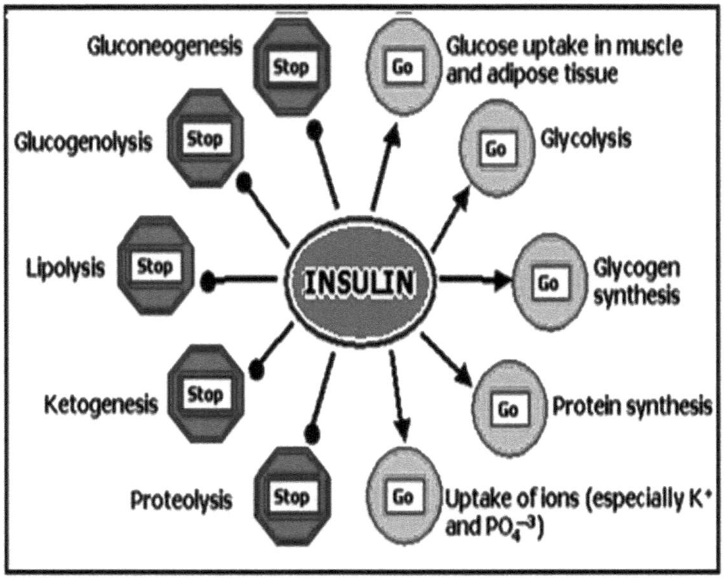

FIGURE 15.7 Action of Insulin

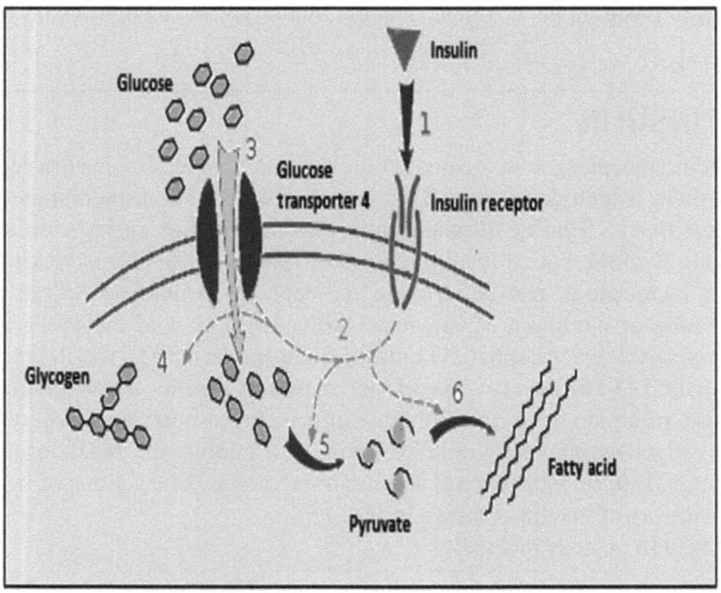

FIGURE 15.8 Pathway of Insulin Action

excrete it into the blood. This is the clinical action of insulin, which is directly useful in reducing high blood glucose levels as in diabetes.

2. *Increased lipid synthesis*: insulin forces fat cells to take in blood lipids, which are converted to triglycerides; lack of insulin causes the reverse.
3. *Increased esterification of fatty acids*: forces adipose tissue to make fats (i.e., triglycerides) from fatty acid esters; lack of insulin causes the reverse
4. *Decreased proteolysis*: decreases the breakdown of protein
5. *Decreased lipolysis*: forces reduction in conversion of fat cell lipid stores into blood fatty acids; lack of insulin causes the reverse
6. *Decreased gluconeogenesis*: decreases production of glucose from nonsugar substrates, primarily in the liver (the vast majority of endogenous insulin arriving at the liver never leaves the liver); lack of insulin causes glucose production from assorted substrates in the liver and elsewhere
7. *Decreased autophagy*: decreases level of degradation of damaged organelles. Postprandial levels inhibit autophagy completely
8. *Increased amino acid uptake*: forces cells to absorb circulating amino acids; lack of insulin inhibits absorption
9. *Increased potassium uptake*: forces cells to absorb serum potassium; lack of insulin inhibits absorption. Insulin's increase in cellular potassium uptake lowers potassium levels in blood. This possibly occurs via insulin-induced translocation of the Na^+/K^+-ATPase to the surface of skeletal muscle cells
10. *Arterial muscle tone*: forces arterial wall muscle to relax, increasing blood flow, especially in micro-arteries; lack of insulin reduces flow by allowing these muscles to contract
11. Increase in the secretion of hydrochloric acid by parietal cells in the stomach
12. Decreased renal sodium excretion

Insulin also influences other body functions, such as vascular compliance and cognition. Once insulin enters the human brain, it enhances learning and memory and benefits verbal memory in particular. Enhancing brain insulin signaling by means of intranasal insulin administration also enhances the acute thermoregulatory and gluco-regulatory response to food intake, suggesting that central nervous insulin contributes to the control of whole-body energy homeostasis in humans. Insulin also has stimulatory effects on gonadotropin-releasing hormone from the hypothalamus, thus favoring fertility.

9 ORAL DELIVERY OF INSULIN

As we know, all peptide and protein drugs must be administered via subcutaneously or intravenously. Since the discovery of insulin, many noninvasive methods of insulin delivery have been thoroughly investigated. However, to date no feasible alternative modes of delivery have achieved enough success to replace subcutaneous administration. The investigation of alternative routes for insulin delivery is driven by the desire to simplify the administration of insulin and potentially reduce the levels of noncompliance currently seen with insulin therapy. The greatest challenge for all approaches to oral insulin delivery is achieving a high bioavailability when compared to subcutaneous delivery. Insulin, if administered via the oral route, will help eliminate the pain caused by injection, psychological barriers associated with multiple daily injections such as needle anxiety and possible infections (Ahmad et al., 2012). Insulin, 51 amino acid protein, can get deteriorated by gastric pH and intestinal enzymes, and even intestinal epithelial cell membranes serve as absorption barrier for intact peptide structure resulting in less than 1% bioavailabity of total insulin taken orally (Sharma et al., 2015). In addition, oral insulin is advantageous because it is delivered directly to the liver, its primary site of action, via the portal circulation, a mechanism very similar to endogenous insulin; subcutaneous insulin treatment, however, does not replicate the normal dynamics of endogenous insulin release, resulting in a failure to achieve a lasting glycemic control in patients. In light of the earlier mentioned distinct benefits, pharmaceutical technologists have been trying to design an oral delivery system for insulin.

10 PROBLEMS IN ORAL DELIVERY

Insulin is one of the most active and robust hypoglycemic agents known for diabetes medication. Oral insulin administration is regarded as a better route of administration because it is delivered directly to the liver (Ramineni et al., 2013), its primary site of action, in such a way that the dosage form replicates the natural route of insulin secretion and absorption via the portal movement. Oral insulin administration can impersonate the physiological fate of insulin and thus is able to provide better glucose homeostasis. Other significant benefits include avoiding use of needles and other injection materials, cost-effectiveness property, deep-rooted acceptability, and fostering patient compliance (Su et al., 2012).

Despite these potential advantages, due to its various barriers like chemicals, such as pH inactivation, or enzymatic or cellular, such as poor drug penetration, as well as absorption, an so on, oral formulations is still a challenge to the scientists (Mukhopadhyaya et al., 2012). These barriers may, in turn, result in low bioavailability. Most peptides and proteins exhibit bioavailability of less than 1%. Thus, it remains an elusive goal to achieve the dream of administering insulin orally, and hence reproducing biological patterns of insulin secretion with associated advantages (Herrero et al., 2012).

To increase the bioavailability of insulin in the bloodstream its oral formulation should be able to bypass the chemical and enzymatic barriers in the GI tract (Hosseininasab et al., 2014), such as the harsh pH conditions in the stomach, the enzymatic activity, the presence of mucus layer of the enterocytes, and the poor absorbance across the intestinal epithelium. A schematic representation of barriers faced on oral delivery of insulin is depicted in Fig. 15.9.

Thus, to achieve effective oral administration of insulin it is vital to understand the characteristics and functions of these barriers. The main barriers in oral insulin therapy are discussed in the next sections.

10.1 CHEMICAL BARRIER

The GI pH varies from the highly acidic in the stomach (pH 1.2 and 5 in the fasted and fed state, respectively) to the slightly basic in the intestine (pH 5.5–7.0 in the fasted and fed state, respectively). The stomach produces gastric juice, which chiefly contains hydrochloric acid (HCl), potassium chloride (KCl), and sodium chloride (NaCl). This acidic environment in the pH range of 1.5–3.5 encourages proteolysis of proteins and peptides into constituent amino acids, peptides for absorption. Strong acidic conditions are also responsible for accelerating the breaking of disulfide bonds between the chains of the insulin molecule, causing chemical degradation of the molecule. This pH variation is unfavorable for polypeptides especially insulin because it is responsible for bond cleavage due to oxidation, hydrolysis, or deamidation, which would cause the protein to lose its activity (Chaturvedi et al., 2013).

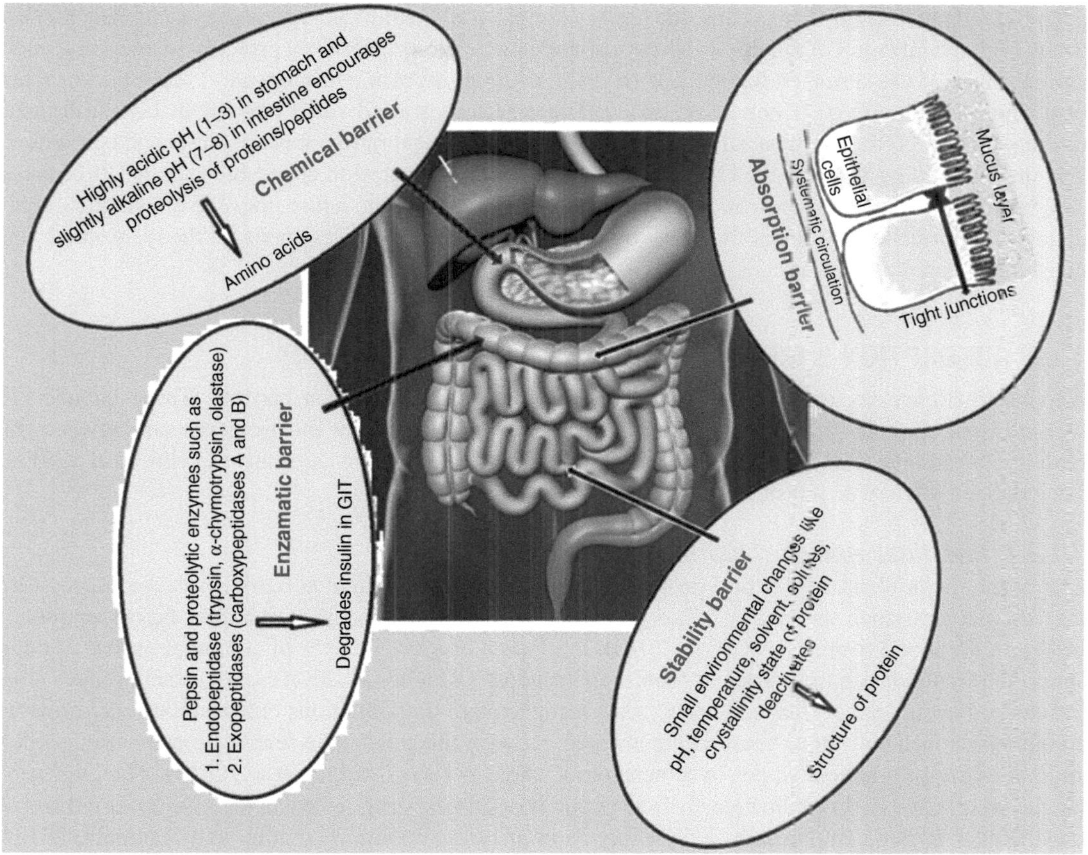

FIGURE 15.9 Different Barriers Coming Across in Oral Insulin Therapy

Deamidation involves hydrolysis of side-chain amide present on glutamine and asparagine residues yielding carboxylic acid, whereas tryptophan, methionine, cysteine, histidine, and tyrosine (present in insulin) are susceptible to oxidation and hydrolysis (occurs in aspartic acid residues), which have been implicated in the cleavage of peptide bonds, leading to a decrease in biological activity. In other words, electrostatic interactions result in increased positive charge on the protein due to low pH in the stomach, whereas higher pH in the intestine leads to repulsion of negative charges. These repulsive forces generate structural incompatibility causing destruction in the functional portion of the therapeutic entity.

10.2 ENZYMATIC BARRIER

The rapid luminal enzymatic degradation is observed as a major blockage for low bioavailability of proteins. In gastrointestinal tract (GIT) gut enzymes possess natural defense mechanism so as to break-down proteins and peptides into smaller peptide fragments and amino acids to prevent the body from potentially dangerous proteins. Thus, the orally administered insulin undergoes enzymatic degradation in the GIT (Najafzadeh et al., 2012) by pepsin and pancreatic proteases or proteolytic enzymes, such as consisting of the serine endopeptidase (trypsin, α-chymotrypsin, elastase) and exopeptidases (carboxypeptidases A and B) (Chen et al., 2011). These proteases are divided into three classes: luminal (secreted by pancreas and released into the small intestine), membrane-bound, and cytosolic proteases. The insulin passing through the GIT gives minimal (<1%) absorption (Kinesh et al., 2010) because GIT insulin is exposed to acid-catalyzed degradation in the stomach, while intestine enzymatic degradation takes place at the lumen, the brush border, the cytosol of the enterocytes, in the lysosomes, and other cell organelles (Pawar et al., 2014).

10.3 ABSORPTION BARRIER

Another challenge for orally administered protein drugs are absorption barriers. These include (1) the intestinal epithelial layer containing intestinal microvilli and tight junctions present between the closely packed intestinal epithelial cells, and (2) the mucus and glycocalyx layer (Vehik et al., 2011). They collectively create physical hindrance to absorption.

10.3.1 Intestinal epithelial barrier

One of the major obstacles for oral protein delivery is intestinal epithelial barrier because hydrophilic macromolecules, such as insulin, is not able to circulate through cell membranes across epithelial cells into the blood torrent (Iyer et al., 2010). It consists of a single layer of columnar epithelial cells supported by the lamina propria and muscularis mucosa (Khulbe, 2015). Across epithelial layer drug penetration occurs either transcellularly (movement through tight junctions between the cells) or paracellularly (movement through intercellular space between the cells). The fenestrate present in porous structure of tight junctions varies in dimension of range 3–10 Å (Pridgen et al., 2014). The large molecular weight and hydrophilic nature of the peptides or clinical drugs restricts their absorption through transcellular or paracellular pathway, leading to poor bioavailability (Lemmer and Hamman, 2013). The epithelial cells are tightly bound to one another by three types of junctions namely desmosomes (zonulae adherens), tight junctions (zonulae occludens), and gap junctions. Tight junctions bind cells with one another strongly providing mechanical strength to cell lining thus these tight junctions inhibits passage between cells (paracellular transport) (Renukuntla et al., 2013).

The epithelium folds to form villi and microvilli (the brush border) of uniform thickness on the apical surface of the epithelial cells. These structures increase the surface area of the intestinal tract, however, these structures contains digestive enzymes, which degrades ingested proteins thus hindering the absorption of proteins in intestine.

10.3.2 Mucus and glycocalyx barrier

Another absorption hurdle is a special insoluble mucus blanket that forms a protective gel-like coating over the entire surface of the gastric mucosa. The mucus layer made up glycoproteins, enzymes, electrolytes, and water and Glycocalyx collectively forms the microenvironment lies on top of the intestinal epithelium (Dorothee and Yu, 2013). Glycocalyx is a fuzzy and fibrous coat that is weakly acidic and consists of sulfated mucopolysaccharides lined by mucus secreted by goblet cells. The mucus layer protects epithelial surfaces by trapping pathogens and foreign particulates and clears them immediately (Alai et al., 2015).

Biologically, mucus has been continuously secreted and detached from the epithelial layer, and renewed in every 4–5 h. This semipermeable hurdle permits for the invasion of nutrients, water and small molecules but remain impervious to bacteria and pathogens. It blocks the circulation of protein molecules into the epithelial cells due to the molecules in the mucus forming a barricading net and the electrostatic repulsion between the negatively charged mucus layer and protein molecules.

10.4 STABILITY BARRIER

The final major barrier toward oral protein drug administration includes fabrication methodology of the protein drug formulation. Proteins have a complex internal structure that helps define their biological activity rather than other low molecular weight drugs. Insulin is a drug containing almost all organizational characteristics of proteins: α-helix, β-sheet, β-turn, high order assembly, allosteric T→R transition, and structural changes in amyloidal fibrillation (Ensign et al., 2012) Either a slight disturbance in the structural sequence of protein (primary, secondary, tertiary, or quaternary structure) results into its deactivation. The smallest change in the environment (or even microenvironment) of the protein can disturb its stability.

The environmental factors affecting protein structure and stability are mainly pH, temperature, solvent, solutes, and crystallinity states of the protein. Thus, any changes in single or combinational factors can lead to denaturation of insulin. Protein drugs are highly sensitive to sunlight, extreme temperatures; humidity, and so on, hence needs to be stored under refrigeration (2–8°C). During storage conditions if insulin is exposed to higher temperatures or sunlight it degrades either by hydrolysis, or transform into higher molecular weight components thus loses its potency and biological activities (Shafaat et al., 2013). Thus, refrigeration is the ideal method of storage of insulin vials.

11 STRATEGIES TO IMPROVE ORAL DELIVERY OF INSULIN

In the pharmaceutical sciences it's a great challenge to design strategies that expand the absorption of insulin through the GIT and significantly boost the treatment of diabetes mellitus. Most peptides are not bioavailable from GIT after oral administration thus most of these systems mainly target GIT barrier by inhibiting acidic and enzymatic degradation, enhancing membrane permeability, or widening tight junctions, chemical modification of insulin, and the formulation of carrier systems (Kalra and Kalra, 2012). Several strategies as shown in Fig. 15.10 have been devised to overcome physiologic and morphologic barriers toward insulin absorption and thus effectively deliver proteins and peptides via the oral route.

Hence, various approaches employed to improve functional characteristics of the biotechnology-derived drug products includes specific excipients, such as absorption enhancers, enzyme inhibitors, mucoadhesive polymers, and formulation vehicles (Park et al., 2010).

11.1 ABSORPTION ENHANCERS

Detergents, surfactants, bile salts, and Ca^{2+}-chelating agents are widely used as penetration enhancers to improve the poor oral bioavailability of protein drugs through the GIT. The absorption enhancers alter the natural framework of the intestinal membrane either by increasing membrane fluidity, varying

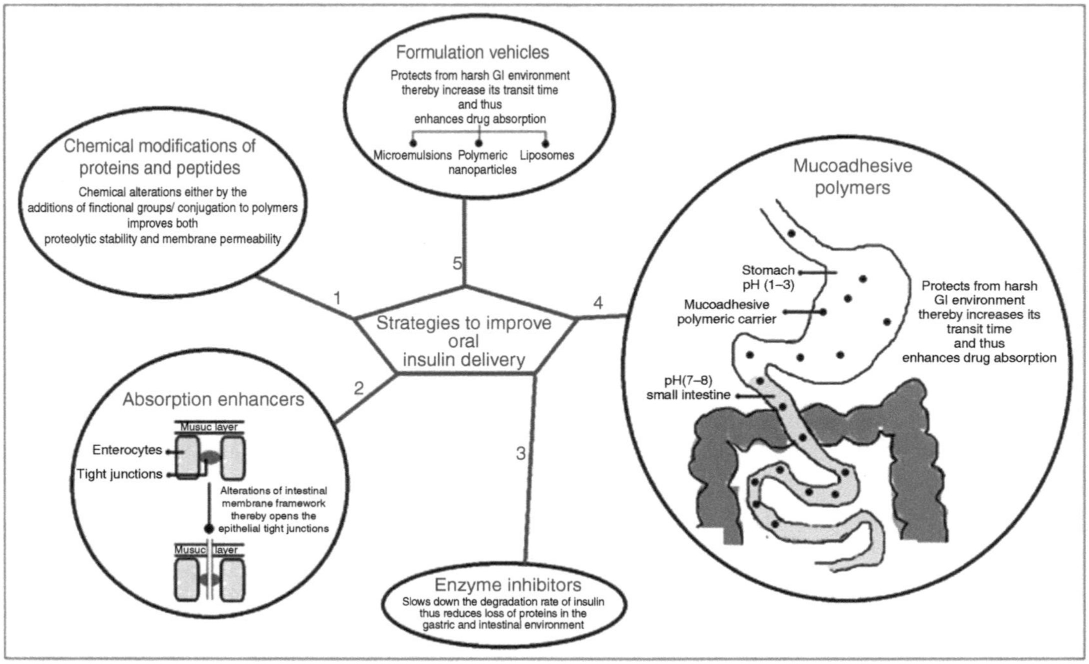

FIGURE 15.10 Strategies to Overcome Physiologic and Morphologic Barriers Towards Insulin Therapy

mucus gumminess, or modifying the structural integrity, thereby opening the epithelial tight junctions to allow drug penetration (Antunes et al., 2013). Permeation enhancers should be harmless, balm, non-allergic, safe, and remain inert either pharmacologically or chemically. Thus its selection and efficiency depends on the physicochemical properties of the protein and peptide drugs, regional differences in intestinal membrane, nature of the vehicle, and other excipients.

Detergents or surfactants enrich transcellular pathway by disrupting the lipid bilayer, rendering the cell membrane more permeable (Aungst, 2012). To enhance paracellular transport of proteins these permeability enhancers, such as chelating agents, forms complexes with calcium ions and rupture tight junctions between intestinal epithelial cells (Anil kumar et al., 2011). In the same manner long alkyl chain enhancers, including fatty acid sodium caprate and acyl carnitines (Hamman and Steenekamp, 2011), shows improved absorption via transient opening of tight junctions.

However, such executed excipients may cause serious side effects, such as damage intestinal epithelial cells and permeate movement of certain toxins, viruses, and pathogens through the intestine (Maher and Brayden, 2012). Nevertheless, these changes in the lipid bilayer of the intestine are reversible in nature. Zonula occludens toxin is a safe and effective enhancer (Song et al., 2015) altering intestinal epithelial tight junctions transiently for the passage of macromolecules, such as insulin, through mucosal barriers.

Carrier molecules can enhance bioavailability of proteins when coadministered with proteins. For example, lipophilic carrier enhancers facilitated the absorption of proteins, such as insulin, human growth hormone, calcitonin, and recombinant parathyroid hormone (rPTH) (Chin et al., 2012). Carrier molecules temporarily stabilize the partially unfolded conformations of proteins exposing their hydrophobic side chains. Thus, the carrier molecules altered lipid solubility, allowing them to gain access to pores of integral membrane transporter. This results in enhanced absorption through lipid bilayers.

11.2 ENZYME INHIBITORS

In oral protein delivery one of main barriers is that protein degrades very rapidly by various proteolytic enzymes. Enzyme inhibitors slow down the rate of degradation of which greatly reduces the loss of proteins in the gastric and intestinal environments, thus increases the amount of insulin available for absorption. Enzyme inhibitors bind target enzymes either reversibly or irreversibly, inactivate it and thereby decreases its activity (Copeland, 2013). Coadministration of enzyme inhibitors provide an effective approach to overcome the enzymatic barrier. The most prominently used enzyme inhibitors were that of trypsin, chymotrypsin, carboxy peptidases, and so on. Aprotinin (inhibitor of trypsin and chymotrypsin), soybean trypsin inhibitor (inhibitor of pancreatic endopeptidases), FK448 (chymotrypsin inhibitor), and chicken ovomucoid (trypsin inhibitor), and so on, are promptly used examples of intestinal protease inhibitors (Gamboa and Leong, 2013).

Enzyme inhibitors are broadly categorized into following groups, namely (1) amino acids and modified amino acids (e.g., boro-leucine, boro-valine), (2) nonamino acids (diisopropyl fluorophosphates, phenylmethylsulfonyl fluoride, camostat mesilate), (3) peptides and modified peptides (bacitracin, pepstatin, antipain, leupeptin, amastatin, chymostatin, and elastatinal), and (4) polypeptides (aprotinin, Bowman–Birk inhibitor, Kunitz trypsin inhibitor, chicken egg white trypsin inhibitor, chicken ovo inhibitor, and human pancreatic trypsin inhibitor).

A novel class of enzyme inhibitor, chicken and duck ovo mucoids has been recently reported, and a formulation has been developed where the insulin and duck ovo mucoids offered 100% protection against the action of trypsin and a-chymotrypsin (Sonia and Sharma, 2014).

Other than enzyme inhibitors, enzymatic activity of GIT can also reduce by usage of pH modifiers. Similarly, a combination of protease inhibitors with permeation enhancers improves oral bioavailability of protein drugs. In combinational strategy the protease inhibitors inhibit the enzymatic degradation of the protein on the one hand while on other side the permeation enhancer facilitates absorption through the intestinal membrane (Araujo et al., 2012).

11.3 MUCOADHESIVE POLYMERS

Another strategy to increase the GI retention time, which would further improve the oral bioavailability of insulin, is the development of delivery devices. These delivery devices are mucoadhesive polymers. Mucoadhesion refers to adhesion between a material and mucosal surface by means of interfacial forces.

The mucoadhesive drug delivery systems possess a number of advantages, like protection of drug the harsh GI environment, enhanced drug absorption across the epithelium, longer transit time in the GIT resulting in extended periods of drug absorption, decreased diffusion barriers thus enhancing bioavailability of drug.

In order to develop a successful adhesive drug delivery system it is necessary to understand the mechanism involved and the adhesive forces that facilitate bond formation. An ideal mucoadhesive polymer should possess the following characteristics, such as presence of chemical groups, which determines the surface charges, high chain flexibility, high molecular weight, hydrophobicity/hydrophilicity, chemically inert, nonirritant, nontoxic, nonallergenic, and optimum surface tension that will induce spreading of polymer onto the mucus layer (Shaikh et al., 2011).

Mucoadhesivity is affected by number of factors, including hydrophilicity, molecular weight, cross-linking, swelling, pH, and the concentration of the active polymer. Mucoadhesive polymers possess numerous hydrophilic groups, such as hydroxyl, carboxyl, amide, and sulfate, through which mucus or the cell membrane interacts either via hydrogen bonding or hydrophobic or electrostatic interactions. These hydrophilic groups also cause polymers to swell in water and, thus expose the maximum number of adhesive sites (Sosnik et al., 2014).

Mucoadhesive polymers are mainly categorized as anionic, cationic, nonionic, amphoteric, thiomers, dendrimers, synthetic glycopolymers (Khutoryanskiy, 2011). In the past few years, a large number of mucoadhesive systems have been developed, including chitosan, methacrylate, alginate, based polymers. In the past 2 decades thiomers are gaining attention toward their mucoadhesivity behavior. Mucoadhesive polymers possess numerous hydrophilic groups, such as hydroxyl, carboxyl, amide, and sulfate. These groups attach to mucus or the cell membrane by various interactions, such as hydrogen bonding and hydrophobic or electrostatic interactions. These hydrophilic groups also cause polymers to swell in water and, thus, expose the maximum number of adhesive sites (Iqbal et al., 2012). These polymers form stronger covalent interactions than other noncovalent bonds, such as hydrogen bonds, van der Waal's forces and ionic interactions of polymer with anionic substructures of the mucus membranes. Anionic/hydrophobic modification of chitosan (LSC) improved the mucoadhesivity of the nanoparticles. The mucoadhesive nanoparticles can either enhance the permeability of insulin across the intestinal cell wall or be absorbed by the transcellular or paracellular route.

11.4 CHEMICAL MODIFICATIONS OF PROTEINS AND PEPTIDES

The chemical alteration of proteins and peptides can be attained by the addition of functional groups or by conjugation to polymers for improving the proteolytic stability and membrane permeability thus enhances bioavailability of drug (Solaro et al., 2010). These chemical modifications may include PEGylation, exchanging of amino acids preferably by substitution of L-amino acid with D-amino acids or by replacement with different amino acids, and hydrophobization. The polymers used for chemical modification of proteins should be water soluble, nonimmunogenic, biocompatible, and devoid of any biological activity.

N-(2-hydroxypropyl) methacrylamide and polyethylene glycol are polymers possess ideal properties protein conjugation. PEGylation offers three advantages as therapeutic perspective: (1) alters the interaction of the therapeutic molecule with the blood proteins, affecting its receptor-binging capacities and filtration through the glomeruli thus decreases the rate of clearance of the therapeutic peptide from the systemic circulation; (2) therapeutic proteins, even of recombinant origin can elicit immune responses on repeated exposure in vivo; and (3) protect proteins from the enzymatic attack due to its steric effect.

Insulin can be modified by PEGylation (adding polyethylene glycol) in parenteral formulations, which enhance the pharmacokinetics of insulin. A PEGylated version of human growth hormone represented similar activity, in terms of growth, as unmodified form but enhanced the sensitivity of insulin in hypophysectomized rats when tested with respect to glucose metabolism (Zhang et al., 2012).

Desmopressin, a vasopressin analog, was developed due to deamination of *N*-terminal amino acid and interchanging of the C-terminal L649 arginine by D-arginine in vasopressin. Metabolic stability, membrane permeability, bioavailability, changes in pharmacokinetic, and pharmacodynamic properties of several peptide are improved due to lipidation (Zhang and Bulaj, 2012). In hydrophobization the surface modification of proteins can occur either by including more hydrophobic amino acids within the peptide backbone or by covalent conjugation with lipophilic moieties, which will enhance the transcellular uptake (Yuan et al., 2011).

11.5 FORMULATION VEHICLES

Numerous strategies suggested the use of formulation vehicles, such as microemulsions, liposomes, nanoparticles, and microspheres to protect the protein formulation against the harsh environment of the GIT (acidic medium and enzymes) so as to achieve successful delivery of proteins and peptides via the oral route (Wu et al., 2012). These formulation vehicles offer a number of advantages, such as enhanced drug delivery across the epithelial mucosa, protecting the encapsulated protein from enzymatic degradation, controlling the release rate, and targeted drug delivery to specific sites in the intestine.

Encapsulation protects the macromolecules against enzymes and pH effects. Thus, the insulin encapsulated in these formulation vehicles are able to cross the intestinal epithelium, either through the Payer's patches (Jung et al., 2010) and/or the regular absorptive epithelium, and travel systemically to other sites including the liver, spleen, and other tissues. These observations suggest three major mechanisms of action of formulation vehicles for improving oral delivery of proteins: (1) the spheres shields proteins from enzymatic degradation, (2) the spheres cross over the intestinal mucosa, and (3) the spheres change the intertissue distribution of the protein in organism (Duraivel et al., 2013). The last proposed mechanism is extremely important for drugs, like insulin, as these carriers are expected to build up in the liver where the loaded insulin can be efficiently used to limit the abnormally elevated hepatic glucose production present in diabetic patients. Different carriers extensively used for encapsulating therapeutic proteins are discussed in the next sections.

11.5.1 Microemulsions

Microspheres (1–1000 μm) can be employed as another strategy for enhancing oral protein or peptide delivery (Prasanth et al., 2011). The major advantage of microspheres is their provision of stability for labile compounds that are easily degraded in the intestinal epithelium. Eudragit S100 microspheres have the potential to act as an oral carrier for peptide drugs, like insulin (Kumar et al., 2014). Alginate–chitosan microcapsules fabricated by layer-by-layer assembly method and provide a simple method for controlling the loading and release of protein molecules with these polysaccharide microcapsules (Sameer, 2014). Insulin loaded in microcapsules of chitosan and sodium alginate fabricated by layer-by-layer assembly has been reported.

11.5.2 Polymeric nanoparticles

The formulation vehicles, such as polymeric nanoparticles, easily entrap and encapsulate therapeutic proteins and peptides and lead to the targeted area. Nanoparticles are colloidal polymeric drug carriers that show uptake by paracellular, transcellular, or intercellular pathways or via the M cells and Peyer's patches, and are stable in the GIT (Chan et al., 2010). The concepts behind these nanocarriers can reduce clearance through the alimentary canal and lead to increased bioavailability of therapeutic protein.

Makhlof et al. (2011) revealed the permeation enhancing properties of the mucoadhesive nanoparticles. Polymers, including poly (lactide-*co*-glycolide) (PLGA), poly(lactide) (PLA) and poly(ε-caprolactone) (PCL), have been used to fabricate NPs for oral insulin delivery.

11.5.3 Liposomes

Liposomes are particulate protein carriers widely used for the oral delivery of proteins (Swaminathan and Ehrhardt 2012). Liposomes are spherical vesicles consisting of an aqueous core entrapped by one or more lipid bilayers. In addition to their size and aqueous core their biocompatible, biologically inert, weakly immunogenic nature and limited toxicity makes it perfect vehicle for protein drug delivery. These are lipid-based delivery systems offering protection for peptides and proteins in GIT (Li et al., 2010). Archaeosomes (liposomal formulations) controlled blood glucose levels more tightly relative to conventional liposomes. Hence archeosomes can be considered as better carriers for protein and peptide delivery due to higher stability. Mohanraj et al. (2010) developed a hybrid silica-liposome nanocapsule (SNLC) system containing insulin. They work out on capabilities of silica-liposome nanocapsule to shield against lipolytic degradation and prolong insulin release in simulated GI conditions, and thus find out that even in presence of digestive enzymes insulin gets protected from enzymatic degradation.

12 RECENT BREAKTHROUGHS IN ORAL INSULIN DELIVERY

Oral delivery of insulin considerably improves the quality of life of diabetic patients who routinely receive insulin by the subcutaneous route. In fact, compared with this administration route, oral delivery of insulin in diabetes treatment offers several advantages. This fact is supported by numerous researchers. Sarmento and his colleague (Fonte et al., 2013) suggested that protein encapsulation into nanoparticles is observed as a promising alternative to administer insulin orally as they have the ability to promote insulin paracellular or transcellular transport across the intestinal mucosa. Also they discussed different delivery systems intended to increase the oral bioavailability of insulin with a special focus on nanoparticulate carrier systems. Similarly Verma et al. (2014) informed that oral insulin avoid the pain associated with subcutaneous administration, anxiety due to needle, and infections developed due to subcutaneous administration. They discussed the use of different types of enzyme inhibitors, like sodium cholate, camostat, mesilate, bacitracin, leupeptin, and so on, which have been used to prevent insulin from enzymatic degradation.

Prusty and Sahu (2013) designed and characterized insulin incorporated nanoparticles by complex coacervation process and did antidiabetic study of orally administered insulin-incorporated nanoparticles in diabetic rats. The results indicate that the prepared nanoparticulate system act as an effective carrier for oral delivery of insulin. Das and Chaudhury (2011) studied lipid nanoparticles based on solid matrix as potential drug carriers to improve GI absorption and oral bioavailability of several drugs, especially lipophilic compounds.

Shan et al. (2015) revealed that orally administered nanoparticles could be absorbed by the epithelial tissue only if they successfully permeate through the mucus that covers the epithelium. The prepared nanoparticles having a nanocomplex core composed of insulin and cell penetrating peptide (CPP), and a dissociable hydrophilic coating of *N*-(2-hydroxypropyl) methacrylamide copolymer (pHPMA) derivatives. After systematic screening using mucus-secreting epithelial cells, nanoparticles exhibit excellent permeation in mucus due to the "mucus-inert" pHPMA coating, as well as high epithelial absorption mediated by CPP. The investigation of nanoparticle behavior shows that the pHPMA molecules gradually dissociate from the nanoparticle surface as it permeates through mucus, and the CPP-rich core is revealed in time for subsequent transepithelial transport through the secretory endoplasmic reticulum/golgi pathway and endocytic recycling pathway. The nanoparticles exhibit 20-fold higher absorption than free insulin on mucus-secreting epithelium cells, and orally administered nanoparticles generate a prominent hypoglycemic response and an increase of the serum insulin concentration in diabetic rats. Their studies provide the evidence of using pHPMA as dissociable "mucus-inert" agent to enhance mucus permeation of nanoparticles, and validate a strategy to overcome the multiple absorption barriers using nanoparticle platform with dissociable hydrophilic coating and drug-loaded CPP-rich core.

13 COMMERCIALIZATION

Oral delivery systems for proteins and peptides are still in development stages. Several oral protein and peptide technologies produced by pharmaceutical companies for application in various clinical situations are still under way. Although most of the formulations based on these technologies are in the developmental stage, many have progressed to the clinical trial stage. The data are summarized in Table 15.1.

The technologies are summarized in the next sections.

13.1 ORAL-LYN

RapidMist, a device as similar to that used by asthmatic patients, is running successfully in the market, launched by Generex Biotechnology Corporation, Canada. The formulation, called Oral-Lyn is delivered to the oral cavity through the RapidMist device (aerosol-type device containing nonchlorofluro-carbon propellant, penetration enhancers, and stabilizers), which permeate across the buccal epithelium and enter the rich network of blood vessels. Due to sufficiently large micellar size (larger than 7 mm) of Oral-Lyn delivery system, it does not enter the respiratory system. Oral-Lyn is claimed to be safe formulation in which Oral-Lyn without insulin formulation was administrated to 40 dogs or nearly 1000 patients and did not show any abnormalities in the buccal mucosa (Muheem et al., 2014). Oral-Lyn is in various stages of clinical development around the world, and now, this product is under development for treatment of types 1 and 2 diabetes. These formulations showed effective results in type 2 diabetes, whose patients were resistant to conventional treatments such as diet, exercise, metformin, sulfonylureas, and thiazolidenes. In recent times, Oral-Lyn has also been permitted in India for import, commercialization, and sale for the treatment of both types 1 and 2 diabetes and the product has been made available with the name Oral Recosulin since January 2009.

13.2 ELIGEN

Eligen technology developed by Emisphere Technologies, Inc. is a macromolecule-delivering platform, which enriches the transcellular permeation across intestinal cell membranes. Emisphere delivery agents, based on sodium *N*-[8-(2-hydroxybenzoyl) aminocaprylate] (SNAC), interact with the drug molecule by forming weak, noncovalent complexes that enable transcellular absorption, without altering the tight junctions (Brayden and Mrsny, 2011). The synthesized lipophilic drug/carrier complex is

Table 15.1 Protein Oral Delivery Technologies Under Development by Companies

S.No.	Product Name	Company	Technology	Formulation	Developmental Stage	Products
1.	Oral-Lyn	Generex	Penetration enhancers	Spray devices and aerosol particles	Phase IV	Insulin, macrotonin
2.	Eligen	Emisphere	Penetration enhancers	Tablet	Phase II	Calcitonin, insulin, PTH, heparin, calcitonin, enzymes (lipases, esterases, proteases)
3.	Capsulin	Diabetology	Penetration enhancers	Capsule	Phase II	Insulin
4.	Octreolin	Chiasma (Israel)	Penetration enhancers [sodium caprylate(C8)]	Suspension	Phase I (phase I completed, phase III)	Enrolling octreotide
5.	ORMD-0801	Oramed	Salts of EDTA (enteric coated +PE)	Capsule	Phase I	Insulin/exenatide
6.	AI-401	Eli–lily	Enzyme inhibitor	Oral formulation	Phase II	Insulin
7.	Oradel	Apollo Life Science	Nanoparticles	Tablet	Clinical phase I b Insulin,	TNF-blocker
8.	Macrulin	Provalis PLC	Lipid-based microemulsion	Emulsion	Phase II	Insulin, salmon calcitonin

capable of diffusing across the epithelial membrane. This method shows reversible and safe delivery of macromolecules without altering the drug, damaging epithelial membranes, or creating new toxicities, due to the weak association between carrier and drugs. Eligen technology was used to develop various types of oral formulations including solutions, tablets, and capsules.

Eligen technology has served as a hydrophobic carrier for the delivery of various macromolecules ranging in size from 500 to >150,000 daltons, such as heparin, insulin, and calcitonin, thereby increasing their lipophilicity. In a study it was found that in patients with type 2 diabetes, the insulin capsule containing 10 mg of insulin and 200 mg of SNAC, when administered 30 min prior to the standardized meal, reduced postprandial excursion, produced a marked increase in systemic insulin levels (Park et al., 2011). The practicality of Eligen technology has been studied in animals and humans for the development of various formulations for improved oral delivery of therapeutic macromolecules. Although this technology shows promising results in oral delivery of protein and peptides, its clinical application is problematic because it causes nausea in patients and it requires a higher amount delivery agent as compared to the protein or peptide, which ultimately decreases patient compliance (Choonara et al., 2014).

13.3 CODES (COLON SPECIFIC DELIVERY SYSTEM) TECHNOLOGY

Yamanouchi Pharmaceutical Co., Ltd. (Tokyo, Japan) formulated CODES technology for colonic-specific delivery. It consists of lactulose-containing tablet coated with two acrylic films that display pH-dependent solubility. Principally, this technology proposes that the enteric soluble polymer protects the tablet from the harsh gastric conditions within the stomach but rapidly dissolves within the intestine (Kotla and Shivapooja, 2014) whereas the Eudragit E coating subsequently protects the tablet within the alkaline environment of the small intestine (Marais et al., 2013). As the drug enters the colon, the existing bacteria enzymatically degrade the lactulose into an organic acid resulting in lowering the surrounding pH of the tablet thus enabling the dissolution of the acid soluble polymer and initiating drug release. This technology significantly proved itself as studies showed reduced blood glucose levels with supplementary use of absorption enhancers and sustained delivery.

13.4 CAPSULIN

Capsulin is under clinically trial by UK-based company Diabetology, which shows the beginning of action within 30 min and duration of action up to 4–6 h. During the fasting condition, higher doses (300 I.U.), given to healthy volunteers with T1DM (type 1 diabetes mellitus), showed a sudden fall in blood glucose level (1.6 mmol/L) and minimum doses (150 I.U.), which represented lowering of blood glucose levels (0.02 mmol/L) (Luzio et al., 2010). Thus, the results obtained supported the fact that Capsulin has the ability to control the progression of diabetic conditions.

13.5 OCTREOLIN

Transient permeability enhancer (TPE) system is an enteric coated formulation produced by Chiasma, Inc. (Jerusalem, Israel), which facilitates intestinal absorbance of drug molecules with limited intestinal bioavailability. It is formulated from sodium caprylate in hydrophobic microparticles and is agitated with castor oils or medium-chain glycerides, yielding emulsions (oily suspension) (Salama et al., 2010). TPE functions by opening of tight junctions between adjacent epithelial cells through paracellular pathway. It was able to improve the bioavailability of the peptide octreotide in a study by Tuvia et al. (2012). Octreolin is an oral octreotide capsule produced by Chiasma for the treatment of acromegaly that is based on their proprietary TPE system. Octreolin is currently in phase 3 trials for acromegaly and if successful is expected to be launched in 2015.

13.6 ORMD-0801

Oramed Pharmaceuticals, Inc. (Jerusalem, Israel) have developed a POD technology to improve the absorption of proteins and peptides across the GIT epithelium by exploiting the action of protease inhibitors and absorption enhancers. Effect of oral insulin was determined by studies in eight volunteers in the fasted condition and demonstrated reduced glucose levels (7%–35%) and also a decline in the C-peptide level (13%–87%) in all formulations (Eldor et al., 2013). When the studies were conducted on fed volunteers, release of insulin was found to be adversely affected by meal and GIT motility. The onset and duration of action from time of administration was found to be 2 h and 5–6 h, respectively (Walsh et al., 2011). However, this approach has numerous shortcomings, such as the increased paracellular permeability of possible toxic molecules and the safety considerations associated with EDTA salts used orally.

13.7 AI-401

AI-401 is an oral formulation of recombinant human insulin and is used for the prevention and treatment of type 1 diabetes. This technique follows the principal of oral-tolerance therapy. Small drug dosages (too small to affect blood glucose levels) are claimed to result in suppression of auto reactive T cells that have been attacking the pancreas. Autoimmune Inc. has agreements with Eli–Lilly to develop and market the product. The interim results of the US phase II trial of AI-401 for the treatment of diabetes showed the feasibility of oral insulin therapy, delaying beta-cell destruction in the pancreas, and thus preserving endogenous insulin secretion in newly diagnosed type 1 diabetic patients (Pandey, 2010).

13.8 ORADEL

Apollo Life Sciences, Australia, has developed Oradel drug delivery technology, which have the ability to protect and transport insulin molecules by encapsulating them in the matrix of nanoparticles made up of a sugar-based (carbohydrate) protective polymer coated with vitamin B12 molecules (Sharma et al., 2015). Nanoparticles containing insulin were claimed to be pulled through the wall of the intestine using an active transport mechanism in the GIT for the absorption and cellular uptake of the relative large vitamin B12 molecules. Vitamin B12 molecules are suggested to protect insulin from enzymatic digestion in the stomach as well as facilitate the transport of insulin across the small intestine into the blood stream. Apollo Life Sciences (Yun et al., 2013) has indicated that it has a unique production method whereby up to 100% of insulin molecules are entrapped in the matrix of the Oraldel nanoparticles and blood glucose levels could be lowered to within normal range for up to 12 h in diabetic rats.

13.9 MACRULIN

The lipid-based microemulsions can load protein drugs, and have been proposed to enhance the bioavailability of orally administered protein drugs (Severino et al., 2014). This system prevents proteins from proteolysis or acidic degradation to enhance the protein absorption in the GIT. Provalis PLC has developed an insulin-based oral pill (Macrulin). The technology uses a water-in-oil (w/o) microemulsion in which the aqueous phase contains insulin and the oil phase contains cholesterol, lecithin (a biological surfactant and a major component of membrane lipids) and nonesterified fatty acids. The lipid phase acts as a vehicle to enhance the uptake of insulin from the small intestine following oral administration (Lopes et al., 2014). Macrulin is in phase II clinical trials for the management of type 2 diabetes.

14 RECENT CHALLENGES AND FUTURE PROSPECTS

Oral delivery of proteins and peptides is the most efficient way to replace the invasive route as well as a very interesting and promising area for research. Major problems arising in oral insulin administration are the high molecular weight, chemical or enzymatic degradation, low intestinal permeability, as well as the low oral bioavailability. Various efficient approaches were discussed to overcome these problems. The research progress on finding a system to deliver insulin orally has concentrated more in academic research, rather than in developments in the pharmaceutical industry.

The technologies proposed that have been used to improve bioavailability of orally delivered proteins are based on specific approaches like chemical modification of proteins and adding absorption enhancers, which prevents degradation of protein by acid and enzymes in the GIT or increasing the permeability of proteins through the epithelial layer of the GIT. A variety of manipulations, such as chemical modification of proteins and adding absorption enhancers, have shown some increase in oral bioavailability. This has been achieved by using different carriers such as synthetic polymeric carriers (e.g., polymeric nanoparticles, polymer-drug conjugates, micelles, and dendrimers), natural polymers, stimuli (e.g., pH, temperature, and enzyme) responsive polymers, liposomes, SLNs, drug nanoparticles, drug emulsions, peptides. The growing field of nanotechnology has allowed cost-effective and pilot-scale production of proteins and peptides used for oral delivery such as Oral Recosulin, Octreolin, Macrulin and Orasome, in which a few are in clinical stage of development.

REFERENCES

Ahmad, A., Othman, L., Zaini, A., Chowdhury, E.H., 2012. Oral nano-insulin therapy: current progress on nanoparticle-based devices for intestinal epithelium-targeted insulin delivery. J. Nanomed. Nanotechnol. S4 (007), 1–10.

Alai, M.S., Lin, W.J., Pingale, S.S., 2015. Application of polymeric nanoparticles and micelles in insulin oral delivery. J. Food Drug Anal. 23, 351–358.

Anil kumar, P., Badarinath, A.V., Naveen, N., Prasad, K., Reddy, R.S.B., Hyndhavi, M., Nirosha, M., 2011. Rationalized description on study of intestinal barrier, drug permeability and permeation enhancers. J. Global. Trends. Pharm. Sci. 2 (4), 431–449.

Antunes, F., Andrade, F., Ferreira, D., Nielson, H.M., Sarmento, B., 2013. Models to predict intestinal absorption of therapeutic peptides and proteins. Curr. Drug. Metab. 14, 4–20.

Araujo, F., Fonte, P., Sarmento, B., 2012. Oral delivery of glucagon-like peptide-1 and analogs: alternatives for diabetes control. J. Diabetes Sci. Technol. 6, 1486–1497.

Aungst, B.J., 2012. Absorption enhancers: applications and advances. AAPS J. 14, 10–18.

Bassam, A., Rasool, H., 2013. Overview on diabetes mellitus (Type 2). J. Chromat. Separ. Techniq. 4 (2), 1000–1114.

Brayden, D.J., Mrsny, R.J., 2011. Oral peptide delivery: prioritizing the leading technologies. Thera. Deliv. 2 (12), 1567–1573.

Chan, J.M., Valencia, P.M., Zhang, L., 2010. Polymeric nanoparticles for drug delivery. Methods Mol. Bio. 624, 163–175.

Chaturvedi, K., Ganguly, K., Nadagouda, M.N., Aminabhavi, T.M., 2013. Polymeric hydrogels for oral insulin delivery. J. Control. Release 165, 129–138.

Chen, L., Magliano, D.J., Zimmet, P.Z., 2012. The worldwide epidemiology of type 2 diabetes mellitus—present and future perspectives. Nature 8, 228–236.

Chen, M.C., Sonaje, K., Chen, K.J., Sung, H.W., 2011. A review of the prospects for polymeric nanoparticle platforms in oral insulin delivery. Biomaterials 32, 9826–9838.

Chin, J., Mahmud, K.A.F., Kim, S.E., Park, K., Byun, Y., 2012. Insight of current technologies for oral delivery of proteins and peptides. Drug Discov. Today 9 (2), e105–e112.

Choonara, B.F., Choonara, Y.E., Kumar, P., kumar, D.B., du Toit, L.C., Pillay, V., 2014. A review of advanced oral drug delivery technologies facilitating the protection and absorption of protein and peptide molecules. Biotechnol. Adv. 32, 1269–1282.

Copeland, R.A., 2013. Evaluation of enzyme inhibitors in drug discovery: a guide for medicinal chemist and pharmacologists. John Wiley and Sons, New Jersey, pp. 57–59.

Das, S., Chaudhury, A., 2011. Recent advances in lipid nanoparticle formulations with solid matrix for oral drug delivery. AAPS Pharm. Sci. Tech. 12 (1), 62–76.

Ding, W.Y., Tanaka, Y., Zhang, W., 2014. Risk factors contributing to type 2 diabetes and recent advances in the treatment and prevention. Int. J. Med. Sci. 11 (11), 1185–1200.

Dorothee, G., Yu, A.S.L., 2013. Claudins and the modulation of tight junction permeability. Physiol. Rev. 93 (2), 525–569.

Duraivel, S., Venkatesh, V., Rajalakshmi, A.N., Gopinath, H., 2013. Peroral administration of peptides and proteins. Elixir Pharm. 58, 14535–14545.

Eldor, R., Arbit, E., Corcos, A., Kidron, M., 2013. Glucose-reducing effect of the ORMD-0801 oral insulin preparation in patients with uncontrolled type 1 diabetes: a pilot study. PLoS One 8 (4), e59524, (1–4).

Ensign, L.M., Cone, R., Hanes, J., 2012. Oral drug delivery with polymeric nanoparticles: the gastrointestinal mucus barriers. Adv. Drug Deliv. Rev. 64 (6), 557–570.

Fonte, P., Araújo, F., Reis, S., Sarmento, B., 2013. Oral insulin delivery: how far are we? J. Diabetes Sci. Technol. 7 (2), 520–531.

Gamboa, J.M., Leong, K.W., 2013. In vitro and in vivo models for the study of oral delivery of nanoparticles. Adv. Drug Deliv. Rev. 65 (6), 800–810.

George, S., Roy, A., 2013. Novel approaches in insulin drug delivery. Int. J. Drug Dev. Res. 5 (4), 25–29.

Hamman, J.H., Steenekamp, J.H., 2011. Oral peptide drug delivery: strategies to overcome challenges. In: Castanho, M., Santos, N.C. (Eds.), Peptide Drug Discovery and Development: Translational Research in Academia and Industry. Wiley-VCH Verlag GmbH & Co. KGaA, Weinheim, Germany.

Herrero, E.P., Alonso, M.J., Csaba, N., 2012. Polymer-based oral peptide nanomedicines. Ther. Deliv. 3, 657–668.

Hilgenfeld, R., Seipke, G., Berchtold, H., Owen, D.R., 2014. The evolution of insulin glargine and its continuing contribution to diabetes care. Drugs 74 (8), 911–927.

Hosseininasab, S., Pashaei-Asl, R., Khandaghi, A.A., Nasrabadi, H.T., Nejati-Koshki, K., Akbarzadeh, A., et al., 2014. Synthesis, characterization, and in vitro studies of PLGA-PEG nanoparticles for oral insulin delivery. Chem. Biol. Drug Des. 84, 307–315.

Iqbal, J., Shahnaz, G., Dunnhaupt, S., Muller, C., Hintzen, F., Bernkop-Schnurch, A., 2012. Preactivated thiomers as mucoadhesive polymers for drug delivery. Biomaterials 33, 1528–1535.

Iyer, H., Khedkar, A., Verma, M., 2010. Oral insulin: a review of current status. Diabetes. Obes. Metab. 12, 179–185.

Jung, C., Hugot, J.P., Barreau, F., 2010. Peyer's patches: the immune sensors of the intestine. Int. J. Inflam. 2010, 1–12.

Kalra, S., Kalra, B., 2012. Storage of insulin in rural areas. J. Acad. Med. Sci. 2 (2), 88–89.

Khulbe, A., 2015. A Review on oral insulin. World J. Pharm. Sci. 3 (5), 884–889.

Khutoryanskiy, V.V., 2011. Advances in mucoadhesion and mucoadhesive polymers. Macromol. Biosci. 11, 748–764.

Kinesh, V.P., Neelam, D.P., Punit, B.P., Bhavesh, S.B., Pragna, K.S., 2010. Novel approaches for oral delivery of insulin and current status of oral insulin products. Int. J. Pharm. Sci. Nanotechnol. 3 (3), 1057–1064.

Kotla, N.J., Shivapooja, A., 2014. Recent developments in colon specific drug delivery systems approaches promising in targeting colon. Int. J. Pharm. Clin. Res. 6, 101–106.

Krishnan, Sudhakar, A.U., 2015. Oral insulin: a challenging approach in insulin delivery. World J. Pharm. Pharm. Sci. 2, 404–420.

Kumar, R., Majumdar, D.K., Panda, A.K., Pathak, D.P., 2014. Eudragit polymers in colon targeted oral delivery of insulin. Int. J. Pharm. Chem. Biol. Sci. 4 (4), 968–979.

Lemmer, H.J., Hamman, J.H., 2013. Paracellular drug absorption enhancement through tight junction modulation. Exp. Opin. Drug Deliv. 10 (1), 103–114.

Li, Z., Chan, J., Sun, W., Xu, Y., 2010. Investigation of archaeosomes as carriers for oral delivery of peptides. Biochem. Biophys. Res. Commun. 394, 412–417.

Lopes, M.A., Abrahim, A.A., Cabral, L.M., Rodrigues, C.R., Seica, R.M.F., 2014. Intestinal absorption of insulin nanoparticles: contribution of M cells. Nanomed. Nanotechnol. Biol. Med. 10 (6), 1139–1151.

Luzio, S.D., Dunsealth, G., Lockett, A., et al., 2010. Glucose lowering effect of an oral insulin (capsulin) during an isoglycaemic clamp study in persons with type 2 diabetes. Diabetes Obes. Metab. 12, 82–87.

Maher, S., Brayden, D.J., 2012. Overcoming poor permeability: translating permeation enhancers for oral peptide delivery. Drug Discov. Today Technol. 9, 113–119.

Makhlof, A., Werle, M., Tozuka, Y., Takeuchi, H., 2011. A mucoadhesive nanoparticulate system for the simultaneous delivery of macromolecules and permeation enhancers to the intestinal mucosa. J Control Release 149 (1), 81–88.

Marais, E., Hamman, J., du Plessis, L., Lemmer, R., Steenekamp, J., 2013. Eudragit L100/*N*-trimethylchitosan chloride microspheres for oral insulin delivery. Molecules 18, 6734–6747.

Mohanraj, V.J., Barnes, T.J., Prestidge, C.A., 2010. Silica nanoparticle coated liposomes: a new type of hybrid nanocapsule for proteins. Int. J. Pharm. 392, 285–293.

Muheem, A., Shakeel, F., Jahangir, M.A., Mohammed, A., Mallick, N., Jain, G.K., Warsi, M.H., Ahmad, F.J., 2014. A review on the strategies for oral delivery of proteins and peptides and their clinical perspectives. Saudi Pharm. J. 24 (4), 413–428.

Mukhopadhyaya, P., Mishra, R., Rana, D., Kundu, P.P., 2012. Strategies for effective oral insulin delivery with modified chitosan nanoparticles: a review. Prog. Polym. Sci. 37, 1457–1475.

Najafzadeh, H., Kooshapur, H., Kianidehkordi, F., 2012. Evaluation of an oral insulin formulation in normal and diabetic rats. Ind. J. Pharmacol. 44 (1), 103–105.

Negrato, C.A., Junior, M.M., Kostrisch, L.M.V., Guedes, M.F., Mattar, R., Gomes, M.B., 2012. Insulin analogues in the treatment of diabetes in pregnancy 56 (7), 405–414.

Olya, S., Khorvash, M., Rahmani, H.R., Esmaeilkhanian, S., Olya, B., Sadri, H., 2014. Oral delivery of insulin-loaded nanoparticles in diabetic rabbits and in sheep. Czech. J. Anim. Sci. 59 (6), 251–256.

Ozougwu, J.C., Obimba, K.C., Belonwu, C.D., Unakalamba, C.B., 2013. The pathogenesis and pathophysiology of type 1 and type 2 diabetes mellitus. J. Physio. Pathophy. 4 (4), 46–57.

Pandey, S., 2010. Non-invasive insulin delivery systems: challenges and needs for improvement. Int. J. Pharm. Tech. Res. 2 (1), 603–614.

Pandit, N., Joshi, T., 2015. Novel approaches for oral delivery of insulin. J. Drug Deliv. Thera. 5 (4), 61–70.

Park, J.W., Kim, S.K., Al-Hilal, T.A., Jeon, O.C., Moon, H.T., Byun, Y., 2010. Strategies for oral delivery of macromolecule drugs. Biotechnol. Biopro. Eng. 15, 66–75.

Park, K., Kwon, I.C., Park, K., 2011. Oral protein delivery: current status and future prospect. React. Funct. Polym. 71, 280–287.

Pawar, V.K., Meher, J.G., Singh, Y., Chaurasia, M., Reddy, B.S., Chourasia, M.K., 2014. Targeting of gastrointestinal tract for amended delivery of protein/peptide therapeutics: strategies and industrial perspectives. J. Control Release 196, 168–183.

Piero, M.N., Nzaro, G.M., Njagi, J.M., 2014. Diabetes mellitus—a devastating metabolic disorder. Asian J. Biomed. Pharm. Sci. 4 (40), 1–7.

Prasanth, V.V., Moy, A.C., Matthew, S.T., Mathapan, R., 2011. Microspheres—an overview. Int. J. Res. Pharm. Biomed. Sci. 2, 332–338.

Pridgen, E.M., Alexis, F., Farokhzad, O.C., 2014. Polymeric nanoparticle technologies for oral drug delivery. Clin. Gastroenterol. Hepatol. 12, 1605–1610.

Prusty, A.K., Sahu, S.K., 2013. Development and evaluation of insulin incorporated nanoparticles for oral administration. ISRN Nanotechnol. 2013, 1–6.

Rajeswari, T.S., 2011. Non invasive insulins: advanced insulin therapy over this decade. J. Appl. Pharm. Sci. 01 (08), 12–20.

Ramachandran, A., Das, A.K., Joshi, S.R., Yajnik, C.S., Shah, S., Kumar, K.M.P., 2010. Current status of diabetes in India and need for novel therapeutic agents. Suppl. Japi. 58, 7–9.

Ramineni, S.K., Cunningham, L.L., Dziubla, T.D., Puleo, D.A., 2013. Competing properties of mucoadhesive films designed for localized delivery of imiquimod. Biomater. Sci. 1 (7), 753–762.

Renukuntla, J., Vadlapudi, A.D., Patel, A., Boddu, S.H.S., Mitra, A.K., 2013. Approaches for enhancing oral bioavailability of peptides and proteins. Int. J. Pharm. 447, 75–93.

Roy, T., Lloyd, C.E., 2012. Epidemiology of depression and diabetes: a systematic review. J. Affect. Disord. 6, 8–21.

Saima, K., Razique, A., Tasneem, F., 2012. Insulin like antigen: sources other than pancreas. Int. J. Curr. Pharm. Res. 4 (2), 24–28.

Salama, P., Mamluk, R., Marom, K, Weinstein, I., Tzabari, M. (2010). Pharmaceutical compositions and related methods of delivery. US2012/0,093,886 A1.

Sameer, N., 2014. Novel noninvasive techniques in management of diabetes. Asian J. Pharm. 8, 141–160.

Seshiah, V., 2012. Premixed insulin analogues in diabetes and gestational diabetes mellitus. Med. Update 22, 231–234.

Severino, P., Andreani, T., Jager, A., Chaud, M.V., Santana, M.H.A., Silva, A.M., Souto, E.B., 2014. Solid lipid nanoparticles for hydrophilic biotech drugs: optimization and cell viability studies (Caco-2 & HEPG-2 cell lines). Euro. J. Med. Chem. 81, 28–34.

Shafaat, K., Hussain, A., Kumar, B., Hasan, R.U., Prabhat, P., Yadav, V.K., 2013. An overview: storage of pharmaceutical products. World J. Pharm. Pharm. Sci. 2 (5), 2499–2515.

Shaikh, R., Singh, T.R., Garland, M.J., Woolfson, A.D., Donnelly, R.F., 2011. Mucoadhesive drug delivery systems. J. Pharm. Bio. Sci. 3, 89–100.

Shaikh, S., Nazim, S., Khan, T., Shaikh, A., Zammiruddin, M., Quazi, A., Allana, A., 2010. Recent advances in pulmonary drug delivery system. Int. J. Appl. Pharm. 2 (4), 27–31.

Shan, W., Zhu, X., Liu, M., Li, L., Zhong, J., Sun, W., Zhang, Z., Huang, Y., 2015. Overcoming the diffusion barrier of mucus and absorption barrier of epithelium by self-assembled nanoparticles for oral delivery of insulin. ACS Nano 9 (3), 2345–2356.

Sharma, G., Sharma, A.R., Nam, J.S., Priya, G., Doss, C., Lee, S.S., Chakraborty, C., 2015. Nanoparticle based insulin delivery system: the next generation efficient therapy for type 1 diabetes. J. Nanobiotechnol. 13 (74), 1–13.

Shivashankar, M., Mani, D., 2011. A brief overview of diabetes. Int. J. Pharm. Pharm. Sci. 3 (4), 22–27.

Solaro, R., Chiellini, F., Battisti, A., 2010. Targeted delivery of protein drugs by nanocarriers. Materials 3 (3), 1928–1980.

Song, K.H., Kim, S.B., Shim, C.K., Chung, S.J., Kim, D.D., Rhee, S.K., Kim, K., 2015. Paracellular permeation-enhancing effect of AT1002 C-terminal amidation in nasal delivery. Drug Des. Devel. Thera. 9, 1815–1823.

Sonia, T.A., Sharma, C.P., 2014. Oral Delivery of Insulin. Elsevier, New York.

Sosnik, A., Neves, J.D., Sarmento, B., 2014. Mucoadhesive, polymers in the design of nano-drug delivery systems for administration by non-parenteral routes: a review. Prog. Polym. Sci. 39 (12), 2030–2075.

Su, F.-Y., Lin, K.-J., Sonaje, K., Wey, S.-P., Yen, T.-C., Ho, Y.-C., Panda, N., Chuang, E.-Y., Maiti, B., Sung, H.-W., 2012. Protease inhibition and absorption enhancement by functional nanoparticles for effective oral insulin delivery. Biomaterials 33, 2801–2811.

Swaminathan, J., Ehrhardt, C., 2012. Liposomal delivery of proteins and peptides. Exp. Opin. Drug Deliv. 9 (12), 1489–1503.

Tuvia, S., Atsmon, J., Teichman, S.L., Katz, S., Salama, P., Pelled, P.D., et al., 2012. Oral octreotide absorption in human subjects: comparable pharmacokinetics to parenteral octreotide and effective growth hormone suppression. J. Clin. Endocrinol. Metab. 97, 1–8.

Ünal, D., Kara, A.D., Aksak, S., Altunkaynak, B.Z., Yıldırım, S., 2012. Insulin hormone: mechanism and effects on the body and relationship with central nervous system. Dicle Med. J. 39 (2), 310–315.

Varewij, A.J., Janssen, J.A., 2012. Insulin and its analogues and their affinities for the IGF1 receptor. Endocr.-Relat. Cancer 19, 63–75.

Vehik, K., Cuthbertson, D., Ruhlig, H., Schatz, D.A., Peakman, M., Krischer, J.P., 2011. Long-term outcome of individuals treated with oral insulin—diabetes prevention trial-type 1 (DPT-1) oral insulin trial. Diabetes Care 34 (7), 1585–1590.

Verma, A., Kumar, N., Malviya, R., Sharma, P.K., 2014. Emerging trends in noninvasive insulin delivery. J. Pharm. 2014, 1–9.
Walsh, E., Maher, S., Adamczyk, B., et al., 2011. Oral delivery of macromolecules: rationale underpinning gastrointestinal permeation enhancement technology (GIPET). Thera. Deliv. 2 (12), 1595–1610.
Wu, Z.M., Zhou, L.Y., Guo, X.D., Jiang, W., Ling, L., Qian, Y., Luo, K.Q., Zhang, L.J., 2012. HP55-coated capsule containing PLGA/RS nanoparticles for oral delivery of insulin. Int. J. Pharm. 425, 1–8.
Yuan, L., Yu, Q., Li, D., Chen, H., 2011. Surface modification to control protein/surface interactions. Macromol. Biosci. 11, 1031–1040.
Yun, Y., Cho, Y.W., Park, K., 2013. Nanoparticles for oral delivery: targeted nanoparticles with peptidic ligands for oral protein delivery. Adv. Drug Deliv. Rev. 65 (6), 822–832.
Zhang, L., Bulaj, G., 2012. Converting peptides into drug leads by lipidation. Curr. Med. Chem. 19, 1602–1618.
Zhang, Z.X., Liu, Y.K., Pan, H., Pan, L., Zhang, Q., Su, H.M., Zhao, Q.L., Li, H., He, C., 2012. The effect of polyethylene glycol recombinant human growth hormone on growth and glucose metabolism in hypophysectomized rats. Growth Horm. IGF. Res. 22 (1), 30–35.

FURTHER READING

Clardy, S.M., Allis, D.G., Fairchild, T.J., Doyle, R.P., 2011. Vitamin B12 in drug delivery: breaking through the barriers to a B12 bioconjugate pharmaceutical. Exp. Opin. Drug Deliv. 8, 127–140.
Yaturu, S., 2013. Insulin therapies: current and future trends at dawn. World J. Diabetes 4 (1), 1–7.

CHAPTER

16

NANOTECHNOLOGY FOR ORAL DELIVERY OF ANTICANCER DRUGS: AN INSIGHT POTENTIAL

Madhu Gupta*, Vikas Sharma*, Nagendra S. Chauhan**

**Shri Rawatpura Sarkar Institute of Pharmacy, Datia, India*

***Government Ayurvedic College Campus, Raipur, India*

CHAPTER OUTLINE

1 Introduction 467

2 Remarkable Addressing Issues Concerns for Challenging of Oral Chemotherapy 469

2.1 Patient's Adherence 475

2.2 Economic Factor 475

3 Specific Consideration of Anticancer Drugs Appropriate for Oral Delivery 476

3.1 Poor Drug Solubility 476

3.2 Intestinal Transit Time 477

3.3 Drug Metabolism and Intestinal Efflux Pumps 477

4 Nanocarrier System for Oral Delivery of Cancer Therapeutics 478

4.1 Drug Nanocrystals 480

4.2 Polymeric Nanosystems 481

4.3 Lipidic Nanoplatform Systems 499

4.4 Dendrimers 504

5 Concluding Remarks 505

References 505

1 INTRODUCTION

Anticancer drugs are the most significant leaders for treatment of complex cancer disease in regard to other therapeutic modalities. In general, the anticancer drug–based therapeutic management of cancer is referred to as chemotherapy. Chemotherapy is a basic customized approach for therapy of human malignancies, but this therapy is directed to cell killing or apoptosis in cancerous cells, as well as normal cells that tend to be toxic for healthy cells (Bromberg, 2008).

Currently, most of the chemotherapeutic drugs are administered by the i.v. route. It is the direct approach for reaching the drug at affected site that surmounts the variability pattern of gastrointestinal tract (GIT). This route offers 100% bioavailability, so dosage accuracy is significant. The higher drug concentration is delivered to normal tissue, hence, it can be a risky route. The i.v. chemotherapy is designed to achieve the maximal drug extent to kill tumor cells within shorter time period followed by several weeks without any i.v. drug regimen administration (Terwogt et al., 1999). For instance, docetaxel is an antineoplastic agent categorized in the second generation of the taxoid family having the therapeutic activity as twice as compared to paclitaxel. It is a microtubule depolymerization inhibitor at in vitro level and only licensed for the treatment of locally advanced or metastatic breast cancer with a first-line chemotherapy regimen (Yin et al., 2009). Presently, the parenteral formulations are available for its clinical use but they suffer serious drawbacks, such as acute hypersensitivity reactions, accumulative fluid retention, peripheral oedema, and occasionally pleural or pericardial effusions (Gupta et al., 2014). Most of the injectable formulations need to use specific pharmaceutical excipients, owing that 40% of anticancer bioactive is generally hydrophobic in nature. The pharmaceutical excipients selected address the solubility issues and facilitate the parenteral administration, but may add toxicity of the final formulation. The one commercialized Cremophor EL (CrEL)–based formulation of PTX by Bristol Myers Squibb as Taxol. CrEL is a low molecular weight surfactant and achieved by the complexation with 1:35 molar ratio of castor oil with ethylene oxide. It is self-assembled into surfactant micelles in aqueous media and widely accepted for enhancing solubility of poorly water-soluble compounds, such as immunosuppressants (cyclosporine A) and anticancer drugs (teniposide, ixabepilone). Taxol contains a higher amount of CrEL and cause toxicological side effects. Serious side effects, such as neuro- and nephrotoxicity, acute hypersensitivity reactions, are produced. To enhance the safety profile that reduces their toxicity, premedication with corticosteroids and antihistamine is suggested by clinical experts. However, documented reports elucidated that one problem arises with these additives and they might be modified the kinetics of the drug (Gaucher et al., 2010).

With all the formulation-based problems, some problems also existed in regard to the patient's comfort zone, such as quality of life and necessity for follow-up therapy after the diagnosis of the disease. All these issues need developing of the most satisfactory dosage forms and routes of administration that could overcome the problems. Oral route is the most abundant, comfortable, and convenient way for drug administration because it has several benefits, such as safety concern, convenience, and patient compliance with acceptability, properties of the drug, access to a disease site, and effectiveness in dealing with the specific disease. The versatility of therapeutics, boundaries for choice of excipients and principally, physiological barriers create a great concern for designing and developing of oral drug delivery system (Agrawal et al., 2014). In this context, the area of oral chemotherapy altered the current scenario of chemotherapy and ultimately improves the survival of patients. Oral chemotherapy manages a sustained release profile of drug concentration in blood level to achieve a greater exposure to cancerous cells and avoids the high peak level above maximum tolerable concentration. Hence, all these facts enhance their usefulness by adding up the therapeutic outcome with decreasing the side effects. In the opinion of patients themselves, oral chemotherapy made them feel less sick and able to fight their disease. The most important issue is to feel freedom, with less interfere with medications, as well as more economical because it confines the cost of hospitalization and the infusion equipment supplies (Fig. 16.1) (Batlle et al., 2004; Feng et al., 2011).

Presently, oral chemotherapy has occupied its position up to 10%, according to the National Comprehensive Cancer Network (Halfdanarson and Jatoi, 2010). Previous reviews stated that almost 78.7%

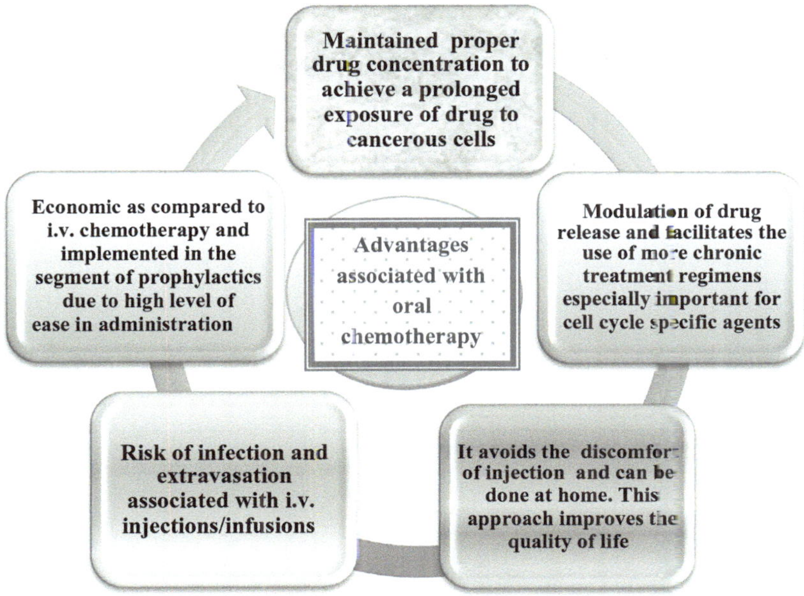

FIGURE 16.1 Schematic Illustration of Advantages for Oral Chemotherapy

patients are interested in oral chemotherapy, while 2.7% patients opted for parenteral route, and the remaining 18.6% showed no choice for recurring breast cancer disease, landed with no preference (Thanki et al., 2013). More than 20 products are available in the market and few of them are in development stage, and play a significant role in oral chemotherapy Table 16.1 and Table 16.2 presented the current oral anticancer drugs in development phase (O'neill and Twelves, 2002).

2 REMARKABLE ADDRESSING ISSUES CONCERNS FOR CHALLENGING OF ORAL CHEMOTHERAPY

Nowadays, oral chemotherapy is the most important option in the medicine world especially for cancer patients. It completely changes the current regimen of chemotherapy and becomes a significant tool for recognizing the cancer patient's dream of chemotherapy at home with themselves. It is a ray of hope for late-stage patients, who are too weak to tolerate any therapy at the clinical level.

If we pay attention and observe Table 16.1, it depicts the various drugs used for oral chemotherapy. The dosage form tablets or capsules are designed but suffer some problems, such as solubility and bioavailability issues. Most of the anticancer drugs are categorized in the class IV of the biopharmaceutical classification system (BCS) that has both low solubility in aqueous fluids and low apparent permeability. This issue can create the complex formulation problems in their oral delivery, such as limited aqueous solubilities, degradation in gastrointestinal fluids, showing affinity toward intestinal

Table 16.1 List of Oral Chemotherapy Drugs Already in Market or in Development Stage

Drug	Trade Name	Form	Strength	Use	Side Effects
Altretamine	Hexalen	Capsules	50 mg	Ovarian cancer	Bone marrow depression, dizziness and fainting
Bexarotene	Targretin	Capsules	75 mg	Cutaneous T cell lymphoma	Hypothyroidism, headache, tiredness, nausea, vomiting, dry skin, or trouble sleeping
Busulfan	Myleran	Coated tablets	2 mg	Chronic myeloid leukemia	Darkening skin or dry mouth, other cancers (such as acute leukemia, tumors
Capecitabine	Xeloda	Coated tablets	150 and 500 mg	Metastatic breast and colorectal and Stage III (Dukes' C) colon	Weakness, back/joint/muscle pain, dizziness, trouble sleeping, skin darkening
Chlorambucil	Chloramino-phene	Capsules	2 mg	Chronic lymphocytic leukemia, malignant lymphomas and Waldensto m's disease	Bruising or bleeding, anemia, mouth sores, ulcers, diarrhea
Cyclophosphamide	Endoxan	Coated tablets	25 and 50 mg	Different cancer diseases (breast, ovarian cancer and leukemia)	Absence of menstrual periods, color change in skin, skin rash, stomach discomfort or pain, texture change in nails
Erlotinib hydrochloride	Tarceva	Coated tablets	25, 100 and 150 mg	Nonsmall cell lung cancer, pancreatic cancer	Interstitial lung disease, gastrointestinal perforation, serious skin condition
Estramustine	Estracyt	Capsules	140 mg	Hormone-resistant prostate cancer	Leukopenia, hypersensitivity reaction, fluid retention, ischemic heart disease
Etoposide	Vepesid	Capsules	50 and 100 mg	Lung cancer, leukemia and cancer of the lymph glands	Diarrhea, dizziness, alopecia, hypotension
Fludarabine	Fludara	Coated tablets	10 mg	Chronic lymphocytic leukemia	Myelosuppression, nausea, vomiting, malaise, peripheral neuropathy, visual disturbance
Gefitinib	Iressa	Tablets	250 mg	Nonsmall lung carcinoma	Eye disorders, respiratory disorders, thoracic, and mediastinal disorders, hepatobiliary disorders
Ibandronic acid	Ibandronic acid Sandroz	Tablets	50 and 150 mg	Breast neoplasm	Anaphylactic reaction/shock, gastrointestinal irritation, and ocular inflammation

Idarubicin	Zavedos	Capsules	5 mg	Acute myelogenous leukemia	Bone marrow suppressant, emetigenic, mucositis, hepatic, and/or renal function impairment
Imatinib	Glivec	Capsules or tablets	50 and 100 mg	Leukemia, gastrointestinal stromal tumor	Weight increase, neutropenia, anaemia, vomiting, diarrhea, dyspepsia, muscle spasm and cramps
Lapatinib	Tyverb	Tablets	250 mg	Breast neoplasm	Skin rash, unusual wheezing, or difficulty in breathing, swollen eyelids, lips or tongue
Lenalidomide	Revlimid	Capsules	2.5, 5, 10, 15, 20 and 25 mg	Multiple myeloma	Dry skin, constipation, nausea, diarrhea, tiredness, easy bruising
Lomustine	Belustine	Capsules	40 mg	Brain tumors, Hodgkin's disease and non-Hodgkin's lymphoma, melanoma, lung and colon cancer	Loss of appetite, nausea, temporary hair loss, mouth or lip sores, diarrhea, unusual fatigue, or weakness
Melphalan	Alkeran	Coated tablets	2 mg	Multiple myeloma	Severe allergic reactions, starry stools, blood in urine or stools, chest pain, unusual tiredness or weakness, weight loss, yellowing eyes, and skin.
Mercaptopurine	Purinethol	Tablets	50 mg	Acute lymphoblastic leukemia	Temporary hair loss, easy bruising/bleeding, dizziness/fainting, joint pain/swelling, tongue
Methotrexate	Methotrexate bellon	Tablets	2.5 mg	Different kinds of cancers	Bone marrow suppression and mucosal damage, ulcerative stomatitis, leucopaenia, nausea, and other gastrointestinal disorders
Mitotane	Lysodren	Tablets	500 mg	Adrenal carcinoma	High fever, chills, uncontrolled shaking, blurred or double vision, problems with speech, balance
Nilotinib	Tasigna	Capsules	150 and 200 mg	Chronic myeloid leukemia	Rash, pruritus, headache, nausea, fatigue, alopecia, muscle spasms, diarrhea, arthralgia, abdominal pain, hematological toxicities
Procarbazine	Natulan, Matulan	Capsules	50 mg	Hodgkin's lymphoma and other malignant lymphomas	Blood and lymphatic systems disorders, hepatobiliary disorders, skin, and subcutaneous tissue disorders

(*Continued*)

Table 16.1 List of Oral Chemotherapy Drugs Already in Market or in Development Stage (*cont.*)

Drug	Trade Name	Form	Strength	Use	Side Effects
Sorafenib tosylate	Nexavar	Coated tablets	200 mg	Renal cell carcinoma, advanced primary liver cancer	Diarrhea, fatigue, alopecia, infection, and rash, myocardial infarction/ischaemia, gastrointestinal perforation, hand–foot syndrome
Sunitinib malate	Sutent	Capsules	12.5, 25, 37.5and, 50 mg	Renal cell carcinoma and imatinib-resistant gastrointestinal stromal tumor	Headache with chest pain and severe dizziness, sudden numbness, or weakness, numbness
Tegafur-uracil	UFT	Capsules	100 mg tegafur, a prodrug of 5-fluorouracil and 224 mg uracil	Metastatic colorectal tumors	Abdominal pain, asthenia, leucopenia neutropenia, paresthesia, neuralgia, neuropathy
Tegafur-gimeracil-oteracil	Teysuno	Capsules	20 mg/5.8 mg/15.8 mg	Stomach cancer neoplasm	Neutropaenia, anemia, and fatigue, diarrhea, nausea, vomiting
Temozolomide	Temodal	Capsules	5, 20, 100, 140, 180, or 250 mg	Brain tumors	Uncontrolled bleeding, seizures (convulsions), severe headache that does not go away
Thioguanine	Lanvis	Tablets	40 mg	Acute leukemia and chronic granulocytic leukemia	Hepato-biliary disorders, intestinal necrosis, and perforation
Topotecan	Hycamtic	Capsules	0.25 and 1 mg	Small lung cancer, uterine, cervical and ovarian neoplasm	Wheezing, feeling short of breath, pale skin, feeling light-headed, rapid heart rate, easy bruising, unusual bleeding (nose, mouth, vagina, or rectum)
Toremifene	Fareston	Tablets	60 mg	Breast neoplasm	Fatigue, oedema (swelling), nausea, headache, weight increase, sleep disorders, changes in the lining of the uterus (endometrium), endometrial cancer
Vinorelbin	Navelbine	Capsules	30 mg	Nonsmall cell lung cancer	Thrombocytopenia, gastrointestinal toxicity with nausea, bacterial, viral or fungal infections resulting from bone marrow depression, severe hyponatraemia

Table 16.2 Some Examples of Oral Anticancer Drugs in Under-Development Phase

Drug/Moiety	Newer Approach	Dosing Schedule	Comments
Paclitaxel (Ptx)	Ptx/cyclosporine A	Paclitaxel dose level of 180, 360, or 540 mg. Cyclosporin A was administered at a dose of 5 mg/kg p.o. 1 h before and concurrently with oral paclitaxel in patients.	P-gp inhibitor; similar bioavailability with daily dosing schedule
Paclitaxel (Ptx)	Ptx/GF120918	GF120918 (25 mg/kg) was administered p.o. by gavage 15 min or 2 h before oral or i.v. dosing of paclitaxel (10 mg/kg), respectively in wild-type and mdr1ab knockout mice.	P-gp inhibitor; bioavailability approximately 40%
Paclitaxel (Ptx)	Ptx-HP-β-CD-nanoparticles	Paclitaxel (10 mg/kg) in different formulations by oral administration in male wistar rats.	Bioavailability approximately 80%
Docetaxel (Dtx)	Dtx/ritonavir	Wild-type mice and Mdr1a/1 b mice (n = 6 per group) with a cannulated jugular vein received oral docetaxel at dose levels of 10, 20, and 30 mg/kg.	CYP450 inhibitor; increment of systemic exposure by 50-fold
Docetaxel (Dtx)	Dtx/cyclosporine (CsA)	Patients received one course of oral docetaxel 75 mg/m^2 with or without a single oral dose of CsA 15 mg/kg CsA preceded oral docetaxel by 30 min.	A P-gp inhibitor; increment of bioavailability
Docetaxel (Dtx)	Dtx/PLA-TPGS/MMT NPs	Nanoparticles of 10 mg/kg DTX dose in form in male Sprague-Dawley (SD) rats.	Bioavailability approximately 78%
Docetaxel (Dtx)	ModraDoc001	ModraDoc001 15 mg capsules were administered orally with 100 mL tap water in patients	Oral solid dispersion formulation
Satraplatin	JM216 (prodrug)	In the initial phase I trial, satraplatin was administered at doses ranging from 60 to 700 mg/m^2 as a single oral dose in patients.	Milder toxicity; lack of cross resistance with cisplatin; same efficiency of other platinum drugs

and liver cytochrome P450 (CYP3A4) and P-glycoprotein (P-gp) in the intestinal barrier with poor intestinal permeabilities (Terwogt et al., 1999). All the issues significantly considered execution of oral delivery for chemotherapy, as well as the need for special concerns while designing the formulation of oral dosage forms.

Instead, different areas should be important and taken in consideration. First, the necessary aspect is to consider the toxicity issue of chemotherapy. After oral administration, the higher amount of chemotherapeutic drugs near the intestinal wall could damage the intestinal cells and impair their physiological functions. Second, a considerable factor is the patient adherence and economic issues that are also important for the development of a new medicine (Fig. 16.2). Adherence is an equivalent of patient compliance, but it is preferred because of simplicity and less judgmental connotation. To provide the knowledge about the medications, such as when and how to take their medications, as well as concern with new techniques for measuring adherence and persistence are equally important (Mazzaferro et al., 2013; Ruddy et al., 2009).

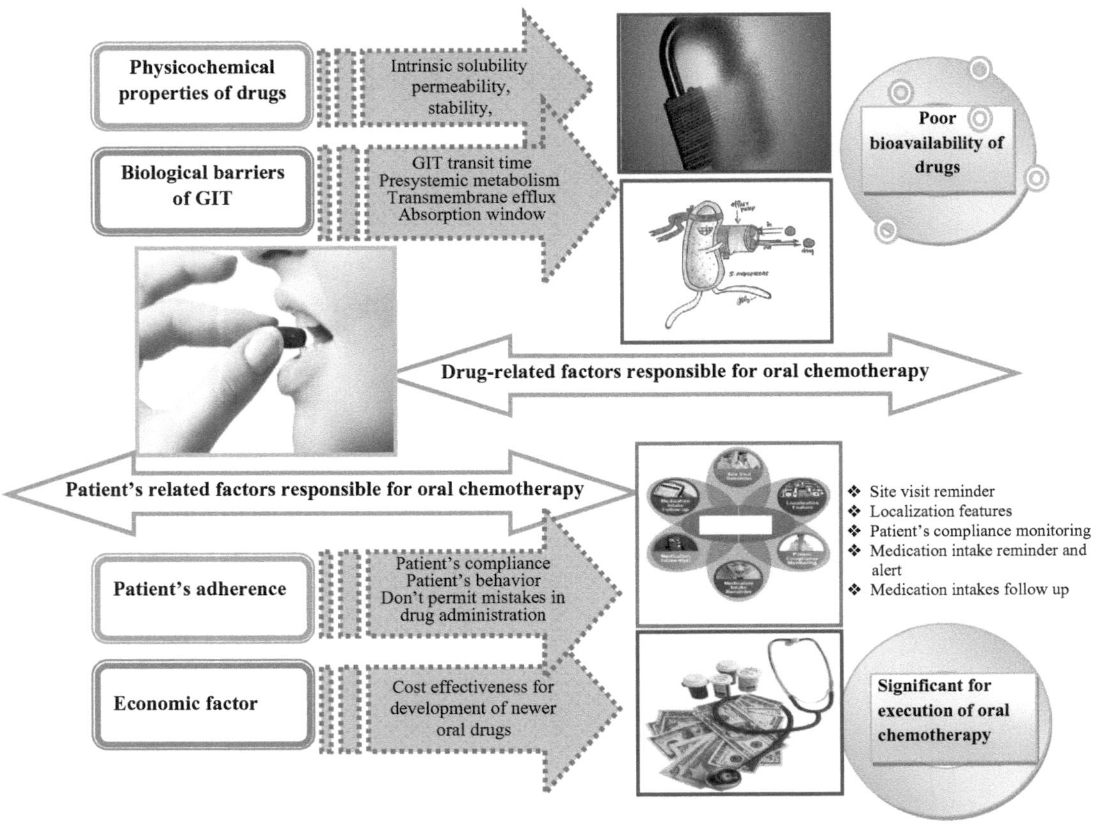

FIGURE 16.2 Remarkable Addressing Issue Concerns for Challenging of Oral Chemotherapy

2.1 PATIENT'S ADHERENCE

The greater availability and increasing use of oral chemotherapy, the more concern about adherence to prescribed clinical regimens by experts would become an increasingly important issue in oncology. The patient is more excited about oral chemotherapy; however, it holds great problems related to the patient's adherence, especially in chronic conditions. Adherence is is related to the patient's behavior that totally coincides with medical care or health advice. A patient is adherent when he or she has not made any mistakes for administration of medication, such as not any dose is missed, no extra dose is received, and no dose is taken in the wrong quantity or at the wrong time (Regnier Denois et al., 2010; Ruddy et al., 2009). There are basically two different methods for monitoring adherence: direct and indirect methods. In direct methods, the direct observation is done during therapy and pharmacokinetic parameters are determined. This method seems to be complicated in the case of oral. Drug or metabolite levels in serum or urine investigation might be a variable for oral chemotherapy and nonadherent patients could manipulate the results, so that they would be counted as adherent. On the other hand, indirect method deals the (1) patient self-report, (2) pill counts and microelectronic monitoring system (MEMS). The former one is the traditional way and suffers some drawbacks, such as inaccuracy because of patient poor memory or reluctance to report nonadherence, while MEMS devices have the potential to record each time after opening. The data are compiled and evaluated by a computer to generate a graphic representation of the number of doses taken daily, the number of missed or extra doses, and the dosing intervals. This method also has some crisis, such as the opening of a pill counter does not necessary signify that the patient really administered the pill, and their expensive cost (Mazzaferro et al., 2013). The documented reports stated that for oral chemotherapy, patient adherence is variable and not easily predicted. The adherence ranged from less than 20% –100% has been reported in certain populations, such as adolescents, can create particular challenges. Future studies should focus on improving measurement and prediction of adherence and on developing technology to improve adherence for both patients in clinical trials or patients taking treatment outside of the research setting (Partridge, 2002).

2.2 ECONOMIC FACTOR

Pharmaco-economic analyses were performed at clinical level to assess the cost effectiveness of newer oral drugs with regard to the cost of infusion administration. At present, pharmaco-economic analyses were carried out on various anticancer drugs, including capecitabine (Cassidy et al., 2006), capecitabine/cisplatin (Irshad and Maisey, 2010), ibandronate (De Cock et al., 2005), and uracil-ftorafur (UFT) (Eng et al., 2001).

For instance, in 2006 a study was conducted in which the comparison between oral capecitabine versus i.v. 5-fluorouracil/leucovorin (5-FU/LV). Researchers evaluated the cost effectiveness of the adjuvant capecitabine from payer [UK National Health Service (NHS)] and societal perspectives. The study revealed that a 5FU/LV administration cost was higher than 57% lower chemotherapy costs for capecitabine, yet acquisition cost was more in the case of capecitabine. The total cost of therapy, including the cost of chemotherapy drugs, cost of visits for drug administration, cost of hospital use, cost of physician consultations for adverse events and for treating them, and cost of ambulance trips. After including the societal cost, the total cost was approximately for £3500 for the oral capecitabine versus £8500 of 5-FU/LV. Hence, this is the main point for capecitabine as a dominant treatment option in terms of economics with the total saving per patient over £5000 (Cassidy et al., 2006).

3 SPECIFIC CONSIDERATION OF ANTICANCER DRUGS APPROPRIATE FOR ORAL DELIVERY

Some critical issues also addressed for designing of oral chemotherapy, such as therapeutic index, dosage adjustment, their pharmacokinetics parameter, inter- and intrasubject variability, and with its potential local toxicity. Literature suggests that anticancer drugs have shown unique and peculiar pharmacokinetic and pharmacodynamic profiles with a narrow therapeutic window. Thus, a small variation in the administered dose may be harmful and causes severe and life-threatening toxicity in some subjects.

The dose adaptation is significant in the case of oral therapy, and this would be directly linked with dosing strategies and inter- to intraindividual variability that is ultimately related to the toxicity aspects. If there are concerns about GIT toxicity with oral administration, it is not an easy task to determine a universal pattern for all existing anticancer drugs. For example, the methotrexate administration–induced focal vacuolization and ultrastructural damage of the immature intestinal crypt cells and ultimately reduction of the intestinal mucosal mass, mucosal protein and DNA content, body weight loss, hence, significant intestinal transit time and permeability is increased. Therefore, this enterotoxicity can be compensated with the oral coadministration of glutamine (Bajin-Katic´ et al., 2006). Published reports predicted that vinorelbine, topotecan, and etoposide cause severe hematological toxicities, neutropenia and/or leucopenia in oral and i.v. treatments. The gastrointestinal toxicities (i.e., nausea, diarrhea, and vomiting) were only slightly more after oral administration as compared to i.v. administration (Mazzaferro et al., 2013).

3.1 POOR DRUG SOLUBILITY

The oral chemotherapy is dependent on the drug's physicochemical properties, such as solubility and permeability. According to Lipinski, it was supported that a classical relationship exists between quantity-sufficient solubility of the drug, and can be determined on the basis of the potency and permeability of the drug. The pharmacokinetics and pharmacodynamic parameter of drug determines its in vivo therapeutic profile (Thanki et al., 2013). Literature assessed that more than 65% of anticancer drugs are commercialized as oral drug delivery systems for clinical use but very a small number is used, because of limited oral bioavailability, and these types of drugs are categorized as poorly water soluble, permeability limited.

Most anticancer drugs are hydrophobic in nature and difficult to assign in formulation form, so various strategies have been implemented to improve anticancer drug solubility by employing pharmaceutical excipients, drug delivery vehicles, or making prodrugs. Water miscible cosolvents namely ethanol, methanol, methylene chloride, or acetonitrile and water-insoluble organic solvents, such as oils, triglycerides, vitamin E, or their associations might be used as pharmaceutical excipients (Mazzaferro et al., 2013). Oral drug delivery systems are designed to overcome the solubility issues, such as micelles, nanoparticles, liposomes, or nanocapsules (Bromberg, 2008) and their detailed discussions are included in the next segment of this chapter. The prodrug strategy is used for improving the bioavailability and to decrease the side effects of the lead compound. Generally, it is inactive but at in vivo level, it is converted into active cytotoxic compounds via different mechanisms, particularly by the action of specific enzymes. For instance, irinotecan (CPT-11) is a prodrug of SN-38 (7-ethyl-10-hydroxy-camptothecin) that has higher potency up to 100–1000-fold more cytotoxic than the parent drug. It represents a broader spectrum of antitumor activity in preclinical and clinical levels for various cancers. In biological system, CPT-11 is converted to SN-38 with the action of various enzymes with major contributions of the carboxylesterases. The presence of this enzyme in the human liver and the GIT make it possible for further use of irinotecan by the oral route (Mathijssen et al., 2001). The apparent water solubility of drugs might be increased with the use of cyclodextrins, which are macromolecules, composed of cyclic oligosaccharides. This option makes the drugs more water-soluble and creates further possibilities for pharmaceutical formulations ranging from conventional to colloidal dispersions. The lipophilic drug moiety complexed with guest molecules of cyclodextrins and results in the formation of drug and/or cyclodextrin complexes and confirmed more aqueous solubility than their parent one and ultimately enhance their water solubility (Mazzaferro et al., 2011). The inclusion complexes can reduce or prevent gastrointestinal irritation, as well as enhance their dissolution rate and consequently in oral absorption. Besides, cyclodextrins work as a permeability enhancer upon oral administration for those drugs that experienced lower intestinal permeability (Calleja et al., 2012). A recent study showed that hybrid cyclodextrin nanocapsules loaded with camptothecin might be successfully explored for the oral delivery in order to maintain drug stability in the body, and subsequently improve its eventual bioavailability.

3.2 INTESTINAL TRANSIT TIME

The intestinal drug absorption is important for the oral drug delivery system and its potential is associated with enzymatic and physical environment of the GIT. After drug administration, the drug reaches the harsh environment of the stomach or entire length of GIT secreted the digestive enzymes, where it can be degraded so stability is compromised (Mei et al., 2013). This can be overcome with the help of various polymeric systems to enhance the intestinal absorption of hydrophilic or hydrophobic drugs. They have the potential to pass through the stomach and ultimately, after reaching the intestine release the drugs. Various polymers, such as polysaccharides (chitosan or dextran), acrylic copolymers (Eudragit), phospholipids, or cellulosic derivatives, represented the adequate gastroresistant properties and easily pass intact through the stomach, reaching the intestines and releasing the drug (Mazzaferro et al., 2013; Sajeesh et al., 2010). Some drugs have a short transit time, leading to incomplete absorption with poor bioavailability. The researchers paid attention to the fact that by using mucoadhesive dosage forms, the duration of time can be improved. The mucoadhesive systems have potential benefits, such as increased local drug concentrations, which is favorable to absorption, improving the drug effectiveness by maintaining their plasma drug concentration, and in some cases specially restricting absorption to a particular site in the intestine (Bravo-Osuna et al., 2007). One study stated that the intestinal permeation of docetaxel based nanosystems (consisted of methyl-β-cyclodextrin, Me-β-CD) combined with poly(isobutylcyanoacrylate) and coated with thiolated chitosan was higher than the ethanol control solution of drug. Interestingly, when mucoadhesive interactions between nanoparticles and the mucosa were avoided, the intestinal permeation of docetaxel was significantly reduced, confirming that the mucoadhesion of the nanoparticles was a mandatory condition to enhance the intestinal permeation of docetaxel (Mazzaferro et al., 2012).

3.3 DRUG METABOLISM AND INTESTINAL EFFLUX PUMPS

Improving the bioavailability of orally administered drugs has been an ongoing research with limited success. Another major problem for oral delivery of anticancer drug is their metabolism or intestinal efflux. The intestinal efflux pump, such as P-gp, is situated on the apical membranes of intestinal absorptive cells. By the drug efflux property of efflux pump, the drug bioavailability furthermore will be reduced and cancer cells would be resistant (Constantinides and Wasan, 2006). The ATP-dependent multidrug transporters belong to the ubiquitous superfamily of ATP-binding cassette (ABC) proteins, have been linked in the resistance to infectious diseases, such as cancer, AIDS, and malaria. However, the presence of ABC transporters and the metabolic CYP450 enzyme in GIT could ultimately impair the bioavailability of anticancer drugs. Physiologically, ABC transporters serve a drastic role in body defense and are able to modulate the normal absorption, distribution, metabolism, excretion, and toxicity of xenobiotics. To increase the oral bioavailability of drugs, researchers must also focus on the concomitant use of ABC transporters and CYP450. But, inhibition of the metabolism mechanisms may affect the distribution and the bioavailability of other xenobiotics, leading to unwanted side effects (Leslie et al., 2005; Schinkel and Jonker, 2003).

4 NANOCARRIER SYSTEM FOR ORAL DELIVERY OF CANCER THERAPEUTICS

Oral chemotherapy is a very promising issue in 21st-century medicine. It can radically change the current scenario of chemotherapy in terms of its sustained and mild plasma drug concentration, convenient and flexible administration, and lower cost with resulting improved compliance. At present, most antitumor drugs, such as taxanes (paclitaxel and docetaxel) are not sufficiently orally bioavailable. This problem can be surmounted by the emerging applications of nanomedicines for anticancer drugs. Nanomedicine has been considered as the utilization of nanotechnology to solve difficulty in medicines. Nanoplatform systems successfully change way that we make drugs or formulations and the way we take medicines, and thus provide an ideal solution for oral chemotherapy (Gong et al., 2012). Numerous examples exist and research for the application of the various nanocarriers, such as conjugated chitosan, polymeric micelles, liposomes, solid lipid nanoparticles (SLNs), and biodegradable polymers-based NPs providing the controlled, sustained and targeted drug delivery across the various physiological drug barriers, including the gastrointestinal barrier for oral chemotherapy (Fig. 16.3). The application of nanomedicine can be best exemplified by considering a published report. For instance, paclitaxel formulated with D-α-tocopheryl polyethylene glycol 1000 succinate (vitamin E TPGS) emulsified poly(lactic-*co*-glycolic acid) (PLGA) NPs presented sustained release up to 168 h than with 22 h for Taxol at the same dose range 10 mg/kg dose and 400% maximum tolerance dose for rats (Feng et al., 2007). In another report, the paclitaxel loaded NPs composed of poly(lactide)/vitamin E TPGS (PLA-TPGS) copolymer for controlled and sustained delivery and the sustained delivery was possible up to 240 h with 1.6-fold AUC as compared with Taxol (Zhang et al., 2008).

Nanoengineered systems were explored in various ways to improve the oral delivery of various anticancer drugs. The nanosystems easily achieved the optimum amount of drug concentration in plasma, as well as in the accumulated maximum drug concentration in a tumor and its surrounding area. The nanomedicines are associated with several benefits, such as increasing solubilizing potential of drug, higher drug loaded, altered absorption pathways, prevention of metabolic degradation within the GIT, inhibition of P-gp mediated efflux pump, flexibility of chosen materials suitable for nanomedicines, easy surface modification, drug and disease specific design alteration, site-specific targeting ability, and the ability to incorporate a wide variety of drug substances. In the same manner, it reduced the

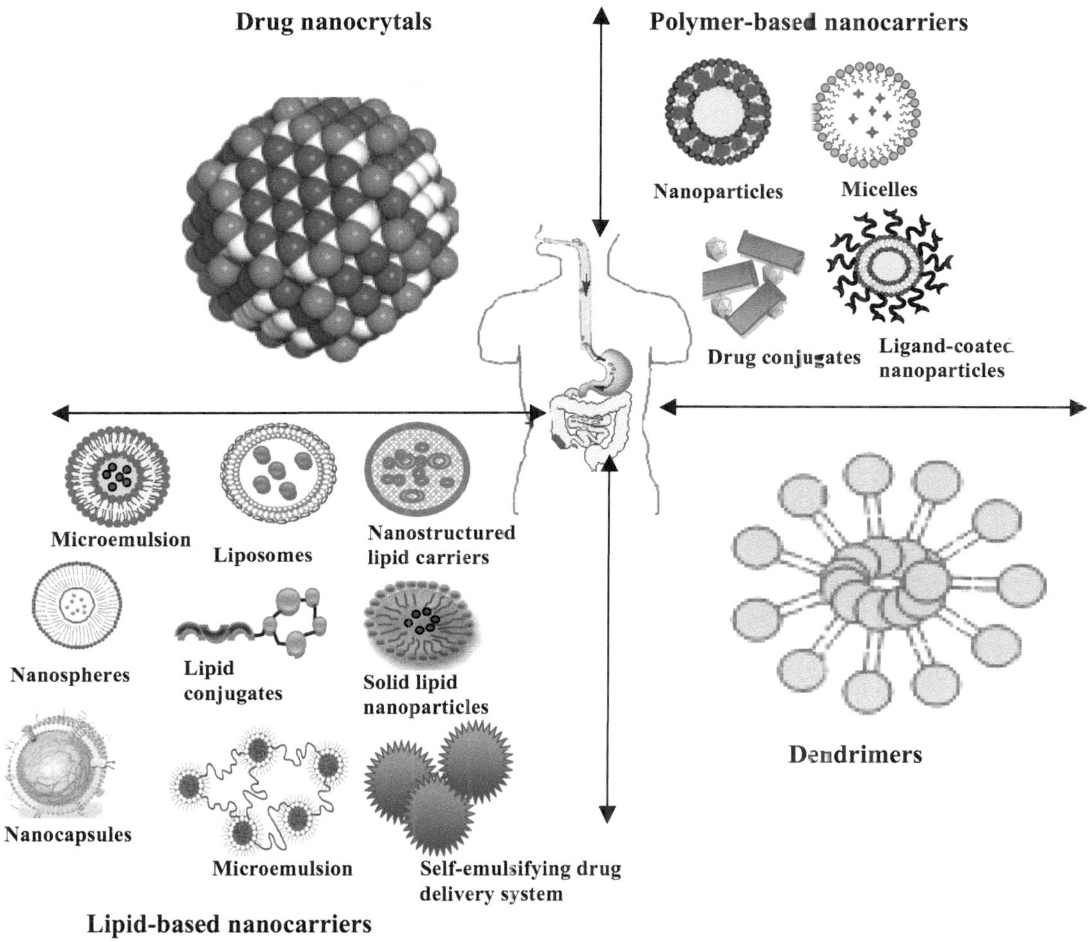

FIGURE 16.3 Schematic Representation of Nanocarriers Administered via Oral Route for Oral Cancer Chemotherapy

toxic effects of the drug to the GIT that is very crucial for patients receiving chronic cancer therapy via oral route (Fig. 16.4). The ease and convenience of flexibility of designing such nanosystems in regard to particle size, shape, and surface properties that serve an important role in the uptake across the GIT membrane and affect the absorption profile. The previous published reports documented that nanomedicine with 50–300 nm size ranges, positive zeta potential and hydrophobic surfaces are favorable biopharmaceutical parameters for preferential uptake from the GIT. Nanocarriers followed the various identified absorption mechanism for enhancing the oral bioavailability of drugs, including higher absorption from enterocytes (with increased solubilization and dissolution), mucoadhesion (interaction between the positively charged nanocarrier with negatively charged mucin), altered the tight junction (potential of nanocarriers to interact with the tight junction proteins), receptor-mediated

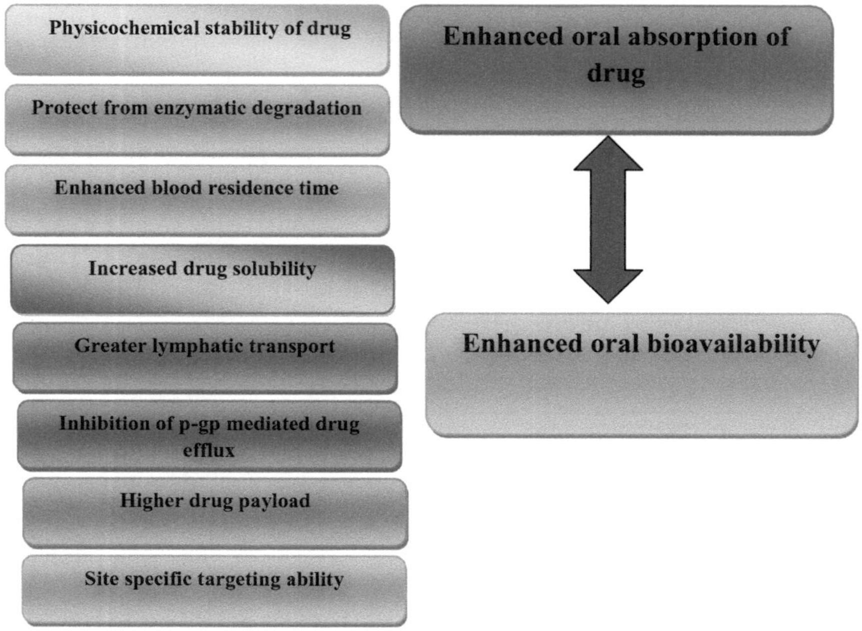

FIGURE 16.4 Advantages of Nanocarriers Based Oral Cancer Chemotherapy

endocytosis and transcytosis (clathrin and caveolae-dependent and independent endocytosis), phagocytosis through microfold cells (M cells) of the Peyer's patches, and other mucosa-associated lymphoid tissues (MALT), as well as chylomicron uptake mechanism from the enterocytes or lymphatic absorption (Thanki et al., 2013).

4.1 DRUG NANOCRYSTALS

The development of nanocrystals emerged and drew the attention of researchers owing to their ease of preparation and not requiring any solubilizing and/or encapsulating chemicals, hence circumventing side effects that may have occurred in many existing delivery systems. Nanocrystals referred to as drug crystals having sizes ranging from dozens to a few hundred nanometers. Drug nanocrystals-based formulations had the potential for increasing dissolution rates due to size reduction to the nanometer range, ultimately improving the oral bioavailability in the case of oral anticancer drug, reducing the fluctuations at pharmacokinetic manner in the fed and fasted conditions, and variability among patient populations. The staying period of nanocrystals in the GIT might be increased by employing the mucoadhesive polymers for improving the adhesiveness of nanocrystals to lumen. Studies further investigate the various approaches that have been imparted for mucoadhesive properties to the nanocrystals, such as suspension layering, spray drying, and so on (Gao et al., 2012). Apart from this, the nanocrystals linked with several benefits that are used for cancer therapy, such as higher drug loading (as high as 100%), experienced greater stability, protected drug molecules from enzymatic degradation, easy surface manipulation, improved pharmacokinetics and biodistribution, less toxic effects, bioimaging becomes possible with hybrid nanocrystals, high level of scalability, and widespread industrial adaptability. The drug release mechanism could be explained as less than 5% of the total administered dose is delivered to the tumor site. Once nanocrytals reached to the tumor site, the dissolution rate or drug release of drug molecule mainly depends on its solubility in the surrounding environment and concentration gradient. The killing of cancer cells in the delivery site starts when the drug concentrations reached their threshold value or exceed the IC_{50} value (i.e., half the maximum inhibitory concentration).

Presently, various marketed nanocrystal-based drug products are available for oral administration. Although the marketed nanocrystal-based products are mainly for oral delivery, they are formulated mostly by the top–down approach. The one product used for oral cancer chemotherapy has entered in phase II, approved by FDA is 2-methoxyestradiol based under the trade name of Panzem, manufactured by EntreMed by top–down, media milling-based method (Lu et al., 2015). Thes formulations that have excellent antimitotic and antioangiogenic activities demonstrated an improved pharmacokinetic performance and antitumor activity as compared to the conventional capsule formulation when given orally to 16 patients with advanced solid malignancies and 14 patients with taxane-refractory, metastatic castrate resistant prostate cancer (Harrison et al., 2011; Tevaarwerk et al., 2009). Although nanocrystal formulation was maximal tolerated and showed biologic activity, the active moiety had not presented sufficient clinical activity and the study was terminated at phase II.

Few published reports are documented for oral drug nanocrystals. Among them, one report investigated the development of novel nanocrystal formulation of paclitaxel and camptothecin using three-phase nanoparticle engineering technology (3PNET technology) that combined the amorphous precipitate; hydrated amorphous aggregate and finally stabilized nanocrystal utilizing F127 as polymer stabilizer. The oral formulations administered at 20 mg/kg and presented the comparable antitumor efficacy as that of intravenous administered taxol at same dose range and significantly had maximum therapeutic efficacy at higher doses of 60 mg/kg and showed 3-fold greater effect in regard to free drug suspension, and clearly showed the evidence of the advantage of nanocrystals in the drug absorption (Liu et al., 2010). In the same series system, porous quaternized chitosan nanoparticles encapsulated with paclitaxel nanocrystals was developed and presented higher intracellular drug retention with enhanced in vitro cytotoxicity as compared to free drugs when experimental studies performed on Caco-2 cell lines and lewis lung carcinoma (LLC) cell lines, respectively. The in vivo results are fully agree with in vitro results, and showed 5-fold increase drug accumulation at the tumor site along with higher antitumor efficacy with maximum safety concerns as compared to standard formulation (Lv et al., 2011).

4.2 POLYMERIC NANOSYSTEMS

4.2.1 *Polymeric nanoparticles*

The nanoparticuluate drug delivery systems provide an unprecedented opportunity for improved oral drug delivery to tumors. Polymeric nanoparticles have a significant impact in the field of cancer therapy and possess remarkable features for imbedding the drugs and diagnostic agents. The existence of the particulate carriers arises since the early 1980s, with the progress in polymer chemistry that allowed the designing of biodegradable and biocompatible materials. Polymeric nanoparticles are submicron in size range 10–1000 nm in diameter and are formed by the assembly of a wide variety of polymers. The size range varies depending on the kinetic control variables, as well as the polymer characteristics.

A large number of polymers and copolymers, including natural polymers, such as gelatin, dextran, albumin, chitosan, and alginate among which chitosan and its derivatives have been widely explored are available. Several synthetic biodegradable polymers, such as polylactic acid (PLA), polyglycolic acid (PGA), copolymers of lactic and glycolic acid (PLGA), poly(ε-caprolactone) (PCL), poly-alkyl cyanoacrylate (PACA), polyethylene imine (PEI), poly(L-lysine), poly(aspartic acid), have been exploited for formulation of polymeric nanoparticles (Gupta et al., 2013).

Nanoparticles are recognized as submicron colloidal systems that contain both nanospheres and nanocapsules. Nanospheres must be considered the matrix-based systems while nanocapsules consist of core–shell structures having an inner liquid core (which can be either aqueous or oily depending on the core material) surrounded by a polymeric wall. Several studies confirmed the potential role of polymeric nanocapsules, including improving poor water solubility, protecting the drug molecule from the environment and maintaining the stability, providing controlled release, and improving the desired pharmacokinetic profile. Moreover, based on the structure of nanocapsules, the oily core enabled an improvement in the solubility of hydrophobic drugs and the polymeric wall exterior to the core, which is able to protect the molecules against the harsh environment (Ünal et al., 2015a,b). A previous study established that after oral administration of docetaxel nanocapsules embedded in microparticles, they presented higher bioavailability with regard to i.v.-based commercial docetaxel solution (taxotere) in rats. The designed nanosystem was also evaluated in animal studies and reported that docetaxel nanocapsules elicited a marked enhanced absorption more than oral Taxotere in minipigs with relative bioavailability and C_{max} values 10- and 8.4-fold higher, respectively. Orally absorbed nanosystems modulated the elimination and distribution pattern of docetaxel, as evidenced in the organ biodistribution rat study, owing to their reinforced coating, while transiting through the enterocytes by surface adsorption of apoproteins and phospholipids. The study concluded that significant enhancement of oral absorption might be due to the unique nanoplatform, as well as lymphatic route, that is involved in nanocapsule absorption. However, this route avoided first-pass effects and protected the drug from CYP3A4 degradation either in the gut or systemic circulation showing higher drug blood levels than i.v.-administered free drug, which was markedly metabolized by CYP3A4 in the circulation (Attili-Qadri et al., 2013). In one study, nanocapsules prepared from amphiphilic cyclodextrins (CDs) or poly-ε-caprolactone (PCL) for the effective oral delivery of an anticancer agent, camptothecin (CPT). The drug-loaded anionic and chitosan (CS)-coated cationic nanocapsules were designed and characterized for various parameters. The results showed that both anionic and cationic formulations were in the range of 180–220 nm with narrow size distribution. The CPT-loaded nanosystems were found to be stable in simulated gastrointestinal media and nanoparticles and mucin interaction was confirmed by turbidimetric methods. In vivo animal studies concluded that intestinal uptake of CS-coated nanocapsules was significantly higher than cationic nanocapsules. The results supported that CPT-loaded positively charged CD nanocapsules might be a fasicinating option for oral chemotherapy (Ünal et al., 2015a,b). In the same series, one of the published reports documented the hybrid cyclodextrin nanocapsules intended for the oral delivery of camptothecin for maintaining drug stability in the body and improving its eventual bioavailability (Ünal et al., 2015a,b).

The polymeric nanoparticles especially used for sustained drug release profile have significant consideration in terms of oral cancer delivery and ensuring that not any traces of drug releasing or leaching was observed from the formulation till it reaches systemic circulation, thus bypassing various physiological barriers to oral delivery of difficult to deliver drugs. Several studies concluded the quick absorption of nanoparticulate system from the GIT after oral ingestion, and consequent drug release from longer periods of time (exceeding GIT transit time). This can be possible with the preferential uptake of polymeric nanoparticles by specialized Peyer's patches (M cells) or the isolated follicles of the gut-associated lymphoid tissue present in the GIT (Thanki et al., 2013).

Jain et al. (2011) actively studied the PLGA nanoparticles for oral delivery of doxorubicin. Doxorubicin loaded PLGA nanoparticles (Dox-NPs) presented greater intracellular uptake and were localized in the site of action in mouse breast cancer, C127I cell lines as compared to free drug. This effect in Caco-2 cell uptake mainly depends on the time and concentration dependent The designed nanosystem showed ~49.06% reduction in the tumor burden upon oral administration of Dox-NPs in 30 days; on the other hand, 158% increase in tumor burden was observed in the untreated group. Considering the toxicity aspects, the developed system posed lower significantly lower cardiotoxicity (Jain et al., 2011). In one report, didodecyl dimethylammonium bromide (DMAB) stabilized PLGA nanoparticles might be a promising approach for oral paclitaxel (PTX) delivery system in a female SD rat tumor xenograft model. It was concluded that oral administration of PTX-loaded nanoparticles at 3.75 mg/kg was significantly more effective in tumor inhibition as compared to oral or i.v. administration of PTX at a dose of 7.5 mg/kg in cremophor EL (Bhardwaj et al., 2009). Lee et al. (2008) suggested that low molecular weight chitosan-PTX conjugate was also effective for oral chemotherapy to inhibit tumor growth. In normal ICR mice, the bioavailability of chitosan-PTX was about 42% and 27%, respectively, at 5 and 10 mg/kg dose after oral administration (Lee et al., 2008).

Recent reports documented that multifunctional polyelectrolyte complex nanoparticles designed by hyaluronic acid (HA) grafted polycaprolactone (PCL) nanoparticles (HA-*g*-PCL NPs), then coated with chitosan (CS) as electrostatically for improving the oral delivery of lipophilic paclitaxel and doxorubicin (DOX). The results indicated that the core–shell nanosystems retained intact and stable in the pH from 3.0 to 6.8, but the CS layer depleted after increasing pH 7.4 and then accelerated drug release that was triggered by hyaluronidase-1. The cell uptake studies reported that HA-*g*-PCL NPs reached into cancer cells (EC109) via receptor-mediated endocytosis and release the drug intracellularly, but were rarely taken up by normal fibroblasts (NIH3T3). Therefore, higher cytotoxicity was achieved in case of cancer cells EC109 but lower cytotoxicity against normal cells NIH3T3 and all results were also confirmed by in vivo studies. Finally, CS/HA-*g*-PCL multifunctional nanoparticles presented higher potential for enhancing the oral delivery efficiency of anticancer drugs (Huang et al., 2014).

One study developed a novel bioadhesive drug delivery system for oral delivery of paclitaxel composed of poly(D,L-lactide-*co*-glycolide)/montmorillonite (PLGA/MMT) nanoparticles. The results concluded that PLGA/MMT nanoparticles enhanced the cellular uptake in presence of MMT ranging 57%–177% for Caco-2 cells and 11%–55% for HT-29 cells as compared to plain PLGA-NPs. The cellular uptake efficiency was dependent on the extent of MMT and the number of particles in incubation. The designed system possibly extends the residence time in the GIT, which promotes oral delivery of paclitaxel (Dong and Feng, 2005).

Jiang et al. (2013) developed three types of nanoparticles from commercial PCL and self-synthesized PLA-PCL-TPGS random copolymer for oral delivery of anticancer agents, including thiolated chitosan-modified PCL nanoparticles, unmodified PLA-PCL-TPGS nanoparticles, and thiolated chitosan-modified PLA-PCL-TPGS nanoparticles. The results showed that thiolated chitosan greatly enhances the mucoadhesiveness and permeation properties. Cell uptake studies favored the thiolated chitosan-modified PLA-PCL-TPGS nanoparticles that had potential for higher cell uptake than the other one. The cell cytotoxicity studies supported the thiolated chitosan-modified PLA-PCL-TPGS nanoparticles over commercial Taxol in regard to cytotoxicity in A549 cells. Moreover, these mucoadhesive nanoparticles enhanced the paclitaxel transport through opening tight junctions and bypassing the efflux pump of P-gp. This research finally established that thiolated chitosan-modified PLA-PCL-TPGS nanoparticles are a remarkable option to the treatment of lung cancer via oral chemotherapy (Jiang et al., 2013). Reported in the same year, the doxorubicin loaded apotransferrin and lactoferrin nanoparticles (apodoxonano or lactodoxonano) for oral treatment of hepatocellular carcinoma in rats. The results concluded that the maximum therapeutic efficacy of apodoxonano and lactodoxonano with the number of neoplastic nodules was significantly lower approximately >93% than that of rats administered with saline or plain drug solution. The apotransferrin and lactoferrin nanoformulation improved the efficacy and increased the oral bioavailability of doxorubicin in a target-specific manner, and hence it is an effective and safer oral drug delivery approach (Golla et al., 2013). Zabaleta et al. explored the capability of PEGylated poly(anhydride) nanoparticles as novel platforms for the oral delivery of paclitaxel. Their results revealed that this system successfully enhanced the oral permeability and increased the drug residence time in plasma at least 48 h. These effects would be probably due to the diffusion abilities of these carriers through the mucus layer and inhibitory effect of PEG on P-gp and P450 cytochrome.

4.2.2 Polymeric micelles

Polymeric micelles represent fascinating options for drug delivery systems for anticancer drugs that are hydrophobic nature and exhibit poor bioavailability. The unique core–shell structure of micelles has able to incorporate poorly soluble drugs into the inner hydrophobic core thus improving their stability and bioavailability (Gupta et al., 2013). Polymeric micelles formed by amphiphilic copolymers and able to provide a controlled and targeted way to deliver the therapeutics, modulate the pathways of drug biocirculation and cellular drug distribution, simultaneously increase the drug amount at targeted sites and afford a high drug efficacy and low side effects. Polymer segments that self-assembled into aggregates and form micelles are an inherent property of copolymers that possess both water-soluble, as well as water-insoluble segments at a certain temperature. The self-assembled nature is driven by a gain in entropy of the solvent molecules with the hydrophobic polymer part tends to aggregate and withdraw from the aqueous medium. Polymeric micelles experienced significantly lower aggregation concentration values and greater shelf life than those of low-molecular weight surfactants (Bromberg, 2008). Amphiphilic polymers are self-associated in aqueous mediums to form polymeric micelles (10–100 nm), however, depending on the delivery site, the physicochemical properties, such as size, surface potential, and surface properties of polymeric micelles, can be easily manipulated. Polymeric micelles showed promising options for poorly water-soluble drugs and various poly(ethylene glycol)-block-poly(D,L-lactic-acid) (PEG-*b*-PLA) micelles can solubilize multiple poorly water-soluble drugs, such as paclitaxel, etoposide, and docetaxel (Shin et al., 2009). Polymeric micelles are emerging nanoconstructs for successful developments of oral cancer chemotherapy. The oral formulations entangled with higher molecular weight, nondegradable polymers and gels are not excretable and cannot find their way into the systemic circulation. The important thing concerning the design of oral formulations is its mucoadhesivity with improved adhesion to the intestinal walls for enhancing the residence time (Bromberg, 2008). Research-based evidence supported that oral chemotherapy-based polymeric micelles are in the exploratory stage with some fields for under investigation, especially the polymeric micelles mechanism and drug release across the gastrointestinal membrane. The enhanced absorption might be included in membrane permeating behavior, inhibition of efflux transports, as well as bioadhesive properties (Gong et al., 2012). Various mechanisms are reported to explain the mechanism of polymeric micelles forms across the gastrointestinal membrane. One research group described that intact nanomicelles form across the Caco-2 cell monolayer and the everted gut sac with transmission electron microscope observation. This process has been mediated by an endocytosis pathway through nonspecific interactions between the polymer and the intestinal membrane (Yao et al., 2011). The amphiphilic micelles-reservoir–based mechanism was confirmed by its high apparent permeability coefficient (Papp) at lower polymer concentration with different Papp of polymer and extent of entrapped drug (Mathot et al., 2006).

P-gp is the most common efflux transport in intestine and several studies were reported. The P-gp inhibitor, such as poloxamer, D-α-tocopheryl PEG succinate (TPGS), and amphiphilic polymers, are another commonly used P-gp inhibitor used for oral administration with polymeric micelles. Some other amphiphilic polymers mPEG-block-polycaprolactone, PEG phosphatidylethanolamine, PEG-*b*-PLA, mPEG-poly(caprolactone-trimethylene carbonate), *N*-octyl-O-sulfate chitosan have been reported as P-gp inhibitors. Bioadhesive polymers had the potential to improve drug absorption by increasing the residence time at absorption site (Gong et al., 2012). In the PTX-nanomicelles system, TPGS1000 as the P-gp inhibitor and dequalinium (cationic amphiphilic compound) as the bioadhesive polymer were used and electrostatic interaction with the negative intestinal membrane. The study concluded that functional PTX nanomicelles showed the strongest inhibitory effect on resistant breast cancer MCF-7/Adr cells at ex vivo and in vivo manner (Yao et al., 2011).

Various blockgraft copolymers of the poly(ethylene oxide) (PEO) and poly(propylene oxide) (PPO) pluronic polyethers and poly(acrylic acid) (PAA) bound by carbon—carbon bonds emerged, however both polymeric components are considered as safe. The experimental studies on animals revealed that pluronic-PAA copolymers are excreted without absorbing into the systemic circulation after orally ingestion. At physiological pH, these micelles are surface-active and self-assembled with hydrophobic cores of dehydrated PPO and multilayered shells of hydrophilic PEO and partially ionized PAA segments. Polymeric micelles have the potential to solubilize hydrophobic drugs, such as paclitaxel and steroids, and protect the molecules, such as camptothecin from the hydrolytic reactions. The pluronic-PAA copolymer-based micelles easily interacted with cell membranes and simultaneously suppressed membrane pumps, such as P-gp. The ionizable carboxyls group in the outer part of micellar facilitates mucoadhesion, ultimately increase the staying period of micelles and solubilized drugs in the GIT. Higher amount of drugs (weakly basic and water-soluble drugs, such as doxorubicin and its analogs, mitomycin C, mitoxantrone, fluorouracil, and cyclophosph) can be encapsulated into pluronic-PAA micelles via electrostatic interactions with the micellar corona (Bromberg, 2008).

Poly(ethylene oxide)–*b*-poly-(propylene oxide)–*b*-poly(ethylene oxide) (PEO–PPO–PEO) amphiphilic triblock copolymers were employed for micelles preparation and successfully explored as drug and gene delivery systems, as well as in diagnostic imaging as carriers for various contrasting agents (Kabanov et al., 2002). PEO–PPO–PEO micelles composed from commercially available copolymers pluronics and presenting various attractive properties, such as resistance of the PEO shell to protein adsorption, temperature responsive PPO core for solubilizing water-insoluble compounds, and availability of hydroxyl groups to which receptor-specific ligands can be attached. However, the PEO–PPO–PEO micelles also linked with micellar disaggregation when they are diluted at physiological pH. Rapoport (1999) and his research group explored various strategies for stabilizing the PEO–PPO–PEO micelles including radical crosslinking of micelle core using benzoyl peroxide as radical initiator, introduction of vegetable oil to increase hydrophobic interactions within the core, formation of hydrogel inside the micelle core (Plurogel). The pluronic copolymers have been able to inhibit drug efflux transporters activity, such as P-gp, MRPs, and BCRP. All these facts addressed that polymeric micelles provided a promising platform for improving oral bioavailability and loading efficiency of drug (Huang et al., 2008). In a study, the PEO–PPO–PEO micelles was stabilized by UV-induced formation of an interpenetrating network of poly(pentaerythritol tetraacrylate), in which the polyether chains were physically entrapped, has been reported (Petrov et al., 2006). Pluoronic-based micelles displayed the extended the GIT residence time of entrapped drug. The orally ingested micelles in mice revealed a high area under the curve of micellar paclitaxel, longer mean residence time (9-times more than i.v. Taxol) and high distribution volume (2-fold higher than i.v. Taxol) indicating an efficient oral absorption (Yoncheva et al., 2012).

One research group reported the *N*-octyl-O-sulfate chitosan (NOSC)–based micelles successfully enhanced the oral bioavailability of paclitaxel as compared with commercially available Taxol at in vivo level. This effect can be illustrated by enhancing the absorption of paclitaxel due to P-gp inhibition by NOSC, and the transport of paclitaxel as a P-gp-independent way. The mechanism of P-gp inhibition by NOSC micelles is interfering with the P-gp ATPase than reducing the P-gp expression and exhibiting a potential to improve oral bioavailability of P-gp substrates (Mo et al., 2011). Table 16.3 reflects the list of anticancer drugs benefiting by the polymeric micelle approach for improving their oral deliverability.

The research study and various clinical trials studies displayed a very opportunistic drug delivery for oral cancer chemotherapy; however, several issues are under observation, such as GIT stability, possibility of mixed micelles in presence of bile acid, large-scale manufacturing process, packaging and storage and patient compliance and so on. The amphiphilic polymers are in unstable form in acidic conditions, but some progression in technology may be useful, such as enteric coating were employed after solidification of polymeric micelles formulation (Gong et al., 2012) (Table 16.4)

4.2.3 Polymer–drug conjugates

Oral chemotherapy can be successfully done on conjugates between drugs and water soluble carriers or polymers to improve solubility and pharmacokinetic profiles of water-insoluble drugs.

The polymer backbone can also be linked with other excipients, such as diagnostic agent, targeting ligand, PEG chains to enhance hydrophilicity, drug efflux inhibitors, hence formulation of such nanosystems that is highly specialized to the disease site. The resulting conjugate had the potential for active transport from the GIT to further facilitate drug absorption. The targeting potential is very important in the case of cancer therapy to reduce the toxicity among normal cells. Other benefits associated with polymer-drug conjugates, such as increased solubilization, greater degree of plasma half-life, higher bioavailability, protection with metabolic or degrading enzymes, and prevention of immune responses, reduce the adverse effects and targeting potential. The design of polymer–drug conjugates has progressed over the past 2 decades, with an increasing number of candidates currently under clinical investigation (Pasut and Veronese, 2007). In the case of polymer drug conjugates drug release is significantly concerned about achieving optimum therapeutic efficacy. For this, the type of linkage should be carefully chosen and various types of linkages include amide linkages, ester linkages, sulfhydryl linkages, hydrazone linkages, enzymatically degradable linkages, and so on. The research studies basically focused on the fact that ester and hydrazone linkages are acid labile and should be less favored while designing oral drug delivery system. The polymer–drug conjugates can be classified, that is, PEGylation, in which covalent binding of PEG chains to drug molecules. PEGylated paclitaxel displayed the 4-fold higher potential in the oral bioavailability as compared to the free drug, which could

Table 16.3 List of Experimented Polymeric Micelles Used for Oral Anticancer Chemotherapy

Drug	System Used	Entrapment Efficiency/ Drug Loading	Therapeutic Outcome	References
Methotrexate	PF-127, PF-127/SDS and PF-127/PC	69.80 ± 9%, 77.75 ± 6% and 81.30 ± 6%, respectively	Micelles of PF127/SDS, exhibited highest cell uptake (84.25%) and strong in vitro cytotoxicity (0.511 ± 0.437 μM; 6.04-fold than free MTX) in MCF-7 cells. In vivo acute toxicity studies in Swiss mice ensured safety (no major abnormalities in liver and kidney tissues) of the developed micelles at 10 mg/kg dose.	Kumar et al. (2014)
Quercetin	Soluplus	>90%	The stable Qu-PMs showed sustained release of entrapped quercetin for up to 10 days in vitro, and, more importantly, a sustained plasma level and enhanced systemic bioavailability of quercetin in vivo.	Dian et al. (2014)
Docetaxel	PEG–PCL micelles in pH responsive hydrogels	Drug loading = 7.76%	DTX-micelle–hydrogel significantly improved the oral bioavailability of docetaxel (75.6%) about 10 times compared to DTX micelles, and this increase in bioavailability was probably due to the small intestine targeting release of the pH-responsive hydrogel.	Wang et al. (2014)
PTX	*N*-deoxycholic acid-*N*, O-hydroxyethyl chitosan (DHC)		After oral administration of PTX-loaded DHC micelles, the bioavailability was 3-fold compared with that of an orally dosed Taxol. The single-pass intestinal perfusion studies (SPIP) showed that the intestinal absorption of micelles was via endocytosis involving a saturable process and a P-gp independent way	Li et al. (2010)
PTX	*N*-octyl-O-sulfate chitosan (NOSC)		The oral bioavailability of PTX loaded in NOSC micelles (PTX-M) was 6-fold improved in comparison with that of an orally dosed Taxol. The effect of NOSC micelles on enhancing the absorption of PTX was resulted from P-gp inhibition by NOSC, and the transport of PTX-M as a P-gp-independent way.	Mo et al. (2011)

(Continued)

Table 16.3 List of Experimented Polymeric Micelles Used for Oral Anticancer Chemotherapy (*cont.*)

Drug	System Used	Entrapment Efficiency/ Drug Loading	Therapeutic Outcome	References
PTX	Dequalinium-Vit TP-GS1000-PEG2000-DSPE		The permeability studies revealed about 36.4-fold higher transport of these functional micelles as compared to free drug when evaluated against Caco-2 cell monolayers.	Yao et al. (2011)
PTX	PEG-DSPE-TPGS	—	Solubility increased to 5 mg/mL. The micelles were stable in vitro under various conditions modeling the physiological ones, in particular, at low pH values and in the presence of bile acids, which is especially important for their possible oral administration.	Dabholkar et al. (2006)
Doxorubicin	PLGA-dextran	Drug content 7.6%–12.3%	Significantly higher in vitro cytotoxicity against doxorubicin resistant HuCC-T1 cell lines and polymeric micelles incorporating doxorubicin are promising vehicles for antitumor drug targeting. Keywords: dextran, polym.	Jeong and Kang (2011)
Doxorubicin	Stearic acid-g-chitosan	—	Increased oral bioavailability and CS-SA with higher amino-substituted degree (SD) was a promising vehicle for oral drugs	Yuan et al. (2011)
PTX	PEG-b-VBODENA-Acrylic acid	—	Complete drug release in 12 h. The pH-dependent release of PTX from HP micelles can be used to increase the bioavailability of PTX upon oral delivery.	Kim et al. (2008)
α-Mangostin	PVP	—	About 13,000-fold increase in solubility of α-mangostin. Improved solubility, self-assembly of nanomicelles, and intracellular delivery through endocytosis may enhance the pharmacological properties of α-mangostin, particularly antitumor efficacy.	Aisha et al. (2011)

Table 16.4 List of Nanocarriers Used for Chemotherapy Via Oral Administration

Biomaterials Used	Delivery System	Biotherapeutics	Disease/Cell Line/ Animal Model	Concluding Remarks	References
Amphiphilic cyclo-dextrins (CDs) or poly-ε- caprolactone (PCL)	Core–shell hybrid anionic and chitosan (CS)-coated cationic nanocapsule	Camptothecin	Caco-2	The results CPT-loaded nanosystems were found to be stable in simulated gastrointestinal media and nanoparticles and mucin interaction was confirmed by turbidimetric methods. Invivo animal studies concluded that intestinal uptake of CS-coated nanocapsules was significantly higher than cationic nanocapsules. The results supported that CPT-loaded positively charged CD nanocapsules might be fasicinating option for oral chemotherapy.	Ünal et al. (2015a,b)
Eudragit L and HPMC solution	Nanocapsules embedded with microparticles	Docetaxel	Rats/minipigs	The study concluded that significant enhancement of oral absorption might be due to the unique nanoplatform, as well as lymphatic route that is involved in nanocapsule absorption. However, this route avoided first-pass effects and protecting the drug from CYP3A4 degradation either in the gut or systemic circulation showing higher drug blood levels than i.v.-administered free drug, which was markedly metabolized by CYP3A4 in the circulation.	Attili-Qadri et al. (2013)
Chitosan-coated polycaprolactone-grafted hyaluronic acid nanoparticles	Multifunctional nanopartilces	Paclitaxel and doxorubicin	EC109/NIH3T3	The ex vivo and in vivo studies showed that CS/HA-*g*-PCL CNPs via oral administration was able to preferentially deliver drugs into tumor tissue with commendable antitumor efficiency and few side effects. CS/HA-*g*-PCL CNPs showed great potential for improving oral delivery efficiency of lipophilic anticancer drugs.	Huang et al. (2014)

(Continued)

Table 16.4 List of Nanocarriers Used for Chemotherapy Via Oral Administration (*cont.*)

Biomaterials Used	Delivery System	Biotherapeutics	Disease/Cell Line/ Animal Model	Concluding Remarks	References
Soluthin MD,	Lipid drug conjugates	Curcumin	—	In vivo anticancer study performed with optimized C-LPNCs showed significant increase in efficacy compared with pure CUR. Thus, lipopolysaccharide nanocarriers show potential delivery strategy to improve oral bioavailability and anticancer efficacy of CUR in the treatment of colorectal cancer.	Chaurasia et al. (2015)
Lyotropic liquid crystalline	Lipid drug conjugates	Phytanyl prodrug analog of capecitabine	Mouse 4T1 breast tumor	The study findings suggested that lipid drug conjugates had significantly higher antitumor efficacy at cell line studies and in vivo tumor model as compared to parent compound.	Gong et al. (2012)
—	Lipid nanospeheres	Tripterine	—	The oral bioavailability of TRI was increased up to 224.88% in regard to TRI-based oral suspensions. The main reason for enhancing oral absorption is improved intestinal permeability and postenterocyte lymphatic transport.	Zhang et al. (2014)
TPGS1000 and dequalinium	Polymeric micelles	PTX	MCF-7/Adr cells	The study findings concluded that functional PTX nanomicelles showed the strongest inhibitory effect on resistant breast cancer MCF-7/Adr cells at ex vivo and in vivo manner.	Yao et al. (2011)
PLGA/MMT	Polymeric nanoparticles	PTX	Caco-2 cells and HT-29 cells	PLGA/MMT nanoparticles enhanced the cellular uptake in presence of MMT ranged 57%–177% for caco-2 cells and 11%–55% for HT-29 cells as compare to plain PLGA. The cellular uptake efficiency was dependent on the extent of MMT and the number of particles in incubation. The designed system is possibly extended the residence time in the GIT, which promotes oral delivery of paclitaxel.	Dong and Feng (2005)

PCL and PLA-PCL-TPGS random copolymer	Thiolated chitosan-modified PCL nanoparticles, un-modified PLA-PCL-TPGS nanoparticles, and thiolated chitosan-modified PLA-PCL-TPGS nanoparticles	PTX	A549 cells	Cell uptake studies favored the thiolated chitosan-modified PLA-PCL-TPGS nanoparticles that had potential for higher cell uptake and cell cytotoxicity studies supported the thiolated chitosan-modified PLA-PCL-TPGS nanoparticles over commercial Taxol in regard to cytotoxicity in A549 cells. Moreover, these mucoadhesive nanoparticles enhanced the paclitaxel transport through opening tight junctions and bypassing the efflux pump of P-gp. This research finally established that thiolated chitosan-modified PLA-PCL-TPGS nanoparticles are a remarkable option to the treatment of lung cancer via oral chemotherapy.	Jiang et al. (2013)
N-octyl-O-sulfate chitosan (NOSC)	Polymeric micelles	PTX	Caco-2 cells/SD rats	The oral bioavailability can be enhanced by the absorption of paclitaxel due to P-gp inhibition by NOSC, and the transport of paclitaxel as a P-gp-independent way. The mechanism of P-gp inhibition by NOSC micelles is interfering with the P-gp ATPase, reducing the P-gp expression and exhibiting a potential to improve oral bioavailability of P-gp substrates.	Mo et al. (2011)
Apotransferrin and lactoferrin	Apodoxonano or lactodoxonano) nanoparticles	DOX	Wistar rats	The results concluded the maximum therapeutic efficacy of apodoxonano and lactodoxonano with the number of neoplastic nodules was significantly lower, approximately >93% than that of rats administered with saline or plain drug solution. The apotransferrin and lactoferrin nanoformulation improved the efficacy and increased the oral bioavailability of doxorubicin in a target-specific manner, and hence it is an effective and safer oral drug delivery approach.	Golla et al. (2013)

(Continued)

Table 16.4 List of Nanocarriers Used for Chemotherapy Via Oral Administration (*cont.*)

Biomaterials Used	Delivery System	Biotherapeutics	Disease/Cell Line/ Animal Model	Concluding Remarks	References
Pluoronics	PEO-PPO-PEO micelles	PTX	Female C57BL/6J mice	Pluoronic-based micelles displayed the extended the GIT residence time of entrapped drug with high area under curve of micellar paclitaxel, longer mean residence time (9-times more than i.v. Taxol) and high distribution volume (2-fold higher than i.v. Taxol) indicating an efficient oral absorption.	Yoncheva et al. (2012)
PLGA-TPGS random copolymer	DMAB-modified PLGA nanoparticles, unmodified PLGA-TPGS nanoparticles, and DMAB-modified PLGA-TPGS nanoparticles	Docetaxel	MCF-7 cells	Oral chemotherapy by DMAB-modified PLGA-TPGS nanoparticle formulation is an attractive and promising treatment option for patients.	Chen et al. (2010)
Pegylated poly(anhydride	Polymeric nanoparticles	PTX	Male Wistar rats	Their results findings revealed that this system successfully enhance the oral permeability and increase the drug residence time in plasma at least 48 h. These effects would be probably due to the diffusion abilities of these carriers through the mucus layer and inhibitory effect of PEG on P-gp and P450 cytochrome.	
Pluronic copolymers and low molecular weight heparin-all-transretinoid acid (LHR) conjugate	Pluronic/LHR mixed polymeric micelles	PTX	MCF-7 cells/ Sprague–Dawley rats	The designed system has higher cytotoxic activity in MCF-7 cells. In situ effective permeability of PTX through rat small intestine was 5- to 6-fold higher with mixed micelles than that of Taxol. Moreover, pluronic/LHR MPMs achieved significantly higher AUC and C_{max} level and this enhancement might be due to the inhibition of both P-gp efflux system and cytochrome P450 metabolism by pluronic copolymers.	Dahmani et al. (2012)

Low molecular weight chitosan	Polymer drug conjugate	DTX	NCI-H358 and brain glioblastoma (U87MG)/nude mice	The orally ingested conjugate presented comparable antitumor efficacy for both NCI-H358 and U87MG models, however, lesser subacute toxicity as seen in body weight loss and hematological toxicity.	Lee et al. (2009)
Capryol 90/cremophor EL/transcutol	Microemulsion	DTX	Caco-2 cell monolayer/male Sprague–Dawley rats	The apical to basolateral transport of drug across the caco-2 cell monolayer from the M-3 formulation was significantly higher improved (0.624 μg/cm^2) as compared to commercial product Taxotere (0.025 μg/cm^2). The oral bioavailability is also increased (34.42%) in case of microemulsion system as compared to that of the orally administered Taxotere (6.63%). All the remarkable results make it an attractive delivery system for further development of docetaxel microemulsions as an oral drug delivery system.	Yin et al. (2009)
PAMAM	Dendrimers	Short hairpin RNA	Tca8113 cells/BALB/c nude mice	These studies suggest that RNAi-mediated hTERT gene silencing, coupled with dendrimer delivery, may provide a promising approach for the treatment of oral cancer, in which hTERT is abundantly expressed.	Liu et al. (2011)
G3.5 PAMAM	Poly(amido amine) dendrimers; dendrimers	SN38	Caco-2 cells and HT-29/female CD-1	The designed system showed significant cytotoxicity and transport of G3.5-glycine-SN38 was highly concentration-dependent, whereas transport of G3.5-βalanine-SN38 was concentration-independent. The results show that PAMAM dendrimers have the potential to improve the oral bioavailability of potent anticancer drugs.	Goldberg et al. (2011)

(*Continued*)

Table 16.4 List of Nanocarriers Used for Chemotherapy Via Oral Administration (*cont.*)

Biomaterials Used	Delivery System	Biotherapeutics	Disease/Cell Line/ Animal Model	Concluding Remarks	References
TPGS, MCT, and HS-15	Self-assembled nanosystem	Teniposide	MCF-7, caco-2 cell monolayer/Balb'c nude mice	The intestinal absorption and cellular uptake in caco-2 cell monolayer was obviously improved by teniposide and cytotoxicity was significantly increased. The oral bioavailability of teniposide was greatly improved over 5.41-fold than that of teniposide solution. In particular, the teniposide concentration in tumor could specifically accumulate in tumor site in a xenograft model after oral administration.	Zhang et al. (2013)
Transcutol-HP , Labrafil-M1944 CS, Labrafac and Lipoïd-S75-3	Lipid nanocapsules	SN 38	Caco-2 cell mono-layers	Sn38-loaded LNCs improved permeability of Sn38 across caco-2 cells ($5.69 \pm 0.87 \times 10^6$ cm/s at 6 h vs. $0.31 \pm 0.02 \times 10^6$ cm/s) and intracellular concentration compared with free Sn38. In conclusion, Sn38 nanocarriers have been developed and display a strong potential for oral administration.	Roger et al. (2011)
Soya lecithin, cho-lesterol and stearyl amine	Polyelectrolyte sta-bilized multilayered liposomes	Paclitaxel	A549 cell lines/fe-male SD rats	The designed system presented the higher uptake, lower IC50 value. The in vivo pharmacokinetics studies revealed about 4.07-fold increase in the overall oral bioavailability of PTX and significant reduction in the tumor growth as compared to the control and comparable to that of i.v. Taxol.	Jain et al. (2012)
Ethylene diamine and methyl acrylate	Poly(amid o amine) dendrimers	Camptothecin	CD-1 mice	This study demonstrates that cationic and anionic PAMAM dendrimers were equally effective in enhancing the oral absorption of camptothecin. Results suggest that drug inclusion in PAMAM interior controlled drug solubilization in SGF and SIF, and increased oral bioavailability.	Sadekar et al. (2013)

Galactosemodified trimethyl chitosan-cysteine (GTC) Conjugates with various galactose grafting densities	Multifunctional polymeric nanoparticles	Sh RNA and Si RNA	Caco-2 monolayers/ tumorbearing mice	In tumor-bearing mice, orally delivered GTC2 NPs could effectively accumulate in the tumor tissues and silence the expression of Survivin and vascular endothelial growth factor (VEGF), evoking increased apoptosis, inhibited angiogenesis and thus the most efficient tumor regression. This study could serve as an effective approach for synergistic cancer therapy via oral gene delivery, and highlighted the importance of ligand grafting density in the rational design of targeted nanocarriers.	Han et al. (2014)
—	Lipid nanocapsules	PTX	Caco-2 cells	The results suggested that the presence of nanosystem on the basolateral side of the caco-2 cell monolayers when LNCs were applied on the apical side and uptaken by caco-2 cells mainly through active endocytic processes and more particularly via clathrin-dependent and caveolae-dependent transport mechanisms to allow transcytosis with improved gastrointestinal crossing of the drug.	Roger et al. (2009)
—	Drug nanocrystals	PTX and camptothecin	—	The oral formulations administered at 20 mg/kg and presented the comparable antitumor efficacy as that of intravenous administered taxol at same dose ranged and significantly maximum therapeutic efficacy at higher doses of 60 mg/kg and showed 3-fold greater effect in regard to free drug suspension, clearly showed the evidence of advantage of nanocrystal in the drug absorption.	Liu et al. (2010)
Chitosan	Porous quaternized chitosan nanoparticles encapsulated with paclitaxel nanocrystals	PTX	Caco-2 and lewis lung carcinoma	The porous quaternized chitosan nanoparticles encapsulated with paclitaxel nanocrystals was developed and presented higher intracellular drug retention with enhanced in vitro cytotoxicity and showed 5 fold increase drug accumulation at tumor site along with higher antitumor efficacy with maximum safety concerns as compared to standard formulation.	Lv et al. (2011)

(Continued)

Table 16.4 List of Nanocarriers Used for Chemotherapy Via Oral Administration (*cont.*)

Biomaterials Used	Delivery System	Biotherapeutics	Disease/Cell Line/ Animal Model	Concluding Remarks	References
PLGA	Polymeric nanoparticles	DOX	Caco-2 and C127I cell	The designed nanosystem showed ~49.06% reduction in the tumor burden upon oral administration of Dox-NPs in 30 days; on the other hand, 158% increase in tumor burden was observed in untreated group. Considering the toxicity aspects, developed system posed significantly lower cardiotoxicity.	Jain et al. (2011)
Pluronic grafted chitosan (Pl-g-CH) copolymer	Nanocrystals	PTX	Caco-2 cell	Nanocrystals rendered higher PTX accumulation inside the cell with significant decrease in transepithelial electrical resistance of caco-2 cell monolayers was observed with nanocrystals. This supports the role of the stabilizer in reversible opening of tight junctions between cells that can allow paracellular transport of drug. Nanocrystals resulted in much enhanced absorption with 12.6-fold improvement in relative bioavailability to that of Taxol. Investigating the role of Pluronic-g-cationic polyelectrolyte as functional stabilizer for nanocrystals: Impact on Paclitaxel oral.	
Mixture of ethyl oleate, tween-80 (T-form) or cremophor EL (C-form), and PEG-400/ethanol	SMEDDS	9-Nitrocamptothecin (9-NC)	SKOV-3 cells/nude mice	The SMEDDS formulations (T-form and C-form) represented significant enhancement of oral bioavailability (37.03% and 37.91%) in rats compared with 9-NC suspension and solution. Hence, SMEDDS is a promising drug delivery system to increase the oral bioavailability and antitumor effects and represents a novel 9-NC therapy for cancer patients.	

Oil, surfactant, and cosurfactant mixture using labrafil and Tween 80	SEDDS	DIM-14	H1650 stem cells/dog	The pharmacokinetic evaluation in dogs showed increased C_{max} (39.18 ± 7.34 vs. 21.68 ± 6.3 μg·dL-1), higher AUC0–t (34,481.34 ± 1125.46 vs. 14,159.53 ± 702.20 μg·min·dL-1) and improved absorption with 3 times more bioavailability of SEDDS compared to DIM-14 solution.	
GMS and soya lecithin	SLN	Quercitin	Male Waster rats	The experimental study concluded that the in vitro and in vivo results were fully agreed and very promising with the relative increasing the bioavailability of quercitin–SLN to quercitin suspension of 571.4%.	Li et al. (2010)
Monostearin, soybean oil, soya lecithin	NLC	Etoposide	Male Sprague–Dawley rats	The relative bioavailability of VP16-DSPE-NLCs were enhanced about 3.5-fold compared with VP16 suspension. Hence, VP16-DSPE-NLCs presented the highest cytotoxicity and improved the oral bioavailability of VP16.	
Phospholipid mixture	Polymeric core enclosed in liposomes	Vincristine, 5-FU, and MTX	Caco-2 and TC7 cell lines	The results suggested that polymeric particles that were the core of lipidic vesicles resulted in a slight increase of their bioavailability by the oral route.	
G3.5 dendrimers using EDC (1-ethyl-3-(3-dimethylaminopropyl)-carbodiimide)/NHS	G3.5 PAMAM dendrimers	SN-38	HT-29 cells	G3.5-glycine-SN38 and G3.5-β-alanine-SN38 displayed the higher apoptotic activity in colorectal cancer cells. SN-38 based both of conjugates increased SN38 transepithelial transport with regard to the free drug and disposed a powerful option for improving the oral bioavailability of potent anti-cancer drugs.	Goldberg et al. (2011)
—	Polyamidoamine (PAMAM) dendrimer	Short hairpin RNA (shRNA)		These studies suggest that RNAi-mediated human telomerase reverse transcriptase (hTERT) gene silencing, coupled with dendrimer delivery, may provide a promising approach for the treatment of oral cancer, in which hTERT is abundantly expressed.	Liu et al. (2011)

(Continued)

Table 16.4 List of Nanocarriers Used for Chemotherapy Via Oral Administration (*cont.*)

Biomaterials Used	Delivery System	Biotherapeutics	Disease/Cell Line/ Animal Model	Concluding Remarks	References
Anionic polyacrylic acid (PAA) followed by cationic chitosan	Layerosomes	PTX	HeLa cell line	The lyophilized chitosan–PAA–PTX-based liposome presented greater stability in simulated gastrointestinal fluids, and shelf life would be increased up to 6 months. The designed system revealed enhanced drug cytotoxicity in human cervical cancer cell culture experiments as compared to plain PTX-liposomes.	
Anionic polymer PAA followed by coating of cationic polymer PAH	Layerosomes	PTX	A549 cell lines/ female Sprague Dawley (SD) rats	The in vivo pharmacokinetics studies revealed about 4.07-fold increase in the overall oral bioavailability of PTX as compared to that of free drug.	Jain et al. (2012)

also be associated with increased solubility, permeability, and reduced presystemic metabolism. In the similar phase, an oral colon-specific drug delivery was developed for 9-aminocamptothecin conjugated with *N*-(2-hydroxypropyl)methacrylamide (HPMA) and presented a versatile option for colon cancer chemotherapy. Scientists focused their attention on designing the oral delivery of anticancer agents by employing the same principal.

Recently, chitosan polymer is a biodegradable, biocompatible, mucoadhesive, and nontoxic and increasingly novel platform for oral delivery of drugs. As contrast, high molecular weight chitosan, the low molecular weight chitosan (≤1 lac Dalton) showed higher solubility, lower toxicity profile with narrower molecular weight distribution (Kumar et al., 2014). The drug conjugates was designed for oral chemotherapy of a paclitaxel prodrug in the form of a conjugate with low molecular weight chitosan (LMWC), in which LMWC acted as a carrier. The systems have several benefits, such as increasing water solubility of paclitaxel by conjugation to water-soluble LMWC, the higher drug retention in the GIT due to the mucoadhesive property of LMWC, having potential to bypass the P-gp mediated efflux, an ability to bypass cytochrome P450-mediated metabolism. Hence, all these features make it attractive for drug delivery systems for oral chemotherapy that showed enhanced bioavailability and comparable antitumor efficacy at in vivo manner (Lee et al., 2008). In the same series, docetaxel was conjugated with low molecular weight chitosan (LMWC) via a cleavable linker for oral administration. The cytotoxicity study showed that conjugates have similar IC_{50} values to the parent drug. The pharmacokinetic parameters estimated that the half-life of conjugate was increased by 15 times and AUC was 3.8–6.2 times greater in comparison with the intravenously injected drug. The orally ingested conjugate presented comparable antitumor efficacy for both NCI-H358 and U87MG models, however, lesser subacute toxicity as seen in body weight loss and hematological toxicity (Lee et al., 2009).

4.3 LIPIDIC NANOPLATFORM SYSTEMS

Several studies signify the various lipids that have been used as delivery devices in several forms, offering the potential of endless opportunities in the area of drug delivery owing to their ability to enhance gastrointestinal solubilization, stablity, and absorption via selective lymphatic uptake of poor bioavailable drugs. All these features can be utilized to raise the therapeutic outcome of the drugs with lower bioavailability and to reduce their effective dose requirement. Numerous lipids are investigated for this purpose, such as endogenous of exogenous to date (Chakraborty et al., 2009). Lipids possess special amphiphilic properties able to form various structures in aqueous media showing great potential for constructing the possibilities of liquid, semi-solid, and solid lipid-based formulations for oral anticancer delivery. The principal lipidic drug delivery systems are present, such as micro- and nanoemulsions, liposomes, SLNs, and self-emulsifying drug delivery systems (SEDDS).

The term lipid refers to triglycerides, partial glycerides, fatty acids, steroids, and waxes. The choice of lipids is based on their features, such as polymorphic, crystallinity, miscibility, and physicochemical structure. The research-based experimental studies evidenced that the combination of the lipidic matrix, such as mono- and diglycerides, probably increases drug solubility as compared to pure lipids. Yet, the chemical nature of lipids is also important due to lipids, which form highly crystalline particles with a perfect lattice showing drug leaching during storage period. To the contrary, the lipids that formed the imperfect crystalline lattice presented stable lipidic nanoparticles.

4.3.1 Lipid–drug conjugates

A lipid–drug conjugate is one of the most significant approaches for lipid-based drug delivery systems in which the concerned drug is covalently attached with lipids that subsequently improves the drug payload into the system up to as high as 40%–50% w/w. Lipid–drug conjugates have the potential to provide sustained and targeted drug delivery (Thanki et al., 2013). The lipid–drug conjugates generally are spherical shape and confirm their potential and good therapeutic efficacy at in vitro levels but they failed during in vivo studies showing the limited oral bioavailability. In lipidic conjugate systems, the drug is covalentely attached with lipid molecules for improving lipiphilicity and stabilized with surfactant molecules. This could lead to optimum penetration and absorption through biological membrane furthermore, enhances the bioavailability of the therapeutic moiety. The lipids used for such a type of system can be phospholipids, fatty acids, such as docosahexaenoic acid, stearic acid, oleic acid, and so onand lipoamino acids.

Lipid–drug conjugates were developed with Soluthin MD, a phosphatidylcholine-maltodextrin-based hydrophilic lipopolysaccharide, displaying superior biocompatibility and bioavailability enhancing properties. In one study, the curcumin-loaded lipopolysaccharide nanocarriers (C-LPNCs) prepared to improve oral bioavailability and anticancer efficacy in colon-26 tumor-bearing mice at in vitro and in vivo manner. Ex vivo cytotoxicity studies showed the 23 fold lesser IC_{50} value as compared to free curcumin, while the pharmacokinetic parameters presented the 130-fold increase in oral bioavailability. The designed lipopolysaccharide nanosystems revealed the potential delivery option for enhancing oral bioavailability and anticancer efficacy of drug in colorectal cancer therapy (Chaurasia et al., 2015). In another set line, phytanyl pro-drug analog of capecitabine synthesized and characterized for various in vitro and in vivo studies. The study findings suggested that lipid–drug conjugates had significantly higher antitumor efficacy at cell line studies and in vivo tumor models as compared to parent compound (Gong et al., 2012). Importantly, the lipid–drug conjugate showed amphiphilic property, able to self-assemble with definite structures, which could further assist in the stabilization and superior absorption of drug in the GIT.

4.3.2 Self-emulsifying drug delivery system

SEDDS is a novel approach, in which the drug is dissolved in a stable liquid isotropic mixture of oils and surfactants. SEDDS has potential in spontaneous ways to emulsify and produce fine oil-in-water emulsions when it is combined with the aqueous GIT, leads to rapid absorption, lesser variability. Coabsorption of excipients in this formulation can be limited presystemic metabolism but not sufficient at effective manner to bypass all restrictions (Sachan et al., 2010).

The delivery system is supposed to self-emulsify rapidly after mixing the GIT contents of stomach in the form of (1) microemulsions (transparent dispersed systems having size less than 30 nm), (2) fine opaque nanoemulsions of droplet size of 50–200 nm, (3) or coarse emulsions with size greater than 500 nm. The oil droplets are much lesser as compared to obtained by in vitro level owing to bile salts would be implicated with surfactant layers of the emulsion droplets. This emulsifying system has remarkable features for dissolving or dispersing the hydrophobic drugs including anticancer drugs and P-gp inhibitors and to improve the absorption through the intestine (Mazzaferro et al., 2013). Nonionic self-emulsifying oil-in-water formulation that is almost free of cremophore was formulated by Lo et al. for paclitaxel delivery. The formulation based on oils, Tween 180 and L-α-phosphatidylcholine solubilizes near 500 mg/L of drug. A recent study was proposed to improve the oral bioavailability of 3,3-Diindolylmethane-14 (DIM-14) via SEDDS delivery system in dogs and evaluate pharmacodynamic characteristics of SEDDS against H1650 stem cell tumor models. The pharmacokinetic evaluation in dogs showed increased C_{max} (39.18 ± 7.34 vs. 21.68 ± 6.3 μg dL-1), higher AUC0–*t* (34,481.34 ± 1125.46 vs. 14,159.53 ± 702.20 μg min dL-1) and improved absorption with 3 times more bioavailability of SEDDS compared to DIM-14 solution

Self-microemulsifying drug delivery system (SMEDDS) provided the small droplet size with higher surface area for easy absorption. To select a suitable self-emulsifying vehicle, it is mandatory to assess the droplet size following self-emulsification. In one study, two SMEDDS formulations of 9-Nitrocamptothecin (9-NC) developed from a mixture of ethyl oleate, Tween80 (T-form), or cremophor EL (C-form), and PEG-400/ethanol were formed as microemulsions and with aqueous phase. The resulting systems were checked for their efficacy at in vitro and in vivo, including the kinetics and antitumor effects in SKOV-3 human ovarian cancer xenograft in nude mice. The results supposed that droplet sizes of resulting microemulsions were (30.8 ± 4.6) nm and (39.8 ± 8.2) nm for SMEDDS T-form and C-form, respectively. The SMEDDS formulations (T-form and C-form) represented significant enhancement of oral bioavailability (37.03% and 37.91%) in rats compared with 9-NC suspension and solution. Hence, SMEDDS is a promising drug delivery system to increase the oral bioavailability and antitumor effects and represents a novel 9-NC therapy for cancer patients. Although several SEDDS are under development (Chaurasiya et al., 2012; Oostendorp et al., 2010), no registered SEDDS formulation of anticancer drugs is known yet.

4.3.3 Lipid nanospheres

Lipid nanospheres are a colloidal therapeutic system and suitable for the treatment of various diseases. They can be obtained from warm oil-in-water microemulsions, having the advantage of being constituted with biocompatible lipids. They are capable of avoiding the drug precipitation, which is the main problem in commercial dosage forms when diluted before administration. Lipid nanospheres are recognized as artificial lipoprotein-like particles, which consist of phospholipids and simple lipids as soybean oil with sizes ranged (25–50 nm) and lipid compositions to endogenous plasma lipoproteins.

The oral route for anticancer drug delivery remains very challenging due to poor water-solubility and/or poor permeability. To enhance the oral bioavailability, recent research reported the tripterine (TRI, a plant-derived anticancer compound) using lipid nanospheres (LNs). The obtained particles were 150 nm in size with higher entrapment potential (99.9%). The designed systems were stable with negligible drug release (<0.2%) in simulated physiological fluid. The oral bioavailability of TRI was increased up to 224.88% in regard to TRI-based oral suspensions. The main reason for enhancing oral absorption is improved intestinal permeability and postenterocyte lymphatic transport (Zhang et al., 2014).

4.3.4 Microemulsion

There is a constantly increasing interest in microemulsion systems, which are considered an ideal system for poorly water-soluble drugs for oral chemotherapy. Microemulsion is a lipidic system having absorption enhancement, as well as high solubilizing capacity, thermodynamic stability, improved dissolution of lipophilic drugs and surfactant-induced permeability enhancement (Araya et al., 2005). In the microemulsion preparation, various excipients commonly added in these systems, namely cremophor, Tween 80, labrasol, and transcutol, all these showed the inhibition of P-gp mediated functions making the microemulsion system an attractive delivery system for oral delivery (Lin et al., 2007). The microemulsion system of docetaxel resulted in an improvement of oral bioavailability by increasing the drug's solubility and permeability with inhibiting the P-gp efflux system. The solubility study and pseudo ternary phase diagrams depicted that microemulsion of about 30 nm in size with improved solubilizing capacity toward the hydrophobic drug, docetaxel. It is composed (M-3 formulation) of capryol 90 (oil), cremophor EL (surfactant), and transcutol (cosurfactant), all these raised solubilization capacity of docetaxel up to 30 mg/mL. The apical to basolateral transport of drugs across the caco-2 cell monolayer from the M-3 formulation (capryol 90/cremophor EL/transcutol = 29.4:24.9:12.4, w/w) was significantly higher improved (0.624 μg/cm^2) as compared to commercial product Taxotere (0.025 μg/cm^2). The oral bioavailability is also increased (34.42%) in the case of microemulsion systems as compared to that of the orally administered Taxotere (6.63%). The increase in bioavailability was mainly due to the combined effect of the solubility enhancement, the inhibition of P-gp efflux system and the increase in permeability. All the remarkable results make it an attractive delivery system for further development of docetaxel microemulsions as an oral drug delivery system (Yin et al., 2009).

4.3.5 Lipid nanocapsules

Lipid nanocapsules are considered as novel systems having a core–shell arrangement of the constituents with a liquid reservoir (usually oils) as core that can be coated by a protective membrane. The protective membrane is mainly made up of polymer chains that are present in the surfactant (Thanki et al., 2013). Several methods were explored for formulation of lipid nanocapsules; however, the phase inversion temperature method is most favorable. The lipid nanocapsules associated with several beneficial features, such as biocompatibility, biodegradability, higher encapsulation, very high drug payloads, specific absorption mechanism across the GIT, sustained release, membrane efflux transporter inhibition, targeting potential and oral deliverability. The aqueous-core lipid nanocapsules came in to existence due to the ability to incorporate both hydrophilic and lipophilic drugmolecules.

Roger et al. (2009) explored the potential of lipid nanocapsules (LNC) and investigated its improvement in oral bioavailability of paclitaxel. The drug-loaded LNCs raised paclitaxel permeability across intestinal epithelium approx. 3.5-fold higher and uptake by Caco-2 cells compared with free drug or Taxol. Furthermore, experiments were performed on transport in the presence of various endocytosis inhibitors. Results showed the drug transport was reduced up to 30 ± 6% when cell cholesterol was depleted. It was found that inhibition of transport was reduced 65 ± 12% and 20 ± 8% by caveolae-mediated endocytosis and clathrin-mediated endocytosis, respectively. The electron microscopy study suggested that the presence of nanosystems on the basolateral side of the Caco-2 cell monolayers when LNCs were applied on the apical side and uptaken by Caco-2 cells mainly through active endocytic processes and more particularly via clathrin-dependent and caveolae-dependent transport mechanisms to allow transcytosis with improved gastrointestinal crossing of the drug.

4.3.6 Solid lipid nanoparticles and nanostructured lipid nanocarriers

Oral delivery of drugs loaded within SLNs and NLC has gained increasing interest from last two decades. They are basically made up from physiologically biocompatible and biodegradable lipids with a wide range of methodologies, showing their safe and effective potential in comparison to the conventional polymeric nanoparticles (Chakraborty et al., 2009). SLNs were first developed in the early 1990s and were principally composed from solid lipid only, while the second generation NLCs were prepared from a blend of a solid lipid with a liquid lipid (oil) leading to certain benefits compared to

SLN. A blend of a liquid and solid lipid forms a less ordered crystalline structure showing many imperfections having more space available for drug accommodation. By controlling the amount of liquid lipids added to the formulation, the NLCs maintained at solid form at body temperature and finally the controlled drug release can be achieved (Gupta et al., 2013). The lipidic nanosystem generally used for all routes of administration, but their use for oral drug delivery has been widely accepted. They showed high drug loading for lipophilic compounds and improvement of gastrointestinal absorption and oral bioavailability of several drugs, with avoid the first-pass metabolism leading to lymphatic transport, sustained and controlled release and minimizes the bioavailability fluctuations. Therefore, lipid-based systems overcome the inherent limitations of slow and incomplete dissolution of poorly soluble drugs and promote the formation of solubilized phases from which absorption may occur. The facilitation of absorption and adhesiveness of lipid nanosystems leads to GIT absorption enhancement's special concern for hydrophobic drugs. After reaching the absorption site, drug release was started after adhesion to the gut wall. The principal of drug release effect is explained as follows: the lipids are degraded by the presence of enzymes in the gut wall and create the formation of surface active mono- and diglycerides on the surface of the lipid droplets or solid lipid particles and after detaching these molecules they form micelles. These micelles interact with bile salts and form mixed micelles. The drug is absorbed together with the micelles.

In vitro studies suggested that P-gp-overexpressing human cancer cells treated with doxorubicin and paclitaxel-loaded lipid nanoparticles are able to overcome multidrug resistance via P-gp inhibition and ATP depletion and confirmed their potential role in improving the therapeutic efficacy of drugs. In vivo studies performed with orally administered lipid nanoparticles containing drugs, such as camptothecin and idorubicin have shown successful enhancing of the oral bioavailability of drugs. Li et al. (2010) suggested the quercitin-loaded SLN due to quercitin is a flavonoid compound and implicated for cancer therapy butshowed poor water solubility and low bioavailability. The experimental study concluded that the in vitro and in vivo results were fully agreed and very promising with the relative increasing the bioavailability of quercitin–SLN to quercitin suspension of 571.4%. The nanostructured lipid carrier formulation for etoposide (VP16) estimated the potential of NLCs as oral delivery system and pharmacokinetic study showed the relative bioavailability of VP16-NLCs, VP16-PEG40-NLCs, and VP16-DSPE-NLCs were enhanced about 1.8-, 3.0-, and 3.5-fold, respectively, compared with VP16 suspension. Hence, VP16-DSPE-NLCs presented the highest cytotoxicity and improved the oral bioavailability of VP16.

4.3.7 Liposomes

Liposomes confirmed their establishment to improve the therapeutic efficacy of both existing and new drugs. The nanoplatform system is enclosed vesicles consisted of one or more phospholipid based bilayers enclosing an aqueous phase. They are categorized into large multilamellar vesicles (MLV), small unilamellar vesicles (SUV) or large unilamellar vesicles (LUV), depending on their size and the number of lipid bilayers. They had potential to accommodate either hydrophilic drug in their inner aqueous phase or hydrophobic drug in their hydrophobic lipid bilayers. The principal advantage associated with their composition is similar like to the biomembranes, they are biodegradable and nontoxic (Chakraborty et al., 2009). Literature suggested that liposomes could be successfully explored for oral administration of drug/protein with improving their solubility or stability and absorption of poorly water-soluble drugs. Liposomes are able to protect the drugs sensitive to the harsh surroundings of the GIT. In contrast, other study concluded that enzymatic activity in the duodenum and bile salts destroyed the lipidic layers of liposomes integrity and subsequently releasing the drug (Chakraborty et al., 2009). In one study, β-sitosterol-containing orally administered liposomes were developed for preventing tumour metastasis and confirmed that increasing host defence play a key role for preventing tumour cells from metastasis.

Liposomes would be stabilized and allow their importance in oral delivery in the form of polymerized liposomes. Liposomal coated by various polymers by polyethylene glycol, carrageenans, collagen, and gangliosides GM1 and GM type III, and showing an interesting approach for enhancing their stability in the GIT. Moutardier et al. designed a system composed of a polymeric core enclosed in liposomes vesicles for encapsulating vincristine, 5-fluorouracil (5-FU), and methotrexate (MTX). Their results suggested that polymeric particles that were the core of lipidic vesicles resulted in a slight increase of their bioavailability by the oral route.

4.3.8 Layerosomes

Layersomes have shown potential for drug delivery that consists of the layer-by-layer (LbL) coating of the polyelectrolytes agents over liposomes to improve the stability purpose in the hostile atmosphere in GIT with facilitated their shelf life (Thanki et al., 2013). The layerosomes have been prepared by a positive charge to the liposomes, coated with anionic polymer polyacrylic acid (PAA) and subsequently coating cationic polymer polyallylamine hydrochloride (PAH). The lyophilized formulation presented higher stability as GIT media and improve the 4.07-fold increase in the overall oral bioavailability of PTX (Jain et al., 2012). In the same line, paclitaxel (PTX)-loaded multilayered liposomes were prepared using LbLassembly via positive surface potential of PTX-liposomes and subsequently coated with PAA followed by cationic chitosan. The lyophilized chitosan–PAA–PTX-based liposome presented greater stability in simulated gastrointestinal fluids, and the shelf life was increased up to 6 months. The designed system revealed enhanced drug cytotoxicity in human cervical cancer cell culture experiments as compared to plain PTX-liposomes.

4.4 DENDRIMERS

Dendrimers may represent a significant approach for oral anticancer drug delivery. A dendrimer is considered a macromolecule that is characterized by a controlled branch, like structure radiating, from the core to the periphery. They can be easily translocated across the intestinal barrier owing to their compact structural integrity. Authors suggested that polyamidoamine (PAMAM) dendrimers crossed the epithelial barrier with the combined effort of transcellular and paracellular routes and rapidly opened the tight junctions, hence facilitating their own transport through the paracellular pathway. They had many properties with traditional polymeric drug delivery systems, such as water solubility, greater drug loading, and improved biodistribution. PAMAM dendrimers conjugated with various anticancer drugs and made into the formulations that are orally bioavailable. Dendrimers can be successfully encapsulated in the camptothecins, paclitaxel, and doxorubicin with enhancing drug aqueous solubility.

SN-38 (7-ethyl-10-hydroxy-campothecin) is a potent topoisomerase-1 poison, effectively used for colorectal cancer and hepatic metastases therapy. The anchorage of SN38 to G3.5 PAMAM dendrimers with ester-linked glycine and β-alanine spacers efficiently improved oral bioavailability while minimizing gastrointestinal toxicity. G3.5-β-alanine-SN38 was more stable than G3.5-glycine-SN38, and released the contents at 10%, 20%, and 56% in simulated gastric, intestinal, and liver environments for up to 6, 24, and 48 h, respectively. G3.5-glycine-SN38 and G3.5-β-alanine-SN38 displayed IC50 values of 0.60 and 3.59 μM, respectively, in HT-29 cells treated for 48 h, indicating the higher apoptotic activity in colorectal cancer cells. SN-38-based conjugates increased SN38 transepithelial transport with regard to the free drug and disposed a powerful option for improving the oral bioavailability of potent anticancer drugs (Goldberg et al., 2011).

5 CONCLUDING REMARKS

This chapter combined or discussed the oral delivery of anticancer chemotherapy with the maximum emphasis on the nanoplatform systems. These ideal nanoengineered structured carriers confirmed their significant importance via various ways, such as flexibility for designing the dosage forms, selection of biomaterials used for the development of nanocarriers, enhancing the oral bioavailability so to maximize the therapeutic response, improvement of the physicochemical properties of the drugs with low aqueous solubility, and increasing the intestinal absorption of the drugs. The oral chemotherapy via novel systems improves the quality of life of patients suffering cancer. Simultaneously, the pharmacoeconomic gain with oral delivery will gain the attention of health-care agencies by reducing the overall cost incurred in the proper health-care management.

Proof-of-concept in vitro and in vivo results for the drug delivery of anticancer drugs by the various nanocarriers after oral administration, which can be found so far from the literature, are best exemplified for the reader's benefit. Moreover, the dream of cancer patients is chemotherapeutic treatment at home, which is almost possible and realized by oral delivery of anticancer drugs pharmaceutical nanotechnology. In addition, these nanoengineered vehicles can be considered as part of an interdisciplinary approach for integrated drug formulations that might be useful for being multifunctional by characteristics, occupying the various stages of clinical evaluation and may provide hope as soon as entering in the market with their appealing approach.

REFERENCES

Agrawal, U., Sharma, R., Gupta, M., Vyas, S.P., 2014. Is nanotechnology a boon for oral drug delivery? Drug Discov Today 19 (10), 1530–1546.

Aisha, A., Ismail, Z., Abu-salah, K., Majid, A., 2011. Solid dispersions of α-mangostin improve its aqueous solubility through self-assembly of nanomicelles. J. Pharm. Sci. 101 (2), 815–825.

Araya, H., Tomita, M., Hayashi, M., 2005. The novel formulation design of O/W microemulsion for improving the gastrointestinal absorption of poorly water soluble compounds. Int. J. Pharm. 305 (1–2), 61–74.

Attili-Qadri, S., Karra, N., Nemirovski, A., Schwob, O., Talmon, Y., Nassar, T., Benita, S., 2013. Oral delivery system prolongs blood circulation of docetaxel nanocapsules via lymphatic absorption. Proc. Natl. Acad. Sci. 110 (43), 17498–17503.

Bajin-Katic´, K., et al., 2006. Intestinal alkaline phosphatase activity as a molecular marker of enterotoxicity induced by single dose of 5-fluorouracil and protective role of orally administered glutamine. Arch. Oncol. 14, 101–105.

Batlle, J.F., et al., 2004. Oral chemotherapy: potential benefits and limitations. Clin Transl. Oncol. 6, 335–340.

Bhardwaj, V., Ankola, D., Gupta, S., Schneider, M., Lehr, C., Kumar, M., 2009. PLGA nanoparticles stabilized with cationic surfactant: safety studies and application in oral delivery of paclitaxel to treat chemical-induced breast cancer in rat. Pharm. Res. 26 (11), 2495–2503.

Bravo-Osuna, I., Vauthier, C., Farabollini, A., Palmieri, G., Ponchel, G., 2007. Mucoadhesion mechanism of chitosan and thiolated chitosan-poly(isobutyl cyanoacrylate) core–shell nanoparticles. Biomaterials 28 (13), 2233–2243.

Bromberg, L., 2008. Polymeric micelles in oral chemotherapy. J. Control. Release 128 (2), 99–112.

Calleja, P., Huarte, J., Agüeros, M., Ruiz-Gatón, L., Espuelas, S., Irache, J., 2012. Molecular buckets: cyclodextrins for oral cancer therapy. Thera. Deliv. 3 (1), 43–57.

Cassidy, J., Douillard, J., Twelves, C., McKendrick, J., Scheithauer, W., Bustová, I., Johnston, P., Lesniewski-Kmak, K., Jelic, S., Fountzilas, G., Coxon, F., Díaz-Rubio, E., Maughan, T., Malzyner, A., Bertetto, O., Beham, A., Figer, A., Dufour, P., Patel, K., Cowell, W., Garrison, L., 2006. Pharmacoeconomic analysis of adjuvant oral capecitabine versus intravenous 5-FU/LV in Dukes' C colon cancer: the X-ACT trial. Br. J. Cancer 94 (8), 1122–1129.

Chakraborty, S., Shukla, D., Mishra, B., Singh, S., 2009. Lipid: an emerging platform for oral delivery of drugs with poor bioavailability. Eur. J. Pharm. Biopharm. 73 (1), 1–15.

Chaurasia, S., Patel, R., Chaubey, P., Kumar, N., Khan, G., Mishra, B., 2015. Lipopolysaccharide based oral nanocarriers for the improvement of bioavailability and anticancer efficacy of curcumin. Carbohyd. Polym. 130, 9–17.

Chaurasiya, A., Singh, A., Jain, G., Warsi, M., Sublet, E., Ahmad, F., Borchard, G., Khar, R., 2012. Dual approach utilizing self microemulsifying technique and novel P-gp inhibitor for effective delivery of taxanes. J. Microencapsul. 29 (6), 583–595.

Chen, H., Zheng, Y., Tian, G., Tian, Y., Zeng, X., Liu, G., Liu, K., Li, L., Li, Z., Mei, L., Huang, L., 2010. Oral Delivery of DMAB-modified docetaxel-loaded PLGA-TPGS nanoparticles for cancer chemotherapy. Nano. Res. Lett. 6, 4.

Constantinides, P., Wasan, K., 2006. Lipid formulation strategies for enhancing intestinal transport and absorption of P-glycoprotein (P-gp) substrate drugs: in vitro/In vivo case studies. J. Pharm. Sci. 96 (2), 235–248.

Dabholkar, R., Sawant, R., Mongayt, D., Devarajan, P., Torchilin, V., 2006. Polyethylene glycol–phosphatidylethanolamine conjugate (PEG–PE)-based mixed micelles: some properties, loading with paclitaxel, and modulation of P-glycoprotein-mediated efflux. Int. J. Pharm. 315 (1–2), 148–157.

Dahmani, F., Yang, H., Zhou, J., Yao, J., Zhang, T., Zhang, Q., 2012. Enhanced oral bioavailability of paclitaxel in pluronic/LHR mixed polymeric micelles: preparation, in vitro and in vivo evaluation. Eur. J. Pharm. Sci. 47 (1), 179–189.

De Cock, E., et al., 2005. Cost-effectiveness of oral ibandronate versus IV zoledronic acid or IV pamidronate for bone metastases in patients receiving oral hormonal therapy for breast cancer in the United Kingdom. Clin. Ther. 27, 1295–1310.

Dian, L., Yu, E., Chen, X., Wen, X., Zhang, Z., Qin, L., Wang, Q., Li, G., Wu, C., 2014. Enhancing oral bioavailability of quercetin using novel soluplus polymeric micelles. Nano. Res. Lett. 9 (1), 684.

Dong, Y., Feng, S., 2005. Poly(D,L-lactide-*co*-glycolide)/montmorillonite nanoparticles for oral delivery of anticancer drugs. Biomaterials 26 (30), 6068–6076.

Eng, C., et al., 2001. Oral fluoropyrimidine treatment of colorectal cancer. Clin. Colorect. Cancer 1, 95–103.

Feng, S., Zhao, L., Zhang, Z., Bhakta, G., Yin Win, K., Dong, Y., Chien, S., 2007. Chemotherapeutic engineering: Vitamin E TPGS-emulsified nanoparticles of biodegradable polymers realized sustainable paclitaxel chemotherapy for 168 h in vivo. Chem. Eng. Sci. 62 (23), 6641–6648.

Feng, S., Zhao, L., Tang, J., 2011. Nanomedicine for oral chemotherapy. Nanomedicine 6 (3), 407–410.

Gao, L., Liu, G., Ma, J., Wang, X., Zhou, L., Li, X., Wang, F., 2012. Application of drug nanocrystal technologies on oral drug delivery of poorly soluble drugs. Pharm. Res. 30 (2), 307–324.

Gaucher, G., Marchessault, R., Leroux, J., 2010. Polyester-based micelles and nanoparticles for the parenteral delivery of taxanes. J. Control. Release 143 (1), 2–12.

Goldberg, D., Vijayalakshmi, N., Swaan, P., Ghandehari, H., 2011. G3.5 PAMAM dendrimers enhance transepithelial transport of SN38 while minimizing gastrointestinal toxicity. J. Control. Release 150 (3), 318–325.

Golla, K., Bhaskar, C., Ahmed, F., Kondapi, A., 2013. A target-specific oral formulation of doxorubicin-protein nanoparticles: efficacy and safety in hepatocellular cancer. J. Cancer 4 (8), 644–652.

Gong, J., Chen, M., Zheng, Y., Wang, S., Wang, Y., 2012. Polymeric micelles drug delivery system in oncology. J. Control. Release 159 (3), 312–323.

Gupta, M., P. Agrawal, G., P. Vyas, S., 2013. Polymeric nanomedicines as a promising vehicle for solid tumor therapy and targeting. Curr. Mol. Med. 13(1), 179–204.

Gupta, M., Chashoo, G., Sharma, P., Saxena, A., Gupta, P., Agrawal, G., Vyas, S., 2014. Dual targeted polymeric nanoparticles based on tumor endothelium and tumor cells for enhanced antitumor drug delivery. Mol. Pharm. 11 (3), 697–715.

Halfdanarson, T.R., Jatoi, A., 2010. Oral cancer chemotherapy: the critical interplay between patient education and patient safety. Curr. Oncol. Rep. 12, 247–252.

Han, L., Tang, C., Yin, C., 2014. Oral delivery of shRNA and siRNA via multifunctional polymeric nanoparticles for synergistic cancer therapy. Biomaterials 35 (15), 4589–4600.

Harrison, M.R., Hahn, N.M., Pili, R., Oh, W.K., Hammers, H., Sweeney, C., Kim, K., Perlman, S., Arnott, J., Sidor, C., Wilding, G., Liu, G., 2011. A phase II study of 2-methoxyestradiol (2ME2) NanoCrystalA (R) dispersion (NCD) in patients with taxane-refractory, metastatic castrate-resistant prostate cancer (CRPC). Invest. New Drugs 29, 1465–1474.

Huang, J., Si, L., Jiang, L., Fan, Z., Qiu, J., Li, G., 2008. Effect of pluronic F68 block copolymer on P-glycoprotein transport and CYP3A4 metabolism. Int. J. Pharm. 356 (1–2), 351–353.

Huang, P., Yang, C., Liu, J., Wang, W., Guo, S., Li, J., Sun, Y., Dong, H., Deng, L., Zhang, J., Liu, J., Dong, A., 2014. Improving the oral delivery efficiency of anticancer drugs by chitosan coated polycaprolactone-grafted hyaluronic acid nanoparticles. J. Mater. Chem. B 2 (25), 4021–4033.

Irshad, S., Maisey, N., 2010. Considerations when choosing oral chemotherapy: identifying and responding to patient need. Eur. J. Cancer Care 19, 5–11.

Jain, A., Swarnakar, N., Das, M., Godugu, C., Singh, R., Rao, P., Jain, S., 2011. Augmented anticancer efficacy of doxorubicin-loaded polymeric nanoparticles after oral administration in a breast cancer induced animal model. Mol. Pharm. 8 (4), 1140–1151.

Jain, S., Kumar, D., Swarnakar, N., Thanki, K., 2012. Polyelectrolyte stabilized multilayered liposomes for oral delivery of paclitaxel. Biomaterials 33 (28), 6758–6768.

Jeong, Y., Kang, D., 2011. Doxorubicin-incorporated polymeric micelles composed of dextran-*b*-poly(D L-lactide-*co*-glycolide) copolymer. Int. J. Nano. 6, 1415–1427.

Jiang, L., Li, X., Liu, L., Zhang, Q., 2013. Thiolated chitosan-modified PLA-PCL-TPGS nanoparticles for oral chemotherapy of lung cancer. Nano. Res. Lett. 8 (1), 66.

Kabanov, A., Batrakova, E., Alakhov, V., 2002. Pluronic block copolymers as novel polymer therapeutics for drug and gene delivery. J. Control. Release 82 (2–3), 189–212.

Kim, S., Kim, J., Huh, K., Acharya, G., Park, K., 2008. Hydrotropic polymer micelles containing acrylic acid moieties for oral delivery of paclitaxel. J. Control. Release 132 (3), 222–229.

Kumar, B., Meher, J., Gupta, A., Pawar, V., Singh, Y., Chourasia, M., 2014. Delivery of methotrexate executed by engineered polymeric micelles intended to improve chemotherapy. J. Biomater. Tissue Eng. 4 (9), 700–709.

Lee, E., Lee, J., Lee, I., Yu, M., Kim, H., Chae, S., Jon, S., 2008. Conjugated chitosan as a novel platform for oral delivery of paclitaxel. J. Med. Chem. 51 (20), 6442–6449.

Lee, E., Kim, H., Lee, I., Jon, S., 2009. In vivo antitumor effects of chitosan-conjugated docetaxel after oral administration. J. Control. Release 140 (2), 79–85.

Leslie, E., Deeley, R., Cole, S., 2005. Multidrug resistance proteins: role of P-glycoprotein, MRP1, MRP2, and BCRP (ABCG2) in tissue defense. Toxicol. Appl. Pharmacol. 204 (3), 216–237.

Li, H., Huo, M., Zhou, J., Dai, Y., Deng, Y., Shi, X., Masoud, J., 2010. Enhanced oral absorption of paclitaxel in *N*-deoxycholic acid-*N*, O-hydroxyethyl chitosan micellar system. J. Pharm. Sci. 99 (11), 4543–4553.

Lin, Y., Shen, Q., Katsumi, H., Okada, N., Fujita, T., Jiang, X., Yamamoto, A., 2007. Effects of labrasol and other pharmaceutical excipients on the intestinal transport and absorption of rhodamine123, a P-glycoprotein substrate, in rats. Biol. Pharm. Bull. 30 (7), 1301–1307.

Liu, F., Park, J., Zhang, Y., Conwell, C., Liu, Y., Bathula, S., Huang, L., 2010. Targeted cancer therapy with novel high drug-loading nanocrystals. J. Pharm. Sci. 99 (8), 3542–3551.

Liu, X., Huang, H., Wang, J., Wang, C., Wang, M., Zhang, B., Pan, C., 2011. Dendrimers-delivered short hairpin RNA targeting hTERT inhibits oral cancer cell growth in vitro and in vivo. Biochem. Pharmacol. 82 (1), 17–23.

Lu, Y., Chen, Y., Gemeinhart, R., Wu, W., Li, T., 2015. Developing nanocrystals for cancer treatment. Nanomedicine 10 (16), 2537–2552.

Lv, P., Wei, W., Yue, H., Yang, T., Wang, L., Ma, G., 2011. Porous quaternized chitosan nanoparticles containing paclitaxel nanocrystals improved therapeutic efficacy in non-small-cell lung cancer after oral administration. Biomacromolecules 12 (12), 4230–4239.

Mathijssen, R., et al., 2001. Clinical pharmacokinetics and metabolism of irinotecan (CPT-11). Clin. Cancer Res. 7, 2182–2194.

Mathot, F., van Beijsterveldt, L., Préat, V., Brewster, M., Ariën, A., 2006. Intestinal uptake and biodistribution of novel polymeric micelles after oral administration. J. Control. Release 111 (1–2), 47–55.

Mazzaferro, S., et al., 2011. Bivalent sequential binding of docetaxel to methyl-bcyclodextrin. Int. J. Pharm. 416, 171–180.

Mazzaferro, S., Bouchemal, K., Skanji, R., Gueutin, C., Chacun, H., Ponchel, G., 2012. Intestinal permeation enhancement of docetaxel encapsulated into methyl-β-cyclodextrin/poly(isobutylcyanoacrylate) nanoparticles coated with thiolated chitosan. J. Control. Release 162 (3), 568–574.

Mazzaferro, S., Bouchemal, K., Ponchel, G., 2013. Oral delivery of anticancer drugs I: general considerations. Drug Discov. Today 18 (1–2), 25–34.

Mei, L., Zhang, Z., Zhao, L., Huang, L., Yang, X., Tang, J., Feng, S., 2013. Pharmaceutical nanotechnology for oral delivery of anticancer drugs. Adv. Drug Deliv. Rev. 65 (6), 880–890.

Mo, R., Jin, X., Li, N., Ju, C., Sun, M., Zhang, C., Ping, Q., 2011. The mechanism of enhancement on oral absorption of paclitaxel by *N*-octyl-O-sulfate chitosan micelles. Biomaterials 32 (20), 4609–4620.

O'neill, V.J., Twelves, C.J., 2002. Oral cancer treatment: developments in chemotherapy and beyond. Br. J. Cancer 87, 933–937.

Oostendorp, R., Buckle, T., Lambert, G., Garrigue, J., Beijnen, J., Schellens, J., van Tellingen, O., 2010. Paclitaxel in self-micro emulsifying formulations: oral bioavailability study in mice. Invest. New Drugs 29 (5), 768–776.

Partridge, A., 2002. Adherence to therapy with oral antineoplastic agents. CancerSpect. Know. Environ. 94 (9), 652–661.

Pasut, G., Veronese, F., 2007. Polymer–drug conjugation, recent achievements and general strategies. Prog. Polym. Sci. 32 (8–9), 933–961.

Petrov, P., Bozukov, M., Burkhardt, M., Muthukrishnan, S., Muller, A., Tsvetanov, C., 2006. Stabilization of polymeric micelles with a mixed poly(ethylene oxide)/poly(2-hydroxyethyl methacrylate) shell by formation of poly(pentaerythritol tetraacrylate) nanonetworks within the micelles. J. Mater. Chem. 16 (22), 2192.

Rapoport, N., 1999. Stabilization and activation of Pluronic micelles for tumor-targeted drug delivery. Colloids Surf. B 16 (1–4), 93–111.

Regnier Denois, V., Poirson, J., Nourissat, A., Jacquin, J., Guastalla, J., Chauvin, F., 2010. Adherence with oral chemotherapy: results from a qualitative study of the behaviour and representations of patients and oncologists. Eur. J. Cancer Care 20 (4), 520–527.

Roger, E., Lagarce, F., Garcion, E., Benoit, J., 2009. Lipid nanocarriers improve paclitaxel transport throughout human intestinal epithelial cells by using vesicle-mediated transcytosis. J. Control. Release 140 (2), 174–181.

Roger, E., Lagarce, F., Benoit, J., 2011. Development and characterization of a novel lipid nanocapsule formulation of Sn38 for oral administration. Eur. J. Pharm. Biopharm. 79 (1), 181–188.

Ruddy, K., Mayer, E., Partridge, A., 2009. Patient adherence and persistence with oral anticancer treatment. A Cancer J. Clin. 59 (1), 56–66.

Sachan, R., Khatri, K., Kasture, S.B., 2010. Self-eumlsifying drug delivery system a novel approach for enhancement of bioavalibility. Int. J. Pharm.Tech. Res. 2 (3), 1738–1745.

Sadekar, S., Thiagarajan, G., Bartlett, K., Hubbard, D., Ray, A., McGill, L., Ghandehari, H., 2013. Poly(amido amine) dendrimers as absorption enhancers for oral delivery of camptothecin. Int. J. Pharm. 456 (1), 175–185.

Sajeesh, S., Bouchemal, K., Sharma, C., Vauthier, C., 2010. Surface-functionalized polymethacrylic acid based hydrogel microparticles for oral drug delivery. Eur. J. Pharm. Biopharm. 74 (2), 209–218.

Schinkel, A., Jonker, J., 2003. Mammalian drug efflux transporters of the ATP binding cassette (ABC) family: an overview. Adv. Drug Deliv. Rev. 55 (1), 3–29.

Shin, H., Alani, A., Rao, D., Rockich, N., Kwon, G., 2009. Multi-drug loaded polymeric micelles for simultaneous delivery of poorly soluble anticancer drugs. J. Control. Release 140 (3), 294–300.

Terwogt, J., Schellens, J., Huinink, W., Beijnen, J., 1999. Clinical pharmacology of anticancer agents in relation to formulations and administration routes. Cancer Treat. Rev. 25 (2), 83–102.

Tevaarwerk, A.J., Holen, K.D., Alberti, D.B., Sidor, C., Arnott, J., Quon, C., Wilding, G., Liu, G., 2009. Phase I trial of 2-methoxyestradiol nanocrystal dispersion in advanced solid malignancies. Clin. Cancer Res. 15, 1460–1465.

Thanki, K., Gangwal, R., Sangamwar, A., Jain, S., 2013. Oral delivery of anticancer drugs: challenges and opportunities. J. Control. Release 170 (1), 15–40.

Ünal, H., Öztürk, N., Bilensoy, E., 2015a. Formulation development, stability and anticancer efficacy of core–shell cyclodextrin nanocapsules for oral chemotherapy with camptothecin. Beilstein J. Organ. Chem. 11, 204–212.

Ünal, H., d'Angelo, I., Pagano, E., Borrelli, F., Izzo, A., Ungaro, F., Quaglia, F., Bilensoy, E., 2015b. Core–shell hybrid nanocapsules for oral delivery of camptothecin: formulation development, in vitro and in vivo evaluation. J. Nano. Res. 17 (1).

Wang, Y., Chen, L., Tan, L., Zhao, Q., Luo, F., Wei, Y., Qian, Z., 2014. PEG–PCL based micelle hydrogels as oral docetaxel delivery systems for breast cancer therapy. Biomaterials 35 (25), 6972–6985.

Yao, H., Ju, R., Wang, X., Zhang, Y., Li, R., Yu, Y., Zhang, L., Lu, W., 2011. The antitumor efficacy of functional paclitaxel nanomicelles in treating resistant breast cancers by oral delivery. Biomaterials 32 (12), 3285–3302.

Yin, Y., Cui, F., Mu, C., Choi, M., Kim, J., Chung, S., Shim, C., Kim, D., 2009. Docetaxel microemulsion for enhanced oral bioavailability: preparation and in vitro and in vivo evaluation. J. Control. Release 140 (2), 86–94.

Yoncheva, K., Calleja, P., Agüeros, M., Petrov, P., Miladinova, I., Tsvetanov, C., Irache, J., 2012. Stabilized micelles as delivery vehicles for paclitaxel. Int. J. Pharm. 436 (1–2), 258–264.

Yuan, H., Lu, L., Du, Y., Hu, F., 2011. Stearic Acid-*g*-chitosan polymeric micelle for oral drug delivery: in vitro transport and in vivo absorption. Mol. Pharm. 8 (1), 225–238.

Zhang, Z.P., Lee, S.H., Gan, C.W., Feng, S.S., 2008. Pharmacokinetics, biodistribution and xenograft tumor model of PLA–TPGS nanoparticles prepared by dialysis method for controlled and sustained small molecule chemotherapy. Pharm. Res. 25 (8), 1925–1935.

Zhang, Z., Ma, L., Jiang, S., Liu, Z., Huang, J., Chen, L., Yu, H., Li, Y., 2013. A self-assembled nanocarrier loading teniposide improves the oral delivery and drug concentration in tumor. J. Control. Release 166 (1), 30–37.

Zhang, X., Zhang, T., Zhou, X., Liu, H., Sun, H., Ma, Z., Wu, B., 2014. Enhancement of oral bioavailability of tripterine through lipid nanospheres: preparation, characterization, and absorption evaluation. J. Pharm. Sci. 103 (6), 1711–1719.

FURTHER READING

Johnsen, K., Gudbergsson, J., Skov, M., Pilgaard, L., Moos, T., Duroux, M., 2014. A comprehensive overview of exosomes as drug delivery vehicles: endogenous nanocarriers for targeted cancer therapy. Biochim. Biophy. Acta 1846 (1), 75–87.

Sharma, S., Verma, A., Pandey, G., Mittapelly, N., Mishra, P., 2015. Investigating the role of pluronic-*g*-cationic polyelectrolyte as functional stabilizer for nanocrystals: impact on Paclitaxel oral bioavailability and tumor growth. Acta Biomater. 26, 169–183.

Sugano, K., 2011. Fraction of a dose absorbed estimation for structurally diverse low solubility compounds. Int. J. Pharm. 405 (1–2), 79–89.

CHAPTER

17

NANOMATERIALS: PROMISING STRUCTURES FOR THE MANAGEMENT OF ORAL CANCER

Görkem Eskiizmir*, Aylin T. Ermertcan*, Kerim Yapici**

**Manisa Celal Bayar University, Manisa, Turkey*

***Cumhuriyet University, Sivas, Turkey*

CHAPTER OUTLINE

1 Oral Cancer 512
- 1.1 Oral Cancer: Overview 512
- 1.2 Epidemiology 512
- 1.3 Etiopathogenesis 512
- 1.4 Clinical Features and Diagnosis 513
- 1.5 Treatment 514
- 1.6 Prognosis 515

2 Nanoparticles in Cancer Treatment and Diagnosis 516
- 2.1 Magnetic Nanoparticles 516
- 2.2 Gold Nanoparticles 517
- 2.3 Graphene Oxide and Graphene NPs 519
- 2.4 Carbon Nanotubes 520

3 Nanomaterials in Early Diagnosis of Oral Cancer 523

4 Nanotherapeutics in Oral Cancer 524
- 4.1 Nanomaterials for Drug Delivery Systems 525
- 4.2 Nanocarriers in Oral Cancer 529
- 4.3 Nanomaterials for Combination Treatment Modalities in Oral Cancer 529

5 Nanomaterials in Tumor Imaging of Oral Cancer 531
- 5.1 Optical Imaging 531
- 5.2 Magnetic Resonance Imaging 531
- 5.3 Positron Emission Tomography 532
- 5.4 Nanotheranostics in Oral Cancer 533

6 Conclusions 533

References 533

1 ORAL CANCER

1.1 ORAL CANCER: OVERVIEW

Oral cancer is the sixth most common cancer worldwide, and one of the major reasons for cancer related morbidity and mortality. A male predominance in the incidence of oral cancer has been reported. In addition, it is more common in developing than developed countries. The major risk factors are tobacco use, alcohol drinking, betel quid chewing, chronic viral infections such as human papilloma virus (HPV), and poor oral hygiene. The most frequent histopathological type is squamous cell carcinoma (SCC). Oral carcinogenesis is a multistep, multifactorial process that involves several alterations of protooncogenes and inactivation of tumor suppressor genes. The most common symptoms of oral cancer are ulcerated, unhealing and hemorrhagic oral lesions with or without pain and cervical lymphadenopathy. Surgery, radiotherapy, and chemotherapy are the contemporary treatment modalities. Unfortunately, survival rates are suboptimal particularly for patients with advanced stage oral cancer. Furthermore, the frequency of locoregional recurrence and distant metastasis are remarkably high even after combined treatment modalities. Therefore, innovative treatment alternatives are urgently required.

1.2 EPIDEMIOLOGY

Globally, approximately 500,000 new cases with oral and pharyngeal cancers are diagnosed annually. In 2013, age standardized incidence rate per 100,000 population was 6.4 worldwide (Global Burden of Disease Control Collaboration, 2015). There is a geographical difference in the incidence of oral cancer, and three quarters of new cases are diagnosed in developing countries with a male predominance. The highest incidence of oral cancer is reported in India (men 16.4/100,000, women 10.6/100,000) (Chaturvedi et al., 2013). Oral cancer accounts for approximately 30% of all diagnosed cancers in India, and is remarkably high when compared to developed countries, such as United Kingdom (3%) and France (6%) (Brocklehurst et al., 2013). However, the incidence of oral cancer is also on the rise in the Western world. Recent epidemiological studies have demonstrated that the incidence of oral cancer has steadily increased over the past two decades particularly in Western Europe (Olaleye et al., 2015; Warnakulasuriya, 2009). The highest male incidence rate was reported in France (15.6/100,000) and Slovakia (15.2/100,000) among the European Union countries, whereas the lowest incidence was determined in Poland (6.6/100,000), Italy (6.3/100,000), and Netherlands (5.6/100,000) (Chaturvedi et al., 2013). In addition, the male incidence rate was relatively low in North America (Canada 5.7/100,000 and United States 6.6/100,000) (Chaturvedi et al., 2013). Currently, the incidence remains higher in older compared with younger groups and the majority of cases occur in people age 50 or over (Scully and Bagan, 2007).

1.3 ETIOPATHOGENESIS

In oral cancer, the most common histopathological type is SCC (more than 90% of all oral cancer) (Das and Nagpal, 2002; Tsantoulis et al., 2007), which is generally developed by a multistep, multifactorial process that involves several alterations of protooncogenes and inactivation of tumor suppressor genes, termed as oral carcinogenesis. It is triggered by several carcinogens, such as tobacco use, alcohol drinking, betel quid chewing, chronic HPV infection, and poor oral hygiene. Meat and red chili powder are anticipated as potential risk factors for the development of oral cancer (Das and Nagpal, 2002), whereas, fruits and vegetables particularly high in vitamins A and C may have a protective effect (Edefonti et al., 2015). In addition, subgingival plaque and synthetic dental materials might also have a genotoxic effect on mucosa of oral cavity. Emdogain, an enamel matrix derivative that is widely used for periodontal treatment, may induce the production of matrix metalloproteinase-2 and matrix metalloproteinase-9, both of which may cause the initiation and progression of oral cancer (Bagan and Scully, 2009). Furthermore, genetic susceptibility may also have a synergic role in oral carcinogenesis (Das and Nagpal, 2002). A correlation between tumor necrosis factor (TNF)-α polymorphism and development of oral cancer (an increase in risk of oral cancer with allele G and the GG genotype of TNF-α-238G > A and a decrease in risk of oral cancer with allele G and the GG + GA genotype of TNF-α-308G > A) was reported (Chen et al., 2013a). A microarray-based gene expression analysis found 601 genes that are significantly regulated in cancer tissue when compared with adjacent healthy mucosa. FMO2, CPA6, TNC, and SIAT1 genes were significantly upregulated in early stages of oral cancer (Fialka et al., 2008).

Leukoplakia (leuko-white and plakia-patch) and erythroplakia (leuko-red and plakia-patch), both of which are clinical definitions of premalignant lesions of oral cavity, and are frequently observed during oral carcinogenesis. They may involve hyperplasia, dysplasia (an early stage of oral carcinogenesis), and invasive carcinoma. The prevalence of leukoplakia is 0.2%–4.9% (Lee et al., 2008; Mishra et al., 2005; Schepman et al., 1996), and the risk of malignant transformation is ranging 0.13%–17.5% (Amagasa et al., 2011). On the other hand, the risk of malignant transformation is remarkably high (44.9%) (Villa et al., 2011) in erythroplakia, even though its prevalence is lower than leukoplakia (0.02%–0.2%) (Hashibe et al., 2000; Mehta and Hamner, 1971). Several molecular changes, such as overexpression of epithelial growth factor receptor (EGFR), a cell-surface receptor, and a member of epithelial growth factor superfamily, are very characteristic for leukoplakia and erythroplakia. Christensen et al. (1992) demonstrated high expression of EGFR even in oral dysplasia. Furthermore, Taoudi Benchekroun et al. (2010) determined an increased EGFR gene copy in oral premalignant lesions with a higher risk of malignant transformation. Consequently, EGFR overexpression is very frequent particularly in advanced stage oral cancers despite intratumoral heterogeneity (Huang et al., 2012; Rössle et al., 2013; Szabo et al., 2011), and generally have a correlation with poor prognosis, which is detailed in the next section.

1.4 CLINICAL FEATURES AND DIAGNOSIS

The most common symptom of oral cancer is ulcerated, unhealing, and hemorrhagic oral lesion with or without pain. In addition, cervical lymphadenopathy, ear pain, mobility of teeth, problems in breathing, halitosis, difficulty in speaking, dysphagia, problems using prosthesis, trismus, and paresthesia are the other symptoms. Patients may also present with skin fistula, cachexia, severe bleeding, and anemia particularly in advanced stage oral cancer. Oral cancer is most frequently detected on certain areas of oral cavity, such as tongue, floor of the mouth, and buccal region (Bagan et al., 2010) (Fig. 17.1A).

Oral SCC is generally characterized as exophytic or endophytic, or a combination of both. Exophytic tumors with a warty appearance are less common, slower growing, and less infiltrative than ulcerative lesions, and therefore generally have a better prognosis. On the other hand, endophytic lesions are more common and often appear as red or grayish ulcers with heaped up edges that have a tendency to bleed easily. In fact, the incidence of deep infiltration, a risk factor for lymphatic and hematogenous dissemination of cancerous cells, is remarkably high in endophytic lesions (Das and Nagpal, 2002).

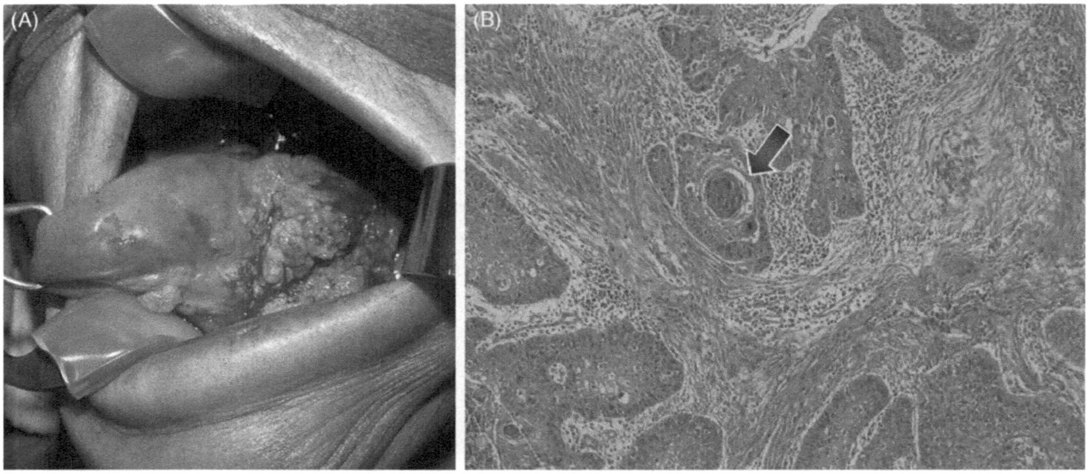

FIGURE 17.1

A view of an oral cancer on the left ventral side of tongue (A) and histopathological view of tumor with a keratinization pearl (*arrow*) (B).

An histopathological examination is a must for the confirmation of diagnosis. Therefore, a biopsy should always be performed to all lesions even to high-volume tumors in which the clinical characteristics are typical. In addition, a detailed description of tumor size and depth of invasion should be reported after total resection. Oral SCCs are graded histologically as (1) low-grade tumors with more extensive keratinization, infrequent mitosis and little nuclear pleomorphism and (2) high-grade tumors showing little keratin, much more mitosis and extreme nuclear pleomorphism (Fig. 17.1B). There are several variants of SCC of which basaloid squamous cell carcinoma and verrucous carcinoma are the most common histopathological types. Basaloid SCC is an aggressive variant of SCC. On the other hand, verrucous carcinoma (accounts less than 5% of oral cancer) is considered a low-grade variant of SCC and generally has a more favorable prognosis.

The staging of oral cancer is essential and necessary for treatment strategy. The staging of oral cancer mainly depends on physical examination, palpation of cervical lymph nodes and radiological assessment and follows the TNM system (da Silva et al., 2011). Vital staining with toluidine blue, different optical techniques, cytology, molecular biomarkers in tissues and saliva are the other diagnostic tools; however, they are not superior to histopathological examination and have not validated yet (Scully and Bagan, 2009).

1.5 TREATMENT

The successful management of oral cancer mainly depends on the appropriate treatment of both oral lesion and neck lymph nodes. The primary treatment modality for oral cancer is surgery (Chinn and Myers, 2015); however, a combination with radiotherapy and chemotherapy is generally required particularly for advanced stage oral cancer. The surgical plan is dictated by the site of primary tumor, tumor size (T stage), involvement of surrounding tissues, such as mandibula or maxilla, presence and number of nodal metastasis (N stage), and/or extranodal spread, and presence of perineural invasion or lymphovascular metastasis. An interdisciplinary approach between head and neck surgeons, facial reconstructive surgeons, dental oncologists, physical therapists, speech pathologists, and maxillofacial prosthodontists is generally required in order to have a satisfactory oncological, functional, and cosmetic outcome (Das and Nagpal, 2002). In addition, a combination of radiotherapy and targeted therapy particularly with anti-EGFR drugs may have a therapeutic effect in oral cancers and improve overall and progression free survival, and locoregional control (Chan et al., 2015). Photodynamic therapy may also be an alternative treatment choice for superficial lesions (Scully and Bagan, 2007). Gene therapies and targeted therapies, including tyrosine kinase inhibitors and cycline dependent kinase inhibitors, are promising treatment modalities for oral cancers (Das and Nagpal, 2002; Scully and Bagan, 2009).

The risk of cervical micrometastasis was relatively high even in patients with early stage oral cancer (Fasunla et al., 2011). Therefore, patients with oral cancer may require sentinel lymph node (SLN) biopsy or prophylactic neck dissection except selected patients with low risk of regional metastasis (Yuen et al., 2009). A large-scale study from Europe demonstrated that SLN mapping, a surgical procedure in which the first draining lymph nodes of a solid tumor were collected in order to determine the absence or presence of lymphatic metastasis, is a highly sensitive and specific diagnostic method for the detection of micrometastasis in patients with early stage (T1/T2) oral cancer (Schilling et al. 2015).

Primary, secondary, and tertiary prevention for oral SCC is of utmost important. Primary cancer prevention includes avoiding known carcinogens, such as tobacco smoking and alcohol use. Secondary prevention includes increased public awareness of early symptoms of premalignant disease and early diagnosis of oral cancer. Unfortunately, studies have reported an alarming lack of awareness about oral cancer; these gaps in knowledge should be addressed by further public education particularly to high-risk populations (Warnakulasuriya, 2009). As tertiary prevention, regular posttreatment follow-up is highly recommended to patients who had a curative treatment for oral SCC. The aim of tertiary prevention is early detection of locoregional recurrence, distant metastasis a second primary cancer, such as head and neck cancer or lung cancer, and treatment related complications (Foy et al., 2013).

1.6 PROGNOSIS

In oral cancer, the five-year survival rate is approximately 50%–55% (Messadi et al., 2009). Of note, the five-year survival rate of early stage oral cancer is approximately 75%, whereas, this rate may dramatically decrease to 16% in advanced stage oral cancer (Silverman and Gorsky, 1990). The most important prognostic factors for oral cancer are tumor size and characteristic, depth of invasion, involvement of regional lymph nodes, and distant metastasis. The prognosis of oral cancer may also vary according to the location of oral cancer. Patients with advanced stage tumors and poorly differentiated tumors have a dismal prognosis. Moreover, the survival may reduce by half when lymph node involvement occurs.

A variety of molecular factors, such as altered expression of Bcl-2 (Mallick et al., 2009) and loss of Bax expression (Bose et al., 2012) are also reported for poor prognosis. Overexpression of *cyclin D1* (Capaccio et al., 2000) and inactivation of p16 expression (Perez-Sayans et al., 2015) are correlated with aggressive tumors and influenced disease free- and overall survival. Moreover, disruptive p53 mutations in tumor cellsmay also lead to poor prognosis, increase in locoregional recurrence and decrease in survival rates (Poeta et al., 2007). Rubin Grandis et al. (1998) reported a decrease in disease free- and cause specific survival when tumors had a high level of EGFR and its ligand, TGF-α. Furthermore, Chung et al. (2006) demonstrated that high EGFR gene copy number is a poor prognostic factor in head and neck cancer.

2 NANOPARTICLES IN CANCER TREATMENT AND DIAGNOSIS

Nanoscience and nanotechnology are defined as the design fabrication and application of the nanomaterials that have at least one dimension smaller than 100 nm. Nanomaterial and nanoparticles (NPs) are umbrella terms for the definition of nanoscale structures. Recent developments demonstrated that NPs may offer new, effective, and innovative strategies in several aspects of cancer management (Fig. 17.2). In this respect, an overview of NPs used in biomedical applications is presented with a major focus on their synthesis, characterization, and using in treatment and imaging of cancer cells.

2.1 MAGNETIC NANOPARTICLES

Magnetic nanoparticles (MNPs), such as magnetite (Fe_3O_4) or its oxidized form of maghemite (γ-Fe_2O_3) exhibit strong magnetic properties; therefore they have a wide range of applications and can be utilized in gas sensors (Belle et al., 2011), adsorbent (Ai et al., 2011; Nethaji et al., 2013), and biomedical fields, such as contrast agents in magnetic resonance imaging (MRI) (Jin et al., 2014; Li et al., 2013; Weinstein et al., 2010; Zhao et al., 2014b), drug delivery system (Chen et al., 2010; Hu et al., 2012), hyperthermic treatment of various cancers (Gonzales-Weimuller et al., 2009; Hervault and Thanh, 2014; Salunkhe et al., 2014). For biomedical applications, MNPs should have some special properties, such as monodisperse size distribution, high surface functionality, biocompatibility, and

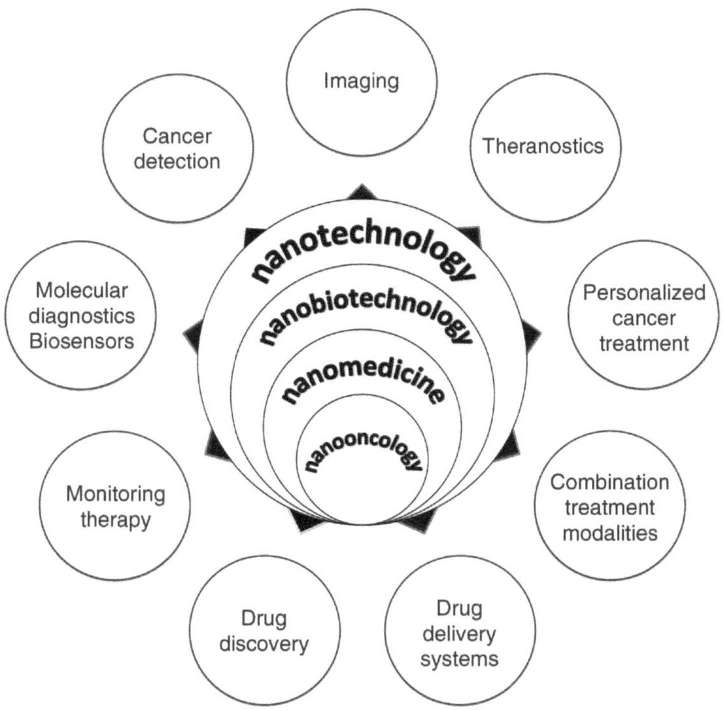

FIGURE 17.2 The Potential Role of Nanotechnology in the Field of Oncology

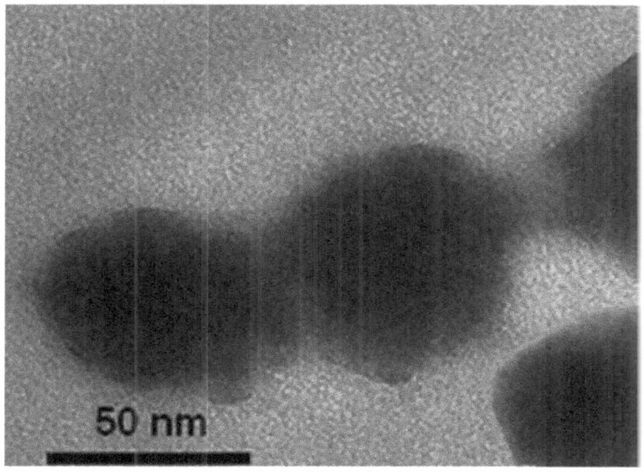

FIGURE 17.3 TEM Image of Fe_3O_4 NPs

nontoxicity (Stanicki et al., 2015). In addition, the stability of MNPs in various based fluids is one of the most important challenges. To date, numerous different approaches have been developed to synthesize MNPs with controllable particle size, shape, and narrow size distribution. These methods are coprecipitation (Hui et al., 2008; Jolivet et al., 2002; Lu et al., 2007), thermal decomposition (Wu et al., 2008), hydrothermal (Cai et al., 2003), microemulsion (Malik et al., 2012), sol–gel (Albornoz and Jacobo, 2006), and sonochemical reactions (Kima et al., 2005). Recently, Lima-Tenórica et al. (2015) and Stanicki et al. (2015) have extensively discussed all of these methods incorporating both advantages and disadvantages. Fig. 17.3 shows the transmission electron microscope (TEM) images of Fe_3O_4 NPs that are synthesized by coprecipitation technique based on the reduction of MNPs from the mixture of ferrous and ferric salts by adding NaOH solution. Among all substantial methods, coprecipitation is considered as the simplest and most efficient technique to produce large amounts of NPs (Stanicki et al., 2015).

Recent advances in the development of magnetic nanosystems for the combination of hyperthermia and chemotherapy have been reviewed by Hervault and Thanh (2014). In this review, they emphasized that even though magnetic NP suspension has shown great promise for heat mediator agents, it is still far from being used in clinical settings. Moreover, they stated that combinations of magnetic hyperthermia with chemotherapy for the treatment of cancer may be more effective than either hyperthermia or chemotherapy alone. The progress related to synthesis, functionalization and in vivo application of MNPs have also been reviewed by Lima-Tenórioa et al. (2015). They pointed out that MNPs may have a great potential in MRI as a contrast agent.

2.2 GOLD NANOPARTICLES

Gold (Au) NPs are good candidates for drug delivery platforms, diagnostic biomedical agents, hyperthermia treatment, and radiotherapy due to their unique physicochemical characteristics including easy functionalization and good biocompatibility, large surface-to-volume ratio, and relatively simple preparation (Butterworth et al., 2013; Jain et al., 2012; Liang et al., 2015). Among the various biomedical applications, most efforts have been devoted to explore potential effectiveness of Au NPs as radiosensitizer for radiotherapy. This is due to strong photoelectric absorption and release secondary

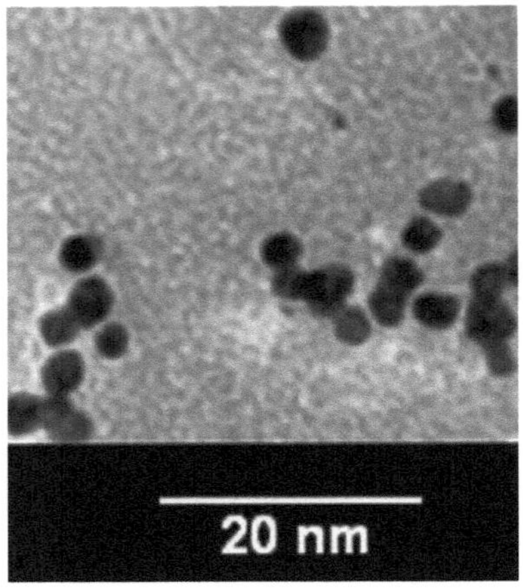

FIGURE 17.4 TEM Image of Au NPs

electron resulted in by gamma or X-ray irritation (Zhang et al., 2012). Radiotherapy, an anticancer treatment modality which is mainly based on bombardment of high-energy ionizing radiation to tumor sides, thereby killing the cancerous cells by direct cleavage or fragmentation of DNA and/or generation of free radicals (H, OH, O_2^-) and charged species (H_2O^+, H_3O^+) (Kwatra et al., 2013). Unfortunately, the hazardous effects of ionizing radiation may also impair the surrounding healthy tissues (Babaei and Ganjalikhani, 2014). In head and neck radiotherapy, a decrease in radiation dose is frequently required in order to protect the healthy vital tissues, such as spinal cord, eye, and salivary glands. In addition, the radioresistance may emerge in cancerous cells, which may lead to locoregional recurrence and poor prognosis (Kwatra et al., 2013). Therefore, an enhancement in radiosensitivity of tumor cells is required in order to overcome these shortcomings. The recent published studies demonstrated that metallic NPs with high atomic numbers, such as Au NPs ($Z = 79$), silver (Ag) NPs ($Z = 47$), and platinum NPs (Pt) ($Z = 78$) can be used as radiosensitizers to enhance the susceptibility of cancerous cells, thereby increasing the antitumor activity of radiotherapy. In a recent review, Babaei and Ganjalikhani (2014) have assessed the radiosensitizer role of different NPs in radiotherapy. They outlined six major factors that may affect the radiosensitivity of Au NPs; these are: (1) accumulation of Au NPs inside the cells and intracellular localization, (2) size, (3) concentration, (4) radiation dose, (5) cell type, and (6) surface modification of Au NPs.

Au NPs can be fabricated in various shapes, such as nanospheres, nanorods, nanocubes, nanocages, and nanostars (Liang et al., 2015) for different application fields. In biomedical applications, the most commonly used Au NPs are nanospheres and nanorods. In literature, various methods have been proposed to synthesize uniform size and shape of Au NPs. These are chemical reduction (Storhoff et al., 1998), electrochemical (Ying et al., 1997), and photochemical (Nidome et al., 2003). The citrate reduction method is the most common preparation technique as it is easy to apply, with low pollution reactant and nontoxic water solvent (Li et al., 2011). This method depends on the reduction of boiling solution of chloroauric acid by adding certain amounts of sodium citrate solution. Besides, the size of the Au NPs and monodispersity are mainly controlled by adjusting ratio between the Au salt and reducing agent and temperature. Fig. 17.4 demonstrates the TEM image of Au NPs fabricated by the reduction of chloroauric acid by sodium citrate solution.

2.3 GRAPHENE OXIDE AND GRAPHENE NPs

Graphene is composed of one atom thick planar sheet of sp^2-bounded carbon atoms densely packed in honey comp crystal lattice (Yin et al., 2015). The fundamental physical properties of graphene especially superior mechanical strength with a Young's modulus of 1,100 GPa (Lee et al., 2008), high planar surface area, 2,630 m^2/g (Stoller et al., 2008), high thermal conductivity, 5,000 W/m//K (Balandin et al., 2008), high carrier mobility, 10,000 $cm^2/V/s$ and capacity (Gomez-Navarro et al., 2007) make it a potential candidate for various applications including energy, electronics, molecular sensing, and catalysis. Apart form the superior mechanical and electronic structures, a recent study, in which the toxicity of graphene was examined and compared with carbon nanotubes (CNTs), demonstrated better results in favor of graphene (Zhang et al., 2010). This finding supports the great promise for implementation of graphene into the biomedical applications, such as biomarker sensors, drug delivery systems for anticancer agents, and hyperthermia treatment of several cancers (Markovic et al., 2011). There are mainly three methods employed to fabricate graphene nanosheets. These are mechanical exfoliation, chemical vapor deposition on the metal substrates and chemical or thermal reduction of graphene oxide (GO). Yin et al. (2015) analyzed and compared all of these fabrication techniques in detail. In their study, they emphasized that the reduction method is more beneficial for mass production of few-layered graphene, whereas the chemical vapor deposition is more efficient for mass production of high-quality graphene.

Graphene oxide is a single atomic layered material and can be produced from the oxidizing graphite powder with a well-known and widely employed method of Hummers proposed by Hummers and Offeman (1958) and improved by Marcano et al. (2010). Briefly, this method is based on the oxidation of graphite crystals in the presence of oxidants ($KMnO_4$ and $NaNO_3$) and acids (H_2SO_4 and H_3PO_4). Graphene sheets and GO have similar structures, both of which contain oxygen groups (Satish et al., 2013). The improved methods proposed by Marcano et al. (2010) have more advantages over classical Hummer's method, such as higher fractions of well-oxidized hydrophilic carbon material, more regular carbon framework, and no toxic gas evaluation during preparation. Therefore, they suggested that improved methods can be preferred for large-scale production of GO. Fig. 17.5 shows the TEM image of exfoliated GO sheet produced by improved Hummer's method. The large surface area and strong light absorbing properties make the graphene and GO an ideal candidate for drug loading and delivery, and photothermal therapy (PTT), in which near infrared (NIR) laser radiation is converted to heat in order to destroy the cancerous cells. Near infrared laser irradiation generally operates in the range of 700–1100 nm, which is accepted as the most appropriate wavelength for the penetration of light into deep tissues (Wu et al., 2013a). In an experimental study, Markovic et al. (2011) compared the photothermal effect and anticancer activity of NIR laser excited graphene and CNT NPs in vitro. They demonstrated that antitumor activity of graphene is superior to CNT NPs under the same operational conditions and concentrations. The superior photothermal sensitivity of graphene is described by its better dispersivity and optical and geometrical properties. In addition, they conclude that large surface area, low toxicity, and easy preparation make graphene-based NPs promising candidates for PTT. Zhang et al. (2011) have conducted an experimental study in order to investigate the tumor ablation by combining the PTT and chemotherapy using polyethylene glycol (PEGylated) GO in vivo and in vitro. They emphasized that combination of chemotherapy and PTT may show synergistic effect and enhance antitumor activity without weight loss and tumor recurrence. Yang et al. (2013) reported the design, synthesis, and evaluation of an active targeted delivery system consisting of PEGylated GO loaded with epirubicin. Their results demonstrated that synergistic targeted treatment enhances the local drug concentration and performs superior results for antitumor activity in U87 mouse xenograft.

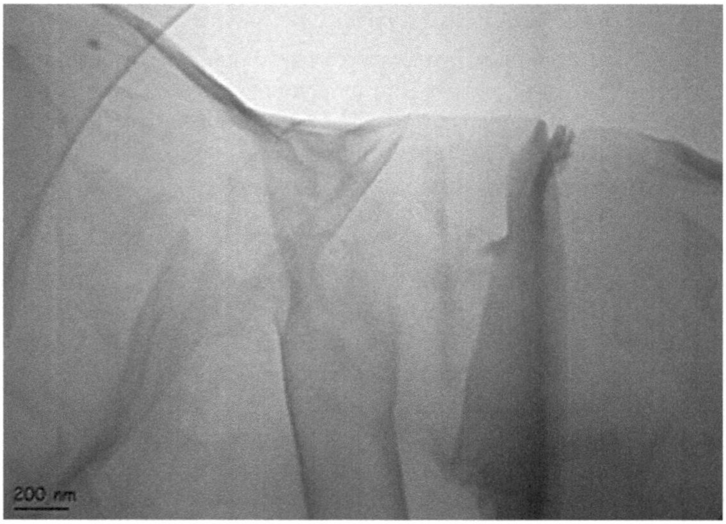

FIGURE 17.5 TEM Image of Graphene Oxide

The preclinical studies in which graphene-based nanomaterials were used for the ablation of various cancerous cell lines are summarized in Table 17.1.

2.4 CARBON NANOTUBES

Carbon nanotubes are allotropes of the carbon and can be considered as the rolled up single layer graphite (Jan Prasek et al., 2011). A CNT can be fabricated using various methods, such as chemical vapor deposition (CVD), plasma-enhanced CVD, laser ablation, arc discharge, electrolysis, sonochemical, or hydrothermal. The advantages and disadvantages of them are discussed in detail by Jan Prasek et al. (2011). Single-walled CNT (SWCNT) and multiwalled CNT (MWCNT) are the two basic forms of the CNT. Single-walled CNTs have diameter differing from 0.4 to 3 nm, while their length is in micrometer range (Eatemadi et al., 2014). On the other hand, MWCNTs are composed of several concentric tubes. Hence, their inner and outer diameters depend on the number of the tubes. In general, inner diameter is in the range of 0.4 nm to few nanometers and outer diameter differs from 2 to 30 nm (Eatemadi et al., 2014). Due to their unique physical, chemical, electrical, and mechanical properties, several practical applications, such as sensors, catalyst support, electronic devices, supercapacitors, hydrogen storage, and biomedicine fields have been reported (Paradise and Goswami, 2007). Like graphene-based NPs, CNTs have also strong light absorption properties in the NIR wavelengths. This unique property makes both SWCNTs and MWCNTs promising agents for PTT (Hashida et al., 2014; Lin et al., 2015; Moon et al., 2009; Neves et al., 2013). Lin et al. (2015) investigated the photothermal efficacy of MWCNTs in a xenograft model of bone metastasis of breast cancer. They concluded that MWCNTs plus NIR laser irradiation is an effective method for the management of bone metastasis of breast cancer. In another study, the photothermal activity of SWCNTs functionalized with a peptide was evaluated both in vitro and in vivo (Hashida et al., 2014). Authors reported that NIR laser irradiation plus SWCNTs/$(KFKA)_7$ showed a rapid increase in temperature in the subcutaneously inoculated colon 26 tumor cells, thus leading to a remarkable reduction in tumor size when compared with the SWCNTs/$(KFKA)_7$ or NIR laser irradiation alone.

Table 17.1 The Preclinical Studies Conducted in the Literature for the Treatment of Cancer Cells Using Graphene Based Nanomaterials

References	Graphene-Based Nanomaterial	Functionalization	Active Targeting	Type Treatment	Cell Line	Anticancer Drug	Thermal Therapy	Photodynamic Therapy
Yang et al. (2010b)	GO	6 arm PEG	NA	in vivo	4T1&KB&U87MG	NA	Yes	
Zhang et al. (2011)	GO	$PEG-NH_2$	NA	in vitro and in vivo	EMT6	DOX	Yes	
Markovic et al. (2011)	GO&CNT	NA	NA	in vitro	U251	NA	Yes	
Robinson et al. (2011)	rGO	$PEG-NH_2$	Arg-Gly-Asp (RGD)	in vitro	U87GM	NA	Yes	
Yang et al. (2012)	GO&rGO	$PEG-NH_2$	NA	in vivo	4T1	NA	Yes	
Yang et al. (2013)	GO	$PEG-NH_2$	EGFR	in vitro and in vivo	U87GM	EPI	Yes	
Lu et al. (2014)	GORs	PL-PEG	NA	in vitro and in vivo	U87GM	DOX	Yes	
Jung et al. (2014)	GO	NA	HA	in vitro and in vivo	Melanoma	NA	Yes	
Tian et al. (2011)	GO	6 arm PEG	NA	in vitro	KB	NA	Yes	Yes
Miao et al. (2013)	GO	$PEG-NH_2$	NA	in vitro and in vivo	SCC7	DOX	No	Yes
Wu et al. (2014)	GO	NA	NA	in vitro	MM RPMI-8226	DOX	No	Yes
Li et al. (2014)	rGO	$PEG-NH_2$	Transferrin	in vitro	AGS	NA	No	Yes
Zhou et al. (2014)	GO	NA	NA	in vitro	A549	SN-38	No	Yes
Wu et al (2014)	GO	PLL	NA	in vitro	HeLa & MCF7 & B16	DOX	No	Yes

(*Continued*)

Table 17.1 The Preclinical Studies Conducted in the Literature for the Treatment of Cancer Cells Using Graphene Based Nanomaterials (*cont.*)

References	Graphene-Based Nanomaterial	Functionalization	Active Targeting	Type Treatment	Cell Line	Anticancer Drug	Thermal Therapy	Photodynamic Therapy
Liu et al. (2008)	GO	6 arm PEG	NA	in vitro	HCT-116	SN-38	No	
Liu et al. (2013)	GO	6 arm PEG	Transferrin	in vitro and in vivo	C6 Glioma	DOX	No	
Tian et al. (2014)	GO	$PEG-NH_2$	NA	in vitro	MCF-7 & CAL-27	CDDP	No	
Zhou et al. (2014)	GO	Polyelectrolyte	Integrin αvβ3	in vitro and in vivo	U87GM& MCF-7	DOX	No	
Wang et al. (2014b)	GO	NA	Chlorotoxin	in vitro	C6 Glioma	DOX	No	
Zhu et al. (2014)	GO	6 arm PEG	NA	in vitro	HeLa & SMMC-7721	PPT	No	
Some et al. (2014)	DGO	NA	NA		HCT-116	Curcumin	No	

CDDP, *Cisplatin;* DGO, *double oxide graphene oxide;* DOX, *doxorubicin;* EPI, *epirubicin;* GO, *graphene oxide;* GOR, *graphene oxide ribbon;* HA, *hyaluronic asit;* PL-PEG, *phospholipid polyethylene glycol;* $PEG-NH_2$, *metoksifenetilamin;* PLL, *poly-L-lysine;* PPT, *podophyllotoxin;* NA, *not available;* rGO, *reduced graphene oxide.*

2.4.1 Quantum dots

Quantum dots (QDs) are nanometer-scale semiconductors that exhibit discrete energy levels related to nanocrystal size and composition (Valizadeh et al., 2012). Their unique optical and electronic properties, such as high brightness and stability, wide and continuous absorption spectra, and tunable wavelengths make them a potential candidate in cancer imaging (Zhang et al., 2008). Various synthesis methods have been proposed for the fabrication of QDs. These are sol–gel, γ irradiation, microwave assist, polyol, and high-temperature methods. All these fabrication techniques are discussed in detail by Obonyo et al. (2010). The carbon-based QDs, including graphene QDs and carbon QDs, have been reviewed by the well-documented study of Wang and Hu (2014) which was mainly focused on the synthesis methods,biomedical applications, optronics, catalysis, and sensors. Gao et al. (2004) have carried out an experimental in vivo study to develop probes based on the semiconductor QDs for cancer targeting and imaging. They reported that QD probes may accumulate at tumors and produce sensitive and multicolor fluorescence imaging of cancer cells.

3 NANOMATERIALS IN EARLY DIAGNOSIS OF ORAL CANCER

In the realm of oncology, early diagnosis is of utmost important for the success of cancer treatment and one of the major contributing factors for better prognosis and survival. Unfortunately, most of the symptoms and signs are recognized in advanced stages, thereby leading to a delay in diagnosis and treatment. In oral cancer, Seoane et al. (2016) analyzed the relationship between mortality and diagnostic delay, and demonstrated a risk increase of more than two-fold in mortality. Hence, they emphasized the survival advantage of early detection. Ideally, a *cancer detection and screening system* may provide early and rapid detection of tumor cells with a very high sensitivity, specificity, and accuracy. Unfortunately, unlike several aspects of oncology, current *cancer detection systems* are far from perfect. Recently, Macey et al. (2015) systemically analyzed the diagnostic value of substantial minimally invasive methods, including vital staining, such as toluidine blue, methylene blue, and rose bengal, cytological techniques, light-based detection systems, molecular analysis of blood and saliva, for the early detection of premalignant and malignant lesions of the oral cavity. They determined that none of the abovementioned methods had an accuracy of biopsy and histopathological examination, an invasive surgical procedure. Similarly, Carreras-Torras and Gay-Escoda (2015) emphasized that biopsy and histopathological examination is the gold standard method for the diagnosis of oral cancer among all noninvasive and invasive diagnostic methods.

NPs regarding their versatile properties may serve several options for the advancement of substantial *cancer detection systems*. For instance, specially designed nanowires, nanostructures that can precisely detect cancer related molecules from body fluids, such as blood, urine, saliva, and so on, may improve the sensitivity of substantial sensor test chips. They may also provide an excellent platform for multiplexed systems, which are useful for the detection of several biomarkers including proteins, genes, and so on. Furthermore, they can enable the development of novel, highly selective, and specific nanodiagnostic techniques with lower expenses. Briefly, Jain (2007) classified the nanodiagnostic techniques as: (1) nanoscale visualization, (2) NP biolabels, (3) nanotechnology-based biochips, (4) nanoproteomic-based diagnostics, (5) biobarcode assays, (6) nanopore technology, (7) DNA nanomachines for molecular diagnostics, (8) NP-based immunoassays, (9) nanobiosensors, and (10) combination of multiple diagnostic techniques. However, the major limitation of the nanodiagnostic techniques is optimization. Therefore, to date, very few of them are commercially available, even though promising results have been reported in preclinical and preliminary studies.

Gold NP, an excellent plasmonic NP with high light absorption capacity especially at NIR wavelengths (650–900 nm), was widely investigated for the early detection of cancerous cells due to its unique optoelectronic properties (Huang et al., 2007b; Jain et al., 2008; Lee and El-Sayed, 2008). However, as emphasized by Yasun et al. (2013), Au NPs need to be targeted to a specialized ligand, such as a receptor or protein depending on the cellular characteristic of cancerous cells. El-Sayed et al. (2005) conjugated gold to anti-EGFR monoclonal antibody in order to improve its detection capacity for oral SCC in vitro. A sixfold increase in binding of cancerous cells was demonstrated when compared with noncancerous cells, and highly specific and homogenous binding of anti-EGFR conjugated Au NPs gave a relatively sharper surface plasmon resonance absorption band. Hence, the authors indicated the potential role of anti-EGFR conjugated Au NPs for molecular biosensor techniques. Kah et al. (2007) also developed a surface-enhanced Raman scattering-based saliva assay using anti-EGFR conjugated colloidal Au NPs in surface-enhanced Raman scattering monolayer film in order to biosense the oral cancer related biochemical changes in saliva. They reported a sensitivity of 70% and a specificity of 100% irrespective of cancer stage, and mentioned that surface-enhanced Raman scattering-based saliva assay is a simple, cost-effective, highly specific and moderately sensitive diagnostic technique for the early detection of oral cancer.

Exfoliative cytology, a simple and minimally invasive technique, is particularly used for the diagnosis of advanced stage oral cancer. Even though it is also considered as an attractive method for early diagnosis of premalignant and malignant lesions of oral cavity, the sensitivity and specificity is variable. Therefore, the clinical applicability of exfoliative cytology for early detection and screening of oral cancer is controversial (Gupta et al., 2014; Perez-Sayansm et al., 2010; Sciubba, 1999; Sousa et al., 2014; Ye et al., 2015). Weigum et al. (2010) aimed to increase the diagnostic accuracy of exfoliative cytology and designed a nanobiochip sensor, an automated, electronic device that can integrate several laboratory procedures on a single chip, in order to analyze cellular biomarkers, such as EGFR and detect several cytomorphological changes that are generally emerged during oral carcinogenesis. In a preliminary study, promising results with a sensitivity value of 97% and a specificity value of 93% were determined for the detection of oral cancer. However, further clinical trials with a large-scale study populationshould be performed before clinical application.

4 NANOTHERAPEUTICS IN ORAL CANCER

Most of the conventional chemotherapeutic agents have several shortcomings that significantly limit their use in clinical settings and lead to a wide range of variability in treatment outcomes. These are: (1) low bioavailability due to their pharmacokinetic properties, (2) nonspecific distribution throughout the body, (3) notorious toxic effects on healthy tissues, (4) metabolic adverse effects especially for patients with comorbidities, such as diabetes mellitus, liver and kidney failures, (5) short blood circulation life span due to their physicochemical instability, quick elimination and renal clearance, (6) high rate of opsonization, (7) uncontrolled drug release kinetics, (8) insufficient accumulation at tumor microenvironment mainly due to biologic barriers, and (9) multidrug resistance.

Nanotechnology is likely to confer a myriad of benefits in the field of cancer treatment. Hence, a variety of NPs may overcome the aforementioned drawbacks of conventional chemotherapeutics and enable development of safer chemotherapeutics with higher therapeutic efficacy, drug delivery systems, targeted *smart* drugs with controlled release, prolonged blood circulation life span and lesser adverse effects, and multifunctional NP-based drugs (Fig. 17.6). Thereby, it may yield several new *ammos*

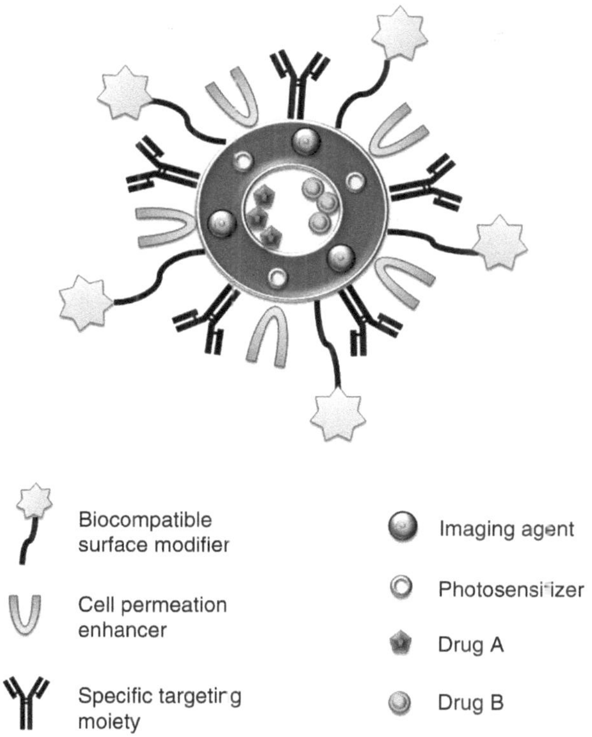

FIGURE 17.6 An Illustration of a Multifunctional NP

for the battle against cancer. In the past two decades, a few *first generation* liposomal- [PEGylated liposomal doxorubicin, Doxil (Alza Corporation, 1995); liposomal daunorubicin, DounoXome (NeXstar Pharmaceutical, 1996); liposomal doxorubicin, Myocet (Cephalon)], polymer- [neocarzinostatin, Zinostatin Stimaler (Yamanuchi); PEG-L-asparaginase, Oncaspar (Sigma-Tau Pharmaceuticals, Inc.)] and protein-based nanodrugs [albumin bound paclitaxel, Abraxane (Celgene Corporation, 2005)] have been approved for the clinical applications particularly for the treatment of breast cancer and Kaposi sarcoma; more importantly, they pave the way for other innovative NPs in the field of drug discovery (Fonseca et al., 2014; Frank et al., 2014).

Current literature is replete with a myriad of preclinical researches, in which appreciable success with promising outcomes has been reported with newly designed nanotherapeutic agents for the treatment of oral cancers.

4.1 NANOMATERIALS FOR DRUG DELIVERY SYSTEMS

NPs have several unique properties for the optimization of drug delivery systems (Allen and Cullis, 2004). They can be individually designed and tailored according to pharmacological and therapeutic profile of drugs. In addition, nanocarriers may improve the biodistribution and bioavailability of active drug, shield it from degradation, prolong its blood circulation life span, regulate its release, enable selective and active targeting, and reduce toxic side effects.

In blood circulation, NPs with a size of less than 6 nm are quickly extravasated and excreted by glomerular filtration. In addition, NPs with a size of more than 10 nm are prone to be opsonized by phagocytes or captured by the cells of reticuloendothelial system (RES) as they are recognized as foreign molecules, thereby leading to rapid degradation and elimination via biliary system (Longmire et al., 2008). Therefore, surface modification of NPs is generally essential in order to enhance their bioavailability. Polyethylene glycol, a widely used, highly efficient surface modifier agent, sterically stabilizes the NPs and generates stealth properties by increasing its hydrophilicity, constructing a three-dimensional structure with an impermeable layer, thereby preventing the recognition and elimination process either by immune system or RES (Peracchia et al., 1999). However, PEG moiety may impair the transport of NPs and disturb its cellular internalization, and cause *accelerated blood clearance* phenomenon, unforeseen immunogenic response particularly emerged after several applications. These potential obstacles are denominated as *PEG dilemma* (Hatekayama et al., 2013). Of note, Hatekayama et al. (2011) presented a multifunctional envelope-type nanodevice in order to overcome PEG dilemma.

Conventional chemotherapeutics are internalized by passive diffusion, a process in which molecules enter the cell according to concentration difference, and generally are deposited around the cell membrane (Fig. 17.7). Therefore, drug accumulation in cellular cytosol and intracellular concentration can be compromised by the action of efflux pumps and transporters, member of ATP-binding cassette transporter superfamily. In several solid tumors, overexpression of efflux transporters, such as P-glycoprotein, multidrug resistance associated protein 1–3, and breast cancer resistance protein may frequently emerge inherently or during the treatment period, thereby leading to an increase in

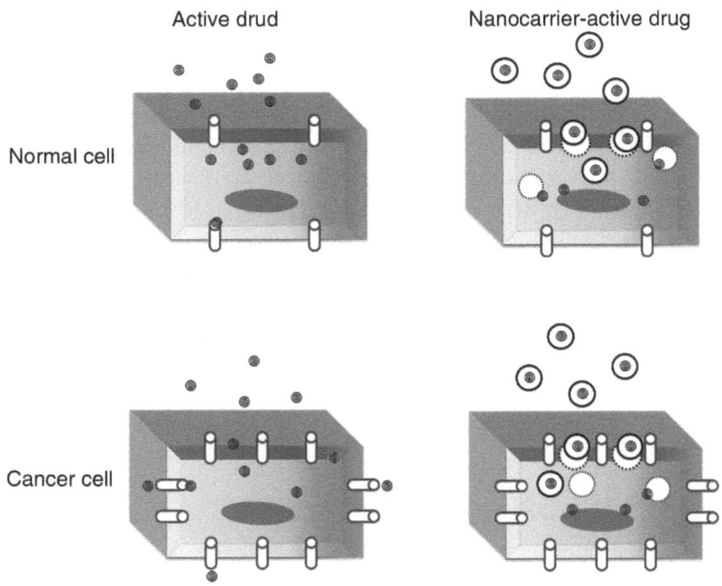

FIGURE 17.7 An Illustration Representing the Cellular Uptake of Chemotherapy and NP-Based Nanodrugs in Cancerous and Noncancerous Cells

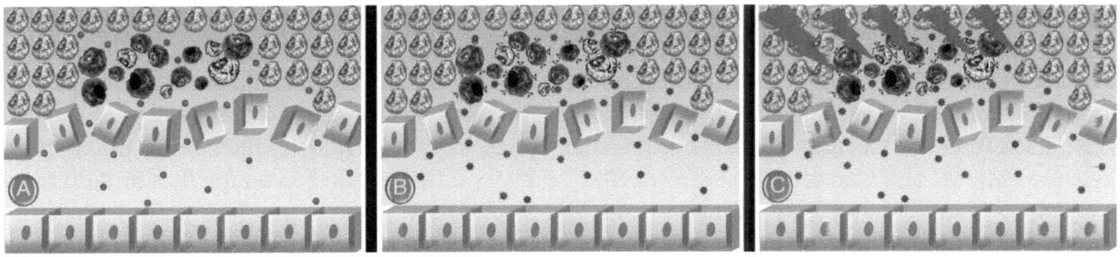

FIGURE 17.8

An illustration representing passive (A), active (B), and stimuli responsive targeting (C).

removal of chemotherapeutic agents and a decrease in drug concentration inside the cancerous cells that may cause treatment failure, tumor recurrence, and dismal prognosis (Brigger et al., 2002; Kirtane et al., 2013; Nobili et al., 2006; Thomas and Coley, 2003; Yang et al., 2014). On the other hand, the major internalization mechanism of NP-based drugs is endocytosis, a mechanism of cellular uptake in which molecules are internalized by a small vesicle (Canton and Battaglia, 2012). Thereafter, the active drug is delivered within the cytosol and deposited in the perinuclear region and can avoid from the action of efflux pump; thereby, chemotherapy resistance might be overcome (Fig. 17.7).

Nanomaterials have three strategies for drug targeting: (1) passive targeting, (2) active targeting, and (3) physical stimuli responsive targeting (Fig. 17.8).

Passive targeting: It is the main route of delivery for conventional chemotherapeutic agents and first-generation NP-based drugs (Barenholz, 2012; Maeda, 2001; Maeda et al., 2001; Rink et al., 2013). In solid tumors, the anarchic nature of tumor microenvironment may facilitate passive targeting (Fig. 17.8A). Tumor microenvironment is a battleground with a dynamic, disorganized, and corrupted mixture of a variety of different molecules (lymphokines, cytokines, matrix metalloproteinase, collagen fibers, proteoglycans, glycosaminoglycans, etc.), cancerous and noncancerous cells (lymphocytes, macrophages, cancer related fibroblasts, endothelial cells, adipocytes, etc.) (Curry et al., 2014; Eskiizmir, 2015). In addition, the blood supply is also variable at tumor microenvironment, and generally impaired due to irregular architecture of tumor vasculature, which has amorphous and leaky vessels, dense and heterogeneous extracellular matrix and abnormal lymphatic network. The morphology of vascular and lymphatic system at tumor microenvironment is dramatically different than healthy tissues. Tumor vessels are chaotic, dilated, and tortuous with an abnormal vessel-wall structure due to poorly aligned, disorganized endothelial cells. In normal blood vessels, the size of interendothelial gaps are approximately 7–10 nm; whereas, these gaps may increase up to 100 nm in solid tumors (Hashizume et al., 2000; McDonald and Foss, 2000). The pore sizes of normal vessels are generally less than 10 nm in diameter, whereas, the pore cutoff size may range 100–780 nm in diameter depending on the tumor size, heterogeneity, vascularization, and location (Li et al., 2012b; Sarin, 2010; Yuan et al., 1995). In addition, extensive production of various vascular permeability mediators, such as bradykinin, nitric oxide, and vascular endothelial growth factors may stimulate the permeability characteristic of tumor vessels (Fang et al., 2003; Maeda et al., 2003). Furthermore, the mean drainage velocity of extracellular fluid, which is approximately 0.1–2 μm/s in normal tissues, is significantly compromised at the tumor microenvironment of solid tumors due to inadequate amount of defective and nonfunctional lymphatic capillaries (Rutkowski and Swartz, 2007; Swartz and Fleury, 2007). Hence, both hyperpermeability of vascular network and insufficient clearance of lymphatic system is conceptualized as enhanced permeability and retention (EPR) effect, which was initially termed by Maeda and Matsumura (1989) approximately 3 decades ago. In fact, EPR effect is the major transport dynamic for passive targeting. In normal tissues, hydrostatic pressure and plasma colloidal osmotic pressure are the driving force for the fluid exchange between capillaries and tissues. Therefore, the extravasation of fluid exchange can theoretically be explained by the Starling equation (Michel and Curry, 1999): $J_v = (L_p S)[P_c - P_i] - \delta(\pi_c - \pi_i)]$ [J_v: volume flux of fluid (mL/min), L_p: hydraulic conductivity (cm/min mmHg), S: capillary surface area (cm^2), P_c: hydrostatic pressure of capillary (mmHg), P_i: hydrostatic pressure of interstitial fluid (mmHg), δ: osmotic reflection coefficient of vessel wall (0–1, 0 means hyperpermeability, whereas, 1 means impermeability), π_c: colloidal pressure of capillary (mmHg), and π_i: colloidal pressure interstitial fluid (mmHg)]. In healthy tissues, interstitial fluid pressure is generally low (approximately 0–5 mmHg), whereas, it is dramatically elevated in solid tumors (Jain, 1994; Jain et al., 2007). For instance, in an experimental model of oral SCC, Raju et al. (2008) demonstrated a significant increase in interstitial fluid pressure (29.1 ± 2.9 mmHg) with defective lymphatic system at tumor microenvironment when compared with healthy tissues (3.1 ± 0.3 mmHg). Therefore, as Nichols and Bae (2014) indicated, EPR effect is unreliable and unpredictable at tumor microenvironment and may be specifically interfered with the volume, composition and pressure of interstitial fluid (Baronzio et al., 2012), which may remarkably vary depending on several factors, such as vascularization, histopathological type, location, and spatial and temporal heterogeneity of tumors (Lunt et al., 2008; Wu et al., 2013b). Moreover, the dense architecture of extracellular matrix, including proteoglycans, glycosaminoglycans, and interconnected collagen matrix, may also impair the EPR effect and decrease passive targeting of NP-based drugs. Ultimately, interstitial hypertension due to the combination of impaired blood flow and fluid drainage and anarchic nature of extracellular matrix generally lead to impairment in transvascular transport and may impede the drug delivery at tumor microenvironment (Heldin et al., 2004), thereby leading to chemotherapy resistance and treatment failure. Therefore, amelioration in drug targeting is essential in order to enhance the efficiency of drug delivery.

Active targeting: It is also termed as ligand-conjugated targeting (Fig. 17.8B). The idea of targeted therapy was initially termed by Paul Ehrlich as *magic bullet*. The rationale of targeted therapy in cancer is to increase the therapeutic efficacy of anticancer drugs while preserving the noncancerous cells and healthy tissues. Upon incorporation of NPs with cancerous cell-specific, overexpressed molecules or receptors, such as transferrin, folate receptor, EGFR, CD95, transmembrane tyrosine kinase receptor, the EPR effect–related limitation and diversity in drug delivery can be overcome (Akhter et al., 2013; Bertrand et al., 2014; Brigger et al., 2002; Byrne et al., 2008; Fonseca et al., 2014; Zwicke et al., 2012).

Stimuli responsive targeting: Even though NP-based drug transport and delivery can be efficiently achieved by passive and active targeting, promoting the release of NP-based drug at the targeted area is still challenging and unpredictable. Recently, physical and/or chemical stimuli responsive targeting systems have drawn great attention to overcome this shortcoming by controlling the drug release (Fig. 17.8C). These NP-based drug systems mainly triggered by: (1) internal stimuli, such as pH, enzymatic activities, and reactive oxygen radicals, and (2) external stimuli, such as temperature, light, irradiation, magnetic field, and ultrasound waves (Fleige et al., 2012; Muthu et al., 2009; Taghizadeh et al., 2015; Wicki et al., 2015). Moreover, Cheng et al. (2013) suggested multistimuli-responsive polymeric NPs with combinations of aforementioned stimuli for a programmed site-specific drug delivery system.

4.2 NANOCARRIERS IN ORAL CANCER

4.2.1 In vitro and in vivo studies

In an experimental oral SCC model, a biodegradable thermosensitive hydrogel, a drug carrier for tailoring drug release kinetics, was designed for the delivery of cisplatin, a highly cytotoxic anticancer drug, which is widely used for lung cancer, ovarian cancer, head and neck cancer, unfortunately, with severe adverse effects, such as nephrotoxicity, neurotoxicity, and ototoxicity, and suberoylanilide hydroxamic acid, a histone deacetylase inhibitor, in order to control the drug delivery (Li et al., 2012a). They reported significant inhibition of cell proliferation and tumoral angiogenesis, decrease in tumor size without systemic side effects after intratumoral injection. In addition, Endo et al. (2013) examined the therapeutic efficacy and adverse reactions of cisplatin incorporated polymeric micellar NP and demonstrated significant inhibition of cellular proliferation with lesser renal injury in an orthotopic tongue cancer model. Recently, Zhao et al. (2015) developed a monomeric self-assembled nucleoside nanoparticle (SNNP) for drug delivery system and loaded with 5-fluorouracile (5-FU), a pyrimidine analog that inhibits thymidylate synthase irreversibly. The administration of 5-FU–loaded SNNP showed high blood stability and significant antitumor activity with acceptable side effects.

4.2.2 Clinical trials

In literature, a paucity of clinical trials was determined about NP-based drugs in oral cancer. Damascelli et al. (2003) intraarterially administered paclitaxel, a member of taxane family, which interferes with the normal breakdown of microtubules during cell division, incorporated human albumin NP to 23 patients with advanced stage oral cancer. They demonstrated complete response in four patients in whom microscopic residual tumors with a size of 1–10 mm were detected after surgical excision. In addition, authors reported that the side effects of the drug were minimum. Hence, they emphasized the potential role of paclitaxel incorporated human albumin NP particularly for induction chemotherapy in the treatment of oral cancers.

4.3 NANOMATERIALS FOR COMBINATION TREATMENT MODALITIES IN ORAL CANCER

In oral cancers, NPs may augment the antitumor activity of contemporary treatment modalities, such as PTT and radiotherapy, and minimize treatment-related toxic effects. Therefore, they may increase therapeutic efficacy without jeopardizing the surrounding healthy tissues

4.3.1 Photothermal therapy

Photothermal therapy is a minimally invasive, local treatment modality with minimal toxicity. It mainly depends on triggering a photosensitizer by an electromagnetic radiation, such as radio frequency, microwaves, NIR, or visible light in order to convert this energy into heat. In clinical settings, this process is called *hyperthermia*. Hyperthermia may trigger several hazardous phenomena, such as denaturation of proteins, lysis of cell membrane, evaporation of cytosol, thereby leading to cell death (Lapotko, 2009). Photothermal therapy, unlike photodynamic therapy which has an antitumor activity relying on the generation of radical oxygen species, hence, presence of oxygen is essential, mainly exerts its effect by increasing the temperature of milieu; therefore, it does not require oxygen in order to generate its cytotoxic effect on cancerous cells. Initially, the idea of NP-based laser-activated PTT was conceptualized by Lin and coworkers (Pitsillides et al., 2003). In oral cancers, El-Sayed et al.

(2006, 2007); Huang et al. (2006a, 2007a) and Huang and El-Sayed (2010) reported the effective role of anti-EGFR conjugated Au NPs in imaging and plasmonic PTT. Moreover, they also demonstrated that a threshold temperature ranging 70–80°C was required in order to destroy cancerous cells in vitro (Huang et al., 2006b). In a xenograft oral SCC model, PEGylated Au nanorods were administered intratumorally (with NIR radiation of 0.9–1.1 W/cm^2 for 10 min) and intravenously (with NIR radiation of 1.7–1.9 W/cm^2 for 10 min) in order to examine and compare their tumor suppressive capacity in plasmonic PTT (Dickerson et al., 2008); and significant decrease in tumor size was observed in both groups (>57% after intratumoral injection and 25% after intravenous injection). El-Sayed and coworkers (Mackey et al., 2014) aimed to determine the optimum size of Au nanorods for plasmonic PTT of oral cancers. Theoretical calculations and in vitro experimental study showed that the most effective photosensitizer agent was Au nanorods with a size of 28 × 8 nm. Furthermore, Wang et al. (2014a) examined the potential role of Rose Bengal, a hydrophilic photosensitizer, conjugated Au nanorods in photodynamic and PTT of oral cancer both in vitro and in vivo, and demonstrated highly efficient cytotoxic effects with production of reactive oxygen species under 532 nm light irradiation and remarkable enhancement in temperature under 810 nm NIR laser irradiation. Of note, a preliminary clinical study using Au nanoshells (AuroShell) and NIR laser irradiation for the PTT of refractory and/or recurrent head and neck cancers has been completed recently (Clinicaltrials.gov, 2010). Hopefully, this study will show high therapeutic efficacy of PTT with Au NPs in oral cancer, even though it has not been published yet.

4.3.2 Radiotherapy

Radiotherapy is widely applied for curative and adjuvant treatment of head and neck cancers with favorable impact on locoregional control and survival (Chinn and Myers, 2015; Corvo, 2007; Gregoire et al., 2015). However, the main drawback of radiotherapy is treatment related toxicities, which may eventually lead to dose limitations, treatment failure, and radiotherapy resistance. Therefore, the quest for increasing therapeutic efficacy without damaging surrounding healthy tissues or dose reduction without compromising treatment success is still an ongoing field of research.

In NP assisted-radiotherapy, encouraging outcomes have been reported especially with dense inorganic NPs, such as Au, bismuth, and silver (Cooper, 2014; Retif et al., 2015). However, Au, as it has several benefits, such as biocompatibility, delivering high-Z material, and high-dose enhancement factor, is the most popular NP for combining with radiotherapy. Hainfeld et al. (2008) intensively examined the efficacy of Au assisted-radiotherapy in oral cancer (Hainfeld et al., 2008, 2010, 2014). In a xenograft oral SCC model, Au NPs (1.9 nm in size) were injected intravenously, and hyperthermia (44°C of water bath for 20 min) was applied thereafter (Hainfeld et al., 2010). They demonstrated an enhancement in therapeutic efficacy of radiotherapy when applied with Au NPs, which is more evident particularly at 42 Gy (68 keV). In addition, they observed a synergistic effect when hyperthermia and Au NP assisted-radiotherapy were combined. They demonstrated that tumor control was obtained at 23 Gy with a radiation-dose reduction factor of 2.4. In another study, the surface plasmonic resonance effect of Au NPs was exploited in order to produce heat, and radiotherapy was applied after intratumoral injection of Au NPs with a size of 15 nm (Hainfeld et al., 2014). Even though infrared radiation was used instead of NIR irradiation, a dose reduction factor of more than 3 (55 Gy was reduced to <15 Gy for 50% tumor control) was achieved. Therefore, they emphasized that Au NPs are promising agents for NP assisted-radiotherapy. However, these results need to be confirmed by clinical studies.

5 NANOMATERIALS IN TUMOR IMAGING OF ORAL CANCER

NPs may revolutionize the current imaging techniques, such as optical imaging, MRI, and positron emission tomography (PET), and may be helpful in increasing their selectivity for the detection of small size tumors and metastasis with fewer side effects.

5.1 OPTICAL IMAGING

QDs have attracted remarkable interest for several medical aspects of optical imaging, a bioimaging application, which utilizes nonionizing radiation especially between visible to NIR wavelengths in order to differentiate pathological and nonpathological tissues. In fact, it may be suitable particularly for in vivo tumor detection and operative applications, such as optical image guided surgery and SLN mapping.

Early tumor detection is one of the major challenges in cancer imaging. As a matter of fact, current imaging techniques, such as computerized tomography or MRI, have a capacity of detecting the tumors involving at least millions of cancerous cells, hence leading to a delay in cancer diagnosis. Yang et al. (2010a) tailored a peptide-conjugated NIR QDs with an emission wavelength of 800 nm (QD800) for early detection of oral cancer. They demonstrated that 10^4 cancerous cells were sufficient for visual detection using the fluorescence properties of QD, even though its intensity faded as tissue depth increased. Thereafter, same group conjugated anti-EGFR antibody to QD800 and reported promising results for the optical detection of cancerous cells in vivo (Yang et al., 2011).

In surgical oncology, total excision of a solid tumor is crucial for treatment success. Indeed, insufficient resections may lead to a high rate of recurrence, whereas, overresection may cause unnecessary violation of surrounding healthy tissues. Therefore, surgical margin is a cornerstone for the successful management of solid tumors, and a *razor's edge* for an oncology surgeon who generally designs a surgical strategy relying on preoperative imaging and also intraoperative visual assessment and palpation. In head and neck surgery, optical image-guided surgery using NPs particularly QDs, as they have several useful properties, such as long photobleaching period and high performance of deep tissue infrared imaging, may provide precise delineation of surgical margins and facilitate removal of the tumor without leaving residual tumor behind and avoiding aggressive surgeries (Ballou et al., 2005; Keereweer et al., 2012; Merian et al., 2012). In addition, QDs are considered an ideal candidate for SLN mapping (Ballou et al., 2007). However, the selection of ideal QD is critical and influenced by its size irrespective of its charge or surface chemistry. Thus, smaller QDs (especially <10 nm) may flow rapidly and surpass the target lymph nodes, thereby compromising the reliability of SLN mapping (Cassette et al., 2013). On the other hand, the flow of large NPs may be hindered and cause accumulation of NPs at the injection site. Of note, one of the major concerns for QD-based agents is toxicity particularly due to release of cadmium ions (Ambrosome et al., 2012). In order to overcome this drawback, QDs were loaded into polymersomes made of poly(dimethylsiloxane)-poly(2-methyloxazoline) diblock copolymers (Camblin et al., 2014) or cadmium-free QDs (Soenen et al., 2014) that are designed for biomedical applications, and surface coatings, such as PEG, are used in order to enhance their biocompatibility.

5.2 MAGNETIC RESONANCE IMAGING

Magnetic resonance imaging, a minimally invasive imaging technique that uses magnetic fields, is elusive for the detection of solid tumors with high spatial and temporal resolution. Unlike computerized tomography, ionizing radiation is not a matter of concern. It is highly effective for the determination of tumor size, depth, and volume, therefore invaluable for staging in oral cancer (Boland et al., 2013; Kwon et al., 2016; Park et al., 2011). A variety of contrast agents involving gadolinium-based, hyperpolarized, and iron oxide magnetic NPs are also administered in order to improve the visualization of targeted tissue by enhancing the image contrast (Brigger et al., 2002). Iron oxide magnetic NPs are good candidate for MRI, as they have a high magnetization capacity (Bulte and Kraitchman, 2004). They are classified according to their size as follows: (1) superparamagnetic (>50 nm) and (2) ultrasmall superparamagnetic (<50 nm) iron oxide NPs (SPIONs and USPIONs). In addition, they are generally coated in order to avoid agglomeration, mainly resulting from dipole interaction, and increase the bioavailability. Therefore, a typical SPION has a monomer- or polymer coated inorganic core [magnetite (Fe_3O_4) or maghemite (Fe_2O_3)] of iron oxide (Bennett et al., 2014; Kainz and Reiser, 2014). Superparamagnetic iron oxide NPs can cause signal reductions in T2-weighted sequences of MRI. However, a nontargeted SPION is generally entrapped by RES, therefore, it is only suitable for the imaging of lymphatic system, liver and spleen. Therefore, they are mainly used for the detection of hepatocellular carcinoma, cholangiosarcoma or liver metastases (Bonnemain, 1998; Corot et al., 2006; Ferrucci and Stark, 1990; Tanimoto and Kuribayashi, 2006). Currently, a couple of commercially available nontargeted SPIONs, such as Endorem or Feridex (dextran-coated iron oxide NPs with a diameter of 80–150 nm) and Resovist or Cliavist (carboxy-dextran coated iron oxide NPs with a diameter of 50–150 nm), and USPIONs, such as Sinerem or Combidex (dextran-coated iron oxide NPs with a diameter of 20–40 nm) are on the market (Ittrich et al., 2013).

Recently, targeted SPION-based imaging has a tremendous interest in preclinical settings. It is considered that diagnostic accuracy of MRI can be maximized and better visualization of different neoplasms can be achieved, as a SPION is targeted by a conjugation of a specific ligand, such as aptamer, antibody, or peptide. In current literature, successful conjugation of anti-EGFR to SPIONs has been reported and described as promising multifunctional contrast agent for different EGFR overexpressing cancers, such as brain tumors (Hadjipanayis et al., 2010; Mu et al., 2015), breast cancer (Chen et al., 2013b), cervix cancer (Yuan et al., 2013), and epidermoid carcinoma (Tseng et al., 2015) in vitro and in vivo.

5.3 POSITRON EMISSION TOMOGRAPHY

Positron emission tomography is a contemporary, minimally invasive imaging technique that examines the metabolic activity of tissues or organs using a specially tailored radioactive tracer generally fluorodeoxyglucose, an analog of glucose. The diagnostic value of PET has been documented in lymphoma (Pyo et al., 2013), brain tumors (Zhao et al., 2014a), lung cancer (Chien et al., 2013), renal cell carcinoma (Wang et al., 2012a), head and neck cancers (Kyzas et al., 2008; Gupta et al., 2011) and nasopharyngeal carcinoma (Shen et al., 2014). NPs may improve the diagnostic efficacy of PET and may be helpful for the combination of PET and therapy. In a xenograft head and neck cancer model, Xie et al. (2010) conjugated ^{64}Cu radionuclide to Au nanoshells and administered in order to determine the biodistribution in vivo. A high correlation between PET images and neutron activation analysis was detected. Therefore, they mentioned that Au nanoshells conjugated ^{64}Cu radionuclide may be useful for the guidance of PTT in head and neck cancer. Furthermore, the application of ^{64}Cu-porphysome NPs in PET imaging and intraoperative real-time visualization of solid tumors in head and neck cancer in an orthotopic head and neck cancer model (Muhanna et al., 2015).

5.4 NANOTHERANOSTICS IN ORAL CANCER

Theranosis is a relatively new term that refers to any agent that enables the integration of diagnosis and therapy simultaneously (Sumer and Gao, 2008; Warner, 2004). NPs may serve as an efficient platform for the development of a dual-purpose nanotheranostic agent regarding their aforementioned versatilities and unique physicochemical properties, such as nanoscale size, high surface-to-volume ratio, and optical character. Wang et al. (2012b) succinctly defined nanotheranostic as a *see and treat* strategy for the management of cancer by using biocompatible NPs. Therefore, an ideal nanotheranostic agent should have good biocompatibility and bioavailability, high tissue penetration and cellular uptake, longer blood circulation life span, selective targeting with minimal toxicity. Shapira et al. (2011) emphasized that a novel, multifunctional nanotheranostic must have four components: (1) a moiety for active targeting, (2) an imaging segment of primary and metastatic lesions, (3) anticancer drugs as payloads, and (4) a special chemosensitizing molecule for the neutralization of chemotherapy resistance. However, it is noteworthy that optimization of these factors is remarkably challenging, and a trade-off between the advantages and disadvantages of NPs should be performed rationally. A nanotheranostic that is specifically tailored for oral cancers has not been investigated in a clinical study yet. Melancon et al. (2014) developed an anti-EGFR (C225)-conjugated hollow gold nanospheres and successfully applied for imaging and selective thermal ablation of head and neck cancers. Lukianova-Hleb and Lapotko (2015) designed nanoquadrapeutics, which offered integration of four treatment modality simultaneously: (1) anti-EGFR conjugated colloidal gold, (2) liposomal drug, (3) laser pulse therapy, and (4) radiotherapy. Nanoquadrapeutics were highly efficient for the detection of microscopic residual tumors after tumor resection in vivo. In addition, tumor activity was increased 17-fold despite decreasing the drug (3% of clinical applications) and radiation doses (6% of clinical applications).

6 CONCLUSIONS

Oral cancer remains a lethal disease for more than 50% of cases diagnosed annually. Therefore, innovative diagnostic techniques with high specificity and sensitivity, and new treatment modalities with high therapeutic efficacy are required. Currently, several NPs have shown promising outcomes in preclinical settings. Hopefully, they will be used in clinical settings in the near future.

REFERENCES

Ai, L., Zhang, C., Chen, Z., 2011. Removal of methylene blue from aqueous solution by a solvothermal-synthesized graphene/magnetite composite. J. Hazard Mater. 192, 1515–1524.

Akhter, S., Ahmad, I., Ahmad, M.Z., Ramazani, F., 2013. Nanomedicines as cancer therapeutics: current status. Curr. Cancer Drug Target. 13, 362–378.

Albornoz, C., Jacobo, S.E., 2006. Preparation of a biocompatible magnetic film from an aqueous ferrofluid. J. Magn. Magn. Mater. 305, 12–15.

Allen, T.M., Cullis, P.R., 2004. Drug delivery systems: entering the mainstream. Science 303, 1818–1822.

Amagasa, T., Yamashiro, M., Uzawa, N., 2011. Oral premalignant lesions: from a clinical perspective. Int. J. Clin. Oncol. 16, 5–14.

Ambrosome, A., Mattera, L., Marchesano, V., Quarta, A., 2012. Mechanisms underlying toxicity induced by CdTe quantum dots determined in an invertebrate model organism. Biomaterials 33, 1991–2000.

Babaei, M., Ganjalikhani, M., 2014. The potential effectiveness of nanoparticles as radio sensitizers for radiotherapy. Bioimpacts 4 (1), 15–20.

Bagan, J.V., Scully, C., 2009. Recent advances in oral oncology 2008: squamous cell carcinoma aetiopathogenesis and experimental studies. Oral Oncol. 45, 45–48.

Bagan, J., Sarrion, G., Jimenez, Y., 2010. Oral cancer: clinical features. Oral Oncol. 46, 414–417.

Balandin, A.A., Ghosh, S., Bao, W., Calizo, I., Teweldebrhan, D., Miao, F., Lau, C.N., 2008. Superior thermal conductivity of single-layer graphene. Nano Lett. 8 (3), 902–907.

Ballou, B., Ernst, L.A., Waggoner, A.S., 2005. Fluorescence imaging of tumors in vivo. Curr. Med. Chem. 12, 795–805.

Ballou, B., Ernst, A., Andreko, S., Harper, T., 2007. Sentinel lymph node imaging using quantum dots in mouse tumors. Bioconjug. Chem. 18, 389–396.

Barenholz, Y., 2012. Doxil: the first FDA-approved nano-drug: lessons learned. J. Control. Release 160, 117–134.

Baronzio, G., Schwartz, L., Kiselevsky, M., Guais, A., 2012. Tumor interstitial fluid as modulator of cancer inflammation, thrombosis, immunity and angiogenesis. Anticancer Res. 32, 405–414.

Belle, C.J., Bonamin, A., Simon, U., Santoyo-Salazar, J., 2011. Size dependent gas sensing properties of spinel iron oxide nanoparticles. Sens. Actuators B 160, 942–950.

Bennett, K.M., Jo, J., Cabral, H., Bakalova, R., 2014. MR imaging techniques for nano-pathophysiology and theranostics. Adv. Drug Deliv. Rev. 74, 75–94.

Bertrand, N., Wu, J., Xu, X., Kamaly, N., 2014. Cancer nanotechnology: the impact of passive and active targeting in the era of modern cancer biology. Adv. Drug Deliv. Rev. 66, 2–25.

Boland, P.W., Watt-Smith, S.R., Hopper, C., Golding, S.J., 2013. Magnetic resonance imaging predicts survival and occult metastasis in oral cancer: a dual-centre, retrospective study. Br. J. Oral Maxillofac. Surg. 51, 696–701.

Bonnemain, B., 1998. Superparamagnetic agents in magnetic resonance imaging: physicochemical characteristics and clinical applications. J. Drug Target. 6, 167–174.

Bose, P., Klimowicz, A.C., Kornaga, E., Petrillo, S.K., 2012. Bax expression measured by AQUAnalysis is an independent prognostic marker in oral squamous cell carcinoma. BMC Cancer 12, 332.

Brigger, I., Dubernet, C., Couvreur, P., 2002. Nanoparticles in cancer therapy and diagnosis. Adv. Drug Deliv. Rev. 54, 631–651.

Brocklehurst, P., Kujan, O., O'Malley, L.A., Ogden, G., 2013. Screening programmes for the early detection and prevention of oral cancer. Cochrane Data. Syst. Rev. 11, CD004150.

Bulte, J.W., Kraitchman, D.L., 2004. Iron oxide MR contrast agents for molecular and cellular imaging. NMR Biomed. 17, 484–499.

Butterworth, K.T., McMahon, S.J., Taggart, L.E., Prise, K.M., 2013. Radiosensitization by gold nanoparticles: effective at megavoltage energies and potential role of oxidative stress. Transl. Cancer Res. 2 (4), 269–279.

Byrne, J.D., Betancourt, T., Brannon-Pappas, L., 2008. Active targeting schemes for nanoparticle systems in cancer therapeutics. Adv. Drug Deliv. Rev. 60, 1615–1626.

Cai, H., An, X., Cui, J., Li, J., 2003. Facile hydrothermal synthesis and surface functionalization of polyethyleneimine-coated iron oxide nanoparticles for biomedical applications. ACS Appl. Mater. Interf. 5, 1722–1731.

Camblin, M., Detampel, P., Kettiger, H., Wu, D., 2014. Polymersomes containing quantum dots for cellular imaging. Int. J. Nanomedicine 9, 2287–2298.

Canton, I., Battaglia, G., 2012. Endocytosis at the nanoscale. Chem. Soc. Rev. 41, 2718–2739.

Capaccio, P., Pruneri, G., Carboni, N., Pagliari, A.V., 2000. Cyclin D1 expression is predictive of occult metastases in head and neck cancer patients with clinically negative cervical lymph nodes. Head Neck 22, 234–240.

Carreras-Torras, C., Gay-Escoda, C., 2015. Techniques for early diagnosis of oral squamous cell carcinoma: systematic review. Med. Oral. Patol. Oral. Cir. Bucal. 20, e305–e315.

Cassette, E., Helle, M., Bezdetnaya, L., Marchal, F., 2013. Design of new quantum dot materials for deep tissue infrared imaging. Adv. Drug Deliv. Rev. 65, 719–731.

Chan, K.K., Glenny, A.M., Weldon, J.C., Furness, S., 2015. Interventions for treatment of oral and oropharyngeal cancers: targeted therapy and immunotherapy. Cochrane Data. Syst. Rev. 12, CD010341.

Chaturvedi, A.K., Anderson, W.F., Lortet-Tieulent, J., Curado, M.P., 2013. Worldwide trends in incidence rates for oral cavity and oropharyngeal cancers. J. Clin. Oncol. 31, 4550–4559.

Chen, F.H., Zhang, L.M., Chen, Q.T., Zhang, Y., 2010. Synthesis of a novel magnetic drug delivery system composed of doxorubicin-conjugated Fe_3O_4 nanoparticle cores and a PEGfunctionalized porous silica shell. Chem. Commun. 46, 8633–8635.

Chen, F.C., Zhang, F., Zhang, Z.J., Meng, S.Y., 2013a. Tumor necrosis factor-α gene polymorphisms and risk of oral cancer: evidence from a meta-analysis. Asian Pac. J. Cancer Prev. 14, 7243–7249.

Chen, H., Wang, L., Yu, Q., Qian, W., 2013b. Anti-HER2 antibody and ScFvEGFR-conjugated antifouling magnetic iron oxide nanoparticles for targeting and magnetic resonance imaging of breast cancer. Int. J. Nanomedicine 8, 3781–3794.

Cheng, R., Meng, F., Deng, C., Klok, H.A., 2013. Dual and multi-stimuli responsive polymeric nanoparticles for programmed site-specific drug delivery. Biomaterials 34, 3647–3657.

Chien, C.R., Liang, J.A., Chen, J.H., Wang, H.N., 2013. Fluorodeoxyglucose-positron emission tomography screening for lung cancer: a systematic review and meta-analysis. Cancer Imag. 13, 458–465.

Chinn, S.B., Myers, J.N., 2015. Oral cavity carcinoma: current management, controversies, and future directions. J. Clin. Oncol. 33, 3269–3276.

Christensen, M.E., Therkildsen, M.H., Hansen, B.L., Albeck, H., 1992. Epidermal growth factor receptor expression on oral mucosa dysplastic epithelia and squamous cell carcinomas. Eur. Arch. Otorhinol. 249, 243–247.

Chung, C.H., Ely, K., McGavran, L., Varella-Garcia, M., 2006. Increased epidermal growth factor receptor gene copy number is associated with poor prognosis in head and neck squamous cell carcinomas. J. Clin. Oncol. 24, 4170–4176.

Clinicaltrials.gov, 2010. Pilot study of AuroLaser therapy in refractory and/or recurrent tumors of the head and neck. National Institute of Health. Available from: http://clinicaltrials.gov/ct2/show/NCT00848042.

Cooper, D.R., 2014. Gold nanoparticles and their alternatives for radiation therapy enhancement. Front Chem. 2, 86.

Corot, C., Robert, P., Idee, J.M., Port, M., 2006. Recent advances in iron oxide nanocrystal technology for medical imaging. Adv. Drug Deliv. Rev. 58, 1471–1504.

Corvo, R., 2007. Evidence-based radiation oncology in head and neck squamous cell carcinoma. Radiother. Oncol. 85, 156–170.

Curry, J.M., Sprandio, J., Cognetti, D., Luginbuhl, A., 2014. Tumor microenvironment in head and neck squamous cell carcinoma. Semin. Oncol. 41, 217–234.

da Silva, S.D., Ferlito, A., Takes, R.P., Brakenhoff, R.H., 2011. Advances and applications of oral cancer basic research. Oral Oncol. 47, 783–791.

Damascelli, B., Patelli, G.L., Lanocita, R., Di Tolla, G., 2003. A novel intraarterial chemotherapy using paclitaxel in albumin nanoparticles to treat advanced squamous cell carcinoma of the tongue: preliminary findings. AJNR Am. J. Roentgenol. 181, 253–260.

Das, B.R., Nagpal, J.K., 2002. Understanding the biology of oral cancer. Med. Sci. Monit. 8, 258–267.

Dickerson, E.B., Dreaden, E.C., Huang, X., El-Sayed, I.H., 2008. Gold nanorod assisted near-infrared plasmonic photothermal therapy (PPTT) of squamous cell carcinoma in mice. Cancer Lett. 269, 57–66.

Eatemadi, A., Daraee, H., Karimkhanloo, H., Kouhi, M., 2014. Carbon nanotubes: properties, synthesis, purification, and medical applications. Nano. Res. Lett. 9, 393.

Edefonti, V., Hashibe, M., Parpinel, M., Turati, F., 2015. Natural vitamin C intake and the risk of head and neck cancer: a pooled analysis in the International Head and Neck Cancer Epidemiology Consortium. Int. J. Cancer 137, 448–462.

El-Sayed, I.H., Huang, X., El-Sayed, M.A., 2005. Surface plasmon resonance scattering and absorption of anti-EGFR antibody conjugated gold nanoparticles in cancer diagnostics: applications in oral cancer. Nano Lett. 5, 829–834.

El-Sayed, I.H., Huang, X., El-Sayed, M.A., 2006. Selective laser photo-thermal therapy of epithelial carcinoma using anti-EGFR antibody conjugated gold nanoparticles. Cancer Lett. 239, 129–135.

El-Sayed, I., Huang, X., Macheret, F., Humstoe, J.O., 2007. Effect of plasmonic gold nanoparticles on benign and malignant cellular autofluorescence: a novel probe for fluorescence based detection of cancer. Technol. Cancer Res. Treat. 6, 403–412.

Endo, K., Ueno, T., Kondo, S., Wakisaka, N., 2013. Tumor-targeted chemotherapy with the nanopolymer-based drug NC-6004 for oral squamous cell carcinoma. Cancer Sci. 104, 369–374.

Eskiizmir, G., 2015. Tumor microenvironment in head and neck squamous cell carcinoma. Turk. Arch. Otorhinolaryngol. 53, 120–127.

Fang, J., Sawa, T., Maeda, H., 2003. Factors and mechanism of "EPR" effect and the enhanced antitumor effects of macromolecular drugs including SMANCS. Adv. Exp. Med. Biol. 519, 29–49.

Fasunla, A.J., Greene, B.H., Timmesfeld, N., Wiegand, S., 2011. A meta-analysis of the randomized controlled trials on elective neck dissection versus therapeutic neck dissection in oral cavity cancers with clinically node-negative neck. Oral. Oncol. 47, 320–324.

Ferrucci, J.T., Stark, D.D., 1990. Iron oxide-enhanced MR imaging of the liver and spleen: review of the first 5 years. AJR Am. J. Roentgenol. 155, 943–950.

Fialka, F., Gruber, R.M., Hitt, R., Opitz, L., 2008. CPA6, FMO2, LGI1, SIAT1 and TNC are differentially expressed in early- and late-stage oral squamous cell carcinoma—a pilot study. Oral. Oncol. 44, 941–948.

Fleige, E., Quadir, M.A., Haag, R., 2012. Stimuli-responsive polymeric nanocarriers for the controlled transport of active compounds: concepts and applications. Adv. Drug Deliv. Rev. 64, 866–884.

Fonseca, N.A., Gregorio, A.C., Valerio-Fernandes, A., Simoes, S., 2014. Bridging cancer biology and the patients' needs with nanotechnology-based approaches. Cancer Treat. Rev. 40, 626–635.

Foy, J.P., Bertolus, C., William, Jr., W.N., Saintigny, P., 2013. Oral premalignancy: the roles of early detection and chemoprevention. Otolaryngol. Clin. North Am. 46, 579–597.

Frank, D., Tyagi, C., Tomar, L., Choonara, Y.E., 2014. Overview of the role of nanotechnological innovations in the detection and treatment of solid tumors. Int. J. Nanomed. 9, 589–613.

Gao, X., Cui, Y., Levenson, R.M., Chung, L.W.K., 2004. In vivo cancer targeting and imaging with semiconductor quantum dots. Nat. Biotechnol. 22, 969–976.

Global Burden of Disease Control Collaboration, 2015. The global burden of cancer 2013. JAMA Oncol. 1, 505–527.

Gomez-Navarro, C., Weitz, R.T., Bittner, A.M., Scolari, M., Burghard, M., Kern, K., 2007. Electronic transport properties of individual chemically reduced graphene oxide sheets. Nano Lett. 7, 3499–3503.

Gonzales-Weimuller, M., Zeisberger, M., Krishnan, K.M., 2009. Size-dependant heating rates of iron oxide nanoparticles for magnetic fluid hyperthermia. J. Magn. Magn. Mater. 321, 1947–1950.

Gregoire, V., Langendijk, J.A., Nuyts, S., 2015. Advances in radiotherapy for head and neck cancer. J. Clin. Oncol. 33, 3277–3284.

Gupta, T., Master, Z., Kannan, S., Agarwal, J.P., 2011. Diagnostic performance of post-treatment FDG PET or FDG PET/CT imaging in head and neck cancer: a systematic review and meta-analysis. Eur. J. Nucl. Med. Mol. Imag. 38, 2083–2095.

Gupta, S., Shah, J.S., Parikh, S., Limbdiwala, P., 2014. Clinical correlative study on early detection of oral cancer and precancerous lesions by modified oral brush biopsy and cytology followed by histopathology. J. Cancer Res. Ther. 10, 232–238.

Hadjipanayis, C.G., Machaidze, R., Kaluzova, M., Wang, L., 2010. EGFRvIII antibody-conjugated iron oxide nanoparticles for magnetic resonance imaging-guided convection-enhanced delivery and targeted therapy of glioblastoma. Cancer Res. 70, 6303–6312.

Hainfeld, J.F., Dilmanian, F.A., Slatkin, D.N., Smilowitz, H.M., 2008. Radiotherapy enhancement with gold nanoparticles. J. Pharm. Pharmacol. 60, 977–985.

Hainfeld, J.F., Dilmanian, F.A., Zhong, Z., Slatkin, D.N., 2010. Gold nanoparticles enhance the radiation therapy of a murine squamous cell carcinoma. Phys. Med. Biol. 55, 3045–3059.

Hainfeld, J.F., Lin, L., Slatkin, D.N., Dilmanian, F.A., 2014. Gold nanoparticle hyperthermia reduces radiotherapy dose. Nanomedicine 10, 1609–1617.

Hashibe, M., Mathew, B., Kuruvilla, B., Thomas, G., 2000. Chewing tobacco, alcohol, and the risk of erythroplakia. Cancer Epidemiol. Biomark. Prev. 9, 639–645.

Hashida, Y., Tanaka, H., Zhou, S., Kawakami, S., 2014. Photothermal ablation of tumor cells using a single-walled carbon nanotube–peptide composite. J. Control. Release 173, 59–66.

Hashizume, H., Baluk, P., Morikawa, S., McLean, J.W., 2000. Openings between defective endothelial cells explain leakiness. Am. J. Pathol. 156, 1363–1380.

Hatekayama, H., Akita, H., Harashima, H., 2011. A multifunctional envelope type nano device (MEND) for gene delivery to tumours based on the EPR effect: a strategy for overcoming the PEG dilemma. Adv. Drug Deliv. Rev. 63, 152–160.

Hatekayama, H., Akita, H., Harashima, H., 2013. The polyethyleneglycol dilemma: advantage and disadvantage of PEGylation of liposomes for systemic genes and nucleic acids delivery to tumors. Biol. Pharm. Bull. 36, 892–899.

Heldin, C.H., Rubin, K., Pietras, K., Ostman, A., 2004. High interstitial fluid pressure: an obstacle in cancer therapy. Nat. Rev. Cancer 4, 806–813.

Hervault, A., Thanh, N.T.K., 2014. Magnetic nanoparticles-based therapeutic agents for thermo-chemotherapy treatment of cancer. Nanoscale 6, 11553–11573.

Hu, J., Qian, Y., Wang, X., Liu, T., 2012. Drug-loaded and superparamagnetic iron oxide nanoparticle surface-embedded amphiphilic block copolymer micelles for integrated chemotherapeutic drug delivery and MR imaging. Langmuir 28, 2073–2082.

Huang, X., El-Sayed, M.A., 2010. Gold nanoparticles: optical properties and implementations in cancer diagnosis and photothermal therapy. J. Adv. Res. 1, 13–28.

Huang, X., El-Sayed, I.H., Qian, W., El-Sayed, M.A., 2006a. Cancer cell imaging and photothermal therapy in the near-infrared region by using gold nanorods. J. Am. Chem. Soc. 128, 2115–2120.

Huang, X., Jain, P.K., El-Sayed, I.H., El-Sayed, M.A., 2006b. Determination of the minimum temperature required for selective photothermal destruction of cancer cells with the use of immunotargeted gold nanoparticles. Photochem. Photobiol. 82, 412–417.

Huang, X., Qian, W., El-Sayed, I.H., El-Sayed, M.A., 2007a. The potential use of the enhanced nonlinear properties of gold nanospheres in photothermal cancer therapy. Lasers Surg. Med. 39, 747–753.

Huang, X., Jain, P.K., El-Sayed, I.H., El-Sayed, M.A., 2007b. Gold nanoparticles: interesting optical properties and recent applications in cancer diagnostics and therapy. Nanomedicine 2, 681–693.

Huang, S.F., Cheng, S.D., Chien, H.T., Liao, C.T., 2012. Relationship between epidermal growth factor receptor gene copy number and protein expression in oral cavity squamous cell carcinoma. Oral Oncol. 48, 67–72.

Hui, C., Shen, C., Yang, T., Bao, L., 2008. Large-scale Fe_3O_4 nanoparticles soluble in water synthesized by a facile method. J. Phys. Chem. C 112, 11336–11339.

Hummers, W.S., Offeman, R.E., 1958. Preparation of graphitic oxide. J. Am. Chem. Soc. 80, 1339.

Ittrich, H., Peldschus, K., Raabe, N., Kaul, M., 2013. Superparamagnetic iron oxide nanoparticles in biomedicine: applications and developments in diagnostics and therapy. Rofo 185, 1149–1166.

Jain, R.K., 1994. Barriers to drug delivery in solid tumors. Sci. Am. 271, 58–65.

Jain, K.K., 2007. Applications of nanobiotechnology in clinical diagnostics. Clin. Chem. 53, 2002–2009.

Jain, R.K., Tong, R.T., Munn, L.L., 2007. Effect of vascular normalization by antiangiogenic therapy on interstitial hypertension, peritumor edema, and lymphatic metastasis: insights from a mathematical model. Cancer Res. 67, 2729–2735.

Jain, P.K., Huang, X., El-Sayed, I.H., El-Sayed, M.A., 2008. Noble metals on the nanoscale: optical and photothermal properties and some applications in imaging, sensing, biology, and medicine. Acc. Chem. Res. 41, 1578–1586.

Jain, S., Hirst, D.G., O'Sullivan, J.M., 2012. Gold nanoparticles as novel agents for cancer therapy. Br. J. Radiol. 85, 101–113.

Jan Prasek, J., Drbohlavova, J., Chomoucka, J., Hubalek, J., 2011. Methods for carbon nanotubes synthesis—review. J. Mater. Chem. 21, 1587–15884.

Jin, R., Lin, B., Li, D., Ai, H., 2014. Superparamagnetic iron oxide nanoparticles for MR imaging and therapy: design considerations and clinical applications. Curr. Opin. Pharmacol. 18C, 18–27.

Jolivet, J.P., Tronc, É., Chaneac, C., 2002. Synthesis of iron oxide-based magnetic nanomaterials and composites. CR Chim. 5, 659–664.

Jung, H.S., Kong, W.H., Sung, D.K., Lee, M.Y., 2014. Nanographene oxide-hyaluronic acid conjugate for photothermal ablation therapy of skin cancer. ACS Nano 8, 260–268.

Kah, J.C., Kho, K.W., Lee, C.G., James, C., 2007. Early diagnosis of oral cancer based on the surface plasmon resonance of gold nanoparticles. Int. J. Nanomed. 2, 785–798.

Kainz, Q.M., Reiser, O., 2014. Polymer- and dendrimer-coated magnetic nanoparticles as versatile supports for catalysts, scavengers, and reagents. Acc. Chem. Res. 47, 667–677.

Keereweer, S., Sterenborg, H.J., Kerrebijn, J.D., Van Driel, P.B., 2012. Image-guided surgery in head and neck cancer: current practice and future directions of optical imaging. Head Neck 34, 120–126.

Kima, E.H., Leea, H.S., Kwakb, B.K., Kimc, B.K., 2005. Synthesis of ferrofluid with magnetic nanoparticles by sonochemical method for MRI contrast agent. J. Magn. Magn. Mater. 289, 328–330.

Kirtane, A.R., Kalscheuer, S.M., Panyam, J., 2013. Exploiting nanotechnology to overcome tumor drug resistance: challenges and opportunities. Adv. Drug Deliv. Rev. 65, 1731–1747.

Kwatra, D., Venugopal, A., Anant, S., 2013. Nanoparticles in radiation therapy: a summary of various approaches to enhance radiosensitization in cancer. Transl. Cancer Res. 2, 330–342.

Kwon, M., Moon, H., Nam, S.Y., Lee, J.H., et al., 2016. Clinical significance of three-dimensional measurement of tumour thickness on magnetic resonance imaging in patients with oral tongue squamous cell carcinoma. Eur. Radiol. 26 (3), 858–865.

Kyzas, P.A., Evangelou, E., Denaxa-Kyza, D., Ioannidis, J.P., 2008. 18F-fluorodeoxyglucose positron emission tomography to evaluate cervical node metastases in patients with head and neck squamous cell carcinoma: a meta-analysis. J. Natl. Cancer Inst. 100, 712–720.

Lapotko, D., 2009. Therapy with gold nanoparticles and lasers: what really kills the cells? Nanomedicine 4, 253–256.

Lee, K.S., El-Sayed, M.A., 2008. Gold and silver nanoparticles in sensing and imaging: sensitivity of plasmon response to size, shape, and metal composition. J. Phys. Chem. B 110, 19220–19225.

Lee, C., Wei, X.D., Kysar, J.W., Hone, J., 2008. Measurement of the elastic properties and intrinsic strength of monolayer graphene. Science 321, 385–388.

Li, C., Li, D., Wan, G., Xu, J., 2011. Facile synthesis of concentrated gold nanoparticles with low size-distribution in water: temperature and pH controls. Nanoscale Res. Lett. 6, 440.

Li, J., Gong, C., Feng, X., Zhou, X., 2012a. Biodegradable thermosensitive hydrogel for SAHA and DDP delivery: therapeutic effects on oral squamous cell carcinoma xenografts. PLoS One 7, e33860.

Li, Y., Wang, J., Wientjes, M.G., Au, J.L., 2012b. Delivery of nanomedicines to extracellular and intracellular compartments of a solid tumor. Adv. Drug Deliv. Rev. 64, 29–39.

Li, L., Jiang, W., Luo, K., Song, H., 2013. Superparamagnetic iron oxide nanoparticles as MRI contrast agents for non-invasive stem cell labeling and tracking. Theranostic 3, 595–615.

Li, J.L., Hou, X.L., Bao, H.C., Sun, L., 2014. Graphene oxide nanoparticles for enhanced photothermal cancer cell therapy under the irradiation of a femtosecond laser beam. J. Biomed. Mater. Res. A 102, 2181–2188.

Liang, H., Tian, H., Deng, M., Chen, X., 2015. Gold nanoparticles for cancer theranostics. Chin. J. Chem. 33, 1001–1010.

Lima-Tenórioa, M.K., Pineda, E.A.G., Ahmad, N.M., Fessi, H., 2015. Magnetic nanoparticles: In vivo cancer diagnosis and therapy. Int. J. Pharm. 493, 313–327.

Lin, Z., Liu, Y., Ma, X., Hu, S., 2015. Photothermal ablation of bone metastasis of breast cancer using PEGylated multi-walled carbon nanotubes. Sci. Rep. 5, 11709.

Liu, G., Shen, H., Mao, J., et al., 2013. Transferrin modified graphene oxide for glioma-targeted drug delivery: in vitro and in vivo evaluations. ACS Appl. Mater. Interfaces. 5, 6909–6914.

Liu, Z., Robinson, J.T., Sun, X., Dai, H., 2008. PEGylated nanographene oxide for delivery of water-insoluble cancer drugs. J. Am. Chem. Soc. 130, 10876–10877.

Longmire, M., Choyke, P.L., Kobayashi, H., 2008. Clearance properties of nano-sized particles and molecules as imaging agents: considerations and caveats. Nanomedicine 3, 703–717.

Lu, A.H., Salabas, E.L., Schüth, F., 2007. Magnetic nanoparticles: synthesis, protection, functionalization, and application. Angew. Chem. Int. Ed. 46, 1222–1244.

Lu, Y.J., Lin, C.W., Yang, H.W., Lin, K.J., 2014. Biodistribution of PEGylated graphene oxide nanoribbons and their application in cancer chemo-photothermal therapy. Carbon 74, 83–95.

Lukianova–Hleb, E.Y., Lapotko, D.O., 2015. Rapid detection and destruction of squamous cell carcinoma of the head and neck by nano-quadrapeutics. Head Neck 37, 1547–1555.

Lunt, S.J., Kalliomaki, T.M., Brown, A., Yang, V.X., 2008. Interstitial fluid pressure, vascularity and metastasis in ectopic, orthotopic and spontaneous tumours. BMC Cancer 8, 2.

Macey, R., Walsh, T., Brocklehurst, P., Kerr, A.R., Lingen, M.W., 2015. Diagnostic tests for oral cancer and potentially malignant disorders in patients presenting with clinically evident lesions. Cochrane Data. Syst. Rev. 5, CD010276.

MacKey, M.A., Ali, M.R., Austin, L.A., Near, R.D., 2014. The most effective gold nanorod size for plasmonic photothermal therapy: theory and in vitro experiments. J. Phys. Chem. B 118, 1319–1326.

Maeda, H., 2001. SMANCS and polymer-conjugated macromolecular drugs: advantages in cancer chemotherapy. Adv. Drug Deliv. Rev. 46, 169–185.

Maeda, H., Matsumura, Y., 1989. Tumoritropic and lymphotropic principles of macromolecular drugs. Crit. Rev. Ther. Drug Carrier Syst. 6, 193–210.

Maeda, H., Sawa, T., Konno, T., 2001. Mechanism of tumor-targeted delivery of macromolecular drugs, including the EPR effect in solid tumor and clinical overview of the prototype polymeric drug SMANCS. J. Control. Release 74, 47–61.

Maeda, H., Fang, J., Inutsuka, T., Kitamoto, Y., 2003. Vascular permeability enhancement in solid tumor: various factors, mechanisms involved and its implications. Int. Immunopharmacol. 3, 319–328.

Malik, M.A., Wani, M.Y., Hashim, M.A., 2012. Microemulsion method: a novel route to synthesize organic and inorganic nanomaterials: 1st nano update. Arab. J. Chem. 5, 397–417.

Mallick, S., Patil, R., Gyanchandani, R., Pawar, S., 2009. Human oral cancers have altered expression of Bcl-2 family members and increased expression of the anti-apoptotic splice variant of Mcl-1. J. Pathol. 217, 389–407.

Marcano, D.C., Kosynkin, D.V., Berlin, J.M., Sinitskii, A., 2010. Improved synthesis of graphene oxide. ACS Nano 4, 4806–4814.

Markovic, Z.M., H-Trajkovic, L.M., T-Markovic, B.M., Kepic, D.P., 2011. In vitro comparison of the photothermal anticancer activity of graphene nanoparticles and carbon nanotubes. Biomaterials 32, 1121–1129.

McDonald, D.M., Foss, A.J., 2000. Endothelial cells of tumor vessels: abnormal but not absent. Cancer Metastasis. Rev. 19, 109–120.

Mehta, F.S., Hamner, J.E. III, 1971. Report on investigation of oral cancer and precancerous conditions in Indian rural populations, 1966–69. Munksgaard, Copenhagen.

Melancon, M.P., Zhou, M., Zhang, R., Xiong, C., 2014. Selective uptake and imaging of aptamer- and antibody-conjugated hollow nanospheres targeted to epidermal growth factor receptors overexpressed in head and neck cancer. ACS Nano 8, 4530–4538.

Merian, J., Gravier, J., Navarro, F., Texier, I., 2012. Fluorescent nanoprobes dedicated to in vivo imaging: from preclinical validations to clinical translation. Molecules 17, 5564–5591.

Messadi, D.V., Wilder-Smith, P., Wolinsky, L., 2009. Improving oral cancer survival: the role of dental providers. J. Calif. Dent. Assoc. 37, 789–798.

Miao, W., Shim, G., Lee, S., Lee, S., 2013. Safety and tumor accumulation of pegylated graphene oxide nanosheets for co-delivery of anticancer drug and photosensitizer. Biomaterials 34, 3402–3410.

Michel, C.C., Curry, F.E., 1999. Microvascular permeability. Physiol. Rev. 79, 703–761.

Mishra, M., Mohanty, J., Sengupta, S., Tripathy, S., 2005. Epidemiological and clinicopathological study of oral leukoplakia. Indian J. Dermatol. Venereol. Leprol. 71, 161–165.

Moon, K.H., Lee, S.H., Choi, H.C., 2009. In vivo near-infrared mediated tumor destruction by photothermal effect of carbon nanotubes. ACS Nano 3, 3707–3713.

Mu, K., Zhang, S., Ai, T., Jiang, J., 2015. Monoclonal antibody-conjugated superparamagnetic iron oxide nanoparticles for imaging of epidermal growth factor receptor-targeted cells and gliomas. Mol. Imag. 14, 14.

Muhanna, N., MacDonald, T.D., Chan, H., Jin, C.S., 2015. Multimodal nanoparticle for primary tumor delineation and lymphatic metastasis mapping in a head-and-neck cancer rabbit model. Adv. Health. Mater. 4, 2164–2169.

Muthu, M.S., Rajesh, C.V., Mishra, A., Singh, S., 2009. Stimulus-responsive targeted nanomicelles for effective cancer therapy. Nanomedicine 4, 657–667.

Nethaji, S., Sivasamy, A., Mandal, A.B., 2013. Preparation and characterization of corn cob activated carbon coated with nano-sized magnetite particles for the removal of Cr(VI). Biores. Technol. 134, 94–100.

Neves, L.V.V., Krais, J.J., Van Rite, B.D., Ramesh, R., 2013. Targeting single-walled carbon nanotubes for the treatment of breast cancer using photothermal therapy. Nanotechnology 24, 375104.

Nichols, J.W., Bae, Y.H., 2014. EPR: evidence and fallacy. J. Control. Release 190, 451–464.

Nidome, Y., Nishioka, K., Kawasaki, H., Yamada, S., 2003. Rapid synthesis of gold nanorods by the combination of chemical reduction and photoirradiation processes; morphological changes depending on the growing processes. Chem. Commun. 18, 2376–2377.

Nobili, S., Landini, S., Giglioni, B., Mini, E., 2006. Pharmacological strategies for overcoming multidrug resistance. Curr. Drug Targets. 7, 861–879.

Obonyo, O., Fisher, E., Edwards, M., Douroumis, D., 2010. Quantum dots synthesis and biological applications as imaging and drug delivery systems. Crit. Rev. Biotechnol. 30, 283–301.

Olaleye, O., Ekrikpo, O., Lyne, O., Wiseberg, J., 2015. Incidence and survival trends of lip, intra-oral cavity and tongue base cancers in South-East England. Ann. R. Coll. Surg. Engl. 97, 229–234.

Paradise, M., Goswami, T., 2007. Carbon nanotubes: production and industrial applications. Mater. Des. 28, 1477–1489.

Park, J.O., Jung, S.L., Joo, Y.H., Jung, C.K., 2011. Diagnostic accuracy of magnetic resonance imaging (MRI) in the assessment of tumor invasion depth in oral/oropharyngeal cancer. Oral Oncol. 47, 381–386.

Peracchia, M.T., Fattal, E., Desmaele, D., Besnard, M., 1999. Stealth PEGylated polycyanoacrylate nanoparticles for intravenous administration and splenic targeting. J. Control. Release 60, 121–128.

Perez-Sayans, M., Suarez-Penaranda, J.M., Padin-Iruegas, M.E., Gayoso-Diz, P., 2015. The loss of p16 expression worsens the prognosis of OSCC. Appl. Immunohistochem. Mol. Morphol. 23, 724–732.

Perez-Sayansm, M., Somoza-Martin, J.M., Barros-Anqeira, F., Renoiras-Lopez, M.D., 2010. Exfoliative cytology for diagnosing oral cancer. Biotech. Histochem. 85, 177–187.

Pitsillides, C.M., Joe, E.K., Wei, X., Anderson, R.R., 2003. Selective cell targeting with light-absorbing microparticles and nanoparticles. Biophys. J. 84, 4023–4032.

Poeta, M.I., Manola, J., Goldwasser, M.A., Forastiere, A., 2007. TP53 mutations and survival in squamous-cell carcinoma of the head and neck. N. Engl. J. Med. 357, 2552–2561.

Pyo, J., Won Kim, K., Jacene, H.A., Sakellis, C.G., 2013. End-therapy positron emission tomography for treatment response assessment in follicular lymphoma: a systematic review and meta-analysis. Clin. Cancer Res. 19, 6566–6577.

Raju, B., Haug, S.R., Ibrahim, S.O., Heyeraas, K.J., 2008. High interstitial fluid pressure in rat tongue cancer is related to increased lymph vessel area, tumor size, invasiveness and decreased body weight. J. Oral Pathol. Med. 37, 137–144.

Retif, P., Pinel, S., Toussaint, M., Frochot, C., 2015. Nanoparticles for radiation therapy enhancement: the key parameters. Theranostics 5, 1030–1044.

Rink, J.S., Plebanek, M.P., Tripathy, S., Thaxton, C.S., 2013. Update on current and potential nanoparticle cancer therapies. Curr. Opin. Oncol. 25, 646–651.

Robinson, J.T., Tabakman, S.M., Liang, Y., Wang, H., 2011. Ultrasmall reduced graphene oxide with high near-infrared absorbance for photothermal therapy. J. Am. Chem. Soc. 133, 6825–6831.

Rössle, M., Weber, C.S., Züllig, L., Graf, N., 2013. EGFR expression and copy number changes in low T-stage oral squamous cell carcinomas. Histopathology 63, 271–278.

Rubin Grandis, J., Melhem, M.F., Gooding, W.E., Day, R., 1998. Levels of TGF-alpha and EGFR protein in head and neck squamous cell carcinoma and patient survival. J. Natl. Cancer Inst. 90, 824–832.

Rutkowski, J.M., Swartz, M.A., 2007. A driving force for change: interstitial flow as a morphoregulator. Trends Cell. Biol. 17, 44–50.

Salunkhe, A.B., Khot, V.M., Pawar, S.H., 2014. Magnetic hyperthermia with magnetic nanoparticles: a status review. Curr .Top Med. Chem. 14, 572–594.

Sarin, H., 2010. Physiologic upper limits of pore size of different blood capillary types and another perspective on the dual pore theory of microvascular permeability. J. Angiogen. Res. 2, 14.

Satish, B., Venkateswara, R.K., Shilpa, C.C.H., Tejaswi, T., 2013. Synthesis and characterization of graphene oxide and its antimicrobial activity against *Klebseilla* and *Staphylococcus*. Int. J. Adv. Biotechnol. Res 4, 142–146.

Schepman, K.P., van der Meii, E.H., Smeele, L.E., van der Waal, I., 1996. Prevalance study of oral white lesions with special reference to a new definition of oral leucoplakia. Eur. J. Cancer B. Oral Oncol. 32B, 416–419

Schilling, C., Stoeckli, S.J., Haerle, S.K., Broglie, M.A., 2015. Sentinel European Node Trial (SENT): 3-year results of sentinel node biopsy in oral cancer. Eur. J. Cancer 51, 2777–2784.

Sciubba, J.J., 1999. Improving detection of precancerous and cancerous oral lesions. Computer-assisted analysis of the oral brush biopsy. U.S. Collaborative OralCDx Study Group. J. Am. Dent. Assoc. 130, 1445–1457.

Scully, C., Bagan, J.V., 2007. Recent advances in oral oncology. Oral Oncol. 43, 107–115.

Scully, C., Bagan, J., 2009. Oral squamous cell carcinoma overview. Oral Oncol. 45, 301–308.

Seoane, J., Alvarez-Novoa, P., Gomez, I., Takkouche, B., 2016. Early oral cancer diagnosis: the Aarhus statement perspective. A systematic review and meta-analysis. Head Neck 38 (Suppl. 1), E2182–E2189.

Shapira, A., Livney, Y.D., Broxterman, H.J., Assaraf, Y.G., 2011. Nanomedicine for targeted cancer therapy: towards the overcoming of drug resistance. Drug Resist. Updat. 14, 150–163.

Shen, G., Zhang, W., Jia, Z., Wang, Q., 2014. Meta-analysis of diagnostic value of 18F-FDG PET or PET/CT for detecting lymph node and distant metastases in patients with nasopharyngeal carcinoma. Br. J. Radiol. 87, 20140296.

Silverman, S., Gorsky, M., 1990. Epidemiologic and demographic update in oral cancer: California and national data, 1973–1985. J. Am. Dent. Assoc. 120, 495–499.

Soenen, S.J., Manshian, B.B., Aubert, T., Himmelreich, U., 2014. Cytotoxicity of cadmium-free quantum dots and their use in cell bioimaging. Chem. Res. Toxicol. 27, 1050–1059.

Some, S., Gwon, A.R., Hwang, E., Bahn, H., 2014. Cancer therapy using ultrahigh hydrophobic drug-loaded graphene derivatives. Sci. Rep. 4, 6314.

Sousa, M.C., Alves, M.G., Souza, L.A., Brandao, A.A., 2014. Correlation of clinical, cytological and histological findings in oral squamous cell carcinomas. Oncol. Lett. 8, 199–802.

Stanicki, D., Elst, L.V., Muller, R.N., Laurent, S., 2015. Synthesis and processing of magnetic nanoparticles. Curr. Opt. Chem. Eng. 8, 7–14.

Stoller, M.D., Park, S.Z., Yanwu, A., Jinho, R., 2008. Graphene-based ultracapacitors. Nano Lett. 8, 3498–3502.

Storhoff, J.J., Elghanian, R., Mucic, R.C., Mirkin, C.A., 1998. One-pot colorimetric differentiation of polynucleotides with single base imperfections using gold nanoparticle probes. J. Am. Chem. Soc. 120, 1959–1964.

Sumer, B., Gao, J., 2008. Theranostic nanomedicine for cancer. Nanomedicine 3, 137–140.
Swartz, M.A., Fleury, M.E., 2007. Interstitial flow and its effects in soft tissues. Annu. Rev. Biomed. Eng. 9, 229–256.
Szabo, B., Nelhubel, G.A., Karpati, A., Kennessey, I., 2011. Clinical significance of genetic alterations and expression of epidermal growth factor receptor (EGFR) in head and neck squamous cell carcinomas. Oral Oncol. 47, 487–496.
Taghizadeh, B., Taranejoo, S., Monemian, S.S., Salehi Moghaddam, Z., 2015. Classification of stimuli-responsive polymers as anticancer drug delivery systems. Drug Deliv. 22, 145–155.
Tanimoto, A., Kuribayashi, S., 2006. Application of superparamagnetic iron oxide to imaging of hepatocellular carcinoma. Eur. J. Radiol. 58, 200–216.
Taoudi Benchekroun, M., Saintigny, P., Thomas, S.M., El-Nagger, A.K., 2010. Epidermal growth factor receptor expression and gene copy number in the risk of oral cancer. Cancer Prev. 3, 800–809.
Thomas, H., Coley, H.M., 2003. Overcoming multidrug resistance in cancer: an update on the clinical strategy of inhibiting P-glycoprotein. Cancer Control. 10, 159–165.
Tian, B., Wang, C., Zhang, S., Feng, L., Liu, Z., 2011. Photothermally enhanced photodynamic therapy delivered by nano-graphene oxide. ACS Nano 5, 7000–7009.
Tian, L., Pei, X., Zen, Y., et al., 2014. Functionalized nanoscale graphene oxide for high efficient drug delivery of cisplatin. J. Nanopart. Res. 16, 2709–2723.
Tsantoulis, P.K., Kastrinakis, N.G., Yourvas, A.D., Laskaris, G., Gorgoulis, V.G., 2007. Advances in the biology of oral cancer. Oral Oncol. 43, 523–534.
Tseng, S.H., Chou, M.Y., Chu, I.M., 2015. Cetuximab-conjugated iron oxide nanoparticles for cancer imaging and therapy. Int. J. Nanomed. 10, 3663–3685.
Valizadeh, A., Mikaeili, H., Samiei, M., Farkhani, S.M., 2012. Quantum dots: synthesis, bioapplications, and toxicity. Nanoscale Res. Lett. 7, 480.
Villa, A., Villa, C., Abati, S., 2011. Oral cancer and oral erythroplakia: an update and implication for clinicians. Aust. Dent. J. 56, 253–256.
Wang, Y., Hu, A., 2014. Carbon quantum dots: synthesis, properties and applications. J. Mater. Chem. C 2, 6921–6939.
Wang, H.Y., Ding, H.J., Chen, J.H., Chao, C.H., 2012a. Meta-analysis of the diagnostic performance of FDG-PET and PET/CT in renal cell carcinoma. Cancer Imag. 12, 464–474.
Wang, S., Kim, G., Lee, Y.E., Hah, H.J., 2012b. Multifunctional biodegradable polyacrylamide nanocarriers for cancer theranostics: a "see and treat" strategy. ACS Nano 6, 6843–6851.
Wang, B., Wang, J.H., Liu, Q., 2014a. Rose-Bengal-conjugated gold nanorods for in vivo photodynamic and photothermal oral cancer therapies. Biomaterials 35, 1954–1966.
Wang, H., Gu, W., Xiao, N., Ye, L., 2014b. Chlorotoxin-conjugated graphene oxide for targeted delivery of an anticancer drug. Int. J. Nanomedicine 9, 1433–1442.
Warnakulasuriya, S., 2009. Global epidemiology of oral and oropharyngeal cancer. Oral Oncol. 45, 309–316.
Warner, S., 2004. Diagnostics plus therapy = theranostics. Scientist 18, 38–39.
Weigum, S.E., Floriano, P.N., Redding, S.W., Yeh, C.K., 2010. Nano-bio-chip sensor platform for examination of oral exfoliative cytology. Cancer Prev. Res. 3, 518–528.
Weinstein, J.S., Varallyay, C.G., Dosa, E., Gahramanov, S., 2010. Superparamagnetic iron oxide nanoparticles: diagnostic magnetic resonance imaging and potential therapeutic applications in neurooncology and central nervous system inflammatory pathologies, a review. J. Cereb. Blood Flow Metab. 30, 15–35.
Wicki, A., Witzigmann, D., Balasubramnian, V., Huwyler, J., 2015. Nanomedicine in cancer therapy: challenges, opportunities, and clinical applications. J. Control. Release 200, 138–157.
Wu, W., He, Q., Jiang, C., 2008. Magnetic iron oxide nanoparticles: synthesis and surface functionalization strategies. Nanoscale Res. Lett. 3, 397–415.

Wu, M.C., Deokar, A.R., Liao, J.H., Shih, P.Y., 2013a. Graphene-based photothermal agent for rapid and effective killing of bacteria. ACS Nano 7, 1281–1290.

Wu, M., Frieboes, H.B., McDougall, S.R., Chaplain, M.A., 2013b. The effect of interstitial pressure on tumor growth: coupling with the blood and lymphatic vascular systems. J. Theor. Biol 320, 131–151.

Wu, C., He, Q., Zhu, A., Li, D., 2014. Synergistic anticancer activity of photo- and chemoresponsive nanoformulation based polylysine-functionalized graphene. ACS Appl. Mater. Interf. 6, 21615–21623.

Xie, H., Wang, Z.J., Bao, A., Goins, B., 2010. In vivo PET imaging and biodistribution of radiolabeled gold nanoshells in rats with tumor xenografts. Int. J. Pharm. 395, 324–330.

Yang, K., Cao, Y.A., Shi, C., Li, Z.G., 2010a. Quantum dot-based visual in vivo imaging for oral squamous cell carcinoma in mice. Oral Oncol. 46, 864–868.

Yang, K., Zhang, S., Zhang, G., Sun, X., 2010b. Graphene in mice: ultrahigh in vivo tumor uptake and efficient photothermal therapy. Nano Lett. 10, 3318–3323.

Yang, K., Zhang, F.J., Tang, H., Zhao, C., 2011. In-vivo imaging of oral squamous cell carcinoma by EGFR monoclonal antibody conjugated near-infrared quantum dots in mice. Int. J. Nanomed. 6, 1739–1745.

Yang, K., Wan, J., Zhang, S., Tian, B., 2012. The influence of surface chemistry and size of nanoscale graphene oxide on photothermal therapy of cancer using ultra-low laser power. Biomaterials 33, 2206–2214.

Yang, H.W., Lu, Y.J., Lin, K.J., Hsu, S.C., 2013. EGRF conjugated PEGylated nanographene oxide for targeted chemotherapy and photothermal therapy. Biomaterials 34, 7204–7214.

Yang, X., Yi, C., Luo, N., Gong, C., 2014. Nanomedicine to overcome cancer multidrug resistance. Curr. Drug. Metab. 15, 632–649.

Yasun, E., Kang, H., Erdal, H., Cansiz, S., 2013. Cancer cell sensing and therapy using affinity tag-conjugated gold nanorods. Interf. Focus 3, 20130006.

Ye, X., Zhang, J., Tan, Y., Chen, G., 2015. Meta-analysis of two computer-assisted screening methods for diagnosing oral precancer and cancer. Oral Oncol. 51, 966–975.

Yin, P.T., Shah, S., Chhowalla, M., Lee, K.B., 2015. Design, synthesis, and characterization of graphene – nanoparticle hybrid materials for bioapplications. Chem. Rev. 115, 2483–2531.

Ying, Y., Chang, S.S., Lee, C.L., Wang, C.R.C., 1997. Gold nanorods: electrochemical synthesis and optical properties. J. Phys. Chem. B 101, 6661–6664.

Yuan, F., Dellian, M., Fukumura, D., Leunig, M., 1995. Vascular permeability in a human tumor xenograft: molecular size dependence and cutoff size. Cancer Res. 55, 3752–3756.

Yuan, Y., Chen, S., Paunesku, T., Gleber, S.C., 2013. Epidermal growth factor receptor targeted nuclear delivery and high-resolution whole cell X-ray imaging of Fe_3O_4@TiO_2 nanoparticles in cancer cells. ACS Nano 7, 10502–10517.

Yuen, A.P., Ho, C.M., Tang, L.C., Cheung, W.Y., 2009. Prospective randomized study of selective neck dissection versus observation for N_0 neck of early tongue carcinoma. Head Neck 31, 765–772.

Zhang, H., Yee, D., Wang, C., 2008. Quantum dots for cancer diagnosis and therapy: biological and clinical perspectives. Nanomedicine 3, 83–91.

Zhang, Y., Ali, S.F., Dervishi, E., Xu, Y., 2010. Cytotoxicity effects of graphene and single-wall carbon nanotubes in neural phaeochromocytoma derived PC12 cells. ACS Nano 4, 3181e6.

Zhang, W., Guo, Z., Huang, D., Liu, Z., 2011. Synergistic effect of chemo-photothermal therapy using PEGylated graphene oxide. Biomaterials 32, 8555–8561.

Zhang, X.D., Wu, D., Shen, X., Chen, J., 2012. Size-dependent radiosensitization of PEG-coated gold nanoparticles for cancer therapy. Biomaterials 33, 6408–6419.

Zhao, C., Zhang, Y., Wang, J., 2014a. A meta-analysis on the diagnostic performance of (18)F-FDG and (11) C-methionine PET for differentiating brain tumors. AJNR Am. J. Neuroradiol. 35, 1058–1065.

Zhao, X., Zhao, H., Chen, Z., Lan, M., 2014b. Ultrasmall superparamagnetic iron oxide nanoparticles for magnetic resonance imaging contrast agent. J. Nanosci. Nanotechnol. 14, 210–220.

Zhao, H., Feng, H., Liu, D., Ji, N., 2015. Self-assembling monomeric nucleoside molecular nanoparticles loaded with 5-FU enhancing therapeutic efficacy against oral cancer. ACS Nano 9, 9638–9651.

Zhou, L., Zhou, L., Wei, S., Ge, X., Zhou, J., Jiang, H., Li, F., Shen, J., 2014. Combination of chemotherapy and photodynamic therapy using graphene oxide as drug delivery system. J. Photochem. Photobiol. B 135, 7–16.

Zhu, S., Zhen, H., Li, Y., Wang, P., 2014. PEGylated graphene oxide as a nanocarrier for podophyllotoxin. J. Nanopart. Res. 16, 2530.

Zwicke, G.L., Mansoori, G.A., Jeffery, C.J., 2012. Utilizing the folate receptor for active targeting of cancer nanotherapeutics. Nano Rev. 3, 1–9.

CHAPTER

THERAPEUTIC POTENTIAL OF SOLUBILIZED NANOLIGNIN AGAINST ORAL DISEASES

18

Hiroshi Sakagami*, Hong Sheng*, Toshikazu Yasui*, Kunihiko Fukuchi, Takaaki Oizumi[†], Hirokazu Ohno[‡], Masaji Yamamoto[‡], Toshiyuki Fukuda[§], Kayoko Kotohda[¶], Hidenobu Yoshida[¶], Taisei Kanamoto[††], Shigemi Terakubo[††], Hideki Nakashima[††]**

*Meikai University School of Dentistry, Saitama, Japan

**Showa University, Tokyo, Japan

[†]Daiwa Biological Research Institute Co., Ltd., Kanagawa, Japan

[‡]Maruzen Pharmaceuticals Co. Ltd., Tokyo, Japan

[§]Satoen Co., Ltd., Shizuoka, Japan

[¶]Cascade Resources Laboratory Co., Ltd., Nagano, Japan

[††]St. Marianna University School of Medicine, Kanagawa, Japan

CHAPTER OUTLINE

1 Studies of Nanostructure of Lignin 546
1.1 Structure of Lignin 546
1.2 Encounter to LCC 547
2 Biological Activity 549
2.1 In Vivo Study 549
2.2 In Vitro Study with Cultured Cells 550
2.3 Antiviral Activity 557
2.4 Synergism with Vitamin C 560
2.5 Anti-UV Activity 561
3 Efficient Utilization of Plant Resources by Alkaline Extraction 562
3.1 Alkaline Extraction of Leaf of *Sasa senanensis* Rehder 562
3.2 Alkaline Extraction of Licorice Root 563
3.3 Alkaline Extraction of Green Tea Leaves 563
3.4 Effect of $FeSO_4$ on Anti-HIV Activity of LCC 564
4 Utilization of Lignin and Its Components 564
4.1 Industrial Use: Dental Application 564
4.2 Utilization of Cellulose 566
4.3 Manufacture of Metal Nanoparticles 566
5 Clinical Application 567
5.1 Effect of lignin-vitamin C tablet on HSV-1-infection 567

5.2 Effect of SE on Oral Lichenoid Dysplasia 567
5.3 Antihalitosis Effect of SE-Containing Toothpaste 567
6 Future Direction 568
Acknowledgments 569
References 569

There are three representative polyphenols in the natural kingdom: lignin, tannin, and flavonoid. Lignin is linked to polysaccharides to form lignin–carbohydrate complex (LCC). Due to the structural complexity, compete structural determination of LCC has not been achieved yet. Surveys through PubMed revealed that the number of publications related to the nanostructure of lignin was only 0.1% of that of total nanostructures publications. The publication of biological processes of lignin, were few. In this chapter, we summarize previous publications of the structure and function of nanostructures of lignin, and discuss their applicability to oral diseases.

1 STUDIES OF NANOSTRUCTURE OF LIGNIN

1.1 STRUCTURE OF LIGNIN

Lignins, formed through phenolic oxidative coupling processes in the plant, are the major class of natural products present in the natural kingdom (Davin et al., 1997). Lignins are formed by the dehydrogenative polymerization of three monolignols: *p*-coumaryl, *p*-coniferyl, and sinapyl alcohols (Lewis and Yamamoto, 1990), produced from L-phenylalanine by general phenylpropanoid pathway (Emiliani et al., 2009). Some polysaccharides in the cell walls of lignified plants are linked to lignin to form lignin-carbohydrate complexes (LCCs).

The carbohydrate portion of LCC is composed of heterologous sugars: glucose, arabinose, mannose, galactose, fucose, and occasionally uronic acids, depending on the plant species. Varying ratios of polysaccharide to phenylpropenoids produces heterogeneity in the acidity, water solubility, ethanol insolubility, and molecular weight. The molecular weight of LCCs from bald cypress, birch, and rice straw distributed from 1.5 to 85 kDa (Azuma and Tetsuo, 1988). Lignins help provide mechanical support and defend against pathogens, as integral cell wall components.

Lignocellulosic biomass is a complex biopolymer that is primary composed of cellulose, hemicellulose, and lignin. Lignin, the main natural aromatic polymer, has a very complex and complicated structure, which depends on the separation method and plant species. There are several lignin isolation methods that use sulfuric acid (final 3%) (Klason) (Abdullah et al., 2013), thioglycolic acid (Moreira-Vilar et al., 2014), phosphoric acid (Zeng et al., 2014), or sodium hydroxide (Si et al., 2015). Acid-ferrous ion ($FeSO_4$)-assisted pretreatment increases solubilization and enzymatic digestion of both cellulose and xylan (Wei et al., 2011).

Depending on the types of extraction solvents used, the size and reactivity of the lignin fragments were considerably changed (Ten et al., 2014). Approximately 90% of lignin is burned for energy.

The kraft lignin particles showed a broad distribution ranging from large micron- to nanosized particles. Shearing reduced the size of nanolignin particles to less than 100 nm (Nair et al., 2014).

The molecular weight of the nanoparticles was reduced by exposure of lignins (wheat straw and Sarkanda grass) to ultrasonic irradiation, without any significant compositional and structural changes (Gilca et al., 2015).

When Kraft lignin was subjected to complete ethanol over an α-MoC_{1-x}/AC catalyst at 280°C, 25 most abundant liquid products (C6-C10 esters, alcohols, arenes, phenols, and benzyl alcohols) were produced without formation of oligomers nor char (Ma et al., 2014).

The chemical structure of untreated solid wood polymers (cellulose, hemicellulose, and lignin) prepared from different habitats can be distinguished with each other by solid-state NMR and IR spectroscopies (Santoni et al., 2015). Since the cell wall is highly resistant to enzymatic degradation, an efficient degradation of lignocellulosic plant material is crucial in plant conversion to energy. Scanning near-field optical microscopy revealed a segmented circumferential nanostructure of 100 nm range in the secondary cell walls of a softwood (spruce), a hardwood (beech), and a grass (bamboo). Changes in lignin distribution or composition, reflecting variations in lignification within the secondary cell wall significantly influence the SNOM signal (Keplinger et al., 2014).

1.2 ENCOUNTER TO LCC

Paying attention to folklore that the intake of hot water extract of pine cone *Pinus parviflora* Sieb. et Zucc. effectively cured gastroenterological tumors, Sakagami et al. (1986) has partially purified the substance that induced the differentiation of human myelogenous leukemic cells into maturing macrophages with gel filtration chromatography, using NBT-reducing activity as differentiation marker. When the active fraction was applied to DEAE-cellulose column, differentiation-inducing substance was tightly bound to the column. The differentiation-inducing substance was not be eluted from the column with 1 M NaCl, but could be eluted with alkaline solution (0.15 M NaOH) with the purification fold of 327, suggesting that it is highly acidic substance (Ikeda et al., 1988) (Fig. 18.1A). When this was applied to polyacrylamide gel electrophoresis, it migrated as single bands. Due to the lower yield (11%), further study of differentiation-inducing substance has been closed.

Using this information, we next isolated the antitumor polysaccharides in large scale. The pine cone was treated with ethanol to remove the sticky cytotoxic resin, extracted with hot water, and then alkaline solution (1% NaOH) (Sakagami et al., 1987). Polysaccharides in the hot water extract were precipitated by 86% ethanol, dialyzed against distilled water to remove low molecular-weight substances and then applied to DEAE-cellulose column chromatography. Neutral polysaccharide fraction (Fr. I) passed through the column, and then acidic polysaccharide fractions (Fr. II and III) rich in uronic acid were eluted from DEAE cellulose column chromatography with 0.5 and 2 M NaCl, respectively. The most acidic polysaccharides (Fr. VI and Fr. V) were eluted with 0.15 M NaOH (Fig. 18.1B). Fr. V was further separated into Fr. V-1 and Fr. V-2 on sepharose CL-4B gel filtration chromatography. Mice-bearing sarcoma-180 ascites tumor cells were treated with each fraction and the survival ratio was determined. The antitumor activity [evaluated by the survival ratio of the treated group to control group (T/C%)] increased with the acidity: Fr. I (T/C = 98%) < Fr. II (T/C = 118%) < Fr. III (T/C = 175%), Fr. IV (T/C = 151%) < Fr. V-2 (T/C = 172%) (Fig. 18.1B) (Sakagami et al., 1987). Higher antitumor activity was recovered by extraction of the residue with 1% NaOH and subsequent precipitation under acidic condition (pH 5) (Fr. VI) (T/C = 227%), and also by ethanol precipitation of the resultant supernatant (Fr. VII) (T/C = 282%). Both Frs. VI and VII showed broad distribution of molecular weight with a peak approximately at 10 kDa. Elemental analysis demonstrated the presence of carbon, hydrogen, and oxygen in all fractions. Gas chromatography demonstrated the presence of neutral sugar [arabinose (or fucose), mannose, galactose, and glucose] (11%–22% of dried weight) and uronic acid (1.7%–5.2%) in Fr. VI and Fr. VII

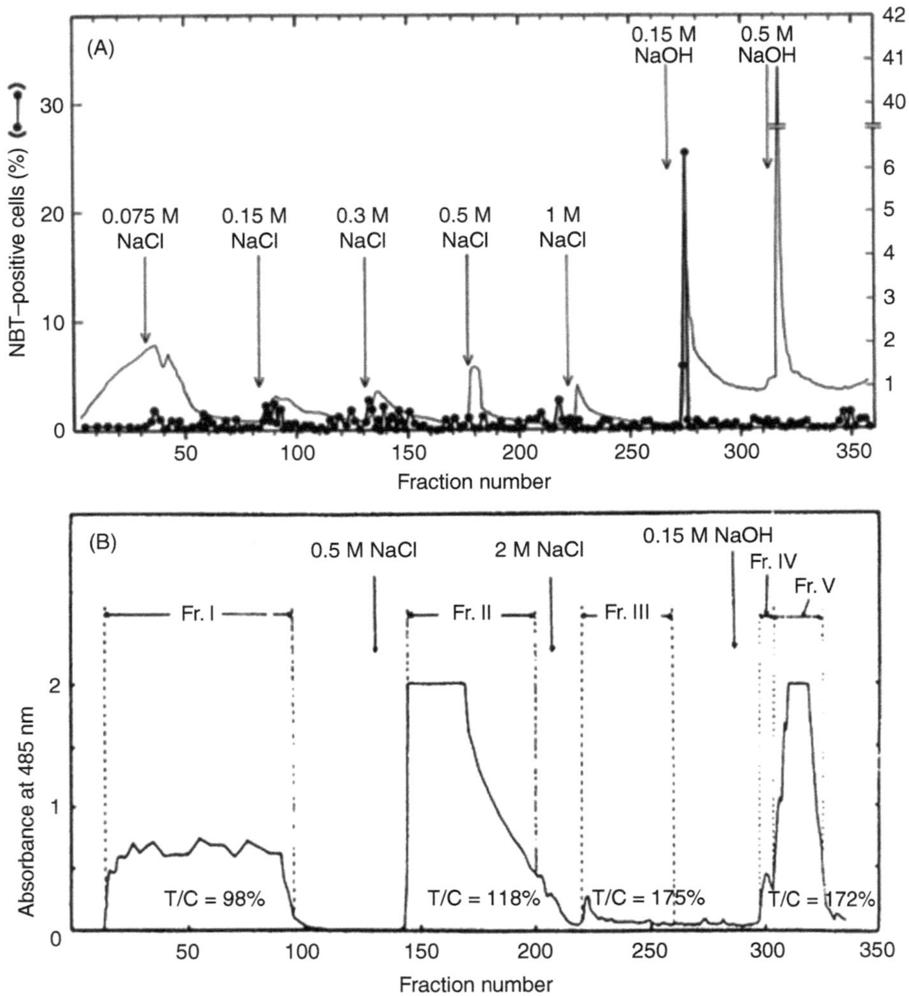

FIGURE 18.1 Separation of differentiation-inducing substances (A) and antitumor substances (B) on the DEAE Sepharose chromatography.

Cited from Ikeda, M., Sakagami, H., Konno, K. , 1988. Purification of differentiation-inducing substance present in pine cone extract (in Japanese). Showa Med. J. 48, 357–361; Sakagami, H., Ikeda, M., Unten, S., Takeda, K., Murayama, J., Hamada, A., Kimura, K., Komatsu, N., Konno, K., 1987. Antitumor activity of polysaccharide fractions from pine cone extract of Pinus parviflora *Sieb. et Zucc. Anticancer Res. 7, 1153–1159 with permission.*

(Sakagami et al., 1987). Frs. VI and VII, prepared from the pinecones of *Pinus parviflora* Sieb. et Zucc. showed higher antitumor activity in mice, extending their survival rates at much higher yield than those prepared from *Pinus densiflora* Sieb. and Zucc., *Pinus thunbergii* Parl., *Pinus elliottii* var. Elliotti, *Pinus taeda* L., *Pinus caribaea* var. Hondurenses, *Pinus sylvestris* L., or from the pine seed shells of *Pinus parviflora* Sieb. et Zucc., and *Pinus armandii Franch* (Harada et al., 1988). These data suggest that acidic polysaccharides (Frs. VI and VII) are responsible for the legendary antitumor potential of the pinecone of *Pinus parviflora* Sieb. et Zucc.

The most active acidic polysaccharide (Fr. IV) was identified as lignin-carbohydrate complex (LCC), based on (1) UV absorption spectra (minimum absorption at 260 nm, maximum absorption at 280 nm, broad maximum absorption at 500 nm), (2) IR spectra (3400 cm^{-1} for hydroxyl group with hydrogen bonding, 2700 cm^{-1} for aliphatic C–H, 1700~1600 cm^{-1} for carbonyl group conjugated to *p*-electron system, 1600, 1500 cm^{-1} for aromatic double bond, 1400~1000 cm^{-1} for C–O expansion and no ester bonding), (3) ESR (one strong signal at $g = 2.003$ under solid state at room temperature, signal intensity was significantly reduced by oxidation and reduction), and ^{1}H-NMR (in 0.2%NaOD-D_2O, δ 6.5~7.5 ppm for hydrogens in aromatic CH, 4.5~5.5 ppm for >C = C <(δ) and δ 3.0~4.0 ppm for –O–CH–CH; in $CDCl_3$ after acetylation by pyridine-acetic acid anhydride, δ 2.3 ppm for acetyl group bound to phenolic OH and δ 2.1 ppm for acetyl group bound to alcoholic OH, and (4) thin layer chromatograph (same Rf value of Fr. VI with commercial alkali-lignin in various solvent systems) (Oh-Hara et al., 1990; Sakagami et al., 1989).

2 BIOLOGICAL ACTIVITY

2.1 IN VIVO STUDY

2.1.1 Pharmacodynamics

Bioavailability of LCC was investigated by administration of the radioactive LCC to the mice. Radioactive LCC was prepared by incubating of pine cones LCC with Na[^{125}I], dialyzed against excess water, and lyophilized. Only 6% of orally administered [^{125}I] LCC appeared in the blood of mice after 3 h, and thereafter excreted into the urine and feces, indicating low levels of absorption of LCC through the stomach and intestine. Approximately 40% of intravenously administered the [^{125}I] LCC was accumulated in the liver with a maximum at 3–5 h after administration, and then mostly excreted into the urine (Sakagami et al., 1999). Intravenously administered LCC maintained its anti-HIV activity for a certain period of time (Shimizu et al., 1993).

2.1.2 Endogenous TNF production

Endogenous production of TNF was well established in mice. TNF thus produced has been reported to induce resistance against microbial infection (Nakane et al., 1988; Shemer-Avni et al., 1988). The optimal condition of endogenous production is significantly affected by concentrations of both priming and eliciting agents, and the interval between priming and eliciting injections. When mice were intravenously treated with eliciting agents, such as OK-432 (Picibanil) (Sakagami, 1990) or *Lactobacillus casei* (Sakagami et al., 1992c), TNF was induced in the blood, reaching a maximum level after 2–3 h, and then rapidly declined. When mice were primed with LCC (pine cone Frs. VI or VII) before elicitation, much higher level of endogenous TNF production and increased numbers of Kupffer cells (specialized macrophages located in the liver lining the walls of the sinusoids) were observed. The LCC-primed endogenous TNF production decreased during the aging of mice and upon the tumor implantation into the mice (Hanaoka et al., 1989).

2.1.3 Antimicrobial activity

Pretreatment of mice with LCC, prepared from cone or seed shell from various pine trees, induced antimicrobial activity against various microorganisms (*Staphylococcus aureus* SH10, *Escherichia coli*

GN2411, *Pseudomonas aeruginosa* H7, *Klebsiella pneumonia* ST101, *Candida albicans* YA2), but not *Salmonella enteritidis* 116-54 (Harada et al., 1988; Oh-Hara et al., 1990). Intraperitoneal administration of LCC transiently induced differentiation-inducing factor (DIF) with a peak at 1–2 h. The rapid decay of DIF activity from the peritoneal cavity was followed by polymorphonuclear cell (PMN) accumulation and enhancement of superoxide generation (detected by luminol-dependent chemiluminescence, LDCL) by peritoneal exudate cells. On the other hand, a neutral polysaccharide fraction from *Pinus parviflora* Sieb. et Zucc. that lacked any of these activities, did not induce PMN accumulation, DIF activity or LDCL generation. The results suggest a significant role of PMN activation for the expression of antimicrobial activity induced by LCC.

When the sugar moiety of LCC was destroyed by treatment with sulfuric acid or trifluoroacetic acid, the antimicrobial activity was significantly reduced. Furthermore, dehydrogenation polymers of phenylpropenoids, which do not contain sugar, showed much lower antimicrobial activity. These data suggest the importance of carbohydrate moiety of LCC for the induction of antimicrobial activity. We found that intraperitoneal administration of water-insoluble β-1,3-glucan (TAK) induces antimicrobial activity whereas water-soluble derivative (carboxymethyl TAK) was inactive, suggesting that insolubility nonspecifically induces antimicrobial activity. Therefore, it is necessary to dissolve completely to assess the biological activity of LCC.

Self-assembled poly(*N*-methylaniline)-lignosulfonate (PNMA-LS) composite spheres having reactive silver-ion adsorbability showed antibacterial activity against *E. coli* and *S. aureus* cells (Lu et al., 2013).

2.1.4 Antiparasite activity

Hymenoleptis nana (Cestoda) is a tiny intestinal tapeworm and the most common human cestoda. It requires only one host but can also cycle through two. Its larvae migrate only within the gut wall, and its life span is relatively short (4–6 weeks). *Hymenoleptis nana* is more frequent in populations living in conditions of poverty and poor hygiene, particularly when fleas are present.

Pretreatment with LCCs (pine cone Fr. VI and VII) protected the infant mice from *Hymenoleptis nana* (Cestoda) infection. Subcutaneous administration of LCCs (10 mg/kg) to 1-week-old mice evoked strong protective immunity against oral infection by *Hymenolepis nana* eggs. Significant antiparasite effects were also induced in 4-week-old mice by intraperitoneal or oral administration of LCCs. Injection of pine cone LCCs by any root (*s.c, i.p., p.o.*) induced higher antiparasite activity as compared with neutral- and uronic acid-enriched pine cone polysaccharide (Frs. I, II), PSK (protein-bound polysaccharide), Schizophyllan (water-soluble glucan), TAK (water-insoluble glucan), carboxymethyl TAK (Abe et al., 1989). Potent antiparasite activity may be derived from the conjugation of carbohydrate and lignin.

2.2 IN VITRO STUDY WITH CULTURED CELLS

2.2.1 Tumor-specific cytotoxicity against oral squamous cell carcinoma

It is believed that oral environment reflects the condition of whole body. Therefore, we used human oral squamous cell carcinoma cell line as target cells. More than 90% of malignant neoplasms of the oral cavity and oropharynx are squamous cell carcinoma of the lining mucosa with relatively rare neoplasms arising in minor salivary glands and soft tissues, causing various symptoms including mucosal growth and ulceration, pain, malodor, difficulty of speaking, opening the mouth, chewing, difficulty and pain with swallowing, bleeding, weight loss, and neck swelling.

Prescreening tests with animals are essential for the development of new anticancer drugs. However, the data obtained from animals cannot be directly applied to humans, due to different metabolic pathways between humans and animals. We have established in vitro assay methods for tumor-specificity, using human oral squamous cell carcinoma cell lines (HSC-2, HSC-3, HSC-4, Ca9-22) and human oral normal mesenchymal cells [gingival fibroblasts (HGF), pulp cell (HPC), periodontal ligament fibroblast (HPLF)]. These cells were incubated for 48 h with increasing concentrations of test samples, and the relative viable cell number was determined by MTT method. Tumor-selectivity index (TS value) was determined by dividing the mean CC_{50} against normal cells by the mean CC_{50} against tumor cells (left panel, Table 18.1). Thus obtained TS values seem to reflect the in vivo antitumor activity even though these cells were different types of cells (epithelial oral squamous cell carcinoma cell lines versus mesenchymal normal oral cells), because (1) antitumor agents, such as doxorubicin (TS = >26.0) and 5-FU (TS = >55.6) showed higher TS values (Sakagami, 2014), and (2) among 24 plant extracts, *Camptotheca acuminate* leaf, a well-known source for camptothecin, showed the highest TS value (88.3) (Suzuki et al., 2014). On the other hand, lignin-carbohydrate complexes (TS = 2.7), flavonoids [flavonols, isoprenyl flavonoids (TS = 2.1 ± 0.4), 2-arylbenzofurans (TS = 1.2 ± 0.2), benzophenones (TS = 1.7 ± 0.4), xanthones (TS = 1.2 ± 0.3), anthraquinones (TS = 3.8 ± 4.9), phenylbutanone glycosides (TS = 1.5~3.3), stilbene glucosides (TS = 2.4 ± 0.8), naphthalene glucosides (TS = 1.1~1.4), luteolin glucosides] (TS = 1~3.8), tannins (TS = 1~4.8), terpenoids and their glycosides (TS = 1.1~2.5) (Sakagami, 2014) and licorice flavonoids (liquiritin apioside, neoliquiritin apioside, liquiritin, neoliquiritin, liquiritigenin, isoliquiritin apioside, neoisoliquiritin apioside, isoliquiritin, neoisoliquiritin, isoliquiritigenin) (TS = 1~4.4) (Ohno et al., 2013) showed much lower TS values.

Because natural polyphenols showed disappointing lower tumor-specificity, we explored new compounds that retained the partial structures of natural products attached by various substituents or functional groups. We found that newly synthesized compound, (*E*)-3-(4-hydroxystyryl)-6-methoxy-4*H*-chromen-4-one (69.0) (Shimada et al., 2014c) and (*E*)-3-(4-chlorostyryl)-7-methoxy-2*H*-chromene (TS = 59.9) (Uesawa et al., 2015) showed very high tumor-specificity, without showing high cytotoxicity against human oral keratinocyte (HOK), in contrast to doxorubicin that showed very strong cytotoxicity to HOK. Multivariate statistics with chemical descriptors for the location of substituted group, molecular shape and electrostatic interaction may be useful for designing the most favorable compound with higher tumor selectivity.

We next investigated how (*E*)-3-(4-hydroxystyryl)-6-methoxy-4*H*-chromen-4-one disturbed the cellular function of HSC-2 cells, the most sensitive cells tested. This compound induced the mitochondrial vacuolization; autophagy suppression followed by apoptosis induction, and changes in the metabolites involved in amino acid and glycerophospholipid metabolisms (increase of diethanolamine and cytidine diphosphate-choline, and decrease of choline). Since the C5~7 position of chromon ring of this compound is open to introduction of various functional groups, chemical modification may be a potential choice for designing new types of anticancer drugs (Sakagami et al., 2015a). It is interesting to state that doxorubicin is very much cytotoxic to human oral keratinocyte (HOK), inducing apoptosis in HOK (Sakagami et al., 2017), whereas (*E*)-3-(4-hydroxystyryl)-6-methoxy-4*H*-chromen-4-one showed very weak cytotoxicity. The reason why HOK is highly sensitive to doxorubicin, and how to protect the HOK from doxorubicin cytotoxicity are underway.

Table 18.1 Antitumor (In Vitro), Anti-HIV Activity and Anti-UV Activity of Natural Products and Synthetic Compounds

	Antitumor Activity (TS)		Anti-HIV Activity (SI)		Anti-UV Activity (SI)	
Lignin-carbohydrate complex						
Pine cone of *Pinus parviflora* Sieb. et Zucc.	>4.1	Sakagami (2014)	14	Lai et al. (1990)	37.1	Nanbu et al. (2013)
Pine cone of *Pinus elliottii* var. Elliottii	>2.7	Sakagami (2014)	28	Satoh et al. (1999b)	>38.1	Nanbu et al. (2013)
Pine seed shell of *Pinus parviflora* Sieb. et Zucc	>1.7	Sakagami (2014)	12	Nakashima et al. (1992a)	25.6	Nanbu et al. (2013)
Bark of *Erythroxylum catuaba* Arr. Cam.			43	Manabe et al. (1992)		
Husk of cacao beans of Theobroma			311	Sakagami et al. (2008b)		
Mass of cacao beans of Theobroma			46	Sakagami et al. (2011)		
Lentinus edodes mycelia extract (LEM)			94	Kawano et al. (2010a)	41.9	Nanbu et al. (2013)
Precipitating fiber fraction of murberry juice			7	Sakagami et al. (2007)		
Alkaline extract of leaves of *Sasa senanensis* Rehder			86	Sakagami et al. (2010b)	38.5	Nanbu et al. (2013)
Dehydrogenation polymers of phenylpropenoids ($n = 23$)			105	Nakashima et al. (1992a)		
Nanoligno (LCC treated with Fe^{2+})	4.0		49~150			
Chemically modified glucans						
Neutral polysaccharides of pine cone of *P. parviflora* Sieb. et Zucc.			1	Sakagami et al. (2010a)		
Uronic acid-containing polysaccharides of pine cone			1	Sakagami et al. (2010a)		
N,N-dimethylaminoethyl paramylon (substituion ratio: 5%)			<1	Koizumi et al. (1993)	<1	Nanbu et al. (2013)
N,N-diethylaminoethyl paramylon (substitution ratio: 10%)			<1	Koizumi et al. (1993)		
2-Hydroxy-3-trimethylammoniopropyl paramylon chloride			<1	Koizumi et al. (1993)		

Sodium paramylon sulfate (substitution ratio: 4%)			>274	Koizumi et al. (1993)	<1	Nanbu et al. (2013)
Dimethylaminoethyl curdlan (substitution ratio: 5%)			<1	Koizumi et al. (1993)		
PSK (protein-bound polysaccharide)			1	Koizumi et al. (1993)		
Lower molecular weight polyphenols						
Hydrolyzable tannins (monomer) (MW: 484–1,255) (*n* = 21)	1.5 ± 0.5	Sakagami (2014)	1.8 ± 2.8	Nakashima et al. (1992b)		
Hydrolyzable tannins (dimer) (MW: 1,571–2,282) (*n* = 39)	1.4 ± 0.2	Sakagami (2014)	2.3 ± 3.2	Nakashima et al. (1992b)		
Hydrolyzable tannins (trimer) (MW: 2,354–2,658) (*n* = 4)			3.4 ± 3.7	Nakashima et al. (1992b)		
Hydrolyzable tannins (tetramer) (MW: 3,138–3,745) (*n* = 3)			7.3 ± 6.5	Nakashima et al. (1992b)		
Condensed tannins (MW: 290–1,764) (*n* = 8)			1.1 ± 0.4	Nakashima et al. (1992b)		
Flavonoids (MW: 84–648) (*n* = 92)	1.2 ± 0.6	Sakagami (2014)	1.5 ± 1.9	Fukai et al. (2000)		
Epigallocatechin gallate	4.1	Sakagami (2014)			7.7	Nanbu et al. (2011)
Gallic acid	1.1	Sakagami (2014)			5.4	Nanbu et al. (2011)
Luteolin 6-C-β-D-glucoside					>2.8	Matsuta et al. (2011)
Luteolin 7-O-β-D-glucoside					7.3	Matsuta et al. (2011)
Luteolin 6-C-α-D-glucoside					>7.1	Matsuta et al. (2011)
Tricin			24	Matsuta et al. (2011)	23.5	Matsuta et al. (2011)
Antioxidants						
Vanillin					63.8	Nanbu et al. (2013)
Sodium ascorbate (vitmin C)					42.4	Nanbu et al. (2011)
Curcumin					<1	Nanbu et al. (2011)
Ar-turmerone					<1	Nanbu et al. (2011)
N-Acetyl-L-cysteine					<1	Nanbu et al. (2011)

(*Continued*)

Table 18.1 AntiTumor (In Vitro), Anti-HIV Activity and Anti-UV Activity of Natural Products and Synthetic Compounds (cont.)

		Antitumor Activity (TS)		Anti-HIV Activity (SI)		Anti-UV Activity (SI)	
Synthetic compounds							
Water-soluble azulenes ($n = 8$)		2.3 ± 0.6	Ueki et al. (2013)			35.5 ± 21.6	Ueki et al. (2013)
(*E*)-3-(4-hydroxystyryl)-6-methoxy-4*H*-chromen-4-one		69	Shimada et al. (2014c)	<1	Shimada et al. (2014c)		
(*E*)-3-(4-Chlorostyryl)-7-methoxy-2*H*-chromene		59.9		<1	Shimada et al. (2014c)		
Alkaline extract versus hot-water extract							
Green tea leaves	Hot-water extraction			<0.022	Sakagami et al. (2013)	3.4	Nanbu et al. (2011)
	Alkaline extraction			3		>10.4	Sakagami et al. (2013)
Oolong tea leaves	Hot-water extraction			<0.033	Sakagami et al. (2013)	<1	Nanbu et al. (2011)
	Alkaline extraction			13		>7.4	Sakagami et al. (2013)
Orange flower	Hot-water extraction			<0.5	Sakagami et al. (2013)		
	Alkaline extraction			>15			
Licorice root	Hot-water extraction			4	Ohno et al. (2014)		
	Alkaline extraction			42			
Kampo medicines ($n = 10$)				1.0 ± 0.0	Kato et al. (2012)	2.4 ± 1.8	Kato et al. (2012)
Constituent plant extracts of Kampo medicines ($n = 25$)				1.3 ± 0.8	Kato et al. (2012)	1.4 ± 1.6	Kato et al. (2012)

Positive Controls						
Dextran sulfate (molecular mass, 5 kDa)			2,956			
Curdlan sulfate (molecular mass, 79 kDa)			11,718			
Azidothymidine			23,261			
2′,3′-Dideoxycytidine (ddC)			2,974			
Anthracyclines (*n* = 4)	181 ± 100					
5-FU	>66					
Docetaxel	>128					
Camptothecin	2,961					

The tumor-selectivity index (TS) was calculated by the following equation: TS = mean CC_{50} against normal cells/mean CC_{50} against tumor cells. Anti-HIV activity was determined by selectivity index (SI) that was determined by the ratio of CC_{50} (mock-infected) and EC_{50} (HIV-infected). Anti-UV activity was determined by SI that was determined by the ratio of CC_{50} (without UV irradiation) and EC_{50} (UV-irradiation). The details of protocols are described in the text.

2.2.2 Mitogenic activity against splenocyte

LCCs (pine cone Frs. V, VI, VII) showed mitogenic activity, stimulating the proliferation ([^{3}H]-tymidine incorporation into acid-insoluble fraction) of isolated splenocytes from both mice and rats, more potently than plant lectins. In mice with subcutaneously transplanted sarcoma-180, responses of splenocytes against concanavalin A declined in the terminal stage of tumor development, whereas responses against LCC remained relatively constant throughout the tumor progression. Acetylation or methylation of LCC suppressed the splenocyte stimulation activity of Fr. V, suggesting the importance of the hydroxyl group of the LCC in splenocyte stimulation (Kurakata et al., 1989).

2.2.3 Stimulation of iodination of myeloperoxidase-positive cells

Monocytes and polymorphonuclear leukocytes (PMN) contain higher levels of myeloperoxidase (MPO). The function of MPO-positive cells, such as antibacterial and phagocytic activity can be evaluated by iodination, that is incorporation of [^{125}I] into acid-insoluble fraction of the cells. The iodination is determined by ingestion, H_2O_2 production, or the content or release of MPO (Klebanoff, 1967) and played a role in the process of bactericidal activity. LCC significantly stimulated the iodination of MPO-positive cells, but not that of MPO-negative cells (Unten et al., 1989). The iodination was also temperature-dependent, reaching a maximum at 37°C. This stimulation was significantly inhibited by MPO inhibitors (sodium azide, aminotriazole). Limited digestion of lignin structure by $NaClO_2$ resulted in the significant loss of stimulation activity, whereas removal of the monosaccharide residues by acid-catalyzed hydrolysis did not significantly affect the stimulation activity. Both hydrolysable and condensed tannins stimulated the iodination of these PMN (Sakagami et al., 1990a) and monocytes (Sakagami et al., 1992a,b). On the other hand, glucans (Schizophyllan, Paramylon, TAK) and their chemically modified derivatives (dimethylaminoethyl-paramylon, paramylon sulfate, carboxymethyl TAK), and polysialoglycoproteins [H-PSGP (200 kD), L-PSGP (9 kD)] did not significantly showed little or no activity (Sakagami et al., 1991). These data indicate the importance of a polymerized phenolic structure for the stimulation of PMN iodination that may relate to their antibacterial activity.

2.2.4 Macrophage activation

We performed DNA microarray analysis to see how LCC [Fr4 from cultured extract of *Lentinus edodes* mycelia (LEM)] affect the mammalian gene expression, using mouse macrophage-like cells J774.1 stimulated with either Fr4 or LPS (positive control) (Kawano et al., 2010b). The authenticity of Fr4 as lignin was confirmed by the presence of lignin precursors (vanillic acid, syringic acid, *p*-coumaric acid and ferulic acid), with undetectable levels (<0.1 μg/g) of flavonoids (fisetin, daizein, genistein, quercetin, kaempferol, apigenin, and chrysin), tannins-related compounds (gallic acid, gallo catechin, catechin, epicatechin, epigallocatechin gallate, gallocatechin gallate, epicatechin gallate, catechin gallate, ellagic acid, theaflavin-3-gallate, theaflavin-3′-gallate, and theaflavin-3,3′-gallate) and chlorogenic acid and eugenol (Kawano et al., 2010b). Fr4 affected the expression of innate immune response-related genes, but did not affect the expression of so many genes that had been affected by LPS. LEM (Fr4) enhanced the expression of dectin-2 (4.2-fold) and TLR-2 (2.5-fold) prominently, but only slightly modified the expression of dectin-1 (0.8-fold). On the other hand, LPS did not affect the expression of dectin-2 nor TLR-2. The expression of the following genes were not apparently changed by Fr4 treatment: complement receptor 3 (0.9-fold), TLR1, 3, 4, 9, and 13 (0.8- to 1.7-fold), Sykb, Zap70, Jak2 (1.0- to 1.2-fold), Nfκb1, NFκb2, Rela, Relb (1.0- to 1.6-fold), Nfκbia, Nfκbib, Nfκbie,

Nfκbi12 Nfκbiz (0.8- to 2.3-fold). These data suggest the significant role of the activation of the dectin-2 signaling pathway in the action of LCC on macrophages (Kushida et al., 2011).

2.3 ANTIVIRAL ACTIVITY

2.3.1 Anti-HIV activity

We found that LCCs from pine cone extract from *Pinus parvifilara* Sieb. and Zucc. inhibited the multiplication of human immunodeficiency virus (HIV) in the cells (Lai et al., 1990). We have established the in vitro assay method of anti-HIV activity using HIV-infected and mock-infected human T-cell lines MT-4 (Nakashima et al., 1992b). Mock- and HIV-infected MT-4 cells (3×10^4 cells/96-microwell) were incubated for 5 days with different concentrations of samples and the relative viable cell number was determined by MTT assay, to yield the CC_{50} and EC_{50} (that increased the viable cell number of the HIV-infected cells to 50%), respectively. The anti-HIV activity (selectivity index, SI) was determined by dividing the CC_{50} by EC_{50} (center panel, Table 18.1). LCCs from pine cone from *Pinus parviflora* Sieb. et Zucc. and *Pinus elliottii* var. Elliotti (SI = 14, 28) (Satoh et al., 1999b), from the pine seed shell of *Pinus parviflora* Sieb. et Zucc. (Sakagami et al., 1992e), bark of *Erythroxylum catuaba* Arr. Cam. (SI = 43) (Manabe et al., 1992), husk and mass of *Theobroma cacao* (SI = 311, 46) (Sakagami et al., 2008b), and cultured extract of *Lentinus edodes* mycelia (LEM) (SI = >94) (Kawano et al., 2010a) and mulberry juice (SI = 7) (Sakagami et al., 2007) showed potent anti-HIV activity. The anti-HIV activity of the hot-water extract of LEM has been also reported by other groups (Suzuki et al., 1990).

The anti-HIV activity of LCCs (SI = 14~311) (Nakashima et al., 1992a) was generally higher than that of lower molecular weight polyphenols, such as tannins (SI = 1-11) (Nakashima et al., 1992b) and flavonoids (SI = 1) (Fukai et al., 2000), and natural and chemically modified glucans (*N,N*-dimethylaminoethyl paramylon, *N,N*-diethylaminoethyl paramylon, 2-hydroxy-3-trimethylammoniopropyl paramylon, sodium caroboxymethyl paramylon, carboxymethyl-TAK) (SI = 1) except for sulfated polysaccharide with higher substitution ratio of sulfate group (SR >4.0%) (paramylon sulfate and dextran sulfate) (Koizumi et al., 1993), phenylpropanoid amides (SI <1) (Shimada et al., 2014b), piperic acid amides (SI <1) (Shimada et al., 2014a), 3-styrylchromones (SI <1) (Shimada et al., 2014c), 3-styryl-2H-chromenes (*S* <1) (Uesawa et al., 2015), and oleoylamides (*S* <1) (Sakagami et al., 2015b).

Ten Kampo medicines (Byakkokaninjinto, Hangesyashinto, Hotyuekkito, Juzentaihoto, Kikyoto, Ninjinyoeito, Rikkosan, Saireito, Shosaikoto, and Unseiin) (*S* = 1) and its constituent plant extracts (*Alisma* rhizoma and *Asiasarum* root, *Astragalus* root, *Atractylodes Lancea* rhizoma, *Bupleurum* root, *cimicifuga* rhizoma, cinnamon bark, *Cnidium* Rhizoma, *Coptis* rhizoma, gardenia fruit, ginger, ginseng, glycyrrhiza, Japanese angelica root, Japanese gentian, jujube, peony root, phellodendron bark, pinellia tuber, platycodon root, polyporus sclerotium, poria sclerotium, *Rehmannia* root, *Saposhnikovia* root, and *Scutellaria* root) (SI = 1~4.4) showed relatively lower anti-HIV activity (Kato et al., 2012).

Limited digestion of lignin structure by $NaClO_2$ resulted in significant loss of anti-HIV activity (from SI = 14 to 3), whereas removal of the carbohydrate residues by acid-catalyzed hydrolysis did not significantly affect the anti-HIV activity (Lai Pk et al., 1992), suggesting the importance of phenylpropenoid polymer, but not carbohydrate moiety, for anti-HIV activity induction. Dehydrogenation polymers of phenylpropenoids without carbohydrate showed higher anti-HIV activity (SI = 105) than LCCs, whereas phenylpropenoid monomers were inactive, suggesting the importance of highly polymerized lignified structure (Nakashima et al., 1992a).

Conventional anti-HIV drugs are classified largely into the following five groups: nucleoside analog RT Inhibitor, which enters into the cells, phosphorylated, used as RT substrate and then inhibits viral replication), nonnucleic acid reverse transcriptase inhibitor (that binds to p66 subunit of HIVRT, modify the conformation of three-dimensional-structure and then noncompetitively inhibit RT activity), protease inhibitor (PI) (peptide-like chemicals that competitively inhibit viral aspartate protease), integrase strand transfer inhibitor: INSTI (that inhibits the enzyme that integrates viral DNA into human DNA), and CCR5 inhibitor (that inhibits HIV bind to lymphocyte surface CCR receptor). For the HIV-treatment, it is most common to adopt the HAART: highly active antiretroviral therapy that uses the combinations of these anti-HIV drugs, but leaving problems such as drug tolerance and adverse effects. LCC that is contained at approximately 40% in alkaline extracts of the plants rapidly binds to virus, inactivating the virus. Due to this unique property, LCC is expected to synergistically enhance the anti-HIV activity of these antiretrovirus drugs. Furthermore, it has broad antiviral spectrum, suggesting its efficacy against virally induced stomatitis. As compared with tannins and flavonoids, LCC can be obtained at three orders higher yields, making it suitable for the large-scale production at the factory. If synergism with conventional chemotherapy was confirmed, the amount of anti-HIV can be reduced, thus making it possible to distribute the antiretrovirus drugs at much lower prices.

2.3.2 Antiinfluenza virus activity

LCCs inhibited the plaque formation (resulting from cytopathic effect of influenza virus infection) and RNA polymerase activity of influenza virus (Nagata et al., 1990). Limited digestion demonstrated the importance of phenylpropenoid polymer, but not sugar moiety, for antiinfluenza virus activity (Harada et al., 1991). Dehydrogenation polymers of phenylpropenoids inhibited both the plaque formation and RNA polymerase more effectively than LCCs (Sakagami et al., 1990b). Sucrose gradient centrifugation demonstrated that without virus, [^{125}I]-LCC floated up to the top of sucrose gradient, whereas its sediment migrated to the band where the virus protein was detected, demonstrating strong affinity of LCC with the influenza virus. Contact to LCC abrogated the infectivity of the influenza virus in mouse infection models (Sakagami et al., 1992d).

When MDCK cells were incubated with [^{125}I]-LCC, the radioactivity was bound to the cells. After 1 h, the binding reached the maximum (about 3.4% initially added radioactivity. Further analysis with Scatchard analysis is needed to identify the receptor (Fig. 18.2).

2.3.3 Anti-HSV activity

Antiherpes simplex virus (HSV) activity was determined by the ability to inhibit the plaque formation of HSV-1, HSV-2 strains in African green monkey kidney cells and human adenocarcinoma cells. LCCs, isolated from the cones of various pine trees (*Pinus parviflora* Sieb. et Zucc, *Pinus densiflora* Sieb. et Zucc., *Pinus thunbergii* Parl., *Pinus elliottii* var. *Elliotti, Pinus* taeda L., *Pinus caribaea* var. Hondurenses, *Pinus sylvestris* L., and the pine seed shells of *Pinus parviflora* Sieb. et Zucc., and *Pinus armandii Franch*) (Fukuchi et al., 1989a) and tannins (oenothein B, coriariin A, rugosin D, cornusiin A, tellimagrandin I, casuarictin, penta-O-galloyl-β-D-glucose, geraniin, 4,8-tetramer of epicatechin gallate) (Fukuchi et al., 1989b) showed comparable anti-HSV activity. However, due to the lack of chemotherapeutic index (safety margin), the direct comparison of the anti-HSV activity between LCC and tannins could not be achieved.

On the other hand, neutral polysaccharide (pine cone Fr. I) (no inhibition at 10 μg/mL) and acidic polysaccharide (rich in uronic acid) (pine cone Fr. II) (no inhibition), glucans (paramylon,

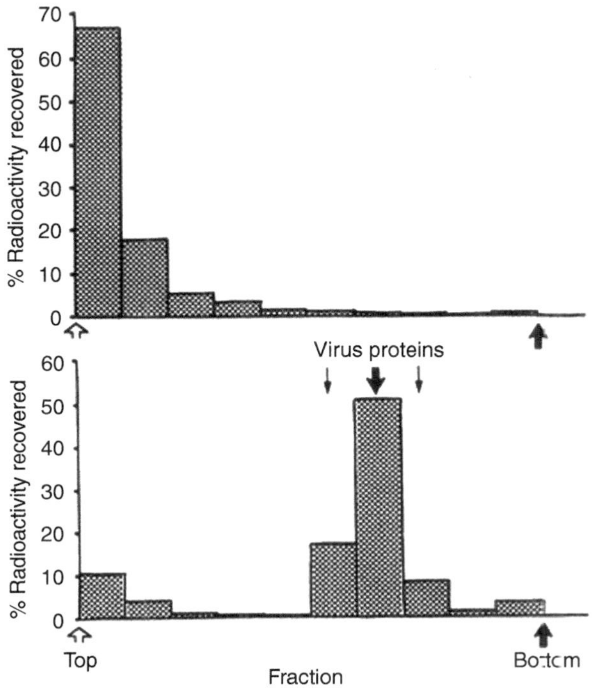

FIGURE 18.2 Binding of [^{125}]-LCC to Influenza Virus

[^{125}]-LCC was mixed without or with influenza virus, and then applied to sucrose gradient centrifugation. Radioactivity distributed into each fraction was determined. The *arrows* indicate the fraction from which the presence of virus proteins demonstrated by Western blot analysis.

Cited from Sakagami, H., Takeda, M., Kawazoe, Y., Nagata, K., Ishihama, A., Ueda, M. & Yamazaki, S., 1992d. Anti-influenza virus activity of a lignin fraction from cone of Pinus parviflora Sieb. et Zucc. In Vivo, 6, 491–495 with permission.

Schizophyllan, TAK), and chemically modified derivatives (*N,N*-dimethylaminoethyl paramylon, sodium paramylon sulfate, carboxymethyl-TAK) and protein-bound polysaccharide (PSK) were inactive. Using [^{3}H]-thymidine labeled virus, we found that lignin-carbohydrate complex (Fukuchi et al., 1989a) and tannins (Fukuchi et al., 1989b) inhibited adsorption to these cells rather than the inhibition of virus penetration into the cells. We recently found that SE enhanced the anti-HSV activity of acyclovir by plaque assay (Sakagami et al., 2016b).

Recently, carboxylated lignins, synthesized using 4-hydroxy cinnamic acid scaffold by enzymatic oxidative coupling inhibited the entry of HSV-1 entry into the cells (Thakkar et al., 2010). LCC isolated from Prunella vulgaris (PPS-2b) (MW8500) having glucose, galactose, mannose, galacturonic acid, rhamnose, xylose, and arabinose with glucose as the major sugar also showed anti-HSV activity possibly by inhibiting the viral binding and penetration into host cells. Prunella cream formulated with a semipurified fraction significantly reduced the skin lesion and mortality induced by HSV-1 infection in Guinea pigs (Zhang et al., 2007). The anti-HSV activity of sulfated lignins depended on their molecular weight, with the maximum at 39.4 kDa (Raghuraman et al., 2007).

2.4 SYNERGISM WITH VITAMIN C

2.4.1 Radical scavenging activity

Electron spin resonance (ESR) spectroscopy has been used for more than 50 years to study a variety of paramagnetic species. Superoxide anion (O_2^-), produced by hypoxanthine and xanthine oxidase, and hydroxyl radical, produced by Fenton reaction, are unstable, being rapidly decayed. However, in the presence of spin-trapping agent such as 5,5-dimethyl-1-pyrroline-*N*-oxide (DMPO), they form stable adducts such as DMPO-OOH and DMPO-OH, respectively, and therefore can be quantified by ESR. Lower concentration of LCC (pine cone Fr. VI) and sodium acorbate showed radical scavenging activity. LCC stimulated the superoxide anion (O_2^-) scavenging activity of sodium ascorbate (Sakagami et al., 2005). LCCs from *Ceriops decandra* (Griff.) Ding Hou (Sakagami et al., 1998) and cacao husk (Sakagami et al., 2008b) scavenged O_2^- and hydroxyl radical, and synergistically enhanced the radical scavenging activity of sodium ascorbate.

LCC (Fr. VI) dose-dependently scavenged the hypochlorite (a product of MPO-catalyzed reaction), monitored by luminol chemiluminescence (Sakagami et al., 1995).

2.4.2 Radical generation and cytotoxicity

Vitamin C exhibited both antioxidant and prooxidant activity, depending on the concentration (Sakagami and Satoh, 1997b; Sakagami et al., 2000). We have reported that ascorbate derivatives that produced the doublet signal of ascorbate radical (sodium-L-ascorbate, L-ascorbic acid, D-isoascorbic acid, 6-β-D-galactosyl-L-ascorbate, sodium 5,6-benzylidene-L-ascorbate) induced apoptotic cell death, characterized by internucleosomal DNA fragmentation (evaluated by agarose gel electrophoresis) and an increase in the intracellular Ca^{2+} concentration) (detected by confocal experiment with Fluo-3 fluorescence) in HL-60 cells. On the other hand, ascorbate derivatives that did not produce radicals (L-ascorbic acid-2-phosphate magnesium salt, L-ascorbic acid 2-sulfate and dehydroascorbic acid) did not induce apoptosis markers (Sakagami and Satoh, 1997a). This suggests the possible involvement of the ascorbate radical in apoptosis-induction by ascorbate analogs. Exposure of HL-60 cells to any active ascorbate derivatives rapidly elevated the intracellular Ca^{2+} concentration, which might serve as the initial signal leading to the cell death pathway (Sakagami and Satoh, 1997a).

LCCs from the pine cone of *Pinus parviflola* Sieb. et Zucc. (Satoh et al., 1999a), pine cone of *Pinus elliottii* var. Elliotti (Satoh et al., 1999b), leaf of *Ceriops decandra* (Griff.) Ding Hou (Sakagami et al., 1998) and, thorn apple of *Crataegu Cuneata* Sieb. et Zucc. (Satoh et al., 1998) modulated the radical intensity of ascorbate biphasically depending on the concentrations. At higher concentration, LCCs strongly enhanced the radical intensity of sodium ascorbate, which rapidly decayed, possibly due to the breakdown of ascorbic acid or to the consumption of ascorbyl radical. LCCs, derived from pinecones (Fr. VI) (Satoh et al., 1999a), Catuaba bark (Manabe et al., 1992), pine seed shell (Sakagami et al., 1992e), A. nikoense maxim. (Sakagami et al., 1997a) and *C. Cuneata* Sieb. et Zucc. (Satoh et al., 1998) enhanced the radical intensity and cytotoxic activity of sodium ascorbate. On the other hand, tannins such as gallic acid, epigallocatechin gallate, and tannic acid counteracted the radical intensity and cytotoxic activity of sodium ascorbate (Satoh et al., 1999a). These results also substantiated the relationship between radical intensity and cytoxicity.

Sodium ascorbate rapidly reduced the oxygen concentration in the culture medium, possibly due to oxygen consumption via its prooxidation action. Simultaneous addition of LCCs further enhanced the ascorbate-stimulated consumption of oxygen (Sakagami et al., 1997b). These data suggest that the synergistic enhancement of the cytotoxic activity of LCCs and ascorbate might be due at least in part to the stimulated induction of hypoxia.

Nakamura et al. (2012, 2015) reported that photoirradiated gallic acid could generate hydroxyl radical and enhance the antimicrobial activity. They suggest that the bactericidal action of hydroxyl radical is a result of the photooxidation of polyphenolic compounds, followed by photolysis of H_2O_2 (Tsukada et al., 2016). The confirmation of this hypothesis needs additional experiments.

2.5 ANTI-UV ACTIVITY

We have established a simple assay method for anti-UV activity (Kantoh et al., 2010). The cells were trypsinized, inoculated onto 96-microwell plate, and incubated for 48 h to achieve the complete attachment to the plate. After replacing with PBS(−), cells were then exposed to UV irradiation (6 $J/m^2/min$) or not for 1 min. The media were replaced with fresh fetal bovine serum (FBS) and cells were cultured for a further 48 h. The viable cell number was determined by MTT method, and expressed as a percentage that of control unirradiated and untreated cells. This method provides the 50% cytotoxic concentration (CC_{50}) (without UV irradiation) and 50% effective concentration (EC_{50}) that returned the viability of UV-induced cells to 50% of that of control, and the anti-UV activity (selectivity index, SI) can be defined as the following equation: $SI = CC_{50}/EC_{50}$. Target cells we have initially used were human oral squamous cell carcinoma (OSCC) cell line HSC-2, which was the most sensitive to UV irradiation (right panel, Table 18.1).

Among the natural polyphenols, LCCs (prepared by sequential alkaline extraction and acid precipitation) showed the highest anti-UV activity. LCC, prepared from pine cone (*Pinus parviflora* Sieb. et Zucc., *Pinus taeda* L and *Pinus elliottii* var. Elliottii), pine seed shell (*Pinus parviflora* Sieb. et Zucc.), *Lentinus edodes* mycelia and *Sasa senanensis* Rehder leaves, showed the excellent anti-UV activity (SI = >3.7.0, >38.1, 25.6, 41.9, 38.5). Anti-UV activity was also recovered from the ethanol (50%) precipitable fraction (Fr. VII) that contains higher amounts of polysaccharides (SI = >12.3, >21.1, >10.3, >7.8, >7.1), but their activity was slightly lower than that of Fr. VI. This suggests that the highly polymerized lignin structures are necessary to express anti-UV activity.

On the other hand, lower molecular weight polyphenols, such as resveratrol, a stilbenoid known for its antiaging effect (SI <1), epigallocatechin gallate (a major component of green tea) (SI = 7.7), gallic acid (a component unit of tannin) (SI = 5.4), luteolin glycosides (luteolin 6-C-β-D-glucoside, luteolin 7-O-β-D-glucoside, luteolin 6-C-α-D-glucoside, tricin) (SI = >2.8, 7.3, >7.1, 23.5) showed lower anti-UV activity (Nanbu et al., 2013).

Ten Kampo medicines (Byakkokaninjinto, Hangesyashinto, Hotyuekkito, Juzentaihoto, Kikyoto, Ninjinyoeito, Rikkosan, Saireito, Shosaikoto, and Unseiin) (S = 1~4.9) and its constituent plant extracts (*Alisma* rhizoma and *Asiasarum* root, *Astragalus* root, *Atractylodes Lancea* rhizoma, *Bupleurum* root, *cimicifuga* rhizoma, cinnamon bark, *Cnidium* rhizoma, *Coptis* rhizoma, gardenia fruit, ginger, ginseng, glycyrrhiza, Japanese angelica root, Japanese gentian, jujube, peony root, phellodendron bark, Pinellia tuber, Platycodon root, Polyporus sclerotium, Poria sclerotium, Rehmannia root, Saposhnikovia root, and Scutellaria root) (SI = 0.75~4.3) (Kato et al., 2012, 2014) and tea extracts [green tea (SI = 1.6~3.4), black tea (SI <1), jasmine tea (SI <1), koh-ki tea (SI <1), burley tea (SI <1)]. LCC showed additive or synergistic anti-UV activity with vitamin C (Nanbu et al., 2011).

Alkaline extract of *Sasa senanensis* Rehder leaves (SE), an LCC-rich OTC drug, similarly showed anti-UV activity.

Among anti-oxidants, vanillin and sodium ascorbate (vitamin C) showed potent anti-UV activity (SI = 63.8 and 42.4, respectively) (Nanbu et al., 2013). On the other hand, curcumin, Ar-turmerone, *N*-acetyl-L-cysteine and catalase showed no anti-UV activity (SI <1).

Among synthetic compounds, water-soluble azulenes (3-methylazulene-1-sulfonate, sodium 3-ethylazulene-1-sulfonate, sodium 3-propylazulene-1-sulfonate, sodium 7-isopropyl-3-methylazulene-1-sulfonate, sodium 7-isopropyl-3-ethylazulene-1-sulfonate, sodium 7-isopropyl-3-propylazulene-1-sulfonate, sodium 3-hexylazulene-1-sulfonate, sodium 7-isopropyl-3-heptylazulene-1-sulfonate, disodium azulene 1,3-disulfonate) showed the highest anti-UV activity (SI = 45.0~65.6) (Ueki et al., 2013), suggesting the importance of water solubility as the determinant of anti-UV activity.

We recently found that human skin derived cells (a spontaneously transformed aneuploid immortal keratinocyte cell line HaCaT) and human sarcoma HT1080 produced higher SI value as compared with HSC-2 cells (Sakagami et al., 2016c).

Deltamethrin (photolabile compound) was encapsulated into corn oil nanoemulsions (NE) by a hot high-pressure homogenization technique followed by coating with chitosan as the first coating layer (CH-NE). The photodegradation of deltamethrin was significantly reduced by including lignosulfonate as the second coating layer to form a double-layer coated NE (L-CH-NE). This novel nanocarrier system may protect the photolabile compounds against photodegradation (Nguyen et al., 2013). When using the appropriate weight ratio of cellulose nanocrystal/lignin (either synthetic or fractionated from spruce and corn stalks), these nanocomposite systems effectively blocked the UV spectrum. Dislocation of the π–π aromatic aggregates increases the extinction coefficient and decreases the transmittance in the UV region (Hambardzumyan et al., 2012).

3 EFFICIENT UTILIZATION OF PLANT RESOURCES BY ALKALINE EXTRACTION

We found that alkaline extracts show similar biological properties with LCC.

3.1 ALKALINE EXTRACTION OF LEAF OF *SASA SENANENSIS* REHDER

Alkaline extract of leaf of *Sasa senanensis* Rehder or its closely related Sasa albino-marginata Makino et Shibata (SE) showed several lignin-like activities: potent anti-HIV activity and antiinfluenza virus activity, scavenging activity of O_2^- and hydroxyl radical in a synergistic fashion with vitamin C (Sakagami et al., 2008a).

SE inhibited the nitric oxide (NO) and prostaglandin E_2 (PGE_2) production by LPS-activated mouse macrophage-like cells RAW264.7, by suppressing the inducible NO synthase (iNOS) and cyclooxygenase-2 (COX-2) expression at both protein and mRNA levels and scavenging NO (Zhou et al., 2009). Chlorophyllin, that composed of 0.25% SE, may be involved in part for the inhibition of NO production. This may be due to the inhibitory action of chlorophyllin present in SE, since chlorophyllin inhibited the LPS-stimulated NO production more efficiently than chlorophyll. Solvent fractionation of SE demonstrated that (1) chlorophyllin in SE was recovered from the water layer, that contains more than 81% of compounds and inhibited the NO production by macrophages more potently than other *n*-hexane, diethyl ether and ethylacetate layers. Furthermore, LCC isolated from SE showed the unique greenish color of chlorophyllin (absorption maximum = 452 nm) (Sakagami et al., 2010b), suggesting the possible association of chlorophyllin with LCC in the native state or during extraction by alkaline solutions.

We used IL-1β-stimulated human gingival fibroblast (HGF) as human gingivitis model. When HGF cells were stimulated with IL-1β (5 ng/mL), one or two orders higher concentrations of inflammatory cytokines (IL-6, IL-8, MCP-1) and PGE_2, but not tumor necrosis factor (TNF)-α and NO, were released into culture medium (Ono et al., 2011). SE significantly inhibited the IL-1β-induced PGE_2 production via inhibition of cyclooxygenase (COX)-2 protein and also reduced the COX-2 activity by approximately 50%.

Metabolomics, the measurement of all intracellular metabolites, has become a powerful tool to gain insight into cellular function and site of action. Using this technology, we recently demonstrated that IL-1β reduced the intracellular concentrations of overall primary metabolites especially those of amino acid, urea cycle, polyamine, *S*-adenosylmethione, and glutathione synthetic pathways. The addition of titanium dioxide nanoparticles further augmented these IL-1β-induced metabolic changes (Garcia-Contreras et al., 2015). We recently found that SE reversed these metabolic changes (Sakagami, manuscript in preparation).

SE inhibited CYP3A activity [IC_{50} = 0.99% (576 μg/mL)], but much weaker than grape juice (IC_{50} = 0.49%) and chlorophyllin (IC_{50} = 109 μg/mL) (Sakagami et al., 2010b), suggesting that the incidence of side effects by the combination with other drugs may be lower than that induced by grapefruit juice. Chlorophyllin present in SE seems to be a potential culprit of CYP3A inhibition.

Other bioactive substances isolate from SE and related Sasa plants are: (1) 4′,5,7-trihydroxy-3,5′-dimethoxylflavone (Tricin) that potently inhibited the cytopathic effect of cytomegalovirus (Sakai et al., 2008), (2) apigenin 6-C-β-D-xylopyranosyl-8-C-β-D-glucopyranoside (Park et al., 2007), kurilensin A and B (Hasegawa et al., 2008) that scavenged the DPPH radical, (3) tricin and (−)-syringaresinol that inhibited P-glycoprotein, enhancing the cytotoxic activity of daunomycin against human breast cancer cells (Jeong et al., 2007). It should be noted that LCC could be obtained at much higher yield than Tricin, showed higher anti-UV activity than Tricin.

3.2 ALKALINE EXTRACTION OF LICORICE ROOT

Alkaline extracts from licorice root showed more favorable effects, as compared with hot-water extracts. Licorice root was successively extracted first with water or alkaline solution (pH 9 or 12), and the alkaline (pH 12.0) extract was further separated into 50% ethanol-soluble and -insoluble fractions. Alkaline extraction yielded slightly higher amounts of dried materials as compared with water extraction. Alkaline extract showed higher anti-HIV and antibacterial activities, and similar magnitudes of CYP3A4 inhibitory and superoxide and hydroxyl radical-scavenging activities, as compared with water extract. Anti-HIV activity was recovered from the 50% ethanol-insoluble high-molecular weight fraction that represents approximately 3% of the alkaline extract, whereas antibacterial activity was concentrated in the soluble fraction rich in glycyrrhizid acid, flavanones and chalcones. Alkaline extractions were found to be superior over water extraction for preparing anti-HIV and antibacterial agents at a higher yield from licorice root (Ohno et al., 2014).

3.3 ALKALINE EXTRACTION OF GREEN TEA LEAVES

Alkaline extraction of green tea leaf, oolong tealeaf, and orange flower could be recovered twice as much as dried materials as compared with water extraction. Water extracts showed higher antibacterial, CYP3A4 inhibitory and superoxide scavenging activities, whereas alkaline extract showed higher anti-HIV and hydroxyl radical scavenging activity. Both water and alkaline extracts showed comparable anti-UV activity (Sakagami et al., 2013).

3.4 EFFECT OF $FeSO_4$ ON ANTI-HIV ACTIVITY OF LCC

Concomitant treatment of biomass with dilute acid and ferrous ion resulted in increased hydrolysis of sugars, and delamination and fibrillation of the cell wall. This was accompanied by attenuation of the major peaks representing the C—O—C and C—H bonds in cellulose, and the association of Fe^{2+} ions with both cellulose/xylan and lignin (Wei et al., 2011). Iron and hydrogen peroxide are capable of oxidizing a wide range of substrates and causing biological damage. The reaction, referred to as the Fenton reaction, is complex and capable of generating both hydroxyl radicals and higher oxidation states of the iron.

We tested the effect of Fe^{2+} concentration and higher temperature on the anti-HIV activity of LCC. Pine cone LCC was incubated with 0~1 mg/mL $FeSO_4$ at room temperature or heating up to 90°C and cooling down. Our preliminary experiment demonstrated that anti-HIV activity of pine-cone LCC was dependent on the concentration of $FeSO_4$ added exogenously. At 0.01 or 0.03 mg/mL $FeSO_4$, the anti-HIV activity was increased by 26% and 45%, respectively, although there was no statistical significance between $FeSO_4$ concentration and CC_{50} in both nonheated and heated samples ($P = 0.985$ and 0.841), between $FeSO_4$ concentration and EC_{50} ($P = 0.676$ and 0.730) (Table 18.2). Heating significantly reduced anti-HIV activity (SI value) ($P = 0.0016$) without affecting the cytotoxicity. Drastic condition of heating may extensively degrade the LCC to monomers that have no anti-HIV activity.

$FeSO_4$-treated LCC showed slightly higher cytotoxicity against human oral squamous cell lines (Ca9-22, HSC-2, HSC-3, and HSC-4) (CC_{50} = 5.4, 7.3, 6.7, 8.0, mean 6.85 ± 1.1%) as compared with normal oral cells (human gingival fibroblast, periodontal ligament fibroblast, pulp cells) (CC_{50} = 12.5, 19.0, 50.0, mean 27.2 ± 20.0%), yielding tumor-specificity index of 4.0. These results indicate that once the optimal condition settled, $FeSO_4$ treatment without heating is a good choice for increasing the anti-HIV activity of LCC, keeping the reasonable tumor-specific cytotoxicity.

4 UTILIZATION OF LIGNIN AND ITS COMPONENTS

4.1 INDUSTRIAL USE: DENTAL APPLICATION

Shearing lignin increased the higher thermal stability when blended with polyvinyl alcohol (Nair et al., 2014). Embedding cellulose nanocrystals improved the thermomechanical performance of lignin-based electrospun fibers and spin-coated thin films. The addition of cellulose nanocrystals to the multicomponent fiber mats increased the reinforcing effect and humidity stability, thus controlling the yield behavior in the composites (Ago et al., 2013).

Incorporation of lignosulfonates to chitosan nanoparticles increased the stability to lysozyme degradation, biocompatibility with human cells, and antimicrobial activity (Kim et al., 2013).

Lignin nanocrystals confined on reduced graphene oxides (RGOs) used for renewable energy-storage materials showed pseudocapacitive behavior. Lignin nanocrystals synergized the fast and reversible redox charge transfer of surface-confined quinone and the interplay with electron-conducting RGOs. This property may be useful to develop the advanced renewable electrodes for energy-storage applications and to understand the redox chemistry of electroactive biomaterials (Kim et al., 2014).

Confocal microscopy imaging revealed that all lignin nanotubes formulations entered HeLa cells, penetrated the cell nucleus and sorbed to DNA. Exposure of HeLa cells to lignin nanotubes coated with

Table 18.2 Effect of $FeSO_4$ Treatment on Anti-HIV Activity of Pine Cone LCC

Sample	$FeSO_4$ (mg/mL)	CC_{50} (μg/mL)	EC_{50} (μg/mL)	SI
Pine cone LCC	0	496	6.1	82
Pine cone LCC	0.0001	416	8.5	49
Pine cone LCC	0.0003	517	5.0	103
Pine cone LCC	0.001	619	5.2	119
Pine cone LCC	0.003	393	5.7	69
Pine cone LCC	0.01	316	11.3	28
Pine cone LCC	0.03	445	12.7	35
Pine cone LCC	0.1	460	7.9	58
Pine cone LCC	0.3	438	5.5	79
Pine cone LCC	1	562	3.7	150
Pine cone LCC (heated)	0	394	14.6	27
Pine cone LCC (heated)	0.0001	446	15.0	30
Pine cone LCC (heated)	0.0003	543	15.3	35
Pine cone LCC (heated)	0.001	533	14.7	36
Pine cone LCC (heated)	0.003	452	13.4	34
Pine cone LCC (heated)	0.01	452	11.4	40
Pine cone LCC (heated)	0.03	435	14.6	30
Pine cone LCC (heated)	0.1	410	15.8	26
Pine cone LCC (heated)	0.3	453	12.5	36
Pine cone LCC (heated)	1	436	18.2	24
DS (μg/mL)		144	0.634	228
CRDS (μg/mL)		573	0.565	1,015
AZT (μM)		243	0.017	13,958
ddC (μM)		2,713	0.645	4,205

Pine cone LCC was treated with the indicated concentrations of $FeSO_4$, without or with heating up to 90°C and cooling down. Anti-HIV activity was then determined. Each value represents mean ±S.D. of triplicate determinations. There was statistical significance in EC_{50} between nonheated and heated groups ($P = 0.0000024$). There was also statistical significance in SI between nonheated and heated groups ($P = 0.0016$). However, there was no statistical significance between $FeSO_4$ concentration and CC_{50} in both nonheated and heated samples ($P = 0.985$ and 0.841), between $FeSO_4$ concentration and EC_{50} ($P = 0.676$ and 0.730) in both nonheated and heated groups, and between the CC_{50} values of nonheated and heated groups ($P = 0.74$).

DNA encoding the green fluorescent protein leads to transfection and expression of green fluorescent protein. The highest transfection efficiency was obtained with lignin nanotubes made from NaOH-lignin having high DNA binding capacity and DNA delivery directly into the nucleus. Lignin nanotubes can be useful as smart delivery vehicles of DNA without the cytotoxicity or the immunogenicity of viral vectors (Ten et al., 2014).

Cationic amphiphilic lignin-based graft copolymers, synthesized from kraft lignin, showed excellent in vitro transfection efficiency that was comparable or even higher than that of branched PEI (25 K), suggesting their application as nonviral gene vectors for future gene delivery application.

Low molecular weight lignin yielded highly porous carbon with more graphitization through direct carbonization without additional activation processes or templating agents. This product may be a good candidate for supercapacitor electrode materials (Jeon et al., 2015).

4.2 UTILIZATION OF CELLULOSE

Cellulose in biomass provides biocomposites, pharmaceutical carriers, and electronic substrates. However, strong lignin layers cause low cellulose accessibility to chemicals, and high-cellulose crystallinity, which inhibit the digestibility of the biomass for cellulose extraction. It is thus necessary to deconstruct noncellulosic content in lignocellulosic biomass, while maintaining cellulose product for further hydrolysis into nanocellulose material (Lee et al., 2014). Lignocellulose was obtained from lignin by alkaline hydrogen peroxide, and used as feeding material for microarthropods *Folsomia candida* (Collembola) (Moreno-Murguia et al., 2015).

Cellulose nanofibers (diameter = 10–70 nm) were obtained by alkali treatment and bleaching, and high-pressure homogenization from depectinated sugar beet pulp, and separated from the hemicellulose and lignin. The thermal degradation temperature elevated 47.3°C by alkali treatment. Cellulose nanofibers may be applicable as reinforcement in the biocomposite material at high temperature (Li et al., 2014).

Cellulose nanowhiskers (CNW) was extracted from unripe coconut husk fibers (CHF), by delignification using organosolv process, followed by alkaline bleaching and subsequent hydrolysis with sulfuric acid. The process yielded a partially delignified acetosolv cellulose pulp and acetic black liquor as a stable aqueous colloidal suspension, from which the lignin was recovered (Nascimento et al., 2014).

Cellulosic fibers from kenaf bast were isolated by three steps: alkali pulping process, bleaching, and mechanical shearing, and properties in fibers were progressively enhanced during the reduction of the size from micro- to nanoscale. Lignin and hemicellulose were almost entirely removed by alkaline pulping process (Karimi et al., 2014).

When lignin was treated with microemulsions, formulated with aqueous solutions of urea or ethylenediamine, interfibril hydrogen bonding was disrupted to produce cellulose nanofibrils. This allowed energy savings during microfluidization, and facilitated higher degrees of deconstruction of cellulose nanofibril structures, with higher water retention value and surface area. Urea-containing microemulsions may reduce energy consumption and weaken the cellulosic matrix (Carrillo et al., 2014).

4.3 MANUFACTURE OF METAL NANOPARTICLES

By increasing concentration of rice straw (lignocellulosic waste material) used as a soft biotemplate, the size, surface area, and porosity of anatase titanium dioxide nanoparticles (TiO_2-NPs) synthesized by sol-gel method was decreased to a 13.0 ± 3.3 nm (Ramimoghadam et al., 2014).

Lignins with spatial three-dimensional structure act as both reducing and stabilizing agents. Lignin accelerated the preparation of silver nanoparticles (AgNPs) from Tollens′ reagent under microwave irradiation. During the formation of AgNPs, lignin was oxidized, and the side chains of lignin were partly disrupted into small molecules, such as hydrocarbon and alcohol. AgNPs thus synthesized can be utilized to detect the toxic ion Hg^{2+} selectively and sensitively (Shen et al., 2014).

Lignin was used effectively as a reducing agent to access nanoparticles of four different transition metals (Au, Pd, Ru, Re). Mechanochemical grinding with lignin more efficiently encapsulated the nanoparticles (Rak et al., 2014).

5 CLINICAL APPLICATION

5.1 EFFECT OF LIGNIN-VITAMIN C TABLET ON HSV-1-INFECTION

Due to the synergistic effect of lignin and vitamin C, we have manufactured the LCC-vitamin C tablet and evaluated its anti-HSV activity by a small-scale clinical trial with 48 healthy patients of both genders (4–61 years old, mean: 31 years) who have active lesions of HSV-1. LCC-ascorbic acid tablet (mixture of 50 mg pine cone extract powder JS, 50 mg ascorbic acid, 83 mg maltitol, 13 mg potato starch and 13 mg calcium stearate) or placebo was orally administered three times daily for a month. The patients who began the LCC-ascorbic acid treatment within the first 48 h did not develop HVI-1 characteristic lesions. Even patients who started the treatment later shortened the duration of cold sore lesions and reduced the symptoms. The majority of the patients reported the reduction in both the severity of symptoms and the recurrence episodes after the LCC-ascorbic acid treatment, suggesting its possible applicability for the prevention and treatment of HSV-1 infection (Lopez et al., 2009). The effect of ascorbic acid-free tablet is under investigation.

5.2 EFFECT OF SE ON ORAL LICHENOID DYSPLASIA

Lichen planus is a chronic mucocutaneous disease that affects the skin, tongue, and oral mucosa. Oral lichen planus (OLP) most commonly presents the reticular form manifested as white lacy streaks on the mucosa. The lesions tend to be bilateral and are asymptomatic. The lacy streaks may also be seen on other parts of the mouth, including the gingiva (gums), the tongue, palate, and lips. The cause of lichen planus is not known, possibly allergic reactions, drugs, stress, and virus infection

Long-term (10 months) treatment of SE progressively reduced both the area of white streaks and the base-line levels of interleukin-6 and 8 in the saliva of oral lichenoid dysplasia patient. His periodontal condition (sway of the molar, swelling of the ligament tissues, irritation) was much improved by unknown reason (Matsuta et al., 2012). In the saliva, IL-8 concentration was two orders higher than that of IL-6, and SE treatment kept the lower level of IL-8 concentration for longer periods. Repeated clinical study with larger numbers of patients is needed to confirm this phenomenon.

5.3 ANTIHALITOSIS EFFECT OF SE-CONTAINING TOOTHPASTE

A total of 12 volunteers participated. They brushed their teeth three times after taking each meal with SE containing toothpaste (SETP) (that contains 26.2% SE) or placebo toothpaste. Halitosis in the breath (expressed as volatile sulfur concentration estimated by Brethtron) and bacterial number on the tongue (measured by bacterial counter) were measured by commercially available portable apparatuses at the fixed time in the morning. Some relationship was observed between halitosis and bacterial number from each individual, especially when severe halitosis subjects were included. Repeated experiments demonstrated that SETP significantly reduced the halitosis but not that of bacterial number on the tongue (Sakagami et al., 2016a).

6 FUTURE DIRECTION

This chapter demonstrate that LCC showed excellent anti-HIV activity and anti-UV activity, in contrast to their weak tumor-selective cytotoxicity. Alkaline extracts contain various sizes of LCC and its degradation products, making it difficult to investigate the relationship between the molecular weight and biological activity. It is conceivable that with the increase in the molecular weight, the water solubility of LCC may decline, and make it difficult to be sterilized by millipore filteration. Even if LCC was purified, the aggregates may be formed when it is dissolved in water, and insolubility may reduce the biological activity.

There is a possibility that bacterial endotoxin (LPS) in the soil and air may contaminate LCC during the isolation step, because LPS is similarly extracted with alkaline solution and precipitated with acid. Alkaline extraction steps inactivate LPS, but also degrade the LCC into its smaller size. Therefore, the conditions for alkaline extraction should be optimized to maximize the LPS inactivation and minimize the loss of biological activity.

It is generally accepted that lignin is linked by various amounts of polysaccharides, yielding LCCs having broad ranges of molecular weights. Phenylpropenoid polymer portion of LCC have potent antiviral activity, whereas carbohydrate moiety have immunopotentiation activity, possibly via activation of TLR pathway and chemokine expression. When LCC is prepared from the leaves, chlorophyllin may be associated during the extraction step with alkaline solution. This association may enhance antiinflammation activity and inhibition of CYP. When the washing procedure is incomplete, bacterial LPS in the soil and air may associate with LCC, considering that both LPS and LCC are precipitated by acid. This association may trigger the background level of immune response–related genes expression. These associations may produce diverse biological activities, depending on the plant species and preparative procedures.

LCC is separated into two distinct parts. Lignin moiety is involved in the antiviral activity, anti-UV activity, iodination of MPO-positive cells and radical scavenging activity, whereas carbohydrate moiety may be involved in immunopotentiation, such as antimicrobial activity and antiparasite activity via dectin-2 (Fig. 18.3). Action points of LCC may be multiple. At present, whether LCC acts outside of the cells or inside of the cells is not clear. We found that *p*-coumaric acid, one of degradation product of SE, was detected from the cells after incubation of cells with SE, suggesting that *p*-coumaric acid is also involved in the expression of biological activities of SE. We have recently found that antistomatitis effect of SE was significantly reduced when it was added to the cells, which are already committed to the inflammatory cells, expressing higher level of inflammatory cytokines (Sakagami, unpublished data). Therefore, it possible that LCC may nonspecifically inhibit the binding of proinflammatory substances, such as LPS and IL-1β to their specific receptors. Further investigations are necessary to confirm these possibilities.

In conclusion, LCCs showed broad and potent antiviral activity and synergism with vitamin C. The factors that affect the biological activity of LCC are molecular weight, solubility, and LPS contamination. Because virus is one of the major risk factors of oral cavity cancer (Sakagami, 2010), antiviral action of LCC may reduce the incidence of virus-triggered diseases, such as cancer. These unique properties of LCCs suggest their functionality as alternative medicine. Excellent anti-UV activity of LCC may suggest its utilization in the cosmetic industry.

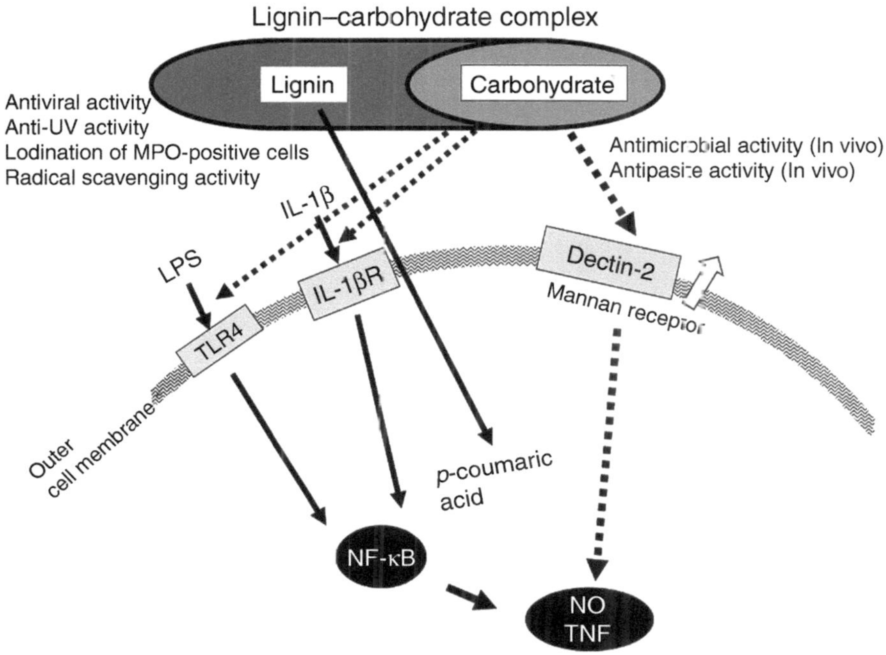

FIGURE 18.3 Putative Mode of Action of Lignin-Carbohydrate Complex

ACKNOWLEDGMENTS

This study was supported in part by a Grant-in-Aid from the Ministry of Education, Science, Sports, and Culture of Japan (Sakagami No.14370607, 2002–2004, Sakagami No. 19592156, 2007–2009, Sakagami No. 16K11519, 2016-2018), and Grant-in-Aid for Challenging Exploratory Research from The Ministry of Education, Culture, Sports, Science, and Technology (Sakagami H. 25670897).

REFERENCES

Abdullah, H.M., Abdul Latif, M.H., Attiya, H.G., 2013. Characterization and determination of lignin in different types of Iraqi phoenix date palm pruning woods. Int. J. Biol. Macromol. 61, 340–346.

Abe, M., Okamoto, K., Konno, K., Sakagami, H., 1989. Induction of antiparasite activity by pine cone lignin-related substances. In Vivo 3, 359–362.

Ago, M., Jakes, J.E., Rojas, O.J., 2013. Thermomechanical properties of lignin-based electrospun nanofibers and films reinforced with cellulose nanocrystals: a dynamic mechanical and nanoindentation study. ACS Appl. Mater. Interf. 5, 11768–11776.

Azuma, J.-I., Tetsuo, K., 1988. Lignin-carbohydrate complexes from various sources. Methods Enzymol. 161, 12–18.

Carrillo, C.A., Laine, J., Rojas, O.J., 2014. Microemulsion systems for fiber deconstruction into cellulose nanofibrils. ACS Appl. Mater. Interf. 6, 22622–22627.
Davin, L.B., Wang, H.B., Crowell, A.L., Bedgar, D.L., Martin, D.M., Sarkanen, S., Lewis, N.G., 1997. Stereoselective bimolecular phenoxy radical coupling by an auxiliary (dirigent) protein without an active center. Science 275, 362–366.
Emiliani, G., Fondi, M., Fani, R., Gribaldo, S., 2009. A horizontal gene transfer at the origin of phenylpropanoid metabolism: a key adaptation of plants to land. Biol. Direct 4, 7.
Fukai, T., Sakagami, H., Toguchi, M., Takayama, F., Iwakura, I., Atsumi, T., Ueha, T., Nakashima, H., Nomura, T., 2000. Cytotoxic activity of low molecular weight polyphenols against human oral tumor cell lines. Anticancer Res. 20, 2525–2536.
Fukuchi, K., Sakagami, H., Ikeda, M., Kawazoe, Y., Oh-Hara, T., Konno, K., Ichikawa, S., Hata, N., Kondo, H., Nonoyama, M., 1989a. Inhibition of herpes simplex virus infection by pine cone antitumor substances. Anticancer Res. 9, 313–317.
Fukuchi, K., Sakagami, H., Okuda, T., Hatano, T., Tanuma, S., Kitajima, K., Inoue, Y., Inoue, S., Ichikawa, S., Nonoyama, M., et al., 1989b. Inhibition of herpes simplex virus infection by tannins and related compounds. Antiviral Res. 11, 285–297.
Garcia-Contreras, R., Sugimoto, M., Umemura, N., Kaneko, M., Hatakeyama, Y., Soga, T., Tomita, M., Scougall-Vilchis, R.J., Contreras-Bulnes, R., Nakajima, H., Sakagami, H., 2015. Alteration of metabolomic profiles by titanium dioxide nanoparticles in human gingivitis model. Biomaterials 57, 33–40.
Gilca, I.A., Popa, V.I., Crestini, C., 2015. Obtaining lignin nanoparticles by sonication. Ultrason. Sonochem. 23, 369–375.
Hambardzumyan, A., Foulon, L., Chabbert, B., Aguie-Beghin, V., 2012. Natural organic UV-absorbent coatings based on cellulose and lignin: designed effects on spectroscopic properties. Biomacromolecules 13, 4081–4088.
Hanaoka, A., Sakagami, H., Konno, K., 1989. Pine cone antitumor substances stimulate cytotoxic factor production in young mice, but not in aged or tumor-bearing mice. Showa Univ. J. Med. Sci. 1, 57–63.
Harada, H., Sakagami, H., Konno, K., Sato, T., Osawa, N., Fujimaki, M., Komatsu, N., 1988. Induction of antimicrobial activity by antitumor substances from pine cone extract of *Pinus parviflora* Sieb. et Zucc. Anticancer Res. 8, 581–587.
Harada, H., Sakagami, H., Nagata, K., Oh-Hara, T., Kawazoe, Y., Ishihama, A., Hata, N., Misawa, Y., Terada, H., Konno, K., 1991. Possible involvement of lignin structure in anti-influenza virus activity. Antiviral Res. 15, 41–49.
Hasegawa, T., Tanaka, A., Hosoda, A., Takano, F., Ohta, T., 2008. Antioxidant C-glycosyl flavones from the leaves of *Sasa kurilensis* var. gigantea. Phytochemistry 69, 1419–1424.
Ikeda, M., Sakagami, H., Konno, K., 1988. Purification of differentiation-inducing substance present in pine cone extract (in Japanese. Showa Med. J. 48, 357–361.
Jeon, J.W., Zhang, L., Lutkenhaus, J.L., Laskar, D.D., Lemmon, J.P., Choi, D., Nandasiri, M.I., Hashmi, A., Xu, J., Motkuri, R.K., Fernandez, C.A., Liu, J., Tucker, M.P., Mcgrail, P.B., Yang, B., Nune, S.K., 2015. Controlling porosity in lignin-derived nanoporous carbon for supercapacitor applications. ChemSusChem 8, 428–432.
Jeong, Y.H., Chung, S.Y., Han, A.R., Sung, M.K., Jang, D.S., Lee, J., Kwon, Y., Lee, H.J., Seo, E.K., 2007. P-glycoprotein inhibitory activity of two phenolic compounds, (−)-syringaresinol and tricin from *Sasa borealis*. Chem. Biodivers. 4, 12–16.
Kantoh, K., Ono, M., Nakamura, Y., Nakamura, Y., Hashimoto, K., Sakagami, H., Wakabayashi, H., 2010. Hormetic and anti-radiation effects of tropolone-related compounds. In Vivo 24, 843–851.
Karimi, S., Tahir, P.M., Karimi, A., Dufresne, A., Abdulkhani, A., 2014. Kenaf bast cellulosic fibers hierarchy: a comprehensive approach from micro to nano. Carbohydr. Polym. 101, 878–885.
Kato, T., Horie, N., Matsuta, T., Naoki, U., Shimoyama, T., Kaneko, T., Kanamoto, T., Terakubo, S., Nakashima, H., Kusama, K., Sakagami, H., 2012. Anti-UV/HIV activity of Kampo medicines and constituent plant extracts. In Vivo 26, 1007–1013.

Kato, T., Hino, S., Horie, N., Shimoyama, T., Kaneko, T., Kusama, K., Sakagami, H., 2014. Anti-UV activity of Kampo medicines and constituent plant extracts: re-evaluation with skin keratinocyte system. In Vivo 28, 571–578.

Kawano, M., Sakagami, H., Satoh, K., Shioda, S., Kanamoto, T., Terakubu, S., Nakashima, H., Makino, T., 2010a. Lignin-like activity of *Lentinus edodes* mycelia extract (LEM). In Vivo 24, 543–551.

Kawano, M., Thet, M.M., Makino, T., Kushida, T., Sakagami, H., 2010b. DNA microarray analysis of signaling pathway in macrophages stimulated by lignin-carbohydrate complex from lentinus edodes mycelia (LEM) extract. Anticancer Res. 30, 2567–2576.

Keplinger, T., Konnerth, J., Aguie-Beghin, V., Ruggeberg, M., Gierlinger, N., Burgert, I., 2014. A zoom into the nanoscale texture of secondary cell walls. Plant Methods 10, 1.

Kim, S., Fernandes, M.M., Matama, T., Loureiro, A., Gomes, A.C., Cavaco-Paulo, A., 2013. Chitosan-lignosulfonates sono-chemically prepared nanoparticles: characterisation and potential applications. Colloids Surf. B 103, 1–8.

Kim, S.K., Kim, Y.K., Lee, H., Lee, S.B., Park, H.S., 2014. Superior pseudocapacitive behavior of confined lignin nanocrystals for renewable energy-storage materials. ChemSusChem 7, 1094–1101.

Klebanoff, S.J., 1967. Iodination of bacteria: a bactericidal mechanism. J. Exp. Med. 126, 1063–1078.

Koizumi, N., Sakagami, H., Utsumi, A., Fujinaga, S., Takeda, M., Asano, K., Sugawara, I., Ichikawa, S., Kondo, H., Mori, S., et al., 1993. Anti-HIV (human immunodeficiency virus) activity of sulfated paramylon. Antiviral Res. 21, 1–14.

Kurakata, Y., Sakagami, H., Takeda, M., Konno, K., Kitajima, K., Ichikawa, S., Hata, N., Sato, T., 1989. Mitogenic activity of pine cone extracts against cultured splenocytes from normal and tumor-bearing animals. Anticancer Res. 9, 961–966.

Kushida, T., Makino, T., Tomomura, M., Tomomura, A., Sakagami, H., 2011. Enhancement of Dectin-2 gene expression by lignin-carbohydrate complex from Lentinus edodes mycelia extract (LEM) in a mouse macrophage-like cell line. Anticancer Res. 31, 1241–1248.

Lai, P.K., Donovan, J., Takayama, H., Sakagami, H., Tanaka, A., Konno, K., Nonoyama, M., 1990. Modification of human immunodeficiency viral replication by pine cone extracts. AIDS Res. Hum. Retrov. 6, 205–217.

Lai Pk, O.-H.T., Tamura, Y., Kawazoe, Y., Konno, K., Sakagami, H., Tanaka, A., Nonoyama M, 1992. Polymeric phenylpropenoids are the active components in the pine cone extract that inhibit the replication of type-1 human immunodeficiency virus in vitro. J. Gen. Appl. Microbiol. 38, 303–323.

Lee, H.V., Hamid, S.B., Zain, S.K., 2014. Conversion of lignocellulosic biomass to nanocellulose: structure and chemical process. Sci. World J. 2014, 631013.

Lewis, N.G., Yamamoto, E., 1990. Lignin: occurrence, biogenesis and biodegradation. Annu. Rev. Plant Physiol. Plant Mol. Biol. 41, 455–496.

Li, M., Wang, L.J., Li, D., Cheng, Y.L., Adhikari, B., 2014. Preparation and characterization of cellulose nanofibers from de-pectinated sugar beet pulp. Carbohydr. Polym. 102, 136–143.

Lopez, B.S., Yamamoto, M., Utsumi, K., Aratsu, C., Sakagami, H., 2009. A clinical pilot study of lignin—ascorbic acid combination treatment of herpes simplex virus. In Vivo 23, 1011–1016.

Lu, Q.F., Zhang, J.Y., Yang, J., He, Z.W., Fang, C.Q., Lin, Q., 2013. Self-assembled poly(*N*-methylaniline)-lignosulfonate spheres: from silver-ion adsorbent to antimicrobial material. Chemistry 19, 10935–10944.

Ma, R., Hao, W., Ma, X., Tian, Y., Li, Y., 2014. Catalytic ethanolysis of Kraft lignin into high-value small-molecular chemicals over a nanostructured alpha-molybdenum carbide catalyst. Angew. Chem. Int. 53, 7310–7315.

Manabe, H., Sakagami, H., Ishizone, H., Kusano, H., Fujimaki, M., Wada, C., Komatsu, N., Nakashima, H., Murakami, T., Yamamoto, N., 1992. Effects of Catuaba extracts on microbial and HIV infection. In Vivo 6, 161–165.

Matsuta, T., Sakagami, H., Satoh, K., Kanamoto, T., Terakubo, S., Nakashima, H., Kitajima, M., Oizumi, H., Oizumi, T., 2011. Biological activity of luteolin glycosides and tricin from *Sasa senanensis* Rehder. In Vivo 25, 757–762.

Matsuta, T., Sakagami, H., Tanaka, S., Machino, M., Tomomura, M., Tomomura, A., Yasui, T., Itoh, K., Sugiura, T., Kitajima, M., Oizumi, H., Oizumi, T., 2012. Pilot clinical study of *Sasa senanensis* rehder leaf extract treatment on lichenoid dysplasia. In Vivo 26, 957–962.

Moreira-Vilar, F.C., Siqueira-Soares Rde, C., Finger-Teixeira, A., De Oliveira, D.M., Ferro, A.P., Da Rocha, G.J., Ferrarese Mde, L., Dos Santos, W.D., Ferrarese-Filho, O., 2014. The acetyl bromide method is faster, simpler and presents best recovery of lignin in different herbaceous tissues than Klason and thioglycolic acid methods. PLoS One 9, e110000.

Moreno-Murguia, B., Soto-Mercado, J.R., Morales-Malacara, J.B., Castano, V.M., 2015. Production of micro- and nanosilica from soil inhabiting *Folsomia candida* fed with treated rice husk. IET Nanobiotechnol. 9, 215–219.

Nagata, K., Sakagami, H., Harada, H., Nonoyama, M., Ishihama, A., Konno, K., 1990. Inhibition of influenza virus infection by pine cone antitumor substances. Antiviral Res. 13, 11–21.

Nair, S.S., Sharma, S., Pu, Y., Sun, Q., Pan, S., Zhu, J.Y., Deng, Y., Ragauskas, A.J., 2014. High shear homogenization of lignin to nanolignin and thermal stability of nanolignin-polyvinyl alcohol blends. ChemSusChem 7, 3513–3520.

Nakamura, K., Yamada, Y., Ikai, H., Kanno, T., Sasaki, K., Niwano, Y., 2012. Bactericidal action of photoirradiated gallic acid via reactive oxygen species formation. J. Agric. Food Chem. 60, 10048–10054.

Nakamura, K., Ishiyama, K., Sheng, H., Ikai, H., Kanno, T., Niwano, Y., 2015. Bactericidal activity and mechanism of photoirradiated polyphenols against gram-positive and -negative bacteria. J. Agric. Food Chem. 63, 7707–7713.

Nakane, A., Minagawa, T., Kato, K., 1988. Endogenous tumor necrosis factor (cachectin) is essential to host resistance against *Listeria monocytogenes* infection. Infect. Immun. 56, 2563–2569.

Nakashima, H., Murakami, T., Yamamoto, N., Naoe, T., Kawazoe, Y., Konno, K., Sakagami, H., 1992a. Lignified materials as medicinal resources. V. Anti-HIV (human immunodeficiency virus) activity of some synthetic lignins. Chem. Pharm. Bull. (Tokyo) 40, 2102–2105.

Nakashima, H., Murakami, T., Yamamoto, N., Sakagami, H., Tanuma, S., Hatano, T., Yoshida, T., Okuda, T., 1992b. Inhibition of human immunodeficiency viral replication by tannins and related compounds. Antiviral Res. 18, 91–103.

Nanbu, T., Matsuta, T., Sakagami, H., Shimada, J., Maki, J., Makino, T., 2011. Anti-UV activity of *Lentinus edodes* mycelia extract (LEM). In Vivo 25, 733–740.

Nanbu, T., Shimada, J., Kobayashi, M., Hirano, K., Koh, T., Machino, M., Ohno, H., Yamamoto, M., Sakagami, H., 2013. Anti-UV activity of lignin-carbohydrate complex and related compounds. In Vivo 27, 133–139.

Nascimento, D.M., Almeida, J.S., Dias, A.F., Figueiredo, M.C., Morais, J.P., Feitosa, J.P., De, F.R.M., 2014. A novel green approach for the preparation of cellulose nanowhiskers from white coir. Carbohydr. Polym. 110, 456–463.

Nguyen, M.H., Hwang, I.C., Park, H.J., 2013. Enhanced photoprotection for photo-labile compounds using double-layer coated corn oil-nanoemulsions with chitosan and lignosulfonate. J. Photochem. Photobiol. B 125, 194–201.

Oh-Hara, T., Sakagami, H., Kawazoe, Y., Kaiya, T., Komatsu, N., Ohsawa, N., Fujimaki, M., Tanuma, S., Konno, K., 1990. Antimicrobial spectrum of lignin-related pine cone extracts of *Pinus parviflora* Sieb. et Zucc. In Vivo 4, 7–12.

Ohno, H., Araho, D., Uesawa, Y., Kagaya, H., Ishihara, M., Sakagami, H., Yamamoto, M., 2013. Evaluation of cytotoxiciy and tumor-specificity of licorice flavonoids based on chemical structure. Anticancer Res. 33, 3061–3068.

Ohno, H., Miyoshi, S., Araho, D., Kanamoto, T., Terakubo, S., Nakashima, H., Tsuda, T., Sunaga, K., Amano, S., Ohkoshi, E., Sakagami, H., Satoh, K., Yamamoto, M., 2014. Efficient utilization of licorice root by alkaline extraction. In Vivo 28, 785–794.

Ono, M., Kantoh, K., Ueki, J., Shimada, A., Wakabayashi, H., Matsuta, T., Sakagami, H., Kumada, H., Hamada, N., Kitajima, M., Oizumi, H., Oizumi, T., 2011. Quest for anti-inflammatory substances using IL-1beta-stimulated gingival fibroblasts. In Vivo 25, 763–768.

Park, H.S., Lim, J.H., Kim, H.J., Choi, H.J., Lee, I.S., 2007. Antioxidant flavone glycosides from the leaves of *Sasa borealis*. Arch. Pharm. Res. 30, 161–166.

Raghuraman, A., Tiwari, V., Zhao, Q., Shukla, D., Debnath, A.K., Desai, U.R., 2007. Viral inhibition studies on sulfated lignin, a chemically modified biopolymer and a potential mimic of heparan sulfate. Biomacromolecules 8, 1759–1763.

Rak, M.J., Friscic, T., Moores, A., 2014. Mechanochemical synthesis of Au, Pd, Ru and Re nanoparticles with lignin as a bio-based reducing agent and stabilizing matrix. Faraday Discuss. 170, 155–167.

Ramimoghadam, D., Bagheri, S., Abd Hamid, S.B., 2014. Biotemplated synthesis of anatase titanium dioxide nanoparticles via lignocellulosic waste material. Biomed. Res. Int. 2014, 205636.

Sakagami, H., 1990. Endogenous production of cytotoxic factor in normal mice by PSK combined with OK-432. Showa Univ. J. Med. Sci. 2, 159–164.

Sakagami, H., 2010. Oral cavity cancer. In: Murat Tuncer, A. (Ed.), Cancer Report 2010 Asian Pacific Organization for CANCER PREVENTION (APOCP). "New Hope in Health" Foundation, MN Medical&Nobel Publishing Company, Chapter 5.1.8, pp. 222–226.

Sakagami, H., 2014. Biological activities and possible dental application of three major groups of polyphenols. J. Pharmacol. Sci. 126, 92–106.

Sakagami, H., Satoh, K., 1997a. Modulating factors of radical intensity and cytotoxic activity of ascorbate (review). Anticancer Res. 17, 3513–3520.

Sakagami, H., Satoh, K., 1997b. Prooxidant action of two antioxidants: ascorbic acid and gallic acid. Anticancer Res. 17, 221–224.

Sakagami, H., Takeda, K., Makino, Y., Konno, K., 1986. Partial purification of novel differentiation-inducing substances(s) from hot water extract of Japanese pine cone. Jpn. J. Cancer Res. 77, 59–64.

Sakagami, H., Ikeda, M., Unten, S., Takeda, K., Murayama, J., Hamada, A., Kimura, K., Komatsu, N., Konno, K., 1987. Antitumor activity of polysaccharide fractions from pine cone extract of *Pinus parviflora* Sieb. et Zucc. Anticancer Res. 7, 1153–1159.

Sakagami, H., Oh-Hara, T., Kaiya, T., Kawazoe, Y., Nonoyama, M., Konno, K., 1989. Molecular species of the antitumor and antiviral fraction from pine cone extract. Anticancer Res. 9, 1593–1598.

Sakagami, H., Hatano, T., Yoshida, T., Tanuma, S., Hata, N., Misawa, Y., Ishii, N., Tsutsumi, T., Okuda, T., 1990a. Stimulation of granulocytic cell iodination by tannins and related compounds. Anticancer Res. 10, 1523–1531.

Sakagami, H., Nagata, K., Ishihama, A., Oh-Hara, T., Kawazoe, Y., 1990b. Anti-influenza virus activity of synthetically polymerized phenylpropenoids. Biochem. Biophys. Res. Commun. 172, 1267–1272.

Sakagami, H., Kawazoe, Y., Oh-Hara, T., Kitajima, K., Inoue, Y., Tanuma, S., Ichikawa, S., Konno, K., 1991. Stimulation of human peripheral blood polymorphonuclear cell iodination by lignin-related substances. J. Leukoc. Biol. 49, 277–282.

Sakagami, H., Asano, K., Hara, Y., Shimamura, T., 1992a. Stimulation of human monocyte and polymorphonuclear cell iodination and interleukin-1 production by epigallocatechin gallate. J. Leukoc. Biol. 51, 478–483.

Sakagami, H., Asano, K., Tanuma, S., Hatano, T., Yoshida, T., Okuda, T., 1992b. Stimulation of monocyte iodination and IL-1 production by tannins and related compounds. Anticancer Res. 12, 377–387.

Sakagami, H., Kuroiwa, Y., Takeda, M., Ota, H., Kazama, K., Naoe, T., Kawazoe, Y., Ichikawa, S., Kondo, H., Yokokura, T., et al., 1992c. Distribution of TNF endogenously induced by various immunopotentiators and *Lactobacillus casei* in mice. In Vivo 6, 247–253.

Sakagami, H., Takeda, M., Kawazoe, Y., Nagata, K., Ishihama, A., Ueda, M., Yamazaki, S., 1992d. Anti-influenza virus activity of a lignin fraction from cone of *Pinus parviflora* Sieb. et Zucc. In Vivo 6, 491–495.

Sakagami, H., Yoshihara, M., Fujimaki, M., Wada, C., Komatsu, N., Nakashima, H., Murakami, T., Yamamoto, N., 1992e. Effect of pine seed shell extract on microbial and viral infection. In Vivo 6, 13–16.

Sakagami, H., Sakagami, T., Yoshida, H., Omata, T., Shiota, F., Takahashi, H., Kawazoe, Y., Takeda, M., 1995. Hypochlorite scavenging activity of polyphenols. Anticancer Res. 15, 917–921.

Sakagami, H., Anzai, S., Goto, S., Takeda, M., 1997a. Enhancement of cytotoxic activity of ascorbate by Acer nikoense Maxim. Ext. Anticancer Res. 17, 4453–4456.

Sakagami, H., Satoh, K., Aiuchi, T., Nakaya, K., Takeda, M., 1997b. Stimulation of ascorbate-induced hypoxia by lignin. Anticancer Res. 17, 1213–1216.
Sakagami, H., Kashimata, M., Toguchi, M., Satoh, K., Odanaka, Y., Ida, Y., Premanathan, M., Arakaki, R., Kathiresan, K., Nakashima, H., Komatsu, N., Fujimaki, M., Yoshihara, M., 1998. Radical modulation activity of lignins from a mangrove plant, *Ceriops decandra* (Griff.) Ding Hou. In Vivo 12, 327–332.
Sakagami, H., Asano, K., Yoshida, T., Kawazoe, Y., 1999. Organ distribution and toxicity of lignin. In Vivo 13, 41–44.
Sakagami, H., Satoh, K., Hakeda, Y., Kumegawa, M., 2000. Apoptosis-inducing activity of vitamin C and vitamin K. Cell. Mol. Biol. 46, 129–143.
Sakagami, H., Hashimoto, K., Suzuki, F., Ogiwara, T., Satoh, K., Ito, H., Hatano, T., Takashi, Y., Fujisawa, S., 2005. Molecular requirements of lignin-carbohydrate complexes for expression of unique biological activities. Phytochemistry 66, 2108–2120.
Sakagami, H., Asano, K., Satoh, K., Takahashi, K., Kobayashi, M., Koga, N., Takahashi, H., Tachikawa, R., Tashiro, T., Hasegawa, A., Kurihara, K., Ikarashi, T., Kanamoto, T., Terakubo, S., Nakashima, H., Watanabe, S., Nakamura, W., 2007. Anti-stress, anti-HIV and vitamin C-synergized radical scavenging activity of mulberry juice fractions. In Vivo 21, 499–505.
Sakagami, H., Amano, S., Kikuchi, H., Nakamura, Y., Kuroshita, R., Watanabe, S., Satoh, K., Hasegawa, H., Nomura, A., Kanamoto, T., Terakubo, S., Nakashima, H., Taniguchi, S., Oizumi, T., 2008a. Antiviral, antibacterial and vitamin C-synergized radical-scavenging activity of *Sasa senanensis* Rehder extract. In Vivo 22, 471–476.
Sakagami, H., Satoh, K., Fukamachi, H., Ikarashi, T., Shimizu, A., Yano, K., Kanamoto, T., Terakubo, S., Nakashima, H., Hasegawa, H., Nomura, A., Utsumi, K., Yamamoto, M., Maeda, Y., Osawa, K., 2008b. Anti-HIV and vitamin C-synergized radical scavenging activity of cacao husk lignin fractions. In Vivo 22, 327–332.
Sakagami, H., Kushida, T., Oizumi, T., Nakashima, H., Makino, T., 2010a. Distribution of lignin carbohydrate complex in plant kingdom and its functionality as alternative medicine. Pharmacol. Ther. 128, 91–105.
Sakagami, H., Zhou, L., Kawano, M., Thet, M.M., Tanaka, S., Machino, M., Amano, S., Kuroshita, R., Watanabe, S., Chu, Q., Wang, Q.T., Kanamoto, T., Terakubo, S., Nakashima, H., Sekine, K., Shirataki, Y., Zhang, C.H., Uesawa, Y., Mohri, K., Kitajima, M., Oizumi, H., Oizumi, T., 2010b. Multiple biological complex of alkaline extract of the leaves of *Sasa senanensis* Rehder. In Vivo 24, 735–743.
Sakagami, H., Kawano, M., Thet, M.M., Hashimoto, K., Satoh, K., Kanamoto, T., Terakubo, S., Nakashima, H., Haishima, Y., Maeda, Y., Sakurai, K., 2011. Anti-HIV and immunomodulation activities of cacao mass lignin carbohydrate complex. In Vivo. 25, 229–236.
Sakagami, H., Amano, O.E., Satoh, S., Kanamoto, K., Terakubo, T., Nakashima, S., Sunaga, H., Otsuki, K., Ikeda, T., Fukuda T, H., 2013. Efficient utilization of plant resources by alkaline extraction. Altern. Integr. Med. 2, 133.
Sakagami, H., S., C., Kanda, Y., Amano, O., Sugimoto, M., Ota, S., Soga, T., Tomita, M., Sato, A., Tanuma, S., Takao, K., Sugita, Y., 2015a. Effects of 3-styrylchromones on metabolic profiles and cell death in oral squamous cell carcinoma cells. Toxocol. Rep. 2, 1281–1290.
Sakagami, H., Uesawa, Y., Ishihara, M., Kagaya, H., Kanamoto, T., Terakubo, S., Nakashima, H., Takao, K., Sugita, Y., 2015b. Quantitative structure-cytotoxicity relationship of oleoylamides. Anticancer Res. 35, 5341–5351.
Sakagami, H., Sheng, H., Ono, K., K., Y., Miyadai, T., T., Y., Nakadai, D., Tanaka, S., Matsumoto, M., Yasui, T., Watanabe, K., Junye, Jia, Natori, T., Suguro-Kitajma, M., Oizumi, H., T., O., 2016a. Anti-halitosis effect of toothpaste supplemented with alkaline extract of the leaves of Sasa senanensis Rehder. In Vivo 30, 107–112.
Sakagami, H., Fukuchi, K., Kanamoto, T., Terakubo, S., Nakashima, H., Natori, T., Suguro-Kitajima, M., Oizumi, H., Yasui, T., Oizumi, T., 2016b. Synergism of Alkaline extract of the leaves of *Sasa senanensis* Rehder and antiviral agents. In Vivo 30, 421–426.
Sakagami, H., Sheng, H., Okudaira, N., Yasui, T., Wakabayashi, H., Jia, J., Natori, T., Suguro-Kitajima, M., 2016c. Oizumi cosmetic potential of lignin-carbohydrate complex. Review. In Vivo 30, 331–339.
Sakagami, H., Okudaira, N., Masuda, Y., Amano, O., Yokose, S., Kanda, Y., Suguro, M., Natori, T., Oizumi, H., Oizumi, T., 2017. Induction of apoptosis in human oral keratinocyte by doxorubicin. Anticancer Res. 37.

Sakai, A., Watanabe, K., Koketsu, M., Akuzawa, K., Yamada, R., Li, Z., Sadanari, H., Matsubara, K., Muroyama, T., 2008. Anti-human cytomegalovirus activity of constituents from Sasa albo-marginata (Kumazasa in Japan). Antivir. Chem. Chemother. 19, 125–132.

Santoni, I., Callone, E., Sandak, A., Sandak, J., Dire, S., 2015. Solid state NMR and IR characterization of wood polymer structure in relation to tree provenance. Carbohydr. Polym. 117, 710–721.

Satoh, K., Anzai, S., Sakagami, H., 1998. Enhancement of radical intensity and cytotoxic activity of ascorbate by *Crataegus cuneata* Sieb et. Zucc. extracts. Anticancer Res. 18, 2749–2753.

Satoh, K., Ida, Y., Ishihara, M., Sakagami, H., 1999a. Interaction between sodium ascorbate and polyphenols. Anticancer Res. 19, 4177–4186.

Satoh, K., Kihara, T., Ida, Y., Sakagami, H., Koyama, N., Premanathan, M., Arakaki, R., Nakashima, H., Komatsu, N., Fujimaki, M., Misawa, Y., Hata, N., 1999b. Radical modulation activity of pine cone extracts of *Pinus elliottii* var Elliottii. Anticancer Res. 19, 357–364.

Shemer-Avni, Y., Wallach, D., Sarov, I., 1988. Inhibition of *Chlamydia trachomatis* growth by recombinant tumor necrosis factor. Infect. Immun. 56, 2503–2506.

Shen, Z., Luo, Y., Wang, Q., Wang, X., Sun, R., 2014. High-value utilization of lignin to synthesize Ag nanoparticles with detection capacity for Hg(2)(+). ACS Appl. Mater. Interf. 6, 16147–16155.

Shimada, C., Uesawa, Y., Ishihara, M., Kagaya, H., Kanamoto, T., Terakubo, S., Nakashima, H., Takao, K., Miyashiro, T., Sugita, Y., Sakagami, H., 2014a. Quantitative structure-cytotoxicity relationship of piperic acid amides. Anticancer Res. 34, 4877–4884.

Shimada, C., Uesawa, Y., Ishihara, M., Kagaya, H., Kanamoto, T., Terakubo, S., Nakashima, H., Takao, K., Saito, T., Sugita, Y., Sakagami, H., 2014b. Quantitative structure-cytotoxicity relationship of phenylpropanoid amides. Anticancer Res. 34, 3543–3548.

Shimada, C., Uesawa, Y., Ishii-Nozawa, R., Ishihara, M., Kagaya, H., Kanamoto, T., Terakubo, S., Nakashima, H., Takao, K., Sugita, Y., Sakagami, H., 2014c. Quantitative structure-cytotoxicity relationship of 3-styrylchromones. Anticancer Res. 34, 5405–5411.

Shimizu, N., Naoe, T., Kawazoe, Y., Sakagami, H., Nakashima, H., Murakami, T., Yamamoto, N., 1993. Lignified materials as medicinal resources. VI. Anti-HIV activity of dehydrogenation polymer of p-coumaric acid, a synthetic lignin, in a quasi-in-vivo assay system as an intermediary step to clinical trials. Biol. Pharm. Bull. 16, 434–436.

Si, S., Chen, Y., Fan, C., Hu, H., Li, Y., Huang, J., Liao, H., Hao, B., Li, Q., Peng, L., Tu, Y., 2015. Lignin extraction distinctively enhances biomass enzymatic saccharification in hemicelluloses-rich Miscanthus species under various alkali and acid pretreatments. Bioresour. Technol. 183, 248–254.

Suzuki, H., Iiyama, K., Yoshida, O., Yamazaki, S., Yamamoto, N., Toda, S., 1990. Structural characterization of the immunoactive and antiviral water-solubilized lignin in an extract of the culture medium of Lentinus edodes mycelia (LEM). Agric. Biol. Chem. 54, 479–487.

Suzuki, R., Matsuno, S., Sakagami, H., Okada, Y., Shirataki, Y., 2014. Search of new cytotoxic crude materials against human oral squamous cell carcinoma using 1H NMR-based metabolomics. Anticancer Res. 34, 4117–4120.

Ten, E., Ling, C., Wang, Y., Srivastava, A., Dempere, L.A., Vermerris, W., 2014. Lignin nanotubes as vehicles for gene delivery into human cells. Biomacromolecules 15, 327–338.

Thakkar, J.N., Tiwari, V., Desai, U.R., 2010. Nonsulfated, cinnamic acid-based lignins are potent antagonists of HSV-1 entry into cells. Biomacromolecules 11, 1412–1416.

Tsukada, M., Sheng, H., Mokudai, T., Oizumi, S., Kamachi, T., Niwano, Y., 2016. Bactericidal action of photo-irradiated aqueous extract from residue of crushed grape for wine preparation. Biocontrol Sci. 21 (2), 113–121.

Ueki, J., Sakagami, H., Wakabayashi, H., 2013. Anti-UV activity of newly-synthesized water-soluble azulenes. In Vivo 27, 119–126.

Uesawa, Y., Sakagami, H., Ishihara, M., Kagaya, H., Kanamoto, T., Terakubo, S., Nakashima, H., Yahagi, H., Takao, K., Sugita, Y., 2015. Quantitative structure-cytotoxicity relationship of 3-styryl-2H-chromenes. Anticancer Res. 35, 5299–5307.

Unten, S., Sakagami, H., Konno, K., 1989. Stimulation of granulocytic cell iodination by pine cone antitumor substances. J. Leukoc. Biol. 45, 168–175.

Wei, H., Donohoe, B.S., Vinzant, T.B., Ciesielski, P.N., Wang, W., Gedvilas, L.M., Zeng, Y., Johnson, D.K., Ding, S.Y., Himmel, M.E., Tucker, M.P., 2011. Elucidating the role of ferrous ion cocatalyst in enhancing dilute acid pretreatment of lignocellulosic biomass. Biotechnol. Biofuel. 4, 48.

Zeng, J., Tong, Z., Wang, L., Zhu, J.Y., Ingram, L., 2014. Isolation and structural characterization of sugarcane bagasse lignin after dilute phosphoric acid plus steam explosion pretreatment and its effect on cellulose hydrolysis. Bioresour. Technol. 154, 274–581.

Zhang, Y., But, P.P., Ooi, V.E., Xu, H.X., Delaney, G.D., Lee, S.H., Lee, S.F., 2007. Chemical properties, mode of action, and in vivo anti-herpes activities of a lignin-carbohydrate complex from *Prunella vulgaris*. Antiviral Res. 75, 242–249.

Zhou, L., Hashimoto, K., Satoh, K., Yokote, Y., Kitajima, M., Oizumi, T., Oizumi, H., Sakagami, H., 2009. Effect of Sasa senanensis Rehder extract on NO and PGE_2 production by activated mouse macrophage-like RAW264. 7 cells. In Vivo 23, 773–777.

CHAPTER

19

NOVEL LIPID NANOSTRUCTURES FOR DELIVERY OF NATURAL AGENTS WITH ANTIOXIDANT, ANTIINFLAMMATORY AND ANTISTROKE POTENTIAL: PERSPECTIVES AND OUTCOMES

Hafsa Ahmad, Abhishek Arya, Satish Agrawal, Anil K. Dwivedi
CSIR-Central drug Research Institute, Lucknow, India

CHAPTER OUTLINE

1 Introduction 578
2 Stroke 579
2.1 Types of Stroke on the Basis of Pathogenesis 580
2.2 Risk Factors 580
2.3 Pathophysiology of Stroke: (Mechanism of Ischemia) 581
2.4 Cell Death in Cerebral Ischemia 581
3 Oxidative Stress Inflammation, Antioxidants, and Their Mechanism of Action 582
3.1 Mechanisms of Antioxidants 584
3.2 Natural Antioxidants 584
4 Bioactives with Potential Antioxidant, Antiinflammatory, and Antistroke Effects 584
5 Limitations of Bioactives to Elicit the Desired Therapeutic Response 589
6 Novel Lipid Nanostructures 589
6.1 Advantages of LNs-Based Delivery 590
6.2 Mechanism of Drug Delivery via LNs 591
6.3 Basic Ingredients of Lipid Nanocarriers 592
6.4 Types of LNs 592
6.5 Insights into SLNs and Liposomes with Their Methods of Preparation 595
6.6 Characterization of LNs 597
6.7 Challenges of LNs-based delivery systems 599
6.8 Role of LNs in Delivery of Bioactives for the Brain 599
7 Conclusions: Way Ahead with LNs 600
Acknowledgement 601
References 601

1 INTRODUCTION

The term "stroke" describes the clinical situation in which acute vascular and brain tissue changes affects focal neurologic deficits lasting for 24 h or more. These deficits result from either inadequate blood flow or hemorrhage (Mackey, 2008). Ischemic stroke accounts for 85% of all strokes. The r-tPA was introduced in 1996, and currently the only treatment available with a narrow therapeutic window (Iadecola and Anrather, 2011). The pathophysiological changes that occur in response to stroke might be free radical production, excitotoxicity, sodium and calcium influx disruption, enzymatic changes, stimulation of the inflammatory processes, endothelin release, platelet and leukocyte activation, delayed coagulation and endothelial dysfunction. These events contribute to ischemic injury that follows after stroke onset (Gupta et al., 2010). It is therefore believed that therapeutic agents that can counter oxidative stress and associated inflammation could be beneficial in brain injury associated with cerebral stroke. Literature suggests that there are several candidates of natural origin like Resveratrol (Shin et al., 2010), Cyclosporine A (Shin et al., 2010), and several plant candidates like *Withania somnifera* (Ahmad et al., 2016), *Panax ginseng* (Nabavi et al., 2015), *Scutellaria baicalensis* (Tang et al., 2004; Zhang et al., 2006), and *Uncaria sinensis* (Yokoyama et al., 2004) that have potential neuroprotective effects in advent of focal ischemia. However, their use for therapy is often restricted due to the biodegradability, poor aqueous solubility, and low bioavailability. Hence, here the aim of the therapy is not merely neuroprotection but administration and delivery of these neuroprotectants in such a manner where they could be put to use in a controlled site-specific mechanism to provide optimum benefit and overcome the demerits of solubility, bioavailability, and degradation. This is probably where the concept of nanoparticulate drug delivery plays a significant role. Fig. 19.1 presents a schematic representation of the lipid nanoparticulate delivery of antioxidant and antiinflammatory agents of natural origin in providing protection against cerebral stroke and alleviating the pathophysiological changes that occur during I/R brain injury.

A nanoparticulate drug delivery system is largely a nanosized formulation that allows introduction of an active in the body and controls its release thereby regulating its therapeutic outcome. Among a wide range of these nanostructures, lipid nanoparticles are the most popular delivery systems owing to associated merits like biocompatibility, reduced toxicity and versatility, thus making them attractive candidates for formulation of drugs and biologicals. These lipid nanocarriers (LNs) have been used primarily to improve the aqueous solubility of drugs (BCS Classes II and IV) and consequently their bioavailability. They also enhance potency by protecting the therapeutic agent from biodegradation or biotransformation. Due to their characteristic size dependent properties LNs provide a means of controlled secondary/tertiary level site-specific drug targeting. LNs also address the toxicity issues associated with drugs by altering their biodistribution and thus contributing to the success of the therapy. Solid lipid nanoparticles (SLN) and liposomes are the most conventional forms of LNs, however newer modifications of SLNs like nanostructured lipid carriers (NLC), lipid drug conjugate (LDC), and liquid crystal delivery systems were introduced in the past decades. LNs exploit the properties of lipids to deliver active drug candidates. Most LNs incorporate one or more of the following lipids: homolipids (cerides, glycerides, and sterides), heterolipids (phospholipids, glycolipids, and sulfolipids) and complex lipids (lipoproteins, chylomicrons). LNs may be produced by high-pressure homogenization, or by solvent emulsification/evaporation techniques. Characterization of LNs involves evaluation of particle size characteristics, zeta potential,

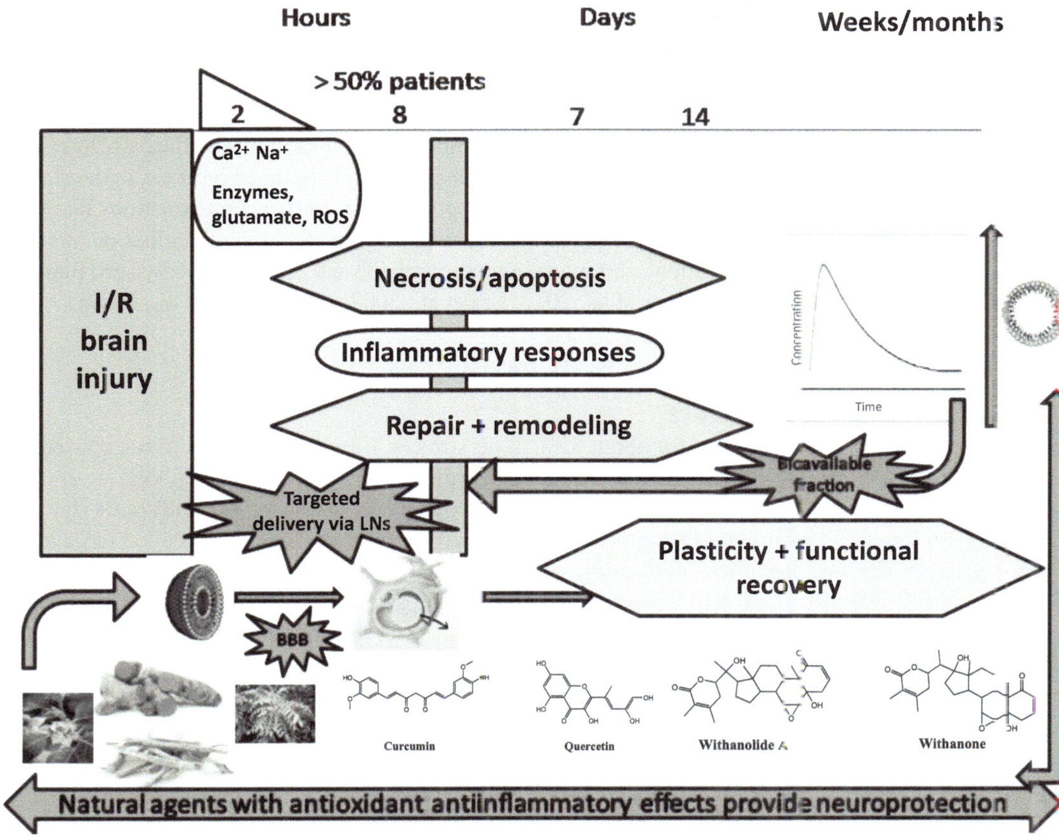

FIGURE 19.1

Lipid nanoparticulate delivery of natural antioxidants and antiinflammatory agents in providing neuroprotection against cerebral stroke and alleviating pathophysiological changes in i/r brain injury.

crystallinity, and polymorphism, presence of additional colloidal structures, drug content, and entrapment efficiency (Attama et al., 2012).

2 STROKE

Stroke is one of the most common disabling neurological disorders and is a leading cause of death worldwide. It involves sudden loss of neurologic function resulting from focal disturbance of cerebral blood flow due to ischemia or hemorrhage. Stroke can cause terminal damage of neurological functions, disability, or mortality (NINDS, 2011; Heron et al., 2009). In ischemic brain tissue: the deprivation of oxygen and glucose supplies and blocking of synthesis of adenosine triphosphate (ATP) causes failure of energy-dependent processes necessary for tissue cell survival. Consequently, a number of cellular and extracellular elements are damaged. A perturbed homeostasis of metal ions is recognized in stroke (Chen et al., 2012).

An event where functions of the temporal and focal brain are disturbed momentarily and symptoms are manifested for less than one hour is termed as a transient ischemic attack (TIA). This cerebral dysfunction, of vascular origin, may not cause neurologic damage, but it increases the risk to a subsequent stroke within three months. Stroke is the third major cause of death in the United States, especially in geriatrics. Statistics shows that 8%–12% of individuals with strokes might result in mortality within 30 days in the age group of 45–64 years. More than 67% deaths due to stroke globally have been reported in developing countries where stroke mortality is higher (World Health Report, 2004). In this context nations like India and China have to face a huge socioeconomic burden in terms of costs involved in rehabilitation of stroke victims and/or survivors as the population is now surviving through the age bracket 55–65 years (Banerjee et al., 2005; Das and Banerjee, 2008; Kaur et al., 2011; Liu et al., 2007; Lloyd-Jones et al., 2009).

2.1 TYPES OF STROKE ON THE BASIS OF PATHOGENESIS

2.1.1 Occlusive/ischemic disease

Occlusive/ischemic disease accounts for about 85% of all strokes. Infarction of brain tissue caused due to occlusion of a cerebral blood vessel results in an ischemic stroke. It can be thrombotic or embolic in nature and can cause permanent damage. Thrombotic: occlusion of the vessel with hemodynamic effects (>70%) resulting from growth/rupture of atherosclerotic plaque. Hence, it is called large artery or atherosclerotic stroke. Embolic: heart, aortic arch, and major cervical vessels might be the site of origin of emboli, which travels across vessels to produce the stroke. The most common emboli are formed due to atrial fibrillation. Small subcortical or lacunar infarcts/lacunar disease of small arteries is a subtype of thrombotic stroke, which is caused due to lipohyalinosis of small brain vessels (<15 mm diameter) in thalamus, pons, and basal ganglia, in elderly patients with hypertension or diabetes and involves a series of tiny ischemic strokes which presents symptoms like clumsiness, weakness and emotional variability (Mackey, 2008).

2.1.2 Transient ischemic attacks

Transient ischemic attacks (TIA) are small strokes due to temporarily reduced blood supply, secondary to small platelet emboli. Symptoms disappear within a few minutes to 24 h and these strokes rapidly dissociate.

2.1.3 Hemorrhagic stroke/hemorrhagic cerebrovascular disease

Hemorrhagic stroke/hemorrhagic cerebrovascular disease involves acute bleeding in the brain parenchyma or subarachnoid space. About 15% of all strokes are hemorrhagic in nature.

2.1.4 Pervasive

Pervasive is global cerebral ischemia and includes cardiovascular dysfunction and pervasive vascular lesions.

2.1.5 Prolonged ischemic neurological deficiency

Prolonged ischemic neurological deficiency (PRIND) is characterized by restoration of clinical symptoms within the first 3 weeks.

2.2 RISK FACTORS

The risk factors associated with cerebral stroke is given in Table 19.1

Table 19.1 Risk Factors Associated with Cerebral Stroke

Risk Factors	
Nonmodifiable	**Modifiable**
• Age: High risk in the elderly (doubles each successive decade of life after 45 years) • Gender: 30% higher incidence in males compared to females • Neonatal weight: High vulnerability in individuals with low birth weight (<5 lbs) • Race: Asians, African Americans, and Hispanics at higher risk than whites • Genetic factors • History of stroke: A previous episode of TIA or stroke is a risk factor; 40% survivors are vulnerable to a second attack within next 5 years	• Hypertension • Diabetes mellitus • Hyperlipidemia • Obesity (central body fat distribution) • High alcoholism • Cigarette smoking • Drug abuse • Physical activity • Diet and nutrition • Use of oral contraceptives (>50mcg estrogen) • Hormone replacement therapy • Atrial fibrillation • Valvular heart disease • Asymptomatic carotid stenosis

2.3 PATHOPHYSIOLOGY OF STROKE: (MECHANISM OF ISCHEMIA)

The mechanisms of ischemia can be understood as follow:

1. *Thrombosis*: Thrombosis is the in situ formation of a clot within an artery that remains there for a sufficiently long time to cause ischemic insult. Pathological mechanisms in the local endothelium often elicit a thrombotic episode.
2. *Embolism*: Besides the heart being a common source of a thrombo-embolus, cerebral circulation might be the carrier of several materials to the brain, which lodge in vessels and precipitate stroke.
3. *Systemic hypoperfusion*: This occurs due to loss of arterial pressure. Cardiac arrest due to myocardial infarction or arrhythmia is a leading cause of systemic hypoperfusion.
4. *Obliteration of the arterial lumen*: Noninflammatory vasculopathy, inflammatory vasculitis, vasospasm, or compression might result in luminal narrowing, which causes ischemia.
5. *Cerebral venous thrombosis (CVT)*: Thrombi develop because of several predisposing conditions. CVT is often multifactorial and can lead to vascular congestion, impairment of forward flow, and eventually infarction.

2.4 CELL DEATH IN CEREBRAL ISCHEMIA

An energy failure occurs in ischemia due to an imbalance between energy demand and supply. Depleted energy reserves result in inhibition of protein synthesis and expression of genes required for programmed cell death. Infarction is synonymous with necrosis and is the predominant histological feature of stroke involving cytoplasmic swelling, organelle, and cell membranes dissolution, and inflammatory responses. In rat MCAO (middle cerebral artery occlusion) model, cell death is primarily affected through necrotic processes; however, neuronal death in cerebral ischemia may be executed as programmed cell death under specific circumstances. In transient ischemia models, because energy failure is only transient, the neuronal cell death occurs slowly and is apoptotic in nature. Here the genomic DNA is cleaved and gels into fragments of different sizes as in programmed cell death. The duration and severity of ischemia in stroke directs the nature of cell death (necrotic or apoptotic). It has, however, been indicated from several studies that mechanisms supporting programmed cell death are activated in ischemic neurons. Programmed cell death via expression and upregulation of Fas receptor, the role of TNF signaling pathways, activation of Caspase 8 and changes in expression and activity of ERK and JNK kinase pathways have all been implicated after ischemia (Graham and Hickey, 2002; Maas and Safdieh, 2009; Mackey, 2008).

3 OXIDATIVE STRESS INFLAMMATION, ANTIOXIDANTS, AND THEIR MECHANISM OF ACTION

Oxidative stress is characterized by increased generation of oxygen free radicals, which result in generations of lipid peroxidation products like malondialdehyde (MDA) that increase vascular inflammation. Inflammation and oxidative stress together act as hallmarks of life-threatening events like myocardial infarction and stroke. Besides a deranged lipid profile, oxidative stress resulting from a perturbation in reactive oxygen species (ROS) and antioxidant defenses deregulates a series of cellular functions and contributes to various pathological conditions like acute ischemic stroke and TIA. Intracellular protection mechanisms that counter oxidative stress function via enzymatic and nonenzymatic antioxidant enzymes like superoxide dismutase (SOD) and vitamin E. Strategies adopted for attenuation of oxidation and vascular inflammation includes administration of antioxidants and secretory phospholipase A 2 (PLA2) inhibitors that reduce oxidative stress, and inhibits production of proinflammatory prostaglandins and leukotrienes (Kaur et al., 2011; Rosenson and Stafforini, 2012; Sindhi et al., 2013). Small amounts of ROS are continuously generated in the plasma membranes and in the vicinity of cellular organelles in the body. They act at the site of their generation and can damage cell structures, including membrane lipids, proteins, enzymes, and nucleic acids. ROS (hydrogen peroxide and hypochlorous acid) and free radicals (hydroxyl radical and superoxide anion) are produced as normal products of metabolism, endogenously by mitochondria during respiratory chain reaction and exogenously by sources like ultraviolet radiations, pollutants, and inflammation. Fig. 19.2A demonstrates the events that occur in oxidative stress. Rapid free radical generation causes oxidative damage to bioactives. These free radicals are atoms or molecules in singlet state, that is, they have unpaired electron, which makes them highly reactive. ROS and RNS are the main source of free radicals that have been implicated in the pathogenesis of neurological disorders like Alzheimer's disease, Parkinson's disease, and strokes (Sharma et al., 2013; Subedi et al., 2014). Most phytomedicinals are redox active molecules and are often referred to as antioxidants, capable of eliminating free radicals, ROS, and RNS, which contribute to most chronic diseases. Synthetic antioxidants available currently include gallic acid esters, butylated hydroxyl anisole, and butylated hydroxyl toluene and these are known to cause several side effects. Plant phytochemicals counteract free radical induced oxidative stress while overcoming the side effects of synthetic antioxidants (Subedi et al., 2014). Antioxidants of natural origin are largely known to have greater antioxidant potential as compared to their synthetic counterparts (Krishnaiah et al., 2007). Phytochemical antioxidants not only provide antioxidant defense, but also foster longevity by management of cell maintenance and DNA repair. Maintaining a comprehensive system of multiple antioxidants like vitamins (C and E) and selenium, nutritionally are essential for good health. These antioxidants provide immunostimulant, antiinflammatory, and anticarcinogenic effects and protect the structural integrity of ischemic or hypoxic tissues (Sindhi et al., 2013). Several dietary and endogenous natural agents, like carotene, tocopherol, ascorbic acid, ferritin, transferrin, and SOD, are found to be free radical scavengers

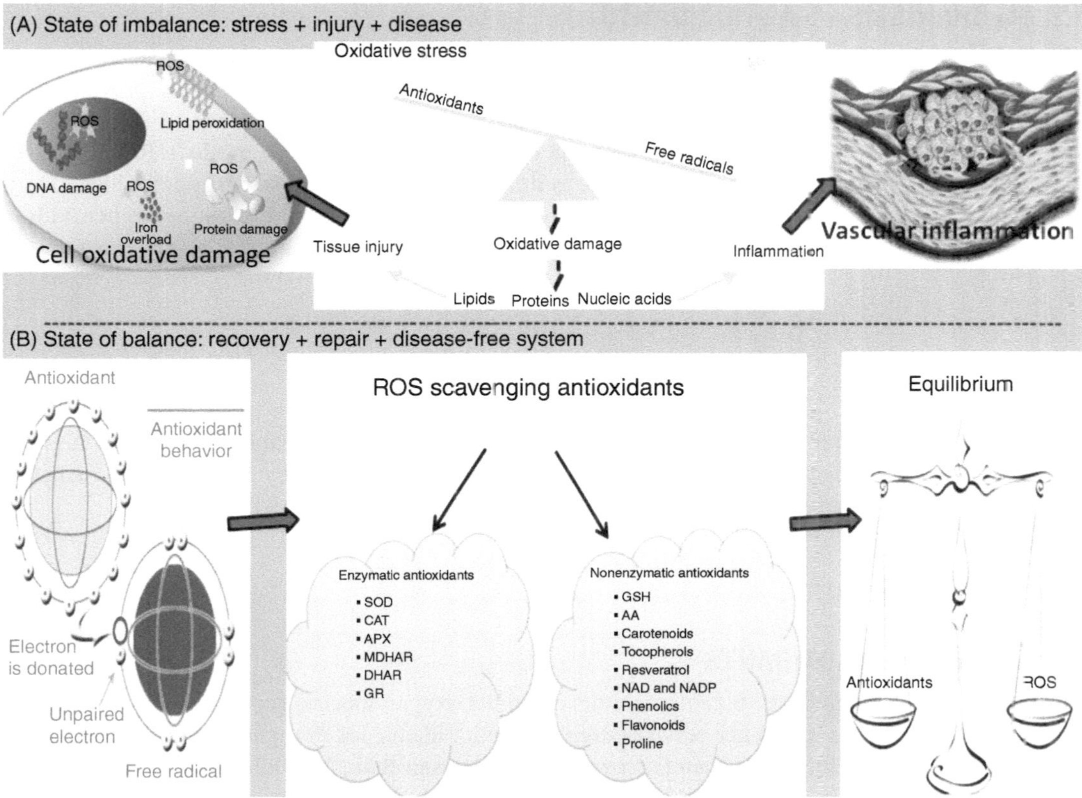

FIGURE 19.2

Demonstrates the state of (A) imbalance and the events that occur in oxidative stress and (B) balance brought about by the mechanism of antioxidant action.

with significant antioxidant properties (Sharma et al., 2013). Antioxidant action is a measure of the decrease in overall oxidation rate and represents the lag phase duration (Krishnaiah et al., 2007). Several assays used to quantify total antioxidant content have been described (Carlsen et al., 2010). An ideal antioxidant quenches free radicals, chelates metal redox, and absorbed readily, at physiologically relative levels (Subedi et al., 2014). Antioxidants are secondary constituents or metabolites occurring naturally, which at low concentration delays the oxidation of proteins, carbohydrates, lipids, and DNA. They inhibit or prevent the oxidation of a susceptible substrate. Fig. 19.2B elucidates the mechanism of antioxidant action to bring about a state of balance in the body. Three main categories of antioxidants are:

1. SOD, catalase (CAT), glutathione reductase (GR), and minerals, like Se, Cu, and Zn, and so on, as the first line defense.
2. Glutathione (GSH), vitamin C, albumin, vitamin E, carotenoids, flavonoids, and so on as the second line defense.
3. A complex group of enzymes that repair damaged DNA and proteins, oxidized lipids, and peroxides (lipase, protease, DNA repair enzymes, transferases, etc.) as the third line of defense (Subedi et al., 2014; Tang et al., 2004).

3.1 MECHANISMS OF ANTIOXIDANTS

Concepts of oxidation retarders, coantioxidants synergism and antagonism are further beneficial for antioxidant action. On the basis of their mechanisms, antioxidants can be either preventive or chain-breaking antioxidant. Preventive antioxidants are peroxide decomposers and work by suppressing oxidation by decreasing the rate of chain initiation. Hydroperoxide product, ROOH causes the initiation process in most cases. Preventive antioxidants transform these hydroperoxides to molecular products, which do not serve as free radical sources. Enzymes like glutathione peroxidase can reduce H_2O_2 to H_2O and lipid hydroperoxides to the corresponding alcohol as demonstrated in Eq. (19.1):

$$\underset{\text{Glutathione peroxides}}{\overset{[2H]}{\text{ROOH} \rightarrow \text{ROH} + \text{H}_2\text{O}}} \tag{19.1}$$

Commercially available chain-breaking antioxidants are mainly phenols or aromatic amines, which exert their effect by trapping peroxyl radicals, as demonstrated by Eq. (19.2) (Krishnaiah et al., 2007).

$$O_{2^-} + \text{antioxidant} \rightarrow O_2 + \text{antioxidant} + \text{heat} \tag{19.2}$$

3.2 NATURAL ANTIOXIDANTS

Many bioactives are found to be potent inhibitors of singlet oxygen species, superoxide radical, and lipid peroxidation products and hence serve as excellent antioxidants, for example, tetrandrine, a bioscoclaurine alkaloid isolated from *Stephania tetrandra* (Ezhilarasan et al., 2014). Flavonoids neutralize all kinds of oxidizing radicals like superoxide and hydroxyl radicals through chelation thereby exerting their antioxidant action. Flavonoids act as free radical scavengers exerting a protective role in ischemic tissue damage and provide beneficial antiinflammatory effects. Their structural features together with their ability to interact with and penetrate the lipid bilayers enable them to modify membrane dependent process like free radical induced membrane lipid peroxidation. Bioflavnoids act as chelating agents as they bind to metal ions in our bodies to prevent them from being available for oxidation. Anthocyanins present in berry nectars of grapes (*Vitis vinifera*), bilberries, blueberries (*Vaccinium myrtillus*), elderberries (*Origanum syriacum*), cranberries (*Vaccinium macrocarpon*), prunes (*Prunus domestica*), and condensed tannins (proanthocyanidins) of grapes, bilberry, blueberry, cranberry, elderberry, prunes, and apples are potent antioxidants (Al-Jaber et al., 2011; Gupta and Sharma, 2006; Krishnaiah et al., 2007; Sindhi et al., 2013).

4 BIOACTIVES WITH POTENTIAL ANTIOXIDANT, ANTIINFLAMMATORY, AND ANTISTROKE EFFECTS

Several actives of natural origin serve as excellent leads to be developed as neuroprotecting agents by virtue of their antioxidant and antiinflammatory properties either directly or through indirect action. A brief review of relevant literature presents the leads in Table 19.2.

Table 19.2 Leads of Natural Origin with Antiinflammatory, Antioxidant, and Antistroke Potential

Description of Bioactives/Botanicals/Plant Extracts with Potential Activity	Reported By
Antiinflammatory and antioxidant therapeutics	
Successively extracted samples of *Cissus quadrangularis*, *Plumbago zeylanica*, *Terminalia bellarica*, and *Terminalia chebula* in water, ethanol, and hexane, inhibited COX-2 activity as compared to COX-1. The ethanol extract demonstrated effective DPPH, OH, and superoxide radical scavenging activity. In vivo antiinflammatory study shows that *T. bellarica* and *T. chebulla* had a significant impact on inhibition of edema formation.	Shaikh et al (2015)
Kang 601 heji (K-601), a traditional Chinese remedy comprising five herbs, *Lonicera japonica*, *Isatis indigotica*, *Rheum palmatum*, *Phellodendron chinense*, and *Scutellaria baicalensis* reduced the level of inflammations in two animal models in dose proportionate manner.	Alolga et al. (2015)
The ethanolic ciplukan (*Physalis angulata* L.; Solanaceae) extract demonstrated significant higher antioxidant activity compared to its aqueous extract.	Kusumaningtyas et al. (2015)
Methanolic extracts of *Premna corymbosa* bark (Verbenaceae), *Premna mucronata* bark (Verbenaceae), *Mallotus philippnensis* bark (Euphorbiaceae), *Trichosanthus bracteata* stem (Euphorbiaceae), *Maharanga bicolor* roots (Euphorbiaceae), *Hibiscus rosa-sinensis* leaves (Malvaceae), *Astilbe rivularis* bark (Saxifragaceae), and *Syzygium operculatus* bark (Myrtaceae) were evaluated for prediction of their antioxidant activity. They showed significant free radical scavenging activity in dose-dependent fashion that was due to the presence of phenols and flavonoids.	Subedi et al. (2014)
The methanol and aqueous extracts of aerial parts of *Cyperus rotundus* showed 88% and 19% inhibition of xanthine oxidase activity with 61.5% and 42.0% inhibition of lipid peroxidation, respectively.	Soumaya et al. (2014)
Polysaccharide obtained from *Astragalus membranaceus* roots and *Bletilla striata* corms, essential oil from *Cleistocalyx operculatus* buds and aerial parts of *Houttuynia cordata*, alkaloids from *Fritillaria wabuensis bulbus* and ripe fruits of *Evodia rutaecarpa*, flavonoids from *Rabdosia rubescens* and fruits of *Citrus aurantium*, terpenoids of *Daphne aurantiaca* stem bark and *Heteroplexis micocephala*, phenolics from *Camellia sinensis* stroma and *Laggera alata* were found to be active antiinflammatory agents.	Wang et al. (2013)
Dichloromethane and petroleum ether extracts of barks of *Adansonia digitata*, *Bolusanthus speciosus*, *Pterocarpus angolensis*, and *Pappea capensis* leaves and ethanolic and aqueous extracts of *Bolusanthus speciosus* stem and *Ekebergia capensis* bark showed the antiinflammatory activity in both COX-1and-2 assays at 250 mg/mL.	Mulaudzi et al. (2013)
Castilleja tenuiflora extracts provided topical antiinflammatory activity in a mouse ear edema model. It could be attributed to verbascoside, thus supporting *its* use in inflammatory disorders.	Sanchez et al. (2013)
A study demonstrated the antiinflammatory effects and strong inhibitory effect against secretion of IL-8 exhibited by *Polygonum bistorta*, *Myrtus communis, Rosa damascena*, *Oligochaeta ramose*, *Syzygium aromaticum*, *Ruta graveolens*, *Terminalia chebula*, and *Tamarix dioica*.	Zaidi et al. (2012)
Ethanolic extract of aerial parts of *Alhagi maurorum*, *Conyza dioscoridis*, *Convolvulus fatmensis*, *Diplotaxis acris*, and *Origanum syriacum* showed significant antiinflammatory activity on carrageenan-induced paw edema in rats.	Al-Jaber et al. (2011)

(Continued)

Table 19.2 Leads of Natural Origin with Antiinflammatory, Antioxidant, and Antistroke Potential (*cont.*)

Description of Bioactives/Botanicals/Plant Extracts with Potential Activity	Reported By
Water and ethanloic extracts of leaves of *Alpina calcarata* (Zingiberaceae), *Ocimum basillicum* (Lamiaceae), *Acorus calamus* (Araceae), *Verbascum thapsus* (Scrophulariaceae), *Hyptis suaveolens* (Lamiaceae), *Solanum indicum* (Solanaceae), *Jatropa gossipifolia* (Euphorbiaceae), *Strebilis aspera* (Moraceae), *Strebilis aspera* (Moraceae), *Sauropus androgynous* (Euphorbiaceae), and flowers of *Jatropa multifida* (Euphorbiaceae), and fruits of *Passiflora edulis* (Passifloraceae) were found to possess high phenolic content and subsequent antioxidant activity.	Narayanaswamy and Balakrishnan (2011)
An investigation into antiinflammatory and antioxidant studies revealed that *Gynura pseudochina* var. *hispida* and *Oroxylum indicum* showed efficient NF-B inhibitory effects with lowest IC50, 41.96, and 47.45 g/mL, respectively while *Muehlenbeckia platyclada* supressed TNF, IL-1, and IL-6 release with IC50 values of 0.28–8.67 g/mL. *Oroxylum indicum* demonstrated significant antioxidant activity via lipid-peroxidation inhibition at IC50 of 0.08 g/mL.	Siriwatanametanon et al. (2010)
The ethnobotanical use of several plants as an antiinflammatory remedy have been reported in a tribe of Arunachal Pradesh in India; a few of them include hot water extract of *Ageratum conyzoides*, *Artemisia nilagirica*, *Blumea fistulosa*, *Eupatorium adenophorum*, *Eupatorium odoratum*, fleshy young stem juice of *Centella asiatica* Linn. (Asclepiadaceae), the fresh paste of bark of *Bombax ceiba* Linn. (Bombacaceae).	Namsa et al. (2009)
Extracts of *Agapanthus campanulatus*, *Cyperus textilis*, *Vernonia natalensis*, *Antidesma venosum*, *Cucumis hirsutus*, *Diospyros lycioides*, *Dissotis princeps*, *Gladiolus dalenii*, *Haworthia limifolia*, *Protea simplex*, *Becium obovatum*, and *Watsonia tabularis* showed dose dependent activity against COX-1 and/or COX-2 enzymes and dichloromethane extract of *Agapanthus campanulatus* root exhibited the highest COX-2 inhibitory activity (83.7%) at 62.5 g/mL.	Fawole et al. (2009)
A pilot study of 70 Fiji grown herbs revealed their tremendous antioxidant potential. Antioxidant capacity was observed as: green leafy vegetables>fruits > root crops. *Ipomoea batata* leaves rich in TPP, quercetin and β-carotene; *Moringa oleifera* leaves rich in TPP, quercetin, kaempferol, β-carotene; *Curcuma longa* rich in TPP, fisetin, quercetin, myricetin; *Zingiber officinate* rich in TPP and *Zingiber zerumbet*; the richest source of kaempferol exhibited high antioxidant capacity.	Lako et al. (2007)
The flavonoids and catecholic xanthones from *Cudrania tricuspidata* root bark demonstrated significant antioxidant activity against LDL oxidation as evident from TBARS assay.	Park et al. (2006)
Mauritian black tea and its fresh leaves have high content of (+)-catechin, (–)-epicatechin, (–)-epicatechin 3-gallate, epigallocatechin, (–)-epigallocatechin 3-gallate, gallic acid, and procyanidin dimers B1 and B2 are rich sources of polyphenolics and antioxidants. Consumption of tea; related to an improved antioxidant status leading to decreased risk of coronary heart disease and stroke.	Luximon-Ramma et al. (2005)
Compounds and antioxidant properties of fruit sweetie Oroblanco, pummelo-grapefruit hybrid (*Citrus grandis*, *C. paradisi*) were compared with that of common white grapefruit. Both fruits were rich in natural antioxidants with higher phenol content and antioxidant activity in sweetie as compared to white grapefruit.	Gorinstein et al. (2004)
Ethanolic extracts of the following Chinese medicinal vine plants were found to have potent antiinflammatory activity: the stems of *Spatholobus suberectus*, *Trachelospermum jasminoides*, *Sinomenium acutum*, *Piper kadsura*, *Polygonum multiflorum*, and *Clematis chinensis*, roots of *Tripterygium wilfordii*, *Tinospora sinensis*, and the root and stem of *Tinospora sagittata*.	Li et al. (2003)

Table 19.2 Leads of Natural Origin with Antiinflammatory, Antioxidant, and Antistroke Potential (*cont.*)

Description of Bioactives/Botanicals/Plant Extracts with Potential Activity	Reported By
Antiinflammatory activity was reported in the following plants with their methanol extract demonstrating greater inhibition than their water and hexane extracts: *Ajuga remota* (Lamiaceae), *Maytenus senegalensis* (Celastraceae), *Galinsoga parviflora* (Asteraceae), *Conyza schimperiana* (Asteraceae), *Oxygonum sinuatum* (Polygonaceae), *Croton macrostachyus* (Euphorbiaceae), *Croton megalocarpus* (Euphorbiaceae), *Mondia whitei* (Periplocaceae), *Plectranthus barbatus* (Lamiaceae), *Spiranthes mauritianum* (Asteraceae), *Zanthoxylum chalybeum* (Rutaceae), and *Zanthoxylum usambarense* (Rutaceae).	Matu and Van Staden (2003)
Antistroke agents	
PEGylated liposome encapsulating FK506 (Tacrolimus, a macrolide lactone derived from a soil bacterium) was prepared and injected i.v. in MCAO rats 1 h after occlusion. Expansion in oxidative stress and brain cell damage was significantly reduced following this treatment and significant amelioration in motor function deficits was observed on administration before reperfusion.	Fukuta et al. (2015)
A study reported that a single injection of curcumin (1 and 2 mg/kg, i.v.) demonstrated significant protective effects 30 min after focal cerebral ischemia/reperfusion in rats. Another recent investigation revealed that i.p. administration of curcumin (100 and 300 mg/kg) 1 h after MCAO significantly reduced infarct volume, and improved neurological deficit in a dose-dependent manner. Curcumin significantly decreased the expression of caspase-3 protein. Yet another study gave insights on effectiveness of curcumin when given 4 h after ischemia in the rat thrombo-embolic stroke model. Also, when curcumin packaged in a suitable novel delivery system, such as solid lipid nanoparticles can provide beneficial effects against cerebral ischemic insult and can cross the blood brain barrier for an efficient targeting. A new study showed that curcumin exerts a neuroprotective effect by regulating the expression of various proteins in focal cerebral ischemia which were successfully identified using a proteomic approach.	Jiang et al. (2007); Lan et al. (2008); Zhao et al. (2008); Dohare et al. (2008); Kakkar et al. (2013); Shah et al. (2016)
A study revealed that an ethanolic extract of *Lavendula officinalis* flowers significantly reduced the blood brain barrier permeability in experimental groups when compared to ischemic group. It reduced MDA levels of plasma and brain tissue in intact group when compared with control group.	Rabiei and Rafieian-Kopaei (2014)
In an investigation conceptualized to impart tissue repair in an ischemic brain on a mice model sequential delivery of two agents was attempted. PEGylated poly(lactic-*co*-glycolic acid) nanoparticles loaded with epidermal growth factor and erythropoetin biphasic microparticles with a PLGA core and a poly(sebacic acid) coating were prepared and dispersed in a hyaluronan methylcellulose hydrogel. The sequential delivery fashioned in such a way resulted in significant tissue repair and minimized damage compared to an ICV infusion.	Wang et al. (2013)
It was found from an investigation that prophylactic treatment with naringenin had significant effects against ischemic brain injury. Another study on hesperidin demonstrated similar effects.	Raza et al. (2011, 2013)
A study undertaken to evaluate black tea consumption and vulnerability to stroke revealed that daily consumption of black tea (four cups or more) is inversely associated with risk of stroke.	Larsson et al. (2013)
Naodesheng, a prescription in traditional Chinese medicine composed of safflower (*Flos Carthami*), sanchi (*Panax notoginseng*), Kudzu root (*Radix Puerariae*), chuanxiong (*Rhizoma Ligustici chuanxiong*), and stoned hawkthorn (*Crataegus pinnatifida* Bge). It is known to promote blood flow to overcome stasis. It has clinical relevance in cerebral arteriosclerosis, ischemic stroke, and hemorrhage. An investigation in rats based on quantification of platelet aggregation, and in mice measurement of clotting time and protection rate revealed its antithrombotic and anticoagulation potential.	Chen et al. (2012)

(*Continued*)

Table 19.2 Leads of Natural Origin with Antiinflammatory, Antioxidant, and Antistroke Potential (*cont.*)

Description of Bioactives/Botanicals/Plant Extracts with Potential Activity	Reported By
It was reported that pretreatment with hydroalcoholic extract of *Mimusops elengi* flowers rich in polyphenolics at 100 and 200 mg/kg afforded significant neuronal improvement and protection against I/R injury with good antioxidant effect.	Nagakannan et al. (2012)
Alcoholic extract of *Passiflora foetida* aerial parts at 25 mg/kg was significantly found to decrease brain infarct volume both in cortical and sub cortical structures of rats while at higher doses (100 and 400 mg $kg^{-1)}$ significant reduction was observed only in cerebral cortex. It improved neurological deficits and sensory response to mechanical and temperature stimuli.	Wattanathorn et al. (2012)
Ten flavonoids including fustin, fisetin, sulfuretin, butein, butin, eriodictyol, morin hydrate, quercetin, kaempferol, and isoliquiritigenin isolated from the methanolic extract of *Rhus verniciflua* Stokes (Anacardiaceae) were found to serve as potential therapeutics for stroke.	Cho et al. (2012)
A codelivery system of dexamethasone and plasmid DNA formed by conjugation with polyethylenimine demonstrated significantly reduced infarct volumes and inflammatory cytokines on a TIA animal model. Also, gene therapy was attempted with heme oxygenase-1 gene delivered to the focal ischemic brain using PEI2k-Dexa. In another study the same carrier (Dexmethasone) was utilized for gene delivery of heme oxygenase-1 to the ischemic brain by formulation of carrier peptide micelles loaded with the desired gene.	Hyun et al. (2011); Lee et al. (2012)
Total oligomeric flavonoids (TOFs) prepared from *Cyperus rotundus* (Cyperaceae) in rat model of cerebral ischemia and reperfusion was found to significantly reduce the neurological deficits and reverse the anxiogenic behavior. It also significantly altered the markers of peroxidation in favor of neuroprotection.	Sunil et al. (2011)
Tanshinone II A is one of the major diterpenes extracted from *Salvia miltiorrhiza*. It was revealed that it protected the brain from damage caused by ischemic insult in pMCAO model on Sprague Dawley rats possibly through antiinflammatory responses.	Wang et al. (2010)
Transsodium crocetinate (TSC), a carotenoid related to crocetin obtained from saffron. Administration of TSC provided beneficial effect either alone or on concomitant delivery with tPA in acute ischemic stroke on a small clot embolic model in rabbits.	Lapchak (2010)
A study suggested the possible role of caffeionol as a neuroprotectant in patients with acute stroke when given along with hypothermia and tPA administration.	Martin-Schild et al. (2009)
The major active constituent of *Ligusticum wallichii*, 2, 3, 5, 6-tetramethylpyrazine (TMP) is known to be used for stroke treatment. A derivative of TMP was synthesized 2-[[(1,1-dimethylethyl)oxidoimino]-methyl]-3,5,6-trimethylpyrazine (TBN), which retained TMPs thrombolytic activity, and had high antioxidant potential and demonstrated significant effects in MCAO stroke model in rats.	Sun et al. (2008)
Radix puerariae extract has shown to demonstrate amelioration of behavior response in post-stroke depression mice, which is closely related with the protective effects of catecholamine synthesized against cerebral ischemia reperfusion brain damage.	Lan et al. (2008)
A study reported that the free radical scavenging potential of flavonoids like wogonin, baicalin, baicalein, catechin, and procyanidin B2 using electron spin resonance spectrometry for superoxide, 1,1-diphenyl-2picrylhydrazyl and hydroxyl radical and their neuroprotective effects on a chick cardiomyocyte model of ischemia.	Chang et al. (2007)

5 LIMITATIONS OF BIOACTIVES TO ELICIT THE DESIRED THERAPEUTIC RESPONSE

Most therapeutic agents of natural origin that have tremendous potential as agents of choice in several disease conditions suffer from drawbacks that limit their clinical use. Many bioactive with antioxidant, antiinflammatory and neuroprotective effects discussed exhaustively in this chapter that might work wonders for the alleviation of neuronal and tissue damage following stroke and in bringing about the reversal of pathophysiological changes in ischemia-reperfusion injury fail clinically due to degradation, poor aqueous solubility, and subsequent bioavailability. Poor bioavailability mostly results from low intrinsic activity, reduced absorption, increased metabolism, inactivity of metabolized products along with rapid elimination and clearance (Anand et al., 2007). Bioavailability of flavonoids varies greatly due to variation in subgroups and differences in chemical properties and polarity. Bioavailability of flavonoids like quercetin, hesperetin, and naringenin have been studied by investigators (Erlund, 2004). It was observed that the disposition of quercetin and its glycosides like rutin in humans depended largely on sugar moiety and to some extent the plant matrix could affect the rate and extent of absorption (Graefe et al., 2001). Curcumin is a bioactive with a wide spectrum of biological activities found to be safe even at a high dose of 12 g/day; however, it exhibited poor bioavailability. Poor bioavailability of curcumin results from its poor absorption, rapid metabolism, and fast systemic elimination. Hence, several approaches have been undertaken to improve the bioavailability of bioactives and thus to validate the use of lipid nanocarriers like liposomes and nanoparticles.

6 NOVEL LIPID NANOSTRUCTURES

Recent times have witnessed many advances in the field of medicine as nanoparticles—often referred to as "controlled release reservoirs"—have revolutionized drug delivery research (Naahidi et al., 2013). Nano is derived from a Greek word that means dwarf and the term nanotechnology finds the first mention in the year 1974 when a Japanese scientist by the name of Norio Taniguchi referred to such materials in nanometers (nm). One nm is equivalent to one billionth of a meter. The size range of interest today goes from 100 nm downward to about 0.2 nm (atomic levels), as in this range material properties change drastically. The National Nanotechnology Initiative defines nanoparticles as structures of sizes in the range 1–100 nm; however nano is generally used for structures up to a few hundred nanometers. Nanostructures <100 nm are promising drug carriers. This small size provides for a much higher surface area per unit mass leading to higher reactivity and altered optical, electrical, and magnetic properties of materials. Also, these nanostructures possess unique physicochemical-biological features. By their large reactive area and ability to cross physiological barriers, a larger cell uptake results, which consequently modifies in vivo behavior. Hence, they can be successfully used as delivery tools for many bioactive (Deleers and Pathak, 2007; Wilczewska et al., 2012). The high surface-to-volume ratio of these nanostructures provides a large functional area for drugs, probes, or proteins to bind or get adsorbed, thus making these nanostructures suitable carriers. Drugs can be incorporated into the carrier matrix or attached to the carrier surface. The large surface area also has implications for improving the dissolution and aqueous solubility of drugs (Athar and Das, 2014).

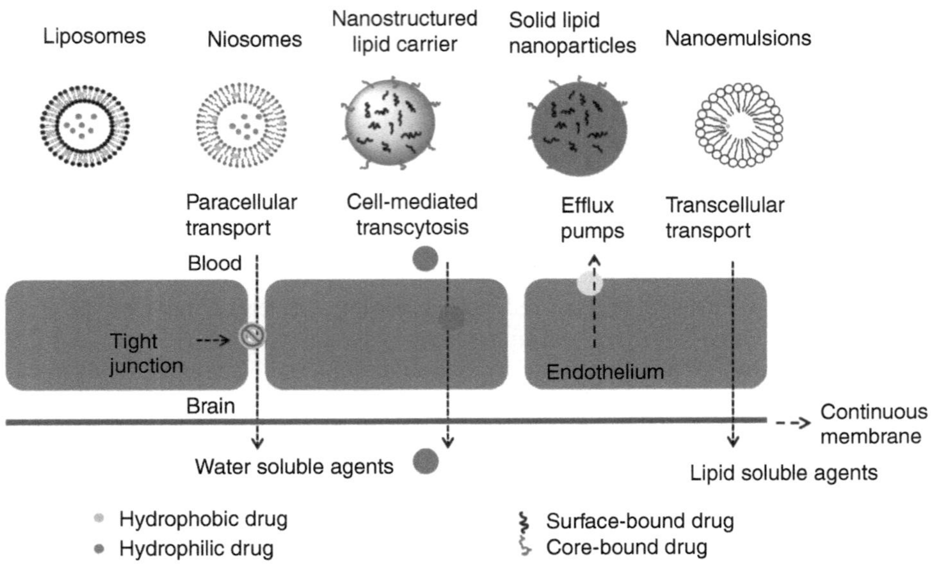

FIGURE 19.3 Schematic Representation of Various Types of Lipid Based Nanocarriers and Their Possible Mechanism of Transport Across Blood Brain Barrier

The capabilities to target the drug and to control its release are the two elements of an ideal delivery system. Targeting ensures high efficiency with minimal side effects (Deleers and Pathak, 2007). For a carrier to be suitable for a delivery system it should satisfy the prerequisites of drug incorporation, it should be able to reduce toxicity and sustain therapeutic outcomes, enhance safety and compatibility, and allow for better targeting and biodistribution. These prerequisites can only be met by materials that are biodegradable, biocompatible, and satisfy the conditions for a biodegradable speed of the carrier and the delivery speed of the drug (Athar and Das, 2014). Issues on biocompatibility have gained ground in recent times and calls for a thorough evaluation of the interaction between a material and the biological system of interest to ensure the effective and safe use of nanocarriers for therapeutic purposes. Biocompatibility of a material must be evaluated with emphasis on the environment in which it is to be placed as it is crucial to safe drug release and reducing cytotoxicity (Naahidi et al., 2013). Use of biodegradable nanoparticles has found increased utility recently because of the potential they hold to act as a viable carriers for target-oriented gene delivery and delivery of biologicals and drugs with advantages of high biocompatibility, better encapsulation, and improved release profiles (Mahapatro and Singh, 2011). Various types of lipid-based nanocarriers include solid lipid nanoparticles, liposomes, liquid drug conjugates, nanoemulsions, transfersomes, niosomes, and liquid crystal drug delivery systems to name a few (Attama et al., 2012). Fig. 19.3 shows a schematic representation of various types of lipid-based nanocarriers and their possible mechanism of transport across blood brain barrier (BBB).

6.1 ADVANTAGES OF LNs-BASED DELIVERY

1. Particle size and surface properties can be easily exploited for targeting.
2. Choice of matrix constituents can easily modulate the control release and particle degradation characteristics.

3. The release of the drug could be controlled or sustained in transit and at the site of localization.
4. Surface modification is possible by attaching various ligands or use of magnetic guidance.
5. High drug loading and superior encapsulation of drugs into the systems is attainable.
6. They can be prepared without the use of organic solvents.
7. Can be administered through different routes like parenteral, oral, nasal, and intraocular.
8. Enhanced physical stability of pharmaceuticals by avoidance of coalescence.
9. Leakage of incorporated active ingredients can be minimized by controlling their mobility.
10. Both lipophilic and hydrophilic drugs can be formulated with ease.
11. Most lipid carriers being nonallergenic, nonirritating, and nontoxic offer the advantages of biodegradability and biocompatibility.
12. They can be easily scaled-up, sterilized, and validated and are less expensive (Attama et al., 2012; Bagul et al., 2012; Mohanraj and Chen, 2007).

6.2 MECHANISM OF DRUG DELIVERY VIA LNs

A drug delivery system utilizing nonmaterials usually consists of two components: an active ingredient and the architectured nanoparticle. This system in totality is on some occasions called smart drugs or theranostics and serves a specific function related to diseases management. However, in the drug delivery via LNs along with the engineered nanoscale particles being used as carriers, the drug on it is own might be formulated at a nanoscale working independently as its carrier

How the drug is conjugated to the nanocarrier and how it's being targeted are two crucial factors in a target-based therapeutic approach. A drug adsorbed or covalently attached to the surface of the nanocarriers could be encapsulated in it. However, covalent linkage offers a more precise control of the quantity of bioactive being delivered as it could control the number of drug molecules connected to the nanocarrier. LNs demonstrate their site-specific delivery of drugs through avoidance of reticulo-endothelial system (RES) and by increasing localization through enhanced vascular permeability and retention effect (EPR). Here two types of approaches are possible.

1. *Surface bound*: the drug particles are attached to nanoparticle surface.
2. *Core bound*: drug particles are incorporated in the nanoparticle matrix.

Cell-specific targeting may be achieved through active or passive mechanisms. Active targeting largely involves surface modification through recognition ligands, bound to conjugated antibodies, low molecular ligands like folic acids and peptides to reach the intended site. This way of targeting may also involve manipulation of physical stimuli like temperature, pH, or magnetism. Selective and passive targeting can be achieved due to EPR effect. On reaching the intended diseased target site/tissues, the drug-nanocarrier conjugates needs to release the therapeutic agents in a controlled manner. It is well documented that a competent nanostructure has the potential to transport or carry the guest (drug) molecules whose early release might cause the confronting problem. Specifically designed biodegradable LN formulations can be used for this purpose. A controlled release of drugs from nanocarriers can be achieved via an enzymatic activity or through alterations in physiological environments, such as temperature, pH, and osmolality (Athar and Das, 2014; Wilczewska et al., 2012). Fig. 19.3 illustrates the mechanism of brain delivery of biologicals via LNs.

6.3 BASIC INGREDIENTS OF LIPID NANOCARRIERS

6.3.1 Emulsifiers

Emulsifiers are required to provide stability to LN dispersions and to prevent agglomeration of particles. Their suitability depends on their surfactant properties, like chemical structure, molecular weight, charge, and hydrophilic-lipophilic balance (HLB). A few emulsifiers and coemulsifiers and their corresponding HLBs includes Lecithin (4–9), Polysorbate 65 (10.5), Cremophor EL (12–14), Polysorbate 80, Solutol HS15 (15), Polysorbate 20 (16.7), Poloxamer 407 (21.5), and Poloxamer 188 (29).

6.3.2 Lipids

A matrix for LNs may be composed of natural, semisynthetic, or synthetic lipids that could be used either alone or in combination. These lipids can further be either bilayer or nonbilayer lipids.

6.3.2.1 Bilayer lipids

Phospholipids and similar compounds with heads and tails (similar in size); that form lipid bilayers in aqueous systems. Here two monolayers exhibit tail–tail association hence reduces contact of hydrophobic parts with water while enhancing hydrophilic associations. These molecules move freely in half of their bilayers; however, a huge energy barrier restricts movement on the bilayer's second side. Additionally, cholesterol helps to regulate the membrane fluidity. Phospholipids like phosphatidylcholine are capable of altering the lipid particles surface chemistry, and their chain packing affects their stretching, swelling, bending, or capability to get deformed (mechanical characteristics). These properties can be manipulated for the design of novel carriers for drug delivery.

6.3.2.2 Nonbilayer lipids

The major lipids in some biological systems are nonbilayer lipids, which cannot be packaged as a lamellar structure in presence of water based systems in purified form. In drug delivery, such lipids are largely utilized as matrix-forming lipids. They have varying functional features based on their properties like crystallinity, melting, and polymorphism. They might promote absorption of lipophilic drugs. Hard fats like stearic acid, palmitic acid, natural hard fats like theobroma oil, triglycerides like trimyristin and tristearin, waxes like bees wax, carnauba wax and cetyl palmitate, mono di and triglyceride mixtures like glyceryl monostearate (Imwitor 900), glyceryl behenate (Compritol 888 ATO) are a few examples of such lipids used in making LNs (Attama et al., 2012).

6.4 TYPES OF LNs

6.4.1 Solid lipid nanoparticles (SLN)

SLNs were developed in the 1990s as an alternative to the conventional carriers, such as emulsions, liposomes, and polymeric nanoparticles. SLNs are new generation colloidal carriers sometimes known as lipospheres. SLNs are biodegradable, biocompatible, and nontoxic particles prepared with safe physiological lipids (GRAS) and may be prepared without organic solvents. They are composed of lipids, waxes, and surfactants for emulsification. These lipids are tolerated well in vivo. The lipoidal matrix may be made up of homolipids like fats and waxes, which not only modify the drug release profile but also protect the incorporated bioactives against physical and chemical degradation. Commonly used lipids may be fatty acids, triglycerides, steroids, and so on. Typical SLNs utilize lipids like paraffin wax or biodegradable glycerides (Compritol 888 ATO) to form the core construct. Emulsifiers are used for making stable lipid dispersion and include sodium cholate and sodium glycocholate (Attama et al., 2012; Ranghar et al., 2014). They can encapsulate both lipophilic and hydrophilic actives under optimized conditions. These SLNs are submicron colloidal particles (50–1000 nm) made from physiological lipids that are dispersed in water/aqueous surfactant solution. SLNs provide all advantages of traditional carrier systems and allow for solutions for their demerits (Attama et al., 2012; Bagul et al., 2012). Their advantages include the solid state of particle matrix, improved bioavailability, protection of chemically labile ingredients against chemical decomposition and controlled release characteristics. Hence, they find widespread use in systems for parenteral use and delivery systems for oral, ocular, brain, topical, rectal applications. However, certain drawbacks of SLNs are particle growth, tendency to gel, dynamics of polymorphism, and poor entrapment because of crystallinity of the solid lipids (Attama et al., 2012; Lembo and Cavalli, 2010).

6.4.2 Nanostructured lipid carriers

Nanostructured lipid carriers (NLCs) are modifications of lipid-based nanoparticles with advantages over conventional SLNs. These are colloidal carriers with average size of particles in nm range. NLCs possess solid lipid matrices having some oil content (liquid lipid phase) and conventionally obtained by mixing and blending of solid lipids with liquid lipids, which results in a unique nanostructure with improved drug loading and better drug retention properties during storage. The resulting lipid matrix although solid at body temperature, exhibits lower melting than the initial solid lipid. For NLC production, the capability to elicit and control drug release is an important parameter that should be considered while mixing lipids. Membrane instability and drug leaching related with liposomes and emulsions and demerits like low payload, drug expulsion, and large water content of SLNs can be overcome via (Lembo and Cavalli, 2010; Wilczewska et al., 2012). Assuming generalization most methods used to prepare SLNs can be used for NLC preparation (Attama et al., 2012). Based on the method of preparation and lipid blend composition, three different kinds of NLCs are obtained. Imperfect crystal model (Type I), Amorphous model (Type II), and Multiple model (Type III) (Eliana and Souto, 2007).

6.4.3 Lipid drug conjugates (LDC)

LDCs were architectured as modifications of SLNs to overcome their limitations of poor loading of hydrophilic bioactives due to partitioning effects during production. With SLNs only low dose and potent lipophobic drugs could be encapsulated in lipoidal matrix. LDCs with up to 33% drug-loading capacities were developed to expand the applicability of lipid-based carriers to lipophobic drug molecules. These insoluble drug-lipid conjugates can be produced by salt formation (with a fatty acid) or by covalent linking (to esters or ethers). High-pressure homogenization is then used for preparation of nanoparticles using aqueous surfactant. These nanoparticles are especially useful in targeting the brain with hydrophilic drugs for treatment of protozoa infections (Attama et al., 2012; Wilczewska et al., 2012).

6.4.4 Liposomes

Liposomes are closed vesicular structures proposed by Gregoriadis as first vesicle based carriers. These are concentric lipid vesicles in which an aqueous volume is completely entrapped in a bilayer made of hydrated phospholipids and cholesterol (Bagul et al., 2012). Liposomes can range 20–30 nm to a few microns depending on their chemical composition and method of preparation used. The bilayers in a liposome are separated from each other by aqueous channels/domains hence this alternating hydrophilic and hydrophobic structure of liposomes makes them suitable delivery systems for compounds of varying solubilities. The liposomal lipid bilayers fuse with cell membranes and deliver their contents (bioactives). The bilayer construct of a liposome can be suitably tailored in some ways to meet different purposes. Functionalizing of their surfaces has also been undertaken to achieve site-specific targeting. Owing to their versatile nature liposomes find widespread applications in radiology, cosmetology, and vaccinology. Liposomes can be further classified depending on lamellarity and size as small unilamellar vesicles (SUV), large unilamellar vesicles, and large multilamellar vesicles. SUVs with nanometric particles are useful for drug and antigen delivery (Attama et al., 2012; Lembo and Cavalli, 2010). Drug release from liposomes depends on their composition, pH, osmotic gradient, and the surrounding environment. Also, an increased residence time can increase the duration of action, but it decreases the number of particles. Liposome-cells interactions can be realized by one or a combination of adsorption, fusion, endocytosis, and lipid transfer (Wilczewska et al., 2012). An introduction of PEGylation in the lipid bilayers of liposomes prevents its interactions with plasma proteins and hence restricts and limits their recognition and removal by RES. However, they have some drawbacks like poor in vitro and in vivo stability, low encapsulation, and expensive production process (Lembo and Cavalli, 2010).

6.4.5 Transfersomes

Transfersomes came into existence with the aim of providing a carrier for dermal delivery of bioactives. These are typically ultradeformable liposomes, made from phospholipids and additional amphiphilic substances (biles), which can severely distort the shape of the vesicle. Drugs and antigens are incorporated into them as in liposomes and these deformable and flexible vesicular systems permeate intact across the intact dermis to deliver these bioactives. Their permeation across the dermal barrier is due to hydrostatic gradient forces existing in the skin. When dispersed in buffer their vesicle diameter is reported to be in the order of 100 nm.

6.4.6 Niosomes

Niosomes are structural analogs of liposomes, the difference being that their vesicle constructs are mainly composed of nonionic bilayer forming surfactants. Like liposomes, they also improve the bioavailability of drugs by prolonging their residence time and reducing their clearance. They are also being used for targeted drug delivery and even antigens, and small molecules have been successfully delivered via niosomes as with liposomes. Their properties depend on bilayer composition and preparation method. Niosomes are obtained when synthetic nonionic surfactants are hydrated with or without using other lipids and cholesterol. The use of synthetic surfactants in niosomes makes them better delivery systems as they are inexpensive and provide better chemical stability than natural phospholipids being used in liposomes (Attama et al., 2012).

6.4.7 Liquid crystal drug delivery systems

A novel class of sustained release matrix systems can be achieved due to the spontaneous self-assembly of certain lipids to produce liquid crystalline structures that are highly stable to dilution. These liquid crystals can potentially act as reservoirs for controlled drug release in gastrointestinal tract fluids and subcutaneous space. They can also be dispersed into nanoparticulate form without compromising the initial (parent) architecture. The drug release rate from these liquid crystals is a direct property of their nanostructure matrix. These are further classified into lamellar, cubic, hexagonal mesophases, and so on. Reversed cubic and hexagonal mesophases types of lyotropic systems have been extensively investigated for their versatility as they can provide sustain release characteristics to a broad category of actives including drugs with low molecular weight, peptides, proteins, nucleic acids, and others (Attama et al., 2012).

6.4.8 Nanoemulsions

Nanoemulsions (NE) typically have sizes <1 μm and recently have gained ground as novel lipid-based delivery systems along with microemulsions. These NEs can enhance drug (especially hydrophobic drugs) bioavailability by increasing their solubility and are now a delivery system of choice for delivery of BCS class II and class IV drugs. Oral NEs utilize safe ingredients, like edible oils and GRAS-grade excipients and are hence regarded as safe. By their minute droplet size and increased kinetic stability, NEs demonstrate high encapsulation of bioactives (Attama et al., 2012).

6.5 INSIGHTS INTO SLNs AND LIPOSOMES WITH THEIR METHODS OF PREPARATION

6.5.1 SLNs

These submicron particles in the range of 52–100 nm consisting of biocompatible physiological lipids have unique properties as potent drug carriers. Salient features of SLNs include improved stability and tolerability, protection of actives from degradation, and controlled release (Ranghar et al., 2014; Wilczewska et al., 2012). SLNs have widely been employed for brain delivery of therapeutic agents because of their ability to cross BBB, for example, camptothecin-loaded SLNs were shown to have higher affinity to the porcine brain capillary endothelial cells compared to macrophages and in vivo studies in rats demonstrated detection of their high amounts localized in the brain post intravenous administration of fluorescent labeled SLNs (Joseph and Saha, 2013). The type of SLNs depends on various factors, like the chemical nature of the bioactive and lipid being used, solubility of the bioactive in the lipid melt, method of preparation used (hot or cold), the production temperature and nature and concentration of surfactants used. Literature finds mention of three different models that describe how actives can be incorporated. The three incorporation models include:

1. Homogeneous matrix model (Type I) obtained from a solid solution of lipid and an active constituent, which in turn is obtained when cold homogenization method is used for SLN production.
2. Drug-enriched shell model (Type II) attained when hot homogenization method is used for SLN production, and the concentration of the active in the melted lipid is usually low. A classical example of this model is the incorporation of coenzyme Q10.
3. Drug-enriched core model (Type III) is obtained when high active ingredient concentration is present in the lipid melt as it is close to its saturation solubility. Cooling of hot oil droplets results in reduced solubility of the active ingredient in the melt; on exceeding saturation solubility, the active ingredient precipitate out resulting in a drug-enriched core (Eliana and Souto, 2007).

6.5.1.1 Preparation methods for SLN production

- High pressure or high shear homogenization (HPH)
  HPH method can be suitably used for the preparation of SLNs and its modifications by melt emulsification and can be performed at high temperature (hot process) or at and below room temperature (cold process). A reduced particle size is attained by impact, shear, cavitation, and turbulence. The cold homogenization method reduces exposure of the bioactive to high temperatures. Both techniques afford very narrow particle size distributions and are suitable for up to 40% lipid concentrations.

- Production of SLN via microemulsions
  A warm microemulsion can be prepared in excess cold water under stirring using a thermostatic syringe in a typical ratio of about 1:50. Furthermore, ultrafiltration or lyophilization can be used to remove excess water to increase the particle concentration.
- Solvent emulsification/evaporation
  The lipophilic drugs are dissolved in an organic solvent, which is not water miscible (like cyclohexane) and then emulsified in an aqueous phase. As the solvent evaporates, nanoparticulate dispersion is produced by lipid precipitation in the aqueous medium (Attama et al., 2012; Bagul et al., 2012).
- Spray drying method
  An aqueous SLN dispersion can be transformed into a drug product by this technique as an alternative to lyophilization as it is comparatively economical. However, its use often leads to aggregation of particles because of high temperature, shear forces, and slight melting of particles. Lipids with melting points >70°C have been used for spray drying. The concentrations of 1% trehalose solution in water or 20% trehalose solution in ethanol-water (10/90 v/v) have been known to give the best results (Bagul et al., 2012).
- Production of SLN's using supercritical fluid technology
  Supercritical fluid technology can produce biodegradable nanoparticles by using environmentally safe supercritical fluids without the use of any organic solvents. A supercritical fluid is a solvent at a temperature above its critical temperature, at which the fluid remains a single phase irrespective of pressure. Two common processing techniques with supercritical fluids are supercritical antisolvent and rapid expansion of critical solution (Mohanraj and Chen, 2007).

6.5.2 Liposomes

Liposomes are vesicles composed of amphiphilic double layered phospholipids with a volume of entrapped aqueous solute and based on lamellarity and diameter can be further categorized as multilamellar vesicles (>400 nm), unilamellar vesicles (100–400 nm), and small unilamellar vesicles (<100 nm). FDA-approved Doxil (doxorubicin liposomes) was the first liposomal delivery system to treat the AIDS-associated Kaposi's sarcoma. Liposomes can be suitably engineered by selection of the appropriate lipid ingredients and surface modification to cater to different properties like specificity of the target, prolonged residence in circulation pH, reduced environmental, and temperature sensitivity (Ranghar et al., 2014). They can deliver hydrophilic (encapsulated in the aqueous core) as well as hydrophobic compounds (encapsulated through phospholipids membrane). Phosphatidylcholine and phosphatidylethanolamine are used to make the vesicle whereas cholesterol is a commonly used additive that modifies the rigidity of the lipid membranes (Athar and Das, 2014). Reduced drug permeability is observed in the "gel" state compared to "fluid" state. Hence, the selection of high-phase-transition lipids like phosphatidylcholines (PC), those with long saturated fatty acyl chains like distearoyl PC and hydrogenated soy PC, which at physiological temperature remains in a gel state, can maximize the stability of liposomal entrapment. The addition of cholesterol (30–50 mol. %) further improves stability by filling in gaps between PC molecules in the lipid bilayers resulting in a tight bilayer and thereby minimizing any chances of insertion of plasma proteins in the lipid bilayer and reducing RES clearance of liposomes (Jun and Lee, 2007). The mechanism of drug delivery via liposomes either involves active extrusion or passive diffusion (Athar and Das, 2014).

6.5.2.1 Types of liposomes based on formulation strategies

Depending on intended use and formulation strategy, adopted liposomes can be of the following types:

- Sterically stabilized liposomes: Include PEGylated systems with increased circulation time due to steric hindrance on the bilayer surface.
- pH-sensitive liposomes: pH-sensitive liposomes are intended to destabilize at mild acidic pH present in solid tumors/endosomal compartments and are composed of dioleoyl phosphatidylethanolamine (DOPE), which have a cone-shaped structure and promote changes to HII state, along with a mildly acidic amphiphiles like cholesteryl hemisuccinate or oleic acid, which provides stability to the bilayer construct at neutral pH but not at slight acidic pH.
- Fusogenic and endosomolytic liposomes: These are produced by reconstitution of viral envelope proteins into liposomes or entrapment of bacterial hemolysins with different degrees of pH dependence.
- Temperature-sensitive liposomes: Liposomal contents can be triggered for local release by hyperthermia by utilizing a bilayer composition with phase-transition temperature >37°C.
- Cationic liposomes: They form electrostatic complexes with plasmid DNA and promote gene transfer. Cationic liposomes can be formed with either multivalent cationic lipids or monovalent cationic lipids.
- Targeted liposomes: Liposome targeting to specific cell populations can be achieved by incorporation of a targeting moiety/ligand that might consist of a lipid-anchored antibody like antitransferrin receptor, anti-CD20, anti-HER2, and anti-CD19 or antibody fragments like Fab and scFv besides folate and carbohydrate (Jun and Lee, 2007). Targeted brain delivery via liposomes resulted in significant improvement of drug concentration in the brain in vitro cell lines (Chiang, 2007).

6.5.2.2 General methods of liposome preparation

There are several established methods of preparations based on the scale of preparation and other factors like encapsulation efficiency. Initially, the lipid ingredients are dissolved in a suitable solvent (chloroform/methanol), dried on a rotary evaporator to obtain a thin film, followed by lipid hydration (in the aqueous medium at a temperature >phase-transition temperature) and lastly particle size reduction is accomplished. This process yields heterogeneous MLVs. Repeated cycles of freezing and thawing can reduce the lamellarity of these MLVs. When phospholipids are dissolved in a solvent like ethanol (water miscible), and quickly diluted into aqueous medium (buffer or distilled water), it resulted in direct formation of liposomes with relatively reduced particle sizes. Subsequently, the solvent is removed from the liposomal formulation with further size modifications. Reduction in particle size can be brought about by several methods like sonication, extrusion, and homogenization (Jun and Lee, 2007).

6.6 CHARACTERIZATION OF LNs

- Particle characteristics, surface properties, and zeta potential
  Quantitative evaluation of particle characteristics and their morphological features are informative tools to assess the biophysical properties that affect biological activity, distribution, and toxicity. Various tools and techniques are used to evaluate lamellarity, size, shape, and polydispersity of LNs. Particle size and size distribution are salient features of nanoparticles. It is known that

submicron nanoparticles perform better than microparticles as drug delivery systems; hence appropriate control of size presumes significance. Photoncorrelation spectroscopy or dynamic light scattering is the most routinely used technique to measure particle size. It determines the diameter of particles through Brownian motion and light scattering when the viscosity of the medium is known. These findings can be further validated by scanning or transmission electron or atomic force microscopic examinations (SEM, TEM, and AFM), freeze fracture electron microscopy (FFEM), cryo-electron microscopy (Cryo-EM), and scanning tunneling microscopy (STM). These microscopic techniques also give information about the morphology of nanoparticles. AFM is largely employed to obtain information on the size, shape, and surface morphology of nanoparticles. It can resolve to as low as 0.01 nm and produce a three-dimensional contrast image of the specimen being investigated. The zeta potential (size distribution kinetics) of LNs is an indicator of their surface charge property and reflects their electrical potential. A zeta potential above (+/−) 30 mV indicates stable systems, as surface charge prevents particle aggregation and fusion. The zeta potential also indicates whether the charged bioactive is incorporated in the core or adsorbed onto the surface (Attama et al., 2012; Bagul et al., 2012; Deleers and Pathak, 2007)

- Solid state characteristics
  Solid state characteristics detect the possible physicochemical modifications of the drug encapsulated in LNs. LNs exhibit alterations in the matrix component crystallinity due to use of emulsifiers, method of preparation adopted and high shear that result in the formation of amorphous, liquid or only semicrystallized metastable systems. Changes due to polymorphism lead to physical and chemical changes including changes in solubility, shape, and melting of the active and auxiliary materials. The solid state evaluation of LNs may utilize the following techniques: differential scanning calorimetry, Fourier-transform infrared spectroscopy, X-ray diffraction, Raman spectroscopy, and hot stage microscopy. Useful information on the extent of crystallization and polymorphism can be obtained from X-ray diffraction and DSC analysis.
- Drug loading
  High drug-loading capacity represents ideal systems as it reduces the requirement of matrix substances. Drug loading can be achieved by incorporation or adsorption/absorption technique. Drug loading and entrapment efficiency are dependent on the solid-state drug solubility in the matrix material. The PEG moiety does not have much effect on drug loading. For small molecules, ionic interaction between the drug and matrix can effectively increase drug loading.
- Drug release
  The rates of drug release are governed by several factors, like its solubility, desorption of the surface-bound/adsorbed drug, its diffusion through the nanoparticle matrix and erosion or degradation of nanoparticle matrix. When drug diffusion is faster than matrix erosion, the release mechanism is controlled by the diffusion process. The initial rapid release (burst release) is due to drugs weakly bound/adsorbed on the surface. The method of drug incorporation affects its release profile. Several methods to study in vitro drug release are known; however, the dialysis technique is the most frequently used method.

  Methods used to evaluate in vitro drug release include:

1. Dialysis bag diffusion technique.
2. Reverse dialysis bag technique.
3. Side-by-side diffusion cells with artificial or biological membranes.

4. Agitation followed by ultracentrifugation/centrifugation.
5. Ultrafiltration or centrifugal ultrafiltration methods.

Others: Presence of additional colloidal structures (liposomes, micelles, supercooled melts, and nanoparticles) and distribution process time frames are other parameters of interest that should be evaluated. The interaction of LNs with cells can be studied with confocal laser scanning microscopy (CLSM). Additional knowledge on qualitative nature and size of nanoparticles can be achieved by nuclear magnetic resonance (NMR) (Attama et al., 2012; Bagul et al., 2012).

6.7 CHALLENGES OF LNs-BASED DELIVERY SYSTEMS

Despite having many advantages LNs-based drug delivery needs to address a few issues that pose a challenge to its clinical and commercial application. Small size and large surface areas result in aggregation of particles, difficult handling in dry form, and restricted drug loading and burst release. Another major challenge in parenteral delivery of LNs is their probable interaction with the cells of RES (Athar and Das, 2014; Bagul et al., 2012).

Some issues of concern are addressed here:

1. The stability of LNs is governed by their sizes. LNs face instabilities like Ostwald ripening and particle aggregation. Colloidal state leads to several additional instabilities, like the presence of big interfaces causes' sorption-desorption, chemical instabilities, and stabilizer layer interactions.
2. Triglycerides crystallize in α, γ, β′, and β-polymorphic forms. α-Polymorph (metastable) transforms to β-polymorph (stable) having a ordered and rigid structure but poor drug loading capacity. Drug expulsion and inability to prolong drug release are the demerits of the stable β-form.
3. Morphological change of LNs from spheres to platelets causes gelation of SLN dispersions. Gelation of the otherwise liquid dispersions upon storage has been observed depending on composition, including the emulsifiers and quantity of matrix used. It poses a problem in parenteral delivery.
4. Self-assembly forms certain additional structures (like micelles/liposomes/vesicles) due to emulsifiers used in large amounts for stabilizing the LNs that tend to redistribute from the particles back to the aqueous medium. Pharmaceuticals can get solublized in these additional structures, and alter drug release along with drug loading capacity.
5. High water content (70%–95%) is a major problem with LNs as it could lead to degradation of the incorporated bioactive and high cost of energy input is required for lyophilization.
6. Selection of a suitable dosage form to package LNs is difficult task, often requiring an extra step in packaging that might require additional technology making the production process more expensive (Attama et al., 2012).

6.8 ROLE OF LNs IN DELIVERY OF BIOACTIVES FOR THE BRAIN

Targeting the drug by conventional approaches in the central nervous system is a challenging task because of the BBB. This barrier, otherwise a life-supporting protection, restricts delivery of many drugs and pharmaceuticals to the brain as they are not capable of crossing the BBB in amounts relevant to produce the therapeutic effect. It consequently results in poor therapeutic efficiency to treat brain diseases as a result of noneffective brain concentrations of these bioactives (Athar and Das, 2014). Hence, there is a need for specialized mechanisms of drug delivery for administering drugs and pharmaceuticals across the brain. Drug targeting via LNs overcomes physiological barriers and facilitates drug transport to desired cellular compartments via passive or ligand-mediated mechanisms. LNs provide high efficacy of the systemically administered drugs thereby minimizing any unwanted effects on healthy tissues and reducing the need of repeated drug administration (Mohanraj and Chen, 2007). Several studies have demonstrated that surfactant-coated SLNs can enhance brain uptake of loaded drugs that exhibit poor diffusion in the brain when given as solutions. SLNs loaded with a prodrug named dioctanoyl-fluoro-deoxyuridine were found to show higher brain distribution of the drug as compared to that from its solution form. The empirical approaches currently being investigated for triggering LN delivery systems translocation across the BBB includes Absorptive mediated transcytosis (AMT) and receptor mediated transcytosis (RMT).

AMT: Absorptive-mediated endocytosis is induced by electrostatic interaction between proteins that are positively charged and plasma membranes having a negative charge; hence, this mechanism holds relevance to brain as well as other organs rather cationized proteins are known to accumulate in other tissues like in liver, kidney, and lung tissues. Cationized proteins like albumin, immunoglobulins have been found to translocate through AMT.

RMT: Due to its ubiquity comparatively high brain specificity may be attained by targeting of receptors that are overexpressed (like receptors of insulin, insulin-like growth factor, and transferrin or low density lipoproteins) at the luminal side of brain endothelial cells in comparison to other organs. LNs of 100 nm or less in diameter, surface modified with receptor recognizing ligand (natural substrate or antibody) can trigger transcytosis. Transferrin ligand was often used for liposomal brain delivery of therapeutic agents (Oliveira, 2007).

7 CONCLUSIONS: WAY AHEAD WITH LNs

The use of LNs as drug carriers is a favored concept as most ingredients and excipients used have GRAS status thus making them suitable for delivery of food and pharmaceutical products. The small molecule, drugs, proteins, and peptides can be suitably delivered via LNs. The main concerns of formulation designs include shelf-life stability and preventing endogenous enzymatic degradation of the therapeutic agent along with its intended delivery to the target. Lipid formulations can control drug release and minimize variations in absorption that results in effective target concentrations of the drug. LNs can be orally administered as water-based dispersions or packaged as conventional dosage forms, like capsules, tablets, pellets powders. LNs are suitable both for hydrophilic and lipophilic drugs that can be incorporated for active as well as passive targeting mechanisms. LNs for topical use can be formulated with high lipid content or formulated into gels and creams. LNs may offer potential merits for delivery of botanicals and natural bioactives through oral and topical routes of administration. Hence, they serve as promising delivery systems that promote clinical development of botanicals for mitigation and treatment of diseases considering their limitations on account of their physicochemical properties. As discussed in the chapter, phytomedicines suffer from the drawbacks of poor aqueous solubility, bioavailability, and degradation, and hence, they do not reach the target sites in clinically significant amounts to produce the desired therapeutic effects. Thus, their true potentials could not be utilized optimally. Phytomedicinals exhibit a wide spectrum of pharmacological activities hence LNs can be used to deliver them to the desired site of action. Targeted delivery via LNs prevents undesired effects and wastage of bioactives. From the point of view of the formulation scientist; lipid particulate drug delivery systems envisage a positive futuristic market trend, and it is likely to have several successful formulations out there in the market.

ACKNOWLEDGEMENT

The authors are thankful to CSIR, New Delhi and knowledge resource centre CSIR-CDRI, Lucknow. CDRI communication no 9449.

REFERENCES

Ahmad, H., Khandelwal, K., Samuel, S.S., Tripathi, S., Mitra, K., Sangwan, R.S., Shukla, R., Dwivedi, A K., 2016. Neuro-protective potential of a vesicular system of a standardized extract of a new chemotype of Withania somnifera Dunal (NMITLI118RT +) against cerebral stroke in rats. Drug Deliv. 23 (7), 2630–2641.

Al-Jaber, N.A., Awaad, A.S., Moses, J.E., 2011. Review on some antioxidant plants growing in Arab world. J. Saudi Chem. Soc. 15, 293–307.

Alolga, R.N., Amadi, S.W., Onoja, V., Assanhou, A.G., Muyaba, M., Kassim, S.A., 2015. Anti-inflammatory and antipyretic properties of Kang 601 heji, a traditional Chinese oral liquid dosage form. Asian Pac. J. Trop. Biomed. 5, 921–927.

Anand, P., Kunnumakkara, A.B., Newman, R.A., Aggarwal, B.B., 2007. Bioavailability of curcumin: problems and promises. Mol. Pharm. 4, 807–818.

Athar, M., Das, A.J., 2014. Therapeutic nanoparticles: state-of-the-art of nanomedicine. Adv. Mater. Rev. 1, 25–37.

Attama, A.A., Momoh, M.A., Builders, P.F., 2012. Lipid Nanoparticulate Drug Delivery Systems: A Revolution in Dosage Form Design and Development. INTECH Open Access Publisher, Croatia.

Bagul, R., Mahajan, V., Dhake, A., 2012. New approaches in nanoparticulate drug delivery system: a review. Int. J. Curr. Pharm. Res 4, 29–38.

Banerjee, T.K., Roy, M.K., Bhoi, K.K., 2005. Is stroke increasing in India: preventive measures that need to be implemented. J. Ind. Med. Assoc. 103, 162, 164, 166 passim-162, 164, 166 passim.

NINDS. (2011). Brain Basics: preventing stroke. National Institute of Neurological Disorders and Stroke (NINDS). NIH Publication No. 11-3440b. Prepared by: Office of Communications and Public LiaisonNINDS, National Institutes of HealthBethesda, MD 20892. Retrieved 2009-10-24.

Carlsen, M.H., Halvorsen, B.L., Holte, K., Bohn, S.K., Dragland, S., Sampson, L., Willey, C., Senoo, H., Umezono, Y., Sanada, C., 2010. The total antioxidant content of more than 3100 foods, beverages, spices, herbs and supplements used worldwide. Nutr. J. 9, 1–11.

Chang, W.-T., Shao, Z.-H., Yin, J.-J., Mehendale, S., Wang, C.-Z., Qin, Y., Li, J., Chen, W.-J., Chien, C.-T., Becker, L.B., 2007. Comparative effects of flavonoids on oxidant scavenging and ischemia-reperfusion injury in cardiomyocytes. Eur. J. Pharmacol. 566, 58–66.

Chen, C., Li, S.-X., Wang, S.-M., Liang, S.-W., 2012. Investigation into the anti-thrombosis effect and contents of total saponins and flavonoids in the bioactive fraction of Naodesheng prescription. J. Ethnopharmacol. 144, 208–212.

Chiang, J.F., 2007. Biological requirements for nanotherapeutic applications. In: Deepak, T., Deleers, M. (Eds.), Nanoparticulate Drug-Delivery Systems. Informa Healthcare, New York, pp. 159–172.

Cho, N., Choi, J.H., Yang, H., Jeong, E.J., Lee, K.Y., Kim, Y.C., Sung, S.H., 2012. Neuroprotective and anti-inflammatory effects of flavonoids isolated from Rhus verniciflua in neuronal HT22 and microglial BV2 cell lines. Food Chem. Toxicol. 50, 1940–1945.

Das, S.K., Banerjee, T.K., 2008. Stroke Indian scenario. Circulation 118, 2719–2724.

Deleers, M., Pathak, Y., 2007. Nanoparticulate Drug Delivery Systems. Drugs and the Pharmaceutical Sciences Informa Healthcare, New York.

Dohare, P., Garg, P., Jain, V., Nath, C., Ray, M., 2008. Dose dependence and therapeutic window for the neuroprotective effects of curcumin in thromboembolic model of rat. Behav. Brain Res. 193, 289–297.

Eliana, B., Souto, R.H.M., 2007. Lipid nanoparticles (solid lipid nanoparticles and nanostructured lipid carriers) for cosmetic, dermal, and transdermal applications. In: Thassu, D., Deleers, M. (Eds.), Nanoparticulate Drug-Delivery Systems. Informa Healthcare, New York, pp. 213–233.

Erlund, I., 2004. Review of the flavonoids quercetin, hesperetin, and naringenin. Dietary sources, bioactivities, bioavailability, and epidemiology. Nutr. Res. 24, 851–874.

Ezhilarasan, D., Sokal, E., Karthikeyan, S., Najimi, M., 2014. Plant derived antioxidants and antifibrotic drugs: past, present and future. J. Coast. Life Med. 2, 738–745.

Fawole, O., Ndhlala, A., Amoo, S., Finnie, J., Van Staden, J., 2009. Anti-inflammatory and phytochemical properties of twelve medicinal plants used for treating gastro-intestinal ailments in South Africa. J. Ethnopharmacol. 123, 237–243.

Fukuta, T., Ishii, T., Asai, T., Sato, A., Kikuchi, T., Shimizu, K., Minamino, T., Oku, N., 2015. Treatment of stroke with liposomal neuroprotective agents under cerebral ischemia conditions. Eur. J. Pharm. Biopharm. 97, 1–7.

Gorinstein, S., Cvikrova, M., Machackova, I., Haruenkit, R., Park, Y.-S., Jung, S.-T., Yamamoto, K., Ayala, A.L.M., Katrich, E., Trakhtenberg, S., 2004. Characterization of antioxidant compounds in Jaffa sweeties and white grapefruits. Food Chem. 84, 503–510.

Graefe, E.U., Wittig, J., Mueller, S., Riethling, A.-K., Uehleke, B., Drewelow, B., Pforte, H., Jacobasch, G., Derendorf, H., Veit, M., 2001. Pharmacokinetics and bioavailability of quercetin glycosides in humans. J. Clin. Pharmacol. 41, 492–499.

Graham, S.H., Hickey, R.W., 2002. Molecular pathophysiology of stroke. In: Kenneth, L., Davis, D.C., Joseph, T.C., Nemeroff, C. (Eds.), Neuropsychopharmacology: The Fifth Generation of Progress. Nature Publishing Group (NPG), New York, pp. 1317–1326.

Gupta, Y., Briyal, S., Gulati, A., 2010. Therapeutic potential of herbal drugs in cerebral ischemia. Indian J. Physiol. Pharmacol. 54 (2), 99–122.

Gupta, V.K., Sharma, S.K., 2006. Plants as natural antioxidants. Nat. Prod. Rad. 5, 326–334.

Heron, M., Hoyert, D., Murphy, S., Xu, J, Kochanek, K.D., Tajeda-Vera, B, 2009. Deaths: final data for 2006. Nat. Vital Stat. Rep. 57, 1–134.

Hyun, H., Lee, J., Kim, S., Hyun, D.K., Choi, J.S., Lee, J.-K., Lee, M., 2011. Combinational therapy of ischemic brain stroke by delivery of heme oxygenase-1 gene and dexamethasone. Biomaterials 32, 306–315.

Iadecola, C., Anrather, J., 2011. Stroke research at a crossroad: asking the brain for directions. Nature Neurosci. 14, 1363–1368.

Jiang, J., Wang, W., Sun, Y.J., Hu, M., Li, F., Zhu, D.Y., 2007. Neuroprotective effect of curcumin on focal cerebral ischemic rats by preventing blood–brain barrier damage. Eur. J. Pharmacol. 561, 54–62.

Joseph, E., Saha, R.N., 2013. Advances in brain targeted drug delivery: nanoparticulate systems. AAPS PharmaSciTech 3 (1), 1–8.

Jun Wu, X.Z., Lee, Robert J., 2007. Lipid-based nanoparticulate drug delivery system. In: Thassu, D., Deleers, M. (Eds.), Nanoparticulate Drug-Delivery Systems. Informa Healthcare, New York, pp. 89–98.

Kakkar, V., Muppu, S.K., Chopra, K., Kaur, I.P., 2013. Curcumin loaded solid lipid nanoparticles: an efficient formulation approach for cerebral ischemic reperfusion injury in rats. Eur. J. Pharma. Biopharm. 85, 339–345.

Kaur, J., Verma, I., Narang, A., Chinna, R., Singh, P., Aggarwal, S., 2011. Int. J. Biol. Med. Res. 2, 575–576.

Krishnaiah, D., Sarbatly, R., Bono, A., 2007. Phytochemical antioxidants for health and medicine: a move towards nature. Biotechnol. Mol. Biol. Rev. 1, 97–104.

Kusumaningtyas, R., Laily, N., Limandha, P., 2015. Potential of Ciplukan (*Physalis Angulata* L.) as source of functional ingredient. Procedia Chem. 14, 367–372.

Lako, J., Trenerry, V.C., Wahlqvist, M., Wattanapenpaiboon, N., Sotheeswaran, S., Premier, R., 2007. Phytochemical flavonols, carotenoids and the antioxidant properties of a wide selection of Fijian fruit, vegetables and other readily available foods. Food Chem. 101, 1727–1741.

Lan, J., Yan, B., Zhao, Y.N., Wang, D., Hu, J., Xing, D., Du, L., 2008. Cerebral ischemia reperfusion exacerbates and pueraria flavonoids attenuate depressive responses to stress in mice. Tsinghua Sci. Technol. 13, 485–491.

Lapchak, P.A., 2010. Efficacy and safety profile of the carotenoid trans sodium crocetinate administered to rabbits following multiple infarct ischemic strokes: a combination therapy study with tissue plasminogen activator. Brain Res. 1309, 136–145.

Larsson, S.C., Virtamo, J., Wolk, A., 2013. Black tea consumption and risk of stroke in women and men. Annals Epidemiol. 23, 157–160.

Lee, J., Hyun, H., Kim, J., Ryu, J.H., Kim, H.A., Park, J.H., Lee, M., 2012. Dexamethasone-loaded peptide micelles for delivery of the heme oxygenase-1 gene to ischemic brain. J. Control. Release 158, 131–138.

Lembo, D., Cavalli, R., 2010. Nanoparticulate delivery systems for antiviral drugs. Antiviral Chem. Chemothera. 21, 53–70.

Li, R.W., Lin, G.D., Myers, S.P., Leach, D.N., 2003. Anti-inflammatory activity of Chinese medicinal vine plants. J. Ethnopharmacol. 85, 61–67.

Liu, M., Wu, B., Wang, W.-Z., Lee, L.-M., Zhang, S.-H., Kong, L.-Z., 2007. Stroke in China: epidemiology, prevention, and management strategies. Lancet Neurol. 6, 456–464.

Lloyd-Jones, D., Adams, R., Carnethon, M., De Simone, G., Ferguson, T.B., Flegal, K., Ford, E., Furie, K., Go, A., Greenlund, K., 2009. Heart disease and stroke statistics: 2009 update a report from the American Heart Association Statistics Committee and Stroke Statistics Subcommittee. Circulation 119, e21–e181.

Luximon-Ramma, A., Bahorun, T., Crozier, A., Zbarsky, V., Datla, K.P., Dexter, D.T., Aruoma, O.I., 2005. Characterization of the antioxidant functions of flavonoids and proanthocyanidins in Mauritian black teas. Food Res. Int. 38, 357–367.

Maas, M.B., Safdieh, J., 2009. Ischemic Stroke: Pathophysiology and Principles of Stroke Localization. In: Atri, A., Milligan, T. (Eds.), Hospital Physician Neurology Board Review Manual. Turner White Communications, Wayne, PA, USA.

Mackey, D.B., 2008. Prevention, Pathophysiology and Prevention of Stroke.

Mahapatro, A., Singh, D.K., 2011. Biodegradable nanoparticles are excellent vehicle for site directed in-vivo delivery of drugs and vaccines. J. Nanobiotechnol. 9, 1–11.

Martin-Schild, S., Hallevi, H., Shaltoni, H., Barreto, A.D., Gonzales, N.R., Aronowski, J., Savitz, S.I., Grotta, J.C., 2009. Combined neuroprotective modalities coupled with thrombolysis in acute ischemic stroke: a pilot study of caffeinol and mild hypothermia. J. Stroke Cerebrovascul. Dis. 18, 86–96.

Matu, E.N., Van Staden, J., 2003. Antibacterial and anti-inflammatory activities of some plants used for medicinal purposes in Kenya. J. Ethnopharmacol. 87, 35–41.

Mohanraj, V., Chen, Y., 2007. Nanoparticles: a review. Trop. J. Pharma. Res. 5, 561–573.

Mulaudzi, R., Ndhlala, A., Kulkarni, M., Finnie, J., Van Staden, J., 2013. Anti-inflammatory and mutagenic evaluation of medicinal plants used by Venda people against venereal and related diseases. J. Ethnopharmacol. 146, 173–179.

Naahidi, S., Jafari, M., Edalat, F., Raymond, K., Khademhosseini, A., Chen, P., 2013. Biocompatibility of engineered nanoparticles for drug delivery. J. Control. Release 166, 182–194.

Nabavi, S.F., Sureda, A., Habtemariam, S., Nabavi, S.M., 2015. Ginsenoside Rd and ischemic stroke: a short review of literatures. J. Ginseng Res. 39 (4), 299–303.

Nagakannan, P., Shivasharan, B., Thippeswamy, B., Veerapur, V., Bansal, P., 2012. Protective effect of hydroalcoholic extract of *Mimusops elengi* Linn. flowers against middle cerebral artery occlusion induced brain injury in rats. J. Ethnopharmacol. 140, 247–254.

Namsa, N.D., Tag, H., Mandal, M., Kalita, P., Das, A., 2009. An ethnobotanical study of traditional anti-inflammatory plants used by the Lohit community of Arunachal Pradesh, India. J. Ethnopharmacol. 125, 234–245.

Narayanaswamy, N., Balakrishnan, K., 2011. Evaluation of some medicinal plants for their antioxidant properties. Int. J. Pharm. Tech. Res. 3, 381–385.

Oliveira, J.-C.O.A.M.P.D., 2007. Nanoparticulate systems for central nervous system drug delivery. In: Thassu, D., Deleers, M. (Eds.), Nanoparticulate Drug-Delivery Systems. Informa Healthcare, New York, pp. 281–290.

Park, K.H., Park, Y.-D., Han, J.-M., Im, K.-R., Lee, B.W., Jeong, I.Y., Jeong, T.-S., Lee, W.S., 2006. Anti-atherosclerotic and anti-inflammatory activities of catecholic xanthones and flavonoids isolated from Cudrania tricuspidata. Bioorgan. Med. Chem. Lett. 16, 5580–5583.

Rabiei, Z., Rafieian-Kopaei, M., 2014. Neuroprotective effect of pretreatment with *Lavandula officinalis* ethanolic extract on blood-brain barrier permeability in a rat stroke model. Asian Pac. J. Trop. Med. 7, S421–S426.

Ranghar, S., Sirohi, P., Verma, P., Agarwal, V., 2014. Nanoparticle-based drug delivery systems: promising approaches against infections. Braz. Arch. Biol. Technol. 57, 209–222.

Raza, S.S., Khan, M.M., Ahmad, A., Ashafaq, M., Khuwaja, G., Tabassum, R., Javed, H., Siddiqui, M.S., Safhi, M.M., Islam, F., 2011. Hesperidin ameliorates functional and histological outcome and reduces neuroinflammation in experimental stroke. Brain Res. 1420, 93–105.

Raza, S., Khan, M., Ahmad, A., Ashafaq, M., Islam, F., Wagner, A., Safhi, M., 2013. Neuroprotective effect of naringenin is mediated through suppression of NF-κB signaling pathway in experimental stroke. Neuroscience 230, 157–171.

Rosenson, R.S., Stafforini, D.M., 2012. Modulation of oxidative stress, inflammation, and atherosclerosis by lipoprotein-associated phospholipase A2. J. Lipid Res. 53, 1767–1782.

Sanchez, P.M., Villarreal, M.L., Herrera-Ruiz, M., Zamilpa, A., Jiménez-Ferrer, E., Trejo-Tapia, G., 2013. In vivo anti-inflammatory and anti-ulcerogenic activities of extracts from wild growing and in vitro plants of *Castilleja tenuiflora* Benth.(Orobanchaceae). J. Ethnopharmacol. 150, 1032–1037.

Shah, F.-A., Gim, S.-A., Sung, J.-H., Jeon, S.-J., Kim, M.-O., Koh, P.-O., 2016. Identification of proteins regulated by curcumin in cerebral ischemia. J. Surg. Res. 201, 141–148.

Shaikh, R.U., Pund, M.M., Gacche, R.N., 2015. Evaluation of anti-inflammatory activity of selected medicinal plants used in Indian traditional medication system in vitro as well as in vivo. J. Trad. Compl. Med. 6 (4), 355–361.

Sharma, S., Singh, L., Singh, S., 2013. A review on medicinal plants having antioxidant potential. Int. J. Pharm. Bio. Sci. 1, 404–409.

Shin, J.A., Lee, H., Lim, Y.-K., Koh, Y., Choi, J.H., Park, E.-M., 2010. Therapeutic effects of resveratrol during acute periods following experimental ischemic stroke. J. Neuroimmunol. 227, 93–100.

Sindhi, V., Gupta, V., Sharma, K., Bhatnagar, S., Kumari, R., Dhaka, N., 2013. Potential applications of antioxidants: a review. J. Pharm. Res. 7, 828–835.

Siriwatanametanon, N., Fiebich, B.L., Efferth, T., Prieto, J.M., Heinrich, M., 2010. Traditionally used Thai medicinal plants: in vitro anti-inflammatory, anticancer and antioxidant activities. J. Ethnopharmacol. 130, 196–207.

Soumaya, K.-J., Zied, G., Nouha, N., Mounira, K., Kamel, G., Genviève, F.D.M., Leila, G.C., 2014. Evaluation of in vitro antioxidant and apoptotic activities of *Cyperus rotundus*. Asian Pac. J. Trop. Med. 7, 105–112.

Subedi, L., Timalsena, S., Duwadi, P., Thapa, R., Paudel, A., Parajuli, K., 2014. Antioxidant activity and phenol and flavonoid contents of eight medicinal plants from Western Nepal. J. Trad. Chin. Med. 34, 584–590.

Sun, Y., Jiang, J., Zhang, Z., Yu, P., Wang, L., Xu, C., Liu, W., Wang, Y., 2008. Antioxidative and thrombolytic TMP nitrone for treatment of ischemic stroke. Bioorgan. Med. Chem. 16, 8868–8874.

Sunil, A., Kesavanarayanan, K., Kalaivani, P., Sathiya, S., Ranju, V., Priya, R.J., Pramila, B., Paul, F.S., Venkhatesh, J., Babu, C.S., 2011. Total oligomeric flavonoids of *Cyperus rotundus* ameliorates neurological deficits, excitotoxicity and behavioral alterations induced by cerebral ischemic: reperfusion injury in rats. Brain Res. Bull. 84, 394–405.

Tang, W., Sun, X., Fang, J.-S., Zhang, M., Sucher, N., 2004. Flavonoids from Radix scutellariae as potential stroke therapeutic agents by targeting the second postsynaptic density 95 (PSD-95)/disc large/zonula occludens-1 (PDZ) domain of PSD-95. Phytomedicine 11, 277–284.

Wang, Y., Cooke, M.J., Sachewsky, N., Morshead, C.M., Shoichet, M.S., 2013. Bioengineered sequential growth factor delivery stimulates brain tissue regeneration after stroke. J. Control. Release 172, 1–11.

Wang, L., Zhang, X., Liu, L., Cui, L., Yang, R., Li, M., Du, W., 2010. Tanshinone II A down-regulates HMGB1, RAGE, TLR4, NF-(B expression, ameliorates BBB permeability and endothelial cell function, and protects rat brains against focal ischemia. Brain Res. 1321, 143–151.

Wattanathorn, J., Sattroopinat, N., Tong-Un, T., Muchmapura, S., Wannanond, P., Sirisa-Ard, P., 2012. Neuroprotective effect against cerebral ischemia of *Passiflora foetida*. Am. J. Appl. Sci. 9, 600.

Wilczewska, A.Z., Niemirowicz, K., Markiewicz, K.H., Car, H., 2012. Nanoparticles as drug delivery systems. Pharmacol. Rep. 64, 1020–1037.

World Health Report, 2004. Deaths by cause, sex and mortality stratum in WHO regions, estimates for 2002. Annex Table 2.

Yokoyama, K., Shimada, Y., Hori, E., Nakagawa, T., Takagi, S., Sekiya, N., Kouta, K., Nishijo, H., Yokozawa, T., Terasawa, K., 2004. Effects of Choto-san and hooks and stems of *Uncaria sinensis* on antioxidant enzyme activities in the gerbil brain after transient forebrain ischemia. J. Ethnopharmacol. 95, 335–343.

Zaidi, S.F., Muhammad, J.S., Shahryar, S., Usmanghani, K., Gilani, A.-H., Jafri, W., Sugiyama, T., 2012. Anti-inflammatory and cytoprotective effects of selected Pakistani medicinal plants in Helicobacter pylori-infected gastric epithelial cells. J. Ethnopharmacol. 141, 403–410.

Zhang, Y., Wang, X., Wang, X., Xu, Z., Liu, Z., Ni, Q., Chu, X., Qiu, M., Zhao, A., Jia, W., 2006. Protective effect of flavonoids from *Scutellaria baicalensis* Georgi on cerebral ischemia injury. J. Ethnopharmacol. 108, 355–360.

Zhao, J., Zhao, Y., Zheng, W., Lu, Y., Feng, G., Yu, S., 2008. Neuroprotective effect of curcumin on transient focal cerebral ischemia in rats. Brain Res. 1229, 224–232.

CHAPTER

CHITOSAN-BASED NANOPARTICULATE SYSTEMS FOR ORAL DRUG DELIVERY

20

Sougata Jana, Sabyasachi Maiti

Gupta College of Technological Sciences, Asansol, India

CHAPTER OUTLINE

1 **Introduction** 607
2 **Novel Technologies for Chitosan-Based Nanoparticles** 611
 2.1 Synthesis of CS-NPs by Ionic Gelation 611
 2.2 Solid–Liquid Phase Separation Method 613
 2.3 Solvent Injection Method for CS-SLNs 613
 2.4 Emulsification and Ultrasonication Method for TMC-*g*-PA SLNs 613
 2.5 In Situ Precipitation Method for Magnetic Nanohydrogels 614
 2.6 Self-Assembled CS-Tocol Nanoparticles 616
 2.7 Spinning Disk Processor for CS–PMA NPs 617
 2.8 Dialysis Method for Chitosan-Based Nanohydrogels 618
 2.9 Amide Coupling Reaction for CS-Drug Nanoconjugates 618
 2.10 Sol–Gel Method for TiO_2-Coated CS-Based Nanohybrid Composites 619
3 **Oral Chitosan Nanosystems: Key Findings** 619
4 **Conclusions** 631
References 631

1 INTRODUCTION

The polysaccharide-based materials have gained importance as drug delivery carriers over last few decades due to their excellent physical and biological properties (Liu et al., 2008). Cationic polysaccharide, chitosan (CS) has been widely studied as a drug delivery vehicle for mucosal, ocular, topical, and controlled delivery of a variety of therapeutic agents and also as targeted drug-delivery systems (Na et al., 2012).

The amine and alcohol groups along CS chains enable this polysaccharide for chemical modification in order to extend the pharmaceutical application of CS (Croisier and Jerome, 2013). Crosslinking by glutyraldehyde, epichlorohydrin, and glyoxal (Coelho et al., 2007; Gupta and Jabrail, 2007; Yang et al., 2005) and surface decoration is a possible way of overcoming the limitations of native CS, such as swelling and dissolving under gastric conditions thereby enhancing the mechanical strength of the polymer. Grafting of CS has a wide range of applications in the field of drug delivery. Hollow CS-*g*-poly(acrylic acid) nanospheres were used as carriers for doxorubicin (DOX) by Hu et al. (2007). Grafting of hydrophilic vinyl monomers onto CS chain causes the formation of interchain bridges between hydrogel polymer and functional groups that increases the bioadhesive and mucoadhesive characteristics of the carrier. As a consequence, residence time and tissue permeability of drug is increased leading to site-specific binding (Bhattarai et al., 2010; Peppas and Sahlin, 1996; Salamat-Miller et al., 2005). CS-*g*-β-cyclodextrin (CD) nanoparticles have been investigated as a carrier for controlled release of ketoprofen (KTP) (Yuan et al., 2013).

The CS, a linear polymer having (1,4)-linked 2-amino-2-deoxy-β-D-glucopyranose units on its structure, is a biodegradable, biocompatible, nontoxic, and nonimmunogenic mucoadhesive polymer, which offers certain advantages over other natural polymers in formulating nanoparticles for oral administration of drug molecules (Nagpal et al., 2010). The cationic nature of CS (due to protonation of amine groups in acidic pH) facilitates effective encapsulation of biomolecules like proteins and drugs and prolongs the drug residence time in the gastrointestinal tract (Luessen et al., 1996). Apart from the endowment of positive charge on the surfaces of nanoparticles, CS also elevates the contact time with intestinal epithelium enhancing permeation through tight junction of epithelial cells via paracellular transport (Schipper et al., 1997). Most importantly, CS is digested by chitosanase enzymes secreted by microorganisms at the intestine after oral ingestion (Nagpal et al., 2010).

Sodium alginate (ALG) is another water-soluble natural linear polysaccharide, containing varying amounts of 1,4-linked β-D-mannuronic acid (M) and α-L-guluronic acid (G) residues, extracted from brown seaweed. It is a very popular pH-responsive polymer due to its shrinkage in lower pH (Dusseault et al., 2005), enabling encapsulated drug retention in the stomach while protecting it against enzymatic deactivation. Properties of biodegradability, biocompatibility, lowtoxicity, low immunogenicity, and good mucoadhesion (Esposito et al., 1996; George and Abraham, 2006) facilitate its application in oral drug delivery.

CS can form polyelectrolyte complexes with anionic biopolymers, such as sodium ALG (Finotelli et al., 2010). CS-ALG complexes are prepared by ionotropic gelation (Sarmento et al., 2007), in which the electrostatic interaction of carboxylic acid groups of ALG with amino groups of CS results in gel formation (Motwani et al., 2008).

CS/ALG nanoparticles were reported to be efficient in protecting insulin from the aggressive environment of the stomach and sustained insulin release was observed under intestinal milieu (Sarmento et al., 2007).

Hydrophilic drugs can be encapsulated in nanoparticles for improving their oral bioavailability, minimizing side effects, facilitating controlled release and permeation across the blood–brain barrier. It has earlier been reported that CS helps in augmenting the brain uptake potential of various drugs for treatment of several neurological disorders (Nagpal et al., 2013; Wilson et al., 2011). Orally administered CS nanoparticles could help the drug in reaching the brain through absorptive transcytosis (Bowman and Leong, 2006; Sarvaiya and Agrawal, 2015).

Various nanotechnologies have been employed for improving oral drug delivery. Natural polymer CS and their derivatives have been widely used for the development of self-assembled nanoparticles, which could increase the drug oral bioavailability (Chung et al., 2013; Lee et al., 2011a,b; Nogueira et al., 2013). Due to their good mucoadhesive character, they are used as pharmaceutical carriers for sustained drug release (Hirano, 1996). CS derivatives are used to enhance the bioavailability and dissolution rates of hydrophobic drugs (Miyazaki et al., 1981). This nanoparticle drug-delivery system could enhance stability of labile drugs, drug bioavailability, and controlled drug release owing to the fact that particles in the nanosize ranges are efficient in crossing permeability barriers (Sharma et al., 2004).

CS-based smart hydrogels has also drawn attention for its potential to achieve site-specific delivery to the colon. Carboxymethyl CS is one of the most important derivatives of CS polysaccharide. Because of carboxylate (COO^-) and amine (NH_2) pendants, the use of carboxymethyl CS has been extended for drug delivery. By introducing superparamagnetic iron oxide nanoparticles (SPIONs) into pH-sensitive hydrogel, there is possibility of obtaining dual sensitive hydrogels, which can simultaneously respond to external magnetic field and change of pH (Philippova et al., 2011). Under this condition, drug-targeting carrier based-SPIONs can target cell or tissue directly with the application of an external magnetic field, leading to decreased circulation time, drug dosage, and thereby reduced side effects (Kayal and Ramanujan, 2010). By applying an external magnetic field, the release of drug from magnetic hydrogels can also be controlled. Magnetic carboxymethyl CS hydrogels have been synthesized and tested for the pharmaceutical and biomedical applications (Shen et al., 2012).

CS has inherent antibacterial activity due to strong interaction between its positively charged amide groups and negatively charged phospholipid components of bacterial plasma membranes. CS molecules kill bacteria either by creating pores in cell membranes, which results in leakage of vital intracellular contents, or by its interaction with negatively charged DNA (Piras et al., 2014). CS nanoparticles (NPs) exhibit superior antibacterial activity due to their higher surface area and charge density as compared to macroscale CS polymer (de Pinho Neves et al., 2014). Jamil et al. (2016) fabricated and characterized CS-based nanocarriers for the enhanced efficacy of cefazolin antibiotics.

The use of biocompatible and biodegradable coatings in medical applications may add various structural and mechanical properties to the drug composites, such as protection from oxidation and degradation of drug, delay in the drug release, changing route of administration, and so on (Cultepe et al., 2010; Maroni et al., 2013). Some of them are titanium oxide (Signoretto et al., 2011), silicon oxide (Wang et al., 2013), polyvinyl pyrrolidone (PVP) (Archana et al., 2013), and pectin (Nguyen et al., 2011), and so on.

Oral delivery of new drug entities is becoming ever more difficult since they exhibit poor water solubility, poor dissolution rate and low oral bioavailability in the gastrointestinal tract (Gaucher et al., 2010; Pinnamaneni et al., 2002). Thus, there is a need of novel technologies for drug solubilization and effective oral administration of sparingly soluble drugs. Besides classical strategies, like drug micronization or use of surfactants, cosolvents, complexing agents and lipids, innovative strategies using nanocarriers are being studied. Among these, polymeric micelles are of particular interest. The amphiphilic block copolymers self-assemble above their critical micelle concentration (CMC) and form micelles. The hydrophilic moiety of the copolymer forms the outer layer of the micelle (the corona) while the hydrophobic block forms the core. The hydrophobic domain interacts with poorly soluble drugs resulting in their solubilization within the micelle. Moreover, its cationic properties below pH 6.5 favor mucoadhesion in the gastrointestinal tract. Hydrophobically modified CSs have been successfully synthesized via derivatization of the primary hydroxyl and amine groups located on the backbone of CS. Nevertheless, due to the insolubility of CS observed at pH values above its pKa (6.4) in water, micelles of amphiphilic CS rapidly precipitate in biological solution (pH 7.4). Therefore, water-soluble CS derivatives are being considered for the development of drug-delivery systems like glycol chitosan (GC), which exhibits good solubility over a broad range of pH (Hyung et al., 2006; Nam et al., 2009).

Nanoparticles delivery system offers great promise to administer anticancer drugs by oral route, because nanoparticles can reduce drug toxicities and control the release of encapsulated or associated drug (Sun et al., 2010). CS-based nanoparticles have attracted numerous interests because of their ability to open tight junctions (TJs) between intestinal epithelial cells and facilitate paracellular transport of drugs (Derakhshandeh and Fathi, 2012; Vllasaliu et al., 2012). Previous studies showed that the electrostatic interaction between CS and integrin receptors on cell membranes was prerequisite to initiate the cascade of TJs disruption and promote intestinal adsorption (Hsu et al., 2012). Since the pKa value of the amino groups on CS is approximately 6.5, CS is protonated only at acidic pH and deprotonated at neutral pH. This means that CS could open TJs merely confined to the duodenum locally (pH 6.0–7.0) where the pH value are below or close to its pKa (Su et al., 2012). Expanding the range of its application in the gastrointestinal tract is thus essential to further increase the efficacy of CS-based nanoparticles. O-carboxymethyl chitosan (CMCS) is a water-soluble CS derivative and the carboxyl group on CMCS makes its capacity high to bind Ca^{2+} (Gujarathi et al., 2012). This property could deprive the divalent ions from extracellular matrix and increase the parallular permeability of the epithelium (Bromberg and Alakhov, 2003). On the other hand, the pKa of CMCS is 2.0–4.0. Negative charged CMCS is able to form polyelectrolyte complex with positive charged CS via electrostatic interaction and maintain the stability of nanostructures in the gastrointestinal tract, which may provide a great potential to overcome the limitation of CS and expand the drug absorption site beyond duodenum.

Recently, it has been reported that paclitaxel-conjugate with low molecular weight CS could favor oral delivery of the anticancer agent. The conjugate showed increased water solubility, prolonged gastrointestinal retention, bypassed the P-glycoprotein-mediated efflux, and bypass cytochrome P450-mediated metabolism. Overall, the conjugate dramatically improved oral bioavailability and in vivo antitumor efficacy (Lee et al., 2008).

Colon-targeted drug-delivery systems have been designed, in particular nanosized particles that accumulate in inflamed colonic tissues due to their predisposition to mucoadhesion, which in turn promote cellular uptake and intracellular drug accumulation (Lamprecht et al., 2002; Moulari et al., 2008). One such mucoadhesive carrier is CS, a popular polymer for the development of particulate colon-targeted drug-delivery systems because of its ready digestion by the endogenous colonic flora in inflammatory bowel disease (IBD) patients (Friend, 1991). To limit CS dissolution only to the colon and not at low pH, which could lead to premature drug release in the upper gastrointestinal tract, an enteric coat of poly(methyl acrylates) (PMA, e.g., Eudragit S100 and L100) can be applied.

CS is an aminopolysaccharide composed of randomly distributed β-1,4-linked glucosamine and *N*-acetyl-D-glucosamine units. It is obtained from deacetylation of chitin, a natural polysaccharide found in the outer skeleton of insects, crabs, shrimp, and lobsters and in the internal structure of other invertebrates (Kumar et al., 2004; Liu et al., 2008). The sugar backbone of CS is similar to cellulose; except that the hydroxyl group of cellulose on C-2 position is replaced by an amino group in CS. Unfortunately, CS is only soluble in an aqueous solution at pH higher than its pKa (~pH 5.5–6.5) as the consequence of protonation of amine D-glucosamine monomeric units while remains insoluble at organic solvents, neutral and alkaline pH values. This is considered as a major negative characteristic for CS, which limits its use in some applications (Casettari et al., 2012; Jia et al., 2001). However, the presence of reactive functional groups in the CS structure provides great opportunity for its chemical modification, which offers a wide range of derivatives to induce specific quality. Grafting or copolymerization of CS with different molecular weights of polymers, such as poly(ethylene glycol) (PEG) is considered an attractive strategy to improve its solubility. PEGylated CS has been widely applied in pharmaceutical fields in the forms of microspheres, nanoparticles, or polymers to enhance stability, prolong half-life of active agents, and alter pharmacokinetics, tissue distribution, and pharmacological properties (Chiu et al., 1998; Du et al., 2009; Greenwald et al., 2003; Hu et al., 2007). These have been correlated to PEG hydrophilic flexibility, nonionic property, biodegradability, biocompatibility, low toxicity, antigenicity, and immunogenicity (Germershaus et al., 2008). However, modification of CS amino groups causes a change in the skeleton of CS, which results in the loss of original biochemical and physiochemical properties (Sugimoto et al., 1998). In addition, the presence of amine groups in CS makes it an excellent bioadhesive material that creates possibility of interaction with mucus and mucosal surface through interaction of positive amine groups and negative sugar groups, such as silaic acid (Deacon et al., 1999). It is therefore essential to enhance water solubility of CS without losing its biochemical and physiochemical properties by derivatizations methods, such as modifications of hydroxyl groups with PEG (Gorochovceva and Makuška, 2004).

Thus, it is understood that CS and its derivatives have been considered to overcome the poor solubility, intestinal permeation, and bioavailablity characteristics of therapeutic entities. Furthermore, superparamagnetic nanoparticles, polyelectrolyte complex nanoparticles, inorganic coatings on nanoparticles, nanoconjugates of CS, or its derivatives have been designed for oral Drug-Delivery Applications. Native CS has been chemically modified in order to make it suitable for Drug-Delivery Applications, too. The delivery of small molecular therapeutics via oral route often leads to patient compliance due to ease administration.

This chapter discusses the novel technologies employed for the design of CS-based nanocarriers, the potential of different CS derivatives for oral drug delivery, and the recent findings on CS-based nanosystems for oral applications.

2 NOVEL TECHNOLOGIES FOR CHITOSAN-BASED NANOPARTICLES

2.1 SYNTHESIS OF CS-NPs BY IONIC GELATION

CS-NPs were synthesized by ionic gelation method using sodium tripolyphosphate (TPP) as a gelating agent (Aktaş et al., 2005; Calvo et al., 1997; Elzatahry and Mohy Eldin, 2008). CS was dissolved in 1% (v/v) acetic acid and the pH was adjusted to 5.0 using 1 M NaOH. The drug, ellagic acid (EA) dissolved in Dimethyl sulfoxide (DMSO) was then added to the CS dispersion. Finally, TPP was added dropwise to the CS-EA solution under mild stirring 1 h to form EA-encapsulated CS nanoparticles (CSNPs). The nanoparticles were isolated by centrifugation, and freeze-dried.

For the preparation of Bay41-4109-loaded CSNPs, an aqueous solution of sodium TPP containing Bay41-4109 was dropped into acetic acid solution of CS under magnetic stirring (1000 rpm). The ionic interaction between the positively charged amino groups of CS and negatively charged sodium TPP resulted in instantaneous formation of CSNPs.

For the preparation of chitosan-*g*-β-cyclodextrin (CD-*g*-CS) nanoparticles, it is of utmost need to gain an idea about the synthestic procedure of this graft copolymer. First of all, mono-6-deoxy-6-(*p*-toluenesulfonyl)-β-cyclodextrin (6-OTs-β-CD) was synthesized as described previously (Petter et al., 1990). Briefly, β-CD was suspended in water, and NaOH solution was added dropwise. The suspension became homogeneous and slightly yellow before the addition was complete. Paratoluenesulfonyl chloride in acetonitrile was added dropwise, causing immediate formation of white precipitate. The resulting precipitate was recovered by filtration and dried for 12 h. CD-*g*-CS was synthesized using Gonil's method (Gonil and Sajomsang, 2011) with slight modification. Briefly, CS was dissolved in 1% (v/v) acetic acid (pH 4.0). The solution of 6-OTs-β-CD in *N*,*N*-dimethylformamide (DMF) was added into the CS solution. The reaction mixture was refluxed at 100°C for 16 h, dialyzed with deionized water, and last, freeze-dried to obtain CD-*g*-CS powder. The reaction scheme is illustrated in Fig. 20.1. The mild ionic gelation procedure developed by Aktaş et al. (2005) was adopted for CD-*g*-CS NPs

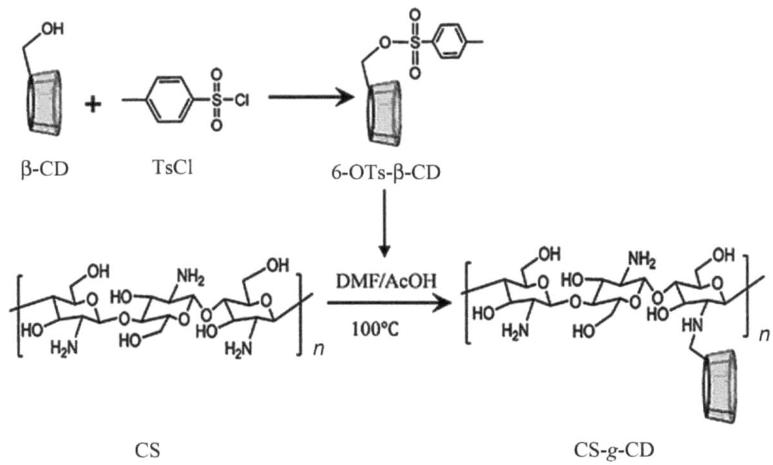

FIGURE 20.1 Synthetic Pathway of CD-*g*-CS Polymer

Reprinted from Yuan, Z., Ye, Y., Gao, F., Yuan, H., Lan, M., Lou, K., Wang, W., 2013. Chitosan-graft-β-cyclodextrin nanoparticles as a carrier for controlled drug release. Int. J. Pharm. 446, 191–198, Copyright (2013), with permission from Elsevier.

preparation. Nanoparticles were spontaneously generated due to ionic gelation between the positively charged amino groups of CD-*g*-CS and negatively charged TPP. Briefly, CD-*g*-CS (DS = 9.6, 14.0, and 20.0%) solutions (0.05%–0.25%, w/w) containing KTP (KTP: CD-*g*-CS = 1:2) was adjusted to pH 3.5–6.0. Crosslinking solution of TPP was added dropwise to CD-*g*-CS solution with magnetic stirring at 800 rpm for 30 min. The nanoparticles thus obtained were freeze-dried with 3% mannitol as lyophilized protection agent.

Before the synthesis of PEG–CS nanoparticles, chemoselective PEGylation of CS was done by Najafabadi et al. (2014). It follows three steps: protection of the CS amine group with sodium dodecylsulfate (SDS) and then chlorination of protected CS, activation of OH–PEG–OCH_3 (mPEG) with NaH, and grafting of activated mPEG onto the chlorinated CS followed by removal of the protecting group.

The SDS/CS was complexed by the procedure described by Cai et al. (2009). Briefly, (SDS)/CS complex (SCC) was prepared by mixing acidic solutions of CS and SDS. Predetermined amounts of CS and SDS were dissolved in 2% acetic acid solution, respectively. Then, the SDS solution was poured into it and gently stirred for 2 h. The resulting precipitate was filtered off, washed, and freeze-dried to obtain a white product. The amine groups at the C2 position of CS were protected using SDS. To prepare chlorinated CS, oxalyl chloride was added in 10-fold excess compared to SCC in pyridine and stirred under N_2 atmosphere. After half an hour, the reaction mixture was precipitated in ice-cold water, filtered, and vacuum-dried to yield chlorinated-SCC. The methoxy PEG was activated by adding OH–PEG–OCH_3 (mPEG) to a suspension of NaH in anhydrous THF. The mixture was stirred at 60°C for 2 h under inert atmosphere. Then, chlorinated-SCC was added to the reaction mixture and stirred for another 24 h. Finally, the reaction was allowed to cool down at room temperature and precipitated in methanol, filtered, washed, and vacuum-dried to yield product. The reaction mixture of SCC and PEG was directly precipitated into 15% Tris (hydroxyl methyl) amino methane aqueous solution (pH 8.0). The precipitate was collected via centrifugation, washed with 15% Tris solution and distilled water, and freeze-dried to yield CS-*g*–PEG. CS-*g*–PEG was dissolved in 1% (v/v) acetic acid solution (pH 5.0) and predetermined amount of ibuprofen was added. TPP was dissolved in water and the pH was adjusted to 3.0. The dropwise addition of TPP to the PEGylated–CS solution produced nanoparticles under constant magnetic stirring for 1 h.

For the production of DOX-loaded CS-CMCS NPs, an aqueous solution of DOX was premixed with CS solution (pH 6.0). Subsequently, CMCS and TPP solutions were blended with the mixture under constant stirring for 1 h, and the nanoparticles were formed and subsequently processed for freeze-drying.

CS-ALG NPs containing curcumin diethyl disuccinate (CDD) were prepared using oil-in-water emulsification and ionotropic gelation method (Lertsutthiwong et al., 2009). Briefly, acetonic CDD solution was added dropwise into ALG solution containing a nonionic surfactant (Pluronic F127) under mechanical stirring at 1000 rpm for 30 min, followed by $CaCl_2$ solution. The mixture was sonicated at a frequency of 45 kHz and sonic power of 80 W for 15 min. Then, CS solution in 1% (v/v) acetic acid was added dropwise into the resultant calcium-ALG pregel with continuous stirring at 1000 rpm for 30 min. The CS-ALG nanoparticles containing CDD were obtained as dispersion in aqueous solution.

2.2 SOLID–LIQUID PHASE SEPARATION METHOD

CS hollow nanospheres (CHNs) were prepared by Liu et al. (2015). 0.05% (w/v) aqueous solution of CS, dissolved in 0.025% (v/v) acetic acid was filled in a plastic tube of 12 mm (d) × 50 mm (h) dimension. The tube was immediately immersed into liquid nitrogen for 2–24 h and the frozen samples were then freeze-dried at −80°C. This process led to the formation of CHNs. Due to freezing of the CS solution, constitutional supercoiling at CS molecular chains and the water interface occurred, and led to Mullins–Sekerka instability and ice cell growth into a well-defined structure. This was a good method for preparation of CHNs, because it is a template-free and surfactants-free process. The freezing time also affects the morphology of the nanospheres. Too short or too long freezing leads to nano/microfibrous structure. This method was also reported earlier (Im et al., 2005; Zhang et al., 2005).

2.3 SOLVENT INJECTION METHOD FOR CS-SLNs

Hansraj et al. (2015) formulated sumatriptan succinate-loaded CS-solid lipid nanoparticles (CS-SLNs) by solvent injection method as described by Nair et al. (2012) with minor modification. Here, ethanol was used as organic solvent to dissolve tripalmitin lipid. The lipid phase, warmed at 75°C was dropped into a prewarmed aqueous phase containing sumatriptan succinate, polysorbate 80, soya lecithin, and acetic acid solution of CS under mechanical stirring at 4000 rpm. The nanoparticles dispersion was then cool-centrifuged at 15,000 rpm for 20 min. The precipitate was washed with HPLC grade water, redispersed under sonication and lyophilized after addition of a cryoprotectant, D-mannitol to have NPs

2.4 EMULSIFICATION AND ULTRASONICATION METHOD FOR TMC-*g*-PA SLNs

Before describing the preparation of trimethyl chitosan (TMC)-grafted-palmitic acid (TMC-*g*-PA)-coated SLNs, it is desirable to disclose the protocol for the synthesis of this graft copolymer (Ramalingam and Ko, 2014). Briefly, CS and sodium iodide were dissolved in *N*-methyl-2-pyrollidinone (NMP) and NaOH solution on water bath at 60°C. After the addition of iodomethane, the product was precipitated with ethanol, centrifuged, and washed with diethyl ether. Then, the product was dissolved in sodium chloride to exchange the iodide with chloride, precipitated in ethanol and centrifuged. Finally, water-soluble TMC product was dialyzed against water for 24 h and lyophilized.

The amidation of the primary amino group of TMC and the carboxyl group of palmitic acid (PA) in presence of EDC HCl resulted in the formation of TMC-*g*-PA copolymer (Du et al., 2009; Lee et al., 2011a). The ethanolic PA solution was added to an aqueous solution of TMC at 80°C. EDC HCl was then added to the reaction mixture and stirred for 6 h. The final reaction mixture was dialyzed against ethanol and then against distilled water to remove unreacted palmitic acid and any trace of ethanol, respectively. Finally, the dialyzed product was lyophilized to obtain TMC-*g*-PA polymer.

Resveratrol-loaded SLNs were produced by the emulsification and ultrasonication. Briefly, the lipid phase consisting of Precirol ATO 5, palmitic acid, and Gelucire 50/13 was melted to a temperature of 5–10°C above the melting point of the constituents. Resveratrol was then dispersed in the molten lipid phase, after which the preheated aqueous solution containing Tween 80 was added to the lipid phase. The mixture was probe sonicated for 5 min by maintaining the same temperature. The resulting SLNs was ice-cooled and then lyophilized. The surface modified SLNs were prepared by simple mixing of SLNs and TMC-*g*-PA at mass ratio of 50:1. The equal volume of TMC-*g*-PA solution and the SLNs dispersion were mixed and stirred for 2 h, centrifuged at 13,000 rpm for 30 min and then freeze-dried.

2.5 IN SITU PRECIPITATION METHOD FOR MAGNETIC NANOHYDROGELS

Saboktakin et al. (2010) prepared CS and dextran sulfate (DS) interpenetrating network (IPN) nnaohydrogels as follows: CS–DS and $FeCl_3$ hexahydrate were dissolved in water under nitrogen flush. $FeCl_2{\cdot}4H_2O$ and a small amount of aqueous ammonia were added. The black color solution was heated to 80°C for 100 min. The mixture was cooled to room temperature, and the ammonia was removed by N_2 flushing over 10 min. Then, the solid mixture was dissolved in acetic acid (1%, w/v) and 5-aminosalicylic acid (5-ASA) was dissolved in DMF. Then, the magnetic CS/DS and the drug solution (1:3) were mixed together and the solution was dropped through 22-gauge needle into sodium-saturated Tris–HCl buffer solution containing glutaraldehyde-saturated toluene (GST) under magnetic stirring (200 rpm). The hydrogel nanoparticles formed spontaneously, filtered, washed with Tris–HCl buffer, and dried in vacuum oven at 40°C.

The magnetic CMCS/κ-carrageenan (mCMCS-CG) IPN nanohydrogels were prepared via in situ precipitation of iron salts in the presence of a mixture of *N*,O-CMCS, and CG (Mahdavinia et al., 2015). The preparation of mCMCS-CG particles is shown in Fig. 20.2. Carrageenan (CG) and CMCS aqueous solution were mixed together at 70°C and gradually dropped into aqueous solutions of iron salts ($FeSO_4{\cdot}7H_2O$ and $FeCl_3{\cdot}6H_2O$). The spherical particles were immediately formed. The resulting particles were cured for 4 h, washed with water to remove iron salts from surface of beads. The iron salt-crosslinked beads were transferred into ammonia solution (1 M) and allowed for 4 h to form the magnetic nanohydrogels. The magnetic nanohydrogels were washed with distilled water.

The carboxylate and amine groups on CMCS as well as the sulfate pendants on CG can interact electrostatically with Fe^{2+} and Fe^{3+} ions. Therefore, the solution of biopolymers was dropped into Fe^{2+}/Fe^{3+} solution for formation of nanohydrogels through ionic crosslinking. The (Fe^{2+}/Fe^{3+})-crosslinked hydrogel particles were kept in ammonia solution to form magnetic Fe_3O_4 nanoparticles in situ. The apparent strength of magnetic beads was not sufficient. Hence, they were immersed into crosslinking solution (0.2 M KCl and 0.2 M $CaCl_2$) and stirred slowly for 1 h. The K^+ and Ca^{2+} cations can interact electrostatically with sulfate pendants on CG and carboxylate groups on CMCS, respectively (Luo et al., 2013; Morris et al., 1980). The treatment of magnetic hydrogels with K^+/Ca^{2+} solutions produced hard gels with high apparent strength. The mCMCS-CG nanoydrogels were air-dried. The nonmagnetic nanoydrogels were formed without dropping them into iron ions.

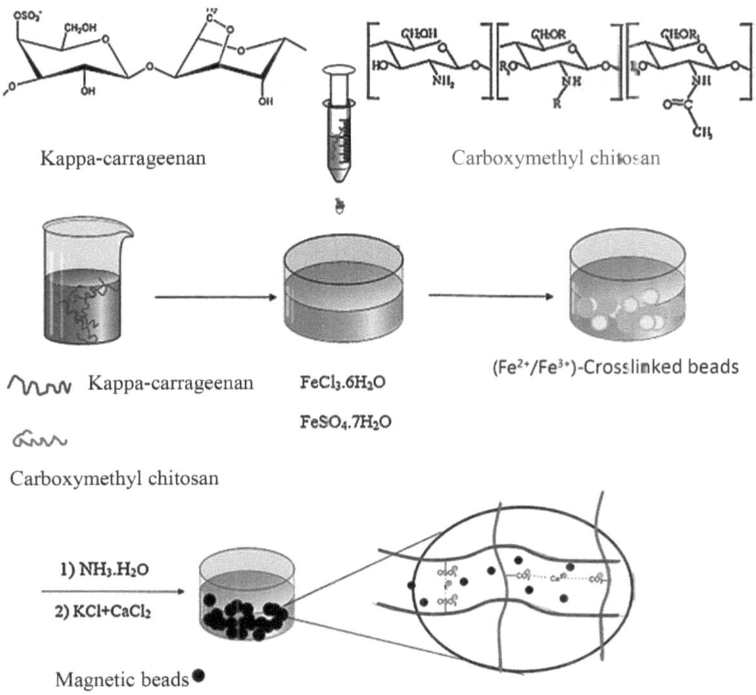

FIGURE 20.2 A Simple Scheme for the Synthesis of Magnetic Carboxymethyl Chitosan-κ-Carrageenan (mCMCS-CG) Particles

Reprinted from Mahdavinia, G.R., Etemadi, H., Soleymani, F.. 2015. Magnetic/pH-responsive beads based on caboxymethyl chitosan and κ-carrageenan and controlled drug release. Carbohydr. Polym. 128, 112–121, Copyright (2015), with permission from Elsevier

For the preparation of magnetite drug-loaded CMCD nanoparticles, carboxymethyl β-cyclodextrin (CMCD) polymer was synthesized following the procedure reported by Fernández et al. (2004). A mixture of β-CD and NaOH in water was treated with monochloroacetic acid solution at 50°C for 5 h. Then the reaction mixture was cooled. The neutral solution was then poured to excess methanol and white precipitate was produced. The solid precipitate was then filtered and dried under vacuum to give CMCD polymer. Then, CMCD-coated Fe_3O_4 magnetic nanoparticles (mCMCD NPs) were synthesized by chemical coprecipitation of Fe^{2+} and Fe^{3+} ions by NaOH in presence of CMCD polymer, followed by hydrothermal treatment.

Typically, a mixture of $FeCl_3$, $FeSO_4$, and CMCD was dissolved in deionized water in a three-necked flask and N_2 was imported for 30 min to extrude the air. The pH value of the mixed solution was adjusted to 9–10. The resulting mixture was heated to 60°C under the protection of N_2 by stirring at 1000 rpm for 30 min. The resulting solution was then transferred to a teflon-lined autoclave and treated at 160°C for 6 h. The obtained products were separated with the help of a magnet, washed with deionized water and ethanol. Finally, the particles were dried in vacuum at 50°C for further processing.

An anticancer drug, 5-flurouracil (5-Fu) was loaded by emulsion crosslinking method with modification (Ding et al., 2015). 5-Fu was dissolved in acetic acid solution of CS under magnetic stirring. Subsequently, mCMCD hybrids were added into the CS–drug solution and the reaction system was stirred for 60 min. The resulting mixture was then added dropwise in paraffin oil containing of Span 80. After 30 min of emulsification, glutaraldehyde was added to crosslink CS. The system was stirred in a water bath at 40°C for 1 h. After that, the pH value of the reaction system was adjusted to 9–10 by using 1 M NaOH and kept in a water bath at 70°C for 2 h. Then, the resulting black products (CS-mCMCD NPs) were washed with ethanol and petroleum ether, alternately. Finally, the resultant nanocomposites were collected with the aid of an external magnet and dried in a vacuum oven at 40°C.

2.6 SELF-ASSEMBLED CS-TOCOL NANOPARTICLES

The tocol derivatives of CS were synthesized via amide bond between the primary amine groups of CS and carboxyl groups of tocols as reported previously by Kwon et al. (2003). The synthesis procedure is shown in Fig. 20.3. The reaction occurs in aqueous/methanol medium in the presence of

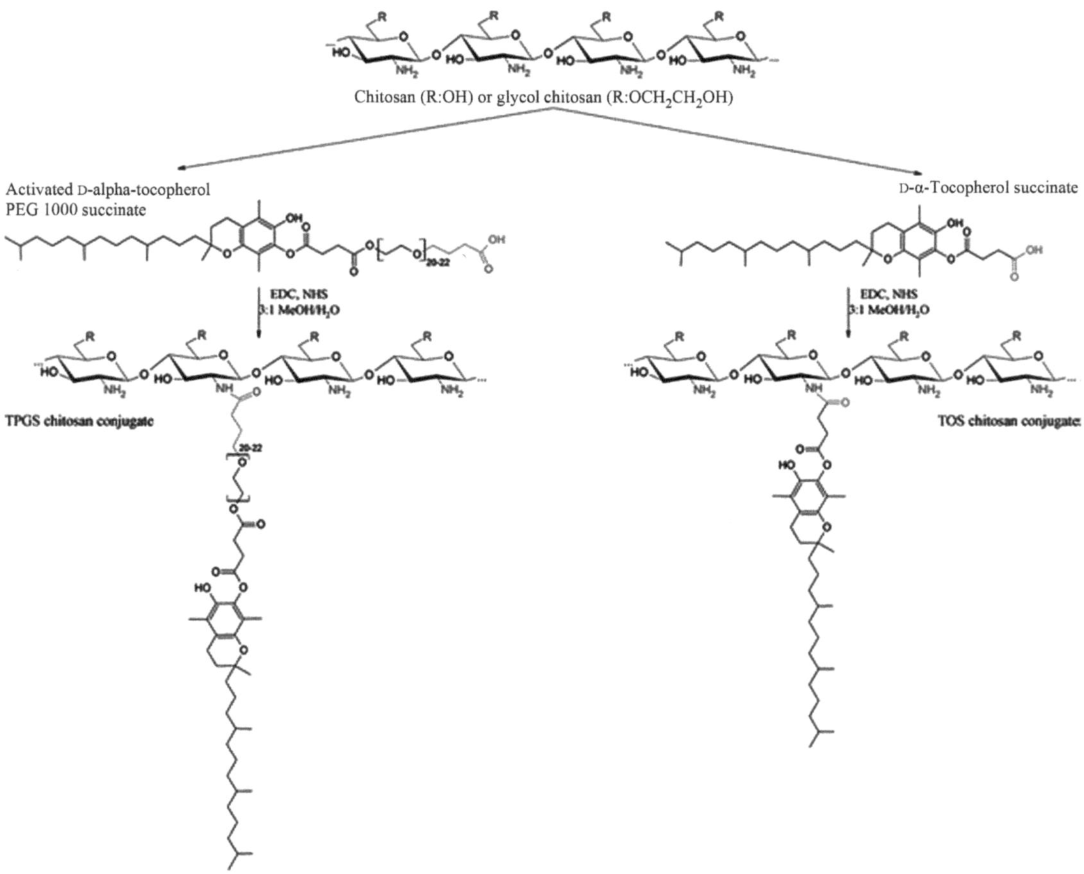

FIGURE 20.3 Synthesis of Tocol Derivatives of Chitosan

Reprinted from Duhem, N., Rolland, J., Riva, R., Guillet, P., Schumers, J.-M., Jérome, C., Gohy, J.-F., Préat, V., 2012. Tocol modified glycol chitosan for the oral delivery of poorly soluble drugs. Int. J. Pharm. 423, 452–460, Copyright (2012), with permission from Elsevier.

N-(3-dimethylaminopropyl)-*N*′-ethylcarbodiimide hydrochloride (EDC) and *N*-hydroxyl succinimide (NHS). Prior to the coupling, D-α-tocopherol polyethylene glycol 1000 succinate (TPGS) was functionalized with a carboxylic acid group. Briefly, TPGS, succinic anhydride, and triethylamine were stirred 48 h in dichloromethane at 30°C under strict anhydrous conditions. The reaction mixture was washed with NaCl solution. The organic phase was dried over Na_2SO_4, filtered. The solvent was evaporated under vacuum.

CS was dissolved in 1% (v/v) acetic solution (pH 4.0) and glycol chitosan (GC) in water before adding EDC. D-α-Tocopherol succinate (TOS) and NHS were dissolved in methanol whereas activated TGPS and NHS were dissolved in water before drop wise addition to the aqueous solution. The resulting solution was stirred for 24 h at room temperature. Then, the solution was dialyzed (MWCO 12 kDa) against water: methanol, and pure water. The ratio between the volume of solvent outside the dialysis membrane and the volume of the solution inside was at least 30. Finally, the dialyzed solution was lyophilized (Kwon et al., 2003). Substitution degree was ~3 times higher in the case of the TPGS derivatives than TOS derivatives to get the same quantity of hydropbic moiety on the derivatives. The dispersion of CS-Tocol derivatives in water caused the formation of self-assembled nanoparticles.

2.7 SPINNING DISK PROCESSOR FOR CS–PMA NPs

The stepwise fabrication of poly(methyl acrylates) (PMA)-coated CS–diclofenac sodium nanocores (CS–PMA NPs) using the spinning disk processor (SDP) is illustrated in Fig. 20.4. To initiate the

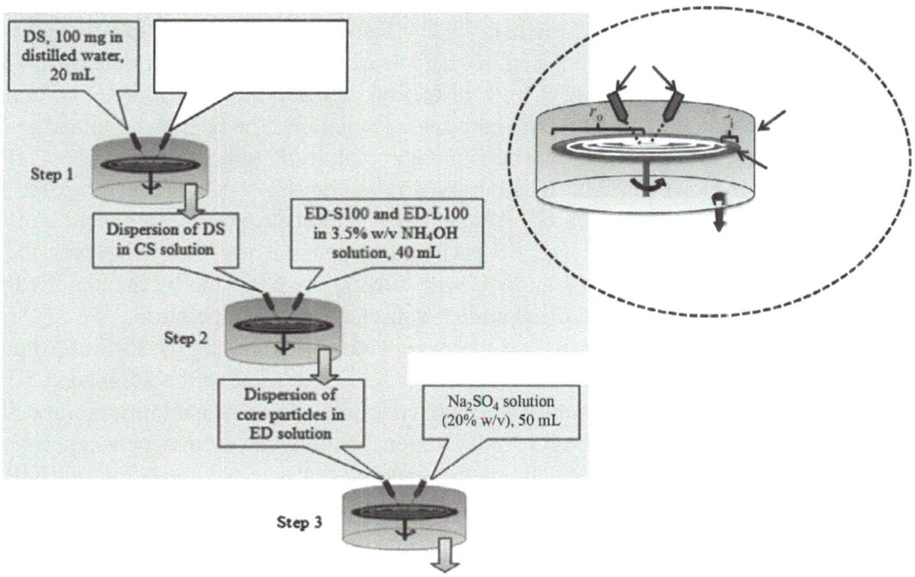

FIGURE 20.4

Schematics of a Spinning Disk Processor (SDP) and the SDP Processing Steps Used to Fabricate CS–PMA NPs.

Reprinted from Huanbutta, K., Sriamornsak, P., Luangtana-Anan, M., Limmatvapirat, S., Puttipipatkhachorn, S., Lim, L.-Y., Terada, K., Nunthanid, J., 2013. Application of multiple stepwise spinning disk processing for the synthesis of poly(methyl acrylates) coated chitosan–diclofenac sodium nanoparticles for colonic drug delivery. Eur. J. Pharm. Sci. 50, 303–311, Copyright (2013), with permission from Elsevier.

process, drug solution was simultaneously introduced with CS solution in 0.5% v/v aqueous acetic acid at drug: CS weight ratio of 1:2, onto SDP disk. The dispersion was collected as it flowed off the edge of the disk, and reintroduced onto the rotating disk for mixing with ED-S100 and ED-L100 ammonia solution. The eluant was again collected for a third round of mixing on the SDP with sodium sulfate solution to salt out the enteric polymers. The final dispersion of CS–PMA NPs was separated from the medium by centrifugation at 7000 rpm for 45 min. Residual sodium sulfate was removed by redispersing CS–PMA NPs in excess distilled water. Purified CS–PMA NPs were obtained by freeze-drying for 24 h (Huanbutta et al., 2013).

2.8 DIALYSIS METHOD FOR CHITOSAN-BASED NANOHYDROGELS

Chitosan-methyl methacrylate (CS-M) was prepared by free radical polymerization of CS and methyl methacrylate (Liu et al., 2016). Briefly, CS was dissolved in 2.0% acetic acid solution and then 0.5 mL of methyl methacrylate was added into the flask. After degassing, the flask was sealed and the solution was bubbled with dried nitrogen for 10 min prior to polymerization. The polymerization reaction was triggered by potassium persulfate and the mixture was heated to 60°C under stirring for 12 h. The resultant suspension was dialyzed in ultrapure water for 24 h through the semipermeable membrane (10 kDa) to remove the unreactive materials and then dried under vacuum at room temperature.

Chitosan-acrylic acid-methyl methacrylate (CS-AM) nanohydro-gel was obtained by graft polymerization of CS, acrylic acid and methyl methacrylate. CS and potassium persulfate were dissolved in acetic acid solution under stirring. Then 0.2 mL of acrylic acid was added and the mixture was heated to 60°C under nitrogen stream. After 1 h, 0.3 mL of methyl methacrylate was added. The graft polymerization was allowed to proceed for 12 h with continuous agitation. The resultant nanosuspension was dialyzed in ultrapure water for 24 h and dried under vacuum at room temperature.

CS-acrylic acid-methyl methacrylate-*N*-isopropylacrylamide (CS-AMNP) was prepared by grafting *N*-isopropylacrylamide (NIPAM) on CS-AM. CS-AM nanohydrogel was prepared as mentioned earlier. NIPAM was added into the dialyzed CS-AM nanosuspension, and then MBA was added as the crosslinker. The reaction was carried out at 25°C with stirring for 6 h. Finally, the nanohydrogel was dialyzed in ultrapure water for 24 h and dried under vacuum at room temperature.

The drug-loaded nanohydrogel suspensions were prepared by incubating the PBS of 5-Fu for 4 h at 25°C. The solution was then dialyzed in ultrapure water for 6 h to remove the nonloaded 5-Fu. Then, the 5-Fu-loaded nanohydrogel suspension was obtained in the semipermeable membrane. The 5-Fu was loaded on the CS-based nanohydrogel by hydrogen bonding interaction between 5-Fu and the nanohydrogels. Additionally, van der Waals interactions existed between the 5-Fu and NIPAM side chains (Rejinold et al., 2011) in the CS-AMNP nanohydrogel.

2.9 AMIDE COUPLING REACTION FOR CS-DRUG NANOCONJUGATES

Chitosan-atorvastatin (CS–AT) conjugate was prepared by amide coupling reaction. AT solution in methanol was activated by EDC for 4 h to form an AT ester. This was then added dropwise to the aqueous acidic CS solution under continuous magnetic stirring (pH 5–6). After 24 h, the excess reagent and the corresponding acylisourea byproduct was removed by washing with distilled water. The reaction mixture was purified by ultrafiltration, and the CS–AT conjugate was lyophilized.

Nanosizing of CS–AT conjugate was achieved using wet milling by high-pressure homogenization (HPH) technique since the thermal energy generated during wet milling is lower than that generated by dry-mills as the drug is suspended in aqueous media. Briefly, the synthesized conjugate was dispersed in deionized water at a concentration of 0.1% (w/v). The suspension was allowed to pass through a high-pressure homogenizer to obtain nanoconjugates. CS–AT nanoconjugates were collected after lyophilization.

2.10 SOL–GEL METHOD FOR TIO_2-COATED CS-BASED NANOHYBRID COMPOSITES

The nanohybrid composites of dopamine (DOP)-CS-titanium dioxide (TiO_2) were synthesized by Safari et al. (2015) using the sol–gel method. CS powder was dissolved in acetic acid to ensure 1% (w/v) concentration in a round flask and required amount of DOP (10%–40%, w/w) was added under continuous stirring. DOP-CS composites were dried in rotary evaporator at 70°C.

DOP-CS composites were coated with different amounts of TiO_2 (10%–50%, w/w) by using titanium tetra *n*-butoxide, as precursor. To prepare DOP-CS-TiO_2 nanohybrid composites, DOP-CS composite was dispersed in *n*-hexane. Then the required amount of titanium tetra *n*-butoxide (10%–50%, w/w) dissolved in *n*-hexane was added dropwise to the mixture. After 30 min, titanium tetra *n*-butoxide was hydrolyzed by slow addition of deionized water (minimum water/butoxide mol% = 2). The reaction mixture was left for 24 h at room temperature. Thereafter, the solid was filtered, dried, and powdered. The intercalated DOP-CS-TiO_2 nanohybrid composites were renamed hereafter as DOP-CS-TiO_2 followed by percent of DOP and TiO_2, respectively in parenthesis. For example, in DOP-CS-TiO_2 (10%, 30%), 10% indicated the mass ratio of DOP to CS, and 30% indicated mass ratio of TiO_2 to DOP-CS composite.

3 ORAL CHITOSAN NANOSYSTEMS: KEY FINDINGS

Considering the biodegradable and nontoxic nature of natural polymers, they have been investigated as drug delivery carriers for small therapeutic entities. Among them, CS is the most widely investigated natural polymer so far. The key findings on particle size, morphology, drug entrapment efficiency, in vitro and in vivo performance, cytotoxic potential of some CS-derived oral nanoparticulate systems developed in last few years are discussed herein.

The clinical application of EA, a naturally occurring anticancer polyphenolic compound has been halted due to its poor water solubility and oral bioavailability. In view of this, Arulmozhi et al. (2013) reported EA-encapsulated CS nanoparticles (EA-CS-NPs) for oral cancer treatment. The synthesis procedure is described in Section 2.1. The nanoparticles (176 nm) had spherical morphology with drug encapsulation efficiency of 94%.

Increased drug loading dramatically affected the encapsulation efficiency of the drug. They speculated that excess drug was loosely adsorbed onto the surface of CS and finally separated during centrifugation. The bare CS-NPs were spherical in shape (~50 nm) and aggregated. It was likely that their high specific surface energy due to reduction of particle size favored aggregation (Venkatesan et al., 2011). However, a distinct, spherical morphology was observed with a relatively higher average diameter around 30–120 nm. The in vitro drug release rate in simulated body fluid (PBS medium, pH 7.4) was rapid up to 3 h, followed by a slow release of about 80% drug in 48 h. It was supposed that the surface-adsorbed drug particles dissolved rapidly in the release medium whereas; the drug encapsulated in the polymer matrix released at a slow rate. The results corroborated well to that reported by other groups for CS-encapsulated polyphenolic compounds (Liang et al., 2011; Marcel-Ionel et al., 2000). Further, they evaluated therapeutic efficacy of EA-CS-NPs in human oral cancer cell line (KB). EA-CS-NPs exhibited significant cytotoxicity in KB cells in a dose-dependent manner (0.238–100 μg/mL). The estimated half-maximal inhibitory concentration (IC_{50}) values for EA and EA-CS-NPs were 3.125 and 0.953 μg/mL, respectively. Higher cytotoxicity at very low concentration suggested the possibility of greater intracellular uptake of EA-CS-NPs due to nonspecific interaction with negatively charged cell membranes, thereby facilitating diffusion of the drug into cell membrane (Venkatesan et al., 2011). It was worthy to note that bare CS-NPs had limited cytotoxic effect in KB cells (IC_{50} ~25 μg/mL) Hence, EA-CS-NPs effectively delivered EA into the cancer cells.

Bay41-4109, a heteroaryl dihydropyrimidine (HAP), usually destabilizes capsid assembly and inhibits hepatitis B virus (HBV) replication. Xue et al. (2015) focused on developing CSNPs (see Section 2.1) that could overcome the solubility and bioavailability problems, thus making Bay41-4109 suitable for HBV treatment. Bay41-4109 nanoparticles (168 nm) showed an encapsulation efficiency of 73.5%. CSNPs released about 60% drug in PBS (pH 7.4) at the end of 70 h. The biphasic drug release pattern, that is, initial fast release followed by slow release was noted. The drug release became slower as a consequence of CS hydration and swelling (Sadeghi et al., 2008).

Cytotoxicity and in vivo pharmacokinetics were also studied to examine the effectiveness of this delivery system. The cytotoxicity on HepDE19 cells seemed negligible even up to 10 μM concentration. Oral administration of Bay41-4109 NPs achieved plasma drug concentration of 1439 mg/L, which was significantly higher than Bay41-4109 suspension (297 mg/L). The C_{max} after administration of nanoparticles was 3.3-fold higher than that of Bay41-4109 suspension, in equivalent doses. The bioavailability of Bay41-4109 was 70.28%, nearly fourfold higher than that obtained by suspension. Thus, the intestinal absorption of Bay41-4109 was significantly enhanced when administered in the form of nanoparticles.

CS is insoluble at neutral and alkaline media. In acidic medium, the amine groups are positively charged, thus conferring polysaccharide a high charge density. As charge density depends on solution pH, the formation process of CS nanoparticles is also pH-responsive (Quan et al., 2005). The formation of nanoparticles as a function of pH has been demonstrated with CD-*g*-CS polymer having different degree of substitution of CD (DS). The synthesis of CD-*g*-CS is illustrated in Section 2.1. It was reported that CD-*g*-CS with lower DS (9.6%) could form nanoparticles in the range of pH 4.0–5.5 (Yuan et al., 2013). The nanoparticle formed at solution pH 4.5 was smaller among the other solution pH. The nanoparticles with higher DS (14.0 and 20.0%) CD-*g*-CS was obtained in the pH range of 3.8–5.5. The optimum pH for smallest nanoparticles was pH 5.0. The particles aggregated at pH ≥ 6.0. The electrostatic repulsion between nanoparticles decreased due to lower protonisation over the particle surface and probably induced particle aggregation.

Because the amino group reacted with active group in 6-OTs-β-CD, the crystallinity of CD-*g*-CS decreased due to deformation of the strong hydrogen bonds pertaining to CS backbone. Low water soluble KTP was incorporated into the drug delivery carriers. The size of CD-*g*-CS nanoparticles was dependent on the weight ratio of CD-*g*-CS to TPP. Under a certain weight ratio range, a strong electrostatic interaction between CD-*g*-CS and TPP possibly consolidated the polymer matrix. The size of spherical particles ranged from 150 to 200 nm, and was smaller than that determined by DLS. They had 202.0–589.0 nm size under DLS measurement. TEM measurement is usually carried out in dry state, while DLS in solution state. Hence, the particle size was bigger in DLS measurements. The encapsulation efficiency of CD_{20}-*g*-CS NPs (75%) was significantly higher than KTP-loaded CS NPs (55.4%). The drug entrapment efficiency decreased with lower substitution, reaching up to 58%. The drug entrapment efficiency obtained with 20% DS was as high as 1.36-fold than that of CS NPs. Therefore, β-CD effectively held low water soluble drugs into its cavity and increased the drug encapsulation efficiency. By regulating the DS of CD-*g*-CS, the physiochemical properties and drug encapsulation efficiency of CD-*g*-CS nanoparticles could be controlled.

Under mild acidic medium (pH 4.0), the drug release from the highly substituted polymer particles was faster than CS nanoparticles. Perhaps, the grafting induced weaker hydrogen bonds among CS and reduced the number of CS amine groups for further interaction with carbonyl groups of KTP, thus caused slight faster drug release (Guerrero et al., 2010). However, CD-*g*-CS nanoparticles displayed sustained-release behavior at pH 6.8 compared to CS NPs. Higher DS of CD-*g*-CS led to slower diffusion of KTP from the nanoparticles. More CD within CD-*g*-CS nanoparticles could load more KTP by a series of weak intermolecular forces, thus enabling sustained-release ability (Maestrelli et al., 2006).

Najafabadi et al. (2014) explored the potential of CS–PEG nanoparticles (Section 2.1) to deliver ibuprofen to the colon. The ibuprofen-loaded CS–PEG nanoparticles had a mean size of ~80 ± 1.2 nm. Increasing polymer concentration (0.1%–1%, w/v) augmented ibuprofen encapsulation from 41% to 99%. Increase in polymer concentration is expected to increase the number of hydrophobic groups available for interaction with the hydrophobic drug (ibuprofen), and consequently resulted in good drug entrapment efficiency. Due to limited number of hydrophobic groups in the nanoparticles available for interaction with the drug, an increase in the drug concentration resulted in low drug entrapment efficiency. In simulated gastrointestinal fluids, CS-NPs released complete load in 12 h. However, CS–PEG nanoparticles did not complete drug release even after 48 h in phosphate buffer (pH 7.4). The hydrophobic–hydrophobic interactions between drug and PEGylated CS might be stronger than CS-NPs and thus, exhibited sustained release profile suitable for colon-specific drug delivery.

Feng et al. (2013) suggested that the ionic interaction between positively charged CS, DOX and negatively charged CMCS could produce nanoparticles (Section 2.1) TPP was used as cross-linker to stabilize nanostructures further. CMCS significantly increased the drug-loading efficiency (72.87%) at DOX:CS:CMCS:TPP ratio of 1:3:4:0.5 and produced the smallest CS-CMCS NPs (279.3 ± 8.3 nm). Due to entanglements of CMCS molecules a negative surface charge was generated (−33.8 ± 1.6 mV). After 60 h, the release of DOX was 44.99, 38.36, and 79.36% in pH 1.2, 6.0, and 7.0, respectively. The pH-responsive stability of the DOX-loaded CS-CMCS NPs was lower in acidic pH than the neutral. The drug release rate at acidic pH was relatively slower than that of neutral pH. This hinted that DOX release from CS-CMCS NPs could be suppressed in stomach, promoting drug release in small intestine. Further, CS-CMCS NPs was found to promote absorption of DOX throughout the entire rat small intestine, especially in jejunum and ileum. Fig. 20.5 displays the CLSM images of absorption process of different formulations in excised rat ileum. For DOX-loaded CS/FITC-CMCS NPs, the green (FITC-CMCS) and red (DOX) fluorescence overlapped and appeared yellow, indicating integrity of the complexes. Following incubation with DOX for 2 h, no apparent red fluorescence was observed in myenteron. However, a relatively weak red fluorescence was observed in myenteron of ileum treated with DOX + CsA and DOX-loaded CS-NPs. Cyclosporine A (CsA) used as P-glycoprotein/cytochrome P450 inhibitor. The intensity of red and green fluorescence was strong in myenteron for DOX-loaded CS/FITC-CMCS NPs, which was indicative of more efficient intestinal permeation of the drug.

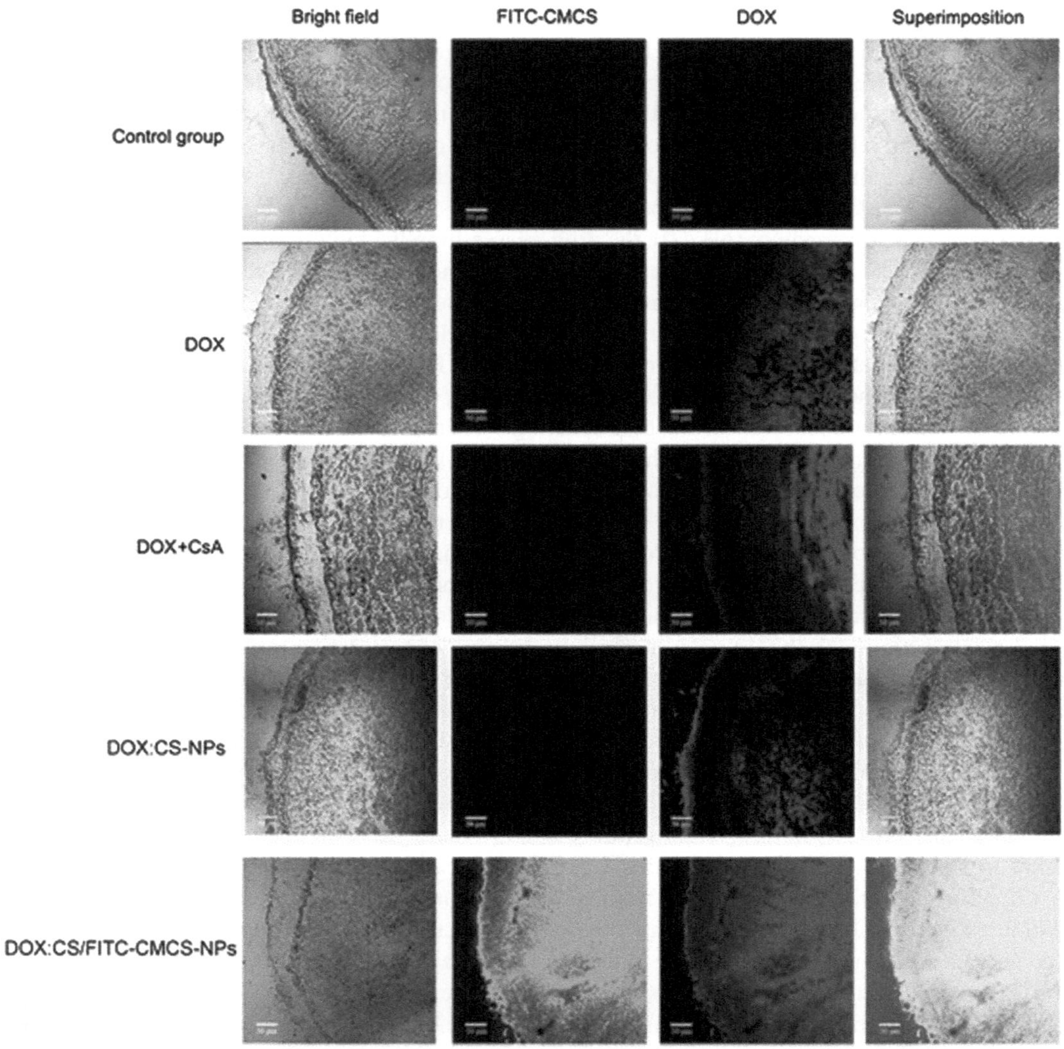

FIGURE 20.5 Ex Vivo Confocal Images of DOX, DOX+ CsA, DOX-Loaded CS-NPs, and DOX-Loaded CS/FITC-CMCS-NPs Absorption in Excised Rat Ileum

Reprinted from Feng, C., Wang, Z., Jiang, C., Kong, M., Zhou, X., Li, Y., Cheng, X., Chen, X., 2013. Chitosan/O-carboxymethyl chitosan nanoparticles for efficient and safe oral anticancer drug delivery: in vitro and in vivo evaluation. Int. J. Pharm. 457, 158–167, Copyright (2013), with permission from Elsevier.

CS-CMCS NPs was effectively delivered DOX into blood, giving an absolute bioavailability of 42%. This was nearly 6-fold and 1.75-fold higher than that of DOX and DOX-loaded CSNPs. The C_{max} value was nearly 4-fold and 2.3-fold higher than that of DOX and DOX-loaded CS-NPs, respectively. The tissue distribution and toxicity data of CS-CMCS NPs showed low level of DOX in heart and kidney, and consequently lowered the chances of cardiac and renal toxicities. Overall, CS-CMCS NPs could be considered safe and effective oral delivery system for DOX.

Bhunchu et al. (2015) encapsulated CDD in CS-ALG (ALG) nanoparticles (Section 2.1) and suggested that Pluronic F127 was the best surfactant in producing lowest particle size (414 ± 16 nm), and highest drug entrapment efficiency (54.9 ± 1.3%) than the others (Cremophor RH40, Tween 80). According to Das et al. (2010), Pluronic F127 could entrap more drugs into its hydrophobic core due to its amphiphilic and self-assembling nature in the aqueous phase. However, the entrapment efficiency decreased with the use of a high level of Pluronic F127 (>1%) possibly due to the surfactant-mediated increase in system viscosity that influenced the large particles production (Arora et al., 2011). In vitro cellular internalization study demonstrated higher cellular uptake of CDD-NPs in Caco-2 cells than free CDD.

Jamil et al. (2016) investigated the activity of cefazolin-loaded CSNPs against multidrug resistant pathogens to develop better therapeutic solutions. The drug encapsulation efficiency of CSNPs (<100 nm) was proportional to the concentration of antibiotic and ranged 28%–62%. Previous studies reported that many β-lactams might exhibit poor intestinal absorption due to efflux system (Zaki and Hafez, 2012). CSNPs at higher initial drug concentration, might improve bioavailability of this kind drug. Cefazolin-loaded CSNPs demonstrated excellent antimicrobial potential against multidrug resistant *Klebsiella pneumoniae*, *Pseudomonas aeroginosa*, and extended spectrum beta lactamase (ESBL) positive *Escherichia coli* over the drug concentration range of 200–2000 μg/mL. However, the free drug could not inhibit the growth of pathogens. CS possesses inherent antibacterial activity due to strong interaction between its positively charged amine groups and negatively charged phospholipid components of bacterial plasma membranes. CSNPs had about +50 mV zeta potential and thus positive surface charge of NPs interacted with bacterial membranes, killed bacteria either by creating pores in cell membranes, or by binding to negatively charged DNA (Piras et al., 2014). The antibacterial activity of CS is superior in nanoscale level due to accompanying higher surface area and charge density than that observed at macroscale (de Pinho Neves et al., 2014). They proposed that CS-based nanocarrier system would be useful against resistant bacterial infections caused by major Gram-negative pathogens

CHNs of 500–1000 nm diameter were investigated by Liu et al. (2015). The detailed preparative procedure is given in Section 2.2. SEM images revealed hollow interiors of CHNs. The bowl shape morphology with an opening of 300–500 nm in diameter on their surface was reported. The pore size ranged from 3 to 120 nm, which indicated there were mesopores and macropores on the interiors and outer surfaces. The curcumin entrapment efficiency was 63.93% at an equal CS: curcumin mass ratio. Low cytotoxic potential of CHNs was revealed on osteoblast-like cells. CHNs released their 48.67% load, mostly the surface adsorbed drug in PBS buffer (pH 7.4) solution in the first 12 h. When the CHNs started releasing the drug from the interiors, the drug release rate decreased over the period of 12–48 h. About 87% curcumin escaped from the nanospheres in 50 h. Obviously, the release of curcumin became more sustained than CS solid nanoparticles, where complete drug release occurred within 2 h (O'Toole et al., 2012). The surface pores on hollow nanospheres could be beneficial in terms of drug adsorption and accommodate high amount of drug in large empty interiors. Subsequently, the shells could restrict free diffusion of encapsulated molecules into cancerous cells.

CS-SLNs were also investigated as possible oral delivery carriers of sumatriptan succinate (Hansraj et al., 2015). The method of SLNs production is discussed in Section 2.3. TEM image of drug-loaded CS-SLNs showed the presence of segregated, uniform-sized spherical nature of the particles. Overall, the size, zeta potential, and drug entrapment efficiency for the SLNs were in the range of 192–301.4 nm, 30.2–51.4 mV, and 76.3%–91.1%, respectively.

The phosphate and amine groups of amphiphilic soya lecithin preferentially attached onto CS-SLNs surface and resulted in higher zeta potential compared to CS and tripalmitin alone. Soya lecithin adversely affected the drug entrapment efficiency at high concentration because of its amphiphilic nature, which allowed drug leakage from the CS-SLNs. CS rigidized the CS-SLNs and favored high drug entrapment. A proportional relationship existed between particle size and lipid (tripalmitin) concentration. Tripalmitin and soya lecithin, in combination, significantly contributed to about 4.54-fold drug accumulation in brain of male Wistar rats, probably by surpassing the blood–brain barrier owing to lipophilic and surfactant properties.

Only 17% drug released in 2 h due to presence of high concentration of lipids in CS-SLNs and this figure reached to almost 50% over 24 h in phosphate buffer (pH 7.4). The solvation of CS-tripalmitin-soya lecithin in dissolution media, diffusion of entrapped drug through swollen matrices and erosion of gel boundary layer imparted slow drug release pattern.

In vivo study suggested that the SLNs successfully permeated blood–brain barrier and exhibited their antimigraine activity. The positively charged CS-SLNs interacted with the negatively charged cytomembrane cells, and entered rapidly into the cells and further, bypassed the endosomes after internalization. By virtue of absorptive transcytosis, a significant amount of CS-SLNs reached the brain to elicit pharmacological action (Kou et al., 2013).

Resveratrol (3,5,4′-trihydroxystilbene), a natural nonflavonoid polyphenolic compound found in peanuts, grapes, red wine, and various berries, offers potential health benefits (Pandita et al., 2014). However, its therapeutic application is very limited owing to its photosensitivity, poor aqueous solubility at physiological pH (30 μg/L), low absorption, rapid first pass metabolism, and enterohepatic recirculation (Summerlin et al., 2015). Following oral administration, very little free compounds have been identified in plasma/tissues due to metabolic breakdown of the compounds (Singh and Pai, 2014).

Ramalingam and Ko (2015) also designed SLNs-based drug-delivery systems for resveratrol according to the procedure described in Section 2.4. However, the poor stability and burst release behavior in acidic pH conditions of SLNs resulted in increased aggregation of the particles in the gastrointestinal environment, limiting the success of these particles as an oral delivery system for hydrophobic drugs. Hence, the surface of SLNs was modified by TMC-*g*-PA copolymer. The mean size of SLNs and TMC-*g*-PA SLNs was found to be 100 nm and 258 nm, respectively. The drug entrapment efficiency of SLNs was 95.45%. The solid lipid core of Precirol ATO 5 and palmitic acid accommodated high levels of resveratrol than the previous reports on resveratrol entrapment efficiency (<80%) (Jose et al., 2014; Neves et al., 2013). The drug was retained by the surface-modified SLNs without appreciable loss during surface modification. The polymer or the SLNs did not exhibit significant toxicity against the NIH/3T3 mouse embryonic fibroblast cells over concentration range of 20–1000 μg/mL. TMC-*g*-PA copolymers and SLNs exhibited excellent biocompatibility and nontoxicity. The drug release rate from SLNs was higher in simulated gastric fluid (SGF, pH 1.2) than in simulated intestinal fluid (SIF, pH 6.8). However, the release of resveratrol from TMC-*g*-PA SLNs was significantly less in SGF. About 43% drug release occurred in SIF at the end of 24 h, revealing sustained drug release behavior. SLNs yielded higher plasma concentrations and $AUC_{0–8h}$ values over resveratrol suspension. TMC-*g*-PA SLNs had 3.8-fold higher relative bioavailability than suspension. TMC-*g*-PA SLNs delayed the attainment of peak plasma drug concentration in comparison to SLNs and suspension. The mucoadhesive nature of polymer and the protection against acidic and enzymatic degradation perhaps allowed the sustained release of the drug. Overall, their findings supported TMC-*g*-PA SLNs as a potential oral drug-delivery system for resveratrol.

Colon-specific drug delivery could be advantageous for treating diseases like amoebiasis, Crohn's disease, ulcerative colitis, and colorectal cancer. Saboktakin et al. (2010) examined CS–DS-coated iron oxide nanoparticles for colon-specific delivery of 5-aminosalicylic acid (5-ASA). The hydrogel nanoparticles were designed by the procedure described in Section 2.5. The average drug entrapment was found to be 81.21 ± 1.86%. FE-SEM images of superparamagnetic CS–DS hydrogel revealed that the incorporation of 5-ASA into the hydrogel produced smooth surface and compact structures. The particle size decreased from 2 μm to about 250 nm with corresponding increase of CS: DS ratio from 1.0 to 8.0. The concentration of CS–DS, GST, and crosslinking time affected drug release in simulated fluids. The effect of CS–DS on drug release was found to be negligible. The drug release decreased at higher crosslinking time due to promotion of crosslinks between CS–DS chains and GST. Superparamagnetic CS–DS hydrogel did not permit premature release of drug in simulated upper conditions (<15% in 5 h). The hydrogels delivered most of the drug in colon (~90% in 9 h), an environment rich in bacterial enzymes that degrade the CS–DS and allowed drug release at the colon. The drug release in simulated colon fluid (SCF, pH 7.0) was less than 65% in 9 h, demonstrating their potential for colon delivery of 5-ASA.

Mahdavinia et al. (2015) highlighted the fact that the incorporation of magnetic nanoparticles into CMCS/κ-CG beads significantly altered the bead morphology. An undulant and coarse structure was obtained for magnetic beads. The procedure of hydrogel synthesis is described under Section 2.5. Solid-shape sections on crosslinked magnetic CMCS/κ-CG might be attributed to the rod-shape of magnetic hydrogel particles. The presence of magnetic nanoparticles and K^+/Ca^{2+} crosslinking induced a coarse microstructure on magnetic hydrogel particles.

TEM micrographs of mCMCS-CG1 (CG:CMCS = 1:1) and mCMCS-CG3 (CG:CMCS = 1.5:0.5) revealed that the magnetic Fe_3O_4 exhibited nanoscale particles with <100 nm size. The mCMCS-CG1 nanoparticles were either rod-shaped or relatively spherical in stabilized or aggregated forms. The varying morphology was attributed to the variation in pore size of hydrogels. At higher ratio, mCMCS-CG3 nanoparticles were in cluster-like structure and spherical. Thus, the size and shape of Fe_3O_4 nanoparticles could be affected by the mass ratio of biopolymers.

The swelling capacity of magnetic beads was lower than the magnetic-free nanohydrogels. The less swelling degree of magnetic nanogels might be due to (1) decrease in ratio of hydrophilic and anionic functional groups (due to nonionic Fe_3O_4 nanoparticles), and (2) more crosslink points as a consequence of interaction between magnetic nanoparticles and polymeric chains (Philippova et al., 2011). In case of mCMCS-CG3 nanohydrogels, the swelling capacity was higher than that other magnetic hydrogel particles in both pH 1.2 and 7.4 buffer solution. Irrespective of the composition, the swelling degree was always less in acidic medium than that in alkaline medium. This might be due to higher weight ratio of sulfate groups containing CG component in mCMCS-CG3 sample. They agreed with the views of Muhamad et al. (2011), where the swelling capacity of carboxymethyl cellulose-CG hydrogels was found to increase at higher CG/cellulose ratio. Based on pKa of sulfate and carboxylate pendants (pKa of methane sulfonic acid is −2.5, while that of acetic acid is 4.8), the sulfate groups on the CG could dissociate in aqueous media more readily than carboxylate groups (Hosseinzadeh et al., 2005; Mahdavinia et al., 2013). Therefore, CG behaved like a strong polyelectrolyte. Consequently, a high swelling force originated from strong electrostatic repulsion between sulfate groups on CG backbone at higher κ-CG content in the hydrogel composition (Mistsumata et al., 2003).

Under acidic medium, the drug release was found <12% for nonmagnetic and <8% for magnetic hydrogel particles at pH 1.2 in 3 h. In addition, the low solubility of diclofenac sodium (pKa ~4) under acidic medium further worsened the situation. At pH 7.4, the drug release remarkably enhanced as compared to those at pH 1.2. The drug release profiles were consistent with the swelling behavior of hydrogels. The maximum drug release was around 82% for nonmagnetic hydrogel particles and about 60% for magnetic hydrogel particles at pH 7.4. The presence of magnetite nanoparticles certainly influenced the drug release patterns. The released amount became higher after the application of external AMF with the strengths of 150 and 300 G. The result was quite similar to that reported by Likhitkar and Bajpai (2012) on controlled release of cisplatin from starch-based magnetic nanoparticles. An appropriate reason for the increased drug release upon applying alternative magnetic field (AMF) may be assigned to the alignment of magnetic nanoparticles with the applied AMF. Due to constant motion of magnetic nanoparticles, the fluctuation of applied AMF could act as stimulus to agitate the magnetic nanoparticles. This agitation resulted in a motion of magnetic nanoparticles and led to relaxation of polymer backbones, which caused higher release of content (Gupta and Bajpai, 2011; Hu et al., 2008). Increasing strength of AMF from 150 to 300 G caused further release of drug. The polymer chain relaxation led to mechanical deformation and subsequent tensile stresses, resulting in enhanced drug release.

Ding et al. (2015) studied CS crosslinked carboxymethyl-β-cyclodextrin (CMCD) polymer modified Fe_3O_4 magnetic nanoparticles for delivering anticancer agent 5-Fu (CS-mCMCD NPs) (see Section 2.5). CMCD-grafting on the Fe_3O_4 nanoparticles (mCMCD NPs) enhanced drug holding capacities in its hydrophobic cavity through host–guest interactions.

SEM micrographs revealed that Fe_3O_4 nanoparticles bonded with CMCD polymers had well-shaped spherical form with smooth surface. The mean particle was approximately 16 ± 3 nm with a narrow size distribution. The mCMCD NPs were slightly bigger than mNPs, attributing to the CMCD polymer coating on the surface of Fe_3O_4 particles (thickness of CMCD layer ~4 nm). The size distributions of CS-mCMCD NPs (~100 nm) were bigger than the mCMCD NPs, revealing that mCMCD NPs were successfully encapsulated into the CS nanocarriers. The magnetic CS composites prepared by crosslinking with glutaraldehyde had a well-distribution pattern with relatively glossy spherical surface. Meanwhile, the average particle sizes of CS-mCMCD NPs gradually increased accompanying with the additive amounts of crosslinking agent at the fixed CS content. The formation of spherical shaped nanocarriers was due to covalent bonding between amino groups of CS and carboxyl groups of glutaraldehyde. During crosslinking and hardening process, the crosslinking became intensified at higher glutaraldehyde concentration with the exclusion of water from the nanospheres (80–200 nm).

They speculated that the reaction time had a significant influence on the size and features of CS-mCMCD hybrids. A prolonged reaction time was favorable for adsorption of CS molecules on the Fe_3O_4 nanoparticle surfaces; however, excessive CS free radicals also might induce the aggregation of nanocarriers. The CS-mCMCD nanocomplexes exhibited maximum entrapment efficiency of 97.6%. The saturation magnetization of the CS-mCMCD NPs was approximately 43.8 emu/g, while the mCMCD NPs were around 62.5 emu/g. The low saturation magnetization could be due to excessive incorporation of CS into the CS-mCMCD NPs (Nara et al., 1996). The existence of CS on the surface of mCMCD particles decreased the uniformity due to the quenching of surface moments, resulting in the reduction of the magnetic moment in such nanoparticles.

CS-mCMCD NPs exhibited pH-dependent, swelling- and diffusion-controlled drug release behaviors in simulated gastrointestinal fluids. The slower drug release profiles in weakly alkaline media were obtained. A lower pH value led to swelling effect and electrostatic repulsion contributing to the protonation amine impact of NH_3^+, and thus resulted in a higher release of 5-Fu.

Additionally, the swelling effects of nanocarriers were pH-sensitive under different pH environments. When nanocarriers were dispersed at pH 1.2, which was below its pKa (8.01) of 5-Fu, thus, it was protonated and positively charged. Meanwhile, the amino groups on CS (pKa ≈ 6.5) and the free carboxymethyl groups (pKa ~ 4.7) around CMCD cavity remained positively charged at pH 1.2 (Badruddoza et al., 2011). Therefore, there was a strong electrostatic repulsive interaction between 5-Fu and CS-mCMCD NPs, which was favorable for the drug diffusion from the carrier. By contrast, when nanocarriers were dispersed at pH 6.8, which was in close proximity to its pKa of CS, higher than that of CMCD and lower than that of 5-Fu, the electrostatic interaction between 5-Fu and CS-mCMCD NPs was weak and cause relatively slow release of drug.

Duhem et al. (2012) developed tocol derivatives of CS (Section 2.6), which could self-assemble in the gastrointestinal tract and enhance the solubility of poorly soluble drugs. The tocols was biocompatible and able to solubilize a variety of hydrophobic compounds, named tocophilic compounds (Constantinides et al., 2006). It was hypothesized that the grafting of TPGS onto CS would decrease the CMC of TPGS (0.02 mM) due to greater thermodynamic stability of the polymeric micelles (Sadoqi et al., 2009; Torchilin, 2007). Further, TPGS was selected because of its ability to inhibit P-glycoprotein efflux. TOS allowed simple synthesis scheme compared to TPGS.

Among TOS-modified polymers, only glycol chitosan (GC) with tocopherol succinate (GC-TOS) (DS ≈ 6% mol/mol) self-assembled into nanosized particles in phosphate buffer at pH 7.0. GC-TOS led to the formation of large aggregates at 20% tocopherol substitution level. Due to low solubility of CS under these conditions, the solubilization of CS-TOS was not possible at pH 7.0. GC imparted nanoparticle stability at pH 7.0 since it had higher water solubility than CS. CS-TPGS and GC-TPGS formed jelly micronic structures and thus were not appropriate as micellar vehicles. Hence, research was focused on the GC-TOS amphiphilic derivative.

The CMC of GC-TOS (2.0 μg/mL) was sufficiently high to withstand dilution and ensure the delivery of carried drug to its absorption site (Gaucher et al., 2010). Consequently, GC-TOS displayed sufficient thermodynamic stability for oral administration of poorly soluble drugs. A hydrodynamic radius of 101 nm for GC-TOS particles was obtained in DLS measurements. TEM analysis showed spherical shape of nanoparticles with a diameter of about 90 nm.

By increasing the concentration of GC-TOS to 1.0 mg/mL, micellar aggregates were realized from DLS measurement. The hydrodynamic radius was equal to 283 ± 30 nm in PBS (pH 7.0) and 353 ± 25 nm in 0.1 N HCl. The presence of such aggregates could be explained by the high molecular weight of the GC (~250 kDa). The influence of molecular weight of CS derivatives on size and structure of self-assembled particles were reported (Kim et al., 2001; Noh et al., 2011; Wang et al., 2001). It was suggested that increasing the micelle concentration of CS-stearic derivative with MW >40 kDa induced further aggregation into poorly organized bird nest–like structures that was assigned to the higher rigidity of the CS chain (Kim et al., 2001). It was hypothesized that GC-TOS might also form bird nest–like structures at higher GC-TOS concentration due to inter- and intrachain interactions with particles containing several hydrophobic domains. As GC-TOS are intended for oral delivery, the large aggregates in concentrated GC-TOS solutions were, however, not regarded as a limitation. Lowering the molecular weight of GC appeared to be a valuable approach for further investigations. In addition, quaternization of some of the amino groups of low-molecular weight GC could be another interesting way to force repulsion between particles and to avoid further aggregation of the GC-TOS nanoparticles. GC-TOS increased ketoconazole solubility by 27 times compared to the free drug solubility in phosphate buffer (pH 7.4). The micelles exhibited a sustained release profile up to 48 h in PBS (pH 7.4). Over the first 12 h, 47% of loaded ketoconazole was released, reached to 84% after 48 h. Similar drug release profile from amphiphilic CS particles was reported, especially with glycol chitosan-5β cholanic acid (Min et al., 2008). Drug delivery vehicles must be biocompatible for their successfully exploitation in the pharmaceutical field.

GC-TOS showed no cytotoxicity at concentration varying from 0.05 to 10 mg/mL. Transepithelial electrical resistance (TEER) measurements at the beginning and at the end of the transport assay showed a reduction of 23% of the initial value after 2 h of incubation with 1% (w/v) GC-TOS probably due to the ability of GC-TOS to open tight junctions (Smith et al., 2004). A 3.4-fold increase of ketoconazole absorption was observed in presence of GC-TOS (1%, w/v) than ketoconazole suspension at 245 μg/mL. Thus, the tocol polymer conjugates might be promising vehicles for the oral delivery of poorly soluble drugs.

Huanbutta et al. (2013) described the formation of uniform-sized poly (methyl acrylates)-coated CS-diclofenac sodium nanocores (CS–PMA NPs) by spinning disk processing (SDP) for colonic drug delivery as given in Section 2.7. SEM and TEM micrographs of the optimized CS–PMA NPs indicated the presence of nanostructural units within dense aggregated particles. Most of the aggregates were about 10 μm in diameter, while the spherical drug-loaded CS nanocores were each about 10 nm in diameter. The nanocores had zeta potential of +68.90 mV, suggesting effective coating of the anionic drug by cationic CS (Harris et al., 2011). Dispersion of the core drug–CS particles in the PMA (or ED) solution caused charge reversal to negative (−42.11 mV). This indicated that the final CS–PMA NPs product comprised of drug-loaded CS nanocores were effectively coated by the anionic PMA polymers. CS–PMA NPs ranged from 7.91 to 68.05 μm and that of drug entrapment efficiency from 47.42% to 101.57%, depending on the coating: core ratio.

The in vitro drug dissolution profile of the optimized formulation showed <20% drug leak in the simulated gastric (pH 1.2) and small intestinal media (pH 6.8) of IBD patients. In contrast, more than 90% of the drug load was released into the simulated colonic fluid (pH 5.0) within 8 h. This implied that the CS–PMA NPs could successfully deliver drug to the colon of IBD patients. The addition of colonic enzyme, β-glucosidase, in pH 5.0 Tris–HCl buffer further accelerated the drug release in the colonic phase. Beta-glucosidase could erode CS coats and enhance the release rate of the encapsulated drug (Nunthanid et al., 2008).

CS–PMA NPs prepared by solvent evaporation method suggested that CS–PMA NPs was larger (100–150 μm), and had higher drug entrapment efficiency (>95% with drug: CS:PMA (ED-S100 + ED-L100) at 1:2:15) under similar formulation composition. However, more than 40% of the drug load was prematurely released in intestinal fluid, limiting its application as colon targeted drug-delivery system. Thus, the CS–PMA NPs generated by SDP processing was found superior. The enzymes in the colonic microflora may further enhance the drug release from the nanocores. In addition, the protonation of CS-nanocores at pH 5.0 would yield positive surface charges that could encourage bioadhesion to the negatively charged mucous layer of the inflamed colonic epithelium cells (Lamprecht et al., 2001). The resultant close proximity between the nanocores and the absorptive epithelium would enhance drug uptake and absorption, with cellular uptake of the nanocores via clathrin-mediated endocytosis pathway (Desai et al., 1997; Ma and Lim, 2003). Drug uptake from CS–PMA NPs into Caco-2 cells was threefold higher than that from a control drug solution, with no apparent cytotoxicity at 0.10 mg/mL. Thus, pH- and enzyme-controlled delivery system with favorable cellular uptake of nanocores in intestinal epithelium supported their use for colon drug trageting.

Liu et al. (2016) investigated the potential of CS-based nanohydrogels (see Section 2.8 for preparative conditions) for controlling the release of 5-Fu in artificial simulated gastric (pH 1.2) and intestinal fluids (pH 6.8). CS-based nanohydrogel particles were of 140–190 nm size. The drug loading efficiencies were about 97.4%–99.5%. The nanohydrogels exhibited stable and consecutive release of 5-Fu at pH 1.2 and 6.8. Nearly all drug loads was quickly released from CS-M nanohydrogels in 6 h. For CS-AM and CS-AMNP nanohydrogels, about 50% of 5-Fu was released quickly in 5 h, and then the release of 5-Fu was slow and consecutive. CS-AM nanohydrogels showed sustained release of 5-Fu for 40 h at pH 1.2. Compared with pH 6.8, the nanohydrogels released 5-Fu quickly. This was because the grafted AA responded to pH value, which promoted hydrogen bonding between 5-Fu and CS-AM nanohydrogels. However, there was little difference in release behaviorsof 5-Fu from the CS-M and CS-AMNP nanohydrogels at pH 1.2 and 6.8. Thus, 5-Fu-loaded CS-M, CS-AM, and CS-AMNP nanohydrogels could deliver 5-Fu under variable pH conditions of gastrointestinal tract.

Atorvastatin (AT) is an orally administered drug used to lower elevated total cholesterol, low-density lipoprotein and triglyceride levels, and elevate high-density lipoprotein cholesterol (Law et al., 2003). Its low oral bioavailability (~12%) is attributable to low aqueous solubility, crystalline nature, and high hepatic first-pass metabolism (Cilla et al., 1996; Lennernas, 2003). Furthermore, the bioavailability is highly variable due to instability in acidic milieu of the stomach (Shah et al., 2008). Poor oral bioavailability of AT results in administration of high doses and engenders dose-related undesirable adverse effects, such as liver abnormalities, rhabdomyolysis, arthralgia, and kidney failure.

Anwar et al. (2011) developed CS–AT nanoconjugate by HPH technique for improving bioavailability of AT according the procedure described in Section 2.9. The particle size had an inverse relationship with both homogenization pressure and number of cycles. Conjugates with lowest particle size 215.3 ± 14.2 nm were obtained after two homogenization cycles at 40,000 psi pressure. Increasing the number of homogenization cycles from 2 to 3 did not reduce the particle size further, indicating the attainment of saturation levels. FT-IR spectra of CS–AT nanoconjugate were superimposable to the corresponding spectra of CS–AT conjugate, suggesting that no chemical change occurred during homogenization. The formation of CS–AT conjugate resulted in a scaffold-like structure while on homogenization, the formation of CS–AT nanoconjugate presented a smooth surface morphology of nanoconjugates. X-ray diffraction (XRD) spectra demonstrated amorphous nature of nanoconjugate. The aqueous solubility of pure AT, CS–AT conjugates, and CS–AT nanoconjugate was found to be 23.5, 589.2, and 2410.2 μg/mL, respectively. The solubility of AT increased by about 25-fold after conjugation with CS. The solubility of CS–AT nanoconjugate was ~4-fold greater than the CS–AT conjugate and nearly 100-fold higher than that of pure AT. This improved water solubility of AT for CS–AT nanoconjugate could be attributed to the collective effect of formation of water soluble conjugate, amorphous AT in CS–AT nanoconjugate and reduced particle size, which offered higher surface area for drug dissolution.

Complete release of AT was attained in SGF (pH 1.2) within 6 h; whereas 92.5 ± 6.2% AT was released from nanoconjugate in 10 h in phosphate buffer (pH 7.4). Approximately, 20% of AT was released upon incubation in SGF within 2 h. Considering the transit time of 1–2 h in stomach, AT release was seemed insubstantial. Hence, acid catalyzed degradation of AT, responsible for variable bioavailability, would be significantly reduced. AT suspension showed plasma half-life ($t_{1/2}$) of 15.8 h, however CS–AT nanoconjugate exhibited delayed $t_{1/2}$ value of 19.3 h suggesting that AT was released from CS–AT nanoconjugate in a sustained manner over prolonged period of time. In addition, the C_{max} of the nanoconjugate (2574 ± 95.4 ng/mL) was greater than that of AT suspension (583 ± 55.5 ng/mL). Oral administration of CS–AT nanoconjugate to rat conferred about fivefold higher bioavailability than AT suspension. This unprecedented high absorption was attributed to enhanced solubility of amorphous AT in CS–AT nanocomplex, mucoadhesive property of CS and the ability to open tight junctions in intestinal epithelial cells.

The polymeric coatings are not always chemically stabile, and may cause local inflammatory reactions and so on during conventional delivery of drugs. Therefore, it would be interesting to embed bioactive compounds and biomolecules within inorganic coating, such as TiO_2, ZrO_2, and SiO_2. This type of coating might increase drug passage time through its pores into simulated fluids (Signoretto et al., 2011) and eliminate the chances of stimuli-responsive (temperature, pH) removal of surface coating of the drug carrier systems (Alvarez-Lorenzo et al., 2013; Ribeiro et al., 2014). If such inorganic coatings have nanostructure properties, they could improve cellular adhesion, osteoblast proliferation, and biomineralization.

Intravenous DOP (3,4-dihydroxyphenylethylamine), commonly used to treat nervous diseases, such as Parkinson's disease, schizophrenia was tested for oral application after incorporation into DOP-CS-TiO_2 nanocomposites. The nanocomposites were prepared by the method as described in Section 2.10 (Safari et al., 2015). Incorporation of the TiO_2 enhanced the drug entrapment considerably, and reduced the drug release rate. The DOP-CS composites without TiO_2 coating released the entire amount of drug after 10 min in pH 7.4 solution, while TiO_2-coated composites released the entrapped drug after 16 h. The presence of TiO_2 in the composite improved the thermal stability of the DOP-CS composite. FE-SEM morphology of the best nanohybrid composite, that is, DOP-CS-TiO_2 (20%, 30%) revealed a uniform coating of the CS/drug composite. The porous nature of the composite was also seen. By coating the DOP-CS composite with a porous TiO_2 shell, one would expect reduction in pore diffusion rate of the drug. Formation of amorphous TiO_2 around the DOP-CS composite was also confirmed in the XRD pattern of the DOP-CS-TiO_2 nanohybrid composite. The nanohybrid composite had no tendency of coagulation or flocculation in water.

Release behavior was strongly influenced by the amount of TiO_2 coating. During hydrolysis of titanium tetra *n*-butoxide, it was likely that the —OH and —NH_2 groups of CS participated in the polymerization of TiO_2 precursor, and formed covalent bonds between the DOP-CS matrixes with TiO_2. Moreover, the strong hydrogen bonding interactions between these groups with OH groups on the surface of TiO_2, held TiO_2 coating strongly on the DOP-CS composite. DOP remained intact during the formation of 30% TiO_2 coating. The drug dissolution from the composites without TiO_2 coating was very fast, and the entire drug was released in 10 min. They thought that the nanohybrid composite DOP-CS-TiO_2 (20, 30) was good enough, but TiO_2 concentrations (10%–50%, w/w) still might have role in controlling drug release.

It was found that 30% TiO_2 coating had much better performance than the 10 and 20% TiO_2 coatings, while the performance of 40 and 50% TiO_2 coatings were very similar to the 30% of TiO_2 coating. Therefore, nanohybrid composite DOP-CS-TiO_2 (20, 30) was selected as the optimum formulation. The drug release rate followed the trend: pH 1.2 > pH 5.3 > pH 7.4. They reasoned that higher solubility of CS at low pH caused faster separation of DOP from the composite. In addition, the amorphous TiO_2 could be hydrolyzed to some extent in acidic conditions (Song et al., 2012). Therefore, by losing the TiO_2 coating, one would expect dramatic increase in drug release rate. It must be noted that even at pH 1.2; the hybrid liberated only 90% of DOP in 4 h. It was noticed that around 70 and 87% of the drug was released within 5 h and 12 h in pH 7.4 phosphate buffer, respectively from DOP-CS-TiO_2 (20, 30) composites. The total DOP was released after 16 h. The results might transform DOP utilization from injection to oral.

4 CONCLUSIONS

CS is the most widely studied biopolymer both in native and modified forms to design different types of drug delivery vehicles including nanoparticles, hydrogels, polyelectrolyte complex with oppositely charged biopolymers, self-assembled micelles, and so on. Research focus has been given to its drug delivery as well as biomedical applications. Current developments on CS-based nanosystems are very promising toward oral delivery of different kinds of small therapeutic molecules. The investigators have developed new systems to circumvent different problems associated with the drugs and their delivery devices. However, the literature lacks the toxicity profile of modified CS materials and their nanodevices. Unless sufficient information is gathered on the toxicity profile of the modified polymers and their nanosystems, their clinical development will definitely take much more time. Overall, the key findings on CS-based nanoparticulate systems are encouraging for oral delivery of low molecular weight drug molecules.

REFERENCES

Aktaş, Y., Andrieux, K., Alonso, M.J., Calvo, P., Gürsoy, R.N., Couvreur, P., Çapan, Y., 2005. Preparation and in vitro evaluation of chitosan nanoparticles containing a caspase inhibitor. Int. J. Pharm. 298, 378–383.

Alvarez-Lorenzo, C., Blanco-Fernandez, B., Puga, A.M., Concheiro, A., 2013. Crosslinked ionic polysaccharides for stimuli-sensitive drug delivery. Adv. Drug Deliv. Rev. 65, 1148–1171.

Anwar, M., Warsi, M.H., Mallick, N., Akhter, S., Gahoi, S., Jain, G.K., Talegaonkar, S., Ahmad, F.J., Khar, R.K., 2011. Enhanced bioavailability of nano-sized chitosan–atorvastatin conjugate after oral administration to rats. Eur. J. Pharm. Sci. 44, 241–249.

Archana, D., Singh, B.K., Dutta, J., Dutta, P.K., 2013. In vivo evaluation of chitosan–PVP–titanium dioxide nanocomposite as wound dressing material. Carbohydr. Polym. 95, 530–539.

Arora, S., Gupta, S., Narang, R.K., Budhiraja, R.D., 2011. Amoxicillin loaded chitosanalginate polyelectrolyte complex nanoparticles as mucopenetrating delivery system for *H. Pylori*. Sci. Pharm. 79, 673–694.

Arulmozhi, V., Pandian, K., Mirunalini, S., 2013. Ellagic acid encapsulated chitosan nanoparticles for drug deliverysystem in human oral cancer cell line (KB). Colloids Surf. B Biointerfaces 110, 313–320.

Badruddoza, A.Z.M., Tay, A.S.H., Tan, P.Y., Hidajat, K., Uddin, M.S., 2011. Carboxymethyl-β-cyclodextrin conjugated magnetic nanoparticles as nanoadsorbents for removal of copper ions: synthesis and adsorption studies. J. Hazard. Mater. 185, 1177–1186.

Bhattarai, N., Gunn, J., Zhang, M., 2010. Chitosan-based hydrogels for controlled, localized drug delivery. Adv. Drug Deliv. Rev. 62, 83–99.

Bhunchu, S., Rojsitthisak, P., Rojsitthisak, P., 2015. Effects of preparation parameters on the characteristics of chitosanealginate nanoparticles containing curcumin diethyl disuccinate. J. Drug Deliv. Sci. Technol. 28, 64–72.

Bowman, K., Leong, K.W., 2006. Chitosan nanoparticles for oral drug and gene delivery. Int. J. Nanomed. 1, 117–128.

Bromberg, L., Alakhov, V., 2003. Effects of polyether-modified poly(acrylic acid) microgels on doxorubicin transport in human intestinal epithelial Caco-2 cell layers. J. Control. Release 88, 11–22.

Cai, G., Jiang, H., Tu, K., Wang, L., Zhu, K., 2009. A facile route for regioselective conjugation of organo-soluble polymers onto chitosan. Macromol. Biosci. 9, 256–261.

Calvo, P., Remuñan-López, C., Luis Vila-Jato, J., José Alonso, M., 1997. Chitosan and chitosan:ethylene oxide-propylene oxide block copolymer NPs as novel carriers for proteins and vaccines. Pharm. Res. 14, 1431–1436.

Casettari, L., Vllasaliu, D., Castagnino, E., Stolnik, S., Howdle, S., Illum, L., 2012. PEGylated chitosan derivatives: synthesis, characterizations and pharmaceutical applications. Prog. Polym. Sci. 37, 659–685.

Chiu, H.-C., Chern, C.-S., Lee, C.-K., Chang, H.-F., 1998. Synthesis and characterization of amphiphilic poly(ethylene glycol) graft copolymers and their potential application as drug carriers. Polymer 39, 1609–1616.

Chung, C.W., Chung, K.D., Jeong, Y.I., Kang, D.H., 2013. 5-Aminolevulinic acid-incorporated nanoparticles of methoxypoly(ethylene glycol)-chitosan copolymer for photodynamic therapy. Int. J. Nanomed. 8, 809–819.

Cilla, D.D., Whitfield, J.L.R., Gibson, D.M., Sedman, A.J., Posvar, E.L., 1996. Multiple dose pharmacokinetics, pharmacodynamics, and safety of atorvastatin, an inhibitor of HMG-CoA reductase, in healthy subjects. Clin. Pharmacol. Ther. 60, 687–695.

Coelho, T.C., Lausa, R., Mangrich, A.S., de Favere, V.T., Laranjeira, M.C.M., 2007. Effect of heparin coating on epichlorohydrin cross-linked chitosan microspheres on the adsorption of copper (II) ions. React. Funct. Polym. 67, 468–475.

Constantinides, P.P., Han, J., Davis, S.S., 2006. Advances in the use of tocols as drug delivery vehicles. Pharm. Res. 23, 243–255.

Croisier, F., Jerome, C., 2013. Chitosan-based biomaterials for tissue engineering. Eur. Polym. J. 49, 780–792.

Das, R.K., Kasoju, N., Bora, U., 2010. Encapsulation of curcumin in alginate-chitosanpluronic composite nanoparticles for delivery to cancer cells. Nanomedicine 6, 153–160.

de Pinho Neves, A.L., Milioli, C.C., Müller, L., Riella, H.G., Kuhnen, N.C., Stulzer, H.K., 2014. Factorial design as tool in chitosan nanoparticles development by ionic gelation technique. Colloids Surf. A Physicochem. Eng. Asp. 445, 34–39.

Deacon, M., Davis, S., White, R., Nordman, H., Carlstedt, I., Errington, N., Rowe, A., Harding, S., 1999. Are chitosan–mucin interactions specific to different regions of the stomach? Velocity ultracentrifugation offers a clue. Carbohydr. Polym. 38, 235–238.

Derakhshandeh, K., Fathi, S., 2012. Role of chitosan nanoparticles in the oral absorption of Gemcitabine. Int. J. Pharm. 437, 172–177.

Desai, M.P., Labhasetwar, V., Walter, E., Levy, R.J., Amidon, G.L., 1997. The mechanism of uptake of biodegradable microparticles in Caco-2 cells is size dependent. Pharm. Res. 14, 1568–1573.

Ding, Y., Shen, S.Z., Sun, H., Sun, K., Liu, F., Qi, Y., Yan, J., 2015. Design and construction of polymerized-chitosan coated Fe_3O_4 magnetic nanoparticles and its application for hydrophobic drug delivery. Mater. Sci. Eng. C Mater. Biol. Appl. 48, 487–498.

Du, Y.-Z., Wang, L., Yuan, H., Wei, X.-H., Hu, F.-Q., 2009. Preparation and characteristics of linoleic acid-grafted chitosan oligosaccharide micelles as a carrier for doxorubicin. Colloids Surf. B Biointerfaces 69, 257–263.

Duhem, N., Rolland, J., Riva, R., Guillet, P., Schumers, J.-M., Jérome, C., Gohy, J.-F., Préat, V., 2012. Tocol modified glycol chitosan for the oral delivery of poorly soluble drugs. Int. J. Pharm. 423, 452–460.

Dusseault, J., Leblond, F.A., Robitaille, R., Jourdan, G., Tessier, J., Ménard, M., Henley, N., Hallé, J.P., 2005. Microencapsulation of living cells in semi-permeable membranes with covalently cross-linked layers. Biomaterials 26, 1515–1522.

Elzatahry, A.A., Mohy Eldin, M.S., 2008. Preparation and characterization of metronidazole-loaded chitosan nanoparticles for drug delivery application. Polym. Adv. Technol. 19, 1787–1791.

Esposito, E., Cortesi, R., Nastruzzi, C., 1996. Gelatin microspheres: influence of preparation parameters and thermal treatment on chemico-physical and biopharmaceutical properties. Biomaterials 17, 2009–2020.

Feng, C., Wang, Z., Jiang, C., Kong, M., Zhou, X., Li, Y., Cheng, X., Chen, X., 2013. Chitosan/O-carboxymethyl chitosan nanoparticles for efficient and safe oral anticancer drug delivery: in vitro and in vivo evaluation. Int. J. Pharm. 457, 158–167.

Fernández, M., Villalonga, M.L., Fragoso, A., Cao, R., Baños, M., Villalonga, R., 2004. α-Chymotrypsin stabilization by chemical conjugation with O-carboxymethyl-poly-β-cyclodextrin. Process Biochem. 39, 535–539.

Finotelli, P.V., Da Silva, D., Sola-Penna, M., Rossi, A.M., Farina, M., Andrade, L.R., Takeuchi, A.Y., Rocha-Le~ao, M.H., 2010. Microcapsules of alginate/chitosan containing magnetic nanoparticles for controlled release of insulin. Colloids Surf. B Biointerfaces 81, 206–211.

Friend, D.R., 1991. Colon-specific drug delivery. Adv. Drug Deliv. Rev. 7, 149–199.

Gaucher, G., Satturwar, P., Jones, M.C., Furtos, A., Leroux, J.C., 2010. Polymeric micelles for oral drug delivery. Eur. J. Pharm. Biopharm. 76, 147–158.

George, M., Abraham, T.E., 2006. Polyionic hydrocolloids for the intestinal delivery of protein drugs: alginate and chitosan—a review. J. Control. Release 114, 1–14.

Germershaus, O., Mao, S., Sitterberg, J., Bakowsky, U., Kissel, T., 2008. Gene delivery using chitosan, trimethyl chitosan or polyethylenglycol-graft-trimethyl chitosan block copolymers: establishment of structure-activity relationships in vitro. J. Control. Release 125, 145–154.

Gonil, P., Sajomsang, W., 2011. Novel quaternized chitosan containing β-cyclodextrin moiety: synthesis, characterization and antimicrobial activity. Carbohydr. Polym. 83, 905–913.

Gorochovceva, N., Makuška, R., 2004. Synthesis and study of water-soluble chitosan-O-poly(ethylene glycol) graft copolymers. Eur. Polym. J. 40, 685–691.

Greenwald, R.B., Choe, Y.H., McGuire, J., Conover, C.D., 2003. Effective drug delivery by PEGylated drug conjugates. Adv. Drug Deliv. Rev. 55, 217–250.

Guerrero, S., Teijón, C., Enriqueta, M., José, M.T., Blanco, M.D., 2010. Characterization and in vivo evaluation of ketotifen-loaded chitosan microspheres. Carbohydr. Polym. 79, 1006–1013.

Gujarathi, N.A., Rane, B.R., Patel, J.K., 2012. pH sensitive polyelectrolyte complex of O-carboxymethyl chitosan and poly (acrylic acid) cross-linked with calcium for sustained delivery of acid susceptible drugs. Int. J. Pharm. 436, 418–425.

Gultepe, E., Nagesha, D., Sridhar, S., Amiji, M., 2010. Nanoporous inorganic membranes or coatings for sustained drug delivery in implantable devices. Adv. Drug Deliv. Rev. 62, 305–315.

Gupta, R., Bajpai, A.K., 2011. Magnetically guided release of ciprofloxacin from superparamagnetic polymer nanocomposites. J. Biomater. Sci. 22, 893–918.

Gupta, K.C., Jabrail, F.H., 2007. Glutyraldehyde crosslinked chitosan microspheres for the controlled release of centchroman. Carbohydr. Res. 342, 2244–2252.

Hansraj, G.P., Singh, S.K., Kumar, P., 2015. Sumatriptan succinate loaded chitosan solid lipid nanoparticles forenhanced anti-migraine potential. Int. J. Biol. Macromol. 81, 467–476.

Harris, R., Lecumberri, E., Mateos-Aparicio, I., Mengíbar, M., Heras, A., 2011. Chitosan nanoparticles and microspheres for the encapsulation of natural antioxidants extracted from Ilex paraguariensis. Carbohydr. Polym. 84, 803–806.

Hirano, S., 1996. Chitin biotechnology applications. Biotechnol. Annu. Rev. 2, 237–258.

Hosseinzadeh, H., Pourjavadi, A., Mahdavinia, G.R., Zohuriaan-Mehr, M.J., 2005. Modified carrageenan. 1. H-CarragPAM, a novel biopolymer-based superab-sorbent hydrogel. J. Bioact. Compat. Polym. 20, 475–490.

Hsu, L.-W., Lee, P.-L., Chen, C.-T., Mi, F.-L., Juang, J.-H., Hwang, S.-M., Ho, Y.-C., Sung, H.-W., 2012. Elucidating the signaling mechanism of an epithelial tight-junction opening induced by chitosan. Biomaterials 33, 6254–6263.

Hu, Y., Ding, Y., Sun, M., Zhang, L., Jiang, X., Yang, C., 2007. Hollow chitosan/poly(acrylic acid) nanospheres as drug carriers. Biomacromolecules 8, 1069–1076.

Hu, S.H., Liu, T.Y., Huang, H.Y., Liu, D.M., Chen, S.Y., 2008. Magnetic-sensitive silica nanospheres for controlled drug release. Langmuir 24, 239–244.

Huanbutta, K., Sriamornsak, P., Luangtana-Anan, M., Limmatvapirat, S., Puttipipatkhachorn, S., Lim, L.-Y., Terada, K., Nunthanid, J., 2013. Application of multiple stepwise spinning disk processing for the synthesis of poly(methyl acrylates) coated chitosan–diclofenac sodium nanoparticles for colonic drug delivery. Eur. J. Pharm. Sci. 50, 303–311.

Hyung, P.J., Kwon, S., Lee, M., Chung, H., Kim, J.H., Kim, Y.S., Park, R.W., Kim, I.S., Bong, S.S., Kwon, I.C., Young, J.S., 2006. Self-assembled nanoparticles based on glycol chitosan bearing hydrophobic moieties as carriers for doxorubicin: in vivo biodistribution and anti-tumor activity. Biomaterials 27, 119–126.

Im, S.H., Jeong, U., Xia, Y., 2005. Polymer hollow particles with controllable holes in their surfaces. Nat. Mater. 4, 671–675.

Jamil, B., Habib, H., Abbasi, S., Nasir, H., Rahman, A., Rehman, A., Bokhari, H., Imran, M., 2016. Cefazolin loaded chitosan nanoparticles to cure multi drug resistant Gram-negative pathogens. Carbohydr. Polym. 136, 682–691.

Jia, Z., shen, D., Xu, W., 2001. Synthesis and antibacterial activities of quaternary ammonium salt of chitosan. Carbohydr. Res. 333, 1–6.

Jose, S., Anju, S., Cinu, T., Aleykutty, N., Thomas, S., Souto, E., 2014. In vivo pharmacokinetics and biodistribution of resveratrol-loaded solid lipid nanoparticles for brain delivery. Int. J. Pharm. 474, 6–13.

Kayal, S., Ramanujan, R.V., 2010. Doxorubicin loaded PVA coated iron oxidenanoparticles for targeted drug delivery. Mater. Sci. Eng. C 30, 484–490.

Kim, Y.H., Gihm, S.H., Park, C.R., Lee, K.Y., Kim, T.W., Kwon, I.C., Chung, H., Jeong, S.Y., 2001. Structural characteristics of size-controlled self-aggregates of deoxycholic acid-modified chitosan and their application as a DNA delivery carrier. Bioconjug. Chem. 12, 932–938.

Kou, L., Sun, J., Zhai, Y., He, Z., 2013. The endocytosis and intracellular fate of nanomedicines: implication for rational design. Asian J. Pharm. Sci. 8, 1–10.

Kumar, M.N., Muzzarelli, R.A., Muzzarelli, C., Sashiwa, H., Domb, A.J., 2004. Chitosan chemistry and pharmaceutical perspectives. Chem. Rev. 104, 6017–6084.

Kwon, S., Park, J.H., Chung, H., Kwon, I.C., Jeong, S.Y., Kim, I.S., 2003. Physicochemical characteristics of self-assembled nanoparticles based on glycol chitosan bearing 5 beta-cholanic acid. Langmuir 19, 10188–10193.

Lamprecht, A., Schäfer, U., Lehr, C.-M., 2001. Size-dependent bioadhesion of micro and nanoparticulate carriers to the inflamed colonic mucosa. Pharm. Res. 18, 788–793.

Lamprecht, A., Stallmach, A., Kawashima, Y., Lehr, C.-M., 2002. Carrier systems for the treatment of inflammatory bowel disease. Drugs Future 27, 961–971.

Law, M.R., Wald, N.J., Rudnicka, A.R., 2003. Quantifying effect of statins on low density lipoprotein cholesterol, ischaemic heart disease, and stroke: systematic review and meta-analysis. Br. Med. J. 326, 1423–1427.

Lee, C.-M., Jang, D., Kim, J., Cheong, S.-J., Kim, E.-M., Jeong, M.-H., Kim, S.-H., Kim, D.W., Lim, S.T., Sohn, M.-H., 2011a. Oleyl-chitosan nanoparticles based on a dual probe for optical/MR imaging in vivo. Bioconjug. Chem. 22, 186–192.

Lee, S.J., Koo, H., Jeong, H., Huh, M.S., Choi, Y., Jeong, S.Y., Byun, Y., Choi, K., Kim, K., Kwon, I.C., 2011b. Comparative study of photosensitizer loaded and conjugated glycol chitosan nanoparticles for cancer therapy. J. Control. Release 152, 21–29.

Lee, E., Lee, J., Lee, I.H., Yu, M., Kim, H., Chae, S.Y., Jon, S., 2008. Conjugated chitosan as a novel platform for oral delivery of paclitaxel. J. Med. Chem. 51, 6442–6449.

Lennernas, H., 2003. Clinical pharmacokinetics of atorvastatin. Clin. Pharmacokinet. 42, 1141–1160.

Lertsutthiwong, P., Rojsitthisak, P., Nimmannit, U., 2009. Preparation of turmeric oilloaded chitosan-alginate biopolymeric nanocapsules. Mater. Sci. Eng. C 29, 856–860.

Liang, J., Li, F., Fang, Y., Yang, W., An, X., Xin, Z., Cao, L., Hu, Q., 2011. Synthesis, characterization and cytotoxicity studies of chitosan-coated tea polyphenols nanoparticles. Colloids Surf. B Biointerfaces. 82, 297–301.

Likhitkar, S., Bajpai, A.K., 2012. Magnetically controlled release of cis-platin from superparamagnetic starch nanoparticles. Carbohydr. Polym. 87, 300–308.

Liu, Z., Jiao, Y., Wang, Y., Zhou, C., Zhang, Z., 2008. Polysaccharides-based nanoparticles as drug delivery systems. Adv. Drug Deliv. Rev. 60, 1650–1662.

Liu, M., Yang, J., Ao, P., Zhou, C., 2015. Preparation and characterization of chitosan hollow nanospheres for anticancer drug curcumin delivery. Mater. Lett. 150, 114–117.

Liu, S., Zhang, J., Cui, X., Guo, Y., Zhang, X., Hongyan, W., 2016. Synthesis of chitosan-based nanohydrogels for loading and release of 5-fluorouracil. Colloids Surf. A Physicochem. Eng. Asp. 490, 91–97.

Luessen, H.L., de Leeuw, B.J., Langemëyer, M.W., de Boer, A.B., Verhoef, J.C., Junginger, H.E., 1996. Mucoadhesive polymers in peroral peptide drug delivery. VI. Carbomer and chitosan improve the intestinal absorption of the peptide drug buserelin in vivo. Pharm. Res. 13, 1668–1672.

Luo, Y., Wang, T.T.Y., Teng, Z., Chen, P., Sun, J., Wang, Q., 2013. Encapsulation of indole-3-carbinol and 3,3'-diindolylmethane in zein/carboxymethyl chitosan nanoparticles with controlled release property and improved stability. Food Chem. 139, 224–230.

Ma, Z., Lim, L.Y., 2003. Uptake of chitosan and associated insulin in Caco-2 cell monolayers: a comparison and chitosan nanoparticles. Pharm. Res. 20, 1812–1819.

Mahdavinia, G.R., Aghaie, H., Sheykhloie, H., Vardini, M.T., Etemadi, H., 2013. Synthesis of CarAlg/MMt nanocomposite hydrogels and adsorption of cationiccrystal violet. Carbohydr. Polym. 98, 358–365.

Mahdavinia, G.R., Etemadi, H., Soleymani, F., 2015. Magnetic/pH-responsive beads based on caboxymethyl chitosan and κ-carrageenan and controlled drug release. Carbohydr. Polym. 128, 112–121.

Marcel-Ionel, P., Nicolae, A., Valentin, I.P., Daniela, A., 2000. Study of the interactions between polyphenolic compounds and chitosan. React. Funct. Polym. 45, 35–43.

Maroni, A., Del Curto, M.D., Zema, L., Foppoli, A., Gazzaniga, A., 2013. Film coatings for oral colon delivery. Int. J. Pharm. 457, 372–394.

Maestrelli, F., González-Rodríguez, M.L., Rabasco, A.M., Mura, P., 2006. Effect of preparation technique on the properties of liposomes encapsulating ketoprofen–cyclodextrin complexes aimed for transdermal delivery. Int. J. Pharm. 312, 53–60.

Min, K.H., Park, K., Kim, Y.S., Bae, S.M., Lee, S., Jo, H.G., Park, R.W., Kim, I.S., Jeong, S.Y., Kim, K., Kwon, I.C., 2008. Hydrophobically modified glycol chitosan nanoparticles-encapsulated camptothecin enhance the drug stability and tumor targeting in cancer therapy. J. Control. Release 127, 208–218.

Mistsumata, T., Suemitsu, Y., Fujii, K., Fujii, T., Taniguchi, T., 2003. pH respond ofchitosan, κ-carrageenan, carboxymethyl cellulose sodium salt complex hydro-gels. Polymer 44, 7103–7111.

Miyazaki, S., Ishii, K., Nadai, T., 1981. The use of chitin and chitosan as drug carriers. Chem. Pharm. Bull. 29, 3067–3069.

Morris, E.R., Rees, D.A., Robinson, G., 1980. Cation-specific aggregation of car-rageenan helices: domain model of polymer gel structure. J. Mol. Biol. 138, 349–362.

Motwani, S.K., Chopra, S., Talegaonkar, S., Kohli, K., Ahmad, F.J.A., Khar, R.K., 2008. Chitosan-sodium alginate nanoparticles as submicroscopic reservoirs for ocular delivery: formulation, optimisation and in vitro characterization. Eur. J. Pharm. Biopharm. 68, 513–525.

Moulari, B., Pertuit, D., Pellequer, Y., Lamprecht, A., 2008. The targeting of surface modified silica nanoparticles to inflamed tissue in experimental colitis. Biomaterials 29, 4554–4560.

Muhamad, I.I., Fen, L.S., Hui, N.H., Mustapha, N.A., 2011. Genipin-cross-linkedkappa-carrageenan/carboxymethyl cellulose beads and effects on beta-carotenerelease. Carbohydr. Polym. 83, 1207–1212.

Na, J.H., Lee, S.Y., Lee, S., Koo, H., Min, K.H., Jeong, S.Y., Yuk, S.H., Kim, K., Kwon, I.C., 2012. Effect of the stability and deformability of self-assembled glycol chitosan nanoparticles on tumor-targeting efficiency. J. Control. Release 163, 2–9.

Nagpal, K., Singh, S.K., Mishra, D.N., 2010. Chitosan nanoparticles: a promising system in novel drug delivery. Chem. Pharm. Bull. 58, 1423–1430.

Nagpal, K., Singh, S.K., Mishra, D.N., 2013. Evaluation of safety and efficacy of brain targeted chitosan nanoparticles of minocycline. Int. J. Biol. Macromol. 59, 20–28.

Nair, R., Kumar, A.C.K., Priya, V.K., Yadav, C.M., Raju, P.Y., 2012. Formulation and evaluation of chitosan solid lipid nanoparticles of carbamazepine. Lipids Health Dis. 11, 72–79.

Najafabadi, A.H., Abdouss, M., Faghihi, S., 2014. Synthesis and evaluation of PEG-O-chitosan nanoparticles for delivery of poor water soluble drugs:ibuprofen. Mater. Sci. Eng. C 41, 91–99.

Nam, H.Y., Kwon, S.M., Chung, H., Lee, S.Y., Kwon, S.H., Jeon, H., Kim, Y., Park, J.H., Kim, J., Her, S., Oh, Y.K., Kwon, I.C., Kim, K., Jeong, S.Y., 2009. Cellular uptake mechanism and intracellular fate of hydrophobically modified glycol chitosan nanoparticles. J. Control. Release 135, 259–267.

Nara, M., Torii, H., Tasumi, M., 1996. Correlation between the vibrational frequencies of the carboxylate group and the types of its coordination to a metal ion: an *ab initio* molecular orbital study. J. Phys. Chem. 100, 19812–19817.

Neves, A.R., Lúcio, M., Martins, S., Lima, J.L.C., Reis, S., 2013. Novel resveratrol nanodelivery systems based on lipid nanoparticles to enhance its oral bioavailability. Int. J. Nanomed. 8, 177.

Nguyen, S., Alund, S.J., Hiorth, M., Kjøniksen, A.L., Smistad, G., 2011. Studies on pectincoating of liposomes for drug delivery. Colloids Surf. B Biointerfaces 88, 664–673.

Nogueira, D.R., Tavano, L., Mitjans, M., Pérez, L., Infante, M.R., Vinardell, M.P., 2013. In vitro antitumor activity of methotrexate via pH-sensitive chitosan nanoparticles. Biomaterials 34, 2758–2772.

Noh, S.M., Han, S.E., Shim, G., Lee, K.E., Kim, C.W., Han, S.S., Choi, Y., Kim, Y.K., Kim, W.K., Oh, Y.K., 2011. Tocopheryl oligochitosan-based self assembling oligomersomes for siRNA delivery. Biomaterials 32, 849–857.

Nunthanid, J., Huanbutta, K., Luangtana-Anan, M., Sriamornsak, P., Limmatvapirat, S., Puttipipatkhachorn, S., 2008. Development of time-, pH-, and enzyme-controlled colonic drug delivery using spray-dried chitosan acetate and hydroxypropyl methylcellulose. Eur. J. Pharm. Biopharm. 68, 253–259.

O'Toole, M.G., Henderson, R.M., Soucy, P.A., Fasciotto, B.H., Hoblitzell, P.J., Keynton, R.S., Ehringer, W.D., Gobin, A.S., 2012. Curcumin encapsulation in submicrometer spray-dried chitosan/Tween 20 particles. Biomacromolecules 13, 2309–2314.

Pandita, D., Kumar, S., Poonia, N., Lather, V., 2014. Solid lipid nanoparticles enhance oral bioavailability of resveratrol, a natural polyphenol. Food Res. Int. 62, 1165–1174.

Peppas, N.A., Sahlin, J.J., 1996. Hydrogels as mucoadhesive and bioadhesive materials: a review. Biomaterials 17, 1553–1561.

Petter, R.C., Salek, J.S., Sikorski, C.T., 1990. Cooperative binding by aggregated mono-6-(alkylamino)-β-cyclodextrins. J. Am. Chem. Soc. 112, 3860–3868.

Philippova, O., Barabanova, A., Molchanov, V., Khokhlov, A., 2011. Magneticpolymer beads: recent trends and developments in synthetic design and applications. Eur. Polym. J. 47, 542–559.

Pinnamaneni, S., Das, N.G., Das, S.K., 2002. Formulation approaches for orally administered poorly soluble drugs. Pharmazie 57, 291–300.

Piras, A.M., Maisetta, G., Sandreschi, S., Esin, S., Gazzarri, M., Batoni, G., Chiellini, F., 2014. Preparation, physical–chemical and biological characterization of chitosan nanoparticles loaded with lysozyme. Int. J. Biol. Macromol. 67, 124–131.

Quan, G., Tao, W., Colette, C., Paul, M.C., 2005. Modulation of surface charge, particle size and morphological properties of chitosan-TPP nanoparticles intended for gene delivery. Colloids Surf. B Biointerfaces 44, 65–73.

Ramalingam, P., Ko, Y.T., 2014. Enhanced oral delivery of curcumin from N-trimethyl chitosan surface-modified solid lipid nanoparticles: pharmacokinetic and brain distribution evaluations. Pharm. Res. 32, 389–402.

Ramalingam, P., Ko, Y.T., 2015. Improved oral delivery of resveratrol from N-trimethyl chitosan-g-palmitic acid surface-modified solid lipid nanoparticles. Colloids Surf. B Biointerfaces 139, 52–61.

Rejinold, N.S., Chennazhi, K.P., Nair, S.V., Tamura, H., Jayakumar, R., 2011. Biodegradable and thermo-sensitive chitosan-g-poly(N-vinylcaprolactam)nanoparticles as a 5-fluorouracil carrier. Carbohydr. Polym. 83, 776–786.

Ribeiro, L.N.M., Alcântara, A.C.S., Darder, M., Aranda, P., Araújo-Moreira, F.M., Ruiz-Hitzky, E., 2014. Pectin-coated chitosan–LDH bionanocomposite beads as potentialsystems for colon-targeted drug delivery. Int. J. Pharm. 463, 1–9.

Saboktakin, M.R., Tabatabaie, R., Maharramov, A., Ramazanov, M.A., 2010. Synthesis and characterization of superparamagnetic chitosan–dextran sulfate hydrogels as nano carriers for colon-specific drug delivery. Carbohydr. Polym. 81, 372–376.

Sadeghi, A.M., Dorkoosh, F.A., Avadi, M.R., Saadat, P., Rafiee-Tehrani, M., Junginger, H.E., 2008. Preparation, characterization and antibacterial activities of chitosan, TMC and DEMC nanoparticles loaded with insulin using both the ionotropic gelation and polyelectrolyte complexation methods. Int. J. Pharm. 355, 299–306.

Sadoqi, M., Lau-Cam, C.A., Wu, S.H., 2009. Investigation of the micellar properties of the tocopheryl polyethylene glycol succinate surfactants TPGS 400 and TPGS 1000 by steady state fluorometry. J. Colloid Interface Sci. 333, 585–589.

Safari, M., Ghiaci, M., Jafari-Asl, M., Ensafi, A.A., 2015. Nanohybrid organic–inorganic chitosan/dopamine/TiO_2 composites with controlled drug-delivery properties. Appl. Surf. Sci. 342, 26–33.

Salamat-Miller, N., Chittchang, M., Johnston, T.P., 2005. The use of mucoadhesive polymers in buccal drug delivery. Adv. Drug Deliv. Rev. 57, 1666–1691.

Sarmento, B., Ribeiro, A.J., Veiga, F., Ferreira, D.C., Neufeld, R.J., 2007. Insulin-loaded nanoparticles are prepared by alginate ionotropic pre-gelation followed by chitosan polyelectrolyte complexation. J. Nanosci. Nanotechnol. 7, 2833–2841.

Sarvaiya, J., Agrawal, Y.K., 2015. Chitosan as a suitable nanocarrier material for anti-Alzheimer drug delivery. Int. J. Biol. Macromol. 72, 454–465.

Schipper, N.G., Olsson, S., Hoogstraate, J.A., deBoer, A.G., Vårum, K.M., Artursson, P., 1997. Chitosans as absorption enhancers for poorly absorbable drugs 2: mechanism of absorption enhancement. Pharm. Res. 14, 923–929.

Shah, R.P., Kumar, V., Singh, S., 2008. Liquid chromatography/mass spectrometric studies on atorvastatin and its stress degradation products. Rapid Commun. Mass Spectrom. 22, 613–622.

Sharma, A., Pandey, R., Sharma, S., Khuller, G.K., 2004. Chemotherapeutic efficacy of poly(DL-lactide-co-glycolide) nanoparticle encapsulated antitubercular drugs at sub-therapeutic dose against experimental tuberculosis. Int. J. Antimicrob. Agents 24, 599–604.

Shen, J.M., Tang, W.J., Zhang, X.L., Chen, T., Zhang, H.X., 2012. Anovel carboxymethyl chitosan-based folate/Fe_3O_4/CdTe nanoparticlefor targeted drug delivery and cell imaging. Carbohydr. Polym. 88, 239–249.

Signoretto, M., Ghedini, E., Nichele, V., Pinna, F., Crocellà, V., Cerrato, G., 2011. Effect of textural properties on the drug delivery behaviour of nanoporous TiO_2 matrices. Micropor. Mesopor. Mater. 139, 189–196.

Singh, G., Pai, R.S., 2014. Recent advances of resveratrol in nanostructured based delivery systems and in the management of HIV/AIDS. J. Control. Release 194, 178–188.

Smith, J., Wood, E., Dornish, M., 2004. Effect of chitosan on epithelial cell tight junctions. Pharm. Res. 21, 43–49.

Song, H., Liang, B., Lü, L., Wu, P., Li, C., 2012. Effect of hydrolysis conditions on hydrousTiO_2 polymorphs precipitated from a titanyl sulfate and sulfuric acid solution. Int. J. Miner. Metall Mater. 19, 642–650.

Su, F.-Y., Lin, K.-J., Sonaje, K., Wey, S.-P., Yen, T.-C., Ho, Y.-C., Panda, N., Chuang, E.-Y., Maiti, B., Sung, H.-W., 2012. Protease inhibition and absorption enhancement by functional nanoparticles for effective oral insulin delivery. Biomaterials 33, 2801–2811.

Sugimoto, M., Morimoto, M., Sashiwa, H., Saimoto, H., Shigemasa, Y., 1998. Preparation and characterization of water-soluble chitin and chitosan derivatives. Carbohydr. Polym. 36, 49–59.

Summerlin, N., Soo, E., Thakur, S., Qu, Z., Jambhrunkar, S., Popat, A., 2015. Resveratrol nanoformulations: challenges and opportunities. Int. J. Pharm. 479, 282–290.

Sun, W., Mao, S., Wang, Y., Junyaprasert, V.B., Zhang, T., Na, L., Wang, J., 2010. Bioadhesion and oral absorption of enoxaparin nanocomplexes. Int. J. Pharm. 386, 275–281.

Torchilin, V.P., 2007. Micellar nanocarriers: pharmaceutical perspectives. Pharm. Res. 24, 1–16.

Venkatesan, P., Nagaprasad, P., Rupesh Dash, B.N., Prashanth, K., Devanand, S., Belal, A., Amita, P., Subhas, C.K., Paul, B.F., Mahitosh, M., 2011. The potential of celecoxib-loaded hydroxyapatite-chitosan nanocomposite for the treatment of colon cancer. Biomaterials 32, 3794–3806.

Vllasaliu, D., Casettari, L., Fowler, R., Exposito-Harris, R., Garnett, M., Illum, L., Stolnik, S., 2012. Absorption-promoting effects of chitosan in airway and intestinal cell lines: a comparative study. Int. J. Pharm. 430, 151–160.

Wang, L., Liao, R., Li, X., 2013. Layer-by-layer deposition of luminescent polymeric microgel films on magnetic Fe_3O_4@SiO_2 nanospheres for loading and release of ibuprofen. Powder Technol. 235, 103–109.

Wang, W., McConaghy, A.M., Tetley, L., Uchegbu, I.F., 2001. Controls on polymer molecular weight may be used to control the size of palmitoyl glycol chitosan polymeric vesicles. Langmuir 17, 631–636.

Wilson, B., Samanta, M.K., Muthu, M.S., Vinothapooshan, G., 2011. Design and evaluation of chitosan nanoparticles as novel drug carrier for the delivery of rivastigmine to treat Alzheimer's disease. Ther. Deliv. 2, 599–609.

Xue, M., Hu, S., Lu, Y., Zhang, Y., Jiang, X., An, S., Guo, Y., Zhou, X., Hou, H., Jiang, C., 2015. Development of chitosan nanoparticles as drug delivery system for a prototype capsid inhibitor. Int. J. Pharm. 495, 771–782.

Yang, Q., Dou, F., Liang, B., Shen, Q., 2005. Studies of cross-linking reaction on chitosan fiber with glyoxal. Carbohydr. Polym. 59, 205–210.

Yuan, Z., Ye, Y., Gao, F., Yuan, H., Lan, M., Lou, K., Wang, W., 2013. Chitosan-graft-β-cyclodextrin nanoparticles as a carrier for controlled drug release. Int. J. Pharm. 446, 191–198.

Zaki, N.M., Hafez, M.M., 2012. Enhanced antibacterial effect of ceftriaxone sodium-loaded chitosan nanoparticles against intracellular Salmonella typhimurium. AAPS PharmSciTech 13, 411–421.

Zhang, H., Hussain, I., Brust, M., Butler, M.F., Rannard, S.P., Cooper, A.I., 2005. Aligned two- and three-dimensional structures by directional freezing of polymers and nanoparticles. Nat. Mater. 4, 787–793.

FURTHER READING

Cao, H.N., He, J., Deng, L., Gao, X.Q., 2009. Fabrication of cyclodextrin-functionalized superparamagnetic Fe_3O_4/amino-silane core–shell nanoparticles via layer-by-layer method. Appl. Surf. Sci. 255, 7974–7980.

Dhudhani, A.R., Kosaraju, S.L., 2010. Bioadhesive chitosan nanoparticles: preparation and characterization. Carbohydr. Polym. 81, 243–251.

Du, J., Hsieh, Y.-L., 2007. PEGylation of chitosan for improved solubility and fiber formation via electrospinning. Cellulose 14, 543–552.

Simeonova, M., Velichkova, R., Ivanova, G., Enchev, V., Abrahams, I., 2003. Poly(butylcyanoacrylate) nanoparticles for topical delivery of 5-fluorouracil. Int. J. Pharm. 263, 133–140.

CHAPTER

PHYTONANOCONJUGATES IN ORAL MEDICINE

21

L. Bharathi Priya*, R. Baskaran, V. Vijaya Padma*[†,‡]**

*Bharathiar University, Coimbatore, India

**National Health Research Institutes, Zhunan, Taiwan

[†]China Medical University, Taichung, Taiwan

[‡]Asia University, Taichung, Taiwan

CHAPTER OUTLINE

1 Introduction 640
2 Mechanism of Polyphenol Protection 641
3 Nanostructures in Oral Delivery 642
3.1 Liposomes 642
3.2 Micelles/Polymeric Micelles 644
3.3 Dendrimers 644
3.4 Self-Emulsifying Drug-Delivery Systems 645
3.5 Inorganic Nanoparticles 645
3.6 Organic Polymeric Nanoparticles 646
3.7 Solid Lipid Nanoparticles and Nanostructured Lipid Carriers 646
3.8 Nanoemulsions 647
3.9 Emulsomes 647
3.10 Cerasomes 648
3.11 Nanocrystals 648
3.12 Multifunctionalized Carrier Systems 648
4 Mechanism of Absorption of Phytonanoconjugates in Epithelial Cells Via Endocytotic Cellular Uptake 648
4.1 Paracellular Pathway 649
4.2 Transcellular Pathway 649
5 Biofate and Pharmacokinetics of the Nanocarriers in the GI Tract After Oral Administration 651
6 Phytonanoparticles in Oral Medicine 651
6.1 Epigallocatechin-3-gallate 651
6.2 Resveratrol 653
6.3 Curcumin 654
6.4 Silymarin 656
7 Other Phytonanoconjugates 657
8 Remarks and Future Perceptions 657
References 658

1 INTRODUCTION

Several natural compounds especially phytochemicals have preventive and therapeutic potential for diseases (Kim et al., 2010, p. 1415). A variety of phytochemicals have been reported to reduce the incidences of coronary heart diseases, cancers, diabetes, neurodegenerative and other chronic diseases (Pandey and Rizvi, 2009, p. 270). These plant-derived compounds are most often the polyphenols possessing less/negligible side effects with excellent antioxidant properties (Curti et al., 2014, p. 1776). The major component of phenolic phytochemical is flavonoids. Flavonoid includes flavones, flavonols, flavanols, isoflavones, and anthocyanidins (Dai and Mumper, 2010, p. 7314). Phenolic acids, hydroxycinnamates, and their derivatives, and stilbenes are other important phenolic phytochemicals (Rio et al., 2013, pp. 1824–1825). A single compound of polyphenolic class may possess varying pharmacological properties at various sites/tissues/cells based on the pathological conditions. For example EGCG (epigallocatechin 3-gallate), a green tea polyphenol was reported to possess modulatory activity over insulin resistance, diabetes, obesity, hypertension, dyslipidemia, chemopreventive, immunomodulatory, metal chelating properties besides being an excellent antioxidant (Legeay et al., 2015, pp. 5445–5457). Protective effect of EGCG over various organs like heart, kidney, liver, lung, testes, ovaries, was proved in several in vivo experiments exposed to pathological conditions/models generated (Adak and Gabar, 2011, pp. 645–664). Some phytochemicals have been shown to reverse epigenetic mutations in cancer cells, inhibits tumor progression and premalignant lesions to cancer (Lee et al., 2013, p. 154). The major drawback with phytochemicals is that most of these compounds have poor aqueous solubility, stability, target specificity, and bioavailability in the body, which makes them unfeasible to be present at their effective levels in the target tissues. The most common administration of the preventive and therapeutic phytochemicals is by oral mode (Wang et al., 2014, p. 365). Oral method of drug administration is the predominant and widely accepted method of drug delivery on behalf of its simple, convenient, noninvasive, safest, and most economical aspect (Bala et al., 2013, pp. 67–72). Oral drug delivery has impact in therapy based on the knowledge of highly absorptive mechanism of human intestinal epithelium composed of villi that increase the total absorptive surface area in the gastrointestinal (GI) tract to 300–400 m^2. Absorptive enterocytes and mucus secreting goblet cells cover the villi, which are intermingled with follicle associated epithelium. These lymphoid regions, Peyer's patches, are covered with M cells or microfold cells specialized for antigen sampling. M cells play a vital role in drug delivery, because they possess high transcytotic ability with comparatively less protection by mucus (Ensign et al., 2012, p. 558). In spite of these advantages, the challenging problems in oral delivery includes, difficulty in swallowing pill, irritant and unpalatable drugs being not suitable for administration by this route, poor stability in the gastric environment, low solubility and/or bioavailability, slow onset of action, less/no control over drug release, nonspecific delivery site and systemic side effects (Gupta et al., 2009, p. 162). Harsh condition of the GI tract, primarily due to physical barriers like the intestinal epithelium and degradation by various enzymes and gastric juices are the major obstacles hindering the clinical availability of naturally derived compounds (Ensign et al., 2012, pp. 558–561). The key factors determining the absorption of these compounds to pass the physiological barrier of GI tract in oral administration include: solubility/aqueous state of the drug to facilitate a suitable concentration gradient, ionization state of drug in lipids, oral dosage form, motility of GI, gut microbiota, blood flow and ensuing pathological processes, food-matrix, other medicines in GI tract (Nunes et al., 2015, pp. 3–24). The protecting mucus layers at the epithelial surfaces act as significant barriers for the penetration of drugs and thereby decreases its efficiency (Ensign et al., 2012, p. 559).

Altogether, the systemic organization of the GI contributes for the low permeation efficiency of phenolic phytochemical due to passive diffusion and active efflux to lumen with no specific receptors present specifically for the uptake of phenolic phytochemicals (Li et al., 2015b, p. 154).

Targeting or specificity toward diseased cells/tissues is the important criteria in determining the success rate of therapy. The lacunae in the conventional therapeutic strategies paved way to merge the addictive effects of plant derived compounds and nanoparticles (NPs) in oral medicine. Recently, nanoparticle-based approach has become popular to overcome these challenges. These NPs are classified based on their size with a diameter of 100–2500 nm, and size between 1 and 100 nm, termed as ultrafine NPs (Bankar, 2012, p. 191). Several reports with a range of phytochemicals have proved enhanced efficiency in treatment of cancers when conjugated with NPs (Doggui et al., 2012, p. 377; Ghosh et al., 2012, p. 206; Lee et al., 2002, p. 1025; Liang et al., 2011, p. 297; Lin et al., 2015, p. 3021). This intensified efficiency can be attributed to the properties of NPs, such as biocompatibility, reduced dimensions, high surface chemical interactions, stability in biological fluids and during storage (Yallapu et al., 2010, pp. 19–29). Enhanced permeation retention (EPR) mechanism explains the accumulation of NPs in solid tumors by escaping the vasculature through leaky endothelial tissues surrounding the tumor (Yallapu et al., 2010, p. 20). This chapter describes a range of effective phytochemicals with nanoconjugates in oral medicine.

2 MECHANISM OF POLYPHENOL PROTECTION

The potential protection offered by polyphenols toward amelioration of several diseases, was attributed to the amount of health promoting properties of the phenolic components present in them. Their potential beneficial effects on human health can be associated with their antimicrobial, antioxidant, antiinflammatory, antimutagenic, anticarcinogenic properties (Pandey and Rizvi, 2009, pp. 270–278), in addition to the capacity to stimulate the digestive process. The human body naturally produces reactive oxygen species (ROS) via many enzymatic systems/reactions upon the consumption of oxygen (Rahman, 2007, p. 220). The optimized amount of these species play a beneficial role in signal transduction by acting as signal transducers/or growth regulators, but when goes beyond the normal range leads to a condition called oxidative stress. Oxidative stress mediated by ROS is the main factor involved in several pathological conditions (Schulze-Osthoff et al., 1997, p. 336). Exogenous consumption of polyphenols at this stage disrupts the chain reaction initiated by ROS and protects the cellular components susceptible for oxidative damage by ROS (Pandey and Rizvi, 2009, p. 273). Their excellent beneficial properties make them suitable for utilization in therapies. A number of polyphenols have been reported to protect the cellular components, such as DNA, proteins, and membrane lipids, from ROS attacks by preventing/curbing the chain reaction initiated by ROS (Upadhyay and Dixit, 2015, pp. 1–15). Thus, polyphenols protects the system by reducing susceptibility to damage and limits/cures the symptoms of damage caused by toxicants or diseased conditions (Shobha and Andallu, 2013, p. 351). The perspective of using polyphenolic compounds in medicine has substantially increased, but faces constraints pertaining to their stability, solubility, oral bioavailability, nonspecific side effects and storage properties (Mignet et al., 2013, pp. 458–459). Nanoencapsulation of bioactive polyphenols within appropriately structured carriers may overcome these problems without adverse effect on sensory characteristics and appearance of final product, which indeed is the major challenge in this area of knowledge.

3 NANOSTRUCTURES IN ORAL DELIVERY

Nanostructures for oral delivery can be prepared based on the material selection from proteins, polysaccharides and synthetic polymers. Factors determining the selection of material include the size of NPs, inherent properties of the drug, surface characteristics such as charge and permeability, degree of biodegradability, etc. Dispersion of preformed polymers, polymerization of monomers and ionic gelation of hydrophilic polymers, supercritical fluid technology and particle replication are the most often exploited methods of choice in preparation of these NPs (Nikam et al., 2014, pp. 1121–1127). NPs synthesized with any of these methods are said to be ideal when possessing a high drug-loading capacity with minimal matrix materials which makes them suitable for administration. Particle size and size division of nanoparticle systems determines the biodistribution, biological fate, toxicity and the targeting ability influencing the drug loading, drug release and stability of NPs. More often, the drug is incorporated or absorbed/ adsorbed to the nanoparticle systems. Solid-state of the drug solubility in matrix material, polymer composition, molecular weight, and the drug polymer interaction determines the efficiency of drug loading and entrapment (Mohanraj and Chen, 2006, pp. 561–570). Nanoparticle-based oral delivery is preferred due to the advantageous properties of improved serum solubility, sustained and controlled release of the carried drug, persistence in systemic circulation, and simultaneous delivery of multiple therapeutic drugs to target tissues/cells one at a time (Singh et al., 2011, pp. 177–179). Fabrication of NPs is an important criterion for the efficient delivery and action of phytochemicals (Bharali et al., 2011, p. 4024). Nanostructure-mediated drug delivery facilitates direct drug delivery into cells and can target tumors inside the healthy tissue (Sutradhar and Lutful Amin, 2014, pp. 1–7). NPs as delivery systems for phenolic phytochemicals have been shown to enhance phenolic phytochemical absorption and bioavailability in the GI tract. Hydrogen bonds and hydrophobic interactions of NPs with phenolic phytochemicals help in the encapsulation of phenolic phytochemicals within NPs which protect the encapsulated phenolic phytochemicals from degradation/ oxidation in the GI tract (Li et al., 2015b, p. 158). Nanoparticle system has been shown to enhance the solubility of low-watersoluble phenolic phytochemicals, prevent phenolic phytochemicals against degradation in the GI tract, enhance absorption of phenolic phytochemicals in epithelial cells via endocytotic cellular uptake with increased diffusion rate and facilitates direct uptake by intestinal lymphatic system (Nunes et al., 2015, p. 11). This suggests the potential role of nano-based phytoconjugates, because, the efficiency of oral medicine lies when there is an increased absorption and bioavailability of the drug in the GI tract. The commonly used nanostructures with promising results in oral drug delivery of phytochemicals are liposomes, micelles, dendrimers, inorganic and organic NPs. Recently developed and successful nanostructures that could enhance oral bioavailability include solid-lipid NPs (SLNs), nanostructured lipid carriers, nanoemulsions, emulsomes, cerasomes and nanocrystals (Ahmad et al., 2015, pp. 809–815) (Fig. 21.1). Cellular uptake of NPs utilizes the mechanisms of clathrin- and caveoli-mediated endocytosis and pinocytosis exempting phagocytosis because of the nanoscale size of particles (Patel et al., 2012, p. 16) (Fig. 21.2).

3.1 LIPOSOMES

Liposomes were first introduced for drug delivery in the year 1965 (Allen and Cullis, 2013, p. 36). They are artificial phospholipid vesicles that can be unilamellar or multilamellar ranging from 50 nm to 1 μm in size, enclosing aqueous compartments. The valuable pharmacological property of liposomes is attributed to its overcoming nature in the rapid uptake by the reticuloendothelial system (RES) and decreased rate of drug leakage, which aids in long circulating liposomal drug systems upon surface

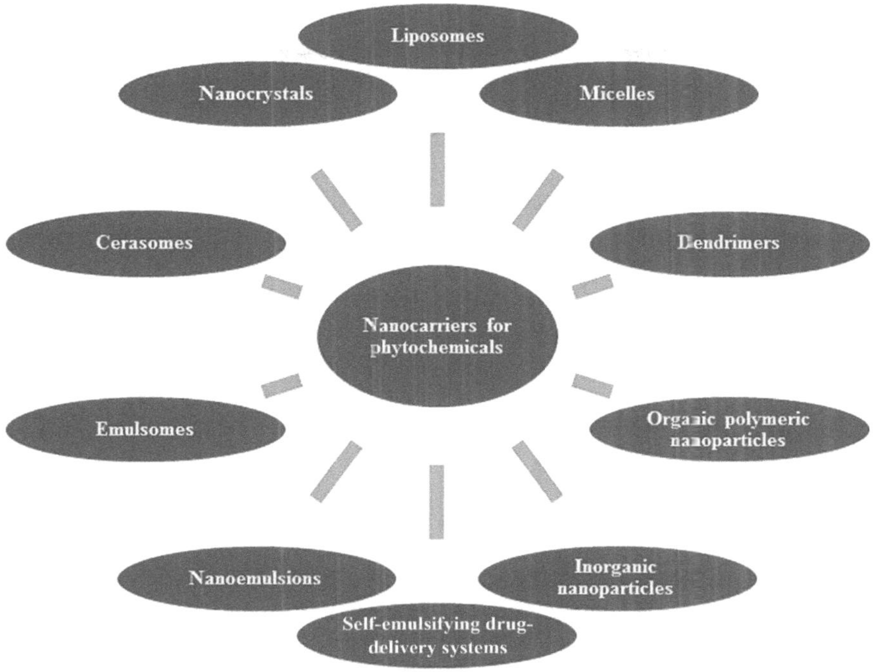

FIGURE 21.1 The Picture Represents the Commonly Used Nanostructures for the Oral Delivery of Phytochemicals

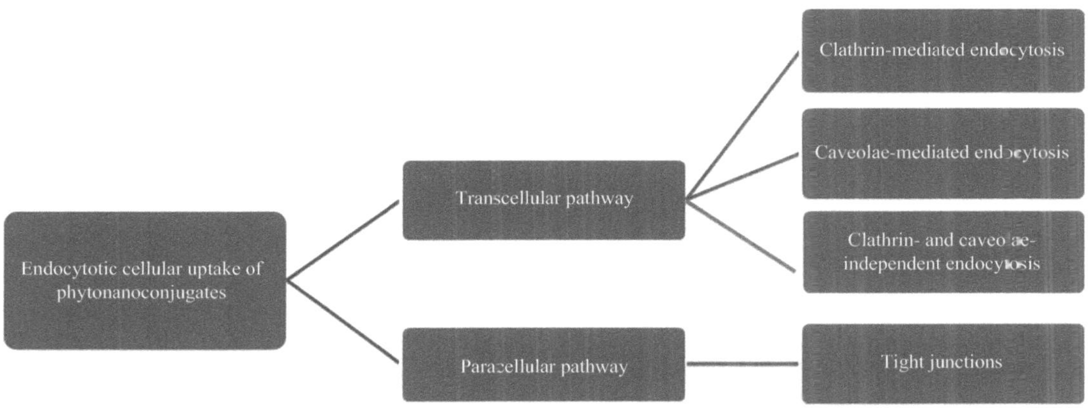

FIGURE 21.2 Mechanistic Flow Chart for the Uptake of Phytonanoconjugates in Gastrointestinal Tract

modifications (Mattheolabakis et al., 2012, pp. 1580–1581). Modifications of liposomes with hydrophilic components like polymers, carbohydrates, or glycolipids lead to changes in pharmacokinetic properties like half-life, clearance rate, and volume of distribution (Allen et al., 1995, pp. 268–274). Size, route of administration, liposomal composition, and dose plays a critical role in the absorption rate of liposomes (Oussoren et al., 1997, p. 261). Long circulating liposomes have the ability to accumulate in transplantable tumors, which makes it suitable as an efficient carrier system in cancer therapy (Deshpande et al., 2013, pp. 1509–1528). Application of liposomes has been extended to a variety of areas, such as drug delivery, cosmetic formulations, diagnostic agents, and food industry (Vishvakrama and Sharma, 2012, pp. 1074–1084). Due to their extensive use, some liposomal formulations were approved by the Food and Drug Administration (FDA) for availability in market for treating different diseases. Liposomes are the most frequently used antimicrobial drug-delivery system (Singh et al., 2011, p. 178). Opsonins and ligands that activate endocytosis in other cell types can also be incorporated into liposome membranes (Al-Jumaily and Jasim, 2015, pp. 101–102). The lipid bilayer structure of liposomes, which mimics the cell membranes of infectious microbes, is exploited for the destruction of microbes by directly releasing the drug inside the bacteria (Ranghar et al., 2014, p. 215). The smaller size of nanoliposomes with larger interfacial area of the encapsulated bioactive compounds, can accumulate well especially in solid tumor treatment, because of the enhanced permeation and retention (EPR) effect increasing the bioavailability of the encapsulated compound (Wang et al., 2014, p. 363).

3.2 MICELLES/POLYMERIC MICELLES

Micelles/polymeric micelles are composed of amphiphilic macromolecules with a hydrophobic core and hydrophilic shell with the size ranging from 20 to 80 nm (Lu and Park, 2013). Traditional micelles are lipid-based. Basically, the hydrophilic heads of phospholipids forms the shell and the fatty acid tails of phospholipids forms the hydrophobic core of the micelle that can accommodate phytochemicals (Wang et al., 2014, p. 365). Polymeric micelles are emerging as a powerful tool in nanomedicine platform for cancer therapy due to their tunable size, in vivo stability, and efficiency in solubilizing water insoluble anticancer drugs with increased blood circulation times (Sutton et al., 2007, p. 1029). Other potent pharmacokinetic benefits include reduced systemic toxicity and patient morbidity. Functionalized micelles as a recent advancement play an important role in cancer therapy by site-specific drug release at the target site (Sutton et al., 2007, p. 1029). Micelles are more advantageous than liposomes owing to its rapid accumulation into tumors due to its smaller size (Weissig et al., 1998, p. 1552). Like liposomes, micelles can be conjugated or functionalized with carbohydrates, peptides, mAbs, aptamers, and polymers for targeting and improving the pharmacological behaviors of payloads (Liu et al., 2012, p. 2521).

3.3 DENDRIMERS

Dendrimers are a class of artificial, highly branched, globular macromolecules with tree-like topological structure with defined molecular weights and entrapment properties (Shishu and Maheshwari, 2009, p. 493). Dendrimers comprises a core, layers of branched repeat units emerging from the core, and functional end groups on the outside layer of repeat units (Gor et al., 2014, p. 264). Altering the molecular properties of dendrimers with respect to size, shape, dimension, density, polarity, flexibility, and solubility by using different building/branching units and surface functional groups as a controllable modality makes dendrimers a well characterized drug-delivery system (Nazemi and Gillies, 2013, p. 15). Site-specific release of dendrimers was achieved by pH- and enzyme-dependent mechanisms (Mishra, 2011, pp. 72–73). Complexation or encapsulation is the common method of drug inclusion into dendrimers. Dendrimers can carry both hydrophobic and hydrophilic agent (Jain and

Bharatam, 2014, p. 2476). The hydrophobic core can carry the hydrophobic drugs, and hydrophilic drugs can be attached to the multivalent surfaces of dendrimers by means of covalent conjugation or electrostatic interaction (Singh et al., 2011, pp. 178–179). Dendrimers have the advantages over other linear polymers as it can be conjugated with several drug molecules, targeting ligands, and solubilizing groups at the terminals or periphery of the dendrimers, thereby making them suitable for not only oral mode of delivery but also for transdermal, intravenous and ocular deliveries (Martinho et al., 2011, p. 513). Functionalizing dendrimers with biocompatible components or by acetylation, glycosylation, PEGylation, and amino acid functionalization have been proved efficient in dendrimers-based cancer therapy (Cheng et al., 2011, p. 2673). The branched nature of dendrimers affords large surface area to size ratio allowing great reactivity with microorganisms in vivo (Sagadevan and Periasamy, 2014, p. 115). Dendrimers have been proved to cross cell barriers via both paracellular and transcellular pathways (Zhou et al., 2015, p. 2642). The widely explored and used oral dendrimers is poly-amidoamine dendrimers. Poly-amidoamine dendrimers are good carriers of oral delivery as it facilitates transport through epithelial barrier (Sadekar and Ghandehari, 2012, p. 571).

3.4 SELF-EMULSIFYING DRUG-DELIVERY SYSTEMS

Self-emulsifying drug-delivery systems (SEDDS) hold unmatched potential in improving oral bioavailability of poorly water-soluble drugs (Sarpal et al., 2015, p. 42). Orally administered SEDDS gets rapidly dispersed in GI fluids, yields micro- or nanoemulsions containing the solubilized drug. The miniscule globular size of SEDDS aids in the rapid absorption of micro/nanoemulsifed drug through lymphatic pathways, bypassing the hepatic first-pass effect (Sarpal et al., 2015, pp. 42–45). SEDDS have received societal impact due to their ability of improving oral absorption of highly hydrophobic compounds (Wang et al., 2014, p. 365). SEDDS formulations are isotropic mixture of oils, surfactants, nutrients (or drugs), usually with one or more coemulsifier or solubilizer/cosurfactants (Weerapol et al., 2014, p. 456). The mechanism behind the dispersion of SEDDS is that orally administered SEDDS (solid, liquid, or semiliquid form), the mixture gets dispersed into GI fluids and yields fine O/W emulsions comprising hydrophobic compounds upon gentle agitation in GI tract (Wang et al., 2014, p. 365). Even dispersion of SEDDS is due to the small size and large surface areas, which enhance the aqueous solubility of hydrophobic compounds, and, in turn, contributes to improved oral bioavailability (Mahapatra et al., 2014, p. 545).

3.5 INORGANIC NANOPARTICLES

Inorganic NPs have received significant impact in cancer therapy due to its optoelectric properties attributed to their size and shape (Bhattacharyya et al., 2011, p. 237). Their versatile property of extensive availability, site-specfic delivery, good biocompatibility, and rich functionality and controlled release of the carried drugs makes it as an effective drug-delivery system (Gupta et al., 2013, p. 268). The commonly used inorganic NPs in conjugation with phytochemicals in therapeutics includes, gold (Sreelakshmi et al., 2013, p. 1258), silver (Varaprasad et al., 2011, p. 1863), selenium (Ramamurthy et al., 2013, p. 1131), silica (Aras et al., 2014, p. 3865), iron oxide (Jayaprabha and Joy, 2014, p. 1), carbon materials (Li et al., 2014, p. 379), zinc oxide (Othman and Hussein, 2015, p. 105), titanium dioxide (Sawant and Kupwade, 2015, p. 37), and layered double hydroxide (LDH) NPs (Khorsandi et al., 2015, p. 93987).

3.6 ORGANIC POLYMERIC NANOPARTICLES

Organic polymeric NPs are commonly solid NPs made of organic compounds (lipids or polymers) ranging from 10 to 1 µm. Polymeric NPs are usually prepared from biodegradable and biocompatible polymers to facilitate controlled drug release at the desired target site (Hillaireau and Couvreur, 2009, p. 2874). Generally, polymeric NPs appear in the form of nanocapsules or nanospheres. Nanocapsules have a polymeric outer membrane surrounding an oily core; the drug or the active constituent can either be adsorbed to the polymeric membrane or can be dissolved in the oily core. Nanospheres comprise only the polymeric structure and drug/active constituent is retained or adsorbed (Ajazuddin and Saraf, 2010, p. 682). The polymeric NPs generally used are poly(lactide) (PLA), poly(lactide-*co*-glycolide) (PLGA), poly(epsilon-caprolactone) (PCL), poly(isobutylcyanoacrylate) (PICBA), poly(isohexylcyanoacrylate) (PIHCA), poly(n-butylcyanoacrylate) (PBCA), poly(acrylate) and poly(methacrylate) (Eudragit) as synthetic polymers; poly(lactide)-poly(ethylene glycol) (PLA-PEG), poly(lactide-*co*-glycolide)-poly(ethylene glycol) (PLGA-PEG) as copolymers; chitosan, alginate, gelatin, albumin and fibroin as natural polymers and biopolymers; polypyrrole (PPy), polyaniline (PANI), polyacetylene, poly[3,4-(ethylenedioxy)thiophene] (PEDOT), poly(dialkylfluorene), poly(*p*-phenyleneethynylene) (PPE), poly(*p*-phenylenevinylene) (PPV) as conjugated polymers (Brayner et al., 2013, pp. 29–33). SLNs, the other form of organic NPs, are made up of lipid matrices derived from glycol esters of fatty acids. They possess high stability, good biocompatibility, less toxicity, and serve as an alternative to polymer-based NPs but shows poor solubility of the carried drugs due to its lipid composition (Brayner et al., 2013, pp. 27–74). Polymeric NPs have been reported for the successful delivery of drugs to cancer cells by active or passive targeting (Magadala et al., 2008, p. 40).

3.7 SOLID LIPID NANOPARTICLES AND NANOSTRUCTURED LIPID CARRIERS

SLNs constitute a structure parallel to O/W nanoemulsion containing a hydrophilic shell and a hydrophobic lipid core (Puri et al., 2009, p. 3). SLNs are a substituting unique carrier system to conventional nanostructures, such as polymeric NPs, nanoliposomes, and nanoemulsions (Muller et al., 1996). The solid lipid core can encapsulate the hydrophobic polyphenol compounds with higher stability, lessened degradation, sustained release, which could minimize toxicity and improves target specificity of the compounds to target tissues (Muller et al., 2000, pp. 161–174). SLNs are normally composed of solid lipids, surfactants, and water, with or without cosurfactants (Mehnert and Mader, 2001). Fatty acids, triglycerides, waxes or its mixture are the commonly used lipids (Mehnert and Mader, 2001, pp. 167–188; Uner, 2006, pp. 375–378). Surfactants are used to stabilize the lipid dispersion. Commonly used surfactants include phospholipids, fatty acid ethoxylates, sorbitan esters, bile salts, or a combination of these components (Mehnert and Mader, 2001, pp. 165–193). Low compound loading capacity and leakage during storage are the main drawbacks in lipid-based nanocarriers inspite of their potential use in pharmaceutical research. Nanostructured lipid carriers (NLCs) were developed to overcome the drawbacks of SLNs (Uner, 2006, p. 376). NLCs are more promising than SLNs due to their small size, stability, biodegradability, and biocompatibility with increased loading capacity (Puri et al., 2009, p. 5). The use of multiplex lipid mixtures boosts the loading capacity of the drug in the hydrophobic core. NLCs serve as a unique carrier system for low soluble phytochemicals (Aditya et al., 2014; Gokce et al., 2012, p. 1841). The hydrophobic phytochemicals, such as resveratrol and quercetin were reported to get readily encapsulated into the lipid core with increased stability inside the lipid core (Teeranachaideekul et al., 2007, p. 198).

Wide investigation on SLNs and NLCs shows its varied application in cosmetics, pharmaceutics, food and agricultural products (Choudhary et al., 2015, p. 521). Both SLNs and NLCs can be used for various routes of administration, including oral, parenteral, percutaneous, nasal, dermal, ocular, rectal, and pulmonary in the pharmaceutical field. The preferable mode of SLNs and NLCs administration is by oral route (Iqbal et al., 2012, p. 813). Together, SLNs and NLCs were reported to increase bioavailability and to attain continuous release for hydrophobic phytochemicals. The encapsulated hydrophobic phytochemicals within SLNs or NLCs after oral administration are absorbed into lacteals in the villi of the small intestine (Bargoni et al., 1998, p. 745). These compounds are then transported from intestinal lymphatic vessels to the thoracic lymph duct and eventually into the systemic circulation at the junction of the jugular and left subclavian vein (Cavalli et al., 2003, p. 1085). The transport evades the first-pass hepatic effect of the encapsulated/carried compound in liver before it gets released into the circulation system, and therefore improves bioavailability.

3.8 NANOEMULSIONS

Nanoemulsions (NEs) can also be referred to as miniemulsions, ultrafine emulsion, and microemulsions with the droplet size ranging between 20 and 200 nm. They are thermodynamically stable isotropic system, comprising transparent dispersions of oil and water stabilized by an interfacial film of surfactant molecule. The transparency of the nanoemulsion systems is attributed to its maximum droplet size, which is less than the one-fourth of the wavelength of visible light. These nanosized droplets are responsible for the enormous interfacial areas associated with NEs in influencing the transport properties of the drug with other advantages like enhanced solubility and effective targeting of lipophilic drugs to cancer cells. Nanoemulsions can be of the type of oil in water (O/W) or water in oil (W/O). Solans et al. (2005, p. 102) reported the use of nanoemulsions in the oral administration of drugs and correlated the absorption of the emulsion to their droplet size in the GI tract. Nanoemulsions are advantageous over the lipid-based systems in terms of high solubility, decreased intersubject variability in terms of GI–fluid volume, rapid onset of action with prolonged self-life (Shafiq-un-Nabi et al., 2007, pp. 12–15), toxicological safety, and capability of large-scale production by high-pressure homogenization (Mehnert and Mader, 2001, pp. 165–190).

3.9 EMULSOMES

Emulsomes carry the features and advantages of liposomes and emulsions. It also bypasses the disadvantages faced by several nanodelivery systems in oral delivery (Kumar et al., 2013, p. 133) The internal cores of emulsomes are made of fats and triglycerides stabilized in form of o/w emulsion with lecithin at higher concentrations. The solidified or semisolidified internal oily core of emulsomes facilitates to load lipophilic drugs at high concentrations and provides a controlled drug release. It can also act as a good carrier for poor watersoluble drugs by encapsulating drugs in aqueous compartments of surrounding phospholipid layers. The mechanism of drug release to the target site happens via fusion of the cationic emulsomes with the endosomal membrane due to charge–charge interaction. After this the drug was released in the cytoplasm before lysosomal degradation and achieves a sustained release over a prolonged period. Emulsomes protects the drug from harsh gastric environment of the GI tract. Emulsomes can be used for tissue-specific drug targeting (Gill et al., 2012, pp. 87–92; Vyas et al., 2006, p. 321). Reduced toxicity issues with prolonged action at lower doses of the drug at the desired target makes them a suitable delivery system for drugs (Paliwal et al. 2009). Emulsomes may play a potential role in the effective treatment against various life-threatening viral infections, such as hepatitis, HIV, and Epstein-Barr virus infection (Vyas et al., 2006, p. 321). Targeted drug delivery was obtained with surface modified emulsomes by S-layer proteins (self-assembling proteins present on the surface of many prokaryotic organisms). Cells can be protected from oxidative stress and membrane damage by modifying the emulsomes surface with this crystalline S-layer lattice. The future perspective with emulsomes is to achieve highly specific targeted delivery of drug by utilizing the capability of recrystallizing S-layer fusion proteins loaded lipidic nanoformulations for the presentation of binding functions or homing protein domains (Ucisik et al., 2015, pp. 392–403). Tripalmitin emulsomes were developed for the treatment of experimental visceral leishmaniasis (Pal et al., 2012, p. 62).

3.10 CERASOMES

Cerasomes are novel organic–inorganic hybrid NPs possessing a liposomal bilayer structure with a siloxane network on their spherical external surface. Siloxane network present on the surface boosts the morphological rigidity and thereby is advantageous in overcoming the inherent morphological instability of liposomes. Cerasomes are alternative therapeutic system to the commonly existing chemotherapeutics. The characteristics and applications of cerasomes are similar to that of liposomes in biocompatibility and drug delivery application. Modified cerasomes in their external surface can be used for targeting specific cancer cells. Localization of vehicles in the cytoplasm determines the control over effective drug release at pharmacologically effective levels (Ahmad et al., 2015, pp. 813–814).

3.11 NANOCRYSTALS

Nanocrystals (NCs) are NPs with a crystalline character. They have a solid, generally a crystalline drug core in the nanometer-size range with an outer layer of stabilizer. The reduced nanoscale size of the drug particles facilitates its aqueous solubility with rapid dissolution velocity, enabling bioavailability after oral administration (Ravichandran, 2012, pp. 52–57). In addition, NCs facilitate sustained drug release, dose reduction, targeting to various tissues. NCs can also be prepared in aqueous solution or nonaqueous solvent medium, which leads to the development of nanosuspensions (nanocrytals). These nanosuspensions are submicron colloidal dispersion of pure drug particles stabilized by surfactants, polymers, or both (Rabinow, 2004, p. 785).

3.12 MULTIFUNCTIONALIZED CARRIER SYSTEMS

Multifunctionalized carrier systems are of recent interest to improve the efficiency of the carried drug with various pharmacological attributes at a single time for the effective curation of disorders. Multifunctionalized carrier systems could help in curbing the limitations faced several conventional methods.

4 MECHANISM OF ABSORPTION OF PHYTONANOCONJUGATES IN EPITHELIAL CELLS VIA ENDOCYTOTIC CELLULAR UPTAKE

The majority of the phenolic phytochemicals has been reported to enter the systemic circulation via passive diffusion by endocytotic cellular uptake, which is responsible for the low permeation efficiency of phytochemicals and their active efflux to lumen. The uptake of NPs has been reported to happen via the penetration across the small intestinal epithelium by either (1) paracellular or (2) transcellular pathway. Generally, these transport ways enhance absorption of carried/encapsulated phenolic phytochemicals in the GI tract. Li et al. (2015b, pp. 153–164) have clearly described the mechanism of nanoparticle-mediated dug delivery via these transport pathways with examples. The mechanism is described in the sections later.

4.1 PARACELLULAR PATHWAY

Tight junctions play a predominant role in the transport of phenolic phytochemicals by paracellular transport/passive diffusion through intercellular spaces among the epithelium. These tight junctions are formed by the junctional proteins of the epithelium and are responsible for the epithelium's integrity (Sonaje et al., 2012, p. 1271). Controlled disruption of tight junctions by NPs or by using the junctional proteins mediators can facilitate the enhanced transport of polyphenolic compounds through paracellular pathway. Fabricated/coated NPs with chitosan, thiolated chitosan, chitosan derivatives, and polyacrylate derivatives were reported as mediators of tight junctions enhancing paracellular transport (Bravo-Osuna et al., 2008, p. 436; des Rieux et al., 2006, pp. 1–6; Kudsiova and Lawrence, 2008, pp. 3998–3999; Martien et al., 2008, p. 1; Sadeghi et al., 2008, p. 270; Vllasaliu et al., 2010, p. 183). Yeh et al. (2011, p. 6164) reported chitosan-mediated modulation of tight junction protein CLDN4 in controlled disruption of tight junctions. Controlled disruption of tight junctions for the delivery of bioactive drugs to capillary blood was achieved by using NPs of size < 20 nm. This is a reversible disruption process and the function goes back to normal after delivery (Acosta, 2009, pp. 3–13; Jevprasesphant et al., 2003, pp. 1543–1544; Zha et al., 2008, pp. 135–136). Similar disruption of tight junction was achieved with polyacrylate derivatives which formed poly(acrylic acid)-Ca^{2+} complexes by interaction with Ca^{2+} from intestinal epithelial cells (Borchard et al., 1996, pp. 131–132; Junginger and Verhoef, 1998, pp. 373–374).

4.2 TRANSCELLULAR PATHWAY

Nanoparticle uptake in cells by transcellular pathway is an energy-dependent process. An example is the uptake of poly(D,L-lactic acid) NPs determined by a fluorescent dye labeling where enhanced uptake was observed with the rise in the incubated temperature from 4 to 37°C (Leroux et al., 1994, pp. 471–479). Similar reports regarding the uptake of nanoparticle were obtained with protein-based NPs, PLGA NPs, and certain types of NPs (Benfer and Kissel, 2012, p. 247; Chang et al., 2009, p. 285; Kim et al., 2007, p. 356; Ofokansi et al., 2010, p. 1; Panyam and Labhasetwar, 2003, p. 212; Tseng et al., 2007, p. 3996).

Generally, NPs uptake in cells happens via macropinocytosis or endocytosis containing clathrin-mediated endocytosis, caveolae-mediated endocytosis, and clathrin- and caveolae-independent endocytosis (Acosta, 2009, pp. 3–13; Roger et al., 2010, p. 293; Wang et al., 2011, p. 1920). Internalization of NPs by macropinocytosis is an actin-driven and receptor-independent process. Particles less than 2 mm in size can be enclosed by macropinosomes formed on the surface of enterocytes. Internalized NPs are then translocated to endolysosome (Mercer and Helenius, 2009, p. 510; Sahay et al., 2010, p. 182). Inhibition of clathrin- and caveolae-mediated endocytosis and uptake via macropinocytosis were reported by cyclodextrin NPs containing a lipid tail in Caco-2 cells (O'Neill et al., 2011, p. 174).

Clathrin-mediated endocytosis needs clathrin-coated pits formed by assembly of clathrin. Coated pits progress to form vesicles of 120 nm, a GTPase-dependent process. The formed vesicles then take up or engulf NPs into cells through low-density lipoprotein receptor or by iron-loaded transferring receptors that interact with clathrin adaptor proteins (Ogunkoya et al., 2009, pp. 109–114). NPs internalized by clathrin-mediated endocytosis are translocated into lysosome for degradation (Roger et al., 2010, p. 297; Wang et al., 2011, p. 1920). A quantum dot clathrin associated study performed by Ogunkoya et al. (2009, pp. 109–114) showed the smooth location of a clathrin adaptor AP-2 to coated pits and vesicles where clathrin and AP-2 cooperated to generate coated pits. This is due to the clathrin recruiting activity of AP-2 bound to the regions of plasma membrane, facilitating the polymerization of clathrin into lattice for membrane invagination until the coated pit get separated from the rest of the membrane to form the coated vesicle (Ogunkoya et al., 2009, pp. 109–114). Other clathrin adaptors have been described to perform the similar activities on the clathrin-mediated endocytosis (Rappoport et al., 2006, pp. 1072–1080; Traub, 2003, p. 203). Here, NPs with usual sizes < 200 nm have the ability to stabilize the clathrin coated pits, and the process of internalizing NPs was mainly by clathrin-mediated endocytosis (Ehrlich et al., 2004, pp. 591–600). For example, LDH NPs were taken up via clathrin-mediated endocytosis by escaped endosomal accumulation possibly by the deacidification (Xu et al., 2008, p. 86). Similarly, the uptake of positive charged poly(lactic acid) NPs in HeLa cells were reported to happen via internalization of clathrin-mediated endocytosis with significantly increased efficiency of uptake compared to the negatively charged NPs facing repulsion with negative surface charges of HeLa cells (Harush-Frenkel et al., 2007, p. 26).

Caveolae-mediated endocytosis happens via the assembly of caveolin protein, choloestrol, and sphingolipids, which facilitates the formation of caveolae on cell membrane (Kiss, 2012, p. 14; Reilly et al., 2012, p. 1280; Roger et al., 2010, p. 296). When caveolae-mediated transport was inhibited by the presence of filipin by its binding with sterol, the clathrin-mediated transport stayed effective (Schnitzer et al., 1994, p. 1217). NPs with an average size < 60 nm are usually internalized by caveolae-mediated endocytosis and the process get enhanced with the smaller size of NPs. Gold NPs of size approximately 5 nm were internalized by HeLa cells via caveolae-mediated endocytosis. Sucrose, which is disruptor to the formation of clathrin-mediated endocytosis, showed no inhibitory on clathrin-mediated endocytosis (Hao et al., 2012, p. 1).

Conversely, a report by Rejman et al. (2004, p. 159) showed the NPs uptake with 500 nm size via caveolae-mediated pathway escaping lysosomal accumulation. Caveolae-mediated endocytosis can help in the escapism of NPs to the degradation of endolysosome, and could deliver NPs to the endoplasmic reticulum and the nucleus of cells.

Clathrin- and caveolae-independent endocytosis is believed to be associated with cholesterol-rich microdomains on cell membrane. Polymeric NPs with size less than 25 nm were reported to happen via a novel endocytotic mechanism that led to the trafficking outside the endo/lysosomal pathway without relying on mediation by cholesterol, clathrin, or caveolae. The internalized bioactive compounds in epithelial cells are then translocated to endo/lysosome and released in the lysosome after lysosomal degradation. Metabolic process of the bioactive compounds may then continue in epithelial cells (Jin et al., 2011, p. 33; Li et al., 2008, p. 1; Ma et al., 2011, p. 8629).

Exocytotic process can also facilitate the uptake of NPs to enter blood circulation when they remain intact in endolysosome. Moreover, NPs can also escape the accumulation in endo/lysosome and may get translocated to the endoplasmic reticulum and Golgi complex.

5 BIOFATE AND PHARMACOKINETICS OF THE NANOCARRIERS IN THE GI TRACT AFTER ORAL ADMINISTRATION

It is clear now that the NPs pass the GI tract by means of passive and active transport. This is evident by its elimination through feces and urine specifying their absorption in the GI tract and their access into the systemic circulation. Water solubility, membrane permeability, and metabolic stability are the key determinants of oral bioavailability of a small molecule drug (Chan and Stewart, 1996, pp. 461–470). Size, composition, surface characteristics, and architecture determines the uptake of NPs by enterocytes or M cells (Nikam et al., 2014, p. 1125). Generally, nanocarriers are absorbed by means of initial adsorption. Physiochemical properties, such as molecular weight, conformation, and hydrophobicity determine the adsorption of the polymer. Water-soluble nanostructures made up of polymers are stable and are adsorbed as particle carriers, but the low/less stable nanocarriers. For example, polyeclectrolyte complexes or polymeric micelles may not be completely dissolved due to partial dissociation. Oral delivery also facilitates to the exposure of NPs to gut associated bacterial population in the GI tract. By means of chemical degradation, the complex polymeric drug-delivery systems may get converted to less complex products. The encapsulated drug may get released by means of bioerosion/biodegradation/swelling of the nanostructures, which ultimately results in the drug diffusion. Commonly, biodegradable polymers get degraded completely after drug release and thereby leaves no residue. An example is PLGA NPs, which degrade to lactic and glycolic acid upon hydrolytic cleavage. This is due to the dissolution of the polymer upon drug release. Modifications in the polymer side groups/surface modification/the destruction of macromolecular backbone influences the size, charge of the nanocarrier systems and pharmacokinetic behavior of nanostructures in its uptake by the RES, half-life and diffusion in the tissues.

6 PHYTONANOPARTICLES IN ORAL MEDICINE

6.1 EPIGALLOCATECHIN-3-GALLATE

Epigallocatechin-3-gallate (EGCG) is the major component of green tea polyphenol which has been investigated as subject in a number of studies for numerous health promoting effects. Its therapeutic potential against a broad range of disorders includes HIV, cancer, Sjögren's syndrome, neurodegeneration, chronic fatigue syndrome, endometriosis, spinal muscular atrophy, CB1 receptor, cannabinoid 1 receptor activity, periapical lesions, cerebrovascular insult. EGCG is unstable in the GI tract and is not readily absorbed in humans and research animals (Chen et al., 1997, p 1045; Lee et al., 2002, p. 1025; Warden et al., 2001, p. 1731). It gets rapidly degraded in both acidic (pH below 2) and neutral conditions. Chitosan encapsulated EGCG NPs, significantly delayed its degradation in simulated digestive fluids (Ponnuraj et al., 2015, p. 387). Differential effects of EGCG in cancer cells versus normal cells also necessities the targeted therapy in order to protect normal cells and chemoprevent the cancer cells (Han, 2014, p. 1). Encapsulated EGCG in polylactic acid–polyethylene glycol NPs (nano-EGCG) increased the biological effectiveness of EGCG more than 10-fold dose advantage for exerting proapoptotic and angiogenic inhibitory effects, which are the critically important determinants of chemopreventive effects of in both in vitro and in vivo systems (Siddiqui et al., 2009, p. 1712; Siddiqui et al., 2010, p. 883). Selenium nanopartilcels conjugated with EGCG and coated with Tet-1 peptide (Tet-1-EGCG@Se) prevented Aβ fibrillation and helped in disaggregation of Aβ fibrils.

Conversely, a study by Zhang et al. (2014, p. 8475) showed that EGCG-dispersed selenium NPs did not mediate notable anticancer activity in vivo. Also, both EGCG@Se and Tet-1-EGCG@Se labeled Aβ fibrils with a high affinity, and Tet-1 peptides enhanced the cellular uptake of Tet-1-EGCG@Se in PC12 cells (Zhang et al., 2014, p. 8475). In a study by Hu and coworkers, encapsulated EGCG within the fabricated food-grade polymer nanoparticle by the electrostatic interaction between chitosan and milk caseinophosphopeptides (CPP) proved the sustained release of EGCG through oral delivery. The food-grade polymer NPs enhanced the intestinal absorption of EGCG significantly in Caco-2 cells. Also, the chitosan–CPP NPs were proved to withstand the harsh pH and enzymatic conditions in GI fluids. Chitosan NPs encapsulating epigallocatechin-3-gallate (Chit-nanoEGCG) showed slow release of EGCG in simulated gastric juice with acidic pH. Chit-nanoEGCG inhibited the tumor growth and secreted prostate-specific antigen levels in subcutaneously implanted 22Rv1 tumor xenografts in athymic nude mice to a significant level (Khan et al., 2014, p. 415). EGCG attached to nanogold particles (EGCG-pNG) improved its anticancer efficacy in B16F10 murine melanoma cells than its individual counterpart (Chen et al., 2014, p. 459). EGCG conjugated nanoparticle promoted DNA damage in lymphocytes but with increasing concentrations it reduced DNA damage in a concentration-dependent manner (Alotaibi et al., 2013, p. 389). Targeted delivery of the phytochemical/therapeutic agent rather than single/combinatorial approach, targeted delivery of EGCG was proved efficient in restricting growth of tumor cells in mice (Ray et al., 2013, p. 3064). Shukla et al. (2012, p. 12426) have reported targeted therapy with EGCG conjugated gold NPs against laminin receptor in PC-3 cells xenografted SCID mice. Block copolymer-based composition of EGCG improved the oral bioavailability and therapeutic activity of EGCG in female Sprague-Dawley rats (Klinski et al., 2013, p. 1). Encouraging reports were obtained with EGCG-nanogold in suppressing the growth of tumor cells in mice model implanted with MBT-2 cells with increased stability over wide range of pH conditions (Tan et al., 2014). Site-specific delivery of a potential target-activated nanosizer comprising a epigallocatechin-3-gallate-loaded fucose-chitosan/polyethylene glycol-chitosan/gelatin NPs, in gastric cancer cells reduced drug release in a controlled manner within gastric acids and inhibited gastric cancer cell growth, induced cell apoptosis, and reduced vascular endothelial growth factor protein expression. It also significantly affected gastric tumor activity and reduced gastric and liver tissue inflammatory reaction in an orthotopic gastric tumor mouse model (Lin et al., 2015, p. 3021). Effective removal of the extracellular collagen caused by CCl4 intoxication in the hepatic fibrosis rat liver by GTE-CS NPs (green tea extract encapsulated into CS-NPs) were reported by Safer et al. (2015, p. 6452). Formulation of EGCG in a novel self-double-emulsifying drug-delivery system (SDEDDS) solid form improved its oral absorption and bioavailability in Caco-2 cells and male Sprague-Dawley rats (Hu et al., 2015, p. 1). Tea polyphenols loaded in CS-NPs showed antitumor activity towards HepG2 cancer cells (Liang et al., 2011, p. 297). It has been reported that chitosan-based NPs are capable of enhancing the bioavailability of EGCG present in tea extract (Dube et al., 2010, p. 219; Hu et al., 2012, p. 2421). The plasma concentrations of EGCG encapsulated CS-NPs increased the plasma concentration of EGCG to 1.5 factor compared to free EGCG (Dube et al., 2011, p. 422). Encased EGCG into gelatin-based 200 nm NPs comprising of a soft gel-like interior with or without a surrounding layer-by-layer shell of polyelectrolytes (dextran sulfate/protamine sulfate, polyglutamic acid/poly-L-lysine, polystyrene sulfonate/polyallylamine hydrochloride, carboxymethyl cellulose/gelatin, type A) retained the biological activity of EGCG and blocked hepatocyte growth factor (HGF)-induced intracellular signaling in the breast cancer cell line MBA-MD-231 similar to free EGCG (Shutava et al., 2009, p. 1877).

6.2 RESVERATROL

Resveratrol (*trans*-3,4′,-5-trihydroxystilebene) is a natural polyphenol from the stilbens family abundant primarily in the skin of red grapes, red wine, some berries, peanuts, dark chocolate, and other cocoa products. Resveratrol has been shown to possess antioxidant, cardioprotective, neuroprotective, antiinflammatory, antiaging, and anticarcinogenic properties (Amri et al., 2012, p. 182; Baur, 2006, p. 337; Baur, 2006a, 493, Jang et al., 1997, p. 218; Wu et al., 2013, p. 1). Generally, resveratrol occurs as two structural isomers: *cis*- and *trans*-resveratrol (*t*-RVT), *t*-RVT is converted into *cis*-resveratrol under UV exposure (Orallo, 2006, pp. 87–89). *t*-RVT is a potent antioxidant but has very low bioavailability (almost zero) due to enterohepatic recirculation and extensive first-pass metabolism by CYP3A4 in liver (Singh and Pai, 2014, p. 647). Poor oral bioavailability, rapid metabolism, and photosensitivity limit the clinical applications of resveratrol (Pandita et al., 2014, p. 1165). It has been reported that nanoencapsulation increases its stability of *t*-RVT offering protection against light-exposure degradation (Sanna et al., 2012, p. 5501). The potential protective role of NPs including liposomes, SLNs, nanospheres, and polymeric lipid-core nanocapsules on increasing the photostability of different *t*-RVT was compared in the study by (Detoni et al., 2012, pp. 913–920). The liposomes maintained *t*-RVT for the longest time and photostability of *t*-RVT were increased with the tested nanoparticle concentrations. *t*-RVT is extensively used in the studies as it is more stable than *cis*-resveratrol (Orallo, 2006, p. 87) and hence, the subject of this chapter describes *t*-RVT as resveratrol. Fabrication of SLNs and NLCs encapsulated with resveratrol resulted in enhanced aqueous solubility (Detoni et al., 2012, pp. 913–920; Neves et al., 2013, p. 177). Encapulated resveratrol in NPs also exhibited a sustained release pattern than from free resveratrol (Lu et al., 2009, p. 89; Neves et al., 2013, 177; Shao et al., 2009, p. 40). Circulating resveratrol is rapidly metabolized similar to EGCG and quercetin with low micromolar levels of free resveratrol in blood (Baur and Sinclair, 2006, p. 493). Resveratrol encapsulated into lipid-core NPs showed more than twofold of resveratrol concentrations in the kidney, brain, and liver tissues with higher GI safety than free resveratrol in male Wistar rats compared to free resveratrol administration (Frozza et al., 2010, p. 694). Nanoresveratrol decreased ROS production with increased antioxidant status in cell culture and research animal model distinguishing its significant efficiency from free resveratrol and void NPs (Gokce et al., 2012, p. 1841; Lee et al., 2012, p. 4662; Lu et al., 2009, p. 89). A study by Lee et al. (2012, p. 4662) showed the hepatoprotective role of nanoresveratrol were doubled in CCl4-induced hepatotoxicity of rats compared to free resveratrol with decreased oxidative stress and reduced inflammatory cytokine production. Surface modification of palmitic acid (PA)-decorated resveratrol-loaded SLNs with *N*-trimethyl chitosan (TMC) graft palmitic acid (PA) (TMC-g-PA) mucoadhesive copolymer circumvents the poor stability and burst release behavior in stomach acidic pH conditions of SLNs with negligible release of resveratrol in simulated gastric and sustained release in simulated intestinal conditions. The surface modified TMC-g-PA SLRNs lead to a 3.8-fold increase in relative bioavailability of resveratrol than that from resveratrol suspension (Ramalingam and Ko, 2015a, p. 52). Optimized PLGA NPs enhanced the oral bioavailability of *t*-RVT with significant increase in absorptivity and permeability parameters of the optimized NPs to transport through the Peyer's patches confirmed by in situ SPIP studies (Singh and Pai, 2014, p. 647). Penalva et al. (2015, p. 5603) has reported that Zein-based NPs improved the oral bioavailability of resveratrol to 19.2-fold higher than the free polyphenol solution with prolonged plasma levels of the resveratrol for at least 48 h. RSV-loaded polyethylene glycol–polylactic acid (PEG–PLA; M.W. 5000–5000) polymer NPs for improved stability showed a significant anticancer activity reducing the cell number of CT26 colon cancer cells significantly to 5.6% and colony forming capacity to 6.3% to that of controls upon 72 h treatment with 40 and 20 μM of RSV-NP, respectively (Jung et al., 2015, p. 251). Resveratrol-loaded carboxymethyl chitosan NPs (RES-CMCSNPs) showed better in vitro DPPH radical scavenging abilities than resveratrol raw powder. Carboxymethyl chitosan NPs also improved the solubility of resveratrol improving the antioxidant activity of the drug. Also, in vitro release experiments with RES and RES-CMCSNPs under simulated human GI tract showed RES-CMCSNPs rendered better releasing effects than resveratrol alone. RES-CMCSNPs exhibited increased in vivo absorption, prolonged duration of action and increased relative bioavailability by 3.516 times more than that of the raw resveratrol (Zu et al., 2015, p. 1). Soy lecithin-based self-microemulsifying drug-delivery system (SMEDDS) of resveratrol showed a complete release of resveratrol from the SMEDDSin approximately 180 min with improved stability of the formulation for 12 weeks confirmed by in vitro release test (Liu et al., 2014, p. 561). Crosslinked poly(acrylic acids) microgels and agarose as semiinterpenetrating networks were found to release resveratrol to about 80% after 7 days of the loaded mass (Tunesi et al., 2015, p. 1). Seljak et al. studied that analysis of the effect of SMEDDS containing resveratrol showed an inhibitory effect on the efflux transporters multidrug resistance-associated protein and P-gp on Caco-2 cells. Also, the formulation allowed for high resveratrol loading (122.5 mg/g SMEDDS) with excellent self-emulsifying properties and very rapid release. Formulation of resveratrol in SMEDDS significantly reduced its metabolite efflux. The formulated SMEDDS (without incorporated resveratrol and its individual components showed no significant effect on cell vitality and integrity (Seljak et al., 2014, p. 3491). Resveratrol bound to caseinate or caseinate–dextran with a zein core protected *t*-RVT from isomerization when exposed to UV light. Caseinate–dextran NPs were more stable to aggregation under simulated GI conditions than those coated by caseinate alone, apparently due to greater steric repulsion. The biopolymer NPs and complexes enhanced the bioaccessibility of resveratrol (Davidov-Pardo et al., 2015, p. 3915). Encapsulated resveratrol into lipid-core NPs improved the preventive effect of resveratrol's on Alzheimer's disease (Frozza et al., 2010, p. 694). Lu et al. (2009, p. 89) also proved the similar effect in their study using resveratrol-loaded polymeric micelles.

6.3 CURCUMIN

Curcumin is a hydrophobic curcuminoid present in the Indian spice turmeric extracted from the herb curcuma longa. Curcumin has been reported to possess several pharmacological properties, including antiangiogenic, antitumorigenic, and antiinflammatory properties and may potentially aid in preventing/treating some chronic diseases, such as diabetes, cancer, and cardiovascular disease (Anand et al., 2008, p. 1590). In spite of these advantages, curcumin has poor aqueous solubility and low bioavailability with low micromolar concentrations in blood (Shoba et al., 1998, p. 353; Wahlstrom and Blennow, 1978, p. 86). It has been reported to get quickly metabolized by hepatic enzymes in humans and research animals (Bansal et al., 2011, p. 1158). To overcome these limitations, many biocompatible and biodegradable NPs have been developed. Curcumin encapsulated in SLNs, nanoemulsions and PLGA NPs have successfully improved the aqueous solubility and chemical stability (Kim et al., 2011, p. 285; Mohanty and Sahoo, 2010, p. 6597; Nair et al., 2012, p. 44; Xie et al., 2011, p. 9280; Yallapu et al., 2010, p. 19). A sustained release pattern was observed with nanocurcumin (Kakkar et al., 2011, p. 495; Nair et al., 2012, p. 44; Shaikh et al., 2009, p. 223; Xie et al., 2011, p. 9280). It has been shown that there was more than two fold increase in oral bioavailability of curcumin when encapsulated into SLNs, nanoemulsions and PLGA NPs, (Gao et al., 2012, p. 659; Kakkar et al., 2011, p. 495; Khalil et al., 2013; Shaikh et al., 2009, p. 223; Xie et al., 2011, p. 9280; Yu and Huang, 2012,

p. 5373) with increased bioactivity (Doggui et al., 2012, p. 377; Kim et al., 2011, p. 285; Mohanty and Sahoo, 2010, p. 6597; Wang et al., 2012, p. 3667; Yallapu et al., 2010, p. 19; Yekollu et al., 2011, p. 2928). Nanocurcumin significantly decreased the cell viability of curcumin on cancer cells, with increased uptake in curcumin by cancer cells (Kim et al., 2011, p. 285; Mohanty and Sahoo, 2010, p. 6597; Nair et al., 2012, p. 44; Yallapu et al., 2010, p. 19). In a study by Punfa et al. (2012, p. 823) Curcumin encapsulated in PLGA NPs conjugated to anti-*P*-glycoprotein (antibody) significantly augmented the binding and targeting specificity of NPs to cancer cells with enhanced uptake by cancer cells. Curcumin PLGA NPs protected neurons against oxidative damage in vitro and increased the accumulation of curcumin in rat brain were by decreasing the ROS accumulation in neurons suggesting the potential application of curcumin NPs in preventing Alzheimer's disease and other neuron degenerative diseases (Doggui et al., 2012, p. 377; Tsai et al., 2011, p. 331s). Furthermore, curcumin liposomes decreased the inflammatory response and reversed insulin resistance in an animal model (Yekollu et al., 2011, p. 2928). Curcumin-loaded SLNs was developed with the aim of using them in formulations of asthmatic therapy showed increased curcumin concentrations in lung with double fold inhibitory effect on the inflammatory responses in the lung to that of the free compound Wang et al. (2012c, p. 3667). Ramalingam and Ko (2015b, p. 389) modified the surface of the SLNs by capping with chitosan and *N*-trimethyl chitosan (TMC, a quaternized chitosan derivative), The modified SLNs enhanced the oral bioavailability of curcumin. Also, *N*-trimethyl chitosan coated SLNs (TMC-SLCNS), showed prolonged stability and improved the TMC-mediated protection of SLNs in acidic conditions, preventing the burst release of the drug in the gastric environment. Liu et al. (2006, p. 720) showed a significant increase in bioavailability with curcumin–phospholipid complex compared to free curcumin. In this, oral administered of curcumin–phospholipid complex to Sprague-Dawley male rats showed a maximum plasma curcumin level with 1.5-fold increase in curcumin half-life to that of free curcumin. Moreover, the efficacy of curcumin as a chemopreventive agent was augmented when incorporated into nanocapsule system. Ghosh et al. (2012, pp. 206–214) reported the enhanced efficiency of curcumin in chemoprevention against diethylnitrosamine-induced rat hepatocellular carcinoma model compared to free curcumin. Curcumin NCs prepared by high-pressure homogenization showed improved solubility with enhanced bioavailability (Rachmawati et al., 2013, p. 204). Curcumin self-nanophospholipic dispersion prepared with Phosal 53 and Miglycl 812 enhanced the drug dissolution and oral bioavailability of curcumin (Allam et al., 2015, p. 117). Li et al. (2015a, p. 202) had developed an amphiphilic curcumin derivative by modifying curcumin with hyaluronic acid and loaded the modified curcumin into nanomicelles. The prepared nanomicelles showed better antioxidant potentials examined by DPPH radical scavenging and reducing power assay. A study by Anwar et al. (2015, p. 162) showed enhanced oral bioavailability of curcumin at 11.6-fold when curcumin was converted to nanosized curcumin by supercritical antisolvent process compared to free curcumin. Amorphous curcumin–chitosan nanoparticle complex was developed as a cost-effective alternative to nanoencapsulation with high chemical stability and supersaturation to enhance the bioavailability of curcumin. The prolonged supersaturation of the curcumin nanoplex owes to the presence of hydroxypropyl methylcellulose (HPMC) at five times of the saturation solubility of curcumin (Nguyen et al., 2015, p. 1). Self-assembled mixed micelles have been showed to improve the oral absorption of curcumin in in vitro and in vivo conditions (Wang et al., 2015, p. 108). 222. D'Souza and Devarajan (2016, p. 1251) showed that galactose polysaccharides adsorbed curcumin NPs played a vital role in enhancing the absorption of drugs through galectin-mediated absorption and absolute bioavailability of greater than 25% with galactose polysaccharides arabinogalactan and kappa-carrageenan.

Optimized formulation of curcumin-loaded lipopolysaccharide nanocarriers (C-LPNCs) enhanced the oral bioavailability at ~130-fold and anticancer efficacy of curcumin in colon-26 tumor-bearing mice in vitro and in vivo compared to free curcumin. The anticancer studies conducted in vitro and in vivo with optimized C-LPNCs exhibited significant increase in efficacy compared with free curcumin in the colorectal cancer model (Chaurasia et al., 2015, p. 9). Hazzah et al. (2015, p. 3913) had developed gelucire-based curcumin NPs with higher entrapment efficiency and stability for targeting oral mucosa infections. The study found a significant retained amount of curcumin onto the mucosal tissue signifying preferential mucosal uptake and exhibited higher antimicrobial activity compared with free cucumin concentrations. Curcumin-loaded lipid cubic liquid crystalline NPs showed a sustained release property in vitro release with enhanced oral absorption of curcumin when compared to crude curcumin (He et al., 2015, p. 5559). Curcumin-loaded chitosan-coated NPs developed with the aim for the local treatment of oral cavity cancer proved to be vital with its activity over significant reduction of SCC-9 human oral cancer cell viability in a concentration and time-dependent manner (Mazzarino et al., 2015, p. 781). In vivo pharmacokinetics studies showed enhanced oral bioavailability of Curcumin to 12-folds in Swiss albino rats revealed when encapsulated into SLNs when compared with marketed formulation of raw curcumin (Adcumin) (Shelat et al., 2015, p. 278). Curcumin NPs possess remarkable antibacterial, antiviral, and antiprotozoan activity. Hence, curcumin nanoparticle-loaded nanogel, microemulsion, and nanocream can be used for drug delivery (*Nature Reviews*).

6.4 SILYMARIN

Silymarin is a flavonolignans extracted from the milk thistle *Silybum marianum* (L.) gaernt. Silymarin has been shown to possess various pharmacological properties like hepatoprotective, antioxidant, antiinflammatory, anticancer, and cardioprotective activities. Although it has been proved by clinical trials that silymarin is safe at high doses (>1500 mg/day) in humans, pharmacokinetic studies revealed poor absorption, rapid metabolism, and ultimately poor oral bioavailability of silymarin (Javed et al., 2011, p. 239). El-Sherbiny et al. (2011, p. 1345) had developed a biodegradable pH-responsive alginate-poly (lactic-*co*-glycolic acid) nano/microhydrogel matrix for oral delivery of silymarin conferring sustained oral release of silymarin with enhanced overall dissolution and oral bioavailability. Nanosilymarin showed protection against γ-radiation-induced oxidative stress evaluated by the reduction in micronucleus frequency and ROS generation, microscopy analysis, and cell-cycle estimation in cultured human embryonic kidney cells (Adhikari and Arora, 2015, p. 792). Oil-based nanocarrier for silymarin improved the oral efficacy of silymarin in vitro and in vivo studies exhibiting hepatoprotection (Parveen et al., 2011, p. 245). Silymarin-loaded SLNs by cold homogenization method than hot homogenization was found to highly efficient with the relative bioavailability at 2.79-fold higher compared to free silymarin solution (He et al., 2007, p. 195). Lu and Park (2013, p. 198) improved oral bioavailability of silymarin with binary lipids-based NLCs. Chen et al. (2015, p. 2403) successfully developed silymarin-loaded SNEDDS, which improved the dissolution, permeability, and oral bioavailability of silymarin. Oral delivery of silymarin loaded SLNs in Beagle dogs showed enhanced bioavailability. Bile-salt containing liposomes loaded with silymarin developed by supercritical fluid technology improved the dissolution and bioavailability compared of silymarin powder (Yang et al., 2015). Maryana et al. (2015) developed silymarin-loaded phytosomes with the aim of increasing the stability and oral bioavailability of silymarin. SNEDDS developed was found to improve the drug malabsorption in rats with roux-en-Y gastric bypass surgery (Chen et al., 2015).

Gupta et al. (2014, p. 156) showed enhanced activity of silymarin when loaded in chitosan NPs in passive targeted therapy in its hepatoprotective action against CCl4 induced hepatotoxicity in Swiss albino mice. Silymarin-loaded SLNs enhanced the oral bioavailability of silymarin in Beagle dogs as experimental model compared to free silymarin (Jun et al., 2005, p. 19).

7 OTHER PHYTONANOCONJUGATES

Synthesized iron-based NPs using oolong tea extracts removed 75.5% of malachite green (50 mg/L) discovering a promising role in environmental remediation (Huang et al., 2014, p. 801). Campos et al. (2014, p. 109) developed an optimized SLNs carrying rosmarinic acid and their herbal extracts for oral administration by changing the concentrations of surfactant—Tween 80, and lipid–witepsol wax, which enabled the association of rosmarinic acid to the lipid at high efficiencies and produced highly stable NPs for storage to a time period of 28 days. Puerarin is a cardioprotective agent and has been used in the treatment of cerebro-vascular diseases. But its use is greatly limited due to poor solubility and short half-life. High doses of intravenous administration is the preferred mode of delivery as no oral formulation of puerarin is available. This may lead to severe and acute side effects. Single intragastric administration puerarin-loaded SLNs showed 3.1-fold increase in oral bioavailability (Luo et al., 2011, p. 238). In the study by Reddy et al. (2014, p. 115), silver NPs were synthesized using aqueous Piper longum fruit extract and confirmed by UV-visible spectroscopy.

8 REMARKS AND FUTURE PERCEPTIONS

Over the past years several studies were conducted in association with the development of delivery systems of polyphenolic compounds due to its health promoting, disease preventive, and therapeutic properties, especially to overcome the challenges of oral absorption and bioavailability of the drug allowing a greater effect associated with these compounds. Bioaccessibility, food matrix effect, and the gut microbiota can also influence the efficacy of polyphenolic compounds during digestion. Nanostructures paved the way to overcome several of these possible challenges related to solubility and oral bioavailability of polyphenolic compounds with the knowledge of GI barriers. Translational efficiency of the phytonanoconjugates needs further transformation for exhibiting a successful clinical trial. Improved nanocarrier efficiency can be achieved by multifunctional nanocarrier system, which could specifically protect the carried polyphenolic compounds from acidic environment, enzymatic degradation in the GI tract, improved drug absorption, prolonged retention in GI, and by conjugating disease–specific diagnostic agents. Moreover, polyphenolic compounds in conjugation and complexation with the drug/proteins that maintains their activity and stability against proteolytic degradation can be in sighted for improved efficiency. Connecting the advantages of such nanocarrier delivery systems for polyphenolic compounds and development of multifunctional nanosystems, the merits associated with the oral route of administration can be maximized with the targeted delivery and minimized side effects rather than treating the symptoms of the disease. In conclusion, the strategy of using phytonanoconjugates would advance potential phytochemicals as a smart therapeutic agent and may certainly promise for future multifunctional therapeutic application with enhanced bioavailability and limited toxicity.

REFERENCES

Acosta, E., 2009. Bioavailability of nanoparticles in nutrient and nutraceutical delivery. Curr. Opin. Colloid. Interface Sci. 14, 3–15.

Adak, M., Gabar, M.A., 2011. Green tea as a functional food for better health: a brief review. RJPBCS 2 (2), 645–664.

Adhikari, M., Arora, R., 2015. Nano-silymarin provides protection against gamma-radiation-induced oxidative stress in cultured human embryonic kidney cells. Mutat. Res. Genet. Toxicol. Environ. Mutagen. 792, 1–11.

Aditya, N.P., Macedo, A.S., Doktorovova, S., Souto, E.B., Kim, S., Chang, P.S., et al., 2014. Development and evaluation of lipid nanocarriers for quercetin delivery: a comparative stud y of solid lipid nanoparticles (SLN), nanostructured lipid carriers (NLC), and lipid nanoemulsions (LNE). LWT Food Sci. Tech. 16 (2), 1321–1330.

Ahmad, U., Hussain, M.T., Faiyazuddin, M., 2015. Nanotechnology-based chemopreventive approaches to improve the oral delivery of silymarin in liver cancer. Nanotechnology: Novel Perspectives and prospects. McGraw Hill Education, 809–817.

Ajazuddin, Saraf, S., 2010. Applications of novel drug delivery system for herbal formulations. Fitoterapia 81, 680–689.

Al-Jumaily, E.F., Jasim, B.H., 2015. A review: liposomes as pharmaceutical carriers. IJAP 5 (2), 101–106.

Allam, A.N., Komeil, I.A., Fouda, M.A., Abdallah, O.Y., 2015. Preparation, characterization and in vivo evaluation of curcumin self-nano phospholipid dispersion as an approach to enhance oral bioavailability. Int. J. Pharm. 489, 117–123.

Allen, T.M., Cullis, P.R., 2013. Liposomal drug delivery systems: from concept to clinical applications. Adv. Drug. Deliv. Rev. 65, 36–48.

Allen, T.M., Hansen, C.B., Demenezes, D.E.L., 1995. Pharmacokinetics of long-circulating liposomes. Adv. Drug. Deliv. Rev. 16, 267–284.

Alotaibi, A., Bhatnagar, P., Najafzadeh, M., Gupta, K.C., Anderson, D., 2013. Tea phenols in bulk and nanoparticle form modify DNA damage in human lymphocytes from colon cancer patients and healthy individuals treated in vitro with platinum-based chemotherapeutic drugs. Nanomedicine 8, 389–401.

Amri, A., Chaumeil, J.C., Sfar, S., Charrueau, C., 2012. Administration of resveratrol: what formulation solutions to bioavailability limitations? J. Control. Release 158 (2), 182–193, 10.

Anand, P., Thomas, S.G., Kunnumakkara, A.B., Sundaram, C., Harikumar, K.B., Sung, B., Tharakan, S.T., Misra, K., Priyadarsini, I.K., Rajasekharan, K.N., Aggarwal, B.B., 2008. Biological activities of curcumin and its analogues (Congeners) made by man and mother nature. Biochem. Pharmacol. 76, 1590–1611.

Anwar, M., Ahmad, I., Warsi, M.H., Mohapatra, S., Ahmad, N., Akhter, S., Ali, A., Ahmad, F.J., 2015. Experimental investigation and oral bioavailability enhancement of nano-sized curcumin by using supercritical anti-solvent process. Eur. J. Pharm. Biopharm. 96, 162–172.

Aras, A., Khokhar, A.R., Qureshi, M.Z., Silva, M.F., Sobczak-Kupiec, A., Pineda, E.A.G., Hechenleitner, A.A.W., Farooqi, A.A., 2014. Targeting cancer with nano-bullets: curcumin, EGCG, resveratrol and quercetin on flying carpets. Asian. Pac. J. Cancer 15, 3865–3871.

Bala, R., Pawar, P., Khanna, S., Arora, S., 2013. Orally dissolving strips: a new approach to oral drug delivery system. Int. J. Pharm. Investig. 3, 67–76.

Bankar, V.B., 2012. Nanotechnology or nano particles as drug carriers. Ph. Tech. Med. 1 (6), 191–196.

Bansal, S.S., Goel, M., Aqil, F., Vadhanam, M.V., Gupta, R.C., 2011. Advanced drug delivery systems of curcumin for cancer chemoprevention. Cancer Prev. Res. 4, 1158–1171.

Bargoni, A., Cavalli, R., Caputo, O., Fundaro, A., Gasco, M.R., Zara, G.P., 1998. Solid lipid nanoparticles in lymph and plasma after duodenal administration to rats. Pharm. Res. 15, 745–750.

Baur, J.A., Sinclair, D.A., 2006. Therapeutic potential of resveratrol: the in vivo evidence. Nat. Rev. Drug Discov. 5, 493–506.

Benfer, M., Kissel, T., 2012. Cellular uptake mechanism and knockdown activity of siRNA-loaded biodegradable DEAPA-PVA-g-PLGA nanoparticles. Eur. J. Pharm. Biopharm. 80 (2), 247–256.

Bharali, D.J., Siddiqui, I.A., Adhami, V.M., Chamcheu, J.C., Aldahmash, A.M., Mukhtar, H., Mousa, S.A., 2011. Nanoparticle delivery of natural products in the prevention and treatment of cancers: current status and future prospects. Cancers 3, 4024–4045.

Bhattacharyya, S., Kudgus, R.A., Bhattacharya, R., Mukherjee, P., 2011. Inorganic nanoparticles in cancer therapy. Pharmaceut. Res. 28, 237–259.

Borchard, G., Lue b en, H.L., de Boer, A.G., Verhoef, J.C., Lehr, C.M., Junginger, H.E., 1996. The potential of mucoadhesive polymers in enhancing intestinal peptide drug absorption. III: effects of chitosan-glutamate and carbomer on epithelial tight junctions in vitro. J. Control. Release 39 (2–3), 131–138.

Bravo-Osuna, I., Vauthier, C., Chacun, H., Ponchel, G., 2008. Specific permeability modulation of intestinal paracellular pathway by chitosan-poly(isobutylcyanoacrylate) core-shell nanoparticles. Eur. J. Pharm. Biopharm. 69 (2), 436–444.

Brayner, R., Fiévet, F., Coradin, T., 2013. Synthesis of organic and bioorganic nanoparticles: an overview of the preparation methods. In: Allouche, J. (Ed.), Nanomaterials: A Danger or a Promise? A Chemical and Biological Perspective. Springer, London, UK, pp. 27–74.

Campos, D.A., Madureira, A.R., Gomes, A.M., Sarmento, B., Pintado, M.M., 2014. Optimization of the production of solid Witepsol nanoparticles loaded with rosmarinic acid. Colloids Surf. B 115, 109–117.

Cavalli, R., Bargon, A., Podio, V., Munton, E., Zara, C.P., Gasco, M.R., 2003. Duodenal administration of solid lipid nanoparticles loaded with different percentages of tobramycin. J. Pharm. Sci. 92, 1085–1094.

Chan, O.H., Stewart, B.H., 1996. Physicochemical and drug-delivery considerations for oral drug bioavailability. Drug. Discov. Today 1, 461–473.

Chang, J., Jallouli, Y., Kroubi, M., Yuan, X.B., Feng, W., Kang, C.S., et al., 2009. Characterization of endocytosis of transferrin-coated PLGA nanoparticles by the blood-brain barrier. Int. J. Pharm. 379 (2), 285–292.

Chaurasia, S., Patel, R.R., Chaubey, P., Kumar, N., Khan, G., Mishra, B., 2015. Lipopolysaccharide based oral nanocarriers for the improvement of bioavailability and anticancer efficacy of curcumin. Carbohyd. Polym. 130, 9–17.

Chen, C.H., Chang, C.C., Shih, T.H., Aljuffali, I.A., Yeh, T.S., Fang, J.Y., 2015. Self-nanoemulsifying drug delivery systems ameliorate the oral delivery of silymarin in rats with Roux-en-Y gastric bypass surgery. Int. J. Nanomed. 10, 2403–2416.

Chen, C.C., Hsieh, D.S., Huang, K.J., Chan, Y.L., Hong, P.D., Yeh, M.K., Wu, C.J., 2014. Improving anticancer efficacy of (-)-epigallocatechin-3-gallate gold nanoparticles in murine B16F10 melanoma cells. Drug Des. Devel. Ther. 8, 459–474.

Chen, L., Lee, M.J., Li, H., Yang, C.S., 1997. Absorption, distribution, elimination of tea polyphenols in rats. Drug Metab. Dispos. 25, 1045–1050.

Cheng, Y., Zhao, L., Li, Y., Xu, T., 2011. Design of biocompatible dendrimers for cancer diagnosis and therapy: current status and future perspectives. Chem. Soc. Rev. 40, 2673–2703.

Choudhary, S., Pandey, A., Kamran Khan, M., Khan, S., Hakki, E.E., Akkaya, M.S., et al., 2015. Chitin nanofibrils and its application in healing the skin. Int. J. Sci. Nat. 6 (3), 521–523.

Curti, V., Capelli, E., Boschi, F., Nabavi, S.F., Bongiorno, A.I., Habtemariam, S., Nabavi, S.M., Daglia, M., 2014. Modulation of human miR-17-3p expression by methyl 3-O-methyl gallate as explanation of its in vivo protective activities. Mol. Nutr. Food Res. 58, 1776–1784.

D'Souza, A.A., Devarajan, P.V., 2016. Bioenhanced oral curcumin nanoparticles: role of carbohydrates. Carbohydr. Polym. 136, 1251–1258.

Dai, J., Mumper, R.J., 2010. Plant phenolics: extraction, analysis and their antioxidant and anticancer properties. Molecules 15, 7313–7352.

Davidov-Pardo, G., Perez-Ciordia, S., Marin-Arroyo, M.R., McClements, D.J., 2015. Improving resveratrol bioaccessibility using biopolymer nanoparticles and complexes: impact of protein-carbohydrate maillard conjugation. J. Agr. Food Chem. 63, 3915–3923.

des Rieux, A., Fievez, V., Garinot, M., Schneider, Y.J., Préat, V., 2006. Nanoparticles as potential oral delivery systems of proteins and vaccines: a mechanistic approach. J. Control. Release 116 (1), 1–27.

Deshpande, P.P., Biswas, S., Torchilin, V.P., 2013. Current trends in the use of liposomes for tumor targeting. Nanomedicine 8, 1509–1528.

Detoni, C.B., Souto, G.D., da Silva, A.L., Pohlmann, A.R., Guterres, S.S., 2012. Photostability and skin penetration of different E-resveratrol-loaded supramolecular structures. Photochem. Photobiol. 88, 913–921.

Doggui, S., Sahni, J.K., Arseneault, M., Dao, L., Ramassamy, C., 2012. Neuronal uptake and neuroprotective effect of curcumin-loaded PLGA nanoparticles on the human SK-N-SH cell line. J. Alzheimers Dis. 30, 377–392.

Dube, A., Nicolazzo, J.A., Larson, I., 2010. Chitosan nanoparticles enhance the intestinal absorption of the green tea catechins (+)-catechin and (-)-epigallocatechin gallate. Eur. J. Pharm. Sci. 41, 219–225.

Dube, A., Nicolazzo, J.A., Larson, I., 2011. Chitosan nanoparticles enhance the plasma exposure of (−)-epigallocatechin gallate in mice through an enhancement in intestinal stability. Eur. J. Pharm. Sci. 44, 422–426.

Ehrlich, M., Boll, W., van Oijen, A., Hariharan, R., Chandran, K., Nibert, M.L., et al., 2004. Endocytosis by random initiation and stabilization of clathrin-coated pits. Cell 118 (5), 591–605.

El-Sherbiny, I.M., Abdel-Mogib, M., Dawidar, A.A.M., Elsayed, A., Smyth, H.D.C., 2011. Biodegradable pH-responsive alginate-poly (lactic-*co*-glycolic acid) nano/micro hydrogel matrices for oral delivery of silymarin. Carbohyd. Polym. 83, 1345–1354.

Ensign, L.M., Cone, R., Hanes, J., 2012. Oral drug delivery with polymeric nanoparticles: the gastrointestinal mucus barriers. Adv. Drug Deliv. Rev. 64, 557–570.

Frozza, R.L., Bernardi, A., Paese, K., Hoppe, J.B., da Silva, T., Battastini, A.M., Pohlmann, A.R., Guterres, S.S., Salbego, C., 2010. Characterization of trans-resveratrol-loaded lipid-core nanocapsules and tissue distribution studies in rats. J. Biomed. Nanotechnol. 6, 694–703.

Gao, Y., Wang, C., Sun, M., Wang, X., Yu, A.H., Li, A.G., Zhai, G.X., 2012. In vivo evaluation of curcumin loaded nanosuspensions by oral administration. J. Biomed. Nanotechnol. 8, 659–668.

Ghosh, D., Choudhury, S.T., Ghosh, S., Mandal, A.K., Sarkar, S., Ghosh, A., Saha, K.D., Das, N., 2012. Nanocapsulated curcumin: oral chemopreventive formulation against diethylnitrosamine induced hepatocellular carcinoma in rat. Chem. Biol. Interact. 195, 206–214.

Gill, B., Singh, J., Sharma, V., Kumar, S.L.H., 2012. Emulsomes: an emerging vesicular drug delivery system. Asian J. Pharmacol. 6, 87–94.

Gokce, E.H., Korkmaz, E., Dellera, E., Sandri, G., Bonferoni, M.C., Ozer, O., 2012. Resveratrol-loaded solid lipid nanoparticles versus nanostructured lipid carriers: evaluation of antioxidant potential for dermal applications. Int. J. Nanomed. 7, 1841–1850.

Gor, R.M., Sonvane, B.B., Shah, V.H., 2014. Dendrimer:- an emerging approach for drug delivery. Int. J. Pharm. Sci. Rev. Res. 26 (1), 264–270.

Gupta, H., Bhandari, D., Sharma, A., 2009. Recent trends in oral drug delivery: a review. Recent Pat. Drug. Deliv. Formul. 3, 162–173.

Gupta, S., Bansal, R., Gupta, S., Jindal, N., Jindal, A., 2013. Nanocarriers and nanoparticles for skin care and dermatological treatments. Indian Dermatol. Online J. 4, 267–272.

Gupta, S., Kumar Singh, S., Girotra, P., 2014. Targeting silymarin for improved hepatoprotective activity through chitosan nanoparticles. Int. J. Pharm. Investig. 4 (4), 156–163.

Han, D.W., 2014. Advances of green tea catechins towards smart anticancer agents. Biochem. Pharmacol. 3, e153.

Hao, X., Wu, J., Shan, Y., Cai, M., Shang, X., Jiang, J., et al., 2012. Caveolae-mediated endocytosis of biocompatible gold nanoparticles in living Hela cells. J. Phys. Condens. Matter. 24 (16), 164207.

Harush-Frenkel, O., Debotton, N., Benita, S., Altschuler, Y., 2007. Targeting of nanoparticles to the clathrin-mediated endocytic pathway. Biochem. Biophys. Res. Commun. 353 (1), 26–32.

Hazzah, H.A., Farid, R.M., Nasra, M.M.A., Hazzah, W.A., El-Massik, M.A., Abdallah, O.Y., 2015. Gelucire-based nanoparticles for curcumin targeting to oral mucosa: preparation, characterization, and antimicrobial activity assessment. J. Pharm. Sci. 104, 3913–3924.

He, J., Hou, S.X., Lu, W.G., Zhu, L., Feng, J.F., 2007. Preparation, pharmacokinetics and body distribution of silymarin-loaded solid lipid nanoparticles after oral administration. J. Biomed. Nanotechnol. 3, 195–202.

He, X.L., Li, Q.H., Liu, X.J., Wu, G.S., Zhai, G.X., 2015. Curcumin-loaded lipid cubic liquid crystalline nanoparticles: preparation, optimization, physicochemical properties and oral absorption. J. Nanosci. Nanotechnol. 15, 5559–5565.
Hillaireau, H., Couvreur, P., 2009. Nanocarriers' entry into the cell: relevance to drug delivery. Cell. Mol. Life Sci. 66, 2873–2896.
Hu, B., Ting, Y., Yang, X., Tang, W., Zeng, X., Huang, Q., 2012. Nanochemoprevention by encapsulation of (-)-epigallocatechin-3-gallate with bioactive peptides/chitosan nanoparticles for enhancement of its bioavailability. Chem. Commun. 48, 2421–2423.
Hu, C., Wang, Q., Zhao, G., Yao, W., Xia, Q., 2015. Improved oral absorption of (−)-epigallocatechin-3-gallate via self-double-emulsifying solid formulation. Eur. J. of Lipid Sci. Technol. 117, 1–10.
Huang, L.L., Weng, X.L., Chen, Z.L., Megharaj, M., Naidu, R., 2014. Synthesis of iron-based nanoparticles using oolong tea extract for the degradation of malachite green. Spectrochim. Acta A. 117, 801–804.
Iqbal, M.A., Md, S., Sahni, J.K., Baboota, S., Dang, S., Ali, J., 2012. Nanostructured lipid carriers system recent advances in drug delivery. J. Drug Target. 20, 813–830.
Jain, V., Bharatam, P.V., 2014. Pharmacoinformatic approaches to understand complexation of dendrimeric nanoparticles with drugs. Nanoscale 6, 2476–2501.
Jang, M., Cai, L., Udeani, G.O., Slowing, K.V., Thomas, C.F., Beecher, C.W., et al., 1997. Cancer chemopreventive activity of resveratrol, a natural product derived from grapes. Science 275, 218–220.
Javed, S., Kohli, K., Ali, M., 2011. Reassessing bioavailability of silymarin. Altern. Med. Rev. 16, 239–249.
Jayaprabha, K.N., Joy, P.A., 2014. Curcumin encapsulated superparamagnetic iron oxide based nanofluids for possible multifunctional applications. J. Nanofluids 3, 1–7.
Jevprasesphant, R., Penny, J., Attwood, D., McKeown, N.B., D' Emanuele, A., 2003. Engineering of dendrimer surfaces to enhance transepithelial transport and reduce cytotoxicity. Pharm. Res 20 (10), 1543–1550.
Jin, H., Lovell, J.F., Chen, J., Lin, Q., Ding, L., Ng, K.K., et al., 2011. Mechanistic in-sights into LDL nanoparticle-mediated siRNA delivery. Bioconjug. Chem. 23 (1), 33–41.
Jun, H.E., Jian-fang, F., Le-le, Z., Wei-gen, L., Shi-xiang, H., 2005. Bioavailability of silymarin-loaded solid lipid nanoparticles. Chinese Traditional Patent Medicine 27 (1), 19–21.
Jung, K.H., Lee, J.H., Park, J.W., Quach, C.H., Moon, S.H., Cho, Y.S., Lee, K.H., 2015. Resveratrol-loaded polymeric nanoparticles suppress glucose metabolism and tumor growth in vitro and in vivo. Int. J. Pharm. 478, 251–257.
Junginger, H.E., Verhoef, J.C., 1998. Macromolecules as safe penetration enhancers for hydrophilic drugs: a fiction? Pharm. Sci. Technol. Today 1 (9), 370–376.
Kakkar, V., Singh, S., Singla, D., Kaur, I.P., 2011. Exploring solid lipid nanoparticles to enhance the oral bioavailability of curcumin. Mol. Nutr. Food Res. 55, 495–503.
Khalil, N.M., do Nascimento, T.C., Casa, D.M., Dalmolin, L.F., de Mattos, A.C., Hoss, I., Romano, M.A., Mainardes, R.M., 2013. Pharmacokinetics of curcumin-loaded PLGA and PLGA-PEG blend nanoparticles after oral administration in rats. Colloids Surf. B. 101, 353–360.
Khan, N., Bharali, D.J., Adhami, V.M., Siddiqui, I.A., Cui, H., Shabana, S.M., et al., 2014. Oral administration of naturally occurring chitosan-based nanoformulated green tea polyphenol EGCG effectively inhibits prostate cancer cell growth in a xenograft model. Carcinogenesis 35, 415–423.
Khorsandi, K., Hosseinzadeh, R., Fateh, M., 2015. Curcumin intercalated layered double hydroxide nanohybrid as a potential drug delivery system for effective photodynamic therapy in human breast cancer cells. RSC Adv. 5, 93987–93994.
Kim, H., Gil, S., Andrieux, K., Nicolas, V., Appel, M., Chacun, H., et al., 2007. Low-density lipoprotein receptor-mediated endocytosis of PEGylated nanoparticles in rat brain endothelial cells. Cell. Mol. Life Sci. 64 (3), 356–364.
Kim, J., Lee, H.J., Lee, K.W., 2010. Naturally occurring phytochemicals for the prevention of Alzheimer's disease. J. Neurochem. 112, 1415–1430.

Kim, T.H., Jiang, H.H., Youn, Y.S., Park, C.W., Tak, K.K., Lee, S., Kim, H., Jon, S., Chen, X., Lee, K.C., 2011. Preparation and characterization of water-soluble albumin-bound curcumin nanoparticles with improved antitumor activity. Int. J. Pharm. 403, 285–291.

Kiss, A.L., 2012. Caveolae and the regulation of endocytosis. Adv. Exp. Med. Biol. 729, 14–28.

Klinski, E., Semov, A., Yan, X., Alakhov, V., Muyzhnek, E., Kiselev, V., 2013. Block copolymer based composition of epigallocatechin-3-gallate with improved oral bioavailability as a way to increase its therapeutic activity. J. Nanomed. Biother. Discov. 3, 1–5.

Kudsiova, L., Lawrence, M.J., 2008. A comparison of the effect of chitosan and chitosan-coated vesicles on monolayer integrity and permeability across Caco-2 and 16HBE14o-cells. J. Pharm. Sci. 97 (9), 3998–4010.

Kumar, R., Seth, N., Kumar, S.L.H., 2013. Emulsomes: an emerging vesicular drug delivery system. J. Drug Deliv. Ther. 3 (6), 133–142.

Lee, J.H., Khor, T.O., Shu, L., Su, Z.Y., Fuentes, F., Kong, A.N., 2013. Dietary phytochemicals and cancer prevention: Nrf2 signaling, epigenetics, and cell death mechanisms in blocking cancer initiation and progression. Pharmacol. Ther. 137 (2), 153–171.

Lee, M.J., Maliakal, P., Chen, L., Meng, X., Bondoc, F.Y., Prabhu, S., et al., 2002. Pharmacokinetics of tea catechins after ingestion of green tea and (−)-epigallocatechin-3-gallate by humans: formation of different metabolites and individual variability. Cancer Epidemiol. Biomark. Prev. 11, 1025–1032.

Lee, C.W., Yen, F.L., Huang, H.W., Wu, T.H., Ko, H.H., Tzeng, W.S., Lin, C.C., 2012. Resveratrol nanoparticle system improves dissolution properties and enhances the hepatoprotective effect of resveratrol through antioxidant and anti-inflammatory pathways. J. Agr. Food Chem. 60, 4662–4671.

Legeay, S., Rodier, M., Fillon, L., Faure, S., Clere, N., 2015. Epigallocatechin gallate: a review of its beneficial properties to prevent metabolic syndrome. Nutrients 7, 5443–5468.

Leroux, J.C., Gravel, P., Balant, L., Volet, B., Anner, B.M., Allémann, E., et al., 1994. Internalization of poly(D,L-lactic acid) nanoparticles by isolated human leu-kocytes and analysis of plasma proteins adsorbed onto the particles. J. Biomed. Mater. Res. 28 (4), 471–481.

Li, W., Chen, C., Ye, C., Wei, T., Zhao, Y., Lao, F., et al., 2008. The translocation of fullerenic nanoparticles into lysosome via the pathway of clathrin-mediated endocytosis. Nanotechnology 19 (14), 145102.

Li, Z., Jiang, H., Xu, C.M., Gu, L.W., 2015b. A review: using nanoparticles to enhance absorption and bioavailability of phenolic phytochemicals. Food Hydrocoll. 43, 153–164.

Li, J., Shin, G.H., Chen, X.G., Park, H.J., 2015a. Modified curcumin with hyaluronic acid: Combination of pro-drug and nano-micelle strategy to address the curcumin challenge. Food Res. Int. 69, 202–208.

Li, H.X., Zhang, N., Hao, Y.W., Wang, Y.L., Jia, S.S., Zhang, H.L., Zhang, Y., Zhang, Z.Z., 2014. Formulation of curcumin delivery with functionalized single-walled carbon nanotubes: characteristics and anticancer effects in vitro. Drug Deliv. 21, 379–387.

Liang, J., Li, F., Fang, Y., Yang, W.J., An, X.X., Zhao, L.Y., Xin, Z.H., Cao, L., Hu, Q.H., 2011. Synthesis, characterization and cytotoxicity studies of chitosan-coated tea polyphenols nanoparticles. Colloid Surf. B 82, 297–301.

Lin, Y.H., Chen, Z.R., Lai, C.H., Hsieh, C.H., Feng, C.L., 2015. Active targeted nanoparticles for oral administration of gastric cancer therapy. Biomacromolecules 16, 3021–3032.

Liu, A.C., Lou, H.X., Zhao, L.X., Fan, P.H., 2006. Validated LC/MS/MS assay for curcumin and tetrahydrocurcumin in rat plasma and application to pharmacokinetic study of phospholipid complex of curcumin. J. Pharmaceut. Biomed. 40, 720–727.

Liu, G.Q., Zhou, H.F., Zhang, J., Yan, Z.M., Duan, M.X., Long, Y.F., Xia, Q., 2014. Preparation and characterisation of a soy lecithin-based self-microemulsifying drug delivery system of resveratrol. Micro. Nano. Lett. 9, 561–565.

Liu, T., Li, X., Qian, Y., Hu, X., Liu, S., 2012. Multifunctional pH-disintegrable micellar nanoparticles of asymmetrically functionalized β-cyclodextrin-based star copolymer covalently conjugated with doxorubicin and DOTA-Gd moieties. Biomaterials 33, 2521–2531.

Lu, X., Ji, C., Xu, H., Li, X., Ding, H., Ye, M., Zhu, Z., Ding, D., Jiang, X., Ding, X., Guo, X., 2009. Resveratrol-loaded polymeric micelles protect cells from Abeta-induced oxidative stress. Int. J. Pharm. 375, 89–96.

Lu, Y., Park, K., 2013. Polymeric micelles and alternative nanonized delivery vehicles for poorly soluble drugs. Int. J. Pharm. 453 (1), 198–214.

Luo, C.F., Yuan, M., Chen, M.S., Liu, S.M., Zhu, L., Huang, B.Y., Liu, X.W., Xiong, W., 2011. Pharmacokinetics, tissue distribution and relative bioavailability of puerarin solid lipid nanoparticles following oral administration. Int. J. Pharm. 410, 138–144.

Ma, X., Wu, Y., Jin, S., Tian, Y., Zhang, X., Zhao, Y., et al., 2011. Gold nanoparticles induce autophagosome accumulation through size-dependent nanoparticle uptake and lysosome impairment. ACS Nano 5 (11), 8629–8639.

Magadala, P., van Vlerken, L.E., Shahiwala, A., Amiji, M.M., 2008. Multifunctional polymeric nanosystems for tumor-targeted delivery. In: Torchilin, V.P. (Ed.), Multifunctional Pharmaceutical Nanocarriers. Springer Publishing, New York, pp. 33–64.

Mahapatra, A.K., Narasimha Murthy, P., Swadeep, B., Prasad swain, R., 2014. Self-emulsifying drug delivery systems (SEDDS): an update from formulation development to therapeutic strategies. Int. J. Pharm. Tech. Res. 6 (2), 545–568.

Martien, R., Loretz, B., Sandbichler, A.M., Schnürch, A.B., 2008. Thiolated chitosan nanoparticles: transfection study in the Caco-2 differentiated cell culture. Nanotechnology 19 (4), 045101.

Martinho, N., Damgé, C., Pinto Reis, C., 2011. Recent advances in drug delivery systems. J. Biomater. Nanobiotechnol. 2, 510–526.

Maryana, W., Rahma, A., Mudhakir, D., Rachmawati, H., 2015. Phytosome containing silymarin for oral administration: formulation and physical evaluation. JBBBE 25, 54–65.

Mattheolabakis, G., Rigas, B., Constantinides, P.P., 2012. Nanodelivery strategies in cancer chemotherapy: biological rationale and pharmaceutical perspectives. Nanomedicine 7, 1577–1590.

Mazzarino, L., Loch-Neckel, G., Bubniak, L.D., Mazzucco, S., Santos-Silva, M.C., Borsali, R., Lemos-Senna, E., 2015. Curcumin-loaded chitosan-coated nanoparticles as a new approach for the local treatment of oral cavity cancer. J. Nanosci. Nanotechnol. 15, 781–791.

Mehnert, W., Mader, K., 2001. Solid lipid nanoparticles: production, characterization and applications. Adv. Drug Deliv. Rev. 47, 165–196.

Mercer, J., Helenius, A., 2009. Virus entry by macropinocytosis. Nat. Cell. Biol. 11 (5), 510–520.

Mignet, N., Seguin, J., Chabot, G.G., 2013. Bioavailability of polyphenol liposomes: a challenge ahead Pharmaceutics 5, 457–471.

Mishra, I., 2011. Dendrimer: a novel drug delivery system. JDDT 1 (2), 70–74.

Mohanraj, V.J., Chen, Y., 2006. Nanoparticles: a review. Trop. J. Pharm. Res. 5 (1), 561–573.

Mohanty, C., Sahoo, S.K., 2010. The in vitro stability and in vivo pharmacokinetics of curcumin prepared as an aqueous nanoparticulate formulation. Biomaterials 31, 6597–6611.

Muller, R.H., Maassen, S., Weyhers, H., Mehnert, W., 1996. Phagocytic uptake and cytotoxicity of solid lipid nanoparticles (SLN) sterically stabilized with poloxamine 908 and poloxamer 407. J. Drug Target. 4, 161–170.

Muller, R.H., Mader, K., Gohla, S., 2000. Solid lipid nanoparticles (SLN) for controlled drug delivery: a review of the state of the art. Eur. J. Pharm. Biopharm. 50, 161–177.

Nair, K.L., Thulasidasan, A.K.T., Deepa, G., Anto, R.J., Kumar, G.S.V., 2012. Purely aqueous PLGA nanoparticulate formulations of curcumin exhibit enhanced anticancer activity with dependence on the combination of the carrier. Int. J. Pharm. 425, 44–52.

Nazemi, A., Gillies, E.R., 2013. Dendritic surface functionalization of nanomaterials: controlling properties and functions for biomedical applications. Braz. J. Pharm. Sci. 49, 15–32.

Neves, A.R., Lucio, M., Martins, S., Lima, J.L.C., Reis, S., 2013. Novel resveratrol nanodelivery systems based on lipid nanoparticles to enhance its oral bioavailability. Int. J. Nanomed. 8, 177–187.

Nguyen, M.H., Yu, H., Kiew, T.Y., Hadinoto, K., 2015. Cost-effective alternative to nano-encapsulation: amorphous curcumin–chitosan nanoparticle complex exhibiting high payload and supersaturation generation. Eur. J. Pharm. Biopharm. 96, 1–10.

Nikam, A.P., Mukesh, Ratnaparkhiand, P., Shilpa, Chaudhari, P., 2014. Nanoparticles: an overview. Int. J. Res. Dev. Pharm. L. Sci. 3 (5), 1121–1127.

Nunes, S., Madureira, R., Campos, D., Sarmento, B., Gomes, A.M., Pintado, M., Reis, F., 2015. Solid lipid nanoparticles as oral delivery systems of phenolic compounds: overcoming pharmacokinetic limitations for nutraceutical applications. Crit. Rev. Food Sci. Nutr., doi: 10.1080/10408398.2015.1031337.

Ofokansi, K., Winter, G., Fricker, G., Coeter, C., 2010. Matrix-loaded biodegradable gelatin nanoparticles as new approach to improve drug loading and delivery. Eur. J. Pharm. Biopharm. 76 (1), 1–9.

Ogunkoya, Y., Nickel, B., Gay, V., Murray, S., 2009. Using quantum dots to visualize clathrin associations. Biotech. Histochem. 84 (3), 109–115.

O'Neill, M.J., Guo, J., Byrne, C., Darcy, R., O' Driscoll, C.M., 2011. Mechanistic studies on the uptake and intracellular trafficking of novel cyclodextrin transfection complexes by intestinal epithelial cells. Int. J. Pharm. 413 (1–2), 174–183.

Orallo, F., 2006. Comparative studies of the antioxidant effects of *cis*- and transresveratrol. Curr. Med. Chem. 13, 87–98.

Othman, S.A., Hussein, M.A., 2015. In vitro antibacterial, antioxidant and hepatoprotective effect of curcumin-zinc oxide nano-particles in combination. Int. J. Pharm. Bio. Sci. 6 (4), 105–118.

Oussoren, C., Zuidema, J., Crommelin, D.J.A., Storm, G., 1997. Lymphatic uptake and biodistribution of liposomes after subcutaneous injection. 2. Influence of liposomal size, lipid composition and lipid dose. BBA Biomembranes 1328, 261–272.

Pal, A., Gupta, S., Jaiswal, A., Dube, A., Vyas, S.P., 2012. Development and evaluation of tripalmitin emulsomes for the treatment of experimental visceral leishmaniasis. J. Liposome Res. 22, 62–71.

Paliwal, R., Paliwal, S.R., Mishra, N., Mehta, A., Vyas, S.P., 2009. Engineered chylomicron mimicking carrier emulsome for lymph targeted oral delivery of methotrexate. Int. J. Pharm. 380, 181–188.

Pandey, K.B., Rizvi, S.I., 2009. Plant polyphenols as dietary antioxidants in human health and disease. Oxid. Med. Cell. Longev. 2, 270–278.

Pandita, D., Kumar, S., Poonia, N., Lather, V., 2014. Solid lipid nanoparticles enhance oral bioavailability of resveratrol, a natural polyphenol. Food Res. Int. 62, 1165–1174.

Panyam, J., Labhasetwar, V., 2003. Dynamics of endocytosis and exocytosis of poly(D,L-lactide-*co*-glycolide) nanoparticles in vascular smooth muscle cells. Pharm. Res. 20 (2), 212–220.

Parveen, R., Baboota, S., Ali, J., Ahuja, A., Vasudev, S.S., Ahmad, S., 2011. Oil based nanocarrier for improved oral delivery of silymarin: in vitro and in vivo studies. Int. J. Pharm. 413, 245–253.

Patel, D.J., Mistri, P.A., Prajapati, J.J., 2012. Treatment of cancer by using nanoparticles as a drug delivery. Int. J. Drug Dev. Res. 4 (1), 14–27.

Ponnuraj, R., Janakiraman, K., Gopalakrishnan, S., Senthilnathan, K., Meganathan, V., Saravanan, P., 2015. Formulation and characterization of epigallocatechin gallate nanoparticles. IAJPR 5 (01), 387–399.

Punfa, W., Yodkeeree, S., Pitchakarn, P., Ampasavate, C., Limtrakul, P., 2012. Enhancement of cellular uptake and cytotoxicity of curcumin-loaded PLGA nanoparticles by conjugation with anti-P-glycoprotein in drug resistance cancer cells. Acta Pharmacol. Sin. 33, 823–831.

Puri, A., Loomis, K., Smith, B., Lee, J.H., Yavlovich, A., Heldman, E., et al., 2009. Lipid-based nanoparticles as pharmaceutical drug carriers: from concepts to clinic. Crit. Rev. Ther. Drug 26, 523–580.

Rabinow, B.E., 2004. Nanosuspensions in drug delivery. Nat. Rev. Drug Discov. 3, 785–796.

Rachmawati, H., Al Shaal, L., Muller, R.H., Keck, C.M., 2013. Development of curcumin nanocrystal: physical aspects. J. Pharm. Sci. 102, 204–214.

Rahman, K., 2007. Studies on free radicals, antioxidants, and co-factors. Clin. Interv. Aging 2, 219–236.

Ramalingam, P., Ko, Y.T., 2015a. Enhanced oral delivery of curcumin from N-trimethyl chitosan surface-modified solid lipid nanoparticles: Pharmacokinetic and brain distribution evaluation. Pharm. Res. 32 (2), 389–402.

Ramalingam, P., Ko, Y.T., 2015b. Improved oral delivery of resveratrol from *N*-trimethyl chitosan-gpalmitic acid surface-modified solid lipid nanoparticles. Colloids Surf. B 139, 52–61.

Ramamurthy, C., Sampath, K.S., Arunkumar, P., Kumar, M.S., Sujatha, V., Premkumar, K., Thirunavukkarasu, C., 2013. Green synthesis and characterization of selenium nanoparticles and its augmented cytotoxicity with doxorubicin on cancer cells. Bioproc. Biosyst. Eng. 36, 1131–1139.

Ranghar, S., Sirohi, P., Verma, P., Agarwal, V., 2014. Nanoparticle-based drug delivery systems: promising approaches against infections. Braz. Arch. Biol. Techn. 57, 209–222.

Rappoport, J.Z., Kemal, S., Benmerah, A., Simon, S.M., 2006. Dynamics of clathrin and adaptor proteins during endocytosis. Am. J. Physiol. Cell. Physiol. 291 (5), C1072–C1081.

Ravichandran, R., 2012. Development of an oral curcumin nanocrystal formulation. J. Nanotechnol. Eng. Med. 3 (4), 041007.

Ray, L., Kumar, P., Gupta, K.C., 2013. The activity against Ehrlich's ascites tumors of doxorubicin contained in self-assembled, cell receptor targeted nanoparticle with simultaneous oral delivery of the green tea polyphenol epigallocatechin-3-gallate. Biomaterials 34, 3064–3076.

Reddy, N.J., Nagoor Vali, D., Rani, M., Rani, S.S., 2014. Evaluation of antioxidant, antibacterial and cytotoxic effects of green synthesized silver nanoparticles by Piper longum fruit. Mater. Sci. Eng. C 34, 115–122.

Reilly, M.J., Larsen, J.D., Sullivan, M.O., 2012. Polyplexes Traffic through caveolae to the Golgi and endoplasmic reticulum en route to the nucleus. Mol. Pharm. 9 (5), 1280–1290.

Rio, D.D., Rodriguez-Mateos, A., Spencer, J.P., Tognolini, M., Borges, G., Crozier, A., 2013. Dietary (poly) phenolics in human health: structures, bioavailability, and evidence of protective effects against chronic diseases. Antioxid. Redox. Signal. 18, 1818–1892.

Roger, E., Lagarce, F., Garcion, E., Benoit, J.P., 2010. Biopharmaceutical parameters to consider in order to alter the fate of nanocarriers after oral delivery. Nanomedicine 5 (2), 287–306.

Sadeghi, A.M.M., Dorkoosh, F.A., Avadi, M.R., Weinhold, M., Bayat, A., Delie, F., et al., 2008. Permeation enhancer effect of chitosan and chitosan derivatives: comparison of formulations as soluble polymers and nanoparticulate systems on insulin absorption in Caco-2 cells. Eur. J. Pharm. Biopharm. 70 (1), 270–278.

Sadekar, S., Ghandehari, H., 2012. Transepithelial transport and toxicity of PAMAM dendrimers: implications for oral drug delivery. Adv. Drug Deliv. Rev. 64, 571–588.

Safer, A.M.A., Hanafy, N.A., Bharali, D.J., Cui, H.D., Mousa, S.A., 2015. Effect of green tea extract encapsulated into chitosan nanoparticles on hepatic fibrosis collagen fibers assessed by atomic force microscopy in rat hepatic fibrosis model. J. Nanosci. Nanotechnol. 15, 6452–6459.

Sagadevan, S., Periasamy, M., 2014. A review on role of nanostructures in drug delivery system. Rev. Adv. Mater. Sci. 36, 112–117.

Sahay, G., Alakhova, D.Y., Kabanov, A.V., 2010. Endocytosis of nanomedicines. J. Control. Release 145 (3), 182–195.

Sanna, V., Roggio, A.M., Siliani, S., Piccinini, M., Marceddu, S., Mariani, A., et al., 2012. Development of novel cationic chitosan-and anionic alginate-coated poly(D,L-lactide-*co*-glycolide) nanoparticles for controlled release and light protection of resveratrol. Int. J. Nanomedicine 7, 5501–5516.

Sarpal, K., Pawar, Y.B., Nansal, A.K., 2015. Self-emulsifying drug delivery systems: a strategy to improve oral bioavailability. CRIPS 11, 42–49.

Sawant, V.J., Kupwade, R.V., 2015. Functionalization of TiO_2 nanoparticles and curcumin loading for enhancement of biological activity. Der. Pharmacia. Lett. 7 (6), 37–44.

Schnitzer, J.E., Oh, P., Pinney, E., Allard, J., 1994. Filipin-sensitive caveolae-mediated transport in endothelium: reduced transcytosis, scavenger endocytosis, and capillary permeability of select macromolecules. J. Cell. Biol. 127 (5), 1217–1232.

Schulze-Osthoff, K., Bauer, M.K.A., Vogt, M., Wesselborg, S., 1997. Oxidative stress and signal transduction. Int. J. Vitam. Nutr. Res. 67, 336–342.

Seljak, K.B., Berginc, K., Trontelj, J., Zvonar, A., Kristl, A., Gasperlin, M., 2014. A self-microemulsifying drug delivery system to overcome intestinal resveratrol toxicity and presystemic metabolism. J. Pharm. Sci. 103, 3491–3500.

Shafiq-un-Nabi, S., Shakeel, F., Baboota, S., Ali, J., Ahuja, A., Khar, R.K., Ali, M., 2007. Development and optimization using nanoemulsion technique: a technical note. AAPS PharmSciTech 8, 12–17.

Shaikh, J., Ankola, D.D., Beniwal, V., Singh, D., Kumar, M.N.V.R., 2009. Nanoparticle encapsulation improves oral bioavailability of curcumin by at least 9-fold when compared to curcumin administered with piperine as absorption enhancer. Eur. J. Pharm. Sci. 37, 223–230.

Shao, J.F., Li, X.L., Lu, X.W., Jiang, C., Hu, Y., Li, Q.P., et al., 2009. Enhanced growth inhibition effect of resveratrol incorporated into biodegradable nanoparticles against glioma cells is mediated by the induction of intracellular reactive oxygen species levels. Colloids Surf. B 72, 40–47.

Shelat, P., Mandowara, V.K., Gupta, D.G., Shetul, V.P., 2015. Formulation of curcuminoid loaded solid lipid nanoparticles in order to improve oral bioavailability. Int. J. Pharm. Pharm. Sci. 7 (6), 278–282.

Shishu, Maheshwari, M., 2009. Dendrimers: the novel pharmaceutical drug carriers. Int. J Pharm. Sci. Nanotech. 2 (2), 493–502.

Shoba, G., Joy, D., Joseph, T., Majeed, M., Rajendran, R., Srinivas, P.S., 1998. Influence of piperine on the pharmacokinetics of curcumin in animals and human volunteers. Planta. Med. 64, 353–356.

Shobha, R.I., Andallu, B., 2013. Oxidative Stress and cancer: role of anticarcinogenic herbs and spices. AJPCT 1 (3), 351–369.

Shukla, R., Chanda, N., Zambre, A., Upendran, A., Katti, K., Kulkarni, R.R., et al., 2012. Laminin receptor specific therapeutic gold nanoparticles ((AuNP)-Au-198-EGCg) show efficacy in treating prostate cancer. P. Natl. Acad. Sci. USA 109, 12426–12431.

Shutava, T.G., Balkundi, S.S., Vangala, P., Steffan, J.J., Bigelow, R.L., Cardelli, J.A., O'Neal, D.P., Lvov, Y.M., 2009. Layer-by-layer-coated gelatin nanoparticles as a vehicle for delivery of natural polyphenols. ACS Nano 3, 1877–1885.

Siddiqui, I.A., Adhami, V.M., Bharali, D.J., Hafeez, B.B., Asim, M., et al., 2009. Introducing nanochemoprevention as a novel approach for cancer control: proof of principle with green tea polyphenol epigallocatechin-3-gallate. Cancer Res. 69, 1712–1716.

Singh, G., Pai, R.S., 2014. Optimized PLGA nanoparticle platform for orally dosed trans-resveratrol with enhanced bioavailability potential. Expert. Opin. Drug Deliv. 11, 647–659.

Singh, S., Pandey, V.K., Tewari, R.P., Agarwal, V., 2011. Nanoparticle based drug delivery system: advantages and applications. Indian J. Sci. Technol. 4 (3), 177–180.

Solans, C., Izquierdo, P., Nolla, J., Azemar, N., Garcia-Celma, M.J., 2005. Nano-emulsions. Curr. Opin. Colloid. Interf. Sci. 10, 102–110.

Sonaje, K., Chuang, E.Y., Lin, K.J., Yen, T.C., Su, F.Y., Tseng, M.T., et al., 2012. Opening of epithelial tight junctions and enhancement of paracellular permeation by chitosan: microscopic, ultrastructural, and computed-tomographic observations. Mol. Pharm. 9 (5), 1271–1279.

Sreelakshmi, C., Goel, N., Datta, K.K.R., Addlagatta, A., Ummanni, R., Reddy, B.V.S., 2013. Green synthesis of curcumin capped gold nanoparticles and evaluation of their cytotoxicity. Nanosci. Nanotech. Lett. 5, 1258–1265.

Sutradhar, K.B., Lutful Amin, Md., 2014. Nanotechnology in cancer drug delivery and selective targeting. ISRN Nanotechnology 2014, 12 pp..

Sutton, D., Nasongkla, N., Blanco, E., Gao, J.M., 2007. Functionalized micellar systems for cancer targeted drug delivery. Pharm. Res. 24, 1029–1046.

Tan, S.W., Yeh, M.K., Hsieh, D.S., Wu C.J., 2014. Physical nano-complexes for preventing and treating cancer and method for manufacturing the same. US Patent 8,759,546. Google Patents.

Teeranachaideekul, V., Muller, R.H., Junyaprasert, V.B., 2007. Encapsulation of ascorbyl palmitate in nanostructured lipid carriers (NLC): effects of formulation parameters on physicochemical stability. Int. J. Pharm. 340, 198–206.

Traub, L.M., 2003. Sorting it out: AP-2 and alternate clathrin adaptors in endocytic cargo selection. J. Cell. Biol. 163 (2), 203–208.

Tsai, Y.M., Chien, C.F., Lin, L.C., Tsai, T.H., 2011. Curcumin and its nano-formulation: the kinetics of tissue distribution and blood-brain barrier penetration. Int. J. Pharm. 416, 331–338.

Tseng, C.L., Wang, T.W., Dong, G.C., Yueh-Hsiu Wu, S., Young, T.H., Shieh, M.J., et al., 2007. Development of gelatin nanoparticles with biotinylated EGF conjugation for lung cancer targeting. Biomaterials 28 (27), 3996–4005.

Tunesi, M., Prina, E., Munarin, F., Rodilossi, S., Albani, D., Petrini, P., Giordano, C., 2015. Cross-linked poly(acrylic acids) microgels and agarose as semi-interpenetrating networks for resveratrol release. J. Mater. Sci. Mater. M 26 (1), 5328.

Ucisik, M.H., Sleytr, U.B., Schuster, B., 2015. Emulsomes meet S-layer proteins: an emerging targeted drug delivery system. Curr. Pharm. Biotechnol. 16, 392–405.

Uner, M., 2006. Preparation, characterization and physico-chemical properties of solid lipid nanoparticles (SLN) and nanostructured lipid carriers (NLC): their benefits as colloidal drug carrier systems. Pharmazie 61, 375–386.

Upadhyay, S., Dixit, M., 2015. Role of polyphenols and other phytochemicals on molecular signaling. Oxid. Med. Cell. Longev. 2015, 504253.

Varaprasad, K., Vimala, K., Ravindra, S., Reddy, N.N., Reddy, G.V.S., Raju, K.M., 2011. Fabrication of silver nanocomposite films impregnated with curcumin for superior antibacterial applications. J. Mater. Sci. Mater. Med. 22, 1863–1872.

Vishvakrama, P., Sharma, S., 2012. Liposomes: an overview. IJRPBS 3 (3), 1074–1084.

Vllasaliu, D., Exposito-Harris, R., Heras, A., Casettari, L., Garnett, M., Illum, L., et al., 2010. Tight junction modulation by chitosan nanoparticles: comparison with chitosan solution. Int. J Pharm. 400 (1-2), 183–193.

Vyas, S.P., Subhedar, R., Jain, S., 2006. Development and characterization of emulsomes for sustained and targeted delivery of an antiviral agent to liver. J. Pharm. Pharmacol. 58, 321–326.

Wahlstrom, B., Blennow, G., 1978. A study on the fate of curcumin in the rat. Acta Pharmacol. Toxicol. 43, 86–92.

Wang, J., Byrne, J.D., Napier, M.E., DeSimone, J.M., 2011. More effective nanomedicines through particle design. Small 7 (14), 1919–1931.

Wang, J.L., Ma, W.Z., Tu, P.F., 2015. The mechanism of self-assembled mixed micelles in improving curcumin oral absorption: In vitro and in vivo. Colloids Surf. B 133, 108–119.

Wang, S., Su, R., Nie, S.F., Sun, M., Zhang, J., Wu, D.Y., Moustaid-Moussa, N., 2014. Application of nanotechnology in improving bioavailability and bioactivity of diet-derived phytochemicals. J. Nutr. Biochem. 25, 363–376.

Wang, W.R., Zhu, R.R., Xie, Q., Li, A., Xiao, Y., Li, K., Liu, H., Cui, D.X., Chen, Y.H., Wang, S.L., 2012. Enhanced bioavailability and efficiency of curcumin for the treatment of asthma by its formulation in solid lipid nanoparticles. Int. J. Nanomed. 7, 3667–3677.

Warden, B.A., Smith, L.S., Beecher, G.R., Balentine, D.A., Clevidence, B.A., 2001. Catechins are bioavailable in men and women drinking black tea throughout the day. J. Nutr. 131, 1731–1737.

Weerapol, Y., Limmatvapirat, S., Nunthanid, J., Sriamornsak, P., 2014. Self-nanoemulsifying drug delivery system of nifedipine: impact of hydrophilic-lipophilic balance and molecular structure of mixed surfactants. AAPS Pharmscitech. 15, 456–464.

Weissig, V., Whiteman, K.R., Torchilin, V.P., 1998. Accumulation of protein-loaded long-circulating micelles and liposomes in subcutaneous Lewis lung carcinoma in mice. Pharmaceut. Res. 15, 1552–1556.

Wu, C.F., Yang, J.Y., Wang, F., Wang, X.X., 2013. Resveratrol: botanical origin, pharmacological activity and applications. Chin. J. Nat. Med. 11, 1–15.

Xie, X., Tao, Q., Zou, Y., Zhang, F., Guo, M., Wang, Y., Wang, H., Zhou, Q., Yu, S., 2011. PLGA nanoparticles improve the oral bioavailability of curcumin in rats: characterizations and mechanisms. J. Agric. Food Chem. 59, 9280–9289.

Xu, Z.P., Niebert, M., Porazik, K., Walker, T.L., Cooper, H.M., Middelberg, A.P.J., et al., 2008. Subcellular compartment targeting of layered double hydroxide nano-particles. J. Control. Release 130 (1), 86–94.

Yallapu, M.M., Gupta, B.K., Jaggi, M., Chauhan, S.C., 2010. Fabrication of curcumin encapsulated PLGA nanoparticles for improved therapeutic effects in metastatic cancer cells. J. Colloids Interf. Sci. 351, 19–29.

Yang, G., Zhao, Y.P., Zhang, Y.T., Dang, B.L., Liu, Y., Feng, N.P., 2015. Enhanced oral bioavailability of silymarin using liposomes containing a bile salt: preparation by supercritical fluid technology and evaluation in vitro and in vivo. Int. J. Nanomed. 10, 6633–6644.

Yeh, T.H., Hsu, L.W., Tseng, M.T., Lee, P.L., Sonjae, K., Ho, Y.C., et al., 2011. Mechanism and consequence of chitosan-mediated reversible epithelial tight junction opening. Biomaterials 32 (26), 6164–6173.

Yekollu, S.K., Thomas, R., O'Sullivan, B., 2011. Targeting curcusomes to inflammatory dendritic cells inhibits NF-kappaB and improves insulin resistance in obese mice. Diabetes 60, 2928–2938.

Yu, H.L., Huang, Q.R., 2012. Improving the oral bioavailability of curcumin using novel organogel-based nanoemulsions. J. Agr. Food Chem. 60, 5373–5379.

Zha, L.Y., Xu, Z.R., Wang, M.Q., Gu, L.Y., 2008. Chromium nanoparticle exhibits higher absorption efficiency than chromium picolinate and chromium chloride in Caco-2 cell monolayers. J. Anim. Physiol. Anim. Nutr. 92 (2), 131–140.

Zhang, J.N., Zhou, X.B., Yu, Q.Q., Yang, L.C., Sun, D.D., Zhou, Y.H., Liu, J., 2014. Epigallocatechin-3-gallate (EGCG)-stabilized selenium nanoparticles coated with Tet-1 peptide to reduce amyloid-beta aggregation and cytotoxicity. ACS Appl. Mater. Inter. 6, 8475–8487.

Zhou, Y.F., Li, J.M., Lu, F., Deng, J.J., Zhang, J.H., Fang, P.J., Peng, X.S., Zhou, S.F., 2015. A study on the hemocompatibility of dendronized chitosan derivatives in red blood cells. Drug Des. Dev. Ther. 9, 2635–2645.

Zu, Y., Zhang, Y., Wang, W., Zhao, X., Han, X., Wang, K., Ge, Y., 2015. Preparation and in vitro/in vivo evaluation of resveratrol-loaded carboxymethyl chitosan nanoparticles. Drug. Deliv. 23 (3), 1–11.

FURTHER READING

Baur, J.A., Pearson, K.J., Price, N.L., Jamieson, H.A., Lerin, C., Kalra, A., et al., 2006. Resveratrol improves health and survival of mice on a high-calorie diet. Nature 444, 337–342.

Penalva, R., Esparza, I., Larraneta, E., Gonzalez-Navarro, C.J., Gamazo, C., Irache, J.M., 2015. Zein-based nanoparticles improve the oral bioavailability of resveratrol and its anti-inflammatory effects in a mouse model of endotoxic shock. J. Agric. Food Chem. 63, 5603–5611.

Rejman, J., Oberle, V., Zuhorn, I., Hoekstra, D., 2004. Size-dependent internalization of particles via the pathways of clathrin- and caveolae-mediated endocytosis. Biochem. J. 377, 159–169.

Sanad, R.A., AbdelMalak, N.S., Elbayoomy, T.S., Badawi, A.A., 2010. Formulation of a novel oxybenzone-loaded nanostructured lipid carriers (NLCs). AAPS PharmSciTech 11, 1684–1694.

Siddiqui, I.A., Adhami, V.M., Ahmad, N., Mukhtar, H., 2010. Nanochemoprevention: sustained release of bioactive food components for cancer prevention. Nutr. Cancer 62, 883–890.

CHAPTER

22

DESIGN AND DEVELOPMENT OF PHARMACEUTICAL MICROPROCESSES IN THE PRODUCTION OF NANOMEDICINE

Norma A. Noguez Méndez*, Carlos T. Quirino Barreda*, Abraham F. Vega*, Jorge E. Miranda Calderon*, César G. Urioste*, Xochitl C. Palomec*, Alejandro R. Martínez, Máximo P. Díaz****

**Laboratory of Molecular Pharmacy and Controlled Release, México D.F., Mexico*

***Instituto Tecnológico de Querétaro, Santiago de Querétaro, Mexico*

CHAPTER OUTLINE

1 Microfluidics: Origins, Present, and Future 670
- 1.1 Physical Foundations of Microfluidics 672
- 1.2 Applications of Microfluidics in Pharmaceutical Processes and Nanomedicine 673

2 Lyotropic Liquid Crystals as Drug Delivery Systems 676
- 2.1 Introduction 676
- 2.2 Liquid Crystals (LCs) 676
- 2.3 Thermotropic Liquid Crystals (TLCs) 676
- 2.4 Lyotropic Liquid Crystals (LLCs) 677

3 Lipid-Carrying Systems 679
- 3.1 Introduction 679
- 3.2 Solid Lipid Nanoparticles 680

4 Microspheres 683
- 4.1 Introduction 683
- 4.2 Criteria for the Design and Development of Microspheres 683
- 4.3 Physicochemical Characteristics of Chitosan 684
- 4.4 Microsphere Preparation Methods 685
- 4.5 Application of Microspheres in the Transport and Vectorization of Pharmaceutical Compounds 686

5 Polymerosomes 687
- 5.1 Introduction 687
- 5.2 Requirements for the Production of Polymerosomes 687
- 5.3 Synthesis of a Polymerosome from Hydrophilic-Hydrophobic Copolymers 688
- 5.4 Polymerosome Applications 689

6 Conclusions 690

References 691

Nanostructures for Oral Medicine

1 MICROFLUIDICS: ORIGINS, PRESENT, AND FUTURE

Microfluidics is a recent research area, and its main characteristics are the manipulation of fluids (gases and liquids) in conduits (from 10^{-9} to 10^{-8} L) with dimensions of tenths of microns, as well as the multidisciplinary complexity of the physical principles and of the fabrication of the devices used (Fernández Rivas, 2011).

The impact of microfluidics on technological development is such that a new branch of fluid dynamics has emerged, possessing great potential in very diverse fields, such as chemical synthesis, biological analysis, optics, and information technology. Recently, microfluidics has also greatly influenced the pharmaceutical production processes. Research in many of these fields, however, is still in the initial phases.

A lecture delivered by Feynman in 1959 (Feynman, 2002) was the starting point regarding micro- and nanoscales. Then, in 1983, Feynman anticipated many of the devices used in microelectromechanical systems (MEMS) (Trimmer, 1997).

Similar to basic sciences, in which the demonstration of concepts is important, importance in technology resides in the development and potentiation of new applications (Whitesides, 2006).

The first application of microfluidics technology originated in chemical analysis, where this technology has been demonstrated to have several advantages, for example, the use of small amounts of sample with which detection and separation processes can be performed with high resolution, sensitivity, low cost, and short analysis times in very small analytic devices (Squires and Quake, 2005; Stone et al., 2004).

Microfluids exploit the two most important characteristics of this type of system. The most obvious characteristic is the small size; the less obvious characteristic is the flow inside small conducts with very low Reynolds numbers, which allows work to be performed under laminar flow (McClain, 2003). These characteristics allow better control of concentration in time and space (Cho, 2003; Lee et al., 2003; Glasgow, 2001; Jensen, 2006; Lucchetta et al., 2005; Walters et al., 2004).

This technology seems very promising, offering several advantages and few disadvantages (at least for the main applications analyzed). However, microfluidics technology is not yet been widely used because many of the high-impact potential applications are limited to the laboratories' levels of familiarity with this technology.

Currently, the prospects for microfluidics are good, particularly in the analysis and synthesis of new molecules, providing the time and conditions needed for microfluidics to develop to its full potential and generate new technologies in the long term, as long as the field of action does not restrict itself to those technologies that have shown the highest potential.

It can be considered that microfluidics originated from the following well-defined branches of science and technology:

1. molecular analysis,
2. biodefense,
3. molecular biology, and
4. microelectronics.

Chronologically, the use of microfluidics originated in molecular analysis (Ng et al., 2002) due to the need for more efficient analytical methods, such as gas chromatography (GPC), high-performance liquid chromatography (HPLC), and capillary electrophoresis (CE), all of which are based on the use of microfluidics, generating a true revolution in the chemical analysis field. These methods, combined with high-potency lasers, have managed to increase the sensitivity and resolution of optical detection instruments when using small amounts of samples at low molecular concentrations.

Due to the success obtained with microanalysis methods, the development of new technologies has followed this line, with new, more compact and versatile processes being developed simultaneously with new applications in emerging fields of chemistry and biochemistry.

The second branch of microfluidics originated from a totally different motivation. At the end of the Cold War, countries that possessed chemical and biological weapons had an advantage in the arms race. However, the possibility that terrorist groups would obtain these weapons made them a threat to all of humanity, not only to the countries involved. This risk still exists; therefore there is a need to consider a less-positive outlook for the countries that developed most of these weapons.

In this context, the Defense Advanced Research Projects Agency (DARPA) of the United States Department of Defense financed a series of projects in the 1990s aiming to develop efficient and compact detection systems using microfluidic devices. This program helped promote the rapid growth of microfluidics in the research and development of this technology, as well as in the increase in the number of academic groups focused on this activity.

The third influence on the development of microfluidics originated in the field of molecular biology, beginning with the explosion in genomics research in the 1980s. This use of microfluidics was followed by its use in other microanalysis areas related to molecular biology, such as the sequencing of deoxyribonucleic acid (DNA), which required analytical methods with high performance, resolution, and sensitivity that had been previously validated by solving problems in biology. Researchers found that microfluidics successfully fulfilled all of these requirements.

The last contribution came from microelectronics. The original idea was to develop microfluidics from soft photolithography associated with the technology used to print microcircuits in silicon wafers, also known as silicon microelectronics, and from MEMS, which allowed the development of more-complex systems.

In the beginning of the development of microfluidics, devices were manufactured from silicon and glass. Although these materials are still used, they have been replaced by systems manufactured from polymers.

These systems are mostly used in the analysis of biological samples in water, where glass- or silicon-based devices are unnecessary or inappropriate. Silicon is usually more expensive and is opaque to visible and UV light, preventing the use of common optical detection methods. Moreover, it is easier to manufacture the materials required for microanalysis systems, such as pumps and valves, from elastomers than from rigid materials. Another important aspect is that glass and silicon do not possess the properties (especially gas permeability) required to work with live cells from animals.

Most research in microfluidic systems has been based on the science of polymers because the properties of microfluidic systems are completely different from those of silicon (Mijatovic et al., 2005; Whitesides and Stroock, 2001). Polydimethylsiloxane (PDMS) was a significant material in the development of these devices because it is an optically transparent elastomer, is permeable to oxygen and has demonstrated great adaptability for the development of new technologies in early stages. However, devices made from glass, silicon, and metal have reemerged as indispensable materials in the development of microprocesses where chemical and thermal stability is required.

1.1 PHYSICAL FOUNDATIONS OF MICROFLUIDICS

The study of microfluids is a multidisciplinary field including physics, chemistry, engineering, and biotechnology; it involves the study of the behavior of fluids at the micro- and mesoscales. This field also includes the design of systems in which small amounts of fluids will be used. The behavior of fluids at the microscale greatly differs from what is observed at the macroscale: the surface tension and energy dissipation are completely different. In microchannels with a diameter of 10–500 nm, the Reynolds number is extremely low. Therefore, the flow is always laminar and no turbulence occurs; only diffusion influences the mixture of the fluids (Liu et al., 2000).

Most microfluidic devices function at low Reynolds numbers, but the flow is still not linear due to interactions at the interface of the fluids. The important effects observed are (1) the limit is not static and (2) the movement of one fluid can slow that of the other fluid. The resulting instability, which promotes the formation of droplets, creates a struggle between the surface tension and the shear force (Atencia and Beebe, 2005; Kim et al., 2014; Thorsen et al., 2001).

The ratio of the viscosity forces to the capillarity is given by the capillarity number (Ca), which represents the relative effects of the viscosity (viscous forces) and the surface tension. Ca acts through an interface between a liquid and a gas, or between two immiscible liquids (Tice et al., 2004).

$$Ca = \frac{\mu v}{\gamma}$$

Where μ is the liquid's dynamic viscosity, v is the characteristic velocity, and γ is the surface tension between both phases.

The flow of a fluid is classified into two flow regimes: laminar and turbulent. Laminar flow is characterized by the smooth and constant movement of the fluid, while turbulent flow is characterized by vortices and flow fluctuations.

Physically, these regimes differ in terms of the relative importance of the viscosity and inertia forces. The relative importance of these two types of forces for a given flow condition is measured by the Reynolds number (Re), which is calculated as follows:

$$\mathrm{Re} = \frac{\text{inertial forces}}{\text{viscous forces}} = \frac{\rho v_s D}{\mu}$$

Where ρ = the density of the fluid, v_s = the characteristic velocity of the fluid, D = the diameter of the conduct through which the fluid passes or characteristic length, and μ = the dynamic viscosity of the fluid.

At low Reynolds numbers, the viscosity effects dominate over the inertial effects and a completely laminar flow is produced. In a laminar flow system under constant environmental conditions, the currents of the fluid flow in parallel and the velocity at any point inside the current does not change over time. This causes the transference of mass by convection only in the direction of the flow, and a mixture can be achieved by molecular diffusion only. On the contrary, at high Re values, the flow is dominated by inertial forces and is characterized by a turbulent flow. This flow is dominated by inertial forces, the fluid's movement is random, and convective mass transport occurs in all directions.

Among the defined regimes of laminar and turbulent flow, a transition interval occurs at certain Reynolds numbers. The exact values of this interval are a function of several parameters, such as the shape of the channel and its internal surface roughness.

In most situations, the numeric value of the transition Re is between 1500 and 2500. The Re of microfluidic systems is lower than 100, and the flow is considered essentially laminar. This characteristic has direct consequences on the mixture inside the microfluidic device.

1.2 APPLICATIONS OF MICROFLUIDICS IN PHARMACEUTICAL PROCESSES AND NANOMEDICINE

Current manufacturing methods and ranges of components are sufficient for the use of microfluidic systems, such as protein crystallization and bioanalysis. The manipulation of multiphase flows is another possible application of microfluidics, allowing the generation of monodisperse bubbles (gas–liquid mixtures). These dispersions suggest new pathways for the production of polymer vesicles, emulsions, and foams.

In chemical synthesis (especially in organic and medical chemistry), the drops formed by microfluids can serve as microreactors for organic synthesis and kits for analytical determination, such as the detection of disease or the concentration of substances in bodily fluids.

In cellular biology research, microfluidic systems have allowed the manufacturing of microchambers in which cells grow and are observed. Moreover, these microchambers can be used to observe the cytoskeleton or the forces exerted by cells on a substrate, or they can be used to separate moving cells from static ones (e.g., sperm and embryos) (Pihl et al., 2005; Sia and Whitesides, 2003; Wheleer, 2004). Table 22.1 shows examples of these applications.

Table 22.1 Microfluidic in Pharmaceutical Processes and Nanomedicine

Year	Application
2000	New platform with which to study phenomena that depend on gradients in concentration, dynamic phenomena in cell biology (Jeon et al., 2000)
2004	Emulsions with two types of internal droplets were prepared by varying geometry and wettability in microchannels (Okushima et al., 2004)
2005	This article provides an overview in the area of continuous microfluidics-based synthesis of polymer colloids (Seo et al., 2005)
	The present research synthesized biodegradable monodisperse microgels by photopolymerization of monodisperse prepolymer droplets (De Geest et al., 2005)
2006	This paper describes a novel method of manufacturing shape-controlled calcium alginate gel microparticles in a microfluidic device (Liu et al., 2006)
	These microfluidic devices allow one to generate monodisperse spherical and nonspherical droplets by appropriately controlling the flow conditions and channel geometries (Kobayashi et al., 2006)
	This letter presents a simple way to prepare monodisperse o/w and w/o emulsions in T-junction microfluidic device (Xu et al., 2006)
	This paper describes microfluidic assembly of colloid-filled hidrogel granules of varying shape and composition (Shepherd et al., 2006)
2007	A new method to tailor liposome size and size distribution in a microfluidic (Jahn, 2007)

(*Continued*)

Table 22.1 Microfluidic in Pharmaceutical Processes and Nanomedicine (*cont.*)

Year	Application
2008	In this article, the authors discuss and analyze the design of a versatile nozzle used to development droplet pairs, based on hydrodynamic coupling of two spatially separated nozzles (Frenz et al., 2008)
	This article presents the use of controlled microfluidic transport methods to study the experimental parameters for fabricating nanoparticles utilizing phase inversion (Laulicht et al., 2008)
	The fabrication of monodisperse mesoporous silica particles are based on the formation of well-defined equally sized emulsion droplets using a microfluidic approach (Carroll, 2008)
	A novel technique called the "lipid-coated ice droplet hydration method" is used for the preparation of giant vesicles with a controlled size (Sugiura et al., 2008)
	This paper reports a novel microfluidic method for the production of crosslinked alginate microparticles and nanoparticles (Rondeau and Cooper-White, 2008)
2009	In this paper, the authors introduce a technique of topographical features, which is applicable on different substrates by modifying their surface energy (Cortese et al., 2009)
	Emusification in microdevices (microfluidic emulsification) involves micrometer-sized droplets and fast interface expansion rates (Steegmans et al., 2009)
	The microfluidic fabrication of macroporous polymer microspheres via the simultaneous reactions within single droplets, induced by UV irradiation (Gong et al., 2009)
	In this paper, the authors introduce a flexible microfluidic device to characterize the mechanical properties of soft viscoelastic solids, such as bacterial biofilms (Hohne et al., 2009)
2010	This work suggests that microfluidics may facilitate the development and optimization of biomimetic colloidal systems for nanomedice applications (Lo et al., 2010)
	This article presents a microfluidic method to direct the self-assembly of temperature-sensitive liposome-hydrogel hybrid nanoparticles (Hong et al., 2010)
	This article demonstrates rapid microfluidic fabrication of hybrid microparticles composed of functionalized viral nanotemplates directly embedded in polymeric hydrogels (Lewis et al., 2010)
	This paper presents a simple fabrication of photo- and thermoresponsive microparticles with a narrow size distribution in the PDMS-based microfluidic device (Lone et al., 2010)
2011	The authors present the relationship between oxygen transport and shear in a membrane bilayer microfluidic device, in which soluble factors are delivered to a cell population by means of flow through a proximated channel separated from the culture channel by membrane (Inamdar et al., 2011)
	The authors show that the production and the geometrical shape of complex polymersomes can be predicted by varying the flow rates of a simple microdevice using an empirical law that predicts the droplet size (Perro et al., 2011)
	Droplet microfluidics performed in poly(methyl methacrylate) PMMA microfluidic devices resulted in significant wall wetting by water droplets formed in a liquid–liquid segmented flow when using a hydrophobic carrier fluid, such as perfluorotripropylamine (Subramanian et al., 2011)
	This work describes a novel, simple microfluidic method for fabricating titania–silica core–shell microspheres (Lan et al., 2011)

Table 22.1 Microfluidic in Pharmaceutical Processes and Nanomedicine *(cont.)*

Year	Application
2012	The authors investigate the influence of microfluidic experimental parameters on the sizes and shell thicknesses of double emulsion templated microcapsules (Chen et al., 2012)
	Microfluidic approaches for controlled generation of colloidal clusters have been used for the synthesis of functional material including drug delivery carriers (Wan et al., 2012)
2013	A new method is described that combines a microfluidic device for the controlled formation of liposomes with instantaneous immobilization by means of ultrarapid cooling (Jahn et al., 2013)
	The authors report the microfluidic production of spherical and nonspherical fat particles from crystallizable oils (Kim and Vanapalli, 2013)
	This article describes a versatile and facile route to the continuous production of monodisperse polylactide (PLA) microcapsules with controllable structures (Watanabe et al., 2013)
	The author introduces a facile approach for the production of gas-filled microcapsules designed to withstand high pressures (Abbaspourrad, 2013)
	Three major nanofabrication strategies, including nanolithography, microelectromechanical system–based techniques, and methods using various nanomaterials (Duan et al., 2013)
	This paper describes the microfluidic assembly of soft dimer capsules by the fusion of individual capsules with distinct properties (Lu et al., 2013)
	The authors investigated the assembly of colloidal silica crystals inside double emulsion drops generated in microcapillary microfluidic devices (Shirk et al., 2013)
2014	This paper reports the controlled formation of internally porous polyelectrolyte particles with diameters ranging from tens to hundreds of micrometers through selective solvent extraction using microfluídos (Watanabe et al., 2014)
	Isotropic microparticles prepared from a suspension that undergoes polymerization have long been used for a variety of applications (Kim et al., 2014)
	This paper proposes new methods to encapsulate amines using double-emulsion templates made in a microfluidic device (Chen et al., 2014a,b)
2015	A new microfluidic process was used to generate unique micrometer-sized hierarchical lipid particles having spherical lipid core and multilamellar shell structures (Mizuno et al., 2015)
	Semipermeable microcapsules are appealing for controlled release of drugs, study of cell-to-cell communication, and isolation of enzymes or artificial catalysts (Kim et al., 2015a,b)
	Thermotropic liquid crystals possess superior foaming capability without the aid of surfactants because of the anisotropic molecular structures (Shi and Yokoyama, 2015)
	Polymeric multicompartmental microparticles have significant potential in many applications due to the capability to hold various functions in discrete domain within a single particle (Choi et al., 2015)

2 LYOTROPIC LIQUID CRYSTALS AS DRUG DELIVERY SYSTEMS

2.1 INTRODUCTION

During the first 15 years of the 21st century, several innovative systems applying micro- and nanoparticles in pharmaceutical technology were designed and commercialized, such as those described in the next sections. These nanoparticles have dimensions from 1 to 1000 nm, and they consist of simple particles or complex structures that usually exhibit colloidal behavior and are used as adjuvants or transporters for one or more drugs that are dissolved, dispersed, encapsulated, absorbed, and/or ligated. Regarding their physical or molecular aggregation state, these nanostructured systems can be in a solid, liquid, semisolid, or mesophasic state, as in the case of liquid crystals and their dispersions. Liquid crystals show a non-Newtonian flow, whereas the flow of their dispersions is Newtonian.

The goal of pharmaceutical nanotechnology is to provide, using nanostructured systems and topical and transdermal administration methods, higher safety, and therapy efficiency compared to conventional means of drug administration.

The physicochemical properties of the active compound and the excipient matrix, as well as their interactions, are important for the release of the active substance and its passive diffusion from the vehicle or topical composition containing it, as well as from the skin's surface and mucosa. This allows the drug to overcome the barrier properties conferred by the molecular and cellular structures of the corneal strata and other epidermal layers, as well as the intercellular lipid matrix, also considering the system's fluidity.

2.2 LIQUID CRYSTALS (LCs)

Liquid crystals (LCs) are mesophases or states of matter that are between an isotropic liquid and a crystalline solid; hence, they combine the properties of both states. Specifically, molecules in LCs tend to point in a certain direction while possessing translational freedom (position). Molecules that can form mesophases are called mesogens. Depending on their molecular form, rod-shaped mesogens form calamitic mesophases, and disk-shaped mesogens form discotic mesophases (Mueller-Goymann, 2002; Patel and Patel, 2010; Tadwee et al., 2012).

LCs are mainly classified according to the physicochemical parameters responsible for the phase transition, such as thermotropic (TLCs) orlyotropic LCs (LLCs). Thus, from their crystalline state, the formation of a mesophase begins when the temperature increases or when a solvent is added, which corresponds to the difference between TLCs and LLCs, respectively. Therefore, LLCs are formed by mesogens that are not molecules per se but are their hydrates or solvates, as well as by associations with hydrated or solvated molecules. On the other hand, with both LLCs and TLCs, a variation in temperature can also cause a phase transition between different mesophases (Mueller-Goymann, 2002; Pasquali et al., 2006; Patel and Patel, 2010).

Although they are better known for applications in television screens, LCs are also a vital part of all forms of life, such as essential organic substances (e.g., DNA), lipids in cell membranes and proteins (Tadwee et al., 2012).

2.3 THERMOTROPIC LIQUID CRYSTALS (TLCs)

As mentioned earlier, the transition phase of a TLC depends on the temperature, with two main types of TLC: smectic and nematic. As the temperature increases, the first LC phase is smectic, which involves a laminar arrangement and the translational and rotational movement of the molecules.

The increase in temperature leads to the nematic phase, where the molecules diffuse toward the exterior of the initial mesh-arranged structure and the laminar-type arrangement. At higher temperatures, the material transforms into an isotropic liquid, and the movement of the molecules changes again (Patel and Patel, 2010).

Calamitic mesophases (formed by rod-shaped mesogens) exhibit a parallel orientation of their molecules, which are also arranged in laminae; these are called smectic LCs. The plane of the lamina can either be oriented perpendicularly or inclined with respect to the longitudinal axis of the molecule.

Discotic LCs (formed by disk-shaped mesogens) originate as nematic or cholesteric mesophases. Their structural characteristics are similar to those of their respective calamitic mesophases. Instead of smectic mesophases, columnar discotic LCs originate from the stacking of the disks one on top of each other. The columns of the columnar discotic phase form a mesh that assumes either a hexagonal or rectangular form. Additionally, the columns can be inclined or not. In contrast, cholesteric LCs do not exclusively exhibit a parallel orientation; their orientation can rotate from laminae in either a right- or left-oriented fashion. The distance of the lamina where there is a 360 degree rotation is called the "pitch," which can include several hundred nanometers. Nematic LCs are similar to cholesteric LCs in that they possess an orientation order. However, they are different in that there is no rotation, as previously described (Mueller-Goymann, 2002; Patel and Patel, 2010).

2.4 LYOTROPIC LIQUID CRYSTALS (LLCs)

In LLCs, the phase transition depends on the temperature and the concentration; thus, LLCs are mixtures of amphiphilic molecules and solvents at given temperatures and concentrations. Their mesomorphic properties change with temperature, pressure, and the relative concentrations of the different components of the mixture.

An important characteristic of LLCs is the self-assembly of the amphiphilic molecules as supramolecular structures, which are the basic units of the mesophases. The level of hydration or solvation depends on the amphiphilic properties of the molecule. The hydration and solvation of most rod-shaped molecules produces different geometric figures, such as cones or cylinders. The cylinders are arranged in laminae, resulting in laminar phases with alternating polar and nonpolar layers.

An increase in the concentration of the solvent transforms the solvated rod-shaped molecule into a conic form, leading to a change of phase. Depending on the polar or nonpolar characteristics of the solvating agent and the molecule itself, the transition results in a hexagonal or inverse hexagonal phase. In the hexagonal phase, the amount of water or nonpolar solvent that can be added is limited. As the molecular geometry changes during solvation, another phase transformation occurs toward the cubic (type I) or inverse cubic phase (type IV), which consist of spherical or ellipsoidal micelles, inverse micelles or both.

In addition to the cubical forms described earlier, other forms exist between the laminar and hexagonal (cubic type II) or inverse hexagonal (cubic type III). In contrast to type I and IV discontinuous phases, type II and III cubic mesophases belong to the bicontinuous phases (Hegmann et al. 2007; Kim et al., 2015a; Mueller-Goymann, 2002; Müller-Goymann, 2004; Pasquali et al., 2005, 2006; Patel and Patel, 2010).

2.4.1 Formation and pharmaceutical applications of LLCs

The lyotropic phases frequently found in pharmaceutical and cosmetic formulas are laminar, hexagonal (normal and inverse), and cubic (Amar-Yuli et al., 2007; Pasquali et al., 2006).

Cubic and hexagonal LLCs are usually formed spontaneously due to the addition of certain amphiphilic lipids in aqueous environments. The phases are usually viscous; therefore, they are difficult to manage and their applications are limited. Moreover, the viscous phase can cause irritation when in contact with biological epithelium. To overcome this, an innovative strategy has been used to disperse the original phase in water in the form of small particles (Guo et al., 2010; Lodha et al., 2014; Yaghmur and Glatter, 2009). However, it is very important to ensure that the intact structure of the initial mesophasic system of interest (e.g., hexagonal or cubic) is maintained when diluted (dispersed) in the aqueous medium (Amar-Yuli et al., 2007).

When cubic or hexagonal LLCs are dispersed in excess water and with the addition of stabilizing agents (such as copolymers of polyoxyethylene/polypropylene: poloxamers or Pluronic), they form colloidal dispersions known as cubosomes and hexosomes, respectively. In general, the methods of preparation of cubic and hexagonal LLCs are simpler than those of their dispersions. For example, cubic and hexagonal LLCs can be prepared by mixing amphiphilic lipid and water phases by vortexing. In contrast, the production of cubosomes or hexosomes is more complex.

Two techniques are typically used to prepare cubosomes or hexosomes; these techniques coincide with the techniques that are generally applied in nanotechnology. The first technique is known as a top-down approach or miniaturization in which a mixture of the lipids forming the structure and stabilizing agents is hydrated to allow the self-assembly of a viscous material. Then, the material is dispersed in an aqueous solution by adding high-level energy (homogenization at high pressure and ultrasonication) to form cubosomes or hexosomes.

The second technique for the preparation of cubosomes and hexosomes is known as the bottom-up approach, or amplification. The key factor is the presence of a hydrotrope, which can dissolve lipids that are insoluble in water to create a liquid precursor and prevent the formation of LLCs at high concentrations. The controlled addition of aqueous medium into the previous solution rapidly reduces the lipid's solubility, resulting in the formation of cubosomes or hexosomes. This method requires less energy and has the advantage of producing nanoparticles with smaller particle sizes and better stability (Chen et al., 2014b; Guo et al., 2010; Lodha et al., 2014).

LLCs are considered a new method for drug release due to their unique physicochemical properties, which confer the following advantages: (1) effective solubilization compared with other traditional carriers; (2) high load capacity for pharmaceutics with a large range of polarities and sizes; (3) 20- to >100-fold improved bioavailability of water-soluble peptides; (4) promise as a vehicle for the protection of drugs sensitive to enzymatic degradation; (5) great flexibility in terms of composition; (6) biodegradability; and (7) promotion of the slow, sustained release of the drugs incorporated in LLCs (Kim et al., 2015a).

In recent years, the cubic and hexagonal phases have received considerable attention due to their potential as drug release systems because they provide a slow release matrix and protect peptides, proteins, and nucleic acids from physical and chemical degradation. Additionally, because the lipids forming LLCs are nontoxic and biodegradable, they can be used for several forms of administration (Chen et al., 2014b; Guo et al., 2010). On the other hand, the large-scale manufacturing methods for the preparation of well-defined and stable dispersions represent a considerable obstacle for the commercialization of LLC-type release systems (Amar-Yuli et al., 2007). Moreover, limited amounts of adequate lipids are available for the preparation of these systems (Kim et al., 2015b). However, several potential forms of administration have been studied (Chen et al., 2014a; Guo et al., 2010).

2.4.2 Stability of lyotropic liquid-crystal systems

Due to their physicochemical and structural properties, lyotropic liquid crystal systems are more stable compared to other types of dispersal systems (Amar-Yuli et al., 2007). Similar observations have been reported for emulsions developed in recent years based on liquid crystals (Kim et al., 2009; Zhang and Liu, 2013) because the formation of an interphase with liquid crystalline characteristics provides greater stability, which is attributed to the increase in the mechanical strength of the oil-water interphase and to the fixation of the emulsion's drops to the liquid-crystal structure. This multilayer surrounding the emulsion's drops reduces the Van der Waals interactions between the oil drops and acts as a barrier against coalescence (Lavaselli et al., 2012).

As has been previously reported, LLCs are usually formed with amphiphilic molecules and/or polar lipids that, at different proportions and in the presence of water, form colloidal dispersions or direct or reverse cubic and hexagonal mesophases with high viscosity. The proportions of water, lipids, and other coparticipant amphiphilic molecules interfere with the formation of such mesophases and their stability.

One of the most studied amphiphilic compounds is the glycerol monooleate (Amar-Yuli et al., 2007; Ganem-Quintanar et al., 2000), which can form at least four well-characterized mesophases: one lamellar phase, two reverse bicontinuous cubic phases at room temperature, and one reverse hexagonal phase at a relatively higher temperature. With monooleate and other compounds, it has been demonstrated that the stability of each phase in LLCs and their dispersions (hexosomes and cubosomes) is a function of the relation between the amphiphilic lipids, the presence of a polymeric costabilizer and the proportion of water, among other factors, where the polymer usually acts as a steric-type stabilizer by adhering (in some cases) to the particle's surface. However, this mechanism has not been clearly elucidated.

Another determining factor in the formation of each phase of the LLC is the proportion of water, which affects the structuration, order or crystallinity of the system, and its particle size. There have been reports of hydration and dehydration phenomena in which the competition between the hydration of the structure and the adsorption of the polymer by the LLC system plays an important role (Amar-Yuli et al., 2007). Other factors governing the formation of the different mesophases for a given composition of lipid-copolymer-water are the agitation-dispersion system and the energy it provides during the formation of the mesophase, as well as changes in temperature during and after system formation.

Other factors that influence the stability of LLCs and their dispersions, although to a lesser degree, are the presence of electrolytes and molecules that favor hydrotropy phenomena and changes in pH that affect hydration and the gelification of the system (Amar-Yuli et al., 2007; Guo et al., 2010).

3 LIPID-CARRYING SYSTEMS

3.1 INTRODUCTION

Nanometric vectors are colloidal systems of submicrometric sizes (<1 μm) that can be produced from a large variety of materials and compositions. These nanometric vectors include liposomes, micelles, dendrimers, solid lipid nanoparticles, metallic nanoparticles, semiconductor nanoparticles, polymeric nanoparticles, quantum dots, and nanotubes, among others (Mishra et al., 2010).

In recent decades, nanometric carriers have received considerable interest as effective release systems for therapeutic agents for many reasons, including (Ram et al., 2011):

1. Physical and biological stability, which can facilitate the encapsulation and/or controlled release of the compound.
2. Good tolerance to components.
3. Simplicity of the formulation and production processes.
4. Possibility to scale up the production process.

A large proportion of the nanoparticles developed to contain therapeutic agents (pharmaceutics and biopharmaceutical agents) by encapsulation or absorption have demonstrated a capacity to direct the administration, minimize the collateral effects, and protect the active substance during storage, release, and transport in the organism. The release of the therapeutic agent only at the site of disease allows a low dose to be administered because only a small fraction of the active substance is lost in other parts of the body. Thus, toxic compounds, such as cytotoxic agents, only affect the target tissue and therefore minimize collateral effects (Amin et al., 2011; Assadi et al., 2011; Bhowmik et al., 2010; Caldorera-Moore et al., 2010; Fernandez-Fernandez et al., 2011; Marcato and Durán, 2008).

3.2 SOLID LIPID NANOPARTICLES

Solid lipid nanoparticles (SLNs) are stable colloidal carrier systems obtained by dispersing a lipid or mixture of lipids melted in an aqueous dilution of the surfactant using different methods.

SLNs were first proposed in early 1991 and combine the advantages of traditional systems. On the one hand, nontoxic molecules can be used in the body similar to liposomes but with higher stability; on the other hand, nanometric sizes can be achieved (Müller et al., 2000; Wissing et al., 2004).

The techniques used to manufacture SLNs are high-pressure homogenization (in cold or heat), microemulsion, solvent evaporation, and double emulsification (w/o/w).

In the high-pressure homogenization method, the compound is dissolved in the chosen lipid, which is melted at temperatures between 5 and 10°C above its melting point, and the mixture is dispersed by agitation in a hot dilution (at the same temperature as the lipid phase) of the surfactant. This preemulsion is homogenized at high pressure. The emulsion is cooled, and the resolidification of the lipid leads to the formation of nanometric structures. It is also possible to encapsulate compounds that are sensitive to temperature or hydrophilic molecules using cold mixing. This way, the migration of the compound to the aqueous phase is avoided during the initial process of the formation of the preemulsion.

In the second method, a microemulsion is formed, with the internal phase being the lipid melted with the incorporated compound. This mixture is added, underagitation, to an aqueous phase with surfactants and cosurfactants at the same temperature. The microemulsion formed is added to a cold aqueous dilution, and the precipitation of the SLN is immediate. The latter method has the disadvantage that the amount of lipid phase compared to the aqueous one can be very small; therefore, to obtain a significant amount of SLN, it is necessary to remove an excessive amount of water.

In the solvent evaporation method, the lipid is dissolved via an organic method and is dispersed in water to form an emulsion o/w. Then, the solvent is evaporated; it leaves the lipid in contact with the water. Due to its insolubility, it precipitates and produces the SLN. However, this method presents a similar problem to that encountered with conventional nanoparticles: the use of organic solvents (Müller et al., 2000).

In the double emulsification (w/o/w) method, the compound is dissolved in an aqueous internal phase, which is emulsified at high speeds (e.g., using an Ultra Turrax) with a lipid previously melted or dissolved in an organic solvent. This first emulsion w/o is then dispersed in an aqueous phase containing stabilizers, producing a double emulsion w/o/w, which is maintained under agitation. Subsequently, the nanoparticles are isolated by filtration. This technique is useful for pharmaceutical compounds that present high solubility in water.

For a release system to be effective, it must be able to efficiently encapsulate the compound, which depends on the following factors:

1. *The solubility of the pharmaceutical compound in the lipid*: if it has low solubility, it will migrate toward the aqueous phase during the preparation process.
2. *Chemical nature and crystalline structure*: lipids can form crystalline particles, which will cause the formation of a perfect network. Therefore, foreign particles (in the case the pharmaceutical compound) will be excluded from the lipid matrix. To avoid this phenomenon, mixtures of lipids are used to avoid the recrystallization of the lipid.
3. Lipid polymorphism: if there is the possibility of different crystallization methods, the lipid may crystallize differently in the dilution core than in the nanoparticle manufacturing process. Therefore, during the production of the SLN, the compound maybe excluded from the lipid core to form a more perfect crystal matrix (Heurtault et al., 2003).

Once the release system is formed, it must be able to release the pharmaceutical compounds in a controlled fashion. The release will depend on the particle size, the lipid matrix, the surfactant concentration, and the parameters of the process (Manjunath et al., 2005).

3.2.1 Release of the pharmaceutical compound from nanoparticles

In most cases, nanoparticles release the compound via their direct interaction with the medium. This can occur through the following mechanisms:

- Desorption of the surface.
- Diffusion through a matrix.
- Diffusion through the polymer or lipid wall.
- Matrix erosion.
- Combination of the diffusion-erosion processes (Soppimath et al., 2001).

The diffusion coefficient and biodegradation are the main factors to consider in the release mechanism of the compound.

The release of the compound from nanoparticles is strongly influenced by the biological environment in which it is found. The mean effect is greater in the case of large doses because a higher degree of bonding to plasmatic proteins can occur, accompanied by a delay in the release of the active compound.

3.2.2 Applications of solid lipid nanoparticles

SLNs are appropriate for systemic supply because their components are physiologically well tolerated and because they have good storage capacity after lyophilization and/or sterilization (Das, 2015). When administered intravenously, SLNs are sufficiently small to circulate through the microvascular system and can avoid being captured by macrophages if they are covered by hydrophilic agents.

Therefore, these lipid nanoparticles have been suggested for viral and nonviral gene administration. It has been demonstrated that cationic SLNs can directly bind to genes through electrostatic interactions, and they have potential benefits in directed gene therapy for the treatment of cancer. The load of particles can be regulated through its composition, which allows the binding of molecules of opposite charges.

The treatment of central nervous system diseases, such as brain tumors, AIDS, and psychiatric and neurological disorders are usually limited by the inability of powerful pharmaceutical agents to pass through the blood–brain barrier (BBB). Coating these agents with hydrophilic colloids can improve their transport through the BBB and hence improve their distribution in tissues (Kreuter, 2001; Wang et al., 2002). Fundaro et al. (2000) prepared doxorubicin loaded into coated and noncoated SLNs. They observed, 24 h after intravenous administration, that the particles coated with the hydrophilic colloid were present in blood at higher concentrations than the noncoated SLNs.

SLNs have been proposed as carriers of anticancer compounds or peptides for the treatment of lung cancer. In a recent study, antituberculosis compounds (rifampicin, isoniazid, and pyrazinamide) were incorporated in several SLN formulations with varied particle sizes ranging from 1.1 to 2.1 μm. These formulations were orally nebulized into guinea pigs for direct lung administration (Mura et al., 2015; Pandey and Khuller, 2005).

The ocular administration of pharmaceutical compounds using SLNs has been studied. The microadhesive and biocompatible properties of SLNs improve their interactions with the eye mucosa and can prolong the residence time of the compound in the cornea to direct the treatment in the eye. Cavalli et al. (2002) evaluated SLNs as vehicles for the ocular administration of tobramycin in rabbit eyes. They found that SLNs significantly increased the bioavailability of the compound in the aqueous humor. Cavalli et al. (1996) also studied the delivery of pilocarpine included in SLNs. This pharmaceutical compound is commonly used in glaucoma treatment. The results showed that the bioavailability of the compounds also improved in the eye (Kearns and Williams, 2009; Puglia et al., 2015).

SLNs are colloidal carrier systems that are attractive for topical use due to their multiple desirable effects in skin in addition to the features of a colloidal vehicle. SLNs are adequate for use in damaged or swollen skin because they are based on nonirritating and nontoxic lipids (Wissing and Müller, 2003). Researchers have reported the frequent topical use of SLNs. During recent years, lipid nanoparticles have been loaded with active compounds, such as vitamin E (Dingler et al., 1999), tocopheryl acetate (Wissing and Müller, 2001), retinol (Jenning et al., 2000), ascorbyl palmitate (Uner et al., 2005), clotrimazole (Souto et al., 2004), triptolide (Mei et al., 2003), podophyllotoxin (Chen et al., 2006), and a nonsteroid antiandrogen, RU 58841, for topical use (Muenster et al., 2005). One novel area has been the use of SLNs in sunscreen creams (Wan et al., 2012).

SLNs can be easily produced with low-cost, safe, stable, and biocompatible materials. They can also be loaded with different classes of therapeutic agents for different administration forms. SLNs also offer a highly versatile platform that should be considered for applications involving molecules whose solubility and/or bioavailability represent significant challenges.

In addition to their components (physiological compounds), the advantages of SLNs include their rapid and efficient production time, the possibility of large-scale production, the avoidance of organic solvent usage, and the creation of high-encapsulation-efficiency carriers.

4 MICROSPHERES

4.1 INTRODUCTION

Microspheres are small spherical particles with diameters from 1 to 1000 μm (Jyothi et al., 2010; Singh et al., 2010). In some cases, microspheres are also known as microparticles. Microspheres can be produced from several natural and synthetic polymeric materials or even from inorganic materials (Jyothi et al., 2010). For example, microspheres can be produced from commercially available polymers or ceramics. Depending on the method, solid or porous microspheres can be obtained for specific intended applications.

The variety of methods for the production of microspheres offers a myriad of opportunities to control the aspects of administration of the pharmaceutical compound. This focus facilitates the precise release of the desired amount of a component at the site of action and its reduction at non-target sites. Similarly, this factor guarantees the protection of the compounds before and after administration (Jyothi et al., 2010). Additionally, the vectorization of pharmaceutical compounds can be manipulated by coupling a recognition molecule to the microsphere. The exploitation of these changes in pharmacokinetic behavior can lead to an improved therapeutic effect. The aim of any pharmaceutical compound administration system is to provide a therapeutic amount of the compound at the correct site in the body to quickly achieve an effective concentration and then maintain it for a given time. A well-designed modified release system for the compound can overcome some of the problems of conventional therapy and improve the therapeutic efficiency, thus improving the patient's quality of life.

4.2 CRITERIA FOR THE DESIGN AND DEVELOPMENT OF MICROSPHERES

There are several approaches for the design of a modified release system for a pharmaceutical compound. One approach involves the use of microspheres, which are spherical particles smaller than 200 μm that are characteristically free-flow powders.

The design of a microsphere must satisfy certain criteria, such as:

1. The capacity to reasonably incorporate the necessary concentrations of the pharmaceutical compound to achieve therapeutic doses.
2. The stability of the preparation after synthesis, with a clinically acceptable lifespan.
3. Controlled particle size and, preferably, dispersibility in aqueous vehicles.
4. The release of the compound with a good control in an adequate time scale.
5. Biocompatibility and biodegradability (Mao et al., 2012).

One of the most important factors affecting the final properties of a microsphere is the selection of the type of polymer, which must be based on the physicochemical characteristics of the pharmaceutical compounds, the loading requirements, and the speed and duration of the release.

A wide variety of polymers are used for the formation of microspheres; these are categorized as synthetic vesus natural and biodegradable versus nonbiodegradable (Mao et al., 2012). The most frequently used synthetic polymers include poly-(methyl methacrylate) (PMMA), glicycidyl-epoxy-methacrylate, lactic acid and glycolic acid polymers and their copolymers, such as poly-(alkyl cyanoacrylate). Commonly used natural polymers include albumin, gelatin, collagen, and carbohydrates, such as chitosan, alginate, and carrageenan.

The selection of a material for the fabrication of microspheres must include a consideration of the properties of the polymer, such as its chemical composition, physical and mechanical properties, mechanism and rate of biodegradation, toxicological profile and even regulatory aspects. In addition, the physicochemical properties of the pharmaceutical compound must be evaluated, such as its solubility in water, compatibility with the polymer, storage, and in vivo transformation (Mao et al., 2012).

4.2.1 Methodologies used for the production of microparticles

The methodologies used to produce microparticles include (Jyothi et al., 2010):

- complex coacervation,
- polyelectrolyte complexes,
- ionotropic gelation,
- emulsification-crosslinking,
- coprecipitation and spray-drying.

It is important to highlight that, despite the diverse documented methods and available chemical agents for producing microparticles, only a few methodologies offer real potential for the development of pharmaceutical applications due to the strict regulatory framework that excipients must fulfill for the release of such compounds. As mentioned, a critical aspect of the formulation of microspheres is the load of the compound, which is related to the dose of the final product and its release regime for the duration of the effect, which is quite difficult to achieve in some cases. Due to these limitations, one of the most frequently used and widely recognized polymers for the production of microspheres is chitosan.

4.3 PHYSICOCHEMICAL CHARACTERISTICS OF CHITOSAN

Chitosan is an abundant biopolymer in nature, and the cost of its production is low. Chemically, the term "chitosan" defines a family of linear heteropolysaccharides that are formed by monomers of 2-amino-2-deoxy-β-D-glucose and 2-acetamide-2-deoxy-β-D-glucose linked by β-1→4 bonds. These compounds include a series of polymers with molecular weights of 50–2000 kDa and deacetylation percentage of 75%–98% (Fig. 22.1). Chitosan has many advantages, especially for the development of microparticles. For example, chitosan has the capacity to control the liberation of active agents, allowing the use of organic solvents to be avoided during manufacturing. Chitosan is a polymer that contains a number of free amino groups that are available for crosslinking, its cationic nature allows ionic reticulation with multivalent ions, and it has a mucoadhesive character that increases time at the absorption site, among other benefits (Prabaharan and Mano, 2005).

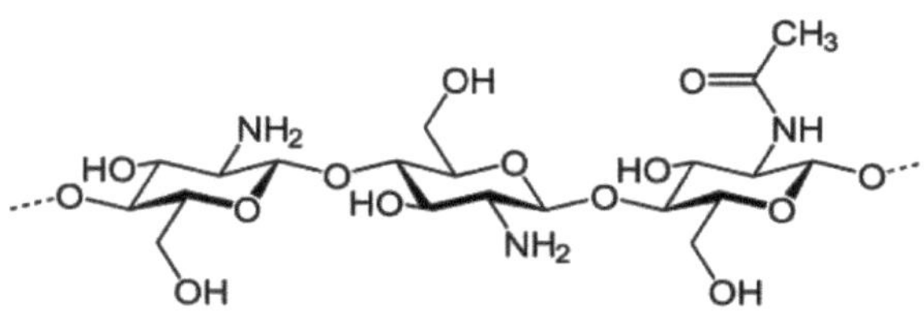

FIGURE 22.1 Structure of Chitosan

4.4 MICROSPHERE PREPARATION METHODS

One of the commonly used methods of microsphere preparation is based on the formation of an emulsion of water in oil (w/o), followed by the later addition of a crosslinking agent (glutaraldehyde), which coagulates the microbeads formed by the dispersed phase (Cerchiara et al., 2015). The stabilization of the emulsion is a critical step for the efficient encapsulation of a dissolved drug in an aqueous phase that is subsequently is dispersed in an organic phase. Surfactants are the most commonly used stabilizers, including nonionic, anionic, and cationic surfactants. The addition of a surfactant results in the reduction of the surface tension of the continuous phase, which facilitates a reduction in the size of the globule; this effect occurs when the concentration of surfactant reaches the critical micelle concentration (CMC). The choice of stabilizing agent is based on the type of polymer. For example, polyvinyl alcohol (PVA) is the most commonly used stabilizer in the production of poly(lactic-*co*-glycolic acid) (PLGA) microspheres (Mao et al., 2012).

Jeong et al. (2003) showed that the microsphere particle size decreased with increasing concentrations of PVA in the aqueous phase. For the preparation of microspheres with chitosan, the surfactant of choice is Span 80. Using this method, it is possible to encapsulate hydrophilic compounds, such as vancomycin hydrochloride (Cerchiara et al., 2015). Research conducted in the Laboratory of Molecular Pharmacy and Controlled Release of UAM-X has shown that with the use of this method of production, it is possible to obtain microspheres with sizes of 1–5 μm and a spherical appearance (Fig. 22.2).

Another method that has gained attention in recent years is the production of microparticles with the spraying process (Rege et al., 2003). Technological advances in spraying have been successfully applied to produce pharmaceutical products, with the two main techniques being (1) spray-drying and (2) freezing. The unitary operation in spray drying is mainly evaporation, whereas in spraying–freezing, it is the phase change from liquid to solid. Both processes are similar, except for the energy flow.

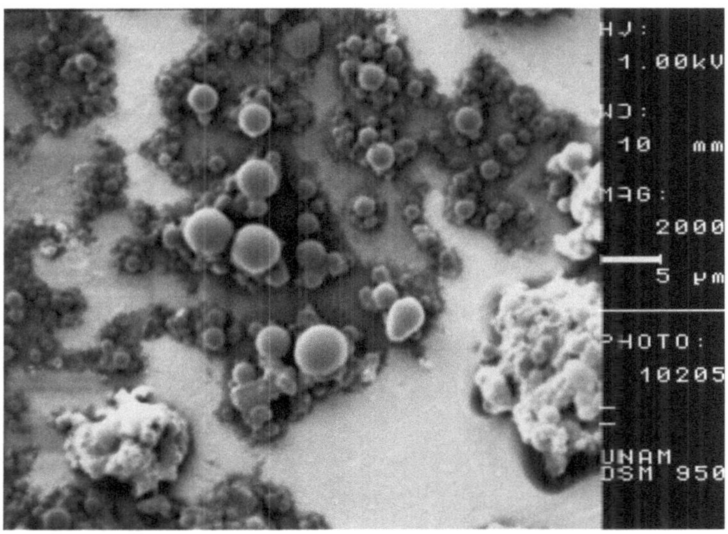

FIGURE 22.2 Scanning Electron Microscopy of Chitosan Microspheres (LFMLC-UAMX)

In the case of spray drying, the energy is applied to the drop, forcing the evaporation of the medium and resulting in the transfer of mass and energy through the drop.

Use of the spray-drying technique to synthesize chitosan microspheres has been reported, and the results show that it is possible to obtain microspheres with diameters between 2 and 5 μm (De La Torre et al., 2003; Rege et al., 2003). The chitosan microspheres appear as free-flowing powder and are particularly suitable for drug delivery applications. However, in this method, the chemical behavior of chitosan must be controlled to reduce the effects of certain limitations, including its insolubility at pH 6.5 and its reactivity under thermal conditions. Chitosan can be sprayed for the preparation of microspheres at neutral pH if a colloidal suspension prepared with sodium hydroxide is achieved.

4.5 APPLICATION OF MICROSPHERES IN THE TRANSPORT AND VECTORIZATION OF PHARMACEUTICAL COMPOUNDS

An important application of microspheres is the oral administration of vaccines. However, this form of administration causes the degradation of antigens in the gastrointestinal tract and shows inefficient specificity for the site of action when administered in a free form. Several studies have shown that incorporating the vaccine into chitosan microspheres significantly improved the capture of antigens by M cells and prevented the degradation of the vaccine in the gastrointestinal tract (Islam et al., 2012). One example of this type of research is a study conducted by Ahire et al. (2007), who examined the systemic and local immune responses after the oral administration of the tetanus toxoid. Chitosan microspheres showed a local immune response, improving the dose-dependent pharmacokinetics 3 weeks after the oral administration of the vaccine compared to free-form administration. The formulation was evaluated in terms of encapsulation efficiency and stability as a function of pH, which produced the following results. The microparticle showed stability in simulated gastric fluid and in intestinal conditions; similarly, the microsphere system was demonstrated to be effective at inducing IgA in the intestine and IgG in the systemic circulation (Ahire et al., 2007).

Microspheres have also been applied as a release system in anticancer therapy to improve the efficiency and reduce the side effects of cytotoxic drugs. Vectorization is achieved through chemoembolization (physical entrapment in small blood vessels) and can be controlled by careful selection of the administration form. Microspheres have proven very effective for chemoembolization in preclinical models and preliminary clinical trials (Tam et al., 2011).

Other applications of microspheres include administration by ophthalmic, topical, nasal, or intramuscular routes. A very wide variety of compounds have been used for encapsulation in microspheres as a modified release system, which has led to fruitful research on the use of chitosan as support and as an important component of the design of pharmaceutical forms for modified release.

Microspheres are a good choice for the transport of low- and high-molecular-weight compounds and of cellular components, such as DNA fragments, peptides, and proteins. The design of transport matrices for microspheres offers several advantages. Different types of materials may be used, depending on the application system. This versatility allows microspheres to be used for drug vectorization, regardless of the physicochemical characteristics. For example, for hydrophobic compounds, the use of poly lactic-glycolic materials is recommended; for water-soluble compounds, chitosan has been recommended. Finally, the choice of drug delivery system based on microspheres depends on the goal of the treatment.

5 POLYMEROSOMES

5.1 INTRODUCTION

Polymerosomes are artificial vesicles formed with copolymers that contain hydrophilic and hydrophobic blocks (amphiphilic). The most common form are hollow spheres that contain an aqueous solution whose coat forms a membrane-like bilayer (Lee and Feijen, 2012).

The use of copolymer blocks for the generation of polymeric vesicles is a new strategy for the development of encapsulation. Depending on the length and chemical nature of the copolymer, the resulting polymerosome can be more robust than a liposome of comparable size.

The thickness of the coating is controlled by varying the polymerization degree of the blocks of the individual molecules, and its fluidity and permeability can be adjusted altering the glass transition temperature (Tg) of the hydrophobic block (Lorenceau et al., 2005).

Polymerosomes can be designed in such a way that the molecules bind to each other very tightly. Their atoms can bind not only inside a simple polymer but also to the atoms of nearby polymers. This is known as crosslinking, and it greatly increases the resistance of these polymeric vesicles.

5.2 REQUIREMENTS FOR THE PRODUCTION OF POLYMEROSOMES

Several important aspects must be consideredin the production of polymerosomes.

1. Apparently, not all copolymer blocks can form polymerosomes, that is, certain composition characteristics must be met (Xu et al., 2005).
2. For the formation of copolymers in water, the copolymer block must have a hydrophilic fraction between 20% and 40%, the same proportion that is exhibited by natural lipids. When the fraction is between 45% and 55%, cylinder-shaped micelles are formed; between 55% and 70%, sphere-shaped micelles predominate (Meng et al., 2003).
3. With a copolymer block in which one of the polymers has a low glass transition temperature, polymerosomes can be directly formed in an aqueous medium (Afsaneh et al., 2002).

5.2.1 Polymerosome production

A polymerosome can be formed from copolymer blocks using an organic or aqueous solvent, or a mixture of both (Discher et al., 2001). The production can be performed in two steps: the first consists of an amphipathic bilayer, and the second consists of the formation of the vesicle (Förster and Borchert, 2005).

Polymerosomes can assume several shapes, depending on the conditions of the production process: the copolymer composition, the nature of the solvent and the system's temperature (Soo and Eisenberg, 2004)

5.2.2 Copolymers

A copolymer is a macromolecule formed by two or more monomers (from the Greek mono, or one, and meros, or part) or different repetitive units, which can form different chemical bonds.

The monomers can be randomly or periodically distributed. A block copolymer consists of long sequences of two alternating polymers. If the change in composition occurs in ramifications, it is known as a branched copolymer. Different copolymerization techniques for the synthesis of copolymers include random synthesis and MADIX radical reactions (Anseth et al., 1996).

In the molecular characterization of a polymer, information is obtained regarding the following:

1. The chemical structure of the macromolecules.
2. The molecular weight of the polymer.
3. The distribution of molecular weights.
4. The sizes of the macromolecules.
5. The molecular movements created by the heating of the system.
6. The arrangement of the macromolecular chains to produce a three-dimensional structure.

The final objective of a characterization process must be to correlate the characteristics and properties of the molecules of a polymer with its macroscopic characteristics, such as chemical structure, molecular weight, distribution of molecular weights, and molecular movements. When these types of correlation are achieved, the three areas that form the production process of a material can be adequately integrated (Palacios et al., 1982).

5.3 SYNTHESIS OF A POLYMEROSOME FROM HYDROPHILIC-HYDROPHOBIC COPOLYMERS

Some methods used for the production of liposomes can be used for the production of polymerosomes, including:

1. *Membrane rehydration*: In this simple process, a copolymer block is rehydrated with an aqueous solution. The amphipathic part mixes with an organic solvent. A glass surface is covered with this solution, which is evaporated using a vacuum (Bermudez et al., 2002). Then, an aqueous solution is added, which causes the formation of the vesicle (Förster and Borchert, 2005).
2. *Electroformation*: Electroformation was invented by Angelova and Dimitrov in 1986, and it involves the use of two 0.5-mm-diameter platinum wires separated by a space of 0.5 cm, where the copolymer solution is deposited. An electric field is used, which, in combination with an aqueous solution, causes a decrease in the surface tension (Förster and Borchert, 2005) and a difference in the dielectric constant of the copolymer blocks, producing a reorientation that induces the formation of microspheres. This method is used in the formation of vesicles with a single layer 10–100 μm in diameter (Hamley, 2003).
3. *Organic solvents*: In this method, the copolymer block is first dissolved in a water-soluble solvent (e.g., dioxane, tetrahydrofuran [THF] or dimethylfuran [DMF]). Water is then added, increasing the surface tension and causing the formation of the vesicles (Meng et al., 2003). Once the polymerosomes are obtained, the solvent is evaporated to make these structures more stable.
4. *Microfluids*: With the help of microchannels, the formation of microcapsules is induced using immiscible fluids, which allows the generation of perfect spherical encapsulation structures. The diameter of these particles depends on the capillary size, which can range 10–100 μm; this can also be modified by varying the flow in each phase. With this technique, monodispersed emulsions can be obtained. It is possible to produce this type of vesicles in a single step through the formation of a double emulsion consisting of one or more small drops suspended in a larger drop of a second, immiscible liquid suspended in a third phase. In this manner, two types of emulsions can be formed: water/oil/water (w/o/w) and its inverse, oil/water/oil (o/w/o).

5.4 POLYMEROSOME APPLICATIONS

5.4.1 Encapsulation of pharmaceutical compounds

In the past decade, a wide variety of active compounds have been studied. There are approximately 12 publications on polymerosomes containing doxorubicin in free form or as one of its salts (chlorhydrate) using different monomer combinations along with different synthesis methods to optimize its load and release. One example is poly(trimethylene carbonate)/polyglutamic acid (PTMC-PGA) (Sanson et al., 2010). Some of these studies have been conducted in animal models and obtained promising results, such as decreases in size and increases in the concentration of the pharmaceutical compound in tumors (Ahmed et al., 2006; Chen et al., 2010; Levine et al., 2008).

Paclitaxel has been studied in ethylene oxide/lactic acid (PEO-PLA) copolymers, in ethylene/butadiene oxide (PEO-PBD), and in combination with doxorubicin. These compounds are the most frequently used agents for the treatment of a wide variety of metastatic aggressive tumors of the breast (Ahmed et al., 2006).

The copolymerpolyethylene glycol/poly(glycerol monomethacrylate) has been used to encapsulate indomethacin, which is an analgesic and antiinflammatory agent (Meng et al., 2009).

Other types of agents that have been encapsulated in polymeric vesicles include pravastatin (hydrophilic statin), which is used for the control of dyslipidemias (Onaca et al., 2009); human erythropoietin (HEP) (a glycoproteic hormone), which is involved in erythropoiesis regulation (Zhang et al., 2010); antibiotics, such as amphotericin B in a copolymer of polyethylene glycol/polylactic acid (PEG-PLA) (Jain and Kumar, 2010); and hydroalcoholic extracts, such as coumarin, in polymerosomes of polyethylene glycol/polyglycolic lactic acid (PEG-PLGA) (Yu et al., 2012).

5.4.2 Polymerosomes for the design of oxygen carriers

Arifin (2005) proposed, for the first time, the use of polymerosomes (PEH) as oxygen carriers by combining a lipid conjugated to polyethylene glycol to encapsulate bovine hemoglobin. Even after PEH was shown to be more stable than lipid vesicles and to possess more affinity to bind oxygen, dispersions of these vesicles were shown to be a precipitated mixture of polymerosomes, micelles, and copolymers. Thus, a solution is to study the minimum molecular weight of a copolymer required for the complete formation of the polymerosome.

Lorenceau et al. (2005) proposed the generation of polymerosomes by the double emulsion method from copolymers of poly(acrylic acid)-poly(butyl acrylate) (PAA-PBA), wherein the synthesis of the copolymer is controlled by the use of a catalyzer (MADIX reaction). This process would allow one to determine the composition of the hydrophilic and hydrophobic blocks. The work of Lorenceau et al. (2005) used a copolymer in which the molecular weight (MW) of the hydrophilic block was 1.5 kDa and the MW of the hydrophobic block was 4 kDa. The membrane formed from this copolymer of PAA-PBA for the development of the polymerosome was not uniform in thickness, and the total evaporation of the solvent used in the synthesis of these polymerosomes was not achieved. These factors had an impact on their permeability.

Based on the earlier research, the Laboratory of Molecular Pharmacy and Controlled Release performed the random synthesis of the PAA-PBA copolymer to generate a flexible and permeable polymeric membrane, which allows the design of a polymerosome containing human hemoglobin to create an artificial oxygen-carrying human cell as a prototype of a red blood cell.

The synthesis of the polymerosome was performed in a T-shaped glass device. The vesicles were observed under an optical microscope, and the size of the polymeric vesicle was determined to range

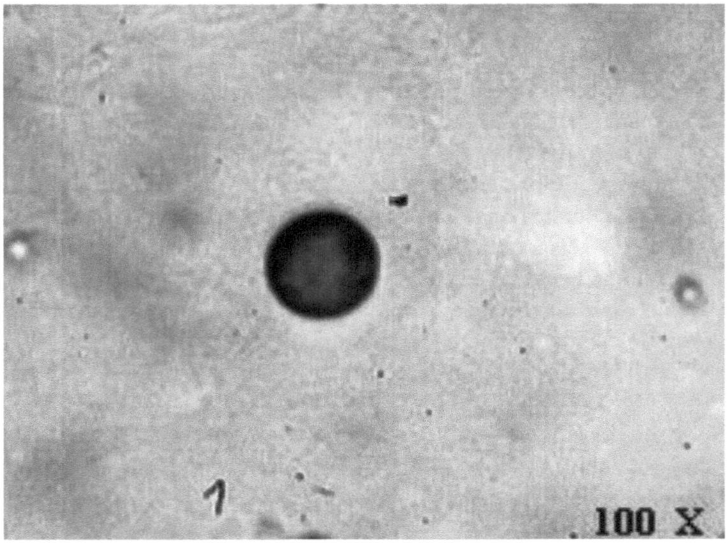

FIGURE 22.3 Polymersome with Hemoglobin (Hb)

from 27 μm to 29 μm, with an encapsulation of hemoglobin (Hb) of 59% to 60%. Compared to the erythrocyte, there was some similarity in physiological characteristics. The formation of the amphiphilic membrane and oxygen diffusion may have been somewhat similar to a red blood cell; see Fig. 22.3 (Noguez, 2015).

Finally, it should be noted that there are different ways to release a pharmaceutical compound from the polymerosome at the site of action by modifying its physicochemical characteristics, including the following: (1) varying pH (active release), (2) degradation (sustained release), (3) release by solubility modification (instant release), (4) release by temperature change, (5) release by change in redox potential, (6) release by change in electrostatic force and other types of stimuli (changes in magnetic field and ultrasound) (Meng et al., 2009).

6 CONCLUSIONS

In this chapter we understand that the study of microfluidic was a multidisciplinary field including physics, chemistry, engineering, and biotechnology; it involves the study of the behavior of fluids at the micro- and mesoscales. This area also includes the design of systems in which small amounts of fluids was involved. The manipulation of multiphase flows is another possible application of microfluidics; allowing the generation of monodisperse bubbles (gas–liquid mixtures). These dispersions suggest new pathways for the production of polymer vesicles, emulsions, in nanoparticles. Other ways to the application of nanotechnology are the design of lyotropic liquid crystals. We knew the cubic and hexagonal phases had a potential as drug release systems because they provide a slow release matrix and protect peptides, proteins, and nucleic acids from physical and chemical degradation. Lipid-carrying systems include nanometric vectors as liposomes, micelles, dendrimers, solid lipid nanoparticles, metallic nanoparticles, semiconductor nanoparticles, polymeric nanoparticles, quantum dots, and nanotubes, among others. Microspheres had the variety of methods for the production and offered a myriad of opportunities to control the aspects of the administration of the pharmaceutical compound. Polymersomes designed in such a way that the molecules bind to each other very tightly. Finally, it should be noted that there are different ways to release a pharmaceutical compound from the polymersome at the site of action by modifying its physicochemical characteristics.

REFERENCES

Abbaspourrad, A., 2013. Microfluidic fabrication of stable gas-filled microcapsules for acoustic contrast enhancement. Langmuir 29, 12352–12357.

Afsaneh, L., Samuel, J., Kwon, G.S., 2002. Poly(ethylene oxide)-block-poly(L-amino acid): micelles for drug delivery. Adv. Drug Deliv. Rev. 54, 169–190.

Ahire, V.J., Sawant, K.K., Doshi, J.B., Ravetkar, S.D., 2007. Chitosan microparticles as oral delivery system for tetanus toxoid. Drug Dev. Ind. Pharm. 33, 1112–1124.

Ahmed, F., Pakunlu, R.I., Brannan, A., Bates, F., Minko, T., Discher, D.E., 2006. Biodegradable polymersomes loaded with both paclitaxel and doxorubicin permeate and shrink tumors, inducing apoptosis in proportion to accumulated drug. J. Control. Release 116, 150–158.

Amar-Yuli, I., Wachtel, E., Shoshan, E.B., Danino, D., Aserin, A., Garti, N., 2007. Hexosome and hexagonal phases mediated by hydration and polymeric stabilizer. Langmuir 23 (7), 3637–3645.

Amin, R., Hwang, S., Park, S.H., 2011. Nanobiotechnology: an interface between nanotechnology and biotechnology. Nano 6, 101–111.

Anseth, K.S., Scott, R.A., Peppas, N.A., 1996. Effects of ionization on the reaction behavior and kinetics of acrylic acid polymerizations. Macromolecules 29, 8308–8312.

Arifin, D.R., 2005. Cellular hemoglobin-based oxygen carriers as potential artificial blood substitutes. Ph. D. Thesis. University of Notre Dame, Indiana, pp 1-173.

Assadi, M., Afrasiabi, K., Nabipour, I., Seyedabadi, M., 2011. Nanotechnology and nuclear medicine; research and preclinical applications. J. of Nucl. Med. 14, 149–159, 197–198.

Atencia, J., Beebe, D.J., 2005. Controlled microfluidic interfaces. Nature 437, 29.

Bermudez, H., Brannan, A.K., Hammer, D.A., Bates, F.S., Discher, D.E., 2002. Molecular weight dependence of polymersome membrane structure, elasticity, and stability. Macromolecules 35 8203–8208.

Bhowmik, D., Chiranjib, Margret Chandira, R., Tripathi, K.K., Sampath Kumar, K.P., 2010. Nanomedicine-an overview. Int. J. PharmTech Res. 2, 2143–2151.

Caldorera-Moore, M., Guimard, N., Shi, L., Roy, K., 2010. Designer nanoparticles incorporating size, shape and triggered release into nanoscale drug carriers. Exp. Opin. Drug Deliv. 7, 479–495.

Carroll, J.N., 2008. Droplet based microfluidics for emulsion and solvent evaporation synthesis of monodisperse mesoporous silica microspheres. Langmuir 24, 658–661.

Cavalli, R., Marengo, E., Rodriguez, L., Gasco, M.R., 1996. Effects of some experimental factors on the production process of solid lipid nanoparticles. Eur. J. Pharm. Biopharm. 42, 110–115.

Cavalli, R., Gasco, M.R., Chetoni, P., Burgalassi, S., Saettone, M.F., 2002. Solid lipid nanoparticles (SLN) as ocular delivery system for tobramycin. Int. J. Pharm. 238, 241–245.

Cerchiara, T., Abruzzo, A., Di Cagno, M., Bigucci, F., Bauer-Brandl, A., Parolin, C., Vitali, B., Gallucci, M.C., Luppi, B., 2015. Chitosan based micro- and nanoparticles for colon-targeted delivery of vancomycin prepared by alternative processing methods. Eur. J. Pharm. Biopharm. 92, 112–119.

Chen, H.B., Chang, X.L., Du, D.R., Liu, W., Liu, J., Weng, T., Yang, Y.J., Xu, H.B., Yang, X.L., 2006. Podophyllotoxin-loaded solid lipid nanoparticles for epidermal targeting. J. Control. Release 110, 296–306.

Chen, W., Meng, F., Cheng, R., Zhong, Z., 2010. pH-Sensitive degradable polymersomes for triggered release of anticancer drugs: a comparative study with micelles. J. Control. Release 142, 40–46.
Chen, P.W., Erb, R.M., Studar, A.R., 2012. Designer polymer-based microcapsules made using microfluidics. Langmuir 28, 144–152.
Chen, P.W., Cadisch, G., Studart, R.A., 2014a. Encapsulation of aliphatic amines using microfluidics. Langmuir 30, 2346–2350.
Chen, Y., Ma, P., Gui, S., 2014b. Cubic and hexagonal liquid crystals as drug delivery systems. BioMed. Res. Int. 815981, 1–12.
Cho, B.S., 2003. Passively driven integrated microfluidic system for separation of mobile sperm. Anal. Chem. 75, 1671–1675.
Choi, C.H., Kang, S.M., Jin, S.H., Yi, H., Lee, C.S., 2015. Controlled fabrication of multicompartmental polymeric microparticles by sequential micromolding via surface-tension-induced droplet formation. Langmuir 31 (3), 1328–1335.
Cortese, B., et al., 2009. Engineering transfer of micro- and nanometer-scale features by surface energy modification. Langmuir 25 (12), 7025–7031.
Das, S., 2015. Lipid-based nanoformulations for active delivery. Curr. Pharm. Biotechnol. 16, 289–290.
De Geest, B.G., Urbanski, J.P., Thorsen, T., Demeester, J., Smedt, S.C.D, 2005. Synthesis of monodisperse biodegradable microgels in microfluidic devices. Langmuir 21, 10275–10279.
De La Torre, P.M., Enobakhare, Y., Torrado, G., Torrado, S., 2003. Release of amoxicillin from polyionic complexes of chitosan and poly(acrylic acid) study of polymer/polymer and polymer/drug interactions within the network structure. Biomaterials 24, 1499–1506.
Dingler, A., Blum, R.P., Niehus, H., Muller, R.H., Gohla, S., 1999. Solid lipid nanoparticles (SLN (TM)/ Lipopearls (TM)): a pharmaceutical and cosmetic carrier for the application of vitamin E in dermal products. J. Microencapsul. 16, 751–767.
Discher, D., Bermudez, H., Lee, J., Discher, B., Hammer, D., Bates, F., 2001. Polymersomes: thermomecanical properties through in vivo testing. Bioeng. Conf. 50, 543–544.
Duan, C., Wang, W., Xie, Q., 2013. Review article: fabrication of nanofluidic devices. Biomicrofluidics 7, 026501(1–45).
Fernández Rivas, D., 2011. Microfluidos: nuevas fronteras. Rev. Cubana de Física 28 (1), 60–67.
Fernandez-Fernandez, A., Manchanda, R., Mcgoron, A.J., 2011. Theranostic applications of nanomaterials in cancer: drug delivery, image-guided therapy, and multifunctional platforms. Appl. Biochem. Biotechnol. 165, 1628–1651.
Feynman, R., 2002. There's plenty of room at the bottom. J. Microelectromech. Syst. 1, 60–66.
Förster, S., Borchert, K., 2005. Polymer vesicles. In: Mark, H.E. (Ed.), Encyclopedia of Polymer Science and Technology. Third ed. John Wiley & Sons, New York, pp. 1–52.
Frenz, L., Blouwolff, J., Griffiths, D.A., Baret, J.C., 2008. Microfluidic production of droplet pairs. Langmuir 24, 12073–12076.
Fundaro, A., Cavalli, R., Bargoni, A., Vighetto, D., Zara, G.P., Gasco, M.R., 2000. Non-stealth and stealth solid lipid nanoparticles (SLN) carrying doxorubicin: pharmacokinetics and tissue distribution after i.v. administration to rats. Pharmacol. Res. 42, 337–343.
Ganem-Quintanar, A., Quintanar-Guerrero, D., Buri, P., 2000. Monoolein: a review of the pharmaceutical applications. Drug Dev. Ind. Pharm. 26 (8), 809–820.
Glasgow, I.K., 2001. Handling individual mammalian embryos using microfluidics. IEEE Trans. Biomed. Eng. 48, 570–578.
Gong, X., Wen, W., Sheng, P., 2009. Microfluidic fabrication of porous polymer microspheres:dual reaction in single droplets. Langmuir 25 (12), 7072–7077.
Guo, C., Wang, J., Cao, F., Lee, R., Zhai, G., 2010. Lyotropic liquid crystal systems in drug delivery. Drug Discov. Today 15 (23/24), 1032–1040.

Hamley, I.W., 2003. Nanostructure fabrication using block copolymers. Nanotechnology 14, R39–R54.
Hegmann, T., Qi, H., Marx, V., 2007. Nanoparticles in liquid crystals: synthesis, self-assembly, defect formation and potential applications. J. Inorg. Organomet. Polymer. Mater. 17 (3), 483–508.
Heurtault, B., Saulnier, P., Pech, B., Proust, J.-E., Benoit, J.-P., 2003. Physico-chemical stability of colloidal lipid particles. Biomaterials 24, 4283–4300.
Hohne, D.N., Younger, J.G., Solomon, M.J., 2009. Fexible microfluidic device for mechanical property characterization of soft viscoelastic solids such as bacterial biofilms. Langmuir 25 (13), 7743–7751.
Hong, J.S., Stavis, S.M., DePaoli Lacerda, S.H., Locascio, L.E., Raghavan, S.R., Gaitan, M., 2010. Microfluidic directed self-assembly of liposome-hydrogel hybrid nanoparticles. Langmuir 26 (13), 11581–11588.
Inamdar, K.N., Griffith, L.G., Borenstein, J.T., 2011. Transport and shear in a microfluidic membrane bilayer device for cell culture. Biomicrofluidics 5, 022213-(1–16).
Islam, M.A., Firdous, J., Choi, Y.J., Yun, C.H., Cho, C.S., 2012. Design and application of chitosan microspheres as oral and nasal vaccine carriers: an updated review. Int. J. Nanomed. 7, 6077–6093.
Jahn, A., 2007. Microfluidic directed formation of liposomes of controlled size. Langmuir 23, 6289–6293.
Jahn, A., Lucas, F., Wepf, R.A., Dittrich, P.S., 2013. Freezing continuous-flow self-assembly in a microfluidic device: toward imaging of liposome formation. Langmuir 29, 1717–1723.
Jain, J.P., Kumar, N., 2010. Development of amphotericin B loaded polymersomes based on (PEG)3-PLA co-polymers: factors affecting size and in vitro evaluation. Eur. J. Pharm. Sci. 40, 456–465.
Jenning, V., Gysler, A., Schäfer-Korting, M., Gohla, S.H., 2000. Vitamin A loaded solid lipid nanoparticles for topical use: occlusive properties and drug targeting to the upper skin. Eur. J. Pharm. Biopharm. 49, 211–218.
Jensen, K.F., 2006. Silicon-based michrochemical systems characteristics and applications. MRS Bull. 31, 101–107.
Jeon, N.L., Dertinger, K.W., Chiu, D.T., Choi, I.S., Stroock, D.A., Whitesides, G.M., 2000. Generation of solution and surface gradients using microfluidic systems. Langmuir 16, 8311–8316.
Jeong, Y.I., Song, J.G., Kang, S.S., Ryu, H.H., Lee, Y.H., Choi, C., Shin, B.A., Kim, K.K., Ahn, K.Y., Jung, S., 2003. Preparation of poly (DL-lactide-*co*-glycolide) microspheres encapsulating all-trans retinoic acid. Int. J. Pharm. 259, 79–91.
Jyothi, N.V.N., Prasanna, P.M., Sakarkar, S.N., Prabha, K.S., Ramaiah, P.S., Srawan, G.Y., 2010. Microencapsulation techniques, factors influencing encapsulation efficiency. J. Microencapsul. 27, 187–197.
Kearns, V.R., Williams, R.L., 2009. Drug delivery systems for the eye. Exp. Rev. Med. Dev. 6, 277–290.
Kim, J., Vanapalli, A.S., 2013. Microfluidic production of spherical and nonspherical fat particles by thermal quenching of crystallizable oils. Langmuir 29, 12307–12316.
Kim, I.-Y., Nakagawa, S., Ri, K., Hashimoto, S., Masaki, H., 2009. Liquid crystal o/w emulsions to mimic lipids and strengthen skin barrier. Cosmet. Toiletries, 1–10.
Kim, J.H., Jeon, T.Y., Choi, T.M., Shim, T.S., Kim, S.H., 2014. Droplet microfluidics for producing functional microparticles. Langmuir 30, 1473–1488.
Kim, D.-H., Jahn, A., Cho, S.-J., Kim, J., Ki, M.-H., Kim, D.-D., 2015a. Lyotropic liquid crystal systems in drug delivery: a review. J. Pharm. Invest. 45, 1–11.
Kim, B., Jeon, T.Y., Oh, Y.K., Kim, S.H., 2015b. Microfluidic production of semipermeable microcapsules by polymerization induced phase separation. Langmuir 31, 6027–6034.
Kobayashi, I., Uemura, K., Nakajima, M., 2006. Controlled generation of monodisperse discoid droplets using microchannel arrays. Langmuir 22, 10893–10897.
Kreuter, J., 2001. Nanoparticulate systems for brain delivery of drugs. Adv. Drug Deliv. Rev. 47, 65–81.
Lan, W., Li, S., Xu, J., Luo, G., 2011. Synthesis of titania-silica core-shell microspheres via a controlled interface reaction in a microfluidic device. Langmuir 27, 13242–13247.
Laulicht, B., Cheifetz, P., Mathiowitz, E., Tripathi, A., 2008. Evaluation of continuous flow nanosphere formation by controlled microfluidic transport. Langmuir 24, 9717–9726.

Lavaselli, S.A., Pedemonte, C.I., Mazon, J.I., Lillini, G.J., Pasquali, R.C., Riquelme, B., 2012. Rheological behavior of liquid-crystalline emulsion of topic application with econazole nitrate. Ser. Biomech. 27 (3–4), 34–38.

Lee, J.S., Feijen, J., 2012. Polymersomes for drug delivery: design, formation and characterization. J. Control. Release 161, 473–483.

Lee, J.N., Park, C., Whitesides, G.M., 2003. Solvent compatibility of poly(dimethysiloxane)-based microfluidic devices. Anal. Chem. 75, 6544–6554.

Levine, D.H., Ghoroghchian, P.P., Freudenberg, J., Zhang, G., Therien, M.J., Greene, M.I., Hammer, D.A., Murali, R., 2008. Polymersomes: a new multi-functional tool for cancer diagnosis and therapy. Methods 46, 25–32.

Lewis, C.L., Lin, Y., Yang, C., Manocchi, A., Yuet, K.P., Doyle, P.S., Yi, H., 2010. Microfluidic fabrication of hydrogel microparticles containing functionalized viral nanotemplates. Langmuir 26 (16), 13436–13441.

Liu, R.H., Stremler, M.A., Sharp, K.V., Olsen, M.G., Santiago, J.G., Adrian, R.J., Aref, H., Beebe, D.J., 2000. Passive mixing in a three-dimensional serpentine microchanel. J. Microelect. Syst. 9, 190–196.

Liu, K., Ding, H.J., Liu, J., Chen, Y., Zhao, X.Z., 2006. Shape-controlled production of biodegradable calcium alginate gel microparticles using a novel microfluidic device. Langmuir 22, 9453–9457.

Lo, C.T., Jahn, A., Locascio, L.E., Vreeland, N.W., 2010. Controlled self-assembly of monodisperse niosomes by microfluidic hydrodynamic focusing. Langmuir 26 (11), 8559–8566.

Lodha, A., Jadhav, G., Pande, V., 2014. Liquid crystals as a cubo-hexagonal topical controlled drug delivery system. Pharmacophore 5 (3), 430–441.

Lone, S., Kim, S.H., Nam, S.W., Park, S., Cheong, I.W., 2010. Microfluidic preparation of dual stimuli-responsive microparticles and light-directed clustering. Langmuir 26 (23), 17975–17980.

Lorenceau, E., Utada, A.S., Link, D.R., Cristobal, G., Joanicot, M., Weitz, D.A., 2005. Generation of polymerosomes from double-emulsions. Langmuir 21, 9183–9186.

Lu, A.X., Jiang, K., DeVoe, L.D., Raghavan, R., 2013. Microfluid assembly of janus-like dimer capsules. Langmuir 29, 13624–13629.

Lucchetta, E.M., Lee, J.H., Fu, L.A., Patel, N.H., Ismagilov, F.R., 2005. Dynamics of Drosophila embryonic patterning network pertube in space and time using microfluidics. Nature 434, 1134–1138.

Manjunath, K., Ready, J.S., Venkateswarlu, V., 2005. Solid lipid nanoparticles as drug delivery systems. Meth. Find. Exp. Clin. Pharmacol. 27, 127–144.

Mao, S., Guo, C., Shi, Y., Li, L.C., 2012. Recent advances in polymeric microspheres for parenteral drug delivery part 1. Exp. Opin. Drug Deliv. 9, 1161–1176.

Marcato, P.D., Durán, N., 2008. New aspects of nanopharmaceutical delivery systems. J. Nanosci. Nanotechnol. 8, 2216–2229.

McClain, M.A., 2003. Microfluidic devices for the high-throughput chemical analysis of cells. Anal. Chem. 75, 5646–5655.

Mei, Z.N., Chen, H.B., Weng, T., Yang, Y.J., Yang, X.L., 2003. Solid lipid nanoparticle and microemulsion for topical delivery of triptolide. Eur. J. Pharm. Biopharm. 56, 189–196.

Meng, F., Hiemstra, C., Engbers, G.H.M., Feijen, J., 2003. Biodegradable polymersomes. Macromolecules 36, 3004–3006.

Meng, F., Zhong, Z., Feijen, J., 2009. Stimuli-responsive polymersomes for programmed drug delivery. Biomacromolecules 10, 197–209.

Mijatovic, D., Ejkel, J.C., van den Berg, A., 2005. Technologies for nanofluidic systems top-down vs bottom-up. Lab. Chip 5, 492–500.

Mishra, B., Patel, B.B., Tiwari, S., 2010. Colloidal nanocarriers: a review on formulation technology, types and applications toward targeted drug delivery. Nanomed. Nanotechnol. Biol. Med. 6, 9–24.

Mizuno, M., Toyota, T., Konishi, M., Kageyama, Y., Yamada, M., Seki, M., 2015. Formation of monodisperse hierarchical lipid particles utilizing microfluidic droplets in a nonequilibrium state. Langmuir 31, 2334–2341.

Mueller-Goymann, C., 2002. Drug delivery-liquid crystals, Second ed Marcel Dekker, New York, 834–851.

Muenster, U., Nakamura, C., Haberland, A., Jores, K., Mehnert, W., Rummel, S., Schaller, M., Korting, H.C., Zouboulis, C.C., Blume-Peytavi, U., Schaefer-Korting, M., 2005. RU 58841-myristate—prodrug development for topical treatment of acne and androgenetic alopecia. Pharmazie 60, 8–12.

Müller, R.H., Mäder, K., Gohla, S., 2000. Solid lipid nanoparticles (SLN) for controlled drug delivery: a review of the state of the art. Eur. J. Pharm. Biopharm. 50, 161–177.

Müller-Goymann, C., 2004. Physicochemical characterization of colloidal drug delivery systems such as reverse micelles, vesicles, liquid crystals and nanoparticles for topical administration. Eur. J. Pharm. Biopharm. 58, 343–356.

Mura, S., Bui, D.T., Couvreur, P., Nicolas, J., 2015. Lipid prodrug nanocarriers in cancer therapy. J. Control. Release 208, 25–41.

Ng, J.M., Gitlin, I., Stroock, A.D., Whitesides, G.M., 2002. Components for integrated poly(dimethylsiloxane) microfluidic systems. Electrophoresis 23, 3461–3473.

Noguez, N.A., 2015. Estudio Comparativo de la capacidad de trnasporte de oxígeno de una célula artificial a partir de polimerosomas. Universidad Nacional Autónoma de Mexico. (Tesis doctoral.) México, D.F.

Okushima, S., Nisisako, T., Torii, T., Higuchi, T., 2004. Controlled production of monodisperse double emulsion by two-step droplet breakup in microfluidic devices. Langmuir 20, 9905–9908.

Onaca, O., Enea, R., Hughes, D.W., Meier, W., 2009. Stimuli-responsive polymersomes as nanocarriers for drug and gene delivery. Macromol. Biosci. 9, 129–139.

Palacios, A.J., Flores, R.H., Guillot, J., 1982. Caracterización de copolímeros. Sociedad Química de México, México, 15–25.

Pandey, R., Khuller, G.K., 2005. Solid lipid particle-based inhalable sustained drug delivery system against experimental tuberculosis. Tuberculosis 85, 227–234.

Pasquali, R., Bregni, C., Serrao, R., 2005. Estructura de las principales fases líquido-cristalinas liotrópicas. Acta Farm. Bonaer. 24 (3), 453–457.

Pasquali, R., Bregni, C., Serrano, R., 2006. Características e identificación de los cristales líquidos liotrópicos. Rev. Mex. Cienc. Farm. 37 (2), 38–53.

Patel, R., Patel, T., 2010. Liquid crystals and their application in the field of drug delivery. Fanun, M. (Ed.), Colloids in drug delivery. Surfactant Science Series, vol. 150, CRC Press Taylor & Francis Group, Florida, pp. 311–336.

Perro, A., Nicolet, C., Angly, J., Lecommandoux, S., Le Meins, J.F., 2011. Mastering a double emulsion in a simple co-flow microfluidic to generate complex polimerosomes. Langmuir 27, 9034–9042.

Pihl, J., Karlsson, M., Chiu, D.T., 2005. Microfluidic technologies in drug discovery. Drug Discov. Today 10, 1377–1383.

Prabaharan, M., Mano, J.F., 2005. Chitosan-based particles as controlled drug delivery systems. Drug Deliv. 12, 41–57.

Puglia, C., Offerta, A., Carbone, C., Bonina, F., Pignatello, R., Puglisi, G., 2015. Lipid nanocarriers (LNC) and their applications in ocular drug delivery. Curr. Med. Chem. 22, 1589–1602.

Ram, Y.M., Yadav, H.K.S., Singh, M.N., Shivakumar, H.G., 2011. Nanoparticles, promising carriers in drug targeting: a review. Curr. Drug Ther. 6, 87–96.

Rege, P.R., Garmise, R.J., Block, L.H., 2003. Spray-dried chitinosans: part I: preparation and characterization. Int. J. Pharm. 252, 41–51.

Rondeau, E., Cooper-White, J.J., 2008. Biopolymer microparticle and nanoparticle formation within a microfluidic device. Langmuir 24, 6937–6945.

Sanson, C., Schatz, C., Le Meins, J.-F., Soum, A., Thévenot, J., Garanger, E., Lecommandoux, S., 2010. A simple method to achieve high doxorubicin loading in biodegradable polymersomes. J. Control Release 147, 428–435.

Seo, M., Nie, Z., Xu, S., Mok, M., Lewis, P.C., Graham, R., Kumacheva, E., 2005. Continuous microfluidic reactors for polymer particles. Langmuir 21, 11614–11622.

Shepherd, R.F., Conrad, J.C., Rhodes, S.K., Link, D.R., Marquez, M., Weitz, D.A., Lewis, J.A., 2006. Microfluidic assembly of homogeneous and Janus colloid filled hydrogel granules. Langmuir 22, 8618–8622.

Shi, S., Yokoyama, H., 2015. Liquid crystal foams generated by pressure-driven microfluidic. Langmuir 31, 4429–4434.

Shirk, K., Steiner, C., Kim, J.W., Marquez, M., Martinez, J.C., 2013. Assembly of colloidal silica crystals inside double emulsion drops. Langmuir 29, 11849–11857.

Sia, S.K., Whitesides, G.M., 2003. Microfluidic devices fabricated in poly(dimethylsiloxano) for biological studies. Electrophoresis 24, 3563–3576.

Singh, M.N., Hemant, K.S.Y., Ram, M., Shivakumar, H.G., 2010. Microencapsulation: a promising technique for controlled drug delivery. Res. Pharm. Sci. 5, 65–77.

Soo, P.L., Eisenberg, A., 2004. Preparation of block copolymer vesicles in solution, journal of polymer science. Polym. Phy. B 42, 923–938.

Soppimath, K.S., Aminabhavi, T.M., Kulkarni, A.R., Rudzinski, W.E., 2001. Biodegradable polymeric nanoparticles as drug delivery devices. J. Control. Release 70, 1–20.

Souto, E.B., Wissing, S.A., Barbosa, C.M., Muller, R.H., 2004. Development of a controlled release formulation based on SLN and NLC for topical clotrimazole delivery. Int. J. Pharm. 278, 71–77.

Squires, T.M., Quake, S.R., 2005. Microfluidics fluid physics at the nanoliter scale. Rev. Mod. Phys. 77, 977–1026.

Steegmans, M., Warmerdam, A., Schroen, K., Boom, M.R., 2009. Dynamic interfacial tension measurements with microfluidic Y-junctions. Langmuir 25 (17), 9751–9758.

Stone, H.A., Stroock, A.D., Ajdari, E., 2004. Engineering flows in small devices microfluidics toward a lab-on-a chip. Annu. Rev. Fluid Mech. 36, 381–411.

Subramanian, B., Kim, N., Lee, W., Spivak, D.A., Nikitopoulos, D.E., McCarley, R.L., Soper, S.A., 2011. Surface modification of droplet polymeric microfluidic devices for the stable and continuous generation of aqueous droplets. Langmuir 27, 7949–7957.

Sugiura, S., Kuroiwa, T., Kagota, T., Nakajima, M., Sato, S., Mukataka, M., Walde, P., Ichikawa, S., 2008. Novel method for obtaining homogeneous giant vesicles from a monodisperse water in oil emulsion prepared with a microfluidic device. Langmuir 24, 4581–4588.

Tadwee, I: Shahi, S., Ramteke, V., Syed, I., 2012. Liquid crystals pharmaceutical application: a review. IJPRAS 1 (2), 06–11.

Tam, K.Y., Leung, K.C.F., Wang, Y.X.J., 2011. Chemoembolization agents for cancer treatment. Eur. J. Pharm. Sci. 44, 1–10.

Thorsen, T., Roberts, R.W., Arnold, F.H., Quake, S.R., 2001. Dynamic pattern formation in a vesicle- generating microfluidic device. Phys. Rev. Lett. 86 (18), 4163–4166.

Tice, J.D., Somg, H., Lyon, A.D., Ismagilov, R.F., 2004. Effects of viscosity on droplet formation and mixing in microfluidic channels. Anal. Chim. Acta 507, 73–77.

Trimmer, W., 1997. I of Electrical and E Engineers Micromechanics and MEMS: classic and seminal papers to 1990. Primera ed. USA: IEEE.

Uner, M., Wissing, S.A., Yener, G., Muller, R.H., 2005. Solid lipid nanoparticles (SLN) and nanostructured lipid carriers (NLC) for application of ascorbyl palmitate. Pharmazie 60, 577–582.

Walters, E.M., Clark, S.G., Beebe, D.J., Wheeler, M.B., 2004. Mammalian embryo culture in a microfluidic device. Meth. Mol. Biol. 254, 375–382.

Wan, J., Shi, L., Benson, B., Bruzek, M.J., Anthony, J.E., Sinko, P.J., Prudhomme, R.K., Stone, H.A., 2012. Microfluidic generation od droplets with a high loading of nanoparticles. Langmuir 28, 13143–13148.

Wang, J.X., Sun, X., Zhang, Z.R., 2002. Enhanced brain targeting by synthesis of 3′,5′-dioctanoyl-5-fluoro-2′-deoxyuridine and incorporation into solid lipid nanoparticles. Eur. J. Pharm. Biopharm. 54, 285–290.

Watanabe, T., Kimura, Y., Ono, T., 2013. Microfluidic fabrication of monodisperse polylactide microcapsules with tunable structures through rapid precipitation. Langmuir 29, 14082–14088.

Watanabe, T., Lopez, C.G., Douglas, J.F., Ono, T., Cabral, J.T., 2014. Microfluidic approach to the formation of internally porous polymer particles by sovent extraction. Langmuir 30, 2470–2479.

Wheleer, A.R., 2004. A microfluidic device to confine a single cardiac myocyte in a subnanoliter volumen on planar microelectrodes to extracellular potential recordings. Lab. Chip 4, 357–362.

Whitesides, G.M., 2006. The origins and the future of microfluidics. Nature 442, 368–373.

Whitesides, G.M., Stroock, A.D., 2001. Flexible methods for microfluidics. Phys. Today 54, 42–48.

Wissing, S.A., Müller, R.H., 2001. A novel sunscreen system based on tocopherol acetate incorporated into solid lipid nanoparticles. Int. J. Cosmet. Sci. 23, 233–243.

Wissing, S.A., Müller, R.H., 2003. Cosmetic applications for solid lipid nanoparticles (SLN). Int. J. Pharm. 254, 65–68.

Wissing, S.A., Kayser, O., Müller, R.H., 2004. Solid lipid nanoparticles for parenteral drug delivery. Adv. Drug Deliv. Rev. 56, 1257–1272.

Xu, J.-P., Ji, J., Chen, W.-D., Shen, J.-C., 2005. Novel biomimetic polymersomes as polymer therapeutics for drug delivery. J. Control. Release 107, 502–512.

Xu, J.H., Li, S.W., Tan, J., Wang, J.Y., Luo, G.S., 2006. Controllable preparation of monodisperse O/W and W/O emulsions in the same microfluidic device. Langmuir 19, 7943–7946.

Yaghmur, A., Glatter, O., 2009. Characterization and potential applications of nanostructured aqueous dispersions. Adv. Colloid. Interf. 147–148, 333–342.

Yu, Y., Pang, Z., Lu, W., Yin, Q., Gao, H., Jiang, X., 2012. Self-assembled polymersomes conjugated with lactoferrin as novel drug carrier for brain delivery. Pharm. Res. 29, 83–96.

Zhang, W., Liu, L., 2013. Study on the formation and properties of liquid crystal emulsion in cosmetic. J. Cosmet. Dermatol. Sci. Appl. 3, 139–144.

Zhang, Y., Wu, F., Yuan, W., Jin, T., 2010. Polymersomes of asymmetric bilayer membrane formed by phase-guided assembly. J. Control. Release 147, 413–419.

FURTHER READING

Meng, F., Engbers, G.H.M., Feijen, J., 2005. degradable polymersomes as a basis for artificial cells: encapsulation, release and targeting. J. Control. Release 101, 187–198.

Pata, V., Dan, N., 2003. The effect of chain length on protein solubilization in polymer-based vesicles (polymersomes). Biophys. J. 85, 2111–2118.

Pratten, M.K., Lloyd, J.B., Hörpel, G., Ringsdorf, H., 1985. Micelle-forming block copolymers: pinocytosis by macrophages and interaction with model membranes. Die Makromol. Chem. 186, 725–733.

CHAPTER

MESOPOROUS MATERIALS AND TECHNOLOGIES FOR DEVELOPMENT OF ORAL MEDICINE

23

Shou-Cang Shen, Yuan-Cai Dong, Kumaran Letchmanan, Wai Kiong Ng
*Institute of Chemical and Engineering Sciences, A*STAR, Jurong Island, Singapore*

CHAPTER OUTLINE

1 Introduction 700
2 Development of Mesoporous Materials as Potential Drug Carriers 701
2.1 Mesoporous Silica 701
2.2 Mesoporous Carbon 703
2.3 Mesoporous Composites 705
2.4 Other Mesoporous Materials 706
3 Drug-Loading Techniques 708
3.1 Solution Immersion 708
3.2 Impregnation and Solvent Evaporation 709
3.3 Cospray Drying 710
3.4 Other Loading Methods 711
4 Characterization and Stabilities Studies of Poorly Soluble Oral Drug Formulation 712
4.1 Effect of Pore Size of Mesoporous Drug Carriers 713
4.2 Stability of Amorphous Formulation and Supersaturation 716
4.3 Sustained Release of Poorly Soluble Drugs 718
4.4 Enhancement of In Vivo Bioavailability 719
4.5 Cytotoxicity of Mesoporous Drug Delivery System 720
5 Application of Functional Mesoporous Materials in Smart Oral Dosage Form 721
5.1 Slow Sustained Release of APIs 721
5.2 Stimuli Responsive Smart Drug Delivery 723
5.3 Development of Tablet Form 731
6 Summary and Outlook 734
References 735

1 INTRODUCTION

Since the discovery of M41S family of ordered mesoporous silica materials (Kresge et al., 1992), the research and development in periodic mesoporous structured materials has attracted tremendous attention for their emerging applications in biomedicine, nanotechnology, sensor, catalysis, and adsorption (Han et al., 1999; Fan et al., 2003; Lai et al., 2003; Yan et al., 2004). Among these applications, the mesoporous materials with tunable pore size at the range of 2–50 nm, large surface area, and pore volume provide a broad range of scaffolds as drug carriers for effective delivery of active ingredients with various objectives (Anglin et al., 2008; Carino et al., 2007; Nunes et al., 2007; Pasqua et al., 2009; Prokopowicz and Przyjazny, 2007; Wang, 2009). Because of their superior properties of stable mesoporous structures, uniform pore sizes, modifiable surface, nontoxic nature, and good biocompatibility, mesoporous silica carriers have exhibited distinct advantages over conventional excipients for controlled drug delivery, especially in oral medicine (Argyo et al., 2014; Tang et al., 2012; Yang et al., 2012).

The development of new drug delivery systems is gaining momentum in overcoming health care and facing challenges through higher efficiency. Among these difficulties, the poor solubility of biopharmaceutics classification system class-II (BCS-II) active pharmaceutical ingredients (API) is a major challenge for the pharmaceutical industry. Among various approaches to increase the dissolution rate and enhance bioavailability, the utilization of inorganic mesoporous materials as drug carriers has emerged as a highly effective and feasible technique. The key advantage lies in their large surface area and pore volume, which enable the carrier to host large quantities of payload molecules, protects them from premature degradation, and promotes immediate release. The nanoscaled pore structures have been found to entrap poorly soluble drugs in the amorphous or nanocrystalline form with significantly enhanced dissolution rate and increased apparent solubility in comparison with raw crystalline APIs. Another advantage feature is the rigid pore scaffold that ensures the stability of the amorphous or nanocrystalline form during storage even under stressed test conditions, which cannot be readily, achieved by in polymer-based solid dispersion formulations.

Due to the high concentration of hydroxyl group on the large surface areas of mesoporous silica based materials, the surface can be modified and functionalized to interact with drug molecules. This is used to achieve sustained release of APIs for maintaining a drug concentration in plasma for extended durations at targeted therapeutic level (Muñoz et al., 2003; Yang et al., 2007; Zeng et al., 2006). Stimuli-responsive polymers or guest particles have been combined to control the pore opening of mesoporous drug carriers for controlled delivery of drugs, enzymes, or proteins to certain part of digestion system through oral administration. Furthermore, composite matrixes of mesoporous materials functionalized with pH-, thermal-, or bioresponsive polymers were developed as drug carriers for smart on demand delivery. Various methods for drug loading in mesoporous materials have been investigated to achieve efficient loadings, such as in situ absorption, impregnation, cospray drying, and supercritical fluid processing.

Mesoporous silica microparticles and nanoparticles have been extensively explored for effective parenteral drug delivery (He et al., 2010a; Manzano et al., 2009; Shen et al., 2013). However, the slow degradation and excretion of mesoporous silica particles from the body circulation system remains challenging for application in intravascular administration. As a comparison, silica materials has been approved and widely used as excipients to manufacture granules or tablets for oral administration. Therefore, the practical application of mesoporous silica in oral medicine is much relatively safer and more feasible due to lower cytotoxicity to digestion system and ease of excretion. In this chapter, the recent progress in development of mesoporous materials as drug carriers for applications in oral medicine formulation are summarized according to the types of mesoporous excipients, drug-loading techniques, characterization, and stability studies and applications in oral medicine.

2 DEVELOPMENT OF MESOPOROUS MATERIALS AS POTENTIAL DRUG CARRIERS

2.1 MESOPOROUS SILICA

Among various mesoporous materials, silica is the most extensively studied as drug carriers with high delivery efficiency. According to pore size, mesoporous silica is generally divided into two categories: MCM (mobile composite matter) and SBA (Santa Barbara amorphous material). The MCM materials are generally fabricated by using quaternary ammonium salts, such as cetyltrimethylammonium bromide (CTAB), as a soft template to direct the formation of mesoporous structure during the hydrothermal synthesis process. To synthesis MCM mesoporous silica, in brief, an alkyltrimethylammonium surfactant (e.g., CTAB) is first dissolved in water at a given concentration. The formation of hexagonal array of cylindrical micelles follows in which the hydrophobic hydrocarbon chains are gathered in the center and the polar groups are aligned on the surface, in contact with a continuous region of water surrounding the micelles. These quaternary ammonium micelles then interact with the silicate species (e.g., sodium silicate) and precipitate. After subsequent hydrothermal treatment, the resulting solid materials comprise organic template in the composite. When the surfactant is removed by calcination or extraction with an appropriated solvent, the hollow structured silica with uniform pore size is obtained. The resultant MCM-41 is amorphous in nature, having ordered hexagonal arrays of cylindrical pores with a very narrow pore size distribution. By adjusting the synthesis condition, the formed pore structure can be tuned and arranged as a two-dimensional (2D) hexagonal (MCM-41), three-dimensional (3D) cubic (MCM-48), or lamellar (MCM-50) structures. As the synthesis of MCM-41 is based on a cylindrical micelle, pore diameter of the materials can be tailored to a desired size by the choice of surfactants with appropriate alkyl chain length. The pore size can be controlled at the range of 2 to 10 nm. Alternatively, micelles can be expanded by dissolving organic substances (e.g., trimethylbenzene) into the fatty regions of micelles. Studies have also shown that factors, such as surfactant chemistry, surfactant-to-silica ratio, and different silica sources can influence the formation of MCM-41, particularly the geometry of the pores being cylindrical or cubic in shape (Wan and Zhao, 2007).

Another type of mesoporous silica SBA with pore size tunable at the range of 5 to 50 nm was synthesized by using nonionic pluronic poly(ethylene oxide)–poly(propylene oxide)–poly(ethylene oxide) (PEO_x–PPO_y–PEO_z) triblock polymer, such as P123, as soft template and tetraethoxysilane (TEOS) is used as silica source. SBA-15 with 2D arranged uniform hexagonal arrays has been intensively studied due to its mesostructural diversity as well as the larger pore. The pore walls are also found to be thicker, which may be an advantage in tableting process where material could resist the compression without pore structure collapse. The pore size can be tailored by using the triblock pluronic polymer surfactant with different block numbers or tuning the synthesis conditions (Johnson et al., 1999; Zhao et al., 1998). The feasibility to obtain different pore sizes and geometries offers wide potential for hosting molecules larger than the ones exhibited for classic microporous materials.

Among a wide variety of mesoporous materials with different mesostructures, MCM-41 was firstly applied as a drug carrier for controlled drug delivery in tablet form for potential oral administration (Vallet-Regi et al., 2001). This work triggered the rapid growth of research and development activity using mesoporous silica and other mesostructured inorganics materials for drug delivery (Heikkila et al., 2007a; Salonen et al., 2005; Ukmar and Planinsek, 2010). The ordered mesoporous matrices like MCM-41 and SBA-15 exhibited very promising results in the application of dissolution-enhancing formulations for poorly soluble drugs with a high degree of physicochemical diversity. Mesoporous materials have been used as an amorphous drug delivery system and achieved solubility enhancement for poorly soluble drugs owing to their nanosized confinement and large surface areas (Qian and Bogner, 2012).

The particle size of mesoporous silica could be several to hundreds of micrometers. The most common morphology of SBA-15 is noodle-like fiber bundles of tens of micrometers in length and coexisting with short rod-like particles of several micrometers in length (Schmidt-Winkel et al., 1999). Zhao et al. (2000) have reported that SBA-15 with various morphologies of rope-, doughnut-, and discoid-like shapes synthesized by self-assembly in the presence of organic additives and inorganic salts. Short rod-like SBA-15 particles with length of 1–2 μm have been obtained by either static hydrolysis of tetraethyl orthosilicate (TEOS) without stirring during the self-assembly process (Sayari et al., 2004), or hydrothermal synthesis in the assistance of high concentration of KCl (KCl/TEOS >1.0) in the mixture (Yu et al., 2002). Submicron-size monodispersed SBA-15 particles were synthesized by static hydrolysis and followed by dry gel conversion (Shen et al., 2006).

Recently, mesoporous silica nanoparticles (MSNs) with a particle size at about 50 to several hundred nm have attracted a great deal of attention for drug delivery application (He and Shi, 2011; Li et al., 2015a). Several researchers have used additional agents or special treatments to control the particle size or morphology of MSNs. Park et al. (1998) used microwave to synthesize MSNs with sizes ranging from 200 nm to 2 μm by adding ethylene glycol (EG). The Okubo group synthesized colloidal mesoporous silica with particle sizes of 70–110 nm by the addition of EG (Gu et al., 2007). Qiao et al. (2009) controlled particle sizes from 25 to 200 nm by adding inorganic bases and alcohols. MSNs have been intensively investigated for drug delivery with superior performance drug carriers could improve drug loading and release efficiency (Wang et al., 2015). MSNs offer the potential to achieve enhanced oral bioavailability of poorly soluble drugs via improved drug dissolution rate and enhanced drug permeability (Zhang et al., 2012). Moreover, MSNs can be modified by functional polymers for controlled release of active ingredients. Zhang et al. (2010a) reported that spherical MSNs were synthesized using an organic template method in an oil/water phase, and large pore diameter MSNs were functionalized with aminopropyl groups through postsynthesis treatment. This kind of MSN was loaded with a poorly water-soluble telmisartan at high loading up to 60 wt.% for intended oral administration. The release rate of telmisartan from MSNs was found to be effectively increased and the release rate of telmisartan from the functionalized MSNs could also to be effectively controlled compared with that from the unmodified MSNs (Li et al., 2014a). Li et al. (2014a) reported that poly(ethyleneimine)s modified MSNs could be synthesized by biomimetic method, and used for controlled release of propranolol hydrochloride (PNH). Furthermore, MSNs were also functionalized with stimuli-responsive polymer or incorporated with nanovalves for smart drug delivery (Dong et al., 2013; Hwang et al., 2015; Tarn et al., 2014).

A novel drug delivery system of hollow mesoporous silica spheres (HMSS) have also been developed for oral administration to improve the drug loading capacity and effective delivery (Guo et al., 2011; Zhao et al., 2009a). Chen et al. (2013) synthesized HMSS with an average diameter of 900 nm and shell thickness of 100 nm by using a soft template. The significant improvement of drug loading and dissolution rate demonstrated that HMSS could provide good reservoirs for drug solutions to enhance the dissolution of poorly water-soluble drugs. Zhu et al. (2005) found that HMSS with uniform size and morphology could be successfully synthesized in a facile route using poly(vinylpyrrolidone) (PVP) and CTAB as cotemplates at room temperature. The hexagonally ordered pore channels were formed on the shell layer. Ibuprofen (IBU) as a model drug was used to examine the storage capacity and the release behavior of drug molecules. The storage capacity of HMSS could reach 1133 ± 52.4 mg/g for IBU, which was 3-fold higher than that of conventional mesoporous materials previously reported (~337 mg/g), and the storage capacity was adjustable in a wide range. Furthermore, this system also showed a sustained-release behavior. In addition, novel core-shell dual-mesoporous silica nanospheres with a tunable pore size were also successfully synthesized using a styrene monomer as a channel template for the core and CTAB as a channel template for the shell in order to improve the dissolution rate of poorly water-soluble drugs (Wu et al., 2014).

2.2 MESOPOROUS CARBON

Mesoporous-activated carbon has been synthesized via chemical activation by using coconut shell as carbon source and $ZnCl_2$ as activating agent with appropriate ratio (Hu et al., 2000; Hu and Srinivasan, 2001). Nevertheless, the pore size of the resulted mesoporous carbon was less uniform and pore channels were in disordered arrangement. The ordered mesoporous carbon materials have been synthesized by nanocasting technology using appropriate ordered mesoporous silica as hard templates (Fuertes, 2003; Liang et al., 2008). Joo et al. (2001) reported the synthesis of CMK-1 mesoporous carbon. The harvested mesoporous carbon generally preserved the morphology of mesoporous silica. Most mesoporous carbon materials usually exhibited the common fiber morphology by inheriting from the parent template of mesoporous silica although various pore structures could be fabricated (Che et al., 2003; Li et al., 2013a; Ting et al., 2010; Vinu et al., 2007). Recently, various ordered mesoporous carbon materials have been synthesized and used as oral drug carriers (Gao et al., 2015; Saha et al., 2015; Zhang et al., 2015a). A hierarchical porous carbon monolith with 3D-ordered macropores (~400 nm) and uniform accessible mesopores (~5.2 nm) was synthesized via a facile dual-templating technique using colloidal silica nanospheres and Poloxamer 407 as templates (Zhang et al., 2014a). The resulted mesoporous carbon exhibited excellent capability for inclusion of poorly soluble drug, such as Valsartan (VAL) in pore channels in amorphous and showed amorphous for at least 6 months. In vitro tests showed that the dissolution rate of mesoporous carbon-based formulation was increased significantly compared with that of crystalline VAL. Furthermore, a pharmacokinetic study in rats demonstrated about 2.4-fold increase in oral bioavailability as compared with the commercially available VAL formulation. Zhu et al. (2014a) synthesized mesoporous carbon with spherical pores by using mesoporous silica as template and used as drug carrier for poorly soluble celecoxib to improve dissolution rate and oral bioavailability with approximately 1.59-fold enhancement as comparison with commercial celecoxib. Shen et al. (2014b) synthesized monodispersed mesoporous submicron carbon particles by cospray drying of SBA-15 submicron particles and sucrose, followed by carbonization in a nitrogen environment and removal of the hard SBA-15 template in NaOH solution. The ratio of sucrose/SBA-15 used for the cospray drying affected the integration and dispersibility of the obtained mesoporous carbon particles and sucrose/SBA-15 ratios at 1.0 ~1.5 resulted in monodispersed mesoporous carbon submicron particles. The monodispersed mesoporous carbon particles were used as drug carrier to enhance the dissolution rate of a poorly soluble drug, indomethacin (IDMC). The morphology of mesoporous submicron particles was not changed after drug loading as most IDMC molecules were encapsulated into its pore channels in amorphous state, contributed significantly to the enhancement in dissolution rate of IDMC. Due to the ordered mesoporous structure, the pore walls separate the IDMC particles and prevent recrystallization. The amorphous IDMC in solid dispersion could exhibit excellent stability under stress storage conditions of 40°C/75%RH for 6 months. Furthermore, Zhang et al. (2014b) reported that a novel mesoporous carbon matrix, with a large pore size (about 26 nm) and a 3D porous structure was successfully prepared by an inverse replica templating method using mesocellular silica template and applied as a drug delivery vehicle for the oral delivery of poorly water-soluble drugs. Characterization demonstrated the outstanding features of mesoporous carbon foam, including a high surface area, large pore volume and a well-defined pore size distribution, making it an excellent drug carrier. Celecoxib incorporation into the prepared large pore mesoporous carbon resulted in an approximately 9-fold increase in the aqueous solubility in comparison with crystalline Celecoxib.

As the hard template synthesis route involves more steps and additional cost of mesoporous silica, soft templates have recently been used to synthesize mesoporous carbon materials (Chang et al., 2013a; Choma et al., 2011, 2013). The new synthesis stratagem involves organic soft-templating or formation of a self-assembly of surfactant micelles in the carbon precursor matrix. This method avoids several difficulties associated with the silica-based hard templating technique. By using soft surfactant template techniques, mesoporous carbon could be synthesized by crosslinking a proper organic resin in the presence of some sacrificial surfactants, such as pluronic triblock polymer F127, which yield negligible carbon upon pyrolysis in N_2 environment leading to removal of soft template and form mesoporous carbon matrix. Phenol (Fang et al., 2010; Meng et al., 2005), resorcinol (Lee and Oh, 2002), phloroglucinol (Liang and Dai, 2006), and hexaphenol (Saha et al., 2012) have been successfully used as carbon precursors for synthesis of mesoporous carbon. Saha et al. (2014a) reported the synthesis of mesoporous carbons by soft template for potential oral drug delivery. The mesoporous carbons, synthesized from phloroglucinol and lignin, a synthetic and a sustainable precursor, respectively, exhibited specific surface area of 200–400 m^2/g and pore volume of 0.2–0.6 cm^3/g. The obtained carbon has narrower pore widths (4–6 nm) and higher pore volume. The mesoporous carbons have excellent potential for the controlled-release of the specific drugs through oral delivery (Saha et al., 2014b). By using soft template synthesis route, the particle size of mesoporous carbon could be feasibly reduced to nanoscale and potentially be used for drug delivery (Schuster et al., 2012; Xu et al., 2014; Zhai et al., 2011; Zhang et al., 2015a). The ordered mesoporous carbon nanospheres with an average diameter of about 170 nm and a smooth surface were synthesized using Pluronic F127 as a soft-template and phenolic resin as a carbon precursor. The mesoporous carbon nanospheres were incorporated with lipid for application of improved oral delivery of the poorly water-soluble drug (Zhang et al., 2015b). Liu et al. (2015) reported that mesoporous carbon nanoparticles with particle size at 100–126 nm with large and adjustable pores have been synthesized by using poly(ethylene oxide)-*b*-polystyrene (PEO-*b*-PS) as a template and resorcinol–formaldehyde (RF) as a carbon precursor. The pore size could be tuned in the range of 13–32 nm. This carbon material with large pores and small particle sizes may have potential application in drug/gene delivery. Moreover, mesoporous carbon nanoparticles have been prepared from glucose and stabilized with phospholipid for delivery of antitumor drug (Zhang et al., 2015a). Mesoporous carbon was also partially oxidized with carboxyl groups on surface for loading a oral poorly water-soluble drug, and it was found to be more efficient to enhance the drug loading capacity and improve the oral bioavailability (Zhang et al., 2013a).

2.3 MESOPOROUS COMPOSITES

To achieve enhanced drug delivery efficiency, mesoporous silica based composites have been developed as drug carriers (Rana et al., 2011; Song et al., 2012; Zhao et al., 2009b; Zhu et al., 2011a). Mesoporous silica could be functionalized with other inorganic materials to form "inorganic-inorganic" composites. Recently, Fang et al. (2014) reported that dual-pore mesostructured carbon@silica composite core–shell nanospheres for multidrug delivery. Monodispersed mesoporous carbon was synthesized by using F127 as a soft template and phenol/formaldehyde as carbon source. On the surface of this kind mesoporous carbon core with pore size about 5.8 nm, a shell layer of silica was coated and exhibited pore size of 3.1 nm. The hierarchical mesostructure and amphiphilicity of the hydrophobic carbon cores and hydrophilic silica shells led to distinct benefits in multidrug combination therapy with anticancer drugs, such as cisplatin and paclitaxel. Moreover, nanocomposites of mesoporous silica and calcium phosphate are of great interest for applications to controlled drug delivery as they have demonstrated excellent bioactivity or biodegradability. The nanoporous composite of silica containing calcium phosphate has been found to have high porosity and provided a sustained release of antibiotics (El-Ghannam et al., 2005). A composite of mesoporous SBA-16 and hydroxyapatite (HAP) was fabricated by the growth of calcium phosphate particles within an organized silica structure (SBA-16) for evaluation of its application as a drug delivery system (Andrade et al., 2014). Yang et al. (2011) reported a novel luminescent mesoporous silica/apatite composite for controlled drug release. A bifunctional composite, combining the mesoporous structure of MCM-48 and the self-activated strong blue luminescence of HAP, was synthesized for drug delivery and in vitro drug release could be monitored in the simulated body fluid (SBF) system. Studies have shown that this spherical composite exhibited ordered mesoscale pores, which were suitable to be a drug carrier for controlled drug release. Drug storage/release test indicated that this composite showed high drug loading capacity and well controlled release behavior for captopril. Most importantly, the composite loaded with captopril still shows strong blue emission centered at 410 nm under UV excitation, and the luminescence intensity kept increasing when drug molecules are gradually released from the composite. Moreover, composites of mesoporous silica based MCM-48 and HAP have been synthesized by in-situ synthesis procedure (Aghaei et al., 2014). MCM-48/HAP nanocomposite was also fabricated by in situ synthesis and investigated as bioactive ceramic and used a drug delivery agent to achieve higher dissolution rate of IBU.

MgO incorporated SBA-15 mesoporous silica was prepared by in situ coating (Wang et al., 2005; Wei et al., 2003). Shen et al. (2007) reported that the submicron particles with modified surface were synthesized by a simple one-pot synthesis approach and used as a drug carrier for controlled release. The delayed release profile from MgO-coated SBA-15 was attributed to the interaction of IBU molecules with MgO on drug carrier surface. In addition, the release rate of IBU could be modulated by varying the content of MgO, and was found to decrease with higher amounts of MgO coated on the surface of SBA-15 submicron particles. Mesoporous silica was also functionalized with magnetic Fe_3O_4 as composites, which can be potential vectors for drug delivery and bioadsorption. Huang et al. (2012) reported that magnetic Fe_3O_4@mesoporous silica composites were synthesized by generating Fe_3O_4 nanoparticles in the mesoporous silica matrix using the sol–gel method in nitrogen atmosphere. The composites possessed suitable mesoporous structure for the adsorption of both small-molecular drug and large biomolecules. The in vitro test indicated that these Fe_3O_4@mesoporous silica composites show sustained release properties for aspirin. The release of the aspirin molecules from the pores of the Fe_3O_4@mesoporous silica composites was basically a diffusion controlled process. In addition, magnetic silica-pillared clay with ordered layered structure and uniform pore size distribution were synthesized by generating magnetic Fe_3O_4 nanoparticles onto the mesoporous silica-pillared clay hosts using molecular self-assembly method (Mao et al., 2014). These magnetic mesoporous silica-pillared clays could exhibit sustained release profiles of hosted drugs.

Composites of polymer and mesoporous silica-based materials were of great interest to be used as drug carriers for high drug loading and controlled delivery (Xue and Shi, 2004; Zhu et al., 2013). Recently, polyelectrolytes-coated mesoporous silica particles was synthesized by postsynthesis coating through electrostatic interactions between a positively charged polyethylenimine and a negatively charged poly(sodium-4-styrenesulfonate) (Varga et al., 2015). The synthesized mesoporous silica exhibited a well-ordered hexagonal structure, high surface area (1254 m^2/g) and relatively large pore diameter, which enabled the high capacity for adsorption of IBU up to 1666 mg/g. After loading with IBU, the drug-loaded particles was immersed with positive charged polyethylenimine solution and was stirred for one more hour, then centrifuged and followed by treatment in negative charged poly(sodium-4-styrenesulfonate) solution to form polymer coating in external surface. With the increase of the number of the polymer shell layers, the release rate was slowed, more than with 30 percentages remained in the composites. The sustained and the controlled release were realized by forming the core-shell composites. The core-shell structured mesoporous particles were suitable for delivery of poorly soluble drugs in aqueous biological systems. Moreover, Islam et al. (2015) investigated polyelectrolyte-mediated hierarchical mesoporous calcium silicates as a platform for drug delivery carrier with ultrahigh loading capacity and controlled release behaviour. A facile method for the poly(allylamine hydrochloride) (PAH)-assisted synthesis of mesoporous calcium silicate hydrates (PAH-CS) with a large specific surface area and pore volume has been developed. IBU and FITC-labeled bovine albumin (FITC-albumin) were chosen as a model drug and model protein, respectively, to test the loading and delivery efficiencies of the PAH-CS carriers. The ultra-high drug loading capacities (IBU/carrier at 3.5 g/g or FITC-Albumin/carrier at 1 g/g) and their release patterns were investigated under controlled pH conditions. The PAH-CS showed no cytotoxic effect on MC3T3-E1 cell lines evaluated by the LDH (lactate dehydrogenase) assay in supernatant medium, indicating good biocompatibility of the PAH-CS drug carriers. A number of "organic-inorganic" mesoporous composite have been synthesized and explored the potential for oral drug delivery (Qian et al., 2014; Zhang et al., 2015b). By incorporation with stimuli-responsive polymer, a smart drug delivery system could be obtained (Wen et al., 2013; Zhang et al., 2014c) and this kind of mesoporous composites for application in oral medicine will be reviewed in controlled drug delivery section.

2.4 OTHER MESOPOROUS MATERIALS

In addition to mesoporous silica and carbon, many other biocompatible mesoporous materials have been synthesized and explored as oral drug carriers. Mesoporous silicon also attracted attention for drug delivery (Anglin et al., 2008; Lehto et al., 2005; Roine et al., 2012). Mesoporous silicon was generally obtained by electrochemical anodization of silicon wafer where the porosity and pore size can easily be tuned by anodization parameters (Salonen et al., 2000). The surface of mesoporous silicon was modifiable and could form a stable hydrophilic matrix with a nanoporous structure (Song and Sailor, 1999). Wang et al. (2010) investigated oxidized mesoporous silicon microparticles for improved oral delivery of poorly soluble drugs. The mesoporous silicon microparticles were reported as a solid dispersion carrier for improving dissolution and enhancing the orally administered pharmacokinetics of IDMC by encapsulation in nanopores as amorphous state. The stability of the amorphous state has been demonstrated for several months under accelerated storage conditions. The mesoporous silicon carrier facilitated the immediate release of IDMC and enhanced oral delivery performance in comparison with crystalline IDMC. Salonen et al. (2005) investigated five model drugs (antipyrine, IBU, griseofulvin, ranitidine, and furosemide) into mesoporous silicon microparticles and found their obviously improved release behavior for oral delivery.

Mesoporous nanoHAP (mn-HAP) was successfully synthesized via a novel micelle-templating method using lauryl dimethylaminoacetic acid as zwitterionic surfactant (Amer et al., 2013) or use yeast as soft template (He et al., 2010b). The use of such a surfactant in combination with microwave energy input enabled the precise control of pore size in a narrow-size distribution range (~36 nm). Liang et al. (2013) reported that mesoporous HAP could be synthesized by using a template-free sonochemistry-assisted microwave method. Due to the rapid formation of mesostructure in the presence of microwave, a continuous microwave flow synthesis technique has been successfully developed for potentially large scale fabrication of mesoporous HAP (Akram et al., 2015). The pore size of mesoporous HAP could be tunable by using 1-dodecanethiol as the pore expander and large pore was achievable (Bakhtiari et al., 2015). The mesoporous HAP has been applied as drug carrier for drug delivery (Gu et al., 2014a). Ibrahim et al. (2015) synthesized pure and mesoporous HAP nanopowder with a large pore volume (1.4 cm^3/g) and surface area (284.1 m^2/g) from raw eggshells at room temperature using a simple two-step procedure. The precursor droplets was controlled by the size of the needle connected to the syringe pump used for precursor additions and the pH value of the reaction solution was stabilized, leading to production of HAP with high surface area and large pore size. The as-produced HAP revealed high IBU (as a model drug) loading (1.38 g/g HAP) and enhanced its dissolution properties. He et al. (2015) reported that mesoporous HAP rod-like nanocrystals with large and tunable mesopores and small particle sizes were synthesized using a novel bioinspired route where MSNs acted as effective nucleating agents, and fetuin (an abundant serum protein) inhibited the growth of HAP. The large pore HAP was investigated as drug delivery vehicles and exhibited a high loading capacity and sustained release of large molecule protein. Demonstration of effective intracellular uptake and low cytotoxicity confirmed the potential of the mesoporous HAP as carriers for loading large molecular pharmaceutics. The similar carbonated HAP was also synthesized and used as drug delivery systems with high drug-loading capacity (Guo et al., 2012, 2013; Long et al., 2013). Zhao et al. (2012a) prepared nanosized hydroxycarbonate apatite as a drug carrier to improve the dissolution rate and increase the bioavailability of poorly soluble oral drugs. Uniform mesoporous hydroxycarbonate apatite nanoparticles were synthesized using $CaCO_3$ as a sacrificial template by the hydrothermal method in the presence of CTAB as a surfactant. The prepared mesoporous hydroxycarbonate apatite was used as a drug carrier to investigate the drug uptake and release of poorly soluble carvedilol. In vitro drug dissolution tests showed that mesoporous hydroxycarbonate apatite could produce burst release of carvedilol as comparison with micronized carvedilol in simulated gastric fluid and intestinal fluid. Moreover, mesoporous HAP has been synthesized with other functional materials to form composite as drug carriers enabled controlled or stimuli-responsive release (Gu et al., 2014b; Guo et al., 2014a; Li et al., 2013b, 2014b; Lin et al., 2013).

Mesoporous zirconium phosphonate has been synthesized and applied for oral medicine delivery study. Tang et al. (2013) reported that the bifunctional mesoporous zirconium phosphonates could be synthesized through surfactant-assisted cocondensation of $ZrCl_4$ with two different phosphoric acids, both 1-phosphomethylproline (H_3PMP) and 1,4-bis(phosphomethyl)-piperazine (BPMP), in an one-pot synthesis procedure. The L-proline group of H_3PMP and piperazine group of BPMP in the frameworks rendered zirconium phosphonates with pH-controllable release function and high cell penetration capability. Further functionalization with biological modifiers was achieved by the carboxyl groups in L-proline groups on the external surface of mesoporous zirconium phosphonates. On the basis of pH-sensitive zirconium phosphonates carriers and assisted by lag-time films coating, the time- and pH-controlled oral colon-targeted nucleic acid delivery systems have been developed, which can carry most of the loaded nucleic acids to the colon under dual time and pH value control. Shi et al. (2010) also investigated the pH-sensitive mesoporous zirconium diphosphonates for controllable oral colon-targeted delivery of antitumor medicine. The drug molecules will be trapped without release when passing through the stomach, and be released completely in a sustained-release kinetics in the colon followed by a minimal drug release in the small intestines.

Moreover, mesoporous calcium carbonate materials have been prepared and applied for oral drug formulation to stabilize the amorphous state of active pharmaceutical ingredients. Forsgren et al. (2013) reported that, after incorporation of the poorly soluble celecoxib in the porous calcium carbonate (vaterite) particles, a 5- to 6-fold enhancement in apparent solubility of celecoxib could be achieved due to pore-induced amorphization. Synthesis of the carrier particles and the incorporation of the active substances were carried out simultaneously in one-step procedure, enabling easy fabrication. Furthermore, an improvement in solubility was also confirmed for poorly soluble ketoconazole when formulated with the mesoporous vaterite.

A number of other mesoporous materials have been developed and used for oral drug application. Mesoporous alumina with uniform pore distribution was reported for application in controlled drug release (Das et al., 2009; Kapoor et al., 2009). Zhang et al. (2014d) reported that Upsalite, a mesoporous magnesium carbonate, could be used to stabilize amorphous IBU as an approach to increase the aqueous solubility of poorly soluble drugs. Mesoporous ZnO nanospheres have been synthesized by simple soluble starch-insertion method and used as a biocompatible carrier for the controlled delivery of captopril via oral administration (Bakrudeen et al., 2013). In addition, bioactive glass has been developed to mesoporous structure and applied as drug carrier for effective oral delivery (Javadzadeh et al., 2014; Li et al., 2013c; Zhao et al., 2008; Zhu et al., 2012).

3 DRUG-LOADING TECHNIQUES

3.1 SOLUTION IMMERSION

The solid dispersion of mesoporous materials with drug molecules have been formulated by using different methods. The model drug could be loaded onto mesoporous silica by solution immersion procedure. The high loading of drugs in mesoporous solid could be achieved due to its large surface area for physically adsorption of drug molecules. Although a large number of poorly soluble drugs encounter the difficulty to be dissolved in aqueous-based medium, most of these drugs could be well dissolved in a proper organic solvent. In 2001, mesoporous silica was first loaded with poorly soluble IBU by soaking MCM-41 powder or tablet in solution of IBU dissolved in hexane (Vallet-Regi et al., 2001). The amount of IBU loaded onto MCM-41 was determined by the IBU concentration change in solution measured by UV spectrophotometer and drug loading was also measured by thermosgravimetry as a reference. IBU was also loaded onto mesoporous silica by soaking in solution of IBU dissolved in ethanol (concentration could be up to 700 mg/mL). Ethanol could be used as the loading solvent as it is safe, nontoxic, and dissolves IBU in large quantities. Heikkila et al. (2007b) performed that the loading of IBU onto TUD-1 in a closed container to prevent evaporation of ethanol for the total loading time of 20 h at ambient conditions, whereupon the sample was vacuum filtered through a Teflon membrane filter with a 1 μm nominal pore size. The IBU adsorbed on the exterior surfaces of the drug-loaded microparticles was removed by washing the sample with 3 mL of ethanol. Finally, the sample was dried for analysis. Salonen et al. (2005) studied five different model drugs onto mesoporous silicon microparticles by the solution adsorption using different solvents, such as purified water, absolute ethanol (99.5%), dimethylformamide (DMF), and dimethylsulfoxide (DMSO). Surface modification of mesoporous silica could result in more drug molecules to be loaded by immersion adsorption. When the mesoporous SBA-15 was modified by alkaline MgO element, the loading of IBU was increased by using the same adsorption procedure (Shen et al., 2007). The loading of IBU on pure silica SBA-15 was only 15.8 wt.%; as a comparison, MgO/SBA-15 (Si/Mg = 20) submicron particles adsorbed 28.6 wt.% of IBU although its surface area was lower than pure silica SBA-15. With increase of MgO content to Si/Mg = 5, the IBU loading was increased to 32.2%. MgO coated SBA-15 has been proven to create basic sites on the large surface of SBA-15 materials, and thus significantly improved its adsorption of acidic model drugs and resulted in higher drug loading. Recently, Li et al. (2015b) functionalized the surface of mesoporous silica with amino and carboxyl groups. It was found that IDMC loading capacity was obviously enhanced when using amino- and carboxyl-modified mesoporous silica as drug carriers due to the stronger hydrogen bonding force formed between surface modified carrier and drug. As the solution adsorption was associated with a washing step to remove surface-free drug particles, the final drug loading could be lower than designed value and lack of consistence due to variation in washing speed and batch scale in different labs (Kiwilsza et al., 2015).

3.2 IMPREGNATION AND SOLVENT EVAPORATION

To achieve the drug loading to be consistent with the active ingredient in designed formula, the solvent evaporation method was used (Zhao et al., 2012a). Mellaerts et al. (2008a) loaded poorly water-soluble itraconazole onto ordered mesoporous silica by suspending the powder into an itraconazole solution in methylene chloride (5 mg/mL). The mixture was agitated for 24 h using a rotary mixer. Subsequently, the solvent was removed by evaporation and the powder was dried overnight at 35°C. Mesoporous silica loaded with itraconazole was heated to 100°C for 5 min and placed under reduced pressure (103 bar) at 40°C for another 48 h to ensure the complete removal of methylene chloride. The final loading in solid dispersion was very closely to the designed loading as the loss of active ingredient was negligible by omitting the washing step used in adsorption method. Thus, solvent evaporation is widely used method for loading API onto mesoporous drug carriers. Itraconazole was also loaded to mesoporous SBA-15 by the incipient wetness impregnation method to use dichloromethane as solvent (Mellaerts et al., 2008b, 2010). During the addition of itraconazole solution, the powder and solution mixture was intensively mixed with a spatula. The drug-loaded powders were obtained by solvent evaporation at 35°C in air for 24 h and subsequently placed under reduced pressure at 40°C for 48 h. The impregnation method was compared with melt method and it was found that drug distribution could be affected by using different loading procedure. The effectiveness of the loading method was found to be strongly compound dependent. Incipient wetness impregnation using a concentrated itraconazole solution in dichloromethane followed by solvent evaporation was most efficient for dispersing itraconazole in SBA-15. The itraconazole molecules were located on the mesopore walls and inside micropores of the mesopore walls. When SBA-15 was loaded by slurrying it in a diluted itraconazole solution from which the solvent was evaporated, the itraconazole molecules ended up in the mesopores. At a loading of 30 wt.%, itraconazole exhibited intermolecular interactions inside the mesopores revealed by UV spectroscopy and endothermic events traced with DSC. The physical mixing of itraconazole and SBA-15 powder followed by heating above the itraconazole melting temperature resulted in formulations in which glassy itraconazole particles were deposited externally on the SBA-15 particles. As a comparison, IBU was preferably positioned inside the micropores either by impregnation or melt loading procedure. Zhang et al. (2010a) studied that Telmisartan, a well-known antihypertensive drug, could be loaded onto MSNs by using acetic acid as the loading solvent because it is also safe, nontoxic, and can dissolve large amounts of Telmisartan. Due to the large pore size and pore volume, high drug loading up to 60 wt.% could be successfully achieved and preserved the drug in solid dispersion in no-crystalline form. A large number of poorly soluble APIs was formulated with mesoporous materials by solvent evaporation methods. Carvedilol was loaded onto MCM-41 and SBA-16 by soaking in methanol/carvedilol solution and followed by evaporation of methanol (Hu et al., 2012). Lovastatin was dissolved in chloroform and loaded onto porous silica monolith by solvent evaporation (Wu et al., 2012). Van Speybroeck et al. (2009) applied incipient wetness method to load 10 model BCS-II poorly soluble drugs (carbamazepine, cinnarizine, danazol, diazepam, fenofibrate, griseofulvin, IDMC, ketoconazole, nifedipine, phenylbutazone) onto mesoporous SBA-15 and evaluated the applicability of the ordered mesoporous silicate SBA-15 as an excipient to enhance dissolution. All model drugs were dissolved in methylene chloride and mixed with SBA-15 under stirring with a spatula until seemingly dry, followed by drying at 40°C in a vacuum oven. In all cases, the formulated drug was found to be amorphous, as evidenced by DSC results. This loss of crystallinity resulted in a significant improvement in dissolution rate as compared to the crystalline drugs, indicating that SBA-15 can be applied as a dissolution-enhancing formulation approach for a very wide variety of poorly soluble drugs. Limnell et al. (2011a) investigated the effect of drug loading methods on performance of poorly soluble drug formulated with mesoporous silica. IDMC was loaded on to MCM-41 by solvent evaporation method and compared with immersion absorption process. High concentrations of drug in solution is required for absorption method in order to achieve higher drug loading, thus the selection of solvent is critical for drug loading by solution absorption process. The drug-loading efficiency of solvent evaporation is higher, and the formation of amorphous form or crystalline phase may also be related with evaporation process. Generally, to harvest amorphous formation by impregnation and solvent evaporation, the drug loading is kept to be below 30 wt.%, the high drug loading could result in partial crystal form on external surface. The concentration of drug is increased while solvent is evaporating, crystallization of drug molecules could occur before being driven to pore channels.

3.3 COSPRAY DRYING

Spray drying is a matured process that has been widely used for manufacture of pharmaceutical products and development of new formulations (Paudel et al., 2013). Shen et al. (2010, 2011) successfully loaded poorly soluble model drugs onto mesoporous SBA-15 by cospray drying. IBU was dissolved in ethanol and SBA-15 submicron particles were dispersed in the solution under stirring overnight. The spray drying was performed on a BÜCHI B-290 minispray dryer (BÜCHI Labortechnik AG, Switzerland) operated in inert loop mode. The poorly soluble was loaded onto porous drug carriers with high efficiency. Amorphous formulation of poorly soluble drug in mesoporous silica was obtained at high loading up to the pore channels being fully filled with guest molecules. Cospray-drying IBU with mesoporous

SBA-15 submicron particles exhibited excellence in production of stable amorphous IBU with significantly enhanced dissolution rate. With drug loading of IBU/SBA-15 ratio being 50:50 (wt./wt.) or below, most drug molecules were entrapped inside the straight mesoporous channels via the cospray drying and the morphology of SBA-15 submicron particles remained unchanged. IBU confined inside the mesoporous structure was in the amorphous state shown by PXRD and DSC measurements. The amorphous state of IBU in the solid dispersion showed remarkable stability when subject to stress test condition of 40°C/75%RH in open pans for 12 months. The uniform pore walls were believed to prevent the recrystallization of the homogeneously dispersed drug molecules inside the mesoporous channels with confined nanospace. The dissolution rate of IBU from the cospray dried solid dispersion was significantly enhanced to achieve a rapid release. Even after the accelerated stability test, the rapid drug release property was well preserved. In addition to IBU, a number of poorly soluble models, such as fenofibrate, IDMC, carbamazepine, and ursodeoxycholic acid, could be effectively formulated with mesoporous silica in amorphous form at high drug loading and showed significantly enhancement in dissolution properties (Shen et al., 2014a). Hong et al. (2016) compared solvent evaporation and cospray-drying methods to load poorly soluble fenofibrate to SBA-15 and MCM-41. It was found that cospray drying showed better efficiency to entrap drug molecules into mesoporous channels than solvent evaporation method. For the formulation was designed with high drug loading of 50 wt.%, impregnation method led to formation of particle crystalline form of fenofibrate, showing lower dissolution rate than the amorphous formulation harvested by cospray drying. Solvent impregnation samples possessed more drug molecules deposited on the external surface of silica particles. The deposited fenofibrate on particle surface may have tendency to aggregate silica particles to larger particle size. As comparison, samples prepared by cospray drying showed smaller particle size and better homogeneity of drug distribution in mesoporous particles. Recently, Letchmanan et al. (2015) enhanced the solubility of poorly water-soluble antimalaria drug, artemisinin, by encapsulating the drug particles inside pore channels of ordered mesoporous SBA-15 via cospray drying. The cospray-dried ART/SBA-15 samples demonstrated significantly improved dissolution rates and supersaturation compared to the untreated artemisinin. The amorphous formulation of artemisinin with SBA-15 can exhibit supersaturation with 2-fold higher than the counterpart crystalline artemisinin. Moreover, poorly soluble IDMC was also efficiently loaded onto monodispersed mesoporous carbon carrier by cospray drying (Shen et al., 2014b). The morphology of mesoporous carbon was not changed after drug loading as most IDMC molecules were encapsulated into its pore channels in amorphous state. The amorphous formulation contributed to the significantly enhanced dissolution rate of IDMC from solid dispersion. This one-step straight spray-drying process possesses high potential for application in formulation of poorly soluble ingredients with mesoporous materials to enhance their dissolution properties. The application is not limited in pharmaceutical industry only, and it could be benefit for agrochemical, nutrient, and food industries.

3.4 OTHER LOADING METHODS

For solution immersion and evaporation method loading into the pores of the mesoporous silica, migration of active ingredients from the inside to the external surface or near the outside surface is inevitable during the process of drying. To prepare poorly water-soluble drug mesoporous silica microparticles with higher drug-loading efficiency and longer sustained-release time, other loading procedures have been attempted to load poorly soluble drugs onto mesoporous carriers. Due to the high diffusibility and the high dissolving capacity of the supercritical carbon dioxide (SCF-CO_2), the poorly water-soluble drugs, IBU, was loaded into the pores of the mesoporous silica by supercritical technique. Ni et al. (2012) studied the solubility of IBU under supercritical conditions and deposited IBU into ordered mesoporous silica SBA-15 using supercritical carbon dioxide as solvent at 17 MPa and 310.15 K. The maximum drug loading of IBU/SBA-15 was as high as 41.96 wt.%, and results of scanning electron microscopy, and nitrogen adsorption-desorption isotherms characterizations indicated that most adsorbed drugs were inside the nanoscale channels. Due to differences in the drug quantity adsorbed inside the pores by using different loading conditions, the samples with similar drug loading may exhibit different release rates. Li-Hong et al. (2013) found that the amount and the depth of IBU entered the pores of the mesoporous silica by supercritical fluid (SCF) technique were both larger than those by the solution immersion method. It was found that IBU loaded into the mesoporous silica by SCF technique was in amorphous form and the highest loading could reach 386 mg/g (w/w, IBU/SiO_2), it was much more than that by the solution immersion method. An in vitro releasing study showed that the dissolution was slower than that prepared by the solution immersion method. Similarly, mesoporous HAP was loaded with IBU via super critical procedure (Ibrahim et al., 2015). High drug loading could be achieved at high pressure supercritical condition and about 1.38 g/g (IBU/HAP) was loaded onto mesoporous HAP. The loaded IBU exhibited enhanced dissolution and controlled release due to the adjustable partition of the IBU on the surface and in the pores of the HAP by loading under different supercritical conditions. The high-pressure supercritical condition loading would squeeze IBU deep into the pores of HAP, which may result in a slower release.

Moreover, many other methods have been investigated. Waters et al. (2013) attempted the inclusion of fenofibrate in a series of mesoporous silica materials using microwave irradiation procedure. This procedure applied microwaves to get sample to be heated with homogeneity and the formulation was similar to conventional melt method (Mellaerts et al., 2008b). Solid state reaction was applied for loading poorly soluble drug mesalazine onto mesoporous MCM-41 silica modified with amino and/or carboxylic groups (Popova et al., 2014). Mesalazine was mixed with the carriers at room temperature in a DDR-GM 9458-type high energy, vibrating ball mill for 3 min. In addition, fluidized bed loading method was used to formulate poorly water soluble drug with mesoporous silica (Limnell et al., 2011a). IDMC was dissolved in ethanol and MCM-41 was added to solution to form the suspension that was shaken for at least 1.5 h while protected from light. The solvent was evaporated by spraying the suspension with the fluid bed equipment: the suspension was pumped through a silicone tube and a 0.5 mm diameter nozzle. The atomizing air was fed to the system at 0.8 bar pressure and was heated to 90°C to ensure fast evaporation of ethanol. The inlet air was fed to the system at 0.1 bar pressure and at room temperature to protect mass from heating. This procedure is similar to spray drying process, however, this fluidized bed loading keep the harvested powder in fluidized bed and provided platform to perform further processes of drying, granulating and fluidized bed coating.

4 CHARACTERIZATION AND STABILITIES STUDIES OF POORLY SOLUBLE ORAL DRUG FORMULATION

More than one third of the drugs listed in the US Pharmacopoeia and half of the new chemical entities (NCEs), or new active ingredients are poorly soluble or insoluble in water. Many NCEs fail to be commercialized due to their insolubility or poor solubility in water. Upon oral administration, these poorly soluble drugs usually showed a very low bioavailability because of their poor solubility in the digestive fluid, causing erratic and incomplete absorption that may lead to utilization of high dosage with extra cost and a loss in therapeutic effect. Much attention has been attracted to improve the dissolution rate and solubility of poorly water-soluble drugs to increase bioavailability. Many techniques have been developed to improve dissolution rates through specific formulation methods, the most common being particle size reduction (Jinno et al., 2006; Shukla and Price, 1989), cocrystallization (Shiraki et al., 2008), cyclodextrins inclusion (Palmieri et al., 1997; Xiang and Anderson, 2003), solid solutions or dispersions, nanocrystalline (Wunderlich et al., 1999), or amorphization. Among various endeavors, producing drugs in the amorphous form is one of most promising approaches, as amorphous API possesses inherent property of high dissolution rate and improved solubility. Due to the confined space of nanosized pore channels, APIs loaded to mesoporous silica materials could be in amorphous or nanocrystalline form. When drug-loaded mesoporous silica is dispersed in a dissolution medium as a powder form, immediate release can be achieved as entrapped drugs easily diffuse out of the uniform mesopores (Cavallaro et al., 2004; Charnay et al., 2004; Du and He, 2010). The rapid dissolution of drugs from mesoporous silica has attracted research interest in the formulation of poorly water-soluble drugs to enhance the dissolution rate and bioavailability of such kind of drugs (Mellaerts et al., 2007; Salonen et al., 2005; Van Speybroeck et al., 2009; Wang et al., 2010; Wu et al., 2014). The powder form of mesoporous loaded with BCS II class drugs can be directly administrated through oral delivery or further developed to granules or tablet form as oral medicine (Santos et al., 2013). Although mesoporous-loaded drugs have been investigated for various objectives, the formulation of poorly soluble drugs with enhanced bioavailability for oral delivery is the most promising system toward practical application and this area has been extensively explored (Santos et al., 2013; Xu et al., 2013). Mesoporous materials have been investigated as a stable amorphous drug delivery system owing to their nanosized mesoporous structures and large surface areas (Qian and Bogner, 2012). Molecules interacting with porous media behave very differently from those in bulk phase. Their changed dynamics and thermodynamics properties are believed to play an important role in the properties and product performance of the amorphous system. There are several factors affecting the performance of mesoporous silica loaded poorly soluble drugs and are discussed in the next sections.

4.1 EFFECT OF PORE SIZE OF MESOPOROUS DRUG CARRIERS

Generally, poorly soluble drugs loaded within bare mesoporous silica resulted faster dissolution rate in comparison with their counterpart pure crystalline form of drugs (Heikkilä et al., 2007a). According to a reported investigation of crystallization processes in confined spaces, crystallization can only occur in channels with diameter to molecular size ratios above 20 (Sliwinska-Bartkowiak et al., 2001). For cylindrical pores with diameters to guest molecule size ratio in the range 12~20, the confined phase at low temperature was a heterogeneous phase with partially crystalline domains mixed with amorphous regions. For cylindrical pores with pore diameter smaller than 12 times of guest molecular size, the confined phase at low temperature consisted of an orientationally ordered contact layer and an amorphous inner region. Most of small molecular pharmaceutical ingredients confined in pore channels of mesoporous materials could be amorphous, which represents the most energetic solid state with increased molecular motion and reduced binding energy. This formulation possessed a reduction in the lattice energy of the drug molecules compared with the crystalline forms and resulted in faster dissolution in liquid medium. The pore size of mesoporous materials at nanoscale was the major factor contributing to the enhanced dissolution rate. When large pore mesoporous materials were used as drug carriers, nanocrystal form could be formed. The pore size of mesoporous drug carrier has been found to determine the physical state of APIs formulated in solid dispersion and affect the performance of formulated poorly soluble drugs.

Heikkila et al. (2007b) studied the dissolution of poorly soluble IBU from siliceous mesoporous material TUD-1 with pore diameter about 5 nm and compared with MCM-41 with pore size of 2.6 nm. Drug release studies in vitro demonstrated a fast and unrestricted liberation of IBU, with 96% released in 210 min of the dissolution assay. The drug dissolution from TUD-1 material with the random, foam-like 3Dmesopore network and high accessibility to the dissolution medium was found to be much faster (kinetic constant $k = 10.7$) and more diffusion based (release constant $n = 0.64$) compared to a mesoporous MCM-41 material with smaller, unidirectional mesopore channels ($k = 4.7$, $n = 0.71$). Also, the mesoporous carriers were found to significantly increase the dissolution rate of IBU, when compared to the pure crystalline form of the drug ($k = 0.6$, $n = 0.96$). Cauda et al. (2009) studied the IBU uptake and release rates through different morphologies and pore diameters (4 and 6 nm) of confined mesoporous silica synthesized within the channels (120–200 nm in diameter, 60 μm long) of anodic alumina membranes. It was found that the presence of mesoporous silica increased the drug adsorption capacity with respect to the anodic alumina membranes. The drug release was found to be affected by pore size and morphology. Concerning the columnar-oriented silica samples, the drug release rate decreased upon reducing the pore diameter. From the mesoporous silica particles with circular morphology in alumina membrane, the IBU was released with fast kinetics; it was proposed that the drug did not deeply diffuse into the structure. Jia et al. (2012) modulated the pore size of MSNs from 3 nm to 10 nm by controlling the etching time of MSNs with the $NaBH_4$ solution and the as-synthesized MSNs were investigated as carriers for loading and delivery of the poorly soluble paclitaxel (PTX). It was found that the drug release rate was faster from the larger pore MSNs.

Shen et al. (2011) investigated the effect of pore size and particles size of mesoporous silica drug carriers on the physical state and dissolution of poorly soluble IBU formulated in solid suspension. Three mesoporous silica materials with different pore diameters [MCM-41 (2.3 nm), SBA-15 (6.0 nm) and large pore SBA-15 (SBA-15-LP (20 nm)] were used as drug carriers to load IBU by cospray-drying process at a high drug loading of 50:50 (w/w). The pore sizes were found to affect the physical state and particle size of IBU in mesoporous structures, which influenced the dissolution profiles. When IBU was cospray dried with MCM-41 and SBA-15 with pore size smaller than 10 nm, amorphous state of IBU was obtained due to nanospace confinement. In contrast, nanocrystals were obtained when IBU was cospray dried with large pore SBA-15-LP with pore size above 20 nm. The physical state of IBU played a key role in affecting the dissolution of IBU from the solid dispersion. As shown in Fig. 23.1, IBU in the amorphous state exhibited a higher dissolution rate than nanocrystalline IBU, even though the larger pore size could facilitate diffusion from the host matrix. As IBU entrapped in both MCM-41 and SBA-15 are in the amorphous form with rapid dissolution property, the pore size appeared to affect the dissolution profiles because the pore size determined the diffusion rate from the pore channels. As the pore size of SBA-15 (~6 nm) was larger than MCM-41 (~2.3 nm), the faster diffusion rate from mesoporous channels of SBA-15 resulted in the higher dissolution rate (Izquierdo-Barba et al., 2009). However, although SBA-15-LP has much larger pore size than MCM-41 and SBA-15, the dissolution rate of IBU from IBU/SBA-15-LP was slower. The lower dissolution rate of IBU/SBA-15-LP was due to the slower dissolution kinetics of nanocrystalline IBU compared to its amorphous form. Although the large pore size facilitated the diffusion of dissolved molecules from internal pore channels to the dissolution medium, the intrinsic stable crystalline form of IBU in SBA-15-LP exhibited a slower

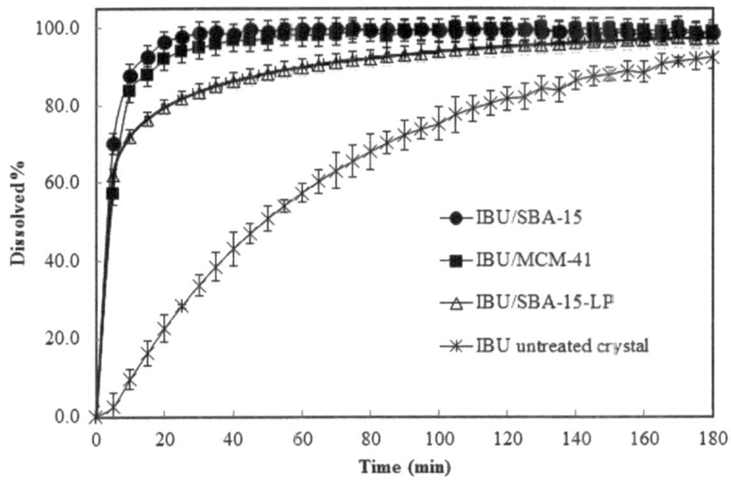

FIGURE 23.1 Dissolution Profiles of Ibuprofen from IBU/MCM-41, IBU/SBA-15 and IBU/SBA-15-LP (Shen et al., 2011)

dissolution rate and controls the overall dissolution profile. Nevertheless, the IBU nanocrystal entrapped in mesoporous SBA-15-LP showed a much higher dissolution rate than untreated IBU crystal. The particle size of mesoporous silica on dissolution of IBU was also studied and found that it showed a less pronounced effect on the dissolution of IBU. Thus, the amorphous/nanocrystalline state of IBU was the most important influence on drug dissolution followed by the diffusion kinetics, particle size of IBU and path length from host matrix to dissolution medium. Similar effect of pore size on dissolution profile was observed for fenofibrate formulated with mesoporous silica with different pore diameter of drug carrier (Hong et al., 2016).

Recently, Zhu et al. (2014b) investigated the influence of pore size on release rate of poorly soluble drugs from 3D face-centered cubic mesoporous silica carrier. 3D face-centered cubic mesoporous silica (FMS) with different pore sizes (16.0, 6.9, and 3.7 nm) was successfully synthesized by using Pluronic block copolymer F127 as a template and adjusting the reaction temperatures. Poorly soluble Celecoxib (CEL), which is a BCS class II drug, was used as a model drug and loaded into FMS with different pore sizes by the solvent deposition method at a drug/silica ratio of 1:4. The results obtained showed that CEL was in a noncrystalline state after incorporation of CEL into the pores of FMS-15 with pore size of 16.0 nm. In vitro dissolution was carried out to demonstrate the effects of FMS with different pore sizes on the release of CEL. The results indicated that the dissolution rate of CEL from FMS-15 with pore size of 16 nm was significantly enhanced as compared with pure CEL. This could be explained by supposing that CEL encountered less diffusion resistance and its crystallinity decreased due to the nanopore channels of FMS-15. Moreover, drug loading and pore size both played an important role in enhancing the dissolution properties for the poorly water-soluble drugs. Comparing the pore size of 3.7 and 16.0 nm, the dissolution rate of CEL from FMS was increased with the increase of pore size. As the active ingredient in this formulation was in amorphous form, the larger pore size facilitated the diffusion through pore channels. It was worth noting that FMSC-15 had the fastest dissolution rate among all the CEL-FMS formulation samples, which was due to its special face-centered cubic pore structure, large pore size of 16.0 nm and uniform drug loading. For formulation with FMS with pore size at 18 and 20 nm, the dissolution rate was decreased, due to the formation of partial crystalline CEL in large pore structures. Riikonen et al. (2015) investigated 16 batches of CEL-loaded mesoporous silicon particles with pore sizes ranging from 17 to 58 nm and CEL content from 5 to 36 wt.%. Dissolution rate of CEL was improved in vitro by lowering the drug-loading degree, which hindered the recrystallization of CEL on the external surface of the particles. The fastest permeation of loaded CEL through the coculture monolayer as well as the highest bioavailability in rats was observed with the particles with small pore size and low loading degree. To study the effect of pore size at a wide range, Dwyer et al. (2015) confined the poorly soluble pharmaceutical fenofibrate in controlled pore glass (CPG) with 10 different pore sizes between 12 and 300 nm. Nanocrystalline fenofibrate was formed in pore sizes greater than 20 nm and showed characteristic melting point depressions following a Gibbs–Thomson relationship as well as enhanced dissolution rates. By comparing the nanocrystal form fenofibrate within nanoporous structures, the dissolution testing showed enhanced dissolution profiles for the nanocrystalline materials confined in different porous matrices, with fenofibrate confined in 70 nm CPG showing the greatest improvement in dissolution. Wang et al. (2013) compared mesoporous MCM-41 and SBA-15 as well as 3D-ordered macroporous (3DOM) silica. It is not surprising to notice that IDMC-loaded mesoporous silica formulations with evidence of the successful inclusion into nanopores, the reduced crystallinity, and finally accelerated dissolution of IDMC. It was worth mentioning that, in comparison to mesoporous silica, 3DOM silica displayed a more rapid release profile, which may be ascribed to the 3D-interconnected pore networks and the highly accessible surface areas.

4.2 STABILITY OF AMORPHOUS FORMULATION AND SUPERSATURATION

The stability of the amorphous state is often a challenge for conventional formulation using polymers as drug carriers because recrystallization could occur due to the low surface area and glass transition of polymer materials (Marsac et al., 2008). Stability of amorphous formulation of mesoporous materials as drug carriers upon storage is one of the crucial factors to determine whether the new formulation can reach market. A key question that remains to be addressed is whether the dispersed state of drug molecules in ordered mesoporous silica carrier can be maintained for longer periods and under different humidities and temperatures. Mesoporous silica could potentially stabilize amorphous formulation due to stability of mesoporous structure and wall thickness of about 4 nm. Shen et al. (2010) found that the cospray drying with mesoporous SBA-15 not only created amorphous IBU, but also stabilized the amorphous state of IBU under severe storage condition for at least 1 year. For a sample of cospray-dried IBU with SBA-15 at high loading of 50 wt.%, the amorphous state could be well preserved after 12 months of storage test under 40°C/75%RH in open pans, as confirmed by XRD and DSC analysis. The nanospace of pore channels of SBA-15 confined IBU molecules within solid dispersion in the amorphous state. In addition, the uniform silica pore walls perfectly separate the fine amorphous particles of IBU and prevent the recrystallization and crystal growth even under severe storage conditions. By cospray drying, only a small amount IBU stayed on the surface SBA-15 submicron particles. As the surface of SBA-15 is rich in hydroxyl groups, the hydrogen bonds between IBU and surface of SBA-15 could also prevent the movement and crystallization of IBU on surface. This shows a distinct advantage of mesoporous SBA-15 as drug carriers for the formulation of poorly water-soluble drugs. Due to the well-preserved amorphous state, the rapid dissolution profile of IBU from cospray-dried solid dispersion was almost not changed as compared with fresh sample. Van Speybroeck et al. (2009)

also observed the excellent amorphous stability of 10 model poorly soluble drugs formulated with mesoporous SBA-15. After storage at 25°C/52%RH for 6 months, no crystallization was detected and the high pharmaceutical performance of all formulations was retained during the 6 months storage period. Mellaerts et al. (2010) investigated aging behavior of pharmaceutical formulations of itraconazole on SBA-15 ordered mesoporous silica carrier material by storage test for 3, 6, and 12 months at 4 and 25°C and relative humidity of 0%, 52%, and 97%. It was interesting to notice that long-term storage at a relative humidity of 52% or 97% did not harm the rapid release profiles of itraconazole from SBA-15, and even showed a beneficial effect on the release. Itraconazole phase analysis using DSC and analysis of the presence of itraconazole concentration gradients using XPS revealed that storage did not cause migration, aggregation, or crystallization of itraconazole. The surface of the pores of SBA-15 was found to undergo hydroxylation at high relative humidity. Such hydroxylation rendered the SBA-15 surface more hydrophilic, which facilitated the liberation of poorly water-soluble itraconazole molecules upon competitive adsorption of water molecules from simulated gastric fluid. The excellent amorphous stability and well preserved dissolution properties were also reported for fenofibrate encapsulated with MCM-41 and SBA-15, regardless the pore size of drug carriers and loading methods applied (Hong et al., 2016). Poorly soluble antimalaria drug artemisinin encapsulated in SBA-15 also exhibited excellent amorphous stability upon storage test for 6 months (Letchmanan et al., 2015). The amorphous artemisinin showed chemical stable under mild humidity condition by preserved more than 95% of active ingredient after storage under condition of for 6 months although high humidity may cause degradation. Similar results were observed for IDMC formulated in silica based mesoporous MCM-41 and SBA-15, where physical stability was satisfied and chemical stability may be subjected to storage conditions (Limnell et al., 2011b). In addition to mesoporous silica, mesoporous carbon formulated amorphous ingredients also exhibited excellent stability upon storage test. Due to the confined space between the uniform pore walls, the amorphous solid dispersion of IDMC with mesoporous carbon exhibited excellent stability under stress storage conditions (40°C/75%RH) as the rigid pore wall effectively prevented any recrystallization to occur (Shen et al., 2014b). The results imply that poorly soluble drugs encapsulation in mesoporous material drug carriers possess good shelf-life and this dissolution-enhancing formulation is a promising approach for a very wide variety of poorly soluble therapeutic compounds.

The rapid release of poorly soluble drugs in amorphous form within mesoporous silica is often associated with the generation of supersaturation, which may face the risk of drug precipitation and reduced availability for absorption (Augustijns and Brewster, 2012; Bevernage et al., 2013; Brouwers et al., 2009). Mesoporous silica loading itraconazole could achieve the extent of supersaturation exceeded 9.6 during at least 4 h (Mellaerts et al., 2008c). To stabilize the supersaturation and prevent the precipitation is another challenge for some poorly soluble drugs. Jia et al. (2011) investigated the amorphous fenofibrate with mesoporous system and supersaturation stabilizing effects of six different polymers. It was found that drug precipitated very quickly in phosphate buffer solution (PBS) solution without any polymer. The addition of polymers resulted in an increased drug concentration in solution. It was found that Eudragit displayed better stabilizing effect than PVP and HPMC, while Eudragit L100-55 was the best to stabilize the supersaturation by inhibit nucleation or crystal growth by adsorption on the crystal interface thereby blocking crystal growth. Van Speybroeck et al. (2010) applied precipitation inhibitors combined with mesoporous silica formulation to enhance the oral absorption of poorly soluble itraconazole. Hydroxypropylmethylcellulose (HPMC) and hydroxypropylmethylcellulose acetate succinate (HPMCAS) were used as precipitation inhibitors and physically blended with itraconazole-loaded SBA-15. In vitro release experiments implementing a transfer from simulated gastric fluid to simulated intestinal fluid were used to evaluate the pharmaceutical performance. With the presence of precipitation inhibitor, the percentage release of itraconazole and supersaturation concentration of itraconazole was enhanced. For in vivo study, when high enough amounts of HPMC were coadministered with itraconazole-loaded SBA-15 (itraconazole:SBA-15:HPMC 1:4:6), the extent of absorption was increased by more than 60% when compared to itraconazole/SBA-15 without precipitation inhibitors (AUC 14,937 ± 1617 vs. 8987 ± 2726 nM h). The results demonstrated that the pharmaceutical performance of SBA-15 was enhanced through addition of an appropriate precipitation inhibitor.

4.3 SUSTAINED RELEASE OF POORLY SOLUBLE DRUGS

The enhanced solubility and oral bioavailability of a poorly water-soluble drug is often inadequate for the desired therapeutic effect. To enhance the total absorption, the initial burst release is needed to be suppressed to an effective therapeutic level. The poorly soluble drugs are also expected to be delivered at a sustained release rate (Deng et al., 2014; Yu and Zhai, 2009). To establish a new oral sustained drug delivery system for the poorly water-soluble drug felodipine, Hu et al. (2015) applied a layer-by-layer (LbL) coating to MSNs loaded with drug molecules in porous structures. Firstly, the model drug was loaded into MSN, and then the loaded MSN were repeatedly encapsulated by chitosan and acacia via LBL self-assembly method. After multilayer coating, the drug release rate was effectively controlled. A sustained drug delivery system for felodipine was successfully obtained, although the drug release behavior was slightly affected by the state of the encapsulated multilayer under different pH conditions. In addition, approaches of shielding the drug from the gastric metabolism and targeted drug release to obtain optimal drug absorption attracted attention for some poorly soluble drugs. Van Speybroeck et al. (2014) found that mesoporous silica loaded poorly soluble IDMC or glibenclamide exhibited very slow release in the acidic environment of the stomach. In in vitro experiments simulating the gastric-to-intestinal transition, the release of both compounds from SBA-15 remained under 1% in simulated gastric fluid, whereas both drugs were completely released within 10 min after transfer to fasted state simulated intestinal fluid (pH 6.5). Roine et al. (2015) studied the controlled release of a poorly water-soluble model drug, griseofulvin, by loading with thermally oxidized mesoporous silicon nanoparticles and dispersed to Eudragit L 100–55 enteric polymer micromatrix particles, which were produced by electro-spraying. The in vitro drug dissolution at pH 1.2 and 6.8, and permeation at pH 7.4 across Caco-2/HT29 cell monolayers from the micromatrix particles, were investigated. The micromatrix particles were found to be gastroresistant, while at pH 6.8 the griseofulvin was released rapidly. Compared to free griseofulvin, the permeability of encapsulated griseofulvin across the intestinal cell monolayers was greatly improved. This formulation could provide oral drug delivery solution for improving the bioavailability of poorly water-soluble drugs to the intestines and avoiding gastric metabolism under acidic condition. The poorly soluble drugs are also expected to achieve a slow sustained release. A core/shell structured nanocomposite, mesoporous nanomatrix-supported lipid bilayer was developed as an oral drug nanocarrier for oral sustained delivery of the water-insoluble drug, lovastatin (Zhang et al., 2015c). The amorphous formulated lovastatin showed significant enhancement in dissolution and the initial burst release was suppressed by the lipid shell layer in composite. Thus animal in vivo test indicated that this formulation enhanced the oral bioavailability by 2-fold. Moreover, the surface of mesoporous silica was modified to carry out controlled release of active ingredients

(Vafaee et al., 2012). For effective delivery of poorly soluble drug, MSNs were functionalized with aminopropyl groups (Xu et al., 2012). MCM-41 was grafted with amino and/or carboxylic functional groups to provide the opportunity for further polymer coating after loading with poorly soluble mesalazine (Popova et al., 2014), and the release rate of mesalazine from the functionalized MSNs was effectively controlled compared with that from the unmodified MSNs. The functionalized, polymer coated mesoporous systems are design of oral drug delivery systems for the dissolution enhancement and/or sustained release of poorly water-soluble drugs. In addition, the drug release rate was also found to be modulated by using super critical loading procedure at high pressure (Ibrahim et al., 2015; Li-Hong et al., 2013) or loading with core-shell structured dual-mesoporous silica nanospheres with small pore at shell layer (Wu et al., 2014) to achieve sustained release of poorly soluble drugs (Li et al., 2015a).

4.4 ENHANCEMENT OF IN VIVO BIOAVAILABILITY

To confirm the enhanced bioavailability of poorly soluble drugs by using mesoporous silica as carriers, animal in vivo tests are required for moving forward to preclinical trial (Riikonen et al., 2015). Compared with crystal counterpart drugs, the bioavailability of mesoporous silica formulated poorly soluble drug was obviously enhanced (Wang et al., 2014). Kumar et al. (2014) evaluated the in vitro release of poorly soluble drug aceclofenac loaded on mesoporous silica and their in vivo performances. The dissolution efficiency of the mesoporous formulation was almost 2 times more than that of plain drug and marketed tablets in acidic discriminating media. A significant enhancement of 189% in oral bioavailability (AUCO-8) was observed for optimized aceclofenac loaded mesoporous formulation as compared to plain aceclofenac in male Wistar rats. Hong et al. (2016) studied the bioavailability of fenofibrate loaded with mesoporous silica in dogs and found the cospray dried formulations could exhibit more than 30-fold increment in fenofibrate bioavailability compared to the reference crystalline fenofibrate. Although mesoporous silica formulation could increase the oral bioavailability of some poorly soluble drugs, it is not feasible to exceed the performance of commercial marketed products. Mellaerts et al. (2008a) used rabbits and dogs for animal in vivo tests to compare the oral bioavailability of itraconazole in mesoporous silica formulation with the crystalline drug and the marketed product Sporanox. After administration of crystalline itraconazole in dogs (20 mg), no systemic itraconazole could be detected. As comparison, by using mesoporous silica as a carrier, the AUC0–8 was boosted to 681 ± 565 nM h. In rabbits, the AUC0–24 increased significantly from 521 ± 159 nM h after oral administration of crystalline itraconazole (8 mg) to 1069 ± 278 nM h when this dose was loaded into mesoporous silica. T_{max} decreased from 9.8 ± 1.8 to 4.2 ± 1.8 h. No significant improvement (AUC, C_{max}, and T_{max}) could be observed when comparing itraconazole/mesoporous silica formulation with commercial Sporanox in both species. The oral bioavailability of itraconazole formulated with ordered mesoporous silica as a carrier was similar to the marketed product Sporanox, in rabbits as well as in dogs. Furthermore, some drug formulation with mesoporous silica cases showed higher extent of absorption than corresponding commercial products. Kiekens et al. (2012) also investigate the bioavailability enhancement of the poorly soluble ezetimibe loaded in ordered mesoporous silica using dogs for animal test. The dog study was designed to determine the oral bioavailability of ezetimibe administrated in capsules and tablets format, and compared in vitro and in vivo with the commercially available tablet Ezetrol. The tablet form of ezetimibe in mesoporous silica showed similar results to that of Ezetrol. However, the capsule formulation demonstrated a faster absorption into the blood circulation, including a superior metabolization of ezetimibe into the active glucuronide conjugate, indicated that ezetimibe absorption was improved by using mesoporous silica as drug delivery technology. Moreover, for poorly soluble glibenclamide loaded with SBA-15 formulation, stable supersaturation could be achieved and animal in vivo study using Wistar rats revealed that extent of absorption showed more than 4-fold higher than that of the commercial glibenclamide product Daonil (van Speybroeck et al., 2014). The poorly soluble carvedilol loaded with mesoporous carbon resulted in 179.3% improvement in oral bioavailability as compared with the marketed product when tested in dogs (Zhang et al., 2013a). In addition to mesoporous silica, mesoporous carbon formulated simvastatin was also found to exhibit enhanced oral bioavailability in beagle dogs by 163% higher relative to commercial formulation (Zocor) (Zhang et al., 2013b). Surface functionalized mesoporous silicon microparticles have also been reported as a solid dispersion carrier for improving dissolution and enhancing the orally administered pharmacokinetics (fasted rat model) of IDMC. The mesoporous silicon carrier facilitates accelerated immediate release of IDMC and enhanced oral delivery performance in comparison with crystalline IDMC and commercial Indocid, that is, a 4-times reduction on T_{max}, a 200% increase on C_{max} and a significant increase in bioavailability (Wang et al., 2010). The results indicated that mesoporous material formulation is a promising oral drug delivery system for enhanced oral bioavailability of poorly water-soluble drugs.

4.5 CYTOTOXICITY OF MESOPOROUS DRUG DELIVERY SYSTEM

The cytotoxicity of mesoporous materials is a crucial factor for development of new formulation while moving forward to clinical trial. The hazards of crystalline silica inhalation are well known and associated with several pulmonary pathologies due to long-term existence in pulmonary. As comparison, mesoporous silica is amorphous form and the toxicity associated with oral administration could be substantially different from inhalation. To study potential biological effects of the mesoporous materials in the gastrointestinal (GI) tract, the cytotoxicity of the mesoporous materials on heterogeneous human epithelial colorectal adenocarcinoma cells (Caco-2) has been widely investigated due to the similarities in morphology and physiology of cells in GI internal membrane. Heikkila et al. (2010) have studied the cytotoxicity of mesoporous silica MCM-41 and SBA-15 microparticles on Caco-2 cells. It was found that when particles tested at the threshold concentration of 2 mg/mL, mesoporous silica materials may induce significant ATP depletion in the cells after 3 h of exposure. However, Letchmanan et al. (2015) investigated the cytotoxicity of mesoporous silica submicron particles SBA-15 on Caco-2 cells and viable rate of cells was determined by flow cytometeric analysis. It was found that SBA-15 submicron particles showed very weak cytotoxicity to Caco-2 cells and viability of Caco-2 cells was more than 90% after 24 h exposures to SBA-15 even at 4 mg/mL. The cytotoxicity of mesoporous carbon was also assayed by in vitro cell culture investigation. Mesoporous carbon exhibited no apparent cytotoxicity to Caco-2 cells in distinct concentrations ranging from 0.01 to 100 μg/mL (Zhu et al., 2014a). Zhao et al. (2012b) found that mesoporous carbon exhibited a weak cytotoxicity at tested concentrations (10–800 μg/mL). When the concentration of carriers increased, cell survival was slightly reduced, and this weak toxic effect observed at high concentrations might be attributed to carrier agglomeration in aqueous culture medium, as a result of strong van der Waals interactions. Even at a concentration of 800 μg/mL, the morphology of cells was not changed and the cell survival remained above 80% after 24 h exposure with mesoporous carbon, confirming the weak cytotoxicity of the mesoporous carbon matrix.

Correlating in vitro results to in vivo situation is notoriously difficult, especially in the case of a complex and fluctuating physiological system as the GI tract. In vivo biological effects including the distribution, clearance and the interactions with tissues and cell types of mesoporous materials exposed through different routes is essential before contemplating their clinical applications. Fu et al. (2013) invested the absorption, distribution, excretion, and toxicity of MSNs in mice following different exposure routes. A systematic investigation of the absorption, distribution, excretion and toxicity of MSNs with the average size of 110 nm after four different exposure routes including intravenous, hypodermic, intramuscular injection, and oral administration to mice were performed. Exposing by oral administration, MSNs were absorbed into the intestinal tract and persisted in the liver, and MSNs administrated by intravenous injection were mainly present in the liver and spleen. In addition, MSNs could cause inflammatory response around the injection sites after intramuscular and hypodermic injection. It was also found that MSNs were mainly excreted through urine and feces after different exposure routes and it has been proven that MSN mainly accumulated in mononuclear phagocytic cells in the liver or spleen, and could be excreted from the body after 4 weeks (Liu et al., 2011a). Pathological examinations illustrated that MSNs possessed good tissue biocompatibility after oral and intravenous injection. These findings revealed that oral and intravenous administration were relatively safe routes for possible biomedical application. Mariano-Neto et al. (2014) investigated the application of mesoporous silica as immunological adjuvant and revealed that silica was only presented in spleen after 35 days and was completely eliminated from all mouse organs after 10 weeks. Kupferschmidt et al. (2013) evaluated the in vivo oral toxicology of mesoporous silica particles by a dose stepwise procedure or range finding test was followed by a consequent confirmatory test including daily administrations of 2000 and 1200 mg/kg doses of mesoporous silica particles. The maximum tolerated dose of different mesoporous silica materials was not reached for Sprague Dawley rats even at the high dosage, although adverse effects could be observed in some animals that were most probably due to the administration by oral dosage of the formulated particles. The results are promising for the use of mesoporous silica materials as drug-delivery systems through oral administration.

5 APPLICATION OF FUNCTIONAL MESOPOROUS MATERIALS IN SMART ORAL DOSAGE FORM

5.1 SLOW SUSTAINED RELEASE OF APIs

Mesoporous silica was initially studied to achieve slow release or sustained release of soluble active ingredients absorbed in tablet form (Doadrio et al., 2004). This kind of formulation exhibited a pronounced initial burst release, followed by a very slow release pattern controlled by diffusion (Lopez et al., 2009; Zhang et al., 2010b). To further control the release rate of active ingredients, the effect of geometric structure of mesoporous silica on drug release profiles was investigated (El-Toni et al., 2012). Recently, Bian et al. (2013) controlled the release of IBU on multilayer mesoporous vesicle. A concentric multilayer vesicle with mesoporous channels in each layer was synthesized and used as drug carrier. The average size of the vesicles was about 100–300 nm, with the cavities of an average distance of about 7 nm between every two layers, and the mesoporous channels of an average diameter of 3.4 nm in each layer. The vesicle showed a fast loading with the IBU capacity up to 30 wt.% within 2 h, and a slow controllable slowly and stably release during about 15 h.

As the bared silica showed less controlling effect to retard the undesired release, much effort was focused on the surface modification of mesoporous silica. Due to the large surface areas with rich hydroxyl groups, various functional groups could be grafted on surface or created by cocondensation during synthesis to provide interaction with drug molecules loaded onto mesoporous carriers (Asefa et al., 2009; Tang et al., 2006). Lin et al. (2009) found that hollow spheres of periodic mesoporous organosilica ethylene groups in its framework could be prepared by using 1,2-bis (trimethoxysilyl) ethane with TEOS as silica sources. When this kind of organosilica mesoporous material was used as drug carriers, it exhibited higher loading capacity and slower release rate compared to the conventional periodic mesoporous silica. Moreover, mesoporous silica was feasibly modified with positive charged amino-groups as drug carriers to achieve a sustained release (Ghasemnejad et al., 2015; Jabbari Zahir Abadi et al., 2012; Yang et al., 2009). Lee et al. (2008) produced a positive surface charged MSN by incorporating trimethylammonium (TA) functional groups into the framework of MSN-TA via direct cocondensation of a TA-silane and TEOS in the presence of a base as a catalyst. Anionic drug molecules, Orange II (a fluorescent tracing molecule), and sulfasalazine (an antiinflammatory prodrug used for bowel disease), were effectively loaded into these MSN–TA samples. When these drug-loaded MSN–TA nanoparticles were placed in physiological buffer solution (pH 7.4), a partial negative surface charge on the MSN–TA was generated due to the deprotonation of silanol groups, and the strong electrostatic repulsion triggered a sustained release of the loaded molecules. MSN–TA as a nanovehicle for pH-dependent loading and controllable release of anionic drug molecules can be used as an oral delivery drug systems targeting at intestine. Yoncheva et al. (2014) compared bared silica and amino functionalized MSN to load with antiinflammatory drug budesonide and additionally postcoated with bioadhesive polymer(carbopol). The combination of amino-functionalization of the carrier with the postcoating of the nanoparticles was found to exhibit a sustained release of budesonide. Gao et al. (2011) reported postgrafting modification of mesoporous silica through the weak chemical interaction or hydrogen bonding in between OH groups of the mesopore surface and 3-aminopropyltriethoxysilane (N-TES) or 3-(2-aminoethylamino) propyltrimethoxysilane (NN-TES). Subsequently, the aspirin adsorption experiments indicated that the adsorption capacities of modified BMMs were improved with the increasing amount of N-TES groups. The in vitro tests exhibited that aspirin kinetic release constant k reduced with the increase amount of the amino groups on the modified mesoporous silica, suggesting that the aspirin release rate from the pores of mesoporous silica was influenced by interaction with organic groups on the surface. On the other hand, mesoporous silica was also modified with carboxyl groups to created negative charge on surface for achieving slower release of drug molecules with opposite charge. Mesoporous SBA-15 was bifunctionalized trimethylsilyl-carboxyl groups to control the release of famotidine (Xu et al., 2008). It was found that the increase of carboxyl content could effectively enhance the loading capacity of famotidine. Meanwhile, the grafting trimethylsilyl groups on the surface of COOH/SBA-15 with hexamethyldisilazane reduced the size of diffusion channels, thus the release of famotidine was greatly delayed with the increasing content of trimethylsilyl groups.

Mesoporous materials combined with functional polymers is another effective route to sustain the drug release (Bhattacharyya et al., 2012; Hartono et al., 2014; Kar et al., 2013; Zhang et al., 2014e). Tamanna et al. (2015) investigated polymeric networks encompassing gentamicin loaded MSNs to tune the release kinetics. Embedding of drug preloaded MSNs was performed by an efficient LbL self-assemble strategy using polystyrene sulfonate (PSS) and poly (allylaminehydrochloride) (PAH). The mean release time from uncoated MSNs was 3.6 days at pH 7.4 and 0.4 days at pH 1.4. A further slower release was achieved by diffusion through one or two PSS/PAH bilayer(s), which had a mean transit time of 6.0 days at pH 7.4 and 3.5 days at pH 1.4. The developed nanodrug carriers combined with the self-assembled polyelectrolyte coating enabled sustained slow release by controlling pH and the number of bilayers. Moreover, poly(tetrafluoroethylene) grafted mesoporous silica was reported to exhibit controlled release of heparin to maintain long-term bioactivity and improve blood compatibility

(Li et al., 2012). Polyamine was anchored on surface of mesoporous silica to function as amine-bearing gate-like scaffoldings to control the release of vitamin B_2 (Bernardos et al., 2008). Similarly, MSNs were functionalized with an 18-crown-6 groups attached at surface as a gate molecule and the releasing kinetic could be controlled by the exchange reaction of "host–guest complex" with K^+ ions at proper concentration (Choi et al., 2011).

In addition to surface modification and combination with polymers, slow sustained release of drug from mesoporous silica was also investigated by surfactant-assisted controlled release (Tsai et al., 2011). Biocompatible surfactant was applied as soft template for self-assembly of mesoporous silica and drug molecules were loaded by conjunction with surfactant. Botella et al. (2012) developed sugar surfactant–silica platforms under biomimetic condition to avoid degradation of potential drugs and used as drug delivery carriers for IBU, by encapsulating drug molecules within the hydrophobic core of the template micelles. A well-controlled sustain release of IBU in 72 h could be achieved. The obtained surfactant-assisted delivery systems constitute a new approach towards controlled release of small therapeutic molecules.

Many other approaches were explored to achieve sustained release of drug from mesoporous carriers, such as coating with lipid on antibiotic-loaded MSN (Mudakavi et al., 2014), dual-capillary electro-encapsulation in Eudragit polymer via electro-spraying (Roine et al., 2012), selective immobilization of curcumin onto the internal surface of mesoporous hollow silica particles by covalent bonding (Jin et al., 2011) and chiral functionalized mesoporous silica materials for slow sustained release of chiral drug (Guo et al., 2010).

5.2 STIMULI RESPONSIVE SMART DRUG DELIVERY

5.2.1 pH-responsive drug delivery

The pH-sensitive drug delivery system is important for oral administration to treat intestine or colon related diseases, such as irritable bowel syndrome, Crohn's disease, ulcerative colitis, and colon cancer. The pH values vary through gastric-intestine tract, which increases from the stomach (pH 1.0–3.0) to the small intestine (pH 6.5–7.0), and to the colon (pH 7.0–8.0). pH-responsive delivery enables the target delivery of drug molecules to colon by protecting the sensitive drug molecules to overcome the dissociation by the enzymes (for example, pepsin) and the low pH of gastric acid in the stomach. A feasible, reversible delivery system could be envisaged by grafting a responsive polymer on the exterior surface of the nanocontainer and the responsive polymer nanoshell could be opened and closed by changing the surrounding environment. Hong et al. (2009) reported that the MSNs are used as drug reservoir and reversible addition-fragmentation chain transfer (RAFT) process was applied to graft poly(acrylicacid) (PAA) onto the exterior surface of MSNs. A novel smart nanocontainers with a MSN as the container drug delivery system with a pH-responsive PAA nanoshell as a smart nanovalve was fabricated. The pH-responsive nanoshell can act as a nanovalve to regulate the uptake and release of guest molecules from the nanocontainer. Filling guest molecules into the nanocontainer was carried out via stirring MSN-PAA in drug solution in the open nanovalve (pH of solution is 8.0) and closed the nanovalve by washing in solution with pH of 4.0. The release of active ingredients was found to be dependent of pH of dissolution medium. At low pH of 4.0, the release of active ingredients was very limited, while the release of model molecules was significantly faster at higher pH of 8.0. In addition to surface grafting, mesoporous drug carrier could be coated with pH-responsive polymer for target delivery (Ma et al., 2014). Choi et al. (2014) developed polymer-coated spherical mesoporous silica for pH-controlled delivery of insulin. Mesoporous silica with adsorbed insulin was freeze-dried to prevent desorption and denaturation of insulin before and then was coated with the pH-sensitive polymer polymethacrylates, Eudragit L30D-55 (methacrylic acid–ethyl acrylate copolymer) and Eudragit L100 (methacrylic acid–methyl methacrylate copolymer). As shown in Fig. 23.2, in an acidic environment of pH 1.2, the pH-sensitive polymer effectively inhibited the release of insulin due to its insolubility of coating layer. However, the controlled delivery system showed sustained release of insulin in medium with pH of 6.8. This result opens up great potential for the effective oral delivery system of many other protein drugs as well as insulin to intestines avoiding to be denatured in acidic condition of stomach.

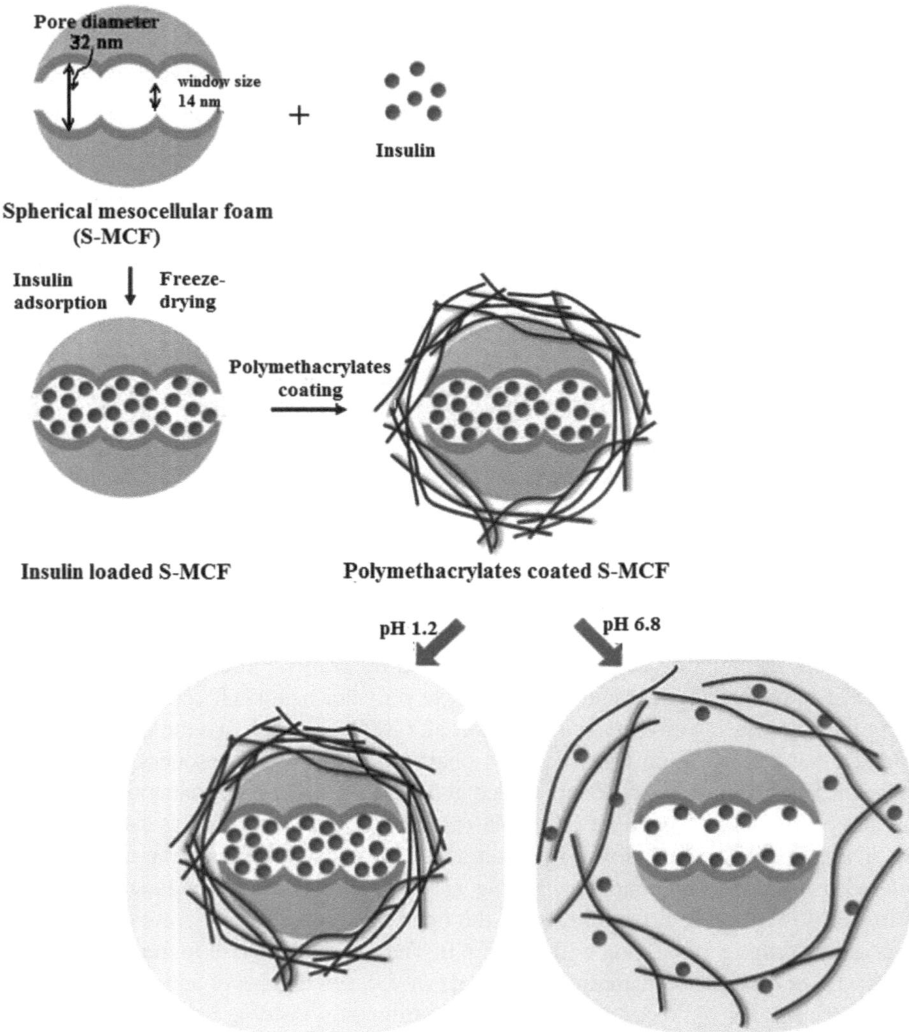

FIGURE 23.2 Schematic for the pH-Controlled Insulin Delivery System Based on Ph-Sensitive Polymer Coating of Insulin-Loaded S-MCF and its pHresponsive Release (Choi et al., 2014)

In addition to use pH sensitive polymer, Cheng et al. (2011) designed a novel oral colon-specific drug delivery system using a modification of MSN surfaces with pH-sensitive trimethylammonium (TA) groups through a pH-sensitive hydrazone bond. The pH-sensitive TA groups can efficiently increase the loading amounts of anionic drugs by a strong electrostatic attraction. After oral administration, the acidic pH of gastric juice can fully hydrolyze the TA–hydrazone bonds and further eliminate the positive charges of TA groups from MSN surfaces. When the hydrolyzed complexes were further delivered to the colon's pH of 7–8, a rapid and complete release of adsorbed drugs was observed. Moreover, nitrogen-doped mesoporous carbon could also exhibit pH-responsive controlled release to allow a sustained drug level of 10 wt.% at pH = 2 and of almost 40 wt.% at pH = 7.4 (Sánchez-Sánchez et al., 2015). Moreover, Xue and Findenegg (2012) found that lysozyme molecules can act as a pH-responsive nanovalve to block and unlock the pore entrances of MCM-41 nanoparticles for guest molecules. It was indicated that pore blocking was due to a pH-induced conformational change by which the effective size of the protein was changed in a reversible manner and inhibited the release under acidic conditions.

Besides the target delivery to intestine by pH-responsive control by suppressing release at acidic condition, MSNs were also functionalized to facilitate the drug release at lower pH for targeted delivery of anticancer drugs to tumor cells. pH-responsive drug delivery is of special interest as the pH value of normal tissues and blood stream is 7.4, but in most tumor tissues the pH values will decrease to a lower extracellular pH values (6.0–7.0), and the pH will drop to pH 4.5–5.5 inside cells especially in endosome (Schmaljohann, 2006). To enable pH-responsive drug release at acidic condition and inhibit the release at neutral condition, Rim et al. (2011) applied inorganic CaP as a novel pore blocker by enzyme-mediated mineralization on the MSN surfaces provides a novel route for smart porous nanocarriers, which are capable of releasing guest drugs from the CaP blocked pore under pH control. The CaP coatings can hold the encapsulated anticancer drugs under extracellular conditions, whereas they can be dissolved within intracellular endosomes as nontoxic ions to initiate drug release within tumor cells. Moreover, this kind of smart drug delivery with pH-responsive release was also developed by coating responsive polymer on MSN. Du et al. (2015) fabricated core/shell composites by the layer-by-layer (LbL) assembly of the fluorescein isothiocyanate modified chitosan and sodium alginate onto the carboxyl modified MSNs, followed by PEGylation. The anticancer drug loaded in the pore channels of MSN showed a low leakage quantity at the neutral environment but significantly enhanced release at lower pH media mimicking the tumor environments. Various pH sensitive polymers, such as poly (methacrylic acid) (Wen et al., 2013), poly(4-vinyl pyridine) (Liu et al., 2011b), poly(2-(diethylamino) ethyl methacrylate) (Sun et al., 2010; Yu et al., 2013), poly(allylamine hydrochloride) and sodium poly(styrene sulfonate) (Yang et al., 2010), poly(methacrylic acid) (Wen et al., 2013), polyadenine (Choi et al., 2012), polydopamine (Zheng et al., 2014), amphiphilic diblock polymer [poly(PDM-*b*-PEGMA)] (Mei et al., 2012) and polyethylene glycol-*co*-polyvinyl pyridine (Pourjavadi et al., 2015) have been coated or grafted on mesoporous silica particles to achieve pH triggered release under acid conditions. Moreover, this kind of pH-responsive release could be achieved by surface modification with thiol- or amino groups (Wani et al., 2012). It was also found that formation of *N*-[*N*′-(*N*′-phenyl)-2-aminophenyl]-3-aminopropyl groups on the external surface of mesoporous silica particles near the pore orifices caused retention of biologically active amino acid at pH ~6.86 and its unhindered release at pH ~1.00 (Roik and Belyakova, 2014). The low pH triggered release was caused by protonation of nitrogen atoms in acidic media leads to repulsion of *N*-[*N*′-(*N*′-phenyl)-2-aminophenyl]-3-aminopropyl groups and rapid release of entrapped biologically active molecules from mesoporous channels. Similar acidic condition triggered release was also using HAP (Yang et al., 2014), ZnO (Wu et al., 2015), or layered double hydroxide (Zheng et al., 2013) as nanovalve capped on mesoporous particles as these cap materials could be degraded under acidic conditions.

It is noted that although the lower pH triggered release offer potential to target deliver active anti-cancer ingredients to tumor cell, the loaded in formulated MSNs may suffer significant loss in acidic stomach condition via oral administration. It is necessary to develop the pH-responsive drug-loaded vehicles in tablet format and coated with polymers to avoid the leakage of active ingredients in stomach acidic fluid, then disintegrated to individual MSNs in the intestines to be target delivered to tumor cell for chemotherapy. Xu et al. (2009) developed a simple pH-controlled drug-release system by coating pH-sensitive polymer hydroxypropyl methylcellulose phthalate (HPMCP) on drug-loaded mesoporous SBA-15 tablet. This formulation can avoid active ingredients within core mesoporous silica tablet to be release in acidic gastric fluid. As comparison, the model drug famotidine could be completely released in only 2 h from mesoporous silica tablet without HPMCP coating. For the tablet with twice coating of HPMCP, only 4.0 wt.% of famotidine could be released within 4 h, the release of active ingredients from core tablet were greatly delayed by HPMCP coating. This coating technology is potential useful to develop formulation in tablet or granules format to deliver acidic triggered release anticancer drug vehicle to intestines via oral administration. Furthermore, to selectively deliver active ingredient to stomach and intestine with different pH, smart dual-cargo selectively controlled release based on a pH-responsive mesoporous silica system was developed (Gui et al., 2015). Two responsive gate-like ensembles were modified onto the surface of MSNs for preparation of the selectively controlled release system. One type of gatekeeper was lysozyme protein molecules, which could lock or unlock the pores by changing the effective size by adjusting the pH and concentration of PBS buffer solution in a reversible manner. The other type of gatekeeper was boronic acid-functionalized gold nanoparticles (B-AuNPs), which could lock and unlock the pores of MSNs by virtue of the reversible reactions between polyalcohols and boronic acid. At the lower pH value (pH 2.0) only one kind of cargo capped by B-AuNP was released from the MSNs, whereas at a higher pH value (pH 7.0) only the other kind of cargo capped with lysozyme was released from the MSNs.

5.2.2 Thermal responsive drug delivery

Among various stimuli-responsive drug delivery systems, temperature-sensitive system also attracted attention as body temperature may be changed due to inflammatory reaction. The construction of thermo-sensitive delivery system was based on an organic–inorganic hybrid nanocomposite consisting of a mesoporous drug reservoir grafted a temperature-responsive polymer to control drug release. The most frequently used thermos-responsive polymer is a thermosensitive poly(*N*-isopropyl acrylamide) (PNIPAAm) derivative, exhibiting volumetric expansion/contraction behavior at 306 K, which can be used to develop a thermosensitive drug delivery system. A lower critical solution temperature (LCST) is associated with thermos-responsive polymers, which shows obvious swelling at the temperatures above LCST. PNIPAAm have been grafted onto mesoporous materials as temperature-responsive gates to control drug delivery (Fu et al., 2007; Zhou et al., 2007; Zhu et al., 2011b). The PNIPAAm layer on the surface of the MSNs could be reversibly opened and closed as triggered by changes in temperature and, thus, could regulate the uptake and release of active ingredient from MSNs grafted with PNIPAAm. The release profiles exhibited a pronounced transition at around 25°C. Since the LCST of PNIPAAm is lower than physiological temperature of 37°C, the researchers tried to increase the LCST by copolymerization with other monomers. For a better way to comply with physiological temperature change, paraffin capped MSNs were developed as finely tuned temperature-controlled drug delivery (Aznar et al., 2011). Silica mesoporous nanoparticles (ca. 100–150 nm) have been selected as the inorganic scaffold. The mesoporous MCM-41 support was loaded with a suitable guest (Safranine O), and the external surface was functionalized with octadecyltrimethoxysilane to be noted as solid S1. These alkyl chains were not able to close the pores but were suitable to interact with paraffins through London forces; the paraffins set up a hydrophobic layer that could block the pores and inhibit guest release. An increase of the temperature above the paraffin melting point would result in melting of the paraffin and the subsequent uncapping of the pores. By selecting proper paraffin, the thermal responsive release could be finely controlled from 37 to 42°C. For instance, as shown in Fig. 23.3, when paraffin heneicosane (H, melting point 39°C) was used, the release of model molecules below 37°C was very limited, however, when temperature was increased to 39°C, most active ingredient encapsulated MSNs could be released. The finely tuned tailor-made temperature triggered release offered potential for applications that demand zero release before stimulus implementation and render them important for delivery triggered by fever or local temperature changes.

To implement local thermos-responsive drug delivery, PNIPAAm was combined with NIR-stimuli gold nanoparticles gatekeepers. The Au nanocage in formulation can effectively absorb and convert light to heat upon irradiation with an NIR laser, resulting in the expansion of the PNIPAAm shell covering the exterior of mesoporous silica, and exposed the pores of mesoporous silica shell, realizing the triggered release of entrapped anticancer drugs (Al-Kady et al., 2011; Yang et al., 2013). Although the thermos-responsive drug delivery has been widely studied, the temperature-responsive drug delivery system has not exhibited as much feasibility as that of pH-trigged drug delivery. Despite these reported examples, temperature-controlled release of molecules from mesoporous materials is still in its early exploration stage, and there are no reported examples where the release temperature could be easily customized depending on target requirements. Moreover, the temperature of warm-blooded animals is generally constant over the whole body, unless the temperature may be slightly changed during a fever period.

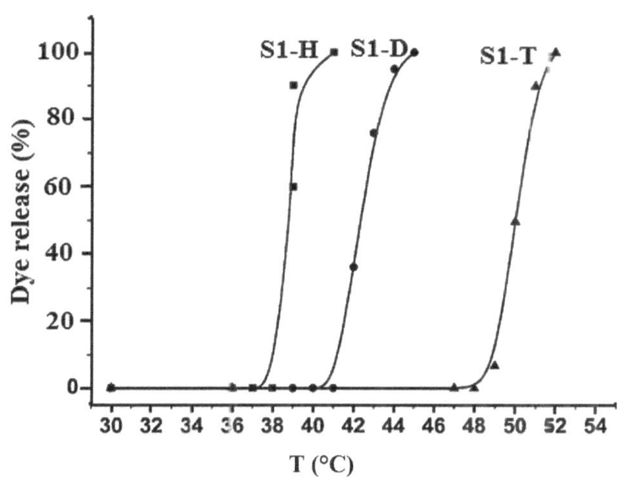

FIGURE 23.3 Release Profile of Solids S1-H, S1-D, and S1-T at Different Temperatures Release data were normalized to the maximum delivery obtained for each solid (Aznar et al., 2011)

5.2.3 Glucose responsive drug delivery

Currently, the treatment for insulin-dependent diabetes is mainly limited to insulin injection, which brings significant suffering to patients. To replace the frequent insulin injections for diabetes patients to control glucose level is a worldwide medical challenge. The emerging glucose-responsive-controlled insulin release system based on mesoporous silica is expected to be a new and highly promising therapy stratagem that can control the insulin dosage smartly, avoiding overdose leading to serious hypoglycemia. Zhao et al. (2011) reported a novel glucose-responsive-controlled release of insulin system through coating enzyme multilayers on mesoporous silica particles. The feasible stratagem applied a glucose-sensing ingredient, such as glucose oxidizing enzyme cooperated with pH-responsive controlling nanovalve to regulate the release of insulin. As shown in Fig. 23.4, the mesoporous silica core served as the reservoir of insulin (Step I), and the enzyme multilayers crosslinked with glutaraldehyde act as a valve to control the release of insulin in response to the external glucose level (Step II). Contacting with glucose, the glucose-oxidizing enzyme could convert glucose into gluconic acid, which decreases the pH value of the microenvironment around the particles and resulted in the volume expansion of the enzyme multilayer shells (Step III) and the subsequent release of insulin from the pore channels of mesoporous silica particles (Step IV). The in vitro insulin release study indicated that glucose triggered release of insulin was achieved, and the amount of insulin released was dependent on the concentration of glucose. Without the presence of glucose in PBS medium, only 5% insulin could be released in 24 h, whereas more than 70% insulin was released in 24 h when 200 mg/dL glucose was in PBS buffer. Furthermore, a similar glucose sensitive delivery system was designed by using MSN core coated with polymer shell of 3-acrylamidophenylboronic acid (AAPBA) and *N*-isopropylacrylamide (NIPAM) as glucose sensor moieties (Sun et al., 2013). In vitro experiments showed that the cumulative release of insulin was dependent on glucose concentration, and the glucose sensitivity could be adjusted simply by different pH values. Simultaneously, compared with the noncrosslinked shell, the crosslinked shell, using dextran-maleic acid as a macromolecule crosslink, enables insulin to release more sustainably. Wu et al. (2013) reported glucose- and pH-responsive controlled release systems by protein-gated carbohydrate-functionalized mesoporous silica nanocontainers. The protein nanogates could be opened either by decreasing the pH value of the buffer or by competitive binding of glucose

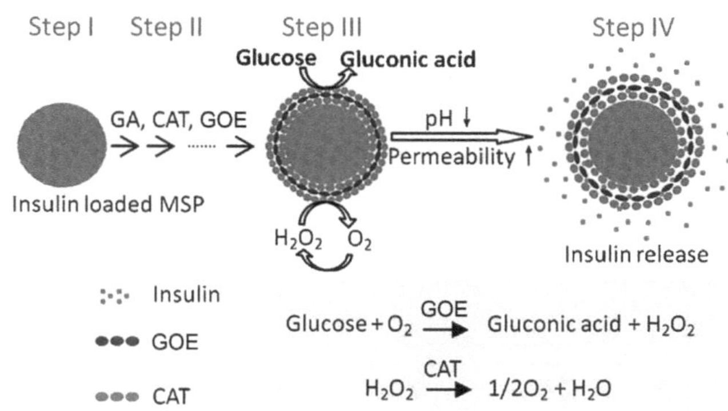

FIGURE 23.4 Schematic Representation of the Glucose-Responsive Controlled Release of Insulin Based on MSPs Coated by Enzyme Multilayers (*GOE*, Glucose Oxidizing Enzyme; *CAT*, Catalase) (Zhao et al., 2011)

at high concentrations, releasing the loaded insulin from the MSN pores on demand. Moreover, Zhao et al. (2009c) designed a glucose-responsive MSN-based double delivery system for both insulin and cyclic adenosine monophosphate (cAMP) with precise control over the sequence of release. Gluconic acid-modified insulin (G-Ins) proteins were immobilized on the exterior surface of MSN and also serve as caps to encapsulate cAMP molecules inside the mesopores of MSN. The release of both G-Ins and cAMP from MSN can be triggered by the introduction of saccharides, such as glucose. Nevertheless, the development of glucose-responsive drug delivery through oral administration still faces the challenge of pH change in GI tract and to inhibit the release of active ingredient in stomach with acidic fluid (Ksenofontova, 2013).

5.2.4 Bio-responsive drug delivery

It is known that there are changes in the pH environment and different enzymes in different parts of the digestive system. The responsive controlled-release system based on different enzyme is another feasible stratagem of great interest as pH-responsive delivery. The target therapeutic to the colon by oral administration is critical for treatment of inflammatory bowel disease, intestinal worms, constipation, and colon cancer, and so on. Guo et al. (2014b) reported a novel enzyme-triggered controlled-release system based on Konjac oligosaccharide coating on MSNs. MCM-41 nanoparticles were selected as the host, and Rhodamine B was utilized as a model drug for convenient detection. After the drug loading, Konjac oligosaccharide was connected to MSNs to close the pores using an amino silane coupling agent (3-aminopropyltriethoxysilane). In vitro studies indicated that when β-mannanase was presented in the medium, the enzyme caused the degradation of Konjac oligosaccharide and made the "gate" open to release ingredients from pore channels up to 80% in 4 h. As comparison, the model ingredient stored in the core mesoporous particles could be well protected and less than 10% to be released in 48 h without the presence of β-mannanase. As β-mannanase is a typical enzyme in the colon of human body, this system exhibited excellent potential to be used for treatment for colon diseases. Popat et al. (2014) developed a programmable drug release system using bioresponsive MSNs for site-specific oral drug delivery. Novel MSNs showing responsive to multiple biological stimuli (pH and enzymes) were prepared through conjugation with a structure modified soy protein isolate. Drug carriers were first functionalized with primary amino groups (MSN–NH_2) with positive charge for increased loading of anionic prodrug, sulfasalazine through ionic interaction. The composite was further coated with a rationally designed succinylated soy protein isolate with the carboxylic groups to bind with amino functionalized MSNs. The succinylation could increase the hydrolysis of capped proteins in the presence of enzymes present in the small intestine. The choice of amino functionalized MSNs provided high drug loading with zero release at pH 1.2 while succinylated soy protein isolate coating restricted the release of sulfasalazine in the stomach (pH 1.2) and duodenum (a medium high pH of 5). The novel drug delivery system featured a high prodrug loading (25% of Sulfasalazine), and site-specific and controlled sulfasalazine release in the presence of pancreatin in the small intestine. The prodrug was further broken down by the in situ enzyme azo-reductase to release active metabolite 5-aminosalycilic acid and sulfapyridine in the colon for disease treatment. This unique and precisely programmed oral drug delivery system could be useful for targeting the small intestine or the colon, which is the absorption site of most orally delivered drugs. Moreover, biotin-avidin was applied as protease-responsive cap system to control release of active ingredients from mesoporous silica (Schlossbauer et al., 2009). Bioresponsive drug delivery system with mesoporous drug host has also been investigated by using lactose (Bernardos et al., 2009), aptamer-based molecule (Zhu et al., 2011c), peptide sequence (Coll et al., 2011), and Prokopowicz and Przyjazny (2007) uril (Liu et al., 2011c) as the controlling gate to be opened by certain type of enzymes. By using trialcoxysilane derivative as a nanovalve grafted on the pore opening of mesoporous host to control the release of active ingredients loaded in pore channels, the release characters showed dual enzyme-triggered controlled release. In the presence of either urease or amidase, the nanovalve could be opened and enable a sustained release of guest molecules (Agostini et al., 2012). In addition to the stratagem of grafting nanovalve on mesoporous silica, smart bioresponsive polymer was applied to coat onto MSNs to form a reductively sheddable polymer shell for controlled drug delivery. Chang et al. (2013b) designed a novel concept of a bioresponsive controlled drug release nanocarrier using the MSN-COOH core applied as the storage platform where entrapped high doses of therapeutic drugs, disulfide-crosslinked thermo/pH-sensitive and redox sheddable P(VCL-s-s-MAA)-PEG shell acted as gatekeepers. In physiological conditions, the P(VCL-s-s-MAA)-PEG shell underwent a distinct transition from a swollen state in pH 7.4 to a collapsed state in pH 5.0. Though sufficiently stable in water, composite nanoparticles were prone to fast dissociation and rupture when subjected to 10 mM glutathione, due to the shedding of polymer walls through reductive cleavage of intermediate disulfide bonds, so that the polymer shell was active in moderating the diffusion of embedded drugs in-and-out of MSN channels. These bioresponsive mesoporous composites drug delivery systems offer potential smart target delivery to specific location of the digestive system via convenient oral administration.

5.2.5 Other responsive and multifunctional delivery

The stimuli-responsive drug delivery systems have been explored in a wide scope of varied objectives. It is noted that tumor tissues and tumor cytosol are also characterized by a reductive environment with concentration of GSH at least 4-fold higher than that in the normal tissues. Stimuli-responsive controlled release systems can be accomplished by using redox-responsive molecules as linkers and grafting suitable gatekeepers onto the surface of silica to prevent the premature release of the cargo before reaching the target. Recently, Chen et al. (2014) reported that mesoporous SBA-15 end-capped by PEG via L-cystine based linker exhibited redox responsive to reductive agent for the controlled target drug delivery. The disulfide bond in L-cystine ester can be effectively cleaved by redox reaction under glutathione of cancer cell level, and then resulting in the release of guest molecules encapsulated in the pores channels (Cui et al., 2012). In addition, the redox responsive controlled release could be achieved by grafting the surfaces of MSNs with (mercaptopropyl)trimethoxysilane (Mendez et al., 2012), ferrocenyl group@b-cyclodextrin attached to Au nanoparticles (b-CDS-AuNPs) (Silveira et al., 2015), or using the copolymer of two oligo(ethylene glycol) macromonomers crosslinked by the disulfide linker *N,N′*-bis-(acryloyl)cystamine as a cap to mesoporous silica (Jiao et al., 2013).

Multifunctional stimuli-responsive drug delivery was investigated based on mesoporous silica as the drug reservoir. Both pH- and redox-triggered synergistic controlled release could be achieved by covalent attachment of ZnO quantum dots to the external surface of MSNs as gatekeepers via disulfide-conjugated two amide linkages encapsulated the anticancer drugs doxorubicin within the MSN cavities and pores, and minimized premature release of the drug (Silveira et al., 2015). The controlled release of the drug from the ZnO-gated mesoporous silica delivery system was realized via the dissolution of ZnO gate triggered by a decrease in pH and cleavage of the disulfide bonds in the presence of glutathione, which was synergically stimulated by tumor cells with weakly acidic environments and high-expressed glutathione. The stimuli-responsive controlled release have also been incorporated with magnetic nano particles or fluorescent dye for magnetically oriented release or enhanced bioimaging (Chen et al., 2011; Xie et al., 2013; Xing et al., 2012) Moreover, multifunctional nanocomposite nanoparticles for simultaneous fluorescence and magnetic resonance imaging, and pH sensitive drug release were fabricated by immobilizing pH-responsive hydrazone bonds, magnetite nanoparticles and fluorescent dyes in MSNs (Lee et al., 2011). The target controlled release could be monitored by fluorescence and magnetic resonance imaging.

5.3 DEVELOPMENT OF TABLET FORM

Tablets dosage format is one of the most convenient and widely used for oral administration. When mesoporous MCM-41 was first applied as a drug carrier for controlled delivery, the MCM-41 powder was compressed to a tablet form disc and then loaded with active ingredients by soaking it in the drug solution and the tablet was disintegrated during dissolution testing (Andersson et al., 2004; Vallet-Regi et al., 2001). For pharmaceutical manufacturing, granulation is usually primarily used to process powder for tableting. Granulated materials result in reduced dustiness and improved flowability. Vialpando et al. (2012) assessed the effect of wet granulation on the performance of mesoporous silica loaded with poorly soluble model drugs. The fine powder of drug-loaded mesoporous silica was granulated using a PVP binder solution. As comparison with fine powder form, the granulation led to particle-size enlargement and enhanced powder flow based on the Carr Index and Hausner Ratio results. When itraconazole was tested as model drug, the release profile of granules was not obvious changed as compared with the powder form. Risk of premature drug release during wet granulation was primarily compound dependent and it could be reduced by decreasing the initial drug load of the material and binder solution addition rate or by increasing the granulation temperature as well as the binder solution concentration. Following by compression to tablet, it was found that the loss in the drug release rate by 120 MPa compression was fully recovered by incorporating croscarmellose sodium to facilitate the disintegration of the compacted granules. No release difference before and after compression was observed. Similar result was observed by Shen et al. (2010) that the dissolution profiles poorly soluble drug loaded with mesoporous silica were almost unchanged by compressing with corn starch to tablet. Once the tablet was dropped in dissolution medium, it was disintegrated quickly and the drug-loaded particles were well dispersed in the medium to show similar release character as drug-loaded powder directly tested. As the pore channels were filled with drug molecules, the compression did not cause major alteration in porous structures of mesoporous silica at various compression forces (Limnell et al., 2011a). Vialpando et al. (2011) tested the effect of the compression force on the pore structure of mesoporous silica and compared with drug loaded mesoporous silica. The compression of pure mesoporous silica resulted in decreasing pore volume and surface area with increasing pressure, due to partial collapsing and wall breaking of pores at the outer layers, and some pores remaining intact. It was found that 20 wt.% drug-loaded mesoporous silica exhibited higher resistance to compression, indicating that drug molecules loaded in pore channels served as a structure support and buffered the impact of the applied pressure. The high compression force at 480 MPa caused an overall decrease in drug release. The addition of microcrystalline cellulose before compression of drug-loaded mesoporous silica materials could result in partial recovery of drug release due to energy dissipation as a consequence of plastic deformation of the additives. For rapid drug release, proper disintegrant may also be required to facilitate the disintegration of compressed tablet and enabled the immediate release as that in the powder form. To avoid high pressure compression, Wang et al. (2012) developed an immediate-release pellet formulation with improved drug dissolution and adsorption. Poorly soluble carbamazepine was adsorbed into mesoporous silica via a wetness impregnation method and then processed by extrusion/spheronization into pellets. Carbamazepine-loaded SBA-15 mesoporous silica was thoroughly mixed with microcrystalline cellulose, sorbitol, and sodium carboxymethyl starch, followed by wetting with some water and extruded at room temperature using a single screw extruder equipped with a dome-shaped extrusion screen (1.2 mm in thickness, 1 mm in perforation diameter). The extrudates were spheronized for 3 min at 800 rpm in a spheronizer and dried at 40°C. Carbamazepine-loaded SBA-15 pellets showed markedly improved dissolution compared with crystalline carbamazepine in vitro, and significantly enhanced the oral bioavailability of carbamazepine in vivo as compared with commercially available carbamazepine tablets. This result indicates that preparing drug-loaded silica into immediate-release pellets may represent a new approach to development of rapidly acting and better absorbed oral formulations for poorly soluble drugs.

To achieve sustained release, the tablet form of drug loaded mesoporous silica was processed surface coating. Wu et al. (2007) investigated the controlled release of vitamin B1 from mesoporous silica tablets coated with a series of sol–gel coatings. Vitamin B1 loaded mesoporous silica was pressed to tablet and then followed by sol–gel coating with TEOS, *n*-propyltriethoxysilane (PTES), *n*-octyltriethoxysilane (OTES), bis(trimethoxysilyl)hexane (TSH), bis(triethoxysilyl) octane (TSO), bis(trimethoxysilylpropyl)amine (TSPA), and bis[(3-trimethoxysilyl)propyl]ethylenediamine (TSPED) were used as precursors for the coatings. Due to the denser structure of the surface coating layer, vitamin B1 release from the tablet was decreased with the increase in the cycles of coatings and drying temperature as coating layer was even denser at higher temperatures. More vitamin B1 was released in 0.01 M HCl solution than in 0.05 M phosphate buffer solution at pH 7.4 because of the electrostatic repulsions between the protonated vitamin B1 cations and the positively charged mesoporous silica carrier/coatings. The experimental results indicated that drug release could be easily controlled by changing the organosilane type and the cycles of coating. Besides of surface coating, recently, Zhao et al. (2015) developed tablet form oral push–pull osmotic pump for sustain release of fenofibrate from MSNs. Fenofibrate, the selected model drug, was first loaded into the MSNs, followed by a suspending agent consisting of a drug layer of push–pull osmotic pump. Polyethylene oxide of molecular weight (MW) 100,000 and polyethylene oxide of MW 6,000,000 were selected as the suspending agent and the expanding agent, respectively. Cellulose acetate was used as the semipermeable membrane, along with polyethylene glycol 6,000 to increase the flexibility and control the membrane permeability. After oral administration, polyethylene oxide in the push layer swelled after absorbed water passed through the semipermeable membrane and pushed drug loaded MSN out of the release hole. The in vitro dissolution studies indicated that the osmotic pump tablet combined with MSNs was able to deliver fenofibrate in an approximately zero-order manner in 24 h.

For controlled target delivery, tablet forms of mesoporous materials loaded with the drug was coated with a pH-responsive polymer (Li and Shi, 2011; Xu et al., 2009). A lag-time combined with pH-sensitive colon-targeted drug delivery system was investigated by coating lag-time films on drug loaded pH-sensitive mesoporous carriers. Ren et al. (2013) developed a time- and pH-controlled oral colon-targeted delivery of bovine insulin by dip-coating lag-time ethylcellulose films on tablet comprising bovine insulin-loaded pH-sensitive mesoporous phosphonate (ZrBPMP) as the tablet core and

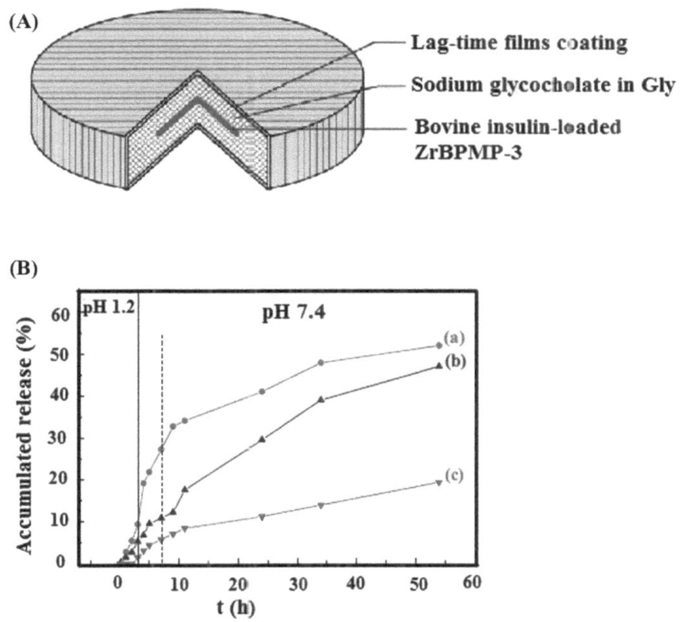

FIGURE 23.5

(A) Schematic illustration of time- and pH-controlled oral colon-targeted delivery system for bovine insulin based on pH-sensitive ZrBPMP and assisted by lag-time films; and (B) the bovine insulin release profiles of tablet with different lag-time film coating for pH transit from 1.2 to 7.4: (a) three times coating; (b) four times coating; (c) five times coating (Ren et al., 2013).

absorption enhancer of insulin and sodium glycocholate at the outer layer as indicated in Fig. 23.5. The lag time can be controlled via adjusting coating cycles of lag-time films and the release of bovine insulin in the colon can be triggered by the pH-sensitive ZrBPMP. For tablets with four cycles of coating with ethylcellulose, the release amount of bovine insulin within 3 h in the simulated gastric fluid and within 4 h in the simulated intestinal fluid decreased to 5.4% and 5.5%, respectively. It was followed by a much higher release rate in medium with pH 7.4 of simulated colon condition. Thus, bovine insulin could be carried passing through the stomach and small intestine with a minimum release and liberated in the colon specifically under dual time and pH value control. Furthermore, the released bovine insulin in vitro release experiment could still retain its native conformation as evidenced by CD spectra, suggesting that the delivery system could preserve bovine insulin with its bioactive conformation from denaturation during the loading, lag time films coating, delivery and release. Moreover, Xu et al. (2011) coated drug-loaded mesoporous silica tablet with pH-responsive Eudragit polymer and enabled potential specific target drug delivery. The drug release was suppressed at acidic condition and increased rapidly with medium pH change from 1.2 to 7.4. The excellent retardation was further enhanced with the increase of coating concentration and thickness of the coating layer. Therefore, the release target position depended on the pH value of the release medium and the coating state. The tablet system could prevent the irritant drug from leakage in the stomach and make it only release at the target position (intestine). It is a suitable oral drug delivery system with a convenient tablet dosage form for intestinal absorption and avoiding irritant drug leakage in stomach.

6 SUMMARY AND OUTLOOK

The development of the ordered mesoporous materials as drug carriers for effective drug delivery has grown rapidly over the last decade due to their great versatility and feasibility to be functionalized. Among various potential applications, the oral delivery system is the most promising. Mesoporous silica has been intensively investigated due to its ease of synthesis, tunable pore structure and particle size, and functionalizable hydroxyl group-enriched surface. In addition, many other mesoporous materials, such as mesoporous structured carbon, HAP, zirconium phosphonate and calcium carbonate have also been explored as oral drug carriers. The application of mesoporous materials as carriers for oral delivery of poorly water-soluble drugs has attracted great attention as it has high potential for practical applications. By using mesoporous materials as drug carrier, it is feasible to develop new formulations and processes that can make poorly soluble APIs in a stable amorphous form with significantly enhanced dissolution rates and apparent solubility. Poorly soluble APIs have been formulated with mesoporous materials via solution absorption, impregnation, solvent evaporation, cospray drying and melt processing. Besides to the enhanced solubility and dissolution results demonstrated by in vitro measurement, animal in vivo test have revealed significant improved bioavailability of poorly soluble drugs by formulation with mesoporous carriers, providing strong evidence for advancing to clinical trials. Due to the stable mesoporous structures, the amorphous formulation has been proven to exhibit excellent stability and can preserve its enhanced solubility after storage under stressed test conditions. Therefore, good shelf life is expected by using mesoporous materials as drug carrier for poorly soluble drugs.

Functionalized stimuli-responsive mesoporous drug delivery system enabled "smart" oral drug delivery, where by drug release could be controlled by changing pH, temperature or biological environment. Especially for the pH-responsive oral drug delivery system, the release under stomach acidic conditions could be suppressed and this enables the target drug delivery to intestine or colon for related disease treatment. This formulation could avoid the drug or protein to be denatured or metabolized by enzyme under strong acidic conditions. Although some ground-breaking results demonstrated that functionalized MSN nanovehicles effectively delivered drugs to cancer cells and suppressed the growth of tumors, further investigations are still required for practical applications. As the release of target anticancer drug in MSN formulation is designed to be triggered by the acid environment within cancer cells, the encapsulated anticancer drug via oral administration also needs to avoid the release in acidic stomach fluid and a primary protection layer on tablet is required. Due to low cytotoxicity and the stable nanostructure protecting active ingredients, mesoporous materials showed high potential for development of specific oral delivery, such as glucose-responsive delivery of insulin for diabetic treatment. To date, most of the reported mesoporous silica based target-specific drug delivery achieved successful results in vitro, and few systems have been tested in vivo using small animals. Therefore, the development of mesoporous drug formulation for oral medicine is still at the preclinical stage and promising breakthrough performance still await translational efforts toward commercialisation.

REFERENCES

Aghaei, H., Nourbakhsh, A.A., Karbasi, S., JavadKalbasi, R., Rafienia, M., Nourbakhsh, N., Bonakdar, S., Mackenzie, K.J.D., 2014. Investigation on bioactivity and cytotoxicity of mesoporous nano-composite MCM-48/hydroxyapatite for ibuprofen drug delivery. Ceram. Int. 40, 7355–7362.

Agostini, A., Mondragon, L., Coll, C., Aznar, E., Marcos, M.D., Martinez-Manez, R., Sancenon, F., Soto, J., Perez-Paya, E., Amoros, P., 2012. Dual enzyme-triggered controlled release on capped nanometric silica mesoporous supports. Chem. Open 1, 17–20.

Akram, M., Alshemary, A.Z., Goh, Y.F., Wan Ibrahim, W.A., Lintang, H.O., Hussain, R., 2015. Continuous microwave flow synthesis of mesoporous hydroxyapatite. Mater. Sci. Eng. C 56, 356–362.

Al-Kady, A.S., Gaber, M., Hussein, M.M., Ebeid el, Z.M., 2011. Nanostructure-loaded mesoporous silica for controlled release of coumarin derivatives: a novel testing of the hyperthermia effect. Eur. J. Pharm. Biopharm. 77, 66–74.

Amer, W., Abdelouahdi, K., Ramananarivo, H.R., Zahouily, M., Fihri, A., Coppel, Y., Varma, R.S., Solhy, A., 2013. Synthesis of mesoporous nano-hydroxyapatite by using zwitterions surfactant. Mater. Lett. 107, 189–193.

Andersson, J., Rosenholm, J., Areva, S., Lindén, M., 2004. Influences of material characteristics on ibuprofen drug loading and release profiles from ordered micro- and mesoporous silica matrices. Chem. Mater. 16, 4160–4167.

Andrade, G.F., Gomide, V.S., da Silva Junior, A.C., Goes, A.M., de Sousa, E.M., 2014. An in situ synthesis of mesoporous SBA-16/hydroxyapatite for ciprofloxacin release: in vitro stability and cytocompatibility studies. J. Mater. Sci. Mater. Med. 25, 2527–2540.

Anglin, E.J., Cheng, L., Freeman, W.R., Sailor, M.J., 2008. Porous silicon in drug delivery devices and materials. Adv. Drug Deliv. Rev. 60, 1266–1277.

Argyo, C., Weiss, V., Bräuchle, C., Bein, T., 2014. Multifunctional mesoporous silica nanoparticles as a universal platform for drug delivery. Chem. Mater. 26, 435–451.

Asefa, T., Otuonye, A., Wang, G., Blair, E., Vathyam, R., Denton, K., 2009. Controlling adsorption and release of drug and small molecules by organic functionalization of mesoporous materials. Adsorption 15, 287–299.

Augustijns, P., Brewster, M.E., 2012. Supersaturating drug delivery systems: fast is not necessarily good enough. J. Pharm. Sci. 101, 7–9.

Aznar, E., Mondragon, L., Ros-Lis, J.V., Sancenon, F., Marcos, M.D., Martinez-Manez, R., Soto, J., Perez-Paya, E., Amoros, P., 2011. Finely tuned temperature-controlled cargo release using paraffin-capped mesoporous silica nanoparticles. Angew. Chem. Int. Ed. Engl. 50, 11172–11175.

Bakhtiari, L., Rezaie, H.R., Javadpour, J., Erfan, M., Shokrgozar, M.A., 2015. The effect of synthesis parameters on the geometry and dimensions of mesoporous hydroxyapatite nanoparticles in the presence of 1-dodecanethiol as a pore expander. Mater. Sci. Eng. C Mater. Biol. Appl. 53, 1–6.

Bakrudeen, H., Tsibouklis, J., Reddy, B.R., 2013. Facile fabrication of mesoporous ZnO nanospheres for the controlled delivery of captopril. J. Nanopart. Res. 15, 1–13.

Bernardos, A., Aznar, E., Coll, C., Martinez-Manez, R., Barat, J.M., Marcos, M.D., Sancenon, F., Benito, A., Soto, J., 2008. Controlled release of vitamin B2 using mesoporous materials functionalized with amine-bearing gate-like scaffoldings, J. Control. Release 131, 181–189.

Bernardos, A., Aznar, E., Marcos, M.D., Martinez-Manez, R., Sancenon, F., Soto, J., Barat, J.M., Amoros, P., 2009. Enzyme-responsive controlled release using mesoporous silica supports capped with lactose. Angew. Chem. Int. Ed. Engl. 48, 5884–5887.

Bevernage, J., Brouwers, J., Brewster, M.E., Augustijns, P., 2013. Evaluation of gastrointestinal drug supersaturation and precipitation: strategies and issues. Int. J. Pharm. 453, 25–35.

Bhattacharyya, S., Wang, H., Ducheyne, P., 2012. Polymer-coated mesoporous silica nanoparticles for the controlled release of macromolecules. Acta Biomater. 8, 3429–3435.

Bian, Z., Tang, J., Hu, J., Li, J., Xu, S., Liu, H., 2013. Controlled-release of ibuprofen on multilayer mesoporous vesicle. Colloids Surf. A 436, 1021–1026.

Botella, P., Corma, A., Quesada, M., 2012. Synthesis of ordered mesoporous silica templated with biocompatible surfactants and applications in controlled release of drugs. J. Mater. Chem. 22, 6394–6401.
Brouwers, J., Brewster, M.E., Augustijns, P., 2009. Supersaturating drug delivery systems: the answer to solubility-limited oral bioavailability? J. Pharm. Sci. 98, 2549–2572.
Carino, I.S., Pasqua, L., Testa, F., Aiello, R., Puoci, F., Iemma, F., Picci, N., 2007. Silica-based mesoporous materials as drug delivery system for methotrexate release. Drug Deliv. 14, 491–495.
Cauda, V., Mühlstein, L., Onida, B., Bein, T., 2009. Tuning drug uptake and release rates through different morphologies and pore diameters of confined mesoporous silica. Micropor. Mesopor. Mater. 118, 435–442.
Cavallaro, G., Pierro, P., Palumbo, F.S., Testa, F., Pasqua, L., Aiello, R., 2004. Drug delivery devices based on mesoporous silicate. Drug Deliv. 11, 41–46.
Chang, B., Fu, J., Tian, Y., Dong, X., 2013a. Soft-template synthesis of sulfonated mesoporous carbon with high catalytic activity for biodiesel production. RSC Adv. 3, 1987–1994.
Chang, B., Chen, D., Wang, Y., Chen, Y., Jiao, Y., Sha, X., Yang, W., 2013b. Bioresponsive controlled drug release based on mesoporous silica nanoparticles coated with reductively sheddable polymer shell. Chem. Mater. 25, 574–585.
Charnay, C., Begu, S., Tourne-Peteilh, C., Nicole, L., Lerner, D.A., Devoisselle, J.M., 2004. Inclusion of ibuprofen in mesoporous templated silica: drug loading and release property. Eur. J. Pharm. Biopharm. 57, 533–540.
Che, S., Garcia-Bennett, A.E., Liu, X., Hodgkins, R.P., Wright, P.A., Zhao, D., Terasaki, O., Tatsumi, T., 2003. Synthesis of large-pore Ia3d mesoporous silica and its tubelike carbon replica. Angew. Chem. Int. Ed. Engl. 42, 3930–3934.
Chen, P.J., Hu, S.H., Hsiao, C.S., Chen, Y.Y., Liu, D.M., Chen, S.Y., 2011. Multifunctional magnetically removable nanogated lids of Fe_3O_4-capped mesoporous silica nanoparticles for intracellular controlled release and MR imaging. J. Mater. Chem. 21, 2535–2543.
Chen, B., Quan, G., Wang, Z., Chen, J., Wu, L., Xu, Y., Li, G., Wu, C., 2013. Hollow mesoporous silicas as a drug solution delivery system for insoluble drugs. Powd. Technol. 240, 48–53.
Chen, L., Zheng, Z., Wang, J., Wang, X., 2014. Mesoporous SBA-15 end-capped by PEG via l-cystine based linker for redox responsive controlled release. Micropor. Mesopor. Mater. 185, 7–15.
Cheng, S.H., Liao, W.N., Chen, L.M., Lee, C.H., 2011. pH-controllable release using functionalized mesoporous silica nanoparticles as an oral drug delivery system. J. Mater. Chem. 21, 7130–7137.
Choi, Y.L., Jaworski, J., Seo, M.L., Lee, S.J., Jung, J.H., 2011. Controlled release using mesoporous silica nanoparticles functionalized with 18-crown-6 derivatives. J. Mater. Chem. 21, 7882–7885.
Choi, Y.L., Lee, J.H., Jaworski, J., Jung, J.H., 2012. Mesoporous silica nanoparticles functionalized with a thymidine derivative for controlled release. J. Mater. Chem. 22, 9455–9457.
Choi, S.R., Jang, D.J., Kim, S., An, S., Lee, J., Oh, E., Kim, J., 2014. Polymer-coated spherical mesoporous silica for pH-controlled delivery of insulin. J. Mater. Chem. B 2, 616–619.
Choma, J., Jedynak, K., Górka, J., Jaroniec, M., 2011. Soft-templating synthesis and adsorption properties of mesoporous carbons with embedded silver nanoparticles. Adsorption 17, 461–466.
Choma, J., Jedynak, K., Marszewski, M., Jaroniec, M., 2013. Organic acid-assisted soft-templating synthesis of ordered mesoporous carbons. Adsorption 19, 563–569.
Coll, C., Mondragon, L., Martinez-Manez, R., Sancenon, F., Marcos, M.D., Soto, J., Amoros, P., Perez-Paya, E., 2011. Enzyme-mediated controlled release systems by anchoring peptide sequences on mesoporous silica supports. Angew. Chem. Int. Ed. Engl. 50, 2138–2140.
Cui, Y., Dong, H., Cai, X., Wang, D., Li, Y., 2012. Mesoporous silica nanoparticles capped with disulfide-linked PEG gatekeepers for glutathione-mediated controlled release. ACS Appl. Mater. Interf. 4, 3177–3183.
Das, S.K., Kapoor, S., Yamada, H., Bhattacharyya, A.J., 2009. Effects of surface acidity and pore size of mesoporous alumina on degree of loading and controlled release of ibuprofen. Micropor. Mesopor. Mater. 118, 267–272.
Deng, Z., Zhang, W., Lu, Y., Zhang, Y., Lu, Z., 2014. Fabrication of hydrogel beads entrappted by mesoporous silicates nanoparticles for pH-sensitive and sustained release of indomethacin. Asian J. Chem. 26, 3609–3614.

Doadrio, A.L., Sousa, E.M., Doadrio, J.C., Perez Pariente, J., Izquierdo-Barba, I., Vallet-Regi, M., 2004. Mesoporous SBA-15 HPLC evaluation for controlled gentamicin drug delivery. J. Control. Release 97, 125–132.

Dong, J., Xue, M., Zink, J.I., 2013. Functioning of nanovalves on polymer coated mesoporous silica nanoparticles. Nanoscale 5, 10300–10306.

Du, X., He, J., 2010. Regulation role of ibuprofen toward the morphology of porous silica nanospheres during its in situ encapsulation. J. Colloid Interf. Sci. 345, 269–277.

Du, P., Zhao, X., Zeng, J., Guo, J., Liu, P., 2015. Layer-by-layer engineering fluorescent polyelectrolyte coated mesoporous silica nanoparticles as pH-sensitive nanocarriers for controlled release. Appl. Surf. Sci. 345, 90–98.

Dwyer, L.M., Michaelis, V.K., O'Mahony, M., Griffin, R.G., Myerson, A.S., 2015. Confined crystallization of fenofibrate in nanoporous silica. CrystEngComm 17, 7922–7929.

El-Ghannam, A., Ahmed, K., Omran, M., 2005. Nanoporous delivery system to treat osteomyelitis and regenerate bone: gentamicin release kinetics and bactericidal effect. J. Biomed. Mater. Res. B 73, 277–284.

El-Toni, A.M., Khan, A., Ibrahim, M.A., Al-Hoshan, M., Labis, J.P., 2012. Fabrication of mesoporous silica shells on solid silica spheres using anionic surfactants and their potential application in controlling drug release. Molecules 17, 13199–13210.

Fan, J., Yu, C., Gao, F., Lei, J., Tian, B., Wang, L., Luo, Q., Tu, B., Zhou, W., Zhao, D., 2003. Cubic mesoporous silica with large controllable entrance sizes and advanced adsorption properties. Angew. Chem. Int. Ed. Engl. 42, 3146–3150.

Fang, Y., Gu, D., Zou, Y., Wu, Z., Li, F., Che, R., Deng, Y., Tu, B., Zhao, D., 2010. A low-concentration hydrothermal synthesis of biocompatible ordered mesoporous carbon nanospheres with tunable and uniform size. Angew. Chem. Int. Ed. Engl. 49, 7987–7991.

Fang, Y., Zheng, G., Yang, J., Tang, H., Zhang, Y., Kong, B., Lv, Y., Xu, C., Asiri, A.M., Zi, J., Zhang, F., Zhao, D., 2014. Dual-pore mesoporous carbon@silica composite core–shell nanospheres for multidrug delivery. Angew. Chem. Int. Ed. Engl. 53, 5366–5370.

Forsgren, J., Andersson, M., Nilsson, P., Mihranyan, A., 2013. Mesoporous calcium carbonate as a phase stabilizer of amorphous celecoxib: an approach to increase the bioavailability of poorly soluble pharmaceutical substances. Adv. Health. Mater. 2, 1469–1476.

Fu, Q., Rao, G.V., Ward, T.L., Lu, Y., Lopez, G.P., 2007. Thermoresponsive transport through ordered mesoporous silica/PNIPAAm copolymer membranes and microspheres. Langmuir 23, 170–174.

Fu, C., Liu, T., Li, L., Liu, H., Chen, D., Tang, F., 2013. The absorption, distribution, excretion and toxicity of mesoporous silica nanoparticles in mice following different exposure routes. Biomaterials 34, 2565–2575.

Fuertes, A.B., 2003. Template synthesis of mesoporous carbons with a controlled particle size. J. Mater. Chem. 13, 3085–3088.

Gao, L., Sun, J., Li, Y., Zhang, L., 2011. Bimodal mesoporous silicas functionalized with different level and species of the amino groups for adsorption and controlled release of aspirin. J. Nanosci. Nanotechnol. 11, 6690–6697.

Gao, Y., Zhu, W., Liu, J., Di, D., Chang, D., Jiang, T., Wang, S., 2015. A geometric pore adsorption model for predicting the drug loading capacity of insoluble drugs in mesoporous carbon. Int. J. Pharm. 485, 25–30.

Ghasemnejad, M., Ahmadi, E., Mohamadnia, Z., Doustgani, A., Hashemikia, S., , 2015. Functionalized silica nanoparticles as a carrier for betamethasone sodium phosphate: drug release study and statistical optimization of drug loading by response surface method. Mater. Sci. Eng. C 56, 223–232.

Gu, J., Fan, W., Shimojima, A., Okubo, T., 2007. Organic-inorganic mesoporous nanocarriers integrated with biogenic ligands. Small 3, 1740–1744.

Gu, L., He, X., Wu, Z., 2014a. Mesoporous hydroxyapatite: preparation, drug adsorption, and release properties. Mater. Chem. Phys. 148, 153–158.

Gu, L., He, X., Wu, Z., 2014b. Mesoporous Fe_3O_4/hydroxyapatite composite for targeted drug delivery. Mater. Res. Bull. 59, 65–68.

Gui, W., Wang, W., Jiao, X., Chen, L., Wen, Y., Zhang, X., 2015. Dual-cargo selectively controlled release based on a pH-responsive mesoporous silica system. Chem. Phys. Chem. 16, 607–613.

Guo, Z., Du, Y., Liu, X., Ng, S.C., Chen, Y., Yang, Y., 2010. Enantioselectively controlled release of chiral drug (metoprolol) using chiral mesoporous silica materials. Nanotechnology 21, 1–11.
Guo, H., Qian, H., Sun, S., Sun, D., Yin, H., Cai, X., Liu, Z., Wu, J., Jiang, T., Liu, X., 2011. Hollow mesoporous silica nanoparticles for intracellular delivery of fluorescent dye. Chem. Cent. J. 5, 1–10.
Guo, Y.P., Yao, Y.B., Guo, Y.J., Ning, C.Q., 2012. Hydrothermal fabrication of mesoporous carbonated hydroxyapatite microspheres for a drug delivery system. Micropor. Mesopor. Mater. 155, 245–251.
Guo, Y.J., Wang, Y.Y., Chen, T., Wei, Y.T., Chu, L.F., Guo, Y.P., 2013. Hollow carbonated hydroxyapatite microspheres with mesoporous structure: hydrothermal fabrication and drug delivery property. Mater. Sci. Eng. C 33, 3166–3172.
Guo, Y., Shi, X., Fang, Q., Zhang, J., Fang, H., Jia, W., Yang, G., Yang, L., 2014a. Facile preparation of hydroxyapatite–chondroitin sulfate hybrid mesoporous microrods for controlled and sustained release of antitumor drugs. Mater. Lett. 125., 111–115.
Guo, W., Yang, C., Cui, L., Lin, H., Qu, F., 2014b. An enzyme-responsive controlled release system of mesoporous silica coated with Konjac oligosaccharide. Langmuir 30, 243–249.
Han, Y.J., Stucky, G.D., Butler, A., 1999. Mesoporous silicate sequestration and release of proteins. J. Am. Chem. Soc. 121, 9897–9898.
Hartono, S.B., Phuoc, N.T., Yu, M., Jia, Z., Monteiro, M.J., Qiao, S., Yu, C., 2014. Functionalized large pore mesoporous silica nanoparticles for gene delivery featuring controlled release and co-delivery. J. Mater. Chem. B 2, 718–726.
He, Q., Shi, J., 2011. Mesoporous silica nanoparticle based nano drug delivery systems: synthesis, controlled drug release and delivery, pharmacokinetics and biocompatibility. J. Mater. Chem. 21, 5845–5855.
He, Q., Zhang, J., Chen, F., Guo, L., Zhu, Z., Shi, J., 2010a. An anti-ROS/hepatic fibrosis drug delivery system based on salvianolic acid B loaded mesoporous silica nanoparticles. Biomaterials 31, 7785–7796.
He, W., Li, Z., Wang, Y., Chen, X., Zhang, X., Zhao, H., Yan, S., Zhou, W., 2010b. Synthesis of mesoporous structured hydroxyapatite particles using yeast cells as the template. J. Mater. Sci. Mater. Med. 21, 155–159.
He, Q., Pan, L., Wang, Y., Meldrum, F.C., 2015. Bioinspired synthesis of large-pore, mesoporous hydroxyapatite nanocrystals for the controlled release of large pharmaceutics. Cryst. Growth Des. 15, 723–731.
Heikkila, T., Salonen, J., Tuura, J., Kumar, N., Salmi, T., Murzin, D.Y., Hamdy, M.S., Mul, G., Laitinen, L., Kaukonen, A.M., Hirvonen, J., Lehto, V.P., 2007a. Evaluation of mesoporous TCPSi, MCM-41, SBA-15, and TUD-1 materials as API carriers for oral drug delivery. Drug Deliv. 14, 337–347.
Heikkila, T., Salonen, J., Tuura, J., Hamdy, M.S., Mul, G., Kumar, N., Salmi, T., Murzin, D.Y., Laitinen, L., Kaukonen, A.M., Hirvonen, J., Lehto, V.P., 2007b. Mesoporous silica material TUD-1 as a drug delivery system. Int. J. Pharm. 331, 133–138.
Heikkila, T., Santos, H.A., Kumar, N., Murzin, D.Y., Salonen, J., Laaksonen, T., Peltonen, L., Hirvonen, J., Lehto, V.P., 2010. Cytotoxicity study of ordered mesoporous silica MCM-41 and SBA-15 microparticles on Caco-2 cells. Eur. J. Pharm. Biopharm. 74, 483–494.
Hong, C.Y., Li, X., Pan, C.Y., 2009. Fabrication of smart nanocontainers with a mesoporous core and a pH-responsive shell for controlled uptake and release. J. Mater. Chem. 19, 5155–5160.
Hong, S., Shen, S., Tan, D.C., Ng, W.K., Liu, X., Chia, L.S., Irwan, A.W., Tan, R.B., Nowak, S.A., Marsh, K., Gokhale, R., 2016. High drug load, stable, manufacturable and bioavailable fenofibrate formulations in mesoporous silica: a comparison of spray drying versus solvent impregnation methods. Drug. Deliv. 23, 316–327.
Hu, Z., Srinivasan, M.P., 2001. Mesoporous high-surface-area activated carbon. Micropor. Mesopor. Mater. 43, 267–275.
Hu, Z., Srinivasan, M.P., Ni, Y., 2000. Preparation of mesoporous high-surface-area activated carbon. Adv. Mater. 12, 62–65.
Hu, Y., Zhi, Z., Zhao, Q., Wu, C., Zhao, P., Jiang, H., Jiang, T., Wang, S., 2012. 3D cubic mesoporous silica microsphere as a carrier for poorly soluble drug carvedilol. Micropor. Mesopor. Mater. 147, 94–101.

Hu, L., Sun, H., Zhao, Q., Han, N., Bai, L., Wang, Y., Jiang, T., Wang, S., 2015. Multilayer encapsulated mesoporous silica nanospheres as an oral sustained drug delivery system for the poorly water-soluble drug felodipine. Mater. Sci. Eng. C 47, 313–324.

Huang, S., Li, C., Cheng, Z., Fan, Y., Yang, P., Zhang, C., Yang, K., Lin, J., 2012. Magnetic Fe_3O_4@mesoporous silica composites for drug delivery and bioadsorption. J. Colloid Interf. Sci. 376, 312–321.

Hwang, A.A., Lu, J., Tamanoi, F., Zink, J.I., 2015. Functional nanovalves on protein-coated nanoparticles for in vitro and in vivo controlled drug delivery. Small 11, 319–328.

Ibrahim, A.R., Li, X., Zhou, Y., Huang, Y., Chen, W., Wang, H., Li, J., 2015. Synthesis of spongy-like mesoporous hydroxyapatite from raw waste eggshells for enhanced dissolution of ibuprofen loaded via supercritical CO_2. Int. J. Mol. Sci. 16, 7960–7975.

Islam, M.S., Choi, H.N., Choi, W.S., Lee, H.-J., 2015. Polyelectrolyte-mediated hierarchical mesoporous calcium silicates: a platform for drug delivery carrier with ultrahigh loading capacity and controlled release behavior. J. Mater. Chem. B 3, 1001–1009.

Izquierdo-Barba, I., Sousa, E., Doadrio, J., Doadrio, A., Pariente, J., Martínez, A., Babonneau, F., Vallet-Regí, M., 2009. Influence of mesoporous structure type on the controlled delivery of drugs: release of ibuprofen from MCM-48, SBA-15 and functionalized SBA-15. J. Solgel Sci. Technol. 50, 421–429.

Jabbari Zahir Abadi, I., Sadeghi, O., Lotfizadeh Zhad, H., Tavassoli, N., Amani, V., Amini, M., 2012. Novel modified nanoporous silica for oral drug delivery: loading and release of clarithromycin. J. Solgel Sci. Technol. 61, 90–95.

Javadzadeh, Y., Bairami-Atashgah, R., Barzegar-Jalali, M., Soleimani, F., Mohammadi, G., Sabzevari, A., Adibkia, K., 2014. Inclusion of piroxicam in mesoporous phosphate glass-ceramic and evaluation of the physiochemical characteristics. Colloids Surf. B 116, 751–756.

Jia, Z., Lin, P., Xiang, Y., Wang, X., Wang, J., Zhang, X., Zhang, Q., 2011. A novel nanomatrix system consisted of colloidal silica and pH-sensitive polymethylacrylate improves the oral bioavailability of fenofibrate. Eur. J. Pharm. Biopharm. 79, 126–134.

Jia, L., Shen, J., Li, Z., Zhang, D., Zhang, Q., Duan, C., Liu, G., Zheng, D., Liu, Y., Tian, X., 2012. Successfully tailoring the pore size of mesoporous silica nanoparticles: exploitation of delivery systems for poorly water-soluble drugs. Int. J. Pharm. 439, 81–91.

Jiao, Y., Sun, Y., Chang, B., Lu, D., Yang, W., 2013. Redox- and temperature-controlled drug release from hollow mesoporous silica nanoparticles. Chemistry 19, 15410–15420.

Jin, D., Park, K.W., Lee, J.H., Song, K., Kim, J.-G., Seo, M.L., Jung, J.H., 2011. The selective immobilization of curcumin onto the internal surface of mesoporous hollow silica particles by covalent bonding and its controlled release. J. Mater. Chem. 21, 3641–3645.

Jinno, J., Kamada, N., Miyake, M., Yamada, K., Mukai, T., Odomi, M., Toguchi, H., Liversidge, G.G., Higaki, K., Kimura, T., 2006. Effect of particle size reduction on dissolution and oral absorption of a poorly water-soluble drug, cilostazol, in beagle dogs. J. Control. Release 111, 56–64.

Johnson, A.S., Ollivier, P.J., Mallouk, T.E., 1999. Ordered mesoporous polymers of tunable pore size from colloidal silica templates. Science 283, 963–965.

Joo, S.H., Jun, S., Ryoo, R., 2001. Synthesis of ordered mesoporous carbon molecular sieves CMK-1. Micropor. Mesopor. Mater. 44-45, 153–158.

Kapoor, S., Hegde, R., Bhattacharyya, A.J., 2009. Influence of surface chemistry of mesoporous alumina with wide pore distribution on controlled drug release. J. Control. Release 140, 34–39.

Kar, M., Tiwari, N., Tiwari, M., Lahiri, M., Gupta, S.S., 2013. Poly-L-arginine grafted silica mesoporous nanoparticles for enhanced cellular uptake and their application in DNA delivery and controlled drug release. Part. Part. Syst. Charact. 30, 166–179.

Kiekens, F., Eelen, S., Verheyden, L., Daems, T., Martens, J., Van Den Mooter, G., 2012. Use of ordered mesoporous silica to enhance the oral bioavailability of ezetimibe in dogs. J. Pharm. Sci. 101, 1136–1144.

Kiwilsza, A., Milanowski, B., Drużbicki, K., Coy, L.E., Grzeszkowiak, M., Jarek, M., Mielcarek, J., Lulek, J., Pajzderska, A., Wąsicki, J., 2015. Mesoporous drug carrier systems for enhanced delivery rate of poorly water-soluble drug: nimodipine. J. Porous Mat. 22, 817–829.

Kresge, C.T., Leonowicz, M.E., Roth, W.J., Vartuli, J.C., Beck, J.S., 1992. Ordered mesoporous molecular sieves synthesized by a liquid-crystal template mechanism. Nature 359, 710–712.
Ksenofontova, O., 2013. Development of the new insulin analogues suitable for the oral delivery in mesoporous silicon devices. Diabetes Technol. Ther. 15, A87–A187.
Kumar, D., Sailaja Chirravuri, S.V., Shastri, N.R., 2014. Impact of surface area of silica particles on dissolution rate and oral bioavailability of poorly water soluble drugs: a case study with aceclofenac. Int. J. Pharm. 461, 459–468.
Kupferschmidt, N., Xia, X., Labrador, R.H., Atluri, R., Ballell, L., Garcia-Bennett, A.E., 2013. In vivo oral toxicological evaluation of mesoporous silica particles. Nanomedicine 8, 57–64.
Lai, C.Y., Trewyn, B.G., Jeftinija, D.M., Jeftinija, K., Xu, S., Jeftinija, S., Lin, V.S., 2003. A mesoporous silica nanosphere-based carrier system with chemically removable CdS nanoparticle caps for stimuli-responsive controlled release of neurotransmitters and drug molecules. J. Am. Chem. Soc. 125, 4451–4459.
Lee, K.T., Oh, S.M., 2002. Novel synthesis of porous carbons with tunable pore size by surfactant-templated sol-gel process and carbonisation. Chem. Commun. 22, 2722–2723.
Lee, C.H., Lo, L.W., Mou, C.Y., Yang, C.S., 2008. Synthesis and characterization of positive-charge functionalized mesoporous silica nanoparticles for oral drug delivery of an anti-inflammatory drug. Adv. Funct. Mater. 18, 3283–3292.
Lee, J.E., Lee, D.J., Lee, N., Kim, B.H., Choi, S.H., Hyeon, T., 2011. Multifunctional mesoporous silica nanocomposite nanoparticles for pH controlled drug release and dual modal imaging. J. Mater. Chem. 21, 16869–16872.
Lehto, V.P., Vaha-Heikkila, K., Paski, J., Salonen, J., 2005. Use of thermoanalytical methods in quantification of drug load in mesoporous silicon microparticles. J. Therm. Anal. Calorim. 80, 393–397.
Letchmanan, K., Shen, S.C., Ng, W.K., Tan, R.B., 2015. Enhanced dissolution and stability of artemisinin by nano-confinement in ordered mesoporous SBA-15 particles. J. Microencapsul. 32, 390–400.
Li, J., Shi, X., 2011. Diclofenac sodium loaded pH-sensitive mesoporous zirconium diphosphonate coated with lag-time films for controlled colon-targeted release. Acta Chim. Sinica 69, 2439–2444.
Li, K., Zhou, Y., Yang, J.Y., Zhu, J.H., Liu, C.J., 2012. In vitro biocompatibility evaluation of ePTFE graft with controlled release of heparin from mesoporous material. Appl. Surf. Sci. 258, 4041–4047.
Li, J., Liang, Y., Dou, B., Ma, C., Lu, R., Hao, Z., Xie, Q., Luan, Z., Li, K., 2013a. Nanocasting synthesis of graphitized ordered mesoporous carbon using Fe-coated SBA-15 template. Mater. Chem. Phys. 138, 484–489.
Li, D., Zhu, Y., Liang, Z., 2013b. Alendronate functionalized mesoporous hydroxyapatite nanoparticles for drug delivery. Mater. Res. Bull. 48, 2201–2204.
Li, Y., Liu, Y.Z., Long, T., Yu, X.B., Tang, T.T., Dai, K.R., Tian, B., Guo, Y.P., Zhu, Z.A., 2013c. Mesoporous bioactive glass as a drug delivery system: fabrication, bactericidal properties and biocompatibility. J. Mater. Sci. Mater. Med. 24, 1951–1961.
Li, J., Xu, L., Liu, H., Wang, Y., Wang, Q., Chen, H., Pan, W., Li, S., 2014a. Biomimetic synthesized nanoporous silica@poly(ethyleneimine)s xerogel as drug carrier: characteristics and controlled release effect. Int. J. Pharm. 467, 9–18.
Li, D., He, J., Cheng, W., Wu, Y., Hu, Z., Tian, H., Huang, Y., 2014b. Redox-responsive nanoreservoirs based on collagen end-capped mesoporous hydroxyapatite nanoparticles for targeted drug delivery. J. Mater. Chem. B 2, 6089–6096.
Li, J., Xu, L., Yang, B.X., Bao, Z.H., Pan, W.S., Li, S.M., 2015a. Biomimetic synthesized chiral mesoporous silica: structures and controlled release functions as drug carrier. Mater. Sci. Eng. C Mater. Biol. Appl. 55, 367–372.
Li, J., Xu, L., Yang, B., Wang, H., Bao, Z., Pan, W., Li, S., 2015b. Facile synthesis of functionalized ionic surfactant templated mesoporous silica for incorporation of poorly water-soluble drug. Int. J. Pharm. 492, 191–198.
Liang, C., Dai, S., 2006. Synthesis of mesoporous carbon materials via enhanced hydrogen-bonding interaction. J. Am. Chem. Soc. 128, 5316–5317.

Liang, C., Li, Z., Dai, S., 2008. Mesoporous carbon materials: synthesis and modification. Angew. Chem. Int. Ed. Engl. 47, 3696–3717.

Liang, T., Qian, J., Yuan, Y., Liu, C., 2013. Synthesis of mesoporous hydroxyapatite nanoparticles using a template-free sonochemistry-assisted microwave method. J. Mater. Sci. 48, 5334–5341.

Li-Hong, W., Xin, C., Hui, X., Li-Li, Z., Jing, H., Mei-Juan, Z., Jie, L., Yi, L., Jin-Wen, L., Wei, Z., Gang, C., 2013. A novel strategy to design sustained-release poorly water-soluble drug mesoporous silica microparticles based on supercritical fluid technique. Int. J. Pharm. 454, 135–142.

Limnell, T., Santos, H.A., Makila, E., Heikkila, T., Salonen, J., Murzin, D.Y., Kumar, N., Laaksonen, T., Peltonen, L., Hirvonen, J., 2011a. Drug delivery formulations of ordered and nonordered mesoporous silica: comparison of three drug loading methods. J. Pharm. Sci. 100, 3294–3306.

Limnell, T., Heikkila, T., Santos, H.A., Sistonen, S., Hellsten, S., Laaksonen, T., Peltonen, L., Kumar, N., Murzin, D.Y., Louhi-Kultanen, M., Salonen, J., Hirvonen, J., Lehto, V.P., 2011b. Physicochemical stability of high indomethacin payload ordered mesoporous silica MCM-41 and SBA-15 microparticles. Int. J. Pharm 416, 242–251.

Lin, C.X., Qiao, S.Z., Yu, C.Z., Ismadji, S., Lu, G.Q., 2009. Periodic mesoporous silica and organosilica with controlled morphologies as carriers for drug release. Micropor. Mesopor. Mater. 117, 213–219.

Lin, K., Chen, L., Liu, P., Zou, Z., Zhang, M., Shen, Y., Qiao, Y., Liu, X., Chang, J., 2013. Hollow magnetic hydroxyapatite microspheres with hierarchically mesoporous microstructure for pH-responsive drug delivery. CrystEngComm 15, 2999–3008.

Liu, T., Li, L., Teng, X., Huang, X., Liu, H., Chen, D., Ren, J., He, J., Tang, F., 2011a. Single and repeated dose toxicity of mesoporous hollow silica nanoparticles in intravenously exposed mice. Biomaterials 32, 1657–1668.

Liu, R., Liao, P., Liu, J., Feng, P., 2011b. Responsive polymer-coated mesoporous silica as a pH-sensitive nanocarrier for controlled release. Langmuir 27, 3095–3099.

Liu, J., Du, X., Zhang, X., 2011c. Enzyme-inspired controlled release of cucurbit[7]uril nanovalves by using magnetic mesoporous silica. Chemistry 17, 810–815.

Liu, C., Yu, M., Li, Y., Li, J., Wang, J., Yu, C., Wang, L., 2015. Synthesis of mesoporous carbon nanoparticles with large and tunable pore sizes. Nanoscale 7, 11580–11590.

Long, T., Guo, Y.P., Liu, Y.Z., Zhu, Z.A., 2013. Hierarchically nanostructured mesoporous carbonated hydroxyapatite microspheres for drug delivery systems with high drug-loading capacity. RSC Advances 3, 24169–24176.

Lopez, T., Ortiz, E., Alexander-Katz, R., Basaldella, E., Bokhimi, X., 2009. Cortisol controlled release by mesoporous silica. Nanomedicine 5, 170–177.

Ma, J., Lin, H., Xing, R., Li, X., Bian, C., Xiang, D., Guo, W., Qu, F., 2014. Synthesis of pH-responsive mesoporous silica nanotubes for controlled release. J. Solgel Sci. Technol. 69, 364–369.

Manzano, M., Colilla, M., Vallet-Regi, M., 2009. Drug delivery from ordered mesoporous matrices. Expert Opin. Drug Deliv. 6, 1383–1400.

Mao, H., Liu, X., Yang, J., Li, B., Chen, Q., Zhong, J., 2014. Fabrication of magnetic silica-pillared clay (SPC) nanocomposites with ordered interlayer mesoporous structure for controlled drug release. Micropor. Mesopor. Mater. 184, 169–176.

Mariano-Neto, F., Matos, J.R., da Silva, L.C.C., Carvalho, L.V., Scaramuzzi, K., Sant'Anna, O.A., Oliveira, C.P., Fantini, M.C.A., 2014. Physical properties of ordered mesoporous SBA-15 silica as immunological adjuvant. J. Phys. D 47, 1–10.

Marsac, P.J., Konno, H., Rumondor, A.C., Taylor, L.S., 2008. Recrystallization of nifedipine and felodipine from amorphous molecular level solid dispersions containing poly(vinylpyrrolidone) and sorbed water. Pharm. Res. 25, 647–656.

Mei, X., Chen, D., Li, N., Xu, Q., Ge, J., Li, H., Lu, J., 2012. Hollow mesoporous silica nanoparticles conjugated with pH-sensitive amphiphilic diblock polymer for controlled drug release. Micropor. Mesopor. Mater. 152, 16–24.

Mellaerts, R., Aerts, C.A., Van Humbeeck, J., Augustijns, P., Van den Mooter, G., Martens, J.A., 2007. Enhanced release of itraconazole from ordered mesoporous SBA-15 silica materials. Chem. Commun., 1375–1377.
Mellaerts, R., Mols, R., Jammaer, J.A., Aerts, C.A., Annaert, P., Van Humbeeck, J., Van den Mooter, G., Augustijns, P., Martens, J.A., 2008a. Increasing the oral bioavailability of the poorly water soluble drug itraconazole with ordered mesoporous silica. Eur. J. Pharm. Biopharm. 69, 223–230.
Mellaerts, R., Jammaer, J.A., Van Speybroeck, M., Chen, H., Van Humbeeck, J., Augustijns, P., Van den Mooter, G., Martens, J.A., 2008b. Physical state of poorly water soluble therapeutic molecules loaded into SBA-15 ordered mesoporous silica carriers: a case study with itraconazole and ibuprofen. Langmuir 24, 8651–8659.
Mellaerts, R., Mols, R., Kayaert, P., Annaert, P., Van Humbeeck, J., Van den Mooter, G., Martens, J.A., Augustijns, P., 2008c. Ordered mesoporous silica induces pH-independent supersaturation of the basic low solubility compound itraconazole resulting in enhanced transepithelial transport. Int. J. Pharm. 357, 169–179.
Mellaerts, R., Houthoofd, K., Elen, K., Chen, H., Van Speybroeck, M., Van Humbeeck, J., Augustijns, P., Mullens, J., Van den Mooter, G., Martens, J.A., 2010. Aging behavior of pharmaceutical formulations of itraconazole on SBA-15 ordered mesoporous silica carrier material. Micropor. Mesopor. Mater. 130, 154–161.
Mendez, J., Monteagudo, A., Griebenow, K., 2012. Stimulus-responsive controlled release system by covalent immobilization of an enzyme into mesoporous silica nanoparticles. Bioconjug. Chem. 23, 698–704.
Meng, Y., Gu, D., Zhang, F., Shi, Y., Yang, H., Li, Z., Yu, C., Tu, B., Zhao, D., 2005. Ordered mesoporous polymers and homologous carbon frameworks: amphiphilic surfactant templating and direct transformation. Angew. Chem. Int. Ed. Engl. 44, 7053–7059.
Mudakavi, R.J., Raichur, A.M., Chakravortty, D., 2014. Lipid coated mesoporous silica nanoparticles as an oral delivery system for targeting and treatment of intravacuolar *Salmonella* infections. RSC Adv. 4, 61160–61166.
Muñoz, B., Rámila, A., Pérez-Pariente, J., Díaz, I., Vallet-Regí, M., 2003. MCM-41 organic modification as drug delivery rate regulator. Chem. Mater. 15, 500–503.
Ni, M., Xu, Q.Q., Yin, J.Z., 2012. Preparation of controlled release nanodrug ibuprofen supported on mesoporous silica using supercritical carbon dioxide. J. Mater. Res. 27, 2902–2910.
Nunes, C.D., Vaz, P.D., Fernandes, A.C., Ferreira, P., Romao, C.C., Calhorda, M.J., 2007. Loading and delivery of sertraline using inorganic micro and mesoporous materials. Eur. J. Pharm. Biopharm. 66, 357–365.
Palmieri, G.F., Giovannucci, G., Antonini, I., Martelli, S., 1997. Inclusion complexation of fenofibrate with β-cyclodextrin and hydroxypropyl-β-cyclodextrin. Evaluation of interactions in solution and solid complex characterization. STP Pharm. Sci. 7, 174–181.
Park, S.E., Kim, D.S., Chang, J.S., Kim, W.Y., 1998. Synthesis of MCM-41 using microwave heating with ethylene glycol. Catal. Today 44, 301–308.
Pasqua, L., Cundari, S., Ceresa, C., Cavaletti, G., 2009. Recent development, applications, and perspectives of mesoporous silica particles in medicine and biotechnology. Curr. Med. Chem. 16, 3054–3063.
Paudel, A., Worku, Z.A., Meeus, J., Guns, S., Van den Mooter, G., 2013. Manufacturing of solid dispersions of poorly water soluble drugs by spray drying: formulation and process considerations. Int. J. Pharm. 453, 253–284.
Popat, A., Jambhrunkar, S., Zhang, J., Yang, J., Zhang, H., Meka, A., Yu, C., 2014. Programmable drug release using bioresponsive mesoporous silica nanoparticles for site-specific oral drug delivery. Chem. Commun. 50, 5547–5550.
Popova, M., Szegedi, A., Yoncheva, K., Konstantinov, S., Petrova, G.P., Aleksandrov, H.A., Vayssilov, G.N., Shestakova, P., 2014. New method for preparation of delivery systems of poorly soluble drugs on the basis of functionalized mesoporous MCM-41 nanoparticles. Micropor. Mesopor. Mater. 198, 247–255.
Pourjavadi, A., Tehrani, Z.M., Bennett, C., 2015. PEG-*co*-polyvinyl pyridine coated magnetic mesoporous silica nanoparticles for pH-responsive controlled release of doxorubicin. Int. J. Polym. Mater. 64, 570–577.
Prokopowicz, M., Przyjazny, A., 2007. Synthesis of sol-gel mesoporous silica materials providing a slow release of doxorubicin. J. Microencapsul. 24, 682–701.

Qian, K.K., Bogner, R.H., 2012. Application of mesoporous silicon dioxide and silicate in oral amorphous drug delivery systems. J. Pharm. Sci. 101, 444–463.

Qian, X., Wang, W., Kong, W., Chen, Y., 2014. Organic-inorganic hybrid hollow mesoporous organosilica nanoparticles for efficient ultrasound-based imaging and controlled drug release. J. Nanomater. 2014, 1–8.

Qiao, Z.A., Zhang, L., Guo, M., Liu, Y., Huo, Q., 2009. Synthesis of mesoporous silica nanoparticles via controlled hydrolysis and condensation of silicon alkoxide. Chem. Mater. 21, 3823–3829.

Rana, V.K., Park, S.S., Parambadath, S., Kim, M.J., Kim, S.H., Mishra, S., Singh, R.P., Ha, C.S., 2011. Hierarchical mesoporous bio-polymer/silica composites co-templated by trimethyl chitosan and a surfactant for controlled drug delivery. MedChemComm 2, 1162–1166.

Ren, Y., Shi, X., Sun, Q., Sun, L., 2013. Dual-controlled oral colon-targeted delivery of bovine insulin based on mesoporous phosphonate. Mater. Res. Bull. 48, 4850–4855.

Riikonen, J., Correia, A., Kovalainen, M., Nakki, S., Lehtonen, M., Leppanen, J., Rantanen, J., Xu, W., Araujo, F., Hirvonen, J., Jarvinen, K., Santos, H.A., Lehto, V.P., 2015. Systematic in vitro and in vivo study on porous silicon to improve the oral bioavailability of celecoxib. Biomaterials 52, 44–55.

Rim, H.P., Min, K.H., Lee, H.J., Jeong, S.Y., Lee, S.C., 2011. pH-tunable calcium phosphate covered mesoporous silica nanocontainers for intracellular controlled release of guest drugs. Angew. Chem. Int. Ed. Engl. 50, 8853–8857.

Roik, N.V., Belyakova, L.A., 2014. Chemical design of pH-sensitive nanovalves on the outer surface of mesoporous silicas for controlled storage and release of aromatic amino acid. J. Solid State Chem. 215, 284–291.

Roine, J., Murtomaa, M., Myllys, M., Salonen, J., 2012. Dual-capillary electroencapsulation of mesoporous silicon drug carrier particles for controlled oral drug delivery. J. Electrostat. 70, 428–437.

Roine, J., Kaasalainen, M., Peurla, M., Correia, A., Araujo, F., Santos, H.A., Murtomaa, M., Salonen, J., 2015. Controlled dissolution of griseofulvin solid dispersions from electrosprayed enteric polymer micromatrix particles: physicochemical characterization and in vitro evaluation. Mol. Pharm. 12, 2254–2264.

Saha, D., Contescu, C.I., Gallego, N.C., 2012. Bimodal mesoporous carbon synthesized from large organic precursor and amphiphilic tri-block copolymer by self-assembly. Micropor. Mesopor. Mater. 155, 71–74.

Saha, D., Warren, K.E., Naskar, A.K., 2014a. Soft-templated mesoporous carbons as potential materials for oral drug delivery. Carbon 71, 47–57.

Saha, D., Warren, K.E., Naskar, A.K., 2014b. Controlled release of antipyrine from mesoporous carbons. Micropor. Mesopor. Mater. 196, 327–334.

Saha, D., Moken, T., Chen, J., Hensley, D.K., Delaney, K., Hunt, M.A., Nelson, K., Spurri, A., Benham, L., Brice, R., Azoro, M., 2015. Micro-/mesoporous carbons for controlled release of antipyrine and indomethacin. RSC Adv. 5, 23699–23707.

Salonen, J., Björkqvist, M., Laine, E., Niinistö, L., 2000. Effects of fabrication parameters on porous p + -type silicon morphology. Phys. Stat. Sol. A 182, 249–254.

Salonen, J., Laitinen, L., Kaukonen, A.M., Tuura, J., Bjorkqvist, M., Heikkila, T., Vaha-Heikkila, K., Hirvonen, J., Lehto, V.P., 2005. Mesoporous silicon microparticles for oral drug delivery: loading and release of five model drugs. J. Control. Release 108, 362–374.

Sánchez-Sánchez, Á., Suárez-García, F., Martínez-Alonso, A., Tascón, J.M.D., 2015. pH-responsive ordered mesoporous carbons for controlled ibuprofen release. Carbon 94, 152–159.

Santos, H.A., Peltonen, L., Limnell, T., Hirvonen, J., 2013. Mesoporous materials and nanocrystals for enhancing the dissolution behavior of poorly water-soluble drugs. Curr. Pharm. Biotechnol. 14, 926–938.

Sayari, A., Han, B.H., Yang, Y., 2004. Simple synthesis route to monodispersed SBA-15 silica rods. J. Am. Chem. Soc. 126, 14348–14349.

Schlossbauer, A., Kecht, J., Bein, T., 2009. Biotin-avidin as a protease-responsive cap system for controlled guest release from colloidal mesoporous silica. Angew. Chem. Int. Ed. Engl. 48, 3092–3095.

Schmaljohann, D., 2006. Thermo- and pH-responsive polymers in drug delivery. Adv. Drug Deliv. Rev. 58, 1655–1670.

Schmidt-Winkel, P., Yang, P.D., Margolese, D.I., Chmelka, B.F., Stucky, G.D., 1999. Fluoride-induced hierarchical ordering of mesoporous silica in aqueous acid-syntheses. Adv. Mater. 11, 303–307.

Schuster, J., He, G., Mandlmeier, B., Yim, T., Lee, K.T., Bein, T., Nazar, L.F., 2012. Spherical ordered mesoporous carbon nanoparticles with high porosity for lithium-sulfur batteries. Angew. Chem. Int. Ed. Engl. 51, 3591–3595.

Shen, S.C., Chen, F.X., Chow, P.S., Phanapavudhikul, P., Zhu, K.W., Tan, R.B.H., 2006. Synthesis of SBA-15 mesoporous silica via dry-gel conversion route. Micropor, Mesopor. Mater. 92, 300–308.

Shen, S.C., Chow, P.S., Chen, F.X., Tan, R.B., 2007. Submicron particles of SBA-15 modified with MgO as carriers for controlled drug delivery. Chem. Pharm. Bull. 55, 985–991.

Shen, S.C., Ng, W.K., Chia, L., Dong, Y.C., Tan, R.B., 2010. Stabilized amorphous state of ibuprofen by co-spray drying with mesoporous SBA-15 to enhance dissolution properties. J. Pharm. Sci. 99, 1997–2007.

Shen, S.C., Ng, W.K., Chia, L., Hu, J., Tan, R.B., 2011. Physical state and dissolution of ibuprofen formulated by co-spray drying with mesoporous silica: effect of pore and particle size. Int. J. Pharm. 410, 188–195.

Shen, S.C., Ng, W.K., Chia, L.S., Dong, Y.C., Tan, R.B., 2013. Applications of mesoporous materials as excipients for innovative drug delivery and formulation. Curr. Pharm. Des. 19, 6270–6289.

Shen, S.C., W.K. Ng, Chia, L., 2014. Mesoporous material excipients for poorly aqueous soluble ingredients, US Patent No. 8,778,401 B2.

Shen, S.C., Ng, W.K., Chia, L.S.O., Dong, Y.C., Tan, R.B.H., 2014b. Preparation and application of monodispersed mesoporous submicron carbon particles as a drug carrier. Powd. Technol. 261, 241–249.

Shi, X., Li, J., Tang, Y., Yang, Q., 2010. pH-sensitive mesoporous zirconium diphosphonates for controllable colon-targeted delivery. J. Mater. Chem. 20, 6495–6504.

Shiraki, K., Takata, N., Takano, R., Hayashi, Y., Terada, K., 2008. Dissolution improvement and the mechanism of the improvement from cocrystallization of poorly water-soluble compounds. Pharm. Res. 25, 2581–2592.

Shukla, A.J., Price, J.C., 1989. Effect of drug (core) particle size on the dissolution of theophylline from microspheres made from low molecular weight cellulose acetate propionate. Pharm. Res. 6, 418–421.

Silveira, G.Q., da Silva, R.S., Franco, L.P., Vargas, M.D., Ronconi, C.M., 2015. Redox-responsive nanoreservoirs: the effect of different types of mesoporous silica on the controlled release of doxorubicin in solution and in vitro. Micropor. Mesopor. Mater. 206, 226–233.

Sliwinska-Bartkowiak, M., Dudziak, G., Gras, R., Sikorski, R., Radhakrishnan, R., Gubbins, K.E., 2001. Freezing behavior in porous glasses and MCM-41. Colloids Surf. A 187–188, 523–529.

Song, J.H., Sailor, M.J., 1999. Chemical modification of crystalline porous silicon surfaces. Comments Inorg. Chem. 21, 69–84.

Song, B., Wu, C., Chang, J., 2012. Controllable delivery of hydrophilic and hydrophobic drugs from electrospun poly(lactic-*co*-glycolic acid)/mesoporous silica nanoparticles composite mats. J. Biomed. Mater. Res. B 100, 2178–2186.

Sun, J.T., Hong, C.Y., Pan, C.Y., 2010. Fabrication of PDEAEMA-coated mesoporous silica nanoparticles and pH-responsive controlled release. J. Phys. Chem. C 114, 12481–12486.

Sun, L., Zhang, X., Zheng, C., Wu, Z., Li, C., 2013. A pH gated, glucose-sensitive nanoparticle based on worm-like mesoporous silica for controlled insulin release. J. Phys. Chem. B 117, 3852–3860.

Tamanna, T., Bulitta, J.B., Yu, A., 2015. Controlling antibiotic release from mesoporous silica nano drug carriers via self-assembled polyelectrolyte coating. J. Mater. Sci. Mater. Med. 26, 1–7.

Tang, Q., Xu, Y., Wu, D., Sun, Y., Wang, J., Xu, J., Deng, F., 2006. Studies on a new carrier of trimethylsilyl-modified mesoporous material for controlled drug delivery. J. Control. Release 114, 41–46.

Tang, F., Li, L., Chen, D., 2012. Mesoporous silica nanoparticles: synthesis, biocompatibility and drug delivery. Adv. Mater. 24, 1504–1534.

Tang, Y., Ren, Y.B., Shi, X., 2013. Bifunctional mesoporous zirconium phosphonates for delivery of nucleic acids. Inorgan. Chem. 52, 1388–1397.

Tarn, D., Ferris, D.P., Barnes, J.C., Ambrogio, M.W., Stoddart, J.F., Zink, J.I., 2014. A reversible light-operated nanovalve on mesoporous silica nanoparticles. Nanoscale 6, 3335–3343.

Ting, C.C., Wu, H.Y., Vetrivel, S., Saikia, D., Pan, Y.C., Fey, G.T.K., Kao, H.M., 2010. A one-pot route to synthesize highly ordered mesoporous carbons and silicas through organic–inorganic self-assembly of triblock copolymer, sucrose and silica. Micropor. Mesopor. Mater. 128, 1–11.

Tsai, C.H., Vivero-Escoto, J.L., Slowing, I.I., Fang, I.J., Trewyn, B.G., Lin, V.S., 2011. Surfactant-assisted controlled release of hydrophobic drugs using anionic surfactant templated mesoporous silica nanoparticles. Biomaterials 32, 6234–6244.

Ukmar, T., Planinsek, O., 2010. Ordered mesoporous silicates as matrices for controlled release of drugs. Acta Pharm. 60, 373–385.

Vafaee, M., Amini, M., Najafi, E., Sadeghi, O., Amani, V., 2012. Modified nanoporous silicas for oral delivery of the water insoluble organotin compound: loading and release of methylphenyltin dichloride as an anti-tumor drug model. J. Solgel Sci. Technol. 64, 411–417.

Vallet-Regi, M., Rámila, A., del Real, R.P., Pérez-Pariente, J., 2001. A new property of MCM-41: drug delivery system. Chem. Mater. 13, 308–311.

Van Speybroeck, M., Barillaro, V., Thi, T.D., Mellaerts, R., Martens, J., Van Humbeeck, J., Vermant, J., Annaert, P., Van den Mooter, G., Augustijns, P., 2009. Ordered mesoporous silica material SBA-15: a broad-spectrum formulation platform for poorly soluble drugs. J. Pharm. Sci. 98, 2648–2658

Van Speybroeck, M., Mols, R., Mellaerts, R., Thi, T.D., Martens, J.A., Van Humbeeck, J., Annaert, P., Van den Mooter, G., Augustijns, P., 2010. Combined use of ordered mesoporous silica and precipitation inhibitors for improved oral absorption of the poorly soluble weak base itraconazole. Eur. J. Pharm. Biopharm. 75, 354–365.

van Speybroeck, M., Mellaerts, R., Thi, T.D., Martens, J.A., Van Humbeeck, J., Annaert, P., Van den Mooter, G., Augustijns, P., 2014. Preventing release in the acidic environment of the stomach via occlusion in ordered mesoporous silica enhances the absorption of poorly soluble weakly acidic drugs. J. Pharm. Sci. 100, 4864–4876.

Varga, N., Benkő, M., Sebők, D., Bohus, G., Janovák, L., Dékány, I., 2015. Mesoporous silica core–shell composite functionalized with polyelectrolytes for drug delivery. Micropor. Mesopor. Mater. 213, 134–141.

Vialpando, M., Aerts, A., Persoons, J., Martens, J., Van Den Mooter, G., 2011. Evaluation of ordered mesoporous silica as a carrier for poorly soluble drugs: influence of pressure on the structure and drug release. J. Pharm. Sci. 100, 3411–3420.

Vialpando, M., Backhuijs, F., Martens, J.A., Van den Mooter, G., 2012. Risk assessment of premature drug release during wet granulation of ordered mesoporous silica loaded with poorly soluble compounds itraconazole, fenofibrate, naproxen, and ibuprofen. Eur. J. Pharm. Biopharm. 81, 190–198

Vinu, A., Hossian, K.Z., Srinivasu, P., Miyahara, M., Anandan, S., Gokulakrishnan, N., Mori, T., Ariga, K., Balasubramanian, V.V., 2007. Carboxy-mesoporous carbon and its excellent adsorption capability for proteins. J. Mater. Chem. 17, 1819–1825.

Wan, Y., Zhao, D., 2007. On the controllable soft-templating approach to mesoporous silicates. Chem. Rev. 107, 2821–2860.

Wang, S., 2009. Ordered mesoporous materials for drug delivery. Micropor. Mesopor. Mater. 117, 1–9.

Wang, Y.M., Wu, Z.Y., Wei, Y.L., Zhu, J.H., 2005. In situ coating metal oxide on SBA-15 in one-pot synthesis Micropor. Mesopor. Mater. 84, 127–136.

Wang, F., Hui, H., Barnes, T.J., Barnett, C., Prestidge, C.A., 2010. Oxidized mesoporous silicon microparticles for improved oral delivery of poorly soluble drugs. Mol. Pharm. 7, 227–236.

Wang, Z., Chen, B., Quan, G., Li, F., Wu, Q., Dian, L., Dong, Y., Li, G., Wu, C., 2012. Increasing the oral bioavailability of poorly water-soluble carbamazepine using immediate-release pellets supported on SBA-15 mesoporous silica. Int. J. Nanomed. 7, 5807–5818.

Wang, Y., Zhao, Q., Hu, Y., Sun, L., Bai, L., Jiang, T., Wang, S., 2013. Ordered nanoporous silica as carriers for improved delivery of water insoluble drugs: a comparative study between three dimensional and two dimensional macroporous silica. Int. J. Nanomed. 8, 4015–4031.

Wang, Y., Sun, L., Jiang, T., Zhang, J., Zhang, C., Sun, C., Deng, Y., Sun, J., Wang, S., 2014. The investigation of MCM-48-type and MCM-41-type mesoporous silica as oral solid dispersion carriers for water insoluble cilostazol. Drug Dev. Ind. Pharm. 40, 819–828.
Wang, Y., Zhao, Q., Han, N., Bai, L., Li, J., Liu, J., Che, E., Hu, Zhang, L.Q., Jiang, T., Wang, S., 2015. Mesoporous silica nanoparticles in drug delivery and biomedical applications. Nanomedicine 11, 313–327.
Wani, A., Muthuswamy, E., Savithra, G.H., Mao, G., Brock, S., Oupicky, D., 2012. Surface functionalization of mesoporous silica nanoparticles controls loading and release behavior of mitoxantrone. Pharm. Res. 29, 2407–2418.
Waters, L.J., Hussain, T., Parkes, G., Hanrahan, J.P., Tobin, J.M., 2013. Inclusion of fenofibrate in a series of mesoporous silicas using microwave irradiation. Eur. J. Pharm. Biopharm. 85, 936–941.
Wei, Y.L., Wang, Y.M., Zhu, J.H., Wu, Z.Y., 2003. In-situ coating of SBA-15 with MgO: direct synthesis of mesoporous solid bases from strong acidic systems. Adv. Mater. 15, 1943–1945.
Wen, H., Guo, J., Chang, B., Yang, W., 2013. pH-responsive composite microspheres based on magnetic mesoporous silica nanoparticle for drug delivery. Eur. J. Pharm. Biopharm. 84, 91–98.
Wu, Z., Jiang, Y., Kim, T., Lee, K., 2007. Effects of surface coating on the controlled release of vitamin B1 from mesoporous silica tablets. J. Control. Release 119, 215–221.
Wu, C., Wang, J., Hu, Y., Zhi, Z., Jiang, T., Zhang, J., Wang, S., 2012. Development of a novel starch-derived porous silica monolith for enhancing the dissolution rate of poorly water soluble drug. Mater. Sci. Eng. C 32, 201–206.
Wu, S., Huang, X., Du, X., 2013. Glucose- and pH-responsive controlled release of cargo from protein-gated carbohydrate-functionalized mesoporous silica nanocontainers. Angew. Chem. Int. Ed. Engl. 52, 5580–5584.
Wu, C., Sun, X., Zhao, Z., Zhao, Y., Hao, Y., Liu, Y., Gao, Y., 2014. Synthesis of novel core–shell structured dual-mesoporous silica nanospheres and their application for enhancing the dissolution rate of poorly water-soluble drugs. Mater. Sci. Eng. C 44, 262–267.
Wu, S., Huang, X., Du, X., 2015. pH- and redox-triggered synergistic controlled release of a ZnO-gated hollow mesoporous silica drug delivery system. J. Mater. Chem. B 3, 1426–1432.
Wunderlich, J.C., Schick, U., Werry, J., Freidenreich, J., 1999. Gelatin or collagen hydrolysate containing drug formulation that provides for immediate release of nanoparticle drug compounds, US Patent No. US5,932,245 A.
Xiang, T.X., Anderson, B.D., 2003. Pharmaceutical formulation for poorly water soluble camptothecin analogues, US Patent No. US6,653,319 B1.
Xie, M., Shi, H., Li, Z., Shen, H., Ma, K., Li, B., Shen, S., Jin, Y., 2013. A multifunctional mesoporous silica nanocomposite for targeted delivery, controlled release of doxorubicin and bioimaging. Colloids Surf. B 110, 138–147.
Xing, R., Lin, H., Jiang, P., Qu, F., 2012. Biofunctional mesoporous silica nanoparticles for magnetically oriented target and pH-responsive controlled release of ibuprofen. Colloids Surf. A 403, 7–14.
Xu, W., Gao, Q., Xu, Y., Wu, D., Sun, Y., Shen, W., Deng, F., 2008. Controlled drug release from bifunctionalized mesoporous silica. J. Solid State Chem. 181, 2837–2844.
Xu, W., Gao, Q., Xu, Y., Wu, D., Sun, Y., 2009. pH-Controlled drug release from mesoporous silica tablets coated with hydroxypropyl methylcellulose phthalate. Mater. Res. Bull. 44, 606–612.
Xu, Y., Qu, F., Wang, Y., Lin, H., Wu, X., Jin, Y., 2011. Construction of a novel pH-sensitive drug release system from mesoporous silica tablets coated with Eudragit. Solid State Sci. 13, 641–646.
Xu, Y., Wang, C., Zhou, G., Wu, Y., Chen, J., 2012. Improving the controlled release of water-insoluble emodin from amino-functionalized mesoporous silica. Appl. Surf. Sci. 258, 6366–6372.
Xu, W., Riikonen, J., Lehto, V.P., 2013. Mesoporous systems for poorly soluble drugs. Int. J. Pharm. 453, 181–197.
Xu, G., Liu, S., Niu, H., Lv, W., Wu, R.A., 2014. Functionalized mesoporous carbon nanoparticles for targeted chemo-photothermal therapy of cancer cells under near-infrared irradiation. RSC Adv. 4, 33986–33997.
Xue, M., Findenegg, G.H., 2012. Lysozyme as a pH-responsive valve for the controlled release of guest molecules from mesoporous silica. Langmuir 28, 17578–17584.

Xue, J.M., Shi, M., 2004. PLGA/mesoporous silica hybrid structure for controlled drug release. J. Control. Release 98, 209–217.

Yan, X., Yu, C., Zhou, X., Tang, J., Zhao, D., 2004. Highly ordered mesoporous bioactive glasses with superior in vitro bone-forming bioactivities. Angew. Chem. Int. Ed. Engl. 43, 5980–5984.

Yang, P.P., Quan, Z.W., Lu, L.L., Huang, S.S., Lin, J., Fu, H.G., 2007. MCM-41 functionalized with YVO 4: Eu3 + : a novel drug delivery system. Nanotechnology 18, 1–10.

Yang, Y.J., Tao, X., Hou, Q., Chen, J.F., 2009. Fluorescent mesoporous silica nanotubes incorporating CdS quantum dots for controlled release of ibuprofen. Acta Biomater. 5, 3488–3496.

Yang, Y.J., Tao, X., Hou, Q., Ma, Y., Chen, X.L., Chen, J.F., 2010. Mesoporous silica nanotubes coated with multilayered polyelectrolytes for pH-controlled drug release. Acta Biomater. 6, 3092–3100.

Yang, P., Yang, P., Teng, X., Lin, J., Huang, L., 2011. A novel luminescent mesoporous silica/apatite composite for controlled drug release. J. Mater. Chem. 21, 5505–5510.

Yang, P., Gai, S., Lin, J., 2012. Functionalized mesoporous silica materials for controlled drug delivery. Chem. Soc. Rev. 41, 3679–3698.

Yang, J., Shen, D., Zhou, L., Li, W., Li, X., Yao, C., Wang, R., El-Toni, A.M., Zhang, F., Zhao, D., 2013. Spatially confined fabrication of core–shell gold nanocages@mesoporous silica for near-infrared controlled photothermal drug release. Chem. Mater. 25, 3030–3037.

Yang, C., Guo, W., Cui, L., Xiang, D., Cai, K., Lin, H., Qu, F., 2014. pH-responsive controlled-release system based on mesoporous bioglass materials capped with mineralized hydroxyapatite. Mater. Sci. Eng. C 36, 237–243.

Yoncheva, K., Popova, M., Szegedi, A., Mihaly, J., Tzankov, B., Lambov, N., Konstantinov, S., Tzankova, V., Pessina, F., Valoti, M., 2014. Functionalized mesoporous silica nanoparticles for oral delivery of budesonide. J. Solid State Chem. 211, 154–161.

Yu, H., Zhai, Q.Z., 2009. Mesoporous SBA-15 molecular sieve as a carrier for controlled release of nimodipine. Micropor. Mesopor. Mater. 123, 298–305.

Yu, C., Fan, J., Tian, B., Zhao, D., Stucky, G.D., 2002. High-yield synthesis of periodic mesoporous silica rods and their replication to mesoporous carbon rods. Adv. Mater. 14, 1742–1745.

Yu, F., Tang, X., Pei, M., 2013. Facile synthesis of PDMAEMA-coated hollow mesoporous silica nanoparticles and their pH-responsive controlled release. Micropor. Mesopor. Mater. 173, 64–69.

Zeng, W., Qian, X.F., Yin, J., Zhu, Z.K., 2006. The drug delivery system of MCM-41 materials via co-condensation synthesis. Mater. Chem. Phys. 97, 437–441.

Zhai, Y., Dou, Y., Liu, X., Park, S.S., Ha, C.S., Zhao, D., 2011. Soft-template synthesis of ordered mesoporous carbon/nanoparticle nickel composites with a high surface area. Carbon 49, 545–555.

Zhang, Y., Zhi, Z., Jiang, T., Zhang, J., Wang, Z., Wang, S., 2010a. Spherical mesoporous silica nanoparticles for loading and release of the poorly water-soluble drug telmisartan. J. Control. Release 145, 257–263.

Zhang, H., Li, Z., Xu, P., Wu, R., Jiao, Z., 2010b. A facile two step synthesis of novel chrysanthemum-like mesoporous silica nanoparticles for controlled pyrene release. Chem. Commun. 46, 6783–6785.

Zhang, Y., Wang, J., Bai, X., Jiang, T., Zhang, Q., Wang, S., 2012. Mesoporous silica nanoparticles for increasing the oral bioavailability and permeation of poorly water soluble drugs. Mol. Pharm. 9, 505–513.

Zhang, Y., Zhi, Z., Li, X., Gao, J., Song, Y., 2013a. Carboxylated mesoporous carbon microparticles as new approach to improve the oral bioavailability of poorly water-soluble carvedilol. Int. J. Pharm. 454, 403–411.

Zhang, Y., Wang, H., Gao, C., Li, X., Li, L., 2013b. Highly ordered mesoporous carbon nanomatrix as a new approach to improve the oral absorption of the water-insoluble drug, simvastatin. Eur. J. Pharm. Sci. 49, 864–872.

Zhang, Y., Che, E., Zhang, M., Sun, B., Gao, J., Han, J., Song, Y., 2014a. Increasing the dissolution rate and oral bioavailability of the poorly water-soluble drug valsartan using novel hierarchical porous carbon monoliths. Int. J. Pharm. 473, 375–383.

Zhang, Y., Wang, H., Li, C., Sun, B., Wang, Y., Wang, S., Gao, C., 2014b. A novel three-dimensional large-pore mesoporous carbon matrix as a potential nanovehicle for the fast release of the poorly water-soluble drug, celecoxib. Pharm. Res. 31, 1059–1070.

Zhang, G., Yang, M., Cai, D., Zheng, K., Zhang, X., Wu, L., Wu, Z., 2014c. Composite of functional mesoporous silica and DNA: an enzyme-responsive controlled release drug carrier system. ACS Appl. Mater. Interf. 6, 8042–8047.

Zhang, P., Forsgren, J., Stromme, M., 2014d. Stabilisation of amorphous ibuprofen in Upsalite, a mesoporous magnesium carbonate, as an approach to increasing the aqueous solubility of poorly soluble drugs. Int. J. Pharm. 472, 185–191.

Zhang, C., Zhao, Q., Wan, L., Wang, T., Sun, J., Gao, Y., Jiang, T., Wang, S., 2014e. Poly dimethyl diallyl ammonium coated CMK-5 for sustained oral drug release. Int. J. Pharm. 461, 171–180.

Zhang, X., Zhang, T., Ye, Y., Chen, H., Sun, H., Zhou, X., Ma, Z., Wu, B., 2015a. Phospholipid-stabilized mesoporous carbon nanospheres as versatile carriers for systemic delivery of amphiphobic SNX-2112 (a Hsp90 inhibitor) with enhanced antitumor effect. Eur. J. Pharm. Biopharm. 94, 30–41.

Zhang, Y., Zhao, Q., Zhu, W., Zhang, L., Han, J., Lin, Q., Ai, F., 2015b. Synthesis and evaluation of mesoporous carbon/lipid bilayer nanocomposites for improved oral delivery of the poorly water-soluble drug, nimodipine. Pharm. Res. 32, 2372–2383.

Zhang, Y., Zhang, H., Che, E., Zhang, L., Han, J., Yang, Y., Wang, S., Zhang, M., Gao, C., 2015c. Development of novel mesoporous nanomatrix-supported lipid bilayers for oral sustained delivery of the water-insoluble drug, lovastatin. Colloids Surf. B 128, 77–85.

Zhao, D., Feng, J., Huo, Q., Melosh, N., Fredrickson, G.H., Chmelka, B.F., Stucky, G.D., 1998. Triblock copolymer syntheses of mesoporous silica with periodic 50 to 300 angstrom pores. Science 279, 548–552.

Zhao, D., Sun, J.Y., Li, Q.Z., Stucky, G.D., 2000. Morphological control of highly ordered mesoporous silica SBA-15. Chem. Mater. 12, 275–279.

Zhao, L., Yan, X., Zhou, X., Zhou, L., Wang, H., Tang, J., Yu, C., 2008. Mesoporous bioactive glasses for controlled drug release. Micropor. Mesopor. Mater. 109, 210–215.

Zhao, W., Lang, M., Li, Y., Li, L., Shi, J., 2009a. Fabrication of uniform hollow mesoporous silica spheres and ellipsoids of tunable size through a facile hard-templating route. J. Mater. Chem. 19, 2778–2783.

Zhao, Y.F., Loo, S.C., Ma, J., 2009b. Co-synthesis and drug delivery properties of mesoporous hydroxyapatite-silica composites. J. Nanosci. Nanotechnol. 9, 3720–3727.

Zhao, Y., Trewyn, B.G., Slowing, I.I., Lin, V.S., 2009c. Mesoporous silica nanoparticle-based double drug delivery system for glucose-responsive controlled release of insulin and cyclic AMP. J. Am. Chem. Soc. 131, 8398–8400.

Zhao, W., Zhang, H., He, Q., Li, Y., Gu, J., Li, L., Li, H., Shi, J., 2011. A glucose-responsive controlled release of insulin system based on enzyme multilayers-coated mesoporous silica particles. Chem. Commun. 47, 9459–9461.

Zhao, Q., Wang, T., Wang, J., Zheng, L., Jiang, T., Cheng, G., Wang, S., 2012a. Fabrication of mesoporous hydroxycarbonate apatite for oral delivery of poorly water-soluble drug carvedilol. J. Non Cryst. Solids 358, 229–235.

Zhao, P., Wang, L., Sun, C., Jiang, T., Zhang, J., Zhang, Q., Sun, J., Deng, Y., Wang, S., 2012b. Uniform mesoporous carbon as a carrier for poorly water soluble drug and its cytotoxicity study. Eur. J. Pharm. Biopharm. 80, 535–543.

Zhao, Z., Wu, C., Zhao, Y., Hao, Y., Liu, Y., Zhao, W., 2015. Development of an oral push–pull osmotic pump of fenofibrate-loaded mesoporous silica nanoparticles. Int. J. Nanomed. 10, 1691–1701.

Zheng, Q., Hao, Y., Ye, P., Guo, L., Wu, H., Guo, Q., Jiang, J., Fu, F., Chen, G., 2013. A pH-responsive controlled release system using layered double hydroxide (LDH)-capped mesoporous silica nanoparticles. J. Mater. Chem. B 1, 1644–1648.

Zheng, Q., Lin, T., Wu, H., Guo, L., Ye, P., Hao, Y., Guo, Q., Jiang, J., Fu, F., Chen, G., 2014. Mussel-inspired polydopamine coated mesoporous silica nanoparticles as pH-sensitive nanocarriers for controlled release. Int. J. Pharm. 463, 22–26.

Zhou, Z., Zhu, S., Zhang, D., 2007. Grafting of thermo-responsive polymer inside mesoporous silica with large pore size using ATRP and investigation of its use in drug release. J. Mater. Chem. 17, 2428–2433

Zhu, Y.F., Shi, J.L., Shen, W.H., Chen, H.R., Dong, X.P., Ruan, M.L., 2005. Preparation of novel hollow mesoporous silica spheres and their sustained-release property. Nanotechnology 16, 2633–2638.

Zhu, M., Wang, H., Liu, J., He, H., Hua, X., He, Q., Zhang, L., Ye, X., Shi, J., 2011a. A mesoporous silica nanoparticulate/beta-TCP/BG composite drug delivery system for osteoarticular tuberculosis therapy. Biomaterials 32, 1986–1995.

Zhu, S., Chen, C., Chen, Z., Liu, X., Li, Y., Shi, Y., Zhang, D., 2011b. Thermo-responsive polymer-functionalized mesoporous carbon for controlled drug release. Mater. Chem. Phys. 126, 357–363.

Zhu, C.L., Lu, C.H., Song, X.Y., Yang, H.H., Wang, X.R., 2011c. Bioresponsive controlled release using mesoporous silica nanoparticles capped with aptamer-based molecular gate. J. Am. Chem. Soc. 133, 1278–1281.

Zhu, M., Shi, J., He, Q., Zhang, L., Chen, F., Chen, Y., 2012. An emulsification–solvent evaporation route to mesoporous bioactive glass microspheres for bisphosphonate drug delivery. J. Mater. Sci. 47, 2256–2263.

Zhu, M., Zhu, Y., Zhang, L., Shi, J., 2013. Preparation of chitosan/mesoporous silica nanoparticle composite hydrogels for sustained co-delivery of biomacromolecules and small chemical drugs. Sci. Technol. Adv. Mater. 14, 1–9.

Zhu, W., Zhao, Q., Sun, C., Zhang, Z., Jiang, T., Sun, J., Li, Y., Wang, S., 2014a. Mesoporous carbon with spherical pores as a carrier for celecoxib with needle-like crystallinity: improve dissolution rate and bioavailability. Mater. Sci. Eng. C 39, 13–20.

Zhu, W., Wan, L., Zhang, C., Gao, Y., Zheng, X., Jiang, T., Wang, S., 2014b. Exploitation of 3D face-centered cubic mesoporous silica as a carrier for a poorly water soluble drug: influence of pore size on release rate. Mater. Sci. Eng. C 34, 78–85.

CHAPTER

NANODENTISTRY: NOVEL APPROACHES

24

Nehi Sinha*, Niha M. Kulshreshtha*, Meenal Dixit*, Indrani Jadhav*, Divya Shrivastava*, Prakash S. Bisen*,,†**

**Jaipur National University, Jaipur, India*
***Jiwaji University, Gwalior, India*
†Tropilite Foods Pvt Ltd., Gwalior, India

CHAPTER OUTLINE

1 **Introduction** 752
1.1 The Bottom-Up Approach 752
1.2 The Top-Down Approach 755
1.3 The Functional Approach 757
1.4 The Biomimetic Approach 757
2 **Types and Properties of Nanostructures Used in Dentistry** 757
2.1 Nanocomposites 758
2.2 Nanoclusters 760
2.3 Nanorobots 761
2.4 Nanosolutions 761
2.5 Nanolight Curable Glass Ionomers 762
2.6 Nanoimpression Materials 762
2.7 Nanoencapsulated Materials 762
2.8 Other Nanostructures 763
3 **Nanotechnology in Dentistry** 763
3.1 Nanodiagnostics 763
3.2 Nanotechnology in Oral and Maxillofacial Surgery 767
3.3 Nanotechnology in Orthodontic Treatment 768
3.4 Nanotechnology in Prosthodontics 768
3.5 Nanotechnology in Periodontics 768
3.6 Nanotechnology in Conservative Dentistry and Endodontics 769
4 **Preventive Nanodentistry** 770
4.1 Dental Caries 770
4.2 Antibacterial Nanotherapy 770
4.3 Biomimetic Remineralization—Reversing an Incipient Caries 770
4.4 Biomimetic Remineralization—Recurrent Decay 770
4.5 Caries Vaccine 771
4.6 Periodontal Diseases 771

5 Challenges Faced by Nanodentistry....771
6 Conclusions....772
References....772

1 INTRODUCTION

Nanotechnology is a fascinating field of science where all the disciplines of natural sciences intertwine at nanoscale, such as physics combines with life sciences, engineering, chemistry, materials science, and computational approaches, which are closely intertwined. This intrinsic multidisciplinary nature of nanotechnology offers vast potential for interdisciplinary applications.

Oral medicine is a field of dentistry that is constantly evolving. Multidisciplinary approach has led to advancement of oral medicine in prevention, diagnostics, and therapeutics.

Nanotechnology is the exploitation of matter at molecular and atomic levels, which has revolutionized the field of dentistry as "Nanodentistry". Nanodentistry is based on four approaches (Fig. 24.1) and can be defined as a promising discipline that provides nanotechnology based on next generation devices and tools for use in oral health care.

1.1 THE BOTTOM-UP APPROACH

Bottom-up approach includes self-assembly to nanofabrication by use of chemical or physical forces operating at nanoscale to assemble smaller units forming larger structures (Das et al., 2007). Chemical synthesis, self-assembly, and molecular fabrication are a few examples of bottom-up approach. In simple language it is like building a house, by taking lots of building blocks and putting them together to produce the final bigger structure (Fig. 24.2). Another common example found in nature is that of cells that use enzymes to produce DNA by using various component molecules and assembling them to make the final molecular structure. Bottom-up approaches in nanodentistry include polymerization, precipitation, and micellar arrangements (Fig. 24.2).

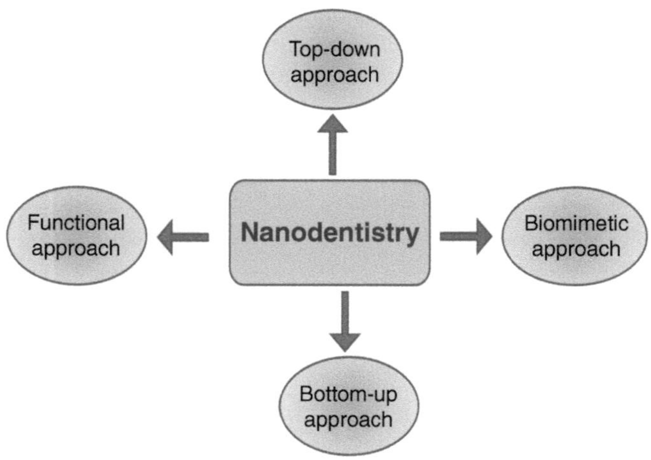

FIGURE 24.1 Different Approaches of Nanodentistry

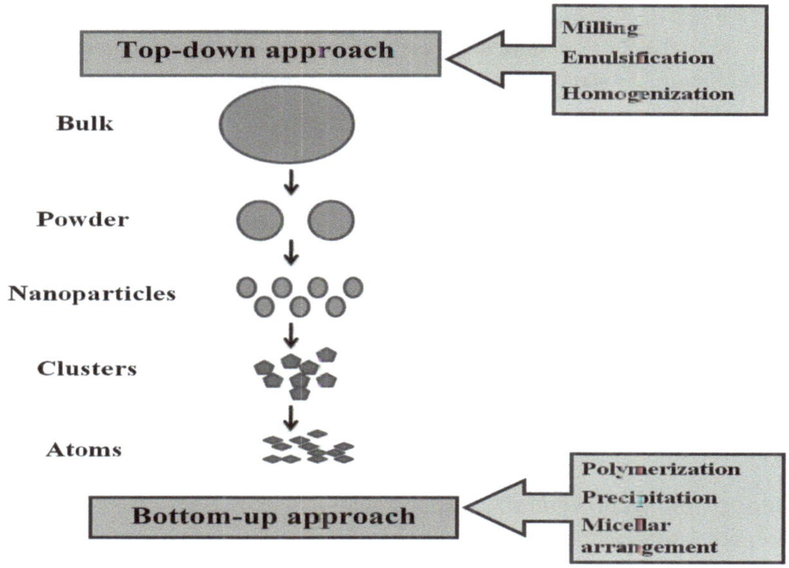

FIGURE 24.2 Top-Down and Bottom-Up Approach

Various applications based on top-down and bottom-up approaches are summarized in Table 24.1.

1.1.1 Nanoanaesthesia

Nanorobots are used for imparting oral anaesthesia. Although nanorobots use a hybrid of top-down and bottom-up approach, in oral anaesthesia, nanorobots are self-assembled from nanobuilding blocks into structure, tools, sensors, and actuators. It consists of colloidal suspension of millions of active analgesic micrometer-sized dental nanorobots. After reaching the surface of the crown or mucosa, the nanorobots

Table 24.1 Applications Based on Top-Down and Bottom-Up Approaches

Top-Down Approach	Bottom-Up Approach
Nanoanaesthesia	Nanocomposites
Major tooth repair	Nanolight-curing glass ionomer restorative
Tooth repositioning	Impression materials
Hypersensitive cure	Nanocomposite denture teeth
Dental durability and appearance	Nanosolution
Dentifrobots	Nanoencapsulation
Cancer diagnosis	Laser plasma application for periodontia
Therapeutic aid in oral disease	Prosthetic implant
	Nanoneedles
	Bone replacement material

enter the dentin by migrating into a patient's gingivae. Nanorobots, on reaching the dentin, proceed toward the pulp through the dentinal tubular holes, which are 1–4 μm in diameter. They are carried by a combination of temperature and chemical gradients, and even the position of navigation by control of the nanocomputer according to the dentist (Freitas, 2000).

1.1.2 Major tooth repair

Tooth repair involves manufacture of tooth in vitro by organizing nanorod-like calcium hydroxyapatite crystals roughly parallel to each other that lead to biomineralization, which is followed by its installation, collectively called the complete dentition therapy. The nanorobotic manufacture of the tooth involves both mineral and cellular components and techniques like genetic engineering, tissue engineering, and tissue regeneration are vital (Chandki et al., 2012). Biomaterials are used to make the scaffold then stem cells are transplanted and/or endogenous stem cells are attached. With the reasonable desktop manufacture facility, which has a potential to fabricate a new tooth at low cost and time, this therapy may become commercially viable. The same technology might be applied for tooth maintenance (Shellhart and Oesterle, 1999).

1.1.3 Tooth repositioning/orthodontic nanorobots

Orthodontic wires are coated with inorganic fullerene, such as tungsten disulfide (WS_2) nanoparticles. Alumina nanostructures with polysulfone are used to prepare orthodontic brackets. Nanorobots, molecular sorting rotors, and robot arms are made up of biocompatible glycocalyx-coated diamondoid material. In contrast to the conventional molar uprighting techniques, which are used currently to achieve rotation, straightening, and repositioning of teeth that take a long time to complete (weeks to months), the use of nanorobots may achieve this feat in a matter of minutes to hours (Sahoo et al., 2007). These orthodontic nanorobots can manipulate periodontal tissues directly like periodontal ligament, alveolar bone, cementum, and gingivae and can vertically reposition teeth rapidly and effortlessly.

1.1.4 Hypersensitivity cure

Nanorobots can be used to relieve hypersensitivity in patients by providing a quick and permanent treatment. This is done by manipulating the pressure transmitted hydrodynamically to the pulp by selective occlusion of selected tubules using natural biomaterials (Nagpal et al., 2011). This can be carried out precisely by reconstructive dental nanorobots in minutes, offering a quick relief.

1.1.5 Dental durability and appearance

The replacement of the upper layer of enamel with artificial materials imparts strength, increases durability, and improves the appearance of teeth. Naturally hard and tough materials, such as sapphire (Fartash et al., 1996) or diamond, may be covalently bonded at the surface for this purpose. However, in their pure form, these materials are brittle and fracture-prone. Incorporation of nanomaterials, such as carbon nanotubes, to form nanocomposites to overcome this drawback and can therefore act as effective replacement of upper enamel (Yunshin et al., 2005).

1.1.6 Dentrifrobots

Daily use products, such as toothpaste and mouthwash, can be enriched with dentifrobots, which can perform debridement of the plaque and calculus. These nanorobotic dentifrices reach to the supragingival and subgingival sites and metabolize organic matter. The organic matter is converted into odorless and harmless vapors and pathogenic bacteria are specifically identified and destroyed whereas natural oral flora is not harmed (Freitas, 2000).

1.1.7 Cancer diagnosis

Nanotechnology can aid in medical diagnosis by providing convenient and noninvasive techniques to identify and/or quantify disease markers. For example, in the case of cancer, early diagnosis, localization, typing, malignancy, and recurrence can be easily monitored by physicochemical modification at the nanoscale called "Nanotexturing" of surfaces (Sunder Rajan et al., 2013).

1.1.8 Therapeutic aid in oral diseases

1.1.8.1 Drug delivery

Limited bioavailability of many therapeutic agent limits their utility in oral formulations. In some cases, reduction in particle size of the drugs has shown to improve the bioavailability (Shabouri, 2002). Nanoparticle formulation can also offer protection from unfavorable pH, can sustain the drug exposure or increase retention of the drug (bioadhesion) for improved stability. All these factors achieve higher effective concentrations of the drug at the intended target site, ensuring better therapeutic effects.

1.1.8.2 Gene therapy

The aim of gene therapy is to treat genetic disorders by introducing a functional copy of the defective genes (Goverdhana et al., 2005). Gene delivery can be carried out by either direct injection using gene guns or via the use of viral or nonviral vectors (Wells, 2004). Among the nonviral vectors, application of nanoparticles for gene delivery has been widely commended (Davis, 1997). The prime advantage of nanoparticles as gene delivery agent is due to their less immunogenic than the viral vectors and therefore has a potential to replace viral vector for gene delivery.

1.2 THE TOP-DOWN APPROACH

1.2.1 Nanocomposites

In the field of restorative dentistry, nanomaterials can be used to refine the resin-based composite systems. Nanocomposites share favorable properties of nanorange particle size (<100 nm), strength, microhardness, flexural strength, smoothness, translucency, and precision in terms of shade characterization. They also offer reduced filling shrinkage and excellent handling properties. Nanocomposites include two types of composite materials: nanohybrid and nanofilled resin-based composites. Nanofilled composites have been reported to have better physical properties like higher elastic modulus than hybrid composites (Beun et al., 2007).

1.2.2 Nanolight-curing glass ionomer restorative

This technology is the combination of "Blends Nanotechnology" and "Fluoraluminosilicate (FAS)" technology and can be used for primary teeth restorations, small Class I, Class III and IV restorations, transitional and sandwich restorations, and core build-ups. The major advantages are improved resistance to wearing, pleasing aesthetics with magnificent polish (Chandki et al., 2012).

1.2.3 Impression materials

Siloxane impression materials are produced by integrating nanofillers with vinylsiloxanes. The impression material improves hydrophilic properties and better flow characteristics thus allowing better pouring with enhanced precision, minimizing voids at the margin (Patil et al., 2008). The additional advantages include high tear, heat and distortion resistance, and snap set that lowers the errors generated due to micromovements.

1.2.4 Nanocomposite denture teeth

Composite denture teeth blend the favorable properties of porcelain denture teeth, such as wear resistance and acrylic resin denture teeth like easy recontouring and polishing (Freitas, 2005). Nanocomposite denture is composed of uniformly dispersed nanosized filler particles in polymethyl-methacrylate (PMMA). The major advantages include enhanced hardness and durability, polishable, impact with stain resistant material.

1.2.5 Nanosolution

Nanosolutions are bonding agents ensuring homogeneous mixing of adhesive used in dentistry. Nanosolutions are composed of solutions of highly dispersible nanoparticles (Patil et al., 2008).

1.2.6 Nanoencapsulation

Nanoparticles can be used to generate nanocapsules for better drug delivery with reduced side effects. These targeted release systems encapsulating novel antibiotics and vaccines were developed by South West Research Institute (SWRI). Similarly, the targeted delivery of drugs and genes to the human liver has been developed at Osaka University in Japan in 2003. This delivery exploited a display peptide that is used by the hepatitis B virus for gaining entry in liver. Hollow nanoparticles displaying the peptide were generated using engineered hepatitis B virus envelope L particles to achieve this. The same principle can be applied in the future to target oral tissues, such as cells derived from the periodontium.

1.2.7 Laser plasma application for periodontia

Certain nanoparticles can be optically broken down upon irradiation with laser pulses for specific applications. For example, TiO_2 nanoparticles of dimensions 20–50 nm can be applied as gel-based emulsion on human skin and irradiated with laser pulses for possible applications in melanin removal, cutting of enamel and dentin, incision of soft tissue, caries preparation, and periodontal treatment (Freitas, 2005).

1.2.8 Prosthetic implant

Nanotechnology can be used to generate specialized surfaces with known chemical properties and controlled topography to help in generating implant surfaces with tissue integrative properties, in gaining better understanding of biological interactions and conferring the ability to manipulate the nature of periimplant tissue by controlling the cell differentiation into specific lineages. Biologically active drugs, such as antibiotics and growth factors can also be incorporated during precipitation of certain coatings, namely, CaP coatings on Ti implants to achieve better cell growth, for example, Nanotite-nanocoated implant achieves 60% higher growth of new cells (Lavenus et al., 2010; Le Guéhennec et al., 2007).

1.2.9 Nanoneedles

Nanoneedles consisting of nanorange stainless steel crystals have been developed with the ultimate aim of cell level surgery (Sandvik Bioloine, RK 91 needles). Nanotweezers are also being developed for this purpose.

1.2.10 Bone-replacement materials

Arranged in parallel, calcium hydroxyapatite crystals in the form of nanorods have been generated as a simulation of natural biomineralization of dental enamel (Chen et al., 2005). Some of the commercially available hydroxyapatite nanoparticles used to treat bone defects include VITOSS (orthovita Inc, USA)

(HA+ TCP), Ostim (Osartis GmbH Germany) (HA), and NanOssTM (Angstrom Medica, USA) (HA). However, there are some challenges, such as economical feasibility of mass-scale nanorobots, accurate positioning and assembly, biocompatibility, coordination between large number of narobots and public acceptance.

1.3 THE FUNCTIONAL APPROACH

The ultimate aim of this approach is to generate desired functionality components irrespective of the actual process of assembly. The properties of interest might be electronic properties in the case of molecular scale electronics or mechanical properties in case of synthetic molecular motors.

1.4 THE BIOMIMETIC APPROACH

Biomimetics, also known as bionics, biognosis, or biomimicry, is the use and implementation of concepts and principles from nature for creating new materials, devices, and systems (Ghalanbor et al., 2005). This approach mimics the biological system to form nanoscale devices. The technology is still under research and has not yet reached to the stage of commercialization.

2 TYPES AND PROPERTIES OF NANOSTRUCTURES USED IN DENTISTRY

The various types of nanostructures used in dentistry are shown in Fig. 24.3.

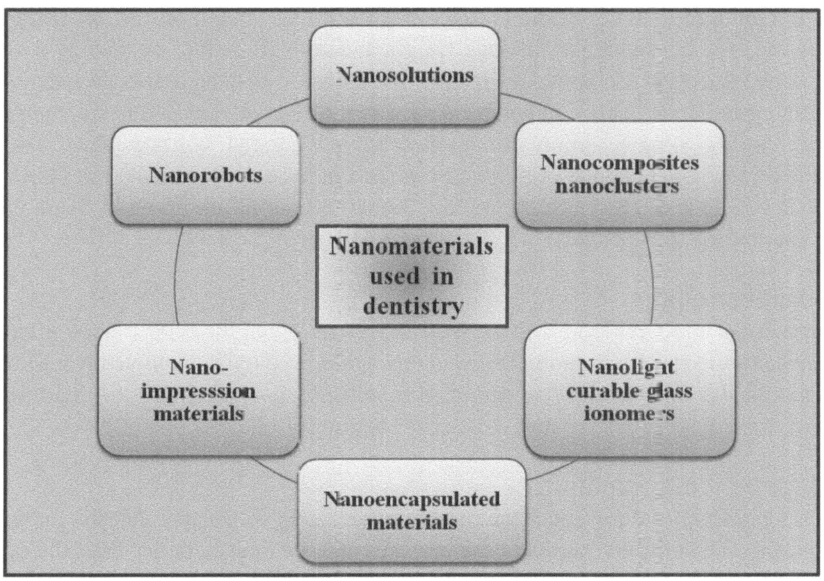

FIGURE 24.3 Types of Nanostructures Used in Dentistry

2.1 NANOCOMPOSITES

The term nanocomposite is used for composite materials composed of nanomaterials integrated with ceramic, metal of polymer matrix. Formation of a composite enhances the properties of the material by achieving better synergy between the components. The properties of the composites depend on the type of matrix used, attributes of nanoscale phase including shape, size, dispersion, orientation, and the interplay between the two phases (Raffaele and Janne, 2009).

2.1.1 Properties of nanocomposites

Nanocomposites blend the beneficial properties of hybrid composites (resistance to wear and mechanical strength) with those of microfill composites (polish and gloss retention).

2.1.1.1 Polymerization shrinkage

The composite resin shows the polymerization shrinkage in the range of 1.4%–1.6%. The reasons for low shrinkage value include strong interaction between resin and nanoparticles and use of epoxy resins, which have low shrinkage characteristics. In addition, the amount of organic matrix of composites also correlates with shrinkage characteristics. Higher organic matrix content leads to higher shrinkage characteristics. This is evident from the shrinkage characteristics of nanohybrid composites (Grandio) organic matrix (13 wt.%), which shows very low shrinkage as compared to nanofill composites (Filtek-Supreme Translucent) having 30 wt.% organic matrix. The chemistry of organic matrix also affects shrinkage (Boaro et al., 2010; Chen et al., 2006; Fortin and Vargas, 2000).

2.1.1.2 Water sorption

Uptake of water by the polymer of composite can either lead to shrinkage or swelling. Shrinkage results when solvent extract unreacted components like monomers from the composite, resulting in a decrease in polymer volume. The consequences include weight and strength reduction. On the other hand, if the solvent diffuses inside polymer chains, this results in expansion with an associated increase in volume and weight (Ortengren et al., 2001). In some cases, a part of the solvent can be accommodated in microvoids created during polymerization with an increase in the actual volume. Thus, it is very difficult to anticipate the dimensional change of polymer matrix due to solvent. The overall effect is dependent on the chemical structure of the polymer matrix. Usually, nanofill composites display higher water sorption than nanohybrid composites.

2.1.1.3 Flexural strength

Flexural strength has a direct relation with the type and chemistry of the filler. Nanocomposites normally show higher flexural strength than microfill, microhybrid, and hybrid composites. Flexural strength also correlates with the filler loading so that nanohybrid composites (lower fill loading) have lower flexural strength than nanofill composites (higher fill loading, Lohbauer et al., 2006).

2.1.1.4 Wear and gloss retention

Nanoclusters are resistant to wear and retain gloss better owing to the fact that the particles that wear off from the surface are primary particles that are in nanosize range, rather than the secondary particles from the resin, which are larger in size. In contrast, hybrid and microhybrid fillers have larger average particle size and are therefore prone to wear and easily lose glossy appearance (Palaniappan et al., 2009).

2.1.2 Types of nanocomposites

On the basis of matrix materials used to design these nanocomposites, they can be categorized as in the following sections.

2.1.2.1 Metal matrix nanocomposites (MMNC)/reinforced metal matrix composites

Metal matrix composites consist of a metal (Al, Mg, and Cu) matrix filled with nanoceramic particles (carbides, oxides, nitrides, and carbon nanotubes) featuring physical and mechanical properties very different from those of the matrix (Riccardo and Maurizio, 2014). The properties that can be improved by nanoparticles include damping, mechanical strength, and wear resistance to the base material.

The MMNCs can be further categorized into continuous reinforced materials and noncontinuous reinforced materials. MMNC based on carbon nanotubes exploiting the properties of high electrical conductivity and high tensile strength of the carbon nanotubes has particularly gained interest. These nanocomposites not only allow homogeneous dispersion of nanotubes inside the metal matrix leading to strong intercomponent adhesion, but also are economical to produce on mass scale. Other than carbon nanotubes, carbon nitride, and boron nitride metal matrix composites are also being explored for their applications in nanodentistry (Bakshi et al., 2010).

2.1.2.2 Ceramic matrix nanocomposites (CMNC)

These are defined as nanocomposites in which the major volume is taken up by a ceramic. Ceramic compounds include nitrides, oxides, silicides, or borides. In most of the CMNCs, a metal is also dispersed in the ceramic component to improve electric, magnetic (Kruis et al., 1998), tribological, and optical properties of the composite. This also makes the composite corrosion resistant and imparts other protective properties (Zhang et al., 2003). However, these composites need to be designed carefully, taking into account the binary phase diagram of the mixture. It should be ensured that the metal does not lose its metallic character in the composite after reacting with the ceramic component in spite of the high process temperatures required for ceramic preparation. This can be easily carried out by selecting ceramic and metal phases that are immiscible with each other. An example of such immiscible CMNC is TiO_2 and Cu mixture.

2.1.2.3 Polymer matrix nanocomposites (PMNC)

In these types of nanocomposites, nanoparticles or nanofillers are dispersed in a polymer or a copolymer matrix (Manias, 2007). The nanoparticles or nanofillers have one dimension in the nanorange of 1–50 nm, however, they may vary in shape. They can be present in the form of fibers, platelets, or spheroids. The production of these multiphase systems require precise complexing to form a homogeneous dispersion, which should be correctly oriented and stabilized. The conversion to nanoscale dimension increases the surface area-to-volume ratio of the particles, simultaneously increasing the interaction with other particles in the mixture and imparting the properties of high heat resistance and strength enhancement. Silicon nanospheres represent one of the examples of PMNCs with outstanding hardness as a unique property in the range between sapphire and diamond. This characteristic is imparted due to the nanorange structure ranging 40–100 nm. PMNCs are further catergorized into: (1) biohybrid nanopolymers and (2) biohybrid polymer nanotubes.

2.1.2.3.1 Biohybrid nanopolymers

These are nanopolymers produced by electrospinning, wherein an electric field of very high strength 103 V/cm is used to convert the polymer solution droplets into a fluid jet. The jet is accelerated toward counter electrodes and simultaneously subjected to powerful circular bending motion driven by electricity, which leads to elongation, thinning, and solvent evaporation leading to the deposition of nanoscale fibers on the opposite electrode (Greiner et al., 2006).

2.1.2.3.2 Bio-hybrid polymer nanotubes

These are produced by wetting technique. The polymer solution in this technique is allowed to come into contact with pores located in high-energy surfaces like silicon and aluminum. This leads to the coating of the walls of the pores with thin films (of the order of nanometers) of polymer solution within minutes at temperatures higher than 50 K above the transition temperature of glass. To retrieve the nanotubes, the template may be selectively dissolved or the system is cooled down followed by mechanical removal of nanotubes. The process parameters can be varied to get nanotubes of desired length, diameter and homogeneity (Greiner et al., 2006).

2.2 NANOCLUSTERS

Nanoclusters represent a group of nanoparticles having at least one nanoscale dimension and size distribution within a narrow range. They may be composed of the single atom of an element or combinations of atoms of different atom in stoichiometric ratios. The forces that hold these atoms together can be covalent, ionic, metallic, van der Waals forces, or hydrogen bonds. This difference in types of bonding forces involved is the basis of differentiation among the different clusters. Transition metal atom clusters are of special interest as they are characterized by unique electronic, magnetic, and chemical properties. These clusters can be produced from the bulk using both top-down and bottom-up approaches. With the advancements in the field of nanotechnology, it is now possible to manipulate size and dispersion characteristics, as well as chemical composition and orientation during the synthesis of nanoclusters.

2.2.1 Properties of nanoclusters

Nanoclusters represent a bridge between atomic and molecular structures and the bulk materials. Unlike the bulk materials, which have constant physical properties, the physical properties of nanoclusters vary according to their size. Electrons are present in continuous energy bands within bulk materials whereas in individual atoms and molecules the energy bands are discrete. The electron distribution in nanocluster or nanocrystal is intermediate between these two extremes. Moreover, the capacitance of these structures is very small having significant charging energy even upon single electron transfer to the cluster (Philip, 2001).

2.2.2 Sources of nanocluster synthesis

Nanoclusters can be synthesized using different sources. Two of the frequently used sources are: (1) supersonic nozzle source, and (2) gas aggregation source/condensation cell type source.

2.2.2.1 Supersonic nozzle source

The nanoclusters are generated from a supersonic beam consisting of vaporized metal mixed in the carrier gas in this method. Briefly, the metal is allowed to evaporate in an oven set at 75–1500 K and pressure of several atmospheres. The metallic vapors are mixed with a carrier gas and passed through a nozzle into a high vacuum that generates the supersonic beam. This is followed by seeding of the beam to generate nanoclusters. The size of the clusters is determined by presence or absence of carrier gas. If small-sized nanoclusters are desired then carrier gas can be omitted (Wilcoxon and Abrams, 2006).

2.2.2.2 Gas aggregation source/condensation cell-type source

The source clustering is achieved by incorporation of supersaturated vapor of cluster material in a flow of inert gas at low temperature and high pressure in this method. The vaporization may be carried out by magnetic sputtering. At low temperatures, the cluster is generated by the successive addition of single atoms. The cluster size and abundance is dependent on the collision statistics, which are independent of the thermodynamic properties of the clusters. Factors that affect size distribution of the clusters include pressure and density of the metal vapor, the rate of sputtering, type, temperature, and flow rate of the gas, the dimension of apertures and the drift region (Wilcoxon and Abrams, 2006).

2.3 NANOROBOTS

Nanorobotics is the technology used to manufacture nanorobots/nanomachines/nanites having micro-dimensional structures that can interact and manipulate activity of nanoscale objects with high precision (10^{-9}). Applications of nanorobots in oral medicine are diverse ranging from rapid orthodontic realignments, hypersensitivity cure, local anaesthesia, oral health maintenance by selective killing of caries causing bacteria, dentition renaturalization, and repair of tooth blemishes.

Nanorobots are primarily composed of nanoscale parts made up of sulfur, carbon, oxygen, hydrogen, and fluoride. The major components of a nanorobot are payload, camera, swimming tail, and a capacitor (Wang et al., 2011). The manufacturing process is complicated involving interfacing between biotic/abiotic and organic/inorganic systems. Apart from these, components for power, control, communication, sensing, and so on are also incorporated in these devices.

2.3.1 Mechanism of action of nanorobots

Nanorobots must use energy efficiently. The energy could be supplied from external source in the form of acoustic energy or it can be generated in vivo by metabolizing glucose and oxygen. Efficient onboard computers will track, control, and manipulate their activity by communicating via acoustic signal navigation network. This system will also ensure that after its desired task is complete, nanorobots are excreted out of the body through normal waste excretion systems or by active scavenging mechanisms (Babel and Mathur, 2011).

2.3.2 Nanorobots in dentistry

In the field of dentistry, nanorobots can be used for restoring teeth structure, curing and preventing periodontal diseases. They can be specifically used to administer local anaesthesia, treat hypersensitivity for rapid realignment and straightening of teeth, and to improve teeth durability. The applications of nanorobots in dentistry are illustrated in Fig. 24.4.

2.4 NANOSOLUTIONS

Nanosolutions represent the new generation bonding agents having single-step application. These solutions are composed of uniformly dispersed nanoparticles, such as nanofillers and impart properties like increased bond strength and reduced particle settling. These solutions are used in bonding agents to ensure perfect mixing of adhesive upon each use (Rybachuk and Cekman, 2009).

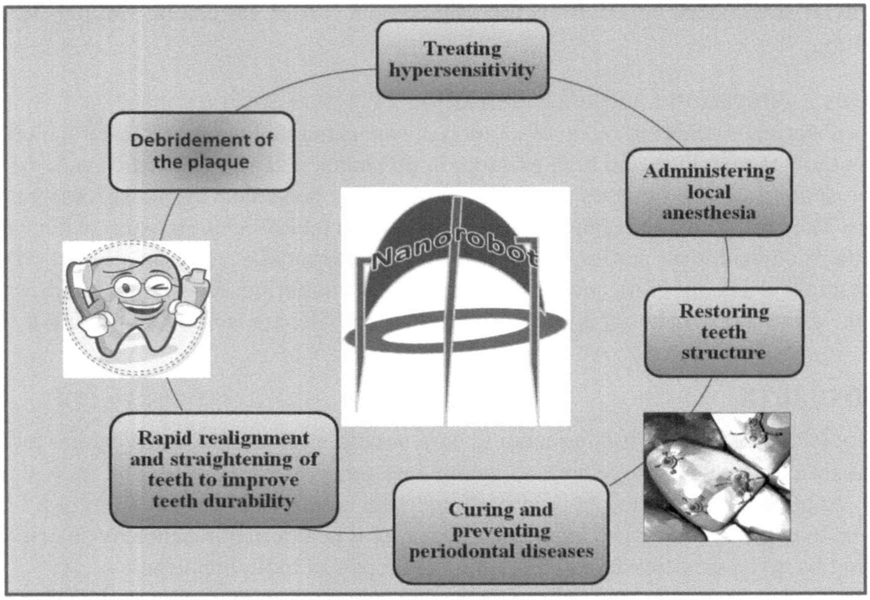

FIGURE 24.4 Applications of Nanorobots in the Field of Dentistry

2.5 NANOLIGHT CURABLE GLASS IONOMERS

Nanolight curable glass ionomers combine the nanostructures like nanofillers and nanocluster fillers with fluoro alumino silicate (FAS) technology. By bonding these nanostructures with FAS, properties like fluoride release, improved aesthetics, excellent polish, and wear resistance are incorporated in the glass ionomers, allowing easy mixing and dispensing, which make them ideal restorative material. These glass ionomers thus satisfy a range of clinical indications like Class I, II, V, and core build-up (Bayne, 2005).

2.6 NANOIMPRESSION MATERIALS

Impression materials, that is, the materials used to get an idea about the shape of teeth and the alveolar ridges, can be categorized as nonelastic and elastic materials. This classification is based on the capability of the material to withdraw over the undercuts it has set. Different types of impression materials are available, which vary in their properties. One of the most commonly used nanoimpression materials are polyvinylsiloxanes (PVS). PVS offer advantages of improved hydrophilic nature and thus better flow precision and accurate detailing (Kumar and Vijayalakshmi, 2006). These are generated by mixing a base paste consisting of nanofillers, polymers of silicone, cross-linker, and inhibitor with a catalytic paste, consisting of platinum catalyst and plasticizer. Fillers, as well as silicone polymers can be blended in different combinations as per requirement (Verma et al., 2010).

2.7 NANOENCAPSULATED MATERIALS

Nanoscale coating of substances on another material is called nanoencapsulation. The material encapsulated is called filler or fill or core material or internal phase whereas the material used for encapsulation is called coating or shell or external phase. The appearance of the encapsulated particle varies with the variation in shape, size, and composition of encapsulated material. Different techniques can be used for encapsulation, including wax and lipid coating, spray congealing, melt extrusion, fluid bed coating, and hydrogel encapsulation.

Nanocapsules are nanodimensional shells made up of nontoxic polymeric material with an inner liquid core and offer advantages of stability, accurate targeting, and controlled sustained release with reduced side effects (Kovvuru et al., 2012). They have been used as nutraceuticals for enhancement of foods after delivery of drugs and self-healing of materials. A specific study related to targeted drug delivery to treat periodontal disease using triclosan-laden nanoparticles has revealed that this type of system results in lower inflammation (Pinon et al., 2005). Arestin consisting of minocycline encapsulated into microspheres has also been used for targeted delivery of drugs to periodontal pocket.

2.8 OTHER NANOSTRUCTURES

Other types of nanostructures that find applications in the field of dentistry are nanospheres, nanotubes, nanorods, and nanoshells.

The use of nanorods in restorative dentistry has been used frequently. Specific enamel-prism-like hydroxyapatite nanorods resembling natural enamel in structure with self-assembly characteristics have been synthesized (Chen et al., 2005; Saunders, 2009). Nanospheres are exploited for the restorative dentistry (Saunders, 2009) and nanotubes are being utilized in dentistry for several applications including coatings that accelerate growth of bone on the implants (titanium oxide nanotubes), to improve flexural strength of RBC (modified single walled carbon nanotubes) (Oh et al., 2005; Saunders, 2009) and for promotion of rapid osseointegration of titanium implants (nanoporous surfaces) (Lavenus et al., 2011). Nanoshells are gold-coated nanoscale beads that can absorb near infrared light in a size dependent fashion to generate intense heat that can be used to selectively kill target cells, for example, cancer cells (Kanaparthy and Kanaparthy, 2011).

3 NANOTECHNOLOGY IN DENTISTRY

3.1 NANODIAGNOSTICS

Nanodiagnostics exploit nanostructures for diagnostic purposes. Nanotechnology-based devices find application in diagnosis of diseases in initial stages of disease development. The optical and magnetic properties of nanoparticles can be manipulated in a size-dependent manner to attain a detectable signal. The binding of nanoparticle probe to its respective target leads to generation of a quantifiable signal proportional to the target molecule. Large number of different probes are being widely used for the purpose of diagnostics, nanoshells, QDs, and metal nanoparticles (West and Halas, 2003). A probe must be water-soluble and stable with least interaction to function in a biological system with the surrounding environment. The most promising nanotechnologies for clinical diagnosis include cantilevers, quantum dots (QDs) and gold nanoparticles (Jain, 2005).

3.1.1 Quantum dots

Structurally, quantum dot is made up of a core enclosed within a shell, both made of different semiconductor materials and they exhibit quantum mechanical properties due to their size, which can vary

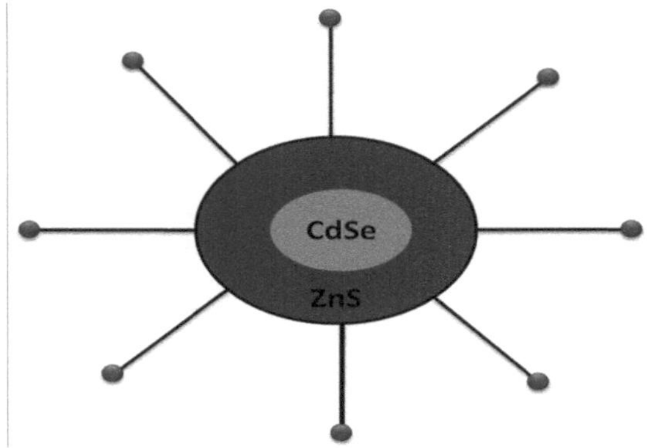

FIGURE 24.5 A Typical CdSe/ZnS QD

from 2 to 10 nm in diameter. QD cores are either composed of elements from groups II and VI, (CdSe) (Fig. 24.5) or groups III and V (InP).

CdSe is the most commonly used for QD cores. The shell is made up of semiconductor material operating at high voltages like ZnS (Bruchez et al., 1998; Chan and Nie, 1998). Types of QDs and their emission wavelengths are summarized in Table 24.2.

3.1.1.1 Preparation of quantum dots

Application-based strategies are used to synthesize QDs (Michalet et al., 2005). One of the approach is the use of organic solvents, such as toluene or chloroform at high temperature in the presence of surfactants. But these QDs are insoluble in water due to surfactant layer. Surfactant-coated QDs are made water-soluble by either replacing the surfactant with the ligand molecules carrying QD binding functional groups at one end and hydrophilic groups at the other end or additionally coating them with amphipathic molecules, phospholipid micelles, and polymer beads. QDs are also encapsulated with phospholipid micelles to impart water-soluble properties (Larson et al., 2003; West and Halas, 2003).

3.1.1.2 Applications of quantum dots

QDs can be labeled by fluorescent molecules, antibodies, oligonucleotide, avidin, or aptamer, or coated with streptavidin, which facilitate the QD toward the target analyte. Semiconductor nanocrystals absorb

Table 24.2 QD Types and Emission Wavelengths

QD	Emission Wavelengths
CdSe/ZnS	550–630 nm
CdSe core	525–655 nm
InP	Near-infrared
InAs	Near-infrared
PbS core	850 and 950 nm
PbSe core	1100, 1310, 1550, and 1900 nm

Table 24.3 Examples of QD Linked Recognition Components (QD assay) and Their Target Molecules

Recognition Component Coupled to QD	Target Molecule	References
Streptavidin	Her2 breast cancer marker or biotinylated PSA	Wu et al. (2003)
Oligonucleotide probes	Target DNA	Hassan et al. (2006)
p-Sulfonatocalixarene	Acetylcholine	Jin et al. (2005)
Primary antibodies	Pgp	Jin et al. (2005)

wider spectrum of light than molecular flourophores, thus they use single excitation laser wavelength to excite a broad spectrum of quantum dot laser wavelength. This can help in probing several markers at the same time in biosensing and bioassays. The recognition component linked to QD and target molecules are summarized in Table 24.3.

The spectrum of applications for quantum dots are summarized in the Fig. 24.6.

Their stability under complex in vivo conditions make them appropriate for imaging at cellular and molecular level and also for drug delivery. They are also resistant to photobleaching in contrast to conventionally used molecular flurophores, making them suitable for DNA detection. Quantum dots can also be used for early diagnosis of oral cancer (Hassan et al., 2006; West and Halas, 2003). Currently, up to 40,000 assays can be run simultaneously with present QD technology.

3.1.2 Gold nanoparticles

Gold nanoparticles can be synthesized in varying shapes and sizes (Fig. 24.7).

Two different systems have been studied in relation to their utility in early diagnosis of oral cancer. They can be used in SERS (surface-enhanced Raman scattering)-based saliva assay for the detection of oral cancer. When gold nanoparticles are used in SERS, an enhanced Raman spectroscopy signal is detected, which can be used to follow the chemical changes occurring in saliva in case of oral cancer (Huang et al., 2007). Gold nanoparticles can also be used as optical probes to provide optical discrimination between normal cells and cancer cells.

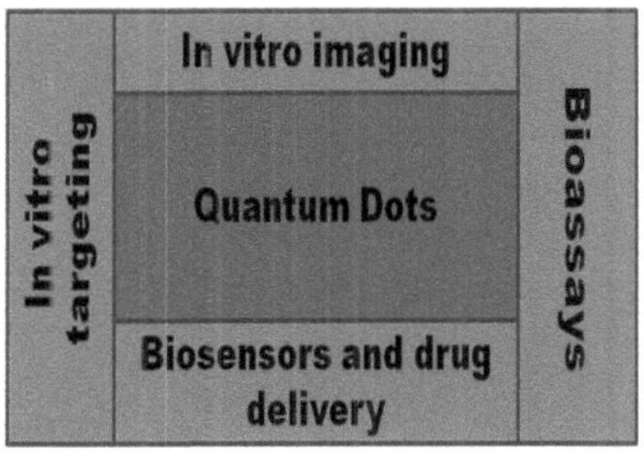

FIGURE 24.6 Broad Applications of Quantum Dots

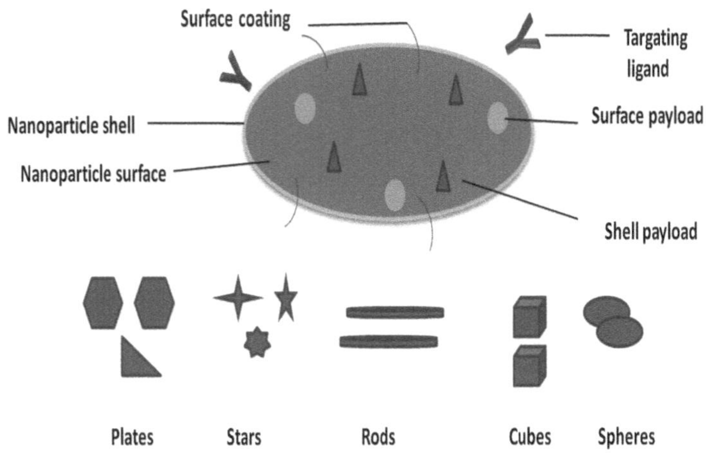

FIGURE 24.7 Schematic Representations of Gold Nanoparticles Used in Clinical Practice

3.1.2.1 Synthesis of gold nanoparticles

AuNPs synthesis can be carried out by employing different methods (Grzelczak et al., 2008). This is mainly performed by reducing Au(III) precursor electrochemically or by chemical methods. For example, reduction of gold can be carried out by citrate using a single-phase water-based reduction method. Factors that determine the size and shape include the reaction temperature and time, the type of reducing agent used and incorporation of capping agent that limits the growth of nanoparticle by selective binding (Baptista et al., 2008). This method used the initiation of particle growth using UV light to achieve better shape- and size-distribution characteristics in case of larger (9–120 nm) AuNPs (Kimling et al., 2006). The absoption peak of the nanoparticles can be manipulated as per requirement by altering their dimensions. The most suitable size range for nanodiagnostic applications is between 700 and 800 nm coinciding with optical window of tissue.

3.1.2.2 Applications of gold nanoparticles

Gold-based nanoparticles can be used to label protein or DNA molecules, which can be specifically targeted for detection or diagnosis of diseases with sensitivity of gold nanoparticles for detection of protein and DNA (West and Halas, 2003). The other benefits include enhanced contrast, biocompatibility, and absence of photobleaching. However, the optical signal might not be comparable to quantum dot analysis.

Diagnosis of cancer cells with gold nanoparticles is based on the detection of a cell surface receptor overexpressed only in epithelial cancer—the epidermal growth factor receptor. Gold nanoparticles can be conjugated with EGF receptor specific antibodies, which will bind specifically to the cancer cells expressing the receptor. The presence of gold nanoparticles on the surface of cancer cells will impart an optical contrast that can be used to easily discriminate the cancerous cell from noncancerous one.

3.1.3 Cantilevers

Cantilevers are nanoscale beams, the surface of which can be modified to impart specificity. The changes in surface stress cause nanoscale bending of the beams upon binding of the target to the cantilever,

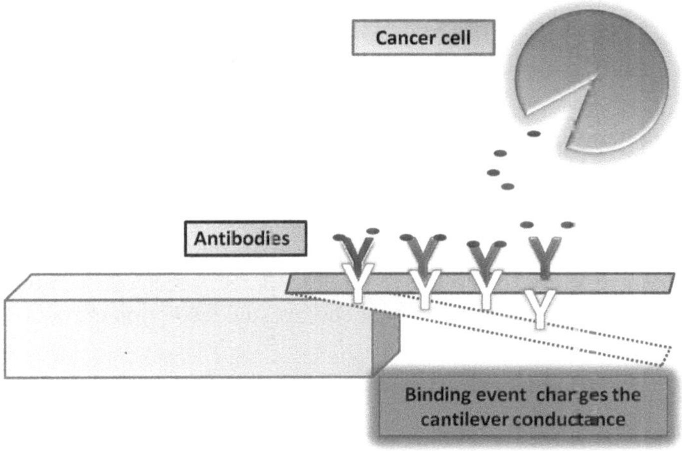

FIGURE 24.8 Cantilevers Detect Biomarkers of Oral Cancer

which can be recorded. For example, a DNA-coated cantilever beam can be used to detect presence of specific genetic sequences in the biological samples. The advantages include rapid and simultaneous detection of analytes at nanomolar concentrations without the requirement of labeling (Jain, 2003). Cantilever-based nanosensing devices are being used to sense analytes for rapid diagnosis of oral diseases (Fig. 24.8).

A comparison of the applications and toxicity of the three nanostructures: QDs, gold nanoparticles, and cantilevers are summarized in Table 24.4.

3.2 NANOTECHNOLOGY IN ORAL AND MAXILLOFACIAL SURGERY

3.2.1 Local anesthesia

Nanotechnology can be used to administer painless anesthesia during dental procedures. Dental robots can be introduced in a patient's gingiva. The nanorobots will be guided to the pulp by using the chemical- and temperature-gradient information using a nanocomputer. Once they reach their target, they can cut off sensitivity. Once the dental procedures are over, the sensitivity can be restored and then the nanorobots can be made to egress from the tooth. The activity of nanorobots will be under the control of the dentist (Sivaramakrishnan and Neelakantan, 2014).

Table 24.4 Applications and Toxicity of Nanostructures Used in Nanodiagnostics

Features	QDs	Cantilevers	Gold Nanoparticles
Applications	Diagnostics; immunoassays; detection; imaging	Qualitative and quantitative estimation of protein and DNA	Detection of DNA by hybridization method
Toxicity	Toxic in case of leakage of semiconductor	Not toxic	Not toxic

3.2.2 Nanoneedles

Nanosized needles made up of nanodimension stainless steel crystals and nanotweezers are under development. These devices help to make cell-surgery a reality in near future (Reyaz, 2014).

3.3 NANOTECHNOLOGY IN ORTHODONTIC TREATMENT

Tooth rotation, repositioning, or straightening can be carried out with the aid of orthodontic nanorobots. The advantages include painless and fast manipulation compared to conventional molar uprighting techniques (Reyaz, 2014). Dry lubrication can be achieved upon coating orthodontic wire with tungsten disulfide nanoparticles thereby reducing the excessive frictional force, which resists movement of tooth along the wire in order to reduce the chances of loss of anchorage and resorption of root (Redlich et al., 2008).

3.4 NANOTECHNOLOGY IN PROSTHODONTICS

3.4.1 Impression materials

Nanotechnology-based impression materials are composed of vinyl polysiloxanes with embedded nanofillers to improve hydrophillicity as well as flow and binding properties. These unique impression materials allow better model pouring and precision with fewer marginal voids (Sivaramakrishnan and Neelakantan, 2014).

3.4.2 Implants

The success of a dental implant relies on the efficient biointegration of the implant with the surrounding tissue. Implant material can be coated with biocompatible nanoparticles to improve the adhesion and integration characteristics (Sivaramakrishnan and Neelakantan, 2014).

3.4.3 Nanocomposite denture teeth

Nanocomposite denture teeth offer the advantage of being stress, stain, impact and wear resistant. They also have superb polishing characteristics giving them better aesthetics. These composite denture teeth are composed of nanosized filler particles homogeneously dispersed in PMMA (Arora and Kapoor, 2014).

3.5 NANOTECHNOLOGY IN PERIODONTICS

3.5.1 Dentinal hypersensitivity

Dental nanorobots could be used for permanent and quick relief from hypersensitivity. This method relies on the fact that naturally hypersensitive teeth are characterized by larger diameter and higher surface density of dentinal tubules. Nanorobots can be programmed to reach tubules and occlude them within minutes to relieve sensitivity. The movement of nanorobots can be controlled by a nanocomputer, which will guide the robots to the target utilizing the chemical and temperature differential data (Sivaramakrishnan and Neelakantan, 2014).

3.5.2 Drug delivery

Delivery systems utilizing nanoparticles have been used for local, targeted delivery of drugs. For periodontal therapy, nanoparticles loaded with triclosan have been produced by emulsification-diffusion (Pinon et al., 2005). Similarly, microspheres based on tetracycline are also being studied for direct delivery to periodontal pockets (Sivaramakrishnan and Neelakantan, 2014).

3.5.3 Materials to induce bone growth

Nanobone mimics the natural bone in the design and consists of nanocrystallites made up of hydroxyapatite with interspersed pores. The smaller particle size achieves larger surface area and the presence of pores leads to formation of a rough interface between the biomaterial and cells which allows rapid cell growth. The size of the pores is smaller as compared to the size of the cells, thus the interstices usually contain the essential plasma proteins. These nanobones can be used to treat bone defects. Some examples of nanoparticles used for the purpose include VITOSSO (Orthovita, Inc, USA) HA + TCP, Ostium (Osartis GmbH, Germany) and HA, NanOSSTM (Angstrom Medica, USA) HA (Reyaz, 2014)

3.6 NANOTECHNOLOGY IN CONSERVATIVE DENTISTRY AND ENDODONTICS

3.6.1 Tooth durability and appearance

Replacement of damaged enamel layers with artificial material is known to improve appearance and durability. Biocompatible materials like diamond and sapphire nanoparticles can be covalently bonded for this purpose, which are many times stronger than the conventional materials (Yunshin et al., 2005).

3.6.2 Nanorobotic dentifrice (dentrifrobots)

Nanoscale hydroxyapatite crystals can be interspersed in toothpaste or mouthwash to restore teeth that are damaged by forming a shielding layer on enamel surface. Dentifrices like microhydrin and microbrite (Patel et al., 2014) are nanoscale molecular cages that can degrade trapped organic matter and convert it into harmless and odorless vapors. Nanoparticle-based dentifrices can also be developed to debride calculus and prevent its accumulation supra and subgingivally. These nanoparticles can be programmed to get deactivated when swallowed (Sivaramakrishnan and Neelakantan, 2014).

3.6.3 Nanocomposite resins

Nanoparticles can be uniformly distributed in resins to enhance properties like flexural strength, hardness, reduce polymerization shrinkage, and improve modulus of elasticity (Sivaramakrishnan and Neelakantan, 2014). A nanocomposite resin is successfully produced by Nanoproducts Corporation used as nanofiller made up of aluminosilicate powder (alumina to silica ration of 1:4 M) with a refractive index of 1.5.

3.6.4 Nanoparticulate-based disinfection in endodontics

Nanoparticles have also found application in endodontics for the disinfection of root canals. The idea is to utilize the antimicrobial potential of these nanoparticles, which is primarily due to the ability to disrupt cell wall and to disintegrate the biofilms. The most commonly used nanoparticles for this purpose are silver nanoparticles. These nanoparticles have been reported to disrupt *Enterococcus faecalis* biofilms when administered in 0.02% gels. Apart from silver and zinc oxide, chitosan nanoparticles have also shown encouraging antibacterial activity against *E. faecalis*.

The second promising nanomaterial for use in endodontics is bioactive glass. This material (SiO_2—Na_2O—CaO—P_2O_5) also exerts antimicrobial effect due to elevation in pH and can also be used for root canal disinfection. Studies on 45S5 bioactive suspension–nanometric/micrometric hybrid have been reported to show up to 10-fold higher silica release and up to a 3-unit increase in pH highlighting its potential as antimicrobial agent (Sivaramakrishnan and Neelakantan, 2014).

4 PREVENTIVE NANODENTISTRY

Preventive dentistry refers to cumulative measures taken to prevent rather than to cure periodontal diseases and dental caries. Nanotechnology provides next generation methods for preventive dentistry.

4.1 DENTAL CARIES

Dental caries result from demineralization of teeth (loss of calcium and phosphate) in response to acid generation by bacteria inhabiting tooth surfaces in oral biofilms or dental plaque. The major bacterial species involved are *Streptococcus sobrinus*, *S. mutans*, and *Lactobacillus* spp. Thus the usual routes to prevent dental caries are to inhibit bacterial activity, to reverse demineralization and to promote remineralization. Anticaries vaccine is also under development.

4.2 ANTIBACTERIAL NANOTHERAPY

Nanoparticles, such as silver nanoparticles (Blecher et al., 2011; Li et al., 2008), zinc oxide nanoparticles (Gu et al., 2012; Xie et al., 2011), and polyethylenimine nanoparticles (Shvero et al., 2010) are incorporated into dental adhesives (Chen et al., 2014) or dental composites (Kasraei et al., 2014) to impart antibacterial properties. Alternately, these can be added in nanocoatings for inhibiting the growth and adhesion of bacteria (Besinis et al., 2014). The mechanisms of action of these antibacterial nanoparticles may vary and include a generation of ROS (Foldbjerg et al., 2009), active transport, and metabolism of sugars (Kawabata and Nishiguchi, 1998), disruption of cell envelope (Kawabata and Nishiguchi,1998), disturbance in membrane electron transport (Li et al., 2008), inhibition of biofilm formation due to reduced enzymatic activity and DNA replication inhibition (Rai et al., 2009). The action of these nanoparticles varies with the size (Lu et al., 2013). Nanoparticles have been found to be effective against biofilms composed of *Lactobacillus acidophilus* and *S. mutans* in in vitro model (das Neves et al., 2014; Li et al., 2014).

4.3 BIOMIMETIC REMINERALIZATION—REVERSING AN INCIPIENT CARIES

Nanotechnology has a role to play in reversing caries in early phases. Colloidal calcium carbonate (CC) nanoparticles can be used for delivery and sustained release of calcium ions at high concentrations and these nanoparticles adhere easily on oral surfaces. CC nanoparticles are reported to elevate the surrounding pH thereby promoting remineralization of early phase enamel lesions (Nakashima et al., 2009). Calcium fluoride nanoparticles act as fluoride reservoirs and are highly soluble. CaF2 reacts with dicalcium phosphate dehydrate to produce apatite, which is easily deposited at enamel surface and can act as an effective anticaries agent in mouth rinse in place of sodium fluoride.

4.4 BIOMIMETIC REMINERALIZATION—RECURRENT DECAY

Nanomaterials can also be used for preventing secondary/recurrent decay that tends to occur around restorations. Nanoparticles can be used as Ca^{+2} and PO_4^{-3} releasing fillers or as high-stress bearing components in combination with other fillers, such as nanoscale silica. Examples of the former category include tetracalciumphosphate (Xu et al., 2009), carbonate hydroxyapatite (Roveri et al., 2008), anhydrous dicalcium phosphate (Xu et al., 2006, 2007a,b), and monocalcium phosphate monohydrate

(Xu et al., 2007a,b). These nanocalcium fillers can bind to eroded dental enamel and can form a protective layer preventing further tooth decay and initiating tooth remineralization (Roveri et al., 2008). The calcium and phosphate released from these nanoparticles is dependent on volume fraction of CaP form and its degradability and it can also be released when required, for example, upon decrease in pH to cariogenic level (Xu et al., 2009).

4.5 CARIES VACCINE

Anticaries vaccines are designed to develop resistance against characteristic bacteria involved in dental caries, such as *S. mutans*. A vaccine can target surface protein antigen or glucosyl transferases (Kt et al., 2013), both of which are involved in adhesion of *S. mutans* to the teeth surface and is responsible for the virulence (Koga et al., 1990; Yamashita et al., 1993). Several DNA vaccines are being developed on this principle, for example, pCIA-P (Su et al., 2014), pcDNA3-Pac (Wang et al., 2014), pGLUA-P (Guo et al., 2004), and pGJGLU/ VAX (Liu et al., 2014). These DNA vaccines can induce both humoral and cellular immune response and offer advantages, such as being stable, effective, safe, and inexpensive (Talwar et al., 1999). However, one of the serious limitations is that their immunogenicity is low in human beings. This limitation can be overcome by employing nanotechnology based delivery vehicles, for example, anionic liposomes in chitosan (Chen et al., 2013), which enhance the immunogenicity. The delivery vehicle can also be programmed to release vaccine when required, for example, when the pH is reduced to cariogenic levels (Chen et al., 2013)

The progress in anticaries vaccine development is impeded by high salivary flow, high diversity of oral flora, enzymatic degradation of vaccine, difficult antigen delivery, and poor internalization.

4.6 PERIODONTAL DISEASES

Periodontal diseases can be prevented by the use of nanorobots/dentifrobots in mouthwash or toothpaste. These nanorobots can move rapidly across supra- and subgingival surfaces and are able to remove organic residues and prevent the accumulation of plaque and calculus. Nanostructured doxycycline gel has been studied in this regard in experimental periodontal disease (Botelho et al., 2010). These nanorobots can be programmed to get deactivated if they are swallowed (Saravana and Vijayalakshmi, 2006).

5 CHALLENGES FACED BY NANODENTISTRY

The potential of nanodentistry has not been realized yet in spite of significant advancement in the field (Freitas, 1999). The major issues range from basic engineering problems to biocompatibility, public acceptance, and insufficient financial support.

The design limitations, such as tools for precise positioning of nanoscale gears and their assembly impede development of nanorobots. Economically viable mass production techniques need to be developed for these nanocomponents. Development of proper techniques for coordinating the activities of large number of nanorobots simultaneously is also a bottleneck.

There are issues related to biocompatibility and public acceptance. The human safety issues need to be addressed before these nanodentistry products come in the market. Lack of FDA regulations regarding control of nanotechnology products highlights the serious lacunae in the field. FDA along with EPA and nuclear protection agencies need to analyze major health risks associated with these products.

Nanomaterials: both nanotechnology-based products and delivery vehicles need to be characterized for their safety and impact on the environment. Nanomaterials might also impact workers exposed to these agents during manufacturing processes, therefore, research is to be carried out to study the interaction of nanomaterials with human systems. The various routes of entry of nanoparticles in the human body include nasal, oral, skin, or arterial route (in case of medical procedures). They can be efficiently absorbed via gastrointestinal tract to reach the circulatory system. Owing to their nanosize, they are highly mobile and may potentially cross the blood–brain barrier. They may also result in overload on phagocytic cells, leading to inflammation. Nanoparticles have a large surface area-to-volume ratio; therefore, they can efficiently adsorb macromolecules inside the body and may interfere and adversely affect nervous and cardiovascular system. Certain nanoparticles, for example, carbon black has been reported to interfere with cell signaling. Nanoparticles may also affect DNA and lead to genetic defects.

In addition, the impact assessment study, upon release of these nanoparticles in the environment, should also be properly analyzed as environmental factors like pH, temperature, and presence of pollutants might alter the properties of nanoparticles in an unanticipated manner. These effects might be long-term effects and might not present themselves for many years.

The future advancements in the field of nanodentistry will rely on the public acceptance, which in turn will be decided by proper address of environmental, health, and safety concerns. The nanotechnology products developed, using green nanotechnologies, have addressed the safety issues to some extent. In spite of all the caveats, nanotechnology still holds promise in the field of nanodentistry.

6 CONCLUSIONS

Nanotechnology can potentially transform the field of oral medicine with its myriad uses, diagnosis, prevention and management of oral diseases and deformities. Nanomaterials can be incorporated in restorative material and implants, as well as in bone replacement material to enhance their properties. However, to be able to integrate nanostructures with clinical medicine, the pathophysiology of disease need to be understood in more detail in order to generate more opportunities for diagnosis, prevention, and treatment of periodontal diseases. This involves multiple scientific disciplines; the proficiency in these disciplines is a prerequisite for effective translation of ideas into products. In presence of proper expertise, the novel nanotechnology-based method can revolutionize nanodentistry by enhancing accuracy, efficiency, and speed of the procedures, reducing the overall cost.

However, the potential risks involved with the use of nanoscale structures in nanodentistry should be carefully analyzed. With proper regulatory guidelines and strategy formulation in place, this field is expected to benefit greatly with nanomaterials in near future.

REFERENCES

Arora, R., Kapoor, H., 2014. Nanotechnology in dentistry—hope or hype. OHDM 13 (4), 928–933.

Babel, S., Mathur, S., 2011. Nanorobotics: headway towards dentistry. Int. J. Res. Sci. Technol. 1 (3), 1–9.

Bakshi, S.R., Lahiri, D., Argawal, A., 2010. Carbon nanotube reinforced metal matrix composites: a review. Int. Mater. Rev. 55, 41–64.

Baptista, P., Pereira, E., Eaton, P., Doria, G., Miranda, A., Gomes, I., Quaresma, P., Franco, R., 2008. Gold nanoparticles for the development of clinical diagnosis methods. Anal. Bioanal. Chem. 391 (3), 943–950.

Bayne, S.C., 2005. Dental biomaterials: where are we and where are we going? J. Dent. Educ. 69, 71–85.
Besinis, A., De Peralta, T., Handy, R.D., 2014. Inhibition of biofilm formation and antibacterial properties of a silver nano-coating on human dentine. Nanotoxicology 48 (7), 745–754.
Beun, S., Glorieux, T., Devaux, J., Vreven, J., Leloup, G., 2007. Characterization of nanofilled compared to universal and microfilled composites. Dent. Mater. 23 (1), 51–59.
Blecher, K., Nasir, A., Friedman, A., 2011. The growing role of nanotechnology in combating infectious disease. Virulence 2 (5), 395–401.
Boaro, L.C., Goncalves, F., Guimaraes, T.C., Ferracane, J.L., Versluis, A., Braga, R.R., 2010. Polymerization stress, shrinkage andelastic modulus of current low-shrinkage restorativecomposites. Dent. Mater. 26, 144–150.
Botelho, M.A., Martins, J.G., Ruela, R.S., Queiroz, D.B., Ruela, W.S., 2010. Nanotechnology in ligature-induced periodontitis: protective effect of a oxycycline gel with nanoparticules. J. Appl. Oral. Sci. 18 (4), 335–342.
Bruchez, M.J., Moronne, M., Gin, P., Weiss, S., Alivisatos, A.P., 1998. Semiconductor nanocrystals as fluorescent biological labels. Science 281, 2013–2016.
Chan, W.C., Nie, S., 1998. Quantum dot bioconjugates for ultrasensitive nonisotopic detection. Science 281, 2016–2018.
Chandki, R., Kala, M., Kumar, N.K., Brigit, B., Banthia, P., Banthia, R., 2012. Nanodentistry: exploring the beauty of miniature. J. Clin. Exp. Dent. 4 (2), 119–124.
Chen, M.H., Chen, C.R., Hsu, S.H., Sun, S.P., Su, W.F., 2006. Low shrinkage light curable nanocomposite for dental restorative material. Dent. Mater. 22, 138–145.
Chen, H., Clarkson, B.H., Sun, K., Mansfield, J.F., 2005. Self assembly of synthetic hydroxyapatite nanorods into enamel prism like structure. J. Colloid Interface Sci. 288 (1), 97–103.
Chen, C., Weir, M.D., Cheng, L., 2014. Antibacterial activity and ion release of bonding agent containing amorphous calcium phosphate nanoparticles. Dent. Mater. 30 (8), 891–901.
Chen, L., Zhu, J., Li, Y., 2013. Enhanced nasal mucosal delivery and immunogenicity of anti-caries DNA vaccine through incorporation of anionic liposomes in chitosan/DNA complexes. PLoS One 8 (8), 71913–71951.
das Neves, P., Agnelli, J., Kurachi, C., De Souza, C., 2014. Addition of silver nanoparticles to composite resin: effect on physical and bacterial properties *in vitro*. Braz. Dent. J. 25, 141–145.
Das, S., Gates, A.J., Abdul, H.A., Rose, G.S., Picconatto, C.A., Ellenbogen, J.C., 2007. Designs for ultra-tiny, special-purpose nanoelectronic circuits'. IEEE Trans. Circ. Syst. I (11), 2528–2540.
Davis, S.S., 1997. Biomedical applications of nanotechnology—implications for drug targeting and gene therapy. Trends Biotechnol. 15 (6), 217–224.
Fartash, B., Tangerud, T., Silness, J., Arvidson, K., 1996. Rehabilitation of mandibularedentulism by single crystal sapphire implants and overdentures: 3–12 year results in 86 patients. A dual center international study. Clin. Oral Implants Res. 7, 220–229.
Foldbjerg, R., Olesen, P., Hougaard, M., Dang, D.A., Hoffmann, H.J., Autrup, H., 2009. PVP-coated silver nanoparticles and silver ions induce reactive oxygen species, apoptosis and necrosis in THP-1 monocytes. Toxicol. Lett. 190, 156–162.
Fortin, D., Vargas, N.A., 2000. The spectrum of composites: new techniques and materials. JADA 131 (6), 26S–30S.
Freitas, Jr., R.A., 2000. Nanodentistry. J. Am. Dent. Assoc. 131, 1559–1566.
Freitas R.A., 1999. Nanomedicine, vol. 1. Basic capabilities. Landes Bioscience, Georgetown, Texas.
Freitas, R.A., 2005. Nanotechnology, nanomedicine and nanosurgery. Int. J. Surg. 3 (4), 243–246.
Ghalanbor, Z., Marashi, S.A., Ranjbar, B., 2005. Nanotechnology helps medicine: nanoscaleswimmers and their future applications. Med. Hypotheses (1), 198–199.
Goverdhana, S., Puntel, M., Xiong, W., Zirger, J.M., Barcia, C., Curtin, J.F., 2005. Regulatable gene expression systems for gene therapy applications: progress and future challenges. Mol. Ther. 12 (2), 189–211.
Greiner, A., Wendorff, J.H., Yarin, A.L., Zussman, E., 2006. Biohybrid nanosystems with polymer nanofibers and nanotubes. Appl. Microbiol. Biotechnol. 71 (4), 387–393.

Grzelczak, M., Perez, J., Mulvaney, P., Liz, L.M., 2008. Shape control in gold nanoparticle synthesis. Chem. Soc. Rev. 37 (9), 1783–1791.

Gu, H., Fan, D., Gao, J., 2012. Effect of $ZnCl_2$ on plaque growth and biofilm vitality. Arch. Oral Biol. 57 (4), 369–375.

Guo, J.H., Jia, R., Fan, M.W., Bian, Z., Chen, Z., Peng, B., 2004. Construction and immunogenic characterization of a fusion anti-caries DNA vaccine against PAc and glucosyltransferase I of *Streptococcus mutans*. J. Dent. Res. 83 (3), 260–270.

Hassan, M.E., Mai, M.H., Steven, C., 2006. Nanodiagnostics: a new frontier for clinical laboratory diagnosis. Clin. Chem. 52 (7), 1238–1246.

Huang, X., Sayed, I.H., Qian, W., 2007. Cancer cells assemble & align gold nanorods conjugated to antibodies to produce highly enhanced, sharp, and polarized surface raman spectra: a potential cancer diagnostic marker. Nano Lett. 7 (6), 1591–1597.

Jain, K.K., 2003. Nanodiagnostics: application of nanotechnology in molecular diagnostics. Expert Rev. Mol. Diagn. 3, 153–161.

Jain, K.K., 2005. Nanotechnology in clinical laboratory diagnostics. Clin. Chim. Acta 358, 37–54.

Jin, T., Fujii, F., Sakata, H., Tamura, M., Kinjo, M., 2005. Amphiphilic *p*-sulfonatocalix arene-coated CdSe/ZnS quantum dots for the optical detection of the neurotransmitter acetylcholine. Chem. Commun., 4300–4302.

Kanaparthy, R., Kanaparthy, A., 2011. The changing face of dentistry: nanotechnology. Int. J. Nanomed. 6, 2799–2804.

Kasraei, S., Sami, L., Hendi, S., Alikhani, M.Y., Rezaei, L., Khamverdi, Z., 2014. Antibacterial properties of composite resins incorporating silver and zinc oxide nanoparticles on *Streptococcus mutans* and *Lactobacillus*. Restor. Dent. Endod. 39 (2), 109–114.

Kawabata, N., Nishiguchi, M., 1998. Antibacterial activity of soluble pyridiniumtype polymers. Appl. Environ. Microbiol. 54, 2532–2535.

Kimling, J., Maier, M., Okenve, B., Kotaidis, V., Ballot, H., Plech, A., 2006. Turkevich method for gold nanoparticle synthesis revisited. J. Phys. Chem. 110 (32), 15700–15707.

Koga, T., Okahashi, N., Takahashi, I., Kanamoto, T., Asakawa, H., Iwaki, M., 1990. Surface hydrophobicity, adherence, and aggregation of cell surface protein antigen mutants of *Streptococcus mutans* serotype c. Infect. Immun. 58 (2), 289–296.

Kovvuru, S.K., Mahita, V.N., Manjunatha, B.S., Babu, B.S., 2012. Nanotechnology: the emerging science in dentistry. J. Orofac. Res., 233–236.

Kruis, F.E., Fissan, H., Peled, A., 1998. Synthesis of nanoparticles in the gas phase for electronic, optical and magnetic applications—a review. J. Aerosol Sci. 29 (5-6), 511–535.

Kt, S., Kmk, M., Jimson, S., 2013. Dental caries vaccine—a possible option? J. Clin. Diagn. Res. 7 (6), 1250–1253.

Kumar, S.R., Vijayalakshmi, R., 2006. Nanotechnology in dentistry. Ind. J. Dent. Res. (2), 62–65.

Larson, D.R., Zipfel, W.R., Williams, R.M., Clark, S.W., Bruchez, M.P., Wise, F.W., 2003. Water-soluble quantum dots for multiphoton fluorescence imaging in vivo. Science 300, 1434–1436.

Lavenus, S., Ricquier, J.C., Louarn, G., Layrolle, P., 2010. Cell interaction with nanopatterned surface of implants. Nanomedicine 5 (6), 937–947.

Lavenus, S., Berreur, M., Trichet, V., Pilet, P., Louarn, G., Layrolle, P., 2011. Adhesion and osteogenic differentiation of human mesenchymalstem cells on titanium nanopores. Eur. Cell. Mater. 22, 84–96.

Le Guéhennec, L., Soueidan, A., Layrolle, P., Amouriq, Y., 2007. Surface treatments of titanium dental implants for rapid osseointegration. Dent. Mater. 23 (7), 844–854.

Li, Q.M.S., Lyon, D.Y., Brunet, L., Liga, M.V., Li, D., Alvarez, P.J., 2008. Antimicrobial nanomaterials for water disinfection and microbial control: potential applications and implications. Water Res. 42 (18), 4591–4602.

Li, F., Weir, M.D., Fouad, A.F., Xu, H.H.K., 2014. Effect of salivary pellicle on antibacterial activity of novel antibacterial dental adhesives using a dental plaque microcosm biofilm model. Dent. Mater. 30 (2), 182–191.

Liu, G., Fan, M., Guo, J., 2014. Efficacy of immune responses induced by anticaries DNA vaccine-loaded bacterial ghost in mice. Zhonghua Kou Qiang Yi Xue Za Zhi 49 (1), 37–41.

Lohbauer, U., Frankenberger, R., Kramer, N., Petschelt, A., 2006. Strength and fatigue performance versus filler fraction of different types of direct dental restoratives. J. Biomed. Mater. Res. 76B, 114–120.

Lu, Z., Rong, K., Li, J., Yang, H., Chen, R., 2013. Size-dependent antibacterial activities of silver nanoparticles against oral anaerobic pathogenic bacteria. J. Mater. Sci. Mater. Med. 24 (6), 1465–1471.

Manias, E., 2007. Nanocomposites: stiffer by design. Nat. Mater. 6 (1), 9–11.

Michalet, X., Pinaud, F.F., Bentolila, L.A., Tsay, J.M., Doose, S., Li, J.J., 2005. Quantum dots for live cells, in vivo imaging, and diagnostics. Science 307, 538–544.

Nagpal, A., Kaur, J., Sharma, S., Bansal, A., Sachdev, P., 2011. Nanotechnology: the era of molecular dentistry. Indian J. Dent. Sci. 3 (5), 80–82.

Nakashima, S., Yoshie, M., Sano, H., Bahar, A., 2009. Effect of a test dentifrice containing nano-sized calcium carbonate on remineralization of enamel lesions in vitro. J. Oral Sci. 51 (1), 69–77.

Oh, S.H., Finõnes, R.R., Daraio, C., Chen, L.H., Jin, S., 2005. Growth of nano-scalehydroxyapatite using chemically treated titanium oxide nanotubes. Biomaterials 26, 4938–4943.

Ortengren, U., Wellendorf, H., Karlsson, S., Ruyter, I.E., 2001. Watersorption and solubility of dental composites and identification of monomers released in an aqueous environment. J. Oral Rehabil. 28, 1106–1115.

Palaniappan, S., Bharadwaj, D., Mattar, D., Peumans, M., VanMeerbeek, B., Lambrechts, P., 2009. Three-year randomized clinical trial to evaluate the clinical performance and wear of a nanocomposite versus a hybrid composite. Dent. Mater. 25, 1302–1314.

Patel, I., Solanki, K., Patel, B., 2014. Nano technology—the future of dentistry: a review article. J. Res. Adv. Dent. 3, 51–54.

Patil, M., Mehta, D.S., Guvva, S., 2008. Future impact of nanotechnology on medicine and dentistry. J. Indian Soc. Periodontol. 12 (2), 34–40.

Philip, M., 2001. Nanostructured materials. Rep. Prog. Phys. 64, 297–381.

Pinon, S.E., Ganem, Q.A., Alonso, P.V., Quintanar, G.D., 2005. Preparation and characterization of triclosan nanoparticles for periodontal treatment. Int. J. Pharm. 294, 217–232.

Raffaele, M., Janne, R., 2009. Nanocomposites: nanoparticles in the right place. Nature Mater. 8, 926–928.

Rai, M., Yadav, A., Gade, A., 2009. Silver nanoparticles as a new generation of antimicrobials. Biotechnol. Adv. 27, 76–83.

Redlich, M., Katz, A., Rapoport, L., Wagnerb, H.D., Feldmanb, Y., 2008. Improved orthodontic stainless steel wires coated with inorganic fullerene-like nanoparticles of WS2 impregnated in electroless nickel–phosphorous film. Dent. Mater. 24, 1640–1646.

Reyaz, A.M., 2014. Role of nanotechnology in dentistry. Sch. J. App. Med. Sci. 2 (2D), 785–789.

Riccardo, C., Maurizio, V., 2014. Metal matrix composites reinforced by nano-particles: a review. Metals 4, 65–83.

Roveri, N., Foresti, E., Lelli, M., 2008. Synthetic biomimetic carbonate hydroxyapatite nanocrystals for enamel remineralization. Adv. Mater. Res. 4 (7–50), 821–824.

Rybachuk, A.V., Cekman, I.S., 2009. Nanotechnology and nanoparticles in dentistry. Pharmocol. Pharm. 1, 18–21.

Sahoo, S.K., Parveen, S., Panda, J.J., 2007. The present and future of nanotechnology in human health care. Nanomedicine 3, 20–31.

Saravana, K.R., Vijayalakshmi, R., 2006. Nanotechnology in dentistry. Indian J. Dent. Res. 17, 62–65.

Saunders, S.A., 2009. Current practicality of nanotechnology in dentistry Part 1. Focus on nanocomposite restoratives and biomemetics. Clin. Cosmet. Invest. Dent. 1, 47–61.

Shabouri, M.H., 2002. Positively charged nanoparticles for improving the oral bioavailability of cyclosporine-A. Int. J. Pharm. 249, 101–108.

Shellhart, W.C., Oesterle, L.J., 1999. Uprighting molars without extrusion. J. Am. Dent. Assoc. 130, 381–385.

Shvero, D.K., Davidi, M.P., Weiss, E.I., Srerer, N., Beyth, N., 2010. Antibacterial effect of polyethyleneimine nanoparticles incorporated in provisional cements against *Streptococcus mutans*. J. Biomed. Mater. Res. 94 (2), 367–371.

Sivaramakrishnan, S.M., Neelakantan, P., 2014. Nanotechnology in dentistry—what does the future hold in store. Dentistry 4, 198.

Su, L.K., Yu, F., Li, Z.F., Zeng, C., Xu, Q.A., Fan, M.W., 2014. Intranasal co-deliveryof IL-6 gene enhances the immunogenicity of anti-caries DNA vaccine. Acta Pharmacol. Sin. 35 (5), 592–598.

Sunder Rajan, S., Acharya, S.R., Saraswathi, V., 2013. Nanodentistry. Indian J. Sci. Res. 4 (2), 233–238.

Talwar, G.P., Diwan, M., Razvi, F., Malhotra, R., 1999. The impact of new technologies on vaccines. Natl. Med. J. India 12 (6), 274–280.

Verma, S.K., Prabhat, K.C., Goyal, L., Rani, M., Jain, A., 2010. A critical review of the implication of nanotechnology in modern dental practice. Natl. J. Maxillofac Surg. 1, 41–44.

Wang, J., Hartmann, F.K., Fedorov, R., 2011. Can man-made nanomachines compete with nature biomotors. ACS Nano 3 (1), 4–9.

Wang, J., Liu, J.G., Jiang, J., Wang, W., Bai, G.H., Guan, X.Y., 2014. Study on gene vaccine pcDNA3-PAc against dental caries by intranasal immunization in rabbits. Shanghai Kou Qiang Yi Xue 23 (2), 133–137.

Wells, D.J., 2004. Gene therapy progress and prospects: electroporationand other physical methods. Gene Ther. 11 (18), 1363–1369.

West, J.L., Halas, N.J., 2003. Engineered nanomaterials for biophotonics applications: improving sensing, imaging, and therapeutics. Annu. Rev. Biomed. Eng. 5, 285–292.

Wilcoxon, J.P., Abrams, B.L., 2006. Synthesis, structure and properties of metal nanoclusters. Chem. Soc. Rev. 35, 1162–1194.

Wu, X., Liu, H., Liu, J., Haley, K.N., Treadway, J.A., Larson, J.P., 2003. Immunofluorescent labeling of cancer marker Her2 and other cellular targets with semiconductor quantum dots. Nat. Biotechnol. 221, 41–46.

Xie, Y., He, Y., Irwin, P.L., Jin, T., Shi, X., 2011. Antibacterial activity and mechanism of action of zinc oxide nanoparticles against *Campylobacter* jejuni. Appl. Environ. Microbiol. 77 (7), 2325–2331.

Xu, H.H., Sun, L., Weir, M.D., 2006. Nano DCPA-whisker composites with high strength and Ca and PO4 release. J. Dent. Res. 85 (8), 722–727.

Xu, H.H., Sun, L., Weir, M.D., Takagi, S., Chow, L.C., Hockey, B., 2007a. Effects ofincorporating nanosized calcium phosphate particles on properties of whisker-reinforced dental composites. J. Biomed. Mater. Res. 81 (1), 116–125.

Xu, H.H., Weir, M.D., Sun, L., 2009. Calcium and phosphate ion releasing composite: effect of pH on release and mechanical properties. Dent. Mater. 25 (4), 535–542.

Xu, H.H., Weir, M.D., Sun, L., Takagi, S., Chow, L.C., 2007b. Effects of calcium phosphate nanoparticles on $CaPO_4$ composite. J. Dent. Res. 86 (4), 378–383.

Yamashita, Y., Bowen, W.H., Burne, R.A., Kuramitsu, H.K., 1993. Role of the *Streptococcus mutans* gtf genes in caries induction in the specific pathogen- free rat model. Infect. Immun. 61 (9), 3811–3817.

Yunshin, S., Park, H.N., Kim, K.H., 2005. Biologic evaluation of chitosan: nanofiber membrane for guided bone regeneration. J. Periodontol. 76, 84–85.

Zhang, S., Sun, D., Fu, Y., Du, H., 2003. Recent advances of superhard nanocomposite coatings: a review. Surf. Coat. Technol. 167 (2-3), 113–119.

FURTHER READING

Agnelli, J.A., Kurachi, C., de Souza, C.W., 2014. Addition of silver nanoparticles to composite resin: effect on physical and bactericidal properties in vitro. Braz. Dent. J. 25 (2), 141–145.

CHAPTER

25

THE ROLE OF NANOMEDICINE, NANOTECHNOLOGY, AND NANOSTRUCTURES ON ORAL BONE HEALING, MODELING, AND REMODELING

Mohamadreza Baghaban-Eslaminejad*, Ahmad Oryan, Amir Kamali**, Ali Moshiri†**

**Royan Institute for Stem Cell Biology and Technology, ACECR, Tehran, Iran*

***Shiraz University, Shiraz, Iran*

†Iran University of Medical Sciences, Tehran, Iran

CHAPTER OUTLINE

1 Introduction 778
2 Nanomedicine 779
3 TERM and Nanomedicine 780
4 Fabrication Methods for Nanomaterials and Nanostructures 781
4.1 Physical and Chemical Synthesis of Nanomaterials 782
4.2 Biological Synthesis of Nanoparticles 782
5 Application of Nanotechnology in Dentistry 788
5.1 Nanomaterials 788
5.2 Nanodiagnostics 805
5.3 Nanorobotics 807
6 Application of Nanomedicine in Oral Bone Regeneration 808
6.1 Nanoceramics 808
6.2 Nanometals 811
6.3 Nanopolymers 811
6.4 Nanocomposites and Nanoscaffolds 811
7 Nanotechnology and Cell Therapy in Oral Bone Regeneration 814
8 Current Challenges and Future Directions 816
References 818

1 INTRODUCTION

The oral cavity is a specialized structure and composed of soft and hard tissues that have uniquely developed in order to initiate and facilitate oral functions particularly chewing, prehension, swallowing, speaking, and so on (Oredugba, 2006). Many reasons can cause oral diseases and injuries to develop. Regardless of the etiology, treatment of oral diseases and injuries is technically demanding and the most interfering factor is that the function of the oral cavity should be preserved during the course of the treatment (Brennan et al., 2007). In addition, despite the passing decades of modern medicine, we do not have proper modalities in order to successfully regenerate the lost tissues in the oral cavity and treat several disorders, such as dental loss, periodontal cysts, and bone loss, large bone defects both in the mandible and maxilla, oral tumors, soft tissue degenerations and infections, congenital disorders (such as cleft palate), massive traumatic injuries, and burns (Elsalanty and Genecov, 2009). As a general rule, the most important diseases and injuries in the oral cavity can be categorized as those causing: (1) tissue loss (e.g., dental loss, bone loss, perialveolar cyst), (2) infections (e.g., candidiasis, osteomyelitis), (3) tumors (odontoblastoma, squamous cell carcinoma, melanoma), and (4) degenerative changes (osteoporosis, jaw osteonecrosis, degeneration of the periodontal ligament) (Baldi et al., 2009; Epstein et al., 2015; Li et al., 2000) (Fig. 25.1). Tissue engineering and regenerative medicine (TERM) is a newer option in managing various oral diseases and injuries and, fortunately, oral medicine and surgery are pioneers in this field. In another meaning, dentists, maxillofacial and head and neck surgeons are one of the first groups of practitioners that used modern TERM strategies for treatment of various diseases of oral cavity. TERM uses scaffolds, healing promotive factors, cells and genes in order to manage various diseases of the oral cavity (Galler and D'souza, 2011).

Today, with increasing demand for advances in diagnosis and treatment modalities for oral health, new treatment methods and applications, such as nanotechnology, have been considered as viable and interesting options in TERM based strategies (Kim et al., 2014). Nanotechnology, which deals with matter in dimensions between approximately 1 and 100 nm, has been used in many studies to

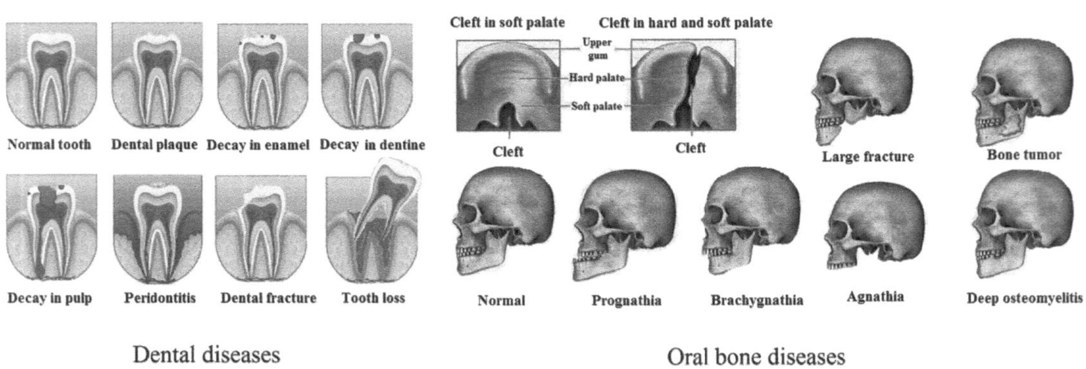

FIGURE 25.1 Dental and Oral Bone Abnormalities and Disorders

enhance health issues and provide novel means of diagnosis and treatment (Jha et al., 2014; Ozak and Ozkan, 2013; Saha, 2009; Saini et al., 2010; Walker and Mouton, 2006). Researchers in the field of oral health and medicine have explored the nanoscale biomaterials that have a potential in therapeutic modalities with moderate success (Balshaw et al., 2005; Mitragotri et al., 2015). The implementation of nanotechnology in the field of oral medicine includes diagnosis of tumors, novel routes of drug delivery, new bone graft materials, modification of implant surface (Abiodun-Solanke et al., 2014; Gu et al., 2013). This chapter illustrates current developments of nanotechnology in the field of oral health and dentistry, and discusses potential future uses of nanotechnology, nanobiotechnology, and nanostructures in oral bone medicine.

2 NANOMEDICINE

"Nano" is derived from the Greek word meaning "dwarf" (Adhikari et al., 2013). The term "nanotechnology" was first used in 1974 by Norio Taniguchio, which refers to the science of manipulating matter on an atomic, molecular, and supramolecular scale. This unit is approximately equal to the long size of two or three atoms of gold (Abiodun-Solanke et al., 2014; Kumar et al., 2011). The Nobel Prize–winning physicist, Richard Feynman, envisioned nanotechnology in 1959 by his seminal lecture. After discussing his ideas with a coworker, Feynman offered the first known proposal for a nanomedical procedure to cure heart disease (Deb et al., 2015; Samir et al., 2015). Today, nanotechnology is a developing branch of science that is being applied in various fields of science, including medicine (Wilkinson, 2003). Like TERM, nanomedicine, as an interdisciplinary field of science, brings physicians, engineers, physicists, biologists, chemists, and mathematicians together to improve the conventional treatments (Moghimi et al., 2005; Porter and Youtie, 2009). Nanomedicine attempts to control, repair, and reconstruct the human biological systems under the supervision of nanotechnology standards (Wong and Liu, 2012). The base of nanomedicine utilizes the nanotechnology to reduce the biomaterial scale in order to increase the specific surface area of a material to increase the bioactivity and chemical activity of a particular substance in viable environments directly or indirectly (Abou Neel et al., 2015).

The application of nanomedicine ranges from the medical uses of nanomaterials, to different types of nanostructures and even development of the future molecular nanotechnology (Basavaraj, 2012). The researchers believe that molecular nanotechnologies can start a revolution in the future of medical sciences (Patil et al., 2008). However, using nanotechnology in medical applications may be a double-edged sword: (1) the safety benefits of nanotechnological innovations in treating many diseases and (2) the potential health hazards posed by exposure to nanomaterials (Schmidt, 2009; Teow et al., 2011).

Nanomaterials are defined, in principle, as materials that deal with sizes 100 nm or smaller (1–100 nm) (Contado, 2015). The spatial arrangements of nanomaterials can create different types of nanostructures. According to the number of dimensions on the nanoscale, nanostructures can be categorized into zero-dimensional (0D), one-dimensional (1D), two-dimensional (2D), and three-dimensional (3D) structures (Sajanlal et al., 2011). Until now, many nanostructures, including nanoparticles, nanotubes, nanosheets, nanosphers, nanofibers, nanofilms, nanofoams, nanomeshes, nanoshells, andothers, have been designed and introduce by nanotechnology experts (Sajanlal et al., 2011). Various types of nanostructures have been illustrated in Fig. 25.2.

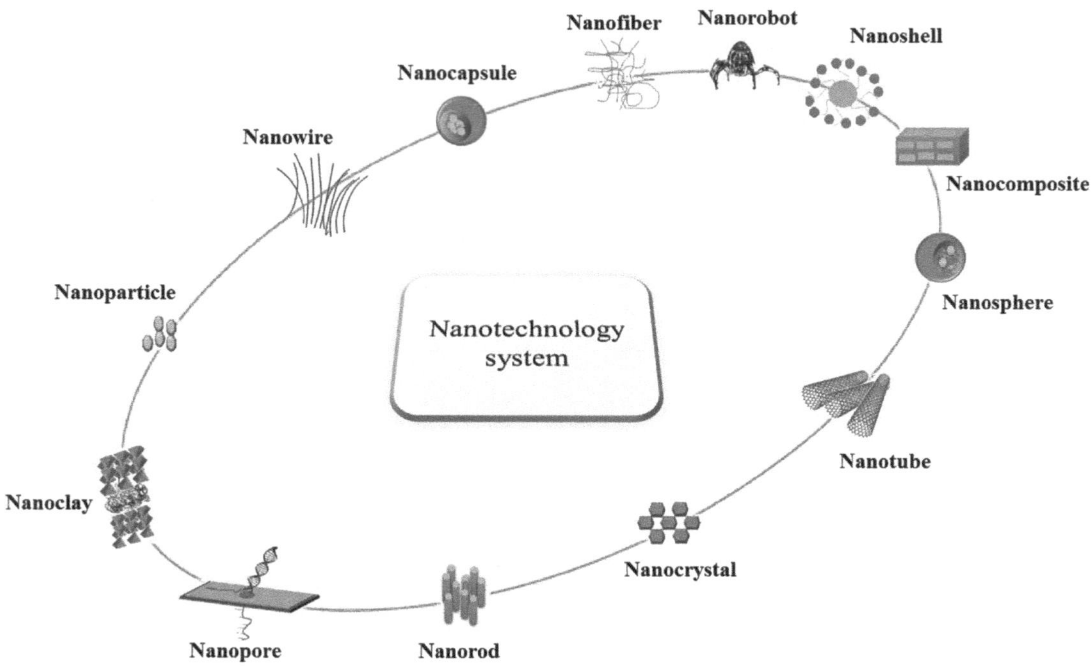

FIGURE 25.2 Various Types of Nanostructures

3 TERM AND NANOMEDICINE

Bone tissue consists of two main osseous structures, including cortical and trabecular bone. Cortical bone forms the cortex or compact shell of most bones, which is composed of repeating osteon units, whereas the cancellous bone is made up of an interconnecting framework of trabeculae that gives more flexible and weaker characteristics (Clarke, 2008). The physiological properties of bone tissues are strongly associated with the organization of the extracellular matrix (ECM) and resident cells, where the organization of the ECM is hierarchical with different levels (nm to cm scale) (Gong et al., 2015) (Fig. 25.3). Therefore, regeneration of large bone defects need novel and creative nanotechnology-based TERM strategies that account for the nano- to macroscale hierarchical assembly of tissue (Gong et al., 2015). In consideration of several limitations in the traditional bone regeneration strategies, nanomedicine provides newer options for acceleration of bone healing process (Chaudhury et al., 2014).

In oral medicine, nanotechnology has a wide meaning so that it presents itself through the use of nanomaterials, biotechnology, genetic engineering, molecular machine systems, and even nanorobots (Bhardwaj et al., 2014). It is well established that nanomaterials have superior stability and hardness that significantly exceed the corresponding characteristics of microsized materials. This may be

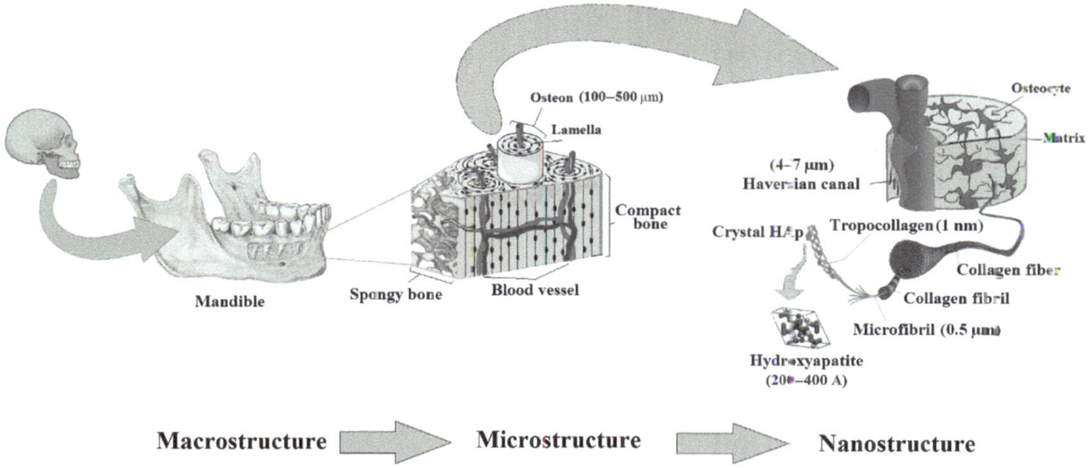

FIGURE 25.3 Hierarchical Structure of Oral Bones

rooted in reduction of impurities (bubbles and defects) in nanocrystalline grain materials (El-Atwani et al., 2014; Gao et al., 2014; Yu and Zhan, 2014). In addition, the effect of conventional lattice dislocation slip on fracture toughness enhancement significantly decline by decreasing in grain size. Simultaneously, enhancing the grain boundary sliding and migration can improve the fracture toughness (Gao et al., 2014). Three characteristics of the nanomaterials, including small grain size, large surface area, and strong interfacial interactions are the most important factors, which can lead to the great torsion and tensile modulus (Stassi et al., 2012). As the result of the aforementioned topics, it is clear that the nanostructured scaffolds can be an excellent alternative to the ECM and achieve full integration with the normal tissues (Motamedian et al., 2015). Regardless of the beneficial role of nanotechnology in increasing the mechanical strength of the bone scaffolds, nanotechnology can be used to produce nanospheres for controlled drug delivery (Park, 2007), to produce nanoparticles in order to enhance the efficacy of the detergents, drugs, and solvents and also several other applications that have been discussed in detail in this chapter.

4 FABRICATION METHODS FOR NANOMATERIALS AND NANOSTRUCTURES

One of the most challenging issues in producing nanomaterials and nanostructures is the method of fabrication. While several methodologies have been utilized to produce different nanostructures, most of them are still under investigation and the reliability of the methods is improving day by day (Daraio and Jin, 2012). Each of these several available protocols are discussed below in order to find out which of the many available protocols may have value in oral medicine.

4.1 PHYSICAL AND CHEMICAL SYNTHESIS OF NANOMATERIALS

There are two general approaches in preparation of nanoparticulate materials; these include "top-down" and "bottom-up" techniques. The former seeks to create smaller pieces by using a bulk material to direct their assembly (top-down). Bottom-up approach, termed as the process of synthesis of the material from atomic or molecular species via chemical reactions, results in the precursor particles to grow in size. Both approaches can be done in gas, liquid, supercritical fluids, solid states, or in vacuum (Phelippeau et al., 2003). There are several processes for producing nanoparticles, including sol–gel processing, chemical vapor deposition (CVD), plasma or flame spraying synthesis, laser pyrolysis, lithography, high-energy ball milling, single or double emulsion and atomic or molecular condensation (Phelippeau et al., 2003). Different precursor particles determine the type of synthesis processes. The protocols and applications of the different physical and chemical synthesis methods are listed in Table 25.1.

4.2 BIOLOGICAL SYNTHESIS OF NANOPARTICLES

Although there are many chemical and physical techniques for synthesis of nanoparticles, these methods have many limitations, including the use of toxic solvents, generation of hazardous substances for human health, and high-energy consumption (Nadagouda and Varma, 2008). Biosynthesis of nanoparticles by biological resources is effective, nontoxic, and eco-friendly (Hulkoti and Taranath, 2014). A wide range of microorganisms, including prokaryotes and eukaryotes are used to synthesize nanoparticles, such as nanometals (e.g., silver, gold, platinum, zirconium, palladium, iron, and cadmium), metal oxides (e.g., titanium oxide, zinc oxide, and iron oxide) and others (e.g., lead carbonate, cadmium carbonate, strontium carbonate, zinc phosphate, and cadmium selenide) (Thakkar et al., 2010). Some of these nanoparticles produced by biological resources are summarized in Table 25.2. These biological resources include bacteria, plants, yeasts, fungi, and algae. The synthesis of nanoparticles may be intracellular or extracellular according to the location where nanoparticles are formed (Kulkarni and Muddapur, 2014). In principle, the mechanism of the nanoparticles synthesis by biological approaches varies based on the type of material and biological resources. For example, in production of silver nanoparticles, the mechanism involves reduction of Ag ions by the electron shuttle enzymatic metal reduction process (bioreduction mechanism) (Chauhan et al., 2013).

Since the regulation of particle morphology in chemical and physical methods is still a continuous area of research, biological synthesis approach with the ability to strictly control particle shape would be an interesting option to overcome the limitations of other methods (Singh et al., 2015). In addition, it might be possible to control the particle size by varying parameters, such as biological resources, growth stages, synthesis conditions, reaction time, substrate concentrations, growth media, source compound of target nanoparticle, reaction conditions (temperature, pH), and addition of non-target ions (Gurunathan et al., 2013). Biosynthesis approach has more benefit because nanoparticles are sometimes covered with a lipid layer, which provides better stability and solubility in physiological conditions and the nanoparticles have superior biocompatibility and less toxicity, which is critical for medical applications and is the bottleneck of other synthetic methods (Li et al., 2011; Zhang et al., 2012). Research is currently being conducted on the manipulation of cells at the genomic and proteomic levels. With a better understanding of the cellular and molecular processes of synthesis, including isolation and identification of the nanoparticle reduction factors, it is hoped that this work can be done in less time and with high synthesis efficiency.

Table 25.1 Protocols and Applications of Different Physicochemical Fabrication Methods

Fabrication Methods	Protocols	Applications
Sol–gel	Solid nanoparticles dispersed in a liquid phase (a sol) and create mostly a colloidal suspension, forming a continuous gelatinous network (a gel).	This technique is a well-known admitted industrial process in production of colloidal nanoparticles from liquid phase. Oxide nanoparticles and composites nanopowders are particularly synthesized through this process.
Spraying	The reactants break down and the particles are formed in a flame by homogeneous nucleation and growth. In the next step formation of nanoparticles occurs by rapid cooling. In the simplest way of spraying method, the reactant is dissolved in liquid phase and then pressurized by compressed air or mechanically compressed through a small nozzle to spray on the hot surface, which leads to form many tiny droplets (100 nm).	Spraying processes using high-energy flame (for example, by plasma spraying equipment or carbon dioxide laser) have been utilized to manufacture some of the nanoparticles, such as ZnO and quantum dots.
Atomic or molecular condensation	A bulk piece is heated and melted in vacuum to produce a stream of vaporized and atomized matter, which is cooled rapidly after transmission into a chamber containing either inert or reactive gaseous phase. The cooled metallic atoms are condensed to create liquid nanoparticles, which then unify in a controlled environment, preserving a spherical shape with smooth surfaces. Further cooling of the liquid nanoparticles solidifies them and stops growing. Surfactants usually added to keep the so-formed nanomaterials separated because they are very reactive and have high tendency to form agglomerates.	Condensation is used mainly for metal-based nanoparticles. If a reactive gas, such as oxygen is used, the reaction results in formation of the metal oxide nanoparticles.
Chemical vapor deposition (CVD)	The process consists of decomposing hydrocarbon vapor that attaches and accumulates onto a substrate in presence of a metal catalyst. It is also called thermal or catalytic CVD to be distinguished from many other kinds of CVD used for various purposes.	The CVD process is the most popular method to produce carbon nanotubes (CNTs) and semiconductor nanowires, such as Si, GaN, and ZnO.
Supercritical fluid synthesis (SCF)	Supercritical fluids are materials forced into supercritical state by regulating their temperature and pressure, which are used to form nanoparticles by a rapid expansion of a supercritical solution. They are obtained by vaporizing a liquid in a closed space and are condensed to reach liquid phase. This supercritical state is very sensitive to the parameters of the process and is adjustable. Supercritical fluids are also used for coating and filling nanostructures.	Supercritical carbon dioxide and water are the most common reactants used in this process because they are inert, nonflammable, inexpensive, and environmentally benign and have higher reaction rates and selectivity. The SCF synthesis of nanoparticles like phosphorus, magnetic materials, carbon nanotubes.

(Continued)

Table 25.1 Protocols and Applications of Different Physicochemical Fabrication Methods (*Cont.*)

Fabrication Methods	Protocols	Applications
Lithography	Lithography is a very common approach in fabricating nanometer-scale structures; this method involves several major and minor techniques based on the manufacturing processes, including optical lithography, electron-beam lithography, nanoimprint lithography, multiphoton lithography, scanning probe lithography.	The main application of lithography is generally for multigate devices, such as field effect transistors, quantum dots, nanowires, gratings, zone plates, and photomasks.
Laser pyrolysis	This method is usually classified as a vapor-phase synthesis process to produce nanoparticles and nanopowders with small-diameter and low agglomeration level. When nucleation occurs, particle growth happens by coalescence/coagulation rather than further nucleation. By reaching to optimum temperature, particle coalescence is faster than coagulation and eventually spherical nanoparticles are formed.	Different nanopowders suchas Si, SiC, Si_3N_4, and Si/C/N composite have been prepared by this method. Laser ablation method, a similar technique to laser pyrolysis, is also used in production of CNTs and nanowires.
High-energy ball milling	Ball mill is a type of grinder, a cylindrical device, used in grinding or mixing the materials. Microstructurally, the mechanism of this process can be divided into four stages: (1) initial stage, (2) intermediate stage, (3) final stage, and (4) completion stage. Briefly, in this technique, the powder macroparticles are trapped between the highly kinetic colliding balls and the inner surface of the vial, resulting in frequent deformation, revulcanization, and fragmentation of premixed powders, which causes formation of very fine and dispersed nanoparticles.	In particular, mechanical attributed contamination by the milling tools (Fe), as well as ambient gas (O_2, N_2) can be some limitations for high-energy ball milling technique. Nanocrystalline powders and oxide nanoparticles can be fabricated through this method.
Templates	In this method, various molds containing regular nanosized pores or voids are used as a template to form nanoparticles. Contrary to the common opinions, the template was not to be a three-dimensional (3D) object, but they can also be created on a plane surface or a gas-liquid interface by forming self-assembled monolayers.	Nanoparticles including porous alumina, zeolites, diblock copolymers, dendrimers, and other molecules have been manufactured by this approach.

Single or double emulsion	Single (oil-water) or double (water-oil-water) emulsion is an approach by which polymers can be used to encapsulate both hydrophobic and hydrophilic drugs in micro- or nanoscale form. In summary, polymer (e.g., PLGA) is dissolved into an oil phase that is emulsified with a surfactant or stabilizer (water and poly vinyl alcohol). Hydrophobic drugs are added directly to the organic solvent (oil), whereas hydrophilic drugs may be first emulsified with the polymer solution prior to formation of the nanoparticles. High-intensity sonication or high-power homogenizer can help to facilitate formation of the small nanospheres. The resulting emulsion is then added to a lot of aqueous phase and stirred for several hours. This allows the organic solvent to evaporate. Finally, the nanoparticles are separated and washed by centrifugation.	These methods are cost-effective and applicable methodsin fabrication of nanospheres, which have been used in delivery systems.
Electrospinning	In electrospinning process, a suspended drop of polymer is charged with thousands of volts. The electric field is used to overcome the surface tension of a polymer solution to shoot a jet of liquid out of a needle toward a conducting collector resulting in fabricating the polymer fibers. Many factors affect this process and also determine the size of the nanofibers. These include the polymer properties and concentration, solvent properties, solution flow rate, voltage, distance from the needle to the collector, type of the collector and method of collection.	This method is used in fabrication of nanofibers from the natural and synthetic polymers.

Table 25.2 Some Selected Nanoparticles Synthesized by Different Biological Resources

Resource Type	Products	Culturing Temperature (°C)	Size (nm)	Shape	Location	References
(A) Bacteria						
AMB-1	Fe_3O_4	28	20	Cubo-octahedral	Intracellular	Amemiya et al. (2007)
Lactobacillus sp.	$BaTiO_3$	25	20–80	Tetragonal	Extracellular	Jha and Prasad (2010)
Fusariumoxysporum	TiO_2	30	6–13	Spherical	Extracellular	Bansal et al. (2005)
Rhodococcus sp.	Au	37	5–15	Spherical	Intracellular	Ahmad et al. (2003)
Corynebacteriumglutamicum	Ag	30	5–50	Irregular	Extracellular	Sneha et al. (2010)
Shewanellaoneidensis MR-1	Fe_2O_3	25	30–43	Pseudohexagonal/ irregular or rhombohedral	Intracellular	Bose et al. (2009)
Lactobacillus sp.	TiO_2	25	8–35	Spherical	Extracellular	Jha et al. (2009)
Brevibacteriumcasei	PHB	37	100–125	Not available	Intracellular	Pandian et al. (2009)
(B) Algae						
Sargassumwightii	Au	Not available	8–12	Planar	Extracellular	Singaravelu et al. (2007)
Shewanella algae	Pt	25	5	Not available	Intracellular	Konishi et al. (2007)
Shewanella sp.	Se	30		Spherical	Extracellular	Lee et al. (2007)
(C) Yeast						
Saccharomyces cerevisiae	Sb_2O_3	25–60	2–10	Spherical	Intracellular	Li et al. (2011)
Candida utilis	Au	37	Not available	Not available	Intracellular	Gericke and Pinches (2006)
Schizosaccharomycespombe	CdS	Not available	1–1.5	Hexagonal lattice	Intracellular	Kowshik et al. (2002)
Candida glabrata	CdS	Not available	2	Hexagonal lattice	Intracellular	Dameron et al. (1989)
(D) Plants						
Avena sativa (Oat)	Au	30–60	5–85	Rod shape	Extracellular	Armendariz et al. (2004)

Iresineherbstii	Ag	30–60	44–64	Cubic	Intracellular	Dipankar and Murugan (2012)
Aloe vera	Ag	30–60	15.2	Spherical	Extracellular	Medda et al. (2015)
Cinnamomumcamphora	Au and Ag	30–60	55–80	Triangular or spherical	Extracellular	Ponarulselvam et al. (2012)
(E) Fungi						
Aspergillusflavus	Ag	25	5–25	Spherical	Extracellular	Vigneshwaran et al. (2007)
Fusariumoxysporum	Au–Ag alloy	25	8–14	Spherical	Extracellular	Senapati et al. (2005)
Fusariumoxysporum	$PbCO_3$, $CdCO_3$	27	120–200	Spherical	Extracellular	Sanyal et al. (2005)
Coriolusversicolor	CdS	25	100–200	Spherical	Extracellular	Sanghi and Verma (2009)
Verticillium sp.	Ag	18–20	25 ± 12	Spherical	Intracellular	Senapati et al. (2004)
Trichodermaviridae	Ag	10–40	2–4	Single crystals	Extracellular	Fayaz et al. (2009)

5 APPLICATION OF NANOTECHNOLOGY IN DENTISTRY

Nanodentistry is a subfield of nanomedicine, which makes possible the maintenance of near-perfect teeth health through the use of nanomaterials and nanobiotechnology. Basically, nanodentistry improves the oral health care by three major routes including nanomaterials, nanodiagnostics, and nanorobotics (Aeran et al., 2015). Various nanomaterials have been produced for tissue engineering of oral bone and dental tissues including natural (organic and inorganic), synthetic (nanoceramics and nanopolymers), and hybrid nanomaterials (Chiara et al., 2012). Over the past few years, resin composite restorative materials have been appealing due to more safety and aesthetic quality in comparison to the traditional dental fillers, such as amalgam (Ilie and Hickel, 2011). Calcium phosphate salts are the main inorganic component of all mineralized tissues in human body. Also, other inorganic materials,such as calcium carbonates and sulfates are present in smaller quantities (Chiara et al., 2012). Hydroxyapatite (HAp) is responsible for 60%–70%, and 90% of the dry matter content of bone and enamel respectively (Dorozhkin, 2011b). Tooth enamel is the most mineralized tissue of human body. More than 96% and less than 4% of the dry matter content of the enamel are composed of inorganic and organic materials respectively. Also, more than 70% of the dry matter content of the dentin is composed of inorganic materials, which are almost comparable to bone. This inorganic material is mainly composed of a calcium phosphate related to the hexagonal HAp, whose chemical formula is $Ca_{10}(PO_4)_6\ 2(OH)$ (Sabel, 2012). HAp is a white powder and natural form of calcium apatite that is the hydroxyl end member of the complex apatite group. The OH^- ion is routinely replaced by fluoride, chloride, or carbonate in order to produce fluorapatite or chlorapatite (Zhou and Lee, 2011). As HAp has formed the most parts of the inorganic compounds of bone and dental tissues, the nanostructured or nanoscaled HA is routinely used in oral and maxillofacial regenerative medicine in order to promote remineralization of the enamel and dentine (Pepla et al., 2014).

Restorative dentistry has traditionally focused on the use of different materials in order to restore physical function following tissue loss. Today, various types of engineered nanomaterials (ENMs) have been used in different medicinal goals in dentistry, including infection controls, dental fillers and implants, targeted delivery, and preventive care productions (Mantri and Mantri, 2013).

5.1 NANOMATERIALS

5.1.1 Antimicrobial nanoagents

There are a wide range of cements, composites, adhesives, and sealants in dentistry in order to inhibit dental decay. One of the most important features of these materials is antibacterial properties (Utneja et al., 2015). Antibacterial properties of dental cements and luting adhesives are very important features because bacteria may still be present in gaps between cavity of the walls and the restorative material. It is also a good opportunity for bacteria to gain access to the dental cavity through microleakages or micropores or microcanals present after cementation (Boeckh et al., 2002). Generally, in endodontic infections, elimination of bacteria and microorganisms from root canal system is totally impossible (Vijayaraghavan et al., 2012). Alternatively, using filler and restorative material with antimicrobial activity is considered necessary to reduce the load of remaining bacteria and to eradicate the infection (Cheng et al., 2015). Endodontic infection has a polymicrobial nature with predominance of anaerobic bacteria, so the antimicrobial spectrum of action in these materials should be investigated (Carrotte, 2004). Many studies have reported that glass ionomer cements (GICs) are the most cariostatic and somehow antibacterial material due to release of fluoride (Cabral et al., 2015; Davidovich et al., 2007; Vahid Dastjerdie et al., 2012; Vermeersch et al., 2005; Yesilyurt et al., 2009). Although numerous studies have been conducted to improve the antimicrobial activity of restorative materials, most of them have focused on sustain release of various incorporated low-molecular-weight (LMW) antibacterial agents including antibiotics, nanometals (silver, zinc, cooper, gold, and titanium), iodine and chlorhexidine (Boeckh et al., 2002; Chen et al., 2012; Farrugia and Camilleri, 2015; Imazato, 2009; Tobias et al., 1985). Table 25.3 lists some of the antibacterial nanomaterials and their mechanisms of action on bacteria.

Silver (Ag) and silver compounds are used as an antimicrobial agent for many years and thus they have several applications (Kim et al., 2007). For instance, they are used for water purification, wound care, as bone implants, and in reconstructive orthopedic surgery, cardiac devices, catheters, and surgical appliances (Lansdown, 2006). The antimicrobial activity of silver is related to the amount of released bioactive silver ions and their ability to contact with microbial cell membranes (Saengmee-anupharb et al., 2013). Usually, silver is used in ionized (Ag +) or elementary forms (silver zeolites and nanoparticles) (Lansdown, 2006). Silver nanoparticles can be constructed in various shapes depending on the application at hand (e.g., spherical, diamond, octagonal, and thin sheets). They can be synthesized by several physical and chemical methods and basically can be divided into three major categories: wet chemistry, ion implantation, or biogenic synthesis (Bhattacharya and Mukherjee, 2008; Rycenga et al., 2011; Wan et al., 2013). Similar to other nanoparticles, very large surface area allows silver to coordinate a vast number of ligands (De Jong and Borm, 2008). When silver nanoparticles are exposed to the bacteria, they bound to the cell membrane. Once connected, some of the nanoparticles pass through the membrane, expose to the cytoplasm and interact with phosphate-containing compounds (DNA and RNA), while other portion sticks to the sulfur-containing proteins on the membrane. The silver-sulfur interactions result in formation of pits and pores at the cell membrane. Then after, the cellular components release into extracellular environment through these pores, which is due to the difference in the intra- and extracellular osmotic pressures. Within the cell, the integration of silver creates a low molecular weight (LMW) region and shrinks the DNA structure. The condensed

Table 25.3 Antibacterial Nanomaterials and Their Mechanisms of Action on Bacteria

Nanomaterial	Mechanism of Action
Silver	Bacterial cell wall damage, impairing the electron transfer system, disrupting the hydrogen bonding between the two antiparallel strands and denaturing the DNA molecule
Gold	Reacts with bacterial cell wall, strong electrostatic effects, weakening DNA replication of bacteria
Titanium dioxide	Production of reactive oxygen species (ROS), damages the bacterial cell membrane and cell wall
Copper oxide	Generation of ROS and hydroxyl free radicals, enhancement of the intracellular oxidative stress, disruption of bacterial cell membrane as the result of lipid oxidation
Carbon nanotube (CNT)	Disruption of cell membrane by ROS, oxidation of proteins and lipids within the cell membrane
Zinc oxide	Accumulation of nanoparticles inside the cells, H_2O_2 generation, enhanced cell permeability, cell wall damage, disruption of cell membrane

state of DNA inhibits the DNA polymerasecontact with the target adhesion site. The introduction of silver nanoparticles block the bacterial replication and thus cell death occurs (Lee et al., 2014; You et al., 2012). The bactericidal activity increases when silver comes in contact with fluids leading to ionization (Hussmann et al., 2013). This has been suggested to be associated with the suppression of enzymes and inhibition of the expression of proteins causing interference in the process of cell metabolism (ATP production) (Lamsal et al., 2011). Many investigations evaluated the antimicrobial activity of silver nanoparticles based on their size (Ajitha et al., 2015; Lu et al., 2013; Pal et al., 2007). Based on the recent investigations and as a general rule, nanoparticles with an average size of 10 nm or less show electronic effects that considerably increase their bactericidal effect (Saion et al., 2013). Recently, synergistic activity of silver nanoparticles with common antibiotics, such as penicillin-G, ampicillin, erythromycin, clindamycin, and vancomycin against *Escherichia coli* and *Staphylococcus aureus* has been noted by some studies (Durán et al., 2010; Rai et al., 2012).

Silver nanoparticles are used in coatings of medical equipment (Ge et al., 2014). It has been shown that the bacterial count is significantly lower when compared to the uncoated one (Furno et al., 2004; Sussman et al., 2015). However, the antibacterial effect of these nanoparticles gradually decreases, because in the process of cleaning and washing for new procedure, a large portion of the silver nanoparticles become less effective due to the loss of silver ions (Jain and Pradeep, 2005).

Recently, a new design of toothbrushes has been developed in which the nanosilver particles are coated over the toothbrushes in order to increase the effectiveness of silver nanoparticles during toothbrushing (Hanke and Guggenbichler, 2004). Silver or gold nanoparticles can also be used in between toothbrush bristles to prevent gingivitis, periodontitis, or dental caries (Hamal et al., 2014). In addition, a new generation of toothpaste has been formulated with silver nanoparticles in order to use their antibacterial effect (Mohamed Hamouda, 2012). In the field of dental restoratives, some studies have been done to investigate the antibacterial effect of materials (composite resin, endodontic materials, acrylic resin, and implants) incorporating nanosilver particles (Abe et al., 2004; Cheng et al., 2011; Zhang et al., 2013). It has been shown that silver nanoparticles possess an inhibitory effect against many microorganisms, including *Streptococcus mutans*, *Candida albicans*, *Pseudomonas aeruginosa*, *Enterococcus faecalis*, and *S. aureus* (Flores et al., 2010). Although, it has been stated that silver nanoparticles are biocompatible with mammalian cells and their applications on dental restoratives does not represent a threat to human health, it is well-documented that the released silver ions undergo dissolutionand have a great cytotoxic effect (Jiao et al., 2014). Silver nanoparticle depositions in the membrane can disrupt regulation of solutes and alter proteins structure and cell recognition. Inflammation, genotoxicity, free radical production, and cytotoxicity have been shown to be associated with long-term exposure to nanosilver particulates. Silver nanoparticles accumulate primarily in the liver but it has also been shown that such particles are toxic in other organs particularly the brain (Sukirtha et al., 2012). Although silver nanoparticles are a promising antimicrobial agent in dentistry, their application on some areas, such as endodontics and dental implants, remains conflicting, and further researches are needed to be done in this area. It is also necessary to illustrate the best ways of incorporation of silver nanoparticles, as well as the possible side effects of their addition in dental materials and devices. Taken together, each strategy that uses silver nanoparticles in oral medicine should be aware that silver nanoparticles should only be available for a short period of time during each treatment strategy and long-term exposure to silver is toxic for the cells, tissues, and organs.

Zinc (Zn) and zinc oxide (ZnO) nanoparticles have also been suggested for medical application in dentistry due to their antibacterial effects (Sevinç and Hanley, 2010). However, the number of Zn applications in dentistry is limited. For instance, Zn nanoparticles have been shown to fail in obtaining the sanitary approval for use in the food industry (De Jong and Borm, 2008). Copper (Cu), another metal element with antimicrobial properties, is cheaper than silver and has the advantage of being a stable chemical element. In addition, its physicalproperties allow it to be easily mixed with other polymers to produce dental implants (Thneibat et al., 2008). However, Cu nanoparticles have not been fully investigated as a potential antibacterial agent in dental materials. Similar arguments have been suggested for gold (Au) and titanium (Ti) nanoparticles (Besinis et al., 2014; Kasraei et al., 2014; Mohamed Hamouda, 2012). The size and shape of ENMs are important factors that affect their antimicrobial activity. Engineered nanoparticles with a particle size of less than 10 nm have been shown to be the most effective sizes against microorganisms. It has been also suggested that the triangular nanoparticles have more bactericidal activity compared to those with spherical or needle-like morphology (Pal et al., 2007).

Metals nanoparticles are not the only type of ENMs known to provide antimicrobial properties for dental health care. Quaternary ammonium polyethylenimine (QA-PEI), chitosan, silicon dioxide (SiO_2), and bioactive glass nanoparticles are other options that have been suggested for this matter (Chen and Chung, 2012; Farah et al., 2015; Prabhakar and Kumar, 2010). The choice of antibacterial nanoparticles is mainly associated to the type of dental application. Nanometals are commonly used in dental nanocomposites or are utilized as antibacterial coatings; while others, such as bioactive glass nanopowders, are mostly used as root canal disinfectants (Mohamed Hamouda, 2012). SiO_2 nanoparticles have been reported to prevent adherence of microorganisms flora and control and inhibit formation of oral biofilm. Although SiO_2 nanoparticles have no inherent antibacterial effect, it is believed that it inhibits or delays microbial adherence and growth rate of biofilm (Chen et al., 2013). The antimicrobial effect of some ENMs synergistically increases with ultraviolet light (e.g., TiO_2 nanoparticles) (Yuan et al., 2010). Certain metallic materials (e.g., Ag and Cu nanoparticles) have been proved effective against abroad spectrum of bacterial strains, including a wide range of Gram negative and positive bacteria, while others are morestrain-specific depending on their mechanism ofaction (e.g., ZnO nanoparticleshave more efficacy in killing the Gram positive bacteria) (Yah and Simate, 2015). In order to replace the traditional antibacterial agents with the new generation of ENMs, the health surveillance systems must first be satisfied that the new therapeutic agents are safe, or safer than the existing products, and also are more effective. Overall, nanotechnology can be helpful to control and prevent oral infections by using novel ENMs in different routes, such as using nanoparticles as carriers for traditional antibacterial agents (chlorhexidine) or impregnation of antimicrobial nanoparticles into resin-based composites to have a long-lasting effect against microorganisms without affecting the physical andmechanical properties of the dental tissue. Table 25.4 represents the important antimicrobial ENMs that are used in dentistry.

5.1.2 Nanofillers

Recently, nanofillers have been introduced to dental health as a promising alternative option to classic fillers (Ozak and Ozkan, 2013). Dental composites typically consist of three major components that are chemically different from each other: the resin-based matrix (organic matrix), the filler (inorganic matrix), and a crosslinker (usually silane, to enhance the chemical bonds between the filler and the organic matrix) (Cramer et al., 2011). Each of these phases can be modified. Different characteristics of composites, including the physical, chemical, clinical, and mechanical properties can be changed according to clinician's requirement. In the past few years, introduction of new materials, such as

Table 25.4 Applications of Antibacterial Nanomaterials in Dental Medicine

Type of Nanomaterials	Size of Nanoparticles	Aim of Application	Results	References
Ag and ZnO NPs	50 nm (ZnO)–20 nm (Silver)	Used in resin composite to inhibit the overgrowth of oral biofilm microorganisms	Colony forming units measurement formed on the surface of blood agar plate showed the significant effect of resin composite incorporated with NPs compared to control group	Kasraei et al. (2014)
Ag NPs	2.7 nm	Used in adhesives for enhancing antimicrobial activity	The agar plate diffusion test showed that the primer adhesive containing 0.1% silver NPs had significantly larger inhibition zone in comparison with controls	Cheng et al. (2013)
QPEI NPs	10 nm	Used in resin composite with antibacterial property	The remaining visible bacteria on the samples containing QPEI NPs significantly decreased compared to the control samples	Beyth et al. (2010)
CSRB	60 ± 20 nm	Used to introduce a novel antibiofilm material	Complete elimination of the biofilm bacteria has been achieved by modification of CSRB dosage without any significant toxic effects	Shrestha et al. (2014)
QPEI NPSs	—	Used in a commercial provisional dental cement	No bacterial growth was observed on the provisional cement samples	Shvero et al. (2010)
Bioactive glass nanopowders (72S, 63S, 58S)	20–90 nm	Used in nanocomposites for disinfection of root canal	The minimum bactericidal concentration for 63Sand 58S was 100 mg/ml (broth concentration of material). The 72S had no antibacterial effect	Mortazavi et al. (2010)
45S5 bioglass NPs	<5 μm	Used in nanocomposites for disinfection of root canal	The bacterial growth in specimens, which were treated by 45S5 NPs, were higher in comparison with micro 45S5 group.	Waltimo et al. (2009)
ZnO and TiO_2 NPs	40–100 nm	Used in resin composite with antibacterial property	There was a 20% reduction in bacterial biofilm growth. The antibacterial effect of TiO2 was similar to the control group	Aydin Sevinc and Hanley (2010)
Ag NPs	100–120 nm	Used in oral tissue conditioner with antimicrobial effect	The antimicrobial inhibitory effect was observed against all bacteria	Nam (2011)
Ag NPs	—	Used as an antibacterial coating for dental implants	The colony forming units of the pathogenic bacteria significantly decreased in implants coated by Ag NPs compared to the uncoated implants	Huang et al. (2010)

CSRB, *Chitosan NPs functionalized with rose Bengal;* QPEI, *Quaternary ammonium polyethylenimine;* NPs, *nanoparticles.*

Table 25.5 Applications of Nanomaterials as Dental Filler

Type of Nanomaterial	Size of Nanoparticle	Aim of Application	Results	References
Zr NPs	20–50 nm	Used in resin composite to improve the properties	Incorporation of Zr NPs with specimen increased the tensile bond strength of the resin composite	Lohbauer et al. (2010)
CaF_2 NPs	56 nm	Used in resin composite to improve mechanical properties and also fluoride release	The composite specimens containing CaF_2 NPs have significantly showed better flexural strength after 48 days immersion in NaCl compared to the control samples	Xu et al. (2010)
n-HAp particles	20–70 nm	Used in adhesive systems to reinforce composites	Incorporation of n-HAp with specimen significantly increased the microtensile bond strength of the adhesives compared to the control group	Wagner et al. (2013)
Hybrid acrylic and Si NPs	75–100 nm	Used in dental resin composites to increase the mechanical resistance	The composites containing 30 wt.% silica NPs showed the highest flexure strength	Canche-Escamilla et al. (2014)
HAp-nanorods	—	Used in adhesive adhesives to improve it	Diametric tensile strength was higher when n-Hap had been incorporated compared to the control	Sadat-Shojai et al. (2010)
Si NPs	20–50 nm	Used in dental resin composites to increase the mechanical resistance	The composites containing 40 wt.% silica NPs showed the highest fracture roughness and flexure strength	Hosseinalipour et al. (2010)
CNT	≤10 nm	Used in resin-based restorative materials to improve the mechanical properties	The flexure strength of CNT composite was significantly higher than the controls	Zhang et al. (2008)
$BaSO_4$ NPs	≤10 nm	Used to control the mechanical properties and reaction time of the ionomer cement	$BaSO_4$ NPs initially reduced the working time but by increasing its concentration in cement, the effect was reversed. The mechanical strength also reduced by adding NPs to the cement	Prentice et al. (2006)

NPs: nanoparticles, CNT: carbon nanotube.

phosphine oxide initiators and monomethacrylate diluents, has led the production of new generation of dental composites (Richards and Antonucci, 2007). However, introducing the nanofillers in the field of dentistry hasmade a significant change in creation of the advanced dental fillers and composites. The physical and chemical properties and also the size of filler particles, which have been used in dental composites, are significantly variable. However, optimization of the final restorative materials can be done simultaneously by incorporation of nanofillers with resin-based materials. Generally, addition of nanofiller particles can increase the mechanical properties of composites, reduce the rate of shrinkage during polymerization process, modify the thermal expansion coefficient of the composite as close as possible to the tooth, offer more uniform particle distribution, admit a higher filler load, attenuate viscosity, improve handling and finally provide radio-opacity (Cramer et al., 2011; Fu et al., 2014).

Conventional dental composites contain a range of fillers, vary from material to material, but often include silica, quartz, and silicate particles based on the metal oxides (e.g., barium, strontium, zinc, aluminum, and zirconium) (Drummond, 2008). These composites are classified based on filler particle size, including macrofiller composites (0.1–100 μm), microfiller composites (0.04–0.1 μm), and hybrid composites (fillers of different sizes). Most of the current typical composites have a range of size from 0.04 to 0.7 μm (De Oliveira et al., 2012). The concern in this area is that these composites may not optimally interact with the nanoscalic structural elements of dental tissues, such as the enamel rods, HAp crystals, dentinal tubules, and collagen fibers. The average filler size in dental composites is 40 nm, and the real innovation of them is their ability to increase the load of the inorganic phase in 80 wt.% when compared to the microfilled composites, 50 wt.%, for example (Subramani and Ahmed, 2011). The first dental nanocomposite that introduced to the market was Filtek Supreme (3 M ESPE, St. Paul, MN, USA) (Gogna et al., 2011). Other commercially available nanocomposites include Ketac (3M ESPE, St. Paul, MN, USA), Ketac N100; nanoionomers (3M ESPE, St. Paul, MN, USA), Fuji IX GP (GC, Leuven, Belgium), nanoprimer, Premise (Kerr/Sybron, Orange, CA, USA), Adper single bond plus adhesive (3M ESPE), Ceram X (DENTSPLY International, Milford, CT, USA), GrandiO (Voco, Cuxhaven, Germany), 4 Seasons (Ivoclar Vivadent, Schaan, Liechtenstein), and Palfique (Tokuyama Dental Corp., Tokyo, Japan).

Natural silica (nanoclays) and their modified varieties are also commonly used to produce dental composites. These materials are nontoxic and nonirritant, especially when used P.O. (Mousavinasab, 2011). In the past decade, silica nanoparticles (e.g., available under the trade names Aerosil [Evonik comp.], Cab-O-Sil [Cabot Corporation], HDK [Wacker Chemie], and Konasil [OCI]), have been suggested as nanofillers to enhance the dental resin-based composites (Karabela and Sideridou, 2011). They can improve the physical properties of the final products, such as surface roughness, abrasion resistance, and rheological behavior. Occasionally, incorporating the nanofillers is an advantageous strategy to increase the physical and mechanical properties of resin-based composites. However, filler-loading in dental composites should be carefully controlled because there are some limitations.

It has been shown that elevating the concentration of Zr nanofillers in dental composites gradually increases the microtensile bond strength. Although the analysis has shown that the highest values of mechanical properties for 20 wt.% filler content, composites with a lower filler content (5 wt.%) had increased indexes compared to the control specimens (Lohbauer et al., 2010). The optimal filler load in dental composites was about 0.5 wt.% for sodium montmorillonite, 10 wt.% for $CaF2$, 0.2 wt.% for HA and 40 wt.% for silica nanofillers (Karabela and Sideridou, 2011). Nevertheless, regarding the optimal filler load, these data are not entirely reliable, because the studies did not adhere to the standardized methodologies when preparing their composites. It should also be highlighted that the nanofillers may not always enhance the mechanical properties compared to the microfillermaterials. It has been known that solubility and shrinkage of both types of compositesare similar, but the evidences displayed that the nanofiller composites have higher water absorption and opacity (Althues et al., 2007).

Among all shapes of nanofiller structures, spherical types are popular because the stress is more uniformly distributed across the bulk volume of the composite and consequently crack formationis inhibited due to the round shape of particles (Zou et al., 2008). Carbon nanotubes (CNTs) are other types of nanofillers, which have been investigated in many researches (Akasaka et al., 2009; Sasani et al., 2014; Terada et al., 2009; Zhang et al., 2008). Zhang et al. (2008) synthesized a dental composite based on CNTs that contained an improved mechanical performance. CNTs are well-known to have exceptional strength and their applications in production of dental nanocomposites is conceivable due to their unique dimensional distribution (aspect ratio >1000).

One of the most important critical issues for patients is the aesthetics feature of restoration materials. For instance, dental composites with a color that closely matches to that of patient's natural tooth is always a demand (Bagheri et al., 2007). Nanoparticles have been employed to modify the shade and translucency for providing desired aesthetics. On the other hand, the optical properties are also important in clinical settings and the nanofillers have been used in this regard to match the radiopacity of dental adhesive (Besinis et al., 2015). Collectively, application of nanofillers for production of nanocomposites may improve some properties while compromising some other characteristics. Eventually, the clinicians should know all the available dental composites in the context of clinical considerations to select an appropriate one according to the position of the cavity and the aesthetic requirements. Table 25.5 lists some of the most important nanofillers for dental application.

5.1.3 Implants

All the ongoing efforts in production of newer dental implants have been summarized in a novel product with a high success rate after the initial implantation and durability. At present, the success rates for dental implanting are 90%–95%, and the failure of this procedure is mainly due to poor osseointegration, infection, or rejection (Tomsia et al., 2011). The future of dental implants is significantly associated to the intrinsic inflammatory response and the quality of integration in implant-tissue interface. Consequently, the types of implant materials, surface chemistry, physical and topographical structures are fundamental aspects that should be considered when designing a desired implant (Sartoretto et al., 2015). The diameter of a cell is typically variable from 2 to 10 μm, but the intracellular components are much smaller, even in the nanoscale. The cells secrete special kind of proteins during osseointegration with a typical size of just 5 nm, which are even smaller than any manmade nanoparticles (Abou Neel et al., 2015). In principle, the aim of producing new generation of dental implants with basic biological components (nanoscale) is to achieve a perfect strategyfor designing the materials that interact with cells and tissues at a molecular level. Table 25.6 displays some of the selected ENMs, which have been used in the application of dental implants.

Modifying the surface of metallic implants is one way to increase the chance for a successful implantation. Today, using ceramic materials to cover the surfaces of metallic implants is a popular modality in order to reduce inflammatory response and increase the biocompatibility and osseointegration of the dental implants (Saini et al., 2015). The ceramic nanomaterials that have been selected for covering implant surfaces include HAp, bioactive glasses, and other CaP compounds (Xuereb et al., 2015). The n-HAp is probably the most desirable material used as a coating biomaterial for stainless steel, titanium, and its alloys implants because of its similarity to the inorganic component of enamel, dentine,

Table 25.6 Applications of Nanomaterials in Dental and Bone Tissue Implants

Type of Nanomaterial	Size	Aim of Application	Results	References
n-HAp particles	70–80 nm	Used in titanium implants as a coating material to stimulate proliferation and differentiation of osteoblasts	The alkaline phosphatase activity and osteocalcin production of cells in n-Hap group were significantly higher than the control group	Sasani et al. (2014)
CNT	—	Used in biodegradable implants (PLA) to increase mechanical properties	The tensile strength, as well as cell growth of PLA-CNT implant was significantly higher compared to the PLA alone	Lee et al. (2011)
HApnanocrystal	5 nm	Used in titanium implants as a coating material	New bone formation after 9 weeks of healing was higher in the single and double coated implants by HAp nanocrystals	Svanborg et al. (2011)
n-HAp particles	1–150 nm	Used in titanium implants to promote osseointegration	Osseointegrationwas significantly higher in the n-HAp coated group after 2 weeks in comparison with the uncoated implant	Lin et al. (2009)
HAp-nanorods	70–80 nm	Used in titanium implants to improve osseointegration	The mean, bone to implant contact increased in the coated implants by HAp-nanorods compared to the control	Yang et al. (2009a)
Diamond NPs	1–50 nm	Used in bone implants to regulate new bone formation	Proliferation of osteoblast was significantly higher in the Diamond NPs coated implants compared to the control	Yang et al. (2009b)
CaP NPs	20–100 nm	Used titanium implants to enhance bone formation and healing	The histomorphometric results showed that the bone to implant contact and bone volume significantly were higher in the CaP NPs coated implants compared to the control	Goene et al. (2007)
YSZ NPs	50 nm	Used in biomedical implants as a bioactive durable coating	The hardness and elastic modulus of the 45S5 bioactive glass implant were improved by YSZ NPs coating	Radice et al. (2007)

CaP, *calcium phosphate;* CNT, *carbon nanotube;* HAp, *hydroxyapatite;* NPs, *nanoparticles;* YSZ, *Yttria-stabilized zirconia.*

and bone (Marchi et al., 2013). Recent investigations have shown that n-HAp coatedmetallic implants have more biocompatible surfaces and encourages cell adherence in the implant-tissue interface compared to the uncoated metallic implants (Pachauri et al., 2014; Rosa et al., 2012). In vivo and in vitro experiments have noted that titanium implants coated with n-HA ph as enhanced osseointegration and accelerated new bone formation when compared to the controls (Kobayashi et al., 2007; Kuroda and Okido, 2012). However, the surface characteristics of metallic implants are not the only factors that determine the human biological responses. It has been proposed that size of nanoparticle and nanostructural morphology (e.g., spherical, fiber, rod-shaped, and crystals) also play important roles in cellular adhesion rate and their bone-forming capacity (Pachauri et al., 2014). Implants coated with CaP nanocrystals with 20–100 nm in diameter have demonstrated similar enhanced osseointegrative behavior in vivo (Yang et al., 2010).

Alumina, another nanoceramic option with promising biocompatible and osseointegrative properties, showed a significant capacity to increase osteoblast proliferation and adhesion when coated on implants (with the size of 24 nm) (Webster et al., 2000a). Its nanofibers morphology (2 nm in diameter, 5 nm long) has also been found to induce osteoblast differentiation and mineralization (Webster et al., 2005). It has been found that osteoblast adhesion increased on nanocrystalline diamond coating (with grain sizes less than 100 nm) whereas microcrystalline diamond coating with greater particle size inhibited osteoblastic cell adhesion rate (Yang et al., 2009b). This latter finding could be the foundation for production of coated implants with variable particle sizes to control the rate of bone apposition and desired bone growth based on the specific anatomical locations. Other types of nanoparticles, such as yttria-stabilized zirconia (YSZ) and CNT-CaP have also been investigated as means to amplifymetallic and polymeric-based implants (Coelho et al., 2009; Radice et al., 2007). Numerous scientific reports have suggested that the nanoparticulate-coated dental implants are more biocompatible so that they have the ability to enhance new bone formation (He et al., 2009; Lee et al., 2011; Sasani et al., 2014; Shi et al., 2013). However, these reports remain conflicting because other results have shown no beneficial effects from implants that have been coated with nanoparticles (Al-Hamdan et al., 2011; Coelho et al., 2009; Schouten et al., 2010). Actually, these conflicting reports may be due to the fact that the optimal bone quality of the implant bed overshadowed the superior osseointegrative properties of the nanocoated implants. In the experimental dental disease models, healthy bone tissues have been selected as implantation site, which does not mimic the poor state of bony base in edentulous patients.

Despite existing extensive research in literature, it is still unclear whether the improved function of the nanoparticulate coated implants are derived from the chemical properties of the coatings or it is due to the modification of surface roughness of the implants by nanoparticles. However, some investigations by focusing on the latter one suggest that improvement of the osseointegration of nano-CaP particles coated implants arises from the complexity of the yield modified surface rather than the CaP chemistry itself (Mendes et al., 2007). In fact, increases in surface roughness improve the degree of osseointegration. This conclusion could be a possible explanation for the role of nanotopography in the success of implantation. However, further studies are required to either support or reject this hypothesis (Wennerberg and Albrektsson, 2009).

Unfortunately, all efforts in this regard have failed to fully characterize the implant surfaces, and in some cases, scanning electron microscopy (SEM) is the only method used to describe these surfaces and compare them together. Thus, more investigation is needed to assess the nanostructure of coated implants and provide quantitative data by more advanced surface characterization techniques, such as 3D measurements for SEM, profilometry, atomic force microscopy (AFM), X-ray photoelectron spectroscopy (XPS), and mechanical testing. Nanoindentation is also a mechanical testing system that perfectly measures the surface roughness and elastic modulus, which can explain the nanotopographical features of various surface coatings. Analysis of these data may then provide a better understanding on how the cell response is determined by the surface nanotopography.

5.1.4 Delivery systems

In order to effectively treat a kind of bone and dental disease, it is often necessary to provide an optimum concentration of a particular substance (e.g., drug, growth factor, biomaterial) for a desirable period of time, which can be achieved by control delivery systems. Nanostructures particularly those produced as nanocapsules, nanoscaffold, nanosell, nanotube, and quantum dots, have been suggested as valuable delivery systems in controlling the dental caries and restoration, local anesthesia, dental remineralization, tooth sensitivity management, dental caries vaccine, regulation of tooth biofilm, root canal disinfection, and periodontal infection (Goldberg et al., 2007).

5.1.4.1 Nanocapsules and nanospheres

Nanospheres and nanocapsules are two forms of the nanostructures that differ in their morphology and architecture. Nanospheres are formed by dense polymeric matrix whereas nanocapsulesare nanoscale shells made up of natural or synthetic polymers. Nanocapsules consist of a liquid/solid core in which the drug, gene, protein, and other substances are incorporated into the interior cavity that is surrounded by a distinctive polymeric membrane (Guterres et al., 2007). In the drug delivery system, encapsulation of the drugs into nanoscale shells is one of the methods in protecting the drugs from destruction, degradation, and rapid metabolism during the desired delivery protocol and also protecting the body against damages of the toxic drugs (rapid release of agents than can be toxic). Encapsulation of drugs into nanocapsules or loading them onto the nanospheres enables controlled drug release and improves local drug delivery. During a specific time period, these nanostructures undergo slow enzymatic degradation and following disintegration of the membrane sustains release of the drug. Usually, the initial burst release occurs at the early hours of administration and then the drug or substance gradually releases over the time (De Jong and Borm, 2008). It has been shown that capsules made up of nanoparticles have higher surface-area-to-volume ratio, smaller membrane porosities and consequently more efficient solubility in comparison with microparticles (microspheres or microcapsules) (Uskoković, 2013). Such properties significantly affect the enzymatic biodegradation time and diffusion of drug through the membrane. In addition, compared to microparticles, nanoparticles have the ability to penetrate into extracellular and intracellular areas that may be unreachable to other drug delivery systems due to their nanoscalic diameter, including the periodontal pocket areas below the gum line (Hubbs et al., 2013). In 2005, a group of researchersproduced and characterized nanoparticles loaded with triclosan (TCS) (a noncationic antibacterial agent with a well-established efficacy against several plaque-forming bacteria) to obtain a novel gingival intrapocket delivery approach adequate for the treatment of periodontitis. Additionally, they performed a preliminary experimental study on dogs in which the TCS-loaded PLGA nanoparticles (9.09% of TCS) were injected in the bottom of the experimental model of pockets (Pinon-Segundo et al., 2005). They showed that TCS nanoparticles significantly diminished the inflammation at the experimental sites. Natural herbal extracts have also been encapsulated into polymeric nanostructures for the treatment of periodontal infections and tooth decays. In a study, *Harungana madagascariensis* leaf extract (HLE) (a herbal extract, which is known for its antibacterial, antifungal, and antiviral properties) has been incorporated with PLGA polymer nanocapsules to investigate its in vitro bactericidal activity (Bonifácio et al., 2014). Interestingly, a reduction in the bactericidal concentration was observed in HLE/PLGA nanoparticle group compared to HLE alone. The bioadhesive property of the PLGA biopolymer may be a possible explanation for these results, which remains on the bacterial cell wall for a prolonged period of time, thus extending the drug action.

As an ongoing research, novel electrically charged nanoparticles have been designed, which were loaded with farnesol (a natural 15-carbon acyclic sesquiterpene alcohol with proved antibacterial effect). The positively charged outer layer of the nanoparticles was able to stay in place at the surface of the dental enamel because it was made up, in part, of HAp, which had been negatively charged. As a general rule, as oppositely charged magnets are attracted to each other, the oppositely charged nanoparticles and HA are also attracted to the dental enamel. Because teeth are usually covered by saliva, the researchers weren't sure that the nanoparticles would bind to dental enamel. But not only did the particles stay in place, they were also able to adhere to the polymeric matrix and stick to dental plaque. A key property of the inner nanoparticle material is that it is unstable at acidic pH (e g., 4.5) allowing the antibacterial drug to escape more rapidly (Horev et al., 2015). This process was exactly what happenes to the pH level in plaque when it is exposed to food products that cause dental caries. In the other words, the drug releases from the nanoparticles when exposed to cavity precisely when it is most needed to quickly stop acidic pH producing bacteria. Until now, the results of this investigation showed that when farensol was delivered by the nanoparticle carriers, both the number and severity of the caries were reduced. Overall, nanospheres and nanocapsules are attractive candidates for delivery systems, not only for controlled drug delivery, but also they can be utilized for future dental regenerative modalities, such as delivery of gene and growth promotive factors.

5.1.4.2 Nanofibrous scaffolds

Previously, nanofibrous scaffolds were only used for drug and hormone delivery purposes but later on, it has been found that these scaffolds have more capabilities so that they can be used to preserve and deliver stem cells into human target organs (Perán et al., 2012). At present, such scaffolds are used for local drug delivery to the target sites and their biocompatibility (scaffold disintegration without any serious immune response) is among their main advantages. The fabrication methods of nanofibrous scaffolds and their components vary depending on the target site and include electrospinning, self-assembly, and phase separation.

5.1.4.2.1 Electrospinning

A newfound manufacturing technology of nanopolymer fibers is based on the principle of spinning dilute polymer solutions in a high voltage electric field. The electrospinning process was first patented in the 1900s. Electrospinning can be used to produce many synthetic and natural nanofiberous polymers and can also fabricate nanofibrous scaffolds in which many cell types can adhere to the nanofiber surface (James et al., 2011). In dentistry, this method has been used to form a gelatin membrane in order to regenerate periodontal tissues. When periodontal ligament cells were seeded on such product it has been shown that cell attachment and proliferation well occurred over 7 days (Zhang et al., 2009). Such fabrication method also supports production of nanofibers from copolymers or polymer composites with biological molecules or minerals (Binulal et al., 2010; Jose et al., 2010). However, similar to other techniques, electrospinning also has some limitations to produce nanoscaffold for example, it cannot produce complex 3D scaffolds or designed pore geometry. In addition, it is often hard to spin some of the biologic polymers, such as collagen and the method is much more valuable for spinning the synthetic polymers, such as PLGA.

5.1.4.2.2 Self-assembly

Self-assembly is defined as the spontaneous organization of components into a specific structure (Whitesides et al., 1991). This important process occurs naturally with the self-assembly of natural intracellular components, such as collagen (main ECM protein in various connective tissues). These natural processes can self-assemble into nanofibrous structures from engineered self-assembling peptides (Hartgerink et al., 2001). These peptides can be designed to form a nanofiber structure through autonomous organization of molecules due to noncovalent interactions. Huang et al. (2010) used a nanofibrous self-assembling peptide amphiphile to induce enamel formation. This peptide injected into a mouse incisor and then was cultured in a kidney capsule for eight weeks. Ordered crystalline HAp was found in "pearls" near the incisor. Although, the results cannot confirm nanofibrous organization following peptide injection, self-assembling peptides have been suggested as an optional initiator in tooth development. In most cases, self-assembly method result in hydrogel formation containing nanofibers of much smaller size than the fibers, which have been produced by electrospinning (Hartgerink et al., 2001). Since the self-assembly process can naturally occur following injection in vivo, these hydrogels can also be useful in injection applications. However, the self-assembled hydrogels have poor mechanical strength and their structural peptides are also susceptible to enzymatic biodegradation, which can affect the degradation rate of scaffold in an uncontrolled fashion.

5.1.4.2.3 Phase separation

Phase separation is a process where a single-phase homogenous solution is critically quenched, causing separation into a two phases; polymer-rich region and a solvent-rich region. Recently, a new method has been developed that uses thermally induced phase separation to produce a synthetic nanofibrous-biodegradable scaffold. In biological applications, it is beneficial for fabrication of nanofibrous scaffolds, which have interconnected porous structure. This latter can facilitate the cell migration, nutrient/waste exchange, and uniform cell and nutrient distribution (Ma, 2008). High porous scaffolds can be manufactured through this method by adding porogen materials. Progens, such as sugar, salt, or paraffin are often added within a bulk material and then leached out following phase separation, leaving open pores, as well as nanofibers. Because of this complex process, phase separation method is more time-consuming than other fabrication procedures. Nonetheless, nanofibers with an average diameter of 100 nm can be produced by such method (Wei and Ma, 2006), which can be difficult when is compared to other techniques. In addition, the phase separation beneficially allows fabrication of interconnected porous scaffolds with complex geometries (Chen et al., 2006). Thus, phase separation is a valuable method for nanofibrous scaffold preparation for bone tissue engineering applications.

5.1.4.3 Nanoshells

Nanoshell is a type of nanoparticle with a dielectric (e.g., silica) core and a thin metal coating (usually gold) (Fuchigami et al., 2012). In the near future, this type of nanoparticle can be used in dentistry for different applications, including treatment of severe gingival and periodontal infections, getting rid of oral and dental cancer cells, and possibly for soft tissue welding of inaccessible area. The nanoshell can be stimulated with infrared light in which the nanoshell absorbs the energy of the light and consequently generates heat. Through this mechanism, nanoshells have a potential photothermal therapeutic ability to destroy bacteria and tumor cells. This method minimizes the risk of trauma to the adjacent tissues (Pattani and Tunnell, 2012). For this purpose, nanoshells are usually shuttled into tumors by the use of phagocytosis, where phagocytes engulf the nanoshells through the cell membrane to form an internal phagosome. In the next step, the cellular lysosomes fuse with the phagosome and form a phagolysosome in order to metabolize it. These nanoshells are not metabolized. Thus, in order to be effective, the nanoshells need to be within the tumor cells and photo-induction is needed (as described earlier) to terminate the tumor cells' life. Additionally, the nanoshells, which are loaded with both cell-targeting agents (antibodies, proteins, etc.) and specific drugs (e.g., antibacterial or chemotherapeutic agents) can be used for targeted drug delivery (Lal et al., 2008). Another interesting application of the nanoshells is the tissue welding capability. Nanoshells may represent a rapid way for treatment of deep lacerations in an emergency room. Recently, a nanoshell-based solder (nanoshells incorporating with bovine serum albumin) has been applied to full thickness incisions, after which the incision areas were irradiated with infrared light for several minutes to start tissue welding and the results of wound healing was better than the control group and was comparable to the suture group (Loo et al., 2004). These results suggest the nanoshells as a novel effective ligature/cautery method in deep gingival laceration or difficult accessible bleeding sites in dental tissues.

5.1.5 Individual care products

The use of oral personal care products and dentifrices (toothpastes, tooth powders, mouthwashes, etc.) is recommended to the public health to manage oral health and generally help to prevent teeth decays. The interest for using nanotechnology in these products is continuously growing over a time. In principle, nanomaterials could be used in dental care products for various purposes, such as controlling the microbial count and plaques as parts of dental biofilm, aiding the mineralization process of the enamel or dentin, providing nanoscale minerals to regulate optimal pH (Hannig and Hannig, 2010).

Fluorides are still the most favored remineralizing agent, which has been used in dentifrices. Topical fluorides enhance the formation of flourapatite in the presence of Ca and *P* ions produced during enamel demineralization. However, for every 2 fluoride ions, 10 Ca ions, and 6 *P* ions are required to form one-unit cell of fluorapatite [Ca_{10} $(PO_4)_6$ F_2] (Swarup and Rao, 2012). Thus, the availability of Ca and *P* ions is the limiting factor for net remineralization to occur. Moreover, based on the mechanism of remineralization that occurs by topical application of fluorides, they attempts to reduce apatite dissolution only rather than promote mineralization of apatite crystals or replacement of the lost minerals (Zero, 2006). Therefore, CaP-based materials are needed to combine with fluoride in order to provide an acceptable solution. As stated earlier, microstructured CaP-based biomaterials have limited value in this regard because they have low biological properties, which are majorly due to their low affinity to do biologically active chemical reactions with cells and host tissues.

Employing artificial nanosized CaP-based biomaterials, such as n-HAp in dentifrices has been considered as a wise strategy in order to restore the lost mineral content of dental tissue. The n-HAp is a revolutionary material and is widely used in dentistry. The n-HAp is considered as one of the most biocompatible and bioactive materials in dentistry and is getting worldwide acceptance these days (Najibfard et al., 2011). Although preliminary attempts tried to clinically applicate HAp in bone regenerative medicine, the results were not encouraging. In recent years and by developing nanotechnology and translational medicine, the n-HAp has been found to have superior bone regenerative properties with acceptable outcomes in maxillofacial surgery and also dentistry (Najibfard et al., 2011). The biomimetic hydroxyapatite function is to protect the teeth by creation a new layer of synthetic enamel around the tooth, rather than hardening the existing layer with fluorine, which chemically changes into calcium halophosphate [$Ca_5(PO_4)_3F$]. Nanoparticles of HAp have similar characteristics when compared to crystalline structure of tooth enamel and several reports have confirmed the potential benefit of n-HAp application in dental enamel repair and regeneration.

Until now, n-HAp has been used in different forms (spherical, rod shape, crystal) (Skrtic and Antonucci, 2011). The nanostructured HAp crystals have high levels of biomimetic properties due to their composition, structure, morphology, bulk and surface physical–chemical properties. They are normally synthesized through various models but the manufacturers are strict to preserve the scale limit, which is routinely selected between the values of 20 and 50 nm in diameter. However, several commercially available products exist that used higher scales, including 100 to up to 200 nm in diameter (Dorozhkin, 2011a). The most valuable range for n-HAp is below 30 nm in order to preserve the most bioactive characteristics of such biomaterial. When the scale is decreased the specific surface area of the n-HAp particles increases, which is the major responsible factor for the superior biologic activity and cell-particle interaction of n-HAp. The n-HAp contained in toothpastes has a strong ability to bond with proteins, as well as with fragments of plaque and bacteria (Pepla et al., 2014). Many studies have found that the n-HAp added in toothpaste not only increased the structural strength of enamel and improved remineralization but also reduced bacterial colonization of the tooth surfaces. Moreover, as previously discussed, it also acts as filler because it repairs small holes and depressions on enamel surface, a function enhanced by the small size of the particles.

The n-HAp can be used alone or in combination particularly with various agents, such as sodium fluoride and strontium in order to provide a solution for such purpose. It has been suggested that the 100 nm carbonated HAp particles are able to cover the demineralized enamel surfaces more effectively when compared with fluoridated toothpaste (Lelli et al., 2014). The fall in the surface Ca/P ratio following demineralization does not practically increase after remineralizing with 2% sodium fluoride application. Suggesting only structural modification of apatite occurs and is restricted to a partial hydroxyl group replacement by fluoride ions without influencing the Ca and phosphate content. It has been shown that application of synthetic n-HAp exhibit surface Ca/P ratio close to that of the biological enamel (Swarup and Rao, 2012). Although n-HAp is an acceptable and beneficial nanostructured biomaterial for remineralization of the tooth surfaces and subsurfaces, different formulations have been developed to remineralize the subsurface lesions by n-HAp-based products. Unfortunately, limited data exist showing whether the n-HAp containing products have superiority over the classic fluoride-based products for such purpose (Tschoppe et al., 2011). It has recently been shown that toothpastes containing n-HAp has higher remineralizing effects compared to amine fluoride toothpastes with bovine dentine, and comparable trends were obtained for enamel (Tschoppe et al., 2011).

For producing n-HAp containing toothpastes several factors exist that may alter the efficacy of n-HAp for remineralization of the tooth surfaces. These majorly include pH of the toothpaste, concentration of the n-HAp in the formulation, the formulation of the toothpaste, the complementary ingredients of the toothpaste, time of application, time of exposure, treatment of the enamel with n-HAp, and so on. Perhaps, pH is the most determinant factor in this regard. It has been claimed that only pH values higher than 5.5 to 6 are able to promote lesion arrest and facilitate remineralization (Peters, 2010). Unfortunately, limited data exist showing which of the many available pH values in the n-HAp toothpaste slurries are able to effectively promote remineralization; however, it can be emphasized that n-HAp has more efficacy in pH with higher ranges than 6 to 7 and this means that the patients should not dilute n-HAp-based toothpastes with water and the best diluting agent during toothpaste application is the saliva. This fact could be extended not only for n-HAp but also for other calcium phosphate based nanomaterials as it has been suggested that higher mineral gain could be observed with a pH of 6.5 compared to 5.5 (Tschoppe et al., 2011). This is basically due to two reasons: (1) n-HAp and other nanostructured calcium phosphate–based biomaterials solve in the solutions that have lower pH values. (2) Tap water has lower viscosity and pH than saliva and for this reason washing the tooth with tap water significantly dilutes the n-HAp-based toothpastes; therefore, when the concentration of the n-HAp is reduced its efficacy is declined, too. Although n-HAp may have higher benefit in higher pH values (e.g., ranging 6 to 7), conflicting issues are seen in the literature. In a study, it was claimed that the largest increase in mineralization was observed in the group with pH 4.0. The group with pH 7.0, however, showed the lowest degree of mineralization. An accumulation of many minerals in the lesions and a corresponding reduction of their depth were also observed. Unlike with this latter study and in line with the former study, another study showed that the effect of n-HAp is better than the effect of micro-HA at pH 7.0 and at the same concentrations (Haghgoo et al., 2014). The concentration of calcium in solutions containing nano-HAp was greater than that detected in solutions containing micro-HAp. Increases in the Ca concentration lead to growth in the saturation of oral fluids with HAp, favoring the deposition of apatite minerals in the lesions and eventually promoting remineralization (Huang et al., 2009).

Regardless of pH, concentration of n-HAp is an important factor that has a great role in the treatment success. To date, there are few studies showing which of the many available used concentrations of n-HAp has the best results for protection of the dental tissues from erosions and to increase, enhance, and promote enamel and dentine remineralization. In a study, the effect of n-HAp concentrations on initial enamel lesions under dynamic pH-cycling conditions has been evaluated and determined in detail (Tay and Pashley, 2008). Initial enamel lesions were prepared in bovine enamel with an acidic buffer. NaF (positive control), deionized water (negative control) and four different concentrations of n-HAp (1%, 5%, 10%, and 15 wt.%) were selected as the treatment agents. The concentration of 10% n-HAp was reported to be optimal for remineralization of early enamel caries. However, based on this result, it can be suggested that n-HAp therapy is an expensive protocol at least with the current technologies of synthesis but by developing in the methods of n-HAp synthesis, the price of n-HAp is continuously decreasing, which supports its application in tooth enamel and dentine protection (Huang et al., 2009). As stated, n-HAp is reasonable to be combined with sodium fluoride in order to increase the efficacy of the treatment strategy. Nanohydroxy and fluorapatite have been synthesized using a sol–gel technique in an ethanol base. Glass ionomers, containing nanobioceramics with improved mechanical properties and strong binding to dentin are very promising restorative dental materials (Moshaverinia et al., 2008). Also, the role of n-HAp and other Ca-F-based materials has been investigated as a treatment for dentine hypersensitivity by occluding the dentinal tubules.

Dental erosion, known as an oral health problemnowadays, is increasing in young children and adolescents in developed and developing countries. The main external cause of increased dental erosion is a higher consumption of acid-containing beverages. In particular the use of acidic sports/water drinks has recently increased, and this may cause erosion and subsequently dental decay according to acid content of beverages (Ehlen et al., 2008). The potential risk of dental erosion in a sports drink containing n-HA as an additive have been examined recently. The experimental solutions were Powerade (PA) alone and PA with 0.05%, 0.10%, and 0.25% n-HAp. Twelve sound bovine enamel specimens for each group were placed in the mentioned diluted solutions for 60 min per day for 7 days and the results were compared with the control group. The pH and dental enamel indices increased with enhancing n-HA concentration in the drinks, whereas the titratable acidity decreased (Min et al., 2011). With regard to the latter results, it may be a very good and also a promising option to add small percentages of n-HAp (0.25%) in sports drinks, in order to prevent dental erosion.The n-HAp has also been usedwidely as an experimental mineral supplement, in order to improve the bone and tooth mineral density. It is used as an experimental addition to commercial glass ionomer cement (GIC), a procedure that has led to significant enhancement in the biological and mechanical properties of these substances. Taken together, n-HAp is, in fact, a valuable and better source of free Cato provide a suitable remineralization, protect enamel against dental caries and is an excellent candidate for oral bone regeneration and repair (Pepla et al., 2014).

Another interesting product that is made up of CaP-based materials is casein phosphopeptide–amorphous calcium phosphate (CPP-ACP). The casein phosphopeptide (CPP) is derived from the milk protein casein by tryptic digestion (Reynolds, 1998). The CPP can bond to Ca-P amorphous nanomaterials, which is commercially known as Recaldent and is available as a product under the brand name GC Tooth Mousse (GC Ltd.). As a rich source of calcium and phosphate ions, the CPP-ACP nanocomplexes have been proposed to promote enamel regeneration. When CPP-ACP nanocomplexes are delivered to the oral cavity (via sugar-free gum, medicated tooth cream, and fortified dairy milk), they adhere to the enamel, pellicle, plaque and soft tissues, providing bioavailable calcium and phosphate ions. The free calcium and phosphate ions then enter the enamel rods and reform into apatite crystals, contributing to the dental remineralization.

Other nanomaterials, including nanoparticulate bioactive glass, nanosized carbonated apatite alone, or in combination with silica, nanosized calcium fluoride, and nanoprecursors of amorphous calcium phosphates have been suggested by several experimental studies to promote enamel and dentine remineralization and can be utilized in toothpastes and other dentifrice components in the near future (Earl et al., 2009). In addition to CaP nanostructures, several types of nanoparticulates have also been found that are used in dental sensitivity. The silica nanoparticles (40 nm) have the ability to form a physical barrier at the entrance of dentinal tubules, preventing the movement of the fluid within the tubules, and subsequently reducing sensitivity problems (Tian et al., 2014). Fifty years ago, strontium chloride ($SrCl_2$) introduced to the market as the first tubule-blocking agent (Sensodyne toothpaste). The findings suggest that 10% $SrCl_2$ dentifrices reduce dentine hypersensitivity (Markowitz, 2009). Nonetheless, the incompatibility of $SrCl_2$ with fluoride leads to replacement of $SrCl_2$ with strontium acetate. Until now, many attempts have been made in order to overcome dental sensitivity and consequently many commercial toothpastes and other dentifrices have been introduced to the global marketing. However, it has to be noted that using the special toothpaste for sensitive teeth has an adverse effect causing dentine erosion and the tubules to open rather than resolving the problem (Miglani et al., 2010).

Whitening toothpastes and mouthwashes have found a special place in the commercial products due to the aesthetics aspect. Recently, one study investigated the discoloration effect of mouthrinses containing different metal nanoparticles (nano-ZnO, nano-CuO, nano-Ag, and nano-TiO_2). The results showed that the mouthwashes containing metal nanoparticles produced the same or even greater enamel discoloration compared to that of chlorhexidine. Moreover, it was found that TiO_2 nanoparticles caused the mildest and the ZnO nanoparticles produced the more severe color change among the tested nanoparticles (Eslami et al., 2015). This indicates that nano-ZnO is not a suitable additive for mouthwashes, whereas nano-TiO_2 should be further investigated for such application, because of two reasons: (1) it caused the least amount of discoloration compared to other nanoparticles; (2) it has unique and inherent whitening potential. Generally, TiO_2 is an effective whitener in powder form, where it is employed as a pigment to provide whiteness and opacity to many medical and industrial products, as well as most toothpastes. Komatsu et al. (2014) suggested the potential application of TiO_2 nanotubes to tooth whitening. They also found that TiO_2 nanotubes facilitated whitening with H_2O_2 at a low concentration and addition of TiO_2 nanotubes to H_2O_2 increased responsiveness to visible light.

Taken together, there is an emerging market for the use of nanotechnology in oral health-care products, and like applications of other nanomaterials in the food or personal care sectors, the safety and efficacy of these products need to be proven.

5.2 NANODIAGNOSTICS

Nanotechnology presents a great opportunity for clinicians to perform tests with higher sensitivity, specificity, simplicity, accuracy, and more affordable without affecting the cells. However, to date, few contributions have been reported on the impact of nanotechnology on the diagnostic procedures of dental conditions. Dental practitioners still use the traditionally routine diagnostic methods for dental conditions, including physical examination and diagnostic imaging (often radiology) and these techniques have not changed basically for decades. Essentially, the introduction of new technologies or novel approaches for clinical use is very difficult when the traditional treatment plans have been established for patients, while ensuring best outcomes (Laroui et al., 2013; McCarroll et al., 2014). Thus, unlike in medicine, dentistry has not yet seen considerable transition from bench to bedside. However, over the past few years, the contributions ofnanodiagnostics in dentistry have been made mostly in atomic force microscopy, nanoparticle contrast enhancers and other miscellaneous approaches:

5.2.1 Atomic force microscopy

Atomic force microscopy (AFM) is a very high-resolution type of scanning probe microscopy with demonstrated resolution on the order of fractions of a nanometer, which allows studying the surface of materials and organisms in detail (Kirmizis and Logothetidis, 2010). In dental biofilm, understanding of bacterial adhesion (as a complex nanomechanics involved in bacterial pathogenesis) is essential for prevention or treatment of dental caries. With the AFM and its capability, it is possible now to characterize the bacteria and the measurement of their adhesion to different substrates (implant, plaque, enamel) through a high-resolution imaging without any disruption of the bacterial morphology and properties. AFM cantilever is a nanomechanical biosensor comprising flexible, small rods with endings that are able to attach to cell, microorganism and so on (Steffens et al., 2012) This biosensor provides a real-time scanning of a live bacterial cell with high sensitivity. The obtained observations from this procedure also provide precise information about the cell's elasticity, membrane-molecules' properties, and even biomechanical interactions of antibacterial agents with a bacterial cell. For instance, vancomycin binds to D-Ala–D-Ala terminal of peptidoglycan precursors of bacterial cell walls. Thus, this approach could help to develop new antibacterial agents against drug-resistant bacteria.

5.2.2 Nanoparticle contrast enhancers

One of the most remarkable contributions from nanoparticles in dentistry relates to the diagnosis of oral cancer lesions. Diagnostic imaging technologies, such as computed tomography (CT), 3D-CT scan, CT-angioscan, PET scan, magnetic resonance imaging (MRI), ultrasonography and radiology, often play a critical role in overall tumor management and designing the treatment plan. These diagnostic techniques are able to identify the anatomical patterns and provide basic information regarding tumor location, size, and spread based on endogenous contrast. The success rate of cancer treatment, especially those of the oral cavity, highly depends on their accurate and prompt diagnosis (Chung et al., 2012). Unfortunately, in the case of tumors and metastases smaller than 5 mm, the routine imaging approaches become less reliable to distinguish tumors between benign and a malignant one (Sureka et al., 2013).

Recent progress toward nanotechnology-based CT imaging has been made by using gold nanoparticles. Gold is an element that has strong absorption of X-ray and can be used as a contrast material (Reuveni et al., 2011). In order to track the tumor cells, it is possible to increase the specificity of gold nanoparticles by functionalizing their surface with antibodies, which are able to bind with overexpressed tumor antigens, such as EGF, HER2, and folate. In this regard, a team of researchers succeeded to produce HER2-targeted gold nanoparticles by coating 15 nm particles with polyethylene glycol and covalently coupling them to anti-HER2 antibodies. Popovtzer et al. (2008) also reported that attenuation coefficient for the oral squamous-cell carcinoma, which molecularly targeted (bioconjugation of the gold nanoparticles to the UM-A9 antibody), is more than five times higher than untracked tumor cells.

Quantum dot (QD) is another nanoparticle that is used in detection of tumor cells. It is a semiconductor nanocrystal and acts as an inorganic fluorophore when exposed to an excitation wavelength (Cormode et al., 2014). Just like the gold nanoparticles, QDs can also be conjugated to different biomolecules, such as biotin, transferrin, avidin, streptavidin, immunoglobulin G (IgG), nucleic acids, peptides. Selectively targeting theoral squamous tumors is a long-standing problem because the chemotherapeutic agents that are used to kill the tumor cells lack specificity and cause serious side effects (Pisanic et al., 2014). Yang et al. (2011) successfully used cell-penetrating peptides to conjugate QDs for tumor cell tracking. In their in vivo (mouse model) study, they managed to engulf QDs conjugates into oral squamous cell carcinoma through endocytosis in order to track the cancer cells by diagnostic imaging. The QDs have also the capability to be used in dental caries and periodontal infections. The first report of using QDs for bacterial labeling was given by Kloepfer et al. (2003). Since this pioneered work, the QDs are widely used for labeling, detection, and quantification of *Mycobacterium* species, *E. coli* species, and *Salmonella* species.

5.2.3 Other miscellaneous approaches

5.2.3.1 Nanopores

A nanopore is a very small pore in a thin membrane, typically just big enough to pass through a single strand of DNA. Thus, shape and electrical properties of each base on a DNA strand may be assessed and consequently provide an opportunity for reading the genetic codes (Venkatesan and Bashir, 2011). These valuable data will help researchers in detecting errors in genes that may develop to cancer in future.

5.2.3.2 Nanotubes

Nanocarbon tubes (CNT) have a cylindrical nanostructure with a diameter half of a DNA. The structural changes in DNA sequence (point mutation, deletion, duplication, insertion, etc.) and more importantly, the exact location of mutation can be detected by using CNT. In dentistry, in addition to oral cancer diagnostics, these tubes can be used for evaluation of dentin collagen network, dentine, and enamel sensitivity, oral biofilm and plaque, surface of dental implants, and comparison of postoperative results (Naderi et al., 2013).

5.2.3.3 Dendrimers

Dendrimers have a central core with polymer branching radially around the central core. Dendrimer nanoparticles have the ability to be used for both diagnostic and therapeutic purposes (Guo and Shi, 2012). One property of these star-like polymers is their wide surface area and thus a single dendrimer has enough places to carry a molecule that recognizes tumor cell and an anticancer drug to kill those cells. Dendrimer nanoparticles have the ability to be an ideal carrier for target drug delivery vehicles capable of targeting oral tumors.

5.2.3.4 Nanowire sensors

The oncology scientists have designed a sensor test chip that contains thousands of semiconductor nanowires (2–20 nm), and is able to detect biomarkers that are released by tumor cells at the early stages of tumor progression (Zheng et al., 2005). These semiconductor nanowires composed of silicon and other materials can also act as field-effect transistor devices. In oral medicine, these devices can detect proteins, DNA, viral particle, and so on, in tiny amounts of the salivary sample where conventional sensors are inefficient.

5.3 NANOROBOTICS

With the help of nanotechnology, scientists have designed amazing novel nanosized robots playing many roles in dentistry. Nanorobots can be used for inducing the analgesia, desensitizing the tooth, manipulation of the biofilm, and improving durability of teeth (Cavalcanti and Freitas, 2005). These intelligent nanorobots can make exact copies and replace or repair the damaged tissues. The function of nanorobotics can be controlled by the nanocomputers that have been previously programmed via acoustic signals (as are used in ultrasonography) (Fig. 25.4). Usually, gingiva is the suggested site for the injection of colloidal suspension containing millions of robots. Upon entry, they begin their journey to reach dental pulp via dentinal tubules, gingival sulcus, and lamina dura or perform their function in other areas as ordered by the clinician (Shetty et al., 2013). For example, in order to apply anesthesia, the colloidal suspension containing millions of active analgesic nanorobots is injected into the gingival tissue that responds to input supplied by the dentist (Agarwal, 2012). Once in the pulp cavity, they apply the analgesia by blocking the nerve impulses in any tooth that requires treatment. After

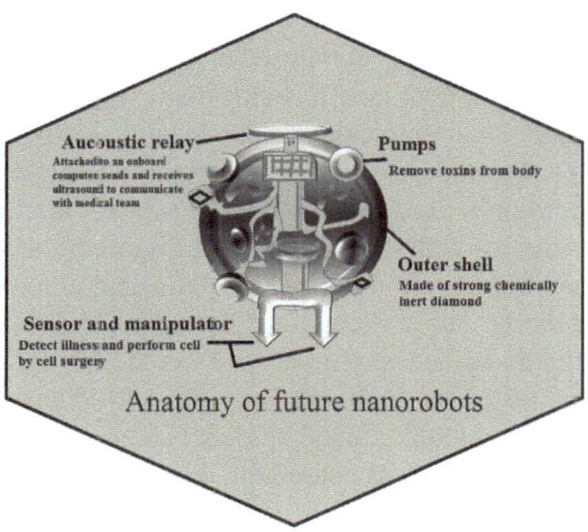

FIGURE 25.4 Schematic Picture of a Future Nanorobot

completion of treatment, the dentist shuts down the analgesic effect of the nanorobots and subsequently the dental sense will return. Thus, this anesthetic approach is fast-acting and reversible, with no side effects associated with its use. Nanorobots can also be used in the future mouthwashes or toothpastes (dentifrobots) (Ozak and Ozkan, 2013). When these products are used and left on the occlusal surfaces of teeth, nanorobots can remove the organic residues and odorless vapors and prevent the calculus accumulation by continuous debridement. These nanorobots, which can move as fast as 1–10 mμ/s, are simultaneously deactivated after swallowing.

6 APPLICATION OF NANOMEDICINE IN ORAL BONE REGENERATION

From past to present, there is a big concern for maxillofacial surgeons to repair and reconstruct the critical-sized oral bone defects caused by trauma, infection, tumor resection, congenital deformity, as well as physical and pathological degeneration (Oryan et al., 2014b). In the last decades, many efforts have been made to promote oral bone regeneration by using bioactive nanomaterials and nanobiodegradable scaffolds (Bigham-Sadegh et al., 2013a; Parizi et al., 2013). Typically, there are several nanomaterilas that can be utilized to create bioactive nanostructured scaffolds fororal bone regeneration, including nanoceramics, nanopolymers, nanometals, and nanocomposites (Mazaheri et al., 2015). The ideal nanomaterial used in bone tissue engineering should be one that exhibits the following properties: induces osteoprogenitor cells to differentiate into osteoblasts and then osteocytes (osteoinductive), allows for the ingrowth of neovasculature and the infiltration of osteogenic precursor cells (osteoconductive), stimulates the mature bone cells to produce new bone (osteogenic), facilitates the connection between living bone and the surface of a load-bearing implant (osseointegrative), does not stimulate an inflammatory response inappropriate to its beneficial effect (biocompatible), degrades after completing its purpose and leaves no harmful components (biodegradable), be easily processed into the final product form, has an acceptable shelf-life and be easily sterilized (Bigham-Sadegh et al., 2013b; Oryan et al., 2014b).

6.1 NANOCERAMICS

Nanoceramic is a type of nanoparticle that is composed of ceramics. Nanoceramics are generally classified as inorganic, heat-resistant, and nonmetallic solids made of both metallic and nonmetallic compounds. Ceramic-based materials, including calcium phosphates (HAp, tricalcium phosphate [TCP]), metallic oxides (aluminum, Zr, Sr, and Ti), and glass ceramics (Bioglass, Ceravital, Cerabone, and Bioverit) are among those commonly used in bone TERM (Gao et al., 2014). These ceramics create strong bonds to hard and soft tissues with degradation and dissolving in physiological fluids, thereby improving osseointegration and bone graft incorporation. The degradation and solubility of these materials depend on their crystallinity and particle size. Therefore, degradation of many ceramics is a controllable function during the manufacturing process (Lv et al., 2013).

Application of nanoceramic materials, such as n-HAp and n-TCP have been shown to increase new bone formation in many experimental and clinical studies (Cao et al., 2012; Fukui et al., 2008; Pang et al., 2015). Since the n-HAp has excellent osteoinductive capacity and improves osseointegration in bone-implant interface, it is used in the field of implantology in recent years. Continuing interest in the n-HAp has prompted many TERM experts to look for new combinations that could improve existing materials or produce new ones that could regenerate oral bone completely. This has led to new complex compounds, such as combinations of n-HAp and zinc carbonate (Pepla et al., 2014) or the n-HAp associated with the polyamide 66 (Huang et al., 2011), which seems to be an excellent material for remineralization in the early dental lesions, and to reproduce a new periodontal membrane with improved properties, respectively. Because of the unique ability of n-HAp in chemically binding to bone, stimulating osteoblasts, and encouraging them for new bone formation without any toxicity or hazardous effect, it has been widely used in combination with other materials for oral and maxillofacial surgery (Pepla et al., 2014).

Up to now, n-HAp particles have been proved superior to their microsized particles by several investigation and findings (Eslaminejad et al., 2010, 2011; Heo et al., 2009; Mao et al., 2015; Nicoletti et al., 2015), representing the n-HAp as an ideal candidate to be applied in bone tissue engineering. In this regard, in vitro studies have shown that n-HAp particles (67 nm) significantly increase osteoblast adhesion and inhibit competitive adhesion of fibroblast compared to conventional 179 nm size HAp (Rezaie et al., 2015). Further examination revealed that nanoceramics significantly adsorbed greater quantities of vitronectine (a protein that enhances osteoblast adhesion, on the nanophase ceramic) while conventional ceramics adsorbed greater concentration of laminin (a protein that increases endothelial cells and fibroblasts adhesion). On this subject, an important question is that why the vitronectine has more affinity to attach to nanoceramics, while laminin is immobilized on conventional ceramics? Webster et al. (2000b) clarified this question. The pore size of the nanoceramics influences the interactions of serum proteins and subsequently affects the type of cell adhesion. It would be more accurate to say that the vitronectine, which is a 15 nm linear protein, may preferentially have been adsorbed to the tiny pores present in the nanoceramics, whereas laminin, a 70 nm cruciform shape protein, would have rather been adsorbed into the larger pores of conventional ceramics (macrosized bioceramics). Additionally, it has been shown that osteoclastic activities, including the synthesis of tartrate-resistant acid phosphatase (TRAP) and formation of resorption pits have also increased on n-HAp compared to micro-HAp (Zhang et al., 2014). The mentioned mechanisms illustrated two critical points in bone repair and regeneration: (1) nanoceramics, such as n-HAp have more capability than microsized ceramics to direct the regeneration mechanism into new bone formation, as well as inhibiting the fibrous callus formation. (2) Osteoclastic activity is enhanced in the following regeneration process by using nanoceramics, which has been known as one of the most substantial factors affecting the bone remodeling process. The in vivo studies have also confirmed these results. The experimental studies on rats demonstrated that n-HAp coated titanium and tantalum scaffolds accelerated new bone formation when used as osteoconductive coatings compared to the uncoated or conventional microsized HAp coated scaffold (Jiang et al., 2015; Pang et al., 2015).

In its granular form, HAp is currently used in clinical practice to reconstruct oral and periodontal bone defects. As a commercial product, HAp is used to fill oral bone defects after cystectomy, apicoectomy, and the loss of dental implants and also used toenhance the thickness of atrophic alveolar ridges. The porous HAp blocks are also used in maxillofacial surgery, as well as granular powder (Pepla et al., 2014). Actually, although bioceramics, such as HAp with grain size more than 100 nm (microparticle) are currently used for oral bone and dental implants, these materials sometimes exhibit inadequate apposition of bone, leading to implant failure. In summary, when the macrosized-bioceramicsare used in oral bone reconstruction, new bone formationis rarely found (Pilia et al., 2013). Fortunately, this problem has been solved by introduction of nanobioceramics, such as n-HAp and nanosized β-TCP powder and granules into the commercial products. The HAp nanoparticles that are especially used to treat the bone defects are available under the brand names including Ostim (Osartis GmbH & Co. KG, Ohernburg, Germany) and NanOSS (Angstrom Medica, USA) (Al Machot et al., 2014). These products can be used in cleft lip-cleft palate patients, osseous defect in periodontal surgeries and massive maxillary and mandibular injuries requiring bone graft.

Recently, a new promising finding regarding HAp nanoparticles, suggests this type of material for filling the defects that develop after tumor resection. In an in vitro study, when these two cultures (culture #1: bone marrow mesenchymal stem cells [BMMSC] and culture 2: osteosarcoma cells) were exposed to n-HAp in the range of 20–80 nm, greater cell viability and proliferation of mesenchymal stem cell wereseen especially in the case of 20 nm nanoparticle. In contrast, the greatest effect on osteosarcoma cell growth inhibition was happened by the smallest n-HAp particles (20 nm). Therefore, small-sized n-HAp can be beneficial for normal stem cells but can also be an inhibiting factor for abnormal cancerous cells (Chen et al., 2011). However, more studies are needed to elucidate such an important finding in animal models.

TCP, another CaP-based bioceramic, has commonly been used in bone regeneration with or without the combination of HAp. It is a calcium salt of phosphoric acid with the chemical formula $Ca_3\ (PO_4)_2$ and also known as bone phosphate of lime (BPL) (Podaropoulos et al., 2009). Recently, the nanosized β-TCP granules have been prepared from the wet chemically precipitated β-TCP in order to use in oral bone regeneration (Lee et al., 2016). The results showed that the osteoconductivity were higher at 75% porosity and 2.50 m^2/g specific surface area due to the nanosized particles of porous β-TCP granules. Furthermore, the histological analysis of dog mandibular defect showed that β-TCP granules demonstrated remarkable bone regeneration effect compared to that of the nontreated group at 12 weeks after implantation, indicating the increased new bone formation and rapid resorption without toxicologically significant changes. Combination of TCP and HAp nanophase ceramics is currently available in the market for dental and oral bone regeneration under the trade name VITOSSO (Orthovita, Inc., Great Valley Parkway Malvern, PA19355, USA) (Verma et al., 2010).

Bioactive glasses are the most applicable bioceramic agents in bone TERM.These popular bioceramics are considered as important osteoconductive and osteostimulative materials valuable for bone reconstruction, healing, and regeneration (Jones et al., 2010). The mechanism behind new bone formation on bioactive glass is closely related to the release of sodium and calcium ions and deposition of a carbonated HAp layer. A strong chemical bond between bioactive glass and host bone is formed by the apatite layer (Hench and Jones, 2015). Further studies have also revealed that the bioactive glass products containing calcium and silicon have a better bioactivity, and both Ca and Si ions have been found to stimulate osteoblast proliferation and differentiation. Among various bioactive glass products, the 45S5 (bioglass®) is still considered as the gold standard type, producing through the traditional melt method at high temperatures (Li et al., 2015). Almost 10 years ago, a new generation of bioactive glass, nanosized mesoporous bioglass (MBG) was developed. MBGs have a highly ordered mesoporous channel structure and a highly specific surface area. In general, MBGs offer a suite of characteristics that are important for oral bone regeneration. The typical characteristics include: (1) they have a great in vitro and in vivo bioactivity. Particularly, they have shown the stimulatory effect to increase in vivo new bone formation, compared with conventional bioactive glasses. (2) They are excellent carriers in delivery of several agents, such as antimicrobial drugs, bioactive small molecules, and growth factors, which have been provided a great applicable option for utilizing in osteomyelitis, as well as stimulating the growth and differentiation of the new bone (Wu and Chang, 2012; Wu et al., 2011).

6.2 NANOMETALS

Typically, orthopedic implants, such as screws, plates, rods, and nails are mainly fabricated using metals and metallic materials (e.g., stainless steel and titanium alloys). To date, in oral bone TERMs, although many works have been conducted on nanoceramics and their composites, several recent studies have focused to assessregeneration efficacy of nanophase metals (McMahon et al., 2013; Zhang et al., 2011). Due to their high surface area, roughness, energy, and load-bearing ability, nanophase metals can be used in bone tissue engineering. In this regard, nanophase titanium, Ti6Al4V and CoCrMo, have been found to significantly enhance osteoblast adhesion, and proliferation and additionally synthesis of ECM proteins (alkaline phosphates and collagen) when compared to conventional metals (McMahon et al., 2013). An in vitro study speculated that increased osteoblast adhesion may be due to the more grain boundaries at the surface of nanophase metals than conventional ones (Webster and Ejiofor, 2004). It is expected to enhance the results when carbon nanofibers are incorporated with nanophase metals.

6.3 NANOPOLYMERS

Synthetic (e.g., polyglycolic acid (PGA), poly(lactide-*co*-glycolide) (PLGA), poly-L-lactide (PLLA), polylactic acid (PLA), polycaprolactone (PCL), and natural (e.g., gelatin, silk, collagen, chitosan) polymers can be easily fabricated (Meimandi-Parizi et al., 2013a,b,c; Moshiri et al., 2015). The nanostructure of such polymers let them to have more biocompatibility and biodegradability compared to the microscale polymers and thus these nanopolymers may be valuable options for bone TERM (Moshiri et al., 2014a,b; Oryan et al., 2014c; Oryan et al., 2015b). Furthermore, these polymers can be modified chemically to improve their properties for bone regenerative mechanisms.Generally, the nanopolymers have been used for fabrication of bioactive nanocomposite scaffolds in order to promote bone regeneration via the following routes: first, nanopolymers can increase the chemical and mechanical properties of final products through providing a basic platform forcarrying the bioactive materials, such as nanoceramics and nanophase metals. Second, nanopolymers, particularly the synthetic forms can be used as a delivery vehicle for bone TERM (e.g., PLGA nanosphere and nanocapsule).

6.4 NANOCOMPOSITES AND NANOSCAFFOLDS

In fact, bone tissue is a biological nanocomposite with a hierarchical structure ranging from the microscale to the nanoscale and composed of organic (collagen type I) and inorganic (n-HAp crystal) components. Thus, the fabricated nanocomposites represent promising platforms in bone TERM with a capacity to mimic the architecture of natural ECM and bone (biomimetic effect). The inherent features of nanocomposites, such as increased toughness, wettability, roughness, and enhanced surface area can also provide a mechanically reinforced lightweight bone implant through increased protein adsorption (e.g., vitronectine and fibronectin), nutrient exchange, and porosity. Until now, several studies have been conducted to produce a reliable and efficient nanocomposite for bone regeneration (Nejati et al., 2009; Zandi et al., 2010). These nanocomposites have been made by several types of nanomaterials via different fabrication procedures. These include polymer/polymer (e.g., gelatin-PLGA) (Thomson et al., 1995), polymers/ceramics (e.g., PLLA-n-HAP) (Wei and Ma, 2008), polymers/metals (e.g., PLGA-TiO_2) (Castillo-Dali et al., 2014), metals/ceramics (e.g., titanium-45S5 bio-glass) (Jurczyk et al., 2011), ceramic/ceramic (e.g., TCP-n-HAp) (Pilia et al., 2013), metal/metal (e.g., SiO_2-TiO_2) (Naga et al., 2014), and polymers/metals/ceramics (e.g., n-HAp-TiO_2-PCL) (Gao et al., 2014) nanocomposite models.

Fabrication of nanocomposite scaffolds, actually provide a structural support for the cells, as well as the possibility to provide adelivery system for drugs (antibacterial, anticancer, antiresorptive, etc.), growth factors (TGF-β, IGF-1, BMPs, and PDGF), and genes (BMPs, RUNX2, osterix, and TGF-β),while changes to the nanoscale level of tissue hierarchy may have significant effects on cell adhesion, function, and integrin-triggered signaling pathways (Oryan et al., 2014a; Oryan et al., 2015a).

Current modalities employed in fabrication of the nanocomposite scaffolds include solvent casting, electrospinning, melt molding, gas foaming, freeze drying, phase-separation, porogen leaching, rapid prototyping, fiber mesh, and self-assembly (Smith et al., 2009). The application, advantages, and disadvantages of these techniques have been listed in Table 25.7.

Table 25.7 The Protocols, Applications, Advantages and Disadvantages of Different Scaffold Fabrication Techniques

Fabrication Method	Protocol and Application	Advantage	Disadvantage
Electrospinning	It uses electric force to draw charged threads of polymer melts up to nanofiber. This method is used for fabrication of the nanofibrous scaffolds from natural or synthetic polymers	Controlling the porosity, pore size, and fiber diameter	Limited mechanical property, pore size decrease with fiber thickness. Cannot produce 3D scaffolds
Self-assembly	The self-assembly process is as an autonomous organization of biological molecules, which can be induced by noncovalent bonds or weak covalent interactions, such as electrostatic, van der Waals, hydrophobic interactions, ionic, and hydrogen bonds. It has been utilized in fabrication of various nanofiberous scaffolds	Control over porosity, pore size, and fiber diameter	Expensive material, complex design parameters
Gas foaming	This process involves exposing the highly porous polymers with CO_2 at 800 psi pressure to saturate the polymer with gas. Under this condition, the dissolved CO_2 becomes unstable and in order to minimize the free energy, becomes cluster; and this causes pore nucleation. After completion of foaming process, the final scaffold is formed	Free of harsh organic solvents, control over porosity, and pore size	Limited mechanical property, inadequate pore interconnectivity, may yield a scaffold with nonporous external surface
Solvent casting	This technique is totally based on evaporation of some solvent in order to form scaffolds. In this method, the polymeric solution is added into a mold and provides sufficient time to evaporate the solvent that create a layer of polymeric membrane	Control over porosity, pore size, and crystallinity	Limited mechanical property, residual solvents, and porogen material, only is used to produce thin membranes up to 3 mm thick

Table 25.7 The Protocols, Applications, Advantages and Disadvantages of Different Scaffold Fabrication Techniques (*Cont.*)

Fabrication Method	Protocol and Application	Advantage	Disadvantage
Phase separation	The phase separation process takes place when a homogeneous polymer solution becomes thermodynamically unstable under certain temperature conditions and tends to separate into a multiphase system domain, comprising a polymer-lean phase and a polymer-rich phase. Subsequently, the polymer-rich phase solidifies to form a matrix (scaffold base) while the polymer-lean phase turns into pores as the result of solvent removal	No decrease in the activity of the molecule	Difficult to precisely control the scaffold morphology
Melt molding	Melt molding involves filling a mold with polymer powder and a porogen and then heating them while applying pressure to the mixture. During the fabrication process, the polymer/ceramic or metals and porogen materials are bind together in order to form a scaffold. Once the mold was removed and the porogen component was leached out, the porous scaffold forms by providing sufficient time to dry	This approach is developed to control the pore size, pore geometry, pore interconnectivity, and spatial distribution of pores, which are important for cells to exchange the nutrient/waste from pore to pore	Require high temperature that preclude the ability of incorporation of the bioactive molecules, possibility of residual porogen
Rapid prototyping(RP)	It is a computer controlled fabrication technique that can rapidly produce 3D objects in a layer by layer fashion. RP technique generally comprises the design of scaffold model by using the computer aided design (CAD) software and computer aided manufacturing (CAM) programs, which is then expressed as a series of cross sections. The RP processes that are used in fabrication of the tissue scaffold include stereo lithography (SLA), selective laser sintering (SLS), fused deposition modeling (FDM), 3D printing, and 3D plotting	Excellent control over geometry, porosity, no supporting material required	Limited polymer type, highly expensive equipment
Fiber mesh	This method consists of individual fibers either woven or knitted into 3D patterns of variable pore sizes. They can be prepared by deposition of polymer solution over a nonwoven mesh of another polymer followed by subsequent evaporation	Large surface area for cell attachment, rapid nutrient diffusion	Lack the structural stability

(*Continued*)

Table 25.7 The Protocols, Applications, Advantages and Disadvantages of Different Scaffold Fabrication Techniques (*Cont.*)

Fabrication Method	Protocol and Application	Advantage	Disadvantage
Freeze drying	Polymer (natural or synthetic or their combination (hybrid form) is first dissolved in a solvent to form a desired polymer concentration. The solution is then frozen and the solvent is removed through sublimation process under the vacuum, and a high porous scaffold is fabricated	Use of ice crystals instead of an organic solvent in the scaffold production process, is more suitable for biomedical applications, high temperature and separate leaching step not required	Cannot fabricate hierarchical structures (e.g., vascularized systems), small pore size, and long processing time
Porogen leaching	This approach is commonly used in fabrication of scaffolds with controlled porosity. The porogen leaching method is totally based on the dispersion of porogen (salt, sugar and wax) either in liquid particulates or powdered materials. Through the process of evaporation, cross linking occurs and the porogens are removed by leaching leading to formation of pores in the porogen domain	Increased the porosity, controlled over porosity and pore geometry	Inadequate pore size and pore interconnectivity

7 NANOTECHNOLOGY AND CELL THERAPY IN ORAL BONE REGENERATION

ECM is a collection of extracellular molecules secreted by the cells that span over different length scales, with strands of collagen fibrils ranging from 35 to 250 nm and a length that can extend over the microscale. Actually, ECM have a 3D nanomicrofiber structure, which is composed of nano- to microscale fibers, holes, and protuberances (Goldberg et al., 2007). At the nano- and microscale interactions with the ECM components, expose the cell to a variety of chemical moieties and passive mechanical forces that can influence cell behaviors. This complex process especially can control the cell behavior within stem cell niche and the cell fate determination is achieved via such well-choreographed spatiotemporal dialog between stem cells and their ECM (as a micro/nanoenvironment) (Gong et al., 2015).

Within a few decades, the nanotechnology has changed the shape of cell therapy in different medical fields. Recently, cell-based therapies are promising new therapeutic methods in bone TERM (Baghaban Eslaminejad and Malakooty Poor, 2014; Ghahramanpoor et al., 2011; Zomorodian and Baghaban Eslaminejad, 2012). The mesenchymal stem cells (MSCs) require a structural framework because of the following reasons: (1) the framework supports the MSCs and provide a suitable platform for exchanging the nutrients and waste products, as well as essential biomaterials in order to facilitate cellular growth, proliferation, and differentiation. In general, the progenitor and stem cells are undifferentiated biological cells that can differentiate into various types of cells (e.g., myocyte, adipocyte, osteocyte, chondrocyte, osteoblast) offering hopes not only for the treatment of diseases but also for complex tissue engineered procedures, such as surgical treatments or tissue/organ transplantation (Eslaminejad and Nadri, 2009; Khojasteh et al., 2008; Jafarian et al., 2008). As the field of TERM progresses, the need for novel synthetic ECMs (bioscaffolds) inspired by tissue-specific niches for programmed stem cell fate has become of paramount importance. The advent of nanotechnology in the field of cell therapy has now made it possible for researchers to design and fabricate a suitable platform (nanobased scaffold) to support the stem cells and to gain valuable insights into their biomechanics. Moreover, different bioinspired and biomimetic nanostructures (nanofibers, nanotubes, nanoparticles, nanospheres, nanopores, nanocomposites, and hydrogel) have recently been employed to control stem cell differentiation specifically into chondroblast and osteoblast for bone tissue regeneration (Malakooty Poor et al., 2014).

In bone injuries, TERM with the help of nanotechnology has designed many scaffolds to regenerate bone tissue in a different approach by using stem cells. These attempts for fabrication of nanobased scaffolds include selecting biomaterials that are present in the native bone (e.g., n-HAp and collagen fibers), seeding various sources of stem cells on the bioscaffolds and incorporating them with growth promotive factors (e.g., BMPs) to provide a biomimetic niche for stimulating bone repair. The effects of nanotopographical density on the morphology and differentiation of hMSCs has previously been demonstrated (Faghihi and Baghaban Eslaminejad, 2014). The formation of cytoskeleton is necessary for the shape effect on the stem cell differentiation through the Rho-associated protein kinase (ROCK)-pathway. ROCK is a subtype of the serine-threonine kinase, involving mainly in the regulation of cell shape and movement by acting on the cytoskeleton. Synthetic bioscaffolds have different physical structures, subsequently exerting different forces to cells. Thus, these various cell tensions stimulate ROCK expression and affect the cell differentiation into osteoblast (Zhao et al., 2014).

Recently, the effect of nanotopography geometrics have been examined on MSCs differentiations and many studies have proven the role of various nanotopographical cues, such as feature size and degree of ordering to direct stem cell fate. In the recent years, the elongation of cell shape has been recognized as a crucial factor for the stem cell differentiation (Gong et al., 2015). Regarding this matter, a study used a nanogrooved platform for stem cells in three nanotopographical patterns (1:1, 1:3, and 1:5 width/spacing ratio, width = 550 nm). Interestingly, it has been found that the hMSCs with the sparser nanogrooved pattern elongated and orientated along the direction of nanogrooves. Furthermore, the effect of nanotopographical density on the osteogenesis of hMSCs was significant at the 1:1 and 1:3 nanogrooved patterns, but not significant at 1:5. Similar to nanogroove pattern, it was found that the symmetry and order of the nanopits significantly affect the osteopontin and osteocalcin expression (two bone-specific ECM proteins). An increased osteogenicity has been evident in hMSCs cultured on this pattern with either completely ordered or completely random features (You et al., 2012).

Collectively, a common point has been highlighted by previous studies, which have been conducted on mechano-sensitivity of stem cells. All studies has explicitly or implicitly suggested that the cell shape and cytoskeleton contractility are involved in regulation of the mechanosensitivity of stem cells and subsequently controlling the direction of the cell fate (Gong et al., 2015; Guilak et al., 2009; Sun et al., 2012). It has also been suggested that the force balance is very important along the mechanical axis of the ECM–integrin–cytoskeleton linkage. Moreover, strong evidence suggested that the stem cell differentiation into the distinct lineages could be maximized if the stem cells were cultured in the mechanical micro- and nanoenvironment mimicking their tissue elasticity in vivo. Additionally, modification of surface properties at the nanometer scale including surface texture, geometry, roughness, spatial position, and height, might potentially change clustering of the integrins, development of focal adhesions, and cytoskeletal structure, thus influencing the osteogenic differentiation to the surface (Pachauri et al., 2014).

8 CURRENT CHALLENGES AND FUTURE DIRECTIONS

Over the past decade, nanotechnology has taken big steps forward in fabrication of bioscaffolds, which have been manufactured from nanofibers, nanotubes, nanoparticles, and other nanostructures in order to better mimic the natural ECM. These novel bioactive scaffolds have recently emerged as promising candidates that resemble the ECM and efficiently regenerate dental tissues or replace bone defects (Fig. 25.5). However, combination of these nanomaterials for fabrication of nanobased bioscaffolds is still under investigation. The design of stronger and stiffer scaffold materials, especially for bone TERM applications, requires incorporation of a hierarchical design encompassing many length scales from the nanolevel to generate strength (i.e., to mimic composite deformation of n-HAp and collagen fiber), as well as microscale structures to influence the crack path and generate toughness (e.g., to mimic osteons and cement lines). Thus, nanotechnology cannot be the final answer for these purposes alone and several limitations in processing techniques have hampered the progress of bioscaffold fabrication. The real challenge is finding a way to use different technologies in combination with nanomaterials.

Incorporation of nanoparticles in bioscaffolds can increase the osteoblast functions, such as adhesion, proliferation, and differentiation. Nanoceramics, as well as coatings and other filler materials are popular types of bone substitutes due to their beneficial abilities to promote bone regeneration. Synthetic and natural nanopolymers are also known as excellent candidates to be applied in dental and bone tissue engineering because of their biodegradability-, biocompatibility- and substance delivery properties and also due to the ease of fabrication. However, hybrid biomaterials have enhanced structural and functional properties and the nanoscale designing let to improve the effects of such biomaterials in order to effectively improve bone regeneration.

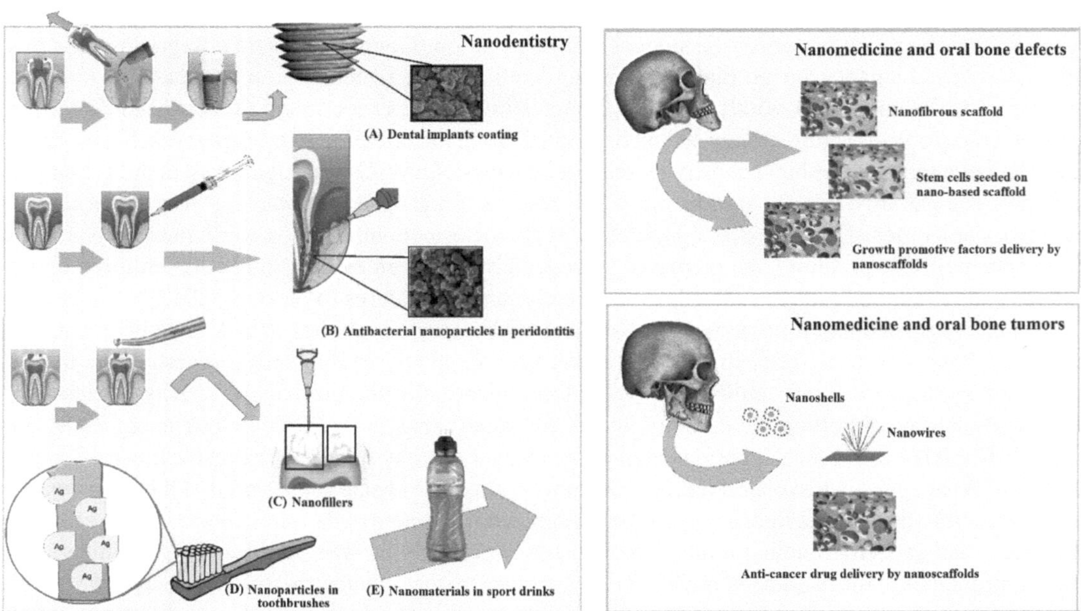

FIGURE 25.5 Some Applications of Nanomedicine in Dental and Oral Bone Disorders

Features or components of the nanostructured bioscaffolds varies from micro- to nanoscales, so that a wide range of interactions can be stimulated at the cellular level. In some instances, some of those interactions can also cause serious side effects, such as toxicity, which can raise serious concerns. In particular, at some instances degraded byproducts of the implanted nanomaterials or their aggregations may result in cell toxicity, such as the toxicity of heavy metals (iron, nickel, and cobalt catalysts) that has been repeatedly reported to date (Sharifi et al., 2012). Nonetheless, limited data exist concentrating on the toxicity of the nanomaterials on viable cells, tissues, and organs. Further investigations need to illustrate the role of nanomaterials on human health and environment and finally to elucidate whether nanoparticles should be used in bioengineering applications without fearing the side effects. In addition, future trends should try to better biomimic the natural bone healing process by fabricating new biomimetic nanostructured bioscaffolds that can encourage stem cell recruitment into the defect site. Although it is so difficult to resemble natural processes, recent scientific findings show potential to recruit endogenous stem cells and osteoblasts by using different factors (angiogenic/hypoxic inductive factors, growth factors, small molecules, etc.). Proper choosing of scaffold nanomaterials, as well as their geometry, pore size, pore interconnectivity, distribution pattern, nanomaterial size, and drug delivery properties at a desired rate will play critical roles in future development of the oral bone scaffolds.

Through the growing emphasis on nanomedicine, numerous possibilities have been opened now to modify the chemical and structural properties of bioscaffold with an unprecedented degree of control. For the first time, nanotechnology provides tools to manipulate the cellular micro-nanoenvironment and to monitor key cellular events at molecular level. However, the final role of nanotechnology in oral bone tissue engineering and its integration in design and production of an ideal oral bone scaffold, implant is still unclear. To reach such purposes, the following scientific questions should be addressed in the future investigations.

(1) What are the most effective nanomaterials in dental and oral bone regeneration and which of the many available composites has the best properties in this matter? (2) What are the exact biodegradation rates of different nanomaterials and their composites in vivo? Which situations and factors particularly affect these rates? (3) What are the optimum biomechanical features (toughness, stiffness, surface elasticity, etc.) of the synthetic dental and oral bone implants? (4) Which of the synthetic methods (chemical, physical, biological, or combination of them) of the nanomaterials are more applicable, cost effective, and less hazardous for biomedical applications? (5) What is the best scaffold design for oral bone regenerative applications (porous, dense, biphasic)? (6) What is the optimum scaffold permeability (pore size, shape, and interconnectivity)? Does it depend on the composition and specific application? (7) What are the best nanostructures for delivery of various agents? Which of the many available agents is more practical and cost benefit for using in these structures in oral bone regeneration? (8) How the methodologies can be adopted for recruiting endogenous stem cells or osteoblastic/odontoblastic precursor cells into the target site. (9) What is the best approach to improve the osseointegration of surrounding tissues with nanofabricated products (implants or scaffolds)? (10) What is the exact mechanism of surface nanotopography and chemistry in cell proliferation and differentiation and how they control cell responses? (Particularly, the differentiation of stem cells towards odontoblastic or osteoblastic lineages.) (11) What processes must be developed to shift from generating small amounts of bone to treating critical-sized bone defects?

Answers to these critical questions will yield information needed to prepare an important guideline for the clinicians who work on oral health care. The final goal of any effort in this regard is to fabricate a "real active bio-product." Such productions will completely integrate with oral bone tissues while they are actively resorbed or remodeled in a programmed manner with controlled osteogenic activity. The real active bioproducts finally result in faster oral bone formation, reduced healing time, and rapid restoration of normal function.

REFERENCES

Abe, Y., Ishii, M., Takeuchi, M., Ueshige, M., Tanaka, S., Akagawa, Y., 2004. Effect of saliva on an antimicrobial tissue conditioner containing silver-zeolite. J. Oral Rehab. 31, 568–573.

Abiodun-Solanke, I.M.F., Ajayi, D.M., Arigbede, A.O., 2014. Nanotechnology and its application in dentistry. Ann. Med. Health Sci. Res. 4, 171–177.

Abou Neel, E.A., Bozec, L., Perez, R.A., Kim, H.-W., Knowles, J.C., 2015. Nanotechnology in dentistry: prevention, diagnosis, and therapy. Int. J. Nanomed. 10, 6371–6394.

Adhikari, U., Ghosh, A., Chandra, G., 2013. Nano particles of herbal origin: a recent eco-friend trend in mosquito control. Asian Pac. J. Trop. Dis. 3, 167–168.

Aeran, H., Kumar, V., Uniyal, S., Tanwer, P., 2015. Nanodentistry: is just a fiction or future. J. Oral Biol. Cranio. Res. 5, 207–211.

Agarwal, A., 2012. The future of anaesthesiology. Ind. J. Anaesth. 56, 524–528.

Ahmad, A., Senapati, S., Khan, M.I., Kumar, R., Ramani, R., Srinivas, V., Sastry, M., 2003. Intracellular synthesis of gold nanoparticles by a novel alkalotolerant actinomycete, *Rhodococcus* species. Nanotechnology 14, 824.

Ajitha, B., Ashok Kumar Reddy, Y., Sreedhara Reddy, P., 2015. Enhanced antimicrobial activity of silver nanoparticles with controlled particle size by pH variation. Powder Technol. 269, 110–117.

Akasaka, T., Nakata, K., Uo, M., Watari, F., 2009. Modification of the dentin surface by using carbon nanotubes. Biomed. Mater. Eng. 19, 179–185.

Al-Hamdan, K., Al-Moaber, S.H., Junker, R., Jansen, J.A., 2011. Effect of implant surface properties on peri-implant bone healing: a histological and histomorphometric study in dogs. Clin. Oral Implant. Res. 22, 399–405.

Al Machot, E., Hoffmann, T., Lorenz, K., Khalili, I., Noack, B., 2014. Clinical outcomes after treatment of periodontal intrabony defects with nanocrystalline hydroxyapatite (ostim) or enamel matrix derivatives (emdogain): a randomized controlled clinical trial. BioMed Res. Int. 2014, 786353.

Althues, H., Henle, J., Kaskel, S., 2007. Functional inorganic nanofillers for transparent polymers. Chem. Soc. Rev. 36, 1454–1465.

Amemiya, Y., Arakaki, A., Staniland, S.S., Tanaka, T., Matsunaga, T., 2007. Controlled formation of magnetite crystal by partial oxidation of ferrous hydroxide in the presence of recombinant magnetotactic bacterial protein Mms6. Biomaterials 28, 5381–5389.

Armendariz, V., Herrera, I., Peralta-Videa, J., Jose-Yacaman, M., Troiani, H., Santiago, P., Gardea-Torresdey, J., 2004. Size controlled gold nanoparticle formation by Avena sativa biomass: use of plants in nanobiotechnology. J. Nano. Res. 6, 377–382.

Aydin Sevinc, B., Hanley, L., 2010. Antibacterial activity of dental composites containing zinc oxide nanoparticles. J. Biomed. Mater. Res. B 94, 22–31.

Baghaban Eslaminejad, M., Malakooty Poor, E., 2014. Mesenchymal stem cells as a potent cell source for articular cartilage regeneration. World J. Stem. Cells 6, 344–354.

Bagheri, R., Burrow, M.F., Tyas, M.J., 2007. Surface characteristics of aesthetic restorative materials: an SEM study. J. Oral Rehabil. 34, 68–76.

Baldi, D., Izzotti, A., Bonica, P., Pera, P., Pulliero, A., 2009. Degenerative periodontal-diseases and oral osteonecrosis: the role of gene-environment interactions. Mutat. Res. 667, 118–131.

Balshaw, D.M., Philbert, M., Suk, W.A., 2005. Research strategies for safety evaluation of nanomaterials. Part III: nanoscale technologies for assessing risk and improving public health. Toxicol. Sci. 88, 298–306.

Bansal, V., Rautaray, D., Bharde, A., Ahire, K., Sanyal, A., Ahmad, A., Sastry, M., 2005. Fungus-mediated biosynthesis of silica and titania particles. J. Mater. Chem. 15, 2583–2589.

Basavaraj, K.H., 2012. Nanotechnology in medicine and relevance to dermatology: present concepts. Ind. J. Dermatol. 57, 169–174.

Besinis, A., De Peralta, T., Handy, R.D., 2014. The antibacterial effects of silver, titanium dioxide and silica dioxide nanoparticles compared to the dental disinfectant chlorhexidine on *Streptococcus mutans* using a suite of bioassays. Nanotoxicology 8, 1–16.

Besinis, A., De Peralta, T., Tredwin, C.J., Handy, R.D., 2015. Review of nanomaterials in dentistry: interactions with the oral microenvironment, clinical applications, hazards, and benefits. ACS Nano 9, 2255–2289.

Beyth, N., Yudovin-Farber, I., Perez-Davidi, M., Domb, A.J., Weiss, E.I., 2010. Polyethyleneimine nanoparticles incorporated into resin composite cause cell death and trigger biofilm stress in vivo. Proc. Natl. Acad. Sci. USA 107, 22038–22043.

Bhardwaj, A., Bhardwaj, A., Misuriya, A., Maroli, S., Manjula, S., Singh, A.K., 2014. Nanotechnology in dentistry: present and future. J. Int. Oral Health 6, 121–126.

Bhattacharya, R., Mukherjee, P., 2008. Biological properties of "naked" metal nanoparticles. Adv. Drug Deliv. Rev. 60, 1289–1306.

Bigham-Sadegh, A., Karimi, I., Alebouye, M., Shafie-Sarvestani, Z., Oryan, A., 2013a. Evaluation of bone healing in canine tibial defects filled with cortical autograft, commercial-DBM, calf fetal DBM, omentum and omentum-calf fetal DBM. J. Vet. Sci. 14, 337–343.

Bigham-Sadegh, A., Oryan, A., Mirshokraei, P., Shadkhast, M., Basiri, E., 2013b. Bone tissue engineering with periosteal-free graft and pedicle omentum. ANZ J. Surg. 83, 255–261.

Binulal, N.S., Deepthy, M., Selvamurugan, N., Shalumon, K.T., Suja, S., Mony, U., Jayakumar, R., Nair, S.V., 2010. Role of nanofibrous poly(caprolactone) scaffolds in human mesenchymal stem cell attachment and spreading for in vitro bone tissue engineering: response to osteogenic regulators. Tissue Eng. Part A 16, 393–404.

Boeckh, C., Schumacher, E., Podbielski, A., Haller, B., 2002. Antibacterial activity of restorative dental biomaterials in vitro. Caries Res. 36, 101–107.

Bonifácio, B.V., Da Silva, P.B., Ramos, M.A.D.S., Negri, K.M.S., Bauab, T.M., Chorilli, M., 2014. Nanotechnology-based drug delivery systems and herbal medicines: a review. Int. J. Nanomed. 9, 1–15.

Bose, S., Hochella, M.F., Gorby, Y.A., Kennedy, D.W., Mccready, D.E., Madden, A.S., Lower, B.H., 2009. Bioreduction of hematite nanoparticles by the dissimilatory iron reducing bacterium *Shewanella oneidensis* MR-1. Geochim. Cosmochim. Acta 73, 962–976.

Brennan, M.T., Kent, M.L., Fox, P.C., Norton, H.J., Lockhart, P.B., 2007. The impact of oral disease and nonsurgical treatment on bacteremia in children. J. Am. Dent. Assoc. 138, 80–85.

Cabral, M.F.C., Martinho, R.L.D.M., Guedes-Neto, M.V., Rebelo, M.A.B., Fontes, D.G., Cohen-Carneiro, F., 2015. Do conventional glass ionomer cements release more fluoride than resin-modified glass ionomer cements? Rest. Dent. Endo. 40, 209–215.

Canche-Escamilla, G., Duarte-Aranda, S., Toledano, M., 2014. Synthesis and characterization of hybrid silica/PMMA nanoparticles and their use as filler in dental composites. Mater. Sci. Eng. C 42, 161–167.

Cao, L., Duan, P.G., Wang, H.R., Li, X.L., Yuan, F.L., Fan, Z.Y., Li, S.M., Dong, J., 2012. Degradation and osteogenic potential of a novel poly(lactic acid)/nano-sized beta-tricalcium phosphate scaffold. Int. J. Nanomed. 7, 5881–5888.

Carrotte, P., 2004. Endodontics: part 9 calcium hydroxide, root resorption, endo-perio lesions. Br. Dent. J. 197, 735–743.

Castillo-Dali, G., Castillo-Oyague, R., Terriza, A., Saffar, J.L., Batista, A., Barranco, A., Cabezas-Talavero, J., Lynch, C.D., Barouk, B., Llorens, A., Sloan, A.J., Cayon, R.V., Gutierrez-Perez, J.L., Torres-Lagares, D., 2014. In vivo comparative model of oxygen plasma and nanocomposite particles on PLGA membranes for guided bone regeneration processes to be applied in pre-prosthetic surgery: a pilot study. J. Dent. 42, 1446–1457.

Cavalcanti, A., Freitas, JR., R.A., 2005. Nanorobotics control design: a collective behavior approach for medicine. IEEE Trans. Nanobiosci. 4, 133–140.

Chaudhury, K., Kumar, V., Kandasamy, J., Roychoudhury, S., 2014. Regenerative nanomedicine: current perspectives and future directions. Int. J. Nanomed. 9, 4153–4167.

Chauhan, R., Kumar, A., Abraham, J., 2013. A biological approach to the synthesis of silver nanoparticles with streptomyces sp jar1 and its antimicrobial activity. Scient. Pharm. 81, 607–621.

Chen, C.Y., Chung, Y.C., 2012. Antibacterial effect of water-soluble chitosan on representative dental pathogens *Streptococcus mutans* and *Lactobacilli brevis*. J. Appl. Oral Sci. 20, 620–627.

Chen, L., Mccrate, J.M., Lee, J.C.M., Li, H., 2011. The role of surface charge on the uptake and biocompatibility of hydroxyapatite nanoparticles with osteoblast cells. Nanotechnology 22, 105708.

Chen, L., Shen, H., Suh, B.I., 2012. Antibacterial dental restorative materials: a state-of-the-art review. Am. J. Dent. 25, 337–346.

Chen, V.J., Smith, L.A., Ma, P.X., 2006. Bone regeneration on computer-designed nano-fibrous scaffolds. Biomaterials 27, 3973–3979.

Chen, M., Yu, Q., Sun, H., 2013. Novel strategies for the prevention and treatment of biofilm related infections. Int. J. Mol. Sci. 14, 18488–18501.

Cheng, L., Weir, M.D., Zhang, K., Arola, D.D., Zhou, X., Xu, H.H., 2013. Dental primer and adhesive containing a new antibacterial quaternary ammonium monomer dimethylaminododecyl methacrylate. J. Dent. 41, 345–355.

Cheng, Y.J., Zeiger, D.N., Howarter, J.A., Zhang, X., Lin, N.J., Antonucci, J.M., Lin-Gibson, S., 2011. In situ formation of silver nanoparticles in photocrosslinking polymers. J. Biomed. Mater. Res. B 97, 124–131.

Cheng, L., Zhang, K., Weir, M.D., Melo, M.A.S., Zhou, X., Xu, H.H.K., 2015. Nanotechnology strategies for antibacterial and remineralizing composites and adhesives to tackle dental caries. Nanomedicine 10, 627–641.

Chiara, G., Letizia, F., Lorenzo, F., Edoardo, S., Diego, S., Stefano, S., Eriberto, B., Barbara, Z., 2012. Nanostructured biomaterials for tissue engineered bone tissue reconstruction. Int. J. Mol. Sci. 13, 737–757.

Chung, W.J., Chung, H.W., Shin, M.J., Lee, S.H., Lee, M.H., Lee, J.S., Kim, M.J., Lee, W.K., 2012. MRI to differentiate benign from malignant soft-tissue tumours of the extremities: a simplified systematic imaging approach using depth, size and heterogeneity of signal intensity. Br. J. Radiol. 85, e831–e836.

Clarke, B., 2008. Normal bone anatomy and physiology. Clin. J. Am. Soc. Nephrol. 3, S131–S139.

Coelho, P.G., Cardaropoli, G., Suzuki, M., Lemons, J.E., 2009. Early healing of nanothickness bioceramic coatings on dental implants: an experimental study in dogs. J. Biomed. Mater. Res. B V 88, 387–393.

Contado, C., 2015. Nanomaterials in consumer products: a challenging analytical problem. Front. Chem. 3, 48.

Cormode, D.P., Naha, P.C., Fayad, Z.A., 2014. Nanoparticle contrast agents for computed tomography: a focus on micelles. Cont. Media Mol. Imag. 9, 37–52.

Cramer, N.B., Stansbury, J.W., Bowman, C.N., 2011. Recent advances and developments in composite dental restorative materials. J. Dent. Res. 90, 402–416.

Dameron, C., Reese, R., Mehra, R., Kortan, A., Carroll, P., Steigerwald, M., Brus, L., Winge, D., 1989. Biosynthesis of cadmium sulphide quantum semiconductor crystallites. Nature 338, 596–597.

Daraio, C., Jin, S., 2012. Synthesis and patterning methods for nanostructures useful for biological applications. In: Silva, G.A., Parpura, V (Eds.), Nanotechnology for Biology and Medicine. Springer, New York.

Davidovich, E., Weiss, E., Fuks, A.B., Beyth, N., 2007. Surface antibacterial properties of glass ionomer cements used in atraumatic restorative treatment. J. Am. Dent. Assoc. 138, 1347–1352.

De Jong, W.H., Borm, P.J.A., 2008. Drug delivery and nanoparticles: applications and hazards. Int. J. Nanomed. 3, 133–149.

De Oliveira, G.U., Mondelli, R.F.L., Charantola Rodrigues, M., Franco, E.B., Ishikiriama, S.K., Wang, L., 2012. Impact of filler size and distribution on roughness and wear of composite resin after simulated toothbrushing. J. Appl. Oral Sci. 20, 510–516.

Deb, S., Ghosh, K., Shetty, S.D., 2015. Nanoimaging in cardiovascular diseases: current state of the art. Indian J. Med. Res. 141, 285–298.

Dipankar, C., Murugan, S., 2012. The green synthesis, characterization and evaluation of the biological activities of silver nanoparticles synthesized from *Iresine herbstii* leaf aqueous extracts. Colloids Surf. B 98, 112–119.
Dorozhkin, S.V., 2011a. Biocomposites and hybrid biomaterials based on calcium orthophosphates. Biomatter 1, 3–56.
Dorozhkin, S.V., 2011b. Calcium orthophosphates: occurrence, properties, biomineralization, pathological calcification and biomimetic applications. Biomatter 1, 121–164.
Drummond, J.L., 2008. Degradation, fatigue and failure of resin dental composite materials. J. Dent. Res. 87, 710–719.
Durán, N., Marcato, P.D., Conti, R.D., Alves, O.L., Costa, F.T.M., Brocchi, M., 2010. Potential use of silver nanoparticles on pathogenic bacteria, their toxicity and possible mechanisms of action. J. Braz. Chem. Soc. 21, 949–959.
Earl, J.S., Wood, D.J., Milne, S.J., 2009. Nanoparticles for dentine tubule infiltration: an in vitro study. J. Nanosci. Nanotechnol. 9, 6668–6674.
Ehlen, L.A., Marshall, T.A., Qian, F., Wefel, J.S., Warren, J.J., 2008. Acidic beverages increase the risk of in vitro tooth erosion. Nutr. Res. 28, 299–303.
El-Atwani, O., Hinks, J.A., Greaves, G., Gonderman, S., Qiu, T., Efe, M., Allain, J.P., 2014. In-situ TEM observation of the response of ultrafine- and nanocrystalline-grained tungsten to extreme irradiation environments. Sci. Rep. 4, 4716.
Elsalanty, M.E., Genecov, D.G., 2009. Bone grafts in craniofacial surgery. Cran. Trauma Recon. 2, 125–134
Epstein, J.B., Kish, R.V., Hallajian, L., Sciubba, J., 2015. Head and neck, oral, and oropharyngeal cancer: a review of medicolegal cases. Oral. Surg. Oral Med. Oral Pathol. Oral Radiol. 119, 177–186.
Eslami, N., Ahrari, F., Rajabi, O., Zamani, R., 2015. The staining effect of different mouthwashes containing nanoparticles on dental enamel. J. Clin. Exp. Dent. 7, e457–e461.
Eslaminejad, M.B., Bagheri, F., Zandi, M., Nejati, E., Zomorodian, E., Mivehchi, H., 2010. Comparison of proliferation and osteoblast differentiation of marrow-derived mesenchymal stem cells on nano-and micro-hydroxyapatite contained composite scaffolds. Iran J. Biotechnol. 8, 234–242.
Eslaminejad, M.B., Bagheri, F., Zandi, M., Nejati, E., Zomorodian, E., 2011. Study of mesenchymal stem cell proliferation and bone differentiation on composite scaffolds of PLLA and nano hydroxyapatite with different morphologies. Cell Journal (Yakhteh) 12, 469–476.
Eslaminejad, M.B., Nadri, S., 2009. Murine mesenchymal stem cell isolated and expanded in low and high density culture system: surface antigen expression and osteogenic culture mineralization. In Vitro Cell. Devel. Biol. 45, 451–459.
Faghihi, F., Baghaban Eslaminejad, M., 2014. The effect of nano-scale topography on osteogenic differentiation of mesenchymal stem cells. Biomed. Pap. Med. Fac. 158, 5–16.
Farah, S., Aviv, O., Laout, N., Ratner, S., Beyth, N., Domb, A.J., 2015. Quaternary ammonium polyethylenimine nanoparticles for treating bacterial contaminated water. Colloids Surf. B 128, 614–619.
Farrugia, C., Camilleri, J., 2015. Antimicrobial properties of conventional restorative filling materials and advances in antimicrobial properties of composite resins and glass ionomer cements: a literature review. Dent. Mater. 31, 10.
Fayaz, A.M., Balaji, K., Kalaichelvan, P., Venkatesan, R., 2009. Fungal based synthesis of silver nanoparticles: an effect of temperature on the size of particles. Colloids Surf. B 74, 123–126.
Flores, C.Y., Diaz, C., Rubert, A., Benítez, G.A., Moreno, M.S., Fernandez Lorenzo, D.E., Mele, M.A., Salvarezza, R.C., Schilardi, P.L., Vericat, C., 2010. Spontaneous adsorption of silver nanoparticles on Ti/TiC_2 surfaces: antibacterial effect on *Pseudomonas aeruginosa*. J. Colloid Interf. Sci. 350, 402–408.
Fu, J., Liu, W., Hao, Z., Wu, X., Yin, J., Panjiyar, A., Liu, X., Shen, J., Wang, H., 2014. Characterization of a low shrinkage dental composite containing bismethylene spiroorthocarbonate expanding monomer. Int. J. Mol. Sci. 15, 2400–2412.
Fuchigami, T., Kitamoto, Y., Namiki, Y., 2012. Size-tunable drug-delivery capsules composed of a magnetic nanoshell. Biomatter 2, 313–320.

Fukui, N., Sato, T., Kuboki, Y., Aoki, H., 2008. Bone tissue reaction of nano-hydroxyapatite/collagen composite at the early stage of implantation. Biomed. Mater. Eng. 18, 25–33.
Furno, F., Morley, K.S., Wong, B., Sharp, B.L., Arnold, P.L., Howdle, S.M., Bayston, R., Brown, P.D., Winship, P.D., Reid, H.J., 2004. Silver nanoparticles and polymeric medical devices: a new approach to prevention of infection? J. Antimicrob. Chemother. 54, 1019–1024.
Galler, K.M., D'souza, R.N., 2011. Tissue engineering approaches for regenerative dentistry. Regen. Med. 6, 111–124.
Gao, C., Deng, Y., Feng, P., Mao, Z., Li, P., Yang, B., Deng, J., Cao, Y., Shuai, C., Peng, S., 2014. Current progress in bioactive ceramic scaffolds for bone repair and regeneration. Int. J. Mol. Sci. 15, 4714–4732.
Ge, L., Li, Q., Wang, M., Ouyang, J., Li, X., Xing, M.M.Q., 2014. Nanosilver particles in medical applications: synthesis, performance, and toxicity. Int. J. Nanomed. 9, 2399–2407.
Gericke, M., Pinches, A., 2006. Biological synthesis of metal nanoparticles. Hydrometallurgy 83, 132–140.
Ghahramanpoor, M.K., Hassani Najafabadi, S.A., Abdouss, M., Bagheri, F., Baghaban Eslaminejad, M., 2011. A hydrophobically-modified alginate gel system: utility in the repair of articular cartilage defects. J. Mater. Sci. Mater. Med. 22, 2365–2375.
Goene, R.J., Testori, T., Trisi, P., 2007. Influence of a nanometer-scale surface enhancement on de novo bone formation on titanium implants: a histomorphometric study in human maxillae. Int. J. Period. Rest. Dent. 27, 211–219.
Gogna, R., Jagadis, S., Shashikal, K., 2011. A comparative in vitro study of microleakage by a radioactive isotope and compressive strength of three nanofilled composite resin restorations. J. Conserv. Dent. 14, 128–131.
Goldberg, M., Langer, R., Jia, X., 2007. Nanostructured materials for applications in drug delivery and tissue engineering. J. Biomater. Sci. Polym. Ed. 18, 241–268.
Gong, T., Xie, J., Liao, J., Zhang, T., Lin, S., Lin, Y., 2015. Nanomaterials and bone regeneration. Bone Res. 10, 15029.
Gu, W., Wu, C., Chen, J., Xiao, Y., 2013. Nanotechnology in the targeted drug delivery for bone diseases and bone regeneration. Int. J. Nanomed. 8, 2305–2317.
Guilak, F., Cohen, D.M., Estes, B.T., Gimble, J.M., Liedtke, W., Chen, C.S., 2009. Control of stem cell fate by physical interactions with the extracellular matrix. Cell. Stem Cell 5, 17–26.
Guo, R., Shi, X., 2012. Dendrimers in cancer therapeutics and diagnosis. Curr. Drug Metab. 13, 1097–1109.
Gurunathan, S., Raman, J., Malek, S.N.A., John, P.A., Vikineswary, S., 2013. Green synthesis of silver nanoparticles using Ganoderma neo-japonicum Imazeki: a potential cytotoxic agent against breast cancer cells. Int. J. Nanomed. 8, 4399–4413.
Guterres, S.S., Alves, M.P., Pohlmann, A.R., 2007. Polymeric nanoparticles, nanospheres and nanocapsules, for cutaneous applications. Drug Targ. Insights 2, 147–157.
Haghgoo, R., Rezvani, M.B., Salehi Zeinabadi, M., 2014. Comparison of nano-hydroxyapatite and sodium fluoride mouthrinse for remineralization of incipient carious lesions. J. Dent. 11, 406–410.
Hamal, J.D., Hensley, D.M., Maller, S.C., Palazzolo, D.J., Vandewalle, K.S., 2014. An in vitro comparison of antimicrobial toothbrushes. Gen. Dent. 62, e24–e27.
Hanke, B., Guggenbichler, P.J., 2004. Anti-microbial body care product. Google Patents.
Hannig, M., Hannig, C., 2010. Nanomaterials in preventive dentistry. Nat. Nanotechnol. 5, 565–569.
Hartgerink, J.D., Beniash, E., Stupp, S.I., 2001. Self-assembly and mineralization of peptide-amphiphile nanofibers. Science 294, 1684–1688.
He, F., Yang, G., Wang, X., Zhao, S., 2009. Effect of electrochemically deposited nanohydroxyapatite on bone bonding of sandblasted/dual acid-etched titanium implant. Int. J. Oral Maxillofac. Impl. 24, 790–799.
Hench, L.L., Jones, J.R., 2015. Bioactive glasses: frontiers and challenges. Front. Bioeng. Biotechnol. 3, 194.
Heo, S.J., Kim, S.E., Wei, J., Hyun, Y.T., Yun, H.S., Kim, D.H., Shin, J.W., Shin, J.W., 2009. Fabrication and characterization of novel nano- and micro-HA/PCL composite scaffolds using a modified rapid prototyping process. J. Biomed. Mater. Res. A 89, 108–116.

Horev, B., Klein, M.I., Hwang, G., Li, Y., Kim, D., Koo, H., Benoit, D.S., 2015. pH-activated nanoparticles for controlled topical delivery of farnesol to disrupt oral biofilm virulence. ACS Nano 9, 2390–2404.

Hosseinalipour, M., Javadpour, J., Rezaie, H., Dadras, T., Hayati, A.N., 2010. Investigation of mechanical properties of experimental Bis-GMA/TEGDMA dental composite resins containing various mass fractions of silica nanoparticles. J. Prosthodont. 19, 112–117.

Huang, H.-L., Chang, Y.-Y., Lai, M.-C., Lin, C.-R., Lai, C.-H., Shieh, T.-M., 2010. Antibacterial TaN-Ag coatings on titanium dental implants. Surf. Coat. Technol. 205, 1636–1641.

Huang, S.B., Gao, S.S., Yu, H.Y., 2009. Effect of nano-hydroxyapatite concentration on remineralization of initial enamel lesion in vitro. Biomed. Mater. 4, 034104.

Huang, D., Zuo, Y., Zou, Q., Zhang, L., Li, J., Cheng, L., Shen, J., Li, Y., 2011. Antibacterial chitosan coating on nano-hydroxyapatite/polyamide66 porous bone scaffold for drug delivery. J. Biomater. Sci. Polym. Ed. 22, 931–944.

Hubbs, A.F., Sargent, L.M., Porter, D.W., Sager, T.M., Chen, B.T., Frazer, D.G., Castranova, V., Sriram, K., Nurkiewicz, T.R., Reynolds, S.H., Battelli, L.A., Schwegler-Berry, D., Mckinney, W., Fluharty, K.L., Mercer, R.R., 2013. Nanotechnology: toxicologic pathology. Toxicol. Pathol. 41, 395–409.

Hulkoti, N.I., Taranath, T.C., 2014. Biosynthesis of nanoparticles using microbes: a review. Colloids Surf. B 121, 474–483.

Hussmann, B., Johann, I., Kauther, M.D., Landgraeber, S., Jäger, M., Lendemans, S., 2013. Measurement of the silver ion concentration in wound fluids after implantation of silver-coated megaprostheses: correlation with the clinical outcome. BioMed Res. Int. 2013, 763096.

Ilie, N., Hickel, R., 2011. Resin composite restorative materials. Aust. Dent. J. 1, 59–66.

Imazato, S., 2009. Bio-active restorative materials with antibacterial effects: new dimension of innovation in restorative dentistry. Dent. Mater. J. 28, 11–19.

Jafarian, M., Eslaminejad, M.B., Khojasteh, A., Abbas, F.M., Dehghan, M.M., Hassanizadeh, R., Houshmand, B., 2008. Marrow-derived mesenchymal stem cells-directed bone regeneration in the dog mandible: a comparison between biphasic calcium phosphate and natural bone mineral. Oral Surg. Oral Med. Oral Pathol. Oral Radiol. Endodontol. 105, e14–e24.

Jain, P., Pradeep, T., 2005. Potential of silver nanoparticle-coated polyurethane foam as an antibacterial water filter. Biotechnol. Bioeng. 90, 59–63.

James, R., Toti, U.S., Laurencin, C.T., Kumbar, S.G., 2011. Electrospun nanofibrous scaffolds for engineering soft connective tissues. Method. Mol. Biol. 726, 243–258.

Jha, R.K., Jha, P.K., Chaudhury, K., Rana, S.V.S., Guha, S.K., 2014. An emerging interface between life science and nanotechnology: present status and prospects of reproductive healthcare aided by nano-biotechnology. Nano Rev. 5, 10.

Jha, A.K., Prasad, K., 2010. Ferroelectric BaTiO 3 nanoparticles: biosynthesis and characterization. Colloids Surf. B 75, 330–334.

Jha, A.K., Prasad, K., Kulkarni, A., 2009. Synthesis of TiO 2 nanoparticles using microorganisms. Colloids Surf. B 71, 226–229.

Jiang, Q.H., Gong, X., Wang, X.X., He, F.M., 2015. Osteogenesis of rat mesenchymal stem cells and osteoblastic cells on strontium-doped nanohydroxyapatite-coated titanium surfaces. Int. J. Oral Maxillofac. Impl. 30, 461–471.

Jiao, Z.-H., Li, M., Feng, Y.-X., Shi, J.-C., Zhang, J., Shao, B., 2014. Hormesis effects of silver nanoparticles at non-cytotoxic doses to human hepatoma cells. PLoS One 9, e102564.

Jones, J.R., Lin, S., Yue, S., Lee, P.D., Hanna, J.V., Smith, M.E., Newport, R.J., 2010. Bioactive glass scaffolds for bone regeneration and their hierarchical characterisation. Proc. Inst. Mech. Eng. H 224, 1373–1387.

Jose, M.V., Thomas, V., Xu, Y., Bellis, S., Nyairo, E., Dean, D., 2010. Aligned bioactive multi-component nanofibrous nanocomposite scaffolds for bone tissue engineering. Macromol. Biosci. 10, 433–444.

Jurczyk, M.U., Jurczyk, K., Miklaszewski, A., Jurczyk, M., 2011. Nanostructured titanium-45S5 Bioglass scaffold composites for medical applications. Mater. Des. 32, 4882–4889.

Karabela, M.M., Sideridou, I.D., 2011. Synthesis and study of properties of dental resin composites with different nanosilica particles size. Dent. Mater. 27, 825–835.

Kasraei, S., Sami, L., Hendi, S., Alikhani, M.-Y., Rezaei-Soufi, L., Khamverdi, Z., 2014. Antibacterial properties of composite resins incorporating silver and zinc oxide nanoparticles on *Streptococcus mutans* and *Lactobacillus*. Rest. Dent. Endodon. 39, 109–114.

Khojasteh, A., Eslaminejad, M.B., Nazarian, H., 2008. Mesenchymal stem cells enhance bone regeneration in rat calvarial critical size defects more than platelete-rich plasma. Oral Surg. Oral Med. Oral Pathol. Oral Radiol. Endodontol. 106, 356–362.

Kim, J.S., Kuk, E., Yu, K.N., Kim, J.H., Park, S.J., Lee, H.J., Kim, S.H., Park, Y.K., Park, Y.H., Hwang, C.Y., Kim, Y.K., Lee, Y.S., Jeong, D.H., Cho, M.H., 2007. Antimicrobial effects of silver nanoparticles. Nanomedicine 3, 95–101.

Kim, E.-S., Ahn, E.H., Dvir, T., Kim, D.-H., 2014. Emerging nanotechnology approaches in tissue engineering and regenerative medicine. Int. J. Nanomed. 9, 1–5.

Kirmizis, D., Logothetidis, S., 2010. Atomic force microscopy probing in the measurement of cell mechanics. Int. J. Nanomed. 5, 137–145.

Kloepfer, J.A., Mielke, R.E., Wong, M.S., Nealson, K.H., Stucky, G., Nadeau, J.L., 2003. Quantum dots as strain- and metabolism-specific microbiological labels. Appl. Environ. Microbiol. 69, 4205–4213.

Kobayashi, T., Itoh, S., Nakamura, S., Nakamura, M., Shinomiya, K., Yamashita, K., 2007. Enhanced bone bonding of hydroxyapatite-coated titanium implants by electrical polarization. J. Biomed. Mater. Res. A 82, 145–151.

Komatsu, O., Nishida, H., Sekino, T., Yamamoto, K., 2014. Application of titanium dioxide nanotubes to tooth whitening. Nano. Biomed. 6, 63–72.

Konishi, Y., Ohno, K., Saitoh, N., Nomura, T., Nagamine, S., Hishida, H., Takahashi, Y., Uruga, T., 2007. Bioreductive deposition of platinum nanoparticles on the bacterium *Shewanella algae*. J. Biotechnol. 128, 648–653.

Kowshik, M., Deshmukh, N., Vogel, W., Urban, J., Kulkarni, S.K., Paknikar, K., 2002. Microbial synthesis of semiconductor CdS nanoparticles, their characterization, and their use in the fabrication of an ideal diode. Biotechnol. Bioeng. 78, 583–588.

Kulkarni, N., Muddapur, U., 2014. Biosynthesis of metal nanoparticles: a review. J. Nanotechnol. 2014, 8.

Kumar, P.S., Kumar, S., Savadi, R.C., John, J., 2011. Nanodentistry: a paradigm shift-from fiction to reality. J. Indian Prosthod. Soc. 11, 1–6.

Kuroda, K., Okido, M., 2012. Hydroxyapatite coating of titanium implants using hydroprocessing and evaluation of their osteoconductivity. Bioinorgan. Chem. Appl. 2012, 7.

Lal, S., Clare, S.E., Halas, N.J., 2008. Nanoshell-enabled photothermal cancer therapy: impending clinical impact. Acc. Chem. Res. 41, 1842–1851.

Lamsal, K., Kim, S.W., Jung, J.H., Kim, Y.S., Kim, K.S., Lee, Y.S., 2011. Application of silver nanoparticles for the control of colletotrichum species in vitro and pepper anthracnose disease in field. Mycobiology 39, 194–199.

Lansdown, A.B., 2006. Silver in health care: antimicrobial effects and safety in use. Curr. Probl. Dermatol. 33, 17–34.

Laroui, H., Rakhya, P., Xiao, B., Viennois, E., Merlin, D., 2013. Nanotechnology in diagnostics and therapeutics for gastrointestinal disorders. Dig. Liver Dis. 45, 995–1002.

Lee, J.-H., Han, J., Choi, H., Hur, H.-G., 2007. Effects of temperature and dissolved oxygen on Se (IV) removal and Se (0) precipitation by *Shewanella* sp. HN-41. Chemosphere 68, 1898–1905.

Lee, W., Kim, K.J., Lee, D.G., 2014. A novel mechanism for the antibacterial effect of silver nanoparticles on *Escherichia coli*. Biometals 27, 1191–1201.

Lee, D.S., Pai, Y., Chang, S., Kim, D.H., 2016. Microstructure, physical properties, and bone regeneration effect of the nano-sized beta-tricalcium phosphate granules. Mater. Sci. Eng. C 58, 971–976.

Lee, H.H., Sang Shin, U., Lee, J.H., Kim, H.W., 2011. Biomedical nanocomposites of poly(lactic acid) and calcium phosphate hybridized with modified carbon nanotubes for hard tissue implants. J. Biomed. Mater. Res. B 98, 246–254.

Lelli, M., Putignano, A., Marchetti, M., Foltran, I., Mangani, F., Procaccini, M. Roveri, N., Orsini, G., 2014. Remineralization and repair of enamel surface by biomimetic Zn-carbonate hydroxyapatite containing toothpaste: a comparative in vivo study. Front. Physiol. 5, 333.

Li, X., Kolltveit, K.M., Tronstad, L., Olsen, I., 2000. Systemic diseases caused by oral infection. Clin. Microbiol. Rev. 13, 547–558.

Li, X., Xu, H., Chen, Z.-S., Chen, G., 2011. Biosynthesis of nanoparticles by microorganisms and their applications. J. Nanomater. 2011, 16.

Li, W., Ding, Y., Yu, S., Yao, Q., Boccaccini, A.R., 2015. Multifunctional chitosan-45s5 bioactive glass-poly(3-hydroxybutyrate-*co*-3-hydroxyvalerate) microsphere composite membranes for guided tissue/bone regeneration. ACS Appl. Mater. Interf. 7, 20845–20854.

Lin, A., Wang, C.J., Kelly, J., Gubbi, P., Nishimura, I., 2009. The role of titanium implant surface modification with hydroxyapatite nanoparticles in progressive early bone-implant fixation in vivo. Int. J. Oral Maxillofac. Impl. 24, 808–816.

Lohbauer, U., Wagner, A., Belli, R., Stoetzel, C., Hilpert, A., Kurland, H.D., Grabow, J., Muller, F.A., 2010. Zirconia nanoparticles prepared by laser vaporization as fillers for dental adhesives. Acta Biomate. 6, 4539–4546.

Loo, C., Lin, A., Hirsch, L., Lee, M.H., Barton, J., Halas, N., West, J., Drezek, R., 2004. Nanoshell-enabled photonics-based imaging and therapy of cancer. Technol. Cancer Res. Treat. 3, 33–40.

Lu, Z., Rong, K., Li, J., Yang, H., Chen, R., 2013. Size-dependent antibacterial activities of silver nanoparticles against oral anaerobic pathogenic bacteria. J. Mater. Sci. Mater. Med. 24, 1465–1471.

Lv, Q., Deng, M., Ulery, B.D., Nair, L.S., Laurencin, C.T., 2013. Nano-ceramic composite scaffolds for bioreactor-based bone engineering. Clin. Ortho. Rel. Res. 471, 2422–2433.

Ma, P.X., 2008. Biomimetic materials for tissue engineering. Adv. Drug Deliv. Rev. 60, 184–198.

Malakooty Poor, E., Baghaban Eslaminejad, M., Gheibi, N., Bagheri, F., Atyabi, F., 2014. Chitosan-pDNA nanoparticle characteristics determine the transfection efficacy of gene delivery to human mesenchymal stem cells. Artif. Cells Nanomed. Biotechnol. 42, 376–384.

Mantri, S.S., Mantri, S.P., 2013. The nano era in dentistry. J. Nat. Sci. Biol. Med. 4, 39–44.

Mao, L., Liu, J., Zhao, J., Chang, J., Xia, L., Jiang, L., Wang, X., Lin, K., Fang, B., 2015. Effect of micro-nano-hybrid structured hydroxyapatite bioceramics on osteogenic and cementogenic differentiation of human periodontal ligament stem cell via Wnt signaling pathway. Int. J. Nanomed. 10, 7031–7044.

Marchi, J., Ribeiro, C., Bressiani, A.H.D.A., Marques, M.M., 2013. Cell response of calcium phosphate based ceramics, a bone substitute material. Mater. Res. 16, 703–712.

Markowitz, K., 2009. The original desensitizers: strontium and potassium salts. J. Clin. Dent. 20, 145–151.

Mazaheri, M., Eslahi, N., Ordikhani, F., Tamjid, E., Simchi, A., 2015. Nanomedicine applications in orthopedic medicine: state of the art. Int. J. Nanomed. 10, 6039–6054.

McCarroll, J., Teo, J., Boyer, C., Goldstein, D., Kavallaris, M., Phillips, P.A., 2014. Potential applications of nanotechnology for the diagnosis and treatment of pancreatic cancer. Front. Physiol. 5, 2.

McMahon, R.E., Wang, L., Skoracki, R., Mathur, A.B., 2013. Development of nanomaterials for bone repair and regeneration. J. Biomed. Mater. Res. B 101, 387–397.

Medda, S., Hajra, A., Dey, U., Bose, P., Mondal, N., 2015. Biosynthesis of silver nanoparticles from Aloe vera leaf extract and antifungal activity against *Rhizopus* sp. and *Aspergillus* sp. Appl. Nanosci. 5, 875–880.

Meimandi-Parizi, A., Oryan, A., Moshiri, A., 2013a. Role of tissue engineered collagen based tridimensional implant on the healing response of the experimentally induced large Achilles tendon defect model in rabbits: a long term study with high clinical relevance. J. Biomed. Sci. 20, 28.

Meimandi-Parizi, A., Oryan, A., Moshiri, A., 2013b. Tendon tissue engineering and its role on healing of the experimentally induced large tendon defect model in rabbits: a comprehensive in vivo study. PLoS One 8, e73016.

Meimandi-Parizi, A., Oryan, A., Moshiri, A., Silver, I.A., 2013c. Novel application of a tissue-engineered collagen-based three-dimensional bio-implant in a large tendon defect model: a broad-based study with high value in translational medicine. Tissue Cell 45, 282–294.

Mendes, V.C., Moineddin, R., Davies, J.E., 2007. The effect of discrete calcium phosphate nanocrystals on bone-bonding to titanium surfaces. Biomaterials 28, 4748–4755.

Miglani, S., Aggarwal, V., Ahuja, B., 2010. Dentin hypersensitivity: recent trends in management. J. Conserv. Dent. 13, 218–224.

Min, J.H., Kwon, H.K., Kim, B.I., 2011. The addition of nano-sized hydroxyapatite to a sports drink to inhibit dental erosion: in vitro study using bovine enamel. J. Dent. 39, 629–635.

Mitragotri, S., Anderson, D.G., Chen, X., Chow, E.K., Ho, D., Kabanov, A.V., Karp, J.M., Kataoka, K., Mirkin, C.A., Petrosko, S.H., Shi, J., Stevens, M.M., Sun, S., Teoh, S., Venkatraman, S.S., Xia, Y., Wang, S., Gu, Z., Xu, C., 2015. Accelerating the translation of nanomaterials in biomedicine. ACS Nano 9, 6644–6654.

Moghimi, S.M., Hunter, A.C., Murray, J.C., 2005. Nanomedicine: current status and future prospects. Faseb J. 19, 311–330.

Mohamed Hamouda, I., 2012. Current perspectives of nanoparticles in medical and dental biomaterials. J. Biomed. Res. 26, 143–151.

Mortazavi, V., Nahrkhalaji, M.M., Fathi, M.H., Mousavi, S.B., Esfahani, B.N., 2010. Antibacterial effects of sol-gel-derived bioactive glass nanoparticle on aerobic bacteria. J. Biomed. Mater. Res. A 94, 160–168.

Moshaverinia, A., Ansari, S., Movasaghi, Z., Billington, R.W., Darr, J.A., Rehman, I.U., 2008. Modification of conventional glass-ionomer cements with *N*-vinylpyrrolidone containing polyacids, nano-hydroxy and fluoroapatite to improve mechanical properties. Dent. Mater. 24, 1381–1390.

Moshiri, A., Oryan, A., Meimandi-Parizi, A., 2015. Synthesis, development, characterization and effectiveness of bovine pure platelet gel-collagen-polydioxanone bioactive graft on tendon healing. J. Cell. Mol. Med. 19, 1308–1332.

Moshiri, A., Oryan, A., Meimandi-Parizi, A., Koohi-Hosseinabadi, O., 2014a. Effectiveness of xenogenous-based bovine-derived platelet gel embedded within a three-dimensional collagen implant on the healing and regeneration of the Achilles tendon defect in rabbits. Exp. Opin. Biol. Ther. 14, 1065–1089.

Moshiri, A., Oryan, A., Meimandiparizi, A., Maffulli, N., 2014b. Collagen implants in experimental tendon injury in rabbits: a clinical, ultra-structural and biomechanical investigation. J. Biol. Reg. Homeost. Agents 28, 381–397.

Motamedian, S.R., Hosseinpour, S., Ahsaie, M.G., Khojasteh, A., 2015. Smart scaffolds in bone tissue engineering: a systematic review of literature. World J. Stem Cells 7, 657–668.

Mousavinasab, S.M., 2011. Biocompatibility of composite resins. Dent. Res. J. 8, S21–S29.

Nadagouda, M.N., Varma, R.S., 2008. Green synthesis of ag and pd nanospheres, nanowires, and nanorods using vitamin : catalytic polymerisation of aniline and pyrrole. J. Nanomat. 2008, 8.

Naderi, N., Madani, S.Y., Ferguson, E., Mosahebi, A., Seifalian, A.M., 2013. Carbon nanotubes in the diagnosis and treatment of malignant melanoma. Anticancer Agents Med. Chem. 13, 171–185.

Naga, S.M., El-Kady, A.M., El-Maghraby, H.F., Awaad, M., Detsch, R., Boccaccini, A.R., 2014. Novel porous Al_2O_3-SiO_2-TiO_2 bone grafting materials: formation and characterization. J. Biomater. Appl. 28, 813–824.

Najibfard, K., Ramalingam, K., Chedjieu, I., Amaechi, B.T., 2011. Remineralization of early caries by a nano-hydroxyapatite dentifrice. J. Clin. Dent. 22, 139–143.

Nam, K.Y., 2011. In vitro antimicrobial effect of the tissue conditioner containing silver nanoparticles. J. Adv. Prosthodont. 3, 20–24.

Nejati, E., Firouzdor, V., Eslaminejad, M.B., Bagheri, F., 2009. Needle-like nano hydroxyapatite/poly(L-lactide acid) composite scaffold for bone tissue engineering application. Mater. Sci. Eng. C 29, 942–949.

Nicoletti, A., Torricelli, P., Bigi, A., Fornasari, P., Fini, M., Moroni, L., 2015. Incorporation of nanostructured hydroxyapatite and poly(*N*-isopropylacrylamide) in demineralized bone matrix enhances osteoblast and human mesenchymal stem cell activity. Biointerphases 10, 041001.

Oredugba, F.A., 2006. Use of oral health care services and oral findings in children with special needs in Lagos, Nigeria. Spec. Care Dent. 26, 59–65.

Oryan, A., Alidadi, S., Moshiri, A., Bigham-Sadegh, A., 2014a. Bone morphogenetic proteins: a powerful osteoinductive compound with non-negligible side effects and limitations. Biofactors 40, 459–481.

Oryan, A., Alidadi, S., Moshiri, A., Maffulli, N., 2014b. Bone regenerative medicine: classic options, novel strategies, and future directions. J. Ortho. Surg. Res. 9, 1–27.

Oryan, A., Moshiri, A., Meimandi-Parizi, A., 2014c. In vitro characterization of a novel tissue engineered based hybridized nano and micro structured collagen implant and its in vivo role on tenoinduction, tenoconduction, tenogenesis and tenointegration. J. Mater. Sci. Mater. Med. 25, 873–897.

Oryan, A., Kamali, A., Moshiri, A., 2015a. Potential mechanisms and applications of statins on osteogenesis: current modalities, conflicts and future directions. J. Control. Release 215, 12–24.

Oryan, A., Moshiri, A., Meimandi-Parizi, A., Maffulli, N., 2015b. Role of xenogenous bovine platelet gel embedded within collagen implant on tendon healing: an in vitro and in vivo study. Exp. Biol. Med. 240, 194–210.

Ozak, S.T., Ozkan, P., 2013. Nanotechnology and dentistry. Eur. J. Dent. 7, 145–151.

Pachauri, P., Bathala, L.R., Sangur, R., 2014. Techniques for dental implant nanosurface modifications. J. Adv. Prosthod. 6, 498–504.

Pal, S., Tak, Y.K., Song, J.M., 2007. Does the antibacterial activity of silver nanoparticles depend on the shape of the nanoparticle? A study of the Gram-negative bacterium *Escherichia coli*. Appl. Env. Microbiol. 73, 1712–1720.

Pandian, S.R.K., Deepak, V., Kalishwaralal, K., Muniyandi, J., Rameshkumar, N., Gurunathan, S., 2009. Synthesis of PHB nanoparticles from optimized medium utilizing dairy industrial waste using Brevibacterium casei SRKP2: a green chemistry approach. Colloids Surf. B 74, 266–273.

Pang, K.M., Lee, J.K., Seo, Y.K., Kim, S.M., Kim, M.J., Lee, J.H., 2015. Biologic properties of nano-hydroxyapatite: an in vivo study of calvarial defects, ectopic bone formation and bone implantation. Biomed. Mater. Eng. 25, 25–38.

Parizi, A.M., Oryan, A., Shafiei-Sarvestani, Z., Bigham-Sadegh, A., 2013. Effectiveness of synthetic hydroxyapatite versus Persian Gulf coral in an animal model of long bone defect reconstruction. J. Orthop. Traumatol.V 14, 259–268.

Park, K., 2007. Nanotechnology: what it can do for drug delivery. J. Control. Release 120, 1–3.

Patil, M., Mehta, D.S., Guvva, S., 2008. Future impact of nanotechnology on medicine and dentistry. J. Indian Soc. Periodontol. 12, 34–40.

Pattani, V.P., Tunnell, J.W., 2012. Nanoparticle-mediated photothermal therapy: a comparative study of heating for different particle types. Lasers Surg. Med. 44, 675–684.

Pepla, E., Besharat, L.K., Palaia, G., Tenore, G., Migliau, G., 2014. Nano-hydroxyapatite and its applications in preventive, restorative and regenerative dentistry: a review of literature. Ann. Stomatol. 5, 108–114.

Perán, M., García, M.A., López-Ruiz, E., Bustamante, M., Jiménez, G., Madeddu, R., Marchal, J.A., 2012. Functionalized nanostructures with application in regenerative medicine. Int. J. Mol. Sci. 13, 3847–3886.

Peters, M.C., 2010. Strategies for noninvasive demineralized tissue repair. Dent. Clin. North Am. 54, 507–525.

Phelippeau, A., Pommier, S., Prioul, C., Clavel, M., 2003. Heavily drawn eutectoid steel: a nanostructured material. In: Tsakalakos, T. et al., (Ed.), Nanostructures: Synthesis, Functional Properties and Applications. Springer, Netherlands.

Pilia, M., Guda, T., Appleford, M., 2013. Development of composite scaffolds for load-bearing segmental bone defects. BioMed. Res. Int. 2013, 458253.

Pinon-Segundo, E., Ganem-Quintanar, A., Alonso-Perez, V., Quintanar-Guerrero, D., 2005. Preparation and characterization of triclosan nanoparticles for periodontal treatment. Int. J. Pharm. 294, 217–232.

Pisanic, 2ND, T.R., Zhang, Y., Wang, T.H., 2014. Quantum dots in diagnostics and detection: principles and paradigms. Analyst 139, 2968–2981.

Podaropoulos, L., Veis, A.A., Papadimitriou, S., Alexandridis, C., Kalyvas, D., 2009. Bone regeneration using beta-tricalcium phosphate in a calcium sulfate matrix. J. Oral Implantol. 35, 28–36.
Ponarulselvam, S., Panneerselvam, C., Murugan, K., Aarthi, N., Kalimuthu, K., Thangamani, S., 2012. Synthesis of silver nanoparticles using leaves of *Catharanthus roseus* Linn. G. Don and their antiplasmodial activities. Asian Pac. J. Trop. Biomed. 2, 574–580.
Popovtzer, R., Agrawal, A., Kotov, N.A., Popovtzer, A., Balter, J., Carey, T.E., Kopelman, R., 2008. Targeted gold nanoparticles enable molecular CT imaging of cancer. Nano Lett. 8, 4593–4596.
Porter, A.L., Youtie, J., 2009. How interdisciplinary is nanotechnology? J. Nano. Res. 11, 1023–1041.
Prabhakar, A.R., Kumar, S., 2010. Antibacterial effect of bioactive glass in combination with powdered enamel and dentin. Indian J. Dent. Res. 21, 30–34.
Prentice, L.H., Tyas, M.J., Burrow, M.F., 2006. The effect of ytterbium fluoride and barium sulphate nanoparticles on the reactivity and strength of a glass-ionomer cement. Dent. Mater. 22, 746–751.
Radice, S., Kern, P., Burki, G., Michler, J., Textor, M., 2007. Electrophoretic deposition of zirconia-Bioglass composite coatings for biomedical implants. J. Biomed. Mater. Res. A 82, 436–444.
Rai, M.K., Deshmukh, S.D., Ingle, A.P., Gade, A.K., 2012. Silver nanoparticles: the powerful nanoweapon against multidrug-resistant bacteria. J. Appl. Microbiol. 112, 841–852.
Reuveni, T., Motiei, M., Romman, Z., Popovtzer, A., Popovtzer, R., 2011. Targeted gold nanoparticles enable molecular CT imaging of cancer: an in vivo study. Int. J. Nanomed. 6, 2859–2864.
Reynolds, E.C., 1998. Anticariogenic complexes of amorphous calcium phosphate stabilized by casein phosphopeptides: a review. Spec. Care Dent. 18, 8–16.
Rezaie, H.R., Bakhtiari, L., Öchsner, A., 2015. Biomaterials and Their Applications. Springer, Berlin.
Richards, N.D., Antonucci, J.M., 2007. The effects of two monofunctional diluent monomers and two photoinitiator systems on the properties of UDMA-based composites. Am. J. Dent. 20, 147–152.
Rosa, M.B., Albrektsson, T., Francischone, C.E., Schwartz Filho, H.O., Wennerberg, A., 2012. The influence of surface treatment on the implant roughness pattern. J. Appl. Oral Sci. 20, 550–555.
Rycenga, M., Cobley, C.M., Zeng, J., Li, W., Moran, C.H., Zhang, Q., Qin, D., Xia, Y., 2011. Controlling the synthesis and assembly of silver nanostructures for plasmonic applications. Chem. Rev. 111, 3669–3712.
Sabel, N., 2012. Enamel of primary teeth: morphological and chemical aspects. Swed. Dent. J. Suppl. 222, 1–77.
Sadat-Shojai, M., Atai, M., Nodehi, A., Khanlar, L.N., 2010. Hydroxyapatite nanorods as novel fillers for improving the properties of dental adhesives: synthesis and application. Dent. Mater. 26, 471–482.
Saengmee-anupharb, S., Srikhirin, T., Thaweboon, B., Thaweboon, S., Amornsakchai, T., Dechkunakorn, S., Suddhasthira, T., Kamaguchi, A., 2013. Antimicrobial effects of silver zeolite, silver zirconium phosphate silicate and silver zirconium phosphate against oral microorganisms. Asian Pac. J. Trop. Biomed. 3, 47–52.
Saha, M., 2009. Nanomedicine: promising tiny machine for the healthcare in future: a review. Oman Med. J. 24, 242–247.
Saini, R., Saini, S., Sharma, S., 2010. Nanotechnology: the future medicine. J. Cutan. Aesth. Surg. 3, 32–33.
Saini, M., Singh, Y., Arora, P., Arora, V., Jain, K., 2015. Implant biomaterials: a comprehensive review. World J. Clin. Cases 3, 52–57.
Saion, E., Gharibshahi, E., Naghavi, K., 2013. Size-controlled and optical properties of monodispersed silver nanoparticles synthesized by the radiolytic reduction method. Int. J. Mol. Sci. 14, 7880–7896.
Sajanlal, P.R., Sreeprasad, T.S., Samal, A.K., Pradeep, T., 2011. Anisotropic nanomaterials: structure, growth, assembly, and functions. Nano Rev. 2, 10.
Samir, A., Elgamal, B.M., Gabr, H., Sabaawy, H.E., 2015. Nanotechnology applications in hematological malignancies: a review. Oncol. Rep. 34, 1097–1105.
Sanghi, R., Verma, P., 2009. A facile green extracellular biosynthesis of CdS nanoparticles by immobilized fungus. Chem. Eng. J. 155, 886–891.
Sanyal, A., Rautaray, D., Bansal, V., Ahmad, A., Sastry, M., 2005. Heavy-metal remediation by a fungus as a means of production of lead and cadmium carbonate crystals. Langmuir 21, 7220–7224.

Sartoretto, S.C., Alves, A.T.N.N., Resende, R.F.B., Calasans-Maia, J., Granjeiro, J.M., Calasans-Maia, M.D., 2015. Early osseointegration driven by the surface chemistry and wettability of dental implants. J. Appl. Oral Sci. 23, 279–287.

Sasani, N., Vahdati Khaki, J., Mojtaba Zebarjad, S., 2014. Characterization and nanomechanical properties of novel dental implant coatings containing copper decorated-carbon nanotubes. J. Mech. Behav. Biomed. Mater. 37, 125–132.

Schmidt, C.W., 2009. Nanotechnology-related environment, health, and safety research: examining the national strategy. Env. Health Persp. 117, A158–A161.

Schouten, C., Meijer, G.J., Van Den Beucken, J.J., Leeuwenburgh, S.C., De Jonge, L.T., Wolke, J.G., Spauwen, P.H., Jansen, J.A., 2010. In vivo bone response and mechanical evaluation of electrosprayed CaP nanoparticle coatings using the iliac crest of goats as an implantation model. Acta Biomater. 6, 2227–2236.

Senapati, S., Mandal, D., Ahmad, A., 2004. Fungus mediated synthesis of silver nanoparticles: a novel biological approach. Indian J. Phys. 78, 101–105.

Senapati, S., Ahmad, A., Khan, M.I., Sastry, M., Kumar, R., 2005. Extracellular biosynthesis of bimetallic Au–Ag alloy nanoparticles. Small 1, 517–520.

Sevinç, B.A., Hanley, L., 2010. Antibacterial activity of dental composites containing zinc oxide nanoparticles. J. Biomed. Mater. Res. B 94, 22–31.

Sharifi, S., Behzadi, S., Laurent, S., Forrest, M.L., Stroeve, P., Mahmoudi, M., 2012. Toxicity of nanomaterials. Chem. Soc. Rev. 41, 2323–2343.

Shetty, N.J., Swati, P., David, K., 2013. Nanorobots: future in dentistry. Saudi Dent. J. 25, 49–52.

Shi, J., Dong, L.L., He, F., Zhao, S., Yang, G.L., 2013. Osteoblast responses to thin nanohydroxyapatite coated on roughened titanium surfaces deposited by an electrochemical process. Oral Surg. Oral Med. Oral Pathol. Oral Radiol. 116, 26.

Shrestha, A., Hamblin, M.R., Kishen, A., 2014. Photoactivated rose bengal functionalized chitosan nanoparticles produce antibacterial/biofilm activity and stabilize dentin-collagen. Nanomedicine 10, 491–501.

Shvero, D.K., Davidi, M.P., Weiss, E.I., Srerer, N., Beyth, N., 2010. Antibacterial effect of polyethyleneimine nanoparticles incorporated in provisional cements against *Streptococcus mutans*. J. Biomed. Mater. Res. B 94, 367–371.

Singaravelu, G., Arockiamary, J., Kumar, V.G., Govindaraju, K., 2007. A novel extracellular synthesis of monodisperse gold nanoparticles using marine alga, *Sargassum wightii* Greville. Colloids Surf. B 57, 97–101.

Singh, P., Kim, Y.J., Wang, C., Mathiyalagan, R., Yang, D.C., 2015. The development of a green approach for the biosynthesis of silver and gold nanoparticles by using Panax ginseng root extract, and their biological applications. Artif. Cells Nanomed. Biotechnol. 14, 1–8.

Skrtic, D., Antonucci, J.M., 2011. Bioactive polymeric composites for tooth mineral regeneration: physicochemical and cellular aspects. J. Funct. Biomater. 2, 271–307.

Smith, I.O., Liu, X.H., Smith, L.A., Ma, P.X., 2009. Nano-structured polymer scaffolds for tissue engineering and regenerative medicine. Nanomed. Nanobiotechnol. 1, 226–236.

Sneha, K., Sathishkumar, M., Mao, J., Kwak, I., Yun, Y.-S., 2010. Corynebacterium glutamicum-mediated crystallization of silver ions through sorption and reduction processes. Chem. Eng. J. 162, 989–996.

Stassi, S., Canavese, G., Cauda, V., Marasso, S.L., Pirri, C.F., 2012. Evaluation of different conductive nanostructured particles as filler in smart piezoresistive composites. Nano. Res. Lett. 7, 327–1327.

Steffens, C., Leite, F.L., Bueno, C.C., Manzoli, A., Herrmann, P.S., 2012. Atomic force microscopy as a tool applied to nano/biosensors. Sensors 12, 8278–8300.

Subramani, K., Ahmed, W., 2011. Emerging Nanotechnologies in Dentistry: Processes, Materials and Applications. William Andrew; Elsevier, USA.

Sukirtha, R., Priyanka, K.M., Antony, J.J., Kamalakkannan, S., Thangam, R., Gunasekaran, P., Krishnan, M., Achiraman, S., 2012. Cytotoxic effect of green synthesized silver nanoparticles using *Melia azedarach* against in vitro HeLa cell lines and lymphoma mice model. Proc. Biochem. 47, 273–279.

Sun, Y., Chen, C.S., Fu, J., 2012. Forcing stem cells to behave: a biophysical perspective of the cellular microenvironment. Ann. Rev. Biophys. 41, 519–542.

Sureka, B., Thukral, B.B., Mittal, M.K., Mittal, A., Sinha, M., 2013. Radiological review of pleural tumors. Indian J. Radiol. Imag. 23, 313–320.

Sussman, E.M., Jayanti, P., Dair, B.J., Casey, B.J., 2015. Assessment of total silver and silver nanoparticle extraction from medical devices. Food Chem. Toxicol. 85, 10–19.

Svanborg, L.M., Hoffman, M., Andersson, M., Currie, F., Kjellin, P., Wennerberg, A., 2011. The effect of hydroxyapatite nanocrystals on early bone formation surrounding dental implants. Int. J. Oral Maxillofac. Surg. 40, 308–315.

Swarup, J.S., Rao, A., 2012. Enamel surface remineralization: using synthetic nanohydroxyapatite. Contemp. Clin. Dent. 3, 433–436.

Tay, F.R., Pashley, D.H., 2008. Guided tissue remineralisation of partially demineralised human dentine. Biomaterials 29, 1127–1137.

Teow, Y., Asharani, P.V., Hande, M.P., Valiyaveettil, S., 2011. Health impact and safety of engineered nanomaterials. Chem. Commun. 47, 7025–7038.

Terada, M., Abe, S., Akasaka, T., Uo, M., Kitagawa, Y., Watari, F., 2009. Multiwalled carbon nanotube coating on titanium. Biomed. Mater. Eng. 19, 45–52.

Thakkar, K.N., Mhatre, S.S., Parikh, R.Y., 2010. Biological synthesis of metallic nanoparticles. Nanomedicine 6, 257–262.

Thneibat, A., Fontana, M., Cochran, M.A., Gonzalez-Cabezas, C., Moore, B.K., Matis, B.A., Lund, M.R., 2008. Anticariogenic and antibacterial properties of a copper varnish using an in vitro microbial caries model. Oper. Dent. 33, 142–148.

Thomson, R.C., Yaszemski, M.J., Powers, J.M., Mikos, A.G., 1995. Fabrication of biodegradable polymer scaffolds to engineer trabecular bone. J. Biomater. Sci. Polym. Ed. 7, 23–38.

Tian, L., Peng, C., Shi, Y., Guo, X., Zhong, B., Qi, J., Wang, G., Cai, Q., Cui, F., 2014. Effect of mesoporous silica nanoparticles on dentinal tubule occlusion: an in vitro study using SEM and image analysis. Dent. Mater. J. 33, 125–132.

Tobias, R.S., Browne, R.M., Wilson, C.A., 1985. Antibacterial activity of dental restorative materials. Int. Endod. J. 18, 161–171.

Tomsia, A.P., Launey, M.E., Lee, J.S., Mankani, M.H., Wegst, U.G.K., Saiz, E., 2011. Nanotechnology approaches for better dental implants. Int. J. Oral Maxillof. Impl. 26, 25–49.

Tschoppe, P., Zandim, D.L., Martus, P., Kielbassa, A.M., 2011. Enamel and dentine remineralization by nano-hydroxyapatite toothpastes. J. Dent. 39, 430–437.

Usković, V., 2013. Entering the era of nanoscience: time to be so small. J. Biomed. Nanotechnol. 9, 1441–1470.

Utneja, S., Nawal, R.R., Talwar, S., Verma, M., 2015. Current perspectives of bio-ceramic technology in endodontics: calcium enriched mixture cement—review of its composition, properties and applications. Rest. Dent. Endodon. 40, 1–13.

Vahid Dastjerdie, E., Oskoui, M., Sayanjali, E., Tabatabaei, F.S., 2012. In-vitro comparison of the antimicrobial properties of glass ionomer cements with zinc phosphate cements. Iranian J. Pharm. Res. 11, 77–82.

Venkatesan, B.M., Bashir, R., 2011. Nanopore sensors for nucleic acid analysis. Nat. Nanotechnol. 6, 615–624.

Verma, S.K., Prabhat, K.C., Goyal, L., Rani, M., Jain, A., 2010. A critical review of the implication of nanotechnology in modern dental practice. Nat. J. Maxillof. Surg. 1, 41–44.

Vermeersch, G., Leloup, G., Delmee, M., Vreven, J., 2005. Antibacterial activity of glass-ionomer cements, compomers and resin composites: relationship between acidity and material setting phase. J. Oral Rehab. 32, 368–374.

Vigneshwaran, N., Ashtaputre, N., Varadarajan, P., Nachane, R., Paralikar, K., Balasubramanya, R., 2007. Biological synthesis of silver nanoparticles using the fungus *Aspergillus flavus*. Mater. Lett. 61, 1413–1418.

Vijayaraghavan, R., Mathian, V.M., Sundaram, A.M., Karunakaran, R., Vinodh, S., 2012. Triple antibiotic paste in root canal therapy. J. Pharm. Bioallied Sci. 4, S230–S233.

Wagner, A., Belli, R., Stotzel, C., Hilpert, A., Muller, F.A., Lohbauer, U., 2013. Biomimetically- and hydrothermally-grown HAp nanoparticles as reinforcing fillers for dental adhesives. J. Adhes. Dent. 15, 413–422.

Walker, B., Mouton, C.P., 2006. Nanotechnology and nanomedicine: a primer. J. Nat. Med. Assoc. 98, 1985–1988.

Waltimo, T., Mohn, D., Paque, F., Brunner, T.J., Stark, W.J., Imfeld, T., Schatzle, M., Zehnder, M., 2009. Fine-tuning of bioactive glass for root canal disinfection. J. Dent. Res. 88, 235–238.

Wan, Y., Guo, Z., Jiang, X., Fang, K., Lu, X., Zhang, Y., Gu, N., 2013. Quasi-spherical silver nanoparticles: aqueous synthesis and size control by the seed-mediated Lee-Meisel method. J. Colloid Interf. Sci. 394, 263–268.

Webster, T.J., Ejiofor, J.U., 2004. Increased osteoblast adhesion on nanophase metals: Ti, Ti_6Al_4V, and CoCrMo. Biomaterials 25, 4731–4739.

Webster, T.J., Ergun, C., Doremus, R.H., Siegel, R.W., Bizios, R., 2000a. Enhanced functions of osteoblasts on nanophase ceramics. Biomaterials 21, 1803–1810.

Webster, T.J., Ergun, C., Doremus, R.H., Siegel, R.W., Bizios, R., 2000b. Specific proteins mediate enhanced osteoblast adhesion on nanophase ceramics. J. Biomed. Mater. Res. 51, 475–483.

Webster, T.J., Hellenmeyer, E.L., Price, R.L., 2005. Increased osteoblast functions on theta + delta nanofiber alumina. Biomaterials 26, 953–960.

Wei, G., Ma, P.X., 2006. Macroporous and nanofibrous polymer scaffolds and polymer/bone-like apatite composite scaffolds generated by sugar spheres. J. Biomed. Mater. Res. A 78, 306–315.

Wei, G., Ma, P.X., 2008. Nanostructured biomaterials for regeneration. Adv. Funct. Mater. 18, 3566–3582.

Wennerberg, A., Albrektsson, T., 2009. Effects of titanium surface topography on bone integration: a systematic review. Clin. Oral Impl. Res. 4, 172–184.

Whitesides, G.M., Mathias, J.P., Seto, C.T., 1991. Molecular self-assembly and nanochemistry: a chemical strategy for the synthesis of nanostructures. Science 254, 1312–1319.

Wilkinson, J.M., 2003. Nanotechnology applications in medicine. Med. Dev. Technol. 14, 29–31.

Wong, K.K.Y., Liu, X.L., 2012. Nanomedicine: a primer for surgeons. Ped. Surg. Int. 28, 943–951.

Wu, C., Chang, J., 2012. Mesoporous bioactive glasses: structure characteristics, drug/growth factor delivery and bone regeneration application. Interf. Focus 2, 292–306.

Wu, C., Chang, J., Xiao, Y., 2011. Mesoporous bioactive glasses as drug delivery and bone tissue regeneration platforms. Ther. Deliv. 2, 1189–1198.

Xu, H.H., Moreau, J.L., Sun, L., Chow, L.C., 2010. Novel CaF(2) nanocomposite with high strength and fluoride ion release. J. Dent. Res. 89, 739–745.

Xuereb, M., Camilleri, J., Attard, N.J., 2015. Systematic review of current dental implant coating materials and novel coating techniques. Int. J. Prosthodont. 28, 51–59.

Yah, C.S., Simate, G.S., 2015. Nanoparticles as potential new generation broad spectrum antimicrobial agents. DARU J. Pharm. Sci. 23, 43.

Yang, G.L., He, F.M., Hu, J.A., Wang, X.X., Zhao, S.F., 2009a. Effects of biomimetically and electrochemically deposited nano-hydroxyapatite coatings on osseointegration of porous titanium implants. Oral Surg. Oral Med. Oral Pathol. Oral Radiol. Endod. 107, 782–789.

Yang, L., Sheldon, B.W., Webster, T.J., 2009b. Orthopedic nano diamond coatings: control of surface properties and their impact on osteoblast adhesion and proliferation. J. Biomed. Mater. Res. A 91, 548–556.

Yang, G.L., He, F.M., Song, E., Hu, J.A., Wang, X.X., Zhao, S.F., 2010. In vivo comparison of bone formation on titanium implant surfaces coated with biomimetically deposited calcium phosphate or electrochemically deposited hydroxyapatite. Int. J. Oral Maxillofac. Impl. 25, 669–680.

Yang, K., Zhang, F.J., Tang, H., Zhao, C., Cao, Y.A., Lv, X.Q., Chen, D., Li, Y.D., 2011. In-vivo imaging of oral squamous cell carcinoma by EGFR monoclonal antibody conjugated near-infrared quantum dots in mice. Int. J. Nanomed. 6, 1739–1745.

Yesilyurt, C., Er, K., Tasdemir, T., Buruk, K., Celik, D., 2009. Antibacterial activity and physical properties of glass-ionomer cements containing antibiotics. Oper. Dent. 34, 18–23.

You, C., Han, C., Wang, X., Zheng, Y., Li, Q., Hu, X., Sun, H., 2012. The progress of silver nanoparticles in the antibacterial mechanism, clinical application and cytotoxicity. Mol. Biol. Rep. 39, 9193–9201.

Yu, X., Zhan, Z., 2014. The effects of the size of nanocrystalline materials on their thermodynamic and mechanical properties. Nano. Res. Lett. 9, 9–516.

Yuan, Y., Ding, J., Xu, J., Deng, J., Guo, J., 2010. TiO_2 nanoparticles co-doped with silver and nitrogen for antibacterial application. J. Nanosci. Nanotechnol. 10, 4868–4874.

Zandi, M., Mirzadeh, H., Mayer, C., Urch, H., Eslaminejad, M.B., Bagheri, F., Mivehchi, H., 2010. Biocompatibility evaluation of nano-rod hydroxyapatite/gelatin coated with nano-HAp as a novel scaffold using mesenchymal stem cells. J. Biomed. Mater. Res. A 92, 1244–1255.

Zero, D.T., 2006. Dentifrices, mouthwashes, and remineralization/caries arrestment strategies. BMC Oral Health 6, S9.

Zhang, F., Xia, Y., Xu, L., Gu, N., 2008. Surface modification and microstructure of single-walled carbon nanotubes for dental resin-based composites. J. Biomed. Mater. Res. B 86, 90–97.

Zhang, S., Huang, Y., Yang, X., Mei, F., Ma, Q., Chen, G., Ryu, S., Deng, X., 2009. Gelatin nanofibrous membrane fabricated by electrospinning of aqueous gelatin solution for guided tissue regeneration. J. Biomed. Mater. Res. A 90, 671–679.

Zhang, Z.-G., Li, Z.-H., Mao, X.-Z., Wang, W.-C., 2011. Advances in bone repair with nanobiomaterials: mini-review. Cytotechnology 63, 437–443.

Zhang, X.-Q., Xu, X., Bertrand, N., Pridgen, E., Swami, A., Farokhzad, O.C., 2012. Interactions of nanomaterials and biological systems: implications to personalized nanomedicine. Adv. Drug Deliv. Rev. 64, 1363–1384.

Zhang, K., Li, F., Imazato, S., Cheng, L., Liu, H., Arola, D.D., Bai, Y., Xu, H.H., 2013. Dual antibacterial agents of nano-silver and 12-methacryloyloxydodecylpyridinium bromide in dental adhesive to inhibit caries. J. Biomed. Mater. Res. B 101, 929–938.

Zhang, B.G.X., Myers, D.E., Wallace, G.G., Brandt, M., Choong, P.F.M., 2014. Bioactive coatings for orthopaedic implants: recent trends in development of implant coatings. Int. J. Mol. Sci. 15, 11878–11921.

Zhao, L., Yang, G., Zhao, X., 2014. Rho-associated protein kinases play an important role in the differentiation of rat adipose-derived stromal cells into cardiomyocytes in vitro. PLoS One 9, e115191.

Zheng, G., Patolsky, F., Cui, Y., Wang, W.U., Lieber, C.M., 2005. Multiplexed electrical detection of cancer markers with nanowire sensor arrays. Nat. Biotechnol. 23, 1294–1301.

Zhou, H., Lee, J., 2011. Nanoscale hydroxyapatite particles for bone tissue engineering. Acta Biomater. 7, 2769–2781.

Zomorodian, E., Baghaban Eslaminejad, M., 2012. Mesenchymal stem cells as a potent cell source for bone regeneration. Stem Cells Int. 2012, 980353.

Zou, H., Wu, S., Shen, J., 2008. Polymer/silica nanocomposites: preparation, characterization, properties, and applications. Chem. Rev. 108, 3893–3957.

CHAPTER

THERAPEUTIC APPLICATIONS OF NANOTECHNOLOGY IN DENTISTRY

26

Muhammad S. Zafar*,, Zohaib Khurshid†, Shariq Najeeb‡, Sana Zohaib†, Ihtesham Ur Rehman§**

**Taibah University, Al Madinah, Saudi Arabia*

***Riphah International University, Islamabad, Pakistan*

†King Faisal University, Al-Hofuf, Saudi Arabia

‡Al-Farabi Colleges, Riyadh, Saudi Arabia

§The University of Sheffield, Sheffield, United Kingdom

CHAPTER OUTLINE

1 Introduction 833
2 Origin of Nanotechnology Concept and History 835
3 Nature and Timeline of Nanomaterials 835
4 Synthesis, Characteristics, and Classifications of Nanomaterials 836
5 Nanobiomaterials in Dentistry 838
5.1 Nanobiomaterials in Restorative Dentistry 839
5.2 Nanobiomaterials for Preventive Dentistry 847
5.3 Nanobiomaterials for Endodontics 848
5.4 Nanobiomaterials in Oral Implantology 852
6 Conclusive Remarks 855
References 855

1 INTRODUCTION

The word "Nano" is derived from the Greek word for "dwarf," combined with a noun to develop words, for example, nanotechnology, nanometer, nanodevices, or nanorobots and it's an emerging branch of science, engineering, and technology. "*Nanotechnology*" is defined as the study, development, and application of extremely small (which is about 1–100 nm) materials, devices, and systems exhibiting physical, chemical, and biological properties that are different from those found on a larger scale (Dunn et al., 2002). The other name of this technology is reported as molecular nanotechnology or molecular engineering (Drexler, 1992). Nanotechnology is a meaningful engineering technology today, helping to make biomaterials, electronics, biomedical instruments, and consumer products by controlling the individual atoms and molecules (Moldovan and Ferrari, 2002; Roszek et al., 2005; West and Halas, 2003).

Before jumping to advances of nanotechnology need to understand the nanoscale measurement. Nanoscale is usually defined as smaller than one-tenth of a micrometer in at least one dimension, though this term is sometimes also used for materials smaller than1 μm. At nanoscale, 1 nm is 1 billionth or 10^{-9} of a meter. It's difficult to imagine just how small that is, so, as shown in Fig. 26.1, here are some examples; a strand of human DNA is 2.5 nm in diameter: a human hair is approximately 80,000–100,000 nm wide, a sheet of paper is about 100,000 nm thick, a single gold atom is about a third of a nanometer in diameter, a single-walled carbon nanotube is around 1 nm in diameter (Agarwal et al., 2005b).

Nanotechnology is one revolutionizing field helping to improve many technology and industry sectors, for example, electronics and information technology, sustainable energy application, environmental remediation application, medicine, dentistry, food safety, and transportation, among many others (Balshaw et al., 2005; Roco and Bainbridge, 2002; Thome, 2006), According to United States (US) president's 2016 budget request submitted to congress on February 2, 2015, provides $1.5 billion for the national nanotechnology initiatives (NNI), a continued investment in support of the president's priorities and innovation strategy. Cumulatively totaling more than $22 billion since the inception of the national nanotechnology initiatives (NNI) in 2001 (including the 2016 request), this support reflects nanotechnology's potential to significantly improve our fundamental understanding and control of matter at the nanoscale and to translate that knowledge into solutions for critical national needs. See further details on the report at http://www.nano.gov/node/1326. There are four main types of materials (metals, polymers, ceramics, and composites), and nanomaterials have been developed in all four categories for practical applications. Due to unique properties, nanomaterials always remain a focus of interest for biomaterials scientists to get benefits for medical and dental applications to improve the quality of life (Gupta, 2011). To date, nanomaterials have been developed and are being used practically for a range of medical applications, such as drug delivery, gene delivery, imaging tools, molecular diagnostics (Tinkle et al., 2009). Therapeutic applications of nanotechnology in dentistry are discussed in this chapter.

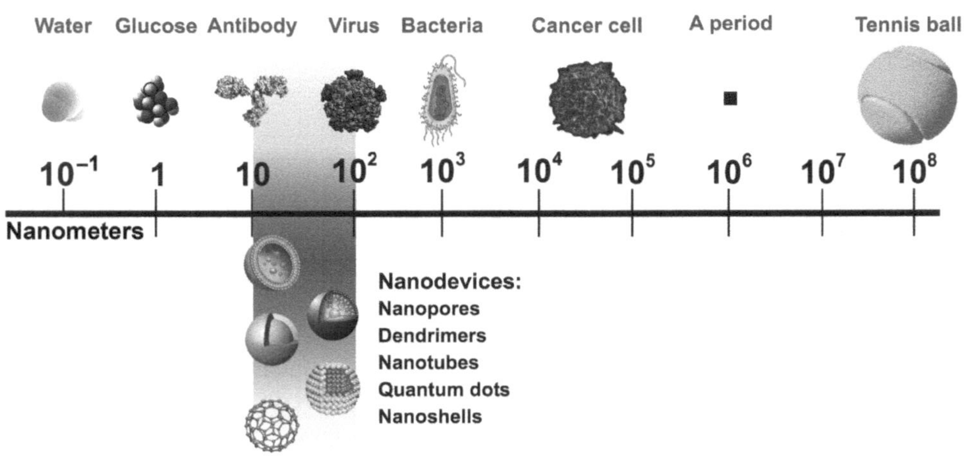

FIGURE 26.1 Illustration Representing Nanoscale With Examples

2 ORIGIN OF NANOTECHNOLOGY CONCEPT AND HISTORY

The aroma of a small particle was in late 1867 smelled by James Clerk Maxwell during his experiment on handling of individual molecules called Maxwell's Demon (Bennett, 1987; Maddox, 2002). After few years, Richard Adolf Zsigmondy first observed gold solution and other materials with a size around 10 nm; he also introduced the first classification of a particle based on the size in a book published in 1914 (Zsigmondy, 1909a,b). Physicist Richard Feynman presented the idea entitled *There's Plenty of Room at the Bottom* at an American Physical Society meeting at the California Institute of Technology in 1959 (Toumey, 2009). The transcript of this talk was published in *Engineering and Science* in 1960. Although Feynman did not apply the term "Nanotechnology or Nanosciences," he did, however, describe the novel process in which scientists can manipulate materials at atomic or molecular levels (Gammaitoni, 2012). Tokyo Science University Professor Norio Taniguchi used the term nanotechnology for the first time in 1974 to describe the processing and properties of materials at single atom and molecule level (Taniguchi, 1974). The idea of nanotechnology was further probed in depth and promoted by Dr. K. Eric Drexler and published a book titled *Engines of Creation-The Coming Era of Nanotechnology* in the late 1980s (Drexler, 2003). In 1991, the publication by *Sumio Lijima* "Helical microtubules of graphitic carbon" introduced the concept of nanotubes and boosted nanomaterials research (Eliseev et al., 2009).

At such a small scale, most of the atoms are surface atoms and it is impossible to see the materials using the naked eye, and even not possible using ordinary microscopes. The development of scanning microscopes that can be used to see images at atomic or molecular level advanced the field of nanotechnology. The scanning tunneling microscope was developed in 1981 (Binnig et al., 1982). Once scientists had better equipment with the invention of scanning tunneling microscopes and atomic force microscope (about 30 years ago) the modern science of nanotechnology was born. An active research is going on to take advantages of nanotechnology and develop materials with enhanced properties and better control over their properties (Roco, 2003).

3 NATURE AND TIMELINE OF NANOMATERIALS

In nature, many examples of nanotechnology are present. From DNA (2.5 nm), to structural architecture of bone tropocollagen (300 nm), and dental enamel crystal size, which can be as small as 30–60 nm in thickness and 100–1000 nm in length (Hannig and Hannig, 2010; Rho et al., 1998; Ritchie, 2011). There is no question about the structure–function relationship and the best performance regarding natural biomaterials.

> Nature is a master chemist with incredible talent.
> Nature is and will continue to be the best material scientist ever.

We can learn many lessons from natural materials and follow the principles of nature for making these materials. All natural nanomaterials are fabricated under ambient conditions and perform well in the environment. Natural biological materials show many levels of hierarchical structures from the macroscopic to the microscopic length scale with the smallest building blocks being designed with nm-sized hard insertions implanted into a soft protein matrix (Wickson, 2008). Biomimetic techniques are the

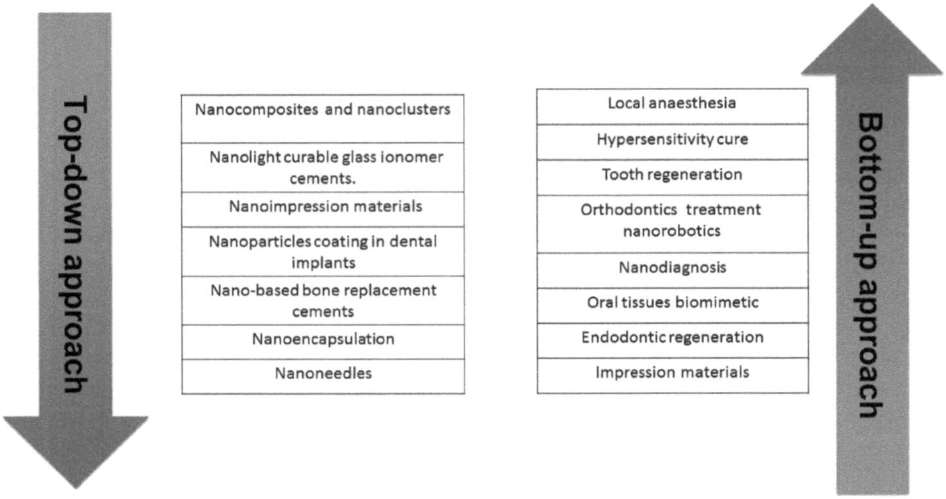

FIGURE 26.2 Name of Technologies Utilizes in Nanofabrication of Materials by Top-Up and Bottom-Up Approaches

way to unlock nature's methodology to optimize nanolevel materials fabrication. Biomimetic is the science of understanding, how nature has designed, processed, and assembled these high-performance materials and then applying those principals to make synthetic materials (Bhushan, 2009).

4 SYNTHESIS, CHARACTERISTICS, AND CLASSIFICATIONS OF NANOMATERIALS

Nanomanufacturing is the collective term used to describe the synthesis of nanosize materials, systems, structure, and devices. Two different approaches for nanomanufacturing are *top-down*, *bottom-up*, and *self-assembling approach*. Top-down fabrication reduces large templates of materials into nanoscale but it requires large quantity of materials and this approach has a major drawback of wasting unused materials. In bottom-down approach development of products by building atomic or molecular component at nanoscale assembling, making process cheaper as compare to top-down approach (Biswas et al., 2012; Majumder et al., 2007). Fig. 26.2 illustrates the nanofabrication of materials by top-down and bottom-up approaches. These nanomanufacturing approaches are more advanced within last few years by introducing chemical vapor deposition, atomic layer epitaxy, molecular beam epitaxy, dip pinned lithography, roll-to-roll processing, and nanoimprint lithography (Carneiro et al., 2011; Faustini et al., 2013; Yao and He, 2014).

Nanomaterials are expected to be stronger, durable, lighter, easy to handle, and manipulate. They can also be water repellent, self-assembled, antifog, antibacterial, antiviral, radiation resistant, scratch free, and biocompatible (Carneiro et al., 2011; Yao and He, 2014). Modern nanofabrication techniques face problems with surface contamination, and as a result the fabrication of micron-scale devices requires very expensive clean-room technology. We hope that understanding the biological system will lead to building bioinspired devices for cleaning on micro- and nanoscales. Hackmann et al. (2015)

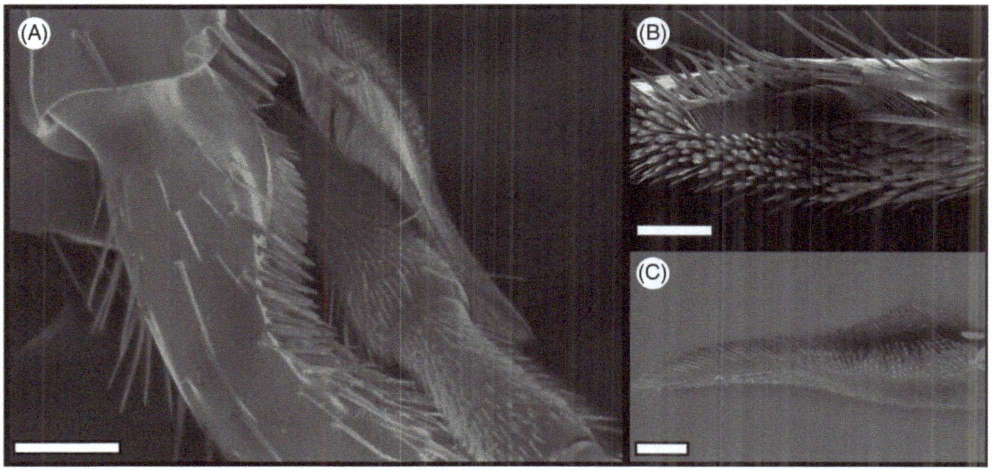

FIGURE 26.3 Antenna Cleaner of C. Rufifemur Ants

(A) Scanning electron micrograph of the antenna clamped by the cleaner (view from the outer=posterior side). (B) Tarsal notch. (C) Tibial spur. In image (B) bristles are shown in red (light gray in top of image in print version); the comb is shown in blue (dark gray in center of image in print version). In image (C) the comb is again shown in blue (dark gray in print version).

have undertaken the first biomechanical investigation of the different types of hair used by ants to clear dirt from their antennae (Fig. 26.3).

It is hoped that mimicking this technology could help us develop technology to clean material at a nanometer level (Hackmann et al., 2015). Properties of nanomaterial may vary depending on a number of factors, such as type of material (metal, polymer, or ceramic), type of nanostructure, and macroscoping morphology. Some key properties of nanomaterials are:

1. Increased reactivity. At nanoscale, most atoms are surface atoms thus increasing the number of sites available for reaction and improved surface functionalities.
2. A large surface area/volume ratio that effects on reactivity and physical properties.
3. Better mechanical interlocking of polymer matrix with nanoparticles due to high surface area.
4. Improved mechanical properties. Inorganic nanoparticles are hard and dense.
5. Better resistance to the crack propagation under cyclic loads.
6. Decrease in grain size. Important for surface reactions, such as catalysis.
7. Nanofillers can improve optical properties and more smooth surface finish.
8. Nanocomposites have also shown better control over biodegradability.

The characterization of nanomaterials involves the use of different techniques and methods to probe their physical, mechanical, chemical, and biological properties. Regarding nanomaterials characterization, the first question should come in mind is, what kind of information are we looking for and, at what resolution? For example:

1. Morphology: Nanostructure and architecture
2. Crystallography: Details of atomic arrangements

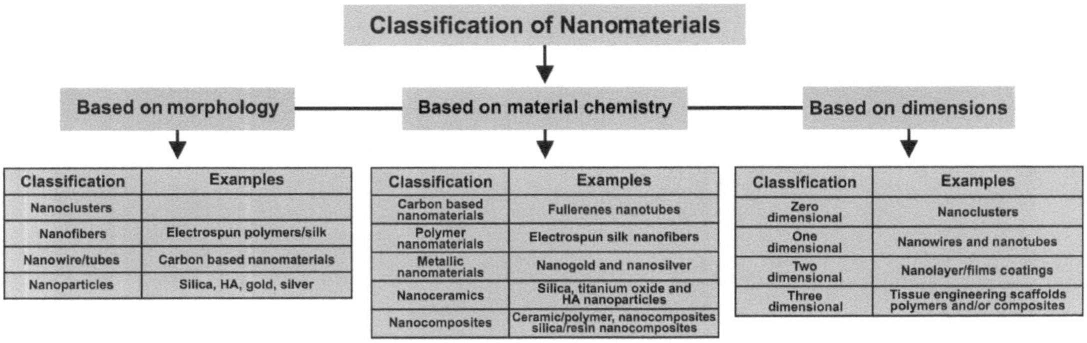

FIGURE 26.4 Classification of Nanomaterials on the Basis of Morphology, Chemistry and Dimensions

3. Chemical structure: Elements present and chemistry of molecules
4. Electronic structure: Nature of bonding in atoms
5. Thermal properties: Reaction to temperature related changes
6. Mechanical properties: Reaction to external stress and forces applied

The data obtained from characterization help to correlate nanostructure, and properties and determine the criteria for fabrication. Before implementing it in clinical practice, a detailed understanding and analysis of nanomaterials is required for a complete understanding of materials reaction both in vitro and in vivo (Najeeb et al., 2015). As shown in Fig. 26.4, nanomaterials are classified on the basis of morphology of basic microscopic unit, materials chemistry and dimensional existence.

5 NANOBIOMATERIALS IN DENTISTRY

There has been remarkable research conducted on nanobiomaterials in recent years that has transitioned it from theoretical foundation to clinical practice. Nanobiomaterials are typically nanometer-sized particles and features. Nanopores, nanotubes, nanosolutions, nanoshells, quantum dots, dendrimers, liposomes, nanorods, fullerenes, nanospheres, nanowires, nanorings, and nanocapsules are all examples of nanometer-sized particles and features that can be used to produce new materials or modify existing ones to improve their performance in the dental practice. Nanotechnology has brought forth many improvements to visualize materials architecture on nanoscale resolution, for example, micro- and nanoleakages develop in resin-based composites (RBC) (Al-Boni and Raja, 2010). Due to the forth mentioned research conducted on applications of nanoscience in dentistry, a new field in dentistry has emerged (Freitas, 2000; Kanaparthy and Kanaparthy, 2011). Nanodentistry is the art of treating oral diseases and improving the oral health by means of nanometer-level materials (nanomaterials). The chapter is divided into various sections according to the specific applications and uses of nanobiomaterials in various dental specialties, for example, restorative dentistry, paediatric dentistry, preventive dentistry, orthodontics, prosthodontics, periodontics, implant dentistry, regenerative dentistry, and endodontics. A summary of nanotechnology and its uses in dentistry is provided in Table 26.1.

Table 26.1 Application of Nanobiomaterials in Dentistry

Nanobiomaterials in Dentistry	Features
Nanocomposites	Enhance the polishability of resin and high filler loading increases hardness and wear resistance, reduced polymerization shrinkage, and less staining. Superior hardness, translucency, and esthetic appeal. Superior flexural strength
Nanoglass ionomers	Surface finish more closely approximates hybrid composite and improved wear resistance
Nanobonding agent	Silica nanofiller technology and higher bond strength performance prevents particle settling. Homogenous coat of bonding agent can be applied. Treatment time is minimal
Nanocoating agent	Used as a final coating over esthetic restorations and uniformly dispersed nanofillers for higher wear resistance. Preventing abrasion and discoloration. Smooth high luster finish retained over time Better wear and stain resistance
Nanopolishing agent	Protect tooth surfaces against the damage caused by cariogenic bacteria as the bacteria can be removed easily from such polished surfaces and this also leads to less staining of the teeth and better aesthetics
Nanoendodontic Sealer	Improved convenience and delivery method.
Nanoimpression materials	Nanofillers are integrated in vinyl-poly-siloxanes producing a unique addition silicone impression material. The material has better flow, improved hydrophilic properties and enhanced detail precision

5.1 NANOBIOMATERIALS IN RESTORATIVE DENTISTRY

Dental caries is the demineralization of dental hard tissues (enamel, dentine, and cementum) by acidogenic bacteria present in dental plaque. The dental hard tissues have almost similar composition with little difference but structurally different architecture. Dental caries is considered as a dynamic process in which enamel, dentine, cementum, and pulp are damaged due to imbalance in mineral ions present in the tooth structure and dental plaque fluid delineated by demineralization and remineralization processes. Demineralization can cause a tooth to lose apatite of calcium and phosphate from enamel, dentine, and cementum due to drop in the pH (<5.5) of the oral biofilm by the action of bacterial acids. In remineralization process damaged apatite of calcium and phosphate from enamel redeposit to tooth surface from dental plaque fluid or saliva. Fluoride is the main ingredient that promotes remineralization and prevents teeth from decay. In the past decade extensive studies have been done to bring out tooth replacement materials containing calcium and phosphate ions on it for remineralize tooth lesion by help of nanotechnology.

Restorative dentistry is a division of dentistry dealing with those dental materials that can replace dental hard tissues in oral environment level to maintain physical function and aesthetic loss by disease processes. The ideal restorative material should be biocompatible, bond permanently to tooth structure or bone, match the natural appearance of tooth structure and other visible tissues, and be capable of initiating tissue repair or regeneration of missing or damaged tissues. For these available materials are amalgam, gold fillings, glass ionomers cements, and resin-based composites. In Table 26.2 represents historical background of dental materials and recent advances. In past decades, dental materials have changed their morphology from macro, micro, and hybrid then nano. These drastic changes bring better improvements to treat dental diseases.

Before the introduction of nanotechnology application toward dentistry, dental surgeon used amalgam as a posterior tooth filling material and resin based composite as an anterior filling material. More recently, nanomodified restorative materials have been used clinically to restore teeth. To date, significant advancements of nanotechnology in restorative dentistry are the development of resin-based composite technology and also glass ionomers cements.

5.1.1 Resin-based dental composites

Resin-based composite technology was introduced in 1960s as a conservative, tooth-adherent material with higher mechanical properties when compared to other plastic-based material (acrylic). Old resin composite system were chemically activated but the new generation introduced photo-light activated composites, which are initiated by ultraviolet (UV) light. This light-cured resin system provides more chair side time for manipulate in tooth for mimicking natural anatomy of tooth.

Table 26.2 History of Dental Materials and Related Applications

Year	Materials With Their Uses
3000 BC	Gold bands and wires were used by the Phoenicians
600 AD	The Mayans used implants consisting of sea shell segments that were placed in anterior teeth sockets
1678–1761	The father of modern dentistry used tin foil and lead cylinders for filling the tooth cavities
1756	Philip Pfaff of Germany described a method for making impressions
1774	Duchateau a French pharmacist designed a process for producing hard, decay proof porcelain dentures
1800	Introduction of amalgam as tooth filling material
1817	Planteau, a French dentist first introduced porcelain teeth
1839	Charles Goodyear has invented vulcanized rubber denture bases
1935	Polymerized acrylic resin was introduced as a denture base material
1960	Introduction of resin based dental composite
1972	New translucent chemically bonded to tooth surface dental cement develop at the Laboratory of the Government Chemist in England by Wilson and Kent

Resin-based composites (RBC) consist of two active ingredient resin and filler with other components, like silane-coupling agent, initiators and accelerators, radiopaque material with pigments. This composition varies slightly in different commercially available material brands of RBCs. Anterior fillings, posterior fillings, losing diastemas, or camouflaging dental malpositions, veneers, treatment of dysmorphias or discolorations, incisal edge restoration, inlay, onlays, and core build-ups are some of the several uses RBCs in dentistry. Depending on the requirements, such as occlusal load and aesthetics, the filler content can vary in each type of composite.

RBCs can be classified according to the size of the filler particles from macrometer to nanometer-size. The resin composite system has four main components; organic polymer matrix, inorganic filler particles, coupling agent, and initiator/activator (Ferracane, 2011). Before the advent of aromatic dimethacrylates monomer introduction as organic polymer resin matrix in contemporary RBCs, use of bisphenol A-glycidyl dimethacrylates (Bis-GMA) was common (Lovelh et al., 1999). But their hydrogen bonding interaction with hydroxyl groups and monomer molecules create materials more viscous. This viscosity can be controlled by the manufacturer by adding low molecular compound with difunctional carbon double bonds, for example, triethylene glycol dimethacrylate (TEGDMA) and ethoxylated-bisphenol-A-dimethacrylate (Bis-EMA) (Ilie and Hickel, 2011). A new silorane-based monomer, introduced as *oxirane* (also known as *epoxy*), reportedly reduced polymerization shrinkage with no internal stress build during curing (Lien and Vandewalle, 2010). The structure of silorane shown in Fig. 26.5, undergoes ring-opening cross-linking through cationic polymerization by a special initiator system, iodonium salts (Ilie and Hickel, 2006). This technology also improves depth of cure, hardness, and strength as compare to conventional Bis-GMA-based resins (Weinmann et al., 2005).

In resin-based dental composites, inorganic fillers are ground quartz or glass, barium, strontium, zirconium or ytterbium, sol-gel–derived ceramics, and silica (Zandinejad et al., 2006). This inorganic filler technology makes the major bulk of dental composite. Its function is to reinforce the resin matrix and improve mechanical properties, like improve tensile strength, compressive strength, decrease abrasion, toughness, dimensional stability, thermal stability (Ferracane, 1995). Also, by increasing filler content, it can reduce polymerization shrinkage, water absorption, and linear expansion coefficient (Zandinejad et al., 2006). Different methods can be utilized to synthesize and

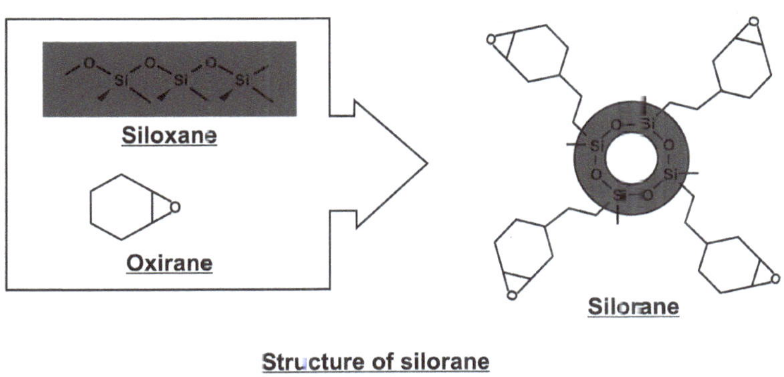

FIGURE 26.5 Schematic Presentation of Silorane Chemical Structure

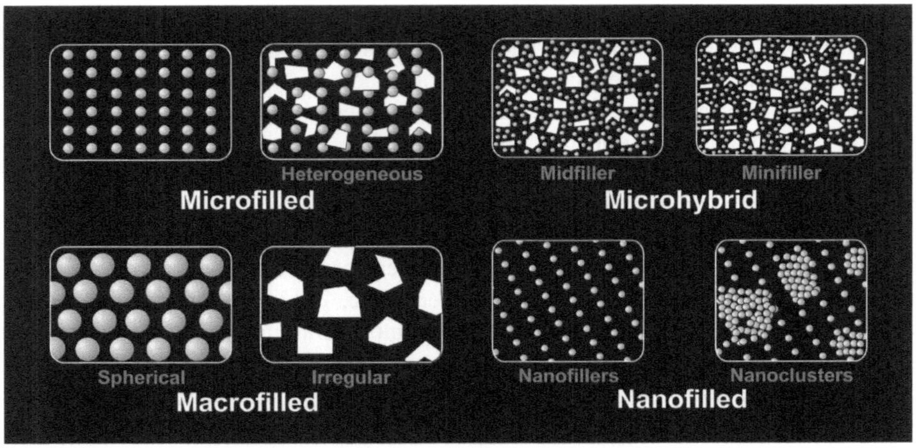

FIGURE 26.6 Classification of Dental Composite on the Basis of Particle Size

control filler particles and are sol-gel, flame pyrolysis, and flame spray pyrolysis (Chen, 2010). Their loading efficiency greatly depend on size. Fig. 26.6 represents the classification of dental composite according to particle size.

According to classification of dental composite on particle size, *macrofilled* dental composite system contains large, irregular, or spherical shaped filler particles (diameter 20–30 μm) with excellent strength, low resistance to wear, but are difficult to polish and their aesthetics are compromised. To overcome the limitations of macrofilled composites, manufacturers can introduce micrometer-sized fillers to the composite system to produce *microfilled* resin composites. The particle size in this system ranged 0.04–0.2 μm that are polishable imparting good aesthetics and suitable for restoring anterior teeth. Furthermore, refinements in the particle by milling and grinding develop new system of composite called *Microhybrid* contains 60%–70% filler by volume. They have good clinical wear resistance and mechanical properties also considered as universal dental composite for anterior and posterior application (Chan et al., 2007; Mitra et al., 2003). Due to the introduction of nanotechnology in dental resin composites, several drawbacks of traditional resin composites can be overcome (Khurshid et al., 2015). As shown in Fig. 26.7 nanomers (NM) and nanoclusters (NCs) particles can be made part of the resin composite system to produce nanomodified materials.

The NM are synthesized from aqueous colloidal silica sols by drying powders of nanosized silica (20–75 nm in diameter) treated with 3-methacryloxypropyltrimethoxysilane (MPTS), as a coupling agent. This coupling agent is bifunctional material because it contains a silica ester function on one end for bonding to inorganic surface and a methacrylate group on the other end to bond with resin matrix when cured. The NCs are like cluster of grapes that are loosely bound agglomerates of nanosized particles. They are synthesized in two forms; the first forms are synthesized by particles of silica (75 nm) and the size of agglomerated particles around 0.6 μm. The second forms are developed from a colloidal solution of silica and a zirconyl salt-producing zirconia silica particles sizes that range from 2 to 20 nm in size. Following synthesis, both the types of particles are treated with a silane-coupling agent to improve bonding between filler and resin thereby improving mechanical properties of the set material (Chen et al., 2006).

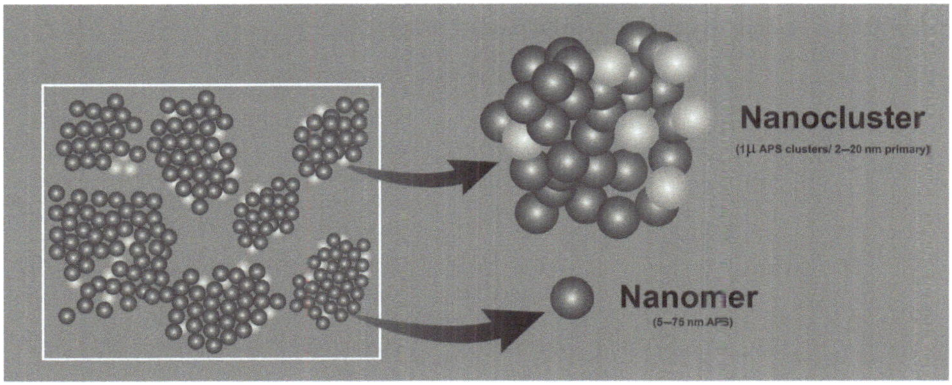

FIGURE 26.7 Illustration Representing Size and Morphology of Nanoclusters and Nanomers

Other ingredients of resin-based dental composites are minor quantities of photosensitizers, such as camphoquinone, present in light-activated resin. In chemically activation systems organic amine works as and accelerator (Chen, 2010; Siang Soh et al., 2006). Table 26.3 provides a brief description and composition of commercially available dental composites.

Recently, the viscoelastic properties of commercially available dental resin-based composites (Filtek P60, Filtek Supreme, and Filtek Z250; 3M ESPE, Bracknell, UK) have been measured to determine the effect of beverages on their storage moduli and damping ratios and concluded Filtek P60 showed better results compared to other restorative materials and its related to pH environment, hydrophilicity, and the chemical composition of composites. (Kaleem et al., 2015). One group performed a comparison of color stability and degree of conversion under different beverages on nanocomposites and microhybrid and concluded the capability of good color stability in nanocomposites samples (Kheraif et al., 2013).

5.1.2 Dental glass-ionomers system

Dental glass-ionomers cements (DGICs) have played an important role in restorative dentistry because of they have unique properties, like direct chemical bonding with natural tooth structure, fluoride release, and fluoride recharge. Fluoride is well documented as an anticariogenic restorative material (Ullah et al., 2015). A variety of mechanisms is involved in the anticariogenic effects of fluoride release from glass ionomers, including the reduction of demineralization, the enhancement of remineralization, the interference of pellicle and plaque formation, and the inhibition of microbial growth and metabolism. These properties have made DGICs a material of choice for luting, lining, and filling materials (Zafar, 2013). In short, glass-ionomers materials are bioactive materials that are fully adhesive to the natural tooth structure forming chemical reaction at tooth cement interface (Nicholson, 1998). Additionally, they release fluoride in the oral cavity for long period of time and recharge by replenishing lost fluoride from topical applications like mouthwashes, toothpastes, and fluoridated varnishes (Gonzalez, 2004). These all properties produced a major shift in the practice of conservative dentistry from "extension for prevention" to maximum conservative approach. Moreover, DGICs made the concept of atraumatic restorative treatment (ART) technique possible and performing art of restorative dentistry without use of any special equipment. This has made the practice of restorative dentistry possible in backward and remote areas possible because no special equipment is required for manipulation

Table 26.3 Commercially Available Dental Resin Composites

Composite	Types	Particle Size	Manufacturer	Composition	Batch No.
P-60	Packable-composite	0.6 μm 0.19–3.3 μm	3M ESPE, St Paul, MN, USA	Filler: 61% vol. and 83% wt. silica/ zirconia filler (mean particle size of 0.6 μm) polymeric matrix: Bis-GMA, Bis-EMA, UDMA, TEGDMA	4MM 0400701 9099 IKY 2004-04
Admira	Microhybrid	0.7 μm	Voco, Anton-Flettner-St., Cuxhaven, Germany	Filler: 3% apatite-sulphate-phosphate and 77% inorganic glass particles ormocer matix: polysiloxane-Si-O + methacrylate monomer	36012 651555
Alert		0.8 μm	Jeneric-Pentron Wallingford CT, USA	Filler: 54% by vol. and 83.5% Ba-B-Al silicate. Resin matrix: Ethoxylated, TEGDMA	860.71
Supreme	Nanofilled composite	0.6–1.4 μm	3M ESPE, St. Paul, MN, USA	Filler: 78.5% wt. of nonagglomerated silica nanoparticles with mean size of 20 nm and nonagglomerations composed of zircon-silica particles with size varying from 5 to 20 nm. Mean agglomeration size varied from 0.6 to 1.4 μm. organic matrix: Bis-GMA, Bis-EMA, TEGDMA, UDMA	6AL 5BW
A-110	Microfilled	0.01–0.09 μm	3M ESPE, St. Paul, MN, USA	Filler: 56% wt. or 40% in volume of silica particles with mean size of 0.04 μm (varying 0.01–0.09 μm). Organic matrix: Bis-GMA and TEGDMA	
Fillmagic	Microhybrid composite	0.5 μm	Vigodent S.A. Ind. Com. Bonsucesso, RJ, Brazil	Filler: 80% by wt. Barium and aluminum silicate glass and pyrogenic silicon (0.5 μm), organic matrix: Bis-GMA, Bis-EMA, TEGDMA and UDMA	129/08
Glacier	Condensable microhybrid composite	0.4–1.0 μm	SDI Limited Bayswater, Victoria, Australia	Filler: 81% wt. strontium glass and amorphous silica, organic matrix: multifunctional methacrylate ester	07/0507
TE-Econom	Hybrid-composite	1.0 μm	Ivoclar Vivadent, Schaan, Liechtenstein	Filler: 72% by wt. Barium glass, aluminum and barium fluorosilicate glass, high/ dispersed silicon dioxide and mutual spheroid oxide (1.0 μm). Organic matrix: Bis-GMA, TEGDMA, and UDMA	16366

Composite	Types	Particle Size	Manufacturer	Composition	Batch No.
Natural look	Microhybrid composite	0.5 µm	DFL Ind. Com, RJ, Brazil	Filler: 77% wt. Barium crystal glass and amorphous silica (0.5 µm), organic matrix: Bis-GMA, Bis-EMA, TEGDMA	08050692
Prisma-APH	Hybrid composite	1.0 µm	Dentsply Ind. Com. Petrópolis, RJ, Brazil	Filler: Barium and aluminum silicate, silanized barium, pyrogenic silicon. Organic matrix: modified urethane	0913918
Filtek-Z250	Microhybrid composite	0.19–3.3 µm	3M ESPE, St Paul, MN, USA	Filler: 60% in vol. and 77.6% wt. of zircon silicate particles. Mean size of 0.6% µm. Organic matrix: Bis-GMA, Bis-EMA, and UDMA	0-EF/1370A2
Sure-Fil	High-density hybrid packable composite	0.8 µm	Dentsply De Trey GmbH, Konstanz, Germany	Filler: 65% vol. and 82% by wt. silanized baium, boron-silicate-aluminum. Resin matix: Bis-GMA, UDMA	040706
Esthet–X	Microhybrid composite Nano hybrid composite	0.04–1 µm	Dentsply/ Caulk, Milford, DE, USA	Filler: 60% vol. and 77% wt. containing Ba-B-fluoroalumino silicate glass, amorphous silicate, photoinitiater and stabilizers. Resin matrix: Bis-GMA, Bis-EMA, TEGDMA	0510082 0510281
Belle Glass		0.7 µm	Kerr Lab Orange CA, USA	Filler: 78.7% by vol. and wt. 74.78% Ba-Si glass with mean particle size of 0.5 µm. Resin matrix Bis-GMA, UDMA, Bis-EMA	P900-181
Art Glass		0.8 µm	Heaeus Kulzer GmbH Wehrheim Germany	Filler: 52% vol. and 69% wt. Ba-Al-S glass. Polymeric matrix. Bis-GMA, UDMA	020107

of these materials (Frencken et al., 1994, 1996). Applications of glass ionomers are in all the specialties of dentistry and mainly are class I, III, IV, and V restorations, cavity liners, root caries, retrograde filling material, luting material for posts, crowns, and bridges, core build-up material, orthodontic band cementation, cementation of orthodontic brackets, with composites in sandwich technique, pits and fissure sealants, cementation of fixed prosthesis, temporary covering of exposed dentine, ART, bone cement, splinting of avulsed teeth (Hes et al., 1999; Lohbauer, 2010).

Like all other biomaterials, DGICs cannot be an ideal material for every application. Indeed it is true that glass ionomers are the only class of materials that can fight against carious attack on tooth and has a base line in the restorative dentistry. However, still there are few limitations with DGICs, for example low strength, long setting reaction, and the need to be reinforced or laminated depending on the requirement (Xie et al., 2000). Since the invention of glass ionomers in 1972 by Wilson and Kent, substantial research has been conducted, which has eventually led to the production of resin modified glass ionomers (RMGICs) with improved quality and properties (Nicholson, 1998). There is a lot of research work to be carried out for the full understanding of action, setting reaction, and mechanism of fluoride release from these materials. Weakness of tensile strength and poor physical properties is another area of research for these materials. Different elements and modifications are needed to investigate to improve the radiopacity, strength, physical properties, and antibacterial properties of these materials. Hydroxyapatite has excellent biological behavior, its composition and crystal structure are similar to those of the apatite in found human dental structures and the skeletal system. Many studies have tried to evaluate the effect of the addition of HA powders on the properties of DGICs (DGICs). Lucas et al. (2003) in their studies added 0.3–50 mm spherical HA particles to the powder of a capsulated GIC (Fuji IX GP) with a particle size of 0.3–200 mm (8% m/m). Samples were immersed in distilled water for 1 day or 1 week. Seventy-four results showed that HA-ionomers are promising filling for dental materials and the incorporation of HA particles into the powder of glass-ionomers cements increased the mechanical properties of the set cement (Lucas et al., 2003). A timeline for the reported modification of DGICs is represented in Fig. 26.8.

Giomers are relatively new members of the glass-ionomer group developed in an attempt to improve the physical, mechanical, aesthetic, and biological properties of existing DGICs (Mousavinasab and Meyers, 2009). This is a hybrid material and modification has been made in the filler particles by prereacted treatment with acid before blending them in resin matrix. Manufacturer of these materials has claimed to show better physical and mechanical properties, biocompatibility, and more affective fluoride release (Chitnis et al., 2006).

Resin-modified nano-GIC, namely Ketac nano by 3M ESPE (paste/paste system), based on methacrylate-modified polyalkenoic acid, is the first glass ionomer to be made available commercially. However, to date, there is no sufficient clinical evidence and systematic reviews indicating any significant improvement of nanoionomers over conventional resin-modified GICs. Nevertheless, manufacturer recommends the material to be used for small class 1 restoration, class III and IV, transitional

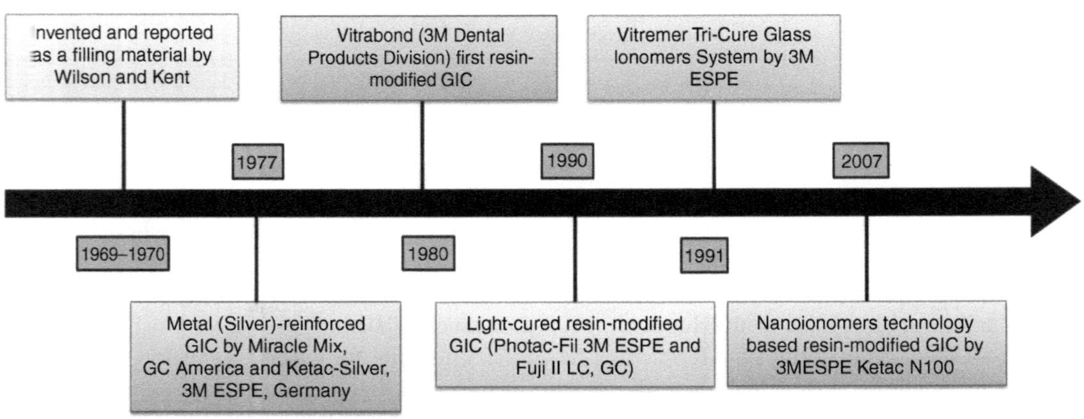

FIGURE 26.8 Representation of Dental Glass-Ionomers Cements (Dgics) Developments

restoration, laminates and core build ups (> 50% coronal tooth intact). One study suggested that shear bond strength (SBS) of Ketac nano is acceptable for bonding orthodontic brackets (Uysal et al., 2010). Equia, another nanoionomer, produced by GC Corporation, has been made available for clinical application. The typical composition of this cement is typically composed of high viscosity glass particles, inorganic nanofillers (represents 15% by weight and 80% by volume), adhesive monomer, functional methacrylate, methyl methacrylate (MMA), and photochemical initiator called as Equia (GC Corp., Japan) (Basso, 2011). Manufacturers of Equia system claims 3 min for complete restoration with several layer coating of resin guarantee lasting protection of the restoration and margins, increasing resistance to both flexion and wear (Friedl et al., 2011). Different groups incorporating particles of microhydroxyapatite (5–10 μm), nanohydroxyapatite (100–500 nm), nanofluroapatite (around 100–200 nm) in conventional-GIC and resin-modified glass ionomers cents to developed compressive strength (177–179 MPa), higher diametral tensile strength (19–20 MPa), and higher biaxial flexural strength (26–28 MPa) (Lee et al., 2010a; Moshaverinia et al., 2008).

5.2 NANOBIOMATERIALS FOR PREVENTIVE DENTISTRY

There are several ways in which nanosized particles can be used to prevent tooth decay. Nanoparticles have been shown to induce remineralization of enamel. For instance, it has been observed that nanocrystals of casein phosphopeptide amorphous calcium phosphate promote not only remineralization but also demonstrate anticarious activity in vivo and in vitro (Dunn et al., 2002). Additionally, the remineralizing effect of rinsing with nanocarbonate-hydroxyapatite nanocrystals on dentin results in a reduced dentin sensitivity (Drexler, 1992). Indeed, TEM analysis shows that these nanocrystals can block dentinal tubules for up to 10 min (Drexler, 1992). Moreover, a double-blinded randomized clinical trial among 70 patients has also concluded that using a dentifrice containing nanocarbonate hydroxyapatite can produce statistically significant reduction in dentine hypersensitivity and discomfort when compared with conventional antisensitivity toothpastes (Moldovan and Ferrari, 2002). Biomimetic remineralization of enamel by using dentifrices incorporated with nanohydroxyapatite crystals has been well documented (Roszek et al., 2005; Zafar and Ahmed, 2015). This has been supported by Nakashima et al. who have observed statistically significant remineralization of tooth structure in artificially induced dental caries after application of dentifrice containing nanohydroxyapatite for 20 days (Agarwal et al., 2005b). However, it has been suggested that the remineralization effect of nanohydroxyapatite crystals is generally limited to the surface of the initial carious lesion (Roco and Bainbridge, 2002; Roszek et al., 2005).

Another mode of delivering anticaries therapy to carious lesions and cavities is incorporation of nanosized bioactive fillers like dicalcium phosphate anhydrous to resin composites without undermining the mechanical properties (Balshaw et al., 2005; Thome, 2006). Resin composites filled with nanocrystals of CaF_2 have been seen to release fluoride comparable to or higher than resin-modified glass-ionomers cements in addition to increased mechanical properties (Gupta 2011; Tinkle et al., 2009). Incorporation of nanofluorohydroxyapatite to conventional and resin-modified glass ionomers cements not only could increase the fluoride release but also enhance their mechanical properties (Moshaverinia et al., 2008). Another way of modifying glass ionomers cements is addition to nanohydroxyapatite (Moshaverinia et al., 2011). Since nanohydroxyapatite is known to inhibit dental caries and promote remineralization, these types of restorations could prevent recurrent caries demineralization around restoration margins. Combining bioactive silver nanoparticles and glass ionomer with antibacterial agents such as chlrohexidine and quaternary ammonium dimethacrylate have also been observed to inhibit the growth of *Streptococcus mutans*, a major causative bacterial agent of dental caries (Siqueira et al., 2015). Glass-ionomers cements modified by addition of TiO_2 nanoparticles have enhanced antibacterial effect and superior mechanical properties but their biocompatibility is still debatable (Garcia-Contreras et al., 2014, 2015). Amelogenin, an extracellular matrix protein, known to stimulate crystallization and organization of enamel crystals (Yang et al., 2010). Apatite crystals as small as 35 nm have been observed to have been formed following application of amelogenin on demineralized enamel (Fan et al., 2012).

Due to the lack of randomized control trials and clinical data, it is too early ascertain whether or not using nanotechnology is superior to conventional preventive measures. However, most of the in vitro and in vivo studies conducted to date show promising results and suggest that in the initial stages, nano-sized particles of particles, such as hydroxyapatite and calcium phosphate could not only prevent dental caries but also promote remineralization to an extent. Nevertheless, more clinical data is definitely vital to determine the future of nanotechnology in preventative dentistry.

5.3 NANOBIOMATERIALS FOR ENDODONTICS

Endodontics is the study of biology, morphology, physiology, aetiology, and diagnosis of pathology, and treatment of dental pulp and periradicular tissues (Siddiqui et al., 2012). The aim of endodontics (surgical and/or nonsurgical) includes: the differential diagnosis and management and treatment of oral pain from pulpal and/or periradicular origin; vital pulp therapy procedures (paedodontics and adult endodontics); nonsurgical treatment of root canal systems and their filling (obturation); surgical removal of diseased tissues resulting from pulpal pathology; repair procedures related to the removed tissues and iatrogenic defects; intentional replantation and replantation of avulsed teeth; hemi-sections, root-end resections, and root resections; root-end fillings; bleaching of discolored teeth (internal bleaching); retreatment of previously endodontically treated teeth; and treatment with posts and cores for coronal restorations. The success of the treatment primarily depends on expatiating of pulp, making the pulp chamber and canals free from microbes, shaping of canals and good sealing (coronal and apical) of canals to prevent microleakage. Endodontic materials and instruments can be broadly categorized as those utilized to cleaning and shaping of pulp canals (reamers and files), those for maintain pulp vitality and used in pulp canal therapy for disinfection of the pulp space (irrigants and intracanal medicaments) and root canal filling (solid materials and sealers) (Rizvi et al., 2014).

5.3.1 Endodontic instruments surface coatings

Stainless steel instrument fracture during cleaning and shaping of canals create permanent endodontic pain and disease required surgical intervention. Therefore, nickel-titanium (Ni-Ti) rotary instrument were introduced because of their superelasticity, superb shape memory, and excellent corrosion resistance capability. Due to complexity in root canal morphology, Ni-Ti instruments get fatigue and failure of that alloy lead to possible canal instrument fracture. With the help of nanotechnology Adini et al. (2011a,b) reported coating of cobalt with impregnated fullerene-like WS_2 nanoparticles on nickel-titanium (Ni-Ti) file and analyzed improvement in fatigue resistance and breakage time of coated files. Another group of researchers, using electrodeposition method for coating nanoparticles of cobalt with impregnated fullerene-like (IF-WS_2), observed 66% reduction in friction coefficient with low piercing rate (Samorodnitzky-Naveh et al., 2009). Medical application of impregnated-fullerene (IF-WS_2)

coated via dry wet method or physical vapor deposition (PVD) method possess lesser lower friction and wear when compared to conventional cobalt files which could aid cleaning and shaping of canals (Adini et al., 2011b). Schäfer (2002a) report a higher cutting efficacy of titanium nitride (TiN) coated and noncoated nickel-titanium (Ni-Ti) K-files. Another study reported there is no effect of exposure to sodium hypochlorite (NaOCl) or repeated sterilization effect on cutting efficiency PVD-coated NiTi K-files (Schäfer, 2002b).

5.3.2 Modifications of endodontic sealer

Due to unavailability of chemical bonding of root obturation materials to teeth, another material can be added to enhance bonding of root dentine and obturation material, also provide seal to canal irregularities or minor discrepancies between root canal walls and gutta-percha, and antibacterial effects as well. This material is called root canal sealer (Cohen et al., 2006). Sealers also serve as lubricant, closed lateral or accessory canals, and improve canal sealing (Chng et al., 2005). In Table 26.4, a list of available root canal sealer discussed in detailed.

According to Gutmann (1998), the requirements of a root canal sealer are:

1. Penetration of sealer to patent dentinal tubules
2. Provide antibacterial effect against endodontic microbes
3. Bind strongly to organic and inorganic phases of dentine
4. Capable of inducing the cementogenesis around apical foramen
5. Strengthen root canals

Different kinds of sealers are present, each with advantages and disadvantages, but no current sealer satisfies all the characteristics needed. The introduction of nanoparticles in dentistry helped in designing the nanoparticles-based root canal sealers that can penetrate dentinal tubules and ensure sealing of all tubules smartly (Mohan et al., 2013). Some nanotechnology-based root canal sealers are discussed in the next sections (Javidi et al., 2014).

5.3.3 EndoSequence BC

EndoSequence BC sealer has been developed using bioceramic nanotechnologies and is composed of calcium silicate, calcium phosphate, and calcium hydroxide, zirconia, and thickening agents (Zoufan et al., 2011). Nanoparticles used in BC sealer has improved its handling and physical properties. During the hydration reaction in the root canal, nanocomposite structure of calcium silicate and hydroxyapatite is formed. The hydration reaction and setting time is affected by the availability of water and the setting time may be prolonged in overly dried canals (Loushine et al., 2011).

5.3.4 Guttaflow sealer

Guttaflow sealer is one of the latest products introduced for endodontic applications. It is a silicon-based sealer with the addition of gutta-percha powder and silver nanoparticles (Anantula and Ganta, 2011). This material is available in capsule form that can be mixed and injected.

5.3.5 Endodontic sealer modified using antibacterial nanoparticles

For infection points of view, the antibacterial activity of endodontic sealers can be very beneficial. Antibacterial quaternary ammonium polyethylene mine (QPEI) nanoparticles have been incorporated into existing sealers, such as AH plus, epiphany, and guttaflow (Bouillaguet et al., 2008). Resin composites

Table 26.4 List of Available Root Canal Sealers

Materials	Description	Advantages	Disadvantages	Examples
Zinc oxide Eugenol cement sealer	- Most commonly used sealer - Modification of the regular ZnOE - The mixing vehicle is mostly eugenol - Powder contains ZnO that has been finely sifted to enhance the flow - Setting process is a chemical reaction with physical embedding of ZnO in a matrix of zinc eugenolate - Free Eugenol always remains in the mass as an irritant	- Absorbed if extruded into periradicular tissue - Antimicrobial effect through ZnO or additives such as rosin and Canada balsam or corticosteroids	- Shrinks on setting - stains the tooth (Roth's Sealer was developed to be as a nonstaining ZnOE sealer)	- Pulp Canal Sealer (SybronEndo) - Pulp Canal Sealer EWT - Proscol (Procosol, Inc., Philadelphia, PA) - Roth (Roth's Pharmacy, Chicago, IL) - Tubliseal (Sybron Endo/ Kerr) - Wach's Sealer Balas Dental, Chicago, IL)
Calcium hydroxide sealers	- It was developed for two main reasons: 1. Antimicrobial effect 2. Osteogenic-Cementogenic potential - However, these two effects were not demonstrated		- Needs to be soluble in order to leach out the hydroxyl group responsible for its activity but sealers shouldn't exhibit solubility - Poor cohesive strength - Poor dentin adhesion	- Calciobiotic Root Canal Sealer (CRCS) (Coltene Whaledent, Altstätten, - Switzerland) - Sealapex (Sybron Endo) - Apexit (Ivoclar Vivadent)
Noneugenol sealers	- Developed from periodontal dressings	No irritating effect of eugenol		Nogenol (GC America, Alsip, IL)
Glass ionomer sealers	- Developed to take advantage of their bond to dentin, fluoride release, antimicrobial activity, and biocompatibility	- Dentin binding - Minimal tissue - Irritation - Low toxicity	- Solubility concern - Inadequate bonding with gutta-percha - Minimal antibacterial effect	- Ketac-Endo (3M ESPE, Minneapolis, MN) - Activ GP (Brassler USA, Savannak, GA), which is a gutta-percha cone impregnated with GI coating.

Materials	Description	Advantages	Disadvantages	Examples
Resin	- Introduced because it provides good adhesion - AH-26 and AH-plus are epoxy-resins. While Epiphany is a dual curable dental resin composed of Bis-GMA, ethoxylated Bis-GMA, UDMA, hydrophilic difunctional methacrylates and fillers	- Good adhesion ability - Does not contain eugenol	AH- 26 contains formaldehyde, which is toxic when freshly mixed but this toxicity declines rapidly after 24 h. This was modified by the introduction of AH- plus that does not release formaldehyde	- AH-26 (Dentsply International Maillfer) - AH-Plus (Dentsply International) - Epiphany (Pentron Clinical Technologies) - EndoREZ (Ultradent Products, South Jordan, UT) - Diaket (3M ESPE)
Silicone-based sealers	- Two materials are available: 1. RoekoSeal (Langenau, Germany): which is a polyvinylsiloxane or polydimethylsiloxane root canal sealer 2. GuttaFlow (Cuyahoga Falls, OH): a. Cold, flowable, self-curing obturation material for root canals that combines gutta-percha and polydimethylsiloxane sealer (RoekoSeal) into one injectable system. b. Fills canal irregularities. c. Inject the material in the canal and then place a single master cone. d. Working time is 15 min and cures in 25–30 min.			
Chloropercha	- White gutta-percha (Alba) mixed with chloroform - Not used anymore	The addition of chloroform allows the gutta-percha cone to fit better in the canal	- No adhesive ability - Toxicity of chloroform	
Bioceramics (calcium silicate) sealers	- Composed of zirconium oxide and calcium silicates	- Hydrophilic - Sets by utilizing moisture - No shrinkage on setting - Antimicrobial effect during setting - Biocompatible	Hardens on setting so retreatment might difficult to perform if required	- iRoot SP or Endosequence BC Sealer (Innovative BioCeramix Inc., Vancouver, Canada) - MTA Fillapex (Angelus) - Tech Biosealer Endo (TBE, Isasan, Rovello Porro, Italy)
Sealers containing formaldehyde	Absolutely contraindicated		- Toxic - Overextension causes irreversible neurotoxicity dysesthesia	- Endomethasone (Septodont, Paris, France) - Rieber's paste (Amubarut; Wera Karl, Biesingen, Germany) - N2 (Indrag-Agsa, Losone, Switzerland) - SPAD

containing QPEI nanoparticles resulted in prolonged antibacterial activity without compromising the mechanical properties (Barros et al., 2014). In order to obtain similar antibacterial effect in endodontic sealers, 0–2 wt.% QPEI nanoparticles were added in to the commercially available sealers.

5.3.6 Nanohydroxyapatite (nano-HA) gutta percha

Nanohydroxyapatite-based endodontic sealer has been developed recently. Nano-HA crystals (40–60 nm) were synthesized using wet chemical methods and utilizing calcium hydroxide and phosphoric acid as precursors. This nanosealer is composed of nano-HA, bismuth oxide (for radiopacity), hexamethylenetetramine as an activator and bisphenol-A-diglycidyl ether as liquid (Masudi et al., 2010). The key advantages of this technology are sealing capabilities that are as good as of epiphany and AH 26 endodontic sealers and no adverse effect of heat application used during root filling.

5.3.7 Bioactive glass nanomaterials as root canal disinfectant

Bioceramics are well known for excellent biocompatibility and bioactivity (Agarwal et al., 2005a; Sheikh et al., 2015). The nanosized bioactive glass particles prepared in specific composition (58S) have demonstrated antibacterial activity (Waltimo et al., 2007). The bioactive glass (58S, 63S, and 72S) have Sio2-CaO-P2O5 system were evaluated for antibacterial activity (Mohn et al., 2010). The nanosized particles (20–90 nm) for these bioactive glasses were prepared using sol-gel technique (Hong et al., 2009). The optimal concentration of 58S and 63S for antibacterial affect was described at 100 mg/mL. At this concentration (100 mg/mL), antibacterial activity was effective and no bacterial growth was observed on agar plates. The rise in the pH of nanoglass materials is also a contributory factor for antibacterial activity (Kishen, 2010). The pH of fresh solution (~7.3) rises up to 8.8 within 1 day and further rises to more than 9.0 in 5 days. The smaller (nanosized) particle bioactive glasses can result in the enhancement of antibacterial activity by leaching of alkaline species (Xia and Chang, 2007). These bioactive glass nanomaterials are considered to be a good potentials candidate for root canal disinfection applications (Waltimo et al., 2007).

5.3.8 Nanobiomaterials for regenerative endodontics

The applications of nanoscaffold materials for endodontic regeneration are well established (Murray et al., 2007). For pulp regeneration, pulp stem cells were purified in the laboratory and grown in sheets on scaffolds (Samant et al., 2012). The scaffolds used were composed of nanofibers of biodegradable collagen type I or fibronectin (Khurshid et al., 2016; Zafar et al., 2015, 2016). Self-assembling polypeptide hydrogels have been used for pulp tissue regeneration. There is formation of a nanofiber mesh that supported the growing cells (Nakashima and Iohara, 2011). Puramatrix (contain amino acids repeats of alanine, arginine, aspartate) has proven to enhance cell growth. Injectable self-assembly collagen I scaffold loaded with exfoliated teeth stem cells resulted in the formation of pulp-like tissue and functional odontoblasts (Zhang et al., 2005b). Collagen type I is the most abundant fibrous protein found in the form of nanofibers in Dentin (~80%–90% of organic matrix) and bone. Odontogenic differentiation and mineralization was promoted in the presence of type I collagen scaffolds (Zhang et al., 2005a).

5.4 NANOBIOMATERIALS IN ORAL IMPLANTOLOGY

A dental implant is a fixture that is placed surgically in the alveolar bone and has the ability to support a removable or fixed prosthesis (Albrektsson et al., 1981; Branemark, 1977). A direct contact should exist between the bone and the dental implant along with stability. This direct contact is known as osseointegration (Albrektsson and Johansson, 2001). However, if a fibrous connective tissue forms between the implant and bone, the implant can fail (Hutton et al., 1994). Successful osseointegration requires a suitable implant material, an adequate surgical technique and a sufficient healing period (Albrektsson et al., 1986). Not only should the dental implant have a suitable design, at the surface level there are many features that must be present to enhance osseointegration (Esposito et al., 1998). Increasing the surface roughness, enhancing the hydrophilicity, and coating the surface with materials that enhance osseointegration (i.e., osseoconductive materials) have all been seen to improve osseointegration (Le Guéhennec et al., 2007). Currently, most dental implants are made of grade 4 commercially pure titanium (cpTi) or grade 5 titanium alloy (Ti6Al4V) (McCracken, 1999). Recently, composites of materials such as zirconia and poly-ether-ether-ketone (PEEK) have also been tested in vivo as dental implants (Najeeb et al., 2015, 2016). Zirconium composite implants made of zirconia and titanium has been approved for clinical use (Berner et al., 2009; Chiapasco et al., 2012; Karl et al., 2013). Table 26.5 lists some nanomodifications of dental implants along with their average reported surface roughness.

5.4.1 Surface hydrophilicity and microlevel surface roughness

At the chemical level, highly hydrophilic implant surfaces interact more favorably with the human tissue owing to their high water content. Implants possessing a low water-contact angle, an indicator of high hydrophilicity, have been shown to have a higher bone-implant contact when compared to those possessing a high water-contact angle (Buser et al., 2004; Zhao et al., 2005). However, prior studies have not observed any significantly higher osseointegration around hydrophilic surfaces (Carlsson et al., 1988). Another important determinant of osseointegration is the implant surface roughness. Surface roughness can be categorized according to size of the roughness. While macrolevel roughness includes the presence of threads and features of more than 10 μm, microlevel and nanolevel characteristics are those in the range of 1–10 μm and nanometers respectively (Le Guéhennec et al., 2007). Plasma-spraying is a process by which highly rough layers of titanium, having a surface roughness of around 7 μm, can be deposited. This is done by spraying a powder of the material through a plasma torch. After contacting the implant surface and fusing upon cooling, a thick layer of the material is deposited onto the implant (Liu et al., 2004). However, a major concern of these layers is the dislodgment of titanium into the blood stream. Titanium particles dislodged from implant sites have been

Table 26.5 Some Nanosized Modifications of Titanium Implants and Their Surface Geometry

Surface Modification	Geometry	Reference(s)
Lithography and laser-pitting	R_a = 40–700 nm	Zhao et al. (2005)
TiO_2 blast + HF immersion	R_a = 50–100 nm	Abron et al. (2001); Cooper et al. (2006)
Sol-deposition of CaP	R_a = 50–100 nm	Liu et al. (2004)
Induction plasma sprayed HAp	Grain size = 15–20 nm	Roy et al. (2011a)
Anodized Ti (nanotubes)	Pore size = 25–60 nm	Gong et al. (2001)
Ion beam–assisted deposition	27 nm thickness	Coelho and Suzuki (2005)

Ra, Arithmetic average of absolute values of surface roughness.

found in liver, spleen, and lymph nodes (Urban et al., 2000). Plasma-sprayed hydroxyapatite coatings have been reported to de-bond, possibly due to the different phases formed because of the heat involved during the spraying procedure (Le Guéhennec et al., 2007). Additionally, implants possessing surface roughness in the micrometer range produced by grit blasting with Al2O3 have also been reported to release the embedded grit particles into the surrounding environment (Mohammadi et al., 2007). Perhaps, a major issue of implants possessing highly rough surfaces is the possibility of ion-leakage and implantitis (Becker et al., 2000).

5.4.2 Nanostructured titanium implant surfaces

To overcome the drawbacks of micrometer-level implant surfaces, studies have attempted to produce controlled surface topography at a nanometer-level (Le Guéhennec et al., 2007; Mendonça et al., 2008). Laser pitting and lithographic printing can be employed to impart a surface with nanofeatures. Both processes have the advantage of producing more controlled features on implant surfaces as opposed to the more random features produced by processes, such as plasma spraying, grit-blasting, and biomimetic deposition of calcium phosphates (Bigerelle et al., 2002; Zhu et al., 2004). Indeed, the average surface roughness (Ra) of 700, 400, 60, and 40 nm on acid-etched, porous anodized, mechanically polished and electro-polished surfaces, respectively, by Zhu et al. (2004) have been observed to increase the proliferation of MG63 osteoblast-like cell. It has been suggested that nanolevel surface roughness aids cellular attachment by making it easier for the filopodia to enter them, providing cell guidance for selective attachment (Klymov et al., 2013).

OsseoSpeed (Astra Tech AB, Molndal, Sweden) implants possess a TiO_2-blasted surface, which is immersed in hydrofluoric acid (HF) to produce a roughness of 50–100 nm (Abron et al., 2001; Cooper et al., 2006). Moreover, this nanotopography has been observed to increase the level of the genes Runx2, osterix, bone sialoprotein, and alkaline phosphatase all of which indicate higher osteoblastic activity (Guo et al., 2007). Similar up regulation of IGF-2, BMP6, and BMP2 has been observed when these implants are tested with human mesenchymal stem cells (Mendonça et al., 2008). Indeed, OsseoSpeed implants have demonstrated higher bone growth, torque removal, and a higher BIC in vivo (Ellingsen et al., 2003; Zandinejad et al., 2006). A multicenter trial involving 1860 OsseoSpeed implants has had favorable results with just 4% of them showing inflammation and just 25 implants failing after one year of placement (Stanford et al., 2006).

Sol-gel deposition of calcium phosphate (CaP) crystals, having an approximate mean size of 20 nm, can be deposition on titanium implants to provide a nanotopography having a surface roughness of 50–100 nm (Mendes et al., 2007). It has been observed that this surface produces superior bone ingrowth when compared to untreated titanium samples. Niobium oxide coatings have a surface roughness of 7–15 nm on titanium discs produced by sol-gel deposition, has also been seen to increase cellular adhesion and collagen production (Eisenbarth et al., 2007).

Induction (supersonic) plasma sprayed hydroxyapatite (HAp) coatings have a nanotopography in the range of 15–20 nm grain-size and a crack-free structure when compared to the often cracked microlevel topography of coatings produced by conventional plasma spraying (Roy et al., 2011a). The main difference between normal plasma spray and induction plasma spray is the high speed of up to 510 m/s in the latte (Le Guéhennec et al., 2007). Induction plasma spray can also be used to produce nanometer-coatings of titanium oxide on titanium implants (Boulos, 1992). More recently, strontium doped HAp nanocoatings produced by induction plasma spraying have shown superior stimulation of cellular activity when compared to nondoped HAp coatings (Roy et al., 2011b). Nevertheless, more in vivo studies, and subsequent clinical trials, are needed before these implant coatings can be used clinically.

When titanium is anodized in an acidic solution, nanotubes of TiO_2, having a size of 25–60 nm, are formed on its surface (Gong et al., 2001). The nanotubular structure has exhibited induction of enhanced cellular proliferation and attachment in vitro (Bjursten et al., 2010; Sjöström et al., 2011). Immersing nanotubular surfaces of titanium in NaOH leads to the formation of sodium tianate, which has been to induce HAp growth when immersed in SBF (Oh et al., 2005) Furthermore, these anodized and nano-CaP coated surfaces have the potential to induce accelerated attachment and proliferation of MG63 osteoblast-like cells when tested in vitro (Lee et al., 2010b; Sjöström et al., 2011). Indeed, when tested in vivo, TiO_2 nanotubes have been observed to induce bone formation almost nine times of that produced by grit-blasted titanium implants after 4 weeks of implantation (Bjursten et al., 2010). Hence, it is not surprising that titanium implants with nanotubes require a significantly higher removal torque than untreated implants 6 weeks after implantation (Son et al., 2003).

Other coating processes like ion beam assisted deposition (IBAD) can also be used to deposit a thin layer of bioactive ceramics like CaP (Coelho and Suzuki, 2005). When compared to acid etched/sand-blasted implants have a micrometer topography, IBAD-coated implants have been to induce higher bone growth when tested in canine models (Coelho et al., 2009; Ehrenfest et al., 2010). Moreover, when compared to plasma-sprayed HAp coated implants, IBAD-coated implants have exhibited lower degradation rates in vitro and in vivo (Luo et al., 2000).

5.4.3 Nanomanufacturing of other implant materials

There has a significant amount of research in modifying the surface of implant reinforced polymers like PEEK (Najeeb et al., 2016). Indeed, processes like spin-coating, electron-beam (e-beam) deposition, plasma gas etching and plasma immersion ion implantation (PIII) have been used to impart nanometer-level coatings and roughness on PEEK's surface (Najeeb et al., 2015). Moreover, nanosized crystals of TiO_2, HAp, and fluorohydroxyapatite can be combined with PEEK to produce bioactive, stronger and antibacterial composites (Sanpo et al., 2009; Tran et al., 2009). However, to date there have many in vivo studies and human trials to ascertain the future of these implant surfaces.

6 CONCLUSIVE REMARKS

The application of nanomaterials is the most challenging in the field of medicine and biomedical purposes. This is an area of very active research all around the globe involving a lot of research funding. It can be expected in future that the science of dental materials may change significantly with better understanding and the introduction of new nanobiomaterials.

REFERENCES

Abron, A., Hopfensperger, M., Thompson, J., Cooper, L.F., 2001. Evaluation of a predictive model for implant surface topography effects on early osseointegration in rat tibia model. J. Prosthetic. Dent. 85, 40–46.

Adini, A.R., Feldman, Y., Cohen, S.R., Rapoport, L., Moshkovich, A., Redlich, M., Moshonov, J., Shay, B., Tenne, R., 2011a. Alleviating fatigue and failure of NiTi endodontic files by a coating containing inorganic fullerene-like WS 2 nanoparticles. J. Mater. Res. 26, 1234–1242.

Adini, A., Redlich, M., Tenne, R., 2011b. Medical applications of inorganic fullerene-like nanoparticles. J. Mater. Chem. 21, 15121–15131.

Agarwal, A., Gupta, N.D., Jain, A., Páll, E., Florea, A., Soriţău, O., Cenariu, M., Petruţiu, A.S., Roman, A., Li, W., 2005a. Position paper: periodontal regeneration. J. Periodontol. 76, 1601–1622.

Agarwal, A., Paul, B.C., Mahmoodi, H., Datta, A., Roy, K., 2005b. A process-tolerant cache architecture for improved yield in nanoscale technologies. VLSI 13, 27–38.

Al-Boni, R., Raja, O.M., 2010. Microleakage evaluation of silorane based composite versus methacrylate based composite. J. Conserv. Dent. 13, 152.

Albrektsson, T., Brånemark, P.I., Hansson, H.-A., Lindström, J., 1981. Osseointegrated titanium implants: requirements for ensuring a long-lasting, direct bone-to-implant anchorage in man. Acta Orthopaedica. 52, 155–170.

Albrektsson, T., Johansson, C., 2001. Osteoinduction, osteoconduction and osseointegration. Eur. Spine J. 10 (2), S96–S101.

Albrektsson, T., Zarb, G., Worthington, P., Eriksson, A., 1986. long-term efficacy of currently used dental implants: a review and proposed criteria of success. Int. J. Oral Maxillofac. Implants 1, 11–25.

Anantula, K., Ganta, A.K., 2011. Evaluation and comparison of sealing ability of three different obturation techniques: lateral condensation, Obtura II, and GuttaFlow: an in vitro study. J. Conserv. Dent. 14, 57.

Balshaw, D.M., Philbert, M., Suk, W.A., 2005. Res. strategies for safety evaluation of nanomaterials, part III: nanoscale technologies for assessing risk and improving public health. Toxicol. Sci. 88, 298–306.

Barros, J., Silva, M., Rodrigues, M., Alves, F., Lopes, M., Pina-Vaz, I., Siqueira, J., 2014. Antibacterial, physicochem and mechanical properties of endodontic sealers containing quaternary ammonium polyethylenimine nanoparticles. Int. Endodon. J. 47, 725–734.

Basso, M., 2011. Teeth restoration using a high-viscosity glass ionomer cement: Equia system. J. Min. Intervent. Dent. 4, 74–76.

Becker, W., Becker, B.E., Ricci, A., Bahat, O., Rosenberg, E., Rose, L.F., Handelsman, M., Israelson, H., 2000. A prospective multicenter clinical trial comparing one-and two-stage titanium screw-shaped fixtures with one-stage plasma-sprayed solid-screw fixtures. Clin. Implant. Dent. Res. 2, 159–165.

Bennett, C.H., 1987. Demons, engines and second law. Sci. Am. 257, 108–116.

Berner, S., Dard, M., Gottlow, J., Molenberg, A., Wieland, M., 2009. Titanium-zirconium: a novel material for dental implants. Eur. Cell. Mater. 17, 189–205.

Bhushan, B., 2009. Biomimetics: lessons from nature: an overview. Philos. Trans. R. Soc. 367, 1445–1486.

Bigerelle, M., Anselme, K., Noel, B., Ruderman, I., Hardouin, P., Iost, A., 2002. Improvement in morphology of Ti-based surfaces: a new process to increase in vitro human osteoblast response. Biomaterials 23, 1563–1577.

Binnig, G., Rohrer, H., Gerber, C., Weibel, E., 1982. Surface studies by scanning tunneling microscopy. Physical. Rev. Lett. 49, 57.

Biswas, A., Bayer, I.S., Biris, A.S., Wang, T., Dervishi, E., Faupel, F., 2012. Advances in top–down and bottom–up surface nanofabrication: techniques, applications future prospects. Adv. Colloid Interface Sci. 170, 2–27.

Bjursten, L.M., Rasmusson, L., Oh, S., Smith, G.C., Brammer, K.S., Jin, S., 2010. Titanium dioxide nanotubes enhance bone bonding in vivo. J. Biomed. Mater. Res. 92, 1218–1224.

Bouillaguet, S., Shaw, L., Barthelemy, J., Krejci, I., Wataha, J., 2008. Long-term sealing ability of pulp canal sealer, AH-plus, guttaflow and epiphany. Int. Endodon. J. 41, 219–226.

Boulos, M., 1992. RF induction plasma spraying: state-of-the-art review. J. Thermal. Spray. Technol. 1, 33–40.

Branemark, P.-I., 1977. Osseointegrated implants in treatment of edentulous jaw: experience from a 10-year period. Scand. J. Plast. Reconstr. Surg. 16, 1–132.

Buser, D., Broggini, N., Wieland, M., Schenk, R., Denzer, A., Cochran, D., Hoffmann, B., Lussi, A., Steinemann, S., 2004. Enhanced bone apposition to a chemically modified SLA titanium surface. J. Dent. Res. 83, 529–533.

Carlsson, L.V., Alberktsson, T., Berman, C., 1988. Bone response to plasma-cleaned titanium implants. Int. J. Oral Maxillofac. Implants 4, 199–204.

Carneiro, J., Teixeira, V., Carvalho, P., Azevedo, S., Manninen, N., 2011. Self-Cleaning Smart Nanocoatings. Nanocoatings and Ultra-Thin Films. Woodhead Publishing Ltd, Cambridge, UK, 397-413.

Chan, K., Lee, Y.-D., Nicolella, D., Furman, B., Wellinghoff, S., Rawls, R., 2007. Improving fracture toughness of dental nanocomposites by interface engineering and micromechanics. Eng. Fract. Mech. 74, 1857–1871.

Chen, M.-H., 2010. Update on dental nanocomposites. J. Dent. Res. 89, 549–560.

Chen, M.-H., Chen, C.-R., Hsu, S.-H., Sun, S.-P., Su, W.-F., 2006. Low shrinkage light curable nanocomposite for dental restorative material. Dent. Mater. 22, 138–145.

Chiapasco, M., Casentini, P., Zaniboni, M., Corsi, E., Anello, T., 2012. Titanium–zirconium alloy narrow-diameter implants, Straumann Roxolid for rehabilitation of horizontally deficient edentulous ridges: prospective study on 18 consecutive patients. Clin. Oral Implants Res. 23, 1136–1141.

Chitnis, D., Dunn, W.J., Gonzales, D.A., 2006. Comparison of in-vitro bond strengths between resin-modified glass ionomer, polyacid-modified composite resin, and giomer adhesive systems. Am. J. Orthodon. Dent. Ortho. 129, e11–e16.

Chng, H.K., Islam, I., Yap, A.U.J., Tong, Y.W., Koh, E.T., 2005. Properties of a new root-end filling material. J. Endodon. 31, 665–668.

Coelho, P.G., Granjeiro, J.M., Romanos, G.E., Suzuki, M., Silva, N.R., Cardaropoli, G., Thompson, V.P., Lemons, J.E., 2009. Basic Res. methods and current trends of dental implant surfaces. J. Biomed. Mater. Res. 88, 579–596.

Coelho, P.G., Suzuki, M., 2005. Evaluation of an IBAD thin-film process as an alternative method for surface incorporation of bioceramics on dental implants: a study in dogs. J. Appl. Oral Sci. 13, 87–92.

Cohen, S., Burns, R.C., Hargreaves, K.M., Berman, L.H., 2006. Pathways of Pulp. Elsevier Mosby St. Louis, USA.

Cooper, L.F., Zhou, Y., Takebe, J., Guo, J., Abron, A., Holmén, A., Ellingsen, J.E., 2006. Fluoride modification effects on osteoblast behavior and bone formation at TiO_2 grit-blasted cp titanium endosseous implants. Biomaterials 27, 926–936.

Drexler, K.E., 1992. Nanosystems: Molecular Machinery, Manufacturing, and Computation. John Wiley & Sons, UK.

Drexler, K.E., 2003. Drexler and Smalley make case for and against "molecular assemblers". Chem. Eng. News 81, 1.

Dunn, S., Whatmore, R. Chambers, G., 2002. Nanotechnology advances in Europe. European Parliament—STOA Working Papers, STOA 108. Europäisches Parliament, Luxembourg.

Ehrenfest, D.M.D., Coelho, P.G., Kang, B.-S., Sul, Y.-T., Albrektsson, T., 2010. Classification of osseointegrated implant surfaces: materials, chemistry and topography. Trends Biotechnol. 28, 198–206.

Eisenbarth, E., Velten, D., Breme, J., 2007. Biomimetic implant coatings. Biomol. Eng. 24, 27–32.

Eliseev, A.A., Chernysheva, M.V., Verbitskii, N.I., Kiseleva, E.A., Lukashin, A.V., Tretyakov, Y.D., Kiselev, N.A., Zhigalina, O.M., Zakalyukin, R.M., Vasiliev, A.L., 2009. Chem. reactions within single-walled carbon nanotube channels. Chem. Mater. 21, 5001–5003.

Ellingsen, J.E., Johansson, C.B., Wennerberg, A., Holmén, A., 2003. Improved retention and bone-to-implant contact with fluoride-modified titanium implants. Int. J. Oral Maxillofac. Implants 19, 659–666.

Esposito, M., Hirsch, J.M., Lekholm, U., Thomsen, P., 1998. Biological factors contributing to failures of osseointegrated oral implants. II. Etiopathogenesis. Eur. J. Oral Sci. 106, 721–764.

Fan, Y., Wen, T., Liao, S., Lallier, T., Hagan, J.L., Twomley, J.T., Xu, X., 2012. Novel amelogenin-releasing hydrogel for remineralization of enamel artificial caries. J. Bioact. Comp. Polym. 27, 585–603.

Faustini, M., Boissiere, C., Nicole, L., Grosso, D., 2013. From Chem. solutions to inorganic nanostructured mater: a journey into evaporation-driven processes. Chem. Mater. 26, 709–723.

Ferracane, J., 1995. Current trends in dental composites. Crit. Rev. Oral Biol. Med. 6, 302–318.

Ferracane, J.L., 2011. Resin composite: state of art. Dent. Mater. 27, 29–38.

Freitas, R.A., 2000. NanoDent. JADA 131, 1559–1565.

Frencken, J.E., Pilot, T., Songpaisan, Y., Phantumvanit, P., 1996. Atraumatic restorative treatment, ART.: rationale, technique, and development. J. Pub. Health Dent. 56, 135–140.

Frencken, J.E., Songpaisan, Y., Phantumvanit, P., Pilot, T., 1994. An atraumatic restorative treatment, ART. technique: evaluation after one year. Int. Dent. J. 44, 460–464.

Friedl, K., Hiller, K.-A., Friedl, K.-H., 2011. Clinical performance of a new glass ionomer based restoration system: a retrospective cohort study. Dent. Mater. 27, 1031–1037.

Gammaitoni, L., 2012. There's plenty of energy at bottom, micro and nano scale nonlinear noise harvesting. Contemp. Phys. 53, 119–135.

Garcia-Contreras, R., Scougall-Vilchis, R.J., Contreras-Bulnes, R., Kanda, Y., Nakajima, H., Sakagami, H., 2014. Effects of TiO_2 nano glass ionomer cements against normal and cancer oral cells. In Vivo 28, 895–907.

Garcia-Contreras, R., Scougall-Vilchis, R.J., Contreras-Bulnes, R., Sakagami, H., Morales-Luckie, R.A., Nakajima, H., 2015. Mechanical, antibacterial and bond strength properties of nano-titanium-enriched glass ionomer cement. J. Appl. Oral Sci. 23, 321–328.

Gong, D., Grimes, C.A., Varghese, O.K., Hu, W., Singh, R., Chen, Z., Dickey, E.C., 2001. Titanium oxide nanotube arrays prepared by anodic oxidation. J. Mater. Res. 16, 3331–3334.

Gonzalez, E.D.H., 2004. Long-term caries inhibitory effects of fluoride releasing tooth-colored restorative material. MSc thesis submitted to National University Singapore.

Guo, J., Padilla, R., Ambrose, W., De Kok, I., Cooper, L., 2007. Modification of TiO_2 grit blasted titanium implants by hydrofluoric acid treatment alters adherent osteoblast gene expression in vitro and in vivo. Biomaterials 28, 5418–5425.

Gupta, J., 2011. Nanotechnology applications in medicine and dentistry. J. Invest. Clin. Dent. 2, 81–88.

Gutmann, J.L., 1998. Obturation of cleaned and shaped root canal system. Path. Pulp, 258–361.

Hackmann, A., Delacave, H., Robinson, A., Labonte, D., Federle, W., 2015. Functional morphology and efficiency of antenna cleaner in *Camponotus rufifemur* ants. R. Soc. Open Sci. 2, 150129.

Hannig, M., Hannig, C., 2010. Nanomatererials in preventive dentistry. Nat. Nanotechnol. 5, 565–569.

Hes, K., Leung, S., Wei, S., 1999. Resin-ionomer restorative materials for children: a review. Aust. Dent. J. 44 (1), 1–11.

Hong, Z., Liu, A., Chen, L., Chen, X., Jing, X., 2009. Preparation of bioactive glass ceramic nanoparticles by combination of sol–gel and coprecipitation method. J. Non-crystal. Solids 355, 368–372.

Hutton, J.E., Heath, M.R., Chai, J.Y., Harnett, J., Jemt, T., Johns, R.B., McKenna, S., McNamara, D.C., van Steenberghe, D., Taylor, R., 1994. Factors related to success and failure rates at 3-year follow-up in a multicenter study of overdentures supported by Branemark implants. Int. J. Oral. Maxillofac. Implants 10, 33–42.

Ilie, N., Hickel, R., 2006. Silorane-based dental composite: behavior and abilities. Dent. Mater. J. 25, 445–454.

Ilie, N., Hickel, R., 2011. Resin composite restorative mater. Aust. Dent. J. 56, 59–66.

Javidi, M., Zarei, M., Naghavi, N., Mortazavi, M., Nejat, A.H., 2014. Zinc oxide nano-particles as sealer in endodontics and its sealing ability. Contemp. Clin. Dent. 5, 20.

Kaleem, M., Khan, A.S., Rehman, I.U., Wong, F.S., 2015. Effect of beverages on viscoelastic properties of resin-based dental composites. Materials 8, 2863–2872.

Kanaparthy, R., Kanaparthy, A., 2011. Changing face of dentistry: nanotechnology. Int. J. NanoMed. 6, 2799.

Karl, M., Krafft, T., Kelly, J.R., 2013. Fracture of a narrow-diameter roxolid implant: clinical and fractographic considerations. Int. J. Oral Maxillof. Implants 29, 1193–1196.

Kheraif, A.A., Qasim, S.B., Ramakrishnaiah, R., Rehman, I.U., 2013. Effect of different beverages on color stability and degree of conversion of nano and microhybrid composites. Dent. Mater. J. 32, 326–331.

Khurshid, Z., Naseem, M., Sheikh, Z., Najeeb, S., Shahab, S., Zafar, M.S., 2016. Oral antimicrobial peptides: types and role in oral cavity. Saudi Pharma. J. 24 (5), 515–524.

Khurshid, Z., Zafar, M., Qasim, S., Shahab, S., Naseem, M., AbuReqaiba, A., 2015. Advances in nanotechnology for restorative dent. Materials 8, 717–731.

Kishen, A., 2010. Advanced therapeutic options for endodontic biofilms. Endodon. Topics 22, 99–123.

Klymov, A., Prodanov, L., Lamers, E., Jansen, J.A., Walboomers, X.F., 2013. Understanding role of nano-topography on surface of a bone-implant. Biomater. Sci. 1, 135–151.

Le Guéhennec, L., Soueidan, A., Layrolle, P., Amouriq, Y., 2007. Surface treatments of titanium dental implants for rapid osseointegration. Dent. Mater. 23, 844–854.

Lee, J.-J., Lee, Y.-K., Choi, B.-J., Lee, J.-H., Choi, H.-J., Son, H.-K., Hwang, J.-W., Kim, S.-O., 2010a. Physical properties of resin-reinforced glass ionomer cement modified with micro and nano-hydroxyapatite. J. Nanosci. Nanotechnol. 10, 5270–5276.

Lee, J., Lee, J., Lim, Y., 2010b. In vitro investigation of anodization and CaP deposited titanium surface using MG63 osteoblast-like cells. Appl. Surf. Sci. 256, 3086–3092.

Lien, W., Vandewalle, K.S., 2010. Physical properties of a new silorane-based restorative system. Dent. Mater. 26, 337–344.

Liu, X., Chu, P.K., Ding, C., 2004. Surface modification of titanium, titanium alloys, and related materials for biomedical applications. Mater. Sci. Eng. 47, 49–121.

Lohbauer, U., 2010. Dental glass ionomer cements as permanent filling matter?—properties limitations and future trends. Materials 3, 76–96.

Loushine, B.A., Bryan, T.E., Looney, S.W., Gillen, B.M., Loushine, R.J., Weller, R.N., Pashley, D.H., Tay, F.R., 2011. Setting properties and cytotoxicity evaluation of a premixed bioceramic root canal sealer. J. Endodon. 37, 673–677.

Lovelh, L., Newman, S., Bowman, C., 1999. Effects of light intensity, temperature, and comonomer composition on polymerization behavior of dimethacrylate dental resins. J. Dent. Res. 78, 1469–1476.

Lucas, M.E., Arita, K., Nishino, M., 2003. Toughness, bonding and fluoride-release properties of hydroxyapatite-added glass ionomer cement. Biomaterials 24, 3787–3794.

Luo, Z., Cui, F., Feng, Q., Li, H., Zhu, X., Spector, M., 2000. In vitro and in vivo evaluation of degradability of hydroxyapatite coatings synthesized by ion beam-assisted deposition. Surf. Coat. Technol. 131, 192–195.

Maddox, J., 2002. Maxwell's demon: slamming door. Nature 417, 903–1903.

Majumder, D.D., Banerjee, R., Ulrichs, C., Mewis, I., Goswami, A., 2007. Nanomaterials: science of bottom-up and top-down. IETE Tech. Rev. 24, 9–25.

Masudi, S., Luddin, N., Mohamad, D., Alkashakhshir, J., Adnan, R. Ramli, R., 2010. In vitro study on apical sealing ability of nano-hydroxyapatite-filled epoxy resin based endodontic sealer. International Conference on Advancement of Materials and Nanotechnology, ICAMN—2007. AIP Conference Proceedings. AIP Publishing, Melville, New York.

McCracken, M., 1999. Dental implant materials: commercially pure titanium and titanium alloys. J. Prosthodon. 8, 40–43.

Mendes, V.C., Moineddin, R., Davies, J.E., 2007. effect of discrete calcium phosphate nanocrystals on bone-bonding to titanium surfaces. Biomaterials 28, 4748–4755.

Mendonça, G., Mendonça, D.B., Aragao, F.J., Cooper, L.F., 2008. Advancing dental implant surface technology–from micron- to nanotopography. Biomaterials 29, 3822–3835.

Mitra, S.B., Wu, D., Holmes, B.N., 2003. An application of nanotechnology in advanced dental materials. JADA 134, 1382–1390.

Mohammadi, Z., Ziaei-Moayyed, A., Mesgar, A.S.-M., 2007. Grit blasting of Ti–6Al–4V alloy: optimization and its effect on adhesion strength of plasma-sprayed hydroxyapatite coatings. J. Mater. Process. Technol. 194, 15–23.

Mohan, S., Gurtu, A., Dixit, K., Mehrotra, A., 2013. Nanotechnology: Its implications in conservative dentistry and endodontics. J. Dent. Sci. Oral Rehabil 4, 9–14.

Mohn, D., Zehnder, M., Imfeld, T., Stark, W., 2010. Radio-opaque nanosized bioactive glass for potential root canal application: evaluation of radiopacity, bioactivity and alkaline capacity. Int. Endodon. J. 43, 210–217.

Moldovan, N.I., Ferrari, M., 2002. Prospects for microtechnology and nanotechnology in bioengineering of replacement microvessels. Arch. Pathol. Lab. Med. 126, 320–324.

Moshaverinia, A., Ansari, S., Moshaverinia, M., Roohpour, N., Darr, J.A., Rehman, I., 2008. Effects of incorporation of hydroxyapatite and fluoroapatite nanobioceramics into conventional glass ionomer cements, GIC. Acta Biomater. 4, 432–440.

Moshaverinia, A., Roohpour, N., Chee, W.L., Schricker, S.R., 2011. A review of powder modifications in conventional glass-ionomer dental cements. J. Mater. Chem. 21, 1319–1328.

Mousavinasab, S.M., Meyers, I., 2009. Fluoride release and uptake by glass ionomer cements, compomers and giomers. Res J. Biol. Sci. 4, 609–616.

Murray, P.E., Garcia-Godoy, F., Hargreaves, K.M., 2007. Regenerative endodontics: a review of current status and a call for action. J. Endodon. 33, 377–390.

Najeeb, S., Khurshid, Z., Matinlinna, J.P., Siddiqui, F., Nassani, M.Z., Baroudi, K., 2015. Nanomodified PEEK dental implants: bioactive composites and surface modification: a review. Int. J. Dent., 2015.

Najeeb, S., Zafar, M.S., Khurshid, Z., Siddiqui, F., 2016. Applications of polyetheretherketone, PEEK. in oral implantology and prosthodontics. J. Prosthodont. Res. 60, 12–19.

Nakashima, M., Iohara, K., 2011. Regeneration of dental pulp by stem cells. Adv. Dent. Res. 23, 313–319.

Nicholson, J.W., 1998. Chemistry of glass-ionomer cements: a review. Biomaterials 19, 485–494.

Oh, S.-H., Finones, R.R., Daraio, C., Chen, L.-H., Jin, S., 2005. Growth of nano-scale hydroxyapatite using chemically treated titanium oxide nanotubes. Biomaterials 26, 4938–4943.

Rho, J.-Y., Kuhn-Spearing, L., Zioupos, P., 1998. Mechanical properties and hierarchical structure of bone. Med. Eng. Phys. 20, 92–102.

Ritchie, R.O., 2011. Conflicts between strength and toughness. Nat. Mater. 10, 817–822.

Rizvi, A., Zafar, M.S., Farid, W.M., Gazal, G., 2014. Assessment of antimicrobial efficacy of MTAD, sodium hypochlorite, EDTA and chlorhexidine for endodontic applications: an in vitro study. Middle-East J. Sci. Res. 21, 353–357.

Roco, M.C., 2003. Nanotechnology: convergence with modern biology and medicine. Curr. Opin. Biotechnol. 14, 337–346.

Roco, M.C., Bainbridge, W.S., 2002. Converging technologies for improving human performance: integrating from nanoscale. J. Nano. Res. 4, 281–295.

Roszek, B., De Jong, W., Geertsma, R., 2005. Nanotechnology in medical applications: state-of-the-art. Mater. Dev.

Roy, M., Bandyopadhyay, A., Bose, S., 2011a. Induction plasma sprayed nano hydroxyapatite coatings on titanium for orthopaedic and dental implants. Surface. Coat. Technol. 205, 2785–2792.

Roy, M., Bandyopadhyay, A., Bose, S., 2011b. Induction plasma sprayed Sr and Mg doped nano hydroxyapatite coatings on Ti for bone implant. J. Biomed. Mater. Res. 99, 258–265.

Samant, P.S., Tewari, T., Gupta, K., 2012. Revascularization regenerative endodontics: a review of current status. Pub. Health Res. Dev. 3, 27.

Samorodnitzky-Naveh, G.R., Redlich, M., Rapoport, L., Feldman, Y., Tenne, R., 2009. Inorganic fullerene-like tungsten disulfide nanocoating for friction reduction of nickel-titanium alloys. Nanomedicine 4, 943–950.

Sanpo, N., Tan, M.L., Cheang, P., Khor, K., 2009. Antibacterial property of cold-sprayed HA-Ag/PEEK coating. J. Thermal. Spray Technol. 18, 10–15.

Schäfer, E., 2002a. Effect of physical vapor deposition on cutting efficiency of nickel-titanium files. J. Endodont. 28, 800–802.

Schäfer, E., 2002b. Effect of sterilization on cutting efficiency of PVD-coated nickel–titanium endodontic instruments. Int. Endodont. J. 35, 867–872.

Sheikh, Z., Najeeb, S., Khurshid, Z., Verma, V., Rashid, H., Glogauer, M., 2015. Biodegradable material for bone repair and tissue engineering applications. Materials 8, 5744–5794.

Siang Soh, M., Sellinger, A., UJ Yap, A., 2006. Dental nanocomposites. Curr. Nanosci. 2, 373–381.

Siddiqui, F.A., Sheikh, A., Rashid, S., Khurshid, Z., Najeeb, S., Akhtar, S.A., 2012. Radiographic evaluation of dentin thickness and pulp space width for different age groups. JPDA 21, 02.

Siqueira, C., Magalhães, R., Pires, W., Pereira, C., Silveira-Lacerda, P., Carrião, M.S., Estrela, C., 2015. Cytotoxicity of glass ionomer cements containing silver nanoparticles. J. Clin. Exp. Dent. 7, e622–e627.

Sjöström, T., Lalev, G., Mansell, J.P., Su, B., 2011 Initial attachment and spreading of MG63 cells on nanopatterned titanium surfaces via through-mask anodization. Appl. Surf. Sci. 257, 4552–4558.

Son, W., Zhu, X., Shin, H., Ong, J.L., Kim, K., 2003. In vivo histological response to anodized and anodized/hydrothermally treated titanium implants. J. Biomed. Mater. Res. 66, 520–525.

Stanford, C., Johnson, G., Fakhry, A., Gartton, D., Mellonig, J., Wagner, W., 2006. Outcomes of a fluoride modified implant one year after loading in posterior-maxilla when placed with osteotome surgical technique. Appl. Osseointeg. Res. 5, 50–55.

Taniguchi, N., 1974. On basic concept of nanotechnology. Proc. Intl. Conf. Prod. Eng. Tokyo, Part II, Japan Society of Precision Engineering.

Thome, J.R., 2006. New frontier in heat transfer: microscale and nanoscale technologies. Heat Trans. Eng. 27, 1–3.

Tinkle, S., Poster, D., Cairns, C., Castner, D., Gause, C., Kojola, W., Kulkarni, A. Roberts, S., 2009. Nanomaterials and Human Health Instrumentation, Metrology, and Analytical Methods: 2009–2010 NNI Series of EHS Workshops and Reports.

Toumey, C., 2009. Plenty of room, plenty of history. Nat. Nanotechnol. 4, 783–784.

Tran, P.A., Sarin, L., Hurt, R.H., Webster, T.J., 2009. Opportunities for nanotechnology-enabled bioactive bone implants. J. Mater. Chem. 19, 2653–2659.

Ullah, R., Zafar, M.S., Al-Munawwarah, A.-M., Arabia, S., 2015. Oral and dental delivery of fluoride: a review. Fluoride 48, 195–204.

Urban, R.M., Jacobs, J.J., Tomlinson, M.J., Gavrilovic, J., Black, J., Peoc'h, M., 2000. Dissemination of wear particles to liver, spleen, and abdominal lymph nodes of patients with hip or knee replacement. J. Bone Joint Surg. 82 (4), 457–1457.

Uysal, T., Yagci, A., Uysal, B., Akdogan, G., 2010. Are nano-composites and nano-ionomers suitable for orthodontic bracket bonding? Eur. J. Orthodont. 32, 78–82.

Waltimo, T., Brunner, T., Vollenweider, M., Stark, W., Zehnder, M., 2007. Antimicrobial effect of nanometric bioactive glass 45S5. J. Dent. Res. 86, 754–757.

Weinmann, W., Thalacker, C., Guggenberger, R., 2005. Siloranes in dental composites. Dent. Mater. 21, 68–74.

West, J.L., Halas, N.J., 2003. Engineered nanomaterials for biophotonics applications: improving sensing, imaging, and therapeutics. Ann. Rev. Biomed. Eng. 5, 285–292.

Wickson, F., 2008. Narratives of nature and nanotechnology. Nature Nanotechnol. 3, 313–315.

Xia, W., Chang, J., 2007. Preparation and characterization of nano-bioactive-glasses, NBG. by a quick alkali-mediated sol–gel method. Mater. Lett. 61, 3251–3253.

Xie, D., Brantley, W., Culbertson, B., Wang, G., 2000. Mechanical properties and microstructures of glass-ionomer cements. Dent. Mater. 16, 129–138.

Yang, X., Wang, L., Qin, Y., Sun, Z., Henneman, J., Moradian-Oldak, J., Nancollas, G.H., 2010. How amelogenin orchestrates the organization of hierarchical elongated microstructures of apatite. J. Phy. Chem. 114, 2293–2300.

Yao, L., He, J., 2014. Recent progress in antireflection and self-cleaning technology: from surface engineering to functional surfaces. Prog. Mater. Sci. 61, 94–143.

Zafar, M.S., 2013. Effects of surface pre-reacted glass particles on fluoride release of dental restorative materials. World Appl. Sci. J. 28, 457–462.

Zafar, M.S., Ahmed, N., 2015. Therapeutic roles of fluoride released from restorative dental materials. Fluoride 48, 184–194.

Zafar, M.S., Khurshid, Z., Almas, K., 2015. Oral tissue engineering progress and challenges. Tissue Eng. Reg. Med. 12, 387–397.

Zafar, M., Najeeb, S., Khurshid, Z., Vazirzadeh, M., Zohaib, S., Najeeb, B., Sefat, F., 2016. Potential of electrospun nanofibers for biomedical and dental applications. Materials 9, 73.

Zandinejad, A., Atai, M., Pahlevan, A., 2006. effect of ceramic and porous fillers on mechanical properties of experimental dental composites. Dent. Mater. 22, 382–387.

Zhang, W., Walboomers, X., Wolke, J., Bian, Z., Fan, M., Jansen, J.A., 2005a. Differentiation ability of rat postnatal dental pulp cells in vitro. Tissue Eng. 11, 357–368.

Zhang, S., Zhao, X., Spirio, L., 2005b. PuraMatrix: self-assembling peptide nanofiber scaffolds. Scaffold. Tissue Eng., 217–238.

Zhao, G., Schwartz, Z., Wieland, M., Rupp, F., Geis-Gerstorfer, J., Cochran, D., Boyan, B., 2005. High surface energy enhances cell response to titanium substrate microstructure. J. Biomed. Mater. Res. 74, 49–58.

Zhu, X., Chen, J., Scheideler, L., Altebaeumer, T., Geis-Gerstorfer, J., Kern, D., 2004. Cellular reactions of osteoblasts to micron-and submicron-scale porous structures of titanium surfaces. Cells Tissues Organs 178, 13–22.

Zoufan, K., Jiang, J., Komabayashi, T., Wang, Y.-H., Safavi, K.E., Zhu, Q., 2011. Cytotoxicity evaluation of Gutta flow and endo sequence BC sealers. Oral Surg. Oral Med. Oral Pathol. Oral Radiol. Endodontol. 112, 657–661.

Zsigmondy, R., 1909a. Colloids and ultra microscope. J. Am. Chem. Soc. 31, 951–952.

Zsigmondy, R., 1909b. Colloids and ultramicroscope: a manual of colloid chemistry and ultramicroscopy. Wiley Sons, UK.

FURTHER READING

Subramani, K., Ahmed, W., Hartsfield, J.K. (Eds.), 2012. Nanobiomaterials in clinical dentistry. William Andrew, USA.

CHAPTER

ORAL NANOMEDICINE AND THE EMERGENT PROCESS OF CLINICAL TRANSLATION

27

Gabriel H. Hawthorne*, Marcelo P. Bernuci*, Mariza Bortolanza, Ana C. Issy**, Elaine Del-Bel****

**University Center of Maringa, Maringá, Brazil*

***University of Sao Paulo, São Paulo, Brazil*

CHAPTER OUTLINE

1 Introduction 863
2 Understanding Clinical Trials 864
3 Overview of Oral Nanomedicine and Clinical Trials 867
4 Analyzing Selected Trials 869
4.1 Bleaching Efficacy 869
4.2 Desensitizing Efficacy 871
4.3 Dental Implants 873
4.4 Response to Pulpotomy 875
4.5 Alveolar Socket Healing 876
4.6 Gingival Invaginations 876
4.7 Antimicrobial Effect 877
4.8 Ongoing Trials 878
4.9 A Note on Indirect Approaches 879
5 Final Considerations 879
6 Appendix 880
References 880

1 INTRODUCTION

Nanotechnologies are a group of technologies inside the nanometric scale; one nanometer corresponds to a billionth of a meter. Some agencies classify nanotechnology as any technology below 100 nm, as the European Medical Organization, while others use higher values, sometimes up to a micron (1000 nm). Although the principles of the field started to be theoretically discussed in 1959 (Feynman, 1959), the term was coined in the 1980 decade by Eric Drexler (Drexler, 1986). Specially designed nanoparticles and nanostructures may ensure desired proprieties. For example, to enhance the contrasting properties of imaging exam nanoparticles are manufactured with specific material and size. Researchers can use nanotechnology approaches on virtually any structure and area of knowledge, using these manipulated proprieties to improve outcomes. Since its rise three decades ago, nanotechnology is growing in importance, improving many different areas, including the environmental sciences, the many branches of engineering, the development of computers, among others. The aim of nanotechnology is to create harder, lighter, and more efficient materials and systems, like semiconductors or energy producers. Also, there are many different approaches in the nanomedicine field, ranging from the area of diagnosis (imaging, biosensors) to therapeutic applications, where new drugs and interventions are created. In this context, many specific areas are affected, and oral medicine has not been an exception.

In dentistry, nanotechnologies approaches have been used in many different scenarios, originating the emergent nanodentistry field. In fact, there are so many different applications that dozens of reviews would be necessary to cover them all. For instance, scientists have been studying this technology to improve esthetics and mechanical proprieties of different materials, including resin composites and implants (e.g El-Safty et al., 2012). Some studies address the inhibition of biofilm growth and the adhesive prevention of caries (Bhardwaj et al., 2009), the increase of bleaching efficacy (Bortolatto et al., 2014), and the treatment of root canal using nanotechnology (Lee et al., 2015). In a short time span, biochips, consisting of small diagnostic devices with sensitivity down to single-molecule level, may come into use to diagnostic practice, together with the introduction of contrasting agents for improvement of imaging diagnosis (Abou Neel et al., 2015). Therefore, nanotechnology applied to oral medicine is in constant movement and innovation, since potential new approaches are constantly discovered and tested. Considering projections for the future and the physical constraints of the field (Drexler, 2013),, as exposed in a broader way by Drexler (Drexler, 2013), and more specifically by Abiodun-Solanke and colleagues (Abiodun-Solanke et al., 2014), the future of oral nanomedicine appears to be very promising.

The translation from laboratory to clinical practice is dependent on results of clinical trials. Understanding that the thousands of approaches created in laboratories around the world are being applied to clinical practice is of vital interest to the credibility of a field in the health area. A bias may be generated when we discuss potential new approaches, but do not emphasize the clinical studies that follow the initial hypothesis, when these studies exist. If clinical practice is not correctly evaluated, the impression that the real impact of the field is reserved to a future setting may arise, which is actually a bias. When the discussion addresses innovative technologies, such as nanomedicine, this bias appears to be even more present. Nanotechnology is a promising science for the future, but also has many applications in the present. As novel options arise from basic studies, there is a tendency for clinical translation. Therefore, in this study we are going to discuss the translational research, from laboratory approaches to clinical applications, in the oral nanomedicine field. First, the essential aspects of clinical trials are reviewed, so this discussion can be better supported.

2 UNDERSTANDING CLINICAL TRIALS

Clinical trials are the studies that generate solid evidence for new drugs or interventions application in the health area. They are the standard in testing novelties, the foundation of evidence-based medicine. These studies test the generated hypothesis in humans, in a very safe manner, so its efficacy can be finally evaluated. Not only innovations are tested, but different applications of a commonly used intervention may also be explored. After solid evidence has been generated, it is reported and enables peer actualization, ultimately guiding clinical decisions (usually when combined with other similar trials and systematically reviewed).

Clinical trials are divided into phases, numbered from I to IV. However, clinical trials are preceded by a preclinical approach (Fig. 27.1A). In this moment, novel hypotheses are generated, and studies are conducted in vitro and in animal preparations, providing evidence for clinical testing in humans.

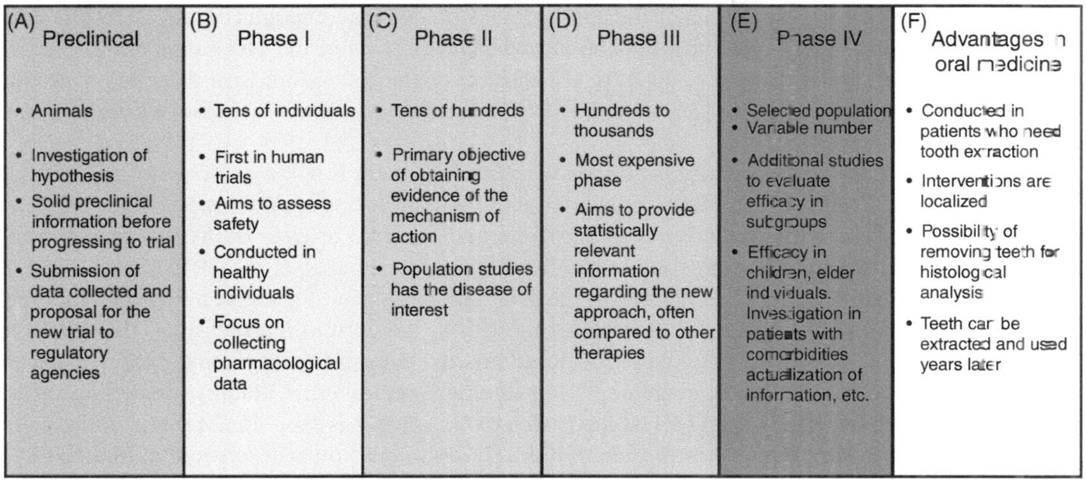

FIGURE 27.1 Summary of Key Information About Clinical Trials

(A) Animals are used to investigate a novel hypothesis in preclinical phase, collecting pharmacological information and data regarding safety. (B) After the approval by a regulatory agency, Phase I is conducted, focusing on safety and enrolling a few tens of individuals. (C) Phase II is done if the approach is proven safe. A larger group of persons is enrolled, and evidence of the mechanisms of action of the given intervention is the goal. (D) After that, if results are promising, proof of statistical superiority of the approach compared other used therapies is required. This is done in Phase III trials, which enroll hundreds or thousands of patients. (E) Subgroups of population (e.g., elders, children) are enrolled in Phase IV trials, where additional investigations are conducted. (F) Trials in oral medicine have some advantages, which enable more trials to be conducted.

In the preclinical phase, efficacy, toxicity, pharmacological parameters, and safety can be estimated (Watt et al., 2013), so the intervention is well characterized. From those parameters, established in animals, an initial dosage and the best route of administration is extrapolated to humans. For instance, the characterization of the nanoparticle, including methods of its production and physical proprieties are part of this phase. Furthermore, the approximate concentration of the nanoparticles, in a given dosage, their safety parameters (in animals, preferentially two different species, one being mammalian) and their mechanism of action would be studied in this step. If the approach has potential, the clinical trials should be conducted, and local regulatory agencies will evaluate a proposal for its conduction.

An extensive ethical analysis is conducted along with a rigid evaluation of safety. If approved, the trial starts at Phase I.

Characteristically, Phase I trials (often described as first in human trials) are conducted with a small group of individuals. The objective is to produce first evidence of safety in humans and achieve a safe dose or mode of application to be used in humans, since the approach has never been tested before (Watt et al., 2013) (Fig. 27.1B). For that goal, a small number of patients is necessary, and the selected individuals are usually healthy humans, so the pharmacological assessment is not heavily influenced by metabolic changes imposed by potential diseases. Conservative doses, which are extrapolated from preclinical studies, are used initially together with the application of a safety factor (often dividing initial dosage by 10). From then, small increases on the dosage are considered if the patients present no significant alterations.

Phase II trials follow these studies if the approach has proven safe enough (Fig. 27.1C). The goal of this new phase is to investigate proof of the mechanism of action of the given intervention in humans, and therefore, individuals tend to have the disease of interest. More than that, larger groups are necessary, often a few hundred patients, so statistical significance can be achieved (Guptill and Chiswell, 2013).

At the end of Phase II, a safe dose is already selected, and there is some evidence of a benefit of the new approach. Therefore, studies can be conducted with still larger populations. This happens in Phase III trials, where the objective is to enroll several hundred or even thousands of individuals to compare the new strategy to the current used approach (Hafley et al., 2013) (Fig. 27.1D). For instance, comparing a novel drug against the drug currently used for a given disease. The new strategy tested should be superior, or at least not inferior to the current intervention. This phase is expensive. As mentioned, they need to enroll hundreds to thousands of patients, prepared teams to deal with data collection and management, and often are multicentric studies (providing information in different populations and lifestyles). However, since this step can show a statistically relevant outcome, it may imply that the intervention has enough power to change guidelines and conducts around the world.

From then on, Phase IV trials may be carried out to evaluate the efficacy of a new drug or intervention in subgroups, such as children, elderly people, or people with different ethnicities (Hafley et al., 2013) (Fig. 27.1E). More than that, other safety parameters and interactions with other drugs or interventions may be evaluated. Although there appears to be a chain of steps, and this is often true, the process of conduction of a clinical trial is not always linear. Sometimes, one phase may be conducted together with other, mainly when it is desirable to speed up the process.

We described the general clinical trial format; this information is summarized in Fig. 27.1. However, some advantages appear to exist in clinical trials in dentistry (Fig. 27.1F), since oral medicine trials may be different from classical trials, designed for the testing of new drugs or other interventions in medicine overall. One example is the possibility of removing teeth for histological analysis or comparison of healing patterns, like in the study conducted by Canuto et al. (2013), where alveolar healing was addressed. Of course, patients that submit themselves to these procedures would already need tooth extraction. Another positive factor in dentistry is that the interventions tend to be localized, and therefore not extensively dangerous or invasive. In contrast, in general medicine, interventions are usually systemic, pose a greater risk for general health, and thus need to be more conservative. Such advantages facilitate the conduction of clinical trials in oral medicine, enabling evidence to be generated and catalyzing the progress of the field. This likely contributes to the fact that oral medicine represents the third most researched area, regarding number of clinical trials conducted, in nanotechnology or nanomedicine overall. Most widely trials are conducted in imaging and cancer research. This also indicates the importance given to oral nanomedicine.

Another advantage of trials in oral medicine is that clinical trials, by definition, can be conducted directly with people or with material of human origin, such as teeth, blood, or DNA. Therefore, studies conducted with teeth are considered clinical trials, even if the tissue has been extracted at some point and used years later. There are some exceptions to this rule, as the United States legislation, for instance, does not consider analysis on tissues of dead subjects as clinical trials (Wong and Schulman, 2013). We did not exclude this kind of trial. However, we found that trials on biological materials alone have not been the majority in the field (better exposed in Appendix). Therefore, using our research methods, clinical trials involving live subjects were most numerous anyway.

Generally, institutes, as the academic research organizations or clinical research organizations, conduct complex clinical trials, such as Phase III or IV (Reist et al., 2013). Clinical research organizations are profit-driven and sometimes hired by pharmaceutical companies to speed up the process of the conduct of a clinical trial. Academic research organizations, in contrast, conduct trials desired for their medical relevance. However, big organizations are not always in charge of clinical studies. Small independent groups may be able to conduct small Phase I and II trials, and generate the first pieces of evidence in a novel field. This behavior appears to be frequently found in oral nanomedicine, where groups of researchers conduct independent and isolated studies. It seems reasonable to expect this pattern in novel fields, especially given the mentioned advantages of conducting clinical trials in oral medicine. These studies provide the first pieces of evidence in a novel field, and then, may eventually open space for the conduction of other ones.

Apart from the years and resources invested in pre-clinical tests, when a clinical trial is prepared, there is a need for funding, capable professionals, organization, and time to conduct and collect the data. Therefore, clinical trials are expensive and time consuming studies, especially Phase III ones, where thousands of individuals are enrolled. After the end of the study, a report is produced, describing the trial in detail. Multiple clinical trial results are necessary to guarantee that a new intervention has the potential to reach the market, and finally, being profitable for the investors. Eventually, multiple trials are necessary. Because of those bureaucratic and expensive steps, many appearing promissory approaches may find resistance to be clinically tested. Very novel approaches, like nanomedicine, where close to none previous evidence is available for the evaluation of safety and efficacy, appear to be under even more resistance. At the same time, it seems inevitable to attribute potential to a field based on clinical trials outcomes. Therefore, clinical trials strike as important study types. The clinical test of hypothesis may cheer researchers to invest in novel approaches, catalyzing the advancement of the field.

3 OVERVIEW OF ORAL NANOMEDICINE AND CLINICAL TRIALS

To understand the magnitude of a group of, we should analyze it in comparison to other studies in the same field.To position the clinical trials on oral medicine inside the group of trials on nanomedicine, we conducted a superficial scientometric analysis of nanomedicine and clinical trials. As mentioned before, oral medicine is the third most clinically studied subject in nanomedicine overall. The most studied topic is imaging, where researchers improve imaging exams using nanomedicine. This can be achieved, for instance, by the use of nanoparticles as contrast in magnetic resonance or position emission tomography tests. There are excellent reviews covering this topic (e.g., Srivatsan and Chen, 2014), and the majority of these studies aim at better cancer detection. The second most studied field is the improvement of pharmacological characteristics of cancer therapy using nanomedicine; most approaches involved the chemotherapy drug paclitaxel (as reviewed by Deepa et al., 2012). Therefore, we can consider that cancer is the condition most clinically studied in nanotechnology overall, followed, then, by oral medicine. The fact that more than 30 clinical trials on oral medicine exist (see Appendix 1), is an indicator of investment and level of activity of the field.

Most clinical trials conducted in oral nanomedicine still cannot serve as the basis to clinical orientation. To be able to do so, trials need to enroll a high number of patients (hundreds to thousands—Phase III trials). Because the field is new, most of the studies aimed to show viability and efficacy, being mostly Phase I and II studies, and, therefore, enrolling a small number of subjects.

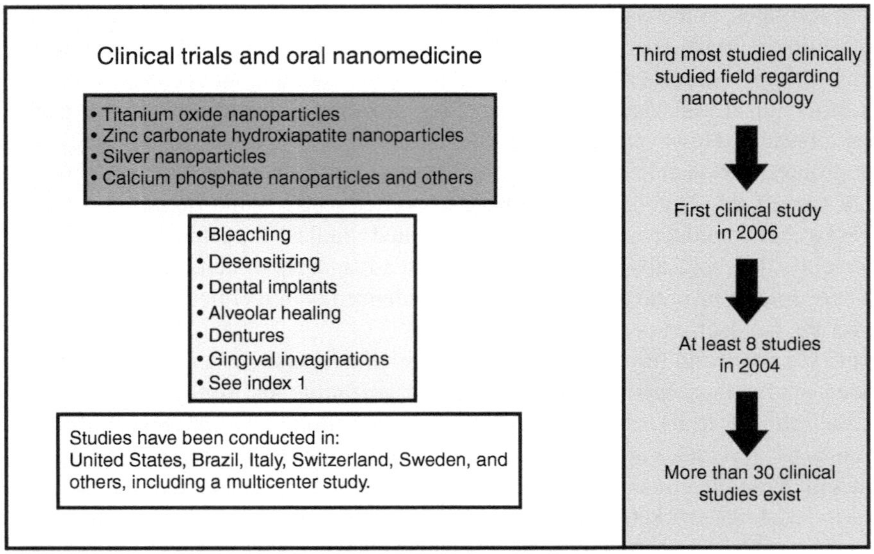

FIGURE 27.2 Overview of the Clinical Studies of Oral Nanomedicine

Left panel: summary of the particles used (top box), topics addressed (middle box) and places where the studies were conducted (bottom box). Right panel: information on the progress and situation of the field.

The first study we could find appeared in 2006 (Hadassah Medical Organization, 2006a). There appears to exist a trend in favor of the conduction of more studies each year. For instance, in 2014 at least 8 clinical trials on the matter have being conducted (Fig. 27.2, right panel). As discussed earlier, because studies on stored human tissues fall into the classification of clinical trials, some studies were conducted in vivo, and some in vitro. Most of the research in the field involves in vivo approaches.

Trials were conducted in many countries (Fig. 27.2, bottom box). Studies were carried out in United States, Mexico, Italy, Switzerland, Denmark, Germany, Sweden, Brazil, Bolivia, Chile, Egypt, Iran, India, and Japan. A huge investment was made in the multicenter study conducted by Ostman et al. (2010), involving 185 patients in Australia, Belgium, Italy, Germany, The Netherlands, Sweden and the United States. This trial is discussed in detail later.

Regarding the topics studied, several different matters were addressed (Fig. 27.2, middle box), including bleaching potential, desensitizing efficacy, antibacterial proprieties, new fluoride formulations, dental implants. These are better discussed in the next section, where we conduct an individual study analysis of some of the clinical trials on the area. At Appendix 1, we provide a list with clinical trials in the field and their addressed topics.

In oral nanomedicine, many different nanoparticles are being used in clinical testing (Fig. 27.2, top box). Titanium oxide nanoparticles, zinc carbonate hydroxyapatite nanoparticles (and other hydroxyapatite nanopastes), silver nanoparticles, and calcium phosphate nanoparticles are illustrations. Because no specific particle dominated the field and the mechanisms of action are still under investigation, we consider that a specific evaluation of these particles is out of the ambitions of our study. However, we provide information regarding some of these nanoparticles in context as selected clinical trials are discussed in the following section.

4 ANALYZING SELECTED TRIALS

To offer insight of the intersection with oral nanomedicine and clinical trials, a few interesting studies were selected to be analyzed in this section. We discuss the objectives, populations, methods, and outcomes of some of the clinical trials in oral medicine, based on what has been already discussed about clinical trials. Our goal, in this session, is to present a picture of how researchers think and what problems they may anticipate when they delineate a clinical trial. For instance, data regarding population is important because one cannot extrapolate results from findings on a small population, with a limited number of subjects or little genetic variability, to clinical practice, similarly, exclusion criteria are important to reduce bias.

4.1 BLEACHING EFFICACY

Many problems may affect teeth, leaving them with unpleasant color or pigmentation. Usually, extrinsic stains occur due to bad oral hygiene, ingest of chromatogenic aliments, and use of tobacco. The causes of intrinsic stains include enamel defects, aging, tetracycline medication intake, severe jaundice, excessive fluoride ingestion (Alqahtani, 2014). The cosmetic procedure of bleaching is very common. Currently, it is available in many different chemical formulas, but the most used chemical in bleaching is hydrogen peroxide (H_2O_2), usually in concentrations of 35%–38%. Although the mechanism by which bleach happens is not entirely understood, it is suspected that as the agent diffuses into the tooth, dissociation of peroxide into free radicals occurs. As hydroxyl radicals, perhydroxyl radicals, perhydroxyl anions, and superoxide anions are formed, attacking the bond between pigmented molecules and tooth tissues (Alqahtani, 2014). The amount of free radicals generated are proportional to the concentration of the bleaching agent and the and the time in contact with enamel. Bleaching can be made in office, in home, or with over-the-counter products, being considered one of the most requested procedures by patients (Bortolatto et al., 2014). The major side effect of bleaching is sensitivity, arising both during and after the treatment. This symptom represents a biological damage, which sometimes is irreversible (Costa et al., 2010).

To investigate improvement on bleaching, Bortolatto et al. (2014) tested the efficacy of a novel bleaching agent containing titanium oxide nanoparticles plus 15% H_2O_2 and the resultant sensitivity from the procedure. The approach was compared with a control bleaching agent, 35% H_2O_2 (Fig. 27.3). This randomized clinical trial was conducted between 2010 and 2011. Forty volunteers, ranging from 18 to 25 years old, were randomly distributed in two groups of 20 participants each. But first, screening was done to reduce bias. Excluding criteria, such as pregnancy or breastfeeding, presence of tetracycline stains, periodontal disease, orthodontic treatment, antiinflammatory intake or tooth malformation, were applied. Patients also received instructions concerning oral hygiene 1 week before the procedure beginning to create more controlled conditions.

In-office bleaching was realized, conducted in the Dental School of Araraquara (UNESP, Brazil). The nanoparticle approach was applied in 3 sessions of 48 min, divided in 3 applications of 16 min. Titanium oxide releases free radicals in the presence of light, so the gel was photocatalyzed via a hybrid light (total time 144 min over the three bleaching sessions). The control groupreceived 35% hydrogen peroxide applied in 3 sessions of 45 min each, divided into 3 applications of 15 min (total time 135 min). Seven women dropped out from the control group and eight from the titanium oxide nanoparticles group, due to tooth sensitivity or unexplained reasons. Sometimes the follow-up is not complete due to many reasons, and this is a liability when conducting clinical trials. Anticipating this problem,

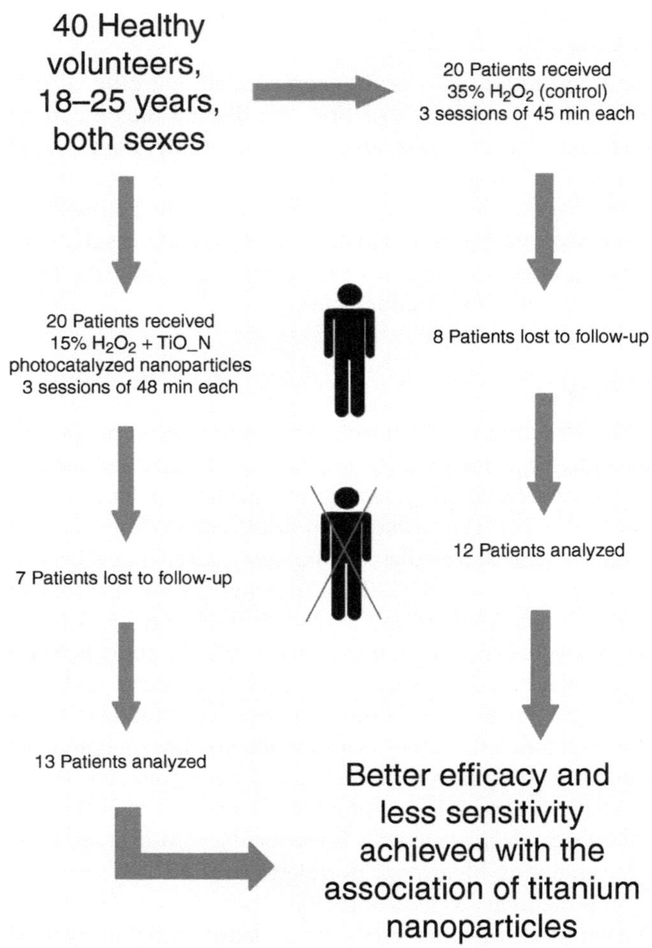

FIGURE 27.3 Bleaching Efficacy of Titanium Nanoparticles Associated with 15% H_2O_2

The diagram shows the study conducted by Bortolatto et al. (2014).

researchers may recruit an additional percentage of patients before the beginning of the study, or are forced to provide results with less significance or to start a new study. In this case, besides the higher dropout rate, the statistical power of analysis was not affected.

Regarding efficacy, the nanoparticle associated protocol had better results, although this outcome could not be seen in the first session. On the following sections, the efficacy became clear. One main constraint of this study was the absence of long-term follow-up and measurements, so the information regarding longevity of the treatment was not available.

Sensitivity was analyzed through a visual analog scale, were patients could classify it as low, average, high, and very high. The scale ranged from no pain, at the left side, to very severe pain, on the right.Patients marked the point along the scale that better represents the pain intensity. The analysis with this scale was conducted 4 times. Lower occurrence of sensitivity was reported with the use of the titanium oxide nanoparticle approach, providing promising results.

Therefore, the novel bleaching agent, containing nanoparticles, promoted less sensitivity and presented greater efficacy compared to control. As described by researchers, regarding sensitivity reduction, at least part of the effect was due to a reduction in the concentration of hydrogen peroxide (from 35% to 15%). What is better, no ultraviolet radiation was necessary, because of the nature of the formulation used; as authors described, titanium oxide doped with nitrogen enabled the efficacy of the approach using wavelengths in the band of visible light. Combining efficacy and safety, the use of titanium oxide nanoparticle approach may be consolidated as a novel strategy in bleaching. Posteriorly, in 2016, Bortolatto and coworkers published another study, a triple-blind trial where 6% hydrogen peroxide solution with nanoparticles of nitrogen-doped titanium oxide was compared with the standard 35% hydrogen peroxide solution in terms of bleaching and tooth sensitivity (Bortolatto et al., 2016). Two sessions of applications were made in 48 volunteers (24 in each intervention arm). In short, they found that the use of nanoparticle solution produced less tooth sensitivity, despite reducing efficacy.

Another clinical trial regarding bleaching was conducted in the end of 2014, sponsored by University of Chile (2014). In this randomized, double-blind clinical trial, a 6% hydrogen peroxide compound associated with titanium oxide nanoparticles was used in the upper hemiarcade, while the control, used on the lower arcade of the same patient, was still a 35% hydrogen peroxide solution.Results were published in 2015 (Martín et al., 2015). Thirty-one patients of 18 years or older were recruited, while the excluding and including criteria were essentially the same from the last trial. The intervention was conducted once a week for three weeks. There were no significant differences in color or sensitivity between the 6% nanoparticle compound and the 35% peroxide solution one month after the intervention. Collectively, these results are a great indicative of titanium oxide nanoparticles efficacy in improving bleaching and reducing collateral effects, since peroxide concentrations can be lowered using nanoparticle associations. These studies present outcomes in different environments, with different population and context, which help assuring that results were not biased. The field of bleaching associated with nanoparticles is under active clinical investigation.

4.2 DESENSITIZING EFFICACY

Dentine hypersensitivity is a condition related to dentine exposal. The clinical expression is, commonly, a sharp and short pain that arises in response to different stimuli, be it thermal, osmotic, or chemical. It is one of the most common dental problems, although its incidence has not been entirely clarified to this point. While in some studies the incidence reaches 74% of patients, this number has been reported to be around 15% in clinical studies (Miglani et al., 2010).

Regarding this problematic, Orsini et al. (2010) studied the desensitizing efficacy of a dentifrice that contained zinc-carbonate hydroxyapatite nanocrystals. Potassium nitrate/fluoride dentifrice, an agent commonly used to reduce bleaching sensitivity, was used as a control. This trial enrolled 70 patients, 36 receiving the novel nanocrystal formulation and 34 receiving the control one. This clinical trial has built to be a superiority test, not a noninferiority one. However, the new approach was considered equal to the conventional clinical approach. In order to reach enough statistical significance, the authors came to the number of 35 subjects per group after calculations. There were several exclusion criteria (chipped teeth, defective restorations, deep dental caries, orthodontic appliances, dentures or bridgework that could interfere with the hypersensitivity analysis, ongoing treatment with antibiotics or antiinflammatory drugs, ongoing treatment for hypersensitivity, heavy smoking, drug or alcohol abuse, and more) to reduce biases; initial 197 patients were screened and excluded if those criteria were achieved. Furthermore, the trial was conducted as a double-blind, randomized study.

Both approaches were successful in reducing sensitivity. In comparison to the control approach, the nanocrystal one had a greater effect in reducing sensitivity on the air blast test, where a blast of air is directed to the affected area of the tooth for 1 s, while the other teeth are isolated. The same was true for the subjective perception of sensitivity. Both differences were already significant 4 weeks after the beginning of the treatment. Results on the cold water test, where ice-cold water was applied selectively to the specific tooth with a syringe, and the tactile test, where a sharp dental explorer was passed across the facial area of the tooth (repeated 3 times) were similar between the two approaches.

Authors quote that the desensitizing effect of the nanocrystal dentifrice is due to the progressive closure of the tubular openings of dentine, starting in few minutes after administration. A few hours later, a mineralized layer is generated, described as the deposition of carbonate hydroxyapatite on enamel or dentin surfaces. The study concluded that the new dentifrice containing zinc nanocrystals significantly reduced hypersensitivity and may be useful in clinical practice.

As discussed earlier, sensitivity also happens after bleaching procedures. Bleaching-related tooth sensitivity is common and is facilitated by enamel defects, such as surface defects or subsurface pores. It has been related that the rapid diffusion of bleaching molecules insult pulp, leading to an inflammatory reaction that releases factors, such as prostaglandins and neuropeptides. In turn, pulpal nociceptors become hypersensitive, and thus the clinical symptom of sensitivity may arise (Cartagena et al., 2015). Addressing this matter, Browning et al. (2011) conducted a clinical trial. A paste that contained nanohydroxyapatite crystal was applied after a bleaching procedure. The hypothesis was essentially simple: the authors wanted to assess if the nanohydroxyapatite crystals paste had efficacy in reducing sensitivity after the use of a whitener. The theory was that correcting enamel defects, by the repairing potential of the nanopaste, sensitivity could be reduced.

The trial was conducted in 42 randomly assigned patients, divided into two groups of 21 people each: one received placebo, and the other the nanopaste. Three different measures of tooth sensitivity were assessed: the percentage of those enrolled in the study that experienced sensitivity, the days required for sensitivity to happen, and the level of pain from daily activities (measured by a visual analog scale or pain rating scale). Note that all these parameters are subjective. Authors expected that at least one of those might change. Patients had their sensitivity evaluated before the beginning of the trial, so a change in this attribute could be realistically calculated at the end of the study. Therefore, each patient acted as its own control.

Regarding the population of the study, all of the participants were adults, in general good health with a similar shade of teeth (equal or darker than shade LV A2, Vita Classical shade guide). This information is important, as the bleaching effect was also evaluated. All of the participants were also examined regarding oral health to reduce potential biases. The enrolled individuals had their maxillary teeth polished and were instructed against using any other oral health products that contained any desensitizer. They were also given specific toothpaste to use throughout the study. A 7% hydrogen peroxide product was used as bleaching agent in each one of the participants, manipulated so it did not contain a desensitizer. The bleaching agent was used twice daily for 2 weeks, while the nano desensitizing agent was used during this period and 1 week after it ended. The desensitizing paste was applied for 5 min after the bleaching sessions. Both participants and researchers were unaware of which group was receiving placebo until the end of the study.

As said, this study was based on a subjective analysis. Patients checked a "yes" or "no" box at the end of each day to report the presence of sensitivity. Pain scales were also completed from then in a 10 cm line that ranged from none to worst imagined. Tooth color was investigated by a research assistant and an investigator, to confirm that bleaching had really occurred. Fifty-one percent of the patients in the placebo group reported sensitivity during active bleaching while only 29% of the nanopaste group had the same outcome. However, this finding proved not to be statistically significant, because of the small enrollment. In contrast, statistically, significant reduction in the number of days of sensitivity was reported. While the placebo group together had 76 sensitivity days, nanopaste group had only 36. Although a reduction in pain was also related, it was not statistically significant, like the first finding. In conclusion, all parameters were better with the use of nanopaste, but the only statistically relevant outcome was a reduction in the number of days where sensitivity was felt. Given the limitations of the study, a larger clinical trial is required to clarify these findings. Furthermore, no relevant differences existed regarding bleaching power between these two groups. This observation is vital because the main objective of this intervention in a clinical setting is the whitening potential. Therefore, the initial conclusion is that nanopaste did not interfere quality of bleaching procedure.

The trial we just analyzed illustrates an important concept. Public perception of clinical trials may not correspond to reality because the limitations and findings of the trial are not correctly interpreted. Some may be inclined to think that clinical trials necessarily provide a short and clear answer, expecting a dualistic result: the approach works, or it does not. In reality, this result is true for the minority of the clinical trials. The safety, organizational and financial barriers to overcome are many, and thus many clinical trials are designed to obtain small pieces of evidence, not definitive, statistical proof. Note that, again, this is especially true in a novel field like nanomedicine. Therefore the "yes or no" answer may not be present. The trials that enable this clear, statistically relevant level of evidence, guiding clinical decisions, are big clinical trials. Those, as discussed, need a high investment, and are conducted by large organizations around the world, with thousands of subjects. To reach this point, many smaller trials likely have been conducted before. Often, multiple large trials need to be analyzed in a systematic review so a clinical decision is suggested. The pivotal concept to understand is that clinical trials may raise more questions than answers

4.3 DENTAL IMPLANTS

Dental implants are extremely common procedures nowadays, conducted all over the world. However, dental implant survival rate is variable, depending on factors, such as quality and quantity of bone in the dental arch (Goiato et al., 2014). Efforts to raise this rate are desired, and nanotechnology approaches have been tested regarding this matter.

Ostman et al. (2010) designed a multicenter, international trial to investigate functional outcomes of implants with nanoparticles. This studyis summarized in Fig. 27.4. The approach consisted of discrete crystalline depositions of calcium phosphate nanoparticles onto a dual acid-etched surface. These nanocrystals ranged from 20 to 100 nm. The product was called NanoTite, and positive preclinical results were obtained when compared to dual acid-etched surfaces alone. Authors described that several animal models and histomorphometric outcomes in humans were obtained, perfectly illustrating how preclinical phase is pivotal in the conduction of clinical trials. For instance, in the preclinical phase, using the nanoparticle approach, 189% greater force was needed for the implant to detach from bone, after 2 weeks of healing, in rabbit tibiae. More than that, greater bone-implant-contact at

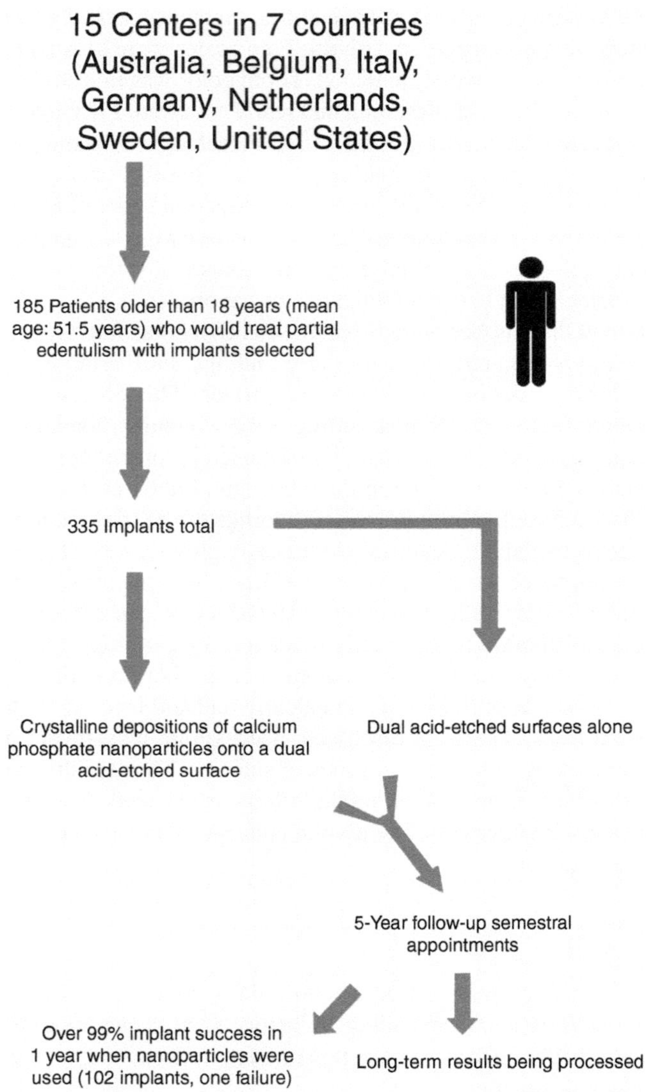

FIGURE 27.4 Crystalline Depositions of Calcium Phosphate Nanoparticles in Dental Implants

The multicenter clinical trial conducted by Ostman et al. (2010) is summarized.

4 weeks was observed in other studies. It was postulated that calcium phosphate nanoparticles might, therefore, improve osseointegration.

More than 15 centers were involved, in 7 countries (Australia, Belgium, Italy, Germany, Netherlands, Sweden, and the United States), and a total of 335 implants were made. Patients were followed for 5 years. These characteristics give a perspective of the size and investment of this study.

Patients of any sex, race, and older than 18 years, who had already made a decision to treat partial edentulism with implants, were selected. Several criteria of exclusion were applied, like infection or inflammation in the areas designed for the implant, uncontrolled diabetes, or metabolic bone disease, pregnancy, etc. One hundred eighty-five patients (56% females, 44% males) with a mean age of 51.5 years were enrolled; in some cases, patients received more than one implant, so a total of 335 implants were placed. In most cases (almost 60% of the restorations) the implant was a single-tooth restoration; multiple unit fixed prostheses were the other 40%.

Follow-ups were scheduled 6 months following implant placement,and posteriorly, at annual intervals during 5 years. Patients were assessed for implant mobility, gingival health, clinical symptoms, and radiographic exams were conducted. Data is still being processed regarding long-term results. However, at one center, in 1 year, of the 102 implants where the nanoparticles were used, only one failed (success rate over 99%). NanoTite surface performs well compared to implants with other surface textures, possibly having a significant effect on primary fixation soon after implant.

This study is one of the most complexes and complete clinical trials regarding oral nanomedicine included in our search criteria. Multicenter analysis helps in the reduction of bias when the same intervention is made in several different countries, odds are raised that subjects with different genetic characteristics will be evaluated. On the other hand, if 100 patients of the same region were analyzed, the study may be biased as genetic or regional factors may impact the outcome. Using distinct regions, the possible influence of different patterns of alimentation, lifestyle, and health habits are also verified. The same is true regarding the number of subjects enrolled; higher enrollment is better for statistical purposes. Therefore, the clinical studies seen as most significant are those conducted on multiple sites around the world and enroll high number of patients.

4.4 RESPONSE TO PULPOTOMY

Bioceramics are materials used for tissue regeneration consisting of nanosphere particles that have the ability to penetrate dentinal tubules. In dentistry, these materials are used as root and perforation repair tools (Azimi et al., 2014). Although some advantages have been reported over time regarding bioceramics efficacy compared to classical approaches, no clinical trial had been performed to assess clinical effects of bioceramics in human vital pulps.

A trial was conducted by the Islamic Azad University Tehran (2012) adressing pulp reactions after partial pulpotomy on healthy human premolars. The use of bioceramics (containing nanosphere particles) was compared to the use of mineral trioxide aggregate, a commonly used pulp-covering material. The pulp response to the procedure was measured using the two interventions. Both clinical and histological parameters were analyzed. Reports were published by Azimi et al. (2014).

Twenty-four premolars from 12 volunteers aged from 12 to 16 years were orthodontic extracted in the same clinic. Several exclusion criteria were applied (premolars with caries, restorations, lingering pain after the stimulus had been removed, etc.), and the 24 premolars were randomly allocated into two groups. A cavity was prepared with a diamond bur, and then patients were treated either with the nanobioceramics or with mineral trioxide aggregate. After 6-weeks follow-up, no significant differences between the two approaches were reported regarding inflammation, formation of hard tissue bridge or appearance. Although the response to pulpotomy was favorable using the two approaches, the difference between the two interventions did not reach statistical significance. In fact, teeth treated with the common mineral material approach had less sensitivity to cold stimuli.

Therefore, although the nanoapproach was not clinically inferior, it could not top the commonly used approach regarding sensitivity. In these cases, it is rational to evaluate if there is a benefit in repeating the clinical study or conducting the intervention in bigger populations using the same formulation. Thus, clinical trials are important studies to signal for liability in the product that could not be seen before. Complementary studies may be conducted to try to correct undesirable parameters on the formulation before the conduction of another clinical trial. And, of course, if the approach is seen as not promissory, the line of research may be discontinued.

4.5 ALVEOLAR SOCKET HEALING

Following tooth extraction, alveolar healing is a complex process. It is estimated that bone production begins 10 days after the removal, although factors, such as disease, can slow down this process (Ahn and Shin, 2008). The use of materials with the goal of maintenance of space helps bone regeneration, which is important to prevent jaw atrophy via osteoclast stimulation (Canuto et al., 2013).

To improve alveolar healing, a nanocrystaline hydroxyapatite paste can be used. This paste, namely Ostim, resembles human bone, has a large bioactive surface area and is quickly replaced by bone tissue. However, the precise mechanisms by which those healing effects happen remain unknown. Canuto et al. (2013) conducted a trial in 2011 where this paste was clinically tested. The small sample was comprised of 14 patients, aged from 45 to 60 years, who had each two monoradicular teeth to be extracted because of problems, such as fractures, nonrestorable caries, or residual roots. Criteria of exclusion including tooth with active infection, smokers of more than 10 cigarettes a day, irradiation in the previous 12 months, pregnancy or breastfeeding, and poor oral hygiene.

In one of the patients' site of extraction the socket was filled with the nanocrystalline paste. The other site of extraction was left to natural healing by clot formation. Both the sockets were not sutured, which means that there was no surgical attempt to close the wound. Also, no antibiotic or analgesic therapy was initiated. Patients were screened at 1, 7, and 14 days, and samples of soft tissues around the tooth were collected. It is important to say that no postsurgical complications existed in any of the patients, providing initial information that the nanoapproach is not prejudicial. In the sockets where nanocrystalline hydroxyapatite was used, at 7 days, a complete epithelialization happened. This effect did not occur in the nonintervention site. This finding is possiblly the most important finding of the study, because no other graft material has presented this result before. More than that, a subjective score for pain was obtained asking patients to classify their pain in a scale that ranged from 0 (no pain) to 10 (worst possible pain). In the second day, the score was higher using the nanopaste, but in all the other times, the score was lower.

Authors also found that vascular endothelial growth factor (VEFG) was significantly increased in the first day with the use of Ostim. This finding indicates that the angiogenesis process,likely necessary for optimal tissue healing, was boosted where the nanopaste was used. Also, proosteogenic factors, such as BMP-4, BMP-7, ALP, and OCN had improved synthesis, being, possibly, the way by which healing is improved.

Because of the small sample size, posterior studies with larger populations are needed. Even so, it seems that Ostim, a hydroxyapatite nanopaste, is a promising formulation that increases angiogenesis, epithelialization, and osteogenesis, improving healing of the alveolar socket.

4.6 GINGIVAL INVAGINATIONS

Gingival invaginations consist of infoldings of at least 1 mm on the gingiva, being frequent after tooth extraction and orthodontic space closure, affecting a range of more than 35% of the patients to virtually

100%. Despite frequency, according to a systematic review conducted, its etiology remains unclear (Gölz et al., 2011). The authors described some of the long-term consequences of gingival invagination, such as inadequate orthodontic space closure, instability of bone closure, and higher chance of plaque and bacterial accumulation. Therefore, approaches that overcome this problem are desired. The present recommended intervention for treatment is the removal of the excess tissue in the invagination, but prevention would be ideal.

Hydroxyapatite is the principal inorganic bone component. Techniques can be used to synthetically produce nanohydroxyapatite, resulting in particles that have structural features which resemble biological bone component. When implanted, these particles show good biocompatibility and are not toxic to the surrounding tissue (Venkatesan and Kim, 2014). Reichert et al. (2014) developed a study to explore how synthetic nanocrystalline hydroxyapatite influences the incidence and degree of gingival invaginations after tooth extraction. Twenty-eight premolars of 10 patients were removed for orthodontic reasons, and then the nanocrystalline hydroxyapatite bone replacement material was inserted. Fourteen premolars received treatment, where another 14 were used as control. Patients had more than 10 years, no metabolic diseases, immune suppression, endocrine bone disorders, and were not taking steroids or other drugs that interfere with calcium homeostasis.

After the implants of nanocrystalline hydroxyapatite, a minimal healing phase of 6 weeks was respected. In both arms of the trial, there were no clinical signs of inflammation or other complications that interfere with the study plan; therefore, the new intervention also appears to be safe. In the clinical evaluation, invaginations deeper than 1 mm were defined as gingival invaginations. Although there was no numeric outcome difference between the treatment arms, with 13 sites developing gingival invaginations in the treatment arm and 14 in the control one (100%), the degree of invaginations varied significantly between the two approaches. Several parameters were evaluated to confirm this finding, but the intervention arm had vertical invagination measurements up to 2 mm smaller.

Therefore, it was postulated that the nanoparticles had the capability to be reabsorbed and participated in remodeling processes, and importantly, were partially successful as a preventive approach. After posterior investigative studies, as authors suggest, this procedure could be conducted in high-risk patients, to enhance the prevention against gingival invaginations.

4.7 ANTIMICROBIAL EFFECT

Composite materials that possess antibacterial activity are desired for oral applications. Although positive results had been reported in vitro, a real assessment of the effect of these nanocompounds is hard to be realistic measured in these studies, since the oral environment presents many different microorganisms and this condition is hard to be replicated in vitro. Therefore, clinical trials have been conducted to investigate the efficacy of cationic polymers with quaternary ammonium groups nanoparticles against microorganisms (Hadassah Medical Organization, 2006b). This trial was reported in 2010.

Bacterial biofilm growth was assessed in two acrylic appliances, one of them containing two normal resin composites and the other one two composites with nanoparticles. These materials were intraoral placed in 10 health volunteers. This trial was reported in 2010, and showed that after 4 h there was a statistically relevant reduction of viable bacteria in the biofilm where nanoparticles were used. This effect was replicated in all the voluntary samples. Even in the most remote parts of the structure, more than 50% of the entire bacterial population on the surface of the biofilm was eliminated. Also, interestingly, the biofilms that had formed in the composite where nanoparticles were used tended to be thicker than the biofilms in the normal resin composites. Therefore, this intervention should potentially extend the service life of dental restorations, reducing costs related to repeated restorative procedures. Adverse events were monitored, with no significant differences between experimental groups. Discomfort or pain was also not felt by the volunteers. This intervention has the potential to extend the service life of dental restorations, reducing costs related to repeated restorative procedures.

Dental caries are a prevalent condition that affects mainly children in undeveloped countries. To prevent caries, fluoride preparations are commonly used worldwide, but because the treatment requires multiple applications per year and have a high cost, the intervention may be not viable in poor regions (Santos et al., 2014). To help in this problem, in the past few years, other formulations have been studied and developed. The antibacterial use of silver compounds is well described and includes currently clinical applications, such as the prevention and treatment of ocular, surgical and dental infections. For this reason, silver fluoride approaches have been tested regarding caries prevention. However, they stain caries dark, an unpleasant esthetical outcome. Therefore, nanosilver fluoride, a product that contains silver nanoparticles, was selected as a possible solution to avoid this black staining inconvenience. Authors suggest that as nanosilver does not have the tendency to form oxides in contact with oxygen, and therefore dark staining may be reduced (Santos et al., 2014). Other advantage of fluoride compounds with silver nanoparticles is the superior antibacterial effect, which is inversely correlated to particle size (Targino et al., 2014).

To evaluate this hypothesis, Santos et al. (2014) tested nanosilver fluoride on caries prevention. It should be noted that their work inspired critics (Burns and Hollands, 2015), mainly regarding biases and part of the methodology. The population was comprised of 60 schoolchildren of a poor region of Brazil. A total of 130 teeth were treated either with nanosilver fluoride for 2 min or with water. Treatment was performed once in a year. After follow-up losses, 51 teeth in the intervention group and 49 teeth in the control one (water) were evaluated, enough for a statistically significant outcome based on authors aspirations. The children were clinically evaluated after 1 week, 5 months, and 12 months. In 12 months, the fluoride approach had a success rate of 66.7% in hardening and arresting dentine caries, while the control group had a success rate of 34.7%. On top of that, the carious lesions did not stain black, and the taste of the solution was not unpleasant.

4.8 ONGOING TRIALS

A study that addresses prevention of enamel demineralization adjacent to brackets is currently being conducted (Universidad Autonoma de San Luis Potosi, 2014). In this intervention, researchers are using silver nanoparticles for the reduction of enamel demineralization with the primary purpose of prevention. Forty patients have being enrolled, aged 10–35 years. Silver nanoparticles are compared to placebo; half of the mouth is applied with nanosilver and half with placebo, both incorporated into the primer orthodontic Transbond XT. Therefore, each patient acts as its own control. This trial can be accessed through the identifier NCT02400957.

Besides that, another study involving nanocrystalline hydroxyapatite gel is being conducted, regarding management of periodontal intrabony defects in patients with chronic periodontitis (Cairo University, 2015). The aim of the researchers is to compare clinical and radiographic patterns of nanohydroxyapatite implants and bone allograft. The estimate enrollment is 30 patients (35–50 years), and evaluations are going to be done at 3 and 6 months. Further actualizations in this trial can be followed searching for the identifier NCT02507596.

Antimicrobial properties of alkylated polyethylenimine nanoparticles are being evaluated in root canal treatment (Hadassah Medical Organization, 2010). Treatment outcome will be assessed at 2 weeks, 3 months, 6 months, and 1 year, with clinical and radiological examinations. Two hundred patients older than 18 years, who needed root canal treatment, were selected for this study. The identifier for this study is NCT01167985.

4.9 A NOTE ON INDIRECT APPROACHES

Some methods cannot be considered direct oral medicine approaches, but impact oral medicine with significance. As an illustration, consider the clinical testing of new drugs. As is well known in clinical settings, diclofenac is a nonsteroidal antiinflammatory drug prescribed for the relief of mild and moderate pain that has a proven efficacy (Bakshi et al., 1994). Acute pain caused by dental procedures is one of the indications to prescribe diclofenac. However, several adverse effects do exist with the use of the drug, especially involving the gastrointestinal tract, liver, and kidneys. Nanotechnology new approaches are being used to develop new diclofenac formulations, and some studies have been able to make the particles one-tenth the original size. Manvelian et al. (2012) conducted a Phase II trial in 2009 (ref) with 331 patients to investigate whether a nano-formulated diclofenac was better than placebo to alleviate postoperative dental pain. The approach was indeed better than placebo and could relieve pain faster compared to the placebo effect. Also, no safety issues were found. As the authors described, nanotechnology approaches may enhance the surface area and enable the use of lower doses, resulting in better tolerability and less collateral effects. Therefore, addressing the whole spectrum of evolution of a field as broad as oral medicine, we should consider improvements on areas not directly related to it.

5 FINAL CONSIDERATIONS

In conclusion, many different areas inside dentistry have been clinically investigated concerning nanotechnology. This fact indicates that the field overall is receiving investment, pointing out that not only isolated, specific subareas of knowledge are being investigated. Currently, not yet in the entire nanomedicine field, but surely in the field of oral medicine, a chain of connection is in the process of being perpetuated, where basic studies are converted to clinical trials. The disseminated investment also implies that there is much space for growth in the field. Promising approaches already initially tested will undergo more elaborate clinical trials, to generate robust statistics in bigger populations. Furthermore, as basic studies in this field bring new ideas, clinical trials addressing brand new technologies will keep being conducted.

In the future, the revolutionary potential of nanomedicine approaches is ample. This potential exists not exclusively in a distant future, where precise atomic manipulation may become a reality, but in a medium and short span as well. Consider, for instance, technologies that improve imaging exams or novel nanobiosensors, and how they may positively influence the field. Today, several clinical trials are being conducted, catalyzing the clinical testing of new options and boosting virtually all processes in the area of oral nanomedicine, from the very basic research to marketing new interventions. Promising approaches will, hopefully, encounter a field with a structured clinical foundation basis for the conduction of studies in their particular subjects.

Analyzing the clinical trials conducted in a field is the only way of predicting the potential for clinical impact of new approaches, and oral nanomedicine stands in a promissory position. Therefore, although nanomedicine can bring improvements on many areas, the field of oral medicine may be one of the first ones where nanomedicine significantly affects the clinical practice.

6 APPENDIX

In this section, studies are presented according to the topic addressed. When the name of the primary author is underlined, it indicates that the study has been conducted in vitro. As discussed, those are considered clinical trials when biological tissue of living humans is used.

Restorations with antibacterial nanoparticles: Hadassah Medical Organization (2006a,b)

Cavities restorations: Andrade et al. (2010, 2011, 2012, 2014)

Root canal treatment: Hadassah Medical Organization (2006a,b, 2013), Waltimo et al. (2009)

Fluorescence measurements: Diniz et al. (2008)

Dental implants fixation: Orsini et al. (2007), Ostman et al. (2010)

Bond strength: Hoshika et al. (2010), Nirupama et al. (2012)

Bleaching efficacy: Bortolatto et al. (2014, 2016), University of Chile (2014), Martín et al. (2015)

Desensitizing efficacy: Orsini et al. (2010, 2013), Federal University of São Paulo (2013), Browning et al. (2011)

Enamel lesion remineralization: Nakashima et al. (2009), Tay and Pashley, 2009 (2009), Universidad Autonoma de San Luis Potosi (2014), Universidad Autonoma de San Luis, Potosí (2015)

Chromatic effect of silver nanoparticle: Hernandez-Sierra et al. (2010)

Socket preservation technique following tooth extraction: Checchi et al. (2011)

Fracture resistance: Taha et al. (2011)

Pulp response to partial pulpotomy: Islamic Azad University Tehran (2012)

Acrylic nanohybrid denture: University of California and Ivoclar Vivadent (2012)

Alveolar healing: Canuto et al. (2013)

Nanoparticle retention in periodontal defects: Gamal and Iacono (2013)

Coating material: Han and Okiji (2013)

Gingival invagination: Reichert et al. (2014)

Pulpal response to pulp capping: Swarup et al. (2014)

Fluoride Formulation: University of Pernambuco (2014), Santos et al., 2014)

Occluding dentin tubules: Amaechi et al. (2015)

Management of periodontal intrabony defects: Cairo University (2015)

REFERENCES

Abiodun-Solanke, I., Ajayi, D., Arigbede, A., 2014. Nanotechnology and its application in dentistry. Ann. Med. Health Sci. Res. 4 (Suppl. 3), S171–S177.

Abou Neel, E.A., Bozec, L., Perez, R.A., Kim, H.W., Knowles, J.C., 2015. Nanotechnology in dentistry: prevention, diagnosis, and therapy. Int. J. Nanomed. 10, 6371–6394.

Ahn, J.J., Shin, H.I., 2008. Bone tissue formation in extraction sockets from sites with advanced periodontal disease: a histomorphometric study in humans. Int. J. Oral Maxillofac. Implants 23 (6), 1133–1138.

Alqahtani, M.Q., 2014. Tooth-bleaching procedures and their controversial effects: a literature review. Saudi Dent. J. 26 (2), 33–46.

Amaechi, B.T., Mathews, S.M., Ramalingam, K., Mensinkai, P.K., 2015. Evaluation of nanohydroxyapatite-containing toothpaste for occluding dentin tubules. Am. J. Dent. 28 (1), 33–39.

Andrade, A.K., Duarte, R.M., Silva, F.D., Batista, A.U., Lima, K.C., Pontual, M.L., Montes, M.A., 2010. Efficacy of composites filled with nanoparticles in permanent molars: six-month results. Gen. Dent. 58 (5), e190–e195.

Andrade, A.K., Duarte, R.M., Medeiros e Silva, F.D., Batista, A.U., Lima, K.C., Pontual, M.L., Montes, M.A., 2011. 30-Month randomised clinical trial to evaluate the clinical performance of a nanofill and a nanohybrid composite. J. Dent. 39 (1), 8–15.

Andrade, A.K., Duarte, R.M., Silva, F.D., Batista, A.U., Lima, K.C., Pontual, M.L., Montes, M.A., 2012. Clinical evaluation of nanofill and nanohybrid composite in class I restorations: a 12-month randomized trial. Gen. Dent. 60 (4), e255–e262.

Andrade, A.K., Duarte, R.M., Medeiros e Silva, F.D., Batista, A.U., Lima, K.C., Monteiro, G.Q., Montes, M.A., 2014. Resin composite class I restorations: a 54-month randomized clinical trial. Oper. Dent. 39 (6), 588–594.

Azimi, S., Fazlyab, M., Sadri, D., Saghiri, M.A., Khosravanifard, B., Asgary, S., 2014. Comparison of pulp response to mineral trioxide aggregate and a bioceramic paste in partial pulpotomy of sound human premolars: a randomized controlled trial. Int. Endod. J. 47 (9), 873–881.

Bakshi, R., Frenkel, G., Dietlein, G., Meurer-Witt, B., Schneider, B., Sinterhauf, U., 1994. A placebo-controlled comparative evaluation of diclofenac dispersible versus ibuprofen in postoperative pain after third molar surgery. J. Clin. Pharmacol. 34 (3), 225–230.

Bhardwaj, S.B., Mehta, M., Gauba, K., 2009. Nanotechnology: role in dental biofilms. Indian J. Dent. Res. 20 (4), 511–513.

Bortolatto, J.F., Pretel, H., Floros, M.C., Luizzi, A.C., Dantas, A.A., Fernandez, E., Moncada, G., de Oliveira, Jr., O.B., 2014. Low concentration H(2)O(2)/TiO_N in office bleaching: a randomized clinical trial. J. Dent. Res. 93 (7 Suppl.), 66s–71s.

Bortolatto, J.F., Trevisan, T.C., Bernardi, P.S., Fernandez, E., Dovigo, L.N., Loguercio, A.D., Batista de Oliveira Junior, O., Pretel, H., 2016. A novel approach for in-office tooth bleaching with 6% H2O2/TiO_N and LED/laser system-a controlled, triple-blinded, randomized clinical trial. Lasers Med. Sci. 31 (3), 437–444.

Browning, W.D., Cho, S.D., Deschepper, E.J., 2011. Effect of a nano-hydroxyapatite paste on bleaching-related tooth sensitivity. J. Esthet. Restor. Dent. 24 (4), 268–276.

Burns, J., Hollands, K., 2015. Nano Silver Fluoride for preventing caries. Evid. Based Dent. 16 (1), 8–9.

Cairo University, 2015. Evaluation of nano-crystalline hydroxyapatite silica gel in management of periodontal intrabony defects. ClinicalTrials.gov Identifier: NCT02507596.

Canuto, R.A., Pol, R., Martinasso, G., Muzio, G., Gallesio, G., Mozzati, M., 2013. Hydroxyapatite paste Ostim, without elevation of full-thickness flaps, improves alveolar healing stimulating BMP- and VEGF-mediated signal pathways: an experimental study in humans. Clin. Oral Implants Res. 24 (Suppl. A100), 42–48.

Cartagena, A.F., Parreiras, S.O., Loguercio, A.D., Reis, A., Campanha, N.H., 2015. In-office bleaching effects on the pulp flow and tooth sensitivity: case series. Braz. Oral Res. 29 (1), 1–6.

Checchi, V., Savarino, L., Montevecchi, M., Felice, P., Checchi, L., 2011. Clinical-radiographic and histological evaluation of two hydroxyapatites in human extraction sockets: a pilot study. Int. J. Oral Maxillofac. Surg. 40 (5), 526–532.

Costa, C.A., Riehl, H., Kina, J.F., Sacono, N.T., Hebling, J., 2010. Human pulp responses to in-office tooth bleaching. Oral Surg. Oral Med. Oral Pathol. Oral Radiol. Endod. 109 (4), e59–e64.

Deepa, G., Ashwanikumar, N., Pillai, J.J., Kumar, G.S., 2012. Polymer nanoparticles: a novel strategy for administration of Paclitaxel in cancer chemotherapy. Curr. Med. Chem. 19 (36), 6207–6213.

Diniz, M.B., Rodrigues, J.A., Hug, I., Cordeiro, R.C., Lussi, A., 2008. The influence of pit and fissure sealants on infrared fluorescence measurements. Caries Res. 42 (5), 328–333.
Drexler, E., 1986. Engines of Creation: the Coming Era of Nanotechnology. Anchor, New York.
Drexler, E., 2013. Radical Abundance: How a Revolution in Nanotechnology Will Change Civilization. PublicAffairs, New York.
El-Safty, S., Akhtar, R., Silikas, N., Watts, D.C., 2012. Nanomechanical properties of dental resin-composites. Dent. Mater. 28 (12), 1292–1300.
Federal University of São Paulo, 2013. Single application of desensitizing pastes as dentin sensitivity treatment.
Feynman, R., 1959. RE: Plenty of room at the bottom.
Gamal, A.Y., Iacono, V.J., 2013. Mixed nano/micro-sized calcium phosphate composite and EDTA root surface etching improve availability of graft material in intrabony defects: an in vivo scanning electron microscopy evaluation. J. Periodontol. 84 (12), 1730–1739.
Goiato, M.C., dos Santos, D.M., Santiago, J.F., Moreno, A., Pellizzer, E.P., 2014. Longevity of dental implants in type IV bone: a systematic review. Int. J. Oral Maxillofac. Surg. 43 (9), 1108–1116.
Gölz, L., Reichert, C., Jäger, A., 2011. Gingival invagination: a systematic review. J. Orofac. Orthop. 72 (6), 409–420.
Guptill, J.T., Chiswell, K., 2013. Phase II clinical trials. In: Lopes, Renato D., Harrington, Robert A. (Eds.), Understanding Clinical Research. 1st ed. Mc Graw Hill Education, Lange, United States.
Hadassah Medical Organization, 2006a. Clinical evaluation of esthetic restorations placed in primary molars with composite resin enriched with insoluble anti bacterial nano particles.
Hadassah Medical Organization, 2006b. Clinical Study of Antibacterial Nanoparticles Incorporated in Composite Restorations.
Hadassah Medical Organization, 2010. Antibacterial properties of silicon incorporated with quaternary ammonium polyethylenimine nanoparticles.
Hafley, G.E., Leonardi, S., Pieper, K.S., 2013. Phase III and IV clinical trials. In: Renato, D., Lopes, Robert, A., Harrington (Eds.), Understanding Clinical Research. first ed. McGraw-Hill Education, Lange, United States.
Han, L., Okiji, T., 2013. Effects of a novel fluoride-containing aluminocalciumsilicate-based tooth coating material (Nanoseal) on enamel and dentin. Am. J. Dent. 26 (4), 191–195.
Hernandez-Sierra, J.F., Ruiz, F., Castanedo-Cazares, J.P., Martinez-Ruiz, V., Mandeville, P., Pierdant-Perez, M., Gordillo-Moscoso, A., Pozos-Guillen Ade, J., 2010. In vitro determination of the chromatic effect of a silver nanoparticles solution linked to the gantrez S-97 copolymer on tooth enamel. J. Clin. Pediatr. Dent. 35 (1), 65–68.
Hoshika, S., Nagano, F., Tanaka, T., Ikeda, T., Wada, T., Asakura, K., Koshiro, K., Selimovic, D., Miyamoto, Y., Sidhu, S.K., Sano, H., 2010. Effect of application time of colloidal platinum nanoparticles on the microtensile bond strength to dentin. Dent. Mater. J. 29 (6), 682–689.
Islamic Azad University Tehran, 2012. Partial pulpotomy on caries free human teeth.
Lee, D.K., Kim, S.V., Limansubroto, A.N., Yen, A., Soundia, A., Wang, C.Y., Shi, W., Hong, C., Tetradis, S., Kim, Y., Park, N.H., Kang, M.K., Ho, D., 2015. Nanodiamond-Gutta percha composite biomaterials for root canal therapy. ACS Nano 9 (11), 11490–11501.
Manvelian, G., Daniels, S., Gibofsky, A., 2012. A phase 2 study evaluating the efficacy and safety of a novel, proprietary, nano-formulated, lower dose oral diclofenac. Pain Med. 13 (11), 1491–1498.
Martín, J., Vildósola, P., Bersezio, C., Herrera, A., Bortolatto, J., Saad, J.R., Oliveira, O.B., Fernández, E., 2015. Effectiveness of 6% hydrogen peroxide concentration for tooth bleaching: a double-blind, randomized clinical trial. J. Dent. 43 (8), 965–972.
Miglani, S., Aggarwal, V., Ahuja, B., 2010. Dentin hypersensitivity: recent trends in management. J. Conserv. Dent. 13 (4), 218–224.
Nakashima, S., Yoshie, M., Sano, H., Bahar, A., 2009. Effect of a test dentifrice containing nano-sized calcium carbonate on remineralization of enamel lesions in vitro. J. Oral Sci. 51 (1), 69–77.

Nirupama, C., Kavitha, S., Jacob, J., Balaji, K., Srinivasan, B., Murugesan, R., Krishnaswamy, N.R., 2012. Comparison of shear bond strength of hydrophilic bonding materials: an in vitro study. J. Contemp. Dent. Pract. 13 (5), 637–643.

Orsini, G., Piattelli, M., Scarano, A., Petrone, G., Kenealy, J., Piattelli, A., Caputi, S., 2007. Randomized, controlled histologic and histomorphometric evaluation of implants with nanometer-scale calcium phosphate added to the dual acid-etched surface in the human posterior maxilla. J. Periodontol. 78 (2), 209–218.

Orsini, G., Procaccini, M., Manzoli, L., Giuliodori, F., Lorenzini, A., Putignano, A., 2010. A double-blind randomized-controlled trial comparing the desensitizing efficacy of a new dentifrice containing carbonate/hydroxyapatite nanocrystals and a sodium fluoride/potassium nitrate dentifrice. J. Clin. Periodontol. 37 (6), 510–517.

Ostman, P.O., Hupalo, M., del Castillo, R., Emery, R.W., Cocchetto, R., Vincenzi, G., Wagenberg, B., Vanassche, B., Valentin, A., Clausen, G., Hogan, P., Goene, R., Evans, C., Testori, T., 2010. Immediate provisionalization of NanoTite implants in support of single-tooth and unilateral restorations: one-year interim report of a prospective, multicenter study. Clin. Implant Dent. Relat. Res. 12 (Suppl. 1), e47–e55.

Reichert, C., Wenghoefer, M., Kutschera, E., Gotz, W., Jager, A., 2014. [Ridge preservation with synthetic nanocrystalline hydroxyapatite reduces the severity of gingival invaginations: a prospective clinical study. J. Orofac. Orthop. 75 (1), 7–15.

Reist, C.J., Rorick, T.L., Berdan, L.G., Lopes, R.D., 2013. The role of academic research organizations in clinical research. In: Renato, D., Lopes, R.A.H. (Eds.), Understanding Clinical Research. Lange (Mc Graw Hill Education), United States, pp. 27–37.

Santos, V.E., Vasconcelos Filho, A., Targino, A G., Flores, M.A., Galembeck, A , Caldas, A.F., Rosenblatt, A., 2014. A new "silver-bullet" to treat caries in children--nano silver fluoride: a randomised clinical trial. J. Dent. 42 (8), 945–951.

Srivatsan, A., Chen, X., 2014. Recent advances in nanoparticle-based nuclear imaging of cancers. Adv. Cancer Res. 124, 83–129.

Swarup, S.J., Rao, A., Boaz, K., Srikant, N., Shenoy, R., 2014. Pulpal response to nano hydroxyapatite, mineral trioxide aggregate and calcium hydroxide when used as a direct pulp capping agent: an in vivo study. J. Clin. Pediatr. Dent. 38 (3), 201–206.

Taha, D.G., Abdel-Samad, A.A., Mahmoud, S.H., 2011. Fracture resistance of maxillary premolars with class II MOD cavities restored with Ormocer, nanofilled, and nanoceramic composite restorative systems. Quint. Int. 42 (7), 579–587.

Targino, A.G., Flores, M.A., dos Santos Junior, V.E., de Godoy Bené Bezerra, F., de Luna Freire, H., Galembeck, A., Rosenblatt, A., 2014. An innovative approach to treating dental decay in children. A new anti-caries agent. J. Mater. Sci. Mater. Med. 25 (8), 2041–2047.

Tay, F.R., Pashley, D.H., 2009. Biomimetic remineralization of resin-bonded acid-etched dentin. J. Dent. Res. 88 (8), 719–724.

Universidad Autonoma de San Luis Potosi, 2014. Addition of silver nanoparticles to a orthodontic primer in preventing enamel demineralization adjacent brackets.

University of California, LA, Ivoclar Vivadent, AG, 2012. Wear characteristics of denture teeth.

University of Chile, 2014. Safety and efficacy of a 6% hydrogen peroxide tooth bleaching agent.

Venkatesan, J., Kim, S.K., 2014. Nano-hydroxyapatite composite biomaterials for bone tissue engineering: a review. J. Biomed. Nanotechnol. 10 (10), 3124–3140.

Waltimo, T., Mohn, D., Paque, F., Brunner, T.J., Stark, W.J., Imfeld, T., Schatzle M., Zehnder, M., 2009. Fine-tuning of bioactive glass for root canal disinfection. J. Dent. Res. 88 (3), 235–238.

Watt, K.M., Chiswell, K., Cohen-Wolkowiez, M., 2013. Phase I Trials: First in Human. In: Lopes, Renato D., Harrington, Robert A. (Eds.), Understanding Clinical Research. first ed. Lange, McGraw Hill Education, United States.

Wong, Y.W., Schulman, K.A., 2013. Ethics of clinical research: an overview and emerging issues. In: Lopes, Renato D., Harrington, R.A. (Eds.), Understanding Clinical Research. 1st ed. Lange, McGraw Hill Education, United States.

FURTHER READING

Alcock, J.P., Barbour, M.E., Sandy, J.R., Ireland, A.J., 2009. Nanoindentation of orthodontic archwires: the effect of decontamination and clinical use on hardness, elastic modulus and surface roughness. Dent. Mater. 25 (8), 1039–1043.

Beyth, N., Yudovin-Farber, I., Perez-Davidi, M., Domb, A.J., Weiss, E.I., 2010. Polyethyleneimine nanoparticles incorporated into resin composite cause cell death and trigger biofilm stress in vivo. Proc. Natl. Acad. Sci. USA 107 (51), 22038–22043.

Fawzy, A.S., Farghaly, A.M., 2009. Probing nano-scale adhesion force between AFM and acid demineralized intertubular dentin: moist versus dry dentin. J. Dent. 37 (12), 963–969.

Hadassah Medical Organization, 2013. A clinical study: the antibacterial effect of insoluble antibacterial nanoparticles (IABN) incorporated in dental materials for root canal treatment.

Lewis, G., Nyman, J.S., 2008. The use of nanoindentation for characterizing the properties of mineralized hard tissues: state-of-the art review. J. Biomed. Mater. Res. B 87 (1), 286–301.

Noorlander, C.W., Kooi, M.W., Oomen, A.G., Park, M.V., Vandebriel, R.J., Geertsma, R.E., 2015. Horizon scan of nanomedicinal products. Nanomedicine 10 (10), 1599–1608.

Orsini, G., Procaccini, M., Manzoli, L., Sparabombe, S., Tiriduzzi, P., Bambini, F., Putignano, A., 2013. A 3-day randomized clinical trial to investigate the desensitizing properties of three dentifrices. J. Periodontol. 84 (11), e65–e73.

Reis, A.F., Giannini, M., Pereira, P.N., 2007. Long-term TEM analysis of the nanoleakage patterns in resin-dentin interfaces produced by different bonding strategies. Dent. Mater. 23 (9), 1164–1172.

Seguier, S., Souza, S.L., Sverzut, A.C., Simioni, A.R., Primo, F.L., Bodineau, A., Correa, V.M., Coulomb, B., Tedesco, A.C., 2010. Impact of photodynamic therapy on inflammatory cells during human chronic periodontitis. J. Photochem. Photobiol. B 101 (3), 348–354.

Universidad Autonoma de San Luis Potosi, 2012. Fluor varnish with silver nanoparticles for dental remineralization in patients with trisomy 21.

University of Pernambuco, 2014. Nanosilver fluoride to prevent dental biofilms growth.

CHAPTER

28

CARBON-BASED NANOSTRUCTURES FOR ELECTROCHEMICAL ANALYSIS OF ORAL MEDICINES

Sevinc Kurbanoglu, Bengi Uslu, Sibel A. Ozkan

Ankara University, Ankara, Turkey

CHAPTER OUTLINE

1 Introduction 886

2 Oral Drugs 887

2.1 Solids and Semisolid Oral Dosage Forms 887

2.2 Liquid Oral Dosage Forms 891

3 Voltammetric Methods for Electrochemical Analysis of Oral Medicines Using Carbon-Based Nanostructures 892

3.1 Potential Sweep Voltammetry 893

3.2 Alternating Current Voltammetry 895

3.3 Stripping Voltammetry 896

4 Carbon-Based Nanostructures in Electrochemical Analysis of Oral Medicines 898

4.1 The Unique Element Carbon 898

4.2 Graphene and Graphene Oxide 899

4.3 Carbon Nanotubes 901

4.4 Carbon Nanofibers 905

4.5 Carbon Nanohorns 915

4.6 Fullerenes 917

4.7 Carbon Nanodiamond 919

5 Conclusions 923

Abbreviations 924

References 925

Nanostructures for Oral Medicine

1 INTRODUCTION

From the discovery of C_{60} in 1985, researched about carbon nanotechnology has become remarkable importance and new carbon nanoforms have been identified. Discovery of carbon-based nanostructures is long-standing related to their synthesis, usage, chemical, and physical properties and geometries (Emsley, 2011; Musameh et al., 2002). Carbon and their allotropes can be used widely in electrochemical nanosensor design and the electronic properties, surface microstructure, surface chemistry, microstructure, and electronic properties are slightly different. This different properties can enhance the modification and effect the redox system, kinetics, and mechanisms of the oral drug (Wang et al., 2001). Nanostructured carbon-based nanomaterials, such as nanocrystalline diamond, carbon nanotubes (CNTs) with different properties, fullerenes, and graphene are still in their initial stages, and they are promising research materials. Diamond is well-known carbon-based material with excellent physical qualities. Diamond has strong covalent bonding between its atoms, enables hardness and high thermal conductivity. Graphite is a semimetal that is mechanically soft due to its weak van der Waals interlayer bonds. Graphite has layers of carbon, called graphene. On the other hand, carbon fibers or carbon nanotube are also carbon-based nanostructures, typically 5–10 μm in diameter. Carbon fibers consist carbon atoms that are bonded together in crystals. They are aligned parallel to the long axis of the fiber leading to high strength-to-volume ratio. With this property, carbon fibers have some unique properties, such as high stiffness, high tensile strength, low weight, high chemical resistance, and low thermal expansion. They have also high surface-to-volume ratio and enhanced reactivity, enhanced electrical properties (Emsley, 2011; Falcao and Wudl, 2007; Streeter et al., 2008; Zhang et al., 2013).

Drugs are called for any chemicals that show effects on the body after taken it is with different ways. A drug must be able to pass from the body into the brain. Once inside the brain, drugs can change the messages in brain cells are sending to each other, and to the whole body (Lachman et al., 1986). The physical form of the drugs can be divided as solid, semisolid, liquid, and gaseous. Drugs can be taken into the body as different dosage forms. Dosage forms are the means by which drug molecules are delivered to sites of action within the body. The need for dosage forms are accurate dose, protection, such as coated tablets (protection from gastric juice, masking taste, and odor), sealed ampules, placement of drugs within body tissues, optimal drug action, insertion of drugs into body cavities (rectal, vaginal), use of desired vehicle for insoluble drugs, sustained release medication, and controlled release medication (Brittain, 1995; Lachman et al., 1986; Sahajwalla, 2004). Oral path is the most favored way of medication administration because of the convenience and low cost of this route.

Modern electroanalytical techniques provide reliable, sensitive, and reproducible data for the quantification of electroactive analytes, such as oral drug active compounds. The electroanalytical methods used for analyzing oral drugs are effective for detecting and determining drug active compounds in their dosage forms in a variety of matrices, such as biological or environmental down to very low concentration levels, generally using stripping techniques (Bagotsky 2006; Koryta et al., 1993; Zoski 2007).

In this chapter, electrochemical assay of oral drugs using carbon-based nanostructures, such as fullerene polymers, CNTs, carbon nanohorns, carbon nanodiamonds, graphene, advantages and disadvantages of producing techniques, are discussed as details, and selected literatures are presented with tables using separate classes of all oral medicines.

2 ORAL DRUGS

Drugs are introduced into the body using several routes. It can be ordered as oral, topical, rectal, parenteral, vaginal, inhaled, ophthalmic, and otic ways. They may be taken by mouth, which is called orally, given by injection, such as intravenously, intramuscularly, by the ocular route or the ear, inserted in the rectum or vagina, by sprayed into the nose, applied to the skin (Martin et al., 1983; Wermuth, 2003; Williams et al., 2002).

Each route has specific purposes, advantages, and disadvantages. Many drugs can be administered orally as liquids, capsules, tablets, or chewable tablets. Because the oral route is the most convenient and usually the safest and least expensive way, it is the most often used way (Martin et al., 1983; Shargel and Kanfer, 2005). However, it has some limitations because of the way a drug typically moves through the digestive tract. When a drug is taken orally, food and other drugs in the digestive tract may affect how much of and how fast the drug is absorbed. Thus, some drugs should be taken on an empty stomach, others should be taken with food, others should not be taken with certain other drugs, and still others cannot be taken orally at all. For drugs administered orally, absorption may begin in the mouth and stomach (Junginger and Horwood, 1992; Shargel and Kanfer, 2005).

Throughout the drug formulation history, oral drug formulations have gained great applications becausee the oral route in form of solids, powders, pellets, and solutions was commonly used. As the technology developed, tablets and capsules were also introduced to oral drug family. Now in the manufacture, generally oral drugs are tablets, capsules, or powder for syrups. However, when preparing oral dosage form many factors should be considered. Inadequate solubility of the ingredients can occur, volatile excipient, such as alcohol can be decreased. On the other hand, oral dosage forms are easier to administration and they have more economical properties. These advantages make them a reason to compare to other types of drugs. Moreover, from the patient point of view, drugs taken orally let the patient feel in complete control (Jithan, 2007; Martin et al., 1983; Shargel and Kanfer, 2005; Wermuth, 2003; Williams et al., 2002).

In this part of the chapter, the details of the different types of oral drugs, classified as tablets, capsules, liquids, powders, and granules, solids, are discussed in detail.

2.1 SOLIDS AND SEMISOLID ORAL DOSAGE FORMS

Oral solid dosage forms are some of the most popular and convenient methods of drug delivery. They can be produced in a nonsterile environment and the process, equipment, and technology is well defined. Solids are the most commonly used, well-ordered, and therefore reason to choose dosage forms in the pharmaceutical field (Carstensen, 1993). Tablets, capsules, and triturates are the main solid oral dosage forms that are commonly prepared and used (Fig. 28.1).

2.1.1 Tablets

A tablet is a hard, compressed medication in round, oval, or square shape. Solid dosage form generally contains a unit dose of one or more drug active compounds. Tablets are usually consisted of drug active compound combined with disintegrant, diluent, binder, lubricant, and some colorant. Step by step, kinds of disintegrant, diluent, glidants, binder, lubricants, and colorant select the oral dosage forms quality. By changing these steps, many modifications can be done and this allows to prepared different kinds of tablets, therefore, generally, tablets are easier to manufacture. Tablets can be prepared in different

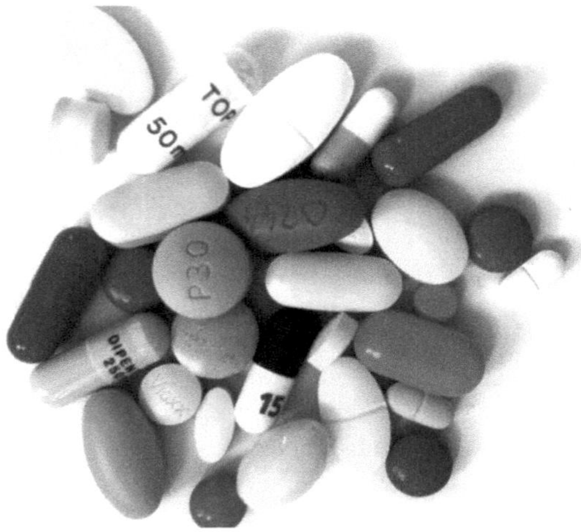

FIGURE 28.1 Some Examples of the Shape of Oral Dosage Forms

sizes, weights, hardness, thickness, and disintegration and dissolution characteristics (Fig. 28.2). Moreover, related to stability and physicochemical properties of drug active compound and the excipients tablets can be prepared with different methods. Four different types of methods, which are named as wet granulation, dry granulation, slugging, and direct compression, can be used for the preparation of tablet dosage forms. Tablets are prepared by mold or compression methods (Ansel et al., 1999; Heinrich Stahl and Wermuth, 2002; William, 1991).

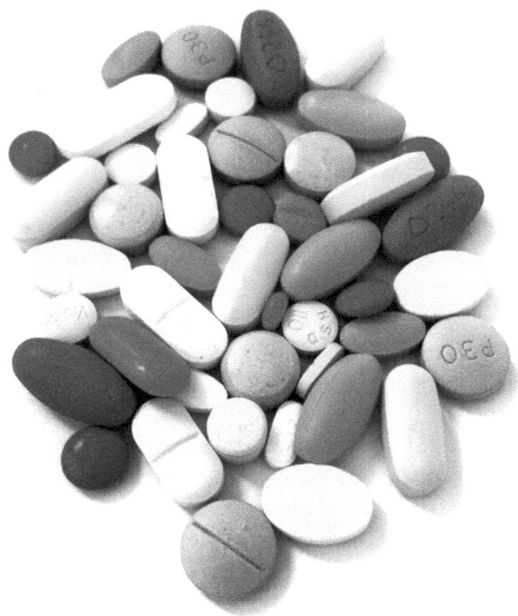

FIGURE 28.2 The Shape and the Examples of the Tablet Dosage Forms

FIGURE 28.3 The Shape and the Examples of the Capsules Dosage Forms

Tablets also allow the preparation of two or more drug active compounds together. Combined drugs can also be prepared to increase the effect of the drugs. Among antihypertensive drugs, hydrochlorothiazide has lots of combined formulations, such as benazepril, captopril, enalapril, lisinopril, moexipril, losartan.

There are several different kinds of tablet dosage forms depending on their healing purpose and physical properties, such as multiple compressed tablets, compressed tablets, film-coated tablets, enteric-coated tablets, sublingual tablets, chewable tablets effervescent tablets, dispensing tablets, sustained-release tablets, extended-release tablets, controlled-release tablets, instant-release tablets (Jithan, 2007; Umesh and Banakar, 1991; Zheng, 2009).

2.1.2 Capsules

A capsule is a pharmaceutical dosage form in a gelatin container. Powders or granules can be capsuled in gelatin shells called capsules. Capsules are generally made of gelatin, sugar, and water and they are generally tasteless. The advantage of it is masking the unpleasant taste of its contents (Fig. 28.3).

Drug active compounds that are not soluble in the gastrointestinal system or that should be administered extemporaneously can be prepared in capsules. Hard gelatin capsules and soft gelatin capsules are two kinds of capsules. Liquid drugs can be encapsulated in soft gelatin capsules. In clinical drug trials, to compare the effects of an investigational drug with another drug product or placebo hard gelatin capsules are generally used (Heinrich Stahl and Wermuth, 2002; Jithan, 2007; William, 1991).

2.1.3 Pellets

Pellets are tablets or capsules like solid dosage forms manufactured by using a variety of novel techniques. They are the small sterile solid mass consisting of a highly purified drug (with or without excipients) made by the formation of granules, or by compression, moldings, or in capsules. Also coated and extended release pellets can be described as "a solid dosage form" in which the drug itself is in the form of granules to which varying amounts of coating have been applied, and which releases a drug (or drugs) in such a manner to allow a reduction in dosing frequency as compared to that drug (or drugs) presented as a conventional dosage form (Martin et al., 1983; Wermuth, 2003; Williams et al , 2002).

Pelletization has some advantages, such as simplification of controlled- and sustained-release formulations, which their enhancement, benefits in mixing to overcome unsuitable formulations,

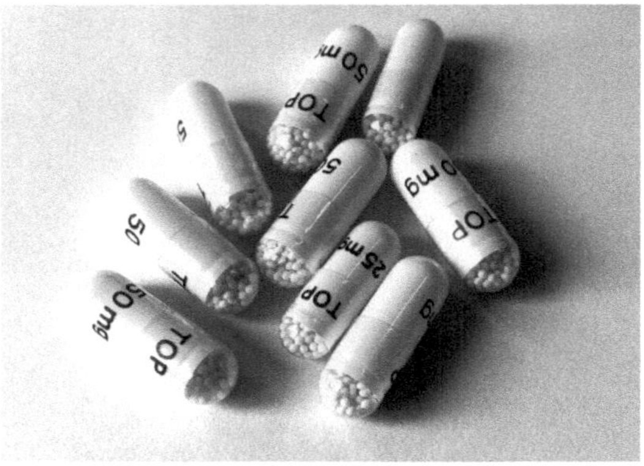

FIGURE 28.4 The Shape and the Examples of the Pellet Dosage Forms

benefits in homogeneity of the oral drug, more patient-friendly in terms of toxic effects compared to others, more environmental friendly related to hazards (Fig. 28.4).

Often, pellets are used for sustained or controlled release systems by subcutaneous or subdermal routes. Several polymers can be used to prepare pellets. Pellets can be used for drugs like proteins, genes, peptides for oral administration.

2.1.4 Lozenges

Lozenges are solid dosage forms that are intended to be dissolved or disintegrated slowly in the mouth. They contain one or more active ingredients and are flavored and sweetened so as to be pleasant tasting. It is generally used for their topical effect, but may also have ingredients that produce a systemic effect. Lozenge is a solid preparation consisting of sugar and gum, the latter giving strength and cohesiveness to the lozenge and facilitating slow release of the medicament (Fig. 28.5).

It is used to medicate the mouth and throat for the slow administration in digestion or cough remedies. Lozenges may contain an anesthetic, a demulcent, or an antiseptic.

Lozenges provide a pleasant dosage form for patients who are unable to swallow other types of solid dosage forms. Because lozenges are formulated to taste good, they must be kept out of the reach

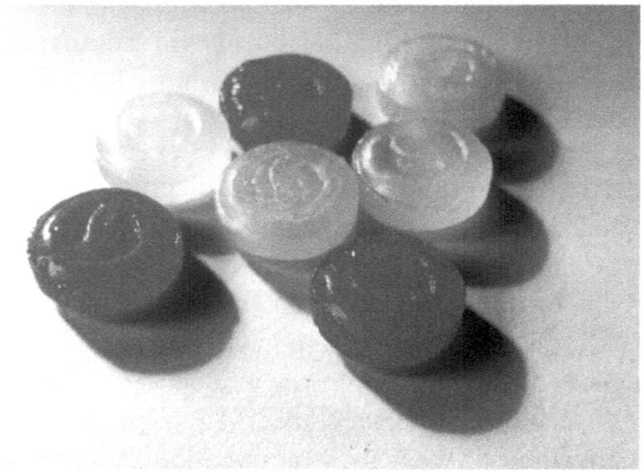

FIGURE 28.5 The Shape and the Examples of the Lozanges

FIGURE 28.6 The Examples of Effervescent Dosage Form

of children, who may view them as candy. Types of lozenges can be ordered as hard candy, chewable gummy gel, and hand-rolled lozenges. Lozenges are produced mostly for the children (Jithan, 2007; Martin et al., 1983; Shargel and Kanfer, 2005).

Pastilles are solid medicated preparations designed to dissolve slowly in the mouth like lozenges. They are softer than lozenges and their bases are either glycerol or gelatin, or acacia and sugar.

2.1.5 Powders and granules

Powder dosage form technology is very challenging because proper delivery of the drug active compound to powder is required. If not, this dosage form can be very harmful. Compared to other types of oral dosage forms very few amounts of powder formulations find a place in the market. Knowing the physical properties of the powders is very essential when preparing the oral dosage form.

Generally, granules are prepared by powders rather than capsule or a tablet because granules can increase the fluidity, delivery, and homogeneity of the dosage forms (Fig. 28.6).

The powder is usually dispersed in water or, in the case of effervescent powders, dissolved before taking (Jithan, 2007; Martin et al., 1983; Shargel and Kanfer, 2005; Williams et al., 2002).

2.2 LIQUID ORAL DOSAGE FORMS

Poorly soluble and low-dose drugs are generally chosen to be liquid oral dosage forms. Liquid preparations provide easy manufacturing and preparation procedures compared to solid forms. These groups can be oral solutions, emulsions, suspensions, elixirs, linctuses, oral drops, gargles, mouthwashes, and syrups are some types of liquid oral dosage forms (Cole, 1989). For poorly soluble compounds, cosolvents, such as ethanol, glycerol, can be used to increase solubility of the drug active compound. By this way, oral drug types can be prepared called "solution." When the drug active compound is acid or base, conjugate pairs of the drug can be used and this can be a solvent in a solvent as a salt. There are some disadvantages of liquid dosage forms. Shelf life of the liquid dosage forms is limited due to the change in the character of drugs (Cole, 1989; Hauss, 2007; Stegemann, 1999).

Suspended and dispersed types of liquid are also used as oral drug formulations. When the solid drugs are unstable in solution it can be prepared as suspended form. Swallowing is easer sometimes compared to solid form and with liquid form of the same drug. Using convenient suspending/dispersing agents, dry powders, which are a mixture of drug active compound and suspending/dispersing agents, can also be prepared.

Oral solutions are clear liquid preparations for oral use containing one or more active ingredients dissolved in a suitable vehicle.

Oral emulsions are stabilized oil-in-water dispersions, either or both phases of which may contain dissolved solids. Emulsions are the most common forms of liquid oral dosage forms. Emulsion is a heterogeneous system that contains oil droplets dispersed in an aqueous media. Water-soluble parts of the drugs go into the water part of the emulsion called aqueous phase, oil soluble ingredients can be in the oil part of the emulsion called lipid phase. Generally, surfactants can be used to stabilize the emulsion (Ansel et al., 1999; Tenjarla, 1999; William, 1991).

Oral suspensions are liquid preparations for oral use containing one or more active ingredients suspended in a suitable vehicle.

Syrup is a concentrated aqueous solution of a sugar, usually sucrose. Some flavors are used for masking disagreeable tastes. Syrups are the liquid oral drugs that are prepared by sucrose and/or little amount of alcohol. Compared to other liquid drugs syrups are viscous, stabilized, and flavored. Due to helping in taste masking and solubility of the drugs, syrups are very communal in pediatric and geriatric treatments. Depending on the physical and chemical characteristics of the drug and ingredients, syrups can be prepared with simple mixture procedures or heating. Since the sugar-based platforms are very good mediums for the growth of microorganisms, special care should be taken in formulations (Jithan, 2007; Martin et al., 1983; Shargel and Kanfer, 2005; Stegemann, 1999).

Elixir is a pleasantly flavored clear liquid oral preparation of potent or nauseous drugs.

Today, newly synthesized drug active compounds are insoluble in water. Crystalline nanosuspensions may help to solve this problem by supplying high weight-per volume-loading, high surface areas that can increase the dissolution rate. Surfactants can also be added to the medium, to increase the pharmaceutical stability by preventing agglomeration (Ansel et al., 1999; Heinrich Stahl and Wermuth, 2002; Tenjarla, 1999; William, 1991).

3 VOLTAMMETRIC METHODS FOR ELECTROCHEMICAL ANALYSIS OF ORAL MEDICINES USING CARBON-BASED NANOSTRUCTURES

In electrochemical analyses of oral drugs, voltammetry is the commonly used electroanalytical technique. Voltammetry records the current as a function of applied potential. It is divided into subgroups called linear sweep, cyclic, step and pulse, and alternating current voltammetric methods and stripping techniques (Bockris and Khan, 1993; Brett and Oliveira-Brett, 1993; Gileadi et al., 1975).

In voltammetric analyses the rate at which material arrives from the bulk of solution to the electrode, mass transport, the rate of electron transfers across the electrode-solution interface, the charge transfer kinetics affects the current obtained. Ions or molecules from bulk solution to or from the electrode surface determine the rate of the electrochemical reaction and the mass transport is achieved by convection, migration, and diffusion. Polarization occurs when the transport is not enough (Frumkin et al., 1980; Smyth and Vas, 1992). In order to increase transportation, agitation or stirring can be performed to create convection, which decreases polarization. In order to achieve mass transport, electrolytes are added with high ionic strength, such as sodium dodecyl sulfate. Diffusion is generally occurring spontaneously as the concentration of electroactive species at the electrode surface decreases, molecules or ions moves from higher concentration to lower concentration (Brett and Oliveira-Brett, 1998; Kissinger and Heineman, 1996).

In voltammetry, a plot of potential versus current is followed and the resulting curve is called as voltammogram. Analytical, kinetic, thermodynamic, chemical (adsorption or diffusion) information can be obtained from the voltammograms, which are obtained by applying potential recording the current. In oral drug analysis, toward voltammetric methods, the following methods are generally used (Bagotsky 2006; Trasatti 1977):

- potential sweep voltammetry
- step and pulse voltammetry
- alternating current voltammetry
- stripping voltammetry

3.1 POTENTIAL SWEEP VOLTAMMETRY

Linear sweep voltammetry (LSV) and cyclic voltammetry (CV) are commonly used and primary experiments that can give information about the electrode reactions and behavior of oral drugs. LSV and CV were first announced in 1938 by Randles and Sevcik, they found place theoretically in 1948 (Christian et al., 2013; Gosser, 1988; Wang, 2006). LSV and CV methods have unique advantages, such as giving qualitative information about electrochemical reactions, they enable rapid determination of formal potentials, and they give insight information about electrochemical mechanism of the oral drugs. Using CV, information about the analytes properties can be gained in terms of electron transfer kinetics and thermodynamics. Using LSV, current can be monitored as a function of potential, as the potential changed the observed current is also changed as a function of time (Brett and Oliveira-Brett 1998; Gosser, 1988; Kissinger and Heineman, 1996; Wang, 2006).

Basically, LSV and CV depend on application of a potential that is decreasing/increasing within time. The following equation is obeyed where the E_i is the starting potential value, t is time, v is the scan rate:

$$E = E_i + vt$$

The degree of peak currents, $I_{p,a}$ and $I_{p,c}$, the potentials where the peaks are observed, $E_{p,a}$ and $E_{p,c}$ are important parameters in CV and LSV (Fig. 28.7).

3.1.1 Step and pulse voltammetry

Pulse voltammetric techniques, such as normal-pulse, alternating current, a differential pulse, square wave, and staircase are presented by Barker et al. in the 18th century (Barker and Gardner, 1960; Barker and Jenkins, 1952; Gosser, 1988). Pulse voltammetric techniques are based on the application of pulse changes of potential and depending on the time of the pulse, the current response is measured (Christian et al., 2013; Ozkan 2012). Applying pulses results better sensitivity due to the measuring process. The current is measured at the end of the pulse. First, a pulse of potential applied, leading to decrease in capacitive current faster than the decrease in faradaic current. Since current is measured at the end of the pulse due to the difference in capacitive and faradaic current, applying pulses results in better sensitivity (Brett and Oliveira-Brett, 1998; Kissinger and Heineman, 1996; Wang, 2006).

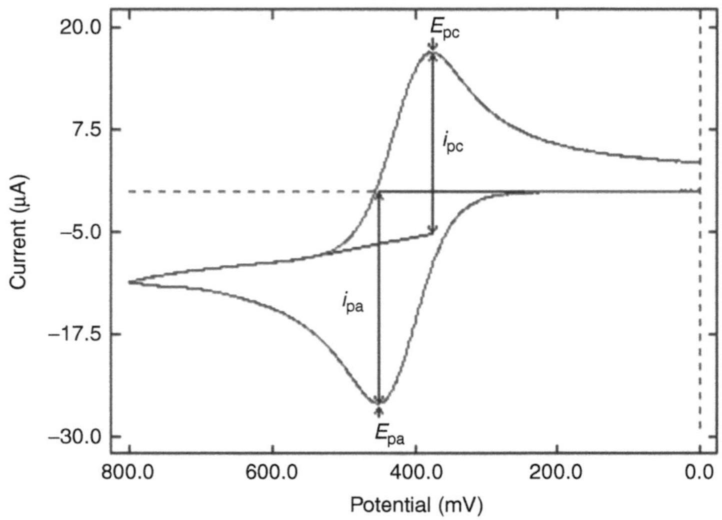

FIGURE 28.7 A Typical Cyclic Voltammogram Showing the Important Peak Parameters

Reprinted with permission from BAS Inc.

3.1.1.1 Normal pulse voltammetry

Normal pulse voltammetry is generally used to increase sensitivity of the measurements by minimizing the charging current effect via applying increased amplitude of the pulses. In normal pulse voltammetry applied pulses are generally start from the initial potential, and the current response is measured at the end of each pulse as shown in Fig. 28.8. (Bard and Faulkner, 2001; Brett and Oliveira-Brett, 1993; Gosser, 1988; Wang, 2006).

3.1.1.2 Differential pulse voltammetry

Differential pulse voltammetry (DPV) measures the difference between the current just before the ends of the pulse and just before application of the pulse as shown in the Fig. 28.9.

The peak-shaped wave form occurs when the difference between these currents versus applied potential graph is plotted (Fig. 28.9). DPV was suggested by Barker and Gardner and became a valuable technique for measuring oral drugs with low limits of detection values. Wave form resulting from the measurement of difference in two currents differs DPV from NPV. Moreover, twice measurements of the current and increase in base potential between pulses are general properties of DPV and differs from other electrochemical techniques that are used in oral drug analyzes. Since the ratio between faradaic current and charging current is large, DPV is one of the most sensitive and applicable electrochemical methods (Bard and Faulkner, 2001; Brett and Oliveira-Brett, 1993; Gosser, 1988; Wang, 2006).

3.1.1.3 Square wave voltammetry

In square wave voltammetry (SWV), the difference between the current at the end of the forward and reverse pulses is measured. This is called net current. SWV is faster than other methods due to the shorter pulses and higher frequencies (Fig. 28.10) (Brett and Oliveira-Brett, 1998; Kissinger and Heineman, 1996; Smyth and Vas, 1992).

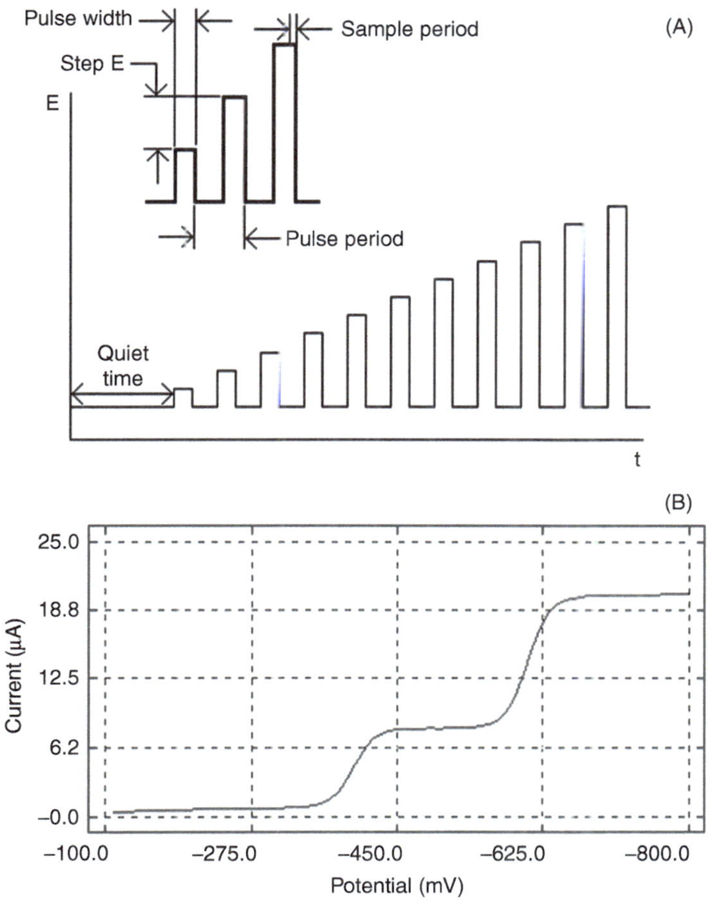

FIGURE 28.8

(A) Potential wave form for normal pulse voltammetry. (B) A typical normal pulse voltammogram.

Reprinted with permission from BAS Inc.

3.2 ALTERNATING CURRENT VOLTAMMETRY

When alternating voltage is applied with a gradually and linearly changing in positive or negative potentials, alternating current occurs. This voltage can be sinusoidal, triangular, or square wave. Based on the wave form, the amplitude can be changed.

Depending on the nature of the electroactive oral drugs and the composition of the supporting electrolyte, the resulting alternating current-potential peak can be changed. The form of the curves gives information about the kinetics and analytical issues. The peak current is directly related with the composition, and the concentration of the electroactive oral drug (Wang, 2006; Trasatti, 1977).

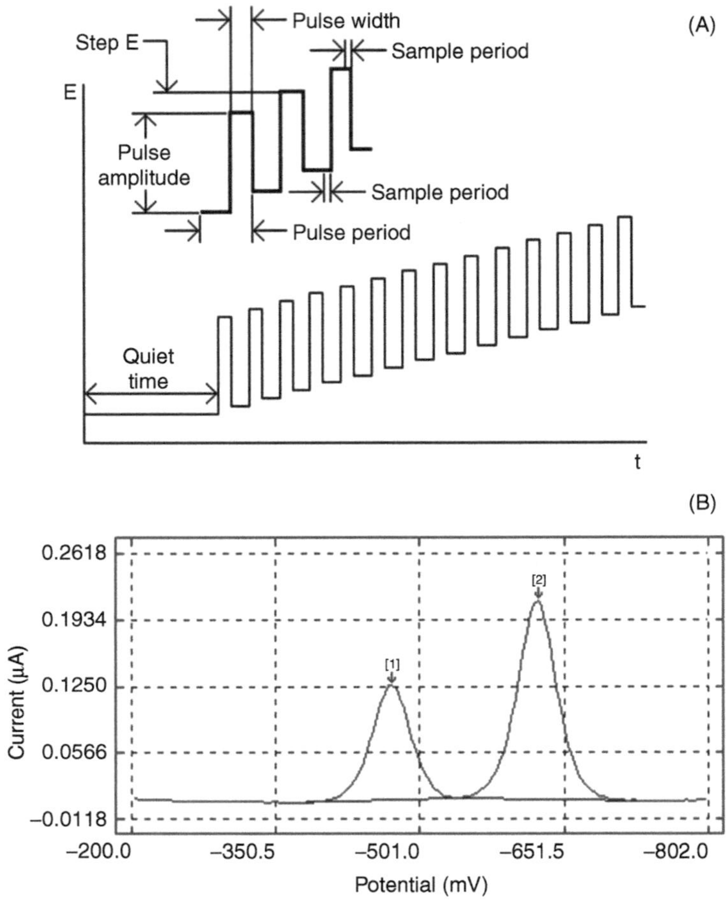

FIGURE 28.9

(A) Potential wave form for differential pulse voltammetry. (B) A typical differential pulse voltammogram.

Reprinted with permission from BAS Inc.

3.3 STRIPPING VOLTAMMETRY

In order to decrease detection limits and increase sensitivity, conditions can be optimized. Preconcentration of the oral drugs on the electrode surface before applying an electrochemical technique can increase the sensitivity of the analytical method. These methods are called stripping methods where the accumulation of the oral drugs at the electrode surface followed by "stripping" from the electrode to the solution. Since the electroactive oral drugs are closer to the electrode surface the detection limits decreases. The preconcentration can generally be achieved by applying potential where the highest adsorption is occurred or close to the limiting current region or at open circuit. The quality of analysis directly depends on the accumulation step and constant stirring is also used to increase sensitivity,

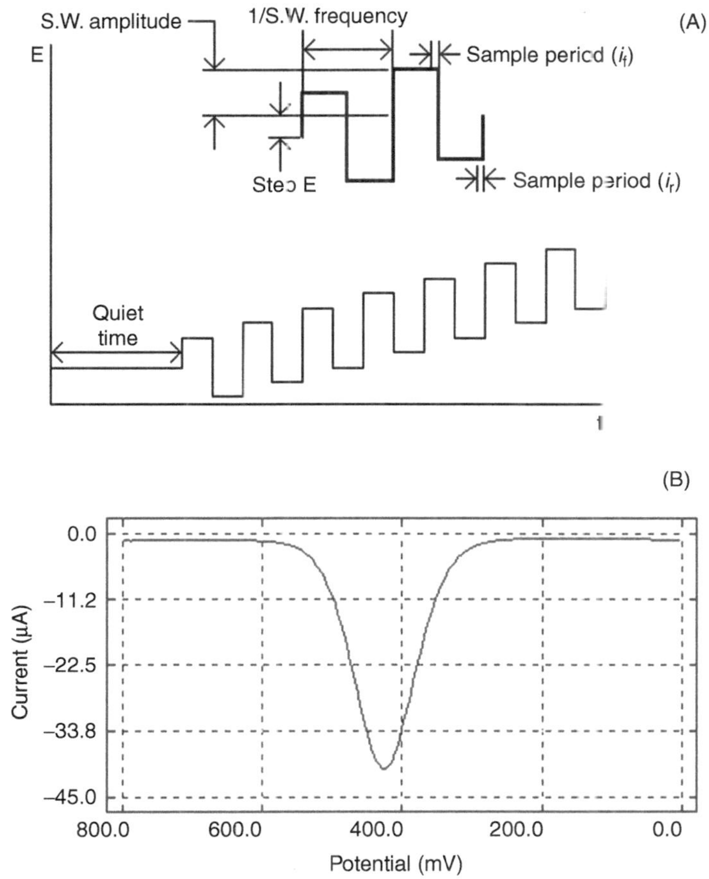

FIGURE 28.10

(A) Potential wave form for square wave voltammetry. (B) A typical square wave voltammogram.

Reprinted with permission from BAS Inc.

reproducibility, and repeatability of analysis. The stripping step can be achieved by controlling the potential, current or at open circuit. In stripping voltammetry, the potential is controlled and the current is proportional to the concentration directly (Brainina and Neyman, 1993; Ozkan, 2012; Wang, 1985). Anodic stripping voltammetry is generally used to analyze cations in the solution by applying negative during the preconcentration step. Cathodic stripping voltammetry, which is inverse of ASV, is also used to analyze halide ions and penicillin. Adsorptive stripping voltammetry is also based on the adsorption accumulation of the oral drug to the electrode surface, with the help of a potential. DPV and SWV can be combined with stripping methods and named, such as adsorptive stripping DPV, AdSDPV, and adsorptive stripping square wave voltammetry, AdSSWV (Brainina and Neyman, 1993; Ozkan, 2012; Wang, 1985).

4 CARBON-BASED NANOSTRUCTURES IN ELECTROCHEMICAL ANALYSIS OF ORAL MEDICINES

4.1 THE UNIQUE ELEMENT CARBON

Carbon, which derived it name from the Latin "carbo," meaning coal, is one of the most abundant elements. It is an extraordinary element, found in all living things, also found in fossilized remains in the form of hydrocarbons (natural gas, crude oil, oil shales, coal, etc.) and carbonates (chalk, limestone, dolomite, etc.). It occurs naturally in pure forms, including graphite, diamond, fullerenes, and graphene (Fig. 28.11). It is essential to life and after hydrogen, helium, and oxygen; it is the mostly found element in the universe. Moreover, carbon is the second most abundant element by mass (about 18.5%) after oxygen in human body (Falcao and Wudl, 2007; Zhang et al., 2013). The family of carbon-based materials now extends from C60, diamond and graphite to graphene, nanohorns, fullerenes, and nanotubes (Falcao and Wudl, 2007; Zhang et al., 2013).

Carbon is one of the most "multipurpose" elements due to its capability to create different compounds related to its electronic configuration. The electronic configuration of carbon ($1s^2\ 2s^2\ 2p^2$) enables carbon to form a huge variety of chains with different length. It can form three different types of bonding, that is, single, double, and triple bonds. In its ground state, two unpaired electrons are presents in 2p, these enable C to form two bonds (Fig. 28.12, Falcao and Wudl, 2007; Uslu and Ozkan, 2007; Zhang et al., 2013).

Carbon has the ability to rearrange the configuration of these valence electrons with the rearrangement process called hybridization. By this hybridization, carbon has ability to form four bonds. As 2s electron promotes into an empty 2p orbital, it goes into its excited state. One 2s orbital + three 2p orbitals, leading to the formation of four sp^3 hybrid orbitals as shown in Fig. 28.12. Another hybridization consisting of one 2s orbital + two 2p orbitals leading to three sp^2 hybrid orbitals, each has one electron, again enabling C to form four bonds two sp hybrid orbitals + $2p^2$. The last possible hybridization is with one 2s orbital + one 2p orbital leading to form two sp hybrid orbitals. Carbon has the ability to form monoatomic chains with double bonds, called carbines (Fig. 28.12). Carbon has the capability to form multiple stable covalent bonds with different atoms, forming almost 10 million different compounds. It has unique chemical and physical properties; at atmospheric pressure it has no melting point. Its triple

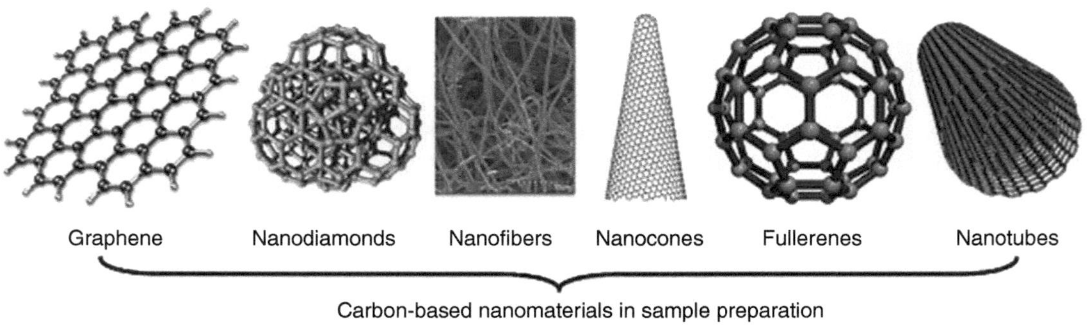

FIGURE 28.11 Carbon-Based Nanostructures

Reproduced from Zhang, B.T., Zheng, X., Li, H.F., Lin, J.M., 2013. Application of carbon-based nanomaterials in sample preparation: a review. Anal. Chim. Acta, 2013, 784, 1–17.

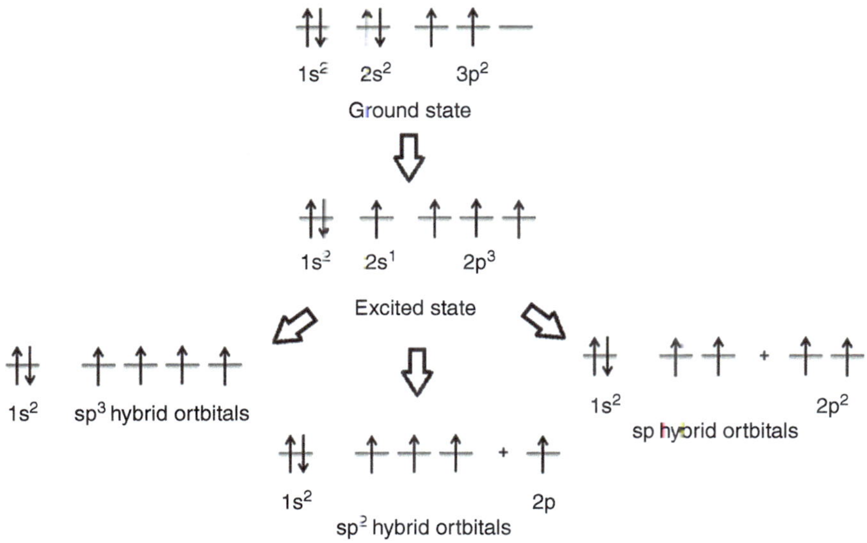

FIGURE 28.12 Electronic Configurations of Carbon Atom

point is at 10.8 ± 0.2 MPa and 4600 ± 300 K (~4330°C or 7820°F) (Bundy et al., 1978; Emsley 2011; Haaland 1976). Carbon has a nonordered atomic structure, that is, "amorphous," and most carbons have small crystallites of graphitic structure. Most of the allotropes of carbon have the same conducting properties in all directions; therefore they are isotropic (Ozkan et al., 2015).

For analyses of oral drugs with electrochemical techniques, nanoforms of carbon, such as graphene, CNTs, carbon nanodiamond, carbon nanohorns found place. It is believed that more platforms will be developed for the diagnostic and analytical applications based on carbon, an extraordinary element, to continue an area of great attention in the field of nanomedicine (Bae et al , 2010; Savvatimskiy, 2005; Whittaker, 1977).

4.2 GRAPHENE AND GRAPHENE OXIDE

Graphene, which is the simplest form of carbon, consist of a single layer of carbon atoms. Simply, a single carbon layer of graphite is called graphene, which is a near planar sheet of sp^2-bonded carbon atoms distributed in a hexagonal network (Hernandez et al., 2008; Novoselov et al., 2004; Singh et al., 2011). After 2004, when the graphene was produced, studies about graphene have gradual increases and in 2010, the Nobel Prize for physics was given a graphene-related research (Novoselov et al., 2004; Tung et al., 2009).

In graphene, all carbon atoms are sp^2-hybridized. The planarity of the layer comes from the π-bonded cloud above the double-bonded C atoms, other carbon atoms form strong directional sigma bonds. Graphene is a "rising-star" carbon-based structure, which is a two-dimensional sheet of carbon atoms bonded through sp^2-hybridization. When graphene is compared with other carbon allotropes, it has some advantages, such as having a high surface area that allows more active sites to the used surfaces, having mobility at room temperature and having more electrical conductivity (Berger et al., 2004, 2006; Geim and Novoselov, 2007).

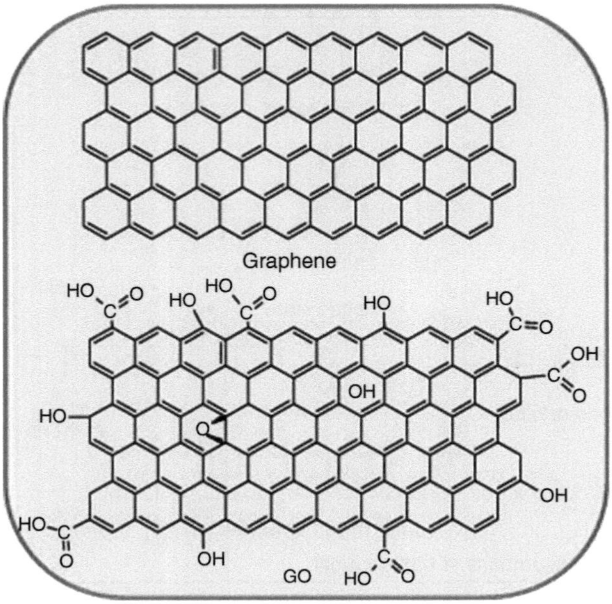

FIGURE 28.13 Chemical Structure of Graphene and Graphene Oxide (GO)

Reprinted with permission from Iwan, A., Chuchmała, A., 2012. Perspectives of applied graphene: polymer solar cells. Prog. Polym. Sci. 37, 1805–1828.

Graphene can be synthesized by different methods:

- Layer removing from graphite by mechanical exfoliation (scotch tape method), dispersing layers of graphite with surfactants, then by reducing graphene oxide, graphene is formed (Liu et al., 2011; Novoselov et al., 2004; Su et al., 2011).
- Chemical vapor deposition (CVD) method is used to produce large-scale graphene. Metallic catalysis, such as ruthenium, iridium, nickel, copper can be used (Reina et al., 2009; Sutter et al., 2008).
- Chemical and electrochemical methods can also be used to produce graphene, such as Hummers method (Hummers and Offeman, 1958; Kosynkin et al., 2010).

Graphene is always a single layer, hence graphene-related forms can be formed, such as graphene nanowalls and graphene nanoribbons. Graphene nanoribbons are a strip of graphene of less than 100 nm width. Graphene and graphene oxide nanostructured materials have gained great attention for the fabrication of several electrochemical sensors for oral drug analysis due to their low cost, high catalytic ability, and good stability. Moreover, the presence of large number of oxygen functional groups on GO, allows GO to hybridize via p–p interaction (Fig. 28.13). This has advantages using GO toward other carbon-based nanostructures (Kukovitsky et al., 2000; Pletikosic et al., 2009; Ray, 2015; Torres et al., 2014).

The individual sheets of carbon atoms that exist in parallel stacked layers are connected to each other by van der Waals forces to form crystallites. The graphene sheet has strong covalent bonding between the carbon atoms. The conductivity depends directly to the parallelism. If it is parallel to the carbon sheets, the conductivity is high. If it is perpendicular, the conductivity is low. Since the van der

Waals forces are weak, the sheets can easily slide past each other. Therefore, graphite is exceptionally soft (Meyyapan, 2005; Ozkan et al., 2015; Pierson, 1993).

In graphene, every carbon atom is bonded to three adjacent carbon atoms that lie at the vertices of equilateral triangles by four valence electrons. These carbon atoms are ready to be involved in the formation of chemical bonds. They are used in forming strong covalent bonds with the neighboring carbon atoms packed in a hexagonal lattice in the graphene sheet. This covalent bond has a short length and high strength due to electrons shared between atoms. The fourth electron is free to move over the surface of the sheet. This property makes the graphene an electrical conductor (Meyyapan, 2005; Ozkan et al., 2015; Pierson, 1993).

In electrochemical oral drug analysis graphene-based platforms found place in recent years. In their work, Asadian and Shahrokhian (2016) fabricated graphene nanosheets decorated with Ag nanoparticles pyrolytic graphite electrode for the voltammetric determination of Azathioprine, which is used to treat an array of autoimmune diseases including rheumatoid arthritis. The electrochemical behavior of Azathioprine was investigated and between 7×10^{-7} and 1×10^{-4} mol L^{-1} with the detection limit of 68 nM Azathioprine detection was achieved by CV. In this research, the prepared modified electrode based on graphene nanosheets decorated with Ag nanoparticles resulted in a remarkable increase toward the electroreduction of Azathioprine. The proposed method was well applied to pharmaceutical formulations and clinical preparations that contain Azathioprine (Asadian and Shahrokhian, 2016).

In another work, Wang et al. (2013) used graphene nanosheets to modify glassy carbon electrode for the simultaneous detection of L-dopa ((*S*)-2-Amino-3-(3,4-dihydroxyphenyl) propanoic acid) and carbidopa ((2*S*)-3-(3,4-dihydroxyphenyl)-2-hydrazino-2-methylpropanoic acid). The authors obtained graphene nanosheets by chemical reduction of graphene oxide using hydrazine. DPV method was utilized in the range of 1–16 μM for both L-dopa and carbidopa. The detection limit of this method was calculated as 0.8 μM for L-dopa and 1.8 μM for carbidopa (Wang et al., 2013). Kalambate et al. (2015) developed graphene nanosheet-gold nanoparticle carbon paste electrode and determination of rivastigmine ((*S*)-3-[1-(dimethylamino)ethyl]phenyl-*N*-ethyl-*N*-methylcarbamate), used for the treatment of Alzheimer's disease, was achieved. Electrochemical characterization of the modified electrode was achieved by CV, electrochemical impedance spectroscopy, chronocoulometry. Graphene was synthesized by exfoliation method from graphite. Adsorptive stripping DPV method was used for the analytical characterization of the designed nanosensor based on graphene nanosheet-gold nanoparticle. Adsorptive stripping DPV peak current was linear in between 2.0×10^{-7} and 6.0×10^{-4} M with a detection limit of 5.3×10^{-8} M (Kalambate et al., 2015).

Some recent studies about the use of graphene for the oral drug detections are summarized in Table 28.1.

4.3 CARBON NANOTUBES

CNTs, discovered in 1991 by Iijima, have diameters from fractions to tens of nanometers and lengths up to several micrometers. CNTs can be considered as a graphene sheet in the shape of a cylinder capped by fullerene-like structures (Iijima, 2002). CNTs can be created by rolling up a single layer of graphite or graphene along a certain direction into a tiny cylinder. A CNT is a tubular hollow-core sp^2-bonded carbon nanostructure with no axially oriented edges. In CNTs, the tube walls are approximately parallel to the tube axis at all times. The number of walls defines the

Table 28.1 Electrochemical Analysis of Some Oral Drugs Using Graphene-Based Platforms

Oral Drug Group	Oral Drug Name	Carbon Nanostructured Platform	Technique	LOD/LOQ	Applications	References
Analgesic and antipyretic	Paracetamol	Pd/GO	DPV	2.2 nM	Tablets, urine	Li et al. (2014a)
Analgesic and antipyretic	Paracetamol	CNT-graphene nanosheet nanocomposite modified GCE	DPV	0.10 μM	Tablets, blood serum	Arvand and Gholizadeh (2013)
Analgesic and antipyretic	Paracetamol	ERGO/NdHCF/GCE	CV, EIS, LSV	14.22 μM	Tablets	Devadas et al. (2012)
Analgesic	Ascorbic acid	PdNPs-graphene oxide/GCE	CV, AMP	0.02–2.28 mM	Tablets	Wu et al. (2012)
Antiallergic	Promethazine	rGO-GCE	AMP	1.99×10^{-7} mol L^{-1}	Tablets	Felix et al. (2015)
Antiarrhythmic	Quinidine	Nanographene/MWCNT/GCE	CV	0.573 nM	Tablets	Jain (2013)
Antiepileptic	Carbamazepine	Graphene–AuNPs-AuE	CV, LSV, EIS	3.03 μM	Bulk form	Pruneanu et al. (2011)
Antilipemic	Rosuvastatin	VACNT-GO	SWAdASV	0.06 μmol L^{-1}	Tablets, urine, human serum	Silva et al. (2015)
Antihelmintic	Mebendazole	GNS/CNS/chitosan modified GCE	DPAdSV	10.5 nM	Tablets, human serum	Ghalkhani and Shahrokhian (2013)
Antihypertensive	Acetazolamide	Graphene paste electrode	CV, DPV	2.2 nM	Human serum, urine	Shakibaian and Parvin (2012)
Antiparkinson's	L-dopa; carbidopa	Graphene/GCE	DPV	0.8 μM; 1.8 μM	Bulk form	Wang et al. (2013)
Antipsychotic	Chlorpromazine	Graphene paste electrode	CV, DPV	6.0 nM	Tablets	Parvin (2011)
Antibiotic	Levofloxacin	Au/PDDA/rGO/GCE	LSV	3.9×10^{-6} mol L^{-1}	Tablets	Borowiec et al. (2015)
Neuroprotective agents	Rivastigmine	GNS–AuNP–CPE	AdSDPV	5.3×10^{-8}	Tablets, blood serum, urine	Kalambate et al. (2015)

Table 28.1 Electrochemical Analysis of Some Oral Drugs Using Graphene-Based Platforms (cont.)

Oral Drug Group	Oral Drug Name	Carbon Nanostructured Platform	Technique	LOD/LOQ	Applications	References
NMDA receptor antagonist	Dextromethorphan	RGO-SPCE	DPV	1.5 ng mL^{-1}	Plasma and urine	Fakhari et al. (2014)
NSAID	Diclofenac	DNA/GO/GCE	DPV	NS	Bulk form	Wei et al. (2012)
Respiratory system	Theophylline	GNP/lcys/Graphene/Nf/GCE	CV, DPV	0.4 nM	Tablets	Zi et al. (2012)
Respiratory system	Theophylline	Graphene/Nf/GCE	CV, EIS, DPV	6 nM	Tablets	Li et al. (2012)
β_2-adrenergic agonists	Clenbuterol	SGS/MWCNTs-COOH/GCE	DPV	4.6 nM	Liver	Zhai et al. (2015)

name of CNTs. CNTs can be single-walled CNTs (SWCNTs), double-walled CNTs, and so on. If there exist few or more they are called as multiwalled carbon nanotubes (MWCNT). All consist concentric cylinders with spacing between nanotube walls approximately the interlayer distance in turbostratic graphite, that is, 0.34 nm. CNTs are most famous nanomaterial among other nanomaterials. They are the choice due to their unique properties. SWCNTs generally form ropes due to van der Waals interactions between other tube walls (Meyyapan, 2005; Pierson, 1993). Moreover, SWCNTs have different chirality, such as arm chair, zigzag, chiral (Fig. 28.14), the rolling-up direction and diameter or the chirality of an SWCNT determines its fundamental properties (Ebbesen and Ajayan, 1992; Fecht and Brühne, 2014; Iijima and Ichihashi, 1993; Saito et al., 1998). To make the ideal ends of an open tube give a zigzagged edge the graphene layer is rolled up and called zigzag type. To give armchair-type nanotubes, the graphene sheet is turned by 30 degree before rolling up. When the angle of turning the graphene layer before rolling up is between 0 and 30 degree, chiral nanotubes can be gained. Armchair nanotubes are electrical conductors whereas zigzag type and chiral nanotubes are semiconducting. Because they are five-membered rings, CNTs are generally closed at each end (Gogotsi, 2006; Grobert, 2007; Iijima and Ichihashi, 1993). The inner layer tubes of the MWCNT have π-orbitals. The interactions between closed tubes within the MWCNT lead to additional stabilization because the space between individual tubes is usually ~0.34 nm. CNTs bonding structure is stronger than the sp^3 bonds found in diamond. This property offers CNT unique strength, however, using hexagons alone, carbon cannot give in closed three-dimensional structures (Barsan et al., 2012; Gogotsi, 2006; Grobert, 2007; Iijima and Ichihashi, 1993; Ozkan et al., 2015).

CNT production methods directly determine the properties of nanotubes, such as diameter, number of walls, and length. Arc-discharge, laser ablation, electrolysis, CVD and plasma-enhanced CVD are the well-known nanotube procedures (Fecht and Brühne, 2014; Grobert, 2007; Iijima and Ichihashi, 1993; Tagmatarchis, 2012).

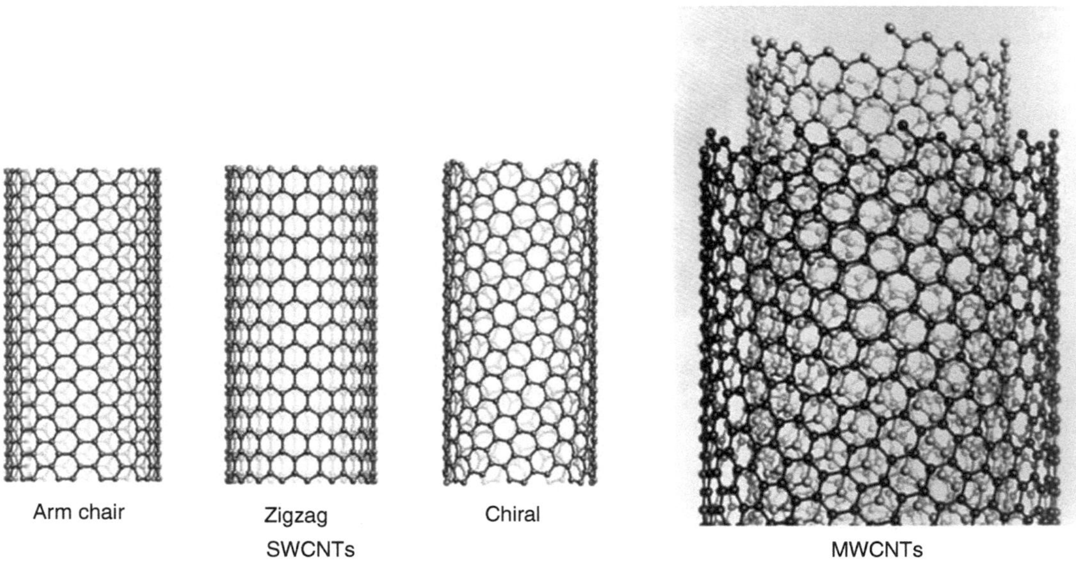

FIGURE 28.14 Computer-Generated Images of Single-Wall Carbon Nanotubes (SWCNTs) with Different Chirality and Multiwalled Carbon Nanotubes (MWCNTs)

Reprinted with permission from Iijima, S. 2002. Carbon nanotubes: past, present, and future. Phys. B Condens. Matter 323, 1–5.

CNTs are the most famous nanomaterials among other nanomaterials. They are the choice due to their unique mechanical, thermal, electrical, and physical properties. Due to active sites associated with CNT defects, enhanced electrocatalytic effects can be followed. Nitrogen-doping during CNT synthesis, nitrogen atoms substituting carbon atoms, can cause extra defects (Barsan et al., 2012; Fecht and Brühne, 2014; Tagmatarchis, 2012; Zhang, 2012). The chemical and electrochemical properties of CNTs enable them to be suitable for the design of a variety of sensors, and CNTs have been used to construct original electrodes with interest for electroanalytical applications for oral drug analysis. CNTs are one of the most commonly used building blocks in nanotechnology. Their effective usage in electrochemical nanosensor design is due to their small size, with larger surface areas, high sensitivity, and fast and reproducible responses, good conductivity, and high chemical stability (Meyyapan, 2005; Pierson, 1993; Torrens and Castellano, 2013). In the literature there exists a huge amount of studies related to CNTs-based modified electrodes for the detection of oral drugs. In their work, Mao et al. (2015) developed an electrochemical sensor for paracetamol based on MWCNT and chitosan–copper complex (Fig. 28.15). The modified electrode was performed to determine paracetamol and using this electrode high electrocatalytic activity toward paracetamol was obtained. In the linear range of 0.1–200 μmol L^{-1} paracetamol was detected with a detection limit of 0.024 μmol L^{-1}. Moreover, the suggested sensor based on MWCNT and chitosan–copper complex, was successfully applied for the determination of paracetamol in tablet and human serum samples (Mao et al., 2015).

In another work, Silva et al. (2015) developed an electrochemical method for the determination of rosuvastatin calcium ((3*R*, 5*S*, 6*E*)-7 [4-(4-fluorophenyl)-2-(*N*-methylmethane sulfonamido) 6-(propan-2-yl) pyrimidin-5-yl] 3,5-dihydroxyhept-6-enoic acid), which is used for the treatment of hypercholesterolemia. Composite of vertically aligned CNTs and graphene oxide was prepared and

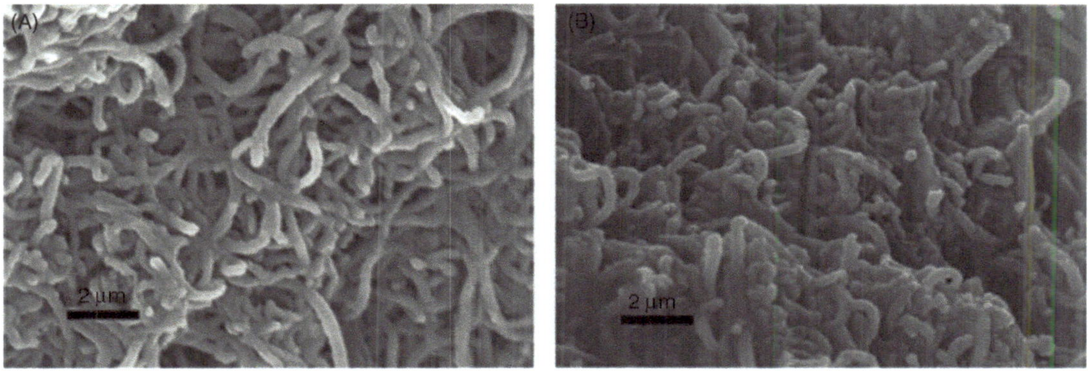

FIGURE 28.15 SEM Images of Multiwalled Carbon Nanotubes and Chitosan–Copper Complex

Reprinted with the permission Mao, A., Li, H., Jin, D., Yu, L., Hu, X. 2015. Fabrication of electrochemical sensor for paracetamol based on multi-walled carbon nanotubes and chitosan–copper complex by self-assembly technique. Talanta 144, 252–257.

square-wave adsorptive anodic stripping voltammetric method was used for the determination of rosuvastatin calcium. In the linear range of 0.48 to 46.72 µmol L^{-1} rosuvastatin the proposed method was successfully used with limit of detection of 0.06 µmol L^{-1} in phosphate buffer solution pH 2. The developed modified electrode was applied to determine rosuvastatin in commercial pharmaceutical samples, human body fluid samples of urine and human serum (Silva et al., 2015).

Valezi et al. (2014) used MWCNT paste electrodes for the simultaneous determination of amlodipine ((4*R*,*S*)-3-ethyl 5-methyl 2-(2-amino-ethoxymethyl)-4-(2-chlorophenyl)-1,4-dihydro-6-methyl pyridine-3,5-dicarboxylate) and enalapril ((2*S*)-1-[(2*S*)-2-{[(2*S*)-1 ethoxy-1-oxo-4-phenylbutan-2-yl] amino}propanoyl] pyrrolidine-2-carboxylic acid), which are generally used antihypertensives. The effect of cationic surfactant cetyltrimethylammonium bromide was also further studied. All the experimental data were optimized in terms of concentration and type of cationic surfactant, pH of Britton–Robinson buffer, and amount of CNTs were all optimized. Square-wave voltammetric method was used for the analyzing amlodipine and enalapril with the MWCNT paste electrode (Valezi et al., 2014).

In Table 28.2, SWCNTs and MWCNTs-based platforms were summarized for the analyses of oral drugs.

4.4 CARBON NANOFIBERS

Carbon nanostructures with diameters above 100 nm are referred to as carbon nanofibers (CNFs) and carbon nanorods. CNFs are higher in size than SWCNTs, behave like them but differ from CNTs in the absence of a hollow cavity. CNFs have a relatively high specific surface, and uniform mesoporous magnitude and distribution. CNFs have the highest specific surface and are above all nanostructured materials. CNFs could be easily available on a large scale using catalytic CVD or CVD methods (Bui et al., 2009; Guzmán et al., 2002; Tagmatarchis, 2012).

CNFs have some advantages toward other carbon nanostructures depending on their high aspect ratio in both electrical conductivity and specific surface area, so their shielding effectiveness is superior. Hence, because of their significantly larger dimensions, CNFs can overcome aggregation without functionalization, coating, or decoration. This is the main disadvantage of CNFs (Adabi et al., 2015; Kang et al., 2006; Lee et al., 2001; Li et al., 2005).

Table 28.2 Electrochemical Analysis of Some Oral Drugs Using SWCNTs and MWCNTs–Based Platforms

Oral Drug Group	Oral Drug Name	Carbon Nanostructured Platform	Technique	LOD/LOQ	Applications	References
Adrenergic agonists	Pseudoephedrine	MWCNT/GCE	DPAdSV, CAdSV	0.82 μM	Tablets, urine	Ahmar and Fakhari (2012)
Analgesic and antipyretic	Paracetamol; ascorbic acid	SWCNT/CCE	DPV	0.12 μM; 3.0 μM	Syrup and tablets, human serum	Habibi et al. (2011)
Analgesic and antipyretic	Paracetamol	MWCNT/GCE	CV	0.02 μM	Tablets	Hou et al. (2011)
Analgesic and antipyretic	Paracetamol	MWCNT/GCE	SWV	9.0×10^{-11} M	Tablets	Kutluay and Aslanoglu (2013)
Analgesic and antipyretic	Levodopa; acetaminophen; tryptophan	DTDF/CNPE	DPV	0.46 μM; 1.11 μM; 1.16 μM	Human blood serum sample	Mazloum-Ardakani et al. (2015)
Analgesic and antipyretic	Paracetamol	MWCNTs/CTS–Cu/GCE	DPV	0.024 μmol L^{-1}	Tablets, human serum	Mao et al. (2015)
Analgesic and antipyretic	Acetaminophen; dipyrone	MWCNT/PTH/CFE	DPV	13.4 μM; 17.8 μM	Tablets	Ghica et al. (2015)
Analgesic and antipyretic	Paracetamol	MWCNT–CTAB/GCE	DPV	4.82×10^{-9} M	Tablets, urine, plasma	Gowda et al. (2015)
Analgesic and antipyretic	Acetaminophen	MWCNT–ACS/GCE	CV, SWV	0.05 μM	Tablets	Lu and Tsai (2011)
Analgesic and antipyretic	Paracetamol	CoNPs/MWCNTs/GCE	SWV	1.0×10^{-9} M	Tablets, human blood serum	Kutluay and Aslanoglu (2014a)
Analgesic and antipyretic	Pyridoxine; acetaminophen	PMG/fCNT/GCE	DPV	9.4 μM; 4.3 μM	Tablets	Barsan et al. (2015a)
Analgesic and antipyretic	Ascorbic acid; paracetamol	GCE/MWCNT–polyhis	CV, DPV	0.76 μM; 32 nM	Tablets	Dalmasso et al. (2012)
Analgesic and antipyretic	Acetaminophen; ascorbic acid	AuNPs/MWCNT/GCE	DPV	0.03 μM; 0.76 μM	Tablets, blood serum	Madrakian et al. (2014)

Analgesic and antipyretic	Ascorbic acid	SDS-MWCNTs/GCE	DPV	3.0×10^{-6} M	Bulk form	Zhang et al. (2014)
Analgesic and antipyretic	Acetaminophen; tryptophan	DMBQ/MCNTPE	SWV	1.0 μM; 4.0 μM	Tablets, human urine, serum	Karimi-Maleh et al. (2014a)
Analgesic and antipyretic	Paracetamol	PtMWCNTs/ TritonX-100/CPE	AMP	17.71 nM	Tablets, blood serum	D'Souza et al. (2015)
Analgesic and antipyretic	Paracetamol	AuNP/MWCNT/ ITO	SWV	4.6 nM	Human urine	Goyal et al. (2011)
Analgesic and antipyretic	Levodopa	HRP/MWCNT–pPDA	SWV	40 nM	Bulk form	Shoja et al. (2016)
Analgesic	Methadone	MWCNT–PGE	DPV	87 nM	Human serum, urine	Alipour et al. (2015)
Analgesic and antipyretic	Levodopa	ZnO/CNTs/CPE	DPV	0.08 μmol L^{-1}	Tablets, urine	Afkhami et al. (2015)
Analgesic and antipyretic drug	Caffeine; acetaminophen; ascorbic acid	$MnFe_2O_4$/CNT–N/GCE	SWV	0.83×10^{-6} mol.dm^{-3}; 0.83×10^{-6} mol.dm^{-3}; 1.8×10^{-6} mol dm^{-3}	Bulk form	Fernandes et al. (2015)
Analgesic and antiepileptic	Phenobarbital, paracetamol	nPt-MWCNTPE	CV, DPV	0.1 μM; 0.17 μM	Tablets, human urine	Raoof et al. (2012)
Analgesic and antipyretic	Paracetamol	CMWCNTs–$NHCH_2CH_2NH)_6$/GCE	DPV	0.092 μM	Human plasma	Li et al. (2014b)
Anesthetic	Phenazopyridine	ds-DNA/CHIT–MWCNTs/PGE	DPV	0.003 μg mL^{-1}	Human plasma and urine	Ensafi et al. (2015a)
Antiallergic	Promethazine	MWCNT/SiO_2/Al_2O_3/ Nb_2O_5/DNA /CPE	SWV	5.9×10^{-6} M	Tablets	Marco et al. (2013)
Antianginal	Nifedipine	MWCNT–β-CD modified CPE	CV, DPV	0.025 μM	Capsules, blood serum	Kor and Zarei (2013)
Antiarrhythmia	Verapamil	MWCNT/GCE	CV; DPV	2.0 μM; 1.6 μM	Tablets, human serum	Chamjangali et al. (2015)
Antibacterial	*R*-ofloxacin; *S*-ofloxacin	MWCNT/poly(L-lysine)/ multi-HRP–AuNPs–Ab_2	AMP	0.15 ng mL^{-1}	Bulk form	He et al. (2015)

(Continued)

Table 28.2 Electrochemical Analysis of Some Oral Drugs Using SWCNTs and MWCNTs–Based Platforms (*cont.*)

Oral Drug Group	Oral Drug Name	Carbon Nanostructured Platform	Technique	LOD/LOQ	Applications	References
Antibiotic	Tetracycline	The MWCNT–COOH–GO/CPE	AdSDPV	3.6×10^{-7} $molL^{-1}$	Tablets, urine	Wong et al. (2015)
Antibiotic	Sulfamethoxazole	MWCNTPE	CV; CA	0.395 nM	Tablets	Arvand et al. (2011)
Antibiotic	Azithromycin	MWCNT decorated with $MgCr_2O_4$	DPV	0.07 μM	Capsules, urine, plasma	Ensafi et al. (2013)
Antibiotic	Norfloxacin	CuO nanoleaves MWCNTs/GC	CV, DPV	0.321 μM	Tablets	Devaraj et al. (2013)
Antibiotic	Amoxicillin	$FeCr_2O_4$/MWCNTs/GCE	EIS, DPV	0.05 μM	Capsules, urine, blood plasma	Ensafi et al. (2012)
Antibiotic	Midecamycin	GNP-MWCNT/GCE	CV	0.5 μM	Tablets	Wan et al. (2011a)
Antibiotic	Piroxicam; cefixime	MWCNTs–NHNPs–MCM–41/GCE	DPV	0.04 μM; 0.05 μM	Human urine and blood serum	Babaei and Afrasiabi (2015)
Antibiotic	Piroxicam	NHPDA/FePt/CNTs/CPE	SWV	0.1 μM	Tablets, urine	Karimi-Maleh et al. (2014b)
Antibiotic	Faropenem	MWCNT–NH_2–IL/GCE	DPV	3.1×10^{-8} mol L^{-1}	Urine	Yang et al. (2015)
Antibiotic	Roxithromycin	SWCNT/GCE	CV, CC	0.5 μM	Capsules	Wan et al. (2011b)
Antibiotic	Meloxicam	MWCNT–CP	DPV	3.22×10^{-5} mg L^{-1}	Human plasma	Azodi-Deilami et al. (2015)
Antibiotic	Levofloxacin	DNA/PoAP/MWCNT/GCE	DPV	1.0 μM	Capsules	Wen et al. (2012)
Antibiotic	Sulfadiazine	DNA/MWCNT–GCE	CV	0.12 μM	Human blood plasma sample	Fotouhi et al. (2013)
Anticancer	6-Mercaptopurine	p-APMCNTPE-TiO_2	CV, DPV	0.065 μM	Tablets, human urine	Ensafi and Karimi-Maleh (2012)

Anticoagulant	Warfarin	CdS-QDs/CS/MWCNTs	DPSV	8.5 nM	Urine, serum	Gholivand and Mohammadi-Behzad (2015)
Anticonvulsant	Diazepam	MWCNTs–MIP	SWV	3.7×10^{-9} M	Tablets, human serum	Hosseini and Motaharian (2015)
Anticonvulsant	Clonazepam	AgNPs/MWCNTs/GCE	DPV	6.0×10^{-9} M	Tablets, human blood serum	Habibi and Jahanbakhshi (2014)
Antidepressant	Venlafaxine	MWCNTs–RTIL/GCE	SWV	1.69×10^{-6} mol L^{-1}	Capsules	Ding et al. (2015)
Antidepressant	Fluvoxamine	HgNPs/MWCNTs/GCE	DPV	0.01 μmol L^{-1}	Tablets, urine	Madrakian et al. (2015)
Antidepressant	Venlafaxine desvenlafaxine	Nf-CNT–GCE	CV, CC, EIS, DPAdSV	12.4 nM; 21.1 nM	Tablets, human urine, human serum	Sanghavi and Srivastava (2011)
Antihistaminic	Cetirizine dihydrochloride	MWCNT/GCE	CV	0.071 μM	Tablets, human urine	Patil et al. (2011)
Antihypertensive	Furosemide	MWCNTPE	CV, DPV	0.29 μM	Tablets, human urine	Malode et al. (2012a)
Antihypertensive	Amlodipine; enalapril	MWCNTPE	SWV	0.049 μM; 0.81 μM	Tablets	Valezi et al. (2014)
Antihypertensive	Methyldopa	p-chloranil/CNTPE	CV, EIS, SWV	0.2 μM	Tablets, human serum, urine	Karimi-Maleh et al. (2012)
Antihypertensive	Captopril	*N-N*-(3,4-dihydroxyphenethyl)-3,5-dinitrobenzamide/CNTPE	CV, CA, EIS	0.034 μM	Tablets, human urine	Ensafi et al. (2011)
Antihypertensive	Captopril	MWCNT–$Zn_3(PO_4)_2$/CPE	SWAdASV	0.1 μM	Tablets, urine	Janegitz et al. (2014)
Antihypertensive	Captopril	FDADO/IPC/MCNTPE	DPV	0.08 μM	Tablets, urine	Karimi-Maleh et al. (2014c)
Antihypertensive	Nicardipine	MWCNTs–β-CD film	sDPV	1×10^{-8} M	Blood serum	Zarei et al. (2015)
Antihypertensive	Captopril	BFCNPEs	SWV	3.0×10^{-8} M	Tablets, urine	Beitollahi et al., 2014

(*Continued*)

Table 28.2 Electrochemical Analysis of Some Oral Drugs Using SWCNTs and MWCNTs–Based Platforms (*cont.*)

Oral Drug Group	Oral Drug Name	Carbon Nanostructured Platform	Technique	LOD/LOQ	Applications	References
Antihypertensive	Furosemide	Carboxyl-MWCNT/GCE	DPV	0.007 μg mL^{-1}	Urine and plasma	Heidarimoghadam and Farmany (2016)
Antihypertensive	Atenolol	GNPS/MWCNTS/GCE	DPV; AMP	0.5 μM; 59 nM	Tablets, urine	Shamsipur et al. (2014)
Antihypertensive	Amlodipine	o-MWCNT/AuE	CV, SWASV	2.65 μM	Tablets	Stoiljković et al. (2012)
Antihypertensive	Methyldopa	$NiFe_2O_4$–MWCNTs–GCE	DPV	0.08 μmol L^{-1}	Tablet, urine, plasma	Ensafi et al. (2015b)
Antihypertensive	Captopril	*N*-4,4'-azodianiline/MWCNT/CPE	DPV	0.03 μM	Urine	Taei et al. (2015)
Antihypertensive	Hydrochlorothiazide; enalapril	MWCNTsPE	SWV	1.4×10^{-8} mol L^{-1}; 4.1×10^{-8} mol L^{-1}	Tablets	Salamanca-Neto et al. (2014)
Antihypertensive	Captopril	PDPP/CdS–QDs/MWCNTs	AMP	15 nM	Tablets, urine	Paimard et al. (2015)
Antihypertensive	Atenolol propranolol	PtNPs/MWCNTs/GCE	DPV	1.17 μM; 0.15 μM	Tablets, serum	Kun et al. (2015)
Antihypertensive	Propranolol	γ-CD–CNT–CME	DPAdSV	4.01×10^{-8} M	Tablets, urine, blood serum	Gaichore and Srivastava (2014)
Antihypertensive	Amlodipine	PLM–GNPs/MWCNTs/GCE	DPV	1 nM	Tablets, plasma, urine	Emami et al. (2014)
Antiinflammatory	Mesalazine	CNTs–N/GCE	SWV	1.2×10^{-8} M	Tablets, human serum	Barsan et al. (2015b)
Antiinflammatory	Naproxen	MWCNT–Gr-IL|CCE	DPV	1.25×10^{-7} mol L^{-1}	Human blood plasma	Sarhangzadeh (2015)
Antilipemic	Bezafibrate	MWCNT–DHP/GCE	SWAdSV	16 nmol L^{-1}	Tablets	Ardila et al. (2014)
Antimalarial	Hydroxychloroquine	MWCNTs/CPE	DPV	6.0 nM	Tablets, human blood serum	Ghoreishi et al. (2015)

Antimigraine	Sumatriptan	Me-CoSal/ MWCNT/ GCE	DPV	0.3 μM	Tablets, Human serum	Amiri et al. (2011)
Anti Parkinson's	Carbidopa	VFCNPE	SWV	0.2 μM	Urine	Keshtkar et al. (2015)
Anti Parkinson's	Levodopa	MWCNT/GCE	DPV	0.57 μM	Tablets	Amiri-Aref et al. (2015)
Anti Parkinson's	Ropinirole	CNT–NAF/GCE	SWV; AdSWV	2.6×10^{-7}; 1.6×10^{-8}	Tablets	Sadiković et al. (2014)
Anti Parkinson's	Levodopa	CNT–NAF/GCE	SWV; AdSWV	1.6×10^{-7}; 5.2×10^{-8}	Tablets	Sadiković et al. (2014)
Anti Parkinson's	Carbidopa	Ferrocene dicarboxylic acid/CNTPE	CV, LSV, SWV	65 nM	Urine	Esfandiari Baghbamidi et al. (2012)
Anti parkinson's	Levodopa	$Co(DMG)_2ClPy$/ MWCNT/BPPGE	CV, SWV	0.86 μM	Tablets	Leite et al. (2012)
Anti parkinson's	Levodopa; carbidopa; tryptophan	2,2′-[1,2-ethanediylbis (nitriloethylidyne)]-bis-hydroquinone/CNTPE	CV, DPV	0.094 μM; 7.2 μM; 12.3 μM	Human urine, water	Mazloum-Ardakani et al. (2011)
Antiprotozoal	Furazolidone	MWCNT/GCE	CV, CA	2.30 μM	Tablets	Fotouhi et al. (2011)
Antiprotozoal	Furazolidone	dsDNA/MWCNT/GCE	CV	0.11 μM	Serum	Fotouhi and Bahmani (2013)
Antiprotozoal	Metronidazole	MWCNT/GCE	AMP	4.92×10^{-5} mg L^{-1}	Tablets	Yuan et al. (2015)
Antipsychotic	Trifluoperazine	MWCNT/GCE	DPAdSV	0.749 nM	Tablets	Dogan-Topal (2013)
Antipsychotic	Chlorpromazine hydrochloride	DNA/MWCNTs–modified AuE	PQCI	NS	Bulk form	Guo et al. (2004)
Antipsychotic	Clozapine	MWCNT–SDS/CPE	CV	1 mM	Bulk form	Manjunatha et al. (2011)
Antipsychotic	Olanzapine	Amine-functionalized TiO_2-MWCNT/GCE	CV, SWV, CA	0.09 μM	Tablets, blood serum	Arvand and Palizkar (2013)

(*Continued*)

Table 28.2 Electrochemical Analysis of Some Oral Drugs Using SWCNTs and MWCNTs–Based Platforms (*cont.*)

Oral Drug Group	Oral Drug Name	Carbon Nanostructured Platform	Technique	LOD/LOQ	Applications	References
Antithyroid	Methimazole	MWCNTs–Salen-Co(III)-MIP/GCE	CV, DPV	0.420 μM	Tablets, human urine	Pan et al. (2012)
Antithyroid	Methimazole	MWCNT/TiO_2NP/CPE	CV	0.17 μmol L^{-1}	Tablets, blood serum	Fouladgar and Mohammadzadeh (2014)
Antitrombotic	Cilostazol	MWCNTs–Bi_2O_3/GCE	CV, SWV	2.06 μM	Tablets, human plasma	Jain and Sharma (2012)
Antiviral drug	Acyclovir	OPPY/MWCNT/GCE	LSV	10.0 nM	Tablets, clinical preparations	Shahrokhian et al. (2015)
Anxiolytic	Buspirone	MWCNT/GCE	CV; DPV; AMP	0.23 μM; 0.18 μM; 0.22 μM	Tablets	Cheemalapati et al. (2014)
Arterial hypertension	Tadalafil	TiO_2-MWCNTPE	AdSSWV	0.11 μM	Tablets, human serum	Demir et al. (2014)
β_2-adrenergic receptor agonist	Salbutamol	CNT/Si	AMP	1 μM	Tablets, syrup	Karuwan et al. (2009)
Beta-adrenergic agonist	Metaproterenol	ITONPs/MWCNTs/GCE	DPV	1.2×10^{-8} M	Tablets, urine	Baytak and Aslanoglu (2015)
Cephalosporins antibiotic	Cefixime	MWCNT decorated $NiFe_2O_3$ magnetic nanoparticles	LSV	0.02 μM	Tablets	Ensafi et al. (2013)
Corticosteroids	Triamcinolone	SWCNT/EPPGE	SWV	8.9×10^{-10} M	Tablets, urine	Goyal et al. (2009a)
Diuretic	Triamterene	MWCNTs/PGE	DPV	3.35 nM	Tablets, serum	Nezhadali and Mojarrab (2015)
Fluoroquinolone antibiotic	Gatifloxacin	MWCNTPE	CV, DPV	4.5 nM	Biological fluids	Brahman et al. (2012)
Hyperlipidemic	Gemfibrozil	Dihexadecyl hydrogen phosphate film/MWCNT/GCE	SWAdSV	0.053 μM	Tablets, urine	Ardila et al. (2013)
Mucolytic	Bromhexine	NiNPs/MWCNT/PtE	SWV	3.0×10^{-6} M	Tablets	Kutluay and Aslanoglu (2014b)

Myorelaxant	Methocarbamol	MWCNT/GCE	DPV	0.104 nM	Human serum	Gholivand and Khodadadian (2011)
NSAID	Diclofenac	MWCNTs/$Cu(OH)_2$/IL/GCE	CV, DPV	0.04 μM	Tablets, ampoule, human and fish blood serum,	Arvand et al. (2012)
NSAID	Diclofenac	IL/CNTPE	SWV	0.09 $\mu mol\ L^{-1}$	Tablets, human urine	Goodarzian et al. (2014)
NSAID	Diclofenac	AuNP/MWCNT/GCE	SVW	0.02 μM	Tablets, urine	Afkhami et al. (2016)
Opioid analgesic	Methadone	MWCNT/GCE	CV, DPV	0.25 μM	Tablets, saliva, human urine	Amiri-Aref et al. (2013)
Psychostimulant	Caffeine; ascorbic acid	MWCNT/GCE	SWV	3.52 nM; 10 nM	Tablets, urine, tea leaves, coffee, cold drink	Gupta et al. (2013)
Peripheric vasodilator	Pentoxifylline	MWCNTPE	CV, DPV	0.169 μM	Tablets, human urine	Abbar et al. (2012)
Respiratory system	Theophylline	ACNTFE	DPV	16 nM	Tablets	Sun and Hu (2013)
Respiratory system	*N*-acetylcysteine	FCMCNTPE	CV, DPV, CA	0.6 μM	Effervescent tablets, human urine, Human serum	Karimi-Maleh et al. (2011)
Respiratory system	Theophylline	MWNT/MnO/GCE	DPV	1.0×10^{-8} M	Tablets	Yang and Li (2015)
Respiratory system	Theophylline	MWCNTPE	CV, DPV	19.7 nM	Tablets, urine	Malode et al. (2012b)
Treatment of liver	Oleanic acid	MIP/CoHCNFe/MWCNTs/PSS/CTS/CE	CV, EIS, SWV	2 nM	Capsules	Hu et al. (2012)
Vitamin	Biotin (B8)	Streptavidin-modified magnetic microbeads	SWV	81 μM	Commercial dietary supplement, infant formula	Kergaravat et al. (2012)
Vitamin	Vitamin B12	SWCNT–chitosan PGE	CV, EIS, SWV	0.89 nM	Tablets	Kuralay et al. (2011)
Vitamin	B1	dsDNA/PMWCNTPE	DPV; DPASV	0.44 $\mu g\ mL^{-1}$; 1.1 $ng\ mL^{-1}$	Serum, plasma, urine	Brahman et al. (2013)

In the literature are studies related to carbon fiber usage in oral drug analyses. Generally, CNFs are filled with nanoparticles to enhance the response of the oral drug one more level. In their work, Bartlett et al. (2005) designed a carbon fiber microdisk electrode for the determination of metronidazole, (1-(hydroxyethyl)-2-methyl-5-nitroimidazole) by SWAdSV. The carbon fiber microdisk electrodes were prepared by sealing short lengths of 7.5 μm diameter carbon fiber directly into glass and metronidazole was detected with a detection limit of 5×10^{-7} mol dm^{-3}. The proposed procedure is successfully achieved to detect metronidazole in urine samples (Bartlett et al., 2005). Antithyroid drugs were also detected with carbon fiber cylindrical microelectrodes by Guzmán et al. (2002). The amperometric detection was well achieved at a potential of +1.6 V with limit of detection of 2.6×10^{-7} mol L^{-1}. Moreover, the proposed method was well applied to spiked feed samples (Guzmán et al., 2002).

Huang et al. (2008) prepared palladium nanoparticle-loaded CNFs for the electrochemical determination of dopamine, uric acid, and ascorbic acid. Using DPV, dopamine, uric acid, and ascorbic acid separation was achieved with the modified electrode. DPV results at bare carbon paste electrode, palladium nanoparticles modified carbon paste electrode, palladium nanoparticle-loaded CNFs modified carbon paste electrode in 0.1M PBS (pH 4.5) containing 2 mM ascorbic acid, 50 μM dopamine, and 100 μM uric acid are shown in Fig. 28.16. Well-separated DPVs can be seen in this figure (Huang et al., 2008).

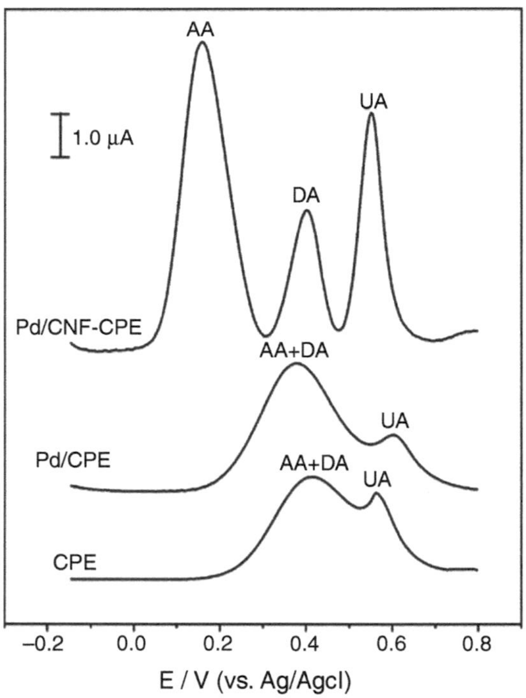

FIGURE 28.16 Differential Pulse Voltammograms of 2 mM Ascorbic Acid, 50 μM Dopamine, and 100 μM Uric Acid Using Pd/CNF/-CPE, Pd/CPE, CPE

Reprinted with permission from Huang, J., Liu, Y., Hou, H., You, T., 2008. Simultaneous electrochemical determination of dopamine, uric acid and ascorbic acid using palladium nanoparticle-loaded carbon nanofibers modified electrode. Biosens. Bioelectron. 24, 632–637.

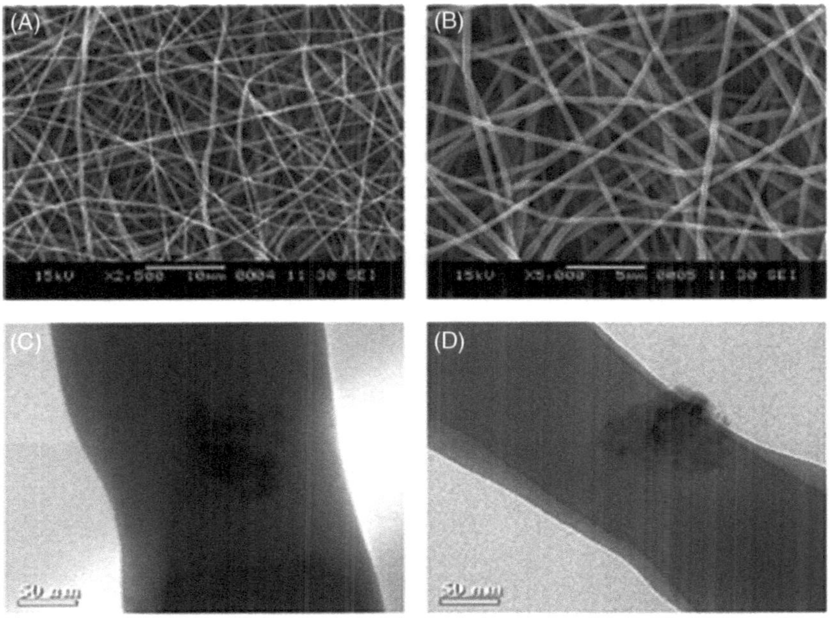

FIGURE 28.17

SEM images of the G/PANI/PS nanofibers with 2500× magnification (A) and 5000× magnification (B), and TEM images of random G distributions in the G/PANI/PS nanofibers (C and D).

Reprinted with permission from Rodthongkum, N., Ruecha, N., Rangkupan, R., Vachet, R. W., Chailapakul, O., 2013. Graphene-loaded nanofiber-modified electrodes for the ultrasensitive determination of dopamine. Anal. Chim. Acta 804, 84–91.

In another work, Rodthongkum et al. (2013) developed electrospun graphene/polyaniline/polystyrene nanofiber-modified screen-printed carbon electrodes for the determination dopamine (Fig. 28.17). With this novel and highly sensitive electrochemical system 9-time increase in the response of dopamine was obtained. Using SWV, wide dynamic range of 0.1 nM to 100 μM, and a very low limit of detection (0.05 nM) and a limit of quantification (0.30 nM) were obtained (Rodthongkum et al., 2013).

4.5 CARBON NANOHORNS

Carbon nanohorns, a new carbon allotrope within the family of elongated CNTs, can be described as disclinations in graphene. When wedges of graphene are removed, reconnect the dangling bonds, which forms a cone. Single-walled carbon nanohorns (SWCNHs) are spherical superstructures of aggregated nanosized graphitic tubes (Cioffi et al., 2007; Rotas et al., 2008).

SWCNHs can be synthesized by laser ablation of pure graphite without using metal catalysts with high production rates, high yields, and high purity. SWCNHs are principally metal-free CNTs and they are used in electrochemical studies for oral drug analyses due to the high surface area, excellent porosity, and considerable internal nanospaces. SWNHs are linked to each other to form the dahlia-flower-like structured aggregate with an average diameter of nearly 80 nm. Therefore, they have inherent nanopores at interstitial sites (Murata et al., 2002a; Pagona et al., 2008; Tagmatarchis, 2012; Tao et al., 2007).

SWCNHs differ from the SWCNTs in terms of their structure. The first SWNH particle is a single graphene tube with 2 nm in diameter and 40 nm in length. One end of a SWNH tube is closed with a

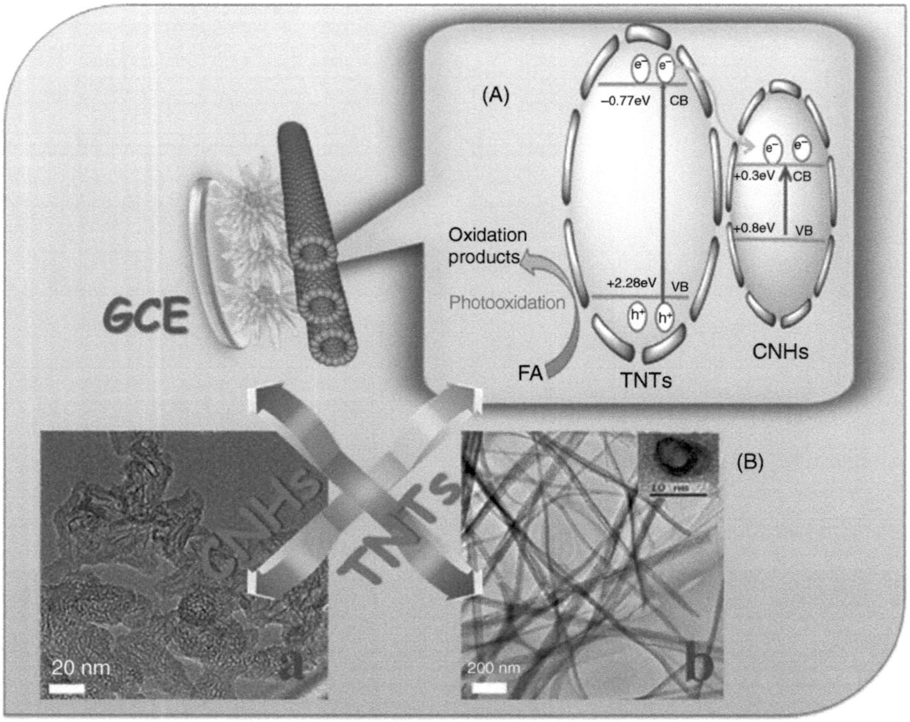

FIGURE 28.18 Sensing Platform Using Interwoven Titanate Nanotubes and Carbon Nanohorns

Reprinted with permission from Dai, H., Li, Y., Zhang, S., Gong, L., Li, X., Lin, Y., 2016. Delicate photoelectrochemical sensor for folic acid based on carbon nanohorns supported interwoven titanate nanotubes. Sensor. Actuat. B Chem. 222, 120126.

conical cap. Compare to SWCNTs, SWCNHs are promising material as adsorbents, catalyst supports, carriers of drug delivery due to their "unique holes" called as "nanowindows." Oxidation treatment can contribute to these holes on carbon walls. By this property, SWCNHs can easily accept molecules inside. Having these considerable internal nanospaces, nanopores and nanowindows are the main advantages of SWCNHs. On the other hand, in SWCNHs can easily aggregate through not only the van der Waals force but also chemical bonding. Compared to CNTs, which can be dispersed in surfactant solution by sonication, SWCNHs cannot (Jiménez-Soto et al., 2010; Murata et al., 2002b; Yudasaka et al., 2008).

Carbon nanohorns are one of the fastest growing areas of research. In the literature, SWCNHs-based studies have novelty, only three studies exist based on oral drug detection by electrochemical methods. In their work, Zhu et al. (2009) developed SWCNHs-modified glassy carbon electrodes for the electrochemical determination of uric acid, dopamine, and ascorbic acid. LSV method used among electrochemical methods, for the simultaneous determination of uric acid, dopamine, and ascorbic acid and calibration curves were obtained in the range of 0.06–10 μM, 0.2–3.8 μM, and 30–400 μM, respectively. Moreover, detection limits were also calculated as 20 nM, 60 nM, and 5 μM for uric acid, dopamine, and ascorbic acid, respectively. The proposed method was well applied to determination of uric acid, dopamine, and ascorbic acid in urine samples (Zhu et al., 2009).

Dai et al. (2016) used carbon nanohorns to determine folic acid. A signal-amplified photoelectrochemical sensing platform based on interwoven titanate nanotubes and carbon nanohorns conjugates was developed as shown in Fig. 28.18. Folic acid was detected in the wide range from 1×10^{-10} to

FIGURE 28.19 Structure of Fullerene C_{60}

Reprinted with permission from Knupfer, M., 2001. Electronic properties of carbon nanostructures. Surf. Sci. Rep. 42, 1–74.

5×10^{-5} M with a detection limit of $(2.5 \pm 0.005) \times 10^{-11}$ M. The developed nanosensor was applied to pharmaceutical folic acid tablets (Dai et al., 2016).

In another work by Tu et al. (2009) a sandwich nanohybrid of SWCNHs–TiO_2–porphyrin toward amperometric detection of chloramphenicol, which is a bacteriostatic antibiotic for the treatment of life-threatening infections was developed. An amount of SWNHs and TiO_2 nanoparticles were optimized and SWNHs–TiO_2–porphyrin nanohybrid was prepared. This modified electrode was used for amperometric detection of chloramphenicol at an applied potential of −0.56 V. With detection limit of 0.9 nM chloramphenicol detection was successfully achieved (Tu et al., 2009).

4.6 FULLERENES

Fullerenes, which were discovered in 1985, are closed single-walled cage molecules exclusively made of carbon. They contain 12 pentagons and varying numbers of hexagons. Buckminsterfullerene, C_{60}, is the well-known fullerene where the atoms and bonds delineate a truncated icosahedrons (Fig. 28.19; Cozzi et al., 2005; Rao et al., 1995; Vávrová et al., 2012).

Fullerene-C_{60} is one of the allotropes of carbon. It is made up of 60 closely packed carbon atoms into fused hexagons and pentagons, it has no edges (Baena et al., 2002; Yang et al., 2011). Fullerene-modified electrodes have an advantage of long stability and wide potential window. A focused pulsed laser of ~30 mJ onto a graphite target in a He atmosphere is the first technique to synthesize fullerenes. Arc-discharge is now the standard production route for fullerenes In 2002, a rational chemical synthesis route was developed for C_{60} (Dresselhaus et al., 1996; Hedberg et al., 1991; Kroto and McKay, 1988; Suarez-Martinez et al., 2012).

Fullerenes are carbon nanoclusters with unusual spherical molecules and unique electron structure. Fullerene films are porous insulators in their neutral state hence, partially reduced C_{60} films are conductive because the film is charged. When the C_{60} film is fully reduced, porosity character is disappearing therefore partially reduced form of C_{60} is generally used where partial reduction does not change the porosity (Fig. 28.20). Partially reducing the C_{60} consists of the modified electrode with as high electroactive surface area, excellent electronic conductivity, and good biocompatibility (Kroto et al., 1985; Scott et al., 2002; Suarez-Martinez et al., 2012; Tan et al., 2009).

Goyal et al. (2009a,b,c) developed a fullerene-C_{60}-modified edge plane pyrolytic graphite electrode for the determination of dexamethasone (9-fluoro-11β, 17,21-trihydroxy- 16 - methylpregna-1,4-diene-3,20-dione), which is a synthetic adrenocortical steroid. They used square wave voltammetric

FIGURE 28.20 A Typical Fesem Image of C_{60}-Modified Edge Plane Pyrolytic Graphite Electrode After Partial Reduction in KOH

Reprinted with permission from Goyal, R.N., Gupta, V.K., Chatterjee, S., 2009b. Fullerene-C60-modified edge plane pyrolytic graphite electrodefor the determination of dexamethasone in pharmaceutical formulations and human biological fluids. Biosens. Bioelectron. 24, 1649–1654.

method and they compared a bare pyrolytic graphite electrode, fullerene-C_{60}-modified glassy carbon electrode, the fullerene-C_{60}-modified edge plane pyrolytic graphite electrode with each other toward dexamethasone detection. Using fullerene-C_{60}-modified edge plane PGE in the concentration range of 0.05–100 μM dexamethasone detection was successfully achieved with a detection limit of 5.5×10^{-8} M. To prove the applicability of the method, proposed modified electrode was applied for quantitative determination of dexamethasone in several commercially available pharmaceutical formulations and human blood plasma (Goyal et al., 2009b). In another work of Goyal et al., C_{60} modified glassy carbon electrode for the voltammetric determination of paracetamol was also achieved. A significant increase in electrode response was observed GCE toward oxidation of paracetamol at fullerene-C_{60}-modified as fullerene anion (C_{60}) catalyzes redox reaction by it makes deprotonation of substrate easier. In the range 0.05–1.5 mM paracetamol was detected and the modified electrode was used to detect paracetamol in different tablets and urine samples (Goyal and Singh, 2006b).

Antiviral drug acyclovir [9(2-hydroxymethyl guanine)] was detected using fullerene-C_{60}-modified glassy carbon electrode by Shetti et al. (2012). By DPV, using fullerene-C_{60}-modified glassy carbon electrode remarkably enhanced electrocatalytic activity toward the oxidation of acyclovir was followed between 9.0×10^{-8} and 6.0×10^{-6} M. The detection limit of the proposed method for the detection of acyclovir was calculated as 1.48×10^{-8} M. The fullerene-C_{60}-modified glassy carbon electrode was further used to determine acyclovir in tablets, human urine, and serum samples (Shetti et al., 2012).

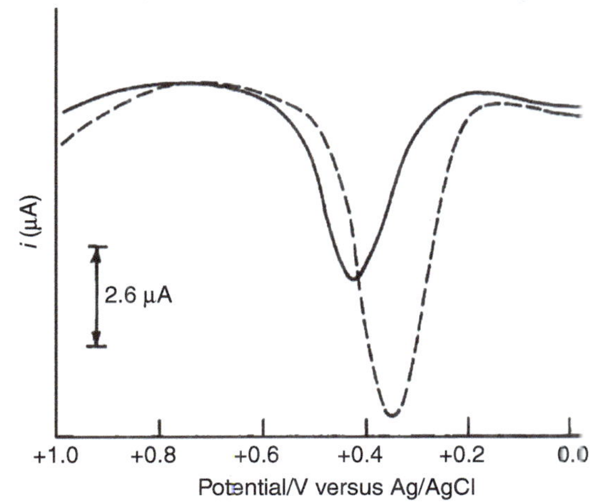

FIGURE 28.21 Voltammograms for 0.5 mM Paracetamol at Bare (—) and C_{60}-Modified Glassy Carbon Electrode

Reprinted with permission from Goyal, R.N., Singh, S.P., 2006b. Voltammetric determination of paracetamol at C60-modified glassy carbonelectrode. Electrochim. Acta 51, 3008–3012.

Stefan-van Staden and Bokretsion (2012) achieved enantioanalysis of antiinflammatory oral drug ketoprofen ((*R*,*S*)-2(3-benzoylphenyl)-propionic acid), using different types of potentiometric membrane electrodes based on C_{60} fullerene and its derivatives. Using fullerene-C_{60}-modified potentiometric membrane electrodes ketoprofen was detected with a limit of detection value of 3.16×10^{-8} M. Using tert-Butyl (1,2-methanofullere C_{60})-61-carboxylate modified potentiometric membrane electrodes ketoprofen was detected with a limit of detection value of 7.08×10^{-8} M and using (1,2-methnofullerene C_{60})-61-carboxylic acid modified potentiometric membrane electrodes ketoprofen was detected with a limit of detection value of 2.73×10^{-7} M. The propose method was also used to determine ketoprofen in capsules (Stefan-van Staden and Bokretsion, 2012).

In another work, Thirumalraj et al. (2016) prepared highly stable fullerene C_{60} decorated graphene oxide nanocomposite and used this nanocomposite for the detection of dopamine in rat brain and pharmaceutical samples. They synthesized graphite oxide from natural graphite by using Hummers' method and the GO–C_{60} nanocomposite was prepared by simple sonication method. The designed nanosensor detects the dopamine in the linear response range of 0.02–73.5 µM with a detection limit of 0.008 µM. In Fig. 28.21, voltammograms for 0.5 mM paracetamol at bare and C_{60}-modified glassy carbon electrode are shown, and enhancement in the response can be followed (Thirumalraj et al., 2016)

4.7 CARBON NANODIAMOND

Diamond, sp^3-hybridized carbon, is an allotrope of carbon. Due to its chemical inertness, hardness, high-thermal conductivity, biocompatibility, and optical transparency, diamond is ideal material (Chen et al., 2008; Zang et al., 2007). Diamond can be prepared by high-pressure high-temperature method (HPHT), producing hundreds of tons each year. Due to its restriction in shape and size and infeasibility diamond can go in mechanical abrasion applications, such as cutting, grinding, and polishing. From the 20th century to now, diamond was synthesized by a low-pressure CVD technique. Nanodiamonds as a novel carbon material can be synthesized by denotation. Nanodiamonds have some unique redox properties that would never

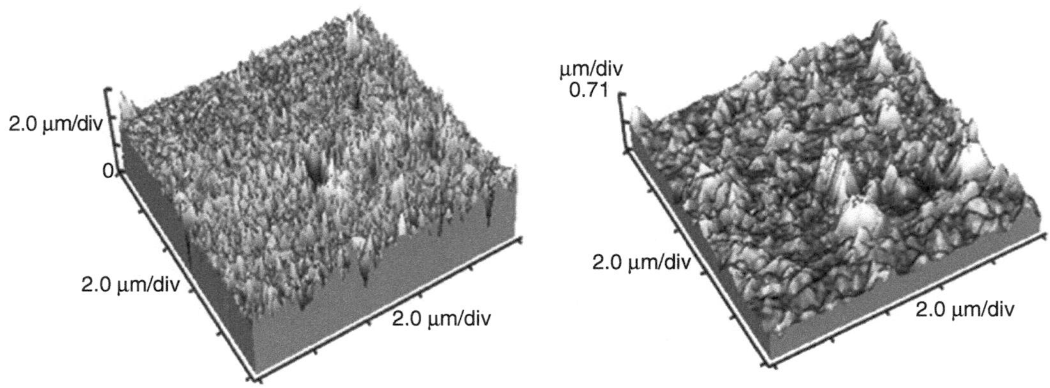

FIGURE 28.22 AFM Images of the Surface of NDG/CS/GCE

Reprinted with permission from Shahrokhian, S., Ghalkhani, M., 2010. Glassy carbon electrodes modified with a film of nanodiamond–graphite/chitosan: Application to the highly sensitive electrochemical determination of Azathioprine. Electrochim. Acta 55, 3621–3627.

been observed for diamond. Nanodiamond particles can have functional groups, such as —OH, —CO, —COOH, —C—O—C, and —CN (Liu et al., 2004; Raina et al., 2008; Schrand et al., 2009). These groups can make nanodiamonds more biocompatible and nontoxic. Some advantages using nanodiamonds as modification agents are NDs that have chemically inert cores with very high surface areas.

- NDs have smooth surface with excellent homogeneous mechanical and thermal properties.
- NDs provide a wide working potential window, low and stable background current.
- NDs have excellent corrosion resistance.
- NDs have microstructural stability at extreme anodic and cathodic potentials.

Modified electrodes using combining different carbon nanostructures, such as graphene oxide together with CNTs, are also a reason to enhance the electrode stability, lower the detection limit of the sensor. In their work, Shahrokhian and Ghalkhani (2010) constructed a film of nanodiamond–graphite/chitosan for the sensitive voltammetric determination of azathioprine (6-[(1-methyl-4-nitro-1H-imidazol-5-yl) sulfanyl]-7H-purine) an immunosuppressive. With the film of nanodiamond–graphite/chitosan, a notable decrease in reduction over potential was obtained with efficient catalytic role toward azathioprine detection. Moreover, with this modified electrode enhancement of the kinetics of the electrode reaction with a significant increase of peak current related to azathioprine was achieved. Linear response was obtained in the range of 0.2–100 μM azathioprine with a detection limit of 65 nM (Shahrokhian and Ghalkhani, 2010) (Fig. 28.22).

In another work, Shahrokhian and Nassab (2013), developed nanodiamond-graphite decorated with Ag nanoparticles-based nanosensor for the voltammetric determination of thioridazine (10-[2-(1-methyl-2-piperidyl) ethyl]-2-methylthiophenothiazine), which is a neuroleptic drug and belongs to the phenothiazine drug group. The modified electrode showed a wide linear response toward thioridazine between 0.08 and 100 μM with a detection limit of 0.01 μM. The modified nanosensor was well applied to determine thioridazine in pharmaceutical and clinical preparations (Shahrokhian and Nassab, 2013).

Electrochemical analysis of some oral drugs using carbon nanohorns, carbon nanodiamond, fullerenes, and other carbon nanostructure–based platforms were summarized in Table 28.3.

Table 28.3 Electrochemical Analysis of Some Oral Drugs Using Fullerenes, Carbon Nanohorns, Carbon Nanodiamond, Carbon Nanofiber–Based Platforms

Oral Drug Group	Oral Drug Name	Carbon Nanostructured Platform	Technique	LOD/LOQ	Applications	Ref.
Adrenocortical steroids	Dexamethasone	Fullerene-C_{60}-modified PGE	SWV	5.5×10^{-8} M	Tablets, human blood plasma	Goyal et al. (2009b)
Analgesic	Ascorbic acid	CNF-CPE	DPV	2 μM	Bulk form	Liu et al. (2008a,b)
Analgesic	Ascorbic acid	Pd/CNF-CPE	DPV	15 μM	Bulk form	Huang et al. (2008)
Analgesic	Ascorbic acid	SWCNH-modified GCE	LSV	5 μM	Urine	Zhu et al. (2009)
Analgesic and antipyretic	Paracetamol	Fullerene-C_{60}-modified GCE	DPV	0.05 mM	Tablets and syrup	Goyal and Singh (2006b)
Adrenal corticosteroid	Prednisolone	Fullerene-C_{60}-modified AuE	DPV	26 nM	Human urine and whole blood sample	Goyal et al. (2009c)
Antibiotic	Chloramphenicol	SWCNHs–TiO_2–porphyrin modified GCEs	AMP	0.9 nM	Bulk form	Tu et al. (2009)
Antiepileptic	Carbamazepine	Fullerene-C_{60}-modified GCE	DPV	16.2 nM	Tablets, spiked human serum, and urine sample	Kalanur et al. (2011)
Antihypertensive	Atenolol	Fullerene-C_{60}-modified GCE	DPV	0.16 mM	Tablets, human urine	Goyal and Singh (2006a)
Antiinflammatory	Prednisone	Fullerene-C_{60}-modified EPPG	SWV	4.8×10^{-8} M	Human urine	Goyal et al. (2010)
Antiinflammatory	Ketoprofen	Fullerene-C_{60}-modified EPMEs; tert-Butyl(1,2-methanofullere; C_{60})-61-carboxylate modified EPMEs; (1,2-methnofullerene C_{60})-61-carboxylicacid modified EPMEs	DP	3.16×10^{-8} M; 7.08×10^{-8} M; 2.73×10^{-7} M	Capsules	Stefan-van Staden and Bokretsion (2012)
Antiinflammatory	Methylprednisolone	Fullerene-C_{60}-modified GCE	DPV	5.6 nM	Tablets, human serum and urine	Goyal et al. (2007)

(*Continued*)

Table 28.3 Electrochemical Analysis of Some Oral Drugs Using Fullerenes, Carbon Nanohorns, Carbon Nanodiamond, Carbon Nanofiber–Based Platforms (*cont.*)

Oral Drug Group	Oral Drug Name	Carbon Nanostructured Platform	Technique	LOD/LOQ	Applications	Ref.
Antithyroid	6-methyl-2-thio-uracil	CFCME	AMP	2.6×10^{-7} mol L^{-1}	Animal feed	Guzmán et al. (2002)
Antiprotozoal	Metronidazole	CFMDE	SWAdSV	5×10^{-7} mol dm^{-3}	Urine	Bartlett et al. (2005)
Antiviral drug	Acyclovir	Fullerene-C_{60}-modified GCE	CV, DPV	14.8 nM	Tablets, human urine, serum	Shetti et al. (2012)
Immunosup-pressive	Azathioprine	NDG/CS/GCE.	CV	65 nM	Tablets, human blood serum	Shahrokhian and Ghalkhani (2010)
Neuroleptic	Thioridazine	AgNPs-NDG	CV	0.01 μM	Tablets, Human Blood Serum	Shahrokhian and Nassab (2013)
Vitamin	Folic acid	Carbon nanohorns supported interwoven titanate nanotubes	PEC	2.5×10^{-11} M	Tablets	Dai et al. (2016)
β_2-adrenergic receptor agonist	Salbutamol	Fullerene-C_{60}-modified GCE	SWV	40 ng mL^{-1}	Tablets and syrup, blood and urine	Goyal et al. (2008)

5 CONCLUSIONS

The unique element carbon has the capability to fill the sp, sp^2, and sp^3 hybrid orbitals. Therefore, carbon allotropes have successful application in electroanalytical studies of electroactive pharmaceutical compounds and is due to the high chemical and electrochemical stability of carbon materials, relatively high hydrogen, and oxygen over potential in different electrolytes, a broad working potential range, simplicity of mechanical renewal of the electrode surface and availability of different carbon materials. Moreover, a high degree of delocalization of π electrons, together with weak van der Waals forces, provide good electrical conductivity, and influence the electrical double-layer, molecular adsorption, and electrode polarity. CNTs, graphene, vitreous carbon, and carbon fibers are in the sp^2 atomic structure whereas diamond and fullerene are in sp^3 atomic structure forms. Among all carbon electrode materials, graphite is one the softest and diamond is one of the hardest known materials, carbon fibers are the highest strength fibers. Carbon fibers are a type of high-strength material and fullerenes are a new class of carbon-only molecules. The well-known and commonly used is 60 carbon atoms (C_{60}) arranged in a soccer ball structure. On the other hand CNTs are generally used in electrochemical oral drug analysis due to their large surface area, stability, and wide positive potential range. CNTs are completely formed of sp^2 bonds and can be SWCNTs with one graphene cylinder. If the many nested cylinders having an interlayer spacing approximately that of graphite it is called MWCNT.

In electrochemical analyses of oral drugs, voltammetry records the current as a function of applied potential, is the commonly used electroanalytical technique. LSV, cyclic, step, and pulse and alternating current voltammetric methods and stripping techniques are the methods generally used in oral drug analyses. LSV and CV are the primary experiment that can give information about the electrode reactions and behavior of oral drugs. LSV and CV methods give qualitative information about electrochemical reactions, they allow performing rapid determination of formal potentials, and they can offer insight information about electrochemical mechanism of the oral drugs. Pulse voltammetric techniques such as normal-pulse voltammetry, alternating current voltammetry, a DPV, square-wave voltammetry are also used in oral drug analyses. In these methods pulse changes of potential applied and depending on the time of the pulse, the current response is measured. Stripping voltammetry is also used where the accumulation of the oral drugs at the electrode surface followed by "stripping" from the electrode to the solution, can increase the sensitivity of the analytical method. Preconcentration of the oral drugs on the electrode surface before applying an electrochemical technique results in stripping voltammetry. The oral route in the form of solids and semisolid oral dosage forms, such as tablets, capsules, pellets lozenges, powders, and granules; liquid oral dosage forms, such as solutions, emulsions, suspensions, syrup, and elixir were generally used drug formulations. In the market nowadays, oral drugs are tablets, capsules, or powder for syrups due to easier administration and economical properties. Oral dosage formulation is very important, because many factors should be considered, such as inadequate solubility of the ingredients, volatile excipients, such as alcohol. Drugs taken orally let the patient feel in complete control, therefore oral drugs are preferred. In this chapter, electrochemical assay of oral drugs using carbon-based nanostructures, such as fullerene polymers, CNTs, carbon nanohorns, carbon nanodiamonds, graphene, and selected literatures are presented with tables using separate classes of all oral medicines. It can be concluded that use of carbon-based nanostructures in oral drug analyses will be increased in the following years, due to their promoting properties in electrochemical analyses. Moreover, new carbon allotropes could also be found and used in electrochemical nanosensor design because this unique element is one of the most "multipurpose" elements due to its capability to create different compounds related to its electronic configuration.

ABBREVIATIONS

Ab	Antibody
ACNTFE	Aligned carbon nanotubes thin film electrode
AdSDPV	Adsorptive stripping differential pulse voltammetry
AgNPs	Silver Nanoparticles
AuNFs	Gold nanoflowers
AMP	Amperometry
CCE	Carbon ceramic electrode
CdS quantum dots	CdS-QDs
CFCME	Carbon fiber cylindrical microelectrodes
CFMDE	Carbon fiber microdisk electrode
CNF	Carbon nanofiber
CNS	Carbon nanospheres
CNT-N	*N*-doped carbon nanotubes
CPE	Carbon-paste electrode
CTS	Chitosan–copper complex
DMQB	8,9-Dihydroxy-7-methyl-12H-benzothiazolo [2,3-b] quinazolin-12-one
DNA	Deoxyribonucleic acid
DTDF	(7,8-Dihydroxy-3, 3,6-trimethyl-3,4-dihydrodibenzo [b, d] furan-1(2H)-one)
DP	Direct potentiometry
DPSV	Differential pulse stripping voltammetry
EPPGE	Edge plane pyrolytic graphite electrode
EPMEs	Potentiometric membrane electrodes
FCMCNTPE	Ferrocene-modified carbon nanotubes paste electrode
GE	Gold electrode
GNS	Graphene nanosheets
HRP	Horseradish peroxidase
IL	Ionic liquid
ITONP	Indium tin oxide nanoparticles
LSV	Linear sweep voltammetry
$MnFe_2O_4$	$MnFe_2O_4$ nanoparticles
NDG	Nanodiamond-graphite
NHNP	Nickel hydroxide nanoparticles
NHPDA	*N*-(4-hydroxyphenyl)-3,5-dinitrobenzamide
NiFe2O4	Ferrite nickel nanoparticles
NS	Not stated
OPPY	Tiron-doped polypyrrole
Pd	Palladium nanoparticle
PDPP	1,4-Phenylene-N,N_0-bis (O,O-diphenylphoramidate)
PEC	Photoelectrochemical current
PGE	Pencil graphite electrode
PLM	Poly-L methionine
PTH/MWCNT/CFE	Poly(thionine)-carbon nanotube modified carbon film electrodes
rGO	Reduced graphene oxide
SDS	Sodium dodecyl sulfate
SGS	Sulfonated graphene sheets
SWAdSV	Square wave adsorptive stripping voltammograms
VFCNPE	Vinylferrocene modified carbon nanotube paste electrode
β-CD	β-Cyclodextrin

REFERENCES

Abbar, J.C., Malode, S.J., Nandibewoor, S.T., 2012. Electrochemical determination of a hemorheologic drug, pentoxifylline at a multi-walled carbon nanotube paste electrode. Bioelectrochemistry 83, 1–7.

Adabi, M., Saber, R., Naghibzadeh, M., Faridbod, F., Faridi-Majidi, R., 2015. Parameters affecting carbon nanofiber electrodes for measurement of cathodic current in electrochemical sensors: an investigation using artificial neural network. RSC Adv. 5, 81243–81252.

Afkhami, A., Kafrashi, F., Madrakian, T., 2015. Electrochemical determination of levodopa in the presence of ascorbic acid by polyglycine/ZnO nanoparticles/multi-walled carbon nanotubes-modified carbon paste electrode. Ionics 21, 2937–2947.

Afkhami, A., Bahiraei, A., Madrakian, T., 2016. Gold nanoparticle/multi-walled carbon nanotube modified glassy carbon electrode as a sensitive voltammetric sensor for the determination of diclofenac sodium. Mater. Sci. Eng. C 59, 168–176.

Ahmar, H., Fakhari, A.R., 2012. Electro-oxidation and adsorptive stripping voltammetric determination of ephedrine and pseudoephedrine at carboxylated multi-walled carbon nanotube-modified electrode. Anal. Method. 4, 812–818.

Alipour, E., Majidi, M.R., Hoseindokht, O., 2015. Development of simple electrochemical sensor for selective determination of methadone in biological samples using multi-walled carbon nanotubes modified pencil graphite electrode. Journal of the Chinese Chemical Society 62, 461–468.

Amiri, M., Pakdel, Z., Bezaatpour, A., Shahrokhian, S., 2011. Electrocatalytic determination of sumatriptan on the surface of carbon-paste electrode modified with a composite of cobalt/Schiff-base complex and carbon nanotube. Bioelectrochemistry 81, 81–85.

Amiri-Aref, M., Raoof, J.B., Ojani, R., 2013. Electrocatalytic oxidation and selective determination of an opioid analgesic methadone in the presence of acetaminophen at a glassy carbon electrode modified with functionalized multi-walled carbon nanotubes: application for human urine, saliva and pharmaceutical samples analysis. Colloids Surf. B 109, 287–293.

Amiri-Aref, M., Raoof, J.B., Ojani, R., 2015. Utilization of a bioactive anthocyanin for the fabrication of a novel carbon nanotube-based electrochemical sensor and its electrocatalytic properties for selective determination of l-dopa in the presence of uric acid. Ionics, 1–10.

Ansel, H.C., Allen, L.V., Popovich, N.G., 1999. Pharmaceutical Dosage Forms and Drug Delivery Systems, seventh ed. Lippincott Williams & Wilkins, Philapdelphia.

Ardila, J.A., Oliveira, G.G., Medeiros, R.A., Fatibello-Filho, O., 2013. Determination of gemfibrozil in pharmaceutical and urine samples by square-wave adsorptive stripping voltammetry using a glassy carbon electrode modified with multi-walled carbon nanotubes within a dihexadecyl hydrogen phosphate film. J. Electroanal. Chem. 690, 32–37.

Ardila, J.A., Oliveira, G.G., Medeiros, R.A., Fatibello-Filho, O., 2014. Square-wave adsorptive stripping voltammetric determination of nanomolar levels of bezafibrate using a glassy carbon electrode modified with multi-walled carbon nanotubes within a dihexadecyl hydrogen phosphate film. Analyst 139, 1762–1768.

Arvand, M., Gholizadeh, T.M., 2013. Simultaneous voltammetric determination of tyrosine and paracetamol using a carbon nanotube-graphene nanosheet nanocomposite modified electrode in human blood serum and pharmaceuticals. Colloids Surf. B 103, 84–93.

Arvand, M., Palizkar, B., 2013. Development of a modified electrode with amine-functionalized TiO_2/multi-walled carbon nanotubes nanocomposite for electrochemical sensing of the atypical neuroleptic drug olanzapine. Mater. Sci. Eng. C 33, 4876–4883.

Arvand, M., Ansari, R., Heydari, L., 2011. Electrocatalytic oxidation and differential pulse voltammetric determination of sulfamethoxazole using carbon nanotube paste electrode. Mater. Sci. Eng. C 31, 1819–1825.

Arvand, M., Gholizadeh, T.M., Zanjanchi, M.A., 2012. MWCNTs/$Cu(OH)_2$ nanoparticles/IL nanocomposite modified glassy carbon electrode as a voltammetric sensor for determination of the non-steroidal anti-inflammatory drug diclofenac. Mater. Sci. Eng. C 32, 1682–1689.

Asadian, E., Shahrokhian, S., 2016. Voltammetric studies of Azathioprine on the surface of graphite electrode modified with graphene nanosheets decorated with Ag nanoparticles. Mater. Sci. Eng. C 58, 1098–1104.

Azodi-Deilami, S., Asadi, E., Abdouss, M., Ahmadi, F., Najafabadi, A.H., Farzaneh, S., 2015. Determination of meloxicam in plasma samples using a highly selective and sensitive voltammetric sensor based on carbon paste electrodes modified by molecularly imprinted polymer nanoparticle–multiwall carbon nanotubes. Anal. Methods 7, 1280–1292.

Babaei, A., Afrasiabi, M., 2015. A glassy carbon electrode modified with MCM-41/nickel hydroxide nanoparticle/multiwalled carbon nanotube composite as a sensor for the simultaneous determination of dopamine, piroxicam, and cefixime. Ionics 21, 1731–1740.

Bae, S., Kim, H., Lee, Y., Xu, X., Park, J.S., Zheng, Y., Balakrishnan, J., Lei, T., Kim, H.R., Song, Y.I., Kim, Y.J., Kim, K.S., Özyilmaz, B., Ahn, J.H., Hong, B.H., Iijima, S., 2010. Roll-to-roll production of 30-inch graphene films for transparent electrodes. Nat. Nanotechnol. 5, 574–578.

Baena, J.R., Gallego, M., Valcarcel, M., 2002. Fullerenes in the analytical sciences. TrAC Trends Anal. Chem. 21, 187–198.

Bagotsky, V.S. (Ed.), 2006. Fundamentals of Electrochemistry. second ed. Wiley Interscience, JohnWiley, Sons Pub, New Jersey.

Bard, A.J., Faulkner, L.R., 2001. Electrochemical Methods, Fundamentals and Applications, second ed. John Wiley and Sons, New York.

Barker, G.C., Gardner, A.W., 1960. Pulse polarography. Fresenius' Zeit. Anal. Chem. 173, 79–83.

Barker, G.C., Jenkins, I.L., 1952. Square-wave polarography. Analyst 77, 685–696.

Barsan, M.M., Carvalho, R.C., Zhong, Y., Sun, X., Brett, C.M., 2012. Carbon nanotube modified carbon cloth electrodes: characterisation and application as biosensors. Electrochim. Acta 85, 203–209.

Barsan, M.M., Ghica, M.E., Brett, C.M., 2015a. Electrochemical sensors and biosensors based on redox polymer/carbon nanotube modified electrodes: a review. Anal. Chim. Acta 881, 1–23.

Barsan, M.M., Toledo, C.T., Brett, C.M., 2015b. New electrode architectures based on poly (methylene green) and functionalized carbon nanotubes: characterization and application to detection of acetaminophen and pyridoxine. J. Electroanal. Chem. 736, 8–15.

Bartlett, P.N., Ghoneim, E., El-Hefnawy, G., El-Hallag, I., 2005. Voltammetry and determination of metronidazole at a carbon fiber microdisk electrode. Talanta 66, 869–874.

Baytak, A.K., Aslanoglu, M., 2015. Decorating carbon nanotubes with nanoparticles of indium tin oxide for the voltammetric determination of metaproterenol. J. Electroanal. Chem. 757, 210–215.

Beitollahi, H., Taher, M.A., Ahmadipour, M., Hosseinzadeh, R., 2014. Electrocatalytic determination of captopril using a modified carbon nanotube paste electrode: application to determination of captopril in pharmaceutical and biological samples. Measurement 47, 770–776.

Berger, C., Song, Z., Li, T., Li, X., Ogbazghi, A.Y., Feng, R., De Heer, W.A., 2004. Ultrathin epitaxial graphite: 2D electron gas properties and a route toward graphene-based nanoelectronics. J. Phys. Chem. B 108, 19912–19916.

Berger, C., Song, Z., Li, X., Wu, X., Brown, N., Naud, C., de Heer, W.A., 2006. Electronic confinement and coherence in patterned epitaxial graphene. Science 312, 1191–1196.

Bockris, J., Khan, S.U.M., 1993. Surface Electrochemistry: A Molecular Level Approach. Plenum Press, New York and London.

Borowiec, J., Yan, K., Tin, C.C., Zhang, J., 2015. Synthesis of PDDA functionalized reduced graphene oxide decorated with gold nanoparticles and its electrochemical response toward levofloxacin. J. Electrochem. Soc. 162, H164–H169.

Brahman, P.K., Dar, R.A., Tiwari, S., Pitre, K.S., 2012. Electrochemical behavior of gatifloxacin at multi-walled carbon nanotube paste electrode and its interaction with DNA. Rev. Anal. Chem. 31, 83–92.

Brahman, P.K., Dar, R.A., Pitre, K.S., 2013. DNA-functionalized electrochemical biosensor for detection of vitamin B1 using electrochemically treated multiwalled carbon nanotube paste electrode by voltammetric methods. Sensor. Actuat. B 177, 807–812.

Brainina, Kh., Neyman, E., 1993. Electroanalytical Stripping Methods. Wiley, New York.

Brett, C.M.A., Oliveira-Brett, A.M. (Eds.), 1993. Electrochemistry, Principles, Methods and Applications. Oxford University Press, New York.

Brett, C.M.A., Oliveira-Brett, A.M.C.F., 1998. Electroanalysis. Oxford University Press, Oxford.

Brittain, H.G., 1995. Physical Characterization of Pharmaceutical Solids (Drugs and the Pharmaceutical Sciences: a Series of Textbooks and Monographs), first ed. Marcel Dekker Inc, New York.

Bui, N.N., Kim, B.H., Yang, K.S., Cruz, M.E.D., Ferraris, J.P., 2009. Activated carbon fibers from electrospinning of polyacrylonitrile/pitch blends. Carbon 47, 2538–2539.

Bundy, F.P., Bassett, W.A., Weathers, M.S., Hemley, R.J., Mao, H.U., Goncharovc A.F., 1978. The controversial carbon solid–liquid–vapour triple point, A. Greenville Whittaker. Nature 276, 695–696.

Carstensen, J.T., 1993. Pharmaceutical principles of solid dosage forms, first ed. Technomic Publishing Company, Lancaster.

Chamjangali, M.A., Goudarzi, N., Bagherian, G., Reskety, A.A., 2015. Development of a new electrochemical sensor for verapamil based on multi-walled carbon nanotube immobilized on glassy carbon electrode. Measurement 71, 23–30.

Cheemalapati, S., Palanisamy, S., Chen, S.M., 2014. A simple and sensitive electroanalytical determination of anxiolytic buspirone hydrochloride drug based on multiwalled carbon nanotubes modified electrode. J. Appl. Electrochem. 44, 317–323.

Chen, L.H., Zang, J.B., Wang, Y.H., Bian, L.Y., 2008. Electrochemical oxidation of nitrite on nanodiamond powder electrode. Electrochim. Acta 53, 3442–3445.

Christian, G.D., Dasgupta, P., Schug, K. (Eds.), 2013. Analytical Chemistry. seventh ed. John Wiley & Sons Inc, New Jersey.

Cioffi, C., Campidelli, S., Sooambar, C., Marcaccio, M., Marcolongo, G., Meneghetti, M., Paolucci, D., Paolucci, F., Ehli, C., Rahman, G.M.A., Sgobba, V., Guldi, D.M., Prato, M., 2007. Synthesis, characterization, and photoinduced electron transfer in functionalized single wall carbon nanohorns. J. Am. Chem. Soc. 129, 3938–3945.

Cole, E.T., 1989. Liquid filled hard gelatin capsules. Pharm. Tech. 13, 124–140.

Cozzi, F., Powell, W.H., Thilgen, C., 2005. Numbering of fullerenes. Pure Appl. Chem. 77, 843–923.

D'Souza, O.J., Mascarenhas, R.J., Thomas, T., Basavaraja, B.M., Saxena, A.K., Mukhopadhyay, K., Roy D., 2015. Platinum decorated multi-walled carbon nanotubes/Triton X-100 modified carbon paste electrode for the sensitive amperometric determination of Paracetamol. J. Electroanal. Chem. 739, 49–57.

Dai, H., Li, Y., Zhang, S., Gong, L., Li, X., Lin, Y., 2016. Delicate photoelectrochemical sensor for folic acid based on carbon nanohorns supported interwoven titanate nanotubes. Sens. Actuat. B 222, 120–126.

Dalmasso, P.R., Pedano, M.L., Rivas, G.A., 2012. Electrochemical determination of ascorbic acid and paracetamol in pharmaceutical formulations using a glassy carbon electrode modified with multi-wall carbon nanotubes dispersed in polyhistidine. Sens. Actuat. B 173, 732–736.

Demir, E., Inam, R., Ozkan, S.A., Uslu, B., 2014. Electrochemical behavior of tadalafil on TiO_2 nanoparticles–MWCNT composite paste electrode and its determination in pharmaceutical dosage forms and human serum samples using adsorptive stripping square wave voltammetry. J. Solid State Chem. 18, 2709–2720.

Devadas, B., Rajkumar, M., Chen, S.M., Saraswathi, R., 2012. Electrochemically reduced graphene oxide/neodymium hexacyanoferrate modified electrodes for the electrochemical detection of paracetomol. Int. J. Electrochem. Sci. 7, 3339–3349.

Devaraj, M., Deivasigamani, R.K., Jeyadevan, S., 2013. Enhancement of the electrochemical behavior of CuO nanoleaves on MWCNTs/GC composite film modified electrode for determination of norfloxacin. Colloids Surf. B 102, 554–561.

Ding, L., Li, L., You, W., Gao, Z.N., Yang, T.L., 2015. Electrocatalytic oxidation of venlafaxine at a multiwall carbon nanotubes-ionic liquid gel modified glassy carbon electrode and its electrochemical determination. Croat. Chem. Acta 88, 81–87.

Dogan-Topal, B., 2013. Electrooxidative behavior and determination of trifluoperazine at multiwalled carbon nanotube-modified glassy carbon electrode. J. Solid State Chem. 17, 1059–1066.

Dresselhaus, M.S., Dresselhaus, G., Eklund, P.C., 1996. Science of Fullerenes and Carbon Nanotubes. Academic, New York.

Ebbesen, T.W., Ajayan, P.M., 1992. Large-scale synthesis of carbon nanotubes. Nature 358, 220–222.

Emami, M., Shamsipur, M., Saber, R., 2014. Design of poly-l-methionine–gold nanocomposit/multi-walled carbon nanotube modified glassy carbon electrode for determination of amlodipine in human biological fluids. J. Solid State Chem. 18, 985–992.

Emsley, J., 2011. Nature's Building Blocks: An A-Z Guide to the Elements, second ed. Oxford University Press, New York.

Ensafi, A.A., Karimi-Maleh, H., 2012. Determination of 6-mercaptopurine in the presence of uric acid using modified multiwall carbon nanotubes-TiO_2 as a voltammetric sensor. Drug Test. Anal. 4, 970–977.

Ensafi, A.A., Karimi-Maleh, H., Mallakpour, S., Rezaei, B., 2011. Highly sensitive voltammetric sensor based on catechol-derivative-multiwall carbon nanotubes for the catalytic determination of captopril in patient human urine samples. Colloids Surf. B 87, 480–488.

Ensafi, A.A., Allafchian, A.R., Rezaei, B., 2012. Multiwall carbon nanotubes decorated with $FeCr_2O_4$, a new selective electrochemical sensor for amoxicillin determination. J. Nano. Res. 14, 1–11.

Ensafi, A.A., Allafchian, A.R., Rezaei, B., 2013. A sensitive and selective voltammetric sensor based on multiwall carbon nanotubes decorated with $MgCr_2O_4$ for the determination of azithromycin. Colloids Surf. B 103, 468–474.

Ensafi, A.A., Lesani, S., Amini, M., Rezaei, B., 2015a. Electrochemical ds-DNA-based biosensor decorated with chitosan modified multi wall carbon nanotubes for phenazopyridine biodetection. J. Taiwan Inst. Chem. Eng. 54, 165–169.

Ensafi, A.A., Saeid, B., Rezaei, B., Allafchian, A.R., 2015b. Differential pulse voltammetric determination of methyldopa using MWCNTs modified glassy carbon decorated with NiFe2O4 nanoparticles. Ionics 21, 1435–1444.

Esfandiari Baghbamidi, S., Beitollahi, H., Karimi-Maleh, H., Soltani-Nejad, S., Soltani-Nejad, V., Roodsaz, S., 2012. Modified carbon nanotube paste electrode for voltammetric determination of carbidopa, folic acid, and tryptophan. J. Anal. Method. Chem. 2012, 305872–305880.

Fakhari, A.R., Koruni, M.H., Ahmar, H., Shahsavani, A., Movahed, S.K., 2014. Electrochemical determination of dextromethorphan on reduced graphene oxide modified screen-printed electrode after electromembrane extraction. Electroanalysis 26, 521–529.

Falcao, E.H., Wudl, F., 2007. Carbon allotropes: beyond graphite and diamond. J. Chem. Technol. Biotechnol. 82 (6), 524–531.

Fecht, H.J., Brühne, K. (Eds.), 2014. Carbon-based Nanomaterials and Hybrids: Synthesis, Properties, and Commercial Applications. CRC Press.

Felix, F.S., Ferreira, L.M., Vieira, F., Trindade, G.M., Ferreira, V.S., Angnes, L., 2015. Amperometric determination of promethazine in tablets using an electrochemically reduced graphene oxide modified electrode. New J. Chem. 39, 696–702.

Fernandes, D.M., Silva, N., Pereira, C., Moura, C., Magalhães, J.M., Bachiller-Baeza, B., Freire, C., Rodríguez-Ramos, I., Guerrero-Ruiz, A., Delerue-Matos, C., 2015. $MnFe_2O_4$@ CNT-N as novel electrochemical nanosensor for determination of caffeine, acetaminophen and ascorbic acid. Sens. Actuat. B 218, 128–136.

Fotouhi, L., Bahmani, F., 2013. MWCNT modified glassy carbon electrode: probing furazolidone-DNA interactions and DNA determination. Electroanalysis 25, 757–764.

Fotouhi, L., Nemati, M., Heravi, M.M., 2011. Electrochemistry and voltammetric determination of furazolidone with a multi-walled nanotube composite film-glassy carbon electrode. J. Appl. Electrochem. 41, 137–142.

Fotouhi, L., Hashkavayi, A.B., Heravi, M.M., 2013. Interaction of sulfadiazine with DNA on a MWCNT modified glassy carbon electrode: determination of DNA. Int. J. Biol. Macromol. 53, 101–106.

Fouladgar, M., Mohammadzadeh, S., 2014. Determination of methimazole on a multiwall carbon nanotube titanium dioxide nanoparticle paste electrode. Anal. Lett. 47, 763–777.

Frumkin, A., Petrii, O.A., Damaskin, B.B., 1980. Bockris, J.O'M., Conway, B.E., Yeager, E. (Eds.), Comprehensive Treatise in Electrochemistry, vol. 1, Plenum Press, New York, London.

Gaichore, R.R., Srivastava, A.K., 2014. Electrocatalytic determination of propranolol hydrochloride at carbon paste electrode based on multiwalled carbon-nanotubes and γ-cyclodextrin. J. Incl. Phen. Macrocycl. Chem. 78, 195–206.

Geim, A.K., Novoselov, K.S., 2007. The rise of graphene. Nature Mater. 6, 183–191.

Ghalkhani, M., Shahrokhian, S., 2013. Adsorptive stripping differential pulse voltammetric determination of mebendazole at a graphene nanosheets and carbon nanospheres/chitosan modified glassy carbon electrode. Sens. Actuat. B 185, 669–674.

Ghica, M.E., Ferreira, G.M., Brett, C.M., 2015. Poly (thionine)-carbon nanotube modified carbon film electrodes and application to the simultaneous determination of acetaminophen and dipyrone. J. Solid State Chem. 19, 2869–2881.

Gholivand, M.B., Khodadadian, M., 2011. Rationally designed molecularly imprinted polymers for selective extraction of methocarbamol from human plasma. Talanta 85, 1680–1688.

Gholivand, M.B., Mohammadi-Behzad, L., 2015. An electrochemical sensor for warfarin determination based on covalent immobilization of quantum dots onto carboxylated multiwalled carbon nanotubes and chitosan composite film modified electrode. Mater. Sci. Eng. C 57, 77–87.

Ghoreishi, S.M., Attaran, A.M., Amin, A.M., Khoobi, A., 2015. Multiwall carbon nanotube-modified electrode as a nanosensor for electrochemical studies and stripping voltammetric determination of an antimalarial drug. RSC Adv. 5, 14407–14415.

Gileadi, E., Kirowa-Eisner, E., Penciner, J., 1975. Interfacial Chemistry: An Experimental Approach. Addison-Wesley, USA.

Gogotsi, Y. (Ed.), 2006. Carbon Nanomaterials, CRC. Taylor & Francis Group, New York.

Goodarzian, M., Khalilzade, M.A., Karimi, F., Gupta, V.K., Keyvanfard, M., Bagheri, H., Fouladgar, M., 2014. Square wave voltammetric determination of diclofenac in liquid phase using a novel ionic liquid multiwall carbon nanotubes paste electrode. J. Mol. Liq. 197, 114–119.

Gosser, Jr., D.K. (Ed.), 1988. Cyclic Voltammetry: Simulation and Analysis of Reaction Mechanisms. Wiley-VCH, New York.

Gowda, J.I., Gunjiganvi, D.G., Sunagar, N.B., Bhat, M.N., Nandibewoor, S.T., 2015. MWCNT-CTAB modified glassy carbon electrode as a sensor for the determination of paracetamol. RSC Adv. 5, 49045–49053.

Goyal, R.N., Singh, S.P., 2006a. Voltammetric determination of atenolol at C_{60}-modified glassy carbon electrodes. Talanta 69, 932–937.

Goyal, R.N., Singh, S.P., 2006b. Voltammetric determination of paracetamol at C_{60}-modified glassy carbon electrode. Electrochim. Acta 51, 3008–3012

Goyal, R.N., Bachheti, N., Tyagi, A., Pandey, A.K., 2007. Differential pulse voltammetric determination of methylprednisolone in pharmaceuticals and human biological fluids. Anal. Chim. Acta 605, 34–40.

Goyal, R.N., Kaur, D., Singh, S.P., Pandey, A.K., 2008. Effect of graphite and metallic impurities of C_{60} fullerene on determination of salbutamol in biological fluids. Talanta 75, 63–69.

Goyal, R.N., Gupta, V.K., Chatterjee, S., 2009a. A sensitive voltammetric sensor for determination of synthetic corticosteroid triamcinolone, abused for doping. Biosens. Bioelectron. 24, 3562–3568.

Goyal, R.N., Gupta, V.K., Chatterjee, S., 2009b. Fullerene-C_{60}-modified edge plane pyrolytic graphite electrode for the determination of dexamethasone in pharmaceutical formulations and human biological fluids. Biosens. Bioelectron. 24, 1649–1654.

Goyal, R.N., Oyama, M., Bachheti, N., Singh, S.P., 2009c. Fullerene C_{60} modified gold electrode and nanogold modified indium tin oxide electrode for prednisolone determination. Bioelectrochemistry 74, 272–277.

Goyal, R.N., Singh, S.P., Chatterjee, S., Bishnoi, S., 2010. Electrochemical investigations of prednisone using fullerene-C_{60}-modified edge plane pyrolytic graphite electrode. Indian J. Chem. A 49, 26.

Goyal, R.N., Rana, A.R.S., Aziz, M.A., Oyama, M., 2011. Effect of gold nanoparticle attached multi-walled carbon nanotube-layered indium tin oxide in monitoring the effect of paracetamol on the release of epinephrine. Anal. Chim. Acta 693, 35–40.

Grobert, N., 2007. Carbon nanotubes: becoming clean. Mater. Today 10, 28–35.

Guo, M., Chen, J., Nie, L., Yao, S., 2004. Electrostatic assembly of calf thymus DNA on multi-walled carbon nanotube modified gold electrode and its interaction with chlorpromazine hydrochloride. Electrochim. Acta 49, 2637–2643.

Gupta, V.K., Jain, A.K., Shoora, S.K., 2013. Multiwall carbon nanotube modified glassy carbon electrode as voltammetric sensor for the simultaneous determination of ascorbic acid and caffeine. Electrochim. Acta 93, 248–253.

Guzmán, A., Agüí, L., Pedrero, M., Yáñez-Sedeño, P., Pingarrón, J.M., 2002. Carbon fiber cylindrical microelectrode-based detector for the determination of antithyroid drugs. Talanta 56, 577–584.

Haaland, D.M., 1976. Graphite-liquid-vapor triple point pressure and the density of liquid carbon. Carbon 14, 357–361.

Habibi, B., Jahanbakhshi, M., 2014. Silver nanoparticles/multi walled carbon nanotubes nanocomposite modified electrode: voltammetric determination of clonazepam. Electrochim. Acta 118, 10–17.

Habibi, B., Jahanbakhshi, M., Pournaghi-Azar, M.H., 2011. Differential pulse voltammetric simultaneous determination of acetaminophen and ascorbic acid using single-walled carbon nanotube-modified carbon–ceramic electrode. Anal. Biochem. 411, 167–175.

Hauss, D.J., 2007. Oral Lipid-Based Formulations. Informa Healthcare USA, Inc, New York.

He, Z., Zang, S., Liu, Y., He, Y., Lei, H., 2015. A multi-walled carbon nanotubes-poly (l-lysine) modified enantioselective immunosensor for ofloxacin by using multi-enzyme-labeled gold nanoflower as signal Enhancer. Biosens. Bioelectron. 73, 85–92.

Hedberg, K., Hedberg, L., Bethune, D.S., Brown, C.A., Dorn, H.C., Johnson, R.D., De Vries, M., 1991. Bond lengths in free molecules of buckminsterfullerene, C60, from gas-phase electron diffraction. Science 254, 410–412.

Heidarimoghadam, R., Farmany, A., 2016. Rapid determination of furosemide in drug and blood plasma of wrestlers by a carboxyl-MWCNT sensor. Mater. Sci. Eng. C 58, 1242–1245.

Heinrich Stahl, P., Wermuth, C.G., 2002. Pharmaceutical Salts: Properties, Selection, and Use, first ed. Wiley YCH, Germany.

Hernandez, Y., Nicolosi, V., Lotya, M., Blighe, F.M., Sun, Z., De, S., Coleman, J.N., 2008. High-yield production of graphene by liquid-phase exfoliation of graphite. Nat. Nanotechnol. 3, 563–568.

Hou, X., Shen, G., Meng, L., Zhu, L., Guo, M., 2011. Multi-walled carbon nanotubes modified glass carbon electrode and its electrocatalytic activity towards oxidation of paracetamol. Russian J. Electrochem. 47, 1262–1267.

Hu, Y., Zhang, Z., Li, J., Zhang, H., Luo, L., Yao, S., 2012. Electrochemical imprinted sensor for determination of oleanic acid based on poly (sodium 4-styrenesulfonate-*co*-acrylic acid)-grafted multi-walled carbon nanotubes-chitosan and cobalt hexacyanoferrate nanoparticles. Biosens. Bioelectron. 31, 190–196.

Huang, J., Liu, Y., Hou, H., You, T., 2008. Simultaneous electrochemical determination of dopamine, uric acid and ascorbic acid using palladium nanoparticle-loaded carbon nanofibers modified electrode. Biosens. Bioelectron. 24, 632–637.

Hummers, W.S., Offeman, R.E., 1958. Preparation of graphitic oxide. J. Am. Chem. Soc. 80, 1339–1343.

Iijima, S., 2002. Carbon nanotubes: past, present, and future. Physica B 323, 1–5.

Iijima, S., Ichihashi, T. 1993. Single-shell carbon nanotubes of 1-nm diameter. Nature 363, 603–605.

Jain, R., 2013. Nano graphene based sensor for antiarrhythmic agent quinidine in solubilized system. Colloids Surf. B 105, 278–283.

Jain, R., Sharma, R., 2012. Novel bismuth/multi-walled carbon nanotubes-based electrochemical sensor for the determination of neuroprotective drug cilostazol. J. Appl. Electrochem. 42, 341–348.

Janegitz, B.C., Figueiredo-Filho, L.C., Vicentini, F.C., Ribeiro, M.F.M., Suarez, W.T., Fatibello-Filho, O., Metters, J.P., Banks, C.E., 2014. Development of a carbon nanotube paste electrode modified with zinc phosphate for captopril determination in pharmaceutical and biological samples. Anal. Method. 6, 1324–1329.

Jiménez-Soto, J.M., Moliner-Martínez, Y., Cárdenas, S., Valcárcel, M., 2010. Evaluation of the performance of single-walled carbon nanohorns in capillary electrophoresis. Electrophoresis 31, 1681–1688.

Jithan, A., 2007. Chapter-7 Oral Formulations, Oral Drug Delivery Technology. Pharma Book Syndicate, Hyderabad, India.

Junginger, H.E., Horwood, E., 1992. Drug Targeting and Delivery: Concepts in Dosage Form Design, first ed. Ellis Horwood Series in Pharmacological Sciences, England.

Kalambate, P.K., Biradar, M.R., Karna, S.P., Srivastava, A.K., 2015. Adsorptive stripping differential pulse voltammetry determination of rivastigmine at graphene nanosheet-gold nanoparticle/carbon paste electrode. J. Electroanal. Chem. 757, 150–158.

Kalanur, S.S., Jaldappagari, S., Balakrishnan, S., 2011. Enhanced electrochemical response of carbamazepine at a nano-structured sensing film of fullerene-C_{60} and its analytical applications. Electrochim. Acta 56, 5295–5301.

Kang, I., Heung, Y.Y., Kim, J.H., Lee, J.W., Gollapudi, R., Subramaniam, S., Narasimhadevara, S., Hurd, D., Kirikera, G.R., Shanov, V., Schulz, M.J., Shi, D., Boerio, J., Mall, S., Ruggles-Wren, M., 2006. Introduction to carbon nanotube and nanofiber smart materials. Compos. Part B 37, 382–394.

Karimi-Maleh, H., Keyvanfard, M., Alizad, K., Fouladgar, M., Beitollahi, H., Mokhtari, A., Gholami-Orimi, F., 2011. Voltammetric determination of N-actylcysteine using modified multiwall carbon nanotubes paste electrode. Int. J. Electrochem. Sci. 6, 6141–6150.

Karimi-Maleh, H., Khalilzadeh, M.A., Ranjbarha, Z., Beitollahi, H., Ensafi, A.A., Zareyee, D., 2012. p-Chloranil modified carbon nanotubes paste electrode as a voltammetric sensor for the simultaneous determination of methyldopa and uric acid. Anal. Method. 4, 2088–2094.

Karimi-Maleh, H., Mehdipour-Ataei, S., Hatami, M., Khalilzadeh, M.A., 2014a. Voltammetric determination of captopril using a novel ferrocene-based polyamide as a mediator and multi-wall carbon nanotubes as a sensor. J. Anal. Chem. 69, 162–168.

Karimi-Maleh, H., Moazampour, M., Ahmar, H., Beitollahi, H., Ensafi, A.A., 2014b. A sensitive nanocomposite-based electrochemical sensor for voltammetric simultaneous determination of isoproterenol, acetaminophen and tryptophan. Measurement 51, 91–99.

Karimi-Maleh, H., Tahernejad-Javazmi, F., Ensafi, A.A., Moradi, R., Mallakpour, S., Beitollahi, H., 2014c. A high sensitive biosensor based on FePt/CNTs nanocomposite/*N*-(4-hydroxyphenyl)-3,5-dinitrobenzamide modified carbon paste electrode for simultaneous determination of glutathione and piroxicam. Biosens. Bioelectron. 60, 1–7.

Karuwan, C., Wisitsoraat, A., Maturos, T., Phokharatkul, D., Sappat, A., Jaruwongrungsee, K., Lomas, T., Tuantranont, A., 2009. Flow injection based microfluidic device with carbon nanotube electrode for rapid salbutamol detection. Talanta 79, 995–1000.

Kergaravat, S.V., Gómez, G.A., Fabiano, S.N., Chávez, T.I.L., Pividori, M.I., Hernández, S.R., 2012. Biotin determination in food supplements by an electrochemical magneto biosensor. Talanta 97, 484–490.

Keshtkar, N., Taher, A.M., Beitollahi, H., 2015. Voltammetric determination of carbidopa and folic acid using a modified carbon nanotubes paste electrode. J. Serbian Chem. Soc. 00, 46–146.

Kissinger, P.T., Heineman, W.R. (Eds.), 1996. Laboratory Techniques in Electroanalytical Chemistry. second ed. Marcell Dekker, New York.

Kor, K., Zarei, K., 2013. β-cyclodextrin incorporated carbon nanotube paste electrode as electrochemical sensor for nifedipine. Electroanalysis 25, 1497–1504.

Koryta, J., Dvorak, J., Kavan, L. (Eds.), 1993. Principles of Electrochemistry. second ed. John Wiley & Sons, New York.

Kosynkin, D.V., Berlin, J.M., Sinitskii, A., Sun, Z., Marcano, D.C., Slesarev, A., Alemany, L.B., Lu, W., Tour, J.M., 2010. Improved synthesis of graphene oxide. ACS Nano 4, 4806–4814.

Kroto, H.W., McKay, K., 1988. The formation of quasi-icosahedral spiral shell carbon particles. Nature 331, 328–331.

Kroto, H.W., Heath, J.R., O'Brien, S.C., Curl, R.F., Smalley, R.E., 1985. C60: Buckminsterfullerene. Nature 318, 162–163.

Kukovitsky, E.F., L'vov, S.G., Sainov, N.A., 2000. VLS-growth of carbon nanotubes from the vapor. Chem. Phys. Lett. 317, 65–70.

Kun, Z., Hongtao, C., Yue, Y., Zhihong, B., Fangzheng, L., Sanming, L., 2015. Platinum nanoparticle-doped multiwalled carbon-nanotube-modified glassy carbon electrode as a sensor for simultaneous determination of atenolol and propranolol in neutral solution. Ionics 21, 1129–1140.

Kuralay, F., Vural, T., Bayram, C., Denkbas, E.B., Abaci, S., 2011. Carbon nanotube–chitosan modified disposable pencil graphite electrode for Vitamin B 12 analysis. Colloids Surf. B 87, 18–22.

Kutluay, A., Aslanoglu, M., 2013. Modification of electrodes using conductive porous layers to confer selectivity for the voltammetric detection of paracetamol in the presence of ascorbic acid, dopamine and uric acid. Sens. Actuat. B 185, 398–404.

Kutluay, A., Aslanoglu, M., 2014a. An electrochemical sensor prepared by sonochemical one-pot synthesis of multi-walled carbon nanotube-supported cobalt nanoparticles for the simultaneous determination of paracetamol and dopamine. Anal. Chim. Acta 839, 59–66.

Kutluay, A., Aslanoglu, M., 2014b. Nickel nanoparticles functionalized multi-walled carbon nanotubes at platinum electrodes for the detection of bromhexine. Sens. Actuat. B 192, 720–724.

Lachman, L., Lieberman, H.A., Kanig, J.L., 1986. The Theory and Practice of Industrial Pharmacy, third ed. Lea-Febiger Publications, Philadelphia.

Lee, B.O., Woo, W.J., Kim, M.S., 2001. EMI shielding effectiveness of carbon nanofiber filled poly (vinyl alcohol) coating materials. Macromol. Mater. Eng. 286, 114–118.

Leite, F.R.F., Maroneze, C.M., de Oliveira, A.B., dos Santos, W.T.P., Damos, F.S., Luz, R.D.C.S., 2012. Development of a sensor for L-dopa based on $Co(DMG)_2ClPy$/multi-walled carbon nanotubes composite immobilized on basal plane pyrolytic graphite electrode. Bioelectrochemistry 86, 22–29.

Li, P., Zhao, T.J., Zhou, J.H., Sui, Z.J., Dai, Y.C., Yuan, W.K., 2005. Characterization of carbon nanofiber composites synthesized by shaping process. Carbon 43, 2701–2710.

Li, Y., Wu, S., Luo, P., Liu, J., Song, G., Zhang, K., Ye, B., 2012. Electrochemical behavior and voltammetric determination of theophylline at a glassy carbon electrode modified with graphene/nafion. Anal. Sci. 28, 497–502.

Li, J., Liu, J., Tan, G., Jiang, J., Peng, S., Deng, M., Qian, D., Feng, Y., Liu, Y., 2014a. High-sensitivity paracetamol sensor based on Pd/graphene oxide nanocomposite as an enhanced electrochemical sensing platform. Biosens. Bioelectron. 54, 468–475.

Li, Y., Feng, S., Li, S., Zhang, Y., Zhong, Y., 2014b. A high effect polymer-free covalent layer by layer self-assemble carboxylated MWCNTs films modified GCE for the detection of paracetamol. Sens. Actuat. B 190, 999–1005.

Liu, Y., Gu, Z., Margrave, J.L., Khabashesku, V.N., 2004. Functionalization of nanoscale diamond powder: fluoro-, alkyl-, amino-, and amino acid-nanodiamond derivatives. Chem. Mater. 16, 3924–3930.

Liu, Q., Shi, J., Zeng, L., Wang, T., Cai, Y., Jiang, G., 2011. Evaluation of graphene as an advantageous adsorbent for solid-phase extraction with chlorophenols as model analytes. J. Chroma. A 1218, 197–204.

Lu, T.L., Tsai, Y.C., 2011. Sensitive electrochemical determination of acetaminophen in pharmaceutical formulations at multiwalled carbon nanotube-alumina-coated silica nanocomposite modified electrode. Sens. Actuat. B 153, 439–444.

Madrakian, T., Haghshenas, E., Afkhami, A., 2014. Simultaneous determination of tyrosine, acetaminophen and ascorbic acid using gold nanoparticles/multiwalled carbon nanotube/glassy carbon electrode by differential pulse voltammetric method. Sens. Actuat. B 193, 451–460.

Madrakian, T., Soleimani, M., Afkhami, A., 2015. Electrochemical determination of fluvoxamine on mercury nanoparticle multi-walled carbon nanotube modified glassy carbon electrode. Sens. Actuat. B 210, 259–266.

Malode, S.J., Abbar, J.C., Shetti, N.P., Nandibewoor, S.T., 2012a. Voltammetric oxidation and determination of loop diuretic furosemide at a multi-walled carbon nanotubes paste electrode. Electrochim. Acta 60, 95–101.

Malode, S.J., Shetti, N.P., Nandibewoor, S.T., 2012b. Voltammetric behavior of theophylline and its determination at multi-wall carbon nanotube paste electrode. Colloids Surf. B 97, 1–6.

Manjunatha, J.G., Swamy, B.K., Mamatha, G.P., Gilbert, O., Srinivas, M.T., Sherigara, B.S., 2011. Electrochemical studies of clozapine drug using carbon nanotube-SDS modified carbon paste electrode: a cyclic voltammetry study. Der Pharm. Chem. 3, 236–249.

Mao, A., Li, H., Jin, D., Yu, L., Hu, X., 2015. Fabrication of electrochemical sensor for paracetamol based on multi-walled carbon nanotubes and chitosan–copper complex by self-assembly technique. Talanta 144, 252–257.

Marco, J.P., Borges, K.B., Tarley, C.R.T., Ribeiro, E.S., Pereira, A.C., 2013. Development of a simple, rapid and validated square wave voltametric method for determination of promethazine in raw material and pharmaceutical formulation using DNA modified multiwall carbon nanotube paste electrode. Sens. Actuat. B 177, 251–259.

Martin, A., Swarbrick, J., Cammarata, A., 1983. Physical Pharmacy: Physical Chemical Principles in the Pharmaceutical Sciences, third ed. Lea & Febiger Publications, Philadelphia.

Mazloum-Ardakani, M., Ganjipour, B., Beitollahi, H., Amini, M.K., Mirkhalaf, F., Naeimi, H., Nejati-Barzoki, M., 2011. Simultaneous determination of levodopa, carbidopa and tryptophan using nanostructured electrochemical sensor based on novel hydroquinone and carbon nanotubes: application to the analysis of some real samples. Electrochim. Acta 56, 9113–9120.

Mazloum-Ardakani, M., Zokaie, M., Khoshroo, A., 2015. Carbon nanotube electrochemical sensor based on and benzofuran derivative as a mediator for the determination of levodopa, acetaminophen, and tryptophan. Ionics 21, 1741–1749.

Meyyapan, M. (Ed.), 2005. Carbon Nanotubes, Science and Applications. CRC Press, London.

Murata, K., Hirahara, K., Yudasaka, M., Iijima, S., Kasuya, D., Kaneko, K., 2002a. Nanowindow-induced molecular sieving effect in a single-wall carbon nanohorn. J. Phys. Chem. B 106, 12668–12669.

Murata, K., Kaneko, K., Kanoh, H., Kasuya, D., Takahashi, K., Kokai, F., Yudasaka, M., Iijima, S., 2002b. Adsorption mechanism of supercritical hydrogen in internal and interstitial nanospaces of single-wall carbon nanohorn assembly. J. Phys. Chem. B 106, 11132–11138.

Musameh, M., Wang, J., Merkoci, A., Lin, Y., 2002. Low-potential stable NADH detection at carbon-nanotube-modified glassy carbon electrodes. Electrochem. Comm. 4, 743–746.

Nezhadali, A., Mojarrab, M., 2015. Fabrication of an electrochemical molecularly imprinted polymer triamterene sensor based on multivariate optimization using multi-walled carbon nanotubes. J. Electroanal. Chem. 744, 85–94.

Novoselov, K.S., Geim, A.K., Morozov, S.V., Jiang, D., Zhang, Y., Dubonos, S.A., Firsov, A.A., 2004. Electric field effect in atomically thin carbon films. Science 306, 666–669.

Ozkan, S.A. (Ed.), 2012. Electroanalytical Methods in Pharmaceutical Analysis and Their Validation. first ed. HNB Pub, New York.

Ozkan, S.A., Kauffmann, J.M., Zuman, P., 2015. Electroanalysis in Biomedical and Pharmaceutical Sciences. Springer-Verlag Berlin Heidelberg, Germany.

Pagona, G., Mountrichas, G., Rotas, G., Karousis, N., Pispas, S., Tagmatarchis, N., 2008. Properties, applications and functionalisation of carbon nanohorns. Int. J. Nanotechnol. 6, 176–195.

Paimard, G., Gholivand, M.B., Shamsipur, M., Gholivand, K., Mohammadi-Behzad, L., Gholami, A., Barati, A., 2015. Fabrication of a highly sensitive amperometric sensor using 1,4-phenylene-*N*,*N*′-bis (O,O-diphenylphoramidate)/CdS quantum dots/multi-walled carbon nanotubes for nanomolar detection of captopril. J. Electroanal. Chem. 738, 176–183.

Pan, M., Fang, G., Duan, Z., Kong, L., Wang, S., 2012. Electrochemical sensor using methimazole imprinted polymer sensitized with MWCNTs and Salen-Co (III) as recognition element. Biosens. Bioelectron. 31, 11–16.

Parvin, M.H., 2011. Graphene paste electrode for detection of chlorpromazine. Electrochem. Comm. 13, 366–369.

Patil, R.H., Hegde, R.N., Nandibewoor, S.T., 2011. Electro-oxidation and determination of antihistamine drug, cetirizine dihydrochloride at glassy carbon electrode modified with multi-walled carbon nanotubes. Colloids Surf. B 83, 133–138.

Pierson, H.O. (Ed.), 1993. Handbook of Carbon, Graphite, Diamond and Fullerenes. Properties, Processing and Applications. Noyes Pub, New Jersey.

Pletikosic, I., Kralj, M., Pervan, P., Brako, R., Coraux, J., N'Diaye, A.T., Busse, C., Michely, T., 2009. Dirac cones and minigaps for graphene on Ir (111). Phys. Rev. Lett. 102, 056808/1–4.

Pruneanu, S., Pogacean, F., Biris, A.R., Ardelean, S., Canpean, V., Blanita, G., Dervishi, E., Biris, A.S., 2011. Novel graphene-gold nanoparticle modified electrodes for the high sensitivity electrochemical spectroscopy detection and analysis of carbamazepine. J. Phys. Chem. C 115, 23387–23394.

Raina, S., Kang, W.P., Davidson, J.L., 2008. Nanodiamond film with 'ridge'surface profile for chemical sensing. Diam. Relat. Mater. 17, 896–899.

Rao, C.N.R., Seshadri, R., Govindaraj, A., Sen, R., 1995. Fullerenes, nanotubes, onions and related carbon structures. Mater. Sci. Eng.R 15, 209–262.

Raoof, J.B., Baghayeri, M., Ojani, R., 2012. A high sensitive voltammetric sensor for qualitative and quantitative determination of phenobarbital as an antiepileptic drug in presence of acetaminophen. Colloids Surf. B 95, 121–128.

Ray, S., 2015. Applications of Graphene and Graphene-oxide Based Nanomaterials. William Andrew, Germany.

Reina, A., Xiaoting, J., Ho, J., Nezich, D., Son, H., Bulovic, V., Dresselhaus, M.S., Kong, J., 2009. Large area, few-layer graphene films on arbitrary substrates by chemical vapor deposition. Nano Lett. 9, 30–35.

Rodthongkum, N., Ruecha, N., Rangkupan, R., Vachet, R.W., Chailapakul, O., 2013. Graphene-loaded nanofiber-modified electrodes for the ultrasensitive determination of dopamine. Anal. Chim. Acta 804, 84–91.

Rotas, G., Sandanayaka, A.S., Tagmatarchis, N., Ichihashi, T., Yudasaka, M., Iijima, S., Ito, O., 2008. (Terpyridine) copper (II)-carbon nanohorns: metallo-nanocomplexes for photoinduced charge separation. J. Am. Chem. Soc. 130, 4725–4731.

Sadiković, M., Nigović, B., Jurić, S., Mornar, A., 2014. Voltammetric determination of ropinirole in the presence of levodopa at the surface of a carbon nanotubes based electrochemical sensor in pharmaceuticals and human serum. J. Electroanal. Chem. 733, 60–68.

Sahajwalla, C.G., 2004. New Drug Development: Regulatory Paradigms for Clinical Pharmacology and Biopharmaceutics (Drugs and the Pharmaceutical Sciences: a Series of Textbooks and Monographs), first ed. Marcel Dekker Inc, USA.

Saito, R., Dresselhaus, G., Dresselhaus, M.S., 1998. Physical properties of carbon nanotubes, vol. 35, 73–81. Imperial college press, London.

Salamanca-Neto, C.A.R., Hatumura, P.H., Tarley, C.R.T., Sartori, E.R., 2014. Electrochemical evaluation and simultaneous determination of binary mixture of antihypertensives hydrochlorothiazide and enalapril in combined dosage forms using carbon nanotubes paste electrode. Ionics 21, 1615–1622.

Sanghavi, B.J., Srivastava, A.K., 2011. Adsorptive stripping differential pulse voltammetric determination of venlafaxine and desvenlafaxine employing Nafion–carbon nanotube composite glassy carbon electrode. Electrochim. Acta 56, 4188–4196.

Sarhangzadeh, K., 2015. Application of multi wall carbon nanotube–graphene hybrid for voltammetric determination of naproxen. J. Iranian Chem. Soc. 12, 2133–2140.

Savvatimskiy, A.I., 2005. Measurements of the melting point of graphite and the properties of liquid carbon (a review for 1963-2003). Carbon 43, 1115–1142.

Schrand, A.M., Hens, S.A.C., Shenderova, O.A., 2009. Nanodiamond particles: properties and perspectives for bioapplications. Crit. Rev. Solid State Mater. Sci. 34, 18–74.

Scott, L.T., Boorum, M.M., McMahon, B.J., Hagen, S., Mack, J., Blank, J., Wegner, H., de Meijere, A., 2002. A rational chemical synthesis of C_{60}. Science 295, 1500–1503.

Shahrokhian, S., Ghalkhani, M., 2010. Glassy carbon electrodes modified with a film of nanodiamond–graphite/chitosan: application to the highly sensitive electrochemical determination of Azathioprine. Electrochim. Acta 55, 3621–3627.

Shahrokhian, S., Nassab, N.H., 2013. Nanodiamond decorated with silver nanoparticles as a sensitive film modifier in a jeweled electrochemical sensor: application to voltammetric determination of thioridazine. Electroanalysis 25, 417–425.

Shahrokhian, S., Azimzadeh, M., Amini, M.K., 2015. Modification of glassy carbon electrode with a bilayer of multiwalled carbon nanotube/tiron-doped polypyrrole: application to sensitive voltammetric determination of acyclovir. Mater. Sci. Eng. C 53, 134–141.

Shakibaian, V., Parvin, M.H., 2012. Determination of acetazolamide by graphene paste electrode. J. Electroanal. Chem. 683, 119–124.

Shamsipur, M., Saber, R., Emami, M., 2014. A highly sensitive electrochemical sensor based on gold nanoparticles/multiwall carbon nanotubes-modified glassy carbon electrode for selective determination of traces of atenolol. Anal. Method. 6, 7038–7045.

Shargel, L., Kanfer, I., 2005. Generic Drug Product Development: Solid Oral Dosage Forms (Drugs and the Pharmaceutical Sciences: a Series of Textbooks and Monographs), first ed. Marcel Dekker, Inc, 2005.

Shetti, N.P., Malode, S.J., Nandibewoor, S.T., 2012. Electrochemical behavior of an antiviral drug acyclovir at fullerene-C_{60}-modified glassy carbon electrode. Bioelectrochemistry 88, 76–83.

Shoja, Y., Rafati, A.A., Ghodsi, J., 2016. Glassy carbon electrode modified with horse radish peroxidase/organic nucleophilic-functionalized carbon nanotube composite for enhanced electrocatalytic oxidation and efficient voltammetric sensing of levodopa. Mater. Sci. Eng. C 58, 835–845.

Silva, T.A., Zanin, H., Vicentini, F.C., Corat, E.J., Fatibello-Filho, O., 2015. Electrochemical determination of rosuvastatin calcium in pharmaceutical and human body fluid samples using a composite of vertically aligned carbon nanotubes and graphene oxide as the electrode material. Sens. Actuat. B 218, 51–59.

Singh, V., Joung, D., Zhai, L., Das, S., Khondaker, S.I., Seal, S., 2011. Graphene based materials: past, present and future. Prog. Mater. Sci. 56, 1178–1271.

Smyth, M.R., Vas, J.G., 1992. Analytical Voltammetry. Comprehensive Analytical Chemistry, Amsterdam.

Stefan-van Staden, R.I., Bokretsion, R.G., 2012. Enantioanalysis of ketoprofen based on its interaction with C_{60} fullerene and its derivatives. Anal. Meth. 4, 1492–1497.

Stegemann, S., 1999. Liquid and semisolid formulation in hard gelatin capsules. Swiss Pharm. 21, 21–28.

Stoiljković, Z.Z., Avramov Ivić, M.L., Petrović, S.D., Mijin, D.Z., Stevanović, S.I., Lačnjevac, U.C., Marinković, A.D., 2012. Voltammetric and square-wave anodic stripping determination of amlodipine besylate on gold electrode. Int. J. Electrochem. Sci. 7, 2288–2303.

Streeter, I., Wildgoose, G.G., Shao, L., Compton, R.G., 2008. Cyclic voltammetry on electrode surfaces covered with porous layers: an analysis of electron transfer kinetics at single-walled carbon nanotube modified electrodes. Sens Actuat. B 133, 462–466.

Su, C.Y., Lu, A.Y., Xu, Y., Chen, F.R., Khlobystov, A.N., Li, L.J., 2011. High-quality thin graphene films from fast electrochemical exfoliation. ACS Nano 5, 2332–2339.

Suarez-Martinez, I., Grobert, N., Ewels, C.P., 2012. Encyclopedia of Carbon Nanoforms. Advances in Carbon Nanomaterials: Science and Applications. Pan Stanford Publishing, Singapore.

Sun, W., Hu, J., 2013. Voltammetric determination of theophylline in pharmaceutical formulations using aligned carbon nanotubes (ACNTs) film modified electrode. J. Anal. Chem. 68, 694–699.

Sutter, P.W., Flege, J.I., Sutter, E.A., 2008. Epitaxial graphene on ruthenium. Nature Mater. 7, 406–411.

Taei, M., Hasanpour, F., Zahedi, G., 2015. Application of new ferrocene derivative for electrocatalytic determination of captopril using multiwall carbon nanotube modified carbon paste electrode. B. Chem. Soc. Ethiopia 29, 149–156.

Tagmatarchis, N. (Ed.), 2012. Advances in Carbon Nanomaterials: Science and Applications. CRC Press, Singapore.

Tan, Y.Z., Xie, S.Y., Huang, R.B., Zheng, L.S., 2009. The stabilization of fused-pentagon fullerene molecules. Nature Chem. 1, 450–460.

Tao, Y., Noguchi, D., Yang, C.-M., Kanoh, H., Tanaka, H., Yudasaka, M., Iijima, S., Kaneko, K., 2007. Conductive and mesoporous singlewall carbon nanohorn/organic aerogel composites. Langmuir 23, 9155–9157.

Tenjarla, S., 1999. Microemulsions: an overview and pharmaceutical applications. Crit. Rev. Ther. Drug 16, 1–62.

Thirumalraj, B., Palanisamy, S., Chen, S.M., Lou, B.S., 2016. Preparation of highly stable fullerene C_{60} decorated graphene oxide nanocomposite and its sensitive electrochemical detection of dopamine in rat brain and pharmaceutical samples. J. Colloid. Interf. Sci. 462, 375–381.

Torrens, F., Castellano, G., 2013. Bundlet model for single-wall carbon nanotubes, nanocones and nanohorns. Methodologies and Applications for Chemoinformatics and Chemical Engineering, United states of America, pp. 228-294.

Torres, L.E.F., Roche, S., Charlier, J.C., 2014. Introduction to Graphene-Based Nanomaterials: From Electronic Structure to Quantum Transport. Cambridge University Press, New York.

Trasatti, S., 1977. Gerischer, H., Tobias, C.W. (Eds.), Advances in Electrochemistry and Electrochemical Engineering, vol. 10, Interscience, New York/London.

Tu, W., Lei, J., Ding, L., Ju, H., 2009. Sandwich nanohybrid of single-walled carbon nanohorns–TiO_2–porphyrin for electrocatalysis and amperometric biosensing towards chloramphenicol. Chem. Commun. 28, 4227–4229.

Tung, V.C., Allen, M.J., Yang, Y., Kaner, R.B., 2009. High-throughput solution processing of large-scale graphene. Nat. Nanotechnol. 4, 25–29.

Umesh, V., Banakar, 1991. Pharmaceutical Dissolution Testing (Drugs and the Pharmaceutical Sciences: a Series of Textbooks and Monographs), first ed. Marcel Dekker Publications, USA.

Uslu, B., Ozkan, S.A., 2007. Electroanalytical application of carbon based electrodes to the pharmaceuticals. Anal. Lett. 40, 817–853.

Valezi, C.F., Duarte, E.H., Mansano, G.R., Dall'Antonia, L.H., Tarley, C.R.T., Sartori, E.R., 2014. An improved method for simultaneous square-wave voltammetric determination of amlodipine and enalapril at multi-walled carbon nanotubes paste electrode based on effect of cationic surfactant. Sens. Actuat. B 205, 234–243.

Vávrová, J., ezáčová, M., Pejchal, J., 2012. Fullerene nanoparticles and their anti-oxidative effects: a comparison to other radioprotective agents. J. Appl. Biomed. 10, 1–8.

Wan, H., Zhao, F., Wu, W., Zeng, B., 2011a. Direct electron transfer and voltammetric determination of roxithromycin at a single-wall carbon nanotube coated glassy carbon electrode. Colloids Surf. B 82, 427–431.

Wan, H., Zhao, F., Zeng, B., 2011b. Direct electrochemistry and voltammetric determination of midecamycin at a multi-walled carbon nanotube coated gold electrode. Colloids Surf. B 86, 247–250.

Wang, J., 1985. Stripping Analysis: Principles, Instrumentation and Applications VCH. Deerfield Beach Florida, USA.

Wang, J., 2006. Analytical Electrochemistry, third ed. Wiley-VCH Pub, New Jersey.

Wang, J., Li, M., Shi, Z., Li, N., Gu, Z., 2001. Electrocatalytic oxidation of 3,4-dihydroxyphenylacetic acid at a glassy carbon electrode modified with single-wall carbon nanotubes. Electrochim. Acta 47, 651–657.

Wang, Q., Das, M.R., Li, M., Boukherroub, R., Szunerits, S., 2013. Voltammetric detection of l-dopa and carbidopa on graphene modified glassy carbon interfaces. Bioelectrochemistry 93, 15–22.

Wei, L., Borowiec, J., Zhu, L., Zhang, J., 2012. Electrochemical investigation on the interaction of diclofenac with DNA and its application to the construction of a graphene-based biosensor. J. Solid State Chem. 16, 3817–3823.

Wen, W., Zhao, D.M., Zhang, X.H., Xiong, H.Y., Wang, S.F., Chen, W., Zhao, Y.D., 2012. One-step fabrication of poly (o-aminophenol)/multi-walled carbon nanotubes composite film modified electrode and its application for levofloxacin determination in pharmaceuticals. Sens. Actuat. B 174, 202–209.

Wermuth, C.G., 2003. The Practice of Medicinal Chemistry, second ed Elsevier Publications, USA.

Whittaker, A.G., 1977. Carbon: a new view of its high-temperature behavior (No. TR-0078 (3950-03)-1). Aerospace corp el segundo ca materials sciences lab, Los Angles.

William A. Hanson, 1991. Handbook of Dissolution Testing, second ed., Aster Publication Corporation, Boca Rotan.

Williams, D.A., Lemke, T.L., Williams, L., 2002. Wilkins, Foye's Principles of Medicinal Chemistry, fifth ed. Philedelphia.

Wong, A., Scontri, M., Materon, E.M., Lanza, M.R., Sotomayor, M.D., 2015. Development and application of an electrochemical sensor modified with multi-walled carbon nanotubes and graphene oxide for the sensitive and selective detection of tetracycline. J. Electroanal. Chem. 757, 250–257.

Wu, G.H., Wu, Y.F., Liu, X.W., Rong, M C., Chen, X.M., Chen, X., 2012. An electrochemical ascorbic acid sensor based on palladium nanoparticles supported on graphene oxide. Anal. Chim. Acta 745, 33–37.

Yang, Y.J., Li, W., 2015. High sensitive determination of theophylline based on manganese oxide nanoparticles/multiwalled carbon nanotube nanocomposite modified electrode. Ionics 21, 1121–1128.

Yang, J., Heo, M., Lee, H.J., Park, S.M., Kim, J.Y., Shin, H.S., 2011. Reduced graphene oxide (rGO)-wrapped fullerene (C_{60}) wires. ACS Nano 5, 8365–8371.

Yang, J., Dong, Y., Wang, L., Zhang, D., Zhang, Z., Zhang, L., 2015. Highly sensitive detection of faropenem using amino functionalized carbon nanotube modified electrode with ionic liquid as binder: Insight into the interaction of faropenem with human serum albumin. Sens. Actuat. B 211, 59–66.

Yuan, L., Jiang, L., Hui, T., Jie, L., Bingbin, X., Feng, Y., Yingchun, L., 2015. Fabrication of highly sensitive and selective electrochemical sensor by using optimized molecularly imprinted polymers on multi-walled carbon nanotubes for metronidazole measurement. Sens. Actuat. B 206, 647–652.

Yudasaka, M., Iijima, S., Crespi, V.H., 2008. Single-wall carbon nanohorns and nanocones. Carbon nanotubes. Springer Berlin, Heidelberg, pp. 605–629.

Zang, J.B., Wang, Y.H., Zhao, S.Z., Bian, L.Y., Lu, J., 2007. Electrochemical properties of nanodiamond powder electrodes. Diam. Relat. Mater. 16, 16–20.

Zarei, K., Fatemi, L., Kor, K., 2015. Stripping voltammetric determination of nicardipine using β-cyclodextrin incorporated carbon nanotube-modified glassy carbon electrode. J. Anal. Chem. 70, 615–620.

Zhai, H., Liu, Z., Chen, Z., Liang, Z., Su, Z., Wang, S., 2015. A sensitive electrochemical sensor with sulfonated graphene sheets/oxygen-functionalized multi-walled carbon nanotubes modified electrode for the detection of clenbuterol. Sens. Actuat. B 210, 483–490.

Zhang, Q. (Ed.), 2012. Carbon Nanotubes and Their Applications. CRC Press, USA.

Zhang, B.T., Zheng, X., Li, H.F., Lin, J.M., 2013. Application of carbon-based nanomaterials in sample preparation: a review. Anal. Chim. Acta 784, 1–17.

Zhang, J., Zhu, Z., Zhu, J., Li, K., Hua, S., 2014. Selective determination of dopamine, ascorbic acid and uric acid at SDS-MWCNTs modified glassy carbon electrode. Int. J. Electrochem. Sci. 9, 1264–1272.

Zheng, J., 2009. Formulation and Analytical Development for Low-Dose Oral Drug Products. John Wiley & Sons, Inc, New Jersey.

Zhu, S., Li, H., Niu, W., Xu, G., 2009. Simultaneous electrochemical determination of uric acid, dopamine, and ascorbic acid at single-walled carbon nanohorn modified glassy carbon electrode. Biosens. Bioelectron. 25, 940–943.

Zi, L., Li, J., Mao, Y., Yang, R., Qu, L., 2012. High sensitive determination of theophylline based on gold nanoparticles/L-cysteine/graphene/nafion modified electrode. Electrochim. Acta 78, 434–439.

Zoski, C.G. (Ed.), 2007. Handbook of Electrochemistry. Elsevier, Amsterdam.

FURTHER READING

Branina Kh., Neyman E. (Eds.), 1993. Electroanalytical stripping methods, vol 126. In: Winefordner, J.D. (series ed.), Chemical analysis: a series of monographs an analytical chemistry, its. applications. John Wiley, Sons Pub, New, York.

Brooks, M.A., Tsai, E.W., 1996. Electrochemistry in pharmaceutical analysis. In: Kissinger, P.T., Heineman, W.R. (Eds.), Laboratory Techniques in Electroanalytical Chemistry. second ed. Marcel Dekker Inc, New York

Bundy, F.P., 1963. Melting of graphite at very high pressure. J. Chem. Phys. 38, 618–630.
Cao, L., Wang, X., Meziani, M.J., Lu, F., Wang, H., Luo, P.G., Lin, Y., Harruff, B.A., Veca, L.M., Murray, D., Xie, S.Y., Sun, Y.P., 2007. Carbon dots for multiphoton bioimaging. J. Am. Chem. Soc. 129 (37), 11318–11319.
Chen, Y., Liu, H., Ye, T., Kim, J., Mao, C., 2007. DNA-directed assembly of single-wall carbon nanotubes. J. Am. Chem. Soc. 129 (28), 8696–8697.
Cottrell, T.L., 1954. The Strengths of Chemical Bonds. Butterworth, London.
Ensafi, A.A., Allafchian, A.R., 2013. Multiwall carbon nanotubes decorated with $NiFe_2O_4$ magnetic nanoparticles, a new catalyst for voltammetric determination of cefixime. Colloids Surf. B 102, 687–693.
Feng, L., Xie, N., Zhong, J., 2014. Carbon nanofibers and their composites: a review of synthesizing, properties and applications. Materials 7, 3919–3945.
Hosseini, M.R.M., Motaharian, A., 2015. Electroanalytical determination of diazepam in tablet and human serum samples using a multiwalled carbon nanotube embedded molecularly imprinted polymer-modified carbon paste electrode. RSC Adv. 5, 81650–81659.
Iijima, S., Yudasaka, M., Yamada, R., Bandow, S., Suenaga, K., Kokai, F., Takahashi, K., 1999. Nano-aggregates of single-walled graphitic carbon nano-horns. Chem. Phys. Lett. 309, 165–170.
Jiménez-Soto, J.M., Cárdenas, S., Valcárcel, M., 2009. Evaluation of carbon nanocones/disks as sorbent material for solid-phase extraction. J. Chrom. A 1216, 5626–5633.
Kellner, R., Mermet, J.-M., Otto, M., Valcarcel, M., Widmer, H.M. (Eds.), 2004. Analytical Chemistry, A Modern Approach to Analytical Science. second ed. Wiley-VCH Pub, New York.
Liu, Y., Huang, J., Hou, H., You, T., 2008a. Simultaneous determination of dopamine, ascorbic acid and uric acid with electrospun carbon nanofibers modified electrode. Electrochem. Comm. 10, 1431–1434.
Liu, Z., Li, X., Tabakman, S.M., Jiang, K., Fan, S., Dai, H., 2008b. Multiplexed multicolor Raman imaging of live cells with isotopically modified single walled carbon nanotubes. J. Am. Chem. Soc. 130, 13540–13541.
Novoselov, K.S., Fal, V.I., Colombo, L., Gellert, P.R., Schwab, M.G., Kim, K., 2012. A roadmap for graphene. Nature 490, 192–200.
Patriarche, G.J., Chateau-Gosselin, M., Vandenbalck, J.L., Zuman, P., 1979. Polarography and related electroanalytical techniques in pharmacy and pharmacology. Bard, A.J. (Ed.), Electroanalytical Chemistry, vol. 11, Marcel Dekker Inc, New York.
Sz cs, Á., Tölgyesi, M., Csiszar, M., Nagy, J.B., Novak, M., 1998. Electrochemistry on partially reduced fullerene films. Electrochim. Acta 44, 613–621.
Vire, J.C., Kauffmann, J.M., 1994. Trends in electrochemistry in drug analysis. Curr. Top. Electrochem. 3, 493–515.
Wang, J. (Ed.), 1988. Electroanalytical Techniques in Clinical Chemistry and Laboratory Medicine. VCH Publishers, New York.
Wang, J., 1989. Voltammetry following nanoelectrolytic preconcentration. Bard, A.J. (Ed.), Electroanalytical Chemistry, vol. 16, Marcel Dekker Inc, New York.
Xiao, N., Venton, B.J., 2012. Rapid, sensitive detection of neurotransmitters at microelectrodes modified with self-assembled SWCNT forests. Anal. Chem. 84, 7816–7822.

CHAPTER

SCIENTOMETRIC OVERVIEW REGARDING ORAL CANCER NANOMEDICINE

29

Ozcan Konur

Yildirim Beyazit University, Ankara, Turkey

CHAPTER OUTLINE

1 Overview 939
1.1 Issues 939
1.2 Methodology 940
1.3 The Cancer Research in General: Overview 941
1.4 The Oral Cancer Research in General: Overview 943
1.5 The Nanoresearch in General: Overview 944
1.6 The Research on the Oral Cancernanomedicine in General: An Overview 945
2 The Citation Classics in Oral Cancer Nanomedicine 946
2.1 The Citation Classics in Oral Cancer Nanomedicine:Overview 946
2.2 The Citation Classics in Head and Neck Cancer Nanomedicine 951
2.3 The Citation Classics in Oral Cancer Nanomedicine 952
2.4 The Citation Classics in Pharyngeal Cancer Nanomedicine 954
2.5 The Citation Classics in Thyroid and Tracheal Cancer Nanomedicine 955
3 Conclusions 956
References 956

1 OVERVIEW

1.1 ISSUES

The field of cancer has been one of the most dynamic and vital research fields in recent years with 2,500,000 papers as indexed by the Science Citation Index-Expanded (SCIE) and Social Sciences Index (SSCI) of the Web of Science (WoS) database as of June 2016 and forming 8.3% of all the papers indexed by these core WoS databases (Hanahan and Weinberg, 2000, 2011; Jemal et al., 2011; Lawrence et al., 2014; Siegel et al., 2012, 2014a,b; Therasse et al., 2000).

Similarly, as a subset of the general cancer research, the field of oral cancer has emerged as one of the most dynamic and vital research fields in recent years with more than 187,000 papers as indexed by the same WoS indices as of June 2016 and forming 7.5% of all the cancer papers. (Aerts et al., 2014; Ang et al., 2010, 2014; Bonner et al., 2006; Chaturvedi et al., 2011; Cooper et al., 2009; Davies and Welch, 2014; Pignon et al., 2000; Stransky et al., 2011).

Besides the general and oral cancer research, the field of nanoresearch has been one of the most dynamic and novel research fields in recent years with more than 1,000,000 papers as indexed by the same WoS indices as of June 2016, forming 3.5% of all the papers indexed by these WoS indices (Dreyer et al., 2010; Hagfeldt et al., 2010; Iijima, 1991; Li et al., 2014; Liu and Kelly, 2014; Liu et al., 2014; Novoselov et al., 2004; Oregan and Gratzel, 1991; Yella et al., 2011).

At the intersection of oral cancer research and nanoresearch, a highly specialized research field of oral cancer nanomedicine emerged during the last decade in meeting the public demand for the efficient nanodiagnosis, nanoimaging, and nanotreatment of oral cancer with more than 1100 papers as indexed by the same WoS indices as of June 2016, forming only 0.6% of the oral cancer papers (Bhirde et al., 2009; El-Sayed et al., 2005; Huang et al., 2007a; Jungebluth et al., 2011; Lu et al., 2014; Ma et al., 2014; Malhotra et al., 2010; Wang et al., 2011, 2014)

Due to their public importance, there have been many scientometric studies relating to the cancer research (Cambrosio et al., 2006; Ho et al., 2010; Mela et al., 1999) and relating to the nanoresearch (Konur, 2016c–i, 2017a–c, 2018a–c; Rafols and Meyer, 2010; Schummer, 2004; Zhou and Leydesdorff, 2006), following the teachings of the general scientometric studies (Konur, 2011, 2012a–p, 2013a–m, 2015a–m, 2016a–b, 2017d, 2018d–e).

However, there have been a limited number of scientometric studies in oral cancer research (Glynn et al., 2012; Grajewski et al., 2010; Sun et al., 2013a,b) and in cancer nanomedicine research (Coccia and Wang, 2015; Dong et al., 2013; Huang et al., 2015; Konur, 2016e; Ma et al., 2015).

It is important to note that there have not been any scientometric studies of oral cancer nanomedicine research, including the citation classics in this field of research. Following North's New Institutional theory, it is important to have up-to-date information about the current public policy issues to develop a set of viable solutions to satisfy the needs of all the key stakeholders relating to the oral cancer nanomedicine research (Konur, 2000, 2002a–c, 2006a,b, 2007a,b, 2012r–s, 2013n; North, 1994).

Therefore, the brief information on a selected set of 25 citation classics in oral cancer nanomedicine are presented following a brief scientometric overview of the underlying cancer research, oral cancer research, nanoresearch, and oral cancer nanomedicine to inform the key stakeholders as the first-ever study of its kind.

1.2 METHODOLOGY

A multistep search process based on the key research streams of oral cancer nanomedicine was employed in this study. These key research streams are cancer research, oral cancer research, nanoresearch, and finally oral cancer nanomedicine research.

For the cancer research stream of the study, the following [I] search syntax was employed: TS = (*cancer* or *tumor* or *tumour* or *carcin* or *neoplas* or *oncol* or chemotherapy* or "*chemo* *therapy*" or p53 or "p-53" or *oncogen* or "onco-gen*" or *melanom* or *sarcoma* or *lymphom* or *leukemia* or *leukaemia* or *blastoma* or *adenoma* or *myeloma* or *astrocytoma* or *mesothelioma* or *doxorubicin* or *cisplatin* or *paclitaxel*) OR WC = (oncology). The keywords were determined from the close analysis of the top citation classics.

For the oral cancer research stream of the study, the following [II] search syntax was employed:

[[I] AND TS = (mouth* or *tongue* or gingiva* or "hard palate*" or "soft palate" or "head and neck" or *throat* or *pharyn* or *laryn* or *trachea* or "oral cavity" or *thyroid* or *tonsil* or *nasal* or *maxillary* or *sinus* or *glottic* or "salivary gland*") OR WC = (Otorhinolaryngology)]

OR [SO = ("oral oncology") or TS = ("*oral cancer*" or "*oral carcinoma*" or "*oral neoplas*" or "*oral tumor*" or "*oral tumour*")]].The keywords were determined from the close analysis of the top citation classics.

For the nanoresearch stream of the study, the following [III] search syntax was employed: TS = (*nano* or *graphene* or *fullerene* or *fullerid* or *fullerit* or *fullero* or*peapod* or *buckyball* or *buckytub* or "*quantum* *dot*" or *silicene* or *graphane* or "*metal* *organic* *framework*" or "*metal organic framework*" or "single* *graphite* *layer*" or *mos2* or "**mos(2)*" or "*mos-2*" or "molybdenum disulfide" or "atomically thin") NOT TS = (*nanog* or "nano-gram" or *nanomolar* or *nanomole or *nanomoles or "nano-mol*" or *nanokelvin* or *nanoamper* or *nanolit* or "*nano-lit*" or *nanosec* or "*nano-sec*" or *nano2* or "*nano(2)" or *nano3* or "*nano(3)*" or *nanoplankton* or * nonanonym* or *nonanomal* or *nanoflagel* or *nanoalga* or *nanoprotist* or *nanofauna* or *nanoheterotroph* or *nanophtalm* or *nanomeli* or nanophyto* or *nanobacteri* or *nanosatellite* or *nanos or *nanor or *nanoa or nanoarchaeota or allopregnanolone or "nanoflow HPLC" or "nano-electrospray" or *nanocol or "Tc-99m" or *nanoflow or nanobody or nanostring*). Again, the keywords were determined from the close analysis of the top citation classics.

For the final oral cancer nanomedicine research stream, the following [IV] search syntax was employed: [II AND III].

The search for the relevant materials was carried in the v.5.22 of the Web of Science (WoS) database in June 2016. The SCIE and Social Sciences Citation Index (SSCI) were used in this study for the years between 1980 and 2016 (as of June 2016).

The key bibliometric data were extracted from these search activities focusing on the reviews and articles as the final set of references.

The highly cited 25 papers were located based on the number of citations they received. The summary information about the located citation classics are presented in the order of the decreasing number of citations for the major research fronts, respectively.These major research fronts were head and neck cancer (HNC) research, oral cancer research, and pharyngeal cancer research. There were also papers in thyroid cancer and tracheal cancer among these citation classics.

The information relating to affiliation of the authors, the number and gender of the authors, the country of the authors, the journal where the paper was published, the subject area of the journal where the paper was indexed, the concise topic of the paper, and total number of citations received for the paper for the WoS databases were given in tables for each paper.

1.3 THE CANCER RESEARCH IN GENERAL: OVERVIEW

Using the keywords related to cancer, 3,524,770 references were located where 2,498,500 of these references were articles and reviews. These papers form 8.3% of all the papers indexed by the core WoS databases. Both the actual and relative size of the cancer research provides substantial evidence for the public importance cancer research in our society.

The most prolific country in terms of the number of publications was the United States with 896,902 papers forming 35.9% of the sample. The other most prolific countries were Japan, China, Germany, England, Italy, and France with 9.1, 7.8, 7.3, 6.4, 5.7, and 5.4% of the sample, respectively. 1.8% of the samples did not disclose any affiliation and in total, 257 countries and territories contributed to the cancer research.

There were 2,263,641 articles and 234,888 reviews where reviews formed 9.4% of the sample, showing the importance of reviews in cancer research. National Natural Science Foundation of China, National Institutes of Health (NIH), and National Cancer Institute (NCI) were the major funding bodies in cancer research.

English was the dominant language of scientific communication in this field comprising 95.6% of the sample. German and French were the other major languages with 1.5 and 1.4% of the sample, respectively.

The most prolific national or regional institution was NIH with 63,151 papers forming 2.5% of the sample. NCI, Institut National de la Sante et de la Recherche Medicale (INSERM), VA Boston Healthcare System, and Unicancer were the other most prolific institutions.

Similarly, the most prolific single institution was Harvard University with 69,378 papers forming 2.8% of the sample. UTMD Anderson Cancer Center, University of Toronto, Johns Hopkins University, and Memorial Sloan Kettering Cancer Center were the other most prolific single institutions.

Like the nanoresearch, the research in this field has boomed after 1990. There was a general increasing trend in the number of papers over time starting with 16,113 papers in 1980 and resulting in a peak with 163,590 papers in 2015. There was a significant rise in 1991, possibly due to the inclusion of the abstracts in the abstract pages of the indices for the first time.

The most prolific journal in terms of the number of publications was *Cancer Research* with 36,879 papers forming 1.5% of the sample. *Plos One*, *Cancer*, *Journal of Biological Chemistry*, *International Journal of Cancer*, *Anticancer Research*, and *Oncogene* were the other most prolific journals. Thus, the journals related to oncology subject category topped the list. In total, 14,477 journals contributed to the cancer research.

The most prolific subject category in terms of the number of publications as expected was oncology with 669,576 papers forming 26.8% of the sample. It appears that 73.2% of the cancer papers were indexed under other subject categories. Therefore, this finding justifies the search outside the oncology subject category as well. Biochemistry molecular biology, surgery, cell biology, pharmacology pharmacy, and radiology nuclear medicine medical imaging were the other most prolific subject categories forming 9.2, 7.3, 7.0, 5.8, and 5.2% of the sample, respectively. Otorhinolaryngology and nanoscience nanotechnology, which are closely related to oral cancer nanomedicine form only 1.2 and 0.5% of the sample, respectively. In total, 227 subject categories were related to the cancer research.

The most-cited papers in this field were dominated by the research fronts relating to the cancer statistics, cancer classification, cancer treatment, and cancer formation. In the most prolific paper, Jemal et al. (2011) discuss global cancer statistics with 14,181 citations. Hanahan and Weinberg (2000) discuss cancer research in a review paper with 13,500 citations, while Therasse et al. (2000) discuss cancer treatment in a paper with 10,390 citations.

There was a similar trend for the hottest papers published in the 2010s. For example, in the most prolific paper, Jemal et al. (2011) discuss global cancer statistics with 14,181 citations. Hanahan and Weinberg (2011) discuss cancer research in a review paper with 10,122 citations, while Siegel et al. (2012) discuss the US cancer statistics in a paper with 6,275 citations. It is interesting to note that Jemal et al. (2011) and Hanahan and Weinberg (2011) have already became citation classics by 2016 as they also appear among the top three citation classics listed earlier.

The most cited paper published in 2014 was Siegel et al. (2014a) discussing cancer statistics in a paper with 4129 citations. Siegel et al. (2014b) discuss colorectal cancer statistics in a paper with 522 citations, while Lawrence et al. (2014) discuss cancer genomics in a paper with 454 citations.

1.4 THE ORAL CANCER RESEARCH IN GENERAL: OVERVIEW

Using the keywords related to oral cancer, 226,658 references were located; 187,452 of these references were articles and reviews. These papers form 7.5% of the papers on cancer in general.

The most prolific country in terms of the number of publications was the United States with 60,770 papers forming 32.4% of the sample. Japan, Germany, China, and Italy followed the United States with 9.1, 7.6, 7.1, and 6.5% of the sample, respectively. In total, 183 countries contributed to the research in this area.

National Natural Science Foundation of China, NIH, NCI, and Ministry of Education Culture Sports Science and Technology of Japan were the major funding bodies supporting cancer research; 152,644 papers did not disclose any funding body and the remaining part of the papers disclosed in total 35,297 funding bodies.

English was the dominant language of scientific communication in this field with 94.8% of the papers. German and French were the other key languages with 2.5 and 1.6% of the sample, respectively.

The most prolific national or regional institution was NIH with 3340 papers. The other most prolific institutions were Unicancer, VA Boston Healthcare System, NCI, and INSERM. On the other hand, the most prolific single institution was Harvard University with 4081 papers. The other most prolific single institutions were UTMD Anderson Cancer Center, University of Toronto, Johns Hopkins University, and University of Pittsburgh; 3,065 papers did not include any affiliation and in total, 42,777 institutions contributed to oral cancer research.

Similar to the nanoresearch booming after 1991, the oral cancer research has boomed after the 1980s comprising 94.0% of the sample. There was a general increasing trend in the number of papers over time starting with just 847 papers in 1980, resulting in a peak with 6363 papers in 2015.

The most prolific journal in terms of the number of publications was *Head and Neck Journal for the Sciences* and *Specialties of the Head and Neck* with 3912 papers. The other most prolific journals were *Laryngoscope*, *Oral Oncology*, *Journal of Laryngology and Otology*, *International Journal of Radiation Oncology Biology Physics*, *Cancer*, *Otolaryngology Head and Neck Surgery* with 3324, 2714, 2481, 2360, 2314, and 2261 papers, respectively. These journals were related to the subject categories of otorhinolaryngology and oncology. In total, 5648 journals contributed to the oral cancer research.

The most prolific subject category in terms of the number of publications was oncology with 44,523 papers forming 23.8% of the sample. The other most prolific subject categories were otorhinolaryngology, surgery, radiology nuclear medicine medical imaging, pathology, and endocrinology metabolism with 29,764, 27,942, 15,139, 13,912, and 13,259 papers, respectively. In total, papers were indexed under 208 subject categories.

The most-cited papers in this field were dominated by the treatment, prevalence, imaging, and diagnosis of HNC, oral cancer, thyroid cancer, laryngeal cancer, and pharyngeal cancer. For example, Cooper et al. (2009) discuss treatment of thyroid cancer in a top paper with 2410 citations. Bonner et al. (2006) and Pignon et al. (2000) discuss treatment of HNC squamous cancer cells (HNSCC) in papers with 2291 and 1437 citations, respectively.

There was a similar trend for the hottest papers published in the 2010s. For example, Ang et al. (2010) discuss treatment of oropharyngeal cancer cells in a top paper with 1376 citations. Stransky et al. (2011) discuss genomics of HNSCC in a paper with 738 citations, while Chaturvedi et al. (2011) discuss prevalence of oropharyngeal cancer cells in a paper with 671 citations.

Similarly, the top paper published in 2014 was Davies and Welch (2014) discussing prevalence of thyroid cancer in a paper with 124 citations. Aerts et al. (2014) discuss imaging of HNC cells in a paper with 73 citations, while Ang et al. (2014) discuss treatment of HNC cells in a paper with 68 citations.

1.5 THE NANORESEARCH IN GENERAL: OVERVIEW

Using the keywords related to nanoresearch, 1,127,280 references were located where 1,063,884 of these references were articles and reviews. These papers form 3.5% of all the papers indexed by the core WoS databases.

The most prolific country in terms of the number of publications was China with 276,182 papers forming 26.0% of the sample. The United States closely followed China with 217,746 papers forming 20.5% of the papers. Japan, Germany, South Korea, and India were the other most prolific countries with 7.6, 7.2, 6.0, and 5.6% of the papers, respectively. It is surprising that China has surpassed the United States in this field.

There were 1,025,657 articles and 38,227 reviews where reviews formed 3.6% of the sample. 7.3% of these papers were also presented in the conferences. It is not surprising that the most prolific funding body was National Natural Science Foundation of China funding 9.3% of the papers. The other major funding bodies were National Science Foundation, Fundamental Research Funds for the Central Universities, and National Basic Research Program of China.

English was the dominant language of scientific communication in this field comprising 98.0% of the sample. Chinese was the other major language with 1.5% of the papers.

The most prolific national or regional institution as expected was Chinese Academy of Sciences with 5.0% of the sample. Centre National de la Recherche Scientifique (CNRS), US Department of Energy, Russian Academy of Sciences, and Max Planck Society were the other most prolific national institutions with 3.3, 2.3, 2.1, 2.1, and 1.2% of the sample, respectively.

On the other hand, the most prolific single institution was as expected Tsinghua University with 10,254 papers. Indian Institute of Technology, Jilin University, Nanyang Technological University, Nanjing University, and University of Science Technology of China were the other most prolific single institutions with 10,381, 8,615, 8,602, 8,523, and 8,431 papers, respectively. Thus, four of the top institutions were Chinese showing dominance of China in this field. In total, 82,351 institutions contributed to the nanoresearch.

The nanomaterial research has boomed after 1990 with the seminal paper on the nanomaterials by Iijima (1991) following a decade of limited fullerene research. There was a general increasing trend in the number of papers over time starting with 64 papers in 1980, resulting in a peak with 132,514 papers in 2015. There was a significant rise (nearly 4 times rise) in 1991, possibly due to the inclusion of the abstracts in the abstract pages of the indices for the first time.

The most prolific journals in terms of the number of publications were *Applied Physics Letters*, *Physical Review B*, *Journal of Applied Physics*, *Journal of Physical Chemistry C*, and *RSC Advances* with 24,965, 23,795, 17,164, 16,981, and 16,839 papers, respectively. It is striking to note that these top journals relate to physics and chemistry, not nanoscience and nanotechnology. *Nanotechnology*, *Journal of Nanoscience and Nanotechnology*, *Nano Letters*, *Nanoscale*, and *ACS Nano* were some of the top journals relating to nanoscience and nanotechnology. In total, 8981 journals contributed to the nanoresearch.

The most prolific subject category was materials science multidisciplinary with 315,107 papers forming 29.6% of the sample. Physics applied, chemistry physical, chemistry multidisciplinary, nanoscience nanotechnology, and physics condensed matter were the other most prolific subject categories with 20.4, 20.1, 16.8, 15.0, and 13 2%, respectively. Materials science, including nanoscience and nanotechnology, physics and chemistry emerge as the top subject categories in nanoresearch. In total, 235 subject categories indexed papers in nanoresearch.

Perhaps, the most important finding from these data is that the only 14.9% of the papers were indexed under the subject category of nanoscience and nanotechnology, providing justification for the development of the search strategy outside this subject category through the keyword search.

The most-cited papers in nanomaterials were dominated by the carbon nanotubes, graphene, and solar nanomaterials. For example, Iijima (1991) discusses the carbon nanotubes in a seminal paper with 25,686 citations. This top paper was followed by Novoselov et al. (2004) discussing the development of graphene in a seminal paper with 21,252 citations and by Oregan and Gratzel (1991) discussing nanosolar cells in a seminal paper with 15,988 citations.

There was a similar trend for the hottest papers in nanoresearch published in 2010s. For example, Hagfeldt et al. (2010) discuss nanosolar cells in a paper with 3595 citations. Following this top hot paper, Dreyer et al. (2010) discuss graphene in a paper with 3410 citations and Yella et al. (2011) discuss nanosolar cells in a paper with 3276 citations.

2014 saw the introduction of phosphorene in the nanoresearch. For example, Li et al. (2014) discuss phosphorene in a top paper with 694 citations. Liu et al. (2014) also discuss phosphorene in a paper with 615 citations following the top paper. Additionally, Liu and Kelly (2014) discuss nanosolar cells in a paper with 519 citations.

1.6 THE RESEARCH ON THE ORAL CANCERNANOMEDICINE IN GENERAL: AN OVERVIEW

Using the keywords related to oral cancer nanomedicine, 1177 references were located where 1114 of them were articles and reviews. These papers form 0.6% of the oral cancer papers in general. These findings suggest that this field has been a specialized field of research with relatively a small sample size.

S.A. Mousa was the most prolific author with 23 papers. R. Chen , P.J. Davis , and S.Y. Feng followed the top author with 18, 17, and 15 papers, respectively. In total, 5497 authors contributed to the research in this area.

The most prolific country in terms of the number of publications was the United States with 347 papers forming 31.1% of the sample. China followed the United States closely with 257 papers forming 23.1% of the sample. Japan, India, Germany, and South Korea were the other most prolific countries. In total, 64 countries contributed to the research in this field.

There were 969 articles and 145 reviews while English was the dominant language of scientific communication in this field with 99.0% of the sample.

National Natural Science Foundation of China and NIH were the most prolific funding agencies; 351 papers did not disclose any funding body, while the remaining of them disclosed 1379 funding bodies in total.

The most prolific single institution was Fudan University with 22 papers. Seoul National University, University of North Carolina, and UTMD Anderson Cancer Center closely followed the top institution with 21, 20, and 19 papers, respectively. On the other hand, CNRS was the most prolific national institution with 27 papers. Chinese Academy of Sciences, Unicancer, and NIH closely followed the top institution with 19, 18, and 15 citations, respectively. In total, 1328 institutions contributed tothe research in this area.

Unlike the nanoresearch booming after 1991, the research in this field boomed only during the last decade with only six papers published in the 1990s. These findings suggest that the research in this field has been relatively recent and more importantly will continue to expand significantly in the near future in the steps of cancer nanomedicine.

The most prolific journal in terms of the number of publications was *International Journal of Nanomedicine*, publishing 32 papers. *Biomaterials*, *Plos One*, and *Journal of Controlled Release* closely followed the top journal with 2, 24, and 19 papers, respectively. It is notable that the top journal list was dominated by the journals related to nanoscience and nanotechnology, as well as medicine. In total, 503 journals contributed to the research in this field.

The most prolific subject category in terms of the number of publications was pharmacology pharmacy with 212 papers forming 19.0% of the sample. Nanoscience nanotechnology closely followed the top category with 185 papers forming 16.6% of the sample. Chemistry multidisciplinary, oncology, and materials science multidisciplinary were the other most prolific categories with 127, 122, and 103 papers, respectively. In total, papers were indexed under 105 subject categories highlighting interdisciplinarity of this research field.

The citation impact of the papers published in this field has been significant with 24.4 citations per paper and H-index of 78.

The most-cited papers in this field were dominated by applications of nanotechnology in oral cancer medicine, including HNC, oral cancer, pharyngeal cancer, and thyroid cancer with a focus on the nanodiagnosis, nanoimaging, and nanotreatment of these cancer cells. For example, El-Sayed et al. (2005) focus on the nanodiagnosis of oral cancer in a paper with 994 citations. Bhirde et al. (2009) discuss nanotreatment of HNSCC in a paper with 391 citations, while Huang et al. (2007b) discuss nanodiagnosis of oral cancer cells in a paper with 320 citations.

There was a similar trend for the hottest papers published in the 2010s. These papers in this field were dominated by applications of nanotechnology in oral cancer medicine, including HNC, oral cancer, pharyngeal cancer, and thyroid cancer with a focus on the nanodiagnosis, nanoimaging, and nanotreatment of cancer cells. For example, in the hottest paper with 189 citations, Jungebluth et al. (2011) discuss nanotreatment of tracheal cancer cells. Malhotra et al. (2010) and Wang et al. (2011) discuss nanobiosensing and nanodiagnostics of HNSCC in a paper with 158 and 139 citations, respectively.

The publication pattern was also similar in 2014 with the similar cancer types and research fronts. For example, Lu et al. (2014) discuss nanotreatment of HNC cells, while Ma et al. (2014) discuss nanotreatment of pharyngeal cancer cells in a paper with 35 citations each. Similarly, Wang et al. (2014) discuss nanotreatment of oral cancer cells in a paper with 34 citations.

2 THE CITATION CLASSICS IN ORAL CANCER NANOMEDICINE

2.1 THE CITATION CLASSICS IN ORAL CANCER NANOMEDICINE:OVERVIEW

The field of oral cancer nanomedicine has been one of the most recent niche research areas with 25 citation classics. These citation classics were located and the key emerging issues from these papers were presented under four research fronts in the decreasing order of the number of citations (Table 29.1):

Table 29.1 The Citation Classics in Oral Cancer Nanomedicine

S. No.	Paper Reference	Affiliation	Country	Number of Authors	M/F	Journal	Subject Area	Topic	Citations
1	El-Sayed et al.	University of California–San Francisco; Georgia Institute of Technology	United States	3	M	Nano Lett.	Chem. Mult., Chem. Phys., Nanosci. Nanotech. +3	Nanodiagnostics	994
2	Bhirde et al.	University of Connecticut +2	United States	10	M	ACS Nano	Chem. Mult., Chem. Phys., Nanosci. Nanotech. +1	Nanotreatment	391
3	Huang et al.	Georgia Institute of Technology; University of California–San Francisco +1	United States	4	F	Nano Lett.	Chem. Mult., Chem. Phys., Nanosci. Nanotech. +3	Nanodiagnostics	320
4	Popovtzer et al.	University of Michigan	United States	7	F	Nano Lett.	Chem. Mult., Chem. Phys., Nanosci. Nanotech. +3	Nanoimaging	299
5	Bae et al.	University of Tokyo	Japan	5	M	Mol. Biosyst.	Bioch. Mol. Biol.	Nanotreatment	270
6	Oyelere et al.	Georgia Institute of Technology; University of California–San Francisco	United States	5	M	Bioconjugate Chem.	Bioch. Res. Meth., Bioch. Mol. Biol., Chem. Mult. +1	Nanodiagnostics	226
7	Jungebluth et al.	University Hospital RWTH Aachen, Uppsala Univ. +4	Germany; Sweden +2	28	M	Lancet	Med. Gen. Int.	Nanotreatment	189
8	Choi et al.	University of Pennsylvania	United States	5	M	Acad. Radiol.	Radiol. Nucl. Med. Med. Imag.	Nanoimaging	178
9	Oyewumi et al.	University of Kentucky	United States	5	M	J. Control. Release	Chem. Mult., Phar. Phar.	Nanotreatment	174

(*Continued*)

Table 29.1 The Citation Classics in Oral Cancer Nanomedicine (*cont.*)

S. No.	Paper Reference	Affiliation	Country	Number of Authors	M/F	Journal	Subject Area	Topic	Citations
10	Malhotra et al.	University of Connecticut, National University of Ireland	United States; Ireland	5	F	Anal. Chem.	Chem. Anal.	Nanobiosensors	158
11	Wang et al.	Emory University +1	United States	9	M	Cancer Res.	Oncol.	Nanodiagnostics	139
12	Feng et al.	Fujian Normal University +3	China	10	M	Biosens. Bioelectron.	Biophys., Biot. Appl. Microb., Chem. Anal. +2	Nanobiosensors	132
13	Huang et al.	Georgia Inst Technology; University of California–San Francisco	United States	4	F	Lasers Surg. Med.	Surg.	Nanotreatment	128
14	Muhammed et al.	Indian Institute of Technology +2	India	7	M	Chem.-Eur. J.	Chem. Mult.	Nanoimaging	112
15	Retnakumari et al.	Indian Institute of Technology +1	India	8	F	Nanotechnology	Nanosci. Nanotech., Mats. Sci. Mult., Phys. Appl.	Nanoimaging	112
16	Setua et al.	Amrita Vishwa Vidyapeetham University	India	5	F	Biomaterials	Eng. Biomed., Mats. Sci. Biomats.	Nanoimaging	112
17	Reuveni et al.	Bar Ilan University +2	Israel	5	M	Int. J. Nanomed.	Nanosci. Nanotech., Phar. Phar.	Nanoimaging	93

18	Wang et al.	Emory University +1	United States	13	M	ACS Nano.	Chem. Mult., Chem. Phys., Nanosci. Nanotech. +1	Nanotreatment	89
19	Schroeder et al.	Hebrew University of Jerusalem +1	Israel	5	F	J. Control. Release	Chem. Mult., Phar. Phar.	Nanotreatment	80
20	Manzoor et al.	Amrita Institute of Medical Sciences and Research Center	India	6	M	Nanotechnology	Nanosci. Nanotech., Mats. Sci. Mult., Phys. Appl.	Nanoimaging	78
21	Shen et al.	Donghua University; University of Michigan	China, United States	9	M	J. Phys. Chem. C	Chem. Phys., Nanosci. Nanotechnol. +1	Nanotreatment	78
22	Damascelli et al.	University of Milan; University Piemonte E +1	Italy	18	M	Cancer	Oncol.	Nanotreatment	75
23	Stevens et al.	Ohio State University	United States	3	M	Pharm. Res.	Chem. Mult., Phar. Phar.	Nanotreatment	72
24	Guo et al.	Purdue University; University of Southern Mississippi	United States	3	M	Gene Ther.	Bioch. Mol. Biol., Biot. Appl. Microb. +2	Nanotreatment	69
25	Hakim et al.	Technion Israel Institute of Technology	Israel	9	F	Br. J. Cancer	Oncol.	Nanodiagnostics	66

F, Female; M, male.

HNC nanomedicine, oral cancer nanomedicine, pharyngeal cancer nanomedicine, and finally thyroid and tracheal cancer nanomedicine.

I.H. El-Sayed, M.A. El-Sayed, and S. Nair, were the most prolific authors with four papers each. The other most prolific authors with three papers each were M. Koyakutty, D. Menon, and S. Setua. In total, 162 authors contributed to these 25 citation classics in this area.

The United States was the most prolific country with 14 papers. India, Israel, and China followed the United States with 4, 3, and 2 papers, respectively. In total, 12 countries contributed to these 25 citation classics.

Amrita Vishwa Vidyapeetham University, Georgia Institute of Technology, and University of California–San Francisco were the most prolific individual institutions with four papers each. Emory University, Indian Institute of Technology, University of Connecticut, and University of Michigan followed these top academic institutions with two papers each. In total, 41 institutions contributed to these citation classics.

The most prolific publication year was 2010 with five papers. 2007, 2009, and 2011 saw four papers each. Thus, 17 papers were published in the last decade providing evidence for the recent emergence of this research area. There were no papers published in the 1980s or 1990s.

The most prolific journal was Nano Letters with three papers. *ACS Nano*, *Journal of Controlled Release*, and *Nanotechnology* followed the top journal with two papers each. It is important to note that seven top papers were published by the journals in nanoscience and nanotechnology. In total, 16 journals published these citation classics.

The most prolific subject categories were chemistry multidisciplinary and nanoscience nanotechnology with 10 papers each. Materials science multidisciplinary, chemistry physical, and physics applied were the other prolific subject area with eight, six, and five papers, respectively. In total, these papers were indexed under 22 subject categories. There were only four and three papers in pharmacy and pharmaceutics and oncology subject areas, respectively. These findings suggest that materials science, including nanoscience and nanotechnology, chemistry, and physics were the major subject areas of these citation classics, rather than pharmacy and pharmaceutics, oncology, and otorhinolaryngology.

The citation impact of these classical papers was significant with 186 citations per paper and H-index of 25. The number of citations ranged from 66 to 994 for the WoS database.

There was a significant gender deficit among these papers as there were only eight papers with a female first author out of 25 papers. This finding supports findings of Sotudeh and Khoshian (2014) that although female nanoresearchers are scarce in number, they equally perform in terms of research output and citation impact (see also Villanueva-Felez et al. (2015) for gender deficits in nanoscience and nanotechnology).

All these citation classics were articles showing a need for substantial reviews of the research in this area although there were 145 reviews in oral cancer nanomedicine.

There are three major research fronts among citation classics: HNC nanomedicine, oral cancer nanomedicine, and pharyngeal cancer nanomedicine. There are also two papers in thyroid cancer nanomedicineand tracheal cancer nanomedicine.

In the following parts, only the brief information on the most cited papers are provided under these major topical research fronts as it is not the primary purpose of this study to provide a detailed information on these citation classics.

2.2 THE CITATION CLASSICS IN HEAD AND NECK CANCER NANOMEDICINE

Bhirde et al. (2009) discuss nanotreatment of HNSCC in a paper originating from the United States with 391 citations. They study targeted in vivo killing of these cancer cells using a drug single-wall carbon nanotube (SWNT) bioconjugate, in comparison to nontargeted bioconjugates. They attach first-line anticancer agent cisplatin and epidermal growth factor (EGF) to SWNTs to specifically target squamous cancer. They find in vitro that SWNT-Qdot-EGF bioconjugates internalized rapidly into the cancer cells. They further find in vivo that these bioconjugates injected into live mice were selectively taken up by HNSCC tumors. HNSCC cells treated with SWNT−cisplatin−EGF were also killed selectively. They finally observe that regression of tumor growth was rapid in mice treated with targeted SWNT−cisplatin−EGF relative to nontargeted SWNT-cisplatin.

Popovtzer et al. (2008) discuss nanoimaging of HNC cells in a paper originating from the United States with 299 citations. They describe a targeted molecular imaging platform that enables cancer detection at the cellular and molecular level with standard clinical X-ray-based computed tomography (CT). They observe that gold nanoprobes selectively and sensitively target tumor selective antigens while inducing distinct contrast in CT imaging. They show in vitro that the attenuation coefficient for the molecularly targeted cells is over 5 times higher than for identical but untargeted cancer cells or for normal cells.

Malhotra et al. (2010) discuss nanobiosensing of HNSCC (HNSCC) in a paper originating from Ireland and the United States with 158 citations. They develop an ultrasensitive electrochemical immunosensor for human interleukin-6 (IL-6) and studies of IL-6 detection in HNSCC cells. They use SWNT forests with attached capture antibodies (Ab_1) for IL-6 in an electrochemical sandwich immunoassay protocol using enzyme label horseradish peroxidase (HRP) to measure very low and elevated levels of IL-6. They observe that secondary antibodies (Ab_2) attached to carboxylated multiwall carbon nanotubes with 106 HRP labels per 100 nm gave the highest sensitivity and the best detection limit (DL) for IL-6 in 10 μL of calf serum. They conclude that these immunosensors accurately measured secreted IL-6 in a wide range of HNSCC cells demonstrated by excellent correlations with standard enzyme-linked immunosorbent assays (ELISA).

Wang et al. (2011) discuss nanodiagnosisof HNSCC in a paper originating from the United States with 139 citations. They report using surface-enhanced Raman spectroscopy (SERS) to directly measure targeted circulating tumor cells (CTCs) in the presence of white blood cells. They find that SERS nanoparticles with EGF peptide as a targeting ligand successfully identified CTCs in the peripheral blood of 19 patients with HNSCC, with a range of CTCs per milliliter of whole blood.

Reuveni et al. (2011) discuss nanoimaging of HNSCC in a paper originating from the United States and Israel with 93 citations. They show in vivo the feasibility of cancer diagnosis based on molecular markers using clinical CT. After anti-EGF receptor conjugated gold nanoparticles were intravenously injected into nude mice implanted with HNSCC cells, they observe that a small tumor is enhanced and becomes clearly visible by the molecularly targeted gold nanoparticles. They further observe that active tumor targeting is more efficient and specific than passive targeting.

Wang et al. (2009) discuss nanotreatment of HNC cells in a paper originating from the United States with 89 citations. They synthesize a ternary conjugate heparin−folic acid−paclitaxel (HFT), loaded with additional paclitaxel (T) resulting in nanoparticle, HFT-T to retain the antitumor activity of paclitaxel and specifically target folate receptor (FR)-expressing tumors. They observe in vitro that HFT-T selectively recognizes FR-positive HNC cell line, displaying higher cytotoxicity compared to the free form of paclitaxel. In a subcutaneous HNC xenograft model, they further observe that HFT-T administration enhanced the specific delivery of paclitaxel into tumor tissues and remarkably improved antitumor efficacy of paclitaxel. HFT-T tumors did not have a resurgence of growth after several weeks of treatment. They conclude that HFT-T could be more effective in preventing tumors from developing drug resistance and is a promising strategy to enhance chemotherapy efficacy and minimize adverse effects.

Schroeder et al. (2007) discuss nanotreatment of HNC cells in a paper originating from Israel with 80 citations. They examine nanoparticles containing quantum dots (QDs) entrapped in a lipid shell, and postloaded with a folate-lipid conjugate for targeting these cells expressing the FR. They mix QDs with 1,2-dipalmitoyl-sn-glycero-3 phosphocholine and methoxy–polyethylene–glycol–distearoyl–phosphatidyl–ethanolamine (mPEG-DSPE) generating a nanoparticle called lipodot. Folate-derivatized PEG-DSPE was postloaded into the lipodots at 0.5% lipid molar concentration. They incubate HNC cells, upregulated for their FRss, with folate-targeted and nontargeted lipodots in vitro. They find that only folate-targeted lipodots were taken up by tumor cells with substantial internalization. Confirming the specificity of folate-targeted lipodots, they observe that binding and internalization were inhibited by free folate, and there was no uptake in a folate-receptor negative cell line. They conclude that folate-targeted lipodots represent an attractive approach for tumor cell labeling both in vitro and in vivo.

Damascelli et al. (2001) discuss nanotreatment of HNSCC (HNSCC) in a paper originating from Italy with 75 citations. They examine the feasibility, maximum tolerated dose, and toxicities of intraarterial administration of paclitaxel-albumin nanoparticles (PANs) in 31 patients with advanced HNSCC. They observe that intraarterial administration of PAN by percutaneous catheterization does not require premedication, is easy and reproducible, and has acceptable toxicity. Most dose levels showed considerable antitumor activity.

Hakim et al. (2011) discuss nanodiagnostics of HNC cells in a paper originating from Israel with six citations. They develop a cost-effective, fast, and reliable method for the differential diagnosis of HNC that is based on breath testing with a nanoscale artificial nose (NA-NOSE). They collect alveolar breath from 87 volunteers (HNC and lung cancer patients and healthy controls) in a cross-sectional clinical trial. They test the discriminative power of NA-NOSE based on an array of five gold nanoparticle sensors, using 62 breath samples. They find that the NA-NOSE could clearly distinguish between (1) HNC patients and healthy controls, (2) LC patients and healthy controls, and (3) HNC and LC patients.

2.3 THE CITATION CLASSICS IN ORAL CANCER NANOMEDICINE

El-Sayed et al. (2005) discuss nanodiagnostics of oral cancer cells in a paper originating from the United States with 994 citations. They record surface plasmon resonance (SPR) scattering images and SPR absorption spectra from both colloidal gold nanoparticles and from gold nanoparticles conjugated to monoclonal anti-EGF receptor (anti-EGFR) antibodies after incubation in cell cultures with both non-malignant malignant oral epithelial cell lines. They observe that the anti-EGFR antibody conjugated nanoparticles specifically and homogeneously bind to the surface of the cancer type cells with 600% greater affinity than to the noncancerous cells. This specific and homogeneous binding gives a relatively sharper SPR absorption band with a red shifted maximum compared to that observed when added to the noncancerous cells. They conclude that SPR scattering imaging or SPR absorption spectroscopy generated from antibody conjugated gold nanoparticles can be useful in molecular biosensor techniques for the diagnosis and investigation of oral epithelial living cancer cells in vivo and in vitro.

Huang et al. (2007a) discuss nanodiagnostics of oral cancer cells in a paper originating from the United States with 320 citations. They note that these cells assemble and align gold nanorods conjugated to anti-EGFR antibodies, which exhibit strong Rayleigh (Mie) scattering useful for imaging. They find that molecules near the nanorods on the cancer cells give a Raman spectrum that is greatly enhanced due to the high surface plasmon field of the nanorod assembly in which their extended surface plasmon fields overlap, sharp due to a homogeneous environment, and polarized due to anisotropic alignments.

Oyelere et al. (2007) discuss nanodiagnostics of oral cancer cells in a paper originating from the United States with 226 citations. They conjugate gold nanorods covalently with a nuclear localization signal peptide through a thioalkyl-triazole linker and incubate with immortalized benign and oral cancer cell lines. Dark field light surface plasmon resonance (SPR) scattering images show that nanorods are located in both the cytoplasm and nucleus of both cell lines. Single cell micro-Raman spectra reveal enhanced Raman bands of the peptide, as well as molecules in the cytoplasm and the nucleus. Further, the Raman spectra reveal a difference between benign and cancer cell lines.

Huang et al. (2007b) discuss nanotreatment of oral cancer cells in a paper originating from the United States with 128 citations. They study selective and efficient laser photothermal therapy (PTT) by exploiting the nonlinear optical properties of aggregated spherical gold nanoparticles conjugated to anti-EGFR antibodies using short near-infrared (NIR) laser pulses. They synthesize and conjugate spherical gold nanoparticles to anti-EGFR antibodies to specifically target these cells. They observe that the laser power threshold for the photothermal destruction of cells after the nanoparticle treatment is 20 times lower than that required to destroy these cells in the normal PTT without nanoparticles. The number of destroyed cells is quadratically dependent on the laser power. The number of dead cells shows a nonlinear dependence on the concentration of gold nanoparticles that are specifically targeted to cancer cells. They argue that aggregated nanospheres are responsible for the observed enhanced photothermal destruction of the cells.

Muhammed et al. (2010) discuss nanoimaging of oral cancer cells in a paper originating from India with 124 citations. They synthesize a luminescent quantum cluster (QC) of gold with a quantum yield of ~4%by the core etching of mercaptosuccinic acid protected gold nanoparticles by bovine serum albumin (BSA), Au_{QC}@BSA. Luminescence of the QC is exploited as a "turn-off" sensor for Cu^{2+} ions and a "turn-on" sensor for glutathione detection. They show metal-enhanced luminescence (MEL) of this QC in the presence of silver nanoparticles with a ninefold maximum enhancement. They observe that folic acid (FA) conjugated Au_{QC}@BSA was internalized to a significant extent by oral cancer cells through FA–mediated endocytosis. They use the inherent luminescence of the internalized Au_{QC}@BSA in cancer cell imaging.

Retnakumari et al. (2010) discuss nanoimaging of oral cancer cells in a paper originating from India with 112 citations. They study molecular-receptor-targeted imaging of FR-positive cells using folic-acid-conjugated fluorescent Au_{25} nanoclusters (Au NCs). They synthesize highly fluorescent Au_{25} clusters by controlled reduction of Au^+ ions, stabilized in BSA, using a green-chemical reducing agent, ascorbic acid (vitamin C). They then conjugate these clusters with FA through amide linkage with the BSA shell. They observe that these bioconjugated clusters show excellent stability and fluorescence efficiency in phosphate buffer saline (PBS), indicating effective protection of nanoclusters by serum albumin during the bioconjugation reaction and cell–cluster interaction. They show receptor-targeted cancer detection using Au clusters on FR^{+ve} oral cancer cells where the FA-conjugated Au_{25} clusters were internalized in significantly higher concentrations compared to the negative control cell lines.

Stevens et al. (2004) discuss nanotreatment of oral cancer cells in a paper originating from the United States with 72 citations. They synthesize Paclitaxel-7-carbonyl-cholesterol (Tax-Chol), a prodrug of paclitaxel with increased lipophilicity, and evaluate it for incorporation into a lipid nanoparticle (LN) formulation. This nanostructure contains folate-polyethylene glycol-cholesterol (f-PEG-Chol) as a ligand that targets the tumor marker FR. They observe that FR-targeted LNs showed greater uptake and cytotoxicity in FR^{+ve} oral cancer cells than nontargeted LNs. They conclude that FR-targeted LNs containing Tax-Chol are a promising novel formulation for the treatment of FR^{+ve} oral cancer cells.

2.4 THE CITATION CLASSICS IN PHARYNGEAL CANCER NANOMEDICINE

Bae et al. (2005) discuss nanotreatment of pharyngeal cancer cellsin a paper originating from Japan with 270 citations. They prepare a new type of multifunctional polymeric micelle drug carrier for active intracellular drug delivery. In order to prepare such micelles, they synthesize self-assembling amphiphilic block copolymers, folate–poly(ethylene glycol)–poly(aspartate hydrazone adriamycin) [Fol–PEG–P(Asp-Hyd-ADR)] by installing a molecular promoter to enhance intracellular transport, folate (Fol), at the end of the shell-forming PEG chain and conjugating the anticancer drug, adriamycin (ADR), to the side chain of the core-forming PAsp segment through an acid-sensitive hydrazone bond. They observe that the tetrazolium dye method (MTT assay) using human pharyngeal cancer cells revealed that folate-bound micelles (FMA) significantly improved cell growth inhibitory activity in spite of a short exposure time due to the selective and strong interaction between folate molecules and their receptors. Subsequent flow cytometric analysis showed that cellular uptake of FMA significantly increased.

Choi et al. (2004) discuss nanoimaging of nasopharyngeal cancer cells in a paper originating from the United States with 178 citations. They use receptor-targeting to deliver contrast-producing superparamagnetic iron oxide (IO) nanoparticle to receptor-expressing tumors for in vivo magnetic resonance (MR) imaging. These nanoparticles were tethered with *N*-hydroxysuccinimide-folate and fluorescence isothiocyanate (FITC). They observe that the internalization of nanoparticles into targeted cells only occurred when IO was conjugated to folate and when the FRs are available and accessible on the cells. The endocytosis was efficient and rapid, as 97.5% cancer cells cultured with folate-FITC-IO showed FITC uptake after 1 h of incubation. In in vivo MR imaging, they observe an average intensity decrease of 38% from precontrast to postcontrast images of the tumor, which was about 3 times the intensity decrease observed at a nontumor-bearing muscle.

Oyewumi et al. (2004) discuss nanotreatment of nasopharyngeal cancer cells in a paper originating from the United States with 174 citations. They compare the cell uptake, biodistribution, and tumor retention of folate-coated and PEG-coated gadolinium (Gd) nanoparticles. They observe that Gd nanoparticles did not aggregate platelets or activate neutrophils. Both folate-coated and PEG-coated nanoparticles had comparable tumor accumulation. However, the cell uptake and tumor retention of folate-coated nanoparticles was significantly enhanced over PEG-coated nanoparticles. Thus, folate ligand coating facilitated tumor cell internalization and retention of Gd-nanoparticles in the tumor tissue.

Feng et al. (2010) discuss nanodiagnosis of nasopharyngeal cancer cells in a paper originating from China and Canada with 132 citations. They develop a SERS method for blood plasma biochemical analysis to develop a simple blood test for noninvasive nasopharyngeal cancer diagnosis. They mix silver nanoparticles (Ag NP) as the SERS-active nanostructures with blood plasma to enhance the Raman scattering signals of various biomolecular constituents, such as proteins, lipids, and nucleic acids. They observe interesting cancer specific biomolecular differences, including an increase in the relative amounts of nucleic acid, collagen, phospholipids, and phenylalanine and a decrease in the percentage of amino acids and saccharide contents in the blood plasma of nasopharyngeal cancer patients as compared to that of healthy subjects. They find features differentiated the nasopharyngeal cancer SERS spectra from normal SERS spectra with high sensitivity (90.7%) and specificity (100%). They show great potentials for developing SERS blood plasma analysis into a novel clinical tool for noninvasive detection of nasopharyngeal cancers.

Setua et al. (2010) discuss nanoimaging of nasopharyngeal cancer cells in a paper originating from India with 112 citations. They study targeted cancer imaging using rare-earth oxide nanocrystals, free from heavy metals (Cd, Se, Te, Hg, and Pb), showing bright red-fluorescence and magnetic resonance imaging (MRI). They prepare Y_2O_3 nanocrystals (YO NC) doped in situ with fluorescent (Eu^{3+}) and paramagnetic (Gd^{3+}) impurities and conjugated with a potential cancer targeting ligand, FA. They show cancer targeting capability of FA conjugated NCs on FR-positive FR^{+ve} human nasopharyngeal cancer cells (KB) with FR depressed KB (FRd) and FR negative FR^{-ve} lung cancer cells A549 as controls. They observe highly specific binding and cellular uptake of large concentration of FA conjugated NCs on FR^{+ve} cells compared to the controls. They emphasize unique bimodal contrast imaging capability, nontoxicity, and cancer targeting capability of multiple impurities doped rare-earth oxide nanocrystals.

Manzoor et al. (2009) discuss nanoimaging of nasopharyngeal cancer cells in a paper originating from India with 78 citations. They study a heavy-metal-free luminescent (QD based on doped zinc sulfide (ZnS), conjugated with a cancer-targeting ligand, FA, as a promising biofriendly system for targeted cancer imaging. They test the cytotoxicity of bare and FA conjugated QDs in vitro using normal lung fibroblast cell line (L929), folate-receptor-positive FR^{+ve} nasopharyngeal epidermoid carcinoma cell line (KB), and FR^{-ve} lung cancer cell line (A549). Both bare and FA-conjugated ZnS QDs elicited no apparent toxicity. They find that interaction of FA–QDs with different cell lines showed highly specific attachment of QDs in the FR^{+ve} cancer cell line, leaving others unaffected. They argue that the bright and stable luminescence of the QDs could be used to image both single cancer cells and colonies of cancer cells without affecting their metabolic activity and morphology. They recommend the use of nontoxic, Cd-, Te-, Se-, Pb-, and Hg-free luminescent QDs for targeted cancer imaging.

Guo et al. (2006) discuss nanotreatment of nasopharyngeal cancer cells in a paper originating from the United States with 69 citations. They conjugate the folate into adenosine 5'-monophosphate (AMP) by 1,6-hexanediamine linkages. They incorporate the folate–AMP into the 5'-end of bacteriophage phi29 motor pRNA. Utilizing the engineered left/right interlocking loops, they construct polyvalent dimeric pRNA nanoparticles using RNA nanotechnology to carry folate, a detection marker, and siRNA targeting at an antiapoptosis factor. They observe that incubation of nasopharyngeal epidermal cancer (KB) cells with the dimer resulted in its entry into cancer cells, and the subsequent silencing of the target gene. They conclude that such a protein-free RNA nanoparticle with undetectable antigenicity has a potential for repeated long-term administration for nasopharyngeal cancer as the effectiveness and specificity were confirmed by ex vivo delivery in the animal trial.

2.5 THE CITATION CLASSICS IN THYROID AND TRACHEAL CANCER NANOMEDICINE

Jungebluth et al. (2011) discuss nanotreatment of tracheal cancer cells in a paper originating from England, Germany, Iceland, and Sweden with 189 citations. They study the clinical transplantation of the tracheobronchial airway with a stem-cell-seeded bioartificial nanocomposite. After complete tumor resection the airway of the cancer patient was replaced with a tailored bioartificial nanocomposite previously seeded with autologous bone-marrow mononuclear cells via a bioreactor for 36 h. They observe an extracellular matrix-like coating and proliferating cells in the scaffold after the reseeding and bioreactor process. There were no major complications, and the patient was asymptomatic and tumor free 5 months after transplantation. Postoperatively, they detected a mobilization of peripheral cells displaying increased mesenchymal stromal cell phenotype, and upregulation of epoetin receptors, antiapoptotic genes, and biomarkers. They conclude that the bioreactor reseeding process and pharmacological-induced site-specific and graft-specific regeneration and tissue protection are key factors for successful clinical outcome.

Shen et al. (2009) discuss nanotreatment of thyroid and oral cancer cells in a paper originating from the United States and China with 78 citations. Polyethyleneimine (PEI) was covalently bonded to acid-treated multiwalled carbon nanotubes (MWCNTs) through amide bond formation. The amine groups of PEI on the surface of MWCNTs were then reacted with acetic anhydride or succinic anhydride to form MWCNTs with neutral or negative surface charges, respectively. In vitro cytotoxicity assays using both FRO cells (a human thyroid cancer cell line) and KB cells (a human oral epithelial cancer cell line), they observe that the biocompatibility of these functionalized MWCNTs is largely dependent on their surface potential. They find that neutral and negatively charged MWCNTs are nontoxic to both cell lines at a concentration up to a limit, whereas positively charged MWCNTs are toxic to FRO cells at a threshold. They conclude that that PEI-modified MWCNTs can be chemically modified to alter their surface charges and cytotoxicity, thereby significantly improving the biocompatibility of the materials with cancer cells.

3 CONCLUSIONS

The field of the oral cancer nanomedicine emerged as a distinct research area mostly during the last decade building on the underlying areas of cancer research, oral cancer research, and nanoresearch. Thus, this new research field shows the strong signs of interdisciplinarity and significant supply of research in meeting the increasing public demand in the coming years.

The HNC nanomedicine, oral cancer nanomedicine, and pharyngeal cancer nanomedicine emerged as the key cancer-specific research fronts among the citation classics in oral cancer nanomedicine. There were also signs of research interest in thyroid cancer nanomedicine and tracheal cancer nanomedicine. On the other hand, the nanodiagnosis, nanoimaging and nanotreatment of these cancer cells emerged as the key topical research fronts among these citation classics.

The citation classics presented in this chapter were helpful in highlighting important papers influencing the development of the research field, as well as in determining the key research fronts in this field.

Further research is recommended for the detailed studies including scientometric studies and citation classic studies with a special emphasis on neck cancer nanomedicine, oral cancer nanomedicine, and pharyngeal cancer nanomedicine, as well as the nanodiagnosis, nanoimaging and nanotreatment of these cancer cells. Similar studies are also recommended for cancer nanomedicine and oral cancer medicine regarding the scarcity of the bibliometric studies in these fields.

REFERENCES

Aerts, H.J.W.L., Velazquez, E.R., Leijenaar, R.T.H., Parmar, C., Grossmann, P., Cavalho, S., et al., 2014. Decoding tumour phenotype by noninvasive imaging using a quantitative radiomics approach. Nat. Commun. 5, 4006.

Ang, K.K., Harris, J., Wheeler, R., Weber, R., Rosenthal, D.I., Nguyen-Tan, P.F., et al., 2010. Human papillomavirus and survival of patients with oropharyngeal cancer. N. Engl. J. Med. 363, 24–35.

Ang, K.K., Zhang, Q., Rosenthal, D.I., Nguyen-Tan, P.F., Sherman, E.J., Weber, R.S., 2014. Randomized phase III trial of concurrent accelerated radiation plus cisplatin with or without Cetuximab for stage III to IV head and neck carcinoma: RTOG 0522. J. Clin. Oncol. 32, 2940–2950.

Bae, Y., Jang, W.D., Nishiyama, N., Fukushima, S., Kataoka, K., 2005. Multifunctional polymeric micelles with folate-mediated cancer cell targeting and pH-triggered drug releasing properties for active intracellular drug delivery. Mol. Biosyst. 1, 242–250.

Bhirde, A.A., Patel, V., Gavard, J., Zhang, G.F., Sousa, A.A., Masedunskas, A., et al., 2009. Targeted killing of cancer cells in vivo and in vitro with EGF-directed carbon nanotube-based drug delivery. ACS Nano 3, 307–316.

Bonner, J.A., Harari, P.M., Giralt, J., Azarnia, N., Shin, D.M., Cohen, R.B., et al., 2006. Radiotherapy plus Cetuximab for squamous-cell carcinoma of the head and neck. N. Engl. J. Med. 354, 567–578.

Cambrosio, A., Keating, P., Mercier, S., Lewison, G., Mogoutov, A., 2006. Mapping the emergence and development of translational cancer research. Eur. J. Cancer 42, 3140–3148.

Chaturvedi, A.K., Engels, E.A., Pfeiffer, R.M., Hernandez, B.Y., Xiao, W.H., Kim, E., et al., 2011. Human papillomavirus and rising oropharyngeal cancer incidence in the United States. J. Clin. Oncol. 29, 4294–4301.

Choi, H., Choi, S.R., Zhou, R., Kung, H.F., Chen, I.W., 2004. Iron oxide nanoparticles as magnetic resonance contrast agent for tumor imaging via folate receptor-targeted delivery. Acad. Radiol. 11, 996–1004.

Coccia, M., Wang, L., 2015. Path-breaking directions of nanotechnology-based chemotherapy and molecular cancer therapy. Technol. Forecast. Soc. Chang. 94, 155–169.

Cooper, D.S., Doherty, G.M., Haugen, B.R., Kloos, R.T., Lee, S.L., Mandel, S.J., et al., 2009. Revised American thyroid association management guidelines for patients with thyroid nodules and differentiated thyroid cancer. Thyroid 19, 1167–1214.

Damascelli, B., Cantu, G., Mattavelli, F., Tamplenizza, P., Bidoli, P., Leo, E., et al., 2001. Intraarterial chemotherapy with polyoxyethylated castor oil free paclitaxel, incorporated in albumin nanoparticles (ABI-007): phase I study of patients with squamous cell carcinoma of the head and neck and anal canal: preliminary evidence of clinical activity. Cancer 92, 2592–2602.

Davies, L., Welch, H.G., 2014. Current thyroid cancer trends in the United States. JAMA Otolaryngol Head Neck Surg. 140, 317–322.

Dong, X.F., Qiu, X.C., Liu, Q., Jia, J., 2013. Bibliometric analysis of nanotechnology applied in oncology from 2002 to 2011. Tumor Biol. 34, 3273–3278.

Dreyer, D.R., Park, S., Bielawski, C.W., Ruoff, R.S., 2010. The chemistry of graphene oxide. Chem. Soc. Rev. 39, 228–240.

El-Sayed, I.H., Huang, X.H., El-Sayed, M.A., 2005. Surface plasmon resonance scattering and absorption of anti-EGFR antibody conjugated gold nanoparticles in cancer diagnostics: applications in oral cancer. Nano Lett. 5, 829–834.

Feng, S.Y., Chen, R., Lin, J.Q., Pan, J.J., Chen, G.N., Li, Y.Z., et al., 2010. Nasopharyngeal cancer detection based on blood plasma surface-enhanced Raman spectroscopy and multivariate analysis. Biosens. Bioelectron. 25, 2414–2419.

Glynn, R.W., Lowery, A.J., Scutaru, C., O'Dwyer, T., Keogh, I., 2012. Laryngeal cancer: quantitative and qualitative assessment of research output, 1945-2010. Laryngoscope 122, 1967–1973.

Grajewski, S., Quarcoo, D., Uibel, S., Scutaru, C., Groneberg, D., Spallek, M., 2010. A scientometric analysis of leukoplakia and erythroplakia. Laryngo Rhino Otol. 89, 210–215.

Guo, S., Huang, F., Guo, P., 2006. Construction of folate-conjugated pRNA of bacteriophage phi29 DNA packaging motor for delivery of chimeric siRNA to nasopharyngeal carcinoma cells. Gene Ther. 13, 814–820.

Hagfeldt, A., Boschloo, G., Sun, L.C., Kloo, L., Pettersson, H., 2010. Dye-sensitized solar cells. Chem. Rev. 110, 6595–6663.

Hakim, M., Billan, S., Tisch, U., Peng, G., Dvrokind, I., Marom, O., et al., 2011. Diagnosis of head-and-neck cancer from exhaled breath. Br. J. Cancer 104, 1649–1655.

Hanahan, D., Weinberg, R.A., 2000. The hallmarks of cancer. Cell 100, 57–70.

Hanahan, D., Weinberg, R.A., 2011. Hallmarks of cancer: the next generation. Cell 144, 646–674.

Ho, Y.S., Satoh, H., Lin, S.Y., 2010. Japanese lung cancer research trends and performance in Science Citation Index. Intern. Med. 49, 2219–2228.
Huang, X.H., El-Sayed, I.H., Qian, W., El-Sayed, M.A., 2007a. Cancer cells assemble and align gold nanorods conjugated to antibodies to produce highly enhanced, sharp, and polarized surface Raman spectra: a potential cancer diagnostic marker. Nano Lett. 7, 1591–1597.
Huang, X., Qian, W., El-Sayed, I.H., El-Sayed, M.A., 2007b. The potential use of the enhanced nonlinear properties of gold nanospheres in photothermal cancer therapy. Lasers Surg. Med. 39, 747–753.
Huang, Y., Ma, J., Porter, A.L., Kwon, S., Zhu, D.H., 2015. Analyzing collaboration networks and developmental patterns of nano-enabled drug delivery (NEDD) for brain cancer. Beilstein J. Nanotechnol. 6, 1666–1676.
Iijima, S., 1991. Helical microtubules of graphitic carbon. Nature 354, 56–58.
Jemal, A., Bray, F., Center, M.M., Ferlay, J., Ward, E., Forman, D., 2011. Global cancer statistics. Cancer J. Clin. 61, 69–90.
Jungebluth, P., Alici, E., Baiguera, S., Le Blanc, K., Blomberg, P., Bozoky, B., et al., 2011. Tracheobronchial transplantation with a stem-cell-seeded bioartificial nanocomposite: a proof-of-concept study. Lancet 378, 1997–2004.
Konur, O., 2000. Creating enforceable civil rights for disabled students in higher education: an institutional theory perspective. Disabil. Soc. 15, 1041–1063.
Konur, O., 2002a. Access to employment by disabled people in the UK: is the Disability Discrimination Act working? Int. J. Discrimin. Law 5, 247–279.
Konur, O., 2002b. Access to nursing education by disabled students: rights and duties of nursing programs. Nurs. Educ. Today 22, 364–374.
Konur, O., 2002c. Assessment of disabled students in higher education: current public policy issues. Assess. Eval. High. Educ. 27, 131–152.
Konur, O., 2006a. Participation of children with dyslexia in compulsory education: current public policy issues. Dyslexia 12, 51–67.
Konur, O., 2006b. Teaching disabled students in higher education. Teach. High. Educ. 11, 351–363.
Konur, O., 2007a. A judicial outcome analysis of the Disability Discrimination Act: a windfall for the employers? Disabil. Soc. 22, 187–204.
Konur, O., 2007b. Computer-assisted teaching and assessment of disabled students in higher education: the interface between academic standards and disability rights. J. Comput. Assist. Lear. 23, 207–219.
Konur, O., 2011. The scientometric evaluation of the research on the algae and bio-energy. Appl. Energ. 88, 3532–3540.
Konur, O., 2012a. The evaluation of the global energy and fuels research: a scientometric approach. Ener. Educ. Sci. Tech. A 30, 613–628.
Konur, O., 2012b. What have we learned from the citation classics in energy and fuels: a mixed study. Ener. Educ. Sci. Tech. A 30 (si 1), 255–268.
Konur, O., 2012c. 100 citation classics in energy and fuels. Ener. Educ. Sci. Tech. A 30 (si 1), 319–332.
Konur, O., 2012d. The scientometric evaluation of the research on the production of bioenergy from biomass. Biomass Bioenergy 47, 504–515.
Konur, O., 2012e. The evaluation of the biogas research: a scientometric approach. Ener. Educ. Sci. Tech. A 29, 1277–1292.
Konur, O., 2012f. The evaluation of the research on the biohydrogen: a scientometric approach. Ener. Educ. Sci. Tech. A 29, 323–338.
Konur, O., 2012g. The evaluation of the research on the microbial fuel cells: a scientometric approach. Ener. Educ. Sci. Tech. A 29, 309–322.
Konur, O., 2012h. The evaluation of the research on the bioethanol: a scientometric approach. Ener. Educ. Sci. Tech. A 28, 1051–1064.
Konur, O., 2012i. The evaluation of the research on the biodiesel: a scientometric approach. Ener. Educ. Sci. Tech. A 28, 1003–1014.

Konur, O., 2012. The evaluation of the research on the biofuels: a scientometric approach. Ener. Educ. Sci. Tech. A 28, 903–916.
Konur, O., 2012. Prof. Dr. Ayhan Demirbas' scientometric biography. Ener. Educ. Sci. Tech. A 28, 727–738.
Konur, O., 2012. The evaluation of the biorefinery research: a scientometric approach. Ener. Educ. Sci. Tech. A 30 (si 1), 347–358.
Konur, O., 2012. The evaluation of the bio-oil research: a scientometric approach. Ener. Educ. Sci. Tech. A 30 (si 1), 379–392.
Konur, O., 2012. The evaluation of the educational research: a scientometric approach. Ener. Educ. Sci. Tech. B 4, 1935–1948.
Konur, O., 2012. Evaluation of the research on the social sciences in Turkey: a scientometric approach. Ener. Educ. Sci. Tech. B 4, 1893–1908.
Konur, O., 2012. The evaluation of the research on the arts and humanities in Turkey: a scientometric approach. Ener. Educ. Sci. Tech. B 4, 1603–1618.
Konur, O., 2012. The scientometric evaluation of the research on the deaf students in higher education. Ener. Educ. Sci. Tech. B 4, 1573–1588.
Konur, O., 2012. The scientometric evaluation of the research on the students with ADHD in higher education. Ener. Educ. Sci. Tech. B 4, 1547–1562.
Konur, O., 2013a. The scientometric evaluation of the institutional research. The Marmara Universities: part 4. Ener. Educ. Sci. Tech. B 5, 365–380
Konur, O., 2013b. The scientometric evaluation of the institutional research. The Marmara Universities: part 3. Ener. Educ. Sci. Tech. B 5, 349–364
Konur, O., 2013c. The scientometric evaluation of the institutional research. The Marmara Universities: part 2. Ener. Educ. Sci. Tech. B 5, 333–348.
Konur, O., 2013d. The scientometric evaluation of the institutional research. The Marmara Universities: part 1. Ener. Educ. Sci. Tech. B 5, 317–332.
Konur, O., 2013e. The scientometric evaluation of the institutional research. The Ege universities: part 3. Ener. Educ. Sci. Tech. B 5, 83–98.
Konur, O., 2013f. The scientometric evaluation of the institutional research. The Ege Universities: part 2. Ener. Educ. Sci. Tech. B 5, 67–82.
Konur, O., 2013g. The scientometric evaluation of the institutional research. The Ege Universities: part 1. Ener. Educ. Sci. Tech. B 5, 51–66.
Konur, O., 2013h. The scientometric evaluation of the institutional research. The Akdeniz Universities: part 3. Ener. Educ. Sci. Tech. B 5, 151–166.
Konur, O., 2013i. The scientometric evaluation of the institutional research. The Akdeniz Universities: part 2. Ener. Educ. Sci. Tech. B 5, 135–150.
Konur, O., 2013. The scientometric evaluation of the institutional research. The Akdeniz Universities: part 1. Ener. Educ. Sci. Tech. B 5, 119–134.
Konur, O., 2013. The scientometric evaluation of the institutional research. The Inner Anatolian Universities: part 4. Ener. Educ. Sci. Tech. B 5, 267–282.
Konur, O., 2013. The scientometric evaluation of the institutional research. The Inner Anatolian Universities: part 3. Ener. Educ. Sci. Tech. B 5, 251–266.
Konur, O., 2013. The research on the attitudes toward disabled people in the educational settings: a scientometric evaluation. Ener. Educ. Sci. Tech. B 5, 483–498.
Konur, O., 2013. The policies and practices for the academic assessment of blind students in higher education and professions. Ener. Educ. Sci. Tech. B 5, 463–472.
Konur, O., 2015a. Algal biosorption of heavy metals from wastes. In: Kim, S.K., Lee, C.G. (Eds.), Marine Bioenergy: Trends and Developments. CRC Press, Boca Raton, FL, pp. 597–626.
Konur, O., 2015b. Algal economics and optimization. In: Kim, S.K., Lee, C.G. (Eds.), Marine Bioenergy: Trends and Developments. CRC Press, Boca Raton, FL, pp. 691–716.

Konur, O., 2015c. Algal high-value consumer products. In: Kim, S.K., Lee, C.G. (Eds.), Marine Bioenergy: Trends and Developments. CRC Press, Boca Raton, FL, pp. 653–682.
Konur, O., 2015d. Algal photobioreactors. In: Kim, S.K., Lee, C.G. (Eds.), Marine Bioenergy: Trends and Developments. CRC Press, Boca Raton, FL, pp. 81–108.
Konur, O., 2015e. Algal photosynthesis, biosorption, biotechnology, and biofuels. In: Kim, S.K. (Ed.), Springer Handbook of Marine Biotechnology. Springer, Berlin, pp. 1131–1161.
Konur, O., 2015f. Current state of research on algal biodiesel. In: Kim, S.K., Lee, C.G. (Eds.), Marine Bioenergy: Trends and Developments. CRC Press, Boca Raton, FL, pp. 487–512.
Konur, O., 2015g. Current state of research on algal bioelectricity and algal microbial fuel cells. In: Kim, S.K., Lee, C.G. (Eds.), Marine Bioenergy: Trends and Developments. CRC Press, Boca Raton, FL, pp. 527–556.
Konur, O., 2015h. Current state of research on algal bioethanol. In: Kim, S.K., Lee, C.G. (Eds.), Marine Bioenergy: Trends and Developments. CRC Press, Boca Raton, FL, pp. 217–244.
Konur, O., 2015i. Current state of research on algal biohydrogen. In: Kim, S.K., Lee, C.G. (Eds.), Marine Bioenergy: Trends and Developments. CRC Press, Boca Raton, FL, pp. 393–422.
Konur, O., 2015. Current state of research on algal biomethane. In: Kim, S.K., Lee, C.G. (Eds.), Marine Bioenergy: Trends and Developments. CRC Press, Boca Raton, FL, pp. 273–302.
Konur, O., 2015. Current state of research on algal biomethanol. In: Kim, S.K., Lee, C.G. (Eds.), Marine Bioenergy: Trends and Developments. CRC Press, Boca Raton, FL, pp. 327–370.
Konur, O., 2015. The scientometric study of the global energy research. In: Prasad, R., Sivakumar, S., Sharma, U.C. (Eds.), Energy Science and Technology V. 1. Opportunities and Challenges. Studium Press LLC, Houston, TX, pp. 475–489.
Konur, O., 2015. The review of citation classics on the global energy research. In: Prasad, R., Sivakumar, S., Sharma, U.C. (Eds.), Energy Science and Technology V. 1. Opportunities and Challenges. Studium Press LLC, Houston, TX, pp. 490–526.
Konur, O., 2016a. Algal omics: the most cited papers. In: Kim, S.K. (Ed.), Marine Omics: Principles and Applications. CRC Press, Roca Baton, FL, pp. 9–34.
Konur, O., 2016b. Glycoscience: the current state of the research. In: Kim, S.K. (Ed.), Marine Glycobiology: Principles and Applications. CRC Press, Boca Raton, FL, pp. 7–21.
Konur, O., 2016c. The scientometric overview in nanobiodrugs. In: Holban, A.M., Grumezescu, A.M. (Eds.), Nanoarchitectonics for Smart Delivery and Drug Targeting. Elsevier, Amsterdam, pp. 405–428.
Konur, O., 2016d. The scientometric overview in cancer targeting. In: Holban, A.M., Grumezescu, A.M. (Eds.), Nanoarchitectonics for Smart Delivery and Drug Targeting. Elsevier, Amsterdam, pp. 871–895.
Konur, O., 2016e. Scientometric overview regarding the surface chemistry of nanobiomaterials. In: Grumezescu, A.M. (Ed.), Surface Chemistry of Nanobiomaterials: Applications of Nanobiomaterials. Elsevier, Amsterdam, pp. 463–486.
Konur, O., 2016f. Scientometric overview regarding the nanobiomaterials in antimicrobial therapy. In: Grumezescu, A.M. (Ed.), Nanobiomaterials in Antimicrobial Therapy: Applications of Nanobiomaterials. Elsevier, Amsterdam, pp. 511–535.
Konur, O., 2016g. Scientometric overview regarding the nanobiomaterials in dentistry. In: Grumezescu, A.M. (Ed.), Nanobiomaterials in Dentistry: Applications of Nanobiomaterials. Elsevier, Amsterdam, pp. 425–453.
Konur, O., 2016h. Scientometric overview regarding nanoemulsions used in the food industry. In: Grumezescu, A. (Ed.), Emulsions: Nanotechnology in the Agri-Food Industry. Elsevier, Amsterdam, pp. 689–711.
Konur, O., 2016i. Scientometric overview in food nanopreservation. In: Grumezescu, A.M. (Ed.), Food Preservation: Nanotechnology in the Agri-Food Industry. Elsevier, Amsterdam, pp. 703–729.
Konur, O., 2017a. Scientometric overview in nanopesticides. In: Grumezescu, A.M. (Ed.), New Pesticides and Soil Sensors: Nanotechnology in the Agri-Food Industry. Elsevier, Amsterdam, pp. 719–744.
Konur, O., 2017b. Scientometric overview regarding water nanopurification. In: Grumezescu, A. (Ed.), Water Purification: Nanotechnology in the Agri-Food Industry. Elsevier, Amsterdam, pp. 693–716.

Konur, O., 2017c. Recent citation classics in antimicrobial nanobiomaterials. In: Ficai, A., Grumezescu, A. (Eds.), Nanostructures for Antimicrobial Therapy. Elsevier, Amsterdam.

Konur, O., 2017d. The top citation classics in alginates for biomedicine. In: Venkatesan, J., Anil, S., Kim, S.K. (Eds.), Seaweed Polysaccharides: Isolation, Biological and Biomedical Applications. Elsevier, Amsterdam.

Konur, O., 2018a. Citation classics in nanodrug delivery by dendrimers. In: Sharma, A.K., Keservani, R.K. (Eds.), Dendrimers for Drug Delivery. Apple Academic Press, Waretown, NJ.

Konur, O., 2018b. Citation classics in nanodrug delivery by nanopolymers. In: Keservani, R.K. (Ed.), Nanoconjugate Nanocarriers for Drug Delivery. Apple Academic Press, Waretown, NJ.

Konur, O., 2018c. Citation classics in nanodrug delivery. In: Keservani, R.K. (Ed.), Nanoconjugate Nanocarriers for Drug Delivery. Apple Academic Press, Waretown, NJ.

Konur, O., 2018d. The state of the research in algal drugs. In: Kim, S.K. (Ed.), Encyclopedia of Marine Biotechnology. Wiley-Blackwell, Oxford.

Konur, O., 2018e. The state of the research in algal genomics. In: Kim, S.K. (Ed.), Encyclopedia of Marine Biotechnology. Wiley-Blackwell, Oxford.

Lawrence, M.S., Stojanov, P., Mermel, C.H., Robinson, J.T., Garraway, L.A., Golub, T.R., et al., 2014. Discovery and saturation analysis of cancer genes across 21 tumour types. Nature 505, 495–501.

Li, L.K., Yu, Y.J., Ye, G.J., Ge, Q.Q., Ou, X.D., Wu, H., et al., 2014. Black phosphorus field-effect transistors. Nat. Nanotechnol. 9, 372–377.

Liu, D.Y., Kelly, T.L., 2014. Perovskite solar cells with a planar heterojunction structure prepared using room-temperature solution processing techniques. Nat. Photon. 8, 133–138.

Liu, H., Neal, A.T., Zhu, Z., Luo, Z., Xu, X.F., Tomanek, D., et al., 2014. Phosphorene: an unexplored 2D semiconductor with a high hole mobility. ACS Nano 8, 4033–4041.

Lu, K.D., He, C.B., Lin, W.B., 2014. Nanoscale metal-organic framework for highly effective photodynamic therapy of resistant head and neck cancer. J. Am. Chem. Soc. 136, 16712–16715.

Ma, D., Lin, Q.M., Zhang, L.M., Liang, Y.Y., Xue, W., 2014. A star-shaped porphyrin-arginine functionalized poly(L-lysine) copolymer for photo-enhanced drug and gene co-delivery. Biomaterials 35, 4357–4367.

Ma, J., Porter, A.L., Aminabhavi, T.M., Zhu, D.H., 2015. Nano-enabled drug delivery systems for brain cancer and Alzheimer's disease: research patterns and opportunities. Nanomedicine 11, 1763–1771.

Malhotra, R., Patel, V., Vaque, J.P., Gutkind, J.S., Rusling, J.F., 2010. Ultrasensitive electrochemical immunosensor for oral cancer biomarker IL-6 using carbon nanotube forest electrodes and multilabel amplification. Anal. Chem. 82, 3118–3123.

Manzoor, K., Johny, S., Thomas, D., Setua, S., Menon, D., Nair, S., 2009. Bio-conjugated luminescent quantum dots of doped ZnS: a cyto-friendly system for targeted cancer imaging. Nanotechnology 20, 065102.

Mela, G.S., Cimmino, M.A., Ugolini, D., 1999. Impact assessment of oncology research in the European Union. Eur. J. Cancer 35, 1182–1186.

Muhammed, M.A.H., Verma, P.K., Pal, S.K., Retnakumari, A., Koyakutty, M., Nair, S., et al., 2010. Luminescent quantum clusters of gold in bulk by albumin-induced core etching of nanoparticles: Metal ion sensing, metal-enhanced luminescence, and biolabeling. Chem. Eur. J. 16, 10103–10112.

North, D., 1994. Economic-performance through time. Am. Econ. Rev. 84, 359–368.

Novoselov, K.S., Geim, A.K., Morozov, S.V., Jiang, D., Zhang, Y., Dubonos, S.V., et al., 2004. Electric field effect in atomically thin carbon films. Science 306, 666–669.

Oregan, B., Gratzel, M., 1991. A low-cost, high-efficiency solar-cell based on dye-sensitized colloidal TiO_2 films. Nature 353, 73–740.

Oyelere, A.K., Chen, P.C., Huang, X.H., El-Sayed, I.H., El-Sayed, M.A., 2007. Peptide-conjugated gold nanorods for nuclear targeting. Bioconj. Chem. 18, 1490–1497.

Oyewumi, M.O., Yokel, R.A., Jay, M., Coakley, T., Mumper, R.J., 2004. Comparison of cell uptake, biodistribution and tumor retention of folate-coated and PEG-coated gadolinium nanoparticles in tumor-bearing mice. J. Control. Release 95, 613–626.

Pignon, J.P., Bourhis, J., Domenge, C., Designe, L., NC, M.A.C.H., Collaborative Grp, 2000. Chemotherapy added to locoregional treatment for head and neck squamous-cell carcinoma: three meta-analyses of updated individual data. Lancet 455, 949–955.

Popovtzer, R., Agrawal, A., Kotov, N.A., Popovtzer, A., Balter, J., Carey, T.E., et al., 2008. Targeted gold nanoparticles enable molecular CT imaging of cancer. Nano Lett. 8, 4593–4596.

Rafols, I., Meyer, M., 2010. Diversity and network coherence as indicators of interdisciplinarity: case studies in bio nanoscience. Scientometrics 82, 263–287.

Retnakumari, A., Setua, S., Menon, D., Ravindran, P., Muhammed, H., Pradeep, T., et al., 2010. Molecular-receptor-specific, non-toxic, near-infrared-emitting Au cluster-protein nanoconjugates for targeted cancer imaging. Nanotechnology 21.

Reuveni, T., Motiei, M., Romman, Z., Popovtzer, A., Popovtzer, R., 2011. Targeted gold nanoparticles enable molecular CT imaging of cancer: an in vivo study. Int. J. Nanomed. 6, 2859–2864.

Schroeder, J.E., Shweky, I., Shmeeda, H., Banin, U., Gabizon, A., 2007. Folate-mediated tumor cell uptake of quantum dots entrapped in lipid nanoparticles. J. Control. Release 124, 28–34.

Schummer, J., 2004. Multidisciplinarity, interdisciplinarity, and patterns of research collaboration in nanoscience and nanotechnology. Scientometrics 59, 425–465.

Setua, S., Menon, D., Asok, A., Nair, S., Koyakutty, M., 2010. Folate receptor targeted, rare-earth oxide nanocrystals for bi-modal fluorescence and magnetic imaging of cancer cells. Biomaterials 31, 714–729.

Shen, M.W., Wang, S.H., Shi, X.Y., Chen, X.S., Huang, Q.G., Petersen, E.J., et al., 2009. Polyethyleneimine-mediated functionalization of multiwalled carbon nanotubes: synthesis, characterization, and in vitro toxicity assay. J. Phys. Chem. C 113, 3150–3156.

Siegel, R., Naishadham, D., Jemal, A., 2012. Cancer Statistics, 2012. Cancer J. Clin. 62, 10–29.

Siegel, R., Ma, J.M., Zou, Z.H., Jemal, A., 2014a. Cancer Statistics, 2014. Cancer J. Clin. 64, 9–29.

Siegel, R., DeSantis, C., Jemal, A., 2014b. Colorectal cancer statistics, 2014. Cancer J. Clin. 64, 104–117.

Sotudeh, H., Khoshian, N., 2014. Gender differences in science: the case of scientific productivity in nano science & technology during 2005-2007. Scientometrics 98, 457–472.

Stevens, P.J., Sekido, M., Lee, R.J., 2004. A folate receptor-targeted lipid nanoparticle formulation for a lipophilic paclitaxel prodrug. Pharm. Res. 21, 2153–2157.

Stransky, N., Egloff, A.M., Tward, A.D., Kostic, A.D., Cibulskis, K., Sivachenko, A., 2011. The mutational landscape of head and neck squamous cell carcinoma. Science 333, 1157–1160.

Sun, G.H., Houlton, J., Moloci, N.M., MacEachern, M.P., Bradford, C.R., Prince, M.E., et al., 2013a. Prospective head and neck cancer research: a four-decade bibliometric perspective. Oncologist 18, 584–591.

Sun, G.H., Houlton, J.J., MacEachern, M.P., Bradford, C.R., Hayward, R.A., 2013b. Influence of study sponsorship on head and neck cancer randomized trial results. J. Sci. Spec. Head Neck 35, 1515–1520.

Therasse, P., Arbuck, S.G., Eisenhauer, E.A., Wanders, J., Kaplan, R.S., Rubinstein, L., et al., 2000. New guidelines to evaluate the response to treatment in solid tumors. J. Natl. Cancer Inst. 92, 205–216.

Villanueva-Felez, A., Woolley, R., Canibano, C., 2015. Nanotechnology researchers' collaboration relationships: a gender analysis of access to scientific information. Soc. Stud. Sci. 45, 100–129.

Wang, X., Li, J., Wang, Y.Q., Cho, K.J., Kim, G., Gjyrezi, A., et al., 2009. HFT-T, a targeting nanoparticle, enhances specific delivery of paclitaxel to folate receptor-positive tumors. ACS Nano 3, 3165–3174.

Wang, X., Qian, X.M., Beitler, J.J., Chen, Z.G., Khuri, F.R., Lewis, M.M., et al., 2011. Detection of circulating tumor cells in human peripheral blood using surface-enhanced Raman scattering nanoparticles. Cancer Res. 71, 1526–1532.

Wang, B.K., Wang, J.H., Liu, Q., Huang, H., Chen, M., Li, K.Y., et al., 2014. Rose-bengal-conjugated gold nanorods for in vivo photodynamic and photothermal oral cancer therapies. Biomaterials 35, 1954–1966.

Yella, A., Lee, H.W., Tsao, H.N., Yi, C.Y., Chandiran, A.K., Nazeeruddin, M.K., et al., 2011. Porphyrin-sensitized solar cells with cobalt (II/III)-based redox electrolyte exceed 12 percent efficiency. Science 334, 629–634.

Zhou, P., Leydesdorff, L., 2006. The emergence of China as a leading nation in science. Res. Policy 35, 83–104.

Index

A

Abraxane, 3
Absorption, 624
  enhancers, 95
Absorptive mediated transcytosis (AMT), 599
Absorptive transcytosis, 608, 624
Accelerated blood clearance phenomenon, 526
N-Acetyl-D-glucosamine, 415
N-Acetylglucosamine, 418
Acidic gastric environment, 409
Acidogenic bacteria, 839
Acid-PEG microspheres, 418
Acrylic copolymers (Eudragit), 477
Active pharmaceutical ingredients (API), 700
  slow sustained release of, 721–723
Acyclovir, 176, 352, 425
Adaptive immunity, 297
  second line defense, as, 297
Adeno-associated virus, 132
Adenocarcinoma, 155
Adenosine triphosphate (ATP), 579
Adipic acid dihydrazide (ADH) modified hyaluronic acid, 418
Adjuvants, nanosystems for, 314
AFM. *See* Atomic force microscopy (AFM)
Agglomeration, 50
Aggregatibacter actinomycetemcomitans, 251
Ag-NPs-coated polyurethane condom (PUC), 354
Air blast test, 872
β−Alanine spacers, 504
Albumin, 600
Alginate, 38, 426
Alginate-based nanocarriers, 426
Alginic acid, 367
Alkaline extraction
  efficient utilization of plant resources by, 562–564
    effect of $FeSO_4$ on anti-HIV activity of LCC, 564
    green tea leaves, 563
    licorice root, 563
    Sasa senanensis Rehder leaf, 562–563
ALS. *See* Amyotrophic lateral sclerosis (ALS)
Alternating current voltammetry, 895
Alternative magnetic field (AMF), 625
Alzheimer's disease (AD), 62, 582, 901
Amalgam, 788
Amantadine, 63
Amelogenin, 847
AMF. *See* Alternative magnetic field (AMF)
Amitriptyline, 217
Amorphization, 712
Amphiphilic chitosan derivatives, 416
Amphiphilic copolymeric nanocontainers formation
  synthesis scheme, 13
Amphiphilic PHAs synthesis, 10
Amphiphilic polymers, 484
Amphiphlic block copolymers, 2
Amphotericin B, 2
AMT. *See* Absorptive mediated transcytosis (AMT)
Amyloid-β peptide, 78
β-Amyloid protein, 63
Amyotrophic lateral sclerosis (ALS), 63
Analytical ultracentrifugation (AUC), 344
Angiogenesis, 421
Antenna cleaner of ants, 837
Anthocyanidins, 640
Antibacterial agents, 805
Antibacterial nanomaterials
  applications in dental medicine, 792
Antibiotics, 788
Anticancer drugs
  specific consideration for oral delivery, 476–477
    dose adaptation, 476
    drug metabolism, 477
    intestinal efflux pumps, 477
    intestinal transit time, 477
    poor drug solubility, 476
      pharmaceutical excipients, role of, 476
      water miscible cosolvents, role of, 476
    therapeutic window, 476
Anticancer polyphenolic compound, 619
Anticancer treatments, 352
Anticarcinogenic effects, 582
Anti-EGFR conjugated colloidal gold, 533
Anti-EGFR (C225)-conjugated hollow gold nanospheres, 533
Anti- EGFR monoclonal antibody, 524
Antigen-antibody complexes, 296
Antigenicity, 610
Antigen presenting cells (APCs), 266, 298
  adaptive immune response, role in, 298
Antigen uptake, 266
  M cell mediated antigen uptake, 266
Antiherpesviral microbicide, 352
Anti-HSV activity, 567
Antiinflammatory agents, 578
Antiinflammatory antibodies (IgA), 299
Antioxidants, 578, 582, 640
  mechanism of action, 582–584
  natural, 584
Antiprotozoan activity, 656

Antiretroviral (ARV)
exist barriers, 45
Antiretrovirus drugs, 558
APCs. *See* Antigen presenting cells (APCs)
Apigenin, 240
Apoptotic markers
internalization and activation, 19
schematic representation, 19
Apparent viscosity, 378
Aprotinin, 453
Aqueous compatibility, 379
Aromatic polymer, 546
ART. *See* Atraumatic restorative treatment (ART)
Artificial lipoprotein-like particles, 501
Ascorbic acid, 582, 916
Astragalus membranaceus, 320
Atenolol, 176, 217, 391, 397–400
Atomic force microscopy (AFM), 51, 344, 805, 835
probe, 51
Atomic/molecular condensation, 782
Atorvastatin, 629
ATP. *See* Adenosine triphosphate (ATP)
ATP-binding cassette (ABC) transporters, 78, 477, 526
Atraumatic restorative treatment (ART), 843
Atrial fibrillation, 580
Au nanorods, 529
Au nanoshells, 529
conjugated ^{64}Cu radionuclide, 532
Autohydrolysis, 124
Azathioprine, 901

B

Bacillus Calmette-Guérin (BCG), 299
Bacterial acids, 839
Bacterial biofilm growth, 877
Bacterial cell, 805
Bacterial morphology, 805
Bacterial polyester nanoparticles
drug delivery, role in
drawbacks and challenges, 19
Basaloid squamous cell carcinoma, 514
Basolateral macrophages, 155
BBB. *See* Blood-brain barrier (BBB)
B cells, 296
BCS class II
definition, 217
examples of, 217
Benzoic acid, 383
Beta-glucosidase, 628
Bimetallic nanoparticles (Ag/CuNPs) synthesis, 343–344
copper (II) nitrite, role of, 343
29 nm Ag/CuNPs, 343
63 nm Ag/CuNPs, 344
silver reduction, role of, 343
sodium citrate and tannic acid mixture, role of, 343
Bioactive glasses, 810, 852
nanoparticles, 791
Bioactives, 589
with antioxidant, antiinflammatory, and antistroke effects, 584–585
limitations to elicit desired therapeutic response, 589
Bioactive targeting, 28
Bioavailability, 124, 151, 168, 578, 640
Bioceramics, 809, 810, 852, 875
Biochips, 864
Biocompatibility, 578, 590, 608
thermoplastics, 7
Biodegradable, 124, 578
amphiphilic nanoparticles, 2
bacterial polyesters
poly(R)-hydroxyalkanoic acids (PHA), 4–7
biodegradation, 6
drug carrier, as, 7
general applications, 6–7
properties, 5
coatings, 609
glycerides, 592
micelles
insulin-phospholipid complex, loaded with, 40
polymeric materials, 64–65
natural polymers, 64–65
synthetic polymers, 65
polymerization anhydride, 126
Biofate, 651
Biofilms, 864
Bioflavnoids, 584
Biohybrid nanopolymers, 759
Bio-hybrid polymer nanotubes, 760
Biological host defense system, 296
Biological membranes, 385
interaction with, 385
Biological polymers, 412
Biomaterials
device-based therapies, 7
medical imaging, 7
tissue engineering, use in, 7
Biomimetic approach, 757
Biomolecules
encapsulation, 608
Bionanotechnology
drug delivery application, 3
Biopharmaceutical classification system (BCS), 175, 217, 232, 469
applications of pellet, dissolution rate of drugs in BCS class II, 217
drug classification, 234

Biosensors, 805, 863
Biotherapeutic macromolecules, 302
Biotransformation, 139, 578
Bis-EMA. *See* Ethoxylated-bisphenol-A-dimethacrylate (Bis-EMA)
Bis-GMA. *See* Bisphenol A-glycidyl dimethacrylates (Bis-GMA)
Bisphenol A-glycidyl dimethacrylates (Bis-GMA), 841
Bleaching agent, 869, 871
Bleaching sensitivity, 871
Block copolymers
  micelle-forming, 2
Blood-brain barrier (BBB), 62, 590, 682
B-lymphocytes, 125
  antigen-binding receptor, with, 298
  maturation, 298
BMMSC. *See* Bone marrow mesenchymal stem cells (BMMSC)
Bone allograft, 878
Bone marrow mesenchymal stem cells (BMMSC), 810
Bone phosphate of lime (BPL), 810
Bone regeneration, 811
Bone replacement material, 877
Bone tissue, 780
  cortical bone, 780
  nanomedicine, for bone healing process, 780
  physiological properties of, 780
  regeneration of large bone defects, 780
  trabecular bone, 780
Boronic acid-functionalized gold nanoparticles (B-AuNPs), 726
Bottom-up approach, 752, 753
  applications based on, 753
  cancer diagnosis, 755
  chemical synthesis, 752
  dental durability and appearance, 754
  dentrifrobots, 754
  hypersensitivity cure, 754
  major tooth repair, 754
  micellar arrangements, 752
  molecular fabrication, 752
  nanoanaesthesia, 753
  polymerization, 752
  precipitation, 752
  self-assembly, 752
  therapeutic aid, in oral diseases, 755
    drug delivery, 755
    gene therapy, 755
  tooth repositioning/orthodontic nanorobots, 754
Bovine insulin based on pH-sensitive ZrBPMP
  schematic illustration, of time- and pH-controlled oral colon-targeted delivery system, 733
Bovine serum albumin (BSA), 279, 426
  loaded PLGA NPs, 313
Bovine spongiform encephalopathy (BSE), 131
Bowel syndrome, 723
BPL. *See* Bone phosphate of lime (BPL)
Brownian motion, 50
BSE. *See* Bovine spongiform encephalopathy (BSE)
Buccal films, preparation, 102
Buccal mucosa chemical permeation enhancers
  chemical structure, mechanisms, and biopharmaceutical applications, 94–100
    chelators, 98
    cyclodextrins, 98
    fatty acids, 98
    polymers, chitosan and its derivatives and associations, 99–100
    surfactants and bile salts, 95–97
Buccal topical delivery, 102
Buckminsterfullerene, 917
Bupivacaine, 102
Burst effect, 17

C

Caco-2 cell lines, 155, 242, 481, 484, 502
Calcium-assisted ionic cross-linked nanoparticles, 413
Calcium carbonates, 788
Calcium ions, 426
Calcium phosphate nanoparticles, 413, 868, 873
  crystalline depositions in dental implants, 874
Calcium phosphate salts, 78[illegible]
Camphoquinone, 843
Camptotheca acuminate, 55
Camptothecin (CPT), 77, 242, 551
  loaded PLGA microspheres, 310
Cancer detection and screening system, 523
Cancer management
  see and treat strategy, 533
Cancerous/noncancerous cells
  chemotherapy
    cellular uptake of, 526
  NP-based nanodrugs
    cellular uptake of, 526
  radioresistance, 517
Cancer research, 941–942
Cancer therapeutics
  nanocarrier system for oral delivery, 478–504
    dendrimers, 504
    lipidic nanoplatform systems, 499–504
      layersomes, 504
      lipid-drug conjugate, 500
      lipid nanocapsules, 502
      lipid nanospheres, 501

Cancer therapeutics (*cont.*)
liposomes, 503–504
microemulsion, 501
nanostructured lipid nanocarriers, 502–503
self-emulsifying drug delivery system (SEDDS), 500–501
solid lipid nanoparticles, 502–503
Cancer treatment and diagnosis
nanoparticles, role of, 516–523
carbon nanotubes (CNTs), 520
gold (Au) nanoparticles (NPs), 517–518
graphene oxide and graphene NPs, 519
magnetic nanoparticles (MNPs), 516–517
quantum dots (QDs), 523
Candida albicans, 103, 251, 549, 790
Cantilevers, 766
detection, biomarkers of oral cancer, 767
Capecitabine, 475
Capillarity number (Ca), 672
Capryol 90, 239
Captopril, 176
CAR. *See* Coxsackie and adenovirus receptor (CAR)
Carbamazepine (CBZ), 220
Carbamazepine-loaded SBA-15 mesoporous silica, 731
Carbines, 898
Carbomer-D matrices, 389
drug diffusion front movement in SDPM, 391
kinetics and release mechanism, 391
alginic acid-D matrices, 394–396
carbomer-D SDPM, 391–394
water uptake, 389
Carbon atom, electronic configurations of, 899
Carbon-based nanostructures, 898
in electrochemical analysis of oral medicines, 898
carbon nanodiamond, 919–920
carbon nanofibers, 905, 914–915
carbon nanohorns, 915–917
carbon nanotubes, 901–905
fullerenes, 917–919
graphene and graphene oxide, 899–901
unique element carbon, 898
Carbon element, 898
Carbon nanodiamonds, 886, 919–920
Carbon nanofibers, 905, 914–915
Carbon nanohorns, 886, 915–917
Carbon nanotubes (CNTs), 42, 195, 242, 284, 520, 795, 886, 901–905
fabrication, 520
chemical vapor deposition (CVD), 520
laser ablation, 520
plasma-enhanced CVD, 520
light absorption properties, 520
multiwalled CNT (MWCNT), 520
single-walled CNT (SWCNT), 520
Carboxymethyl-β−cyclodextrin (CMCD), 615, 625
Carboxymethylcellulose-D matrices, 398–399
Carboxymethyl CS, 609
Carrageenan-D matrices, 399–400
Carvedilol, 176
Casein phosphopeptide (CPP), 804
Case study, 221
coating of pellets, 221
comparison, of in vitro release profiles of drug, 222
dissolution tests, 222
reconstitution study, 222
results, 222
Cationic polysaccharide, 607
Caveolae-mediated endocytosis, 650
C-Azithromycin, 379
C57BL/6 mice, HSV-2 infection of, 355
tannic acid-modified AgNPs, effect of, 359
CCR5 antagonist, 353
CD44 receptors, 413, 420
Cefazolin antibiotics, 609
Cefodizime, 320
Celecoxib (CEL), 703, 715
Cell-bound mucins, 30
Cell lysis, 139
Cell penetrating peptide (CPP), 456
Cellular delivery, 147
Cellulose nanofibers, 566
Cellulose nanowhiskers (CNW), 566
Cellulosic fibers isolation, 566
alkali pulping process, 566
bleaching, 566
mechanical shearing, 566
Ceramic matrix nanocomposites (CMNC), 759
Cerasomes, 648
Cerebral paralysis, 63
Cerebral stroke, risk factors associated with, 581
Cerebral venous thrombosis (CVT), 581
Ceriops decandra, 560
Cervical lymphadenopathy, 513
Cervical micrometastasis, 515
Cetyltrimethylammonium bromide (CTAB), 701
Chebulagic acid (CHLA), 354
Chelation agents, 98
Chemical coupling reaction
PHA polymers and mPEG, 12
Chemical stability, 378, 388
Chemical vapor deposition (CVD), 782, 836, 900
Chemotherapeutic drugs, 48
Chemotherapy via oral administration
nanocarriers used, 489
CHF. *See* Coconut husk fibers (CHF)

Chitosan (CS), 148, 415, 607, 684, 791
based nanoparticles, novel technologies, 611–619
chitosan-based nanohydrogels, dialysis method, 618
CS-PMA NPs, spinning disk processor for, 617
emulsification and ultrasonication method, for TMC-*g*-PA SLNs, 613
magnetic nanohydrogels, in situ precipitation method, 614–615
self-assembled CS-tocol nanoparticles, 616–617
solid-liquid phase separation method, 613
solvent injection method, 613
synthesis by ionic gelation, 611–613
diclofenac sodium nanocores, 628
drug nanoconjugates, amide coupling reaction, 618–619
drug solubility, role in, 36
grafting of, 607
hydration, 620
limitations, 607
for oral drug delivery, 611
pharmaceutical application, 607
physicochemical characteristics, 684
structure, 684
swelling, 620
as targeted drug-delivery systems, 607
TIO_2-coated nanohybrid composites, sol-gel method, 619
Chitosan-alginate core-shell nanoparticles, 413
Chitosan chloride, 164
Chitosan (CS)-coated cationic nanocapsules, 482
Chitosan-methyl methacrylate (CS-M), 618
Chitosan nanoparticles, 413
lignosulfonates, incorporation of, 564
Chitosan nanoparticles (CN), 111
Chitosan-PAA-PTX-based liposome, 504
Chlorapatite, 788
Chloroquine, 217
Cholera toxin B (CTB), 125
Chondroitin sulfate, 413, 420, 424
Chondroitin sulfate-α- linolenic acid micelles, 420
Cimetidine, 176
Ciprofloxacin, 384, 386, 403
Cisplatin, 16, 529
Classic microporous materials, 701
Clathrin-mediated endocytosis, 650
Claudin-4, 416
Clean-room technology, 836
Clinical trials, 864–867
double-blind, 871
phase I, 865
phase II, 865
phase III, 865, 866
phase IV, 865, 866
preclinical, 865
Clotrimazole, 251
CLSM. *See* Confocal laser scanning microscopy (CLSM)
CMC. *See* Critical micelle concentration (CMC)
CMCD. *See* Carboxymethyl-β-cyclodextrin (CMCD)
CMK-1 mesoporous carbon, 703
CN. *See* Chitosan nanoparticles (CN)
CNTs. *See* Carbon nanotubes (CNTs)
CNW. *See* Cellulose nanowhiskers (CNW)
Coantioxidants synergism, 584
Coconut husk fibers (CHF), 556
Cocrystallization, 712
Coenzyme Q10 (CoQ10), 240
Cold water test, 872
Colon cancer, 723
Colonic fluid, 628
Colorectal cancer therapy, 500
Combidex, 531
Commercially available oral vaccines, 265
Compatibility, 388
Complexation, 185
Composite materials, 877
Condensed tannins, 336
Confocal laser scanning microscopy (CLSM), 599
Confocal microscopy imaging, 564
Congenital deformity, 808
Controlled release reservoirs, 589
Conventional chemotherapeutic agents, shortcomings, 524
Conventional drugs, 154
Copper alloys, antiviral activity, 355
Copper (Cu) nanoparticles, 790
antimicrobial activity, 350
Escherichia coli, 350
methicillin-resistant *Staphylococcus aureus*, 350
Corona effect, 351
Cortical bone, 780
Corticosteroids, for mouth sores treatment, 105
Cortisone, 300
Corynebacterium parvum, 299
Counterionic condensation, 373
affinity constants for, 374
Coupling agent, 842
Covalent-like association, 373
Coxsackie and adenovirus receptor (CAR), 134
C-Pilocarpine, 379
C-procaine, 379
Cremophor EL (CrEL)-based formulation, 468
Critical micelle concentration (CMC), 1, 40, 70, 609, 685
Crohn's disease, 157, 723
Cryo-electron microscopy (Cryo-EM), 597
Cryo-EM. *See* Cryo-electron microscopy (Cryo-EM)
cryoFESEM. *See* Cryo Field Emission Scanning Electron microscopy (cryoFESEM)
Cryo Field Emission Scanning Electron microscopy (cryoFESEM), 3[illegible]5

CryoTEM. *See* Cryo transmission electron microscopy(CryoTEM)
Cryo transmission electron microscopy(CryoTEM), 315
Crystalline PHA granules, 5
CS. *See* Chitosan (CS)
CS-M. *See* Chitosan-methyl methacrylate (CS-M)
CTB. *See* Cholera toxin B (CTB)
Curcuma longa, 109
Curcumin, 234, 306, 589, 654–656
Curcumin-loaded lipopolysaccharide nanocarriers (C-LPNCs), 500, 656
Curcumin SMEDDS, 322
CVT. *See* Cerebral venous thrombosis (CVT)
Cyclic voltammetry (CV), 893, 894
Cyclodextrins (CD), 88, 134, 423
  complexes, 476
  inclusion, 712
Cyclooxygenase (COX)-2 protein, 562
Cyclosporine, 299, 578, 621
  graft rejection, role in, 299
  immunosuppressant, as, 299
  renal toxicity, 302
Cyclosporine A, 41, 428
Cysteamine, 158
Cystectomy, 809
Cysteine trimethyl chitosan-carboxymethyl dextran nanoparticles, 428
Cytokines, as immunoadjuvants, 299
Cytomegalovirus, cytopathic effect, 563
Cytotoxicity, 140, 149, 590, 619
Cytotoxic T cells, 296

D

Danazol, 176, 217
DDS. *See* Drug-delivery systems (DDS)
DEAE-cellulose column chromatography, 547
DEAE Sepharose chromatography, 548
Degradative enzymes, 151
Deltamethrin, 562
Demineralization process, 839, 843
Dendrimers, 38–39, 62, 71–73, 285, 644
  based technologies, 412
  generation 4 structure, 72
  in oral drug delivery, 241, 242
    drug conjugation and, 249
    host guest interactions, bioactives and, 248
    list of drugs and effect, on solubility and stability with, 253
    as nanocarrier, through oral route, 248
    properties of, 247
      cores of generation 4 dendrimers with primary amines, as terminal functional groups, 247
      EDA Core PAMAM, influence of different terminal functionality of, 247
      enhanced permeability retention (EPR) effect, 247
      physico-chemical properties, 247
      viscosity, 247
    solubilizing potential of, 242
    synthetic approaches of, 243
      click chemistry approach, 244, 245
      convergent approach, 243, 245
      divergent approach, 243, 244
      double exponential approach, 244, 246
      lego chemistry approach, 244, 245
  platinum complex, 39
  synthesis, 38, 72
Dendritic architectures, as drug delivery systems, 39
Dendritic cells (DCs), 131, 266
Dengue virus (DENV), 354
Dental caries, 839, 847, 871, 875, 878
  prevention, 878
Dental composites, 844
  classification, 842
    macrofilled, 842
    microfilled, 842
    microhybrid, 842
Dental erosion, 803
Dental glass-ionomers cements (DGIC), 843, 846
Dental implants, 852, 868, 873
  OsseoSpeed implants, 854
  survival rate, 873
  titanium implants, 853
Dental materials, history and applications, 840
Dental plaque, 839
Dental restorations, 877
Dentifrices, 847
Dentine hypersensitivity, 871
Dentistry, nanostructures used in, 757
  properties of, 757
  types, 757
Dentures, 871
Deoxycholic acid (DOCA), 421
  derivatives, 421
Deoxyribonucleic acid (DNA), 671
Desensitizer, 872
Desmopressin, 455
Dexamethasone, 105, 299, 917
Dexamethasone phosphate, 381
Dextran, 428
Dextran sulfate, 614
DGIC. *See* Dental glass-ionomers cements (DGIC)
D-Glucuronic acid, 418
Diabetes
  diagnosis test for, 440
  plants used in treatment of, 441
  in world scenario, 441
Diabetes mellitus, 436

Dialysis bag diffusion, 598
Diamond, 886
Diblock copolymer
  nanoparticles self-assembly, 12–13
Diclofenac, 401, 879
  adverse effects, 879
Diclofenac sodium, 625
Diethylmethyl chitosan (DEMC), 415
DIF. *See* Differentiation-inducing factor (DIF)
Differential pulse voltammetry, 894, 914
Differential pulse voltammograms, 914
Differential scanning calorimetry (DSC), 320
Differentiation-inducing factor (DIF), 549
Diffusion coefficient, 376, 681
Diffusion-controlled drug release, 626
Dihydroxyphenylethylamine, 630
Diltiazem, 176
Dimethylethyl chitosan, 415
Dimethylformamide (DMF), 708
5,5-Dimethyl-1-pyrroline-*N*-oxide (DMPO), 560
Dimethylsulfoxide (DMSO), 708
Dioleoyl phosphatidylethanolamine, 597
Direct imaging technique, 49
Dissolution, 176
  cubic root law, 178
  definition, 176
  Fick's law of, 177
  mechanism of, 177
  Nernst- Brunner and Hixson-Crowell equation, 177
  Noyes-Whitney equation, 177
  physical phenomena, 177
  rate, 176
  stages, 177
β–D-mannuronic (M-block) acid, 38
DMPO. *See* 5, 5-Dimethyl-1-pyrroline-*N*-oxide (DMPO)
DNA delivery, 126
DNA fragmentation, 560
Docetaxel, 468
Docetaxel based nanosystems, 477
Docosahexaenoic acid, 500
Dopamine, 916
Dose adaptation, 475
Double-distilled water, 239
DOX. *See* Doxorubicin (DOX)
DOX-loaded poly(isohexylcyanoacrylate) nanoparticles, 3
Doxorubicin (DOX), 2, 416, 551, 607
  CNT loaded, 43
  functionalized dendrimer, 39
  loaded lipid nanoparticles, 503
  loaded PLGA nanoparticles (Dox-NPs), 483
  transdrug, 3
Drosophila, 128
Drug absorption, 609
Drug accumulation, 526
Drug bioavailability, 88
Drug/cyclodextrin interactions, 423
Drug-delivery systems (DDS), 28, 62, 88, 94, 174, 592, 607
  carriers, 104
    polymeric micelles as, 113
  controlled targeted release (CTR), 174
  cyclodextrins as carriers in, 102
  devices, 1
  drug release rate, 174
  efficacy of, 174
  hydration rate and wettability of tablets, 174
  nanostructured materials, 174
  nanostructures used, 37
Drug encapsulation, 68, 101
Drug entrapment, 623
Drug expulsion, 599
Drug loaded polymeric nanoparticles
  benefits, 4
Drug-loading techniques, 708
  cospray drying, 710
  impregnation and solvent evaporation, 709
  microwave irradiation procedure, 712
  solution immersion, 708
  spray drying, 710
  supercritical fluid (SCF) technique, 711
Drug nanocrystals-based formulations, 480
Drug oral bioavailability, 608
Drug release, 154, 380, 625
Drug residence time, 608
Drug-resistant bacteria, 805
Drugs
  absorption enhancers, 36
    categories, 36
  bioavailability, measurement of, 175, 176
    approaches for improvement, 183
    definition of, 183
    dissolution, 176–178
    permeability, 178–179
    solubility, 178
  biopharmaceutical classification system (BCS) of, 175
  carrier-mediated transport, 33
  chemical modification, 185
  complexation, 185
  dispersion in carriers, 184
  dissolution, 175
  endocytosis, 33
    phagocytosis, 33
    pinocytosis, 33
    potocytosis, 33
    receptor-mediated, 33
  formulation, of poorly water-soluble, 218
  maximum tolerable concentration (MTC), 47

Drugs (*cont.*)
micronization of, 218
modification, of crystal habit, 184
nanostructured lipid carriers, 41
permeability, 175
physical modification, pharmaceutics approach, 183
particle size reduction, 183–184
homogenizers, 184
micronization, 183
nanosuspension, 184
receptor-mediated transport, 33
retention, 608
solubility, 175
solubilization, 217, 609
by surfactants, 185
stability of oral drugs, improvement, 186
transport mechanism in GIT, 179
Drug-targeting systems, 62
DSC. *See* Differential scanning calorimetry (DSC)
Dual-purpose nanotheranostic agent, 533
Duchenne muscular dystrophy, 136
Dynamic light scattering (DLS), 344, 376, 377
Dysmorphia treatment, 841

E

EA-encapsulated CS nanoparticles (EA-CS-NPs), 619
Echinacea angustifolia, 306
Echinacea purpurea, 306
EE-dexamethasone phosphate-Cl complexes, 381
EE-D matrices, 400
Effector B cells (plasma cells), 298
Effector T cells, 296
EGFR. *See* Epithelial growth factor receptor (EGFR)
Electrical properties, 886
Electroformation, 688
Electrokinetic potentials, 375, 376
Electroneutrality, 387
Electron microscopy, 155, 345
Electron spin resonance (ESR), 560
Electrostatic attraction, 373
Electrostatic interactions, 385
Electrostatic repulsion, 625, 626
Emdogain, 512
periodontal treatment, use in, 512
Emend (aprepitant), 241
Emulsions, 185
Emulsion-solvent evaporation, 305
Emulsomes, 647
Enalapril, 385, 386
Enalapril maleate, 385
Enamel demineralization, 878
Encephalomyelitis virus, 137
Endocytosis, 151, 628
Endodontics, 848
Endogenous adjuvants, 299
histamine, 299
tuftsin, 299
Endorem, 531
Endosomes, 624
Endothelial cells, 809
Endothelial dysfunction, 578
Endothelin release, 578
Energy dispersive X-ray and spectroscopy (EDX), 49
Engineered nanomaterials (ENMs), 788
antimicrobial effect of, 791
size and shape of, 790
Enhanced permeability and retention (EPR), 164
mechanism, 641
Enoxaparin, 413
Enterococcus faecalis, 252, 790
Enterocytes, 134, 151
Enzymatic degradation, 547
Enzyme inhibitors, 453
Epidermoid carcinoma, 532
Epigallocatechin gallate, 561
Epigallocatechin-3-gallate, 651
Epithelial cells, 125, 149, 609
Epithelial growth factor receptor (EGFR), 513
Epithelium, 608
EPR. *See* Enhanced permeability and retention (EPR)
EPR effect, 3, 527
Erythroplakia, 513
malignant transformation, risk of, 513
Erythroxylum catuaba, 557
ESBL. *See* Extended spectrum beta lactamase (ESBL)
Escherichia coli, 252, 264, 549, 623, 789
Ethambutol, 217
Ethoxylated-bisphenol-A-dimethacrylate (Bis-EMA), 841
7-Ethyl-10-hydroxy-camptothecin (SN38), 242
Etoposide, 9, 476
Eudragit, 47, 410, 723
Eudragit S100 (ES), 47
microspheres, 454
Excitotoxicity, 578
Exfoliative cytology, 524
Exophytic tumors, 513
Extended spectrum beta lactamase (ESBL), 623
External magnetic field, 609
Extracellular matrix (ECM), 413, 780
Extraction solvents, 546
Extrusion spheronization process, pellets production, 205, 206
disadvantages of, 207
formulation, 211
excipients, for microcrystalline cellulose, 212, 213

microcrystalline cellulose, as spheronization aid, 211–212
use of other excipients, 212
physicochemical properties of the API, 205
process and equipment, 207
drying, 209
dry mixing and wet granulation, 207
extrusion, 207
friction plate design, 210
pellet quality, 209
photograph (plan view), of spheroniser plate in motion, 208
schematic overview, of extruder types applied in, 208
screening, 210
spheronization, 207
schematic overview of steps in, 206
schematic presentation, of different pellet formation stages, 211
steps of, 206

F

Factor Xa, 421
FAE. *See* Follicle-associated epithelium (FAE)
Familial adenomatous polyposis, 149
Fast-acting insulin analogs, 443
Aspart, 444
Glulisine, 444
Lispro, 444
Fasted state simulated intestinal fluid (FaSSIF), 422
Fc receptor, 47
Fed state simulated intestinal fluid (FeSSIF), 422
Fenofibrate, 176
Fenton reaction, 560, 564
Feridex, 531
Ferritin, 582
$FeSO_4$-treated LCC, 564
FFEM. *See* Freeze fracture electron microscopy (FFEM)
Fibronectin, 852
Fickian diffusion, 381
Fick's law, 177
Filtek P60, 843
Fissure sealants, 843
FITC-BSA-loaded chondroitin sulfate-chitosan nanoparticles, 420
FK448 inhibitors, 453
Flame spraying synthesis, 782
Flame spray pyrolysis, 841
Flavanols, 640
Flavonoids, 640
Flexion, 846
Fluorapatite, 788
Fluoride, anticariogenic effects of, 843
Fluoro alumino silicate (FAS) technology, 762
Fluoroquinolone antimicrobials, 385
Fluorouracil, 485
5-Fluorouracil (5-FU), 428, 504, 529
Folate-bound micelles (FMA), 954
Folic acid, 916
Follicle-associated epithelium (FAE), 33, 149, 156, 180, 266
Folsomia candida, 566
Foot ulcers, 437
Freeze drying, 812
Freeze fracture electron microscopy (FFEM), 597
Fullerenes, 917–919
polymers, 886
structure, 917
Functional approach, 757
Furosemide, 176

G

GALT. *See* Gut-associated lymphoid tissue (GALT)
Gastrointestinal luminal milieu, drug absorption, 179
Gastrointestinal microbiota, 182
Gastrointestinal (GI) system, 409
Gastrointestinal tract (GIT)
drug absorption and. *See also* Drugs
gastric emptying rate, 183
gastrointestinal luminal milieu and, 179–181
barrier, to drug absorption, 181
primary port, for oral drug delivery, 179
gastrointestinal microbiota, 182
pH variation in GIT, 181
colonic pH, 182
gastric pH, 181
intestinal pH, 182
transportation of drugs, mechanism, 179
schematic representation of, 180
villi, 180
transport mechanisms
enterocytes, role of, 33
M cells, role of, 33
paracellular transport, 33
transcellular transport, 33
GBM. *See* Glioblastoma multiforme (GBM)
GC. *See* Glycol chitosan (GC)
Gelatin, 412
Gemcitabine, 43
Gemfibrozil, 240
Gendicine, 134
Gene expression, 130, 58[illegible]
Gene therapy, 147
Genetic engineering, 127
Genetic vaccination, 167
Gibbs-Thomson relationship as, 715

Gingival invaginations, 876
Ginseng phytosome, 306
Giomers, 846
Glass ionomer cement (GIC), 788, 803
Glass ionomers
 applications, 843
Glibenclamide, 217
Glicycidyl-epoxy-methacrylate, 683
Glioblastoma, 62
Glioblastoma multiforme (GBM), 63
Gliomas, 63
Glutamine conjugated oligochitosan, 413
Glutaraldehyde (GSH), 413, 416, 626
Glutaraldehyde-saturated toluene (GST), 614
Glutathione peptide, 423
Glutathione reductase, 583
Glutathione synthetic pathways, 563
Glycocalyx, 34, 151, 450
Glycol chitosan (GC), 609
Glycolic acid polymers, 683
Glycoproteins, 151, 416
Glycosylphosphatidylinositol-linked folate receptor (FR), 321
Glycyrrhizic acid micelles
 taxol loaded, 40
Gold (Au) nanoparticles, 195, 765
 applications of, 766
 biomedical applications, 518
 biomedical research, role in, 350
 citrate reduction method, 518
 cytotoxic effects, 350
 fabrication, 518
 human liver HepG2 cells, effect on, 350
 mean size, for different molar ratios of sodium citrate and tannic acid, 340
 molecular biosensor techniques, use in, 524
 nontoxic, 350
 radiosensitizer for radiotherapy, use as, 517
 schematic representations of, 766
 site-specific drug delivery, role in, 350
 surface-enhanced Raman scattering-based saliva assay, use in, 524
 synthesis, 342, 766
  ammonium citrate tribasic, role of, 342
  citrate concentration, effect of, 339
  with citrate ions
   overall redox reaction, 339
  gold (III) chloride hydrate, role of, 342
  10 nm AuNPs, 342
  29 nm AuNPs, 342
  62 nm AuNPs, 342
  polyvinylidene fluoride (PVDF) membrane, role of, 342
  reagents ratio, effect of, 339
  sodium citrate, role of, 342
  synthesis conditions optimization, effect of, 339
 TEM image, 518
Gonococcal ophthalmia neonatorum, credés prophylaxis of, 352
G/PANI/PS nanofibers, 915
Graft copolymers
 cationic amphiphilic lignin-based, 566
Graft rejection, 300
Gram-negative pathogens, 623
Graphene, 886, 899–901
 applications, 519
 biomarker sensors, as, 519
 biomedical applications, 519
 chemical structure, 900
 composition, 519
 fundamental physical properties, 519
 hyperthermia treatment, role in, 519
Graphene based nanomaterials
 cancer cells treatment
  preclinical studies, 521
Graphene nanosheets
 fabrication method, 519
  chemical vapor deposition, 519
  graphene oxide, chemical or thermal reduction of, 519
  Hummer's method, 519
  mechanical exfoliation, 519
 photothermal therapy (PTT), 519
Graphene oxide, 899–901
 chemical structure, 900
 TEM image, 520
Graphite, 886
 crystals, oxidation, 519
Graphitization, 566
Grave's disease, 296, 300
Green tea leaves
 water extracts
  antibacterial activity, 563
  CYP3A4 inhibitory, 563
  superoxide scavenging activities, 563
Griseofulvin, 2
GST. *See* Glutaraldehyde-saturated toluene (GST)
Gut-associated lymphoid tissue (GALT), 128, 137, 266, 310

H

Halitosis, 513
Harungana madagascariensis, 798
Head and neck cancer nanomedicine, citation classics in, 951–952
HEKs. *See* Human epidermal keratinocytes (HEKs)
HeLa cells, 9, 564
Helicobacter pylori, 426

Helper T cells, 296
Hemagglutinin, 126
Hematopoietic cells, 133
Hemolysins, 597
Hemophilia, 149
Heparin, 420, 421, 424
Hepatic metastases therapy, 504
Hepatitis B antigen (HBsAg), 49, 276
Hepatitis C virus (HCV), 354
Hepatocytes, 133
Herbal drug delivery, 306
Herbal immunomodulators, nanosystems for, 311
Herpes simplex infection
  antiviral drugs, use of, 352
  chronic suppressive therapy, 352
  genital HSV-2 infection, 351
  symptomatic recurrences, 351
  vaccines, 352
Herpes simplex virus (HSV), 335, 558
  HSV-1 infection
    treatment, 567
Hesperetin, 589
Heteropolysaccharides, 413
Hexaphenol, 704
HGF. *See* Human gingival fibroblast (HGF)
High-energy ball milling, 782
High-pressure homogenization, 619
Histocompatibility, 130
Histone methyl transferase enzyme (hSET1), 428
HLB. *See* Hydrophilic-lipophilic balance (HLB)
HL-60 cells, 560
Hodgkin's lymphoma, 322
HOK. *See* Human oral keratinocyte (HOK)
Hollow mesoporous silica spheres (HMSS), 702
  drug loading and dissolution rate, 702
  storage capacity of, 702
Homeostasis, 579
Homolipids, 592
Homopolysaccharides, 413
Horror autotoxicus, 296
Host-guest complex, 722
HPC. *See* Human pulp cell (HPC)
HPLF. *See* Human periodontal ligament fibroblast (HPLF)
HPV. *See* Human papilloma virus (HPV)
HSC-2 cells, 562
HSV. *See* Herpes simplex virus (HSV)
HT-29 cells, 504
Human colon adenocarcinoma cell line, 48
Human cytomegalovirus (HCMV), 354
Human epidermal keratinocytes (HEKs), 350
Human gingival fibroblast (HGF), 551, 562
Human immunodeficiency virus (HIV), 353, 557
  CD4-dependent virion binding, 353
  encoded proteins, 353
  gold NPs (Au NPs), effect of, 353
  polyvinylpyrrolidone-coated silver NPs (PVP-coated AgNPs), effect of, 353
  prevention
    Ag-NPs-coated polyurethane condom (PUC), role of, 354
    condoms, role of, 354
    copper, role of, 354
    nanotechnology-based condoms, need of, 354
  primary targets, 353
Human oral keratinocyte (HOK), 551
Human papilloma virus (HPV), 512
Human periodontal ligament fibroblast (HPLF), 551
Human pulp cell (HPC), 551
Huntington's disease, 63
Hyaluronan receptor for endocytosis (HARE), 418
Hyaluronic acid (HA), 38, 413, 418, 424
Hyaluronic acid-*g*-PCL copolymer, 419
Hyaluronic acid (HA) grafted polycaprolactone (PCL) nanoparticles (HA-*g*-PCL NPs), 483
Hyaluronic acid-SMA polymers, 413
Hydrochlorothiazide, 176
Hydrodynamic diameters, 375, 376
Hydrogel NPs (nanogels), 238
Hydrogels, 371, 409, 413
Hydrogel synthesis, 625
Hydrogen bonding, 373
  interaction, 618
Hydrogenophaga pseudoflava
  accumulated PHA granules
    transmission electron microscopic image, 4
Hydrolysable tannins, 336
Hydrolysis, 630
Hydroperoxide, 584
Hydrophilic drugs, 591, 608
Hydrophilic filler, 425
Hydrophilic-lipophilic balance (HLB), 592
Hydrophilic polysaccharide, 413
Hydrophobic drugs, 426
  encapsulation of, 104, 107
  solubility of, 102
  solubilization, 1
Hydrophobic-hydrophobic interactions, 621
Hydrophobic interactions, 373
Hydrophobicity, 126
Hydroxyapatite (HAP), 705, 845
4-Hydroxybutyrate (4HB) monomers, 5
Hydroxyl radical-scavenging activities, 563
Hydroxypropyl methylcellulose (HPMC), 168, 654
Hydroxypropyl methylcellulose phthalate (HPMCP), 726
N-Hydroxysuccinimide-folate, 954
Hymenolepis nana, 550
Hypertension, 580

Hyperthermia, 529, 597
Hypoxia, 560

I

IBAD. *See* Ion beam assisted deposition (IBAD)
Ibandronate, 475
IBD. *See* Inflammatory bowel disease (IBD)
Ibuprofen (IBU), 176, 428, 621, 702
  dissolution profiles of, 715
Immune system
  components of, 298
  defense, 296
    recognition, 296, 297
    response, 296
  mechanisms, humoral immunity, 296
  summary of, 296–299
Immunogenicity, 135, 413, 608, 610
Immunoglobulins, 600
Immunological memory, 297
Immunomodulation, 299
  cytokines, role of, 299
Immunomodulator oral delivery
  nanosystems used, advantages and disadvantages of, 324
Immunomodulators
  nanosystem, uptake in intestinal lumen, 321
Immunomodulators (IMs), 295
  classification and examples of, 301
  classification of, 299–300
    immunoadjuvants, 299
    immunostimulants, 299
    immunosuppressants, 300
  conventional marketed products, 302
  current routes of administration, 300–301
  fate after oral delivery, 320–323
  need of nanosystems, 302–303
  oral nanosystems for, 303–315
    herbal immunomodulators, 306–310
    synthetic immunomodulators, 305–306
    toxins and venoms, 313–315
    vaccines and adjuvant, 310–313
  oral route for, 301–302
    conventional immunomodulators, 301–302
    drawbacks of, 302
  other routes of administration, 324, 325
  overview of, 299–300
  several advantages, 300
Immunopotentiation, 568
Immunostimulatory antigen, 167
Impregnated-fullerene (IF-WS_2), 848
Indomethacin (IDMC), 426, 689, 703
Infectious diseases, 263
Inflammation, 578
Inflammatory bowel disease (IBD), 149, 301, 610
  mucosal barrier, changes in, 32
Inflammatory diseases of skin
  prevention and treatment, 296
Infrared spectroscopy, 51
Innate immunity, 297
  cellular mechanisms, 297
  molecular mechanisms, 297
Inorganic-based nanocarriers, 194
  carbon nanotubes, 195
  gold nanoparticles, 195
  layer-by- layer structured nanocarriers, 195
  nanoshell, 196
  quantum dots (QDs), 196
  silica-based nanocarriers, 195
Inorganic-inorganic composites, 705
Inorganic nanocarriers, 283
  gold nanocarrier, 283
  silica nanocarrier, 284
  silver, 283
  superparamagnetic iron oxide nanocarriers (SPIONs), 284
Inorganic nanoparticles, 305, 645
Insulin, 442
  action of, 445–447
  analogs, 443
    fast-acting insulin analogs, 443
    long-acting insulin analogs, 444
  formulation vehicles
    liposomes, 456
    microemulsions, 455
    polymeric nanoparticles, 455
  oral delivery, 447
    problems in, 448
      absorption barrier, 450
      chemical barrier, 448
      enzymatic barrier, 449
      stability barrier, 451
    recent breakthroughs, 456
    strategies to improve, 451
      absorption enhancers, 451
      chemical modifications of proteins and peptides, 454
      enzyme inhibitors, 453
      formulation vehicles, 455
      mucoadhesive polymers, 453
  premixed analog, 444
  structure of, 442
  therapy, 440
Insulin delivery system
  schematic for the pH-controlled, 724
Integrin receptors, 426
Intelligent polymers, 67
Interfibril hydrogen bonding, 566
Internal bleaching, 848

Internal vibrational energy, 50
Interpenetrating network nanohydrogels, 614
Intestinal epithelium
drug uptake mechanisms, 35
Intestinal immune response, 266
antigen uptake, 266
enterocytes-mediated antigen uptake, 268
M cell mediated antigen uptake, 266
immune response, 268
nanocarrier uptake mechanism, and intestinal immune response, 267
in vivo tumor detection, role in, 531
Ion beam assisted deposition (IBAD), 855
Ionic strength, 413
Ion-ion interaction, 373
Ion-pair equilibrium, 387
Irinotecan (CPT-11), 476
Iron oxide magnetic NPs
classification, 531
superparamagnetic, 531
ultrasmall superparamagnetic, 531
Ischemia
cerebral, cell death in, 581
mechanisms of, 581
arterial lumen, obliteration, 581
cerebral venous thrombosis, 581
embolism, 581
systemic hypoperfusion, 581
thrombosis, 581
ISCOMs (immunostimulatory complexes), 286
Isoflavones, 640
Isoniazid (INH), 42, 217

K

Kampo medicines, 561
Kaposi sarcoma, 524, 596
Keratinization, 93
Keratinocytes, 90
Ketoconazole, 176, 217, 251, 627
Ketoprofen, 176, 607
Kinetic behavior, 384
Kinetic data, 394
Kinetic energy, 378
Kinetic stability, 595
Klebsiella pneumoniae, 549, 623
Kollagen hydrolizate derivative (KLHT), 219
Konjac glucomannan, 425
Kraft lignin particles, 546
Kupffer cells, 549

L

Lactic acid, 683
Lactobacillus casei, 549
Lactodoxonano, 483
Lamina propria lymphocytes, 266
Laminin, 809
Laminin receptor precursor (LRP), 132
Laser pitting, 854
Laser pulse therapy, 533
Laser pyrolysis, 782
Layer-by- layer structured nanocarriers, 195
LCC. *See* Lignin-carbohydrate complex (LCC)
LCC-ascorbic acid tablet, 567
LDC. *See* Lipid-drug conjugates (LDC)
Lectins, 42
LEM. *See Lentinus edodes* mycelia (LEM)
Lentinus edodes mycelia (LEM), 556, 557
Leukocyte activation, 578
Leukoplakia, 513
prevalence, 513
Leuprolide, 416
Levodopa, 63
Lewis lung carcinoma (LLC) cell lines, 481
Lidocaine, 374, 391
Lidocaine release profile, 382, 383
Ligand-conjugated poly(R)-hydroxyalkanoic acids (PHA) nanoparticles
active targeting, 9–10
target drug delivery, use in, 9–10
Lignin
digestion of, 557
formation, 546
isolation methods, 546
nanostructure, studies of, 546–549
encounter to LCC, 547–549
lignin structure, 546–547
reducing agent, as, 567
utilization of components and, 564–567
cellulose utilization, 566
industrial use, dental application, 564–566
metal nanoparticle manufacture, 566–567
Lignin-carbohydrate complex (LCC), 546
binding to influenza virus, 559
bioavailability, 549
biological activity, 549–562
anti-UV activity, 552, 561–562
antiviral activity, 557–559
anti-HIV activity, 552, 557–558
anti-HSV activity, 558–559
antiinfluenza virus activity, 558
synergism with vitamin C, 560–561
radical generation and cytotoxicity, 560–561
radical scavenging activity, 560
in vitro study with cultured cells, 550–556
macrophage activation, 556
mitogenic activity against splenocyte, 556

Lignin-carbohydrate complex (LCC) (*cont.*)
stimulation of iodination of myeloperoxidase-positive cells, 556
tumor-specific cytotoxicity against oral squamous cell carcinoma, 550–551
in vivo study, 549–550
antimicrobial activity, 549–550
antiparasite activity, 550
endogenous TNF production, 549
pharmacodynamics, 549
carbohydrate portion of, composition, 546
clinical application, 567
antihalitosis effect of SE-containing toothpaste, 567
lignin-vitamin C tablet, effect on HSV-1-infection, 567
SE, effect on oral lichenoid dysplasia, 567
putative mode of action, 569
Lignocellulosic biomass, 546, 566
Linear expansion coefficient, 841
Linear polymethacrylate, 371
Linear sweep voltammetry (LSV), 893
1-4 linked α–L-guluronic (G-block) residue, 38
Lipid-based formulations, 194
liposomes, 194
nanostructured lipid carriers (NLCs), 194
solid lipid nanoparticles (SLNs), 194
Lipid based nanocarriers types, schematic representation, 590
Lipid-carrying systems, 679
overview, 679
Lipid-drug conjugates (LDC), 500, 578
Lipid nanocapsules (LNC), 502
Lipid nanocarriers (LNs), 279, 578
archaeosomes, 282
based delivery systems, challenges, 599
basic ingredients, 592
emulsifiers, 592
lipids, 592
bilayer lipids, 592
nonbilayer lipids, 592
bilosomes, 280–281
characterization, 597–599
drug-loading, 598
drug release, 598
particle characteristics, surface properties, and zeta potential, 597
solid state characteristics, 598
liposomes, 280
nanoemulsion, 282
niosomes, 281–282
role in delivery of bioactives for brain, 599–600
types, 592–595
lipid drug conjugates, 593
liposomes, 593
liquid crystal drug delivery systems, 594
nanoemulsions, 595
nanostructured lipid carriers, 593
niosomes, 594
solid lipid nanoparticles, 592
transfersomes, 594
Lipids
complex lipids, 578
heterolipids, 578
homolipids, 578
nanospheres, 501
peroxidation, 582
polymorphism, 681
Lipid-witepsol wax, 657
Lipofectamine, 159
Lipophilic drugs, 596, 600
carrier complex, 36
Lipopolysaccharide, 156
Liposomes, 40–41, 194, 239, 303, 456, 642
calcitonin-loaded CS-coated, 41
composition, 40
encapsulation property, 40
insights and methods of preparation, 596–597
insulin-loaded, 41
large unilamellar vesicles (LUV), 503
methods of preparation, 40
multilamellar vesicles (MLV), 74, 503
oral DDS, use as, 41
preparation method, 304, 597
sizes, 40
small unilamellar vesicles (SUV), 503
N-trimethyl chitosan (TMC) coated, 41
types, based on formulation strategies, 597
cationic liposomes, 597
fusogenic and endosomolytic liposomes, 597
pH-sensitive liposomes, 597
sterically stabilized liposomes, 597
targeted liposomes, 597
temperature-sensitive liposomes, 597
unilamellar, 74
vs.phytosomes, 303
Liquid crystals (LCs), 676
Liquid oral dosage forms, 891
elixir, 892
oral emulsions, 892
oral solutions, 892
oral suspensions, 892
syrup, 892
Listeria, 264
Lithography, 782
LMWH-TCA-based nanocarriers, 422
LNs. *See* Lipid nanocarriers (LNs)

Long-acting insulin analogs, 444
Detemir, 444
Glargine, 444
Long terminal repeat (LTR), 133
Lou Gehrig's disease, 63
Lovastatin, 46
Low molecular weight chitosan (LMWC), 499
LPS-activated mouse macrophage-like cells RAW264.7, 562
LPS-stimulated NO production, 562
LRP. *See* Laminin receptor precursor (LRP)
LTR. *See* Long terminal repeat (LTR)
Luciferase, 167
Lupron Depot, 234
Lymphocytes, 417
adaptive immune response, role in, 298
B lymphocytes (B cells), 298
formation, 298
T lymphocytes (T cells), 298
Lymphoid organs, 298
Lymphovascular metastasis, 514
Lyophilization, 11, 596, 599
Lyotropic liquid crystals (LLCs), 676, 677
formation and pharmaceutical applications, 677
stability, 679
Lysosome, 127

M

Macrophage, 416
Macrulin, 460
Madin–Darby canine kidney (MDCK) cells, 179
MADIX radical reactions, 687
Magic bullet, 528
Magnetic carboxymethyl chitosan-κ−carrageenan, synthesis of, 615
Magnetic carboxymethyl CS hydrogels, 609
Magnetic nanoparticles (MNPs), 516–517
adsorbent, use as, 516
biomedical fields, use as, 516
contrast agents in magnetic resonance imaging (MRI), use as, 516
drug delivery system, use as, 516
gas sensors, use as, 516
properties
biocompatibility, 516
high surface functionality, 516
monodisperse size distribution, 516
nontoxicity, 516
synthesis, 516
coprecipitation, 516
hydrothermal, 516
microemulsion, 516
sol-gel, 516
sonochemical reactions, 516
thermal decomposition, 516
Magnetic nanosystems
development, 517
Magnetic resonance imaging (MRI), 805
Magnetite (Fe_3O_4) nanoparticles
TEM image, 517
Magnetization capacity, 531
Major histocompatibility complex (MHC), 266, 298
Malignant neoplasms, 550
Malondialdehyde, 582
Malondialdehyde (MDA), 582
MALT. *See* Mucosa-associated lymphoid tissues (MALT)
Mannose, 416
Marketed nanocrystal-based drug products, 481
Materials nanofabrication
bottom-up approaches, 836
self-assembling approach, 836
top-down approaches, 836
Matrix erosion, 598
Matrix metalloproteinase-2, 512
Matrix metalloproteinase-9, 512
Maxillofacial surgery, 809
Maxwell's Demon, 835
MBG. *See* Mesoporous bioglass (MBG)
M cells. *See* Microfold cells (M cells)
MCM-41, as drug carrier, 702
MDA. *See* Malondialdehyde (MDA)
Measles virus (MV), 354
Mebendazole, 217
Mefenamic acid, 217
Megace ES (Megestrol), 241
Melatonin, 38
Memory B cells, 298
Memory cells, 297
Mercaptobenzoic acid coated gold NPs, 353
Mercaptoethanesulfonate (MES)-decorated AgNPs, 353
Mesalazine, 401
Mesenchymal stem cells, 810, 814
Mesoporous-activated carbon, 703
features of, 703
fiber morphology, 703
as oral drug carriers, 703
for potential oral drug delivery, 704
synthesis of, 703
Mesoporous alumina, 708
Mesoporous bioglass (MBG), 810
characteristics, 810
Mesoporous calcium carbonate materials, 708
Mesoporous composites, 705
as drug carriers, 705

Mesoporous drug carriers
effect of pore size of, 713
crystallization, 713
dissolution rate, 713
pore size of, 713
Mesoporous drug delivery system
cytotoxicity of, 720
Mesoporous hydroxycarbonate apatite, 707
Mesoporous magnesium carbonate, 708
Mesoporous nanoHAP (mn-HAP), 707
Mesoporous silica, 700, 701
categories, 701
MCM (mobile composite matter), 701
SBA (Santa Barbara amorphous material), 701
particle size of, 702
Mesoporous silica nanoparticles (MSNs), 195, 702
for drug delivery application, 702
synthesis of, 702
Mesoporous structured materials, 700
development of drug delivery systems, 700
as potential drug carriers, 701
Mesoporous zirconium phosphonate, 707
Metabolomics, 563
Metal-based nanocarriers, 194
carbon nanotubes, 195
gold nanoparticles, 195
layer-by- layer structured nanocarriers, 195
nanoshell, 196
quantum dots (QDs), 196
silica-based nanocarriers, 195
Metal-enhanced luminescence (MEL), 953
Metallic nanoparticles (MNPs)
characterization techniques, 344–345
dynamic light scattering (DLS), 344
electron microscopy, 345
results, 345–348
detailed synthesis, 342–344
synthesis with tannins, 336–341
reactions scheme, 340
Metallic nanoparticles colloids
syntheses and characterization with tannic acid, 336–345
Metallic nanoparticles synthesis
citrate ion oxidation, 338
citrate ion structure, 338
sodium citrate, role of, 337
stable nuclei formation, role of, 341
tannic acid, role of, 337
Metal matrix nanocomposites (MMNC), 759
Metastatic breast cancer, 468
3-Methacryloxypropyltrimethoxysilane (MPTS), 842
Methotrexate (MTX), 299, 504
Methyl methacrylate (MMA), 846
Methylmethacrylate-sulfopropylmethacrylate (MMA-SPM), 46
Metoclopramide, 393
Metoprolol, 176
Metronidazole, 426, 914
Micelles formation process, 70
Micelles/polymeric micelles, 644
Micelle unimolecular, 72
Micellization, 70
Miconazole nitrate, 423
Microarray-based gene expression analysis, 512
Microbial PHB, 10
Microcrystalline cellulose, 211–212
Microelectromechanical systems (MEMS), 670
Microemulsions, 185, 239, 304
Microfluidics, 670
applications of, 673
in pharmaceutical processes and nanomedicine, 673
physical foundations of, 672
Microfluidics technology, 670
Microfluidization, 566
Microfold cells (M cells), 33, 35, 478
Microparticles
absorption in gastrointestinal tract, 125
absorption promotion, 129
applications in oral delivery of drug molecules, 124
characteristics and effect of intake in vivo, 126–128
particles
age of animals, 128
performance of, 126
size of, 127
surface modification and charge of, 127
protein transduction and absorption, 128–129
Micropellets, 305
Microspheres, 305, 683
application, 686
design and development, criteria for, 683
methodologies used to produce, 684
overview, 683
preparation methods, 685–686
Microtubule depolymerization inhibitor, 468
Microvilli, 156
Miniemulsion, 213
Mitomycin C, 485
Mitoxantrone, 485
MMA. *See* Methyl methacrylate (MMA)
MMA-SPM. *See* Methylmethacrylate-sulfopropylmethacrylate (MMA-SPM)
Mobile composite matter (MCM) mesoporous silica, 701
synthesis of, 701
Model drugs, molecular structure, 372

Modern electroanalytical techniques, 886
Molecular engineering, 833
Monocarboxylate transporters (MCTs), 43
Monodispersed mesoporous carbon, 705
Mononuclear phagocyte system (MPS), 66
Monophosphoryl lipid A (MPLA), 276
Mouse fibroblast L929 cell lines, 9
Mouse keratinocyte 291.03C cell culture, 355
Mouse monocyte RAW 265.7 cell line, 348
mPEGylated nanoparticles
  degradation profiles, 14
MPO. *See* Myeloperoxidase (MPO)
MPO inhibitors, 556
MPO-positive cells, 568
MPS. *See* Mononuclear phagocyte system (MPS)
MPTS. *See* 3-Methacryloxypropyltrimethoxysilane (MPTS)
MRI. *See* Magnetic resonance imaging (MRI)
MUC gene family, 30
Mucins, 30, 181
  adhesion phenomena, role in, 30
Mucoadhesion, 99, 416, 608
Mucoadhesive polymers, 451, 454
Mucosa-associated lymphoid tissues (MALT), 478
Mucosal epithelium, 126
Mucosal immune-regulation, 31
Mucosal neck cells, 34
Mucus, 30
  adhesion, 323
  antimicrobial molecules, 30
    defensins, 30
    immunoglobulins, 30
    lysozymes, 30
  clearance mechanisms, 34
  mesh-pore spacing, 34
  orally administered drugs
    absorption, role in, 34
    bioavailability, role in, 34
  protective capability, 30
  viscoelastic properties, 30
Multidrug transporters
  ATP-dependent, 477
Multifunctionalized carrier systems, 648
Multifunctional NP-based drugs, 525
  ammos, 524
  protein-based nanodrugs, 524
Multifunctional stimuli-responsive drug delivery, 730
Multiwalled carbon nanotubes (MWCNTs), 42, 904, 905, 956
  applications, 520
Muscularis mucosa, 151
Mutagenesis, 133
Mycobacterium, 806
Myeloperoxidase (MPO), 158, 162, 556

N

Nanobiochip sensor, 524
Nanobiomaterials
  in dentistry, 838–855
    for endodontics, 848–852
      bioactive glass nanomaterials as root canal disinfectant, 852
      endodontic instruments surface coatings, 848
      endodontic sealer modified using antibacterial nanoparticles, 849
      EndoSequence BC, 849
      guttaflow sealer, 849
      modifications of endodontic sealer, 849
      nanohydroxyapatite (nano-HA) gutta percha, 852
      for regenerative endodontics, 852
    in oral implantology, 852–855
      nanomanufacturing of other implant materials, 855
      nanostructured titanium implant surfaces, 854–855
      surface hydrophilicity and microlevel surface roughness, 853
    in preventive dentistry, 847–848
    in restorative dentistry, 839–846
      dental glass-ionomers system, 843–846
      resin-based dental composites, 840–843
Nanobonding agent, 839
Nanocapsules (NC), 1, 101, 763, 798
  submicroscopic colloidal drug carrier systems, as, 1
Nanocarbon tubes (CNT), 806
Nanocarrier based oral vaccine delivery, 268
  affecting properties, antigen uptake, 270
    hydrophobicity, 271
    nanocarrier surface charge, 270
    particle shapes, 271
    particle size of nanovaccine, 270
  agencies under initiatives for, 288
  antigen-loading mechanism, 269
  classification of, 272
  future prospects, 288–289
  illustration of, 275
  regulatory aspects of, 286
    environmental, health, and safety risk, 286
    ethical, legal, and social issues, 287
    regulatory challenges associated with nanoparticulate systems, 287
    regulatory challenges associated with vaccine systems, 287
  vaccine manufacturing, regulatory bodies for, 288
Nanocarriers, 651
  in oral delivery, 234
    carbon nanotubes (CNTs), 242
    dendrimers, 241
    hydrogel NPs (nanogels), 238

Nanocarriers (*cont.*)
liposomes, 239
nanocrystals, 241
nanoemulsion and self-nanoemulsifying system, 239
nanoemulsion for oral drug delivery, 239
self-nanoemulsifying drug-delivery system, 240
nanospheres, 241
poly(lactic-*co*-glycolic acid) nanoparticles, 234
polymeric micelles, 238
solid lipid nanoparticles, 238
types, conjugated with drug, and effect on pharmacological property, 236
Nanocarriers technology
and modifications to improve specificity, 66–76
dendrimers, 62, 71–73
functional surface building, 75–76
liposomes, 62, 73–74
polymeric micelles, 62, 69–71
polymeric nanoparticles, 62, 68–69
Nanoceramics, 809
Nanoclusters, 760, 842, 843
properties of, 760
sources of synthesis, 760
condensation cell-type source, 761
gas aggregation source, 761
supersonic nozzle source, 760
Nanocoating agent, 839
Nanocomposite, 758, 839, 843
properties of, 758
flexural strength, 758
polymerization shrinkage, 758
water sorption, 758
wear and gloss retention, 758
types of, 759
ceramic matrix nanocomposites (CMNC), 759
metal matrix nanocomposites (MMNC), 759
polymer matrix nanocomposites (PMNC), 759
biohybrid nanopolymers, 759
bio-hybrid polymer nanotubes, 760
reinforced metal matrix composites, 759
Nanocrystal dentifrice
desensitizing effect of, 872
Nanocrystal formulation, 871
Nanocrystalline fenofibrate, 715
Nanocrystalline hydroxyapatite, 878
Nanocrystals (NCs), 241, 648
Nanodentistry, 788, 838
challenges, 771–772
different approaches of, 752
Nanodesensitizing agent, 872
Nanodevices, 833
Nanodiagnostics, 763
cantilevers, 766
gold nanoparticles, 765
quantum dots, 763
typical CdSe/ZnS QD, 764
Nanoemulsion and self-nanoemulsifying system, 239
nanoemulsion for oral drug delivery, 239
self-nanoemulsifying drug-delivery system, 240
Nanoemulsions (NE), 213, 562, 590, 647
definition, 213
excipients used for formulation of, 217
methods of production, 215
high-energy emulsification methods, 215
low-energy emulsification methods, 215–216
materials used in, 216
transparent and semitranslucent appearance, 213
Nanoencapsulated materials, 762
Nanoendodontic sealer, 839
Nanogels, 38
chronic diseases treatment, role in, 38
Nanoglass ionomers, 839
Nanohydroxyapatite, 877
Nanoimpression materials, 762, 839
Nanoimprint lithography, 836
Nanoionomer, 846
Nanolight curable glass ionomers, 762
Nanolignin particles, 546
Nanomaterials, 788
antimicrobial nanoagents, 788–791
endodontic infection, 788
in endodontic infections, 788
low-molecular-weight (LMW) antibacterial agents, 788
mechanisms of action, on bacteria, 789
characterization of, 780, 837
classification of, 838
definition, 779
delivery systems, 798
electrospinning, 799
nanofibrous scaffolds, 799
nanoshell, 800
nanospheres and nanocapsules, 798–799
phase separation, 800
self-assembly, 800
in dental and bone tissue implants, 796
drug targeting strategies, 527
active, 527, 528
passive, 527
physical stimuli-triggered active, 527, 528
fabrication methods for, 781
physical and chemical synthesis of, 782
implants, 795–797
individual care products, 801
medical applications
in drug delivery, 834
in gene delivery, 834

in imaging tools, 834
in molecular diagnostics, 834
nanofillers, 791–795
nature and timeline of, 835
properties of, 837
synthesis, characteristics, and classifications of, 836–838
Nanomedicine, 779, 864, 867, 899
application of, 779
in cancer therapy, 867
nanostructures, category, 779
in oral bone regeneration, application, 808–812
nanoceramics, 808–810
nanocomposites and nanoscaffolds, 811
nanometals, 811
nanopolymers, 811
term and, 780
Nanomers, 842, 843
Nanometer, 863
Nanoparticles (NP), 61, 124, 608, 641
biological synthesis of, 782
characterization, 49–51
atomic force microscopy (AFM), 51
dynamic light scattering, 50
electron spectroscopy, for chemical analysis, 51
fourier transform infrared spectrometry (FTIR), 51
light microscopy, 49
Raman spectroscopy (RS), 50
scanning electron microscopy (SEM), 49
transmission electron microscopy (TEM), 50
x-ray photoelectron spectroscopy, 51
zeta potential analysis, 50
classification, 41
DDS, used in, 41
delivery system, 609
lectin conjugated, 42
surface charge determination, 50
synthesis, 49
bottom-up methods, 49
synthesis, by different biological resources, 786
targeting molecules of surface-modified, 76
types, active molecules transport of, 62
as vehicles for drug transport and release, advantages, 66
Nanoparticulate drug delivery, 578
Nanopolishing agent, 839
Nanoquadrapeutics, 533
Nanoresearch, 944–945
Nanorobotics, 761
Nanorobots, 761, 807, 833
in dentistry, 761, 762
mechanism of action, 761
Nanorods, 763
Nanoscale, 834
Nanoshells, 196, 763, 800
Nanosolutions, 761
Nanospheres (NS), 41, 101, 241, 763, 798
Nanostructured lipid carriers (NLC), 45, 76, 79, 88, 107, 194, 304, 578, 646
benefits, 304
drug loading capacity of, 75
vs. solid lipid nanoparticles (SLN), 304
Nanostructures
category, 779
in oral delivery, 642
types of, 779, 780
Nanosystems
characterization methods, 315–320
drug content stability, 320
imaging, 315
particle size and zeta potential, 315
spectroscopic evaluations, 320
carbon-nuclear magnetic resonance (C^{13} NMR), 320
Fourier transform infrared spectroscopy (FTIR), 320
hydrogen-nuclear magnetic resonance (H^1 NMR), 320
in vitro and in vivo evaluation, 320
surface tension measurement, 320
transition temperature, 320
immunomodulator loaded, 321
preparation, 303
Nanotechnology, 124, 833, 863
and cell therapy in oral bone regeneration, 814–815
in conservative dentistry and endodontics, 769
nanocomposite resins, 769
nanoparticulate-based disinfection in endodontics, 769
nanorobotic dentifrice (dentrifrobots), 769
tooth durability and appearance, 769
current challenges and future directions, 816–817
in dentistry, 763, 788
nanodiagnostics, 763–767
in dentistry, applications, 788–807
nanodiagnostics, 805–807
atomic force microscopy, 805
dendrimers, 806
nanoparticle contrast enhancers, 805
nanopores, 806
nanotubes, 806
nanowire sensors, 807
future perspectives, 79–80
in oral and maxillofacial surgery, 767–768
in oral medicine, 864
origin and history, 835
in orthodontic treatment, 768
overview, 61–63
in periodontics, 768

Nanotechnology (*cont.*)
dentinal hypersensitivity, 768
drug delivery, 768
materials, to induce bone growth, 769
in prosthodontics, 768
implants, 768
impression materials, 768
nanocomposite denture teeth, 768
NanoTite, 873, 875
Nanotopography, 815
Nanotubes, 763, 835, 838
Nano ZS zetasizer system, 344
National nanotechnology initiatives (NNI), 834
Native PHA granules, 5
Natural antioxidants and anti-inflammatory agents
lipid nanoparticulate delivery of, 579
Natural origin with anti-inflammatory, antioxidant, and antistroke potential, leads, 585
Natural polymers, 608
biodegradable nature, 619
nontoxic nature, 619
oral drug delivery, role in, 38
NC. *See* Nanocapsules (NC)
Neomycin B, 176
Neoplastic diseases progression
immune system, role of, 296
Nephropathy, 437
Nernst–Brunner and Hixson-Crowell equation, 177
Neurodegenerative disorders (ND), 63
clinical applications in, 78–79
Neurofibrillary tangles, 63
Neuroleptic drug, 920
Neuropathy, 437
Neuropeptides, 872
Neuroprotectants, 578
Niclosamide, 217
Nifedipine, 217
Niosomes, 105, 590
Nitric oxide (NO), 251
9-Nitrocamptothecin (9-NC), 501
NLC. *See* Nanostructured lipid carriers (NLC)
NMR. *See* Nuclear magnetic resonance (NMR)
NMR spectroscopy, 373
Noncondensing polymers, 148
Nondeliberate immunosuppression, 300
Nonfolate conjugated cells, 9
Nonionic self-emulsifying oil-in-water formulation, 500
Non-mPEG counterpart particles
degradation profiles, 14
Nonself antigens, 296
Nonsteroidal antiinflammatory drug (NSAID), 102
Nonviral gene transmission, 140
Normal pulse voltammetry, 894, 895
Noscapine, 322
Novel lipid nanostructures, 589–600
LNs-based delivery
advantages, 590–591
mechanism, 591
Noyes-Whitney equation, 177
NP. *See* Nanoparticles (NP)
NP assisted-radiotherapy, 530
NP-based drug systems
triggered by, 528
NPs-based antiviral agents
advantages, 352
NS. *See* Nanospheres (NS)
NSAID. *See* Nonsteroidal antiinflammatory drug (NSAID)
Nuclear magnetic resonance (NMR), 599
Nucleic acid administration, oral delivery systems for, 159–168
cationic lipid-based nanostructures, 159–162
inside a polymeric matrix, 162
per se, 159
condensing polymer-based nanostructures, 162–165
inside a micrometric matrix, 165
per se, 163–164
chitosan, 163–164
modified chitosan, 164
micrometric delivery systems and minitablets, 167–168
alginate micrometric formulations, 167
PLGA micrometric formulations, 167
tablets, 168
naked molecules (essentially oligonucleotides), 159
other nanostructures, 166

O

OATP. *See* Organic anion transporting polypeptide (OATP)
Octreolin, 460
N-Octyl-O-sulfate chitosan (NOSC)-based micelles, 486
Odontoblasts, 852
Oil-in-water solvent-evaporation method, 15
Oleyl conjugated trimethyl chitosan, 164
Oligonucleotide, 168
Oncology
nanotechnology, potential role of, 516
Opsonization-promoting IgG antibodies, 299
Optical image-guided surgery, 531
Optical microscope, 49
Oral anticancer chemotherapy
experimented polymeric micelles used, 487
Oral bioavailability, 629
Oral biofilm, 839
Oral bone regeneration, 808
Oral cancer, 512–515
anti-EGFR drugs, effect of, 514
clinical features, 513–514

CPA6 genes, role of, 512
cycling dependent kinase inhibitors, role of, 514
diagnosis, 513–514
cytological techniques, 523
gold standard method, 523
biopsy, 523
histopathological examination, 523
light-based detection systems, 523
epidemiology, 512
age standardized incidence rate, 512
incidence, 512
male incidence rates, 512
etiopathogenesis, 512–513
carcinogens, role of, 512
protooncogenes, 512
risk factors, 512
tumor suppressor genes, inactivation of, 512
five-year survival rate, 515
FMO2 genes, role of, 512
gene therapies, 514
histopathological examination, 514
left ventral side of tongue, on, 514
management, 514
neck lymph nodes, role of, 514
oral lesion, role of, 514
radiotherapy and chemotherapy combination, 514
molecular biomarkers, 514
mortality and diagnostic delay, relationship between, 523
nanodiagnostic techniques, 523
biobarcode assays, 523
biochips, 523
DNA nanomachines, 523
nanobiosensors, 523
nanopore technology, 523
nanoproteomic, 523
nanoscale visualization, 523
NP-based immunoassays, 523
NP biolabels, 523
nanomaterials in early diagnosis, 523–524
nanotherapeutics, role of, 524–530
combination treatment modalities, nanomaterials for, 529–530
photothermal therapy, 529
radiotherapy, 530
nanocarriers, 529
clinical trials, 529
in vitro and in vivo studies, 529
nanomaterials for drug delivery systems, 525–528
nuclear pleomorphism, 514
overview, 512
photodynamic therapy, role of, 514
prognosis, 515
Bax expression loss, role of, 515
Bcl-2 expression, role of, 515
cyclin D1 overexpression, role of, 515
high EGFR gene copy number, role of, 515
prognostic factors, 515
distant metastasis, 515
invasion depth, 515
regional lymph nodes involvement, 515
tumor size and characteristic, 515
radiotherapy and targeted therapy, combination of, 514
SIAT1 genes, role of, 512
staging, 514
surgical plan, 514
symptom, 513
TNC genes, role of, 512
treatment, 514–515
tumor imaging, nanomaterials in, 531–533
magnetic resonance imaging (MRI), 531–532
nanotheranostics, 533
optical imaging, 531
positron emission tomography, 532
tumor necrosis factor (TNF)-α polymorphism, correlation between, 512
tumor with keratinization pearl, histopathological view of, 514
tyrosine kinase inhibitors, role of, 514
vital staining, 514
Oral cancer chemotherapy
nanocarriers administered via oral route
schematic representation, 479
nanocarriers based, advantages of, 480
Oral cancer lesions, 805
Oral cancer nanomedicine, 940
citation classics in, 946, 950, 952–954
research, 940, 945–946
Oral carcinogenesis, 512
genetic susceptibility, role of, 512
Oral cavity, 778
category, diseases and injuries, 778
Oral chemotherapy
advantages, 469
drugs
in development stage, 470
in market, 470
patient compliance, 474
remarkable addressing issues, 469–475
economic factor, 475
patient's adherence monitoring, 475
direct methods, 475
indirect method, 475
toxicity, 474
Oral chitosan nanosystems
findings, 619–630

Oral delivery nanostructures
therapeutic applications, 45–49
arthritis, 46
cancer, 47–48
cerebrovascular disease, 45–46
diabetes, 48
hepatitis B, 49
hypercholesterolemia, 46
transporter, characteristics of, 37–45
carbon structures, 42–43
dendrimers, 38–39
hydrogels, 37–38
liposomes, 40–41
micelles, 40
nanoparticles, 41–42
polymers used, 43–45
Oral delivery systems, 457
AI-401, 460
Capsulin, 459
CODES technology, 459
Eligen technology, 457–458
Macrulin technology, 460
Octreolin, 459
Oradel drug delivery technology, 460
ORAL-LYN, 457
ORMD-0801, 459
Oral drug delivery, 77, 186, 426, 608
challenges, 36
dendrimers, 193
future perspective, 460
gastrointestinal (GI) mucus barriers, 30–34
absorption, 32
mucosal barrier, 34
mucus, 30–31
interactions, 34
transport mechanisms, 33
nanocrystals, 193
nanoemulsions, 193
nanostructures for, 29–30
for nanotechnologies, 608
polymeric formulation, 186
polymeric micelles, 192
polymeric nanoparticles, 186
nanocapsules, 192
nanospheres, 192
polymer drug conjugate, 192
polysaccharide blends (PSB), 429
properties, advantage, disadvantage and application, of nanostructures, 187
self-assembled chitosan nanoparticles, 193
strategies, 35–36
Oral drugs, 887
electrochemical analysis of, 906, 921
Oral gene delivery
digestion circumstance of, 138
feasibility of, 130
vector system, 130–131
Oral ingestion, 608
Oral lichen planus (OLP), 567
Oral medicine, 124, 864, 879
clinical trials in, advantages, 865, 866
Oral mucosa, 88–94
concept and functions, 89
organization and structure, 89–92
permeability, 93
permeation pathways, 94
types
lining, 88, 90, 92
masticatory, 88, 90
specialized, 88, 90, 92
Oral nanomedicine
and clinical trials, 867–868
alveolar socket healing, 876
antimicrobial effect, 877–878
bleaching efficacy, 869–871
dental implants, 873–875
desensitizing efficacy, 871–873
gingival invaginations, 876
indirect approaches, 879
response to pulpotomy, 875
Oral nanosystems, 303
Oral recosulin, 460
Oral route, 232
drug uptake mechanisms, in intestinal epithelium via paracellular pathway, 235
future prospects, 252
intestinal epithelium, as physical barrier, 232
for pediatriatics and geriatric patients, 232
schematics diagrammatic uptake mechanism of drug, 233
Oral SCC
primary prevention, 515
secondary prevention, 515
tertiary prevention, 515
Oral therapy, 132
Oral vaccination, 138
Oral vaccine delivery, 264
challenges in, 268
under clinical trials, 265
commercially available oral vaccines, 265
future prospects, 288–289
nanocarrier based, 268
nanocarrier-based system for, 273
regulatory aspects of, 286
type, of nanocarrier for, 272
carbon nanotube, 284

dendrimers, 285
inorganic nanocarriers, 283
gold nanocarrier, 283
silica nanocarrier, 284
silver (Ag), 283
superparamagnetic iron oxide nanocarrier, 284
ISCOMs, 286
lipid nanocarriers, 279
archaeosomes, 282
bilosomes, 280–281
liposomes, 280
nanoemulsion, 282
niosomes, 281–282
polymeric nanocarriers, 272
chitosan, 278
dextrans, 279
mannan, 279
polyanhydride, 277
poly(D, L-lactic acid), 277
poly(lactic-*co*-glycolic acid), 276
virus-like particles (VLPs), 285
vaccine manufacturing, regulatory bodies for, 288
Oral viral vector
potential, problems, and prospects, 139–140
selection suitable for oral delivery, 132–137
adeno-associated virus, 135–137
adenovirus, 133–135
other viral vector systems, 137
retrovirus, 133
transmission, distribution, and transgenic product pharmacokinetics of, 138–139
transmission system, advantages, 139
uptake, 131–132
Orasome, 460
Organic acids, 371
Organic anion transporting polypeptide (OATP), 44
Organic-inorganic mesoporous composite, 706
Organic polymeric nanoparticles, 646
Organic solvents, 688
Oridonin, 423
Origanum syriacum, 584
Osseointegration, 808, 852, 873
Osseointegrative, 808
Osteoblasts, 808
Osteocalcin expression, 815
Osteoconductive coatings, 809
Osteocytes, 808
Osteogenic precursor cells, 808
Osteomyelitis, 810
Ostim, 876
Ovalbumin antigen (OVA), 41, 276
Oxidation retarders, 584
Oxidative stress, 578, 583
inflammation, mechanism of action, 582–584
Oxirane, 841

P

Paclitaxel (PTX)
delivery system, 3, 234, 415, 468, 483, 689, 714, 867
loaded lipid nanoparticles, 503
loaded multilayered liposomes, 504
loaded NPs, 478
PAMAM dendrimers. *See* Poly(amidoamine) (PAMAM) dendrimers
PAMAM-SN38 complexes, 242
Panax ginseng, 306, 578
Panzem nanocrystal colloidal dispersion, 241
Paracellular transport, 415
Parallel artificial membrane permeability assay (PAMPA), 179
Parathyroid hormone, 166
Parkinson's disease, 62, 582
Particle-coating technology, 124
Pastilles, 891
Patient's adherence monitoring
direct methods
direct observation, 475
pharmacokinetic parameters determination, 475
indirect method
microelectronic monitoring system (MEMS), 475
patient self-report, 475
pill counts, 475
PCA. *See* Poly(ε-caprolactone) (PCA)
PCL. *See* Poly-ε-caprolactone (PCL)
PE-D aqueous dispersions, 378
PE-D complexes, 375, 381
in solid state, 388
wetting of particulate and compacted, 389
PE-D dispersions, 377, 385
PEEK. *See* Poly-ether-ether-ketone (PEEK)
PEG. *See* Polyethylene glycol (PEG)
PEG400, 239
PEG-coated nanoparticles, 954
PEG dilemma, 526
PEGylated chitosan derivatives, 416
PEGylated drug conjugates, 18
PEGylation, 593
Pellets, 203. *See also* Extrusion spheronization process; pellets production
advantages, 205
definition, 203
disadvantages, 205
evaluation of, 213
methods for determination of properties, 214

Pellets (*cont.*)
techniques for manufacturing, 204
cryopelletization, 204
direct powder pelletization, 204
dry powder layering, 204
extrusion-spheronization, 205
freeze pelletization, 204
hot-melt extrusion, 204
solution and suspension layering, 204
Penetration enhancers, 95
chemical structures and mechanisms of action of, 96
fatty acids used as, 98
PEO. *See* Polyethylene oxide (PEO)
Peptidoglycan, 805
Periodontitis, 878
Permeability, 178–179
Permeation promoters, 95
Perturbation, 582
Peyer's patches, 30, 33, 35, 125, 417, 478
PGA. *See* Poly(glycolic acid) (PGA)
P-glycoprotein (P-gp), 469, 526
P-gp inhibitor, 485
P-gp mediated efflux pump, 478
P-gp-overexpressing human cancer cells, 503
PHA. *See* Poly(R)-hydroxyalkanoic acids (PHA) polymers
PhaC genes, 9
PHA depolymerases, 4
Phagocytosis, 73, 125
PHA-mPEG nanoparticles, 12
3-D AFM image, 16
drug loaded
schematic representation, 19
rhodamine-loaded, 18
PHA-mPEG nanoparticles characterization, 13–18
cellular uptake, 18
core-shell topology in aqueous solution, 13–15
drug loading, 16–17
drug release from, 17
morphological characteristics, 15
PhaP genes, 9
Pharmaceutical proteins
production, 36
Pharmacodynamics, 124
Pharmacokinetics, 124, 134, 610, 620, 651
Pharmacopoeia, 154
Pharyngeal cancer nanomedicine
citation classics in, 954–955
Phase-contrast imaging, 51
PHB. *See* Polyhydorxybutyrate (PHB)
P(3HB) depolymerase, 9
PHB4HB. *See* Poly(3-hydroxybutyrate-*co*-4-hydroxybutyrate) (PHB4HB)
PHB-mPEG diblock copolymer
surface modified, synthesis of, 11
PHBV. *See* Poly(3-hydroxybutyrate-*co*-3-hydroxyvalerate) (PHBV)
Phenol, 704
Phenolic phytochemicals, 640
Phenylpropanoid pathway, 546
Phenylpropenoids, 546
Phenytoin, 217
Phloroglucinol, 704
Phosphatidylcholines, 592, 596
Phosphatidyl inositol triphosphate (PIP3), 445
Phospholipids phosphatidylserine (PS), 321
Phospholipid vesicle-based permeation assay (PVPA), 179
Photobleaching period, 531
Photodynamic therapy, 105, 529
Photo-light activated composites, 840
Photosensitivity, 624
PHV. *See* Poly(3-hydroxyvalerate) (PHV)
P(3HV-*co*-4HB)-*b*-mPEG nanoparticles
cellular internalization
microscopic study, 18
Physical vapor deposition (PVD), 848
Physicochemical fabrication methods
protocols and applications of, 783
Phytochemical antioxidants, 582
Phytochemicals, 640
Phytomedicines, 600
Phytonanoconjugates, 648, 657
endocytotic cellular uptake, 648
paracellular pathway, 649
transcellular pathway, 649–650
Phytonanoparticles in oral medicine, 651
curcumin, 654–656
epigallocatechin-3-gallate, 651
resveratrol (*trans*-3, 4′, -5-trihydroxystilebene), 653
silymarin, 656
Picibanil, 549
Pine cone LCC
anti-HIV activity
$FeSO_4$ treatment, effect of, 565
antiparasite activity of, 550
Pinus caribaea, 547
Pinus densiflora, 547
Pinus elliottii, 547, 557
Pinus parviflora, 547
Pinus sylvestris, 547
Pinus taeda, 547
Pinus thunbergii, 547
Piroxicam, 217
PLA. *See* Poly-D, L-lactide (PLA)
Placebo toothpaste, 567
Plantago ovata, 425

Plasma cells, 298
Plasma spraying synthesis, 782
PLGA. *See* Poly-D, L-lactide-*co*-glycolide (PLGA)
Plurogel, 485
Pluronic block copolymer F127, 715
Pluronics, 71
Pluronic triblock polymer F127, 704
PM. *See* Polymeric micelles (PM)
PMA. *See* Poly(methyl acrylates) (PMA)
p53 mutations, 515
PN. *See* Polymeric nanoparticles (PNs)
Poliovirus, 132
Poloxamer 407, 703
Polyacrylamide, 154
Polyacrylamide gel electrophoresis, 547
Polyacrylic acid (PAA), 43
Poly(acrylic acid)-poly(butyl acrylate) (PAA-PBA), 689
Polyalkenoic acid, 846
Poly-alkyl cyanoacrylate (PACA), 481, 683
Poly-amidoamine dendrimers, 644
Poly(amidoamine) (PAMAM) dendrimers, 39, 241, 504
  cationic charge of, 242
  cisplatin, conjugation with, 39
  hydrophilic tendency, 39
  macromolecular nature, 39
  in vitro for potential oral drug delivery applications, 242
  water-soluble polymers, 242
Polyamine, 722
Poly(butyl cyanoacryalate) (PBCA), 45
Poly butyl methacrylate, 426
Polycaprolactone (PCL) conjugated hyaluronic acid (HA) nanoparticles, 416
Polycarbophil, 367
Polydimethylsiloxane (PDMS), 671
Poly(dimethylsiloxane)-poly(2-methyloxazoline) diblock copolymers, 531
Poly D-lactic acid (PDLA), 44
Poly-D, L-lactide (PLA), 110
Poly-D, L-lactide-*co*-glycolide (PLGA), 110
Poly(ε-caprolactone) (PCA), 65
Poly-ε-caprolactone (PCL), 110
Polyelectrolyte-drug complexes, 371, 375, 380
  in aqueous dispersion, 371
Polyelectrolyte-drug complexes, distribution, 373
Polyelectrolyte-drug counterionic condensation, 371
Polyelectrolyte-drug interactions, 378
Polyelectrolytes (PE), 367
Polyester-mPEG block polymers, 10
Poly-ether-ether-ketone (PEEK), 852, 855
Polyethylene glycol (PEG), 2, 36, 66, 113, 134, 410, 454
  drug solubility, role in, 36
Polyethyleneimine (PEI), 956
Polyethylene oxide (PEO), 71
Poly(ethylene oxide)-*b*-poly-(propylene oxide)-*b*-poly(ethylene oxide) (PEO-PPO-PEO) amphiphilic triblock copolymers, 485
Poly(glycolic acid) (PGA), 55, 481
Polyhydorxybutyrate (PHB), 5, 7
  depolymerases, 6
Poly(3-hydroxybutyrate-*co*-4-hydroxybutyrate) (PHB4HB), 5
Poly(3-hydroxybutyrate-*co*-3-hydroxyvalerate) (PHBV), 5
Poly(3-hydroxyvalerate) (PHV), 5
Poly(lactic acid) (PLA), 12, 65, 124, 481
Poly(lactide-*co*-glycolide) (PLGA) nanoparticles, 12, 55, 124, 148, 234, 410
Polylactides (PLAs), 45
PolyL-lactic acid (PLLA), 44
Polymeric micelles (PM), 69, 101, 238
  shielding effect, 4
Polymeric micro/nanoparticles, 322
Polymeric nanocarriers, 272
  chitosan, 278
  dextrans, 279
  mannan, 279
  polyanhydride, 277
  poly(D, L-lactic acid), 277
  poly(lactic-*co*-glycolic acid), 276
  synthetic polymers, in preparation of, 65
  as vehicles for oral-delivery systems, 76–77
Polymeric nanoparticles (NPs), 68, 89, 305
  benefits, 305
  drug-delivery applications, role in, 3–4
  nanocapsules, 101
  nanospheres, 101
  pharmaceutical applications, 3
  polymeric micelles, 101
  types, 68
    nanocapsules, 68
    nanospheres, 68
Polymer matrix nanocomposites (PMNC), 759
  biohybrid nanopolymers, 759
  bio-hybrid polymer nanotubes, 760
Polymerosomes, 687, 688
  applications, 689
    for design of oxygen carriers, 689
    pharmaceutical compounds, encapsulation of, 689
  copolymer, 687
  with hemoglobin (Hb), 690
  methods used for production, 688
  overview, 687
  production, 687
    requirements for, 687
  synthesis from hydrophilic-hydrophobic copolymers, 688
Polymersomes, 113
Polymethacrylic acid derivatives, 376
Poly(methyl acrylates) (PMA), 617, 683

Polymorphism, 598
Polymorphonuclear cell (PMN), 549
Polyoxyethylene (PEO), 103
Poly (allylaminehydrochloride) (PAH), 722
Polyphenol protection, mechanism, 641
Poly-(1, 4-phenyleneacetone dimethylene thioketal) polymers-based nonviral vectors, 412
Polypropylene oxide (PPO), 71
Poly(R)-hydroxyalkanoic acids (PHA) polymers, 4–10
 based drug-loaded nanoparticles, 13
  solvent-displacement method, preparation by, 13
 based materials, degradation of, 4
 based nanoparticles
  drawbacks and challenges, 19
 based polyesters, 4
 biodegradable materials, 6
 chemical modification
  drug delivery, role in, 10
  passive targeting, role in, 10
 copolyesters, 6
 copolymers production, 5
 crystalline PHA, 5
 degrading microorganisms, 6
 depolymerases, 6
 fermentative production, 4
 homopolyesters, 6
 nanoparticles
  diffusion release, 17
  drug delivery system, 8
 synthase, 9
Polysaccharides (PS), 413, 547
 for oral drug delivery, 411
Polystyrene, 124
Polystyrene sulfonate (PSS), 722
Polyvinyl alcohol (PVA), 165, 685
Poly(vinyl pyrrolidone) (PVP), 44, 609
Polyvinylpyrrolidone-poly(acrylic acid)-pilocarpine systems, 379
Polyvinylsiloxanes (PVS), 762
Poorly soluble oral drug
 characterization and stabilities, 712
  effect of pore size, of mesoporous drug carriers, 713–715
  stability of amorphous formulation and supersaturation, 716–717
 enhancement, of in vivo bioavailability, 719
 sustained release of, 718
Porogen leaching, 812
Porphyromonas gingivalis, 105, 251
Posphatidylserine, 128
Postprandial blood glucose levels, 48
Potassium persulfate, 618
Potential sweep voltammetry, 893
 step and pulse, 893
  differential pulse, 894
  normal pulse, 894
  square wave, 894
Povidone, 44
PPO. *See* Polypropylene oxide (PPO)
Prasiquantel, 217
Prednisone, 299
Preventive nanodentistry, 770
 antibacterial nanotherapy, 770
 biomimetic remineralization
  recurrent decay, 770
  reversing, incipient caries, 770
 caries vaccine, 771
 dental caries, 770
 periodontal diseases, 771
Primary immunodeficiency, 299
Procyanidins, 336
Prograf injection, 305
Proline, 30
Proliposomes, 303
Propranolol, 176, 249
Propranolol hydrochloride (PNH), 702
Prostaglandins, 872
Prostate cancer DU145 cells, 18
Protein synthesis, 581
Protein transduction, 128
Protein tyrosine phosphatase (PTP), 416
Proteolytic enzyme
 degradations, 409
Protozoa infections, 593
Prunus domestica, 584
Pseudomonas aeruginosa, 549, 623, 790
Psyllium polysaccharides, 425
PTX-nanomicelles system, 485
Puerarin, 657
Pullulan based nanogels, 38
Pulpal nociceptors, 872
Pulp canal therapy, 848
Pulpotomy, 875
Punicalagin (PUG), 354
PVD. *See* Physical vapor deposition (PVD)
PVP. *See* Poly(vinyl pyrrolidone) (PVP)
PVPA. *See* Phospholipid vesicle-based permeation assay (PVPA)
Pyrazinamide (PZA), 42, 217

Q

QDs. *See* Quantum dots (QDs)
Quantum dots (QDs), 196, 523, 763–765, 806, 952
 bioimaging application, 531
 cadmium-free, 531
 carbon-based, 523

fabrication techniques, 523
operative applications, role in, 531
optical image guided surgery, 531
SLN mapping, 531
preparation of, 764
probes, 523
cancer imaging, use in, 523
cancer targeting, use in, 523
Quaternary ammonium polyethylenimine (QA-PEI), 791
Quercitin-loaded SLN, 503
Quinidine, 217

R

rAAV. *See* Recombinant adeno-associated virus (rAAV)
Radiology, 805
Radiotherapy
antitumor activity, 517
Ralstonia pickettii T1, 13
Raman fluorescence signals, 50
Raman spectroscopy (RS), 50
factors affecting, 50
Ranitidine, 176
Rapamune(rapamycin), 241
Rayleigh (Mie) scattering, 953
RBC. *See* Resin-based composites (RBC)
Reactive oxygen species (ROS), 412, 582
Receptor mediated endocytosis (RME), 321
Receptor mediated transcytosis (RMT), 599
Recombinant adeno-associated virus (rAAV), 135
Recombinant human epidermal growth factor (rhEGF), 9
Recombinant parathyroid hormone (rPTH), 452
Remineralization process, 839, 847
Reovirus, 131
RES. *See* Reticuloendothelial system (RES)
Resin-based composites (RBC), 838, 839
components, 841
initiators and accelerators, 841
inorganic fillers in, 841
radiopaque material with pigments, 841
silane-coupling agent, 841
Resin composites, 864, 877
Resin modified glass ionomers (RMGIC), 845
Resistant bacterial infections, 623
Resorcinol, 704
Resorcinol-formaldehyde (RF), 704
Resovist, 531
Respiratory syncytial virus (RSV), 354
Restorative dentistry, 788, 839
Restorative material, 839
amalgam, 839
glass ionomers cements, 839
gold fillings, 839
resin-based composites (RBC), 839
Resveratrol, 561, 624
Resveratrol (*trans*-3, 4′, -5-trihydroxystilebene), 653
Reticuloendothelial system (RES), 321, 526, 591
Retinopathy, 437
Retrovirus, 132
Reynolds number (Re), 670, 672
rhEGF. *See* Recombinant human epidermal growth factor (rhEGF)
Rheological properties, 377
Rheumatoid arthritis, 296
Rifampicin (RIF), 42, 217
Ring-opening polymerization, 11
RME. *See* Receptor mediated endocytosis (RME)
RMGIC. *See* Resin modified glass ionomers (RMGIC)
RMT. *See* Receptor mediated transcytosis (RMT)
RNAi. *See* RNA interference (RNAi)
RNA interference (RNAi), 147
RNA nanotechnology, 955
Root canals, 864
filling, 848
sealer, 849, 850
treatment, 879
Root-end fillings, 848
ROS. *See* Reactive oxygen species (ROS)
ROS measurement
in monocytes
2′, 7′−dichlorofluorescin diacetate (DCFH-DA)
fluorescence, role of, 350, 351
RS. *See* Raman spectroscopy (RS)

S

Salinity, 388
Salivary film, 89
Saliva wash out phenomenon, 89
Salmonella enteritidis, 277, 549
Santa Barbara amorphous material (SBA) mesoporous silica, 701
synthesis of, 701
Sarkanda grass, 546
Sasa senanensis, 562
Saturation magnetization, 526
SBA-15, as drug carrier, 702
SBA-15 mesoporous silica, 705
SBA-15 submicron particles, 703
Scaffold fabrication techniques
protocols, applications, advantages and disadvantages, 812
Scanning electron microscopy (SEM), 125, 315, 345
Scanning probe microscopes (SPM), 51, 805
Scanning transmission electron microscopy (STEM), 344
Scanning tunneling microscope, 835
SCC. *See* Squamous cell carcinoma (SCC)
Schistosoma mansoni, 273
Schizophyllan, 550

SCID. *See* Severe combined immunodeficiency (SCID)
Science Citation Index-Expanded (SCIE) database, 939
SCNT. *See* Single-wall carbon nanotubes (SCNTs)
Scutellaria baicalensis, 578
SDP. *See* Spinning disk processor (SDP)
SEAP. *See* Secreted alkaline phosphatase (SEAP)
Sebacic acid, 126
SE containing toothpaste (SETP), 567
Secreted alkaline phosphatase (SEAP), 134
Secreted mucin types, 30
SEDDS. *See* Self-emulsifying drug delivery systems (SEDDS)
Self-assembled nucleoside nanoparticle (SNNP), 529
Self-emulsifying drug delivery systems (SEDDS), 212, 219, 240, 499–501, 645
Self-emulsifying (SE) pellets, 219
SEM. *See* Scanning electron microscope (SEM)
Semiconductor nanowires, 807
Semipermeable membrane, 381
Sensing platform, 916
Sentinel lymph node (SLN) mapping, 515
Serine, 30
Seroepidemiology, 135
Severe combined immunodeficiency (SCID), 299
Sexually transmitted diseases (STDs), 335
Sialic acid, 416
Silica-based nanocarriers, 195
Silicon dioxide, 791
Silicone urinary catheters, 252
Silorane, 841
Silver nanoparticles (AgNPs), 868, 878, 954
  as antimicrobial agent, 789
  in coatings of medical equipment, 790
  cytotoxic effect, 790
  cytotoxic effects, 350
  mean size
    for different molar ratios of sodium citrate and tannic acid, 341
  mechanisms of action, on bacteria, 789
  in oral medicine, 790
  synergistic activity of, 789
  synthesis, 343, 789
    citrate ions concentration, effect of, 341
    10 nm AgNPs, 343
    10-69 nm AgNPs, 343
    33 nm AgNPs, 343
    46 nm AgNPs, 343
    polyvinylidene fluoride (PVDF) membrane, role of, 343
    reducing agents mixture, use of, 340
    silver nitrite, role of, 343
    sodium borohydride, role of, 343
    sodium citrate concentration, effect of, 341
    sodium citrate, role of, 343
    tannic acid, role of, 343
  in toothbrushes, 790
Silver sulfadiazine
  antiinfective topical medicine, as, 352
Silymarin microemulsion, 310, 426, 656
Silymarin proliposomes, 306
Silymarin solution
  vs. silymarin microspheres, 310
Sindbis virus, 137
Sinerem, 531
Single/double emulsion, 782
Single-emulsion solvent evaporation method, 238
Single-wall carbon nanotubes (SCNTs), 42, 833, 901, 904
Single-walled carbon nanohorns (SWCNHs), 915, 916
  SWNHs-TiO_2-porphyrin nanohybrid, 917
siRNA nanocomplex, 416
Sirolimus antibiotics, 300
SKOV-3 human ovarian cancer xenograft, 501
SLN. *See* Solid lipid nanoparticles (SLNs)
Smallpox infection, 263
Small unilamellar vesicles (SUV), 593
SN-38 (7-ethyl-10-hydroxy-campothecin), 504
SNNP. *See* Self-assembled nucleoside nanoparticle (SNNP)
Social Sciences Index (SSCI) database, 939
SOD. *See* Superoxide dismutase (SOD)
Sodium alginate, 167, 608
Sodium oleate, 410
Sol-gel processing, 782
Solid lipid nanoparticles (SLNs), 89, 107, 194, 238, 304, 578, 646, 680–681
  applications, 681
  factors, for release system, 681
  methods of preparation, 304, 595–597
    high shear homogenization, 595
    solvent emulsification/evaporation, 596
    spray drying method, 596
    using supercritical fluid technology, 596
    via microemulsions, 596
  release of pharmaceutical compound from, 681
Solids and semisolid oral dosage forms, 887
  capsule, 889
  lozenges, 890
  pellets, 889
  powders and granules, 891
  tablets, 887–889
Solubility, 178, 378, 381, 681
  complexation with carbomer, 379
  enhancer, methyl-βCD as, 103
Soluble silver compounds
  STDs, treatment of, 352
Solvent casting, 812
Solvent emulsification, 578

Sonication, 597
South West Research Institute (SWRI), 756
Soya lecithin, 624
Spinning disk processor (SDP), 617, 628
SPIONs. *See* Superparamagnetic iron oxide nanoparticles (SPIONs)
SPM. *See* Scanning probe microscopes (SPM)
Squamous cell carcinoma (SCC), 512, 550
Square wave voltammetry, 894, 897
Staphylococcus aureus, 549, 789
STDs. *See* Sexually transmitted diseases (STDs)
Stephania tetrandra, 584
Stimuli-responsive hydrogels, 37
Stimuli-responsive polymers, 67
Stimuli responsive smart drug delivery, 723
  bio-responsive drug delivery, 729
  glucose responsive drug delivery, 728
  Other responsive and multifunctional delivery, 730
  pH-responsive drug delivery, 723–726
    for oral administration, 723
    poly(acrylicacid) (PAA), 723
    reversible addition-fragmentation chain transfer (RAFT) process, 723
  schematic representation, of glucose-responsive controlled release of insulin, 728
  thermal responsive drug delivery, 726–727
    lower critical solution temperature (LCST), 726
    organic-inorganic hybrid nanocomposite, 726
    release profile of solids S1-H, S1-D, and S1-T at different temperatures, 727
Stratum basale, 90
Stratum corneum, 92
Stratum granulosum, 92
Stratum intermedium, 92
Stratum spinosum, 91
Stratum superficiale, 92
Streptococcus mutans, 790, 847
Streptococcus sp., 279
Stripping voltammetry, 896
Stroke, 578–581
  onset, 578
  pathophysiology, 581
  risk factors, 580–581
  types, on basis of pathogenesis, 580
    hemorrhagic stroke/hemorrhagic cerebrovascular disease, 580
    occlusive/ischemic disease, 580
    pervasive, 580
    prolonged ischemic neurological deficiency, 580
    transient ischemic attacks, 580
Styrene-*co*-maleic anhydride (SMA), 413
Submicron emulsion, 213
Sulfamethoxazole, 252
Sumatriptan succinate, 623
Superoxide, 426
Superoxide dismutase (SOD), 582
Superparamagnetic CS-DS hydrogel, 625
Superparamagnetic iron oxide nanoparticles (SPIONs), 10, 28, 609, 954
Superparamagnetic nanoparticles, 611
Surface-confined quinone, 564
Surface energy, 619
Surface nanotopography, 817
Surface-to-volume ratio, 886
Surfactants, 185, 892
  kinds of, 185
Surgical oncology
  razor's edge, 531
SUV. *See* Small unilamellar vesicles (SUV)
Swellable drug-polyelectrolyte matrices (SDPM), 388
  drug release events, 395
Synthetic antioxidants, 582
Synthetic biodegradable polymers, 481
Synthetic immunomodulators
  nanosystems for, 307
Synthetic polymers, 410, 426
  carriers, 460

T

Tablets dosage format, 731–732
Tacrolimus, 300
Tacrolimus-loaded NPs, 305
Tactile test, 872
Tamoxifen (TAM), 47
Tannic acid-modified AgNPs/Ag/CuNPs
  HSV-2 attachment, blockage of, 357
  and HSV-2 virion
    interaction between, 356
  inflammatory properties, 356
  in vivo HSV-2 infection, reduction of, 358
Tannic acid modified metallic nanoparticles colloids
  antiviral activities, 354–358
  biological activity
    size and size distribution, effect of, 337
  biological processes of viruses and cell lines, effect on, 337
  preparation of, 337
  toxicity, 345–351
    apoptosis, 345
    caspase-9 activity, role of, 345
    dose-dependent, 345
    necrosis, 345
    reactive oxygen species (ROS), 345
  toxicity and antiviral properties, 345–358
    herpes simplex infection, 351–352
  virus-mediated STDs, treatment of, 352–354

Tannic acid-modified silver NPs
antiviral activity, 355
Tannic acid modified/unmodified NPs
toxicity testing, 348
Tannic acid structure
condensed gallic acid structural fragments, 339
conventional formula, 339
Tannins, 336
antimicrobial activity, 354
Bacillus subtilis, 354
Clostridium perfringens, 354
Salmonella typhimurium, 354
biological activity, 336
biological antioxidant, as, 336
chemical structure, 336
classification, 336
complex tannins, 336
ellagitannines, 336
gallotannines, 336
common features, 336
complexing or precipitating agent, as, 336
metal ion chelating agent, as, 336
toxicity for fungi, bacteria, and yeasts, 354
wound healing, role in, 354
Targeted polymeric nanoparticles (TPNPs), 322
Targeted *smart* drugs, 524
Targeted SPION-based imaging, 532
Tartrate-resistant acid phosphatase (TRAP), 809
Tat protein, 128
Tau protein, 63
Taurocholic acid, 422
Taurocholic acid (TCA)-conjugated heparin derivatives, 422
Taurocholic acid-linked heparin-docetaxel conjugates (HDTX), 423
Taxanes, 47, 478
Taxol, 40, 176
Taxotere, 501
T cell receptor (TCR), 298
TCR. *See* T cell receptor (TCR)
TEER. *See* Transepithelial electrical resistance (TEER)
Teeth
bleaching, 869
hydrogen peroxide (H_2O_2), used in, 869
side effect of, 869
intrinsic stains, causes, 869
pigmentation, 869
sensitivity, 869
TEGDMA. *See* Triethylene glycol dimethacrylate (TEGDMA)
TEM. *See* Transmission electron microscopy (TEM)
Tensile stresses, 625
Terminalia chebula
hydrolysable tannins, isolation of, 354
5, 5′, 6, 6′-tetrachlorol, 1′, 3, 3′-tetraethyl-benzimidazolylcarbocyanine iodide (JC-1), 348
Tetracyclin hydrochloride, 425
Tetraethyl orthosilicate (TEOS), 702
Theobroma cacao, 557
Theranosis, 68, 533
Therapeutic activity, 409
Therapeutic efficacy, 410
Therapeutic nucleic acids, 148
oral delivery of, 149–158
advantages and limitations of experimental methods, 153
in gut, 150
obstacles and potential targets, 150–152
cellular barriers and potential cellular targets, 151–152
physiological barriers, 150–151
oral route of administration, advantages, 149–150
stability and efficiency, experimental models for studying, 152–158
in cellulo models, 155–156
ex vivo models, 156–157
in situ models, 157
in vitro models, 153–155
in vivo models, 157
pathway and sites of action, overview of, 148
Therapeutic proteins
oral bioavailability, 48
Thermal expansion, 886
Thermal stability, 671
Thermosensitive hydrogel, 529
Thermotropic liquid crystals (TLCs), 676
Thioketal nanoparticles, 412
Thiolated chitosan, 416
Three-phase nanoparticle engineering technology (3PNET technology), 481
Threonine, 30
Thymoquinon, 16
Thyroid and tracheal cancer nanomedicine
citation classics in, 955
TIA. *See* Transient ischemic attack (TIA)
Tight junctions (TJs), 609
TiO_2 nanoparticles, 847
Tissue distribution, 610
Tissue engineering and regenerative medicine (TERM), 778
Titanium dioxide nanoparticles (TiO_2-NPs) synthesis, 566
Titanium nanoparticles, 790
bleaching efficacy of, 870
Titanium oxide nanoparticles, 868, 869
in bleaching, 871
TJs. *See* Tight junctions (TJs)
TLR. *See* Toll like receptors (TLR)
TMC. *See* Trimethyl chitosan (TMC)
TMC-*g*-PA. *See* TMC-grafted-palmitic acid (TMC-*g*-PA)
TMC-grafted-palmitic acid (TMC-*g*-PA), 613

Tocopherol, 582
Tocopherol succinate, 617, 627
Toll like receptors (TLR), 296
  in human, 297
Tonofibrils, 91
Tonofilaments, 91
Tooth decay, 847
Tooth enamel, 788
Tooth extraction, 876
Tooth lesion, 839
Top-down approach, 753, 755
  applications based on, 753
  bone-replacement materials, 756
  impression materials, 755
  laser plasma application, for periodontia, 755
  nanocomposite denture teeth, 756
  nanocomposites, 755
  nanoencapsulation, 756
  nanolight-curing glass ionomer restorative, 755
  nanoneedles, 756
  nanosolutions, 756
  prosthetic implant, 756
Topoisomerase-1 poison, 504
Topotecan, 476
Toxins
  examples of, 316
  nanosystems, 318
TPNPs. *See* Targeted polymeric nanoparticles (TPNPs)
Transbuccal drug delivery, 88
  chitosan used in, 99
  development of nanostructured DDS used in, 101
  limitations, 88, 113
  nanoparticles for, 100–113
    lipid carriers, 106–109
    liposomes, 104–106
    polymeric micelles, 113
    polymeric nanoparticles, 110–112
  nanostructures for, 100–113
    cyclodextrins, 102–103
  overview, 88–89
  polymers used as matrices for, 99
Transcytosis, 478, 600
Transepithelial electrical resistance (TEER), 628
Transfection efficiency, 564
Transfersomes, 590
Transgene expression, 133
Transient ischemic attack (TIA), 580
Transmission electron microscopy (TEM), 315, 345
Transplanted organs, 296
TRAP. *See* Tartrate-resistant acid phosphatase (TRAP)
Trelstar, 234
Triamcinolone, 105
Tricalcium phosphate, 808
Triclosan (TCS), 250, 798
TriCor (fenofibrate), 241
Triethyl chitosan (TEC), 415
Triethylene glycol dimethacrylate (TEGDMA), 841
Triglide (fenofibrate), 241
Trimethylammonium (TA), 721
Trimethyl chitosan (TMC), 415, 613
Trimethyl chitosan-TPP nanoparticles, 413
Trinitrobenzene sulfonic acid, 158
Tripalmitin, 624
Tripolyphoshate (TPP), 164, 413
Tripterine (TRI)-based oral suspensions, 501
Tripterygium wilfondil, 310
Triptolide (TP) nanostructures carriers (TP-NLCs), 45
Tumor cells, 468
Tumor microenvironment, 527
Tumor necrosis factor (TNF)-α, 562
Tumor progression, 556
Tungsten disulfide (WS2) nanoparticles, 754
Tween-20, 239
Tween 80, 501, 657
Type 1 diabetes, 437–438
Type 2 diabetes, 439–440
typical CdSe/ZnS QD, 764

U

UEA. *See* Ulex europaeus agglutinin (UEA)
Ulcerative colitis, 723
Ulex europaeus agglutinin (UEA), 129
Ultrafiltration, 373, 596
Ultrafine emulsion, 213
Ultrasonography, 805, 807
Uncaria sinensis, 578
Untreated solid wood polymers, 547
Uracil-ftorafur (UFT), 475
Urea-containing microemulsions, 566
Uric acid, 916
Uronic acids, 546

V

Vaccines, 299
  composition, 271
    adjuvants, 271
    antigen, 271
  development, 264
  nanosystems for, 314
Vaccinia virus (VV), 137
Vaccinium macrocarpon, 584
Valsartan (VAL), 703
van der Waals forces, 900
Van der Waals interactions, 618
VAS. *See* Visual analog scale (VAS)

Vascular endothelial growth factors (VEFG), 527, 876
Vascular permeability mediators, 527
VEFG. *See* Vascular endothelial growth factors (VEFG)
Venoms, 315
  composition, 315
  diagnostic reagents, as, 315
  drug designing, role in, 315
  examples of, 316
  nanoformulations, 315
  nanosystems, 318
  pharmaceutical agents, as, 315
Verapamil, 176
Verrucous carcinoma, 514
Viaclathrin-coated pits, 33
Vibrio cholerae, 132
Vincristine, 504
Vinorelbine, 476
Vinpocetine (VP), 45
Virus encoded proteins, 353
  gp 41, 353
  gp 120, 353
Virus-like particles (VLPs), 285
Viscosity, 672
Visual analog scale (VAS), 104
Vitronectine, 809
Voltammetric methods, 892–893
VV. *See* Vaccinia virus (VV)

W

WAS. *See* Wiskott-Aldrich syndrome (WAS)
Water-soluble azulenes, 562
Water sorption, 413
Water uptake coefficient, 389
Water vapor sorption-desorption isotherm, 389
Web of Science (WoS) database, 939
Wet mass extrusion. *See* Extrusion spheronization process, pellets production
Wiskott-Aldrich syndrome (WAS), 299
Withania somnifera, 578
World Health Organization (WHO), 351

X

Xenograft oral SCC model, 529
X-linked agammaglobulinemia, 299
X-ray diffraction, 629

Y

Y_2O_3 nanocrystals (YO NC), 955
Yttria-stabilized zirconia (YSZ), 797

Z

Zedoary essential oil, 240
Zero order kinetics, 381
Zinc carbonate hydroxyapatite nanoparticles, 868, 871
Ziziphus mauritiana, 310
Zn nanoparticles, 790
  in food industry, 790
Zwitterionic compounds
  acid-base equilibria speciation, 384
Zwitterionic drugs, 379
Zwitterionic surfactants, 185, 707
Zwitterions, 384

Kolibri 360 GmbH
Ferdinand-Jühlke-Straße 7
99095 Erfurt, Deutschland
produktsicherheit@kolibri360.de